Guide to the

Vascular Plants of Central French Guiana

Memoirs of The New York Botanical Garden

The MEMOIRS are published at irregular intervals in volumes of various sizes and are designed to include results of original botanical research by members of The New York Botanical Garden's staff, or by botanists who have collaborated in one or more of The New York Botanical Garden's research programs. Ordinarily only manuscripts of 100 or more typewritten pages will be considered for publication.

Manuscripts should be submitted to the Editor. For further information regarding editorial policy and instructions for the preparation of manuscripts, contact the Editor.

Orders for published and forthcoming issues and volumes should be placed with:

THE NEW YORK BOTANICAL GARDEN PRESS
200th St. & Kazimiroff Blvd.
Bronx, New York 10458-5126 U.S.A.
(718) 817-8721; fax (718) 817-8842
email: nybgpress@nybg.org
http://www.nybg.org/bsci/spub

Guide to the

Vascular Plants of Central French Guiana

Part 2. Dicotyledons

Scott A. Mori
Georges Cremers
Carol A. Gracie
Jean-Jacques de Granville
Scott V. Heald
Michel Hoff
John D. Mitchell

Illustrations by:
Bobbi Angell

Photographs by:
Carol A. Gracie (unless otherwise credited)

The New York Botanical Garden Press

Memoirs of The New York Botanical Garden Volume 76, Part 2

Composition by Eisner/Martin Typographics

Printed in China

Cover photographs by Carol A. Gracie

International Standard Serial Number 0077-8931

The paper used in this publication meets the requirements of ANSI/NISO Z39.48–1992 (R1997). ♾™

The Metropolitan Life Foundation is a leadership funder of The New York Botanical Garden Press.

Library of Congress Cataloging-in-Publication Data

Guide to the vascular plants of central French Guiana / Scott A. Mori . . . [et al.] ; illustrations by Bobbi Angell ; photos by Carol Gracie (unless otherwise credited).
p. cm. — (Memoirs of the New York Botanical Garden ; v. 76)
Summary in English and French.
Includes bibliographical references (v. 2, p.) and indexes.
Contents: pt. 1. Pteridophytes, gymnosperms, and monocotyledons. pt. 2. Dicotyledons
ISBN 0-89327-445-3
1. Botany—French Guiana. 2. Plants—Identification. I. Mori, Scott A., 1941– . II. Series.
QK1.N525 vol. 76
[QK269]
581 s—dc20
581.9882] 96-30166

CIP

02 03 04 05 06 07 08 09 10 9 8 7 6 5 4 3 2 1

The New York Botanical Garden Press
200th St. & Kazimiroff Blvd.
Bronx, NY 10458-5126, USA
nybgpress@nybg.org
http://www.nybg.org/bsci/spub

WWF Guianas
Gravenstraat 63, Suite E
Paramaribo, Suriname
wwf@wwfsuriname.net

This book is dedicated to all those botanists whose collections and knowledge of the Neotropical flora made it possible.

CONTRIBUTORS

Pedro Acevedo-Rodríguez, United States National Herbarium, Botany Department, MRC-166, Smithsonian Institution, Washington, DC 20560-0166, U.S.A.

Lucile Allorge, Museum National d'Histoire Naturelle, Laboratoire de Phanérogamie, 16 rue Buffon,75005 Paris, France.

William S. Alverson, Environmental and Conservation Program, The Field Museum of Natural History, 1400 S. Lake Shore Drive, Chicago, IL 60605-7436, U.S.A.

William R. Anderson, University of Michigan Herbarium, North University Building, Ann Arbor, MI 48109-1057, U.S.A.

Bobbi Angell, P.O. Box 158, Marlboro, VT 05344-0158, U.S.A.

Scott Armbruster, Biology and Wildlife, 211 Irving Building, P.O. Box 756100, Fairbanks, AK 99775-1200, U.S.A.

Daniel Austin, Department of Biological Sciences, Florida Atlantic University, P.O. Box 3091, Boca Raton, FL 33431-0991, U.S.A.

Gerardo A. Aymard, Herbario Universitario, Bio Centro—UNELLEZ, Mesa de Cavacas 3323, Portuguesa, Venezuela.

Rupert C. Barneby†, Institute of Systematic Botany, The New York Botanical Garden, Bronx, NY 10458-5126, U.S.A.

Cornelis C. Berg, Arboretum and Botanical Garden, University of Bergen, 5067 Store Milde, Norway.

Brian M. Boom, Institute of Systematic Botany, The New York Botanical Garden, Bronx, NY 10458-5126, U.S.A.

Bruno Bordenave, Museum National d'Histoire Naturelle, Laboratoire de Phanérogamie, 16 rue Buffon,75005 Paris, France.

John Brandbyge, Hortensiavej 21, 8270 Højberg, Denmark.

John L. Brown, Institute of Systematic Botany, The New York Botanical Garden, Bronx, NY 10458-5126, U.S.A.

Georges Cremers, Antenne IRD (formerly ORSTOM), Museum National d'Histoire Naturelle, Laboratoire de Phanérogamie, 16 rue Buffon, 75005 Paris, France.

Douglas C. Daly, Institute of Systematic Botany, The New York Botanical Garden, Bronx, NY 10458-5126, U.S.A.

Robert A. DeFilipps, United States National Herbarium, Botany Department, MRC-166, Smithsonian Institution, Washington, DC 20560-0166, U.S.A.

Piero G. Delprete, Institute of Systematic Botany, The New York Botanical Garden, Bronx, NY 10458-5126, U.S.A.

Laurence J. Dorr, United States National Herbarium, Botany Department, MRC-166, Smithsonian Institution, Washington, DC 20560-0166, U.S.A.

Stefan Dressler, Forschungsinstitut Senckenberg, Senckenberganlage 25, D-60325 Frankfurt/M., Germany.

Christian Feuillet, United States National Herbarium, Botany Department, MRC-166, Smithsonian Institution, Washington, DC 20560-0166, U.S.A.

Beat Fischer, BAB—Büro für Angewandte Biologie, Depotstrasse 28, 3012 Bern, Switzerland.

Enrique Forero, Facultad de Ciencias, Universidad Nacional de Colombia, Apartado Aéreo 5997, Santafé de Bogotá, Colombia.

Peter W. Fritsch, Department of Botany, California Academy of Sciences, Golden Gate Park, San Francisco, CA 94118-4599, U.S.A.

Lynn J. Gillespie, Research Division, Canadian Museum of Nature, P.O. Box 3443, Station D, Ottawa, ON K1P 6P4, Canada.

Ara Görts-van Rijn, Institute of Systematic Botany, State University of Utrecht, Postbus 80.102, 3508 TC Utrecht, Netherlands.

Carol A. Gracie, Institute of Systematic Botany, The New York Botanical Garden, Bronx, NY 10458-5126, U.S.A.

Jean-Jacques de Granville, IRD (formerly ORSTOM), Centre de Cayenne, B. P. 165, F-97323 Cayenne Cédex, France.

James W. Grimes, Royal Botanic Gardens, Melbourne, Birdwood Ave., South Yarra 3141, Victoria, Australia.

Mats H. G. Gustafsson, Department of Systematic Botany, Institute of Biological Sciences, University of Aarhus, Nordlandsvej 68, 8240 Risskov, Denmark.

Bruce F. Hansen, Herbarium, Department of Biology, University of South Florida, Tampa, FL 33260-5150, U.S.A.

Raymond Harley, Royal Botanic Gardens, Kew, Richmond, Surrey TW9 3AB, England, U.K.

Scott V. Heald, Institute of Systematic Botany, The New York Botanical Garden, Bronx, NY 10458-5126, U.S.A.

Willem H. A Hekking†, Institute of Systematic Botany, University of Utrecht, Postbus 80.102, 3508 TC Utrecht, Netherlands.

Paul Hiepko, Botanischer Garten und Botanisches Museum Berlin-Dahlem, Königin-Luise-Straße 6–8, D-14191 Berlin 33, Germany.

Michel Hoff, IRD (formerly ORSTOM), Service du Patrimoine Naturel, Museum National d'Histoire Naturelle, 57 rue Cuvier, F-75231 Paris Cédex, France.

Noel H. Holmgren, Institute of Systematic Botany, The New York Botanical Garden, Bronx, NY 10458-5126, U.S.A.

Bruce Holst, Herbarium, The Marie Selby Botanical Garden, 811 South Palm Avenue, Sarasota, FL 34236-7726, U.S.A.

Hugh H. Iltis, Herbarium, Botany Department, University of Wisconsin, Birge Hall, Madison, WI 53706-1381, U.S.A.
Ariane Jacobs-Brouwer, Institute of Systematic Botany, State University of Utrecht, Postbus 80.102, 3508 TC Utrecht, Netherlands.
Marion Jansen-Jacobs, Institute of Systematic Botany, University of Utrecht, Postbus 80.102, 3508 TC Utrecht, Netherlands.
Jacquelyn A. Kallunki, Institute of Systematic Botany, The New York Botanical Garden, Bronx, NY 10458-5126, U.S.A.
Maria Lúcia Kawasaki, Herbário, Instituto de Botânica, Caixa Postal 4005, 01061-970 São Paulo-SP, Brasil.
Job Kuijt, Herbarium, Biology Department, University of Victoria, Victoria, BC V8W 2Y2, Canada.
Beat Ernst Leuenberger, Botanischer Garten und Botanisches Museum Berlin-Dahlem, Königin-Luise-Straße 6–8, D-14191 Berlin, Germany.
Lúcia G. Lohmann, Missouri Botanical Garden, P.O. Box 299, St. Louis, MO 63166-0299, U.S.A.
James L. Luteyn, Institute of Systematic Botany, The New York Botanical Garden, Bronx, NY 10458-5126, U.S.A.
Paul Maas, Institute of Systematic Botany, University of Utrecht, Postbus 80.102, 3508 TC Utrecht, Netherlands.
Hiltje Maas-van de Kamer, Institute of Systematic Botany, University of Utrecht, Postbus 80.102, 3508 TC Utrecht, Netherlands.
Santiago Madriñán, Dept. Ciencias Biológicas, Universidad de Los Andes, Apartado Aéreo 4976, Santafé de Bogotá, Colombia.
Shirley L. Maina, United States National Herbarium, Botany Department, MRC-166, Smithsonian Institution, Washington, DC 20560-0166, U.S.A.
John D. Mitchell, Institute of Systematic Botany, The New York Botanical Garden, Bronx, NY 10458-5126, U.S.A.
Scott A. Mori, Institute of Systematic Botany, The New York Botanical Garden, Bronx, NY 10458-5126, U.S.A.
Gilberto Morillo, Herbario, Facultad de Ciencias Forestales, Departamento de Botánica, Codigo Postal 5101-A, Mérida, Edo. Mérida, Venezuela.
Michael Nee, Institute of Systematic Botany, The New York Botanical Garden, Bronx, NY 10458-5126, U.S.A.
Terence D. Pennington, Royal Botanic Gardens, Kew, Richmond, Surrey TW9 3AB, England, U.K.
Thomas Philbrick, Department of Biology, Western Connecticut State University, Danbury, CT 06810-6860, U.S.A.
John J. Pipoly III, Fairchild Tropical Garden, 10935 Old Cutler Rd., Coral Gables (Miami), FL 33156-4299, U.S.A.
Vanessa Plana, Royal Botanic Garden Edinburgh, 20A Inverleith Row, Edinburgh EH3 5LR, Scotland, U.K.
Marcel-Marie Plumel†, Museum National d'Histoire Naturelle, Laboratoire de Phanérogamie, 16 rue Buffon, 75005 Paris, France.
Odile Poncy, Museum National d'Histoire Naturelle, Laboratoire de Phanérogamie, 16 rue Buffon, 75005 Paris, France.
Ghillean T. Prance, The Old Vicarage, Silver Street, Lyme Regis, Dorset DT7 3HS, England, U.K.
John F. Pruski, United States National Herbarium, Botany Department, MRC-166, Smithsonian Institution, Washington, DC 20560-0166, U.S.A.
Susanne Renner, Department of Biology, University of Missouri at St. Louis, Natural Bridge Rd. 8001, St. Louis, MO 63121-4499, U.S.A.
Adrian C. de Roon, Institute of Systematic Botany, University of Utrecht, Postbus 80.102, 3508 TC Utrecht, Netherlands.
Daniel Sabatier, IRD/Cirad-Forêt, Campus de Baillarguet, BP 5035, F-34032 Montpellier Cédex, France.
Claude Sastre, Museum National d'Histoire Naturelle, Laboratoire de Phanérogamie, 16 rue Buffon, 75005 Paris, France.
Laurence E. Skog, United States National Herbarium, Botany Department, MRC-166, Smithsonian Institution, Washington, DC 20560-0166, U.S.A.
Damon A. Smith, 445 W. Center Street, Whitewater, WI 53190-1978, USA.
Nathan P. Smith, Institute of Systematic Botany, The New York Botanical Garden, Bronx, NY 10458-5126, U.S.A.
Bertil Ståhl, Department of Botany, University of Göteborg, P.O. Box 461, SE 40530 Göteborg, Sweden.
Wm. Wayt Thomas, Institute of Systematic Botany, The New York Botanical Garden, Bronx, NY 10458-5126, U.S.A.
Alberto Vicentini, Departamento de Botânica, Instituto Nacional de Pesquisas da Amazônia, Caixa Postal 478, 69083 Manaus, Amazonas, Brazil.
Bruno Wallnöfer, Herbarium, Department of Botany, Naturhistorisches Museum Wien, Burgring 7, 1014, Wien, Austria.
Dieter C. Wasshausen, United States National Herbarium, Botany Department, MRC-166, Smithsonian Institution, Washington, DC 20560-0166, U.S.A.
Henk van der Werff, Missouri Botanical Garden, P.O. Box 299, St. Louis, MO 63166-0299, U.S.A.
John Wurdack†, United States National Herbarium, Botany Department, MRC-166, Smithsonian Institution, Washington, DC 20560-0166, U.S.A.

MEMOIRS OF THE NEW YORK BOTANICAL GARDEN **76(2):** viii + 1–776

Guide to the Vascular Plants of Central French Guiana:

Part 2. Dicotyledons

Scott A. Mori, Georges Cremers, Carol A. Gracie, Jean-Jacques de Granville, Scott V. Heald, Michel Hoff, and John D. Mitchell

Contents

Summary/Résumé 4
Introduction 5
Scientific and Common Names 5
Voucher Specimens 5
Systematic Concepts 5
Aids to Identification 5
Habit 6
Sulcate (fluted) boles 6
Flying buttresses 6
Tree Bark 6
White or off-white exudate 6
Red exudate 6
Aroma 6
Leaves 6
Plants with variegated leaves 6
Trees, shrubs, and lianas with heteromorphic leaves 6
Trees with simple, opposite leaves 7
Shrubs with prickles on leaves 7
Trees, shrubs, and lianas with cavities on the leaf blades 7
Trees, shrubs, and lianas with glands on the leaves 7
Trees, shrubs, and lianas with peltate scales on leaf blades 8
Trees and shrubs with stellate and dendritic hairs on leaf blades 8
Trees, shrubs, and lianas with punctations/striations on leaf blades 8
Inflorescences and Flowers 9
Trees, shrubs, and lianas with glands on inflorescences or flowers 9
Fruits 9
Trees, shrubs, and lianas with fruits having unusual surfaces 9
Seeds 9
Trees and lianas with winged seeds 9
Lianas 9
Special adaptations for climbing 9
Distinctive aromas 9
Notes on Descriptions 9
Group V: Magnoliopsida (dicotyledons) 10
Key to Keys of Magnoliopsida 10
Key 1. Achlorophyllous Plants 10
Key 2. Herbaceous Terrestrial or Aquatic Plants 11
Key 3. Climbing Plants 14
Key 4. Epiphytic Plants 17
Key 5. Trees 18
Key 6. Shrubs 27
Acanthaceae(Acanthus Family) by Dieter C. Wasshausen 31
Amaranthaceae (Amaranth Family) by Robert A. DeFilipps & Shirley L. Maina 40
Anacardiaceae (Cashew Family) by John D. Mitchell 43
Annonaceae (Custard-apple Family) by Paul Maas and Hiltje Maas-van de Kamer 53

Apiaceae (Carrot Family) by Scott A. Mori 67
Apocynaceae (Dogbane Family) by Lucile Allorge (*Forsteronia* by Bruce F. Hansen and *Himatanthus* by Marcel-Marie Plumel) 69
Araliaceae (Ginseng Family) by Scott A. Mori 84
Aristolochiaceae (Birthwort Family) by Christian Feuillet & Odile Poncy 87
Asclepiadaceae (Milkweed Family) by Gilberto Morillo 89
Asteraceae (Aster Family) by John F. Pruski 94
Balanophoraceae (Balanophora Family) by Carol A. Gracie 116
Begoniaceae (Begonia Family) by Scott V. Heald 117
Bignoniaceae (Trumpet-creeper Family) by Lúcia G. Lohmann, John L. Brown & Scott A. Mori 118
Bixaceae (Lipstick-tree Family) by Scott V. Heald 139
Bombacaceae (Kapok-tree Family) by William S. Alverson & Scott A. Mori 139
Boraginaceae (Borage Family) by Christian Feuillet 145
Brassicaceae (Mustard Family) by John D. Mitchell 151
Burseraceae (Frankincense Family) by Douglas C. Daly 151
Cactaceae (Cactus Family) by Beat Ernst Leuenberger 165
Caesalpiniaceae (Caesalpinia Family) by Rupert C. Barneby & Scott V. Heald 167
Campanulaceae (Bellflower Family) by Carol A. Gracie 183
Canellaceae (Canella Family) by John L. Brown & Scott A. Mori 184
Capparaceae (Caper Family) by Hugh H. Iltis & Scott A. Mori 186
Caricaceae (Papaya Family) by Scott A. Mori 188
Caryocaraceae (Souari Family) by Jean-Jacques de Granville 191
Caryophyllaceae (Pink Family) by Robert A. DeFilipps & Shirley L. Maina 193
Cecropiaceae (Cecropia Family) by Cornelis C. Berg 194
Celastraceae (Bittersweet Family) by John D. Mitchell 199
Chrysobalanaceae (Cocoa-plum Family) by Ghillean T. Prance 202
Clusiaceae (Mangosteen Family) by John J. Pipoly III & Mats H. G. Gustafsson 212
Combretaceae (Indian Almond Family) by Maria Lúcia Kawasaki 224
Connaraceae (Connarus Family) by Enrique Forero 227
Convolvulaceae (Morning-glory Family) by Daniel Austin 231
Cucurbitaceae (Squash Family) by Michael Nee 236
Dichapetalaceae (Dichapetalum Family) by Ghillean T. Prance 247
Dilleniaceae (Dillenia Family) by Gerardo A. Aymard & Scott A. Mori 250
Ebenaceae (Ebony Family) by Bruno Wallnöfer & Scott A. Mori 254
Elaeocarpaceae (Elaeocarpus Family) by Scott V. Heald, Alberto Vicentini & Damon A. Smith 258
Ericaceae (Heath Family) by James L. Luteyn 261
Erythroxylaceae (Coca Family) by Scott V. Heald 263
Euphorbiaceae (Spurge Family) by Lynn J. Gillespie (*Dalechampia* by Scott Armbruster) . . . 266
Fabaceae (Bean Family) by Rupert C. Barneby & Scott V. Heald 298
Flacourtiaceae (Flacourtia Family) by Scott A. Mori & Beat Fischer 319
Gentianaceae (Gentian Family) by Hiltje Maas-van de Kamer & Paul Maas 328
Gesneriaceae (Gesneriad Family) by Christian Feuillet & Laurence E. Skog 334
Hernandiaceae (Hernandia Family) by Scott A. Mori & John L. Brown 344
Hippocrateaceae (Hippocratea Family) by Ara Görts-van Rijn 347
Hugoniaceae (Hugonia Family) by Daniel Sabatier 353
Humiriaceae (Humiria Family) by Daniel Sabatier 355
Icacinaceae (Icacina Family) by Adrian C. de Roon & Scott A. Mori 358
Lacistemataceae (Lacistema Family) by Scott A. Mori 362
Lamiaceae (Mint Family) by Raymond Harley 363
Lauraceae (Avocado Family) by Henk van der Werff (*Rhodostemonodaphne* by Santiago Madriñán) 370
Lecythidaceae (Brazil-nut Family) by Scott A. Mori 385
Lentibulariaceae (Bladdernut Family) by Nathan P. Smith 397
Loganiaceae (Logania Family) by Bruno Bordenave 398
Loranthaceae (Showy Mistletoe Family) by Job Kuijt 405
Lythraceae (Loosestrife Family) by Maria Lúcia Kawasaki 408
Malpighiaceae (Malpighia Family) by William R. Anderson 410
Malvaceae (Mallow Family) by Laurence J. Dorr 428
Marcgraviaceae (Shingle Plant Family) by Scott V. Heald, Adrian C. de Roon & Stefan Dressler 431
Melastomataceae (Melastome Family) by John Wurdack & Susanne Renner 437
Meliaceae (Mahogany Family) by Terry Pennington 465

Mendonciaceae (Mendoncia Family) by Dieter C. Wasshausen 472
Menispermaceae (Moonseed Family) by Rupert C. Barneby 474
Mimosaceae (Mimosa Family) by James W. Grimes (*Inga* by Odile Poncy) 484
Monimiaceae (Monimia Family) by John D. Mitchell 510
Moraceae (Mulberry Family) by Cornelis C. Berg 515
Myristicaceae (Nutmeg Family) by John D. Mitchell 525
Myrsinaceae (Myrsine Family) by John J. Pipoly III 532
Myrtaceae (Myrtle Family) by Maria Lúcia Kawasaki & Bruce Holst 539
Nyctaginaceae (Four O'Clock Family) by Robert A. DeFilipps & Shirley L. Maina 551
Nymphaeaceae (Water-lily Family) by Scott A. Mori 554
Ochnaceae (Ochna Family) by Claude Sastre 554
Olacaceae (Olax Family) by Paul Hiepko 559
Onagraceae (Evening Primrose Family) by Maria Lúcia Kawasaki 562
Oxalidaceae (Wood-sorrel Family) by John D. Mitchell 564
Passifloraceae (Passion-flower Family) by Christian Feuillet 566
Phytolaccaceae (Pokeweed Family) by Robert A. DeFilipps & Shirley L. Maina 571
Piperaceae (Pepper Family) by Ara Görts-van Rijn 574
Podostemaceae (River-weed Family) by C. Thomas Philbrick 585
Polygalaceae (Milkwort Family) by Ariane Jacobs-Brouwer 585
Polygonaceae (Buckwheat Family) by John Brandbyge 590
Proteaceae (Protea Family) by Vanessa Plana 592
Quiinaceae (Quiina Family) by Scott A. Mori 594
Rafflesiaceae (Rafflesia Family) by Scott A. Mori 598
Rhabdodendraceae (Rhabdodendron Family) by Scott A. Mori 598
Rhamnaceae (Buckthorn Family) by Scott V. Heald 600
Rhizophoraceae (Red Mangrove Family) by Scott A. Mori 603
Rosaceae (Rose Family) by John D. Mitchell 605
Rubiaceae (Coffee Family) by Brian M. Boom and Piero Delprete 606
Rutaceae (Rue Family) by Jacquelyn A. Kallunki 649
Sapindaceae (Soapberry Family) by Pedro Acevedo-Rodríguez 656
Sapotaceae (Sapodilla Family) by Terence D. Pennington 669
Scrophulariaceae (Figwort Family) by Noel H. Holmgren 683
Simaroubaceae (Quassia Family) by Wm. Wayt Thomas 685
Solanaceae (Nightshade Family) by Michael Nee 689
Sterculiaceae (Cacao Family) by Laurence J. Dorr 700
Styracaceae (Storax Family) by Peter W. Fritsch, Scott A. Mori & John L. Brown 706
Symplocaceae (Sweetleaf Family) by Scott A. Mori & John L. Brown 708
Theophrastaceae (Theophrasta Family) by Bertil Ståhl 709
Thymelaeaceae (Mezereum Family) by Maria Lúcia Kawasaki & Scott A. Mori 711
Tiliaceae (Linden Family) by Laurence J. Dorr 712
Trigoniaceae (Trigonia Family) by Scott A. Mori 716
Turneraceae (Turnera Family) by Michel Hoff 718
Ulmaceae (Elm Family) by Cornelis C. Berg 720
Urticaceae (Nettle Family) by Cornelis C. Berg 722
Verbenaceae (Verbena Family) by Marion Jansen-Jacobs 725
Violaceae (Violet Family) by Wilhem H. A. Hekking & John D. Mitchell 732
Viscaceae (Mistletoe Family) by Job Kuijt 739
Vitaceae (Grape Family) by Scott V. Heald 740
Vochysiaceae (Vochysia Family) by Scott A. Mori 742

Additions and Corrections to Part 1 746
Glossary (supplement to Part 1) 748
Acknowledgments 750
Literature Cited 751
Index to Common Names 754
Index to Scientific Names 762

Summary

Mori, Scott A. (Institute of Systematic Botany, The New York Botanical Garden, Bronx, New York, 10458-5126, U.S.A.), **Georges Cremers** (IRD ex ORSTOM, Laboratoire de Phanérogamie, Museum National d'Histoire Naturelle, 16, rue Buffon, F-7500 Paris, France), **Carol A. Gracie** (Institute of Systematic Botany, The New York Botanical Garden, Bronx, New York, 10458-5126, U.S.A.), **Jean-Jacques de Granville** (IRD ex ORSTOM, Centre de Cayenne, B.P. 165, F-97323 Cayenne Cédex, France), **Scott V. Heald** (Institute of Systematic Botany, The New York Botanical Garden, Bronx, New York, 10458-5126, U.S.A.), **Michel Hoff** (IRD ex ORSTOM, Service du Patrimoine Naturel, Museum National d'Histoire Naturelle, 57, rue Cuvier, F-75231 Paris cédex 05, France), and **John D. Mitchell** (Institute of Systematic Botany, The New York Botanical Garden, Bronx, New York, 10458-5126, U.S.A.). Guide to the vascular plants of central French Guiana. Part 2. Dicotyledons. Mem. New York Bot. Gard. 76(2): viii + 1–776. 2002. — This is the second of a two-part guide to the vascular plants of central French Guiana. Keys and descriptions are provided for the 1483 species in 112 families of dicotyledons (Magnoliophyta class Magnoliopsida) native or naturalized in central French Guiana. An additional 20 species expected to occur, but not yet collected, and 30 cultivated species are also treated. No taxonomic novelties are published, but a new lectotype for *Casearia bractifera* is proposed. In addition, an annotated list of species with notable ecological or morphological features of use in plant identification complements the list presented in Part 1. Three hundred and twenty-six line illustrations and 128 color plates facilitate use of this guide. Additional terms supplement the glossary presented in Part 1. Although based on species from central French Guiana, this guide can be used to help identify families and genera of vascular plants throughout the forests of lowland northeastern South America.

Résumé

Mori, Scott A. (Institute of Systematic Botany, The New York Botanical Garden, Bronx, New York, 10458-5126, U.S.A.), **Georges Cremers** (IRD ex ORSTOM, Laboratoire de Phanérogamie, Museum National d'Histoire Naturelle, 16, rue Buffon, F-7500 Paris, France), **Carol A. Gracie** (Institute of Systematic Botany, The New York Botanical Garden, Bronx, New York, 10458-5126, U.S.A.), **Jean-Jacques de Granville** (IRD ex ORSTOM, Centre de Cayenne, B.P. 165, F-97323 Cayenne Cédex, France), **Scott V. Heald** (Institute of Systematic Botany, The New York Botanical Garden, Bronx, New York, 10458-5126, U.S.A.), **Michel Hoff** (IRD ex ORSTOM, Service du Patrimoine Naturel, Museum National d'Histoire Naturelle, 57, rue Cuvier, F-75231 Paris cédex 05, France), and **John D. Mitchell** (Institute of Systematic Botany, The New York Botanical Garden, Bronx, New York, 10458-5126, U.S.A.). Guide to the vascular plants of central French Guiana. Part 2. Dicotyledons. Mem. New York Bot. Gard. 76(2): viii + 1–776. 2002. — Cet ouvrage est le deuxieme d'un manuel en deux volumes sur les plantes vasculaires de la Guyane française centrale dans la Amerique du Sud. Les clés et les descriptions concernent les 1483 espèces réparties dans 112 familles de dicotylédons (Magnoliophyta class Magnoliopsida) indigènes ou naturalisées en Guyane française centrale. De plus, 20 espèces probablement présentes mais actuellement non encore collectées, ainsi qui 30 espèces cultivées sont également traitées. Aucune nouveauté taxonomique n'a été publiée, mais un nouveau Lectotype pour *Casearia bractifera* est proposé. En supplément, une liste annotée des espèces, avec les caractéristiques écologiques et morphologiques remarquables nécessaires à l'identification des plantes, est incluse. Trois cent vingt-six planches d'illustrations au trait et 128 planches de photos en couleur en facilitent l'identification. Un glossaire supplement les définitions des termes techniques de le premier volume. Bien que basé sur les espèces de la Guyane française centrale, ce manuel peut servir d'aide à l'identification des familles et des genres des plantes vasculaires de basse altitude, partout dans le nord-est de l'Amérique du Sud.

Introduction

This is the second of a two-volume book, and most of the introductory material has already been covered in Part 1 (Mori et al., 1997).

Our long-term goal is to make the information we have gathered about the vascular plants of central French Guiana available in two electronic formats: 1) as a searchable specimen and species database with more information about species than is presently online in our specimen database (http://www.nybg.org/bsci/hcol/ french_guiana) and 2) as an electronic flora illustrated with additional photographs and with multiple entry keys.

As pointed out in the introduction to Part 1, one of our aims is to make central French Guiana a place where biological studies are facilitated because of the ability to identify the plants found there. The next steps are for biologists to carry on with 1) inventories of animals, fungi, lichens, bryophytes (William R. Buck of The New York Botanical Garden is preparing a moss flora), and microorganisms; 2) ecological studies (including descriptions of the different types of vegetation, pollination, dispersal, and other plant/animal interactions); and 3) exploration for interesting plant and animal products of use to mankind. We are already engaged in some of these kinds of studies and invite others interested in investigating this little disturbed rain forest to join us.

Most botanical exploration in central French Guiana has taken place in the vicinity of Saül. Therefore, this guide will work best when used in the trail system surrounding the village and will be less inclusive in more remote areas. For example, our year 2000 expedition to Pic Matécho revealed a number of new taxa previously unknown to central French Guiana. Because this book was already in an advanced stage of the editorial process, it was not possible to add all of the new discoveries to this edition.

We hope that our work in central French Guiana will demonstrate that this area is worthy of preservation as a biological reserve because of its natural beauty and high diversity of fungi, plants, and animals. We argue that the forests of central French Guiana are more valuable if left intact because of potential tourism income (Mori et al., 1998) as well as for the ecosystem services they provide than they would be if they were cut and converted to pasture and agricultural fields.

Scientific and Common Names

In Part 1, Brummitt and Powell (1992) were followed except the authors' names were spelled out in full. In this part, we follow the Brummitt and Powell abbreviations because most databases, including NYpc and its successor, CASSIA, used for this study, follow the Brummitt and Powell abbreviations. The only exception is that a space is placed between the initials of the authors as is done in NYpc but not by Brummit and Powell.

The Index to Common names differs from that provided in Part 1 in that the common name is followed by the scientific name rather than by the page number where the scientific name can be found. The same sources for common names as used for determining the common names of species in part 1 were consulted. However, Georges Cremers made a special effort to search the Aublet database for the common names of species occurring in our area.

Portuguese names are included because of the influence of Brazilians throughout French Guiana. Many of these names have made it into Brazilian Portuguese from various Indian languages, but we have made no effort to identify the original Indian languages from which they are derived.

Voucher Specimens

Since the publication of Part 1, the specimen database of nearly 14,000 entries has been made available on the Internet. The database can be accessed by visiting The New York Botanical Garden web site at <http://www.nybg.org>. At the present time, this database includes only the flowering plants. Checklists with specimen citations have been distributed to the libraries of the Flora of the Guianas Consortium. The pteridophytes are found in the Aublet database of IRD (formerly ORSTOM)/Cayenne but this is not yet available on the World Wide Web. For further information about the pteridophytes and their vouchers, contact Georges Cremers of the Museum National d'Histoire Naturelle at gcremers@mnhn.fr. Specimens recorded in these databases are mostly deposited in the herbaria of The New York Botanical Garden (NY) and IRD/Cayenne (CAY).

Systematic Concepts

The classification of the flowering plants is undergoing refinement as a result of new information obtained from molecular studies and phylogenetic analyses (Judd et al., 1999; Angiospermum Phylogeny Group, 1998). Although we follow the Cronquist (1981) system, we have included comments on changes in taxonomic concepts when appropriate. For example, we still treat *Strychnos* as part of the Loganiaceae, but indicate that there is evidence to support its segregation as a separate family (Struwe et al., 1994). Likewise, although morphological and molecular data support the placement of *Potalia* in the Gentianaceae rather than in the Loganiaceae (Struwe et al., 1994), we continue to place it in the former family as suggested by Cronquist (1981).

We have chosen to continue with the Cronquist system (1981) because an overall system in which molecular data are included was not available during most of the twelve years that this *Guide* has been in preparation. Grayum and Hammel (1999) reached the same conclusion for family and generic concepts in the preparation of *A Manual of the Plants of Costa Rica*.

Our goal is to make it possible for others to put names of species on plants from central French Guiana and this goal is not seriously compromised by the changes in familial classification and by new generic placements taking place because of information derived from molecular studies.

Aids to Identification

With continued botanical exploration and increased botanical knowledge of central French Guiana, the following additions to the "Aids to Identification" presented in Part 1 of this guide have been discovered. This process will continue for many years to come as our knowledge of tropical plants becomes more complete.

Habit

Sulcate (Fluted) Boles

- *Alibertia myrciifolia* (Rubiaceae). A species of small trees with opposite, simple leaves.

Flying Buttresses

- *Sloanea latifolia* (Elaeocarpaceae) should be added to the list of species of this genus with flying buttresses. See Part 1, page 22.

Tree Bark

White or Off-white Exudate

- *Jacaratia spinosa* (Caricaceae). This species of tree with alternate, palmately compound leaves and prickles on the trunk and stems emits an off-white exudate from the cut trunks and broken branches.

Red Exudate

- *Coussapoa angustifolia* (Cecropiaceae). This species of hemiepiphytic shrubs or trees with alternate leaves and obliquely oriented stipules emits a red exudate from the cut stems.
- *Croton matourensis* (Euphorbiaceae). This species of tree of secondary habitats with alternate, simple leaves emits a red exudate from cut trunks.

Aroma

- *Pourouma* spp. (Cecropiaceae). Species of this genus possess alternate, simple, often lobed leaves with basally attached petioles. At least some of the species of this genus emit aromas of spearmint when the trunk is cut. The exact distribution of this feature has not yet been determined.

Leaves

Plants with Variegated Leaves

- *Buforrestia candolleana* (Commelinaceae). The alternate, simple, sheathed leaves of this herb, which roots at the lower nodes of the prostrate part of the stems, possess two arching, silver lines, one on each side of the midrib.
- *Monstera dubia* (Araceae). The alternate, simple, cordate based, shingle leaves of this climbing herb possess silver markings. This species was called *Monstera* sp. nov. in Part 1, p. 179.
- *Nautilocalyx pictus* (Gesneriaceae). The often bullate, subequal leaves of a pair at a node of this species often have green, white, silver, or red markings adaxially.

Trees, Shrubs, and Lianas with Heteromorphic Leaves

- *Aegiphila integrifolia* (Verbenaceae). The opposite, simple leaves of this shrub or small tree are often unequal in size at the same node.
- *Bagassa guianensis* (Moraceae). The opposite, simple leaves of this relatively common, secondary growth tree have distinctly crenate margins. The blades may be nearly cordate and unlobed or display various degrees of lobing. Leaves of juvenile plants are often 3-parted. This is the only species of Moraceae in the Neotropics with opposite leaves.
- *Cuphea carthagenensis* (Lythraceae). This subshrub, common in open, weedy areas, possesses opposite, simple leaves with one leaf of the pair often markedly smaller than the other. Sometimes the smaller leaf is so small that the leaves appear to be alternate. At other nodes, the leaves at a node may be nearly equal in size.

Plates 1–8

Plates 1–5. Contributors to Parts 1 and 2 of the *Guide to the Vascular Plants of Central French Guiana* and their contributions.

Plate 1. Left—Right, Top—Bottom: **Pedro Acevedo-Rodríguez**, Sapindaceae. **Lucile Allorge**, Apocynaceae excepting *Forsteronia* and *Himatanthus*. **William S. Alverson**, Bombacaceae. [Photo by S. Boykin] **Patti Anderson**, Heliconiaceae. **William R. Anderson**, Malpighiaceae. [Photo by D. Bay] **Lennart Andersson**, Marantaceae. [Photo by A. Nilsson] **Bobbi Angell**, artist. **Scott Armbruster**, *Dalechampia* (Euphorbiaceae). **Daniel Austin**, Convolvulaceae. [Photo by P. Maserjian] **Rubert Barneby**, Caesalpiniaceae, Fabaceae, Menispermaceae. **Cornelis C. Berg**, Cecropiaceae, Moraceae, Ulmaceae, Urticaceae. **Brian M. Boom**, Rapateaceae, Rubiaceae. **Bruno Bordenave**, Loganiaceae. **John Brandbyge**, Polygonaceae. [Photo by R. Høerleik] **John L. Brown**, Bignoniaceae, Canellaceae, Hernandiaceae, Styracaceae, Symplocaceae. **Eric A. Christenson**, Orchidaceae.

Plate 2. Left—Right, Top—Bottom: **Georges Cremers**, Pteridophytes. **Thomas B. Croat**, Araceae. **Douglas C. Daly**, Burseraceae. **Robert A. DeFilipps**, Amaranthaceae, Caryophyllaceae, Nyctaginaceae, Phytolaccaceae. **Piero G. Delprete**, Rubiaceae. **Laurence Dorr**, Malvaceae, Sterculiaceae, Tiliaceae. **Christian Feuillet**, Aristolochiaceae, Boraginaceae, Gesneriaceae, Passifloraceae. **Beat Fisher**, Flacourtiaceae. **Enrique Forero**, Connaraceae. **Peter W. Fritsch**, Styracaceae. [Photo by D. Linn, California Academy of Sciences] **Lynn J. Gillespie**, Euphorbiaceae, except *Dalechampia.* **Ara Görts-van Rijn**, Hippocrateaceae, Piperaceae. **Eric Gouda**, Bromeliaceae. **Carol A. Gracie**, Balanophoraceae, Campanulaceae, Commelinaceae, Heliconiaceae; photographer. [Photo by S. Mori] **Jean-Jacques de Granville**, Arecaceae, Caryocaraceae. **James W. Grimes**, Mimosaceae, except *Inga.*

Plate 3. Left—Right, Top—Bottom: **Mats H. G. Gustafsson**, Clusiaceae. **Bruce Hansen**, *Forsteronia* (Apocynaceae). [Photo by F. Essig] **Raymond Harley**, Lamiaceae. [Photo by S. Mayo] **Scott V. Heald**, Begoniaceae, Bixaceae, Elaeocarpaceae, Erythroxylaceae, Fabaceae, Marcgraviaceae, Rhamnaceae, Vitaceae. **Willem H. A. Hekking**, Violaceae. **Paul Hiepko**, Olacaceae. **Michel Hoff**, Turneraceae; database designer. [Photo by C. Hoff] **Noel H. Holmgren**, Scrophulariaceae. **Bruce Holst**, Myrtaceae. [Photo by M. Dean] **Hugh H. Iltis**, Capparaceae. **Arian Jacobs-Brouwer**, Polygalaceae. [Photo by A. Jacobs] **Marion Jansen-Jacobs**, Verbenaceae. **Emmet Judziewicz**, Poaceae. [Photo by L. Clark] **Jacquelyn A. Kallunki**, Rutaceae. **Maria Lúcia Kawasaki**, Combretaceae, Lythraceae, Myrtaceae, Onagraceae, Thymelaeaceae. [Photo by M. Amaral] **Job Kuijt**, Loranthaceae, Viscaceae.

Plate 4. Left—Right, Top—Bottom: **Beat Ernst Leuenberger**, Cactaceae. [Photo by S. Arroyo-Leuenberger] **Lúcia G. Lohmann**, Bignoniaceae. **James L. Luteyn**, Ericaceae. **Hilte Maas**, Annonaceae, Burmanniaceae, Cannaceae, Costaceae, Gentianaceae, Haemodoraceae, Strelitziaceae, Triuridaceae, Zingiberaceae. **Paul Maas**, Annonaceae, Burmanniaceae, Cannaceae, Costaceae, Gentianaceae, Haemodoraceae, Strelitziaceae, Triuridaceae, Zingiberaceae. **Santiago Madriñán**, *Rhodostemonodaphne* (Lauraceae). **Shirley L. Maina**, Amaranthaceae, Caryophyllaceae, Nyctaginaceae, Phytolaccaceae. **John D. Mitchell**, Anacardiaceae, Brassicaceae, Celastraceae, Monimiaceae, Myristicaceae, Oxalidaceae, Rosaceae, Smilacaceae, Violaceae. **Scott A. Mori**, Apiaceae, Araliaceae, Bignoniaceae, Bombacaceae, Canellaceae, Capparaceae, Caricaceae, Dilleniaceae, Ebenaceae, Flacourtiaceae, Heliconiaceae, Hernandiaceae, Icacinaceae, Lacistemataceae, Lecythidaceae, Nymphaeaceae, Quiinaceae, Rafflesiaceae, Rhabdodendraceae, Rhizophoraceae, Styracaceae, Symplocaceae, Thymelaeaceae, Trigoniaceae, Vochysiaceae. **Gilberto Morillo**, Asclepiadaceae.

Michael Nee, Cucurbitaceae, Solanaceae. **Terence D. Pennington**, Meliaceae, Sapotaceae. **Thomas Philbrick**, Podostemaceae. **John J. Pipoly III**, Clusiaceae, Myrsinaceae. **Vanessa Plana**, Proteaceae. [Photo by A. McRobb] **Marcel-Marie Plumel**, *Himatanthus* (Apocynaceae). [Photo by M. Chalopin]

Plate 5. Left—Right, Top—Bottom: **Odile Poncy**, Aristolochiaceae, *Inga* (Mimosaceae). **Ghillean T. Prance**, Chrysobalanaceae, Dichapetalaceae. **John F. Pruski**, Asteraceae. **Susanne Renner**, Melastomataceae. [Photo by R. Ricklefs] **Adrian C. de Roon**, Icacinaceae, Marcgraviaceae. **Daniel Sabatier**, Hugoniaceae, Humiriaceae. [Photo by G. Sabatier] **Claude Sastre**, Ochnaceae. [Photo by C. Sastre] **Laurence E. Skog**, Gesneriaceae. **Bertil Ståhl**, Theophrastaceae. **Dennis Stevenson**, Gnetaceae. **Wm. Wayt Thomas**, Cyperaceae, Simaroubaceae. **Alberto Vicentini**, Elaeocarpaceae. **Bruno Wallnöfer**, Ebenaceae. [Photo by S. Mori] **Dieter C. Wasshausen**, Acanthaceae, Mendonciaceae. **Henk van der Werff**, Lauraceae. [Photo by T. van der Werff] **John Wurdack**, Melastomataceae.

Plate 6. Some Saül Collectors. a. May 1992 French Guiana expedition members at Eaux Claires, Saül. Left—Right, front: Eric Gouda, Carol Gracie, Patti Anderson; back: Scott Mori, Bobbi Angell, Ara Görts-van Rijn, Pedro Acevedo-Rodríguez. **b. June 1988 Flora Neotropica field trip members at ORSTOM field station, Saül**. Left—Right, front: Scott Mori, Iain Prance, Ray Harley, Henrik Balslev; back: Jean-Jacques de Granville, Daniel Katz, Paulo Windisch, Stephan Beck, Al Gentry. **c. February 1993 French Guiana expedition members at Eaux Claires, Saül**. Left—Right, front: Carol Gracie, Bernard Jardin, Frieda Billiet, Jean-Jacques de Granville; middle: Tom Croat, Hiltje Maas, Scott Mori, Georges Cremers; back: Paul Maas, David Read, Terry Pennington. **d. September 1995 at Eaux Claires, Saül**. Scott Mori and Carol Gracie. [Photo by S. Heald] **e. August 1993 at Eaux Claires, Saül**. Scott Mori, Henk van der Werff, and Michael Rothman. **f. February 1993 at Eaux Claires, Saül**. Paul and Hiltje Maas. **g. June 1988 on Mt. Fumée, Saül**. Stephan Beck, Al Gentry, and Scott Mori. **h. August 1987 in forest near Saül**. Scott Mori collecting specimen of *Cordia alliodora*. **i. February 1992 in forest near Saül**. Jette Knudsen and Bertil Ståhl.

Plate 7. Preparing Specimens and Cayenne Support. a. September 1993 in forest near Saül. Scott Mori pressing plants. **b. February 1993 in forest near Saül**. Terry Pennington pressing plants. **c. September 1993 in forest near Saül**. Carol Gracie pressing plants. **d. May 1992 in forest near Saül**. Bobbi Angell sketching. **e. September 1994 at Eaux Claires, Saül**. Scott Mori and Brian Boom preparing plant press for drying. **f. October 1991 at Eaux Claires, Saül**. Eric Christenson preparing orchid collections. **g. September 1994 at Eaux Claires, Saül**. Scott Mori, Ryan Fowler, and Rob Jones carrying plant collections to airstrip. **h. August 1994 at Emerald Village, Matoury, near Cayenne**. Marjke, Bernie, and Joep Moonen. **i. September 1995 at ORSTOM herbarium in Cayenne**. Georges Cremers.

Plate 8. Scenes and People from Saül. a. August 1993 in village of Saül. Eglise St.-Antoine-de-Padoue, built in 1952. **b. June 1986 in village of Saül**. I.R.D. (ex ORSTOM) field station. **c. August 1987 in village of Saül**. Michel Modde (center). **d. August 1993 at Eaux Claires, Saül**. Carbet Manderley on Crique a l'Est. **e. August 1993 at Eaux Claires**. Allinckx family. Left—Right: Andy Allinckx, Jean-Yves "Le Gitan" (friend of family), Betty Allinckx Bourgeot holding two-toed sloth, Ghislaine Allinckx, Michel Combes, Yvan Allinckx with daughter, Mandy in front, and Marie-Claude Allinckx holding daughter, Cindy. **f. May 1992 at Eaux Claires**. Stream at Eaux Claires. **g. September 1994 at Eaux Claires**. Carbet with hammocks. **h. August 1993 at Eaux Claires**. Ghislaine Allinckx with homemade bread. **i. May 1992 at Eaux Claires**. Yvan Allinckx helping to prepare dinner.

CONTRIBUTORS

Pedro Acevedo-Rodríguez

Lucile Allorge

William S. Alverson

Patti Anderson

William R. Anderson

Lennart Andersson

Bobbi Angell

Scott Armbruster

Daniel Austin

Rubert Barneby

Cornelis C. Berg

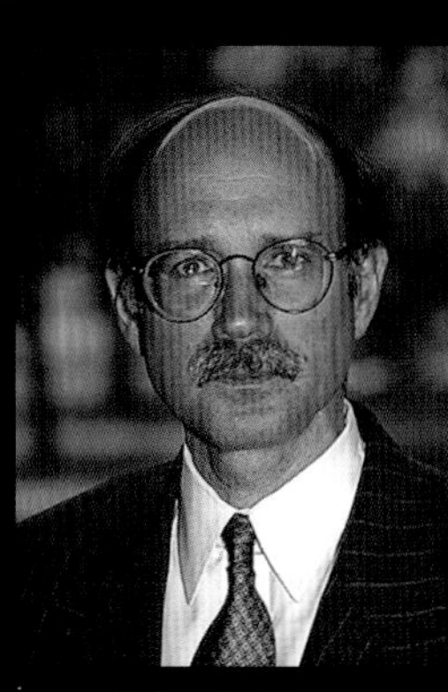
Brian M. Boom

Bruno Bordenave

John Brandbyge

John L. Brown

Eric A. Christenson

CONTRIBUTORS

Georges Cremers

Thomas B. Croat

Douglas C. Daly

Robert A. DeFilipps

Piero G. Delprete

Laurence Dorr

Christian Feuillet

Beat Fisher

Enrique Forero

Peter W. Fritsch

Lynn J. Gillespie

Ara Görts-van Rijn

Eric Gouda

Carol A. Gracie

Jean-Jacques de Granville

James W. Grimes

Plate 2

CONTRIBUTORS

Mats H. G. Gustafsson

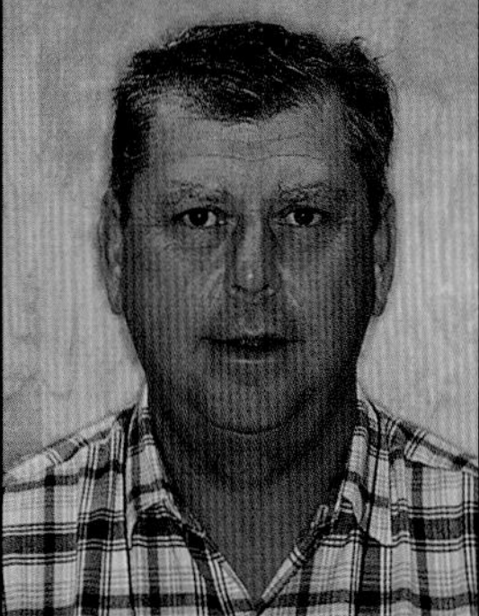
Bruce Hansen

Raymond Harley

Scott V. Heald

Willem H. A. Hekking

Paul Hiepko

Michel Hoff

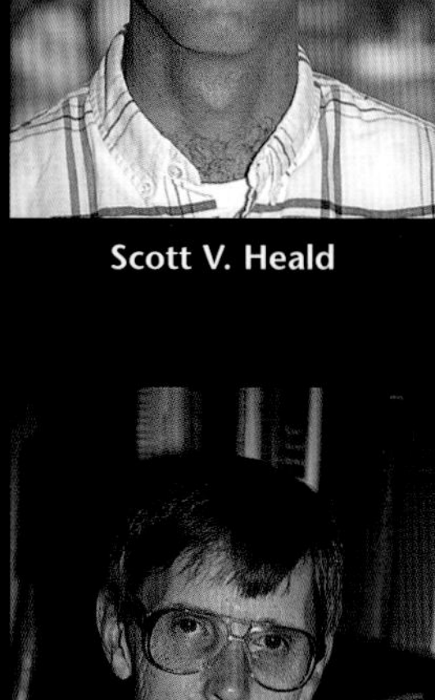
Noel H. Holmgren

Bruce Holst

Hugh H. Iltis

Arian Jacobs-Brouwer

Marion Jansen-Jacobs

Emmet Judziewicz

Jacquelyn A. Kallunki

Maria Lúcia Kawasaki

Job Kuijt

CONTRIBUTORS

Beat Ernst Leuenberger

Lúcia G. Lohmann

James L. Luteyn

Hilte Maas

Paul Maas

Santiago Madriñán

Shirley L. Maina

John D. Mitchell

Scott A. Mori

Gilberto Morillo

Michael Nee

Terence D. Pennington

Thomas Philbrick

John J. Pipoly III

Vanessa Plana

Marcel-Marie Plumel

CONTRIBUTORS

Odile Poncy

Ghillean T. Prance

John F. Pruski

Susanne Renner

Adrian C. de Roon

Daniel Sabatier

Claude Sastre

Laurence E. Skog

Bertil Stahl

Dennis Stevenson

Wm. Wayt Thomas

Alberto Vicentini

Bruno Wallnöfer

Dieter C. Wasshausen

Henk van der Werff

John Wurdack

SOME SAÜL COLLECTORS

a.

b.

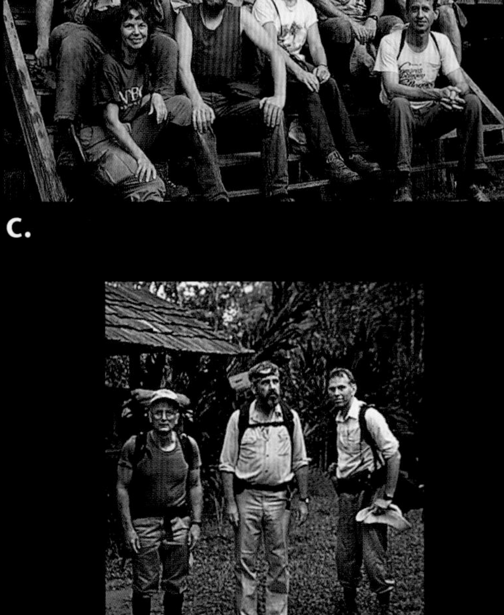

c.

d.

e.

f.

g.

h.

i.

Plate 6

PREPARING SPECIMENS AND CAYENNE SUPPORT

Plate 7

SCENES AND PEOPLE FROM SAÜL

a.

b.

c.

d.

e.

f.

h.

g.

i.

Plate 8

- *Macfadyena unguis-cati* (Bignoniaceae). The abundant, brilliant yellow flowers of this liana with opposite, compound leaves are especially conspicuous because they are present before new leaves have expanded. The juvenile leaves of this species are much smaller than the adult leaves (Gentry, 1993).
- *Roupala montana* (Proteaceae). The alternate leaves of this species are pinnately compound when juvenile and simple when adult.

Trees with Simple, Opposite Leaves

- *Sloanea* spp. (Elaeocarpaceae). Species of this genus with pulvinate-geniculate petioles and irregular trunks may have various combinations of opposite and alternate leaves, e.g., *S. guianensis* has opposite leaves, *S. brevipes* and *Sloanea* sp. A have opposite to subopposite leaves, and *S.* cf. *multiflora* and *S. tuerckheimii* have subopposite to alternate leaves.

Shrubs with Prickles on Leaves

- *Urera laciniata* (Urticaceae). This shrub or treelet possesses alternate, simple, pinnately lobed leaves with long, pointed prickles and stinging hairs on the stems, petioles, and leaf blades.

Trees, Shrubs, and Lianas with Cavities on the Leaf Blades

- *Annona muricata* (Annonaceae). This small tree possesses alternate, simple leaves. The domatia are pocket-shaped hairy pits in secondary vein axils on the abaxial leaf blade surface.
- *Arrabidaea triplinervia* (Bignoniaceae). This liana possesses opposite, compound leaves with simple tendrils derived from the terminal leaflet. The domatia, when present, are located in the axils of the principal veins.
- Lauraceae. A few species of this family (e.g., *Cinnamomum triplinerve*, *Ocotea cinerea*, *O. diffusa*, and *O. subterminalis*) possess slight depressions surrounded by tufts of hairs in the axils of the secondary veins abaxially that suggest domatia.
- *Miconia mirabilis* and *M. sastrei* (Melastomataceae). This tree and shrub, respectively, possess opposite, simple leaves with arcuate primary veins and ladder-like tertiary veins. The domatia-like cavities are located in the axils of the primary veins.
- *Tabebuia serratifolia* (Bignoniaceae). This large tree with opposite, palmately compound leaves possesses abaxial cavities lined by simple hairs in the axils of the secondary veins of the leaflets.

Trees, Shrubs, and Lianas with Glands on the Leaves

- *Aegiphila* spp. (Verbenaceae). Species of this genus of scandent shrubs or lianas with opposite, simple leaves often possess scattered, disc-shaped glands on the abaxial leaf blade surface. The glands are more concentrated near the base and the midrib. They are less conspicuous and perhaps even absent in species with pubescent leaves.
- Bignoniaceae. In addition to glands on the petioles (e.g., *Paragonia pyramidata*) and gland fields between the petioles (especially common in *Arrabidaea* spp. and *Mansoa* spp.), species of lianaceous Bignoniaceae often have glands on the abaxial leaf blade surface, especially toward the base (e.g., *Arrabidaea candicans*, *Schlegelia* spp., and numerous other species). Glands are also found on the pseudostipules of some species (e.g., see Plate IXc of *Cydista aequinoctialis* in part 1).
- *Barnhartia floribunda* (Polygalaceae). This liana with alternate, simple leaves possesses two circular glands at the junction of the petiole and the blade.
- *Bixa orellana* (Bixaceae). This cultivated tree with alternate, simple leaves often possesses glands at the apex of the petiole.
- *Bunchosia decussiflora* (Malpighiaceae). This small tree with opposite, simple leaves has a pair of abaxial yellow glands at the leaf blade base and two small stipules inserted on the adaxial petiole surface. In part 1, this species was identified as *Bunchosia* sp. 1.
- *Cayaponia* spp. (Cucurbitaceae). Species of this genus of lianas with alternate, simple leaves usually possess disc-shaped glands at the base of the blades abaxially.
- *Ficus* spp. (Moraceae). Species of *Ficus* (figs) possess alternate, simple leaves, pronounced stipular scars, and exude white, sometimes caustic, latex when their stems and trunks are cut. Many are hemiepiphytic trees or shrubs, some of which are potential stranglers, and others are free-standing trees. All species have glands at various positions at the base of their leaf blades. These glands have a waxy surface in fresh material. As the leaves become older, the glands may appear only as darkened spots.
- *Hasseltia floribunda* (Flacourtiaceae). This tree with alternate, simple leaves possesses 2 sessile glands on the adaxial leaf surface near the petiole.
- *Ipomoea batatoides* (Convolvulaceae). This liana with alternate, simple leaves possesses microscopic glands on the leaf blade abaxially.
- *Mandevilla rugellosa* (Apocynaceae). This liana with opposite, simple leaves and white latex often has glands along the adaxial surface of the midrib, especially toward the base of the blade. Other members of the genus possess similar glands.
- *Maprounea guianensis* (Euphorbiaceae). This species of tree with alternate, simple leaves usually possesses flat glands on each side of the midrib on the abaxial surface near the base.
- *Maripa glabra* (Convolvulaceae). This liana with alternate, simple leaves possesses numerous sunken oil glands on the leaf blade surface adaxially, especially near the base.
- *Moutabea guianensis* (Polygalaceae). This species of liana with alternate, simple leaves possesses two circular glands on the stem at the base of the petiole and circular glands on the leaf blade abaxially.

- *Polygala* spp. (Polygalaceae). The stems near the base of the petioles of this genus often possess triangular to rhombic or tubular glands.
- *Pothomorphe peltata* (Piperaceae). This small shrub of disturbed areas with distinctive alternate, simple, peltate leaves has scattered glands on the sheathing petiole base (Fig. 254).
- *Securidaca pubescens* and *S. volubilis* (Polygalaceae). These lianas with alternate, simple leaves possess glands at the base of the petiole which are probably stipular in origin. *Securidaca volubilis* was misidentified as *S. diversifolia* on page 38 in Part 1.
- *Tabebuia serratifolia* (Bignoniaceae). This tree with opposite, palmately compound leaves possesses glands on the petioles (Thomas & Dave, 1992) and leaflet blades.

Trees, Shrubs, and Lianas with Peltate Scales on Leaf Blades

- Bignoniaceae. Many species in this family possess peltate scales on their leaf blades. Two examples are *Tabebuia serratifolia*, a tree with opposite, palmately compound leaves, and *Arrabidaea patellifera*, a liana with opposite, bifoliolate leaves.
- *Miconia chrysophylla* and *M. plukenetii* (Melastomataceae). These shrubs or small trees possess opposite, simple leaves with lepidote/stellate hairs, especially on the abaxial leaf blade surface.
- *Rhabdodendron amazonicum* (Rhabdodendraceae). This small tree with alternate, simple leaves has small, scarcely discernable pellucid punctations in the leaf blades, and fruits subtended by a swollen red cupule. The small, scattered peltate scales, found on both leaf surfaces, are seen best with 20× magnification.

Trees and Shrubs with Stellate and Dendritic Hairs on Leaf Blades

It is sometimes difficult to distinguish between stellate and dendritic hairs because they often intergrade (Webster et al., 1996).

- *Clidemia dentata*, *C. octona* subsp. *guyanensis*, and *C. laevifolia* (Melastomataceae). These species of shrubs possess opposite, simple leaves. The stellate hairs are scattered among simple pilose or hispid hairs on the stems, leaves (more common on the abaxial surface) and hypanthia. Other species of *Clidemia* in our area may be pubescent but they lack stellate hairs.
- *Cordia alliodora* (Boraginaceae). This tree, common in disturbed habitats, possesses alternate, simple leaves, swollen ant-inhabited nodes (Plate XIa, b in Part 1), and sparse to dense stellate pubescence on the stems and leaves.
- *Henriettea ramiflora* and *H. succosa* (Melastomataceae). These trees with opposite, simple leaves of disturbed habitats are variously covered with stellate hairs, especially on their abaxial leaf surfaces. The central ray of the hairs is more strongly developed than the radial rays.
- *Mabea salicoides* and *M. subsessilis* (Euphorbiaceae). These trees have abundant, free-flowing white latex and alternate, simple leaves with reddish-brown dendritic hairs on the petioles, midrib, and sometimes scattered over the abaxial leaf blade surfaces.
- *Miconia acuminata*, *M. alata*, *M. argyrophylla*, *M. chrysophylla*, *M. diaphanea*, *M. dispar*, *M. holosericea*, *M. eriodonta*, *M. gratissima*, *M. mirabilis*, *M. plukenetii*, *M. sastrei*, *M. tomentosa*, and *M. trailii* (Melastomataceae). These species of shrubs or small trees have opposite, simple leaves with stellate, stellate/dendritic, or stellate/lepidote hairs, especially on the abaxial leaf blade surface. An especially striking pubescence is that of the stalked stellate hairs of *M. diaphanea*. The hairs of *M. chrysophylla* and *M. plukenetii* are more lepidote than stellate. The pubescence of *M. sastrei* is dominated by simple, spreading hairs. However, smaller stellate/dendritic hairs are scattered among the simple hairs. Other species in our area may be either glabrous or pubescent with simple hairs
- *Minquartia guianensis* (Olacaceae). This common tree species, with a fenestrate trunk, possesses alternate, simple leaves. The olive-like fruits have a small amount of white latex and the seeds have a ruminate endosperm. Because of its rot-resistent wood, it is widely used as posts for telephone lines and in construction. The stellate, sometimes ferruginous, caducous hairs are found mostly on the stems, petioles, and inflorescence rachises and less frequently on the leaf blades.
- *Ryania speciosa* (Flacourtiaceae). This small tree with alternate, simple leaves has stellate hairs on the midrib abaxially.
- *Tabebuia capitata* and *T. serratifolia* (Bignoniaceae). These large trees have opposite, palmately compound leaves. The leaves possess stellate hairs mostly when they are young (Gentry & Morawitz, 1992) so they are difficult to find on most collections.
- *Trichilia euneura* (Meliaceae). The alternate, pinnately compound leaves of this understory tree posseses minute stellate hairs on the leaflet blades abaxially.
- *Virola kwatae* (Myristicaceae). This is a large, buttressed tree with red exudate and alternate, simple leaves. This species possesses a mixture of stellate and dendritic trichomes that are caducous. The adult leaves are mostly glabrous in contrast to the persistently stellate pubescent leaves of the morphologically similar *V. michelii* (see page 41, Part 1).
- *Vismia sandwithii*, *V. guianensis*, *V. latifolia*, and *V. glaziovii* (Clusiaceae). These small to medium-sized trees have colorful (often orange) exudate and opposite, simple leaves. The stellate hairs are most abundant on the abaxial leaf blade surface; those of *V. sandwithii* are stalked.

Trees, Shrubs, and Lianas with Punctations/Striations on Leaf Blades

- *Rhabdodendron amazonicum* (Rhabdodendraceae). This small tree with alternate, simple leaves and small, scattered peltate scales has fruits subtended by a swollen red cupule. The pellucid punctations are small and scarcely discernable.
- *Stizophyllum* spp. (Bignoniaceae). Species of this genus of lianas have hollow stems and opposite, 2-3-foliolate leaves. The latter are pellucid-punctate. This is especially apparent in *S. riparium*.

Inflorescences and Flowers

Trees, Shrubs, and Lianas with Glands on Inflorescences and Flowers

- *Aparisthmium cordatum* (Euphorbiaceae). This species of tree with alternate, simple, cordate leaves possesses biglandular bracts in the pistillate inflorescences.
- Bignoniaceae. Many species of lianaceous Bignonaceae have glands on the calyx (e.g., *Adenocalymma saülense*, *Anemopaegma* spp., *Arrabidaea nigrescens*, *Clytostoma binatum*, *Cydista aequinoctialis*, *Distictella elongata*, *Memora* spp., *Paragonia pyramidata*, *Schlegelia* spp., *Stizophyllum* spp., *Tanaecium nocturnum*) and some even possess glands on the corolla (e.g., *Anemopaegma ionanthum*, *Parabignonia steyermarkii* and *Tanaecium nocturnum*). The glands are not always present on all flowers of a given species.

Fruits

Trees, Shrubs, and Lianas with Fruits Having Unusual Surfaces

- *Clytostoma binatum* (Bignoniaceae). The suborbiculate capsules of this liana with opposite, 2-foliolate leaves are distinctly echinate. The subulate pseudostipules in miniature, bromeliad-like, axillary clusters are characteristic of this species.
- *Pithecoctenium crucigerum* (Bignoniaceae). The thick, woody capsules of this liana with opposite, 2(3)-foliolate leaves are distinctly echinate. The tendrils, which are divided >3 times are unique among the Bignoniaceae of our area.
- *Tynanthus pubescens* (Bignoniaceae). The linear capsules of this liana with opposite, 3-foliolate leaves are distinctly winged. Species of this genus often have a strong clove aroma vegetatively (Gentry, 1992).

Seeds

Trees and Lianas with Winged Seeds

- *Alseis longifolia* (Rubiaceae). The seeds of this tree with fenestrate trunks and opposite, simple leaves are peltate with bilobed wings that are tapered at the ends.

Lianas

Special Adaptations for Climbing

- *Entada polyphylla* (Mimosaceae). The rachis of the alternate, bipinnately compound leaves of this liana is sometimes modified into a tendril (Keller, 1996). This feature is seldom represented in herbarium specimens.
- *Macfadyena unguis-cati* (Bignoniaceae). Juvenile plants of this species climb appressed to tree trunks with the aid of adventitious rootlets. The juvenile leaves are much smaller than the adult leaves (Gentry, 1993). This species also possesses trifid, cat-claw like tendrils and brilliant yellow flowers.
- *Pacouria guianensis* (Apocynaceae). The inflorescences of this liana with opposite, simple leaves and white latex are sometimes modified into tendrils.
- *Parabignonia steyermarkii* (Bignoniaceae). Juvenile plants of this species with opposite leaves climb appressed to tree trunks with the aid of adventitious rootlets. The first leaves are simple with apical teeth; the later ones are bifoliolate with entire leaflets.

Distinctive Aromas

- *Mansoa alliacea* (Bignoniaceae). The vegetative parts of this liana with opposite, compound leaves emit a garlic-like aroma when cut or crushed.
- *Tynanthus polyanthus* and *T. pubescens* (Bignoniaceae). The vegetative parts of species of *Tynanthus* usually emit an aroma of cloves when cut or crushed (Gentry, 1992). The flowers of *T. polyanthus* are much smaller than those of *T. pubescens*.

Notes on Descriptions

Only those species vouchered with specimens are included in the *Guide* unless it is clearly indicated that the species has not been collected from our area but is to be expected from central French Guiana.

At the end of family and generic descriptions we have added references to the *Flora of Venezuelan Guayana*, the *Flora of the Guianas*, *Flora Neotropica*, and the *Flora of Suriname* if the taxon has been treated in one of these Floras. Additional references were added if they are useful in helping to identify plants of central French Guiana. These references should be viewed only as leads into the literature and not as references from which the descriptions were crafted. The *Checklist of the Plants of the Guianas* (Boggan et al., 1997) is useful in helping to determine potential names for species not included in this guide and in the *Flora of the Reserva Ducke* (Ribeiro et al., 1999) of the central Brazilian Amazon provides many illustrations of species representing families and genera also found in French Guiana.

An attempt was made to base most of the descriptions on specimens from within the area of the guide. However, when species are known only from sterile collections in our area the descriptions of flowers and fruits, if provided, are based from specimens gathered from outside the area of the guide. For a local Flora, the goal should be complete descriptions from specimens gathered from within the area under consideration, but this goal, especially in tropical areas, is seldom achieved. Because so many different authors crafted our descriptions, it is often not possible to determine what specimens were used in making the descriptions.

Every attempt was made to illustrate species with collections from central French Guiana, but, when that was not possible, the origin of the collections is indicated in the legends.

Many characters used in our keys are not preserved in herbarium specimens. For example, plant habit, color of exudate, aroma of cut bark, color of fruit, etc. Therefore, collectors must be careful to include all features of a plant lost in collecting for inclusion on their labels.

We emphasize that our phenological indications provide only the months in which fertile specimens have been collected. In order to avoid confusion, each month is listed separately even if fertile specimens have been collected in all months of a range, e.g., fl (Aug, Sep, Oct) not fl (Aug–Oct). Ecological notes only provide approximations of where species might be encountered in our area because label information from different collectors does not provide standard and reliable descriptions of where the plants were collected.

Group V. Magnoliopsida (dicotyledons)

The Magnoliopsida are a class within the phylum Magnoliophyta (the flowering plants). The dicots are distinguished from the other class of flowering plants, the Liliopsida (monocotyledons), by the features listed in the key to groups (Part 1, page 56). Most of the trees (other than palms), shrubs, and lianas, i.e., the woody plants, in our flora belong to the Magnoliopsida. This group possesses approximately 3.5 times as many species as the Liliopsida. The Liliopsida were treated in Part 1 of this guide along with the pteridophytes and the single gymnosperm.

It should be noted that recent molecular data places the Liliopsida within the Magnoliopsida (Angiosperm Phylogeny Group, 1998).

Key to Keys of Magnoliopsida

1. Plants achlorophyllous (saprophytes and parasites; in our area the parasitic Rafflesiaceae produce flowers from the trunk of Flacourtiaceae in our area, which are easily misinterpreted as flowers of the host plant. . . . . *Key 1. Achlorophyllous Plants* (p. 10).
1. Plants with chlorophyll (including hemiparasites).
    2. Plants herbaceous.
        3. Plants not growing on or climbing into trees. . . . . . . *Key 2. Herbaceous Terrestrial or Aquatic Plants* (p. 11).
        3. Plants growing on or climbing into trees.
            4. Plants climbing (vines), rooted in ground. . . . . *Key 3. Climbing Plants* (p. 14).
            4. Plants not climbing, without contact with ground or with contact with ground for only part of life cycle (hemiepiphytes). . . . . *Key 4. Epiphytic Plants* (p. 17).
    2. Plants woody.
        5. Plants with climbing stems (lianas). . . . . *Key 3. Climbing Plants* (p. 14).
        5. Plants without climbing stems.
            6. Plants epiphytic (including hemiepiphytes and hemiparasites). . . . . *Key 4. Epiphytic Plants* (p. 17).
            6. Plants not epiphytic, hemiepiphytic, or hemiparasitic.
                7. Trees (single-stemmed and ≥5 cm diam.) or treelets (single-stemmed, <5 cm diam. and >2 m tall). . . . . *Key 5. Trees* (p. 18).
                7. Shrubs (multiple stemmed or single stemmed and <5 cm diam. and <2 m tall). . . . *Key 6. Shrubs* (p. 27).

Key 1. Achlorophyllous Plants

1. Plants parasitic, growing within and flowering on trunks of Flacourtiaceae in our area. . . . . *Rafflesiaceae*.
1. Plants terrestrial, parasitic or saprophytic, never growing on tree trunks, sometimes growing on tree roots.
    2. Plants mushroom-like. Scale-like leaves alternate. Flowers in globose capitula; petals not fused into tube; ovaries inferior. Fruits 1-seeded achenes. . . . . *Balanophoraceae*.
    2. Plants not mushroom-like. Scale-like leaves opposite. Flowers not in globose capitula; petals fused into tube; ovaries superior. Fruits multi-seeded capsules. . . . . *Voyria* & *Voyriella* (*Gentianaceae*).

Key 2. Herbaceous Terrestrial or Aquatic Plants Key to Subkeys

If your plant does not key out here and is woody at the base, try the key to shrubs, Key 6 (p. 27).

1. Leaves compound. . . . . *Subkey 1. Compound Leaves* (p. 11).
1. Leaves simple.
   2. Leaves opposite or whorled. . . . . *Subkey 2. Simple Opposite or Whorled Leaves* (p. 11).
   2. Leaves alternate. . . . . *Subkey 3. Simple Alternate Leaves* (p. 12).

Key 2. Herbaceous Terrestrial or Aquatic Plants

Subkey 1. Compound Leaves

1. Leaves 3-foliolate. Stamens not tetradynamous.
   2. Stipules present. Flowers zygomorphic. Fruits specialized legumes (loments), i.e., contracted between seeds. . . . . *Desmodium* (*Fabaceae*).
   2. Stipules absent. Flowers actinomorphic. Fruits capsules. . . . . *Oxalidaceae.*
1. Leaves pinnately divided. Stamens tetradynamous. . . . . *Brassicaceae.*

Key 2. Herbaceous Terrestrial or Aquatic Plants

Subkey 2. Simple Opposite or Whorled Leaves

1. Leaves whorled. Flowers in dense spikes, apetalous. . . . . *Peperomia* (*Piperaceae*).
1. Leaves opposite. Flowers not in dense spikes, petals present or absent.
   2. Plants with milky exudate.
      3. Leaves <2 cm long. Flowers unisexual, numerous, congested in leaf axils, in specialized cups called cyathia; peduncle not developed; ovary with 3 carpels. . . . . *Euphorbia*, in part (*Euphorbiaceae*).
      3. Leaves >2 cm long. Flowers bisexual, solitary or, if numerous, not congested, not in cyathia; peduncule developed; ovary with 2 carpels.
         4. Leaves <5 cm long, rounded-apiculate at apex. Corolla fused into long tube; stamens and pistil not fused into column; pollen not in pollinia. . . . . *Apocynaceae.*
         4. Leaves >5 cm long, long attenuate, not apiculate at apex. Corolla not fused into long tube; stamens and pistil fused into column; pollen in pollinia. . . . . *Asclepiadaceae.*
   2. Plants without milky exudate.
      5. Flowers with petals not fused into distinct tube.
         6. Leaves mostly in basal rosette, with pungent aroma when crushed, the margins markedly serrate. Inflorescences cylindric, subtended by whorled bracts. . . . . *Eryngium* (*Apiaceae*).
         6. Leaves not in basal rosette, without pungent aroma when crushed, the margins not markedly serrate. Inflorescences not cylindric nor subtended by whorled bracts.
            7. Leaf blades often orbicular, cordate, or reniform. Placentation free-central. . . . . *Drymaria* (*Caryophyllaceae*).
            7. Leaf blades not orbicular, cordate, or reniform. Placentation not free-central.
               8. Perianth biseriate; petals conspicuous; ovary with >1 ovule.
                  9. Leaf blades with arcuate secondary veins and characteristic ladder-like (scalariform) tertiary veins between the secondary veins. Flowers actinomorphic. . . . . . *Melastomataceae.*
                  9. Leaf blades without venation as described above. Flowers zygomorphic.
                     10. Leaves often heteromorphic, one leaf of node much larger than other. Flowers perigynous; petals magenta; placentation axile. . . . . *Lythraceae.*
                     10. Leaves not heteromorphic. Flowers hypogynous; petals not magenta; placentation parietal. . . . . *Hybanthus* (*Violaceae*).
               8. Perianth uniseriate; tepals inconspicuous; ovary with 1 ovule.
                  11. Leaves without stipules. Flowers bisexual, subtended by scarious bracts; filaments at least partially connate; pollen not released explosively at anthesis. . . . . . . *Amaranthaceae.*
                  11. Leaves with stipules. Flowers unisexual, not subtended by scarious bracts; filaments not connate; pollen released explosively at anthesis. . . . . *Laportea* & *Pilea* (*Urticaceae*).

5. Flowers with petals fused into distinct tube (the lobes of the corolla may sometimes be longer than the tube, e.g., *Coutoubea* in the Gentianaceae).

12. Flowers mostly actinomorphic (zygmorphic flowers may occur at periphery of inflorescence in the Asteraceae).

13. Flowers congested into heads subtended by numerous, overlapping bracts (phyllaries); calyx modified into hairs, scales, or bristles (pappus); ovary inferior. Fruits cypselas. . . . . *Asteraceae.*

13. Flowers not congested into heads or, if in capitula, subtended by only two bracts (e.g., a few species of *Psychotria* that could be interpreted as herbaceous); calyx not modified into pappus; ovary superior or inferior. Fruits not cypselas.

14. Ovary inferior. . . . . *Rubiaceae.*

14. Ovary superior.

15. Leaves often in verticils of 3, the leaf blades glandular-punctate. Corolla with lobes longer than tube; stamens 4. . . . . *Scoparia* (*Scrophulariaceae*).

15. Leaves opposite, the leaf blades not glandular-punctate. Corolla with lobes shorter than tube; stamens 5.

16. Stipules or stipule scars present. Inflorescences with flowers conspicuously on one side of axes. Fruits either bilobed or rough on surface. . . . . *Mitreola* & *Spigelia* (*Loganiaceae*).

16. Stipules absent. Inflorescences not with flowers conspicuously on one side of axes. Fruits neither bilobed nor rough on surface. . . . . *Gentianaceae.*

12. Flowers all zygomorphic.

17. Crushed leaves often aromatic. Ovaries with single ovule per locule, the ovary often lobed at summit. Fruits often splitting into 1-seeded portions.

18. Flowers in capitula or, if not in capitula, the calyx with distinct protuberance; corolla markedly zygomorphic; style gynobasic. Fruits of 4 nutlets. . . . . *Lamiaceae.*

18. Flowers not in capitula and calyx without protuberance; corolla slightly zygomorphic; style terminal. Fruits not of 4 nutlets. . . . . *Verbenaceae.*

17. Crushed leaves not aromatic. Ovaries with 2 or more ovules per locule, the ovary never lobed at summit. Fruits never splitting into 1-seeded portions.

19. Leaves often with cystoliths on adaxial surface, these appearing as white streaks. Flowers often, but not always, subtended by conspicuous bracts. Fruits fiddle- or club-shaped. Seeds without endosperm. . . . . *Acanthaceae.*

19. Leaves without cystoliths. Flowers not subtended by conspicuous bracts. Fruits not fiddle- or club-shaped. Seeds with endosperm.

20. Leaf blades cordate, <15 mm long. Flowers purple, <10 mm long. . . . . . . *Scrophulariaceae.*

20. Leaf blades not cordate, >15 mm long. Flowers not purple, usually >10 mm long.

21. Leaf blade margins entire. Corollas white, green, or blue; stamens 5, the anthers not connivent. Plants of open, sunny areas. . . . . *Chelonanthus* (*Gentianaceae*).

21. Leaf blade margins dentate to serrate. Corollas white or yellow; stamens 4, the anthers connivent. Plants of shaded areas. . . . . *Gesneriaceae.*

Key 2. Herbaceous Terrestrial or Aquatic Plants

Subkey 3. Simple Alternate Leaves

1. Plants with white exudate.

2. Milky exudate sparse, not conspicuous from broken petiole. Flowers congested in heads, subtended by a series of overlapping bracts (involucre). Calyx modified into hairs. Fruits cypselas. . . . . . . *Emilia* (*Asteraceae*).

2. Milky exudate abundant, conspicuous from broken petiole. Flowers not congested in heads, not subtended by a series of overlapping bracts. Calyx not modified into hairs. Fruits berries or capsules.

3. Flowers bisexual, not in cyathium; corolla red or deep pink, with petals fused into tube; anthers connate; ovary inferior, bilocular. Fruits berries. . . . . *Campanulaceae.*

3. Flowers unisexual, situated within a cyathium; corolla not red or deep pink, the petals not fused into tube; anthers not connivent; ovary superior, trilocular. Fruits capsules. . . . . *Euphorbiaceae.*

1. Plants without white exudate.

4. Ovaries inferior.

5. Leaves without stipules. Flowers congested in capitula, subtended by a series of overlapping bracts (involucre). Calyx modified into hairs, bristles, or scales (pappus); corolla with petals fused into tube above ovary. Fruits cypselas, not winged. . . . . *Asteraceae*.
5. Leaves with stipules. Flowers not congested in capitula nor subtended by overlapping bracts. Calyx not modified into pappus; corolla with tepals free above ovary. Fruits capsules, winged. . . . . . . *Begoniaceae*.

4. Ovaries superior.
6. Plants aquatic.
7. Submersed leaves lettuce-like. Flowers covered by sac-like spathella. Plants often associated with rapidly flowing water. . . . . *Podostemaceae*.
7. Submersed leaves not lettuce-like. Flowers not covered by sac-like spathella. Plants associated with still water.
8. Cauline leaves simple, not lobed at base, the basal leaves pinnate. Flowers <10 mm diam.; sepals and petals distinct; petals 4; stamens 6. . . . . *Brassicaceae*.
8. All leaves simple, deeply lobed at base. Flowers >50 mm diam.; sepals and petals (tepals) not distinct; stamens numerous. . . . . *Nymphaeaceae*.
6. Plants not aquatic.
9. Flowers zygomorphic (weakly zygomorphic in *Petiveria*).
10. Roots absent. Leaves often provided with bladders. . . . . *Lentibulariaceae*.
10. Roots present. Leaves not provided with bladders.
11. Corolla with 4 tetals. Fruits with downward curved spines at apex, epizoochorous. . . . . *Petiveria* (*Phytolaccaceae*).
11. Corolla with 5 petals or lobes. Fruits without downward curved spines at apex, not epizoochorous.
12. Flowers with a spurred petal; ovary unilocular, with parietal placentation. . . . . *Noisettia* (*Violaceae*).
12. Flowers without a spurred petal; ovary bilocular, with axile placentation.
13. Leaf blades with entire margin. Corolla with petals free, usually violet. . . . . *Polygala* (*Polygalaceae*).
13. Leaf blades with serrate margin. Corolla fused at base, white. . . . . *Capraria* (*Scrophulariaceae*).
9. Flowers actinomorphic.
14. Petals fused into tube.
15. Inflorescences helicoid. Corolla with narrow tube. Fruits not surrounded by inflated calyx, splitting into 4 parts at maturity. . . . . *Boraginaceae*.
15. Inflorescences not helicoid. Corolla with broadly spreading tube. Fruits surrounded by inflated calyx, not splitting into 4 parts at maturity. . . . . *Physalis* (*Solanaceae*).
14. Petals not fused into tube.
16. Stipules persistent, distinctly ciliate, some cilia nearly as long as stipule. Staminodia dimorphic, the outer numerous, filiform, the inner 5, petaloid. . . . . . . *Sauvegesia* (*Ochnaceae*).
16. Stipules not as above. Staminodia absent or not as above.
17. Leaf blades oblong, <10 × 4 mm. Flowers solitary or clustered in leaf axils. . . . . *Phyllanthus* (*Euphorbiaceae*).
17. Leaf blades not oblong, >10 × 4 mm. Flowers in racemes or spikes.
18. Leaves with stellate hairs. Stamens fused into tube (monadelphous). Fruits splitting into individual units (mericarps), these with pointed apices, epizoochorous. . . . . *Pavonia* (*Malvaceae*).
18. Leaves glabrous or with simple hairs. Stamens not fused into tube. Fruits not as above.
19. Plants with stinging hairs. Leaf blades with markedly dentate margins. . . . . *Laportea* (*Urticaceae*).
19. Plants without stinging hairs. Leaf blades with entire margins.
20. Inflorescences spikes. Flowers apetalous; ovary unilocular. . . . . *Peperomia* (*Piperaceae*).
20. Inflorescences racemes. Flowers with perianth; ovary multilocular or unilocular.
21. Leaf blades obtuse to rounded at base; petioles longer than leaf blades. Tepals papery (scarious). Fruits dehiscent. . . . *Amaranthus* (*Amaranthaceae*).
21. Leaf blades acute or decurrent at base; petioles shorter than leaf blades. Tepals not papery. Fruits indehiscent. . . . . *Phytolaccaceae*.

Key 3. Climbing Plants

Key to Subkeys

Climbers are vines (herbaceous) and lianas (woody) that use other plants for support. Note, however, that climbers may sometimes be found in open areas, e.g., along roadsides, where they may be trailing over the ground rather than climbing into shrubs and trees.

1. Leaves compound. . . . . *Subkey 1. Compound Leaves* (p. 14).
1. Leaves simple.
   2. Leaves opposite. . . . . *Subkey 2. Simple Opposite Leaves* (p. 15).
   2. Leaves alternate. . . . . *Subkey 3. Simple Alternate Leaves* (p. 16).

Key 3. Climbing Plants

Subkey 1. Leaves Compound

One species, *Ipomoea quamoclit,* with simple but deeply pinnatifid leaves is included here.

1. Leaves opposite. . . . . *Bignoniaceae.*
1. Leaves alternate.
   2. Plants climbing with aid of tendrils or hooks (not always included on herbarium specimens).
      3. Plants climbing with aid of short, woody hooks, the stem often flattened and undulate. Leaves of 2 leaflets. Fruits long pods. . . . . *Bauhinia* (*Caesalpiniaceae*).
      3. Plants climbing with aid of tendrils, without woody hooks, the stem not flattened and undulate. Leaves with >2 leaflets. Fruits not long pods.
         4. Leaves bipinnate. Tendrils formed from extension of leaf rachis (seldom seen on herbarium specimens). . . . . *Entada* (*Mimosaceae*).
         4. Leaves pinnate, ternate, or palmate. Tendrils not formed from extension of leaf rachis.
            5. Tendrils opposite leaves. Inflorescences arising opposite leaves. Flowers <2 mm diam. Fruits with 1 seed. . . . . *Vitaceae.*
            5. Tendrils axillary or borne at sides of nodes. Inflorescences axillary. Flowers >2 mm diam. Fruits with >1 seed.
               6. Leaves pinnate or ternate; stipules present. Tendrils axillary, part of modified inflorescence. Flowers with separate petals, these often pubescent and with adaxial appendage. Fruits either with wings or dehiscent. Seeds not embedded in pulp, subtended by white aril. . . . . *Sapindaceae.*
               6. Leaves palmate; stipules absent. Tendrils borne at sides of nodes at right angles to leaf axil, modified branches. Flowers with fused petals, these usually not pubescent nor with adaxial appendage. Fruits not winged nor dehiscent. Seeds embedded in pulp, not subtended by white aril. . . . . *Cucurbitaceae.*
   2. Plants climbing without aid of tendrils or hooks.
      7. Leaves palmate. Plants often with white exudate.
         8. Plants with white exudate. Flowers unisexual; perianth not connate into tube. Fruits with one seed per locule. . . . . *Manihot* (*Euphorbiaceae*).
         8. Plants with or without white exudate. Flowers bisexual; perianth connate into tube. Fruits with >1 seed per locule. . . . . *Convolvulaceae.*
      7. Leaves pinnatifid or pinnate (sometimes reduced to 3 leaflets in some legumes, e.g., *Dioclea*) or bipinnate. Plants usually without white exudate.
         9. Plants armed with recurved prickles along stems or unarmed (*Entada*). Leaves bipinnate. Petals fused, the corolla tubular, white to yellow. . . . . *Mimosaceae.*
         9. Plant not armed or, if armed, the spines in position of stipules. Leaves not bipinnate. Petals free, the corolla not tubular or, if connate, the corolla bright red.
            10. Leaves pinnatifid, the ultimate leaf segments ca. 1 mm wide. Corolla bright red. Introduced species growing in village. . . . . *Ipomoea quamoclit* (*Convolvulaceae*).
            10. Leaves pinnate, the ultimate leaf segments >10 mm wide. Corolla not bright red. Native species.
               11. Leaves without stipules. Flowers actinomorphic, usually <5 mm diam; carpels, when more than one, separate. Fruits developing from one carpel, dehiscent along one side (follicles). Seeds usually black, subtended by yellow aril. . . . . *Connaraceae.*

11. Leaves with stipules. Flowers slightly to markedly zygomorphic, usually >5 mm diam.; carpels 1. Fruits dehiscent along both sides (legumes) or fruits indehiscent and winged. Seeds usually not as above.
    12. Flowers slightly zygomorphic, without pronounced standard and keel, the uppermost petal (standard) internal to lateral petals (best seen in bud); filaments not as below. . . . . *Caesalpiniaceae.*
    12. Flowers markedly zygomorphic, with pronounced standard and keel, the uppermost petal (standard) external to lateral petals (best seen in bud); filaments of 9 fused and 1 free stamen. . . . . *Fabaceae.*

Key 3. Climbing Plants

Subkey 2. Simple Opposite Leaves

1. Petals not fused together above ovary.
   2. Leaves usually with stipules, often with glands on petioles and/or abaxial leaf surface. Calyx with distinct glands; petals clawed. Fruits splitting into winged mericarps. . . . . *Malpighiaceae.*
   2. Leaves with (Trigoniaceae) or without stipules, without glands on petioles or blades. Calyx without distinct glands; petals not clawed. Fruits not winged mericarps (the seeds, however, may be winged in some species of Hippocrateaceae).
      3. Plants often growing tightly appressed to tree trunks, with adventitious roots at nodes (these often not included on herbarium specimens). Leaf blades with arcuate secondary veins and characteristic ladder-like (sclariform) tertiary veins between secondary veins. Anthers poricidal. . . . . *Adelobotrys* (*Melastomataceae*).
      3. Plants not growing tightly appressed to tree trunks, without adventitious roots at nodes. Leaf blades with pinnate secondary veins and reticulate tertiary veins. Anthers not poricidal.
         4. Leaves with small, often caducous stipules. Flowers zygomorphic. Fruits dehiscent, seeds not winged. . . . . *Trigoniaceae.*
         4. Leaves without stipules. Flowers actinomorphic. Fruits indehiscent, or, if dehiscent, the seeds with wings.
            5. Stamens 8 or 10 or more; stigma unlobed; ovary inferior, not surrounded by conspicuous, tightly appressed disc, unilocular, the ovules pendulous from apex of locule. Fruits conspicuously ridged. . . . . *Combretaceae.*
            5. Stamens 3; stigma various but often 3-lobed; ovary superior, surrounded by variously differentiated disc, trilocular, the placentation axile. Fruits not ridged. . . . . *Hippocrateaceae.*
1. Petals fused together, at least at base, around or above ovary.
   6. Plants with abundant, white exudate.
      7. Flowers with stamens strongly adnate to stigma; pollen aggregated in masses (pollinia); corolla tube scarcely developed, shorter than lobes, green or white; corona present. Seeds with white hairs. . . . . *Asclepiadaceae.*
      7. Flowers usually without stamens strongly adnate to stigma, but sometimes stamens weakly adnate to stigma; pollen not aggregated in masses; corolla tube well developed (the corolla itself may be small or large), longer than lobes, yellow or white; corona absent. Seeds either without hairs or, if hairs present, the hairs brown to golden-brown. . . . . *Apocynaceae.*
   6. Plants without colored exudate.
      8. Ovary inferior.
         9. Flowers aggregated into capitula, these subtended by a series of numerous overlapping bracts (involucre); calyx modified into bristles (pappus). Fruits cypselas. . . . . *Asteraceae.*
         9. Flowers not aggregated into capitula or, if in capitula, subtended by two non-overlapping bracts (e.g., species of *Psychotria*); calyx not modified. Fruits berries or capsules. . . . . *Rubiaceae.*
      8. Ovary superior.
         10. Plants with hooklike tendrils (these often not included on herbarium specimens). Flowers actinomorphic. . . . . *Strychnos* (*Loganiaceae*).
         10. Plants without hooklike tendrils. Flowers zygomorphic.
             11. Leaf blades rough to touch. Corollas slightly zygomorphic; calyx lavender to violet. Fruits with persistent, wing-like calyx lobes. . . . . *Petrea* (*Verbenaceae*).

11. Leaf blades smooth to touch. Corollas markedly zygomorphic; calyx green. Fruits without persistent, wing-like calyx lobes.
    12. Bracts red; corolla yellow with red streaks or spots on inside. ...... *Drymonia* (*Gesneriaceae*).
    12. Bracts, if present, green; corolla red, white, or pink, never yellow.
        13. Leaf blades without glands toward base of abaxial surface. Flower buds enclosed in two bracteoles, these often filled with liquid. Fruits with 1–2 seeds. .... *Mendonciaceae*.
        13. Leaf blades with glands toward base of abaxial surface. Flower buds not enclosed in two bracts. Fruits with numerous seeds. .... *Schlegelia* (*Bignoniaceae*).

Key 3. Climbing Plants

Subkey 3. Simple Alternate Leaves

1. Plants climbing with aid of tendrils or hooks.
    2. Plants climbing with aid of short, woody hooks, the stem often flattened and undulate. Fruits long pods. .... *Bauhinia* (*Caesalpiniaceae*).
    2. Plants climbing with aid of tendrils, without woody hooks, the stem not flattened and undulate. Fruits not long pods.
        3. Tendrils coiled in a single plane, shape of butterfly tongue, (not always present on herbarium specimens). Fruits dry, strongly 3-angled, nearly winged. .... *Gouania* (*Rhamnaceae*).
        3. Tendrils not coiled in a single plane. Fruits fleshy, not angled or winged.
            4. Tendrils opposite leaves. Flowers <2 mm diam. Fruits with a single seed. .... *Cissus* (*Vitaceae*).
            4. Tendrils not opposite leaves. Flowers >2 mm diam. Fruits with >1 seed.
                5. Tendrils axillary. Plants often with extrafloral nectaries. Flowers bisexual; petals free; ovary and stamens elevated on stalk (androgynophore); ovary superior. .... *Passifloraceae*.
                5. Tendrils at right angles to leaf axils. Plants without extrafloral nectaries. Flowers unisexual; petals fused; ovary and stamens not elevated on stalks; ovary inferior. .... *Cucurbitaceae*.
1. Plants climbing without aid of tendrils.
    6. Flowers zygomorphic.
        7. Leaves with secondary venation arcuate from base. Inflorescences from along stems well below persistent leaves. Flowers unusual, often large and smoking pipe-shaped; sepals fused into convoluted tube; petals absent; stamens joined into column; ovary inferior. .... *Aristolochiaceae*.
        7. Leaves with secondary venation pinnate. Inflorescences axillary or terminal. Flowers not unusual, not large or smoking pipe-shaped; sepals not fused into convoluted tube; petals present; stamens not joined into column; ovary superior.
            8. Lower petal extended into long spur; anthers with connective extended into conspicuous spur; ovary with parietal placentation. .... *Violaceae* (*Corynostylis* to be expected).
            8. Lower petal not extended into long spur; anthers without connective extended into conspicuous spur; ovary with axile placentation. .... *Polygalaceae*.
    6. Flowers actinomorphic (subtending bracts of Marcgraviaceae are zygomorphic but the flowers are actinomorphic).
        9. Leaves pinnatifid, the ultimate leaf segments ca. 1 mm wide. Corolla bright red. Introduced species growing in village. .... *Ipomoea quamoclit* (*Convolvulaceae*).
        9. Leaves not pinnatifid. Corolla not bright red. Native species.
            10. Flowers unisexual.
                11. Plants with armed stems. .... *Celtis* (*Ulmaceae*).
                11. Plants without armed stems.
                    12. Stipules modified into sheath (ocrea) surrounding stem. .... *Coccoloba* (*Polygonaceae*).
                    12. Stipules absent or not fused into sheath surrounding stem.
                        13. Plants dioecious. Ovary apocarpous. Fruits indehiscent, often with hard, ornamented endocarp (the endocarp of *Disciphania* is cartilaginous). .... *Menispermaceae*.
                        13. Plants monoecious. Ovary not apocarpous. Fruits dehiscent or indehiscent, without hard, ornamented endocarp. .... *Euphorbiaceae*.
            10. Flowers bisexual.
                14. Flowers subtended by modified, cup-like nectar-producing bracts. .... *Marcgraviaceae*.
                14. Flowers not subtended by modified, cup-like nectar-producing bracts.
                    15. Petals fused for ≥1/2 length.

16. Inflorescences helicoid. Corolla tubular. .... *Boraginaceae.*
16. Inflorescence not helicoid. Corolla bell-shaped (campanulate).
17. Some species with milky exudate. Corolla often convolute in bud; calyx with 5 lobes; stamens often of markedly different lengths, the anthers with lateral dehiscence; stigma distinctly capitate. Fruits capsules or 1-seeded berries. .... *Convolvulaceae.*
17. All species without milky exudate. Corolla not convolute in bud; calyx with 10 lobes; stamens not of markedly different lengths, the anthers with poricidal dehiscence; stigma not distinctly capitate. Fruits many-seeded berries. .... *Lycianthes* (*Solanaceae*).
15. Petals absent, or, if present, not fused or fused for <1/2 length.
18. Leaf blades with 2–4 basal secondary veins arching upwards from base, i.e., at least the basal two secondary veins as conspicuous as midrib. Ovary inferior. .... *Sparattanthelium* (*Hernandiaceae*).
18. Leaf blades with pinnate secondary venation, the secondary veins less conspicuous than midrib. Ovary superior.
19. Bark fibrous, easily peeled off in long strips. Flowers purplish-pink; sepals 3; petals in two whorls of 3 each; stamens and pistils both numerous. Seeds with ruminate endosperm. .... *Annona haematantha* (*Annonaceae*).
19. Bark not exceptionally fibrous, not easily peeled off in long strips. Flowers not red; sepals usually >3 (except some Dilleniaceae); petals not 3 or, if 3, then in one whorl; neither stamens nor pistils numerous. Seeds without ruminate endosperm.
20. Leaves often, but not always, scabrous. Stamens numerous (>10); placentation usually parietal, less frequently basal. .... *Dilleniaceae.*
20. Leaves not rough to touch. Stamens usually ≤5; placentation apical, basal, or axile.
21. Stems markedly swollen at nodes. Flowers and fruits in dense spikes. Perianth absent. .... *Piperaceae.*
21. Stems not swollen at nodes. Flowers and fruit not in dense spikes. Perianth present.
22. Stems and leaves with spines. Anthers with poricidal dehiscence. Fruits bilocular berries. .... *Solanum* (*Solanaceae*).
22. Stems and leaves without spines. Anthers without poricidal dehiscence. Fruits not bilocular berries.
23. Plants with stipules. Inflorescences arising from petioles. Petals bifid at apex. .... *Dichapetalaceae.*
23. Plants without stipules. Inflorescences not arising from petioles. Petals not bifid at apex.
24. Calyx persistent in fruit, the sepals fused into conspicuous reflexed "skirt;" stamens 10, the connective not prolonged beyond thecae. .... *Heisteria scandens* (*Olacaceae*).
24. Calyx not conspicuous in fruit; stamens usually 5, the connective prolonged beyond thecae. .... *Icacinaceae.*

Key 4. Epiphytic Plants

This key includes all those dicotyledons that grow upon other plants, including the hemiparasitic Loranthaceae and Viscaceae. Some plants, such as in species of *Ficus*, may start as epiphytes but later develop into free standing trees. The roots of others may (hemiepiphtyes) or may not reach the ground during part of their life cycle. See definitions of epiphyte and hemiepiphyte in Part 1, page 11.

1. Plants herbaceous.
2. Flowers apetalous; ovary unilocular, with single ovule. .... *Peperomia* (*Piperaceae*).
2. Flowers with tepals or petals; ovary multilocular, with >1 ovule per locule.
3. Plants without apparent leaves, the stems fleshy and green. .... *Cactaceae.*
3. Plants with leaves, the stems not fleshy and green.
4. Leaves alternate; stipules conspicuous. Flowers unisexual; corolla not fused into tube; ovary , winged. Seeds dust-like. .... *Begoniaceae.*

4. Leaves opposite; stipules absent. Flowers bisexual; corolla fused into tube; ovary superior, not winged. Seeds not dust-like. . . . . *Gesneriaceae.*
1. Plants woody.
5. Flowers subtended by modified, nectar-producing bracts. . . . . *Marcgraviaceae.*
5. Flowers not subtended by modified, nectar-producing bracts.
6. Leaves simple and alternate.
7. Plants with colored exudate. Flowers located within special branch ends (figs or syconia). . . . . *Ficus* (*Moraceae*).
7. Plants without colored exudate. Flowers not located in special branch ends.
8. Inflorescences not in heads or congested umbels. Flowers >5 mm long; perianth biseriate, the petals fused into tube.
9. Corolla tube >50 mm long; anthers with longitudinal dehiscence; ovary superior. . . . . *Markea* (*Solanaceae*).
9. Corolla tube <30 mm long; anthers with poricidal dehiscence; ovary inferior. . . . . . . *Ericaceae.*
8. Inflorescences in heads or congested umbels. Flowers <5 mm long; perianth uniseriate, or, if biseriate, the petals not fused into tube.
10. Petioles unequal in length. Perianth biseriate; ovary inferior, multilocular; placentation axile. Fruits free from one another, green, not surrounded by fleshy perianth (the fruit itself may be fleshy). . . . . *Oreopanax* (*Araliaceae*).
10. Petioles not markedly unequal in length. Perianth uniseriate; ovary superior, unilocular; placentation basal. Fruits fused to one another, yellow-orange, surrounded by fleshy perianth. . . . . *Coussapoa* (*Cecropiaceae*).
6. Leaves simple and opposite (sometimes some alternate leaves also present on same stems as opposite leaves in some Loranthaceae).
11. Base of plant penetrating tissue of host (hemiparasite).
12. Flowers usually >3 mm long (flowers of *Oryctanthus florentus* may be slightly smaller), in some species very large and showy, bisexual; perianth 6-merous; calyx (as small, rim-like calyculus) and corolla present. Inflorescences with pedicellate flowers. . . . . *Loranthaceae.*
12. Flowers <3 mm long, never showy, unisexual; perianth 3–4-merous; calyx and corolla both not present, i.e., only one perianth whorl present. Inflorescences with sessile flowers. . . . . *Viscaceae.*
11. Base of plant not penetrating tissue of host.
13. Plants with colored exudate, this often yellow. Flowers unisexual, with sticky exudate; stamens >10. . . . . *Clusiaceae.*
13. Plants without colored exudate. Flowers bisexual, without sticky exudate; stamens ≤10.
14. Leaves with interpetiolar stipules. Ovary inferior; petals fused into tube above ovary. . . . . *Rubiaceae.*
14. Leaves without interpetiolar stipules. Ovary inferior or superior; petals free if ovary inferior, fused into tube around ovary if ovary superior.
15. Leaf blades with arcuate secondary veins and characteristic ladder-like (scalariform) tertiary veins between the secondary veins. Flowers with free petals above ovary; stamens 10; ovary bilocular; placentation axile. . . . . *Melastomataceae.*
15. Leaf blades with pinnate venation and reticulate tertiary veins. Flowers with petals fused into tube around ovary; stamens 4; ovary unilocular; placentation parietal. . . . . *Gesneriaceae.*

Key 5. Trees

Key to Subkeys

This key includes treelets (single-stemmed woody plants <5 cm DBH and >2 m tall) as well as trees (single-stemmed >5 cm DBH). However, if your plant is <5 cm DBH and does not key out under tree dicotyledons, try the Key to Shrubs (p. 27).

1. Leaves compound.
2. Leaves mostly palmate or trifoliolate (simple leaves may also be present on same plant, e.g., *Allophyllus* in the Sapindaceae). . . . . *Subkey 1. Palmate Leaves* (p. 19)
2. Leaves pinnate or bipinnate. . . . . *Subkey 2. Pinnate or Bipinnate Leaves* (p. 19).
1. Leaves simple (including deeply lobed leaves, e.g., some species of *Pourouma*, are included here).

3. Leaves opposite or whorled. . . . . *Subkey 3. Simple Opposite or Whorled Leaves* (p. 21).
3. Leaves alternate.
 4. Plants with colored exudate. . . . . *Subkey 4. Simple Alternate Leaves, Colored Exudate Present* (p. 22).
 4. Plants without colored exudate. . . . . *Subkey 5. Simple Alternate Leaves, Colored Exudate Absent* (p. 23).

Key 5. Trees

SUBKEY 1. PALMATE LEAVES

1. Leaves opposite.
 2. Trunks with long, running buttresses. Flowers actinomorphic; corolla not fused into distinct tube; stamens numerous. Fruits with spiny endocarp. . . . . *Caryocaraceae*.
 2. Trunks without long, running buttresses. Flowers zygomorphic; corolla fused into distinct tube; stamens 4. Fruits without spiny endocarp.
  3. Corollas yellow. Fruits dry, dehiscent. Seeds numerous, winged. . . . . *Tabebuia* (*Bignoniaceae*).
  3. Corollas lavender or purple. Fruits fleshy, indehiscent. Seeds solitary, not winged. . . . . . *Vitex* (*Verbenaceae*).
1. Leaves alternate.
 4. Plants with colored exudate.
  5. Leaves with >3 leaflets. Plants dioecious. Flowers zygomorphic; ovary unilocular; placentation parietal. Fruits fleshy throughout, indehiscent. . . . . *Jacaratia* (*Caricaceae*).
  5. Leaves with 3 leaflets. Plants monoecious. Flowers actinomorphic; ovary 3-locular; placentation axile. Fruits dry, explosively dehiscent capsules. . . . . *Hevea* (*Euphorbiaceae*).
 4. Plants without colored exudate.
  6. Leaves with >6 leaflets. Flowers <10 mm diam.
   7. Plants without conspicuous stilt roots or flying buttresses. Stipules not encircling stem. Inflorescences umbels; ovary inferior. Fruits not enclosed by fleshy perianth. . . . . *Schefflera* (*Araliaceae*).
   7. Plants with conspicuous stilt roots or flying buttresses. Stipules encircling stem. Inflorescences in finger-like spikes; ovary superior. Fruits enclosed by fleshy perianth. . . . . *Cecropia sciadophylla* (*Cecropiaceae*).
  6. Leaves usually with ≤ 6 leaflets. Flowers >10 mm.
   8. Leaves with 3 leaflets. Flowers <20 mm long; stamens 5–10, not fused at bases.
    9. Leaflet blades with pits (domatia) in axils of secondary veins. Corolla 17 –24 mm long; ovary with carpels only slightly fused. Fruits dry and dehiscent. . . . . *Ticorea* (*Rutaceae*).
    9. Leaflet blades without cavities in axils of secondary veins. Corolla 2-3 mm long; ovary with carpels fused. Fruits fleshy and indehiscent. *Allophyllus* (*Sapindaceae*).
   8. Leaves with 3–6(8) leaflets. Flowers >20 mm long; stamens fused at bases.
    10. Trees. Leaflets not densely pubescent. Inflorescences not cauliflorous (ramiflorous in *Eriotheca*). Petals not differentiated; staminodia absent. Fruits not markedly ridged. Seeds not surrounded by a slimy pulp. . . . . *Bombacaceae*.
    10. Treelets. Leaflets densely pubescent, especially on abaxial surface. Inflorescences cauliflorous. Petals differented into long, narrow blade distally and cucullate claw proximally; staminodia 5. Fruits markedly ridged. Seeds surrounded by slimy pulp. . . . . *Herrania* (*Sterculiaceae*).

Key 5. Trees

SUBKEY 2. PINNATE OR BIPINNATE LEAVES

1. Leaves opposite or subopposite.
 2. Leaves pinnate.
  3. Leaflet margins serrate, the tertiary veins inconspicuous. Flowers actinomorphic, yellow. . . . . *Touroulia* (*Quiinaceae*).
  3. Leaflet margins entire, the tertiary veins conspicuous. Flowers zygomorphic, rose purple. . . . . *Taralea oppositifolia* (*Fabaceae*).
 2. Leaves bipinnate.
  4. Leaflets <10 mm wide. Flowers grouped into dense, globose heads, nocturnal; corolla actinomorphic, white or yellow. Fruits much longer than wide. . . . . *Parkia decussata* & *P. nitida* (*Mimosaceae*).

4. Leaflets >10 mm wide. Flowers not grouped into dense, globlar heads, diurnal; corolla zygomorphic, lavender. Fruits nearly as wide as long. . . . . *Jacaranda* (*Bignoniaceae*).

1. Leaves alternate.
 5. Leaves bipinnate.
  6. Leafstalks usually with glands. Stamens long exserted, the filaments and petals fused at base into obvious tube (stemonozone). Fruits various, usually longer than wide, sometimes coiled. . . . . *Mimosaceae*.
  6. Leafstalks without glands. Stamens not long exserted, the filaments and petals not fused into stemonozone. Fruits flattened, usually nearly as long as wide. . . . . *Dimorphandra* (*Caesalpiniaceae*).
 5. Leaves pinnate.
  7. Flowers zygomorphic.
   8. Stipules absent. Perianth uniseriate; stamens adnate to perianth. . . . . *Proteaceae*.
   8. Stipules present. Perianth biseriate (uniseriate in a few Caesalpiniaceae, e.g., *Dialium* and *Dicorynia*); stamens not adnate to perianth.
    9. Flowers slightly zygomorphic, without pronounced standard and keel, the standard internal to lateral petals (best seen in bud), or petals reduced to one or three, or petals absent; stamens with free filaments, numerous (then often heterantherous), 10, or reduced to one. . . . . *Caesalpiniaceae* (also *Bocoa* & *Swartzia* of Fabaceae).
    9. Flowers markedly zygomorphic, with pronounced standard and keel, the standard external to lateral petals (best seen in bud); stamens usually with 9 filaments fused into distinct column surrounding ovary and one filament free. . . . . *Fabaceae*.
  7. Flowers actinomorphic.
   10. Plants with white or yellowish, free-flowing exudate, this especially apparent in cut stems and broken leaves. . . . . *Thyrsodium* (*Anacardiaceae*).
   10. Plants without white or yellowish, free-flowing exudate (a whitish exudate may be present in some Burseraceae but it is not free-flowing).
    11. Small, slender treelets. Inflorescences usually cauline. Fruits follicles. Seeds usually black, subtended by yellow aril. . . . . *Connaraceae*.
    11. Usually medium-sized to large trees, infrequently treelets. Inflorescences usually not cauline. Fruits usually not follicles. Seeds variously colored, the arilloid, when present, usually red.
     12. Cut trunk and crushed leaves usually with turpentine aroma. Leaves usually, but not always, with pulvinules, the leaflets always opposite. . . . . *Burseraceae*.
     12. Cut trunk and crushed leaves may be aromatic but usually without turpentine aroma (except some Annacardiaceae, e.g., *Astronium*). Leaves usually without pulvinules (Sapindaceae and some Meliaceae may have pulvinules), the leaflets alternate to subopposite.
      13. Ovaries with >1 carpel, the carpels weakly fused, the individual carpels separate and evident.
       14. Cut trunk with bitter taste, without spines. Leaves without citrus aroma when crushed, without pellucid punctations. Ovary on short gynophore. Fruits indehiscent. . . . . *Simaroubaceae*.
       14. Cut trunk without bitter taste, with spines. Leaves sometimes with citrus aroma when crushed, often with pellucid punctations. Ovary not on short gynophore. Fruits dehiscent. . . . . *Zanthoxylum* (*Rutaceae*).
      13. Ovaries with a single carpel or with carpels strongly fused, the individual carpels not separate and evident.
       15. Leaves with glands between leaflets. Stamens numerous, long exserted. . . . . *Inga* (*Mimosaceae*).
       15. Leaves without glands between leaflets. Stamens usually ≤10, not long exserted.
        16. Leaf rachis often extended beyond distal leaflets, this without indeterminate growth. Petals often hairy and with a hairy appendage; nectary disc extrastaminal; stamens (5)8. . . . . *Sapindaceae*.
        16. Leaf rachis not extended beyond distal leaflets or, if extended, then with indeterminate growth(*Guarea*). Petals not markedly hairy and without an appendage; nectary disk often intrastaminal; stamens usually (7)10.
         17. Leaves sometimes (species of *Guarea*), but not always (remaining genera), with indeterminate growth. Cut trunk often with spicy aroma. Stamens usually fused into tube (the tube very short in species of *Trichilia*); ovary with >1 ovule per locule. Fruits dehiscent. . . . . . . *Meliaceae*.

17. Leaves never with indeterminate growth. Cut trunk without spicy aroma. Stamens not fused into tube; ovary with one ovule per locule. Fruits indehiscent. . . . . *Anacardiaceae*.

Key 5. Trees

Subkey 3. Simple Opposite or Whorled Leaves

1. Plants with colored exudate.
    2. Margins of leaf blades serrate. Flowers unisexual, staminate flowers in catkins, pistillate flowers in capitula. . . . . *Bagassa guianensis* (*Moraceae*).
    2. Margins of leaf blades entire. Flowers bisexual, not in catkins nor in capitula.
        3. Exudate clear. Petioles often pulvinate-geniculate. Perianth uniseriate. Fruits often, but not always, with conspicuous hairs or spines. . . . . *Elaeocarpaceae*.
        3. Exudate white, yellow, or orange. Petioles not pulvinate-geniculate. Perianth biseriate. Fruits without conspicuous hairs or spines.
            4. Exudate white. Petals fused into a tube, the lobes contorted. . . . . *Apocynaceae*.
            4. Exudate yellow or orange (infrequently white in some species of *Clusia*). Petals free, the lobes not contorted. . . . . *Clusiaceae*.
1. Plants without colored exudate.
    5. Flowers with petals or tepals (if only one perianth whorl present) fused into distinct tube, never on or surrounded by a fleshy receptacle.
        6. Leaves with stipules or connate petiole bases.
            7. Petioles touch to form rim extending from one petiole base to next, the stem appearing ensheathed by petiole bases. Flowers subtended by a series of bracts. Ovary superior. Fruits dehiscent. . . . . *Antonia* (*Loganiaceae*).
            7. Stipules not as above. Flowers not subtended by a series of bracts as above. Ovary inferior. Fruits dehiscent or indehiscent. . . . . *Rubiaceae*.
        6. Leaves without apparent stipules or connate petiole bases.
            8. Leaves often opposite, subopposite, and even alternate on same plant. Flowers actinomorphic; perianth uniseriate. . . . . *Nyctaginaceae*.
            8. Leaves strictly opposite. Flowers zygomorphic; perianth biseriate. . . . . *Verbenaceae*.
    5. Flowers with free petals or tepals or flowers small and on or surrounded by fleshy receptacle (Monimiaceae), a distinct tube never present.
        9. Flowers zygomorphic (Malpighiaceae are sometimes nearly actinomorphic).
            10. Calyx with one sepal spurred, without glands on abaxial surface; corolla reduced to 1–3 petals, the petals not clawed. Fruits dry, dehiscent and seeds winged or fruits indehiscent and calyx developed into wings of unequal length. . . . . *Vochysiaceae*.
            10. Calyx not spurred, usually with glands on abaxial surface; corolla 5-merous, the petals clawed. Fruits fleshy, indehiscent, neither fruits nor seeds winged. . . . . *Malpighiaceae*.
        9. Flowers actinomorphic
            11. Cut bark aromatic, the aroma not almond-like. Flowers <5 mm diam.; perianth uniseriate.
                12. Crushed leaves and cut bark with spicy aroma. Leaf blades with secondary veins decurrent along midrib. Anthers with valvate dehiscence; gynoecium not apocarpous. Fruits drupaceous, often subtended by expanded, red cupule. . . . . *Licaria debilis* (*Lauraceae*).
                12. Crushed leaves and cut bark with lemony aroma. Leaf blades without secondary veins decurrent along midrib. Anthers with (*Siparuna*) or without valvate dehiscence (*Mollinedia*); gynoecium apocarpous. Fruits with separate monocarps, the monocarps enclosed within fleshy, enlarged hypanthium (*Siparuna*) or clustered on flattened hypanthium (*Mollinedia*). . . . . *Monimiaceae*.
            11. Cut bark not aromatic or, if aroma present, almond-like. Flowers >5 mm diam.; perianth biseriate.
                13. Ovary inferior.
                    14. Leaf blades with pinnate secondary veins and often with pellucid punctations. Stamens usually >20, the anthers without appendages or oil glands, with lateral dehiscence. . . . . *Myrtaceae*.
                    14. Leaf blades usually with arcuate secondary veins and without pellucid punctations (except for *Mouriri* which has normal pinnate venation). Stamens usually 10, the anthers often with appendages or oil glands, with poricidal dehiscence. . . . . *Melastomataceae*.

13. Ovary superior.
    15. Stipules conspicuous, usually acicular or leaf-like. Leaf blades with tertiary and higher order venation parallel, closely spaced, and often inconspicuous. . . . . *Quiinaceae*.
    15. Stipules inconspicuous (caducous or small). Leaf blades with tertiary and higher order venation reticulate.
        16. Flowers unisexual. . . . . *Tovomita*, in part (*Clusiaceae*).
        16. Flowers bisexual.
            17. Petioles often swollen at base and apex, often geniculate at apex. Stamens >20. Fruits with exocarp often, but not always, with hairs or spines. Seeds usually with conspicuous red aril. . . . . *Elaeocarpaceae*.
            17. Petioles not pulvinate-geniculate. Stamens ≤20. Fruits with exocarp always without hairs or spines. Seeds without conspicuous red aril.
                18. Petals brownish red, dark yellow, or orange; stamens 3; ovary surrounded by nectary disc perforated by 3 staminiferous pockets. . . . . *Cheiloclinium cognatum* (*Hippocrateaceae*).
                18. Petals white; stamens >3; ovary not surrounded by nectariferous disc.
                    19. Petals with fringed margins; stamens ca. 20, the anthers without expanded connective; ovary bilocular; placentation axile. . . . . . *Rhizophoraceae*.
                    19. Petals without fringed margins; stamens 5, the anthers with expanded connective; ovary unilocular; placentation parietal. . . . . *Violaceae*.

Key 5. Trees

Subkey 4. Simple Alternate Leaves, Colored Exudate Present

The colored exudate is sometimes difficult to detect and may be present and obvious only in cut fruit or in the cut petioles.

1. Exudate red or quickly oxidizing to red or reddish-brown upon exposure to air.
    2. Trees with flying buttresses. Leaf blades palmately incised; stipules leaving conspicuous scars encircling stem. . . . . *Pourouma*, in part (*Cecropiaceae*).
    2. Trees without flying buttresses. Leaf blades not palmately incised; stipules not leaving conspicuous scars encircling stem.
        3. Leaf blades cordate or oblanceolate. Perianth free; ovary 3-locular, the style divided. . . . . *Croton draconioides* & *Pausandra fordii* (*Euphorbiaceae*).
        3. Leaf blades elliptic. Perianth (either tepals or petals) fused at least at base; ovary 1–2-locular, the style not divided.
            4. Flowers unisexual; perianth uniseriate, usually 3-merous. Fruits dehiscent. Seeds surrounded by red, laciniate aril. . . . . *Myristicaceae*.
            4. Flowers bisexual; perianth biseriate, 5-merous. Fruits indehiscent or dehiscent. Seeds without red, laciniate aril.
                5. Exudate exuding from junction of inner bark and sapwood, the slash making a hissing sound when first cut. Flowers perigynous; style gynobasic from one side of ovary. Fruits indehiscent. Seeds not winged. . . . . *Licania heteromorpha* (*Chrysobalanaceae*).
                5. Exudate mostly from petioles, not exuding from junction of inner bark and sapwood, the slash not making a hissing sound when first cut. Flowers hypogynous; style terminal. Fruits dehiscent. Seeds winged. . . . . *Aspidosperma*, in part (*Apocynaceae*).
1. Exudate not red, usually white (infrequently brown or slightly reddish in Moraceae or clear in Elaeocarpaceae).
    6. Petioles often swollen at base and apex, often geniculate at apex. Perianth uniseriate. Fruits dehiscent, the exocarp, often but not always, with hairs or spines. Seeds with aril. *Elaeocarpaceae*.
    6. Petioles not pulvinate-geniculate. Perianth uniseriate or biseriate. Fruits indehiscent without exocarp with spines or hairs or, if dehiscent, the seeds without aril.
        7. Flowers bisexual; perianth biseriate.
            8. Trunks clearly fenestrate. Stellate hairs present on young stems, petioles, and inflorescence rachises. Exudate in fruit only. Seeds with ruminate endosperm. . . . . *Minquartia* (*Olacaceae*).
            8. Trunks usually not fenestrate but sometimes sulcate (*Aspidosperma*, in part) or infrequently fenestrate (*Geissospermum* spp.). Stellate hairs not present. Exudate usually from cut bark, but sometimes only in fruit or petioles. Seeds without ruminate endosperm.

9. Corollas white, yellow, or infrequently green, the lobes contorted; stamens opposite sepals; ovary of two separate carpels. Fruits flattened or horn-shaped, dehiscent (*Aspidosperma* and *Himatanthus*) or indehiscent (*Geissospermum*). Seeds without shiny, black or brown testa and without conspicuous hilum, surrounded by wing (*Aspidosperma* and *Himatanthus*) or wing absent (*Geissospermum*). . . . . *Apocynaceae.*
9. Corollas green or infrequently red, the lobes not contorted; stamens opposite petals; ovary of united carpels. Fruits globose, indehiscent. Seeds with shiny, black or brown testa and conspicuous hilum, not surrounded by wing. . . . . *Sapotaceae.*

7. Flowers usually unisexual, but some bisexual flowers may also be present; perianth uniseriate or biseriate.
10. Leaves with conspicuous stipules, the scars entirely or mostly surrounding stem. Flowers in modified branch ends called syconia or in catkins or heads; ovary 1-locular; placentation apical, the ovule solitary. . . . . *Moraceae.*
10. Leaves with or without stipules, the scars, when present, not entirely or mostly surrounding stem. Flowers not in syconia or usually not in catkins (*Mabea* of the Euphorbiaceae has catkin-like inflorescences) or heads; ovary 1- or 3-locular; placentation axile, basal (when 1-locular), or parietal, the ovules 1 to many.
11. Leaf blades deeply dissected. Placentation parietal. Cultivated but sometimes escaped from cultivation. . . . . *Carica* (*Caricaceae*).
11. Leaf blades not deeply dissected. Placentation axile or basal. Native.
12. Perianth usually 3-merous, forming tube; ovary 1-locular; placentation basal. Seeds surrounded by aril; endosperm ruminate. *Myristicaceae* (*Osteophloeum*).
12. Perianth not 3-merous, not forming tube; ovary 3-locular; placentation axile. Seeds not surrounded by laciniate aril; endosperm present but not ruminate. . . . . *Euphorbiaceae.*

Key 5. Trees

Subkey 5. Simple Alternate Leaves, Colored Exudate Absent

1. Flowers zygomorphic, sometimes only gynoecium zygomorphic (style arises from one side of ovary, e.g., Chrysobalanaceae and Rhabdodendraceae) or pistil inserted on one side of hypanthium, or only androecium zygomorphic (anthers offset to one side or 1 or 2 stamens longer than others).
2. Petioles with pulvini at base and apex. Flowers pea-like; stamens with 9 filaments fused together and 1 free. Fruits typical legumes, splitting along 2 sutures. . . . . *Poecilanthe hostmannii* (*Fabaceae*).
2. Petioles without pulvini. Flowers not pea-like; stamens not as above. Fruits not legumes, indehiscent or splitting along more than 2 sutures.
3. Slash of bark soft and fibrous. Flowers markedly zygomorphic, the androecium with flap from one side, this arching over summit of ovary; ovary inferior. Fruits woody, circumscissile capsules. . . . . . . *Lecythidaceae.*
3. Slash of bark hard and friable or soft but not fibrous. Flowers not markedly zygomorphic, the androecium not as above; ovary superior. Fruits indehiscent.
4. Flowers with a single fertile stamen or 1 or 2 stamens much longer than others. Fruits subtended by swollen pedicel (hypocarp) or fruits fleshy, edible drupes with fibrous stones. . . . . *Anacardium* & *Mangifera* (*Anacardiaceae*).
4. Flowers with more than 1 or 2 stamens, all ± same length. Fruits not as above.
5. Stamens with filaments fused into tube (monadelphous), the anthers offset to one side; style terminal. . . . . *Matisia* (*Bombacaceae*).
5. Stamens with filaments not fused into tube; style gynobasic from one side of ovary.
6. Leaves oblanceolate, with inconspicuous tertiary venation, with punctations on abaxial surface. Flowers not perigynous. . . . . *Rhabdodendraceae.*
6. Leaves usually elliptic, with well developed tertiary venation (sometimes covered by hairs), without punctations on abaxial surface. Flowers perigynous. . . . . *Chrysobalanaceae.*

1. Flowers actinomorphic.
7. Ovary inferior.
8. Slash of bark fibrous. Flowers >5 cm diam.; stamens >500, the filaments fused into conspicuous, elevated ring at base, free above; ovary with >10 ovules/locule, these on expanded apical-axile placenta. Fruits not winged (sometimes with 6 ridges), not with expanded calyx surrounding seed, with >1 seed. . . . . *Gustavia* (*Lecythidaceae*).

8. Slash of bark not fibrous. Flowers <2 cm diam.; stamens 3–50, the filaments not fused into conspicuous, elevated ring; ovary with <10 ovules/locule, usually apical but not on expanded placenta. Fruits winged, or with expanded calyx, or with single seed.
  9. Leaf blades ovate to cordate, with three equal veins arising from base. Flowers unisexual; stamens 3. Fruits surrounded by a bladder-like cupule. . . . . *Hernandia* (*Hernandiaceae*).
  9. Leaf blades elliptic to oblanceolate, with pinnate venation throughout. Flowers bisexual; stamens 10–40. Fruits not surrounded by bladder-like cupule.
    10. Stem growth monopodial. Leaf margins crenate. Petals fused for 1/2 length into tube; stamens adnate to corolla; ovary (3)5-locular. Fruits fleshy, never winged, surrounded by tightly appressed calyx. . . . . *Symplocaceae*.
    10. Stem growth sympodial. Leaf margins entire. Petals free; stamens not adnate to corolla; ovary unilocular. Fruits dry, winged or, if not winged, not surrounded by tightly appressed calyx. . . . . *Combretaceae*.

7. Ovary superior.
  11. Perianth with either tepals or petals fused into tube, the tube may be either long (Boraginaceae) or short (Myrsinaceae and some species of Ebenaceae).
    12. Leaf blades with stellate hairs or peltate scales.
      13. Perianth uniseriate; ovary with carpels secondarily apocarpous. Fruits pin-wheel shaped, the carpels free, opening along one suture. . . . . *Sterculiaceae* (*Sterculia*).
      13. Perianth biseriate; ovary with carpels completely fused. Fruits not pin-wheel shaped, indehiscent or opening along more than one suture.
        14. Leaf blades with stellate hairs. Flowers >5 mm diam.; petals without long, attenuated apices; stamens with filaments fused into tube at base; anthers bright yellow. . . . . . *Styracaceae*.
        14. Leaf blades with peltate scales with fringed margins. Flowers <5 mm diam.; petals with long, attenuate apices; stamens without filaments fused into tube at base; anthers not bright yellow. . . . . *Dendrobangia* (*Icacinaceae*).
    12. Leaf blades without stellate hairs or peltate scales.
      15. Stem growth sympodial. Stamens with versatile anthers; style divided twice. . . .*Cordia* (*Boraginaceae*).
      15. Stem growth monopodial. Stamens with basifixed anthers; style simple or only divided once.
        16. Leaf blades often, but not always, serrate. Perianth uniseriate; placentation parietal. Fruits usually dehiscent. . . . . *Casearia* (*Flacourtiaceae*).
        16. Leaf blades entire. Perianth biseriate; placentation axile, apical, or free central. Fruits indehiscent.
          17. Slash of bark with strong turpentine aroma. Petioles with pulvini. Seeds surrounded by white arilloids. . . . . *Protium occultum* (*Burseraceae*).
          17. Slash of bark without turpentine aroma. Petioles with or without pulvini. Seeds not surrounded by white arilloids.
            18. Inflorescences arising from petioles. Petals with bifid lobes. . . . *Tapura* (*Dichapetalaceae*).
            18. Inflorescences not arising from petioles. Petals with entire lobes.
              19. Flowers unisexual. Calyx, especially in fruit, irregularly shaped; style divided. . . . . *Ebenaceae*.
              19. Flowers bisexual. Calyx regularly shaped; style not divided.
                20. Leaves, flowers, and fruits with glandular striations or punctations. . . . . *Myrsinaceae*.
                20. Leaves, flowers, and fruits without glandular striations or punctations.
                  21. Understory treelets or understory trees. Leaf blades oblanceolate; petioles swollen at apices. Flowers large, the petals 43–75 mm long. . . . . *Erythrochiton* (*Rutaceae*).
                  21. Understory or canopy trees. Leaf blades elliptic to oblong; petioles not swollen at apices. Flowers smaller, the petals <7 mm long. . . . .*Olacaceae*.
  11. Perianth without either tepals or petals fused into tube, i.e., a perianth tube absent.
    22. Perianth absent or uniseriate.
      23. Flowers small, usually <3 mm diam. or <3 mm long.
        24. Flowers bisexual.
          25. Inflorescences branched. Stamens usually 8. . . . . *Ampelocera* (*Ulmaceae*).
          25. Inflorescences unbranched. Stamens 1–6.
            26. Stems not swollen at nodes. Stamens bifid; ovary with parietal placentation, >1 ovule per locule. Fruits dehiscent. . . . . *Lacistemataceae*.

26. Stems swollen at nodes. Stamens not bifid; ovary with basal placentation, with 1 ovule per locule. Fruits indehiscent. . . . . *Piperaceae*.
24. Flowers unisexual.
27. Ovary 3-locular. Fruits capsules. . . . . *Euphorbiaceae*.
27. Ovary unilocular. Fruits achenes or drupes.
28. Plants with flying buttresses. Leaf blades palmately incised; stipules leaving scars encircling stem. . . . . *Cecropiaceae*.
28. Plants without flying buttresses. Leaf blades not palmately incised; stipules not leaving scars encircling stem.
29. Leaf blades ovate to widely ovate, not markedly asymmetrical at base. Placentation basal. Fruits achenes, surrounded by fleshy perianth. . . . . *Urera* (*Urticaceae*).
29. Leaf blades narrowly ovate, markedly asymmetrical at base. Placentation apical. Fruits drupes, not surrounded by fleshy perianth. . . . . *Trema* (*Ulmaceae*).
23. Flowers larger, usually >3 mm diam. or 3 mm long.
30. Trunks usually with very thin buttresses, sometimes with flying buttresses. Petioles swollen at base and apex, often geniculate at apex. Ovary usually 4-locular. Seeds often with red aril. . . . . *Sloanea* (*Elaeocarpaceae*).
30. Trunks without thin or flying buttresses. Petioles not markedly swollen at base and apex. Ovary usually 1–3-locular. Seeds usually, but not always (Flacourtiaceae), without red aril.
31. Perianth parts linear, the apices recurving at anthesis; stamens adnate to perianth. Seeds winged. . . . . *Roupala* (*Proteaceae*).
31. Perianth parts not linear, the apices not cucullate; stamens usually free from perianth (adnate in some Flacourtiaceae). Seeds not winged.
32. Leaf blades with 3–5 secondary veins arching upward from base, with no other secondary veins further up midrib, without glands. Flowers with distinct intrastaminal disc. Fruits with single seed. . . . . *Ziziphus* (*Rhamnaceae*).
32. Leaf blades with pinnate venation (infrequently with secondary veins from base), with or without glands. Flowers without distinct intrastaminal disc. Fruits with more than one seed.
33. Ovary unilocular; placentation parietal. . . . . *Flacourtiaceae*.
33. Ovary 3-locular; placentation axile. . . . . *Euphorbiaceae*.
22. Perianth present, biseriate.
34. Stamens united, at least at base.
35. Slash of bark with medicinal aroma (aroma of wintergreen). Leaf blades finely punctate (need to hold against light and use lens to see), aromatic. Placentation parietal. . . . . . *Canellaceae*.
35. Slash of bark without medicinal aroma. Leaf blades not finely punctate, not aromatic. Placentation axile.
36. Stellate hairs present, usually conspicuous on veins of abaxial leaf surface, but sometimes only conspicuous on inflorescences and flowers. Stamens with filaments united into scarcely defined tube. . . . . *Bombacaceae*.
36. Stellate hairs usually not present (sometimes present in Sterculiaceae). Stamens with filaments united, but not into well defined tube.
37. Inflorescences mostly cauline, usually along main trunk and branches, less frequently axillary. Petals highly modified, either with "ant-like" appendages or with basal, saccate claw and apical blade; staminodia alternating with stamens. . . . . *Sterculiaceae*.
37. Inflorescences generally not cauline, either terminal or in leaf axils, infrequently along stems. Petals not modified; staminodia absent.
38. Stems often with conspicuous bracts (cataphylls). Inflorescences in axillary clusters. Petals usually with appendage toward base on adaxial surface. Fruits single-seeded. . . . . *Erythroxylaceae*.
38. Stems without cataphylls. Inflorescences not in axillary clusters. Petals without appendages. Fruits with >1 seed.
39. Leaf blade without minute punctations. Style simple; anthers with thick, prolonged connective. Fruits with markedly sculptured endocarp dehiscing by valves. . . . . *Humiriaceae*.

39. Leaf blade with minute punctations on both surfaces. Styles 5; anthers without well developed connective. Fruits woody, but endocarp not especially sculptured, nor dehiscing by valves. . . . . *Hugoniaceae*.

34. Stamens not united.

40. Ovary with free carpels (apocarpous) or at least with upper part of carpels separate (*Goupia*). Fruits with separate carpels or carpels fused but with sutures conspicuous.

41. Leaf blades with few arcuate secondary veins, the tertiary veins oriented at right angles to midrib. Inflorescences stalked umbels. . . . . *Goupia* (*Celastraceae*).

41. Leaf blades without above venation. Inflorescences various but not in stalked umbels (do not mistake stalked monocarps of Annonaceae as an umbellate inflorescence).

42. Leaf blades distinctly pellucid-punctate. . . . . *Conchocarpus* (*Rutaceae*).

42. Leaf blades not pellucid-punctate.

43. Slash of bark with conspicuous vertical streaks, fibrous, the bark easily peeled off in long strips, often with slightly spicy aroma. Calyx 3-parted; corolla in two whorls of 3 parts each, the petals white, green, or dull yellow; stamens >10; styles numerous, one per carpel, from apex of carpel. Fruits without red, enlarged torus. Seeds with ruminate endosperm. . . . . *Annonaceae*.

43. Slash of bark without conspicuous vertical streaks, not fibrous, the bark not easily peeled off in long strips, without spicy aroma. Calyx 3–5 parted; corolla in single whorl of 5 parts, the petals bright yellow; stamens ≤10; style one, gynobasic, from middle of carpels. Fruits with red, enlarged torus. Seeds without ruminate endosperm. . . . . *Ouratea* (*Ochnaceae*).

40. Ovary with fused carpels or unicarpellate. Fruits with fused carpels or developed from unicarpellate ovaries, the sutures not conspicuous.

44. Slash of bark with conspicuous vertical streaks, fibrous, the bark easily peeled off in long strips. Calyx of 3 sepals; corolla in 2 whorls of 3 petals each; pistils numerous. Fruits globose. Seeds with ruminate endosperm. . . . . *Fusaea* (*Annonaceae*).

44. Slash of bark without conspicuous vertical streaks, not fibrous, the bark not easily peeled off in long strips. Calyx and corolla not as above; pistil 1. Fruits not globose. Seeds without ruminate endosperm.

45. Crushed leaves and cut bark often emitting spicy aroma. Leaf blades often with secondary veins decurrent along midrib. Anthers opening by small flaps. Fruits with a single seed, often subtended by red, enlarged pedicel. . . . . *Lauraceae*.

45. Crushed leaves and cut bark not emitting spicy aroma. Leaf blades without secondary veins decurrent along midrib. Anthers not opening by small flaps. Fruits usually with >1 seed, if 1 seed, then without red, enlarged pedicel.

46. Ovary unilocular; placentation parietal, with >2 ovules/locule.

47. Leaf blades cordate. Fruits covered with prickles. Seeds surrounded by arilloid. . . . . *Bixaceae*.

47. Leaf blades usually elliptic. Fruits usually not covered by prickles. Seeds sometimes arillate, but then aril not completely surrounding seeds.

48. Slash of bark with fetid aroma. Leaves with stellate hairs. Petals 4; ovary on gynophore. . . . . *Capparaceae*.

48. Slash of bark without fetid aroma. Leaves without stellate hairs. Petals usually 5; ovary not on gynophore.

49. Flowers unisexual or bisexual; stamens >10; anthers without prolonged connective. Fruits dehiscent or indehiscent. . . . . *Flacourtiaceae*.

49. Flowers bisexual; stamens <10; anthers with prolonged connective. Fruits indehiscent. . . . . *Violaceae*.

46. Ovary multilocular or, if unilocular, without parietal placentation; placentation axile or pendulous from apex of each locule, if unilocular with ≤2 ovules/locule.

50. Flowers perigynous. . . . . *Prunus* (*Rosaceae*).

50. Flowers not perigynous.

51. Ovary unilocular, with 2 apical, pendulous ovules/locule. . . . . . *Icacinaceae*.

51. Ovary with >1 locule, the placentation axile, apical-axile, or basal-axile.

52. Flowers unisexual. Fruits dehiscent or indehiscent.
    53. Fruits globose, concolorous at maturity, splitting into three segments. . . . . *Euphorbiaceae*.
    53. Fruits oblong, one side white and the other side black at maturity, not splitting into segments. . . . . . *Discophora* (*Icacinaceae*).
52. Flowers bisexual. Fruits indehiscent or, if dehiscent, not splitting into three segments.
    54. Inflorescences pendulous cymes. Anthers with poricidal dehiscence. Fruits bilocular, many-seeded berries. . . . . *Cyphomandra* (*Solanaceae*).
    54. Inflorescences not pendulous cymes. Anthers without poricidal dehiscence. Fruits not bilocular, many-seeded berries.
        55. Calyx circular, much expanded in fruit. . . . . *Chaunochiton* (*Olacaceae*).
        55. Calyx not circular, not expanded in fruit.
            56. Plants with pubescence of stellate hairs. Flowers ≥10 mm diam.; instrastaminal nectary disc absent; stamens ≥10. . . . . *Tiliaceae*.
            56. Plants glabrous. Flowers <10 mm diam.; intrastaminal nectary disc present; stamens usually 5. . . . . *Celastraceae*.

Key 6. Shrubs

Key to Subkeys

If you fail to identify the plant as a shrub (woody plant branched at the base or unbranched but <2 m tall), try the Key to Herbaceous Terrestrial or Aquatic Plants (p. 12) if the plant is small and scarcely woody or the Key to Trees (p. 18) if the plant has a single stem and is woody.

1. Leaves compound or so deeply lobed as to appear compound. . . . . *Subkey 1. Compound Leaves* (p. 27).
1. Leaves simple.
    2. Leaves opposite or whorled. . . . . *Subkey 2. Simple Opposite or Whorled Leaves* (p. 28).
    2. Leaves alternate. . . . . *Subkey 3. Simple Alternate Leaves* (p. 29).

Key 6. Shrubs

Subkey 1. Compound Leaves

1. Leaves palmate or so deeply palmately lobed as to appear compound.
    2. Plants with white exudate. Ovary 3-locular. . . . . *Manihot* (*Euphorbiaceae*).
    2. Plants without white exudate. Ovary not 3-locular.
        3. Leaf blades with pellucid punctations and cavities (domatia) in axils of secondary veins. Flowers not cauline; petals not differentiated into claw and limb. . . . . *Ticorea* (*Rutaceae*).
        3. Leaf blades without pellucid punctations and without cavities in axils of secondary veins. Flowers cauline; petals differentiated into claw and limb, the limb very long. . . . . *Herrania* (*Sterculiaceae*).
1. Leaves pinnate or bipinnate.
    4. Stems armed. Leaves bipinnate, the primary leaflets aggregated at end of rachis, the secondary leaflets sensitive to touch. Inflorescences in heads. . . . . *Mimosa*, in part (*Mimosaceae*).
    4. Stems not armed. Leaves pinnate, the primary leaflets not aggregated at end of rachis, the leaflets not sensitive to touch. Inflorescences not in heads.
        5. Leaves with stipules. Flowers zygomorphic. Fruits legumes (developed from one carpel, opening along two sutures) or modified legumes (loments).
            6. Leaves trifoliolate. Flowers markedly zygomorphic, with pronounced standard and keel, the standard external to lateral petals (best seen in bud), the corolla white to lavender or red or blue violet; 9 filaments fused into distinct column surrounding ovary, one filament free. Fruits loments. . . . . *Desmodium* (*Fabaceae*).

6. Leaves usually not trifoliolate. Flowers slightly zygomorphic, without pronounced standard and keel, the standard internal to lateral petals (best seen in bud), the corolla yellow; filaments not fused into distinct column surrounding ovary. Fruits not loments. . . . . *Senna* (*Caesalpiniaceae*).
5. Leaves without stipules. Flowers actinomorphic. Fruits not legumes.
7. Slash of bark and crushed leaf blades with strong aroma of turpentine. Petiolules with pulvinules. Seeds surrounded by white arilloids. . . . . *Protium pilosum* (*Burseraceae*).
7. Slash of bark and crushed leaf blades without strong aroma of turpentine. Petiolules without pulvinules. Seeds not surrounded by white arilloids.
8. Cut bark not bitter to taste, often with slight spicy aroma. Leaves with indeterminate growth, a small bud at apex. Stamen filaments fused into tube; ovary entire, not on gynophore. Fruits dehiscent. . . . . *Guarea* (*Meliaceae*).
8. Cut bark bitter to taste, without slight spicy aroma. Leaves with determinate growth, without bud at apex. Stamen filaments not fused into tube; ovary lobed, on gynophore. Fruits indehiscent. . . . . *Simaroubaceae* (*Simaba*).

Key 6. Shrubs

Subkey 2. Simple Opposite or Whorled Leaves

1. Ovaries inferior.
2. Petals fused into tube.
3. Leaves without stipules. Flowers aggregated into heads, these subtended by >2 overlapping bracts (phyllaries); calyx modified into bristles or scales (pappus). Fruits cypselas. . . . . *Asteraceae*.
3. Leaves with stipules. Flowers not aggregated into heads nor subtended by >2 overlapping bracts or, if in heads, these subtended by two colorful bracts (some species of *Psychotria*); calyx not modified. Fruits berries, drupes, or capsules. . . . . *Rubiaceae*.
2. Petals not fused into tube.
4. Leaf blades with pellucid punctations, without arcuate secondary veins, the tertiary veins reticulate. Stamens >20; anthers without appendages or oil glands. Fruits berries. . . . . *Myrtaceae*.
4. Leaf blades without pellucid punctations, with arcuate secondary veins, the tertiary veins in ladder-like arrangement between secondary veins. Stamens ≤10; anthers often with appendages or oil glands. Fruits capsules or berries. . . . . *Melastomataceae*.
1. Ovaries superior.
5. Plants with colored exudate.
6. Exudate yellow. . . . . *Clusiaceae*.
6. Exudate white.
7. Shrubs. Leaves without stipules. Flowers without corona; stamens adnate to corolla; pollen not in pollinia. . . . . *Apocynaceae*.
7. Subshrubs (or herbs). Leaves with stipules. Flowers with corona; stamens adnate to gynoecium; pollen in pollinia. . . . . *Asclepiadaceae*.
5. Plants without colored exudate.
8. Flowers actinomorphic.
9. Crushed leaves often with lemony aroma. Flowers inserted within swollen receptacle; anthers valvate. Fruits surrounded by fleshy receptacles to form pseudocarps. . . . . *Monimiaceae*.
9. Crushed leaves without lemony aroma. Flowers not inserted within swollen receptacle; anthers not valvate. Fruits not surrounded by fleshy receptacles.
10. Petals not fused into tube.
11. Stamens 5, not inserted within disc, the anthers with connective extended into appendage; stigma undivided, the ovary with parietal placentation. . . . . *Rinorea* (*Violaceae*).
11. Stamens 3, inserted within disc, the anthers without connective extended into appendage; stigma divided into 3 lobes, the ovary with axile placentation. . . . . *Hippocrateaceae*.
10. Petals or tepals fused into tube.
12. Perianth uniseriate. . . . . *Nyctaginaceae*.
12. Perianth biseriate.
13. Leaves with interpetiolar stipules. Ovary 2–3-locular; placentation axile. . . . . . *Loganiaceae*.
13. Leaves without interpetiolar stipules. Ovary unilocular; placentation usually not axile. . . . . *Gentianaceae*.

8. Flowers slightly to markedly zygomorphic.
 14. Petals not fused into tube.
  15. Flowers perigynous; petals magenta; ovules >1 per locule. Fruits capsules, the seeds released from fruit. . . . . *Lythraceae.*
  15. Flowers hypogynous; petals yellow; ovules 1 per locule. Fruits breaking into mericarps, the seeds retained within mericarps. . . . . *Malpighiaceae.*
 14. Petals fused into tube.
  16. Crushed leaves often with pungent aroma. Ovaries with 1 ovule per locule, the ovary often lobed at summit.
   17. Flowers in heads or, if not in heads, the calyx with distinct protuberance; corolla markedly zygomorphic; style gynobasic. Fruits of 4 nutlets. *Lamiaceae.*
   17. Flowers not in heads and calyx without protuberance; corolla slightly zygomorphic; style terminal. Fruits not of 4 nutlets. . . . . *Verbenaceae.*
  16. Crushed leaves without pungent aroma. Ovaries with 2 to numerous ovules per locule, the ovary never lobed at summit.
   18. Leaf blades often with cystoliths, appearing as light-colored streaks or protuberances. Flowers often, but not always, subtended by conspicuous bracts. Fruits fiddle- or club-shaped. Seeds without endosperm. . . . . *Acanthaceae.*
   18. Leaf blades without cystoliths. Flowers not subtended by conspicuous bracts. Fruits not fiddle- or club shaped. Seeds with endosperm.
    19. Flowers <10 mm long; ovary bilocular; placentation axile. . . . . *Scrophulariaceae.*
    19. Flowers usually >10 mm long; ovary unilocular; placentation parietal. . . . . . *Gesneriaceae.*

Key 6. Shrubs

Subkey 3. Simple Alternate Leaves

1. Plants with white exudate.
 2. Leaves with stipules. Inflorescences with staminate flowers in spikes, pistillate flowers in heads. Flowers actinomorphic, small; perianth uniseriate, not deep pink; ovary superior. . . . . *Clarisia* (*Moraceae*).
 2. Leaves without stipules. Inflorescences solitary, not of separate staminate or pistillate flowers. Flowers zygomorphic, large; perianth biseriate; corolla deep pink; ovary inferior. . . . . . . *Centropogon* (*Campanulaceae*).
1. Plants without white exudate.
 3. Stems with swollen nodes. Flowers and fruits in dense spikes. Flowers without perianth. . . . . *Piper* (*Piperaceae*).
 3. Stems without swollen nodes. Flowers and fruits not in dense spikes. Flowers with perianth.
  4. Flowers with either tepals or petals fused into tube around or above ovary, at least toward base.
   5. Perianth zygomorphic. . . . . *Brunfelsia* (*Solanaceae*).
   5. Perianth actinomorphic.
    6. Flowers congested into dense heads, subtended by a series of overlapping bracts (involucre). Calyx modified into hairs, scales, or bristles (pappus); ovary inferior. Fruits cypselas. . . . . *Asteraceae.*
    6. Flowers not congested into heads nor subtended by overlapping bracts. Calyx not modified as above; ovary superior. Fruits not cypselas.
     7. Perianth uniseriate.
      8. Inflorescences terminal, branched. Fruits orange, ca. 1 cm diam., with a single seed. . . . . *Schoenobiblus* (*Thymelaeaceae*).
      8. Inflorescences clustered in leaf axils, not branched. Fruits light yellow at maturity, 1–4 cm diam., with ca. 7 seeds. . . . . *Casearia bracteifera* (*Flacourtiaceae*).
     7. Perianth biseriate.
      9. Inflorescences helicoid (*Heliotropium*). Style often twice divided into 4 parts (*Cordia*). . . . . *Boraginaceae.*
      9. Inflorescences not helicoid. Style not twice divided into 4 parts.
       10. Flowers arising from petiole. Petals bifid at apex. . . . . *Dichapetalaceae.*
       10. Flowers not arising from petiole. Petals entire at apex.
        11. Leaves, flowers, and fruits with pellucid streaks or punctations (these sometimes present only on petals and ovary). . . . . *Myrsinaceae.*
        11. Leaves, flowers, and fruits without pellucid streaks or punctations.

12. Leaves without pulvinate petioles. Anthers with poricidal dehiscence. Fruits like miniature tomatoes, bilocular with axile placentation. . . . . . *Solanaceae.*
12. Leaves with pulvinate petioles. Anthers with lateral dehiscence. Fruits not like miniature tomatoes, unilocular with free central placentation or of 5 free mericarps.
13. Pulvinus at base of petiole. Calyx inconspicuous; corolla orange; ovary entire. Fruits indehiscent. *Theophrastaceae.*
13. Pulvinus in middle or at apex of petiole. Calyx conspicuous, pink-red; corolla white; ovary lobed, nearly apocarpous. Fruits dehiscent. . . . . *Erythrochiton* (*Rutaceae*).
4. Flowers with neither tepals or petals fused into tube.
14. Ovary lobed, at least upper or lower part of carpels free. Fruits often of distinct monocarps or separating at maturity.
15. Flowers unisexual.
16. Leaf blades with secondary veins arching upward from base, the margins entire. Plants dioecious. Fruits of separate monocarps. . . . . *Menispermaceae.*
16. Leaf blades with secondary veins pinnate, the margins finely serrate. Plants monoecious. Fruits not of separate monocarps. . . . . *Acalypha diversifolia* (*Euphorbiaceae*).
15. Flowers bisexual.
17. Leaf blades with pellucid punctations. Carpels free except at apex where they join style. . . . . *Almeida* (*Rutaceae*).
17. Leaf blades without pellucid punctations. Carpels entirely free or free at least toward apex.
18. Bark fibrous, easily peeled off in long strips, often emitting slightly spicy aroma. Corolla in two whorls of 3 petals each; stamens >10. . . . . *Annonaceae.*
18. Bark not fibrous, not easily peeled off in long strips, not emitting slightly spicy aroma. Corolla in single series; stamens ≤10.
19. Leaf blades with stellate hairs. Flowers bisexual. Filaments fused into conspicuous tube (monadelphous); style not gynobasic. Fruits not on enlarged, red pedicel. . . . . *Malvaceae.*
19. Leaf blades without stellate hairs. Flowers unisexual or bisexual. Filaments not fused into tube; style gynobasic from middle of ovary. Fruits on enlarged, red torus. . . . . *Ochnaceae.*
14. Ovary entire.
20. Stamens with filaments fused into tube, at least for most of lower half.
21. Flowers zygomorphic. . . . . *Polygala* (*Polygalaceae*).
21. Flowers actinomorphic.
22. Stems often with conspicuous bracts. Leaf blades without stellate hairs. Petals with appendages toward base on adaxial surface; style simple. Fruits with a single seed. . . . . *Erythroxylaceae.*
22. Stems without conspicuous bracts. Leaf blades with stellate hairs. Petals without appendages; style divided. Fruits with >1 seed. . . . . *Malvaceae.*
20. Stamens without filaments fused into tube.
23. Ovary unilocular; placentation parietal.
24. Flowers zygomorphic; lower petal gibbous or spurred at base; anthers with connectives prolonged into appendages. . . . . *Hybanthus* & *Noisettia* (*Violaceae*).
24. Flowers actinomorphic; lower petal not gibbous or spurred at base; anthers without connectives prolonged into appendages.
25. Leaf blades oblanceolate, with several glands along petioles and lower part of leaf blades. Petals bright yellow; stigma penicellate.. . . . . *Turnera* (*Turneraceae*)
25. Leaf blades usually elliptic, without glands. Petals usually white or dull yellow; stigma not penicellate. . . . . *Flacourtiaceae.*
23. Ovary multilocular or, if unilocular, the placentation not parietal.
26. Perianth uniseriate.
27. Leaf with distinctly serrate or lobed margins. Flowers unisexual, <2 mm diam. . . . . *Urera* (*Urticaceae*).
27. Leaf blades with entire to subentire, unlobed margins. Flowers bisexual, >3 mm diam. . . . . *Phytolaccaceae.*
26. Perianth biseriate.

28. Corolla yellow; ovary inferior. Fruits dehiscent. .... *Onagraceae.*
28. Corolla usually white; ovary superior. Fruits indehiscent.
    29. Calyx not entire nor enlarged and red; anthers with poricidal dehiscence; ovary bilocular, each locule with numerous ovules. Fruits berries. ..... *Solanaceae.*
    29. Calyx entire or enlarged and red; anthers without poricidal dehiscence; ovary not bilocular, each locule with a single ovule. Fruits drupes. .... *Olacaceae.*

ACANTHACEAE (Acanthus Family)

Dieter C. Wasshausen

Prostrate, erect, or rarely climbing herbs or shrubs. Stems decussate, frequently angled, more rarely terete, often with transverse ridges across node. Stipules absent. Leaves simple, usually decussately arranged; blades usually with cystoliths (intercellular concretions which appear as white streaks on adaxial surface, but lacking in *Gynocraterium* and *Aphelandra*), the margins entire or undulate. Inflorescences axillary or terminal, solitary or arranged in spikes, cymes, or racemes, often densely clustered, frequently in axils of conspicuous bracts, the pedicels with two bracteoles; bracts often leaf-like, sometimes enclosing corolla tube. Flowers zygomorphic to nearly actinomorphic, bisexual; calyx synsepalous at least basally, persistent, the lobes 4 or 5; corolla sympetalous, the tube usually divided into distinct narrow tube basally and wider throat apically, the limb 5-lobed or 2-lipped (rarely 1-lipped), the lobes imbricate or contorted in bud; stamens 4 fertile and 0 or 1 staminode, usually didynamous, or 2 fertile and 0 or 2 staminodia, adnate to corolla tube, the anther locules 1–2, usually parallel, rarely divergent or transverse, muticous (without a projection), obtuse or rounded, or basally spurred; ovary superior, 2-locular, the stigma often 2-lobed, one lobe frequently smaller than other or entire or crateriform; placentation of 2–10 superposed ovules per locule, rarely (*Gynocraterium*) with numerous ovules in 2 rows in each locule. Fruits loculicidal 2-valved capsules, often fiddle- or club-shaped, frequently explosively dehiscent on drying or wetting. Seeds usually borne on hooklike funicles (retinacula), these persistent in capsule after seeds have been discharged, or funicles lacking (*Gynocraterium*), 2 to many, usually flattened, sometimes spheroidal (*Gynocraterium*), the testa smooth or roughened, sometimes with mucilaginous trichomes which expand when moistened; endosperm absent.

The genus *Mendoncia* is often included in the Acanthaceae but is placed in the Mendonciaceae in this Guide.

Bremekamp, C. E. B. 1938. Acanthaceae. *In* A. Pulle (ed.), Flora of Suriname **IV(2):** 166–256. J. H. De Bussy, Amsterdam.
Wasshausen, D. C. 1995. Acanthaceae. *In* J. A. Steyermark, P. E. Berry & B. K. Holst (gen. eds.), Flora of the Venezuelan Guayana **2:** 335–374. P. E. Berry, B. K. Holst & K. Yatskievych (vol. eds.). Missouri Botanical Garden, St. Louis; Timber Press, Portland, Oregon.

1. Stigma crateriform. Seeds irregularly shaped, borne on papilliform placentae, hooklike funicles lacking. .... *Gynocraterium.*
1. Stigma minutely 2-lobed or entire. Seeds subglobose to lenticular, borne on curved hooklike funicles.
  2. Fertile stamens 4.
    3. Anthers all 1-locular. .... *Aphelandra.*
    3. Anthers all 2-locular or those of longer pair 2-locular and shorter pair 1-locular.
      4. Calyx lobes similar or nearly so.
        5. Inflorescences dense, terminal or axillary spikes, these sometimes 4-ranked; bracts conspicuous, green or pale green, closely imbricate, 8–19 × 5.5–14 mm.
          6. Sprawling to erect perennial herbs. Corolla 5-lobed; anthers all 2-locular. Capsules substipitate. Seeds 8–12. .... *Blechum.*
          6. Perennial shrubs to subshrubs. Corolla bilabiate; anthers of longer pair 2-locular, those of shorter pair 1- locular. Capsules stipitate. Seeds 2–4. .... *Herpetacanthus.*
        5. Inflorescences axillary, or loose terminal or axillary cymes, racemes, or panicles (rarely in terminal heads); floral bracts inconspicuous, usually small, not imbricate, 0.8–1.5 × 0.75–1.2 mm.
          7. Inflorescences panicles, cymes, spikes, or racemes, not secund. Corollas bright red, white, white with mauve, or pale rose-colored.
            8. Inflorescences branching, axillary, once or twice dichasially branched, long-pedunculate dichasia or cymes; peduncle 1.5–10 cm long. Corollas funnelform, the limb of 5 equal, spreading lobes 30–45 mm long, white or white with mauve or pale rose; nectary disc annular, without distinct projections. .... *Ruellia.*
            8. Inflorescences narrow, terminal and axillary panicles; peduncle 9–14.5 cm long. Corollas tubular, bilabiate, 52–83 mm long, bright red; nectary disc annular, with projections 1.4–3 mm high. .... *Polylychnis.*
          7. Inflorescences secund racemes. Corollas blue, mauve, or white drying yellow. .... *Asystasia.*
      4. Calyx lobes very dissimilar, posterior and anterior lobes much larger than lateral lobes. ..... *Lepidagathis.*

2. Fertile stamens 2.
   9. Staminodia absent, or if present then rudimentary.
      10. Corollas bright red, sometimes becoming orange-red with age; anther locules parallel or slightly divergent at base, definitely not superposed and anthers not mucronate or appendaged.
         11. Inflorescences dense, solitary, terminal spikes, the flowers not secund; bracts ovate to lanceolate, 15–25 × 4–9 mm. Corollas 55–70 mm long. . . . . *Pachystachys.*
         11. Inflorescences small panicled racemes terminating branches, the flowers secund; bracts narrowly triangular, ca. 2 × 0.5 mm. Corollas 25–35 mm long. . . . . *Anisacanthus.*
      10. Corollas violet, pale violet, or mauve, marked with lighter streaks or maroon markings on the lower lip, or pale red to dull crimson; anthers 2-locular with 1 locule rarely sterile, locules superposed, with lower anther usually spurred or apiculate at base. . . . . *Justicia.*
   9. Staminodia 2.
      12. Shrubs, often spiny in branch axils. Calyx lobes 4; corolla large and conspicuous, 35–45 mm long. . . . . *Barleria.*
      12. Herbs, subshrubs, or shrubs, never spiny in branch axils. Calyx lobes 5; corolla not large and conspicuous, 10–17 mm long. . . . . *Pulchranthus.*

ANISACANTHUS Nees

Branching herbs or subshrubs. Stems (older) covered with brown or gray exfoliating bark. Leaves: petioles present or absent; blades linear to lanceolate, cystoliths present. Inflorescences spicate, racemose, or paniculate, the flowers secund or opposite, borne singly or several at inflorescence node; bracts and bracteoles mostly triangular to linear, usually caducous. Flowers: calyx 3–5-lobed, the lobes triangular to linear; corolla tubular to funnelform, ± arcuate, 2-lipped (lips usually recurved), mostly pilose, usually red, the posterior lip entire or slightly emarginate, the anterior lip 3-lobed; stamens 2, the anthers 2-locular, subequal, not mucronate or appendaged. Capsules subpyriform, slightly beaked. Seeds 2–4, homomorphic, flattened, each supported by curved retinaculum.

Anisacanthus secundus Leonard

Herbs, 1–3 m tall. Stems terete, finely striate. Leaves: petioles to 5 mm long; blades lanceolate, 5–8 × 2–2.5 cm, glabrous except on primary and secondary veins, these puberulent adaxially, white-pilose abaxially. Inflorescences borne in small panicled racemes terminating branches, the flowers secund; bracts narrowly triangular, ca. 2 × 0.5 mm. Flowers: calyx campanulate, the lobes subulate, 2–2.5 × 1 mm; corolla 25–35 mm long, finely pubescent, bright red, the lips ca. 9 mm long. Capsules 20 × 6 × 4 mm, glabrous. Seeds ca. 5 mm diam., white. Fl (Jun); occasional, on dry sandy soils in forests. Not yet collected from our area but to be expected.

APHELANDRA R. Br.

Perennial suffrutescent herbs or shrubs. Stems terete to quadrangular, with swollen nodes. Leaves: petioles present; blades oblong to elliptic, large, cystoliths lacking. Inflorescences terminal or axillary spikes, these conspicuous with showy bracts and flowers; bracts imbricate, the margins toothed; bracteoles 2, laterally subtending calyx. Flowers: calyx 5-lobed, the lobes lanceolate, subequal in length; corolla straight or curved, usually pale to bright red, the limb bilabiate, the upper lip erect, 2-lobed or entire, the lower lip 3-lobed, reflexed or spreading, the lobes subequal to strongly dimorphic; stamens 4, the anthers 1-locular, narrow, often pilose dorsally. Capsules clavate to subglobose. Seeds 4, homomorphic, flattened to subglobose, brown, each supported by curved retinaculum.

Aphelandra aurantiaca (Scheidw.) Lindl.

Unbranched perennial herbs, 1.5 m tall. Stems flattened, glabrous. Leaves: blades ovate to ovate-elliptic to elliptic, 15–20 × 5–7 cm, glabrous. Inflorescences terminal, spikes, 9–11 cm long; bracts lance-ovate to ovate-elliptic, 20–25 × 10–12 mm, glandular-puberulent, green or reddish along margins, the margins dentate; bracteoles subulate, 5–6 × 0.5–1 mm. Flowers: calyx ca. 10 mm long; corolla 55–60 mm long, pubescent, red to reddish-orange, the upper lip 20–22 mm long, entire to emarginate, the lower lip spreading, the middle lobe 20–25 × 12 mm, the lateral lobes reduced. Capsules 15–16 mm long, glandular puberulent. Seeds subcircular, 5 × 4–4.5 mm, covered with trichome-like papillae. Fl (Jul, Aug); occasional, along streams and on slopes in forests.

ASYSTASIA Blume

Perennial herbs or shrubs. Stems erect, procumbent, or clambering. Leaves: blades narrowly lanceolate to ovate, cystoliths present. Inflorescences terminal, simple or branched racemes or spikes, these often lax and unilateral, the flowers opposite or alternate, solitary or glomerate in axils or bracts; bracts linear to narrowly deltoid; bracteoles minute. Flowers: calyx 5-lobed, the lobes linear-setaceous or lanceolate, subequal; corolla funnelform, purplish, blue, yellow, or white, the tube short, the throat campanulate, the limb 5-lobed, the lobes equal; stamens 4, didynamous, the anthers 2-locular, minutely spurred or muticous at base, the locules parallel. Capsules stipitate, pubescent, elliptic. Seeds 2–4, orbicular or angled, compressed, each supported by curved retinaculum.

Asystasia gangetica (L.) T. Anderson

Erect to ascending shrubs, to 1.5 m. Stems subquadrangular, branching. Leaves: petioles present; blades ovate, 4.5–6 × 3.5–4.5 cm, glabrous, the base truncate. Inflorescences axillary or terminal, often 1-sided racemes, 10–15 cm long, the flowers single at nodes, sparse; pedicels present; bracts and bracteoles similar, triangular, 0.8–1.5 × 0.75–1 mm, the margins ciliolate. Flowers: calyx 6–8 mm long, hirsute, gland-dotted; corolla 30–40 mm long, glandular, puberulous, blue, mauve, or white drying yellow, the lobes rounded, 15 mm diam., conspicuously and coarsely reticulate-veined. Capsules 20–30 × 7 × 5 mm, glandular puberulent. Seeds ca. 4 mm diam., roughened, gray. Fl (Aug); native of Old World tropics, introduced and persisting as an escape.

BARLERIA L.

Erect herbs or shrubs. Stems sometimes with spines in axils. Leaves petiolate; blades often with spine-tipped apices, cystoliths present. Inflorescences axillary clusters or terminal spikes, each flower in axil of large bract, the bract sometimes with apical spine; bracteoles usually 2 at base of calyx. Flowers: calyx with 2 unequal pairs of lobes inserted at right angles to each other, the lobes entire or with toothed margins; corolla bilabiate, with tube narrow at base, widening into throat, usually yellow or blue, the lobes 5, the lowest lobe fused to adjacent lobes for shorter length than other 4 lobes, the upper lip 4-lobed, the lower lip 1-lobed; stamens 4, didynamous (shorter pair sometimes minute, but fertile) or 2, then with 0 or 2 staminodia, the anthers 2-locular, the locules equal, parallel to subsagittate; staminodia 2 when only 2 stamens present. Capsules ovoid. Seeds 2–4, homomorphic, lenticular, each supported by curved retinaculum.

Barleria lupulina Lindl.

Prickly shrubs, 0.5–2 m. Stems glabrous, often with downturned spines, ca. 1 cm long at nodes. Leaves: petioles short; blades narrowly elliptic, 5–11 × 0.8–1.2 cm, rigid, glabrous, dark green with red midrib adaxially, the apex spine-tipped. Inflorescences terminal, erect spikes; bracts broadly obovate, imbricate, 15–17 × 12 mm, green with reddish-brown upper half, pubescent with cup-shaped glands basally; bracteoles lanceolate, 4–5.5 mm long. Flowers: calyx lobes lanceolate, wider pair 10 mm long, narrower pair 8 mm long; corolla 35–45 mm long, pale yellow to orange-yellow or salmon-colored, bilabiate, the tube sparingly puberulous, the lobes of upper lip obovate, 20 × 10 mm, the lower lobe oblong, 20 × 5 mm; stamens 2, the anthers 2- locular; staminodia 2. Capsules 15 × 7 × 4 mm, glabrous. Seeds ovoid, 8 × 5 mm, covered with long, appressed trichomes. Fl (Feb, Sep); native of Madagascar, cultivated in the village, escaped in other areas outside our area.

BLECHUM P. Browne

Sprawling to erect perennial herbs. Leaves: petioles present; blades thin, cystoliths present, the margins entire, crenate, or repand-dentate. Inflorescences terminal or axillary, densely spicate, the flowers commonly sessile or short pedicellate in axils of foliaceous bracts; bracts opposite, usually ovate to suborbicular, imbricate, green, the margins entire; bracteoles 2. Flowers: calyx 5-lobed, the lobes linear-subulate; corolla usually lavender, white, or purplish, the tube slender, somewhat enlarged above, the limb subregular, 5-lobed; stamens 4, didynamous, the anthers 2-locular, the locules parallel, equal, rounded or acute at base. Capsules substipitate, ellipsoid. Seeds 8–12, homomorphic, lenticular, each supported by curved retinaculum.

Blechum pyramidatum (Lam.) Urb. [Syn.: *Blechum brownei* Juss.]

Herbs, 0.2–0.9 m tall. Stems quadrate to bisulcate, pubescent. Leaves: blades ovate to elliptic, 2–12 × 1–7 cm, pubescent. Inflorescences terminal or axillary, densely bracteate, 4-sided, dichasiate spikes 3–6 cm long, the dichasia opposite, 1–3-flowered; bracts 8–19 × 5.5–14 mm, pubescent; bracteoles 4.5–11 × 1–4.5 mm. Flowers: calyx 2.5–5 mm long, the lobes 2–3.8 mm long; corolla extending well beyond subtending bracts, 10 –20 mm long, pubescent, the tube 7.5–13 mm long, the lobes broadly subelliptic to subcircular, 1.5–6 × 1–6 mm. Capsules 5.5–7 × 3 × 1.5 mm, pubescent. Seeds 1.5–2 × 1.3–1.7 mm, the margins ringed by hygroscopic trichomes. Fl (Feb, Aug, Sep, Oct, Nov), fr (Feb, Nov); common, in weedy habitats.

GYNOCRATERIUM Bremek.

Herbs or subshrubs. Stems green, quadrangular and articulated, the roots whitish. Leaves with petioles; blades elliptic to ovate, cystoliths absent. Inflorescences dense terminal spikes, the flowers solitary or in groups of 3 at inflorescence nodes, similar in length to bracts; bracts and bracteoles narrow and extremely long, fimbriate. Flowers: calyx deeply 5-lobed, the posterior lobe similar to bracteoles, trinerved, the remaining 4 lobes equal, shorter and much narrower, uninerved; corolla tubular, white or pinkish-white, the lower half cylindrical, the upper half infundibular, the limb short, the lobes subequal, orbicular, the upper pair enclosing lower lobes in flower bud; nectary disc short, cupular; stamens 4, didynamous, inserted at middle of tube, the anthers 2-locular, ovoid, minutely apiculate basally, the locules equal; staminodia 1; stigma included, crateriform. Capsules ellipsoid, seed bearing nearly entire length. Seeds numerous, minute, irregularly shaped, borne on papilliform placentae, retinacula lacking.

Gynocraterium guianense Bremek.

Subshrubs, 0.3–1.2 m tall. Stems 3.5 mm diam., ferruginous-tomentose above, glabrous below. Leaves: petioles 10–25 mm long, ferruginous-pubescent; blades elliptic to ovate, 15–20 × 6–10 cm, rather thin, glabrous on both surfaces, deep-green adaxially, lighter green and shaded reddish abaxially, the margins subentire or undulate. Inflorescences terminal, sessile spikes 3–8 cm long; bracts narrowly linear, 15–17 × 1–1.3 mm, pinkish-white; bracteoles narrowly linear, 17–20 × 0.7–0.8 mm, ciliate. Flowers: calyx lobes linear, pinkish-white, the posterior lobe 13–21 × 1.2 mm, the remaining lobes 8–13 × 0.5–1 mm, ciliate; corolla 18–21 mm long, minutely puberulent adaxially, white, the lobes 2–2.5 × 1.5–2 mm; stamens included, the anthers and filaments glandular pubescent, the anther locules 0.8–0.9 mm long; stigma 1–2 mm diam. Capsules brownish, 4–5 × 2 × 1 mm, glandular puberulous, the apex obtuse. Seeds 0.3–0.4 × 0.3 mm, minutely puberulous, brown. Fl (Mar); occasional, in low wet areas in forest.

HERPETACANTHUS Nees

Perennial shrubs or subshrubs. Stems erect, subquadrangular, articulated. Leaves prominently covered with cystoliths. Inflorescences terminal, spicate, dense, the spikes solitary or in groups; bracts conspicuous, herbaceous; bracteoles small. Flowers: calyx 5-lobed, the lobes equal; corolla small, the tube cylindrical basally, the upper half campanulate and personate, the limb bilabiate, the upper lip erect, emarginate, the lower lip 3-lobed; stamens 4, didynamous, the anthers of longer pair 2-locular, locules superposed, the anthers of shorter pair 1-locular, all obtuse basally. Capsules stipitate, the apex acute. Seeds 2–4, homomorphic, lenticular, glabrous, each supported by curved retinaculum.

Herpetacanthus rotundatus (Lindau) Bremek.

Erect to ascending shrubs, to 1.5 m. Stems subquadrangular, branching. Leaves: petioles 15–25 mm long; blades elliptic to ovate, 7–14 × 4–6 cm, glabrous. Inflorescences terminal, several pedunculate spikes, 3–4 cm long, the flowers in pairs or solitary, subtended by paired bracts; bracts 3-nerved, oblique, ovate-orbicular, 12–13 × 8–10 mm, pale green; bracteoles narrowly triangular, 4 × 0.6 mm. Flowers: calyx lobes similar to bracteoles; corolla 15 mm long, white, the tube 11 mm long, the lips 3.8 mm long, the lower lip reflexed and puberulous. Capsules 10 mm long, puberulous, the stipe 4 mm long. Fl (Aug); rare, in non-flooded forest.

JUSTICIA L.

Herbaceous or shrubby perennials. Leaves: petioles absent or present; blades lanceolate, ovate, or elliptic, cystoliths present, the margins entire. Inflorescences simple spikes, or compound with dichasial or spicate subunits, occasionally reduced to condensed cluster of flowers, the flowers sessile or pedicellate; bracts prominent or inconspicuous, green or brightly colored, the margins entire; bracteoles 2. Flowers: calyx deeply 4- or 5-lobed, the lobes equal, or with posterior lobe sometimes reduced or absent; corolla greenish, white, yellow, orange, pinkish, red, or purplish, usually with white or colored markings on lower lip, the tube cylindric, the limb 2-lipped, the upper lip shallowly 2-lobed (rarely entire), the lower lip 3-lobed; stamens 2, the anthers 2-locular (1 locule rarely sterile), locules often unequal, usually superposed, often oblique, the lower anther usually spurred or apiculate at base; staminodia 0. Capsules stipitate. Seeds 2–4, homomorphic, lenticular, each supported by curved retinaculum.

1. Inflorescences of dense clusters of spikes forming a capitulum; bracts, bracteoles, and calyx lobes with margins long-ciliate. . . . . *J. sprucei.*
1. Inflorescences diffuse, solitary spikes or occasionally groups of 2 or 3 spikes or panicles of dichasiate spikes; bracts, bracteoles, and calyx lobes with margins not distinctively long-ciliate.
    2. Inflorescences of solitary spikes or occasionally groups of 2–3 spikes.
        3. Spikes dense, conspicuously bracteate, the flowers in superposed pairs or solitary, not secund; bracts lanceolate-ovate, 4–5 × 1.5–2 mm. . . . . *J. cayennensis.*
        3. Spikes lax, extremely long, not conspicuously bracteate, the flowers solitary, secund; bracts narrowly linear to spatulate, 5–8 × 0.6–1.5 mm. . . . . *J. potarensis.*
    2. Infloresence of large terminal panicles or panicles of dichasiate spikes.
        4. Decumbent herbs, 0.6–1 m long. Corollas small, 7.5–10 mm long, mauve with maroon markings on lower lip. . . . . *J. pectoralis.*
        4. Herbs or scandent shrubs, 2–3 m tall. Corollas large, 30–35(40) mm long, pale red to dull crimson. . . . . *J. secunda.*

Justicia cayennenis (Nees) Lindau

Erect herbs, 0.3–0.5 m tall. Stems pubescent or ultimately glabrescent, 2-ribbed. Leaves: petioles 2–5 mm long; blades lanceolate or lanceolate-elliptic, (3.5)6–10 × (1.1)1.6–3 cm, glabrous except midrib and lateral veins abaxially, these appressed pubescent, the margins indistinctly revolute. Inflorescences terminal spikes, usually solitary, occasionally in groups of 2–3, subsessile or pedunculate, the flowers borne either in superposed pairs or solitary and accompanied by rudiment of second flower in axils of bracts; bracts decussate, lanceolate-ovate, 4–5 × 1.5–2 mm, green, the margins ciliate; bracteoles linear, 4–5 × 0.5 mm, 1-nerved, the margins ciliate. Flowers: calyx 5-lobed, 4 lobes linear, equal, 4.5 × 0.25–0.4 mm, the posterior lobe reduced, the margin sparingly ciliolate;

corolla 10–12 mm long, puberulous abaxially, violet, marked with lighter streaks on lower lip, the upper lip 5–5.5 × 2 mm, entire, the lower lip obovate, 5–5.5 × 5 mm. Capsules 7 × 2 × 1 mm, puberulent, the stipe 2–2.5 mm long. Seeds 4, lenticular, 1 × 1.25 mm, brown. Fl (Oct, Nov); in non-flooded forest and in flood plain forest.

Justicia pectoralis Jacq. FIG. 1

Decumbent herbs, 0.6–1 m long, often rooting at lower nodes. Stems weak. Leaves: petioles slender, 2–12 mm long; blades narrowly to rather broadly lanceolate, 3–5 × 1 cm, glabrous or hirsutulous adaxially, glabrous abaxially. Inflorescences terminal, pedunculate panicles of dichasiate spikes, the inflorescence axes alternate or opposite, not appearing verticillate, the dichasia alternate or opposite, 1-flowered, 1 per axil, sessile; peduncle 15–30 mm long; pedicels absent to very short; bracts opposite, subulate, 1–2 × 0.2 mm, the margins minutely ciliolate; bracteoles subulate, 1–2 × 0.2 mm, the margins minutely ciliolate. Flowers: calyx 5-lobed, 4 lobes subulate, equal, 1.5–3 × 0.3–5 mm, the posterior lobe reduced, the margin minutely ciliolate; corolla 7.5–10 mm long, mauve, with maroon markings on lower lip, pubescent, the upper lip 2.5–4 mm long, entire, the lower lip 3–4.5 mm long. Capsules 7 × 2.5 × 1.5 mm, puberulent, the stipe 3.5–4 mm long. Seeds 4, lenticular, 1.5 × 1.5 mm, brown. Fl (Aug, Sep, Nov); common, in open, weedy habitats. *Cramentin, herbe chapentier, trevo cumaru* (Portuguese), *zerb charpentier* (Créole).

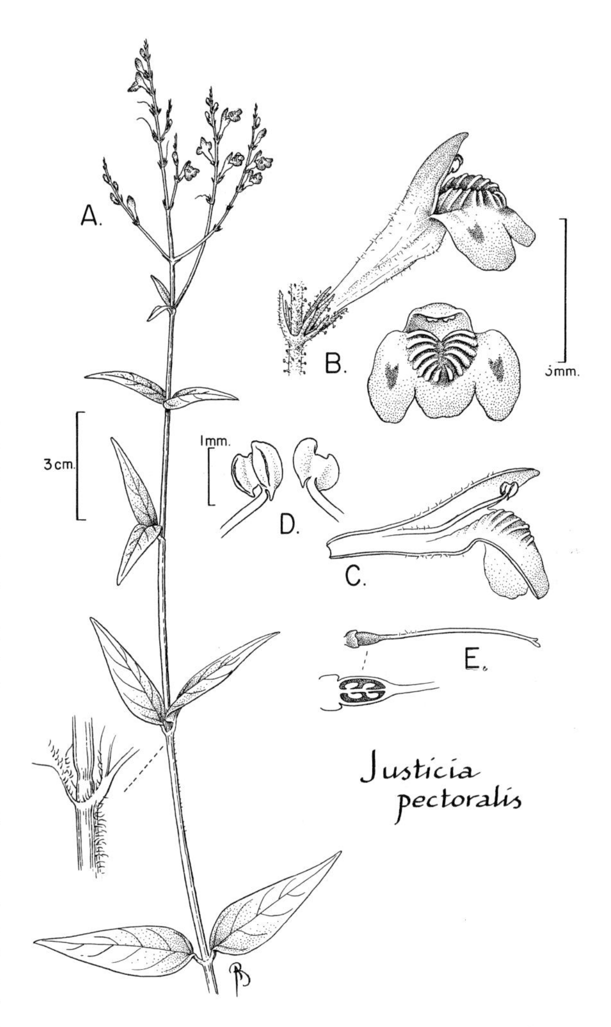

FIG. 1. ACANTHACEAE. *Justicia pectoralis (Mori et al. 20946).* **A.** Stem with leaves and inflorescence (right) and detail of leaf insertions on stem (below). **B.** Lateral (left) and frontal (right) views of flower. **C.** Medial section of corolla tube. **D.** Two views of athers. **E.** Pistil (above) with detail of medial section of ovary (below).

Justicia potarensis (Bremek.) Wassh.

Unbranched, decumbent herbs, (0.07)0.4–0.5 m long, rooting at base. Stems obtusely quadrangular and bisulcate, pubescent. Leaves: petioles 3–5 mm long; blades elliptic to oblong, 10–13 × 4–5 cm, glabrous except midrib and lateral veins below, these puberulent. Inflorescences axillary and terminal spikes, these in groups of 2–3, lax, secund, extremely long, glandular pubescent, the flowers solitary at inflorescence nodes; peduncle ca. 10 mm long; bracts decussate, narrowly linear or spatulate, 5–8 × 0.6–1.5 mm, glandular hirtellous, the margins ciliolate; bracteoles subequal, the margins ciliolate. Flowers: calyx 5-lobed, the lobes subequal, narrowly triangular, 4.5–5 × 0.3 mm, hirtellous, the margins ciliolate; corolla 12 mm long, shortly pubescent, pale violet, the tube cylindric, 5 mm long, the upper lip 7 × 4 mm, triangular, scarcely bilobed, the lobes of lower lip orbicular, 3 mm long, cleft. Capsules 8.5 mm long, pubescent, the stipe 3 mm long. Seeds 4, lenticular, 2.3 × 2 mm, brownish. Fl (Sep), fr (Jan); uncommon, in higher elevation forest.

Justicia secunda Vahl PL. 9a

Herbs or scandent shrubs, 1–3 m tall. Stems obtusely quadrangular, glabrous or sparingly puberulous. Leaves: petioles 10–35 mm long; blades ovate to oblong-ovate, 8–15 × 3–5 cm, firm, glabrous except midrib and lateral veins, these sometimes puberulous, the margins entire. Inflorescences terminal panicles, initially small but becoming large and much branched with age, 5–9 × 2–4 cm (without corollas), flowers secund and crowded on branches of panicles or distant (5–7 mm) with age; bracts subulate or narrowly triangular, 1.5 × 0.5 mm, the margins eciliate; bracteoles subulate, 1.5–2 × 0.25 mm, the margins eciliate. Flowers: calyx 5-lobed, the lobes subequal, 6–7 × 1.25 mm, hirtellous, the margin eciliate; corolla 30–35(40) mm, sparingly puberulous, pale red to dull crimson, the lips subequal, the upper lip erect, narrowly ovate, 22–23 × 4–5 mm, entire, the lower lip spreading, oblong, the lobes ovate, 2 × 1.5–2 mm. Capsules 9–10 × 4–4.5 × 2–2.5 mm, puberulent, the stipe 4–5 mm long. Seeds 4, lenticular, 2.8–3 × 2.5 mm, glabrous, slightly roughened. Fl (Jul, Aug, Sep, Nov, Dec); common, in secondary vegetation, especially along roads in forest. *Radié divin, radié du sang, Saint John, zerb vin* (Créole).

Justicia sprucei V. A. W. Graham

Creeping herbs or subshrubs, 0.45–0.8 m long, often rooting at lower nodes. Stems terete, puberulous. Leaves: petioles 10–20 mm long; blades oblong, 10–15 × 3–5(6) cm, glabrous except midrib and lateral veins abaxially, these puberulent. Inflorescences axillary, sometimes terminal dense clusters of spikes with 2 flowers per inflorescence node forming capitulum; bracts green, lanceolate, 16 × 2 mm, the margins long-ciliate; peduncle short; bracteoles 15 × 1 mm, margin long-ciliate. Flowers: calyx 5-lobed, the lobes subequal, lanceolate, 10 × 1.5 mm, the margins long-ciliate; corolla 35 mm long, white, the tube 25 mm long, puberulous, the upper lip 10 × 6 mm, entire, the lower lip 11 mm long, the lobes obovate, the lateral lobes 6–7 × 6 mm, the middle lobe 7 × 8 mm. Capsules oval, 15 × ca. 5 mm, puberulent, brown, the stipe 6 mm long. Seeds 4, lenticular, 3.5 × 3 mm, nitid, brown. Fl (Jun, Aug), fr (Sep, Nov); occasional, in non-flooded forest.

LEPIDAGATHIS Willd.

Diffuse or rarely erect perennial herbs. Leaves: petioles absent to present; blades ovate to lanceolate, cystoliths present. Inflorescences axillary or terminal, densely bracteate spikes composed of verticillasters, these (subtended by 3-veined bracts or lowermost by a pair of cauline leaves) consisting of 3–7 flowers, the lateral flowers subtended by 1–3-veined, green bracts; bracteoles 1-veined; pedicels scarcely developed. Flowers: calyx 5-lobed, the lobes heteromorphic, the posterior lobe largest, the lateral lobes smallest, the anterior lobes partially connate; corolla small, as long as or slightly exceeding calyx, whitish, lavender, or violet, the limb bilabiate, the upper lip rugulate, emarginate or shallowly 2-lobed, the lower lip 3-lobed; stamens 4, didynamous, included in corolla tube, the anthers 2-locular, the locules equal to unequal, lower or sometimes both minutely appendaged basally; staminodia 0. Capsules estipitate. Seeds 2–4, homomorphic, lenticular, each supported by curved retinaculum.

Lepidagathis alopecuroidea (Vahl) Griseb. [Syn.: *Teliostachya alopecuroidea* (Vahl) Nees]

Erect herbs, to 0.4 m tall. Stems quadrangular, bifariously pubescent. Leaves: petioles 15–30 mm long; blades elliptic to ovate-elliptic, 3–8 × 1.2–2.5 cm, glabrous except midrib and margin, these pubescent, the base attenuate. Inflorescences terminal, cylindric, densely bracteate spikes, 2–6 cm long, the verticillaster usually 3-flowered; peduncle present or absent; bracts obovate, 5–8 × 1.8–3 mm, the apex aristate; pedicels poorly developed; bracteoles lanceolate, 4–6 × 0.5 mm, the apex aristate. Flowers: calyx 5–7 mm long, the posterior lobe obovate, the lateral lobes lance-subulate, the anterior lobes oblanceolate to narrowly elliptic; corolla 4.5–6 mm long, white with irregular pink markings on lower lip, the tube cylindric, 2.5–4 mm long, the upper lip 1.5–2 mm long, 2-lobed, the lower lip 1.8–2 mm long, the lobes subcircular. Capsules 3.5–4.5 mm long. Seeds 4, 1–1.2 × 1 mm. Fl (Aug, Sep); common, in secondary vegetation around Saül.

PACHYSTACHYS Nees

Herbs or subshrubs. Stems terete or subquadragular, puberulous when young. Leaves: petioles present; blades large, cystoliths present. Inflorescences terminal, spicate, dense; bracts conspicuous, herbaceous; bracteoles small or none, the flowers in verticillasters (each consisting of 3–4 flowers). Flowers: calyx 5-lobed, the lobes relatively short; corolla bright red (or white with yellow bract if cultivated), the tube slenderly obconic, curved, the limb bilabiate, the upper lip narrow, 2-lobed, the lower lip 3-lobed, the lobes subequal, oblong or ovate; stamens 2, equaling or slightly exserted beyond tip of upper lip, the anthers 2-locular, deeply sagittate, the locules equal, basally muticous; staminodia absent or rudimentary. Capsules clavate. Seeds 4, homomorphic, flattened, each supported by curved retinaculum.

Pachystachys coccinea (Aubl.) Nees FIG. 2, PL. 9b

Shrubs, 1–2.5 m tall. Stems puberulous when young. Leaves: petioles 2–6 cm long; blades elliptic to oblong-obovate, 18–25 × 6–9 cm, glabrous except primary and secondary veins, these puberulous. Inflorescences solitary, terminal, spikes, 9–22 cm long, bearing conspicuous bracts and flowers; bracts green, ovate to lanceolate, 15–25 × 4–9 mm, puberulent, usually glandular, the margins entire; bracteoles linear-lanceolate, 5 × 0.5 mm. Flowers: calyx campanulate, 5.5–6 mm long, the lobes lanceolate, equal, 5 × 0.5 mm; corolla 55–70 mm long, sparingly hirtellous, bright red becoming orange-red with age, the upper lip erect, 20–25 mm long, emarginate, the lower lip oblong, 17–20 × 4–5 mm. Fl (Jan, Sep, Oct, Nov); locally common, in forest clearings and margins and in moist areas along streams. *Cramentine rouge* (Créole).

POLYLYCHNIS Bremek.

Large herbs or shrubs. Stems quadrangular and articulated. Leaves: petioles present; blades elliptic to narrowly ovate, cystoliths present. Inflorescences narrow, terminal and axillary, panicles, the basal branchlets once or twice dichasially branched, sometimes subtended by very reduced leaf blades, the distal branching monochasial; bracts small; bracteoles none; pedicels present. Flowers: calyx campanulate, basally unequal and shortly 5-lobed or toothed; corolla, tubular, slightly curved, bright red, the limb short, bilabiate, not at all spreading, the upper lobes united into bilobate upper lip; nectaries annular, with distinct projections; stamens 4, didynamous, the anthers 2-locular, the base obtuse, the connective apiculate. Capsules clavate. Seeds 6, homomorphic, flattened to lenticular, each supported by curved retinaculum.

Polylychnis fulgens Bremek. FIG. 3, PL. 9c

Large herbs or scandent shrubs, 1.5–3 m tall. Stems green, glabrous. Leaves: petioles 5–30 mm, canaliculate; blades elliptic-oblong, 13–30 × 4.5–15 cm, glabrous on both surfaces, subentire or irregularly dentate. Inflorescences terminal, sometimes axillary, panicles; peduncle 9–14.5 cm long; basal branchlets 20–55 mm long, subtended by narrow, triangular bracts 3 × 1.2 mm; floral bracts broadly triangular, 1–1.5 × 1.2 mm; pedicels 6–12 mm long. Flowers: calyx green, glabrous, the lobes narrowly triangular, 2.5–3.75 × 1.5 mm; corolla 52–83 mm long, the lips 9–12 mm long, the lobes of upper lip obovate, 3–6 mm long, the lower lip 3-lobed, with lobes obovate, 11 × 7.5 mm; disc 0.7–2 mm high, conical projections 1.4–3 mm high; stamens exserted, the anthers 4 mm long. Capsules 24–26 × 5 × 3 mm, glabrous, apiculate, the stipe 12 × 2.5 mm. Seeds 3–3.5 mm diam., glabrous, the margins ringed by hygroscopic trichomes. Fl (Jan, Feb, Mar, Jun, Sep, Oct, Nov, Dec); common, in low wet areas in forests. Visited and presumably pollinated by hummingbirds, the nectaries visited by ants after corollas fall (Gracie, 1991).

Plates 9–16

Plate 9. ACANTHACEAE. **a.** *Justicia secunda* (*Mori et al. 20887*), inflorescence with open flower and buds. **b.** *Pachystachys coccinea* (*Mori & Gracie 24209*), two inflorescences; note bracts subtending flowers. **c.** *Polylychnis fulgens* (*Mori et al. 23903*), inflorescence with open flower and buds. **d.** *Pulchranthus congestus* (*Mori et al. 20863*), inflorescence with open flower and bud; note lack of conspicuous bracts; compare with *P. variegatus*. **e.** *Ruellia rubra* (*Mori et al. 22085*), flower. **f.** *Pulchranthus variegatus* (*Mori et al. 23079*), inflorescence with open flowers; note conspicuous bracts; compare with *P. congestus*.

Plate 10. ANACARDIACEAE. **a.** *Astronium ulei* (*Mori & Gracie 23859*), staminate flowers and buds. **b.** *Astronium ulei* (*Mori & Gracie 23859*), slash of trunk. **c.** *Astronium ulei* (*Mori et al. 24730*), wind-dispersed fruit with expanded calyx lobes. **d.** *Spondias mombin* (*Mori et al. 21534*), part of inflorescence. **e.** *Tapirira obtusa* (*Mori & Gracie 24213*), part of inflorescence. **f.** *Thyrsodium puberulum* (*Mori et al. 24013*), inflorescence with open flowers and buds. **g.** *Thyrsodium spruceanum* (*Mori & Gracie 24215*), inflorescence with open flowers and buds.

Plate 11. ANNONACEAE. **a.** *Annona prevostiae* (*Mori et al. 22000*), flower. **b.** *Annona prevostiae* (*Mori et al. 22000*), flower with one outer petal bent down to show basal ring of stamens surrounding many carpels. **c.** *Cremastosperma brevipes* (*Mori et al. 22721*), flowers and bud arising from stem. **d.** *Cremastosperma brevipes* (*Mori et al. 22721*), fruit of several monocarps. **e.** *Cymbopetalum brasiliense* (*Mori & Gracie 23995*), fruit with two dehisced monocarps, one revealing arillate seeds. **f.** *Duguetia cadaverica* (*Mori et al. 21519*), flower and bud at apex of long, trailing inflorescence hidden by leaves.

Plate 12. ANNONACEAE. **a.** *Duguetia surinamensis* (*Mori et al. 23178*), apical view of flower. **b.** *Fusaea longifolia* (*Mori et al. 20761*), flower showing staminodes, stamens, and carpels. **c.** *Guatteria punctata* (*Mori et al. 23881*), flowers. **d.** *Guatteria punctata* (*Mori et al. 23881*), young fruits of many monocarps. **e.** *Guatteria foliosa* (*Mori et al. 21559*), young flower.

Plate 13. ANNONACEAE. **a.** *Oxandra asbeckii* (*Mori et al. 22779*), fruit of four monocarps. **b.** *Rollinia elliptica* (*Maas 8062*), flower. **c.** *Rollinia exsucca* (*Maas 8061*), flowers. **d.** *Unonopsis rufescens* (*Maas 8092*), infructescences. **e.** *Unonopsis stipitata* (*Mori et al. 20988*), flower and bud arising from trunk. **f.** *Unonopsis stipitata* (unvouchered), fruits with immature (orange) and mature (red) monocarps. **g.** *Xylopia nitida* (*Mori et al. 22201*), apical view of flower. **h.** *Xylopia nitida* (*Mori et al. 21618*), part of fruit with one monocarp sectioned to show immature arillate seeds.

Plate 14. APOCYNACEAE. **a.** *Ambelania acida* (*Mori et al. 22814*), axillary flower and bud. **b.** *Aspidosperma album* (*Mori & Gracie 23965*), flowers; note elongate, twisted corolla lobes. **c.** *Aspidosperma* sp. (unvouchered), seedling still partially covered by seed coat with winged fruit in background. **d.** *Bonafousia sananho* (*Mori & Gracie 21190*), fruit. **e.** *Bonafousia siphilitica* (*Mori et al. 21503*), apical view of flowers and buds. **f.** *Forsteronia guyanensis* (*Mori et al. 24197*), domatia in axils of secondary veins and midrib on abaxial surface of leaf. **g.** *Forsteronia acouci* (*Mori et al. 23000*), fruits.

Plate 15. APOCYNACEAE. **a.** *Forsteronia guyanensis* (*Mori et al. 21633*), part of inflorescence. **b.** *Himatanthus speciosus* (*Mori et al. 23792*), dehisced fruit showing winged seeds. [Photo by S. Mori] **c.** *Lacmellea aculeata* (*Mori & Gracie 23891*), freshly opened flowers and buds in evening. **d.** *Lacmellea aculeata* (*Mori & Gracie 18895*), immature inflorescence and fruit. **e.** *Lacmellea aculeata* (unvouchered, photographed at Sinnamary River, French Guiana), armed trunk.

Plate 16. APOCYNACEAE. **a.** *Macoubea guianensis* (*Mori et al. 14929*), inflorescence. [Photo by S. Mori] **b.** *Macoubea guianensis* (*Mori et al. 24193*), flowers with contorted petals. **c.** *Macoubea guianensis* (*Mori et al. 22954*), fruit cut open to show oblong seeds and white latex. **d.** *Odontadenia perrottetii* (*Mori & Gracie 21161*), unopened fruit. **e.** *Mandevilla rugellosa* (*Mori et al. 23767*), inflorescence with open flowers and buds. [Photo by S. Mori]

ACANTHACEAE

Justicia secunda **a.**

Pachystachys coccinea **b.**

Polylychnis fulgens **c.**

Pulchranthus congestus **d.**

Ruellia rubra **e.**

Pulchranthus variegatus **f.**

Plate 9

ANACARDIACEAE

Astronium ulei a.

Spondias mombin d.

Astronium ulei b.

c.

Thyrsodium puberulum f.

Tapirira obtusa e.

Thyrsodium spruceanum g.

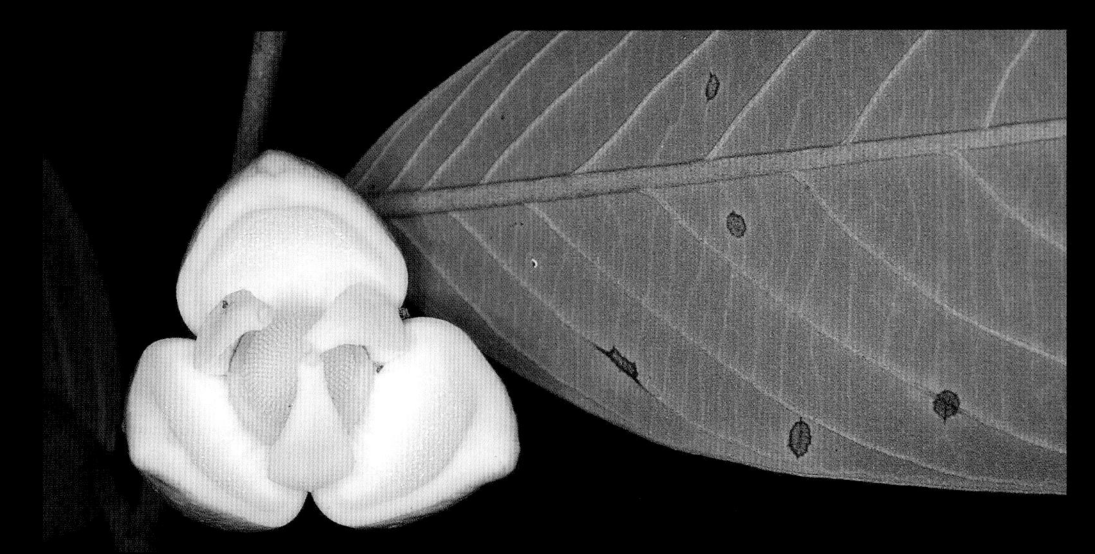

Annona prevostiae a.

Annona prevostiae b.

Cremastosperma brevipes c.

Cymbopetalum brasiliense e.

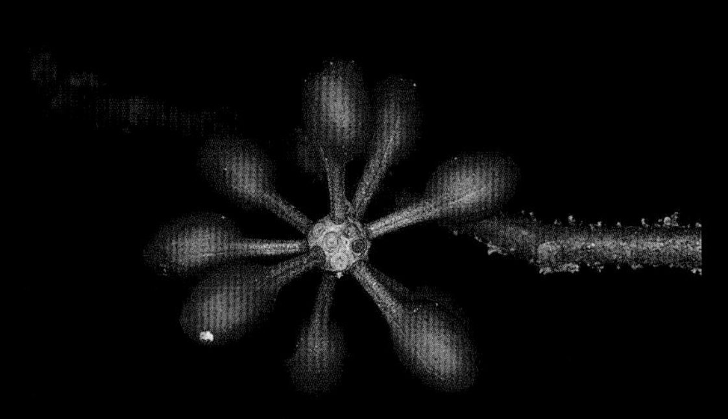

Cremastosperma brevipes d.

Duguetia cadaverica f.

Duguetia surinamensis a.

Fusaea longifolia b.

Guatteria punctata c.

Guatteria punctata d.

Guatteria foliosa e.

Plate 12

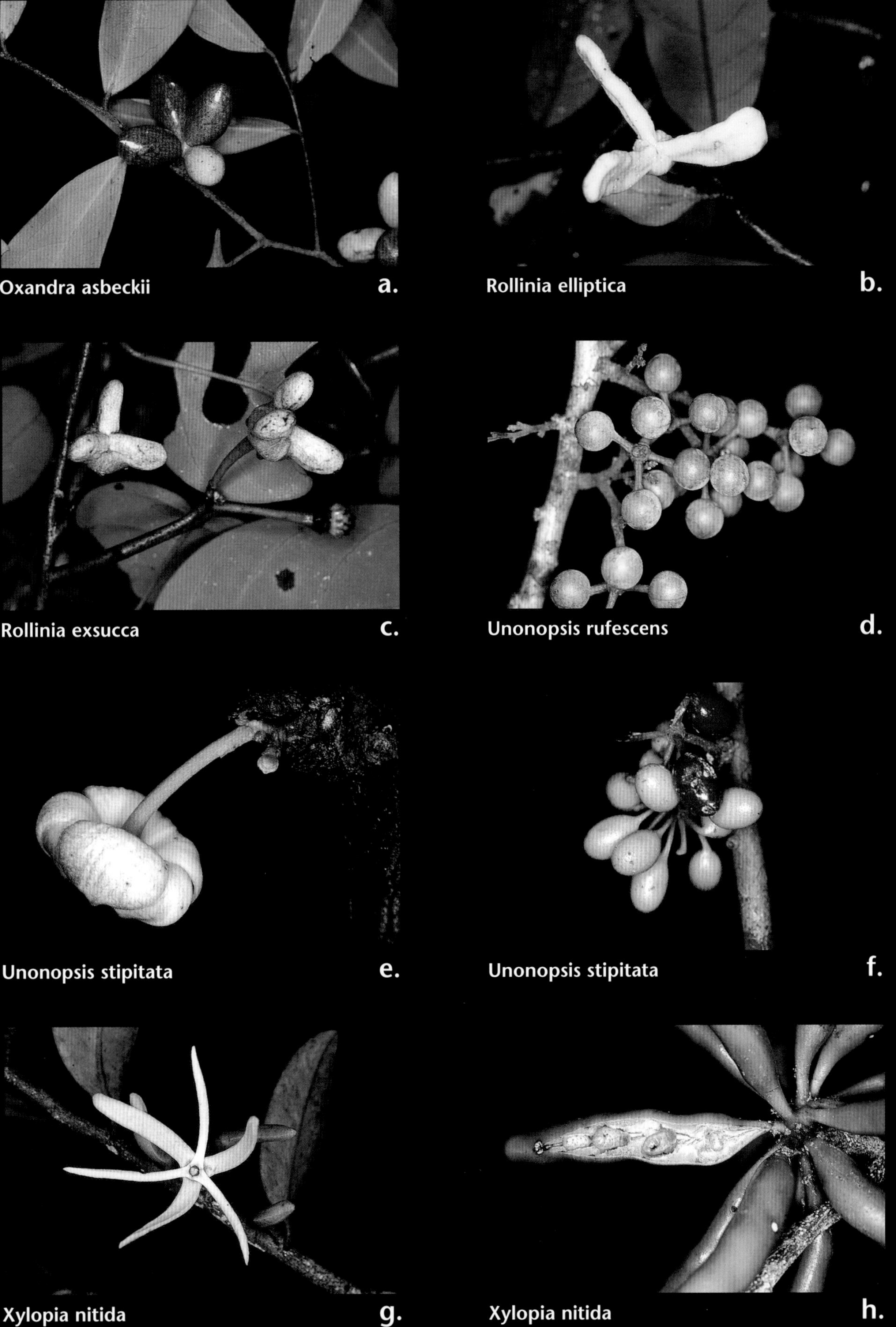

Oxandra asbeckii a.

Rollinia elliptica b.

Rollinia exsucca c.

Unonopsis rufescens d.

Unonopsis stipitata e.

Unonopsis stipitata f.

Xylopia nitida g.

Xylopia nitida h.

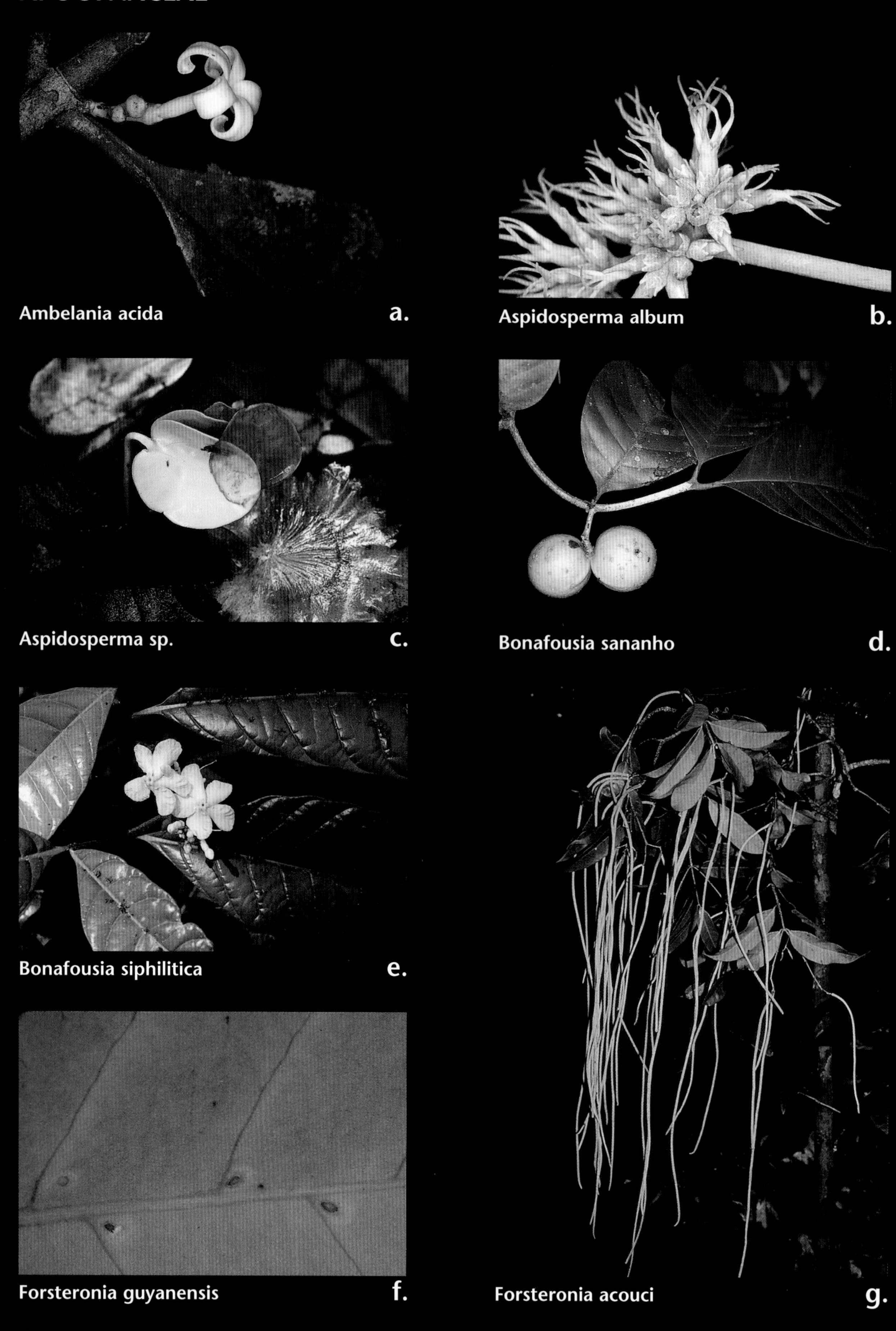

Ambelania acida a.

Aspidosperma album b.

Aspidosperma sp. c.

Bonafousia sananho d.

Bonafousia siphilitica e.

Forsteronia guyanensis f.

Forsteronia acouci g.

Plate 14

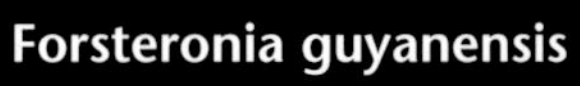

Forsteronia guyanensis a.

Himatanthus speciosus b.

Lacmellea aculeata c.

Lacmellea aculeata d.

Lacmellea aculeata e.

Plate 15

Macoubea guianensis a.

Macoubea guianensis b.

Macoubea guianensis c.

Odontadenia perrotteti d.

e. Mandevilla rugellosa

Plate 16 (Apocynaceae continued on Plate 17)

FIG. 2. ACANTHACEAE. *Pachystachys coccinea* (*Mori et al. 21504*). **A.** Upper part of stem with leaves and inflorescences. **B.** Lateral view of flower with subtending bract. **C.** Medial section of corolla with adnate stamen. **D.** Adaxial (far left) and abaxial (near left) views of anthers. **E.** Lateral view of ovary and base of style in calyx (far left), medial section of ovary (near left), and detail of apex of style and stigma (above).

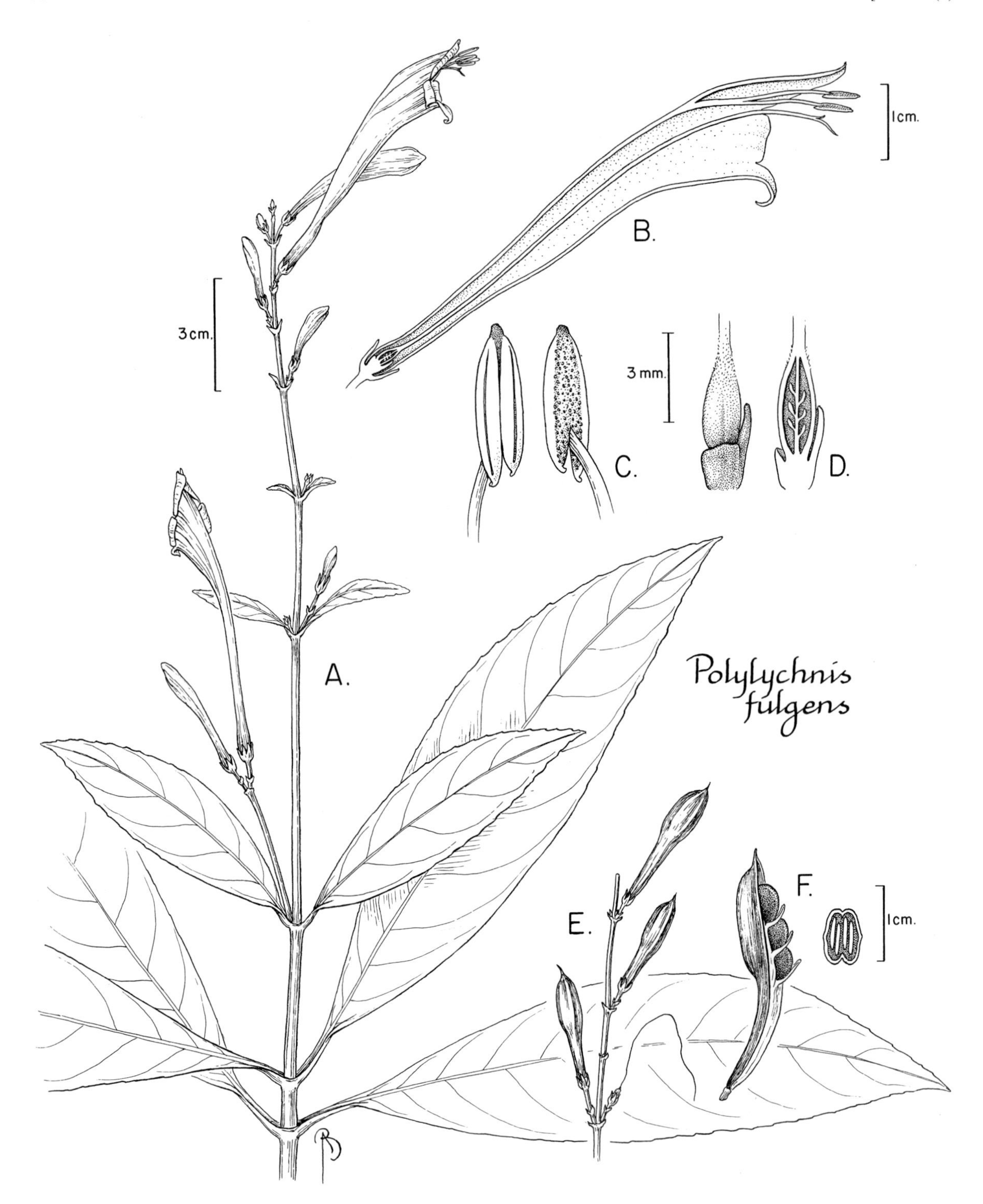

FIG. 3. ACANTHACEAE. *Polylychnis fulgens* (A, B, E, *Mori et al. 20853*; C, D, F, *Mori et al. 8698*). **A.** Stem with inflorescences. **B.** Medial section of flower. **C.** Upper part of stamens. **D.** Intact (left) and medial section (right) of ovary; note nectary. **E.** Part of infructescence. **F.** Open fruit and transverse section of fruit.

PULCHRANTHUS V. M. Baum, Reveal & Nowicke

Herbs or subshrubs. Stems terete or subquadrangular, glabrous. Leaves: petioles present; blades narrowly elliptic to oblong, glabrous or sparsely pubescent, cystoliths present, the margins entire. Inflorescences terminal, spikes, racemes, or well branched terminal panicles with flowers in well defined fascicles, glandular-puberulent. Flowers distinctly bilabiate, curved, white to violet with dark-purplish dots; calyx 5-lobed, the lobes narrowly triangular, glandular-puberulent; corolla glandular-puberulent abaxially, the tube short, broad and curved, the limb bilobed, the upper lip 2-lobed, the lower lip 3-lobed, ciliate, often purplish-dotted; stamens 2, exserted, the filaments curved, the anthers 2-locular, the locules parallel, blunt basally; staminodia 2. Capsules clavate. Seeds 4, lenticular, each supported by curved reticulum.

1. Leaf blades elliptic to oblong, 6–15 × 2–6 cm. Inflorescences open; peduncle 55–60 mm long. Flowers 2–4 per node; corolla white to pink and tinged with purple, the tube white to cream-colored, the upper lip magenta, the lower lip with outer 2 lobes whitish or lavender, the central lobe white with magenta spots. . . . . *P. congestus.*
1. Leaf blades elliptic-lanceolate, 13–26 × 5–9 cm. Inflorescences compact; peduncle 10–15 mm long. Flowers 2 to many per node; corolla pale lavender with purple markings on the limb, the tube white, the lobes of the upper lip lavender with purple markings, the lower lip with lobes all lavender with purple markings, the central lobe more densely so than outer 2 lobes.. . . . . *P. variegatus.*

Pulchranthus congestus (Lindau) V. M. Baum, Reveal & Nowicke Pl. 9d

Herbs or subshrubs, 0.2–0.6 m tall. Stems subquadrangular with red, exfoliating bark. Leaves: petioles 3–25 mm long; blades elliptic to oblong, 6–15 × 2–6 cm, glabrous or sparingly pilose, sometimes reddish, the apex acuminate. Inflorescences terminal, compact, open racemes or panicles, 5–20 cm long, green, the flowers 2–4 per node; peduncles 55–60 mm long; bracts small, narrowly triangular, 1.5–2 × 0.5 mm; pedicels 2–7 mm long, becoming shorter toward apex, glandular-puberulent; bracteoles 0.2–0.5 mm long. Flowers: calyx lobes narrowly triangular, 4–6 × 0.5 mm; corolla 10–13 mm long, glandular, white to pink and tinged with purple, the tube white to cream-colored, 3–4 mm long, the upper lip magenta, the oblong lobes 5–7 × 2–2.5 mm, the lower lip with 3 oblong lobes, the 2 outer ones whitish or lavender, the central one white with magenta spots. Capsules 15 × 5 × 3 mm, pubescent and gland-dotted, tan. Seeds lenticular, 3 × 3.5 mm, covered with trichome-like papillae, brown. Fl (Sep), fr (Sep); occasional, in non-flooded forest.

Pulchranthus variegatus (Aubl.) V. M. Baum, Reveal, & Nowicke Pl. 9f

Subshrubs or shrubs, 0.5–2 m tall. Stems terete with red, exfoliating bark. Leaves: petioles 3–13 mm long; blades elliptic-lanceolate, (8)13–26 × (3)5–9 cm, glabrous, the apex acuminate-cuspidate. Inflorescences terminal, racemes or panicles, 4–15 cm long, green, the flowers 2 to many per node; peduncle 10–15 mm long; bracts small, narrowly triangular, 2.5–3 × 0.5 mm; pedicels lacking to short, <1.5 mm long; bracteoles 1.5–2 mm long. Flowers: calyx lobes narrowly triangular, 2.5–4 × 0.5–1 mm; corolla 10–17 mm, glandular, pale lavender with purple markings on limb, the tube 5–7 mm long, white, the upper lip with elliptic-oblong lobes 5–6 × 1.5–2 mm, with purple markings, the lower lip with elliptic lobes, all lavender with purple markings, the central lobe more densely so than other 2 lobes. Capsules 20 × 5 × 3 mm, glabrous, brown. Seeds lenticular, 3 × 3.5 mm, covered with trichome-like papillae, tan. Fl (Aug), fr (Oct); occasional, in non-flooded forest.

RUELLIA L.

Perennial herbs or shrubs, sometimes with long, thin, tuberous roots. Leaves: petioles absent or present; blades elliptic, ovate-lanceolate or oblong-spatulate, cystoliths present. Inflorescences branching, axillary, long-pedunculate dichasia or cymes, or 1–3 or more, sessile, occasionally long-pedicellate flowers in upper axils combined into terminal racemes or spikes; bracts opposite, green, the margins entire; bracteoles 2. Flowers usually large and showy, regular, cleistogamous flowers often present; calyx deeply 5-lobed, the lobes narrow, mostly equal; corolla blue, blue-purple, mauve, white, occasionally red or yellow, the tube funnelform or salverform, the limb of 5 equal, spreading lobes; nectary disc annular, without distinct projections; stamens 4, didynamous, the anthers 2-locular, the locules equal, muticous; staminodia 0. Capsules substipitate or stipitate. Seeds 4–20, lenticular, each supported by curved retinaculum.

1. Small shrubs or subshrubs, 0.3–0.8 m tall. Leaf blades 5–11.5 × 2.5–4.5 cm, glabrous or sparingly hirsute, green abaxially. Calyx lobes 8–9 × 0.5–1 mm, sparingly glandular pubescent, the trichomes not yellowish; corollas white with mauve or pale rose-colored lobes. . . . . R. rubra.
1. Shrubs, 1 m tall. Leaf blades 15–16 × 4.5–5 cm, densely pilose, purple abaxially. Calyx lobes 10–14 × 0.5 mm, conspicuously glandular-pilose, the trichomes yellowish; corolla white. . . . . R. saülensis.

Ruellia rubra Aubl. Pl. 9e

Small shrubs or subshrubs, 0.3–0.8 m tall. Stems quadrangular, unbranched. Leaves: petioles 2–5 mm long; blades elliptic, 5–11.5 × 2.5–4.5 cm, glabrous or sparingly hirsute. Inflorescences axillary, once or twice dichasially branched, long-pedunculate cymes; peduncle 5–10 cm long; basal bracts linear, to 20 mm long, the others smaller; pedicels of central flowers 1–2 mm long; bracteoles subulate, 1.5–2 × 0.4–0.5 mm. Flowers: calyx lobes linear, 8–9 × 0.5–1 mm, sparingly glandular pubescent; corolla 30–40 mm long, glabrous except upper part of tube, white with mauve or pale rose-colored lobes, the limb suberect, lobes obovate, 7–8 × 5.5 mm, retuse apically; stamens included, the anthers 1.6 mm long. Capsules 15 × 4 × 3 mm, pubescent, apiculate, tinged with red apically, the stipe 5 mm long. Seeds 2 × 1.5 mm, glabrous, dark brown, the margins ringed by hygroscopic trichomes. Fl (Mar, May, Jun, Jul, Aug, Sep, Oct); common, along roadsides in non-flooded forest.

Ruellia saülensis Wassh.

Shrubs, 1 m tall. Stems quadrangular, branching. Leaves: petioles 5–15 mm long; blades elliptic, 15–16 × 4.5–5 cm, densely pilose, purple abaxially. Inflorescences terminal and axillary, once or twice dichasially branched, pedunculate cymes; peduncle 1.5–6 cm long; basal bracts linear, to 10 mm long, the others smaller; pedicels of central flower 3 mm long; bracteoles subulate, 2.5–3 × 0.4–0.5 mm. Flowers: calyx lobes linear, 10–14 × 0.5 mm, conspicuously glandular-pilose, the trichomes yellowish; corolla 35–45 mm long, glabrous except upper part of tube, white, the limb suberect, the lobes obovate, 7–8 × 6–8 mm, the apex retuse; stamens included, the anthers 3 mm long. Capsules 15 × 4 × 3 mm, pubescent, apiculate, tinged with red apically, the stipe 5 mm long. Seeds 2.5 × 2 mm, glabrous, dark brown, the margins ringed by hygroscopic trichomes. Fl (Mar, Jul, Aug, Sep), fr (Jul); common, on granitic outcrops.

AMARANTHACEAE (Amaranth Family)

Robert A. DeFilipps and Shirley L. Maina

Annual or perennial herbs, subshrubs, or shrubs. Stipules absent. Leaves simple, opposite or alternate, entire or nearly so. Inflorescences axillary or terminal spikes, heads, or panicles, often with ultimate 3-flowered cymules, the individual flowers subtended by 1 bract and 2 bracteoles, the lateral flowers in 3-flowered cymules, the flowers sometimes sterile and modified into scales or bristles. Flowers actinomorphic, bisexual or unisexual (plants monoecious); tepals 3 or 5, free, often persistent, scarious; stamens (2)3 or 5, the filaments united below in a cup, the anthers 2- or 4-locular; pseudostaminodia present or absent; ovary superior, 1-locular, the style 1, the stigmas 1–3; placentation usually basal, the ovule solitary. Fruits dry, indehiscent, or irregularly dehiscent capsules (utricles). Seeds small, shiny.

Nee, M. H. 1995. Amaranthaceae. *In* J. A. Steyermark, P. E. Berry & B. K. Holst (gen. eds.), Flora of the Venezuelan Guayana. P. E. Berry, B. K. Holst & K. Yatskievych (vol. eds.) **2:** 384–399. Missouri Botanical Garden, St. Louis; Timber Press, Portland, Oregon.
Scheygrond, A. 1966. Amaranthaceae. *In* A. Pulle (ed.), Flora of Suriname **I(1):** 25–44. E. J. Brill, Leiden.

1. Leaves alternate. . . . . *Amaranthus.*
1. Leaves opposite.
   2. Flowers reflexed, several (2 or more) flowers per glomerule reduced to stiff, hooked (uncinate) spines; bracts and bracteoles modified into herbaceous, hooked spines. . . . . *Cyathula.*
   2. Flowers neither reflexed nor with hooked spines.
      3. Leaves elliptic, ovate, oblanceolate, or spathulate, glabrous or sparsely pubescent with usually simple hairs. . . . . *Alternanthera.*
      3. Leaves narrow, usually lanceolate, densely pubescent.
         4. Hairs branched at base. . . . . *Alternanthera halimifolia.*
         4. Hairs simple. . . . . *Pfaffia.*

ALTERNANTHERA Forssk.

Annual or perennial herbs or subshrubs. Leaves opposite, petiolate, entire. Inflorescences axillary or terminal, sessile or pedunculate heads; bracts and bracteoles present, scarious. Flowers bisexual; tepals 5, equal or outer 3 longer than and enclosing inner 2; stamens 3 or 5, the anthers 2-locular; pseudostaminodia present, often fimbriate. Fruits indehiscent utricles. Seeds cochleate-orbicular or lenticular.

1. Bracts and bracteoles ≤0.8 mm long. Tepals glabrous, 1-veined, to 2 mm long; stamens 3. Fruits obcordate. . . . *A. sessilis.*
1. Bracts and bracteoles >1 mm long. Tepals pubescent, 3-veined, >2.5 mm long; stamens 5. Fruits ovoid or globose.
   2. Leaves narrowly lanceolate or elliptic, green. Tepals 2.6–5 mm long. . . . . *A. halimifolia.*
   2. Leaves very narrowly spatulate, variegated with splotches of red, yellow, orange, and/or purple. Tepals 3.7–4.5(5) mm long. . . . . *A. tenella.*

Alternanthera halimifolia (Lam.) Standl.

Herbs or shrubs. Stems to 2 m tall, densely cinereous or velutinous, with basally branched or stellate hairs, glabrescent with age. Leaves: petioles 2–10 mm long; blades mostly narrowly lanceolate or elliptic, sometimes ovate or orbicular, 1–5 × 0.5–2.5 cm, green, cinereous on veins, densely pubescent, the margins entire. Inflorescences sessile, axillary, ovoid heads 4–12 × 3–5 mm; bracts and bracteoles broadly ovate to elliptic, 1.4–2.3 mm long, pubescent, the apex acuminate to aristate. Flowers: tepals pubescent, the outer 3 tepals ovate to elliptic, 2.6–5 mm long, (1)3-veined, hispidulous abaxially, the apex acuminate; stamens 5; pseudostaminodia apically fimbriate or lacerate. Fruits globose or ovoid, 1–1.5 mm long. Seeds 0.8–1.1 mm diam., reddish-brown. Fl (May); weed in and around village of Saül.

Alternanthera sessilis (L.) DC.

Annual or perennial herbs, terrestrial or shallow water aquatics. Stems rooting at lower nodes, to 1 m tall, glabrous or pubescent in 2 lines. Leaves: petioles short and not well defined, 5 mm long; blades ovate, oblanceolate, or elliptic, 1–6 × 0.5–1.5 cm, glabrous on both sides or sparsely pubescent on midrib abaxially, the margins shallowly serrulate. Inflorescences axillary, congested, sessile, globose heads 2–4 × 3–6 mm; bracts and bracteoles ovate, 0.3–0.8 mm long, glabrous, prominently 1-veined, the apex acuminate. Flowers: tepals glabrous, the outer 3 tepals lance-ovate, narrowly ovate, or elliptic-ovate, 1–2 mm long, 1-veined, the apex acute; stamens 3; pseudostaminodia entire or apically dentate. Fruits obcordate, bifacially compressed, emarginate, 1.5–2 mm long, much exceeding the tepals. Seeds cochleate-orbicular or lenticular, 1.0–1.2 mm diam., brown or reddish-brown. Fl (Jun, Aug), fr (Jun); weed in secondary vegetation near Saül.

Alternanthera tenella Colla [Syns.: *Alternanthera tenella* Colla 'Bettzickiana,' *A. bettzickiana* (Regel) Voss, *A. ficoidea* (L.) R. Br. 'Bettzickiana,' *A. ficoides* (L.) Sm. var. *bettzickiana* (Regel) Backer, *A. paronychoides* A. St.-Hil. var. *bettzickiana* (Regel) Fosberg, *A. ficoidea* (L.) Roem. & Schult. var. *spathulata* (Lam.) L. B. Sm. & Downs]

Herbs or subshrubs. Stems rooting at nodes, to ca. 5 cm tall, glabrous or densely pubescent throughout or only in 2 lines. Leaves: petioles <1 cm long; blades very narrowly spatulate, 2–9 × 0.5–4 cm, glabrous to sparsely pubescent with usually simple or infrequently branched hairs, variegated with blotches of red, yellow, orange, and/or purple, the base cuneate, decurrent on petiole, the margins entire. Inflorescences axillary or terminal, sessile, globose heads 3–10 mm diam.; bracts and bracteoles ovate to elliptic-lanceolate, 1.5–3.5 mm long, spreading, pubescent, the apex aristate. Flowers: tepals pubescent, the outer 3 tepals ovate to elliptic, 3.7–4.5(5) mm long, prominently 3-veined, the apex aristate; stamens 5; pseudostaminodia apically fimbriate or lacerate. Fruits ovoid or suborbicular, ca. 1 mm diam. Seeds 1 mm long, dark reddish-brown. Fr (Aug); cultivated as an ornamental, but sometimes escaping and appearing native.

AMARANTHUS L.

Annual, monoecious or dioecious herbs. Leaves alternate, petiolate, entire. Inflorescences glomerules aggregated into axillary clusters, often also aggregated into terminal spikes. Flowers unisexual; tepals 3–5, apiculate; stamens (2)3 or 5, the anthers 4-locular; pseudostaminodia absent. Fruits dehiscent (circumscissile) or apparently indehiscent. Seeds lenticular or ovoid.

1. Stems glabrous. Leaf blades usually strongly emarginate or bilobed at apex. Inflorescences to 9 cm long. Fruits indehiscent. *A. blitum.*
1. Stems pubescent. Leaf blades obtuse or acute at apex. Inflorescences to 30 cm or more long. Fruits dehiscent (circumscissile). *A. caudatus.*

Amaranthus blitum L.

Annual, monoecious herbs to 1 m tall. Stems often pinkish to deep red, glabrous. Leaves: blades rhomboid, ovate, or elliptic, (0.7)2–6(7.5) × (0.2)2.7–5 cm, glabrous, the apex usually strongly emarginate to bilobed to 3 mm deep. Inflorescences axillary clusters and simple terminal spikes to 9 cm long; bracteoles lanceolate, oblong, or ovate, to 1 mm long, the apex acute or obtuse. Staminate flowers developing far in advance of, and in fewer numbers, than pistillate flowers; tepals 3, oblong, ovate, or elliptic, 0.8–1.3 mm long, incurved, the apex acute; stamens (2)3. Pistillate flowers: tepals 3, 2 linear-oblong to somewhat spatulate, 0.9–2 mm long, 1 shorter and scale-like. Fruits indehiscent, ovoid-globose, smooth or somewhat rugulose, 1.5–2.5 mm long. Seeds lenticular, 0.8–1.2(1.6) mm diam., dark reddish-brown or black, shiny. Fr (Aug); weed in and around village of Saül.

Amaranthus caudatus L.

Annual, monoecious herbs to 2 m tall. Stems often red, pubescent, with tangled white hairs. Leaves: blades lanceolate, elliptic, ovate, or rhomboid-ovate, (2.5)6–20 × (1)2–8 cm, glabrous or somewhat puberulent abaxially, the apex obtuse or acute. Inflorescences small axillary clusters and massive, lax, tail-like, terminal panicles or spikes to 30(50) cm long; bracteoles deltoid-ovate, lanceolate, or lance-acuminate, the apex rigidly aristate. Staminate flowers interspersed throughout inflorescence, fewer than pistillate flowers; tepals 5, ovate to oblong, 2.5–3.2 mm long; stamens 5. Pistillate flowers: tepals 5, the outer one elliptic or oblanceolate, 1.5–2.8 mm long, the inner 4 somewhat shorter than outer, spatulate or oblong-spatulate, recurved. Fruits circumscissile, ovoid-globose, smooth or rugulose, 2–2.5 mm long, as long as or exceeding perianth. Seeds subspherical or lenticular, 0.8–1.5 mm diam., white, tawny, reddish-brown, or black, dull or shiny. Naturalized, a presumed escape from cultivation, the leaves edible. *Ortie* (Créole).

CYATHULA Blume

Annual or perennial herbs. Leaves opposite, petiolate, entire. Inflorescences long, axillary and terminal spikes, the fertile and sterile flowers clustered in pedunculate glomerules along major axes; glomerules comprised of bracteate triads of 1 central bisexual fertile flower and 2 modified sterile flowers formed of uncinate or straight spines, all falling together with fruit; bracts ovate, oblong, the apex acuminate. Flowers bisexual and reduced unisexual, reflexed. Unisexual flowers reduced to perianth segments with 5 rigid, hooked awns or spines (glochidia), developing in axils of bracteoles after bisexual flowers develop; tepals 5, often hairy; stamens 5, the anthers 2-locular; pseudostaminodia present. Fruits indehiscent utricles. Seeds ovoid.

Cyathula prostrata (L.) Blume FIG. 4; PART 1: FIG. 8

Prostrate herbs to 1 m tall. Stems rooting at lower nodes. Leaves: blades elliptic to rhombic or obovate-rhombic, narrowed to base and apex, 2–8.5 × 2.5–4 cm, strigose, the apex obtuse. Inflorescences spiciform, with distant to apically congested glomerules; glomerules broadly ovoid or globoid, 1.5–3 mm long, usually comprised of 2 fertile, 1 sterile, and 4 rudimentary (reduced to glochidia) flowers. Bisexual flowers 2–3 per cluster; tepals oblong-lanceolate, 1.6–2.4 mm long, 3-veined, scarious, pubescent, the apex mucronate. Rudimentary flowers unisexual, greenish-white, about as long as bisexual flowers, reduced to 4–10 uncinate glochidia ca. 1–1.5 mm long. Seeds ovoid, pale brown, smooth, shiny. Fl (Jun), fr (May, Oct); weed in and around village of Saül. *Chaine d'enfant* (Créole).

PFAFFIA Mart.

Perennial herbs or subshrubs. Leaves opposite, petiolate, entire. Inflorescences dense terminal heads, or flowers distant and borne singly in short spikes arranged in diffuse, axillary or terminal, long-pedunculate cymes. Flowers bisexual, sessile; tepals 5, the outer 3 somewhat larger than inner 2, distinctly longitudinally 3-veined, exceeding perianth; stamens 5, the anthers 2-locular (becoming unilocular at anthesis); pseudostaminodia absent. Fruits indehiscent, included. Seeds lenticular or cochleate-orbicular.

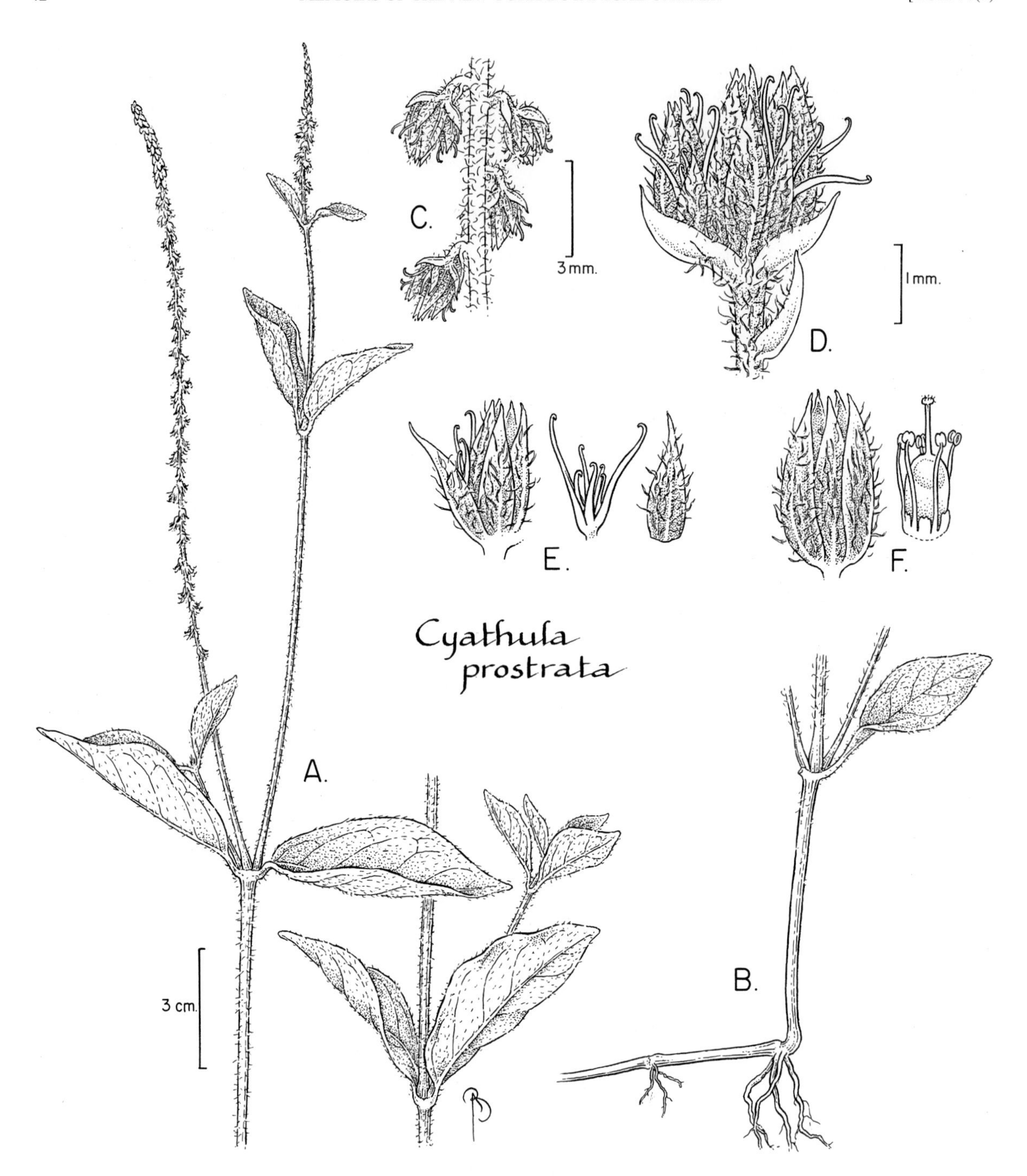

FIG. 4. AMARANTHACEAE. *Cyathula prostrata* (A, B, *Mori et al. 22346*; C–F, *Skog & Feuillet 7291*). **A.** Upper part of stem with leaves and inflorescences (left) and detail of middle section of stem (lower right). **B.** Basal portion of stem showing roots at lower nodes. **C.** Detail of inflorescence. **D.** Glomerule with 3 fertile flowers and 4 rudimentary flowers. **E.** Lateral view of fertile flower with one rudimentary flower attached (left), fertile flower (near right), tepal (far right). **F.** Lateral views of fertile flower (left) and fertile flower with tepals removed (right).

Pfaffia glomerata (Spreng.) Pedersen

Herbs or subshrubs, to 2 m tall. Stems erect, glabrous proximally, pubescent distally, striate, bluish-green (glaucous). Leaf blades narrowly lanceolate, to 10 × 3.2 cm, appressed-pilose adaxially and abaxially, the base somewhat cuneate, the apex acuminate. Inflorescences terminal cymes of peduncled, subglobose heads 4–8 mm diam., the heads elongating to become floriferous for 1 cm; rachis pubescent; bracts broadly ovate, 1/3 as long as tepals, the apex acute. Flowers: tepals oblong, 2–3 mm long, 3-veined, greenish-white, glabrous abaxially, with very small tuft of short hairs much shorter than perianth adaxially, the apex acute. Seeds reddish-brown, shiny. Weed in village of Saül. *L'arbre sensible* (Créole).

ANACARDIACEAE (Cashew Family)

John D. Mitchell

Small to very large trees, sometimes with sap causing contact dermatitis. Stipules usually absent. Leaves simple or pinnately compound, alternate, frequently aggregated toward branch tips; leaflets opposite or alternate. Inflorescences terminal or axillary, thyrsoid, paniculate or racemose; bracts leaflike to sepaloid. Flowers bisexual or unisexual (plants with all bisexual flowers or dioecious, monoecious, andromonoecious, or polygamous); hypanthium sometimes present (*Thyrsodium*); perianth actinomorphic; calyx 4–5-merous, the lobes free or connate, persistent (accrescent in fruit of *Astronium*) or deciduous; corolla 4–5-merous, imbricate or valvate; androecium actinomorphic or zygomorphic (*Anacardium*, *Mangifera*), diplo- or haplostemonous, the stamens usually 5–10, the anthers bilobed, longitudinally dehiscent, the filaments free or basally fused into tube (*Anacardium*); nectary disc intrastaminal, extrastaminal (*Mangifera*), or absent (*Anacardium*); gynoecium 1–5-carpellate, the ovary superior, 1–5-locular, each with 1 apotropous ovule; pistillode usually present in staminate flowers. Fruits drupes, fleshy or dry, indehiscent, the mesocarp thick or very thin, often resinous, a fleshy hypocarp (derived from the receptacle and pedicel) subtending drupe in *Anacardium*. Seeds 1–5; endosperm scanty or absent; embryo curved or straight.

Jansen-Jacobs, M. J. 1976. Anacardiaceae. *In* J. Lanjouw & A. L. Stoffers (eds.), Flora of Suriname **I(2):** 441–444. E. J. Brill, Leiden.
Mitchell, J. D. 1995. Anacardiaeae. *In* J. A. Steyermark, P. E. Berry & B. K. Holst (gen. eds.), Flora of the Venezuelan Guayana **2:** 399–412. P. E. Berry, B. K. Holst & K. Yatskievych (vol. eds.). Missouri Botanical Garden, St. Louis; Timber Press, Portland, Oregon.
———. 1997. Anacardiaceae. *In* A. R. A. Görts-van Rijn & M. J. Jansen-Jacobs (eds.), Flora of the Guianas Ser. A, **19:** 1–47, 68–76. Royal Botanic Gardens, Kew.
Nannenga, E. T. 1966. Anacardiaceae. *In* A. Pulle (ed.), Flora of Suriname **II(1):** 132–145. E. J. Brill, Leiden.

1. Leaves simple.
    2. Leaf blades usually obovate, the apex obtuse to rounded, sometimes shortly acuminate. Flowers with 8–10 stamens and staminodia, the filaments connate, forming tube basally; nectary disc absent. Pericarp woody; hypocarp present. . . . . *Anacardium.*
    2. Leaf blades usually lanceolate, the apex acute to acuminate. Flowers with 4–6 stamens and staminodia, the filaments free; extrastaminal nectary disc present. Pericarp fleshy; hypocarp absent. . . . . *Mangifera.*
1. Leaves compound.
    3. Leaves and inflorescence without milky exudate. Hypanthium absent.
        4. Leaflets with cladodromous secondary venation. Stamens or staminodia 5; styles 3. Fruits with accrescent calyx. . . . . *Astronium.*
        4. Leaflets with brochidodromous secondary venation or secondary veins joining to form intramarginal vein. Stamens or staminodia (8)10; styles 4–5. Fruits without accrescent calyx.
            5. Leaflets with intramarginal veins. Ovary glabrous, (3)4–5-locular. Mature drupes yellow, orange, or red; endocarp bony. . . . . *Spondias.*
            5. Leaflets with typical brochidodromous secondary veins. Ovary pubescent, 1-locular. Mature drupes dark purple-black; endocarp brittle (when dry). . . . . *Tapirira.*
    3. Leaves and inflorescence with milky exudate. Hypanthium present. . . . . *Thyrsodium.*

ANACARDIUM L.

Small to large trees. Leaves simple, aggregated toward branch tips, evergreen or briefly deciduous; petioles present; blades usually obovate, chartaceous to coriaceous, glabrous, the apex obtuse to rounded, sometimes shortly acuminate, the margins entire; venation brochidodromous. Inflorescences terminal or axillary, thyrsoid; pedicels present. Flowers staminate and bisexual (plants andromonoecious); calyx imbricate, 5-merous; corolla imbricate, 5-merous; stamens 8–10, dimorphic, one stamen larger and exserted from corolla, the others much smaller, the filaments basally connate into tube; nectary disc lacking; pistil rudimentary in staminate flowers; ovary 1-locular, the style one, terminal or lateral, the stigma punctiform; placentation basal. Drupes reniform, attached to fleshy hypocarp ("cashew apple") developed from pedicel, the pericarp woody.

Mitchell, J. D. 1992. Additions to *Anacardium* (Anacardiaceae). *Anacardium amapaënse*, a new species from French Guiana and eastern Amazonian Brazil. Brittonia **44:** 331–338.
——— & S. A. Mori. 1987. The cashew and its relatives (*Anacardium*: Anacardiaceae). Mem. New York Bot. Gard. **42:** 1–76.

1. Foliaceous bracts and leaves subtending inflorescence cream-colored to pale green adaxially. Staminal tube 0.3–0.9 mm long. Mature hypocarp yellow or red. Cultivated. . . . . *A. occidentale.*
1. Foliaceous bracts and leaves subtending inflorescence bright white or light pink adaxially. Staminal tube usually (1.2)3.5–4.5 mm long. Mature hypocarp white. Native. . . . . *A. spruceanum.*

Anacardium occidentale L.

Small trees, 2–12 m × 40 cm. Bark gray or brown, the inner bark orange to reddish-brown, resinous. Leaves and foliaceous bracts subtending inflorescences cream-colored to pale green adaxially. Flowers: corolla cylindric, light pink to red after fertilization; staminal tube 0.3–0.9 mm long. Drupes 2–3.5 × 1–2 cm, gray or brown when mature; hypocarp 5–20 × 2–8 cm, yellow or red. Fl (Sep), fr (Jan, fide local inhabitants); cultivated, may occasionally become adventive. *Cajou* (Créole), *caju* (fruit in Portuguese), *cajueiro* (tree in Portuguese), *cashew* (English), *noix de cajou* (French), *pomme-cajou*, *pommier-cajou* (Créole).

Anacardium spruceanum Engler FIG. 5

Trees, to 35 m × 100 cm. Trunks cylindrical, unbuttressed. Bark light gray, smooth, the inner bark reddish-brown, resinous. Leaves and foliaceous bracts subtending inflorescences white or light pink adaxially. Flowers: corolla cylindric, pink to dark purple after fertilization; staminal tube (1.2)3.5–4.5 mm long. Drupes 1.3–1.5 × 1.3–2 cm, black when mature; hypocarp white, with a strong resinous aroma. Fl (May), fr (May), uncommon, in non-flooded forest.

ASTRONIUM Jacq.

Large trees, dioecious. Leaves deciduous, imparipinnate; leaflets opposite, rarely alternate, petiolulate, the margins entire (within our area); secondary venation cladodromous. Inflorescences terminal or axillary, thyrsoid. Flowers: calyx imbricate, 5-merous, enlarging after anthesis in pistillate flowers; corolla imbricate, 5-merous; nectary disc intrastaminal, very thin, 5-lobate; stamens 5, rudimentary or frequently absent in pistillate flowers; ovary 1-locular, the styles 3, terminal, often persistent, the stigmas capitate; placentation apical, with a single ovule. Fruits fusiform, glabrous, subtended by accrescent, chartaceous calyx, the mesocarp resinous, the endocarp thin, brittle when dry; embryo straight.

Astronium ulei Mattick FIG. 6, PL. 10a,b

Trees, to 40 m × 100 cm. Trunks with poorly developed buttresses. Bark gray or dark gray-brown, moderately smooth, sometimes flaky on buttresses and adjacent trunk, lenticels in irregular rows, the outer bark thin, the inner bark light orange or light salmon colored, ca. 15–20 mm thick. Exudate clear, somewhat sticky. Branchlets, leaves, and inflorescences essentially glabrous. Leaves 3–7-foliolate; leaflet blades ovate, obovate, or oblong, chartaceous to subcoriaceous, shiny adaxially, the base truncate, rounded, obtuse, or cuneate, symmetrical or slightly oblique, the apex acuminate. Inflorescences to 31 cm long; pedicels 1.5–2 mm long. Staminate flowers: calyx lobes suborbicular, 0.5 mm long, glabrous except for ciliate margins; petals greenish, ovate, 1.5–2 mm long. Pistillate flowers: calyx lobes orbicular, ca. 1.6 mm long; petals broadly ovate, 1.4–1.6 mm long; ovary ovoid, ca. 1 mm long, the styles 0.5 mm long. Fruits 11–13 mm long, with accrescent calyx lobes 10–15 mm long, the distance between calyx base and articulation of pedicel in fruit 7–15 mm. Fl (Sep), fr (Nov), uncommon, in non-flooded forest.

MANGIFERA L.

Large trees. Leaves evergreen, simple; petioles present; blades lanceolate, the apex acute to acuminate, the margins entire. Inflorescences terminal or axillary, thyrsoid; pedicels present. Flowers staminate and bisexual (plants andromonoecious); calyx 5-merous, imbricate; corolla 5-merous, imbricate; stamens dimorphic, 1(2) fertile, the other 4 sterile, the filaments free; nectary disc extrastaminal; pistil reduced in staminate flowers, the ovary 1-locular, the style one, lateral or eccentric, the stigma punctiform; placentation basal. Drupes ovoid to subreniform, somewhat laterally compressed, the mesocarp very thick, fleshy, the endocarp fibrous.

Mangifera indica L. FIG. 7

Large trees, to 40 m tall. Bark gray or dark brown, longitudinally fissured. Drupes 4–25 × 1.5–10 cm, orange, edible. Fl (Aug); cultivated. *Mango* (English), *manguier* (French), *pied mangue* (Créole).

SPONDIAS L.

Small to large trees. Leaves aggregated toward branch tips, deciduous, imparipinnate. Leaflets opposite or subopposite; blades petiolate, the margins entire or crenulate; intramarginal vein conspicuous. Inflorescences terminal or axillary, open paniculate or pseudospicate. Flowers usually all bisexual or sometimes unisexual (e.g., *S. purpurea*): perianth 5-merous; calyx sometimes imbricate; corolla valvate; stamens (8)10, slightly unequal; nectary disc intrastaminal; ovary (3)5-locular, glabrous, the styles (4)5; placentation with ovules apically suspended. Drupes ovoid to oblong, the mesocarp fleshy, the endocarp bony with fibrous matrix. Seeds 3–5.

Mitchell, J. D. & D. C. Daly. 1998. The "tortoise's cajá" — a new species of *Spondias* (Anacardiaceae) from southwestern Amazonia. Brittonia **50:** 447–451.

1. Leaves mature at time of flowering. Leaflet margins entire. Panicles open, broadly branched. Corolla white or cream-colored. Mature drupe yellow to orange. Native. . . . . *S. mombin.*
1. Leaves absent or very young at flowering. Leaflet margins usually slightly crenulate. Panicles pseudospicate. Corolla usually red. Mature drupe usually red. Cultivated. . . . . *S. purpurea.*

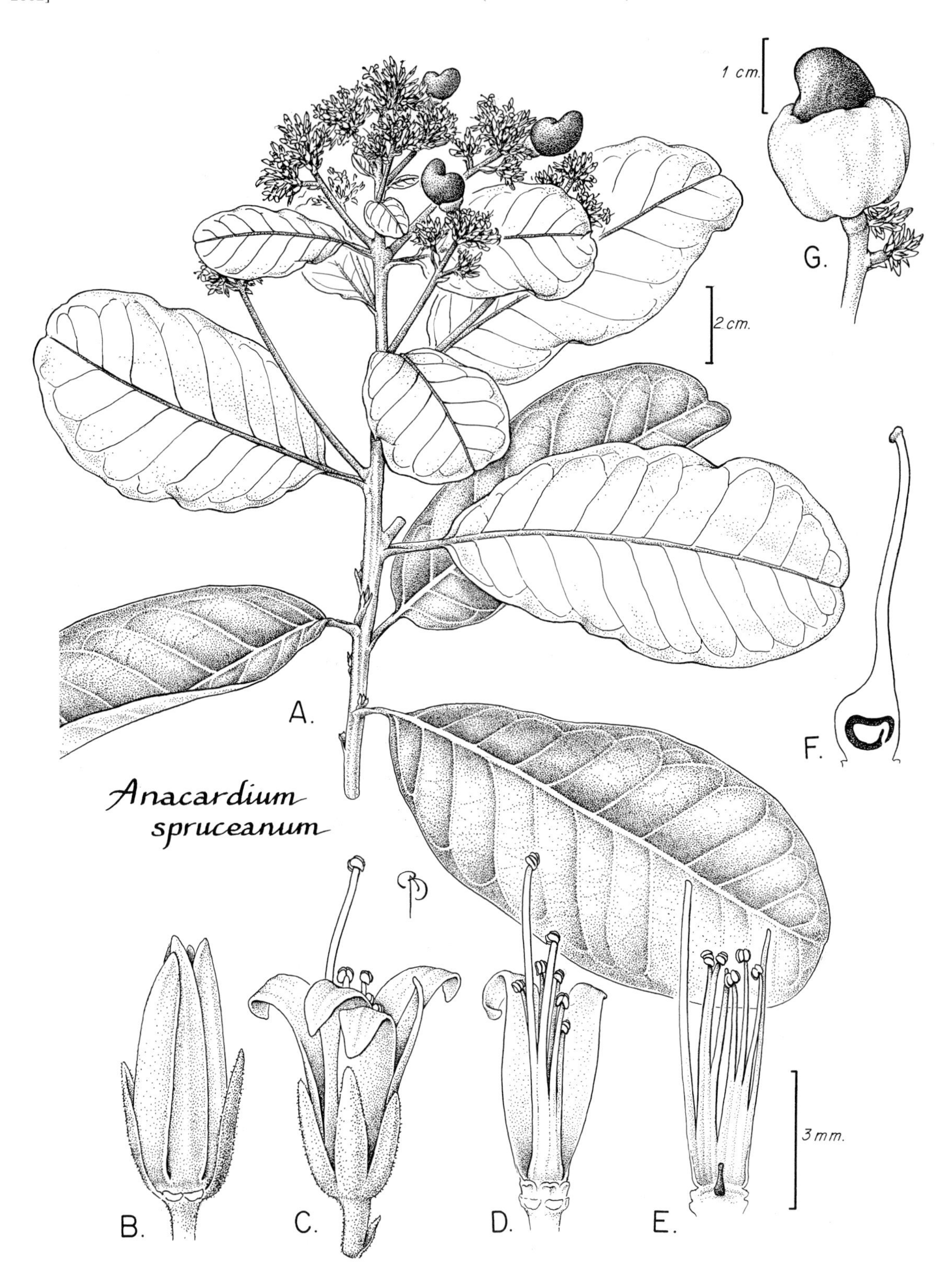

FIG. 5. ANACARDIACEAE. *Anacardium spruceanum* (*Mori et al. 15770*). **A.** Stem with inflorescences bearing three young fruits, in life the bracts subtending the inflorescences have lighter colored adaxial surface than those of the leaves. **B.** Bud just before anthesis. **C.** Staminate flower at anthesis. **D.** Staminate flower with most of perianth removed. **E.** Staminate flower with staminal tube opened to reveal varying degrees of filament coalescence; note the small pistillode. **F.** Pistil. **G.** Fruit, in life the drupe is black and the hypocarp is either white or red. (Reprinted with permission from J. D. Mitchell & S. A. Mori, Mem. New York Bot. Gard. 42, 1987.)

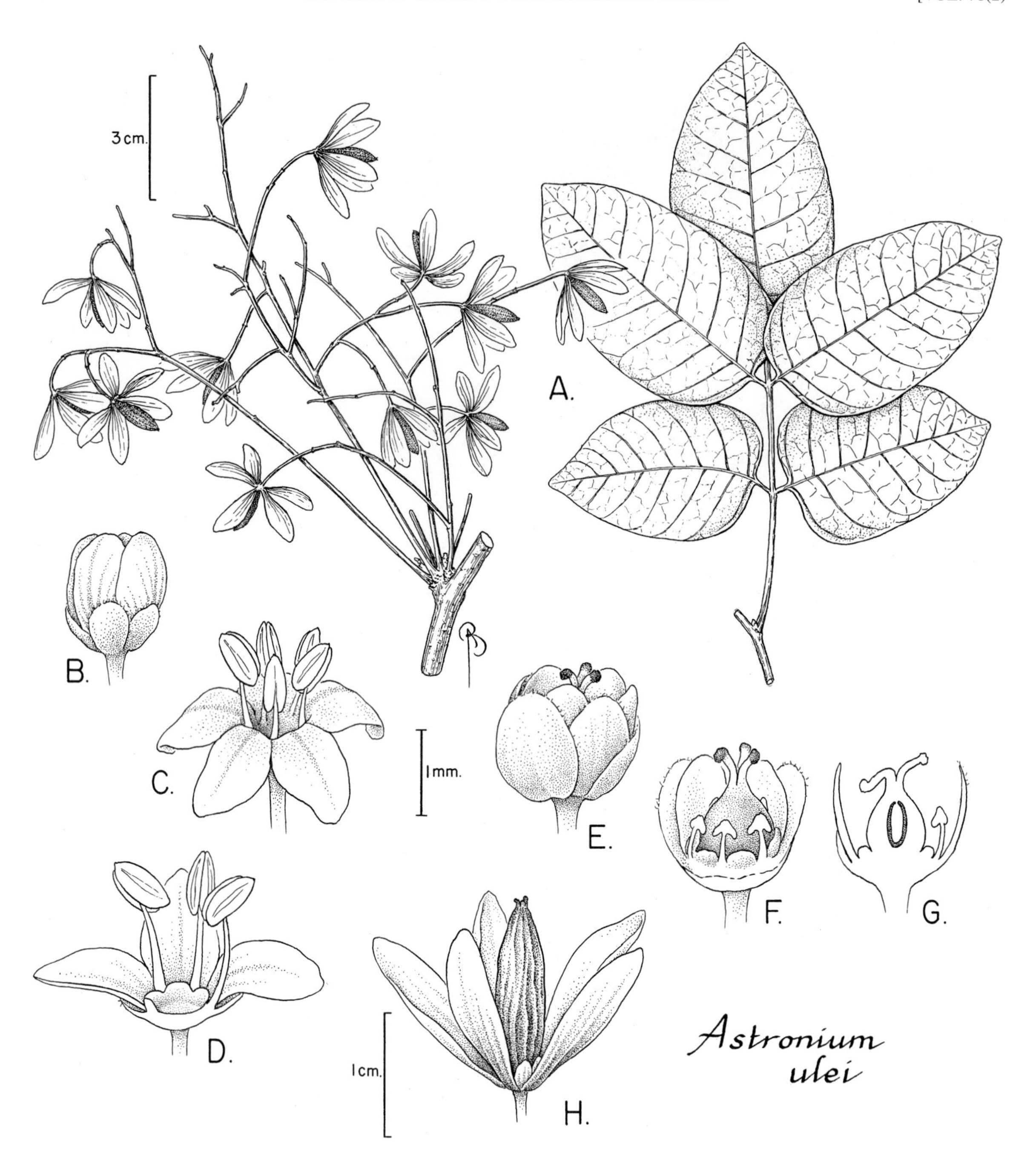

FIG. 6. ANACARDIACEAE. *Astronium ulei* (A, H, *Rodrigues et al. 10237*; B–D, *Ule 7958*; E, G, *Steyermark 86578*). **A.** Leaf (right) and infructescence (left). **B.** Staminate flower just before anthesis. **C.** Staminate flower at anthesis. **D.** Staminate flower with part of perianth removed. **E.** Pistillate flower. **F.** Pistillate flower with part of perianth removed. **G.** Medial section of pistillate flower showing position of ovule. **H.** Fruit showing enlarged sepals.

Spondias mombin L. [Syn.: *Spondias lutea* L.]
FIG. 8A–D, PL. 10b

Large trees, to 30 m × 60 cm. Bark light brown, thick, longitudinally fissured, spines sometimes present, the inner bark orange, resinous. Leaves mature at time of flowering. Leaflets: margins usually entire, the apex acuminate. Inflorescences open, broadly branched panicles. Flowers: calyx not imbricate; corolla white or cream-colored. Drupes 1.5–3.5 × 0.5–2 cm, yellow to orange when mature, edible. Fl (Nov), fr (Dec, Mar); native, primary and secondary non-flooded forest, but also cultivated in village. *Cajá* (Portuguese), *Hog plum* (English), *mombin* (Créole), *tapereba* (Portuguese).

Spondias purpurea L. FIG. 8E–I

Small trees, 3–10 m tall. Bark smooth, gray. Leaves absent or very young at flowering. Leaflets: margins usually slightly crenulate, the apex often mucronate. Inflorescences pseudospicate panicles.

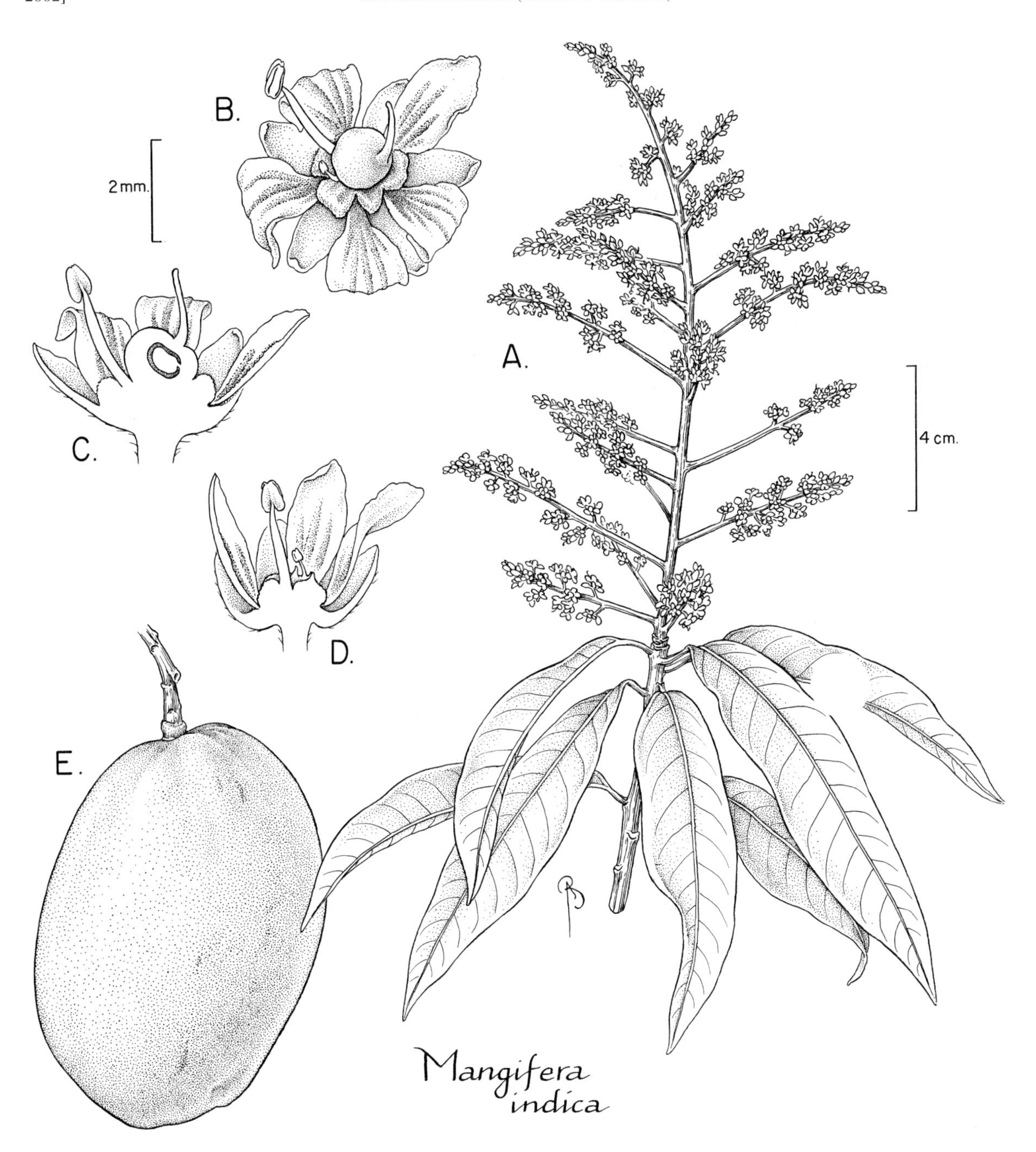

FIG. 7. ANACARDIACEAE. *Mangifera indica* (A, *Mori & Mitchell 18778*; B–D, *De La Cruz 3590*; E, unvouchered photo by Mitchell). **A.** Stem with leaves and inflorescence. **B.** Top view of bisexual flower. **C.** Medial section of bisexual flower. **D.** Medial section of staminate flower. **E.** Fruit.

Flowers: calyx imbricate; corolla often light pink to red. Drupes 1.5–3.5 × 1–2.5 cm, usually red, sometimes yellow at maturity, edible. Fl (Aug), fr (Aug, Sep, Nov); cultivated. *Spanish plum* (English), *mombin* (Créole).

TAPIRIRA Aubl.

Small to very large trees. Leaves evergreen, usually imparipinnate. Leaflets opposite or subopposite; petiolules present; blades with entire margins; secondary venation brochidodromous. Inflorescences terminal or axillary, paniculate; pedicels present. Flowers unisexual, rarely bisexual (plants polygamodioecious); calyx imbricate, 5-merous; corolla imbricate, 5-merous; stamens (8)10, slightly unequal; nectary disc intrastaminal; ovary l-locular, densely pubescent, the styles (4)5; placentation apical or subapical. Drupes oblong-oblique, ellipsoid or globose, dark purple to black when mature, the mesocarp thin, the endocarp cartilaginous (brittle when dry) to bony (outside our area); embryo curved, the cotyledons often with purplish striations.

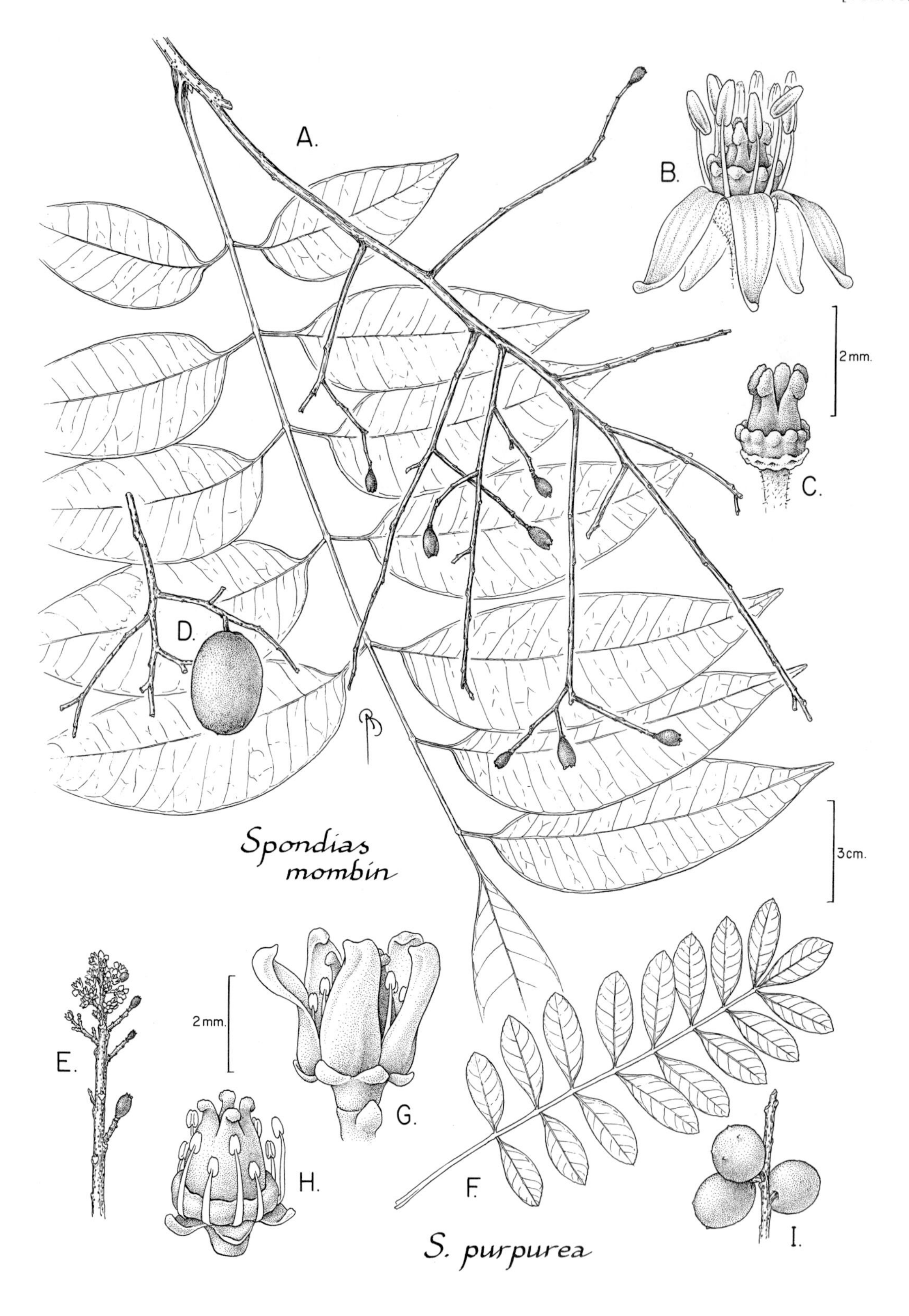

FIG. 8. ANACARDIACEAE. A comparison of species of *Spondias*. **A–D.** *Spondias mombin* (A, *Mori et al. 21534*; B, C, *De La Cruz 3839*; D, *Lanjouw & Lindeman 3152*). **A.** Part of stem with leaf and inflorescence. **B.** Bisexual flower. **C.** Bisexual flower with perianth and stamens removed. **D.** Part of infructescence. **E–I.** *Spondias purpurea* (E, G, H, *Daly 137*; F, I, *Nee 37681*). **E.** Part of inflorescence. **F.** Imparipinnate leaf. **G.** Pistillate flower. **H.** Pistillate flower with petals removed. **I.** Part of infructescence.

Mori, S. A. & J. D. Mitchell. 1990. *Tapirira bethanniana* (Anacardiaceae) and *Eschweilera piresii* subsp. *viridipetala* (Lecythidaceae), two new taxa from central French Guiana. Mem. New York Bot. Gard. **64:** 229–234.

1. Bark rough, very thick, longitudinally fissured, composed of quadrangular plates. Trunks with tall buttresses. Leaflets glabrous abaxially; secondary and tertiary veins scarcely prominent abaxially, the higher order venation usually inconspicuous abaxially. Drupes globose. . . . . *T. bethanniana.*
1. Bark smooth to shallowly fissured, thin. Trunks with low buttresses or buttresses absent. Leaflets sparsely to densely pubescent or occasionally glabrous abaxially; secondary and often tertiary veins prominent abaxially, the higher order venation usually clearly visible abaxially. Drupes ellipsoid or obliquely ovoid to oblong.
   2. Leaflets with appressed hairs only, midrib glabrous adaxially. . . . . *T. guianensis.*
   2. Leaflets with both appressed and erect hairs, midrib sparsely to densely pubescent with erect hairs adaxially. . . . . *T. obtusa.*

Tapirira bethanniana J. D. Mitch. FIG. 9A–D,H,I

Large trees, to 50 m × 80 cm. Trunks with tall buttresses. Bark brown, very thick, longitudinally fissured, broken into quadrangular plates, the inner bark laminated, pinkish to reddish-brown, resinous. Leaves imparipinnate, occasionally paripinnate, with (5)7(11) leaflets, the rachis horizontally flattened. Leaflets: blades usually coriaceous, glabrous ab- and adaxially; higher order venation often inconspicuous abaxially, the midrib and secondary veins often flattened adaxially, the tertiary and higher order venation barely visible adaxially. Drupes globose, 1.9–2 cm diam., dark purple to black. Fl (Nov, Dec), fr (Jan, Mar, Dec); scattered, in primary non-flooded forest.

Tapirira guianensis Aubl. [Syn.: *T. myriantha* Triana & Planch.] FIG. 9J

Small to large trees, to 30 m × 60 cm. Trunks unbuttressed or with low buttresses. Bark brown or gray, thin, smooth or very slightly fissured, the inner bark pinkish to reddish-brown, resinous. Leaves imparipinnate, occasionally paripinnate, with (5)9(13) leaflets, the rachis terete. Leaflets: blades usually chartaceous, occasionally subcoriaceous, sparsely pubescent with appressed hairs only or glabrous abaxially, usually glabrous adaxially; secondary and tertiary veins prominent and conspicuous abaxially, scarcely prominent but clearly visible adaxially. Drupes obliquely ovoid to oblong, 0.8–1.5 × 0.5–1 cm, dark purple to black, edible but rarely eaten by man. Fl (Nov); common, in secondary growth. *Mombin blanc, mombin faux, mombin fou, mombin sauvage* (Créole), *tatapirica* (Portuguese).

Tapirira obtusa (Benth.) J. D. Mitch. [Syns.: *T. peckoltiana* Engl., *T. marchandii* Engl.] FIG. 9K,L, PL. 10e

Small to large trees, to 35 m × 40 cm. Trunks unbuttressed or with low buttresses. Bark light brown or gray, thin usually shallowly fissured, the inner bark pinkish to reddish-brown, resinous. Leaves usually imparipinnate, with 5–13 leaflets, the rachis usually terete. Leaflets: blades chartaceous to subcoriaceous, sometimes rugose, sparsely to very densely pubescent with both appressed and erect hairs, with mostly erect hairs abaxially, the midrib and often the secondary veins with erect hairs adaxially; secondary and tertiary veins prominent and conspicuous abaxially, secondary veins impressed or scarcely prominent adaxially, the tertiary and higher orders of venation frequently inconspicuous adaxially. Drupes ellipsoid, 1–2 × 1–1.7 cm, dark purple to black. Fl (Sep, Oct, Nov), fr (Apr); relatively common, in primary and secondary non-flooded forest.

THYRSODIUM Benth.

Small to large trees. Milky exudate when damaged. Leaves evergreen, pari- or imparipinnate. Leaflets usually alternate; petioles present; blades with entire margins; secondary veins arch upwards near margin. Inflorescences terminal or axillary, thyrsoid, exuding milky sap when damaged; pedicels present. Flowers unisexual (plants dioecious); hypanthium present; calyx valvate, 5-merous, persistent in fruit; corolla imbricate, 5-merous; stamens 5, opposite sepals, the filaments very short; nectary disc intrastaminal; pistil rudimentary in staminate flowers; ovary 1-locular, the style terminal, the stigmas 1–3; placentation lateral, subbasal to subapical. Drupes ovoid, oblong, ellipsoid, or globose, the mesocarp fleshy, the endocarp crustaceous; embryo straight.

Mitchell, J. D. & D. C. Daly. 1993. A revision of *Thyrsodium* (Anacardiaceae). Brittonia **45:** 115–129.

1. Leaflets (sub)glabrous. Pistillode, pistil, and fruits glabrous.
   2. Leaflet apex usually short-acuminate, rounded, or emarginate (rarely long acuminate); tertiary venation slightly prominent adaxially. Petals 2–2.5 mm long; staminodia 0.3–0.4 mm long; stigmas 2–3. . . . . *T. guianense.*
   2. Leaflet apex usually long acuminate; tertiary venation flat adaxially. Petals 3.2–3.6 mm long; staminodia 0.1–0.2 mm long; stigma 1. . . . . *T. puberulum.*
1. Leaflets sparsely to densely pubescent. Pistillode, pistil, and fruits pubescent. *T. spruceanum.*

Thyrsodium guianense Marchand FIG. 11A–C

Medium-sized trees, to 36 m × 50 cm. Trunks with low buttresses. Outer bark smooth, gray, thin, the inner bark reddish-orange. Leaves 6–14-foliolate; petiole pubescent. Leaflets opposite to alternate; blades broadly to narrowly obovate, ovate, elliptic, to rarely oblong, coriaceous, both surfaces glabrous or abaxial surface sparsely pubescent, often shiny adaxially, the base usually cuneate, sometimes acute or obtuse, the apex short acuminate, rounded-truncate, emarginate, or retuse, very rarely long acuminate. Inflorescences densely pubescent. Flowers: petals ovate to narrowly ovate, 2.2–2.5 × 1.3 mm, white when fresh. Staminate flowers: filaments sparsely pubescent; pistillode cylindric, 2–2.5 long, glabrous, crowned by 2–3 stigmatic lobes. Pistillate flowers: staminodia 0.3–0.4 mm long; ovary ovoid, glabrous, the style crowned by 2–3 capitate stigmas. Drupes obovoid, ellipsoid or

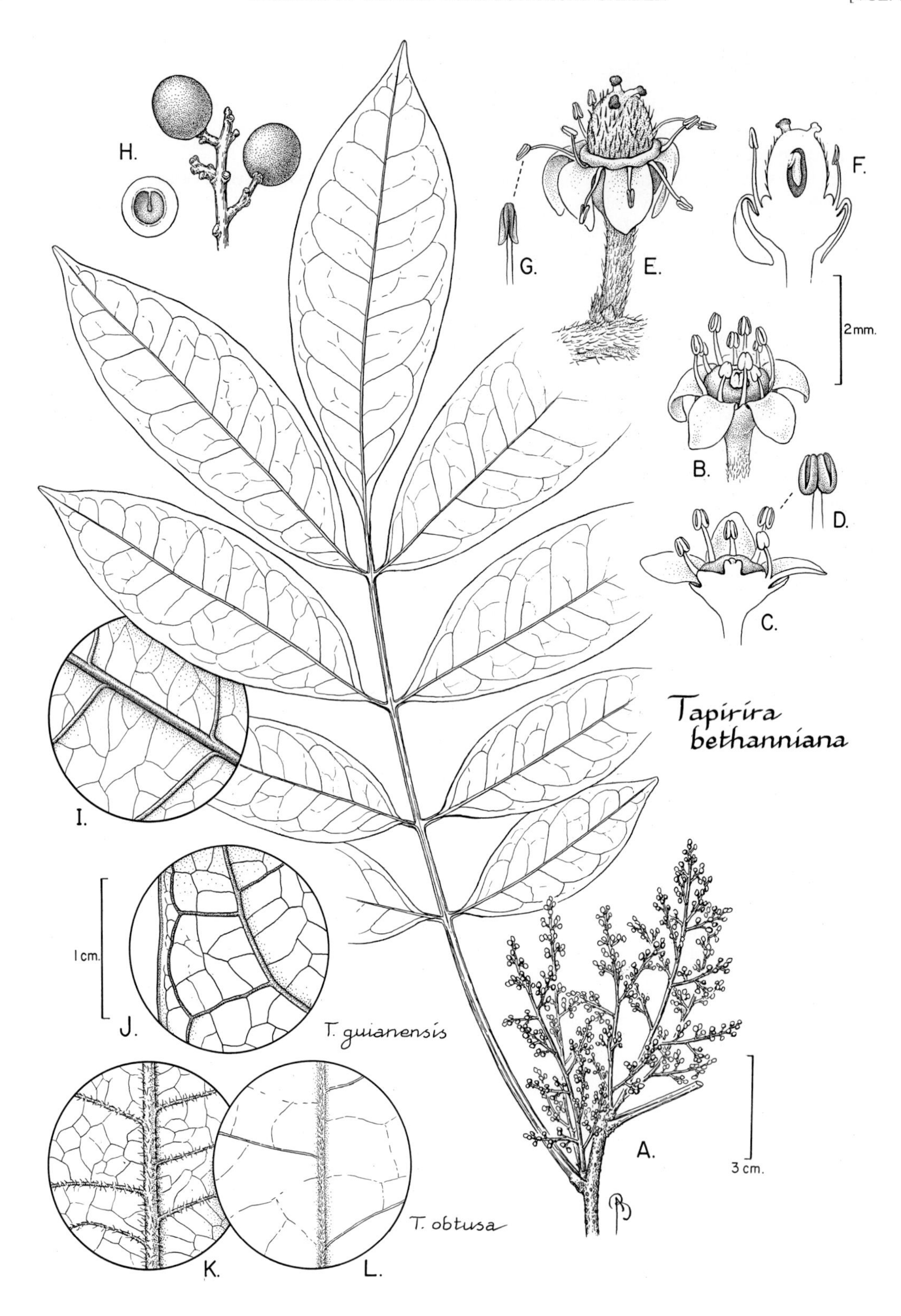

FIG. 9. ANACARDIACEAE. **A–H.** *Tapirira bethanniana* (A–D, *Mori & Boom 15253*; E–G, *Pipoly & Godfrey 7434*; H, *Sabatier & Riera 2046*). **A.** Part of stem with leaf and inflorescence. **B.** Staminate flower. **C.** Medial section of staminate flower. **D.** Enlargement of stamen. **E.** Pistillate flower. **F.** Medial section of pistillate flower. **G.** Enlargement of staminode. **H.** Mature drupes with transverse section on left. **I.** Detail of abaxial leaflet surface. **J.** *T. guianensis* (*Mori & Mitchell 18766*). Detail of nearly glabrous abaxial leaf blade surface. **K, L.** *T. obtusa* (*Mori & Mitchell 18775*) **K.** Detail of abaxial leaflet surface; note the erect trichomes on midrib and secondary veins. **L.** Detail of adaxial leaflet surface; note the densely pubescent midrib. (Reprinted with permission from J. D. Mitchell, Mem. New York Bot. Gard. 64, 1990.)

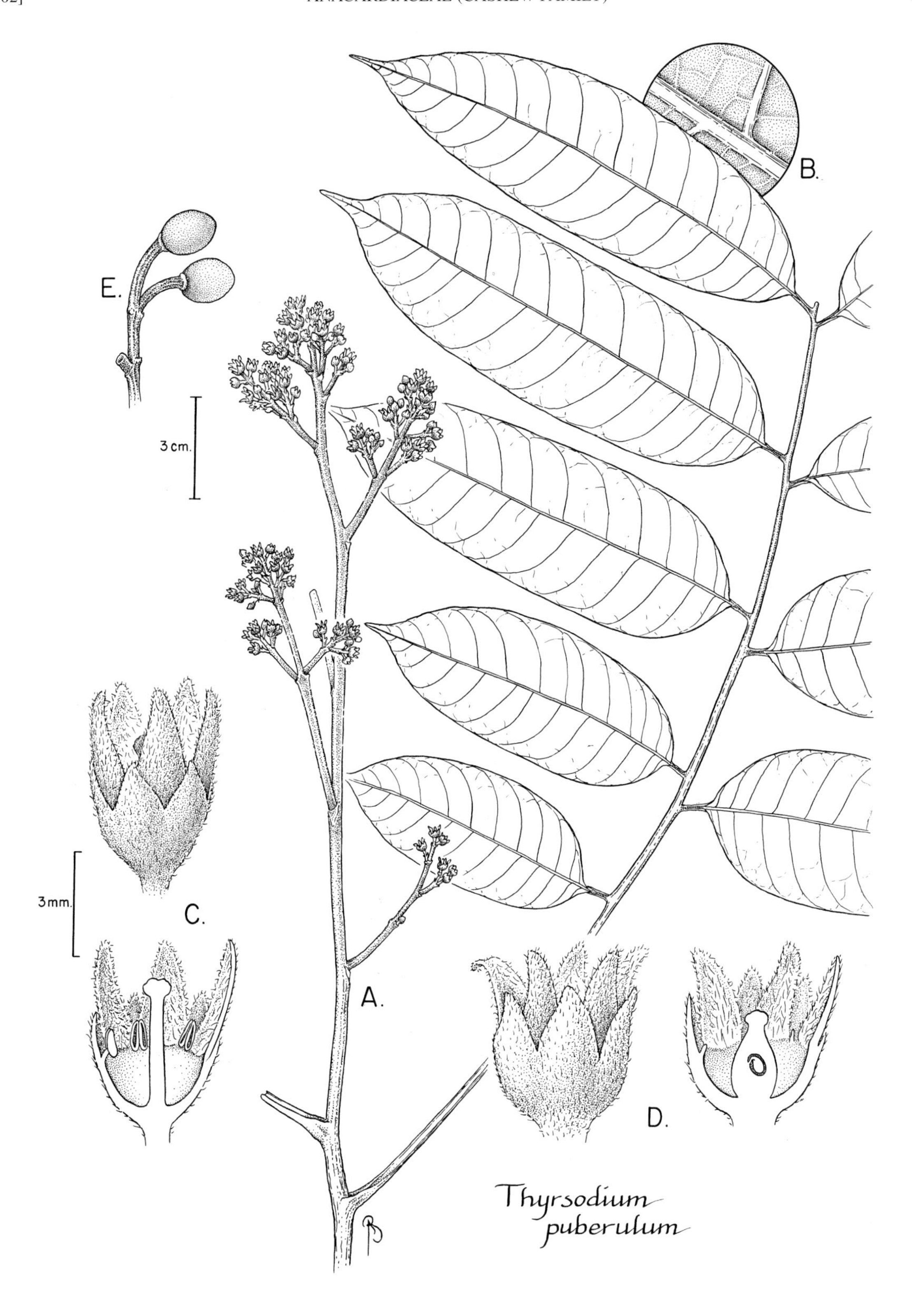

FIG. 10. ANACARDIACEAE. *Thyrsodium puberulum* (A, B, D, *Prance & Silva 58888*; C, *Irwin et al. 48121*; E, *Rabelo et al. 3113*). **A.** Part of stem with leaf and inflorescence. **B.** Detail of abaxial leaflet surface. **C.** External view (top) and medial section (bottom) of staminate flower showing corolla greatly exceeding calyx and subsessile stamens. **D.** External view (left) and medial section of pistillate flower showing corolla only slightly exceeding calyx, the staminodes without rudimentary anthers, and the glabrous ovary. (Reprinted with permission from J. D. Mitchell & D. C. Daly, Brittonia 45(2), 1993.)

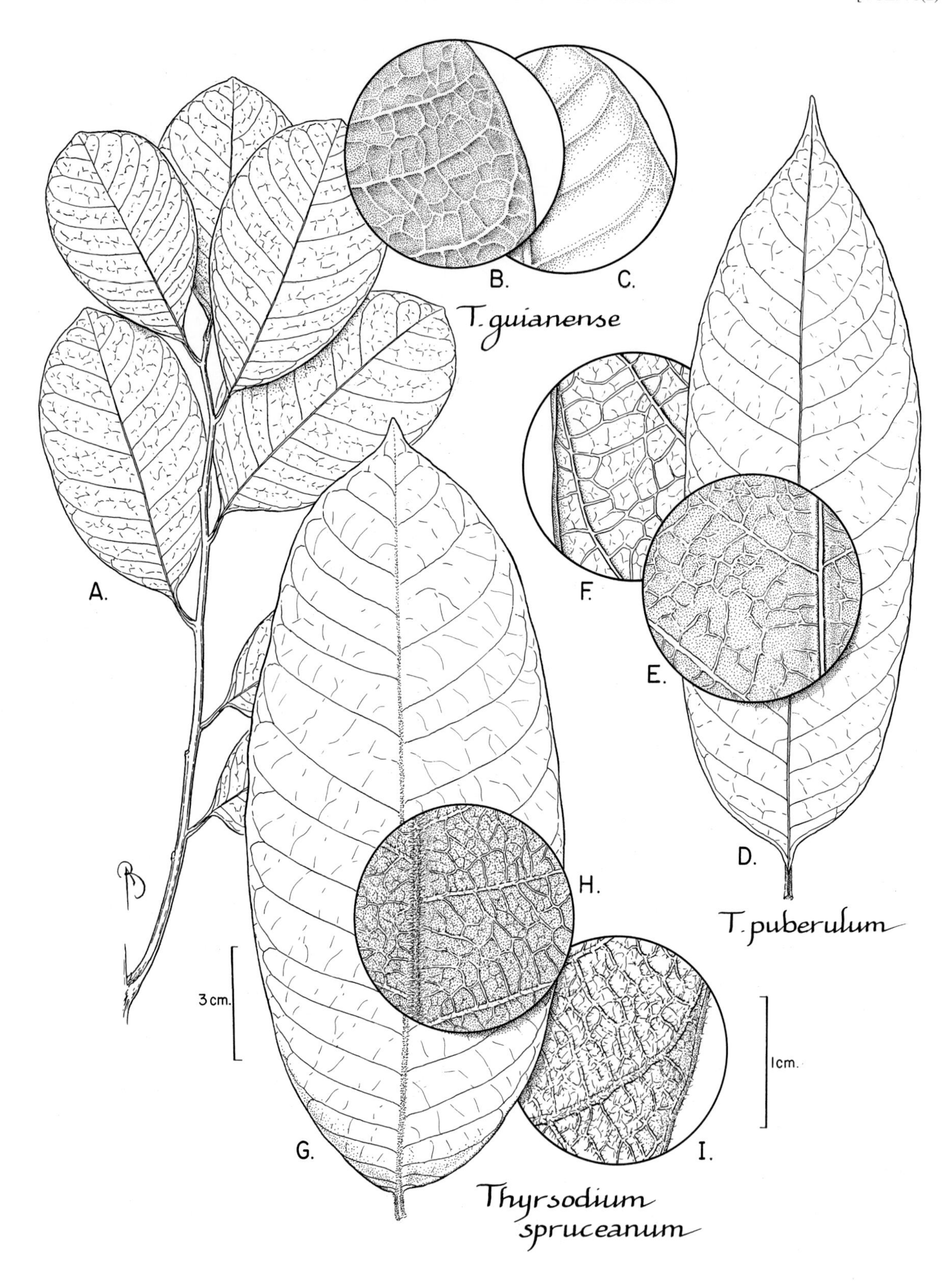

FIG. 11. ANACARDIACEAE. A comparison of the leaves of species of *Thyrsodium*. **A–C.** *Thyrsodium guianense* (*de Granville 3392*). **A.** Pinnately compound leaf. **B.** Detail of adaxial leaflet surface. **C.** Detail of abaxial leaflet surface. **D–F.** *Thyrsodium puberulum* (*Boom 2006*). **D.** Leaflet. **E.** Detail of adaxial leaflet surface. **F.** Detail of abaxial leaflet surface. **G–I.** *Thyrsodium spruceanum* (*Lindeman 4773*). **G.** Leaflet. **H.** Adaxial leaflet surface; note the densely pubescent midrib. **I.** Detail of abaxial leaflet surface.

oblong, 2.5–1.4 cm, bluish-green at maturity, glabrous. Fl (Nov); in non-flooded forest.

Thyrsodium puberulum J. D. Mitch. & Daly
FIGS. 10, 11D–F, PL. 10f

Small to large trees, 8–35 m × 50 cm. Trunks with shallow buttresses to 50 cm tall. Outer bark brown, thin, longitudinally fissured, the inner bark reddish-orange. Leaves 8–11-foliolate; petiole glabrous to sparsely pubescent. Leaflets alternate or subopposite; blades narrowly oblong, lanceolate, or occasionally narrowly elliptic, ovate, or obovate, usually coriaceous, often shiny adaxially, abaxial surface glabrous to sparsely pubescent, the adaxial surface glabrous, the base rounded to obtuse or occasionally cuneate, sometimes oblique, the apex long acuminate. Inflorescences sparsely to densely pubescent. Flowers: petals narrowly ovate, 3.2–3.6 × 1.5–1.7 mm, greenish-white to pale yellow (when fresh). Staminate flowers: corolla greatly exceeding calyx; filaments glabrous; pistillode cylindric, 3–3.5 mm long, glabrous, the rudimentary stigma capitate, sometimes apically notched. Pistillate flowers: corolla not greatly exceeding calyx; staminodia present, without or with rudimentary anthers, 0.1–0.2 mm long; ovary ovoid, glabrous, the style crowned by a single capitate stigma. Drupes ellipsoid to subglobose, 2–3 mm diam., glabrous, color at maturity unknown. Fl (Sep); in non-flooded forest.

Thyrsodium spruceanum Benth.
[Syn.: *T. schomburgkianum* Benth.] FIG. 11G–I, PL. 10g

Small to medium-sized trees, 3–35 m × to 100 cm. Trunks shallowly buttressed or unbuttressed. Outer bark gray-brown, thin, hard, very shallowly fissured, the inner bark dark reddish-orange. Leaves 7–15-foliolate; petiole densely pubescent. Leaflets: blades narrowly oblong, lanceolate, or occasionally narrowly elliptic, ovate, obovate, or oblanceolate, chartaceous to subcoriaceous, shiny or dull adaxially, both surfaces sparsely to densely pubescent, the base obtuse, rounded, truncate, or cordate, sometimes oblique, the apex short to long acuminate. Inflorescences sparsely to densely pubescent. Flowers: petals narrowly ovate, 2.6–3.8 × 1.3–2 mm, cream-colored to pale yellow (when fresh). Staminate flowers: hypanthium more deeply cupular than in pistillate flowers; filaments sparsely pubescent; pistillode cylindric, 2.5–4 mm long, sparsely to densely pubescent, rudimentary stigmas 2–3-lobed or subcapitate. Pistillate flowers: rudimentary anthers 1 mm long; ovary globose, densely pubescent, the style 1 mm long, crowned by capitate or bilobed stigma. Drupes ovoid, ellipsoid, or oblongoid, 1.7–2.5 × 2–3 cm, densely pubescent, green to bluish-green when ripe. No fertile collections seen from our area; apparently rare (one collection only from our area), in non-flooded forest.

ANNONACEAE (Custard-apple Family)

by Paul Maas and Hiltje Maas-van de Kamer

Small to large trees, rarely lianas. Bark fibrous, easily peeled off in long strips, the slash showing conspicuous vertical streaks. Indument of simple, stellate or scale-like hairs, or lacking. Stipules absent. Leaves simple, distichous. Plants sometimes cauliflorous, ramiflorous, or flagelliflorous. Inflorescences axillary or non-axillary, 1- to many-flowered cymes, bracteate or ebracteate. Flowers actinomorphic, bisexual; sepals 3, free or ± connate; petals 6, free or sometimes connate, in 2 whorls of 3, rarely 1 whorl reduced (*Annona*), generally much longer than sepals; stamens numerous, with shield-like apical expansion of connective; carpels free, few to numerous, each with 1 to several ovules. Fruits of free, stipitate monocarps (apocarpous), syncarpous, or pseudosyncarpous (*Duguetia*). Seeds with ruminate endosperm.

Fries, R. E. 1976. Annonaceae. *In* J. Lanjouw & A. L. Stoffers (eds.), Flora of Suriname **I(2):** 341–383. E. J. Brill, Leiden.
Jansen-Jacobs, M. J. 1976. Annonaceae. *In* J. Lanjouw & A. L. Stoffers (eds.), Flora of Suriname **I(2):** 658–687. E. J. Brill, Leiden.
Steyermark, J. A., P. J. M. Maas, P. E. Berry, D. M. Johnson, N. A. Murray & H. Rainer. 1995. Annonaceae. *In* J. A. Steyermark, P. E. Berry & B. K. Holst (gen. eds.), Flora of the Venezuelan Guayana **2:** 413–469. P. E. Berry, B. K. Holst & K. Yatskievych (vol. eds.). Missouri Botanical Garden, St. Louis; Timber Press, Portland, Oregon.

1. Indument of stellate or scale-like hairs (visible with hand lens). Fruits pseudosyncarpous, the lower carpels sterile and often connate, forming a collar. . . . . *Duguetia.*
1. Indument of simple hairs or lacking (rarely furcate hairs in *Annona*; minute stellate hairs in *Anaxagorea*). Fruits not pseudosyncarpous, the lower carpels not connate and not forming a collar.
    2. Flowers and fruits axillary. Fruits of free, 1- to several-seeded monocarps (apocarpous).
        3. Midrib of leaf blades raised adaxially.
            4. Flowers subtended by 4–10 bracts. Petals 4–7 mm long.
                5. Leaf blades with distinct marginal vein very close to margin; secondary veins in 15–20 pairs. Monocarps with stipes 10–15 mm long. . . . . *Pseudoxandra.*
                5. Leaf blades without marginal vein; secondary veins in 6–7 pairs. Monocarps with stipes 1–2 mm long.. . . . . *Oxandra.*
            4. Flowers subtended by 2 bracts. Petals 2–20 mm long.
                6. Midrib of leaf blades distinctly grooved adaxially. Petals ca. 20 mm long, greenish. Seeds irregularly ridged. . . . . *Cremastosperma.*
                6. Midrib of leaf blades not grooved adaxially. Petals 2–6 mm long, whitish. Seeds pitted. . . . *Unonopsis.*
        3. Midrib of leaf blades impressed adaxially.
            7. Petals with brownish indument of stellate to simple hairs, these hardly visible even with handlens; staminodia present between stamens and carpels. Monocarps explosively dehiscent, 2-seeded. . . . *Anaxagorea.*

7. Petals glabrous or with indument of simple hairs clearly visible with handlens; staminodia absent. Monocarps indehiscent or not explosively dehiscent, 1- to many-seeded.
   8. Petals subequal. Monocarps indehiscent, 1-seeded. . . . . *Guatteria.*
   8. Petals unequal. Monocarps dehiscent, 1- to many-seeded. . . . . *Xylopia.*

2. Flowers and fruits terminal, leaf-opposed, or supra-axillary. Fruits of free, 1- to several-seeded monocarps (apocarpous) or monocarps fused (syncarpous).
   9. Bracts absent. Fruits apocarpous.
      10. Flowers pendent, pedicels 40–60 mm long. Petals free, the inner ones boat-shaped. . . . . . *Cymbopetalum.*
      10. Flowers not pendent, pedicels 20–30 mm long. Petals basally connate, the inner ones not boat-shaped. . . . . *Cardiopetalum.*
   9. Bracts present. Fruits syncarpous.
      11. Flowers winged (propeller-shaped). . . . . *Rollinia.*
      11. Flowers not winged (not propeller-shaped).
         12. Leaf blades with distinct marginal vein. Staminodia present between stamens and petals. Fruits with distinct basal, woody collar, the surface smooth (i.e., carpels not elevated). . . . . *Fusaea.*
         12. Leaf blades without distinct marginal vein. Staminodia absent. Fruits without a woody collar, the surface rough (i.e., carpels mostly distinctly elevated). . . . . *Annona.*

ANAXAGOREA A. St.-Hil.

Trees to shrubs. Indument of simple to stellate, minute hairs. Leaf blades with midrib impressed adaxially. Inflorescences axillary, 1- to few-flowered; bracts 2 per flower. Flowers when ripe emitting strong banana-like aroma; sepals valvate, thin; petals valvate, unequal, yellowish, thick, often with brownish indument; innermost stamens staminodial, as long as stamens. Fruits apocarpous, monocarps several, club-shaped with stipe-like basal part and globose, beaked head, explosively dehiscent. Seeds 2 per monocarp, shiny black.

Maas, P. J. M. & L. Y. Th. Westra. 1984. Studies in Annonaceae II. A monograph of the genus *Anaxagorea* A. St. Hil. Bot. Jahrb. Syst. **105(1):** 73–134.

Anaxagorea dolichocarpa Sprague & Sandwith FIG. 12

Trees, 2–8 m tall. Bark brown to black with white spots; wood yellow to white. Young stems covered with brownish hairs, soon becoming glabrous. Leaves: petioles 3–10 × 1–3 mm; blades narrowly elliptic, 11–27 × 4–8 cm, the apex acuminate; secondary veins in 7–14 pairs. Inflorescences 1(2)-flowered, sometimes cauliflorous; pedicels 5–10 mm long. Flowers: sepals 7–11 mm long; petals ovate to obovate, 8–12 mm long, 2–10 mm thick, the outer ones flat adaxially, the inner ones keeled in upper part. Monocarps 3–9, green to reddish, ca. 30 mm long. Fl (Aug, Oct, Nov, Dec), fr (Jan, Mar, Apr, May, Nov); in non-flooded forest and streamside habitats. This species is very widespread and highly variable. *Mamayawé* (Créole).

The specimens found in our area have relatively small flowers with caducous sepals.

ANNONA L.

Trees or lianas. Indument of simple, rarely furcate hairs. Leaf blades with midrib impressed adaxially. Inflorescences terminal, internodal, or leaf-opposed, 1- to several-flowered; bracts 2 per flower. Flowers: sepals valvate; outer petals valvate, the inner ones imbricate or valvate, sometimes absent. Fruits syncarpous, of many carpels, fleshy, mostly areolate and often with a spiny to tuberculate surface.

1. Trees or lianas. Petals reddish to pink, basally connate.
   2. Trees. Leaf blades thick, densely pubescent abaxially; secondary veins in 13–15 pairs. . . . . *A. ambotay.*
   2. Lianas. Leaves thin, sparsely pubescent abaxially; secondary veins in 6–9 pairs. . . . . *A. haematantha.*
1. Trees. Petals white to yellowish white, free.
   3. Leaf blades sparsely hairy, with pocket-shaped, hairy pits in axils of secondary veins (domatia) abaxially; secondary veins in 5–9 pairs. Fruits edible. Cultivated. . . . . *A. muricata.*
   3. Leaf blades densely covered with erect hairs, without pocket-shaped pits in axils of secondary veins (domatia) abaxially; secondary veins in 15–22 pairs. Fruits not edible. Native. . . . . *A. prevostiae.*

Annona ambotay Aubl.

Trees, 10–20 m × to 25 cm. Bark brown, with shallow fissures, the slash emitting slight spicy aroma. Young stems and most other vegetative parts, pedicels, and sepals densely covered with erect, brown, simple to furcate hairs. Leaves: petioles 10 × 2–3 mm; blades ovate to obovate, 15–36 × 7–14 cm, thick, densely covered with erect, brown hairs abaxially, the apex distinctly acuminate; secondary veins in 13–15 pairs. Inflorescences 1–2-flowered; pedicels 10–15 × 1 mm, to 20 × 3 mm in fruit. Flowers: sepals 2–3 mm long; petals basally connate, narrowly ovate, ca. 15 mm long, wine-red. Fruits first green, pale yellow at maturity, subglobose, ca. 4 cm diam., the surface smooth, not areolate. Fl (Sep), fr (Apr); in non-flooded forest and in streamside habitats.

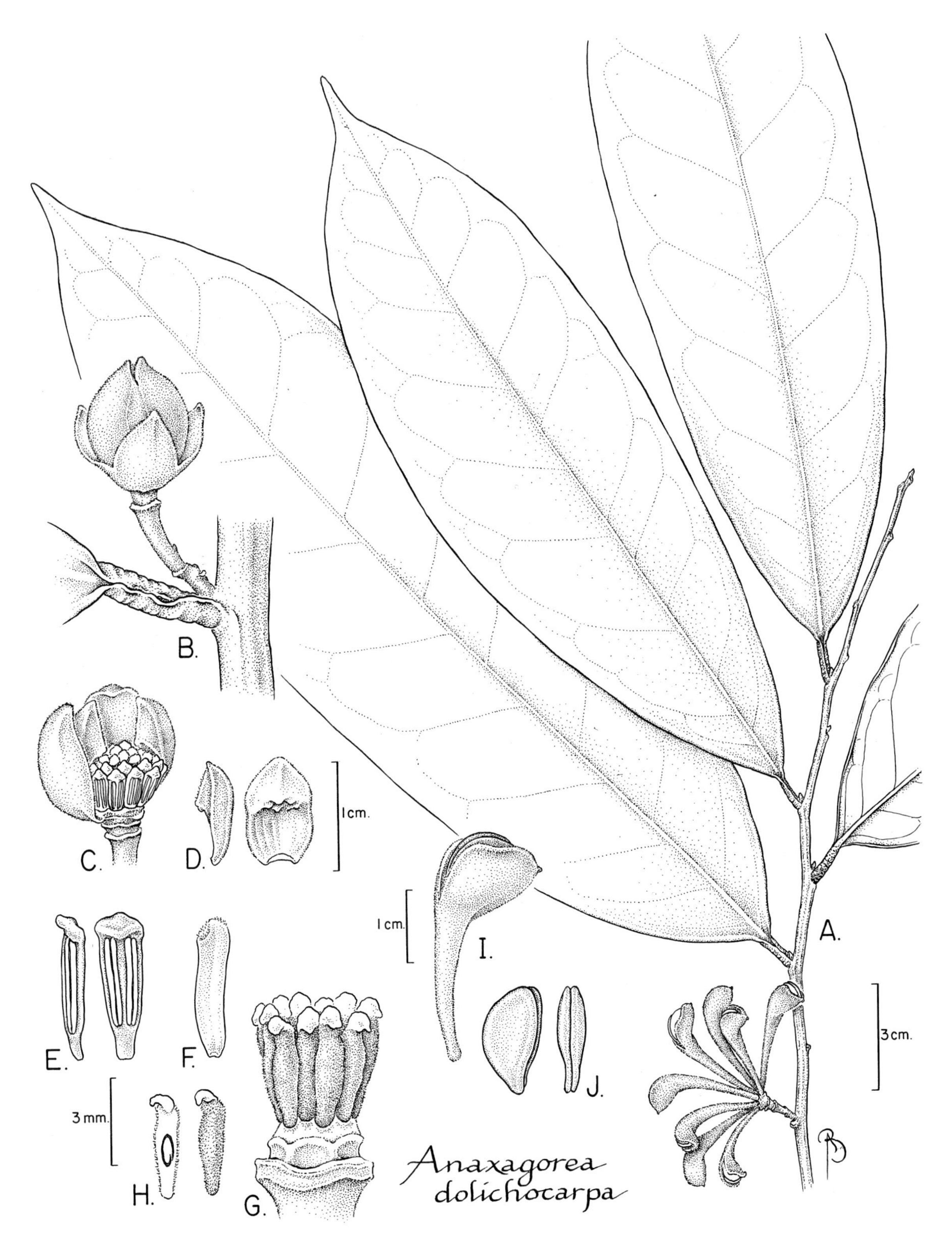

FIG. 12. ANNONACEAE. *Anaxagorea dolichocarpa* (A, *Mori & Pennington 17992*; B, I, J, *Mori et al. 22809*; C–H, *Mori et al. 22783*). **A.** Part of stem with leaves and fruit consisting of several monocarps. **B.** Close-up of part of stem showing inrolled petiole and axillary flower bud. **C.** Close-up of flower with one outer and one inner petal removed to show stamens. **D.** Lateral view (near right) and adaxial view (far right) of inner petals. **E.** Oblique view (near right) and adaxial view (far right) of stamen expanded connectives. **F.** Staminode. **G.** Flower with petals, stamens, and staminodes removed to show free carpels. **H.** Medial section of carpel (near right) and carpel (far right). **I.** A single dehiscent monocarp. **J.** Two views of seeds.

Annona haematantha Miq.

Lianas. Young stems and most other vegetative parts, pedicels, and sepals densely covered with erect, brown, simple hairs. Leaves: petioles 5–10 × 1 mm; blades elliptic-obovate, 12–17 × 5–7 cm, thin, sparsely hairy abaxially, the apex short-acuminate; secondary veins in 6–9 pairs. Inflorescences 1-flowered; pedicels 15–20 × 0.5 mm. Flowers: sepals ca. 3 mm long; petals connate for basal 5–7 mm, the free part narrowly triangular, 15–18 × 5–7 mm long, petals purplish-pink. Fruits not seen from our area (elsewhere subglobose, 1.7–2 cm diam.). Fl (Jun, Aug); in non-flooded forest.

Annona muricata L.

Small trees, to 10 m tall. Young stems densely covered with appressed hairs, soon glabrous. Leaves: petioles 3–5 × 1–2 mm; blades mostly narrowly obovate, 6–15 × 3–7 cm, rather thick, shiny adaxially, sparsely covered with appressed hairs, and with pocket-shaped hairy pits (domatia) in secondary vein axils abaxially, the apex short-acuminate to obtuse; secondary veins in 5–9 pairs. Plants often cauliflorous. Inflorescences 1-flowered; pedicels 15–20 × 2–3 mm. Flowers emitting spicy aroma; sepals 3–5 mm long; petals yellowish white, the outer ones valvate, broadly ovate, 2.5–6 × 2–4 cm, thick (3–4 mm when fresh), the inner ones imbricate, subcircular, to 4 × 3 cm, much thinner than outer ones. Fruits ovoid, 15–20 cm long, with many upcurved spine-like excrescences, green, the pulp white, sweet, edible. Cultivated. Fr (Sep); cultivated. *Cachiman épineux*, *corossol*, *corossolier* (Créole), *graviola* (Portuguese).

Description is based on material from outside our area.

Annona prevostiae H. Rainer FIG. 13, PL. 11a,b

Trees, (4)15–25 m × 20–40 cm. Bark gray, smooth, lenticellate, the inner bark orangish-brown, laminated, emitting strong spicy aroma. Young stems, petioles, pedicels, and sepals densely covered with erect, whitish, simple hairs. Leaves: petioles 5–10 × 2 mm; blades narrowly elliptic to narrowly obovate, 15–22 × 7–10 cm, thick, pale glaucous green; densely hairy abaxially, the apex short-acuminate; secondary veins in 15–22 pairs, strongly prominent, the tertiary veins parallel. Inflorescences 1–3-flowered; pedicels 15–25 × 2–3 mm (to 35 mm long in fruit). Flowers: sepals 3–5 mm long; petals white to cream-colored, the outer ones deltate, 20–25 mm long, densely whitish sericeous, the inner ones narrowly triangular, 15–20 mm long. Fruits ellipsoid, to 7 × 4 cm, with many spine-like excrescences, first green, yellow-orange at maturity, the pulp white. Fl (Mar, Apr, Aug, Sep, Oct, Nov), fr (Mar, Oct, Dec); in non-flooded forest.

CARDIOPETALUM Schltdl.

Trees. Indument of simple hairs. Leaf blades with midrib raised adaxially. Inflorescences internodal, 1-flowered; bracts absent. Flowers: sepals valvate, almost as long as petals; petals imbricate, unequal, connate at base. Fruits apocarpous, monocarps many, constricted, dehiscent, the stipe short to long. Seeds several per monocarp, arillate.

Johnson, D. M. & N. A. Murray. 1995. Synopsis of the tribe Bocageeae (Annonaceae), with revisions of *Cardiopetalum*, *Froesiodendron*, *Trigynaea*, *Bocagea*, and *Hornschuchia*. Brittonia **47:** 248–319.

Cardiopetalum surinamense R. E. Fr.

Trees, 4–19 m × 6–17 cm. Trunks with fluted base. Bark smooth, grid-cracked in rectangles, the inner bark cream-colored. Young stems rather densely covered with whitish, appressed hairs. Leaves: petioles 5–6 × 1 mm; blades narrowly ovate to narrowly elliptic, 10–15 × 4–6 cm, nearly glabrous except for some hairs along midrib abaxially, the apex short-acuminate; secondary veins in 6–8 pairs. Inflorescences: pedicels 25–30 × 1–2 mm. Flowers: sepals 13–15 mm long; petals obovate, ca. 15 × 11 mm, pale yellow. Monocarps 10–15, sickle-shaped, constricted, 0.5–3.5 × ca. 1 cm, yellow-orange, the stipes 5–15 × 1.5–2.5 mm. Fl (May), fr (Jan, Sep); in non-flooded forest.

CREMASTOSPERMA R. E. Fr.

Trees. Indument of simple hairs, or lacking. Leaf blades with midrib raised and grooved adaxially. Inflorescences axillary, 1-flowered; bracts 2 per flower. Flowers: sepals imbricate; petals imbricate, subequal, with thin margins. Fruits apocarpous, the monocarps several, indehiscent, the stipe short to long. Seeds 1 per monocarp, irregularly ridged.

Cremastosperma brevipes (Dunal) R. E. Fr. PL. 11c,d

Trees, 3–5(8) m × 5–10 cm. Young stems and other vegetative parts glabrous or nearly so. Leaves: petioles 5–15 × 2 mm; blades narrowly elliptic to obovate, 20–35 × 8–12 cm, the apex short-acuminate; secondary veins in 11–14 pairs, the tertiary veins subparallel. Plants sometimes cauliflorous. Pedicels 15–30 × 1–2 mm. Flowers with fruity aroma; sepals green, 3–4 mm long; petals ovate, ca. 20 mm long, green with reddish margins, yellow at anthesis. Monocarps 3–16, ellipsoid, 15–16 × 7–10 mm, with lateral apicule less than 1 mm long, reddish pink, becoming purple-black at maturity, the stipes red, 7–15 mm long. Fl (Feb, Apr, Oct, Nov), fr (Sep, Oct, Nov, Dec); in non-flooded forest at all elevations. *Mamayawé* (Créole).

CYMBOPETALUM Benth.

Shrubs. Indument of simple hairs. Leaf blades with midrib raised adaxially. Inflorescences leaf-opposed or internodal, 1-flowered; bracts absent. Flowers pendent on long pedicels; sepals valvate, small; petals valvate, the outer ones thin, the inner ones thick, boat-shaped. Fruits apocarpous, the monocarps several, constricted, dehiscent, the stipe long. Seeds several to many per monocarp, arillate.

Murray, N. A. 1993. Revision of *Cymbopetalum* and *Porcelia* (Annonaceae). Syst. Bot. Monogr. **40:** 1–121.

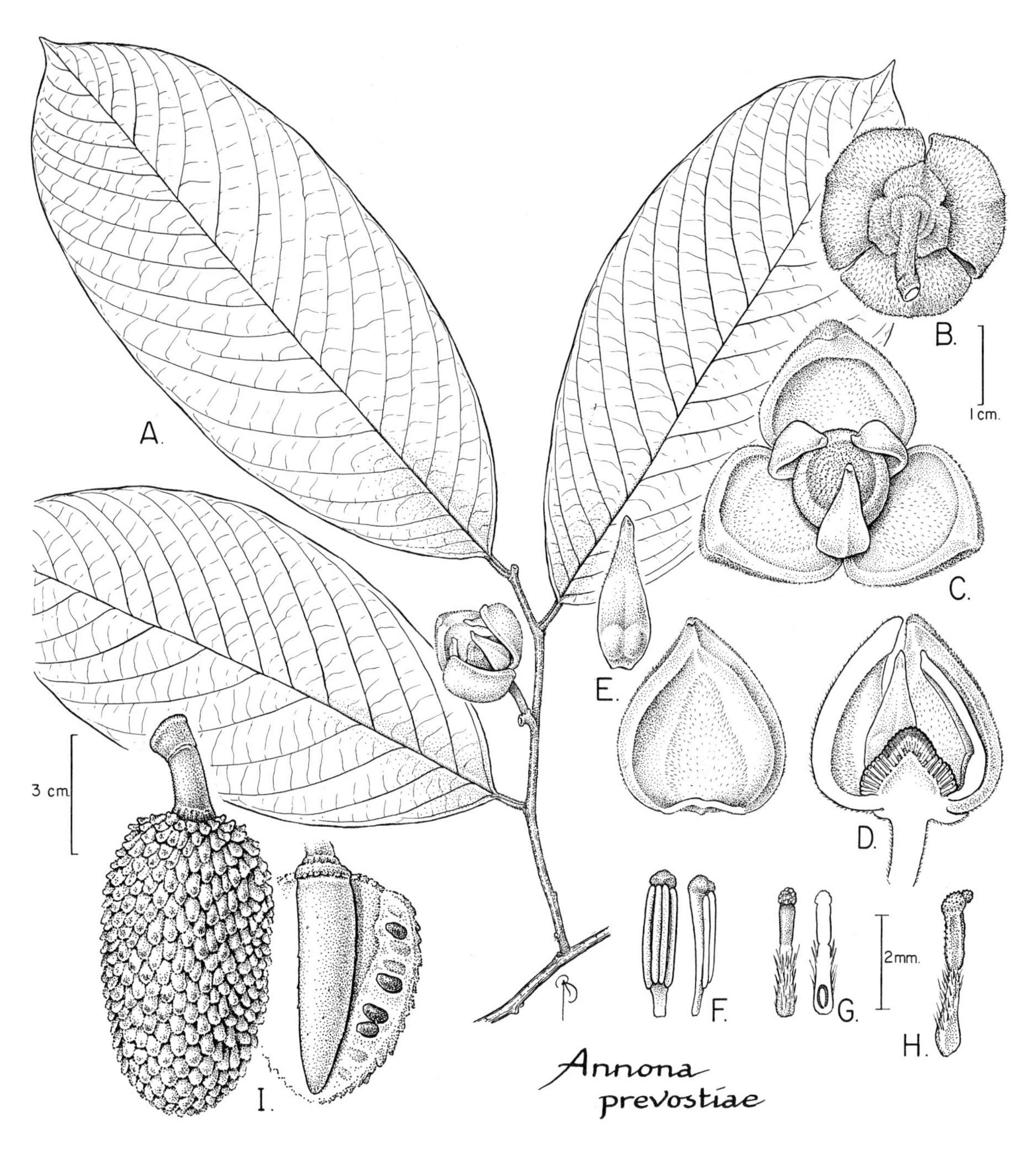

FIG. 13. ANNONACEAE. *Annona prevostiae* (A–H, *Mori et al. 21600*; I, *Mori et al. 22000*). **A.** Part of stem with leaves and partially opened flower. **B.** Basal view of flower. **C.** Apical view of flower showing both whorls of petals. **D.** Medial section of flower. **E.** Inner petal (above) and outer petal (right). **F.** Adaxial view (far left) and lateral view (near left) of stamen showing shield-like apical expansion of connective. **G.** Lateral view (far left) and medial section (near left) of a carpel. **H.** Sterile carpel with thickened stylar region. **I.** Mature fruit (left) and longitudinal section of fruit showing thickened receptacle (right).

Cymbopetalum brasiliense (Vell.) Baill. PL. 11e

Shrubs, 1–2 m tall. Young stems and other vegetative parts sparsely covered with appressed hairs. Leaves: petioles 3–5 × 1–2 mm; blades narrowly elliptic, 16–25 × 5–8 cm, glandular-punctate, the apex acute to acuminate; secondary veins in 8–12 pairs. Inflorescences: pedicels 40–60 × 1–2 mm. Flowers: sepals 4–5 mm long; petals green, cream-colored at anthesis, the outer ones broadly ovate, 15–25 × 15–25 mm, the inner ones very broadly ovate, 15–35 mm long. Monocarps 8–15, sickle-shaped, constricted, 20–35 mm long, green to yellow, tinged with red, aril orange, the stipes well developed. Fl (Sep, Oct), fr (Jan, Feb, Apr, May, Sep, Dec); in non-flooded forest at all elevations.

DUGUETIA A. St.-Hil.

Trees or shrubs. Indument of stellate or scale-like hairs. Leaf blades with midrib impressed adaxially. Inflorescences leaf-opposed or internodal, rarely from base of trunk and flagelliform, 1- to many-flowered; bracts 2 per flower. Flowers: sepals valvate; petals imbricate, subequal. Fruits pseudosyncarpous, of few to many, ± pyramidal carpels, free to rarely connate, the lower carpels sterile and often connate into a collar.

1. Inflorescence flagelliform, to 1.5 m long, arising from base of trunk and trailing over forest floor. Flowers emitting fetid aroma. . . . . *D. cadaverica.*
1. Inflorescence not flagelliform, not arising from base of trunk and trailing over forest floor. Flowers not emitting fetid aroma.
   2. Large trees, to 30 m tall. Flowers salmon-colored. Fruits large, 4–11.5 cm diam., of 125–200 carpels. . . . . *D. surinamensis.*
   2. Small trees, to 10 m tall. Flowers white to yellow. Fruits smaller, 1–6 cm in diam., of 5–150 carpels.
      3. Sepals 3–5 mm long; petals 5–6 times as long as sepals. Fruits of 5–15 carpels, the carpels verrucose. . . . . *D. inconspicua.*
      3. Sepals 8–20 mm long; petals 1–3 times as long as sepals. Fruits of 10–100 carpels, the carpels not verrucose.
         4. Leaf blades 30–40 × 8–12 cm. Inflorescences 3–7-flowered. . . . . *D. eximia.*
         4. Leaf blades 8–30 × 3–8 cm. Inflorescences 1–2-flowered.
            5. Leaf blades discolorous. Sepals basally connate. Fruiting carpels markedly wrinkled. . . . . . *D. yeshidan.*
            5. Leaf blades concolorous. Sepals free. Fruiting carpels smooth.
               6. Petals about as long as sepals. Fruits 1.5–2.7 cm diam., of 10–30 carpels.
                  7. Young twigs distinctly grooved. Leaf blades with secondary veins raised adaxially. Sepals 10–15 mm long. . . . . *D. pycnastera.*
                  7. Young twigs terete. Leaf blades with secondary veins impressed adaxially. Sepals 15–22 mm long. . . . . *D. granvilleana.*
               6. Petals 1.5–2 times as long as sepals. Fruits 2–6 cm diam., of 50–100 carpels.
                  8. Leaf blades bullate, with stellate hairs abaxially. Fruiting carpels long-apiculate, the apiculus to 15 mm long. . . . . *D. riparia.*
                  8. Leaf blades not bullate, with stellate scales abaxially. Fruiting carpels short-apiculate, the apiculus generally <5 mm long. . . . . *D. calycina.*

Duguetia cadaverica Huber FIG. 14, PL. 11f

Shrubs or trees, (2)3–8 m × 8 cm. Young stems rather densely covered with stellate scales. Leaves: petioles 2–8 × 1–3 mm; blades narrowly elliptic to narrowly obovate, 10–30 × 3–7 cm, sparsely covered with stellate scales abaxially, the apex long-acuminate; secondary veins in 10–18 pairs. Inflorescences from base of trunk, flagelliform, trailing over forest floor, many-flowered, to 1.5 m long; pedicels 10–40 mm long. Flowers at anthesis emitting fetid aroma; sepals basally connate, 10–25 mm long; petals ovate, 15–30 mm long, reddish abaxially, white with reddish base adaxially, the inner ones thickened basally. Fruits subglobose, 2.5–4 cm diam., brownish, of 15–25 free carpels. Fl (Jan, Sep, Oct, Nov), fr (Sep); in non-flooded forest and along streams at all elevations.

Duguetia calycina Benoist

Trees, 5–10 m × 10–12 cm. Young stems densely covered with brownish, stellate scales. Leaves: petioles 5–10 × 2 mm; blades narrowly oblong-elliptic, 10–30 × 3–8 cm, sparsely to rather densely covered with stellate scales abaxially, the apex acuminate; secondary veins in 10–12 pairs. Plants sometimes cauliflorous. Inflorescences 1–2(several)-flowered; pedicels 5–10 × 2–3 mm. Flowers: sepals free, 10–17 mm long; petals broadly ovate to obovate, 15–25 mm long, whitish to cream-colored. Fruits ovoid to subglobose, 2–5 cm diam., brownish to yellow, of ca. 100 free carpels. Fl (Aug, Oct, Dec), fr (Apr, Jul); in non-flooded forest at all elevations. *Mamayawé*, *mamayawé-piment* (Créole), *yaoui*.

Duguetia eximia Diels PART 1: FIG. 9

Trees, 3–7 m × 4–10 cm. Young stems densely covered with stellate hairs. Leaves: petioles 5–15 × 2–5 mm; blades narrowly elliptic to narrowly obovate, 30–40 × 8–12 cm, sparsely covered with stellate hairs abaxially; the apex acute to short-acuminate; secondary veins in 12–17 pairs. Inflorescences 3–7-flowered; pedicels 5–10 × 2–4 mm. Flowers: sepals free, 12–15 mm long; petals narrowly oblong-obovate, 18–21 mm long, white to cream-colored, inner base wine red. Fruits globose, 4–6 cm diam., with dark brown velutinous indument, of 100–150 free carpels. Fl (Nov, Dec), fr (Jan, Feb, Mar, May, Oct, Dec); common, in non-flooded forest at all elevations.

Duguetia granvilleana Maas

Trees or shrubs, 1.5–8 m × 1.5–4 cm. Young stems densely covered with stellate hairs. Leaves: petioles 2.5–5 × 1.5–2 mm; blades narrowly obovate to narrowly elliptic, 8–19 × 3–6.5 cm, sparsely covered with persistent, erect, stellate hairs abaxially, the apex acute to acuminate; secondary veins in 9–12 pairs, strongly impressed adaxially. Inflorescences cauliflorous, 1–5-flowered; pedicels 7–11 × 2–2.5 mm. Flowers: sepals free, 15–20 mm long; petals obovate to narrowly elliptic, 20–25 mm long, cream-colored to yellow. Fruits subglobose, 2.3–2.7 cm diam., green to yellowish, of 13–21 free carpels. Fl (Jul); in non-flooded forest.

Duguetia inconspicua Sagot

Shrubs, 1–2(3) m × 1–8 cm. Young stems sparsely covered with stellate scales. Leaves: petioles 1–3 × 1–3 mm; blades narrowly

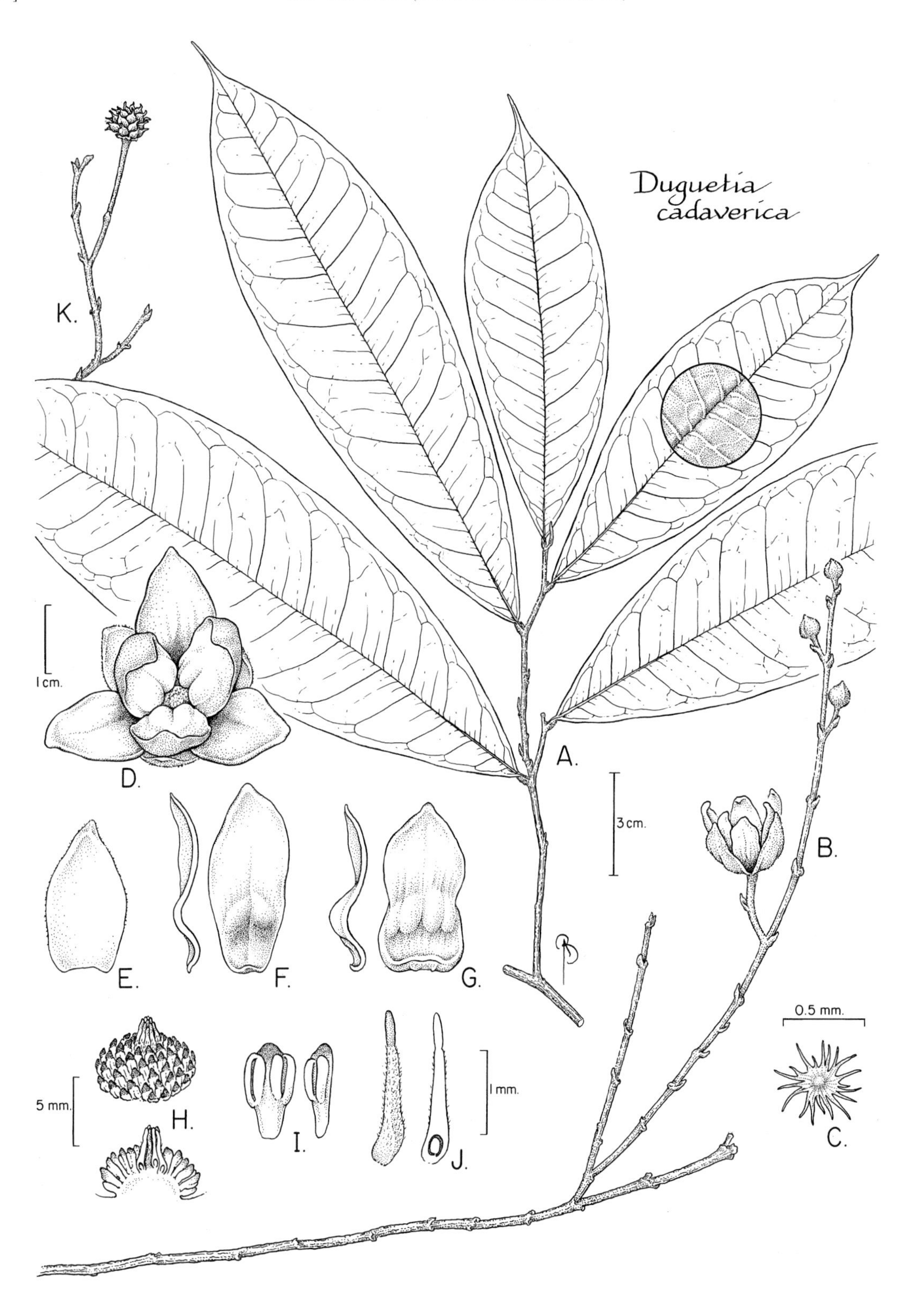

FIG. 14. ANNONACEAE. *Duguetia cadaverica* (A, *Mori et al. 21659*; B–K, *Mori et al. 21519*). **A.** Part of stem with leaves and enlarged detail of adaxial leaf surface. **B.** Part of flagelliform inflorescence with open flower and buds. **C.** Stellate scale from abaxial surface of petal. **D.** Apical view of flower showing two whorls of three petals each and three sepals. **E.** Adaxial view of sepal. **F.** Lateral (far left) and adaxial view (near left) of outer petal. **G.** Lateral (far left) and adaxial view (near left) of inner petal. **H.** Reproductive part of flower showing numerous stamens surrounding several carpels (above) and medial section of same (below). **I.** Adaxial (left) and lateral view (left) of stamens showing small apical expansion of connective. **J.** A carpel (far left) and medial section of same (near left). **K.** Part of infructescence with immature fruit consisting of fused monocarps.

ovate to narrowly obovate, 10–25 × 3–8 cm, sparsely covered with stellate scales and hairs abaxially, the apex acute to acuminate; secondary veins in 10–12 pairs. Inflorescences 1–5-flowered; pedicels 1–5 × 1–2.5 mm. Flowers: sepals basally connate, 3–5 mm long; petals narrowly elliptic, 15–30 mm long, white to yellow. Fruits subglobose, 1–3.5 cm diam., brownish to whitish, of 5–15 verrucose, free carpels. Fl (Sep), fr (Jan, Mar); in non-flooded forest at all elevations. *Mamayawé-piment* (Créole).

Duguetia pycnastera Sandwith

Trees, 3–7 m × 10 cm. Young stems distinctly grooved, densely covered with yellow-brown, stellate hairs. Leaves: petioles 2–4 × 1–3 mm; blades narrowly obovate to narrowly elliptic, 10–25 × 4–8 cm, sparsely covered with caducous, erect, stellate hairs abaxially, the apex long-acuminate; secondary veins in 8–10 pairs. Inflorescences 1–2-flowered, with leafy, caducous bracts to 30 mm long; pedicels 2–20 × 1–4 mm. Flowers: sepals free, 10–15 mm long; petals ovate, 10–20 mm long, cream-colored. Fruits globose, 1.5–2 cm diam., greenish yellow, of 10–30, free carpels. Fr (Feb, Dec); along streams in forest and at all elevations. *Mamayawé* (Créole).

Duguetia riparia Huber

Trees, 3–5(7) m tall. Young stems densely covered with stellate hairs. Leaves: petioles 2–8 × 1–2 mm; blades narrowly elliptic, 10–25 × 3–6 cm, bullate, sparsely covered with stellate hairs abaxially, the apex long-acuminate; secondary veins in 10–15 pairs, the venation often impressed adaxially. Inflorescences 1–2-flowered; pedicels 2–15 × 2–3 mm, to 6 mm in fruit. Flowers: sepals free, 10–20 mm long; petals elliptic to obovate, 20–30 mm long, cream-colored to yellow. Fruit broadly ovoid to broadly ellipsoid, 2–6 cm long, dark ochre-yellow with green carpel tips to completely brown, of 50–100 free carpels. Fl (Aug, Sep), fr (Feb); in forest along streams and in secondary vegetation. *Mamayawé* (Créole).

Duguetia surinamensis R. E. Fr. — PL. 12a

Trees, (10)18–30 m × 20–40 cm. Trunks swollen at base. Bark with reticulate pattern of vertical cracks, the slash brown with lighter streaks. Young stems densely covered with (stellate) scales. Leaves: petioles 3–10 × 1–1.5 mm; blades narrowly elliptic, 10–18 × 3–6 cm, shiny adaxially when dried, sparsely covered with stellate scales abaxially, the apex acute to short-acuminate; secondary veins in 10–12 pairs. Inflorescences 1–4-flowered; pedicels 12–20 × 1 mm, to 30 × 10 mm in fruit. Flowers emitting slightly sweet aroma; sepals free, 20–25 mm long; petals ovate-triangular, 20–30 mm long, salmon-colored, base of inner ones red. Fruits subglobose, 4–11.5 cm diam., brownish maturing red, of 125–200, connate (to 50% fused) carpels. Fl (Aug), fr (Apr, Aug, Sep); in non-flooded forest at lower elevations. *Mamayawé-piment.*

Duguetia yeshidan Sandwith

Trees or shrubs, 1–5 m × 2–8 cm. Young stems densely covered with stellate scales. Leaves: petioles 2–3 × 1–2 mm; blades narrowly obovate, 10–20 × 3–6 cm, discolorous, sparsely covered with stellate scales abaxially, the apex acuminate; secondary veins in 10–15 pairs, the venation distinctly impressed adaxially. Plants sometimes cauliflorous. Inflorescences 1–2-flowered; pedicels 2–10 × 1–3 mm. Flowers: sepals basally connate, 10–15 mm long; petals ovate to elliptic, 20–40 mm long, white to cream-colored. Fruits subglobose, 3–5 cm diam., greenish when young, of 50–100 free, strongly wrinkled carpels. Fl (Feb, Aug, Nov), fr (Jan, Feb, Mar, Apr, Aug, Nov); in non-flooded forest at lower elevations. *Mamayawé-piment* (Créole).

FUSAEA (Baill.) Saff.

Trees. Indument of simple hairs. Leaf blades with secondary veins impressed adaxialy, forming distinct marginal vein. Inflorescences supraaxillary, few-flowered; bracts 2 per flower. Flowers: sepals initially entirely fused, rupturing irregularly at anthesis; petals imbricate, subequal; outer stamens staminodial, much larger than fertile ones. Fruits syncarpous, smooth, of many, areolate, minutely apiculate, scarcely elevated carpels, with a distinct basal, woody collar formed by persistent base of receptacle.

Fusaea longifolia (Aubl.) Saff. — FIG. 15, PL. 12b

Trees, (4)6–12 m × 8–15 cm. Young stems densely covered with brown, appressed hairs. Bark shallowly but conspicuously fissured, black. Leaves: petioles 2–4 × 1–2 mm; blades narrowly elliptic-oblong, 15–30 × 3–7 cm, sparsely covered with appressed hairs abaxially, the apex long-acuminate; secondary veins in 15–18 pairs. Plants ramiflorous. Inflorescences 1–2-flowered; pedicels 15–30 × 2–5 mm. Flowers: sepals ca. 15 mm long, caducous except for persistent connate basal part, the basal part circular in outline and 10–15 mm diam.; petals narrowly oblong-obovate, 30–50 mm long, densely sericeous abaxially, green to white, maturing yellowish brown; staminodia 4–8 mm long. Fruits globose, 5–10 cm diam., glabrous, smooth, green, with edible pulp. Fl (Jan, Feb, Apr, Jul, Aug, Sep), fr (Mar, May); common but scattered, in non-flooded forest at lower elevations. *Malmanmanyaret, mâle mamanyaret, mamayawé* (Créole).

GUATTERIA Ruiz & Pav.

Trees, rarely lianas. Indument of simple hairs. Leaf blades with midrib impressed adaxially. Inflorescences axillary, 1- to several-flowered; pedicels with suprabasal articulation; bracts 2 per flower. Flowers: sepals small, valvate; petals imbricate or valvate, subequal, patent when young, but lengthening and ascending during anthesis, enclosing "pollination chamber," sepals and petals often hairy. Fruits apocarpous, the monocarps many, fleshy, 1-seeded, indehiscent, often apiculate, the stipe short to long.

1. Adaxial leaf blade surface strongly verrucose (use hand lens to see minute warts). Monocarps 15–17 × 6–8 mm. *G. blepharophylla.*
1. Adaxial leaf blade surface not verrucose. Monocarps not exceeding 10 mm in length.

2. Leaf blades 4–7.5 cm wide, the abaxial surface with conspicuous secondary veins. . . . . *G. punctata.*
2. Leaf blades 2–5 cm wide, the abaxial surface with inconspicuous secondary veins.
   3. Adaxial leaf blade surface with raised reticulate, tertiary venation, shiny when dried. . . . . *G. foliosa.*
   3. Adaxial leaf blade surface without raised reticulate tertiary venation, dull when dried.
      4. Leaf blades with attenuate base, the apex acuminate, the margins recurved. . . . . *G.* sp. A.
      4. Leaf blades with acute base, the apex acute, the margins not recurved. . . . . *Guatteria* aff. *oblonga.*

Guatteria blepharophylla Mart. [Syn.: *Guatteriopsis blepharophylla* (Mart.) R. E. Fr.]

Trees, to 6 m tall. Young stems rather densely covered with white, appressed hairs. Leaves: petioles 5–8 × 2–3 mm; blades narrowly elliptic to narrowly obovate, 20–25 × 5–6 cm, verrucose and dull grayish when dried adaxially, sparsely covered with white, appressed hairs abaxially, the apex long-acuminate; secondary veins in 13–15 pairs. Inflorescences 1–2-flowered; pedicels 6–10 × 2 mm. Flowers: sepals patent, 7 mm long; petals obovate-triangular, to 9–15 × 8–9 mm in ripe flowers, cream-colored. Monocarps ellipsoid, 15–17 × 6–8 mm, rather densely hairy, green, maturing red to black, the stipes red to purple, 5–8 × 1–1.5 mm. Fr (Jul); in non-flooded forest at lower elevations.

This species could be placed in *Guatteriopsis* R.E. Fr. which differs in its valvate inner petals vs. the imbricate ones of *Guatteria.*

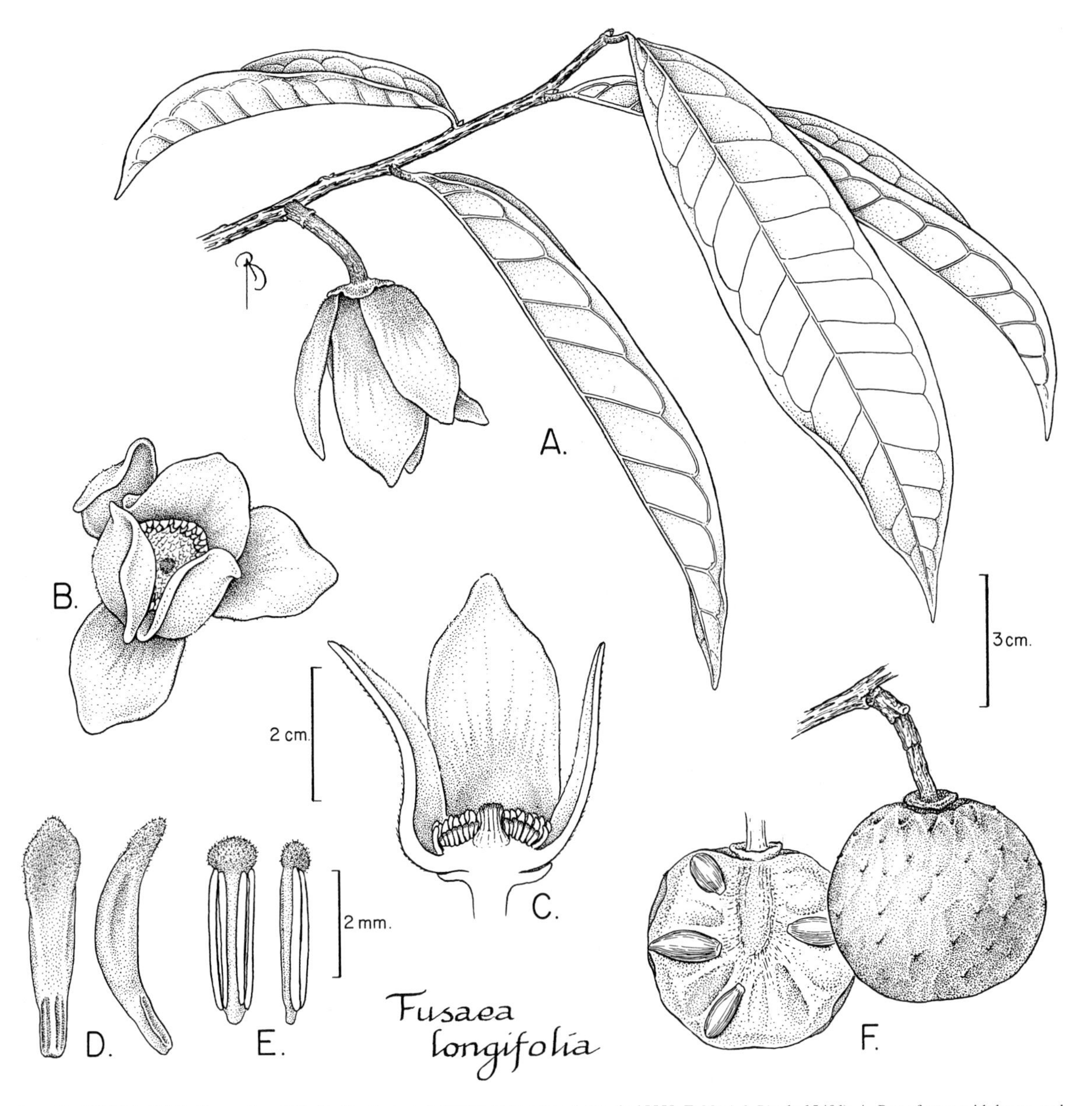

FIG. 15. ANNONACEAE. *Fusaea longifolia* (A–C, *Mori et al. 20987*; D, E, *Mori & Pipoly 15558*; F, *Mori & Pipoly 15401*). **A.** Part of stem with leaves and flower. **B.** Apical view of flower. **C.** Medial section of young flower. **D.** Adaxial view (left) and lateral view (right) of staminodes. **E.** Adaxial view (left) and lateral view (right) of stamens showing apical expansion of connective. **F.** Medial section (left) and fruit of fused monocarps (right).

Guatteria foliosa Benth. PL. 12e

Trees, 20–25 m × 20–28 cm. Young stems rather densely covered with reddish brown, appressed hairs. Leaves: petioles 5–10 × 1 mm; blades narrowly elliptic, 8–15 × 3–5 cm, adaxially shiny when dried, rather densely covered with brown, appressed hairs abaxially, the base narrowly cuneate, the apex distinctly acuminate; secondary veins in 6–8 pairs, with raised, tertiary reticulate venation adaxially, and inconspicuous venation abaxially. Inflorescences 1–2-flowered; pedicels 13–20 × 1–2 mm. Flowers: sepals ca. 3 × 3 mm; petals narrowly oblong-obovate, to 10–20 × 6 mm in ripe flowers, green. Monocarps obovoid to ellipsoid, 5–8 × 3–4 mm, green, maturing black, the stipes 10–20 × 0.5 mm. Fl (Sep, Oct, Nov), fr (Sep, Oct, Nov); in non-flooded forest at lower elevations.

Guatteria aff. oblonga R. E. Fr.

Trees, 30 m × 35 cm. Young stems rather densely covered with brownish, appressed hairs. Leaves: petioles 2–5 × 1 mm; blades narrowly elliptic, 8–10.5 × 2.5–4 cm, midrib densely covered with brown, appressed hairs adaxially, rather densely covered with brownish, appressed hairs abaxially, dull blackish when dried adaxially, the base acute, the apex acute to shortly acuminate; secondary veins in 10–13 pairs, the venation inconspicuous ad- and abaxially. Inflorescences 1–2-flowered; pedicels 10 × 1 mm long. Flowers: sepals ca. 4 × 3 mm; petals obovate, to 10–12 × 6–8 mm in ripe flowers, cream-colored with brown base. Monocarps not seen. Fl (Oct); in non-flooded forest at lower elevations.

Guatteria punctata (Aubl.) R. A. Howard [Syn.: *Guatteria chrysopetala* (Steud.) Miq.] PL. 12c,d

Trees, 2–20 m × 10–30 cm. Trunks with low buttresses. Bark dark green or brown to black, the inner bark white to buff-colored, emitting a strong spicy aroma, the wood white. Young stems densely covered with brownish, appressed hairs. Leaves: petioles 3–10 × 1.5–2.5 mm; blades narrowly obovate to narrowly elliptic, 11–20 × 4–7.5 cm, rather densely to densely covered with reddish brown, appressed hairs abaxially, the apex acuminate; secondary veins in 11–17 pairs, inconspicuous adaxially, but very conspicuous abaxially. Inflorescences 1–2(4)-flowered; pedicels 10–25 × 1–2 mm. Flowers: sepals 4–5 mm long, often reflexed; petals narrowly obovate to narrowly ovate, to 22–25 × 7–10 mm in ripe flowers, green to yellow. Monocarps obovoid to ellipsoid, 7–10 × 3–5 mm, green, maturing black, the stipes red, 14–40 × 0.5–1 mm. Fl (Jan, Mar, Jul, Oct, Sep, Nov, Dec), fr (Jul, Sep, Oct, Nov); in non-flooded forest and in forest near creeks at all elevations. *Mamanyawé, mamayawé* (Créole).

One specimen, *Granville B4561* (CAY), differs from other collections of this species in the dense, brown hairs on the petals.

Guatteria sp. **A**

Trees, 20–30 m × 35–40 cm. Trunks with steep, simple, rounded buttresses. Bark gray, scaly, hard, the inner bark orange, emitting spicy aroma. Young stems rather densely covered with brown, appressed hairs. Leaves: petioles 4–5 × 1–1.5 mm, blades narrowly elliptic, 9–14 × 2–3 cm, stiff, very densely covered with reddish brown, appressed hairs abaxially, dull grayish green to black when dried adaxially, the base attenuate, the apex distinctly acuminate, the margins recurved; secondary veins in 8–13 pairs, the venation inconspicuous adaxially and abaxially. Inflorescences 1-flowered; pedicels 10 × 1–1.5 mm. Flowers not seen. Monocarps ellipsoid, 5–9 × 3–5 mm, dark purple, the stipes 5–15 × 1 mm. Fr (Sep, Dec); in secondary vegetation, blue-gray and silver-beaked tanagers seen eating fruit.

This species, vouchered by *Mori & Boom 15360* (CAY, NY) and *Mori et al. 24197* (NY), probably represents a new species.

OXANDRA A. Rich.

Trees. Indument of simple hairs. Leaf blades with midrib raised adaxially. Inflorescences axillary, 1-flowered; bracts to 4 per flower; pedicels very short. Flowers small; sepals imbricate; petals imbricate, subequal, white. Fruits apocarpous, monocarps few, indehiscent, shortly stipitate. Seeds 1 per monocarp.

Oxandra asbeckii (Pulle) R. E. Fr. PL. 13a

Trees, 7–16 m × 8–15 cm. Trunks angled, not buttressed. Bark smooth, dark brown, the inner bark creamy yellow. Young stems densely covered with appressed hairs. Leaves: petioles 2–5 × 1 mm; blades narrowly elliptic, 7–13 × 2.5–5.5 cm, sparsely covered with long, appressed hairs abaxially, the apex long-acuminate; secondary veins in 6–7 pairs. Inflorescences: pedicels 2–5 mm long. Flowers: sepals 1.5–2 mm long, persistent, the margins ciliate; petals ovate, 6–7 mm long, white, the margins ciliate. Monocarps 2–7, ellipsoid, 15–20 × 8–12 mm, glabrous, green, maturing purple to black, the stipes 1–2 mm long. Fr (Aug, Sep, Oct, Nov, Dec); in non-flooded forest at lower elevations.

PSEUDOXANDRA R. E. Fr.

Trees. Indument of simple hairs. Leaf blades with midrib raised adaxially, the marginal vein distinct, very close to margin. Inflorescences axillary, 1–2-flowered; pedicels very short; bracts to 10 per flower. Flower buds globose; sepals and petals imbricate; petals subequal. Fruits apocarpous, the monocarps several, indehiscent, the stipe long. Seeds 1 per monocarp.

Pseudoxandra cuspidata Maas

Trees, 20 m × 18 cm. Young stems covered with very long, appressed, white hairs, soon glabrous. Leaves: petioles 4–6 × 1–1.5 mm; blades narrowly elliptic, 10–13 × 3–4 cm, glabrous to sparsely covered with very long, appressed hairs abaxially, the apex long-acuminate; secondary veins in 15–20 pairs, hardly distinguishable from tertiary veins, the marginal vein very close to margin (at ca. 1 mm). Inflorescences: pedicels 2–5 × 1–2.5 mm. Flowers: sepals connate for 1/3 of length, 1–1.5 mm long; petals ovate to elliptic, 5–7 mm long, green (at least when young). Monocarps 2–10, subglobose, 10–20 mm diam., with lateral apicule <1 mm long, green, maturing yellow to orange (black when dry), the stipes 10–15 × 2 mm. Fr (Apr, Jul), in non-flooded forest at lower elevations. *Lamoussé* (Créole).

ROLLINIA A. St.-Hil.

Trees. Indument of simple hairs. Leaf blades with midrib impressed adaxially. Inflorescences supraaxillary to leaf-opposed, 1- to several-flowered; bracts 2 per flower. Flowers propeller-shaped; sepals valvate, very small; petals valvate, connate at base, green to yellow, base reddish adaxially, the outer ones prolonged into wing, the inner ones small. Fruit syncarpous, areolate, of many, flat to shortly apiculate carpels.

Maas, P. J. M., L. Y. Th. Westra & collaborators. 1992. *Rollinia*. Fl. Neotrop. Monogr. **57:** 1–188.

1. Sepals without dorsal thickening; wings of outer petals ca. 5 mm long. Fruiting carpels ca. 15. . . . . *R. cuspidata.*
1. Sepals with dorsal thickening; wings of outer petals 7–25 mm long. Fruiting carpels usually >15.
    2. Small trees (to 7 m tall). Leaf blades oblong-elliptic. Wings of outer petals 7–8 mm long, 3–4 mm high. Fruiting carpels 30–50. . . . . *R. exsucca.*
    2. Larger trees (15–18 m tall). Leaf blades narrowly oblong-obovate to narrowly oblong-elliptic. Wings of outer petals 15–25 mm long, 5–6 mm high. Fruits unknown. . . . . *R. elliptica.*

Rollinia cuspidata Mart.

Trees, 2–4 m tall. Bark black. Young stems and petioles densely covered with brown, erect and appressed hairs. Leaves: petioles ca. 5 × 1 mm; blades narrowly elliptic, 9–11 × 3–4 cm, thin, adaxial surface sparsely and abaxial surface rather densely covered with appressed pale brown hairs, blackish green when dried, the apex short-acuminate; secondary veins in ca. 10 pairs. Inflorescences 1- to few-flowered; pedicels 20 × 1 mm. Flowers: sepals ca. 3 mm long, without dorsal thickening; petals yellow, the wings of outer ones horizontal, ca. 5 mm long, ca. 3 mm high, ca. 1 mm thick. Fruits 1.2–3 cm diam., green, maturing yellow to orange, of ca. 15 carpels, the areoles flat. Fl (Jan, Feb, Mar, Aug), fr (Feb, Mar, May); in secondary vegetation and in non-flooded forest. *Abriba sauvage* (Créole).

Rollinia elliptica R. E. Fr. PL. 13b

Trees, 15–18 m × 15–20 cm. Young stems densely covered with brown, appressed hairs, soon glabrous. Leaves: petioles 5–8 × 1–1.5 mm, narrowly oblong-obovate to narrowly oblong-elliptic, 6–15 × 3.5–5 cm, thin, sparsely covered with appressed hairs and glaucous green abaxially, the apex obtuse, acute, or short-acuminate; secondary veins in 10–12 pairs. Inflorescences 1–2-flowered; pedicels 15–25 × 1 mm. Flowers: sepals 1–2 mm long, with dorsal thickening; petals creamy white, the wings of outer petals at angle of ca. 45° with axis of flower, 15–25 mm long, 5–6 mm high, ca. 1 mm thick. Fl (Jan, Feb), fr (Feb, Apr); in somewhat disturbed vegetation along roads in forest at lower elevations.

Rollinia exsucca (Dunal) A. DC. FIG. 16, PL. 13c

Trees 2.5–7 m × 5–8 cm. Young stems rather densely covered with appressed hairs. Leaves: petioles 7–8 × 1 mm; blades oblong-elliptic, 10–15(23) × 4–7(11) cm, glaucous-brown, rather densely covered with appressed hairs abaxially, the apex acute to shortly acuminate; secondary veins in 8–12 pairs. Inflorescences 2–4-flowered; pedicels 15–30 × 1 mm. Flowers: sepals ca. 2 mm long, with dorsal thickening; petals yellow at anthesis, the wings of outer petals slightly ascending, 7–8 mm long, 3–4 mm high, ca. 2 mm thick. Fruit globose, ca. 1.5 cm diam., yellow when ripe, of 30–50 carpels, the areoles slightly protruding, apiculate. Fl (Jan, Feb, Mar), fr (Feb, Mar, Apr, Aug, Dec); in non-flooded forest, in wet area along streams, and in secondary vegetation at lower elevations. *Abriba sauvage*, *corossol sauvage*, *corossol-grand-bois* (Créole).

UNONOPSIS R. E. Fr.

Trees. Indument of simple hairs. Leaf blades with raised midrib adaxially. Inflorescences axillary, 1- to many-flowered, often compound; bracts 2 per flower. Flowers small: sepals valvate; petals valvate, concave, subequal, whitish. Fruits apocarpous, the monocarps several to many, indehiscent, the stipe short to long. Seeds 1 to several per monocarp, pitted.

1. Inflorescence repeatedly branched, many-flowered, to 8 cm long. . . . . *U. rufescens.*
1. Inflorescence not repeatedly branched, 1- to several-flowered, much shorter than 8 cm.
    2. Leaf blades narrowly elliptic, 8–15 × 1.5–4 cm. Monocarps globose, 6–8 mm diam., the stipes 8–10 mm long. . . . . *U. perrottetii.*
    2. Leaf blades generally narrowly obovate, 15–30 × 4–10 cm. Monocarps ellipsoid, 15–20 mm long, the stipes 10–20 mm long. . . . . *U. stipitata.*

Unonopsis perrottetii (A. DC.) R. E. Fr.

Trees, 9–20 m × 10–15 cm. Bark rough, with small boat-shaped fissures, the middle bark red. Young stems covered with appressed hairs. Leaves: petioles 2–7 × 1 mm; blades narrowly elliptic, slightly asymmetric, 8–15 × 1.5–4 cm, thin, glabrous abaxially, the apex long-acuminate; secondary veins in 10–12 pairs. Inflorescences 1–2-flowered; pedicels 20–25 × 1 mm, to 2.5 mm in fruit. Flower buds

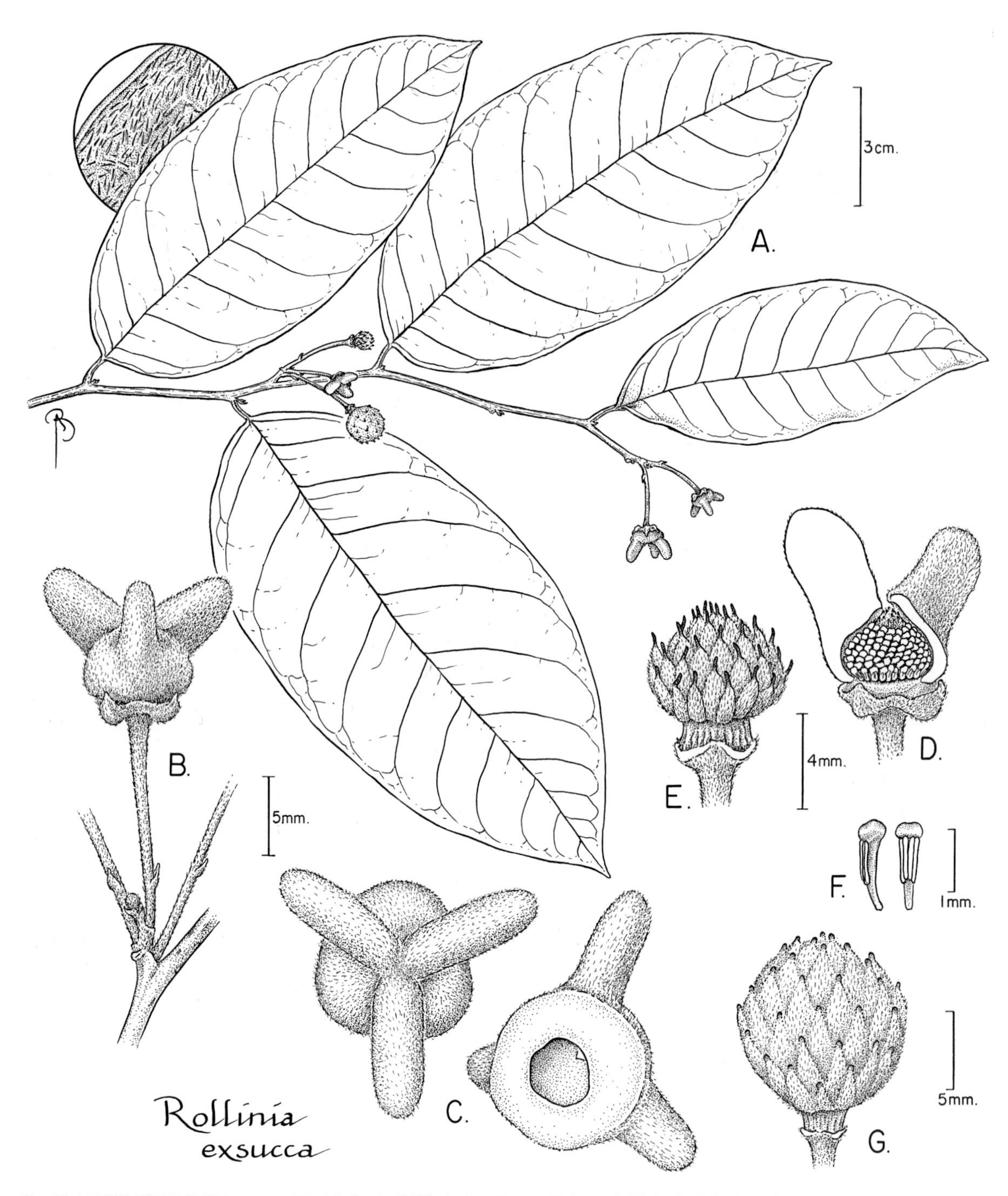

FIG. 16. ANNONACEAE. *Rollinia exsucca* (*Mori & Gracie 21099*). **A.** Part of stem with leaves (with detail of abaxial surface), flowers, and immature fruits. **B.** Inflorescence with lateral view of flower. **C.** Apical (left) and oblique basal (right) views of corolla. **D.** Medial section of flower with one and a half petals removed. **E.** Lateral view of flower with perianth and stamens removed and carpels beginning to mature. **F.** Outermost (near right) and innermost (far right) stamens showing apical expansion of connective. **G.** Lateral view of immature fruit of fused monocarps.

white; sepals free, ca. 1 mm long; petals strongly concave, 3–5 mm high. Monocarps ca. 25, globose, 6–8 mm diam., 1-seeded, the stipes 8–10 × ca. 1 mm. Fl (Nov), fr (Apr); in non-flooded forest at lower elevations. *Mamayawé* (Créole).

Unonopsis rufescens (Baill.) R. E. Fr. FIG. 17, PL. 13d

Trees, 8–20 m × 8–16 cm. Bark shallowly fissured, the slash orange. Young stems rather densely covered with erect and appressed,

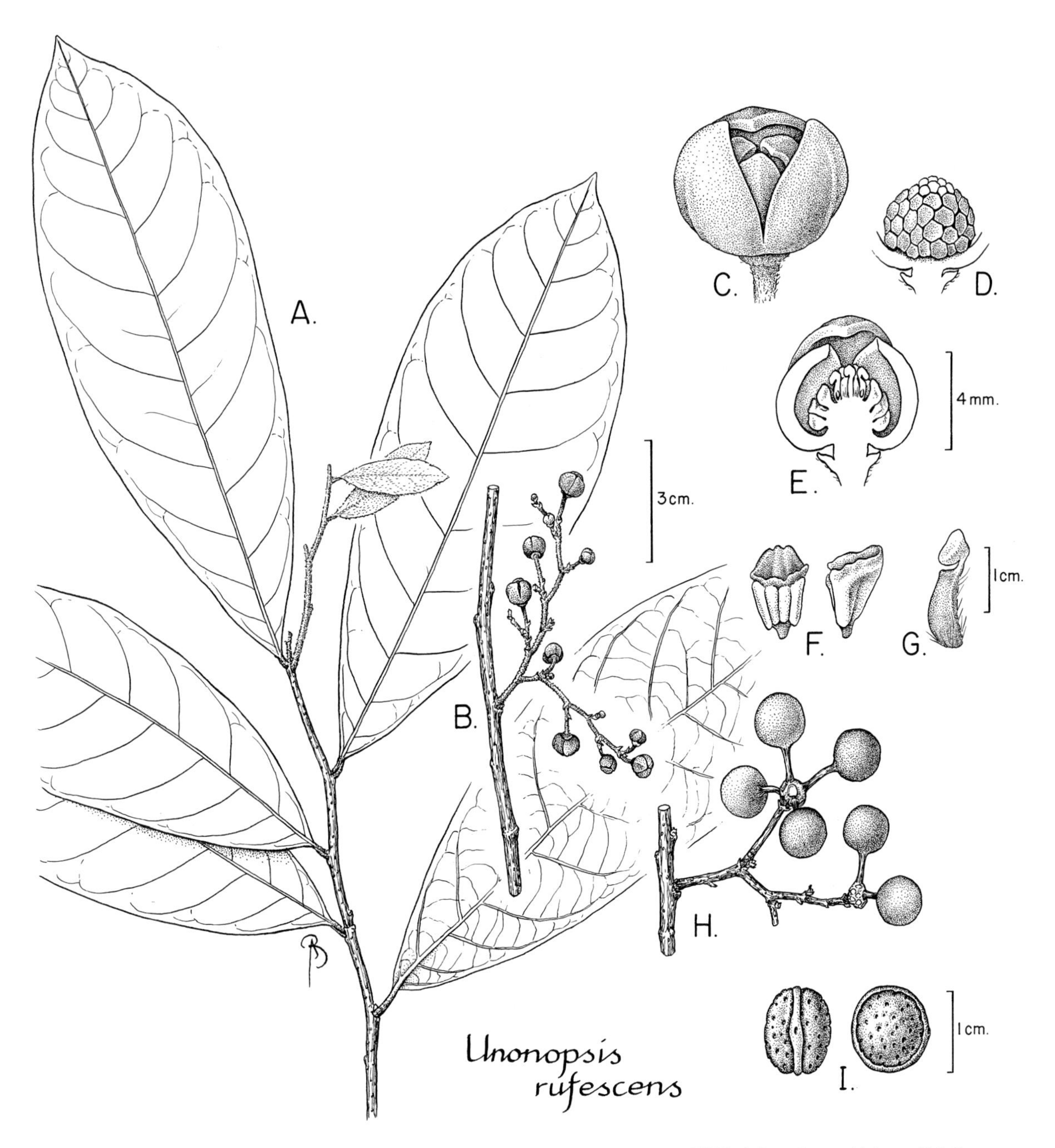

FIG. 17. ANNONACEAE. *Unonopsis rufescens* (A–G, *Mori & Boom 14831*; H, I, *Mori & Pennington 18035*). **A.** Part of stem with leaves. **B.** Inflorescence. **C.** Flower at anthesis. **D.** Detail of flower with perianth removed. **E.** Medial section of flower showing stamens and carpels in the interior. **F.** Adaxial view (left) and lateral view (right) of stamens. **G.** Lateral view of carpel. **H.** Infructescence with fruits consisting of separate monocarps. **I.** Detail of pitted seeds.

pale brown hairs. Leaves: petioles 5–7 × 2 mm; blades narrowly obovate to narrowly elliptic, 8–16 × 3–7 cm, sparsely covered with erect and appressed, pale brown hairs abaxially, the apex short-acuminate; secondary veins in 8–10 pairs. Plants ramiflorous. Inflorescences repeatedly branched, many-flowered, to 8 cm long; pedicels 5–20 × 1–1.5 mm, to 2.5 mm in fruit. Flower buds glaucous. Flowers: sepals ca. 1 mm long, connate into cup-shaped, persistent calyx; petals strongly concave, mostly cream-colored, 3.5–6 high. Monocarps 5–15, globose, 8–12 mm diam., apiculate, glaucous green, 1-seeded, the stipes 10–12 × 1–2 mm. Fl (Sep, Oct, Nov, Dec), fr (Feb, Mar, May, Aug,); in non-flooded forest at lower elevations. *Mamayawé* (Créole).

Unonopsis stipitata Diels PL. 13e,f

Trees, 3–15 m × 5–12 cm. Young stems rather densely covered with erect to appressed hairs. Leaves: petioles 5–10 × 2–3 mm; blades narrowly obovate to narrowly oblong-ovate, 15–30 × 4–

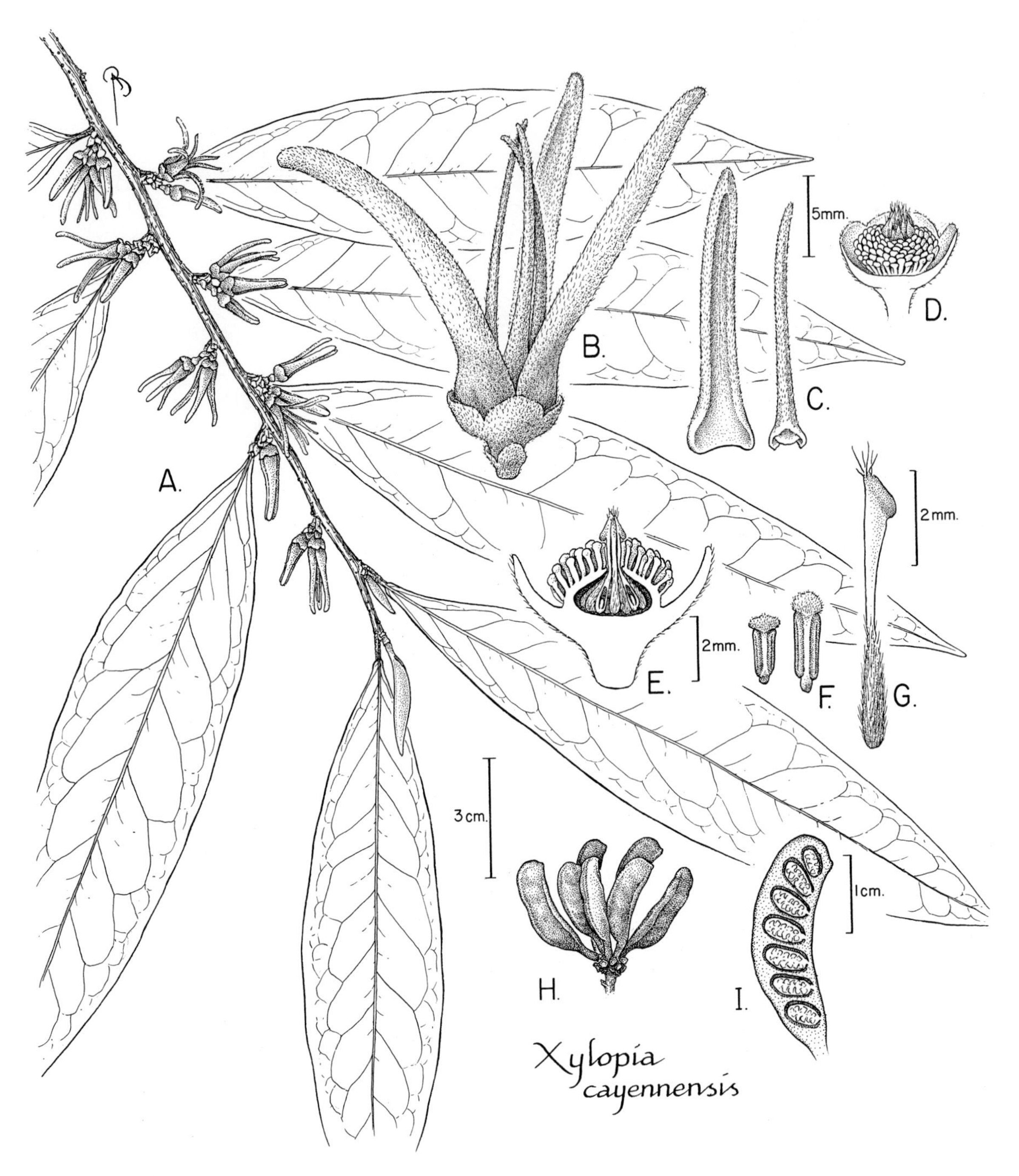

FIG. 18. ANNONACEAE. *Xylopia cayennensis* (A–G, *Mori & Pennington 17985*; H, *de Granville 5318*; I, *Cowan 38950*). **A.** Stem with leaves and inflorescences. **B.** Flower. **C.** Outer (left) and inner (right) petal. **D.** Flower with petals removed. **E.** Medial section of flower. **F.** Adaxial view of stamens showing apical expansion of connective. **G.** One carpel. **H.** Fruit consisting of separate monocarps. **I.** Medial section of monocarp showing ruminate endosperm of seeds.

10 cm, shiny adaxially, subglabrous abaxially, except for the hairy primary vein, the apex distinctly short-acuminate to long-acuminate; secondary veins in 10–12 pairs. Plants cauliflorous or ramiflorous. Inflorescences 1- to several-flowered; pedicels 15–20 × 1–2 mm, to 3 mm in fruit. Flowers: sepals free, ca. 1 mm long; petals strongly concave, cream-colored, rarely orange, 2–5 mm high. Monocarps 20–25, ellipsoid, 15–20 × 8–12 mm, shiny green, maturing pale orange, 1-seeded, the stipes 10–20 × 1–2 mm. Fl (Jun, Aug, Sep, Oct), fr (Jan, Feb, Apr, May, Jul, Aug); in non-flooded forest at lower elevations. *Femelle mamayawé*, *mamayawé* (Créole).

XYLOPIA L.

Trees. Indument of simple hairs or lacking. Leaf blades with midrib impressed adaxially, secondary veins inconspicuous. Inflorescences axillary, 1- to several-flowered; bracts 2 per flower. Flowers: sepals valvate, free or connate into cup-shaped calyx; petals valvate, unequal. Fruits apocarpous, monocarps many to several, not explosively dehiscent, the stipe short. Seeds 1 to several per monocarp, arillate.

1. Leaf blades 15–20 × 4–5 cm, sparsely pubescent abaxially. . . . . *X. cayennensis.*
1. Leaf blades 5–8 × 1–2.5 cm, densely covered with silky hairs abaxially.
   2. Young stems covered with erect hairs. Leaf blades 1–1.5 cm wide. Monocarps 8–12 mm long, 1–3-seeded, the stipes 1–2 mm long. . . . . *X. frutescens.*
   2. Young stems covered with appressed hairs. Leaf blades 2–2.5 cm wide. Monocarps 35–40 mm long, 4–7-seeded, the stipes ca. 10 mm long. . . . . *X. nitida.*

Xylopia cayennensis Maas [Syn.: *Xylopia longifolia* (Sagot) R. E. Fr., nom. illeg.] FIG. 18

Trees, (12)18–20 m × 15–25 cm. Trunks with thin buttresses. Bark smooth. Young stems subglabrous. Leaves: petioles 5–8 × 1–2 mm; blades narrowly elliptic, 15–20 × 4–5 cm, sparsely covered with white, appressed hairs abaxially, the apex acute to short-acuminate; secondary veins in ca. 10 pairs. Inflorescences 2–5-flowered; pedicels 1–2 × 1 mm. Flowers: sepals connate into a cup-shaped calyx 2–3 mm long; petals whitish, dark brown sericeous abaxially, the outer ones linear, 15–17 × 2 mm, the inner ones slightly smaller and narrower. Monocarps 10–15, sickle-shaped, 15–20 mm long, 2–7-seeded, greenish red, with sharp turpentine-like aroma when cut, the stipes 5 × 1 mm. Fl (May), fr (Nov); in non-flooded forest at lower elevations.

Xylopia frutescens Aubl.

Trees, 12–15 m × 15 cm. Young stems densely covered with very long, erect, white hairs. Leaves: petioles 1–3 × 1 mm; blades narrowly ovate, 5–8 × 1–1.5 cm, densely covered with very long, silky, white, appressed hairs abaxially, the apex long-acuminate to acute; secondary veins in ca. 5 pairs. Inflorescences 1–6-flowered; pedicels ca. 1 mm long. Flowers aromatic; sepals ca. 2 mm long, connate at very base; petals white, densely sericeous at abaxial side, the outer ones linear, 12–13 × 2–3 mm, the inner ones slightly smaller. Monocarps ca. 5, ellipsoid to globose, 8–12 × 7–10 mm, 1–3-seeded, red when ripe, the stipes 1–2 × 1 mm. Fl (Sep, Oct, Nov), fr (Sep); in non-flooded forest at lower elevations.

Xylopia nitida Dunal PL. 13g,h

Trees, 20–35 m × 25–50 cm. Trunks with steep buttresses. Young stems densely covered with white, appressed hairs. Leaves: petioles 5–6 × 1 mm; blades narrowly elliptic, 6–8 × 2–2.5 cm, densely covered with silky, long, white, appressed hairs abaxially, the apex acute to short-acuminate. Inflorescences 3–5-flowered; pedicels 2–5 × 1 mm. Flowers very fragrant; sepals connate into cup-shaped calyx ca. 2 mm long; petals linear, dull ochre-orange, deep maroon adaxially, the outer ones 15–17 × 2 mm, sericeous abaxially, the inner ones slightly smaller. Monocarps ca. 25, narrowly oblong, constricted, 35–40 mm long, 4–7-seeded, green maturing dark red, the stipes ca. 10 × 1 mm. Seeds blue-black with blue-gray aril. Fl (May), fr (Sep, Nov); in non-flooded forest at lower elevations.

APIACEAE (Carrot Family)

Scott A. Mori

Herbs, often with hollow stems, rarely woody but never so in our area. Leaves compound, deeply incised or simple, sometimes peltate (*Hydrocotyle*), alternate, cauline, from stolons, or in basal rosettes; petioles often with conspicuous sheath, especially in taxa from outside our area. Inflorescences simple umbels, compound umbels, or heads. Flowers usually small, actinomorphic, bisexual or unisexual; calyx 5-toothed or vestigial; petals 5, free; nectary disc present; stamens 5, inserted on disk alternate with petals; ovary inferior, 2-locular, the styles 2, with or without expanded base called stylopodium; placentation axile, the ovules 1 per locule. Fruits schizocarps, consisting of 2 dry, indehiscent mericarps).

Berry, P. E. 1995. Apiaceae. *In* J. A. Steyermark, P. E. Berry & B. K. Holst (gen. eds.), Flora of the Venezuelan Guayana **2:** 384–399. P. E. Berry, B. K. Holst & K. Yatskievych (vol. eds.
). Missouri Botanical Garden, St. Louis; Timber Press, Portland, Oregon.
Stafleu, F. A. 1951. Umbelliferae. *In* A. Pulle (ed.), Flora of Suriname **III(2):** 252–256. Royal Institute for the Indies.

ERYNGIUM L.

Herbs. Leaves simple, in basal rosettes, the margins distinctly serrate or spiny. Inflorescences solitary to numerous, involucrate heads. Flowers small, bisexual; sepals vestigial; petals 5; ovary often rugose or tuberculate, the style without stylopodium.

Eryngium foetidum L. FIG. 19

Herbs, 0.2–0.5 m tall. Leaves: petioles expanded into narrow sheath; blades narrowly oblanceolate, 9–14 × 2–3.5 cm, aromatic when crushed, the margins distinctly serrate. Inflorescences trichotomously branched, the lower trichotomies subtended by two large, opposite, leaf-like bracts with toothed margins, the flowers in cylindric heads, the heads subtended by whorl of bracts, the individual flowers subtended by single bracteole. Flowers ca. 1 mm diam. Fl (Aug); cultivated as a condiment, also growing in weedy places as a non persistent escape from cultivation. *Chardon-bénit, radié la fièvre* (Créole).

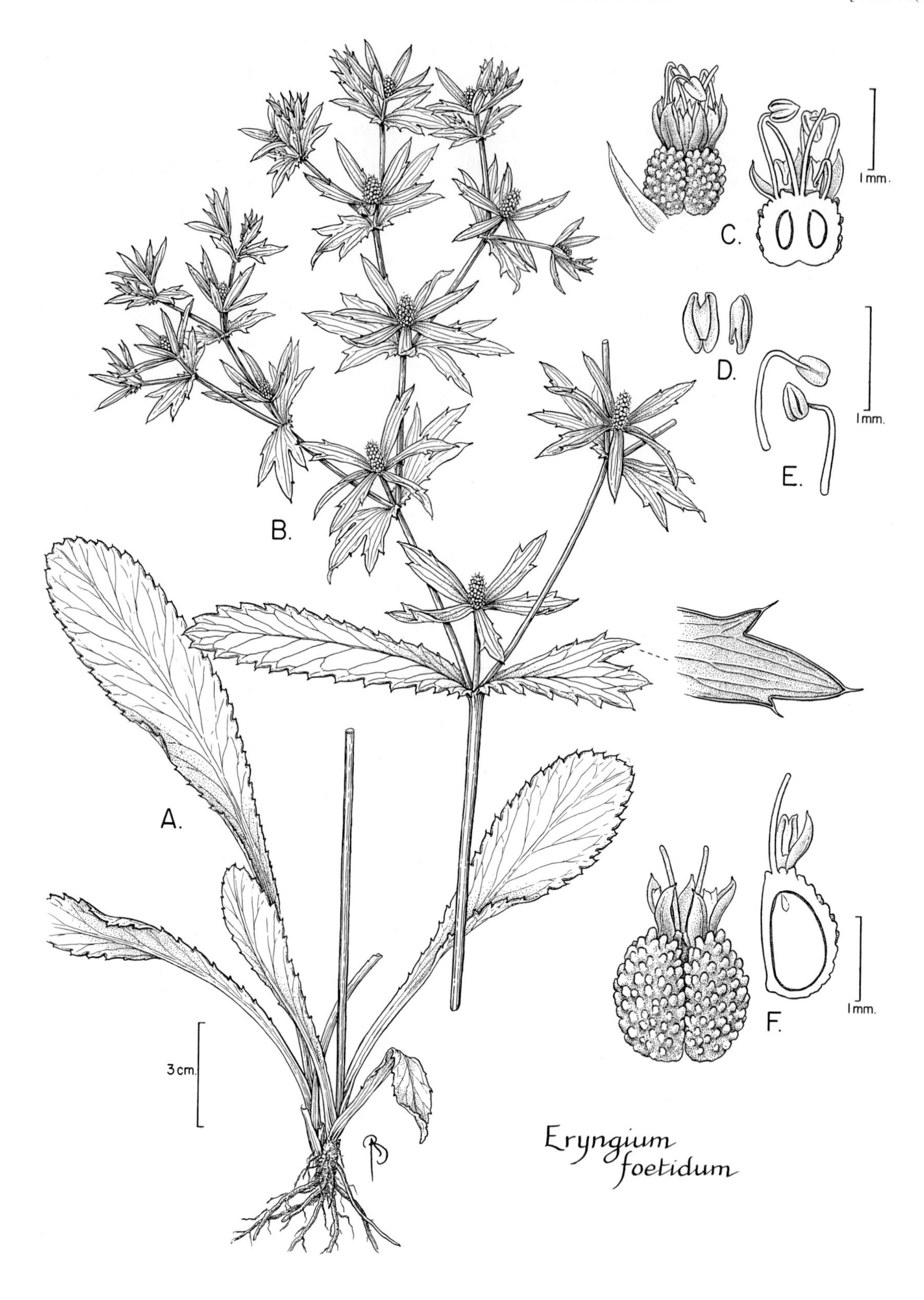

FIG. 19. APIACEAE. *Eryngium foetidum* (A, *Nelson 774* from Brazil; B–E, *Mori et al. 19214*; F, *López-Palacius 1873*). **A.** Basal part of plant. **B.** Upper part of stem and inflorescences with detail of leaf margin (right). **C.** Entire (left) and medial section (right) of flower. **D.** Adaxial view (left) and lateral view (right) of petals. **E.** Stamens. **F.** Whole fruit (left) and medial section (right) through one mericarp.

APOCYNACEAE (Dogbane Family)

Lucile Allorge

Trees, shrubs, lianas, or rarely perennial herbs. Latex often present, white (milky). Stipules absent but intrapetiolar colleters often present. Leaves simple and entire, usually opposite, sometimes spirally arranged (*Plumeria*, *Himatanthus*, *Laxoplumeria*), alternate (*Aspidosperma*, *Geissospermum*), or verticillate (*Allamanda*, *Rauvolfia*, *Couma*). Inflorescences of cymes, thyrses, racemes, panicles, or sometimes of solitary flowers. Flowers actinomorphic, bisexual; sepals usually 5, often with glandular appendages (colleters) adaxially; corolla gamopetalous, usually 5-lobed, salverform to urceolate or campanulate, the aestivation sinistrorse or dextrose; stamens 5, adnate to corolla tube and alternate with corolla lobes, the filaments free or connivent around style head, the anthers 2-celled, the pollen usually in monads, sometimes in tetrads; nectaries often present, these 5 (rarely 2, *Catharanthus*) or forming ring surrounding ovary; ovary superior or half-inferior (*Plumeria*, *Himatanthus*), 1–2-locular, bicarpellate, the carpels free or united, the style thick, the stigmatic surface below apex; placentation parietal, the ovules anatropous. Fruits either indehiscent, then a berry or drupe, or of 1–2 dehiscent follicle-like segments. Seeds unadorned, comose, with a papery wing, or with an aril; endosperm present.

Markgraf, Fr. 1966. Apocynaceae. *In* A. Pulle (ed.), Flora of Suriname **IV(1):** 1–65, 443–467. E. J. Brill, Leiden.

Morillo, G. 1996. Clave genérica de las Apocynaceae (Sf Plumerioideae) del norte de Sudamerica. Pittieria **25:** 43–69.

Zarucchi, J. L., G. N. Morillo, M. E. Endress, B. F. Hansen & A. J. M. Leeuwenberg. 1995. Apocynaceae. *In* J. A. Steyermark, P. E. Berry & B. K. Holst (gen. eds.), Flora of the Venezuelan Guayana **2:** 471–571. P. E. Berry, B. K. Holst & K. Yatskievych (vol. eds.). Missouri Botanical Garden, St. Louis; Timber Press, Portland, Oregon.

1. Trees or shrubs.
  2. Leaves spirally arranged or alternate.
    3. Stems thick. Leaves spirally arranged.
      4. Fruits robust, >10 mm diam. Seeds with membranous wing. . . . . *Himatanthus.*
      4. Fruits very slender, 2–4 mm diam. Seeds surrounded by long hairs. . . . . *Laxoplumeria.*
    3. Stems slender. Leaves alternate.
      5. Corollas with sinistrorse aestivation. Fruits woody, more or less laterally compressed, dehiscent. Seeds surrouded by a membranous wing. . . . . *Aspidosperma.*
      5. Corollas with dextrorse aestivation. Fruits fleshy, not laterally compressed, indehiscent. Seeds without a membranous wing. . . . . *Geissospermum.*
  2. Leaves opposite or verticillate.
    6. Leaves verticillate, 3 at a node. . . . . *Couma.*
    6. Leaves opposite, 2 at a node.
      7. Fruits of 2 follicle-like segments.
        8. New growth composed of 1 pair of leaves. Seeds with white aril. *Bonafousia.*
        8. New growth composed of 2 pairs of leaves. Seeds with red to orange aril. . . . . *Anartia.*
      7. Fruits single berries or 1–2 hollow mericarps.
        9. Trunks black, with prickles. Fruits globose to ovoid, ca. 1.5 cm diam., yellow or orange at maturity. . . . . *Lacmellea.*
        9. Trunks not black, without prickles. Fruits oblong or flattened globose, ≥4 cm diam., yellow or brown at maturity.
          10. Leaf blades with indistinct tertiary veins. Inflorescences axillary or cauline. Fruits yellow, ellipsoid to pyriform at maturity. Seeds not as below, flattened in transverse section, the seed coat finely rugulose but not pitted; aril absent. . . . . *Ambelania.*
          10. Leaf blades with distinct tertiary veins. Inflorescences terminal. Fruits brown, flattened globose at maturity. Seeds easily breaking free from fruit wall to rattle inside hollow fruit, ± round in transverse section, the seed coat pitted; aril present, colorless. . . . . . . *Macoubea.*
1. Herbs, subshrubs, or lianas.
  11. Herbs or subshrubs. Flowers subsessile, pink or white. Escaped from cultivation. . . . . *Catharanthus.*
  11. Lianas or scandent woody plants. Flowers pedicellate, generally cream-colored, white, pale yellow, or yellow. Native species.
    12. Inflorescences sometimes modified into tendrils. Fruits globose. . . . . *Pacouria.*
    12. Inflorescences not modified into tendrils. Fruits much longer than broad.
      13. Leaf blades with dense abaxial pubescence. Fruits articulate, the segments indehiscent, covered with dense hairs oriented at right angles to axis. Seeds glabrous. . . . . *Condylocarpon.*
      13. Leaf blades without dense abaxial pubescence. Fruits not articulate, dehiscent, not covered with dense hairs oriented at right angles to axis. Seeds comose.
        14. Leaf blades with occasional domatia in axils of secondary veins. Inflorescences thyrsiform. Flowers small, ca. 4 mm long × 3 mm diam. at throat; stamens exserted. Fruit segments long and slender (15–50 × 2–7 mm) in our area. . . . . *Forsteronia.*

14. Leaf blades without domatia. Inflorescences not thyrsiform. Flowers large, much larger than 4 mm long × 3 mm diam. at throat; stamens included. Fruit segments shorter and stouter.
 15. Leaf blades cordate at base. Pedicels with numerous, narrowly linear, pubescent bracts. Corolla white or cream-colored in our area. . . . . *Macropharynx.*
 15. Leaf blades not cordate at base. Pedicels without linear bracts. Corolla yellow or orange.
  16. Leaf blades not glandular along midrib adaxially. Inflorescences branched. Sepals large, imbricate. Fruits usually thick-walled. . . . . *Odontadenia.*
  16. Leaf blades glandular along midrib adaxially. Inflorescences not branched. Sepals small, not imbricate. Fruits usually thin-walled. . . . . *Mandevilla.*

AMBELANIA Aubl.

Understory trees. Trunks cylindric. Stems stout, glabrous, opposite, without prickles, cylindric except flattened at nodes. Latex white, abundant. Interpetiolar colleters present as distinct ridge. Leaves opposite; petioles without colleters; blades coriaceous; secondary veins not joined, running almost to margin, the tertiary veins indistinct. Inflorescences axillary or cauline. Flowers: sepals small, unequal, without colleters; corolla white, salverform, the tube with cylindrical portion pubescent inside, the lobes obtuse, the aestivation sinistrorse in bud; stamens inserted at base of corolla tube, free from style head; ovary syncarpous, 2-locular; ovules numerous. Fruits berries, narrowly ellipsoid or pyriform, the pulp white, edible, with copious white latex. Seeds brown, ellipsoid, glabrous, with honeycombed testa; embryo straight, the cotyledons very short.

Ambelania acida Aubl. PL. 14a

Trees, usually 5–20 m tall. Bark finely cracked into rectangles 3–5 mm on a side. Latex white, sticky, abundant. Leaves: petioles 1–2 cm long; blades elliptic, 15–25 × 5.5–9.5 cm, glabrous, the base obtuse to rounded, the apex shortly acuminate; secondary veins in 12–15 pairs, spaced 1–1.5 cm apart. Inflorescences mostly cauline, sessile, corymbose. Flowers white or greenish white, fragrant; corolla ca. 3 cm long at anthesis; anthers all fertile, mucronate, basifixed; ovary conic, 2 mm high, the stigma with 2 appendages as long as head. Fruits oblong, ca. 10 × 4 cm, yellow. Seeds flattened in transverse section, ca. 1 cm long, dark brown. Fl (Oct, Nov), fr (Jan, Apr, May, Oct); in non-flooded forest, flowers probably nocturnal. *Bagasse*, *graine biche*, *papaye biche* (Créole), *pau de leite*, *pepino de mato* (Portuguese).

ANARTIA Miers

Shrubs or understory trees. Trunks cylindric. Stems slender, cylindric, the new growth of two decussate pairs of leaves (the first pair isophyllous, the second pair anisophyllous) and an inflorescence subtended by a pair of bracts. Latex white, abundant. Leaves opposite; petioles with numerous colleters at base; blades often long acuminate, glabrous. Inflorescences simple racemes; pedicels well developed, long. Flowers: sepals 5, free, ca. 0.5 cm long, with colleters in 1–2 series adaxially, the margins not ciliate, the outer sepals smaller than inner ones; corolla with tube expanded above stamens, with infrastaminal indument at least equalling stamens, glabrous above stamens; stamens inserted on upper half of corolla tube, free from style head, included, the pollen with distinctly raised equatorial zone; nectary disc adnate to ovary; ovary long attenuate into style, the style head with short appendages, the involucrum inflexed. Fruits 2 follicles, these longer than wide, acuminate. Seeds entirely enveloped in red to orange aril; embryo inferior-radicular.

Anartia meyeri (G. Don) Miers [Syns.: *Tabernaemontana meyeri* G. Don, *T. undulata* G. Mey.]

Treelets, 1.5–3(6) m tall. Leaves: petioles slender, with the blade narrowly decurrent, 1–2 cm long; blades elliptic to oblong, 8–11.5 × 3–4.5 cm, discolorous, dark green adaxially, pale green abaxially, the base obtuse, the apex long acuminate, the acumen 1–2 cm long; secondary veins in 7–9 pairs, ca. 1–1.5 cm apart. Flowers erect; calyx with colleters in 2 series; corolla tube green, swollen at base, constricted in middle, and expanded at apex, the lobes greenish-white, with yellow at throat. Fruits globose, ca. 3 × 2 cm, dark olive green. Seeds ca. 40 per fruit. Fl (Jan), fr (Jul); in non-flooded forest.

ASPIDOSPERMA Mart. & Zucc.

Canopy to emergent trees. Trunks distinctly fluted from base to top, fenestrate, or cylindric and unbuttressed. Stems slender. Latex obvious in branches, but also sometimes present in trunks, white, often immediately oxidizing orange or red upon exposure to air. Leaves usually alternate, rarely opposite, often pubescent. Inflorescences terminal or axillary. Flowers: sepals 5, without colleters at base; corolla salverform or tubular, 4–5-lobed, often pubescent below insertion of stamens, the aestivation sinistrose; stamens inserted at middle of corolla tube, free from style head, the filaments short, the anthers fertile throughout; ovary 2-carpellate, superior or ± inferior, the stigma globose. Fruits dehiscent, of 2 or often only 1 follicle-like segment, woody, more or less laterally compressed, oblique, with 2 lateral ribs or with echinate outer surface, the inner surface wrinkled. Seeds numerous, orbicular, ovate, or cordate, surrounded by membranous wing (Pl. 14c), anemochorous, the slender funicle centrally attached, albuminous; cotyledons cordate, the radicle short.

Allorge, L. & C. Poupat. 1991. Position systématique et révision du genre *Aspidosperma* (Apocynaceae) pour les trois Guyanes. Le point sur leur étude chimique. Bull. Soc. Bot. France **138(4/5):** 267–301.

1. Sepals linear, the apices acute; corolla lobes 1–1.3 cm long, ca. 2× longer than corolla tube. . . . . *A. schultesii.*
1. Sepals ovate, the apices obtuse; corolla lobes <0.8 cm long, ca. as long as or shorter than corolla tube.

2. Leaf blades densely pubescent abaxially.
   3. Trunks sulcate. Leaves, inflorescences, and flowers with dense, reddish-brown indumentum. Corolla lobes ovate. Fruits echinate. . . . . *A. carapanauba.*
   3. Trunks not sulcate. Leaves, inflorescences, and flowers with dense, white indumentum. Corolla lobes linear. Fruits smooth.
      4. Leaf blades coriaceous, the margins revolute; secondary veins in 14–16 pairs, the secondary veins spaced ca. 10 mm apart. . . . . *A. sandwithianum.*
      4. Leaf blades membranous, the margins not reflexed; secondary veins in 16–26 pairs, the secondary veins spaced ca. 5 mm apart. . . . . *A. album.*
2. Leaf blades sericeous or glabrous abaxially.
   5. Leaf blades elliptic to oblong, the apex obtuse to acuminate.
      6. Corolla glabrous abaxially. Fruits smooth. . . . . *A. cruentum.*
      6. Corolla sericeous abaxially. Fruits echinate.
         7. Sepals subequal, ca. 1 mm long. Fruits dolabriform . . . . *A. oblongum.*
         7. Sepals unequal, ca. 2 mm long. Fruits not dolabriform. . . . . *A. marcgravianum.*
   5. Leaf blades elliptic to obovate, the apex rounded, obtuse, or emarginate.
      8. Corolla sericeous abaxially. Fruits echinate. . . . . *A. excelsum.*
      8. Corolla glabrous abaxially. Fruits smooth. . . . . *A. spruceanum.*

Aspidosperma album (Vahl) Pichon FIG. 20, PL. 14b

Canopy trees, 25 m tall. Trunks cylindric. Bark: outer bark several mm thick, the inner bark 20 mm thick, orange with white streaks. Latex from broken branches oxidizing red. Leaves alternate; petioles 1.5–2.5 cm long; blades oblong, 8–12 × 3.5–4.5 cm, membranous, shiny green adaxially, whitish, with dense covering of short, thick hairs or scales abaxially, the base obtuse to rounded, the apex obtuse to slightly emarginate, the margins not reflexed; secondary veins in 16–26 pairs, ca. 5 mm apart, conspicuous abaxially, less conspicuous adaxially. Inflorescences branched cymes. Flowers: sepals ovate, ca. 1 mm long, densely whitish-pubescent, the apex obtuse; corolla greenish, glabrous, the lobes linear, twisted. Fruits single follicle-like segments, ca. 9 × 5.5 cm. Seeds 8–14, orbicular. Fl (Jul, Sep), fr (Jan, May); uncommon, in non-flooded forest. *Bois macaque, bois patagaie, flambeau rouge* (Créole), *kumãti udu, tyõtiudu* (Bush Negro).

Aspidosperma carapanauba Pichon

Canopy trees, 18–35 m tall. Trunks distinctly sulcate. Young stems densely, reddish-brown pubescent. Bark light brown, very bitter to taste. Latex from stems and inner bark initially white, sometimes appearing reddish when dried. Leaves alternate; petioles 1–2.5 cm long, densely pubescent; blades narrowly ovate, elliptic, or oblong, 9.5–21 × 4.5–6.5 cm, coriaceous, glabrous to subglabrous, often undulate adaxially, dirty white to reddish-brown, with short papillae and dense covering of erect hairs abaxially, the base obtuse to nearly rounded, often asymmetric, the apex obtuse to acuminate, the margins often revolute, especially toward base; secondary veins in 14–17 pairs, ca. 10–15 mm apart, impressed on adaxial surface, salient on abaxial surface. Inflorescences lateral, inserted near apex of stems. Flowers densely pubescent; sepals ovate, ca. 3 mm long; corolla cylindric, ca. 7 mm long, greenish-white, densely pubescent, the lobes ovate, ca. 1 mm long. Fruits 1–2 follicle-like segments, 7–8 × 6–7 cm, the pericarp distinctly echinate, the prickles widened at base, tapered to apex, bent toward suture of segment, arranged in irregular longitudinal lines. Seeds ca. 6 cm diam. Fl (Aug, Sep, Nov), fr (May); relatively common, in non-flooded forest. *Bois pagaïe* (Créole).

Aspidosperma cruentum Woodson

Understory to canopy trees, 15–25 m tall. Trunks cylindric. Bark with finely reticulate, vertically oriented cracks, usually without latex from slash, the inner bark yellow-orange to orange, often mottled with white. Latex from cut stems rapidly oxidizing reddish. Leaves alternate; petioles 2–3 cm long, at first pubescent, later subglabrous; blades usually elliptic, less frequently oblong, 11.5–19 × 4.5–7 cm, coriaceous, both surfaces glabrous, shiny adaxially, dull abaxially, the base acute to obtuse, the apex obtuse to acuminate, the margins finely revolute; secondary veins in 20–24 pairs, obscure and difficult to count, ca. 5 mm apart. Inflorescences cymose, with many flowers, terminal, all axes covered with grayish-white, dense pubescence. Flowers: sepals ovate, ca. 3 mm long, densely grayish-white pubescent, the apex obtuse; corolla ca. 7 mm long, glabrous, light orange to yellow, the tube expanded below stamens and at base, the lobes linear, twisted around themselves, ca. 1–2 mm long; stamens ca. 1 mm long, the filaments short, ca. 0.2 mm long; ovary truncate at apex, surmounted by style, expanded stigma, and 2 pubescent appendages. Fruits smooth, of 1 follicle-like segment, stipitate, densely brown-pubescent, ca. 12 × 10 cm, the stipe 1–1.5 cm long. Seeds ca. 10–12, elliptic, ca. 7 × 2 cm; cotyledons ca. 2.5 × 2 cm, radicle ca. 0.5 cm long; germination epigeous. Fl (Aug, Sep); relatively common, in non-flooded forest.

Aspidosperma excelsum Benth.

Canopy trees, 25–32 m tall. Trunks distinctly sulcate. Latex not conspicuous in trunk, more obvious in stems, white. Leaves alternate; petioles 0.8–1.5 cm long, glabrous; blades elliptic, 7.5–11 × 4–7.5 cm, glabrous, often covered with whitish bloom adaxially, whitish, minutely papillate abaxially, the base acute to obtuse, the apex obtuse to rounded, the margins finely revolute; secondary veins in 10–14 pairs, 7–8 mm apart. Inflorescences terminal or lateral, often poorly developed. Flowers fragrant; sepals ovate, ca. 3 mm long, pubescent, the apex obtuse; corolla tube funnel-shaped, pubescent except at margin, the lobes not incurved, the anthers ca. 0.6 mm long; ovary glabrous. Fruits 2 suborbicular follicle-like segments, 5–7 × 2–2.5 cm, the surface echinate. Seeds disciform, ca. 5–6 cm diam. Fl (Apr, Jul, Sep), fr (May); common, in non-flooded forest.

Aspidosperma marcgravianum Woodson

Canopy trees, 25–45 m tall. Trunks sulcate for entire length. Bark black. Leaves: petiole 0.8–1.2 cm long; blades elliptic, 6–7 × 1.7–2.5 cm, membranous, glabrous adaxially, whitish, with minute appressed, scale-like hairs abaxially, the base acute, the apex acute

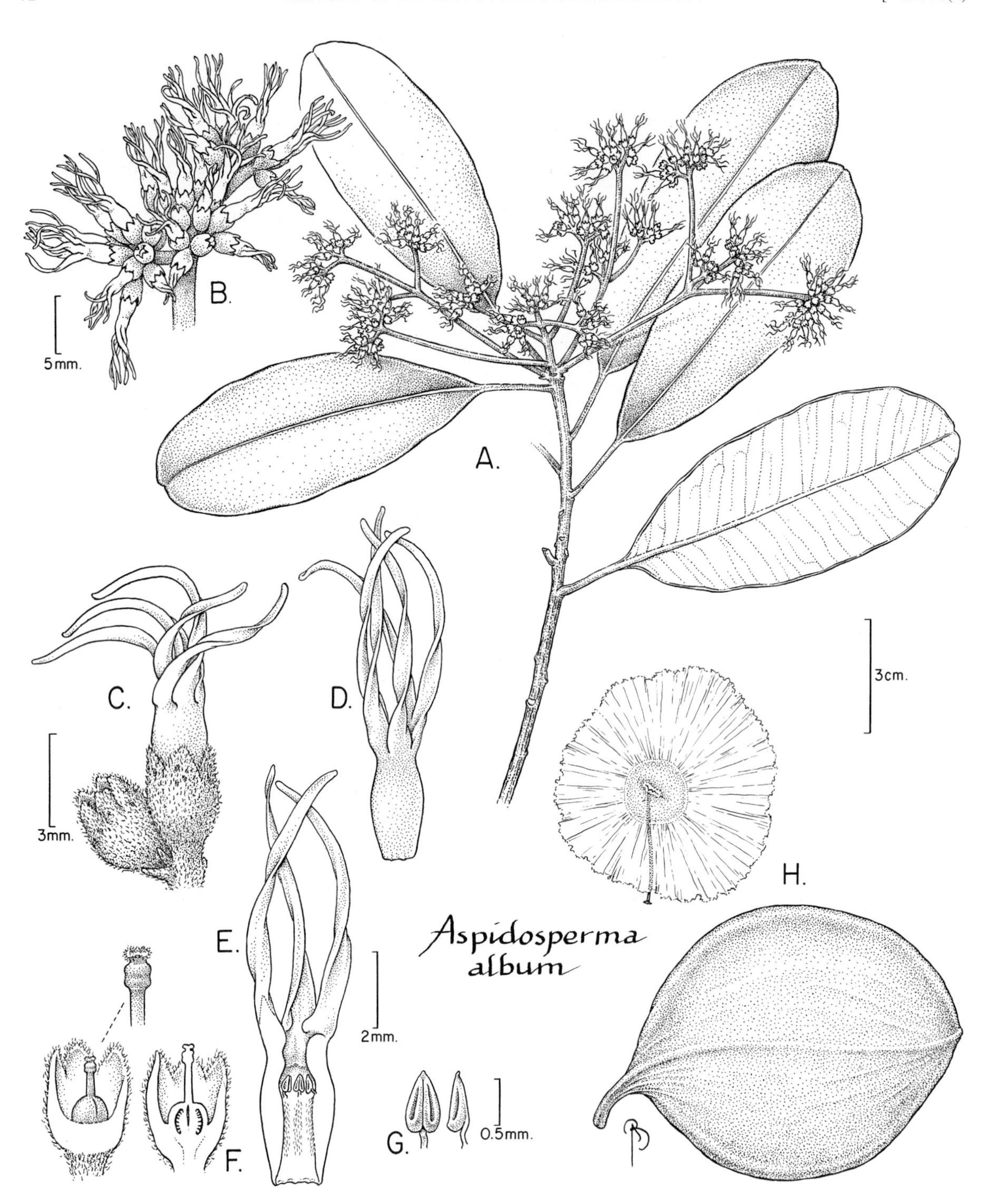

FIG. 20. APOCYNACEAE. *Aspidosperma album* (A–G, *Mori & Gracie 23965*; H, *Mori & Pepper 24297*). **A.** Part of stem with leaves and inflorescences; leaf on bottom right turned to show abaxial surface. **B.** Part of inflorescence. **C.** Lateral view of flower (right) and calyx (below). **D.** Lateral view of corolla. **E.** Medial section of corolla showing adnate stamens. **F.** Section of basal part of flower (far left) with detail of style head (above), and medial section of basal part of flower (near left). **G.** Adaxial (near right) and lateral (far right) views of stamens. **H.** Fruit (below) and seed (above).

to acuminate; secondary veins in 9–11 pairs. Inflorescences thyrsiform, 3–6 cm long, pubescent. Flowers: calyx unequally parted, the sepals ovate, ca. 2 mm long, the 2 outer sepals longer, the base cordiform, the apex obtuse, the margins ciliate; corolla greenish, pubescent except glabrous where covered by calyx, the lobes 1/2 as long as tube; stamens basifixed, inserted at apex of tube; ovary pubescent, with 4 raised areas at apex, the style longer than ovary, the stigma spherical. Fruits 1(2) follicle-like segments, pyriform or obliquely pyriform, ca. 5–6 × 6 × 1.2 cm, rugose, covered with warty, conical tubercles, ca. 1.5–2 × 0.1 cm, yellow-velutinous. Seeds ca. 2 cm diam.; embryo somewhat excentrically placed; funicle stout. Apparently rare in our area, not collected fertile from our area, therefore, flower and fruit descriptions based on material from outside our area, in non-flooded forest. *Bois flambeau, bois pagaïe, citronnelle blanc, païcoussa* (Créole).

Aspidosperma oblongum A. DC.

Canopy trees, to 45 m tall. Trunks buttressed, deeply sulcate, exuding transparent latex when cut. Stems black, lenticellate, pubescent and brown when young. Leaves: petioles 0.5–2 cm long, black when dried; blades elliptic-oblong, 7–14 × 2–4 cm, membranous, dull green adaxially, glaucous abaxially, the base acute, the apex acute, the margins undulate, reflexed from ca. 1 cm above petiole; secondary veins in 16–17 pairs, inconspicuous. Inflorescences erect; peduncles densely appressed-pubescent; pedicels ca. 1 mm long. Flowers mostly white, fragrant; sepals subequal, ovate, ca. 1 mm long, the apex obtuse; corolla sericeous abaxially, the lobes violet-hued, ca. 1 mm long; anthers inserted at ca. middle of corolla tube, the suprastaminal clefts conspicuous; ovary ovate, pubescent, the style twice as long as ovary, the style head tapering. Fruits of 1–2 dolabriform follicle-like segments, 7–7.5 × ca. 4.5 cm, ca. 1.1 cm thick, brown-pilose, coarsely verrucose and furrowed, the base tapering for ca. 0.5 cm, the apex obtuse, the margins with lateral obtuse prickle 1/2 way from base to apex. Seeds ca. 5 cm diam.; funicle centrally attached, ca. 3 cm long; embryo yellow. Fr (Aug); in non-flooded forest.

Aspidosperma sandwithianum Markgr. PART 1: FIGS. 7, 9

Canopy to emergent trees, 30–50 m tall. Trunks cylindric, sometimes with buttresses. Stems thick, ca. 8–10 mm diam. Bark with shallow, vertical cracks. Latex white, oxidizing red-orange upon exposure to air, exuding from cut stems but not apparent in slash. Leaves: petiole 2–4 cm long; blades oblong, 12–27 × 5–7.5 cm, coriaceous, surface glabrous adaxially, whitish, covered with very dense, thick, scale-like hairs abaxially, the base obtuse to rounded, the apex acute to obtuse, the margins slightly revolute; secondary veins in 14–16 pairs, ca. 10 mm apart. Inflorescences with grayish-white pubescence, cymose, the flowers tightly congested; pedicels absent. Flowers: sepals ovate, ca. 2 mm long, densely white pubescent, the apex obtuse; corolla pale yellow, glabrous outside, the lobes linear, twisted, angular, longer than tube. Fruits smooth, of 1 or 2 follicle-like segments, ca. 16 × 11 cm, with straight median line from base to apex but otherwise smooth, white pubescent when young. Seeds ca. 8 cm diam., the wings orbicular; embryo ca. 2.5 cm long. Fl (Sep), fr (May, Aug, old fr Sep); scattered, in non-flooded forest.

Aspidosperma schultesii Woodson [Syns.: *A. morii* (L. Allorge) L. Allorge, *A. macrophyllum* Müll. Arg. subsp. *morii* L. Allorge]

Canopy trees, 20–25 m tall. Trunks irregularly fenestrate, sulcate towards base, the furrows not continuous for entire length of trunk or more markedly sulcate with furrows extending up trunk. Stems often drying black. Bark rough, with vertically oriented reticulate pattern of shallow fissures, the outer bark ca. 3 mm thick, the inner bark orange with white bands, with a bright yellow layer at junction with sapwood. Latex white, apparent in cut branches but not in slash. Leaves widely spaced; petioles 3–5 cm long; blades elliptic, 14–20 × 5.5–8 cm, glabrous, shiny adaxially, glabrous, with whitish caste caused by minute papillae abaxially, the base obtuse, the apex obtuse; secondary veins in 14–18 pairs, these obscure adaxially, 10–15 mm apart. Inflorescences cymose, terminal or axillary, the axes with white pubescence. Flowers showy, 15–20 mm long; sepals linear, not appressed against corolla, ca. 1 mm long, white pubescent, the apex acute; corolla white, pubescent abaxially, the lobes 10–13 mm, longer than tube, pubescent adaxially in lower half, the tube ca. 5 mm long; ovary apocarpous, hirsute, the style ca. 1 mm long, the stigma sleeve-like. Fruits orbicular, 13–14 × 10–11 cm, thickly woody, sulcate with 12–13 ridges and granulate. Seeds orbicular, ca. 7 cm diam.; embryo cordate, ca. 3 cm diam. Fl (Jul, Aug, Sep); uncommon and scattered, in non-flooded forest.

Aspidosperma spruceanum Müll. Arg.

Canopy to emergent trees, 25–40 m tall. Trunks cylindric. Latex oxidizing red. Leaves: petioles ca. 1.5 cm long; blades obovate, 10–20 × 4–8 cm, coriaceous, green-brown and shiny adaxially, squamose abaxially, the apex often emarginate, the margins revolute; secondary veins in 25–30 pairs, perpendicular to primary vein, fine and closely spaced. Inflorescences densely white-pubescent. Flowers: sepals ovate, ca. 2 mm long, densely white pubescent, the apex obtuse; corolla white, pale yellow when dried, glabrous, the tube ca. 3 mm long, sulcate, the lobes as long as tube, linear, twisted in bud; ovary semi-inferior, glabrous, ovate, with a spherical stigma. Fruits smooth, usually of 1 follicle-like segment, ca. 8 × 4.5 cm, densely covered with ferruginous trichomes, with a prominent median vein and numerous secondary veins ± parallel from base to apex. Seeds 9–14, orbicular, ca. 3.5 cm diam. Fl (May); known only by *Mori & Pennington 18054* (NY) from our area, the description is based on collections from outside our area.

BONAFOUSIA A. DC.

Understory trees or shrubs. Trunks cylindric. Stems slender, the new growth of 1 pair of leaves. Bark soft. Latex white. Leaves opposite; petioles with numerous colleters at base. Inflorescences 1 per node (except *B. sananho* and *B. siphilitica*), usually cymose, rarely racemose. Flowers: sepals with colleters at base, the margins often ciliate; corolla white or with tube pink or lilac, exceptionally yellow; stamens usually inserted at middle of and included in corolla tube, the anthers with more than 1/2 of sterile portion adnate to filaments, the staminal tails straight and short; nectary disc prominent and entire; ovary smooth, the style filiform, 3–5 times longer than ovary, the style head with horizontal involucrum entire, the body pentagonal with appendages often 1/3 as long as style head. Fruits of 2 follicle-like segments, these smooth, verrucose, or muriculate. Seeds black, sulcate, with white aril, adaxially incompletely covering seed (except completely covering seed in *B. siphilitica*).

1. Stems terete, triangular, ribbed, or flattened.
   2. Stems distinctly angular.
      3. Sepals <0.5 cm long, the lobes separate. . . . . *B. angulata.*
      3. Sepals >1 cm long, the lobes united at base. . . . . *B. macrocalyx.*
   2. Stems terete or slightly angular.
      4. Corolla basically cream-colored, the tube yellow at apex, mauve-lilac at base. Fruits curved. . . . . *B. disticha.*
      4. Corolla white. Fruits globose. . . . . *B. sananho.*
1. Stems terete, without angles or ribs. . . . . *B. siphilitica.*

Bonafousia angulata (Müll. Arg.) Boiteau & L. Allorge [Syn.: *Tabernaemontana angulata* Müll. Arg.].

Small trees, 1–6 m tall. Stems distinctly angled. Leaves: petioles 1–5 mm long, narrowly winged; blades elliptic, 10–18 × 4–6.5 cm, chartaceous, glabrous, the base acute to obtuse, the apex acuminate; secondary veins in 10–13 pairs. Inflorescences one per node, 3–4-flowered. Flowers: sepals free, ca. 0.5 cm long, membranous, green, with 5–6 colleters adaxially, the margins ciliate; corolla tube pink, ca. 1.5 cm long, with infrastaminal hairs in tube 1/2 as long as stamens, the lobes white, pubescent at overlapping edge in bud; stamens inserted at middle of tube; nectary disc inconspicuous; ovary attenuate into style, the style slender, the style head with an involucrum at base, a pentagonal body, and 2 appendages 1/3 as long as style head. Fruits reniform, follicle-like, 3–4 × 1.2–2 cm, brown, slightly granulous externally, light green internally. Seeds brown; aril white, not covering adaxial side of seed. Fl (Aug), fr (Aug); a poorly known species in our area, in non-flooded forest.

Bonafousia disticha (A. DC.) Boiteau & L. Allorge [Syn.: *Tabernaemontana disticha* A. DC.] FIG. 21

Small trees, 1–3 m tall. Stems slightly trigonous but not markedly angular. Leaves: petioles 5–10 mm long; blades elliptic, 11–20 × 3.5–6.5 cm, glabrous, the base obtuse, decurrent onto petiole, the apex acuminate; secondary veins in 9–13 pairs. Inflorescences 1 per node, cymose, with 2–4 flowers. Flowers diurnal, fragrant; sepals free, ca. 0.4 cm long, membranous, green, ciliate, with 3–4 colleters adaxially, the margins ciliate; corolla 3.5–4 cm long at anthesis, the lobes erect, ca. 1 cm long, yellow, the tube white flushed with pink or mauve at base, sometimes greenish above; stamens inserted at ca. middle of corolla tube; nectary disc adnate to lower half of ovary; ovary ca. 1.5 cm long, the style ca. 1.7 cm long, surmounted by style head ca. 1 mm long with an entire involucrum at base and 2 appendages 1/3 as long as body of style head. Fruits 2 follicle-like segments with lateral ribs, ca. 4.5 × 1 cm, yellow at maturity. Seeds brown, sulcate, the testa with concave cells; aril white, not covering adaxial side of seed. Fl (Jan, Feb, Jul, Aug, Sep, Nov); relatively common, in understory of non-flooded forest. *Bois lézard, couzou* (Créole).

Bonafousia macrocalyx (Müll. Arg.) Boiteau & L. Allorge [Syn.: *Tabernaemontana macrocalyx* Müll. Arg.]

Small trees, 2–4 m tall. Bark brown. Stems dark brown, conspicuously triangular. Leaves: petioles ca. 0.5 cm long; blades elliptic to lanceolate, 18–20 × ca. 6 cm, glabrous, concolorous, dark green, the base obtuse, the apex acute; secondary veins in 10–12 pairs, ca. 1 cm apart. Inflorescences 1 per node, cymose, of ca. 35 flowers. Flowers: calyx 1–1.5 cm long, crassulate, pinkish-white, the sepals united for 1/3 length, with 7–8 colleters at base of sinuses, the margins not ciliate; corolla pinkish-white with golden yellow eye, the lobes triangular, convolute, ca. 1 cm long, the apex curved, retuse, the tube ca. 2.5 cm long, swollen at base, the infrastaminal indument only reaching filaments, the suprastaminal indument scattered; stamens inserted at middle of corolla tube; nectary disc conspicuous, lobed; ovary tapering at apex, the style ca. 4 times as long as ovary, the style head with horizontal involucrum at base, the body pentagonal, the appendages 1.3 times as long as body. Fruits 2 reniform follicle-like segments, ca. 2.5 × 1 cm, the pericarp granulous. Seeds black, sulcate; aril white, not covering adaxial side of seed. Fl (Mar, Apr, Jul), fr (Jul, Nov); in understory of non-flooded forest in open areas.

Bonafousia sananho (Ruiz & Pav.) Markgr. [Syn.: *Tabernaemontana sananho* Ruiz & Pav.] PL. 14d

Small trees, to 6 m tall. Stems triangular, sometimes flattened at apex. Leaves: petioles 7–15 mm long; blades elliptic to widely elliptic, 12.5–20 × 5.5–10.5 cm, glabrous, the base acute to obtuse, the apex acuminate; secondary veins in 8–12 pairs. Inflorescences cymose, with 10–30 flowers; peduncles 2–4 cm long; pedicels slender, as long as calyx. Flowers: sepals free, ca. 0.3 cm long, membranous, green, with 3–5 colleters adaxially, the margins ciliate; corolla with tube 1.2–1.5 cm long, the lobes reflexed, slightly twisted, longer than tube, fringed on margin, white (sometimes with brown spot); nectary disc poorly developed, surrounding ovary for more than half length; style slender, surmounted by style head with well developed horizontal involucrum at base, a pentagonal body, and 2 short appendages 1/4 as long as body. Fruits usually 2 globose follicle-like segments, 5–6 × ca. 5 cm, smooth, with lateral ribs, yellow-orange at maturity. Seeds brown to black, sulcate on dorsal side; aril white, not covering adaxial side of seed, sweet to the taste; embryo with membranous cotyledons with conspicuous nervation. Fl (Oct), fr (Jan, Feb, Apr, May, Jun, Sep, Oct, Dec); in non-flooded forest.

Bonafousia siphilitica (L.f.) L. Allorge [Syn.: *Tabernaemontana siphilitica* (L.f.) Leeuwenb.] PL. 14e

Shrubs, 2–3 m tall. Stems terete, without angles or ribs. Leaves: petioles 10–15 mm long; blades elliptic to widely elliptic, 17–30 × 8.5–14 cm, glabrous, the base acute to obtuse, the apex acuminate; secondary veins in 11–15 pairs. Inflorescences with ultimate segments racemose; peduncles 2.5–5 cm long. Flowers nocturnal, fragrant; sepals free, ca. 0.4 cm long, membranous, green, with 6–8 colleters adaxially; corolla white, the margins ciliate, the lobes ca. 2 × 1.5 cm long; nectary disc prominent around ovary; ovary attenuate into style ca. 1.5 cm long, surmounted by style head with horizontal involucrum at base and 2 short appendages. Fruits 2 follicle-like segments, yellow to orange-brown, oblong, 3–5 × 1.5 cm, with finely granulous pericarp and lateral ribs. Seeds black, the testa rugose with concentric, acute wings and 2 short appendages; aril white, musty-smelling, completely covering seed; embryo with straight radicle twice as long as cotyledons. Fl (Sep, Oct); found only at one locality in dense stand growing in low, wet area. *Bois lait, liane du lait, radié capiaï* (Créole).

CATHARANTHUS G. Don

Herbs or subshrubs. Stems cylindric or tetragonal. Latex white, copious. Leaves opposite. Inflorescences axillary, alternately on right and left of nodes, usually 2-flowered. Flowers 5-merous; sepals without colleters; corolla salverform, the aestivation sinistrorsely contorted; stamens 5, alternating with corolla lobes; nectaries 2, alternate with carpels; ovary 2-carpelar; ovules many in each carpel. Fruits 2 narrowly cylindric, many-seeded follicle-like segments. Seeds minute, brown, without appendages.

Catharanthus roseus (L.) G. Don [Syn.: *Vinca rosea* L.].

Perennials, usually <50 cm tall, finely pubescent. Stems cylindric. Leaves: petioles 3–5 mm long; blades elliptic-oblong, 2.5–4 × 1–2 cm, puberulous on both surfaces, the base acute to obtuse, the apex rounded with minute apiculus. Inflorescences with subsessile flowers. Flowers: sepals linear-subulate, ca. 5 mm long, pubescent; corolla pink, white, or white with pink or yellow eye, the tube ca.

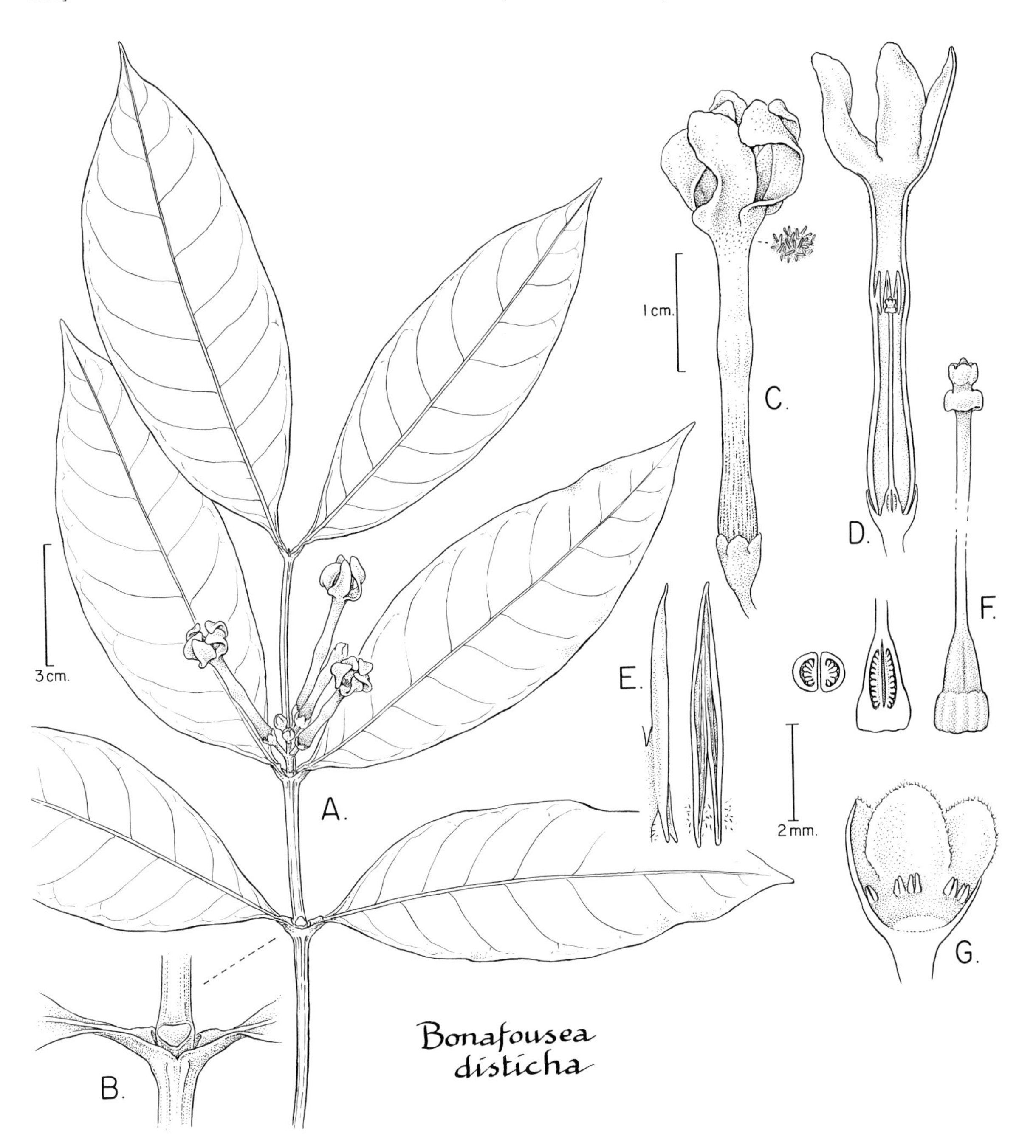

FIG. 21. APOCYNACEAE. *Bonafousia disticha* (*Mori et al. 21025*). **A.** Part of stem with leaves and flowers. **B.** Detail of node. **C.** Lateral view of flower with detail of pubescence on abaxial apex of corolla tube. **D.** Medial section of flower. **E.** Lateral (near right) and adaxial (far right) views of anthers. **F.** Lateral view of pistil (near left) and medial (middle left) and cross (far left) sections of ovary. **G.** Longitudinal section of calyx showing colleters.

20–25 mm long, the lobes 10–15 × 6–8 mm. Fruits follicle-like, ca. 3 cm long. Seeds ca. 2 mm long. Fl (Aug); escaped from cultivation but not persisting, infrequent in weedy areas. *Caca poule* (Créole), *lavandeira* (Portuguese), *pervenche de Madagascar* (French), *roseate periwinkle* (English).

CONDYLOCARPON Desf.

Lianas. Latex white, copious. Leaves opposite in our area. Inflorescences thyrsoid, axillary or terminal, many-flowered. Flowers small, the aestivation sinistrorse; sepals 5, without colleters; corolla often globose in bud, salverform; stamens 5, without sterile basal appendages, free from style-head, inserted near middle of tube, the filaments short, basifixed, the pollen in tetrads; ovary 2-carpelar, apocarpous, ovoid, depressed at top, the style shorter than ovary, the style-head orbicular or turbinate, bilobed at apex. Fruits of 2 pendent segments, articulate into one-seeded, indehiscent units. Seeds fusiform, longitudinally folded, the testa verrucose; embryo straight, the cotyledons narrowly elliptic.

1. Abaxial surfaces of leaf blades very densely pubescent throughout. . . . . *C. amazonicum.*
1. Abaxial surfaces of leaf blades pubescent only on veins. . . . . *C. pubiflorum.*

Condylocarpon amazonicum (Markgr.) Ducke

Stems terete, pilose, the hairs golden-brown. Leaves opposite; petioles 5–10 mm long, canaliculate; blades narrowly elliptic to oblong, 7–18 × 2.5–6.5 cm, adaxial surface ± glabrous or sparsely pubescent, abaxial surface densely pubescent, the hairs golden-brown, the base acute, the apex acuminate; secondary veins in 14–17 pairs. Inflorescences terminal and axillary, puberulous to densely pubescent; peduncle long. Flowers ca. 4 mm long; sepals 5; corolla salverform, the tube ca. 2 mm long, enlarged near middle, the lobes 1–1.8 × ca. 0.7 mm; anthers ovate; ovary conical, ca. 0.6 mm long, the style subsessile, the style head turbinate. Fruits 10–15 cm long, woody, pale green, densely velutinous with spreading golden-brown hairs ca. 3 mm long. Seeds 1 per unit, ca. 1.5 cm × 3 mm, brown, glabrous. Fl (Sep); collected only three times in our area.

Condylocarpon pubiflorum Müll. Arg.

Stems terete, pilose, the hairs golden-brown. Leaves opposite; petioles 8–12 mm long, canaliculate; blades elliptic to narrowly lanceolate, 8–12 × 3–5 cm, both surfaces pubescent on veins only, the hairs golden-brown, the base obtuse, the apex long acuminate; secondary veins in 12–15 pairs. Known only from sterile collections from our area; apparently common, in disturbed areas.

COUMA Aubl.

Shrubs or trees, to 25–30 m tall. Trunks cylindric. Latex white, abundant. Leaves in verticils of 3 (opposite in seedlings); intrapetiolar colleters present. Inflorescences axillary, corymbose, many-flowered. Flowers: sepals 5, small, without colleters; corolla salverform, white, pink, red, or violet, the aestivation sinistrorse in bud; stamens adnate to middle of corolla tube, the anthers ovate, short, not adherent, free from style head; nectary disc adnate or indistinct; ovary syncarpous, 1-locular, the style head cylindric, with 2 free terminal appendages. Fruits globose berries. Seeds ellipsoidal, ca. 10 per fruit; embryo with radicle longer than cotyledons.

Couma guianensis Aubl.

Trees, 20–25 m tall. Trunk cylindrical, the wood soft. Bark finely and regularly grid cracked, the slash red, exuding abundant white latex. Stems brown, trigonous when young, later rough and black. Leaves: petioles 7–10 mm long; blades elliptic to very broadly elliptic, 8.5–18 × 6–10.5 cm, brittle, dark green adaxially, pale green abaxially, the base acute to rounded, the apex acuminate; secondary veins in 10–13 pairs, impressed adaxially, prominent abaxially. Inflorescences in groups of 3 corymbs, the axes brown-violet with light brown lenticels. Flowers: sepals brown-violet; corolla tube pubescent abaxially, long-hairy at throat adaxially, the lobes pink, pubescent; ovary glabrous, the style head barely surpassing calyx. Fruits ca. 4 × 4 cm, red at maturity, the pulp edible. Seeds 4–5 per fruit, flat, elliptic, ca. 1 cm long. Fl (Jul); infrequent in our area but common elsewhere in French Guiana, in non-flooded forest.

FORSTERONIA G. Mey.

Bruce F. Hansen

Lianas. Stems with nodes with colleters mostly caducous, deltoid to subulate. Latex white. Leaves opposite; blades with paired, basal, subulate glands on each side of midrib on adaxial surface, abaxial surface usually with domatia in axils of secondary veins. Inflorescences terminal and lateral, thyrsiform, many-flowered. Flowers small; sepals scarious, with adaxial colleters, usually 1 per sepal but sometimes lacking; corolla with dextrorse aestivation, the tube obconic to subcylindric, with a densely pilose ring at or below throat within, the lobes erect; stamens with filaments free, the anthers closely agglutinated around style head, bilocular, cartilaginous, fertile for upper 1/3, ventrally dehiscent, bases cordate and swollen, apices hyaline, erect to inflexed. Fruits apocarpous, follicle-like segments, pendent, narrowly cylindric. Seeds apically comose, subterete to compressed, linear, longitudinally 5–8-ribbed, with a single wide ventral groove.

1. Larger stems often very corky, with multilayered, conspicuous projections. Leaves 8–18 cm long. Sepals 1.5–2 mm long; anther apices often fimbriate. . . . . *F. acouci.*
1. Larger stems merely lenticellate, without conspicuous projections. Leaves 3–7(9) cm long. Sepals 0.5–0.8 mm long; anther apices entire. . . . . *F. guyanensis.*

Forsteronia acouci (Aubl.) A. DC.
Pl. 14g; Part 1: Pl. XIVd

Bark of larger stems thick and very corky, with thick, multilayered, conspicuous projections. Leaves: petioles 3–5 mm long; blades elliptic to ovate or slightly obovate, 8–15(18) × 3–8 cm, glabrous, the base obtuse or rounded to acute, the apex acuminate to somewhat acute; basal glands caducous, subulate, 0.5–1 mm long; domatia glabrous. Inflorescences terminal and lateral, usually surpassing subtending leaves, open-thyrsiform, usually subglomerulate, many-flowered, puberulent to pubescent, narrowly conic to cylindric or obconic, 5–15 × 3–10 cm. Flowers: sepals lanceolate to narrow-triangular, 1.5–2 mm long, puberulent to pubescent; corolla cylindric to obconic, white, the tube 1.5–2.5 × 2.5–3 mm at throat, puberulent to glabrous abaxially, glabrous with a densely pilose ring below throat adaxially, the lobes erect, sparsely to densely strigose, lanceolate to oblong, 2–3.5 × ca. 1 mm; stamens 0.5 mm long, the anthers quarter to half exserted, glabrous, oblong, 1.5–2 mm long, the bases cordate, the basal tails swollen, the apices hyaline, acuminate to short-attenuate, usually somewhat fimbriate, suberect to slightly inflexed; ovary densely pubescent, ovoid, 0.4–0.6 mm long, the style ca. 0.5 mm long, the style head fusiform, 1–1.5 mm long. Fruits with follicle-like segments long and pendent, 20–35(50) cm × (3)4–7 mm. Seeds subterete to flattened, papillate, fusiform, dorsally 5–8-ribbed with a single wide ventral groove, 14–18(22) × 2–4 mm, the apices truncate, subapically strigose, the coma brownish yellow, 3–5 cm long. Fl (Nov), fr (Feb); common, in non-flooded forest.

Forsteronia guyanensis Müll. Arg. FIG. 22, PLS. 14f, 15a

Bark of larger stems lenticellate but not corky. Leaves: petioles subsessile to 3 mm long; blades obovate or oblanceolate to sometimes elliptic, 3–7(9) × 1.5–3(5) cm, glabrous, the base acute, the apex short acuminate; basal glands caducous, subulate, ≤0.5 mm long; domatia glabrous. Inflorescences terminal and lateral, usually surpassing subtending leaves, open-thyrsiform, many-flowered, puberulent, conic to subhemispheric or subcylindric, 3–10 × 3–8 cm. Flowers: sepals ovate, 0.5–0.8 × 0.8 mm, puberulent; corolla obconic, white, the tube 1.5 × 2 mm at throat, glabrous to puberulent abaxially, mostly glabrous but with densely pilose ring below throat adaxially, the lobes erect, puberulent to pubescent abaxially, puberulent adaxially, lanceolate to ovate, 2 × 1.5 mm; stamens 0.3–0.5 mm long, the anthers included or barely exserted, oblong, glabrous, 1.5–1.8 mm long, the bases cordate, the basal tails swollen, the apices hyaline, acute to acuminate, inflexed; ovary densely pubescent above, ovoid, 0.6 mm long, the style 0.2 mm long. Fruits with follicle-like segments pendent, narrowly cylindric, 30–45 cm × 4–6 mm. Seeds subterete to flattened, glabrate, minutely papillate, narrowly elliptic, longitudinally distinctly 5–8-ribbed with a single wide ventral groove, 15–18 × ca. 2 mm, the coma yellowish brown, ca. 3 cm long. Fl (Aug, Sep, Nov), fr (Feb); common, in non-flooded forest.

GEISSOSPERMUM Allemão

Small to medium-sized trees, to 40 m tall. Trunks fenestrate or sulcate. Stems horizontal, the young stems terete, glabrous or pubescent. Latex very bitter, white or transparent. Leaves alternate, glabrous or densely sericeous or silvery beneath. Inflorescences supra axillary, few- to many-flowered. Flowers pubescent, the aestivation dextrorse; sepals 5, subtended by 2–3 bracts, without colleters; corolla greenish or cream-colored, the lobes 5, pubescent, reflexed, the tube glabrous adaxially except under stamens, the lobes overlapping to right; stamens 5, inserted near throat, subsessile, free from style head; ovary pubescent. Fruits (1)2 horn-like, indehiscent, fleshy, glabrous or tomentous segments. Seeds orbicular, ovate, albuminous; cotyledons cordate or ovate, the radicle straight.

1. Leaf blades glabrous abaxially except when young. Corolla tube 0.9–1 cm long, the lobes 0.4–0.5 cm long. Fruits a single follicle-like segment. . . . . *G. laeve.*
1. Leaf blades densely pubescent abaxially. Corolla tube <0.7 cm long, the lobes <0.3 cm long. Fruits of 2 follicle-like segments.
   2. Floral buds obtuse; sepals acute; corolla tube cylindrical, 0.4–0.5 cm long, the lobes 0.1–0.2 cm long. . . . . *G. argenteum.*
   2. Floral buds acute; sepals obtuse; corolla tube angled, 0.6–0.7 cm long, the lobes 0.2–0.3 cm long. . . . . *G. sericeum.*

Geissospermum argenteum Woodson

Canopy trees, 25–30 m tall. Trunks irregular, fenestrate. Bark: outer bark smooth with paper-like flakes, dark reddish brown. Stems tomentose when young. Latex transparent in trunk, white in branches and fruits. Leaves: blades ovate, 13–16 × 5–6.5 cm, silvery pubescent abaxially, the base symmetric, attenuate onto petiole, the apex long acuminate-mucronate; secondary veins in 5–6 pairs. Flowers: buds subspherical, the apex obtuse; sepals curved, ca. 0.2 cm long, densely tomentose, the apex acute; corolla tube not angled, cylindric, 0.4–0.5 cm long, the lobes 0.1–0.2 cm long. Fruits of 2 follicle-like segments, ca. 4 × 2.5 cm, yellow, tomentose, with abundant milky latex. Seeds ca. 2 × 1 cm.; cotyledons ovate, 0.5 cm long, as long as radicle. Fl (Oct), fr (May); in non-flooded forest. *Acariquara branca* (Portuguese), *bita udu* (Bush Negro), *maria-congo* (Créole).

Geissospermum laeve (Thunb.) Miers FIG. 23

Canopy trees, to 25–30 m tall. Trunks fenestrate. Outer bark brown, rough, with vertical fissures, several mm thick. Stems cylindric, tomentose when young. Latex white in stems. Leaves: blades 13–15 × 4–5 cm, membranous, glabrous when fully expanded, pubescent when young, the base asymmetric, the apex acuminate, the margins undulate; secondary veins in 5–6 pairs, forming angle of ca. 80° with midrib, arcuate. Flowers: sepals curved, 0.4 cm long, not - appressed to corolla; corolla tube 0.9–1 cm long, the lobes 0.4–0.5 cm long. Fruits of 1 follicle-like segment, to 10–12 × 4.5–5 cm, glabrous, green. Seeds ca. 3 × 1.5 cm; cotyledons cordiform, longer than radicle. Fl (Aug), fr (Mar), in non-flooded forest. *Maria-congo* (Créole).

Geissospermum sericeum (Sagot) Benth. & Hook.f.

Canopy trees, to 40 m tall. Trunks cylindric, sulcate, brownish. Bark yellowish-brown, furrowed. Stems pubescent. Latex transparent in trunk, white in branches and stems. Leaves: blades lanceolate, 7–10 × 2.5–3 cm, whitish to yellowish-brown adaxially, densely pubescent abaxially, the hairs appressed, the base asymmetric, cuneate, the apex acuminate; secondary veins in 7–8 pairs, forming angle of ca. 60° with midrib, somewhat arcuate toward margin. Flowers: buds with apex acute; sepals straight, ca. 0.2 cm long, densely pubescent abaxially, sparsely pubescent, dark red adaxially, the apex obtuse; corolla tube 0.6–0.7 cm long, the lobes 0.2–0.3 cm long, silky brown pubescent abaxially. Fruits of 2 follicle-like segments, not seen from our area. Fl (Aug); in non-flooded forest. *Maria-congo* (Créole).

HIMATANTHUS Schultes

Marcel-Marie M. Plumel

Trees or shrubs. Trunks cylindric. Stems terete, thick, sometimes becoming hollow. Latex white, abundant. Leaves spirally arranged; petioles absent or present; blades pinnately nerved, the marginal vein usually well developed. Inflorescences terminal, usually regularly thyrsoid, the secondary peduncles alternate and usually somewhat distant; bracts large, persistent until anthesis, then falling at anthesis.

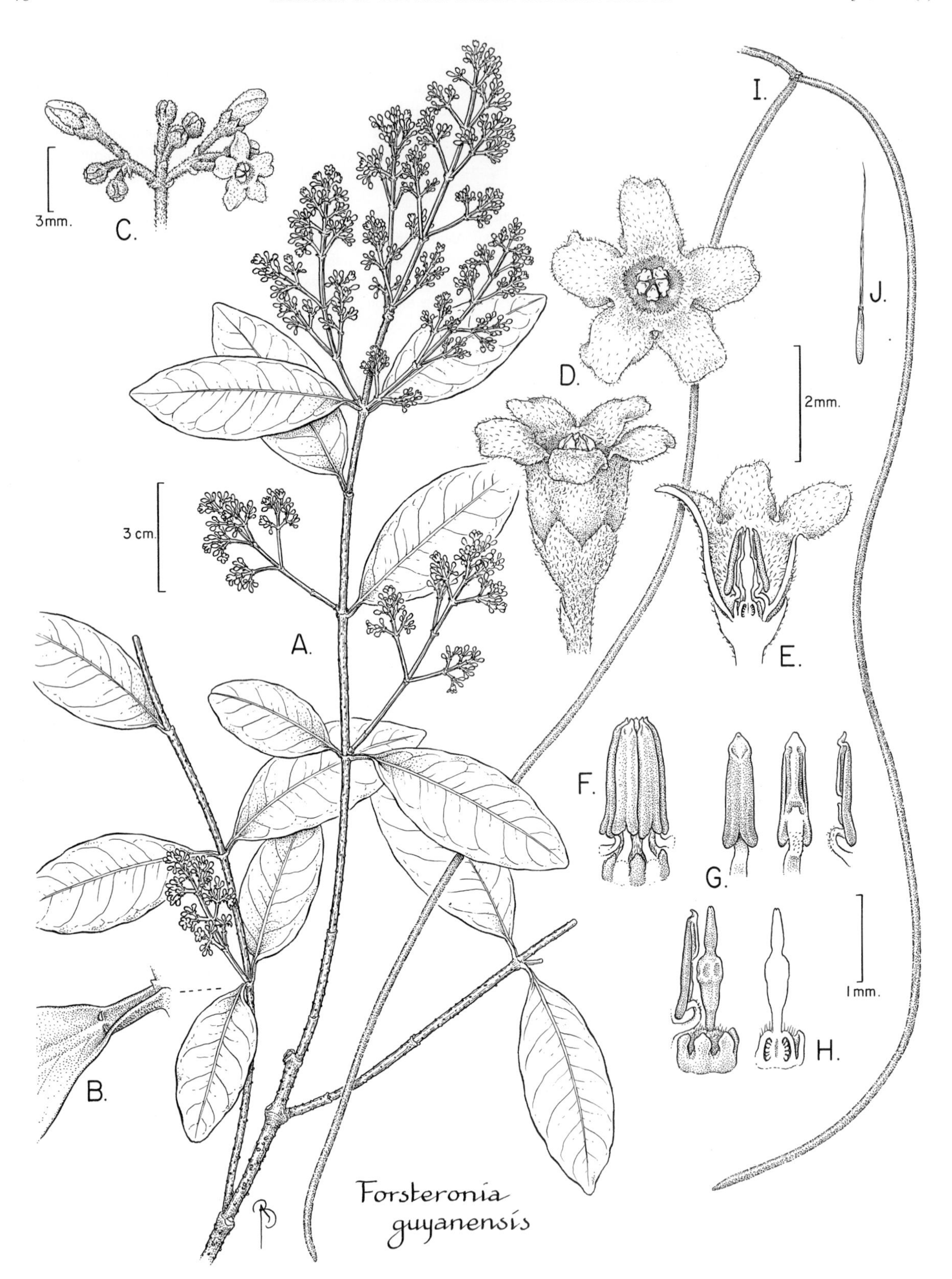

FIG. 22. APOCYNACEAE. *Forsteronia guyanensis* (A–H, *Mori et al. 21633*; I, J, *Leeuwenberg 11787*). **A.** Part of stem with leaves and inflorescences. **B.** Detail of node showing glands on adaxial surface of leaf blade. **C.** Detail of part of inflorescence. **D.** Apical (above right) and lateral (below) views of flower. **E.** Medial section of flower. **F.** Lateral view of androecium showing glandular nectary disc between stamens. **G.** Abaxial (near right), adaxial (middle right), and lateral (far right) views of stamens. **H.** Lateral view of pistil surrounded by glandular nectary disc and one stamen (far left) and medial section of pistil (near left). **I.** Two-parted fruit. **J.** Seed.

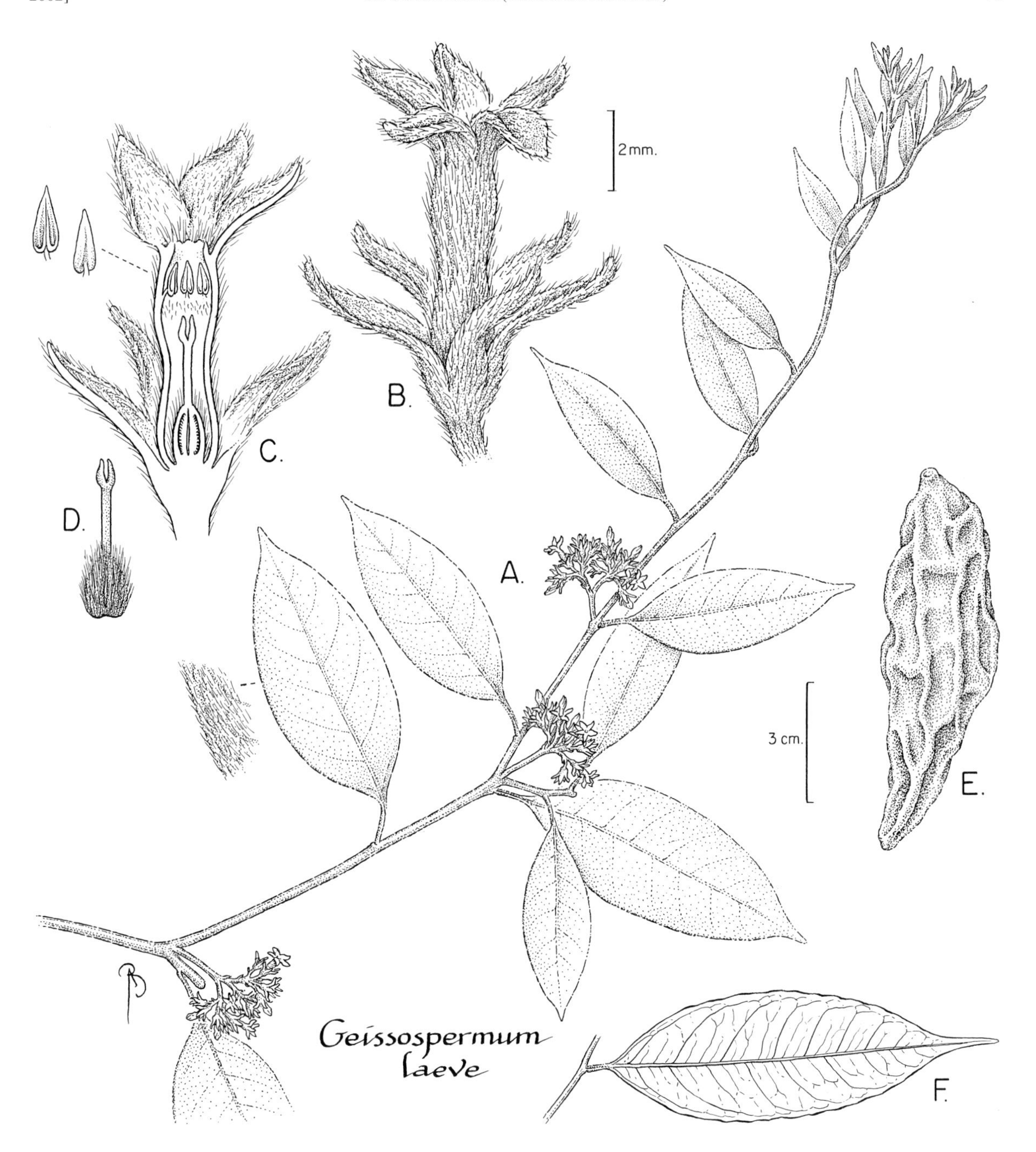

FIG. 23. APOCYNACEAE. *Geissospermum laeve* (A–D, F, *Mori et al. 19177*; E, *Mori & Pepper 24259*). **A.** Part of stem with leaves and inflorescences. **B.** Lateral view of flower. **C.** Medial section of flower (near left) with detail of stamens (far left). **D.** Pistil. **E.** Dried and shriveled fruit. **F.** Less pubescent leaf from older stem.

Flowers large and showy, white, sometimes with yellowish central spot, often fragrant; calyx irregulary developed, without colleters, the sepals often obsolete or of 1–5 acute or lanceolate lobes, these very unequal, distinct almost to receptacle, not imbricate; corolla flaring from a straight or slightly flexuous tube, lacking internal appendices, 5-lobed, the lobes sometimes finely ciliate, the aestivation dextrorse; stamens 5, inserted just above base of corolla tube, the filaments short, the anthers distinct, with 2 thecae entirely fertile, without basal appendices; ovary partially inferior, 2-carpellate, the stigma columnar, the style head with fine rim of papillae and with 2 appendages at apex; placentation axile, the ovules numerous. Fruits 2 follicle-like segments, fusiform, straight or curved, somewhat compressed at apex, ventrally dehiscent. Seeds numerous, flat, surrounded by membranous, irregular wing.

Plumel, M.-M. 1991. Le genre *Himatanthus* (Apocynaceae). Révision taxonomique. Bradea **5(supl.):** 1–118.

1. Inflorescences with bracts 12–15 mm long, the apex acute. Flowers 35–37 mm long; corolla tube ca. 10 mm long, the lobes ca. 25 mm long. . . . . *H. articulatus.*
1. Inflorescences with bracts 18–25 mm long, the apex acuminate. Flowers 70–80 mm long; corolla tube ca. 25 mm long, the lobes 47–48 mm long. . . . . *H. speciosus.*

Himatanthus articulatus (Vahl) Woodson

Understory to canopy trees, to 22 m × 30 cm. Leaves: petioles 2.5–3.5 cm long; blades oblong, 18–20 × 8–8.5 cm, glabrous, the base rounded to cuneate, the apex acute to shortly and obtusely apiculate; secondary veins in 12–15 pairs. Inflorescences fascicles of racemes; bracts 12–15 mm long, the apex acute. Flowers 35–37 mm long; sepals 5, unequal, ovate to oblanceolate; corolla tube ca. 10 mm long, the lobes ca. 25 mm long. Fruits with follicle-like segments slightly incurved, 20–30 cm long, with prominent lateral ribs. Fl (Dec); along rivers.

Himatanthus speciosus (Müll. Arg.) Plumel Pl. 15b

Understory trees, to 20 m tall. Leaves: petioles 2–3.5 cm long; blades narrowly oblanceolate to oblanceolate, 20–38 × 7–10.5 cm, glabrous, the base cuneate, the apex short acuminate; secondary veins in 15–19 pairs. Inflorescences cymes: bracts 18–25 mm long, the apex acuminate, with small colleters in axils. Flowers probably nocturnal, 70–80 mm long; calyx truncate or with one small linear lobe; corolla tube ca. 25 mm long, the lobes 47–48 mm long. Fruits with follicle-like segments 17–22 cm long. Fl (Jul, Sep), fr (Sep); uncommon, in non-flooded forest.

LACMELLEA H. Karst.

Understory trees or shrubs, to 5–20 m tall. Trunks cylindric, often armed with deciduous prickles, the wood orange. Bark dark, often nearly black. Stems with interpetiolar line. Latex white, abundant, sweet-tasting, potable. Leaves opposite. Inflorescences axillary, cymose, lax or congested, with 3–30 flowers, shorter than leaves. Flowers: sepals 5, glabrous or hirtellous; corolla curved, white or greenish-white, swollen above calyx and at insertion of stamens, the lobes rather short, the aestivation sinistrorse; stamens included, the filaments short; ovary unilocular; placentation parietal. Fruits berries, globose to ovoid, with persistent style, glabrous, yellow or orange at maturity, the pulp sweet, yellow, thin, edible. Seeds 1 or 2, brown; cotyledons elliptic, the radicle short.

Lacmellea aculeata (Ducke) Monach. Pl. 15c–e

Understory trees, to 20 m tall. Stems as well as trunk often with prickles. Leaves: petioles 0.5–0.9 cm; blades elliptic, 6–20 × 3–8 cm, the base acute to obtuse, the apex acuminate; secondary veins in 9–17 pairs, the tertiary veins obscure. Inflorescences axillary, 3–7-flowered; peduncle very short. Flowers curved, odorless, probably nocturnal; sepals green; corolla creamy white to pale yellow with age, the tube ca. 3.5 cm long, swollen at base at insertion of stamens, the lobes obtuse, ca. 6 mm long. Fruits 1.5–1.7 × ca. 1 cm. Fl (Aug, Sep), fr (Feb, May, Jun, Oct, Nov); common, in non-flooded forest. *Cumahy, pau de chicle* (Portuguese).

LAXOPLUMERIA Markgr.

Trees. Stems thick, without interpetiolar line. Latex white. Leaves spirally arranged. Inflorescences terminal, many-flowered; bracts minute. Flowers: sepals minute, without colleters; corolla with linear lobes, the aestivation sinistrorse; stamens inserted on upper part of corolla tube, free from style head; ovary apocarpous, with two small scales opposite suture of carpels; ovules numerous. Fruits 2, long, slender follicle-like segments. Seeds surrounded by long hairs.

Laxoplumeria baehniana Monach.

Stems dark brown with elongate lenticels when young, becoming lighter brown with age. Leaves: petioles 1–1.5 cm long; blades narrowly obovate, 8–15 × 3–6.5 cm, glabrous; secondary veins in 11–13 pairs. Inflorescences much branched; peduncles long and slender. Flowers greenish-yellow in bud; sepals ca. 1 mm long; corolla with narrowly cylindric tube, the tube widened at insertion of stamens, glabrous, the lobes longer than tube, the tube and lobes ca. 2.5 cm long; anthers lanceolate, dehiscing along entire length; nectary disc adnate to ovary, with 2 lobes opposite sutures of carpels; ovary glabrous, longer than style, the style head spherical, ca. 1 mm long, with 2 long appendices, annular at base. Fruits 30–40 cm × ca. 2–4 mm, glabrous at maturity. Seeds numerous. Fl (Apr, Sep); apparently rare, in non-flooded forest.

MACOUBEA Aubl.

Canopy trees. Trunks cylindric. Stems with interpetiolar lines and colleters. Latex white. Inflorescences terminal, pedunculate, many-flowered cymes. Leaves opposite; petioles with numerous colleters at base. Inflorescences many-flowered cymes. Flowers white, fragrant; sepals with 5–6 colleters at base; corolla with tube swollen at level of stamens, the aestivation sinistrorse; stamens inserted at base of corolla tube, the anthers lanceolate with divergent tails; ovary pubescent, truncate at apex, the stigma nearly sessile, the style head comprising a horizontal involucrum, a pentagonal body, and 2 long appendages. Fruits of 1–2 segments, hollow at maturity. Seeds numerous, cylindric, long and narrow, sometimes curved, the testa pitted; aril colorless, sweet tasting.

Macoubea guianensis Aubl. Pl. 16a–c

Trees, to 30 m tall. Trunks cylindric to base. Bark smooth, the inner bark yellow to orange. Leaves: petioles 1.5–3 cm long, the base expanded into stipule-like involucrum with axillary colleters; blades elliptic to orbiculate, 10–22 × 6–17 cm, mostly glabrous, the base obtuse to rounded to truncate; secondary veins in 10–17 pairs, 1–2 cm apart. Inflorescences many-flowered; peduncles 2–8 cm long. Flowers: sepals ciliate; corolla with lobes longer than tube, the tube pale yellow to yellowish-white, the lobes white. Fruits usually with only one indehiscent segment developing, flattened globose, to 11 × 12 cm, brown, exuding abundant white latex when cut, hollow at maturity. Seeds ca. 1.5 cm long, easily breaking free from fruit wall and rattling inside hollow fruit when shaken. Fl (Aug, Sep, Oct), fr (Feb, Nov); common, in non-flooded forest. *Poué blanc* (Créole).

MACROPHARYNX Rusby

Lianas. Latex white or clear, oxidizing orange. Leaves opposite, eglandular. Inflorescences axillary, racemes; bracts numerous, linear (epicalyx). Flowers: sepals 5, linear; corolla salverform, the aestivation dextrorse; stamens 5, connivent with stigma head; nectary disc of 5 white nectaries as long as ovary. Fruits cylindric, pubescent follicles. Seeds numerous, with apical coma.

Macropharynx spectabilis (Stadelm.) Woodson FIG. 24

Latex clear, oxidizing orange. Stems densely pubescent, with 2 types of hairs, one long, pointed, white and the other short, matted, ferruginous. Leaves: petioles ca. 6–6.5 cm long, with same pubescence as stems; blades widely elliptic to suborbiculate, 20–23 × 15–17.5 cm, both surfaces with ferruginous, matted hairs, the midrib with a few long, pointed, white hairs abaxially, the base cordate, the apex narrowly acuminate; secondary veins in 9–11 pairs. Inflorescences: bracts conspicuous and abundant, 10 × 0.5 mm, with long, pointed, white hairs. Flowers ca. 4 cm long; corolla white or cream-colored, the lobes ciliate at apex. Fruits ca. 35 cm long (fide label *Phillippe et al. 27022*, NY), the follicle-like segments cylindric, 1.5 cm diam., joined at apices. Seeds ca. 9 mm long, the coma ca. 7 mm long. Fl (Sep), collected only once, in secondary vegetation along airport road.

MANDEVILLA Lindl.

Lianas or infrequently scandent shrubs. Latex white. Leaves opposite (also verticillate outside our area); blades often with glands near base or along midrib. Inflorescences axillary or terminal, racemes. Flowers: sepals 5, small, not imbricate, with many colleters adaxially; corolla funnelform, salverform, or tubular-salverform, white, pink, or yellow, the lobes 5, the aestivation dextrorse; stamens 5, connivent to stigma, the anthers sagittate; nectary disc of 5 nectaries, these ± coalesced at base; ovary 2-carpellate, the style and stigma head fused. Follicles 2. Seeds numerous, with apical coma.

Mandevilla rugellosa (Rich.) L. Allorge PL. 16e

Lianas. Leaves opposite; petioles 4–7 mm long; blades narrowly elliptic to elliptic, 5–10.5 × 1.5–4.5 cm, with glands along midrib adaxially, glabrous adaxially, puberulous, especially along veins, abaxially, the base rounded to slightly lobed, the apex acuminate; secondary veins in 7–9 pairs, the venation slightly impressed adaxially. Inflorescences terminal, 14–17 cm long. Flowers ca. 4 cm long; corolla salverform, yellow with red streaks in throat. Follicles ca. 15 × 0.4 cm, periodically constricted along length. Seeds ca. 0.8 cm long, the coma ca. 1.5 cm long. Fl (Sep); collected only once, on granitic outcrop E of Eaux Claires.

ODONTADENIA Benth.

Lianas. Latex white. Leaves opposite, eglandular. Inflorescences axillary, cymose. Flowers: sepals unequal, imbricate, with 5 colleters adaxially; corolla funnelform, orange, yellow, or white, the lobes symmetric; stamens 5, adnate to stigma, the anthers sagittate; nectary disc coalescent at base, dissected; ovary 2-carpellate, the style head turbinate. Fruits follicles, cylindric or ovoid, the walls often thick. Seeds numerous, cylindric, striate, apiculate, with apical brown coma; embryo straight, the radicle short.

1. Sepals almost as long as corolla tube; corolla minutely puberulent abaxially. . . . . *O. perrottetii.*
1. Sepals 1/5 as long as corolla tube; corolla glabrous except for ciliate margins of lobes (cilia visible only with hand lens). . . . . *O. puncticulosa.*

Odontadenia perrottetii (A. DC.) Woodson FIG. 25, PL. 16d

Leaves: petioles 1–2.5 cm long; blades elliptic to widely elliptic, 9–19 × 5.5–11 cm, coriaceous, glabrous to finely puberulous on veins abaxially, the base rounded to truncate, the apex acute to acuminate; secondary veins in 9–11 pairs, the tertiary venation percurrent. Flowers 5–6 cm long; sepals foliaceous, unequal, markedly imbricate, the inner ones larger, 10–15 mm long; corolla pale yellow, glabrous abaxially, the lower part of tube ca. 15 × 1 mm, the upper part 20 × 5 mm, the lobes ca. 20 mm long. Follicles single or paired, ca. 11–13 × 1.2–1.5 cm, green, with velvet, ferruginous pubescence, subtended by persistent calyx. Fl (Sep, Nov, Dec), fr (Feb); common, as indicated by numerous corollas observed fallen to ground, but difficult to collect, in non-flooded forest.

Odontadenia puncticulosa (Rich.) Pulle PL. 17a; PART 1: FIG. 5

Leaves: petioles ca. 1 cm long; blades ovate to ovate-elliptic, 10–12 × 5–6 cm, coriaceous, glabrous, the base cuneate, the apex acute or acuminate; secondary veins in 6–13 pairs, the tertiary venation forming fine, dense, transverse system. Flowers 7–8 cm long; sepals subequal, not markedly imbricate, 3–5 mm long (5–6 mm when fresh), coriaceous; corolla glabrous except for ciliate margins (cilia visible only with hand lens), pale yellow, the lower part of tube ca. 20 × 3 mm, the upper part ca. 30–50 × 8 mm, the lobes 20–25 mm long. Follicles 2, ca. 22 × 1.5 cm, pubescent, subtended by persistent calyx. Fl (Feb, Oct, Nov); in undisturbed and disturbed non-flooded forest.

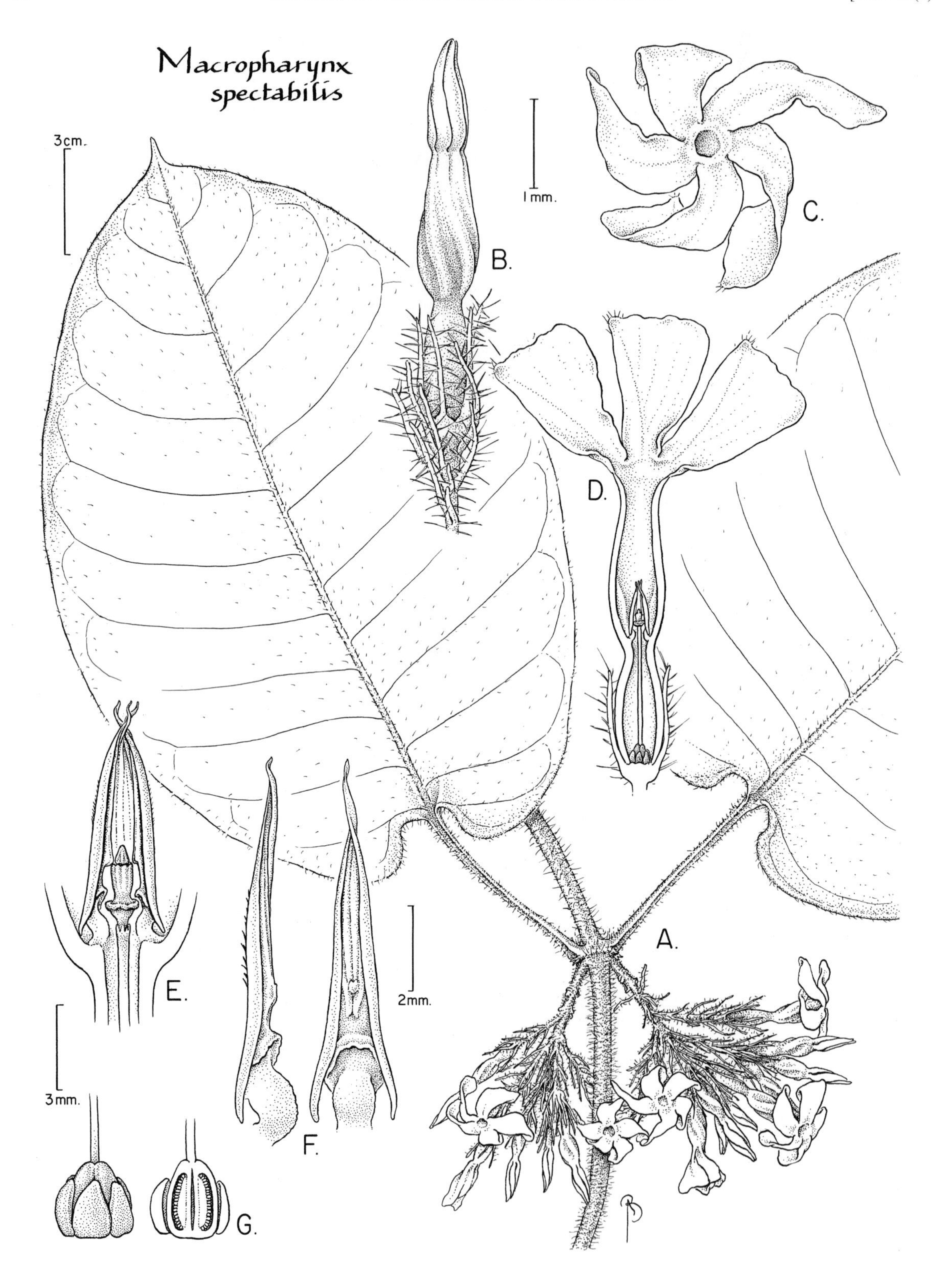

FIG. 24. APOCYNACEAE. *Macropharynx spectabilis* (*Phillippe et al. 27022*). **A.** Part of stem with leaves and inflorescences. **B.** Lateral view of flower bud. **C.** Apical view of corolla. **D.** Longitudinal section of flower. **E.** Longitudinal section of throat of corolla showing stigma surrounded by anthers. **F.** Lateral (left) and adaxial (right) views of anthers. **G.** Ovary surrounded by glands (far left) and medial section of ovary (near left).

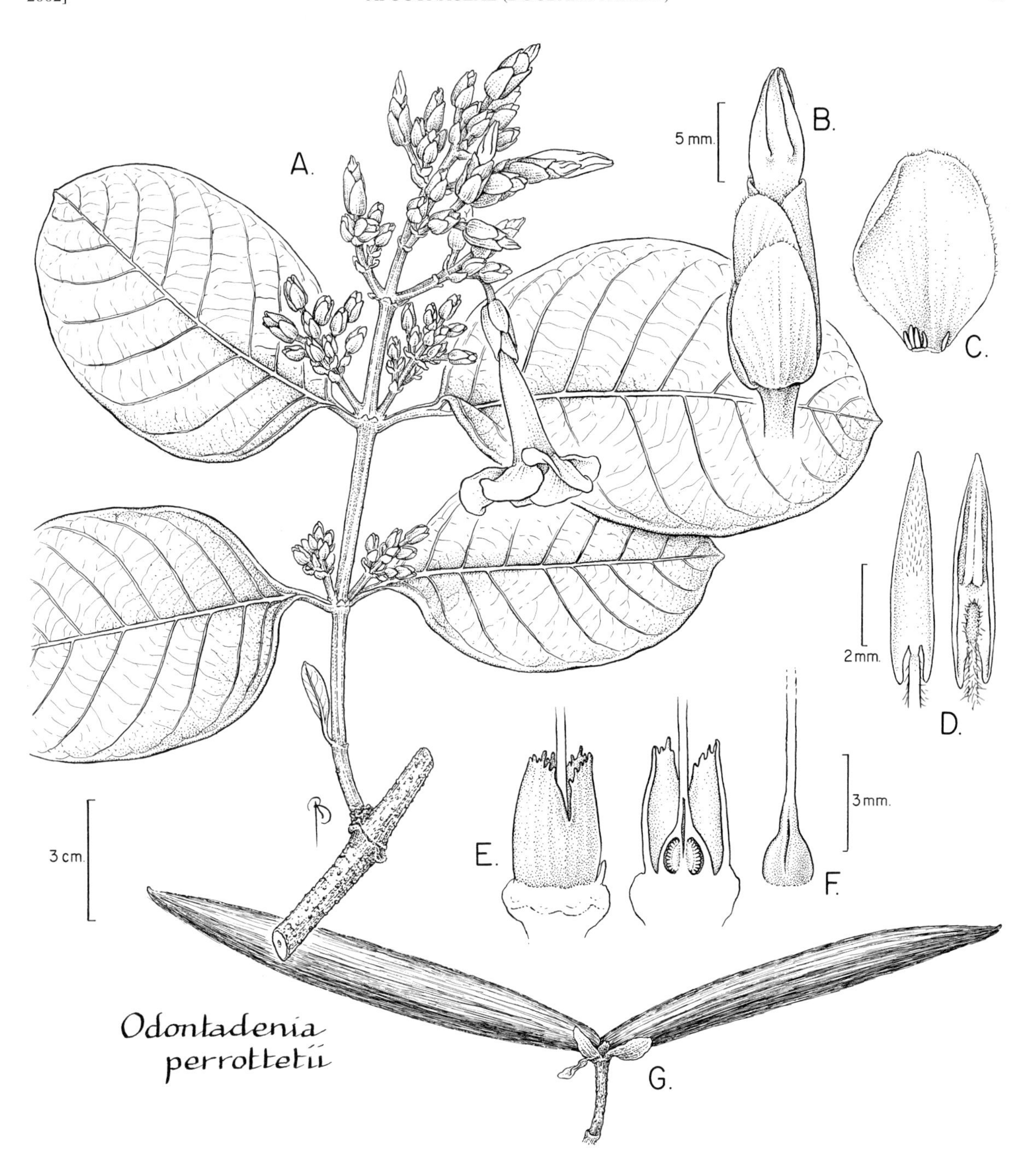

FIG. 25. APOCYNACEAE. *Odontadenia perrottetii* (A, B, *Mori et al. 21594*; C–F, *Mori & Boom 15362*; G, *Mori & Gracie 21161*). **A.** Part of stem with leaves and inflorescences. **B.** Detail of inflorescence with bud subtended by bracts. **C.** Adaxial view of inflorescence bract showing colleters. **D.** Abaxial (left) and adaxial (right) views of anthers. **E.** Lateral view (near right) and medial section (far right) of base of flower showing nectary disc surrounding ovary. **F.** Lateral view of base ovary and base of style. **G.** Fruit.

PACOURIA Aubl.

Lianas. Latex white. Leaves opposite. Inflorescences terminal or axillary, panicles, the flowers clustered, the axes robust, often terminating in tendril. Flowers: stamens 5, the anthers free from style head; ovary syncarpous, 1-locular; placentation parietal, the ovules 10–36 on 2 placentae. Fruits edible berries. Seeds: embryo cordate, the radicle small.

Pacouria guianensis Aubl.

Leaves: petioles ca. 1 cm long, often twisted, with numerous interpetiolar glands; blades widely elliptic to elliptic, 11–22 × 7–9.5 cm, glabrous adaxially, puberulous, especially on veins, abaxially, the base obtuse to rounded, the apex obtuse to acuminate; secondary veins in 9–14 pairs. Flowers probably nocturnal; sepals ca. 2 mm long, brown, pubescent, glabrous, without colleters; corolla salverform, green abaxially, white adaxially, the tube 1–1.3 cm long, the lobes linear-oblong, ca. 1 cm long; stamens inserted at base of corolla tube, the anthers linear, rounded at base; ovary globose, pubescent, ca. 1 mm long, the style 2× as long as ovary, the style head cylindric, the appendices as long as body. Fruits globose, ca. 8 × 8 cm, green to yellow, with whitish lenticels, with abundant latex, the pulp yellowish, sweet-tasting. Seeds ca. 12, ellipsoid, 2.5 cm long, glabrous. Fl (Aug); in disturbed, non-flooded forest. *Liane caoutchouc* (Créole).

ARALIACEAE (Ginseng family)

Scott A. Mori

Trees or shrubs, terrestrial or epiphytic. Ethereal oils, resins, or gums in parenchyma tissue of stem, petioles, and larger veins of leaves. Stipules present. Leaves simple or pinnately or palmately compound, alternate. Inflorescences often umbellate, but sometimes in heads, panicles, or racemes. Flowers usually small, usually bisexual (when unisexual, plants often dioecious); calyx with 4–5 minute teeth; petals 4–5, usually free, infrequently connate or even calyptrate; stamens usually as many as and alternate with petals; ovary inferior, (1)2–5(many)-locular, the styles separate distally, the basal portion of style often swollen to form stylopodium; nectary disc usually present, covering top of ovary; placentation axile. Fruits drupes or berries. Seeds with small embryo, the cotyledons 2, the endosperm abundant, oily.

Frodin, D. G. 1997. Araliaceae. *In* J. A. Steyermark, P. E. Berry & B. K. Holst (gen. eds.), Flora of the Venezuelan Guayana **3:** 1–31. P. E. Berry, B. K. Holst & K. Yatskievych (vol. eds.). Missouri Botanical Garden, St. Louis.

Lindeman, J. 1986. Araliaceae. *In* A. L. Stoffers & J. C. Lindeman (eds.), Flora of Suriname **III(1–2):** 351–353. E. J. Brill, Leiden.

1. Epiphytic shrubs. Leaves simple. Inflorescences with ultimate units cylindric-capitate, the individual flowers connate. . . . . *Oreopanax.*
1. Trees. Leaves palmately compound. Inflorescences with ultimate units umbellate, the individual flowers free from one another. . . . . *Schefflera.*

OREOPANAX Decne. & Planch.

Trees or shrubs, terrestrial or epiphytic. Leaves simple; petioles of varying lengths. Inflorescence terminal, the ultimate units capitate, the flowers connate. Flowers bisexual or functionally unisexual; calyx of 5 inconspicuous teeth; petals 5, valvate; stamens 5, the anthers dorsifixed; summit of ovary recessed, the styles 2–10, separate above, connate and expanded below to form stylopodium, the functionally staminate flowers with fewer styles than pistillate flowers. Fruits berries.

Oreopanax capitatus (Jacq.) Decne. & Planch. [Syn.: *Aralia capitata* Jacq.] FIG. 26

Epiphytic shrubs. Leaves: petioles varying in length, 2–10 cm long, with stellate hairs at base; blades elliptic, variable in size, the smallest 5–9 × 3–5 cm, the largest 10–17 × 5–9 cm. Inflorescences branched three times, the ultimate units cylindric-capitate, all axes with stellate hairs; peduncle 4–10 mm long. Fruits globose, 4.5 mm diam., green. Seeds 4–5 per fruit, elongate, 3 × 2 mm, the integument finely striate. Fl (Jan, Jun, Jul, Aug), fr (Jan, Feb, Aug, Sep, Oct, Dec); common, high in trees of primary forest.

SCHEFFLERA J. R. Forst. & G. Forst.

Freestanding trees in our area, may be epiphytic in other areas. Leaves palmately compound. Inflorescences with ultimate units umbellate, the flowers entirely separate from one another. Flowers bisexual or unisexual; petals 5; stamens 5, the styles 2–5. Fruits distinctly ribbed or laterally compressed.

1. Leaf blades coriaceous, glabrous on both surfaces, concolorous; longest petioles <40 cm long. Fruits with 5 distinct ribs, not flattened. . . . . *S. decaphylla.*
1. Leaf blades membranous, pubescent on abaxial surface, often discolorous; longest petioles >50 cm long. Fruits without 5 distinct ribs (infrequently with 3 distinct ribs), usually flattened. . . . . *S. morototoni.*

Schefflera decaphylla (Seem.) Harms [Syn.: *S. paraensis* Ducke]

Large trees, to 32 m tall. Trunks with swollen base. Leaves with 11 leaflets; petioles ca. 38 cm long; petiolules variable in length, 2.5–9.5 cm long; leaflet blades elliptic to oblong, 8.5–14 × 3–6 cm, coriaceous, glabrous, concolorous, the base obtuse, the apex obtuse, the margins entire; secondary veins in ca. 8–9 pairs. Fruits very strongly 5-ribbed. Fr (Jun); apparently rare, in non-flooded forest.

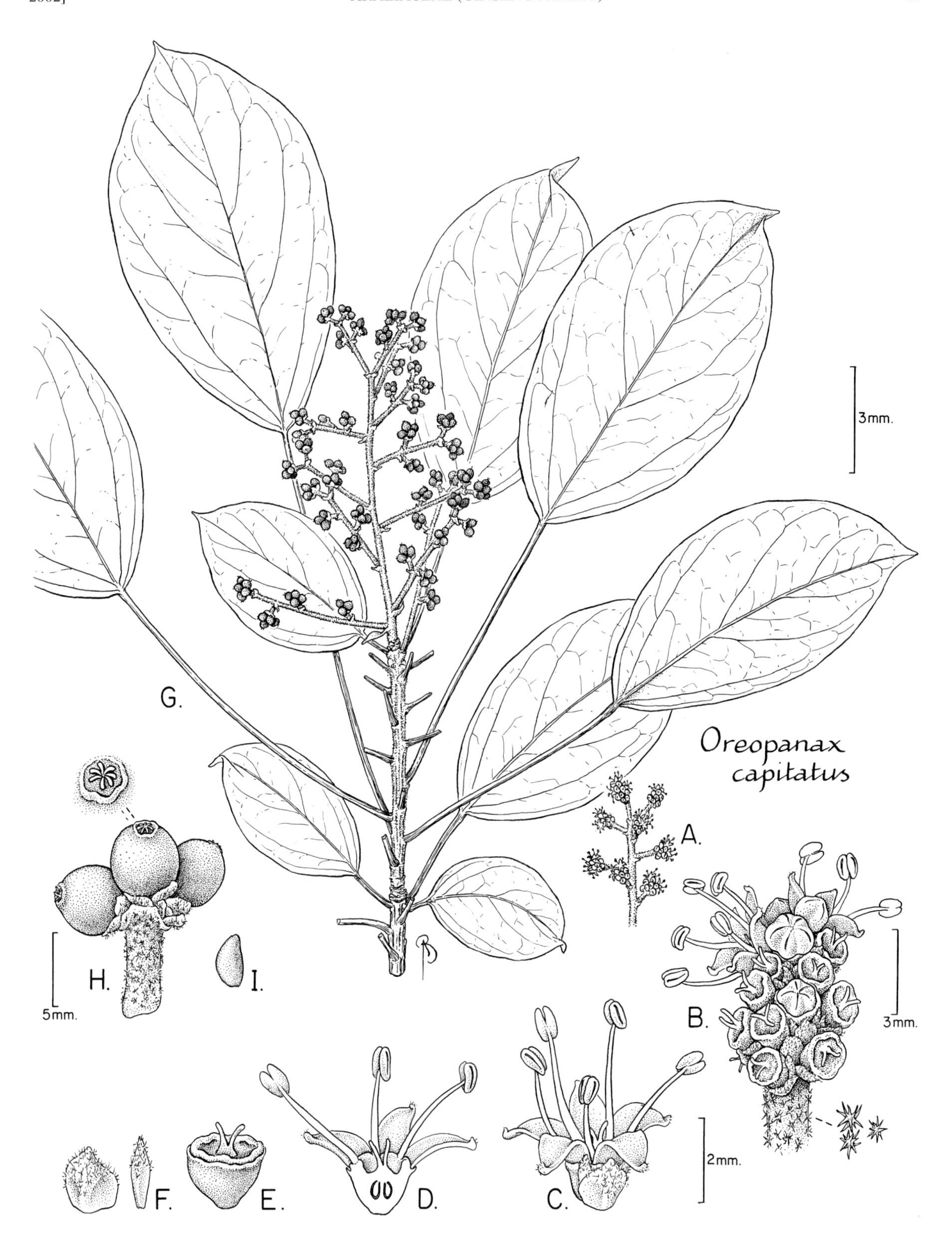

FIG. 26. ARALIACEAE. *Oreopanax capitatus* (A–F, *Mori et al. 23379*; G, *Mori et al. 20746*; H, I, *Granville 2784*). **A.** Part of inflorescence. **B.** Flowers on ultimate stem of inflorescence. **C.** Lateral view of functionally staminate flower. **D.** Medial section of functionally staminate flower. **E.** Ovary of functionally staminate flower with two styles. **F.** Bracteole (near left) and bract (far left). **G.** Stem with leaves and inflorescence. **H.** Part of infructescence with detail of remnant stigma on apex of fruit; note the 7-parted stigma. **I.** Seed.

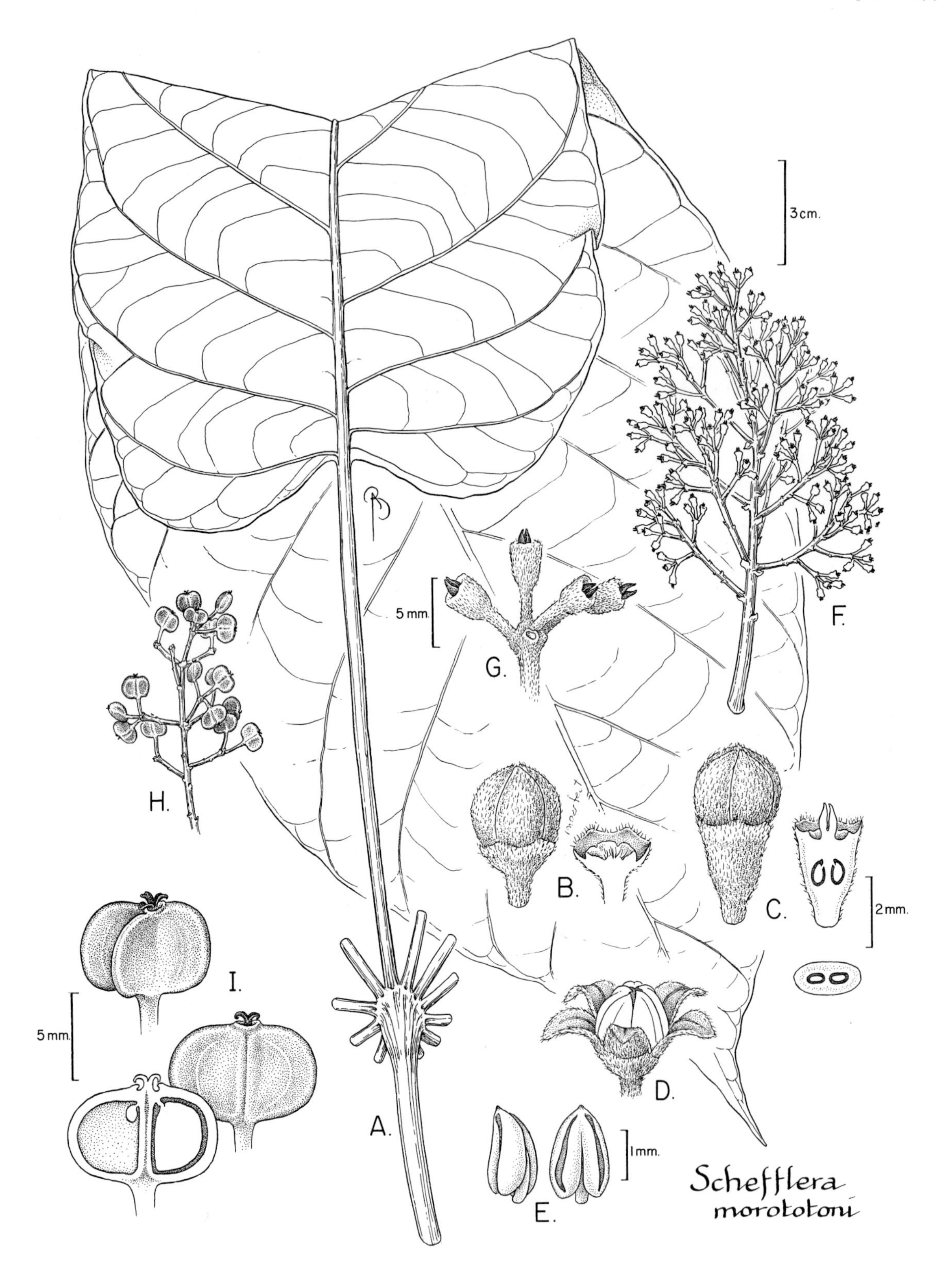

FIG. 27. ARALIACEAE. *Schefflera morototoni* (A, F, G, *Acevedo-Rodríguez 2887* from St. John, U.S. Virgin Islands; B–E, *Br. Hiosam s.n.* from Puerto Rico; H, I, *Mejia 31358* from Dominican Republic). **A.** Leaf with all but one leaflet removed. **B.** Lateral view (left) and medial section with petals removed (right) of imperfect flower bud. **C.** Lateral view (left), medial section with petals removed (right), and transverse section of ovary (below right) of perfect flower bud. **D.** Oblique view of flower at anthesis. **E.** Lateral (left) and adaxial (right) views of anthers. **F.** Infructescence with immature fruit. **G.** Detail of inflorescence with immature fruit. **H.** Detail of infructescence. **I.** Two views (left and below) of fruits and medial section of fruit (below left). (Reprinted with permission from P. Acevedo-Rodríguez, *Flora of St. John, U.S. Virgin Islands*, Mem. New York Bot. Gard. 78, 1996.)

Schefflera morototoni (Aubl.) Maguire, Steyerm. & Frodin [Syn.: *Didymopanax morototoni* (Aubl.) Decne. & Planch.] FIG. 27, PL. 17b

Trees, to 35 m tall. Leaves with 9–13 leaflets; petioles 80–110 cm long; petiolules variable in length, 5.5–22 cm long; leaflet blades oblong, 29–35 × 11.5–13 cm, membranous, pubescent abaxially, discolorous, the base rounded to truncate, the apex acuminate, the margins entire; secondary veins in 10–15 pairs. Fruits usually laterally compressed, infrequently with 3 ribs. Phenology not determined; common, in disturbed habitats. *Bois la Saint-Jean* (Créole), *morototo* (Waypi), *tobitoutou* (Bush Negro).

ARISTOLOCHIACEAE (Birthwort Family)

Christian Feuillet and Odile Poncy

Lianas or vines, herbs or shrubs (outside our area). Stipules usually absent. Leaves alternate, simple, mostly entire, sometimes trilobed. Inflorescences terminal or lateral, racemes, cymes, or of solitary flowers. Flowers bisexual, usually zygomorphic, sometimes actinomorphic, mostly epigynous, less frequently hemi-epigynous or perigynous, often smelling like rotting meat; calyx synsepalous, 3-lobed, 1-lobed, or without lobes, often large and corolla-like; petals wanting or much reduced; stamens 4 to many, free or joined to style to form gynostemium; ovary 4–6-locular; placentation axile or parietal. Fruits capsular, less frequently follicular or indehiscent.

Barringer, K. A. & F. A. González G. 1997. Aristolochiaceae. *In* J. A. Steyermark, P. E. Berry & B. K. Holst (gen. eds.), Flora of the Venezuelan Guayana **3:** 122–129. P. E. Berry, B. K. Holst & K. Yatskievych (vol. eds.). Missouri Botanical Garden, St. Louis.

Feuillet, C. & O. Poncy. 1998. Aristolochiaceae. *In* A. R. A. Görts-van Rijn & M. J. Jansen-Jacobs (eds.), Flora of the Guianas Ser. A, **20:** 1–31. Royal Botanic Gardens, Kew.

ARISTOLOCHIA L.

Lianas or vines. Stems often with thick, fissured bark. Pseudostipules present or absent. Leaves: blades entire (3-lobed in 1 species in our area), the base cordate; venation arching upward from base (campylodromous). Inflorescences axillary or from woody stems, racemes, cymes, or solitary. Flowers zygomorphic; perianth of 3 fused sepals, variable in shape, size and color, with an inflated basal utricle, a narrow trumpet-shaped central tube, and a usually 1–3-lobed distal limb; ovary 6-locular, long and narrow, barely thicker than pedicel, the style enclosed within utricle, fused with androecium and forming gynostemium with 6 anthers and 6 stigmatic lobes. Fruits pendent, 6-locular capsules, each locule with a dorsal keel and beaked apex, the dehiscence septicidal, beginning from base and incomplete at apex. Seeds winged.

1. Pseudostipules present. Leaves trilobed. Superior lobe of perianth limb ending in filament >10 cm long. Cultivated. . . . . *A. trilobata.*
1. Pseudostipules absent. Leaves entire. Lobes of perianth limb emarginate, round, or acute. Native.
    2. Perianth limb divided into a superior and an inferior lobe.
        3. Flowers axillary. Utricle 3–4 cm diam. Perianth limb short fimbriate at margin, the superior lobe cleft more than halfway into two lobules, each 1 cm wide, the inferior lobe 1.5–2 cm long, flat.. . . . . . . *A. didyma.*
        3. Flowers in racemes. Utricle 0.3–0.5 cm diam. Perianth limb not fimbriate at margin, the superior lobe entire, 0.2–0.4 cm wide, the inferior lobe 0.5–0.8 cm long, trough-shaped. . . . . *A. cremersii.*
    2. Perianth limb surrounding throat with an inferior sinus, or consisting of two lateral lobes.
        4. Perianth limb pale to dark whitish pink, with a narrow sinus in lower part, but the two sides in contact with each other. . . . . *A. stahelii.*
        4. Perianth limb brown and yellow, with a broad sinus in the lower part, or bilobed.
            5. Perianth limb horse-shoe-shaped, with a broad sinus in the lower part, the extremities pointing downward or slightly divergent, brown with golden speckles near margin, color fading toward middle, the center and throat yellow. . . . . *A. bukuti.*
            5. Perianth limb divided into two lateral lobes pointing horizontally to obliquely downward, bright yellow with brown markings, center yellow, throat pink. . . . . *A. iquitensis.*

Aristolochia bukuti Poncy FIG. 28, PLS. 17c, 18a,c; PART 1: PL. XIVc

Lianas. Pseudostipules absent. Leaves: blades widely triangular, 11–16 × 15–20 cm, grayish, pubescent abaxially, the base slightly cordate, the apex acute to slightly acuminate. Inflorescences cauliflorous, short, ± 5-flowered racemes in dense clusters. Flowers: perianth whitish, with protruding violet-brown veins abaxially, the utricle ovoid, 2–3 × 1.5 cm, the tube 2 × 0.5 cm, the limb spreading, with one large, suborbicular, 5–6 × 4–5 cm lobe surrounding the throat, the apex acute, lower part with a broad sinus, the extremities pointing downward or slightly divergent, not fimbriate at margin, adaxially brown with golden speckles near margin, the color fading toward middle, the center and throat yellow. Fruits ca. 6 cm long. Fl (Aug, Sep, Nov), fr (Feb, May, Aug); occasional, in non-flooded forest.

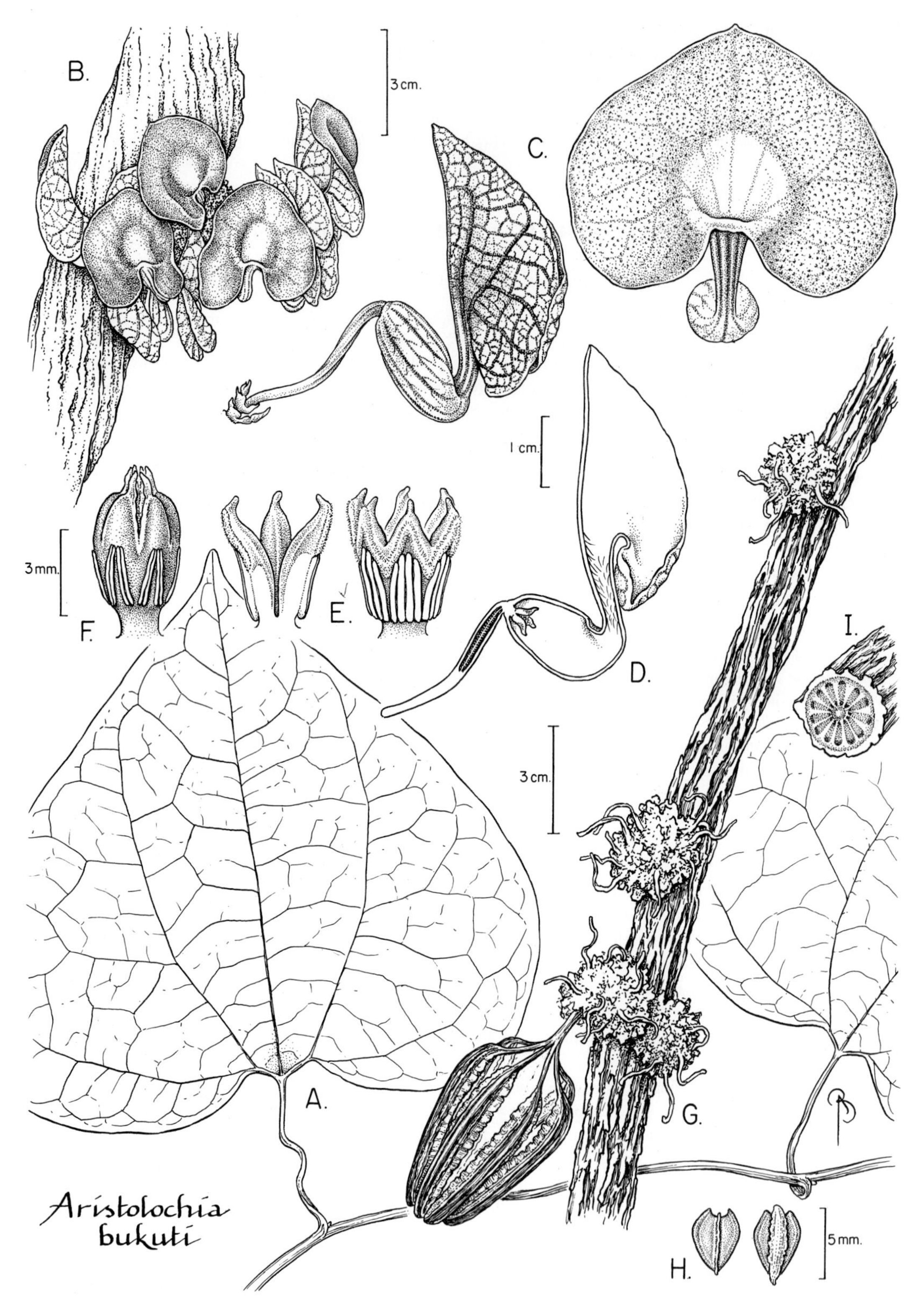

FIG. 28. ARISTOLOCHIACEAE. *Aristolochia bukuti* (A, *Mori et al. 23966*; B, *Mori et al. 23347*; C–F, I, *Mori et al. 21637*; G, H, *Mori et al. 22251*, this collection is of fruit only and, although all of its features agree with those of other collections of *A. bukuti*, it is impossible to be absolutely certain that this voucher actually represents this species. Field observations of the stem and the infructescence support this determination). **A.** Leaves on part of stem. **B.** Cluster of cauliflorous flowers. **C.** Lateral view (left) and frontal view (right) of flower. **D.** Medial section of flower. **E.** Medial section (left) and young gynostemium in pistillate phase (right). **F.** Older gynostemium in staminate phase. **G.** Corky stem of liana with a dehisced fruit. **H.** Seeds, the one on far right showing raphe. **I.** Transverse section of liana.

Aristolochia cremersii Poncy

Lianas. Pseudostipules absent. Leaves: blades widely triangular, 9–14 × 7–14 cm, grayish and slightly tomentose abaxially, the base obtuse to slightly cordate, the apex acute to acuminate. Inflorescences cauliflorous or in leaf axils, 10–15-flowered racemes, 5–45 cm long. Flowers: utricle globose, 3–5 mm diam., the tube 2–2.5 cm long, yellowish, with brown veins, the limb erect, not fimbriate at margin, bilobed, with superior lobe brown, linear, 3–3.5 × 0.2–0.4 cm, entire, acute at apex, the inferior lobe 0.5–0.8 cm long, in shape of a trough, emarginate. Fruits 3–5 cm long. Fl (Sep), fr (Feb, Sep); occasional, in non-flooded forest.

Aristolochia didyma S. Moore

Lianas. Pseudostipules absent. Leaves: blades triangular to widely cordiform, 8–25 × 6–17 cm, glabrous, glaucous abaxially, the base deeply cordate, with acute sinus, the apex triangular, acute to acuminate, the margins undulate. Inflorescences solitary in leaf axils. Flowers: perianth yellow-green outside, the utricle 3–4 × 5–6 cm, ovoid, black spotted, the tube cylindrical, ca. 3 × 1–1.5 cm, the limb erect or oblique, bilobed, the superior lobe 4 cm long, cleft more than halfway, the margin usually violet, purple veined, short fimbriate, each lobule ca. 2.5 × 1 cm, the apex rounded, the inferior lobe 1.5–2 cm long, flat, round, or emarginate at apex, the throat yellow. Fruits ca. 6 cm long, chartaceous. Fl (Jan); in non-flooded forest.

Aristolochia iquitensis O. C. Schmidt

Lianas. Pseudostipules absent. Leaves: blades elongate elliptic with strongly arched to parallel margins, 8–17(20) × 5–7 cm, pubescent abaxially, the base deeply cordate to auriculate, the apex usually acuminate (5–10 mm long). Inflorescences cauliflorous, 1–2 racemes together, 2–6 cm long, bearing ≥10 flowers. Flowers: perianth purple-brown abaxially, the utricle ovoid, 7 × 4.5 mm, the tube funnel-shaped, short, usually <1 cm long, the limb spreading, bilobed, with lateral lobes pointing horizontally to obliquely downward, the apex round, bright yellow with brown markings, the center yellow, the lower margin fleshy-fimbriate, the throat pink. Fruits 6–8 cm long. Fl (Apr), fr (Apr); occasional, in non-flooded forest.

Aristolochia stahelii O. C. Schmidt

Lianas. Pseudostipules absent. Leaves: blades widely cordiform, 7–18 × 10–22 cm, grayish pubescent abaxially, the base slightly cordate, the apex obtuse or rounded to acuminate. Inflorescences cauliflorous, short, 4–5-flowered racemes, to 1 cm long. Flowers: perianth whitish pink, with darker veins abaxially, the utricle ovoid, 3–3.5 cm long, the tube reduced and hidden by limb, the limb surrounding throat, with a narrow sinus in lower part, the two sides touching each other, broadly round, reflexed to utricle, then erect to level of mouth, with red-purple dots nearly masking whitish background color, darker in middle adaxially, the margin of limb recurved, not fimbriate. Fruits 8–10 cm long. Fl (Feb, Nov), fr (Nov); occasional, in non-flooded forest.

Aristolochia trilobata L. Pl. 18b

Lianas. Pseudostipules present. Leaves: blades trilobed, the lobes variable in shape and size, the middle one often diamond-shaped, the lateral ones often falciform, always pointing forward, 5–9 × 6–10 cm, pubescent abaxially, the base cordate, the apex rounded. Inflorescences solitary in leaf axils. Flowers: perianth greenish-yellow, glabrous, dark red veined abaxially, the utricle ovoid, 4–5 × 4 cm, the tube cylindrical, 4–5 × 1.5 cm, the limb not fimbriate, erect, bilobed, the superior lobe ovate, with an apical filament at least 10 cm long, whitish adaxially, the lower lobe short, round. Fruits 6 cm long. Fl (May), fr (May); cultivated as an ornamental in village. *Feuille trèfle, liane trèfle, trèfle* (Créole).

ASCLEPIADACEAE (Milkweed Family)

Gilberto Morillo

Herbs, subshrubs, vines, or lianas. White latex usually present. Stipules absent or, if present, small and inconspicuous, often with colleters along interpetiolar line or in leaf axils (these sometimes mistaken for stipules). Leaves simple, opposite or whorled, sometimes caducous; blades often with two or more glands on adaxial surface toward base; venation brochidodromous. Inflorescences usually subaxillary, sometimes axillary or terminal, usually cymose, the cymes umbelliform or racemiform, but sometimes thyrses, pleiothyrses, or pseudopanicles, or reduced to fascicles or a single flower. Flowers bisexual, usually actinomorphic; calyx of 5 distinct or basally fused lobes, the lobes imbricate or open in bud, generally with 1 or more adaxial glands at base; corolla sympetalous, 5-lobed, the lobes usually contorted, sometimes valvate in bud; stamens 5, inserted at or near base of corolla tube, highly modified, the filaments flat and usually connate to form a tube, the anthers widened, the entire structure united to stigma head to form gynostegium, the anthers 2-celled, sometimes with abaxial laminate or inflated appendages; pollen agglutinated in pollinia, the pollinia connected in pairs by translators and a central hardened structure called the corpusculum; corona usually present, sometimes very complex in structure; ovary superior, 2-carpelar, the styles distinct for most of length but united at apex by common stigma head; placentation marginal or submarginal, the ovules few to many, anatropous, pendulous, imbricate. Fruits of 1–2 follicles, the follicles usually ovoid to fusiform, sometimes inflated and balloon-like (ventricose) or linear-cylindric, smooth or with projections, ridges, or wings. Seeds few to >1000, usually flattened, often with an apical coma of white, silky hairs; embryo flattened, the endosperm hard and oily.

Jonker, F. P. 1940. Asclepidaceae. *In* A. Pulle (ed.), Flora of Suriname **IV(2):** 326–352. J. H. De Bussy, Amsterdam.
Morillo, G. N. 1997. Asclepiadaceae. *In* J. A. Steyermark, P. E. Berry & B. K. Holst (gen. eds.), Flora of the Venezuelan Guayana **3:** 129–177. P. E. Berry, B. K. Holst & K. Yatskievych (vol. eds.). Missouri Botanical Garden, St. Louis.

1. Herbs or subshrubs. Leaf blades lanceolate.
    2. Leaf blades without glands at base. Flowers red and yellow; corona distinct, with an adaxial ligule; pollinia laterally flattened, not concave on one side, pendulous. Fruits usually erect. . . . . *Asclepias.*

2. Leaf blades with 2–5 finger-like glands at base. Flowers usually green; corona scarcely differentiated from column of gynostegium, without an adaxial ligule; pollinia not laterally flattened, concave on one side, usually horizontally oriented. Fruits usually pendent. . . . . *Matelea.*

1. Lianas. Leaf blades usually broadly cordate, broadly elliptic, or ovate-elliptic.
   3. Leaf blades glabrous, with 9–10 glands at base. Flowers usually <10 mm diam.; stigma head ± stellate in shape. . . . . *Metalepis.*
   3. Leaf blades usually with at least some scattered hairs, with 2–5 glands at base. Flowers usually >10 mm diam.; stigma head ± pentagonal in shape. . . . . *Matelea.*

ASCLEPIAS L.

Herbs, subshrubs, or shrubs, usually erect. Leaves usually opposite; blades with or without glands. Inflorescences terminal, subaxillary, or axillary, umbelliform cymes. Flowers small to medium-sized; calyx deeply lobed, the lobes with one gland in each sinus; corolla with tube poorly developed, the lobes reflexed or spreading, valvate in bud; gynostegium usually stipitate; stigma head subpentagonal, usually concave or flat at top; stamens with anthers erect; pollinia pendulous, laterally flattened, uniformly fertile up to point of attachment to translators; translator arms curved, lined by sterile, translucent ribbon, the translators sagittate or oblong; corona usually adnate to base of anthers, of 5 fleshy, cucullate or cylindric-cucullate segments, very often with adaxial ligule. Fruits in pairs, erect, subcylindric or fusiform to broadly ovoid or spherical, usually smooth. Seeds many per fruit, with apical coma.

Asclepias curassavica L.
FIG. 29, PLS. 19, 20a; PART 1: FIG. 5

Erect herbs, to 2 m tall. Leaves: petioles 1–2 cm long, puberulent; blades lanceolate, 5–16 × 1–4 cm, the base narrowly obtuse to cuneate, without glands, the apex acuminate; secondary veins in 12–15 pairs. Inflorescences with 8–9 flowers; peduncle 4–5.5 cm long; pedicel 1.2–1.8 cm long. Flowers red and yellow; calyx lobes oblong-ovate, 2.5–3 × 0.7–1 mm, narrowly obtuse; corolla 8–10 mm diam., the tube 0.5 mm long, glabrous, the lobes reflexed, 4.5–7 × 2–3 mm, red or orange-red; gynostegium 4–5 mm long, the stigma head 1.5–2 mm diam.; pollinia not concave on one side; corona distinct, with adaxial ligule, the segments cucullate, concave, yellow. Fruits narrowly fusiform, 6–10 × 0.6–1.5 cm. Seeds narrowly ovate, 4.8–6 × 3.8–4.1 mm. Fl (Jun, Oct, probably year round); common, in weedy habitats. *Cadrio*, *Codio*, *crodio* (Créole), *oficial de sala*, *suspiro* (Portuguese).

MATELEA Aubl.

Lianas, less frequently erect or prostrate herbs, often pubescent, the pubescence usually consisting of two types of hairs, long multiseptate and short glandular. Leaves opposite; blades usually with 2–5, adaxial, minute, finger-like glands at base. Inflorescences subaxillary, only one per node, helicoid or racemiform cymes, sometimes reduced to a single flower. Flowers small to large, usually green; calyx deeply lobed, the lobes usually with 1 or 2 adaxial glands at base, these rarely absent; corolla usually rotate or campanulate, rarely urceolate or tubular, the lobes often with conspicuous venation of contrasting colors, usually valvate in bud; gynostegium stipitate or sessile; anthers usually horizontal or nearly horizontal, rarely subvertical, without dorsal appendages; pollinia usually horizontally oriented, sometimes pendulous, conspicuously concave on one side, with at least a sterile hyaline zone near attachment to caudicules; corona usually complex, usually surrounding base of gynostegium and often partially adnate to base of gynostegium or to corolla tube. Fruits usually solitary, ovate to narrowly truncate, elliptic, frequently 5-winged, ridged or tuberculate, sometimes smooth. Seeds usually many per fruit, ovate to oblong-ovate, with or without coma.

1. Lianas. Stems with golden-brown pubescence of long septate and short glandular hairs. Leaf blades broadly elliptic to obovate-elliptic, both surfaces pubescent, the base narrowly cordate. Corolla lobes deltoid, 6–6.5 × 6.5–6.7 mm. . . . . *M. gracieae.*
1. Suffrutescent herbs or shrubs. Stems short puberulent, the pubescence not brownish or of two lengths. Leaf blades lanceolate, glabrous except for short hairs along veins, the base obtuse to cuneate. Corolla lobes narrowly ovate or ovate-oblong, 5–5.3 × 1.7–2 mm. . . . . *M. palustris.*

Matelea gracieae Morillo FIG. 30, PL. 20a,b

Lianas. Stems covered with conspicuous golden-brown pubescence of two types, long septate and short glandular. Leaves: petioles 4.5–5 cm long, with pubescence similar to stems; blades broadly elliptic or obovate-elliptic, 20–23.2 × 11.4–12.5 cm, both surfaces with long, septate, erect to appressed hairs, the midrib on abaxial surface with mixture of long and short hairs, the base narrowly cordate, without glands, the apex obtuse to abruptly acuminate; secondary veins in 8–10 pairs, the tertiary venation conspicuous on abaxial surface. Inflorescences subaxillary, 3–4-flowered helicoid cymes, with only one flower in anthesis at a time; peduncle ca. 4.5 cm long, pubescent, the rachis 1.8–2 cm long, pubescent, with conspicuous scars; pedicels 1–1.2 cm long. Flowers usually green and yellow; calyx lobes narrowly elliptic or narrowly ovate-elliptic, 4.7–5 × 2.2–2.4 mm, pubescent abaxially, glabrous adaxially, without glands, the apex narrowly obtuse to subacute; corolla rotate-campanulate, 17–18 mm diam., the lobes deltoid, 6–6.5 × 6.5–7 mm, glabrous except for very short hairs at base adaxially, with short erect hairs covering part of surface abaxially, green with conspicuous reticulate

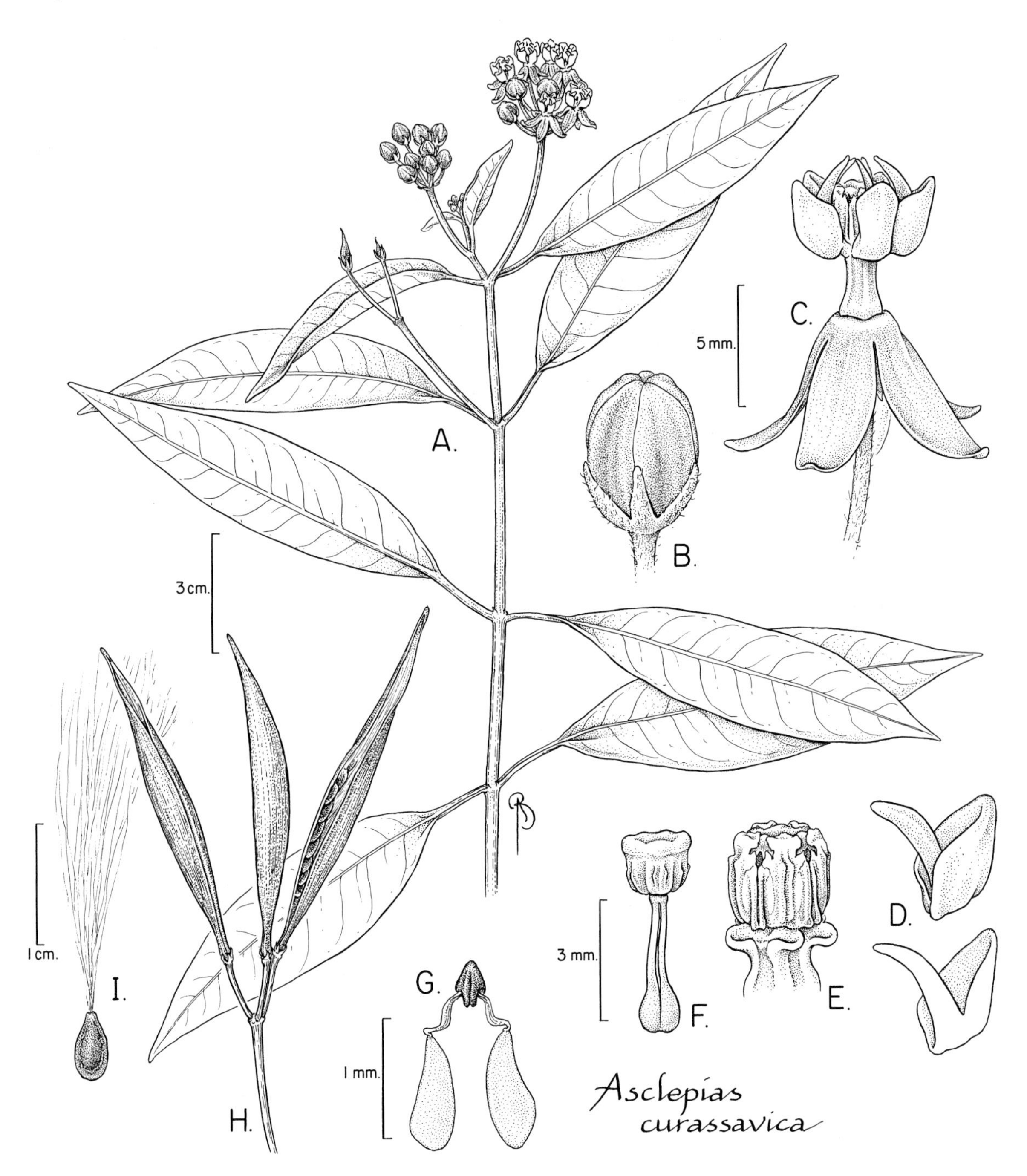

FIG. 29. ASCLEPIADACEAE. *Asclepias curassavica* (A–G, *Mori & Gracie 18446*; H, I, *Killip 11184* from Colombia). **A.** Part of plant with leaves and inflorescences. **B.** Flower bud. **C.** Lateral view of flower. **D.** Intact hood and horn (above) and medial section of hood and horn (below). **E.** Gynostegium. **F.** Pistil with capitate stigma. **G.** Pollinia. **H.** Part of infructescence. **I.** Seed with coma.

venation, the apex obtuse to rounded and slightly emarginate; gynostegium stipitate, 2.4–2.5 mm long, the stigma head white, ± pentagonal, 3–3.1 mm diam.; pollinia subhorizontally oriented, 0.66 × 0.5–0.55 mm, with an upper sterile margin, the translators 0.17 mm long, the corpusculum broadly sagittate, 0.27 × 0.22–0.23 mm; corona fleshy, yellow, deeply 5-lobed, surrounding base of gynostegium, the lobes irregularly trapezoidal in outline, 1.2–1.4 × 2.5–2.7 mm, rugose-caruncululate at base. Fruits ca. 19 × 8 cm. Fl (Aug, Sep), fr (Sep); known from two collections, one from along a stream and the other from non-flooded forest.

Matelea palustris Aubl.

Suffrutescent herbs or shrubs. Stems shortly and densely puberulent. Leaves: petioles 1–1.6 cm long, densely puberulent; blades lanceolate, 12–21 × 3.5–5.1 cm, glabrous except for short

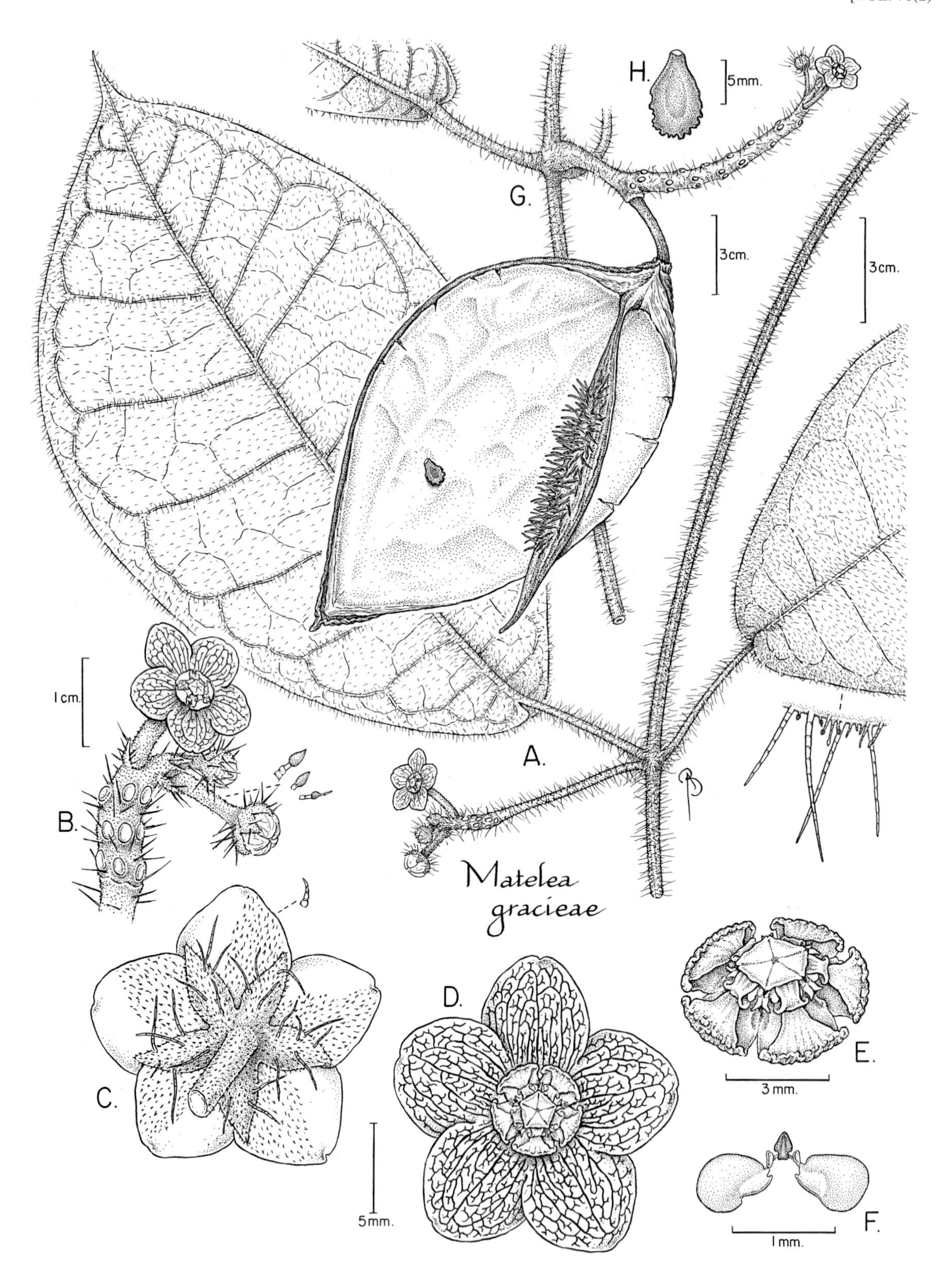

FIG. 30. ASCLEPIADACEAE. *Matelea gracieae* (A–F, *Mori et al. 23142*; G, H, *Phillippe et al. 26496*). **A.** Stem with leaves and inflorescence; note that the apex of the stem is downward. **B.** Part of inflorescence with details of hairs. **C.** Basal view of flower with detail of hair. **D.** Apical view of flower. **E.** Gynostegium. **F.** Adaxial view of pollinia and corpusculum. **G.** Dehisced follicle with single seed. **H.** Seed. (Reprinted with permission from G. Morillo, Brittonia 50(3), 1998.)

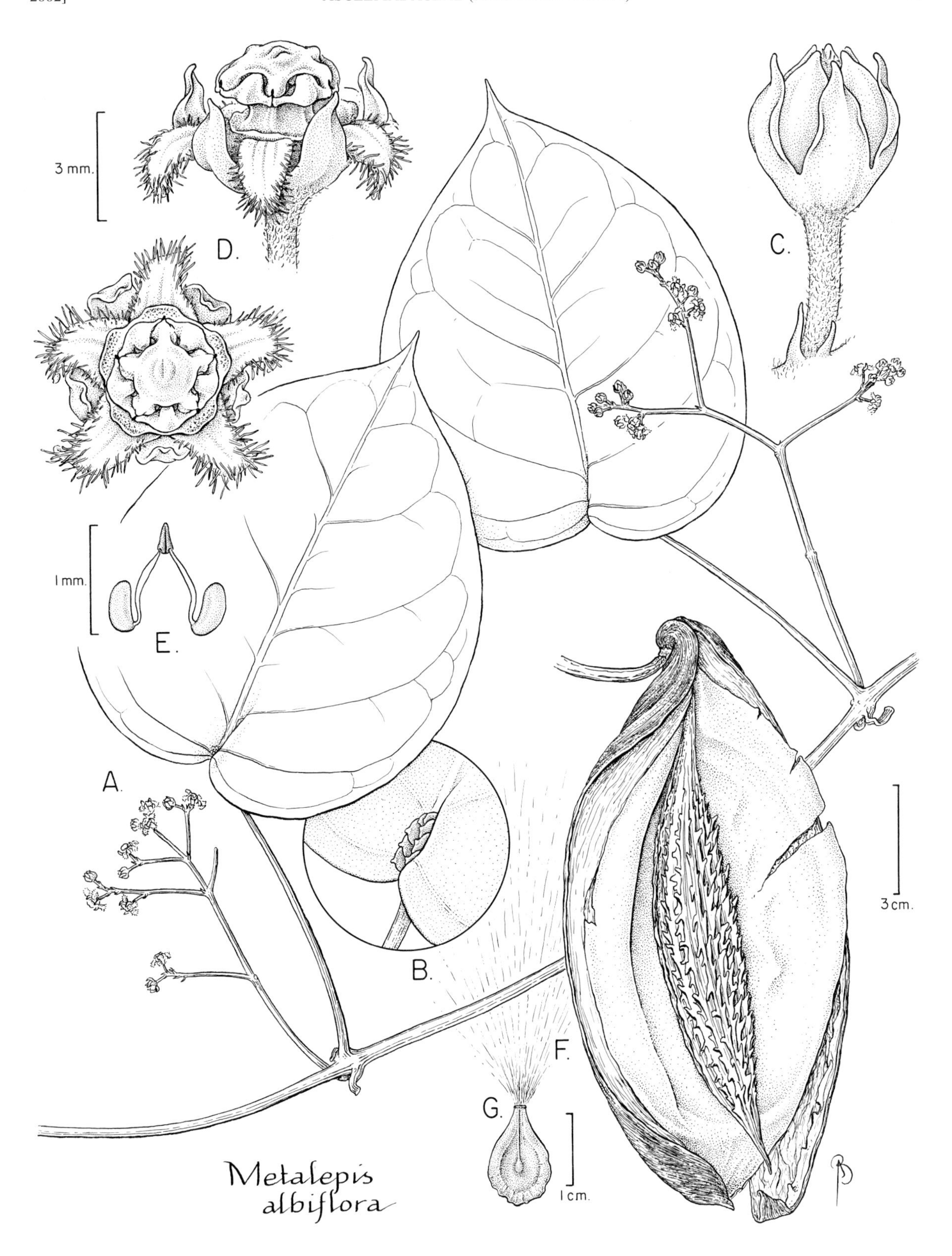

FIG. 31. ASCLEPIADACEAE. *Metalepis albiflora* (A–E, *Mori et al. 23941*; F, *Gentry et al. 11135* from Venezuela; G, *Dumont et al. VE–7710* from Venezuela). **A.** Stem with leaves and inflorescences. **B.** Enlargement of adaxial surface of leaf base showing gland-like structures. **C.** Lateral view of flower bud. **D.** Lateral (above) and apical (below) views of flower. **E.** Pair of pollinia with translators and gland. **F.** Dehisced follicle with seeds already dispersed. **G.** Seed with coma.

hairs along veins, the base obtuse to cuneate, with 2–5 finger-like glands, the apex long acuminate; secondary veins in 6–10 pairs, the tertiary veins usually conspicuous. Inflorescences subaxillary, helicoid cymes, with 6–9 flowers; peduncle 1.5–2 cm long, puberulent, the rachis 1.2–2.6 cm long, with conspicuous scars; pedicels 1.5–2.3 cm long, densely puberulent. Flowers usually green; calyx lobes narrowly ovate or oblong-ovate, usually somewhat reflexed, 1.6–2.1 × 0.7–0.8 mm, abaxial surface with minute appressed hairs, with 1 gland in axil, the apex obtuse; corolla rotate, 12–14 mm diam., the lobes narrowly ovate or ovate-oblong, 5–5.3 × 1.7–2 mm, conspicuously veined, usually glabrous, green, the apex obtuse; gynostegium yellow, stipitate, 1.2–1.3 mm long, the stigma head roundly pentagonal, 1.7–1.8 mm diam.; pollinia horizontally oriented, 0.32–0.33 × 0.25–0.26 mm, the translators ± 0.14 mm long, the corpusculum narrowly sagittate, 0.14 × 0.8 mm; corona scarcely differentiated from column of gynostegium, marginally toothed at base. Fruits pendent, narrowly fusiform, 7–8 × 1–1.2 cm, glabrous. Seeds many, ± ovate, 8–8.5 × 5–5.5 mm, comose. Fl (May, Jun, Aug, Sep, Oct, Nov), fr (Jun); usually in moist areas along streams.

METALEPIS Griseb.

Lianas or vines. Leaves opposite; blades with 3–14 glands. Inflorescences axillary or subaxillary, usually with two, sometimes one, per node, racemose cymes or pseudopanicles. Flowers small; calyx deeply lobed, the lobes with 1 adaxial gland at base; corolla shortly campanulate to subrotate, the lobes inflexed to reflexed, dextro-contorted, usually green, the margins usually revolute; gynostegium shortly stipitate or sessile, the stigma head pentagonal-stellate, shortly mammillate or conical; stamens with anthers subhorizontally oriented to slightly erect; pollinia pendulous, uniformly rounded and fertile to point of attachment of translators, the translators usually very long and narrow, ± curved and without membranes; corpusculum usually narrowly ovate; corona subrotate to openly cyathiform, 5-segmented, the segments alternate with anthers, recurved to erect, subtruncate or somewhat toothed, the abaxial side usually papillose. Fruits single, narrowly ovoid to narrowly ellipsoid-ventricose, glabrous, smooth. Seeds many per fruit, ovate, flat, comose.

Metalepis albiflora Urb. FIG. 31, PL. 20d

Lianas. Stems to 15–20 m long. Leaves: petioles 4–10 cm long; blades broadly cordate, 9.2–21 × 7–15 cm, glabrous, the base with 9–10 glands, the apex obtuse, rounded, or shortly acuminate; secondary veins in 5–7 pairs. Inflorescences often 2 per node, pseudopanicles; peduncle 1–2.8 cm long, densely puberulent; pedicels 2.5–4 mm long. Flowers: calyx lobes narrowly ovate to suboblong, 3–5 × 1–1.3 mm, glabrous, the apex acuminate; corolla subrotate, 5–6 mm diam., the lobes narrowly ovate or ovate, 2.7–3 × 1.4–1.6 mm, with white, flat hairs at apex adaxially and along margins, light green, the apex somewhat reflexed, the margins revolute; gynostegium stipitate, 1.4–1.6 mm long, the stigma head stellate, 2.5–2.7 mm diam., light green; corona segments densely papillate adaxially. Fruits 15–20 × 8–9 cm, obtuse-concave at apex. Fl (Sep); collected only once in our area, in non-flooded forest.

ASTERACEAE (Composite Family)

John F. Pruski

Annual or perennial herbs, shrubs, trees, vines, or lianas, frequently with glandular or non-glandular trichomes. Milky latex sometimes present. Stipules absent. Leaves simple or less commonly ternately or pinnately dissected, alternate, opposite, or whorled; blades commonly concolorous. Capitulescences (the secondary arrangement of capitula) of 1 to many, variously arranged capitula generally free from each other, but occasionally fused (syncephalous); capitula (= heads or primary inflorescence) of 1 to many homogamous or heterogamous flowers on a common receptacle (with or without pales; pales, when present, commonly scarious or membranous) surrounded by an involucre of 2 to many, commonly persistent phyllaries in 1 to several series, phyllaries generally stramineous or the outer ones herbaceous, usually flat and persistent, often spreading with age; the capitula when homogamous (with a single flower type) discoid (only disk flowers present), ligulate (only ligulate flowers present, almost entirely restricted to the Lactuceae), or bilabiate (at least some bilabiate flowers present, found outside our area in the Mutisieae); the capitula when heterogamous (with 2 or more flower types) radiate (ray and disk flowers present) or disciform (all flowers actinomorphic, the central ones commonly bisexual, the marginal ones pistillate). Flowers (florets) unisexual (plants commonly monoecious) or bisexual, actinomorphic (disk flowers, with commonly 5 subequal corolla lobes) or zygomorphic (ray flowers, ligulate flowers, bilabiate flowers), (3)5-merous; calyx (pappus) of scales, awns, setae, bristles (these all commonly of subequal parts), crown-like structure, or absent; corolla sympetalous; zygomorphic flowers sometimes sterile. Bisexual flowers protandrous. Staminate and bisexual flowers: filaments adnate to corolla, commonly cylindrical, the anthers connate into tube (syngenesious); pollen mostly spheroidal and tricolporate. Pistillate and bisexual flowers: ovary inferior, 2-carpellate, unilocular, the style commonly filiform, 2-branched, the style shaft commonly glabrous, the branches with a continuous or 2-banded ventral or ventromarginal stigmatic surface, commonly reaching to near branch apex; placentation basal, the ovule solitary. Fruits cypselae (achene developed from an inferior ovary), very rarely drupaceous, commonly isomorphic, dry (rarely baccate). Seeds lacking endosperm, storing fatty oils.

Koster, J. Th. 1938. Composiate. *In* A. Pulle (ed.), Flora of Suriname **IV(2):** 87–165. J. H. De Bussy, Amsterdam.

Pruski, J. F. 1997. Asteraceae. *In* P. E. Berry, B. K. Holst & K. Yatskievych (eds.), Flora of the Venezuelan Guayana. Volume **3:** 177–393. Araliaceae–Cactaceae. Missouri Botanical Garden, St. Louis.

An understanding of the inflorescence type and of the structure of the flowers of Asteraceae are needed before the keys can be effectively used. See Fig. 32 and the glossaries found at the end of Parts 1 and 2 for help in understanding the terminology used to describe Asteraceae.

Calea solidaginea Kunth subsp. *deltophylla* (R. S. Cowan) Pruski, *Ichthyothere davidsei* H. Rob., and *Riencourtia pedunculosa* (Rich.) Pruski have been collected in savannas and on inselbergs near the periphery of our area and should be looked for in similar habitats in our area. All would represent genera new to our area. Tribes of genera in our area are indicated in the key.

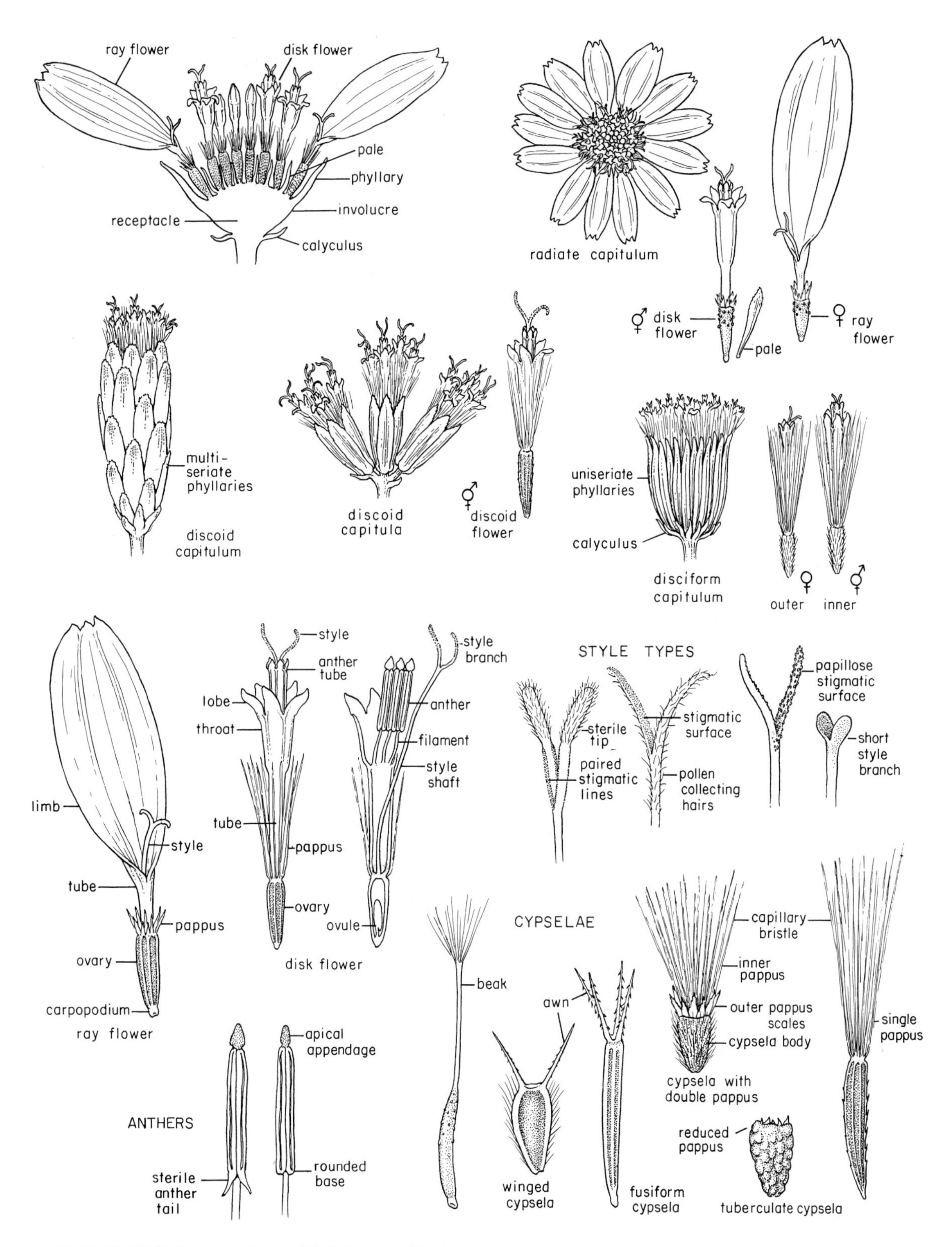

FIG. 32. ASTERACEAE. Capitulum, flower, and fruit characters of the Asteraceae.

1. Leaves commonly ternately or pinnately dissected, a few sometimes simple (Heliantheae, in part).
   2. Ray flowers present or absent, the corollas white or yellow; filaments glabrous. Cypselae without a terminal beak, (2)3–4(6)-awned. . . . . *Bidens.*

2. Ray flowers always present, the corollas pink or rarely white; filaments hirsute. Cypselae terminally beaked, 2-awned. . . . . *Cosmos.*

1. Leaves all simple.
   3. Capitula radiate; receptacles with or without pales.
      4. Leaves alternate. Receptacle without pales. Ray corollas lavender; disk corollas yellow; anthers pale brown; style branches with paired stigmatic lines in proximal 2/3, the apex sterile. Cypselae with pappus of numerous capillary bristles (Astereae). . . . . *Symphyotrichum.*
      4. Leaves opposite. Receptacle with pales. Corollas commonly white or yellow to orangish; anthers commonly black or dark brown; style branches with paired stigmatic lines reaching to apex. Cypselae with a pappus of a few setae, stout awns, short teeth, fimbriae, absent or nearly absent (Heliantheae, in part).
         5. Capitula sessile or often nearly so, commonly axillary. Pappus of few awns or absent.
            6. Disk flowers bisexual, with divided styles. Cypselae dimorphic, those of rays flattened with prominent dentate lateral wings, those of the disks obconical, with a 2(4)-awned pappus.. *Synedrella.*
            6. Disk flowers functionally staminate, with undivided styles. Cypselae isomorphic, ovoid, neither dentate or awned, without pappus. . . . . *Unxia.*
         5. Capitula usually obviously pedunculate, terminal or occasionally axillary. Pappus wanting or reduced to a low crown, sometimes with a few short teeth, aristae, or fimbriae present.
            7. Ray flowers sterile, without styles.
               8. Peduncles 2–5 cm long. Involucres 6–9 mm broad. Ray flower limbs ca. 6.2 mm long, shortly exserted from involucre. Pales scarious. Cypselae not fleshy at maturity, the pappus with 2–4 aristae. . . . . *Elaphandra.*
               8. Peduncles to 7(15) cm long. Involucres 10–15 mm broad. Ray flower limbs 10–17 mm long, well exserted from involucre. Pales rigid. Cypselae with fleshy surface at maturity, the pappus absent. . . . . *Tilesia.*
            7. Ray flowers pistillate, with styles.
               9. Pales filiform. Corollas white, caducous; cypselae caducous. . . . . *Eclipta.*
               9. Pales elliptic-lanceolate. Corollas yellowish, not caducous; cypselae persistent.
                  10. Leaves commonly trilobed, glandular abaxially. Capitula obviously radiate. Disk corollas 5-lobed. Cypselae pyriform, tuberculate at maturity; pappus fimbriate, fimbriae obscured by corky coroniform collar. . . . . *Sphagneticola.*
                  10. Leaves not trilobed, eglandular. Capitula inconspicuously radiate. Disk corollas 4- or 5-lobed. Cypselae strongly compressed, ciliate, the face smooth; pappus of 2–4 fragile setae. . . . . *Acmella.*
   3. Capitula discoid or disciform; receptacle generally without pales.
      11. Leaves alternate; venation pinnate.
         12. Principal or all phyllaries uniseriate, subequal. Stamens with swollen filament collars, the anthers basally rounded; style branches with paired stigmatic lines, apically triangular or truncate, papillose, the shaft glabrous (Senecioneae).
            13. Capitula discoid. Involucre not calyculate. Corollas lavender or purple to pale pink; style branch apex triangular. . . . . *Emilia.*
            13. Capitula disciform. Involucre calyculate. Corollas greenish white; style branch apex truncate. . . . . *Erechtites.*
         12. Phyllaries 2–3(5)-seriate, graduated or sometimes subequal. Stamens with cylindrical filament, the anthers basally spurred; style branches with stigmatic surface continuous over inner surface, slender and gradually attenuate, hispidulous, the distal portion of shaft also hispidulous (Vernonieae).
            14. Leaves discolorous, abaxially white-tomentose. Capitula uniflorous. Phyllaries 2. Corollas 3- or 4- lobed. Pappus a low laciniate crown. . . . . *Rolandra.*
            14. Leaves concolorous, pilose abaxially. Capitula with 4 or more flowers. Phyllaries several. Corollas 5-lobed. Pappus wholly or partly of elongate bristles.
               15. Annual herbs. Capitula pedunculate, distinct, ca. 15-flowered, not bracteate. Corolla lobes subequal. Pappus biseriate, inner series of many fragile bristles. . . . . *Cyanthillium.*
               15. Perennial herbs. Capitula sessile, glomerate, 4-flowered, the glomerules bracteate. Corolla lobes unequal. Pappus uniseriate, of 5 persistent basally enlarged bristles.. . . . . *Elephantopus.*
      11. Leaves opposite; venation commonly palmate or subpalmate.
         16. Receptacle with or without pales. The style branches with stigmatic surfaces in paired lines reaching nearly to branch apex (Heliantheae, in part).
            17. Annual herbs. Capitula discoid. Receptacles obviously paleate. Flowers bisexual; corollas yellowish. Cypselae flattened. . . . . *Acmella.*

17. Shrubs. Capitula disciform. Receptacle without pales or less commonly with inconspicuous pales. The outer flowers pistillate, the limb wanting, represented by 2-4 minute lobes; the inner flowers functionally staminate; corollas white. Cypselae obovoid. . . . . *Clibadium.*

16. Receptacle not paleate. The stigmatic surfaces in paired lines restricted to proximal half of style branches, the apex a long, sterile appendage (Eupatorieae).
18. Vines or lianas. Capitula with 4 flowers and 4 principal phyllaries. . . . . *Mikania.*
18. Herbs to shrubs. Capitula with >4 flowers and >4 principal phyllaries.
19. Phyllaries weakly imbricate, subequal, persistent. Pappus of 5 awns, rarely reduced to a crown. . . . . *Ageratum.*
19. Phyllaries imbricate, graduated, some or all deciduous. Pappus of many bristles.
20. Phyllaries rounded at apex, not spreading with age, all completely deciduous thereby completely exposing glabrous receptacle. . . . . *Chromolaena.*
20. Phyllaries narrowly acute at apex, the outer ones spreading and persistent, the inner ones deciduous. Receptacle pilose, not completely exposed. . . . . *Hebeclinium.*

ACMELLA Rich.

Annual or perennial herbs. Stems erect or decumbent, rooting at nodes, glabrous to pilose. Leaves simple, opposite; petioles present; blades chartaceous, both surfaces eglandular, glabrous to weakly puberulent, the margins subentire to serrulate; venation of 3 main veins from near base. Capitulescences terminal or occasionally axillary, of 1 to few capitula; peduncles long, ascending, glabrous to weakly puberulent. Capitula of marginal ray and central disk flowers (radiate), the ray flowers sometimes inconspicuous, or with only disk flowers (discoid), many-flowered; involucre hemispherical; phyllaries subequal, 1–3-seriate, weakly imbricate, commonly elliptic, herbaceous; receptacle long-conical, with pales, the pales elliptic-lanceolate, conduplicate. Ray flowers (when present) pistillate; corolla commonly yellow or white, the limb sometimes inconspicuous, the tube pilose. Disk flowers bisexual; corollas funnelform or with limb campanulate, usually greenish white or yellow-orange, glabrous, shortly 4- or 5-lobed, the lobes short, deltoid, papillose within; anthers 4 or 5, commonly dark brown; style branches with poorly developed paired broad stigmatic lines. Cypselae dimorphic, the outer ones 3-angled, the inner ones strongly compressed, sometimes with an evident corky, non-winged margin, this commonly ciliate, the face smooth; pappus of 2–4 fragile setae shorter than cypselae and corollas, the cilia and pappus grading into one another.

This genus was recently (Jansen, 1985) resurrected from synonymy within *Spilanthes* Jacq., which differs from *Acmella* by a different base chromosome number, by sessile leaves, and by capitula completely lacking ray flowers. Other species of *Acmella* are known from just outside our area and some of these (e.g., *A. brachyglossa* Cass., *A. ciliata* (Kunth) Cass., and *A. radicans* (Jacq.) R. K. Jansen) may eventually be found in our area.

Acmella uliginosa (Sw.) Cass. [Syn.: *Spilanthes uliginosa* Sw.] FIG. 33, PL. 21a

Low annual herbs, to 75 cm high. Stems erect to ascending or occasionally decumbent, rarely rooting at nodes, sparsely pilose to subglabrous. Leaves: petioles 2–15 mm long, sometimes narrowly winged, sparsely pilose to subglabrous; blades lanceolate to elliptic, rarely broadly elliptic, 2–5 × 0.5–2.5 cm, glabrous to sparsely pilose, the base attenuate, the apex narrowly acute to acuminate, the margins serrulate. Capitulescences several-headed; peduncles 1.5–6.5 cm long, sparsely pilose. Capitula 4.5–8 × 4–6.5 mm, without ray flowers or with inconspicuous ray flowers; phyllaries uniseriate, 5 or 6, 2.2–3.4 mm long; receptacle 3–6 mm long. Ray flowers (when present) 4–7; corolla 2–3.5 mm long, yellow to yellow-orange, the limb apically notched, not much exserted from phyllaries. Disk flowers 4(5)-lobed; corolla 1–1.5 mm long, yellow to yellow-orange. Cypselae 1.2–1.8 mm long, the margins ciliate, not corky; pappus of 2(4) often subequal bristles 0.2–0.7 mm long, ≤2× as long as cilia. Fl (Mar, May, Jun, Jul, Aug, Sep, Oct, Dec); in secondary vegetation in vicinity of Saül and along roads through non-flooded forest.

AGERATUM L.

Herbs (annual or perennial) to shrubs. Stems glabrous, glandular, or pilose. Leaves simple, generally opposite; petioles present; blades commonly 3-veined. Capitulescences generally terminal, in compact or open cymose or corymbiform clusters. Capitula of all disk flowers (discoid), many-flowered; peduncles present; involucre campanulate; phyllaries 2–3-seriate, weakly imbricate, subequal, lanceolate, striate, pilose to glabrous, the apex acuminate to rounded; receptacle conical, generally without pales. Flowers bisexual; corolla funnelform, shortly 5-lobed, pale blue to white, the triangular lobes papillose within; anthers included, appendages large; style branches long, greatly exserted from corolla, the distal half of style branches with a large sterile appendage, the stigmatic lines paired in proximal half of branches, the style base glabrous, not enlarged. Cypselae prismatic, 4–5-angled, glabrous or with bristles on angles, the carpopodium distinct, usually asymmetric; pappus of 5(6) apically tapering scales, coroniform, or absent.

Ageratum conyzoides L. FIG. 34, PL. 21b

Erect, few-branched annual herbs to subshrubs, to 1 m tall. Stems pilose. Leaves: petioles to 3 cm long, pilose; blades ovate, 2–8 × 1–5 cm, chartaceous, both surfaces pilose, the abaxial surface also glandular, the base acute to rounded, the apex mostly obtuse, the margins crenate. Capitulescences irregularly branched panicles in the distal third of the plant, of several cymose clusters to 3 × 4

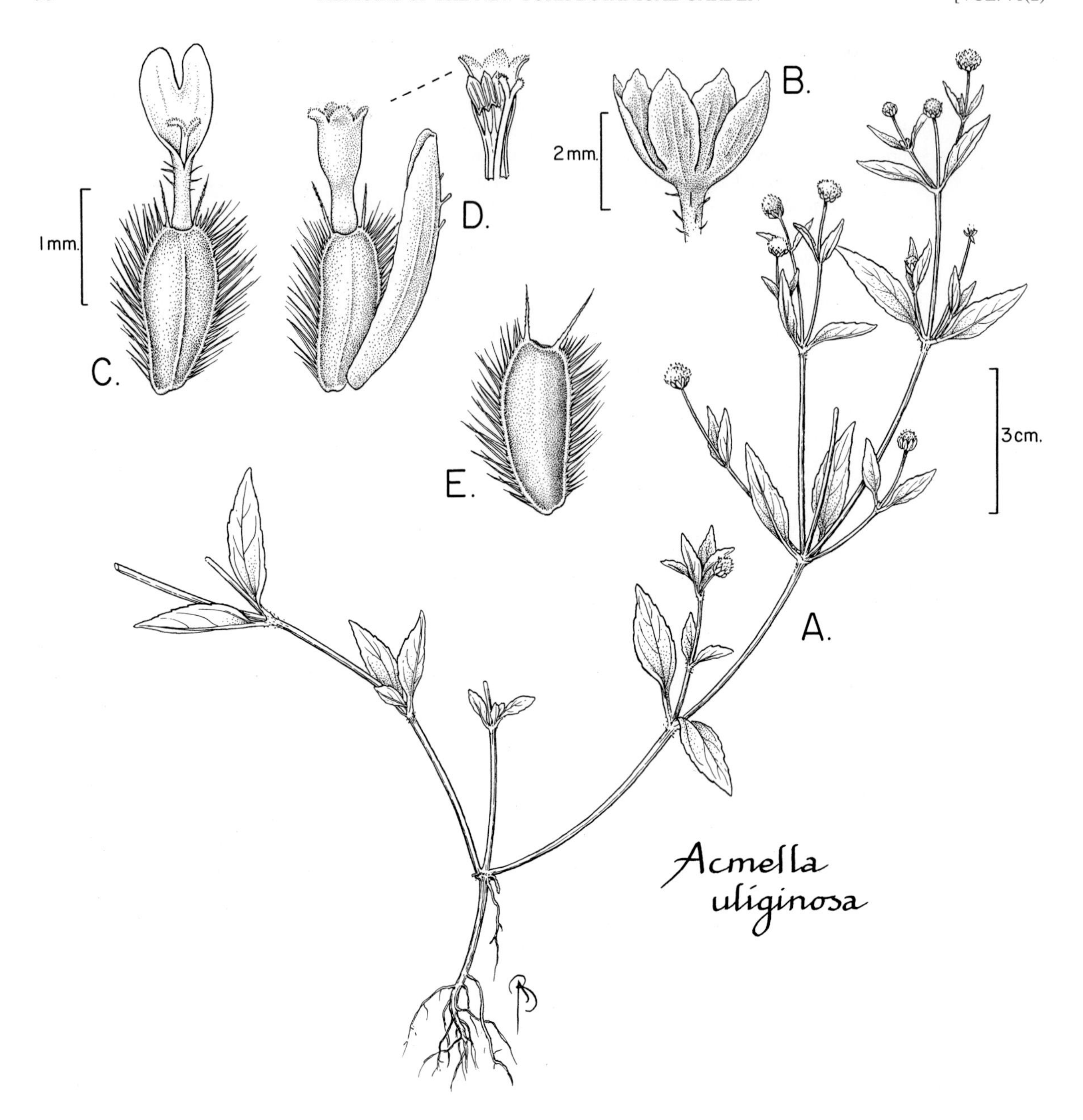

FIG. 33. ASTERACEAE. *Acmella uliginosa* (*Mori & Gracie18455*). **A.** Part of plant with capitulescence. **B.** Lateral view of involucre. **C.** Adaxial view of ray flower. **D.** Disk flower with subtending pale (lower left) and longitudinal section (upper right) of corolla showing stamens and style. **E.** Lateral view of cypsela.

cm, each with to ca. 25 capitula. Capitula 40–70-flowered, to 5.5 mm long and wide; involucre campanulate to hemispherical, to ca. 4 mm long; phyllaries biseriate, lanceolate, 2–3-striate, sparsely pilose to glabrous, the apex acuminate; receptacle ca. 0.8 mm wide; peduncles 3–10 mm long, pilose, bracteolate. Flowers: corolla 1.5–2 mm long, glabrous or sparsely pilose, cream-colored to pale blue. Cypselae 1.2–1.7 mm long, black, the angles with bristles; pappus awns 1.5(2.5) mm long, slightly longer than corollas, the pappus rarely reduced to a crown or lacking. Fl (Feb, Jun, Sep, Oct); in disturbed areas. *Radié François* (Créole).

BIDENS L.

Annual or perennial herbs, glabrous or nearly so. Stems erect, commonly striate or sulcate. Leaves simple to pinnatifid or ternately dissected, opposite or becoming alternate distally; petioles present, frequently narrowly winged; blades ovate or simple ones elliptic, chartaceous, the margins commonly serrate and ciliate; venation pinnate. Capitulescences terminal, of few long-pedunculate capitula. Capitula of marginal ray flowers and central disk flowers (radiate) or occasionally of all disk flowers (discoid), many-flowered; involucre campanulate to subhemispherical; phyllaries biseriate and dimorphic, persistent and reflexed when past fruit, outer series ca. 8, herbaceous, subequal, green, narrow, the inner phyllaries brown with widely hyaline margins, membranous, striate, wider than outer phyllaries; receptacle flat to convex, paleaceous; pales flat. Ray

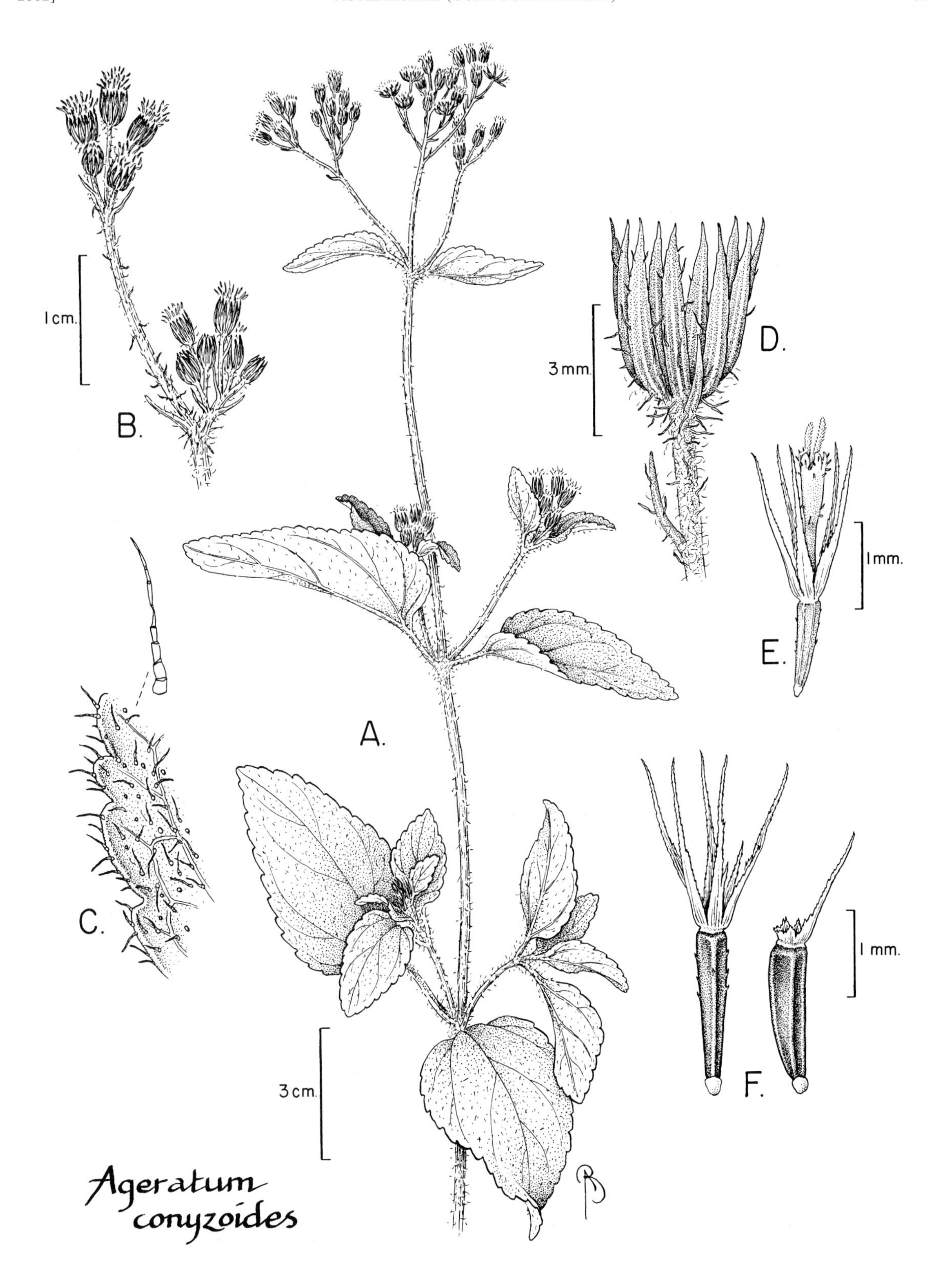

FIG. 34. ASTERACEAE. *Ageratum conyzoides* (A–C, *Eggers s.n.* from St. Thomas, V.I.; D, E, *Fishlock 281* from Tortola, B.V.I.; F, *Correll 49537* from Bahamas). **A.** Distal part of stem with leaves and capitulescences. **B.** Capitulescences. **C.** Detail of leaf margin (right) and non-glandular trichome (above right). **D.** Lateral view of involucre. **E.** Lateral view of disk flower. **F.** Lateral views of disk cypselae. (Reprinted with permission from P. Acevedo-Rodríguez, *Flora of St. John, U.S. Virgin Islands*, Mem. New York Bot. Gard. 78, 1996.)

flowers (when present) with sterile ovaries, but styles sometimes present; corolla white or yellow, glabrous. Disk flowers many, bisexual; corolla funnelform, shortly lobed, usually yellow, glabrous or puberulent, the lobes triangular; anthers black or occasionally yellow, the filaments glabrous; style branches flattened, bearded toward pilose appendage, commonly with weakly paired stigmatic lines. Disk cypselae tangentially flattened (flattened in plane with phyllaries or tangentially compressed) or 4-angled, linear or outer ones clavate, not apically beaked, longitudinally grooved, glabrous or with a few apical or marginal setae; pappus (2)3–4(6)-awned, much shorter than cypselae, the awns retrorsely barbed.

Bidens alba (L.) DC. var. *radiata* (Sch. Bip.) Ballard is to be expected in our area.

1. Stems hexagonal. Leaves 2- or 3-pinnatifid to pinnately compound; leaflets slightly toothed to serrate. Phyllaries linear to lanceolate. Ray flowers (when present) with corolla usually pale yellow. Disk cypselae mostly 4-angled and stoutly (3)4(6)-awned, ca. 3 times as long as the involucre, not tuberculate, outer ones falcately curved. . . . . *B. cynapiifolia.*
1. Stems quadrangular. Leaves simple or ternately divided to 1-pinnately compound; leaflets coarsely serrate. Outer series of phyllaries often oblanceolate. Ray flowers (when present) with white corolla. Disk cypselae ± tangentially flattened and (2)3(5)-awned, <2 times as long as the involucre, occasionally tuberculate, erect. . . . . *B. pilosa.*

Bidens cynapiifolia Kunth FIG. 35; PART 1: FIG. 8

Branched annual herbs, to 2 m tall. Stems hexagonal, glabrous to villous. Leaves 2- or 3-pinnatifid to pinnately compound; petioles to ca. 3.5 cm long, sparsely puberulent to subglabrous, often winged to near base; blades ovate to elliptic, 2–6.5 × 2.5–6 cm, the lobes or leaflets elliptic to ovate, 1–4 × 0.3–1 cm, both surfaces eglandular, the abaxial surface pilose on veins, the base acuminate to cuneate, the apex acuminate to acute, the margins slightly toothed to serrate; rachis (when leaf compound) to 2.5 cm long, often winged toward terminal lobe or leaflet. Capitulescences terminal, of few to several capitula; peduncles 5–11 cm long. Capitula with inconspicuous ray flowers or with all disk flowers, commonly 10–45-flowered; involucre campanulate, 5–8 mm wide, the base setose; phyllaries biseriate, ca. 20, linear to lanceolate, 3.5–6 mm long, greatly spreading in fruit. Ray flowers (when present) inconspicuous, 3–6; corolla pale yellow, less commonly white, the limb apically entire or notched. Disk flowers 10–40; corolla 2.3–3 mm long, pale yellow to yellowish-orange, the lobes sometimes puberulent. Disk cypselae 10–15 mm long, mostly 4-angled and stoutly (3)4(6)-awned, ca. 3 times as long as involucre, not tuberculate, outer ones falcately curved, the awns 2–3.2 mm long. Fl (Oct), fr (Jun); in disturbed areas. *Carrapicho de agulha* (Portuguese), *herbe aiguille*, *persil diable*, *zerb zaiguille* (Créole).

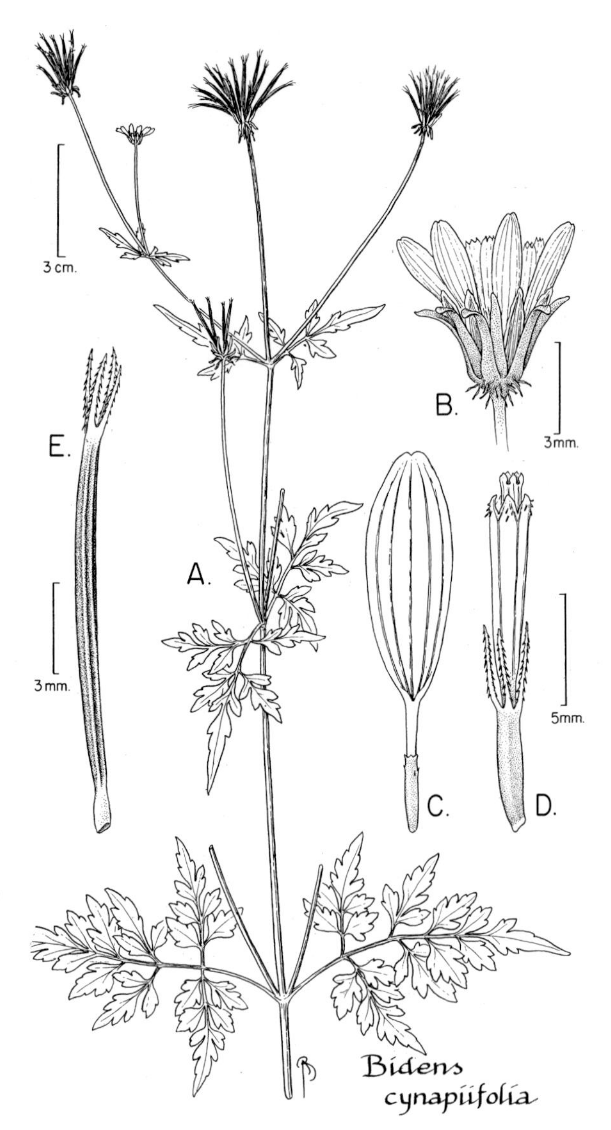

FIG. 35. ASTERACEAE. *Bidens cynapiifolia* (*Britton 1922* from Trinidad). **A.** Distal part of stem with leaves and capitula, some with maturing cypselae. **B.** Lateral view of capitulum. **C.** Adaxial view of ray flower. **D.** Lateral view of disk flower. **E.** Lateral view of disk cypsela. (Adapted with permission from P. Acevedo-Rodríguez, *Flora of St. John, U.S. Virgin Islands*, Mem. New York Bot. Gard. 78, 1996.)

Bidens pilosa L. PART 1: FIG. 8

Branched annual herbs, to 1(1.5) m tall. Stems quadrangular, sparsely pilose. Leaves commonly ternately divided but ranging from simple to 1-pinnately compound; petioles 0.5–2(3) cm long, often sparsely pilose; blades ovate, 2.5–10 × 1–8 cm; leaflets (when leaf compound) ovate or lanceolate, 1–6 mm long, both surfaces sparsely pilose, eglandular, the apex acuminate to narrowly acute, the margins coarsely serrate; rachis (when leaf compound) ca. 1(2) cm long, often winged toward terminal leaflet. Capitulescences terminal, cymose, of few to several capitula; peduncles 2–7 cm long. Capitula of all disk flowers or, less commonly, with inconspicuous ray flowers, commonly 35–75-flowered; involucre hemispherical, 7–15 mm wide; phyllaries 15–20, in 2 dissimilar series, the outer series with phyllaries 4–6 mm long, commonly oblanceolate, herbaceous, the inner series with phyllaries lanceolate, 5–6 mm long, scarious, striate. Ray flowers (when present) inconspicuous; corolla 2–3 mm long, white, the limb apically notched. Disk flowers 35–75; corolla 3–4 mm long, yellow. Disk cypselae to 10(16) mm long, ± tangentially flattened, (2)3(5)-awned, occasionally tuberculate, erect, <2 times as long as involucre, awns 2–3 mm long. Fl (Jun), fr (May); in disturbed areas.

CHROMOLAENA DC.

Erect or sprawling perennial herbs or shrubs. Stems commonly pubescent. Leaves simple, generally opposite; blades commonly 3-veined. Capitulescences mostly corymbiform. Capitula of all disk flowers (discoid), 10–40-flowered; involucre constricted-cylindrical; phyllaries 4–7-seriate, imbricate, graduated, 3–5-veined, not spreading with age, deciduous, the apex widely acute to rounded; receptacle elongate, flat or conical on very top, generally without pales, glabrous. Flowers bisexual; corolla funnelform, shortly 5-lobed, white to blue or purplish, the lobes generally papillose or glandular; anthers largely included within corolla; style branches linear, elongate, greatly exserted from corolla, the distal half of style branches with papillose sterile appendage, the proximal half with paired stigmatic lines. Cypselae narrowly obconical, mostly 5-ribbed, sometimes glandular, the ribs setose or only sparsely so; pappus of ca. 40 bristles, about as long as corollas and cypselae.

Chromolaena odorata (L.) R. M. King & H. Rob. [Syn.: *Eupatorium odoratum* L.] FIG. 36

Herbs to shrubs, 1–3 m tall. Stems branched distally, densely pilose to puberulent. Leaves: petioles 0.4–2.5 cm long, pilose; blades deltoid to rhombic-ovate, 2.5–12 × 1.2–7.5 cm, chartaceous, the adaxial surface puberulent to glabrous, the abaxial surface pubescent, densely red-glandular, the base cuneate-attenuate to subtruncate, the apex acuminate, the margins entire to serrate; 3-veined from near base. Capitulescences terminal, corymbiform-paniculate, to 8 × 4 cm. Capitula ca. 15–25-flowered, 9–12 × 3–4.5 mm; involucre 8–9.5 mm long; phyllaries 5–6-seriate, ca. 15, glabrous to sparsely puberulent, scarious or with subherbaceous apices, the outer ones oblong-elliptic, ca. 2 × 1 mm, the apex rounded, grading into inner ones, the inner ones lanceolate, ca. 9 × 1.3 mm, the apex obtuse to acute; receptacle shortly clavate, ca. 1.3 × 1 mm; peduncle 1–15 mm long, weakly pilose, sparingly glandular. Flowers: corolla 5.3–6 mm long, white to pale lavender. Cypselae 3.5–4 mm long; pappus bristles 5–6 mm long. Fl (Jun, Oct); in disturbed areas. *Radier maringouin*, *raguet maringouin* (Créole).

CLIBADIUM L.

Shrubby herbs to small trees. Stems pubescent. Leaves simple, opposite, petiolate; blades without glands, the margins serrulate to serrate; venation of usually 3 main veins from above base, the veins frequently strongly arching toward leaf apex. Capitulescences terminal, open, corymbiform-paniculate, suboppositely branched, of many capitula, these pedunculate or sessile, sometimes glomerate. Capitula of outer, actinomorphic pistillate and central disk flowers (disciform), small; involucre subglobose; phyllaries mostly biseriate, few, imbricate, pubescent, scarious, the outer ones orbicular, the inner ones enclosing marginal cypselae; receptacle slightly convex, without pales or less commonly with inconspicuous pales. Outer flowers pistillate; corolla tubular, white, the limb wanting, represented by 2–4 small lobes at apex of the tube; style branches linear, with poorly developed paired stigmatic lines. Inner flowers functionally staminate; corolla white, funnelform, 5-lobed, the lobes shortly triangular, pubescent; anthers black, slightly exserted; ovary sterile, the style undivided. Cypselae from outer flowers obovoid, slightly compressed, apically pubescent; pappus none.

Clibadium sylvestre (Aubl.) Baill.

Shrubs, 1–3 m tall. Stems strigose, subterete or sometimes angled. Leaves: petioles slender, (1)2–5(7) cm long, strigose; blades elliptic to ovate, 5–15 × 1.5–10 cm, stiffly chartaceous, adaxial surface slightly scabrous, abaxial surface strigose, the base decurrent onto petiole, then abruptly tapering, the apex narrowly acute or more commonly attenuate, the margins irregularly and remotely serrate; venation of 3 main veins from above base. Capitulescences to ca. 15 × 10 cm, often wider than long. Capitula 12–20-flowered, ca. 4 mm long and wide, short-pedunculate; involucre hemispherical; phyllaries ca. 8, striate, but drying black and obscuring striations, remotely strigose, the outer ones ovate, 2–3 mm long, the apex acute, the inner ones widely ovate, 3–4 mm long, the apex obtuse, in fruit sometimes wider than long; receptacles small. Outer flowers 4–6; corolla ca. 2.1 mm long, minutely 2–4-toothed. Inner flowers 8–14; corolla ca. 3.5 mm long, the lobes triangular, shortly setose. Cypselae from outer flowers ca. 2 mm long, ovate, when young lightly glandular or setose at apex; sterile ovary of disk flowers linear, to ca. 1.5 mm long, apically setose. Fl (Jul, Sep), fr (Mar, Jul, Sep); in disturbed areas or in cultivation; extracts from the leaves and stems used throughout much of its distribution as a fish poison. *Counami*, *topa*, *topa noir* (Créole).

COSMOS Cav.

Annual or perennial herbs, mostly glabrate. Stems erect, striate. Leaves opposite, sometimes becoming alternate toward apex; petioles present; blades pinnately dissected, ovate, chartaceous, eglandular; venation pinnate. Capitulescences terminal, monocephalous or of few long-pedunculate capitula. Capitula of marginal ray flowers and central disk flowers (radiate), showy, many-flowered; involucre campanulate; phyllaries in 2 dissimilar series, ca. 5–8 in each series, lanceolate, the outer ones foliar, the inner ones subtending ray flowers yellow to reddish; receptacle with pales, the pales flat. Ray flowers 5–8(12), sterile; corolla pink, white, or orange. Disk flowers bisexual; corolla funnelform, yellow, orangish, or whitish, the short-triangular lobes papillose adaxially; anthers black, the apical appendage yellowish, the filaments hirsute; style branches with paired stigmatic lines, the apex long-apiculate, papillose. Disk cypselae fusiform, flattened in plane of phyllaries (tangentially flattened), 4-angled, scabridulous, the apex beaked, the faces rounded; pappus commonly stoutly 2-awned, occasionally with a third shorter awn, the awns divergent, retrorsely strigose, much shorter than cypselae.

Cosmos caudatus Kunth FIG. 37

Branched annual herbs, to 1.5(2.5) m tall. Stems subterete. Leaves 2–3 times pinnatifid; petioles 1–5 cm long, occasionally weakly puberulent; rachis to 5 cm long, winged, especially toward terminal lobe; blades ovate, 6–15.5 × 2.5–6.5 cm, glabrous or nearly so, both surfaces smooth, the lobes elliptic-ovate to lanceolate, 1.5–5 cm long, the base attenuate, the apex narrowly acute to attenuate or mucronate, the margins ciliate or entire. Capitulescences of 1 to few capitula; peduncles usually 10–20 cm long. Capitula usually

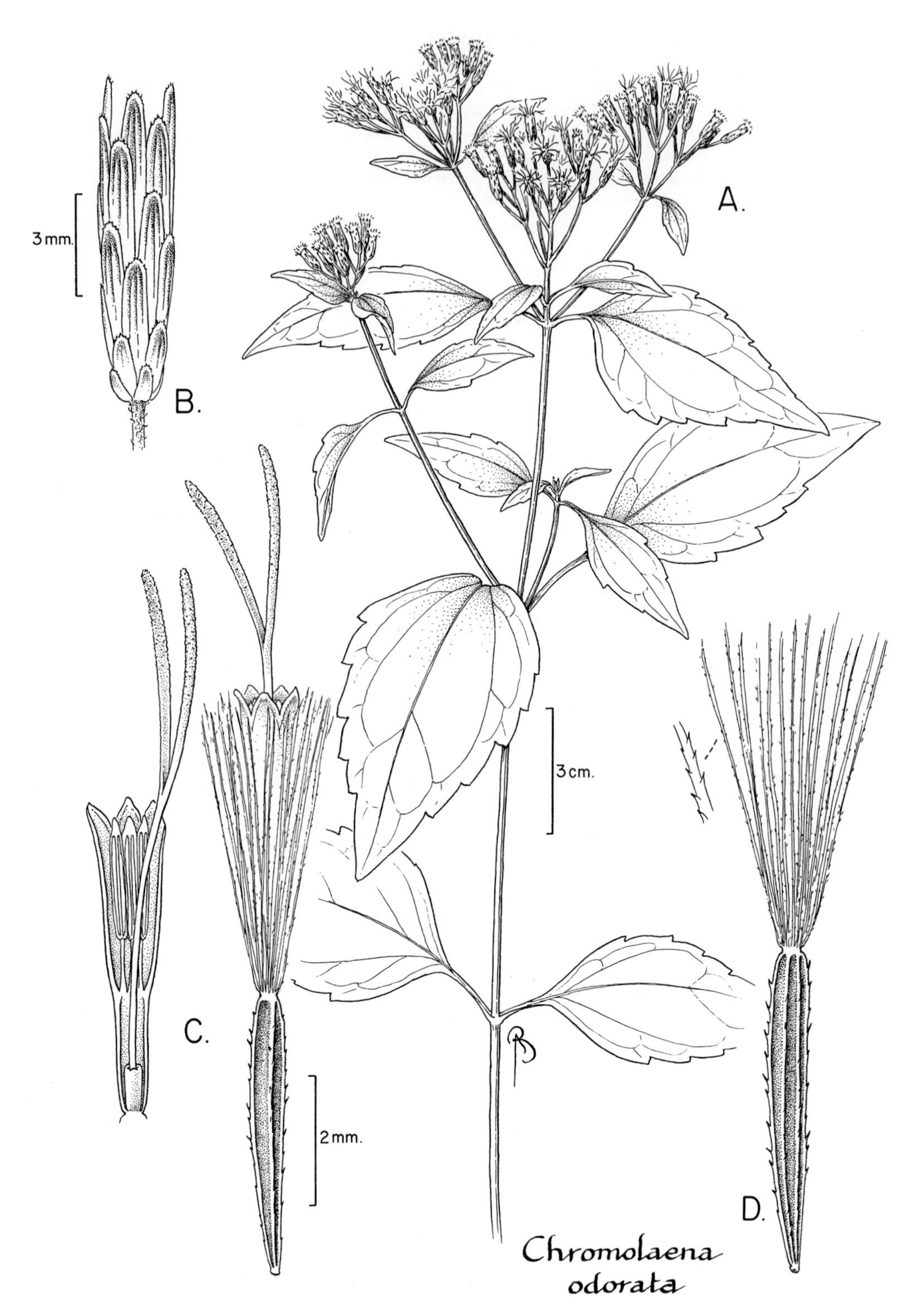

FIG. 36. ASTERACEAE. *Chromolaena odorata* (*Acevedo-Rodríguez 2360* from St. John, U.S. Virgin Islands). **A.** Distal part of stem with leaves and capitulescences. **B.** Lateral view of involucre. **C.** Lateral view of disk flower (right) and longitudinal section (left) of disk corolla showing stamens and style. **D.** Lateral view of disk cypsela (right) and detail of pappus bristle (above left). (Reprinted with permission from P. Acevedo-Rodríguez, *Flora of St. John, U.S. Virgin Islands*, Mem. New York Bot. Gard. 78, 1996.)

30–40-flowered; involucre 1–2 × 1–1.5 cm; outer phyllaries lanceolate, 6–12(20) mm long, as long as or longer than inner phyllaries, the inner phyllaries elliptic-lanceolate, 8–13 mm long, striate. Ray flowers mostly 8; corolla limb pink or rarely white, 15–25 mm long, much broadened, well exserted from involucre, commonly 3-notched at very broadened apex, the tube short. Disk flowers usually 20–30 or

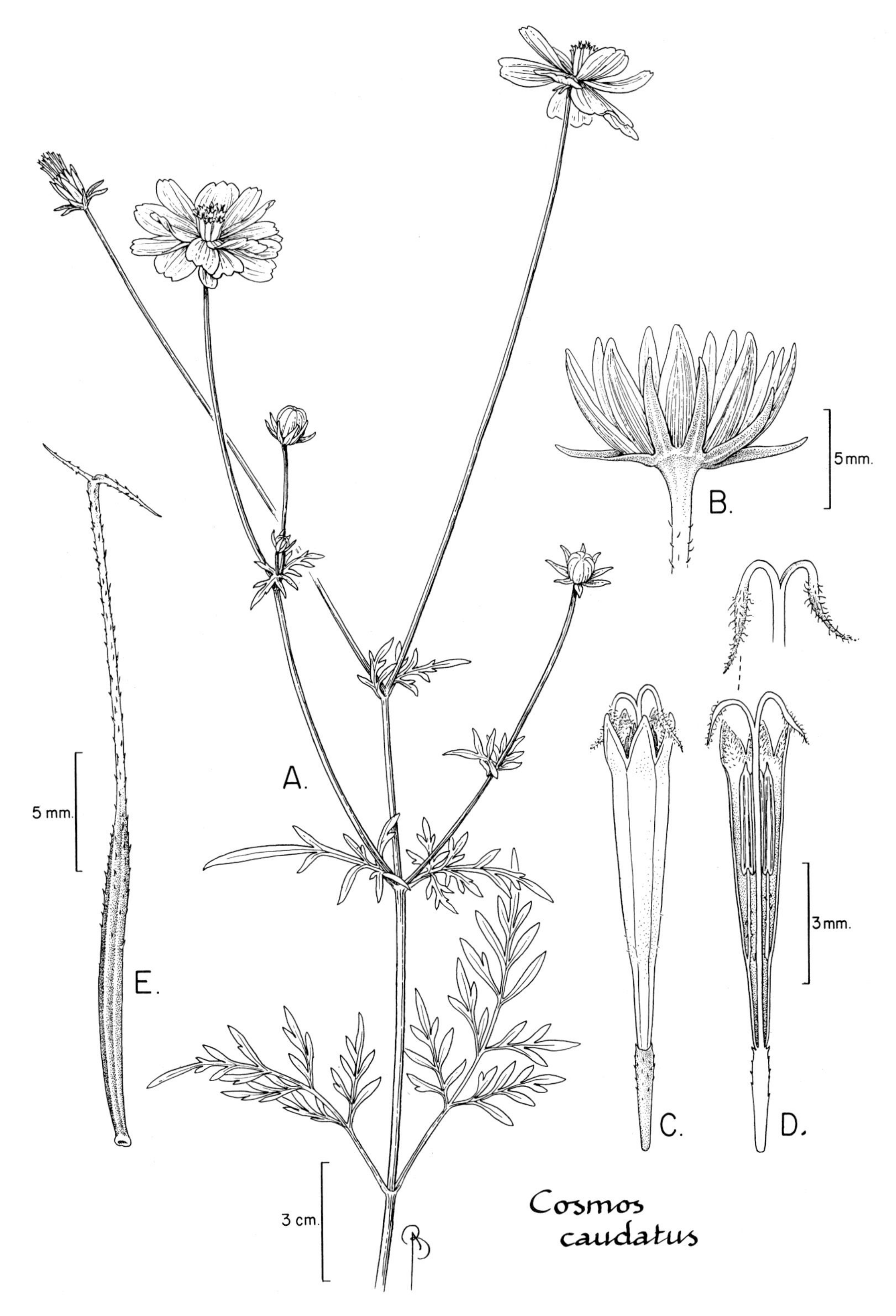

FIG. 37. ASTERACEAE. *Cosmos caudatus* (A–D, *Flores 124* from St. John, U.S. Virgin Islands; E, *Cronquist 10354*). **A.** Distal part of stem with leaves and capitulescence. **B.** Lateral view of involucre. **C.** Lateral view of disk flower. **D.** Medial section of disk flower showing stamens and style and detail of style branches (above). **E.** Lateral view of disk cypsela.(Reprinted with permission from Flora of the Guianas.)

more, the outer ones forming fruits, the innermost often sterile; corolla 5.5–8 mm long, golden-yellow to white, the lobes adaxially cream-colored and papillose. Disk cypselae (including beak) 1.5–3 cm long, (0)2- or 3(5)-awned, dark brown, well exserted from involucre when mature, the awns 2.5–4.5 mm long. Fl (Jun); weedy, escaped ornamental found in vicinity of Saül.

CYANTHILLIUM Blume

Annual or rarely short-lived perennial herbs, to ca. 1 m tall. Stems erect, commonly few-branched, striate, puberulent. Leaves simple, alternate; petioles short, sometimes winged; blades chartaceous; venation pinnate. Capitulescences terminal, corymbiform or corymbiform-paniculate. Capitula of all disk flowers (discoid); involucre campanulate, phyllaries ca. 3(5)-seriate, imbricate to weakly so, graduated, narrowly lanceolate to ovate, pilose, glandular, the apex sharply attenuate, reflexed in fruit, the outer phyllaries much reduced, the inner phyllaries subequal; receptacle without pales. Flowers bisexual; corolla funnelform or with limb campanulate, shortly 5-lobed, the limb short, blue to lavender, the lobes lanceolate, pilose or glandular; anthers included, basally spurred; style shaft upwardly hispidulous, the branches hispidulous, lavender, slender, gradually attenuate, with a continuous stigmatic surface. Cypselae subcylindrical, non-ribbed, appressed-pilose, pustulate; pappus biseriate, white, the outer series of persistent, reduced scales, the inner series of many fragile bristles longer than cypselae.

Cyanthillium cinereum (L.) H. Rob. [Syn.: *Vernonia cinerea* (L.) Less.] PART 1: FIG. 5

Weedy herbaceous annuals, to 0.8 m tall. Stems unbranched or more commonly few-branched. Leaves: petioles winged, to 1.5 cm long; blades elliptic to obovate, tapering into petiole, 1.5–3 × 1–1.5 cm, chartaceous, adaxial surface puberulent to nearly glabrous, abaxial surface pilose, also usually glandular, the base attenuate, the apex acute to obtuse, the margins subentire to serrate. Capitulescences open, held above leaves, of 5–many pedunculate capitula; peduncles 0.5–2 cm long. Capitula ca. 15-flowered; involucre ca. 3.5 mm long; outer phyllaries much reduced, the inner ones subequal. Corolla exserted from involucre, 3–4 mm long, the tube long and narrow, brown or cream-colored, the limb short, lavender, pilose, the lobes also glandular. Cypselae ca. 1.5 mm long; inner pappus bristles to 3 mm long, exserted from involucre and nearly as long as corolla. Fl (Mar, Jun, Nov); in disturbed areas. *Radié albumine*, *vingt quatre heures* (Créole).

ECLIPTA L.

Annual or short-lived perennial herbs. Stems often prostrate or procumbent, usually much branched, strigose. Leaves simple, opposite; petioles nearly absent; blades narrowly lanceolate to elliptic, chartaceous, both surfaces eglandular, the base cuneate to tapering, the apex attenuate to narrowly acute, the margins subentire to serrulate, strigose; venation pinnate. Capitulescences terminal or axillary, thyrsoid-paniculate, solitary or of a few short-pedunculate capitula. Capitula small, with marginal ray flowers and central disk flowers (radiate), many-flowered; involucre hemispherical, of several foliaceous phyllaries later spreading to expose receptacle; receptacle convex or flat, with pales, the pales filiform. Ray flowers many, in several series, pistillate; corolla small, white, caducous. Disk flowers many, bisexual; corolla mostly 5-lobed, yellow or white, caducous, the limb narrowly campanulate; anthers usually black, included; style branches short, puberulent at apex, with poorly developed paired stigmatic lines. Cypselae quickly maturing, caducous, obconical, 4-angled or cypselae from ray flowers trigonous, somewhat flattened, truncate at apex, tuberculate, sometimes weakly winged; pappus absent or of a few short teeth.

Eclipta prostrata (L.) L. [Syn.: *Eclipta alba* (L.) Hassk.] FIG. 38

Weedy annual or short-lived perennial herbs, to 1 m tall. Stems often prostrate proximally and ascending distally, strigose. Leaves: blades 2–11 × 0.4–3 cm, the base narrowly cuneate and clasping stem, the apex acute to acuminate, the margins serrate to crenate; venation of 3 main veins from near base. Capitulescences terminal, of solitary pedunculate capitula; peduncles 1–5 cm long, slender, strigose, commonly shorter than leaves. Capitula with ray flowers; phyllaries mostly biseriate, weakly imbricate, 2.5–7 mm long, elliptic to lanceolate, strigose, the apex acute to caudate. Ray flowers many; corolla white, 1–2.5 mm long, the tube short, glabrous or weakly puberulent, the limb lanceolate, shortly exserted from involucre, glabrous, notched or entire. Disk flowers many; corolla white, 1–2 mm long, the tube short, glabrous, the limb glabrous or sparingly puberulent near apices of the lobes. Cypselae 2–2.5 mm long, essentially glabrous with a few short non-glandular trichomes at apex; pappus of a few inconspicuous teeth or absent. Fl (Jun); in disturbed areas. *Langue-poule* (Créole).

ELAPHANDRA Strother

Scandent to erect, subshrubs or shrubs. Stems terete to somewhat hexagonal, hirsute or less commonly strigose. Leaves simple, opposite; petioles present; blades lanceolate to ovate, both surfaces eglandular, sometimes with abaxial black dots; venation subpalmately 3–5-veined from near base. Capitulescences terminal, in cymose groups of 3 capitula or sometimes monocephalous; peduncles ascending, hirsute to weakly so, less commonly strigose. Capitula with ray and disk flowers (radiate), rarely with all disk flowers (discoid); involucre campanulate; phyllaries imbricate, ca. 2–3-seriate, subequal or with outer ca. 5 longer than inner ca. 5, generally pubescent or inner ones sometimes glabrous, sometimes with black dots or lines, the outer ones herbaceous, the inner ones scarious or with herbaceous apices, much less commonly herbaceous throughout; receptacle with pales, the pales conduplicate. Ray flowers (when present) sterile; corolla yellow, the tube glabrous or weakly puberulous distally, the limb apically notched, eglandular, glabrous or sometimes puberulous. Disk flowers bisexual; corolla funnelform, yellow or less commonly blackish, the tube and throat generally glabrous, the lobes triangular, glabrous to puberulous; anthers black or with appendage rarely tan distally, sometimes glandular; style branches densely papillose, with paired stigmatic lines. Cypselae blackish or less commonly greenish, obpyramidal, somewhat compressed, commonly puberulous, generally smooth, the apex constricted above into a neck, the carpopodia without obvious elaiosomes; pappus with a short corona and commonly 2–4 aristae.

Elaphandra moriana Pruski

Scandent shrubs, to 2 m tall. Stems subterete, strigose. Leaves: petioles 0.5–1.5 cm long, strigose-hispidulous; blades lanceolate to elliptic, 4–12 × 0.8–5.2 cm, chartaceous, surfaces without black dots, scabrous adaxially, surfaces hispidulous abaxially, the base cuneate or sometimes attenuate, the apex acute to acuminate, the margins serrulate; venation subpalmately 3-veined from above

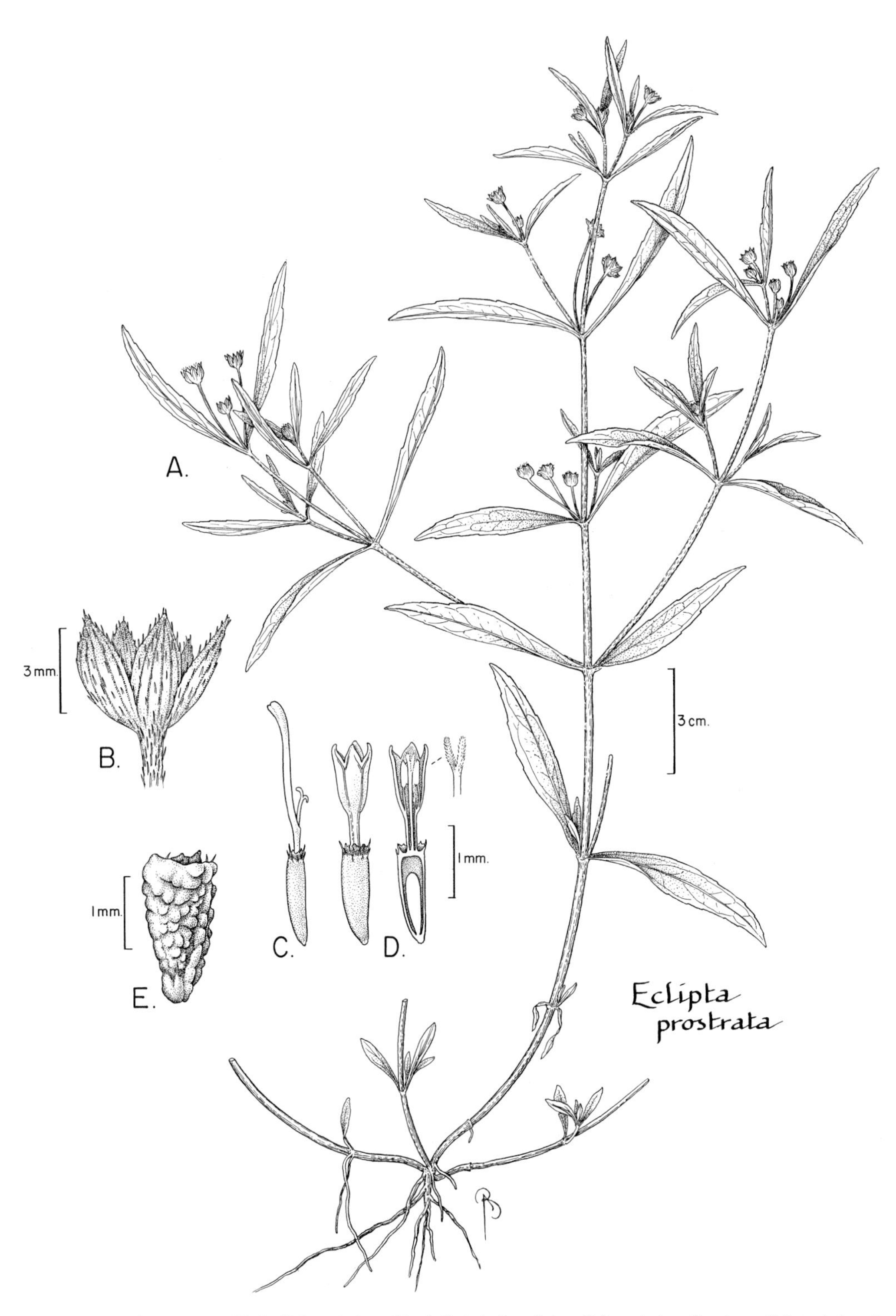

FIG. 38. ASTERACEAE. *Eclipta prostrata* (*Shafer 32* from Antigua, West Indies). **A.** Part of plant. **B.** Lateral view of involucre. **C.** Lateral view of ray flower. **D.** Lateral view (left) and medial section (right) of disk flower with detail of style branches (above right). **E.** Lateral view of cypsela. (Reprinted with permission from P. Acevedo-Rodríguez, *Flora of St. John, U.S. Virgin Islands*, Mem. New York Bot. Gard. 78, 1996.)

base. Capitulescences of 1–3 capitula; peduncles 2–5 cm long, densely strigose or hispid distally. Capitula with ray and disk flowers (radiate); involucres 6–9 × 6–8 mm; phyllaries 10–13, in ca. 2 series, subequal, oblong, 5–9 × 2–2.5 mm, herbaceous or inner ones scarious at base, strigose, non-black-streaked; receptacle flat, 2.5–3 mm wide, the pales glabrous with acuminate apices. Ray flowers (5) 8; corolla yellow, ca. 7.1 mm long, the tube ca. 0.9 mm long, glabrous, the limb ca. 6.2 × 3.1 mm, shortly bilobed at apex,

6–8-veined with 2 veins larger than others, these abaxially puberulent distally; ovaries sterile, 3-angled, ciliate on ribs, 3-aristate, the aristae 1.8–2.4 mm long. Disk flowers 24–29; corolla yellow, non-black-streaked, ca. 4.6 mm long, the tube ca. 1.5 mm long, the throat ca. 2.2 mm long, glabrous, the lobes long-triangular, ca. 0.9 mm long, weakly puberulent; anther appendages tan distally. Cypselae 4–4.2 × 1.5–2 mm, black, weakly puberulent on neck and shoulders; pappus 2–4-aristate, the aristae 1.5–3.2 mm long, arising from low fimbriate crown, ca. 0.6 mm tall. Fl (Sep); in low forest at margin of rock outcrop on Pic Matécho.

ELEPHANTOPUS L.

Perennial rhizomatous herbs. Stems erect, few-branched, pilose. Leaves simple, cauline or basal, alternate; petioles absent; blades clasping stem, both surfaces pilose, the abaxial surface also finely glandular, the margins entire to crenate or dentate, pinnately veined. Capitulescences terminal, of corymbiform panicles, the capitula glomerate, the glomerules dense, cupular, subtended by 2 or 3 leafy bracts. Capitula of all disk flowers (discoid), 4-flowered; involucre cylindrical; phyllaries 8, in 4 decussate pairs in 2 series, glandular, the outer 4 shorter than inner 4, the apex attenuate; receptacle without pales. Flowers bisexual; corolla somewhat zygomorphic, funnelform with a short campanulate limb, deeply and unequally 5-lobed, with a much deeper cut on adaxial side; anthers white, basally spurred; styles hispidulous in distal half, the branches linear, hispidulous, slender, gradually attenuate, with a continuous stigmatic surface. Cypselae obconical, 10-ribbed, hispidulous on ribs; pappus of 5(8) straight subequal bristles commonly basally enlarged.

Elephantopus mollis Kunth FIG. 39, PL. 21d

Few-branched herbs, 1–2 m tall. Stems pilose. Leaves cauline; blades oblanceolate to obovate, 5–27 × 1.5–8(10) cm, chartaceous, adaxial surface pilose, abaxial surface pilose and finely glandular, the base long-attenuate, the apex acute, the margins crenulate. Capitulescences diffuse; main capitulescence stalk 1–6 cm long, pilose. Glomerules 10–40-headed, the leafy bracts 3, cordate to deltoid, to 1.5 × 1 cm, the abaxial surface pilose and glandular. Individual capitula ca. 6–7 × 1–2 mm; outer 4 phyllaries ca. 2–3 mm long, the inner 4 phyllaries ca. 5–7 mm long. Flowers: corolla 5–6.5 mm long, cream-colored, with lavender lobes, glabrous or lobes occasionally glandular. Cypselae 2–2.5 mm long; pappus of 5 bristles, ca. 4 mm long, reaching to base of corolla limb. Fl (May, Jun, Sep, Oct), fr (Sep); in disturbed areas. *Langue de boeuf* (Créole), *lingua de vaca* (Portuguese), *vervine blanc*, *vervine sauvage* (Créole).

EMILIA Cass.

Short-lived, weedy, taprooted herbs. Stems simple or few-branched, glabrous or sparsely pilosulous. Leaves simple, alternate, but often clustered at stem base and reduced distally, the distal ones amplexicaulous, the proximal ones often tapering to narrowly winged petiole; blades with terminal lobe often large; venation pinnate. Capitulescences mostly terminal, loosely corymbiform, 1- to few-headed. Capitula of all disk flowers (discoid), 15–50-flowered; peduncles short to long; involucre not calyculate, cylindrical to weakly campanulate; phyllaries uniseriate, 8–13, subequal, connate but separating distally with age, eventually completely reflexed; receptacle without pales, mostly flat. Flowers bisexual, the inner ones sometimes sterile; corolla long-tubular, shortly 5-lobed, orange, pink, red or purple, rarely white, the tube much longer than limb; filament collar swollen, the anthers basally rounded; style branches weakly exserted, with paired stigmatic lines, the apex papillose, triangular. Cypselae cylindrical, 5-angled, 5–10-ribbed, minutely puberulent; pappus of many deciduous white capillary bristles about as long as corolla and much longer than cypselae.

Emilia fosbergii Nicolson, characterized by dentate leaves and flowers clearly longer than the involucre, is expected to occur in our area.

Emilia sonchifolia (L.) DC. var. **sonchifolia** PL. 21c; PART 1: FIG. 5

Herbs, to 0.5(1) m tall. Leaves: petioles absent or long-attenuate and winged; blades variable, often elliptic-lanceolate or oblanceolate (the proximal ones lyrate), 4–11.5 × 1–5 cm, chartaceous, both surfaces eglandular, glabrous to sparsely puberulent, the base either auriculate and expanded or tapering, the apex acute, the margins serrate to lobed, the terminal lobe commonly larger than others. Capitulescences held above leaves, usually 10–25 cm long, branched portion near apex mostly 5–8 cm wide, few-headed, the capitula not strongly spreading; peduncles 1.5–4 cm long. Capitula 15–30-flowered, 10–12 × 3–4 mm; involucre narrowly cylindrical, 3–4 times longer than wide; phyllaries 8(10), 9–12 mm long, loosely pilosulous, the apex acute; receptacle 2–3.5 mm wide. Flowers: corolla 7–8.5 mm long, not exserted from phyllaries, lavender or purple to pale pink, the lobes 0.5–0.7 mm long, spreading, often papillose. Cypselae 5(10)-ribbed, 3–4 mm long; pappus bristles 5.5–7 mm long. Fl (Feb, Jun); in disturbed areas. *Algodão-do-praia* (Portuguese), *salade Madame Hector*, *taba taba* (Créole).

ERECHTITES Raf.

Short-lived herbs from fibrous roots. Stems sulcate, simple to few-branched. Leaves simple, alternate; petioles nearly absent; blades decurrent or amplexicaulous, both surfaces eglandular, the abaxial surface glabrous to pilose, the margins sharply serrate to variously pinnately lobed or incised; venation pinnate. Capitulescences mostly terminal, corymbiform. Capitula with marginal, actinomorphic pistillate flowers and central disk flowers (disciform), many-flowered; peduncles short; involucre calyculate, cylindrical, often enlarged at base; principal phyllaries uniseriate, several, subequal, linear to lanceolate, connivent to deflexed at maturity; receptacle flat to weakly convex, without pales. Marginal flowers pistillate, 1- or 2-seriate, the corollas filiform, greenish white; central flowers numerous, bisexual, the corollas long-tubular, greenish white, the tube much longer than short lobes; filament collar swollen, the anthers basally rounded; style branch with paired stigmatic lines, apex truncate, with a terminal appendage of fused papillose hairs. Cypselae subcylindrical to subfusiform, with ca. 8 ribs, puberulent in grooves; pappus of many subequal cream-colored bristles about as long as flowers and slightly exserted from involucre.

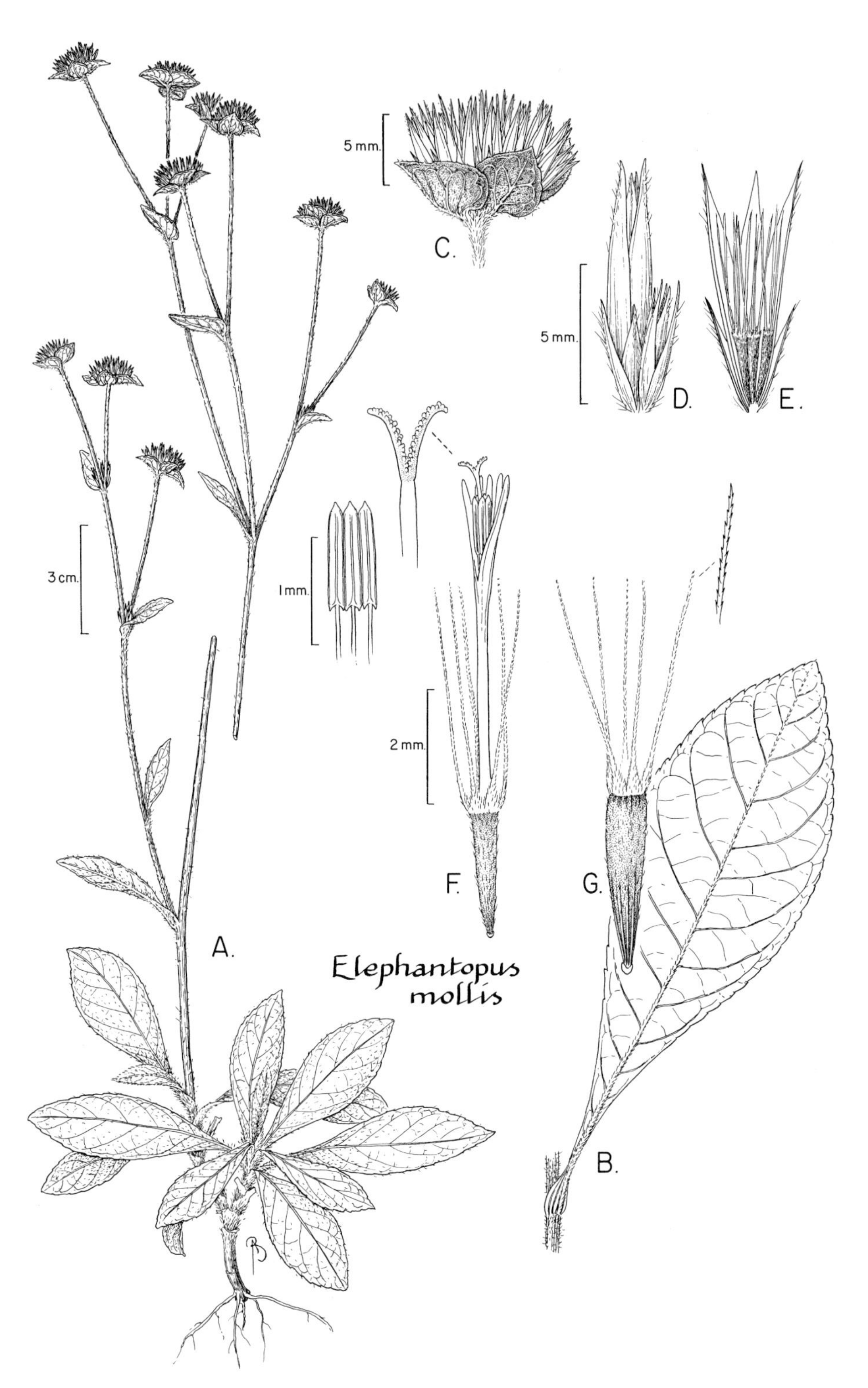

FIG. 39. ASTERACEAE. *Elephantopus mollis* (A, F, *Britton 727* from Cuba; B–E, G, *Boom 6858* from Puerto Rico). **A.** Base and apex of plant. **B.** Abaxial view of lower cauline leaf. **C.** Lateral view of glomerule consisting of several capitula. **D.** Lateral view of capitulum. **E.** Longitudinal section of capitulum showing cypselae. **F.** Lateral view of disk flower (right) with detail of style branches (above, near left) and anthers (above, far left). **G.** Lateral view of disk cypsela (right) and detail of pappus bristle (above right). (Reprinted with permission from P. Acevedo-Rodríguez, *Flora of St. John, U.S. Virgin Islands*, Mem. New York Bot. Gard. 78, 1996.)

Erechtites hieracifolius (L.) DC.

Erect annual herbs, to 1(3) m tall. Stems usually simple, many-striate, glabrous to pilose (especially in leaf axils). Leaves: petioles mostly absent or clasping; blades often oblanceolate, 3–15(35) × 2–4(9) cm, chartaceous, glabrous or abaxial surface less commonly weakly pilose, the apex acute, the margins irregularly dentate to shallowly incised. Capitula 20–150- or more-flowered; involucre cylindrical, about 2 times as long as wide; phyllaries (8)10–12(15) × 0.5–1 mm; calycular bracts usually 1–2 mm long, much shorter than phyllaries, linear-lanceolate. Flowers: corolla of marginal flowers 7–10 mm long; corolla of inner flowers 8–11 mm long, the tube elongate, the throat ca. 1 mm long, the lobes ca. 0.5 mm long; cypselae 2.5–3 mm long; pappus 8–10 mm long, white. Fl (Jul, Sep); in disturbed areas.

HEBECLINIUM DC.

Perennial herbs to subshrubs. Stems erect, moderately branched, sometimes vining, tomentellous. Leaves simple, opposite; petioles long; blades often glandular; venation pinnate or palmate. Capitulescences terminal or axillary from distal nodes, shortly to moderately exserted from leaves, loose, many-headed, corymbiform-paniculate. Capitula of all disk flowers (discoid), many-flowered; peduncles short; phyllaries 4–6-seriate, subimbricate or imbricate, graduated, 3(5)-veined, the outer ones spreading, persistent, the inner ones deciduous; receptacle without pales, convex to shortly conical, pilose, partly hidden by persistent phyllaries. Flowers bisexual; corolla tubular to funnelform, shortly 5-lobed, white to pink, the lobes commonly pilosulous and glandular; anthers included, the apical appendages ovate-triangular; style base not enlarged, glabrous, the branches linear, markedly exserted from corolla, the distal half with a filiform, terete sterile appendage, the proximal half with paired stigmatic lines. Cypselae prismatic, narrowly obconical, 4- or 5-ribbed, glabrous or with a few apical setae, the carpopodium symmetric or nearly so; pappus of many capillary bristles, these about as long as corolla and much longer than cypsela.

Hebeclinium macrophyllum (L.) DC. [Syn.: *Eupatorium macrophyllum* L.] FIG. 40, PL. 21e; PART 1: FIG. 5

Tomentellous herbs or subshrubs, 1–2.5 m tall. Stems subterete, finely striate. Leaves: petioles to 10 cm long; blades widely ovate to cordiform, 5–15 × 2–17 cm, chartaceous, adaxial surface puberulent, abaxial surface tomentellous and glandular, the base obtuse to cordate, the apex acute to acuminate, the margins crenate, palmately 3-veined from base. Capitulescences to 17 × 11 cm; peduncles 1–6(10) mm long. Capitula 50–80-flowered, to 6 × 4 mm; involucre cylindrical to campanulate, to 5 × 4 mm; phyllaries 4–5-seriate, 3–5-costate, puberulent, apically narrowly acute, the outermost ones deltoid, ca. 0.5 mm long, grading to innermost ones lanceolate, to 5 mm long. Flowers: corolla funnelform, 3–3.5 mm long, white to pale pink, the lobes commonly pilosulous. Cypselae to 1.5 mm long, glabrous or with few apical setae, black, the angles and carpopodium stramineous; pappus bristles ca. 3 mm long. Fl (May, Jun, Jul, Aug, Sep, Oct, Dec), fr (Apr, Aug); in disturbed areas. *Herbe à chat, radié maringouin, zerbe chat* (Créole).

MIKANIA Willd.

Twining vines or lianas, rarely erect herbs. Stems subterete or hexagonal. Leaves simple, opposite or less commonly whorled; petioles generally present; blades chartaceous to fleshy or coriaceous, the margins entire or toothed; venation commonly palmate or subpalmate. Capitulescences terminal or axillary, ultimate branches spiciform, racemiform, glomerate, or thyrsoid-corymbiform, commonly many-headed. Capitula of all disk flowers (discoid), 4-flowered; peduncles absent or present; involucre cylindrical; phyllaries 4, subequal, weakly imbricate, persistent, often subtended by a smaller subinvolucral bract; receptacle flat, without pales, glabrous. Flowers bisexual; corolla funnelform or with throat turbinate to widely campanulate, mostly cream-colored, the lobes glabrous, glandular, or setose, commonly shorter than, but sometimes as long as or longer than throat; style branches erect and elongate, linear, cream-colored, the distal half with large sterile appendage, smooth to long-papillose, the proximal half with paired stigmatic lines, the style base enlarged. Cypselae commonly pentagonal, brownish to black; carpopodium symmetric, nonsculptured; pappus of numerous bristles, these about as long as corolla and longer than cypselae.

Mikania congesta DC., *M. cordifolia* (L.f.) Willd., and *M. micrantha* Kunth are to be expected in our area.

1. Capitula short-pedunculate. Style branches smooth or weakly papillose.
   2. Stems commonly subterete, not winged. Blades coriaceous to subcoriaceous, commonly abaxially glabrous, the base cuneate to rounded. Corolla lobes shorter than throat. . . . . *M. gleasonii.*
   2. Stems hexagonal, often winged. Blades chartaceous, abaxially puberulent to glabrate, the base cordate. Corolla lobes about as long as or longer than throat. . . . . *M. microptera.*
1. Capitula sessile, usually ternate in clusters of 3. Style branch appendage long-papillose.
   3. Leaves ovate, eglandular, the base usually conspicuously decurrent on petiole; petioles 1–6 cm long, subglabrous to finely puberulent. Corolla frequently weakly puberulous. Involucre 4–4.5 mm long. Subinvolucral bract, when present, lanceolate, ca. 1 mm long. . . . . *M. guaco.*
   3. Leaves broadly lanceolate to elliptic, often abaxially glandular, base obtuse to rounded; petiole 0.7–2.5 cm long, commonly densely puberulent. Corolla lobes weakly glandular. Involucre 5.5–8 mm long. Subinvolucral bract (when present) oblanceolate to obovate, 3–4 mm long. *M. parviflora.*

Mikania gleasonii B. L. Rob.

Lianas, to 20 m long. Stems commonly subterete, striate, puberulent or glabrate. Leaves: petioles 1–2.5(4) cm long; blades elliptic-ovate to ovate, (2.5)7–20(25) × (1.5)3–9(11) cm, coriaceous to subcoriaceous, adaxial surface glabrous or nearly so, abaxial surface pale green, minutely glandular-punctate or rarely also weakly puberulent with short appressed trichomes, the base cuneate to rounded, the apex acuminate, the margins entire; venation of 3–5 main veins from near base, the tertiary veins immersed or not so.

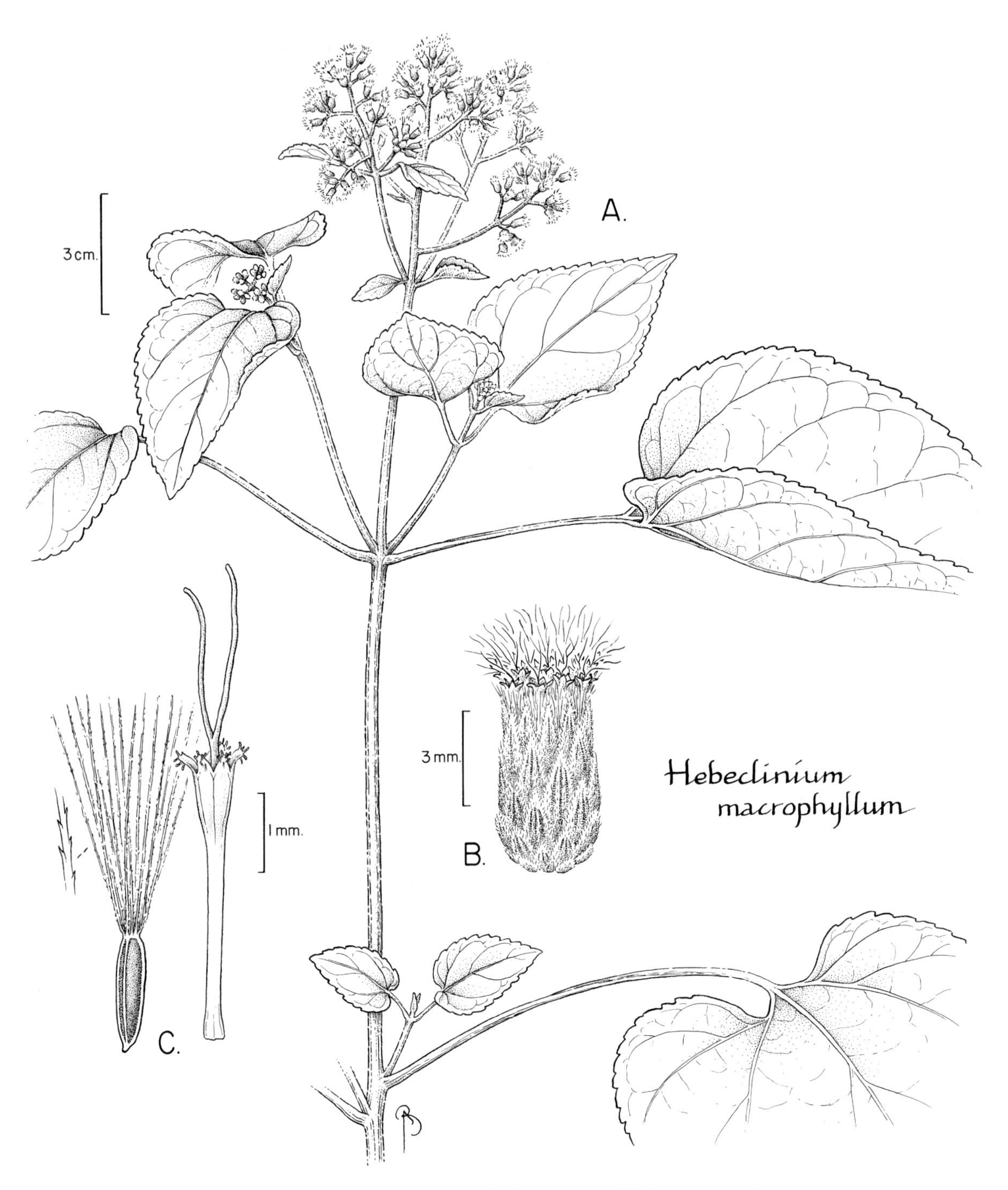

FIG. 40. ASTERACEAE. *Hebeclinium macrophyllum* (A, *Mori et al. 22321*; B, C, *Mori & Gracie 18257*). **A.** Distal part of stem with leaves and capitulescences. **B.** Lateral view of capitulum. **C.** Lateral views of disk corolla and style branches (right) and disk cypsela (near left) with detail of pappus bristle (above, far left).

Capitulescence corymbiform, often wide and diffuse, individual clusters open, often flat-topped; peduncles 0.5–5(6.5) mm long. Capitula 9–11(13) mm long; involucre 6–10 mm long, commonly subtended by a subinvolucral bract; phyllaries 1.1–1.5 mm wide, slightly swollen basally, commonly glabrous. Flowers: corolla funnelform, glabrous or sometimes sparingly glandular, ca. (5)5.5–6.5(7.4) mm long, the lobes (0.5)0.7–1(1.4) mm long, shorter than throat; style branches smooth or weakly papillose. Cypselae 5.5–8.5 mm long, glabrous or less frequently slightly glandular; pappus 6–9 mm long. Fl (Sep), fr (Jul); in non-flooded forest.

Mikania guaco Bonpl. PART 1: FIG. 5

Herbaceous vines, clambering shrubs, or lianas. Stems subterete, puberulent. Leaves: petioles 1–6 cm long, subglabrous to finely puberulent; blades ovate, to 17 × 15 cm, eglandular, adaxial surface glabrous to nearly so, abaxial surface finely puberulent to glabrate, the base usually conspicuously decurrent onto petiole, the apex acute to acuminate, the margins entire to denticulate; venation of 5–7 main veins from near base. Capitulescences corymbiform, often wide and diffuse, individual clusters dense, flat-topped, the branches flattened, or only so at nodes. Capitula 7–8.5 mm long, sessile, ternate; involucre 4–4.5 mm long, often subtended by a lanceolate subinvolucral bract, ca. 1 mm long; phyllaries oblanceolate, ca. 1 mm wide, puberulent, the apex obtuse. Flowers: corolla funnelform, 4–4.5 mm long, much exserted from involucre, frequently weakly puberulous, the lobes deltoid, ca. 0.5 mm long; style branch appendage long-papillose. Cypselae 2–2.5 mm long, puberulent; pappus ca. 4.5 mm long. Fl (Oct); in disturbed areas. *Cipó catinga* (Portuguese), *Radié grage, radié serpent, zerb'grage* (Créole).

Mikania microptera DC.

Herbaceous twining vines. Stems hexagonal, puberulent, angles with narrow subherbaceous wings. Leaves: petioles slender, to 4 cm long, glabrous to puberulent; blades cordiform, not decurrent onto petiole, 5–12 × 3–7 cm, chartaceous, adaxial surface puberulent to glabrate, abaxial surface commonly eglandular, puberulent, rarely glabrous, the base cordate, the apex acute to attenuate, the margins coarsely subentire to serrate; venation palmate, with 3–7 main veins. Capitulescences corymbiform, often wide and diffuse, individual clusters rounded, branches slightly flattened at nodes, with prominent wings. Capitula 5.7–6.5 mm long; peduncles 1–5 mm long; involucre 4.5–5.5 mm long; phyllaries 1–1.4 mm wide, weakly puberulent or glabrous, the apex acute to rounded, often mucronate. Flowers: corolla funnelform, the throat narrowly campanulate, 3.5–4 mm long, largely included within involucre, lightly glandular, the lobes lanceolate, ca. 1 mm long, about as long as or longer than the throat; style branch appendage smooth or finely papillose. Cypselae 2.5–3 mm long, glandular; pappus ca. 3 mm long. Fl (May); in secondary vegetation.

Mikania parviflora (Aubl.) H. Karst.

Twining vines to lianas. Stems subterete, densely puberulent. Leaves: petioles 0.7–2.5 cm long, commonly densely puberulent; blades broadly lanceolate to elliptic, 8–20 × 3–11 cm, the base obtuse to rounded, the apex acute, the margins entire, the adaxial surface glabrous to nearly so, the abaxial surface often puberulent and glandular (but glands destroyed when pressed in alcohol); venation pinnate. Capitulescences terminal or axillary, corymbiform, the branches subterete. Capitula ca. 10 mm tall, sessile, ternate; involucre 5.5–8 mm long, commonly subtended by an oblanceolate to obovate subinvolucral bract, 3–4 mm long; phyllaries narrowly elliptic to oblanceolate, ca. 1 mm wide, the apex obtuse, sometimes pinkish, the outer 2 phyllaries puberulent to glabrate, the inner 2 commonly glabrous. Flowers: corolla funnelform, 4–4.5 mm long, partly exserted from involucre, the lobes deltoid, ca. 0.5 mm long, weakly glandular; style branch appendage long-papillose. Cypselae ca. 4 mm long, mostly glabrous; pappus 4.5–5 mm long. Fl (Jan); in forest and along forest margins.

ROLANDRA Rottb.

Perennial herbs or subshrubs. Stems white-sericeous distally, glabrescent and brown with age. Leaves simple, alternate; petioles short; blades strongly discolorous; venation pinnate. Capitulescences axillary or terminal, sessile, forming secondary non bracteate capitulescences of spherical glomerules, glomerules of numerous capitula. Capitula of a single disk flower (discoid), sessile or nearly so; involucre cylindrical, flattened, of 2 unequal phyllaries, these strongly conduplicate and strongly imbricate (the inner within and opposite the outer), a third smaller bracteole sometimes loosely subtending 2 principal phyllaries; receptacle minute, without pales, weakly pilose. Flowers bisexual; corolla funnelform, deeply 3- or 4-lobed, generally white; anthers basally spurred, the spurs longer than filament collar; styles hispidulous on distal half, the branches hispidulous, very short, slender, gradually attenuate, with a continuous stigmatic surface. Cypselae obconic, weakly 4- or 5-veined, glandular; pappus a low, thin, laciniate crown.

Rolandra fruticosa (L.) Kuntze FIG. 41

Frequently erect, few-branched herbs, to 1(1.5) m tall. Stems subterete, sometimes trailing. Leaves: petioles 2–8 mm long, sericeous; blades elliptic-lanceolate, 3–10 × 1–4 cm, chartaceous, adaxial surface green, sparsely pilose to glabrate, abaxial surface white-tomentose, glandular, the midrib and secondary veins often sericeous, the base cuneate, the apex acute, sometimes mucronulate, the margins entire or serrulate. Glomerules at distal stem nodes, 1–1.5 cm diam; involucre 0.6–1.3 mm wide; phyllaries elliptic-lanceolate, membranous, brownish, weakly sericeous, glandular to glabrate, the outer phyllary aristate, 4.5–6 mm long, the arista to 1.2 mm long, the inner phyllary 3.2–4.5 mm long. Flowers: corollas 3–3.5 mm long, glabrous, the lobes 1–1.5 mm long. Cypselae 1.5–2 mm long; pappus crown to 0.3 mm long. Fl (Jun), fr (Jan, Jun, Aug, Oct, Dec); in disturbed areas. *Radié commandeur, tête-nègre* (Créole).

SPHAGNETICOLA O. Hoffm.

Procumbent perennial, puberulent to pubescent, sometimes succulent herbs. Stems elongating sympodially. Leaves simple, opposite; petioles absent or short; blades lanceolate to trilobed and rhombic or ovate in outline; venation of 3 main veins from above base. Capitulescences terminal (usually appearing axillary when capitula laterally displaced), with a single or a few capitula; peduncles elongate, nonleafy. Capitula with ray and disk flowers (radiate), many-flowered; involucre campanulate to subhemispherical; phyllaries loosely imbricate, subequal, weakly 2(3)-seriate, foliaceous or inner ones only apically so; receptacle with pales, the pales elliptic-lanceolate, conduplicate. Ray flowers pistillate; corolla yellow or orange, the tube short, the limb showy. Disk flowers bisexual; corolla funnelform, yellow or orange; anther thecae black, apical appendages glandular; style branches strongly papillose, with poorly developed paired stigmatic lines, nearly erect or slightly reflexed, weakly exserted. Cypselae often angled, black and tuberculate at maturity, the tubercles tan, lacking a well developed carpopodium; pappus fimbriate, the fimbriae obscured by a corky coroniform collar, this much shorter than cypsela.

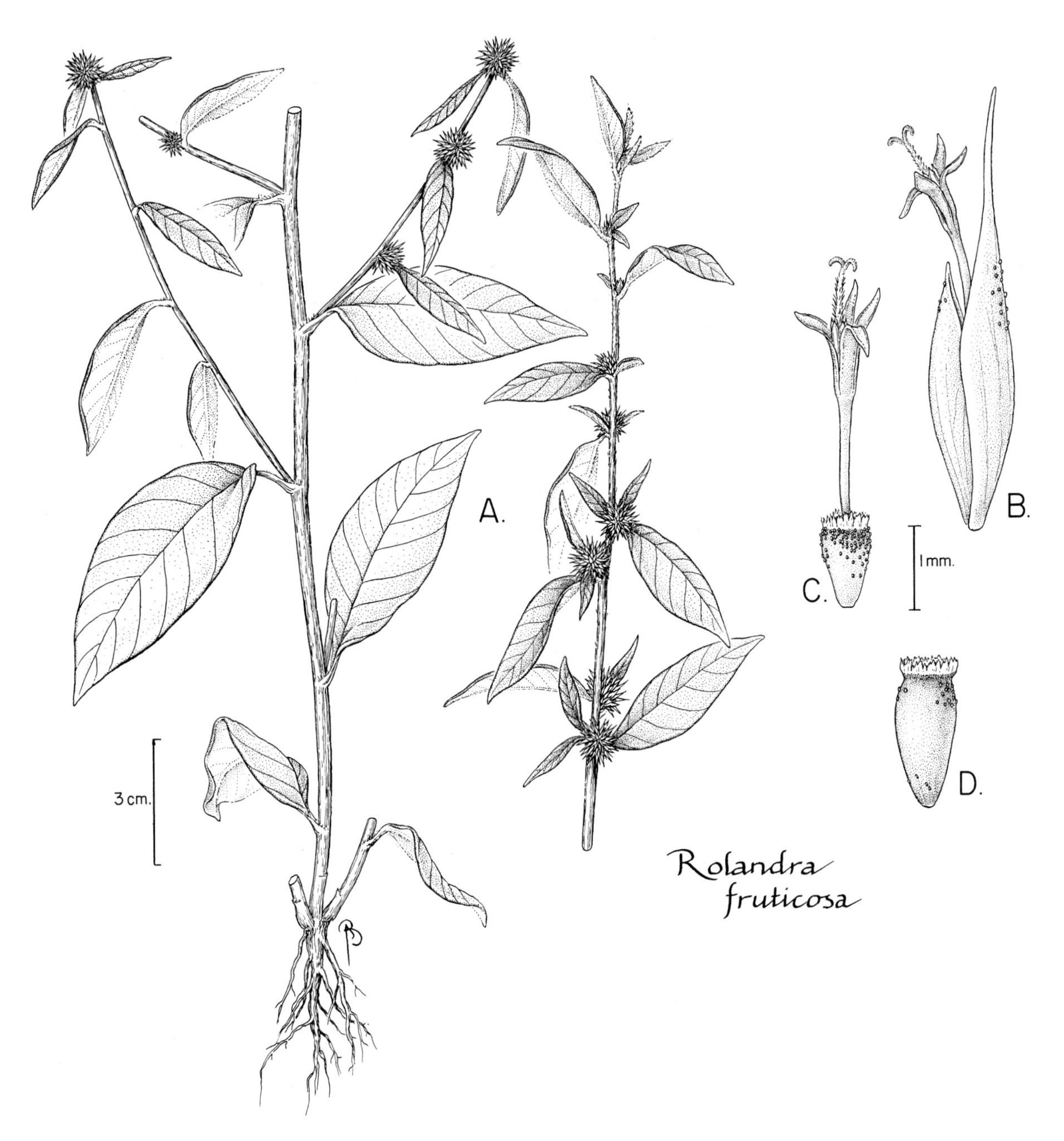

FIG. 41. ASTERACEAE. *Rolandra fruticosa* (*Mori & Gracie 18395*). **A.** Lower (left) and upper (right) parts of plant. **B.** Lateral view of capitulum. **C.** Lateral view of disk flower. **D.** Lateral view of disk cypsela.

Sphagneticola trilobata (L.) Pruski [Syns.: *Wedelia trilobata* (L.) Hitchc., *W. paludosa* DC.] FIG. 42

Procumbent herbs, rooting at the proximal nodes. Stems subterete, to 2 m or more long, the apices ascending. Leaves sometimes inconspicuously basally connate; petioles 0–5 mm long; blades oblanceolate to rhombic, often 3-lobed, 3–10.5 × 2.5–8 cm, chartaceous to fleshy, surfaces hirsute to strigose, glandular on abaxial surface, the base gradually or abruptly tapering, the apex acute to acuminate, the margins subentire or toothed, each margin often with a prominent medial lobe. Capitula solitary; peduncles 3.5–14 cm long; involucre campanulate, 10–14 mm long and wide; phyllaries 12–15, subequal, oblanceolate to oblong, 10–14 × 2.5–4.5 mm, green, strigose, weakly glandular or inner ones merely puberulent; pales oblanceolate. Ray flowers 4–14; corolla yellow, the limb to 15 mm long, ca. 10-veined, the apex 3-lobed, the abaxial surface glandular. Disk flowers many; corolla 4.5–5.5 mm long, shortly 5-lobed, yellow, the lobes 0.6–0.8 mm long, strongly pubescent-papillose adaxially or marginally, occasionally glandular on abaxial surface. Cypselae pyriform, ca. 3 mm long, the collar to ca. 1.1 mm long. Fl (Jun); in disturbed areas. *Vervine crabe* (Créole).

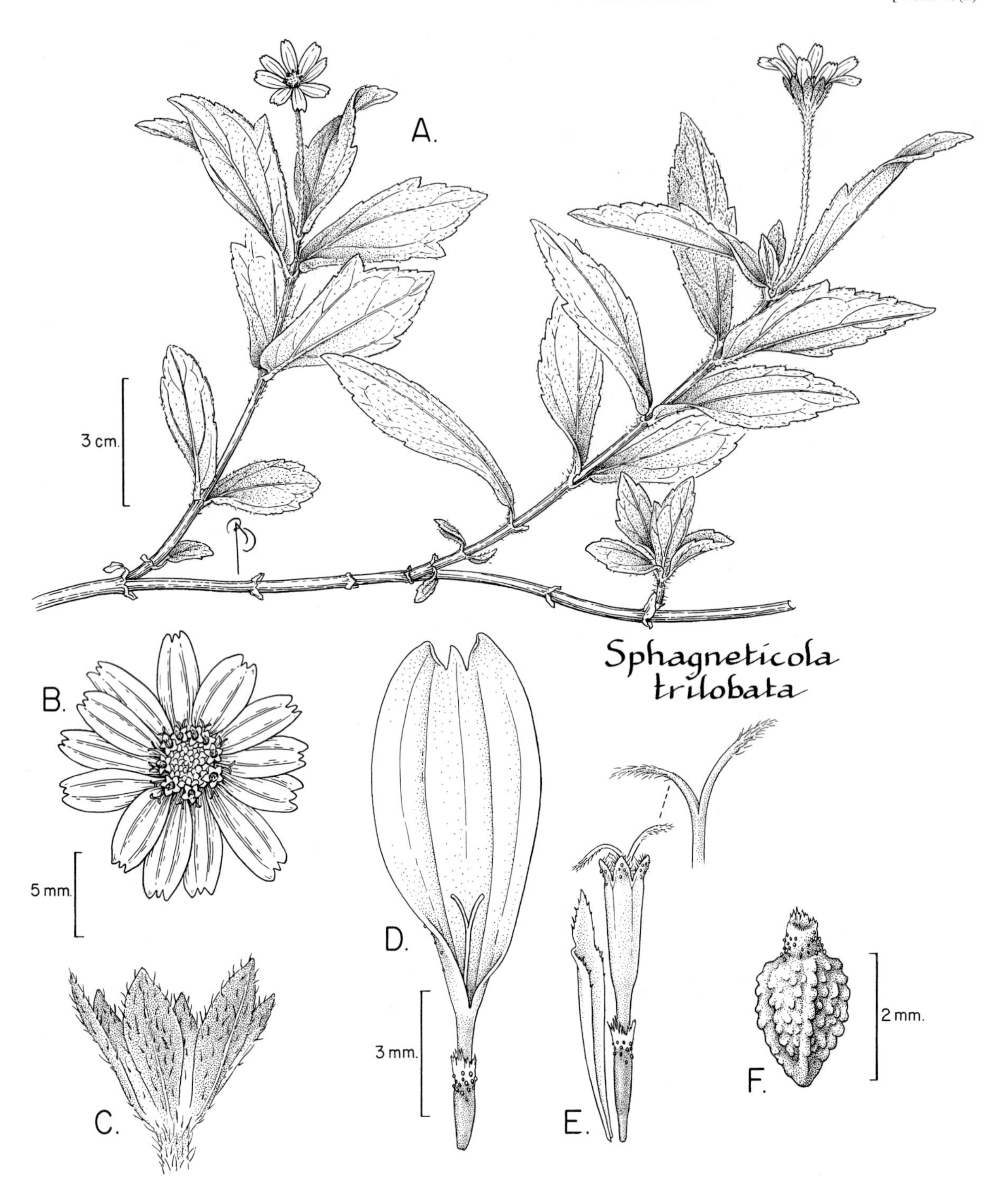

FIG. 42. ASTERACEAE. *Sphagneticola trilobata* (A–E, *Acevedo-Rodríguez 2894* from St. John, U.S. Virgin Islands; F, *Zanoni 10417* from Dominican Republic). **A.** Part of stem with leaves and capitula. **B.** Apical view of capitulum. **C.** Lateral view of involucre. **D.** Adaxial view of ray flower. **E.** Lateral view of disk flower (far right) with subtending pale (near right), and detail of style branches (above right). **F.** Lateral view of cypsela. (Reprinted with permission from P. Acevedo-Rodríguez, *Flora of St. John, U.S. Virgin Islands*, Mem. New York Bot. Gard. 78, 1996.)

SYMPHYOTRICHUM Nees

Perennial, mostly glabrous herbs. Stems erect, few-branched distally, glabrous or occasionally puberulent in lines on axes of capitulescence. Leaves simple, alternate, the distal ones commonly sessile and clasping, rarely cordate, the proximal ones commonly with tapering winged petioles; venation usually pinnate. Capitulescences terminal, open thyrsoid-panicles of few to many capitula. Capitula with ray and disk flowers (radiate), many-flowered; peduncles bracteate; involucre hemispherical or campanulate; phyllaries imbricate, commonly graduated, the outer ones sometimes leafy or involucre sometimes closely subtended by reduced leaves; receptacle without pales. Ray flowers pistillate; corolla lavender (ours), often showy, exserted from involucre. Disk flowers bisexual; corolla funnelform, yellow (ours); anthers slightly exserted, pale brown; style branches dorsally papillose, short, erect, linear-lanceolate, with paired stigmatic lines proximally, the apex sterile, triangular. Cypselae obconic to fusiform, terete, multiveined, eglandular, puberulent when young or more commonly glabrous; pappus of numerous capillary bristles slightly shorter than disk corollas, but longer than cypselae.

This genus was considered to be part of *Aster* as used by New World authors. *Aster* is typified by material from the Old World with flattened, glandular cypselae, and matches only one of the approximately 181 species of *Aster s. lat.* from the New World. The New World species have been removed from *Aster* and placed into 14 genera, including *Symphyotrichum* into which 97 species are accommodated (Nesom, 1994).

Symphyotrichum laeve (L.) Á. Löve & D. Löve [Syn.: *Aster laevis* L.]

Eglandular herbs, to 0.5(1) m tall. Stems subterete, striate. Leaves: petioles absent in distal leaves, present or absent in proximal ones; blades of distal leaves lanceolate to oblanceolate, auriculate-clasping, to ca. 11 × 3 cm, of proximal leaves oblanceolate to spatulate, to ca. 19 × 4 cm, the base often tapered into winged petioliariform, weakly clasping base, ca. 2–7 cm long, all blades chartaceous, both surfaces usually glabrous, glaucous, the apex acute to obtuse, the margins entire or occasionally weakly serrate; venation obscurely pinnate. Capitulescences terminal or axillary from distal nodes, open, several-headed, with leaves reduced and bract-like, the branches and peduncles glabrous or with puberulence in lines; peduncles 0.5–3.5 cm long, often bracteolate. Capitula ca. 1 cm long and wide; involucre 4.5–7 mm long; phyllaries imbricate, strongly graduated, 3–5-seriate, 1.5–2 mm long, with a diamond-shaped, herbaceous patch, the apex acute, the base stramineous, indurate. Ray flowers mostly 20–25; corolla commonly lavender, the tube ca. 2 mm long, the limb 9–13 mm long, minutely 3-notched at apex, 4-veined, exserted from involucre. Disk flowers many; corolla 5–6 mm long, yellow, the lobes 0.6–0.9 mm long; anthers stramineous. Cypselae 1–1.5 mm long, 5-veined, glabrous; pappus bristles 3.5–4.5 mm long, stramineous or sometimes brownish. Fl (Jun); in disturbed areas.

SYNEDRELLA Gaertn.

Annual or short-lived perennial, pubescent, eglandular herbs. Stems strigose. Leaves simple, opposite; petioles winged; blades short-decurrent at base; venation palmate, with 3 veins from near base. Capitulescences of 1 to several capitula in axils of most pairs of stem leaves, these sessile or nearly so and glomerate, or sometimes long-pedunculate. Capitula with ray and disk flowers (radiate); involucre cylindrical; phyllaries few, imbricate, subequal, striate, the outer 1–3 foliaceous, the inner ones scarious, striate; receptacle with pales, the pales flat. Ray flowers pistillate; corolla yellow, glabrous. Disk flowers bisexual, shortly 4- or 5-lobed; corolla narrowly funnelform, yellow, the lobes pubescent; anthers black, included; style branches linear-lanceolate, with poorly developed paired stigmatic lines. Cypselae dimorphic; ray cypselae large, oblong-ovate, tangentially flattened, body black, winged, the wings lateral, stramineous, deeply cut into teeth, the pappus of 2 stout awns, these much shorter than cypsela, grading into wings of cypsela; disk cypselae small, obconical, unwinged, sometimes slightly compressed, with 2(4) stout, divergent awns about as long as cypsela and nearly reaching to base of corolla lobes.

Synedrella nodiflora (L.) Gaertn. FIG. 43, PL. 21f

Herbs, to 1(1.5) m tall. Stems subterete, weakly branched, erect or procumbent. Leaves: petioles obscure or to 2.5 cm long, commonly winged, strigose; blades elliptic to ovate, 3–12 × 1.5–6 cm, chartaceous, adaxial surface weakly strigose, abaxial surface strigose, the base abruptly contracted and attenuate, the apex acuminate to obtuse, the margins subentire to somewhat serrate. Capitulescences with peduncles lacking or to 0.5(4) cm long. Capitula 9–18-flowered, ca. 9 × 4 mm; phyllaries elliptic-lanceolate, the outer ones green, to ca. 9 × 2.5 mm, strigose, the inner ones stramineous, 5–6.5 × 0.7–1.8 mm, glabrous; receptacle convex, the pales ca. 6 × 1 mm, narrowly elliptic, conduplicate. Ray flowers 3–6(7); corolla 3–4 mm long, the limb shortly exserted, sometimes shallowly notched. Disk flowers 6–11, bisexual; corolla to 3.5 mm long, the lobes ca. 0.3 mm long. Ray cypselae 3–5 mm long, the awns and marginal teeth of wings ca. 1 mm long. Disk cypselae ca. 3 mm long, the awns to ca. 3 mm long. Fl (Jun, Aug, Sep, Oct, Nov, Dec); in disturbed areas. *Bouton d'or, radié pisser, razié pisser* (Créole).

TILESIA G. Mey. [Syn.: *Wulffia* Cass.].

Scandent herbs, shrubs, or sometimes climbing shrubs to lianas. Stems weakly angled or striate, puberulent. Leaves simple, opposite; petioles present; blades lanceolate to ovate, both surfaces eglandular; venation pinnate to palmate with 3–5 main veins. Capitulescences terminal from distal nodes, in cymose or subumbellate groups of 3(7) capitula; peduncles stout, pubescent. Capitula with ray and disk flowers (radiate) or with all disk flowers (discoid); involucre hemispherical; phyllaries imbricate, 2–3-seriate, the outer ones herbaceously tipped or sometimes foliar, the inner ones scarious, strongly striate, resembling pales; receptacle with pales, the pales obovate, stoutly apiculate, rigid, conduplicate, strongly striate, protruding in fruit. Ray flowers (when present) sterile; corolla yellow or yellow-orange, the tube glabrous, the limb exserted from involucre, abaxially puberulent. Disk flowers bisexual; corolla funnelform, yellow or orange-yellow, the lobes long-triangular, apically puberulent; anthers included, the thecae black, the connective cream-colored, the appendages deltoid, cream-colored; style branches becoming reflexed, papillose, the paired stigmatic lines weakly discernible. Disk cypselae black, glabrous or apically puberulent, fat and surface fleshy at maturity; pappus absent.

Tilesia baccata (L.) Pruski var. **baccata** [Syns.: *Wulffia baccata* (L.) Kuntze var. *baccata*, *Wulffia stenoglossa* (Cass.) DC.] FIG. 44

Herbs or shrubs, 1–3 m tall. Stems sometimes purple-mottled. Leaves: petioles slender, to 2(3.2) cm long, strigose; blades elliptic to ovate, (5)10–16 × (3)4–8(9) cm, chartaceous to more commonly subcoriaceous, adaxial surface scabrous, abaxial surface strigose, the base cuneate, rounded, or rarely becoming decurrent, the apex acute or acuminate, the margins serrate; venation palmate, with 3–5 main

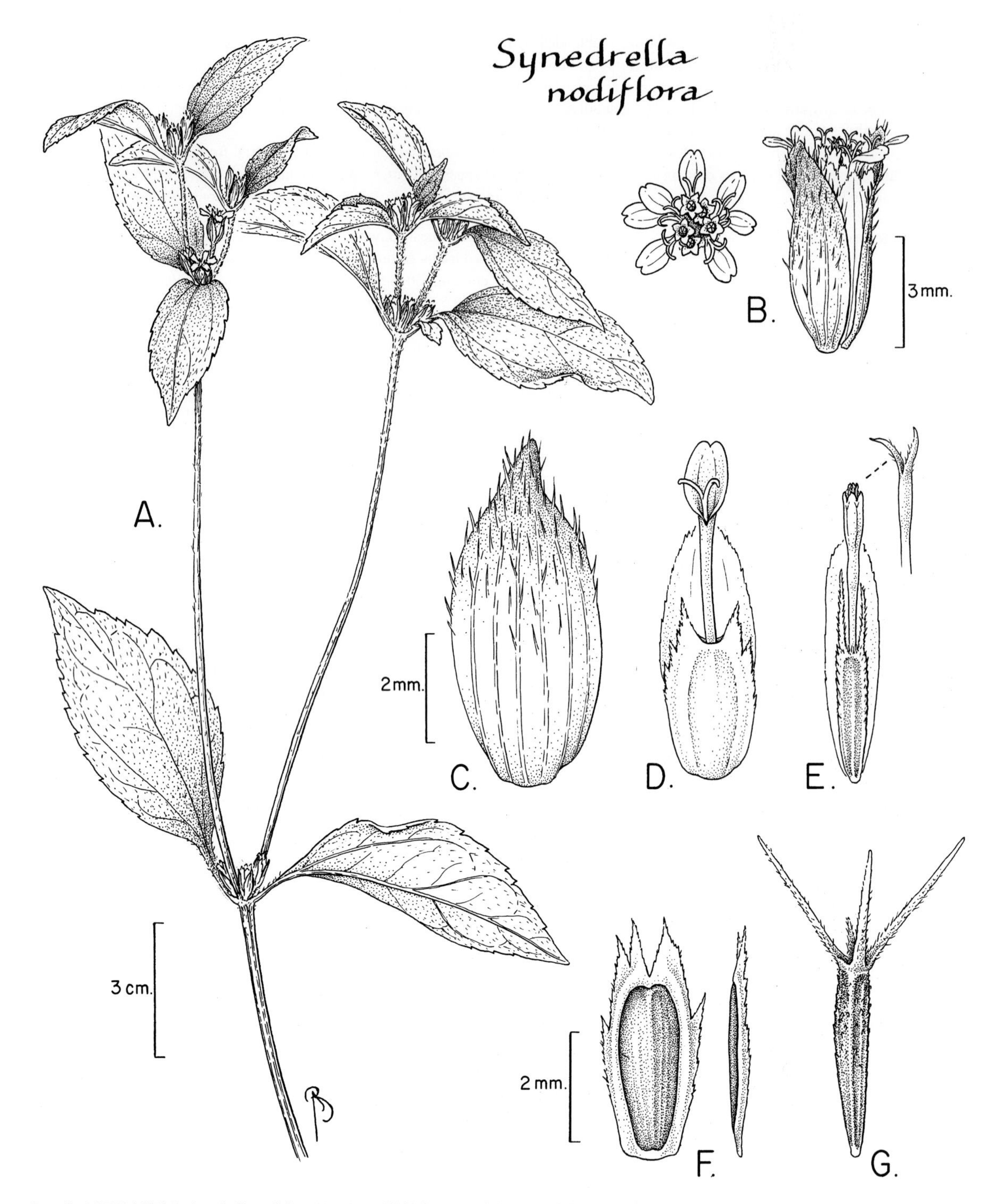

FIG. 43. ASTERACEAE. *Synedrella nodiflora* (A, *Mejia 10293* from Dominican Republic; B, *Acevedo-Rodríguez 4041* from St. John, U.S. Virgin Islands; C–G, *Burch 2522* from Dominican Republic). **A.** Distal part of stem with leaves and capitulescences. **B.** Lateral (right) and apical (left) views of capitula. **C.** Abaxial view of outer phyllary. **D.** Adaxial view of ray flower with subtending phyllary. **E.** Adaxial view of disk flower with subtending pale and detail of style branches (above right). **F.** Two lateral views of flattened winged ray cypselae. **G.** Lateral view of disk cypsela. (Reprinted with permission from P. Acevedo-Rodríguez, *Flora of St. John, U.S. Virgin Islands*, Mem. New York Bot. Gard. 78, 1996.)

veins from near base. Capitulescences with peduncles to 7(15) cm long. Capitula with ray and disk flowers, globose, 1–1.5 cm long and wide; phyllaries lanceolate to elliptic, 2.5–4.5 mm long, strigose; receptacle ca. 1.5 mm wide. Ray flowers 8–15; corolla yellow, the tube ca. 1 mm long, the limb 10–17 × 4 mm, entire or notched. Disk flowers 40–90; corolla 5–6 mm long, yellow, the tube ca. 1 mm long, the lobes 1–1.5 mm long. Disk cypselae oblong or pyriform, 3–5 mm long. Fl (Jan, Feb, Mar, Jun, Aug); in non-flooded forest, along forest edges, and in disturbed areas. *Bouton d'or* (Créole), *cambará amarela* (Portuguese), *manger lapin*, *radié-jaunâtre*, *zerb carême* (Créole).

Plates 17–24

Plate 17. APOCYNACEAE. **a.** *Odontadenia punticulosa* (*Mori et al. 24667*), apical view of flower. [Photo by S. Mori] ARALIACEAE. **b.** *Shefflera morototoni* (unvouchered), tree; note palmately divided leaves. ARISTOLOCHIACEAE. **c.** *Aristolochia bukuti* (*Mori et al. 23347*), liana with many cauliflorous flowers.

Plate 18. ARISTOLOCHIACEAE. **a.** *Aristolochia bukuti* (*Mori et al. 21637*), flower. **b.** *Aristolochia trilobata* (*Mori et al. 22351* grown at NYBG from seed collected in Saül), flower. **c.** *Aristolochia bukuti* (*Mori et al. 22965*), liana showing corky ridges and old fruit. **d.** *Aristolochia trilobata* (*Mori et al. 22351* cultivated in Saül), fruits.

Plate 19. ASCLEPIADACEAE. *Asclepias curassavica* (unvouchered, photographed in Trinidad), inflorescence with flowers and buds.

Plate 20. ASCLEPIADACEAE. **a.** *Asclepias curassavica* (unvouchered, photographed in cultivation in Florida), flowers and buds. **b.** *Matelea gracieae* (*Phillippe 26496*), apical view of flower. **c.** *Matelea gracieae* (*Phillippe 26496*), dehisced fruit. **d.** *Metalepis albiflora* (*Mori et al. 23941*), flowers and buds. [Photo by S. Mori]

Plate 21. ASTERACEAE. **a.** *Acmella uliginosa* (*Mori et al. 22077*), capitula. **b.** *Ageratum conyzoides* (*Mori et al. 24754*), capitula. **c.** *Emilia sonchifolia* var. *sonchifolia* (*Mori et al. 21187*), capitula. **d.** *Elephantopus mollis* (*Mori & Gracie 23928*), two synflorescences with subtending bracts. **e.** *Hebeclinium macrophyllum* (*Mori et al. 23985*), capitulescence. **f.** *Synedrella nodiflora* (unvouchered), capitulum.

Plate 22. BALANOPHORACEAE. **a.** *Helosis cayennensis* (unvouchered), flowering plant. **b.** *Helosis cayennensis* (unvouchered), inflorescence with remnants of peltate scales. **c.** *Helosis cayennensis* (unvouchered), flowering plant with peltate scales. **d.** *Helosis cayennensis* (unvouchered), staminate flowers. BEGONIACEAE. **e.** *Begonia glabra* (*Mori et al. 23064*), winged fruits. **f.** *Begonia prieurii* (*Mori et al. 22073*), inflorescences with pistillate flowers (left) and staminate flowers and buds (right).

Plate 23. BIGNONIACEAE. **a.** *Adenocalymna saülense* (*Mori et al. 24133*), inflorescence and fruit. **b.** *Adenocalymna saülense* (*Mori & Gracie 24211*), flower and buds; note terminal tendril. **c.** *Amphilophium paniculatum* var. *imatacense* (*Mori & Pepper 24261*), inflorescence with one flower; note double calyx, the outer one flared. [Photo by S. Mori] **d.** *Arrabidaea inaequalis* (*Mori & Gracie 21145*), flower and buds. **e.** *Arrabidaea nigrescens* (*Mori & Gracie 24210*), flowers and buds; note deep purple calyces. **f.** *Arrabidaea patellifera* (*Mori et al. 22030*), flower and bud; note constricted corolla tube and cup-shaped calyx. [Photo by S. Mori]

Plate 24. BIGNONIACEAE. **a.** *Arrabidaea trailii* (*Mori et al. 20869*), part of inflorescence. **b.** *Arrabidaea trailii* (*Mori et al. 20869*), flower and buds.

APOCYNACEAE (continued)

Odontadenia punticulosa a.

ARALIACEAE

Shefflera morototoni b.

ARISTOLOCHIACEAE

Aristolochia bukuti c.

Aristolochia bukuti a.

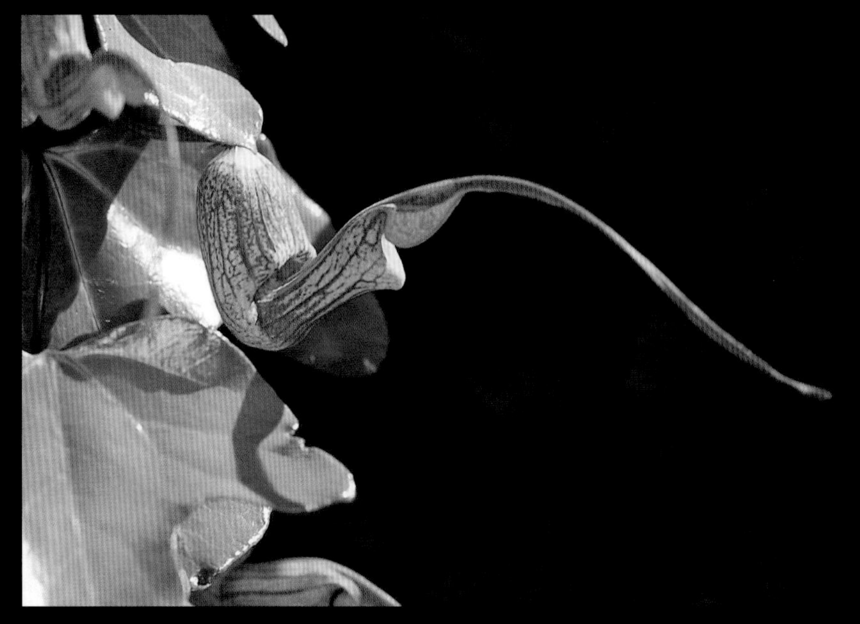

Aristolochia trilobata b.

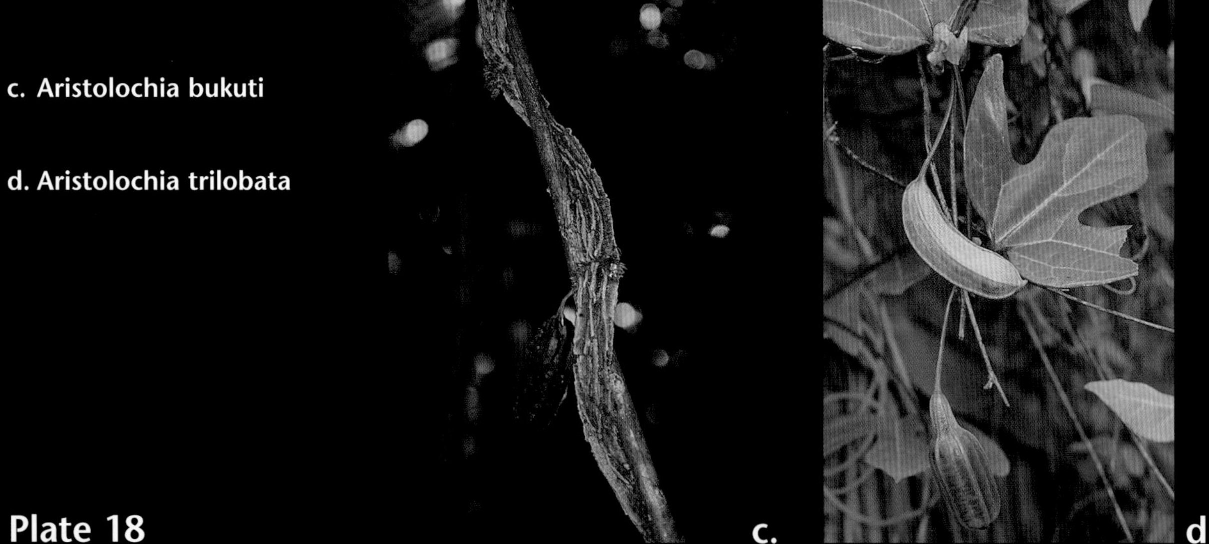

c. Aristolochia bukuti

d. Aristolochia trilobata

Plate 18

Asclepias curassavica

Asclepias curassavica **a.**

Matelea gracieae **b.**

c. *Matelea gracieae*

Metalepis albiflora **d.**

Plate 20

Acmella uliginosa a.

Emilia sonchifolia var. sonchifolia c.

Ageratum conyzoides b.

Elephantopus mollis d.

Synedrella nodiflora f.

Hebeclinium macrophyllum e.

Plate 21

Helosis cayennensis **c.**

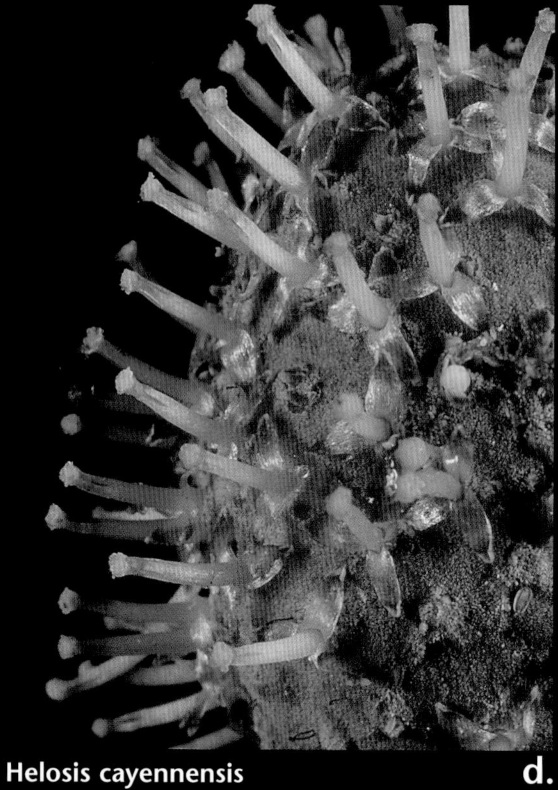

Helosis cayennensis **d.**

BEGONIACEAE

Begonia glabra **e.**

Begonia prieurii **f.**

Plate 22

Adenocalymna saülense a.

Adenocalymna saülense b.

Amphilophium paniculatum var. imatacense c.

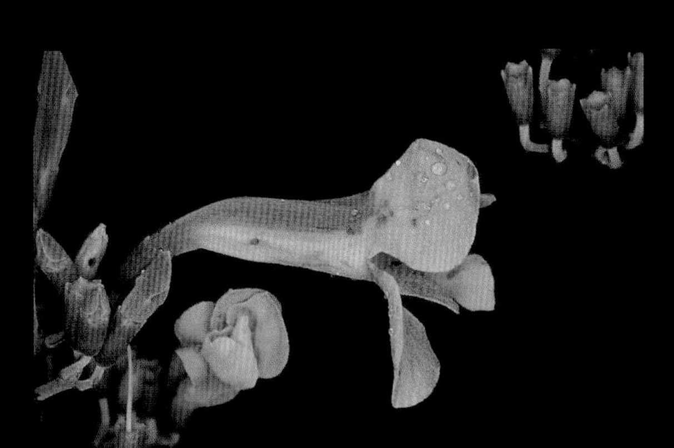

Arrabidaea inaequalis d.

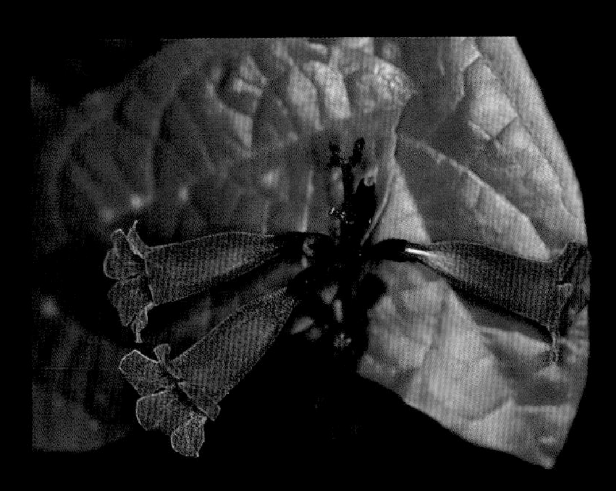

Arrabidaea nigrescens e.

Arrabidaea patellifera f.

Arrabidaea traili **a.**

Arrabidaea traili **b.**

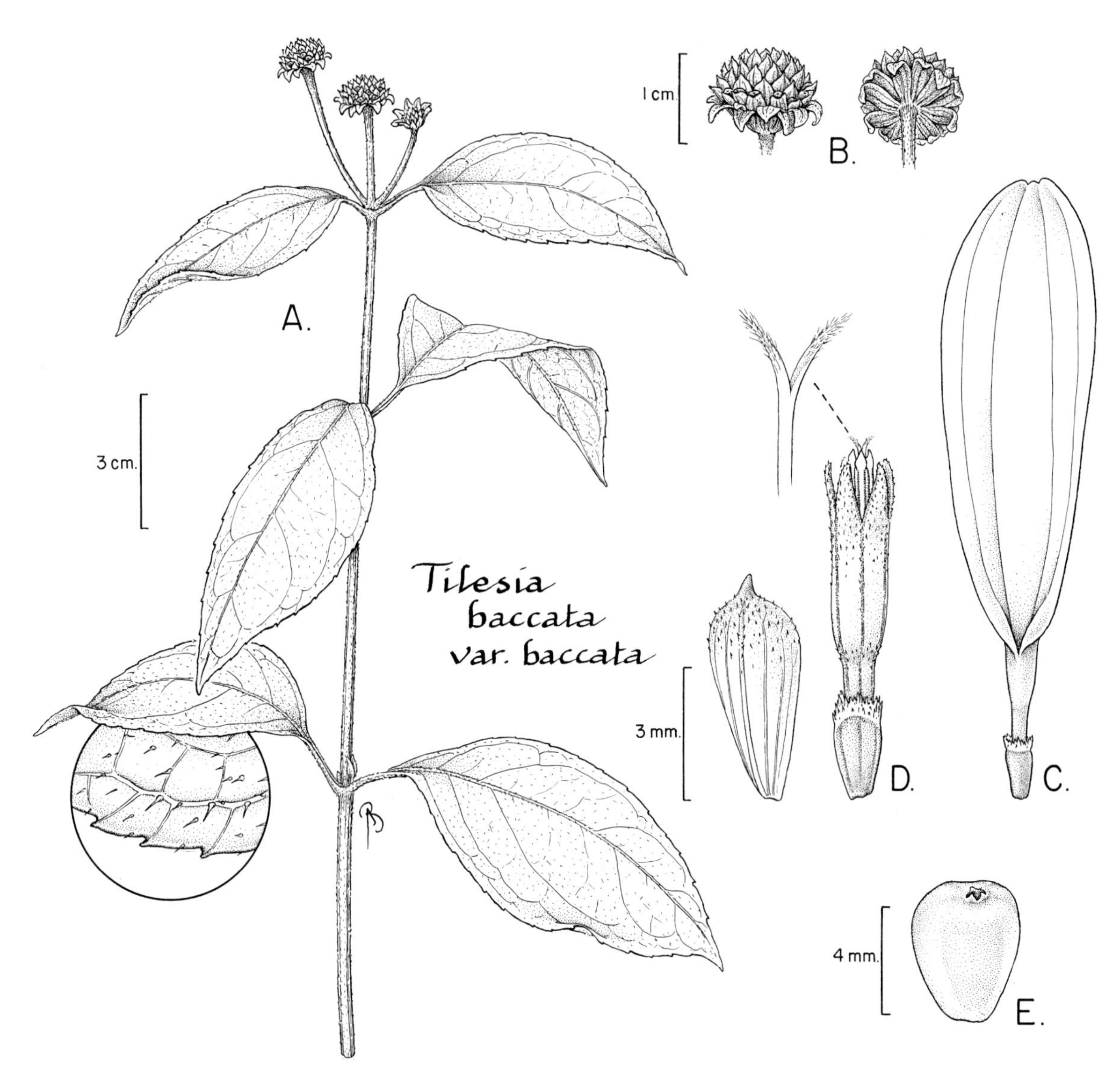

FIG. 44. ASTERACEAE. *Tilesia baccata* var. *baccata* (A, B, *Mori et al. 22320*; C–E, *Lindeman 11757*). **A.** Distal part of plant with capitulescence and detail of adaxial leaf surface (below left). **B.** Lateral (left) and oblique ventral (right) views of capitula. **C.** Adaxial view of ray flower. **D.** Lateral view of disk flower (right) with detail (above left) of style branches, and pale (below left). **E.** Lateral view of cypsela.

UNXIA L.f.

Annual herbs to perennial subshrubs. Stems moderately dichotomously branched, erect, glabrous to pilose, subterete. Leaves simple, opposite; petioles absent or short; blades lanceolate to elliptic; venation of 3(5) subpalmate veins from near base. Capitulescences terminal or axillary in uppermost nodes, of 1 to few subsessile, clustered capitula. Capitula with ray and disk flowers (radiate); involucre hemispherical; phyllaries imbricate, several-seriate, striate, the outer 2 or 3 herbaceous, the inner ones membranous, often yellowish; receptacles with pales, the pales scarious, brittle, narrowly lanceolate, not conduplicate, readily deciduous. Ray flowers pistillate; corollas yellow, the limb shortly exserted from involucre, glandular abaxially; style branches linear with paired stigmatic lines. Disk flowers functionally staminate; corollas yellow-orange, glabrous, glandular, or the tube pilose, the lobes triangular, erect; anthers yellow to black; ovary sterile; style undivided. Cypselae ovoid, glabrous, compressed, black; pappus none.

Unxia camphorata L.f. [Syns.: *Unxia hirsuta* Rich., *Melanpodium camphoratum* (L.f.) Baker]

Herbs or rarely suffrutescent annuals, to 50 cm tall. Stems glabrous to pilose. Leaves: petioles sessile to 4 mm long.; blades lanceolate, 1.5–8(10) × 0.3–1.7(2.7) cm, chartaceous, subglabrous to pilose, also sparsely glandular abaxially, the base commonly rounded to cuneate, the apex acute to acuminate, the margins subentire to serrulate. Capitula 9–15-flowered, not exserted from subtending leaves; peduncles commonly 1–5 mm long, subglabrous to pilose; involucres 3–5 × 3–5 mm; phyllaries elliptic to oblong, 3–5 × 1.5–2.5 mm, the outer ones pubescent or pilose; receptacle convex, ca. 1 mm diam., the pales small. Ray flowers 3–5; corolla ca. 2.3 mm long, the limb ca. 2 mm long, the tube ca. 0.3 mm long, glabrous to weakly pilose. Disk flowers 6–10; corolla funnelform, ca. 2.9 mm long, glandular, otherwise glabrous to weakly pilose, the lobes ca. 0.5 mm long. Cypselae ca. 2 mm long. Fr (Sep); near rock outcrop on Pic Matécho. *Radier camphe* (Créole).

BALANOPHORACEAE (Balanophora Family)

Carol A. Gracie

Achlorophylous, fleshy parasites on roots of various tree species. Leaves, if present, alternate, spiralled, or arising at ± same level, scale-like. Inflorescences terminal, ellipsoid or ovoid, the flowers numerous. Flowers small, unisexual. Staminate flowers: perianth absent or in one whorl; tepals, when present, usually 3; stamens fused, the anthers 3, separate or connate. Pistillate flowers: perianth absent or of minute tepals; gynoecium of 2–3 carpels united to form compound ovary; ovary superior or inferior, usually solid, without locule, the styles 2, less frequently 1; embryo generally one, few-celled, developing in central tissue of ovary. Fruits usually small, indehiscent achenes, the numerous fruits sometimes aggregating to form fleshy multiple fruits. Seeds solitary.

Hansen, B. 1980. Balanophoraceae. Fl. Neotrop. Monogr. **23:** 1–80.

———. 1993. Balanophoraceae. *In* A. R. A. Görts-van Rijn (ed.), Flora of the Guianas, Ser. A, **14:** 40–43, 65–73. Koeltz Scientific Books, Koenigstein.

Lanjouw, J. 1966a. Balanophoraceae. *In* A. Pulle (ed.), Flora of Suriname **I(1):** 45–46. E. J. Brill, Leiden.

Miller, J. S. 1997. Balanophoraceae. *In* J. A. Steyermark, P. E. Berry & B. K. Holst (gen. eds.), Flora of the Venezuelan Guayana **3:** 393–395. P. E. Berry, B. K. Holst & K. Yatskievych (vol. eds.). Missouri Botanical Garden, St. Louis.

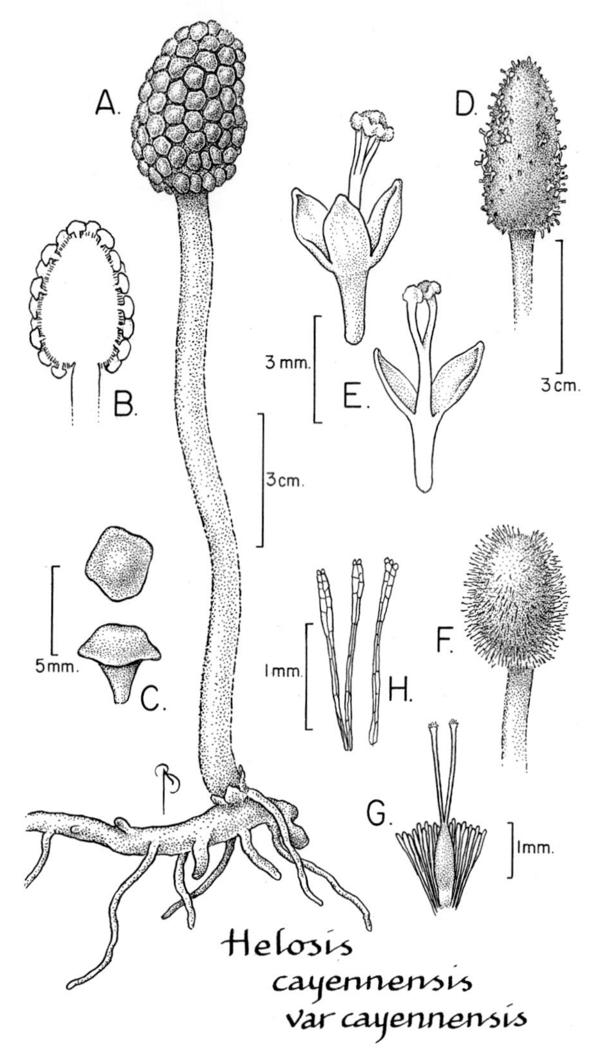

FIG. 45. BALANOPHORACEAE. *Helosis cayennensis* var. *cayennensis* (A–C, *Mori et al. 22276*; D, *Marshall & Rombold 174*; E, *Knapp 6601* from Peru; F, unvouchered photo by C. Gracie; G, H, *Wurdack & Monchino 39757* from Venezuela). **A.** Aerial portion of plant and part of subterranean portion. **B.** Medial section of inflorescence. **C.** Apical (above) and lateral (left) views of peltate bracts from inflorescence. **D.** Inflorescence with a few staminate flowers. **E.** Lateral view (above) and medial section (right) of staminate flower with one anther removed; surrounding hairs not shown. **F.** Inflorescence covered with ciliate hairs during pistillate phase. **G.** Lateral view of pistillate flower with perianth and some of surrounding hairs removed. **H.** Three hairs.

HELOSIS Rich.

A monospecific genus. The generic description same as the description of the species.

Helosis cayennensis (Sw.) Spreng. var. **cayennensis** FIG. 45, PLS. 22a–d

Herbs, the fleshy stalk, 5–15 cm tall, topped by 2–10 × 1–3 cm, ellipsoid-ovoid capitulum, appearing mushroom-like, the above ground portion of plant pinkish-red to orange, sometimes yellowish, the below ground portion with direct connection with xylem of host, producing subterranean starchy tuber containing both host root tissue as well as parasite tissue at site of connection. Inflorescences on stems arising at various intervals from subterranean, horizontal, rhizome-like structures produced by tubers, the stems with 2–6 small, triangular scales, often ± joined to form ring, the globose part of inflorescences covered by adjacent, hexagonal, stalked, early deciduous, peltate bracts, 3.5–4.5 mm diam. Flowers subtended by clavate, white hairs. Staminate flowers: perianth of 3 white lobes with red to purple tips; stamens 3, the filaments united into white column split into three filaments distally, the anthers connate. Pistillate flowers: perianth segments 2, broadly ligulate, adnate to compressed ovary, the styles 2, filiform, extending above surrounding layer of hairs. Fruits small achenes. Fl (Apr, May, Jul, Aug, Dec), fr (Jan, Feb); common, but scattered in non-flooded forest, reported to parasitize *Inga* (Mimosaceae) and *Boehmeria* (Urticaceae) by Hansen, 1980.

BEGONIACEAE (Begonia Family)

Scott V. Heald

Herbs, succulent, erect or scandent. Stems glabrous in our area. Stipules persistent. Leaves simple, alternate. Inflorescences axillary, cymose. Flowers unisexual (plants monoecious); tepals petaloid, white in our area. Staminate flowers lacking bracteoles; tepals 2–4, valvate. Pistillate flowers with bracteoles; tepals 4–5, free, imbricate; ovary inferior, usually 3-locular, with 3 unequal wings, the styles 3, bipartite, the branches with slight helical twisting, entirely covered in papillae in our area; placentation axile, the ovules numerous. Fruits loculicidal capsules. Seeds small, with or without terminal sacs.

Steyermark, J. A. 1997. Begoniaceae. *In* J. A. Steyermark, P. E. Berry & B. K. Holst (gen. eds.), Flora of the Venezuelan Guayana **3:** 397–403. P. E. Berry, B. K. Holst & K. Yatskievych (vol. eds.). Missouri Botanical Garden, St. Louis.

BEGONIA L.

The only genus of the family to occur in our area. Generic description same as the description of the family.

1. Plants scandent, hemiepiphytic, rooting at nodes. Leaf blades glabrous, ± symmetrical. Staminate flowers with 4 unequal tepals. Fruits with three wings, one distinctly elongate, this generally >1 cm wide. . . . . *B. glabra*.
1. Plants erect, terrestrial, not rooting at nodes. Leaf blades sericeous adaxially, asymmetrical. Staminate flowers with 2 equal tepals. Fruits with three wings, one ± triangular, wider distally, this ca. 0.5 cm wide. . . . . *B. prieurii*.

Begonia glabra Aubl. FIG. 46, PL. 22e

Herbs, hemiepiphytic, scandent, rooting at nodes. Stipules entire, 1–2.25 cm long. Leaves: petioles 2–4.5 cm long; blades ovate-suborbiculate, ± symmetrical, 6–12.5 × 3.5–9.0 cm, glabrous, the base often slightly inequilateral, obtuse, the apex acuminate, the margins remotely and bluntly toothed, occasionally with 1–2 irregular, blunt lobes. Inflorescences unisexual in our area. Staminate inflorescences 2-dichotomous; peduncle 2.5 cm long. Pistillate inflorescences 3–6-dichotomous; peduncle 8.5–17 cm long. Staminate flowers: tepals 4, two widely ovate, 5 × 4 mm, two elliptic, 4 × 1.5 mm; stamens ca. 20, the filaments free, the anthers oblong, as long as filaments. Pistillate flowers subtended by minute, elliptical bracteoles; tepals 4, subequal, elliptic, 6 × 2.5 mm; style branches longer than basal stalk. Fruits 8–9 mm long, with three wings, one elongate, triangular, ca. 1.2 cm wide, the other two ca. 0.5 cm wide; bracteoles ± deciduous, not accrescent. Seeds minute, cylindrical, 0.5 mm long, with terminal sacs. Fl (May, Aug, Sep), fr (Jun, Aug, Sep, Nov); usually found growing on bases of tree trunks. *Loseille bois*, *salade tortue*, *salade toti* (Créole).

Begonia prieurii A. DC. PL. 22f

Herbs, terrestrial, erect, not rooting at nodes. Stipules entire, 0.6–1.2 cm long. Leaves: petioles 1.5–2.5 cm long; blades narrowly ovate, oriented obliquely to petiole, 9.5–12 × 3.5–4.5 cm, asymmetrical, sericeous adaxially, glabrous abaxially, the base markedly asymmetrical, one side cuneate, the other semicircular, the apex acute-acuminate, the margins doubly serrate, ciliate. Inflorescences bisexual, 2–3-dichotomous, the staminate flowers caducous; peduncle 2.5–4 cm long. Staminate flowers: tepals 2, very widely ovate, 3 × 3 mm; stamens ca. 20, the filaments fused basally into column, the anthers elliptic-ovate, shorter than filaments. Pistillate flowers subtended by widely ovate, ciliate-serrate bracteoles as long as ovary; tepals 4(5), two suborbiculate, 2 × 1.5 mm, two elliptic, 2 × 0.75 mm (the fifth smaller); style branches shorter than basal stalk. Fruits 6–7.5 mm long, with three wings, one elongate, upswept triangular, 5–5.5 mm wide, the other two 1.5–2.5 mm wide; bracteoles persistent and accrescent, 4–5 mm long. Seeds minute, oblong, 0.2 mm long, lacking terminal sacs. Fl (Feb, Apr, May, Jun, Jul, Aug, Oct, Nov), fr (Feb, Apr, Jun, Jul, Aug, Oct, Nov), common, in disturbed areas, plants exposed to high light often with red pigmentation in leaves and stem.

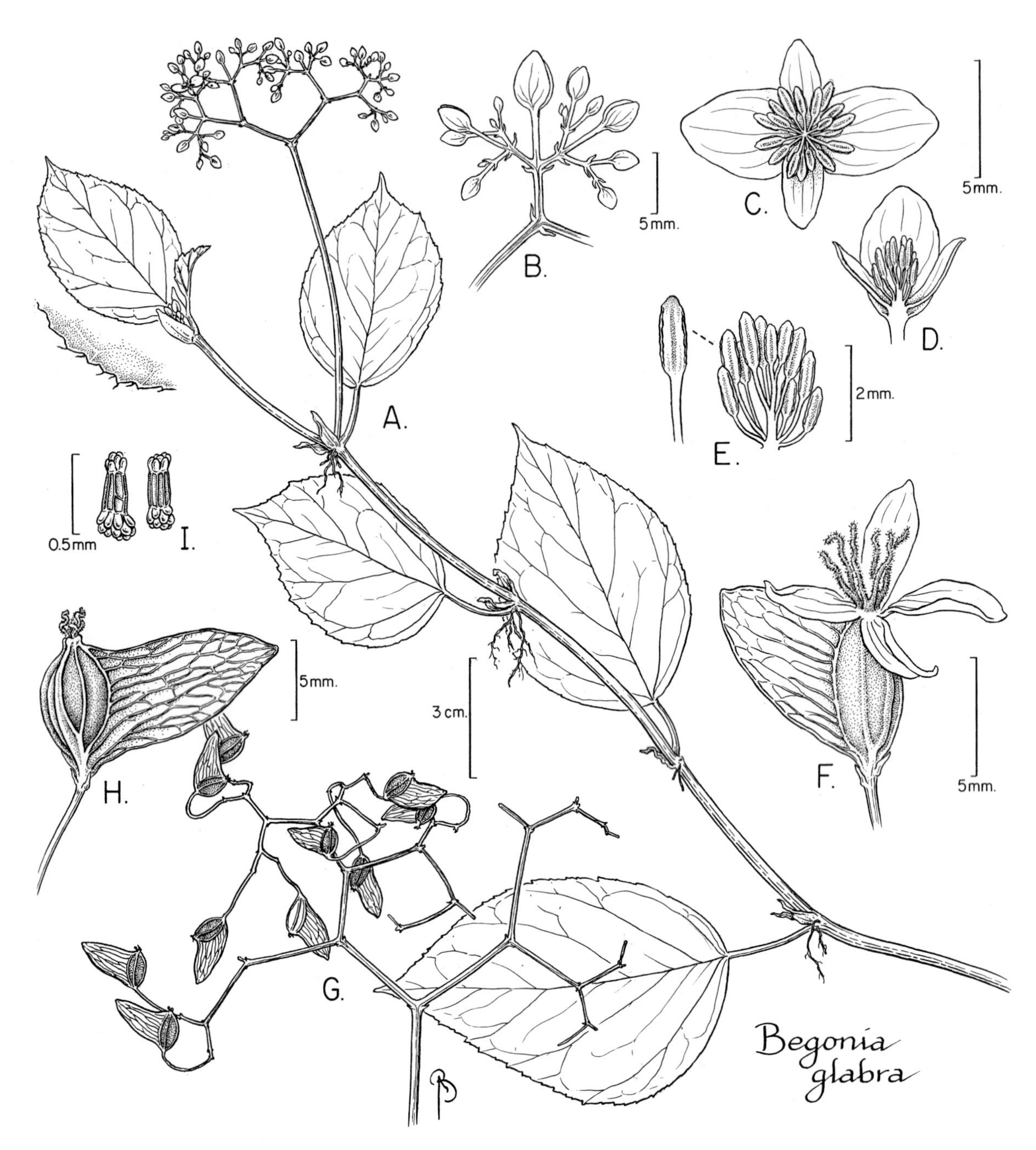

FIG. 46. BEGONIACEAE. *Begonia glabra* (A–E, *Mori et al. 22215*; F, *Granville et al. 8707*; G–I, *Mori et al. 23064*). **A.** Part of plant showing rooting at the nodes, leaves, and immature inflorescence with detail of leaf margin. **B.** Detail of part of inflorescence. **C.** Apical view of staminate flower. **D.** Medial section of staminate flower. **E.** Part of androecium (right) and detail of stamen (left) from bud. **F.** Lateral view of pistillate flower. **G.** Infructescence. **H.** Lateral view of fruit. **I.** Seeds.

BIGNONIACEAE (Trumpet-creeper Family)

Lúcia G. Lohmann, John L. Brown, and Scott A. Mori

Trees, lianas, or hemiepiphytic lianas in our area (also shrubs, vines, or rarely herbaceous outside our area). Stems terete or 4–6-angular, the lianas with anomalous phloem growth, usually with 4 phloem arms, sometimes with 6, 8 (*Cydista*, *Lundia*, *Macfadyena*, *Mansoa*, *Parabignonia*, and *Tanaecium*), or 16 (*Tanaecium* in our area), lacking in the hemiepiphytic lianas (*Schlegelia*), outer scales of axillary buds often pseudostipular, sometimes foliaceous, frequently forming clusters resembling "miniature bromeliads"; glands and ridges frequently present between petioles. Stipules absent. Leaves usually compound, infrequently simple (*Crescentia* and *Schlegelia* in our area), usually opposite, infrequently alternate (*Crescentia* in our area), the terminal leaflet of lianas often replaced by a simple, bifid (*Paragonia*), trifid (e.g., *Distictella*), multifid (*Pithecoctenium*), or "cat-claw" like (*Macfadyena* and *Parabignonia* in our area) tendril; petioles rarely with glands distally (*Paragonia*). Inflorescences terminal or axillary, panicles or racemes, sometimes reduced to fascicles or single flowers. Flowers zygomorphic, bisexual; buds usually rounded at apex, less frequently distinctly pointed (*Lundia* in our area); calyx synsepalous, usually cupular, truncate to variously lobed, or spathaceously split, with a single or double margin (*Amphilophium* in our area), sometimes subtended by a conspicuous pair of bracts (*Memora* in our area), frequently with scattered glands; corolla sympetalous, the tube conspicuous, the lobes usually 5, imbricate, less frequently valvate (*Pyrostegia*, not in our area), frequently with glands arranged in lines on the upper portion; stamens usually inserted on corolla tube, sometimes exserted (*Pyrostegia*, not in our area, and *Tynanthus* in our area), didynamous, usually 4, rarely 2 (*Catalpa* and *Pseudocatalpa*, both not in our area), or 5 (several Old World genera), the anthers with 2 thecae, rarely with 1 (*Jacaranda copaia* in our area), the thecae usually divergent or divaricate, glabrous, rarely villose (*Lundia* in our area); single staminode usually reduced, glabrous or puberulous, rarely glandular-pubescent and exceeding stamens (*Jacaranda*) or lacking; nectary disc usually present surrounding base of ovary, sometimes lacking (*Clytostoma*, *Cydista* and *Lundia* in our area); ovary superior, 1–2-locular, rarely stipitate (*Anemopaegma*), the style slender, the stigma bilamellate; placentation axile in 2-locular and incomplete 2-locular ovaries (*Schlegelia*), parietal in unilocular ovaries (*Crescentia* in our area), the ovules numerous. Fruits loculicidal or septicidal capsules (*Jacaranda* and *Tabebuia* in our area), berries (*Schlegelia*), or hard-shelled pepos (*Crescentia* in our area). Seeds somewhat flattened, without endosperm, usually winged, rarely not winged and embedded in pulp (*Crescentia* in our area); cotleydons foliaceous.

Gentry, A. 1980. Bignoniaceae — Part I (Crescentieae and Tourrettieae). Fl. Neotrop. Monogr. **25(I):** 1–130.
———. 1982. Bignoniaceae. Flora de Venezuela **8(4):** 1–433.
———. 1992. Bignoniaceae — Part II (Tribe Tecomeae). Fl. Neotrop. Monogr. **25(II):** 1–370.
———. 1997. Bignoniaceae. *In* J. A. Steyermark, P. E. Berry & B. K. Holst (gen. eds.), Flora of the Venezuelan Guayana **3:** 403–491. P. E. Berry, B. K. Holst & K. Yatskievych (vol. eds.). Missouri Botanical Garden, St. Louis.
Sandwith, N. Y. 1938. Bignoniaceae. *In* A. Pulle (ed.), Flora of Suriname **IV(2):** 1–86. J. H. De Bussy, Amsterdam.

1. Trees.
   2. Leaves compound. Flowers diurnal. Fruits dehiscent, septacidal capsules. Plants native.
      3. Leaves palmate. Flowers yellow; staminode shorter than stamens, usually glabrous or puberulous, sometimes pubescent at base. Fruits linear, much longer than wide. . . . . *Tabebuia.*
      3. Leaves bipinnate. Flowers purplish-blue; staminode much longer than stamens, conspicuously glandular-pubescent. Fruits oblong, only slightly longer than wide. . . . . *Jacaranda.*
   2. Leaves simple. Flowers nocturnal. Fruits indehiscent, pepos. Plants cultivated.. *Crescentia.*
1. Lianas.
   4. Lianas hemiepiphytic. Stems without anomalous phloem growth. Leaves simple. Tendrils absent. Staminode lacking. Fruits berries.. . . . . *Schlegelia.*
   4. Lianas not hemiepiphytic. Stems with 4–8-phloem arms. Leaves compound. Tendrils present. Staminode present. Fruits loculicidal capsules.
      5. Leaves more than once compound (2–3-ternate or 2–4-pinnate (with >3 leaflets).
         6. Interpetiolar glands present. Leaflets not conspicuously thickened at joint. Corolla lavender. . . . . *Arrabidaea inaequalis.*
         6. Interpetiolar glands absent. Leaflets conspicuously thickened at joint. Corolla white to yellow.
            7. Stems terete. Calyx with glands; corolla bright yellow. . . . . *Memora.*
            7. Stems 4-angular. Calyx without glands; corolla white or cream-colored, sometimes with yellow throat. . . . . *Pleonotoma.*
      5. Leaves once compound (2–3-foliolate).
         8. Stems hollow. Leaflets pellucid-punctate. . . . . *Stizophyllum.*
         8. Stems solid. Leaflets not pellucid-punctate.
            9. Plants with strong aroma of almond, clove, or garlic (even when dried).
               10. Plants with almond (cyanide) aroma. Leaflets with secondary veins palmately arranged. Flowers nocturnal; calyx tubular, with glands arranged in lines on lower portion; corolla tubular, 10.5–14 cm long. Fruits cylindrical. . . . . *Tanaecium.*
               10. Plants without almond aroma. Leaflets with secondary veins pinnately arranged. Flowers diurnal; calyx cupular, without glands; corolla infundibuliform, <6 cm long. Fruits linear.
                  11. Plants with clove aroma. Corolla markedly bent, bilabiate, 1.5–2 cm long; anthers exserted. Fruits never ridged, sometimes conspicuously winged. . . . . *Tynanthus.*
                  11. Plants with garlic aroma. Corolla not bent, not bilabiate, >3 cm long; anthers included. Fruits with conspicuous longitudinal ridge on middle portion, but never winged.. . . . . *Mansoa.*
            9. Plants without aroma of almond, clove, or garlic.

12. Nectary disc surrounding ovary lacking.
    13. Leaflets with secondary veins ± palmately arranged proximally, pinnately arranged distally. Buds distinctly pointed; corolla without nectar guides; anthers villose. . . . . . *Lundia.*
    13. Leaflets with secondary veins pinnately arranged throughout. Buds not distinctly pointed; corolla with nectar guides; anthers glabrous.
        14. Pseudostipules subulate, in clusters, resembling "miniature bromeliads." Abaxial leaflet blade surfaces without glands in axils of veins. Calyx with distinct marginal teeth; ovary finely tuberculate, not lepidote-pubescent. Fruits elliptic, echinate. . . . . *Clytostoma.*
        14. Pseudostipules not as above. Abaxial leaflet blade surfaces with glands in axils of veins. Calyx without distinct marginal teeth, merely 5-denticulate; ovary not finely tuberculate, lepidote-pubescent. Fruits linear, smooth. . . . . *Cydista.*
12. Nectary disc surrounding ovary present.
    15. Corolla tube yellow, cream-colored, or white (the lobes sometimes wine-red or tinged with pink and the throat sometimes with red streaks; in *Amphilophium* the corolla sometimes turning lavendar at anthesis).
        16. Corolla tube bright yellow.
            17. Calyx spongy, inflated, ca. 35 mm long; corolla with red streaks in throat. Fruits oblong-elliptic, woody. . . . . *Callichlamys.*
            17. Calyx not spongy, not inflated, ca. 7–10 mm long; corolla without red streaks in throat. Fruits linear, not woody.
                18. Juvenile plants growing appressed against tree trunks, often with adventitious roots. Tendrils trifid, cat-claw like. Inflorescences in cymes or reduced to a single flower. Calyx without glands. . . . . *Macfadyena.*
                18. Juvenile plants not growing appressed against tree trunks, without adventitious roots. Tendrils simple, not cat-claw like. Inflorescences in racemes. Calyx with glands. . . . . *Memora racemosa.*
        16. Corolla tube pale yellow, cream-colored, or white (in *Amphilophium* the corolla sometimes turning lavender at anthesis).
            19. Corolla membranous, neither bent nor bilabiate.
                20. Ovary stipitate. Fruits elliptic or orbiculate, stipitate. . . . . *Anemopaegma.*
                20. Ovary sessile. Fruits linear or strap-shaped, sessile.
                    21. Calyx with glands. Fruits strap-shaped, tuberculate. . . . . *Adenocalymna.*
                    21. Calyx without glands. Fruits linear, not tuberculate.
                        22. Interpetiolar area with glands, not ridged. Leaflets with 3 veins arising from base, with domatia in axils of secondary veins. Tendrils simple. . . . . *Arrabidaea triplinervia.*
                        22. Interpetiolar area without glands, ridged. Leaflets pinnately veined throughout, without domatia in axils of secondary veins. Tendrils trifid. . . . . *Martinella.*
            19. Corolla coriaceous, markedly bent or, if not bent, then conspicuously bilabiate.
                23. Stems terete, somewhat flattened at nodes. Leaflets elliptic, coriaceous . . . . *Distictella.*
                23. Stems distinctly hexagonal, not flattened at nodes. Leaflets ovate to suborbiculate.
                    24. Tendrils multifid. Calyx with a single margin, with glands; corolla bent, not bilabiate, open at anthesis, pubescent. Fruits echinate (not verrucate). . . . . *Pithecoctenium.*
                    24. Tendrils trifid. Calyx with a double margin, without glands; corolla not bent, bilabiate, closed at anthesis, glabrous. Fruits verrucate (not echinate). . . . . *Amphilophium.*
    15. Corolla tube pink to reddish-purple.
        25. Tendrils bifid. Pseudostipules subulate. Petioles with glands distally. Fruit surfaces scabrous, like sandpaper. . . . . *Paragonia.*
        25. Tendrils simple or trifid. Pseudostipules absent or not subulate. Petioles without glands distally. Fruit surfaces smooth.

26. Juvenile plants growing appressed against tree trunks, often with adventitious roots. Stems tetragonal. Tendrils trifid, cat-claw like. Calyx with 5 distinct lobes. . . . . *Parabignonia.*
26. Juvenile plants not growing appressed against tree trunks, without adentitious roots. Stems terete. Tendrils simple, not cat-claw like. Calyx truncate or with 5 indistinct lobes (denticulate). . . . . *Arrabidaea.*

ADENOCALYMNA Meisn.

Lianas. Stems with 4 phloem arms, terete, solid, without interpetiolar glands, frequently with interpetiolar ridge; pseudostipules small, subulate, sometimes forming clusters resembling "miniature bromeliads," not foliaceous, persistent. Leaves 2–3-foliolate, the terminal leaflet often replaced by a simple tendril; petioles without glands distally; leaflets not pellucid-punctate. Inflorescences axillary, ramiflorous, or terminal, narrow bracteate racemes. Flowers diurnal; calyx cupular or campanulate, usually with conspicuous glands, with a single margin, the apex lobed to ± truncate; corolla infundibuliform to campanulate, 5-parted, not bilabiate, open at anthesis, usually pubescent to subglabrous abaxially, white or yellow (outside our area), without nectar guides, the tube straight, membranous, without glands; stamens inserted, the anthers glabrous; staminode not exceeding stamens; nectary disc present; ovary sessile, lepidote-pubescent, puberulous, or papillate. Fruits loculicidal capsules, elliptic, sometimes linear or strap-shaped, flattened, sometimes distinctly winged or ridged, often tuberculate. Seeds 2-winged or nearly wingless.

Adenocalymna saülense A. H. Gentry Fig. 47, Pl. 23a,b

Stems glabrous to puberulous, with interpetiolar ridge; pseudostipules small, subulate. Leaves 2–3-foliolate; petioles 1–3 cm long, swollen at base; lateral petiolules 0.7–3 cm long, the terminal one 2.5–3 cm; leaflets elliptic to oblong-elliptic, 7–17 × 3–7.7 cm, concolorous, coriaceous, glabrous adaxially, glabrous to puberulous, with scattered glands abaxially, the base acute to truncate, the apex obtuse to acute to slightly acuminate; secondary veins pinnately arranged. Inflorescences ramiflorous racemes, the axes with simple and stellate hairs. Flowers: calyx campanulate, 9–11 × 7–9 mm, pubescent, the hairs dendritic, the apex 2-lobed, with conspicuous glands; corolla 4–4.5 × 1.5–1.6 cm, glabrous except toward apex, the tube white or yellowish-white, the lobes lighter colored; ovary papillate. Fruits strap-shaped, 42 × 2.5 cm, puberulous, tuberculate, thickened along each margin. Seeds with 2 well developed wings. Fl (Aug, Sep), fr (Sep); scattered, in non-flooded forest.

AMPHILOPHIUM Kunth

Lianas. Stems with 4 phloem arms, hexagonal, solid to slightly hollow, without interpetiolar glands, with interpetiolar ridge; pseudostipules small, variable in shape, not forming clusters resembling "miniature bromeliads," foliaceous, caducous. Leaves 2–3-foliolate, the terminal leaflet often replaced by a trifid tendril; petioles without glands distally; leaflets not pellucid-punctate. Inflorescences terminal, racemose panicles. Flowers diurnal; calyx cupular, without conspicuous glands, with a double margin, the outer series undulate and the inner series 2-lobed, the apex irregular; corolla tubular, 5-parted, strongly bilabiate, the lobes not opening and remaining nearly closed at anthesis, glabrous abaxially, white, sometimes turning lavender at anthesis, without nectar guides, the tube straight, membranous to coriaceous, without glands; stamens inserted, the anthers glabrous; staminode not exceeding stamens; nectary disc present; ovary sessile, densely lepidote-pubescent, verruco-papillose. Fruits loculicidal capsules, elliptic, convex, not winged or ridged, verrucose. Seeds 2-winged.

Amphilophium paniculatum (L.) Kunth var. **imatacense** A. H. Gentry Pl. 23c

Stems pubescent, peltate scales and granular hairs both present; pseudostipules inconspicuous. Leaves 2-foliolate; petioles 2–4 cm long, not swollen at base; petiolules 1.5–2.5 cm long; leaflets ovate, 4–8 × 2.5–6.5 cm, discolorous, coriaceous, lepidote-pubescent adaxially, densely pubescent with dendritic hairs abaxially, the base rounded, truncate to slightly cordate, often asymmetric, the apex short acuminate; secondary veins pinnately arranged. Inflorescences terminal, racemose panicles, the axis stellate-pubescent. Flowers: calyx 9–12 × 9–17 mm, pubescent, the outermost series undulate; corolla 2.5–3 × 0.7–0.9 cm, coriaceous, glabrous to puberulous, sticky, white, sometimes turning lavender at anthesis, fading white; ovary with simple, dendritic hairs. Fruits elliptic, 10–12 × 4–5 cm, glabrous (fide Gentry, 1982). Fl (Jan); collected only twice, in non-flooded forest.

ANEMOPAEGMA Meisn.

Subshrubs, shrubs, or lianas. Stems with 4-phloem arms, terete, solid to slightly hollow, with or without interpetiolar glands, without interpetiolar ridge; pseudostipules small or large, orbiculate, not forming clusters resembling "miniature bromeliads," foliaceous, persistent or caducous, sometimes lacking. Leaves 2–5-foliolate, the terminal leaflet often replaced by a trifid, rarely simple, tendril; petioles without glands distally; leaflets not pellucid-punctate. Inflorescences axillary or terminal, racemes, or single flowers. Flowers diurnal; calyx cupular to campanulate, usually with glands, with a single margin, the apex truncate or 5-denticulate; corolla tubular-campanulate, 5-parted, not bilabiate, open at anthesis, glabrous to puberulous abaxially, white or yellow, the lobes sometimes red or bright to pale yellow (fide Gentry 1982), without nectar guides, the tube straight, membranous, with or without glands; stamens inserted, the anthers glabrous; staminode not

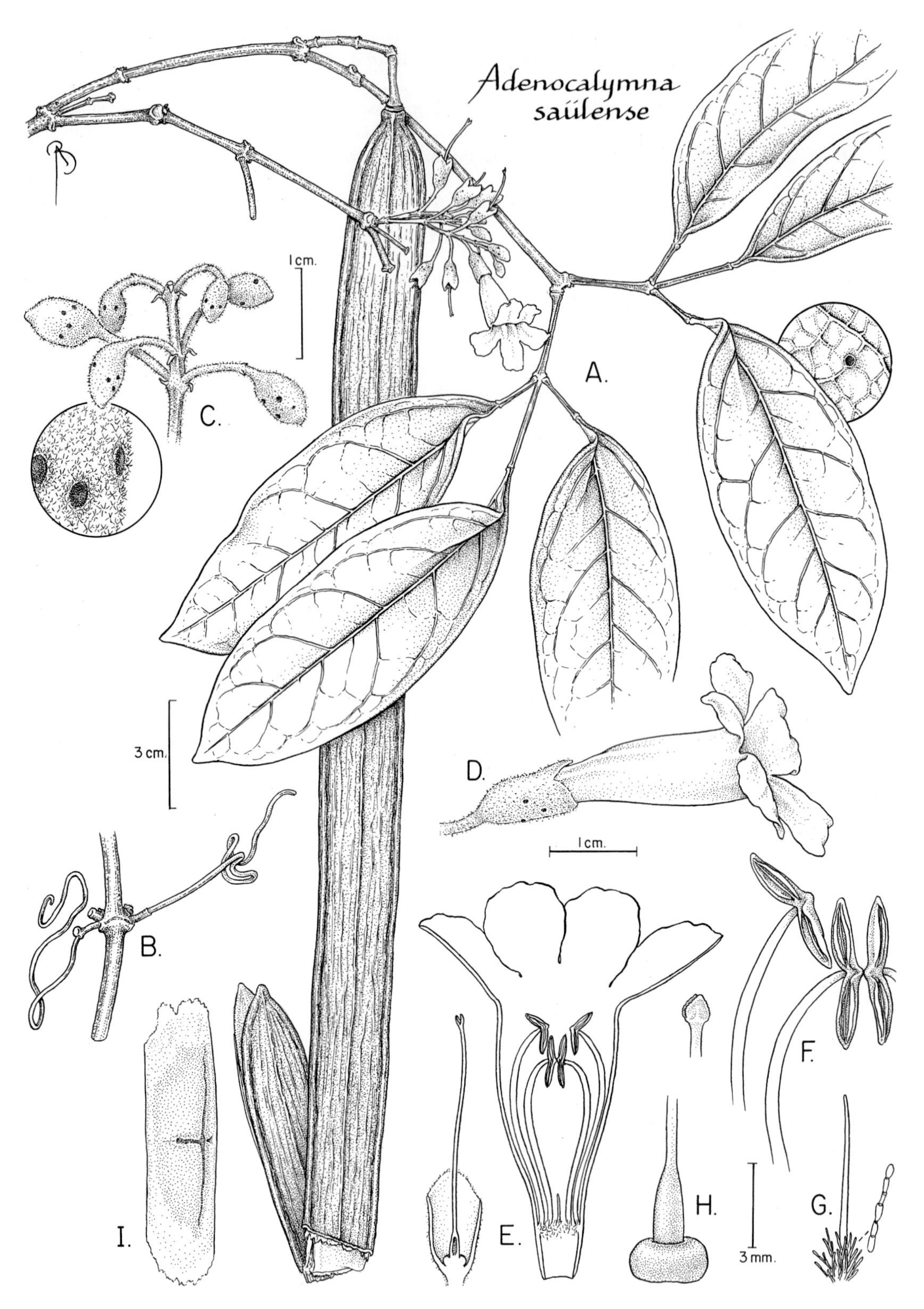

FIG. 47. BIGNONIACEAE. *Adenocalymna saülense* (A, B, I, *Mori & Gracie 24211*; C–H, *Mori et al. 24133*). **A.** Part of stem with leaves, inflorescence, fruit, and detail of abaxial surface of leaf showing gland (far right). **B.** Detail of node with tendrils. **C.** Detail of inflorescence in bud and close-up of glands on bud. **D.** Lateral view of flower. **E.** Medial section of calyx and pistil (left) and section of corolla showing stamens and staminode (right). **F.** Detail of three of the four anthers on upper portion of filaments. **G.** Lateral view of staminode surrounded by basal hairs (near right) and detail of hair (far right). **H.** Lateral view of ovary and basal portion of style surrounded by nectary disc (left) and detail of stigma (above). **I.** Winged seed.

exceeding stamens; nectary disc present; ovary stipitate, lepidote-pubescent. Fruits loculicidal capsules, elliptic to orbiculate, stipitate, convex, not winged or ridged, smooth. Seeds completely surrounded by a broad hyaline wing or wingless.

1. Stems without interpetiolar glands. Leaflets elliptic to ovate, not bullate, glabrous, without scattered glands abaxially. Calyx cupular, coriaceous, ca. 5 mm long, pubescent, with conspicuous clusters of glands distally, the apex truncate; corolla cream-colored with wine-red to maroon lobes, the lobes with conspicuous glands arranged in lines. . . . . *A. ionanthum.*
1. Stems with interpetiolar glands. Leaflets cordate, bullate, puberulent, with scattered glands abaxially. Calyx campanulate, membranous, 12–15 mm long, glabrous, without glands, the apex 2–3-parted; corolla with creamy white lobes, the lobes without conspicuous glands. . . . . *A. oligoneuron.*

Anemopaegma ionanthum A. H. Gentry

Lianas. Stems glabrous or puberulous, without interpetiolar glands. Leaves 2-foliolate; petioles 4–4.5 cm long, not swollen at base; petiolules 1.5–2 cm long; leaflets elliptic to ovate, ca. 15 × 9.5 cm, concolorous, coriaceous, glabrous abaxially, with conspicuous clusters of glands distally, not bullate, the base rounded, sometimes acute or truncate, the apex acute to short acuminate; secondary veins pinnately arranged. Inflorescences axillary racemes, the axis pubescent. Flowers: calyx cupular, ca. 4–5 × 5–6 mm long, pubescent, with conspicuous clusters of glands on the upper part, the apex truncate; corolla 4–5 × 1.5–1.8 cm, puberulous abaxially, with glands arranged in lines, cream-colored, the lobes wine-red, with conspicuous glands arranged in lines; ovary lepidote-pubescent. Fruits unknown. Fl (Oct); collected only twice, in non-flooded forest.

Anemopaegma oligoneuron (Sprague & Sandwith) A. H. Gentry

Lianas. Stems glabrous or puberulous, with interpetiolar glands. Leaves 2-foliolate; leaflets cordate, concolorous, coriaceous, puberulent, bullate (described from *Steege 248*, MO, from Guyana). Flowers: calyx campanulate, 12–15 × 12–20 mm, glabrous, without glands, the apex 2–3-parted; corolla 5–6.5 × 1.7–1.9 cm, glabrous abaxially, without glands, creamy-white except yellow inside tube; ovary lepidote-pubescent. Fruits oblong-elliptic, 13–14 × 5.5–6 cm, stipitate, densely velutinous. Fl (May); known from our area by a single gathering of calyces and corollas from the ground (*Mori et al. 22342*, NY), our description completed from Gentry (1982).

ARRABIDAEA DC.

Lianas, rarely shrubs or subshrubs (outside our area). Stems with 4 phloem arms, terete, solid, with or without interpetiolar glands, usually without interpetiolar ridge; pseudostipules small and inconspicuous, not forming clusters resembling "miniature bromeliads," notfoliaceous, persistent or caducous. Leaves 2–3-foliolate, or infrequently biternate (*A. inaequalis* in our area), the terminal leaflet frequently replaced by a simple tendril or tendrils developing in leaf axils; petioles without glands distally; leaflets not pellucid-punctate. Inflorescences axillary or terminal, often large panicles with many flowers. Flowers diurnal; calyx cupular, campanulate, or tubular, with or without glands, with a single margin, the apex truncate or bilabiate, usually minutely 5-denticulate; corolla tubular-campanulate, tubular-infundibuliform, simply tubular, simply campanulate, or simply infundibuliform, not bilabiate, open at anthesis, usually pubescent abaxially, especially on lobes, purple, lavender, maroon, pale pink, red, or infrequently white, sometimes with white on throat adaxially, usually without nectar guides, the tube usually straight, less frequently curved or bent, usually membranous, with or without glands; stamens inserted, the anthers glabrous; staminode not exceeding stamens; nectary disc present; ovary sessile, minutely lepidote-pubescent. Fruits loculicidal capsules, usually linear, sometimes oblong, flattened, not winged or ridged, smooth or ornamented. Seeds 2-winged or completely surrounded by a broad hyaline wing.

1. Leaves biternate. . . . . *A. inaequalis.*
1. Leaves 2–3-foliolate.
    2. Venation with 3 principal veins arising from base. Calyx tubular; corolla white, cream-colored, or yellow (at least in bud).
        3. Stems and leaves densely pubescent. Corolla yellow in bud, turning pink with age. . . . . *A. japurensis.*
        3. Stems and leaves glabrous to puberulous. Corolla white or cream-colored in bud, not turning pink with age.
            4. Leaflets dark brown when dried, without domatia. Calyx bifid. . . . . *A. oligantha.*
            4. Leaflets light brown when dried, with domatia in axils of secondary veins. Calyx truncate. . . . *A. triplinervia.*
    2. Venation without 3 principal veins arising from base. Calyx cupular; corolla pink, lavender, magenta, or purple in bud and when mature.
        5. Lateral leaflets asymmetric at base; venation impressed adaxially, salient abaxially.
            6. Leaflets bullate, drying olive, the lateral leaflets 14.5–18.2 cm long. Calyx ≥10 mm long, the apex 5-parted. . . . . *A. cinnamomea.*
            6. Leaflets not bullate, drying black, the lateral leaflets 6.5–11 cm long. Calyx ≤4 mm long, the apex truncate. . . . . *A. nigrescens.*
        5. Lateral leaflets symmetric at base; venation ± plane adaxially and abaxially.

7. Corolla tube glabrous abaxially; buds distinctly white at apex; calyx not appressed to corolla tube. . . . . *A. patellifera.*
7. Corolla tube pubescent abaxially; buds always pink at apex; calyx appressed to corolla tube.
   8. Leaflets markedly discolorous, the abaxial surface usually densely covered with whitish hairs (in *A. candicans* leaflets densely whitish-pubescent to nearly glabrous).
      9. Stems without interpetiolar ridges. Calyx cupular; corolla tubular-infundibuliform, lavender. . . . . *A. candicans.*
      9. Stems with interpetiolar ridges. Calyx tubular; corolla tubular, red to reddish-pink. . . . . . *A. trailii.*
   8. Leaflets concolorous.
      10. Stems with conspicuous interpetiolar glands. Corolla tubular-campanulate, the tube curved or bent, 2–3 mm wide. . . . . *A. fanshawei.*
      10. Stems without conspicuous interpetiolar glands. Corolla infundibuliform, the tube ± straight, 5–7 mm wide. . . . . *A. florida.*

Arrabidaea candicans (Rich.) DC.

Lianas. Stems with simple hairs and lepidote scales, with conspicuous interpetiolar glands, without interpetiolar ridge. Leaves 2–3-foliolate, the terminal leaflet often replaced by a simple tendril; petioles 1.5–7 cm long; lateral petiolules 0.3–2.1 cm long, the terminal one 0.6–4.6 cm long; leaflets elliptic, ovate or narrowly ovate, 3.3–12 × 1.3–7 cm, not bullate, chartaceous, discolorous, often drying red, glabrous or with simple white hairs and minute peltate scales adaxially, mostly glabrous to densely white pubescent abaxially, with scattered glands, the base cuneate, truncate, acute or obtuse, the apex acute to short acuminate; venation plane ad- and abaxially, the secondary veins in 6–7 pairs, pinnately arranged, without domatia in vein axils. Flowers: calyx cupular, appressed to corolla tube, 3–5 × 2–5 mm, densely pubescent, without conspicuous glands, the apex 5-denticulate; corolla tubular-infundibuliform, 1.6–4 × 1 cm, pubescent abaxially, lavender. Fruits linear, 12–23 × 0.9–1.2 cm, glabrous to puberulous (fide Gentry, 1982). Fl (Sep), fr (Jun); apparently rare, in non-flooded forest.

Our flowering collection (*Phillippe et al. 26981*, NY) lacks the conspicuous calyx glands seen in other collections from the Guianas.

Arrabidaea cinnamomea (DC.) Sandwith

Lianas. Stems glabrous to puberulous, without interpetiolar glands, without interpetiolar ridge. Leaves always 3-foliolate; tendrils developing in leaf axils; petioles 10.5–11.8 cm long; lateral petiolules 0.8–1 cm long, the terminal one ca. 2.5 cm long; leaflets elliptic or obovate, 14.5–18.2 × 8.2–10.4 cm, bullate, chartaceous, discolorous when dried, often drying olive colored, glabrous to puberulous adaxially, puberulous to pubescent, especially on veins abaxially, the hairs simple or branched, the base asymmetric, acute to rounded, the apex rounded or acuminate; venation impressed adaxially, salient abaxially, the secondary veins in 5–6 pairs, pinnately arranged, without domatia in vein axils. Flowers: calyx cupular, appressed to corolla tube, 10–12 × 5–6 mm, densely pubescent, with scattered glands, the apex 5-parted; corolla infundibuliform, 4–4.5 × 1.5–1.7 cm, pubescent abaxially, pink (based on *Philcox 3882*, NY, from Mato Grosso, Brazil). Fruits linear, 49–55 × 3.5–4 cm, with simple hairs and lepidote scales, with scattered glands. Fr (Jun, Nov); rare, in non-flooded forest.

Arrabidaea fanshawei Sandwith

Lianas. Stems not lepidote-pubescent, with interpetiolar glands, interpetiolar ridge inconspicuous or absent. Leaves 3-foliolate; petioles ca. 3 cm long; lateral petiolules 1–1.5 cm long, the terminal one 0.7 cm long; leaflets elliptic, 7.5–8.5 × 3.5–4 cm, not bullate, chartaceous, concolorous, drying blackish or dark brown, mostly glabrous on both surfaces but lepidote-pubescent on some parts abaxially, the base acute, the apex acute to short acuminate; venation plane ad- and abaxially, the secondary veins in 8 pairs, pinnately arranged, without domatia in vein axils. Flowers: calyx cupular, 2–2.5 × 4–5 mm, whitish-pubescent, the apex 5-parted; corolla tubular-campanulate (the tube sometimes curved or bent), 1.5–1.7 × 0.2–0.3 cm, pubescent abaxially, lavender, sometimes with white spot at throat, the base of tube sometimes deep purple-black. Fruits linear, 12.5 × 0.8–1.1 cm, lepidote-pubescent. Fl (May, Nov); known from our area only by corollas from ground (*Mori et al. 21653*, *22341*, both NY).

Leaves, calyx, and fruit described from *Cremers 10105* (CAY, (from French Guiana but outside our area).

Arrabidaea florida DC.

Lianas. Stems not lepidote-pubescent, without conspicuous interpetiolar glands, without interpetiolar ridge. Leaves 2–3-foliolate; petioles 0.9–6.2 cm; lateral petiolules 0.5–2.7 cm long, the terminal one 0.9–2.6 cm long; leaflets ovate to widely ovate, 5–14 × 4–8.4 cm, not bullate, chartaceous, concolorous, not drying red, puberulous on veins only, the base widely cuneate to truncate, the apex acute to acuminate; venation plane ad- and abaxially, the secondary veins in 4–7 pairs, pinnately arranged, without domatia in vein axils. Flowers: calyx cupular, appressed to corolla tube, 3–4 × 2–3 mm, puberulous, without conspicuous glands, the apex truncate, with 5 small teeth; corolla infundibuliform, 1.2–1.7 × 0.5–0.7 cm, pubescent abaxially, lavender with the tube white adaxially. Fruits linear, 11–22 × 0.9–1 cm, glabrous. Known only from central French Guiana by a sterile collection (*Gentry 63167*, CAY, MO), our description modified from Gentry (1982).

Arrabidaea inaequalis (Splitg.) K. Schum. PL. 23d

Lianas. Stems puberulous, with simple hairs and lepidote scales, conspicuously lenticellate, with conspicuous interpetiolar glands, without interpetiolar ridge. Leaves 2–3-ternate, the terminal division often replaced by a simple tendril; petioles 4–6 cm long; primary petiolules 4 cm long, the secondary petiolules 1–2 cm long, the terminal one 2 cm long; leaflets ovate to elliptic, 6.5–10 × 3–5 cm, not bullate, chartaceous, concolorous, drying brownish, glabrous adaxially, mostly glabrous, only with a few scattered hairs on veins abaxially, the base obtuse to rounded, the apex acuminate;

venation impressed adaxially, salient abaxially, the secondary veins in 5–6 pairs, pinnately arranged, without domatia in vein axils. Flowers: calyx tubular to cupular, appressed to corolla tube, 6 × 4 mm, pubescent, without conspicuous glands, the apex truncate with small marginal teeth; corolla infundibuliform, 4 × 1 cm, pubescent abaxially, lavender to lavender pink. Fruits linear, 7.5–40 × 1.1–1.6 cm, with scattered glands (fide Gentry, 1982). Fl (Feb); rare, in non-flooded forest.

Arrabidaea japurensis (DC.) Bureau & K. Schum.

Lianas. Stems densely pubescent, with simple hairs and lepidote scales, with interpetiolar glands, without interpetiolar ridge. Leaves 2-foliolate, the terminal leaflet often replaced by a simple tendril; petioles 2.5 cm long; petiolules ca. 1.5 cm long; leaflets elliptic, 8.5 × 5.7 cm, not bullate, chartaceous, slightly discolorous, puberulous to pubescent on both surfaces, the base rounded, the apex acute or obtuse; venation impressed adaxially, salient abaxially, the secondary veins in 4–5 pairs, pinnately arranged, without domatia in vein axils. Flowers: calyx tubular, 7 × 4 mm, pubescent, without conspicuous glands, the apex truncate, without conspicuous teeth; corolla infundibuliform, 1.1–1.6 × 5–7 cm, puberulent abaxially, light yellow in bud, turning pink with age. Fruits linear, 3.3–4.4 × 1.2–1.5, with scattered glands. Fl (Dec); collected only once, in non-flooded forest.

Our description completed from Gentry (1982).

Arrabidaea nigrescens Sandwith PL. 23e

Lianas. Stems not distinctly lepidote-pubescent, with interpetiolar glands, without interpetiolar ridge. Leaves 3-foliolate, the terminal leaflet often replaced by a simple tendril; petioles 4–8 cm long; lateral petiolules 0.5–1 cm, the terminal one 1–2.5 cm long; leaflets elliptic to suborbiculate or ovate to widely ovate, 6.5–17 × 3–7 cm, not bullate, chartaceous, slightly discolorous, drying black on both sides, glabrous to nearly glabrous adaxially, puberulous on veins only to pubescent throughout, with occasional glands abaxially, the base rounded, often asymmetric, especially the lateral ones, the apex acute to obtuse or acuminate; venation impressed adaxially, salient abaxially, the secondary veins in 4–5 pairs, pinnately arranged, without domatia in vein axils. Flowers: calyx cupular, appressed to corolla tube, 4 × 4–5 mm, pubescent, the hairs purplish, sometimes branched or even dendritic, with scattered glands, the apex truncate, sometimes with poorly developed teeth; corolla tubular-infundibuliform, 2.5–3 × 0.8–1 cm, pubescent abaxially, magenta or dark purple. Fruits linear, glabrous. Fl (Sep), fr (Jun); rare, in non-flooded forest.

Arrabidaea oligantha Bureau & K. Schum.

Lianas. Stems mostly not lepidote-pubescent but with some peltate scales especially around nodes, with conspicuous interpetiolar glands, without interpetiolar ridge. Leaves 2-foliolate, the terminal leaflet often replaced by a simple tendril; petioles 1.2–1.7 cm long; lateral petiolules 2–2.5 cm long, the terminal one 2.8 cm long; leaflets narrowly ovate, 12–18.5 × 5.5–7.5 cm, not bullate, chartaceous, slightly discolorous, drying dark brown, glabrous on both surfaces, the base obtuse to rounded, the apex acuminate; venation impressed adaxially, salient abaxially, the secondary veins in 5 pairs, pinnately arranged, without domatia in vein axils. Flowers: calyx long, tubular, appressed to corolla tube, 25 × 7 mm, puberulous, without conspicuous glands, the apex bifid; corolla narrowly funnelform, 5–5.5 × 1 cm, puberulous abaxially, maroon with white spot on upper lobe at throat. Fl (Nov), fr (Jun); rare, in non-flooded forest.

Arrabidaea patellifera (Schltr.) Sandwith PL. 23f

Lianas. Stems lepidote-pubescent, without interpetiolar glands, without interpetiolar ridge. Leaves 2-foliolate, the terminal leaflet often replaced by a simple tendril; petioles 0.7–1.5 cm long; petiolules 1–1.5 cm long; leaflets narrowly ovate, ovate, elliptic, or widely elliptic, 5.5–10 × 3–5.5 cm, not bullate, chartaceous, concolorous, minutely and sparsely lepidote-pubescent adaxially, minutely and more densely lepidote-pubescent with long, white hairs, especially on veins toward base abaxially, the base acute, obtuse, or rounded, the apex acuminate; venation impressed adaxially, salient abaxially, the secondary veins in 4–5 pairs, pinnately arranged, without domatia in vein axils. Flowers: calyx widely cupular, not appressed to corolla tube as in other species, 4–5 × 7–8 mm, usually wider than long, glabrous except for minute, sparsely distributed lepidote scales, without conspicuous glands, the apex truncate to undulate; corolla infundibuliform to tubular-infundibuliform, 2.5–3 × 0.9–1.2 cm long, pale lavender or purple, distinctly white at apex in bud, the tube glabrous, the lobes pubescent abaxially, this especially evident in bud. Fruits linear, an immature one 28.5 × 1.3 cm, lepidote-pubescent. Fl (Jan, Feb, Aug, Oct), fr (Feb); common, in non-flooded forest.

Arrabidaea trailii Sprague PL. 24a,b

Lianas. Stems lepidote-pubescent, sometimes with interpetiolar glands, sometimes with interpetiolar ridge. Leaves 2-foliolate, the terminal leaflet often replaced by a simple tendril; petioles 3–3.5 cm long; petioles 2–2.5 cm long; leaflets ovate to widely ovate, 12–18.5 × 6–10.5 cm, not bullate, chartaceous, discolorous, lepidote-pubescent adaxially, densely white pubescent and lepidote-pubescent (under the dense covering of white hairs) abaxially, the base obtuse to rounded, less frequently acute, the apex long acuminate; venation plane ad- and abaxially, the secondary veins in 5–6 pairs, pinnately arranged, without domatia in vein axils. Flowers: calyx tubular, ribbed toward apex, appressed to corolla tube, 8–10 × 3 mm, without conspicuous glands, the apex with conspicuous blunt teeth formed by extension of calycine ribs; corolla tubular, ca. 2 × 0.4 cm, glabrous to pubescent abaxially, red to reddish-pink. Fruits linear, 17–23 × 1.3–1.4 cm, lepidote-pubescent (fide Gentry, 1982). Fl (Feb, Sep, Nov); locally common, in disturbed areas along roads and trails.

Arrabidaea triplinervia (DC.) Bureau

Lianas. Stems puberulous, with interpetiolar glands, without interpetiolar ridge. Leaves 2–3-foliolate, the terminal leaflet often replaced by a simple tendril; petioles 1.5–3 cm long; petiolules ca. 1.5 cm long; leaflets ovate to widely ovate, 8.7–12.2 × 3.4–5.5 cm, not bullate, chartaceous, concolorous, drying light brown, glabrous adaxially, puberulous abaxially, the base truncate, slightly cordate to rounded, the apex long acuminate; venation impressed adaxially, salient abaxially, the secondary veins in 5–6 pairs, with three veins from base, the others pinnately arranged, with domatia in some vein axils. Flowers: calyx tubular, without conspicuous glands, appressed to corolla tube, the apex truncate or biparted; corolla tubular-infundibuliform, 3.3–4.6 × 0.7–1.7 cm, glabrous abaxially, white with pinkish dots on throat adaxially. Fruits linear, 36 × 1.4 cm, lepidote-pubescent. Known only from our area by two sterile collection (*Cremers 1151*, *7151*, CAY)

Our description modified from Gentry (1982).

CALLICHLAMYS Miq.

Lianas. Stems with 4 phloem arms, terete to slightly tetragonal, solid to slightly hollow, without interpetiolar glands, with interpetiolar ridge; pseudostipules absent. Leaves 2–3-foliolate, the terminal leaflet sometimes replaced by a simple tendril; petioles without glands distally; leaflets not pellucid-punctate. Inflorescences axillary (rarely terminal), 1–12-flowered, short racemes. Flowers diurnal; calyx narrowly cupular or tubular, spongy, with scattered glands, with a single margin, the apex bipartite; corolla campanulate-infundibuliform, 5-parted, not bilabiate, open at anthesis, glabrous abaxially, yellow, with nectar guides, the tube straight, thick, without glands; stamens inserted, the anthers glabrous; staminode not exceeding stamens; nectary disc present; ovary sessile, glabrous or lepidote-pubescent. Fruits loculicidal capsules, elliptic, widely elliptic, or oblong, flattened, not ridged or winged, smooth, woody, not ornamented. Seeds 2-winged or not well differentiated from body of seed.

Callichlamys latifolia (Rich.) K. Schum.

Stems glabrous. Leaves 2–3-foliolate; petioles 11–13 cm long; lateral petiolules 1.5–2 cm, pulvinate at base and apex, the terminal one 3–4 cm long; leaflets elliptic to narrowly ovate, 16–22 × 5.5–9, concolorous, chartaceous, glabrous on both surfaces, the base obtuse to rounded, the apex acuminate; secondary veins in 6–8 pairs, pinnately arranged. Flowers: calyx widely tubular, 35 × 20 mm, glabrous to pubescent, with simple hairs and lepidote scales, with scattered glands, the apex lobed; corolla funnelform, ca. 8 × 2 cm, thick, yellow, with red streaks in throat. Fruits elliptic, 24–32 × 6–11.5 cm, flattened, woody (fide Gentry, 1982). Fl (Sep, Oct), fr (Jan, immature); in non-flooded forest and in low, wet areas.

CLYTOSTOMA Bureau

Lianas. Stems with 8 phloem arms, terete to 4-angular, solid, without interpetiolar glands, with or without interpetiolar ridge; pseudostipules various, usually subulate, forming clusters resembling "miniature bromeliads" in leaf axils (Gentry, 1993), not foliaceous, persistent. Leaves simple or 2-foliolate, the terminal leaflet sometimes replaced by a simple tendril; petioles without glands distally; leaflets not pellucid-punctate. Inflorescences axillary, ramiflorous, or terminal, few-flowered panicles or fascicles. Flowers diurnal; calyx cupular to campanulate, glands sometimes present, with a single margin, the apex truncate, usually 5-cuspidate; corolla tubular to infundibuliform, 5-parted, not bilabiate, closed at anthesis, somewhat puberulous to glandular lepidote-pubescent abaxially, white to purple, with nectar guides, the tube straight, membranous, without glands; stamens inserted, the anthers glabrous; staminode not exceeding stamens; nectary disc absent; ovary sessile, finely tuberculate, with glandular hairs. Fruits loculicidal capsules, ellipsoid to suborbiculate, convex, winged or ridged, convex, echinate. Seeds winged or not.

Clytostoma binatum (Thunb.) Sandwith Pl. 25a

Stems terete to 4-angular, glabrous to puberulous, without interpetiolar ridge; pseudostipules forming clusters resembling "miniature bromeliads." Leaves 2-foliolate; petioles 0.8–1.5 cm; petiolules ca. 0.8 cm long; leaflets elliptic, 7–12 × 3–5.3 cm, concolorous, chartaceous, glabrous adaxially, lepidote-pubescent with minute reddish-brown scales (visible only with at least 10× magnification) and a few glands abaxially, the base obtuse, the apex usually acuminate, sometimes acute, rarely rounded; secondary veins in 7–8 pairs, pinnately arranged. Inflorescences ramiflorous or terminal on new shoot, solitary or in few-flowered corymbs. Flowers: calyx cupular, 7 × 6–8 mm, glabrous to puberulous toward pedicel junction, the hairs simple, peltate scales absent, glands present, the apex truncate with distinct marginal teeth; corolla infundibuliform, 7.5–8 × 1.8–2 cm, pubescent in bud, puberulous abaxially at anthesis, lavender with white throat. Fruits ellipsoid to suborbiculate, 4.5–6.5 × 3.2–4.5 cm, glabrous, densely echinate (fide Gentry, 1982). Fl (Sep); rare, in forest in low, wet area.

CRESCENTIA L.

Small to medium-sized trees. Stems without anomalous phloem growth, terete or subterete, solid, without interpetiolar glands, without interpetiolar ridge; pseudostipules absent. Leaves simple (in our area) or 3-foliolate, opposite or alternate; petioles without glands distally; leaflets not pellucid-punctate. Inflorescences cauliflorous or ramiflorous, with 1–2 flowers. Flowers nocturnal; calyx large, usually cupular, with or without glands, with a single margin, the apex usually 2-lobed; corolla tubular, variously parted, not bilabiate, open at anthesis, glabrous or lepidote-pubescent abaxially, whitish, usually with dark red nectar guides, the tube bent, thick, without glands; stamens inserted, the anthers glabrous; staminode not exceeding stamens; nectary disc present; ovary sessile, lepidote-pubescent. Fruits pepos, globose, not winged or ridged, not ornamented. Seeds flat, without wings, scattered throughout pulp of fruit.

Crescentia cujete L.

Small trees. Stems subterete, glabrous. Leaves simple, grouped at branch ends; petioles absent; blades obovate, 3–26 × 1–8 cm, concolorous, chartaceous, with peltate scales on midrib of both surfaces, otherwise mostly glabrous, the base attenuate, the apex acute to obtuse; secondary veins in 5–14 pairs, pinnately arranged. Flowers with musty aroma; calyx cupular, ca. 2 cm long, glabrous, with conspicuous glands, the apex bifid; corolla ca. 4 cm long, glabrous to puberulous, with scattered glands, off-white to yellowish, with purplish venation, especially on lobes. Fruits globose, 13–20 cm diam., glabrous, green with small, scattered white glands especially in immature fruits, these glands visited by ants (Elias & Prance, 1978). Fl (Jun), fr (Jun); cultivated. *Calebassier* (tree), *coui* (fruit) (Créole).

CYDISTA Miers

Lianas. Stems with 8 phloem arms, terete to 4-angular, solid or slightly hollow, usually without interpetiolar glands in adults, sometimes in juveniles, with interpetiolar ridge; pseudostipules present or absent, not forming clusters resembling "miniature bromeliads," sometimes foliaceous, caducous or persistent. Leaves simple (outside our area) or 2-foliolate, the terminal leaflet sometimes replaced by a simple tendril; petioles without glands distally; leaflets not pellucid-punctate. Inflorescences axillary or terminal, racemes or few-flowered panicles. Flowers diurnal; calyx cupular to campanulate, glabrous to lepidote-pubescent, with or without glands, with a single margin, the apex truncate, 5-denticulate; corolla tubular-infundibuliform, 5-parted, not bilabiate, with the tube constricted and closed at anthesis, lepidote-pubescent abaxially, white to reddish-purple, with nectar guides, the tube straight, membranous, without glands; stamens inserted, the anthers glabrous; staminode not exceeding stamens; nectary disc absent; ovary sessile, lepidote-pubescent or somewhat puberulous. Fruits loculicidal capsules, linear, flattened, not winged or ridged, not ornamented. Seeds 2-winged.

Cydista aequinoctialis (L.) Miers

FIG. 48, PL. 25b; PART 1: PL. IXc

Stems terete to ± 4 angular, lepidote-pubescent, interpetiolar glands absent, interpetiolar ridge absent; pseudostipules orbicular with scattered glands or inconspicuous. Leaves 2-foliolate; petioles 2.5–6 cm long; petiolules 0.7–3 cm long; leaflets elliptic to widely elliptic or ovate to widely ovate, 6.5–12.5 × 3.5–8 cm, slightly concolorous, lepidote-pubescent adaxially and abaxially, the peltate scales most common on veins toward base, frequently with conspicuous groups of glands in axils of veins abaxially, other types of hairs absent, the base obtuse to rounded, the apex acute to acuminate; secondary veins in 4–6 pairs, pinnately arranged. Inflorescences axillary, racemes, the rachis 1–3.5 cm long, lepidote-pubescent. Flowers: calyx cupular to campanulate, sometimes flared at margin, 7–9 × 8–10 mm, lepidote-pubescent, usually with small glands arranged in clusters, the apex truncate; corolla campanulate, 4.5–7 × 1.2–1.7 cm, lepidote, white (lavender outside our area), often with purple nectar guides in throat, sometimes variously tinged with yellow. Fruits linear, 21–43 × 1.4–2.4 cm, lepidote (fide Gentry, 1982). Fl (Sep, Oct), fr (Jun); common, in non-flooded forest. *Liane panier* (Créole).

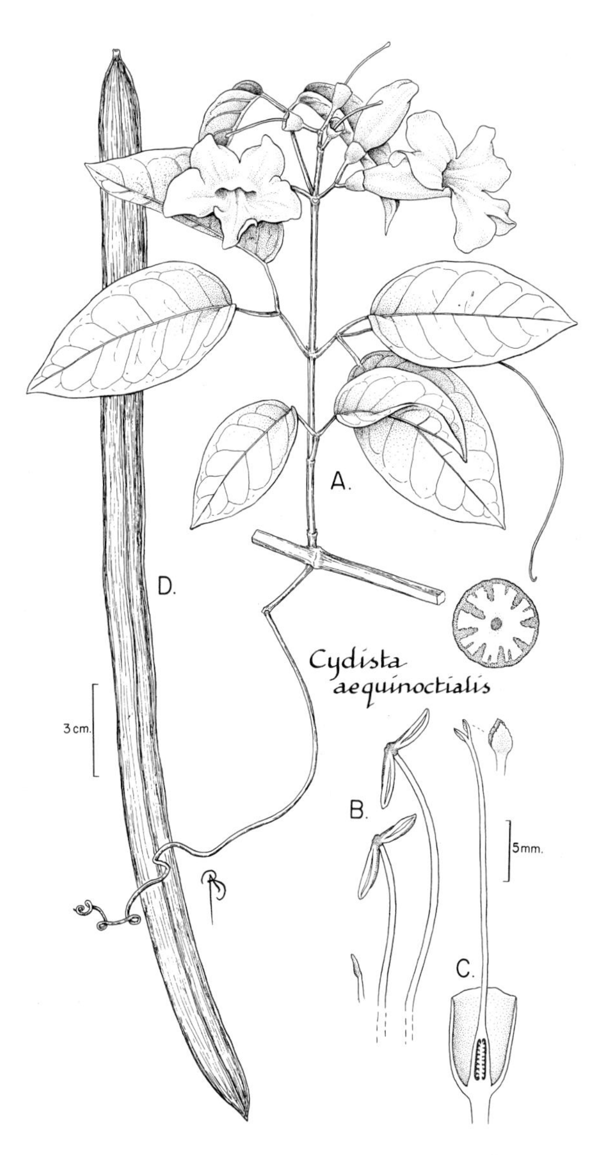

FIG. 48. BIGNONIACEAE. *Cydista aequinoctialis* (A–C, *Acevedo-Rodríguez 2810* from St. John, U.S. Virgin Islands; D, *Questal 575* from Guadaloupe). **A.** Part of stem with leaves, tendrils, flowers in various stages, and cross-section of stem. **B.** Dimorphic stamens (right) and staminode (below). **C.** Medial section of calyx and pistil with detail of stigma. **D.** Fruit. (Adapted with permission from P. Acevedo-Rodríguez, *Flora of St. John, U.S. Virgin Islands*, Mem. New York Bot. Gard. 78, 1996.)

DISTICTELLA Kuntze

Lianas, rarely shrubs. Stems with 4 phloem arms, terete, solid to slightly hollow, without interpetiolar glands, with or without interpetiolar ridge; pseudostipules inconspicuous, subulate, not forming clusters resembling "miniature bromeliads," usually not foliaceous, persistent. Leaves 2–3-foliolate, rarely simple (outside our area), the terminal leaflet often replaced by a trifid tendril;

petioles without glands distally; leaflets not pellucid-punctate. Inflorescences axillary or terminal, racemes, or paniculate racemes. Flowers diurnal; calyx cupular, thick, usually with glands, with a single margin, the apex truncate; corolla tubular-campanulate, 5-parted, not bilabiate, open at anthesis, pubescent abaxially, white, without nectar guides, the tube usually strongly bent, membranous or coriaceous, with or without glands; stamens inserted, the anthers glabrous; staminode not exceeding stamens; nectary disc present; ovary sessile, puberulous. Fruits loculicidal capsules, oblong, flattened, strongly bent, the midrib usually raised, not-ornamented. Seeds winged.

Distictella elongata (Vahl) Urb. PART 1: FIGS. 7, 11

Lianas. Stems sometimes slightly flattened at nodes, densely ferruginous pubescent, with interpetiolar ridge; pseudostipules inconspicuous. Leaves 2-foliolate, densely ferruginous pubescent; petioles 2.5–3 cm long; petiolules 1.5–3 cm long, densely ferruginous pubescent; leaflets elliptic to broadly elliptic, 12–17.5 × 10–14.5 cm, concolorous to slightly discolorous, coriaceous, mostly glabrous with a few scattered crateriform glands adaxially, densely pubescent with simple hairs and peltate scales abaxially, the base rounded to truncate, the apex apiculate; secondary veins in 4–6 pairs, pinnately arranged. Inflorescences axillary or terminal, racemose. Flowers: calyx cupular, 10–12 × 10 mm, pubescent, with conspicuous glands near the margin, the apex truncate; corolla markedly bent, 5.5–6 × 1.8 cm, coriaceous, pubescent ad- and abaxially, white. Fruits oblong, 15.5–18.5 × 5 cm, densely ferruginous pubescent. Fl (Sep), fr (Sep, Nov); the fruits commonly seen on the ground along trails through non-flooded forest, especial in Nov.

Rabelo et al. 2958 (NY) from Amapá, Brazil used to complete description.

JACARANDA Juss.

Trees (in our area), shrubs, or subshrubs. Stems without anomalous phloem growth, terete, solid, without interpetiolar glands, without interpetiolar ridge; pseudostipules absent. Leaves compound, usually bipinnate, occasionally pinnate, without tendrils; petioles without glands distally; leaflets not pellucid-punctate. Inflorescences axillary or terminal, racemes or panicles, few- to many-flowered. Flowers diurnal; calyx short, tubular, cupular, to campanulate, usually without glands, with a single margin, the apex truncate to lobed, usually 5-denticulate, with a single margin; corolla tubular-campanulate, 5-parted, bilabiate, open at anthesis, pubescent to essentially glabrous adaxially, blue, purplish-blue, magenta, or infrequently white, without nectar guides, the tube straight, membranous, without glands; stamens included, the anthers glabrous, sometimes with only 1 theca; staminode exceeding stamens, conspicuously glandular-pubescent at middle and at apex; nectary disc present; ovary sessile, glabrous or pubescent. Fruits septicidal capsules, usually oblong, sometimes elliptic, ± flattened, not ridged or winged, not ornamented. Seeds winged.

Gentry, A. & W. Morawetz. 1992. *Jacaranda*. *In* A. H. Gentry, Bignoniaceae — Part II (Tribe Tecomeae). Fl. Neotrop. Monogr. **25(II):** 51–105.

Jacaranda copaia (Aubl.) D. Don FIG. 49, PL. 25c,d

Trees, 20–30 m tall. Trunks unbuttressed. Bark brown, fissured, the inner bark orange. Stems with simple hairs and lepidote scales. Leaves bipinnate, very large, especially in juvenile trees; ultimate petioles absent to 2–5 mm long; leaflets narrowly elliptic to elliptic or obovate, 3.5–6 × 1.3–3 cm, concolorous, chartaceous, glabrous to lepidote-pubescent on both surfaces, the base ± symmetric to asymmetric, acute to attenuate, the apex acute to obtuse or acuminate. Inflorescences terminal, large, multi-flowered panicles. Flowers: calyx cupular, 0.4–0.7 × 0.3–0.5 cm, densely puberulous to pubescent, with simple hairs and lepidote scales, the apex irregularly dentate, not ribbed to finely ribbed, without glands near margin; corolla 2.4–3.4 × 0.7–1.5 cm, densely puberulous with dendroid hairs, without glands, blue to purple; ovary glabrous. Fruits oblong, 6–13 × 3–10 cm, glabrous to lepidote-pubescent, the valves woody (fide Gentry & Morawetz, 1992). Fl (Aug, Sep, Oct, Nov), fr (Jun, Dec); scattered but common, in non-flooded forest usually in disturbed vegetation, very conspicuous from the air when in flower. *Bois-blanc*, *bois pian* (Créole), *caroba* (Portuguese), *coupaia*, *coupaya*, *faux simarouba* (Créole), *marupa falso* (Portuguese).

In contrast to Gentry and Morawetz (1992), we do not recognize *J. copaia* (Aubl.) D. Don subsp. *spectabilis* (DC.) A. H. Gentry.

LUNDIA DC.

Lianas to lax shrubs. Stems with 4–8 phloem arms, terete to slightly angular, solid to slightly hollow, with or without interpetiolar glands, with or without interpetiolar ridge; pseudostipules small, inconspicuous or absent, not forming clusters resembling "miniature bromeliads," not foliaceous, persistent or caducous. Leaves 2–3-foliolate, the terminal leaflet often replaced by a simple or trifid tendril; petioles without glands distally; leaflets not pellucid-punctate. Inflorescences axillary or terminal, panicles. Flowers diurnal; buds distinctly pointed; calyx often calyptrate, cupular, without glands, with a single margin, the apex truncate, or variously lobed; corolla tubular-infundibuliform to tubular-campanulate, 5-parted, not bilabiate, open at anthesis, pubescent, white to purplish-red, without nectar guides, the tube straight, membranous, without glands; stamens inserted, the anthers villose; staminode not exceeding stamens; nectary disc absent; ovary sessile, pubescent. Fruits loculicidal capsules, linear, flattened, not winged, ridged, or ornamented. Seeds winged.

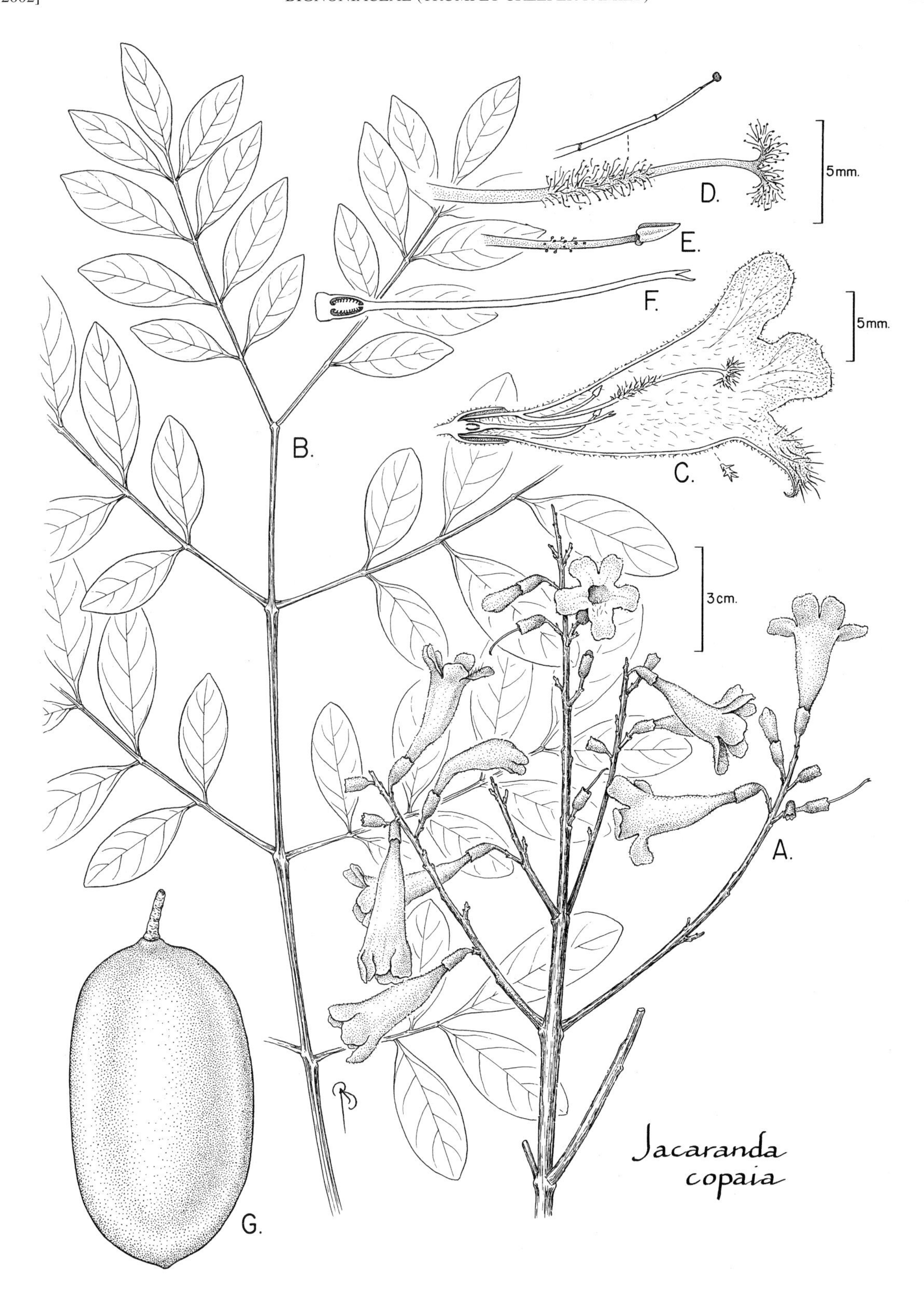

FIG. 49. BIGNONIACEAE. *Jacaranda copaia* (A–C, *Mori et al. 20895*; D–F, *Mori & Mitchell 18785*; G, without voucher). **A.** Part of inflorescence. **B.** Part of bipinnately compound leaf. **C.** Medial section of flower with detail of pubescence (below). **D.** Staminode with detail of glandular pubescence (above). **E.** Part of fertile stamen with glandular pubescence and monothecal anther. **F.** Medial section of pistil. **G.** Fruit.

Lundia corymbifera (Vahl) Sandwith

Lianas. Stems terete, puberulous, with interpetiolar glands, without interpetiolar ridge; pseudostipules inconspicuous. Leaves 2-foliolate; petioles 3.2–5.3 cm; petiolules 1.2–3.2 cm; leaflets ovate to elliptic, 6.5–11.5 × 3.3–7.6 cm, discolorous, chartaceous, puberulous on both surfaces, especially on principal veins and in vein axils toward base abaxially, sparsely lepidote-pubescent on both surfaces, occasionally with glands abaxially, the base cordate to truncate, the apex acute to acuminate; secondary veins in 4–6 pairs, palmately arranged proximally, pinnately arranged distally. Flowers: calyx calyptrate in bud, at anthesis cupular, 3–6 × 4–5 mm, puberulous, without glands, the hairs simple, the apex truncate; corolla 2.5–4.3 × 0.8–1.3 cm, pubescent ab- and adaxially, white. Fruits linear, 29–60 × 1.5–1.9 cm, puberulous. Fl (Apr); known only from *Granville 4421* (CAY), specimen not yet seen by us.

Our description modified from Gentry (1982).

MACFADYENA DC.

Lianas or vines attached to trees by adventitious roots. Stems usually with 8 phloem arms, terete, solid, with or without interpetiolar glands, with or without interpetiolar ridge; pseudostipules large, lanceolate to ovate, not forming clusters resembling "miniature bromeliads," not foliaceous, caducous and therefore usually absent. Leaves 2-foliolate, the terminal leaflet often replaced by a trifid, cat claw-like tendril, of which the terminal divisions uncinate; petioles without glands distally; leaflets not pellucid-punctate. Inflorescences axillary, cymes or panicles, sometimes reduced to a single flower. Flowers diurnal; calyx cupular to campanulate, without glands, with a single margin, the apex truncate to spathaceous, frequently irregularly lobed; corolla tubular-campanulate, 5-parted, not bilabiate, open at anthesis, glabrous abaxially, yellow, without nectar guides, the tube straight, membranous, without glands; stamens inserted, the anthers glabrous; staminode not exceeding stamens; nectary disc present; ovary sessile, minutely lepidote-pubescent to subpuberulous or glabrescent. Fruits loculicidal capsules, narrow, linear, flattened, not winged or ridged, smooth. Seeds not winged.

Macfadyena unguis-cati (L.) A. H. Gentry
FIG. 50, PL. 25e

Lianas. Stems terete, flattened in some places, glabrous, with interpetiolar glands, with interpetiolar ridge; pseudostipules ovate; juvenile plants attached to supporting tree by adventitious rootlets, the leaves much smaller than those of adult plants. Leaves 2–3-foliolate; petioles 0.4–3.9 cm long; lateral petiolules 0.1–0.8 cm long; leaflets elliptic, 1.6–4.8 × 0.5–1.8 cm, discolorous, chartaceous, glabrous, the base rounded, the apex acute; secondary veins pinnately arranged. Flowers: calyx cupular, flared at margin, 7–10 × 8–9 mm, glabrous, without glands, the apex shallowly and irregularly lobed; corolla 4–5 × 1–1.3 cm, glabrous, yellow. Fruits linear, 26–95 × 1–1.9 cm, lepidote-pubescent (fide Gentry, 1982). Fl (Aug); in non-flooded forest, conspicuous because of large number of yellow flowers produced at anthesis. *Griffe chatte*, *griffes de chat* (Créole).

MANSOA DC.

Lianas. Vegetative parts with strong aroma of garlic. Stems with 4–8 phloem arms, terete to 4–6-angular, solid, with or without interpetiolar glands, with or without interpetiolar ridge; pseudostipules usually small, subulate, sometimes forming clusters resembling "miniature bromeliads," never foliaceous, persistent. Leaves 2–3-foliolate, the terminal leaflet often replaced by a trifid or simple tendril, sometimes with a peltate disc at apex; petioles with or without glands distally; leaflets not pellucid-punctate. Inflorescences axillary or terminal, racemes, panicles or corymbs. Flowers diurnal; calyx cupular to tubular campanulate, with or without glands, with a single margin, the apex truncate, lobed, or with 5 teeth, the teeth prolongations of calycine costae; corolla tubular-campanulate to tubular-infundibuliform, 5-parted, not bilabiate, open at anthesis, puberulous and glandular-lepidote-pubescent at least on lobes, pink to reddish purple, with or without nectar guides, the tube straight, membranous, without glands; stamens inserted, the anthers glabrous; staminode not exceeding stamens; nectary disc absent; ovary sessile, ± glandular-papillose and inconspicuously puberulous. Fruits loculicidal capsules, linear-oblong, somewhat flattened or cilindrical, sometimes ridged, not winged, sometimes ornamented. Seeds with or without wings.

1. Stems with interpetiolar ridge; pseudostipules subulate. Petioles 1–1.3 cm long; leaflets 9.3–11 cm long, chartaceous, puberulous; secondary veins 3 from base. Calyx with scattered glands; corolla yellow, the lobes rose-red. . . . . *M. alliacea*.
1. Stems without interpetiolar ridge; pseudostipules conical. Petioles 3–3.5 cm long; leaflets 22–24 cm long, coriaceous, glabrous; secondary veins not 3 from base. Calyx without scattered glands; corolla white, tinged with purple, the lobes purple. . . . . *M. standleyi*.

Mansoa alliacea (Lam.) A. H. Gentry

Stems with 8 phloem arms, terete, glabrous to sparsely puberulous, with interpetiolar glands, with interpetiolar ridge; pseudostipules small, subulate. Leaves 2-foliolate; petioles 1–1.3 cm long; petiolules 0.6–0.9 cm long; leaflets elliptic or obovate, 9.3–11 × 4.7–5.5 cm, concolorous, chartaceous, puberulous on both surfaces, the base rounded to cordiform, the apex acuminate; secondary veins pinnately arranged. Inflorescences axillary or terminal, racemes. Flowers: calyx cupular, 9–21 mm long, puberulous with peltate scales and simple hairs, with scattered glands, the apex 5-lobed; corolla tubular-infundibuliform, 4.3 × 1.5 cm, with branched hairs, rose-red on lobes, the base of tube puberulous, yellow ad- and abaxially. Fruits linear, 35–75 × 2.6–4.2 cm, glabrous. Seeds winged. *Cipó d'alho* (Portuguese), *douvant-douvant*, *liane-ail* (Créole).

Known only by two sterile collections (*Granville B5455* and *Jacquemin 1426*, both CAY) from our area.

Mansoa standleyi (Steyerm.) A. H. Gentry

Stems with 8 phloem arms, terete, glabrous, with interpetiolar glands, without interpetiolar ridge; pseudostipules small, conical. Leaves 2–3-foliolate; petioles 3–3.5 cm long; petiolules 1–1.5 cm long; leaflets elliptic, 22–24 × 9.5–12 cm, concolorous, coriaceous, glabrous, the base acute, the apex acute or acuminate; secondary veins pinnately arranged, 3-veined at base. Inflorescences axillary or terminal, racemes. Flowers: calyx cupular, with scattered simple hairs and scales, without scattered glands, the apex distinctly 5-lobed; corolla tubular-infundibuliform, 3.5–6 × 0.8–1.9 cm, with scattered simple hairs and scales, the tube white tinged with purple, the lobes purple, usually glabrous. Fruits linear, 35–75 × 2.6–3.8 cm, glabrous (fide Gentry, 1982). Seeds winged. Fl (Aug). *Liane-ail* (Créole).

Known from our area by a single flower photographed from the ground and determined as this species by Gentry in 1988. The collection (*Mori & Gracie s.n.*, NY) consists only of a mounted photo at NY. *Aymard 5388* (NY) from Venezuela used to complete description.

MARTINELLA Baill.

Lianas. Stems with 4 phloem arms, terete or tetragonal, solid or slightly hollow, without interpetiolar glands (an occasional gland sometimes in *M. obovata*), with interpetiolar ridge; pseudostipules absent. Leaves 2–3-foliolate, the terminal leaflet often replaced by a trifid tendril; petioles without glands distally; leaflets not pellucid-punctate. Inflorescences axillary, racemes. Flowers diurnal; calyx tubular to campanulate, without glands, with a single margin, the apex truncate, bilabiate or 3–4-lobed; corolla tubular-campanulate, 5-parted, bilabiate, open at anthesis, glabrous to inconspicuously lepidote-pubescent abaxially, wine red to lilac, sometimes white on tube, usually without nectar guides, the tube straight, membranous, without glands; stamens inserted, the anthers glabrous; staminode not exceeding stamen; nectary disc present; ovary sessile, sparsely lepidote-pubescent or puberulous. Fruits loculicidal capsules, linear, flattened, smooth, not winged or ridged. Seeds winged.

Martinella obovata (Kunth) Bureau & K. Schum.

Stems with nodes sometimes thickened, pubescent, with both simple and glandular hairs. Leaves 2-foliolate: petioles 3.5 cm long; petiolules 1 cm long; leaflets ovate to narrowly ovate, 9–10.5 × 4–4.7 cm, concolorous, chartaceous, mostly glabrous on both surfaces except a few scattered glands toward base and punctations throughout abaxially, the base rounded, the apex acuminate; secondary veins in 5 pairs, pinnately arranged. Flowers: calyx cupular to campanulate, 20 × 10 mm, puberulous, clavate, glandular hairs present, the apex irregularly lobed; corolla infundibuliform, 4.5 × 1.3 cm, mostly glabrous, except puberulous on lobes abaxially, the tube white tinged with lilac, the lobes lilac. Fruits linear, 55–130 × 1.2–1.8 cm, glabrous to puberulous (fide Gentry, 1982). Fl (May); in non-flooded forest.

MEMORA Miers

Lianas, subshrubs, or shrubs. Stems of lianas with 4 phloem arms, terete, solid, without interpetiolar glands, with or without interpetiolar ridge; pseudostipules large or small, orbiculate or subulate, not forming clusters resembling "miniature bromeliads", sometimes foliaceous, persistent, or absent. Leaves simple (outside our area), 2-foliolate, pinnate, bipinnate, tripinnate, or biternate, the terminal leaflet often replaced by a simple or trifid tendril; petioles without glands distally; leaflets conspicuously thickened at joint, not pellucid-punctate. Inflorescences axillary, racemes, paniculate arrangements of cymes, or reduced to a single flower. Flowers diurnal; calyx cupular to campanulate or spathaceous, with glands, with a single margin, the apex truncate to bilabiate or minutely 5-denticulate; corolla tubular-infundibuliform, 5-parted, bilabiate, closed at anthesis, glabrous or pubescent abaxially, yellow, without nectar guides, the tube straight, membranous, without glands; stamens inserted, the anthers glabrous; staminode not exceeding stamens; nectary disc present; ovary sessile, glabrous, lepidote-pubescent, or puberulous. Fruits loculicidal capsules, linear to oblong, flattened to subterete, smooth. Seeds slender and winged or thick and corky and not winged.

1. Leaves tripinnate, the ultimate leaflets ≤3 cm long, the terminal most leaflet markedly rhomboid. Calyx spathaceous. . . . . *M. moringiifolia.*
1. Leaves not tripinnate, 2-foliolate, pinnate, bipinnate, or 2–3-ternate, the ultimate leaflets usually >4 cm long, the terminal most leaflet elliptic to ovate, never markedly rhomboid. Calyx truncate.
    2. Inflorescences and calyces densely ferruginous pubescent. . . . . *M. tanaeciicarpa.*
    2. Inflorescences and calyces glabrous, lepidote-pubescent, or puberulous, not ferruginous pubescent.
        3. Leaves bipinnate; leaflets glabrous abaxially; tendrils trifid. Inflorescences paniculately arranged cymes; bracts ≤2 mm long. Calyx straight, lepidote-pubescent, without glands, the apex truncate. . . . . *M. flavida.*
        3. Leaves 2-foliolate to pinnate; leaflets densely glandular-pubescent abaxially; tendrils simple. Inflorescences racemes; bracts ≥6 mm long. Calyx flared, puberulous toward apex, with scattered glands, the apex 5-denticulate. *M. racemosa.*

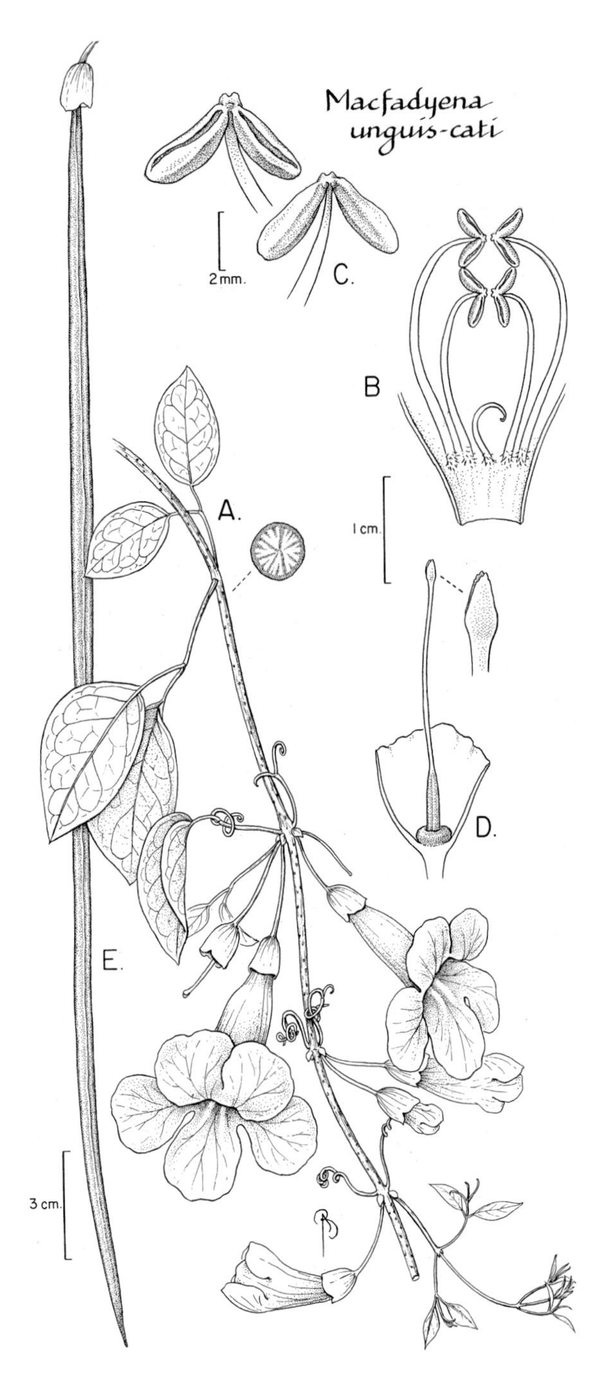

FIG. 50. BIGNONIACEAE. *Macfadyena unguis-cati* (A, *Holdridge 1074* from St. John, U.S. Virgin Islands; B–D, *Mori & Gracie 18641*; E, *Acevedo-Rodríguez 2803* from St. John, U.S. Virgin Islands). **A.** Part of stem with leaves, tendrils, flowers in various stages, and cross-section of the stem. **B.** Base of corolla opened to show arrangement of stamens and staminode. **C.** Adaxial (far left) and abaxial (near left) views of anthers. **D.** Longitudinal section of calyx to show pistil surrounded by nectary disc (left) and detail of stigma (above). **E.** Fruit subtended by calyx.(Adapted with permission from P. Acevedo-Rodríguez, *Flora of St. John, U.S. Virgin Islands*, Mem. New York Bot. Gard. 78, 1996. Note: B and D have been added to the original illustration; C has been adapted from B on the St. John plant and E is equal to C on the St. John plate.)

Memora flavida (DC.) Bureau & K. Schum.

Lianas. Stems terete, with interpetiolar ridge; pseudostipules small, clavate but constricted in middle. Leaves usually bipinnate, the terminal leaflet frequently replaced by a trifid tendril; petioles 3.5–5.5 cm; primary petiolules 3.5–5 cm; secondary lateral petiolules absent to 0.8 cm long, the terminal secondary petiolules longer; leaflets elliptic to narrowly ovate, 6–13.5 × 3–4.5 cm, concolorous, subcoriaceous, glabrous on both surfaces, the base acute, obtuse, or rounded, the apex acute to acuminate; secondary veins in 5–9 pairs, pinnately arranged. Inflorescences paniculately arranged cymes, the axis lepidote-pubescent; bracts 1–2 mm long. Flowers: calyx tubular to cupular, 8–10 × 7–9 mm, lepidote-pubescent, without glands, the apex truncate; corolla 6.5 × 1–1.8 cm, glabrous, yellow. Fruits linear, 17–26 × 2.3–3 cm, lepidote-pubescent (fide Gentry, 1982). Fl (Aug, Oct, Nov); in non-flooded forest.

Memora moringiifolia (DC.) Sandwith PL. 26a

Lianas. Stems terete, without interpetiolar ridge; pseudostipules subfoliaceous. Leaves 3-pinnate, the terminal leaflet frequently replaced by a simple tendril; petioles ca. 3 cm long; ultimate petiolules of lateral leaflets absent to 0.5 cm long; leaflets elliptic, often asymmetrical, 1.5–3 × 0.8–1.5 cm, concolorous, chartaceous, glabrous adaxially, puberulous abaxially, the base asymmetric, the apex acute to obtuse, the terminal leaflet markedly rhomboid; secondary veins of leaflets in 3–7 pairs, pinnately arranged. Inflorescences axillary, solitary, glabrous. Flowers: calyx spathaceous, 15–17 × 6–7 mm, glabrous, with glands arranged in lines, the apex denticulate; corolla 6 × 1.3 cm, glabrous, yellow. Fruits linear, 35 × 2 cm, lepidote. Fl (Aug on the Sinnamary River); abundant, seen commonly sterile in non-flooded forest.

Flowering measurements from *Mori et al. 23422* (NY) from outside our area on the Sinnamary River. Fruit description from *Sabatier 804* (MO) from French Guiana but outside our area.

Memora racemosa A. H. Gentry

PL. 26b; PART 1: FIGS. 7, 9

Lianas. Stems terete, without interpetiolar ridge; pseudostipules inconspicuous, subfoliaceous. Leaves 2-foliolate to pinnate, infrequently bipinnate, the terminal leaflet frequently replaced by a simple tendril; petioles 2–5 cm long; lateral petiolules 0.3–2 cm long, often pulvinate, the terminal petiolule somewhat longer; leaflets narrowly ovate to elliptic, 8–15 × 2.5–5.5 cm, concolorous, subcoriaceous, glabrous adaxially, densely glandular pubescent abaxially, the base obtuse to rounded, the apex acute, acuminate, or attenuate; secondary veins in 5–8 pairs. Inflorescences racemes, the axis glabrous to puberulous; bracts 6–14 mm long. Flowers: calyx cupular to campanulate, flared, 7–10 × 9–11 mm, glabrous to puberulous, with scattered glands toward apex, the apex 5-denticulate; corolla 5.5–6 × 1.3–1.5 cm, glabrous, yellow. Fruits alternately expanded and constricted in conformity with seeds, 38–51 × 1.5–2.5 cm, glabrous to minutely puberulous. Fl (Sep), fr (Nov); abundant, fallen flowers common on trails through non-flooded forest.

Memora tanaeciicarpa A. H. Gentry

Lianas. Stems terete, without interpetiolar ridge; pseudostipules inconspicuous. Leaves 2–3-ternate, the terminal leaflet frequently replaced by a simple tendril; leaflets elliptic to ovate, 4–15 × 1.5–7 cm, concolorous, coriaceous, variously puberulous or lepidote-pubescent on both surfaces, the base cuneate to truncate, the apex acuminate. Inflorescences axillary, racemes, the axis densely ferruginous pubescent. Flowers: calyx campanulate, 10–13 × 6–9 mm, densely ferruginous pubescent, the apex truncate, minutely 5-toothed; corolla ca. 4 cm long, markedly pubescent abaxially, yellow. Fruits elliptic, 9–10 × 5 cm, tomentose. Known only from a single sterile specimen (*Gentry et al. 63138*, CAY).

Our description adapted from Gentry (1982).

PARABIGNONIA K. Schum.

Lianas. Stems with 8 phloem arms, terete or tetragonal, solid, with or without interpetiolar glands, with or without interepetiolar ridge; pseudostipules absent. Leaves (1)2–3-foliolate, the terminal leaflet frequently replaced by a trifid, uncinate cat claw-like tendril; petioles without glands distally; leaflets not pellucid-punctate. Inflorescences axillary, racemes. Flowers diurnal; calyx campanulate, usually with glands, with a simple margin, the apex 5-lobed; corolla tubular-infundibuliform, 5-parted, not bilabiate, open at anthesis, glabrous or puberulous abaxially, reddish-purple, without nectar guides, the tube straight, membranous, without glands; stamens inserted, the anthers glabrous; staminode not exceeding stamens; nectary disc present; ovary sessile, lepidote-pubescent. Fruits loculicidal capsules, linear, flattened, not ridged or winged, smooth. Seeds winged.

Parabignonia steyermarkii Sandwith

Juvenile plants tightly appressed to tree trunks, rooting at nodes. Stems glabrous to puberulous, with interpetiolar glands, with interpetiolar ridge. Leaves 1–2-foliolate, the terminal leaflet modified into trifid, cat claw-like tendril; petioles 0.7–2 cm long; petiolules 0.2–0.5 cm long; leaflets elliptic or ovate, 4–8 × 2–4.5 cm, discolorous, chartaceous, pubescent when young, glabrescent with time, the base rounded, the apex obtuse to acute, the margins entire, sometimes serrate on young leaflets. Inflorescences few-flowered. Flowers: calyx campanulate, 8–10 mm long, densely puberulous, with glands, the apex 5-lobed; corolla 4.5–7.5 × 1.3–2 cm, densely puberulous, reddish-purple abaxially and on lobes, cream-colored on throat adaxially. Fruits linear, an immature one 10 × 0.5 cm. Known only by sterile collections from our area.

Our description adapted from Gentry (1982).

PARAGONIA Bureau

Lianas. Stems with 4 phloem arms, terete to subtetragonal, solid, without interpetiolar glands, with interpetiolar ridge; pseudostipules large, subulate, not forming "miniature-bromeliads," not foliaceous, persistent. Leaves 2-foliolate, the terminal leaflet often replaced by a minutely bifid, or rarely trifid, tendril; petioles with glands distally; leaflets not pellucid-punctate. Inflorescences terminal or axillary, multiflowered panicles or paniculate arrangements of cymes. Flowers diurnal; calyx cupular, lepidote-pubescent, usually with glands, with a single margin, the apex truncate to variously lobed; corolla tubular-infundibuliform, 5-parted, not bilabiate, open at anthesis, pubescent abaxially, usually lavender to purple, without nectar guides, the tube straight, membranous, with glands; stamens inserted, the anthers glabrous; staminode not exceeding stamens; nectary disc present; ovary sessile, lepidote-pubescent. Fruits loculicidal capsules, linear, flattened, not ridged or winged, finely tuberculate, frequently with texture like sandpaper. Seeds winged.

Paragonia pyramidata (Rich.) Bureau

Stems terete to subtetragonal, glabrous to pubescent. Leaves 2-foliolate; petioles 1–2 cm long, with conspicuous gland fields at apex; petiolules 1.5–2 cm; leaflets elliptic, 11.5–14 × 5–6.5 cm, concolorous, chartaceous, mostly glabrous to minutely lepidote-pubescent adaxially, minutely lepidote-pubescent with scattered glands along midrib abaxially, the base acute or rounded, obtuse, the apex short acuminate to acuminate; secondary veins in 4–6 pairs, pinnately arranged. Inflorescences axillary, panicles. Flowers: calyx cupular, 8–9 × 6–12 mm, lepidote-pubescent, with glands, the apex truncate to irregularly lobed; corolla 5.5–7 × 1–1.2 cm, densely pubescent externally, lavender to purple. Fruits linear, 32–61 × 1.2–1.4 cm, glabrous, finely tuberculate (fide Gentry, 1982). Fl (Sep, Oct), in non-flooded forest.

PITHECOCTENIUM Meisn.

Lianas. Stems with 4 phloem arms, hexagonal, solid, without interpetiolar glands, with interpetiolar ridge; pseudostipules small, spatulate, not forming clusters resembling "miniature bromeliads," usually not foliaceous, persistent or caducous. Leaves 2(3)-foliolate, the terminal leaflet often replaced by a 3–15-fid tendril; petioles without glands distally; leaflets not pellucid-punctate. Inflorescences terminal or axillary, racemes or racemose arrangements of cymes. Flowers diurnal; calyx cupular, with or without glands, with a single margin, the apex truncate to 5-denticulate; corolla tubular-infundibuliform, 5-parted, not bilabiate, open at anthesis, densely pubescent abaxially, white, the tube bent, coriaceous, without glands; staminode not exceeding stamens; stamens inserted, the anthers glabrous; nectary disc present; ovary sessile, pubescent. Fruits loculicidal capsules, elliptic, flattened, not winged or ridged, echinate. Seeds winged.

Pithecoctenium crucigerum (L.) A. H. Gentry Pl. 26c

Stems nearly glabrous to pubescent; pseudostipules lanceolate. Leaves mostly 2-(3)-foliolate, the terminal leaflet often replaced by a multifid tendril with tips thickened into a disc; leaflets ovate to suborbiculate, 3.3–18 × 2–14.7 cm, concolorous, chartaceous, lepidote-pubescent and variously pubescent on both surfaces, with small, hyaline, peltate scales, with glands in axils of veins abaxially, the base cordate to truncate, the apex acuminate; secondary veins in 4–6 pairs, palmately arranged proximally, pinnately arranged distally. Inflorescences axillary, racemes. Flowers: calyx cupular, 8–12 × 9–11 mm, densely pubescent, with glands, with teeth thickened below margin; corolla 3.6–6.1 × 1–1.8 cm, densely pubescent, white with yellow throat. Fruits elliptic, 12–31 × 5.2–7.5 cm, conspicuously echinate. Known only by sterile collections from our area.

Our description adapted from Gentry (1982).

PLEONOTOMA Miers

Lianas. Stems with 4 phloem arms, often 4-angular, solid or hollow, without interpetiolar glands, with interpetiolar ridge; pseudostipules variable in shape and size, not forming clusters resembling "miniature bromeliads," sometimes absent. Leaves biternate, triternate, bipinnate, or tripinnate, the terminal leaflet often replaced by a trifid tendril; petioles without glands distally; leaflets conspicuously thickened at joint, not pellucid-punctate. Inflorescences axillary or terminal, racemes. Flowers diurnal; calyx cupular to campanulate, sometimes with glands, with a single margin, the apex truncate or denticulate; corolla tubular-infundibuliform or campanulate, 5-parted, not bilabiate, open at anthesis, glabrous to variously pubescent abaxially, white to pale yellow, the tube straight, membranous, without glands; stamens inserted, the anthers glabrous; staminode not exceeding stamens; nectary disc present; ovary sessile, lepidote-pubescent or pubescent. Fruits loculicidal capsules, linear, narrow to wide, flattened, not ridged or winged, smooth. Seeds winged.

1. Plants glabrous. Inflorescences axillary. Calyx without glands. . . . . *P. albiflora.*
1. Plants covered with dendritic hairs. Inflorescences terminal. Calyx with glands near margin. . . . . *P. dentrotricha.*

Pleonotoma albiflora (DC.) A. H. Gentry

Stems quadrangular, glabrous. Leaves 2–3-ternate; petioles ca. 6.5 cm long; primary petiolules 4.5–7.9 cm long; secondary petiolules 1–2.5 cm long; leaflets ovate to elliptic, 2.5–14 × 1.5–7.5 cm, concolorous, chartaceous, glabrous on both surfaces, the base rounded, the apex obtuse to short acuminate; secondary veins pinnately arranged. Inflorescences axillary, racemes or racemose panicles, the axis glabrous. Flowers: calyx cupular, 6–10 × 7–11 mm, glabrous, the apex 2-parted, without glands; corolla campanulate, ca. 4.5 cm long, glabrous on tube, puberulous at base, white or cream-colored. Fruits linear, 35–51 × 2.1–3 cm, glabrous (fide Gentry, 1982). Known only from sterile collections from non-flooded forest from our area.

Pleonotoma dendrotricha Sandwith

Stems quadrangular, pubescent, interpetiolar ridge covered with dendritic hairs. Leaves 2–3-ternate; petioles 4.2–5 cm long, covered with dendritic hairs; lateral petiolules 1–2.5 cm long, the terminal petiolule 2.5–4.5 cm long; leaflets ovate to elliptic, 4–12 × 2–6 cm, concolorous, chartaceous, pubescent on both surfaces, the hairs dendritic, the base rounded, the apex acute to obtuse; secondary veins pinnately arranged. Inflorescences terminal, racemes, the axis pubescent. Flowers: calyx cupular, 8–9 × 6–8 mm, covered with dendritic hairs, with glands near margin; corolla campanulate, ca. 6.5 cm long, glabrous on tube, pubescent on lobes, white or pale yellow. Fruits linear, 45 × 1.7 cm, glabrous. Fr (Oct); in non-flooded forest.

Flowers not known from our area, our description of them adapted from Gentry (1982).

SCHLEGELIA Miq.

Lianas, hemi-epiphytic shrubs, or lax shrubs (outside our area). Stems without anomalous phloem growth, terete, solid, without interpetiolar glands, without interpetiolar ridges; pseudostipules small or absent, subulate, not forming clusters resembling "miniature-bromeliads," not foliaceous, persistent or caducous. Leaves simple, without tendrils; petioles without glands distally; blades not pellucid-punctate; secondary veins pinnately arranged. Inflorescences axillary or terminal, racemes, fascicles, or racemose panicles. Flowers diurnal; calyx tubular to cupular, often lepidote-pubescent, with glands, with a single margin, the apex truncate to irregularly lobed; corolla tubular, 5-parted, not bilabiate, open at anthesis, glabrous abaxially, densely glandular-pubescent adaxially, white, rose, purple, or red, without nectar guides, the tube straight, thick, without glands; stamens inserted, the anthers glabrous; staminode lacking; nectary disc present; ovary not stiptate, glabrous. Fruits berries, orbicular, globose, or subglobose, not winged or ridged, smooth. Seeds not winged.

1. Pseudostipules subulate, 0.2–0.3 cm long. Leaf blades with distinct gland fields abaxially toward base. Inflorescences multiflowered racemes. Calyx irregularly 3–5-lobed, 5–7 mm long; corolla 1–1.5 cm long, the lobes white tinged with pink. Fruits 0.5–1 cm diam., purple. . . . . *S. fuscata.*
1. Pseudostipules lanceolate, 0.7–0.8 cm long. Leaf blades with scattered glands abaxially toward base. Inflorescences 3-flowered fascicles. Calyx 2-lobed, ca. 1.4 cm long; corolla 3.5–4 cm long, the lobes purplish red. Fruits 1.5–2 cm diam., brown. . . . . *S. paraensis.*

Schlegelia fuscata A. H. Gentry Pl. 26d,e

Lianas. Stems terete; pseudostipules subulate, 0.2–0.3 cm long. Leaves: petioles 1–1.5 cm long, glabrous; blades elliptic to widely elliptic, 9–13 × 4–7.5 cm, coriaceous, drying light brown, glabrous adaxially, glabrous with distinct glands toward base abaxially, the base obtuse to rounded, the apex obtuse. Inflorescences multiflowered racemes; pedicels 4–7 mm long. Flowers: calyx irregularly 3–5-lobed, 5–7 × 3–4 mm; corolla 1–1.5 × 0.6–0.7 cm, the tube white, the lobes white tinged with pink, the throat pink. Fruits depressed globose, 0.5–1 cm diam., purple. Fl (May, Jul, Sep), fr (Jan, Jun, Nov); common, flowers often seen in non-flooded forest in season. Bees observed visiting flowers (*Mori & Gracie 18598*, NY).

Schlegelia paraensis Ducke FIG. 51

Lianas or epiphytic shrubs. Stems terete, glabrous; pseudostipules lanceolate, 0.7–0.8 cm long. Leaves: petioles 1 cm long; blades elliptic, 8–14.5 × 4.5–8 cm, coriaceous, drying reddish-brown, glabrous adaxially, glabrous with scattered glands abaxially, the base obtuse, the apex acute to obtuse. Inflorescences axillary or terminal, 3-flowered fascicles; pedicels ca. 10 mm long. Flowers: calyx 2-lobed, 14 × 8 mm; corolla 3.5–4 × 1–1.5 cm, the tube white, the lobes purplish red. Fruits globose, ca. 1.5–2 cm diam., brown. Fl (Jul, Aug, Sep), fr (Aug, Sep, Nov); apparently less common than *S. fuscata*, in non-flooded forest. Hummingbird observed visiting flowers (*Mori et al. 19210*, NY).

STIZOPHYLLUM Miers

Lianas. Stems with 4 phloem arms, terete, distinctly hollow, without interpetiolar glands, with or without interpetiolar ridge; pseudostipules small and inconspicuous or large, ± spatulate, never forming clusters resembling "miniature-bromeliads," caducous. Leaves 2–3-foliolate, the terminal leaflet often replaced by a simple or trifid tendril; petioles without glands distally; leaflets conspicuously pellucid-punctate (sometimes difficult to see under the pubescence in some species but obvious in glabrous species). Inflorescences axillary or terminal, fascicles or few-flowered racemes. Flowers diurnal; calyx funnelform, cupular, or campanulate, with glands, with a single margin, the apex irregularly lobed; corolla tubular-infundibuliform, 5-parted, bilabiate, open at anthesis, pubescent, often with both simple hairs and lepidote scales, abaxially, white or cream-colored, the lobes sometimes flushed with pink or purple, without nectar guides, the tube straight, membranous, without glands; stamens inserted, the anthers glabrous or pubescent along margins of openings; staminode not exceeding stamens; nectary disc present; ovary sessile, lepidote-pubescent. Fruits loculicidal capsules, narrow linear, subterete ("pencil-like" fide Gentry, 1993) or flattened, not ridged or winged. Seeds winged.

1. Stems and leaves villose. Tendrils densely pubescent. . . . . *S. inaequilaterum.*
1. Stems and leaves pubescent. Tendrils glabrous. . . . . *S. riparium.*

Stizophyllum inaequilaterum Bureau & K. Schum. PL. 27a.

Stems villose, without interpetiolar ridge. Leaves 2–3-foliolate, the terminal leaflet frequently replaced by a trifid, densely pubescent tendril; petioles 2–3.5 cm long, densely pubescent; petiolules 0.7–1 cm long, villose; leaflets ovate to elliptic, 4–7 × 2–5 cm, concolorous, chartaceous, densely pubescent on both surfaces, the hairs simple, surrounded at base by a ring, especially on adaxial surface, the pellucid punctations difficult to see, the base obtuse to rounded, asymmetrical, the apex acute to obtuse; secondary veins in 4–5 pairs, palmately arranged proximally, pinnately arranged distally. Inflorescences terminal racemes. Flowers: calyx cupular, inflated, 6–8 × 7–9 mm, densely pubescent, with glands, the apex irregularly and shallowly lobed; corolla 3.5–4.5 × 0.8–1.2 cm, pubescent with simple hairs and peltate scales abaxially, the tube white flushed with magenta, the lobes solid magenta. Fruits linear, 30–32.5 × 0.4–0.5 cm, villose (fide Gentry, 1982). Fl (Sep); in non-flooded forest.

The leaves of our specimen (*Mori et al. 24242*, NY) do not appear to be fully expanded and therefore our measurements may under represent leaf sizes.

Stizophyllum riparium (Kunth) Sandwith

Stems short pubescent, without interpetiolar ridge. Leaves 2–3-foliolate, the terminal leaflet frequently replaced by a simple, glabrous tendril; petioles 2–11 cm long, pubescent; lateral petiolules 1–2.5 cm long, the terminal petiolule 1.8–5.6 cm long, both pubescent; leaflets ovate, obovate, or oblong-elliptic, 2.9–20 × 1.9–12.5 cm, concolorous, chartaceous, puberulous on both surfaces, the pellucid punctations conspicuous, the base truncate to subcordate, the apex acute to acuminate; secondary veins in 3–6 pairs, pinnately arranged. Inflorescences axillary racemes. Flowers: calyx funnelform, 5–12 × 4–10 mm, pubescent, with simple hairs and peltate scales, with glands, the apex 2–5-lobed; corolla 3.2–5 × 0.7–1.3 cm, pubescent with simple hairs and peltate scales abaxially, uniformly white to cream-colored. Fruits linear, 24–45 × 0.5–0.7 cm, puberulous. Known only by a single sterile specimen (*Gentry 63154*, MO) from our area.

Our description adapted from Gentry (1982).

TABEBUIA DC.

Shrubs to large trees. Stems without anomalous phloem growth, terete, solid, without interpetiolar glands, without interpetiolar ridge; pseudostipules absent. Leaves palmately compound (in our area) to simple or 1-foliolate; petioles without glands distally; leaflets not pellucid-punctate. Inflorescences terminal, panicles, sometimes reduced to a few-flowered raceme or a single flower. Flowers diurnal; calyx cupular, campanulate, or tubular, with or without glands, with a single margin, the apex truncate, 2-lobed, or slightly 5-lobed; corolla tubular-infundibuliform to tubular-campanulate, 5-parted, not bilabiate, open at anthesis, glabrous or puberulous abaxially, yellow in our area, white, lavender, magenta, or red elsewhere, the tube straight, usually membranous, sometimes thickened, without glands; stamens inserted, the anthers glabrous; staminode not exceeding stamens; nectary disc present; ovary sessile, frequently lepidote-pubescent. Fruits septicidal capsules, linear, usually subterete or flattened, not ridged or winged, not ornamented. Seeds usually winged (in our area) or thick and corky and only with vestigial wings.

1. Inflorescences few flowered corymbs. Flowers when leaves present; calyx lepidote, with glands; corolla white. Growing in wet areas along streams or lakes. . . . . *T. insignis.*
1. Inflorescences multiflowered panicles. Flowers when leafless; calyx pubescent, without glands; collolla yellow. Growing in well-drained forest.

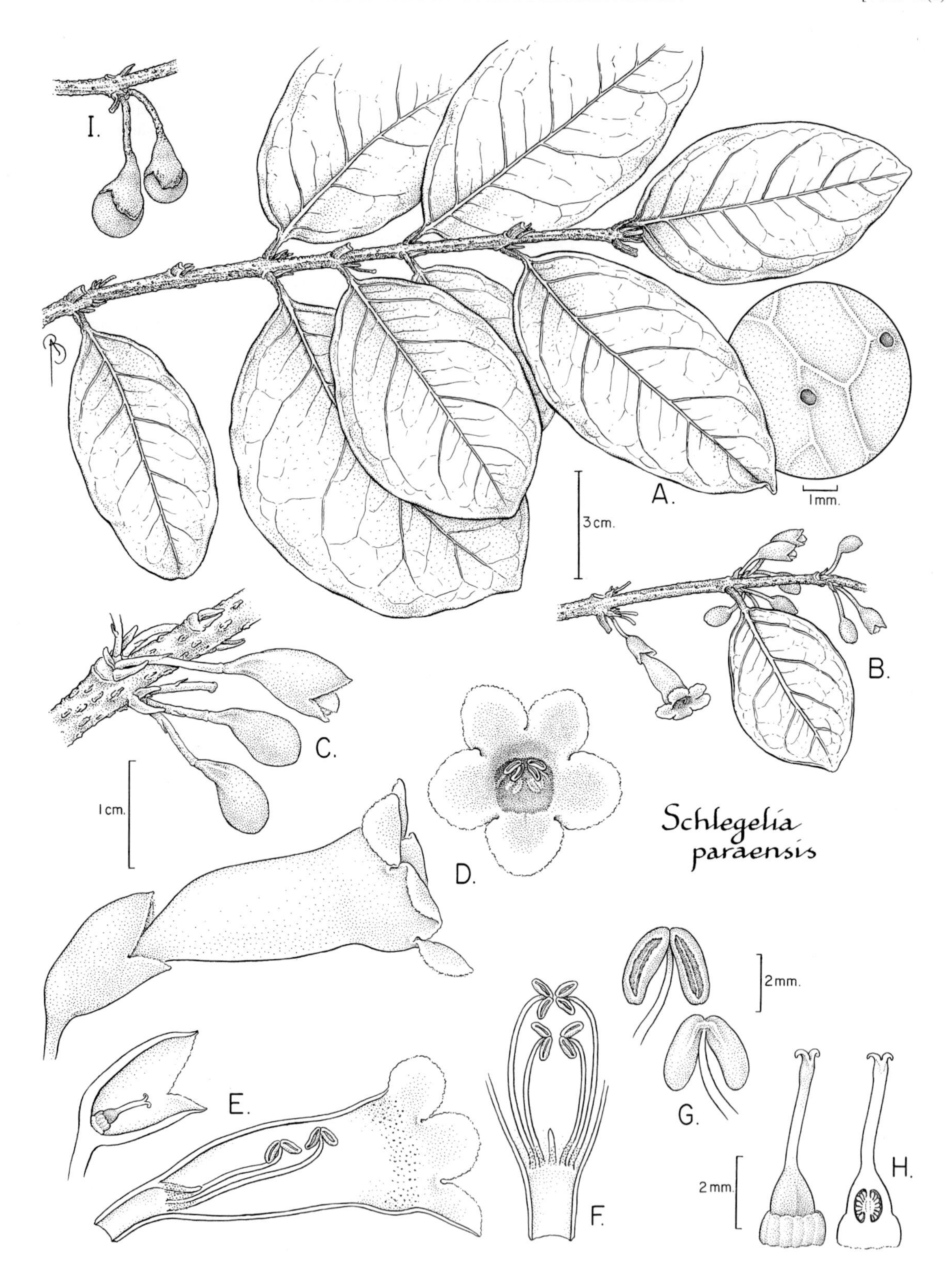

FIG. 51. BIGNONIACEAE. *Schlegelia paraensis* (A–H, *Mori et al. 19210*; I, *Mori et al. 22820*). **A.** Part of stem with leaves and detail of abaxial surface of leaf showing glands. **B.** Part of stem with leaf and flowers in various stages. **C.** Detail of node with flower buds and post-anthesis flower. **D.** Lateral (left) and apical (right) views of flower. **E.** Longitudinal section of calyx with pistil surrounded by nectary disc (left) and longitudinal section of corolla with two stamens and staminode. **F.** Part of corolla base opened to show arrangement of stamens and staminode. **G.** Adaxial (far above) and abaxial (near above) views of anthers. **H.** Lateral view (far left) and medial section (near left) of pistil and surrounding nectary disc. **I.** Fruits on stem.

2. Bark fissured. Lower leaflets not distinctly articulate; axillary domatia absent. Corolla 4.5–7 cm long. . . . *T. capitata.*
2. Bark not fissured. Lower leaflets distinctly articulate; axillary domatia lined by simple hairs present. Corolla 8–12 cm long. . . . . *T. serratifolia.*

Tabebuia capitata (Bureau & K. Schum.) Sandwith

Trees, to 40 m tall. Bark light brown, deeply fissured, the outer bark thick, laminated. Stems stellate-pubescent. Leaves palmately 5-foliolate; petioles 3.5–6 cm long, puberulous; lateral petiolules 1–2 cm long, the terminal one 2.2–3.5 cm long, both puberulous, the lowermost not distinctly articulate; leaflets elliptic to narrowly elliptic, 5.5–12.5 × 3.5–4.5 cm, concolorous, chartaceous, nearly glabrous with a few stellate hairs toward base adaxially, minutely lepidote-pubescent, with a few stellate hairs, and glands toward base abaxially, the base obtuse to rounded, the apex acuminate, the margins entire; secondary veins in 6–9 pairs, pinnately arranged, without axillary domatia. Inflorescences multiflowered panicles. Flowers: calyx campanulate, 8–12 × 6–11 mm, conspicuously pubescent, irregularly lobed, without glands; corolla 4.5–7 × 1–1.8 cm, glabrous adaxially, yellow. Fruits linear, 31–50 × 1.5–1.7 cm, with scattered scales and stellate hairs, smooth (fide Gentry, 1992). Known by a single sterile collection from our area (*Mori et al. 20965*, NY); rare, in non-flooded forest. *Ébène blanc, ébène souffré, ébène verte, poirier* (Créole).

Tabebuia insignis (Miq.) Sandwith

Trees, 10 m tall. Stems lepidote. Leaves palmately 5–7 foliolate; petioles to 20 cm long; lateral petiolules 2–5.5 cm long, the terminal one ca. 6 cm long, both lepidote, not distinctly articulate; leaflets elliptic to oblanceolate, 15–25 × 9.5–12 cm, lepidote ad- and abaxially, the base rounded to cuneate, the apex acute to acuminate, the margins entire; secondary veins in 6–10 pairs, pinnately arranged, without axillary domatia. Inflorescences few-flowered corymbs. Flowers: calyx campanulate, 15–20 × 11–14 mm, lepidote, irregularly 2–3 labiate, with glands; corolla 6.5–11 × 1.4–2.5 cm, glabrous adaxially, white. Fruits linear, 20–24 × 1.2–1.3 cm, lepidote, smooth. Fl (Sep); known by a single collection (*Granville 14193*, CAY, NY) from our area, growing on shore of lake at Pic Matécho. *Poirier* (Créole).

Tabebuia serratifolia (Vahl) G. Nicholson FIG. 52

Trees, to 30 m tall. Bark not fissured, the outer bark thin (<1 mm thick). Stems glabrate to puberulous. Leaves palmately 5–7-foliolate; petioles 4.5–5 cm long, puberulous; lateral petiolules 1–3.5 cm long, the terminal one 3–4 cm long, both puberulous, all distinctly articulate; leaflets elliptic to narrowly elliptic to ovate to narrowly ovate, 7.5–12 × 3–5.5 cm, puberulous with a few minute peltate scales, simple hairs, and glands adaxially, minutely lepidote-pubescent, with a few simple hairs, and glands abaxially, the base obtuse, the apex acuminate, the margins usually serrate; secondary veins in 7–8 pairs, pinnately arranged, with axillary domatia lined by simple hairs. Inflorescences multiflowered panicles. Flowers: calyx campanulate, 8–11 × 6–11 cm, sparsely pubescent, 3–5-lobed, without glands; corolla 8(12) × 2–2.5 cm, glabrous, yellow with red lines on throat adaxially. Fruits linear, (8)12–60 × 1.6–2.4 cm, lepidote (fide Gentry, 1982). Fl (Sep); although collected only twice, this species is scattered but common in non-flooded forest, conspicuous in season because of massive display of yellow flowers in the leafless canopy. *Ebénier de Guyane* (French), *ébène souffré, ébène verte, lébène* (Créole), *pau d'arco* (Portuguese).

TANAECIUM Sw.

Lianas. Vegetative parts often with aroma of almonds (cyanide). Stems with 8 or 16 phloem arms in transverse section, terete, solid, with or without interpetiolar glands, without interpetiolar ridge; pseudostipules small and inconspicuous or absent, not forming clusters resembling "miniature bromeliads," not foliaceous, persistent. Leaves 2–3-foliolate, the terminal leaflet often replaced by a simple tendril; petioles without glands distally; leaflets not pellucid-punctate. Inflorescences axillary or terminal, racemes or paniculate racemose. Flowers nocturnal: calyx tubular to cupular, with glands arranged in lines on lower part, with a single margin, the apex truncate to irregularly lobed; corolla tubular, very long, 5-parted, not bilabiate, open at anthesis, the tube glabrous with lobes granular pubescent abaxially, white, the tube curved, membranous, with glands; stamens inserted, the anthers glabrous; staminode not exceeding stamens; nectary disc present; ovary sessile, lepidote-pubescent. Fruits loculicidal capsules, cylindric, not flattened, not ridged or winged, smooth. Seeds with or without wings.

Tanaecium nocturnum (Barb. Rodr.) Bureau & K. Schum. PL. 27b; PART 1: FIG. 7

Vegetative parts with conspicuous almond aroma. Stems lepidote-pubescent, without interpetiolar glands; pseudostipules lacking. Leaves 2-foliolate; petioles 4–7 cm long, minutely and sparsely lepidote-pubescent; petiolules 1.5–4.5 cm long, minutely and sparsely lepidote-pubescent; leaflets widely ovate, 8–15 × 5–11 cm, concolorous, chartaceous, glabrous to minutely and sparsely lepidote-pubescent adaxially, glabrous to conspicuously lepidote-pubescent abaxially; secondary veins in 1–2 pairs, palmately arranged. Flowers: calyx tubular, 15–17 × 8–10 mm, minutely lepidote-pubescent, with glands, the apex truncate to irregularly lobed; corolla tubular, curved, 10.5–14 × 1–1.2 cm, glabrous on most of tube, pubescent at base of tube, with glands or stalked peltate scales toward apex, the lobes glandular-pubescent. Fruits long cylindric, 18 × 4.5 cm, glabrous to lepidote-pubescent, woody. Seeds winged. Fl (Jan, Feb, May, Aug, Sep, Dec), fr (Nov); common, corollas often seen on ground along trails through non-flooded forest. *Curimbo* (Portuguese), *liane noyau, liane noyo* (Créole).

TYNANTHUS Miers

Lianas. Vegetative parts with aroma of clove when broken. Stems with 4 phloem arms, terete to tetragonal, solid, with or without interpetiolar glands, with or without interpetiolar ridge; pseudostipules small or large, orbiculate, not forming clusters resembling "miniature bromeliads," usually foliaceous and caducous or absent. Leaves 2–3-foliolate, the terminal leaflet often replaced by a simple or trifid tendril; petioles without glands distally; leaflets not pellucid-punctate. Inflorescences axillary or terminal, panicles. Flowers diurnal; calyx

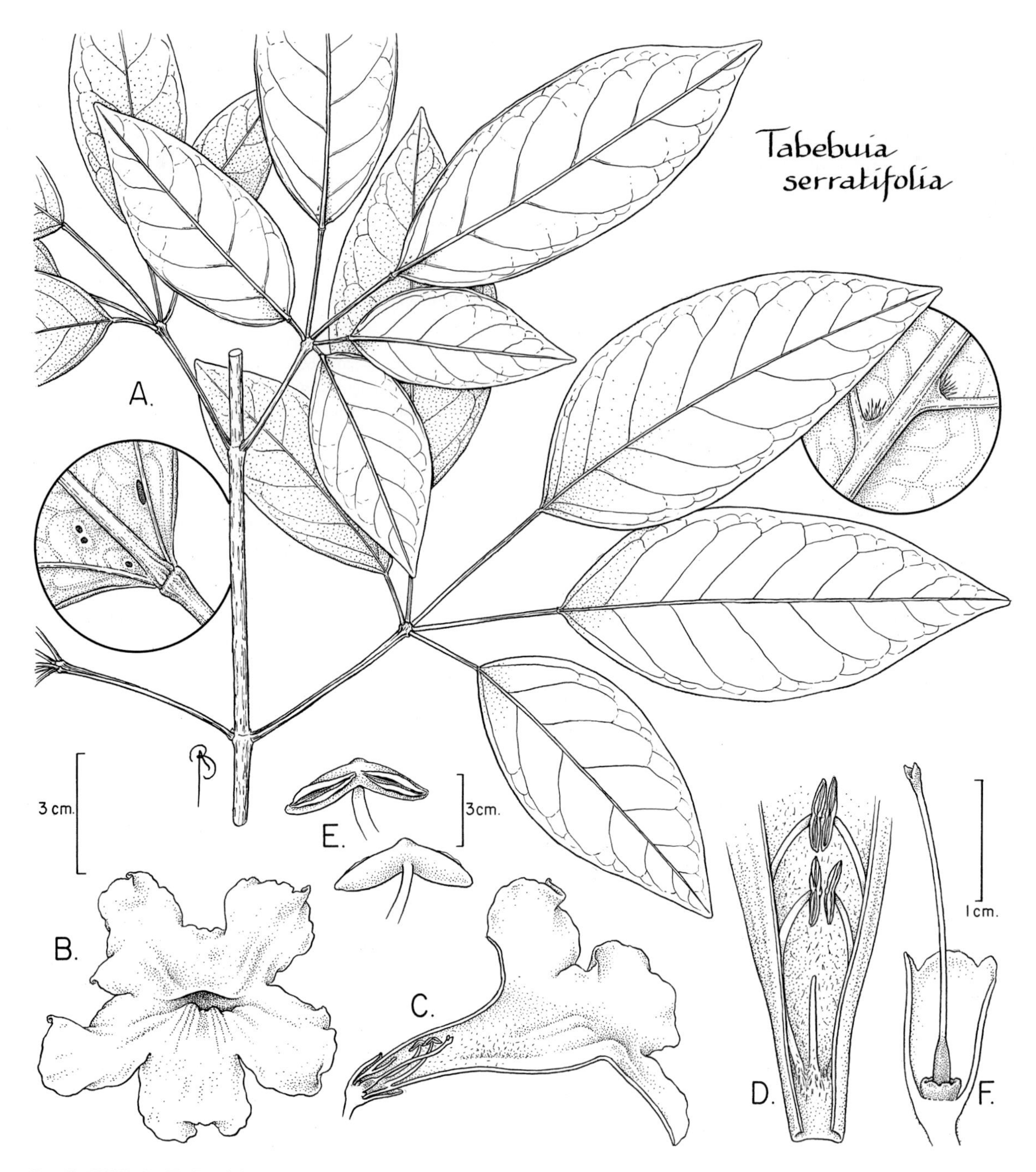

FIG. 52. BIGNONIACEAE. *Tabebuia serratifolia* (A, *Gentry et al. 62996*; B–F, *Mori et al. 21021 A*). **A.** Part of stem with leaves, detail of base of abaxial surface of leaf blade showing glands (left) and detail of abaxial surface of leaf showing tufts of hairs in axils of secondary veins (right). **B.** Apical view of corolla. **C.** Medial section of flower. **D.** Longitudinal section of base of corolla with stamens and staminode. **E.** Adaxial (right) and abaxial (below) views of anthers. **F.** Longitudinal section of calyx showing pistil surrounded by nectary disc.

cupular, small, without glands, with a single margin, the apex truncate or minutely denticulate; corolla infundibuliform, small, 5-parted, bilabiate, open at anthesis, pubescent abaxially, white or pale yellow, the tube bent, membranous, without glands; stamens exserted, the anthers glabrous; staminode not exceeding stamens; nectary disc present but sometimes inconspicuous; ovary sessile, densely pubescent. Fruits loculicidal capsules, linear, flattened, often winged, not ridged, often rough but not echinate. Seeds winged.

1. Stems tetragonal. Tendrils simple. Flowers 1 cm long; corolla glabrous, pale yellow, the ridges leading into throat not differentiated in color from remainder of corolla. Fruits not winged. *T. polyanthus*.
1. Stems terete. Tendrils trifid. Flowers 1.5–2 cm long; corolla pubescent, white, the ridges leading into throat yellow. Fruits conspicuously winged. . . . . *T. pubescens*.

Tynanthus polyanthus (Bureau) Sandwith

Stems tetragonal, glabrous to puberulous, with interpetiolar glands, usually with interpetiolar ridge; pseudostipules orbicular, foliaceous, usually caducous, sometimes persistent. Leaves (2)3-foliolate, the terminal leaflet often replaced by a simple tendril; petioles 2.5–3 cm long, pubescent; petiolules 1.5–2 cm long, pubescent; leaflets obovate, 8.5–9.8 × 5.5–6.5 cm, concolorous, chartaceous. Flowers: calyx cupular, ca. 10 mm long, densely pubescent, without glands, the apex truncate; corolla bent, 1 × 0.3 cm, glabrous, pale yellow, the ridges leading into throat not differentiated in color from remainder of corolla. Fruits linear, ≥30 cm long, ca. 1.5 cm wide, pubescent, not winged. Fl (Aug), fr (Jun); known only from a single gathering of fallen corollas and a fruiting collection.

Vegetative parts described from NY collections from outside our area, especially *Pennell 1602* from Colombia.

Tynanthus pubescens A. H. Gentry PL. 27c

Stems terete, with interpetiolar glands, the glands obscured by pubescence and only seen on youngest growth; pseudostipules ovate, not foliaceous, caducous. Leaves (2)3-foliolate, the terminal leaflet often replaced by a trifid tendril; petioles 6–6.5 cm long, densely white pubescent; lateral petiolules 1.5–1.8 cm long, the terminal one 2–3 cm long, both densely white pubescent; leaflets obovate, 8.5–12.5 × 4.5–7.5 cm, concolorous, chartaceous, pubescent adaxially, densely pubescent abaxially, the base obtuse, the apex rounded to apiculate; secondary veins pinnate, in 6–9 pairs. Flowers: calyx cupular, 3.5 × 3.5 mm, densely white pubescent, without glands, the apex truncate, with minute marginal teeth; corolla bent, 1.5–2 × 0.4 cm, pubescent, white, the ridges leading into throat yellow. Fruits linear, 15–27 × 0.5–0.7 cm, conspicuously winged, puberulous (fide Gentry, 1982). Fl (Aug, Sep); in non-flooded forest.

BIXACEAE (Lipstick-tree Family)

Scott V. Heald

Trees or shrubs. Exudate clear, reddish-orange. Stems with minute, reddish-orange, peltate scales, these stalked or sessile. Stipules deciduous. Leaves simple (in our area), alternate; petioles with distinct but short pulvinus at apex; blades palmately veined. Inflorescences terminal, paniculate or racemose. Flowers actinomorphic to slightly zygomorphic, complete; calyx 5-merous; corolla 5-merous, the petals distinct, imbricate or contorted; stamens numerous; pistil single, the ovary superior, 2–5-carpellate but unilocular or incompletely and falsely bilocular by placental intrusion, the style simple, the stigma two-lobed; placentation parietal, the ovules numerous. Fruits 2–5-valved, loculicidal capsules.

The Bixaceae is treated in the broad sense to include the Cochlospermaceae.

Eichler, A. G. 1871. Bixaceae [including Flacourtiaceae]. *In* C. F. P. von (ed.), Flora Brasiliensis **13(1):** 420–515, pls. 86–102.

BIXA L.

Leaves simple; blades ovate, with scattered reddish-orange, peltate scales abaxially, the base truncate to subcordate, the apex acuminate, the margins entire; palmately 5-veined, the midrib more prominent than others. Inflorescences paniculate. Flowers actinomorphic; calyx with ± apparent basal glands; corolla white to pink; anthers horseshoe-shaped, dehiscing by two apical pores. Fruits 2-valved, usually ovoid, spiny, but sometimes strongly flattened and unarmed. Seeds covered by brilliant red arilloid.

Eyma, P. J. 1966. Bixaceae. *In* A. Pulle (ed.), Flora of Suriname **III(1):** 158–159. E. J. Brill, Leiden.
Steyermark, J. A. & B. K. Holst. 1997. Bixaceae. *In* J. A. Steyermark, P. E. Berry & B. K. Holst (gen. eds.), Flora of the Venezuelan Guayana **3:** 492–495. P. E. Berry, B. K. Holst & K. Yatskievych (vol. eds.). Missouri Botanical Garden, St. Louis.

Bixa orellana L.

Small trees, to 4–5 m tall. Leaves: petioles 3.2–6.5 cm long, the pulvinus apical, distinct, 3.5–5.0 mm long, with glands just below blade; blades 9.5–15.5 cm long. Fruits ovoid, 4–5 × 3.5–4 cm, spiny, usually red at maturity but some varieties outside our area green. Fl (Oct), fr (Oct); cultivated, often persisting around old homesteads, the arilloid is the source of annatto dye. *Roucou*, *roucouyer* (Créole and French), *urucu* (Portuguese).

BOMBACACEAE (Kapok-tree Family)

William S. Alverson and Scott A. Mori

Small to very large trees, usually with simple, stellate, or lepidote hairs. Stipules usually caducous. Leaves simple or palmately compound, alternate. Inflorescences terminal or axillary, often appearing ramiflorous, solitary, racemose, or cymose. Flowers actinomorphic or slightly zygomorphic; pedicels with or without bracteoles; sepals 5, connate except sometimes free at apex; petals 5, free, convolute or imbricate, often large and showy, usually adnate to base of staminal tube; stamens 5 to many, usually connate into well developed or scarcely developed tube, rarely appearing free; ovary superior, 2–5(15)-locular, the style 1, the stigma simple or lobed; placentation axile, the ovules usually ≥2 per locule. Fruits woody capsules or indehiscent and then fleshy or woody. Seeds relatively small and numerous, then surrounded by woolly hairs derived from inner wall of ovary (*Ceiba*, *Eriotheca*, and some *Pachira*), or larger and less numerous and not surrounded by woolly hairs (*Catostemma*, *Matisia*, *Pachira*, *Quararibea*), or distinctly winged (*Huberodendron*).

Concepts of genera of Bombacaceae with palmately compound leaves are not yet clearly established (Alverson, 1994; Steyermark & Stevens, 1988). Recent molecular studdes indicate that this family, as traditionally circumscribed, is not a monphyletic group, and, thus, its nomenclature will have to be revised (Alverson et al., 1999; Baum et al., 1998; Bayer et al., 1999).

Alverson, W. S. & J. A. Steyermark. 1997. Bombacaceae. *In* J. A. Steyermark, P. E. Berry & B. K. Holst (gen. eds.), Flora of the Venezuelan Guayana **3:** 496–527. P. E. Berry, B. K. Holst & K. Yatskievych (vol. eds.). Missouri Botanical Garden, St. Louis.
Jansen-Jacobs, M. J. 1986. Bombacaceae. *In* A. L. Stoffers & J. C. Lindeman (eds.), Flora of Suriname **III(1–2):** 277–282. E. J. Brill, Leiden.
Uittien, H. 1966. Bombacaceae. *In* A. Pulle (ed.), Flora of Suriname **III(1):** 26–33, 436. E. J. Brill, Leiden.

1. Leaves simple or appearing so (unifoliolate).
   2. Canopy to emergent trees, trunks usually with well developed buttresses. Petioles 4.5–9 cm long. Fruits fully or partially dehiscent. Seeds with or without wings.
      3. Leaf blades ovate to elliptic. Inflorescences with flowers secund. Stamens 5, with tube exserted from flower. Fruits much longer than wide. Seeds winged. . . . . *Huberodendron.*
      3. Leaf blades oblong. Inflorescences without flowers secund. Stamens >25, with tube included in flower. Fruits not much longer than wide, usually ± globose. Seeds with arils. . . . . *Catostemma.*
   2. Understory trees, trunks without well developed, buttresses. Petioles 1–2.5 cm long. Fruits indehiscent. Seeds not winged.
      4. Calyx golden-yellow, distinctly rugose. Petals erect at anthesis; anthers disposed to one side; ovary 5-locular. . . . . *Matisia.*
      4. Calyx green, not conspicuously rugose. Petals spreading or reflexed at anthesis; androecium actinomorphic, the anthers not disposed to one side; ovary 2–4-locular. . . . . *Quararibea.*
1. Leaves palmately compound.
   5. Young branches and sometimes bole with conical prickles. Stamens appearing as 5. Seeds surrounded by long, woolly hairs. . . . . *Ceiba.*
   5. Young branches and bole without conical prickles. Stamens >5. Seeds surrounded or not surrounded by long, woolly hairs.
      6. Filaments very long (>5 cm long), in fascicles above staminal tube, the anthers ± linear. Seeds not surrounded by long, woolly hairs in our species.. . . . . *Pachira.*
      6. Filaments shorter (<1.5 cm long), not in fascicles above staminal tube, the anthers globose. Seeds surrounded by long, woolly hairs. . . . . *Eriotheca.*

CATOSTEMMA Benth.

Trees. Leaves: typically unifoliolate, but juvenile leaves sometimes palmately compound; petioles not articulate. Inflorescences axillary fascicles; pedicels bracteolate or ebracteolate. Flowers actinomorphic; sepals enclosing well developed buds, splitting to form irregular lobes; stamens numerous, 25–120, fused at base into short tube included in flower, the anthers dorsifixed, folded downward from summit of filament; ovary 3-locular. Fruits capsules. Seeds 1–2, without wings, at least in some species surrounded by orange arils.

Catostemma commune Sandwith

Canopy trees. Trunks with simple, symmetrical plank buttresses. Leaves: petioles 5–9 cm long, with darkened pulvinus at apex; blades oblong, 32–40 × 7.5–9 cm, the base acute to obtuse, the apex short acuminate, the margins entire; secondary veins in ca. 17 pairs. Fruits not much longer than wide, usually ± globose.

Known from a single sterile collection (*Boom & Mori 2023*, NY) from our area.

CEIBA Mill.

Trees. Leaves palmately compound, mostly glabrous; petiolules articulate. Inflorescences grouped toward ends of branchlets, ramiflorous, usually fasciculate; pedicels bracteolate, the bracteoles caducous. Flowers actinomorphic; sepals connate, not enclosing well developed buds; stamens complex, due to apparent secondary fusion, appearing as 5, forming short tube, the anthers versatile, contorted; ovary 5-locular. Fruits capsules. Seeds without wings, surrounded by long, non-wettable, woolly hairs ("kapok").

Gibbs et al. (1988) have proposed uniting *Chorisia* with *Ceiba*. There are no native species of *Chorisia* and only a single species of *Ceiba* (Boggan et al., 1997) known from the Guianas. Our generic description is based only on Guianan material.

Ceiba pentandra (L.) Gaertn.

FIG. 53, PL. 28; PART 1: FIG. 5

Canopy to emergent trees. Stems and trunk often with conical prickles. Leaves usually with 7 leaflets; petioles 14.5–26.5 cm long; petiolules 1–2.5 cm long; leaflet blades narrowly elliptic, 8.5–17.5 × 2–3.5 cm. Flowers nocturnal, present when tree leafless, 2.5–3.5 × 2–3 cm; petals white; filaments white, the anthers golden yellow. Fruits 12–24 cm long. Infrequent, known by us only from one sterile collection and one other uncollected tree in the vicinity of Saül, in non-flooded forest. *Fromager* (Créole and French), *kktri* (Bush Negro), *sumauma* (Portuguese).

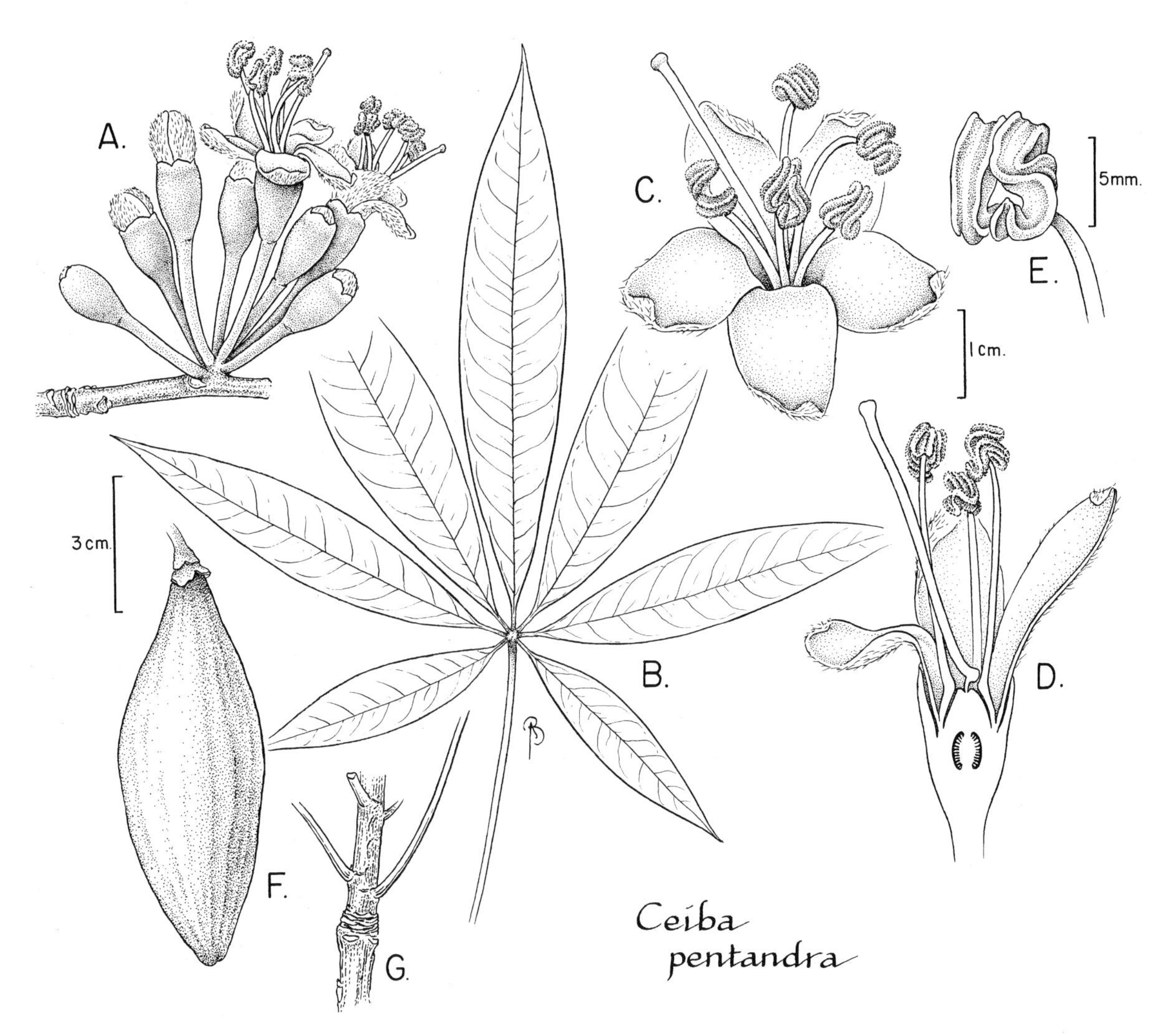

FIG. 53. BOMBACACEAE. *Ceiba pentandra* (A, C–E, *Acevedo-Rodríguez 4231* from St. John, U.S. Virgin Islands; B, *Little 22001* from St. John, U.S. Virgin Islands; F, *Morrow s.n.* from St. John, U.S. Virgin Islands; G, *Mori & Gracie 18379*). **A.** Inflorescence. **B.** Leaf. **C.** Oblique-apical view of flower. **D.** Medial section of flower. **E.** Detail of anther. **F.** Fruit. **G.** Part of stem with petiole bases and leaf scars.(Adapted with permission from P. Acevedo-Rodríguez, *Flora of St. John, U.S. Virgin Islands*, Mem. New York Bot. Gard. 78, 1996. Note that G has been added.)

ERIOTHECA Schott & Endl.

Trees. Leaves palmately compound, mostly glabrous; petiolules articulate. Inflorescences ramiflorous, in fascicles or short racemes; pedicels bracteolate, the bracteoles caducous. Flowers actinomorphic; sepals connate, not enclosing well developed buds; stamens numerous (ca. 100–200), forming well developed tube, the free filaments not in fascicles, the filaments <1.5 cm long, the anthers basifixed, globose; ovary 5-locular. Fruits dehiscent capsules. Seeds without wings, surrounded by conspicuous hairs.

1. Leaflet blades elliptic, with scattered lepidote scales abaxially. Inflorescences usually below leaves; pedicels 1.5–2.5 cm long. Flowers diurnal; staminal tube included. . . . . *E. globosa.*
1. Leaflet blades narrowly obovate to oblanceolate, densely covered with lepidote scales abaxially. Inflorescences usually among leaves; pedicels 3.5–4 cm long. Flowers nocturnal; staminal tube exserted. . . . . *E. longitubulosa.*

Eriotheca globosa (Aubl.) A. Robyns

Understory to canopy trees. Leaves with 3–6 leaflets; petioles 7.5–10 cm to longer in saplings; petiolules 6–15 mm long; blades elliptic, 11.5–18.5 × 4.8–7 cm, the abaxial surface often with scattered lepidote scales. Inflorescences usually below leaves; pedicels 1.5–2.5 cm long. Flowers diurnal, ca. 2 cm long; petals obovate, white abaxially, yellowish adaxially, the apex rounded; staminal tube ca. 0.5 cm long, included. Known from our area only by two sterile specimens (*Gentry et al. 63050*, MO and *Boom & Mori 2279*, NY).

The description taken, in part, from flowering collections *Westra 48554* (NY) and *Irwin et al. 48180* (NY) gathered from the French Guiana/Brazil border along the Oiapoque River.

Eriotheca longitubulosa A. Robyns

Canopy to emergent trees. Trunks with well developed, sometimes branched buttresses. Leaves with 4–7 leaflets; petioles 3–8.5 cm long; petiolules 5–7 mm long; blades narrowly obovate to oblanceolate, 6–8 × 2–5 cm, the abaxial surface densely covered with lepidote scales. Inflorescences usually among leaves; pedicels 3.5–4 cm long. Flowers nocturnal, 4–5 cm long; petals narrowly lanceolate to linear, doubly reflexed at anthesis, yellow-green, often tinged with reddish-pink near middle, the apex acute; staminal tube exserted, 2.5–3 cm long. Fl (Jul, Aug); rare, in non-flooded forest.

HUBERODENDRON Ducke

Trees. Leaves presumabaly unifoliolate (appearing simple), with scattered, simple and stellate hairs; petiolules articulate. Inflorescences terminal, cymose, the flowers secund; pedicels bracteolate, the bracteoles persistent. Flowers actinomorphic; sepals connate, enclosing well developed buds; stamens 5, forming well developed tube exserted from flower, the anthers closely attached to staminal tube; ovary 5-locular. Fruits dehiscent capsules. Seeds with conspicuous unilateral wing, not surrounded by conspicuous hairs.

Huberodendron swietenioides (Gleason) Ducke
FIG. 54, PL. 29; PART 1: FIG. 7

Emergent trees, to 60 m tall. Trunks with buttresses 10–15 m tall, steep, relatively thin, branched. Bark reddish-brown, the slash white with translucent streaks, oxidizing yellowish-brown. Leaves mostly glabrous, pubescent only toward base of blade and on petiole; petioles slender, 4.5–5.5 cm long; blades ovate to elliptic, 10–12.5 × 5.5–6.5 cm, with 8–11 pairs of secondary veins. Flowers ca. 2 cm diam.; petals white. Fruits 5-valved, much longer than wide, 20–25 cm long. Seeds 10–12 cm long, the wing 8–10 cm long. Fl (Oct), fr with seed (May); only known in our area from a small group of trees growing in non-flooded forest on the Sentier Botanique.

MATISIA Humb. & Bonpl.

Trees. Leaves simple, with simple and stellate hairs. Inflorescences axillary or ramiflorous, the flowers solitary; pedicels ebracteolate in our area. Flowers zygomorphic because of staminal orientation; sepals connate, enclosing well developed buds; stamens forming well developed tube, the anthers attached to abaxial surface of apical lobes of staminal tube, disposed to one side; ovary 5-locular. Fruits indehiscent, fibrous-fleshy drupes. Seeds 1–5 per fruit, without wings, not surrounded by conspicuous hairs, dispersed within persistent, bony-fibrous mesocarp.

Matisia is part of a clade that appears to be more closely related to the traditional Malvaceae than to the core Bombacaceae (Baum et al., 1998).

Matisia ochrocalyx K. Schum. FIG. 55, PL. 30a

Understory trees, 10–14 m tall, without buttresses. Bark ± smooth, sometimes with hoop marks, the slash of inner bark yellow, with vertically oriented reticulations. Leaves mostly glabrous, pubescent only toward base of blade, along midrib, and on petiole; petioles 1.7–2.5 cm long; blades elliptic or narrowly to widely obovate, 16–20 × 7.2–10.5 cm, with 5–7 pairs of secondary veins. Flowers ca. 1.5 cm diam.; calyx golden-yellow, distinctly rugose, especially in herbarium specimens; petals white, erect at anthesis. Fruits 3–4 × 3.5 cm, the calyx accrescent, covering ca. 1/2 of fruit. Fl (Apr, May), fr (Jul, Nov); scattered in non-flooded forest.

PACHIRA Aubl.

Trees. Leaves palmately compound, glabrous or with stellate hairs or lepidote scales; petiolules articulate. Inflorescences ramiflorous, mostly solitary, sometimes paired or in clusters of 3; pedicels bracteolate, the bracteoles caducous. Flowers largest of family, actinomorphic; sepals connate, not enclosing well developed buds; stamens numerous (100–700), forming well developed tube, the tube surmounted by fascicles of filaments, the filaments very long (>5 cm long), the anthers versatile, ± linear; ovary 5-locular. Fruits dehiscent woody capsules. Seeds without wings, not surrounded by conspicuous hairs in our species.

Alverson (1994) made a case for uniting *Bombacopsis*, *Pochota*, and *Rhodognaphalopsis* in *Pachira*, a position accepted in this treatment.

1. Leaves with 5–7 leaflets; leaflet blades 11.5–21 cm long. Flowers very large, 20–35 cm long; calyx ca. 3 cm diam. at anthesis. . . . . *P. insignis.*
1. Leaves with 3 leaflets; leaflet blades 3–8.5 cm long. Flowers large, to 19 cm long; calyx ca. 1 cm diam. at anthesis. . . . . *P.* cf. *macrocalyx.*

Pachira insignis (Sw.) Savigny PL. 30b

Canopy trees, the leaf bearing branches stout, 6 mm diam. Leaves with 5–7 leaflets; petioles ca. 9 cm long; petiolules poorly developed; leaflet blades narrowly obovate, 11.5–21 × 2.6–2.9 cm, the abaxial surface glabrous. Flowers very large, 20–35 cm long; calyx ca. 3 cm diam. at anthesis; petals red; anthers yellow. Fruits ca. 22 cm long. Seeds numerous. Fl (Nov), fr (Feb); collected twice from the same tree in low, wet area along stream near Eaux Claires.

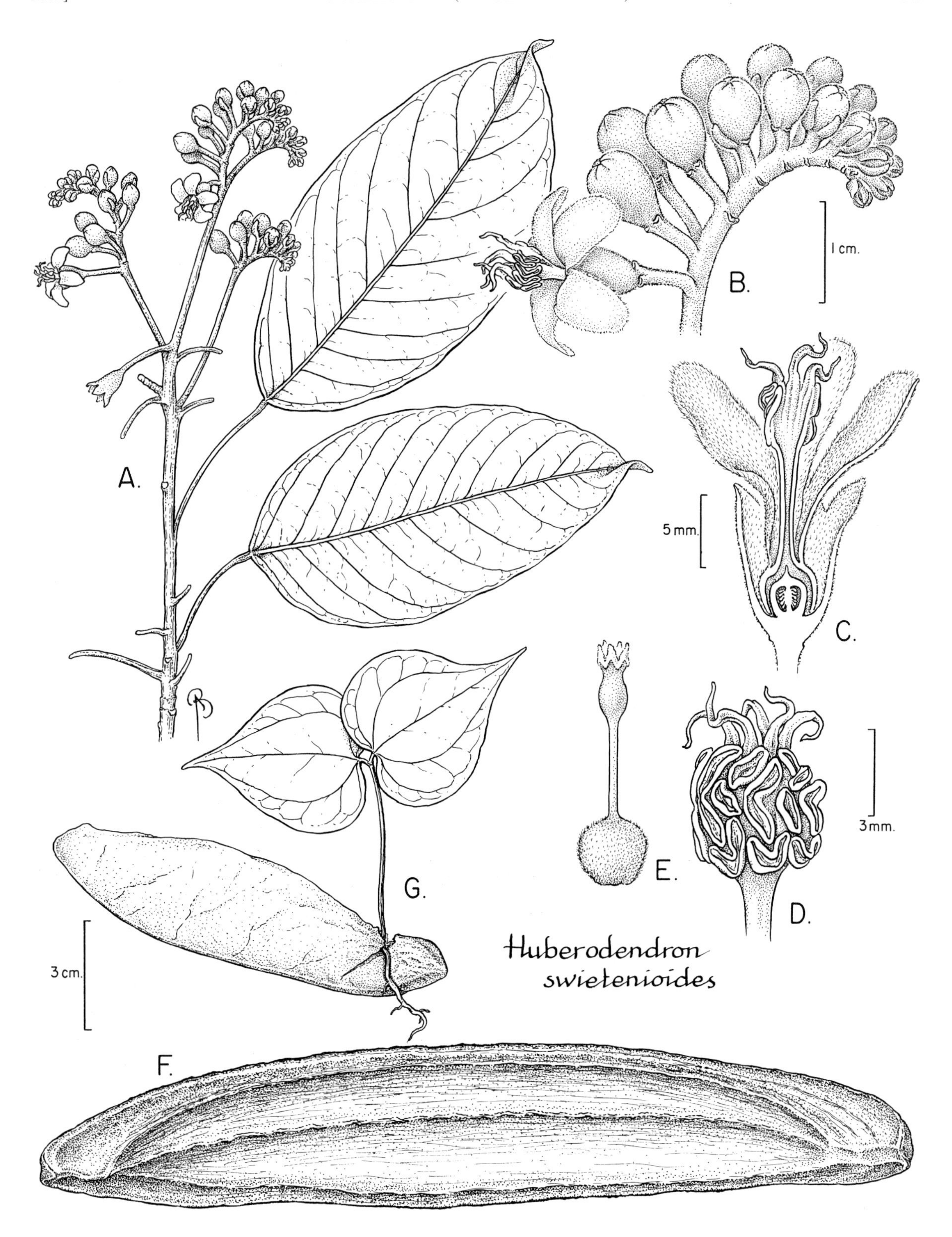

FIG. 54. BOMBACACEAE. *Huberodendron swietenioides* (A–E, *Mori et al. 22065*; F, G, *Mori et al. 22254*). **A.** Part of stem with leaves and inflorescence. **B.** Detail of inflorescence. **C.** Medial section of flower with style removed. **D.** Lateral view of apex of staminal column. **E.** Lateral view of pistil. **F.** Adaxial view of one valve of capsular fruit. **G.** Seedling growing from winged seed.

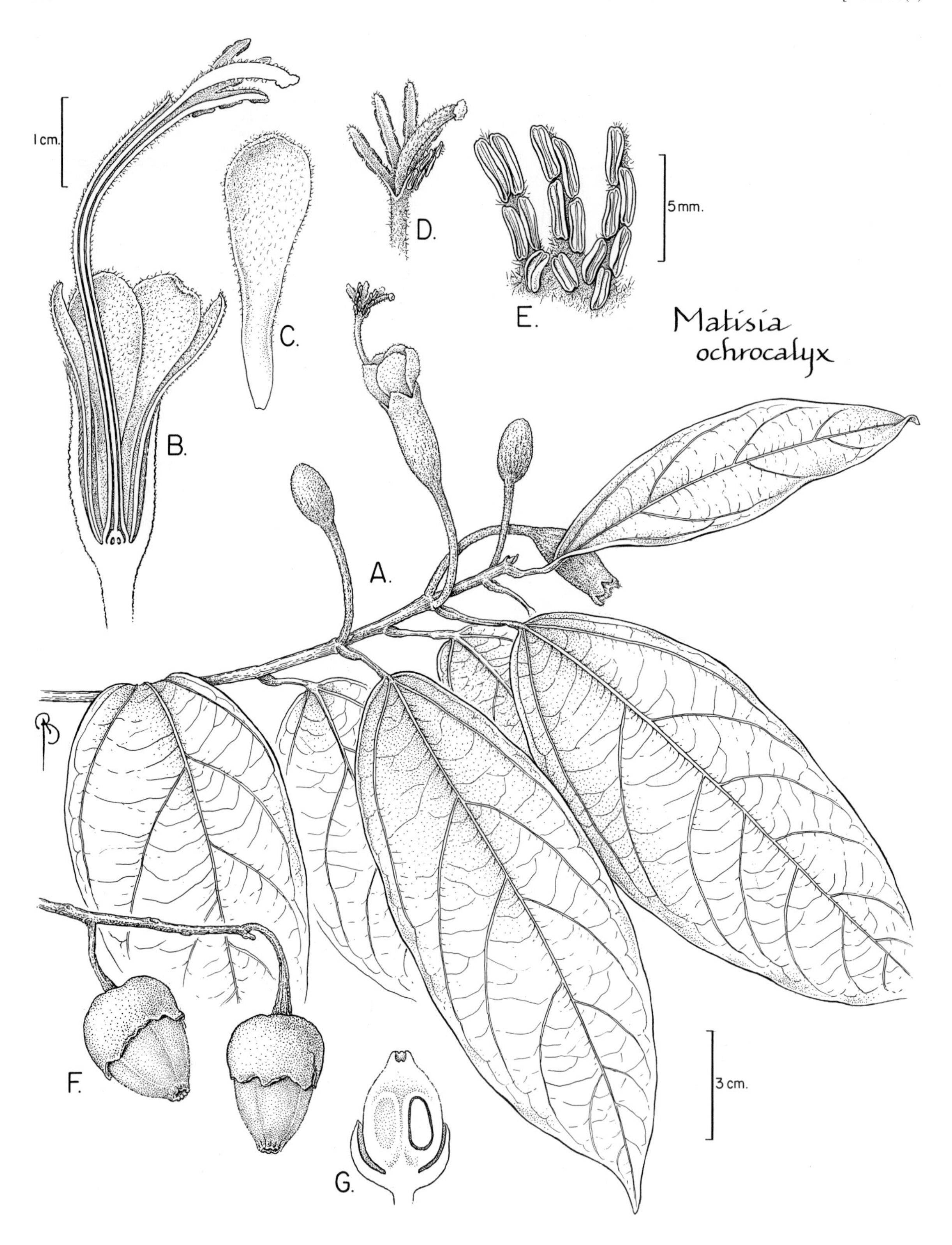

FIG. 55. BOMBACACEAE. *Matisia ochrocalyx* (A–E, *Mori et al. 22240*; F, G, *Mori et al. 22836*). **A.** Part of stem with leaves, axillary buds and flowers. **B.** Medial section of flower. **C.** Adaxial view of petal. **D.** Upper part of androecium surrounding style and stigma. **E.** Detail of upper part of androecium showing anthers. **F.** Part of infructescence with nearly mature fruits. **G.** Medial section of fruit.

Pachira cf. macrocalyx (Ducke) Fern.-Alonso

Understory trees, the leaf bearing branches relatively slender, 3 mm diam. Leaves with 3 leaflets; petioles 2–4 cm long; petiolules 3–5 mm long; leaflet blades widely elliptic to elliptic, 3.5–8.5 × 2–4.8 cm, the abaxial surface glabrous. Flowers large, to ca. 19 cm long; calyx ca. 1 cm diam. at anthesis; petals twisted, brown; staminal tube white. Fruits unknown from our area. Fl (May); rare, collected only once along Antenne Est of the La Fumée Mountain Trail. Our collection deviates from the type by having longer, more slender flowers.

QUARARIBEA Aubl.

Trees. Leaves simple, with simple and stellate hairs. Inflorescences axillary or ramiflorous, the flowers solitary or on short shoots with several flowers; pedicels bracteolate. Flowers actinomorphic; sepals connate, enclosing well developed buds; stamens forming well developed tube, the anthers attached to apical lobes of tube or to tube itself; ovary 2–4-locular. Fruits indehiscent. Seeds without wings, not surrounded by conspicuous hairs, retained within hard, fibrous endocarp during dispersal.

Quararibea is part of a clade that appears to be more closely related to the traditional Malvaceae than to the core Bombacaceae (Baum et al., 1998).

1. Leaf blades with conspicuous stellate hairs abaxially, the base finely auriculate, without domatia in axils of secondary veins. Flowers 4–8 × 3.5–5.5 cm; petals spreading at anthesis; anthers arising from staminal tube itself. . . . . *Q. spatulata.*
1. Leaf blades without conspicuous stellate hairs abaxially, the base obtuse to rounded, with conspicuous domatia in axils of secondary veins. Flowers ca. 2 × 1.5 cm; petals reflexed at anthesis; anthers arising from lobes at apex of staminal tube. . . . . *Q. duckei.*

Quararibea duckei Huber PL. 30c

Understory trees, without buttresses. Bark smooth, the slash of inner bark yellow. Leaf blades elliptic, 12.5–22.5 × 6.5–10.4 cm, without conspicuous stellate hairs abaxially, with well developed domatia in axils of secondary veins, the base obtuse to rounded; secondary veins in 6–8 pairs. Flowers ca. 2 × 1.5 cm; petals pure white, reflexed at anthesis; anthers arising from 5 lobes at apex of tube. Mature fruits unknown from our area. Fl (Feb, May, Sep, Oct, Nov); abundant in non-flooded forest. *Bois gelet, bois lélé, bois lélé blanc* (Créole).

Quararibea spatulata Ducke PL. 30d

Understory trees, usually without buttresses. Bark smooth, the slash of inner bark white, with reticulate network of fibers. Leaf blades widely to narrowly obovate or elliptic, 17–27.5 × 8–12.4 cm, with conspicuous stellate hairs abaxially, without well developed domatia in axils of secondary veins, the base finely auriculate, the secondary veins in 7–8 pairs. Flowers 4–8 × 3.5–5.5 cm; petals white to cream-colored, tinged with yellow, spreading at anthesis; anthers arising from staminal tube itself. Fruits 1–2-seeded fibrous-fleshy drupes. Fl (Apr, May); infrequent in non-flooded forest.

BORAGINACEAE (Borage Family)

Christian Feuillet

Herbs, subshrubs, shrubs, trees, or lianas. Indument often of stiff, unicellular, partly calcified or silicified hairs. Stipules absent. Leaves alternate, simple. Inflorescences cymose to paniculate, the branches commonly sympodial, often helicoid or scorpioid cymes, or reduced and capitate-glomerate, or spicate, rarely of solitary flowers (outside our area). Flowers mostly bisexual, actinomorphic or slightly zygomorphic, gamopetalous; sepals (4)5(6); petals and stamens as many as sepals; stamens adnate to corolla tube; nectary disc either surrounding base of ovary or absent; ovary superior, 2-carpellate, the carpels variously united, often with twice as many segments or locules as carpels (usually not segmented in species in our area), the style terminal or gynobasic (outside our area); placentation axile, ovule 1 per locule. Fruits capsules, 1–4 nutlets, or dry or fleshy, 2–4-stoned drupes.

Johnston, I. M. 1966. Boraginaceae. *In* A. Pulle (ed.), Flora of Suriname **IV(1):** 306–333. E. J. Brill, Leiden.

Miller, J. S., J. Gaviria, R. Gómez & G. Rodríguez. 1997. Boraginaceae. *In* J. A. Steyermark, P. E. Berry & B. K. Host (gen. eds.), Flora of the Venezuelan Guayana **3:** 527–547. P. E. Berry, B. K. Holst & K. Yatskievych (vol. eds). Missouri Botanical Garden, St. Louis.

Uittien, H. 1966. Boraginaceae. *In* A. Pulle (ed.), Flora of Suriname **IV(1):** 496–497. E. J. Brill, Leiden.

1. Flowers with 4 stigmas born on conspicuous 4-branched style. Fruits with 1–4-locular endocarp forming one bony pit. . . . . *Cordia.*
1. Flowers with 1 stigma borne on an undivided style or sessile. Fruits with endocarp split at maturity into 2–4 parts.
   2. Herbs. Mature fruits dry, 2–4 dry nutlets. . . . . *Heliotropium.*
   2. Shrubs or lianas. Mature fruits fleshy, including 2–4 bony stones. . . . . *Tournefortia.*

CORDIA L.

Trees or shrubs. Leaves alternate (crowded at apex of growth unit in *C. nodosa*), simple. Inflorescences loosely paniculate, glomerate, capitate, or spicate cymes. Flowers bisexual or unisexual by abortion; calyx usually 5-toothed or 5–10-lobed, usually persistent, rarely marcescent; corolla small or exserted to large and showy, usually 5-merous, white (greenish or yellow to red-orange outside our area); stamens as many as corolla lobes, included or exserted, extrorse; ovary 4-locular, with 1–4 ovules per locule, the 4 stigmas borne on conspicuous 4-branched style. Fruits unlobed drupes, white or yellow to red, with a 1–4-locular endocarp forming one bony pit, usually one-seeded by abortion, the mesocarp fleshy or dry and papery.

Gaviria, J. 1987. Die Gattung *Cordia* in Venezuela. Mitt. Bot. Staatssamml. München **23:** 1–279.

1. Inflorescences unbranched, flowering basipetally, adnate to petiole. . . . . *C. schomburgkii.*
1. Inflorescences branched, flowering acropetally or simultaneously, free from petiole.
    2. Calyx cylindric to funnel-shaped around fruit. Corolla marcescent, falling as unit with calyx and fruit. Mature fruits dry.
        3. Leaves with stellate hairs. Inflorescences with main axis often swollen. Calyx 4–6 mm long; corolla lobes 5–7 mm long. . . . . *C. alliodora.*
        3. Leaves glabrous. Inflorescences with main axis not swollen. Calyx 8–10 mm long; corolla lobes 12–17 mm long. . . . . *C. goeldiana.*
    2. Calyx flattened or broken under fruit. Corolla deciduous. Mature fruits fleshy.
        4. Stems with bristles, asymmetrically swollen, club-shaped below inflorescences or at nodes. . . . . . . *C. nodosa.*
        4. Stems glabrous or pubescent, but without bristles, terete or angular, not swollen below inflorescences or at nodes.
            5. Leaves glabrous abaxially.
                6. Ovary pubescent. Fruits minutely strigose. . . . . *C. nervosa.*
                6. Ovary glabrous. Fruits glabrous.
                    7. Leaves heteromorphic. Corolla tube 5–7 mm long. Fruits orange to red, the stone erect, ellipsoid. . . . . *C. exaltata.*
                    7. Leaves homomorphic. Corolla tube 2–4 mm long. Fruits yellow, the stone globose or transversely ovoid. . . . . *C. lomatoloba.*
            5. Leaves pubescent abaxially.
                8. Leaves heteromorphic, abaxially strigose.
                    9. Flower buds elongate, obovoid, 4–5 mm long; corolla tube 5–7 mm long. Fruits orange to red. . . . . *C. exaltata.*
                    9. Flower buds subglobose, 2–3 mm long; corolla tube 2–3 mm long. Fruits yellow. . . . *C. naidophila.*
                8. Leaves homomorphic, abaxially with erect, or soft-appressed trichomes.
                    10. Leaves with erect trichomes abaxially. Ovary glabrous. Fruits glabrous, the stone erect, ovoid. . . . . *C. sagotii.*
                    10. Leaves with appressed trichomes abaxially. Ovary pubescent. Fruits pubescent, the stone transversely ovoid. . . . . *C. bicolor.*

Cordia alliodora (Ruiz & Pav.) Oken

PL. 31a; PART 1: PL. XIa,b

Canopy trees. Stems terete, with stellate hairs. Leaves homomorphic, sparingly to densely stellate pubescent; petioles 10–30 mm long; blades oblong to elliptic, usually widest at or above middle, 10–20 × 3–8 cm, stellate pubescent to glabrous on both surfaces, the base acute or obtuse, the apex acuminate, the margins entire. Inflorescences terminal, loosely and widely branched, to 30 cm wide, free from petiole, the flowers crowded on rachises, flowering simultaneously, the main axis commonly inflated, gall-like, irregular, usually serving as ant domatia. Flowers: calyx cylindrical, prominently 10-ribbed, 4–6 mm long, with 5 inconspicuous lobes; corolla white, drying brown, marcescent, the tube 6–7 mm long, the lobes 5–7 mm long, spreading. Fruits sausage-shaped, 3–5 mm long, dry, with fibrous, chartaceous wall, enveloped by corolla and calyx, all 3 falling as unit, the lobes acting as parachute. Fl (Feb, Jul, Aug), fr (Aug), common, in secondary forest. *Cédre-sam* (Créole).

Cordia bicolor DC.

Shrubs to canopy trees. Stems angular, velvety, with very abundant, tawny, short hairs. Leaves homomorphic; petioles 5–10 mm long; blades ovate to broadly lanceolate, widest at or below middle, 8–16 × 2–7 cm, fine, soft, appressed hairs adaxially, the base acute to rounded, the apex acuminate, the margins entire. Inflorescences usually at forks of stem, loosely branched, free from petiole, the main axis not inflated, flowering basipetally. Flowers: calyx covered with abundant, short, tawny hairs abaxially; corolla white, deciduous, the tube ca. 3 mm long, the lobes ca. 2.5 mm long; ovary glabrous or hairy on upper part. Fruits glabrescent, the stone transversely ovoid, >1 cm long. Fl (Jan), fr (Jan, Feb, Nov); common, in non-flooded forest. *Lamoussé* (Créole).

Cordia exaltata Lam.

Shrubs to canopy trees. Stems terete, sparsely strigose throughout. Leaves heteromorphic; petioles 5–10 mm long; blades of larger leaves elliptic to broadly oblanceolate, widest at or above middle,

8–20 × 4–10 cm, scabrous, the base acute or rounded, the apex short acuminate, the margins entire. Inflorescences borne at fork of stems, loosely branched, 10–20 cm wide, free from petiole, the main axis not inflated, flowering basipetally. Flowers: calyx 5-toothed, 4–5 mm long, not ribbed, flat under fruit; corolla white, deciduous, the tube 5–7 mm long, the lobes ca. 2 mm long; ovary glabrous. Fruits fleshy, orange-yellow or red, glabrous, the stone ellipsoid, erect or nearly so, 1–1.5 cm long. Fl (Jun, Aug, Oct), fr (Dec); in secondary vegetation near Saül airstrip. *Bois sip* (Créole).

Cordia goeldiana Huber

Emergent trees. Stems terete, glabrous. Leaves homomorphic, glabrous; petioles 20–35 mm long; blades oblong to elliptic, usually widest at or above middle, 10–20 × 4–7 cm, glabrous on both surfaces, the base acute or obtuse, the apex acute or slightly acuminate, the margins entire. Inflorescences terminal, loosely and widely branched, to 30 cm wide, free from petiole, the flowers crowded on rachises, the main axis not inflated, flowering simultaneously. Flowers: calyx funnel-shaped, obscurely 10-ribbed, 8–10 mm long, with 5 triangular lobes; corolla white, drying brown, marcescent, the tube 9–10 mm long, the lobes spreading, 15–25 mm long. Fruits dry, sausage-shaped, 5–8 mm long, with fibrous, chartaceous wall enveloped by corolla and calyx, all 3 falling as a unit, the corolla lobes acting as parachute. Fl (Nov); occasional, in non-flooded forest.

Cordia lomatoloba I. M. Johnst.

Understory to canopy trees, to 15–25 m tall. Stems terete, glabrescent. Leaves homomorphic; petioles 8–15 mm long; blades lanceolate or lanceolate-elliptic, widest at or just below middle, 8–14 × 2–5.5 cm, coriaceous, glabrous or sparsely and inconspicuously strigose, the base acute, the apex often acuminate, the margins entire. Inflorescences loosely branched and thin, ca. 10 cm long, free from petiole, brown-pubescent, the main axis not inflated, flowering basipetally. Flowers: calyx 2.5 mm long; corolla white, deciduous, the tube 1 mm long, the lobes 3 mm long; ovary glabrous. Fruits fleshy, yellow, glabrous, on flattened calyx, the stone transversely elongate, 15 × 20 mm. Fl (Sep), fr (Mar, Apr, Dec); in forest.

Cordia naidophila I. M. Johnst.

Understory trees. Stems terete, short pubescent. Leaves heteromorphic; petioles 4–7 mm long; blades of largest leaves oblong, widest in upper 1/3 of length, 8–13 × 3.5–5 cm, thin, strigose on both surfaces, the base cuneate from widest point to petiole, the apex widely rounded, then abruptly and shortly acuminate, the margins entire; blades of smaller leaves round, 2–4.5 × 2–5 cm, otherwise similar to larger ones. Inflorescences loosely branched, 4 cm long (on type collection), free from petiole, the main axis not inflated, flowering basipetally. Flowers: calyx 2–3 mm long, strigose; corolla white, deciduous, the tube 2–3 mm long, the lobes ca. 2 mm long; ovary glabrous. Fruits fleshy, yellow, glabrous, the stone erect, obovoid, ca. 12 mm long. Known only by a sterile collection from our area, in non-flooded forest near Crique Limonade.

Cordia nervosa Lam.

Shrubs to understory trees. Stems terete, antrorsely strigose. Leaves homomorphic; petioles, 5–10 mm long; blades oblong to broadly lanceolate, 10–25 × 4–10 cm, coriaceous, glossy, smooth, glabrous to strigose adaxially, the base usually slightly asymmetrical, the apex acuminate, the margins weakly recurved. Inflorescences small and compact, branched, mostly axillary, 1–4 cm long, free from petiole, the main axis not inflated, flowering basipetally. Flowers: calyx 4–5 mm long, minutely short strigose abaxially, not ribbed, enclosing bud, the bud apiculate, opening irregularly; corolla white, deciduous, the tube ca. 5 mm long, the lobes ca. 3 mm long; ovary pubescent toward apex. Fruits fleshy, bright red, minutely and abundantly strigose, the calyx often persisting as irregular fragments, the stone transversely ovoid, 10–13 mm long. Fl (Jul), fr (May, Jul, Nov); in forest on hill slopes and ridges.

Cordia nodosa Lam. FIG. 56; PART 1: PL. XIc

Shrubs to understory trees. Stems terete, bearing stiff, spreading bristles, usually abundant, but sometimes sparse, the stems below each fork abruptly and asymmetrically enlarged to form cavity usually inhabited by ants. Leaves heteromorphic, clustered at apex of new growth; petioles 2–5 mm long; blades lanceolate to elliptic, the large ones 10–40 × 3–30 cm, lustrous, with impressed veins, bullate adaxially, the base obtuse, the apex acuminate, the margins entire. Inflorescences cymose-paniculate, loose or dense, 2–10 cm diam., free from petiole, the main axis not inflated, flowering simultaneously. Flowers: calyx papery, very obscurely ribbed, opening irregularly and forming several irregular lobes, frequently persisting and breaking into fibers, ca. 5 mm long in bud, somewhat puberulent and strigose, bristly; corolla white, deciduous, the tube 4–6 mm long, the lobes 2–3 mm long; ovary hairy. Fruits fleshy, white, yellow, orange, or red, usually bristly, the stone transversely ovoid, 10–17 mm long. Fl (Aug, Sep, Nov), fr (Mar, Apr, May, Jun); common, in understory of non-flooded forest. *Arua peludo* (Portuguese), *Bois fourmis* (Créole), *grão de galo* (Portuguese), *lamoussé, lamoussé formis* (Créole).

Cordia sagotii I. M. Johnst. [Syn.: *C. hirta* I. M. Johnst.] PL. 31b

Shrubs to understory trees. Stems minutely scabrous to brown hirsute, with erect or ascending hairs with enlarged bases. Leaves homomorphic; petioles 4–10 mm long; blades elliptic to oblong or oblanceolate, 8–30 × 3.5–15 cm, coriaceous, with inconspicuous to abundant and short appressed to erect trichomes adaxially, with abundant short to long, erect trichomes abaxially, the base acute to round or oblique, the apex abruptly and shortly acuminate, the margins entire. Inflorescences often in branch forks, rarely axillary, branched, 10–30 cm wide, free from petiole, the main axis not inflated, flowering basipetally. Flowers: calyx ca. 3–4 mm long in bud, sparsely strigulose abaxially, with 5 lobes; corolla white, deciduous, fragrant, the tube 4–6 mm long, the lobes 1.5–2 mm long, rounded, recurved; ovary glabrous. Fruits fleshy, yellow, glabrous, the stone narrowly ovoid, erect, smooth, 15–18 mm long. Fl (Nov), fr (Mar, Jun); in non-flooded forest.

Cordia schomburgkii DC.

Shrubs. Stems terete, with both fine curved pubescence and short, coarse, curved bristles. Leaves homomorphic; petioles 6–9 mm long; blades ovate to elliptic, 5–10 × 2–7 cm, lustrous, distinctly strigose adaxially, the base obtuse to subrounded, the apex acute, the margins entire to sharply dentate. Inflorescences axillary, unbranched, adnate to petiole for 5–8 mm, flowering basipetally; peduncle ascending, to 9 cm long, not swollen, the rachis 5–15 cm long. Flowers: calyx nearly glabrous, bearing numerous resinous granules, with a few bristles near apex; corolla white, deciduous, the tube 5–7 mm long, the lobes very shallow; ovary minutely strigose. Fruits fleshy, minutely strigose, red, ensheathed by calyx, the stone ovoid, 4–5 mm long. Fl (Jan, Feb, Mar); in secondary open vegetation near village.

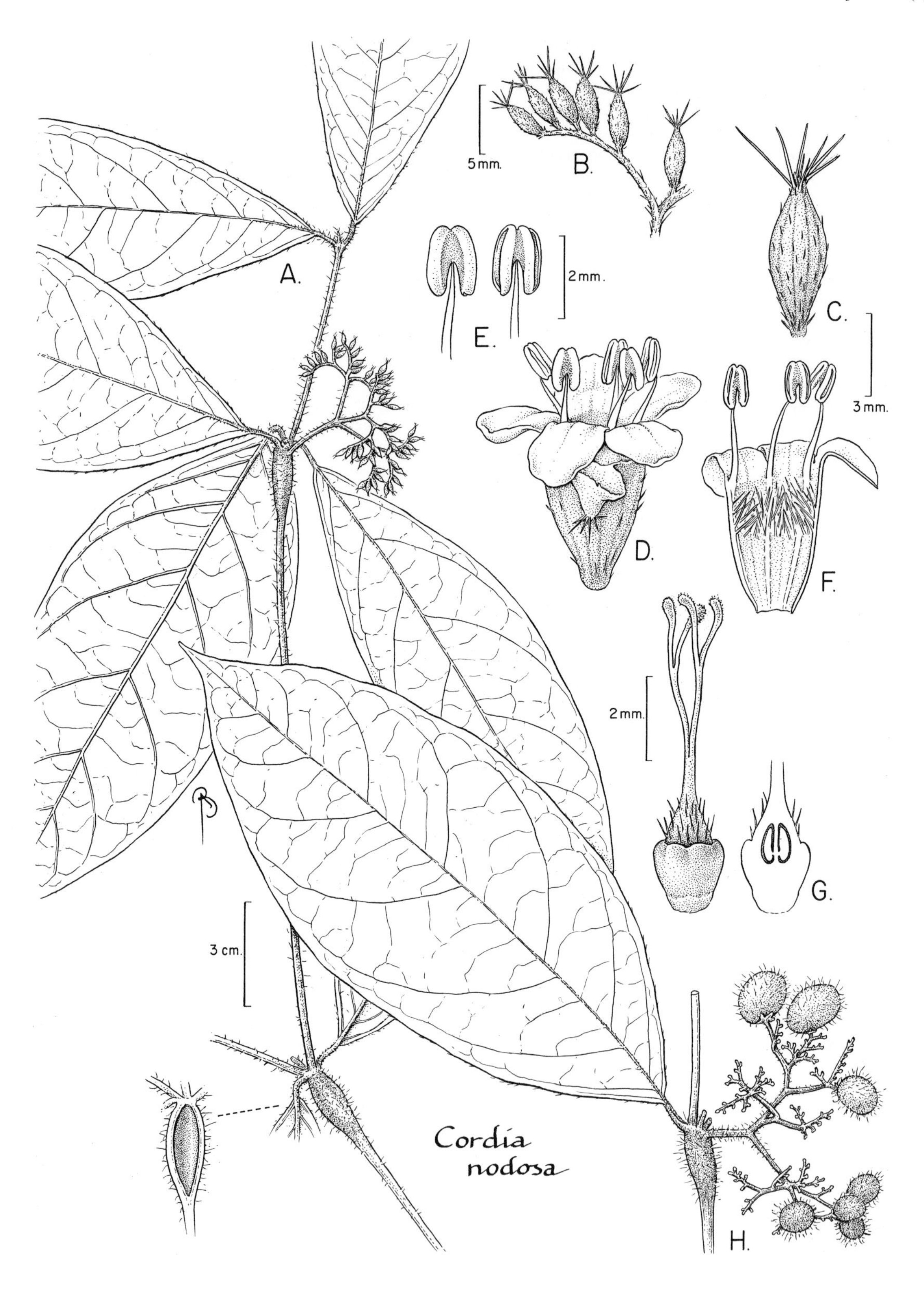

FIG. 56. BORAGINACEAE. *Cordia nodosa* (A–C, *Heald et al. 2*; D–G, *Gleason 583* from Guyana; H, *Acevedo-Rodríguez et al. 4991*). **A.** Part of stem with enlarged nodes, leaves, and inflorescence in bud with detail of medial section of hollow node. **B.** Detail of inflorescence in bud. **C.** Lateral view of flower bud. **D.** Lateral view of flower showing calyptrate calyx. **E.** Abaxial (left) and adaxial (right) views of apical portion of stamens. **F.** Medial section of corolla with adnate stamens. **G.** Pistil surrounded by nectary disc (far left) and medial section of ovary and nectary disc (near left). **H.** Part of stem with leaf, swollen node, and infructescence.

HELIOTROPIUM L.

Herbs or subshrubs (outside our area). Leaves alternate, simple. Inflorescences single, paired, or ternate scorpioid cymes, or of solitary flowers. Flowers bisexual; calyx 5-toothed or 5-lobed, persistent or deciduous; corolla small, 5-merous, yellow, white, or blue, the tube cylindrical; stamens 5, included; ovary 4-locular, the style terminal, undivided, or absent, the stigma 1. Fruits lobed or unlobed, dry, breaking at maturity into 2–4 dry, 1–2-seeded nutlets, frequently with 1–2 sterile locules.

Heliotropium indicum L.

Annual, erect, coarse, weedy herbs, to 1 m tall, mostly branched above middle. Stems, leaves, and calyx with pale, hirsute hairs. Leaves: petioles 4–10 mm long, winged just below blade; blades ovate to elliptic, 5–15 × 3–9 cm, the base obliquely acute to subcordate, the apex acute, the margins repand or undulate. Inflorescences bractless, single, scorpioid cymes to 30 cm long. Flowers: calyx 2–2.5 mm long, somewhat accrescent in fruit, the lobes subulate or cuneate; corolla pale blue, purple to lavender, infrequently white, salverform, the tube 2–5 mm long. Fruits deeply 2-lobed, strongly ribbed, glabrous, breaking into 4-angular nutlets, 2–3 mm long. Fl (Mar, May), fr (May, Sep, Oct); in open, weedy places. This species has a number of medicinal uses (Grenand et al., 1987). *Crêpe denne*, *crête-coq*, *crête d'Inde*, *crèque dinde* (Créole), *crista de galo*, *rabo de galo* (Portuguese).

TOURNEFORTIA L.

Shrubs or lianas (or small trees outside our area.) Leaves alternate, simple. Inflorescences scorpioid cymes borne in dichotomous panicles. Flowers bisexual; calyx usually 5-lobed, persistent; corolla small, usually 5-merous, yellowish or white; stamens usually 5, included; ovary 4-locular, the style terminal, solitary, or absent, the stigma 1. Fruits fleshy at maturity, the endocarp lobed or unlobed, breaking after maturity into 2–4 bony, 1–2-seeded stones, frequently with 1–2 sterile locules, drying later and splitting into 2–4 nutlets.

1. Pedicel thickened under fruit. Corolla lobes long acuminate at apex. Fruits conspicuously lobed, yellow, or yellowish usually spotted with black.
   2. Leaves, young stems, inflorescences, calyx, and corolla sparsely or minutely strigose or glabrous. .... *T. maculata*.
   2. Leaves, young stems, inflorescences, calyx, and corolla conspicuously and usually abundantly strigose or hirsute. .... *T. paniculata* var. *spigeliiflora*.
1. Pedicels thin under fruit or fruit sessile. Corolla lobes rounded to wide mucronate at apex. Fruits unlobed, white.
   3. Leaves short appressed-pubescent adaxially. Flowers sessile; corolla throat constricted; stigma sessile on mature fruit. .... *T. bicolor*.
   3. Leaves glabrous adaxially. Flowers mostly pedicellate even under fruit; corolla throat inflated, bubble-shaped; stigma on well developed, 2–3 mm long style on mature fruit. .... *T. ulei*.

Tournefortia bicolor Sw.

Shrubs or lianas. Stems with short, ascending appressed hairs. Leaves: petioles 5–15 mm long; blades ovate to elliptic, 4–15 × 3–9 cm, shortly appressed-pubescent adaxially, slightly more hairy abaxially, the base obtuse to rounded, the apex acute. Inflorescences dense, branched, 5–20 cm wide, the rachises usually 5 cm long even in fruit; pedicels usually absent even under fruit. Flowers: calyx sparsely strigose at anthesis, the lobes lanceolate to ovate, 1–2.5 mm long; corolla greenish-white, strigose, the tube 4–5 mm long, the limb 6–7 mm broad, spreading, the lobes broad, rounded to wide mucronate at apex. Fruits sessile, unlobed, white, very fleshy, the stigma sessile on fruit; pedicels absent in fruit. Fl (Nov), fr (Nov); in non-flooded forest.

Tournefortia maculata Lam. [Syn.: *T. syringifolia* Vahl] PL. 31c

Shrubs or lianas. Stems inconspicuously short-pubescent when young, the trichomes whitish. Leaves: petioles 7–15 mm long; blades ovate to elliptic or broadly lanceolate, 4–15 × 2–8 cm, both surfaces very sparsely short-strigose, usually abundantly and very minutely tuberculate, the base acute to rounded, the apex acuminate. Inflorescences slender, loosely branched, 5–15 cm wide, the rachises usually 5 cm long even in fruit; pedicels inconspicuous to 1 mm long at anthesis. Flowers: calyx glabrous, 1–1.5 mm long, the lobes subulate to ovate; corolla greenish-white, glabrous, the tube 3–8 mm long, the lobes 1–2.5 mm long, spreading, triangular below middle, coarsely long acuminate at apex. Fruits conspicuously 3–4-lobed, with persistent style at apex, yellow or yellowish, usually spotted with black, breaking into 3–4, 1-seeded nutlets; pedicels accrescent, thickened, to 1–5 mm long under mature fruit. Fl (Jan, Jul, Aug, Dec), fr (Feb, Dec); in non-flooded forest and in the village of Saül.

Tournefortia paniculata Cham. var. **spigeliiflora** (A. DC.) I. M. Johnst. FIG. 57

Shrubs or lianas. Stems conspicuously tawny pubescent, especially when young. Leaves: petioles 7–15 mm long; blades ovate to elliptic or broadly lanceolate, 4–15 × 2–8 cm, both surfaces conspicuously and usually abundantly strigose, usually abundantly and very minutely tuberculate, the base acute to rounded, the apex acuminate. Inflorescences slender, loosely branched, 5–15 cm wide, the rachises usually 5 cm long even in fruit; pedicels inconspicuous to 1 mm long at anthesis. Flowers: calyx 1–1.5 mm long, hirsute, the lobes subulate to ovate; corolla greenish white, hirsute, the tube 3–8 mm long, the lobes 1–2.5 mm long, spreading, triangular below middle, coarsely long acuminate at apex. Fruits conspicuously 3–4-lobed, with persistent style at apex, yellow or yellowish, usually spotted with black, breaking into 3–4, 1-seeded nutlets; pedicels accrescent, thickened, to

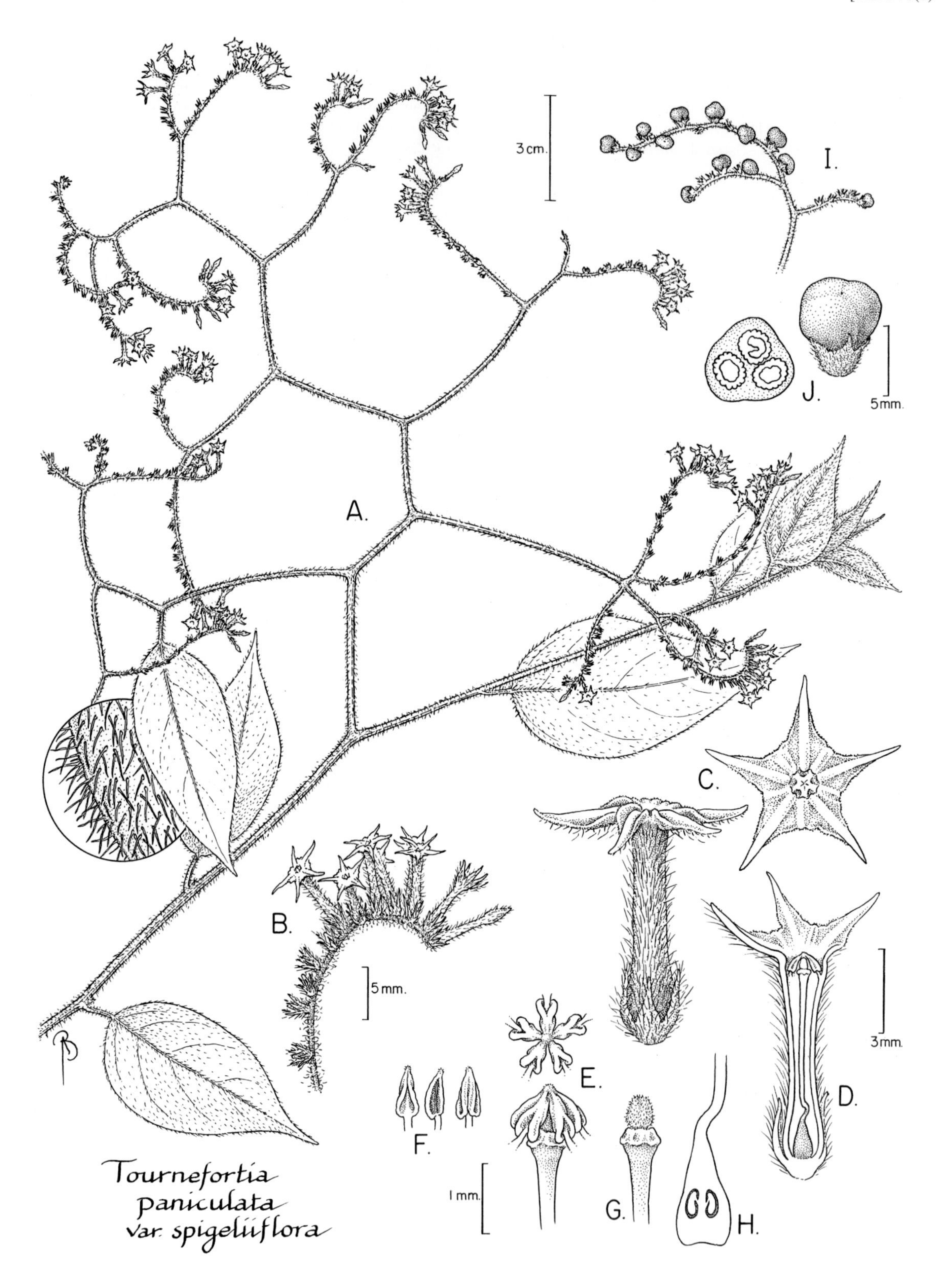

FIG. 57. BORAGINACEAE. *Tournefortia paniculata* var. *spigeliiflora* (A–H, *Mori & Boom 15099*; I, J, *Mori & Gracie 18737*). **A.** Part of stem with leaves and inflorescence; detail of hairs on adaxial surface of leaf. **B.** Part of inflorescence. **C.** Lateral view (below left) and apical view (right) of flower. **D.** Longitudinal section of calyx and corolla with intact pistil. **E.** Apical view of anthers (above left) and lateral view of apex of style with stigma covered by anthers (below left). **F.** Three views of stamens. **G.** Lateral view of upper part of pistil. **H.** Medial section of ovary and base of style. **I.** Part of infructescence. **J.** Lateral view (right) and transverse section (left) of fruit.

1–5 mm long under mature fruit. Fl (Feb, Mar, Oct), fr (Aug); in forest or in secondary vegetation.

Tournefortia ulei Vaupel PL. 31d

Shrubs or lianas. Stems puberulent. Leaves: petioles 8–18 mm long; blades ovate or ovate-elliptic or rarely broadly lanceolate, 6–17 × 3–8 cm broad, rather thin, both surfaces glabrous except for inconspicuous pubescence on veins abaxially, frequently with numerous, scattered, minute, usually pale tubercules, the base obtuse, the apex acuminate. Inflorescences loosely branched, the racemes loosely flowered, becoming 2–10 cm long; pedicels mostly 2–3 mm long even under fruit, neither accrescent nor reduced under mature fruit. Flowers: calyx 1.5–2 mm long, strigose, weakly accrescent, the lobes triangular or subulate; corolla 7–8 mm long, greenish white, short strigose, the tube 3–4 mm long, the throat 1–1.5 mm long, inflated, bubble-shaped, the lobes broadly ovate, rounded to widely mucronate at apex, ca. 1 mm long. Fruits white, unlobed, 4–5 mm diam., not quite as long as in diam., broadest below middle, glabrous. Fl (Sep, Oct); at forest edges.

BRASSICACEAE (Mustard Family)

John D. Mitchell

Annual, biennial, or perennial herbs. Indumentum of diverse types or none. Leaves simple to pinnately dissected, or occasionally pinnately compound (*Nasturtium officinale*), usually alternate. Stipules not present. Inflorescences usually racemose. Flowers bisexual, usually actinomorphic; calyx lobes 4; petals 4, often with elongate claw and abruptly speading limb; stamens usually 6, the 2 outer ones shorter than 4 inner ones (tetradynamous), the anthers longitudinally dehiscent; ovary superior, 2-locular, the style present or absent, the stigma capitate to bifid; placentation parietal. Fruits dry, usually dehiscent siliques.

Al-Shehbaz and Price (1998) have provided evidence for segregating *Nasturtium* from *Rorippa*.

Raechal, L. J. 1997. Brassicaceae. *In* J. A. Steyermark, P. E. Berry & B. K. Holst (gen. eds.), Flora of the Venezuelan Guayana **3:** 547–548. P. E. Berry, B. K. Holst & K. Yatskievych (vol. eds.). Missouri Botanical Garden, St. Louis.

NASTURTIUM R. Br.

Annual or perennial herbs, glabrous or pubescent. Leaves simple, pinnatifid, or pinnately compound; margins entire to toothed. Flowers: petals white or yellow; short stamens flanked at base by pair of minute glands, or glands confluent into annular gland, the long stamens separated by short, cone-shaped gland; ovary cylindric, the style short, the stigma capitate to shortly bifid, the ovules numerous. Fruits somewhat terete, subglobose to cylindric, the persistent style (beak) short, the valves thin, the midrib of valve inconspicuous. Seeds numerous.

Nasturtium officinale R. Br. [Syn.: *Rorippa nasturtium-aquaticum* (L.) Hayek]

Perennial herbs, to 0.6(2) m tall, glabrous. Stems submersed, partly floating or resting on mud, decumbent, branched, rooting at nodes. Leaves imparipinnate: petioles present, the upper ones somewhat narrower than lower ones, to 15 cm long. Leaflets opposite to alternate, in 1–5 pairs; blades with margins entire to crenulate. Inflorescences congested. Flowers: sepals oblong, green, 1.9–2.9 mm long; petals usually twice as long as sepals, 3.5–6.6 × 1.3–2.6 mm, the claw green, the limb white to pale purple; stamens 6, the anthers yellow; stigma entire to shortly bifid. Fruits 11–19 × 1.7–3.2 mm, elliptic, dehiscent, the beak (persistent style) 0.5–2 mm long. Seeds in 2 rows in each locule. An edible weed introduced from the Old World, growing on rocks in streams. *Cresson* (French), *watercress* (English).

BURSERACEAE (Frankincense Family)

Douglas C. Daly

Canopy or understory trees, less often shrubs. Bark thin and smooth to thick and fissured, shed as flakes or variously sized plates, often lenticellate, sometimes hooped. Resin canals associated with virtually all vascularized tissues, the resin clear and drying as white to yellow or blackish crystalline powder or mass, or milky and drying in yellowish globules, flammable in some taxa. Stipules absent. Leaves imparipinnately compound (a few species in some genera simple or unifoliolate), alternate; lateral leaflets (sub)opposite; pulvinulus present below base of terminal and often lateral leaflet blades in many taxa (most Protieae and many Canarieae). Inflorescences panicles of racemes, axillary to subterminal, laxly branched or variously reduced, sometimes pseudospicate (*Protium* sect. *Icicopsis*), structurally indeterminate. Flowers actinomorphic, 3–5(6)-merous, diplostemonous, unisexual (plants dioecious to polygamo-dioecious) or less often functionally bisexual; calyx synsepalous, usually lobed, sometimes irregularly split (*Tetragastris*), the aestivation valvate; corolla apopetalous to variously sympetalous, the aestivation induplicate-valvate; stamens sometimes didynamous but more often the two series of stamens nearly isomorphic, the filaments free or less often connate basally, inserted at base of intrastaminal, annular nectary disc (inserted on disc in *Trattinnickia*). Staminate flowers with reduced pistillode usually present, this sometimes provided with reduced locules and ovules but stigmas not developing, sometimes a solid parenchymatous mass and this sometimes cylindric, in some taxa the pistillode obsolete altogether or ontogenetically fused with nectary disc to form an "ovariodisc." Pistillate flowers: anthers of staminodia devoid of pollen, often some staminodia persisting in fruit; ovary superior, syncarpous, 2–5-locular, the style 1, apical, sometimes slightly branched distally, the stigmas as many as locules or 1 and with as many lobes as locules; placentation axile, the ovules 2, collateral, pendulous, epitropous. Fruits drupaceous to pseudocapsular, dehiscent (all Protieae) or indehiscent (all Canarieae), at most one fertilized ovule developing per locule, the dehiscent fruits with valves as many as developed locules and falling away via acropetal septicidal dehiscence, usually leaving a columellate structure, pyrenes attached to apex of pedicel and variously covered with a red to yellow fleshy arillate structure (Bursereae/Burserinae), or winged and wind-dispersed (Bursereae/Boswellinae), or attached to apex of a columella and the arillate structure usually white, spongy, and sweet (Protieae), (thinly) cartilaginous to bony, the indehiscent fruits mostly (all Canarieae) a single 2–3-locular compound drupe, the exocarp thin, the mesocarp oily (in many cases edible when ripe), the endocarp cartilaginous to bony. Seeds exalbuminous, the testa thin or irregularly thickened, sometimes (e.g., some Protieae) infolded with cotyledons; embryo minute, straight, the cotyledons entire and plano-convex (these sometimes uncinately curved) or palmatifid to lobed and contortuplicate, less often transversely twice-folded.

Daly, D. C. 1997. Burseraceae. *In* J. A. Steyermark, P. E. Berry & B. K. Holst (gen. eds.), Flora of the Venezuelan Guayana **3:** 689–728. P. E. Berry, B. K. Holst & K. Yatskievych (vol. eds.). Missouri Botanical Garden, St. Louis.

Lindeman, J. C. 1986. Burseraceae. *In* A. L. Stoffers & J. C. Lindeman (eds.), Flora of Suriname **III(1–2):** 556–562. E. J. Brill, Leiden.

Swart, J. J. 1951. Burseraceae. *In* A. Pulle (ed.), Flora of Suriname **III(2):** 204–251. Royal Institute for the Indies.

1. Flowers 3-merous; petals not cucullate. Fruits indehiscent, with 1 plurilocular pyrene, 1–3 locules developing (tribe *Canarieae*).
   2. Flowers 3–5 mm long; petals connate, fleshy, provided with retrorse hairs abaxially; stigmas (sub)sessile (style mostly obsolete). Fruits globose to umbonate; endocarp bony, rugose, 2–3-lobed. . . . . *Trattinnickia.*
   2. Flowers 1–2 mm long; petals free, membranous, retrorse hairs absent; style evident. Fruits ovoid to ellipsoid (rarely globose); endocarp cartilaginous, smooth, not lobed.. . . . . *Dacryodes.*
1. Flowers (3)4–5-merous, when trimerous petals cucullate (*Protium* sect. *Sarcoprotium*). Fruits dehiscent, with 1–5 unilocular pyrenes (tribe Protieae).
   3. Lateral leaflets with pulvinulus (sometimes absent in *Protium pilosum*, then acumen serrulate). Petals free (rarely basally connate). . . . . *Protium* (most species).
   3. Lateral leaflets without pulvinulus. Petals free or connate.
      4. Leaflets entire.
         5. All trichomes simple. Resin clear. Petals connate at least 1/2 length; stamens and staminodia (some usually persisting in fruit) usually with anthers continuous with filaments (i.e., the base entire, not sagittate). . . . . *Tetragastris.*
         5. Trichomes of two types, simple hairs and "snail-shaped" glands. Resin milky. Petals free; stamens and staminodia (some usually persisting in fruit) with anthers sagittate. . . . . *Protium* (a few species).
      4. Leaflets serrate (sometimes sparsely).
         6. Resin clear. Leaflets uniformly serrate, the terminal pulvinulus lacking. Petals glabrous adaxially, strongly reflexed at anthesis. Pyrenes papery to cartilaginous. *Crepidospermum.*
         6. Resin milky. Leaflets irregularly serrate, the terminal pulvinulus present. Petals with dense long hairs adaxially, (sub)erect at anthesis. Pyrenes bony. *Protium subserratum* group.

CREPIDOSPERMUM Hook.f.

Understory trees or occasionally shrubs. Trunks sometimes provided with plank buttresses. Bark brown or gray, smooth, sometimes with hoop marks. Fresh resin clear. Trichomes of two types: 1) simple, ferruginous, usually erect to ascending, stiff hairs to 0.7 mm long; and 2) multicellular snail-shaped glands, the latter occurring on most vegetative surfaces and on inflorescence axes, sometimes present on pedicels, calyx, and (rarely) ovary. Leaves usually clustered at ends of branchlets; pulvinuli absent; leaflet margins simply serrate, the teeth acute to obtuse or rarely rounded, rarely incrassate, ascending, to 1 mm long. Inflorescences paniculate but tertiary axes usually poorly developed. Flowers functionally unisexual, 5-merous; petals free, slightly to markedly carinate, the margins long-papillate; stamens 5, exceeding pistillode; staminodia 5 or 10 (when 10 weakly didynamous); pistillode either subcylindric and parenchymatous or ovoid and provided with locules and reduced ovules and contracted into a short, thick, 5-sulcate style. Fruits dehiscent, compound drupes, obliquely ovoid, usually red when mature, dehiscent by a valve (endocarp and mesocarp) falling away from each developed locule, each pyrene suspended from fruit apex by inverted-V-shaped structure; pyrenes papery and brittle to cartilaginous, enveloped in sweet white pulp. Seeds obliquely ovoid, the apex acute, the testa papery; cotyledons fleshy, entire, essentially plano-convex but both distally replicate ca. 1/2 length, nearly U-shaped.

Crepidospermum goudotianum (Tul.) Triana & Planch.

Trees, 3–28 m tall. Bark brown, smooth, with incomplete hoops. Leaves 3–4(5)-jugate; lateral leaflets (narrowly) elliptic, the terminal leaflets oblanceolate to slightly obovate, the base obtuse to acute. Flowers: calyx lobes triangular; petals (narrowly) lanceolate, strongly reflexed >1/2 length at anthesis. Staminate flowers: stamens 5, 1.3–1.7 mm long; nectary disc 0.35–0.45 mm high × 0.5–0.7 mm thick; pistillode subcylindric to narrowly obclavate, 0.55–0.85 mm high, devoid of locules and ovules, glabrous. Pistillate flowers: staminodia 5 or 10, the outer series 0.95–1.25 mm and, when present, the inner series 0.6–1 mm long; nectary disc 0.15–0.2 mm high; pistil 1.6–1.75 mm high; ovary rarely glabrous initially, usually somewhat densely provided with erect to ascending hairs 0.1–0.25(0.45) mm long, pubescence never obscuring surface, the style 0.55–0.75 mm long. Fruits 1–1.5 × 0.5–0.7 cm, glabrescent, maturing red. Rare, known only by sterile collections from our area, in non-flooded forest.

DACRYODES Vahl

Trees, usually of understory. Bark tan to brown or gray, smooth or somewhat rough due to raised lenticels. Resin clear to somewhat milky, drying into opaque, brittle globules. Trichomes of simple, straight or crispate hairs, also sometimes capitate and/or spiralled glands. Leaves: pulvinuli sometimes apparent on lateral and terminal leaflets; leaflets with insertion parallel or perpendicular to plane of leaf, rarely asperous abaxially or on both surfaces, often provided with superficial to slightly sunken minute capitate glands, drying black in some species, the margins entire. Flowers 1–2 mm long, 3-merous, unisexual or less often apparently functionally bisexual; calyx shallowly cupular to urceolate, the 3 lobes continuous or reduced to apicula separated by a flat sinus; petals free; stamens 6, ± equal, the filaments broadly strap-shaped, the anthers continuous with filaments; staminodia in pistillate flowers resembling reduced stamens; ovary 2–3-locular, the style present, the stigma usually with as many lobes as locules; pistillode in staminate flowers reduced to conical parenchymatous structure surmounting nectary disc. Fruits indehiscent, compound drupes, the exocarp thin, usually maturing black or deep purple, drying wrinkled, the mesocarp oily, resinous, often edible, the endocarp (thinly) cartilaginous, smooth, not lobed, with usually only one locule developing and occupying most of pyrene, the other 1–2 usually aborted and vestigial. Seeds: testa smooth; cotyledons palmatifid and folded or contortuplicate; germination epigeal, phanerocotylar, the eophylls opposite and trifoliate.

1. Trichomes of crispate hairs 0.25–0.4 mm long, twisted glands, and erect to ascending hairs. Leaflets usually asperous abaxially or on both surfaces, drying black, the apex narrowly long-acuminate to cuspidate (acumen to 3 cm long); secondary veins acutely prominent abaxially, impressed to flat adaxially. Calyx shallowly and broadly cupular; petals patent at anthesis, red. . . . . *D. cuspidata.*
1. Trichomes of appressed hairs and capitate glands. Leaflets with both surfaces relatively smooth, drying (yellowish-)brown, the apex broadly and abruptly short-acuminate to obtuse or rounded (acumen ≤1.5 cm); secondary veins narrowly prominulous abaxially, broadly prominulous to flat adaxially. Calyx somewhat urceolate; petals suberect at anthesis, pale green. . . . . *D. roraimensis.*

Dacryodes cuspidata (Cuatrec.) Daly [Syn.: *Trattinnickia cuspidata* Cuatrec.]

Trees, 4–10 m tall. Trunks cylindrical, not buttressed. Outer bark gray, smooth, scaly, corky, the inner bark orangish, ca. 1 mm thick. Exudate clear, sticky. Vesture on vegetative parts of crispate hairs 0.25–0.4 mm long, inflorescences also provided with erect to ascending, sharp hairs to 0.1 mm long and with blunt, twisted glandular trichomes, the flowers lacking crispate hairs. Leaves unifoliolate to 4-jugate; pulvinuli conspicuous; lateral petiolules 1–2 cm long; leaflets lanceolate to (narrowly) elliptic, asperous abaxially or on both surfaces, drying black, the base acute, the apex usually narrowly long-acuminate to cuspidate, the acumen narrow, to 3 cm long; midrib and secondary veins acutely prominent and higher-order veins prominulous abaxially, the midrib narrowly prominulous, secondary veins flat to impressed, and higher order veins flat to prominulous adaxially. Inflorescences terminal, (9)13–25 cm long. Flowers unisexual; calyx shallowly and broadly cupular; petals ovate, patent at anthesis, red. Fruits ovoid to subellipsoid, ca. 1.6–1.9 × 1–1.3 cm, drying blackish. Rare, but to be expected from our area.

Dacryodes roraimensis Cuatrec.

Trees, 9–40 m tall. Bark brown, finely fissured, thin, flaky. Leaves (1)2–6-jugate; petioles and rachises usually with dense to sparse appressed hairs and capitate glands; lateral leaflets (oblong-)elliptic to slightly obovate, (markedly) coriaceous at maturity, drying (yellowish-)brown, the apex broadly and abruptly short-acuminate (the acumen ≤1.5 cm) to obtuse or rounded; midrib with scattered capitate glands and scattered appressed hairs to 0.1 mm long, rest of surface often with scattered sunken glands abaxially; secondary veins spreading, the midrib narrowly prominent and secondary veins narrowly prominulous abaxially, the midrib narrowly prominulous and rest of venation broadly prominulous to flat adaxially. Inflorescences axillary, (5)7–13 cm long, robust, the axes with dense to sparse glands and appressed hairs to 0.05 mm long. Flowers apparently functionally bisexual; calyx slightly urceolate, 1–1.4 × 2–2.5 mm, the lobes absent to 0.05(0.1) mm long, with sparse to scattered appressed hairs abaxially; petals oblong-ovate to oblong-lanceolate, 2–3 × 1.1–1.5 mm, usually suberect at anthesis, surface (sub)glabrous abaxially, pale green; stamens 0.8–0.95 mm long (antesepalous ones to 1.05 mm), the filaments broad and flat, 0.25–0.3 mm long; nectary disc 0.15–0.3 mm high × 0.1–0.15 mm thick; ovary ovoid, sessile, glabrous, the style 0.2–0.5 mm long, the stigmatal area 0.2–0.25 mm high. Fruits ovoid to ellipsoid, 1.5–1.7 × 1–1.1 cm, maturing purple. Fr (May, Aug, immature); rare, in non-flooded forest.

PROTIUM Burm.f.

Trees or treelets, rarely shrubs. Trunks often but not always with buttresses, these simple or less often branched, plank or less often flying. Bark gray to brown, usually relatively smooth, thin, brittle, usually lenticellate, often hooped; sapwood white to red. Resin clear and watery or opaque and milky, drying white and powdery, to translucent and hard (in globules), to yellowish and gummy, flammable in some species. Leaves: pulvinulus virtually always visible below base of terminal leaflet blade and usually of laterals, the leaflet margins rarely somewhat revolute, entire (rarely subserrate or sparsely serrate, or acumen sparsely serrate). Inflorescences panicles of racemes, sometimes essentially divaricately branched, sometimes reduced to branched pseudospikes with sessile flowers (sect. *Icicopsis*). Flowers 4–5-merous, unisexual, almost always diplostemonous (isostemonous in pistillate flowers of *P. robustum* and some *P. sagotianum* and *P. tenuifolium*); petals almost always free, rarely petals fused to 2/3 length (but then irregularly divided), lanceolate to broadly ovate, rarely cucullate with clawed base (*P. plagiocarpum*), the apex swollen and inflexed in an often conspicuous apiculum; stamens sometimes didynamous but more often the two series of stamens nearly isomorphic, the anthers sagittate; staminodia present but reduced in pistillate flowers; nectary disc usually annular; pistillode usually of rudimentary ovary with locules and reduced ovules, sometimes solid and parenchymatous, free from nectary disc or sometimes nectary disc and pistillode ontogenetically fused to form convex or less often discoid "ovariodisc," the style usually distinct but stigmas never differentiated; pistil with as many locules as petals, the stigmas distinct, usually sessile, usually globose to discoid, sometimes caudiculate and spiculate. Fruits septicidally dehiscent, compound drupes, usually glabrous, variously shaped but usually slightly to markedly oblique-ovoid, with as many lobes as developed locules (if more than one develops), when mature usually red but occasionally green or brown, drying wrinkled or smooth depending on whether mesocarp is fleshy or lignified, the surface smooth or less often lenticellate or rugose; dehiscent by a valve (endocarp and mesocarp) falling away at maturity from each developed locule, each pyrene suspended from fruit apex by inverted-V-shaped structure clasping just behind apex; columella and valves white or pink to red adaxially; pyrenes thinly cartilaginous to thick and bony, usually smooth-textured, ventral side with funicular scar of varying shape and size, often set in relief from rest of surface, the pyrenes partly or completely enveloped in spongy, sweet, edible, white pulp ("pseudaril"). Seeds: testa papery to slightly fleshy to crustose, often irregularly thickened; cotyledons large, lobed and contortuplicate (then testa infolded and partly lignified), or essentially entire and plano-convex, or transversely folded (sect. *Icicopsis*).

The following keys are meant to be useful for identification of herbarium specimens. The measurements refer to dry fruits and to rehydrated flowers. Rehydration of flowers is necessary for proper dissection. In most instances dissection of flowers is necessary for access to the taxonomically useful characters. The fruit shapes and measurements refer to fruits that have only one pyrene and therefore are not lobed. Fruits of *Protium* tend to reach their mature size and display their mature color rather rapidly, but the pseudaril and embryo develop only shortly before the fruit dehisces.

There are two keys, one for flowering material and one for fruiting material. Their use requires distinguishing staminate and pistillate flowers; a key to gender is provided below.

Some species occur twice in the keys, because they may display two character states. For example, *Protium aracouchini* may have a glabrous or pubescent pistillode.

Key to flower gender

1. Anthers dehiscent and producing pollen; pistillode (when present) always exceeded by stamens; ovules not filling locule when locules and ovules present, the stigmas not developed, instead style apex rounded, lobulate, or acutely winged; pistillode sometimes obsolete, instead forming with nectary disc a continuous, homogeneous, discoid or depressed-conical structure. . . . . staminate flowers.
1. Anthers indehiscent and producing no pollen; true pistil almost always exceeding staminodia; ovules completely filling locules, the stigmas globose to discoid. . . . . pistillate flowers.

Key to flowering material

1. Leaves consistently unifoliolate (0-jugate), the petiole with two pulvinuli. . . . . *P. occultum*.
1. At most a few leaves on any branch unifoliolate, on these the petiole with one pulvinulus.
    2. Inflorescences pseudospicate (flowers (sub)sessile); petals glabrous adaxially; pistillode and nectary disc not distinct (represented by a discoid or depressed-conical structure; section *Icicopsis*).
        3. Leaves 68–110 cm long. Inflorescence axes provided with flexuous appressed hairs to 0.3 mm long. Stamens exceeding petals (exserted); staminodia always 5, reaching stigmas; ovariodisc usually discoid; ovary substipitate, the stigmas globose. . . . . *P. robustum*.
        3. Leaves (12.7)17.5–38.5(53) cm long. Inflorescence axes provided with straight, ascending to appressed hairs to 0.15 mm long, sometimes also with scattered, appressed hairs to 0.15 mm long. Stamens shorter than petals; staminodia 10 (rarely 5), not reaching stigmas; ovariodisc usually depressed-conical; ovary sessile, the stigmas discoid or obdeltoid.
            4. Leaflet apex gradually and broadly short-acuminate. . . . . *P. tenuifolium*.
            4. Leaflet apex abruptly and narrowly long-acuminate. . . . . *P. sagotianum*.
    2. Inflorescences paniculate or glomeriform or, if pseudospicate, the petals lanate adaxially. Pistillode and nectary disc distinct.
        5. Leaflets papillate abaxially. Petals lanate adaxially, with dense, long, ascending hairs 0.3–1.1 mm long.. . . . . *P. subserratum* subsp. *subserratum*.
        5. Leaflets not papillate abaxially. Petals glabrous or, if pubescent adaxially, not lanate and provided with hairs to 0.25 mm long at most.
            6. Resin milky. Leaflet surfaces with scattered, regularly spaced, thick, spiralled, apparently glandular hairs to 0.05 mm long abaxially. Petals not cucullate; stamens didynamous (anthers of two series not overlapping); nectary disc pubescent; stigmas spiculate. . . . . *P. apiculatum*.
            6. Resin translucent or transparent. Leaflet surface without spiralled, apparently glandular trichomes abaxially. Stamens not didynamous (anthers of two series ± level); nectary disc glabrous or, if pubescent, the petals clawed (*P. plagiocarpium*); stigmas not spiculate.

7. Inflorescences longer than petioles of subtending leaves.
  8. Pistillode and pistil glabrous.
    9. Flowers 5-merous; calyx far exceeding disc; anthers ovate in profile; style in staminate flowers markedly 5-winged, in pistillate flowers 5-branched. . . . . *P. pallidum.*
    9. Flowers 4-merous; calyx exceeded by disc; anthers broadly elliptic in profile; style in staminate flowers 4-lobulate, in pistillate flowers unbranched. . . . . *P. aracouchini.*
  8. Pistillode and pistil pubescent.
    10. Leaflets drying reddish, the apex broadly and obtusely acuminate. Petals slightly cucullate, the base clawed; nectary disc pubescent.. . . . . *P. plagiocarpium.*
    10. Leaflets drying green or brown, the apex narrowly and acutely acuminate. Petals not cucullate, the base only slightly narrowed; nectary disc glabrous.
      11. Anthers in staminate flowers 0.15–0.35 mm long, in lateral profile nearly as broad as long, those in pistillate flowers 0.2–0.35 mm long, ovate in dorsiventral view.
        12. Calyx usually exceeding disc, 0.9–1.3 × 1.9–2.5 mm, the calyx and corolla with sparse to dense appressed hairs to 0.15 mm long abaxially, the petals usually sparsely provided with ascending hairs to 0.1 mm long adaxially. . . . . *P. divaricatum* subsp. *fumarium.*
        12. Calyx exceeded by disc, 0.3–0.7 × 1.2–1.7 mm, calyx and corolla glabrous on both surfaces. . . . . *P. aracouchini.*
      11. Anthers in staminate flowers 0.45–0.75 mm long, in lateral profile much longer than broad, those in pistillate flowers (0.35)0.5–0.8 mm long, lanceolate to obtuse-oblong in dorsiventral view.
        13. Petals pubescent or at least provided with few ascending hairs near apex adaxially.
          14. Leaflet midrib with scattered appressed hairs to 0.1 mm long abaxially. Flowers 4-merous; calyx exceeded by disc, (0.6)0.9–1.2 mm long, the lobes 0.2–0.4 mm long. . . . . *P. morii.*
          14. Leaflet midrib and sometimes secondary veins with dense, flexuous hairs 0.2–0.4(0.7) mm long abaxially. Flowers 5-merous; calyx greatly exceeding nectary disc in both sexes, (1.5)1.8–2.4 mm long, the lobes (0.9)1.1–1.6 mm long. . . . . *P. opacum* subsp. *rabelianum.*
        13. Petals glabrous adaxially.
          15. Inflorescences (2.2)4–14(17.5) cm long. Flowers 5-merous.
            16. Leaves (0)1–2-jugate. Calyx and corolla with sparse to scattered hairs to 0.2 mm long abaxially; style in pistillate flowers unbranched (stigmas sessile). . . . . *P. decandrum.*
            16. Leaves (1)2–4(5)-jugate. Calyx and corolla glabrous abaxially; style in pistillate flowers 5-branched (stigmas stipitate).
              17. Leaflets usually glossy, drying (pale) brown abaxially. Calyx exceeded by disc, 0.6–0.8 × 1.5–1.6 mm; style on pistillode laterally 5-lobate; pubescence on pistil dense. . . . . *P. inodorum.*
              17. Leaflets usually dull, drying with a whitish cast abaxially. Calyx far exceeding disc, 1–1.7 × 1.7–2.5 mm; style on pistillode acutely 5-winged; pubescence on pistil sparse to absent. . . . . *P. pallidum.*
          15. Inflorescences 0.4–3 cm long. Flowers 4-merous.. . . . . *P. guianense.*
7. Inflorescences shorter than petioles of subtending leaves.
  18. Pistillode and pistil glabrous.
    19. Flowers 5-merous.
      20. Leaflets inserted perpendicular to plane of leaf, broadly elliptic to broadly oblong, glabrous abaxially, the apex broadly short-acuminate, the acumen obtuse to truncate (rarely slightly notched), to 0.5(0.7) cm long. Calyx usually exceeded by nectary disc in staminate flowers, exceeding disc in pistillate flowers; pistil substipitate, the style laterally 5-branched. . . . . *P. cuneatum.*
      20. Leaflets inserted parallel to plane of leaf, elliptic, with scattered appressed hairs to 0.2 mm long abaxially, the apex abruptly and narrowly long-acuminate, the acumen 0.8–1.8(2) cm long. Calyx level with nectary disc in staminate flowers, exceeded by disc in pistillate flowers; pistil sessile, the style unbranched (stigmas sessile). . . . . *P. demerarense.*

19. Flowers 4-merous.
  21. Leaflets glabrous. Inflorescence bracts not semi-clasping. Petals free, lanceolate, the margins long-papillate; anthers (0.55)0.75–1 mm long; pistil (1.65)1.85–2.25 mm long, the style (0.6)0.8–1.1 mm long. . . . . *P. heptaphyllum* subsp. *heptaphyllum.*
  21. Leaflets pubescent at least on midrib abaxially. Inflorescence bracts semi-clasping. Petals usually basally connate 1/5–3/5 length, ovate, the margins short-papillate; anthers 0.3–0.65 mm long; pistil 1.1–1.4(1.75) mm long, the style (0.15)0.25–0.5 mm long.
    22. Leaves 0–5-jugate; petioles and rachises with dense to sparse erect to ascending hairs to 0.05 mm long; lateral petiolules 0.05–0.3 cm long, the lateral pulvinuli inconspicuous or absent; leaflets narrowly elliptic. Calyx exceeding disc, 1–1.6(1.8) × (1.5)1.7–2.2 mm; anthers 0.3–0.5 long. . . . . *P. pilosum.*
    22. Leaves 0–2(3)-jugate; petioles and rachises with sparse stiff erect hairs 0.3–1(1.3) mm long, also with dense to sparse fine erect hairs to 0.05 mm long; lateral petiolules 0.2–1(1.5) cm long, the lateral pulvinuli usually conspicuous; leaflets (broadly) lanceolate to elliptic. Calyx exceeded by nectary disc in staminate flowers, usually exceeding nectary disc in pistillate flowers, 0.5–1(1.2) × 1.7–2.2(2.4) mm; anthers 0.5–0.65 mm long. . . . . *P. trifoliolatum.*

18. Pistillode and pistil pubescent.
  23. Lateral leaflets elliptic, trichomes appressed abaxially. Flowers 5-merous; petals lanceolate. . . . . *P. demerarense.*
  23. Lateral leaflets ovate to broadly lanceolate or less often elliptic, trichomes erect abaxially. Flowers 4-merous; petals ovate. . . . . *P. guianense.*

Key to fruiting material

[*Note:* Fruits not known for *Protium inodorum.*]

1. Fruits subsymmetric and globose or elliptic (sometimes acuminate).
  2. Infructescences shorter than petioles.
    3. Leaves 0–2(3)-jugate. Fruits smooth, green, glossy, not lenticellate; pseudaril covering only thick lateral band around pyrene. . . . . *P. cuneatum.*
    3. Leaves (1)2–4(5)-jugate. Fruits rough-textured, gray, dull, lenticellate; pseudaril covering most of pyrene. . . . . *P. demerarense.*
  2. Infructescences longer than petioles.
    4. Leaflets with pulvinuli inconspicuous; blades not pallid abaxially. Fruits gray, lenticellate, dull, the calyx forming cupule at base; pseudaril covering most of pyrene surface. . . . . . . *P. opacum* subsp. *rabelianum.*
    4. Leaflets with pulvinuli conspicuous; blades pallid abaxially. Fruits green, smooth, glossy, the calyx not forming cupule at base; pseudaril occurring as band 0.5 cm wide around periphery of pyrene. . . . . *P. pallidum.*
1. Fruits markedly obliquely ovoid, gibbous.
  5. Infructescences paniculate or glomeriform (fruits pedicellate). Pyrenes with site of funicle above middle; cotyledons plano-convex or contortuplicate but not transversely folded.
    6. Infructescences exceeding petioles of subtending leaves.
      7. Fruit base stipitate, cordate above stipe. . . . . *P. plagiocarpium.*
      7. Fruit base substipitate or sessile, truncate or acute.
        8. Fruit base sessile.
          9. Branchlets and infructescences robust. Fruits usually >2 cm long. . . . . *P. apiculatum.*
          9. Branchlets and infructescences slender. Fruits <2 cm long.
            10. Leaflets glabrous. Infructescences 4–11.5 cm long. Fruits glabrous, maturing bicolored, red on dorsal side and green on ventral side. . . . . *P. aracouchini.*
            10. Leaflets with midrib and secondary vein with short erect hairs. Infructescences 0.4–3 cm long. Fruits with short, appressed trichomes persisting at least at base and apex, maturing red only. . . . . *P. guianense.*
        8. Fruit base (sub)stipitate.
          11. Calyx lobes and stigmas 5.
            12. Leaves 1–3-jugate; petioles without fine erect hairs; leaflets glossy. . . . . *P. inodorum.*
            12. Leaves 0–2-jugate; petioles with fine erect hairs to 0.1 mm long; leaflets dull. . . . *P. decandrum.*

11. Calyx lobes and stigmas 4.
    13. Leaves 3–5-jugate. Fruits maturing green to gray-green, often lenticellate, usually drying smooth, the apex abruptly acute or rounded; pyrenes thickly cartilaginous.. . . . *P. morii.*
    13. Leaves (0)2–3-jugate; fruits maturing green or brown, drying wrinkled, the apex (long-)acuminate; pyrenes thin-cartilaginous or papery. . . . . *P. divaricatum* subsp. *fumarium.*
6. Infructescences usually shorter than petioles of subtending leaves.
   14. Resin milky when fresh. Leaflets with pulvinuli lacking on lateral petiolules; blades papillate abaxially. . . . . *P. subserratum* subsp. *subserratum.*
   14. Resin clear when fresh. Leaflets with pulvinuli apparent on all petiolules (sometimes not on *P. pilosum*); blades not papillate abaxially.
       15. Leaflet acumen sparsely serrulate. Fruits markedly trigonous, pale green to whitish at maturity. . . . . *P. pilosum.*
       15. Leaflet acumen entire. Fruits not trigonous, red at maturity.
           16. Leaves consistently unifoliolate, the rachis with two pulvinuli. . . . . *P. occultum.*
           16. Most leaves on all branchlets 1–4-jugate, the rachis with one pulvinulus proximal to terminal leaflet.
               17. Leaflets glabrous. . . . . *P. heptaphyllum* subsp. *heptaphyllum.*
               17. Leaflets pubescent.
                   18. Petioles and rachises with scattered to dense flexuous hairs 0.1–0.6(0.8) mm long, often also with sparse fine erect hairs to 0.1 mm long. Fruits with short appressed hairs persisting at least at base and apex. . . . . *P. guianense.*
                   18. Petioles and rachises with sparse, stiff, erect hairs 0.3–1(1.3) mm long. Fruits glabrous. . . . . *P. trifoliolatum.*
5. Infructescences pseudospicate (fruits sessile). Pyrenes with site of funicle below middle; cotyledons transversely folded (sect. *Icicopsis*).
   19. Branchlets robust, (5)7–12 mm diam at 2 cm from apex. Leaves 68–110 cm long. Fruits strongly oblique, almost perpendicular to pedicels; cotyledons not distally incurved. . . . . *P. robustum.*
   19. Branchlets 1.9–4.4 mm diam at 2 cm from apex. Leaves (12.7)17.5–38.5(53) cm long. Fruits only slightly oblique; cotyledons distally incurved.
       20. Leaflets abruptly and narrowly long-acuminate. . . . . *P. sagotianum.*
       20. Leaflets gradually and broadly short-acuminate. . . . . *P. tenuifolium.*

Protium apiculatum Swart [Syn.: *P. firmum* Swart] PL. 32a

Trees, 5–30 m tall. Trunks often with simple, spreading plank buttresses. Bark gray to beige to brown, relatively smooth, scaly, with short hoop marks and lenticels. Resin milky when fresh. Branchlets ferruginous-pubescent. Leaves 3–5(8)-jugate; leaflets with scattered, regularly spaced, thick, twisted, apparently glandular trichomes to 0.05 mm long abaxially, the apex often apiculate. Inflorescences longer than petioles of subtending leaves, the axes ferruginous-pubescent. Flowers 5-merous; stamens didynamous; staminodia exceeding pistil; nectary disc pubescent. Fruits obliquely ovoid, 1.7–2.5 × 1.2–1.9 cm, base sessile, maturing red, scattered ascending to appressed trichomes to 0.2–0.35 mm long persisting at base and apex, the nectary disc often persistent at base of fruit. Seeds with cotyledons plano-convex. Fl (Dec), fr (Feb, Apr); in non-flooded primary forest.

Protium aracouchini (Aubl.) Marchand PL. 32b

Trees or shrubs, 2–20 m tall. Trunks sometimes with flying buttresses. Bark pale chocolate or usually dark brown, smooth. Leaves 0–4-jugate; leaflets chartaceous, the apex narrowly and acutely acuminate. Inflorescences 4–11.5 cm long, exceeding petioles of subtending leaves, laxly branched, the axes glabrous or rarely with scattered appressed trichomes to 0.15 mm long; pedicels 0.6–2.7 mm long. Flowers 4-merous; calyx 0.3–0.7 × 1.2–1.7 mm, exceeded by nectary disc in both sexes; anthers in staminate flowers 0.15–0.3(0.4) mm long; pistil and pistillode glabrous or pubescent. Fruits strongly obliquely ovoid, 1–1.6 cm long, sessile, maturing bicolored, red dorsally and green ventrally; pyrenes with site of funicle above middle. Seeds with cotyledons plano-convex. Fl (Aug); in non-flooded forest. *Encens tites feuilles* (Créole).

Protium cuneatum Swart PL. 32c; PART 1: PL. IIIb

Trees, 5–20(40) m tall. Trunks consistently with plank and flying buttresses. Bark gray to light brown, smooth, with scattered lenticels. Leaves 0–2(3)-jugate; lateral leaflets inserted perpendicular to plane of leaf, broadly elliptic to broadly oblong, the apex broadly short-acuminate, the acumen obtuse to truncate (rarely slightly notched), 0–0.5(0.7) cm long. Inflorescences shorter than petioles. Flowers 5-merous; calyx usually exceeded by disc in staminate flowers, exceeding disc in pistillate flowers; pistillode and pistil glabrous; style divided into 5 lateral branches terminated by hemispherical stigmas. Fruits subsymmetric, essentially globose, 2.4–2.8 × 1.5–2.3 cm (when 1 pyrene develops), (sub)stipitate, maturing green and glossy; pseudaril occurring only as a thick lateral band ca. 0.5 cm wide around pyrene. Seeds with cotyledons contortuplicate. Fl (Sep), fr (Feb, Sep); in non-flooded forest and by creeks and in swampy areas.

Protium decandrum (Aubl.) Marchand

Trees, (8)15–25(41) m tall. Trunks fluted, with narrow plank buttresses and often with simple or branched flying buttresses. Bark gray, thin, smooth. Leaves (0)1–2-jugate; petioles shorter than interjuga, provided with fine, erect hairs to 0.1 mm long. Inflorescences exceeding petioles of subtending leaves. Flowers 5-merous; calyx usually exceeded by disc, the lobes contiguous, calyx and corolla abaxially with sparse to scattered hairs to 0.2 mm long; pistil and pistillode pubescent. Fruits broadly obliquely ovoid, 1.5–2.1(3.1) × 1.6–2.1 cm (when 1 pyrene develops), substipitate, the apex abruptly acute or rounded, maturing light green; pseudaril covering most of pyrene, the site of funicle above middle. Seeds with cotyledons contortuplicate. Fl (Sep, Oct, Nov), fr (Apr, Nov, Dec); in non-flooded forest.

Protium demerarense Swart PL. 32d

Trees, (7)10–30 m tall. Trunks often with low buttresses. Bark gray to black, scaly. Leaves (1)2–4(5)-jugate; lateral leaflets elliptic, with appressed trichomes to 0.2 mm long abaxially, the apex abruptly and narrowly long-acuminate, the acumen 0.8–1.8(2) cm long. Inflorescences shorter than petioles of subtending leaves. Flowers 5-merous (rarely some 4-merous); calyx level with disc in staminate flowers and exceeded by disc in pistillate flowers; petals lanceolate, cream-colored; filaments cream-colored, the anthers yellow; pistillode sparsely pubescent and pistil glabrous (based on few collections). Fruits mostly solitary in leaf axils of outer branches, essentially globose, (3)3.5–4(5) × (2)2.4–3(4) cm (when 1 pyrene develops), rough-textured, lenticellate, maturing dull gray; pseudaril covering most of pyrene. Seeds with cotyledons contortuplicate. Fl (Jul, Aug, Sep), fr (Feb, May, Jun, Sep, Oct); in non-flooded forest. *Encens*, *encens rouge* (Créole).

Protium divaricatum Engl. subsp. **fumarium** Daly

Trees, 3–15 m tall. Trunks at least occasionally with simple plank buttresses. Bark brown to reddish-brown, smooth, thin, with shallow fissures, lenticellate, hooped. Leaves (0)2–3-jugate; leaflets with apex narrowly and acutely acuminate. Inflorescences much longer than petioles of subtending leaves; bracts semi-clasping. Flowers 4-merous; calyx usually exceeding disc, 0.9–1.3 × 1.9–2.5 mm, with sparse to dense appressed hairs to 0.15 mm long abaxially; petals ovate, usually sparsely provided with ascending hairs to 0.1 mm long adaxially; anthers in staminate flowers 0.25–0.35 mm long, in lateral profile nearly as broad as long, those in pistillate flowers 0.3–0.35 mm long; pistillode and pistil pubescent. Fruits obliquely ovoid, (2.1)2.5–3 (3.3) × (1.4)1.6–2.1(2.5) cm, stipitate, maturing brown or green, drying wrinkled, the apex long-acuminate (*Mori & Pipoly 15590*, NY); pyrenes thin, brittle and easily cracked when dry, the funicular scar on pyrene subapical. Seeds with cotyledons plano-convex. Fl (May, Aug, Sep), fr (Apr, Aug, Sep); in non-flooded forest.

Protium guianense (Aubl.) Marchand [Syn.: *P. hostmannii* (Miq.) Engl.]

Shrubs or trees, 2–25(45) m tall. Trunks often with simple plank buttresses. Bark gray or less often brown, smooth, lenticellate, with incomplete hoops. Leaves 0–4-jugate; petioles and rachises with scattered to dense, flexuous hairs 0.1–0.6(0.8) mm long, often also with sparse, fine, erect hairs to 0.1 mm long; lateral leaflets ovate to broadly lanceolate or less often elliptic, surface with short, erect trichomes abaxially, the base oblique, the apex narrowly and acutely acuminate. Inflorescences short and slender, 0.4–3 cm long, approximately same length as petioles of subtending leaves. Flowers 4-merous; calyx usually exceeded by disc; anthers much longer than broad in lateral profile; pistillode and pistil pubescent. Fruits obliquely ovoid, 1.2–1.6 × 0.5–1.2 cm (when 1 pyrene develops), base sessile, with short appressed trichomes persisting at least at base and apex, maturing red. Seeds with cotyledons plano-convex. Fl (Oct, Dec); in non-flooded forest.

Protium heptaphyllum (Aubl.) Marchand subsp. **heptaphyllum** [Syn.: *P. octandrum* Swart]

Shrubs or trees, 1–25 m tall. Trunks usually with plank buttresses. Leaves (0)1–3-jugate; pulvinuli apparent on all petiolules; leaflets glabrous. Inflorescences usually glomeriform, shorter than petioles of subtending leaves. Flowers 4-merous; petals lanceolate, the margins long-papillate; anthers (0.55)0.75–1 mm long; pistil (1.65)1.85–2.25 mm long, the style (0.6)0.8–1.1 mm long; pistillode and pistil glabrous. Fruits obliquely ovoid, 1.4–2.3 cm long, maturing red, the base sessile, the apex acute. Seeds with cotyledons plano-convex. Fl (Oct, Nov), fr (Feb, Apr, May, Jun, Nov); in non-flooded forest. *Encens* (Créole).

Protium inodorum Daly

Trees, ca. 20 m tall. Trunks with simple plank buttresses. Bark gray, smooth, with scattered white lenticels and hoop marks. Resin clear and sticky. Leaves 1–3-jugate; leaflets often glossy, drying tan abaxially and grayish-brown adaxially, the apex narrowly and acutely acuminate. Inflorescences longer than petioles of subtending leaves, 4.6–8.5 cm long, the secondary axes usually poorly developed. Flowers 5-merous; calyx exceeded by disc, calyx and petals glabrous or rarely with a few scattered trichomes abaxially; anthers in lateral profile much longer than broad; pistillode and pistil densely provided with appressed hairs, also with sparse thick glands <0.05 mm long; style laterally 5-branched. Immature fruits obliquely ovoid, apparently glabrescent, stipitate. Fl (Oct, Dec); fr (Dec, immature); in non-flooded forest.

Protium morii Daly FIG. 58

Trees, (8)20–46 m tall. Trunks with thin, tall plank buttresses. Bark light gray to brown, whitish or reddish-gray, smooth, finely lenticellate. Leaves 3–5-jugate; leaflets with midrib with scattered appressed hairs to 0.1 mm long abaxially. Inflorescences relatively short and robust but longer than petioles of subtending leaves, (3.5)5–9.5(15) cm long. Flowers 4-merous; calyx exceeded by disc, (0.6)0.9–1.2 mm long, the lobes 0.2–0.4 mm long; petals with scattered appressed to ascending hairs 0.1–0.25 mm long distributed along 3 discrete vertical lines adaxially; pistillode and pistil pubescent. Fruits obliquely ovoid, 2.2–2.9 × (1.6)1.8–2 cm (when 1 pyrene develops), substipitate, maturing green or grayish-green, often lenticellate, usually drying smooth, the apex abruptly acute or rounded; pyrenes thickly cartilaginous. Seeds with cotyledons contortuplicate. Fl (Aug, Oct), fr (Feb, Dec); in non-flooded forest.

Protium occultum Daly PL. 33a

Trees, ca. 20 m tall. Trunks with low, relatively thin, spreading buttresses. Bark reddish-brown, smooth. Leaves consistently unifoliolate, the rachis with two pulvinuli; leaflets with apex usually abruptly and narrowly long-acuminate. Inflorescences shorter than petioles of subtending leaves, to 1.3 cm long. Flowers 4-merous; calyx approximately level with nectary disc; stamens white; pistillode and pistil glabrous. Fruits obliquely ovoid, 1.6–2 × 1.3–1.4 cm (when one pyrene develops), maturing red; pyrenes cartilaginous. Seeds with cotyledons plano-convex, curved. Fl (Oct, Nov); in non-flooded forest.

Protium opacum Swart subsp. **rabelianum** Daly PL. 32e

Trees, (5)8–30 m tall. Trunks with simple or branched, plank or flying buttresses. Bark brown to whitish, smooth, with incomplete hoop marks. Leaves (1)2–4-jugate; pulvinuli inconspicuous; leaflets

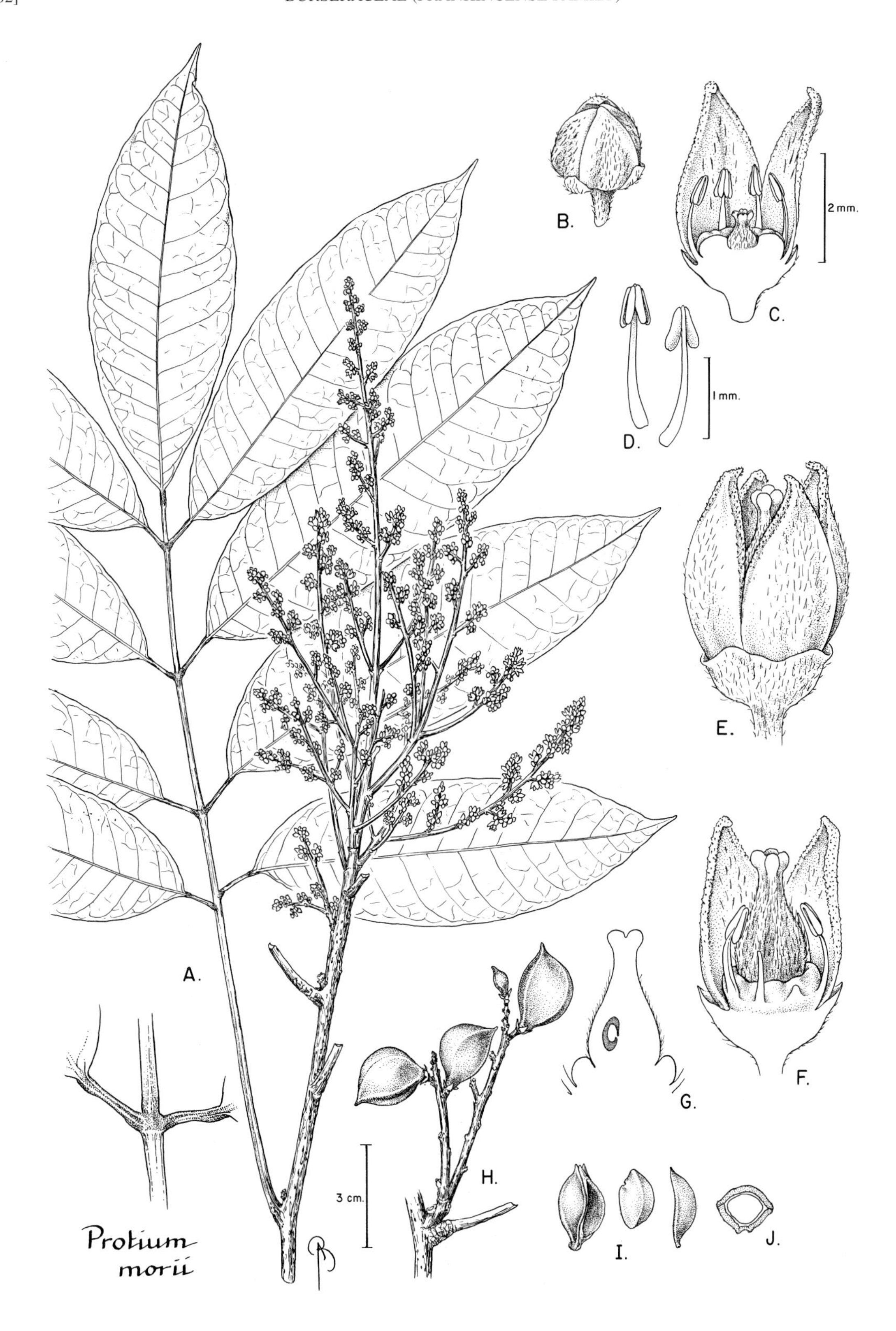

FIG. 58. BURSERACEAE. *Protium morii* (A–D, *Irwin 48119* from Amapá, Brazil; E–G, *Irwin 48474* from Oiapoque River, French Guiana; H, I, *Daly 4049* from Amapá, Brazil). **A.** Branchlet apex with leaf and inflorescence (right) and detail of leaflet pair (left below). **B.** Lateral view of staminate flower bud. **C.** Longitudinal section of staminate flower. **D.** Adaxial (near right) and abaxial (far right) views of stamens. **E.** Lateral view of pistillate flower. **F.** Longitudinal section of pistillate flower showing nectary disc surrounding pistil. **G.** Medial section of pistil. **H.** Infructescence. **I.** Fruit (left), pyrene (above), and valve of fruit (right). **J.** Transverse section of pyrene. (Reprinted with permission from D. C. Daly, Brittonia 44(3), 1992.)

with surface, midrib, and sometimes secondary veins with dense, flexuous hairs 0.2–0.4(0.7) mm long abaxially. Inflorescences longer than petioles of subtending leaves. Flowers 5-merous; calyx greatly exceeding nectary disc in both sexes, (1.5)1.8–2.4 mm long, the lobes (0.9)1.1–1.6 mm long; petals pubescent or at least provided with a few ascending hairs near apex adaxially; pistillode and pistil pubescent. Fruits essentially globose but sometimes slightly obliquely ovoid, (2.4)2.7–3(3.2) × (2.1)2.4–3.1 cm (when 1 pyrene develops), gray and lenticellate, the base sessile, the apex obtusely short-acuminate, the calyx forming cupule at base; pseudaril covering most of pyrene surface. Seeds with cotyledons contortuplicate. Fl (May, Jul, Aug, Sep), fr (Feb, Aug, Sep, Oct, Nov); in non-flooded forest. *Encens, encens blanc, encens grands bois, l'encens* (Créole).

Protium pallidum Cuatrec. PL. 32f

Trees, 10–31 m tall. Trunks usually with simple plank buttresses and (often branched) flying buttresses and prop roots, even on rather young individuals. Bark light brown to gray, smooth, with fine, closely spaced, incomplete hoops and scattered lenticels. Leaves (1)2–4(5)-jugate; lateral pulvinuli conspicuous; leaflets glossy and with whitish cast when fresh or dry abaxially. Inflorescences longer than petioles of subtending leaves. Flowers 5-merous; calyx deeply cupular, 1–1.7 × 1.7–2.5(3) mm, the lobes (0.5)0.7–1.2 mm long, greatly exceeding nectary disc; pistillode glabrous, the style in staminate flowers acutely 5-winged; pistil glabrous or provided with sparse, appressed trichomes to 0.05 mm long, the style with 5 lateral branches to 0.3 mm long. Fruits essentially globose or slightly obliquely ovoid, 2.1–2.8 × (1.4)1.9–2.4 cm, smooth, glossy, maturing green; pseudaril occurring as vertical band 0.5 cm wide around pyrene. Seeds with cotyledons contortuplicate. Fl (Sep, Oct, Nov, Dec), fr (Jan, Mar, Apr, Nov, Dec); in non-flooded forest.

Protium pilosum (Cuatrec.) Daly [Syn.: *Tetragastris pilosa* Cuatrec.] PL. 32g

Small trees or shrubs, 1–12 m tall. Bark light to dark brown, smooth, with raised dark lenticels. Leaves 0–5-jugate; petioles and rachises with dense to sparse, erect to ascending hairs to 0.05 mm long; lateral petiolules 0.05–0.3 cm long, the lateral pulvinuli inconspicuous or absent; leaflets narrowly elliptic, the apex gradually and narrowly long-acuminate and usually sparsely serrulate. Inflorescences shorter than petioles of subtending leaves, 0.4–1.1 cm long, the secondary axes poorly developed. Flowers 4-merous; calyx deeply cupular, 1–1.6(1.8) × (1.5)1.7–2.2 mm, the lobes 0.5–0.9 mm long, greatly exceeding disc; petals fused (30)40–60% length, the margins short-papillate; anthers 0.3–0.4 (staminate) or 0.4–0.5 (pistillate) mm long; pistil 1.25–1.35(1.75) mm long, glabrous, the style 0.3–0.5 mm long; pistillode glabrous. Fruits obliquely ovoid and usually markedly trigonous, 1.5–2.2 × 1.1–1.5 cm (when 1 pyrene develops), substipitate, maturing pale green to white, the base truncate above stipe; pyrenes thickly cartilaginous. Seeds with cotyledons plano-convex, distally uncinately incurved. Fl (Jun, Sep, Oct, Nov, Dec), fr (Jan, Feb, Mar, Dec); in non-flooded forest and secondary forest. *Encens, encens blanc, l'encens petites feuilles* (Créole).

Protium plagiocarpium Benoist

Trees, 9–12 m tall. Trunks with low buttresses. Bark gray-brown, smooth, grid-cracked. Leaves (0)1–2-jugate; lateral leaflets elliptic or less often (broadly) ovate, drying reddish, the apex broadly and obtusely acuminate. Inflorescences relatively slender, 1–7 cm long, longer than petioles of subtending leaves; pedicels clavate, shorter than flowers. Flowers 4-merous; calyx exceeded by nectary disc in both sexes; petals slightly cucullate, clawed at base, the apiculum conspicuous and strongly inflexed; anthers orange; nectary disc pubescent; pistillode depressed-ovoid, pubescent, usually a solid parenchymatous mass; pistil pubescent, the stigmas lateral, microscopically spiculate. Fruits markedly oblique, in dorsiventral view essentially lanceolate, in lateral profile semi-lanceolate, 2.1–2.8 × 0.7–1.3 cm (when 1 pyrene develops), maturing red, the base stipitate and cordate, the apex long-acuminate; pyrenes with site of funicle above middle; cotyledons plano-convex. Fl (Jul, Aug, Sep); in non-flooded forest. *Encens rouge* (Créole).

Protium robustum (Swart) D. M. Porter

Trees, 5–15(25) m tall. Bark dark gray. Branchlets robust, (5)7–12 mm diam at 2 cm from apex. Milky resin sometimes present. Leaves 3–5-jugate, 68–110 cm long; lateral leaflets usually oblong, (19)26–50 cm long, the apex abruptly and narrowly long-acuminate. Inflorescences pseudospicate, robust, shorter than petioles of subtending leaves, the staminate ones 12–21 cm long, the pistillate ones 5–7.2 cm long, the axes provided with dense, flexuous, appressed hairs to 0.3 mm long. Flowers sessile, 5-merous; pistillode and nectary disc not distinct (represented by a usually discoid ovariodisc); stamens exserted at anthesis; staminodia always 5, reaching stigmas; ovary substipitate, the stigmas large and globose. Fruits strongly obliquely ovoid, almost perpendicular to pedicels, 2–2.2 × 0.9–1.3 cm (with one pyrene developing), maturing red; pyrenes with site of funicle below middle. Seeds with cotyledons transversely folded but not distally incurved. Known only by sterile collections from our area; in non-flooded forest.

Protium sagotianum Marchand [Syn.: *P. insigne* (Triana & Planch.) Engl.] PL. 32h

Trees, 4–15(25) m tall. Trunks often with simple plank buttresses to 80 cm high. Outer bark gray or brown, smooth, sometimes with incomplete hoops. Resin whitish, sticky. Branchlets slender, 2–4 mm diam at 2 cm from apex. Leaves 3–4(5)-jugate, 19–38.5 cm long; lateral leaflets usually elliptic, sometimes oblanceolate or oblong, 6–19 cm long, the apex abruptly and narrowly long-acuminate. Inflorescences pseudospicate, usually shorter than petioles of subtending leaves, the staminate ones 4–10.5(17) cm long, the pistillate ones 1.5–4.2 cm long, the axes provided with thick, ascending to appressed hairs to 0.15 mm long, sometimes also with scattered acute appressed hairs to 0.15 mm long. Flowers sessile, 5-merous; stamens didynamous, exceeded by petals; staminodia 10 (rarely 5), not reaching stigmas; ovariodisc usually depressed-conical; ovary sessile, the stigmas discoid or obdeltoid. Fruits only slightly obliquely ovoid, 2.1–2.5 × 1.5–1.7 cm (when 1 pyrene develops), maturing red; mesocarp thick and fleshy. Seeds with cotyledons transversely folded and distally incurved, forming in profile a "6"-shaped figure. Fl (Aug, Oct, Nov), fr (Feb, Apr, May, Jun, Oct, Dec); in non-flooded forest.

Protium subserratum (Engl.) Engl. subsp. **subserratum**

Trees, (5)19–30 m tall. Trunks often angled and often with low plank buttresses. Bark red-brown, thin, smooth or sometimes rough due to raised horizontal lenticels, brittle, scalloped, shed in irregular plates. Resin milky white, gummy. Leaves (3)4–12-jugate; lateral petiolules lacking distal pulvinuli, 0.2–0.6(0.8) cm long; leaflets chartaceous, lacking trichomes but densely papillate abaxially, glossy adaxially, the margins entire; secondary and tertiary veins impressed. Inflorescences usually congested near branchlet apices, usually

Plates 25–32

Plate 25. BIGNONIACEAE. **a.** *Clytostoma binatum* (*Mori et al. 24173*), flowers. **b.** *Cydista aequinoctialis* (*Mori & Gracie 22099*), fallen flower. **c.** *Jacaranda copaia* (*Mori et al. 20895*), flower; note hairy stigma. **d.** *Jacaranda copaia* (*Mori et al. 20895*), lateral view of flower with corolla removed to show hairs on style. **e.** *Macfadyena unguis-cati* (*Mori & Gracie 18641*), liana in flower.

Plate 26. BIGNONIACEAE. **a.** *Memora moringifolia* (unvouchered, photographed at Sinnamary River, French Guiana), liana with newly flushed leaves. **b.** *Memora racemosa* (*Mori et al. 24141*), flower and buds. **c.** *Pithecoctenium crucigerum* (*Mori & Gracie 24228*), ants visiting glands on pseudostipules in leaf axil; note branched tendrils. **d.** *Schlegelia fuscata* (*Mori et al. 24739*), infructescence. [Photo by S. Mori] **e.** *Schlegelia fuscata* (*Mori & Gracie 18598*), flower and buds.

Plate 27. BIGNONIACEAE. **a.** *Stizophyllum inaequilaterum* (*Mori et al. 24242*), flower; note ruffled calyx. **b.** *Tanaecium nocturnum* (*Mori et al. 24765*), nocturnal flower. **c.** *Tynanthus pubescens* (*Mori et al. 23971*), flower.

Plate 28. BOMBACACEAE. **a.** *Ceiba pentandra* (unvouchered, photographed in Venezuela), buds and freshly opened flowers in evening; note nectar in flowers. **b.** *Ceiba pentandra* (unvouchered), crown of emergent tree when leafless. **c.** *Ceiba pentandra* (*Mori & Gracie 18379*), part of buttress of trunk with conical prickles.

Plate 29. BOMBACACEAE. **a.** *Huberodendron swietenioides* (*Mori et al. 22065*), inflorescence with flower and buds. [Photo by S. Mori] **b.** *Huberodendron swietenioides* (*Mori et al. 22254*), dehisced fruits, winged seed, and seedling. **c.** *Huberodendron swietenioides* (unvouchered), base of trunk with branched, plank buttresses.

Plate 30. BOMBACACEAE. **a.** *Matisia ochrocalyx* (*Mori et al. 22240*), flowers. **b.** *Pachira insignis* (*Mori et al. 23009*), fruit valves and seeds. **c.** *Quararibea duckei* (*Mori & Gracie 21090*), flower. **d.** *Quararibea spatulata* (*Mori et al. 22252*), flower.

Plate 31. BORAGINACEAE. **a.** *Cordia alliodora* (*Mori et al. 24798*), flowers. [Photo by S. Mori] **b.** *Cordia sagotii* (*Mori et al. 22748*), flowers and buds. **c.** *Tournefortia maculata* (*Mori et al. 21597*), flowers and buds. **d.** *Tournefortia ulei* (*Mori et al. 24200*), inflorescence.

Plate 32. BURSERACEAE. **a.** *Protium apiculatum* (*Mori & Gracie 21149*), infructescences. **b.** *Protium aracouchini* (*Mori et al. 23301*), inflorescence. **c.** *Protium cuneatum* (*Mori et al. 22922*), flowers and buds. **d.** *Protium demerarense* (*Mori et al. 24011*), flowers and buds. **e.** *Protium opacum* subsp. *rabelianum* (*Mori et al. 24722*), infructescence. **f.** *Protium pallidum* (*Mori et al. 24185*), flowers and buds. **g.** *Protium pilosum* (*Mori et al. 22816*), flowers and buds. **h.** *Protium sagotianum* (*Mori & Pipoly 15560*), fruits. [Photo by S. Mori]

Clytostoma binatum a.

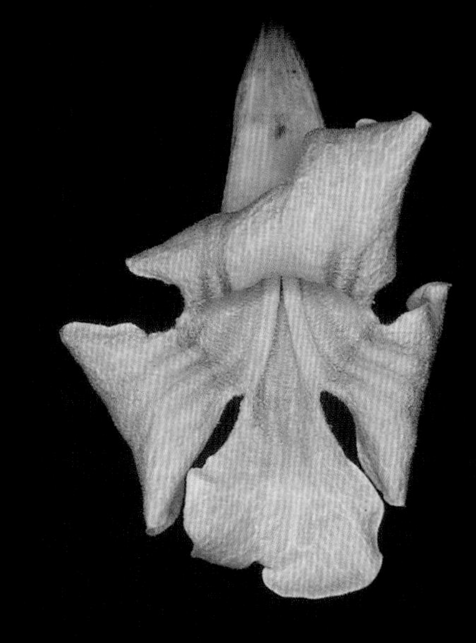

Cydista aequinoctialis b.

Jacaranda copaia c.

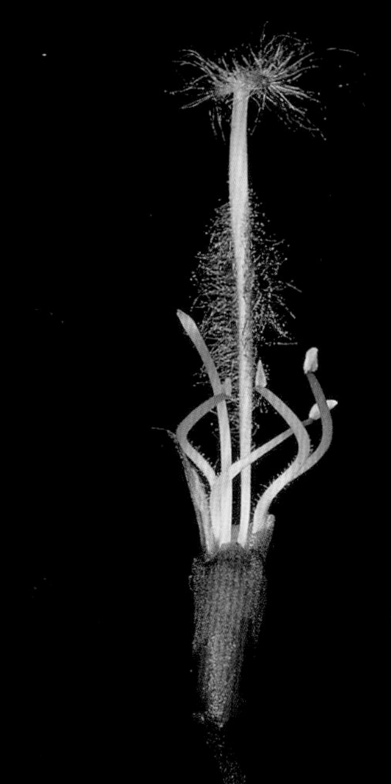

Jacaranda copaia d.

Macfadyena unguis-cati e.

Memora moringifolia 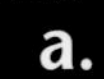

Memora racemosa **b.**

Pithecoctenium crucigerum **c.**

Schlegelia fuscata **d.**

Schlegelia fuscata **e.**

Plate 26

Stizophyllum inaequilaterum a.

Tanaecium nocturnum b.

c.

Tynanthus pubescens

Plate 27

Ceiba pentandra

a.

b.

Ceiba pentandra

c.

Huberodendron swietenioides

a.

b.

Huberodendron swietenioides

c.

Plate 29

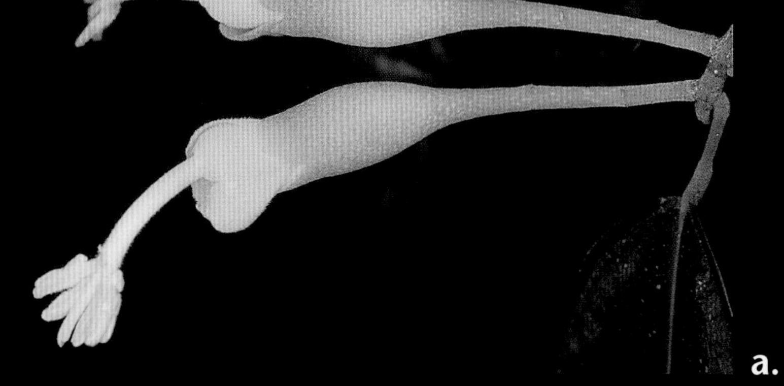

a.

Matisia ochrocalyx

Pachira insignis

b.

c.

Quararibea duckei

d.

Quararibea spatulata

BORAGINACEAE

Cordia alliodora 

b.

Cordia sagotii

Tournefortia maculata

c.

d.

Plate 21

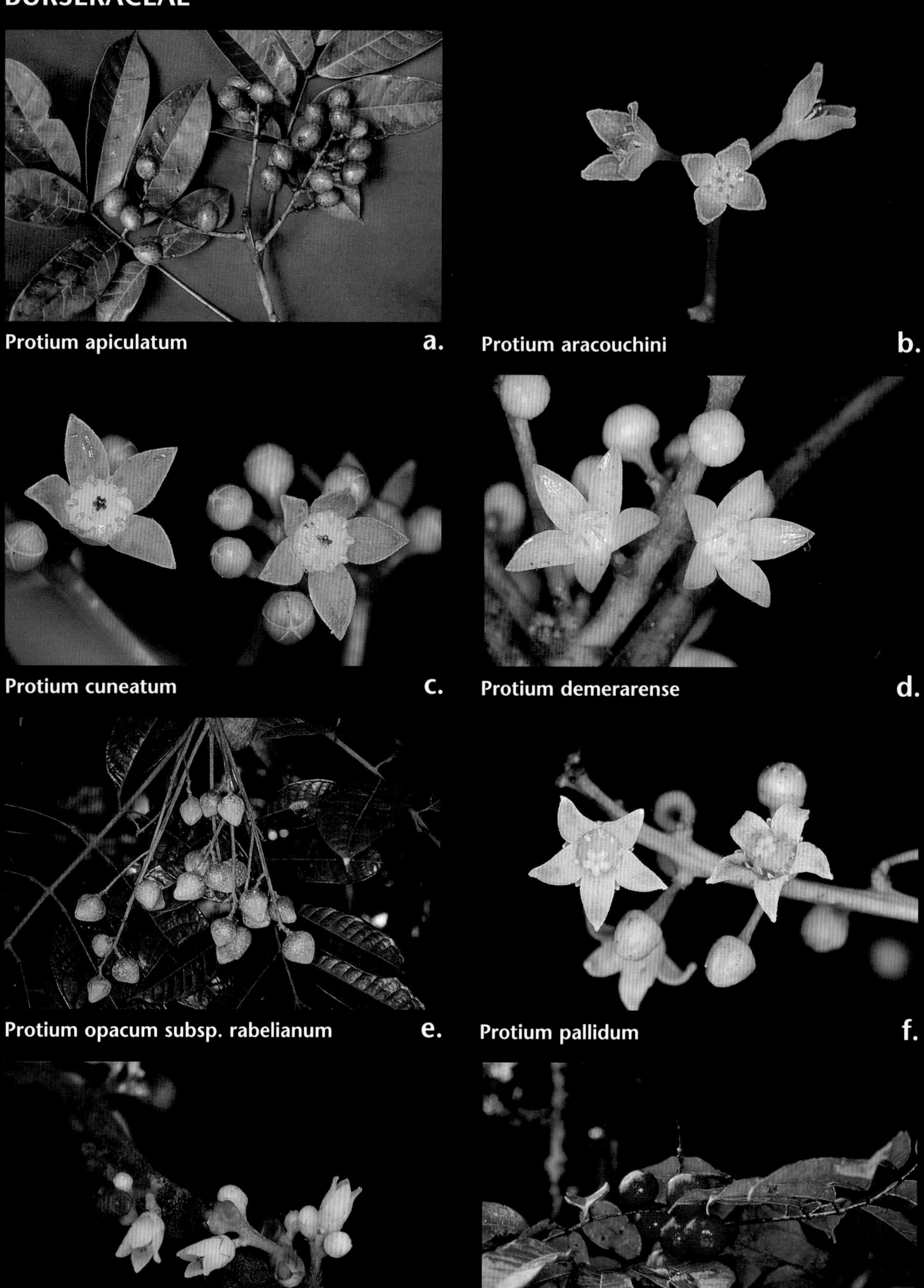

Protium apiculatum a.

Protium aracouchini b.

Protium cuneatum c.

Protium demerarense d.

Protium opacum subsp. rabelianum e.

Protium pallidum f.

Protium pilosum g.

Protium sagotianum h.

Plate 32 (Burseraceae continued on Plate 33)

shorter than petioles of subtending leaves, (2.2)3.5–14 cm long, the secondary axes poorly developed; pedicels 0–1.2(1.5) mm long. Flowers (based on material from outside our area) (4)5-merous, relatively large, 3–4(5) mm long when dry; petals (sub)erect at anthesis, lanate (densely provided with ascending, flexuous hairs 0.3–1 mm long) adaxially, the margins long-papillate; stamens didynamous; nectary disc pubescent; pistillode and pistil pubescent. Fruits slightly to markedly obliquely ovoid, (1.1)1.3–1.9(2.1) × (0.6)1.1–4 cm (when 1 pyrene develops), apparently maturing brown, the base sessile or substipitate; pyrenes bony. Seeds with cotyledons plano-convex. Fl (Oct); in non-flooded forest.

Protium tenuifolium (Engl.) Engl., s.l. [Syn.: *P. neglectum* Swart]

Trees, (5)10–34 m tall. Trunks often with simple or branched, spreading plank buttresses to 140 cm high. Bark reddish-brown, thin, scalloped, somewhat fissured, shed in small plates. Resin reportedly white. Branchlets robust, 1.9–4.4 mm diam at 2 cm from apex. Leaves 2–4(5)-jugate, (12.7)17.5–36.2(53) cm long; lateral leaflets oblong to broadly elliptic, 5.1–16 cm long, the apex gradually and broadly short-acuminate, the acumen 0.3–1.1 cm long. Inflorescences pseudospicate, robust, occasionally longer than petioles of subtending leaves, 4.5–17.8 cm long, the secondary axes often well developed, the axes with scattered to dense, thick, blunt, appressed or twisted trichomes to 0.1 mm long. Flowers 5-merous; stamens didynamous, exceeded by petals; staminodia 10 or 5, not reaching stigmas; ovariodisc usually depressed-conical; ovary sessile, the stigmas discoid or obdeltoid. Fruits only slightly obliquely ovoid, 1.7–2.4 × 1.3–1.7(2) cm (when 1 pyrene develops), maturing red, the base sessile; pyrenes bony. Seeds with cotyledons transversely folded and distally incurved, forming in profile a "6"-shaped figure. Fr (Feb); in non-flooded forest.

Protium trifoliolatum Engl. [Syn.: *Tetragastris trifoliolata* (Engl.) Cuatrec.]

Small trees or shrubs, 1.5–15 m tall. Trunks sometimes with simple plank buttresses. Bark gray to gray-brown, smooth but scaly, with hoop marks and scattered lenticels. Leaves 0–2(3)-jugate; lateral petiolules 0.2–1(1.5) cm long, the pulvinuli usually conspicuous; leaflets (broadly)lanceolate to elliptic; petioles and rachises with sparse stiff erect hairs 0.3–1(1.3) mm long, also with dense to sparse fine erect hairs to 0.05 mm long. Inflorescences usually shorter than petioles of subtending leaves, 0.4–1.8(2) cm long; bracts semi-clasping. Flowers 4-merous; calyx exceeded by nectary disc in staminate flowers (usually exceeding disc in pistillate flowers), 0.5–1(1.2) × 1.7–2.2(2.4) mm; petals usually basally connate, the margins short-papillate; anthers 0.5–0.65 mm long; pistil 1.1–1.4(1.75) mm long, the style (0.15)0.25–0.5 mm long. Fruits obliquely ovoid, (1.4)1.6–2 × 0.9–1.3 cm (when 1 pyrene develops), glabrous, maturing red, the base substipitate; pyrenes cartilaginous. Seeds with cotyledons plano-convex, the distal portion of posterior one often incurved. Fl (Jan, Mar, May, Oct), fr (Mar, Apr, May); in non-flooded forest.

TETRAGASTRIS Gaertn.

Small to large trees. Bark gray to brown, usually thick and fissured on older trees, shed in large irregular plates. Fresh resin clear, resin reportedly drying reddish. Leaves: lateral petiolules lacking pulvinuli, the terminal one distally pulvinulate; leaflet margins entire. Inflorescences paniculate, rarely reduced and almost fasciculate (some *Tetragastris panamensis*). Flowers functionally unisexual (plants dioecious), 4–5-merous; sinuses between calyx lobes sometimes irregularly split; petals (greenish-)yellow, cream-colored, or white, connate at least 1/2 length, only slightly inflexed or reflexed at anthesis, the apex inflexed-apiculate, the sinuses of equal depth; filaments usually strap-shaped, often pubescent abaxially, the anthers usually continuous with filament (base entire, not sagittate). Staminate flowers with nectary disc and non-functional ovary replaced by single, continuous, conical, parenchymatous structure. Fruits septicidally dehiscent, variously shaped but usually slightly to markedly oblique-ovoid, with as many lobes as developed locules, glabrous, usually red but occasionally green or brown at maturity, the surface smooth or less often lenticellate or rugose, drying wrinkled or smooth depending on whether mesocarp is fleshy or lignified, dehiscent by a valve (endocarp and mesocarp) that falls away at maturity from each developed locule, each pyrene suspended from fruit apex by inverted-V-shaped structure clasped just behind apex, the columella and valves pale red within; pyrenes cartilaginous, the ventral side with funicular scar of varying shape and size, often set in relief from rest of surface; pyrenes partly or completely enveloped in spongy, sweet, edible, white pulp ("pseudaril"). Seeds obliquely ovoid, acute, the testa papery; cotyledons fleshy, entire and plano-convex. Seedlings: germination epigeal, phanerocotylar, the eophylls opposite, simple.

1. Lateral leaflets (broadly) elliptic, less often ovate, the terminal one broadly elliptic to obovate (rarely elliptic), the apex abruptly (short-)acuminate or rarely rounded. Inflorescence bracteoles (broadly) ovate with acuminate apex, 1.5–3.5(4) mm long, semi-clasping (broad scars visible on infructescences), often foliose and often enclosing maturing buds. Calyx deeply cupular, 40–60% length of flower, exceeding ovariodisc in staminate flowers, the lobes (0.8)1–1.5(2) mm. Fruits with apex rounded. . . . . *T. altissima.*
1. Lateral leaflets (narrowly) elliptic to (narrowly) lanceolate or (narrowly) oblong-lanceolate, the terminal one oblanceolate to elliptic or rarely obovate, the apex usually long-acuminate. Inflorescence bracteoles lanceolate to ovate with acute (very rarely acuminate) apex, 0.7(1) mm long, occasionally semi-clasping but never nearly covering maturing buds. Calyx more shallowly cupular, 25–38% length of flower, exceeded by ovariodisc in staminate flowers, the lobes (0)0.3–1 mm. Fruits with apex (short-)acuminate or acute, less often rounded. . . . . *T. panamensis.*

Tetragastris altissima (Aubl.) Swart [Syns.: *Icica altissima* Aubl.; *Tetragastris phanerosepala* Sandwith]
FIG. 59, PL. 33b; PART 1: FIG. 9

Trees, to 35 m tall. Trunks sometimes with buttresses. Bark pale gray, somewhat scaly. Leaves 2–4-jugate, 17.8–60 cm long; lateral leaflets (broadly)elliptic, less often ovate, 7.5–18.5 × 2.8–7.7 cm, the terminal one broadly elliptic to obovate (rarely elliptic), 9–21 × 4.5–9 cm, glabrous, the apex abruptly (short-) acuminate or rarely rounded, the acumen 0.2–1.3 cm long. Inflorescences 5–16 cm long, the axes with scattered to dense, ascending to appressed hairs to 0.1 mm; bracteoles (broadly)

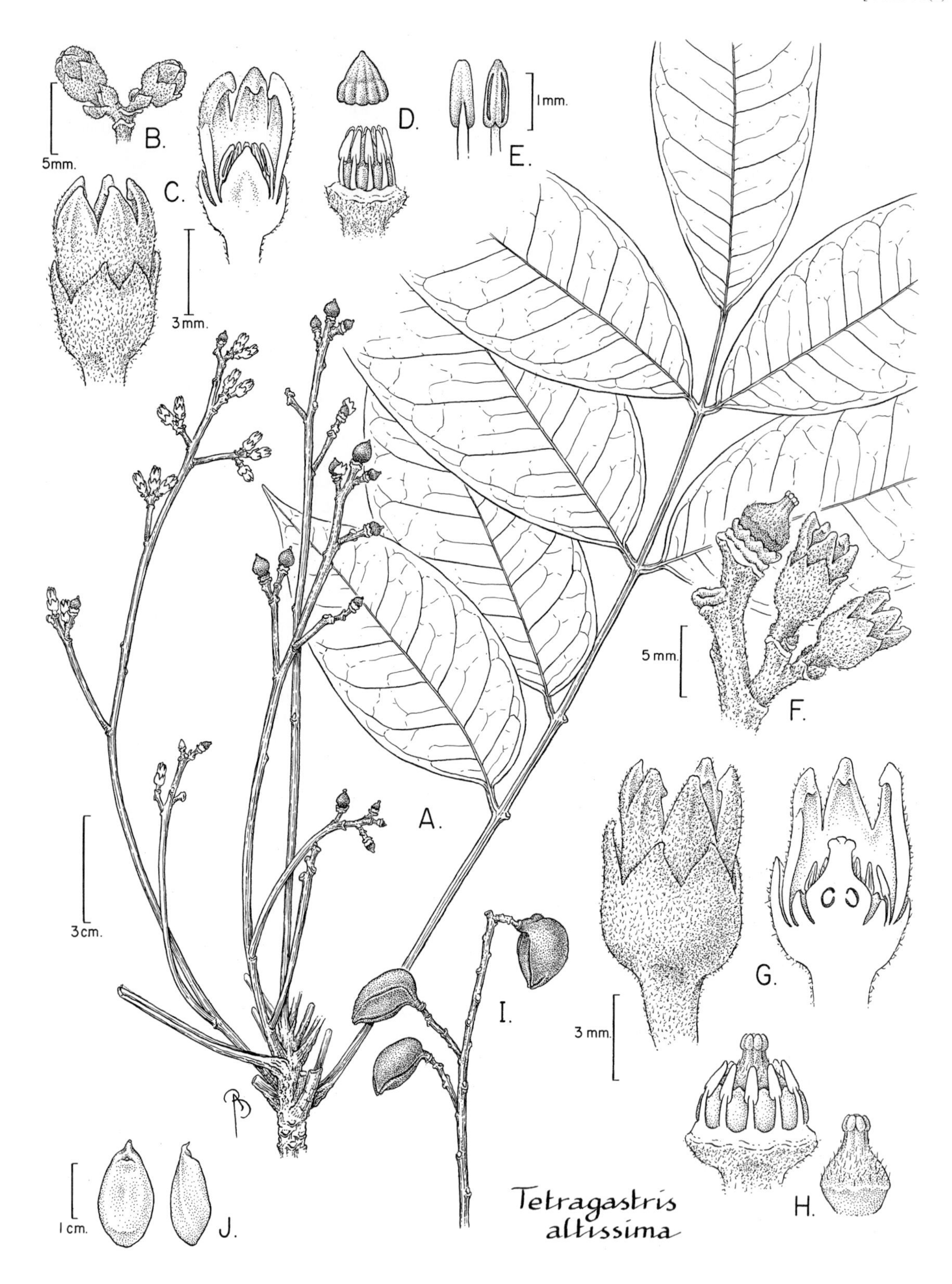

FIG. 59. BURSERACEAE. *Tetragastris altissima* (A, F–H, *Evans et al. 2442* from Suriname; B–E, *Evans et al. 2441* from Suriname; I, J, *Smith 2844* from Guyana). **A.** Branchlet apex with leaf and inflorescence with flowers and young fruits. **B.** Detail of staminate inflorescence with two flower buds showing semi-clasping bracts and their scars. **C.** Lateral view (left) and medial section (right) of staminate flower. **D.** Staminate flower with perianth removed (below left) and ovariodisc (above left). **E.** Abaxial (far left) and adaxial (near left) views of anthers. **F.** Detail of pistillate inflorescence with two flowers at anthesis and young fruit. **G.** Lateral view (left) and medial section (right) of pistillate flower. **H.** Pistillate flower with perianth removed (left) and pistil (right). **I.** Part of infructescence. **J.** Ventral (far left) and lateral (near left) views of pyrenes.

ovate with acuminate apex, semi-clasping, deciduous (broad scars visible on infructescences), often foliose and often covering maturing buds, 1.5–3.5(4) mm long; pedicels often broader than long, 0.8–1.2(2) mm long. Flowers 3.5–4(5.1) mm long; calyx deeply cupular, 40–60% length of flower, the lobes (0.8) 1–1.5 (2) mm long, exceeding ovariodisc in staminate flowers; fertile stamens 1–1.2(1.5) mm long, level with top of ovario-disc, usually provided with fine appressed hairs to 0.2 mm long abaxially, otherwise glabrous or apparently papillate, the staminodes level with stigmas; pistil 1.8–2.2 × 0.8–1.3 mm, glabrous or usually densely provided with ascending hairs to 0.3 mm long, substipitate, the style 0.2–0.5 × 0.4–0.6 mm. Fruits (2)2.2–2.8 × 1.1–1.8(2.4) cm (when 1 pyrene develops), the base usually substipitate, the apex rounded; pyrenes 1.3–1.8 × 0.8–1.2 cm. Fl (Sep, Oct, Dec), fr (Apr, May, Dec); common, in non-flooded forest and along streams.

Tetragastris panamensis (Engl.) Kuntze [Syns.: *Tetragastris stevensonii* Standl., *Tetragastris paraensis* Cuatrec.]

Trees, to 40 m tall. Trunks with simple plank buttresses. Bark gray, smooth, scaly, with hoop marks. Lateral leaflets (narrowly)-elliptic to (narrowly)lanceolate or (narrowly)oblong-lanceolate, the terminal one oblanceolate to elliptic or rarely obovate, the apex usually gradually long-acuminate. Inflorescence bracts lanceolate to ovate, the apex acute (very rarely acuminate), occasionally semi-clasping but never nearly covering maturing buds, those on primary axes to 0.8(1.7) mm long; bracteoles to 0.7(1) mm; pedicels (0.3)0.5–1.5 mm long. Flowers: calyx shallowly cupular, 25–38% length of flower, the lobes (0)0.3–1 mm long, exceeded by usually glabrous ovariodisc (staminate flowers); ovary pubescent (pistillate flowers). Fruits 2–2.8 × 1.5–1.7 cm, the apex (short-)acuminate or acute, less often rounded; pedicels 2.5–4(5) × 2–3.2 mm; pyrenes 1.4–1.7 × 0.8–1 cm. Fl (Sep); in non-flooded forest.

TRATTINNICKIA Willd.

Medium-sized to large trees, less often small trees, rarely shrubby. Trunks sometimes with shallow, thick plank buttresses. Resin clear or white, watery (sometimes reported as oily), slightly sticky, drying yellowish or white and hard, flammable in some species. Leaves: petioles with inverted vascular bundles in medulla; pulvinuli usually apparent on lateral leaflets and almost always on terminal one; leaflets coriaceous, slightly to markedly asperous in most species ab- and/or adaxially, the margins entire. Inflorescence axes tinged red in some species. Flowers 3-merous, unisexual; calyx often irregularly 3-fid, light green (tinged with red in some taxa); petals connate usually <1/2 length, (wine-)red or green, always with at least some retrorsely descending or appressed hairs abaxially, often densely invested with minute capitate glandular trichomes adaxially; stamens 6, ± equal, inserted at base of nectary disc or less often on nectary disc, the filaments much shorter than anthers, the anthers sagittate, sometimes red; pistillode reduced to conical parenchymatous structure surmounting nectary disc and often continuous with it; pistil 2–3-locular, the stigmas (sub)sessile, the style mostly obsolete, with as many lobes as locules. Fruits compound drupes, globose to depressed-ovoid, maturing blue-black to (purplish-)black or brown, the apex often acuminate, the exocarp thin and drying wrinkled, the mesocarp oily and resinous, the endocarp of 2–3 connate pyrenes, bony, rugose, 2–3-lobed at apex. Seeds: usually only one developing per fruit; cotyledons palmatifid, contortuplicate.

1. Lateral leaflets narrowly to broadly lanceolate, less often oblong-lanceolate to oblong-elliptic or rarely ovate, usually glossy adaxially, the base acute to obtuse or less often narrowly truncate, the apex narrowly acute to gradually and narrowly long-acuminate. Primary bracts subtending flowers subulate to ovate, not longer than mature buds. Calyx with dense appressed hairs to 0.2 mm long adaxially; corolla urceolate (lobes somewhat reflexed), red to orange. *T. boliviana.*
1. Lateral leaflets often broadly oblong, less often broadly oblong-lanceolate to oblong-ovate, often dull, the base slightly cordate or less often broadly truncate, the apex abruptly acuminate. Primary bracts subtending flowers cordiform and usually exceeding mature buds. Calyx glabrous adaxially; corolla essentially tubular, green. *T. rhoifolia.*

Trattinnickia boliviana (Swart) Daly [Syn.: *Trattinnickia lawrancei* Standl. var. *boliviana* Swart]

Trees, 10–23 m tall. Bark gray, finely and densely fissured. Leaves: lateral leaflets narrowly to broadly lanceolate, less often oblong-lanceolate to oblong-elliptic or rarely ovate, usually glossy adaxially, both surfaces sometimes with scattered raised glandular punctations, midrib with scattered appressed hairs to 0.15 mm long and sometimes with sparse fine erect hairs abaxially, the base acute to obtuse or less often narrowly truncate, the apex narrowly acute to gradually and narrowly long-acuminate, the margins often slightly revolute, with scattered fine ascending hairs to 0.2 mm long. Inflorescences usually terminal, (8)11–25(30) cm long, the secondary axes well developed, with dense glandular trichomes and fine erect hairs to 0.15 mm long; primary bracts subtending flowers subulate to ovate, not longer than mature buds, 2.6–3 mm long; pedicels (2)2.2–4.5 mm long. Flowers 3–5 mm long; calyx 2.7–3.3 mm long, the lobes 2.5–2.8 mm long, with dense glands and fine erect and sometimes appressed hairs to 0.15 mm long abaxially, also with dense appressed hairs to 0.2 mm adaxially; corolla urceolate, fleshy, red to orange, the lobes 1.5–2.5 mm long, somewhat reflexed, with dense retrorsely appressed hairs to 0.2 mm long abaxially and dense retrorse hairs only near apex adaxially. Fruits broadly ovoid, ca. 1.2 × 0.9–1 cm, maturing black; pedicels 3.5–7 mm long. Known only from sterile collections from our area.

Trattinnickia rhoifolia Willd. [Syn.: *Trattinnickia rhoifolia* Willd. subsp. *sprucei* Engl.] FIG. 60, PL. 33c

Trees, 16–33 m tall. Trunks with low, wide buttresses. Bark brown, scaly, relatively thick, deeply fissured. Leaves: lateral leaflets often broadly oblong, less often broadly oblong-lanceolate to oblong-ovate, both surfaces often dull, markedly asperous, often with raised glandular punctations, the base slightly cordate to truncate, the apex abruptly acuminate. Inflorescences 8–20 cm long, the axes with sparse glandular trichomes and dense suberect

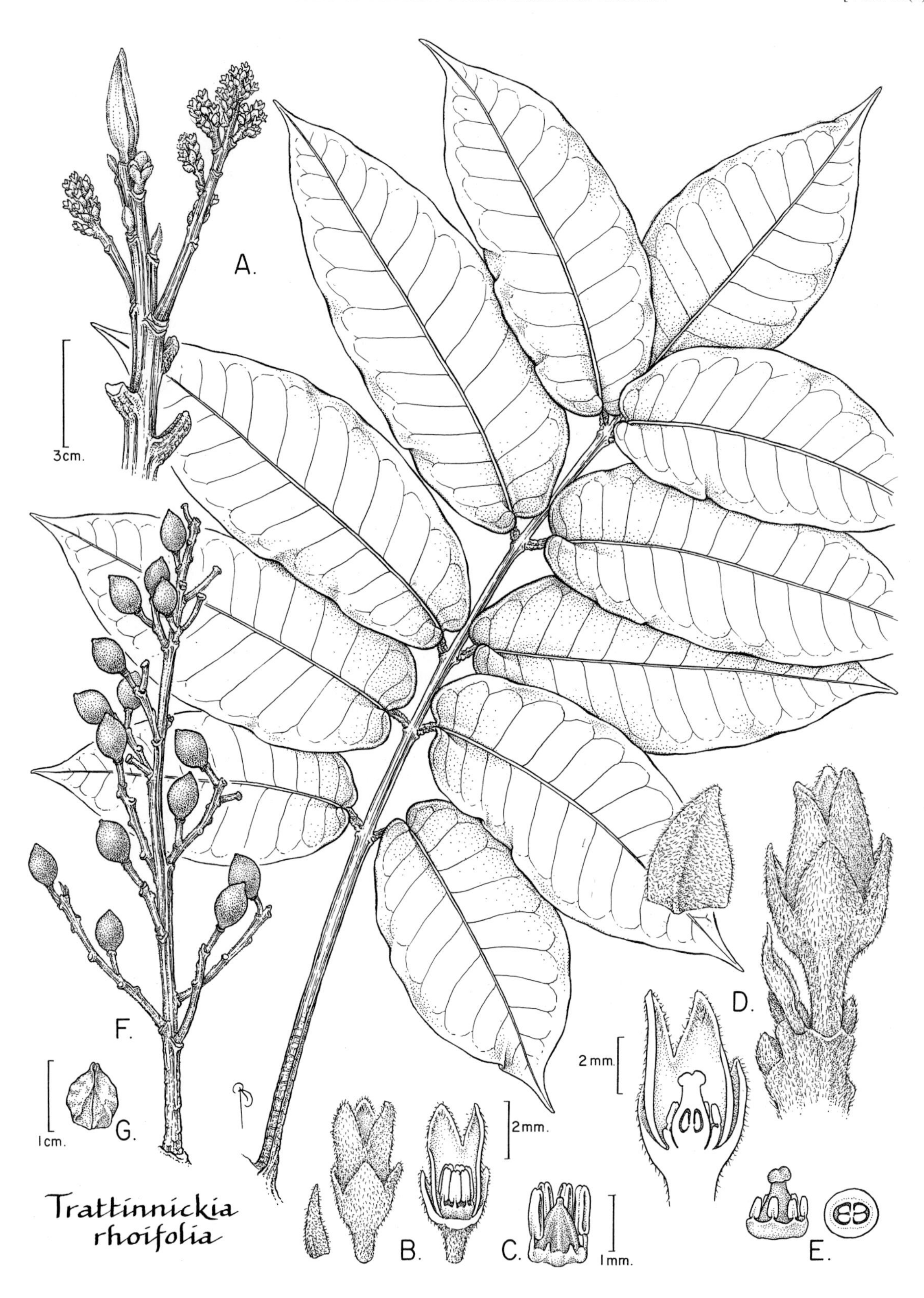

FIG. 60. BURSERACEAE. *Trattinnickia rhoifolia* (A, D–G, *Mori & Gracie 18337*; B, C, *Silva 1912* from Amazonas, Brazil). **A.** Leaf (right) and branchlet apex with inflorescences (left). **B.** Bracteole (far left), lateral view (near left), and longitudinal section (right) of staminate flower. **C.** Androecium with two stamens removed to show ovariodisc. **D.** Lateral view of pistillate flower on inflorescence (right), bracteole (above left), and medial section (below left) of pistillate flower. **E.** Lateral view of pistil and staminodial ring (left) and transverse section of ovary (right). **F.** Infructescence. **G.** Pyrene.

hairs to 0.25 mm long; bracts subulate, to at least 5 mm long, the primary bracteoles subtending flowers cordiform and usually larger than mature buds; pedicels ca. 2.2 mm long. Flowers ca. 4.5 mm long; calyx (2)2.3–2.5 mm long, the lobes 1–1.3 mm long, glabrous abaxially; corolla ± tubular, green, the lobes 1.2–1.4 mm long, with some retrorsely descending hairs to 0.25 mm long adaxially. Fruits ovoid (less often slightly obovoid), 1.3–1.5 × 1 cm, maturing brown, the apex broadly acute to broadly acuminate; pedicels 4.5–10 mm long. Fl (Jun), fr (Jun, immature); in non-flooded forest.

CACTACEAE (Cactus Family)

Beat Ernst Leuenberger

Perennials of diverse habit, usually succulent terrestrial trees, shrubs, climbers, epiphytic shrubs, short cylindric to depressed-globose succulents or woody terrestrial shrubs or climbers. Leaves usually rudimentary or vestigial, rarely large and flattened (*Pereskia*) or terete and caducous (*Opuntia*); axillary buds (termed areoles) usually developing cushion-like indumentum and leaves usually transformed into spines. Inflorescences terminal or arising from areoles. Flowers 0.5–30 × 0.5–30 cm, usually actinomorphic, bisexual; receptacle (pericarpel and epigynous hypanthium) pronounced, enclosing ovary, often bearing bracts or fleshy bract-scales and areoles, often elongated into conspicuous receptacular tube bearing bract-scales which intergrade with tepals; tepals five to numerous in a graded series, the inner ones usually showy; stamens numerous, the anthers 2-locular, dehiscing longitudinally; ovary inferior (half-inferior in some *Pereskia*), unilocular, 3–20-carpellate, the style usually long, the stigma 3–20-lobed; ovules numerous. Fruits indehiscent or variously dehiscent, juicy berries, the outer wall formed by receptacle, naked or bearing bract-scales and areoles with or without tomentum, bristles, or spines. Seeds numerous, black or, in *Opuntia*, covered by bony aril, the testa smooth or variously sculptured; embryo curved to nearly straight, the cotyledons foliaceous, reduced, or vestigial.

Leuenberger, B. E. 1997. Cactaceae. *In* A. R. A. Görts-van Rijn & M. J. Jansen-Jacobs (eds.), Flora of the Guianas Ser. A, **18:** 1–63. Royal Botanic Gardens, Kew.

Trujillo, B. 1997. Cactaceae. *In* J. A. Steyermark, P. E. Berry & B. K. Holst (gen. eds.), Flora of the Venezuelan Guayana **3:** 732–749. P. E. Berry, B. K. Holst & K. Yatskievych (vol. eds.). Missouri Botanical Garden, St. Louis.

1. Stems triangular, armed with spines. . . . . *Selenicereus*.
1. Stems terete or flattened, unarmed.
   2. Terrestrial. Stems with segments obovate, flattened, bearing areoles throughout, the margins not crenate or serrate. Flowers with cupular, green receptacle bearing numerous areoles and glochids. Seeds entirely covered by pale-colored aril, never black. . . . . *Opuntia*.
   2. Epiphytic or epilithic. Stems with segments terete or flattened, not bearing areoles throughout, when flattened, the margins crenate to serrate. Flowers with long and narrow or short and inconspicuous receptacle, not bearing numerous areoles. Seeds without pale colored aril, black or nearly so.
      3. Stems regularly segmented, terete, branching at tip. Flowers <1 cm long, without slender tube. Fruits <1 × 1 cm. . . . . *Rhipsalis*.
      3. Stems not conspicuously segmented, flattened (often with terete base), branching laterally. Flowers to 29 cm long, with slender tube. Fruits to 9 × 5 cm.. . . . . *Epiphyllum*.

EPIPHYLLUM Haw.

Epiphytic or epilithic shrubs with flattened, not conspicuously segmented, leaf-like stems, main shoots terete or rarely triangular at base, the margins crenate, serrate or lobed. Leaves lacking except for thick fleshy cotyledons; areoles minute, spines absent, but weak bristles often present on terete stem bases and on seedlings. Flowers nocturnal, funnelform or salverform; receptacular tube long and slender with few bract-scales and usually naked areoles; tepals shorter than tube, spreading, white to pale yellow or pink-tinged; stamens exserted; stigma-lobes linear. Fruits berries, ovoid or oblong, somewhat ridged or angled, naked or with few (fleshy) bract-scales, spineless. Seeds obovate to reniform, with interstitial pits, black.

Epiphyllum phyllanthus (L.) Haw. PL. 33d–f

Epiphytic or epilithic shrubs. Stems long, flattened, leaf-like, branching from base and from lateral areoles of main stem; main shoots with terete base, the terete part short or up to 100 cm long, middle and apical part of main stems flattened, 30–80(100) × (1.5)4–6(8) cm, ca. 1–2 mm thick when fresh, coriaceous; distal branches ca. 20–50 × (2)3–6(7) cm, the apex obtuse to truncate, the margins shallowly serrate with rounded teeth, or strongly to shallowly crenate, sometimes variable on same plant; areoles in axils of teeth, spines absent, but some weak bristles (5–10 mm long) present on seedlings and terete stem bases. Flowers salverform, 19–29 × 5(11) cm; pericarpel and receptaclar tube very slender, the tube 12–24 cm × 2–4 mm in dried specimens (ca. 4 mm diam. when fresh), green, yellowish-green, reddish to brilliantly purplish red in upper part, with few scattered, triangular bract-scales 5–10 × 1–2 mm, the areoles minute; perianth segments much shorter than tube, 1.5–3 cm long, spreading to reflexed, the outer ones pink-tinged to reddish, the inner ones cream-colored to white. Berries ovoid or oblong, 4–7(9) × 2–5 cm, somewhat ridged or angled, purplish red or partly greenish red, with fine whitish spots, the pericarp thin, purplish, white inside. Seeds 3–4 × 2 mm, black. Fr (Jan, Jun); high in tree canopies and at lower levels on inselbergs.

OPUNTIA Mill.

Plants shrubby to tree-like with trunk formed by stem-segments, or low to creeping. Stems thick and succulent, flattened or cylindric, usually clearly segmented. Leaves linear terete to subulate, succulent, usually small and caducous; areoles with short tomentum; spines, if present, barbed at least at tip, the small barbed bristles (glochids) formed above spines in center or upper part of areole. Inflorescences lateral or rarely terminal. Flowers diurnal; receptacle stout, green, with numerous caducous bracts, the areoles usually numerous and producing glochids; tepals numerous, spreading or erect; stamens included or exserted, the filaments white or colored and petal-like; style stout, sometimes with collar-like rim toward base closing nectar-chamber. Fruits globose to obovoid or clavate, usually fleshy, indehiscent, with numerous areoles and often with glochids, sometimes spiny, apically with broad, often depressed to deep umbilicus left by abscission zone (i.e., where perianth remnants, stamens and style are shed). Seeds numerous, usually compressed, covered by bony aril.

Opuntia cochenillifera (L.) Mill.

Succulent shrubs, to 2 m tall. Stem regularly segmented, the segments obovate to obliquely obovate, flattened, to ca. 22 × 8 cm, thick and fleshy, green. Leaves terete, early deciduous, usually lacking on herbarium specimens, 3–4 × 1 mm; areoles few, prominent, ca. 3–4 cm apart, 2–3(5) mm diam., with dense, short, white to brownish tomentum, spines lacking, glochids inconspicuous, 2–3(5) mm long, in dense fascicles barely emerging from tomentum. Flowers, including exserted stamens and style, ca. 7 × 2 cm; receptacle green, 3 × 2 cm, the areoles numerous, 1–1.5 mm diam., bearing dense tomentum and glochids 2–3 mm long; outer tepals ca. 3–5, triangular-ovate, erect, 6–8 × 4–8 mm, the inner tepals ca. 10–12, triangular-ovate to ovate, acute, erect, 10–15 × 6–10 mm, pink to purplish; stamens numerous, clustered around style, the filaments purplish, the anthers yellow; style green. Fruits (not observed in material from the Guianas) ca. 5 cm long, red. Seeds 5 × 3 mm, cream-colored. Probably cultivated and not part of native flora. *Raquette* (Créole).

RHIPSALIS Gaertn.

Epiphytic, rarely epilithic, subshrubs or shrubs, the plants often pendulous. Stems terete, angled, or flattened and leaf-like, segmented; ramification at nodes, bifurcate to verticillate or subverticillate in whorls; stem-segments with minute, usually fugacious scale-leaves; areoles minute, with short wool and often with 1–2 minute, nectar-secreting spines; regular spines absent or weak and bristly; seedling and juvenile stems often with weak bristles and appearing very different from mature stems. Inflorescences with one to few flowers per areole. Flowers small, 0.5–2 cm long, white or cream-colored; receptacle naked or rarely with 1–2 bract-scales, green; receptacular tube absent or inconspicuous; outer tepals few, scale-like, the inner tepals often ca. 5, ovate to obovate, delicate, thin, spreading to recurved; stamens numerous; style slender, the stigma-lobes ca. 5. Fruits globose or elliptic to oblong, white or pink to purplish, juicy, pulp sticky. Seeds small, few to numerous, ovate to oblique-obovate or oblong, flattened, reddish brown to black, smooth.

Rhipsalis baccifera (J. S. Muell.) Stearn

Epiphytic subshrubs, the plants pendulous, succulent, often 0.5 to 1 m, but sometimes to over 2 m long. Stems terete, sometimes with short aerial roots, the first segment of main branches to 60 cm × 2–4(7) mm; ramification in pairs, or verticillate with 3–9 branches from tip of older branches, to 7 or more whorls in regular or irregular sequence, the ultimate branchlets ca. 5–17 cm × ca. 1 mm; areoles ca. 5–10 mm apart, spineless; scale leaves minute, triangular to broadly ovate, with an appressed apex, 0.2 × 0.5 mm, drying papery, the areoles of seedlings and juvenile plants with numerous straight and spreading bristles 5–8 mm long. Flowers lateral, self fertile, ca. 6 × 6 mm; pericarpel subcylindric, 3 × 1.5–2 mm, usually naked, pale green; outer tepals ca. 5, 0.5 × 0.5 mm, the inner tepals ca. 5, ovate, 2.5–3 × 1.5 mm, white to translucent greenish-white; stamens ca. 20, the anthers yellow; stigma-lobes 3–4. Berries oblong or subglobose, 4–8 × 3–7 mm, smooth, milky white or pale pink (particularly toward apex). Seeds to over 50, obliquely oblong, 1.2–1.4 mm long, brownish black. Fr (Jun); in canopies of tall trees.

SELENICEREUS Britton & Rose

Epiphytic or epilithic, scandent shrubs, often with aerial roots. Stems irregularly segmented, (2)3-winged or 3-angled to many-ribbed; areoles, at least on new growth, tomentose; spines bristly, acicular, or short subulate, often additional bristly spines present. Flowers nocturnal (rarely diurnal), funnelform, rarely salverform, large to very large; pericarpel often tuberculate, bearing inconspicuous or small, fleshy, triangular bract-scales and tomentose areoles with hair-like spines or acicular spines; receptacular tube slender to stout, with small to large bract-scales and conspicuous hair-like spines, bristles, or acicular spines; perianth shorter than tube, the outer tepals greenish, yellow, or brownish-tinged, the inner tepals white; stamens very numerous, the filaments white, the anthers included; stigma lobes numerous, linear to subulate. Fruits medium to large, globose to ovoid, red or yellow, fleshy berries, tuberculate or smooth, the areoles spiny, the pulp fleshy, white. Seeds very numerous, small, black.

Selenicereus extensus (DC.) Leuenb.

Scandent shrubs. Stems to several meters long, irregularly segmented, triangular, rarely some segments 4-angled, 8–30 mm diam., green, often with aerial roots; areoles usually elevated, ca. 2–3 cm apart, with white to grayish brown tomentum, the spines 2–8, short, subulate-conical, to 5 mm long, 2 mm thick at base, brown, the spines on juvenile plants more numerous (to ca. 20) and slender, acicular to bristly, to 10 mm long. Flowers not seen in material from central French Guiana. Fruits unknown. Known from one collection (*Mori et al. 23768*, NY) gathered on a granitic rock outcrop.

CAESALPINIACEAE (Caesalpinia Family)

Rupert C. Barneby and Scott V. Heald

Trees, shrubs, lianas, and infrequently (weedy) herbs. Plants pubescent with simple hairs or glabrous, eglandular, but some furnished with petiolar glands. Stipules usually present, often caducous. Leaves pinnate or rarely bipinnate, either pari- or imparipinnate, alternate, spirally or distichously inserted, pulvinate; leaflets 1 to many pairs, either opposite or alternate. Inflorescences either axillary and lateral or assembled into terminal panicles, rarely cauliflorous, either racemose or spicate, the individual flowers bracteate and often bibracteolate. Flowers bisexual; perianth either hypogynous or perigynous, zygomorphic, irregular, or infrequently actinomorphic (*Acosmium* in our area); sepals 4 or 5, imbricate in bud, either free or partly united into tube; petals commonly 5 or 3, the adaxial (to inflorescence axis) one then interior in bud, but petals sometimes 1 or 0; stamens commonly (9)10, generally all antheriferous with free filaments, but often some stamens staminodial, or diadelphous, or only 2 or 5; ovary with solitary carpel, stipitate or sessile; placentation adaxial, the ovules 2 to many; embryo straight. Fruits legumes, diverse in form, compression, and texture, often longitudinally winged, the dehiscence either inert, or elastic, or none.

Barneby, R. C., B. Stergios, R. S. Cowan, P. E. Berry, J. L. Zarucchi, R. P. Wunderlin, D. M. Kearns, M. F. da Silva A. S. Tavares, D. Velázquez, N. Xena & G. A. Aymard C. 1998. Caesalpiniaceae. *In* J. A. Steyermark, P. E. Berry & B. K. Holst (gen. eds.), Flora of the Venezuelan Guayana **4:** 1–121. P. E. Berry, B. K. Holst & K. Yatskievych (vol. eds.). Missouri Botanical Garden Press, St. Louis.

Cowan, R. S. & J. C. Lindeman. 1989. Caesalpiniaceae. *In* A. R. A. Görts-van Rijn (ed.), Flora of the Guianas Ser. A, **7:** 1–167. Koeltz Scientific Books, Koenigstein.

Refer to glossary for specialized terms used in describing legumes, e.g. leafstalk = petiole + rachis; areolate = with small surface regions bounded by reticulate venation.

1. Leaves bipinnate. Inflorescences of long, dense spikes. Flowers small, almost actinomorphic; stamens 10, alternately fertile and staminodial. . . . . . *Dimorphandra.*
1. Leaves pinnate, 2- to many-foliolate. Inflorescences not as above. Flowers and stamens not as above.
    2. Flowers actinomorphic; petals 5, the vexillum not or scarcely differentiated from other petals; stamens 5. Fruits indehiscent, linear-elliptic in profile, stipitate, planocompressed, inconspicuously winged along ventral service. Seeds mostly 1, inserted near middle of fruit. . . . . . *Acosmium.*
    2. Flowers zygomorphic; petals commonly 10, less frequently 5 or fewer, the vexillum differentiated from other petals. Fruits and seeds not as above, i.e., fruits either dehiscent, sessile, biconvex, or with several seeds.
        3. Leaves paripinnate, the rachises often produced between distal pair of leaflets into appendage, but not into leaflet.
            4. Leaflets of each leaf >2 pairs; blades distinct.
                5. Plant in flower.
                    6. Anthers dorsifixed, versatile.
                        7. Phyllotaxy distichous; petioles lacking; leaflets in ≥20 pairs. Bracteoles united into 2-dentate sheath clasping hypanthium of flower. . . . . . *Elizabetha.*
                        7. Phyllotaxy spiral; petioles present; leaflets in 2–7 pairs. Bracteoles free from each other, deciduous.
                            8. Petal 1. . . . . . *Eperua.*
                            8. Petals 5.
                                9. Leafstalks with adaxial groove bridged at insertion of leaflets, but bridge eglandular. . . . . . *Tachigali.*
                                9. Leafstalks with adaxial groove bearing gland at insertion of leaflets. . . . . . *Batesia.*
                    6. Anthers basifixed, not versatile.
                        10. Glands either at base of petioles or between pairs of leaflets. Androecia with 3 long abaxial stamens simply incurved, their anthers beaked, the beak dehiscent by terminal pore or slit. . . . . . *Senna.*
                        10. Glands absent. Androecia with 3 long abaxial stamens sigmoidally arched, curved downward proximally and incurved distally, their anthers beakless, dehiscent by slits. . . *Cassia.*
                5. Plant in fruit.
                    11. Fruits either cylindric or plumply bifacial, 30–70 cm long, not winged, the cavity septate, divided by horizontal partitions into 1-seeded locules packed with foetid pulp. . . . . . *Cassia.*
                    11. Fruits variously compressed but never more than 30 cm long, winged or not winged, if body subcylindric then 1-locular, i.e., the cavity not divided into 1-seeded locules.
                        12. Phyllotaxy distichous; petioles absent. . . . . . *Elizabetha.*
                        12. Phyllotaxy spiral or irregular; petioles present.
                            13. Petiolar glands present. Fruits 3–4.5 cm long, the valves 2-keeled lengthwise. Seeds red. . . . . . *Batesia.*

13. Petiolar glands absent. Fruits mostly >5 cm long, the valves never laterally keeled. Seeds brown or black.
14. Fruits sessile or subsessile, oblong-elliptic in profile, strongly compressed, indehiscent. Seeds 1–2. . . . . *Tachigali.*
14. Fruits stoutly stipitate and broader than linear, or if sessile, then linear, dehiscent. Seeds usually >2.
15. Leaflets 2–4(7) pairs. Fruits 5–9 cm wide, woody. Seed coats not areolate. . . . . *Eperua.*
15. Leaflets (2)7–14 pairs. Fruits always <3 cm wide. Seed coats areolate. . . . *Senna.*
4. Leaflets of each leaf 1 pair; blades either free or united part-way into bilobed blade (in some *Bauhinia*).
16. Lianas. Stems often flattened and ladder-like, climbing by random tendrils. Leaflet blades palmately 3–6-veined from pulvini. . . . . *Bauhinia.*
16. Trees. Stems not flattened and ladder-like, without tendrils. Leaflet blades not palmately 3–6-veined from pulvini.
17. Petal 1. Fruits planocompressed, ≥7 cm long, woody. . . . . *Macrolobium.*
17. Petals 5. Fruits various, but if planocompressed, then <3.5 cm long.
18. Leaflet blades asymmetrically decurrent onto pulvinule. Styles filiform. Fruits plumply biconvex or subcylindric, indehiscent. Seeds not arillate. *Hymenaea.*
18. Leaflet blades subsymmetrically decurrent onto pulvinule. Styles capped with peltate stigma. Fruits planocompressed, dehiscent. Seeds arillate. *Peltogyne.*
3. Leaves imparipinnate, the rachises terminating in single leaflet.
19. Plant in flower.
20. Stamens 2; anthers basifixed.
21. Inflorescences cymose-paniculate. Petals absent; anther sacs 1-locular. . . . . *Dialium.*
21. Inflorescence racemose-paniculate. Petals 3; anther sacs 4–10-locular. . . . . *Dicorynia.*
20. Stamens ≥4, mostly 10; anthers variously attached.
22. Sepals narrowly lance-acuminate; anthers narrowly flask-shaped, 8.5–12 mm long, basifixed. . . . . *Martiodendron.*
22. Sepals elliptic, ovate, or obovate; anthers long-ellipsoid, 2–6 mm long, dorsifixed.
23. Panicle branches bearing flowers distichously; pedicels lacking. Hypanthia absent; petals absent. . . . . *Copaifera.*
23. Panicle branches bearing flowers spirally; pedicels present. Hypanthia campanulate; petals 5.
24. Petiolar glands absent. Petals ca. 23 mm long, twice as long as sepals. . . . . *Recordoxylon.*
24. Petiolar glands present at insertion of most pairs of leaflets. Petals 4–7.5 mm long, not or scarcely longer than sepals.
25. Trunks not fenestrate. Sepals 5–6 mm long; petals 6–7.5 mm long. . . . . *Batesia.*
25. Trunks fenestrate. Sepals 3.5–4 mm long; petals 4–4.5 mm long. . . . . . *Vouacapoua.*
19. Plant in fruit.
26. Fruits planocompressed, ≥5 cm long and more than twice as wide as long, winged along one or both sutures, the wing ≥2 mm wide.
27. Fruits conventional pods, 3–4 times as long as wide, winged along adaxial surface only, inertly dehiscent. Seeds 4–5. . . . . *Recordoxylon.*
27. Fruits samaroid, not more than twice as long as wide, winged along both sutures, indehiscent. Seeds 1(2).
28. Trees unbuttressed or with poorly developed buttresses. Bark rough, the slash exuding slow-flowing, gelatinous sap after cutting. Fruits 5–8 × 3–4 cm, truly sessile, the cavity extended to very base, and including the sutural wings. . . . . *Dicorynia.*
28. Trees with well developed buttresses. Bark smooth, the slash without sap. Fruits (valves and wings) 12–18 × 4.5–6 cm, technically stipitate, but sutural wings decurrent length of stipe, i.e., proximal to pod-cavity. . . . . *Martiodendron.*
26. Fruits either drupaceous or, if compressed, discoid, if discoid ≤4.5 cm long, not winged.
29. Fruits in profile oblong-ovate, the valves 2-keeled lengthwise, the dehiscence follicular. Seeds 2–3, red. . . . . *Batesia.*
29. Fruits in profile obovate, pyriform, ellipsoid, or subglobose, the valves not keeled, the dehiscence either tardy through both sutures, or none. Seed 1, brown or black.

30. Boles fenestrate. Fruits sessile, the body 6–8.5 cm long. Seeds obese, filling cavity, not arillate.. *Vouacapoua.*
30. Boles not fenestrate. Fruits stipitate, the body 2.5–6 cm long. Seeds either discoid or arillate.
    31. Fruits planocompressed but plump, tardily dehiscent through both sutures. Seed discoid, arillate.. *Copaifera.*
    31. Fruit indehiscent (but valves brittle). Seed plump, not arillate............... *Dialium.*

ACOSMIUM Schott

Trees. Stipules small, caducuous or inconspicuous. Leaves paripinnate, infrequently imparipinnate; leafstalk glands absent; stipels minute; leaflets petiolulate; blades chartaceous and glabrous in our area, occasionally with black dots or pellucid-punctate; secondary veins pinnate. Inflorescences axillary, paniculate. Flowers subactinomorphic; hypanthium campanulate; calyx lobed, occasionally with glands; petals subequal; stamens 5(10 outside our area); ovary sessile or shortly stipitate. Fruits pods or samaroid.

Acosmium is considered by some to be a synonym of *Sweetia* and is usually placed in the Fabaceae rather than the Caesalpiniaceae (Polhill, 1981). We place it in the Caesalpiniaceae because the flowers are distinctly non pea-like and nearly actinomorphic, both features that lead to the Caesalpiniaceae in our key to families.

A complete description of the genus awaits resolution of the differences between *Sweetia* and *Acosmium*.

Acosmium praeclarum (Sandwith) Yakovlev [Syn.: *Sweetia praeclara* Sandwith]

Unarmed trees, 20 m tall in our area. Trunks unbuttressed in our area but sometimes buttressed elsewhere. Stipules fugacious. Leaves 10–20 cm long; leaflets usually in 3 pairs, broadly elliptic, larger ones 6.5–14 × 3–5.5 cm, glabrous, sometimes thinly black speckled, the apex abruptly short acuminate. Inflorescences subterminal panicles, the racemose components 2–7 cm long; pedicels ± 1 mm long, the bracteoles minute, caducous. Flowers 4–6 mm diam.; hypanthium obconic; sepals 5, distally free; petals 5, strap-shaped, pubescent toward base adaxially, white in our collection; stamens 5; ovary densely white pubescent. Fruits (not known from our area) samaroid, stipitate. Seeds 1(2), if 1 inserted near middle of fruit, the ripe valves becoming brown, stiffly papery. Fl (Aug); known from a population of ca. five individuals in non-flooded forest on the Sentier Botanique near Eaux Claires.

BATESIA Benth.

Canopy or emergent trees. Trunks with steep, rounded buttresses. Bark soft, smooth. Young stems finely brownish-tomentulose. Stipules absent. Leaves either pari- or imparipinnate; leafstalks rounded abaxially, openly grooved adaxially, bearing gland at insertion of leaflets; stipels absent; leaflets petiolulate; blades chartaceous, finely brownish-tomentulose abaxially; secondary veins pinnate. Inflorescences racemose-paniculate, finely brownish-tomentulose; bracteoles 2, free, deciduous. Flowers subactinomorphic, nodding in bud; hypanthia campanulate, bearing from its rim sepals, petals, and 10 stamens; sepals 5, elliptic; petals elliptic-obovate, only slightly longer than sepals, pale yellow; stamens with filaments villosulous below middle, the anthers dorsifixed, versatile, dehiscent lengthwise; ovary subsessile at bottom of hypanthium, the style short, the stigma hollow, ciliolate. Fruits sessile, in profile oblong-obovoid, slightly decurved, turgid but laterally compressed and obtusely triangular in section, dehiscence through adaxial suture only, the valves woody, nigrescent, obtusely 2-keeled lengthwise, opening like book. Seeds 2–3, glossy, light red, exareolate.

Batesia floribunda Benth. PL. 34a,b; PART 1: FIG. 9

Trees, fertile at 7 m tall, attaining 50 m tall. Leaves: leafstalks 10–30 cm long; leaflets in 4–6 pairs, with or without petiolulate terminal one; blades oblong-elliptic, larger ones to 10–18 × 4–6.5 cm, the apex shortly acuminate; midrib centric, the secondary veins in 11–19 pairs, widely ascending. Inflorescences paniculate, pyramidal, 10–25 cm long. Flowers with almond-like aroma; sepals 5–6 mm long, abaxially tomentulose; petals 6–7.5 mm long, pilosulous abaxially below middle; anthers long-ellipsoid, 2–6 mm long. Fruits in profile 3–4.5 × 1.6–2 cm. Seeds plumply lenticular, 8–10 mm long. Fl (Feb, Nov), fr (May, Nov); scattered, in non-flooded forest.

BAUHINIA L.

Lianas. Stems either terete or often flattened and ladder-like ("monkey ladder"); tendrils borne randomly, shaped like short, woody hooks. Stipules thinly herbaceous, ephemeral. Leaves finely silky with pallid or coppery hairs; leaflets in 1 pair, either free or partially united by distal margins; blades palmately 3–6-veined from pulvini. Inflorescences terminally racemose or racemose-paniculate, finely silky with pallid or coppery hairs, bracteate; pedicels solitary, 2-bracteolate. Flowers: hypanthia campanulate or shallowly urceolate; sepals 5, narrowly or broadly ovoid, the free teeth either subobsolete or small and herbaceous, these forming rosette at apex of flower buds; petals 5, imbricate, coppery-silky-pilose abaxially, the vexillum interior; stamens 10-merous, the filaments free, subequal, the anthers versatile; ovary subsessile, the stigma obliquely truncate. Fruits oblong-oblanceolate, plano-compressed, the valves crustaceous, elastically dehiscent and coiling, 1-locular.

1. Stems broadly dilated and undulate. Pedicels 1–12 mm long. Calyx at anthesis either campanulate or ovoid, the teeth either depressed-deltate or shortly appendaged.

2. Expanded calyx campanulate, the lobes much wider than long, very shortly appendaged.
 3. Leaflet blades free from each other. Pedicels stout, 1–3 mm long. . . . . *B. guianensis.*
 3. Leaflet blades united through ca. half length. Pedicels slender, 5–12 mm long. . . . . *B. surinamensis.*
2. Expanded calyx urceolate, constricted at orifice, the lobes dilated into lance-ovate herbaceous appendages ca. 1–2 mm long. Leaflets and pedicels as in *B. guianensis*. . . . . *B. outimouta.*
1. Stems terete. Pedicels 20–30 mm long. Calyx at anthesis narrowly ovoid, cleft more than half length into lanceolate lobes. . . . . *B. siqueirae.*

Bauhinia guianensis Aubl. PL. 34c,d

Lianas. Stems broadly dilated, undulate. Leaves variable in size; petioles 1.5–4.5 cm long; blades free above pulvinule, broadly or narrowly semi-ovate, those near inflorescences 5–14 × 2–5.5 cm, shade-leaves to twice as large, glabrous olivaceous adaxially, appressed-silky and often coppery abaxially, the base cordate, the apex abruptly short-acuminate; venation palmately 4–5-veined. Inflorescences racemose, the axes (including short peduncle) 8–15 cm long; pedicels stout, 1–3 mm long; bracts and bracteoles subulate, <2 mm long, caducous. Flowers: buds ovoid, umbonate at apex; calyx campanulate when fully expanded, ca. 6–8 mm long, pallid, dark-ribbed, the rim deltately lobed or subtruncate, the lobes much wider than long, very shortly appendaged; petals 9–12 mm long, white or pinkish adaxially, fulvous-silky abaxially. Fruits ca. 6–8.5 × 1.7–2.4 cm. Fl (Feb, Sep); in non-flooded forest.

Bauhinia outimouta Aubl. PL. 34f; PART 1: FIG. 11

Hardly different from *B. guianensis* except bracts and bracteoles sometimes also foliaceous; calyx urceolate when fully expanded, constricted at orifice, the lobes dilated into lance-ovate, herbaceous appendages, ca. 1–2 mm long. Fl (Jan, Nov, Dec), fr (Mar, Dec); often growing with *B. guianensis*.

Bauhinia siqueirae Ducke

Lianas, high-climbing. Stems terete in transverse section. Leaves: petioles 2–7 cm long; leaflets in 1 pair; blades obliquely ovate, 5–10 × 2.5–3 cm, coriaceous, glabrous except on veins abaxially, the base shallowly cordate, the apex obtuse. Inflorescences racemose, the axes 15–25 cm long, coppery-silky; bracts narrowly lanceolate, 1–17 mm long, caducous; pedicels 2–3 cm long; bracteoles resembling bracts but shorter. Flowers large; buds ovoid, coarsely 15-ribbed, with free sepal-tips <1 mm long; calyx at anthesis narrowly ovoid, cleft more than than 1/2 length into lanceolate lobes, 15–18 mm long, coriaceous; petals oblanceolate, 21–27 mm long, white or greenish adaxially, densely coppery-silky abaxially; filaments ca. 16 mm long. Fruits narrowly oblanceolate, 9–15 × 2.8–3.8 cm, obtuse apiculate, the exocarp dark red, nigrescent. Fl (Jun, Dec); in forest canopy.

Bauhinia surinamensis Amshoff PL. 34e

Essentially like *B. guianensis* except for leaflets united through 1/2 length or more into bilobed blade, and only thinly setulose abaxially; pedicels slender, 5–12 mm long. Flowers: calyx campanulate, the lobes much wider than long, very shortly appendaged. Fl (Sep, Oct); in non-flooded forest.

CASSIA L.

Trees. Stipules either basifixed and simple or laterally attached and bilobed. Leaves either distichous or spiral, paripinnate, pilosulous with simple hairs. Inflorescences of terminal racemes, either on foliate branchlets or on efoliate short-shoots on old wood; pedicels subtended by bracts; bracteoles 2, at base. Flowers: hypanthia turbinate or slightly vase-shaped; sepals 5, reflexed at anthesis, caducous; petals 5, homomorphic or vexillum modified; stamens and staminodia 10, zygomorphic, of a) 3 short adaxial (to raceme axis) staminodia, b) 4 erect median stamens with straight filament and basally dehiscent anther, and c) 3 much longer abaxial stamens with filament sigmoidally arched, curved downward proximally and incurved distally, their anthers beakless, all anthers basifixed, not versatile, not beaked, dehiscing by slits; ovary stipitate; ovules many. Fruits linear, 30–70 cm long, sometimes massive, straight or almost so, either terete or biconvex, not winged, the valves indehiscent, rigidly coriaceous or woody, the seed cavity septate into 1-seeded locules. Seeds horizontal, embedded in foetid pulp, the testa hard, smooth, not areolate.

1. Leaflets in 3–5 pairs; blades to 8–12 × 4–5 cm. . . . . *C. spruceana.*
1. Leaflets in (8)10–29 pairs; blades to 2.7–6 × 0.9–2.4 cm.
 2. Plant in flower.
  3. Petals uniformly golden-yellow; anthers of 3 long stamens either glabrous or thinly pilose abaxially.
   4. Stipules triangular-subulate, <2.5 mm long, caducous. Pedicels 8–16 mm long. Petals 9–11 mm long. . . . . *C. cowanii* var. *cowanii.*
   4. Stipules foliaceous, 2-lobed, the distal lobe larger, 6–14 mm long, persistent. Pedicels 30–60 mm long. Petals 25–38 mm long.. . . . . *C. fastuosa.*
  3. Petals opening white or yellowish-pink and fading deeper pink or apricot, drying yellowish-tan, the vexillum with small yellow eyespot at base of blade; anthers of 3 long stamens woolly. . . . . *C. grandis.*
 2. Plant in fruit.
  5. Fruits massive, 3.6–5 cm diam., bluntly 2-carinate along adaxial suture, 1-carinate along abaxial one, the valves coarsely venose.. . . . . *C. grandis.*
  5. Fruits relatively slender, 1.1–2.5 cm diam., either subcylindric or 2-carinate along both sutures, the valves smooth or almost so.
   6. Fruits 1.1–1.6 cm diam., the valves densely minutely velutinous. . . . . *C. fastuosa.*
   6. Fruits ca. 2.5 cm diam., the valves glabrous. . . . . *C. cowanii* var. *cowanii.*

Cassia cowanii H. S. Irwin & Barneby var. **cowanii**

Trees, to 30 m tall. Trunks steeply buttressed. Stems densely yellowish-pubescent. Stipules triangular-subulate, <2.5 mm long, caducous. Leaves: leafstalks ca. 15–30 cm long; leaflets 12–20 pairs; blades oblong, obtuse or emarginate, to 2.7–5 × 0.9–1.3 cm, minutely yellowish-pubescent. Inflorescences racemose, ascending, the axes becoming 8–16 cm long, 20–50-flowered, densely yellowish-pubescent; pedicels 8–16 mm long. Flowers: hypanthia narrowly turbinate, ca. 1 mm long; sepals oblance-obovate or suborbicular, the inner ones 5–6.5 mm long, all pilosulous abaxially and adaxially; petals obovate or elliptic-obovate beyond short claw, 9–11 mm long, golden-yellow overall; stamens: 3 abaxial stamen filaments 10–11 mm long, the anthers either glabrous or thinly pilose abaxially; ovary densely pilosulous. Fruits pendulous, subcylindric, to 70 × 2.5 cm, the valves obtusely carinate by sutures, becoming woody, the exocarp lustrous green, becoming blackish, the cavity interrupted by horizontal septa 7–11 mm apart. Seeds ca. 13–14 mm long, suspended in pulp. Fl (Oct); in non-flooded forest.

Cassia fastuosa Benth.

Trees, to 40 m tall, but fertile from 4 m upward. Stipules foliaceous, persistent, 2-lobed, attached laterally, the distal lobe 6–14 mm long, the proximal one shorter. Leaves 15–30 cm long, the young foliage sordidly tomentulose but early glabrate; leaflets alternate, 10–29 pairs; larger blades lance-oblong or -ovate, 3–6 × 1–1.7 cm. Inflorescences terminal on short leafless branchlets arising mostly from year-old or older wood, racemose, the axes drooping, 15–35 cm long, loosely 15–35-flowered; bracts ovate-acuminate, 6–12 mm long; pedicels 3–6 cm long; bracteoles 2, about half as long as bracts. Flowers large, fragrant; hypanthia 2.5–5 mm long; sepals ovate or obovate, to 9.5–14 mm; petals subhomomorphic, obovate beyond conspicous claw, the longest 24–38 mm, golden-yellow overall; 3 long stamens with filaments dilated, 34–62 mm long, the anthers either glabrous or thinly pilose abaxially; ovary pilosulous, the stipe 4–7 mm long. Fruits pendulous, narrowly rod-shaped, convex laterally, 40–70 × 1.1–1.6 cm, the valves bluntly 2-carinate by both sutures, thinly woody, velutinous-puberulent. Fl (Oct), fr (Aug, Sep); in disturbed forest.

Cassia grandis L.f.

Trees, 13–30 m tall, but flowering at 5 m upward. Trunks buttressed. Stems foetid when bruised, the new branches pilosulous. Stipules <1.5 mm long, caducous. Leaves distichous; leafstalks 15–40 cm long; leaflets in 8–20 pairs; blades oblong, the longer ones 3.5–6.5 × 1.2–2.4 mm, the apex obtuse-emarginate. Inflorescences arising from defoliate branchlets proximal and often prior to growth of new leaves, racemes, the axes ascending or drooping, 8–25 cm long, pilosulous, 20–45-flowered; bracts 2–5 mm long, caducous; pedicels 8–20 mm long; bracteoles somewhat smaller than bracts, caducous. Flowers: hypanthia turbinate, ≤1 mm long; sepals oblong-obovate, the inner ones 6–9 mm long, all tomentulose; petals hetermorphic, opening white or yellowish-pink and fading deeper pink or apricot, drying yellowish-tan, the vexillum oblong-elliptic, bent forward, 2-callose and yellow-eyed at base of blade, the other petals obovate-flabellate, short-clawed, the longest 9–11 mm long; stamens with filaments of 3 long stamens 14–22 mm long, all anthers woolly; ovary pilose, the stipe 7–11 mm long. Fruits pendulous, massively linear-oblong, nearly straight, when fertile 40–60 × 3.6–5 cm, long persistent on tree, the valves biconvex, woody, coarsely venose, keeled abaxially by 1 and adaxially by 2 obtuse ribs, the ribs 7–25 mm wide, the seed-cavities filled with sweetish, edible, cathartic, malodorous pulp. Seeds ca. 14–16 × 9–10 mm. Fr (Feb); frequently cultivated, may or may not be native.

Cassia spruceana Benth. PL. 35a

Trees, 10–30(40) m tall. Stems densely minutely whitish- or yellow-puberulent when young. Stipules minute, evanescent. Leaves: leafstalks 7–22 cm long; leaflets in 3–5 opposite pairs; pulvinules 3–3.5 mm long; blades broadly ovate or ovate-ellipitic, distal pair 8–12 × 4–5.5 cm, all ample, firmly chartaceous, densely minutely whitish- or yellow-puberulent abaxially, the apex obtuse or very shortly acuminate; venation prominulously pinnate. Inflorescences arising at defoliate annotinous or older nodes, racemose, the axes including short peduncle 7–19 cm long, densely minutely whitish- or yellow-puberulent, loosely 15–30-flowered; bracts and bracteoles small, caducous. Flowers golden-yellow; sepals broadly obtuse, subequal in length; petals subhomomorphic, 11–16 mm long; stamens with filaments of 3 long stamens sigmoid, 15–19 mm long, glabrous, but anthers pilosulous; ovary loosely strigulose, the stipe 3–5 mm long. Fruits pendulous, rod-like, subterete, to 30–60 × 2–2.5 cm, the sutures immersed or nearly so, the adaxial one 2- and the abaxial 1-ribbed longitudinally, the seed-cavities filled with inedible pulp. Fl (Aug, Sep); in non-flooded forest.

COPAIFERA L.

Trees. Trunks without fenestrate boles. Bark coumarin-scented. Stipules small, caducous. Leaves paripinnate; leaflets ample, subopposite, resinous-punctate. Inflorescences paniculate, divaricately branched, the axes bearing flowers distichously, minutely bracteate; pedicels lacking. Flowers small; hypanthia absent; sepals 4, narrowly imbricate in vernation, silky-pilose adaxially; petals absent; stamens 10, the filaments free, spreading-incurved, the anthers long-ellipsoid, dorsifixed, versatile; ovary shortly stipitate, the style about as long as ovary, the stigma capitate; ovules 2, one ovule early abortive. Fruits stipitate, plumply discoid or biconvex, not winged, tardily dehiscent into 2 valves; endocarp fibrous-resinous. Seed partly enveloped in colored aril.

Copaifera guianensis Desf.

Trees, 6–17 or more m tall. Leaves glabrous; leafstalks 9–17 cm long; petiolules 4–6 mm long, twisted but not sulcate; leaflets in 3–4 pairs; blades ovate-oblong, 9–16 × 3–6 cm, carinate abaxially by midrib, the base equilaterally rounded, the apex acuminate; venation pinnate and reticulate. Inflorescences: primary axes 9–18 cm long. Flowers: buds subglobose, gray-puberulous; sepals ovate, 3.5–4.5 mm long, carnosulous; stamens with filaments 6–7 mm long, the anthers 2–2.5 mm long; ovary pilosulous around margin, glabrous facially. Fruits stipitate, the stipe 4–6 mm long, the body in profile obovate, a little declined, 2.5–3.3 × 2–3 cm, the exocarp dark red, glabrous. Seeds lustrous black with red aril. Fl (Sep). *Bois capayou* (Créole), *copahu* (French), *copaïba* (Portuguese), *copalier* (French), *coupawa* (Créole), *panchi mouti* (commercial name), resin from trunk has medicinal use.

DIALIUM L.

Trees. Trunks narrow-buttressed, the boles not fenestrate. Leaves imparipinnate; leaflets alternate, the indumentum of short simple brown hairs. Inflorescences terminal, pyramidal, cymose-paniculate, the cymose axes of inflorescence 2–3 times dichotomous, 5–15-flowered; pedicels ebracteolate. Flowers small, greenish, cream-colored or brownish; hypanthia absent; sepals 5, imbricate in bud; petals absent; stamens 2, inserted at edge of plane nectarial disc, the anthers basifixed, the anther sac 1-locular, dehiscent by apical slit; ovary subsessile, velutinous, the stigma terminal, scarcely dilated; ovules 2. Fruits indehiscent pods, ellipsoid to subglobose, slightly compressed, stipitate, the valves when ripe thin, brittle, the mesocarp spongy, fibrous, the seed-cavity septate, 1- or rarely 2-seeded. Seeds plumply biconvex, the testa lustrous, reticulate.

Dialium guianense (Aubl.) Sandwith PL. 35c

Trees, ca. 25 m tall, exceptionally to 60 m tall. Leaves: leafstalks 4–9 cm long; leaflets 4–7(11) per leaf, alternate along rachis on wrinkled pulvinules, the pulvinules 2–4.5 mm long, longer distally; blades ovate or lanceolate, the terminal one 5.5–12 × 2–5.5 cm, the base rounded or broadly cuneate, the apex acuminate; venation pinnnate and reticulate. Inflorescences with primary axes of panicles 10–30 cm long; bracts small, caducous; pedicels 1–4 mm long. Flowers: sepals ovate, 2–2.5 mm long; nectary disc ca. 2 mm diam.; stamens with filaments nearly 2 mm long, the anthers 0.7–0.9 mm long. Fruits 17–23 × 10–19 mm, contracted at base into stipe-like neck to 3 mm long, indehiscent but valves readily breaking under pressure, the exocarp dark gray-brown, minutely puberulent. Fl (Nov), fr (Mar, Jun); in forest. *Arouma* (commercial name).

DICORYNIA Benth.

Canopy or emergent trees. Trunks without buttresses or with narrow rounded buttresses. Bark rough, the heartwood red. Exudate slow-flowing, gelatinous. Stipules suborbicular, very early caducous. Leaves imparipinnate, the young foliage silky-strigulose with brown-golden hairs. Inflorescences terminal, efoliate, racemose-paniculate, silky-strigulose with brown-golden hairs, the ultimate branches 1–3-flowered cymes; pedicels ebracteate. Flowers white or pinkish; hypanthia absent; sepals 5; petals 3, broadly obovate-emarginate beyond short claw; stamens 2, unequal, the thecae of massive basifixed anthers 4–10-locular; ovary sessile, ± as long as style, the stigma poriform. Fruits samaroid, erect, estipitate, planocompressed, in profile oblong-elliptic, the valves narrowly winged along adaxial suture, stiffly papery. Seeds 1(2), compressed.

Dicorynia guianensis Amshoff PL. 35b

Trees, to 30–50 m tall. Leafstalks ca. 15–30 cm long; leaflets in 3–5 subopposite or alternate pairs and stalked terminal one, the pulvinules 4–7 mm long; blades ovate or oblong-ovate, the longer ones 8–18 × 3.5–6.5 cm, the base rounded or subtruncate, the apex abruptly short-acuminate; venation pinnate and reticulate. Inflorescences with primary axes of panicles 10–35 cm long, the secondary branches 2–16 cm long; pedicels 4–8 mm long. Flowers aromatic; sepals 8–10 mm long; petals 13–16 × 10–13 mm, white; stamens with filaments of shorter stamen 2–3 mm long, of longer one 4–7 mm long, the anthers oblong or obovoid, 4–6 mm long. Fruits in profile 5–8 × 3–4 cm, the valves with adaxial wing 4–7 mm wide, minutely strigulose. Seeds ca. 15–20 × 12–17 mm. Fl (Jan, Dec), fr (Mar); common, in non-flooded forest. *Angelique* (commercial name).

DIMORPHANDRA Schott

Trees. Stipules absent. Leaves ample, abruptly bipinnate, the young parts orange-brown-tomentulose; pinnae (sub)opposite; leaflets opposite, many. Inflorescences panicles of elongate, spiciform racemes, the racemes densely many-flowered; bract minute; bracteoles 1. Flowers small, subactinomorphic; hypanthia obscure, the tube campanulate; sepals shortly 5-toothed; petals 5, subuniform, ca. twice as long as sepals; stamens 10, the 5 opposed to calyx-teeth staminodial, distally dilated and incurved over pistil, the 5 opposed to petals fertile, the anther dorsifixed; ovary centric. Fruits shortly stipitate or nearly sessile, planocompressed, in profile either broad linear or obovate to broad-oblanceolate, the valves woody, either elastically dehiscent or indehiscent.

Freitas da Silva, M. 1986. *Dimorphandra* (Caesalpiniaceae). Fl. Neotrop. Monogr. **44:** 1–128.

1. Leaflets of longer pinnae in ca. 15–20 pairs; blades of larger leaflets 1.6–2.5 × 0.35–0.5 cm. Fruits dimidiately obovate-oblanceolate, ca. 12 × 7 cm, curved, the valves coarsely striate with straight veins ascending from adaxial suture, elastically dehiscent. . . . . *D. macrostachya* subsp. *glabrifolia.*
1. Leaflets of longer pinnae in 6–10 pairs; blades of larger leaflets 4.5–7 × 1.5–3 cm. Fruits broad-linear, ca. 20 × 3–3.5 cm, straight, the valves smooth, indehiscent. . . . . *D. multiflora.*

Dimorphandra macrostachya Benth. subsp. **glabrifolia** (Ducke) M. F. Silva

Trees, to 40 m tall. Trunks with rounded buttresses. Leaves: leafstalks 9–20 cm long; pinnae (3)4–9 pairs; pinnae rachises 8–12 cm long; leaflets of longer pinnae in ca. 15–20 pairs; blades linear, ca. 16–25 × 3.5–5 mm, glabrous adaxially, minutely puberulent abaxially, centrally carinate. Inflorescences appearing spicate, the axes with peduncle 30–45 cm long; pedicels <1 mm long. Flowers: sepals ca. 1.5 × 1.5 mm, the teeth obtuse; petals ca. twice length of sepals; staminodia linear-oblanceolate. Fruits obovate-oblanceolate, ca. 12 × 7 cm, the valves nigrescent, ascending, coarsely striate with straight veins ascending from adaxial suture, elastically dehiscent. Fr (Nov); in non-flooded forest.

Dimorphandra multiflora Ducke [Syn.: *D. pullei* Amshoff]

Trees, to 40 m tall. Bark smooth, gray. Leaves ample; primary leafstalks 2–3.5 dm long; pinnae (4)5–9 pairs; pinnae rachises of distal ones 8–18 cm long; leaflets 6–9(10) pairs, alternate or subopposite; blades symmetrically oblong-elliptic, the larger ones 4.5–7 × 1.5–3 cm, bicolored, lustrous glabrous adaxially, dull gray- or brown-puberulent abaxially, the base rounded, the apex acuminulate. Inflorescences terminal, broad-topped panicles of erect spikes, the primary rachises 8–25 cm long, the individual flower spikes to 4.5–8 cm long, densely flowered. Flowers: sepals including very short hypanthium ca. 1.5 mm long; petals 2.5–3 mm long, ochroleucous; ovary pilosulous; ovules 20–28. Fruits in profile broad-linear, nearly straight, ca. 20 × 3–3.5 cm, the valves thick, coriaceous, puberulent but glabrescent, becoming dark dull brown, not striate, indehiscent. Fr (May, Nov); in non-flooded forest near Saül and on La Fumée Mountain trail.

ELIZABETHA Benth.

Trees. Stems when young brown-pilosulous or glabrate, with perulate resting-buds at stem apices and in some leaf-axils. Stipules at early nodes of new growth lanceolate, caducous. Leaves distichous, paripinnate; petioles pulvinate; leaflets ≥20 pairs in our area. Inflorescences axillary or terminal, racemose, usually densely flowered; pedicels subtended by bract; bracteoles 2, united around hypanthia as 2-dentate sheath. Flowers: hypanthia campanulate; sepals 4; petals 5, 4, or 1; stamens 9 or 10, but only 3 fertile, the filaments shortly united on abaxial (to raceme axis) side of flower, the anthers dorsifixed, versatile; ovary stipitate, the stipe adnate to adaxial wall of hypanthium. Fruits stipitate, in profile oblong or broad-linear, planocompressed, dehiscent elastically by two valves, the valves woody, coiling, the adaxial suture marginate or narrowly winged.

1. Leaves with pulvinus glabrous; leaf stalks glabrous ad- and abaxially, troughlike. *E. leiogyne.*
1. Leaves with pulvinus pubescent; leaf stalks glabrous ad- and pubescent abaxially, scarcely grooved. . . . . *E. princeps.*

Elizabetha leiogyne Ducke

Trees, 18 m tall. Trunks without buttresses. Stems only thinly puberulent. Leaves: pulvinus glabrous; leafstalks 6.5–13 cm long, glabrous ad- and abaxially, troughlike; leaflets in 29–32 pairs; blades narrow, the larger ones 15–18 × 2.5–3.5 mm. Known by one sterile collection from our area; in non-flooded forest.

Elizabetha princeps Benth.

Trees, to 23 m tall. Trunks buttressed. Stems brown-pilosulous. Leaves: pulvinus pubescent; leafstalks 8–18 cm long, glabrous adaxially, pubescent abaxially, scarcely grooved and narrowly winged adaxially; leaflets in 23–32 pairs; blades linear, straight, the longer ones 20–28 × 3–4.5 mm, the distal ones smaller, the base shortly auriculate, the apex obtuse, apiculate. Fl (Sep, Oct), fr (Mar); in non-flooded forest.

EPERUA Aubl.

Forest trees, 20–40 m tall. Stipules either foliaceous, free, and deciduous, or united above leaf-pulvinus and more persistent. Leaves paripinnate; petioles present; glands absent; leaflets opposite, in 3–7 pairs; blades chartaceous or coriaceous, glabrous or almost so; venation reticulate. Inflorescences terminal, rarely axillary, of simple or paniculately branched racemes; pedicels subtended by bracts; bracteoles 2, free, not covering flower bud, caducous. Flowers: hypanthia cupular, glandular adaxially; sepals 4, slightly unequal, imbricate; petal 1, obovate-flabellate, adaxially convex; stamens 10-merous, diadelphous, alternately stronger and weaker, the latter often staminodial, 9 filaments shortly united opposite petal, the anthers dorsifixed, versatile; ovary elevated on free gynophore. Fruits in profile quadrate, narrowly oblong, or scimitar-shaped, planocompressed, the valves woody, without keels, when ripe elastically dehiscent through both sutures and coiling. Seeds several, not areolate.

1. Midrib of leaflet blades incurved. Inflorescences pendulous, simple (few-branched) racemes, the primary axes 60–70 cm long. . . . . *E. falcata.*
1. Midrib of leaflet blades straight. Inflorescences erect, compactly racemose-paniculate; primary axes <15 cm long.
    2. Leaflets in 3–4 pairs. Pedicels 4–5 mm long. Petal lavender-purple. Fruits in profile subquadrate. . . . . . *E. grandiflora.*
    2. Leaflets in 4–7 (mostly 5–6) pairs. Pedicels 11–27 mm long. Petal white. Fruits in profile oblong-oblanceolate. . . . . *E. schomburgkiana.*

Eperua falcata Aubl.

Trees, 20–32 m tall. Trunks cylindric. Stems sordid-puberulent when young. Stipules triangular acute, 2–4 mm long, sometimes connate in front of leaf-pulvini. Leaves: leafstalks 7–14 cm long; pulvinules 4–6 mm long; leaflets in 3–4 pairs; blades inequilaterally ovate-elliptic, 8–15 × 3.5–5.5 cm, gently incurved, chartaceous, glabrous, lustrous, the base rounded or broad-cuneate, the apex acuminate. Inflorescences of simple or few-branched racemes; peduncles long, pendulous, the primary axes 60–70 cm long, the raceme axes 2–4.5 cm long; bracts and bracteoles ovate, 2–4 mm long, deciduous; pedicels 9–15 mm long. Flowers opening at night but persisting into day; hypanthia 4–5 mm long; sepals 16–20 mm long; petal 11–14 mm long, pink or red; stamens with filaments to 3–4.5 cm long. Fruits narrowly oblong, 25–32 × 5–9 cm, attenuate at each end, incipiently falcate, the valves woody, plane, puberulent. Fl (Sep, Oct), fr (Oct);

common, in non-flooded forest. *Apa* (Portuguese), *bĭúdu* (Bush Negro), and *wapa* (Créole and French commercial name).

Eperua grandiflora (Aubl.) Benth.

Trees, to 40 m tall. Stipules foliaceous, lanceolate or falcately elliptic, 4–12 mm long, caducous. Leaves: leafstalks 5–11 cm long; pulvinules 4–9 mm long; leaflets 3–4 pairs; blades symmetrically ovate or broad-elliptic, 6–11.5 × (2)2.5–5 cm, chartaceous, glabrous, plane, the base rounded or cuneate. Inflorescences terminal and randomly axillary, racemose panicles, small, compact, erect, finely puberulous, the primary axes 2.5–7 cm long, the raceme axes 1–1.5 cm long; bracts ovate, 2.5–4 mm long; pedicels 4–5 mm long; bracteoles ovate, 2.5–4 mm long. Flowers: hypanthia 1.5–2 mm long; sepals 8.5–12 mm long; petal 2.2–3.4 cm long, lavender-purple. Fruits in profile subquadrate, ca. 7–8 × 5–5.5 cm, the base abruptly contracted into stipe 10–12 mm long, the exocarp either glabrous or glabrate. Fl (Jul), fr (Aug); in non-flooded forest.

Eperua schomburgkiana Benth.

Trees, to 37 m tall. Trunks with low buttresses. Stipules foliaceous, inequilaterally ovate or semicordate, 1.5–3 cm long, deciduous. Leaves: leafstalks 15–30 cm long, weakly bridged; pulvini 7–10 mm long; leaflets in (4)5–6(7) pairs; blades ovate-elliptic, the longer ones 8–15 × 3.5–5.5 cm, plane, coriaceous, the base broad-cuneate or rounded, the apex short- or long-acuminate. Inflorescences terminal, panicles of racemes, erect, glabrous or rusty-tomentulose, the primary axes 7–12 cm long; bracts obovate, 3.5–5.5 mm long, concave, caducous; pedicels 11–27 mm long; bracteoles obovate, 3.5–5.5 mm long, concave, caducous. Flowers: buds claviform; petal 25–37 × 35–60 mm, white. Fruits with stipe ca. 2 cm long, the body oblong-oblanceolate, to 12–25 × 6–8 cm, the apex acute, the exocarp castaneous, glabrous, rugulose. In non-flooded forest.

Only sterile collections from our area, fertile characters described from specimens collected outside our area.

HYMENAEA L.

Trees. Bark with resiniferous slash. Stipules small, fugacious. Leaves drought-deciduous; leaflets 2; blades lustrous, the base asymmetrically decurrent onto pulvinule. Inflorescences terminal, pyramidal or flat-topped panicles of few-flowered racemes, the axes all abruptly flexuous; pedicels subtended by caducuous bracts; bracteoles 2, caducous. Flowers: hypanthia turbinate-campanulate, thick-walled, the base narrowed into stalk; sepals 4, free to hypanthium-rim; petals 5, subsessile, scarcely or not longer than sepals, subuniform; stamens 10, the filaments free, well exserted, the anthers versatile; ovary centric, shortly stipitate, the style filiform. Fruits subsessile, in profile oblong, obtuse at both ends, plumply biconvex, indehiscent, the valves woody, the seed-cavity continuous. Seeds not arillate.

Hymenaea courbaril L. FIG. 61; PART 1: FIG. 9

Trees, to 40 m tall. Trunks with low buttresses. Leaves: petioles 8–20 mm long; petiolules 2–4.5 mm long; blades ovate-elliptic, 4–10 × 2–5 cm, glabrous, the apex falcate. Inflorescences gray-puberulent, the primary axes 3–15 cm long; pedicels ca. 6–10 mm long. Flowers (described from collections made outside our area): hypanthia ca. 6–7 mm long; sepals: lobes ovate, 12–18 mm long; petals 13–20 mm long. Fruits 8–11 × 3.5–5.5 × 1.5–21.5 cm, the exocarp glabrous, verruculose. Seeds 1 to several per fruit. Fl (Aug), fr (May, already fallen); uncommon in our area, in non-flooded forest. *Caca chien* (Créole), *copal, copal du Brésil* (French), *courbaril* (Créole and commercial name), *jatobá, jutaí* (Portuguese), resin from trunk has medicinal use.

MACROLOBIUM Schreb.

Subshrubs, shrubs, or trees. Stems ± terete, without tendrils. Stipules either foliaceous or rudimentary, deciduous. Leaves paripinnate, glabrous or glabrate; leaflets in 1 to many pairs. Inflorescences axillary or terminal, racemose; pedicels subtended by bracts; bracteoles 2, valvately clasping flower bud, separating at anthesis. Flowers: hypanthia sessile or stipitate, campanulate or cylindric; sepals 4 or 5; petal 1, vexillary, either sessile or long clawed, the blade often undulately crimped; stamens 3-merous, far-exserted; ovary stipitate, inserted either at bottom of hypanthia or ± adnate to adaxial (to raceme axis) side. Fruits in profile obovate or oblong-oblanceolate, either dehiscent or indehiscent, the valves woody.

Macrolobium bifolium (Aubl.) Pers.

Trees, to 15–20 m tall. Trunks blackish, white-spotted, the wood white. Indumentum none or nearly so except for on inflorescence. Stipules early caducous. Leaves: leafstalks 7–14 mm long; pulvinules 3–6 mm long, twisted; leaflets 2; blades obliquely elliptic or ovate-elliptic, mostly 6–16 × 3–6 cm, firm, the base broad-cuneate, proximally decurrent, the apex shortly acuminate; midrib incurved, forwardly displaced from mid-blade, canaliculate adaxially, carinate abaxially. Inflorescences axillary, racemose, either solitary or fasciculate with 2–4 flowers; peduncles short, the axes mostly <10 cm long; bracts ovate, 3–5–6 mm long, densely puberulent abaxially, glabrous adaxially, deciduous, the apex acute; pedicels ca. 3–5 mm long; bracteoles ovate, 3–6 mm long, densely puberulent abaxially, glabrous adaxially, deciduous, the apex acute. Flowers: sepals 4, 3.5–5.5 mm long; petal erect, the claw linear, involute, scarcely shorter than blade, the blade undulately crimped; ovary puberulent overall; ovules 2. Fruits stipitate, the stipe 5–14 mm long, the body in profile ± obovate, 8–14 × 4–8 cm, the valves woody, planocompressed, the ventral suture dilated. Seed 1, discoid, ca. 3–4.5 cm diam. Fl (Oct); in non-flooded forest.

In addition to *M. bifolium*, *M. angustifolium* (Benth.) R. S. Cowan is expected to occur in central French Guiana. It resembles *M. bifolium* in habitat and foliage but differs in the adaxially prominent costa of the leaflets and the adaxially pubescent bracts and bracteoles.

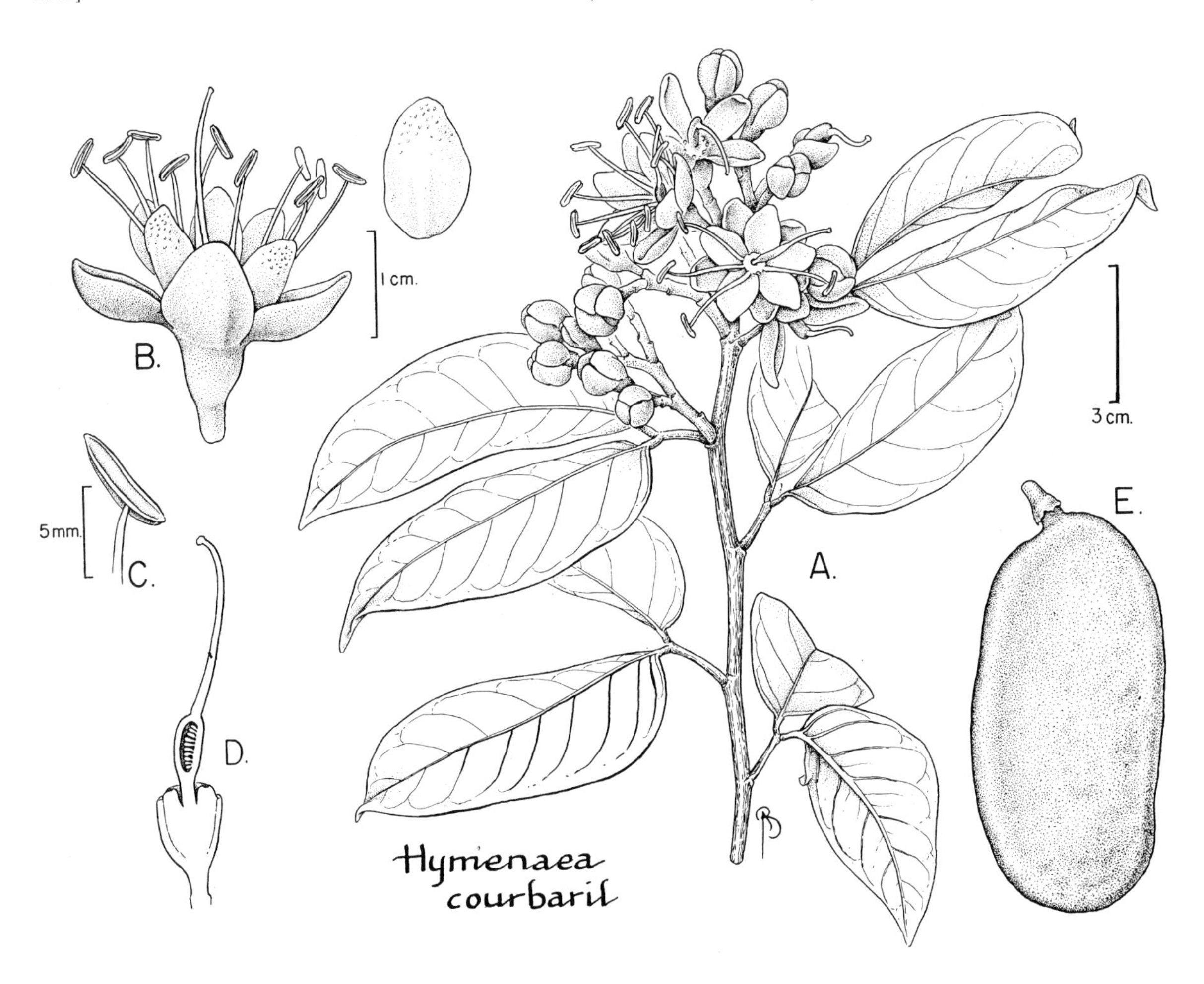

FIG. 61. CAESALPINIACEAE. *Hymenaea courbaril* (A–D, *Acevedo-Rodríguez 5073* from St. John, U.S. Virgin Islands; E, unvouchered photo). **A.** Part of stem with leaves and flowers. **B.** Lateral view of flower (near right) and petal (far right). **C.** Anther. **D.** Medial section of receptacle showing pistil with medial section of ovary. **E.** Fruit.(Reprinted with permission from P. Acevedo-Rodríguez, *Flora of St. John, U.S. Virgin Islands*, Mem. New York Bot. Gard. 78, 1996.)

MARTIODENDRON Gleason

Canopy trees. Trunks buttressed. Bark smooth. Exudate lacking. Stipules absent. Leaves ample, imparipinnate, when young silky-strigulose with golden-bronze hairs, though early glabrate; pulvinules present; leaflets alternate; venation pinnate. Inflorescences terminal, cymose-paniculate, efoliate, silky-strigulose with golden-bronze hairs; pedicels ebracteolate. Flowers zygomorphic, 5-merous; buds narrowly lance-acuminate; hypanthia absent; sepals imbricate in bud, silky-pubescent adaxially; petals equal in length but vexillum broadest, yellow or orange; stamens functionally 4, the filaments short, the anthers narrowly flask-shaped, apically porose, basifixed; ovary sessile, the style as long as ovary, grooved adaxially; ovule 1. Fruits samaras, planocompressed, elliptic or oblong-elliptic, technically stipitate but winged all around sutures, the wing decurrent on stipe, the exocarp reddish-purple when young, the valves stiffly chartaceous. Seed 1, compressed, without pleurogram; dispersal anemochorous.

Martiodendron parviflorum (Amshoff) R. Koeppen
PL. 36a,b; PART 1: FIG. 6

Trees, 25–40 m tall. Trunks with well developed, thin, often fused buttresses. Leaves: leafstalks (4)5–12(15) cm long; leaflets (4)5–6(7) per leaf; blades ovate or oblong-ovate, the longer ones 7–12.5 × 2.7–5 cm. Inflorescences paniculate, the primary axes 5–20 cm long; pedicels 4–16 mm long. Flowers: sepals narrowly lance-acuminate, 10–17 mm long, decurved, the base abaxially gibbous; petals 12–21 mm long, yellow; stamens: the larger anthers 8.5–12 mm long. Fruits in profile 12–18 × 4.5–6 cm, indehiscent, the valves puberulent, the sutural wings at middle of pod 7–20 mm wide. Fl (Jul, Aug), fr (Mar, Apr, Sep, Nov); common, in non-flooded forest.

PELTOGYNE Vogel

Trees. Trunks smooth. Stems ± terete, glabrous. Stipules lacking. Leaves glabrous; leaflets 2; leaflet blades chartaceous; venation reticulate. Inflorescences terminal, paniculate; pedicels subtended by transient bracts; bracteoles 2. Flowers pallid-silky-pilose; hypanthia campanulate; sepals 4, broadly oblong-obovate, free to hypanthium rim, reflexed at anthesis; petals 5, about as long as sepals, nearly uniform, subsessile, gland-dotted; stamens 10-merous, bicyclic, the anthers shortly exserted, versatile; ovary villous, stipitate, the stipe adnate to wall of hypanthium, the style slender, the stigma peltate; ovules 2–3, suspended from near apex of ovary, only one maturing. Fruits plano-compressed, in profile obliquely triangular, dehiscence inertly bivalvate, the adaxial suture thickened or narrowly winged. Seed discoid, cupped at base by fibrous aril, remaining attached to valve during dispersal.

Peltogyne paniculata Benth. subsp. **pubescens** (Benth.) M. F. Silva PART 1: FIG. 6

Trees, to 30 m tall. Bark ferruginous. Leaves: leafstalks 0.7–2.8 cm long, subterete; pulvinules 2.5–8 mm long; blades ovate or ovate-elliptic, 5–14 × 2.5–6 cm, scarcely or strongly incurved, the base subsymmetrical, either obtuse or acuminate, decurrent onto pulvinule. Inflorescences pyramidal, the primary axes 10–25 cm long; pedicels 3–7 mm long; bracts at middle. Flowers: hypanthium 2.5–4 mm long; sepals 4–5.5 mm long, membranous-marginate, the adaxial surface concave, sericeous at middle; petals white or pink. Fruits obtusely triangular-semicordate, the thick adaxial suture straight or shallowly concave, 2.5–3 mm wide, the abaxial suture distended backward below middle. Fl (Nov), fr (Sep); in non-flooded forest. *Amarante* (commercial name), *bois violet* (French), *purple-heart* (English).

RECORDOXYLON Ducke

Canopy trees. Bark exfoliating in scaly plates. Stems glabrous. Stipules lacking. Leaves imparipinnate, glabrous, glands lacking; leaflets subopposite and alternate, pulvinate, the terminal one petiolulate. Inflorescences terminal, racemose-paniculate, thinly brown-puberulent; pedicels articulate below short pseudopedicel, 2-bracteolate below joint. Flowers yellow; hypanthia campanulate, coriaceous; sepals free to hypanthium rim, lance-ovate, membranous-marginate; petals 5, subuniform, clawed, the blades broadly oblong-ovate; stamens 8–9, the filaments glabrous, the anthers long-elliptic, dorsifixed, introrsely dehiscent; ovary free from hypanthium, stipitate, the style linear, truncate. Fruits short-stipitate, the body planocompressed, slightly incurved, narrowly winged along adaxial suture, dehiscence inert, the valves papery reticulate, through both sutures. Seeds 4–5, discoid, lustrous.

Recordoxylon speciosum (Benoist) Barneby

Trees, to 40 m tall. Leaves: leafstalks 9–26 cm long, not bridged; leaflets 9–11 pairs; pulvinules 4–6 mm long; blades oblong, the larger ones 7–11 × 3–4.5 cm, all carinate abaxially by midrib, the base broadly rounded, the apex shortly acuminate; venation finely reticulate adaxially. Inflorescences paniculate, depressed-pyramidal, the primary axes 10–25 cm long; pedicels 6–8 mm long. Flowers showy, abundant; pseudopedicels and hypanthia together 7–8 mm long; hypanthium campanulate; sepals 11–12 mm long; petals ca. 23 mm long, yellow. Fruits 8–14 × 1.8–2.4 cm, the adaxial wing 2–2.5 mm wide. Fl (Apr, Aug); in non-flooded forest. *Wacapou* (Créole and commercial name).

SENNA Mill.

Trees, shrubs, lianas, and herbs, some monocarpic. Indumentum of simple hairs, eglandular. Stipules present. Leaves paripinnate; petioles often with glands, when present, ovoid, globose, or clavate; leaflets in (1)2–18 pairs, opposite, estipellate. Inflorescences axillary or terminal, paniculate racemes; pedicels ebracteate. Flowers: hypanthia solid; perianth 5-merous; sepals free to base, imbricate in vernation; corolla either zygomorphic or nearly actinomorphic, yellow, the two abaxial (to raceme axis) petals either alike or one variously modified; stamens zygomorphic, 10, increasing in length from adaxial (to raceme axis) to abaxial side of flower, the 3 adaxial ones staminodial, the filaments of abaxial stamens simply incurved, the anthers of fertile stamens basifixed, not versatile, modified into sets of 4 shorter median and (2)3 longer abaxial, all variously beaked at apex, dehiscent by 2 separate or confluent pores; pistil either centric or enantiostylous; ovules 5 to many. Fruits polymorphic in length, texture and compression, primitively planocompressed, becoming turgid, or cylindric, or angular, or winged longitudinally, the ripe valves separating along one or both sutures, never coiling, the seed-cavity dry or pulp-filled, either continuous or septate between seeds. Seeds either 1- or 2- seriate, areolate or not, the funicles filiform.

1. Petioles without glands. Leaflets in 7–14 pairs, the longer ones 7–20 cm long. Flower buds enveloped in cone of imbricate, petaloid, yellow bracts.
    2. Petioles (including, or consisting largely of, pulvini) <3.5 cm long. Fruits 4-angular, bicarinate by sutures and winged lengthwise along middle of valves, the wings crenate. . . . . *S. alata.*
    2. Petioles (including pulvini) ≥3.5 cm long. Fruits planocompressed, not winged. . . . . *S. reticulata.*
1. Petioles with glands, either between some leaflet-pairs, or at base of leafstalk. Leaflets either ≤6 pairs when as large as preceding, or at once much smaller and more numerous. Flower buds not enveloped in cone of imbricate, petaloid, yellow bracts.
    3. Plants usually herbaceous, weedy (exception is *S. obtusifolia*). Glands sessile, mounded, at base of petiole, close to pulvini.
        4. Stems, leaflets (except for minute cilia), and fruits appearing glabrous. . . . . *S. occidentalis.*
        4. Stems, leaflets, and fruits hirsute. . . . . *S. hirsuta* var. *hirsuta.*
    3. Plants with habit various but usually shrubs or lianas, less frequently herbs. Glands either sessile or stipitate, between one or several leaflet-pairs.

5. Leaflets of larger leaves in 2 or 3 pairs (only 1 pair in some leaves of *S. lourteigiana*). Corolla zygomorphic about vertical axis, the two abaxial (to raceme axis) petals similar. Fruits biconvex, cylindrical, or tetragonal.
    6. Weedy monocarpic herbs, <1.2 m tall. Leaflets of larger leaves in 3 pairs. Racemes 1–2-flowered. Fruits ≤5.5 mm diam. . . . . *S. obtusifolia.*
    6. Trees or shrubs >2 m tall, or lianas. Leaflets of all leaves (1)2 pairs. Racemes 3- to many-flowered. Fruits 7–20 mm diam.
        7. Leaves with glands between both pairs of leaflets.
            8. Leaflets obliquely elliptic, the midrib curved forward, the margins revolute. Longer sepals 4–6.5 mm long, gray-puberulent externally. Fruits subcylindric, the valves coarsely venulose. . . . . *S. quinquangulata.*
            8. Leaflets symmetrically elliptic, the margins not revolute. Longer sepals 13–18 mm long, glabrous. Fruits tetragonal, papillate. . . . . *S. lourteigiana.*
        7. Leaves with glands between proximal pair of leaflets only.
            9. Blades of distal pair of leaflets 9–17 cm long. Inner sepals 10–20 mm long, glabrous. Fruits ca. 22–31 cm long, glabrous. . . . . *S. latifolia.*
            9. Blades of distal pair of leaflets 2.5–6 cm long. Inner sepals 8–11.5 mm long, yellow pubescent abaxially. Fruits 4–9.5 cm, thinly pubescent when young. . . . . *S. chrysocarpa.*
5. Leaflets in 16–35 pairs. Corolla with no plane of symmetry, one abaxial (to raceme axis) petal longer than its fellow and curved through ± 90° (bomerang-shaped). Fruits linear, planocompressed.. . . . . *S. multijuga.*

Senna alata (L.) Roxb.

Shrubs of rapid growth, in flower when 1–6 m tall, the young growth minutely pilosulous. Stipules herbaceous, deltate-ovate, 6–16 mm long, reflexed. Leaves ample, 30–70 cm long; petioles (including pulvini) <3.5 cm long; glands absent; leaflets in 7–14 pairs; blades broadly oblong-obovate, larger distally, the largest 7–20 × 3–10 cm, the base semicordate. Inflorescences: axes in bud capped with cone of imbricate, petaloid bracts, the racemose axes many-flowered, becoming 15–60 cm long; bracts 1.7–3 cm long, orange-yellow, deciduous; pedicels 5–11 mm long. Flowers: perianth orange-yellow; longest sepals to 11–16 mm; petals, the vexillar petal 16–24 mm long; stamens functionally 2-merous, the anther of fertile, abaxial stamens lanceolate-sagittate, 9–13 mm long. Fruits stiffly ascending, subsessile, in profile broad-linear, 4-angular in transverse section, 11–18 cm long, 2-carinate by sutures, winged lengthwise along middle of valves, ca. 9–12 mm wide between sutures and 20–28 mm from wing to wing, the wings crenate, the valves nigrescent, the seed cavity rhombic in section, septate between seeds. Fl (Oct), fr (Oct), probably flowering and fruiting year round; in low wet places, roadside ditches, and in disturbed forest. *Bois dartre, cassialata, dartrier* (Créole), *matapasto* (Créole), used as a vermifuge.

Senna chrysocarpa (Desv.) H. S. Irwin & Barneby
FIG. 62; PART 1: FIG. 4

Weak shrubs or lianas. Plants either gray- or yellow-pilosulous almost throughout. Stipules setiform, caducous. Leaves 3–8 cm long; petioles 1–2.5 cm long; glands between proximal leaflet pair only; leaflets uniformly 2 pairs; blades: the distal pair ovate-elliptic, 2.5–6 × 1.2–3.2 cm, the base semicordate. Inflorescences thyrsiform-paniculate or pseudoracemose, subumbellate when young, 3–9-flowered, the primary axes and peduncles together 1.5–5.5 cm long; bracts caducous; pedicels 1–2.4 cm long. Flowers: buds subglobose, pilosulous; perianth yellow or orange-yellow; sepals strongly graduated, the longest 8–11.5 mm, the inner sepals yellow-pubescent abaxially; adaxial petals 13–20 mm long; stamens with anthers of 3 abaxial stamens 3.5–5 mm long, shortly beaked; ovary densely yellowish-pilose. Fruits stipitate, the stipe 3.5–7 mm long, the body cylindric, 4–9.5 × 0.7–0.9 cm, the exocarp lustrous, blackish, thinly pubescent when young, indehiscent, the valves chartaceous when ripe. Fl (Aug, Sep), fr (Aug, Oct); in forest margins and in secondary growth. *Galibi* (Amerindian).

Senna hirsuta (L.) H. S. Irwin & Barneby var. **hirsuta**

Herbs, 0.5–2 m tall, becoming softly woody, coarse, malodorous. Plants hirsute throughout (stems, both surfaces of leaflets, and fruits) with straight lustrous hairs 1–2.5 mm long. Stipules linear or linear-elliptic, herbaceous, deciduous. Leaves ca. 10–30 cm long; petioles 1.5–6.5 cm long; glands subglobose or fusiform, 1–2.5 mm long, sessile, mounded, at base of petiole, close to pulvini; rachises 4–12 cm long; leaflets in 3–6 pairs; blades larger distally, the distal pair ovate-, rhombic-, or lanceolate, 4.5–10.5 × 1.5–4 cm, the apex acuminate. Inflorescences axillary, narrow thyrses of 2–8-flowered racemes, at first far surpassed by leaves, later pseudoracemose; peduncles 1–15 mm long, the axes mostly <2 cm long; pedicels 9–25 mm long. Flowers: buds nodding, pilosulous or glabrate; sepals obovate, the abaxial ones ca. 4–7 mm long, the adaxial ones 8–15 mm long; petals subequal, 8–15 mm long, yellow, drying whitish with dark veins; stamens functionally 6-merous, the anthers of 4 median stamens ca. 4–5.5 mm long, those of 2 longest abaxial stamens ca. 5–7 mm long, the base sagittate; ovary hirsutulous. Fruits ascending, in profile linear, 11–15 × 0.4–0.65 cm, gently curved outward, compressed-tetragonal, dehiscent through both sutures, the valves white-setose-hirsute. Fr (Jun), probably flowering and fruiting intermittently year round; weedy in gardens and waste places. *Café zerb pian* (Créole).

Senna latifolia (G. Mey.) H. S. Irwin & Barneby

Lianas. Plants minutely puberulent or glabrous. Stipules herbaceous, ovate, 5–30 mm long. Leaves 12–18 cm long; petioles 3–10 cm long; glands ovoid, 2–5 mm long, sessile, between proximal pair of leaflets; leaflets 2 pairs, stiff, lustrous, the distal pair ovate-elliptic, 9–17 × 4.5–9.5 cm, the apex shortly acuminate; venation prominent on both surfaces; secondary veins 4–6 pairs, venules reticulate. Inflorescences either short or extended, subcorymbose or pseudoracemose, ca. 5–15-flowered; peduncles and axes together 2–9 cm long; pedicels 1.5–4 cm long. Flowers: buds globose, yellow or reddish, glabrous; sepals suborbicular, 10–20 mm long, adaxially concave,

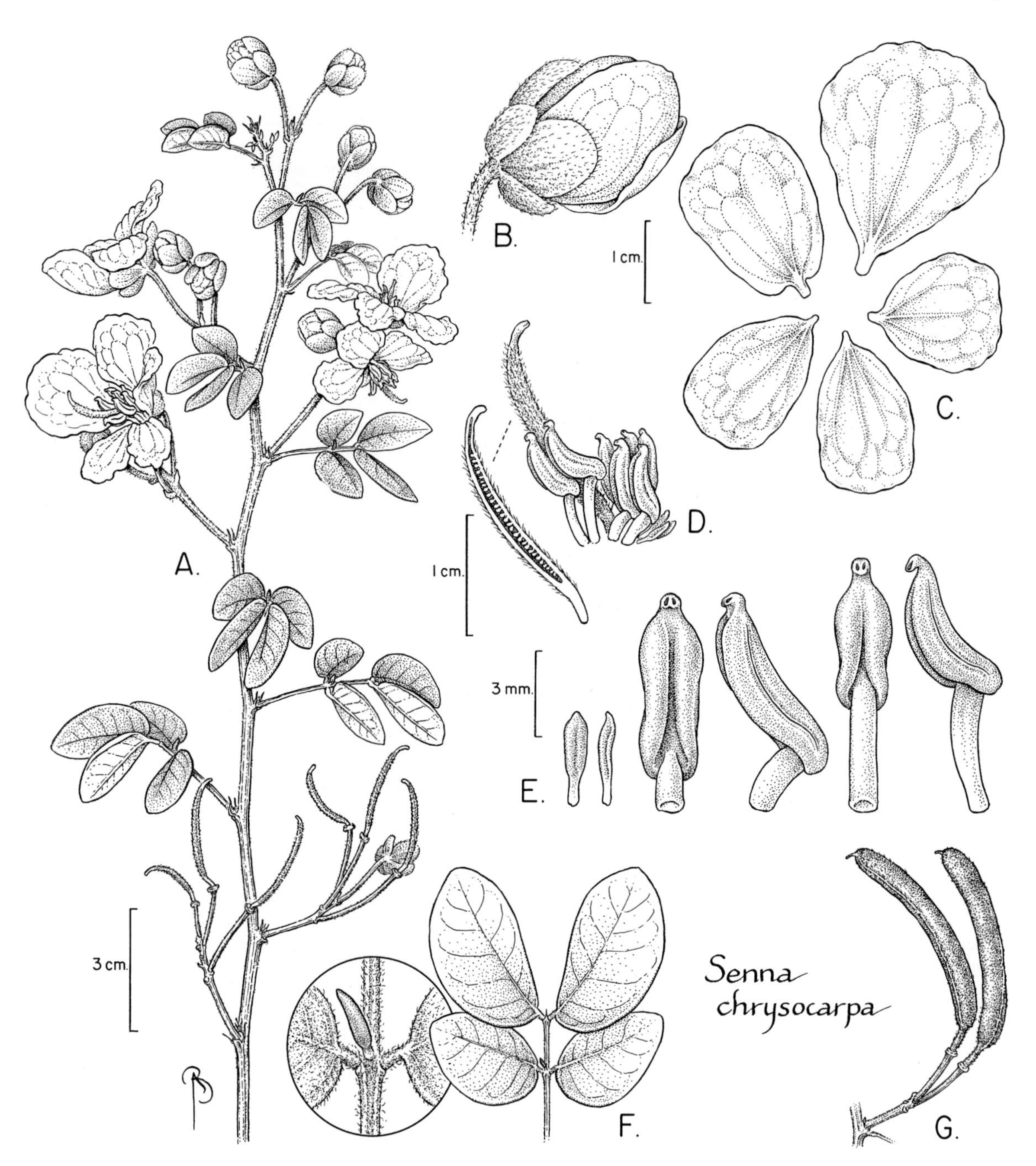

FIG. 62. CAESALPINIACEAE. *Senna chrysocarpa* (A–E, *Mori et al. 20949*; F, G, *Mori et al. 19148*). **A.** Part of stem with leaves, flowers, buds, and immature fruits. **B.** Lateral view of flower bud. **C.** The five petals of a flower. **D.** Lateral view of androecium and pistil (near left) and medial section of pistil (far left). **E.** Adaxial (near right) and lateral (second to right) views of staminodes, adaxial (third to right) and lateral (fourth to right) views of shorter stamens, and adaxial (fifth to right) and lateral (far right) views of longer stamens. **F.** Leaf with detail of gland between leaflets. **G.** Fruits.

connivent over corolla; petals golden- or orange-yellow, the abaxial one 24–32 mm long, the rest narrower but scarcely shorter; stamens with anther of 4 median stamens 6–11 mm long, that of 3 abaxial ones a little shorter but with longer beak; ovary usually glabrous. Fruits pendulous, the stipe 4–8 mm long, the body straight, cylindric, 22–31 × 1–1.3 cm, the exocarp glabrous, coarsely venose, dehiscence follicular through the gaping adaxial suture, the valves coriaceous. Seeds in 2 rows, embedded in blackish pulp, the testa not areolate. Fl (Feb, Apr, Aug), fr (Jun); at forest edges and in disturbed woodland.

Senna lourteigiana H. S. Irwin & Barneby PL. 36c

Lianas. Wood yellow. Plants almost or quite glabrous. Stipules herbaceous, falcate-oblanceolate or linear, 4–13 mm long, caducous. Leaves 9–20 cm long; leafstalks 3.5–6 cm long, widely sulcate adaxially; glands fusiform, 1.5–3 mm long, between each pair of leaflets; leaflets (1)2 pairs, the proximal pair sometimes lacking, but associated gland present; distal blades elliptic, 5.5–13 × 2.5–6.5 cm, the base subequilateral, the margins entire, the apex acuminate, the acumen retuse. Inflorescences axillary or terminal panicles of racemes, the primary axes 2–3.5 cm long, the racemose axes 3–7-flowered; pedicels 2.5–3.5 cm long. Flowers: buds glabrous; sepals obovate, graduated, the adaxial ones 13–18 mm long, glabrous; petals obovate, shortly clawed, golden- or orange-yellow, the longest 2.6–3.2 cm; stamens with anthers of 4 median stamens ca. 9 mm long, straight, those of 3 abaxial ones shorter but with longer beak; ovary strigulose. Fruits pendulous, shortly stipitate, the body straight, ca. 14.5 × 5.5 cm, tetragonal, each suture produced as two longitudinal wings, the exocarp papillate. Seeds biseriate, embedded in thin black pulp. Fl (Jan, Feb, Mar, Sep), fr (May); in disturbed areas and along trails in forest.

Senna multijuga (Rich.) H. S. Irwin & Barneby

Trees, to 25 m tall, though precociously flowering. Stipules lanceolate, the base unilaterally dilated, 1–2 mm wide, deciduous. Leaves 13–35 cm long, glabrous to strigulose to pilosulous; glands ovoid or fusiform, 1.5–4.5 mm long, sessile or stipitate, between first pair of leaflets, similar but smaller ones often at second pair and at some distal leaflet-pairs; leaflets 16–37, opposite, inserted 4–15 mm apart along rachises; blades oblong or lance-oblong, 20–45 × 5–12 mm, paler abaxially, the apex emarginate or mucronulate. Inflorescences pyramidal panicles of racemes, glabrous to strigulose to pilosulous, exserted or basally foliate; peduncles of individual racemes 1–5 cm long, the rachises of more vigorous raceme-branches to 16-flowered, the distal racemes shorter; pedicels 13–32 mm long. Flowers: buds opening long before anthesis; sepals petaloid when mature, graduated, the adaxial ones 4–8 mm long; petals heteromorphic, the 3 adaxial and 1 abaxial petals obovate-oblanceolate, 8–21 mm long, the fifth petal semi-ovate or boomerang-shaped, a little longer than rest; venation coarse; stamens with anthers of 3 abaxial stamens incurved, 4.5–9 mm long, the beak 1–2 mm long, porrect. Fruits: stipe 2–9 mm long, the body broad-linear, 9–20 × 1.3–2 cm, planocompressed, the valves castaneous-nigrescent, papery, venulose, transversely ridged over seeds, inertly dehiscent through both sutures. Seeds narrowly oblong-elliptic, areolate. Collected on upper Mana River just outside of our area, but likely to occur in our area. Fl and fr (probably year round); not vouchered, but in riparian forest and disturbed forest elsewhere. *Marimari* (Portuguese).

Senna obtusifolia (L.) H. S. Irwin & Barneby

Herbs, in age basally lignescent, mostly <1 m tall, monocarpic. Roots yellow-tipped. Stipules linear, 5–16 mm long, deciduous. Leaves 3.5–16 cm long, foetid, either glabrous or thinly pubescent; petioles 1–4 cm long; rachises ca. 1–4 cm long; glands lance-fusiform, 1–3 mm long, between first and often second pair of leaflets; leaflets of almost all leaves in 3 pairs; blades obovate-cuneate or oblanceolate, 2–6.5 × 1–4 cm, the larger ones distal. Inflorescences axillary, racemose, 1–2-flowered; peduncles 1–8 mm long; pedicels at anthesis slender, 8–25 mm long, thickened in fruit. Flowers: buds nodding, glabrous or puberulent; sepals: adaxial ones obovate, 5.5–9 mm long, greenish; petals: vexillar one obcordate, 9–14 mm long, pale yellow; ovary pubescent. Fruits stiffly ascending and arched outward, linear, attenuate at each end, 7–17 × 0.3–0.55 cm, green and 6-ribbed when young, brown and turgid when ripe, dehiscence tardy, inert, the seed-cavity septate. Seeds lustrous, the pleurogram linear. Fl (Jun, Jul, Aug, Oct), fr (Aug, Oct), probably flowering and fruiting year round; in weedy habitats around Saül. *Café zerb pian* (Créole), seeds used to make coffee substitute, leaves and roots have some medicinal value.

Senna occidentalis (L.) Link

Herbs, 0.15–1.4 m tall, either monocarpic or becoming soft-woody at base. Plants coarse, foetid, appearing glabrous. Roots black. Stipules herbaceous, 4–12 mm long, the base auriculate, deciduous. Leaves 10–25 cm long; petioles 2.5–5 cm long; rachises 5–14 cm long; glands globose or hemispherical, 0.7–1.8 mm long, sessile, mounded, at base of petiole, close to pulvini; leaflets in (3)4–5(6) pairs; blades lanceolate or ovate, the distal pair largest, 4.5–10 × 1.3–3.8 cm, the apex acuminate. Inflorescences at first axillary, later thyrsiform, 1–5-flowered racemes; peduncles short to inconspicuous, together with raceme axes ≤1 cm long; pedicels 8–18 mm long. Flowers: buds nodding; sepals submembranous, the adaxial ones 7–10 mm long; petals subequilong, but vexillar one broadest, 12–16 mm long, yellow, drying whitish with brown veins; stamens with anthers of 4 median stamens 3–5 mm long, those of 2 longer, abaxial stamens 3.7–5 mm long; ovary pubescent. Fruits erect-ascending, in profile linear, 8–13.5 × 0.7–0.9 cm, sessile, planocompressed unless distended by seeds, dehiscence inert, the valves carinate by sutures, green-stramineous along sutures, red at middle, the seed-cavity septate. Seeds areolate. Fl (Mar, Jun, Oct), fr (Mar, Jun, Sep, Oct), probably flowering and fruiting year round; weeds in gardens and waste places. *Bois-puant, café zerb pian, digo, indigo* (Créole), uses same as for *S. obtusifolia.*

Senna quinquangulata (Rich.) H. S. Irwin & Barneby

Lianas, climbing into forest canopy, flowering precociously. Plants gray-puberulent or glabrate. Stipules linear-oblanceolate, 3–10 mm long, caducous, the apex falcate. Leaves mostly 10–22 cm long; petioles 2–5.5 cm long; glands tongue-like or clavate, 2–6 mm long, either sessile or stipitate, between each leaflet-pair; leaflets in (1)2 pairs; blades bicolored, the distal pair obliquely elliptic, ovate, or lanceolate, (5)6–16 × (2.5)3–7.5 cm, the apex acuminate, the margins revolute; midrib curved forward. Inflorescences thyrses or panicles of racemes; peduncles and raceme-axes together 1.5–9 cm long, the raceme-axes mostly 7–20-flowered; pedicels 12–32 mm long. Flowers: young buds globose; sepals opening before anthesis, ovate-oblong or suborbicular, the longest 4–6.5 mm, firm, gray-puberulent externally; petals: longest 10–16 mm, either pale yellow or golden-orange, gray-puberulent abaxially; stamens with anthers of 4 median stamens 4.5–9 mm long, those of 3 abaxial stamens 3–5.5 mm long, beaked; ovary pubescent. Fruits pendulous, short-stipitate, the body narrowly cylindric, 11–28 × 0.9–1.5 cm, dehiscence follicular, the valves coriaceous, coarsely venose, gaping to expose biseriate seeds. Seeds sometimes faintly areolate, embedded in foetid pulp. Fl (Jun, Aug, Sep, Oct), fr (Jan, Nov, Dec), the fruit long persistent; in disturbed habitats.

Senna reticulata (Willd.) H. S. Irwin & Barneby

Shrubs, arborescent, 2–9 m tall, closely resembling *S. alata* in habit, foliage, pubescence, and in axillary or later corymbose racemes capped by cone of petaloid bracts, but differing in longer petioles and planocompressed wingless pods. Leaves coarse; petioles, including pulvini, 3.5–14 cm long; glands absent; leaflets in 7–14 pairs; larger blades 7–18 × 3–7 cm. Inflorescences: bracts obovate-acuminate, 14–22 mm long, yellow, caducous; pedicels <1 cm long. Flowers as in *S. alata*. Fruits: stipes 3–6 mm long, the body straight, in profile broad-linear, 9–16 × 1.2–1.7 cm, planocompressed, not winged, the exocarp nigrescent, dehiscence inert through both sutures, the valves stiffly chartaceous, raised over each seed as low transverse ridge. Seeds areolate. Fl (Aug), fr (Aug), probably flowering and fruiting in other months as well; in disturbed habitats.

TACHIGALI Aubl. (including *Sclerolobium* Vogel)

Trees, often of large size. Trunks buttressed. Stems when young with indumentum of gray or rusty, simple or, in one species, stellate hairs. Stipules either small and simple, or foliaceous, then simple or 2–5-lobed, or pectinately decompound into linear lobes, in any case transient, absent from most flowering and from all fruiting specimens. Leaves mostly paripinnate (randomly imparipinnate); leafstalks of some species facultatively or obligately swollen, hollow, and inhabited by *tachi* ants (*Pseudomyrmex* spp.), grooved adaxially, bridged at leaflet insertions, the bridge eglandular; leaflets 3–9 pairs; blades inequilateral, broader below middle on anterior side of midrib, the indumentum of gray or rusty, simple or, in one species, stellate hairs. Inflorescences terminal, efoliate or proximately few-foliate panicles of spiciform racemes, these either solitary or fasciculate; bracts present; pedicels short; bracteoles absent. Flowers either subactinomorphic or distinctly zygomorphic (Pl. 37a); hypanthia hemispherical or obliquely truncate; sepals 5, imbricate; petals 5, subequiform, either expanded or filiform, white or yellow; stamens 10, either of equal length or 3 shorter and stouter, the filaments pilosulous proximally, especially adaxially, the anthers dorsifixed, versatile, laterally dehiscent; ovary shortly stipitate, inserted either on bottom of hypanthia or basally adnate to hypanthium adaxially (to raceme axis); ovules 2–6. Fruits samaras, oblong or elliptic-oblong, (sub)sessile, strongly compressed, stiffly papery or coriaceous, 1–2-seeded, the exocarp often nigrescent, cracking and peeling when ripe to expose mesocarp, the mesocarp tan-stramineous, longitudinally veined. Seeds compressed but plump, without pleurogram.

Flowers are usually required for specific determination.

1. Racemose elements of inflorescences usually many and short, their axes, including peduncle, 2–14 cm long. Flowers actinomorphic or almost so; buds subglobose; hypanthia hemispherical, the rim consequently horizontal; petals <1 mm wide; ovary attached at base of hypanthia.
    2. Hairs stellate, especially on young growth and on abaxial surface of mature leaflets. . . . . *T. melinonii.*
    2. Hairs all simple, basifixed.
        3. Stipules (caducous, but often observed at terminal buds or at lowest branches of inflorescence) either once or twice compound into linear lobes. . . . . *T. guianensis.*
        3. Stipules foliaceous, either entire or 2- to several-lobed, the lobes broader than linear (stipules unknown for *Tachigali* sp. C).
            4. Larger leaflet blades 7–18 × 2.5–7.5 cm, plane, i.e., not bullate.
                5. Larger leaflet blades 3–7.5 cm wide, coriaceous, glabrous abaxially, the base ± symmetric. . . . . *T. paraënsis.*
                5. Larger leaflet blades 2.5–3 cm wide, chartaceous, the base ± asymmetric. . . . . *Tachigali* sp. C.
            4. Larger leaflet blades 15–35 × 9–20 cm, ± bullate, the margins loosely revolute. . . . . *T. amplifolia.*
1. Racemose elements of inflorescences usually few and elongate, their axes, including peduncle, mostly 15–25 cm long. Flowers zygomorphic; buds obliquely pyriform; hypanthia deeply cupulate or campanulate, longer on adaxial side, the rim consequently tilted forward; petals ≥3 mm wide; ovary attached to adaxial wall of hypanthium.
    6. Stipules not seen. Leafstalks slender, not dilated proximally, ≤2 mm diam.; leaflets ca. 7 pairs. . . . . *Tachigali* aff. *bracteolata.*
    6. Stipules foliaceous, 2–5-foliolate. Leafstalks dilated proximally, 4–6 mm diam.; leaflets 3–4 pairs. . . . . *Tachigali* aff. *paniculata.*

Tachigali amplifolia (Ducke) Barneby [Syn.: *Sclerolobium amplifolium* Ducke] PL. 37b; PART 1: PL. XIIIb

Trees, 25–30 m tall cm. Trunks buttressed, the bole smooth. Plants with simple, basifixed pubescence. Stems: young growth densely silky-strigulose. Stipules large and foliaceous, especially in saplings, entire or lobed, caducous. Leaves ample; leaf stalks 20–45 cm long, longer in saplings; leaflets 3–5 pairs; pulvinuli 5–9 mm long; blades amply ovate or ovate-oblong, the distal ones 15–35 × (7)9–20 cm, glabrescent, thinly chartaceous, ± bullate, the base broadly rounded or shallowly cordate, the apex shortly or obscurely acuminate, the margins loosely revolute; secondary veins in 6–8 pairs, immersed or impressed adaxially, sharply prominulous abaxially. Inflorescences terminal, efoliate panicles, 20–40 cm long, commonly well exserted, densely gray-strigulose overall, the individual racemes ca. 6–14 cm long, densely many-flowered, the bracts linear-lance-attenuate, 2.5–5.5 mm long, caducous; pedicels 0.6–1.4 mm long. Flowers fragrant; buds subglobose; hypanthia hemispherical, a little oblique, 1.7–2.2 × 2.2–2.8 mm; sepals broadly or narrowly ovate, 2.6–3.3 × 1.2–1.8 mm, abaxially (to raceme axis) convex, puberulent adaxially; petals linear-spatulate, 2.5–3.5 × <1 mm; stamens with filaments 4.5–6.6 mm long, golden-pilose on proximal half, yellow; ovary attached at base of hypanthium, shortly stipitate, ferruginous-pilose. Fruits not seen from Saül, in Amazonian Brazil 9–13 × 2.5–3 cm. Fl (Aug); common, in secondary forest, often conspicuous at certain times of the year because of large leaflets carpeting ground.

Tachigali aff. bracteolata Dwyer PL. 37a

Trees, ca. 20 m tall. Trunks with thin buttresses. Stems: young growth densely puberulent. Stipules unknown. Leaves: leafstalks slender, ≤21 × <0.2 cm; leaflets in 7 pairs; blades narrowly ovate, ca. 8.5–10 × 2.5–4 cm, microscopically strigulose adaxially, glabrate abaxially except along principal veins, the base shallowly cordate, the apex acuminate; secondary veins in 5–6 pairs. Inflorescences few-branched panicles, the raceme axes ca. 20–25 cm long; bracts to 3 mm long, caducous; pedicels 1.5–2 mm long. Flowers zygomorphic, the buds obliquely pyriform, 9 mm long, dull reddish, gray-silky overall; hypanthia deeply campanulate, 5 mm long on adaxial (to raceme axis) side, 3 mm long on abaxial side, the rim tilted forward; sepals ≤4 mm long; petals silky-puberulent adaxially, light yellow; stamens dimorphic, the 7 longer ones subsigmoid, ca. 12 mm long, the 3 shorter ones erect, deeper yellow; ovary attached to adaxial wall of hypanthium, densely appressed-silky. Fl (Oct); in non-flooded forest.

Tachigali guianensis (Benth.) Zarucchi &Herend. [Syn.: *Sclerolobium guianensis* Benth.]

Trees, 15–40 m tall. Trunks buttressed. Plants with simple, basifixed pubescence. Stems: young growth densely red-pilose. Stipules caducous (but often evident on terminal leaf-buds), 6–15 mm long, either once or twice compound into linear lobes. Leaves: leafstalks 15–40 cm long; leaflets 5–8 (in saplings –11) pairs; blades inequilaterally ovate- or oblong-elliptic, the larger ones 10–23 × 3–8.5 cm, glabrate except along principal veins, the base subcordate, the apex shortly acuminate; secondary veins in 7–12 pairs. Inflorescences dense, the primary axes mostly 15–25 cm long, shorter than uppermost leaves, the spiciform racemes 3–8 cm long; pedicels ≤0.5 mm long. Flowers actinomorphic, gray-pilosulous abaxially; buds subglobose; hypanthia 0.6–1 mm long, hemispherical, the rim horizontal; sepals obovate-elliptic, 2–2.5 mm long; petals linear, <1 mm wide; stamens with filaments densely yellow-pilose below middle; ovary attached at base of hypanthium. Fruits in broad view narrowly oblong-elliptic, 7–9 × 2.4–3 cm, the adaxial wing 3–5 mm wide, the exocarp dull blackish-brown, quickly glabrate. Known from the vicinity of Saül only by sterile specimens; apparently uncommon, in non-flooded forest.

Fertile characters derived from collections made outside our area.

Tachigali melinonii (Harms) Zarucchi & Herend. [Syn.: *Sclerolobium melinonii* Harms]

Trees, ca. 30 m tall. Trunks with flying buttresses, the bole cylindric. Bark brown to orangish. Stems: young growth pubescent with fine, stellately branched, orange or partly gray hairs. Stipules 7–10 mm long, pinnately divided, the linear lobes 2–3 on each side of primary axis, caducous. Leaves: leafstalks 12–30 cm long, pubescent with fine, stellately branched, orange or partly gray hairs; leaflets 6–9 pairs; blades oblong-acuminate, the longer ones 5.5–11 × 2–4 cm long, thinly chartaceous, glabrescent, sometimes pubescent abaxially with fine, stellately branched, orange or partly gray hairs, the base inequilaterally rounded; secondary veins in 6–8 pairs, prominulous abaxially. Inflorescences: primary axes of panicles 10–20 cm long, not or scarcely surpassing distal leaves, the raceme axes ca. 2–4 cm long; pedicels 1–2 mm long. Flowers actinomorphic; buds subglobose; hypanthia hemispherical, 0.8–1.2 mm long, the rim horizontal; sepals obtuse, 1.5–2 mm long, densely gray-puberulent; petals linear-elliptic, <1 mm wide, scarcely exserted; ovary attached at base of hypanthium. Fruits in broad view elliptic-oblong, 6–8 × 1.9–2.6 mm, the adaxial wing 2.5–3.5 mm wide, the exocarp turning lustrous brown, glabrous, smooth or papillate. Known from the vicinity of Saül only by sterile specimens; along streams or in low areas in moist forest. *Cèdre remy, tachi, tassi* (Créole).

Fertile characters derived from collections made outside our area.

Tachigali aff. **paniculata** Aubl.

Trees, ca. 30 m tall. Trunks with well developed, plank-like buttresses. Stipules foliaceous, 2–5-foliolate. Leaves: leafstalks ca. 15 cm long; petioles and first interfoliolar segments of rachis dilated into ant-domatia, ca. 4–6 mm diam.; leaflets in 3–4 pairs; blades broadly oblong-ovate, to ca. 12–22 × 6.5–8 cm, firmly chartaceous, subglabrous, the apex short-acuminate. Known only by sterile specimens in our area; in non-flooded forest.

Tachigali paraënsis (Huber) Barneby [Syn.: *Sclerolobium paraënse* Huber, *S. albiflorum* Benoist]

Trees, 15–45 m tall. Trunks buttressed, the bole cylindric or sometimes fluted. Plants with simple, basifixed pubescence. Stems: young growth densely minutely appressed-silky. Stipules foliaceous, obovate-flabellate or -elliptic, the base narrowed into short petiole-like stalk, the blade 6–29 × 6–19 mm, the margins revolute, caducous (but some usually observable around terminal buds). Leaves: leafstalks 8–26 cm long; leaflets 3–4(5) pairs; blades inequilaterally oblong-elliptic, the longer ones 11–18 × 3–7.5 cm, coriaceous, glabrous or early glabrate, adaxially glossy, the base ± symmetric, either rounded or broad-cuneate, the apex shortly acuminate, the margins entire; secondary veins in 7–10 pairs, nearly immersed on distal adaxial surface. Inflorescences densely minutely appressed-silky, dense panicles of racemes, the primary axes ca. 13–23 cm long, the raceme axes 4–10 cm long; pedicels 0.4–1.4 mm long. Flowers gray-silky abaxially; buds subglobose; hypanthia shallowly cupular, almost symmetrical, 1–1.4 mm long; sepals ovate, ca. 3 mm long; petals linear, <1 mm wide; stamens with filaments densely pilose below middle; ovary attached at base of hypanthium, thinly rusty-pilosulous. Fruits in broad view narrowly elliptic, to 11 × 3 cm. Fl (Jul), fr (Jan, Dec); locally common, in non-flooded forest. *Tachi, tassi* (Créole).

Tachigali sp. **C** (*Moretti 743*, CAY)

Trees. Stipules unknown. Leaves: leafstalks 20–25 cm long, covered with rust-colored pubescence of simple hairs; leaflets in 6 pairs; blades oblong to elliptic, 7–9 × 2.5–3 cm, chartaceous, glabrous adaxially, with sparse, simple hairs abaxially, the base ± asymmetric, the apex acuminate above obtuse to rounded margins. Inflorescences: primary axes of panicles 12–20 cm long, the secondary axes slender, 7–12 cm long, pubescent; bracts acicular, ca. 1.8 mm long; pedicels ca. 2 mm long. Flowers small and delicate, 2.5–3 mm diam.; buds subglobose; hypanthia hemispherical, the rim horizontal; sepals ovate-oblong, ca. 1.4 mm long; petals shorter than sepals, ca. 1.1 × 0.8 mm; stamens with filaments sparsely pilose toward base; ovary attached at base of hypanthia, with long simple hairs, especially along margins. Fruits not known. Fl (Jun).

VOUACAPOUA Aubl.

Canopy trees. Trunks buttressed, the boles usually fenestrate. Stipules lacking. Leaves imparipinnate, ample, glabrous; leafstalks with sessile, convex glands between some pairs of leaflets; leaflets opposite, the terminal one stalked; blades chartaceous, pale abaxially;

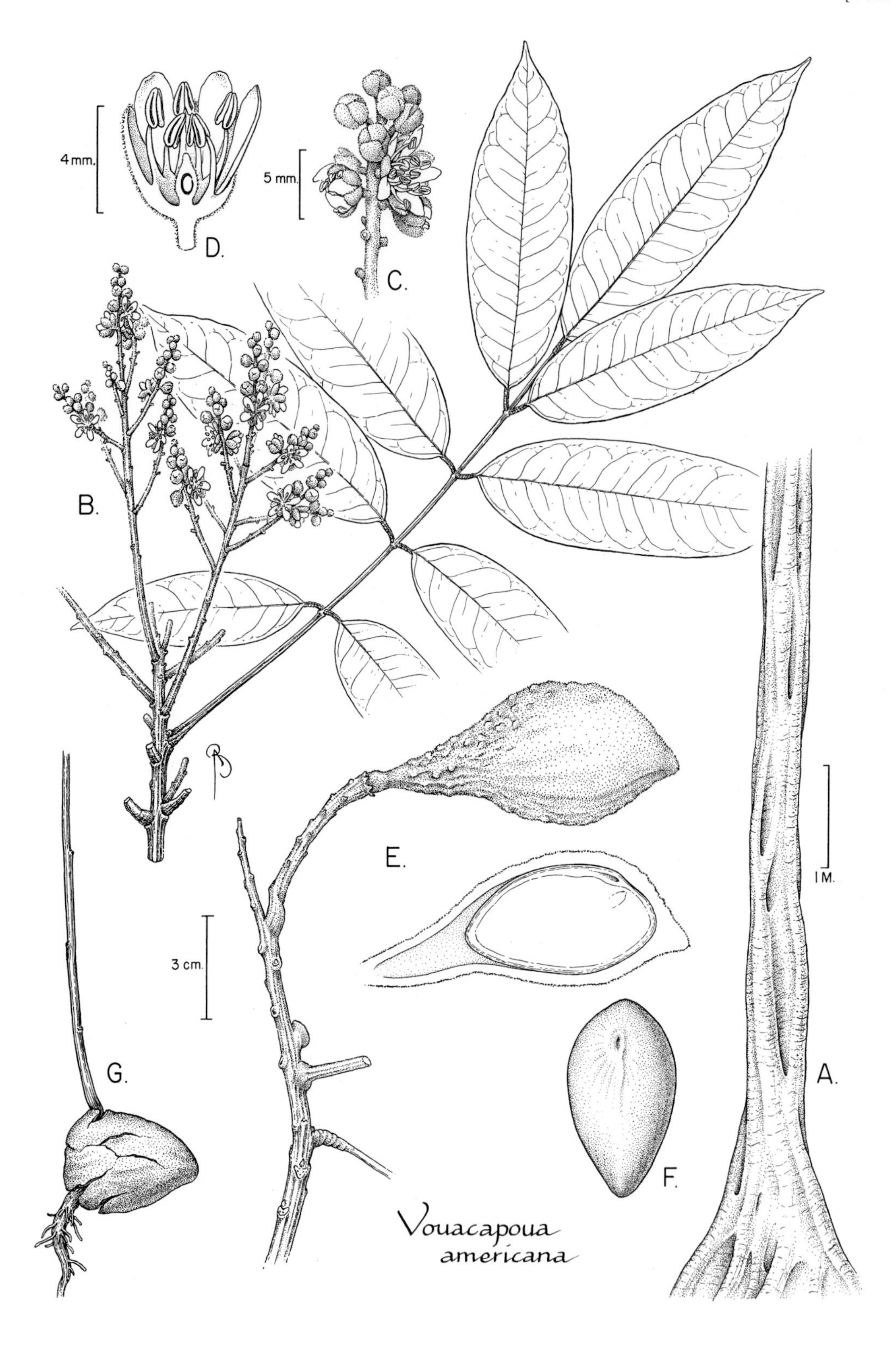

FIG. 63. CAESALPINIACEAE. *Vouacapoua americana* (A, unvouchered field sketch; B–D, *Mori et al. 22757*; E, F, *Mori et al. 22238*; G, *Mori 18521*). **A.** Fenestrate trunk. **B.** Stem with leaf and inflorescences. **C.** Part of inflorescence. **D.** Medial section of flower. **E.** Whole (above) and medial section (below) of fruit. **F.** Seed showing hilar scar. **G.** Base of seedling with attached cotyledons.

secondary veins pinnate, the tertiary venation reticulate. Inflorescences terminal, once or twice branched paniculate arrangement of racemes, densely golden-brown-tomentellous, the racemose axes subtended by caducous bracts; pedicels short; bracteoles 2, caducous. Flowers small, actinomorphic, yellow; hypanthium shallowly campanulate, thickened adaxially; sepals 5, ovate, free to hypanthium rim; petals 5, shortly exserted; stamens 10, the filaments glabrous, the anthers ovate-sagittate, dorsifixed, extrorsely dehiscent; ovary tomentulose, subsessile, the style ≤ ovary in length, glabrescent; ovule 1. Fruits sessile, obovoid-pyriform or broadly claviform, tardily separating along one or both sutures, the valves woody. Seed obese, filling cavity, not arillate.

Vouacapoua americana Aubl.
FIG. 63, PL. 37d,e; PART 1: FIG. 9

Trees, 20–37 m tall. Trunks buttressed, irregular, the boles almost always fenestrate. Leaves: leafstalks 13–31 cm long, the petiolar glands round or transversely ellipsoid, 0.6–1.2 mm long; lateral leaflets in 3–5 pairs; pulvinules black, wrinkled, 6–8 mm long; blades lance-elliptic, graduated, the larger ones 11–17 × 3.5–4.5 cm, the base rounded, the apex shortly acuminate. Inflorescences panicles 12–20 cm long, the raceme axes 1–5.5 cm long; pedicels 1–1.5 mm long. Flowers: hypanthia 1–1.5 × 2.5–4 mm; sepals 3.5–4 mm long; petals 4–4.5 × 2–2.3 mm; stamens with filaments 2.5–3 mm long, the anthers 1.5–1.8 mm long. Fruits ovoid, nearly as thick as broad, 6–8.5 × 2–4 cm, the apex pointed, the exocarp yellowish-brown, glabrate, falling to ground intact. Seeds brown or black. Fl (Mar, Nov), fr (Apr, May, Jul); common, in non-flooded forest. *Wacapou* (Créole and commercial name), the wood is hard, durable, insect-proof, and much utilized in construction.

CAMPANULACEAE (Bellflower Family)

Carol A. Gracie

Herbs or shrubs, rarely trees. White exudate often present. Leaves simple, alternate. Stipules absent. Inflorescences cymose but appearing racemose. Flowers 5-merous, zygomorphic, bisexual, protandrous; calyx 5-merous; corolla tubular, with five lobes at apex; stamens 5, fused into column, adnate to base of corolla; ovary inferior, (1)2–3-locular, the style enclosed in staminal column; placentation axile or, less frequently, parietal with intruding placentae. Fruits berries. Seeds lenticular, minutely foveolate-reticulate.

Lanjouw, J. 1966. Campanulaceae. *In* A Pulle (ed.), Flora of Suriname **IV(1):** 302–305, 494–495. E. J. Brill, Leiden.
Stein, B. A. 1998. Campanulaceae. *In* J. A. Steyermark, P. E. Berry & B. K. Holst (gen. eds.), Flora of the Venezuelan Guayana **4:** 122–129. P. E. Berry, B. K. Holst & K. Yatskievych (vol. eds.). Missouri Botanical Garden Press, St. Louis.

CENTROPOGON C. Presl

Herbs, shrubs, or vines. Leaves: blades with margins usually denticulate or serrulate, rarely deeply laciniate. Inflorescences axillary and solitary or terminal racemes subtended by reduced bracts. Flowers: calyx lobes usually free, occasionally fused into tube; corolla zygomorphic, usually brightly colored, the tube straight or abruptly curved, often constricted near base and expanded toward apex, the limb bilabiate; stamens adnate to corolla, the filaments connate except near base, the anthers connate, the 2 lower anthers tipped with tuft of hairs or with scale-like appendage. Fruits fleshy berries. Seeds numerous.

Centropogon cornutus (L.) Druce FIG. 64, PL. 38a,b

Suffrutescent herbs, to 2.5 m tall. Latex white. Stems grooved, finely glandular. Leaves: petioles grooved, glandular, 0.8–2.8 cm long; blades oblong to ovate-oblong, elliptical-oblong to elliptic, 10–16 × 3–6.5 cm, both surfaces of leaves glandular, the abaxial surface with glandular trichomes tightly appressed to primary and secondary veins, the base rounded to cuneate, sometimes slightly asymmetrical, with one side extending down petiole, the apex acute or acuminate, the margins serrate-dentate; secondary veins in 7–11 pairs, the tertiary veins terminating in glandular marginal teeth. Inflorescences axillary and solitary; pedicels 35–65(80) mm long, with one pair of bracteoles near base. Flowers: calyx with cup-like portion adnate to ovary, the marginal teeth 5, small, glandular, spreading, lanceolate to linear, green; corolla tube ca. 40–50 mm long, curved, with constriction near base, the surface glandular, the lobes 7–9 mm long, deep pink to red; stamens adnate to corolla near base, the staminal tube exserted 7–9 mm from corolla, the 2 lower anthers with tufts of dense white trichomes along connectives; ovary unilocular, the style glabrous, slightly exserted from staminal column at maturity; placentation parietal, with 2 intruding placentae. Fruits berries, 10–12 mm diam., with persistent style and calyx lobes. Seeds numerous, small, with lenticel like markings. Fl (Jan, Mar, May, Jun, Jul, Aug, Sep, Oct), fr (May, Jul, Aug, Oct); mostly in disturbed areas along trails.

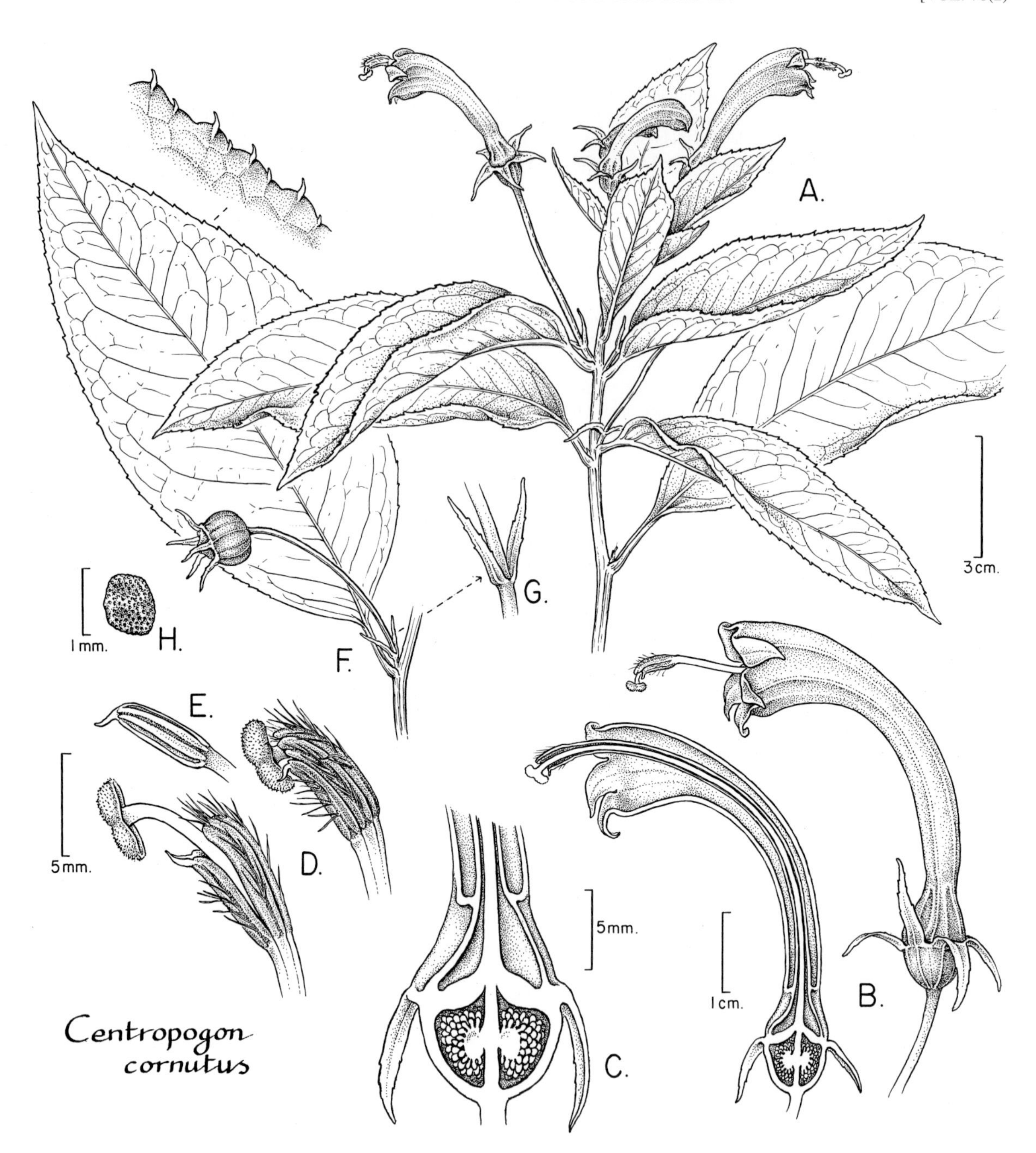

FIG. 64. CAMPANULACEAE. *Centropogon cornutus* (A–D, *Mori et al. 20975*; E, *Mori & Gracie 18352*; F–H, *Mori et al. 22200*). **A.** Apical part of plant with leaves and flowers. **B.** Flower (right) and medial section of flower (left). **C.** Medial section of base of flower. **D.** Two views of apical portion of androecium with style exserted to different degrees. **E.** Adaxial view of lower anther. **F.** Fruit and leaf with detail of leaf margin. **G.** Detail of floral bracts. **H.** Seed.

CANELLACEAE (Canella Family)

John L. Brown and Scott A. Mori

Trees. Leaves simple, alternate, aromatic; blades with glandular dots, the margins entire; pinnately veined. Stipules absent. Inflorescences terminal or axillary, cymes, racemes, or solitary. Flowers actinomorphic, bisexual; sepals 3; petals 5–12 in 2(4) whorls; stamens

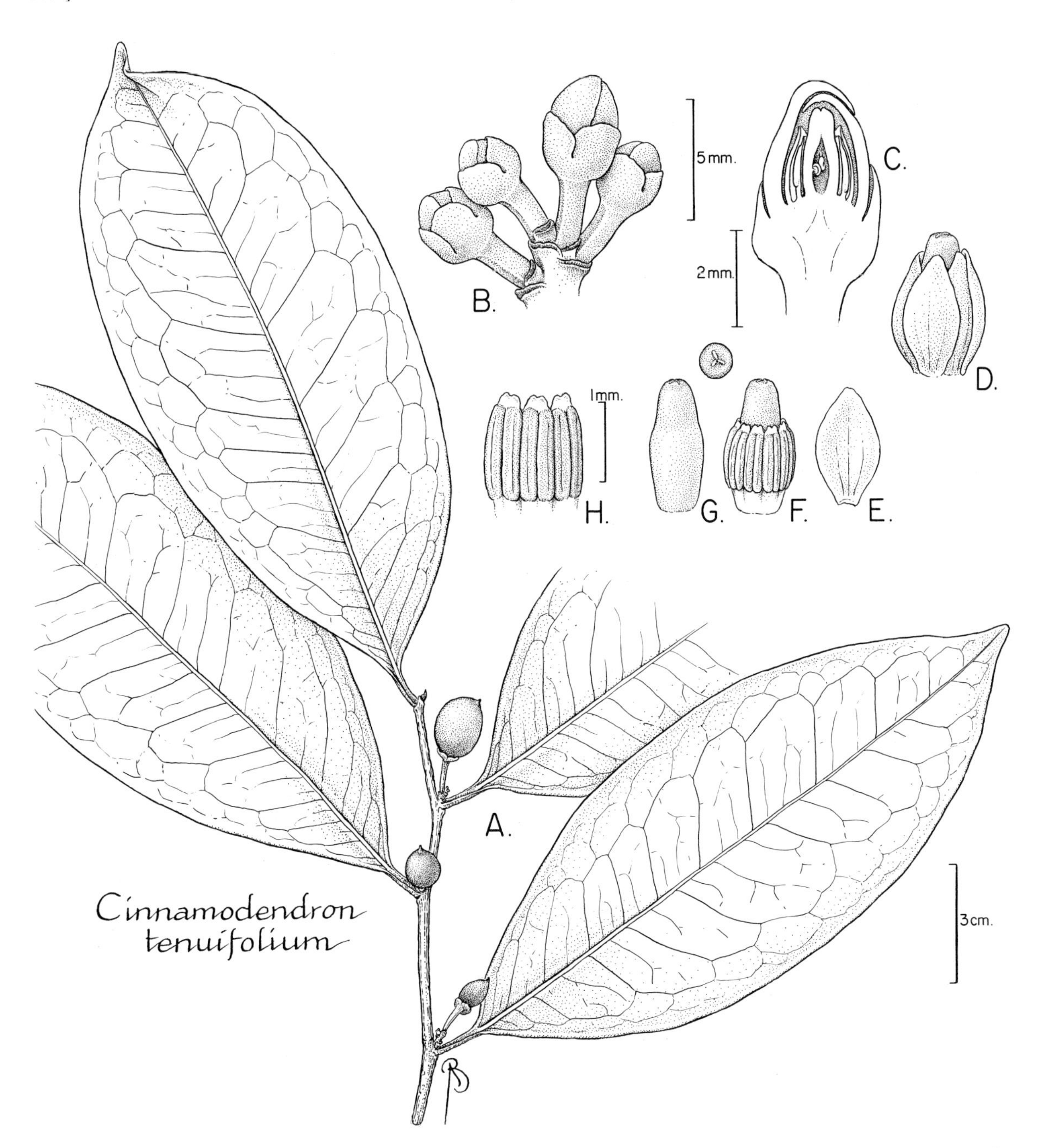

FIG. 65. CANELLACEAE. *Cinnamodendron tenuifolium* (A, *Mori & Pennington 18141*; B–H, *Mori et al. 23002*). **A.** Stem with leaves, fruits. **B.** Part of inflorescence. **C.** Medial section of bud. **D.** Petals surrounding pistil. **E.** Petal. **F.** Androecium surrounding pistil. **G.** Lateral view of pistil (left) and apical view of stigma (above). **H.** Portion of fused anther tube.

6–12, connate into tube, the anthers adnate to outside of tube, extrose; carpels 2–6, united to form unilocular ovary, the style simple, the stigma 2–6-lobed; placentation parietal. Fruits berries.

Uittien, H. 1966. Canellaceae. *In* A. Pulle (ed.), Flora of Suriname **III(1):** 304–305. E. J. Brill, Leiden.

CINNAMODENDRON Endl.

A single genus native to the Guianas. The generic description the same as the family description for our area.

Cinnamodendron tenuifolium Uittien FIG. 65, PL. 38c,d

Small trees, 8–18 m tall. Bark lenticellate, the inner bark red with white streaks and strong, medicinal aroma (aroma of wintergreen). Leaves: petioles 7–9 mm long; blades elliptic to oblong, 10–21 × 5.5–9 cm, finely punctate, aromatic when crushed. Inflorescences in leaf axils, in 1–6 fascicles; pedicels 7–9 mm long. Flowers 3–4 mm long; petals 10, in two whorls, elliptic, densely punctate with oil glands, white, the apex obtuse; staminal tube with 10 small lobes, the linear anthers inserted equidistant around outside of tube; ovary unilocular, the style short, conical; placentae 3, biovulate. Fruits berries, to 1 cm diam., bluish-black, with few seeds embedded in small quantity of thin mucilage. Fl (Jan), fr (May, Jun, Jul, Sep); uncommon, in non-flooded forest.

CAPPARACEAE (Caper Family)

Hugh H. Iltis and Scott A. Mori

Annual or perennial herbs, shrubs, or small to large trees (rarely lianas), with strong, often fetid aroma, all tissues containing glucosinolates (mustard oil precursors evoking pungent sensation when tasted). Stipules minute or lacking (except for pseudostipular spines in some species of *Cleome*). Leaves simple or palmate, alternate; petioles sometimes pulvinate at one or both ends; blades pinnately veined. Inflorescences racemose to corymbose or flowers solitary in axils of leaves. Flowers often zygomorphic, usually bisexual, infrequently unisexual; sepals 4 (to 8 outside our area), free or rarely united; petals usually 4, free, equal and cruciform or posterior pair larger; stamens 6 to ca. 250 (never tetradynamous as in Brassicaceae), as long as petals to long-exserted, the anthers basifixed, introrse, longitudinally dehiscent; receptacle ± conical, often with prominent nectariferous disc, nectariferous gland(s), or scales between calyx and corolla (in species of *Capparis*) or between corolla and stamens; ovary superior, sessile to usually borne on short to elongate stalk called a gynophore (in *Podandrogyne* and some species of *Cleome*, the filaments fused to gynophore to form ± elongate androgynophore), 2-carpellate, 1-locular, the style 1, short, the stigma 1, often sessile, truncate to capitate; placentation parietal, with 2 to many ovules. Fruits various, sessile to long-stipitate on gynophore, either dry to somewhat fleshy, linear-cylindric to globose capsules (siliques) with 2 longitudinal double lines of dehiscence allowing valves to fall from persistent, circular to oblong, ring-shaped, parietal placenta (replum), or globose to obovoid "berries," ± fleshy to sometimes woody-walled, similarly dehiscent to above or often only partially dehiscent or indehiscent, with replum often obscure. Seeds few to many per fruit, ± cochleate-reniform, ovoid to globose, sometimes arillate; embryo white or green, curved or very rarely straight, incumbent, the testa often ± deeply invaginated between radicle and cotyledons, these often much convolute; endosperm very thin.

Jansen-Jacobs, M. J. 1976. Capparaceae. *In* J. Lanjouw & A. L. Stoffers (eds.), Flora of Suriname **I(2):** 512–517. E. J. Brill, Leiden.
Ruiz-Zapata, T. & H. H. Iltis. 1998. Capparaceae. *In* J. A. Steyermark, P. E. Berry & B. K. Holst (gen. eds.), Flora of the Venezuelan Guayana **4:** 132–157. P. E. Berry, B. K. Holst & K. Yatskievych (vol. eds.). Missouri Botanical Garden Press, St. Louis.
Went, J. C. 1937. Capparidaceae. *In* A. Pulle (ed.), Flora of Suriname **II(1):** 397–405. J. H. De Bussy, Amsterdam.
———. 1966. Capparidaceae. *In* A. Pulle (ed.), Flora of Suriname **II(1):** 397–405. E. J. Brill, Leiden.

CAPPARIS L.

Shrubs to large trees (rarely lianas), glabrous or pubescent with tufted, stellate, echinoid, lepidote-peltate, or short, unbranched hairs. Leaves simple; petioles with pulvinus at apex or at both apex and base. Inflorescences racemose, corymbose, compound corymbose-paniculate, or flowers solitary and axillary. Flowers: sepals 4, free (at least at anthesis) or tardily separating, valvate or imbricate, or rarely minute (the flower not enclosed by calyx in bud), equal or in 2 unequal decussate pairs (outer pair completely enclosing smaller inner, or inner pair larger than outer), often enclosing fleshy nectary disc or each sepal subtending nectariferous scale; petals 4, equal (rarely unequal), convolute-imbricate or, rarely, minute and flower not enclosed by calyx in bud, usually white, cream-colored, or purplish- or yellowish-greenish; stamens 6 to >250, the filaments inserted on very short, discoid or conical receptacle (androgynophore); ovary usually borne on short to elongate gynophore. Fruits linear-cylindric to fusiform siliques or oblongoid, obovoid, or ± spherical, ± fleshy "berries," the exocarp coriaceous to hard, dehiscent or tardily dehiscent to indehiscent, the often internally fleshy valves usually persistent on usually thin, often obscure replum (placenta). Seeds 1 to many, usually ± cochleate-reniform, exarillate or with white or red, oily, funicular aril completey surrounding seed, or sometimes with a fibril-suffused clear sarcotesta, embedded in white, orange, or red pulp of fruit wall; embryo white or green, the cotyledons with back flat against radicle but often much enlarged and strongly convoluted, one inside the other and around radicle.

Capparis leprieurii Briq. FIG. 66, PL. 38e

Canopy trees, 15–45 m tall. Trunks usually cylindric to base. Bark smooth, brown to light grayish-brown, the slash yellow, emitting fetid aroma when cut. Leaves when young densely, then sparsely covered with stellate hairs on both surfaces, later becoming glabrescent; petioles 8–22 mm long; blades 11.7–23 × 4–11 cm, the base obtuse to rounded, the apex narrowly to broadly acuminate, sometimes acute, the margins entire. Flowers nocturnal, ca. 3–4 cm diam.; petals creamy white, by morning

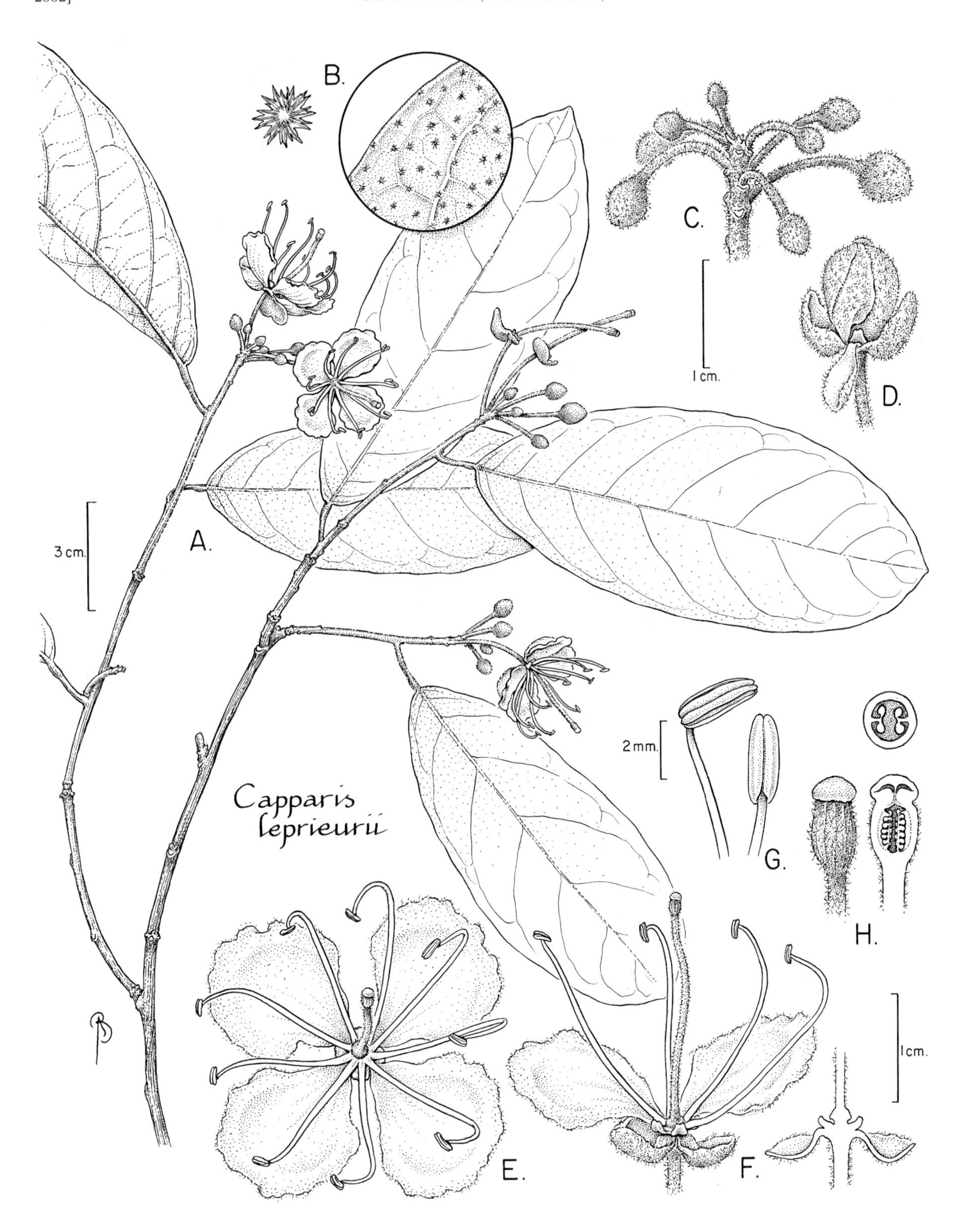

FIG. 66. CAPPARACEAE. *Capparis leprieurii* (*Mori et al. 24021*). **A.** Part of stem with leaves, buds and flowers. **B.** Detail of abaxial surface of nearly fully expanded leaf (right) and enlargement of stellate-echinate hair (left). **C.** Inflorescence in bud. **D.** Mature bud with inner sepal pulled down to show nectar scale. **E.** Apical view of flower. **F.** Lateral view of flower with two petals and five stamens removed (left) and medial section of base of flower, both drawings showing nectar scales. **G.** Lateral (far left) and abaxial (near left) views of anthers. **H.** Lateral view of pistil at apex of gynophore (left), medial section of pistil (right), and transverse section of ovary (above).

caducous; stamens 8(9) in this population, ± 16 in other populations in Guianas. Fruits unknown from our area, oblongoid to subspherical. Fl (May, Sep, Oct); common, scattered in non-flooded forest.

CARICACEAE (Papaya Family)

Scott A. Mori

Trees, soft-stemmed (*Carica*), woody (*Jacaratia*), or sometimes shrubs (some species of *Carica*) or herbs (*Jarilla*) outside of our area. Exudate white. Stipules usually absent. Leaves simple and palmately divided or palmately compound, infrequently simple and entire in some species of *Carica* and *Jarilla* outside our area, alternate. Inflorescences solitary or multi-flowered. Flowers actinomorphic, usually unisexual in our area (plants often dioecious); calyx much reduced in comparison with corolla, usually 5-merous; corolla gamopetalous, the lobes free, usually 5-merous; stamens adnate to corolla, 5 long and 5 short, the anthers with introrse dehiscence; ovary superior, 1-locular; placentation parietal. Seeds numerous. Fruits fleshy berries. Seeds with 2 broad, flat cotyledons, the embryo embedded in soft, fleshy endosperm.

Lindeman, J. 1984. Caricaceae. *In* A. L. Stoffers & J. C. Lindeman (eds.), Flora of Suriname **V(1):** 585–591. E. J. Brill, Leiden.
Miller, J. S. 1998. Caricaceae. *In* J. A. Steyermark, P. E. Berry & B. K. Holst (gen. eds.), Flora of the Venezuelan Guayana **4:** 162–163. P. E. Berry, B. K. Holst & K. Yatskievych (vol. eds.). Missouri Botanical Garden Press, St. Louis.

1. Trunks soft and succulent, the trunk and stem unarmed. Leaves simple but palmately divided. Staminate flowers yellow; ovary 1-locular. . . . . *Carica.*
1. Trunks woody, the trunk and stem armed with prickles. Leaves palmately compound. Staminate flowers green; ovary 5-locular. . . . . *Jacaratia.*

CARICA L.

Trees, soft-stemmed, usually unbranched, unarmed. Trunks with conspicuous leaf scars. Leaves simple, palmately divided. Inflorescences axillary. Staminate inflorescences multi-flowered, much branched. Pistillate inflorescences solitary, appearing cauline with development of fruit. Flowers unisexual (plants dioecious); ovary 1-locular.

Carica papaya L. FIG. 67

Trees, to 10 m tall. Staminate inflorescences branched, multi-flowered. Pistillate inflorescences not branched, few-flowered. Staminate flowers yellow, with alternating long and short stamens, the filaments pubescent. Pistillate flowers yellow, much larger than staminate flowers, the stigma divided into 5 lobes, the lobes in turn divided at apex. Fruits borne along trunk, large (to 20 cm or more long), yellow to yellow-orange, edible berries with numerous black seeds. Fl (year round), fr (year round); cultivated, near homesites but sometimes found in weedy areas and appearing native. *Mamão* (Portuguese), *papaya* (English), *papaye* (Créole and French).

This is one of the most appreciated fruits in the region. The green fruits are eaten as a salad or as a boiled vegetable. The ripe fruits are eaten as a dessert or made into a juice. The latex and other parts of the plant possess many medicinal properties (Grenand et al., 1987) and the latex is the source of the digestive enzyme papain which is commercially used to make meat tenderizers.

JACARATIA A. DC.

Trees, woody, branched, armed. Trunks without conspicuous leaf scars. Leaves palmately compound. Inflorescences axillary. Staminate inflorescences much branched, multi-flowered, unarmed. Pistillate inflorescences of solitary flowers, these on long prickly peduncle, not becoming cauline with development of fruit. Flowers unisexual (plants dioecious); ovary 5-locular.

Jacaratia spinosa (Aubl.) A. DC. FIG. 68, PL. 39a

Trees, to 20 m tall. Trunks and branches armed with prickles. Leaves probably deciduous; petioles 6–23 cm long; leaflets 4–7; blades 6.5–16 × 2–5.5 cm, the abaxial usually much lighter in color than adaxial surface, subtended by conspicuous, triangular stipels ca. 1 mm long; secondary veins in 13–15 pairs. Staminate flowers green; sepals 5, imbricate, ovate, ca. 1 × 1.5 mm; petals 5, fused at base into tube ca. 17 mm long, the lobes spreading, 8 × 2 mm; stamens with filaments extending up abaxial sides of anthers, the anthers apiculate; pistillode ca. 3 mm long. Pistillate flowers unknown from our area, petals nearly free in collections from other areas. Fruits to 8 cm long. Fl (Oct), fr (Dec, Mar); scattered, in non-flooded forest.

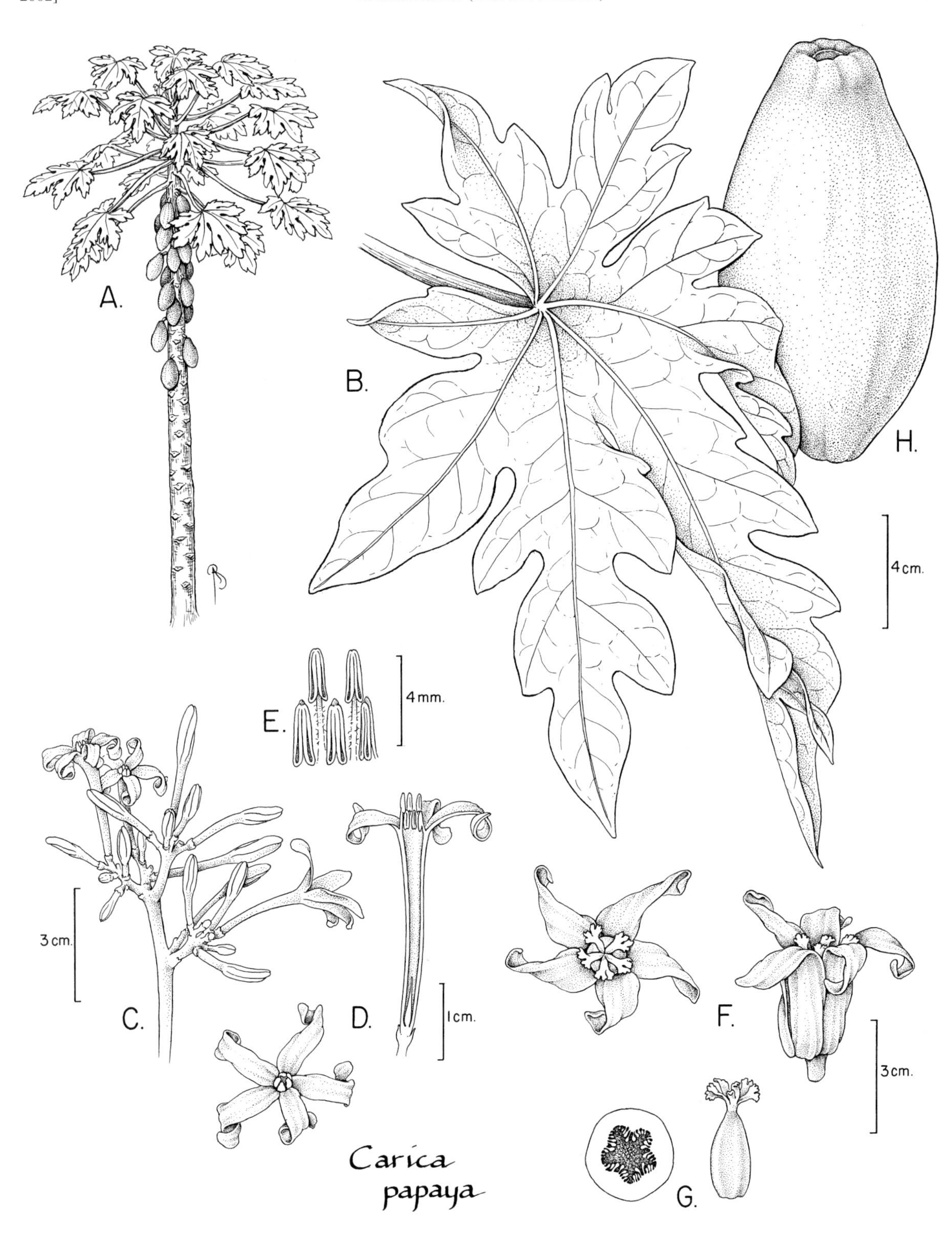

FIG. 67. CARICACEAE. *Carica papaya* (unvouchered, from field sketch in Saül, French Guiana). **A.** Upper part of pistillate plant with stem, leaves, and fruit. **B.** Leaf. **C.** Part of staminate inflorescence. **D.** Apical view (left) and medial section (right) of staminate flower. **E.** Adaxial view of part of androecium. **F.** Apical (left) and lateral (right) views of pistillate flower. **G.** Transverse section of ovary (left) and lateral view (right) of pistil. **H.** Fruit.(Reprinted with permission from P. Acevedo-Rodríguez, *Flora of St. John, U.S. Virgin Islands*, Mem. New York Bot. Gard. 78, 1996.)

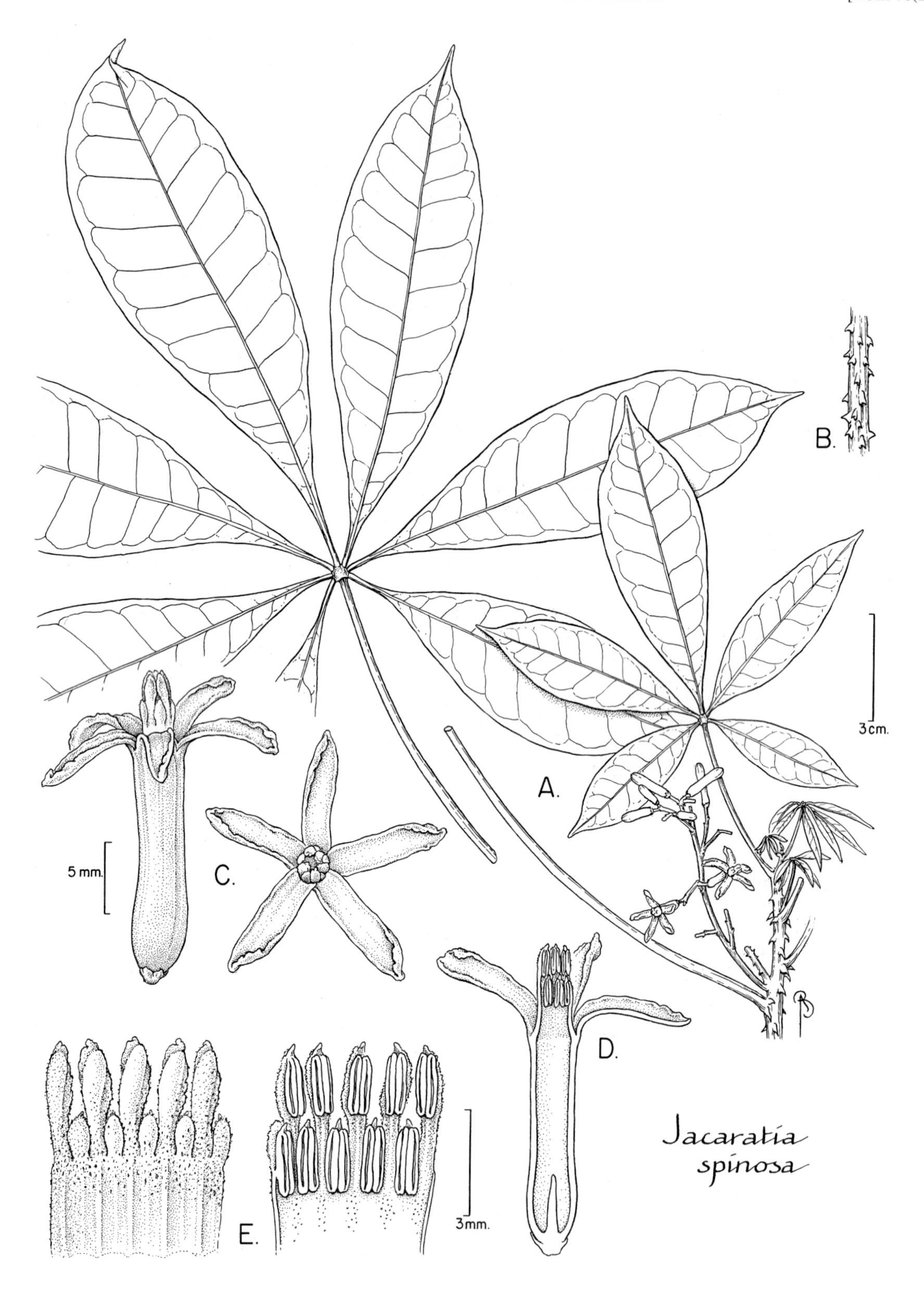

FIG. 68. CARICACEAE. *Jacaratia spinosa* (A, C–E, *Mori et al. 22659*; B, *Mori & Boom 15345*). **A.** Upper part of stem with staminate inflorescence. **B.** Stem with prickles. **C.** Lateral (left) and apical views of staminate flowers (right). **D.** Medial section of staminate flower; note the pistillode. **E.** Exterior (left) and interior (right) of corolla tube.

Plates 33–40

Plate 33. BURSERACEAE. **a.** *Protium occultum* (*Mori et al. 22719*), stem with bases of leaves of single leaflets (note pulvini on petioles) and inflorescences in bud. **b.** *Tetragastris altissima* (*Mori & Gracie 23987*), flowers and buds. **c.** *Trattinnickia rhoifolia* (*Mori & Gracie 18337*), infructescences. CACTACEAE. **d.** *Epiphyllum phyllanthus* (unvouchered from Amazonas, Brazil), flower open at night. **e.** *Epiphyllum phyllanthus* (*Mori et al. 21315* from Amazonas, Brazil), flower. **f.** *Epiphyllum phyllanthus* (*Mori & Gracie 18988*), fruit.

Plate 34. CAESALPINIACEAE. **a.** *Batesia floribunda* (unvouchered), open fruits and seeds on ground. **b.** *Batesia floribunda* (*Mori et al. 22985*), open flower and buds. **c.** *Bauhinia guianensis* (*Mori et al. 24010*), flower. **d.** *Bauhinia guianensis* (*Mori & Gracie 21124*), deeply bifid leaf and tendrils. **e.** *Bauhinia surinamensis* (*Mori et al. 22647*), stem of liana. **f.** *Bauhinia outimouta* (*Mori et al. 24681*), fresh flower and old flowers. [Photo by S. Mori]

Plate 35. CAESALPINIACEAE. **a.** *Cassia spruceana* (*Mori et al. 24150*), flower; note heteromorphic stamens. **b.** *Dicorynia guianensis* (*Mori & Snyder 24340*), flowers, each with one long and one short stamen and pistil with dark brown ovary. [Photo by S. Mori] **c.** *Dialium guianense* (*Mori et al. 24678*), flowers; note two stamens per flower and pistil with dark brown ovary. [Photo by S. Mori]

Plate 36. CAESALPINIACEAE. **a.** *Martiodendron parviflorum* (unvouchered), base of tree with branched, joined buttresses. **b.** *Martiodendron parviflorum* (*Mori et al. 24247*), fruits. **c.** *Senna lourteigiana* (*Mori & Gracie 21104*), pistil and heteromorphic stamens.

Plate 37. CAESALPINIACEAE. **a.** *Tachigali* aff. *bracteolata* (*Mori & Gracie 22100*), flower. **b.** *Tachigali amplifolia* (*Mori et al. 23060*), inflorescence. **c.** *Tachigali* sp. (*Mori et al. 24804*), inflorescence. [Photo by S. Mori] **d.** *Vouacapoua americana* (unvouchered), base of trunk. **e.** *Vouacapoua americana* (*Mori et al. 22757*), part of inflorescence.

Plate 38. CAMPANULACEAE. **a.** *Centropogon cornutus* (*Mori et al. 22200*), flower. **b.** *Centropogon cornutus* (unvouchered), plants with flowers. CANELLACEAE. **c.** *Cinnamodendron tenuifolium* (*Mori et al. 23982*), slash of bark. **d.** *Cinnamodendron tenuifolium* (*Mori et al. 23982*), fruits. CAPPARACEAE. **e.** *Capparis leprieurii* (*Mori et al. 24021*), flower.

Plate 39. CARICACEAE. **a.** *Jacaratia spinosa* (*Mori et al. 22659*), staminate inflorescence. CARYOCARACEAE. **b.** *Caryocar glabrum* (*Mori et al. 24720*), nocturnal flowers and buds. [Photo by S. Mori] CECROPIACEAE. **c.** *Cecropia sciadophylla* (*Mori et al. 21555*), infructescence. **d.** *Coussapoa latifolia* (*Mori et al. 23728*), staminate inflorescence. **e.** *Coussapoa angustifolia* (*Mori et al. 23097*), fruits. **f.** *Pourouma minor* (*Mori et al. 23973*), terminal inflorescence.

Plate 40. CELASTRACEAE. **a.** *Goupia glabra* (*Mori et al. 23369*), flowers. CHRYSOBALANACEAE. **b.** *Hirtella silicea* (*Mori & Gracie 23934*), inflorescence; note whorls of glands on pedicels. **c.** *Licania discolor* (*Mori et al. 23304*), inflorescence. **d.** *Licania alba* (*Mori et al. 22994*), sectioned immature fruit showing developing seed.

Protium occultum a.

Tetragastris altissima b.

Trattinnickia rhoifolia c.

CACTACEAE

Epiphyllum phyllanthus d.

Epiphyllum phyllanthus e.

Epiphyllum phyllanthus f.

Plate 33

Batesia floribunda a.

Batesia floribunda b.

Bauhinia guianensis c.

Bauhinia guianensis d.

Plate 34

Bauhinia surinamensis e.

Bauhinia outimouta f.

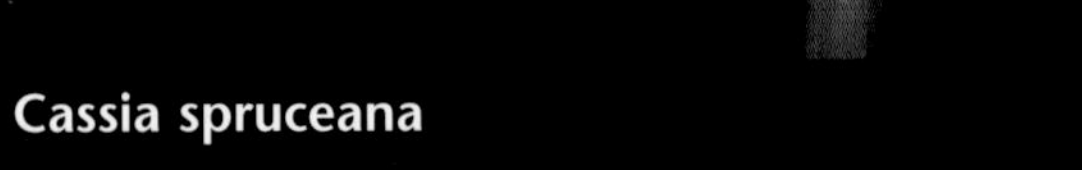

Cassia spruceana **a.**

Dicorynia guianensis **b.**

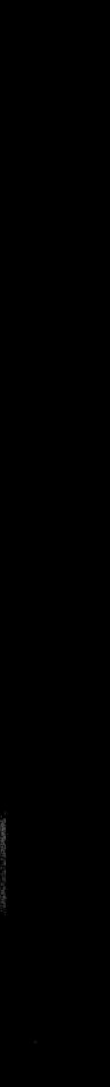

Dialium guianense **c.**

Plate 35

Martiodendron parviflorum a.

Martiodendron parviflorum b.

Senna lourteigiana c.

Plate 36

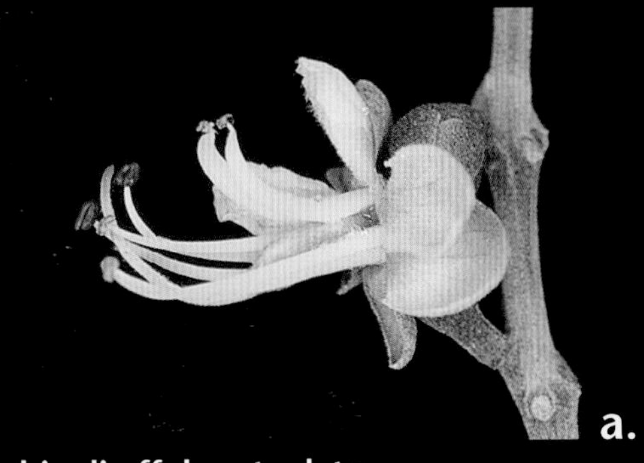

a.

Tachigali aff. bracteolata

c.

Tachigali sp.

b.

Tachigali amplifolia

d.

Vouacapoua americana

e.

Vouacapoua americana

CAMPANULACEAE

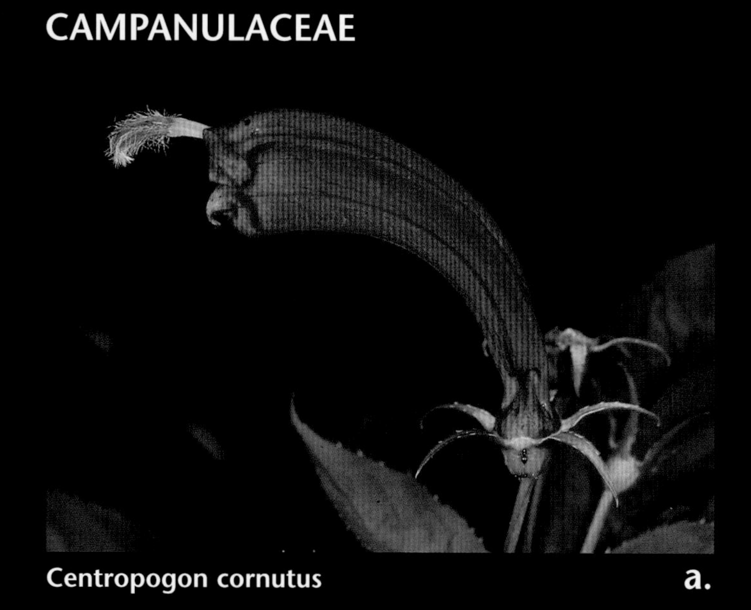

Centropogon cornutus a.

Centropogon cornutus b.

CANELLACEAE

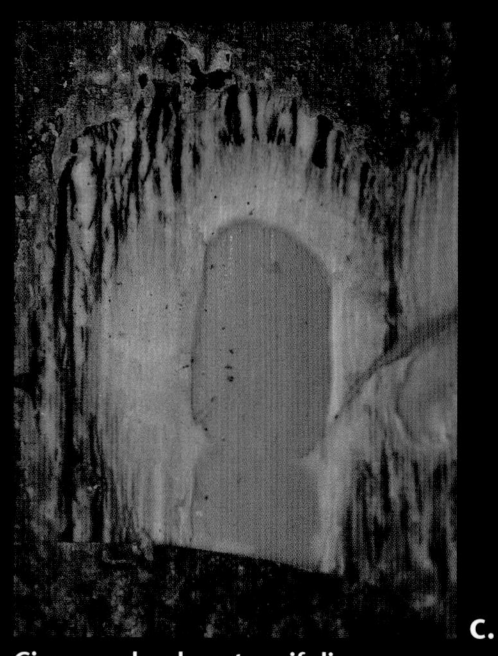

Cinnamodendron tenuifolium c.

Cinnamodendron tenuifolium d.

CAPPARACEAE

Capparis leprieuri e.

Plate 38

CARICACEAE

Jacartia spinosa **a.**

CARYOCARACEAE

Caryocar glabrum **b.**

c.

Cecropia sciadophylla

CECROPIACEAE

Coussapoa angustifolia **e.**

d.

Coussapoa latifolia

f.

Pourouma minor

Plate 39

CELASTRACEAE

Goupia glabra **a.**

CHRYSOBALANACEAE

Hirtella silicea **b.**

Licania discolor **c.**

Licania alba **d.**

Plate 40 (Chrysobalanaceae continued on Plate 41)

CARYOCARACEAE (Soari Family)

Jean-Jacques de Granville

Medium-sized to very large trees. Trunks cylindrical, with or without buttresses, the buttresses often extending long distances from tree. Outer bark usually rough, scaly to deeply fissured, the inner bark fibrous, pale yellow to flesh-colored. Stipules 2, caducous. Leaves palmately compound, opposite (in our area) or alternate (*Anthodiscus*, outside our area); petioles ± long, without persistent stipels at apex. Inflorescences terminal, erect and often at periphery of crown, racemose, with 8–30 flowers; peduncle long; bracts and bracteoles absent or caducous. Flowers actinomorphic, bisexual, large and nocturnal in our area, smaller and diurnal in *Anthodiscus*; sepals 5, imbricate and large in *Caryocar*, small and reduced in *Anthodiscus*; petals 5, imbricate, scarcely united at base; stamens numerous, adnate to base of petals, the filaments long, much exceeding corolla, coiled into S-shape in bud, straight at anthesis, the inner ones shorter, tuberculate distally, the anthers small, oblong, bilocular, the stamens and petals falling together; ovary superior, usually 4-locular in *Caryocar*, to 20 locular in *Anthodiscus*, the styles 4 in *Caryocar*, to 20 in *Anthodiscus*, as long as stamens; placentation axile, the ovules 1 per locule. Fruits drupe-like, eventually splitting into one-seeded, reniform pyrenes, the mesocarp fleshy, the endocarp spiny. Seeds with hypogeal germination; embryo with straight (*Caryocar*) or spirally twisted radicle (*Anthodiscus*).

Görts-van Rijn, A. R. A. 1986. Caryocaraceae. *In* A. L. Stoffers & J. C. Lindeman (eds.), Flora of Suriname **III(1–2):** 473–474. E. J. Brill, Leiden.

Lanjouw, J. & P. F. Baron van Heerdt. 1966. Caryocaraceae. *In* A. Pulle (ed.), Flora of Suriname **III(1):** 366–372. E. J. Brill, Leiden.

Prance, G. T. 1998. Caryocaraceae. *In* J. A. Steyermark, P. E. Berry & B. K. Holst (gen. eds.), Flora of the Venezuelan Guayana **4:** 164–170. P. E. Berry, B. K. Holst & K. Yatskievych (vol. eds.). Missouri Botanical Garden Press, St. Louis.

——— & M. Freitas da Silva. 1973. Caryocaraceae. Fl. Neotrop. Monogr. **12:** 1–75.

CARYOCAR L.

Leaves opposite. Flowers: sepals imbricate and large; ovary usually 4-locular, the styles 4. Fruits large drupes. Seeds: embryo with straight radicle.

Mesocarp and endosperm often edible.

1. Outer bark scaly to shallowly fissured. Blades glabrous to slightly puberulous abaxially; leaflets with bases cuneate to acute; lateral veins in 7–12 pairs. Endocarp with spines 5–15 mm long. . . . . . . *C. glabrum* subsp. *glabrum*.
1. Outer bark deeply fissured. Blades brown tomentose to hirsute abaxially; leaflets with bases rounded to cordate; lateral veins in 12–14 pairs. Endocarp with spines ca. 3 mm long. . . . . *C. villosum*.

Caryocar glabrum (Aubl.) Pers. subsp. **glabrum**
FIG. 69, PL. 39b; PART 1: FIG. 9, PL. IIb

Large to very large trees, to 46 m tall. Trunks cylindrical, to 130 cm diam., with or without buttresses, when present the buttresses usually low, thick, rounded, usually simple, infrequently branched, and spreading. Outer bark smooth in young trees, in older trees rough, scaly to shallowly fissured, grayish to dark brown, lenticellate, the lenticels in vertical rows, the inner bark flesh colored, with yellow inclusions, very fibrous, the fibers oriented in different directions. Young stems glabrous. Leaves without persistent stipels; petioles (2)4–8(15) cm long, glabrous; petiolules 2–10(15) mm long, canaliculate adaxially, glabrous; leaflets elliptic to obovate, 6–11(16) × 2.5–6(8.5) cm, glabrous to slightly puberulous abaxially, the base cuneate to acute, the apex long acuminate, the margins entire to crenate; secondary veins in 7–12(15) pairs, prominent abaxially, somewhat plane adaxially. Inflorescences erect, bearing 8–20(30) flowers, the axes glabrous, lenticellate; peduncle 2–11 cm long; rachis 1–4(6.5) cm long; pedicels 1–2.5 cm long. Flowers: calyx cupuliform, green, the lobes rounded, 4–5 mm wide; petals elliptic-oblong, 2–3 × 0.8–1.5 cm; filaments 4–6 cm long, purple or purplish-red or red to pink at base and gradually grading into white toward apex, or entirely white, the anthers yellow; styles yellowish, grading to red basally. Fruits ellipsoid-globose, 5–6 × 5–10 cm; exocarp tan, glabrous, crustaceous, the mesocarp fleshy, ca. 5 mm thick, enveloping 1–2 subreniform stones 3–4 × 4–5 cm, the endocarp with numerous 5–15 mm long, brown, woody spines. Fl (Aug, Nov, Dec), fr (Feb, May, Jun); fairly common, in non-flooded forest. *Chawari* (Créole).

Caryocar villosum (Aubl.) Pers.

Usually very large trees, to 50 m tall. Trunks cylindrical, to 200(250) cm diam., with or without buttresses, the buttresses, when present thick, low, simple, spreading. Outer bark dark grayish brown, deeply fissured, the inner bark yellowish cream-colored, fibrous. Young stems puberulous to glabrescent. Leaves without stipels; petioles (4)7–10(15) cm long, villose-tomentose; petiolules 2–3 mm long, canaliculate adaxially, brown tomentose; leaflets broadly elliptic to ovoid or obovoid, 7–22 × 5–12 cm, the lateral ones usually slightly smaller than terminal one, brown tomentose to hirsute abaxially, puberulous on veins adaxially, the base rounded to cordate, the apex shortly acuminate, the margins crenulate to crenate, sometimes serrate; secondary veins in 12–14 pairs, prominent abaxially, slightly impressed adaxially. Inflorescences erect, bearing 15–30 flowers, the axes puberulous to glabrescent, lenticellate; peduncle 5–13 cm long; rachis 2–4 cm long; pedicels 2–3.5 cm long. Flowers: calyx campanulate, grayish-green, the lobes 6–7 mm wide, rounded; petals elliptic-oblong to rhomboid, somewhat asymmetric, 2.5–3.5 × 1–1.5 cm; longest filaments 5–7 cm long, white to yellowish white, the anthers yellow; styles yellow. Fruits ellipsoid-globose, 6–7 × 7–8 cm; exocarp tan, glabrous, lenticellate, the mesocarp fleshy, enveloping reniform stone ca. 4 × 5 cm, the endocarp with numerous 3 mm long brown spines. Fl (Sep, Nov); uncommon, in non-flooded forest. *Chawari* (Créole).

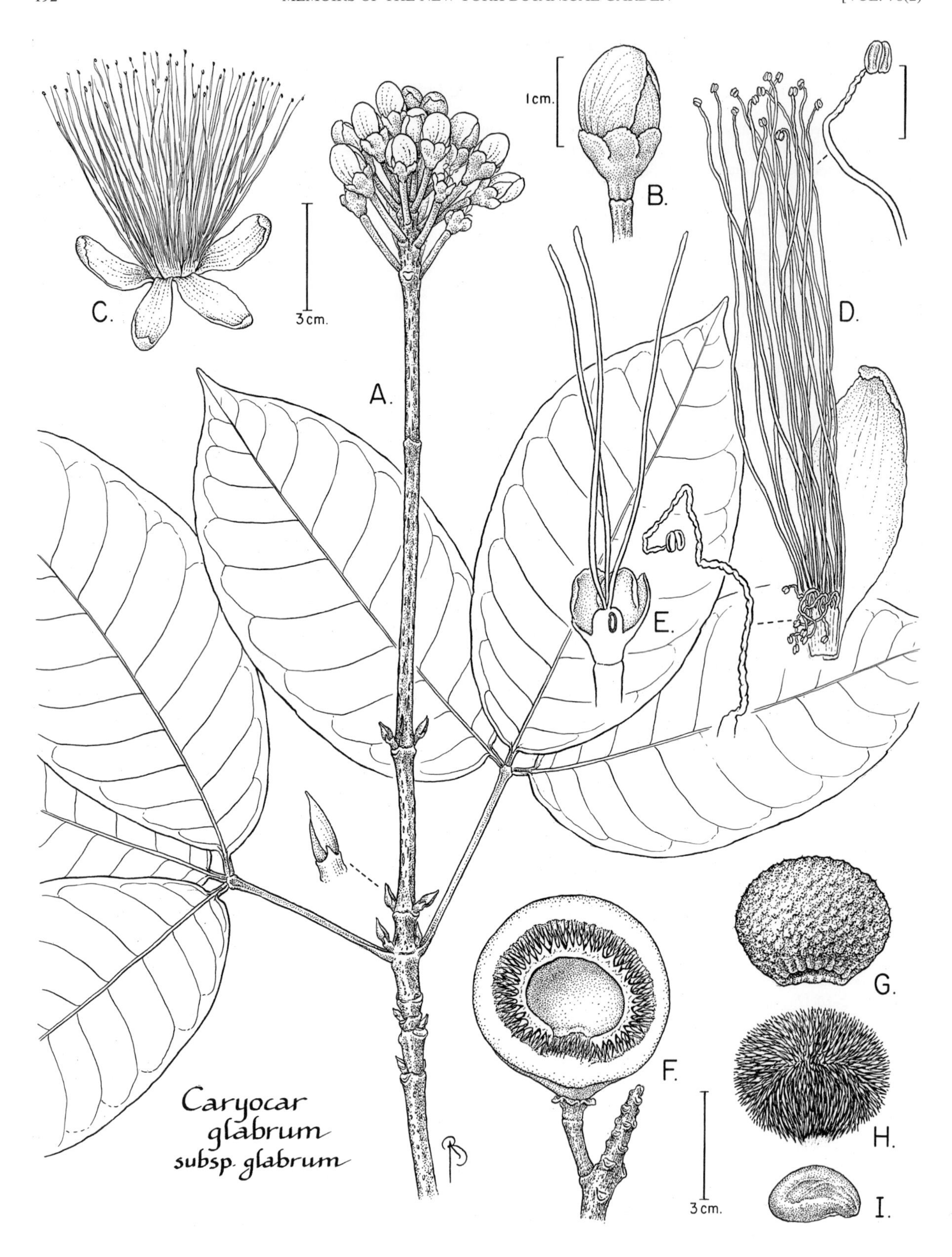

FIG. 69. CARYOCARACEAE. *Caryocar glabrum* subsp. *glabrum* (A, B, E, *Mori & Boom 15295*; C, D, *Mori & Boom 15298*; F, *Mori et al. 22997*; G, *Mori et al. 18181*; H, I, *Mori et al. 18182*). **A.** Part of stem with leaves and inflorescence in bud (right); detail of leaf bud (below). **B.** Lateral view of flower bud. **C.** Lateral view of corolla and stamens as they fall from the tree. **D.** Lateral view of petal and part of the androecium (left) with detail of shorter, inner stamen (below left) and apex of longer, outer stamen (above). **E.** Medial section of pistil and calyx. **F.** Medial section of fruit with seed removed; note spiny endocarp. **G.** Lateral view of thin skin that covers spiny endocarp. **H.** Lateral view of endocarp. **I.** Seed.

CARYOPHYLLACEAE (Pink Family)

Robert A. DeFilipps and Shirley L. Maina

Annual or perennial, often prostrate herbs. Stipules present. Leaves opposite and decussate, simple, sessile or indistinctly petiolate, often connate and amplexicaulous at base, the margins entire. Inflorescences monochasial or dichasial, bracteate cymes. Flowers actinomorphic, bisexual; sepals 4–5, persistent; petals 4–5, free or nearly so, sometimes bifid and clawed; stamens 2–10, free or filaments basally united, the anthers 2-locular; ovary superior, 2- to 5-carpellate and -locular, the style 1, the stigma simple or 2- to 5-lobed, or the styles 2–5 and free or united at base; placentation basal, central, or free-central, the ovules campylotropous. Fruits capsules, valvate or dehiscing by 2–5 apical teeth. Seeds 1-numerous, small.

Aymard, C. & N. L. Cuello A. 1998.Caryophyllaceae. *In* J. A. Steyermark, P. E. Berry & B. K. Holst (gen. eds.), Flora of the Venezuelan Guayana **4:** 171–174. P. E. Berry, B. K. Holst & K. Yatskievych (vol. eds.). Missouri Botanical Garden Press, St. Louis.

Ooststroom, S. J. van. 1966. Caryophyllaceae. *In* A. Pulle (ed.), Flora of Suriname **I(1):** 150–153. E. J. Brill, Leiden.

DRYMARIA Schult.

Annual herbs. Stems usually prostrate, spreading, sometimes rooting at nodes. Stipules small, scarious, ochraceous, persistent or fugaceous. Leaves petiolate, glabrous to villous, often with glandular hairs. Inflorescences few-flowered cymes; bracts scarious. Flowers: sepals 5, free, herbaceous with scarious margins; petals (0–3)5, often deeply 2-cleft (bifid); stamens 2–5, the filaments slightly basally united; ovary sessile or substipitate, 1-locular, the styles 3, basally united; placentation free-central. Fruits capsules, 3-valved. Seeds globose-reniform, cochleate, or hippocrepiform, usually granular or tuberculate, rarely smooth.

Drymaria cordata (L.) Schult. Fig. 70

Annual, erect to prostrate herbs. Stems weakly branched, angled, often rooting at nodes, to 45 cm long, glabrous to densely glandular-puberulent. Stipules membranous, multilacerate, to 2 mm long. Leaves: petioles filiform, 2–15 mm long; blades orbicular, cordate, or reniform, 5–25 × 5–30 mm, glabrous or puberulent, the apex obtuse or acute and mucronulate, the base rounded to cordate. Inflorescences terminal or axillary, laxly few-flowered dichasial cymes; pedicels 2–15 mm long, densely stipitate-glandular. Flowers: sepals ovate, oblong-lanceolate, or lanceolate, 2.5–3.5 mm long, mostly glabrous but with stipitate-glandular puberulent midvein; petals 5, deeply clawed and bifid into linear lobes, white, 2–3 mm long; stamens 2–3(5), the filaments ca. 2 mm long; ovary ovoid, the styles 0.5–1 mm long. Fruits capsules, ovoid, 1.5–2.5 mm long. Seeds 1–12, cochleate, 1–1.5 mm long, dark reddish-brown or black, densely tuberculate. Fl (Aug), fr (Aug); weed in disturbed habitats. *Mignonette*, *petit quinine*, *timignonette* (Créole).

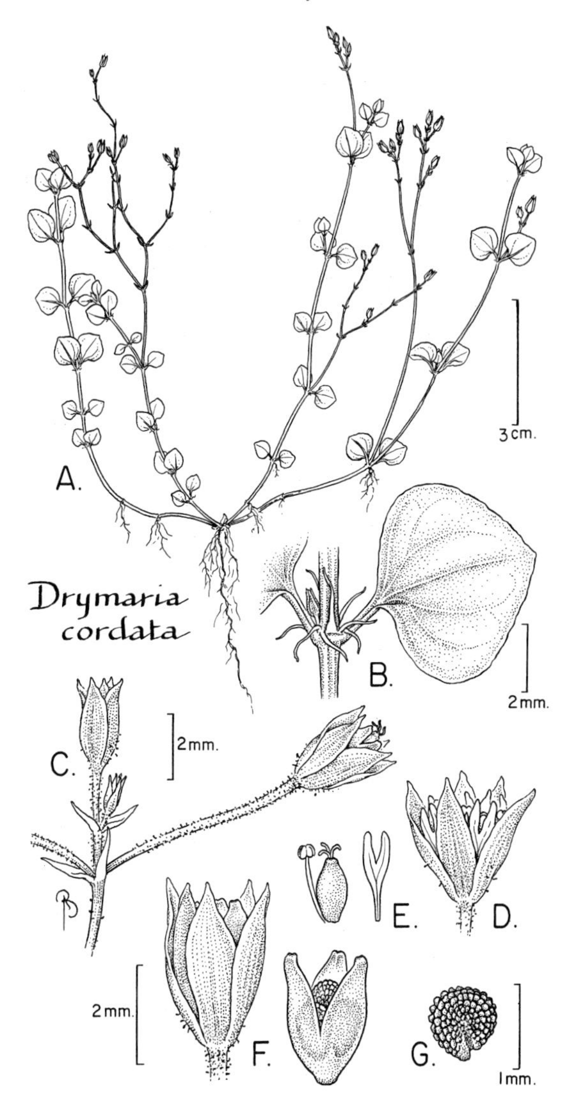

FIG. 70. CARYOPHYLLACEAE. *Drymaria cordata* (*Mori & Gracie 23381*). **A.** Habit. **B.** Detail of node showing leaf and stipules. **C.** Detail of inflorescence showing glandular pubescence. **D.** Lateral view of flower. **E.** Petal (near left) and lateral view of pistil and one stamen (far left). **F.** Lateral views of fruit within calyx (left) and dehisced fruit with calyx removed (right). **G.** Seed.

CECROPIACEAE (Cecropia Family)

Cornelis C. Berg

Trees or shrubs, with adventitious roots. Plants, when cut, exuding watery sap turning black on exposure to air. Stipules fully encircling stem, fused. Leaves spirally arranged. Inflorescences in pairs in leaf axils, usually branched. Flowers actinomorphic, unisexual (plants dioecious). Staminate flowers: tepals 2–4, free or fused; stamens 1–4, free or fused. Pistillate flowers: perianth tubular; ovary 1-locular, the stigma 1, ± peltate to penicillate, infrequently capitate (*Pourouma minor*); placentation basal, the ovules 1 per ovary. Fruits achenes, enclosed by, but free of fleshy perianth. Seeds small or large; endosperm present (in small seeds) or absent (in large seeds).

Berg, C. C., R. W. A. P. Akkermans & E. C. H. van Heusden. 1990. Cecropiaceae: *Coussapoa* and *Pourouma*, with an introduction to the family. Fl. Neotrop. Monogr. **51:** 1–208.

———. 1992. Cecropiaceae. *In* A. R. A. Görts-van Rijn (ed.), Flora of the Guianas Ser. A, **11:** 93–124, 192–222. Koeltz Scientific Books, Koenigstein.

———. 1998. Cecropiaceae. *In* J. A. Steyermark, P. E. Berry & B. K. Holst (gen. eds.), Flora of the Venezuelan Guayana **4:** 174–190. P. E. Berry, B. K. Holst & K. Yatskievych (vol. eds.). Missouri Botanical Garden Press, St. Louis.

1. Free-standing trees. Stipule scars horizontally oriented.
    2. Stems hollow. Leaf blades peltate, radially incised. Flowers in spikes. Seeds small. . . . . *Cecropia.*
    2. Stems not hollow. Leaf blades entire or palmately incised. Flowers not in spikes. Seeds large. . . . . *Pourouma.*
1. Hemiepihytic trees or shrubs. Stipule scars obliquely oriented. . . . . *Coussapoa.*

CECROPIA Loefl.

Free-standing, understory or canopy, pioneer trees, often with stilt-roots and or flying buttresses. Stem and branches with hollow internodes in our area, the internodes inhabited by ants (*C. obtusa*) or not (*C. sciadophylla*). Stipule scars horizontally oriented. Leaves peltate; petioles long, often with patch of brown indumentum ("trichilium") at base, this containing 1–2 mm long ellipsoid bodies ("Muellerian bodies"), these absent in *C. sciadophylla*; blades large, radially incised. Inflorescences usually 4 or more finger-like spikes borne on common peduncle, - enclosed by spathe until flowering. Flowers: perianth tubular. Staminate flowers: stamens 2, free, the anthers detaching from filaments at flowering. Pistillate flowers sessile or subsessile; stigma penicillate. Fruits small (1–3 mm long), enclosed in fleshy perianth, greenish at maturity.

1. Trichilia present. Leaf blades not incised to petiole, with 7–9 lobes, scabrous adaxially. . . . . *C. obtusa.*
1. Trichilia absent. Leaf blades incised to petiole, with 11–15 lobes, smooth adaxially. . . . . *C. sciadophylla.*

Cecropia obtusa Trécul

FIG. 71; PART 1: FIG. 9, PLS. IIIa, VIIb

Trees, to 15 m tall. Stems usually inhabited by ants. Leaves: petioles often whitish because of dense cobwebby hairs, with trichilium at base; blades incised to 8/10 distance between margin and petiole, with 7–9 lobes, ca. 60 cm diam., chartaceous to subcoriaceous, the adaxial surface scabrous, the abaxial surface often whitish because of dense cobwebby hairs. Staminate inflorescences with spikes ca. 0.3 cm thick. Fl (Jun, Sep, Oct), fr (Aug, Oct); common, in secondary vegetation, especially along airport road. *Bois-canon* (Créole).

Cecropia sciadophylla Mart. PL. 39c

Canopy trees, to 30 m tall. Stems not inhabited by ants. Leaves: petioles without cobwebby hairs; blades incised to petiole, with 11–15 lobes, to 100 cm diam., the segments often petiolulate, coriaceous, the adaxial surface smooth, the abaxial surface with very inconspicuous cobwebby hairs. Staminate inflorescences with spikes 0.4–0.8 cm thick. Fl (Nov), fr (Aug, Nov); abundant, in secondary vegetation, especially along airport road. *Bois-canon*, *Bois-canon mâle* (Créole).

COUSSAPOA Aubl.

Hemiepiphytic trees or shrubs, the principal descending root often with pairs of lateral roots clasping trunk of host tree. Stem and branches usually without hollow internodes (hollow in *C. asperifolia* from outside our area). Stipule scars obliquely oriented. Leaves not peltate, the petiole attached to base of blade; petioles not unusually long, without patch of brown indumentum at base; blades medium-sized, entire. Flowers: tepals fused, the perianth urceolate. Staminate and pistillate flowers in globose heads. Fruits small (1–3 mm long), enclosed by fleshy perianth, yellow-orange at maturity.

1. Petioles 1.5–2 cm long; blades broadest at or above middle, usually 2.5–5 cm wide, the base cuneate; secondary veins in 2–3 pairs. Pistillate inflorescences of single heads. . . . . *C. angustifolia.*
1. Petioles 2–7 cm long; blades broadest at or below middle, 6.5–10.5 cm wide, the base obtuse to rounded; secondary veins in 5–6 pairs. Pistillate inflorescences branched. . . . . *C. latifolia.*

Coussapoa angustifolia Aubl. PL. 39e

Leaves: petioles 1.5–2 cm long; largest blades elliptic to narrowly obovate, 6–7.5 × 2.5–5 cm, adaxial surface smooth, the base cuneate, the apex rounded, slightly emarginate; secondary veins in 2–3 pairs. Pistillate inflorescences of single heads. Fr (Aug); uncommon, in large trees usually in non-flooded forest. *Figuier* (Créole).

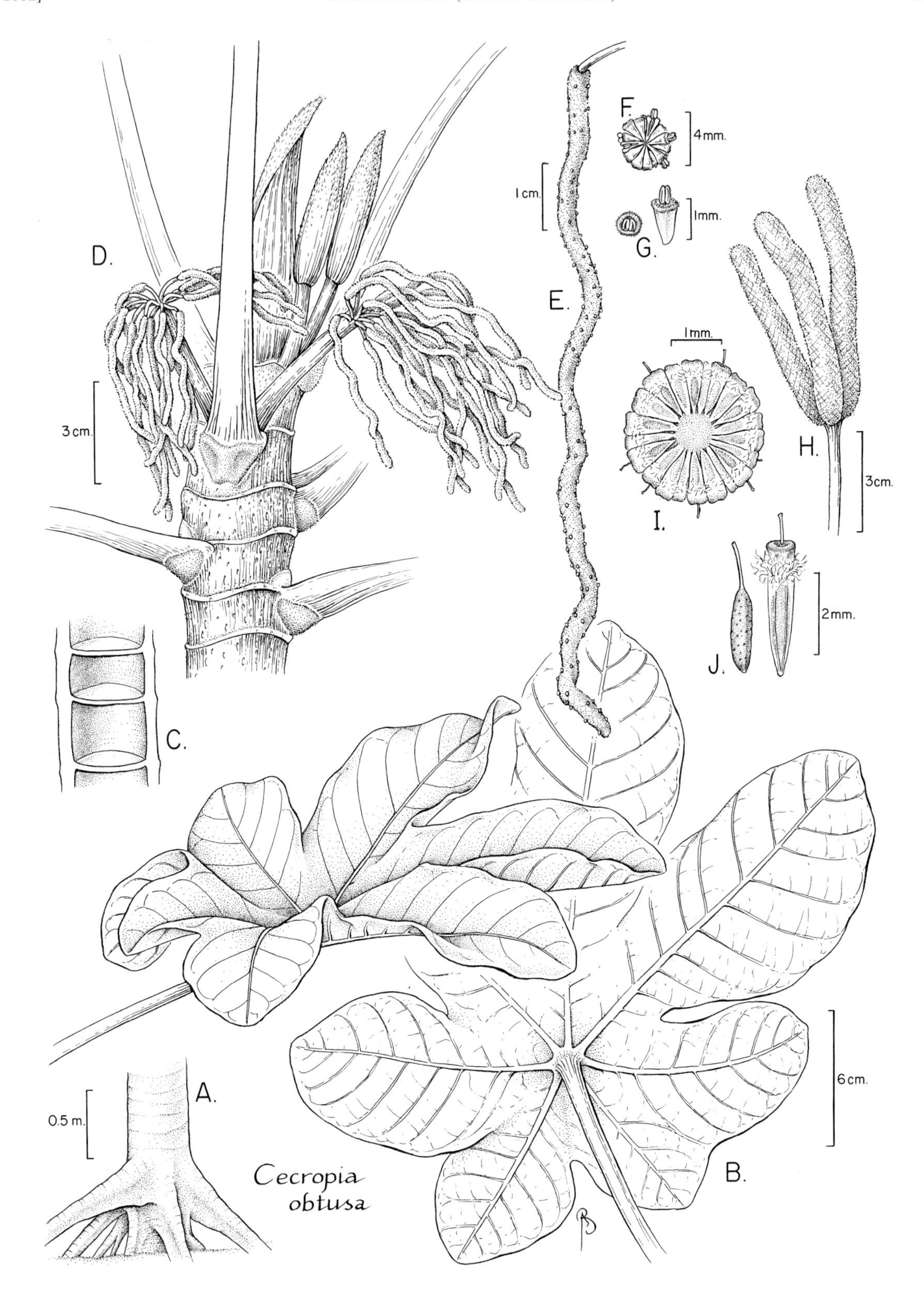

FIG. 71. CECROPIACEAE. *Cecropia obtusa* (A–G, *Mori et al. 20907*; H–J, *Marshall 188*). **A.** Flying buttresses at base of trunk. **B.** Oblique abaxial (far left) and abaxial (near left) views of peltate leaves. **C.** Medial section of stem with hollow internodes. **D.** Apex of stem with sheathing stipule, staminate inflorescences, and petioles with trichilia; note the spathes covering unopened inflorescences (right). **E.** Single spike of staminate inflorescence. **F.** Transverse section of staminate inflorescence. **G.** Apical view (left) and lateral view of staminate flowers with tubular perianths and exserted anthers. **H.** Pistillate inflorescence. **I.** Transverse section of pistillate inflorescence. **J.** Pistil (left) and entire pistillate flower (right) with tubular perianth.

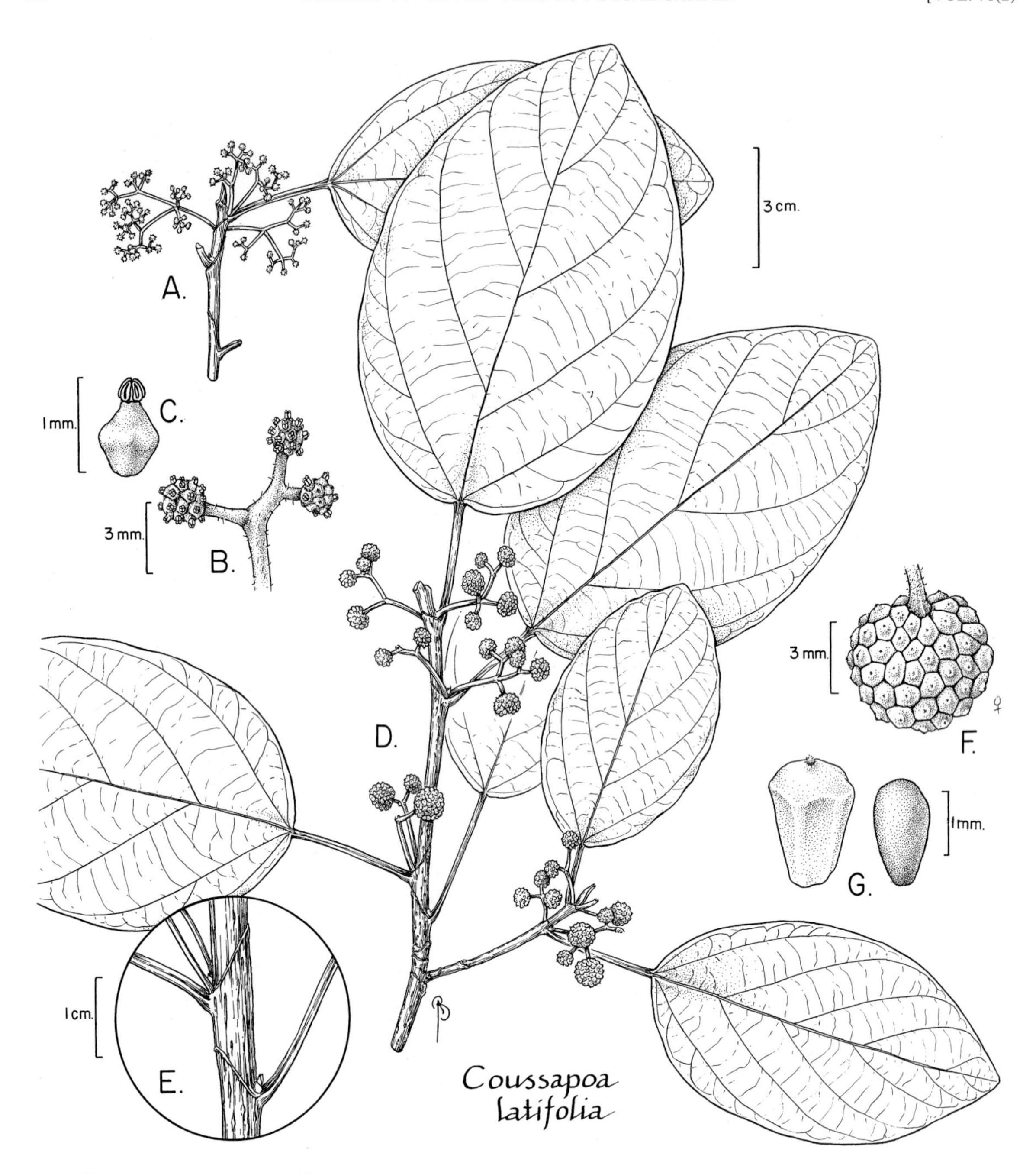

FIG. 72. CECROPIACEAE. *Coussapoa latifolia* (A–C, *Mori et al. 21020*; D–G, *Mori et al. 21506*). **A.** Apex of stem with staminate inflorescences. **B.** Part of staminate inflorescence. **C.** Staminate flower. **D.** Stem with leaves and infructescences. **E.** Detail of stem with oblique stipular scars and bases of infructescences. **F.** Infructescence. **G.** Perianth in fruit (left) and fruit (right).

Coussapoa latifolia Aubl. FIG. 72, PL. 39d

Leaves: petioles 2–7 cm long; largest blades ovate, suborbiculate, or widely elliptic, 9–15 × 6.5–10.5 cm, adaxial surface smooth, the base obtuse to rounded, the apex rounded; secondary veins in 5–6 pairs. Inflorescences branched. Fl (Aug, Sep, Nov), fr (Oct, Nov); common, in large trees usually in non-flooded forest.

POUROUMA Aubl.

Free-standing, understory or canopy trees, often with stilt-roots. Stems and branches without hollow internodes, not inhabited by ants. Stipule scars horizontally oriented on branches. Leaves not peltate, the petiole attached at base of blade; petioles without patch of brown indumentum; blades entire or palmately lobed, sometimes both on same tree. Inflorescences branched or infrequently subumbellate.

Staminate flowers: tepals only basally fused or fused into a tube; stamens (3)4, the filaments shorter (in species with basally fused tepals) or longer than tepals (in species with tepals fused into tube). Pistillate flowers: stigma usually subpeltate. Fruits large (1–2 cm long), enclosed by fleshy perianth, this blackish or brownish at maturity.

1. Leaf blades scabrous adaxially.
   2. Stipules glabrous adaxially. Leaf blades with patent hairs on main veins abaxially. . . . *P. guianensis* subsp. *guianensis.*
   2. Stipules hairy adaxially. Leaf blades with appressed hairs on main veins abaxially.
      3. Leaf blades entire or at most 3-lobed when mature. . . . . *P. bicolor* subsp. *bicolor.*
      3. Leaf blades 5–7-lobed when mature. . . . . *P. bicolor* subsp. *digitata.*
1. Leaf blades smooth adaxially.
   4. Basal secondary veins unbranched.
      5. Stipules 1–3 cm long. Pistillate inflorescences branched. . . . . *P. saulensis.*
      5. Stipules 3–12 cm long. Pistillate inflorescences subumbellate. . . . . *P. minor.*
   4. Basal secondary veins branched.
      6. White cobwebby indumentum on abaxial stipule surfaces, petioles, leafy stems, main veins of blade abaxially, inflorescences, and perianth of pistillate flower. . . . . *P. tomentosa* subsp. *maroniensis.*
      6. White cobwebby indumentum short and (almost) confined to areoles of blades abaxially.
         7. Stipules glabrous adaxially. Leaf blades with short appressed hairs abaxially. . . . . *P. melinonii* subsp. *melinonii.*
         7. Stipules hairy adaxially. Leaf blades with patent hairs abaxially. . . . . *P. villosa.*

Pourouma bicolor Mart. subsp. **bicolor**

Understory trees, to 20 m tall. Stipules 3–15 cm long, hairy adaxially. Leaves: petioles ca. 5–30(45) cm long; blades entire or 3-lobed, the juvenile blades sometimes with more lobes, ca. 10–20 × 5–12 cm, subcoriaceous, scabrous adaxially, with appressed hairs on main veins abaxially. Staminate flowers in loose glomerules. Pistillate inflorescences with ca. 5–20 flowers. Perianth in fruit ca. 1.5 cm long, scabrous. Fr (Feb, Nov); uncommon, in non-flooded forest. *Bois-canon, bois-canon mâle, mâle bois-canon* (Créole).

Pourouma bicolor Mart. subsp. **digitata** (Trécul) C. C. Berg & Heusden

Understory trees, to 15 m tall. Stipules 3–15 cm long, hairy adaxially. Leaves: petioles ca. 5–30(45) cm long; blades 5–7(9)-lobed, 20–35 × 20–35 cm, subcoriaceous, scabrous adaxially, with appressed hairs on main veins abaxially. Staminate flowers in loose glomerules. Pistillate inflorescences with ca. 20–50 flowers. Perianth in fruit ca. 1.5 cm long, scabrous. Fl (Sep), fr (Feb, Oct, Dec); common, in non-flooded forest.

Pourouma guianensis Aubl. subsp. **guianensis**

Understory or canopy trees, to 25 m tall. Stipules ca. 5–15 cm long, glabrous adaxially. Leaves: petioles ca. 5–25 cm long; blades entire and broadly ovate to elliptic or 3-lobed, sometimes 5-lobed, ca. 10–25(40) × 10–25(40) cm, subcoriaceous, scabrous adaxially, with patent hairs on main veins abaxially. Staminate flowers in loose glomerules. Pistillate inflorescences with ca. 10–25 flowers. Perianth in fruit ca. 1.5 cm long, velutinous. Fl (Oct), fr (Jan, Oct, Nov); common, in non-flooded forest. *Bois-canon, bois-canon mâle, mâle bois-canon* (Créole).

Pourouma melinonii Benoist subsp. **melinonii**

Understory or canopy trees, to 30 m tall. Stipules 8–12 cm long, glabrous adaxially. Leaves: petioles ca. 5–15 cm long; blades entire, ovate or 3-lobed, ca. 10–25 × 10–20 cm, coriaceous, smooth adaxially, with appressed hairs on main veins abaxially; basal secondary veins branched. Staminate flowers in globose heads. Pistillate inflorescences with ca. 10–30 flowers. Perianth in fruit 1.5–2 cm long, puberulous. Fl (Oct), fr (Oct, Nov); in non-flooded forest. *Bois-canon* (Créole).

Pourouma minor Benoist Pl. 39f

Understory or canopy trees, to 18 m tall. Stipules 3–12 cm long, glabrous adaxially. Leaves: petioles 1.5–4 cm long; blades entire, obovate to oblong, ca. 5–25 × 2–8 cm, coriaceous, smooth adaxially, with appressed hairs on main veins abaxially; basal secondary veins unbranched. Staminate flowers in loose glomerules. Pistillate inflorescences subumbellate, with 1–4(10) flowers. Flowers: stigma knob-shaped. Perianth in fruit 2–2.5 cm long, subsericeous. Fl (Jan, Aug, Sep), fr (Jan, Feb, Sep); common, in non-flooded forest. *Bois-canon, bois-canon mâle, mâle bois-canon* (Créole).

Pourouma saulensis C. C. Berg & Kooy

Understory or canopy trees, to 30 m tall. Stipules 1–3 cm long, glabrous adaxially. Leaves: petioles 2–9 cm long; blades entire, ovate to elliptic, ca. 5–15 × 3–10 cm, subcoriaceous, smooth adaxially, with appressed hairs on main veins abaxially; basal secondary veins unbranched. Staminate inflorescences unknown. Pistillate inflorescences paniculate, with 7–11 flowers. Perianth in fruit ca. 2 cm long, puberulous to subglabrous. Fl (Oct), fr (Nov, Dec); apparently rare, in non-flooded forest. *Bois-canon* (Créole).

Pourouma tomentosa Miq. subsp. **maroniensis** (Benoist) C. C. Berg & Heusden Fig. 73

Understory or canopy trees, to 35 m tall. Stipules 3–18 cm long, hairy adaxially, with white cobwebby pubescence abaxially. Leaves: petioles ca. 5–15 cm long; blades entire and ovate to elliptic or 3-lobed, ca. 5–40 × 5–40 cm, coriaceous, smooth adaxially, with patent hairs on main veins abaxially; basal secondary veins branched. Staminate inflorescences with flowers in globose heads. Pistillate inflorescences with ca. 10–15 flowers. Perianth in fruit ca. 2 cm long, yellow, puberulous to velutinous, often with cobwebby hairs. Fl (Sep, Oct, Nov), fr (Feb, Dec); common, in non-flooded forest.

Pourouma villosa Trécul

Understory or canopy trees, 10–25 m tall. Stipules ca. 3–15 cm long, pubescent adaxially. Leaves: petioles ca. 5–25 cm long; blades usually 3(5)-lobed, less frequently entire and ovate, ca. 10–30 ×

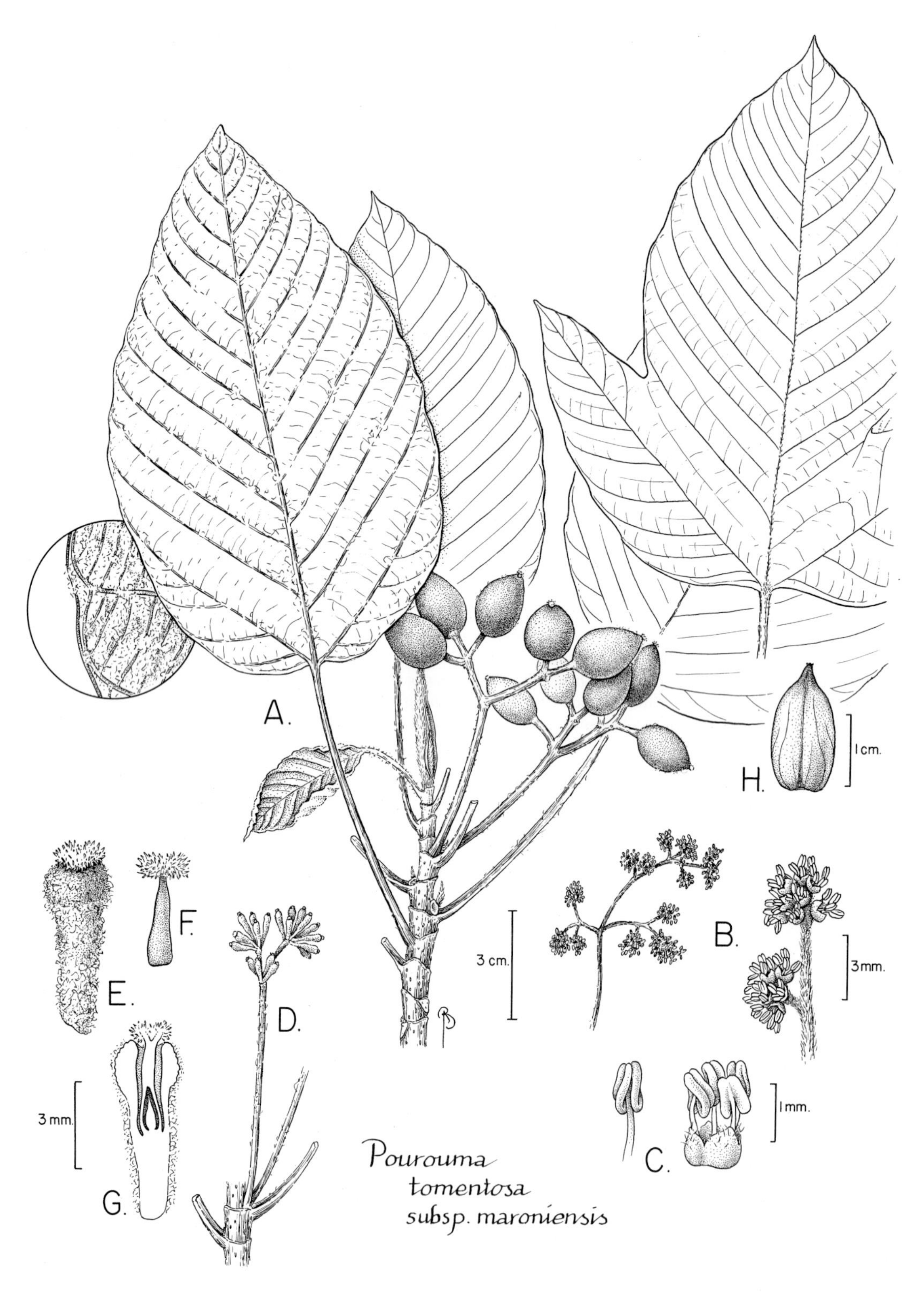

FIG. 73. CECROPIACEAE. *Pourouma tomentosa* subsp. *maroniensis* (A, H, *Mori & Gracie 21125*, lobed leaf *Pires 51193*; B, C, *Boom 1988*; D–G, *Lanjouw 432* from Surinam). **A.** Apex of stem with entire leaves and infructescences; note transverse stipular scars, stipules at apex, detail of abaxial leaf surface (left), and lobed leaf (right). **B.** Staminate inflorescence (left) and detail (right). **C.** Staminate flower (right) and stamen (left). **D.** Pistillate inflorescence. **E.** Pistillate flower with tubular perianth and exserted stigma. **F.** Pistil. **G.** Medial section of pistillate flower. **H.** Fruit.

10–30(40) cm, coriaceous, smooth adaxially; basal secondary veins branched. Staminate inflorescences with flowers in glomerules. Pistillate inflorescences with ca. 15–30 flowers. Perianth in fruit ca. 2 cm long, sparsely puberulous. Fl (Sep, Oct); in non-flooded forest.

CELASTRACEAE (Bittersweet Family)

John D. Mitchell

Trees or shrubs (sometimes lianas outside our area). Stipules mostly small, caducous or absent (*Goupia* with conspicuous linear stipules on young stems). Leaves simple, alternate (opposite in some taxa outside our area). Inflorescences usually axillary, fasciculate, umbellate, or rarely solitary. Flowers unisexual (plants monoecious or dioecious) or bisexual, actinomorphic, 4–5-merous; calyx imbricate, rarely valvate; corolla imbricate or occasionally valvate (e.g., *Goupia*); stamens generally unicyclic, occasionally bicyclic, the anthers introrse, longitudinally dehiscent; nectary disc usually intrastaminal, conspicuous; ovary superior, of 2–5 united carpels (*Goupia* ± free at apex), the style single, terminal, simple with branched or unbranched stigma or 5 distinct styles (*Goupia*); placentation axile, the ovules (1)2 or numerous (*Goupia*) in each locule, erect or pendulous, apotropous. Fruits loculicidal capsules (*Maytenus*) or indehiscent berry-like drupes (*Goupia*) (other fruit types outside of our area).

The Celastraceae include the Hippocrateaceae and excludes *Goupia* as a segregate family, the Goupiaceae, in circumscriptions of other researchers (e.g., Simmons & Hedin, 1999).

Kearns, D. M. 1998. Celastraceae. *In* J. A. Steyermark, P. E. Berry & B. K. Holst (gen. eds.), Flora of the Venezuelan Guayana **4:** 190–197. P. E. Berry, B. K. Holst & K. Yatskievych (vol. eds.). Missouri Botanical Garden Press, St. Louis.

1. Stipules large, conspicuous on young stems. Leaf blades with strongly arcuate secondary veins and parallel tertiary veins oriented at right angles to midrib. Petals valvate in bud; ovary with numerous ovules per locule, the styles 5, separate. Fruits indehiscent berry-like drupes. . . . . *Goupia.*
1. Stipules small, inconspicuous or absent on young stems. Leaf blades with normal pinnately arranged secondary veins and reticulate tertiary veins with no special orientation. Petals imbricate in bud; ovary with 1–2 ovules per locule, the style single. Fruits loculicidal capsules. . . . . *Maytenus.*

GOUPIA Aubl.

Trees. Plants pubescent. Stipules lanceolate, conspicuous on young stems, caducous. Leaves: petioles present; blades with entire to crenate margins; secondary veins strongly arcuate, the tertiary veins parallel, oriented at right angles to midrib. Inflorescences axillary, irregularly umbellate. Flowers bisexual; perianth 5-merous; sepals imbricate in bud; petals valvate in bud, narrow, much longer than sepals, with geniculate, erose, apical appendages; stamens 5, the filaments very short; nectary disc intrastaminal, thin, cupular, sinuate on margin; ovary 5-locular, the carpels fused at base, ± free toward apex, the styles 5, separate; placentation with numerous ascending ovules per locule. Fruits subglobose, 2–3-locular, hard, berry-like drupes. Seeds not arillate.

Goupia glabra Aubl. FIG. 74, PL. 40a

Trees, to 40 m × 120 cm. Trunks cylindrical, with low to steep, thick, running buttresses. Outer bark rough, somewhat fissured, pale grayish-brown, the inner bark hard, 3–6 mm thick, variegated with orange and white streaks. Indument of short to very long trichomes to 2.5 mm long, sparsely to densely covering stems, petioles, leaf blades, and inflorescences, especially evident on young leaves and stems. Stipules lanceolate, caducous, 6–10 mm long. Leaves: petioles 5–10 mm long; blades ovate to lanceolate, elliptic or obovate, 7.7–16 × 3–6.9 cm, chartaceous to subcoriaceous, the base rounded, obtuse, or attenuate, somewhat oblique, the apex usually long acuminate; secondary veins in 3–6 pairs. Inflorescences irregularly umbellate. Flowers: calyx lobes deltoid; petals lanceolate, ca. 4.2 × 1 mm, yellow-orange with red at base, the margins revolute, the apical appendage 1.7–2.5 mm long; stamens with shortly villous anthers; ovary sparsely pubescent, depressed globose, the styles glabrous. Fruits subglobose, ca. 0.6 mm diam., purple to black when ripe. Fl (May, Jun), fr (Aug, Sep, Oct); common, in secondary growth and primary non-flooded forest where it is often associated with gaps. *Cupiúba* (Portuguese), *goupi* (Créole).

MAYTENUS Molina

Shrubs or trees. Plants glabrous. Stipules small, ± persistent or absent. Leaves: blades with margins entire to serrulate or denticulate. Inflorescence axillary, usually fasciculate, occasionally solitary. Flowers unisexual (plants dioecious or polygamodioecious) or bisexual; perianth usually 5-merous, imbricate in bud; stamens 5(6), the filaments subulate, short or elongate; nectary disc intrastaminal, fleshy, partially enclosing base of ovary, 5–10-lobed; ovary 2–3(4)-locular, the style simple, the stigma 2(3–4) branched; placentation with (1)2 erect ovules per locule. Fruits loculicidal capsules. Seeds arillate.

1. Leaf margins always entire; secondary veins in 10–17 pairs. Capsules >2 cm diam. . . . . *M. oblongata.*
1. Leaf margins entire to serrulate or denticulate; secondary veins in 6–10 pairs. Capsules <2 cm diam.
    2. Leaf margins entire to serrulate; tertiary venation inconspicuous on adaxial leaf surface. Pedicels (2) 4–5 mm long. Petals 1.3–1.5 mm long. . . . . *M. floribunda.*
    2. Leaf margins entire to denticulate; tertiary venation conspicuous on adaxial leaf surface. Pedicels 1–1.5 mm long. Petals ca. 0.9 mm long. . . . . *M. myrsinoides.*

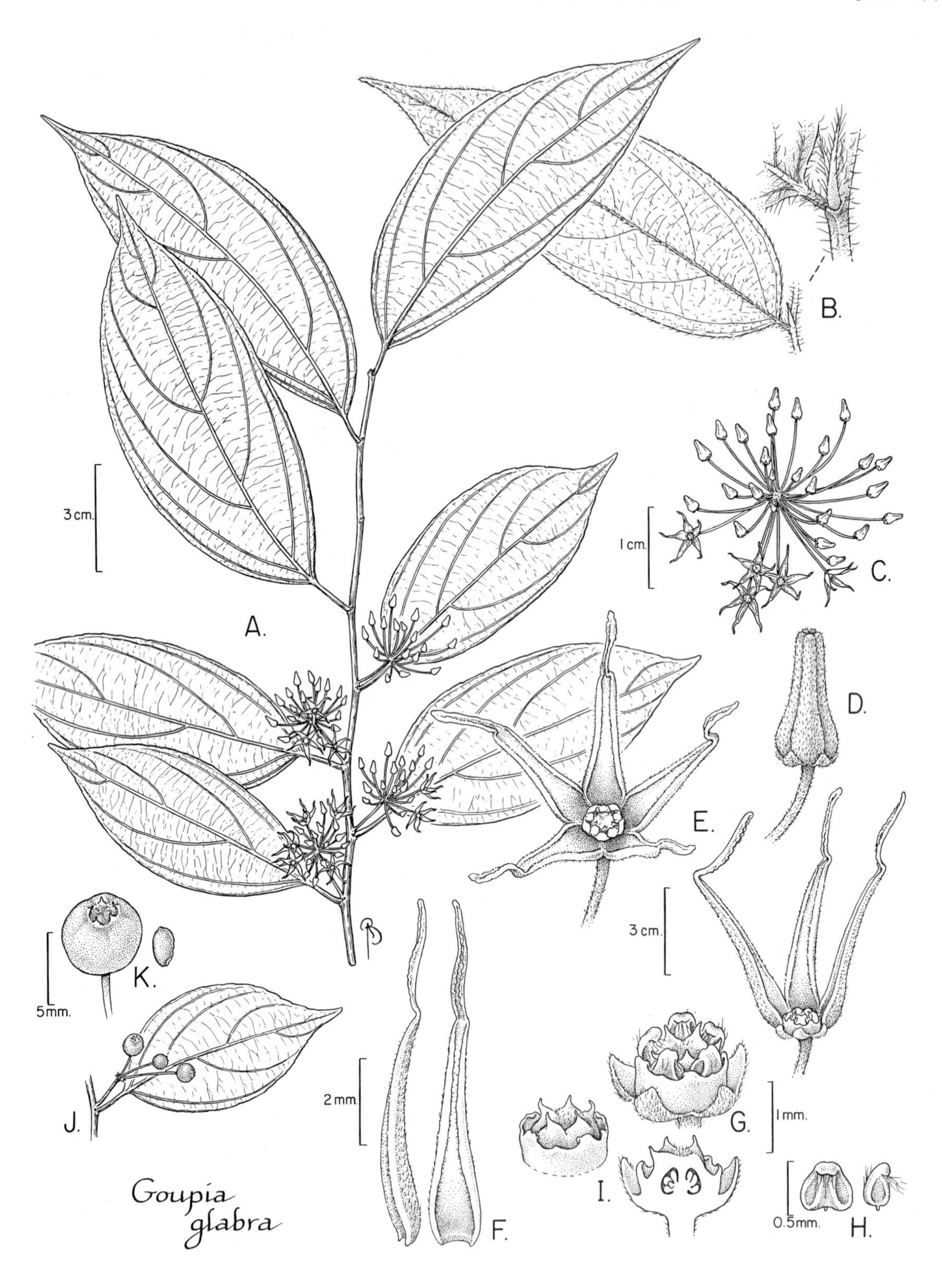

FIG. 74. CELASTRACEAE. *Goupia glabra* (A, *Mori et al. 23364*; B, *Mori et al. 18161*; C–I, *Mori et al. 23369*; J, K, *Mori et al. 21500*). **A.** Stem with leaves and axillary, umbellate inflorescences. **B.** Leaf of young plant (below) with detail of stipule (above); note that the leaves of saplings are pubescent. **C.** Apical view of umbellate inflorescence. **D.** Floral bud. **E.** Apical (above left) and lateral with two petals removed (below right) views of flowers. **F.** Petals. **G.** Flower with all petals removed; note the nectary disc surrounding the stamens. **H.** Adaxial and lateral views of stamens. **I.** Nectary disc surrounding pistil; note separate styles (above left) and medial section of flower with petals removed. **J.** Leaf with axillary, umbellate infructescence. **K.** Fruit (left) and seed (right).

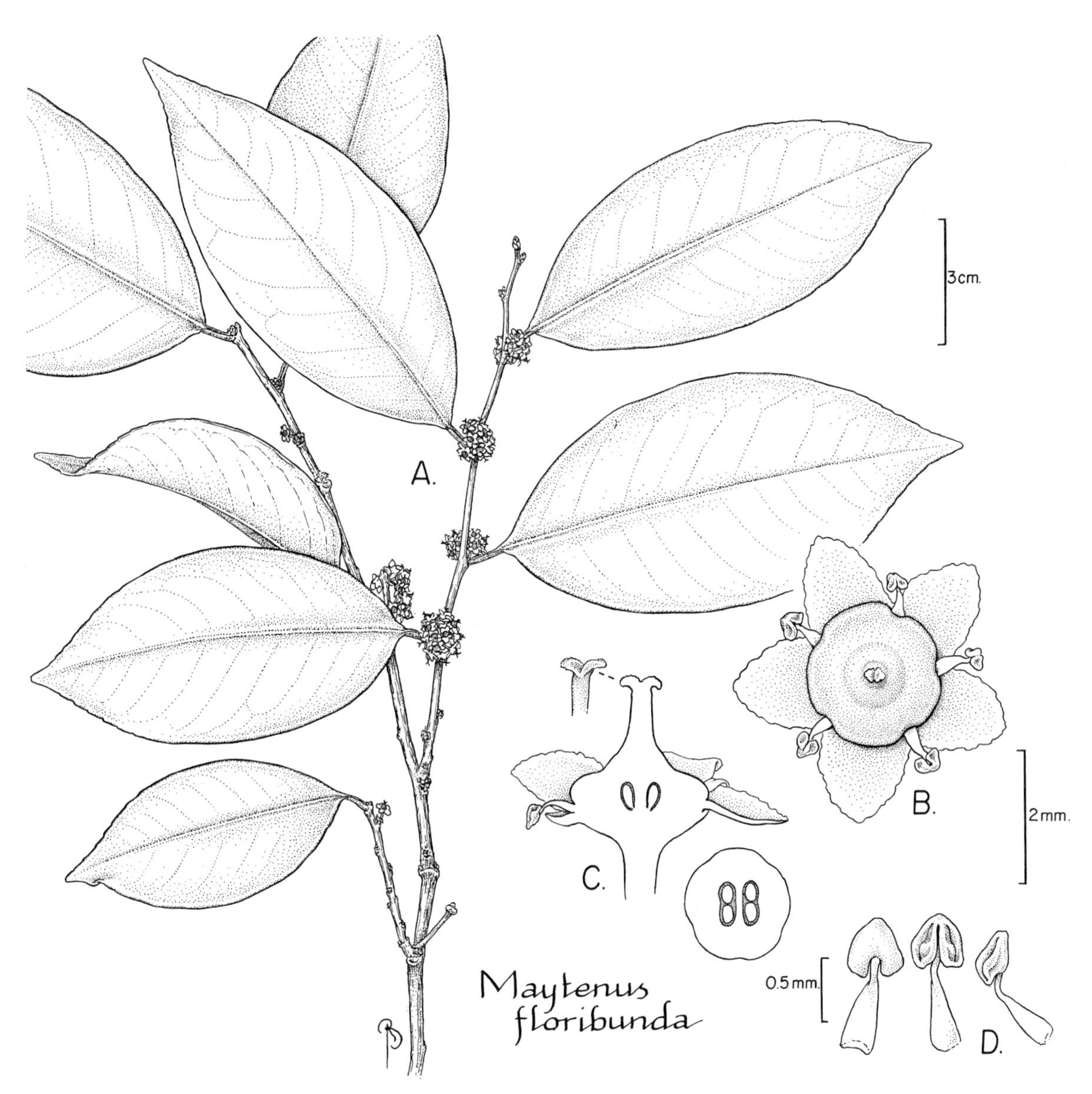

FIG. 75. CELASTRACEAE. *Maytenus floribunda* (*Mori et al. 15104*). **A.** Stem with leaves and axillary inflorescences. **B.** Apical view of flower. **C.** Medial section of flower with detail of stigma (above) and transverse section of ovary (below). **D.** Adaxial (left), abaxial (center), and lateral (right) views of stamens.

Maytenus floribunda Reissek FIG. 75

Small to medium-sized trees, to 25 m × 35 cm. Trunks cylindric. Bark with slash of inner bark thick, dark pink. Leaves: blades elliptic or obovate, 7.6–11.5 × 3–5.7 cm, coriaceous, the margins entire to serrulate; secondary veins in 8–10 pairs, usually flattened on both surfaces, the tertiary venation inconspicuous, particularly on adaxial surface. Inflorescences dense fascicles; pedicels (2)4–5 mm long. Flowers unisexual or bisexual (plants polygamodioecious), pistillate and staminate flowers similar in size and shape; petals reflexed, ovate, 1.3–1.5 × 1.1–1.3 mm. Capsules globose, dehiscing into 2 halves, 1.3–1.5 cm diam., green, becoming yellowish. Seeds brown, partially enclosed by white aril. Fl (Sep, Oct); uncommon, in non-flooded forest.

Maytenus myrsinoides Reissek

Small to medium-sized trees, to 20 m × 115 cm. Trunks sulcate or fenestrate. Bark with outer bark gray, the inner bark dark red; wood pale red. Leaves: blades narrowly elliptic, oblong, or oblanceolate, 10–12.5 × 2.2–4.5 cm, chartaceous, the margins entire to denticulate; secondary veins in 6–10 pairs, flattened to slightly prominent on both surfaces, the tertiary venation usually conspicuous on both surfaces. Inflorescences fasciculate; pedicels 1–1.5 mm long. Flowers unisexual or bisexual; petals ovate, 0.9 × 0.6 mm. Capsules globose, dehiscing into 2 halves, ca. 1.2 cm diam., green, becoming yellow-orange at maturity (black reported outside French Guiana). Seeds partially enclosed in white aril. Fr (May, Jul, Oct, Dec); in non-flooded forest at all elevations.

Maytenus oblongata Reissek

Small trees, to 13 m × 20 cm. Trunks fluted at base, angled. Bark with outer bark appearing smooth, but with cracks 3 mm deep × 5 mm wide, soft, gray-brown, the inner bark bright red, striate, 4 mm thick; exudate a clear sticky resin. Leaves: blades oblong, elliptic, or occasionally ovate, 7.5–18.5 × 4.5–7.2 cm, chartaceous to subcoriaceous, the margins always entire; secondary veins in 10–17 pairs, slightly prominent on both surfaces, the tertiary venation conspicuously prominent on both surfaces. Inflorescences and flowers not seen. Capsules globose, dehiscing into 2 halves, ca. 2.5 cm diam. Known only from a single sterile collection (*Boom 1748*, NY) from our area, apparently in non-flooded forest.

CHRYSOBALANACEAE (Cocoa-plum Family)

Ghillean T. Prance

Trees or infrequently shrubs. Bark: the outer bark thin, often white in slash, the inner bark thick, usually red, often friable. Stipules small and caducous to large and persistent. Leaves simple, alternate, the margins entire. Inflorescences racemose, paniculate or cymose, the flowers bracteate and usually 2-bracteolate. Flowers actinomorphic to zygomorphic, bisexual, markedly perigynous; receptacle short to elongate; nectary disc always present, forming a lining to the receptacle; calyx lobes 5, imbricate, often unequal, erect or reflexed; petals 5, occasionally absent, commonly unequal, imbricate, usually caducous; stamens 3–65, inserted on margin or surface of disc, or basally adnate to it, forming complete circle or, in zygomorphic flowers, unilateral, all fertile or some without anthers and then often reduced to small staminodia, the filaments filiform, free, connate at base, included to far exserted, the anthers small, dorsifixed, longitudinally dehiscent, glabrous; gynoecium of one carpel attached to base, middle, or mouth of receptacle-tube, pubescent or villous, unilocular with two ovules or bilocular (owing to a false partition) with one ovule in each compartment, the style filiform, gynobasic, arising from receptacle at base of ovary, the stigma distinctly or indistinctly 3-lobed; placentation with ovules erect, epitropous, the micropyle directed towards base. Fruits dry or fleshy drupes, the endocarp various, thick or thin, fibrous or bony, with two basal stoppers (obturamenta) for seedling escape in *Parinari*, often densely hairy inside. Seeds erect, almost exalbuminous; cotyledons planoconvex, fleshy.

Prance, G. T. 1972. Chrysobalanaceae. Fl. Neotrop. Monogr. **9:** 1–410.
———. 1986. Chrysobalanaceae. *In* A. R. A. Görts-van Rijn (ed.), Flora of the Guianas Ser. A, [no fascicle no.]: 1–146. Koeltz Scientific Books, Koenigstein.
———. 1989. Chrysobalanaceae. Fl. Neotrop. Monogr. **9S:** 1–267.
———. 1998. Chrysobalanaceae. *In* J. A. Steyermark, P. E. Berry & B. K. Holst (gen. eds.), Flora of the Venezuelan Guayana **4:** 202–246. P. E. Berry, B. K. Holst & K. Yatskievych (vol. eds.). Missouri Botanical Garden Press, St. Louis.
——— & A. R. A. Görts-van Rijn. 1976. Chrysobalanaceae. *In* J. Lanjouw & A. L. Stoffers (eds.), Flora of Suriname **II(2):** 524–555. E. J. Brill, Leiden.

1. Abaxial leaf surface without stomatal crypts (except in *Licania alba, L. albiflora, L. majuscula, L. octandra*). Ovary unilocular. Fruits without basal stoppers, opening by longitudinal slits or indehiscent.
   2. Flowers globose; petals present or absent; stamens included or exserted; ovary inserted at or near base of receptacle. . . . . *Licania.*
   2. Flowers elongate; petals present; stamens always exserted; ovary inserted at mouth of receptacle.
      3. Abaxial leaf surface glabrous, or with stiff appressed hairs. Stamens 3–7, usually purple. fruits with thin endocarp, opening along longitudinal lines. . . . . *Hirtella.*
      3. Abaxial leaf surface lanate or arachnoid pubescent or rarely glabrous. Stamens 9–65, usually white. Fruits with thick granular endocarp, not opening along longitudinal lines. . . . . *Couepia.*
1. Abaxial leaf surface with stomatal crypts. Ovary bilocular. Fruits opening by two basal stoppers (obturamenta). . . . . *Parinari.*

Note: The only other genus likely to occur in the Saül region is the little-collected *Acioa* which differs from all the above genera by the stamens which are united into a strap.

COUEPIA Aubl.

Trees or shrubs. Stipules subulate or deltate, usually persistent or subpersistent, axillary. Leaves: petioles eglandular; blades often with 1 or 2 pairs of glands at base, the abaxial surface glabrous or with lanate or arachnoid indumentum. Inflorescences congested thyrses, often with a few ascending branches, or few-flowered spikes or racemes; pedicels usually shorter than, and often much shorter than, receptacle-tube; bracts and bracteoles usually persistent, rarely enclosing flower buds in small groups, eglandular. Flowers bisexual, slightly zygomorphic; receptacle-tube turbinate to narrowly cylindric, ventricose, rarely longer than calyx, hollow, glabrous inside except at throat or beneath ovary, rarely pubescent inside; annulus well developed at throat; sepals 5, subequal, spreading or reflexed, usually eglandular, the apex acute or rounded; petals 5, orbicular to lingulate, sometimes shortly unguiculate, more or less equalling sepals; stamens 9–65, the filaments undulate in bud with 3 or more bends, inserted on abaxial surface of annulus, usually forming complete circle, less frequently unilateral, far-exserted, but not much longer than combined length of calyx and receptacle-tube, usually white; staminodia absent or short and filiform; ovary 1-carpellate, 1-locular, inserted laterally at mouth of receptacle-tube, the style filiform, far-exserted, indistinctly 3-lobed at apex, hairy for greater part of length. Fruits drupes, 2.5–12 cm or more long, the endocarp hard, thick, granular, shortly hairy inside, on germination breaking up irregularly, surface without longitudinal lines, rough and irregular owing to fusiform anticlinal aggregations of stone cells and fibers penetrating mesocarp more or less deeply.

1. Leaves prominently reticulate abaxially. Receptacles cylindrical. Fruit exteriors pubescent. . . . . *C. parillo.*
1. Leaves not prominently reticulate abaxially. Receptacles campanulate, subcylindrical, or cylindrical. Fruit exteriors glabrous.
   2. Inflorescences much branched panicles. Interior of receptacles filled with hairs to base. . . . . . *C. caryophylloides.*
   2. Inflorescences spikes, racemes, or sparsely branched panicles (*C. guianensis*). Interior of receptacles glabrous except for deflexed hairs at throat.
      3. Flowers with exterior of receptacles densely ferruginous sericeous. . . . . *C. habrantha.*
      3. Flowers with exterior of receptacles glabrous or sparsely gray-appressed pubescent.
         4. Leaf blades pubescent abaxially. Stamens 14–30.
            5. Leaf blades obovate, the apex bluntly acuminate. . . . . *C. obovata.*
            5. Leaf blades oblong to oblong-lanceolate, the apex with finely pointed acumen. . . . . . . *C. guianensis.*
         4. Leaf blades glabrous abaxially. Stamens 9–11. . . . . *C. joaquinae.*

Couepia caryophylloides Benoist

Trees, to 25 m tall. Leaves: blades oblong-elliptic to oblong-lanceolate, 10–21 × 4–7.5 cm, densely gray to brown arachnoid pubescent abaxially, smooth, not reticulate, the apex acuminate, the acumen 4–10 mm long. Inflorescences much branched panicles, the rachis and branches densely gray-pubescent. Flowers: receptacle subcylindrical, gray pubescent on exterior, hairy inside to base within; stamens 22–35 inserted in semicircle. Fruits globose, the exocarp smooth, glabrous. St (Sep); rare, in non-flooded forest. *Gaulette*, *gaulette noir*, *gris-gris*, *gris-gris-rouge* (Créole).

Couepia guianensis Aubl.

Trees, to 30 m tall. Leaves: blades oblong to oblong-lanceolate, 5–13 × 2.5–5.5 cm, densely gray-lanate-arachnoid pubescent abaxially, smooth, not reticulate, the apex acuminate, the acumen finely pointed, 8–18 mm long. Inflorescences racemose or sparsely branched panicles, the rachis and branches sparsely puberulous or glabrous. Flowers: receptacle cylindrical to subcampanulate, sparsely gray-puberulous to glabrous on exterior, glabrous within except for deflexed hairs at throat; stamens 14–30, unilateral. Fruits globose to ovoid, 3–4 × 2.5–3 cm, the exocarp smooth, glabrous. St (Oct); rare, in non-flooded forest.

Couepia habrantha Standl.

Trees, to 12–25 m tall. Leaves: blades oblong-elliptic, 6–13 × 2.5–6 cm, densely reddish brown to gray arachnoid pubescent abaxially, smooth, not reticulate, the apex acuminate, the acumen 4–14 mm long. Inflorescences spikes, the rachis ferruginous-brown pubescent. Flowers: receptacle subcylindrical, densely ferruginous-sericeous on exterior, glabrous within except for deflexed hairs at throat; stamens ca. 25, inserted in a semicircle. Fruits ellipsoid, the exocarp smooth, glabrous. Fl (May); fr (May, immature); rare, in non-flooded forest. *Gaulette gris-gris* (Créole).

Couepia joaquinae Prance

Trees, to 30 m tall. Leaves: blades oblong to oblong lanceolate, 10–15 × 3–5.5 cm, glabrous on both surfaces, smooth not reticulate abaxially, the apex acuminate, the acumen 8–15 mm long, often slightly curved. Inflorescences sparsely branched panicles, the rachis and branches glabrous or sparsely puberulous. Flowers: receptacle narrowly cylindrical, almost glabrous and striate on exterior when dry, glabrous within except for beneath stamens; stamens 9–11, unilateral. Fruits ovoid, 3.5–4.5 × 1.9–2.4 cm, the exocarp smooth, inconspicuously lenticellate. Fl (Jul, Sep); rare, in non-flooded forest.

Couepia obovata Ducke

Trees, to 20 m tall. Leaves: blades obovate, 4–10 × 2–5 cm, the apex bluntly acuminate, the acumen 2–5 mm long, sparsely gray-appressed-pubescent abaxially, smooth, not reticulate. Inflorescences racemose, the rachis sparsely silver-gray puberulous. Flowers: receptacle cylindrical, with sparse, gray appressed hairs on exterior, glabrous within except for deflexed hairs at throat; stamens 16–21, inserted around complete circle. Fruits ovoid, 2–3 × 1.5–2.5 cm, the exocarp smooth, glabrous. Fl (Jul); rare, in non-flooded forest.

Couepia parillo DC. FIG. 76

Trees, to 20 m tall. Leaves: blades oblong to oblong-elliptic, 5–15.5 × 1.7–5.8 cm, abaxial surface prominently reticulate with silver-gray pubescence between reticulations, the apex acuminate, the acumen 5.5–18 mm long. Inflorescences racemose, the rachis ferruginous-brown pubescent. Flowers: receptacle tubular-cylindrical, dense hirtellous pubescent on exterior, glabrous within except for beneath ovary; stamens 45–62, inserted almost around circle. Fruits globose, ca. 2.5 cm diam., the exocarp soft-yellow-brown velutinous. Fl (Aug, Sep); common, in non-flooded forest. *Gaulette*, *gaulette petites feuilles* (Créole).

HIRTELLA L.

Trees and shrubs. Stipules lateral, often subulate or filiform, subpersistent. Leaves: petioles eglandular; blades infrequently with 2 large, bulbous, ant-inhabited domatia at base, otherwise lacking large basal glands, often with many small, submarginal or scattered, discoid glands, the abaxial surface glabrous or with few strigose or strigulose hairs, or hirsute. Inflorescences lax racemes with patent flowers, or panicles or cymes; pedicels usually longer than and often much longer than the receptacle-tube; bracts and bracteoles not enclosing flower buds in small groups, often with stipitate or sessile glands. Flowers bisexual, slightly zygomorphic; receptacle-tube subcampanulate, slightly gibbous, usually shorter than sepals, glabrous inside except near throat; sepals 5, subequal, usually spreading or reflexed, often with sessile or shortly stipitate glands on margin, the apex acute; petals 5, shorter than sepals; stamens 3–7, the filaments usually coiled in bud, inserted on posterior rim of annulus at throat, far-exserted, usually much longer than combined length of calyx and receptacle-tube, usually purple; staminodia short, filiform; ovary 1-carpellate, 1-locular, usually inserted at mouth of receptacle-tube, the style filiform, far-exserted, very shortly 3-lobed at apex. Fruits drupes, usually <2.2 cm long, with poorly developed mesocarp and smooth, thin, hard, non-glandular endocarp with 4–7 longitudinal shallow channels representing lines of weakness permitting seedling to escape.

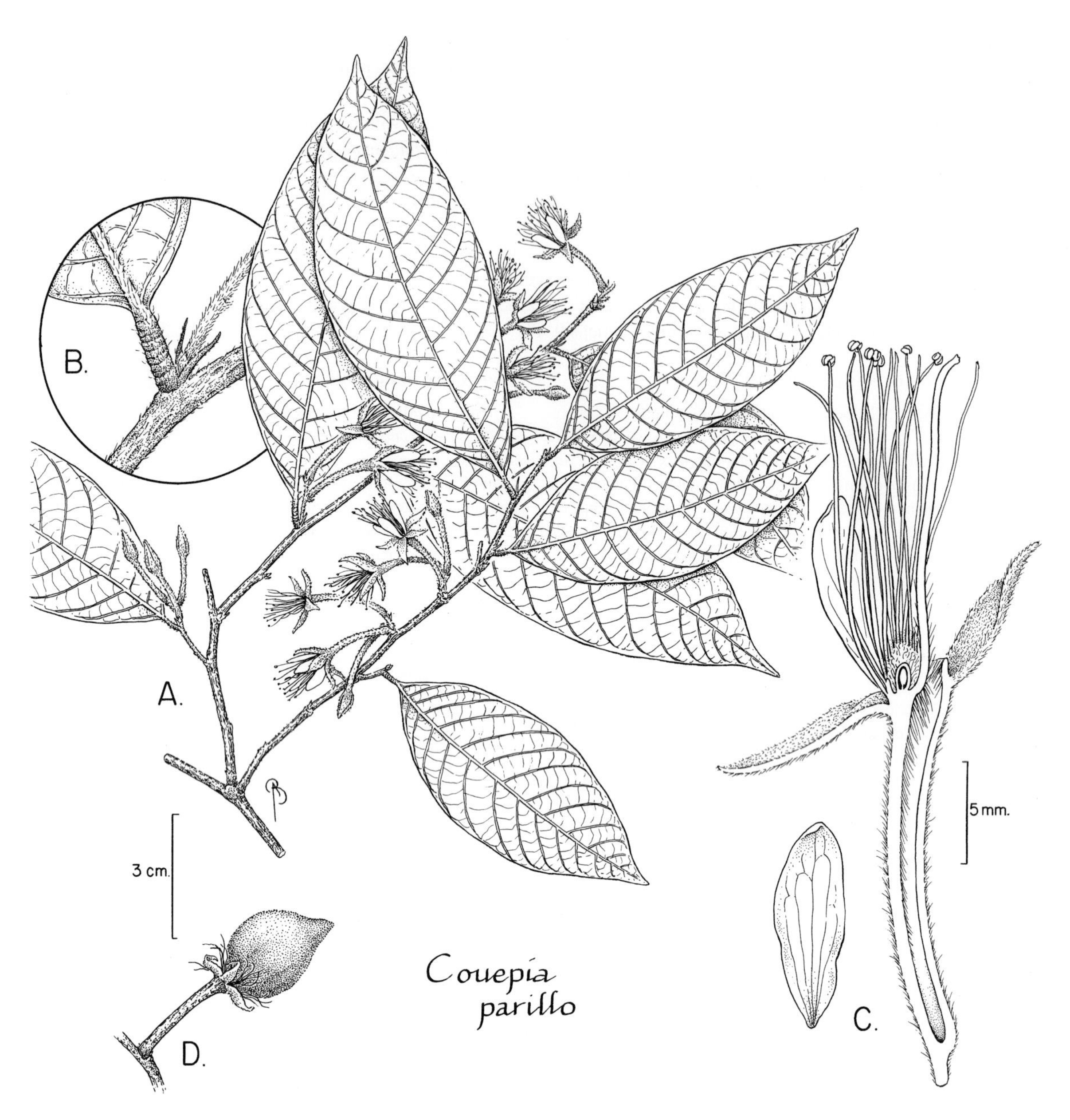

FIG. 76. CHRYSOBALANACEAE. *Couepia parillo* (A–C, *Mori et al. 20759*; D, *Oliveira 4111*). **A.** Part of stem with leaves and inflorescences; note pedicellate flowers. **B.** Detail of node with paired stipules, corrugated petiole, and abaxial surface of leaf base. **C.** Petal (left) and medial section of flower (right); note the tubular hypanthium with gynoecium perched at the apex and gynobasic style. **D.** Immature fruit.

1. Leaf blade bases with two ant-inhabited domatia. . . . . *H. physophora.*
1. Leaf blade bases without ant-inhabited domatia.
   2. Inflorescences panicles.
      3. Bracteoles with numerous stipitate glands. Stamens usually 3 (sometimes to 5 in *H. glandulosa*).
         4. Leaf blades glabrous abaxially; venation plane adaxially. . . . . *H. davisii.*
         4. Leaf blades hirsute abaxially; venation slightly impressed adaxially. . . . . *H. glandulosa.*
      3. Bracteoles eglandular or apex terminating in single gland. Stamens 4–6 (sometimes 3 in *H. bicornis* var. *pubescens*).
         5. Inflorescences with long central rachis, the flowers borne on short branches; bracteoles exceeding receptacle in length, persistent. . . . . *H. suffulta.*
         5. Inflorescences spreading panicles; bracteoles not exceeding receptacle in length, caducous. . . . . *H. bicornis* var. *pubescens.*

2. Inflorescences racemes.
    6. Inflorescences puberulous to glabrescent; bracteoles without glands or with sessile glands.
        7. Leaf blades lanceolate. . . . . *H. tenuifolia.*
        7. Leaf blades elliptic to oblong.. . . . . *H. racemosa.*
    6. Inflorescences tomentellous or hispid; bracteoles with stipitate glands.
        8. Young branches and lower part of inflorescences hispid-setose.. . . . . *H. hispidula.*
        8. Young branches and inflorescences tomentellous. . . . . *H. silicea.*

Hirtella bicornis Mart. & Zucc. var. **pubescens** Ducke

Trees, to 25 m tall. Leaves: blades oblong, 3.5–9.5 × 1.7–4 cm, membranous to subcoriaceous, glabrous or with few appressed hairs only abaxially, without ant-inhabited domatia, the apex cuspidate; secondary veins in 8–13 pairs. Inflorescences spreading panicles, the rachis and branches tomentellous; bracteoles 1–2 mm long, not exceeding receptacle in length, caducous, eglandular. Flowers: stamens (3)5. Fl (May, Jul, Aug, Sep, Nov); common, in non-flooded forest. *Gaulette*, *gaulette petites feuilles* (Créole).

Hirtella davisii Sandwith

Trees, to 30 m tall. Leaves: blades oblong to oblong-elliptic, 6–13.5 × 2.3–5 cm, coriaceous, glabrous abaxially, without ant-inhabited domatia; secondary veins in 9–10 pairs, the veins plane adaxially. Inflorescences panicles, the rachis and branches sparsely pilose; bracteoles 1–3.5 mm long, with numerous stipitate glands. Flowers: stamens 3. Fl (Jun, Oct), fr (Jun); rare, in non-flooded forest.

Hirtella glandulosa Spreng.

Trees, to 25 m tall. Leaves: blades oblong to ovate, 4.5–23 × 2.5–11.5 cm, coriaceous, hirsute abaxially, without ant-inhabited domatia; secondary veins in 8–15 pairs, the veins slightly impressed adaxially. Inflorescences panicles, the rachis and branches hirsute-tomentellous; bracteoles with numerous stipitate glands. Flowers: stamens 3(5). Fl (Nov); rare, in non-flooded forest. *Gaulette blanc*, *gris-gris* (Créole).

Hirtella hispidula Miq.

Trees, to 15 m tall. Leaves: blades elliptic to oblong, 5.5–15 × 2.2–5 cm, coriaceous, hirsute on principal venation abaxially, without ant-inhabited domatia; secondary veins in 8–12 pairs. Inflorescences racemes, the rachis hispid-setose; bracteoles 1.5–3.5 mm long, linear-lanceolate, with several stipitate glands on margins. Flowers: stamens 4–5. Fl (Aug, Sep), fr (Sep); rare, in non-flooded forest. *Gaulette* (Créole).

Hirtella physophora Mart. & Zucc. PART 1: PL. XId

Treelets, to 6 m tall, usually smaller. Leaves: blades oblong-elliptic to oblong, 17–30 × 6–11 cm, membranous, hirsute abaxially on venation, with two swollen ant-inhabited domatia at base; secondary veins in 13–16 pairs. Inflorescences fasciculate racemes, the rachis hispid; bracteoles 2–8 mm long, linear, eglandular. Flowers: stamens 6. Fl (Jan, Feb, Mar, Apr, May), fr (Jan, immature); common, in non-flooded forest. *Gaulette fourmi*, *lamoussé-fourmi*, *petite gaulette*, *petite gaulette fourmi* (Créole).

Hirtella racemosa Lam. FIG. 77

Trees, to 6 m tall. Leaves: blades elliptic to oblong, 7–16.5 × 1.5–7 cm, coriaceous, glabrous or sparsely appressed pubescent abaxially, without ant-inhabited domatia; secondary veins in 6–10 pairs. Inflorescences racemes, the rachis puberulous to glabrescent; bracteoles 0.5–3 mm long, with paired sessile glands toward base. Flowers: stamens 5–7. Fl (Apr, Jul, Aug, Sep, Oct, Dec), fr (Sep); in non-flooded forest. *Dibo*, *dur bois*, *gaulette*, *gris-gris*, *petite gaulette rouge* (Créole).

Hirtella silicea Griseb. PL. 40b

Trees, to 10 m tall. Leaves oblong, 10–22 × 3–9 cm, chartaceous to thin coriaceous, sparsely appressed pubescent abaxially, without ant-inhabited domatia; secondary veins in 6–12 pairs. Inflorescences racemes, the rachis tomentellous; bracteoles 2–6 mm long, oblong, with few to numerous stipitate glands. Flowers: stamens 5–6. Fl (Jan, Mar, Jul, Sep), fr (Sep); rare, in non-flooded forest. *Gaulette*, *gaulette fourmi* (Créole).

Hirtella suffulta Prance

Trees, to 35 m tall. Leaves: blades oblong, 5.5–11.5 × 2.5–3 cm, subcoriaceous, glabrous abaxially or with a few stiff appressed hairs only, without ant-inhabited domatia; secondary veins in 8–9 pairs. Inflorescences panicles, the central rachis long, the flowers borne on short branches, the rachis and branches sparsely puberulous; bracteoles 3–6 mm long, longer than receptacle, persistent, eglandular or apex terminating in a single gland. Flowers: stamens 5–6. Fl (May, Aug, Sep, Nov), fr (Sep); common, in non-flooded forest.

Hirtella tenuifolia Prance

Trees, to 6 m tall. Leaves: blades lanceolate, 9–14 × 2.1–3.5 cm, coriaceous, glabrous abaxially, without ant-inhabited domatia; secondary veins in 11–13 pairs. Inflorescences racemes, the rachis sparsely puberulous; bracteoles 1–2 mm long, with sessile glands on margins. Flowers: stamens 5. Fl (Jul, Aug), fr (Sep); in non-flooded forest. *Gaulette*, *gaulette rouge*, *petite gaulette rouge* (Créole).

LICANIA Aubl.

Trees or shrubs. Stipules small, free, subulate or narrowly deltate, intrapetiolar or inserted on base of petiole, caducous or persistent. Leaves: blades with abaxial surface glabrous, lanate or strigose, or with stomatal crypts filled with densely matted hairs. Inflorescences usually racemose panicles or simple racemes, rarely secondary branches bearing small few-flowered cymules; bracts and bracteoles usually eglandular and not enclosing flower buds. Flowers actinomorphic or weakly zygomorphic; receptacle-tube 0.1–0.8 cm long, variable in shape, usually cupuliform, campanulate or urceolate, hairy inside, but throat without long retrorse hairs; sepals 5, subequal, the apex acute; petals 5, more or less equalling sepals, or absent; stamens 3–40, the filaments forming complete circle or unilateral, included and much shorter than sepals to exserted and about twice as long, usually united at base, usually glabrous; staminodia usually absent; ovary 1-locular, inserted at or near base of

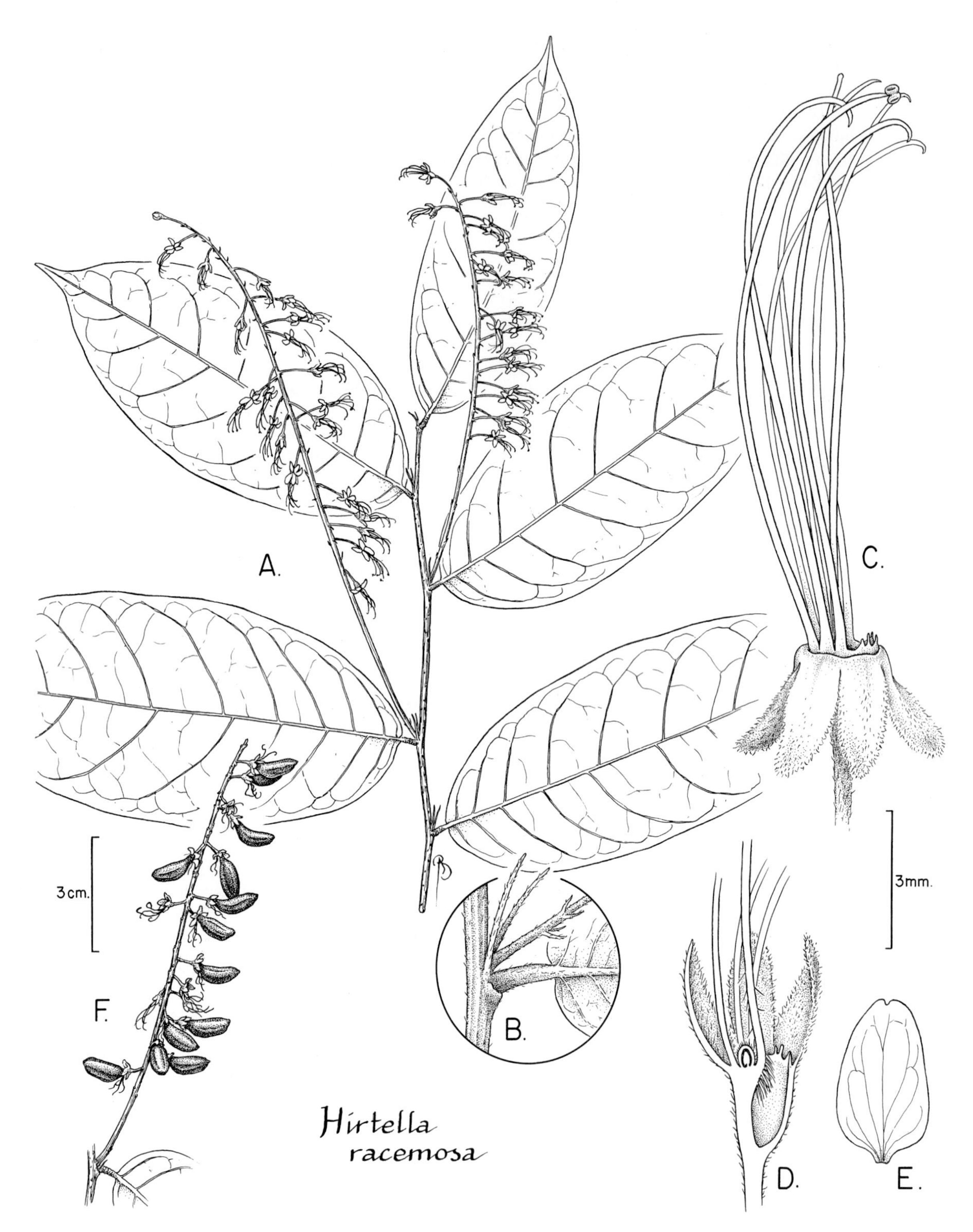

FIG. 77. CHRYSOBALANACEAE. *Hirtella racemosa* (A–E, *Mori & Boom 14870*; F, *Cowan 39357*). **A.** Apex of stem with leaves and inflorescences. **B.** Detail of node with stipules, base of raceme and base of leaf. **C.** Flower with corolla removed; note asymmetrical androecium with staminodes on right. **D.** Medial section of flower with corolla removed; note gynobasic style and gynoecium perched at mouth of shallow tubular hypanthium. **E.** Petal. **F.** Infructescence.

receptacle-tube, the style filiform, indistinctly 3-lobed at apex. Fruits drupes, 1.2–10 cm × 0.8–5 cm, dry or fleshy, the exocarp tomentose, glabrous or verrucose, the endocarp thick, hard and woody, or thin and fibrous, without special mechanisms for seedling escape.

Species of this genus are known as *bois gaulettes* and *gaulettes* in Créole.

1. Stamens (8)10–40, usually exserted.
   2. Petals present.
      3. Stipules small and caducous, <10 mm long. Stamens ca. 35, far exserted. . . . . *L. guianensis.*
      3. Stipules large and persistent, 10–18 mm long. Stamens 10, slightly exserted. . . . . *L. amapaensis.*
   2. Petals absent.
      4. Abaxial leaf blade surface glabrous, without stomatal crypts.
         5. Leaf blades thickly coriaceous, the apex finely attenuate, the acumen 10–25 mm long. Inflorescences and flower exteriors glabrous or sparsely puberulous.. . . . . *L. granvillei.*
         5. Leaf blades chartaceous, the apex acuminate, the acumen 3–15 mm long. Inflorescences and flower exteriors densely gray-puberulous.. . . . . *L. apetala.*
      4. Abaxial leaf blade surface pubescent, with distinct stomatal crypts filled with hairs.
         6. Leaf blades 11–25 × 5–11 cm. Pedicels 3.5–5 mm long.. . . . . *L. albiflora.*
         6. Leaf blades 3–10 × 2–4 cm. Pedicels 0–0.2 mm long. . . . . *L. octandra.*
1. Stamens 3–8 (to 11 in *L. majuscula*), included.
   7. Abaxial leaf blade surface always glabrous. Petals present.
      8. Leaf blade apex caudate. . . . . *L. caudata.*
      8. Leaf blade apex acute or acuminate, not caudate.
         9. Leaf blade apex with well developed acumen 5–9 mm long. Exterior of flowers and inflorescences sparsely hirsute. . . . . *L. glabriflora.*
         9. Leaf blade apex rounded, acute, or bluntly acuminate. Exterior of flowers and inflorescences gray or brown puberulous.
            10. Stipules caducous. Petioles eglandular; blades with base cuneate, confluent onto petiole, the apex acuminate.. . . . . *L. laevigata.*
            10. Stipules subpersistent. Petioles with sessile glands; blades with base rounded to subcordate, not confluent onto petiole, the apex rounded, bluntly acuminate or acute. . . . . *L. hetermorpha* var. *heteromorpha.*
   7. Abaxial leaf blade surface usually pubescent (except glabrous in *L. fanshawei*). Petals absent.
      11. Leaf blades glabrous abaxially.. . . . . *L. fanshawei.*
      11. Leaf blades pubescent abaxially.
         12. Abaxial leaf blade surfaces pulverulent-farinaceous.. . . . . *L. canescens.*
         12. Abaxial leaf blade surfaces lanate pubescent or with stomatal crypts filled with pubescence.
            13. Flowers borne in small cymules on long slender secondary branches of inflorescences. . . . . *L. membranacea.*
            13. Flowers not borne in cymules as described above.
               14. Leaf blades with veins plane adaxially, the abaxial surface not deeply reticulate, without stomatal crypts.
                  15. Inflorescences and exterior of receptacle with dense brown tomentellous to densely puberulous pubescence. Stamens 3. . . . . *L. micrantha.*
                  15. Inflorescences and exterior of receptacle with sparse puberulous pubescence. Stamens 5–6.. . . . . *L. kunthiana.*
               14. Leaf blades with veins impressed adaxially, the abaxial surfaces either deeply reticulate or with stomatal crypts.
                  16. Leaf blades with conspicuous, hair-filled stomatal crypts abaxially.
                     17. Leaf blades brown-lanate abaxially. Receptacles conical, 4–5 mm long. Fruits with velutinous pubescence, the stipe 2–6 mm long.. . . . . *L. majuscula.*
                     17. Leaf blades white-lanate abaxially. Receptacles campanulate, 2.5–3 mm long. Fruits with pulverulent pubescence, the stipe 8–15 mm long. . . . . *L. alba.*
                  16. Leaf blades reticulate, without hair-filled stomatal crypts abaxially.
                     18. Stipules inserted at base of petiole. Stamens 3. . . . . *L. discolor.*
                     18. Stipules axillary. Stamens 6–8.. . . . . *L. laxifora.*

Licania alba (Bernoulli) Cuatrec. PL. 40d

Trees, to 35 m tall. Stipules axillary, elliptic, to 7 mm long, persistent. Leaves: petioles 9–17 mm long, glandular, canaliculate; blades oblong-elliptic to elliptic, 9–27 × 4–10 cm, coriaceous, abaxial surface with deep stomatal crypts obscured by dense white lanate pubescence, the apex acuminate, the acumen 3–10 mm long; midrib and secondary veins impressed adaxially, the secondary veins in 8–12 pairs. Inflorescences terminal and axillary, racemose panicles, the rachis and branches tomentellous; pedicels absent. Flowers: receptacles campanulate, 2.5–3 mm long, tomentellous on exterior; petals absent; stamens 6–8, included, inserted around complete circle. Fruits pyriform, to 9 cm long, the stipe 8–15 mm long, the exocarp sordid-ferruginous-brown pulverulent. Fl (Aug, Sep, Nov), fr (Feb, May); in non-flooded forest. *Gaulette*, *gaulette azon*, *gris-gris* (Créole).

Licania albiflora Fanshawe & Maguire

Trees, to 25 m tall. Stipules caducous. Leaves: petioles 7–14 mm long, terete; blades obovate to oblong-elliptic, 11–25 × 5–11 cm, coriaceous, with stomatal crypts filled with lanate pubescence abaxially, the apex acuminate, the acumen 3–6 mm long; midrib prominulous adaxially; secondary veins in 7–10 pairs. Inflorescences terminal and axillary, racemose panicles, the rachis and branches gray-brown villous pubescent; pedicels 3.5–5 mm long. Flowers: receptacles campanulate, tomentose on exterior; petals absent; stamens 12–13, exserted, inserted around complete circle. Fr (Mar); rare, in non-flooded forest.

Licania amapaensis Prance

Trees, to 15–20 m tall. Stipules axillary, persistent, 10–18 mm long. Leaves: petioles 10–16 mm long, terete; blades oblong, 15–33 × 4.5–9.5 cm, chartaceous, abaxial surface hirsute, the apex acuminate, the acumen 3–15 mm long; midrib plane adaxially, the secondary veins in 13–17 pairs. Inflorescences terminal and subterminal, racemose panicles, the rachis and branches ferruginous-tomentose; pedicels 0.5 mm long. Flowers: receptacles campanulate, tomentose on exterior; petals present; stamens 10, slightly exserted, inserted around three fourths of circle. Fruits ovoid, the exocarp densely ferruginous-velutinous-tomentose. Fl (Oct); rare, in non-flooded forest.

Licania apetala (E. Mey.) Fritsch

Trees, to 40 m tall. Stipules axillary, to 4 mm long, caducous. Leaves: petioles 3–6 mm long, terete; blades oblong to elliptic, 4–12 × 1.8–5 cm, chartaceous, glabrous on both surfaces, without stomatal crypts, the apex acuminate, the acumen 3–15 mm long; midrib prominulous adaxially, the secondary veins in 7–12 pairs. Inflorescences panicles, the flowers in small groups of cymules on short secondary branches, the rachis and branches densely gray-puberulous; pedicels 0–0.5 mm long. Flowers: receptacles campanulate, densely gray-puberulous on exterior; petals absent; stamens ca. 10, exserted, inserted around complete circle. Fruits globose to ovoid, the exocarp smooth and glabrous. Fl (Mar); common, in non-flooded forest. *Gaulette*, *gaulette couepi* (Créole).

Licania canescens Benoist

Trees, to 25 m tall. Stipules adnate to base of petiole, linear, 2–4 mm long, persistent. Leaves: petioles 3–5 mm long, terete to shallowly canaliculate; blades elliptic to oblong-elliptic, 5–15 × 2–5.5 cm, coriaceous, abaxial surface with waxy gray pulverulent-farinaceous pubescence, the apex acuminate, the acumen 4–15 mm long; midrib plane adaxially, the secondary veins in 6–9 pairs. Inflorescences terminal and axillary, racemose panicles, the rachis and branches glabrous or sparsely puberulous; pedicels absent. Flowers: receptacles campanulate, tomentellous on exterior; petals absent; stamens 5, included, unilateral. Fruits pyriform, to 3 cm long, the exocarp glabrous, drying yellow and wrinkled. Fl (Aug, Sep, Oct), fr (Apr, May, Sep); common, in non-flooded forest. *Gaulette*, *gaulette noir*, *gaulette rouge*, *gris-gris* (Créole).

Licania caudata Prance

Trees, to 15 m tall. Stipules caducous. Leaves: petioles 5–7 mm long, weakly canaliculate; blades ovate-elliptic, 8–11 × 3.5–5.5 cm, subcoriaceous, glabrous on both surfaces, the apex caudately acuminate, the acumen 7–9 mm long; midrib prominulous adaxially, the secondary veins in 8–9 pairs. Inflorescences terminal and axillary, racemose panicles, the rachis and branches glabrous or sparsely hirsutulous. Flowers: receptacles campanulate, sparsely hirsute on exterior; petals present; stamens 7–8, included, inserted around complete circle. Fruits oblong, ca. 8 × 25 mm, the exocarp smooth or slightly rugulose, glabrous. Fr (Nov); common, in non-flooded forest. *Gaulette*, *gaulette gris-gris*, *gris-gris* (Créole).

Licania discolor Pilg. PL. 40c

Trees, to 30 m tall. Stipules adnate to base of petioles, to 2 mm long, early caducous. Leaves: petioles 5–10 mm long, terete; blades oblong-ovate, 4–9 × 1.5–6.3 cm, coriaceous, lower surface with brown lanate pubescence obscuring deeply reticulate venation, the apex acuminate, the acumen 2–7 mm long; midrib and secondary veins slightly impressed adaxially, the secondary veins in 5–7 pairs. Inflorescences terminal, racemose panicles, the rachis and branches tomentose; pedicels absent. Flowers 1–2 mm long; receptacles campanulate, tomentose on exterior; petals absent; stamens 3, included, unilateral. Fruits pyriform, exocarp rufous-velutinous. Fl (Aug, Sep); uncommon, in non-flooed forest.

Licania fanshawei Prance

Trees, to 30 m tall. Stipules axillary, to 2 mm long, caducous. Leaves: petioles 3–5 mm long, shallowly canaliculate; blades ovate-elliptic, 4–11 × 2.5–7 cm, coriaceous, glabrous on both surfaces, the apex obtuse; midrib prominulous adaxially, the secondary veins in 7–8 pairs, the venation papillose abaxially. Inflorescences terminal and subterminal, panicles, the rachis and branches brown-pubescent to puberulous; pedicels absent. Flowers: receptacles campanulate, brown-pubescent on exterior; petals absent; stamens 3, included, unilateral. Fruits unknown. Fl (Aug, Sep); rare, in non-flooded forest.

Licania glabriflora Prance

Trees, to 20 m tall. Stipules axillary, lanceolate, to 3.5 mm long, persistent. Leaves: petioles 2–5 mm long, terete; blades oblong to oblong-elliptic, 7–11 × 2.5–4.2 cm, submembranous, glabrous on both surfaces, the apex acuminate, the acumen 5–9 mm long; midrib prominulous adaxially, the secondary veins in 8–10 pairs. Inflorescences axillary, racemose panicles, the rachis and branches sparsely hirsute. Flowers: receptacles campanulate, sparsely hirsutulous on exterior; petals present; stamens 5, included, inserted around three fourths of circle. Fruits unknown. St (Oct); rare, in non-flooded forest.

Licania granvillei Prance FIG. 78

Trees, to 30 m long tall. Stipules <10 mm long, caducous. Leaves: petioles 3–8 mm long, canaliculate; blades oblong, 6–13 × 2.3–6 cm, coriaceous, glabrous on both surfaces, without stomatal crypts, the apex finely attenuate, the acumen 10–25 mm long; midrib prominent adaxially, the secondary veins in 10–16 pairs; Inflorescences racemose panicles, the rachis and branches sparsely appressed puberulous; pedicels absent. Flowers: receptacles campanulate, glabrous or sparsely puberulous on exterior; petals absent; stamens 12, far exserted, inserted around complete circle. Fruits globose to ellipsoid, 5.5 × 3.5–4 cm, the exocarp glabrous and lenticellate. Fl (Aug, Sep), fr (Mar, Apr, May); common, in non-flooded forest.

Licania guianensis (Aubl.) Griseb.

Trees, to 15 m tall. Stipules <10 mm long, caducous. Leaves: petioles 5–8 mm long, canaliculate; blades oblong-elliptic to oblong, 7–18 × 2–5.5 cm, coriaceous, glabrous on both surfaces, the apex

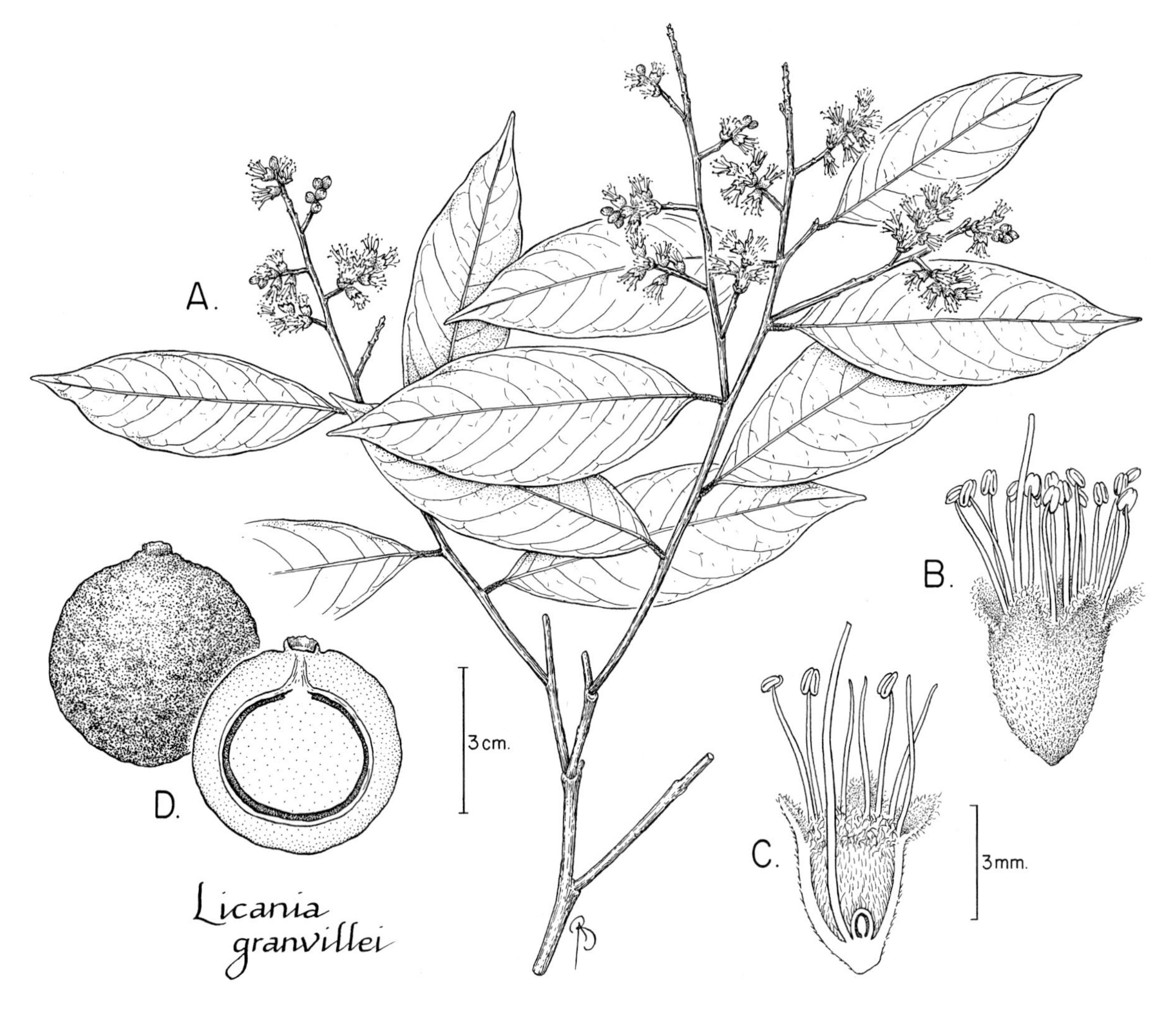

FIG. 78. CHRYSOBALANACEAE. *Licania granvillei* (A–C, *Mori et al. 20789*; D, *Mori & Pennington 18012*). **A.** Apex of stem with leaves and inflorescences; note sessile flowers. **B.** Lateral view of entire flower; note lack of corolla. **C.** Medial section of flower with gynobasally attached pistil and gynoecium located at the base of the cup-shaped hypanthium. **D.** Entire (left) and medial section (right) of fruit with attachment scar at top.

finely pointed, the acumen 8–22 mm long; midrib prominulous adaxially, the secondary veins in 8–12 pairs. Inflorescences terminal and axillary, racemose panicles, the rachis and branches almost glabrous; pedicels 1–3 mm long. Flowers: receptacles cupuliform, almost glabrous on exterior; petals present; stamens ca. 35, far exserted, inserted around complete circle. Fruits oblong to globose, to 5 cm diam., the exocarp smooth and glabrous. Fl (Mar, Apr, May); rare, in non-flooded forest. *Cèdre* (Créole).

Licania heteromorpha Benth. var. **heteromorpha**

Trees, to 30 m tall, usually with low buttresses, these sometimes branched and even prop root like. Bark: slash with red exudate, emitting hissing sound shortly after being cut. Stipules axillary, linear, 1.5–3.5 mm long, subpersistent. Leaves: petioles 2–12 mm long, shallowly canaliculate, with two sessile glands; blades ovate to oblong, usually elliptic, 4.5–11 × 2–6 cm, coriaceous, glabrous on both surfaces, the apex rounded to bluntly acuminate; midrib prominulous adaxially, the secondary veins in 6–11 pairs. Inflorescences terminal and subterminal, racemose panicles, the rachis and branches gray-puberulous; pedicels absent. Flowers: receptacles campanulate, tomentellous to tomentose on exterior; petals present; stamens 5–7, included, inserted around complete circle. Fruits globose, the exocarp glabrous or velutinous pubescent. Fl (Jul, Aug, Sep, Dec), fr (May, Aug, Sep, Oct, Nov); common, in non-flooded forest. *Gaulette*, *gaulette blanc*, *gaulette indien*, *gaulette rouge* (Créole).

Licania kunthiana Hook.f. PL. 41a

Trees, to 25 m tall. Stipules adnate to base of petiole, lanceolate, 2–3 mm long, persistent. Leaves: petioles 2–5 mm long, terete or shallowly canaliculate; blades oblong-ovate to oblong-lanceolate, 3–8.5 × 1.3–5 cm, chartaceous, abaxial surface densely lanate pubescent, the apex acuminate, the acumen 2–13 mm long; midrib and secondary veins plane adaxially, the secondary veins in 7–9 pairs. Inflorescences axillary and terminal, racemose panicles, the rachises and branches sparsely puberulous; pedicels absent. Flowers 1.5–2 mm long; receptacles campanulate, sparsely gray-puberulous on exterior; petals absent; stamens 5–6, included, unilateral. Fruits oblong-elliptic, to 2 cm long, the exocarp yellow-brown, sordid puberulent. Fl (Sep, Nov); rare, in non-flooded forest. *Gaulette* (Créole).

Licania laevigata Prance

Trees, to 25 m tall, with low symmetrical buttresses. Stipules axillary, caducous. Bark light gray, inner bark deep red, the slash hissing when first cut. Leaves: petioles 5–8 mm long, terete, eglandular; blades oblong, 9–18 × 4.2–7 cm, coriaceous, glabrous on both surfaces, the apex acuminate to bluntly acuminate; midrib prominulous adaxially, the secondary veins in 7–10 pairs. Inflorescences terminal and subterminal, racemose panicles, the rachis and branches puberulous; pedicels 0.5 mm long. Flowers: receptacles urceolate, brown-tomentellous on exterior; petals present; stamens 6–7, included, inserted around three fourths of circle. Fruits globose, exocarp densely short-ferruginous-tomentellous. Fl (Oct, Nov), fr (May, Aug); rare, in non-flooded forest. *Gaulette* (Créole).

Licania laxiflora Fritsch

Trees, to 30 m tall. Stipules axillary, elliptic to lanceolate, 2–5 mm long, persistent. Leaves: petioles 4–8 mm long, terete; blades ovate to elliptic, 5–22 × 2.5–9 cm, coriaceous, abaxial surface deeply reticulate with hirsutulous venation, the areas between veins lanate, the apex acuminate, the acumen 2–10 mm long; midrib and secondary veins slightly impressed adaxially, the secondary veins in 7–10 pairs. Inflorescences terminal and subterminal, racemose panicles, the rachis and branches rufous-tomentose; pedicels absent. Flowers: receptacles campanulate, tomentose on exterior; petals absent; stamens 6–8, included, inserted around complete circle. Fruits globose to pyriform, to 10 cm long including stipe, the exocarp rufous-velutinous pubescent. Fl (Mar, Nov), fr (Mar); rare, in non-flooded forest.

Licania majuscula Sagot

Trees, to 30 m tall. Stipules adnate to base of petiole, linear-lanceolate, 3–8 mm long, persistent. Leaves: petioles 8–13 mm long, canaliculate, with two or more pairs of sessile glands; blades elliptic to oblong-ovate, 7.5–18 × 2.8–9 cm, coriaceous, abaxial surface deeply reticulate with stomatal crypts, brown-lanate between venation in mouth of crypts, the apex cuspidate to acuminate, the acumen 2–12 mm long; midrib and secondary veins impressed adaxially, the secondary veins in 8–10 pairs. Inflorescences terminal and axillary, racemose panicles, the rachis and branches tomentose; pedicels absent. Flowers: receptacles conical, 4–5 mm long, tomentose on exterior; petals absent; stamens 8–11, included, inserted around complete circle. Fruits globose, 5–6 mm diam., the stipe 2–6 mm long, the exocarp velutinous pubescent. In non-flooded forest. *Gris-gris*, *gris-gris-noir* (Créole).

Licania membranacea Laness.

Trees, to 35 m tall. Stipules adnate to base of petioles, linear, 3–7 mm long, persistent. Leaves: petioles 8–12 mm long, canaliculate; blades oblong, 8–19 × 3.7–7.8 cm, coriaceous, abaxial surface with short appressed lanate-arachnoid pubescence, the apex finely acuminate, the acumen 10–25 mm long; midrib plane or prominulous adaxially, the secondary veins in 7–10 pairs. Inflorescences terminal and axillary, paniculate, the flowers borne in small cymules on slender secondary branches, the rachis and branches puberulous; pedicels 0.25–1 mm long. Flowers: receptacles campanulate, tomentellous on exterior; petals absent; stamens 3–5, included, unilateral. Fruits pyriform, to 2.5 cm long, the exocarp brown rufous tomentose. Fl (Sep), fr (Nov); immature fr (Nov); rare, in non-flooded forest. *Gaulette*, *gaulette noir* (Créole).

Licania micrantha Miq.

Trees, to 30 m tall. Stipules adnate to extreme base of petiole, lanceolate, to 7 mm long, persistent. Leaves: petioles 4–12 mm long, terete; blades ovate elliptic to oblong, 4.5–15 × 2.3–7 cm, coriaceous, abaxial surface with dense brown lanate pubescence easily removed to reveal smooth surface, the apex acuminate, the acumen 2–14 mm long; midrib and secondary veins plane adaxially, the secondary veins in 5–7 pairs. Inflorescences terminal and axillary, much-branched racemose panicles, the rachis and branches densely brown tomentellous to puberulous; pedicels absent. Flowers 2–3 mm long; receptacles densely brown campanulate, tomentellous on exterior; petals absent; stamens 3, included, unilateral. Fruits pyriform, to 5 cm long including stipe of 5–10 mm, the exocarp with waxy pulverulent indumentum. St (Oct); rare, in non-flooded forest. *Bois gaulette*, *gaulette*, *gaulette-marécage* (Créole).

Licania octandra (Roem. & Schult.) Kuntze PL. 41b

Trees, to 27 m tall. Stipules axillary, to 5 mm long, subpersistent. Leaves: petioles 3–7 mm long, terete to shallowly canaliculate; blades ovate to oblong, 3–10 × 2–4 cm, coriaceous, with stomatal crypts filled with lanate pubescence abaxially, the apex obtuse to bluntly acuminate, the acumen 1–5 mm long; midrib prominulous adaxially, the secondary veins in 8–13 pairs. Inflorescences terminal and axillary, racemose panicles, the rachis and branches gray-brown tomentose; pedicels 0–0.2 mm long. Flowers: receptacles campanulate, tomentose on exterior; petals absent; stamens 8–12, exserted, inserted around complete circle. Fruits globose or ovoid, to 2.5 cm long, the exocarp smooth and glabrous. Fl (Aug, Sep, Nov), fr (Nov); rare, in non-flooded forest.

PARINARI Aubl.

Small or large trees. Stipules 0.5–7 cm long, axillary, caducous or persistent. Leaves: petioles with 2 circular glands on adaxial surface; blades often with several small marginal or submarginal discoid glands along length, the abaxial surface with small stomatal crypts filled with densely matted hairs. Inflorescences many-flowered complex cymes or cymose panicles; bracts and bracteoles eglandular, completely concealing flower buds, both individually and in small groups. Flowers bisexual, slightly zygomorphic; flower buds straight; receptacle-tube longer than sepals, subcampanulate, slightly swollen on one side, hollow, hairy inside throughout; sepals 5, deltate, eglandular, densely hairy on both surfaces, the apex acute; petals 5, as long as or shorter than sepals, caducous; stamens 7–10, the filaments white, shorter than sepals, slightly curved in bud and at anthesis, slightly expanded at base; staminodia ca. 6, minute, subulate; ovary 1-carpellate, 2-locular, inserted on upper half of receptacle-tube below mouth, the style arcuate, included. Fruits fleshy drupes, the exocarp verrucose, the endocarp hard, thick, with a rough, fibrous surface, with 2 basal stoppers.

Species of this genus are known as *gaulette* in Créole.

1. Leaf blades 3–9 cm long; midrib plane to prominulous adaxially. Fruits 2.5–4 cm long. . . . . *P. excelsa.*
1. Leaf blades 9–17 cm long; midrib distinctly impressed adaxially. Fruits 8–10 cm long. . . . . *P. montana.*

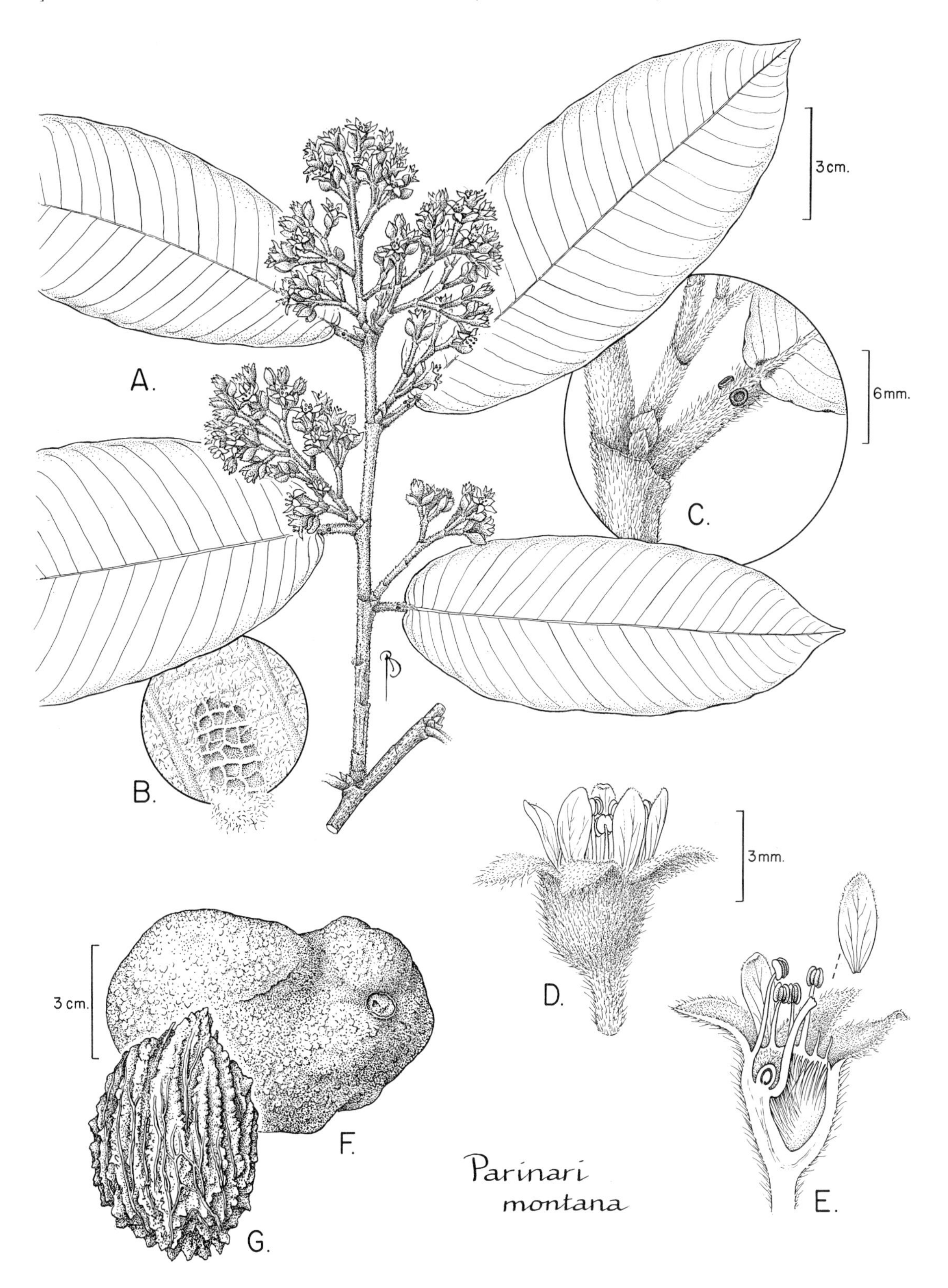

FIG. 79. CHRYSOBALANACEAE. *Parinari montana* (A–E, *Mori et al. 20790*; F, *Mori & Boom 15160*; G, *Prance 58753*). **A.** Apex of stem with leaves and inflorescences. **B.** Abaxial leaf surface with part of woolly pubescence removed to show detail of venation. **C.** Detail of node with stipules, base of panicle, and petiole with distinct glands at distal end. **D.** Lateral view of flower. **E.** Medial section of flower with perched gynoecium, gynobasic style, cup-shaped hypanthium, and petal (above). **F.** Fruit with attachment scar (right). **G.** Seed with corrugated endocarp.

Parinari excelsa Sabine

Trees, to 40 m tall. Stipules ca. 1 mm long, caducous. Leaves: blades ovate to oblong-elliptic, 3–9 × 1.5–5 cm, the base rounded to cuneate, the apex acuminate, the acumen 2–10 mm long; midrib plane to prominulous adaxially, the secondary veins in 13–20 pairs. Fruits oblong, 2.5–4 × 1.8–2.5 cm. Fr (Aug, immature); rare, in non-flooded forest. *Bois gaulette, gris-gris* (Créole).

Parinari montana Aubl. FIG. 79; PART 1: FIGS. 4, 9

Trees, to 40 m tall. Stipules 1–3 mm long, caducous. Leaves: blades oblong to oblong-lanceolate, 9–17 × 3–6.5 cm, the base rounded to subcuneate, the apex acuminate, the acumen 5–9 mm long; midrib distinctly impressed adaxially, the secondary veins in 21–27 pairs. Fruits oblong, 8–10 cm long, the endocarp brain-like. Fl (Sep), fr (Nov); rare, in non-flooded forest.

CLUSIACEAE (Mangosteen Family)

John J. Pipoly III and Mats H. G. Gustafsson

Trees or shrubs, terrestrial or epiphytic, glabrous, or stellate tomentose. Exudate clear, white, yellow or orange. Leaves opposite or alternate (outside our area), simple, entire; petioles at times with an adaxial margined pit at base; blades with linear latex canals or glandular punctations or punctate lineations drying translucent or blackish. Inflorescences terminal or axillary, cymose or rarely in fascicles or flowers solitary; pedicels usually bracteolate. Flowers bisexual or unisexual (plants dioecious or polygamous); sepals 2–14; petals 2–14, free or basally connate, imbricate or contorted; stamens 4 to numerous, of variable shape, at times grouped in fascicles or whorls, or completely fused into synandria; staminodia similar to stamens or connate into resiniferous ring or central mass; ovary superior, of 2–15 carpels, the style single, equal in number to carpels, or absent. Fruits capsules, berries, or drupes (outside our area). Seeds 1 to many per carpel, often arillate.

Tovomitopsis and *Balboa* have been united under Chrysochlamys; *Rheedia* under *Garcinia*; and *Decaphalangium*, *Havetia*, *Havetiopsis*, *Oedematopus*, *Quapoya*, and *Renggeria* under *Clusia*.

Features of leaf venation and latex ducts are described in the present treatment as they appear on dried or pressed material.

Eyma, P. J. 1932. The Polygonaceae, Guttiferae and Lecythidaceae of Surinam. Meded. Bot. Mus. Herb. Rijks Univ. Utrecht **4:** 1–77.

———. 1966. Guttiferae. *In* A. Pulle (ed.), Flora of Suriname **III(1):** 65–118. E. J. Brill, Leiden.

Görts-van Rijn, A. R. A. 1986. Guttiferae. *In* A. L. Stoffers & J. C. Lindeman (eds.), Flora of Suriname **III(1–2):** 306–331. E. J. Brill, Leiden.

Kearns, D. M., P. E. Berry, P. F. Stevens, N. L. Cuello A., J. J. Pipoly, N. K. B. Robson, B. K. Holst, K. Kubitzki & A. L. Weitzman. 1998. Clusiaceae. *In* J. A. Steyermark, P. E. Berry & B. K. Holst (gen. eds.), Flora of the Venezuelan Guayana **4:** 248–329. P. E. Berry, B. K. Holst & K. Yatskievych (vol. eds.). Missouri Botanical Garden Press, St. Louis.

1. Inflorescences axillary fascicles, terminal umbels, or flowers terminal and solitary on pendent, short pedicels, the latter on plants with cruciate (decussate) branching. Style 1 or stigma sessile. Fruits few-seeded berries.
   2. Branching opposite but not decussate. Inflorescences axillary fascicles. Stamens free or connivent, not grouped in fascicles or fused into tube; style obsolete, the stigma 1. Fruits tuberculate (in our area). . . . . . *Garcinia.*
   2. Branching decussate. Inflorescences solitary or umbels. Stamens connate by filaments and forming fascicles or a tube; style elongate, the stigma 4–5 lobed. Fruits not tuberculate.
      3. Flowers several together in umbels. Stamens forming apically 5-lobed tube surrounding pistil. . . . *Symphonia.*
      3. Flowers solitary or 2–3 together but not in umbels. Stamens in separate fascicles, each consisting of 3 to many stamens.
         4. Staminal fascicles of 3–5 stamens, spirally wound around pistil. Fruit usually one-seeded, with spiral markings from staminal fascicles. . . . . *Moronobea.*
         4. Staminal fascicles of >15 stamens, not spirally wound. Fruit usually 5-seeded, without spiral markings. . . . . *Platonia.*
1. Inflorescences terminal cymes or panicles of cymes, or axillary and/or terminal thyrses. Styles absent (stigma sessile) or equal to carpels in number. Fruits berries with numerous seeds or capsules.
   5. Exudate thin, reddish-orange. Petals and fruit with black or orange punctations or lines. Petals pubescent adaxially; stamens in five fascicles. Fruits berries with numerous seeds. Seeds without arils. . . . . *Vismia.*
   5. Exudate thick, usually yellow, cream-colored, white, or clear (rarely orange). Petals and fruit without black or orange punctations or lines. Petals glabrous; stamens variously arranged, but not in five fascicles. Fruits capsules. Seeds with arils.
      6. Sepals at most very slightly reflexed at anthesis; stamens 4, 8, 12 or numerous, the filaments either free and flattened or at least the central ones fused into resiniferous synandria of varying shape; ovules mostly 4 to many per locule; aril not vascularized. . . . . *Clusia.*
      6. Sepals highly reflexed in anthesis; stamens numerous, linear, on concave receptacle, the central ones connivent basally, without resin; ovules one per locule; aril vascularized.
         7. Petiole bases with well developed adaxial margined pits. Sepals not imbricate in bud, the two outer sepals completely enclosing bud. Fruits lenticellate at maturity. . . . . *Tovomita.*
         7. Petiole bases with inconspicuous adaxial margined pits. Sepals imbricate in bud. Fruits smooth at maturity. . . . . *Chrysochlamys.*

CHRYSOCHLAMYS Poepp. [Syn.: *Tovomitopsis* Planch. & Triana]

Glabrous shrubs or small trees. Exudate white or yellow, often sparse. Leaves: petioles with minute adaxial margined pits at base; blades thin; secondary veins widely spaced, usually forming partial submarginal vein. Inflorescences terminal or axillary, loose cymes or cymose panicles or rarely flowers solitary; bracts minute. Flowers unisexual (plants dioecious) or bisexual; sepals 4–6, imbricate in bud; petals 4–6, white, imbricate; stamens numerous (25–200), the filaments free or the innermost fused, the anthers reduced; ovary (4)5(6)-locular with 1 ovule per locule, the stigmas sessile; pistillode absent in staminate flowers. Fruits fleshy, valvular (septifragally) dehiscent capsules with persistent stigmas. Seeds 1–5, arillate.

Chrysochlamys membranacea Planch. & Triana [Syn.: *Tovomitopsis membranacea* (Planch. & Triana) D'Arcy]

Trees, to 10 m tall. Stems terete, 5–8 mm diam. Leaves: petioles 1–2.5 cm, adaxial margined pit minute, to 1.5 mm long; blades membranaceous, oblong, 11–20 × 4–8 cm, the base cuneate, the apex acuminate, the margins slightly raised. Inflorescences terminal or axillary, often below leaves (cauliflorous). Staminate flowers: sepals 5, widely ovate, 5–7 × 4–6 mm, the apex obtuse; petals 5, obovate, 1–1.2 × 3–3.5 mm, the apex rounded; stamens numerous, the outer fertile, the inner sterile and united in column. Pistillate flowers: perianth as in staminate flowers; ovary 5-locular, ovoid, ca. 8–10 × 2–3 mm, the stigmas 5, deltate to circular, ca. 1 mm wide, sessile. Fruits ellipsoid to globose, 1.8–2.5 × 1.5–2 cm, greenish white or reddish. Seeds 1–5, blackish, partly enveloped by white aril. Fr (Feb, Aug, Sep, Nov); fairly frequent, in non-flooded forest along streams, <400 m alt. *Palétuvier grand-bois* (Créole).

CLUSIA L.

Glabrous, free-standing, epiphytic, epipetric, hemiepiphytic or rarely strangling shrubs or trees or sometimes lianas. Exudate clear, translucent, white, cream-colored, yellow or orange. Leaves decussate, petiolate; petioles with adaxial margined pit at base; blades with ± conspicuous latex canals visible adaxially and/or abaxially in dried material. Inflorescences terminal or very rarely lateral cymes or panicles of cymes, flowers rarely solitary; pedicels bracteolate, the bracteoles sometimes grading into sepals. Flowers unisexual or rarely bisexual, sometimes apomictic; sepals 2–9; petals 4–12. Staminate flowers: stamens 4-numerous, free, connate or agglutinated by resin, a pistillode sometimes present. Pistillate flowers: staminodia 4, 5, or numerous, free or connate into ring, sometimes resin-secreting; ovary 4–15-locular, the stigmas 4–15, sessile or on very short styles, sometimes connivent, the ovules few to many per locule. Fruits valvular dehiscent capsules with persistent stigmas. Seeds arillate.

1. Petioles with well defined adaxial margined pits; blades with few or no obvious intersecondary veins. Stamens numerous; stigmas 4–14.
    2. Petioles >1 cm long, the basal adaxial margined pit wider than petiole. Secondary inflorescence bracts persistent. Outer stamens fertile and fused by filaments into ring, the anthers subulate, the central stamens sterile and agglutinated by resin; stigmas >7, the stigmas fusing into raised ring, or stigmatic areas flat.
        3. Stems >(9)10 mm diam. Petioles >5 cm long; foliar latex canals inconspicuous ad- and abaxially. Flowers >5 cm diam.
            4. Stems terete, not winged. Stigmas flat patches, not forming raised ring on mature fruits. . . . *C. platystigma.*
            4. Stems not terete, winged. Stigmas protruding, forming raised ring on mature fruits. . . . . *C. grandiflora.*
        3. Stems <8(10) mm diam. Petioles <5 cm long; foliar latex canals conspicuous ad- or abaxially. Flowers ≤5 cm diam.
            5. Stems terete. Foliar latex canals conspicuous adaxially. Anther connective barely protruding above thecae. Fruits ovoid. . . . . *C. palmicida.*
            5. Stems tetragonal. Foliar latex canals inconspicuous adaxially, somewhat conspicuous abaxially. Anther connective forming filamentous appendage longer than thecae. Fruits oblongoid or ellipsoid. . . . . *C. nemorosa.*
    2. Petioles <1 cm long, basal adaxial margined pit narrower than petiole or absent. Secondary inflorescence bracts caducous. Stamens free or connate (by resin) into conical or pentagonal synandrium, the anthers obovate, oblong, or muticous; the stigmas <10, free, not fusing into ring.
        6. Stems subterete, (3)5–7(12) mm diam. Petioles stout, widely marginate; blades broadly obovate. Stigmas concave or convex. Fruits smooth or transversely costate.
            7. Bracteoles 2. Stamens fused; stigmas 5, 5–10 mm long. Fruits 1.5–7 cm diam.
                8. Leaf blades cartilaginous. Flowers unisexual; petals dehiscent; stamens fused into conical synandrium crowned by pistillode. Fruits globose to ovoid, transversely wrinkled, 3.5–7 cm diam. at maturity. . . . . *C. leprantha.*
                8. Leaf blades thinly coriaceous. Flowers bisexual; petals persistent in fruit; stamens in ring surrounding ovary. Fruits oblongoid, not transversely wrinkled, 1.5–2.5 cm diam.. . . . . *C. scrobiculata.*

7. Bracteoles 4–14. Stamens free; stigmas usually 4, 0.6–0.8 mm long. Fruits 0.3–0.8 cm diam. . . . . *C. melchiori.*
6. Stems terete, 2.5–3.5 mm diam. Petioles thin, slightly canaliculate; blades oblanceolate to obovate. Stigmas convex. Fruits transversely wrinkled. . . . . *C. panapanari.*
1. Petioles with poorly defined adaxial margined pits; blades with numerous obvious intersecondary veins parallel to secondary veins. Stamens 4 or 8; stigmas 4.
9. Stamens 4; stigmas at anthesis on short styles.. . . . . *C. flavida.*
9. Stamens 8; stigmas at anthesis sessile, or rarely short styles..
10. Stems terete to angular, not tetragonal. Leaf blades oblanceolate or oblong; submarginal collecting vein ca. 0.5 mm from margin. . . . . *C. octandra.*
10. Stems tetragonal. Leaf blades obovate, the submarginal collecting vein ca. 1 mm from margin. . . . . *C. obovata.*

Clusia flavida (Benth.) Pipoly [Syns.: *Havetiopsis flavida* (Benth.) Planch. & Triana, *Havetia flavida* Benth., *Havetiopsis flexilis* Planch. & Triana]

Epiphytic or epipetric shrubs or trees, to 8 m tall. Exudate white, often sparse. Stems terete, 4.5–5 mm diam. Leaves: petioles 5–10 mm long, broadly marginate, the adaxial basal pit inconspicuous; blades obovate to oblanceolate, 6–20 × 3–5 cm, membranaceous or chartaceous, rarely subcoriaceous, the base cuneate, decurrent on petiole, the apex broadly rounded, the margins slightly inrolled; secondary and intersecondary veins numerous, parallel, united by submarginal collecting vein 0.5 mm from margins; foliar latex canals sometimes visible ad- and abaxially. Staminate inflorescences 9- to many-flowered. Pistillate inflorescences 1- to few-flowered. Staminate flowers: sepals 4, decussate, the outer oblate to orbicular, 1.5–1.7 × 1.9–2.2 mm, the inner orbicular, 2.5–3 mm long and wide; petals 4, decussate, yellowish, carnose, suborbicular, 2.5–3 mm long, the inner cucullate; stamens 4, the filaments basally connate, pistillode absent. Pistillate flowers: as in staminate but staminodia 4, stamen-like; ovary globose, 3–3.5 mm diam, 4(6)-locular, the styles <1 mm long, the stigmas 4(6), tetragonal, 0.5–1 mm diam. Fruits globose, 10–17 mm diam. Fl (Jul), fr (Jan); in non-flooded forests.

Clusia grandiflora Splitg. PART 1: FIG. 9

Free-standing or epiphytic trees, to 12 m tall. Exudate white, oxidizing yellow. Stems decussately alate, 10–15 mm diam. Leaves: petioles 5.5–8.0 cm long, the basal adaxial margined pit wider than petiole, 1–1.5 cm long; blades obovate or elliptic, 15.5–24.5 × 9–16.8 cm, coriaceous, nitid adaxially only when young, otherwise dull, the base acute, the apex broadly rounded, the margins flat; submarginal veins 0.5–1 mm from margins, the secondary veins numerous, prominulous ad- and abaxially; foliar latex canals linear, numerous, conspicuous adaxially. Inflorescences pendent, 3-flowered cymes or flowers solitary. Staminate flowers: sepals 6, decussate, the outer 2 oblate, 1.3–1.5 × 2–2.2 cm, with scarious margins, the inner 4 oblong, 3.5–4 × 1.5–2 cm, thinly coriaceous, without scarious margins; petals 8, imbricate, basally fused, obovate-spathulate, 6–9 × 3.5–6 cm, white to pink, the apex broadly rounded; stamens numerous, the outer 2.2–3 cm long with filaments fused into tube 1.6–2 cm long, the anthers subulate, 1–1.5 cm long, the connective appendage 3–5 mm long, the inner stamens (staminodia) much shorter, resiniferous, closely connivent into disc-like structure; pistillode absent. Pistillate flowers as in staminate but staminodia 8–10 mm long, resiniferous and fused into ring, anantherous; ovary 10–15-locular, globose, 1.6–1.8 cm diam., the stigmas 10–15, sessile, thick, triangular, ca. 10 × 4 mm, fused into raised ring on mature fruit. Fruits ovoid to depressed-globose, to 8 × 11 cm. Fl (Jan); in non-flooded along creeks. *Zognon sauvage* (Créole).

Clusia leprantha Mart. [Syn.: *Clusia purpurea* (Splitg.) Mart.] PL. 41c

Epiphytic shrubs or trees, to 12 m tall. Exudate white, cream-colored, or yellowish. Stems subterete, 5–8 mm diam. Leaves: petioles 5–10 mm long, the basal adaxial margined pit as wide as petiole; blades obovate to oblanceolate, 10.5–16 × 4.5–9 cm, cartilaginous, the base acute, the apex truncate to very broadly rounded, the margins inrolled to revolute; submarginal veins ca. 1.5 mm from margins, the secondary veins often wavy when dry; foliar latex canals forming fine, non-continuous lines. Staminate inflorescences erect, 1–6-flowered; pedicels 5–10 mm long, the bracteoles 2. Pistillate inflorescences 1–3-flowered. Staminate flowers: sepals 7–9, suborbicular with scarious, hyaline margins, the outer 2 opposite, 8–10 mm long, the inner imbricate, 12–15 mm long; petals 5, imbricate, suborbicular to very widely obovate, 10–12 × 13–15 mm, stiffly membranaceous, dark red; stamens very numerous, connate by resin into conical synandrium, 5–7 × 10 mm, the anthers sessile, minute; pistillode consisting of 5 stigmatic areas at synandrium apex. Pistillate flowers as in staminate but petals to 2 cm long; staminodia ca. 10, to 4 mm long, with rounded antherodes; ovary ovoid, 5-locular, the stigmas 5, sessile, subdeltate, 8–10 × 8–9 mm. Fruits depressed globose or ovoid, 3–5 × 3.5–7 cm, transversely wrinkled, the stigmatic areas often grayish when dry, nearly connivent on apical protrusion of fruit. Fl (Feb, Dec), fr (Nov); in non-flooded forest.

Clusia melchiori Gleason

Terrestrial or epiphytic shrubs or small trees, to 10 m tall. Exudate clear, oxidizing yellowish. Stems subterete, angular, (3)7–12 mm diam. Leaves: petioles broadly marginate, 3–10 mm long, adaxial margined pit as wide as petiole; blades obovate, rarely oblong, (6.5)8–10(17) × (2.8)5.5–8(11) cm, coriaceous, dull ad- and abaxially, the base cuneate, fully decurrent to petiole base, the apex broadly rounded to truncate, the margins flat or inrolled; submarginal veins 1–1.5 mm from margins, the secondary veins numerous, prominulous ad- and abaxially; foliar latex canals inconspicuous. Inflorescences erect, 5- to many-flowered; bractoles 4–14. Staminate flowers: sepals 4, oblate, 3.5–4 × 4–4.5 mm wide, coriaceous, with scarious margin; petals 4–6, obovate, 3–3.5 × 2.5–3 mm, carnose, white, cream-colored or greenish; stamens numerous, 3–3.5 mm long, the filaments flat, 2.8–3 mm long, the anthers muticous, oblong, ca. 0.5 × 0.2 mm, with rounded apex; pistillode absent. Pistillate flowers as in staminate but staminodia 4, linear, 1–1.5 mm long, the antherodes minute; ovary 4-locular, oblong to obovoid, 1.5–3 × 2–4 mm, the styles 4, erect, connivent, 2–2.4 mm long, the stigmas cuneiform, 0.6–0.8 mm long and wide; ovules 4 per locule. Fruits ovoid, 10–13 × 3–8 mm. Fl (Sep), fr (Jan, Nov, Dec); in forest at higher elevations.

Clusia nemorosa G. Mey.

Terrestrial or hemiepiphytic shrubs or trees, to 8 m tall. Exudate white, oxidizing yellow. Stems tetragonal, 6–10 mm diam. Leaves:

petioles 2.2–4 cm long, the basal adaxial margined pit inconspicuous, slightly wider than petiole, 3–5 mm long; blades obovate to elliptic, 8–19.5 × 5–11 cm, coriaceous, the base acute or obtuse, the apex acute to rounded, the margins revolute; submarginal veins to 1 mm from margins, the secondary veins prominent ad- and abaxially; foliar latex canals conspicuous abaxially. Inflorescences pendent, 3–7-flowered. Flowers unisexual in our area, sometimes bisexual outside our area. Staminate flowers: sepals 4, decussate, orbicular, 5–12 × 6–14 mm, stiffly chartaceous, with somewhat scarious margins; petals 4–5, imbricate, obovate, 1.5–2.5 × 1–1.5 cm, coriaceous, white or basally reddish, the apex broadly rounded; stamens numerous, the outer fertile, in 2–3 rows, 5–8 mm long, the filaments basally fused into short tube, the anthers linear, subulate, the connective appendage longer than thecae, the inner stamens (staminodia) much shorter, resiniferous, closely connivent into disc-like structure; pistillode absent. Pistillate flowers as in staminate but staminodia 2–3 mm long, fused into ring, resiniferous; ovary 5–8-locular, 1–1.5 × 0.8–1 cm, the stigmas 5–8, subdeltate, 2.5–3 mm long, fused into raised ring on mature fruit. Fruits oblong or ellipsoid at maturity, 3–4 × 2–2.5 cm. Fl (Sep), fr (Nov); on rocky outcrops and in non-flooded forest. *Zognon sauvage* (Créole).

Clusia obovata (Planch. & Triana) Pipoly [Syn.: *Oedematopus obovatus* Planch. & Triana]

Epiphytic lianous shrubs, to 4 m tall. Stems 3.5–5 mm diam., tetragonal. Leaves: petioles 5–10 mm long, the adaxial margined pits minute; blades obovate, 5–8 × 3–5 cm, thinly to thickly coriaceous, the base acute to obtuse, the apex obtuse to rounded, the margins slightly inrolled to revolute; submarginal collecting veins ca. 1 mm from margins; foliar latex canals conspicuous abaxially. Inflorescences erect, 3–10 cm long. Staminate flowers: sepals 4, decussate, orbicular, the outer pair 3–3.5 mm long, the inner pair widely ovate, 2–4 × 2.5–3 mm; petals 4, decussate, obovate, 4–6 mm long at anthesis, the apex broadly rounded, the inner narrower, cucullate; stamens 8, in two cycles of 4, both equal in length, the filaments 2–2.4 mm long, basally fused, the anthers obovoid, the connective wide; pistillode absent. Pistillate flowers like staminate but ovary 4-locular; staminodia 4, flat, 3 mm long, stamen-like, stigmas 4. Fruits globose, 5–7 mm diam. Fl (Dec); in non-flooded forest.

Collections from our area assigned to this species are very similar to *C. octandra*. It is doubtful whether there are consistent differences between these two species other than leaf shape.

Clusia octandra (Poepp. & Endl.) Pipoly [Syn.: *Oedematopus octandrus* (Poepp. & Endl.) Planch. & Triana]

Epiphytic lianous shrubs, to 4 m tall. Exudate white. Stems terete or angular, not tetragonal, 2.5–3.5 mm diam. Leaves: petioles 5–10 mm long, the adaxial margined pits minute; blades oblanceolate or oblong, 6–8 × 1.5–3 cm, chartaceous to thinly coriaceous, the base cuneate, the apex acute to obtuse, the margins slightly inrolled to revolute; submarginal collecting veins ca. 0.5 mm from margins, the secondary veins fairly prominent above and below; foliar latex canals linear, conspicuous abaxially. Inflorescences pendent, 6–10 cm long. Staminate flowers: sepals 4, decussate, orbicular, the outer 2.5–3 × 2.5–3 mm, the inner widely ovate, 3–3.5 × 2.5–3 mm; petals 4, decussate, orbicular to obovate, 3–4 mm long at anthesis, yellowish white to yellow, the inner pair cucullate; stamens 8, in two cycles of 4, equal in length, the filaments 2.2–2.6 mm long, basally widened, fused at very base, the anthers obovoid to rectanguloid, emarginate at apex; pistillode absent. Pistillate flowers like staminate but staminodia 4, flat, 3 mm long, stamen-like, devoid of pollen; ovary 4-locular, globose, 3–4 mm diam., the stigmas 4, rounded, ca. 1 mm diam, sessile. Fruits globose, ovoid or elliptic, 5–9 × 5–16 mm. Fl (Nov, Dec), fr (Dec); in non-flooded forest.

Clusia palmicida Rich. FIG. 80, PL. 41d,e

Epiphytic, hemiepiphytic or terrestrial shrubs or trees, to 20 m tall. Exudate yellow. Stems terete, 4–8 mm diam. Leaves: petioles (2)3–4.5 cm long, the basal adaxial margined pit much wider than petiole, 1–1.5 cm long; blades obovate to widely obovate, 8–18 × 5–11 cm, thinly to thickly coriaceous, the base cuneate, the apex truncate to very broadly rounded, the margins flat, scarious; submarginal veins 0.5–1 mm from margins; foliar latex canals linear, conspicuous adaxially. Inflorescences pendent, 1–3-flowered. Staminate flowers: sepals 6, decussate, the outer pair orbicular, 1–1.2 cm long and wide, the inner four oblong, 1.2–1.5 × 1–1.2 cm; petals 7–8, imbricate, obovate-spathulate, 4–5 × 2–4 cm, membranaceous, white or pink; stamens numerous, the outer 2–2.5 cm long, the filaments fused into tube 1–1.2 cm long, the anthers subulate, 1–1.3 cm long, the appendage <1 mm long, the inner stamens (staminodia) much shorter, resiniferous, closely connivent into disc-like structure; pistillode absent. Pistillate flowers as in staminate but outer sepals 5–7 × 7–9 mm, the inner 7–9 × 4–5 mm; petals 3–4 × 1.8–2.2 cm; staminodia 5–7 mm long, resiniferous and fused into ring; ovary 10–12-locular, 10–15 × 7–9 mm, the stigmas 10–12, triangular, 5–7 × 1.5–2 mm, fused into raised ring on mature fruit. Fruits ovoid, 3–5 × 2.5–3 cm. Fl (Jun, Jul, Aug, Sep, Oct, Dec), fr (Feb); in non-flooded forest.

Clusia panapanari (Aubl.) Choisy [Syns.: *Clusia colorans* Engl., *Clusia microphylla* Engl.] PL. 42a

Epiphytic or epipetric shrubs or small trees, to 10 m tall. Exudate white, oxidizing yellow. Stems terete, 2.5–3.5 mm diam. Leaves: petioles 5–10 mm long, without basal adaxial margined pit; blades coriaceous, oblanceolate to obovate, 5–10 × 1.8–5 cm, the base cuneate, the apex broadly rounded or acute or rarely subacuminate tip, the margins flat, scarious; submarginal veins absent, the secondary veins inconspicuous; foliar latex canals conspicuous and linear adaxially, inconspicuous abaxially. Inflorescences 3- to many-flowered. Staminate flowers: sepals 6, 3.2–4 × 3.4–4.8 mm, the outer opposite, the inner imbricate, suborbicular to oblate, margins hyaline, scarious; petals 5, imbricate, oblong to oblanceolate-spathulate, 8–10 × 2.5–4 mm, white to yellowish, with conspicuous petal resin canals, the apex broadly rounded; stamens 10-numerous, stout, basally fused, forming pentagonal, disciform synandrium, 0.5–1 mm tall, the anthers dehiscent by 2–4 short slits; pistillode absent. Pistillate flowers as in staminate but outer sepals 1.5–2.5 × 2.2–2.5 mm; petals 4.2–4.5 × 2.3–2.5 mm, the apex broadly rounded; staminodia 4–5, 2.5–3 mm long, the antherodes obovate; ovary 5-locular, ovoid, 2.5–3 × 2.5–3.5 mm, the stigmas 5 sessile, triangular, 1.8–2 × 0.8–1 mm, with acute apex. Fruits ovoid, 1.5–2.5 × 1–1.5 cm, transversely wrinkled, crowned by conical structure formed by connivent stigmas. Fl (Apr, Jun, Jul), fr (Jan, May, Aug, Sep, Nov, Dec), in non-flooded forest and on granitic outcrops.

Clusia platystigma Eyma

Epiphytic or terrestrial shrubs or trees, to 5 m tall. Exudate yellow to orange. Stems terete, 9–15 mm diam. Leaves: petioles (4.5)5–9 cm long, the basal adaxial margined pit as wide as or slightly wider than petiole, 1.5–2 cm long; blades obovate, (15)23–26 × 9.5–15 cm, cartilaginous, the base acute or rarely attenuate, the apex broadly rounded, the margins flat; submarginal veins ca. 1 mm from margins, the secondary veins numerous, visible but not prominently raised ad- or abaxially; foliar latex canals linear, inconspicuous adaxially, few and parallel. Inflorescences pendent, the flower solitary, or rarely

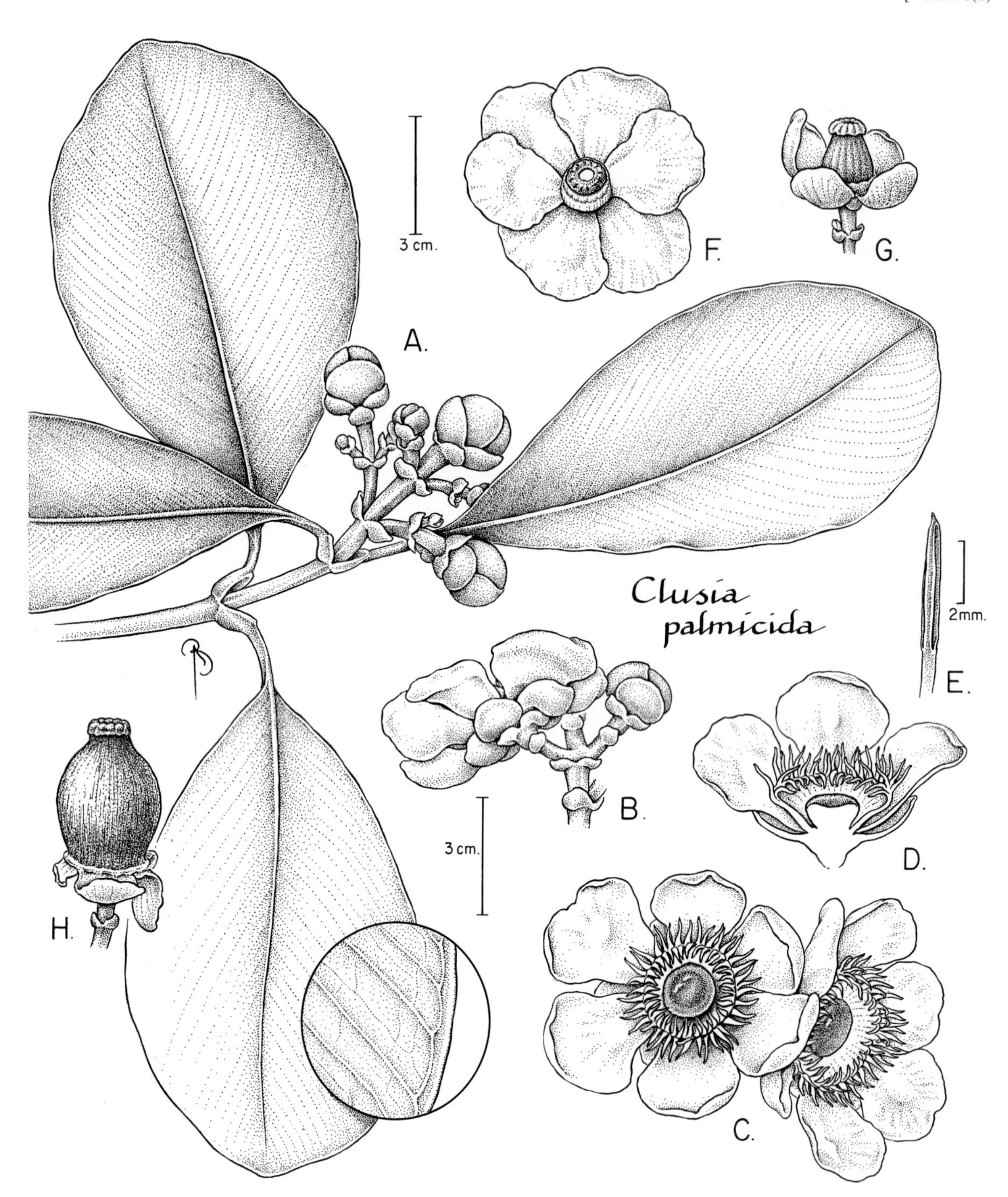

FIG. 80. CLUSIACEAE. *Clusia palmicida* (A, *Mori & Gracie 18959*; B–D, *Mori et al. 21014*; E, *Mori et al. 20847*; F, *Mori 23374*; G, *Mori et al. 20943*; H, *Jansen-Jacobs et al. 1988* from Guyana). **A.** Terminal portion of stem with leaves and flower buds. **B.** Lateral view of flower and flower buds. **C.** Apical (left) and oblique (right) views of staminate flowers. **D.** Medial section of staminate flower. **E.** Anther. **F.** Apical view of pistillate flower. **G.** Lateral view of immature fruit. **H.** Lateral view of immature fruit.

in 2–3-flowered dichasium. Staminate flowers: sepals 6, decussate, suborbicular, thinly coriaceous with scarious margins, the outer pair 2–2.5 × 3–3.5 cm, gradually decreasing in size acropetally; petals 8–10, imbricate, oblanceolate-spathulate, 6.5–7.5 × 3.5–4.5 cm, basally fused, white or pink, the base abruptly contracted, the apex truncate; stamens numerous, the outer 2–2.2 cm long, the filaments fused into tube 1–1.2 cm long, the anthers subulate, 1–1.2 cm long, the connective appendage 3–4 mm, the inner stamens shorter, sterile, resiniferous, connivent into disclike structure; pistillode absent. Pistillate flowers as in staminate but staminodia 8 mm long, resiniferous and fused into ring, antherous; ovary 13–14-locular, 6–8 × 10–12 mm, the stigmas 13–14, sessile, not raised, appearing as stigmatic patches at carpel apices on mature fruit. Fruits ovoid-globose, 5–6 × 6–8 cm. Fl (May, Jun, Aug, Sep), fr (Jan, Sep); in non-flooded forest.

Clusia scrobiculata Benoist Pl. 42b

Hemiepiphytic stranglers, to 6 m tall. Exudate cream-colored, oxidizing red. Stems subterete (3)5–7(12) mm diam. Leaves: petioles widely marginate, 5–10 mm long, the adaxial pits less than petiole width; blades obovate to oblanceolate, (5)8–15(24) × (3.5)5–7(10.5) cm, thinly coriaceous, the apex obtuse or acute, the base acute or gradually narrowed, decurrent to petiole base; secondary veins numerous, conspicuous ad- and abaxially; foliar latex canals numerous, conspicuous adaxially or not. Inflorescences terminal, erect, (1)3-flowered dichasia; bracteoles 2. Flowers bisexual (the only species known to possess bisexual flowers in our area); sepals 4–5, the outer pair opposite, the inner contorted, orbicular, 1–1.5 × 1–1.5 cm, chartaceous, the apex broadly rounded; petals 6, white tinged with pink, imbricate, obovate to orbicular, 1.8–3.2 × 1.5–3 cm, coriaceous, persistent in fruit, the apex broadly rounded; stamens 10, cylindrical, connate to form cupuliform ring, the anthers dehiscing by single terminal pits; ovary 5-locular, oblongoid, 12–20 × 7–9 mm, the stigmas 5 triangular with rounded angles, convex, 5–9 × 7–9 mm. Fruits oblongoid, 2.5–3.5 × 1.5–2.5 cm. Fl (Sep), fr (Jan); in non-flooded forest.

GARCINIA L. [Syn.: *Rheedia* L.]

Glabrous shrubs or trees. Stems opposite but not decussate. Leaves opposite or whorled, with latex canals; petioles transversely wrinkled when dry, the base usually with deep adaxial margined pits. Inflorescences axillary fascicles or flowers solitary. Flowers bisexual and/or unisexual (plants dioecious or polygamous); sepals 2–4(5), imbricate; petals 2–4(6), decussate or imbricate. Staminate flowers: stamens free, numerous, the anthers about as wide as long, central nectary disc present. Pistillate flowers: staminodia stamen-like but fewer in number; ovary 2–4-locular, surrounded by disc, the stigma peltate, dome-shaped or lobed, obscuring short style, the ovules one per carpel. Fruits berries, mostly 1-locular, ellipsoid to globose or ovoid, the exocarp thin or coriaceous, tuberculate in our area, the mesocarp often juicy. Seeds 1(4).

Garcinia madruno (Kunth) Hammel [Syns.: *Calophyllum madruno* Kunth, *Rheedia madruno* (Kunth) Planch. & Triana, *Rheedia acuminata* (Ruiz & Pav.) Planch. & Triana, *Rheedia spruceana* Engl., *Rheedia kappleri* Eyma] Pl. 42c

Trees, 5–15 m tall. Trunks without prop roots. Exudate yellow, copious. Stems terete, 3–4 mm diam. Leaves: petioles 1–2 cm long, adaxial margined pit slightly wider than petiole; blades 10–20 × 4.5–9 cm, thinly coriaceous, the base narrowly acute, the apex acuminate; midrib prominently raised abaxially; foliar latex canals numerous, visible abaxially, oriented at a 45° angle or less to midrib. Staminate inflorescences axillary fascicles, the flowers few to numerous; pedicels 0.8–3 cm. Staminate flowers: sepals 4, orbicular, 1.5–3 × 2–3.5 mm, the apex rounded; petals 4–5, oblong, 5–7 × 2–3 mm; stamens numerous, free; pistillode absent. Pistillate flowers as in staminate but usually solitary, staminodia fewer, with reduced antherodes; ovary ovoid. Fruits ellipsoid, 3–5 × 2–3 cm, the exocarp coriaceous, densely tuberculate, the endocarp fleshy. Seeds 1–2. Fl (Aug); in non-flooded forest.

MORONOBEA Aubl.

Tall, glabrous trees. Stems decussate. Exudate yellow. Leaves with numerous, dense secondary veins. Inflorescences terminal, often on short-shoots, the flowers solitary or rarely 2–3 together on short pedicels. Flowers bisexual; sepals 5, imbricate; petals 5, contorted, ± erect; stamens in 5 fascicles, the fascicles spirally arranged around pistil, each of 3–5 stamens, the anthers long, linear, locellate; ovary 5-locular, the style single, distally 5-branched, with minute terminal stigmatic pores, the ovules 3–10 per locule. Fruits subglobose, soft berries, usually with spiral markings from stamens, green. Seeds 1 to few, large.

Moronobea coccinea Aubl. Pl. 42d,e

Trees, to 50 m × 45 cm. Exudate abundant, free-flowing, yellow, remaining so. Bark peeling in irregular, rectangular plates. Stems terete, 2–5 mm diam. Leaves: petioles 5–10 mm long, the adaxial margined pits absent; blades lanceolate, oblong, oblanceolate or elliptic, (3.5)7–12 × (1.7)2–3(5) cm, coriaceous, the base narrowly acute, the apex acuminate; midrib impressed adaxially, prominently raised abaxially, the secondary veins numerous, dense, prominulous ad- and abaxially, united by submarginal veins 0.5–1 mm from margins; foliar latex canals inconspicuous. Inflorescences: pedicels 1–1.5 cm long. Flowers large and showy; sepals ovate to suborbicular or oblate, 4–8 × 5–9 mm; petals orbicular, enclosing most of stamens and pistil, 10–15 × 11–15 mm, pink, white or red; stamens united by filament base into 5 fascicles, the fascicles 2.5–3.5 cm long, without stalk, spirally contorted or rarely straight, each consisting of 3 stamens; ovary ovoid, ca. 1 cm, the style slightly surpassing stamens. Fruits with spiral markings from staminal fascicles when young, globose at maturity, ca. 5 cm across. Seeds single, globose, ca. 3.5 cm diam. Fl (Oct, Nov, Dec), fr (May, Oct); in non-flooded forest, probably pollinated by passerine birds (Gill et al., 1998). *Manil* (Créole).

PLATONIA Mart.

A monotypic genus. The generic description the same as the description of *P. insignis*.

Platonia insignis Mart. Pl. 42f

Tall, glabrous trees, to 35 m tall × 40 cm. Bark brown, distinctly fissured. Exudate slow-flowing, appearing as droplets, yellow, oxidizing brownish yellow. Stems decussate, terete or angular, 3–10 mm diam. Leaves: petioles 1.4–2.2 cm; blades broadly lanceolate to elliptic or slightly obovate, 5.5–11 × 4.2–6.5 cm, coriaceous, the base obtuse, the apex often acuminate; the midrib impressed adaxially, prominently raised abaxially, the secondary veins numerous, dense, translucent when fresh, prominent when dry, the marginal veins 0.5–1 mm from margins; foliar latex canals not visible. Flowers bisexual; sepals 5, quincuncial, orbicular or wider than long, 7–10 × 10–12 mm; petals 5, contorted, nearly orbicular, slightly oblique, 4.5–5 × 3.5–4.5 cm, enclosing most of pistil and stamens, peach to red; stamens numerous, fused by filaments in 5 fascicles 3.5–4 cm long, with short stalks, not spirally wound, the anthers linear, long, numerous; ovary 5-locular, globose, to 1.5 cm long, surrounded by disc, the style apically 5-branched, with minute, terminal, stigmatic pores, slightly protruding above stamens. Fruits globose berries, to 6 cm diam., green, yellow or orange, the pericarp coriaceous, the pulp white. Seeds usually 5, ellipsoidal, to 3.5 cm long. Fl (Nov); in non-flooded forest, probably pollinated by passerine birds (Gill et al., 1998). *Pacouri* (Créole).

SYMPHONIA L.f.

Glabrous trees. Exudate latex bright yellow. Stems decussate. Leaves with numerous, dense secondary veins. Inflorescences terminal or on axillary short-shoots, few- to many-flowered umbels. Flowers bisexual; sepals 5, quincuncial; petals 5, contorted; stamens connate by filaments forming apically 5-lobed tube surrounding pistil, the anthers 3–4 on each lobe, extrorse; ovary 5-locular, ovules few per carpel, style 1, with five radiating branches, the stigmas forming minute pores at apices. Fruits coriaceous berries. Seeds 1-few.

Symphonia globulifera L.f.

FIG. 81, PL. 43a; PART 1: PL. IIIc

Trees, to 30 m tall. Trunks with prop roots, flying buttresses, or sometimes pneumatophores. Bark almost smooth. Exudate bright yellow, copious. Stems terete, 2.5–4 mm diam. Leaves: petioles 8–10 mm long, without adaxial margined pit; blades lanceolate, oblong or oblanceolate, 6–12 × 2–2.5 cm, coriaceous to chartaceous, the base acute, the apex abruptly acuminate, the margins flat; midrib impressed adaxially, prominently raised abaxially, the secondary veins numerous, prominulous adaxially, prominent abaxially, united by submarginal veins; foliar latex canals not visible. Inflorescences axillary; peduncle obsolete to 2 cm long; pedicels 4–15 mm. Flowers globose, mostly red; sepals suborbicular to widely ovate, 3–10 × 3–9 mm; petals orbicular, enclosing most of stamens and pistil, 10–15 × 8–12 mm; staminal tube as long as pistil, persistent after perianth falls, with 5 spreading lobes, each with 3 anthers; pistil enclosed in staminal tube, the ovary globose to ellipsoid, 2–6 × 2–3 mm, the style-branches spreading, alternating with lobes of staminal tube. Fruits ellipsoid or globose, 3–5 × 2–3 cm, green to purple. Seeds 1 to few, angular. Fl (May, Jun, Aug, Sep), fr (Dec); scattered but common, in non-flooded forest but in other areas individuals of the same or a related species found in wetter habitats, pollinated by passerine birds (Gill et al., 1998). *Anani* (Portuguese), *mani*, *manil*, *manil chêne* (Créole).

TOVOMITA Aubl. [Syn.: *Marialvea* Mart.]

Glabrous trees or shrubs. Exudate white or yellow, sometimes very sparse. Trunks often with prop roots. Leaves: petioles with an adaxial margined pit; blades chartaceous to coriaceous, sometimes with punctations, the margins often strongly raised. Inflorescences terminal or axillary, cymose panicles. Staminate flowers: outer sepals 2–4, decussate, the inner sepals 4–6, cucullate, coriaceous; petals 2–8, decussate or fused in bud; stamens numerous, the filaments free, linear, the anthers 2-locular; short pistillode sometimes present. Pistillate flowers as in staminate but dimensions smaller; staminodia resembling stamens but with smaller anthers; ovary 4–6-locular, conical, the stigmas ovoid to discoid, sessile or on short styles. Bisexual flowers like pistillate but with stamens larger than staminodia. Fruits globose or obovoid, fleshy, valvular dehiscent capsules. Seeds one per locule, mostly enveloped by an aril.

1. Secondary veins of two sizes, the thicker ones 1 cm or more apart from each other.
    2. Tertiary leaf venation prominently raised ad- and abaxially.
        3. Leaf blades obovate, coriaceous, the apex obtuse to rounded, the margins subrevolute. . . . . . . *T. macrophylla.*
        3. Leaf blades elliptic, chartaceous, the apex acute to acuminate, the margins flat. . . . . *T. calodictyos.*
    2. Tertiary leaf venation not prominently raised adaxially.
        4. Pedicels 1.5–2.5 cm long. . . . . *T. obovata.*
        4. Pedicels 0.8–1.2 cm long. . . . . *T. longifolia.*
1. Secondary veins all same size, mostly <0.5 cm apart.
    5. Leaf blades membranaceous or chartaceous, the apex attenuate-acuminate or abruptly acuminate.
        6. Leaf blades mostly <2× as long as wide. Flower buds globose, to 7 mm long. . . . . *T. brevistaminea.*
        6. Leaf blades mostly >2× as long as wide. Flower buds ovoid, to 4.5 mm long.
            7. Petioles with adaxial margined pit slightly narrower than petiole. Sepals oblong. Fruits ellipsoid to obovoid, 2.5–3.5 cm long. . . . . *T. schomburgkii.*
            7. Petioles with adaxial margined pit as wide as petiole. Sepals ovate. Fruits globose to subglobose, 1.3–1.8 cm long. . . . . *T. brasiliensis.*
    5. Leaf blades coriaceous, the apex acute to subacuminate.
        8. Inflorescences 7–10 cm long. Flower buds acute at apex. . . . . *T.* aff. *tenuiflora.*
        8. Inflorescences 2.5–3.5 cm long. Flower buds beaked at apex. . . . . *T. grata.*

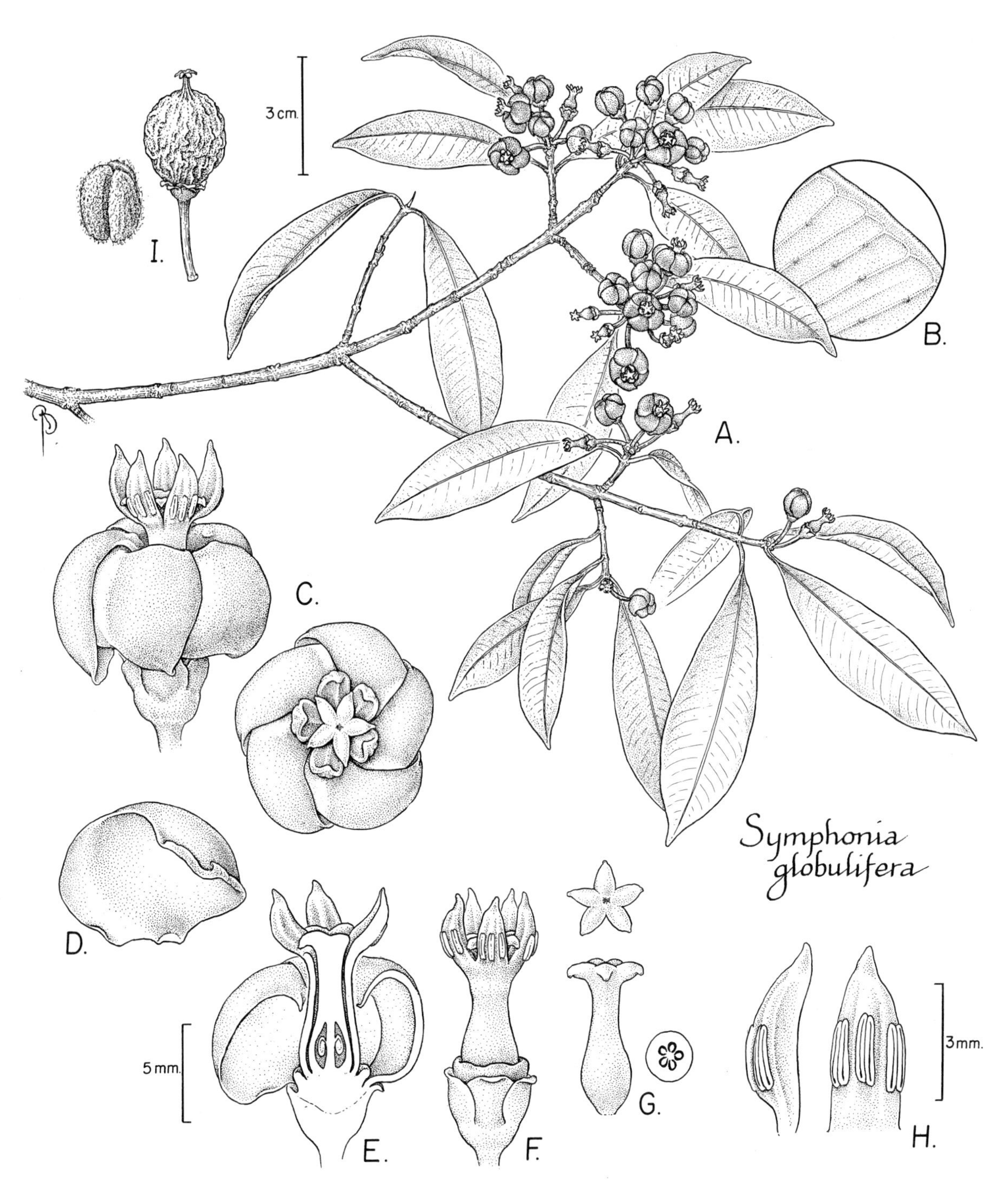

FIG. 81. CLUSIACEAE. *Symphonia globulifera* (A–H, *Mori et al. 23852*; I, *Oldeman 2019A*). **A.** Stem with leaves and inflorescences. **B.** Detail of leaf venation. **C.** Lateral (above left) and apical (below right) views of flowers. **D.** Cucullate petal. **E.** Medial section of flower; note placentation and staminal tube. **F.** Flower with petals removed; note staminal tube, anthers, and stigmatic lobes between lobes of staminal tube. **G.** Pistil (below left), apical view of stigma (above left), and transverse section of ovary (below right). **H.** Lateral (left) and abaxial (right) views of lobes of staminal tube with anthers. **I.** Fruit (right) and seeds (left).

Tovomita brasiliensis (Mart.) Walp.

FIG. 82; PART 1: PL. XVa

Shrubs or trees, to 5 m tall. Trunks without prop roots. Exudate sparse. Stems terete, 1.5–2.5 mm diam. Leaves: petioles 3–7 mm long, the adaxial margined pit 1.5–2.5 mm long, as wide as petiole; blades narrowly elliptic to elliptic to lanceolate, 3.5–10 × 1.2–4 cm, chartaceous, the base acute, the apex long-attenuate to acuminate; punctations not prominent, few, linear, perpendicular to midrib; secondary veins all same size, mostly <0.5 cm apart. Inflorescences terminal, less than half length of subtending leaves; peduncle 6–10 mm; pedicels (3)5–10 mm long. Staminate flowers: buds ovoid; sepals 2,

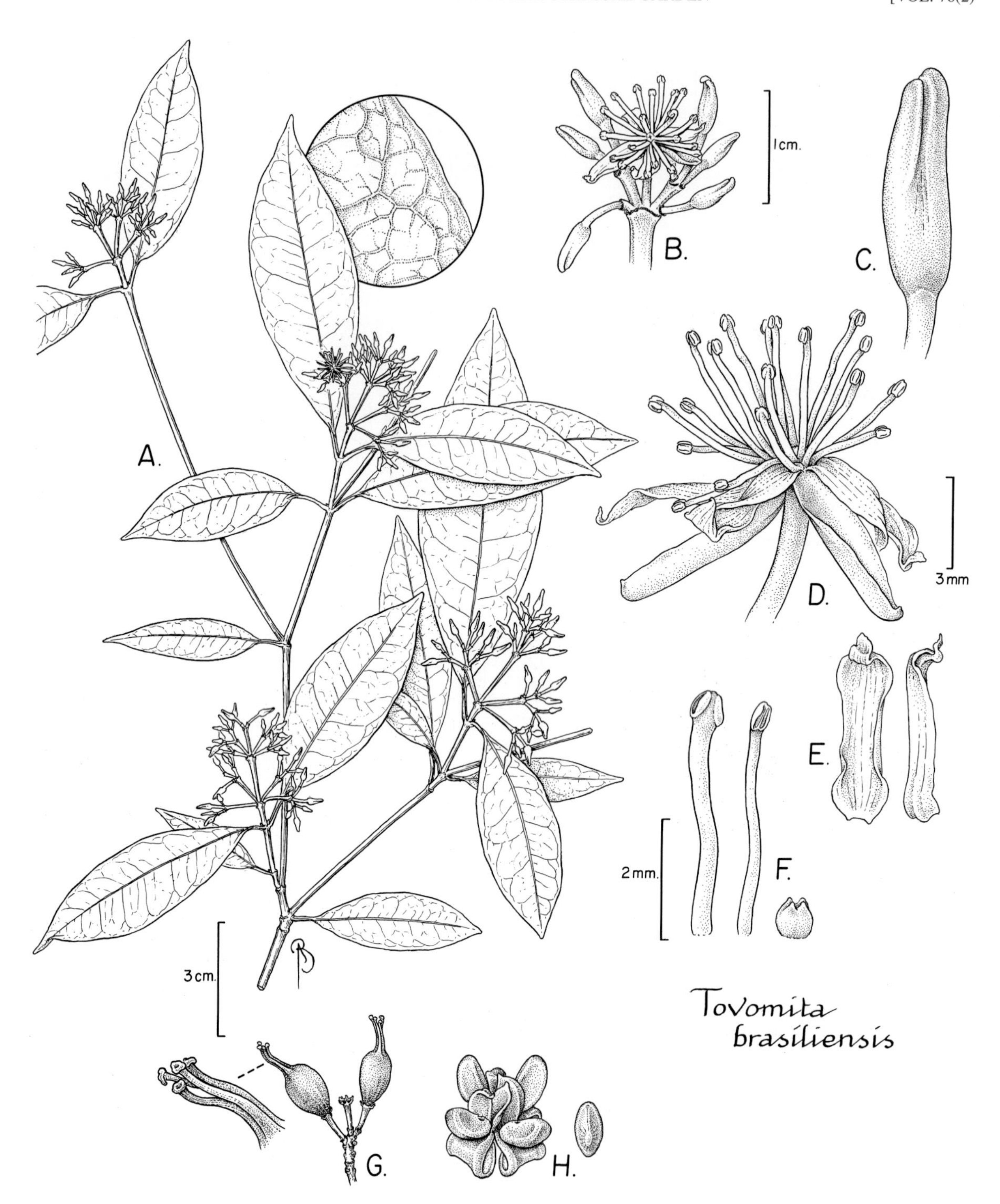

FIG. 82. CLUSIACEAE. *Tovomita brasiliensis* (A–F, *Mori et al. 24002*; G, *Mori & Pipoly 15551* based on the CAY sheet; H, *Mori et al. 22720*). **A.** End of stem showing leaves and inflorescences. **B.** Part of staminate inflorescence with one open flower. **C.** Staminate bud. **D.** Staminate flower. **E.** Adaxial (left) and lateral view of petals from staminate flower. **F.** Stamens (left) and pistillode (right). **G.** Part of infructescence with immature fruits (right) and "beak" formed by remnant stigmas (left). **H.** Open fruit (left) and seed (right).

ovate, 3–4 × 4–5 mm, longitudinally striate medially, the apex obtuse to broadly rounded, the margins scarious; petals 4, oblong, 5–7 mm long, the apex acute; stamens subsessile. Pistillate flowers as in staminate but staminodia with subcapitate callus at apex; ovary 4-locular, oblong, 3.5–4.5 mm long, the stigmas 4, sessile or on short styles, deltoid. Fruits globose to subglobose, often with apical beak, 1.3–1.8 cm long, green, the inside of locules purple. Seeds 4, the arils orange. Fl (Sep), fr (Apr, Oct); in non-flooded forest. *Palétuvier* (Créole).

Plates 41–48

Plate 41. CHRYSOBALANACEAE. **a.** *Licania kunthiana* (*Mori et al. 24172*), flowers. **b.** *Licania octandra* (*Mori et al. 22812*), flowers. CLUSIACEAE. **c.** *Clusia leprantha* (*Mori et al. 24760*), pistillate flower. **d.** *Clusia palmicida* (*Mori et al. 23374*), pistillate flower. **e.** *Clusia palmicida* (*Mori et al. 21014*), staminate flowers.

Plate 42. CLUSIACEAE. **a.** *Clusia panapanari* (*Mori & Gracie 18856*), staminate flower and bud. **b.** *Clusia scrobiculata* (*Mori & Pepper 24274*), immature fruits. [Photo by S. Mori] **c.** *Garcinia madruno* (*Mori et al. 20603* from Amazonas, Brazil), staminate flowers. **d.** *Moronobea coccinea* (*Mori et al. 24698*), slash of bark. [Photo by S. Mori] **e.** *Moronobea coccinea* (*Mori et al. 24698*), buds. [Photo by S. Mori] **f.** *Platonia insignis* (*Mori et al. 24699*), slash of bark. [Photo by S. Mori]

Plate 43. CLUSIACEAE. **a.** *Symphonia globulifera* (*Mori et al. 23365*), flower and bud. **b.** *Tovomita schomburgkii* (*Mori et al. 23370*), pistillate flowers. **c.** *Tovomita schomburgkii* (*Mori et al. 23372*), staminate flowers. **d.** *Vismia sandwithii* (*Mori et al. 23333*), flower and buds; note fascicles of stamens. **e.** *Vismia guianensis* (*Mori et al. 24683*), buds and flowers with some petals and sepals of one flower removed to show fascicles of stamens. [Photo by S. Mori]

Plate 44. COMBRETACEAE. **a.** *Combretum rotundifolium* (*Mori et al. 18625*), flowers. **b.** *Combretum rotundifolium* (*Mori et al. 18625*), buds and flowers. **c.** *Combretum laxum* (*Mori et al. 22642*), flowers. **d.** *Terminalia guyanensis* (*Mori & Gracie 18653*), part of inflorescence. **e.** *Terminalia guyanensis* (*Mori et al. 23905*), apex of stem with terminal bud and leaves with paired glands on petioles.

Plate 45. CONNARACEAE. **a.** *Cnestidium guianense* (*Mori et al. 21628*), flowers. **b.** *Cnestidium guianense* (*Mori & Gracie 21071*), opened fruit showing seed with aril. **c.** *Rourea surinamensis* (*Mori et al. 23318*), flowers. CONVOLVULACEAE. **d.** *Dicranostyles ampla* (*Mori et al. 24184*), flowers. [Photo by S. Mori] **e.** *Ipomoea batatoides* (*Mori et al. 22349*), flowers and buds. [Photo by B. Angell]

Plate 46. CONVOLVULACEAE. **a.** *Ipomoea quamoclit* (unvouchered), flower and bud with delicate, pinnately compound leaves. **b.** *Maripa glabra* (*Mori et al. 23732*), flower and buds. **c.** *Ipomoea squamosa* (*Mori et al. 24008*), dehisced fruits with seeds surrounded by hairs. **d.** *Merremia macrocalyx* (*Mori et al. 21057*), flower. **e.** *Maripa scandens* (*Mori & Pepper 24263*), flowers and buds.

Plate 47. CUCURBITACEAE. **a.** *Cayaponia ophthalmica* (*Mori et al. 24108*), fruits; note leaf-opposed tendrils. **b.** *Gurania bignoniacea* (*Mori & Gracie 21119*), buds and flower. **c.** *Gurania reticulata* (*Mori & Gracie 21135*), inflorescence. **d.** *Cayaponia jenmanii* (*Mori & Gracie 21086*), flowers. **e.** *Gurania lobata* (*Mori & Gracie 21123*), flower and buds; note tephrid fly on open flower.

Plate 48. CUCURBITACEAE. **a.** *Gurania subumbellata* (*Mori & Gracie 21208*), flowers and buds. **b.** *Melothria pendula* (*Mori et al. 21056*), flower and fruit; note tendril. **c.** *Gurania subumbellata* (*Mori & Gracie 21208*), infructescence. **d.** *Momordica charantia* (*Mori et al. 21055*), staminate flower. **e.** *Momordica charantia* (*Mori et al. 21055*), dehisced fruit showing seeds. **f.** *Helmontia leptantha* (*Mori & Gracie 21111*), flower and buds.

CHRYSOBALANACEAE (continued)

Licania kunthiana a.

Licania octandra b.

CLUSIACEAE

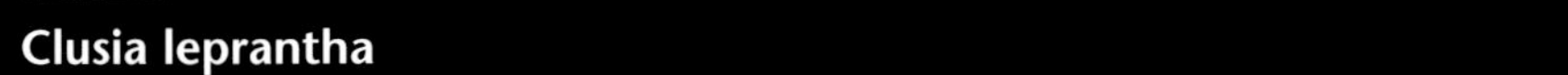

Clusia leprantha c.

Clusia palmicida d.

Clusia palmicida e.

Clusia panapanari a.

Clusia scrobiculata b.

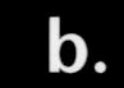

Garcinia madruno c.

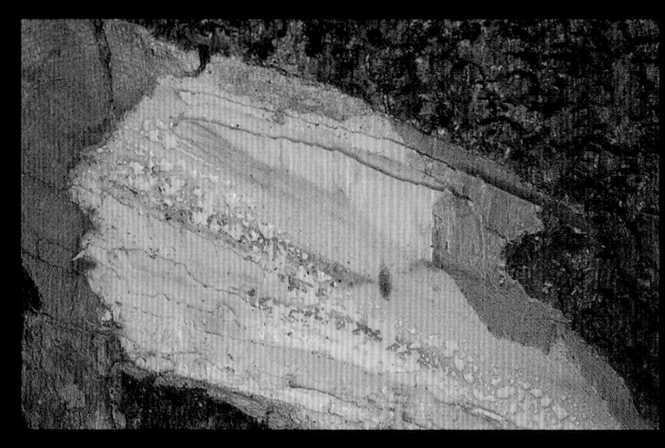

Moronobea coccinea d.

Moronobea coccinea e.

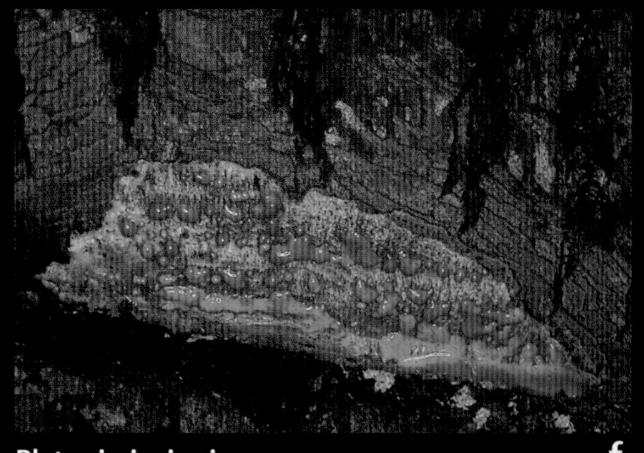

Plate 42

Platonia insignis f.

Tovomita schomburgkii b.

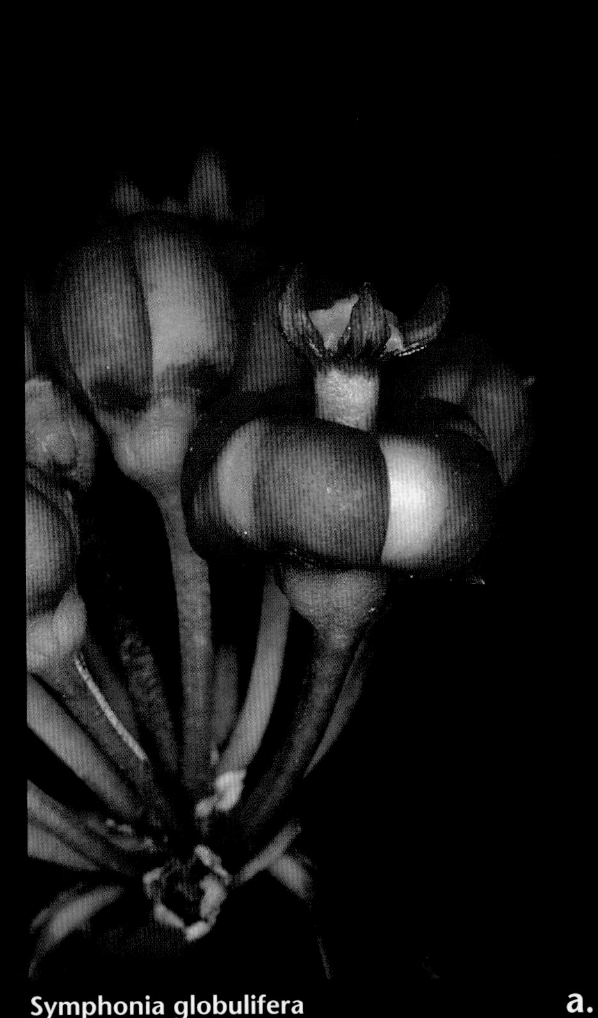

Symphonia globulifera a.

Tovomita schomburgkii c.

Vismia sandwithii d.

Vismia guianensis e.

Plate 43

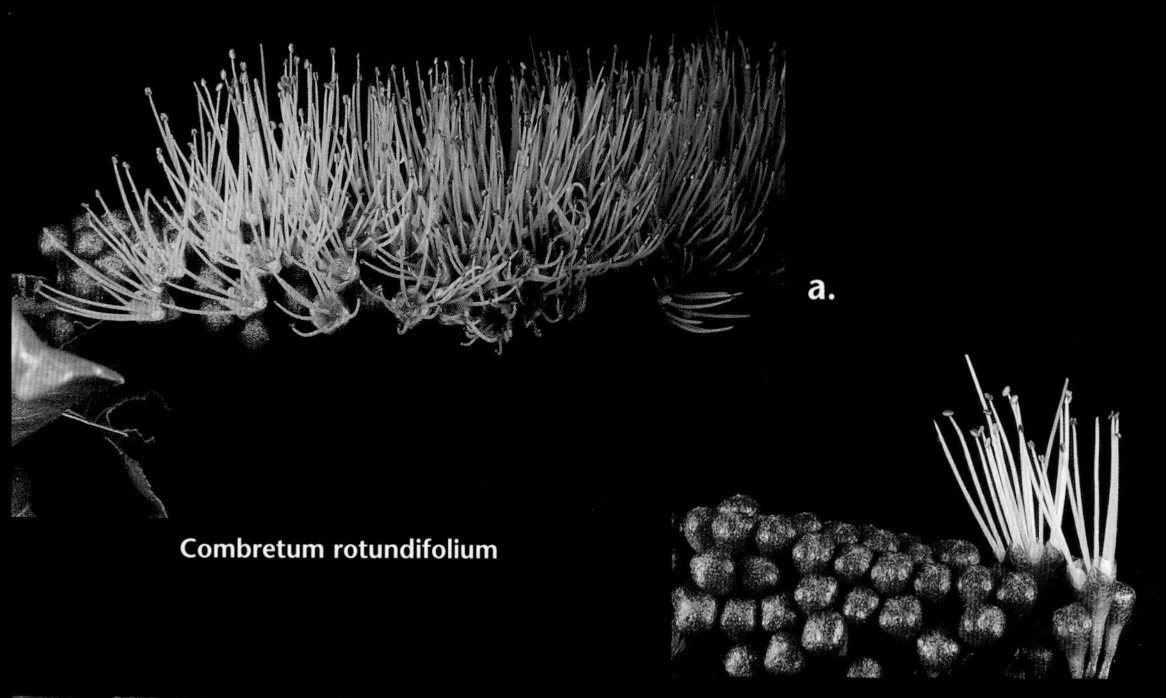

a.

Combretum rotundifolium

b.

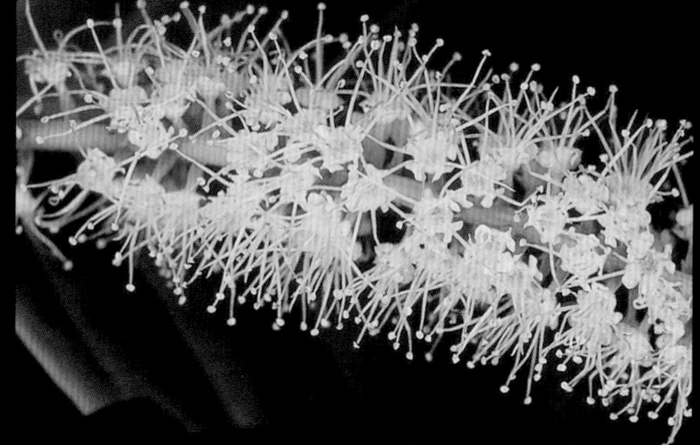

Combretum laxum

c.

Terminalia guyanensis

d.

Terminalia guyanensis

e.

Plate 44

Cnestidium guianense a.

Cnestidium guianense b.

Rourea surinamensis c.

CONVOLVULACEAE

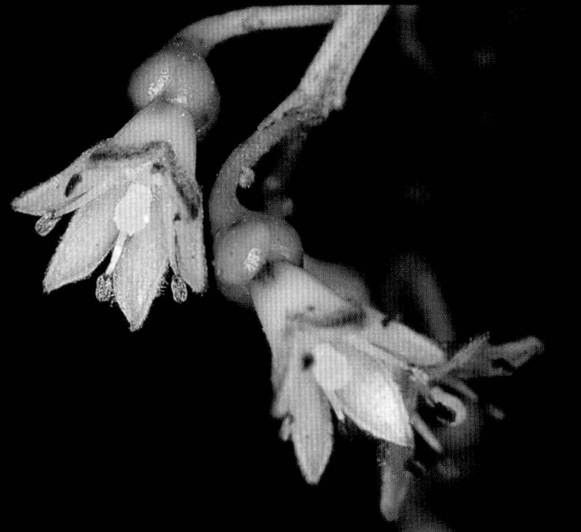

Dicranostyles ampla d.

Ipomoea quamoclit a.

Maripa glabra b.

Ipomoea squamosa c.

Merremia macrocalyx d.

Maripa scandens e.

Plate 46

Cayaponia ophthalmica a.

Gurania bignoniacea b.

Gurania reticulata c.

Cayaponia jenmanii d.

Gurania lobata e.

Gurania subumbellata a.

Melothria pendula b.

Momordica charantia d.

Momordica charantia e.

Gurania subumbellata c.

Helmontia leptantha f.

Plate 48 (Cucurbitaceae continued on Plate 49)

Tovomita brevistaminea Engl.

Shrubs or trees, to 15 m tall. Trunks with prop roots. Exudate sparse, white. Stems terete, 2–4 mm diam. Leaves: petioles 1–2 cm, the adaxial margined pit 1–1.5 mm long, narrower than petiole; blades oblong to elliptic, 6–14 × 3.5–7 cm, the base acute, the apex abruptly acuminate, the margins revolute; punctations inconspicuous; secondary veins all same size, mostly <0.5 cm apart. Inflorescences terminal, to 3/4 length of subtending leaves; peduncle 1–3 cm; pedicels 7–12 mm long. Staminate flowers: buds globose; sepals 2, ovate, 4–6 × 2–3 mm, longitudinally striate medially, the apex obtuse, the margins not scarious; petals 4, oblong, 4–6 × 2–3 mm, the apex acute; stamens on periphery with very short filaments. Pistillate flowers as in staminate but sepals 7–9 × 2.5–3.5 mm; petals 6–8 × 2–2.5 mm; staminodia anantherous, subterete; ovary 4-locular, oblong, 4–6 mm long, the stigmas 4, sessile or on short styles, deltoid. Fruits obovoid, 3.5–5 × 1–2.5 cm. Fl(May); in non-flooded forest.

Tovomita calodictyos Sandwith

Shrubs or trees, to 28 m tall. Trunks without prop roots. Exudate latex yellow, copious. Stems terete, 4–7 mm diam. Leaves: petioles 2–3.7 cm long, the adaxial margined pit slightly wider than petiole, ca. 0.5 mm long; blades elliptic to rarely oblanceolate, 15–32 × 4.5–13.5 cm, chartaceous, the base acute, the apex attenuate to acuminate to rarely acute, the margins entire; punctations not prominent; secondary veins of two sizes, the thicker ones ≥1 cm apart. Staminate inflorescences terminal, slightly longer than petioles to half length of subtending leaves; peduncle 3.5–7 cm long; pedicels 5–10 mm long. Staminate flowers: buds ovoid; sepals 4, ovate, 5–7 × 3.5–4.5 mm, the apex obtuse; petals 4, oblong, 5–7 × 2.5–3 mm, the apex acute; stamens subsessile, the peripheral ones much shorter. Pistillate flowers unknown. Fruits obovoid, 3–5 × 1.5–2.2 cm, the stigmas subdeltate, on styles 2–3 mm long. Fr (May); in non-flooded forest.

Tovomita grata Sandwith

Trees, to 20 m tall. Trunks with prop roots well developed. Exudate copious, thick, bright yellow. Stems terete, 2–4 mm diam. Leaves: petioles 1–1.9 cm long, the adaxial margined pit 1–2 mm long, slightly wider than petiole; blades obovate, elliptic or oblanceolate, 5–12 × 2–6 cm, coriaceous, the base acute, the apex acute or subacuminate, the margins flat; punctations not prominent; secondary veins all same size, mostly <0.5 cm apart. Staminate inflorescences terminal, to half as long as length of leaves; peduncle 0.8–1.5 cm long; pedicels 0.9–1.2 mm long. Pistillate inflorescences as in staminate, but peduncle 6–9 mm; pedicels 7–9 mm long. Staminate flowers: buds ovoid, the apex long-apiculate; sepals 4, widely ovate, 7–8 × 6–8.5 mm, the apex acute; petals 5, linear-oblong, 7–10 × 1.3–4 mm wide, the apex long attenuate; stamens all equal; pistillode absent. Pistillate flowers: sepals 8–10 × 6–8.5 mm; petals 12–15 × 1.5–2 mm; staminodia numerous; ovary 5-locular, ovoid to cylindric, 6–8 × 3–3.5 mm, the stigmas 5, deltate, on short styles 1.5–2 mm long. Fruits ovoid, 3.5–6.5 × 2.5–3 cm. In non-flooded forests.

Tovomita longifolia (Rich.) Hochr. [Syns.: *Tovomita richardiana* Planch. & Triana, *Tovomita choisyana* Planch. & Triana]

Trees, to 20 m tall. Trunks with prop roots. Exudate reddish orange. Stems terete, 3.5–10 mm diam. Leaves: petioles 2–3.5 cm long, the adaxial margined pit as wide or wider than petiole, 3–5 mm long; blades obovate to lanceolate, 14–42 × 6.5–12 cm, chartaceous to thinly coriaceous, the base acute, the apex obtuse to acute, rarely short-acuminate, the margins entire; punctate-lineations somewhat conspicuous; secondary veins of two sizes, the thicker ones ≥1 cm apart. Staminate inflorescences terminal; peduncle 1–3.5 cm; pedicels 8–10 mm. Pistillate inflorescences as in staminate but peduncle 1–1.5 cm long; pedicels 10–12 mm long. Staminate flowers: buds globose, ca. 15 mm long; sepals 2, widely ovate, 3.5–4.5 × 2–3 mm, longitudinally striate throughout, the apex obtuse, the margins opaque, not scarious; petals 4, ovate, 4.5–6 × 2–3 mm, the apex obtuse; stamens centrifugally shorter. Pistillate flowers: sepals 8–10 × 8–10 mm; petals 10–12 × 5–6 mm; staminodia numerous; ovary globose, (4)5–6-locular, 8–10 × 6–8 mm, the stigmas (4)5–6, ovoid, peltate. Fruits obovoid, 4–5 × 2–3 cm. Fr (Oct); in non-flooded forest.

Tovomita macrophylla (Poepp.) Walp.

Trees, to 30 m tall. Trunks with well developed prop roots. Exudate bright yellow, copious. Stems distally angular, 7–12 mm diam. Leaves: petioles (2)3.5–5, the adaxial margined pit as wide or wider than petiole, 3–7 mm long; blades obovate, (11)15–30 × (4.5)9–14 cm, coriaceous, the base subacute to obtuse, the apex obtuse to rounded, the margins subrevolute; punctations inconspicuous; secondary veins of two sizes, the thicker ones ≥1 cm apart. Inflorescences slightly longer than petioles; peduncle 1–3 cm long; pedicels 3–10 mm long. Staminate flowers: sepals 4, ovate, 7–8 mm long; petals 6, oblong, 8–10 mm long; stamens centrifugally smaller; pistillode absent. Pistillate flowers unknown. Fruits obovoid, 5–7 × 3–5 cm diam. Known from our area only by sterile collections; in non-flooded forest.

Tovomita obovata Engl.

Trees, to 30 m tall. Trunks with well developed prop roots. Exudate yellowish-green, moderately abundant. Stems terete, 4.5–10 mm diam. Leaves: petioles (2)3–4.5 cm long, adaxial margined pit less than petiole diameter, 3–5 mm long; blades obovate or rarely oblanceolate, 13.5–38 × 6–13 cm, chartaceous, the base broadly acute, the apex obtuse to subacuminate, the margins entire, flat; punctations not prominent; secondary veins of two sizes, the thicker ones ≥1 cm apart. Staminate inflorescences terminal, ca. twice length of petiole; peduncle ca. 1 cm long; pedicels 1.5–2 cm long. Staminate flowers: buds globose, 13–15 mm long; sepals 4, widely ovate, 8–11 × 5–6 mm, the apex obtuse; petals oblong, 9–12 × 5–7 mm wide, the apex obtuse; stamens roughly equal in size. Pistillate flowers as in staminate but sepals oblate, 10–12 × 13–14 mm; petals 12–15 × 6–8 mm; staminodia numerous, all same length; ovary 6-locular, ovoid, 10–13 × 6–8 mm, the stigmas 6, deltoid, sessile. Fruits obovoid, 4–6 × 2.5–3.5 cm, with short apical beak. Fl (Jun), fr (Jun, Aug); in non-flooded forest.

Tovomita schomburgkii Planch. & Triana PL. 43b,c

Shrubs or small trees, to 12 m tall. Trunks with small prop roots. Exudate bright yellow, copious. Stems terete, 2–3 mm diam. Leaves: petioles 1–1.5 cm, the adaxial margined pit to equal diameter of petiole, ca. 2 mm long; blades elliptic to oblanceolate, 7.5–12 × 3–4.8 cm, chartaceous, the base acute, the apex abruptly acuminate; punctate-lineations inconspicuous except on very young leaves, then conspicuous as lines; secondary veins all same size, mostly <0.5 cm apart. Inflorescences terminal; peduncle 1.5–2 cm; pedicels 5–8 mm long. Staminate flowers: buds ovoid; sepals 2, oblong, 3.5–4.5 × 1.5–2 mm long, the apex acute; petals 4, oblong, 4–5 × 1.5–2 mm, the apex acute; stamens equal in size. Pistillate flowers as in staminate but staminodia fewer, reduced in size; ovary 4-locular, ellipsoid, 15 × 10 mm, the stigmas 4, subsessile. Fruits ellipsoid to obovoid, 2.5–3.5 × 1.5–2 cm. Fl (Apr, May, Jun, Jul, Aug, Oct, Nov), fr (Feb, Mar, Apr, Aug, Nov); in non-flooded forest.

Tovomita aff. **tenuiflora** Planch. & Triana

Trees, to 20 m tall. Trunks with well developed prop-roots or flying buttresses. Exudate white, sparse. Stems 3–4 mm diam. Leaves: petioles 1–2.5 cm, adaxial margined pit smaller than petiole diameter, 1–1.5 mm long; blades oblong or elliptic, 15–25 × 3.8–8 cm, coriaceous, the base acute, the apex acute, rarely subacuminate, the margins slightly inrolled; punctate-lineations inconspicuous, in faint lines abaxially; secondary veins all same size, mostly <0.5 cm apart. Inflorescences terminal, 7–10 cm long; peduncle 1–2 cm long; pedicels 5–10 mm long. Staminate flowers: buds acute; sepals oblong, 5–7 × 2–3 mm, the apex acute; petals linear, 7–9 × 2–3 mm, the apex acute; stamens all equal. Pistillate flowers unknown. Fruits unknown. Fl (Aug, Nov); in non-flooded forests.

VISMIA Vand.

Trees or shrubs. Exudate orange to red, sometimes sparse in vegetative parts. Stems, abaxial leaf blade surface, and inflorescence parts often stellate tomentose. Leaves often with punctations. Inflorescences terminal or rarely axillary, paniculate. Flowers bisexual; perianth usually with dark or orange punctae or lines; sepals with scarious margins; petals white, yellow or green, lanose adaxially; stamens numerous, fused by filaments into five antepetalous fascicles alternating with nectaries, the anthers ovoid or globose, small; ovary 5-locular, the styles 5, the stigmas capitate. Fruits baccate, globose or ovoid, mostly with black or orange punctations or punctate lineations. Seeds numerous, not arillate.

1. Leaf blades glabrous or minutely puberulous at maturity, the base obtuse to rounded.
    2. Leaf blades membranaceous, with black punctations and the absence of papillae. Inflorescences cauliferous, axillary, subterminal, or terminal. Calyx 3–4 mm long; corolla 5–6 mm long. . . . . *V. ramuliflora.*
    2. Leaf blades chartaceous to subcoriaceous, without black punctations but with papillae abaxially. Inflorescences terminal. Calyx 7–8 mm; corolla 9–15 mm long. . . . . *V. cayennensis.*
1. Leaf blades with at least abaxial surface with sessile, rotate, stellate or scale-like trichomes, the base various.
    3. Leaf punctations prominently raised, easily visible without magnification, the base acute. Fruits with persistent stamens. . . . . *V. sandwithii.*
    3. Leaf punctations absent or flat, not visible without aid of magnification, the base various. Fruits without persistent stamens.
        4. Stems 2–3 mm diam. Petioles 1–1.3 cm long; leaf base acute. . . . . *V. guianensis.*
        4. Stems 3.5–5 mm diam. Petioles 1.4–2 cm long; leaf base obtuse to broadly rounded.
            5. Stems, abaxial leaf blade surfaces, and inflorescences densely rufous-tomentose, the tomentum not appressed. . . . . *V. latifolia.*
            5. Stems, abaxial leaf blades surfaces, and inflorescences densely silvery or tawny to ferruginous-tomentose, the tomentum appressed. . . . . *V. gracilis.*

Vismia cayennensis (Jacq.) Pers.

Glabrous or glabrate shrubs or small trees, to 8 m tall. Stems terete, 2–2.5 mm diam. Leaves: petioles 6–10 mm long; blade elliptic or oblong, 8–13 × 3.5–7 cm, chartaceous to subcoriaceous, the base obtuse to rounded, the apex acuminate to cuspidate; punctations black, prominently raised, several per areole. Inflorescences terminal; peduncle 3–4.5 cm long; pedicels 7–10 mm long. Flowers: sepals ovate, 7–8 mm long; petals obovate, rarely oblong, 9–15 mm long, woolly adaxially, sparsely black punctate; anthers glandular; ovary orange. Fruits ovoid to ellipsoid or subglobose at maturity, 8–10 mm diam., without persistent stamens. Fl (Sep, Oct); in non-flooded forest. *Bois-dartre* (Créole), *gomme-gutte de la Guyane* (French), *pau lacre* (Portuguese), *pindia udu* (Bush Negro).

Vismia gracilis Hieron. [Syns.: *V. amazonica* Ewan, *V. glaziovii* Ruhland]

Trees, to 15 m tall. Stems, peduncle, pedicels, petioles, and abaxial leaf blade surfaces with tomentum of sessile, rotate, appressed tawny to ferruginous stellate trichomes, at times appearing scale-like. Stems 3.1–4.5 mm diam. Leaves: petioles 1.4–2 cm long; blades ovate, 12–15 × 5.5–8 cm, coriaceous, the base obtuse to broadly rounded, the apex acuminate to cuspidate, the margins entire; punctations small, not prominently raised, most easily visible on mature glabrescent leaves, several per areole. Inflorescences terminal; peduncle 2–5 cm long; pedicels 4–6 mm long. Flowers: sepals oblong or ovate, 5–6 mm long; petals obovate or ovate, 5–8 mm long, densely tomentose adaxially, punctate-lineate; stamen fascicles densely hairy, the anthers orange. Fruits ovoid, 10–13 × 6–8 mm, without persistent stamens. Fl (Aug, Oct, Nov), fr (Dec); in gaps in non-flooded forest or in secondary areas. *Bois-dartre* (Créole).

Vismia guianensis (Aubl.) Choisy PL. 43e

Trees, to 15 m tall. Stems, peduncle, pedicels, petioles, and abaxial leaf blade surfaces with tomentum of sessile, rotate, rufous, stellate trichomes, at times appearing scale-like. Stems terete, 2–3 mm diam. Leaves: petioles 1–1.3 cm long; blades oblong or lanceolate, 8–13 × 3.2–4.5 cm, subcoriaceous to coriaceous, the base acute, the apex long acuminate to cuspidate, the margins entire; punctations extremely small, inconspicuous, not normally visible at 30×. Inflorescences terminal; peduncle 2.5–3.5 cm long; pedicels 3–5 mm long. Flowers: sepals lanceolate, 5–7 mm long; petals obovate to oblong, 8–10 mm long, densely woolly adaxially, the punctations few, not prominent; anthers orange. Fruits ovoid, then subglobose, 8–12 long, without persistent stamens. Fl (Nov), fr (Feb); in non-flooded forest. *Bois-dartre* (Créole).

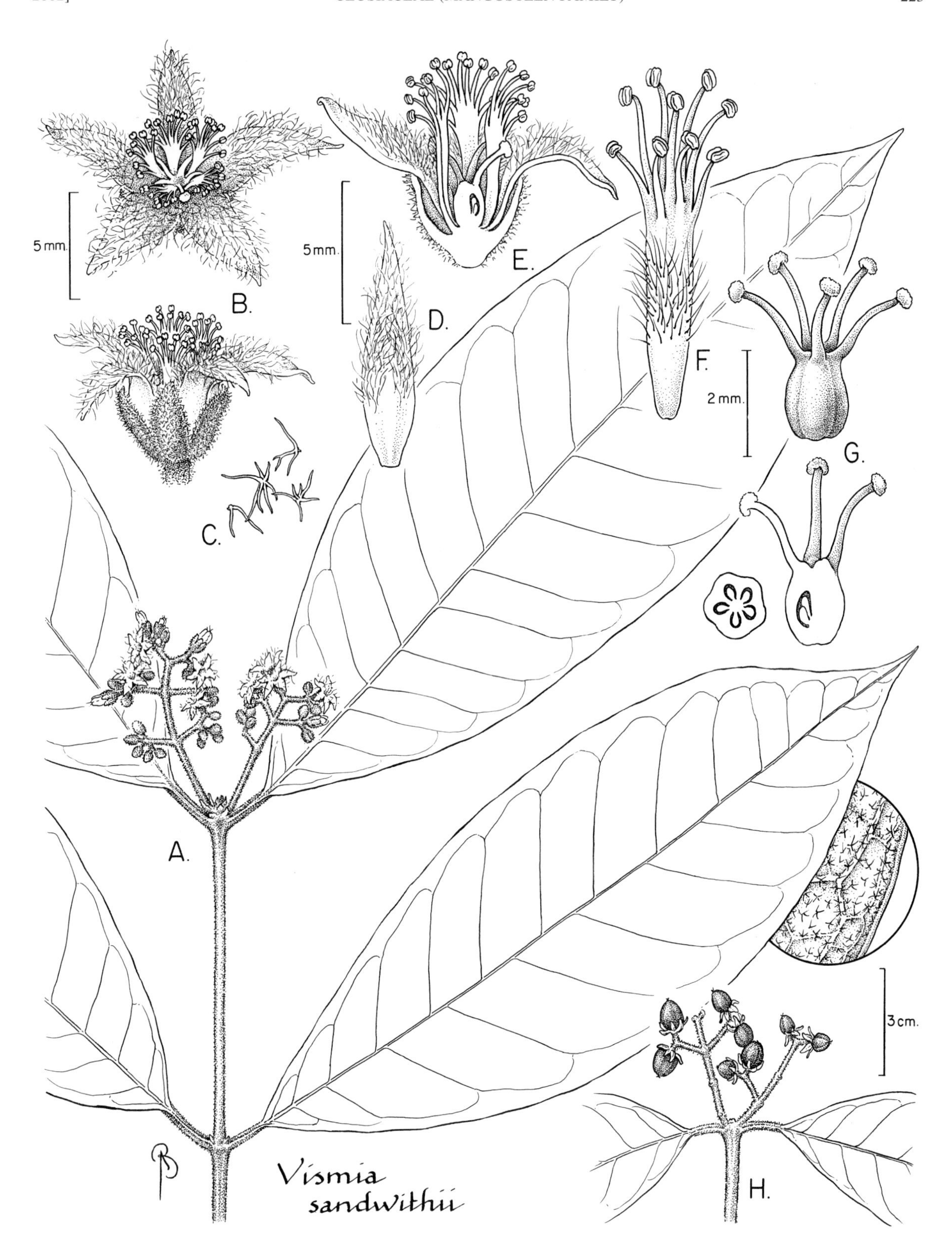

FIG. 83. CLUSIACEAE. *Vismia sandwithii* (A, *Mori & Ek 20741*; B–G, *Mori et al. 23333*; H, *Granville et al. 8895*). **A.** Terminal part of stem with leaves and inflorescences and detail of leaf pubescence. **B.** Apical (above) and lateral (below) views of flower. **C.** Stellate pubescence from calyx. **D.** Adaxial surface of petal. **E.** Medial section of flower. **F.** Staminal fascicle. **G.** Lateral view (above), medial section (below), and transverse section (below left) of pistil. **H.** Infructescence.

Vismia latifolia (Aubl.) Choisy

Trees, to 7 m tall. Stems, peduncle, pedicels, petioles, and abaxial leaf blade surfaces with tomentum of sessile, rotate, rufous, stellate trichomes, at times appearing scale-like. Stems angular, 4–5 mm diam. Leaves: petioles 1.4–2 cm long; blades ovate, rarely oblong, 8–21 × 3.5–7 cm, coriaceous, the base obtuse to broadly rounded, the apex acuminate, the margins entire; punctations black, inconspicuous. Inflorescences terminal; peduncle 3–4 cm long; pedicels 1–3 mm long. Flowers: sepals oblong, 6–8 mm long; petals obovate or rarely oblong, 6–8 mm long, moderately woolly adaxially, the punctations small, few; anthers orange. Fruits ovoid, 10–13 × 6–9 mm, without persistent stamens. Fl (Nov), fr (Feb); in non-flooded forest. *Bois-dartre* (Créole).

Vismia ramuliflora Miq.

Shrubs or small trees, to 5(7) m tall. Glabrous or branchlets, abaxial leaf surface surfaces, and inflorescence remaining minutely puberulent. Stems terete, 1–2 mm diam. Leaves: petioles 5–10 mm long; blades membranaceous, elliptic, oblong or obovate, 5–11 × 2–6 cm, the base obtuse to rounded, the apex caudate to cuspidate, the margins undulate to widely crenulate; punctations large, prominently raised. Inflorescences axillary or subterminal; peduncle 2–4 cm long; pedicels 6–8 mm long. Flowers: sepals oblong, 3.5–4 mm long; petals elliptic to oblong, 8–10 mm, the punctations prominent; anthers black punctate. Fruits subglobose to ovoid, 8–10 × 6–8 mm, without persistent stamens. Fl (Aug, Sep, Nov), fr (Jan, Oct); in gaps in non-flooded forest, at low and higher elevations.

Vismia sandwithii Ewan FIG. 83, PL. 43d

Shrubs to small trees, to 5 m tall. Stems, peduncle, pedicels, petioles, and abaxial leaf blade surfaces with tomentum of stalked, multiangular, rufous, stellate trichomes. Stems terete, 3–4.5 mm diam. Leaves: petioles 8–20 mm long; blades oblanceolate or obovate, 13–22 × 5–10 cm, chartaceous, the base acute, the apex acuminate; punctations prominently raised, several per areole. Inflorescences terminal; peduncle 1–2 cm long; pedicels 3–5 mm long. Flowers: sepals oblong, 5–7 mm long; petals oblong, 7–9 mm long, densely and prominently black punctate abaxially, densely woolly adaxially; anthers black punctate. Fruits globose, 8–10 mm, with persistent stamens. Fl (Aug, Sep), fr (Jan, Oct, Nov, Dec); in non-flooded forest. *Bois-dartre* (Créole).

COMBRETACEAE (Indian Almond Family)

Maria Lúcia Kawasaki

Trees, shrubs, or lianas. Stipules absent. Leaves simple, alternate, in whorls, or opposite. Inflorescences racemes, spikes, panicles, or globular clusters (*Conocarpus*). Flowers actinomorphic, usually bisexual; calyx lobes 4–5; petals 4–5 or absent; stamens 8 or 10; nectary disc present or absent; ovary inferior, unilocular; placentation with 2–6 pendulous ovules. Fruits drupaceous (*Buchenavia*) or samara-like (*Combretum* and *Terminalia* in our area). Seed 1.

Görts-van Rijn, A. R. A. 1986. Combretaceae. *In* A. L. Stoffers & J. C. Lindeman (eds.), Flora of Suriname **III(1–2):** 354–355. E. J. Brill, Leiden.

Stace, C. A. & A. Alwan. 1998. Combretaceae. *In* J. A. Steyermark, P. E. Berry & B. K. Holst (gen. eds.), Flora of the Venezuelan Guayana **4:** 329–352. J. A. Steyermark, P. E. Berry & B. Holst (vol. eds.). Missouri Botanical Garden Press, St. Louis.

1. Lianas. Leaves opposite. Petals present. . . . . *Combretum.*
1. Trees. Leaves alternate or in whorls. Petals absent.
    2. Leaves markedly aggregated at ends of stems; petioles without paired glands toward apex. Anthers basifixed. Fruits not winged. . . . . *Buchenavia.*
    2. Leaves not markedly aggregated at ends of stems; petioles with paired glands toward apex. Anthers dorsifixed. Fruits winged. . . . . *Terminalia.*

BUCHENAVIA Eichler

Large trees. Stem growth sympodial. Leaves alternate or in whorls, often markedly aggregated; petioles without paired glands toward apex. Inflorescences spikes. Flowers 5-merous; petals absent; stamens 10, the anthers basifixed. Fruits drupaceous, 5-ridged.

1. Leaf blades 2–3 cm long. . . . . *B. parvifolia.*
1. Leaf blades 5.5–24 cm long.
    2. Leaf blades 5.5–7 × 2–3 cm, abaxial surface reddish pubescent. . . . . *B. grandis.*
    2. Leaf blades 13–24 × 5.5–8.5 cm, glabrous. . . . . *B. nitidissima.*

Buchenavia grandis Ducke

Large trees, to 30 m tall. Leaves in whorls; petioles ca. 10 mm long; blades oblanceolate, 5.5–7 × 2–3 cm, coriaceous, minutely puberulous adaxially, reddish pubescent on veins abaxially, the base cuneate to attenuate, the apex obtuse or retuse; venation salient on both surfaces. Inflorescences terminal spikes, the axes tomentose. Fruits oblong, ca. 2 cm long, reddish to yellowish tomentose. Fr (Nov, Dec); in non-flooded forest.

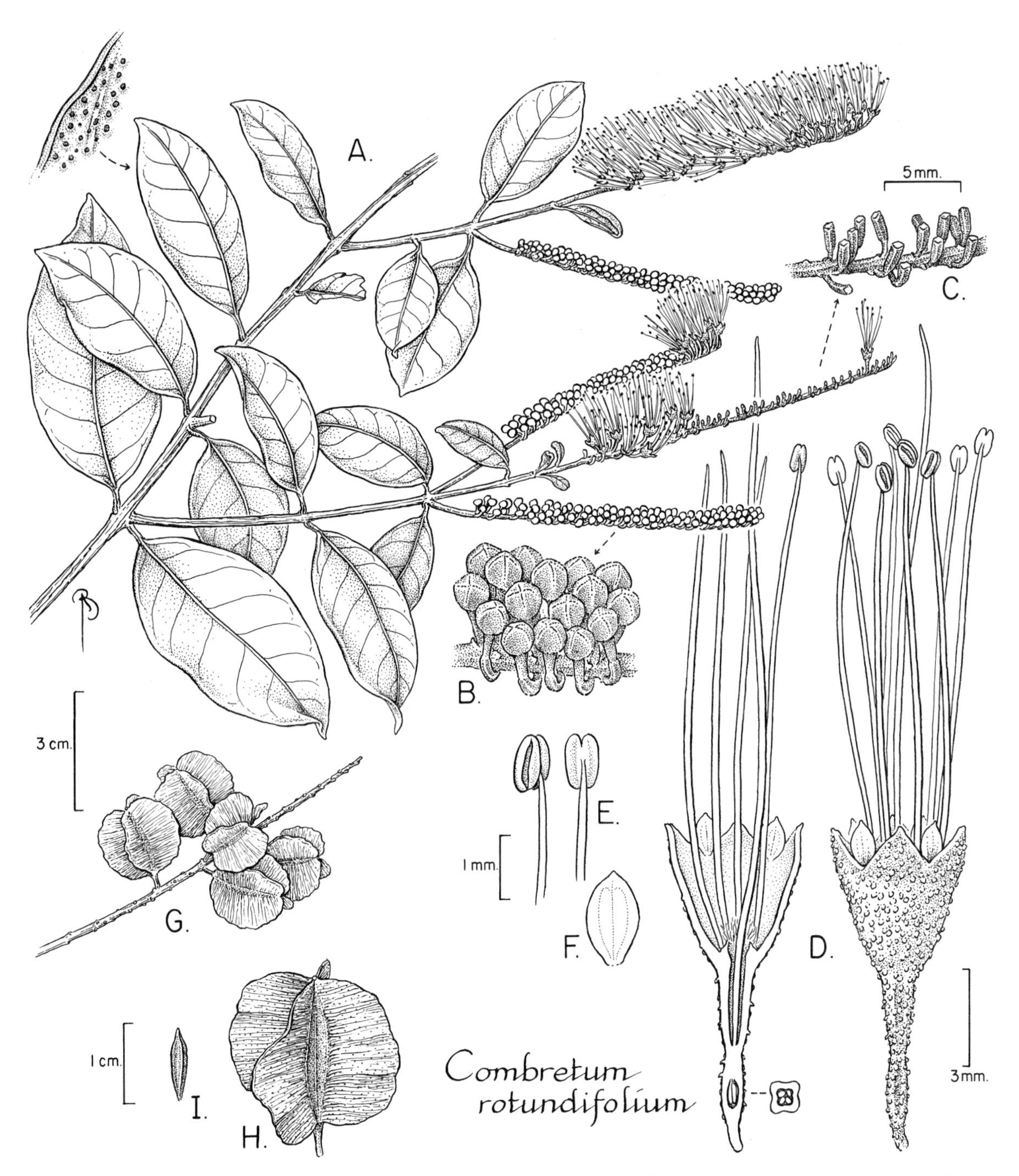

FIG. 84. COMBRETACEAE. *Combretum rotundifolium* (*Mori & Gracie 18625*). **A.** Part of stem with leaves and inflorescences and detail of abaxial surface of leaf (upper left). **B.** Detail of flower buds. **C.** Detail of pedicels of fallen flowers. **D.** Medial section (left) with transverse section of ovary and lateral view (right) of flower. **E.** Two views of stamens. **F.** Petal. **G.** Part of infructescence. **H.** Fruit. **I.** Seed.

Buchenavia nitidissima (Rich.) Alwan & Stace

Large trees, to 30–40 m tall. Outer bark brown, scaly, the middle bark pink, the inner bark cream-colored. Leaves in whorls; petioles to 30 mm long; blades oblanceolate to obovate, 13–24 × 5.5–8.5 cm, coriaceous, glabrous, the base attenuate, the apex shortly acuminate to obtuse, retuse; venation salient on both surfaces. Inflorescences terminal, spikes, the axes reddish-brown tomentose. Fruits pubescent when immature. Fl (Nov), fr (Nov, Dec); in non-flooded forest. *Amandier sauvage* (Créole).

Buchenavia parvifolia Ducke

Trees. Outer bark rough, reddish-brown, the inner bark creamy-yellow, laminated. Leaves in whorls; petioles ca. 1 mm long; blades obovate, 2–3 × 1–1.5 cm, chartaceous, glabrous adaxially, reddish pubescent on veins abaxially, the base attenuate, the apex obtuse, retuse; venation salient on both surfaces. Known by two sterile collections from our area; in non-flooded forest.

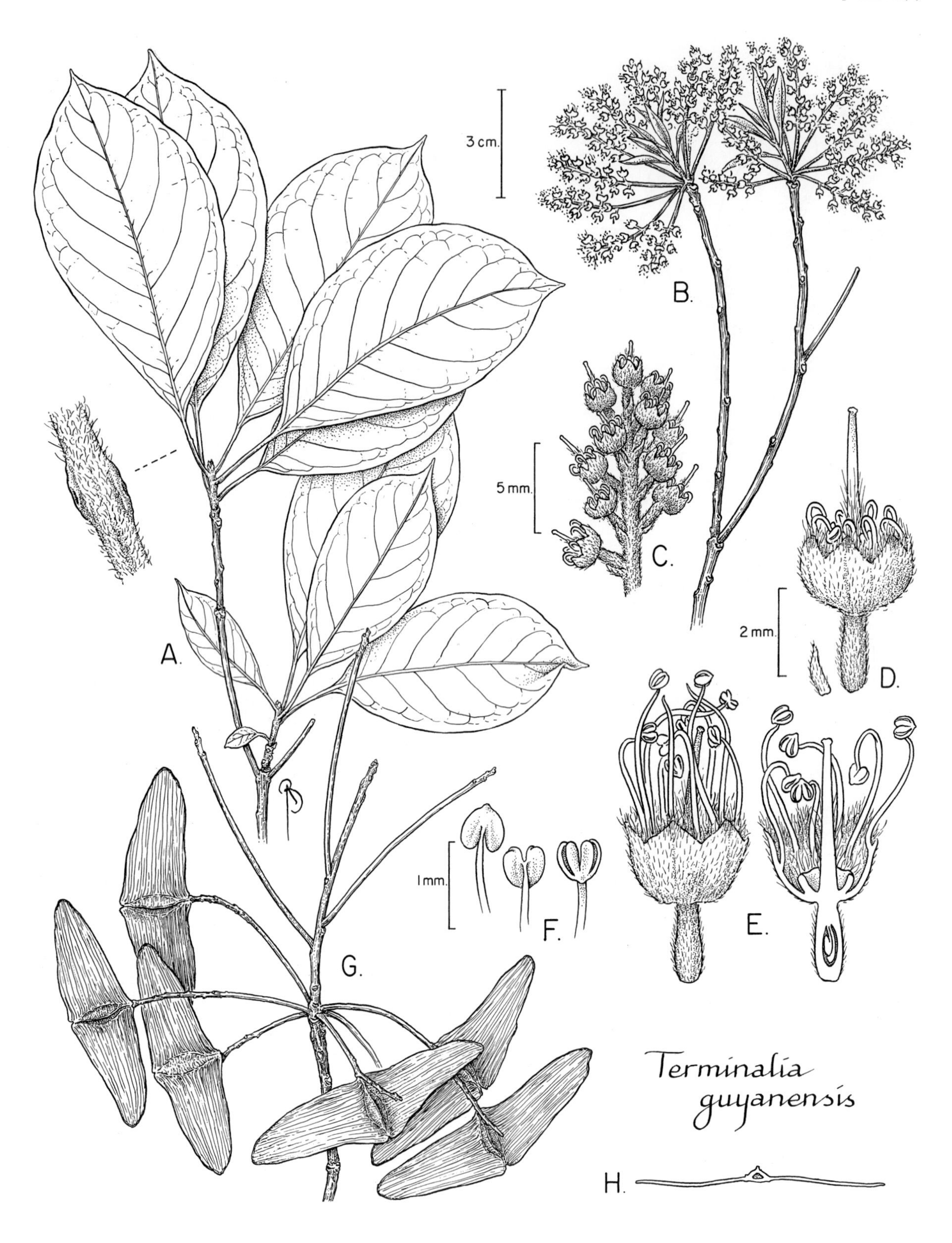

FIG. 85. COMBRETACEAE. *Terminalia guyanensis* (A, *Mori et al. 23905*; B–F, *Mori & Gracie 18653*; G, H, *Mori & Boom 15121*). **A.** Apex of stem with leaves and detail of glands on petiole. **B.** Inflorescences. **C.** Detail of part of inflorescence. **D.** Lateral view of flower in pistillate phase (left) and a subtending bract (far left). **E.** Lateral view (left) and medial section (right) of flower in staminate phase. **F.** Three views of stamens. **G.** Infructescence of winged fruits. **H.** Transverse section of immature fruit.

COMBRETUM Loefl.

Lianas. Leaves opposite, not aggregated; petioles without paired glands toward apex. Inflorescences spikes or panicles of spikes. Flowers 4-merous; petals 4; stamens 8, the anthers dorsifixed. Fruits 4-winged.

1. Inflorescences panicles. Floral receptacle ca. 3 mm long, yellowish tomentose. Fruits oblong. . . . . *C. laxum.*
1. Inflorescences spikes. Floral receptacle ca. 10 mm long, reddish lepidote-pubescent. Fruits suborbicular. . . . . *C. rotundifolium.*

Combretum laxum Jacq. PL. 44c

Lianas. Leaves: petioles to 15 mm long; blades elliptic to narrowly elliptic, 9–16 × 4.5–5.5 cm, chartaceous to coriaceous, minutely pubescent or glabrous, the base obtuse to subcordate, the apex acuminate, apiculate. Inflorescences axillary and terminal panicles of spikes, to 17 cm long. Flowers: receptacle ca. 3 mm long, yellowish tomentose; petals ca. 2 mm long, longer than calyx lobes; nectary disc inconspicuous; filaments ca. 3 mm long. Fruits oblong, puberulous. Fl (Aug, Sep, Oct); in non-flooded forest.

Combretum rotundifolium Rich.
FIG. 84, PL. 44a,b; PART 1: FIG. 6

Lianas. Leaves: petioles ca. 5 mm long; blades elliptic to narrowly elliptic, 6.5–7.5 × 2.5–3.5 cm, coriaceous, reddish lepidote-pubescent abaxially, the base obtuse, the apex acute. Inflorescences axillary and terminal spikes, to 11 cm long. Flowers: receptacle ca. 10 mm long, reddish lepidote-pubescent; petals ca. 1 mm long; nectary disc well developed, pilose at margin; filaments ca. 13 mm long. Fruits suborbicular, lepidote-pubescent. Fl (Aug); in non-flooded forest. *Feuille singe rouge, peigne singe rouge, queue du singe rouge* (Créole).

TERMINALIA L.

Trees. Stem growth sympodial. Leaves alternate or in whorls, aggregated but not markedly so; petioles often with paired glands toward apex. Inflorescences spikes. Flowers 4–5-merous; petals absent; stamens 8 or 10, the anthers dorsifixed. Fruits 2-winged in our area.

1. Inner bark cream-colored. Leaf blades with reddish-brown tuft of hairs in axils of secondary veins abaxially. Fruits golden brown, ca. 2 cm wide. . . . . *T. amazonia.*
1. Inner bark yellow. Leaf blades without reddish-brown tuft of hairs in axils of secondary veins abaxially. Fruits green, 4.5–7.5 cm wide. . . . . *T. guyanensis.*

Terminalia amazonia (J. F. Gmel.) Exell

Trees, 25–30 m tall. Outer bark scaly, reddish-brown, the inner bark cream-colored. Leaves in whorls; petioles 8–12 mm long, densely reddish-brown pubescent; blades elliptic to obovate, 6–10 × 2.5–4.5 cm, chartaceous, densely reddish-brown pubescent on veins abaxially, the base acute to obtuse, the apex short acuminate, the margins entire; secondary veins in 5–7 pairs. Inflorescences not known in our area. Flowers not known from our area. Fruits 0.6 × 2 cm, sparsely pubescent, golden-brown. Fr (Sep); known by separate sterile collection with leaves and by fruits picked from ground, in non-flooded forest. *Angouchi, bois-blanchet, goué-goué, graine-hocco* (Créole).

Terminalia guyanensis Eichler
FIG. 85, PL. 44d,e; PART 1: FIG. 6

Trees, to 56 m tall. Outer bark brown, the inner bark yellow. Leaves in whorls; petioles 15–30 mm long; blades elliptic to narrowly elliptic, or obovate, 7–13 × 3–6.5 cm, chartaceous, puberulous to glabrous, the apex acuminate, the base cuneate; secondary veins in 7–8 pairs. Inflorescences terminal, spikes, ca. 4 cm long, the axes yellowish-tomentose. Flowers: receptacle ca. 3 mm long, yellowish-tomentose; style and filaments ca. 3 mm long, exserted. Fruits 2 × 4.5–7.5 cm, glabrous, green. Fl (Aug, Oct), fr (Oct, Nov); common, one of the tallest trees in our forest, in non-flooded forest.

CONNARACEAE (Connarus Family)

Enrique Forero

Lianas or small trees. Pubescence of simple or dendroid hairs. Stipules absent. Leaves imparipinnate, alternate. Inflorescences axillary, cauliflorous, pseudoterminal, or terminal, usually panicles or racemes. Flowers actinomorphic, bisexual; sepals 5; petals 5; stamens 10, connate at base, the 5 opposite petals distinctly larger than 5 opposite sepals; ovary superior, carpels 1 (*Connarus*) or 2–5 and free; placentation basal, the ovules 2 per carpel, collateral, erect. Fruits follicles, 1 or 5 developing per flower. Seeds one per follicle; arilloid present, incompletely covering basal half, leaving funicle free; endosperm present or absent.

Forero, E. 1983. Connaraceae. Fl. Neotrop. Monogr. **36:** 1–208.
——— & A.R.A. Görts-van Rijn. 1976. Connaraceae. *In* J. Lanjouw & A. L. Stoffers (eds.), Flora of Suriname **I(2):** 654–657. E. J. Brill, Leiden.
——— & E. Santana. 1998. Connaraceae. *In* J. A. Steyermark, P. E. Berry & B. K. Holst (gen. eds.), Flora of the Venezuelan Guayana **4:** 365–365. P. E. Berry, B. K. Holst & K. Yatskievych (vol. eds.). Missouri Botanical Garden Press, St. Louis.
Lanjouw, J. 1976. Connaraceae. *In* J. Lanjouw & A. L. Stoffers (eds.), Flora of Suriname **I(2):** 332–340. E. J. Brill, Leiden.

1. Petals punctate; carpel solitary at anthesis. . . . . *Connarus.*
1. Petals epunctate; carpels 5 at anthesis.
   2. Inflorescences densely pilose. Sepals valvate or only narrowly imbricate. Fruits densely pilose, without accrescent calyx. Endosperm present but scanty. . . . . *Cnestidium.*
   2. Inflorescences usually glabrous or tomentose but not densely pilose. Sepals imbricate. Fruits usually glabrous or tomentose but not densely pilose, with accrescent calyx. Endosperm absent. . . . . *Rourea.*

CNESTIDIUM Planch.

Lianas. Pubescence of simple hairs. Young stems tomentose. Leaves 5–7-foliolate; blades ovate-lanceolate (rarely elliptic), abaxial surface puberulous (hairs appressed), the base attenuate. Inflorescences densely pilose. Flowers: sepals valvate or only narrowly imbricate, densely puberulous; carpels 5, free. Fruits dark brown pubescent (pilose), without accrescent calyx. Seeds: endosperm present but scanty.

Cnestidium guianense (G. Schellenb.) G. Schellenb. FIG. 86, PL. 45a,b

Lianas. Leaflets: blades (2.5)4.5–11(16) × (1.4)2–6(7.8) cm, appressed puberulous, more densely so on veins abaxially, the hairs occasionally shiny; veinlets transverse. Flowers: sepals 2–5 × 1 mm, densely puberulous; petals 2.5–3 × 1 mm, glabrous, white. Fruits with 1–2(4) follicles, the follicles 1.2–2 × 0.4–0.7 cm, striate, dark brown pubescent (in herbarium specimens), orange. Seeds black, with light yellow arilloid. Fl (Sep, Nov), fr (Feb, Apr); in non-flooded forest.

CONNARUS L.

Treelets or lianas. Pubescence of dendroid hairs. Leaves 5–21-foliolate; blades with abaxial surface glabrous or pubescent. Inflorescences axillary, pseudoterminal, or cauliflorous, paniculate or racemose, densely pubescent or tomentose, with simple, dendroid, and sometimes glandular hairs, rarely glabrous. Flowers: sepals imbricate; petals punctate; carpel 1. Fruits densely tomentose or glabrous, sessile or shortly stipitate, without accrescent calyx. Seeds: endosperm absent.

1. Leaflets 11–21, the blades with abaxial surface glabrous. Inflorescences cauliflorous. Fruits glabrous. . . . . *C. fasciculatus* subsp. *fasciculatus.*
1. Leaflets 5–7, the blades with abaxial surface densely ferruginous-tomentose. Inflorescences axillary or pseudoterminal. Fruits densely ferruginous-tomentose. . . . . *C. perrottetii* var. *perrottetii.*

Connarus fasciculatus (DC.) Planch. subsp. **fasciculatus** FIG. 87

Treelets, 2–4 m tall. Leaflets 11–21; blades 9–20 × 3–6 cm, glabrous abaxially. Inflorescences cauliflorous, racemose, 1.5–2 cm long. Flowers: sepals 3.5 mm long, tomentose, densely and conspicuously punctate; petals 4.5–5 × 1.2–1.5 mm, conspicuously punctate. Fruits 2–2.5 × 1–1.2 cm, stipitate, glabrous, burnt orange. Seeds black with whitish arilloid. Fl (Aug, Oct), fr (Jan, Feb, Aug, Sep, Oct, Nov, Dec); in non-flooded forest and secondary forest. *Gaulette* (Créole).

Connarus perrottetii (DC.) Planch. var. **perrottetii**

Lianas. Leaflets 5–7; blades 3.5–7 × 1–2 cm, densely ferruginous tomentose (described as dirty white on *Mori & Granville 8818*, NY) abaxially. Inflorescences axillary or pseudoterminal, paniculate, to 9.5 cm long. Flowers: sepals 2–2.2 mm long, tomentose, sparsely and conspicuously punctate; petals 2.8–3 × 1–1.2 mm, sparsely punctate. Fruits 2.2 × 1.5 cm, sessile or shortly stipitate, densely ferruginous-tomentose (hairs easily wiped off). Fr (Dec); in non-flooded forest. *Mara-sacaca, mata-cachorro* (Portuguese).

ROUREA Aubl.

Lianas. Pubescence of simple hairs. Leaves 3–9-foliolate. Inflorescences axillary, paniculate, usually glabrous or tomentose but not densely pilose. Flowers: sepals imbricate; petals epunctate; carpels 5, free. Fruits sessile, glabrous or glabrescent, with accrescent calyx. Seeds: endosperm absent.

1. Leaflets (8)9, the blades chartaceous, pubescent abaxially. Sepals tomentose. . . . . *R. pubescens* var. *pubescens.*
1. Leaflets (1)3(5), the blades coriaceous to rigid-coriaceous, glabrous abaxially. Sepals mostly glabrous. . . . *R. surinamensis.*

Rourea pubescens (DC.) Radlk. var. **pubescens**

Lianas. Leaflets (8)9; blades 11.5–22 × 5.8–8.8 cm, chartaceous, pubescent adaxially, pale, pubescent abaxially, the hairs white, the apex acuminate, the margins plane. Inflorescences tomentose. Flowers: sepals 2–3 × 1–1.5 mm, densely tomentose, the margins ciliate; petals free, 3.5–6 × 1–1.5 mm, glabrous, white, the apex villous. Fruits 1–1.1 cm long, glabrous or glabrescent, the apex villous. In non-flooded forest. Not known from our flora but to be expected.

Rourea surinamensis Miq. PL. 45c

Lianas. Leaflets (1)3(5); blades 6–13.5 × 2.5–5 cm, coriaceous to rigid-coriaceous, glabrous throughout, the apex acuminate, the margins revolute. Inflorescences glabrous or glabrescent. Flowers: sepals 2 × 1.5 mm, mostly glabrous except apex barbate and margins sometimes ciliate; petals free, 3–5 × 1.5 mm, glabrous, white. Fruits 1–1.2 cm long, glabrous. Fl (Aug), fr (Jan); in non-flooded forest.

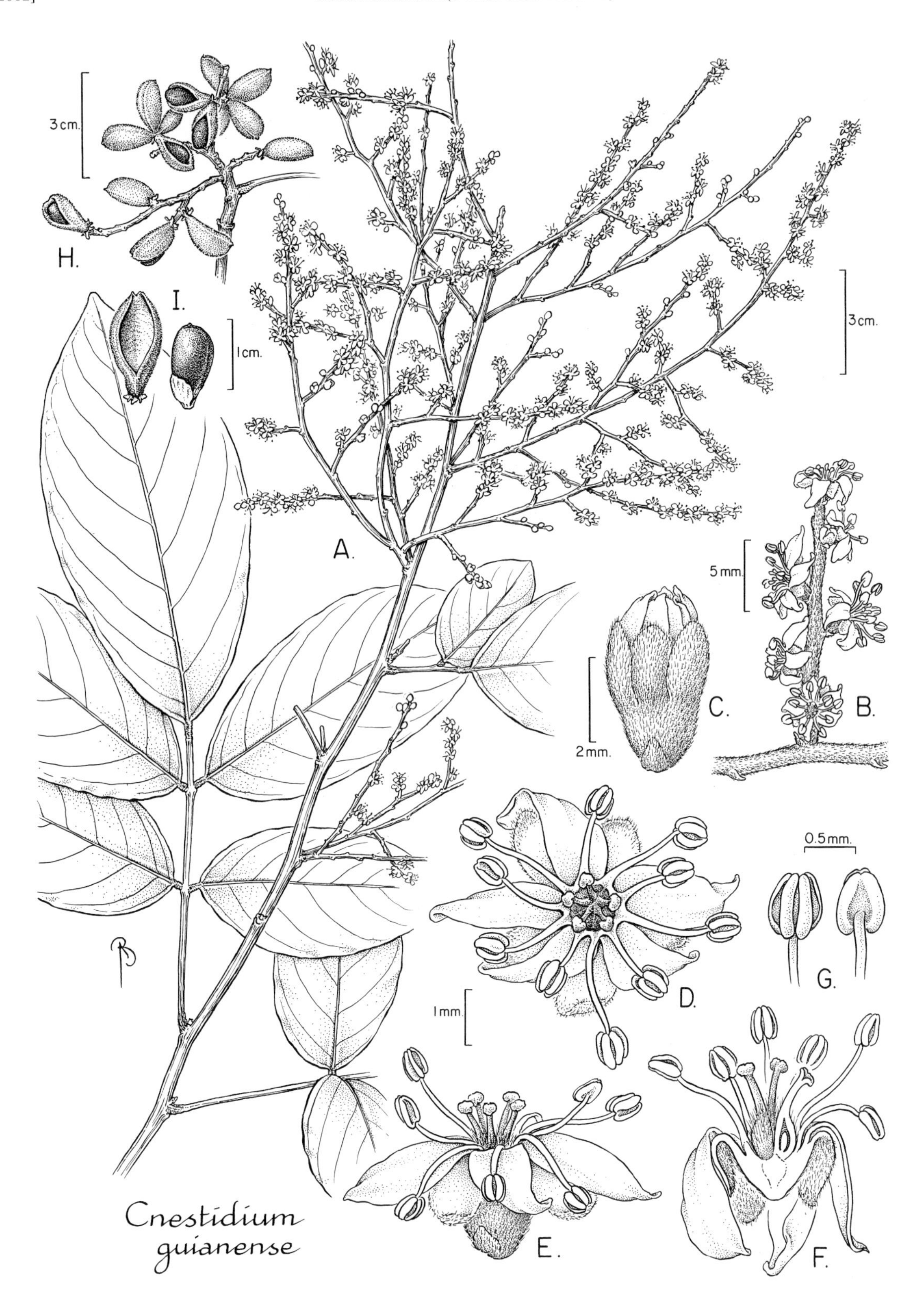

FIG. 86. CONNARACEAE. *Cnestidium guianense* (A–G, *Mori et al. 2415*; H, I, *Mori & Pipoly 15571*). **A.** Part of stem with leaves and inflorescences. **B.** Detail of part of inflorescence. **C.** Lateral view of flower bud. **D.** Apical view of flower. **E.** Lateral view of flower. **F.** Medial section of flower. **G.** Adaxial (left) and abaxial (right) views of anthers. **H.** Infructescence. **I.** Empty follicle and seed with aril.

FIG. 87. CONNARACEAE. *Connarus fasciculatus* subsp. *fasciculatus* (A, *Mori et al. 22952*; B–F, *Irwin & Westra 47701* from Amapá, Brazil; G, H, *Maguire 40781* from Surinam). **A.** Part of stem with pinnate leaf. **B.** Inflorescence. **C.** Lateral view of flower. **D.** Adaxial surface of petals (right) and adaxial surface of part of calyx with detail of hair from abaxial surface (above). **E.** Lateral view of pistil surrounded by staminal ring (left) and medial section of pistil (above right). **F.** Adaxial view of part of staminal ring. **G.** Part of stem with infructescences. **H.** Fruit (near left) and seed with aril (far left).

CONVOLVULACEAE (Morning-glory Family)

Daniel F. Austin

Herbs, vines, lianas, shrubs, or trees. Sap milky in some species. Stipules absent. Leaves mostly simple, pinnately lobed or pectinate, palmately compound in some species. Inflorescences axillary, dichasial, solitary, racemose, or paniculate. Flowers usually actinorphic, usually bisexual, small and inconspicuous to large and showy but mostly wilting quickly; sepals 5, free, imbricate, equal or unequal, persistent, occasionally accrescent in fruit; corolla sympetalous, tubular, funnelform, campanulate, urceolate or salverform, the limb with 5 lobes or teeth or almost entire, with plicae and interplicae; stamens 5, distinct, the filaments inserted on corolla tube base alternate with corolla lobes, the anthers mostly linear or oblong, 2-celled, introrse or extrorse; nectary disc annular or cup-shaped, sometimes 5-lobed, occasionally absent; ovary superior, of 2–4 carpels, usually (1)2-locular, the style filiform, simple or bifid or with 2 distinct styles, the stigma capitate or bilobate or stigmas 2 and linear, ellipsoid, branched, or globose; placentation axile, biovulate. Fruits 1–4-locular, capsular, dehiscent by valves, transversely dehiscent, irregularly dehiscent, or indehiscent and baccate or nut-like. Seeds 1–4, commonly fewer than ovules, glabrous or pubescent.

Austin, D. F. 1982. Convolvulaceae. *In* Z. Luces de Febres & J. A. Steyermark (eds.), Flora de Venezuela **8(3):** 15–226.

———. 1998. Convolvulaceae. *In* J. A. Steyermark, P. E. Berry & B. K. Holst (gen. eds.), Flora of the Venezuelan Guayana **4:** 377–424. P. E. Berry, B. K. Holst & K. Yatskievych (vol. eds.). Missouri Botanical Garden Press, St. Louis.

——— & G. W. Staples. 1981. Convolvulaceae. *In* B. Maguire (ed.), Botany of the Guayana Highland. Mem. New York Bot. Gard. **32:** 309–323.

Ooststroom, S. J. van. 1966. Convolvulaceae. *In* A. Pulle (ed.), Flora of Suriname **IV(1):** 66–102, 468–471. E. J. Brill, Leiden.

1. Corollas pubescent abaxially. Fruits indehiscent, nut-like.
   2. Corollas <10 mm. long. Anthers versatile. . . . . *Dicranostyles.*
   2. Corollas >10 mm. long. Anthers basifixed. . . . . *Maripa.*
1. Corollas glabrous abaxially. Fruits capsular.
   3. Leaves entire to pinnately lobed. Flowers purple, blue or lavender to scarlet. Dehisced anthers straight. Pollen panto-porate and spinulose (visible with 20x magnification, sometimes less). . . . . *Ipomoea.*
   3. Leaves palmately divided to palmately compound. Flowers white, with or without purplish or reddish center. Dehisced anthers spiralled. Pollen 3-colpate, not spinulose. . . . . *Merremia.*

DICRANOSTYLES Benth.

Lianas. Larger stems smooth to slightly fluted. Leaves simple, alternate; the petioles canaliculate; blades chartaceous, occasionally ± coriaceous, densely appressed puberulous or erect pubescent to glabrescent, the base attenuate, acute to obtuse, occasionally rounded, cordate, or truncate. Inflorescences axillary, racemose, racemose-thyrsiform, cylindric-thyrsiform, or thyrsiform, usually ± fasciculate in leaf axils or along internodes, with many flowers. Flowers 5-merous, the aestivation valvate-induplicate; sepals quincuncial, the outer usually ovate, the apex usually acute, the inner mostly broadly ovate to ± rotund, the apex usually rounded, the exposed surfaces usually appressed pubescent, the covered surfaces usually glabrous, the margins mostly ciliate; corollas ± rotate or infundibular, lobed almost to base when ± rotate, or half corolla length or less when infundibular, mostly <10 mm long, white to pinkish; stamens epipetalous, the filaments glandular or glabrous, the anthers dorsifixed, versatile, basally sagittate; the style entire, divided into 2 short branches or completely divided, the stigma 2-lobed, free or closely appressed, variable in shape. Fruits nut-like, indehiscent ellipsoid to ellipsoid-cylindric, 1(4)-seeded through abortion. Seeds glabrous, the seed coat coriaceous.

1. Leaf blades glabrous to glabrescent abaxially. . . . . *D. ampla.*
1. Leaves pubescent abaxially.
   2. Abaxial leaf blade surface reddish with appressed indumentum. Style divided into short branches, these sometimes not visible without dissection, the stigmas with 2 distinct lobes or 2 free stigmas. . . . . *D. guianensis.*
   2. Abaxial leaf blade surface white with appressed indumentum. Style entire, the stigmas with 2 lobes closely appressed. . . . . *D. integra.*

Dicranostyles ampla Ducke PL. 45d

Stems brown to reddish-brown, glabrescent. Leaves: blades oblong, elliptic-oblong, ovate to obovoid, 6–21 × 1.5–8 cm, ± coriaceous, glabrescent, the base attenuate, obtuse to rounded, or ± cordate, the apex acuminate. Inflorescences in axillary groups or terminal, cylindric-thyrsiform to pyramidal-thyrsiform, with brownish pubescence; pedicels 1–2 mm long. Flowers: sepals ovate to narrowly ovate, the outer acute, the inner rounded, 1.5 mm long, with appressed pubescence; corolla ± rotate, 3–4 mm long, the tube 1 mm long, the lobes 2–3 mm long; ovary ovoid, 1–1.3 mm long, the upper half or only apex pubescent, the styles 2, reaching 2–2.5 mm long. Fruits ellipsoid to obpyriform, 15–20 mm long, the pericarp smooth to striate, dark brown to black. Seeds 1, ellipsoid, 15 mm long. Fl (Aug, Sep), fr (Aug); non-flooded forest.

Dicranostyles guianensis Mennega FIG. 88

Stems ferruginous with appressed indumentum when young, glabrescent. Leaves: blades elliptic-ovate to narrowly ovate, 6–11 ×

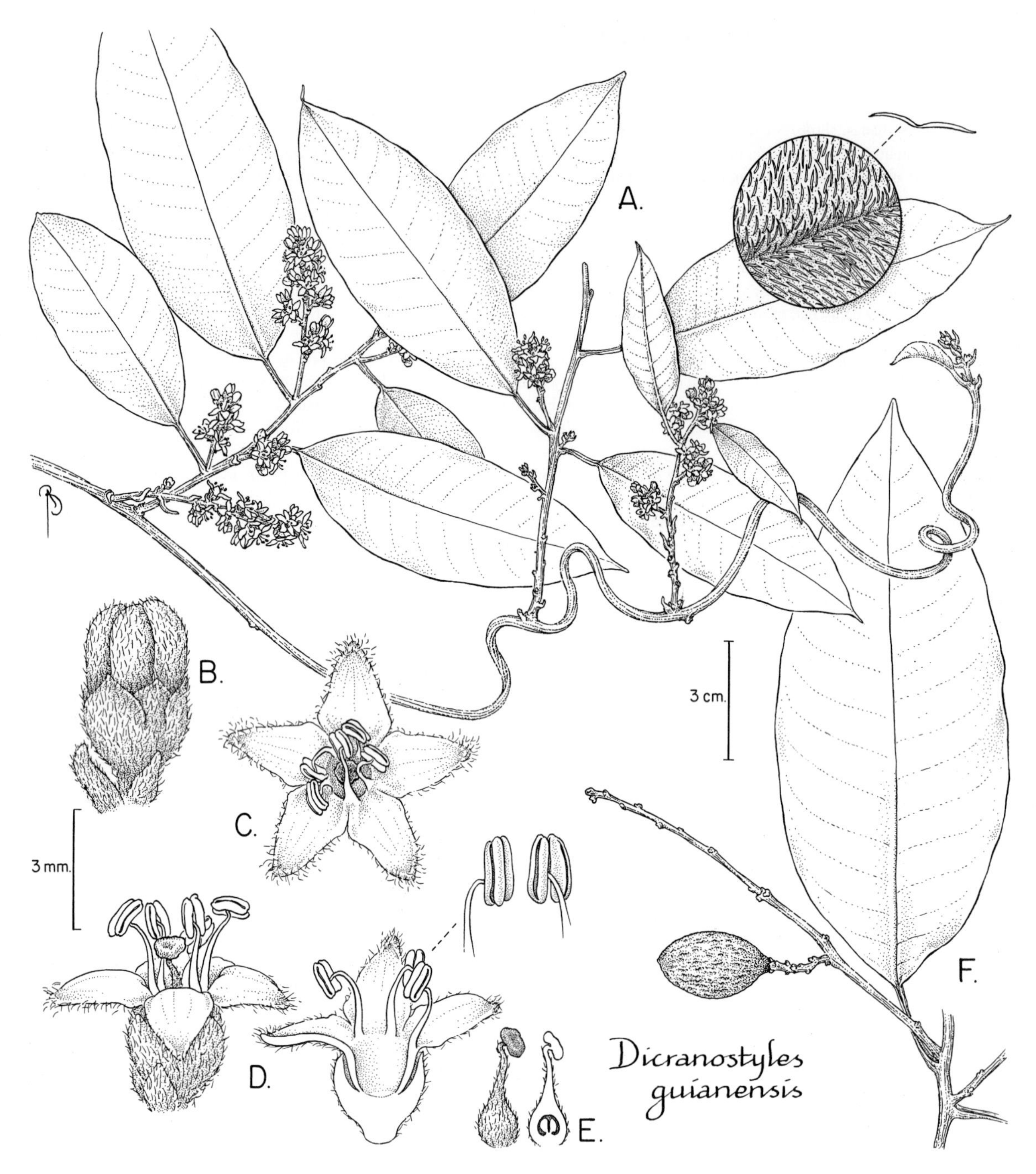

FIG. 88. CONVOLVULACEAE. *Dicranostyles guianensis* (A–E, *Mori et al. 22837*; F, *Mori & Pennington 17957*). **A.** End of stem with leaves, inflorescences, and details of hairs. **B.** Flower bud. **C.** Apical view of flower. **D.** Lateral view of flower (left), medial section of flower with pistil removed (right), and stamens (upper right). **E.** Intact pistil (far left) and medial section of pistil (left). **F.** Part of stem with infructescence.

3–7 cm, chartaceous to ± coriaceous, glabrous except along midrib adaxially, white to reddish setose abaxially, the base rounded to ± rotund, the apex abruptly acuminate. Inflorescences axillary and along internodes of short lateral branches, cymose to paniculate, with reddish pubescence; pedicels 2–3 mm long. Flowers: sepals narrowly ovate to ± deltoid, 3.5 × 2.5–3 mm, the apex acute, with reddish pubescence; corolla ± rotate, deeply lobed, 4–5.5 mm long, pubescent ab- and adaxially, the lobes enlarging during anthesis; stamens exserted, adnate between lobes of corolla; ovary conic, 2–2.5 mm long, villose, the styles 2–3 mm long, shortly branched at apex, pubescent on lower half. Fruits ovoid, 12 mm long, grayish. Seeds unknown. Fl (Nov), fr (May); in canopy, non-flooded forest.

Dicranostyles integra Ducke

Stems grayish-brown with appressed indumentum when young, glabrescent. Leaves: blades oblong, elliptic to ovate, less often obovate, 9–16 × 3–6 cm, chartaceous, glabrescent adaxially, densely pubescent abaxially, the indumentum white, the base obtuse, the apex acuminate. Inflorescences cylindric-thyrsiform, these solitary or fasciculate, axillary, gray-white or pale yellow-white, the indumentum appressed; pedicels 1–2 mm long. Flowers: sepals broadly ovate to rounded, 1.8–2.5 × 2.2–2.5 mm, the apex acute, gray-white pubescent, the indumentum same as inflorescence branches; corolla funnelform, 4–5 mm long, the lobes inflexed at anthesis, pubescent abaxially; stamens usually included, rarely obscurely exserted between corolla lobes; ovary globose, 1–1.5 mm long, villose, the styles 1 mm long, entire, glabrous. Fruits ovoid, 10–15 mm long, black, smooth, shiny. Seeds 1, ellipsoid. Fl (Sep, Nov); in canopy, non-flooded forest.

IPOMOEA L.

Vines, lianas, shrubs, or trees. Stems herbaceous to woody, usually twining, sometimes erect, prostrate, or scandent, glabrous or pubescent. Leaves: petioles present; blades variable in shape and size, entire, lobed, divided or rarely compound, glabrous or pubescent. Inflorescences mostly axillary, flowers 1 to many, in dichasia, rarely paniculate; peduncle long or short; bracts scale-like to foliose. Flowers: sepals ovate to oblong or lanceolate, herbaceous to coriaceous, glabrous to pubescent, often somewhat enlarged in fruit but not markedly accrescent; corolla small to large, regular or rarely slightly zygomorphic, mostly funnelform, less often campanulate, tubular or salverform, purple, blue, lavender, red, pink, white, or less often yellow, the limb shallowly or deeply lobed, the midpetaline bands well defined by two distinct veins; stamens included or rarely exserted, the filaments filiform, often triangular-dilate at base, mostly unequal in length, the anthers straight, the pollen panto-porate, globose, spinulose; ovary usually 2- or sometimes 4-locular, 4-ovulate, rarely 3-locular, 6-ovulate, glabrous or pubescent, the styles simple, filiform, included or rarely exserted, the stigmas capitate, entire or 2- or rarely 3-globose. Fruits globose to ovoid capsules, mostly 4- or rarely 6-valved or splitting irregularly. Seeds 1–4, rarely to 6, glabrous or pubescent.

1. Leaf blades pinnatifid. Corolla salverform; stamens exserted. . . . . *I. quamoclit.*
1. Leaf blades not pinnatifid. Corolla funnelform to campanulate; stamens included.
    2. Sepals unequal, the outer much shorter than inner.
        3. Sepals pubescent or at least ciliate (rarely glabrous), the apex of outer sepals acute to acuminate and cuspidate. . . . . *I. batatas.*
        3. Sepals glabrous, the apex of outer sepals rounded. . . . . *I. squamosa.*
    2. Sepals equal to ± equal.
        4. Sepals 6–9 mm long. . . . . *I. batatoides.*
        4. Sepals 15–18 mm long. . . . . *I. phyllomega.*

Ipomoea batatas (L.) Lam.

Vines. Stems usually somewhat succulent but sometimes slender and herbaceous, erect, procumbent, or twining, reaching 4 m or more long, but often shorter in cultivated plants, glabrous or pubescent. Leaves: blades variable, cordate to ovate, 5–10(15) cm long, chartaceous to fleshy, glabrous or pubescent, the margins entire, dentate to often deeply (3)5–7-lobed. Inflorescences of solitary or few-flowered cymes, glabrous or pubescent; pedicels 3–12 mm long. Flowers apparently absent in some varieties; sepals chartaceous to almost fleshy, unequal, the outer 2 shorter than inner, oblong, (8)10–15 mm long, mostly pubescent or only ciliate, the apex acute to acuminate and cuspidate; corolla funnelform, 4–7 cm long, glabrous, with lavender to purple-lavender limb and darker throat, white in some varieties; stamens included. Fruits rarely formed, 5–8 mm long ovoid, glabrous, brown. Seeds 0–1(4), rotund, glabrous or with wings of short trichomes. Fl (Aug, Sep, Oct, Feb); often cultivated, sometimes found wild in disturbed vegetation. *Patate douce* (Créole and French), *sweet potato* (English).

Ipomoea batatoides Choisy Pl. 45e

Lianas. Stems herbaceous, becoming woody with age, glabrous. Leaves: blades broadly ovate to ± orbicular, 3–10 cm long, glabrous or more commonly obscurely puberulent abaxially, with microscopic glands abaxially, the base shallowly cordate to truncate, the apex acuminate. Inflorescences 1- to several-flowered, in axillary or terminal cymes or thyrses. Flowers: sepals ± equal, elliptic to ± orbicular, cochleate, 6–9 mm long, ± coriaceous; corollas funnelform, 4–5 cm long, glabrous, pale pink or white. Fruits capsular, broadly ovoid, 10–12 mm long, the apex attenuate to base of persistent style. Seeds oblong, 1–4, woolly. Fl (Jan, Apr, May); non-flooded forest.

Ipomoea phyllomega (Vell.) House

Lianas. Stems woody basally, herbaceous toward apex, glabrate. Leaves: blades rounded-cordate, 8–20 × 8–20 cm, glabrate or pilose along veins abaxially, the base cordate, the apex acute or cuspidate-acuminate. Inflorescences compound-cymose or paniculate clusters. Flowers: sepals ± equal, broadly ovate, 15–18 mm long, pilose or glabrescent, often colored purplish after drying, the apex obtuse to ± acute; corolla funnelform, 5–8 cm long, glabrous, purple, the throat darker; stamens included. Fruits capsular, ± globose, 1 cm diam., glabrous. Seeds 1–4, oblong, woolly. Fl (Feb), fr (May); non-flooded forest, often in disturbed areas.

Ipomoea quamoclit L. Pl. 46a

Vines. Stems slender, herbaceous, annual, glabrous. Leaves: blades ovate to elliptic in outline, 1–9 cm long, deeply pinnatifid with 9–10 alternate to opposite pairs of linear lobes, glabrous. Inflorescences solitary or cymes with 2–6 flowers. Flowers: sepals elliptic to oblong, 4–7 mm long, glabrous, the apex obtuse, with 0.25–0.75 mm long mucro, glabrous; corolla salverform, 2–3 cm long, glabrous, scarlet, red, or rarely white; stamens exserted. Fruits capsular, ovoid, 6–9 mm wide, glabrous. Seeds dark to black, with dark patches of short trichomes scattered somewhat irregularly. Fl (Feb, possibly flowers year round); cultivated as an ornamental and escaped in disturbed areas.

Ipomoea squamosa Choisy Pl. 46c

Vines. Stems herbaceous toward apex, suffrutescent at base, glabrous. Leaves: blades variable in shape but usually ± sagittate in outline, 7–8.5 × 7–9 cm, the base cordate or sagittate to hastate, the apex long-acuminate, glabrous to pubescent on both surfaces. Inflorescences simple to 10-flowered cymes; peduncles equalling or longer than petioles. Flowers: sepals unequal, ovate to ± orbicular, the outer 3–5 mm long, the inner 5–10 mm long, glabrous, the apex rounded to truncate, the margins scarious; corollas funnelform, (4)5–8 cm long, glabrous, blue, pink to purple; stamens included. Fruits capsular, ovoid, 10–15 mm long, glabrous, brown. Seeds 1–4, long brown woolly. Fr (Jun, Sep); non-flooded forest and secondary vegetation.

MARIPA Aubl.

Lianas. Larger stems fluted, the younger stems often angled. Leaves: petioles canaliculate, glabrous to glabrescent or occasionally with stellate trichomes; blades mostly elliptic, ovate, or obovate to oblong. Inflorescences terminal, on lateral branches, or axillary, racemose or paniculate-thyrsiform, multiple-flowered, the branches usually pubescent. Flowers often fragrant; sepals quincuncial, ovate, oblong, broadly ovate to rotund, the inner often emarginate, the outer acute to rounded, equal to ± equal in length, the inner usually wider than long, mostly coriaceous with membranaceous margins, ciliate, the pubescence appressed, dibrachiate, stellate or peltate; corolla funnelform, campanulate-funnelform, cylindric-funnelform or campanulate, medium-small (10 mm) to large (60 mm), white to violet or pinkish, the lobes shallow, occasionally deep, mostly rounded, appressed pubescent on interplicae, glabrous on plicae; stamens mostly included, the anthers basifixed; ovary rotund, cylindric to conic, mostly glabrous, the apex occasionally pubescent, the style entire, rarely divided for one-half to one-third length or less, mostly glabrous, occasionally pubescent at apex, the stigma capitate, the lobes 2, closely appressed if style entire, or free, one lobe often slightly below the other. Fruits nut-like, rounded to ellipsoid, the pericarp ligneous, firm to hard, 1–4-seeded through abortion. Seeds glabrous, ovoid to oblong-ellipsoid, rounded, flattened to trigonous if more than one per fruit.

1. Leaves with minute, sunken oil glands adaxially (best seen with 10× magnification). Outer sepals glabrous or with peltate scales; corolla 15–23 mm long. . . . . *M. glabra.*
1. Leaves without oil glands adaxially. Outer sepals with dense cover of trichomes (becoming less obvious with age in fruiting specimens); corolla 15–35 mm long. . . . . *M. scandens.*

Maripa glabra Choisy FIG. 89, PL. 46b

Stems terete when mature, compressed when young. Leaves: blades elliptic to ovate-elliptic, 8–17 × 3–10 cm, coriaceous, with numerous, minute, sunken oil glands adaxially (best seen with 10× magnification), especially prominent near base, the base obtuse, the apex acuminate. Inflorescences terminal, paniculate-thyrsiform, cylindric or ± corymbose; pedicels 3–4(7) mm long. Flowers: sepals ± orbicular, 4–5 × 4–6 mm, the outer lobes glabrous, the inner lobes appressed pubescent abaxially; corollas cylindric-funnelform, 15–23 mm long, white, the lobes rotund at apex; ovary conic, 2 × 1.5 mm, glabrous. Fruits long-cylindric to ellipsoid, 25–27 × 12–18 mm, smooth-rugose to slightly striate, brown, the persistent sepals loosely appressed. Seeds 1, oblong to oblong-ellipsoid or oblong-cylindric to cylindric, 20 × 10–15 mm. Fl (Jul, Aug, Sep, Oct), fr (Mar, May, Nov); common, non-flooded forest.

Maripa scandens Aubl. PL. 46e

Stems terete, brown- to gray-black, often with white lenticels. Leaves: blades broadly ovate to ovate-oblong, (6)10–18 × 4–9 cm, coriaceous, glabrescent, the base cordate, rounded or shortly attenuate, the apex obtuse, acute or acuminate. Inflorescences terminal, paniculate-thyrsiform, yellowish- to whitish-pubescent; pedicels 2.5–8 mm long. Flowers: sepals ovate, ± orbicular, to widely ovate, 5–9 × 4–14 mm, coriaceous, both the inner and outer sepals with appressed dibrachiate trichomes dense enough to impart yellowish or whitish color; corollas funnelform to cylindric-funnelform, 15–35 mm long, white to violet or pink, the lobes rotund; ovary ovoid to conic, 2–5 × 1.5–2 mm, pubescent on upper half to one-third, glabrescent. Fruits ellipsoid to ellipsoid-obovoid, 20–30 × 13–20 mm, smooth to slightly striate, dark brown, often grayish tinged, the persistent calyx tightly appressed around base of fruit. Seeds 1(2), ovoid to ovoid-ellipsoid, 15–22 × 10–15 mm wide. Fl (Jan, Nov, Dec), fr (Feb, May); non-flooded forest.

MERREMIA Dennst.

Vines. Stems twining, glabrous or pubescent. Leaves: blades entire, palmately lobed, or palmately compound with 3–7 leaflets, glabrous or pubescent. Inflorescences axillary, in 1- to few-flowered dichasia; bracts linear or lanceolate; pedicels often 1–3 cm long. Flowers: sepals ± equal, oblong to elliptic, glabrous or pubescent, the apex acute to obtuse; corolla campanulate, white or yellow, some with purple center, rarely other colors; filaments equal or ± equal, mostly glabrous at base, the anthers spirally twisted with complete dehiscence, the pollen 3-colpate or rarely panto-colpate; ovary usually glabrous, 2–3-carpellate, 4–6-ovulate, the style filiform, the stigma globose to 2-globose, included. Fruits capsular, 2–4-celled, longitudinally dehiscent by 4–6 valves, the pericarp thin and fragile. Seeds 1–4(6), glabrous or pubescent.

1. Stems pubescent with hirsute yellowish trichomes, glabrescent. Leaves lobed, cleft, or parted, rarely divided to base; leaflets with sinuate, serrate, or incised margins. . . . . *M. dissecta.*
1. Stems entirely glabrous. Leaves palmately divided to base, leaflets with entire margins.. . . . . *M. macrocalyx.*

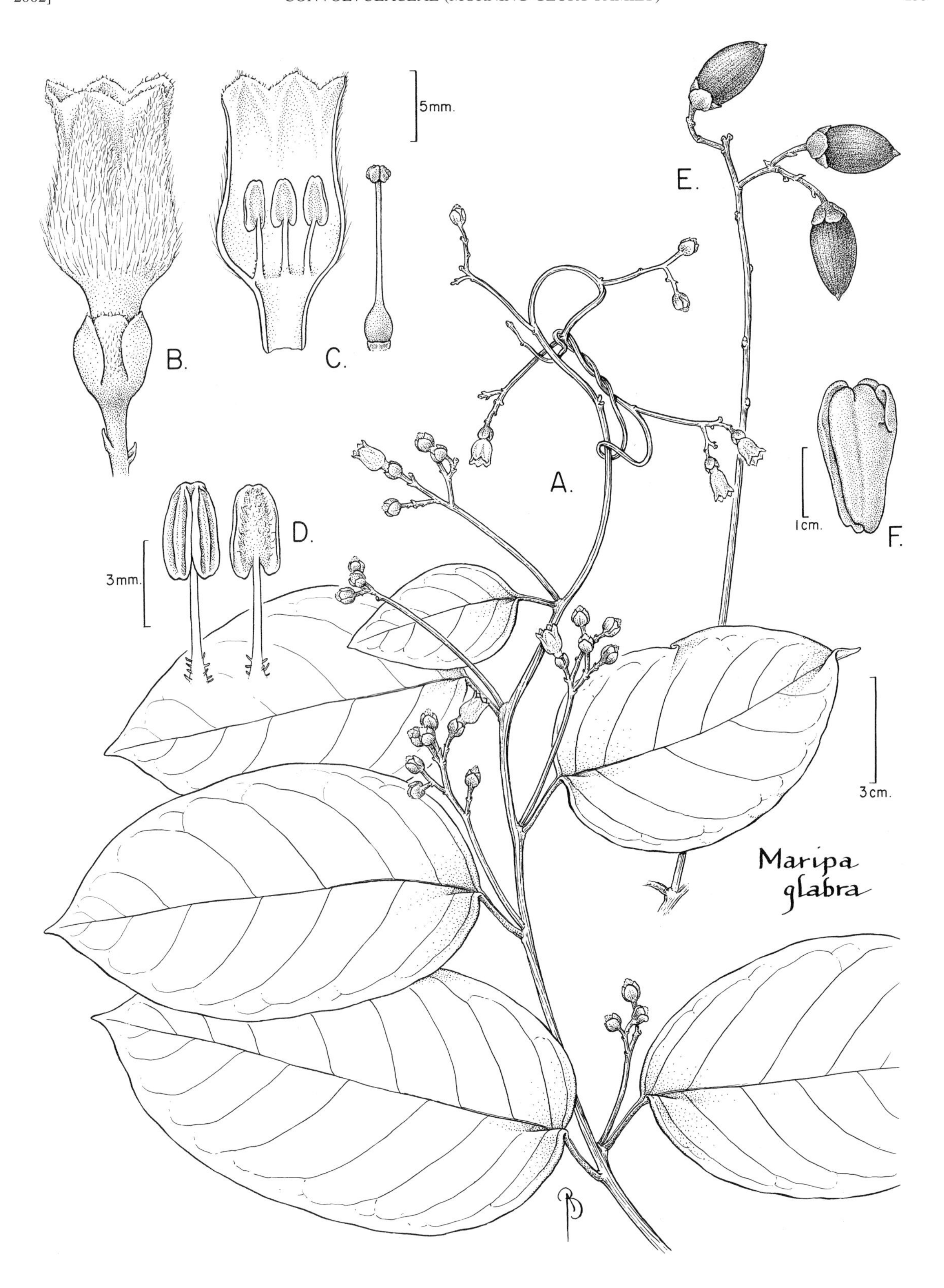

FIG. 89. CONVOLVULACEAE. *Maripa glabra* (A–D, *Mori et al. 21999*; E, F, *Mori & Pennington 18142*). **A.** Part of stem with leaves and inflorescences. **B.** Lateral view of flower. **C.** Medial section of corolla with adnate stamens (left) and lateral view of pistil (right). **D.** Abaxial (left) and adaxial (right) views of stamens. **E.** Part of stem with infructescence. **F.** Cotyledons.

Merremia dissecta (Jacq.) Hallier f.

Stems perennial but herbaceous toward tips, 3–4 m long, rough, hispid to hirsute with simple, erect yellowish trichomes reaching 3 mm long, glabrescent. Leaves: blades palmately lobed, cleft, or parted nearly to base, rarely divided to base, the lobes (4)5–7, lanceolate to elliptic, glabrous, the base acute to attenuate, the apex usually attenuate or sometimes acute, obtuse or rounded, the margins mostly sinuate, serrate, or incised, the median leaflet 9.5–10 × 4–4.5 cm. Inflorescences axillary, cymose, the flowers 1 to many; pedicels clavate, 1.2–1.5 cm long, glabrous; bracts scale-like, <2 mm long, fugacious. Flowers: sepals unequal, ovate-lanceolate, the base rounded to truncate, the apex obtuse to acute, mucronate, the margins entire, membranous, the outer larger, 1.7–2.3 × 0.9–1.1 cm, the inner 1.6–2.1 cm long; corolla campanulate, 3–4 × 3–4 cm, with wine-red, purplish, or brownish throat. Fruits papery, ± globose capsules, 2-locular, dehiscing by irregular valves, surrounded by chartaceous calyx. Seeds often 4, ± globose, 6–8 mm diam., glabrous, black, often with almond taste, the hilum circular. Fl (Jun, Oct), fr (Jun, Oct, Nov, Dec); weed near village and sometimes in cultivation for medicinal properties of seed, at least in other areas. *Pâte d'amande* (Créole).

Merremia macrocalyx (Ruiz & Pav.) O'Donell [Syn.: *Merremia glaber* (Aubl.) Hallier f.] PL. 46d

Stems twining or less often trailing, 5–6 m or more long, glabrous or densely and finely appressed puberulent. Leaves palmately compound; petioles 2–8 cm long; leaflet blades lanceolate to oblong, chartaceous, glabrous or densely appressed pubescent on abaxial side or along main veins, the apex acute to obtuse, the margins entire. Inflorescences compound cymes, often with 10–20 flowers; pedicels 1–1.5 cm long, usually glabrous; bracts lanceolate, 9–10 mm long. Flowers: sepals ± equal or the inner slightly longer, oblong, obtuse to acute, papery, glabrous, the outer 20–25 × 5–7 mm, the inner 2.5–2.9 cm long, the margins scarious; corolla campanulate, 4.5–5 cm long, glabrous, white. Fruits capsular, 4-locular, 4-valvate, depressed-globose, 12–13 mm wide, glabrous, subtended by persistent sepals. Seeds often 4, rotund, 3–4 mm diam., finely pubescent with short trichomes. Fl (Jul, Aug, Sep, Oct), fr (Sep); in disturbed vegetation.

CUCURBITACEAE (Squash Family)

Michael Nee

Herbaceous or soft-woody vines, annuals, or perennials. Indumentum often of spiculate hairs or mineral plaques. Stipules absent. Tendrils almost always present, simple or branched, lateral to petioles. Leaves simple or palmately lobed or compound, often with variable lobing within a single plant, usually cordate at base, usually somewhat fleshy, alternate. Inflorescences axillary, the staminate and pistillate flowers in different axils or different inflorescences in same axil, rarely mixed in single inflorescence. Flowers unisexual (plants monoecious or dioecious, often serially monoecious and plants therefore appearing dioecious); hypanthium present; calyx and corolla generally 5-merous; stamens basically 5, but often reduced to three anthers, one monothecous and the other two dithecous, or anthers connate into one complex structure; ovary inferior, 3-carpellate or reduced to 1–2 carpels, the style 1; placentation axile or parietal, the ovules usually numerous. Fruits dry or fleshy, capsules, or juicy or coriaceous berries (pepos), armed or smooth. Seeds 1-numerous; cotyledons usually large.

The following species have been cultivated for their edible fruits in gardens at Saül or for their use as sponges: *Cucurbita moschata* (Lam.) Poir. (pumpkin), *Luffa acutangula* (L.) Roxb. (loofah sponge of bathrooms).

Jeffrey, C. 1984. Cucurbitaceae. *In* A. L. Stoffers & J. C. Lindeman (eds.), Flora of Suriname **V(1):** 457–518. E. J. Brill, Leiden.

Kearns, D. M. 1998. Cucurbitaceae. *In* J. A. Steyermark, P. E. Berry & B. K. Holst (gen. eds.), Flora of the Venezuelan Guayana **4:** 431–461. P. E. Berry, B. K. Holst & K. Yatskievych (vol. eds.). Missouri Botanical Garden Press, St. Louis.

Key based on flowering material

1. Flowers bright yellow, orange, red, or pink.
    2. Sepals green. Petals bright red. . . . . *Psiguria.*
    2. Sepals green or bright orange. Petals yellow.
        3. Sepals green, inconspicuous and much shorter than petals.
            4. Leaves only shallowly lobed. Ovary smooth. . . . . *Melothria.*
            4. Leaves deeply 3–5(7)-lobed, the lobes dentate. Ovary tuberculate. . . . . *Momordica.*
        3. Sepals bright orange, longer than and more conspicuous than petals. . . . . *Gurania.*
1. Flowers greenish white.
    5. Staminate flowers many, in long-pedunculate racemes. Anthers 2, free. . . . . *Helmontia.*
    5. Staminate flowers few, fasciculate or in short-pedunculate racemes. Anthers 3, coherent in a head or free.
        6. Pistillate flowers short-pedicellate; anthers coherent in a head. . . . . *Cayaponia.*
        6. Pistillate flowers long-pedicellate, the pedicel to 8 cm long in fruit; anthers free. . . . . *Selysia.*

Key based on fruiting material

1. Fruits fusiform, tuberculate, dehiscent in three fleshy valves.. . . . . *Momordica.*
1. Fruits globose, oblong, or flattened-cylindric, smooth, indehiscent or eventually irregularly breaking open.

2. Fruits oblong, slightly flattened, 2-carpellate, the seeds loose in pulpy single locule at maturity. .... *Gurania, Helmontia, Psiguria.*
2. Fruits globose or nearly so, 3-carpellate, the seeds loose in a single locule or embedded in still intact 3-locular structure at maturity.
   3. Fruits soft, 0.8–2 cm diam. .... *Melothria.*
   3. Fruits with coriaceous rind, 1.1–4.5 cm diam.
      4. Fruits usually several per axil, on pedicels ≤2 cm long. .... *Cayaponia.*
      4. Fruits single, born on a pedicel 8 cm long. .... *Selysia.*

CAYAPONIA Silva Manso

Perennial vines. Tendrils simple or 2–5-branched. Leaves simple or palmately lobed to 3–5-foliolate, usually with disc-shaped glands at base of blades abaxially. Plants probably monoecious. Flowers greenish white; corolla campanulate or rotate, 5-lobed. Staminate flowers solitary to paniculate; stamens 3, one anther monothecous, the other two dithecous, usually connate into convoluted head. Pistillate flowers of solitary flowers or in racemes, often of much different size and shape than staminate; ovary 3-carpellate, or reduced to fewer carpels, the stigmas 3. Fruits globose or ellipsoid fleshy, indehiscent coriaceous berries. Seeds 1–30.

1. Leaves trifoliolate. .... *C. rigida.*
1. Leaves simple.
   2. Fruits densely clustered, mixed with prominent foliaceous bracts 15–65 mm long. .... *C. ophthalmica.*
   2. Fruits widely separated, with or without foliaceous bracts.
      3. Hypanthium of staminate flowers tubular, 20–30 × 15–20 mm, densely pilose abaxially. Fruits globose, hard-coriaceous berries, 4.5 cm diam. .... *C. jenmanii.*
      3. Hypanthium of staminate flowers obconic-campanulate, 2.5–4 × 3–3.5 mm, glabrous abaxially, pilose adaxially. Fruits ellipsoid, coriaceous berries, 1.1–2.1 × 0.7–1.2 cm. .... *C. racemosa.*

Cayaponia jenmanii C. Jeffrey — FIG. 90

Vines, drying blackish, nearly glabrous. Leaves: petioles 1–2.5 cm long, pubescent; blades trilobed, 7–13 × 9–14 cm, glabrous adaxially except along main veins, sparsely pubescent abaxially, the lobes apiculate, the base deeply cordate. Staminate inflorescences axillary racemes, to 16 cm long; pedicels 1 cm long, pilose, the foliaceous bracts to 3 cm long. Staminate flowers: hypanthium tubular, 2–3 × 1.5–2 cm, densely pilose abaxially; sepals lanceolate, 12 × 4 mm, pubescent, green; corolla 1.5 cm long, densely puberulent abaxially, white; filaments white, the anthers cream-colored, coherent in fusiform head 9 × 4.5 mm. Pistillate flowers not known. Fruits hard-coriaceous, globose berries, 4.5 cm diam., yellow, the pedicel stout, 2 cm long. Seeds 2, with raised margin, 2 cm diam., gray-brown, surrounded by slimy, bright yellow pulp. Fl (Feb), fr (Aug); in forest.

The description is from Saül plants which may represent a new species; it is somewhat uncertain whether the fruiting specimen is the same species as the flowering ones. The flowers differ somewhat from those of typical *C. jenmanii*, especially in the dense pilosity, in which they resemble those of *C. tessmannii* Harms, a similar, but distinct, species from lowland Peru.

Cayaponia ophthalmica R. E. Schult. — PL. 47a

Robust, soft-woody vines, minutely puberulent to nearly glabrous. Tendrils robust, 4–5-branched. Leaves: petioles 5–15 cm long; blades widely ovate-cordate, 11–28 × 20–35 mm, palmately 5–7-lobed, subcoriaceous, the base deeply cordate. Inflorescences axillary, crowded fascicles; bracts prominent, 15–65 mm long. Staminate hypanthium, 1.8–2 cm long, greenish; sepals yellowish, triangular, 6–12 × 4–5 mm; corolla white, the petals 10 × 6 mm; anthers connate in head 5 mm long. Pistillate flowers 4–7 mm long. Fruits densely clustered, obovoid, 3 × 2.5–2.8 cm, smooth, glabrous, turning red when ripe. Seeds 3, 15–19 × 12.5 × 6 mm. Fr (Feb, Sep); in forest. *Concombre, espèce de concombre* (Créole).

Cayaponia racemosa (Mill.) Cogn.

Herbaceous vines, essentially glabrous. Tendrils 2–3-branched. Leaves very variable on same stem; blades widely ovate, usually 3–5-lobed on lower stems to reduced, rhombic, entire on flowering portion, 4.5–20 × 2.5–25 cm, the base usually abruptly narrowed onto winged petiole. Plants monoecious. Inflorescences: flowering nodes with single pistillate and a few-flowered raceme of staminate flowers, usually with very reduced or bracteal leaves, the whole simulating a panicle. Pistillate inflorescences: pedicels 3–7 mm long. Staminate flowers: hypanthium obconic-campanulate, 2.5–4 × 3–3.5 mm, pilose adaxially, pale green; sepals triangular, 0.5–1.5 mm; petals oblong, 4.5–10 × 2.5–3.5 mm, spreading, pilose adaxially, greenish white; filaments free, 2.8 mm long, the anthers coherent in oblong head. Fruits coriaceous berries, ellipsoid, 11–21 × 7–12 mm, green, turning yellow-orange. Seeds 2–3, 3–11 × 2.2–6.5 mm. Fl (Jul), fr (Sep); common, in secondary vegetation.

Cayaponia rigida (Cogn.) Cogn.

Vines, glabrous. Leaves trifoliolate; petioles 3.5–8.5 cm long; petiolules 5–6 mm long; leaflets elliptic, 9–17 × 3–6 cm, coriaceous, the lateral pair slightly smaller and asymmetric, the base abruptly attenuate, the apex attenuate. Inflorescences of several racemes to 25 cm long, arranged paniculately, the flowers sessile, in clusters of up to 6. Staminate flowers: hypanthium cylindric, 7 mm long; sepals triangular, 2 mm long, the apex acute; corolla lobes ca. 8 mm long, greenish white, pubescent adaxially, greenish white;

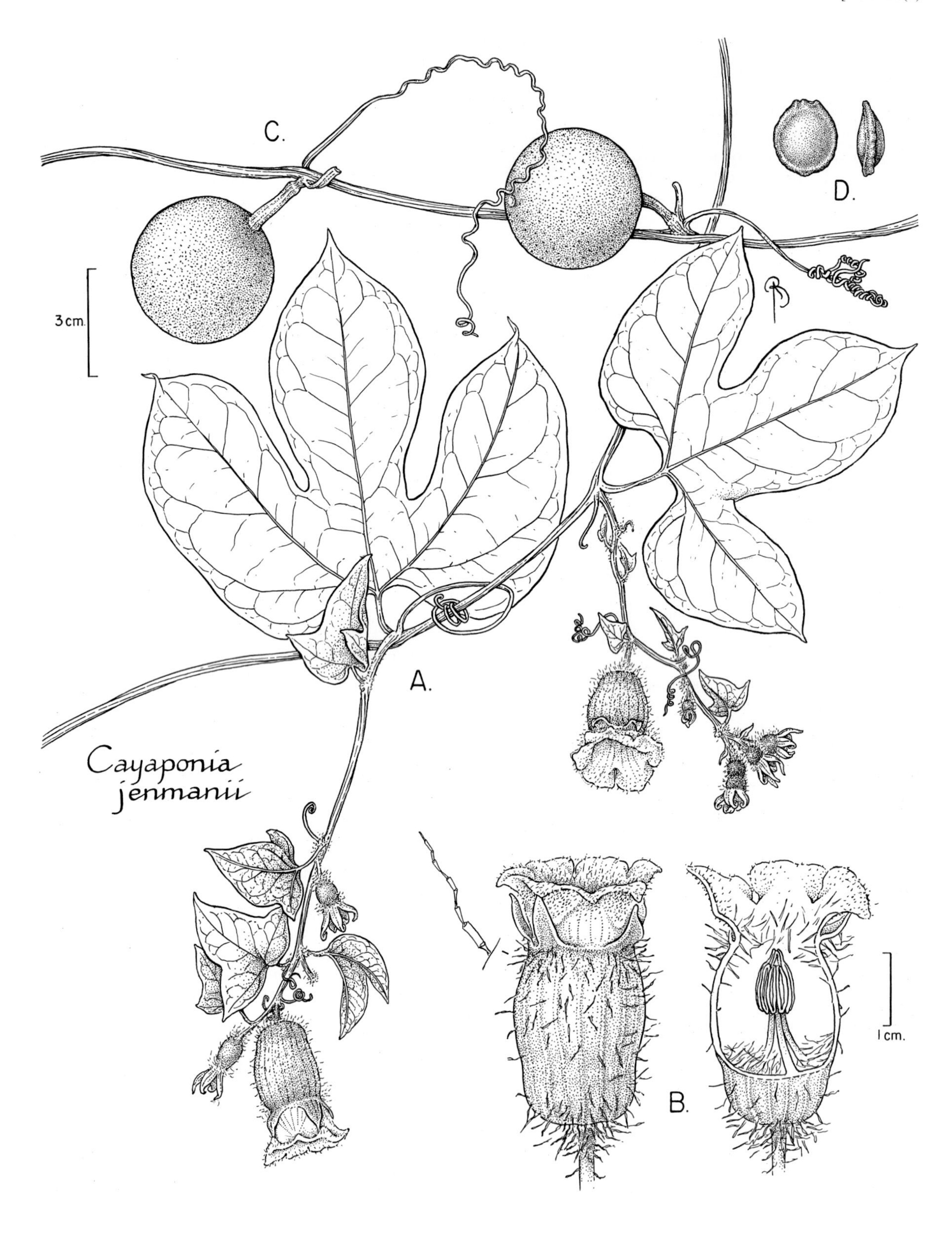

FIG. 90. CUCURBITACEAE. *Cayaponia jenmanii* (A, B, *Mori & Gracie 21086*; C, D, *Mori et al. 23085*). **A.** Part of stem with leaves, tendrils, and staminate and pistillate flowers. **B.** Lateral view (left) with detail of hair (far left) and staminate flower with perianth cut away to show connivant anthers (right). **C.** Part of stem with tendrils and fruits. **D.** Two views of seed.

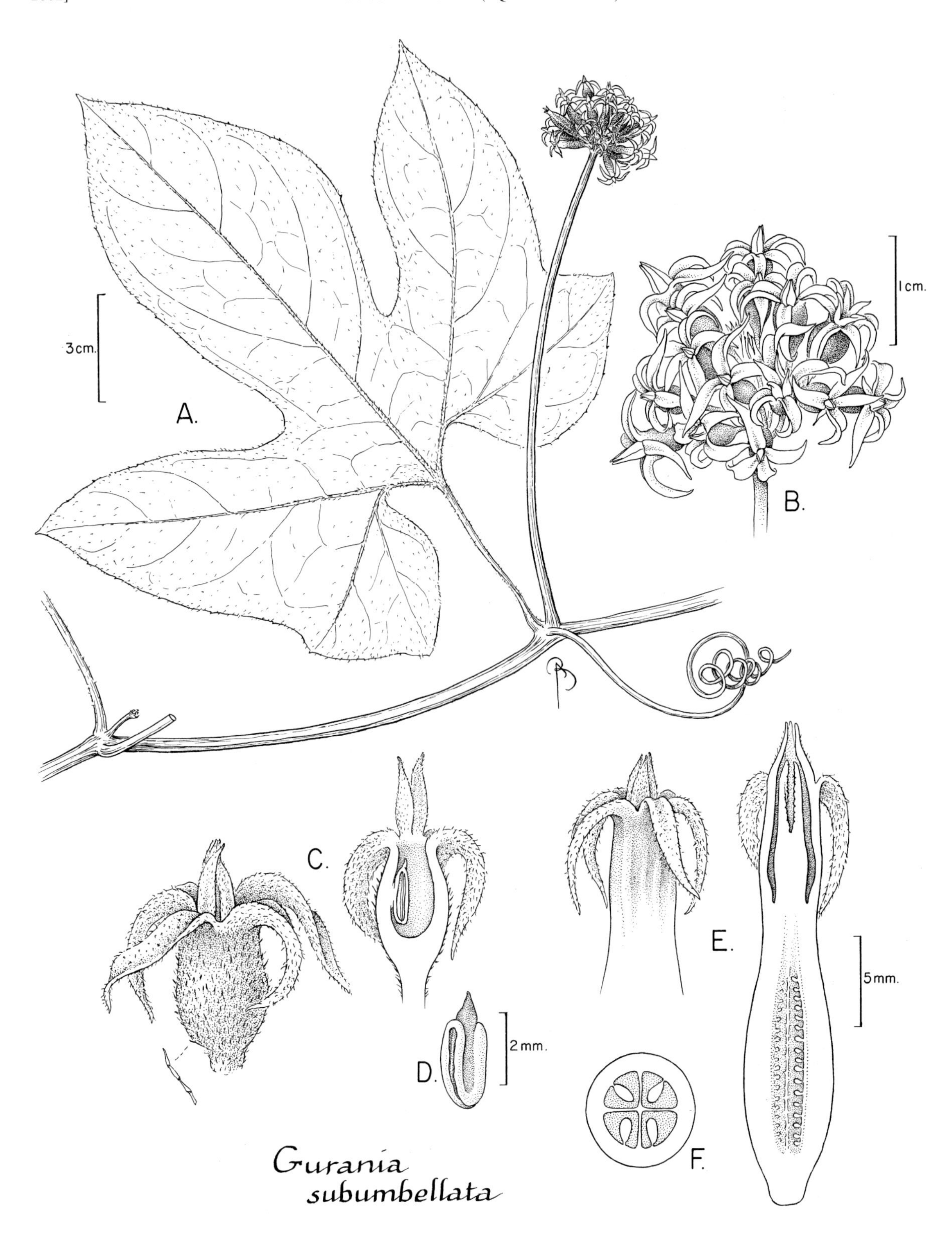

FIG. 91. CUCURBITACEAE. *Gurania subumbellata* (*Mori & Gracie 21208*). **A.** Part of stem with leaf, pistillate inflorescence, and simple tendril; note placement of tendril 90 from leaf. **B.** Staminate inflorescence with calyx more conspicuous than corolla. **C.** Lateral view (left) with detail of multicellular hair (far left) and medial section (right) of staminate flower. **D.** Anther with connective. **E.** Apex (left) and medial section (right) of pistillate flower with inferior ovary. **F.** Transverse section of ovary showing four locules and parietal placentation.

filaments free, the anthers connate in head 2 mm long. Pistillate flowers unknown. Fruits globose berries, 1–1.5 (or more ?) cm diam. Fr (Feb, Apr); in forest.

Collections from Saül are infected by galls on the stems which can easily be mistaken for fruits. The leaves are similar to those of *Gurania bignoniacea*, but lack the pilosity of that species.

GURANIA (Schltdl.) Cogn.

Perennial vines or soft-woody lianas. Tendrils simple. Leaves simple and entire to palmately lobed, or trifoliolate. Staminate inflorescences of fascicled to racemose or subumbellate flowers; peduncles present. Pistillate inflorescences of 1–4 flowers in axils of reduced or absent leaves, pendulous, panicle-like. Plants probably monoecious. Staminate flowers: hypanthium urceolate to cylindric, bright yellow or orange, usually larger in pistillate flowers; sepals 5, usually much longer than petals, bright yellow or orange; petals 5, bright yellow; stamens 2, dithecous, inserted in hypanthium tube. Pistillate flowers like staminate flowers but usually larger; ovary 2-carpellate. Fruits fleshy, oblong-elliptic and slightly flattened berries. Seeds oblong-elliptic, loose in watery pulp.

1. Staminate flowers umbellate, the sepals recurved. . . . . *G. subumbellata.*
1. Staminate flowers in dense racemes or subumbellate, the sepals erect or spreading.
    2. Plants sparsely pilose, the hairs white, 2–3.5 mm long. Staminate flowers with hypanthium urceolate, 4–6 mm long; sepals narrowly triangular, 1.2–5.2 mm long. . . . . *G. bignoniacea.*
    2. Plants nearly glabrous to puberulent. Staminate flowers with hypanthium nearly tubular, 3–17 mm long; sepals linear to lanceolate, 10–25 mm long.
        3. Plants appearing nearly glabrous, minutely puberulent, the hairs scarcely visible at 10x. Staminate flowers with hypanthium 8–17 long. . . . . *G. lobata.*
        3. Plants evidently puberulent. Staminate flowers with hypanthium 3–5 mm long. . . . . *G. reticulata.*

Gurania bignoniacea (Poepp. & Endl.) C. Jeffrey [Syn.: *Anguria bignoniacea* Poepp. & Endl.] PL. 47b

Herbaceous vines, sparsely pilose, the hairs white, 2–3.5 mm long. Leaves simple or trifoliolate; petioles 2–7 cm long; blades ovate to widely reniform, 6–23 × 8–22 cm, membranaceous, the base cordate. Staminate inflorescences densely racemose or subumbellate; peduncle 8–27 cm long; pedicels 0.8–1.5 mm long. Pistillate inflorescences fascicles of 1–4 flowers; pedicels 2–8 mm long. Staminate flowers: hypanthium urceolate, 4–6 × 3–5 mm, bright orange; sepals narrowly triangular, 1.2–5.2 mm, spreading; petals lanceolate, 1.7–6 mm long, yellow. Fruits ellipsoid, 3.5–4.5 × 2 cm, green. Seeds 10 × 5 mm. Fl (Feb, May); in forest openings and along forest edges.

Gurania lobata (L.) Pruski [Syns.: *Gurania spinulosa* (Poepp. & Endl.) Cogn., *Anguria spinulosa* Poepp. & Endl.] PL. 47e

Soft-woody lianas, climbing into small trees, minutely puberulent. Leaves: petioles 5–17 cm long; blades usually deeply 3(5)-lobed, ovate to widely reniform, 10–35 × 11–40 cm, pubescent abaxially. Staminate inflorescences densely racemose; peduncle 10–45 cm long; pedicels 2–26 mm long. Pistillate inflorescences fascicled, the flowers 2–11 per node, clustered in leafless terminal branches. Staminate flowers: hypanthium ± cylindrical, slightly constricted at throat, 8–17 mm long, yellow-orange; sepals lanceolate, spreading, 2.5 cm long, orange with green aristate tips; petals lanceolate, 2–3 mm long, yellow. Fruits ellipsoid-cylindric, 4–7 × 1.5–2.3 cm, green with whitish spots, turning orange. Seeds numerous, 5–9 × 3.5–5 mm, embedded in gelatinous pulp. Fl (Jan, Feb, Mar, Apr, Nov); along forest edges, in clearings, and in secondary vegetation.

Gurania reticulata Cogn. PL. 47c

Herbaceous vines, to 5 m long, densely to sparsely puberulent. Leaves simple; petioles 2–6 cm; blades unlobed, ovate-cordate, 13–23 × 9–19 cm, sparsely puberulent ad- and abaxially. Staminate inflorescences densely racemose; peduncle 5–12 cm long; pedicels to 5 mm long. Staminate flowers: hypanthium tubular, 3–5 mm long, sparsely pubescent, orange; sepals linear, 1–1.5 cm, orange; petals triangular, 1 mm long. Pistillate flowers not known. Fruits not known. Fl (Feb); in forest. *Concombre, kete-poule* (Créole).

Gurania subumbellata (Miq.) Cogn. [Syn.: *Anguria subumbellata* Miq.] FIG. 91, PL. 48a,c

Soft-wooded lianas, high-climbing, villous. Leaves usually 3–5-lobed; petioles 1.5–9 cm long; blades ovate, 11–25 × 8.5–26 cm, the base cordate. Staminate inflorescences of densely umbellate flowers; peduncle 5–40 cm; pedicels 8–40 mm long. Pistillate inflorescences fascicles at nodes with reduced or absent leaves. Staminate flowers: hypanthium urceolate, 4–7 × 2.5–4.5 mm, bright orange; sepals lanceolate, recurved, 5–10 mm long, bright orange; petals narrowly triangular, erect, 2–6 mm long, bright yellow. Fruits ovoid, 5 × 3 cm, glabrous, green. Seeds elliptic, flattened, 6.5 × 4 mm, light tan. Fl (Jan, Feb, Mar, Jun), fr (Aug); along forest edges and in clearings.

HELMONTIA Cogn.

Perennial herbaceous vines. Tendrils simple. Leaves simple, entire to lobed or trifoliolate.Plant dioecy or monoecy not known. Staminate inflorescences racemose or subumbellate. Pistillate flowers in the axils of young leaves. Staminate flowers: hypanthium obconic to cylindric;

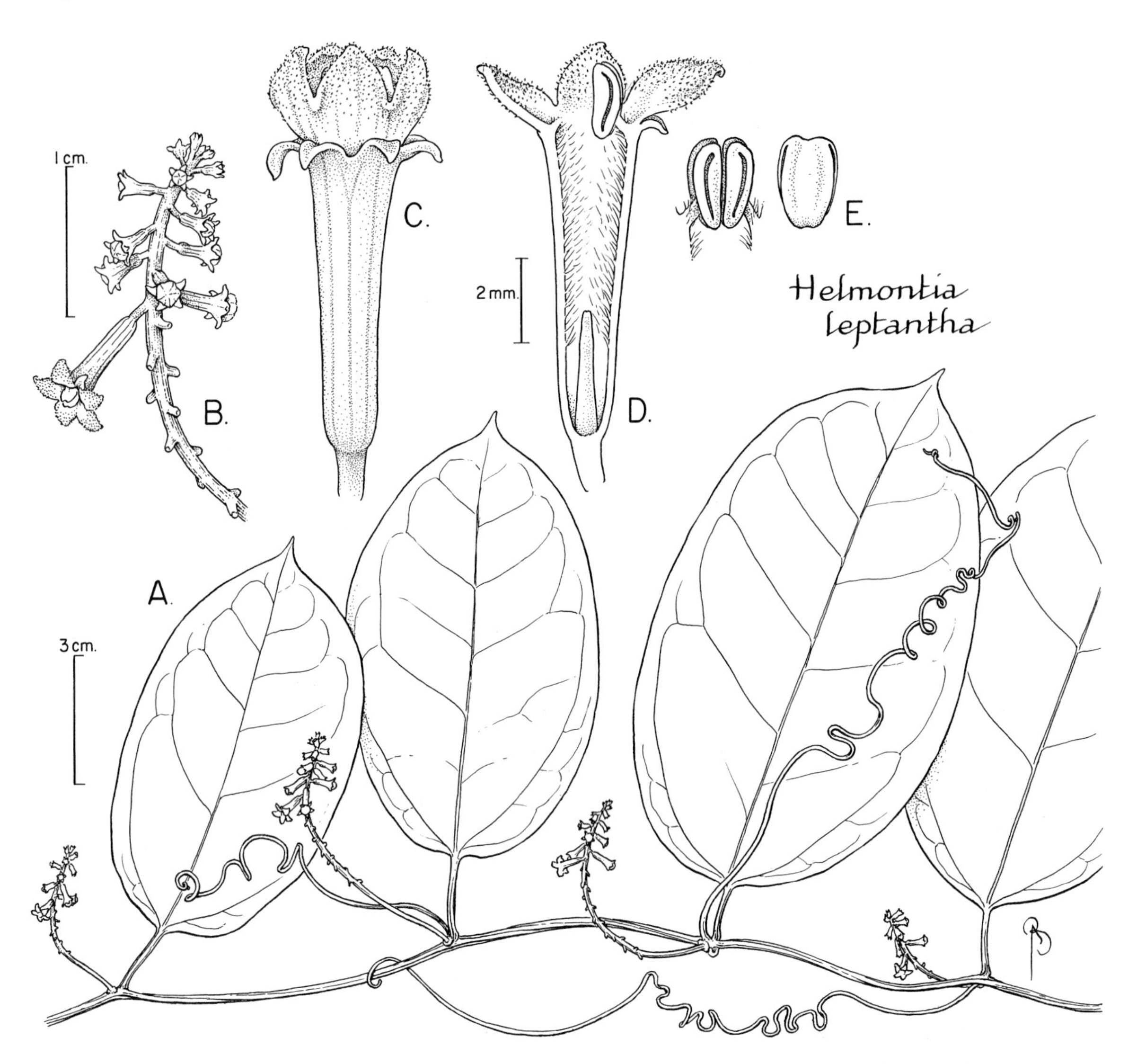

FIG. 92. CUCURBITACEAE. *Helmontia leptantha* (*Mori & Gracie 21111*). **A.** Part of stem with leaves, inflorescences of staminate flowers, and tendrils. **B.** Inflorescence of staminate flowers. **C.** Lateral view of staminate flower. **D.** Medial section of staminate flower with one stamen removed. **E.** Lateral view of pair of stamens (far left) and adaxial view of anther (near left).

sepals 5, small; corolla 5-parted or with 5 free petals; stamens 2, free, the anthers dithecous, exserted. Fruits fleshy, green berries. Seeds numerous, but mature ones poorly known.

1. Plants pilose. Leaves simple, the base deeply cordate. . . . . *H. cardiophylla*.
1. Plants glabrous or minutely puberulent. Leaves trifoliolate, or if simple, the base rounded. . . . . *H. leptantha*.

Helmontia cardiophylla Harms

Vines, pubescent. Tendrils simple. Leaves simple; petioles 2–3 cm long; blades ovate to narrowly triangular, 10–16 × 5–7 cm, the base deeply cordate, the apex attenuate. Staminate inflorescences of racemose flowers; peduncle 1.5–5 cm long, the rachis to 3.5 cm long, with >15 flowers; pedicels 4 mm long. Pistillate inflorescences of solitary flowers in axils; pedicels 7 mm long. Staminate flowers: hypanthium 5 × 0.8 mm, tubular, dilated at summit; sepals triangular, 1 mm long; petals 1.5 mm long, puberulent abaxially, white or greenish. Pistillate flowers: hypanthium tubular, wider than that of staminate flowers; ovary fusiform, 5 mm long, pilose; petals ovate, 2 × 2 mm. Fruits soft berries, pyriform, 2 cm long, beaked by persistent hypanthium. Seeds 6–8 per fruit, elliptic, flattened, 5–6 × 2 mm, white. Fl (Jul), fr (Jul, immature); in secondary vegetation and on rocky outcrops.

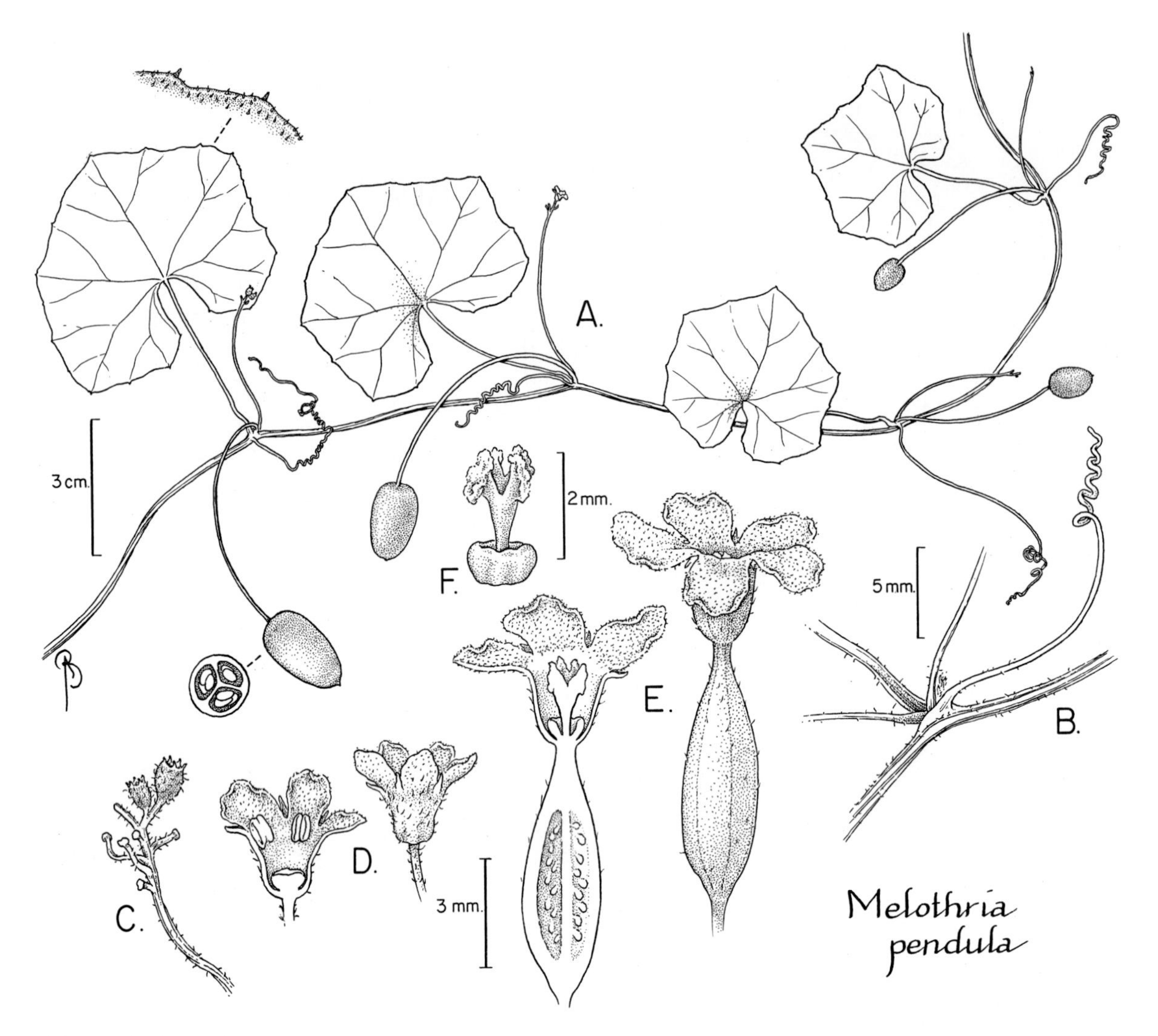

FIG. 93. CUCURBITACEAE. *Melothria pendula* (A, B, *Mori et al. 22352*; C, E, F, *Schunke 4946* from Peru; D, *Stergios 8628* from Venezuela). **A.** Part of stem with leaves and axillary fruits. **B.** Part of stem with tendril. **C.** Part of infructescence with immature fruits. **D.** Lateral view of staminate flower (right) and medial section of staminate flower (left). **E.** Lateral view of pistillate flower (right) and medial section of pistillate flower (left). **F.** Nectary surrounding style and stigma.

The description is taken from French Guiana material; the application of the name, based on a plant from Santarém, Brazil, is uncertain.

Helmontia leptantha (Schltdl.) Cogn. [Syn.: *Anguria leptantha* Schltdl.] FIG. 92, PL. 48f

Vines, glabrous or minutely puberulent. Tendrils simple. Leaves simple (or trifoliolate); petioles 1.5–2.3 cm long; blades elliptic, 13–19 × 6–7.5 cm, the base rounded, the apex apiculate. Staminate inflorescences racemose; peduncle 1.5–2.5 cm long, the rachis to 8 cm long, with 20–120 flowers; pedicels 1 mm long, scarcely differentiated from base of obconic hypanthium. Staminate flowers: hypanthium 5–15 mm long; sepals ovate, 0.5–1.5 mm long; petals ovate, 2–3 × 1–1.7 mm, creamy green. Fruits fleshy ovoid to ellipsoid berries, apparently yellow at maturity. Mature seeds not known. Fl (Jan, Feb, Mar, Jun, Nov); in forest and secondary vegetation.

MELOTHRIA L.

Delicate annual or perennial vines, monoecious. Tendrils simple. Leaves simple; blades entire to lobed, usually scabrous. Staminate inflorescences racemose or subumbellate; peduncles present. Pistillate inflorescences of solitary or few flowers. Flowers small. Staminate

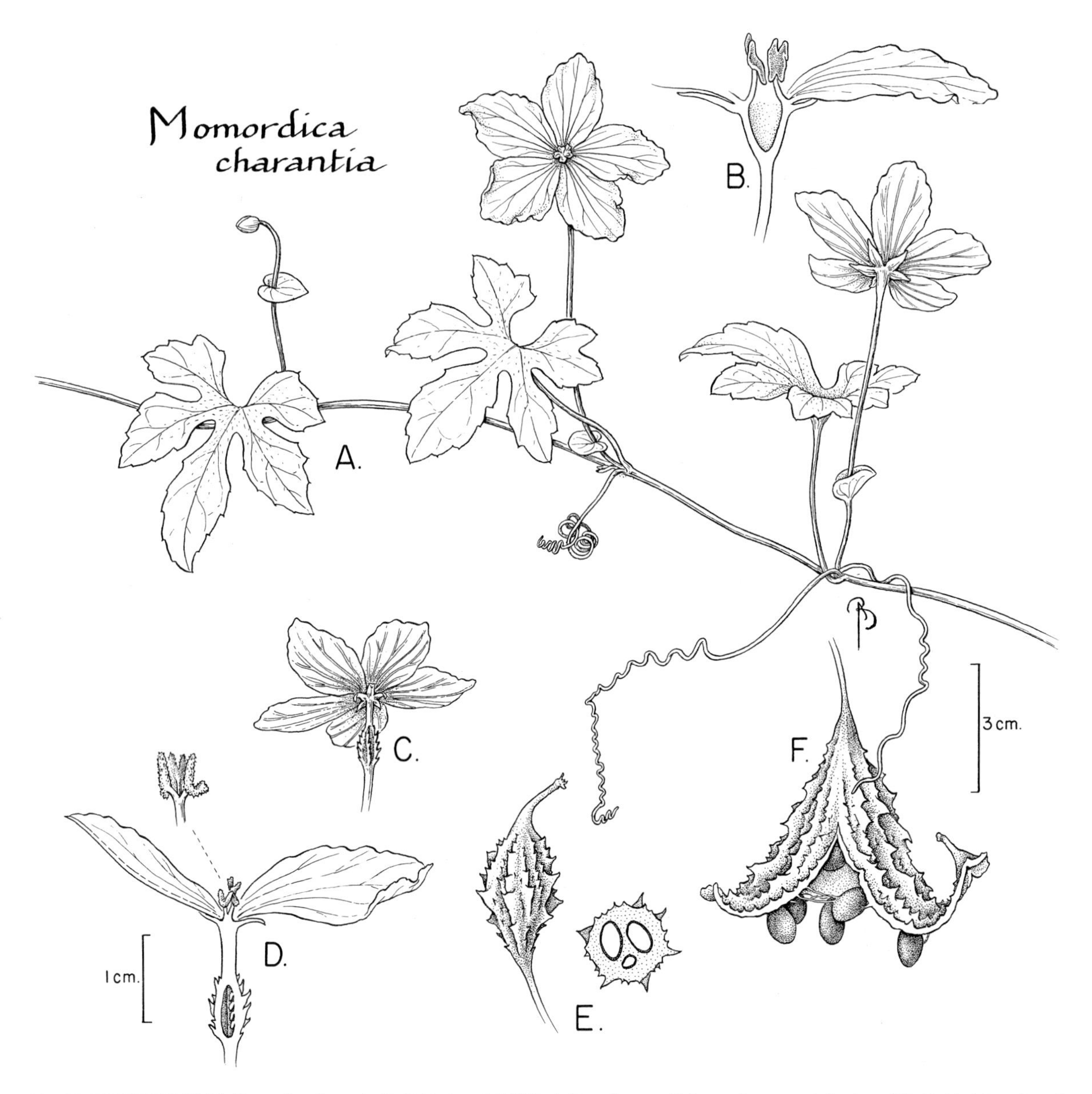

FIG. 94. CUCURBITACEAE. *Momordica charantia* (A–F, *Mori et al. 21055*). **A.** Part of stem with leaves, flowers, and simple tendrils. **B.** Medial section of staminate flower. **C.** Oblique abaxial view of pistillate flower. **D.** Medial section of pistillate flower with detail of stigmas. **E.** Whole (left) and transverse section (right) of immature fruit. **F.** Mature fruit.

flowers: corolla deeply 5-parted, yellow or white; stamens 3, free, one monothecous, the other two dithecous. Pistillate flowers: ovary 3-carpellate, the ovules numerous, the stigmas 3. Fruits small, soft, juicy, globose or ellipsoid berries. Seeds flattened, pilose.

Melothria pendula L. [Syns.: *Melothria fluminensis* Gardner, *Melothria guadalupensis* (Spreng.) Cogn.] FIG. 93, PL. 48b

Delicate vines, to 4 m long climbing over low vegetation, glabrous to hispid. Leaves: petioles 1.5–5 cm long; blades ovate to ovate-triangular, usually shallowly 3-lobed or sagittately 3-lobed, 5–10 × 4–8.5 cm, usually scabrous-pubescent abaxially, the base deeply cordate. Inflorescences axillary, both sexes often found at same node. Staminate inflorescences dense racemes or subumbellate, with 5–15 flowers; peduncle 0.5–2.5 cm long; pedicels 2–5 mm long. Pistillate inflorescences with 1–2 flowers per node; pedicels

0.5–3 cm long. Staminate flowers: hypanthium narrowly campanulate, 1.5–2.5 mm long; sepals 0.2–0.5 mm long; corolla yellow, the lobes 1.5 mm long; anthers 1.5 mm long, sessile in corolla throat. Fruits subglobose or ellipsoid, juicy berries, 0.8–2 × 0.6–1.2 cm, mottled green, then dark purple, almost black at maturity. Seeds 2–4 × 2–2.5 mm. Fl (Feb, May, Jul, Aug, Sep), fr (Feb, May, Jul, Aug, Sep); in disturbed vegetation.

MOMORDICA L.

Delicate vines, probably annuals. Tendrils simple. Leaves 3–7-lobed. Plants monoecious. Staminate inflorescences of solitary flowers, racemose, or corymbose. Pistillate inflorescences of solitary flowers. Staminate flowers: corolla of 5 free petals, yellow; stamens 3, the filaments free, the anthers flexuous, connate, one monothecous, the other two dithecous. Pistillate flowers: stigmas 3, bilobed. Fruits fusiform or ellipsoid fleshy capsules, smooth, sculptured, or tuberculate.

Momordica charantia L.
FIG. 94, PL. 48d,e; PART 1: FIG. 11

Much-branched delicate herbs, forming masses over shrubs or along ground in open areas, pubescent or glabrous. Leaves: petioles 1.5–4 cm long; blades deeply 3–5(7)-lobed, 4–11 × 4–11 cm, the margins dentate. Inflorescences of solitary flowers, axillary. Staminate inflorescences: peduncles 4.5–8.5 cm long, with foliaceous bract 3–6 mm wide, borne 1–3.5 cm above base. Staminate flowers: corolla 7–10 mm long, yellow; anthers connate into head 2.5–3.5 mm long. Pistillate flowers: ovary fusiform, beaked, tuberculate. Fruits fusiform, 4 cm long, fleshy, tuberculate, bright orange, opening explosively into 3 recurved valves to reveal seeds embedded in bright red pulp. Seeds oblong, 8–10 × 4–6 mm, sculptured, black. Fl (Jun, Jul, Sep), fr (Jul, Sep); in disturbed or early secondary vegetation. *Noyau sauvage*, *sorossi* (Créole).

PSIGURIA Arn.

High-climbing, soft-wooded lianas. Tendrils simple. Leaves very variable between and within species, entire to lobed or 3–5-foliolate. Plants said to be dioecious, but probably monoecious. Staminate inflorescences racemose or spicate; peduncle long. Pistillate inflorescences axillary, with 1–3 flowers, frequently on leafless branches and then simulating panicle. Staminate flowers: hypanthium cylindrical or urceolate; sepals short, green; petals 5, free, bright orange, red or rose; stamens 2, free, included in hypanthium, the anthers dithecous. Pistillate flowers: ovary cylindric, 2-carpellate, the ovules numerous. Fruits slightly flattened cylindrical berries, usually dark green at maturity, probably bat-dispersed. Seeds oblong, flattened.

Psiguria triphylla (Miq.) C. Jeffrey [Syn.: *Anguria triphylla* Miq.]
FIG. 95, PL. 49a

Canopy vines, nearly glabrous. Leaves: petioles to 2 cm long; petiolules 2–3 cm long; blades variable, usually 3-lobed but sometimes entire to trifoliolate, when entire the blade ovate, 8–20 × 7–15 cm, fleshy, the apex acuminate. Staminate inflorescences crowded spikes, the flowers opening one per day over long period; peduncle 15–25 cm long; rachis to 4.5 cm long. Pistillate inflorescences axillary, the distal leaves of stem bearing flowers usually with reduced leaves, with 1–3 flowers, usually borne high in canopy; pedicels 2.5–5 cm long in fruit. Staminate flowers: tubular-urceolate, 6–10 mm long; sepals triangular, 2–2.5 mm long; petals ovate to obovate, 13 mm long when fully expanded, densely pubescent, red or red-orange; anthers 2.5 × 2 mm. Fruits flattened-cylindric, 4–6 × 2–2.5 cm, apparently green at maturity. Seeds 7 × 4 mm. Fl (Jan, Feb, Jul, Aug, Dec), fr (Nov); in forest gaps and along forest edges.

SELYSIA Cogn.

Perennial herbaceous vines or soft-woody lianas. Tendrils 2-branched. Leaves simple but often lobed. Plants monoecious. Inflorescences axillary, often both sexes at same node; pedicels present. Staminate inflorescences of solitary flowers or few and fascicled flowers. Pistillate inflorescences of solitary flowers. Staminate flowers: hypanthium cupuliform; sepals 5; corolla campanulate, deeply 5-lobed; stamens 3, one monothecous and two dithecous, the filaments free, the anthers free. Fruits ellipsoid berries, long-pedicellate, red when mature. Seeds 2–15, 3-lobed at apex.

Selysia prunifera (Poepp. & Endl.) Cogn. FIG. 96

Herbaceous vines, sparsely pubescent. Leaves: blades widely ovate, shallowly to deeply 3-lobed, 5–22 × 5–28 cm. Staminate inflorescences axillary, with 2–3 flowers; pedicels 1–2.5 cm long. Pistillate inflorescences of solitary flowers; peduncle to 8 cm long in fruit. Staminate flowers: hypanthium campanulate, 4–5 × 5–7 mm; sepals lanceolate, 3.5–7.5 long; petals elliptic, 13–22 × 7–8 mm, creamy green to white; anthers 4 mm long. Fruits ovoid to ellipsoid, 2.5–5.5 × 1.5 × 2.5 cm, bright orange. Seeds 2–15, triangular, 3-lobed at apex, 12–22 × 7–11 mm, brown, embedded in orange pulp. Fr (Feb, Sep), fr (Jan, May, Jun, Sep); in forest.

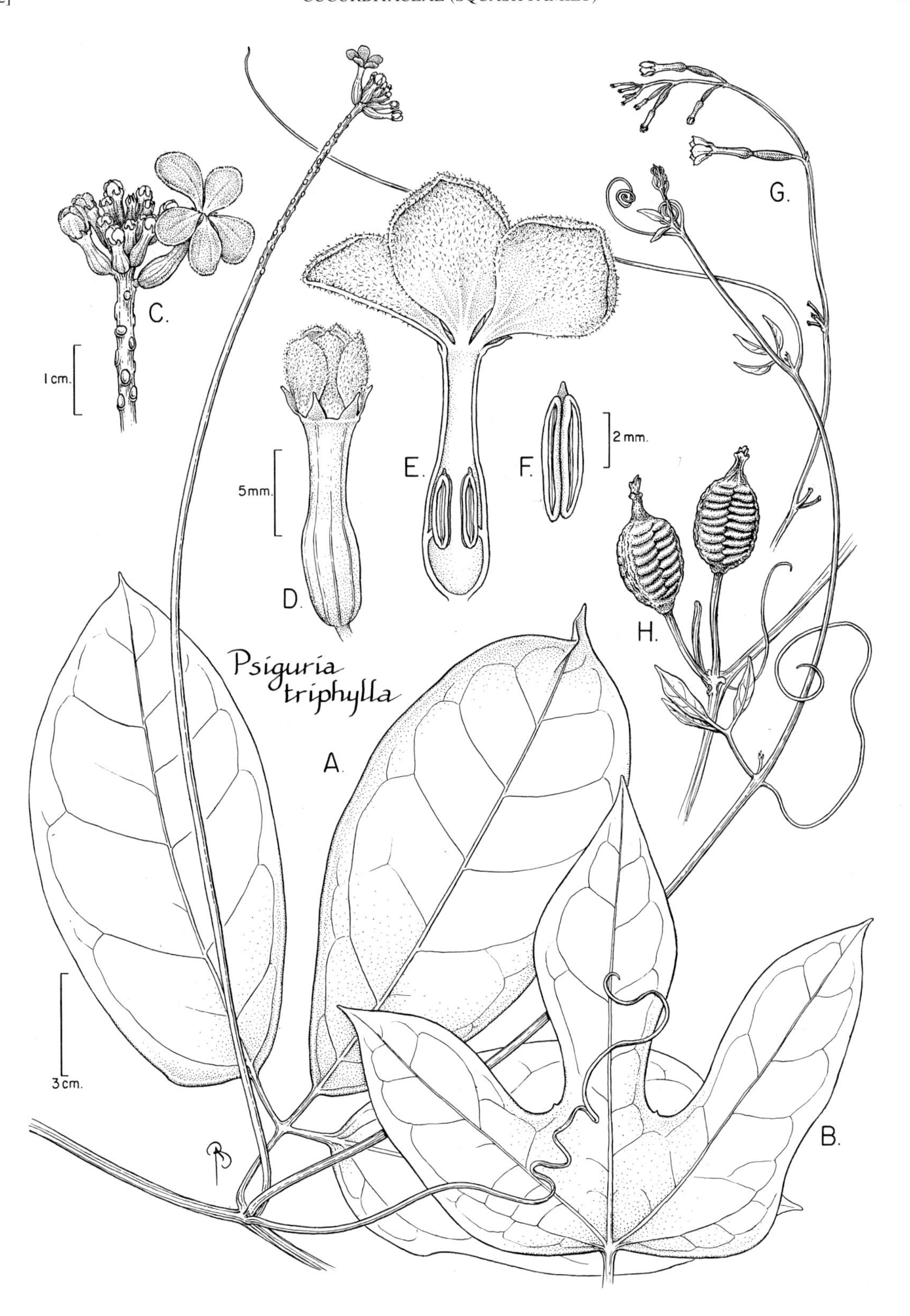

FIG. 95. CUCURBITACEAE. *Psiguria triphylla* (A, *Marshall & Rombold 179*; B, *Hahn 3652*; C, *Mori & Gracie 21082*; D–F, *Mori et al. 19041*; G, *De la Cruz 1297* from Guyana; H, *Bernardi 6249* from Venezuela). **A.** Stem with compound leaf, staminate inflorescence, and tendrils. **B.** Lobed leaf. **C.** Staminate inflorescence with one open flower. **D.** Lateral view of staminate flower bud. **E.** Medial section of staminate flower. **F.** Adaxial view of anther. **G.** Pistillate inflorescence. **H.** Fruits.

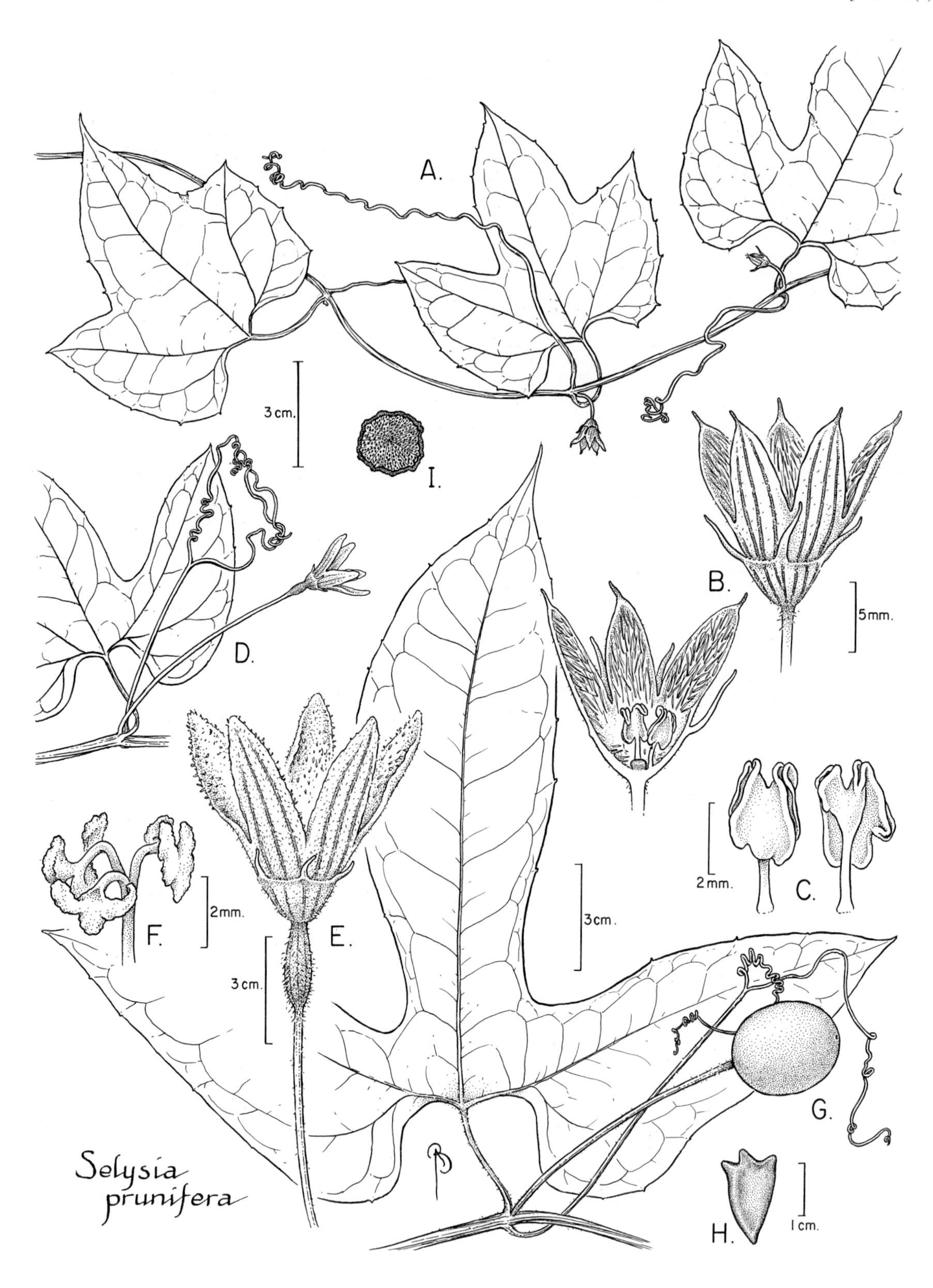

FIG. 96. CUCURBITACEAE. *Selysia prunifera* (A–C, *Fonnegra et al. 2161* from Colombia; D–F, *Dorr et al. 4816* from Venezuela; G, H, *Mori & Gracie 18875*; I, *Gentry et al. 63075*). **A.** Part of stem with leaves, tendrils, and staminate flowers. **B.** Lateral view (right) and medial section (lower left) of staminate flower. **C.** Adaxial (left) and abaxial (right) views of stamens. **D.** Detail of node showing point of origin of petiole, tendril, and flower. **E.** Lateral view of pistillate flower. **F.** Detail of apex of style showing three stigmas. **G.** Part of stem with leaf, tendril, and fruit. **H.** Seed. **I.** Transverse section of stem.

DICHAPETALACEAE (Dichapetalum Family)

Ghillean T. Prance

Trees, shrubs, or lianas. Stipules present but usually caducous. Leaves simple, alternate; blades with margins entire; venation pinnate. Inflorescences axillary or more frequently attached to petiole, corymbose-cymose or subcapitate, or flowers fasciculate; pedicels often articulate. Flowers small, actinomorphic to weakly zygomorphic, bisexual; petals 5, either free, imbricate and almost equal, or connate into tube, the lobes markedly unequal, usually bifid at apex, frequently cucullate or inflexed, often clawed at base; stamens 5, all fertile or only 3 fertile, the other 2 forming staminodia, free or adnate to corolla tube; anthers bilocular, dehiscing longitudinally; nectary disc of 5 equal or unequal hypogynous glands alternating with stamens or united into ring; ovary superior, free, 2–3-locular, the styles 2–3, free or more frequently connate nearly to apex, often recurved, the stigma frequently capitate; placentation axile, the ovules anatropous, pendulous, paired at top of each locule. Fruits dry drupes, the epicarp most frequently pubescent, the mesocarp thin, the endocarp hard. Seeds pendulous, usually solitary in each locule; endosperm absent; embryo large, erect.

Lindeman, J. C. 1986. Dichapetalaceae. *In* A. L. Stoffers & J. C. Lindeman (eds.), Flora of Suriname **III(1–2):** 548–549. E. J. Brill, Leiden.
Prance, G. T. 1972. Dichapetalaceae. Fl. Neotrop. Monogr. **10:** 1–84.
———. 1998. Dichapetalaceae. *In* J. A. Steyermark, P. E. Berry & B. K. Holst (gen. eds.), Flora of the Venezuelan Guayana **4:** 666–685. P. E. Berry, B. K. Holst & K. Yatskievych (vol. eds.) Missouri Botanical Garden Press, St. Louis.
Stafleu, F. A. 1951. Dichapetalaceae. *In* A. Pulle (ed.), Flora of Suriname **III(2):** 166–172. Royal Institute for the Indies.

1. Lianas. Inflorescences with long, distinct peduncles. Petals free; stamens not adnate to corolla tube. . . . *Dichapetalum*.
1. Shrubs or trees. Inflorescences sessile. Petals united; stamens adnate to corolla tube. *Tapura*.

DICHAPETALUM Baill.

Small trees, shrubs, or lianas. Inflorescences axillary or adnate to petiole, branched cymose or corymbose panicles; peduncle long. Flowers: petals 5, free to base, usually bicucullate, the apex 2-lobed, the margins inflexed, surrounding anthers; stamens 5, equal, free, the filaments free; nectary disc of 5 hypogynous glands opposite petals, the glands entire, shallowly lobed, free or united. Fruits dry drupes.

1. Leaf blades coriaceous, bullate, densely hirsute abaxially; venation impressed adaxially. Inflorescences tomentose. . . . . *D. rugosum*.
1. Leaf blades membranous, not bullate, glabrous or with a few appressed hairs on venation abaxially; venation plane adaxially. Inflorescences glabrescent to tomentellous. . . . . *D. pedunculatum*.

Dichapetalum pedunculatum (DC.) Baill.

Lianas. Stipules 5–10 mm long, persistent, puberulous. Leaves: blades oblong to ovate-elliptic, 5–16 × 2–7.5 cm, membranous, not bullate, glabrous or with few stiff appressed hairs beneath; venation plane adaxially. Inflorescences glabrescent to tomentellous. Fl (May); in nonflooded forest.

Dichapetalum rugosum (Vahl) Prance FIG. 97

Lianas. Stipules 2–4 mm long, caducous, densely tomentose. Leaves: blades oblong to ovate-elliptic, 6–32 × 3.5–21 cm, coriaceous, bullate, densely hirsute abaxially; venation impressed adaxially. Inflorescences tomentose. Fl (Feb, Apr); in nonflooded forest.

TAPURA Aubl.

Trees or shrubs. Inflorescences adnate to and sessile on petioles, densely crowded glomerules. Flowers: petals 5, connate at base to form distinct tube, with 2 larger broad lobes, these bifid and bicucullate at apex, and 3 smaller linear-lanceolate entire lobes; stamens 5, all fertile or 3 fertile and 2 reduced to sterile staminodia, the filaments adnate to inside of corolla tube; nectary disc semiannular or 2–3-partite. Fruits dry drupes.

1. Leaves, petioles, and young stems densely pubescent. . . . . *T. amazonica* var. *amazonica*.
1. Leaves, petioles, and young stems glabrous.
    2. Usually small trees >5 m tall. Fertile stamens 5; staminodia absent. . . . . *T. capitulifera*.
    2. Usually shrubs or small tress <5 m tall. Fertile stamens 3; staminodia 2. . . . . *T. guianensis*.

Tapura amazonica Poepp. var. **amazonica**

Small trees, 12 m tall. Leaves, petioles, and young stems densely pubescent. Inner bark ca. 10 mm thick, pure white, with orangish streaks when first cut. Leaves 13–22 × 4–8 cm, densely pubescent abaxially. Flowers: stamens 3, staminodia 2. Fl (May); in nonflooded forest.

Tapura capitulifera Baill.

Trees, to 15 m tall. Leaves, petioles and young stems glabrous. Leaves 5–12 × 1.8–5 cm. Flowers: fertile stamens 5; staminodia absent. In nonflooded forest, not yet collected from our area but to be expected.

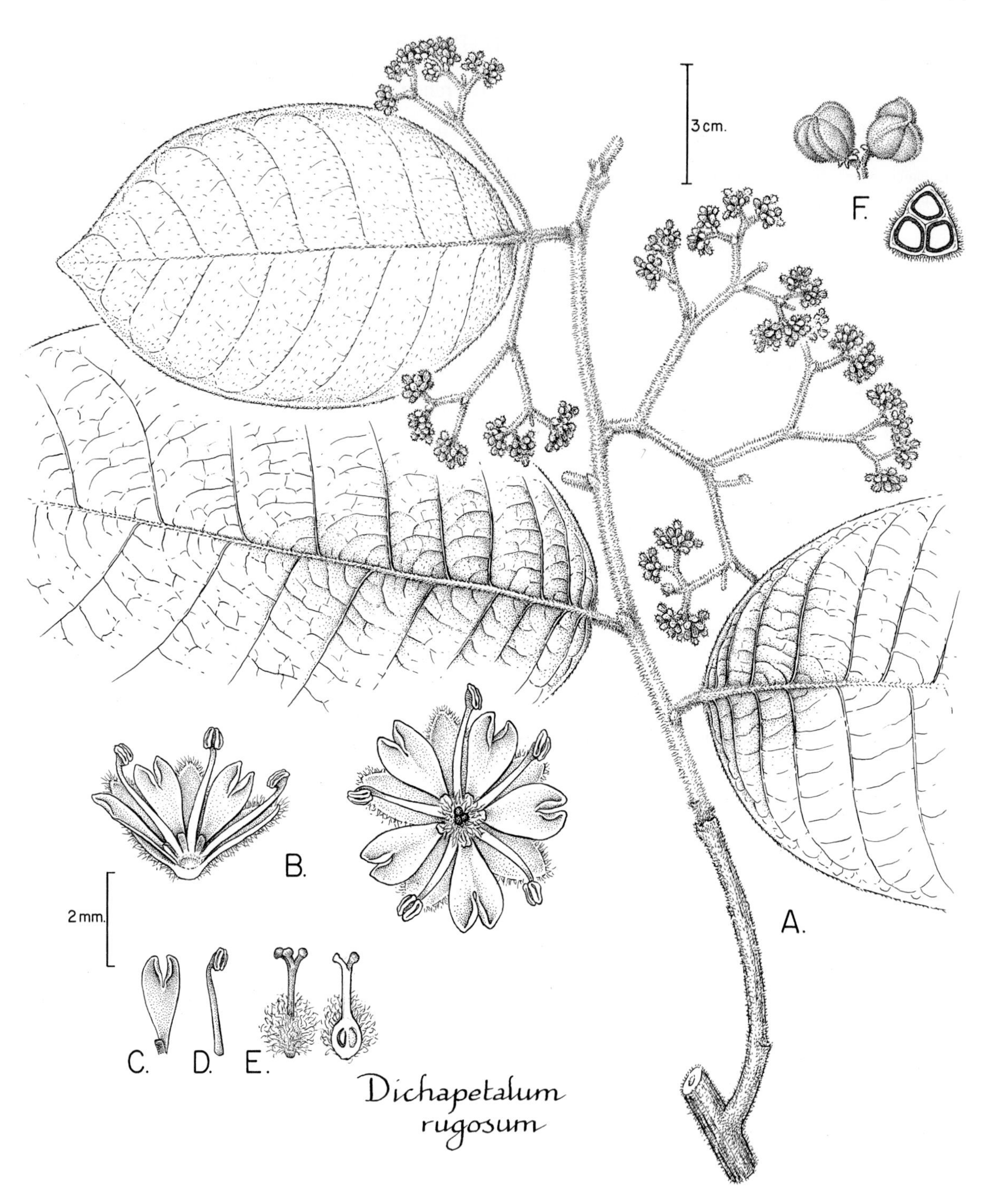

FIG. 97. DICHAPETALACEAE. *Dichapetalum rugosum* (A, E, *Mori & Pipoly 15586*; B–D, *Davidse 18087* from Brazil; F, *Santos 590*). **A.** Part of stem of liana with leaves and generally dichotomously branched inflorescences; note that the inflorescences arise from the petiole of the upper leaf. **B.** Apical view (right) and medial section with pistil removed (left) of staminate flower. **C.** Petal with bifid apex and gland at base. **D.** Stamen. **E.** Pistil. **F.** Part of infructescence (above) and transverse section (below) of trilocular fruit with three seeds.

Tapura guianensis Aubl.

FIG. 98, PL. 49b,c; PART 1: PL. VIIIa

Shrubs or treelets, to ca. 5 m tall. Leaves, petioles and young stems glabrous. Leaves 6–23 × 2.1–9 cm, glabrous. Flowers: fertile stamens 3, staminodia 2. Fl (Jan, Feb, May, Jul, Sep, Oct, Dec), fr (Aug, Sep); common, in understory of nonflooded forest.

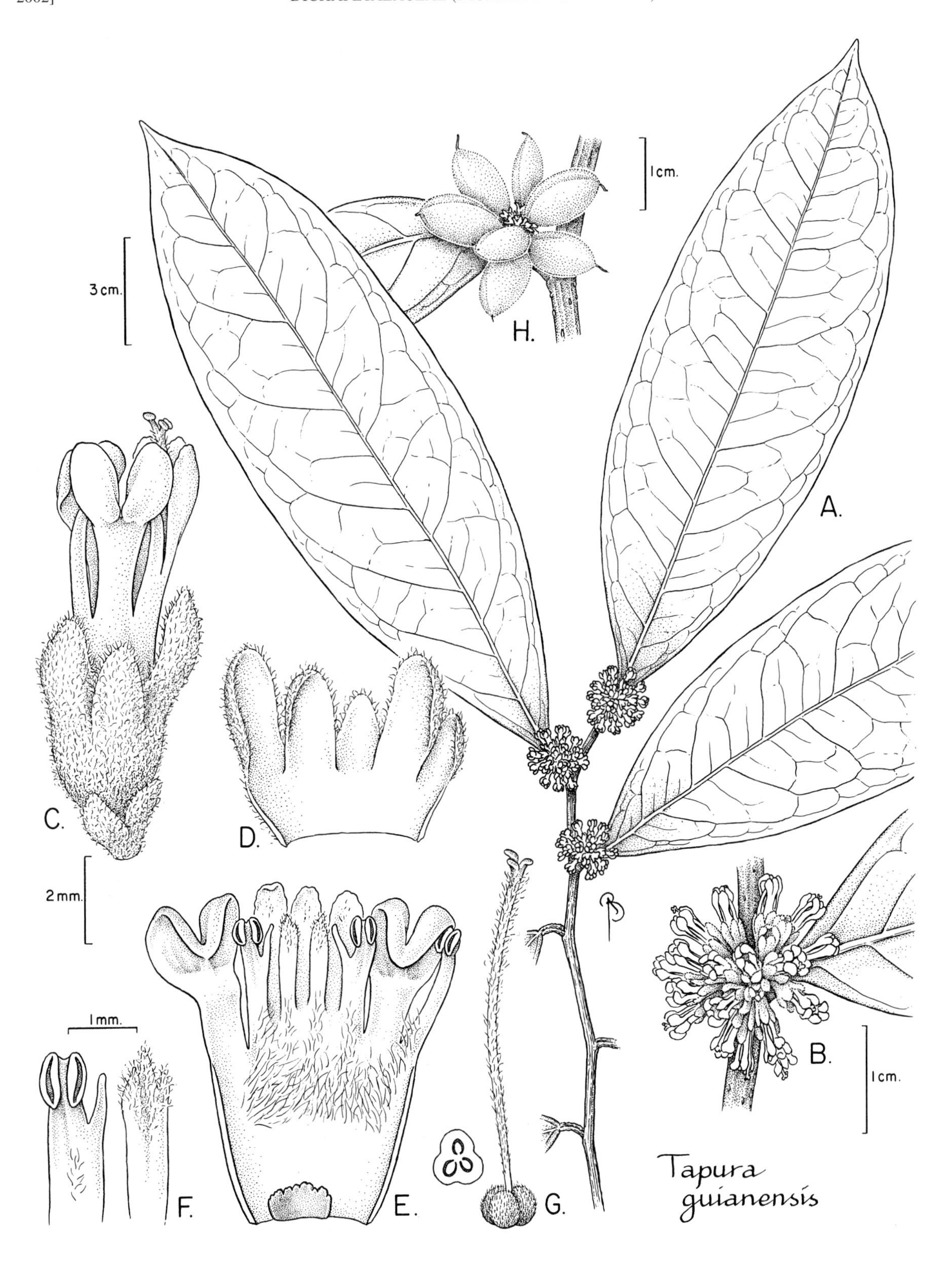

FIG. 98. DICHAPETALACEAE. *Tapura guianensis* (A–G, *Mori et al. 22685*; H, *Mori 24252* photo). **A.** Part of stem with leaves and inflorescences. **B.** Inflorescence on petiole of leaf. **C.** Lateral view of flower. **D.** Adaxial view of calyx opened up. **E.** Adaxial view of corolla opened up showing stamens, staminodes, and gland at base. **F.** Adaxial views of stamen (far left) and staminode (near left). **G.** Lateral view of pistil (near left) and transverse section of ovary (far left). **H.** Cluster of fruits on petiole of leaf.

DILLENIACEAE (Dillenia Family)

Gerardo A. Aymard and Scott A. Mori

Lianas (in our area), small trees, or shrubs. Stems of lianas generally >5 cm diam., the vascular tissue arranged in bands or concentric rings separated by abundant parenchyma (mostly in *Doliocarpus* and *Pinzona*). Stipules absent or present, when present caducous or infrequently wing-like and adnate to petiole. Leaves simple, alternate; blades often smooth or scabrous adaxially because of sclerified or silicified trichomes, the margins entire or serrate. Flowers actinomorphic, usually bisexual, unisexual in *Tetracera* (plants androdioecious); calyx 2–7(to 14 in *Tetracera*)-merous, the sepals usually imbricate at bases except the two inner sepals of *Davilla* cucullate and imbricate for most of length, two larger than others in *Davilla*; corolla polypetalous, usually 3–7-merous, the petals imbricate, caducous, usually white or yellow; stamens usually numerous, free and persistent; ovary superior, the carpels 1–20, free or connate along ventral side (*Curatella*, *Pinzona*); placentation parietal to basal, the ovules 1 or 2 or numerous in each locule. Fruits dry and dehiscent follicles or capsules or berry-like, surrounded by two accrescent sepals in *Davilla*. Seeds often arillate; endosperm well developed; embryo small, straight.

The stems of many species of this family are known as *lianes chasseurs* because, when cut into segments of about a meter in length and held erect, they provide water for hunters. Some stems, however, are irritating to the touch (Grenand et al., 1987).

Aymard C., G. A. 1997. Dilleniaceae Novae Neotropicae IX: *Neodillenia*, a new genus from the Amazon Basin. Harvard Pap. Bot. **10:** 121–131. [Includes key to genera of Dilleniaceae worldwide.]

———. 1998. Dilleniaceae. *In* J. A. Steyermark, P. E. Berry & B. K. Holst (gen. eds.), Flora of the Venezuelan Guayana **4:** 671–685. P. A. Berry, B. K. Holst & K. Yatskievych (vol. eds.). Missouri Botanical Garden Press, St. Louis.

Jansen-Jacobs, M. J. 1986. Dilleniaceae. *In* A. L. Stoffers & J. C. Lindeman (eds.), Flora of Suriname **III(1–2):** 475–484. E. J. Brill, Leiden.

Lanjouw, J. & P. F. Baron von Heerdt. 1966. Dilleniaceae. *In* A. Pulle (ed.), Flora of Suriname **III(1):** 386–408. E. J. Brill, Leiden.

1. Two sepals distinctly larger than others. Fruits surrounded by two accrescent sepals. *Davilla.*
1. All sepals ± equal. Fruits not surrounded by accrescent sepals.
   2. Inflorescences fascicles, glomerules or flowers solitary. Pistil with a single style; carpels 1. Aril white. *Doliocarpus.*
   2. Inflorescences racemes or paniculate arrangements of racemes. Pistil with two distinct styles; carpels 2. Aril orange. *Pinzona.*

DAVILLA Vand.

Lianas or scandent shrubs. Inflorescences axillary or terminal, panicles or paniculate arrangements of clustered cymes. Flowers: sepals 5, unequal, the inner two large, imbricate, cucullate, hardening and covering fruit; petals 3–6, often 5, caducous; stamens 50–200; ovary with 1–2 carpels, free when more than 1, surmounted by single, sublateral style, the stigma peltate or capitate, each carpel with 1–2 ovules. Fruits capsules, surrounded by two persistent sepals, usually with single seed. Seeds surrounded by entire, white aril.

1. Young stems, petioles, and midrib and secondary veins abaxially with golden brown, spreading hairs. Leaf blades ± concolorous, the abaxial surface without markedly salient tertiary veins. . . . . *D. rugosa* var. *rugosa.*
1. Young stems, petioles, and midrib and secondary veins abaxially with whitish, appressed hairs. Leaf blades ± discolorous, the abaxial surface with markedly salient tertiary veins. *D. kunthii.*

Davilla kunthii A. St.-Hil. [Syn.: *Davilla aspera* (Aubl.) Benoist]

Lianas. Young stems with whitish appressed hairs, without short, recurved hairs, the bark not conspicuously peeling. Leaves: petioles 12–33 mm long; blades ovate, elliptic, or widely obovate, 5.5–20 × 2.5–10, ± discolorous, both surfaces scabrous to slightly scabrous, with appressed, scattered, white, acicular hairs, these more common on veins, the base obtuse, narrowly decurrent onto petiole, the apex acute, obtuse, or rounded, the margins revolute, dentate; secondary veins in 6–15 pairs, all orders of venation salient abaxially, impressed to varying degrees adaxially, sometimes finely bullate. Inflorescences axillary and terminal, much branched paniculate arrangements of clusters of cymes; peduncle present; pedicels 4.5–9 mm long, pubescent. Flowers ca. 8 mm diam.; sepals 5, unequal, densely appressed pubescent abaxially; petals 4–6; stamens ca. 50–90; style glabrous, the stigma capitate. Fruits enclosed in two, imbricate, persistent, inner sepals. Seeds 1 per fruit, globose, black, completely surrounded by white aril. Fl (Nov); in nonflooded forest on La Fumée Mountain Trail.

Davilla rugosa Poir. var. **rugosa** [Syn.: *Davilla pilosa* Miq.]

Lianas. Young stems with golden brown, spreading hairs, with short, recurved hairs, these not seen on specimens of this species outside our area, the bark conspicuously peeling. Leaves: petioles 5–15 mm long, narrowly decurrent onto petiole; blades elliptic to widely elliptic, the largest 8.5–15.5 × 2.5–8, ± concolorous, both surfaces scabrous, with erect or spreading, golden-brown acicular hairs over entire surface, the base acute to obtuse, narrowly decurrent onto petiole, the apex very short acuminate, the margins serrate, especially toward apex; secondary veins in 12–14(18) pairs, only midrib and secondary veins salient abaxially, the tertiary veins ± plane, the midrib and secondary veins impressed adaxially. Inflorescences mostly axillary or sometimes terminal, sparsely branched panicles; peduncle present; pedicels 5–7 mm long, these covered with same kinds of hairs as on stems. Flowers ca. 8–10 mm diam.; sepals 5, unequal; stamens ca. 40–50 or even more; style simple, the stigma capitate. Fruits and seeds unknown from our area. Fl (Jan); in nonflooded forest on Plateau La Douane.

Plates 49–56

Plate 49. CUCURBITACEAE. **a.** *Psiguria triphylla* (*Mori et al. 19041*), flower and buds. DICHAPETALACEAE. **b.** *Tapura guianensis* (*Mori et al. 22685*), inflorescences on petioles. **c.** *Tapura guianensis* (*Mori & Gracie 24252*), infructescence on petiole.

Plate 50. DILLENIACEAE. **a.** *Doliocarpus brevipedicellatus* (*Mori et al. 24153*), inflorescence. **b.** *Pinzona coriacea* (*Mori & Gracie 23864*), infructescences with one open fruit. EBENACEAE. **c.** *Diospyros ropourea* (*Mori & Gracie 23885*), pistillate flower. **d.** *Diospyros ropourea* (*Mori et al. 23878*), staminate inflorescences on trunk of tree.

Plate 51. EBENACEAE. **a.** *Diospyros ropourea* (unvouchered), fruits on trunk. **b.** *Diospyros carbonaria* (Mori 24139), fruits; note persistent calyces. **c.** *Diospyros cayennenis* (*Mori et al. 22913*), immature fruits. **d.** *Diospyros cavalcantei* (*Mori et al. 22939*), nearly mature buds. ELAEOCARPACEAE. **e.** *Sloanea latifolia* (*Mori et al. 24165*), flowers and buds.

Plate 52. ELAEOCARPACEAE. **a.** *Sloanea* aff. *synandra* (*Wallnöfer 13494*), flowers, buds, and flowers after petals have fallen. [Photo by M. Gustafsson] **b.** *Sloanea tuerckheimii* (*Mori et al. 23322*), base of buttressed trunk. **c.** *Sloanea tuerckheimii* (*Mori et al. 24716*), flowers. [Photo by S. Mori] **d.** *Sloanea* aff. *brevipes* (*Mori 24748*), flowers. [Photo by S. Mori] **e.** *Sloanea* sp. A (*Mori & Pipoly 15498*), dehisced fruit with seeds. [Photo by S. Mori] **f.** *Sloanea tuerckheimii* (*Mori et al. 24773*), fruits.

Plate 53. ERYTHROXYLACEAE. **a.** *Erythroxylum macrophyllum* (*Mori et al. 22753*), flowers and buds. **b.** *Erythroxylum macrophyllum* (*Mori & Gracie 21153*), fruits. EUPHORBIACEAE. **c.** *Acalypha diversifolia* (*Mori & Gracie 24221*), axillary spicate inflorescence with pistillate flower at base and staminate flowers above. **d.** *Alchorneopsis floribunda* (*Mori et al. 22920*), capsules, one dehisced. **e.** *Aparisthmium cordatum* (*Mori et al. 24015*), base of leaf with stipels at junction of blade and petiole and apical view of pistillate flower. **f.** *Conceveiba guianensis* (*Mori & Gracie 21133*), capsules.

Plate 54. EUPHORBIACEAE. **a.** *Dalechampia stipulacea* (*Mori et al. 23721*), inflorescence with bracts spread to show staminate and pistillate flowers and gland. **b.** *Dalechampia stipulacea* (*Mori et al. 23721*), fruits. **c.** *Dodecastigma integrifolium* (*Mori et al. 22811*), staminate flower and bud. **d.** *Drypetes variabilis* (*Mori et al. 24761*), staminate flowers. **e.** *Euphorbia thymifolia* (*Mori & Gracie 23817*), part of stem with developing capsules. **f.** *Hevea guianensis* (*Mori & Pennington 17954*), capsules. [Photo by S. Mori] **g.** *Hura crepitans* (unvouchered), prickles on young trunk.

Plate 55. EUPHORBIACEAE. **a.** *Hura crepitans* (unvouchered), staminate inflorescence (red) and pistillate flower (black). [Photo by S. Mori] **b.** *Hura crepitans* (unvouchered), fruit. [Photo by S. Mori] **c.** *Mabea speciosa* (*Mori et al. 22676*), pistillate flowers at base (top of photo) and staminate flowers above (bottom of photo). **d.** *Omphalea diandra* (*Mori et al. 23917*), part of inflorescence with pistillate flowers and staminate buds. **e.** *Omphalea diandra* (*Mori et al. 23917*), immature fruits. **f.** *Omphalea diandra* (*Mori et al. 23912*), staminate flowers collected from ground.

Plate 56. EUPHORBIACEAE. **a.** *Phyllanthus urinaria* (*Mori et al. 22076*), part of abaxial surface of stem with leaves, flower, and fruits in various stages. **b.** *Plukenetia polyadenia* (*Mori & Gracie 18668*), staminate buds and flower. **c.** *Sagotia racemosa* (*Mori et al. 23778*), apical view of two pistillate flowers. [Photo by S. Mori] FABACEAE. **d.** *Desmodium axillare* (*Mori et al. 22661*), flowers. **e.** *Desmodium incanum* (*Mori & Gracie 24224*), flowers. **f.** *Desmodium incanum* (*Mori & Gracie 24224*), fruits.

Psiguria triphylla a.

DICHAPETALACEAE

Tapura guianensis b.

Tapura guianensis c.

DILLENIACEAE

Doliocarpus brevipedicellatus a.

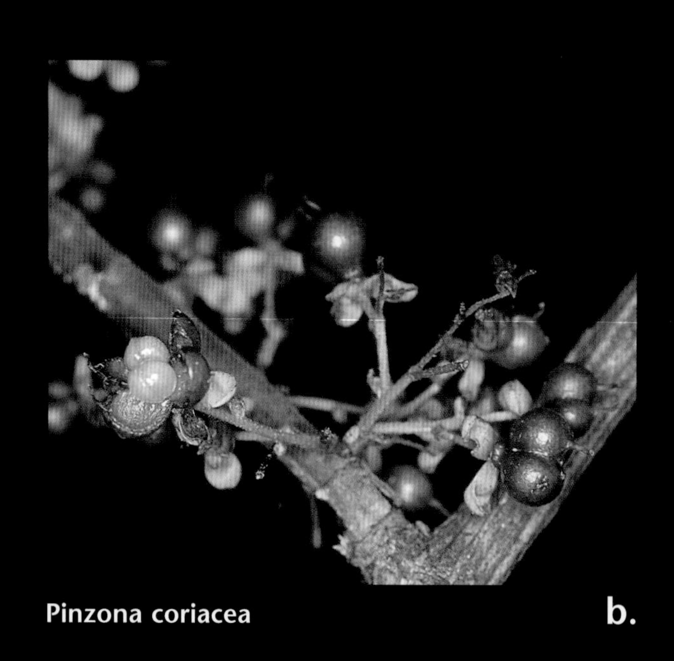

Pinzona coriacea b.

EBENACEAE

Diospyros ropourea c.

Diospyros ropourea d.

EBENACEAE

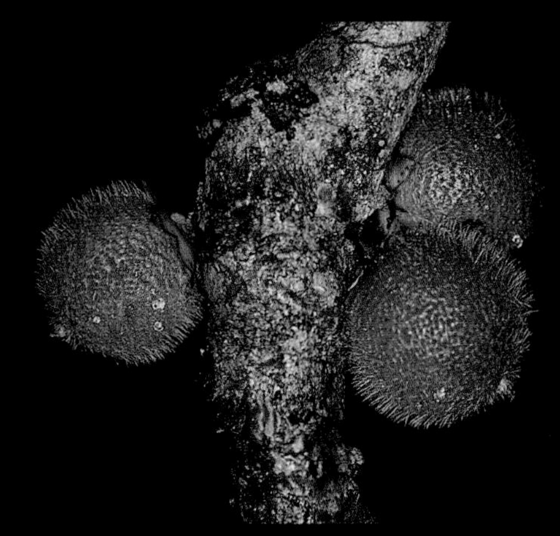

Diospyros ropourea a.

Diospyros carbonaria b.

Diospyros cayennenis c.

Diospyros cavalcantei d.

ELAEOCARPACEAE

Sloanea latifolia e.

Sloanea aff. synandra a.

Sloanea tuerckheimii b.

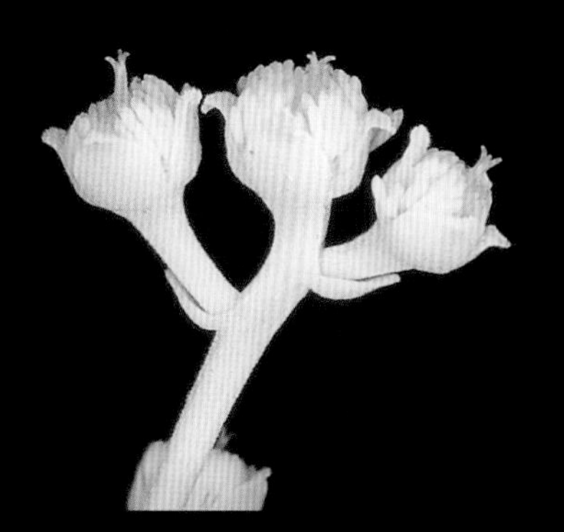

Sloanea tuerckheimii c.

Sloanea aff. brevipes d.

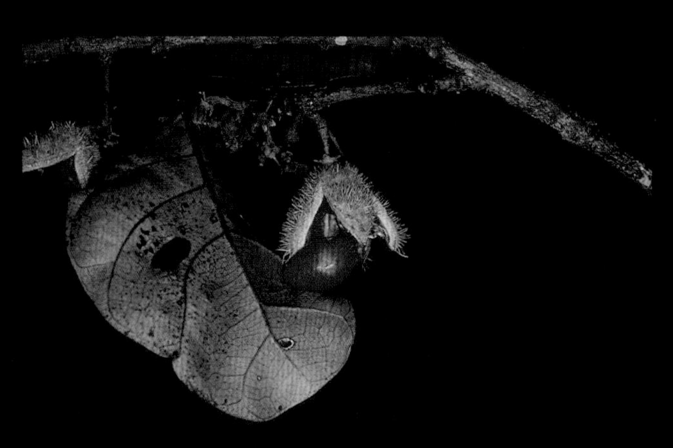

Sloanea sp. A e.

Sloanea tuerckheimii f.

Plate 52

Erythroxylum macrophyllum a.

Erythroxylum macrophyllum b.

EUPHORBIACEAE

Acalypha diversifolia c.

Alchorneopsis floribunda d.

Aparisthmium cordatum e.

Conceveiba guianensis f.

Plate 53

Dalechampia stipulacea a.

Dalechampia stipulacea b.

Dodecastigma integrifolium c.

Euphorbia thymifolia e.

Drypetes variabilis d.

Hevea guianensis f.

Hura crepitans g.

Hura crepitans a.

Hura crepitans b.

Mabea speciosa c.

Omphalea diandra d.

Omphalea diandra e.

Omphalea diandra f.

Plate 55

Phyllanthus urinaria a.

Plukenetia polyadenia b.

Sagotia racemosa c.

FABACEAE

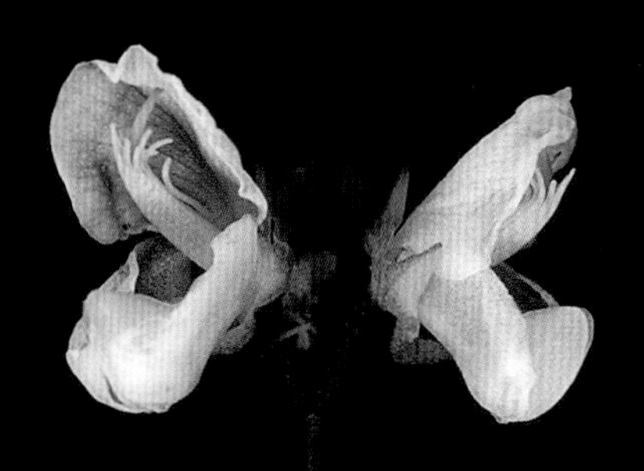

Desmodium axillare d.

Desmodium incanum e.

Desmodium incanum f.

Plate 56 (Fabaceae continued on plate 57)

DOLIOCARPUS Rol.

Lianas. Stems with vascular tissue arranged in bands or cencentric rings separated by abundant parenchyma. Inflorescences axillary or along stems, fascicles, glomerules, or flowers solitary. Flowers: sepals 2–6, ± equal, persistent; petals 2–6, often 3–4, caducous; stamens 20–100, the anthers longitudinally dehiscent; ovary 1-carpellate, 1-locular, surmounted by single style, each carpel with (1)2 basal ovules. Fruits berry-like (in our area) or capsules, not surrounded by persistent sepals, each carpel with 1 or 2 seeds. Seeds surrounded by entire, white aril.

1. Leaf blade margins entire; tertiary venation reticulate. Ovary pubescent. . . . . *D. paraensis.*
1. Leaf blade margins serrate, at least toward apex; tertiary venation subparallel. Ovary glabrous.
    2. Stems without conspicuous annular stipule scars. Leaf blades with appressed, acicular hairs, especially on veins abaxially at maturity; secondary veins usually <1 cm apart. . . . *D. brevipedicellatus* subsp. *brevipedicellatus.*
    2. Stems with conspicuous annular stipule scars. Leaf blades glabrous to puberulous at maturity; secondary veins usually >1 cm apart.
        3. Petioles 5–15 mm long; blades obovate or obovate-oblong. Inflorescences racemes. . . . . *D. guianensis.*
        3. Petioles 30–45 mm long; blades elliptic to widely elliptic. Inflorescences fascicles or flowers solitary. . . . . *D. dentatus* subsp. *latifolius.*

Doliocarpus brevipedicellatus Garcke subsp. **brevipedicellatus** FIG. 99, PL. 50a

Stems without conspicuous annular stipule scars, the new growth pubescent but not densely so. Leaves: petioles 5–10 mm long; blades lanceolate, elliptic to widely elliptic, 5.5–14 × 3–5.5 cm, both surfaces smooth, with appressed, scattered, white, acicular hairs, these more common on veins abaxially at maturity, the base obtuse, the apex acuminate, the margins serrate toward apex, entire toward base; secondary veins in 7–10 pairs, usually <1 cm apart, these and midrib impressed on adaxial surface, salient on abaxial surface, the tertiary veins subparallel, salient on abaxial surface and plane to slightly salient on adaxial surface. Inflorescences racemose; pedicels 1–5 mm long, adpressed pubescent. Flowers ca. 8 mm diam.; sepals (4)5, subequal, cucullate; petals 3, white; stamens ca. 30–40. Fruits globose, ca. 6 mm diam, surmounted by single style, probably splitting open into two valves to expose single seed. Seeds surrounded by white arils. Fl (Sep); in nonflooded forest near entrance to Sentier Botanique at Eaux Claires.

Doliocarpus dentatus (Aubl.) Standl. subsp. **latifolius** Kubitzki

Stems with conspicuous annular stipule scars, the new flushes densely pubescent. Leaves: petioles 30–45 mm long; blades elliptic to widely elliptic, 10–21 × 6–12 cm, both surfaces smooth, glabrous to puberulous, especially along veins and in new leaves, the base obtuse, narrowly decurrent onto petiole, the apex very shortly acuminate, the margins serrate toward apex, entire toward base; secondary veins in 11–18 pairs, usually >1 cm apart, these plane and the midrib impressed on adaxial surface, all orders of venation salient on abaxial surface, the tertiary veins subparallel. Inflorescences fascicles or flowers solitary (based on collections from outside our area). Known only by one sterile specimen from our area.

Doliocarpus guianensis (Aubl.) Gilg

Stems with conspicuous annular stipule scars, the new growth glabrous. Leaves: petioles 5–15 mm long; blades obovate or obovate-oblong, 6–20 × 3–9 cm, both surfaces smooth, glabrous, the base cuneate, the apex rounded or obtuse-apiculate, the margins entire or subsinuate toward apex; secondary veins in 7–10 pairs, usually >1 cm apart, the midrib salient abaxially, the tertiary venation subparallel. Inflorescences racemes, 2–4.5 cm long; pedicels 3–4 mm long. Flowers unknown from our area. Fruits globose, 8–12 mm diam., red, splitting open by two valves. Seeds 1–2, entirely surrounded by white arils. Fr (Jun); in nonflooded forest on Mont Galbao.

Doliocarpus paraensis Sleumer [Syn.: *Doliocarpus surinamensis* Lanj.]

Stems without conspicuous annular stipule scars, the new growth probably glabrous. Leaves: petioles 10–22 mm long; blades narrowly to widely obovate, 7–9 × 3–4.5 cm; both surfaces smooth, glabrous, the base acute, the apex short acuminate, the margins entire; secondary veins in 8–9 pairs, 0.8–1 cm apart, these salient abaxially, ± plane adaxially, the midrib salient on both surfaces, the tertiary veins reticulate. Inflorescences: pedicels ca. 8 mm long. Flowers unknown from our area. Fruits globose, 20–30 mm diam., red, splitting open into two valves to expose single seed. Seeds entirely surrounded by white arils. Fr (May); in nonflooded forest along Antenne Nord of La Fumée Mountain Trail.

PINZONA Mart. & Zucc.

Lianas. Stems with vascular tissue arranged in bands or concentric rings separated by abundant parenchyma. Inflorescences racemes or paniculate arrangements of racemes. Flowers: sepals 3–4 sepals; petals 2–3; stamens 25–40; ovary with 2 carpels, each carpel surmounted by separate style, each carpel with 2 ovules. Fruits paired capsules (at least in our area), not surrounded by persistent sepals, each carpel with 1–2 seeds. Seeds obovate, surrounded by orange aril.

Pinzona coriacea Mart. & Zucc. FIG. 100, PL. 50b

Stems, especially young ones ridged or angulate. Bark shedding in large plates when mature. Leaves: petioles 18–25 mm long; blades elliptic to widely elliptic, oblong, or obovate, 9.5–16 × 5–12 cm, both surfaces slightly scabrous, with appressed, scattered, white, acicular hairs, these more common on veins, the base acute to obtuse, narrowly decurrent onto petiole, the apex rounded into a short apiculus, the margins entire; secondary veins in 7–9 pairs, convergent and joining near margin, these impressed on adaxial surface, salient on abaxial surface, the tertiary veins slightly salient on both surfaces. Inflorescences axillary, racemes or paniculate arrangements of racemes, pubescent, the principal rachis 0.7–7 cm long; pedicels 2–8 mm long. Flowers ca. 3–5 mm diam.; sepals 3–4, cucullate; stamens ca. 15–40, the filaments slender. Fruits globose, bilobed capsules, green, the 2 carpels distinct, the styles persistent, splitting open at right angles to juncture between carpels. Seeds 1 per carpel, obovate, black, shiny, surrounded by orange arils. Fr (Sep); in nonflooded forest along Route de Bélizon.

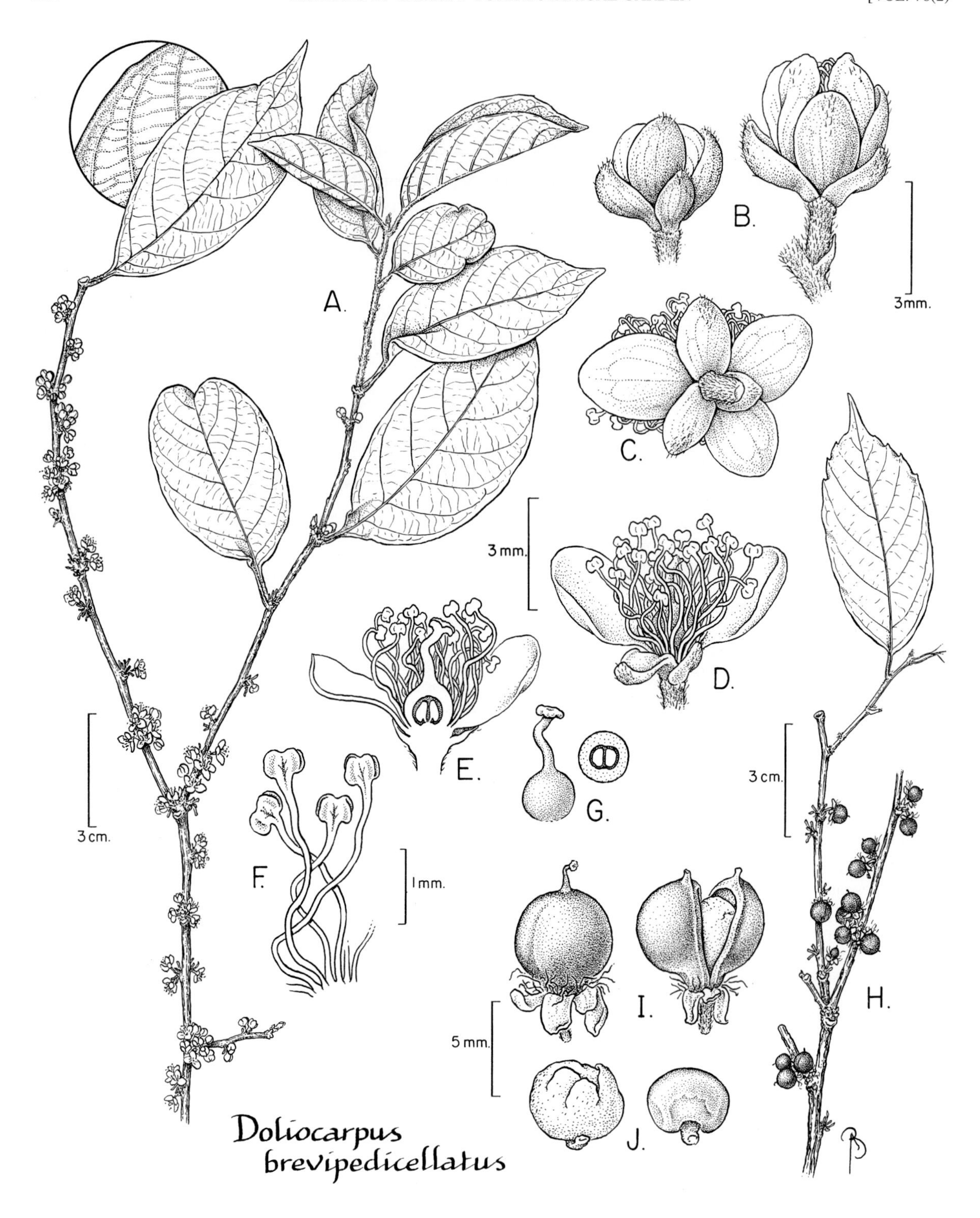

FIG. 99. DILLENIACEAE. *Doliocarpus brevipedicellatus* (A–G, *Mori et al. 24153*; H–J, *Rabelo & Souza 2932* from Brazil). **A.** Part of stem with leaves and inflorescences and detail of abaxial surface of leaf. **B.** Lateral views of two flower buds in different stages. **C.** Basal view of flower. **D.** Lateral view of flower with one petal removed. **E.** Medial section of flower. **F.** Cluster of stamens. **G.** Lateral view (left) and transverse section (above) of pistil. **H.** Part of stem with leaf and infructescences. **I.** Immature (left) and mature (right) fruits. **J.** Seed enclosed in aril (left) and with aril removed (right).

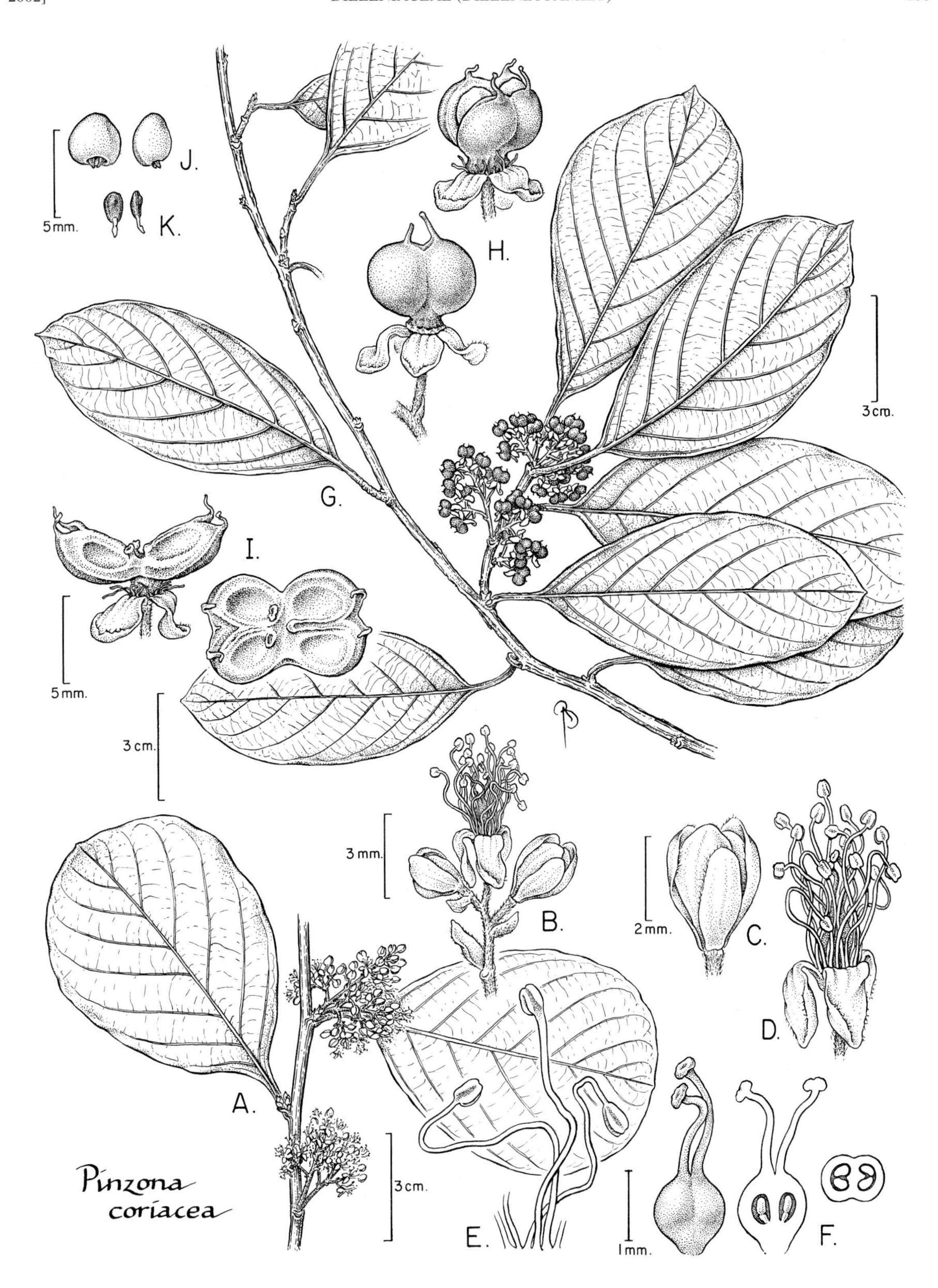

FIG. 100. DILLENIACEAE. *Pinzona coriacea* (A–F, *Steyermark et al. 122371* from Venezuela; G–K, *Mori & Gracie 23864*). **A.** Part of stem with leaves and inflorescences. **B.** Detail of inflorescence. **C.** Lateral view of floral bud. **D.** Lateral view of flower. **E.** Stamens. **F.** Lateral view (far left), medial section (near left), and transverse section (above) of pistil. **G.** Part of stem with leaves and infructescences. **H.** Lateral views of intact fruit (below) and newly dehisced fruit (above). **I.** Lateral (left) and apical (below) views of dehisced fruit. **J.** Two views of arillate seeds. **K.** Two views of seeds with arils removed; note funicles.

EBENACEAE (Ebony Family)

Bruno Wallnöfer and Scott A. Mori

Trees (also shrubs outside our area). Growth architecture of various species of Massart's Model (Hallé et al., 1978). Bark often brittle, often dark, nearly black externally, the slash often yellow at junction of inner bark and sapwood, the heartwood sometimes black. Latex absent. Unicellular trichomes generally present, simple or bifurcate, one arm usually shorter than other, multicellular, club-shaped glandular hairs present in some species, epidermal papillae often present on leaf blades abaxially. Stipules absent. Leaves simple, alternate, spirally or distichously arranged; blades entire, often slightly revolute, most species with ± patelliform extrafloral nectaries on abaxial surface near midrib and especially toward base but often also near apex, these seldom conspicuous. Inflorescences axillary or cauline, cymose (sometimes densely clustered), or the flowers solitary (especially the pistillate flowers); pedicels articulate. Flowers actinomorphic, 3–5(8)-merous, usually unisexual (plants usually dioecious, rarely monoecious or polygamous), rarely bisexual; calyx gamosepalous, persistent and mostly conspicuously accrescent in fruit; corolla gamopetalous, with conspicuous tube or fused only at base, the lobes contorted. Staminate flowers with stamens adnate to base of corolla tube or attached directly to receptacle; stamens (3)12–20(100), pubescent, distinct or often the filaments united, the anthers linear or lanceolate, usually longitudinally dehiscent; pistillode present or absent. Pistillate flowers with superior ovary; staminodia present or rarely absent, adnate to base of corolla tube, the styles free or basally to fully connate; placentation apical-axile. Fruits usually berries, subtended by persistent calyx. Seeds 1–16 per fruit; endosperm hard, often ruminate, the embryo with flat, straight cotyledons.

Sothers, C. & P. E. Berry. 1998. Ebenaceae. *In* J. A. Steyermark, P. E. Berry & B. K. Holst (gen. eds.), Flora of the Venezuelan Guayana **4:** 704–712. P. E. Berry, B. K. Holst & K. Yatskievych (vol. eds.). Missouri Botanical Garden Press, St. Louis.

Wallnöfer, B. 1999. Neue *Diospyros*-Arten (Ebenaceae) aus Südamerika. Ann. Naturhist. Mus. Wien **101B:** 565–592.

———. 2000. Neue *Diospyros*-Arten (Ebenaceae) aus Südamerika. II. Ann. Naturhist. Mus. Wien **102B:** 417–433.

DIOSPYROS L.

The generic description same as description of the family, at least as it applies to species in the New World.

1. Leaf blades densely and minutely glandular hairy at 30× abaxially (in some leaves of *D. vestita*, the glandular hairs are not obvious so several leaves should be studied, especially along the margins and midrib), glaucous when fresh, mostly distinctly paler abaxially when dry, either densely sericeous or only with scattered hairs abaxially.
    2. Petioles absent or to 2(3) mm long; blades ± densely sericeous abaxially, the hairs whitish-gray to golden or brownish when dry.
        3. Stems with distinct hair-free lines running in zig-zag fashion from node to node, the pubescence persistent. Leaf blade length/width ratio >3. . . . . *D. dichroa.*
        3. Stems without distinct hair-free lines running from node to node, the pubescence caducous. Leaf blade length/width ratio <3. . . . . *D. vestita.*
    2. Petioles 4–10 mm long; blades with scattered hairs abaxially.
        4. Leaf blades broadly ovate, (10)15–28 × (4.5)8–12 cm. Fruits barrel-shaped, irregularly gibbose, the apex markedly truncate, the surface without tubercles, densely covered with ± appressed hairs, the hairs to 1 mm long. . . . . *D. martinii.*
        4. Leaf blades broadly lanceolate to elliptic, (3)5–9(12) × 2–3(5) cm. Fruits subglobose, not gibbose, the apex rounded, the surface with tubercules, covered with stiff, patent, golden-brown to ferruginous hairs, the hairs ca. 3 mm long. . . . . *D. capreifolia.*
1. Leaf blades not minutely glandular hairy at 30× abaxially, not glaucous when fresh, mostly not paler abaxially when dry, glabrous or only with scattered hairs abaxially.
    5. Leaf blades glabrous; higher order venation (3rd and 4th order) generally salient adaxially (in some leaves of *D. cayennensis* these ± plane). Fruits glabrous, ± globose.
        6. Leaf blades (especially on distal leaves) often ± shiny adaxially. Flowers generally pedicellate; calyx ca. 10 mm long (use flowers, not fruits for this measurement), the lobes ovate-orbicular; stamens ca. 10. Fruits 4–5 cm diam., brownish when dry, the persistent calyx lobes not strongly undulate-plicate. . . . . *D. cayennensis.*
        6. Leaf blades dull, sometimes slightly shiny adaxially. Flowers ± sessile; calyx 3–4 mm long (use flowers, not fruits for this measurement), the lobes subulate; stamens 4. Fruits ca. 2 cm diam., black when dry, the persistent calyx lobes markedly undulate-plicate. . . . . *D. tetrandra.*
    5. Leaf blades with scattered hairs; higher order venation (3rd and 4th order) plane, impressed, or ± indistinct adaxially. Fruits hairy or glabrescent, not globose or, if ± globose, then densely covered with hispid hairs.

7. Treelets to small trees, to 6 m tall. Inflorescences mostly cauliflorous (ramiflorous in taller plants). Leaf blades (10)15–25 × (3)5–8(10) cm; veins of 2nd and 3rd order markedly impressed adaxially. Fruits with longest hairs to 4 mm long, the hairs stiff, patent, brown to ferruginous. . . . . *D. ropourea.*
7. Large trees. Inflorescences ramiflorous. Leaf blades (3)6–13(15) × (1.7)2–5.5(6) cm; veins of 2nd and 3rd order plane adaxially. Fruits with longest hairs rarely exceeding 1 mm long, the hairs ± appressed to slightly patent, gray to light brown or fruits glabrescent (except proximally and distally).
    8. Stems densely pubescent. Petioles 1–1.5 mm long; blades (3)6–9 × (1.7)2–3 cm, the margins densely pubescent ad- and abaxially; midrib densely pubescent, the secondary veins indistinct adaxially. Fruits wider than long, not covered with tubercles, glabrescent (except proximally and distally). . . . . *Diospyros cavalcantei.*
    8. Stems with scattered hairs. Petioles (3)4–8 mm long; blades (6)9–13(15) × (3.5)4–5.5(6) cm, the margins with scattered hairs; midrib with scattered hairs ad- and abaxially, the secondary veins ± distinct adaxially. Fruits longer than wide, densely covered with tubercles, pubescent. . . . . . *D. carbonaria.*

Diospyros capreifolia Hiern [Syns.: *Maba melinonii* Hiern, originally spelled as "*mellinoni,*" *Diospyros melinonii* (Hiern) A. C. Sm.]

Trees, to 25 m tall. Trunks with or without poorly developed buttresses. Stems pubescent, without hair-free lines. Leaves: petioles 4–5 mm long, pubescent; blades broadly lanceolate to elliptic, (3)5–9(12) × 2–3(5) cm, chartaceous, with scattered, white hairs, densely minutely glandular hairy, glaucous when fresh, paler when dry abaxially, the extrafloral nectaries patelliform, inconspicuous, the base obtuse, the apex acuminate, the margins not revolute; secondary veins in 7–9 pairs. Inflorescences axillary, ramiflorous; pedicels pubescent. Flowers not known from our area. Fruits subglobse, 2–2.7 cm diam., the surface covered with stiff, patent, golden-brown to ferruginous hairs arising singly from tubercles, the hairs ca. 3 mm long, the tubercles covered with minute hairs. Fr (Feb, Apr); in nonflooded forest. *Bois-charbon* (Créole).

Diospyros carbonaria Benoist [Syns.: *D. duckei* Sandwith, *D. capimnensis* Pires & Calvalc.] PL. 51b

Trees, to 20 m tall. Stems glabrous to puberulous with scattered hairs distally, without hair-free lines. Bark dark to nearly black, the slash red with white streaks, yellow at junction of inner bark and sapwood. Leaves: petioles (3)4–8 mm long, pubescent; blades elliptic to widely elliptic, (6)9–13(15) × (3.5)4–5.5(6) cm, chartaceous, sparsely pubescent, not glandular hairy abaxially, the extrafloral nectaries mostly absent, if present then inconspicuous and often removed from midrib, the base acute to obtuse, the apex acuminate, the margins not revolute, with scattered hairs; midrib with scattered hairs ad- and abaxially, the secondary veins in 6–9 pairs, the higher order venation (3rd and 4th order) plane or ± indistinct adaxially. Inflorescences axillary, ramiflorous. Flowers not known from our area. Fruits ovoid to cylindric, ca. 2.5 × 1.7–2 cm, densely tuberculate, pubescent, the hairs whitish, of 2 distinct lengths, the longest not >1 mm long, appressed or slightly patent. Fr (Sep, Oct); in nonflooded forest.

Diospyros cavalcantei Sothers PL. 51d

Trees, to 22 m tall. Trunks with steep buttresses. Bark dark, nearly black externally, the inner bark pink with white streaks. Stems densely pubescent, without hair free lines. Leaves: petioles 1–1.5 mm long; blades elliptic to narrowly elliptic, (3)6–9 × (1.7)2–3 cm, chartaceous, surface ± glabrous adaxially, densely pubescent on midrib ad- and abaxially, surface less pubescent abaxially than midrib, not glandular hairy abaxially, the extrafloral nectaries small, patelliform, the base acute, the apex acute to slightly acuminate, the margins densely pubescent; secondary veins in ca. 7 pairs, indistinct and difficult to count, the higher order venation (3rd and 4th order) ± indistinct adaxially. Inflorescences axillary, ramiflorous, cymose, the axis ca. 5 mm long; pedicels densely pubescent. Flowers nocturnal, collected from ground during day but none open in canopy. Staminate flowers: calyx ca. 9 mm long, pubescent ad- and abaxially; corolla ca. 14 mm long, thick, densely pubescent except for tips of lobes abaxially, white. Fruits (known from outside our area) markedly depressed globose (wider than long), not tuberculate, glabrescent (except on distal and proximal parts), the hairs ca. 1 mm long. Fl (Feb); in nonflooded forest.

Known from a single flowering collection (*Mori et al. 22939*, NY) and from a sterile treelet (*Wallnöfer et al. 13460*, NY).

Diospyros cayennensis A. DC. [Syn.: *Diospyros ierensis* Britton] PL. 51c

Canopy trees, to 30 m tall. Trunks cylindric. Bark nearly black externally, friable when cut, the inner bark light brown to yellowish with white streaks. Stems puberulous, soon glabrescent, without hair-free lines. Leaves: petioles 9–12 mm long, glabrous; blades elliptic to oblong, (6)10–18(20) × (3)5–8 cm, slightly coriaceous, often ± shiny (especially on distal leaves) adaxially, glabrous, not glandular hairy abaxially, the extrafloral nectaries small, patelliform, the base obtuse, the apex acuminate, the margins slightly revolute; secondary veins in 7–10 pairs, the higher order venation (3rd and 4th order) generally distinct and salient on both surfaces, forming a dense reticulum (rarely only distinct and prominent abaxially). Inflorescences axillary, ramiflorous, 1-few-flowered; pedicels generally present, densely pubescent. Flowers 17 mm diam.; calyx ca. 10 mm long (use flowers not fruits for this measurement), the lobes ovate-orbicular; petals fused ca. 2/3 length, yellow-green, the corolla tube ca. 14 × 4 mm, the lobes reflexed. Staminate flowers with ca. 10, pubescent stamens, these loosely adnate to base of corolla tube, the filaments 1.5 mm long, the anthers narrowly sagittate, ca. 3 mm long, attenuate at apex. Fruits globose, 4–5 cm diam., glabrous, drying brownish, the fruiting calyx 4 cm wide, 4-parted, the lobes

1.2 × 1.5 cm, slightly undulate. Fl (May), fr (Feb, Nov); in nonflooded forest.

Diospyros dichroa Sandwith

Canopy to emergent trees, to 45 m tall. Trunks at least sometimes buttressed. Bark dark, nearly black, the slash of outer bark black, the inner bark pinkish to red, with white streaks. Stems of mature plants densely, persistent pubescent except for hair-free lines running in zig-zag fashion from node to node. Leaves: petioles absent to 2 mm long; blades narrowly elliptic, less frequently narrowly ovate, 8.5–10.5 × 2–3 cm, length/width ratio >3, chartaceous, minutely glandular hairy, pale, densely covered with whitish-gray to golden or brownish appressed, sericeous hairs abaxially, the extrafloral nectaries small, hidden below indumentum, the base obtuse to rounded, less frequently acute, the apex long acuminate, the margins slightly revolute; secondary veins barely distinguishable. Known only from one sterile collection (*Granville 3390*, CAY) from our area.

Diospyros martinii Benoist [non *D. martinii* Amshoff]

Trees, usually <10 m tall. Bark gray to black, often mottled with white, the inner bark yellow or possibly the sapwood yellow (not clear from labels). Stems glabrous to pubescent, without hair-free lines. Leaves: petioles 7–10 mm long, pubescent; blades broadly ovate, (10)15–28 × (4.5)8–12 cm, slightly coriaceous, sparsely pubescent, minutely and densely glandular hairy abaxially, glaucous when fresh, usually distinctly paler than adaxially when dry, the extrafloral nectaries inconspicuous, the base obtuse to rounded, the apex acute to acuminate, the margins revolute; secondary veins in 6–11 pairs. Inflorescences axillary, ramiflorous, cymose, the axes densely pubescent; pedicels densely pubescent. Flowers ca. 20 mm diam.; petals fused only at base, white, sometimes tinged with pink, the lobes spreading. Staminate flowers with ca. 50 stamens. Pistillate flowers with ca. 30 staminodia adnate to corolla base, the staminodia covered with acicular hairs. Fruits barrel-shaped, irregularly gibbous, the apex markedly truncate, at least 2.5 cm diam., without tubercules, densely covered with ± appressed hairs, the hairs to 1 mm long. Fl (Aug, Oct, Nov), fr (Feb, Dec); in understory of nonflooded forest.

Diospyros ropourea B. Walln. [= *Ropourea guianensis* Aubl. non Diospyros *guianensis* (Aubl.) Gürke. Syns.: *Diospyros cauliflora* Mart. in Miq. non *D. cauliflora* Blume; *Maba cauliflora* (Mart. in Miq.) Hiern; *Diospyros martinii* Amshoff non *D. martinii* Benoist; *D. matheriana* auct. non *D. matheriana* A. C. Sm.] FIG. 101, PLS. 50c,d, 51a

Treelets to small trees, to 6 m tall. Bark blackish, smooth, the wood yellowish. Stems pubescent, without hair-free lines. Leaves: petioles 5–9 mm long, pubescent; blades narrowly elliptic to elliptic, (10)15–25 × (3)5–8(10) cm, chartaceous, sparsely pubescent, especially along veins, not glandular hairy abaxially, the extrafloral nectaries absent, the base acute, the apex acuminate, the margins not revolute; secondary veins in 6–9 pairs, the 2nd and 3rd order veins distinct and markedly impressed adaxially. Inflorescences mostly cauliflorous (ramiflorous in taller plants), fasciculate; pedicels pubescent. Flowers ca. 15 mm diam.; petals 4–5, connate at base, white, the lobes spreading. Staminate flowers with ca. 35 stamens, these adnate to base of corolla, the anthers densely pubescent, yellow. Pistillate flowers with staminodia long pubescent, entire or bifid at apex; ovary densely pubescent, yellow, the style trifid with each branch in turn bilobed, white. Fruits subglobose, 2–3 cm diam., sparsely covered with stiff, patent, to 4 mm long, brown to ferruginous hairs arising singly from tubercles, the tubercles densely covered with minute hairs, the pericarp yellow to orange at maturity. Seeds ca. 5 per fruit, surrounded by slimy pulp. Fl (Feb, Jul, Aug, Sep, Oct, Nov, Dec), fr (Jan, Feb, Apr, May, Jun, Sep, Oct, Nov, Dec); common, in understory of nonflooded forest. *Marie poil* (Créole).

Diospyros tetrandra Hiern [non *D. tetrandra* Span., nom. nud. in syn. sub *D. maritima* Blume]

Small trees, 5–8 m tall. Bark scaly, greenish-gray, the wood yellow. Stems glabrous. Leaves: petioles 6–10 mm long, glabrous; blades broadly lanceolate to elliptic, 9.5–16 × 4–6 cm, chartaceous, glabrous, dull, sometimes slightly shiny adaxially, not glandular hairy abaxially, dull, the extrafloral nectaries small, the base obtuse to tapering, the apex acuminate, the margins slightly revolute; secondary veins in 8–9 pairs, the higher order venation (3rd and 4th orders) distinct and salient on both surfaces, forming a distinct reticulum. Inflorescences: pedicels ± absent. Flowers: not known from our area; calyx 3–4 mm long, the lobes subulate; stamens 4. Fruits globose, ca. 2 cm diam. at maturity, glabrous, black when dry, the persistent calyx ca. 1.7 cm wide, the lobes markedly undulate-plicate. Fr (Feb, Aug); known only from two fruiting collections from our area (*Mori et al. 24920*, CAY, *Oldeman 1971*, CAY).

Diospyros vestita Benoist [non *D. vestita* Bakh., Syn.: *D. praetermissa* Sandwith]

Trees, to 40 m tall. Trunks cylindric. Bark dark, with poorly developed longitudinal fissures, detaching in irregular, small pieces, the slash of outer bark black, the inner bark pink, reddish with white streaks, or reddish-brown, the sapwood yellowish. Stems densely pubescent, without hair-free lines, becoming glabrous with age. Leaves: petioles inconspicuous, 2–3 mm long, pubescent; blades narrowly ovate, 6–7 × 2–3 cm, length/width ratio <3, chartaceous, with whitish-grey hairs (white when fresh), sericeous, minutely glandular hairy abaxially, the extrafloral nectaries often completely hidden below indumentum, the base obtuse, the apex acuminate, the margins not to slightly revolute; secondary veins in 8–9 pairs, these difficult to count. Inflorescences axillary, ramiflorous, 1- to few-flowered; pedicels pubescent. Flowers ca. 20 mm diam.; petals fused for 1/3 length, the lobes spreading. Staminate flowers with ca. 40 stamens, these adnate to base of corolla, the filaments 2 mm long, the anthers linear, 5 mm long, pubescent; pistillode pubescent. Pistillate flowers with ovary densely covered with long, simple hairs. Fruits cylindric to globose, 2–4 cm diam., pubescent, glabrescent at maturity, the hairs whitish, of two distinct types, long and short. Seeds ca. 5 per fruit, shiny black, covered with mucilaginous pulp, ca. 20 mm long; endosperm not ruminate. Fl (Sep, Nov), fr (Feb, May, Nov); in nonflooded forest.

This species is similar to *D. sericea* A. DC. of *cerrados* and margins of gallery forests of SE Brazil. With further study, the two names may be shown to represent the same species. In that case, the earlier name, *D. sericea*, would have priority.

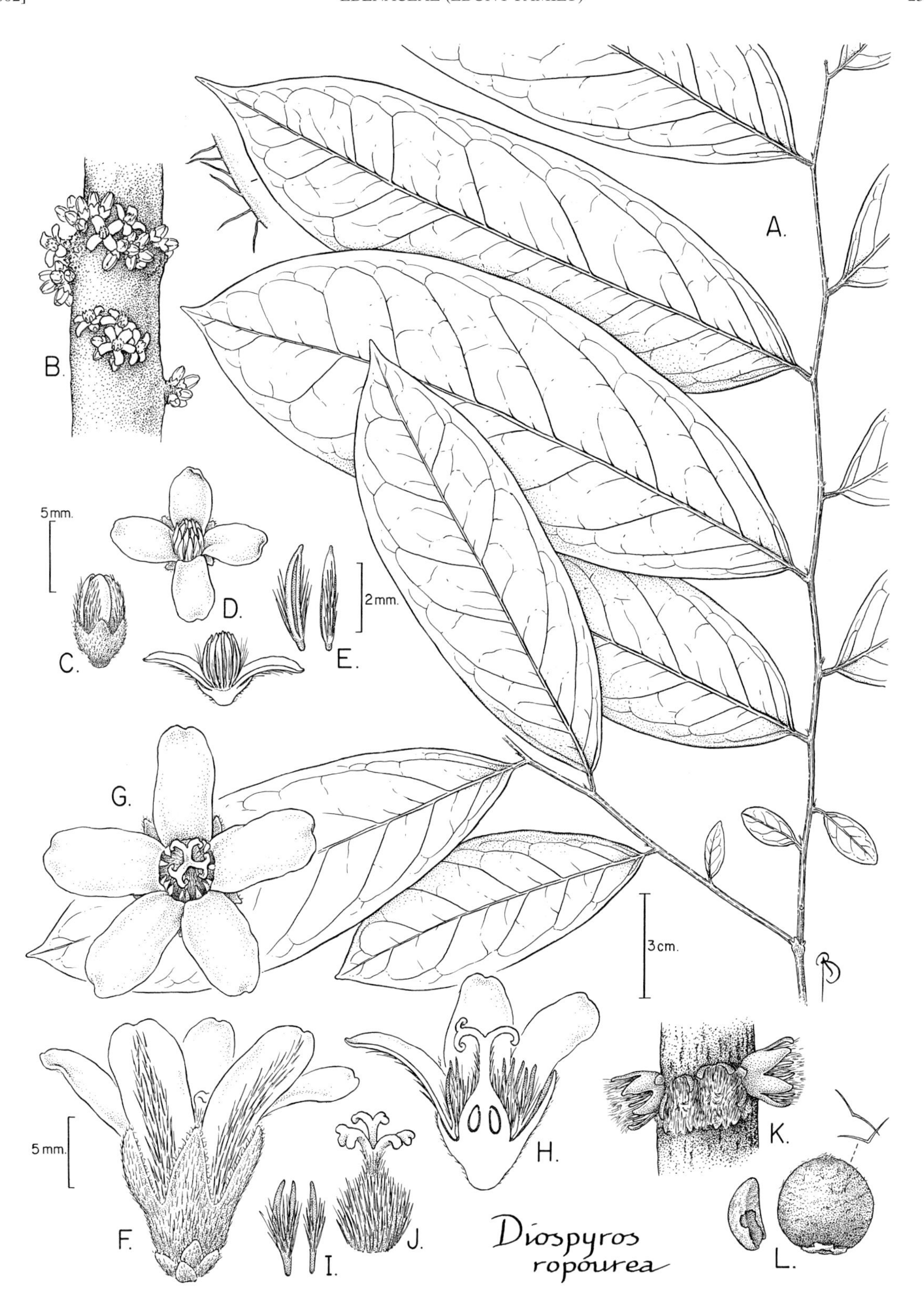

FIG. 101. EBENACEAE. *Diospyros ropourea* (A, F–J, *Mori et al. 21023*; B–E, *Mori et al. 23878*; K, unvouchered photo; L, *Mori & Pennington 18136*). **A.** Part of stem with leaves and detail of ciliate leaf margin. **B.** Part of trunk with cauliflorous flowers. **C.** Flower bud. **D.** Apical view (above) and medial section with intact androecium (below) of staminate flower. **E.** Adaxial view (far left) and abaxial view (near left) of stamen. **F.** Lateral view of flower showing hairs on middle of petals. **G.** Apical view of pistillate flower. **H.** Medial section of pistillate flower. **I.** Staminodes. **J.** Pistil. **K.** Immature fruits on tree trunk. **L.** Fruit (right) with enlargement of hairs (above right) and seed (left).

ELAEOCARPACEAE (Elaeocarpus Family)

Scott V. Heald, Alberto Vicentini, and Damon A. Smith

Trees or shrubs. Pubescence, when present, of simple trichomes. Stipules present, persistent or caducous. Leaves alternate or opposite (or both on same branch in species of *Sloanea*), simple; petioles not thickened or often thickened at both ends (*Sloanea*); blades with venation typically Malvalean (basally 3-veined outside our area) or pinnate (*Sloanea*). Inflorescences racemose, paniculate, or cymose (dichasia). Flowers actinomorphic, bisexual in our area; sepals (3)4–7(9), free or connate, often valvate; petals (3)4–5, free, or absent (in our area); stamens numerous, inserted on surface or margin of disc, the anthers with poricidal or lateral dehiscence; pistil compound, the ovary superior, (1)2–many-locular, the style 1; placentation axile, the ovules 2–many per locule, though often only 1 seed per fruit developing. Fruits loculicidal capsules in our area. Seeds with arils or sarcotesta (in our area).

Smith, C. E., Jr. 1954. The New World species of *Sloanea* (Elaeocarpaceae). Contr. Gray Herb. **175:** 1–114.

Jansen-Jacobs, M. J. 1986. Elaeocarpaceae. *In* A. L. Stoffers & J. C. Lindeman (eds.), Flora of Suriname **III(1–2):** 301–305. E. J. Brill, Leiden.

Uittien, H. 1966. Elaeocarpaceae. *In* A. Pulle (ed.), Flora of Suriname **III(1):** 58–64. E. J. Brill, Leiden.

SLOANEA L.

Trees. Trunks frequently developing narrow, steep, thin, occasionally flying buttresses, the boles often irregular. Leaves alternate or opposite (occasionally both on single branch); petioles usually thickened at both ends; blades with pinnate venation. Flowers: sepals 4(8), free or partly connate, valvate in bud; petals lacking (present in some species outside our area); ovary usually (3)4(5)-carpellate; ovules pendent, few-numerous. Fruits capsules, smooth or echinate. Seeds generally 1–2 per fruit, with aril or sarcotesta, the raphe lateral.

Common names applied to at least some members of this genus include *roucou sauvage* and *châtaignier*.

The last revision of Neotropical *Sloanea* was produced by C. Earl Smith (1954). Despite this effort, more monographic work is needed, and, therefore, names employed for species of this genus must be viewed as provisional.

1. Petioles <3 cm long.
  2. Leaves opposite (sometimes a leaf at a pair failing to develop and giving the impression that the leaves are alternate). . . . . *S. guianensis.*
  2. Leaves alternate to subopposite.
    3. Petioles distinctly pubescent, the individual hairs ≥0.1 mm long. Leaf blades with veins impressed adaxially. . . . . *S. echinocarpa.*
    3. Petioles glabrous or, if puberulous, the hairs <0.1 mm long. Leaf blades with all veins salient or plane adaxially.
      4. Leaf blades (4.5)6–11.5 cm wide.
        5. Cut trunk with strong smell of bitter almond. Petioles and leaf blades entirely glabrous. . . . *Sloanea* sp. B.
        5. Cut trunk without strong smell of bitter almond. Petioles and leaf blades at least somewhat pubescent. . . . . *S.* aff. *synandra.*
      4. Leaf blades 3.5–6(7) cm wide.
        6. Fruits smooth, the fruit valves oblong. . . . . *S. brachytepala.*
        6. Fruits echinate, the fruit valves elliptic.. . . . . *S. laxiflora.*
1. Petioles, at least some, >3 cm long.
  7. Stipules persistent. Fruits with spines curved toward apex. . . . . *S. grandiflora.*
  7. Stipules caducous. Fruits either with spines straight or without spines.
    8. Inflorescences pseudoterminal and axillary, cymose, the peduncle usually well developed. Fruits smooth. . . . . *S. latifolia.*
    8. Inflorescences axillary or from branches, racemose, the peduncle not well developed. Fruits with spines.
      9. Petioles and abaxial leaf blade surfaces glabrous or at most puberulous abaxially. Fruits with spines 3–4 mm long. . . . . *Sloanea* sp. A.
      9. Petioles and abaxial leaf blade surfaces distinctly pubescent abaxially. Fruits with spines 10–25 mm long.
        10. Longest petioles 3–4.5 cm long; blades with pubescence mostly on veins abaxially. . . . *S.* aff. *brevipes.*
        10. Longest petioles 3.5–14.5 cm long; blades with pubescence mostly over entire surface abaxially. . . . . *S. tuerckheimii.*

Sloanea brachytepala Ducke

Trunks with high buttresses. Stipules caducous. Leaves alternate, glabrous throughout; petioles 1–1.5 cm long × 1–1.5 mm diam.; blades widely elliptic, 7–9 × 3.5–6 cm, the base acute, very slightly oblique, the apex acute to bluntly acuminate, the margins entire; midrib and secondary veins plane to slightly salient adaxially, the secondary veins in 5–7 pairs. Inflorescences axillary, cymose; peduncle 13 mm long; pedicels 5–6 mm long. Flowers: sepals 4, probably free, pink abaxially; style simple. Fruits ellipsoid, 2.2 × 1.5 cm, smooth, the valves oblong. Seeds 1 per fruit. Fl (Aug), fr (Aug); apparently uncommon, in nonflooded forest.

Sloanea aff. **brevipes** Benth. PL. 52d

Trunks with steep, narrow, occasionally branched, sometimes flying buttresses, the boles irregular. Bark smooth, lenticellate, the slash pink, the outer bark reddish brown, the inner bark light orange. Stipules caducous. Leaves opposite or subopposite; petioles 3–4.5 cm long × 1.5–2.5 mm diam., pubescent, the hairs 0.1–0.2 mm long; blades suborbicular, obovate, or widely elliptic, 17–22 × 6–13 cm, the base usually obtuse to rounded, infrequently acute, the apex obtuse to acuminate, the margins dentate, especially toward apex, midrib and veins pubescent adaxially, pubescent mostly on veins abaxially; midrib slightly impressed adaxially, the secondary veins in 11–14 pairs. Inflorescences mostly from branches, less frequently axillary, racemose; peduncle 4–6.5 mm long; pedicels 10–14 mm long. Flowers: sepals (4)5–6(8), greenish, basally connate, erect at anthesis; style 3–4 parted at apex. Fruits 2–2.5 × ca. 1.5 cm, echinate, the spines densely covering fruit, straight, to 25 mm long, the valves narrowly elliptic. Seeds 1 per fruit, 1.5 cm long; aril bright red, completely covering seed. Fl (Oct), fr (Apr, Jul, Oct, Nov); in nonflooded forest.

Sloanea echinocarpa Uittien

Trunks with steep, thin, branched buttresses, the bole irregular. Stipules caducous. Leaves alternate; petioles 0.5–2 cm long × 1.5–2.5 mm diam., densely pubescent, the hairs 0.1–0.2 mm long; blades elliptic, 8–14 × 4–7 cm, midrib pubescent adaxially, all veins pubescent abaxially, the hairs 0.1–0.2 mm long, the base acute to rounded, the apex obtuse to acuminate, the margins entire; veins impressed adaxially, the secondary veins in 10–12 pairs. Flowers not known from our area. Fruits 2.5–3 cm long, echinate, the spines straight, 5–7 mm long, the valves narrowly elliptic. Old fr (Jul); apparently uncommon, in nonflooded forest.

Sloanea grandiflora Sm. FIG. 102

Trunks with steep, thin, flying buttresses. Stipules persistent, pulverulent, naviculate, 2–3 cm long, the margins erose to serrate. Leaves alternate; petioles 6–14.5 cm long × 3–4 mm diam., pulverulent; blades elliptic to widely elliptic, 33–40.5 × 18–21.5 cm, midrib and secondary veins puverulent adaxially, glabrous or with a few scattered hairs abaxially, the hairs <0.1 mm long, the base obtuse to cordate, the apex bluntly acuminate, the margins irregularly undulate to dentate; midrib and secondary veins plane to somewhat salient adaxially, the secondary veins in 13–15 pairs. Inflorescences terminal, axillary, or from branches, racemose or panicles of racemes; peduncle nearly absent to 12 mm long; pedicels ca. 10 mm long. Flowers: sepals cupulate, 4 mm long, greenish, with ca. 5 blunt, toothed lobes; style simple. Fruits globose, 2.5–5 × 3–5 cm, densely echinate, the inner surface magenta, the longer spines curved toward apex, 10–18 mm long, the valves elliptic. Seeds ca. 2 per fruit, oblong, 1.5 × 0.6 cm, the seed coat shiny, white; aril orange, completely covering seed. Fl (May, Aug), fr (Jan, Apr, Oct, Nov, Dec); common, at all altitudes in nonflooded forest on well-drained and poorly drained soil.

Sloanea guianensis (Aubl.) Benth.

Trunks with steep, high, occasionally flying, branched buttresses and aerial roots, the bole fluted. Bark brown, smooth, with hoops and small scales, the inner bark thin and green with an inner creamy layer. Stipules caducous. Leaves opposite, though sometimes only 1 developing and then appearing alternate; petioles 0.5–2.5 cm long × 1–1.5 mm diam., puberulent-pulverulent; blades narrowly obovate-widely elliptic, 7–19.5 × 3.6–10 cm, midrib puberulous-pulverulent adaxially, glabrous to puberulous abaxially, the hairs <0.1 mm long, the base usually acute, less frequently obtuse, the apex acute to acuminate, the margins entire; midrib and secondary veins plane to slightly impressed adaxially, the secondary veins in 7–10 pairs. Inflorescences axillary or from branches, racemose; peduncle nearly absent to 5 mm long; pedicels ca. 10 mm. Flowers: sepals 4–5. Fruits ellipsoid to obovoid, 1.1–1.4 × 0.5–0.9 cm, echinate, the outer surface dirty-red, velutinous, the inner surface white, the spines straight, 5–6 mm long, yellowish-green, the valves narrowly elliptic to oblong. Seeds 1 per fruit, ellipsoid, 1–1.3 × 0.5–0.7 cm, white; aril orangish-crimson, almost completely covering seed. Fl (Oct), fr (Jan, Apr, May, Nov); common, in nonflooded forest at all altitudes.

Sloanea latifolia (Rich.) K. Schum. PL. 51e

Trunks with steep, high, narrow, branched, occasionally flying buttresses, sometimes with prop roots, the bole irregular. Stipules caducous. Leaves alternate; petioles 3–10 cm long × 1.5–2 mm diam., pulverulent; blades elliptic to widely obovate, 10–23 × 5.5–13 cm, glabrous adaxially, pulverulent abaxially, especially along veins, the base obtuse to rounded, the apex acute to acuminate, the margins entire; midrib salient adaxially, the secondary veins in 7–10 pairs. Inflorescences pseudoterminal and axillary, compound cymes; peduncle 30–75 mm long; pedicels 15–20 mm long. Flowers strongly and pleasantly aromatic; sepals 4, free, reflexed at anthesis, light pink abaxially, deeper pink adaxially; style simple. Fruits asymmetrically globose, 3.5–5 × 2.5–4.5 cm, smooth, the valves elliptic. Seeds globose, 1 per fruit, 2.7 × 2.3 cm; aril white, covering basal 1/2 of seed. Fl (Sep, Oct), fr (Feb, Dec); scattered but common, in nonflooded forest.

Sloanea laxiflora Benth.

Trunks with high, narrow, thin buttresses. Bark reddish-brown, smooth. Stipules caducous. Leaves alternate; petioles 0.5–2.5 cm long × 1–1.5 mm diam., glabrous to puberulous, the hairs <0.1 mm long; blades elliptic to widely elliptic, 6–17.5 × 3–7 cm, glabrous adaxially, glabrous to puberulous on veins abaxially, the hairs <0.1 mm long, the base acute to obtuse, sometimes ± oblique, the apex acute to acuminate, the margins entire; midrib usually plane, the secondary veins in 5–7 pairs. Inflorescences axillary or terminal, compound cymes; peduncle poorly developed to 20 mm long; pedicels 5–15 mm long. Flowers: sepals 4, reflexed at anthesis, free, cream-colored suffused with pink, deepest pink at base adaxially; style simple. Fruits ellipsoid, ca. 3 × 2 cm, the spines stout, straight, 5 mm long, the valves elliptic. Seeds 1–2 per fruit, ovoid, 1.7 × 0.9 cm; aril orange, completely covering seed. Fl (Sep, Oct, Nov), fr (Apr, Nov); common, in nonflooded forest.

Sloanea aff. **synandra** Benth. PL. 52a

Trees, to 30 m tall. Trunks with steep, narrow, plank buttresses. Stipules caducous. Leaves alternate; petioles 1.5–2.5 cm long × 2–2.5 mm diam., densely pubescent, the hairs <0.1 mm long; blades elliptic to widely elliptic, 13–18 × 8–11 cm, discolorous, midrib and secondary veins pubescent ad- and abaxially, the hairs <0.1 mm long, the base acute to rounded, the apex acute to short acuminate, the margins entire; midrib and secondary veins usually plane adaxially, the secondary veins in 10–14 pairs. Inflorescences axillary in uppermost

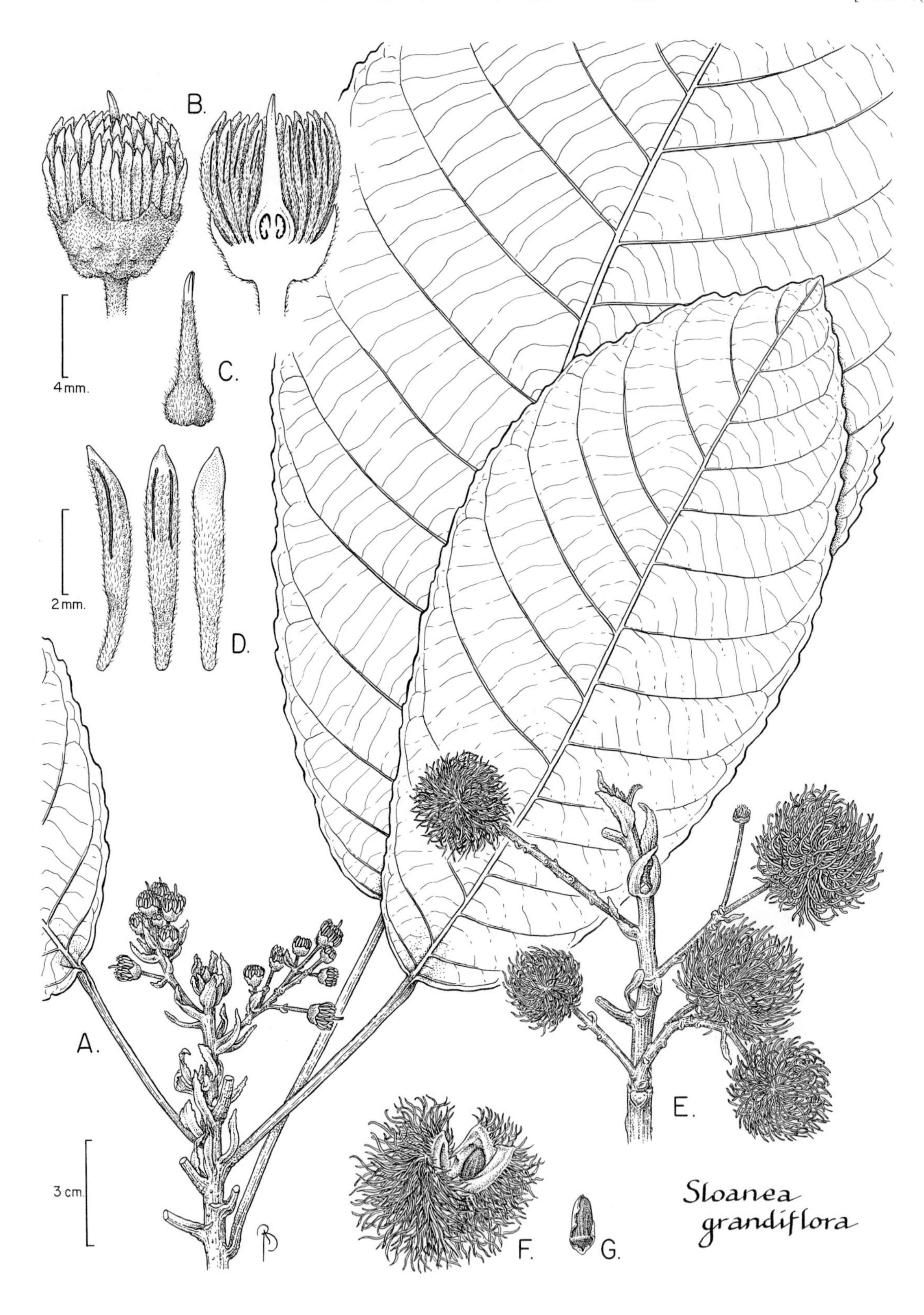

FIG. 102. ELAEOCARPACEAE. *Sloanea grandiflora* (A–D, *Mori et al. 18154*; E, *Mori & Boom 15119*; F, G, *Mori et al. 15697*). **A.** Part of stem with leaves and inflorescences. **B.** Lateral view (left) and medial section (right) of flower. **C.** Pistil. **D.** Lateral (far left), adaxial (center), and adaxial (near left) views of anthers. **E.** Part of infructescence with immature fruits. **F.** Dehisced fruit. **G.** Seed with aril.

leaves, compound cymes; peduncle 25–95 mm long; pedicels 15–30 mm long. Flowers strongly and pleasantly aromatic; sepals 4, free, spreading at anthesis, pink; style simple. Fruits not known from our area. Fl (Nov, Dec); in nonflooded forest at higher elevations.

A specimen, *Mori et al. 15678* (NY), collected at a lower elevation on the La Fumée Mountain Trail, differs in its slightly longer petioles and more reddish abaxial leaf blade surface. This collection was not used in preparing the description.

Sloanea tuerckheimii Donn. Sm. PL. 52b,c,f

Trunks with high, narrow, thin, branched buttresses. Bark reddish-brown, smooth, scaly, the inner bark reddish-brown, 1 mm thick. Stipules caducous. Leaves alternate to subopposite; petioles (2.5)3.5–14.5 cm long × 2–4 mm diam., puberulous-pulverulent; blades suborbiculate, widely obovate to widely elliptic, 19–46 × 8–27 cm, glabrous adaxially, usually densely puberulous over entire surface abaxially, the hairs <0.1 mm long, the base obtuse to rounded, the apex obtuse to rounded, the margins entire to bluntly dentate, especially toward apex; midrib usually plane to slightly salient, the secondary veins in 8–12 pairs. Inflorescences from branches, racemose; peduncle scarcely developed; pedicels 5–8 mm long. Flowers: sepals 6–7, free nearly to base, erect at anthesis; style 4-parted at apex. Fruits ellipsoid, ca. 2 × 1.2 cm, echinate, the spines straight, ca. 8 mm long, first green, then red as fruit matures, the valves elliptic. Fl (Sep, Nov), fr (Feb); in nonflooded forest.

Two collections, *Mori et al 23322* and *24716* (both at NY) do not have the the same abaxial leaf blade pubescence as the other collections. The latter collection is said to come from the same tree as *Mori et al. 24773* (NY) which does have well developed abaxial leaf pubescence. Although provisionally determined as this species, these collections were not used in preparing the description.

Sloanea sp. **A** (*Mori et al. 15498*, NY) PL. 52e

Trunks with steep, high buttresses, the bole fluted. Stipules caducous. Leaves opposite to subopposite; petioles (2.5)3–4 cm long, glabrous to puberulous, the hairs <0.1 mm long; blades widely elliptic, 9–14 × 8–9.5 cm, mostly glabrous, slightly puberulous on lower midrib adaxially, the base obtuse to rounded, the apex rounded to obtuse, the margins entire; midrib and secondary veins plane to slightly impressed adaxially, the secondary veins in 6–9 pairs. Inflorescences from branches, racemose. Flowers not known. Fruits ellipsoid, 2.5–3 × 2 cm, echinate, green on outside, dark pink on inside, the spines ca. 4 mm long, the valves elliptic. Seeds apparently 1 per fruit; aril reddish-orange, surrounding seed. Fr (Apr); in nonflooded forest.

Sloanea sp. **B** (*Boom 1645*, *1664*; *Mori et al. 15681*, *23849*, *24848*, *24850*, NY)

Trunks with steep, branched, thin, flying buttresses and aerial roots, the bole irregular along lower portion. Bark smooth, lenticellate, scaly, the slash pinkish, with strong smell of bitter almond. Stipules caducous. Leaves alternate to subopposite, glabrous throughout; petioles 0.5–2.5 cm long × 2–3 mm diam.; blades narrowly obovate to elliptic, 10–20 × 4.5–11.5 cm, the base cuneate, acute, obtuse, or rounded, the apex acute, shortly acuminate, or rounded, the margins entire; midrib and secondary veins prominent adaxially, the secondary veins in 7–9 pairs. Inflorescences unknown. Flowers unknown. Fruits 4–6 × ca. 4 cm, echinate, the spines robust, ca. 20 mm long, the valves elliptic. Seeds ca. 2 × 2 cm; aril orange. Fr (Apr with seed, May on ground without seed, Jul on ground without seed); in nonflooded forest.

ERICACEAE (Blueberry Family)

James L. Luteyn

Epiphytic shrubs, sometimes lianoid. Indumentum of uni- to multicellular hairs, these sometimes glandular. Stipules absent. Leaves simple, alternate; petioles present; blades evergreen, coriaceous, usually bearing deciduous, minute, glandular trichomes, the margins entire; venation with midvein accompanied by several other nearly equal secondary veins arising at or near the base of the blade (plinerved). Inflorescences axillary, racemose or flowers solitary, the flowers in axils of floral bracts; pedicels 2-bracteolate. Flowers actinomorphic, bisexual, 4–5-merous, obdiplostemonous, without floral aromas, rarely with extrafloral nectaries, the aestivation valvate; hypanthia terete or bluntly angled, the bases sometimes apophysate; perianth biseriate; calyx synsepalous; corolla sympetalous, cylindric to cylindric-urceolate; stamens 4 or 10, equal or alternately unequal, the filaments distinct or connate, equal or unequal, the anthers distally with 2 tubules, with dehiscence introrse by clefts or pores, the pollen grains in tetrahedral tetrads; pistils single, the ovary inferior, 4–5-carpellate, the styles single, the stigmas simple; placentation axile, the ovules numerous per locule. Fruits berries. Seeds small, ca. 1–1.5 mm long, numerous, sometimes enclosed in mucilaginous sheath; endosperm fleshy; embryos straight, white or green.

Most Ericaceae are pollinated by hummingbirds. The fruits are edible but insipid.

Lanjouw, J. 1937. Ericaceae. *In* A. Pulle (ed.), Flora of Suriname **IV:** 492–493.

———. 1966. Ericaceae. *In* A. Pulle (ed.), Flora of Suriname **II(1):** 299–301. E. J. Brill, Leiden.

Luteyn, J. L. 1983. Ericaceae - Part I. *Cavendishia*. Fl. Neotrop. Monogr. **35:** 1–290.

———. (ed.). 1995. Ericaceae — Part II. The superior-ovaried genera (Monotropoideae, Pyroloideae, Rhododendroideae, and Vaccinioideae p.p.). Fl. Neotrop. Monogr. **66:** 1–560.

———. 1998. Ericaceae. *In* J. A. Steyermark, P. E. Berry & B. K. Holst (gen. eds.), Flora of the Venezuelan Guayana **4:** 735–769. P. E. Berry, B. K. Holst & K. Yatskievych (vol. eds.). Missouri Botanical Garden Press, St. Louis.

1. Leaves ca. 0.9–1.8 cm long, the apex obtuse or rounded. Inflorescences of solitary flowers. Pedicels continuous with calyx. Corollas 4–6 mm long. . . . . *Sphyrospermum.*
1. Leaves 6–14(21) cm long, the apex acuminate. Inflorescences racemose. Pedicels articulate with calyx. Corollas 8–20 mm long.
    2. Rachises (2.5)5–10 cm long; bracts 13–25 mm long. Corolla white; staminal filaments distinct. . . . . *Cavendishia.*
    2. Rachises 0.6–1.5 cm long; bracts 1–1.75 mm long. Corolla red; staminal filaments connate. . . . . *Satyria.*

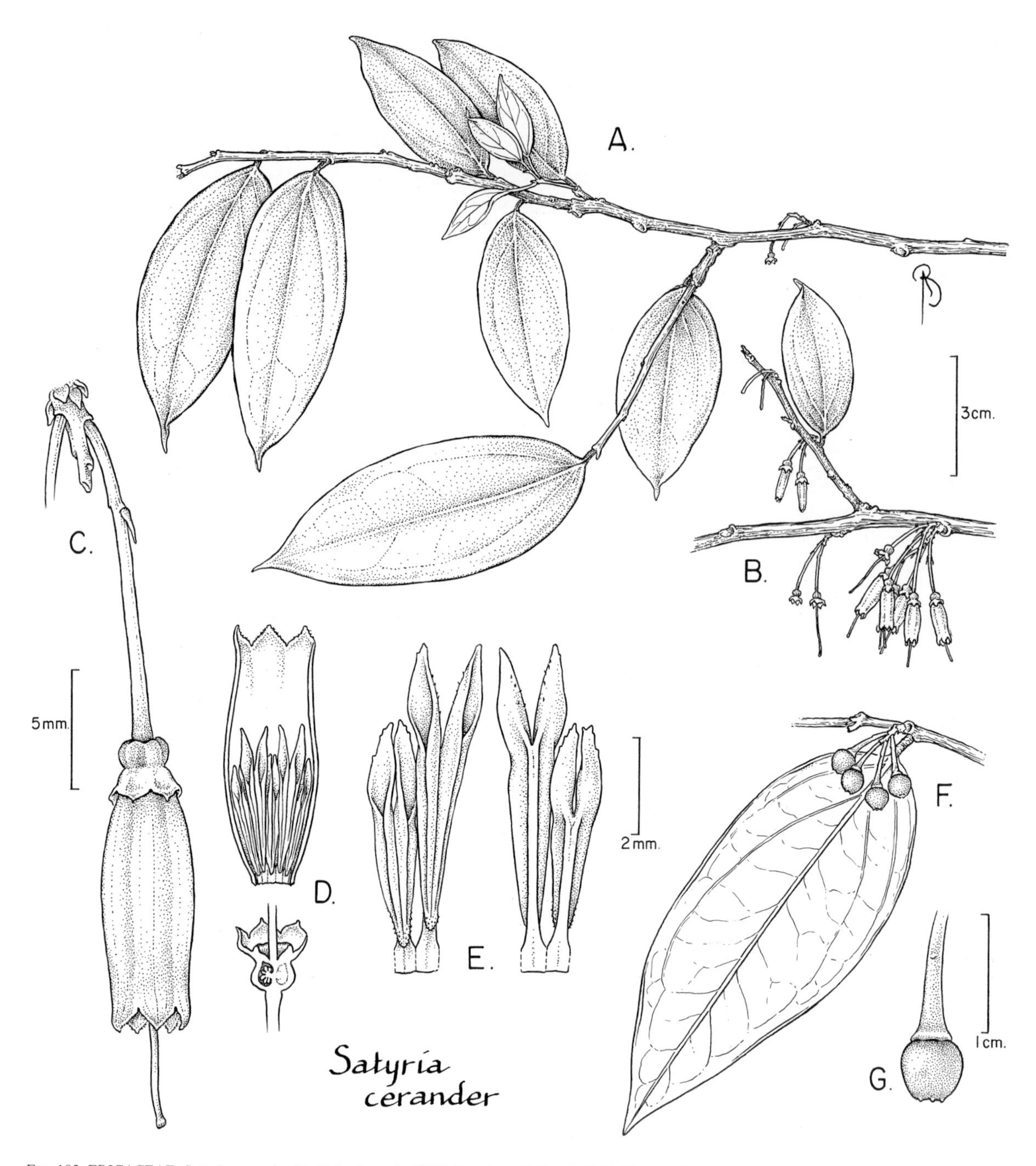

FIG. 103. ERICACEAE. *Satyria cerander* (A–E, *Irwin et al. 47607* from Amapá, Brazil; F, G, *Mori et al. 23656*). **A.** Part of stem with leaves and old inflorescence. **B.** Part of stem with leaf and inflorescences in various stages of maturity. **C.** Lateral view of flower. **D.** Medial section of corolla and androecium (above) and calyx and pistil (below). **E.** Adaxial (left) and abaxial (right) views of two types of stamens. **F.** Part of stem with leaf and fruits. **G.** Lateral view of fruit.

CAVENDISHIA Lindl.

Epiphytic shrubs. Inflorescences racemose; bracts large and showy. Flowers 5-merous; hypanthium base apophysate, the limb and lobes erect after anthesis, the lobes entirely glandular; calyx articulate with pedicel; stamens 10, alternately unequal, subequalling corolla, the filaments distinct and alternately unequal, shorter than anthers, the anthers alternately unequal, the tubules about same width as thecae and about twice as long and dehiscing by elongate clefts; ovaries 5-locular. Fruits dark blue-black. Seeds with white endosperm.

Cavendishia callista Donn. Sm. [Syn.: *Cavendishia duidae* A. C. Sm.]

Stems glaucous, muricate with tiny reddish papillae, glabrous. Leaves: petioles 5–11 mm long, glabrous; blades nitid, ovate to ovate-elliptic, (7)9–14(21) × (2)3–6 cm, glabrous, the base rounded or often slightly cordate, the apex short- to long-acuminate, sometimes abruptly short-acuminate; 3(5)-veined, the inner lateral veins arising to 2 cm above base. Inflorescences viscid, (10)15–25-flowered, the rachises glabrous, (2.5)5–10 cm long; bracts oblong to obovate, 13–25 × 11–20 mm, glabrous, pink to deep rose; pedicels 10–15 mm long, glabrous; bracteoles ovate-lanceolate, 2–3 mm long, glabrous, the apex glandular-callose. Flowers: hypanthium coarsely 10-ribbed, the limb cylindric or campanulate, the lobes triangular, 1–1.2 mm long, erect after anthesis, entirely glandular-callose; calyx 5–10 mm long, glabrous; corolla 18–20 mm long, glabrous, the tube white to grayish pearl-white, the lobes 1–3 mm long, with purple margins; stamens 18–20 mm long, the filaments alternately ca. 3 mm and 8 mm long, the anthers alternately 18 mm and 13 mm long. Fl (Jan, Nov, Dec), fr (Jan); in forest at all elevations.

Staminal measurements taken from corollas from the Venezuelan Gran Sabanna (*Luteyn et al. 6331*, NY).

SATYRIA Klotzsch

Epiphytic shrubs. Inflorescences often ramiflorous, racemose; bracts small and inconspicuous. Flowers 5-merous; hypanthium basally rounded, the limb flaring or spreading, the lobes eglandular; calyx articulate with pedicel; stamens 10, alternately unequal, usually about 1/3 as long as corolla, the filaments connate and equal, shorter than anthers, the anthers alternately unequal, dimorphic in shape, the longer ones with tubules spreading distally and tips recurved, the shorter ones with tubules laterally coherent distally and tips straight, the tubules not sharply differentiated from thecae and dehiscing by subapical clefts; ovaries 5-locular. Fruits dark blue-black. Seeds with white endosperm.

Satyria cerander (Dunal) A. C. Sm. FIG. 103

Stems glabrous. Leaves: petioles 3–6 mm long, glabrous; blades ovate to ovate-elliptic, (6)7–12 × (2.5) 3–5 cm, glabrous, the base rounded, the apex short-acuminate; 3–5-veined, the lateral veins arising from near base. Inflorescences with 1–3 rachises per axis, 4–9-flowered, the rachises 6–15 mm long, glabrous; bracts minute, 1–1.75 mm long, green; pedicels, 7–15 mm long, glabrous; bracteoles ovate, often abruptly acuminate, 1.2–2 mm long, persistent. Flowers: hypanthium cylindric-campanulate, 1.5–2 mm long, basally truncate, the limb campanulate-spreading, 1–1.5 mm long, the lobes apiculate, connivent after anthesis, the sinuses broadly rounded; calyx ca. 3–3.5 mm long, minutely puberulent; corollas apparently 5-angled, 8–11 mm long, glabrous, red; stamens alternately 4.5–5 mm and 5.5–6 mm long, the filaments equal, 1.5–2 mm long, the anthers alternately 4–4.5 mm and 5–5.5 mm long, the longer tubules flaring distally and dehiscing by broad oval clefts ca. 2 mm long; styles long-exserted. Mature berries not seen. Fr (Jan, Dec); in forest on Mont Galbao at 762 m.

Staminal measurements taken from corollas from Amapá, Brazil (*Pires et al. 50786*, NY).

SPHYROSPERMUM Poepp. & Endl.

Slender, epiphytic shrubs. Inflorescences of solitary flowers; bracts small and inconspicuous; pedicels thin, cernuous, distally swollen. Flowers 4-merous; hypanthia subglobose or obconic, the limb campanulate, suberect; calyx continuous with pedicel; stamens 4, equal, about equalling corolla, the filaments distinct and equal, longer than anthers, the tubules distinct, as long as or longer than thecae and dehiscing by oval clefts; ovary 4-locular. Fruits translucent white to lavender or purplish, when immature pericarp dry and brittle, when mature thin and papery. Seeds with green embryo clearly visible through endosperm.

Sphyrospermum buxifolium Poepp. & Endl.

Stems glabrate. Leaves: petioles ca. 1 mm long, puberulous to glabrate; blades suborbicular to oblong-ovate, 9–18 × 6–12 mm, glabrate, the base rounded, the apex rounded or obtuse; obscurely 3-veined from base. Inflorescences with flowers extending beyond subtending leaves; pedicels 10–13 mm long, glabrous to sparingly pilose, slightly and gradually swollen distally. Flowers: hypanthium subglobose, 1–1.5 mm diam., the limb suberect, ca. 0.5 mm long, the lobes apiculate, the sinuses rounded with a tuft of short trichomes or glabrous; calyx pilose to glabrate; corolla 4–6 mm long, sparsely pilose to glabrous, white; stamens 4, equal, slightly shorter than corolla, the filaments 1.8–3 mm long, the anthers 1.8–2.5 mm long. Fruits 4–6 mm diam., pilose or glabrous. Fl (Jan, Sep, Dec), fr (Jan, Sep, Dec); in forest on Mont Galbao.

ERYTHROXYLACEAE (Coca Family)

Scott V. Heald

Shrubs or small trees. Plants glabrous throughout. Stipules intrapetiolar, with or without numerous, distinct, longitudinal striations, one per leaf though often splitting medially from base. Cataphylls keeled, distichous, generally with two or more lateral keels, ± deciduous from base of expanding stems. Leaves simple, alternate, distichous; petioles present; blades elliptic, the margins entire; midrib impressed adaxially toward base, prominent abaxially. Inflorescences axillary, congested into heads; bracts ± numerous, with or without distinct longitudinal striations; pedicels present. Flowers actinomorphic, bisexual, heterostylous; calyx 5-merous, with valvate aestivation, the sepals connate at base, with ± prominent ridges; corolla 5-merous, with imbricate aestivation, the petals free, keeled, variously infolded adaxially to produce

1–2 series of bilobed appendages; stamens 10, the filaments basally connate into tube; ovary superior, (2)3-locular, smooth to tuberculate, the styles three, separate, the stigmas three, capitate; placentation axile, 1(2) ovules in fertile locule, only one developing into seed. Fruits drupes, often subtended by persistent calyx and stamens.

The family is economically important for cocaine manufactured from the leaves of the Andean species, *Erythroxylum coca*.

Plowman, T. & P. E. Berry. 1999. Erythroxylaceae. *In* J. A. Steyermark, P. E. Berry, K. Yatskievych & B. K. Holst (gen. eds.), Flora of the Venezuelan Guayana **5:** 59–71. P. E. Berry, K. Yatskievych & B. K. Holst (vol. eds.). Missouri Botanical Garden Press, St. Louis.

Lindeman, J. C. 1986. Erythroxylaceae. *In* A. L. Stoffers & J. C. Lindeman (eds.), Flora of Suriname **III(1–2):** 501–506. E. J. Brill, Leiden.

Westhoff, V. 1942. Erythroxylaceae. *In* A. Pulle (ed.), Flora of Suriname **III(2):** 1–12. J. H. De Bussy, Amsterdam.

ERYTHROXYLUM P. Browne

The only neotropical genus. The generic description same as the description of the family.

1. Stipules and cataphylls with distinct longitudinal striations.
   2. Petioles ≤5 mm long; blades <16 cm long. Inflorescences in heads ≤4 mm wide. Flowers with ovary ± tuberculate, obovoid.
      3. Stipules and cataphylls 2.5–5.5 mm long. Petioles 2–3 mm long; blades 13–15.5 cm long. Inflorescences in heads 2–3 mm wide. Flowers with ovary distinctly tuberculate. . . . . *E. citrifolium*.
      3. Stipules and cataphylls 7–15 mm long. Petioles 4–5 mm long; blades 11–11.3 cm long. Inflorescences in heads 3–4 mm wide. Flowers with ovary slightly tuberculate. . . . . *E. mucronatum*.
   2. Petioles 8–15 mm long; blades 17–19.5 cm long. Inflorescences in heads 8–15 mm wide. Flowers with ovary smooth, globose. . . . . *E. macrophyllum*.
1. Stipules and cataphylls lacking distinct longitudinal striations. . . . . *E. kapplerianum*.

Erythroxylum citrifolium A. St.-Hil.

Shrubs or small trees, to 6(12) m tall. Stipules attenuate, 2.5–5.5 mm long, with numerous distinct, longitudinal striations, one per leaf though often splitting medially from base. Cataphylls 3.5–4 mm long, deciduous from bases of expanding branches, with distinct longitudinal striations. Leaves: petioles 2–3 mm long; blades 13–15.5 × 5–6.3 cm, coriaceous, ± dull on both surfaces, the base acute to obtuse, the apex acuminate with rounded tip; secondary veins in 9–13 pairs. Inflorescences congested into heads, 2–3 mm wide, with or without subtending cataphyll; bracts numerous, 1 mm long, lacking distinct longitudinal striations; pedicels 6.5–9 mm long. Flowers: sepals cuneate, 1 mm long; petals oblong, 3 mm long; androecium with basal staminal tube slightly lobed, 1.5 mm long; ovary elliptical to obovoid, 3(4)-locular, distinctly tuberculate. Fruits ellipsoid drupes, 1 cm long, turning from white or green to red when mature. Fl (Feb), fr (Mar, Apr); in nonflooded forest. *Pimenta de nambu* (Portuguese).

Erythroxylum kapplerianum Peyr.

Shrubs or small trees, to 8 m tall. Stipules 2–2.5 mm long, lacking distinct, longitudinal striations, one per leaf. Cataphylls 3 mm long, ovate and adaxially concave, few in number and deciduous from bases of expanding branches, lacking distinct, longitudinal striations. Leaves: petioles 2–3 mm long; blades 8.5–11 × 3–4.5 cm, chartaceous, ± dull on both surfaces, the base acute to cuneate, the apex acuminate; secondary veins in 7–9 pairs. Inflorescences congested into few-flowered heads, 2 mm wide, without distinct subtending cataphyll; bracts few, 1–1.5 mm long, lacking longitudinal striations; pedicels 3–4 mm long. Flowers: sepals cuneate, 1.5 mm long; petals ovate, 3.5 mm long; basal staminal tube slightly lobed, 1 mm long; ovary obovoid, 3-locular, smooth. Fruits ellipsoid drupes, 1 cm long, red when mature. Fl (Jan), fr (May); in nonflooded forest.

Erythroxylum macrophyllum Cav. FIG. 104, PL. 53a,b

Small trees, to 7 m tall. Stipules attenuate, 15–20 mm long, with numerous distinct, longitudinal striations, one per leaf though often splitting medially from base. Cataphylls 18–22 mm long, ± persistent at bases of expanding branches, with distinct longitudinal striations. Leaves: petioles 8–15 mm long; blades 17–19.5 × 6–7.5 cm, coriaceous, shiny adaxially, duller abaxially, the base acute to cuneate (obtuse), the apex acute, occasionally with mucro; secondary veins in 9–11 pairs, divided into reticulum toward margins. Inflorescences congested into many-flowered heads, 8–15 mm wide, often with subtending cataphyll; bracts numerous, 1.5–3.5 mm long, with numerous longitudinal striations; pedicels 5–12 mm long. Flowers fragrant; sepals ovate, 4–5 mm long, green with dark midrib; petals oblong, 4 mm long, white with dark midrib; basal staminal tube 10-lobed, 1.5 mm long; ovary globose, 3-locular, smooth. Fruits ovoid drupes, 1–1.2 cm long, yellowish-orange when mature. Fl (Feb, Sep, Oct, Nov, Dec), fr (Jan, Feb, Mar, Jul, Oct); in nonflooded forest.

Erythroxylum mucronatum Benth.

Shrubs or small trees, to 10 m tall. Stipules attenuate, 7–11 mm long, with numerous distinct, longitudinal striations, one per leaf though often splitting medially from base. Cataphylls 12–15 mm long, interspersed in groups of 4–6 along branches and occasionally subtending inflorescences, with distinct longitudinal striations. Leaves: petioles 4–5 mm long; blades 11–11.3 × 4.6–4.9 cm, coriaceous, ± dull on both surfaces, the base acute to obtuse, the apex acute, often with blunt mucro; secondary veins in 11–14 pairs. Inflorescences congested into several-flowered heads, 3–4 mm wide, with subtending cataphyll; bracts numerous, 1 mm long, with distinct longitudinal striations; pedicels 4 mm long. Flowers: sepals cuneate, 1.5–2 mm long; petals elliptic, 3.5 mm long; basal staminal tube 10-lobed, 1.7 mm long; ovary obovoid, 3-locular, slightly tuberculate. Fruits oblong drupes, 1.2 cm long, turning from green to red when mature. Fl (Jan, Oct), fr (Feb, Jul, Oct); in nonflooded forest.

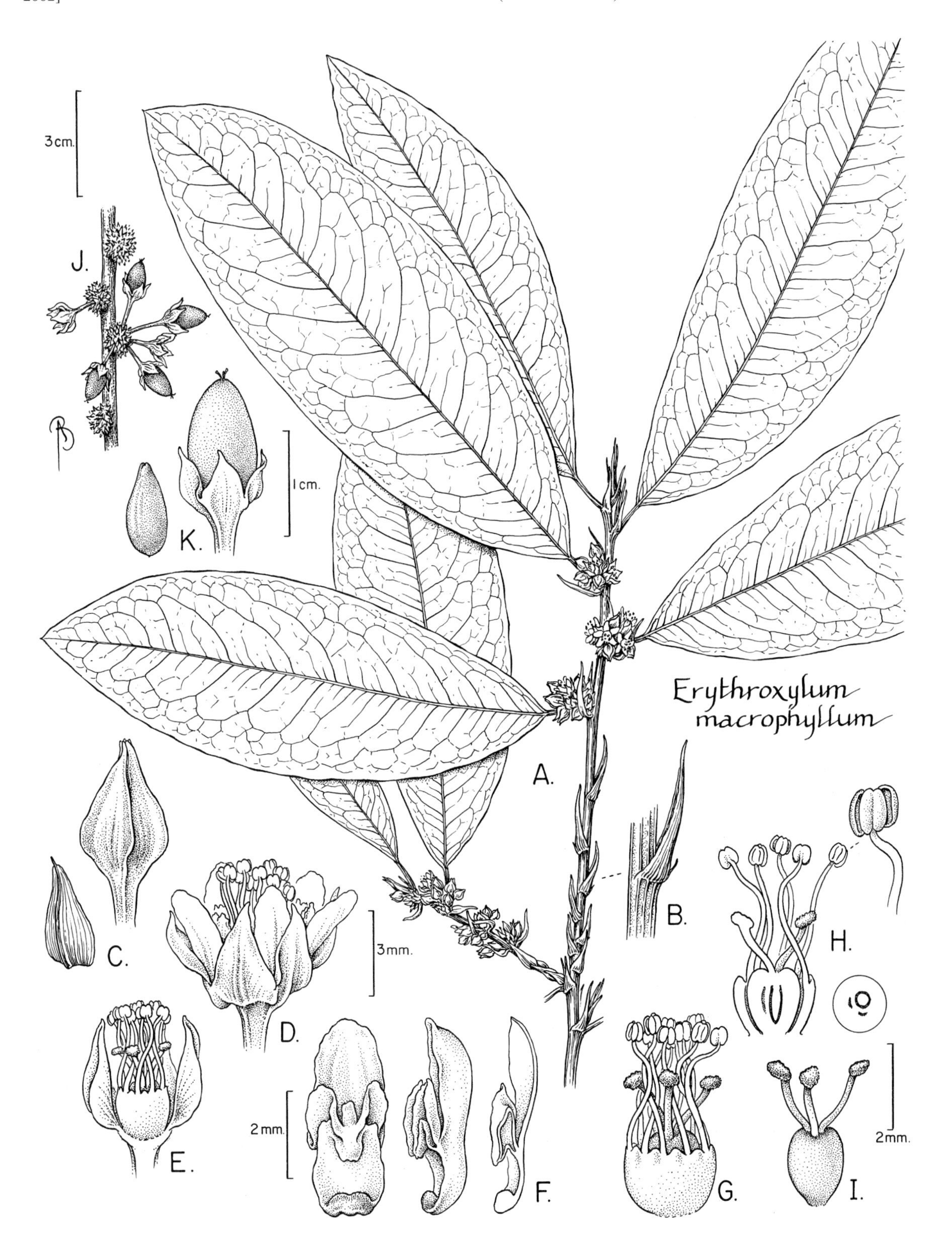

FIG. 104. ERYTHROXYLACEAE. *Erythroxylum macrophyllum* (A–I, *Mori et al. 22753*; J, K, *Mori & Gracie 21153*). **A.** Part of stem with leaves, stipules, cataphylls, and inflorescences. **B.** Cataphyll. **C.** Floral bud (above) and floral bract (left). **D.** Lateral view of flower. **E.** Lateral view of flower with petals and two sepals removed. **F.** Adaxial (far left), lateral (middle left), and medial section (near left) of petal. **G.** Androecium and pistil. **H.** Medial section of androecium and pistil (left), transverse section of ovary (below), and detail of anther (above). **I.** Lateral view of pistil. **J.** Part of stem with infructescences. **K.** Fruit (right) and seed (left).

EUPHORBIACEAE (Spurge Family)

Lynn J. Gillespie

Trees, shrubs, herbs, vines, or lianas. Latex sometimes present, milky white, colored, or clear. Stipules usually present. Leaves simple or compound, alternate or, less commonly, opposite. Inflorescences diverse, sometimes (*Euphorbia*) of specialized structures called cyathia. Flowers mostly actinomorphic, unisexual (plants monoecious or dioecious); calyx of 3–6 lobes or segments, rarely absent; petals 3–6, often reduced or absent; glandular nectary disc often present; stamens (1)3–20(1000); ovary superior, (1)3(20)-locular; placentation axile, the ovules 1–2 per locule, the styles free to connate, entire to multifid. Fruits mostly capsular with 3 explosively dehiscing cocci separating from a persistent columella, sometimes drupaceous or baccate. Seeds 1 or 2 per locule, or fewer by abortion; fleshy caruncle sometimes present.

Gillespie, L. J. 1993. Euphorbiaceae of the Guianas: Annotated species checklist and key to the genera. Brittonia **45(1):** 56–94.

——— & W. S. Armbruster. 1997. A contribution to the Guianan Flora: *Dalechampia*, *Pera*, *Plukenetia*, and *Tragia* (Euphorbiaceae) with notes on subfamily Acalyphoideae. Smithsonian Contr. Bot. **86:** 1–48.

Görts-van Rijn, A. R. A. 1976. Euphorbiaceae. *In* J. Lanjouw & A. L. Stoffers (eds.), Flora of Suriname **II(2):** 387–424. E. J. Brill, Leiden.

Lanjouw, J. 1939. Euphorbiaceae. *In* A. Pulle (ed.), Flora of Suriname **I(1):** 457–470, J. H. De Bussy, Amsterdam.

———. 1966. Euphorbiaceae. *In* A Pulle (ed.), Flora of Suriname **II(1):** 1–101. E. J. Brill, Leiden.

Webster, G. L., P. E. Berry, W. S. Armbruster, H.-J. Esser, L. J. Gillespie, W. J. Hayden, G. A. Levin, R. de S. Secco & S. V. Heald. 1999. Euphorbiaceae. *In* J. A. Steyermark, P. E. Berry, K. Yatskievych & B. K. Holst (gen. eds.), Flora of the Venezuelan Guayana **5:** 72–228. P. E. Berry, K. Yatskievych & B. K. Holst (vol. eds.). Missouri Botanical Garden Press, St. Louis.

1. Vines or lianas.
    2. Lianas, climbing by means of tendrillate shoots. Petioles distinctly biglandular at apex. . . . . *Omphalea.*
    2. Vines or lianas, climbing by twining. Petioles not biglandular at apex.
        3. Leaf blades with pair of flat glands near base. Stinging hairs absent. Fruits fleshy, indehiscent, 6–11 cm diam., 4-seeded. . . . . *Plukenetia.*
        3. Leaf blades eglandular. Stinging hairs present (at least on capsule). Fruits capsular, <2 cm diam., 3-seeded.
            4. Leaf blades lobed or not lobed, with pair of stipels at base. Inflorescences pseudanthial, with flowers subtended by pair of involucral bracts, these often large and colorful, but sometimes reduced. Styles entirely connate into a stout column. . . . . *Dalechampia.*
            4. Leaf blades unlobed, lacking stipels. Infloresences not pseudanthial (racemose, sometimes with single elongate basal branch), not subtended by large bracts. Styles only partly connate, slender. . . . *Tragia.*
1. Herbs, shrubs, or trees (*Manihot* is sometimes a scandent shrub).
    5. Herbs or subshrubs.
        6. Leaf blades deeply 3–5-lobed, with branched glandular hairs. . . . . *Jatropha.*
        6. Leaf blades not lobed, the hairs not glandular.
            7. Erect or sometimes decumbent herbs. Latex or colored exudate absent. Leaves alternate, on deciduous stems with appearance of pinnate leaves. Flowers axillary, single or clustered, mostly pedicellate, not enclosed within cyathia. . . . . *Phyllanthus.*
            7. Prostrate, decumbent, or erect herbs. Latex white. Leaves opposite (or both opposite and alternate in *E. cyathophora*), not on deciduous stems. Flowers enclosed within cup-shaped cyathia. . . . . . *Euphorbia.*
    5. Shrubs or trees.
        8. Leaves mostly palmately compound or very deeply palmately lobed so as to appear palmately compound.
            9. Trees. Leaves 3-foliolate. . . . . *Hevea.*
            9. Shrubs, sometimes scandent shrubs. Leaves 5-foliolate or -lobed. . . . . *Manihot.*
        8. Leaves simple (but sometimes palmately lobed).
            10. Leaf blades deeply palmately lobed.
                11. Leaf blades 7–11-lobed, 10–60 cm long, peltate. Stamens 100–1000, partly and irregularly connate. . . . . *Ricinus.*
                11. Leaf blades 3–5-lobed, <15 cm long, not peltate. Stamens 8–12, free to partly connate.
                    12. Latex clear or colored. Leaves with branched glandular hairs. Inflorescence cymose. Sepals not petaloid; petals present.. . . . . *Jatropha.*
                    12. Latex milky white. Leaves without glandular hairs. Inflorescence racemose or paniculate. Sepals petaloid; petals absent. . . . . *Manihot.*
            10. Leaf blades not lobed.
                13. Hairs lepidote (peltate, scale-like) or stellate.
                    14. Hairs stellate (minutely stellate in *Conceveiba* and *Alchornea*).
                        15. Petioles with pair of conspicuous crateriform glands at apex; blades cordate at base. Inflorescences bisexual, the plants monoecious. . . . . *Croton* in part.

15. Petioles eglandular or glands minute; blades acute to obtuse at base. Inflorescences unisexual, the plants dioecious.
    16. Leaf blades with 3 major veins at base. Styles 2, free, not lobed. Capsules 2-locular. . . . . *Alchornea.*
    16. Leaf blades with one major vein at base (pinnately veined). Styles 3, connate at base, bifid at apex. Capsules 3-locular. . . . . *Conceveiba.*

14. Hairs lepidote.
    17. Plants monoecious. Inflorescences terminal or sometimes axillary, bisexual. Fruits 3-seeded capsules, >6 mm diam. . . . . *Croton* in part.
    17. Plants dioecious. Inflorescences axillary, unisexual. Fruits 1-seeded drupes, 3–6 mm diam. . . . . *Hyeronyma.*

13. Hairs simple, dendritic (*Mabea*), malpighiaceous (*Dodecastigma*, *Pausandra*) or absent.
    18. Latex or colored exudate absent. Leaves eglandular, glabrous; petioles short; blades elliptic, the margins entire. Flowers clustered in leaf axils or on short stems.
        19. Leaf blades chartaceous, the apex acuminate. Fruits irregularly dehiscing, 4–5-loculed capsules. Seeds 2 per locule. . . . . *Margaritaria.*
        19. Leaf blades coriaceous, the apex obtuse to shortly acuminate. Fruits 1-loculed drupes. Seed 1 per fruit. . . . . *Drypetes.*
    18. Latex or colored exudate present or absent. Leaves not as above. Flowers in spicate, racemose, or paniculate inflorescences, or rarely pistillate flowers 1 to several in leaf axils (*Acalypha*, *Hura*).
        20. Leaf blades with 3 major veins from base, i.e., tripliveined.
            21. Leaf blades with small cavities (domatia) in axils of basal lateral veins on abaxial surface, the glands outside basal lateral veins near base on adaxial surface (0)1–2, flat. Inflorescences spicate. Fruits capsular, 0.3–0.4 cm long. Seeds with outer coat fleshy, bright red. . . . . *Alchorneopsis.*
            21. Leaf blades without cavities, the glands at base on adaxial surface 2, distinctly rimmed. Inflorescences racemose or paniculate. Fruits drupaceous, 4.5–5.5 cm long. Seeds with outer seed coat dry, not bright red. . . . . *Glycydendron.*
        20. Leaf blades with single major vein from base, i.e., pinnately veined.
            22. Petioles with pair of glands at apex.
                23. Latex reddish or not evident. Leaf blades oblanceolate, >(20)30 cm long, the base narrowly cuneate. . . . . *Pausandra.*
                23. Latex milky white or somewhat clear white, copious. Leaf blades narrowly to broadly ovate, narrowly elliptic, or narrowly oblong, <20 cm long, the base acute to cordate.
                    24. Trunks with prickles. Petioles >6 cm long; blades broadly ovate, the base cordate. Styles connate into thick column, 2–5 cm long, with lobed stigmatic disc at apex. Capsules 5–20-seeded, 6–10 cm diam. . . . . *Hura.*
                    24. Trunks without prickles. Petioles <3 cm long; blades usually narrowly elliptic, the base acute. Styles <4 mm long. Capsules (2)3-seeded, <2 cm diam. . . . . *Sapium.*
            22. Petioles eglandular at apex (glands may be present at blade base; sometimes minute glandular knobs present at junction of blade and petiole in *Dodecastigma*).
                25. Leaf blades 2–6 cm long, thinly chartaceous. Inflorescences <3 cm long, bisexual. . . . . *Maprounea.*
                25. Leaf blades >7 cm long (except 5–11 cm long in *Mabea subsessilis*), chartaceous to coriaceous. Inflorescences >3 cm long (except (0.5)2–20 cm long in *Richeria*), unisexual or bisexual.
                    26. Hairs dendritic (sometimes present only on inflorescences and capsules). Latex white, copious. Leaf blades narrowly oblong-elliptic to oblong-elliptic, with conspicuous looped secondary veins, without pair of glands at base. Inflorescences with staminate bracts biglandular. . . . . *Mabea.*
                    26. Hairs simple, malpighiaceous, or plant glabrous. Latex absent, clear, or whitish, usually scant. Leaf blades not as above. Inflorescences with staminate bracts eglandular.

27. Leaf blades with pair of glands at base.
 28. Leaf blades ovate or broadly ovate, chartaceous, the base rounded to cordate, with pair of stipels at base on adaxial surface, and pair of flat glands at base on abaxial surface. . . . . *Aparisthmium.*
 28. Leaf blades elliptic or oblanceolate, subcoriaceous, the base acute to obtuse, without stipels, with pair of rimmed glands at base on adaxial surface. . . . . *Glycydendron.*
27. Leaf blades lacking pair of glands at base (sometimes minute knobs present in *Dodecastigma*).
 29. Petioles neither pulvinate nor flexed; blades chartaceous to subcoriaceous, the margins serrate or subentire. Inflorescences spicate, narrowly racemose, or pistillate flowers sessile in leaf axils. Staminate flowers without petals. Capsules <15 mm diam., sessile or on very short pedicels.
  30. Leaf blades elliptic or ovate-elliptic, the margins serrate. Pistillate flowers sessile in leaf axils or at base of staminate spike; styles deeply divided into many filiform branches. Capsules <4 mm diam. . . . . *Acalypha.*
  30. Leaf blades obovate, the margins subentire. Pistillate flowers in unisexual, racemose inflorescences; styles bifid. Capsules >5 mm diam. . . . . *Richeria.*
 29. Petioles pulvinate and flexed at apex; blades subcoriaceous to coriaceous, the margins entire. Inflorescences racemose or paniculate. Staminate flowers with petals. Capsules 15–25 mm diam.
  31. Plants monoecious. Inflorescences 4–15 cm long, the pedicels 10–30 mm long. Pistillate flowers with sepals 7–13 mm long, without petals. Capsules with conspicuous, persisting sepals. . . . . *Sagotia.*
  31. Plants dioecious. Inflorescences 15–35 cm long, the staminate pedicels 3–5 mm long. Pistillate flowers with sepals <5 mm long, with petals. Capsules without conspicuous sepals. . . . . *Dodecastigma.*

ACALYPHA L.

Shrubs, small trees, or herbs (outside our area), mostly monoecious. Latex absent. Hairs simple. Leaves alternate, simple, eglandular, the blade margins usually serrate or dentate; venation pinnate or palmate. Inflorescences unisexual or bisexual, axillary or terminal, spicate, rarely paniculate or rarely with pistillate flowers clustered in leaf axils; pistillate flowers 1–3 per node, usually subtended by a large bract, at basal node(s) in bisexual inflorescences. Flowers unisexual (plants mostly monoecious). Staminate flowers: sepals 4; petals absent; nectary disc absent; stamens 4–8, free or basally connate, the anther-sacs elongate, twisted; pistillode absent. Pistillate flowers: sepals 3–5; petals absent; nectary disc absent; ovary 3-locular, with 1 ovule per locule, the styles free or basally connate, divided into many filiform branches. Fruits capsular. Seeds 3; caruncle mostly reduced or absent.

Acalypha diversifolia Jacq. PL. 53a,b

Shrubs or small trees, 2–4 m tall. Stipules linear-lanceolate, 4–7 mm long. Leaves: petioles 0.6–4 cm long; blades elliptic or ovate-elliptic, 9–20 × 3–7 cm, chartaceous, glabrescent with major veins sparsely pubescent, eglandular, the base acute, obtuse, or rounded, the apex acuminate to long-acuminate, the margins serrate, with serrations closely spaced; venation pinnate, the secondary veins arched. Inflorescences axillary, spicate, 3–8 cm long, bisexual or staminate, green, the staminate flowers densely clustered on spikes, the pistillate flowers in leaf axils or at base of bisexual spike, subtended by 1–2 bracts; pedicels absent. Staminate flowers minute. Pistillate flowers: ovary densely hispidulose, the styles 2–4 mm long. Capsules 3-lobed, ca. 2 × 3 mm, subtended by persistent enlarged bract, 1 to several in leaf axil, usually at base of staminate axis; bracts very broadly ovate, 1–3 × 2–4 mm, the apex obtuse, the margins shallowly dentate. Fl (Sep, Oct, Nov, Dec), fr (Dec); secondary vegetation, especially along roadsides in nonflooded forest.

ALCHORNEA Sw.

Trees. Latex absent. Hairs minutely stellate or simple (outside our area). Leaves alternate, simple; petioles eglandular; blades with a pair of flat glands in axil of basal lateral veins on abaxial surface, the margins usually serrate or dentate; venation pinnate or tripliveined (3 major veins at base). Inflorescences axillary, spicate or paniculate with short spicate branches, unisexual (plants dioecious). Staminate flowers very small, subsessile; sepals 2–5, valvate; petals absent; nectary disc central; stamens 8, in 2 whorls; pistillode absent. Pistillate flowers: sepals 4; petals absent; nectary disc absent; ovary 2-locular, with 1 ovule per locule, the styles 2, free, elongate, unlobed. Fruits capsular, 2-carpelled (2-lobed). Seeds 2, ecarunculate, the outer seed coat dry.

Alchornea triplinervia (Spreng.) Müll. Arg.

Small to medium-sized trees. Leaves: petioles 1–3 cm long; blades elliptic, 6–11 × 3–5 cm, sparsely stellate-pubescent abaxially, glabrescent adaxially with scattered stellate hairs on major veins, the base acute to obtuse, the apex acuminate, the margins serrate, with distinct glandular setae; venation of 3 veins from base and 2–4 pairs of secondary veins diverging from distal half of central primary vein. Inflorescences 3–10 cm long, the axes stellate-pubescent. Pistillate flowers with styles 5–8 mm long. Capsules 2-lobed, globose to depressed globose, 4–6 × 4–6 mm, each lobe subglobose. Fl (Jan); in forest on Mont Galbao.

ALCHORNEOPSIS Müll. Arg.

Trees. Latex absent. Hairs simple. Leaves alternate, simple; petioles eglandular; blades with cavities (pit domatia) in axils of basal lateral primary veins, the opening on abaxial surface, with 1 or 2 flat glands often present near base outside lateral primary veins on adaxial surface; venation tripliveined. Inflorescences unisexual, axillary, spicate, 1 to several per axil. Flowers unisexual (plants dioecious). Staminate flowers very small; sepals 3–4, valvate; petals absent; nectary disc present; stamens 4–7, free; pistillode 3-lobed. Pistillate flowers: sepals 4–5; petals absent; nectary disc annular at ovary base; ovary 3-locular, with 1 ovule per locule, the styles 3, free, unlobed. Fruits capsular. Seeds 3, ecarunculate, the outer seed coat fleshy.

Alchorneopsis floribunda (Benth.) Müll. Arg.
FIG. 105, PL. 53d

Medium-sized to canopy trees. Trunks with branched buttresses. Bark smooth, lenticillate, the inner bark orange. Leaves: petioles 2–4 cm long; blades elliptic or narrowly obovate, 9–12 × 3.5–6 cm, glabrous or glabrescent, with adaxial gland ca. 4–20 cm above base, the base acute, the apex abruptly acuminate, the margins subentire to undulate; venation 3-veined at base, with 2(3) pairs of secondary veins diverging from central primary vein in distal half. Staminate inflorescences 3–12 cm long. Pistillate inflorescences 2–6 cm long, the axes densely appressed puberulous. Pistillate flowers: styles ca. 0.5 mm long, reflexed. Staminate flowers: nectary disc lobed, pubescent; filaments 1.5–2 mm long; pistillode pubescent. Capsules 3-lobed, 3–4 × 5 mm, green, to ca. 20 per infructescence. Seeds often remaining attached to columella after dehiscence, black, irregularly ribbed, the outer seed coat bright red. Fl (Jan, Nov), fr (Jan, Feb); in nonflooded forest.

APARISTHMIUM Endl.

A monotypic genus. The generic description same as the description of the species.

Aparisthmium cordatum (A. Juss.) Baill.
FIG. 106, PL. 53e; PART 1: FIG. 4, PL. Xc

Shrubs or understory trees. Latex absent. Hairs simple. Leaves alternate, simple; petioles 7–15 cm long, eglandular, flexed near apex; blades ovate or broadly ovate, 13–37 × 9–24 cm, sparsely pubescent to glabrescent, with pair of stipels (sometimes glandular) at base adaxially and pair of flat circular glands in axils of basal (and often distal) secondary veins abaxially, the base rounded or shallowly cordate, the apex acuminate, the margins serrate or serrulate; venation pinnate. Inflorescences unisexual (plants dioecious), the axes pubescent. Staminate inflorescences terminal or both terminal and axillary, paniculate with branches spicate, to 35 cm long, often appearing broom-like with many long, slender branches bearing glomerules of very small sessile flowers. Pistillate inflorescences axillary, racemose, to 25 cm long, the flowers single per node, each subtended by biglandular bract; pedicels 3–5 mm long, elongating to ca. 7 mm in fruit. Flowers unisexual. Staminate flowers: calyx 2- to 3-lobed, pubescent; petals absent; nectary disc absent; stamens 3–5, basally connate; pistillode absent. Pistillate flowers: sepals 4–6, pubescent; petals absent; nectary disc absent; ovary 3-locular, pubescent, with 1 ovule per locule, the styles 3, thick, ca. 3 mm long, recurved, connate at base, shortly bifid at apex, the inner surface papillose. Fruits capsules, subglobose, 7–10 × 9–10 mm, green turning red, deeply 3-lobed at apex, with persistent styles; fruiting pedicels ca. 7 mm long. Seeds 3, ecarunculate. Fl (Sep, Oct, Nov, Dec), fr (Mar, Nov); common, in secondary vegetation, especially along roadsides. *Bois di vin* (Créole).

CONCEVEIBA Aubl.

Trees. Latex absent. Hairs minutely stellate. Leaves alternate, simple; petioles eglandular (or rarely minutely biglandular at junction of blade and petiole); blades often with small flat glands scattered on abaxial surface, the margins serrulate, the serrations usually widely spaced; venation pinnate or palmate. Inflorescences terminal, unisexual (plants dioecious), the axes stellate-pubescent. Staminate inflorescences paniculate, the flowers in sessile clusters. Pistillate inflorescences racemose, the bracts biglandular, the flowers single per node; pedicels short. Flowers unisexual. Staminate flowers very small; sepals 3–4; petals absent; nectary disc absent; stamens 15–45(60); pistillode absent. Pistillate flowers: sepals 5–10; petals absent; nectary disc absent; ovary 3-locular, with 1 ovule per locule, the styles 3, connate at base, spreading, the apex bifid. Fruits capsular. Seeds 3, carunculate.

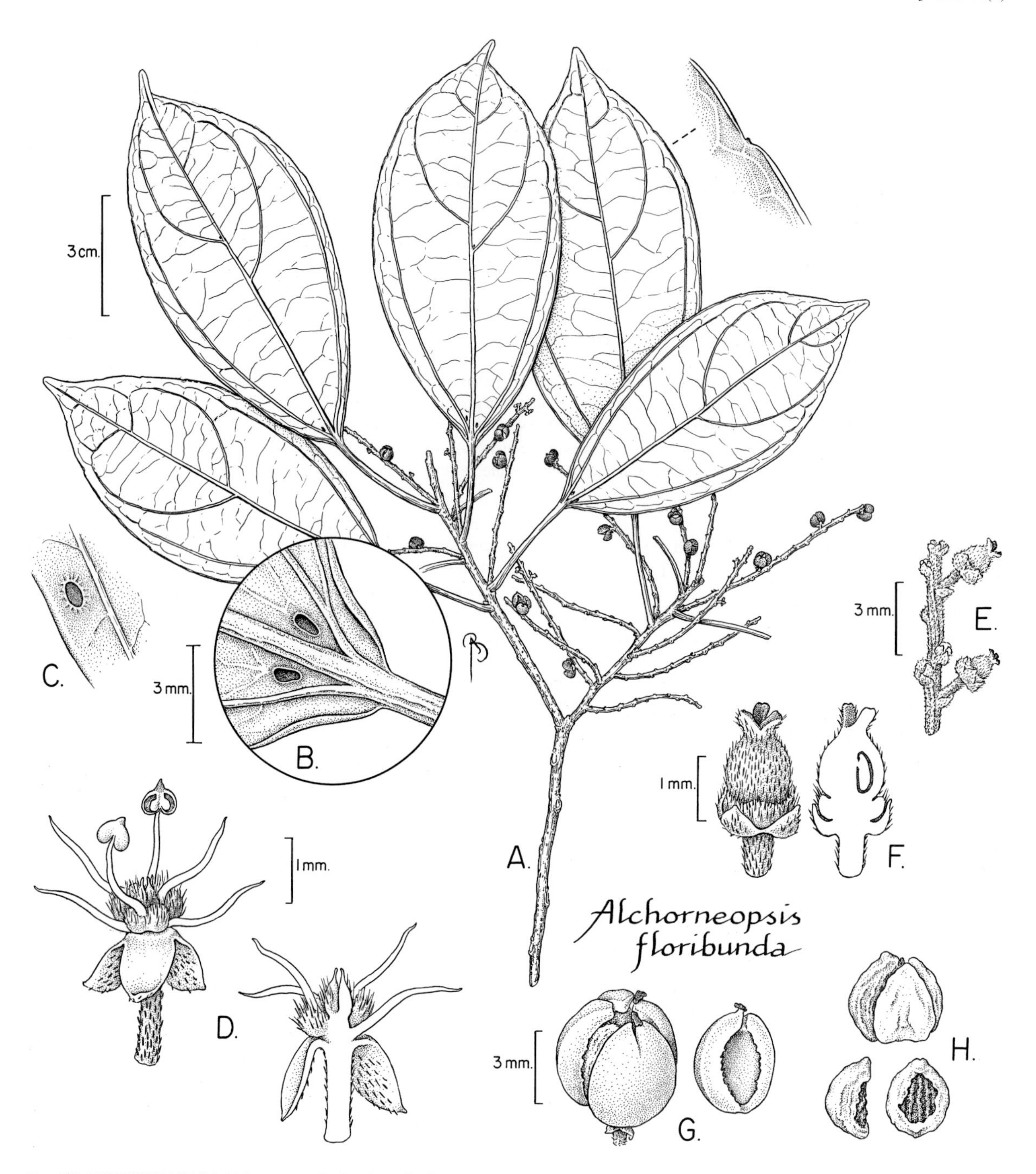

FIG. 105. EUPHORBIACEAE. *Alchorneopsis floribunda* (A, B, G, H, *Mori et al. 22920*; C, *Mori et al. 21587*; D, *Ducke 245* from Brazil; E, F, *Prance et al. 21462* from Brazil). **A.** Stem with leaves and infructescences. **B.** Detail of base of abaxial leaf blade showing domatia. **C.** Detail of adaxial leaf blade margin showing gland. **D.** Lateral view (left) and medial section (right) of staminate flowers. **E.** Part of pistillate inflorescence. **F.** Intact (left) and medial section of pistillate flower (right). **G.** Capsule (left) and one valve of capsule (right). **H.** Cluster of seeds (above) and two views of seeds with fleshy seed coat (below).

Conceveiba guianensis Aubl. FIG. 107, PL. 53f

Understory trees. Leaves: petioles 5–15 cm long, thickened and flexed near apex; blades elliptic, 14–32 × 7–18 cm, glabrescent to sparsely stellate-pubescent, the base obtuse, the apex acuminate; venation pinnate; leaves of saplings sometimes with gland at base of stipules, and pair of small glands at junction of blade and petiole. Staminate inflorescences 7–25 cm long. Pistillate inflorescences 6–12 cm long, elongating to 18 cm in fruit; bracts biglandular. Staminate flowers: sepals glabrous or glabrescent; stamens 15–20, the inner whorl often sterile and elongate. Pistillate flowers: sepals alternating with disciform glands; ovary covered with minute stellate hairs, the styles 6–7 mm long, very shortly bifid, shortly connate at base, the inner surface papillose near apex. Capsules subglobose or broadly ellipsoid, 18–25 × 17–23 mm, very sparsely minutely stellate-pubescent, each carpel carinate, with persistent strongly recurved styles. Fr (Jan, Feb, Mar, Nov); scattered, in nonflooded forest.

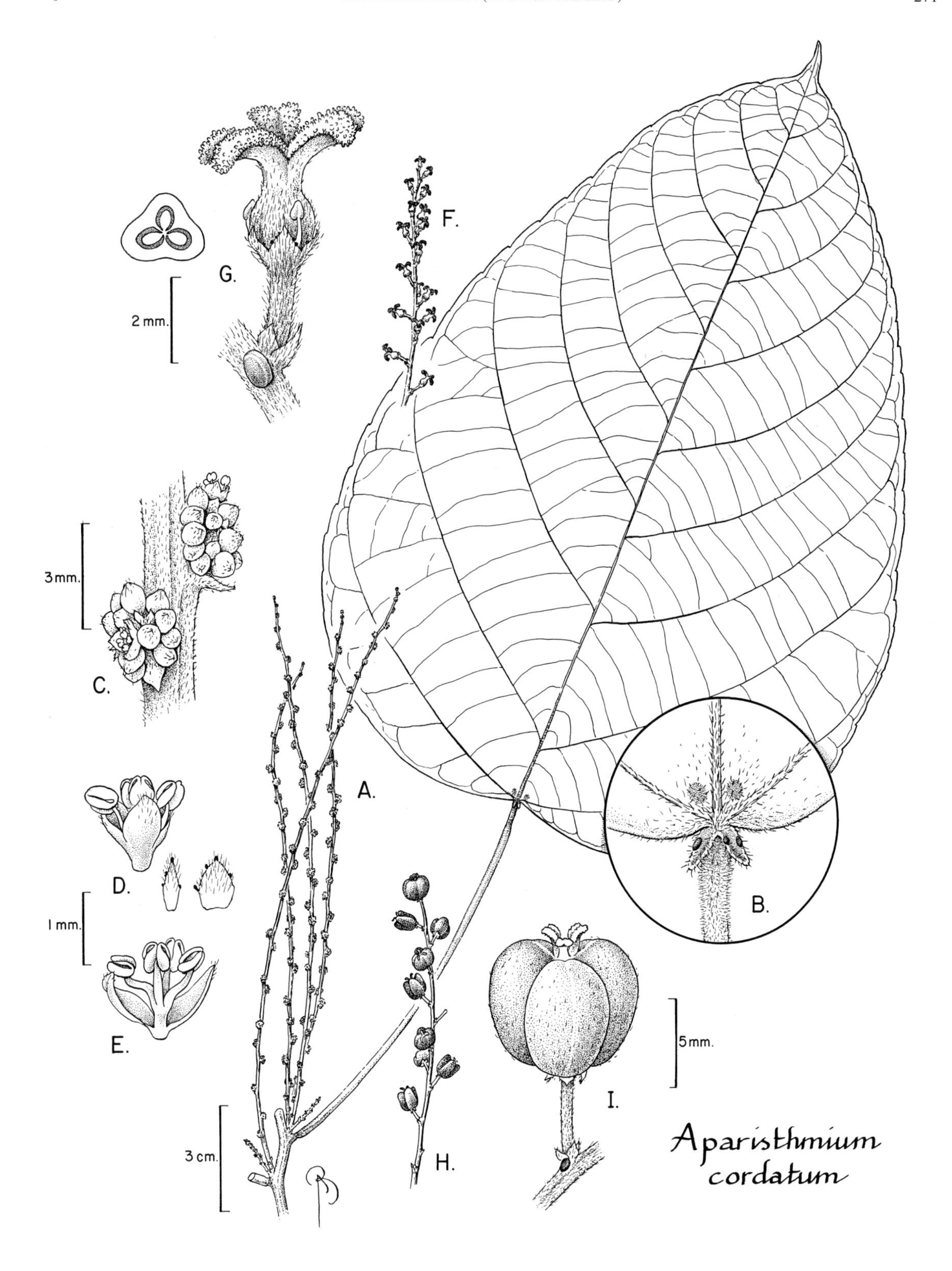

FIG. 106. EUPHORBIACEAE. *Aparisthmium cordatum* (A–C, *Mori & de Granville 8820*; D, E, H, I, *Daly 1163*; F, G, *Boom 4063*). **A.** Part of stem with leaf and staminate inflorescences. **B.** Detail of stipels and associated glands. **C.** Detail of staminate inflorescence. **D.** Staminate flower and associated bracts. **E.** Staminate flower with sepal removed. **F.** Part of pistillate inflorescence. **G.** Pistillate flower with transverse section of ovary (left). **H.** Part of infructescence. **I.** Capsule.

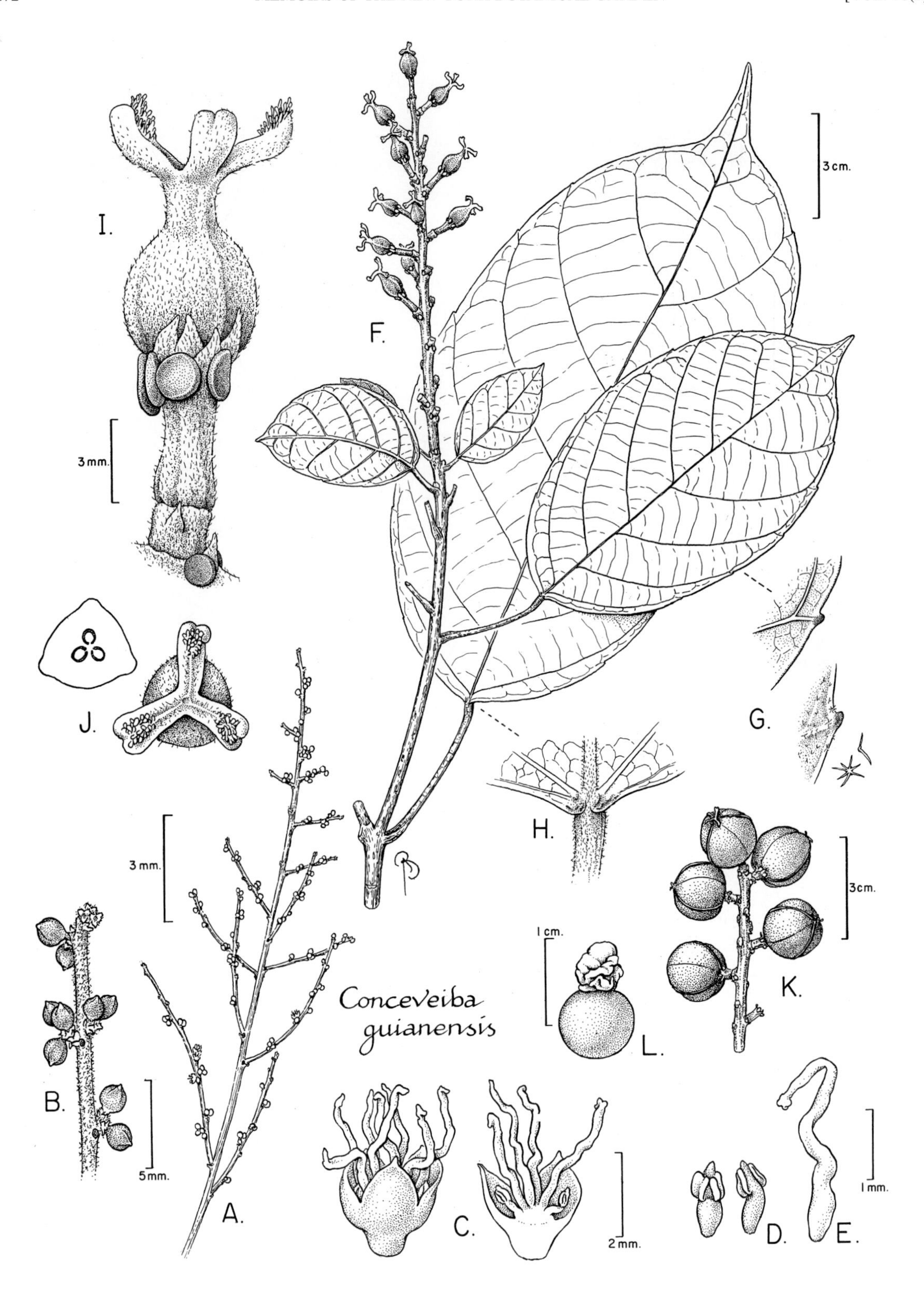

FIG. 107. EUPHORBIACEAE. *Conceveiba guianensis* (A–E, *Ducke 1079* from Brazil; F–J, *Mori & de Granville 8917*; K, L, *Mori & Pipoly 15378*). **A.** Staminate inflorescence. **B.** Detail of staminate inflorescence in bud. **C.** Lateral view (left) and medial section (right) of staminate flower. **D.** Adaxial (far left) and lateral (near left) views of stamen. **E.** Staminode. **F.** Part of stem with leaves and infructescence with young capsules. **G.** Details of adaxial (above) and abaxial (below) leaf surfaces and stellate hairs. **H.** Adaxial view of leaf base. **I.** Pistillate flower. **J.** Apical view (right) and transverse section (above) of pistillate flower. **K.** Part of infructescence. **L.** Seed with caruncle.

CROTON L.

Trees, shrubs, or herbs (outside our area). Latex usually present, colored. Hairs stellate or lepidote (simple hairs sometimes outside our area). Leaves alternate, simple; petioles often biglandular; blades unlobed or palmately lobed, eglandular, the margins entire or serrate; venation pinnate or palmate. Inflorescences bisexual or sometimes unisexual, axillary or terminal, spicate, racemose or paniculate. Flowers unisexual (plants monoecious or sometimes dioecious outside our area). Staminate flowers: sepals 4–6; petals 5, rarely absent; nectary disc present; stamens 8–50+, free, the filaments inflexed in bud. Pistillate flowers: sepals mostly 5–7-lobed; petals 5, often reduced or absent; ovary 3-locular, with 1 ovule per locule, the styles free or basally fused, bifid or divided several times. Fruits capsular. Seeds 3, carunculate.

1. Hairs stellate. Leaf blades ovate or triangular-ovate, the base cordate or rounded.
   2. Leaf blade apex acuminate. Inflorescences solitary. Styles free, slender, bifid. . . . . *C. draconoides.*
   2. Leaf blade apex obtuse to shortly acuminate. Inflorescences solitary to clustered. Styles fused at base, deeply 2–3 divided at apex into flat star-shaped structure. . . . . *C. palanostigma.*
1. Hairs lepidote. Leaf blades narrowly elliptic to elliptic, the base acute or obtuse.
   3. Canopy trees. Petioles with pair of stalked glands at apex; blades glabrous adaxially, with 18–26 pairs of secondary veins. . . . . *C. matourensis.*
   3. Shrubs or understory trees. Petioles eglandular; blades sparsely lepidote-pubescent adaxially, with 7–10 pairs of secondary veins. . . . . *C. schiedeanus.*

Croton draconoides Müll. Arg. FIG. 108; PART 1: FIG. 4

Understory trees, to 22 m tall. Trunks cylindrical, not buttressed. Bark smooth, densely lenticellate, brown, the inner bark cream-colored. Latex red or reddish brown, sticky. Hairs stellate. Leaves: petioles 6–14 cm long, with pair of subsessile crateriform glands at apex; blades triangular-ovate, 12–20 × 10–18 cm, sparsely stellate-pubescent, the base cordate, the apex acuminate; venation weakly palmate. Inflorescences terminal, spicate-racemose, 22–28 cm long, solitary, erect above leaves; cymules mostly bisexual with single pistillate flower (distal cymules often staminate), the axes densely stellate-pubescent. Flowers: sepals densely stellate-pubescent. Pistillate flowers: ovary densely stellate-pubescent, the styles free, slender, deeply bifid. Capsules 6–7 × 9 mm, stellate-pubescent. Fl (Feb, Mar, Aug, Sep), fr (Sep); common, in disturbed nonflooded forest. *Bois-Saint-Michel* (Créole).

The fruits of the central French Guianan population are larger than those of western Amazonia where the type was collected.

Croton matourensis Aubl.

Canopy trees. Trunks cylindrical, not buttressed. Bark with shallow vertical fissures and lenticels, the inner bark brick red, striate. Latex red. Hairs lepidote with irregularly fringed margin. Leaves: petioles 1.5–5 cm long, with pair of stalked glands at apex; blades narrowly elliptic to elliptic, 12–20 × 3.5–8 cm, glabrous adaxilly, lepidote-pubescent and distinctly paler abaxially, the base acute or obtuse; venation pinnate, the secondary veins in 18–26 pairs. Inflorescences terminal, spicate-racemose, 10–22 cm long, in clusters of 2 to 5, with pistillate flowers 8–16 along basal half, the axes densely lepidote-pubescent; pistillate pedicel 2–3 mm long, to 5 mm long in fruit. Flowers: sepals densely lepidote-pubescent. Pistillate flowers: styles 5–6 mm long, fused at base, deeply divided several times into numerous slender style branches. Capsules ca. 6 × 7 mm, densely lepidote-pubescent; sepals persistent, ca. 7 × 3 mm. Fl (Mar), fr (May); in nonflooded forest. *Bois ramier, crotane* (Créole).

Croton palanostigma Klotzsch

Small trees. Hairs stellate. Leaves: petioles 4–10 cm long, with pair of subsessile crateriform glands at apex; blades ovate or triangular-ovate, 15–20 × 9–19 cm, sparsely stellate-pubescent, the base shallowly cordate or rounded, the apex obtuse to shortly acuminate; venation weakly palmate. Inflorescences terminal, spicate-racemose, solitary or in clusters of 2–3; cymules bisexual with single pistillate flower or distal cymules staminate, the axes densely stellate-pubescent. Flowers: sepals densely stellate-pubescent. Pistillate flowers: ovary densely stellate-pubescent, the styles fused at base, deeply 2–3 divided at apex into centrally flattened structure with radiating style branches. Capsules 6–8 × 9–11 mm, densely stellate-pubescent. Fl (Feb, Dec), fr (Mar); in partially disturbed areas.

The Saül population differs from typical *C. palanostigma* of central and western Amazonia by its shorter, less divided style branches and capsules with stellate hairs that are not appressed.

Croton schiedeanus Schltdl.

Shrubs or understory trees. Bark smooth, gray. Hairs lepidote. Leaves: petioles 1–3(4.5) cm long, eglandular; blades elliptic, 9–20 × 4–7 cm, sparsely lepidote-pubescent adaxially, densely lepidote-pubescent abaxially, the base obtuse or acute; venation pinnate, the secondary veins in 7–10 pairs. Inflorescences terminal or axillary in upper leaves, paniculate or racemose, 2–15 cm long; pistillate flowers usually 1(3) per branch, the axes densely lepidote-pubescent; pistillate pedicel 8–14 mm, to 18 mm long in fruit. Flowers: sepals densely lepidote-pubescent. Pistillate flowers: styles 2–3 mm long, free, slender, 2–3 times bifid. Capsules ca. 10 × 10 mm, lepidote-pubescent; sepals persistent, ca. 3 × 2 mm. Fl (Jan, Feb, Mar, Apr, Aug, Sep, Oct, Dec), fr (Mar, Apr, Aug, Oct, Sep, Dec); in secondary vegetation or disturbed areas in nonflooded forest.

DALECHAMPIA L.

Scott Armbruster

Perennial vines, lianas, or shrubs (outside our area). Latex absent. Stems climbing, spreading, or erect, usually covered with minute urticating hairs. Leaves simple, palmately lobed or compound, alternate; petioles present, eglandular; blades with pair of stipels at base (stipels sometimes obscure or deciduous). Inflorescences pseudanthial, subtended by 2 usually large showy involucral bracts, comprising

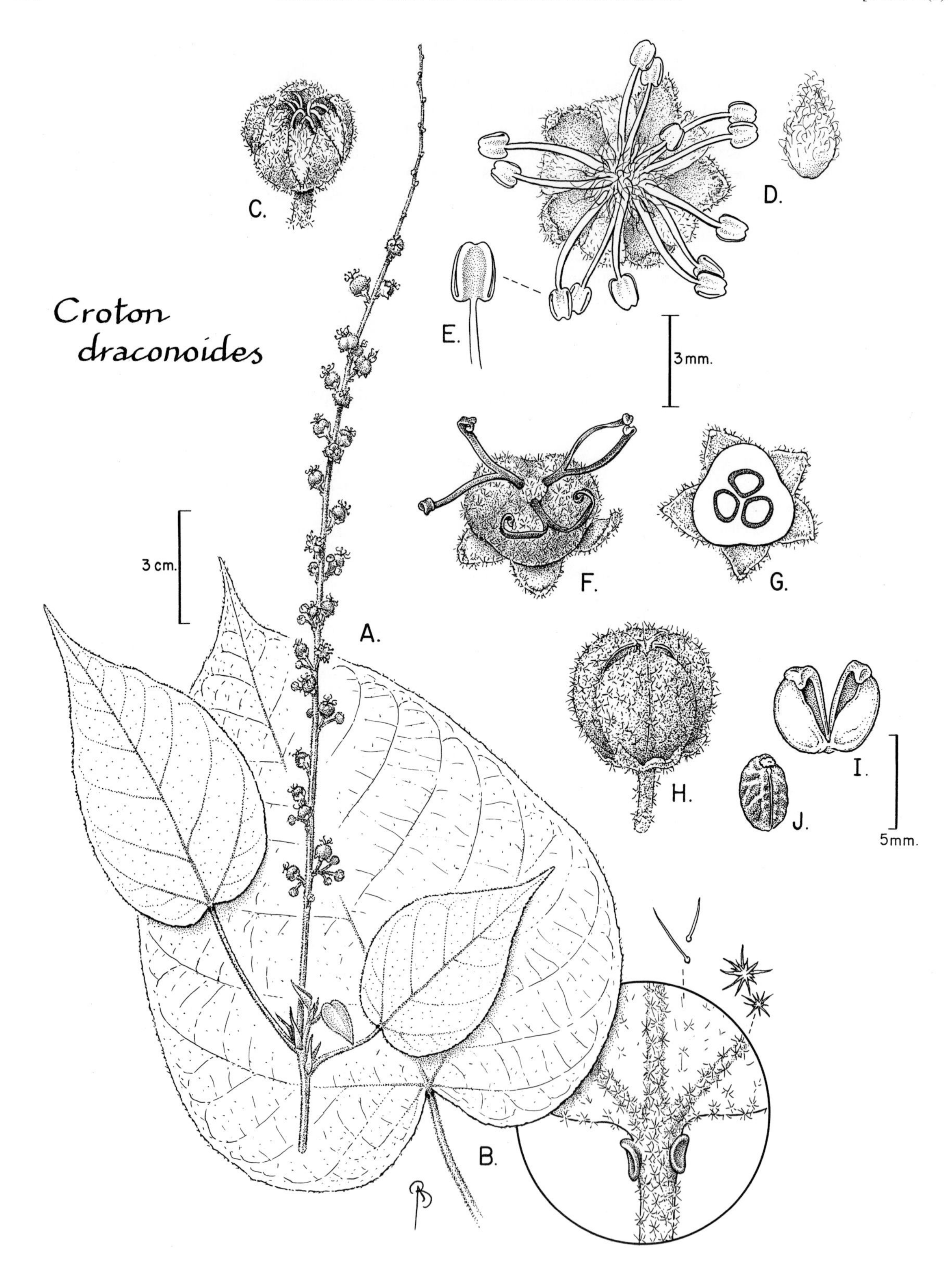

FIG. 108. EUPHORBIACEAE. *Croton draconoides* (A, C–G, *Mori & Mitchell 18762*; H–J, *Phillippe 27019*). **A.** Part of stem with inflorescence and leaves. **B.** Leaf and detail of base of leaf showing glands and stellate hairs. **C.** Bud of staminate flower. **D.** Apical view of staminate flower with calyx lobe on right. **E.** Stamen. **F.** Apical view of pistillate flower. **G.** Transverse section of ovary. **H.** Capsule. **I.** Section of dehisced capsule. **J.** Seed.

a cymule of 3 pistillate flowers, a 3–5-branched pleiochasium of 4 to ca. 15 staminate flowers, and usually a prominent "gland" formed by a cluster of secretory bractlets. Pistillate cymule subtended by an involucel of 1 bract and 0–2 bractlets. Staminate pleiochasium subtended by a cuplike involucellar bract, a bilabiate involucel, or 4(5) free bracts. Staminate flowers: calyx splitting into 3–6 valvate segments at anthesis; petals absent; stamens 5–90 on flat, dome-shaped, or elongate receptacle (column), the anthers subsessile or on short filaments, usually in cluster at apex of column, dehiscing longitudinally. Pistillate flowers: sepals 5–12, entire or lobed; petals absent; ovary of 3 carpels, the styles connate into elongate column, often dilated at apex. Fruits usually small (usually <2 cm diam.), 3-seeded capsules, dehiscing explosively by elastic twisting of dry, ± woody cocci, often with irritating hairs. Seeds subglobose to globose, ecaruncúlate, the surface smooth, roughened, or tuberculate, usually gray-brown mottled.

Armbruster, W. S. 1996. Cladistic analysis and revision of *Dalechampia* sections *Rhopalostylis* and *Brevicolumnae* (Euphorbiaceae). Syst. Bot. **21(2):** 209–235.

1. Leaves unlobed. Inflorescences with involucral bracts equal or unequal, one or both bracts inconspicuous, small (<6 mm long), stipuliform, deep green at anthesis; resin-producing gland absent or present. Pistillate sepals 5–6, unlobed.
    2. Leaf blades narrow, width <1/3 length, the base obtuse; venation with ≥6 arcuate secondary veins branching from basal and distal regions of central vein. Inflorescences on short shoots (peduncles) <2 cm long; resin-producing gland usually present. Staminal column extremely short or absent. Stylar column cylindrical, slightly dilated at apex. . . . . *D. brevicolumna.*
    2. Leaf blades broader, width >1/3 length, the base truncate to deeply cordate or hastate; venation with 1–5 arcuate secondary veins, usually confined to distal half of central vein. Inflorescence on short shoots (peduncles) usually >2 cm long; resin-producing gland absent. Staminal column well developed and conspicuous. Stylar columns greatly enlarged, tapering at both ends, not dilated at apex.
        3. Stipules usually deciduous before or shortly after expansion of leaves. Leaf bases usually shallowly cordate. Inflorescences with involucral bracts subequal, stipuliform, usually <5 mm long, green. Stylar column thick, width ca. 1/3 length; stigma green. . . . . *D. fragrans.*
        3. Stipules usually persistent. Leaf bases with sinus usually broad and U-shaped. Inflorescences with involucral bracts usually dimorphic, the lower one (subtending pistillate flowers) >1 cm long, whitish, the upper one (subtending staminate flowers) <5mm long, stipuliform, green. Stylar column thinner, width ca. 1/4 length; stigma whitish. . . . . *D. heterobractea.*
1. Leaves lobed or unlobed. Inflorescences with involucral bracts equal, conspicuous, >1 cm long, pink, white, or pale green at anthesis; resin-producing gland present. Pistillate sepals (6)8–13, pinnatifid.
    4. Leaf blades unlobed. Inflorescences with involucral bracts pink, the gland secreting brownish-maroon resin; staminate involucel with 4 free involucellar bracts. Styles distinctly peltate at apex, tip ± symmetric. . . . . *D. dioscoreifolia.*
    4. Leaf blades usually 3-lobed on mature plants. Inflorescences with involucral bracts pale green to whitish, the gland secreting yellow resin; staminate involucel with involucellar bracts ± connate (at least at base). Styles moderately to strongly dilated at apex, tip asymmetric.
        5. Stipules lanceolate-linear, margins not glandular capitate. Inflorescences with involucral bracts creamy-white, the apices shallowly three-lobed, the incisions <1/5 of bract length, the three lobes subequal; bracts of staminate involucel free distally, fused at base, forming ± bilabiate involucel. . . . . *D. tiliifolia.*
        5. Stipules lanceolate-ovate, margins glandular capitate. Inflorescences with involucral bracts pale green, the apices deeply 3-lobed, the incisions ca. 1/4 of bract length, the central lobe ca. twice length of lateral lobes; bracts of staminate involucel completely fused into cup-like involucel. . . . . *D. stipulacea.*

Dalechampia brevicolumna Armbr.

Twining lianas reaching low canopy. Stipules lanceolate, 2–4 × 0.8–1 mm. Leaves simple; petioles 0.5–2(6) cm long; blades unlobed, relatively narrow, width <1/3 length, 5–15 × 1.5–5 cm, the base obtuse, the apex acute-mucronate; secondary veins ≥6, diverging from central primary vein throughout its length. Inflorescences on short shoots (peduncles) 0.5–1.5 cm long, with 6–9 staminate flowers and 3 pistillate flowers; involucral bracts subequal, inconspicuous, green, stipuliform, 2.5–3 mm long, the apex unlobed; inflorescence glands secreting whitish resin. Staminate involucel with 3 free bracts. Staminate flowers: stamens 20–30, attached to ± flat receptacle, the column extremely short or absent. Pistillate flowers: sepals 5–6, unlobed; stylar column ± cylindrical, straight, width <1/5 length, with slightly dilated apex, the stigma extending from apex to >3/4 length of style. Seeds spherical, 4–4.3 mm in diameter. Fr (Jul); in disturbed forest between Saül and Bélizon.

Dalechampia dioscoreifolia Poepp. FIG. 109

Twining lianas reaching canopy. Stipules lanceolate, 3–3.5 × 1–1.5 mm. Leaves simple; petioles 2–7 cm long; blades unlobed, ovate, relatively broad, width >1/3 length, 5.5–13.5 × 4–9.5 cm, chartaceous, the base distinctly cordate, the apex acuminate; secondary veins usually <6, diverging from central primary vein along its distal two-thirds. Inflorescences on short shoots (peduncles) usually 2–3 cm long, usually with 10 staminate flowers and 3 pistillate flowers; involucral bracts subequal, showy, pink, ovate, 1.3–4.5 × 1.2–3 cm, with 3–5 major veins, the base clawed, the apex unlobed; inflorescence glands secreting brown-maroon resin. Staminate

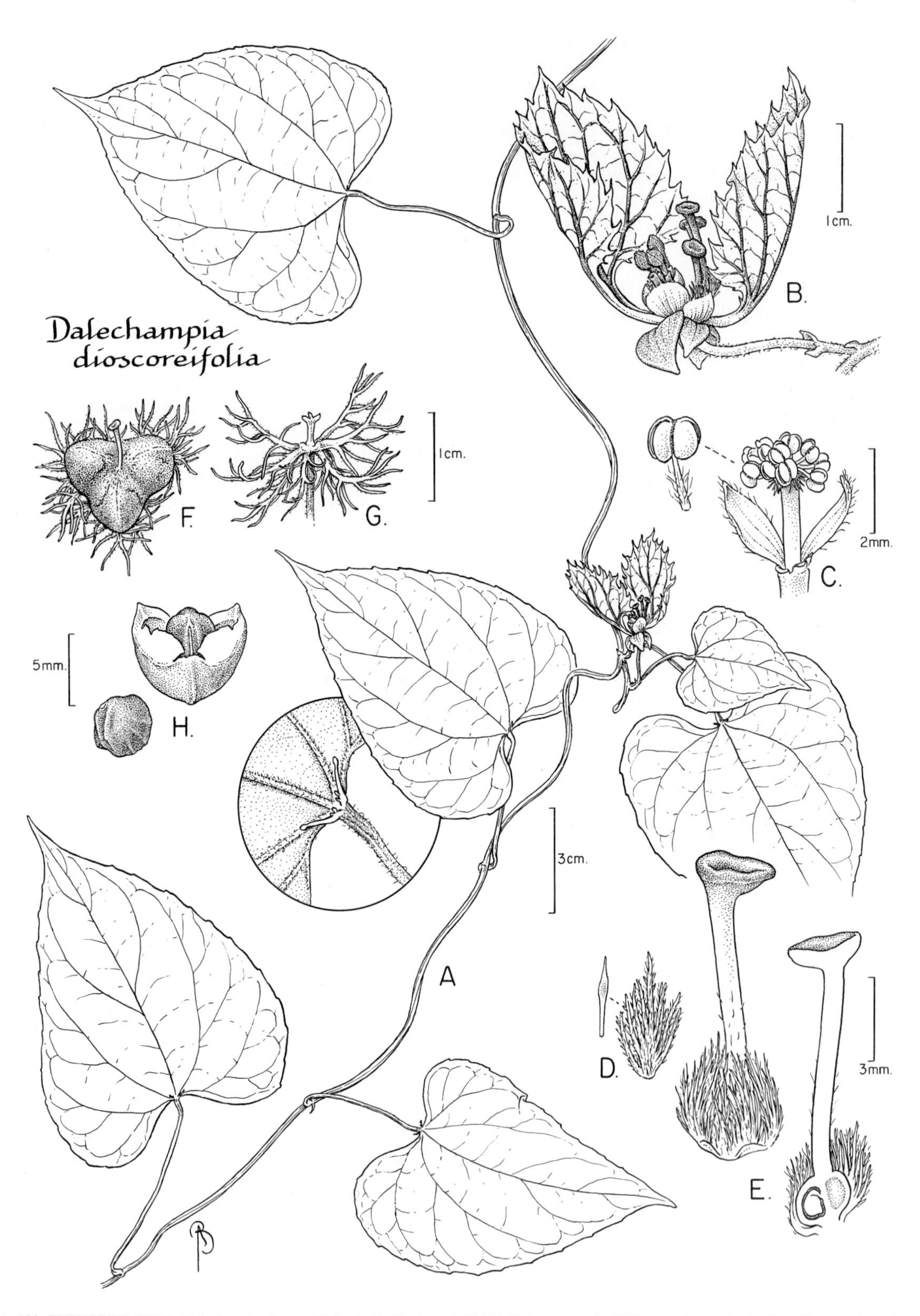

FIG. 109. EUPHORBIACEAE. *Dalechampia dioscoreifolia* (A–E, *Mori et al. 22355*; F, *Luteyn et al. 4896* from Colombia; G, H, *Oldeman 2067*). **A.** Part of stem with leaves and inflorescence; detail of adaxial leaf blade base with stipels. **B.** Lateral view of inflorescence subtended by bracts. **C.** Lateral view of staminate flower with three sepals removed (near left) and detail of stamen (far left). **D.** Sepal surrounding pistillate flower (left) and detail of hair (above). **E.** Lateral view (above) and medial section (right) of pistillate flower. **F.** Oblique-apical view of fruit. **G.** Sepals after fruit has dehisced and fallen. **H.** Fruit segment with seed (above) and seed (left).

involucel with 4 free bracts. Staminate flowers: stamens 20–30, attached to elongate receptacle or column. Pistillate flowers: sepals (6)8–12, pinnatifid; stylar column symmetrically dilated and peltate, the width (excluding apex) <1/5 length, 3–6 mm long, the expanded stylar apex symmetrically peltate-disciform, the stigmatic surface confined to expanded apex. Seeds subspherical and weakly three-angled, ca. 5 mm long. Fl (Feb, May, Jun, Jul, Nov), fr (Feb, Jul, Nov); occasional, at edge of or in canopy of nonflooded forest.

Dalechampia fragrans Armbr.

Twining lianas reaching canopy. Stipules ovate, 3–4 × 1–2 mm, usually deciduous before or shortly after full expansion of leaves. Leaves simple; petioles 1–5 cm long; blades unlobed, ovate, relatively broad, width >1/3 length, 6–14 × 4–12 cm, the base shallowly cordate, usually with broad U-shaped sinus, the apex acuminate; secondary veins usually <6, diverging from central primary vein along its distal half. Inflorescences on short shoots (peduncles) usually 3–10 cm long, with usually 13 staminate flowers and 3 pistillate flowers; involucral bracts subequal, inconspicuous, green, stipuliform, <6 mm long, the apex unlobed; inflorescence glands absent. Staminate involucel with 4 free bracts. Staminate flowers: stamens 4–5, attached to elongate receptacle or column. Pistillate flowers: sepals 6, unlobed; the stylar column thickened, tapering at both ends, straight, width ca. 1/3 length, widest about 2/3 of way to rounded apex, 7–10 mm long, the stigma green, extending from apex to ca. 3/4 length of style. Fl (Sep); in canopy of nonflooded forest.

Dalechampia heterobractea Armbr.

Twining lianas reaching canopy. Stipules lanceolate, (3)5–7 × 1–1.5 mm, persistent. Leaves simple; petioles 3–9 cm long; blades unlobed, relatively broad, width >1/3 length, generally broadest at 1/4 to 1/2 distance from base, 6–19 × 5–15 cm, the base deeply cordate, with broad U-shaped sinus, usually wider than deep (base of blade hastate in young plants), the apex abruptly acuminate; secondary veins usually <6, diverging from central primary vein along its distal half. Inflorescences on short shoots (peduncles) 3–10 cm long, with usually 10 staminate flowers and 3 pistillate flowers; involucral bracts unequal, the distal (upper) inconspicuous, green, stipuliform, 2.5–5 × 1–2 mm, the apex unlobed, the proximal (lower) showy, whitish, 10–16 × 5–6 mm, the apex unlobed; inflorescence glands absent. Staminate involucel with 4 free bracts. Staminate flowers: stamens 10–11, attached to elongate receptacle or column. Pistillate flowers: sepals 6, unlobed, obovate-oblanceolate, 3–4 mm long; stylar column thickened, tapering at both ends, straight, width ca. 1/4 length, widest about 1/2–2/3 of way to rounded apex, 4–9 mm long, the stigma whitish, extending from apex to ca. 2/3 length of style. Seeds subspherical, 3.5–4.1 mm diam.

Dalechampia stipulacea Müll. Arg. PL. 54a,b

Twining lianas sometimes reaching lower canopy. Stipules lanceolate-ovate, 10–20 × 3–5 mm, the margins glandular-capitate. Leaves simple; petioles 2–6 cm long; blades usually 3-lobed, usually ovate, rarely lanceolate, relatively broad, width >1/3 length, 5–12 × 6–14 cm, chartaceous, sparsely pubescent ad- and abaxially, the base deeply cordate, the apex of each lobe narrowly acuminate, the margins usually glandular capitate; secondary veins ≥6, diverging from the central primary vein throughout its length. Inflorescences on short shoots (peduncles) 5–15 cm long, with usually 10 staminate flowers and 3 pistillate flowers; involucral bracts equal, pale green at anthesis, broadly ovate, 4–6 × 2–4 cm, sparsely pubescent, the apex deeply three-lobed, central lobe ca. twice length of lateral lobes, the margins often glandular-capitate; inflorescence gland secreting yellow resin. Staminate involucel of fully connate bracts forming a cup-like involucel. Staminate flowers: stamens 30–50, attached to elongate receptacle or column. Pistillate flowers: sepals 10–12, these pinnatifid, capitate-glandular, hispid; pistil with elongate stylar column slightly curved, width (excluding apex) <1/5 length, 10–15 mm long, the apex moderately dilated, stigma extending from apex to ca. 1/2 length of style. Fruits capsular, the surface glabrescent. Seeds spherical, ca. 4 mm diam., the surface smooth. Fr (Aug); in secondary vegetation and along forest edges.

Dalechampia tiliifolia Lam.

Twining lianas sometimes reaching lower canopy. Stipules lanceolate-linear, 3–7 × ca. 1 mm. Leaves simple; petioles 2–8 cm long; blades usually 3-lobed mixed with unlobed when in bloom, usually ovate, rarely lanceolate, relatively broad, width >1/3 length, 6–15 × 4–13 cm, chartaceous, sparsely pubescent adaxially, velutinous abaxially, the base shallowly to deeply cordate, the apex of each lobe acuminate; secondary veins ≥6, diverging from central primary vein throughout its length. Inflorescences on short shoots (peduncles) usually 5–20 cm long, with usually 10 staminate flowers and 3 pistillate flowers; involucral bracts equal, creamy-white at anthesis, broadly ovate, 3–6.5 × 3–6 cm, velutinous, the apex shallowly 3-lobed, lobes subequal; inflorescence glands secreting yellow resin. Staminate involucel of 4 partially connate bracts forming a bilabiate involucel. Staminate flowers: stamens 40, attached to elongate receptacle or column. Pistillate flowers: sepals 10–13, these pinnatifid, densely hispid; pistil with elongate, strongly curved stylar column, with width (excluding apex) <1/5 length, 6–12 mm long, the apex asymmetrically dilated or umbraculiform, stigma extending from apex to >3/4 length of style. Seeds spherical, ca. 5 mm diam., the surface smooth. Fl (Oct), fr (Mar, Oct); locally common, in open areas, secondary vegetation, and along forest edges.

DODECASTIGMA Ducke

Trees. Exudate clear. Hairs simple or forked at base (malpighiaceous). Leaves alternate, simple; petioles eglandular or often with pair of minute glandular knobs at junction of blade and petiole; blades coriaceous, with very small flat glands scattered on abaxial surface, the margins entire; venation pinnate. Inflorescences unisexual, terminal; pedicels present. Staminate inflorescences paniculate with spicate-racemose branches, the flowers clustered at widely spaced nodes. Pistillate inflorescences racemose or sometimes little-branched, the flowers mostly single at widely spaced nodes. Flowers unisexual (plants dioecious). Staminate flowers: calyx 3- to 4-lobed; petals 3–4; nectary disc annular, pubescent; stamens 7–16, free; pistillode absent. Pistillate flowers: sepals 3–5; petals 3–5; nectary disc absent; ovary 3-locular, densely pubescent, with 1 ovule per locule, the styles 3, divided several times to base, much shorter than ovary. Fruits capsular. Seeds 3, ecarunculate.

Dodecastigma integrifolium (Lanj.) Lanj. & Sandwith Fig. 110, Pl. 54c

Understory trees. Bark dark brown, the inner bark orange-red. Leaves: petioles 3–6 cm long, pulvinate and flexed at apex; blades narrowly elliptic or narrowly oblong-elliptic, 14–20 × 4.5–7 cm, coriaceous, glabrous, often with pair of minute glandular knobs at junction of blade and petiole, the base acute, the apex acuminate. Inflorescences 15–35 cm long, the staminate pedicels 3–5 mm long, the pistillate pedicels 3–10 mm long, elongating in fruit. Flowers: petals 3, broadly ovate, 6–7 × 5–6 mm, rounded at apex. Staminate flowers: calyx 3-lobed; stamens 7–9. Pistillate flowers: sepals 3; ovary densely appressed pubescent, the style branches numerous, 1–1.5 mm long. Capsules 3-lobed, ca. 1.6–2 × 2.5 cm, appressed pubescent, the fruiting pedicels 3–4 cm long. Seeds ellipsoid, ca. 1.5 × 1.1 cm, somewhat flattened, smooth. Fl (Nov); rare, in nonflooded forest.

DRYPETES Vahl

Trees. Latex or colored exudate absent. Hairs, if present, simple. Leaves alternate, simple, eglandular; petioles short; blades with margins entire to serrate (outside our area); venation pinnate. Inflorescences of flowers in axillary clusters; pedicels present. Flowers very small, unisexual (plants dioecious). Staminate flowers: sepals 4–5, imbricate; petals absent; nectary disc intrastaminal; stamens 4–12, free; pistillode absent or reduced. Pistillate flowers: sepals 4–5; petals absent; nectary disc cupular-annular; ovary 1-locular (to 3(6)-locular outside our area), with 2 ovules per locule, the stigmas as many as the carpels, subsessile. Fruits indehiscent, drupaceous. Seeds 1 (or as many as the carpels outside our area), ecarunculate, the surface smooth.

Drypetes variabilis Uittien Fig. 111, Pl. 54d

Canopy trees. Trunks with buttresses. Bark gray, smooth with numerous, conspicuous, horizontal lenticels, the slash yellow-orange, oxidizing darker. Leaves: petioles 7–15 mm long; blades elliptic, 9–20 × 4–11 cm, glabrous, coriaceous, the base obtuse or sometimes acute, decurrent, often somewhat assymetrical, the apex obtuse to shortly acuminate, the margins entire, sometimes slightly undulate; secondary veins looped, in 7–9 pairs, often irregularly spaced, with numerous intersecondary veins, the tertiary and higher order veins reticulate. Staminate flowers fragrant; stamens 4–5, the filaments white; nectary disc lobed, yellow. Pistillate flowers: ovary densely puberulous, the stigma subpeltate. Fruits ellipsoid, 2.5–2.8 × 1.5–1.7 cm, puberulent when immature, becoming glabrous at maturity, pale yellow when mature, the apex obtuse with persistent stigma, the base narrowed; fruiting peduncle 7–20 mm long, sparsely puberulous to glabrescent, with small cupular nectary disc persistent after fruit dispersed. Fl (Feb, Jun, Jul, Aug, Nov), fr (Feb, May, Apr, Jun, Jul, Aug, Oct, Nov); common, in nonflooded forest.

EUPHORBIA L.

Herbs (in our area), shrubs, or trees. Latex milky white, copious in all parts. Hairs simple. Stipules absent or present and intrapetiolar. Leaves alternate, opposite or whorled, simple, eglandular; petioles present or absent; blades with margins entire or serrate; venation pinnate, palmate, or reduced. Inflorescences bisexual, pseudanthial, the flowers contained within cupular cyathium composed of 5 connate bracts, the cyathial rim with bract lobes alternating with (1,2)4 or 5 glands, these often with petaloid appendages, each cyathium with 1 pistillate flower and several to many staminate flowers. Flowers unisexual (plants monoecious). Staminate flowers naked (sepals, petals, and nectary disc absent); stamen 1. Pistillate flowers naked; ovary 3-locular, with 1 ovule per locule, the styles 3, free or connate at base, usually bifid. Fruits capsular, usually exserted from cyathium on elongate pedicel. Seeds 3; ecarunculate or caruncle reduced.

The key includes all species that could occur in our area. These species of *Euphorbia* in our area, with the exception of *E. cyathophora*, are often treated in the segregate genus *Chamaesyce*.

1. Erect herbs. Stipules absent. Leaves alternate, except for those subtending inflorescences; blades symmetrical at base. Cyathial gland 1, without appendage [section *Euphorbia*]. . . . . *E. cyathophora.*
1. Prostrate to erect herbs. Stipules intrapetiolar. Leaves opposite; blades assymetrical at base. Cyathial glands 4, each with petaloid appendage [section *Chamaesyce*].
    2. Erect herbs. Capsules glabrous. . . . . *E. hyssopifolia.*
    2. Prostrate or decumbent herbs. Capsules hairy.
        3. Stem internodes 0.5–5 cm long; leaf blades >8 mm long. Cyathia in densely flowered pedunculate glomerules. . . . . *E. hirta.*
        3. Stem internodes 0.1–1.2 cm long; leaf blades ≤8 mm long. Cyathia solitary or in few-flowered clusters. . . . . *E. thymifolia.*

Euphorbia cyathophora Murray [Syn.: *Poinsettia cyathophora* (Murray) Klotzsch & Garcke]

Erect herbs or subshrubs, to 1 m tall. Stipules absent. Leaves alternate, often opposite just below inflorescence; petioles 0.5–1.5 cm long; blades elliptic, ovate, lanceolate, or pandurate, 3–10 × 1–3 cm, glabrous to sparsely pubescent, the base symmetrical, decurrent onto petiole, the apex acute, the margins entire to dentate or sometimes pinnately lobed, the leaves subtending inflorescence red at base or sometimes throughout; venation pinnate. Cyathia in terminal clusters, each cyathium with single compressed cup-like gland lacking appendage. Capsules ca. 4 × 5 mm, yellow-green, glabrous, exserted. Fl (Jun), fr (Jun); open weedy areas, near village.

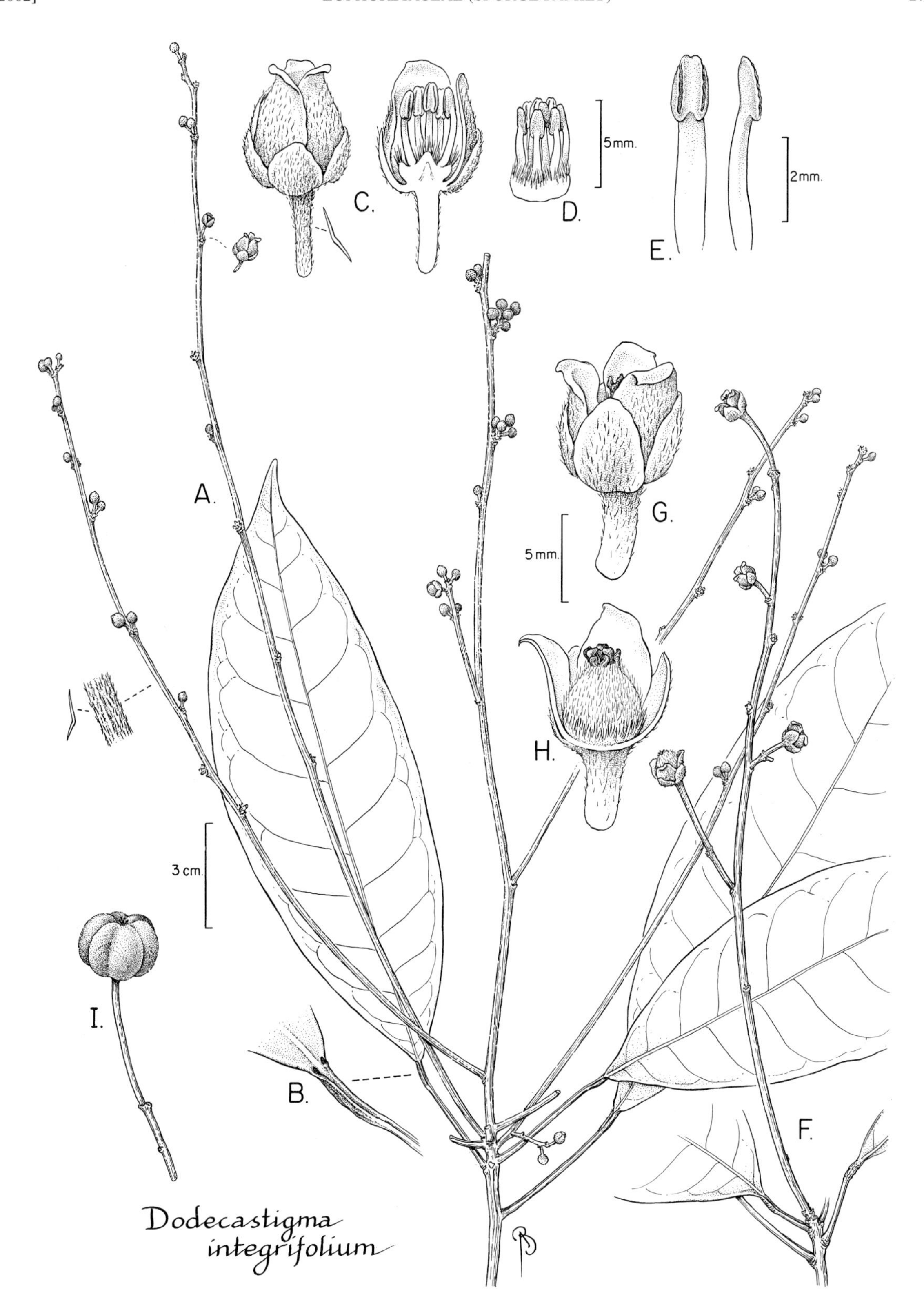

FIG. 110. EUPHORBIACEAE. *Dodecastigma integrifolium* (A–E, *Mori et al. 22811*; F–H, *Silva & Bahia 3546* from Brazil; I, *Ferreira 5944* from Brazil). **A.** Part of stem with leaves and staminate inflorescence with detail of inflorescence axis and malphigiaceous hair. **B.** Base of leaf blade showing glands. **C.** Lateral view of staminate flower with close-up of malphigiaceous hair (left) and medial section of staminate flower (right). **D.** Staminate flower with perianth removed; note the nectary disc with pubescent margin. **E.** Stamens. **F.** Part of stem with pistillate inflorescence. **G.** Lateral view of pistillate flower. **H.** Pistillate flower with part of perianth removed. **I.** Immature capsule.

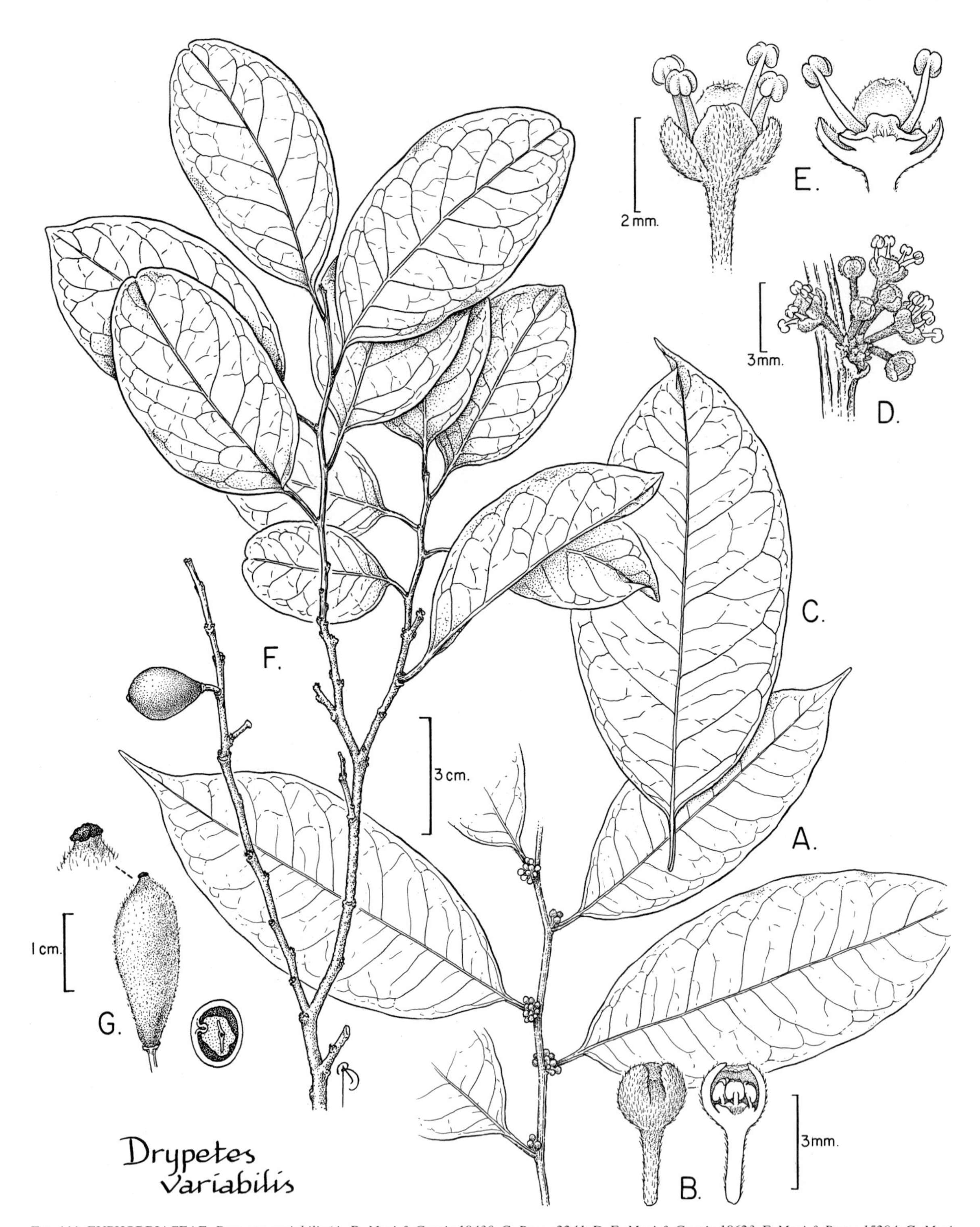

FIG. 111. EUPHORBIACEAE. *Drypetes variabilis* (A, B, *Mori & Gracie 18430*; C, *Boom 2241*; D, E, *Mori & Gracie 18623*; F, *Mori & Boom 15294*; G, *Mori & Gracie 18331*). **A.** Stem with leaves and staminate inflorescences. **B.** Lateral view (left) and medial section (right) of intact staminate flower buds. **C.** Leaf. **D.** Staminate inflorescence on stem. **E.** Lateral view (left) and medial section (right) of staminate flower. **F.** Stem with leaves and fruit. **G.** Lateral view of immature fruit (near right) with detail of remnant stigma (above) and transverse section of fruit (far right).

Euphorbia hirta L. [Syn.: *Chamaesyce hirta* (L.) Small]

Decumbent, sprawling herbs, to 30 cm. Stems puberulent and hirtellous with multicellular hairs. Stipules intrapetiolar. Leaves opposite; petioles 1–2 mm long; blades ovate, 9–40 × 5–15 mm, hirtellous, the base obtuse or rounded, asymmetrical, the apex acute, the margins serrate; venation of 3 veins from base. Cyathia often pinkish-puberulent, densely crowded in mostly axillary, pedunculate,

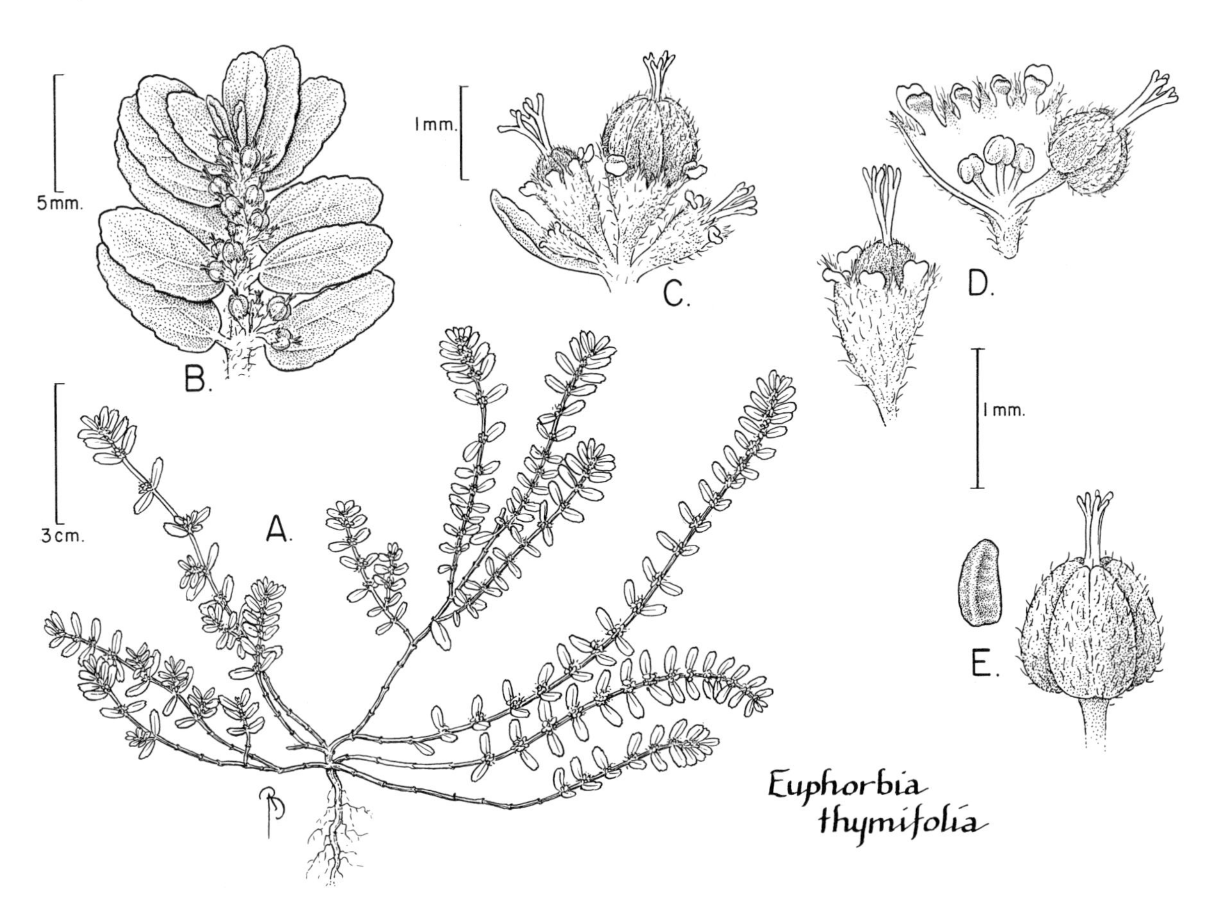

FIG. 112. EUPHORBIACEAE. *Euphorbia thymifolia* (*Mori & Gracie 23961*). **A.** Habit of plant. **B.** Part of stem with leaves and inflorescences. **C.** Detail of inflorescence. **D.** Lateral view of cyathium (left) and cyathium opened to show flowers (above right). **E.** Lateral view of capsule (right) and seed (above).

leafless glomerules; cyathial glands 4, each with petaloid appendage; peduncles 2–9 mm long. Capsules broadly ovoid, 1.1–1.3 mm diam., puberulent, pedicellate and fully exserted from cyathium at maturity. Vouchered by *Luu 103* (probably at MPU) but this collection not confirmed. *Madlomé* (Créole).

Euphorbia hyssopifolia L. [Syn.: *Chamaesyce hyssopifolia* (L.) Small]

Erect or ascending herbs, to 60 cm. Stems often reddish, glabrous. Stipules intrapetiolar. Leaves opposite; petioles 0.5–2 mm long; blades narrowly oblong to narrowly oblong-elliptic, 10–30 × 3–9 mm, glabrous, the base acute to cordate, asymmetrical, the apex obtuse to acute, the margins serrulate; venation of (4)3 veins from base. Cyathia glabrous, on axillary or terminal leafy cymes; cyathial glands 4, each with petaloid appendage. Capsules subglobose, 1.5–2 mm diam., glabrous, pedicellate and fully exserted at maturity. Open weedy areas. Vouchered by *Luu 3* (probably at MPU) but this collection not confirmed.

Euphorbia thymifolia L. [Syn.: *Chamaesyce thymifolia* (L.) Millsp.] FIG. 112, PL. 54e

Prostrate herbs, forming mats to 0.4 m diam. Stems often pinkish, pubescent. Stipules intrapetiolar. Leaves opposite; petioles 0.5–1.5 mm long; blades oblong, 3–8 × 1–4 mm, sparsely pubescent to glabrescent, the base obtuse or truncate, often strongly asymmetrical, the apex acute to rounded, the margins serrate; venation of 3 veins from base. Cyathia often pinkish or purplish, pubescent, axillary, solitary or usually clustered, mostly on leafy short shoots, cyathial glands 4, each with petaloid appendage. Capsules ca. 1 × 1 mm, with scattered hairs, not fully exserted and splitting cyathium at maturity. Fl (Jun, Sep, Oct), fr (Jun, Sep, Oct); open weedy areas, such as airstrip and road edge. *Madlomé*, *madlomé rouge*, *malnommée* (Créole).

GLYCYDENDRON Ducke

Trees. Latex clear or whitish. Hairs simple. Leaves alternate, simple; petioles present; blades biglandular at base, the margins entire; venation tripliveined, the lateral, primary vein pair often indistinct and blades sometimes appearing pinnately veined in our area. Inflorescences unisexual, axillary (in axils of leaves or bracts below leaves on new shoots), racemose or sometimes paniculate. Flowers small, unisexual (plants dioecious). Staminate flowers: sepals 4–5; petals absent; nectary disc annular, lobed; stamens 25–30, free; pistillode absent. Pistillate flowers: sepals 4–5; petals absent; nectary disc cupular-annular, shallowly lobed; ovary 2-locular, with 1 ovule per locule, the styles 2, short, connate at base, deeply bifid. Fruits drupaceous. Seeds 2, ecarunculate.

Glycydendron amazonicum Ducke

Canopy trees. Trunks with low, thick buttresses. Inner bark yellow-orange with reddish streaks. Leaves: petioles 1.5–6(8) cm long, pulvinate and flexed at apex; blades elliptic, narrowly elliptic, or oblanceolate, 7–20(27) × 4–10(13) cm, subcoriaceous, glabrous, sometimes with scattered hairs on abaxial midrib, with 2, flat to crateriform glands with prominent rim at base on adaxial surface, the base acute to obtuse, the apex acuminate; lateral primary vein pair often indistinct, sometimes appearing pinnately veined in our area, the secondary veins in 2–5 pairs, looped. Inflorescences 1–4 cm long. Fruits ellipsoid, 4.5–5.5 × 2–3 cm, yellow, glabrescent; endocarp 4–9 mm thick, woody, irregularly coursely tuberculate and ridged; peduncle to 5 cm long; pedicel 6–8 mm long. Fr (Aug); uncommon, in nonflooded forest. *Saint Martin blanc* (Créole).

The Saül collections differ from typical *G. amazonicum* in their larger fruits (4.5–5.5 vs 2–3.5 cm long) and leaf blades that are less prominently 3-veined with more secondary veins (3–5 vs 2–3 pairs) that are not restricted to the distal half.

HEVEA Aubl.

Trees. Latex copious. Hairs simple. Leaves alternate, palmately compound, 3-foliolate; petioles long, eglandular; blades eglandular, the margins entire or slightly undulate; venation of leaflets pinnate. Inflorescences bisexual, axillary, paniculate; cymules bisexual with central pistillate flower or cymules staminate. Flowers very small, unisexual (plants monoecious). Staminate flowers: calyx (4)5-lobed, the lobes valvate; petals absent; nectary disc lobed, dissected, or absent; stamens 5–10, the filaments connate into a column, the anthers sessile in 1–3 whorls; pistillode apparently present as elongate tip of column. Pistillate flowers: calyx 5-lobed; petals absent; staminodia present or absent; ovary 3-locular, with 1 ovule per locule, the stigmas 3, subsessile. Fruits capsular. Seeds 3, ecarunculate.

Hevea guianensis Aubl. FIG. 113; PL. 54f

Canopy trees. Trunks cylindrical, without buttresses. Bark scaly, brown, the inner bark pale orange. Latex cream-colored or creamy yellow. Leaves deciduous; petioles 8–25 cm long; petiolules 6–13 mm long; leaflet blades oblanceolate, 10–26 × 4–9 cm, glabrous, chartaceous, paler abaxially, the base acute, the apex obtuse-acuminate, the margins entire; venation pinnate, secondary veins in 12–17 pairs. Inflorescences developing after leaf fall, pendent, in axils of bracts below new leaves, 15–40 cm long, the axes pubescent; pedicels present. Flowers yellow; sepals densely pubescent. Staminate flowers: nectary disc absent; anthers 5, in 1 whorl. Pistillate flowers larger than staminate flowers: calyx ca. 4 × 3 mm, tubular, the lobes triangular; staminodia present or reduced; ovary densely pubescent. Capsules ca. 3 × 4 cm, woody. Seeds broadly oblong-ellipsoid, ca. 1.8 × 1.6 cm, pale brown with darker blotches, smooth. Fl (Sep, Oct, Nov), fr (Mar, May); scattered, nonflooded forest. *Hévéa*, *éféa* (Créole).

HURA L.

Trees. Latex milky white, somewhat clear, copious, caustic. Hairs simple. Leaves alternate, simple; petioles long, distinctly biglandular at apex; blades eglandular, the margins serrate to rarely subentire; venation pinnate. Inflorescences unisexual or sometimes bisexual. Staminate inflorescences terminal, spicate, cone-like, pedunculate. Pistillate inflorescences of solitary flowers in axil of upper leaf (leaves) or sometimes at base of staminate spike; staminate flowers sessile, each covered by a thin bract rupturing at anthesis; pistillate pedicels present. Flowers unisexual (plants monoecious). Staminate flowers: calyx cup-shaped; petals absent; nectary disc absent; stamens numerous, the filaments connate into a thick column bearing 2–10 whorls of anthers; pistillode absent. Pistillate flowers: calyx cup-shaped; petals absent; nectary disc absent; ovary 5–20-locular, with 1 ovule per locule, the styles connate into a long column dilated apically into massive lobed nectary disc or funnel, the style branches as many as the locules. Fruits capsular. Seeds 5–20, ecarunculate.

Hura crepitans L. FIG. 114, PLS. 54g, 55a,b; PART 1: FIG. 9

Canopy trees. Trunks cylindrical, with hard conical prickles and simple, rounded, thick buttresses. Outer bark pale brown, smooth, lenticellate, the inner bark cream-colored with orange mottles, fibrous. Leaves: petioles 6–13 cm long, biglandular at apex; blades broadly ovate, 10–15 × 10–15 cm, glabrous adaxially, glabrous to pubescent abaxially, often with midrib sides densely long-pubescent, the base cordate, the apex acuminate to caudate; secondary veins prominent, in 10–13 pairs, ca. perpendicular to midrib near base, becoming less broadly angled toward apex distally. Inflorescences: staminate spikes cylindrical-conical, 2–4.5 cm long, very densely flowered; peduncle to 10 cm long. Pistillate inflorescences with pedicels stout, to 2(5) cm long. Staminate flowers: staminal column with anthers in 2–3 whorls. Pistillate flowers red (turning blackish with age), the stylar column 2–5 cm long, the apical nectary disc or funnel ca. 1.5 cm diam., with 5–20 radiating style branches. Capsules oblate, 3–5 × 8–10 cm, woody, explosively dehiscing into 5–20 cocci. Seeds lenticular, ca. 2 cm diam. Fl (Sep, Oct, Dec); in nonflooded forest. *Assacu* (Portuguese), *bois-diable*, *sablier* (Créole), *sandbox tree* (English).

HYERONIMA Allemão

Trees. Latex or exudate absent. Hairs lepidote. Stipules present or absent. Leaves alternate, simple; petioles eglandular; blades eglandular, or with very small flat glands scattered on abaxial surface, the margins entire; venation pinnate. Inflorescences unisexual, axillary, paniculate with branches racemose, the axes often densely lepidote-pubescent. Flowers very small, unisexual (plants dioecious). Staminate flowers: calyx cup-shaped, shallowly 4–6-lobed; petals absent; nectary disc annular or lobed; stamens (3)4–6, free, opposite calyx lobes; pistillode present. Pistillate flowers: calyx cup-shaped, 4–6-lobed; petals absent; nectary disc cupular-annular; ovary 2-locular, with 2 ovules per locule, the stigmas 4 (probably 2, deeply bifid), subsessile. Fruits drupaceous. Seeds 1, rarely more, ecarunculate.

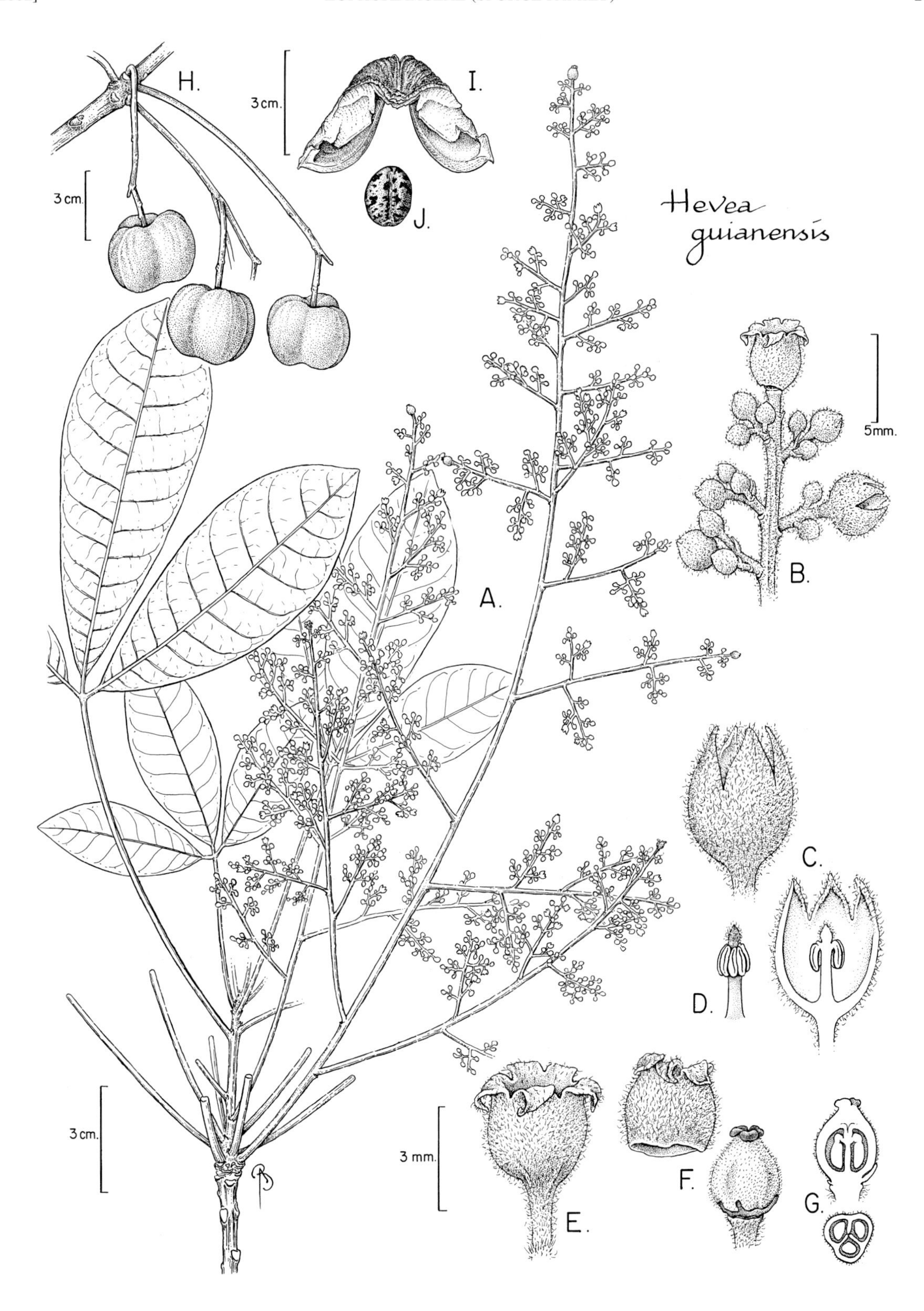

FIG. 113. EUPHORBIACEAE. *Hevea guianensis* (A–G, *Mori & Boom 14917*; H–J, *Mori & Pennington 17954*). **A.** Part of stem with leaves and inflorescences. **B.** Detail of inflorescence with terminal pistillate flower, one open staminate flower, and staminate flower buds in various stages. **C.** Lateral view (left) and medial section (below) of staminate flower. **D.** Lateral view of androecium. **E.** Lateral view of pistillate flower. **F.** Dehiscent corolla (above) and lateral view of pistil (right). **G.** Medial section (above) and transverse section (below) of ovary. **H.** Infructescence. **I.** Dehisced capsule segment. **J.** Seed.

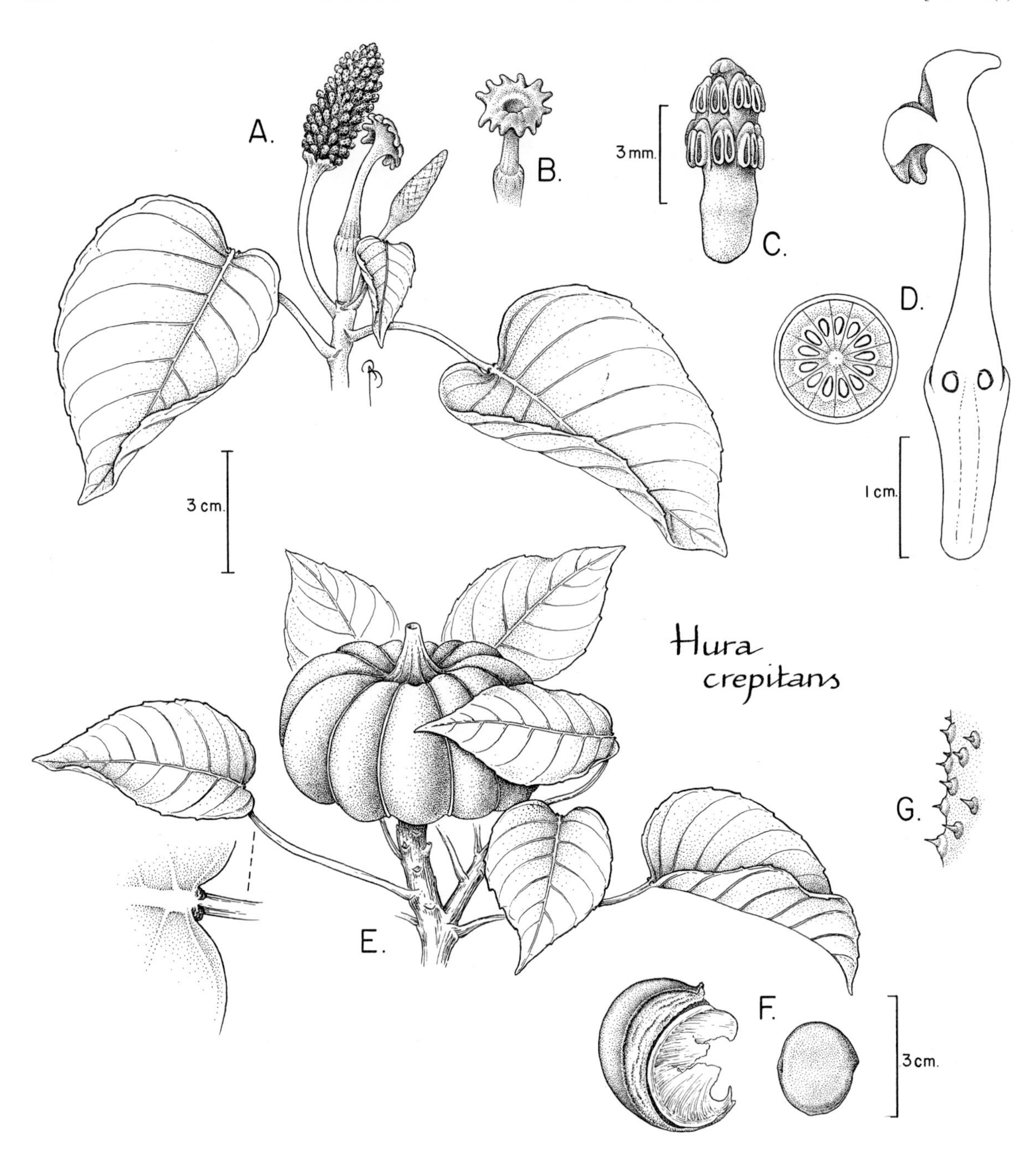

FIG. 114. EUPHORBIACEAE. *Hura crepitans* (A–E, *Acevedo-Rodríguez 4010* from St. John, U.S. Virgin Islands; F, *Zanoni 30602* from the Dominican Republic; G, unvouchered). **A.** Stem with leaves and staminate inflorescence (left) and pistillate flower (right). **B.** Pistillate flower. **C.** Androecium with staminal column bearing 2 rows of anthers. **D.** Ovary in transverse section (left) and medial section of pistillate flower (right). **E.** Stem with capsule. **F.** Segment of capsule (left) and seed (right). **G.** Prickles on trunk. (Reprinted with permission frrm P. Acevedo-Rodríguez, *Flora of St. John, U.S. Virgin Islands*. Mem. New York Bot. Gard. 78, 1996.)

1. Stipules large, leaf-like, usually petiolate. Petioles 3–8 cm long; blades broadly elliptic. Staminate nectary disc annular, extrastaminal. . . . . *H. alchorneoides* var. *alchorneoides*.
1. Stipules absent. Petioles 1–2 cm long; blades obovate. Staminate nectary disc 5-lobed, intrastaminal. . . . . . . *H. oblonga*.

Hyeronima alchorneoides Allemão var. **alchorneoides**
PART 1: PL. XIIId

Canopy trees. Trunks with rounded, simple, running buttresses. Bark gray-brown or brown, rough, scaly, flaky, the inner bark brick red or dark red. Stipules usually leaf-like, petiolate, to 1.5 × 1 cm; saplings often with much broader, sessile stipules. Leaves: petioles 3–8(13) cm long, sparsely lepidote-pubescent; blades broadly elliptic, 11–22(30) × 6–14(17) cm, sparsely lepidote-pubescent on both surfaces, major veins often pilose abaxially, the base obtuse or obtuse-acuminate, the apex obtuse-acuminate; tertiary veins parallel. Staminate flowers: nectary disc annular, extrastaminal. Drupes subglobose, 3–4.5 mm, blackish, sparsely lepidote-pubescent. Fl (Jun, Oct, Nov), fr (Aug, Oct, Nov); in nonflooded forest. *Bois di vin* (Créole).

Hyeronima oblonga (Tul.) Müll. Arg.
FIG. 115; PART 1: FIG. 4

Canopy trees. Trunks with high (steep) narrow, sometimes branched buttresses. Bark reddish brown, rough, scaly, the inner bark dark red or pink, without exudate. Stipules absent. Leaves: petioles 1–3 cm long, lepidote-pubescent; blades obovate, 8–16 × 4–8 cm, very sparsely lepidote-pubescent on both surfaces, midrib more densely lepidote-pubescent, the base acute, the apex obtuse or obtuse-acuminate; tertiary veins mostly reticulate. Staminate flowers: nectary disc 5-lobed, intrastaminal, stellate-lepidote-pubescent. Drupes broadly ellipsoid, 6–8 × 4–6 mm, red, glabrous. Fl (Sep), fr (Jul, Sep, Nov); in nonflooded forest.

JATROPHA L.

Shrubs, trees, or herbs. Latex clear or colored. Hairs simple or sometimes glandular (these sometimes branched). Stipules obsure to well developed, often glandular or spinose, the margins entire to serrate. Leaves alternate, simple or compound, eglandular, except glandular hairs sometimes present; blades often palmately lobed, the margins entire to serrate; venation palmate, venation of leaflets, if present, pinnate. Inflorescences bisexual or unisexual, terminal or axillary, cymose, with the pistillate flowers terminating proximal axes. Flowers unisexual (plants monoecious or dioecious). Staminate flowers: sepals 5, imbricate; petals 5, free to partly connate, red, white, or greenish; nectary disc annular or segmented, extrastaminal; stamens 8–12, the filaments partly connate; pistillode reduced or absent. Pistillate flowers: sepals 5; petals 5, as in staminate flowers; nectary disc annular or segmented; ovary 3-locular, with 1 ovule per locule, the styles free or basally connate, undivided or bifid. Fruits capsular. Seeds 3, carunculate.

Jatropha gossypiifolia L.

Robust herbs or subshrubs, to 2 m tall. Stipules 3–5 mm long, deeply divided into many filiform glandular-tipped lobes. Leaves: petioles 3–9 cm long, with scattered branched glandular hairs; blades deeply 3- or 5-lobed, 5–12 × 7–15 cm, the lobes elliptic or obovate, 1.5–4 cm wide, the apex acuminate, the margins with numerous glandular hairs. Inflorescences: peduncle of cyme 4–5 cm long. Flowers: sepals with glandular hairs along margin; petals dark red. Capsules oblong-globose, 1–1.4 mm diam. Cultivated, persistent plants around homesteads may appear native. *Médecinier béni*, *médecinier rouge* (Créole), *pião roxo* (Portuguese).

MABEA Aubl.

Shrubs or trees. Latex milky white, copious. Hairs dendritic (irregularly branched), usually minute. Stipules mostly absent. Leaves alternate, simple; petioles short, eglandular; blades usually with small flat circular glands often present along abaxial margin, the margins entire to serrulate, often somewhat undulate; venation pinnate, secondary veins diverging at a wide angle and usually conspicuously looped at some distance from margin. Inflorescences bisexual, terminal or axillary, racemose or paniculate, usually pendent, the flowers in pedunculate racemose units with pistillate flowers solitary at basal nodes and staminate flowers in cymules above, each staminate cymule usually subtended by a biglandular bract, the axes and flowers usually covered with minute dendritic hairs; pedicels long. Flowers unisexual (plants monoecious). Staminate flowers globose; calyx 3–5-lobed, not covering anthers in bud; petals absent; nectary disc absent; anthers 10–70, subsessile on convex or subglobose receptacle. Pistillate flowers; sepals 3–6, unequal, acute at apex; petals absent; nectary disc absent; ovary 3-locular, with 1 ovule per locule, the styles 3, connate into a long column, tips unlobed and spreading. Fruits capsular. Seeds 3, carunculate.

1. Stems and leaves glabrous to glabrescent. Inflorescences rarely branched, with racemose units ≥2 cm diam.
    2. Trees, to 20 m tall. Leaf blades 2.5–4.5 cm wide. Inflorescences with racemose unit(s) 6–8 × 2–3.5 cm; peduncle 2.5–9 cm long; glands on bract subtending staminate cymule 1.5–4 mm above rachis; staminate pedicels usually 6–9 mm long; 2–4 pistillate flowers per racemose unit. . . . . *M. piriri.*
    2. Treelets or small trees to 8 m. Leaf blades 3.5–6 cm wide. Inflorescences with racemose unit(s) 9–15 × 3–5 cm; peduncle (4)10–21 cm long; glands on bract subtending staminate cymule touching to 1.5 mm above rachis; staminate pedicels 10–22 mm long; usually 4–10 pistillate flowers per racemose unit.. . . . *M. speciosa.*
1. Stems and leaves with reddish brown dendritic hairs. Inflorescences mostly branched, with racemose units ≤1 cm diam.
    3. Stems, petioles, and leaf blades sparsely covered with dendritic hairs. Inflorescences axillary; pistillate flower pedicels 10–30 mm long. . . . . *M. salicoides.*
    3. Stems and petioles densely covered with dendritic hairs, the leaf blades mostly glabrous with midrib covered with dendritic hairs. Inflorescences terminal; pistillate flower pedicels 5–7 mm long. . . . . . . *M. subsessilis.*

Mabea piriri Aubl.

Understory trees, to 20 m tall. Trunks with very low rounded buttresses. Bark pale brown, smooth with few vertical fissures, the inner bark orange-brown. Stems glabrous to glabrescent. Leaves: petioles 7–11 mm long, glabrous; blades narrowly oblong-elliptic, 7–13 × 2.5–4.5 cm, glabrous, with small circular glands along margin abaxially, the base acute, the apex acuminate to caudate, the margins serrulate to subentire. Inflorescences terminal and in axils of upper leaves, sometimes branched, 1–3(5) at branch end, the racemose

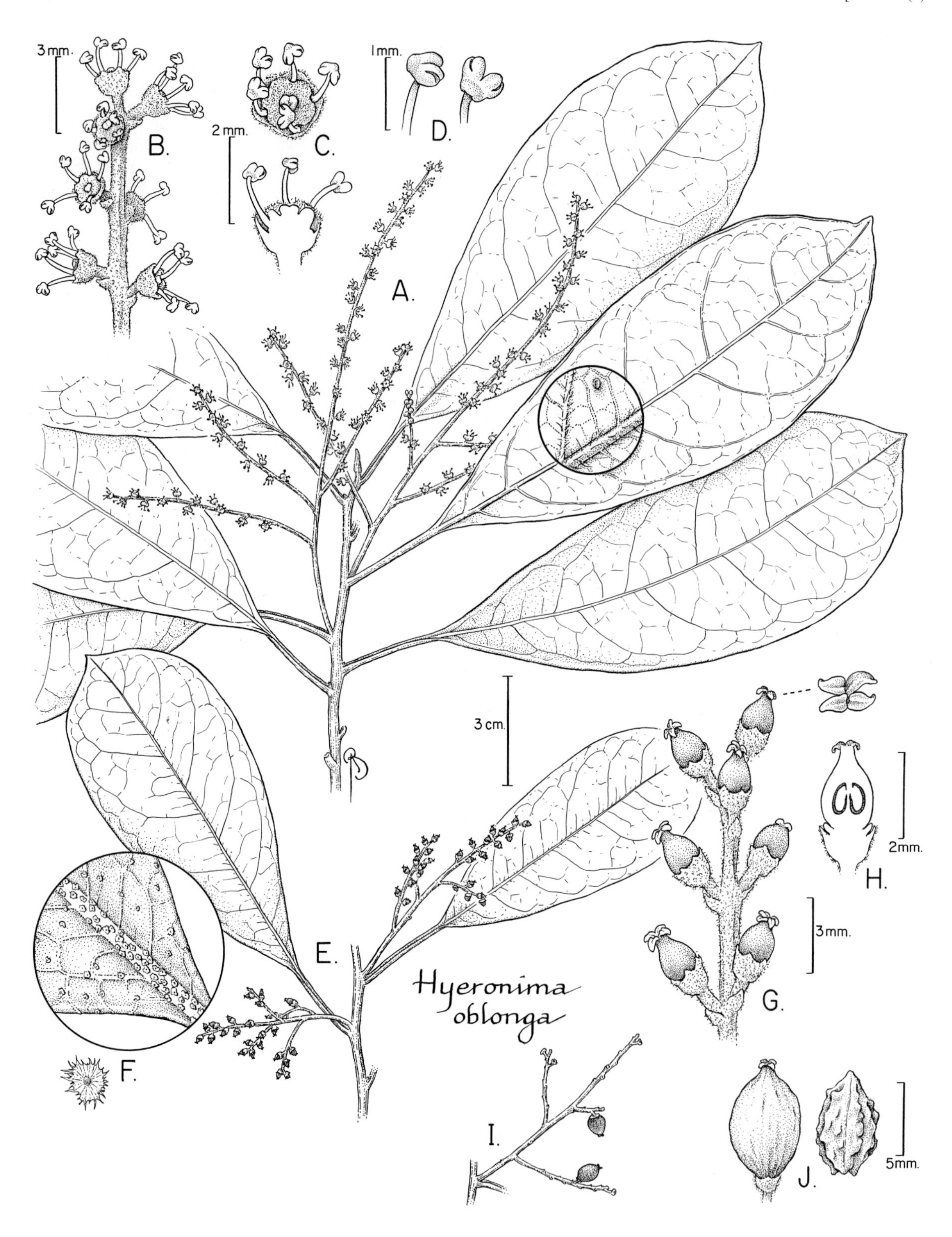

FIG. 115. EUPHORBIACEAE. *Hyeronima oblonga* (A–D, *Mori et al. 20961*; E–H, *Mori et al. 14935*; I, J, *Mori & Boom 15226*). **A.** Part of stem with leaves and staminate inflorescences with enlarged detail of abaxial surface of leaf showing gland. **B.** Detail of staminate inflorescence. **C.** Apical view (above) and medial section (below) of staminate flower. **D.** Two views of anthers. **E.** Part of stem with leaves and pistillate inflorescences with detail (left) of adaxial surface of leaf. **F.** Lepidote hair from leaf. **G.** Detail of pistillate inflorescence (left) with detail of apical view of stigma (above, right). **H.** Medial section of pistillate flower. **I.** Part of infructescence. **J.** Fruit (left) and seed (right).

unit(s) 6–8 × (2)2.5–3.5 cm, pale green, on peduncle 2.5–9 cm long, with 2–4 pistillate flowers per racemose units; staminate bract glands ca. 3 mm long, dark green, inserted 1.5–4 mm above rachis; staminate pedicels 6–9(12) mm long; pistillate pedicels 10–15 mm long. Staminate flowers 1.5–2 mm diam.; stamens 15–25. Pistillate flowers: ovary densely covered with dendritic hairs, the styles 2–2.6 cm long, connate for half of length. Capsules 1.5–2 cm diam., covered with reddish-brown hairs. Fl (Sep, Oct, Nov), fr (Feb, Dec); in nonflooded forest. *Bois lait, bois lélé* (Créole).

Mabea salicoides Esser

Understory trees, ca. 20 m tall. Bark reddish brown. Stems and petioles sparsely covered with minute reddish-brown dendritic hairs. Leaves: petioles 1–1.4 cm long; blades narrowly oblong-elliptic, 12–18 × 4–7 cm, mostly glabrous adaxially, whitish with scattered dendritic hairs abaxially, the base acute to obtuse, the apex obtuse to rounded and shortly caudate, the margins subentire to serrulate, distinctly undulate. Inflorescences axillary, usually branched near base, the racemose units 4–5 × 0.6–1 cm, on peduncle 0.8–1.6 cm long, with ca. 2 pistillate flowers per racemose unit; staminate bracts eglandular (in our area); staminate pedicels 1–3 mm long; pistillate pedicels 10–30 mm long. Staminate flowers ca. 1 mm diam.; stamens 10–20. Pistillate flowers: styles 7–9 mm long, connate ca. one third of length. Capsules ca. 1.6 × 2 cm, covered with dendritic hairs; fruiting peduncle ca. 3 cm long; pedicels ca. 3 cm long. Seeds broadly ovoid-ellipsoid, ca. 9 × 8 mm, dark brown, shiny. Fl (Oct), fr (Apr, Oct); in nonflooded forest.

Mabea speciosa Müll. Arg. FIG. 116, PL. 55c

Understory treelets or small trees, to 10 m tall. Stems glabrous or glabrescent. Leaves: petioles 4–10 mm long, glabrous; blades oblong-elliptic, 8–15 × 3.5–6 cm, glabrous, whitish abaxially, with small circular glands sometimes present along abaxial margin, the base acute or obtuse, the apex rounded or obtuse and distinctly caudate, the margins serrulate to subentire. Inflorescences terminal, sometimes also axillary, usually unbranched, 1(2) at branch end, the racemose unit(s) 9–15 × 3–5 cm, often reddish or purplish green, on peduncle (4)10–21 cm long, with (2)4–10 pistillate flowers per racemose unit; staminate bract glands 3–4 mm long, pale green, touching rachis to 1.5 mm above; pedicels 10–22 mm long. Staminate flowers 2–3 mm diam., pale yellow, often reddish; stamens ca. 30–50. Pistillate flowers pale green, often reddish; ovary densely covered with dendritic hairs, the styles 2–2.8 cm long, connate ca. two thirds of length. Capsules ca. 1.5 cm diam., reddish. Seeds ovoid, ca. 9 × 7 mm. Fl (Aug, Sep, Oct, Nov, Dec), fr (Jan, Feb, May, Sep, Nov); in secondary and mature nonflooded forests at all elevations. *Bois lélé* (Créole).

Mabea subsessilis Pax & K. Hoffm. [Syn.: *M. argutissima* Croizat]

Understory trees, to 15 m tall. Stems and petioles densely covered with reddish-brown dendritic hairs. Stipules linear-triangular, 4–6 mm long; with dendritic hairs. Leaves eglandular; petioles 4–6 mm long; blades narrowly oblong-elliptic, 5–11 × 2–4 cm, mostly glabrous, with dendritic hairs on midrib abaxially, the base rounded, the apex obtuse and distinctly caudate, the margins serrulate with teeth closely spaced. Inflorescences terminal, on peduncle 1–2 cm long, usually branched above, the racemose units ca. 4 × 1 cm, on peduncle, with 2–3 pistillate flowers per racemose unit; staminate bract glands ca. 0.8 mm diam., touching rachis; pistillate pedicels 5–7 mm long. Staminate flowers ca. 1 mm diam.; stamens 10–15. Pistillate flowers: styles 8–9 mm long, connate ca. 2/3 of length. Capsules ca. 1.5 cm diam., sparsely covered with dendritic hairs. Seeds ovoid-ellipsoid, ca. 7–8 × 6 mm diam., dark brown, shiny. Fl (Sep); uncommon, in nonflooded forest.

MANIHOT Mill.

Subshrubs, shrubs, or lianas (in our area), or trees, sometimes scandent. Latex milky white. Hairs simple. Leaves alternate, simple, sometimes appearing palmately compound; petioles mostly long, eglandular; blades often deeply lobed, eglandular, the margins entire or subentire, the lobes sometimes pinnately lobed; venation palmate in lobed leaves, pinnate in unlobed leaves, the venation of each lobe pinnate. Inflorescences bisexual, terminal (or axillary), racemose or paniculate, with pistillate flowers mostly at basal nodes; pistillate pedicel long. Flowers unisexual (plants monoecious). Staminate flowers: calyx 5-lobed, petaloid; petals absent; nectary disc central; stamens 10, free, in 2 whorls of 5 each. Pistillate flowers: sepals 5, petaloid; petals absent; nectary disc annular, fleshy; ovary 3-locular, with 1 ovule per locule, the styles 3, short connate at base, the stigmas dilated. Fruits capsular. Seeds 3, carunculate.

Rogers, D. J. & S. G. Appan. 1973. *Manihot Manihotoides* (Euphorbiaceae). Fl. Neotrop. Monogr. 13: 1–272.

1. Stems distinctly enlarged at nodes. Leaf blades deeply 3–7-lobed, or sometimes unlobed, the lobes >4 mm wide at base (in our area). . . . . *M. esculenta.*
1. Stems not distinctly enlarged at nodes. Leaf blades very deeply 5-lobed, appearing palmately compound, the lobes very constricted and <3 mm wide at base.
    2. Leaf blade lobes entire. . . . . *M.* aff. *quinquepartita.*
    2. Leaf blade lobes pandurate. . . . . *Manihot* sp. A

Manihot esculenta Crantz

Shrubs, to ca. 4 m tall. Stems and leaves glabrous. Leaves: petioles 5–14 cm long; blades deeply 3–7-lobed, 8–12 × 9–19 cm, or sometimes unlobed near inflorescence and then narrowly elliptic and 3–6 cm wide, membranous or thin-chartaceous, the lobes oblanceolate, sometimes sinuate or pandurate, 6–12 × 1.5–3.5 cm, the base narrowed, the apex acuminate. Staminate flowers: calyx campanulate, 6–10 mm long; nectary disc 10-lobed, glabrous; stamens dimorphic, 5 long and 5 short, alternating with nectary disc lobes. Pistillate flowers: sepals narrowly oblong, ca. 8 × 3 mm, the adaxial margins pubescent; ovary ellipsoid, glabrous, with longitudinal ribs, the style tips strongly dilated and multilobed. Capsules broadly oblong-ellipsoid, ca. 1.5 cm diam. Fl (Jun, Aug, Sep, Oct), fr (Oct); cultivated for its edible tubers, sometimes escaped, in open weedy areas. *Maman-manioc* (Créole), *mandioca* (Portuguese), *manioc*, *manioc blanc*, *manioc jaune*, *manioc sauvage* (Créole).

Manihot aff. **quinquepartita** D. J. Rogers & Appan FIG. 117

Scandent shrubs or lianas. Stems and petioles sparsely pubescent. Leaves: petioles 4–14 cm long; blades very deeply 5-lobed and appearing palmately compound, 8–12 × 10–16 cm, membranous or

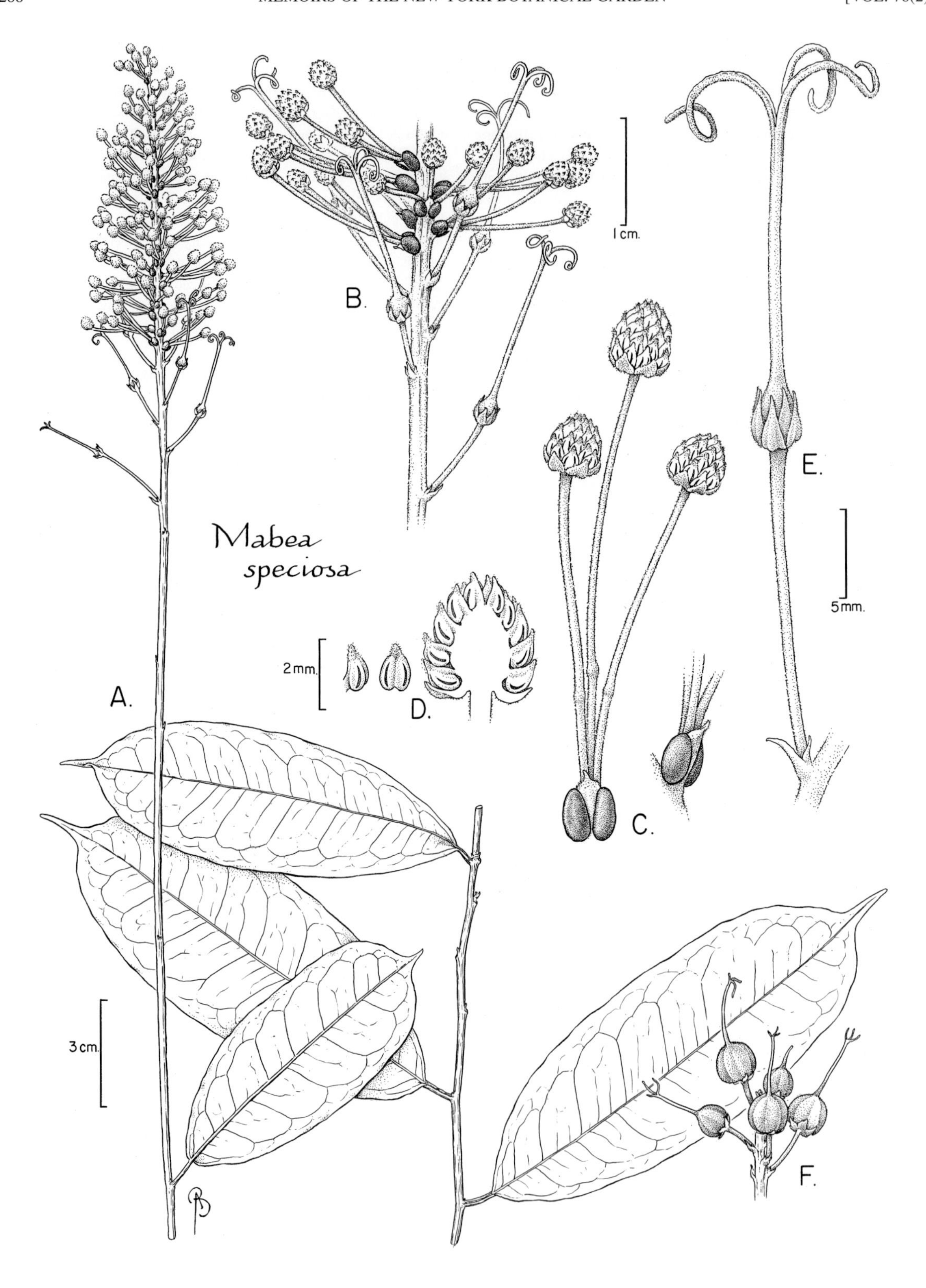

FIG. 116. EUPHORBIACEAE. *Mabea speciosa* (A–E, *Mori et al. 23367*; F, *Mori 8902*). **A.** Part of stem (below, right) and inflorescence with leaf (right); note that inflorescence is pendent on plant. **B.** Detail of proximal portion of inflorescence with pistillate flowers at base and staminate flowers above; note glands on bracts subtending staminate flowers. **C.** Cluster of staminate flowers and subtending biglandular bract (left) and detail of bract (right). **D.** Medial section of staminate flower (right) and two views of anthers (left). **E.** Pistillate flower. **F.** Immature capsules.

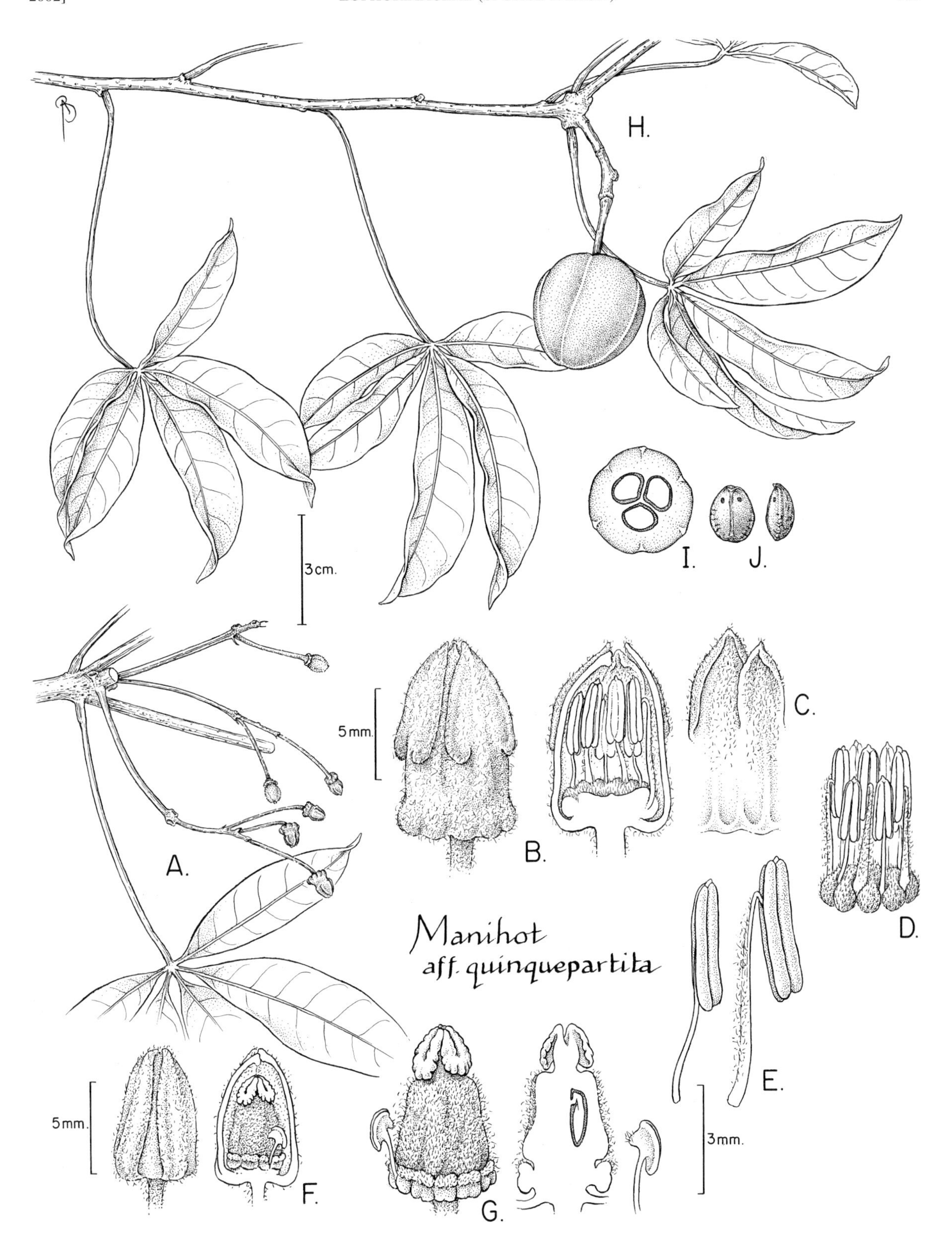

FIG. 117. EUPHORBIACEAE. *Manihot* aff. *quinquepartita* (A–E, *Mori et al. 19063*; F, G, *Oldeman 3196*; H–J, *Mori & Gracie 18622*). **A.** Stem with leaf and inflorescence. **B.** Lateral view of staminate bud (left) and medial section of staminate bud (right). **C.** Detail of abaxial surface of imbricate calyx lobes. **D.** Androecium and lobed disc. **E.** Stamens. **F.** Lateral view (far left) and lateral view with some sepals removed (right) of pistillate bud. **G.** Lateral view of ovary with one staminode (left), medial section of ovary (near right), and staminode (far right). **H.** Stem with leaves and capsule. **I.** Transverse section of capsule. **J.** Seeds.

thin-charctaceous, sparsely pubescent at base and on veins abaxially, the lobes elliptic or narrowly oblong-elliptic, entire, 6–12 × 2–4 cm, the base constricted and very narrow (<3 mm wide), the apex acuminate. Staminate flowers: calyx campanulate, ca. 10 mm long, pubescent; nectary disc 10-lobed, pubescent; stamens dimorphic, alternating with nectary disc lobes. Pistillate flowers: sepals pubescent; ovary ellipsoid, densely pubescent. Capsules broadly ovoid, ca. 2.7 cm diam. Fl (Aug), fr (Feb, Aug); growing into canopy of low disturbed forest.

The Saül plants belong to *Manihot* section *Peruvianae* and most closely resemble *M. quinquepartita* known from eastern Amazonian Brazil, but differs from that species in having small narrow rather than foliaceous inflorescence bracts.

Manihot sp. A

Subshrub, ca. 2 m tall. Stems red. Leaves: petioles 10–13 cm long, red; blades very deeply 5-lobed and glabrous, appearing palmately compound, ca. 18 × 20–30 cm, the lobes obovate, pandurate, 12–18 × 4–8 cm, chartaceous, the base constricted and very narrow (1–2 mm wide), the apex caudate-acuminate. Growing in gap in forest. Known only from a single sterile specimen (*Granville 5070*, CAY).

MAPROUNEA Aubl.

Shrubs or trees. Latex scant, white. Hairs absent. Leaves alternate, simple, glabrous; petioles eglandular; blades glandular or eglandular, the margins entire; venation pinnate. Inflorescences bisexual or sometimes unisexual (outside our area), terminal, racemose, the pistillate flowers 1–4 at basal nodes or sometimes absent, the staminate flowers in dense terminal raceme separated from pistillate flowers by a long internode; bracts biglandular; staminate pedicels very short; pistillate pedicels long. Flowers unisexual (plants monoecious). Staminate flowers: calyx cupular, 3-lobed; petals absent; nectary disc absent; stamens 2, exserted from calyx, the filaments entirely connate. Pistillate flowers: sepals 3; petals absent; nectary disc absent; ovary 3-locular, with 1 ovule per locule, the styles 3, partly connate, the style branches unlobed, spreading. Fruits capsular. Seeds 3, carunculate.

Maprounea guianensis Aubl.

Shrubs or small trees. Leaves: petioles 8–15 mm long; blades ovate or elliptic, 2–6 × 1.2–3.5 cm, thinly chartaceous, with 1–2 flat, elliptic glands usually present on each side of midrib near base on abaxial surface, the base acute to obtuse, the apex acuminate. Inflorescences with staminate part of raceme ovoid, 2.5–5 mm long; peduncle 5–15 mm long; pistillate pedicels 3.5–6 mm long. Staminate flowers: staminal column ca. 1 mm long. Pistillate flowers: styles ca. 1 mm long. Capsules ca. 6 mm diam.; fruiting pedicel to 13 mm long. Seeds with foveolate surface, the caruncle large, cap-like, red. Fl (Nov), fr (Nov); in disturbed areas. *Radié chancre* (Créole), *vaquinha* (Portuguese).

MARGARITARIA L. f.

Shrubs or small trees. Latex absent. Hairs simple or absent. Leaves alternate, simple, deciduous eglandular; petioles short, blades chartaceous, the margins entire; venation pinnate. Inflorescences unisexual, the flowers clustered or rarely solitary, axillary or on short leafless shoots, often below new flush of leaves; staminate flowers clustered, long pedicellate; pistillate flowers 1–5 per node, on stout pedicels. Flowers unisexual (plants dioecious). Staminate flowers very small; sepals 4, imbricate; petals absent; nectary disc annular; stamens 4, free; pistillode absent. Pistillate flowers small; sepals 4; petals absent; nectary disc annular; ovary (3)4–5-locular, with 2 ovules per locule, the styles 4–5, free to partly connate, bifid. Fruits capsular, irregularly dehiscing; pericarp weak, brittle, fragmenting irregularly. Seeds 6–10, usually 2 per locule, ecarunculate, the outer seed coat fleshy.

Margaritaria nobilis L. f. FIG. 118

Shrubs or small understory trees. Stems with conspicuous whitish lenticels. Leaves: petioles 3–8 mm long, canaliculate; blades elliptic, 6–14 × 3–5 cm, glabrous, chartaceous, the base acute, decurrent onto petioles, the apex acuminate, the margins entire; venation with secondary veins arched, indistinctly looped, 8–14 on each side of midrib. Pistillate flowers: styles 2–3 mm long, 4–5, connate at base, bifid at apex, reflexed. Capsules oblate, 8 × 10 mm, 4–5-lobed, green, glabrous, the styles persistent, spreading; fruiting pedicels 2–10(15) mm long. Seeds with outer seed coat dark blue. Fr (Jan, Feb, Mar, Aug); in disturbed areas.

OMPHALEA L.

Lianas (in our area), shrubs, or trees. Latex red or pinkish, scant to copious. Hairs simple. Leaves alternate, simple; petioles biglandular at apex adaxially; blades usually with flat glands scattered abaxially, the margins entire; venation pinnate or palmate. Inflorescences bisexual, terminal, paniculate, the cymules condensed to lax, bisexual with central pistillate flower(s) or sometimes entirely staminate; bracts subtending cymules large, green or yellow-green, usually biglandular. Flowers unisexual (plants monoecious). Staminate flowers: sepals 4, imbricate; petals absent; nectary disc annular, outside stamens; androecium mushroom-shaped, the stamens 2 or 3, connate, the anthers with connectives highly expanded, forming slender staminal column and hemispheroidal cap; pistillode absent. Pistillate flowers: sepals 4; petals absent; nectary disc absent; ovary 3-locular, with 1 ovule per locule, the styles completely connate into a stout column continuous with the ovary. Fruits capsular or berry-like. Seeds 3, ecarunculate.

Omphalea diandra L. FIG. 119, PL. 55d–f

Canopy lianas, climbing by means of terminal tendrillate shoots. Latex dark red, usually scant; exudate clear to pinkish from young stems. Stems cylindrical, to ca. 20 cm diam., with normal secondary growth. Leaves: petioles 2.5–6(12) cm long; blades ovate or elliptic, 2–20 × 6–16 cm, subcoriaceous to coriaceous, glabrescent to velutinous abaxially, the base rounded or shallowly cordate, the apex

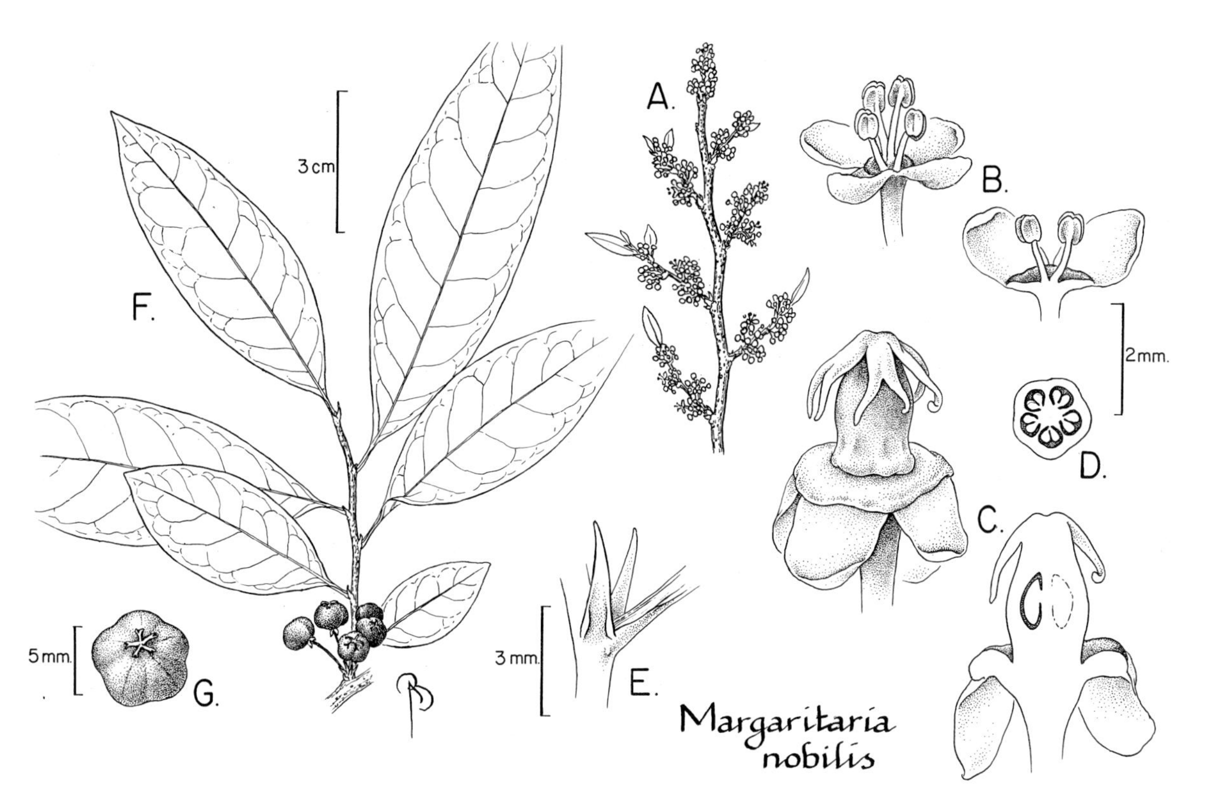

FIG. 118. EUPHORBIACEAE. *Margaritaria nobilis* (A–E, *Harris 3183* from Jamaica; F, G, *Zanoni 16285* from Dominican Republic). **A.** Stems with flowers in clusters and young leaves. **B.** Lateral view (left) and medial section (below right) of staminate flower. **C.** Lateral view (left) and medial section (below right) of pistillate flower. **D.** Transverse section of ovary. **E.** Stipules. **F.** Part of stem with leaves and infructescences. **G.** Apical view of fruit.(Adapted with permission from P. Acevedo-Rodríguez, *Flora of St. John, U.S. Virgin Islands.* Mem. New York Bot. Gard. 78, 1996.)

rounded or obtuse, with short acumen; juvenile leaves or leaves from stump sprouts with blades often much narrower, chartaceous, the base obtuse. Inflorescences to 65 cm long; bracts yellow-green or green, to 6 × 0.8 cm. Staminate flowers: sepals dark red at base adaxially; nectary disc dark red; androecial cap pale green, with 2 anthers. Pistillate flowers: pistil 4–6 mm long, greenish yellow, densely pubescent, the stylar column subglobose, ca. 2 mm long, rounded and often dilated at apex. Fruits subglobose, 8–12 cm diam., indehiscent or tardily dehiscent into 3 mericarps, green or greenish-yellow. Seeds broadly ellipsoid, 3.4–4.8 cm long, rough-surfaced, dark brown. Fl (Jun, Aug, Sep), fr (Dec); in nonflooded forest. *Caiaté, castanha de cotia* (Portuguese), *liane papaye*, *ouabé* (Créole).

PAUSANDRA Radlk.

Trees. Latex reddish or not evident. Hairs simple or malpighiaceous. Leaves alternate, simple, with pair of small glandular knobs at petiole apex; blades with serrate or serrulate margins; venation pinnate. Inflorescences unisexual, axillary, spicate; staminate flowers subsessile, clustered in glomerules; pistillate flowers subsessile, solitary at each node. Flowers unisexual (plants dioecious). Staminate flowers: calyx lobes 5, imbricate; corolla open-tubular, 5-lobed; nectary disc cupular-annular, extrastaminal; stamens 5–7, free; pistillode absent. Pistillate flowers: calyx 5-lobed; petals 5, free or connate; nectary disc annular; ovary 3-locular, with 1 ovule per locule, the styles 3, bifid. Fruits capsular. Seeds 3, carunculate.

1. Leaf blades rigidly coriaceous. Inflorescence internodes mostly ≤1 cm long. Staminate corolla tubes with band of hairs inside (pistillate flowers not seen). Staminate nectary disc glabrous. Capsules 10–12 mm long. . . . . . . *P. fordii.*
1. Leaf blades chartaceous or subcoriaceous. Inflorescence internodes 1–6 cm or more long. Staminate and pistillate corolla tubes glabrous inside. Staminate nectary disc pubescent. Capsules 11–20 mm long. . . . . *P. martinii.*

Pausandra fordii Secco

Understory trees, to ca. 10 m. Leaves: petioles 3–7 cm long, stout, dilated, strongly flexed and with pair of small glandular knobs at apex; blades oblanceolate, 35–55(65) × 8–14 cm, glabrescent, coriaceous, the base narrowly cuneate, the apex acuminate, the margins with glandular setae; secondary veins in 16–26 pairs, the tertiary veins closely parallel. Inflorescences 10–45 cm long; internodes mostly ≤1 cm long. Staminate flowers: calyx glabrescent; corolla tube with band of hairs adaxially; nectary disc glabrous. Capsules subglobose, 3-lobed, ca. 1–1.2 cm long. Seeds ca. 9 × 8 mm, reddish brown. Fr (Apr); locally common, in nonflooded forest.

Pausandra martinii Baill.

Understory trees, to ca. 20 m. Leaves: petioles 1–7 cm long, dilated, flexed and with pair of small, often pointed, glandular knobs at apex; blades oblanceolate, 20–50 × 6–13 cm, glabrescent,

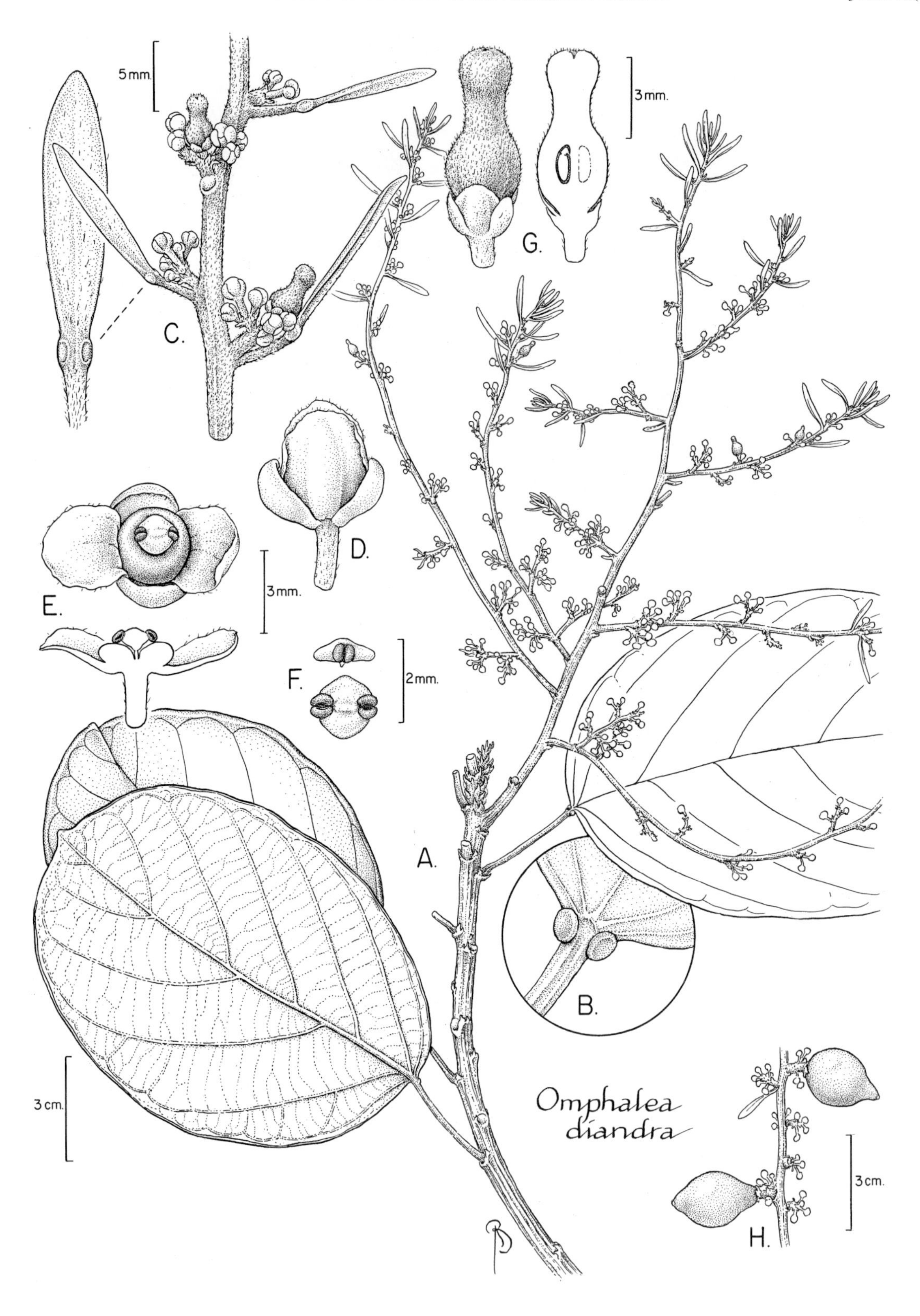

FIG. 119. EUPHORBIACEAE. *Omphalea diandra* (A, B, G, *Mori & Gracie 18926*; C, H, *Mori et al. 23917*; D–F, *Mori et al. 23912*). **A.** Part of stem with leaves and inflorescence. **B.** Apex of petiole and base of leaf blade showing glands. **C.** Part of inflorescence (right) and enlargement of biglandular bract (left). **D.** Staminate flower bud. **E.** Apical view (above) and medial section (below) of staminate flower. **F.** Lateral view (above) and apical view (below) of androecium showing 2 anthers, each with a highly expanded connective. **G.** Lateral view (left) and medial section (right) of pistillate flower. **H.** Part of infructescence with two immature fruits.

chartaceous or subcoriaceous, the base narrowly cuneate or narrowly acute-cuneate, the apex acuminate, the margins with glandular setae; secondary veins in 12–20 pairs, the tertiary veins parallel. Inflorescences (10)20–60 cm long; axis slender; internodes 1–6+ cm long. Staminate flowers: calyx glabrous or glabrescent; corolla tube whitish, glabrous adaxially; nectary disc pubescent. Pistillate flowers: corolla tube glabrous adaxially. Capsules subglobose, 3-lobed, 11–20 mm long. Seeds 9–13 mm long. Known by a single sterile collection in our area (*Mori et al. 23273*, NY).

Identification of this collection as *P. martinii* needs to be confirmed with the collection of fertile material.

PHYLLANTHUS L.

Herbs (in our area), shrubs, or trees. Latex absent. Hairs simple or absent. Stipules often persistent. Leaves often borne on deciduous stems with appearance of pinnate leaves. Leaves alternate, simple; petioles short, eglandular; blades eglandular (in our area) or rarely with flat glands, the margins entire; venation pinnate. Inflorescences axillary, unisexual or bisexual, flowers solitary or clustered in leaf axils; staminate flowers pedicellate; pistillate flowers pedicellate or subsessile. Flowers usually very small, unisexual (plants monoecious in our area or rarely dioecious). Staminate flowers: sepals 4–6; petals absent; nectary disc mostly segmented, extrastaminal; stamens 3(6), the filaments free to connate; pistillode absent. Pistillate flowers: sepals 5 or 6; petals absent; nectary disc annular or segmented; ovary 3-locular, with 2 ovules per locule, the styles 3, variously divided. Fruits capsular or less commonly baccate or drupaceous. Seeds usually 6, 2 per locule, ecarunculate.

1. Flower clusters bisexual, 2-flowered. Capsules on pedicel 1.5–2 mm long. Seeds longitudinally ribbed. . . . *P. amarus*.
1. Flower clusters unisexual, the staminate flowers several per leaf axil, the pistillate flowers one per axil. Capsules subsessile. Seeds transversely ribbed. . . . . *P. urinaria*.

Phyllanthus amarus Schum. & Thonn.

Erect herbs, to ca. 0.5 m tall. Stems glabrous. Leaves borne on deciduous stems; petioles ≤0.5 mm long; blades oblong-elliptic, 4–11 × 2–6 mm, membranous, glabrous, the apex rounded to apiculate. Flower clusters bisexual, each with 1 staminate flower and 1 pistillate flower; pistillate flowers pedicellate. Pistillate flowers: sepals 5, obovate-oblong, ca. 0.8 mm long; ovary smooth, the styles very short, free, bifid at tips. Capsules ca. 1.2 × 2 mm, surface smooth; fruiting pedicel 1.5–2 mm long. Seeds longitudinally ribbed. Fl (Nov), fr (Sep, Nov); open, disturbed areas. *Graine en bas feuille* (Créole), *quebrapedras* (Portuguese).

Phyllanthus urinaria L. PL. 56a

Erect or decumbent herbs, to ca. 1 m tall. Stems sparsely hispidulous. Leaves borne on deciduous stems; petioles ca. 0.5 mm long; blades oblong, 6–15 × 2–5 mm, membranous, hispiduous abaxially, the apex obtuse and often mucronate. Flower clusters unisexual, the pistillate flowers solitary in proximal leaf axils, the staminate flowers in clusters of 3–7 in distal leaf axils; staminate pedicels present, pistillate flowers subsessile. Staminate flowers white. Pistillate flowers: sepals 6, linear-lanceolate, ca. 1 mm long; ovary papillose, the styles very short, connate at base, bifid at tips. Capsules ca. 1.6 × 2.4 mm, yellow, subsessile, surface somewhat verrucose. Seeds transversely ribbed. Fl (Feb, Jun, Jul, Sep, Oct, Nov), fr (Feb, Jun, Jul, Sep, Oct, Nov); open, disturbed areas. *Graine en bas-feuille* (Créole).

PLUKENETIA L.

Twining vines or lianas. Latex absent. Hairs simple. Leaves alternate, simple; petioles eglandular; blades with 1 to several pairs of glands at base, the margins subentire to serrate; venation pinnate or palmate. Inflorescences bisexual with pistillate flower(s) basal or rarely unisexual, racemose. Flowers unisexual (plants monoecious or rarely dioecious). Staminate flowers: sepals 4–5, valvate; petals absent; nectary disc present or absent; stamens 15–40, free; pistillode absent. Pistillate flowers: sepals 4; petals absent; nectary disc absent; ovary 4-locular, with 1 ovule per locule, the styles partly to completely connate, the stylar column thick. Fruits capsular or indehiscent and fleshy berries. Seeds 4.

Plukenetia polyadenia Müll. Arg. PL. 56b

Canopy lianas. Stems twining. Leaves: petioles 1.5–5 cm long; blades elliptic or ovate-elliptic, 7–14 × 4–10 cm, glabrous, the base rounded or obtuse, the margins subentire; venation 3 veined from base, the secondary veins in (1)2–3 pairs, usually diverging from central primary vein in distal half. Inflorescences axillary, unisexual or less often bisexual, when bisexual with pistillate flowers at basal 1–2 nodes, the staminate and bisexual inflorescences 5–25 cm long, the pistillate inflorescences 1.5–3(7) cm long. Staminate flowers: stamens 20–25, the filaments 2–3 mm long. Pistillate flowers: styles connate ca. 2/3 of length. Fruits indehiscent, fleshy, subglobose, 6–11 cm diam., squarish in transverse section, each angle keeled. Seeds ovoid, dull brown, 4.9–5.6 cm long. Fl (Apr, Aug), fr (Mar, Apr, May); in nonflooded forest.

RICHERIA Vahl

Trees. Latex absent. Hairs simple or absent. Leaves alternate, simple; petioles eglandular; blades eglandular or rarely with basal flat glands (outside our area), with margins entire or somewhat crenulate distally; venation pinnate. Inflorescences unisexual, axillary, spicate or racemose; staminate flowers in dense glomerules; pistillate flowers single per node; staminate flowers sessile; pistillate pedicels present. Flowers very small, unisexual (plants dioecious). Staminate flowers: calyx 3–5-lobed, the lobes imbricate; petals absent; nectary disc segments 3–5; stamens 3–6, free, alternating with nectary disc segments; pistillode present. Pistillate flowers: calyx 3–5-lobed; nectary disc cup-shaped; ovary 2-(outside our area) or 3-locular, with 2 ovules per locule, the styles free, bifid. Fruits capsular, somewhat fleshy and tardily dehiscent. Seeds (2)3 (1 per locule), ecarunculate.

Richeria grandis Vahl [Syns.: *R. laurifolia* (Baill.) Müll. Arg., *R. racemosa* (Poepp. & Endl.) Pax & K. Hoffm.]

Small trees. Leaves glabrous or glabrescent: petioles 1–3 cm long, often thickened at base; blades obovate, 11–18 × 4–7 cm, chartaceous or subcoriaceous (coriaceous outside our area), the base cuneate and slightly decurrent on petiole, the apex acuminate or acute (sometimes obtuse or rounded outside our area); midrib conspicuous, the secondary veins in 5–9 pairs, arched and looped. Inflorescences with axes puberulous. Staminate inflorescences 2–5(20) cm long. Pistillate inflorescences (0.5)2–5 cm long. Staminate flowers: calyx puberulous. Pistillate flowers: sepals glabrous or sparsely puberulous, with densely pubescent margin; ovary glabrous, the styles 1–1.5 mm long, recurved. Infructescences 2–10 cm long. Capsules broadly ellipsoid, 1–1.2 × 0.7–0.9 cm, glabrous, tardily dehiscent, splitting irregularly and not separating into separate segments (cocci), with calyx and styles persistent. Seeds 8–7 × 4–5 mm, pale brown. Fl (Dec); in nonflooded forest.

RICINUS L.

A monotypic genus. The generic description same as the description of the species.

Ricinus communis L.

Weak-stemmed shrubs, to 5 m tall. Latex absent. Hairs simple. Leaves alternate, simple; petioles 10–60 cm long, with pair of shortly stalked crateriform glands at apex; blades palmately 7–11-lobed, circular in outline, 10–60+ cm diam, peltate, glabrous, the lobe apices acute to attenuate, the margins irregularly serrate; venation palmate. Inflorescences bisexual, terminal, narrowly paniculate (racemose with lax cymules), the staminate flowers in proximal cymules, the pistillate flowers in distal cymules, the cymules in between often bisexual; pedicels present. Flowers unisexual (plants monoecious). Staminate flowers: calyx spitting into 3–5 valvate lobes; petals absent; nectary disc absent; stamens 100–1000, the filaments irregularly and partly connate, white or cream-colored. Pistillate flowers: calyx spitting into 3–5 lobes; petals absent; nectary disc absent; ovary 3-locular, muricate, with 1 ovule per locule, the styles red, connate at base, spreading, deeply bifid, densely papillose. Fruits capsular, subglobose, 12–20 mm diam., echinate. Seeds ellipsoid, somewhat flattened, smooth, mottled, carunculate. Fl (Jun), fr (Jun); naturalized in disturbed areas, cultivated for the oil extracted from the seeds. *Carrapateira* (Portuguese), castor bean (English), *Grand ricin, huile de ricin* (French), *mamona* (Portuguese), *palma-christi* (Créole).

SAGOTIA Baill.

Trees. Latex absent. Hairs simple. Leaves alternate, simple, eglandular; blades with entire margins; venation pinnate. Inflorescences unisexual or bisexual with pistillate flowers at basal nodes, terminal, racemose or sometimes staminate inflorescences paniculate; pedicels long. Flowers unisexual (plants monoecious). Staminate flowers: sepals 5, imbricate, rounded; petals 5, rounded; nectary disc absent; stamens 20–40, free; pistillode absent. Pistillate flowers: sepals 5, large; petals absent; nectary disc absent; ovary 3-locular, with 1 ovule per locule, the styles 3, free, deeply bifid. Fruits capsular, subtended by persistent, enlarged sepals. Seeds 3, carunculate.

Sagotia racemosa Baill. FIG. 120, PL. 56c

Understory trees. Leaves: petioles 2–6 cm long, pulvinate and flexed at apex and base; blades 10–22 × 4–9 cm, elliptic or oblong-obovate, subcoriaceous, glabrous or glabrescent, the base acute to obtuse, the apex shortly acuminate or acute; secondary veins looped, the tertiary veins reticulate. Inflorescences 4–15 cm long, the axes glabrous; pedicels 1–3 cm long. Staminate flowers: petals obovate, 2–3 mm long, white. Pistillate flowers: sepals linear-oblong or linear-oblanceolate, 7–13 × 2–4 mm, green, glabrous; ovary tomentose, the styles 1.5–2.5 mm long, white. Capsules ca. 1.5 cm diam., puberulous, 3-lobed at apex, the conspicuous, persistent sepals narrowly oblong, to 16 mm long, turning reddish; fruiting pedicel 1.5–4 cm long. Fl (Aug, Sep); infrequent, in nonflooded forest. *Bois lait* (Créole).

SAPIUM P. Browne

Trees. Latex milky white, copious. Hairs absent. Stipules small. Leaves alternate, simple; petioles glabrous, with pair of glands usually present at apex (or sometimes at blade base); blades often with small marginal cupular glands, the margins usually serrulate; venation pinnate. Inflorescences bisexual, terminal, spicate, the axis thick, with small, subsessile flowers; pistillate flowers solitary at basal nodes, the staminate flowers in sessile clusters above, each solitary flower or cluster subtended by a biglandular bract, the glands usually large, disc-shaped. Flowers unisexual (plants monoecious). Staminate flowers: calyx 2–3-lobed, not covering anthers in bud; petals absent; nectary disc absent; stamens 2–3, free or connate at base. Pistillate flowers: calyx 3–5-lobed or subentire; petals absent; nectary disc absent; ovary (2)3-locular, with 1 ovule per locule, the styles (2)3, unlobed, usually connate at base. Fruits capsular. Seeds (2)3, ecarunculate, the outer seed coat fleshy.

1. Leaf blades with 12–20 pairs of secondary veins, the apex acute to shortly acuminate, with tip involute and distinctly thickened. . . . . *S. glandulosum.*
1. Leaf blades with 6–10 pairs of secondary veins, the apex acuminate, with tip flat and not thickened.
    2. Glands sessile at apex of petiole. . . . . *S. argutum.*
    2. Glands stalked, arising from blade margin near base, bent inwards and often hidden under blade. . . *S. paucinervium.*

Sapium argutum (Müll. Arg.) Huber [Syn.: *S. montanum* Lanj.]

Small trees. Leaves: petioles 6–12 mm long, with pair of sessile glands at apex; blades narrowly elliptic or narrowly obovate-elliptic, 8–12 × 2–3.5 cm, chartaceous to sometimes subcoriaceous, the base acute to narrowly cuneate, the apex acuminate, with the tip flat and not thickened, the margins serrulate, with numerous small spinose setae, often widely and irregularly spaced, with conspicuous, cupular glands; secondary veins in 6–10 pairs, arched, looped. Capsules ca. 1

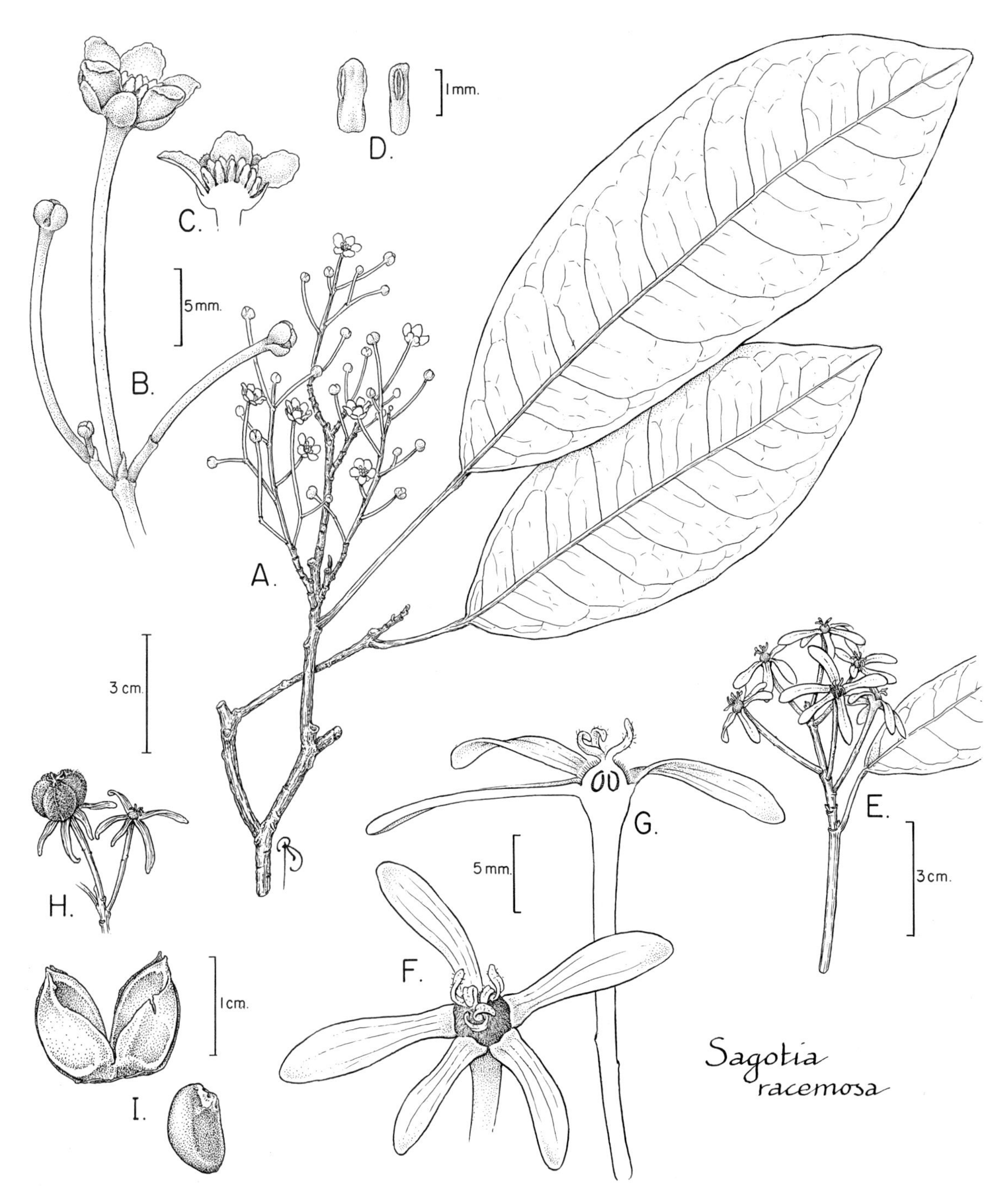

FIG. 120. EUPHORBIACEAE. *Sagotia racemosa* (A–G, *Mori & van der Werff 23275*; H, *Milliken 536* from Roraima, Brazil; I, *Prance et al. 1461* from Pará, Brazil). **A.** Part of stem with leaves and staminate inflorescence. **B.** Detail of staminate inflorescence with flower buds and one open flower. **C.** Medial section of staminate flower. **D.** Two views of anthers. **E.** Pistillate inflorescence. **F.** Oblique-apical view of pistillate flower. **G.** Medial section of pistillate flower. **H.** Part of infructescence. **I.** Dehisced capsule segment (above) and seed (right).

× 1.4 cm, usually 2-seeded. Seeds ca. 7 × 6.5 × 5 mm, black, the outer coat red. Fl (Jan); locally common, in forest bordering open rocky area on hill at ca. 500 m alt.

Sapium glandulosum (L.) Morong [Syns.: *S. klotzschianum* (Müll. Arg.) Huber, *S. lanceolatum* (Müll. Arg.) Huber]

Understory trees. Trunks slightly buttressed at base. Bark smooth, tan, with numerous irregular shallow vertical cracks, the inner bark pink. Leaves: petioles 1–2 cm long, with pair of sessile glands at apex; blades narrowly oblong-elliptic, 6–11(15) × 1.5–3(5) cm, chartaceous to subcoriaceous, the base acute, the apex acute to shortly acuminate, with the tip involute and distinctly thickened, the margins subentire to serrulate; secondary veins in 12–20 pairs, the midrib thick. Capsules ca. 1 cm diam., 3-seeded, purplish red. Seeds 5.5 × 4 × 4 mm, black. Fr (Feb, Aug); in secondary forest.

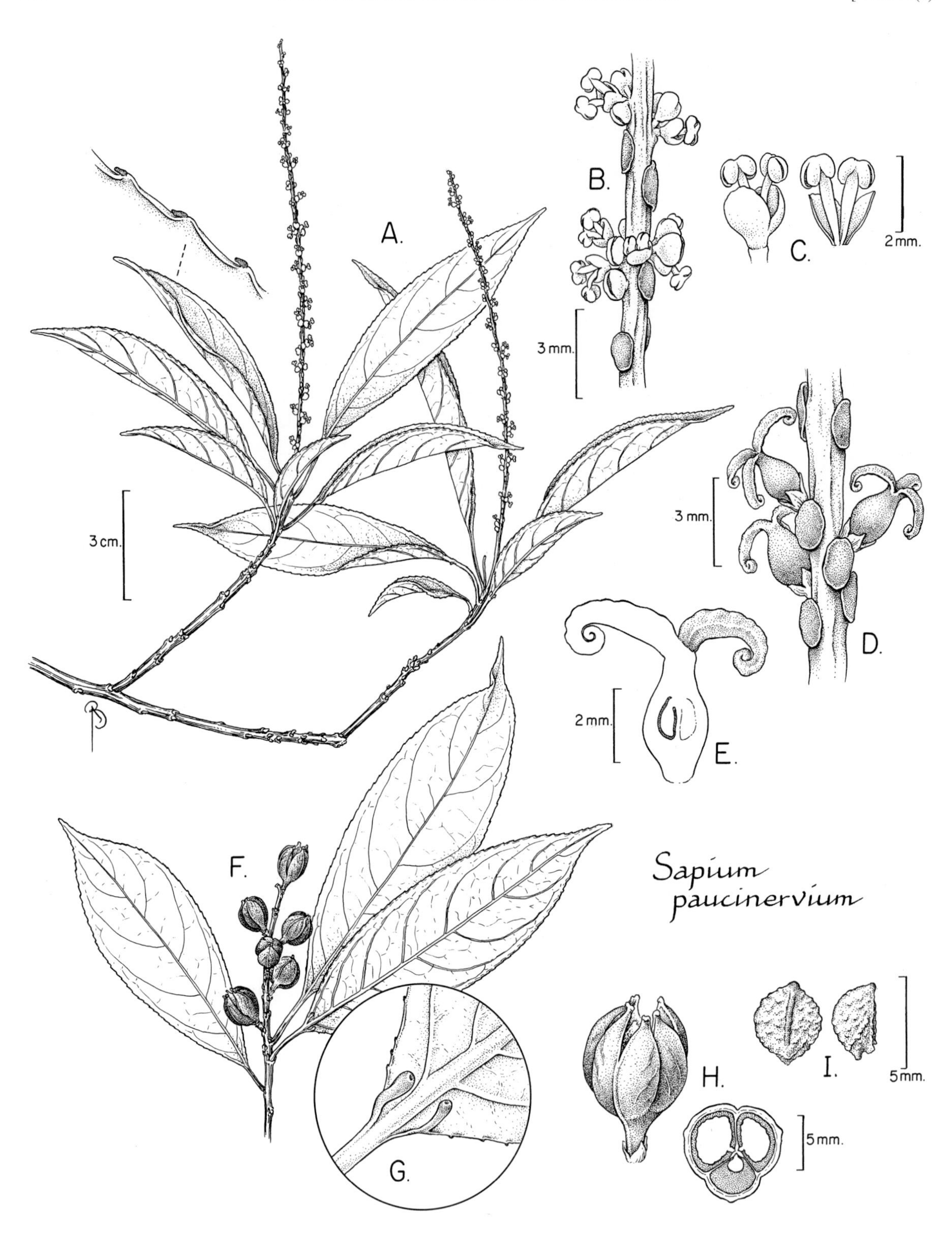

FIG. 121. EUPHORBIACEAE. *Sapium paucinervium* (A–C, *Prévost & Sabatier 2970*; D, E, *Steyermark 88154* from Venezuela; F–I, *Mori et al. 15364*). **A.** Stem with inflorescences, leaves, and detail of leaf blade margin. **B.** Part of inflorescence showing staminate flowers. **C.** Lateral view of staminate flower (left) and staminate flower with part of perianth removed (right). **D.** Part of inflorescence showing pistillate flowers. **E.** Pistil in medial section. **F.** Stem with infructescence. **G.** Base of leaf showing details of glands. **H.** Capsule (left) and transverse section of capsule (right) showing only two seeds developed. **I.** Two views of seeds.

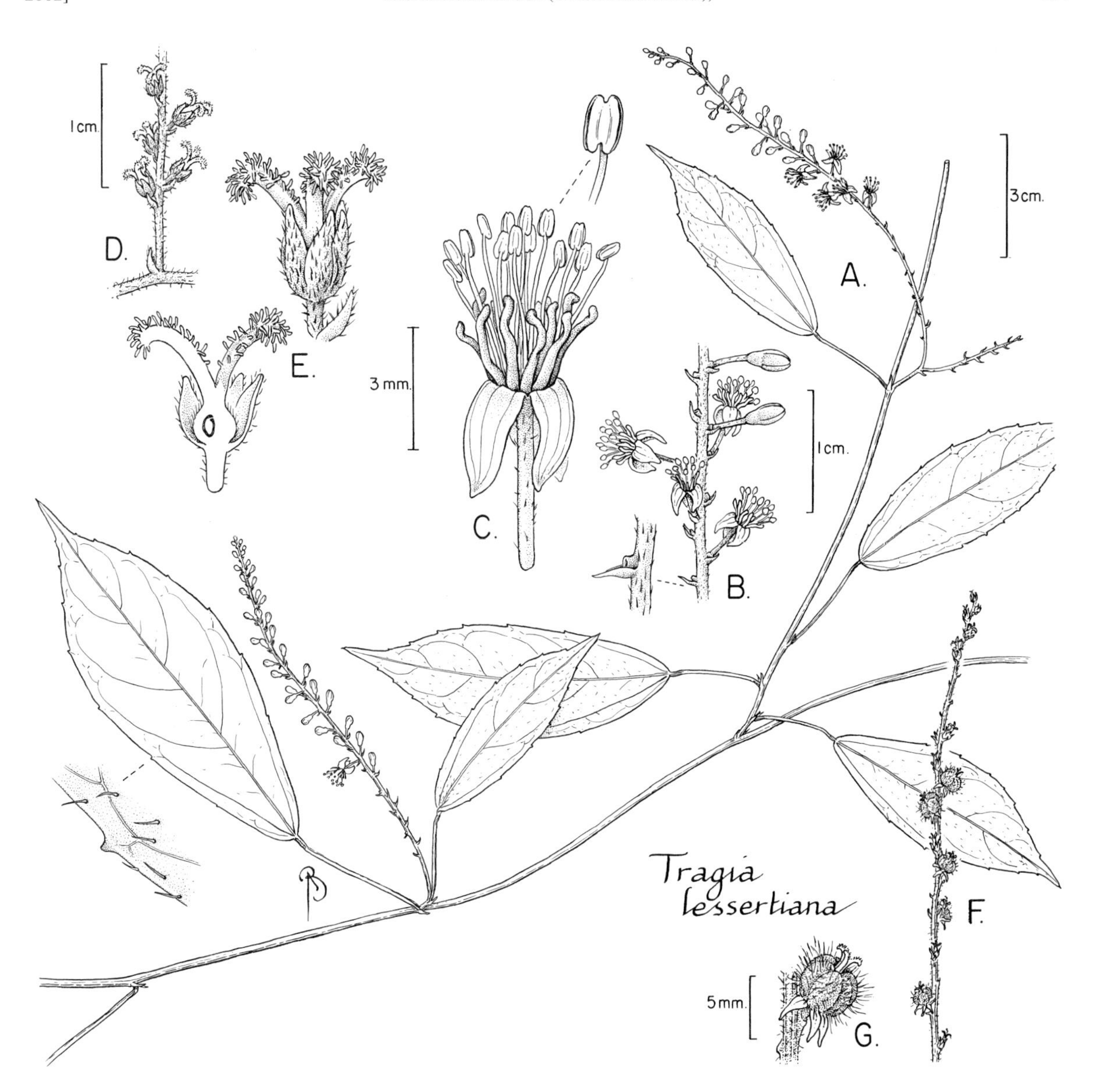

FIG. 122. EUPHORBIACEAE. *Tragia lessertiana* (A–C, *Mori et al. 14874*; D, E, *Daly et al. 475* from Maranho, Brazil; F, G, *Lobo et al. 317* from Maranho, Brazil). **A.** Part of stem with leaves and staminate inflorescences; note close-up of leaf margin showing pubescence. **B.** Part of staminate inflorescence axis with detail of bract on left. **C.** Staminate flower with detail of stamen. **D.** Part of pistillate axis inflorescence. **E.** Lateral view (above) and medial section (below left) of pistillate flower. **F.** Part of infructescence. **G.** Oblique-lateral view of capsule.

Sapium paucinervium Hemsl. FIG. 121

Understory to canopy trees. Trunks unbuttressed or with small spreading buttresses. Bark smooth, hard, brown, with vertical fissures, the middle bark purplish red, the inner bark cream-colored. Leaves deciduous; petioles 1–2 cm long; blades narrowly elliptic, 7–15 × 2–4 cm, chartaceous, with pair of stalked glands 1–2 mm long, arising from blade margin just above base, these bent inwards abaxially and often hidden under blade, the base acute, the apex long-acuminate, with the tip flat and not thickened, the margins serrulate, with numerous closely set glandular setae; secondary veins in 6–10 pairs, strongly arched, mostly not looped. Inflorescences erect, appearing with new growth of leaves. Capsules ca. 1 cm diam., 3-seeded. Seeds 4 × 3.5 × 3 mm, dull yellow, rough-surfaced, the outer coat red. Fl (Jan), fr (Feb, Mar), leafless (Oct); in mature or secondary forest.

TRAGIA L.

Twining vines, herbs, or subshrubs. Hairs simple, at least some stinging. Latex absent. Leaves alternate, simple, eglandular; blades with margins serrulate or serrate; venation pinnate or palmate. Inflorescences mostly bisexual with pistillate flower(s) basal, terminal or axillary, racemose or sometimes with 1 long basal branch. Flowers unisexual (plants monoecious or rarely dioecious). Staminate flowers: sepals 3–6;

petals absent; nectary disc absent or present; stamens 2–50, mostly free; pistillode absent. Pistillate flowers: sepals 6; petals absent; nectary disc absent; ovary 3-locular, with 1 ovule per locule, densely covered with stinging hairs, the styles 3, slender, partly connate, unlobed. Fruits capsular, covered with stinging hairs. Seeds 3, ecarunculate.

Tragia lessertiana (Baill.) Müll. Arg. FIG. 122

Twining vines. Stems and foliage sparsely covered with stinging hairs or sometimes glabrescent. Leaves: petioles 1–5 cm long; blades oblanceolate or narrowly elliptic, 6–11.5 × 2–5 cm, the margins serrate, the base cuneate, acute, or narrowly obtuse; venation pinnate, 3-veined from base. Infloresences consisting of staminate main axis to 20 cm long and pistillate basal branch to 10 cm long, elongating to 25 cm in fruit (sometimes only one axis present). Staminate flowers: sepals 3; nectary disc segments 8–9, slender, alternating with outer whorl of stamens; stamens 14–17. Pistillate flowers: styles to 3 mm long, connate basally 1/3 to 1/2 of length, papillose adaxally. Capsules 3-lobed, ca. 4.5 × 7 mm; infructescence axis to 25 cm long. Fl (Feb, Jun, Sep, Nov); in open, disturbed areas.

FABACEAE (Bean Family)

Rupert C. Barneby and Scott V. Heald

Trees, lianas, shrubs, or herbs, pubescent with basifixed hairs, seldom glandular. Stipules generally present. Leaves alternate, less often opposite, phyllotaxy spiral or rarely distichous, compound, either pari- or imparipinnate, some 3-, 2- or 1-foliolate, pulvinate. Inflorescences simple or branched racemes, pseudoracemes, spikes, or panicles; pedicels often bracteolate. Flowers zygomorphic; calyx mostly synsepalous, sometimes split to base; corolla hypogynous or perigynous, papilionaceous (i.e., petals 5, imbricate in bud, the adaxial one or vexillum exterior, the 2 abaxial or keel innermost), but sometimes petals only 1 (vexillum) or 0; stamens commonly 10, rarely fewer or indefinite, the filaments either free or variously united, the anthers either dorsi- or basifixed; carpel 1, rarely 2; placentation marginal or subapical, the ovules 1 to several. Fruits pods, either elastically or valvately dehiscent, drupes, samaras, or loments. Seeds sometimes arillate; embryo usually curved.

The genus *Acosmium*, sometimes treated as belonging to the Fabaceae, is placed by us in the Caesalpiniaceae because the flowers are distinctly non pea-like and nearly actinomorphic.

Amshoff, J. H. 1976. Papilionaceae. *In* J. Lanjouw & A. L. Stoffers (eds.), Flora of Suriname **II(2):** 1–257.

Aymard C., G. A., N. L. Cuello A., P. E. Berry, V. E. Rudd, R. S. Cowan, P. R. Fantz, R. H. Maxwell, C. H. Stirton, H.-H. Poppendieck, H. Cavalcante de Lima, R. H. Fortunato, B. Stergios, N. Xena de Enrich, D. A. Neill, R. T. Pennington & C. Gil. 1999. Fabaceae. *In* J. A. Steyermark, P. E. Berry, K. Yatskievych & B. K. Holst (gen. eds.), Flora of the Venezuelan Guayana **5:** 231–433. P. E. Berry, K. Yatskievych & B. K. Holst (vol. eds.). Missouri Botanical Garden Press, St. Louis.

1. Leaves 1-foliolate (*Bocoa viridiflora* is mostly 3-foliolate, but may be randomly 1-foliolate). . . . . *Poecilanthe.*
1. Leaves 2- to many-foliolate.
   2. Leaves all 2-foliolate. . . . . *Zornia.*
   2. Leaves 3- to many-foliolate (*Bocoa viridiflora* is mostly 3-foliolate, but may be randomly 1-foliolate).
      3. Calyx entire in bud, at anthesis splitting to base into 3–4 recurved segments; petal 1 or 0; stamens free, >10 per flower. Seeds arillate.
         4. Filaments shorter than anthers, the anthers basifixed, 5–6 mm long. Seeds several, randomly disposed within cavity. . . . . *Candolleodendron.*
         4. Filaments much longer than anthers, the anthers either versatile or basifixed, but if basifixed then <2 mm long. Seeds solitary or, if more than 1, regularly uniseriate.
            5. Foliage not foetid; leaves commonly 5-several-foliolate, if 3-foliolate then rachises green-winged. Anthers dorsifixed. . . . . *Swartzia.*
            5. Foliage foetid; leaves mostly 3-foliolate but randomly 1-foliolate; rachises not winged. Anthers basifixed. . . . . *Bocoa.*
      3. Calyx clearly dentate or lobed prior to anthesis, the limb at anthesis campanulate or tubular; petals 5; stamens either free or united into sheath, but if free then exactly (9)10. Seeds not arillate (except *Cajanus* with rim-aril).
         6. Leaves 3-foliolate.
            7. Leaflets and often also vexillum of flowers with spherical glands abaxially.
               8. Leaflets palmately veined from pulvinule; vexillum 4–7 mm long. Fruits 2-seeded. . . . . *Rhynchosia.*
               8. Leaflets pinnately veined; vexillum 15–18 mm long. Fruits ± 5-seeded. . . . . *Cajanus.*
            7. Leaflets and petals glandless.
               9. Leaves palmately 3-foliolate. Petals yellow. Fruits pods, inflated. . . . . *Crotalaria.*
               9. Leaves pinnately 3-foliolate. Petals whitish, pink, lilac, or purple. Fruits various, but not inflated.
                  10. Lianas, high-climbing with woody stems. Pseudoracemes pendulous. Flowers large; longer petals 3.5–4 cm long. Fruits hispid with brown, stinging hairs. . . . . *Mucuna.*

10. Either lianas or herbaceous twiners, but if stems woody then pseudoracemes incurved-ascending and petals shorter. Fruits either strigose or glabrate, but never with stinging hairs.
    11. Lianas, high-climbing. Style terete, not dilated. . . . . *Dioclea.*
    11. Either herbaceous twiners ≤1.5 m tall, or in *Pueraria* and *Pachyrrhizus* basally woody and taller. Style dilated or, if plants woody at base, then style terete.
        12. Flowers resupinate, the ample vexillum twice or more as long as keel, spurred abaxially at base of blade. Fruits dehiscent pods, the valves with accessory vein running longitudinally near and parallel to abaxial suture. . . . . *Centrosema.*
        12. Flowers not resupinate, the vexillum little longer than inner petals, not spurred abaxially. Fruits various, but if pods, then lacking accessory vein.
            13. Roots tuberous. Style transversely dilated and strongly incurved.. . . . *Pachyrrhizus.*
            13. Roots not tuberous. Style linear, terete, not strongly incurved.
                14. Inflorescences racemose (sometimes branched racemose), the flowers few, crowded at apices of primary axes. Blades of keel-petals incurved through ca. 3/4-circle. . . . . *Vigna.*
                14. Inflorescences either racemose or pseudoracemose, the flowers numerous, scattered along primary axes. Blades of keel-petals not as above.
                    15. Inflorescences pseudoracemose, the secondary axes tuberculate. Vexillum 14–17 mm long. Fruits elastically dehiscent pods, pubescent with straight hairs. . . . . *Pueraria.*
                    15. Inflorescences racemose, the secondary axes not tuberculate. Vexillum 4–6 mm long. Fruits 2–7-seeded loments, densely beset with short, hooked hairs. . . . . *Desmodium.*

6. Leaves usually (4)5- to many-foliolate (3-foliolate in *Clitoria javitensis*).
    16. Stems, leaves, and components of inflorescence opposite or verticillate.
        17. Calyx lobes much modified, the 2 adaxial petaloid, oblong, clasping corolla, the 3 abaxial much shorter, united into 3-denticulate lip; petals rose-purple. Fruits dehiscent pods. . . . . *Taralea.*
        17. Calyx lobes subuniform in shape and length; petals yellow. Fruits planocompressed samaras with central seed. . . . . *Platymiscium.*
    16. Stems, leaves, and components of inflorescence alternate.
        18. Vines or lianas. Flowers resupinate. . . . . *Clitoria.*
        18. Trees, shrubs, or lianas. Flowers not resupinate.
            19. Plants in flower.
                20. Stamen filaments arising separately from hypanthial rim.
                    21. Larger leaflets 7–16 cm long. Calyx symmetrical or, if 2-lipped, the 2 partly united adaxial teeth only a little longer than 3 abaxial ones; keel shorter than wings.
                        22. Secondary veins of leaflets in ± 15 pairs, connected by scalariform tertiary veinlets. Vexillum densely silky-puberulent abaxially. . . . . *Dussia.*
                        22. Secondary veins of leaflets ≤10 pairs, not connected by scalariform tertiary veinlets. Vexillum glabrous abaxially.
                            23. Calyx in profile symmetrical, the hypanthium rim horizontal, the 2 adaxial teeth not partly united; blade of vexillum yellow, red-eyed, unappendaged. . . . . *Ormosia.*
                            23. Calyx in profile asymmetrical, the hypanthium rim oblique, the 2 adaxial teeth partly united; blade of vexillum lavender, appendaged adaxially near base on each side. . . . . *Diplotropis.*
                    21. Larger leaflets 2.5–6 cm long; calyx strongly 2-lipped, the adaxial lobes united into oblong, shallowly emarginate blade, the 3 abaxial ones united into a much smaller, 3-denticulate lip; wings shorter than keel. . . . . *Monopteryx.*
                20. Stamen filaments united into tube, or into sheath open along adaxial side (to inflorescence axes), or 9 united and 1 free, or united into 2 distinct fascicles.
                    24. Leaflets alternate, the leaf rachises produced far beyond distal leaflets. Calyx 2-labiate, the adaxial lip of 2 oblong obtuse lobes free from each other but clasping corolla, the abaxial lip very small, 3-denticulate. . . . . *Dipteryx.*

24. Leaflets either opposite or alternate, but if alternate then rachis terminating in leaflet. Calyx otherwise.
 25. Anthers basifixed, not versatile, terminally dehiscent by pores or transverse slits. . . . . *Dalbergia*.
 25. Anthers dorsifixed, versatile, dehiscent laterally.
  26. Lianas, unarmed. Inflorescences composed of elongate narrow pseudoracemes, the individual flowers biseriate along outer face of short, thickened, abruptly incurved or hooked secondary axes. . . . . . . *Derris*.
  26. Trees and lianas, the latter randomly armed with spinescent stipules. Inflorescences composed of simple or branched racemes or paniculate, the individual flowers borne singly along primary axis.
   27. Lianas, randomly armed with spinescent stipules. . . . . *Machaerium*.
   27. Trees, unarmed.
    28. Inflorescences axillary and supra-axillary racemes. Vexillum ± 6–9 mm long, dark red or red-brown. . . . . *Poecilanthe*.
    28. Inflorescences terminal panicles. Vexillum 10–16.5 mm long, not dark red.
     29. Leaflets in opposite pairs along rachis. . . . . *Hymenolobium*.
     29. Leaflets alternate along rachis.
      30. Petals deep yellow with red eye-spot. . . . . *Pterocarpus*.
      30. Petals anthocyanic (purple, violet, red).
       31. Leafstalks lacking stipels. Calyx in profile subsymmetrically turbinate-campanulate; ovary not appendaged over locule. . . . . *Vatairea*.
       31. Leafstalks with stipels. Calyx in profile asymmetrically turbinate-campanulate; ovary incipiently winged over locule. . . . . *Vataireopsis*.

19. Plants in fruit.
 32. Leaflets alternate, the rachis green-marginate, produced 2–5 cm beyond distal leaflets. Fruits oblong-ellipsoid, subterete, 1-seeded drupes. . . . . *Dipteryx*.
 32. Leaflets either opposite or alternate, the rachis not as above if leaflets alternate. Fruits either dehiscent pods, or compressed indehiscent samaras or nuts.
  33. Fruits pods, the valves woody or stiffly coriaceous, dehiscent through one or both sutures.
   34. Pods narrowly elliptic in profile and narrowly 2-winged along both sutures. Longest leaflets 2.5–6 cm long. . . . . *Monopteryx*.
   34. Pods oblong-oblanceolate, plumbly oblanceolate-ellipsoid, or elliptic in profile, winged on neither suture. Longest leaflets 8–18 cm long.
    35. Secondary veins of leaflets in ± 15 pairs, connected by scalariform tertiary venules. Seeds obese, red. . . . . *Dussia*.
    35. Secondary veins of leaflets in ≤10 pairs, tertiary venulation faint, irregular. Seeds not obese, brown or black.
     36. Inflorescences axillary and supra-axillary, racemes. Pods not constricted, the seed-cavity sometimes divided by papery partitions. . . . . *Poecilanthe*.
     36. Inflorescences terminal, panicles. Pods constricted between seeds, the seed-cavity not divided by papery partitions. . . . . . . *Ormosia*.
  33. Fruits indehiscent, samaroid, or compressed-nuciform.
   37. Fruits mostly water-dispersed, nuciform, firm-walled, in profile round, round-oblong, or reniform, biconvex or convex on one face and concave on reverse, the largest 2–4.5 × 1–2.5 cm. . . . . *Dalbergia*.
   37. Fruits wind-dispersed samaras of papery or submembranous texture, or conspicuously winged, if not winged, >4.5 cm long.
    38. Seed-cavity of samaras basal, surmounted by a long papery wing.
     39. Basal cavity of samaras either smoothly biconvex or obscurely low-carinate longitudinally.
      40. Trees, unarmed. . . . . *Vatairea*.
      40. Lianas, randomly armed with spinescent stipules. . . . . *Machaerium*.

39. Basal cavity of samaras bearing fimbriate longitudinal wing at least 1 cm wide. . . . . *Vataireopsis.*

38. Seed cavity of samaras central.

41. Leaflets opposite. Samaras 1.4–2.8 cm wide.

42. Trees. Inflorescences terminal, paniculate. . . . . *Hymenolobium.*

42. Lianas. Inflorescences axillary-cauliflorous, pseudo-racemose . . . . *Derris.*

41. Leaflets alternate. Samaras 3.4–8 cm wide.

43. Fruits stipitate, in profile oblong-elliptic, narrowly winged along abaxial suture only, the valves membranous, translucent. . . . . *Diplotropis.*

43. Fruits sessile, in profile suborbicular, widely winged around whole periphery, the valves chartaceous, opaque. . . . . . *Pterocarpus.*

ACOSMIUM Schott

See treatment of Caesalpiniaceae.

BOCOA Aubl.

Trees, glabrous except for minutely brown-puberulent inflorescence. Stipules small, ephemeral. Infloresecences short racemes from knots on defoliate stems. Flowers small, greenish, fragrant; calyx entire in bud, globose-apiculate, valvately splitting at anthesis into 3–4 recurving, deciduous lobes; petals 0; stamens ca. 25, free, isomorphic, the anthers basifixed, narrowly oblong, dehiscent by laterals slits, the filaments much longer than anthers; ovary shortly stipitate, the style terminal, the stigma capitellate. Fruits pods, the body inequilaterally obovoid-ellipsoid, plumply biconvex, the valves thin-coriaceous becoming brittle, tardily bivalvate. Seeds 1, arillate.

Bocoa is scarcely different from small-flowered, apetalous *Swartzia* spp. except in basifixed anthers.

Bocoa viridiflora (Ducke) R. S. Cowan

Trees, 10–20(26) m × 10–27 cm. Trunks cylindric. Leaves mostly 3-foliolate, randomly 1-foliolate, pale olivaceous; leafstalks of 3-foliolate leaves 5–9 cm long; lateral leaflets (sub)opposite, smaller than terminal one; pulvinules 5–7 mm; blades broadly ovate- or obovate-elliptic, the terminal leaflet 10–14.5 × 4.5–8 cm, thinly chartaceous, the base broad-cuneate, the apex shortly acuminate. Inflorescences racemose, 1.5–9 cm long; bracts deltate, ≤1 mm long, convex abaxially; pedicels 1.5–4 mm long, ebracteolate. Flowers: sepals ± 3 mm long, glabrous adaxially; stamens with filaments 6–7 mm long, the anthers 1.2–1.4 mm long; gynoecium glabrous, the stipe ± 1 mm long, the style 1.5–2.5 mm long. Fruits pods, in profile 9–17 × 5–13 mm. Fl (Nov), fr (Jul, Sep, Nov, Dec); in nonflooded forest.

CAJANUS DC.

Weakly arborescent shrubs, silky-puberulent or -pilosulous, except for glabrous adaxial surface of leaflets. Younger stems pliantly arching. Leaves pinnately trifoliolate; leaflets stipellate; blades with small yellow glands abaxially; venation pinnate. Inflorescences in distal leaf axils, short racemes, the flowers either solitary or 2–3 at a node. Flowers papilionaceous, yellow, red-streaked; hypanthium lacking; calyx campanulate, 5-toothed, the 2 abaxial teeth partly fused; vexillum glabrous, moderately recurved, the keel obtuse; stamens diadelphous (9 + 1); ovary sessile, the style glabrous, wider at middle, persistent on fruit; ovules ± 5. Fruits pods, undulately narrow-oblong, attenuate at each end, dehiscence elastic, the valves coiling, coriaceous, obliquely sulcate between seeds. Seeds ± 5, subglobose, with rim-aril.

Cajanus cajan (L.) Millsp. FIG. 123

Shrubs, commonly 1–2.5 m tall. Trunks several. Stipules lanceolate, 3–6 mm long, deciduous, the apex acuminate. Leaves: leafstalks 1–5 cm long; blades equilaterally elliptic, the terminal one ± 6–8 × 1.6–3.5 cm, the lateral pair little shorter, pallid abaxially, the base cuneate, the apex shortly acuminate; secondary venation in 5–8 pairs. Inflorescences: peduncles and short raceme-axes together 2–8 cm long; bracts ovate, 3–5 mm long, submembranous, caducous, at 2-flowered nodes a narrower bract at one side; pedicels 8–15 mm long. Flowers: sepals 7–11 mm long, the abaxial tooth 3–5 mm long; vexillum 15–18 mm long. Fruits ± 6–7 × 1–1.2 cm, the coriaceous valves low-convex over each seed, pilosulous, often mottled. Seeds gray or brown. Fl (Apr, Aug, Sep), fr (Sep), but probably flowering and fruiting year round; cultivated for edible seeds but weedy in village and around homesteads. *Haricot pigeon* (Créole), *pigeon pea* (English), *pois congo*, *pois d'Angola*, *pois d'Angole* (Créole).

CANDOLLEODENDRON R. S. Cowan

Slender arborescent shrubs. Bark papery, peeling. Stipules minute. Leaves imparipinnate, glabrous; leaflets opposite. Inflorescences short, cauliflorous; bracts minute; pedicels ebracteolate. Flowers *Swartzia*-like; calyx in bud fusiform, entire, at anthesis splitting to base into 3–4 recurving, coriaceous lobes; petal 1 (adaxial), short-clawed, yellow, the blade adaxially concave; stamens numerous (≤ca. 45), uniform, free,

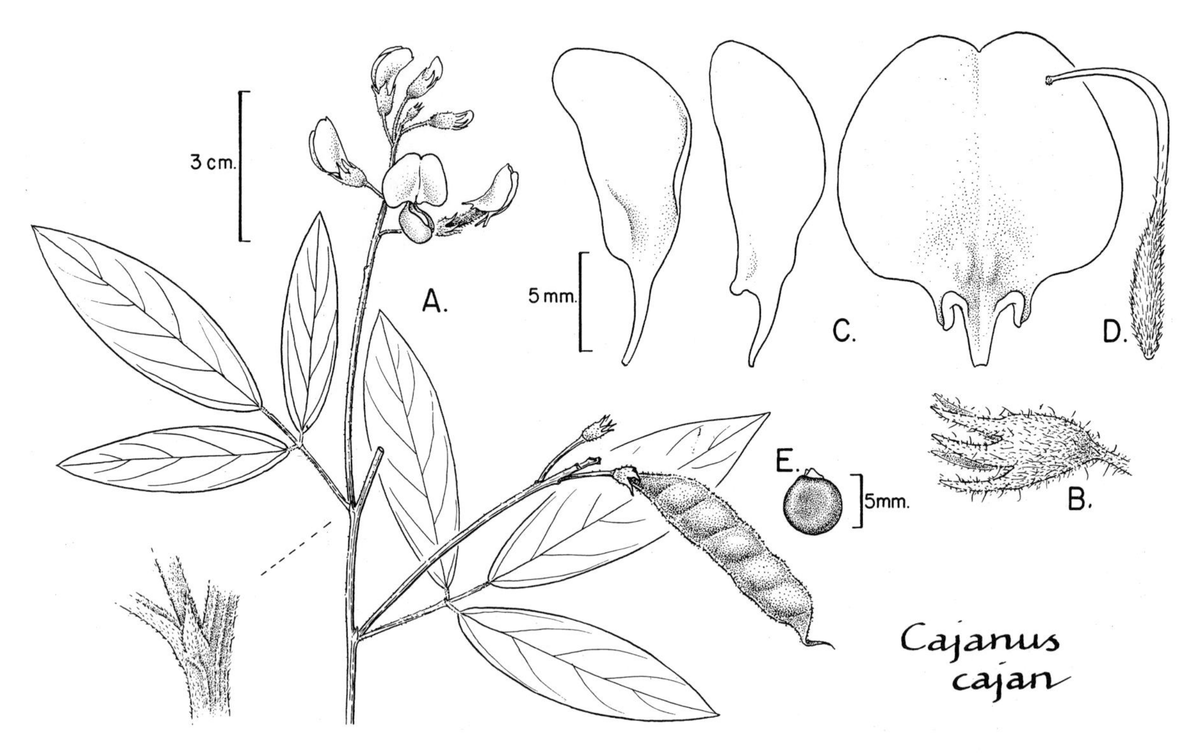

FIG. 123. FABACEAE. *Cajanus cajan* (*Acevedo-Rodríguez 1952* from St. John, U.S. Virgin Islands). **A.** Part of stem with leaves, flowers, and fruit and detail of stem showing stipule at base of petiole (below, left). **B.** Lateral view of calyx. **C.** Wing (far left), keel (near left), and standard (right) petals. **D.** Pistil. **E.** Seed. (Reprinted with permission from P. Acevedo-Rodríguez, *Flora of St. John, U.S. Virgin Islands*, Mem. New York Bot. Gard. 78, 1996.)

the anthers basifixed, erect, linear acute, ± 2–3 times as long as filaments; ovary subsessile, the style linear. Fruits pods, subsessile, obesely short-oblong, plumply 2-convex, the ends rounded, the ripe valves thin-textured, the cavity replete with several randomly oriented seeds. Seeds several, soft, of different shapes, the funicles very long, filiform, expanded at apex into fleshy arils.

Candolleodendron brachystachyum (DC.) R. S. Cowan

Shrubs, fertile at 2 m tall and upward, attaining 10 m × ± 15 cm. Leaves: leafstalks 13–22 cm long; pulvinules 3–7 mm long; leaflets 5–7 per leaf; blades broadly or narrowly elliptic, the larger ones 11–20 × 5–7 cm, pale olivaceous, the base either rounded or cuneate, the apex acuminate or caudate, the margins undulate. Inflorescences racemose, ± 10–20-flowered, the axes 5–9 cm long; pedicels 8–14 mm long. Flowers: sepals 11–15 mm long; petal 17–20 × 10–15 mm, the margins undulate; stamens with filaments 1.5–2 mm long, the anthers 5–6 × 0.6–0.8 mm. Fruits 5.5–6.5 × 2.5–3 cm, glabrous. Fl (Jan, Feb, Dec), fr (Mar); in understory of nonflooded forest.

CENTROSEMA (DC.) Benth.

Low-twining and some high-climbing vines, pubescent in part with hooked hairs. Stipules striately veined, subpersistent. Leaves pinnate; leaflets 3–7 per leaf, stipellate. Inflorescences axillary, solitary or paired pseudoracemes, the flowers borne singly or sometimes in pairs from small tubercle; bracts and bracteoles striate, the latter larger, often covering young flower. Flowers showy, resupinate, pink-purple or whitish; hypanthium lacking; calyx campanulate, 5-toothed, the abaxial teeth often much longer than adaxial pair; vexillum short-clawed, the blade ample, suborbicular, abaxially silky and spurred near base; stamens 10, mostly monadelphous, the anthers uniform; style glabrous, persistent in fruit. Fruits pods, in profile mostly linear, straight or almost so, sometimes round-oblong, elastically dehiscent through both sutures, the valves coiling, coriaceous or stiffly papery, often charged with raised vein parallel and close to adaxial suture, this sometimes narrowly winged, the cavity incipiently septiferous between seeds.

Centrosema vexillatum Benth.

Slender subherbaceous, diffuse, or opportunistically climbing vines. Young stems pilosulous or glabrate. Stipules lance-ovate, 4–9 mm long, persistent. Leaves: leafstalks 2.5–13 cm long, the one interpinnal segment ± 1–2.5 cm long; leaflets 3; blades subsymmetrically ovate or lance-oblong, the terminal one 7–12(17) × 2–5 cm, thinly chartaceous, pilosulous or glabrate, the apex shortly acuminate; midrib and 5–8 pairs of secondary veins slenderly prominulous only abaxially. Inflorescences mostly 2–7-flowered; peduncles 3–9 cm long; bracts small, semiorbicular; bracteoles ovate, 20–34 mm long, the apex acute, clasping and concealing calyx. Flowers: calyx oblique, the ventral and lateral teeth depressed-deltate, the dorsal tooth linear-subulate, 3–5.5 mm long; petals: vexillum 40–48 mm wide, white or bluish-tinged, the wings and keel about 2/3 as long, bluish. Fruits subsessile, linear, straight, 5–10 × 0.6–0.9 cm, the apex contracted into erect linear beak 1–17 mm long, the valves glabrous. Fl (Sep), fr (Sep); at forest margins, collected on trails to Mont Galbao and Crique Limonade.

CLITORIA L.

Trees, lianas, functionally herbaceous subshrubs, or twining herbs, pubescent with simple hairs. Stipules present. Leaves mostly pinnate, 3-foliolate, a few 1-foliolate or imparipinnate; leaflets stipellate; blades pinnately veined. Inflorescences axillary or cauliflorous, either simply racemose or racemose-paniculate, the raceme axes often reduced to 1 to few flowers; pedicels 2-bracteolate. Flowers resupinate; hypanthium well differentiated; calyx deeply campanulate or broad-cylindric, the teeth 5, broad; petals pink or blue, the vexillum geotropic, much exceeding wings and keel; stamens either 1- or 2-adelphous; style barbellate on adaxial side (to raceme axis). Fruits pods, linear or oblong, elastically dehiscent, the valves coiling, either coriaceous becoming brittle, or woody, plane or with accessory rib parallel and approximate to adaxial suture.

1. Lianas. Leaflets 3 per leaf; larger blades 9–26 × 3.5–11 cm. Vexillum lavender-purple. Pods 14–27 × 1.5–2.3 cm. Native. . . . . *C. sagotii* var. *sagotii*..
1. Twining vines. Leaflets 5–7 per leaf; larger blades 2–7 × 1–4.5 cm. Vexillum blue. Pods 7–10 × 0.8–1 cm. Cultivated. . . . . *C. ternatea*.

Clitoria sagotii Fantz var. **sagotii** (Kunth) Benth.

Lianas, flowering in or below forest canopy. Stipules triangular-lanceolate, 2–5 mm long, deciduous. Leaves pedately trifoliolate; leafstalks 4–15 cm long, the petioles 2–11 cm long, the one interfoliolar segment (1)2–5 cm long; pulvinules (2.5)4–11 mm long; leaflets stipellate; blades broadly ovate to elliptic or lanceolate, the larger ones (9)10–26 × 3.5–11 cm, glabrous or abaxially puberulent, the apex acuminate; venation pinnate, the secondary veins in 5–9 pairs, prominulous abaxially. Inflorescences at contemporary foliate or annotinous defoliate nodes, short, often corymbiform, 2–6(12)-flowered racemes; peduncles and raceme axes together 0–2(5) cm; pedicels 4–9 mm long; bracteoles 3–7(10) mm long. Flowers: hypanthium 3–7(10) mm long; calyx broadly cylindric, 22–30 × 6.5–11 mm, the teeth 3–7 mm long; vexillum (4)4.5–7 × 3.5–4 cm, lavender-purple; ovary pubescent. Fruits stipitate, the body broad-linear, planocompressed, in profile 14–27 × (1.5)1.8–2.3 cm, the valves stiffly coriaceous, brown, glabrescent. Fl (Jan); in non-flooded forest.

Clitoria ternatea L.

Slender, herbaceous, twining, thinly strigulose vines. Leaves imparipinnate; leafstalks 3–9 cm long; leaflets 5–7 per leaf; blades elliptic or ovate, the larger ones 2–7 × 1–4.5 cm. Inflorescences axillary racemes, 1–2(3)-flowered; peduncles 2–10(35) mm long; pedicels 2–5 mm long; bracteoles suborbicular, 5–7 mm long, membranous. Flowers large: calyx 17–21 mm long, the teeth ovate, 4.5–7 mm long, the apex acuminate; petals vividly blue-bordered, the vexillum flabellate-obovate, 3.6–5 cm long, puberulent abaxially, the keel scarcely half as long. Fruits linear, slightly declined, 7–10 × 0.8–1 cm, thinly pilosulous glabrescent. Fl (Oct), flowering nearly year round; native of Paleotropics, cultivated in Saül where it is found climbing on walls.

CROTALARIA L.

Herbs and shrubs, glabrous to pubescent with silky hairs, eglandular. Stipules present. Leaves alternate, often digitately 3-foliolate, sometimes simple; petiolules present; leaflets estipellate, pinnately veined. Inflorescences terminal, leaf-opposed racemes; bracts transient; pedicels 2-bracteolate. Flowers papilionaceous; hypanthium obscure; calyx campanulate, often asymmetrically so, obscurely bilabiate, the teeth lanceolate; petals usually yellow, the broad vexillum reflexed through 90°, the keel incurved, rostrate; stamens 10, monadelphous, the anthers alternately larger basifixed and smaller versatile; style dilated and horny at base. Fruits pods, shortly stipitate, the body inflated, 1-locular, valvately dehiscent. Seeds several.

Crotalaria anagyroides Kunth

Coarse, potentially shrubby herbs, ≤3 m tall, silky-pilosulous or -strigulose throughout, except for adaxial surfaces of leaflets and petals. Stipules subulate, always small, sometimes subobsolete. Leaves: leafstalks 2.5–8 cm long; blades elliptic, the middle one 5.5–9 × 1.5–2.7, the outer pair a little smaller, the apex acute or obtuse, mucronate. Inflorescences at first compact, lengthening during anthesis; peduncles and axes together becoming 5–20 cm long; bracts linear-setiform, 5–11 mm long; pedicels 5–10 mm long. Flowers: sepals 9–11.5 mm long, the teeth ± 4.5–6.5 mm long; petals: vexillum 14.5–18 mm long, the keel blades densely barbate along inner margins. Fruits: stipes ca. as long as calyx-tube, the body ellipsoid, 3–4.5 × 1.2–1.5 cm, the valves brown, sericeous. Disturbed woodland, fields, hedgerows, waste places. Expected from central French Guiana, but not yet collected.

DALBERGIA L. f.

Species of this genus are expected in central French Guiana, but they have not yet been collected from our area. The genus has been provisionally entered in the generic key. *Dalbergia* differs from the closely related *Machaerium* by the terminally dehiscent anthers and by fruits with the seed near the middle not at the base of the valves.

DERRIS Lour.

High-climbing lianas. Stipules small, caducous. Leaves imparipinnate; leaflets estipellate, opposite; blades minutely strigulose abaxially; venation pinnate. Inflorescences long, narrow pseudoracemes, the secondary axes very short, incurved, thickened, densely brown-pilosulous; pedicels 2-bracteolate at apex. Flowers small, papilionaceous; calyx campanulate, broadly short-toothed or subtruncate; petals subequal in length, glabrous, pink or ochroleucous; stamens monadelphous; ovary stipitate, brown-pubescent. Fruits samaras, linear-elliptic or plano-compressed, narrowly winged along adaxial suture. Seeds 1–2.

Derris pterocarpa (DC.) Killip

Lianas. Stems terete in transverse section. Sap red. Leaves: leafstalks 10–20 cm long; leaflets 5 per leaf; pulvinules 6–9 mm; blades ovate-, obovate-, or oblong-elliptic, 10–17 × 4–6.5 cm, the base either rounded or cuneate, the apex shortly acuminate; secondary veins in ± 7–9 pairs from each side of centric midrib. Inflorescences axillary-cauliflorous, stiffly incurved-ascending from year-old and older wood, the primary axes 10–20 cm long, the secondary axes 3–5 mm long; pedicels filiform, 3–5 mm long. Flowers: sepals 2.5–3 mm long, the teeth depressed-deltate, <0.5 mm long; petals ochroleucous, the vexillum obovate-cuneate, 6–8 mm long. Fruits in profile linear-elliptic, 5.5–11 × 1.4–1.8 cm, planocompressed, the valves papery, brown-strigulose, stramineous when ripe, the dorsal wing 2.5–3.5 mm wide. Fl (Nov), fr (Jan, Feb); in nonflooded forest.

DESMODIUM Desv.

Herbs, some monocarpic, and subshrubs, pubescent with simple, small hooked hairs. Stipules early dry, papery, or scarious, striately veined, either free or connate into leaf-opposed amplexicaulous sheath. Leaves 3-foliolate; leaflets stipellate. Inflorescences either axillary or terminal racemes or pseudoracemes, bracteate in vernation; bracts, subtending either 1 or 2 flowers, caducous; pedicels slender, either minutely bracteolate at base or naked, the flowers either solitary or mostly paired. Flowers small, papilionaceous; hypanthium lacking; calyx campanulate or turbinate-campanulate, 5-dentate; petals as long or twice as long as sepals, pink, lavender, lilac, purple, red-purple, red-violet, blue-violet, or whitish; stamens 10, either 1- or 2-adelphous. Fruits sessile or stipitate, straight or gently recurved loments consisting of (1)2–7 compressed, 1-seeded, tardily dehiscent joints, the sutures variably indented or continuous between joints, the free-falling articles with short hooked hairs and individually dispersed thereby.

1. Racemes short and dense, often capituliform. Calyx teeth silky-barbate; corolla with petals about as long as sepals. . . . . *D. barbatum.*
1. Racemes not short and dense, not capituliform. Calyx teeth weakly setulose or glabrous; corolla with vexillum at least 1 mm longer than sepals.
    2. Inflorescences mostly simple, scapiform racemes erect from ground, adventitiously rooting stems, the peduncles and raceme axes together 15–60 cm long. Stipes of loments 5–8 mm long, the articles 1–3, 8–12 mm long. . . . . *D. axillare.*
    2. Inflorescences not as above, arising from normally foliate stems, the peduncles and raceme axes together 5–24 cm long. Stipes of loments 0–1 mm long, the articles mostly 4–7, 3–4 mm long, but in *D. wydleriana* as above.
        3. Stems not adventitiously rooting at lower nodes. Stipules fully amplexicaulous and often united opposite petioles into sheath, deciduous from horizontal scar fully girdling stems. Leaflet blades with secondary veins in 7–9 pairs. . . . . *D. incanum.*
        3. Stems sometimes adventitiously rooting at lower base. Stipules ampexicaulous, free, the abscission scar not fully girdling stems. Leaflet blades with secondary veins in 4–7 pairs.
            4. Larger leaflets 1–2.5 cm long. Loments composed mostly of 3–5 articles, the articles ≤3 mm long. . . . . *D. adscendens.*
            4. Larger leaflets 6–9 cm long. Loments composed of (1)2–3 articles, the articles 5–8 mm long. . . . . *D. wydlerianum.*

Desmodium adscendens (Sw.) DC.

Herbs, 15–50 cm tall. Stems erect or incurved, softly pilose, sometimes adventitiously rooting at lowest nodes. Stipules amplexicaulous, free, narrowly lance-attenuate, 4–9 mm long, striate, persistent, the abscission scar not fully girdling stems. Leaves: leafstalks 1–2 cm long; blades obovate, suborbicular, or obtusely elliptic, the terminal one longest, ± 1–2.5 × 1–2 cm, either glabrous or abaxially strigose; secondary veins in 4–7 pairs. Inflorescences: peduncles and raceme axes together ± 5–15 cm long; bracts 5–6 mm long, pilose abaxially, caducous; pedicels 6–10 mm long. Flowers: sepals 2–2.4 mm long; petals lavender with darker eyespots, the vexillum 4.5–6 mm long. Fruits: loments (2)3–5(7)-articulate, 5–24 × 3 mm, the articles ≤3 mm long, the adaxial suture continuous, the abaxial one crenately indented. Fl (Sep, Oct, Dec), fr (Dec), probably flowering and fruiting intermittently year round; along roadsides through forest.

Desmodium axillare (Sw.) DC. PL. 56d

Herbs. Stems rooting at random nodes. Stipules triangular-acuminate, ± 7-veined, deciduous. Leaves: leafstalks 4.5–10 cm long; blades broadly ovate, ± 5–9 × 2.5–5 cm, thinly puberulent or glabrous, the apex acuminate. Inflorescences mostly simple, scapiform racemes, erect from ground, the peduncles with raceme axes together 15–60 cm long; bracts caducous before anthesis; pedicels becoming 6–12 mm long. Flowers: calyx ± 3 mm long; petals lilac or whitish, the vexillum 4–5 mm long. Fruits (1)2(3)-articulate, the stipes 5–8 mm long, the articles broadly lunate, 8–12 × 5–6 mm, the abaxial suture strongly evenly convex, the adaxial one shallowly concave at middle, the valves finely reticulate. Fl (Jan, Feb, Jun, Sep, Oct, Dec), fr (Feb, Jun, Sep), probably flowering and fruiting year round; along trailsides and in forest clearings.

Desmodium barbatum (L.) Benth.

Erect or diffuse, monocarpic herbs, ± 25–50 cm tall, pilose with straight setiform and short uncinate hairs. Stems basally woody with age. Stipules lance-attenuate, 4–10 mm long, papery, striate. Leaves: leafstalks 1–1.5 cm long; blades elliptic or obovate-elliptic, the terminal one ± 1–2.5 × 0.6–1.2 cm, the apex obtuse. Inflorescences terminal, racemes, sessile, short and dense, often capituliform, the axes 1–2.5 cm long; bracts resembling stipules; pedicels 4–7 mm long, spreading-recurved in fruit. Flowers: sepals ± 3.5 mm long, the tube 1 mm long, the teeth lance-attenuate, silky-barbate; petals about as long as sepals, lavender or red-violet, drying bluish. Fruits 3–4-articulate, evenly recurved, the adaxial suture continuous, the abaxial one crenately indented, the articles ± 2.5 × 2 mm. Fl (Jun, Aug); in disturbed habitats. *Pistache sauvage* (Créole).

Desmodium incanum DC. PL. 56e,f; PART 1: FIG. 8.

Herbs, 0.4–0.6 m tall, finely pilosulous throughout, except for adaxial surface of leaflets. Stems erect-ascending from woody base, not adventitiously rooting at lower nodes. Stipules lanceolate, 5–8 mm long, chartaceous, brown, amplexicaulous and often united opposite leaf into sheath, deciduous from horizontal scar fully girdling stems, the apex acuminate. Leaves: leafstalks (1)1.5–4 cm long; blades ovate, broad-rhombic, or oblong-elliptic, the larger ones 3.5–7 × 2–4.5 cm, pallid, dark green adaxially, the base rounded, the apex obtuse or deltately subacute; venation prominulous abaxially, the secondary veins in 7–9 pairs. Inflorescences: peduncles and raceme axes together 9–24 cm long; bracts narrowly lanceolate, submembranous, fugacious; pedicels 6–9 mm long. Flowers: sepals 2.5–3 mm long; petals blue-violet, the vexillum 5–6 mm long. Fruits 18–28 × 2.2–2.8 mm, (4)5–7-articulate, the adaxial suture continuous straight, the abaxial one crenately indented, the articles 3–4 mm long. Fl (Jun, Aug, Sep, Oct, Dec), fr (Jun, Aug, Sep, Dec), probably flowering and fruiting year round; in disturbed habitats.

Desmodium wydlerianum Urb.

Herbs, trailing or weakly ascending, 30–70 cm tall. Stems adventitiously rooting. Stipules amplexicaulous, free, membranous, several-veined, lanceolate or ovate, 3–6 mm long, deciduous from horizontal scar, the apex acuminate, the abscission scar not fully girdling stems. Leaves: leafstalks 4–7.5 cm long; pulvinules 1.7–2.3 mm long; blades rhombic-ovate, the terminal one 6–9 × 4–6 cm, membranous, the base broadly flabellate or subtruncate, the apex acuminate; secondary veins in 4–7 pairs. Inflorescences terminal, laxly few-flowered racemes; pedicels mostly solitary, filiform, 15–19 mm long. Flower: sepals ± 2 mm long; petals scarcely 4 mm long. Fruits subsessile, (1)2–3-articulate, each article in profile lunately elliptic, 5–8 × 2.5–3.3 mm, the adaxial suture of each thickened, concave at middle, the abaxial suture evenly convex, the interseminal sinuses narrow and deep. Fl (Jan, Dec); in moist places.

DIOCLEA Kunth

Lianas, pilosulous with brownish or gray basifixed hairs, eglandular. Stipules either basifixed or produced backward from point of attachment. Leaves pinnately 3-foliolate; leaflets ample, stipellate or not; blades with ventation pinnate. Inflorescences either from axils of current leaves or from old wood, pseudoracemose, the primary axes erect, the secondary axes abruptly incurved, clavate, tubercular, the flowers closely biseriate on outer face; pedicels 2-bracteolate. Flowers: hypanthium and calyx tube inequilaterally turbinate-campanulate, 4-toothed, the adaxial tooth sometimes minutely 2-denticulate; corolla papilionaceous, pink or purple, with eyespot in groove of vexillum, the petals glabrous or vexillum minutely puberulent; stamens monadelphous, the anthers uniform or 5 sterile; ovary subsessile, the style linear or swollen at middle. Fruits sessile pods, in profile linear-oblong, or broad-oblong, or elliptic, either straight or gently incurved, biconvex or planocompressed, dehiscent or indehiscent, the valves coriaceous or woody, variously thickened, either carinate or narrowly winged along sutures, the endocarp papery, intruded between seeds.

1. Stipels subtending leaflets evident, 4–6 mm long. Bracteoles at apex of pedicels ovate, 5–8 mm long, membranous. Inner margins of keel-blades fimbriate. Pods 1.5–2.1 cm wide. . . . . *D. virgata.*
1. Stipels either lacking or 1–4 mm long. Bracteoles at apex of pedicels 1.5 mm long. Inner margins of keel-blades entire. Pods 5–7 cm wide.
   2. Stipules basifixed. Leaflet blades subglabrous. Ripe pods dehiscent, glabrous, 3-ribbed along adaxial suture, not winged. Seeds obese, distorted by mutual pressure, in broad view ± 3–4.5 × 3–4 cm. . . . . *D. macrocarpa.*
   2. Stipules produced backward from point of insertion. Leaflet blades pilosulous on both surfaces, densely so abaxially. Ripe pods indehiscent, persistently brown-pilosulous, 1-ribbed and narrowly 2-winged along adaxial suture. Seeds plumply discoid, ± 3 × 2.2 cm. . . . . *Dioclea* sp. A.

Dioclea macrocarpa Huber

FIG. 124, PL. 57a–d; PART 1: PL. IXa

Robust lianas, young surfaces finely minutely puberulent throughout, except for subglabrous leaflet blades. Stems terete. Bark flaking. Stipules triangular, 1.5–4 mm long, basifixed, persistent. Leaves: leafstalks 5–16 cm long; pulvinules 4–8 mm long; leaflets estipellate; blades ovate or broad-elliptic, the terminal one 9–17 × 5–9 cm, thinly chartaceous, subglabrous, the base either rounded or cuneate, the apex shortly acuminate or caudate; secondary veins in 4–6 pairs. Inflorescences solitary or in fascicles of 2–3 from knots on trunk proximal to current foliage, pseudoracemes, the primary axes 8–50 cm long, stout, virgate; peduncles of tubercular secondary axes 2–4 mm long; pedicels 2–3 mm long; bracteoles suborbicular, 1.5 × 0.8–1.3 mm. Flowers: hypanthium and calyx tube together 6.5–8 × 5–7 mm; sepals 11.5–13.5 mm long, firm, brown, the dorsal tooth 5–7 mm long; petals

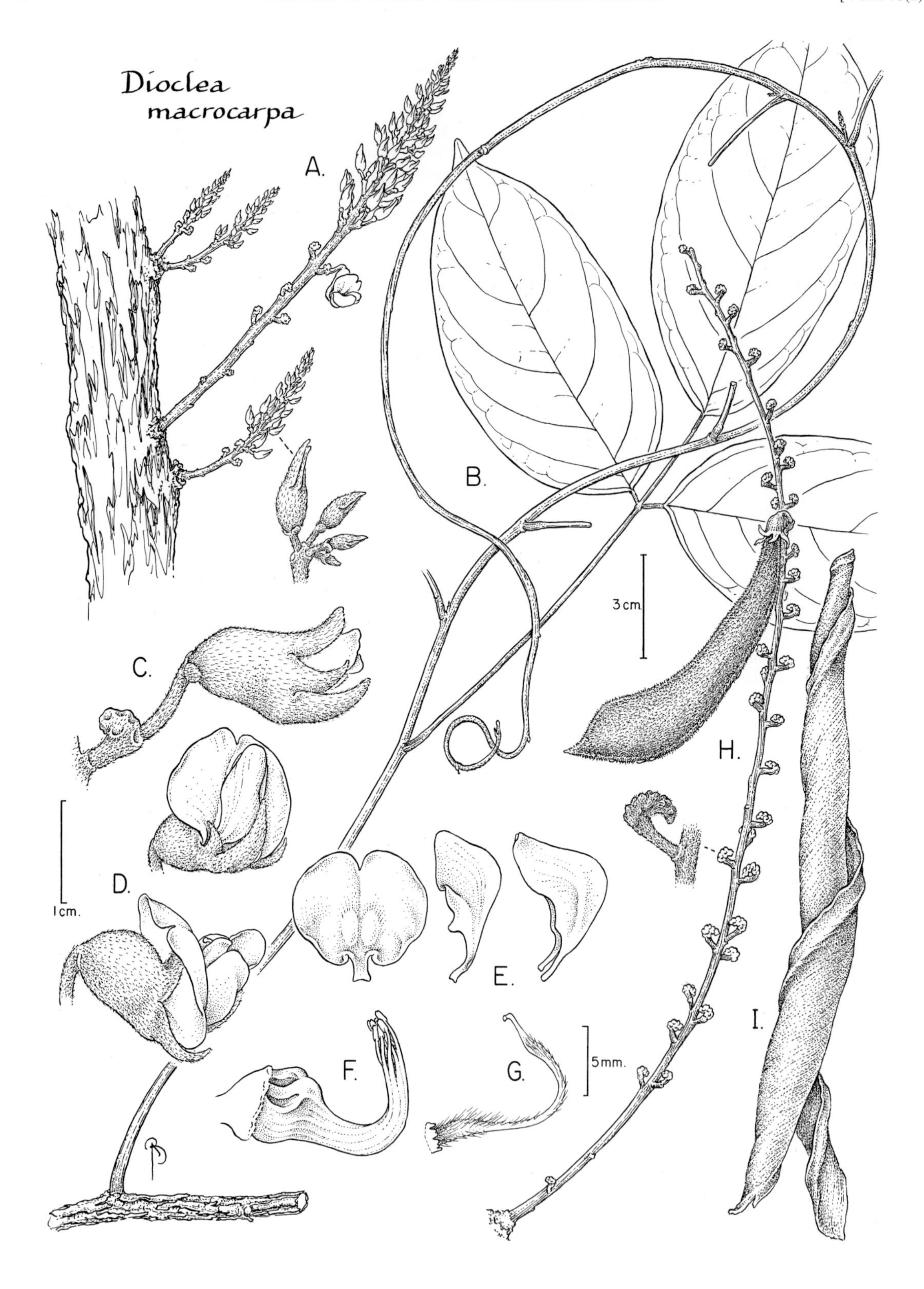

FIG. 124. FABACEAE. *Dioclea macrocarpa* (A, B, *Mori & Gracie 24208*; C–H, *Mori et al. 22825*; I, *Mori et al. 22746*). **A.** Part of stem with cauliflorous inflorescences and detail of flower buds. **B.** Part of young liana and leaf. **C.** Lateral view of bud attached to short shoot. **D.** Oblique (above) and lateral (below) views of flower. **E.** Standard (far left), wing (near left), and keel (right) petals. **F.** Lateral view of androecium. **G.** Lateral view of pistil. **H.** Infructescence with immature fruit and detail of short shoot. **I.** Dehisced fruit with twisted valves.

pink-purple, yellow-eyed, the vexillum strongly recurved, scarcely longer than longest calyx-tooth; stamens subuniform, all fertile. Fruits in profile 19–28 × 5–7 cm, the valves woody, convex, densely puberulent when young, at length brown and glabrous, wrinkled, coarsely 3-ribbed along adaxial suture, coiling after dehiscence. Seeds obese, obtusely quadrate, in broad view ± 3–4.5 × 3–4 × 0.8–1.8 cm. Fl (Sep, Oct, Nov), fr (May, Jul, Nov); in nonflooded forest.

Dioclea virgata (Rich.) Amshoff PL. 57e,f

Lianas, pilosulous nearly throughout. Young stems slender, twining. Stipules triangular deltate, 4–6 mm long, the base corneous. Leaves: leafstalks 4–10 cm long; stipels subulate-setaceous; pulvinules 4–6 mm long; blades ovate, the terminal one 6–11 × 3.5–7.5 cm, the base rounded or shallowly cordate, the apex deltately acuminate. Inflorescences from new growth, the axes of pseudoracemes including peduncles 25–40 cm long, the tubercular floriferous axes sessile, 2.5–4 mm long; pedicels slender, 5–9 mm long; bracteoles ovate, 5–8 mm long, membranous, pallid or purple-tinged, caducous. Flowers: hypanthium together with calyx tube 6.5–8.5 × 6–8 mm; sepals 13–18.5 mm long, glabrous, purplish, the tube abaxially (to inflorescence axis) convex, the abaxial tooth 6–10 mm long; petals pink-purple, the vexillum minutely puberulent, with dark purple aureole and yellow 2-lobed eyespot, the inner margins of keel-blades fimbriate proximal to middle; stamens with fertile anthers 10. Fruits broad-linear, 7–10 × 1.5–2.1 cm, slightly decurved, planocompressed, the valves stiff-coriaceous, densely whitish- or brown-pilosulous, twisting after dehiscence. Seeds 5–9. Fl (Aug, Sep); in disturbed habitats, especially common around the village.

Dioclea sp. **A** (*Jacquemin 1678*, *Oldeman 2094*, both at CAY)

Lianas, brown-gray-pilosulous, the leaflets thinly so adaxially, densely and softly so abaxially. Stipules shortly produced backward from point of attachment. Leaves: stipels 1 mm long; blades broadly obovate, the terminal one 7.5–10.5 × 7–10 cm, the apex obtuse; secondary veins in ± 8 pairs. Fruits 18–21 × 5.5 cm, slightly incurved, compressed but plump, indehiscent, the valves woody, persistently brown-pilosulous, low-convex over seeds, the abaxial suture 2-ribbed, the adaxial suture 1-ribbed, 2-winged, the wings thick, ± 7 mm wide. Seeds plumply discoid, in broad view ± 3 × 2.2 cm. Fr (Feb, Sep); in secondary forest.

DIPLOTROPIS Benth.

Canopy trees, surfaces finely, densely sordid puberulent. Stipules transient. Leaves ample, alternate; leaflets alternate, pulvinulate; blades chartaceous, glabrous or minutely puberulent, sometimes also papillate abaxially. Inflorescences terminal, efoliate or basally foliate panicles of racemes; bracts present; pedicels short; bracteoles present. Flowers fragrant; hypanthium asymmetrically campanulate, the tube longer on adaxial side, the rim consequently oblique; calyx-limb also longer adaxially, 2-lipped, the 2 adaxial lobes broadly triangular, united ± half way, the 3 adaxial ones shorter; petals free, all slenderly clawed, the vexillum blade rhombic, appendaged adaxially near base on each side, the keel-and wing-petals a little longer, the blades obliquely dilated, the keel shorter than wings; stamens 10, the filaments arising separately from hypanthial rim, alternately longer and shorter; ovary subsessile. Fruits samaroid, shortly stipitate, the body planocompressed, in profile oblong-elliptic obtuse, sometimes incipiently twisted near base, the valves stiffly membranous, translucent, coarsely reticulate, the adaxial suture narrowly winged. Seeds obliquely ascending, linear-elliptic, the testa papery.

Diplotropis purpurea (Rich.) Amshoff FIG. 125, PL. 58a

Trees, 35–45 m × 70–100 cm. Trunks with low, thick buttresses. Leaves: leafstalks 6–15 cm long, subterete; pulvinules 3.5–7 mm long; leaflets 5–9 per leaf; blades oblong-obovate, the longer ones 7–12 × 3–5.5 cm, the apex obtuse or shortly acuminate; secondary veins in ≤10 pairs, not connected by scalariform tertiary veinlets. Inflorescences: primary axes 9–22 cm long; bracts deltate, 0.6–1 mm long, persistent; pedicels 2–4 mm long. Flowers: hypanthium 1.8–3.3 mm long; calyx 5.5–6.5 mm long, coriaceous, lavender; petals pink-lavender, the abaxial ones ± 2× as long as calyx; ovary strigulose. Fruits: stipes ± as long as calyx, the body 8–16 × 3.4–4.3 cm, the ripe valves stramineous, glabrate. Seeds 1–4. Fl (Feb, Nov), fr (Sep); in nonflooded forest. *Coeur dehors*, *moutouchi de montagne* (Créole).

DIPTERYX Schreb.

Trees. Trunks with plank buttresses, the wood extremely hard and heavy. Stipules evanescent or lacking. Leaves ample, paripinnate; leafstalks obcompressed, narrowly winged, auriculate at insertion of leaflets; leaflets alternate; blades inequilateral. Inflorescences terminal, racemose-paniculate, rufous-tomentulose; bracts caducous; pedicels short; bracteoles 2, caducous. Flowers: calyx often coriaceous, bilabiate, the 2 adaxial lobes foliaceous, oblong, free to rim of hypanthium, clasping, scarcely shorter than papilionaceous corolla, the 3 abaxial lobes united into much smaller and narrower 3-denticulate lip; petals: vexillum obcordate beyond claw, not appendaged; stamens 10, monadelphous, the anthers versatile, either uniform or dimorphic; ovary substipitate; ovules 1. Fruits drupaceous, plumply oblong-ellipsoid, subterete, the apex obtuse-mucronulate. Seeds 1.

Dipteryx odorata (Aubl.) Willd.

Trees, 20–40 m tall. Leaves: leafstalks ± 10–30 cm long, low-convex abaxially, shallowly concave adaxially, narrowly green-marginate, produced 2–5 cm beyond distal leaflets; leaflets (2)3–4 pairs; pulvinules 3–5 mm long; blades oblong or oblong-elliptic, broader on distal side of straight midrib, the longer ones 10–18 × 5–8 cm, chartaceous, lustrous, the base rounded, the apex shortly deltate-acuminate. Inflorescences short panicles, 10–15 cm long, densely flowered; bracts and bracteoles firm, brown-tomentulose, caducous, the latter valvately clasping young flower. Flowers fragrant; hypanthium 3–4.5 mm long; calyx 15–18 mm long, tomentulose, like bracts, the adaxial lobes oblong, ± 12–15 × 5 mm, the apex obtuse, the abaxial lip 3–5 mm long, 3-denticulate; petals pink, the vexillum scarcely longer than calyx. Fruits subsessile, 5–5.5 × 2.5–3 cm, glabrous. Known only by sterile collections from our area; in nonflooded forest. *Coumarounda* (Créole), *cumaru*, *cumaru roxo* (Portuguese), *faux-gaïac*, *fève tonka* (French), *gagnac*, *gaïac*, *tonka* (Créole).

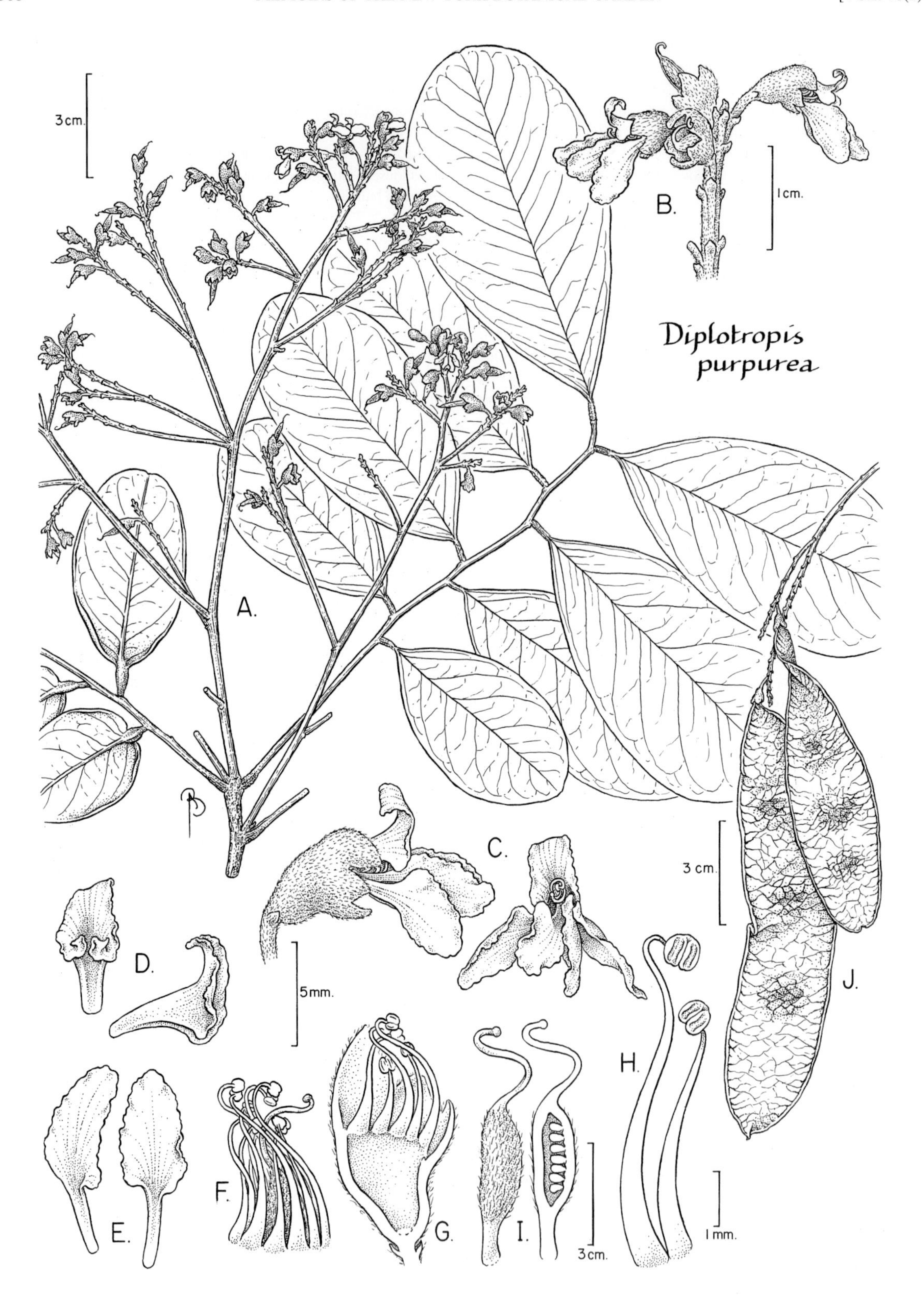

FIG. 125. FABACEAE. *Diplotropis purpurea* (A–C, *Mori & Gracie 21108*; D–I, *Mori et al. 22881*; J, *Granville B.5149*). **A.** Part of stem with leaves, flowers, and very immature fruits. **B.** Part of inflorescence. **C.** Lateral (left) and apical (right) of flowers. **D.** Adaxial (left) and lateral (right) views of standard petal. **E.** Keel (left) and wing (right) petals. **F.** Androecium and upper part of pistil. **G.** Medial section of perianth with adnate stamens. **H.** Two stamens. **I.** Pistil (left) and medial section of pistil (right). **J.** Fruits.

DUSSIA Taub.

Emergent trees. Trunks with tall, plank buttresses. Stipules lacking. Leaves ample, imparipinnate; petioles planocompressed adaxially; leaflets alternate. Inflorescences terminal, panicles, coeval with new foliage; bracts present; bracteoles 2, inserted on decurrent hypanthium. Flowers densely silky-puberulent; hypanthium obliquely turbinate; calyx tubular-campanulate, 5-dentate, obscurely 2-lipped, the adaxial teeth little more distally united than 3 abaxial ones; petals not strongly graduated, the keel-blades imbricately adherent, shorter than wings; stamens 10, the filaments free to hypanthium rim, the anthers dorsifixed, uniform. Fruits pods, sessile, ellipsoid or oblong, plumply biconvex, dehiscence tardy, through both sutures. Seeds 1, then fruits ellipsoid, or 2 and then fruits oblong, obese, red.

Dussia discolor (Benth.) Amshoff PL. 58b

Trees, 30–56 m tall. Leaves: leafstalks 16–28 cm long; leaflets 9–13 per leaf; pulvinules 4–5 mm long; blades ovate, oblong-ovate, or broadly lance-ovate, the larger ones distal, 8–16 × 3–5.5 cm, thinly chartaceous, glabrous adaxially, pallid-velutinous abaxially, the base rounded or subcordate, the apex obdeltate or shortly acuminate; secondary veins in ± 15 pairs, connected by scalariform tertiary veinlets. Inflorescences ≤15 cm long; bracts rhombic-ovate, 4–8 mm long; pedicels stout, 3–4 mm long; bracteoles rhombic-ovate, 4–8 mm long, persistent on fallen flowers. Flowers: hypanthium and calyx tube together 4–5 mm long; calyx 8–9.5 mm long, the teeth deltate; petals pink-lavender, the vexillum including claw 14–17 mm long, green at middle with purple aureole. Fruits 6–11 × 2.8–4.5 cm, the valves woody, densely minutely orange- or tan- velutinous. Seeds: testa white, foam-like, with bright red surface; cotyledons green. Fl (May, Oct), fr (Nov); in nonflooded forest. *Moutouchi, moutouchi de marécage* (Créole).

HYMENOLOBIUM Benth.

Deciduous trees, 15–35 m tall, flowering when leafless, glabrous or nearly so, except for inflorescences. Trunks with buttresses, the wood extremely hard. Stipules lacking or very small, fugacious. Leaves imparipinnate; stipels lacking or very small, fugacious; leaflets opposite. Inflorescences terminal, panicles, ample, densely minutely gray-strigulose; bracts present; pedicels 2-bracteolate. Flowers fragrant; calyx turbinate-campanulate, the rim undulate-truncate; corolla papilionaceous, lilac-pink, the vexillum abruptly bent backward over calyx, unappendaged, with eyespot, the wing- and keel-petals falcate, free; stamens 10, monadelphous, the anthers dorsifixed; ovary stipitate, the style linear, the stigma porose. Fruits samaroid, shed with marcescent calyx, the shortly stipitate body oblong, obtuse at each end, planocompressed but often incipiently twisted toward base, the valves membranous, glabrous, translucent, red or, at maturity, stramineous, intramarginally 2-veined from near base, coarsely openly reticulate. Seeds 1–2, vertically descending, narrowly oblong, the testa thin, brittle.

Hymenolobium petraeum Ducke

Trees, to 45 m tall. Bark exfoliating in plates. Leaves: leafstalks 8–13 cm long, the adaxial sulcus shallow, weakly bridged; leaflets 9–17 per leaf; pulvinules 4–5 mm long; blades oblong-elliptic, the lateral ones larger, 3.2–5 × 1.2–3 cm, chartaceous, the apex emarginate or retuse. Inflorescences effuse, the primary axes 15–30 cm long; bracts <1 mm long; pedicels 2–6 mm long; bracteoles <1 mm long. Flowers: hypanthium ± 1.5–2 mm long, basally attenuate; sepals 4.5–5 mm long; vexillum 10–12 mm long. Fruits in profile 4.5–9 × 1.5–2 cm. Fr (Apr); in nonflooded forest.

MACHAERIUM Pers.

Lianas or trees. Red latex or latex oxidizing red sometimes present. Stems terete or subangular-angular. Stipules hard, spinescent, at random nodes. Leaves imparipinnate, alternate; leaflets either few and large or many and small, often glabrous adaxially. Inflorescences complex terminal or simpler axillary panicles or either solitary or paired racemes, the axes silky-strigulose or -pilosulous with short, sordid or bronze-gold hairs, the secondary axes bracteate; pedicels 2-bracteolate. Flowers small; hypanthium scarcely differentiated or lacking; calyx campanulate, shortly 5-toothed or subtruncate; petals papilionaceous, white, purple, or yellow, the vexillum suborbicular to obcordate beyond claw, silky abaxially, the wings falcate, the keel petals weakly united, these a little shorter than vexillum; stamens 10, either diadelphous (9 + 1; 5 + 5) or monadelphous; ovary stipitate, the style filiform; ovule 1. Fruits samaras, stipitate, the basal seed-cavity biconvex, unappendaged, produced into wing, the wing oblance-oblong, papery, straight or sigmoidally arcuate, the surface reticulately veined. Seeds 1.

1. Trees (infrequently lianas outside our area). . . . . *M. quinata.*
1. Lianas.
    2. Leaflets 3–11(17) per leaf, 6–13 × 2–6 cm.
        3. Secondary veins of leaflet blades many, inconspicuous, closely spaced. Petals golden-yellow. . . . *M. aureiflorum.*
        3. Secondary veins of leaflet blades few, conspicuous, widely spaced. Petals white or lavender.
            4. Leaflets 3–5(7) per leaf, pubescent only on veins abaxially; secondary veins in 5–7 pairs. Calyx mostly glabrous except minutely ciliolate on margin. Inflorescences to 5.5 cm long. Fruits glabrous. . . . . *M. paraënse.*
            4. Leaflets 7–11 per leaf, pubescent throughout abaxially. Calyx brown velutinous. Inflorescences ± 20–30 cm long. Fruits pubescent. . . . . *M. floribundum.*
    2. Leaflets 38–70 per leaf, 0.7–1.6 × 0.2–0.5 cm. . . . . *M. altiscandens.*

Machaerium altiscandens Ducke

Lianas climbing into forest canopy, armed at many, but not all nodes, sordid- or brown-pilosulous throughout, except for glabrous adaxial surface of leaflet blades and glabrescent calyx. Stipules spinescent, 1.5–3 mm long. Leaves: leafstalks 4.5–11 cm long; pulvinules 0.4–1 mm long; leaflets 38–70 per leaf, firm, small; blades narrowly oblong-oblanceolate, the largest 0.7–1.6 × 0.2–0.5 cm, the apex obtuse or emarginate; secondary veins inconspicuous. Inflorescences terminal, effuse, efoliate panicles, the primary axes 2–5 dm long; pedicels 1–2.3 mm long; bracteoles semi-orbicular, 1–1.7 mm long, brown-silky. Flowers: calyx campanulate, 3.2–4.5 × 2.2–3 mm, the tube glabrous, nigrescent, the teeth depressed-deltate, <0.5 mm long, puberulent; petals lavender-violet, the vexillum 9–10 mm long. Fruits 4–5 cm long, the stipe ± 5 mm long, the wing 11–14 mm wide, terminal, decurved, glabrate. Fl (Aug); in nonflooded forest.

Machaerium aureiflorum Ducke

Lianas, randomly armed, subappressed-pilosulous with bronze-gold hairs, except for adaxial surface of leaflets. Stems terete-subangular. Stipules spinescent. Leaves ample; leafstalks 8–18 cm long; pulvinules 2.5–5 mm long; leaflets 9–13(17) per leaf, alternate and subopposite; blades oblong-, oblanceolate-, or obvate-elliptic, the larger ones 7–10 × 3–4.3 cm, chartaceous, carinate abaxially by midrib, the base rounded or broad-cuneate, the apex shortly acuminate, the margins corneous; secondary veins of many, inconspicuous, closely spaced pairs. Inflorescenses terminal, the primary axes ± 10–25 cm long; bracts suborbicular, 0.8–1.5 mm long, caducous; bracteoles suborbicular, 0.8–1.5 mm, persistent. Flowers: sepals 4–4.5 mm long, the teeth depressed-deltate, 0.4–0.9 mm long; petals golden-yellow, the vexillum 9–11 mm long; ovary densely golden-pilosulous. Fruits: stipes ± 8 mm long, the seminiferous body ± 2 × 1.2 cm, coriaceous, the wing 3.5–4 × 2–2.2 cm, almost straight. Fl (Oct), fr (Nov); in nonflooded forest.

Machaerium floribundum Benth.

High-climbing lianas. Stems terete-angular. Stipules absent in our area. Leaves ample; leafstalks 9–26 cm long; pulvinules 2–5.5 mm long; leaflets 7–11 per leaf; blades oblong-obvate or -elliptic, the larger ones 8–12 × (3)3.5–5.5 cm, glabrous adaxially, finely minutely brownish-strigulose throughout abaxially, the base rounded, the apex shortly acuminate; secondary veins in 9–14 pairs, conspicuous, widely spaced. Inflorescences terminal, effuse, the primary axes ± 20–30 cm long (to 40 cm outside our area), finely minutely brownish-strigulose; bracts deltate-ovate, ± 1–2 mm long, deciduous; pedicels ± 1 mm long; bracteoles deltate-ovate, ± 1–2 mm long, deciduous. Flowers: calyx broad-campanulate, 3–4 mm long, brown-velutinous, the orifice crenately denticulate; petals white, purple eyed, the vexillum 5–9 mm long, thinly silky abaxially; ovary velutinous. Fruits pubescent, the stipes 3–7 mm long, the seed-cavity biconvex, 14–17 × 7–10 mm, sometimes minutely crested over seed, the wing 4.5–6 × 1.2–2 cm, slightly decurved. Fr (Feb, Dec); in nonflooded forest.

Machaerium paraënse Ducke

Lianas, randomly armed, glabrous, except sordid-strigulose inflorescence axes and minutely pubescent leafstalks and leaflet veins. Stems terete. Stipules spinescent, ascending or recurved. Leaves: leafstalks 5–13 cm long; pulvinules 4–6 mm long; leaflets 3–5(7) per leaf; blades broadly ovate or oblong-elliptic, the larger ones 6–13 × 3.5–6 cm, pubescent only on veins abaxially, the base either cuneate or rounded, the apex shortly acuminate; secondary veins in 5–7 pairs, conspicuous, widely spaced. Inflorescences axillary and terminal panicles or axillary racemes, compact, the primary axes ± 3–5.5 cm long (to 9 cm outside our area); pedicels 1–2 mm long; bracteoles elliptic-oblong, 4–5 mm long, facially glabrate. Flowers: calyx ± 3.5–5 × 2 mm, glabrous abaxially, minutely ciliolate on margin; petals lavender, the vexillum 7–8 mm long. Fruits glabrous, often drying black, the stipes 10–14 mm long, the seed-cavity 2.5–3 cm long, the wing 5–8 × 2–2.6 cm, nearly straight, or decurved and distally incurved, hence sigmoid. Fl (May), fr (Aug); in nonflooded forest.

Machaerium quinata (Aubl.) Sandwith

Trees (sometimes lianas outside our area). Stems terete when older. Stipular spines absent. Leaves: leafstalks 9–28 cm long; 7–11 leaflets per leaf; blades oblong or oblong-obvate, the longer ones 8–12 × 3–5 cm, glabrous or finely minutely pubescent adaxiallly, finely and densely tan pubescent abaxially, the base rounded to cuneate, the apex acuminate to obtuse; secondary veins in 10–14 pairs, prominent. Inflorescences terminal, the primary axes ± 8–40 cm long, finely minutely brownish pubescent; bracts deltate-ovate, ± 1–2 mm long; bracteoles deltate-ovate, ± 1.5–3.5 mm long. Flowers: calyx campanulate, 4–6 mm long, brownish-velutinous, the orifice denticulate; the vexillum obcordate, ± 9–15 mm long, purple with central white stripe, pubescent abaxially, the keel white; stamens monadelphous; ovary velutinous. Fruits ± 5–9 × 2–3.2 cm, velutinous, often rugose proximally, the stipe 5–8 mm long, the wing 4.5–5.5 × 1.5–2.5 cm. Fl (Apr); in nonflooded forest.

MONOPTERYX Benth.

Trees. Trunks buttressed. Stems glabrous. Stipules ephemeral. Leaves distichous, imparipinnate; leafstalks shallowly, continuously sulcate; stipels lacking; pulvini short; leaflets alternate. Inflorescences both terminal and axillary racemes, immersed in foliage or shortly exserted; bracts present; pedicels present; bracteoles 2. Flowers: hypanthium shallowly turbinate; calyx densely pilosulous adaxially, 2-lipped, the 2 adaxial (to raceme axis) sepals united into oblong, shallowly emarginate blade, the abaxial lip much shorter, 3-denticulate; corolla papilionaceous, green and pink, the vexillum sessile, obcordate, widely gaping at anthesis, the keel-petals oblong, valvately coherent, widely gaping at anthesis, the wings minute, whitish; stamens 10, the filaments free, the anthers almost basifixed; ovary shortly stipitate, the stigma poriform. Fruits pods, shortly stipitate, the body narrowly elliptic, attenuate at both ends, planocompressed, elastically dehiscent, the valves coiling, woody, transversely cracked, narrowly 2-winged along both sutures.

Monopteryx inpae W. A. Rodrigues — Pl. 58c

Trees, to 40–50 m tall. Leaves: leafstalks 7–18 cm long; pulvinules 1–1.5 mm long; leaflets 11–17 per leaf; blades of lateral ones elliptic or oblong-elliptic, the base inequilateral, the terminal one a little larger and symmetrical, the longer ones 2.5–6 × 1–2.4 cm, the base cuneate, the apex emarginate. Inflorescences: peduncle and raceme axes together (2)5–13 cm long; bracts 1.5–3 mm long;

pedicels 3–4.5 mm long; bracteoles 1.5–3 mm long. Flowers: sepals 7–8 mm long, the adaxial lip broadly obtuse, 3.5–4 mm long; gynoecium glabrous. Fruits 10–12 × 2.2–3.1 cm, the sutural wings 2–3.5 mm wide, the valves dull brown, glabrous. Fl (Sep, Oct), fr (Nov); in nonflooded forest.

MUCUNA Adans.

Lianas, nearly or quite glabrous, except for inflorescences. Stipules narrowly ovate, caducous. Leaves pinnately 3-foliolate; leaflets thin-textured, nigrescent when dried; venation palmate, 3–5-veined. Inflorescences axillary, pseudoracemes, silky-strigulose and hispidulous with shining brown, deciduous, stinging hairs; pedicels long, pendulous, the flowers borne 2–3 together on sessile tubercles along abruptly flexuous axes; bracts subtending tubercles foliaceous, caducous; pedicels present; bracteoles lacking. Flowers large; hypanthium turbinate; calyx ellipsoid in bud, at anthesis campanulately expanded beyond hypanthium, the orifice 2-lipped, the 2 adaxial (to inflorescence axis) teeth united to apex or nearly so, the 3 abaxial very short, lanceolate or depressed-deltate; petals ample, lurid-purple, nigrescent, the wings and keel a little surpassing vexillum; stamens 10, monadelphous, the anthers alternately versatile and basifixed. Fruits pods, sessile, broad-linear, biconvex, the valves woody, obliquely ridged, hispid, with caducous, stiff, brown stinging hairs. Seeds 2–4, plumpy discoid, with linear hilum around ± 3/4 periphery.

Mucuna urens (L.) DC.

Lianas. Leaves: leafstalks 6–12 cm long; petioles 1.5–2 times as long as rachises; stipels either setiform or lacking; blades oblong- and ovate-elliptic, the pair postically dilated below middle, the terminal one equilateral and a little longer, 8–14 × 4–6.5 cm, the apex shortly acuminate. Inflorescences: peduncles 5–12 dm long, the floriferous axes at first condensed, early elongating and ultimately 10–25 cm long; bracts ovate-acuminate, submembranous, caducous; pedicels 1.5–4 cm long. Flowers: hypanthium, including calyx, 8–12 × 16–20 mm; wing- and keel-petals 3.5–4 cm long. Fruits straight, ± 8–18 × 5 cm. Fl (Jan, Feb), fr (Feb, May); climbing within and below canopy in nonflooded forest. *Olho de boi* (Portuguese), *zieu bourrique* (Créole).

ORMOSIA Jacks.

Trees, puberulent or silky on young growth with brownish or sordid hairs. Stipules small, caducous. Leaves imparipinnate; stipels lacking. Inflorescences terminal, efoliate panicles; bracts present; pedicels bibracteate at apex, the flowers solitary. Flowers papilionaceous; hypanthium turbinate, basally stipitate, articulate with pedicel; calyx obliquely campanulate, coriaceous, broadly 5-toothed; petals free, the vexillum short-clawed, the blade basally cordate, unappendaged, the keel-blades free, shorter than wings; stamens 10, free beyond hypanthium rim, the anthers subisomorphic; ovary shortly stipitate, the style filiform, the stigma small, latero-terminal. Fruits pods, woody, oblong-oblanceolate in profile, biconvex over seeds and somewhat constricted between them, valvately dehiscent through both sutures. Seeds 1–4, hard, plump, the testa black, red, or particolored, the hilum terminal.

Ormosia flava (Ducke) Rudd

Trees, 25–35 m tall. Trunks with branched buttresses. Young stems silky-pubescent. Leaves ample; leafstalks 9–19 cm long, silky-pubescent; pulvinules 4–7 mm long; leaflets 5–9 per leaf; blades ovate-elliptic, the larger ones 8–15 × 4–7.5 cm, chartaceous, glabrous, the base either cuneate or rounded, the apex shortly acuminate; secondary veins in ≤10 pairs, not connected by scalariform tertiary veinlets. Inflorescences terminal, broad panicles, the primary axes 10–16 cm long; bracts deltate, ≤1 mm long, caducous; pedicels 2.5–3.5 mm long; bracteoles deltate, ≤1 mm long, caducous. Flowers: hypanthium, including stipe, 4.5–5.5 mm long; calyx puberulent abaxially, white-silky adaxially, the tube campanulate, 7–8 × 6–7 mm, the teeth deltate, 2.5–3 mm long; petals yellow, red-eyed, the vexillum 14–17 mm long; ovary silky-puberulent, the style 7–8 mm long. Fruits obovate-cuneate or oblanceolate in profile, 3–5 × 1.2–2 cm, 1–2(3)-seeded, the valves woody, biconvex over each seed. Seeds plump, 9–15 mm long, lustrous black. Fl (Sep), fr (Sep, immature); in nonflooded forest. *Saint-Martin* (Créole).

PACHYRRHIZUS DC.

Herbaceous or basally woody twiners, strigose with subappressed, basifixed, straight hairs. Roots massive, tuberous (edible). Stipules linear-lanceolate, small, caducous. Leaves pinnately trifoliolate; stipels present; blades angular (entire) or potentially lobed. Inflorescences axillary, pseudoracemose, erect, exceeding subtending leaves, the secondary axes very short but not adnate or tuberculate, 2–6-flowered; bracteoles 2, caducous. Flowers: hypanthium lacking; calyx campanulate, the teeth little shorter than tube, the 2 adaxial (to inflorescence axis) united almost to apex; vexillum moderately reflexed, lacking appendages, the keel obtuse, neither coiled nor beaked; stamens diadelphous; style transversally dilated and strongly incurved, the stigma anterior near apex. Fruits pods, linear-oblong, straight, compressed, depressed between seeds, the valves coriaceous, coiled after dehiscence, the cavity narrowly partitioned between seeds. Seeds ± 6–10, compressed but plump, the testa smooth, tan or red-brown, the funicles broad, short.

Pachyrrhizus erosus (L.) Urb.

Stems twining or diffuse and tangled, to 6 m long or more, basally woody with age. Tubers turnip- or carrot-shaped, to 30 cm diam. Stipules 5–10 mm long. Leaves: leafstalks 3–15 cm long, the rachis 1.5–6 cm long; blades ± 3–18 cm wide, the lateral pair ovate-triangular or rhombic, the base broadly flabellate, the margins subentire or angular; venation slenderly prominulous abaxially. Inflorescences:

axes including peduncles ± 4–40 cm long, the secondary axes very short, 2–5-flowered; pedicels ≤5 mm long. Flowers: sepals ± 10 mm long; vexillum with flare near base of blade, ± 20 mm long, lilac-purple; ovary densely pubescent. Fruits 8–14 × 1–1.8 cm, the valves densely strigose. Seeds ± 5–10.5 mm diam. Fl (Feb); in disturbed areas, perhaps escaped from cultivation, persistent in abandoned gardens and sometimes naturalized. Native of tropical lowland Mexico (southward from Jalisco and Tamaulipas) and Central America, extensively cultivated in the New and Old World Tropics for the edible root known as *jicama* (Spanish) or *yam-bean* (English).

PHASEOLUS L.

Subherbaceus or somewhat woody, twining vines, pubescent with simple, straight or evenly curved hairs mixed with small, abruptly hooked hairs, sometimes with gland-tipped setules. Roots fibrous or tuberous. Stipules basifixed, often deflexed, commonly several-veined. Leaves pinnately trifoliolate; stipels present; blades palmately veined from petiolule. Inflorescences pseudoracemose, the secondary axes small, sessile tubercles, the flowers borne solitary or in pairs; pedicels 2-bracteolate. Flowers papilionaceous; calyx campanulate, 5-toothed; vexillum commonly pink or purple with eyespot, strongly recurved, the wing-petals as long or longer, the keel-petal blade rostrate, the beak coiled through ≥ full circle; stamens diadelphous (9 + 1), the anthers uniform; style coiled within keel-beak, barbellate on inner face. Fruits pods, sessile, linear, linear-oblanceolate, or narrow-oblong, usually gently incurved, compressed but becoming biconvex when ripe, dehiscent through both sutures, the valves coiling.

Phaseolus lunatus L.

Herbaceous, erratically twining vines, surfaces thinly setulose, the fine setules mixed with few small hooked hairs. Stipules triangular-lanceolate, 2–6 mm long, submembranous. Leaves: petioles 5–11 cm long; rachis 1.5–2 cm long; stipels 4–9 mm long; blades ovate, the terminal one 4.5–9.5 × 3–5.5 cm, the opposite pair inequilateral, the base broad-cuneate, the apex shortly acuminate. Inflorescences: peduncles and rachises together 3–15 cm long, the flowers either solitary or fasciculate. Flowers small; calyx campanulate, ± 2.5 mm long, the teeth deltate; vexillum ± 5–7 mm long, puberulent abaxially, the keel-petal incurved through nearly 360°. Fruits sessile, lunately oblanceolate in profile, 3–7 × 0.8–1.5 cm, planocompressed, the valves papery, coiled after dehiscence. Seeds 2–4. Fl (Sep), fr (sep, immature); weedy in gardens and waste places, a polymorphic species of wide distribution, cultivated for edible seeds. *Haricot-blanc* (French), *lima bean* (English), *Pois-savon*, *pois-sept-ans* (French).

PLATYMISCIUM Benth.

Trees of medium to large size. Young foliage coumarin-scented. Stems (sometimes hollow, inhabited by ants), leaves and racemose divisions of inflorescence all opposite or verticillate, all glabrous or inflorescences puberulent. Stipules small, evanescent. Leaves ample, imparipinnate; stipels lacking; pulvinules present; blades thinly chartaceous, lustrous, often darkening when dried; venation pinnate, reticulate. Inflorescences terminal, paniculate, the secondary axes racemose; bracts caducous; pedicels filiform, articulate, 2-bracteolate at apex; bracteoles caducous. Flowers small, papilionaceous; hypanthium turbinate; calyx symmetrically campanulate, shortly 5-toothed; petals little graduated, slenderly clawed, glabrous, yellow, the blade of vexillum obcordate, those of keel-petals valvately adherent distally; stamens 10, ordinarily monadelphous, the anthers obcordate, uniform; ovary stipitate, the style linear, glabrous; ovules 1. Fruits samaras, stipitate, planocompressed, the valves papery, glabrous. Seeds 1, reniform, flat.

Platymiscium ulei Harms PL. 58d

Trees, 15–22 m tall. Trunks with or without buttresses. Leaves: leafstalks 16–28 cm long, narrowly grooved; pulvinules 4–8 mm long; leaflets 7–9 per leaf; blades ovate-acuminate, the larger ones 12–17 × 4.5–7 cm, the base rounded. Inflorescences racemes, the axes ± 5 cm long; pedicels 3–6 mm long. Flowers: hypanthium, including stalk and calyx, ± 4 mm long; vexillum ± 11 mm long. Fruits: stipe of samara 5–12 mm long, the body in profile ± 10 × 4 cm, glabrous. Fl (Aug); in nonflooded forest. *Saint-Martin gris* (Créole).

POECILANTHE Benth.

Trees. Stems glabrous. Stipules evanescent or lacking. Leaves 1- to few-foliolate; leaflets either alternate or solitary; stipels minute or lacking; blades chartaceous, glabrous; venation pinnate. Inflorescences axillary and supra-axillary racemes, loose, sordid-puberulent; bracts present; pedicels transiently bibracteate. Flowers small, papilionaceous; calyx turbinate-campanulate, 4-toothed, the tooth opposed to vexillum minutely emarginate; petals dark purple, dark red, red brown, or yellow and purple, little graduated, the vexillum erect, flabellate beyond claw, the wing-petal blades dilated, obovate, the keel-petal blades distally adherent; stamens 10, monadelphous, the anthers versatile, alternately longer and shorter; ovary stipitate or subsessile. Fruits pods, oblanceolate or oblanceolate-ellipsoid, sessile or tapering into stipe, elastically dehiscent from apex downward, the valves stiffly coriaceous, brittle, weakly twisted, nigrescent, the seed cavity divided or not divided by papery partitions. Seeds discoid, the testa thin.

1. Leaflets 5 or 7 per leaf. Pods sessile, 3.5–4.5 × 1.3–1.7 cm, the cavity divided by papery partitions. . . . . *P. effusa*.
1. Leaflets solitary, terminal. Pods stipitate, 12–14 × 2–2.2 cm, the cavity not divided by papery partitions. . . . *P. hostmannii*.

Poecilanthe effusa (Huber) Ducke

Trees, (4)8–20 m × 10–30 cm. Trunks unbuttressed, the wood hard, yellow. Stipules minute, caducous. Leaves ample, imparipinnate, glabrous; leafstalks ± 15–25 cm long; pulvinules 3–6 mm long, wrinkled, discolored; leaflets 5 or 7, the lateral ones usually alternate; blades ovate or lance-elliptic, the distal ones longest, 8–18 × 3.5–8.5 cm, 2–3 times as long as wide, thin-textured, the base rounded or broad-cuneate, the apex acuminate; midrib prominent abaxially, the secondary venation pinnate, in 6–9 pairs, incurved well short of margin. Inflorescences axillary from annotinous or lately fallen leaves, racemose-paniculate or simply racemose, gray-puberulent, the primary axes ≤2 cm long, the secondary axes 3–12 cm long, loosely flowered; bracts minute; pedicels 1–4 mm long; bracteoles minute. Flowers small; hypanthium including calyx tube ± 2 mm long, the whole calyx 5–6 mm long, gray-puberulent; petals dark red, the vexillum ± 6 mm long. Fruits plumply oblanceolate-ellipsoid, 3.5–4.5 × 1.3–1.7 cm, sessile, biconvex, the valves woody, smooth, glabrous, dark brown, the cavity divided by papery partitions. Seeds 2–4. Fl (Aug); in nonflooded forest.

Poecilanthe hostmannii (Benth.) Amshoff

Trees, 4–8 m tall. Leaves unifoliolate; leafstalks 1–3.5 cm long, sulcate adaxially; pulvinules 4–7 mm long; blades broadly ovate-elliptic, 11–27 × 5–10 cm, chartaceous, the apex shortly acuminate; midrib prominent, the secondary veins in ± 10 pairs, the tertiary veins sharply reticulate. Inflorescences racemes, the axes ± 5–10 cm long; pedicels 2–3 mm long. Flowers: calyx ± 7 mm long, the tube turbinate-campanulate, the teeth narrowly lanceolate, little longer than tube; petals dark red, the vexillum ± 9 mm long. Fruits oblanceolate, including stipe 12–14 × 2–2.2 cm, straight, glabrous, the cavity not divided by papery partitions. Fl (Jul), fr (Feb); in nonflooded forest.

PTEROCARPUS Jacq.

Trees and arborescent shrubs, pubescent with simple hairs or subglabrous. Trunks commonly buttressed. Exudate red. Stipules small, caducous. Leaves imparipinnate; stipels lacking; pulvinules slender; leaflets alternate; blades with secondary venation pinnate, divergent at wide angle from midrib, the tertiary venation reticulate. Inflorescences simple or few-branched racemes, the flowers solitary; bracts caducous; pedicels 2-bracteolate in bud. Flowers papilionaceous; hypanthium turbinate; calyx campanulate, 5-toothed, the tube little longer and more convex on adaxial (to inflorescence axis) side, the 2 adaxial teeth united more distally than rest; petals orange-yellow or violaceous, with contrasting eye-spot, the claws slender, the blades broadly dilated; stamens 10, the filaments variously united, in our area monadelphous, the anthers basifixed, dehiscent by lateral slits; ovary sessile or stipitate; ovules 2–6, only 1–2 maturing. Fruits nuts or samaras, indehiscent, in profile obliquely ovate or orbicular, the sutures either carinate or winged.

Pterocarpus rohrii Vahl PL. 58e,f

Trees, to 40 m × 40 cm, glabrous, except for inflorescences. Trunks buttressed. Leaves: leafstalks 6–15 cm long; pulvinules 3–6 mm long; leaflets (4)5–9 per leaf; blades ovate or broad-lanceolate, slightly accrescent distally, the distal ones 6–10(12) × 2.5–4.5(5) cm, the base broadly obtuse or shallowly cordate, the apex bluntly shortly acuminate. Inflorescence axes ≤12 cm long, brown- or golden-brown-pilosulous; pedicels 3–6 mm long. Flowers: hypanthium including calyx 5–7 mm long; petals orange-yellow with red eye-spot, the vexillum 11–14.5 mm long, the wings and keel little shorter. Fruits samaras, in profile suborbicular, 5–8 cm diam., sessile, planocompressed, puberulent but early glabrate, the seed-cavity centric, 1.4–2 cm diam., low-convex and finely venulose, the wing chartaceous, opaque, ± 2–4 cm wide all around, obliquely deltate-apiculate near middle of adaxial edge. Fl (Aug); in nonflooded forest. *Moutouchi*, *moutouchi rubanée* (Créole).

PUERARIA DC.

Vines, sometimes basally woody, eglandular. Stems pilose with spreading or loosely deflexed, straight setiform hairs. Stipules lanceolate, basifixed, striately several-veined, deciduous. Leaves pinnately trifoliolate, distichous; stipels present; petiolules present; blades ovate-flabellate, the terminal one symmetric, the proximal pair broader and more abruptly angular on proximal side, all pilose with spreading or loosely deflexed, straight setiform or subappressed-strigulose hairs; venation 3-veined from base and thence pinnately veined, the secondary veins reaching margins. Inflorescences axillary, pseudoracemose, the primary axes erect, including peduncles mostly longer than subtending leaves, the secondary axes short, adnate, tuberculate, ≥4-flowered; bracts striate, caducous; bracteoles 2, caducous. Flowers: calyx campanulate, the tube longer than teeth, the teeth ovate-acuminate, the 2 abaxial teeth united ± halfway; petals glabrous, lilac-purple, pallid proximally, the vexillum recurved through 50–80°, the wings and keel slightly shorter, laterally adnate, broadly obtuse; stamens 10, the filaments united but vexillary one sometimes shortly so; ovary sessile, strigulose. Fruits pods, narrowly linear, deflexed, sessile, straight or nearly so, laterally compressed but low-convex facially, elastically dehiscent, the valves coiling, stiffly papery, fuscous, with straight hairs, the seed-cavity interrupted by interseminal septa. Seeds uniseriate, the testa fuscous.

Pueraria phaseoloides (Roxb.) Benth.

Vines, 1.5–7 m long. Stems sinuous but not tightly coiling, opportunistically scandent or diffusely tangled. Leaves: leafstalks ± 7–14 cm long; petioles 6–10 cm long, the one interfoliolar segment ± 1.5–4 cm long; blades to 7–14 × 8–12 cm, the margins either entire or shallowly undulate-lobulate. Inflorescences: peduncles and raceme axes together (1)1.5–4 dm long. Flowers: sepals 4.5–6 mm long, pilosulous or strigulose; vexillum 14–17 mm long, the wings and keel only slightly shorter. Fruits 7–11.5 × 0.35–0.45 cm. Seeds ± 2.5 cm long. Fl (Nov); scrambling over herbs and bushes in weedy habits, especially in gardens; native of tropical Asia, sporadically naturalized in tropical South America and West Indies. *Kudzu-vine* (English).

RHYNCHOSIA Lour.

Vines and twining herbs of indefinite length, puberulent or pilosulous with simple hairs intermixed when young, especially on abaxial surfaces of leaflets and vexillum, with small, spherical, yellow, orange or fuscous glands. Stipules ovate or linear-lanceolate, caducous, leaving small, transverse scar. Leaves pinnately 3-foliolate; stipels present or lacking; blades of lateral pair inequilaterally distended on proximal side; venation palmate, 3-veined from pulvinule, the midrib pinnately branched. Inflorescences axillary, racemose; bracts transient; pedicels ebracteolate. Flowers papilionaceous; calyx campanulate, 5-dentate, the abaxial tooth longest; petals yellow, purple- or fuscous-veined, the vexillum erect; stamens 10, diadelphous, the anthers uniform. Fruits pods, sessile, oblong or oblanceolate, valvately dehiscent, the valves stiffly papery, convex over each seed. Seeds 2.

1. Terminal leaflets 2–5 cm long. Fruits oblanceolate, 3–4 mm wide, obscurely constricted between seeds. Seeds brown or black, sometimes mottled. . . . . *R. minima.*
1. Terminal leaflets 8–14 cm long. Fruits oblong, 8–11 mm wide, constricted at middle. Seeds proximal half scarlet, distal half black. . . . . *R. phaseoloides.*

Rhynchosia minima (L.) DC.

Vines, pilosulous throughout or glabrate, densely gland-sprinkled. Stems slender, decumbent or twining from somewhat woody caudex, of indefinite length. Leaves: leafstalks 2–8 cm long; blades rhombic or rhombic-ovate, the terminal one commonly 2–5 cm long, ± as wide, the base broadly cuneate or subtruncate, the apex deltately acute. Inflorescences: peduncles and raceme axes together 2–20 cm long; pedicels 0.7–2.5 mm long. Flowers: calyx 3–6 mm long, the tube 1.5–2 mm long, the teeth lanceolate, nearly as long as corolla; vexillum 4–7 mm long. Fruits in profile falcately oblanceolate, 9–17 × 3–4 mm, decurved, obscurely constricted between seeds, the valves pilosulous and gland-sprinkled. Seeds brown or black, sometimes mottled. To be expected from our area but not yet collected; in waste places, fields, roadsides.

Rhynchosia phaseoloides (Sw.) DC. FIG. 126, PL. 59b,e

Vines. Stems terete in transverse section when young, becoming woody and ribbon-like with age. Leaves: leafstalks 7–15 cm long; blades ovate or obtusely rhombic, the terminal one 8–14 × 7–12 cm, the base rounded or shallowly cordate, the apex acuminate. Inflorescences erect, the axes including short peduncles 10–30 cm long; pedicels ≤1 mm long. Flowers: sepals ± 3.5 mm long; vexillum ± 7 mm long. Fruits in profile oblong, 12–24 × 8–11 mm, straight, constricted at middle, the valves brown, densely puberulent, gland-sprinkled. Seeds ± 4 × 5 mm, the testa with proximal half scarlet, distal half black. Fl (Jun, Aug, Oct), fr (Aug, Oct); in open places, especially at forest margins.

SWARTZIA Schreb.

Trees, pubescent with short simple hairs when young or glabrous. Exudate resinous, often reddish. Stipules present. Leaves (1)3–11-foliolate; rachis either terete or green-marginate; foliar glands lacking; pulvini and pulvinules present; leaflets opposite, the terminal leaflet stalked. Inflorescences arising either from axils of contemporary or year-old and older leaves, or from knots on trunk or stems, simple (then solitary or fasciculate) or paniculate racemes; pedicels either 2-bracteolate or naked. Flowers: calyx in bud subglobose, entire, at anthesis splitting valvately into 4–5 recurved, usually persistent lobes; petal either 1 (adaxial to inflorescence axis) or lacking, either white or yellow, the blade orbicular-flabellate beyond short claw; stamens indefinite in number, 2 or more abaxial ones often larger, the anthers dorsifixed, the filaments much longer than anthers; ovary stipitate. Fruits usually pods, stipitate, the body ovoid, ellipsoid, or oblong-moniliform, dehiscence sometimes tardy or fruits rarely indehiscent. Seeds 1 to few, arillate, the aril white or colored.

Cowan, R. S. 1968. *Swartzia* (Leguminosae, Caesalpinioideae Swartzieae). Fl. Neotrop. Monogr. **1:** 1–228.

1. Leaflets 3 per leaf. Racemes axillary to contemporary leaves, the primary axes ≤3 cm long, 2–5-flowered; pedicels filiform, 10–23 mm long. . . . . *S. arborescens.*
1. Leaflets 5–13 per leaf. Racemes axillary to year-old leaves or cauliflorous, the primary axes 6–20 cm long, ≥10-flowered; pedicels either <10 mm long or stout.
    2. Stipules obovate-semicircular, deciduous, leaving crescentic, semiamplexicaulous scar. . . . *S. panacoco* var. *sagotii.*
    2. Stipules linear-subulate or triangular, deciduous, scar not as above, usually small, round.
        3. Leafstalks with ascending auricle on each side at insertion of leaflets.
            4. Pedicels ebracteolate. Petal lacking. . . . . *S. amshoffiana.*
            4. Pedicels minutely 2-bracteolate at apex. Petal present. . . . . *Swartzia* sp. B.
        3. Leafstalks not auriculate.
            6. Leaflets 9–11 per leaf. Racemes simple. Petal white. Fruits densely lenticellate. . . . . *S. polyphylla.*
            6. Leaflets 5 per leaf. Racemes paniculately branched. Petal and fruits unknown. . . . . *Swartzia* sp. A.

Swartzia amshoffiana R. S. Cowan PL. 59

Trees, 16–32 m × 20–35 cm. Minutely puberulent, except for glabrous leaflets. Stipules linear, 0.5–2.5 mm long, deciduous. Leaves: leafstalks 7–11 cm long, not winged but with auricle at base of each pulvinule, the auricle linear, 1–3 mm long, erect; pulvinules 2–4 mm long; leaflets 7 per leaf; blades ovate-elliptic, the larger ones 8–12 × 3–5 cm, lustrous, chartaceous, the base rounded or broad-cuneate, the apex shortly acuminate; venation reticulate on both surfaces. Inflorescences arising singly or 2–3 together, axillary

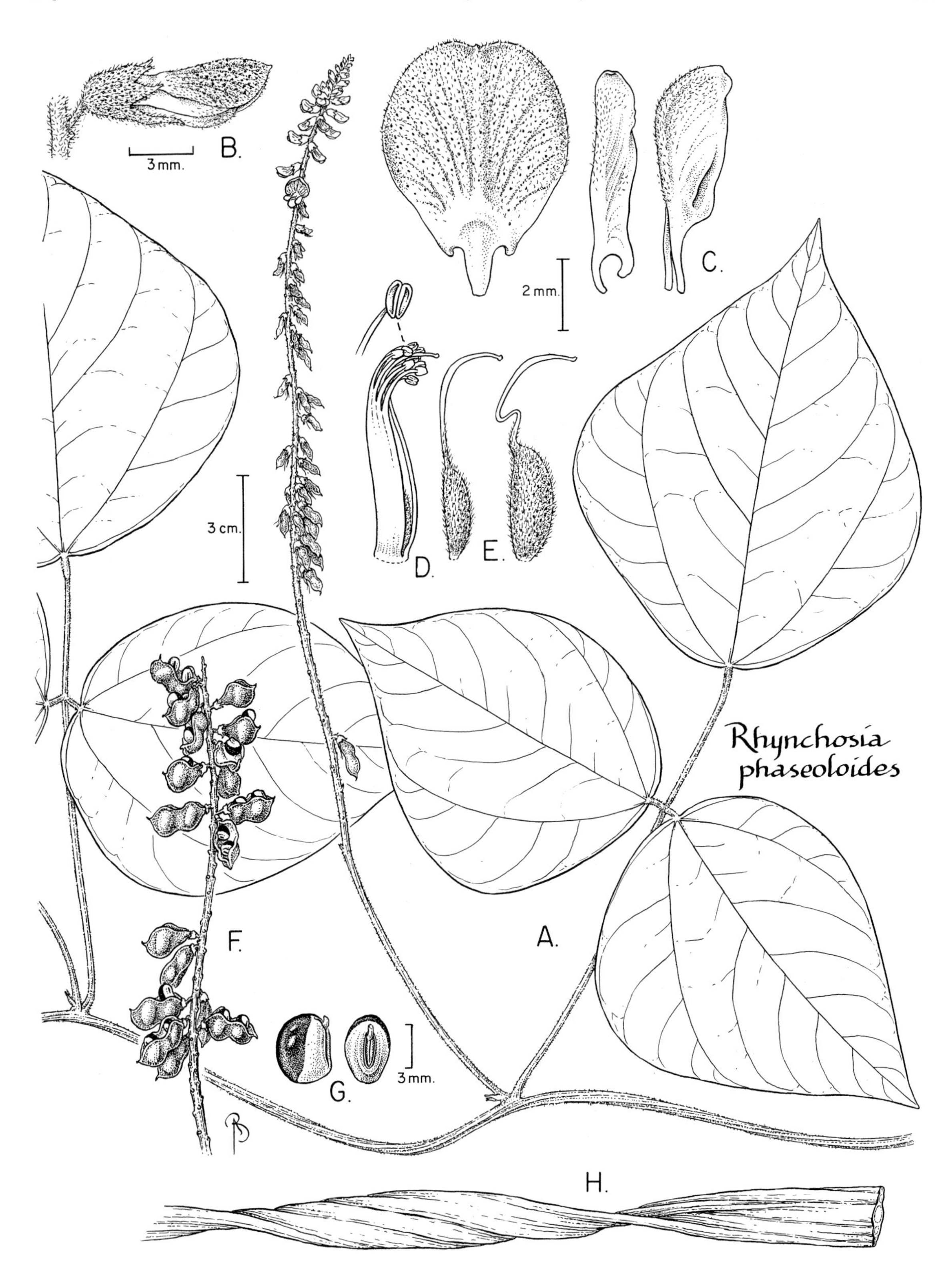

FIG. 126. FABACEAE. *Rhynchosia phaseoloides* (A–E, *Marshall & Rombold 172*; F, G, *Mori & Gracie 19015*; H, *Acevedo-Rodríguez 9365* from Puerto Rico). **A.** Part of stem with leaves and inflorescence. **B.** Lateral view of flower bud. **C.** Three petals: standard (far left), wing (middle left), and keel (near left). **D.** Androecium with detail of anther. **E.** Pistils. **F.** Infructescence. **G.** Two views of seeds. **H.** Part of twisted liana.

to year-old leaves or cauliflorous, racemes, the axes 6–12 cm long; pedicels 3–7 mm long, ebracteolate. Flowers small; buds globose, 4–4.5 mm diam., minutely strigulose; sepals 4.5–6 mm long, glabrous adaxially; petals lacking; larger stamens 2; ovary glabrous, the stipe 1–1.5 mm long, the style infraterminal, 0.4–0.7 mm long. Fruits plumply ovoid-ellipsoid, 17–22 × 10–13 × 5 mm, broadly obtuse. Fl (May, Aug), fr (Aug, immature, Sep, immature, Nov); in nonflooded forest. *Bois-Corbeau* (Créole).

Swartzia arborescens (Aubl.) Pittier

Trees, 15–20 m × 15–20 cm. Trunks buttressed. Stems thinly pilosulous. Stipules linear, 1–2.5 mm long. Leaves: leafstalks ± 1.5–3 cm long, green-winged, shortly auricled at insertion of leaflets; leaflets 3 per leaf; blades ovate, of unequal size, the terminal one ± 5–9 × 3–5 cm, about 2× as long as lateral pair, the apex shortly acuminate. Inflorescences axillary to contemporary leaves, racemes, the axes very slender, 1–3 cm long, 2–5-flowered; pedicels filiform, 10–23 mm long, ebracteolate. Flowers: buds plumply ellipsoid, ± 4.5 mm long; sepals 3–5.5 mm long, deciduous; petal flabelliform, short-clawed, ± 6–12 mm long, yellow; functional stamens 10–15, isomorphic; ovary glabrous, the stipe 2–2.5 mm long, the style terminal, ± 2 mm long. Fruits ellipsoid, ± 3–4.5 × 1.7–2.5 cm, attenuate at each end, compressed but plump, glabrous. Seeds 1. Fl (Jun, Aug, Sep); in nonflooded forest.

Swartzia panacoco (Aubl.) R. S. Cowan var. **sagotii** (Sandwith) R. S. Cowan Pl. 59c

Trees, 20–32 m × 20–35 cm. Trunks sometimes buttressed. Pubescence brown-tomentulose, except for adaxial surface of leaflet blades. Young stems sulcate. Stipules obovate-semicircular, 6–13 mm long, deciduous from narrowly crescentic, semiamplexicaulous scar. Leaves: leafstalks 14–31 cm long, subterete; pulvinules 2–5 mm long; leaflets 7–9(11) per leaf; blades ovate-oblong or oblong-elliptic, larger distally, the distal pair 14–22 × 6–8.5 cm, bicolored, the apex abruptly deltate-acuminate, the margin either plane or revolute; venation prominently pinnate abaxially. Inflorescences cauliflorous, at defoliate nodes on branches or on trunk, racemose, the axes 11–17 cm long; bracts broad-ovate, 2.5–4 mm long; pedicels 15–24 mm long, stout, ebracteolate. Flowers large; buds globose, ± 1 cm diam.; calyx lobes 9–13 mm long, thick-coriaceous; petal including claw 16–27 mm long, pilosulous abaxially, yellow; filaments of larger stamens pilosulous toward base; ovary narrowly ellipsoid, 10–13 mm long, densely velutinous, the gynophore ± 7–9 mm long, glabrescent, the style ± 7–9 mm long, glabrous. Fruits: body either obovoid 1-seeded or oblong several-seeded, in profile ± 6–12 × 3.5 cm, the valves woody, obliquely rugose. Fl (Oct, Nov, Dec), fr (Mar, Jun, Oct, immature, Nov, immature); in nonflooded forest. *Angelique* (French and Créole).

Swartzia polyphylla DC. Fig. 127, Pl. 59d

Trees, 20–35 m × 50–80 cm. Trunks fluted, sometimes buttressed at base. Young stems strigulose with short subappressed, gray or brown hairs. Stipules triangular, ≤1 mm long, caducous. Leaves: leafstalks 10–17 cm long, not marginate, strigulose with short subappressed, gray or brown hairs; pulvinules 3.5–5 mm long; leaflets 9–11 per leaf; blades ovate, the larger ones 6–11 × 2–4.5(5), chartaceous, bicolored, glabrous, the base rounded or broad-cuneate, the apex shortly acuminate. Inflorescences solitary, fasciculate from leaf-axils ≥1 year old, racemose, the axes 6–14 cm long, strigulose with short subappressed, gray or brown hairs; pedicels 4–7 mm long; bracteoles 2, at or distal to middle. Flowers: buds subglobose, 4–5 mm diam., puberulent; sepals becoming 6–7.5 mm long, glabrous adaxially; petal flabellate, including claw 11–17 mm long, white; stamens dimorphic, glabrous; ovary lunately elliptic, 5–8 mm long, glabrous, the stipe 5–12 mm long, the style filiform, 4–8 mm long, incurved. Fruits obovoid or oblong-obovoid, 7–12 × 4–5 cm, the valves woody, densely lenticellate, brown. Seeds 1, then fruits obovoid or 2 and then fruits oblong-obovoid. Fl (Sep); in nonflooded forest. *Bois-corbeau*, *bois pagaïe* (Créole).

Swartzia (ser. *Benthamianae*) sp. **A**

Trees, 25–30 m tall. Subglabrous except for golden-silky racemose-paniculate inflorescences. Stipules linear, 1.5 mm long, caducous. Leaves: leafstalks 4–9 cm long, terete; leaflets 5 per leaf; blades elliptic, 6–11 × 2–4.5 cm, the apex acuminate. Inflorescences axillary to older leaves, paniculately branched racemes; pedicels 2-bracteolate. Known from our area by 2 collections, one sterile, one in young bud. Bud (Dec); in nonflooded forest at all elevations (*Boom & Mori 2388* and *Mori et al. 8773*, both at NY).

Swartzia (subsect. *Terminales*) sp. **B**

Trees, ± 7 m × 10 cm. Bark gray but white- and black-spotted. Stipules deciduous, leaving small round scar, the blade unknown. Leaves ample; leafstalks 13–19 cm long, narrowly green-marginate distal to middle, 4–6 mm wide at distal leaflet-pulvinules and prolonged on each side into ascending auricle; pulvinules ± 1.5 mm long; leaflets 11–13 per leaf; lateral blades oblong or elliptic, larger upward, the distal pair ± 9–15 × 3–5 cm, thin-textured, the base obtuse, the apex acuminate; midrib adaxially prominulous, the secondary veins in 6–10 pairs. Inflorescences cauliflorous, simply racemose, the axes including peduncles 15–17 cm long, densely, shortly brown-pilosulous; pedicels 10–27 mm long, stout, compressed, dilated upward, minutely 2-bracteolate at apex. Flowers: buds ovoid-ellipsoid, angular, rugulose; sepals 4, ≤12 mm long, glabrous adaxially; vexillum (not seen entire) yellow, adaxially pilosulous; stamens glabrous; ovary (in bud) pubescent, the style not seen expanded. Fruits unknown. Fl (May, Oct); in nonflooded forest forest near Saül on Carbet Maïs and Crique Limonade trails (*Granville B.4468, B.4563*, both at CAY), this incomplete material is suggestive of *S. longicarpa* Amshoff.

TARALEA Aubl.

Trees. Trunks buttressed. Stipules lacking, or ephemeral. Leaves ample, opposite; leafstalks dorsiventrally compressed, sometimes narrowly winged on each side proximal to insertion of leaflets; leaflets paripinnate, either alternate or subopposite; blades equilateral, chartaceous, glabrous. Inflorescences terminal, paniculate, pyramidal, finely puberulent, the secondary axes transiently bracteate; pedicels transiently 2-bracteolate. Flowers: hypanthium broadly turbinate, ± as long as calyx tube; calyx bilabiate, the tube shortly campanulate, the lobes of adaxial (to inflorescence axis) lip broadly oblong, obtuse, plane, clasping young corolla, the lower lip minute, 3-denticulate; petals rose-purple, the vexillum strongly recurved, the wings longer, the blade 2-lobed; stamens 10, monadelphous; ovary short-stipitate; ovule 1. Fruits pods, obovate-elliptic in profile, strongly compressed, dehiscent through both sutures, the valves twisting, becoming thin-woody, brittle, the adaxial suture narrowly winged distally. Seeds discoid, the testa brittle. Species of riverine forest.

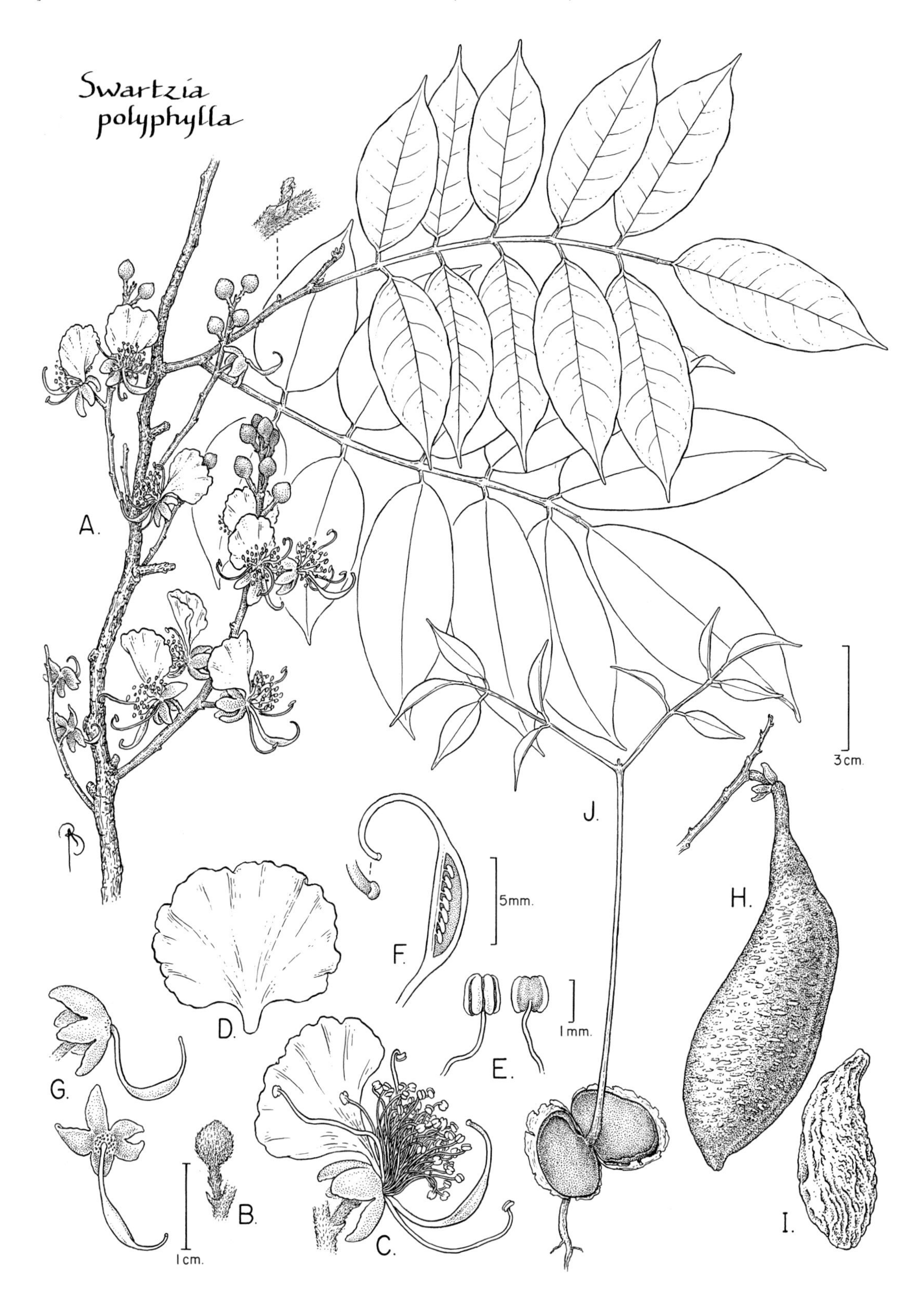

FIG. 127. FABACEAE. *Swartzia polyphylla* (A–G, *Mori et al. 20989*; H, I, *Archer 7854* from Amazonian Brazil; J, *Mori et al. 15663*). **A.** Part of stem with leaves and cauliflorous inflorescences (right) and detail of lateral bud showing stipule (above). **B.** Flower bud. **C.** Lateral view of flower. **D.** Single petal. **E.** Two views of stamens. **F.** Medial section of pistil (right) with detail of tip of style and stigma (above left). **G.** Lateral view (above) and oblique view (below) of calyx and ovary. **H.** Fruit. **I.** Seed. **J.** Seedling.

Taralea oppositifolia Aubl.

Trees, 9–25 m × 30–70 cm. Leaves: leafstalks 8–18 cm long; leaflets (4)6–8 per leaf; pulvinules 4–7 mm long; blades ovate- or oblong-elliptic, the longer ones 9–15 × 3.5–6 cm, the base rounded or broad-cuneate, the apex shortly acuminate; midrib adaxially depressed, abaxially carinate. Inflorescences: primary axes 15–35 cm long; pedicels 2–3 mm long. Flowers: sepals 9–11 mm long; corolla with wings exserted by ± 3 mm; ovary white-silky. Fruits in profile 5–7 × 2.3–3.5 cm, cuneately contracted at each end, the valves dark brown, glabrate. Fl (Nov); seen only once along small stream. *Bois-crapaud, bois violet, gagnac, gagnac rivière, gaïac, gaïac de l'eau* (Créole).

VATAIREA Aubl.

Trees, deciduous, emergent, flowering when leafless. Trunks buttressed. Stipules small, ephemeral. Leaves imparipinnate; stipels lacking; leaflets alternate; blades chartaceous, glabrous. Inflorescences terminal, effuse panicles, finely silky-puberulent; bracts present; pedicels present; bracteoles 2, near base of pedicel, small, caducous. Flowers papilionaceous; hypanthium ribbed, subsymmetrically turbinate-campanulate; calyx turbinate-campanulate, the orifice crenately 5-dentate; petals lavender-violet, the vexillum strongly recurved, not appendaged, the wing-petal blades entire, the keel-petal blades imbricate by exterior margins; stamens 10, monadelphous; ovary stipitate, longitudinally ridged over solitary ovule. Fruits samaras, the wing terminal, the seed-cavity basal, ellipsoid, low-carinate longitudinally.

Vatairea paraensis Ducke

Trees, 25–45 m × ≤150 cm. Leaves: leafstalks 8–14 cm long; leaflets 5–7(9) per leaf; pulvinules 4–7 mm long; blades obovate or oblong-obovate, the larger ones 6–12 × 3–5 cm, the base broad-cuneate or rounded, the apex emarginate. Inflorescences: primary axis 12–25 cm long; pedicels filiform, 4.5–8 mm long. Flowers: hypanthium 1.8–3.5 mm long; calyx 5.5–8.5 mm long, blackish-purple; vexillum 12–19 mm long; ovary brownish-velutinous, glabrescent. Fruits oblanceolate in profile, 7–9 cm long, the semi-niferous locule ± 2.5–3 cm long, the wing 2.5–3 cm wide, papery, transversely venulose, longitudinally 1-veined shortly within adaxial margin. Fl (Feb, Aug, Sep), fr (Sep); in nonflooded forest. *Djãgo* (Bush Negro).

VATAIREOPSIS Ducke

Trees, deciduous, emergent, flowering when leafless. Trunks buttressed. Stipules small, ephemeral. Leaves imparipinnate; stipels present; leaflets alternate; blades chartaceous, glabrous. Inflorescences terminal, effuse panicles, finely silky-puberulent, the secondary axes with bracts; pedicels present; bracteoles 2, near base of pedicel, small, caducous. Flowers papilionaceous; hypanthium ribbed, asymmetric, longer adaxially; calyx turbinate-campanulate, the orifice crenately 5-dentate; petals lavender-violet, the vexillum strongly recurved, not appendaged, the wing-petal blades entire, the keel-petal blades imbricate by exterior margins; stamens 10, monadelphous, the tube relatively short (shorter than in *Vatairea* spp.); ovary stipitate, incipiently winged over solitary ovule. Fruits samaras, the wing terminal, the seed-cavity basal, ellipsoid, winged longitudinally.

Vataireopsis surinamensis H. C. Lima
PL. 60a,b; PART 1: FIG. 6.

Trees, 18–35 m tall. Stems obese, lenticellate. Stipules ephemeral or lacking. Leaves crowded, plurifoliolate; leafstalks 25–55 cm long, finely pilosulous; stipels subulate, ≤1 mm long, inserted 1–2.5 mm proximal to each leaflet; pulvinules 3–4.5 mm long; leaflets alternate, 26–42 per leaf; blades narrowly oblong, those distal to mid-rachis 6–8 × 2–3 cm, thinly chartaceous, either glabrous or minutely strigulose, the base rounded, the apex shallowly emarginate; midrib centric, the secondary venation pinnate, in 8–12 pairs. Inflorescences terminal, paniculate, pyramidal, the primary axes 10–15 cm long; pedicels 3–5.5 mm long; bracteoles 2, at apex of pedicel. Flowers papilionaceous; hypanthium and calyx tube subequal; calyx 6–8 mm long, reddish-brown-tomentulose; petals blue-violet, the vexillum 14–16.5 mm long. Fruits 10–12.5 cm long, the seed-cavity 3.5–4 × 2–2.5 cm, with wing on each side, the wings forwardly appressed, fimbriate, 1–1.4 cm wide, the terminal wing broadly elliptic in profile, 2.7–3.2 cm wide, glabrate. Fl (Feb), fr (Apr); in nonflooded forest. *Djãgo* (Bush Negro).

VIGNA Savi

Vines, pilose with straight setiform, often brownish hairs, those of leafstalks and peduncles retrorse. Stems weak, diffuse or randomly twining. Stipules ovate or lanceolate, thinly herbaceous, the base shortly auriculate. Leaves pinnately 3-foliolate; stipels present. Inflorescences axillary, racemose, pedunculate, the primary axes long, erect, the secondary axes 2–5-flowered at apex; pedicels transiently bracteolate. Flowers: calyx campanulate, the tube submembranous, 15-veined, the teeth 5, lanceolate or triangular, subequal, the apex acuminate; corolla papilionaceous, whitish lavender-veined or bluish, the vexillum unappendaged, little longer than inner petals, the eye-spot yellowish, the keel-petal attenuate, incurved through nearly 3/4 circle; stamens diadelphous; style barbate on inner face, the stigma capitate. Fruits pods, sessile, linear, when mature biconvex, dehiscence through both sutures, the valves tightly coiling, chartaceous.

Vigna vexillata (L.) A. Rich.

Vines of indefinite length. Stipules 5–7-veined. Leaves: leafstalks 3.5–9 cm long, stalk of terminal leaflet ± 1–1.5 cm long; stipels linear-subulate, 1-veined; blades lance-ovate, the pair dilated at base on proximal side, the terminal one symmetrical, the longer blades 6–12 × 1.7–5 cm, the apex obtuse or shortly acuminate. Inflorescences: peduncles 13–26 cm long; pedicels 1–2 mm long. Flowers:

calyx 10–11 mm long, the teeth narrowly lanceolate, 3.5–5.5 mm long; petals ± 20–25 mm long. Fruits 7–9 × 0.35–0.45 cm, the valves dark brown, ascending-pilose. Fl (Jun, Sep), fr (Jun, Sep), probably flowering and fruiting year round; in weedy habitats.

ZORNIA J. F. Gmel.

Herbs, perennial or monocarpic. Stems slender. Stipules elliptic, peltate, attached below middle by adaxial face, deciduous. Leaves bifoliolate, sparsely, minutely gland-punctate. Inflorescences both axillary and terminal, spicate, sessile; bracts present; pedicels lacking; bracteoles 2, conspicuous, resembling stipules in attachment and texture, the anterior lobe appressed to and covering calyx. Flowers small; calyx campanulate, scarious, the teeth 5; corolla papilionaceous, yellow, the vexillum purple-veined, regularly graduated; stamens 10, monadelphous, the filaments alternately longer and shorter, the anthers of different lengths. Fruits loments, straight, narrow, 4–6-articulate, setose, indehiscent. Seeds 1 per article

Zornia latifolia Sm. PL. 60c

Herbs. Stems either erect or diffuse, ≥1 m long, freely branched, glabrous or remotely strigulose. Stipules caducous. Leaves: leafstalks ≤2 cm long; blades of longer leaves lance-ovate, 2–4.5 × 0.8–1.5 cm, those of distal (on stems) leaves linear-lanceolate or -elliptic, shorter and narrower. Inflorescences loosely 10–20-flowered, the axes ≤12 cm long; bracteoles ± 8–10 × 2 mm, 5-veined, ciliolate, gland-punctate. Flowers: sepals 3–3.5 mm long; vexillum 8–9 mm long. Fruits: articles ± 2 × 2 mm, the valves armed with barbellate setae ± 1 mm long. Fl (Aug), fr (Aug), probably flowering and fruiting year round; in weedy habitats. *Herbe canard, zerb canard* (Créole).

FLACOURTIACEAE (Flacourtia Family)

Scott A. Mori and Beat Fischer

Shrubs or trees. Leaves simple, alternate. Stipules usually present, often caducous, rarely absent (*Xylosma*). Inflorescences usually axillary, sometimes terminal, often in clusters, sometimes simple racemes, or paniculate arrangements of racemes. Flowers actinomorphic, bisexual or unisexual (plants dioecious); sepals 3–5, free or fused at base into calyx tube, sometimes petaloid (*Ryania*); petals absent or 3–9; stamens 5 to numerous, often flanked by staminodia; nectaries sometimes present; ovary superior, unilocular, rarely 2-locular by 2 deeply intruding placentae, the style 1 or divided; placentation parietal, the placentae 2–9. Fruits fleshy or dry, indehiscent or capsular, the pericarp sometimes winged or echinate. Seeds often arillate or surrounded by pulp; endosperm fleshy, abundant, the embryo straight, the cotyledons mostly cordate-foliaceous.

Jansen-Jacobs, M. J & A. R. A. Görts-van Rijn. 1986. Flacourtiaceae. *In* A. L. Stoffers & J. C. Lindeman (eds.), Flora of Suriname **III(1–2):** 430–451. E. J. Brill, Leiden.

Olson, M., Paul E. Berry & G. A. Aymard C. 1999. Flacourtiaceae. *In* J. A. Steyermark, P. E. Berry, K. Yatskievych & B. K. Holst (gen. eds.), Flora of the Venezuelan Guayana **5:** 434–474. P. E. Berry, K. Yatskievych & B. K. Holst (vol. eds.). Missouri Botanical Garden Press, St. Louis.

Sleumer, H. O. 1980. Flacourtiaceae. Fl. Neotrop. Monogr. **22:** 1–499.

——— & H. Uittien. 1966. Flacourtiaceae. *In* A. Pulle (ed.), Flora of Suriname **III (1):** 282–303. E. J. Brill, Leiden.

1. Inflorescences terminal, usually branched.
   2. Leaf blades pinnately veined, the margins markedly serrate; glands stalked, often present on leaf blade margins near petiole. Axes of inflorescence not in whorls. Sepals and petals 3. . . . . *Banara.*
   2. Leaf blades with three principal veins arching upward from base, the remaining secondary veins pinnately arranged, the margins entire; glands flattened, on adaxial leaf surface near petiole. Axes of infloresecence in whorls. Sepals and petals 4. . . . . *Hasseltia.*
1. Inflorescences axillary or from stem below current leaves, usually not branched.
   3. Sepals 3; petals 5–9.
      4. Leaf blades not densely pubescent abaxially, the margins entire to serrulate, not glandular. Stigmas 2–3. Fruits not winged, with or without slender bristles. . . . . *Mayna.*
      4. Leaf blades densely pubescent abaxially (in our area), especially when young, the margins with widely spaced, glandular teeth. Stigmas 6–7. Fruits winged. . . . . *Carpotroche.*
   3. Sepals 4–5; petals absent.
      5. Plants with well developed, branched spines on trunk and branches. . . . . *Xylosma.*
      5. Plants without spines.
         6. Sepals usually distinctly fused at bases into short calyx-tube; staminodia present. . . . . *Casearia.*
         6. Sepals not distinctly fused into short calyx-tube; staminodia absent.
            7. Leaf blades glabrous. Inflorescences of fascicles, supra-axillary or from along stem below leaves. Flowers <10 mm diam. . . . . *Laetia.*
            7. Leaf blades (especially on veins abaxially) with stellate hairs. Inflorescences of solitary flowers, axillary. Flowers >20 mm diam. . . . . *Ryania.*

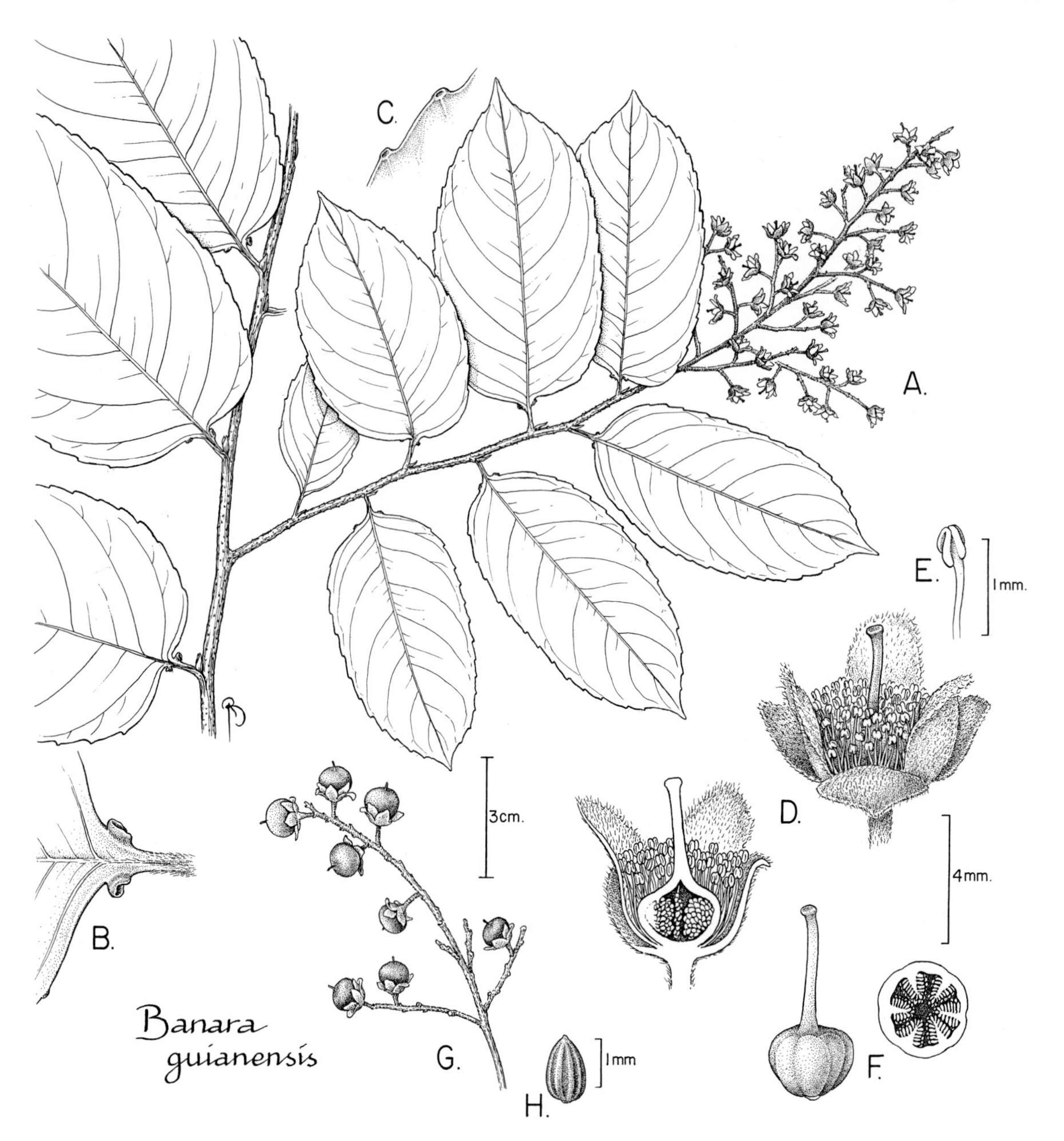

FIG. 128. FLACOURTIACEAE. *Banara guianensis* (A–F, *Hahn 3692*; G, H, *Mori & Pipoly 15618*). **A.** Apex of stem with leaves and terminal panicle. **B.** Base of leaf with glands. **C.** Margin of leaf blade with glandular teeth. **D.** Lateral view (right) and medial section (left) of flower showing parietal placentation of ovules. **E.** Stamen. **F.** Pistil (left) and transverse section (right) of ovary with placenta. **G.** Part of infructescence. **H.** Seed.

BANARA Aubl.

Small trees. Leaf blades with pinnate venation, the margins markedly serrate. Inflorescences terminal, once-branched, the ultimate axes racemose. Flowers bisexual; sepals 3, scarcely united at base; petals 3, scarcely united at base; stamens numerous, the filaments slender, the anthers very small; staminodia and glands absent; placentae 6–7, intruded. Fruits indehiscent. Seeds numerous, embedded in fleshy pulp.

Banara guianensis Aubl. FIG. 128, PL. 60e; PART 1: FIG. 4

Small trees, to 10 m tall. Young stems pubescent. Leaves: petioles 5–10 mm long, pubescent; blades narrowly ovate to oblong, 7–14 × 4.5–6 cm, pubescent abaxially, the base rounded to cordate, the margins distinctly serrate, often with 1–2 stalked glands near junction with petiole; secondary veins in (8)10–12 pairs. Flowers 7–10 mm diam.; sepals gray pubescent; petals gray pubescent, white; stamens with slender, glabrous filaments, the anthers ca. 0.3 mm long; ovary glabrous. Fruits globose, green, ca. 7–10 mm diam. Fl (Jan, Feb, Aug, Dec), fr (Jan, Feb, Apr, Aug); common, in secondary forest, especially along airport road, each plant producing abundant, aromatic flowers for a single day. *Mavévé sucrier* (Créole).

CARPOTROCHE Endl.

Shrubs or small trees. Leaf blades with pinnate venation, the margins entire to serrate. Inflorescences axillary, from stems below leaves, or from trunk. Staminate inflorescences short, generally few-flowered. Pistillate inflorescences usually 1-flowered. Flowers unisexual (plants dioecious); sepals 3; petals 6–9, free, the staminate flowers with numerous stamens, lacking rudiment of ovary, the anthers elongate, the filaments shorter than anthers; ovary usually with 4–8 vestigial ridges or wings, the stigmas 6–7; placentae 4–8. Fruits dehiscent, winged. Seeds generally numerous, embedded in pulp; endosperm copious; embryo straight, with foliaceous cotyledons.

Carpotroche crispidentata Ducke
FIG. 129, PL. 60d; PART 1: PL. XVIb

Shrubs or small trees, to 10 m tall. Young stems densely pubescent. Stipules linear, densely pubescent, ca. 10 mm long. Leaves: petioles 8–15 mm long; blades obovate to narrowly obovate, 13–24 × 6–10 cm, densely pubescent abaxially, especially when young, the margins with widely spaced glandular teeth; secondary veins in 10–12 pairs. Flowers ca. 20 mm diam.; sepals densely pubescent; petals densely pubescent, white; anthers yellow, pilose, 4 mm long. Fruits densely pubescent, with numerous, conspicuous outgrowths oriented in vertical ridges, dehiscing into 4–5 segments. Seeds gray, with red aril. Fl (Apr, May, Jun, Jul, Aug, Sep, Nov), fr (Jan, Feb, Mar, May, Jul, Aug, Sep, Oct, Nov, Dec); scattered, in nonflooded forest.

CASEARIA Jacq.

Shrubs or small trees. Leaf blades with pinnate venation, the margins entire to serrate. Inflorescences usually axillary, less frequently from stems below current leaves, in dense sessile or stalked clusters. Flowers bisexual, small (3–7 mm diam.), green to greenish-white or white; sepals 4–5, usually fused at base into short calyx-tube, infrequently divided to base; petals absent; stamens 5–15, often with longer stamens alternating with shorter ones, the long ones opposite sepal lobes, the short ones alternate sepal lobes, usually alternating with staminodia, sometimes in row separate from staminodia, both stamens and staminodia usually fused to sepal tube; style divided toward apex and stigmas separate (usually 3) or with single stigma, the stigmas usually capitate; ovary 3-locular. Fruits dry to succulent capsules, dehiscing into 3 valves. Seeds glabrous or pubescent, completely to partially enveloped by soft, often colored and fimbriate aril; endosperm fleshy, the cotyledons flat.

1. Style divided at apex, the stigmas separate.
    2. Leaf blades with lateral veins impressed adaxially. Staminodia forming distinct inner row separate from stamens.
        3. Shrubs or treelets <8 m tall at flowering. Leaf blades serrate. Flowers not exserted from inflorescence. Sepals not reflexed; stamens 10. . . . . *C. commersoniana.*
        3. Trees >10 m tall at flowering. Leaf blades entire to serrulate. Flowers exserted from inflorescence. Sepals reflexed; stamens ca. 15. . . . . *C. javitensis.*
    2. Leaf blades with lateral veins plane or salient adaxially. Staminodia not forming distinct inner row separate from stamens.
        4. Leaf blades drying brown, chartaceous, usually ≥9 cm long. Pedicels and sepals glabrous adaxially. . . . . *C. sylvestris* var. *sylvestris.*
        4. Leaf blades drying green, subcoriaceous, usually <9 cm long. Pedicels and sepals pubescent. . . . *Casearia* sp. B.
1. Style not divided at apex, the stigma simple.
    5. Leaf blade margins serrulate to serrate, especially toward apex.
        6. Inflorescences short pedunculate. Fruits more than 1/2 enclosed by calyx at maturity.
            7. Stems and leaf blades densely pubescent abaxially. Flowers not exserted from inflorescence, ca. 7 mm diam. . . . . *C. rusbyana.*
            7. Stems and leaf blades glabrous to puberulous abaxially. Flowers exserted from inflorescence, ca. 3 mm diam. . . . . *C. ulmifolia.*
        6. Inflorescences sessile. Fruits not more than 1/2 enclosed by calyx at maturity.
            8. Leaf blades usually ≥15 cm long. Pedicels and sepals golden sericeous. Fruits 8–10 cm diam. . . . *C. singularis.*
            8. Leaf blades usually <16 cm long. Pedicels and sepals not golden sericeous. Fruits 1.5–3 cm diam.
                9. Leaf blades with secondary veins in 5–7 pairs. Anthers pilose. . . . . *C. pitumba.*
                9. Leaf blades with secondary veins in 3–4(5) pairs. Anthers glabrous. . . . . *C. acuminata.*
    5. Leaf blade margins entire.
        10. Leaf blades drying light yellow-green, especially on abaxial surface. Sepals fused for >1/2 length; stamens 10. . . . . *C. bracteifera.*
        10. Leaf blades drying brown. Sepals fused for ≤1/2 length; stamens <10.
            11. Shrubs or treelets. Petioles usually ≤10 mm long. Flowers markedly exserted from inflorescence. Sepals free to base; fertile stamens 8. . . . . *C. negrensis.*
            11. Treelets. Petioles usually ≥10 mm long. Flowers scarcely exserted from inflorescences. Sepals fused for 1/2 length; fertile stamens 5 (plus five staminodia with reduced anthers). . . . . . *Casearia* sp. A.

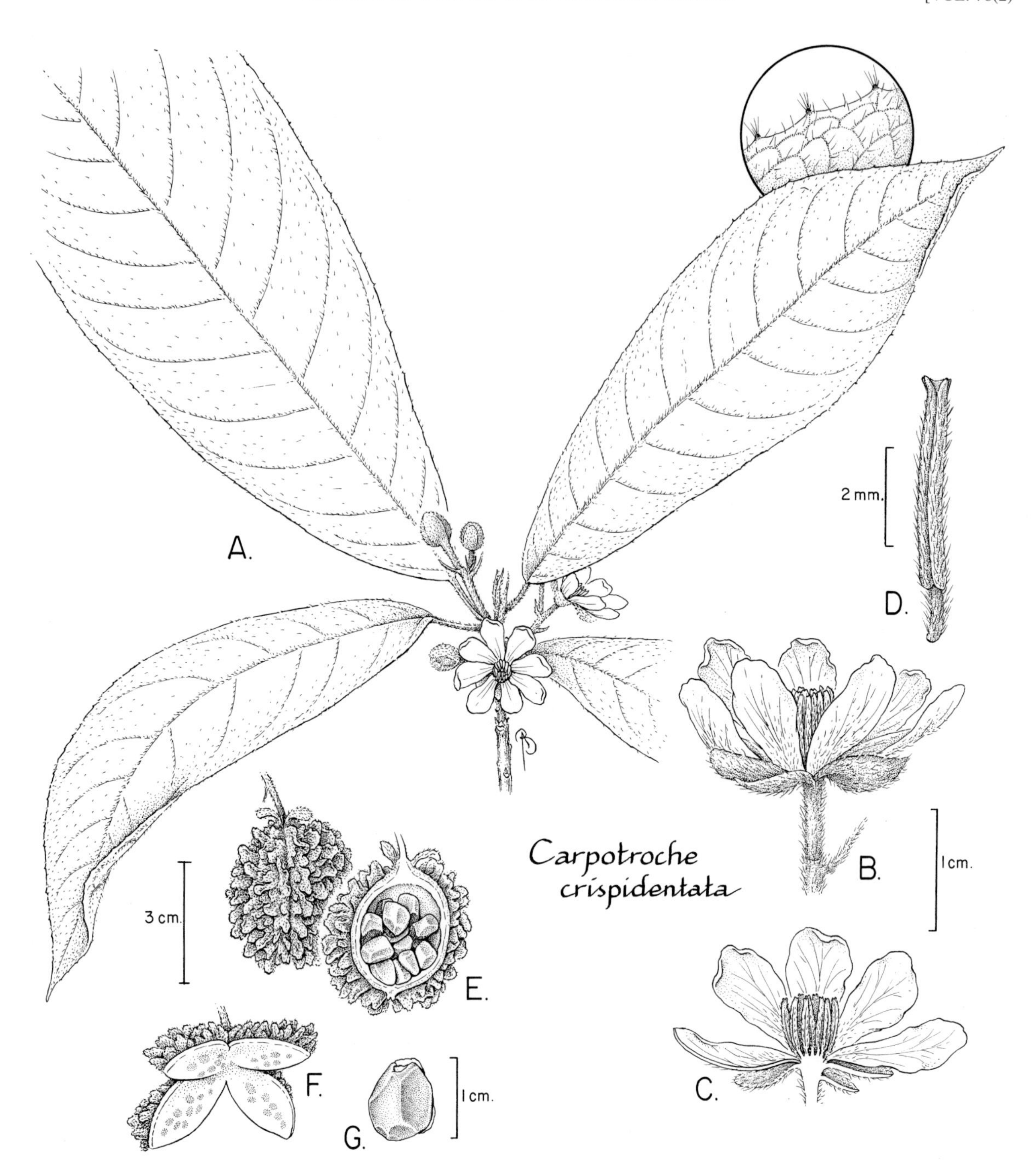

FIG. 129. FLACOURTIACEAE. *Carpotroche crispidentata* (A–D, *Mori et al. 20898*; E, F, *Mori et al. 19191*). **A.** Apex of stem with leaves, inflorescences, and detail of leaf margin. **B.** Lateral view of staminate flower. **C.** Medial section of staminate flower. **D.** Stamen with trichomes covering anther. **E.** Whole (left) and medial section (right) of fruit with seeds and fruit after dehiscence and dispersal of seeds (below). **F.** Seed.

Casearia acuminata DC. PL. 61a,b

Shrubs or treelets, to 7 m tall. Young stems brown, glabrous to puberulous. Leaves: petioles 4–6 mm long; blades narrowly ovate, 7–11 × 3–4.5 cm, glabrous, the margins serrate; secondary veins in 3–4(5) pairs, plane adaxially, the tertiaries percurrent to reticulate. Inflorescences axillary, sessile, with flowers exserted. Flowers white (infrequently green or yellow) except for yellow anthers; sepals connate for less than 1/2 length, not reflexed, glabrous; long stamens 5, short stamens 5, the filaments pilose on lower half, the anthers glabrous; staminodia 10, pilose, both stamens and staminodia attached to sepals; ovary pilose, the style undivided, the stigma capitate. Fruits globose, 1.5–2 cm diam., light orange, not more than 1/2 enclosed by calyx at maturity. Seeds ca.

3–5, surrounded by pulp. Fl (Aug, Sep, Oct, Nov, Dec), fr (Jan, Feb, Mar, May, Jun, Aug, Sep, Oct, Nov); common, in nonflooded forest.

Casearia bracteifera Sagot FIG. 130, PL. 61c

Shrubs or treelets, to 3 m tall. Young stems reddish-brown, glabrous. Leaves: petioles 5–10 mm long; blades narrowly elliptic, 19–31 × 5–9 cm, drying light yellow-green, glabrous, the margins entire; secondary veins in 9–12 pairs, plane to slightly raised adaxially, the tertiaries reticulate. Inflorescences axillary, sessile, with flowers long exserted. Flowers white, drying yellow-green; sepals connate for >1/2 length, not reflexed, glabrous; long stamens 5, short stamens 5; staminodia 10, both stamens and staminodia attached to sepals, glabrous except for pilose long filaments; ovary glabrous, the style undivided, the stigma capitate. Fruits globose, 3–4 cm diam., at first green, then light yellow at maturity. Seeds ca. 7, surrounded by clear pulp. Fl (Jan, Aug, Sep, Oct, Nov), fr (Feb, Mar, May, Jun, Aug, Nov, Dec); common, in nonflooded forest.

Collections of this species have previously been determined by H. O. Sleumer as *Casearia combaymensis*, a species based on *Goudot 106* (holotype at P) collected in Tolima, Colombia. The type (examined by S. A. Mori) differs from collections representing *C. bracteifera* in the sunken vs. salient midrib adaxially and the wider leaf blade width. Moreover, a bud removed from an inflorescence of *Goudot 106* revealed five free tepals, an androecium with five long and five short stamens, the stamens alternating with the staminodia, and the androecium in a single cycle free from the tepals. These features are in conflict with *Casearia* sect. *Casearia* group *Singulares* as defined by Sleumer (1980) as well as with the floral structure of *C. bracteifera*. Following the determinations of R. L. Liesner, we apply *C. bracteifera* to this species. However, an examination of the original syntypes (*Sagot s.n.*, *Leprieur 267*, and *Mélinon 118* at P) indicates that *Sagot s.n.*, designated by Sleumer (1980) as the lectotype, neither matches our material nor the other syntypes. The lectotype proposed by Sleumer (1980) includes three separate sheets, all marked as holotypes. Because it will never be possible to confirm that all three sheets represent the same collection of the same species, the use of *Sagot s.n.* as the lectotype is ambiguous and in conflict with the protologue. We propose that *Leprieur 267* serve as lectotype because the flowers that are needed for the taxonomic placement of species of *Casearia* are present on this collection.

In his monograph of Neotropical Flacourtiaceae, Sleumer (1980) placed *C. bracteifera* Sagot in synonymy under *C. combaymensis*, but the former must now be resurrected as a distinct species because of the flower and leaf features enumerated above. *Casearia combaymensis* no longer applies to any species in our area.

Casearia commersoniana Cambess. PL. 61d

Shrubs to treelets, to 8 m tall. Young stems gray, pubescent. Leaves: petioles absent to 5 mm long; blades widely elliptic to elliptic, 9–17 × 4–7.5 cm, glabrous, the margins serrate; secondary veins in 6–8 pairs, impressed adaxially, the tertiary veins percurrent. Inflorescences axillary, sessile, with flowers not exserted. Flowers white to greenish-white; sepals fused for <1/2 length, not reflexed, pubescent; stamens 10, the filaments glabrous, the anthers glabrous; staminodia 10, separated from stamens in inner row, densely pubescent, both stamens and staminodia attached to sepals; style 3-parted toward apex, pubescent; ovary pubescent. Fruits 1–1.5 cm diam., red, glabrous to sparsely pubescent. Fl (Aug, Sep, Oct), fr (Jan, May, Jul, Aug, Sep, Oct, Nov, Dec); common, in nonflooded forest.

Casearia javitensis Kunth

Trees, to 17 m tall. Trunks sulcate. Young stems pale brown, pubescent. Leaves: petioles 5–8 mm long; blades elliptic to oblong, 12–18 × 6–7 cm, glabrous, the margins entire to serrulate; secondary veins in 5–7 pairs, impressed adaxially, the tertiaries percurrent, glabrous. Inflorescences axillary or from stems below leaves, sessile, with flowers exserted. Flowers with green sepals, white filaments, brown anthers, and staminodia with pink spot at apex; sepals free, pubescent, reflexed; stamens ca. 15, the filaments glabrous, the anthers glabrous; staminodia separated from stamens in inner whorl, densely pubescent, both stamens and staminodia attached to sepals; ovary pubescent, the style 3-parted toward apex. Fruits red, densely pubescent. Seeds pubescent. Fl (Sep, Oct, Dec), fr (Feb); in nonflooded forest.

Casearia negrensis Eichler

Shrubs or treelets, to 7 m tall. Young stems reddish-brown, pubescent. Leaves: petioles 5–10 mm long; blades oblanceolate to narrowly oblong, 13–22 × 4–6 cm, drying brown, glabrous, the margins entire; secondary veins in 7–10 pairs, plane to slightly raised adaxially, the tertiaries percurrent, glabrous. Inflorescences axillary, sessile, with flowers exserted. Flowers white; sepals free, not reflexed, pubescent; stamens 8, of equal length, the filaments glabrous, the anthers glabrous; staminodia 8, fused at bases with bases of stamens, clavate, in same row as stamens, densely short pubesecent, both stamens and staminodia free from sepals; ovary pilose, the style undivided, the stigma capitate, with short but dense pubescence. Fruits oblong, 1–1.5 cm diam. Fl (Jul, Oct, Dec), fr (Jan, Feb, Apr, Jul, Aug, Oct, Dec); in nonflooded forest.

Casearia pitumba Sleumer FIG. 131, PL. 61e,f

Treelets, to 8 m tall, less frequently shrubs. Young stems brown, pubescent. Leaves: petioles 4–6(12) mm long; blades elliptic, 10–16(21) × 4–7 cm, glabrous to sparsely pubescent on midrib abaxially, the margins serrate; secondary veins in 5–7 pairs, plane adaxially, the tertiaries percurrent to reticulate. Inflorescences axillary, sessile, with flowers exserted. Flowers green (sometimes white) except for yellow anthers; sepals fused <1/2 length, not reflexed, pubescent; long stamens 5, short stamens 5, the filaments pilose, the anthers with apical gland crowned with tuft of hairs; staminodia 10, densely pilose, in same row as stamens, both stamens and staminodia attached to sepals; ovary pubescent, the style undivided, the stigma capitate. Fruits 1.5–3 cm diam., yellow to orange, not more than 1/2 enclosed by calyx at maturity. Seeds surrounded by transparent pulp. Fl (Aug, Sep), fr (Jan, Feb, Mar, Apr, May, Aug, Nov, Dec); common, in secondary and primary nonflooded forest.

Casearia rusbyana Briq.

Small trees, to 15 m tall. Young stems with dense, reddish-brown pubescence. Leaves: petioles 6–7 mm long; blades oblong to narrowly oblong, 13–19 × 5–7 cm, sparsely pubescent adaxially, densely pubescent abaxially, the margins serrate; secondary veins in 9–12 pairs, slightly impressed adaxially, the tertiaries distinctly percurrent. Inflorescences axillary, short pedunculate, flowers not exserted. Flowers ca. 7 mm diam.; sepals fused for <l/2 length, not reflexed, pubescent; stamens 10, equal length, the filaments

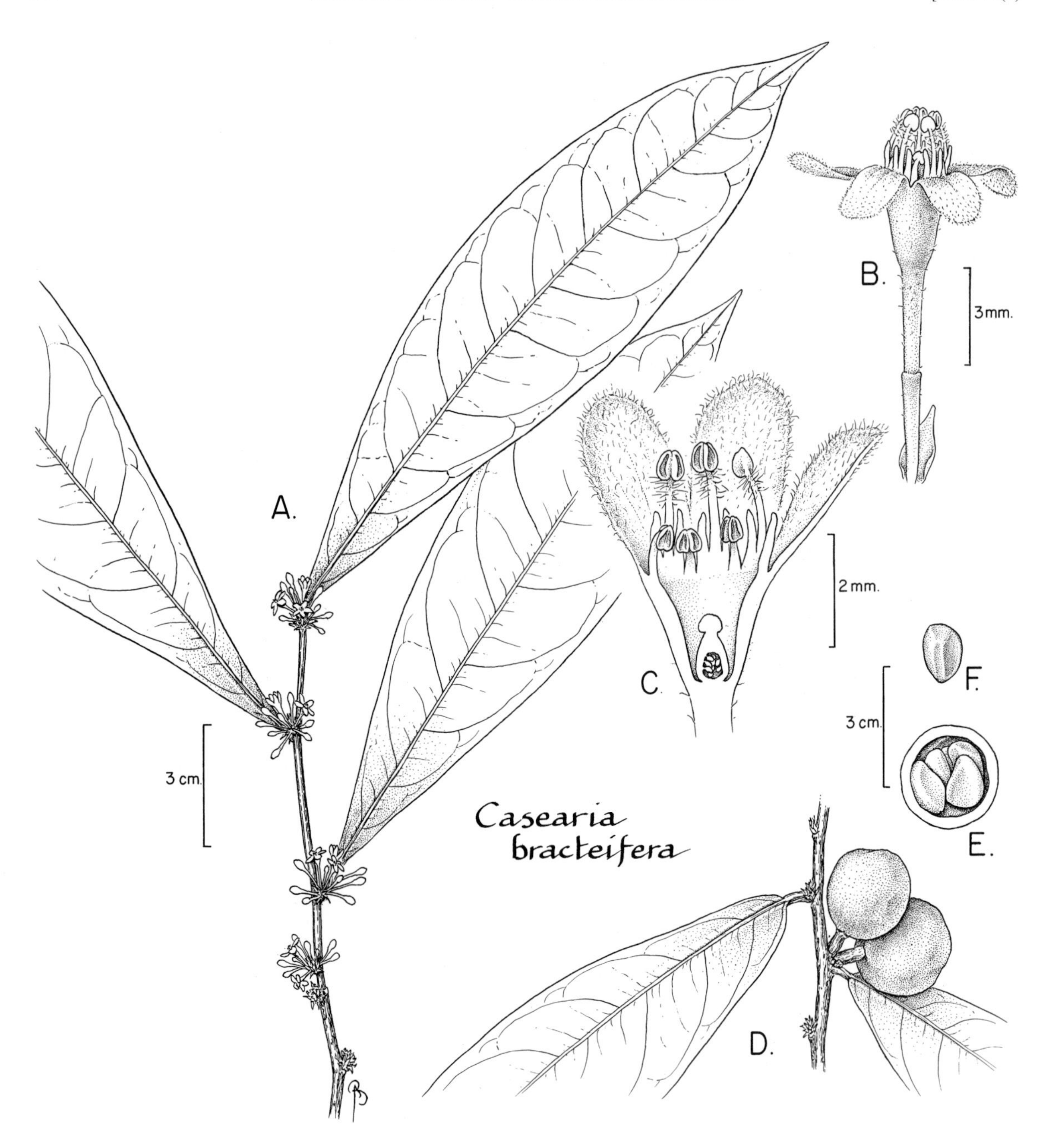

FIG. 130. FLACOURTIACEAE. *Casearia bracteifera* (*Acevedo-Rodríguez 5010*). **A.** Apex of stem with axillary fascicles of flowers. **B.** Lateral view of flower and pedicel. **C.** Medial section of flower showing fertile stamens of two lengths, staminodes, and parietal placentation of ovules. **D.** Part of stem with infructescence. **E.** Transverse section of fruit showing seeds. **F.** Seed.

glabrous, the anthers with a few, scattered hairs; staminodia in same row as stamens, densely pubescent, both stamens and staminodia attached to sepals; ovary pilose, the style undivided, with lower part pilose , the stigma capitate. Fruits more than 1/2 enclosed by calyx at maturity. Fl (Jan, Feb, Aug, Nov, Dec), fr (Nov); scattered, in nonflooded secondary forest.

Casearia singularis Eichler FIG. 132

Trees, to 20 m tall. Bark with yellow inner bark. Young stems reddish-brown, lenticellate, glabrous to puberulous. Leaves: petioles 7–10 mm long; blades elliptic, 15–19 × 7–9 cm, glabrous, the margins serrulate; secondary veins in 6–8 pairs, salient on both surfaces, the tertiaries reticulate-percurrent. Inflorescences axillary and from stems below leaves, sessile, with flowers slightly exserted; pedicels golden sericeous. Flower buds greenish-yellow; sepals fused for <1/2 length, not reflexed, golden sericeous; long stamens 5, short stamens 5, the filaments glabrous, the anthers glabrous; staminodia 10, glabrous, both stamens and staminodia attached to sepals; ovary glabrous, the style undivided, short, the stigma capitate. Fruits globose, 8–10 cm diam., green, falling to ground at maturity, not more than 1/2 enclosed by calyx at maturity. Seeds

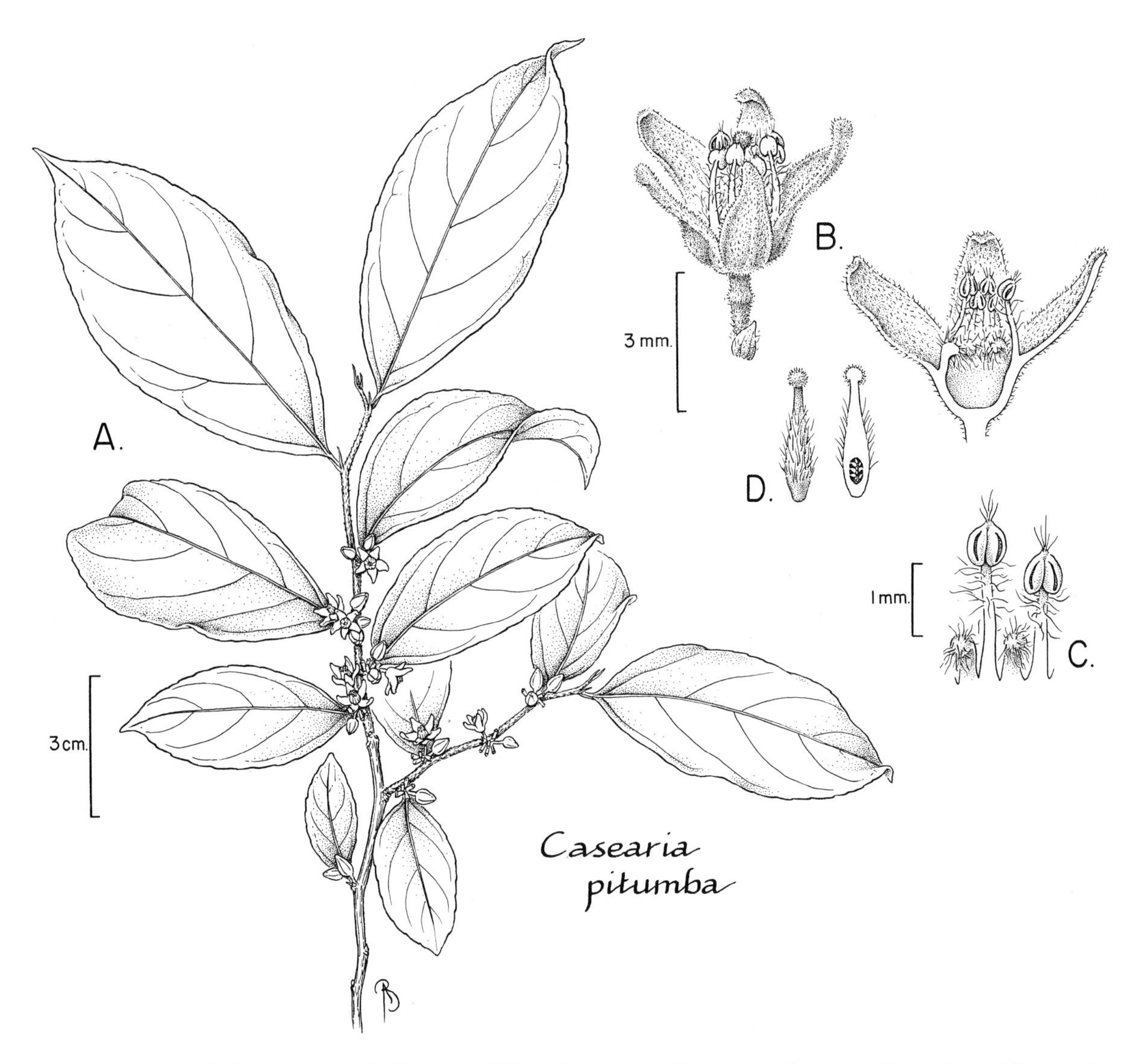

FIG. 131. FLACOURTIACEAE. *Casearia pitumba* (*Mori et al. 20911*). **A.** Apex of stem with leaves and inflorescences. **B.** Lateral view (left) and medial section (right) of flower. **C.** Alternating stamens and staminodes; note pubescence on filament and connective. **D.** Lateral view (left) and medial section (right) of pistil.

surrounded by white pulp. Fl (Sep, Oct, Nov), fr (Feb, Mar, Apr, May, Dec); in nonflooded forest.

Collections of this species have previously been determined by H. O. Sleumer as *Casearia combaymensis*. Sleumer included what we consider to be two separate species (*C. bracteifera* and *C. singularis*) under his concept of *C. combaymensis*.

Casearia combaymensis was based on *Goudot 106* (Holotype, P; photo F 34893) collected in Tolima, Colombia. The type of *C. combaymensis* (examined by S. Mori) differs from collections representing *C. singularis* in the sunken vs. salient midrib adaxially and the sparsely vs. densely pubescent inflorescence pads and floral buds. The flowers of *C. combaymensis* differ from this species in the same features discussed above under *C. bracteifera*. An illustration of the flowers of *C. singularis*, matching the features shown in our illustration (Fig. 132), can be found in Flora Brasiliensis 13(1): t. 95, f. 2 (Eichler, 1871).

In his monograph of Flacourtiaceae, Sleumer (1980) placed *C. singularis* Eichler in synonymy under *C. combaymensis*, but the former must now be resurrected as a distinct species. *Casearia combaymensis* no longer applies to any species in our flora. We are grateful to R. L. Liesner whose determinations of our material as *C. singularis* first made us aware of this problem.

Casearia sylvestris Sw. var. sylvestris

Trees, to 22 m tall. Trunks buttressed. Young stems dark gray, glabrous. Leaves: petioles 6–8 mm long; blades narrowly ovate to elliptic, 9–14 × 4–5 cm, drying brown, chartaceous, glabrous, the margins entire; secondary veins in 6–8 pairs, plane adaxially, salient abaxially, the tertiaries reticulate. Inflorescences axillary, sessile, with flowers exserted; pedicels glabrous. Flowers white; sepals fused <1/2 length, not reflexed, glabrous adaxially; stamens 10,

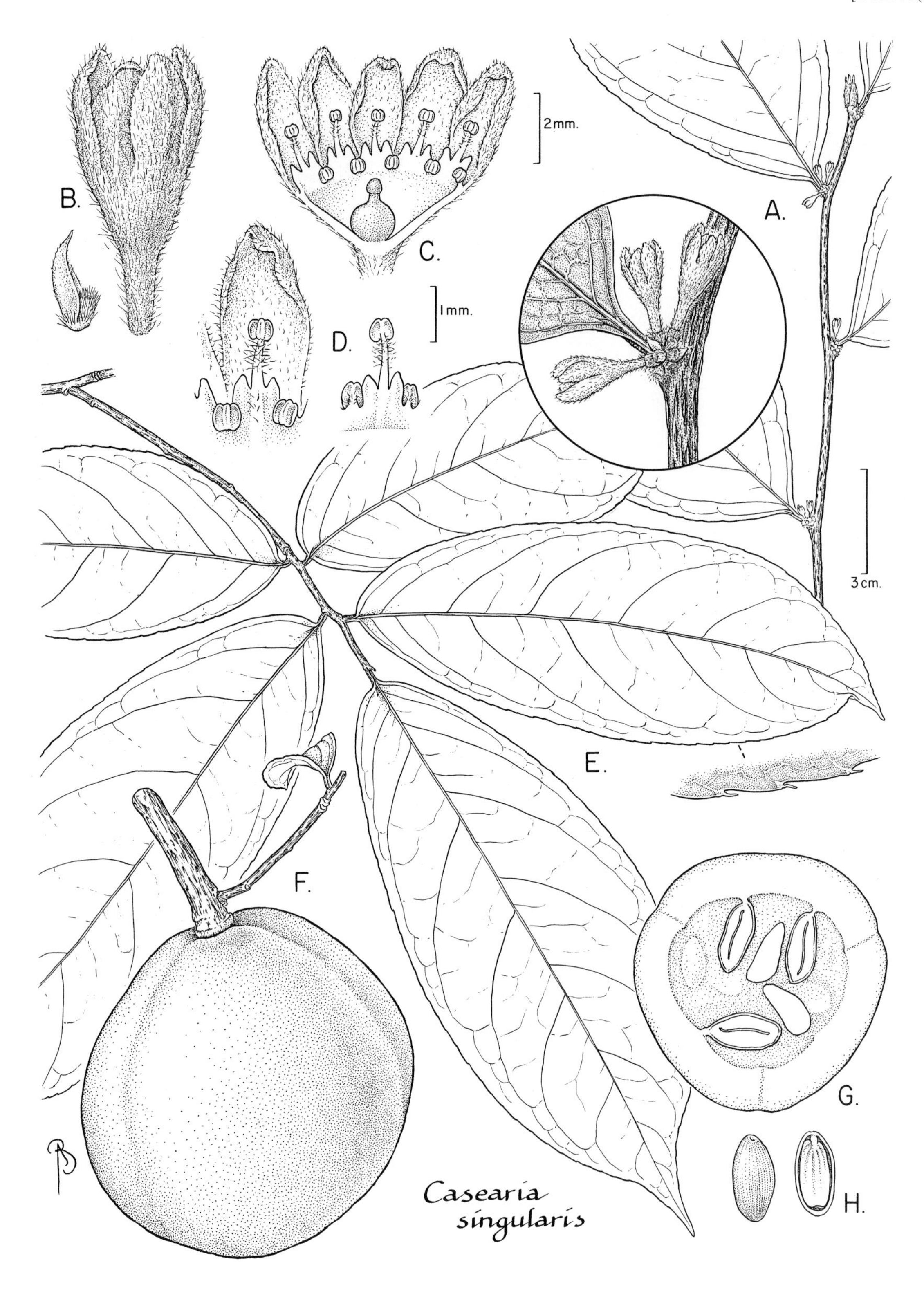

FIG. 132. FLACOURTIACEAE. *Casearia singularis* (A–D, *Wallnöfer et al. 13529*; E–H, *Mori et al. 22210*, holotype). **A.** Part of stem with leaves and axillary inflorescences (right) and detail of node with inflorescence (below left). **B.** Lateral view of flower (right) and floral bract (below). **C.** Flower opened to show adnate stamens and pistil. **D.** Detail of abaxial view of one tepal and part of androecium (left) and detail of adaxial view of part of androecium (right). **E.** Part of stem with leaves and detail of leaf margin. **F.** Fruit. **G.** Transverse section of fruit. **H.** Seed (far left) and medial section of seed (near left).

subequal, the filaments glabrous, the anthers glabrous; staminodia 10, in same row as stamens, pilose, both stamens and staminodia attached to sepals; ovary glabrous, the style short, 3-parted, the stigmas capitate, exserted, slightly longer than stamens. Fl(Apr); collected only once, in nonflooded foreest.

Casearia ulmifolia Vent. PL. 61g

Trees, to 25 m tall. Trunks of larger individuals buttressed. Young stems reddish-brown, puberulous. Leaves: petioles 3–4 mm long; blades elliptic to narrowly ovate or oblong, 7–15 × 3–5.5 cm, young leaves pubescent, especially on petioles and midrib abaxially, the margins serrate; secondary veins in 6–8 pairs, plane to slightly raised adaxially, salient abaxially, the tertiaries reticulate. Inflorescences axillary, short pedunculate, with flowers exserted. Flowers ca. 3 mm diam., green to greenish-white; sepals fused for <1/2 length, not reflexed, pubescent abaxially; long stamens 5, slightly shorter stamens 5, the filaments glabrous, the anthers with scattered hairs to glabrous; staminodia 10, pilose, in same row as stamens, both stamens and staminodia attached to sepals; ovary pubescent, the style undivided, the stigma capitate, pubescent. Fruits more than 1/2 enclosed by calyx at maturity. Fl (Jan, Feb, Apr); in non-flooded forest.

Casearia sp. **A**

Small trees, to 7 m tall. Bark with small depressions, the slash orange, the inner bark laminated. Young stems gray, glabrous. Leaves: petioles 10–15 mm long; blades elliptic to narrowly obovate, 14–14.5 × 6–7 cm, drying brown, glabrous, the margins entire; secondary veins in 5–7 pairs, plane to slightly salient adaxially, the tertiaries reticulate. Inflorescences from stems below leaves, sessile, with flowers slightly exserted. Flowers: sepals fused for ca. 1/2 length, not reflexed, gray pubescent; fertile stamens 5 in inner row, alternate with sepal lobes, the filaments short (0.2–0.3 mm long), glabrous, the anthers glabrous; staminodia of two types, one with vestigial anthers (ca. 1/5th size of anthers of normal stamens) and densely pilose filaments, the other without anthers and glabrous filaments, both opposite sepal lobes, both stamens and staminodia attached to sepals; ovary glabrous, the style undivided, the stigma capitate; ovary glabrous. Fruits 1.3 cm diam. Fl (Aug), fr (Apr, Sep); in nonflooded forest.

This species belongs to group *Decandrae* and is represented in our area by the collections *Mori et al. 14721*, *15570*, and *24026* (all at CAY and NY).

Casearia sp. **B**

Trees, to 10 m tall. Trunks sometimes sulcate toward base. Young stems dark brown, glabrous, slightly zig-zagged. Leaves: petioles 6–8 mm long; blades elliptic to narrowly elliptic, 5–8 × 2–3 cm, drying green, subcoriaceous, glabrous, the margins entire; secondary veins in 6–8 pairs, plane adaxially, the tertiaries reticulate, inconspicuous. Inflorescences axillary, sessile, with flowers exserted; pedicels pubescent. Flowers yellowish-green to white; sepals fused for <1/2 length, not reflexed, pubescent adaxially; stamens 10, subequal, the filaments sparsely pilose, the anthers glabrous; staminodia 10, in same row as stamens, pilose, shorter than stamens, both stamens and staminodia attached to sepals; ovary glabrous at base, sparsely pilose toward apex, the style 3(4)-parted, sparsely pilose, the stigmas scarcely capitate. Fl (May, Aug, Sep), fr (May); in nonflooded forest.

This species belongs to section *Crateria* and is represented in our area by the collections *Mori et al. 18124* (CAY, NY), *19166* (NY), and *20785* (CAY, NY).

HASSELTIA Kunth

Trees. Leaf blades with three principal veins arching upward from base, the remaining secondary veins in 3–4 pinnately arranged pairs, with 2 sessile, flattened glands on adaxial surface near petiole, the margins entire. Inflorescences terminal or axillary, long-peduncled, the axes in whorls. Flowers bisexual; sepals 4, fused at bases; petals 4, fused at bases; stamens numerous, attached to petals, the filaments long and slender, the anthers small; glands small, globose, glabrous, located between bases of filaments; ovary 2-locular by early union of 2 opposite, parietal placentae; placentation parietal but appearing axile because of intruded placentae. Fruits berry-like. Seeds 1(2); endosperm present, the embryo straight, with foliaceous cotyledons.

Hasseltia floribunda Kunth

Trees, to 17 m tall. Leaves: petioles 25–35 mm long; blades elliptic, 12–18 × 4.5–7 cm. Inflorescences with peduncles 3–5 cm long, the secondary and tertiary axes whorled, the ultimate axes racemose. Flowers: buds small (ca. 2–3 mm diam.), globose, densely pubescent; sepals abaxially puberulent. Fl (Dec); collected once at base of Mont Galbao, in nonflooded forest.

LAETIA Loefl.

Trees. Leaf blades with pinnate venation, the margins entire to serrulate. Inflorescences in sessile globose clusters or stalked corymbs. Flowers bisexual; sepals 4–5, free or slightly fused at bases; petals 0; stamens 10 to numerous, inserted on fused part of sepals; staminodia and glands absent; ovary 1-locular, with 3 parietal placentae. Fruits berry-like or tardily dehiscent capsules. Seeds arillate; endosperm copious, the embryo straight, the cotyledons broad, foliaceous.

Laetia procera (Poepp.) Eichler

Trees, to 30 m tall. Trunks often with low, thick buttresses, these sometimes running along ground for several meters from trunk. Bark smooth, conspicuously lenticellate, the inner bark orange with white streaks. Young stems reddish-brown to blackish, glabrous. Leaves: petioles 8–12 mm long; blades elliptic to narrowly elliptic or oblong to narrowly oblong, 12–17 × 4.5–5.5 cm, entirely glabrous, the base obtuse to rounded, the apex short to long acuminate, the margins entire to finely serrulate; secondary veins in 6–10 pairs. Inflorescences fascicles, supra-axillary or along stems below leaves, the flowers long exserted; pedicels with reddish tinge.

Flowers ca. 7 mm diam.; sepals 5, reflexed at flowering, glabrous, white; stamens 15–20, the filaments slender, the anthers 1.5–2 mm long, with filaments attached slightly above base; ovary glabrous. Fruits 1 cm diam., green. Fl (Aug, Sep, Oct), fr (Mar, Sep, Oct); common, in nonflooded forest, its presence probably indicative of past disturbance.

MAYNA Aubl.

Shrubs or small trees. Leaves with pinnate venation, the margins entire to serrate. Inflorescences axillary, the staminate flowers in clusters, the pistillate flowers solitary. Flowers unisexual (plants dioecious); sepals 3; petals 5–8; stamens numerous (20–50), the anthers linear; staminodia and glands absent; ovary 1-locular, usually with bristles, less frequently smooth, the styles 2–3, the stigmas 2–3, each 2-lobed, the lobes laciniate; placentation parietal, the placentae 3. Fruit indehiscent or opening at apex, globose, usually covered with bristles, the bristles sometimes absent. Seeds with copious endosperm, the embryo straight, the cotyledons cordate.

Mayna odorata Aubl. PL. 61h

Shrubs or treelets, to 4 m tall. Young stems gray, with scattered lenticels, pubescent. Stipules acicular. Leaves: petioles 10–25 mm long, pubescent; blades narrowly elliptic to oblanceolate, 12–25 × 3.5–6.5 cm, glabrous to puberulous on veins, the margins entire to serrulate, often with minute glandular hairs, especially toward apex; secondary veins in 7–11 pairs. Flowers ca. 8 mm diam.; buds pubescent; sepals 3, pubescent; petals 5, white; stamens 20–30, the anthers 1.5 mm long, pubescent, orange; ovary pubescent. Fruits 15–20 mm diam., yellow to orange, without bristles. Fl (Jan, Feb, Mar, Apr, Oct, Dec), fr (Jan, Feb, Apr, May, Jun, Aug, Sep, Oct, Nov, Dec); locally common, in nonflooded forest.

The fruits of the Saül population lack the bristles (see plate between pages 256 and 257 in Grenand et al., 1987) found in other populations of this species.

RYANIA Vahl

Shrubs or small trees. Leaves with pinnate venation, the margins entire. Inflorescences axillary, fasciculate or on very short rachises, 1–4-flowered. Flowers >20 mm diam., bisexual, heterostylous; sepals 5, fused at very base, petaloid; petals absent; stamens 30–70, inserted in 2 or 3 rows at apex of calyx-tube, the filaments slender, the anthers oblong to linear; ovary 1-locular, the style undivided or divided at apex; placentation parietal, the placentae 3–9. Fruits indehiscent or ultimately valvately dehiscent. Seeds with membranous basal aril; endosperm copious, the embryo straight, the cotyledons flat, thin.

Ryania speciosa Vahl

Small trees, to 10 m tall. Young stems with rusty colored pubescence. Stipules subulate, persistent. Leaves: petioles 3–5 mm long, pubescent; blades narrowly oblong, 15–19 × 3.5–6.5 cm, pubescent on midrib abaxially, the hairs stellate; secondary veins in 8–9 pairs. Inflorescences of solitary flowers. Flowers large and showy, ca. 4 cm diam.; buds covered with stellate hairs; filaments long, slender, the anthers 7 mm long; style to 25 mm long, undivided. Fl (Jan, Jul, Sep); apparently rare in our area, in nonflooded forest.

XYLOSMA G. Forst.

Shrubs or trees. Leaves with pinnate venation, the margins serrate. Trunks, branches, and stems often armed with simple or branched spines. Stipules absent. Inflorescences axillary or from along stems, in short racemes or sometimes reduced to clusters. Flowers unisexual (plants dioecious); sepals 4, slightly united at base; petals absent. Staminate flowers with numerous stamens, surrounded by globose, separate nectaries, the filaments slender, much longer than sepals, the anthers globose. Pistillate flowers rarely with staminodia present, with nectaries separate or sometimes fused into ring; ovary 1-locular, the style simple or short branched distally; placentation parietal, the ovules 2–3. Fruits with thin-coriaceous pericarp, rarely somewhat fleshy. Seeds generally few; endosperm copious, the embryo large, the cotyledons broad.

Xylosma benthamii (Tul.) Triana & Planch.

Small trees, to 10 m tall. Trunks with well developed, branched spines to 7.5 cm long. Leaves: petioles 3–5 mm long; blades elliptic, (5)7.5–17.5 × (2)4.5–7.5 cm, the margins crenate, with glandular hairs, especially conspicuous on young leaves; secondary veins in 5–9 pairs. Inflorescences in short, bracteate racemes, the rachis 3–5 mm long, pubescent. Staminate flowers ca. 3 mm diam.; sepals 4, pubescent; nectaries globose, separate, extrastaminal; stamens ca. 25–35, the filaments slender, long exserted, the anthers globose, ca. 0.3 mm long. Fruits 0.6 cm diam., rose colored. Fl (Jan), fr (Jan, Feb); uncommon, in secondary or slightly disturbed vegetation in nonflooded forest.

GENTIANACEAE (Gentian Family)

Hiltje Maas-van de Kamer and Paul Maas

Saprophytic (*Voyria*, *Voyriella*) or autotrophic herbs, subshrubs, or treelets, glabrous. Stipules absent. Leaves simple, opposite, small and scale-like or normally developed. Inflorescences terminal or axillary dichasia, racemes, or spikes, or flowers solitary. Flowers actinomorphic or slightly zygomorphic, bisexual, generally 5-merous; sepals connate or free; petals connate, with distinct tube and contorted

Plates 57–64

Plate 57. FABACEAE. **a.** *Dioclea macrocarpa* (*Mori & Gracie 24208*), inflorescences in bud on liana. **b.** *Dioclea macrocarpa* (*Mori & Gracie 24208*), lateral view of flower and bud. **c.** *Dioclea macrocarpa* (*Mori et al. 24679*), flowers. [Photo by S. Mori] **d.** *Dioclea macrocarpa* (*Mori et al. 22211*), opened pod with seeds. **e.** *Dioclea virgata* (*Mori et al. 19179*), flowers and buds. **f.** *Dioclea virgata* (*Mori et al. 19179*), flower.

Plate 58. FABACEAE. **a.** *Diplotropis purpurea* (*Mori & Gracie 21108*), flower and immature fruit. **b.** *Dussia discolor* (*Mori et al. 24728*), opened fruit with seeds, those to left with seed coat partially or totally removed. [Photo by S. Heald] **c.** *Monopteryx inpae* (*Mori et al. 22107*), flowers and buds. [Photo by S. Mori] **d.** *Platymiscium ulei* (*Mori et al. 19039*), flowers beginning to develop fruits. **e.** *Pterocarpus rohrii* (*Mori & Gracie 23209*), slash of trunk showing red sap. **f.** *Pterocarpus rohrii* (*Mori & Gracie 23209*), flowers.

Plate 59. FABACEAE. **a.** *Swartzia amshoffiana* (*Mori et al. 23325*), buds and flowers; this species lacks petals. **b.** *Rhynchosia phaseoloides* (*Mori & Gracie 19015*), flower and buds. **c.** *Swartzia panacoco* var. *sagotii* (*Mori et al. 24710*), flowers and buds. [Photo by S. Heald] **d.** *Swartzia polyphylla* (*Mori et al. 20989*), flower and bud. **e.** *Rhynchosia phaseoloides* (*Mori & Gracie 19015*), dehisced fruits and seeds.

Plate 60. FABACEAE. **a.** *Vataireopsis surinamensis* (*Mori et al. 24759*), leafless crown of flowering tree. **b.** *Vataireopsis surinamensis* (*Mori et al. 24759*), flower and buds. **c.** *Zornia latifolia* (*Mori et al. 23288*), flower; note margin of standard chewed by unknown animal. FLACOURTIACEAE. **d.** *Carpotroche crispidentata* (*Mori et al. 20898*), flower with beetle. **e.** *Banara guianensis* (*Mori et al. 24755*), flowers.

Plate 61. FLACOURTIACEAE. **a.** *Casearia acuminata* (*Mori et al. 22081*), flowers and buds. **b.** *Casearia acuminata* (*Mori et al. 23331*), fruit. **c.** *Casearia bracteifera* (*Mori et al. 21651*), flowers and buds. **d.** *Casearia commersoniana* (*Mori et al. 23902*), dehisced fruit. **e.** *Casearia pitumba* (*Mori et al. 20877*), flowers and buds. **f.** *Casearia pitumba* (*Mori et al. 23093*), close-up of flower and buds. **g.** *Casearia ulmifolia* (*Mori et al. 23012*), flowers and buds. **h.** *Mayna odorata* (*Mori & Gracie 18927*), pistillate flowers and fruits; this species usually has armed fruits in other parts of its range.

Plate 62. GENTIANACEAE. **a.** *Coutoubea spicata* (unvouchered), inflorescences. [Photo by S. Mori] **b.** *Chelonanthus alatus* (*Mori & Gracie 18417*), inflorescence with bud, flower, and immature fruits. **c.** *Tachia grandiflora* (*Acevedo-Rodríguez 4977*), flower. **d.** *Voyria aurantiaca* (*Mori et al. 23911*), two plants with flowers. **e.** *Voyria caerulea* (*Mori et al. 22267*), plant with open flower and bud; corolla tube may be white in some individuals.

Plate 63. GENTIANACEAE. **a.** *Voyria corymbosa* (unvouchered), flowers. **b.** *Voyria flavescens* (*Maas 8105*), flower. **c.** *Voyria rosea* (*Mori et al. 22299*), plant with flowers and buds. **d.** *Voyria rosea* (foreground) and *V. corymbosa* (background), (unvouchered), two species of saprophytes growing in close proximity.

Plate 64. GENTIANACEAE. **a.** *Voyria tenella* (*Mori et al. 22280*), flower. **b.** *Voyria tenuiflora* (*Mori et al. 22241*), flower; note recurved corolla lobes. **c.** *Voyriella parviflora* (*Mori et al. 22283*), plants with inflorescences; note opposite, scale-like leaves. GESNERIACEAE. **d.** *Besleria insolita* (*Mori & Gracie 21176*), buds, open flowers, and fruits subtended by persistent, split calyces. **e.** *Besleria patrisii* (*Mori & Gracie 18726*), flowers and bud.

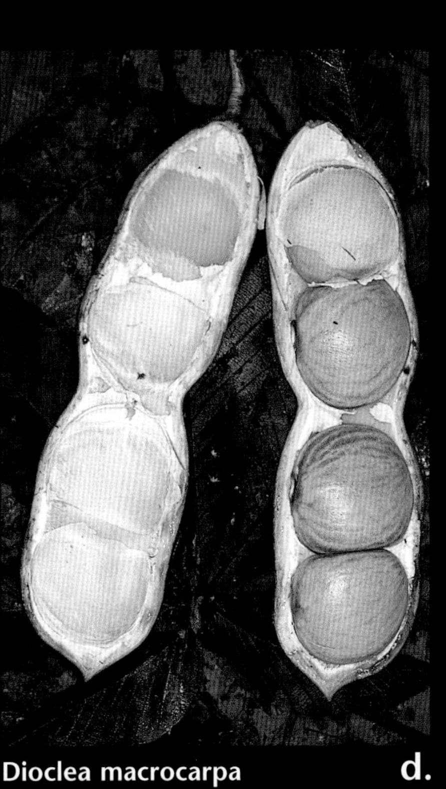

Dioclea macrocarpa d.

Dioclea virgata e.

Dioclea virgata f.

Plate 57

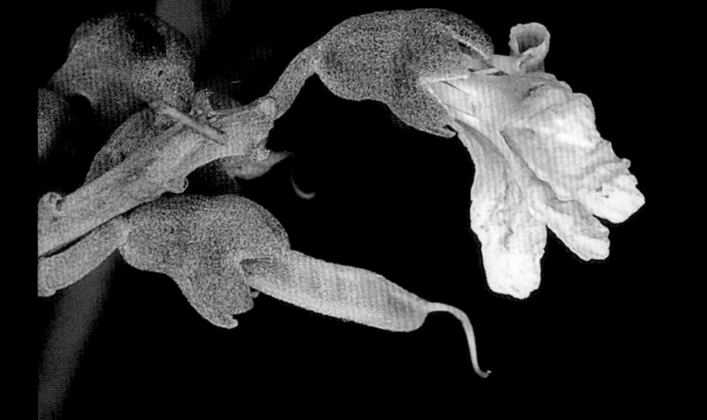

Diplotropis purpurea a.

Dussia discolor b.

Monopteryx inpae c.

Platymiscium ulei d.

Pterocarpus rohrii f.

e.

Plate 58

Swartzia amshoffiana a.

Rhynchosia phaseoloides b.

Swartzia panacoco var. sagotii c.

Swartzia polyphylla d.

Rhynchosia phaseoloides e.

a.

Vataireopsis surinamensis

b.

c.

Zornia latifolia

FLACOURTIACEAE

Carpotroche crispidentata

d.

Banara guianensis

e.

Plate 60

Casearia acuminata a.

Casearia acuminata b.

Casearia bracteifera c.

Casearia commersoniana d.

Casearia pitumba e.

Casearia pitumba f.

Casearia ulmifolia g.

Mayna odorata h.

Plate 61

Coutoubea spicata a.

Chelonanthus alatus b.

Tachia grandiflora c.

Voyria aurantiaca d.

Voyria caerulea e.

Voyria corymbosa a.

Voyria flavescens b.

Voyria rosea c.

Voyria rosea (front)
V. corymbosa (back) d.

Plate 63

Besleria insolita **d.**

Besleria patrisii **e.**

Plate 64 (Gesneriaceae continued on Plate 65)

lobes; stamens usually 5; ovary superior, 1-locular with 2 parietal placentae, or 2-locular with axile placentation. Fruits septicidally dehiscent, 2-valved capsules, or indehiscent. Seeds many, rarely filiform; endosperm present.

Recent morphological and molecular studies by Struwe et al. (1994) support the placement of *Potalia* in the Gentianaceae rather than in the Loganiaceae (Cronquist, 1981) as is done in this *Guide*.

Jonker, F. P. 1966. Gentianaceae. *In* A. Pulle (ed.), Flora of Suriname **IV(1):** 400–427. E. J. Brill, Leiden.

Maas, P. J. M. & P. Ruyters. 1986. *Voyria* and *Voyriella* (saprophytic Gentianaceae). Fl. Neotrop. Monogr. **41:** 1–93.

Struwe, L., P. J. M. Maas, O. Pihlar & V. A. Albert. 1999. Gentianaceae. *In* J. A. Steyermark, P. E. Berry, K. Yatskievych & B. K. Holst (gen. eds.), Flora of the Venezuelan Guayana **5:** 474–542. P. E. Berry, K. Yatskievych & B. K. Holst (vol. eds.). Missouri Botanical Garden Press, St. Louis.

1. Saprophytic herbs.
   2. Inflorescences capitate. Corolla caducous, 6–7 mm long, scarcely exceeding the free sepals. . . . . *Voyriella.*
   2. Inflorescences not capitate, of a single flower or few- to many-flowered dichasia. Corolla persistent, 8–90 mm long, far exceeding the connate sepals. . . . . *Voyria.*
1. Autotrophic herbs, shrubs, or treelets.
   3. Leaf blades amplexicaulous. Inflorescences spikes. Flowers 4-merous. . . . . *Coutoubea.*
   3. Leaf blades not amplexicaulous. Inflorescences of a single flower or dichasia. Flowers 5-merous.
      4. Inflorescences of a single flower in leaf axils. Corolla yellow to yellow-orange.. . . . . *Tachia.*
      4. Inflorescences dichasia, terminal. Corolla greenish, yellowish, white, or blue.. . . . . *Chelonanthus.*

CHELONANTHUS Gilg

Herbs with woody base (subshrubs). Stems often 4-angled to 4-winged. Leaves fully developed, with chlorophyll; petioles absent to present; blades not amplexicaul. Inflorescences terminal, few- to many-flowered dichasia, the branches monochasial. Flowers slightly zygomorphic, 5-merous; sepals basally connate, the lobes abaxially with oblong glandular area; corolla funnelform to salverform, deciduous. Capsules "woody," often nodding, dehiscing by 2 valves, crowned by persistent style. The status of genera of the *Irlbachia* complex, which includes *Chelonanthus*, has not yet been resolved and therefore the name of this genus as applied to our area is subject to change (Struwe & Albert, 1998).

1. Leaves usually sessile. Corolla green, yellow, or white.
   2. Stems 4-angled to basally slightly 4-winged. Corolla greenish or yellow; basal part of style persisting after dehiscence of fruit. . . . . *C. alatus.*
   2. Stems 4-winged throughout. Corolla white; complete style persisting after dehiscence of fruit. . . . . . . *C. longistylus.*
1. Leaves mostly petiolate. Corolla blue (sometimes with a white throat). . . . . *C. purpurascens.*

Chelonanthus alatus (Aubl.) Pulle [Syns.: *Irlbachia alata* (Aubl.) Maas subsp. *alata*, *Lisyanthus alatus* Aubl.] FIG. 133, PL. 62b

Herbs, 0.5–1.5(3.5) m tall. Stems 4-angled to basally 4-winged. Leaves: petioles absent; blades ovate, 6–15 × 4–9 cm, the apex acute. Inflorescences: pedicels 5–10 mm long. Flowers: sepals 4–13 mm long; corolla funnelform, 30–45 mm long, greenish or yellowish. Capsules ellipsoid, 10–20 × 5 mm, crowned by basal part of style. Fl (Feb, Jun, Jul, Sep); common, in secondary vegetation, especially near airport. *Tabaco bravo* (Portuguese)

Chelonanthus longistylus (J. G. M. Pers. & Maas) Struwe & V. Albert [Syn.: *Irlbachia alata* (Aubl.) Maas subsp. *longistyla* J. G. M. Pers. & Maas

Herbs, 1–1.2 m tall. Stems 4-winged throughout. Leaves: petioles absent; blades elliptic to narrowly elliptic, 13–14 × 4–5 cm, the apex acute. Inflorescences: pedicels 7–9 mm long. Flowers: sepals 6–7 mm long; corolla funnelform, 30–35 mm long, white. Capsules ellipsoid, 13 × 6 mm, crowned by complete style until after dehiscence of capsule. Fl (Jan), fr (Jan); only collected from small island in lake 1 km S of Pic Matécho.

Chelonanthus purpurascens (Aubl.) Struwe, S. Nilsson & V. Albert [Syns.: *Irlbachia purpurascens* (Aubl.) Maas, *Lisianthus uliginosus* Griseb., *Lisyanthus purpurascens* Aubl., *Chelonanthus uliginosus* (Griseb.) Gilg]

Herbs, to 2 m tall. Stems slightly 4-angled or terete. Leaves: petioles 0–7 mm; blades (narrowly) elliptic-ovate, 2–9 × 1–4 cm, the apex acute to short-acuminate. Inflorescences: pedicels to 20 mm long. Flowers: sepals 5–7 mm long; corolla broadly funnelform, 50–60 mm long, blue (sometimes with white throat). Capsules ellipsoid, 8–17 × 3–6 mm, crowned by persistent style. Fl (Jan, Dec), fr (Jan); collected in rocky places with low vegetation on Mont Galbao and Pic Matécho.

COUTOUBEA Aubl.

Herbs with woody base (subshrubs). Stems terete, sometimes 4-angled. Leaves fully developed, with chlorophyll; petioles absent; blades amplexicaulous. Inflorescences terminal or axillary, many-flowered spikes. Flowers actinomorphic, 4-merous; sepals basally connate; corolla salverform, marcescent. Capsules dehiscing by 2 valves.

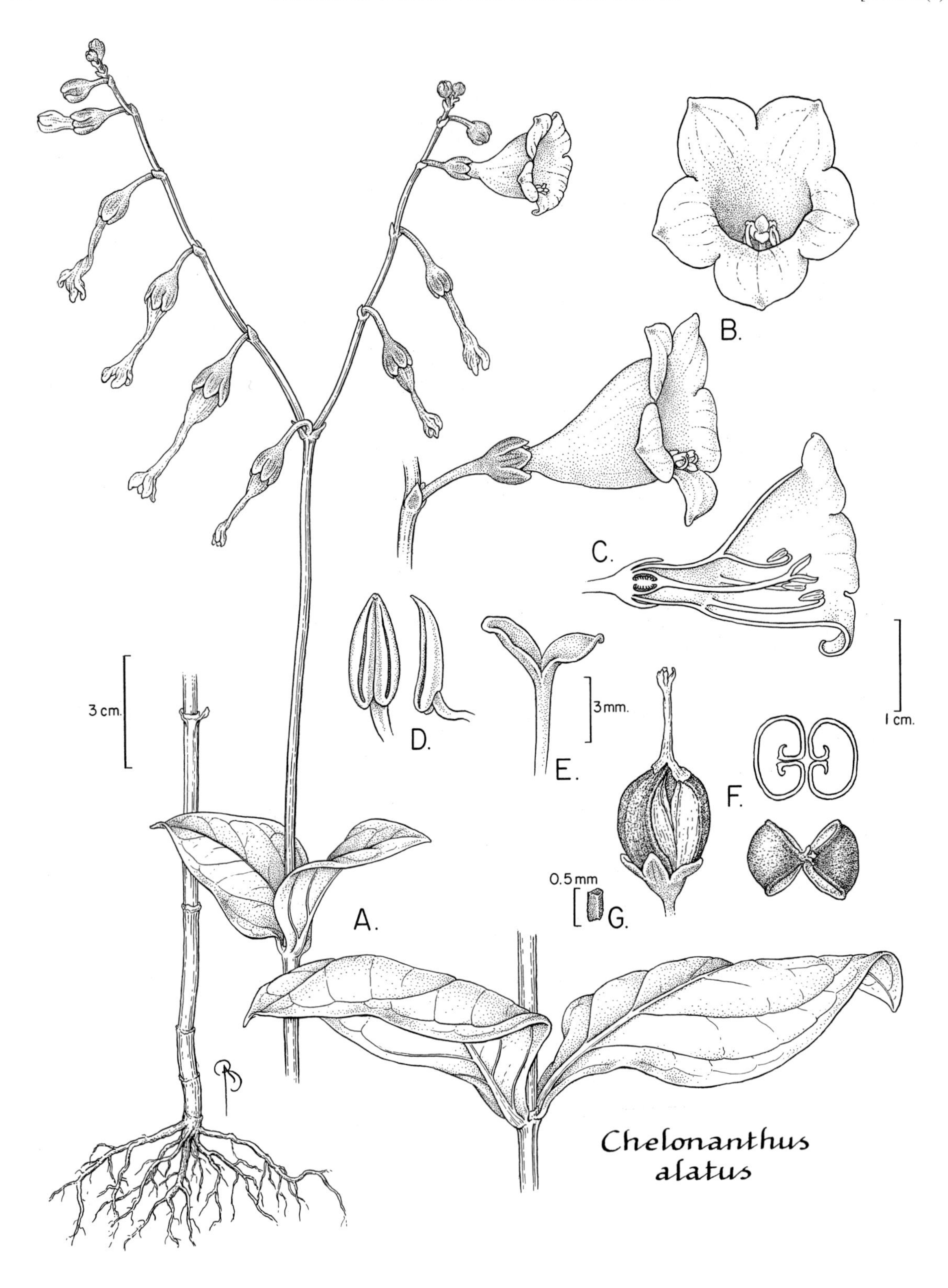

FIG. 133. GENTIANACEAE. *Chelonanthus alatus* (*Mori et al. 20950*). **A.** Apex of stem with terminal inflorescence (near left), base of plant showing leaf scars and roots (far left), and paired leaves from mid-section of stem (right). **B.** Apical (above) and lateral (below) views of flower; note stippling represents glandular areas on calyx. **C.** Medial section of flower. **D.** Adaxial (left) and lateral (right) views of anthers. **E.** Apex of style with bilobed stigma. **F.** Lateral view of fruit with persistent style (left); transverse section of ovary (above right) and apical view (below right) of fruit. **G.** Seed.

Coutoubea spicata Aubl. FIG. 134, PL. 62a

Herbs, 0.25–1.5 m tall. Leaves narrowly elliptic, 6–10 × 1.5–2.5 cm, the apex acute. Inflorescences 5–15 × 1–2 cm. Flowers: sepals narrowly triangular, 5–8 mm long, green with broad hyaline margins; corolla 10–15 mm long, white to pinkish white with dark red to purple throat; stamens and style exserted. Capsules ovoid or ellipsoid, 5–9 × 1.5–3 mm. Fl (Feb, Apr, May, Jun, Sep, Aug, Oct), fr (Jul, Aug, Oct); common, in secondary vegetation, especially roadsides and the airport landing strip, reputed to be poisonous to cattle. *Centorel* (Créole), *diambarana* (Portuguese).

TACHIA Aubl.

Shrubs or treelets. Stems terete, often with resin in upper leaf axils. Leaves fully developed, with chlorophyll; petioles present; blades not amplexicaulous. Inflorescences axillary, the flowers solitary, successively appearing. Flowers actinomorphic, 5-merous; calyx tubular, keeled to winged; corolla salverform, deciduous. Capsules dehiscing by 2 valves.

Tachia grandiflora Maguire & Weaver FIG. 135, PL. 62c

Shrubs, 2–4 m tall. Leaves: petioles 10–15 mm long; blades elliptic, 9–18 × 4.5–7 cm, the apex short-acuminate; midrib raised on both surfaces, secondary venation inconspicuous. Flowers: calyx 20–25 mm long, keeled, green; corolla salverform, 60–70 mm long, yellow to yellow-orange. Capsules narrowly ellipsoid, 20–40 mm long, exceeding persistent calyx, crowned by persistent style. Fl (Mar, May, Jul, Oct), fr (Jul); in nonflooded forest. *Mahot noir* (Créole).

VOYRIA Aubl.

Saprophytic herbs. Stems terete. Leaves small and scale-like, without chlorophyll; petioles absent; blades not amplexicaulous. Inflorescences terminal, solitary or few- to many-flowered dichasia. Flowers actinomorphic, (4)5(6)-merous; calyx tubular; corolla salverform, mostly marcescent; ovary sometimes provided with glands. Capsules septicidally dehiscent or indehiscent. Seeds subglobose or filiform.

1. Corolla usually <25 mm long.
    2. Roots forming dense, stellate clump. Flowers solitary, buds nodding; corolla lobes caducous, the lobes blue to purple; ovary with 2 stalked glands. . . . . . *V. tenella.*
    2. Roots not as above. Flowers solitary or in many-flowered dichasia, buds not nodding; corolla lobes not caducous, the lobes yellow, orange, salmon, cream-colored, or white.
        3. Inflorescences of a single flower.
            4. Stamens with hairy tail at base of thecae. Flowers 4- or 5-merous. . . . . . *V. spruceana.*
            4. Stamens without tails. Flowers 5(6)-merous.
                5. Corolla 12–13 mm long; ovary with two sessile glands at base. . . . . . *V. flavescens.*
                5. Corolla 20–30 mm long; ovary without glands. . . . . . *V. aphylla.*
        3. Inflorescences usually with >1 flower.
            6. Corolla yellow or orange; calyx 5–10 mm long. . . . . . *V. aurantiaca.*
            6. Corolla white; calyx 2–4 mm long. . . . . . *V. corymbosa.*
1. Corolla ≥25 mm long.
    7. Corolla tube constricted in upper part, the lobes distinctly recurved. . . . . . *V. tenuiflora.*
    7. Corolla tube not constricted, the lobes not recurved.
        8. Flowers solitary; corolla yellow to orange. . . . . . *V. aphylla.*
        8. Inflorescences 1- to few-flowered; corolla never yellow.
            9. Calyx distinctly 5-lobed; corolla lobes pink to red. . . . . . *V. rosea.*
            9. Calyx irregularly split, indistinctly 5-lobed; corolla lobes blue, purple, or whitish. . . . . . *V. caerulea.*

Voyria aphylla (Jacq.) Pers.

Herbs, 5–20 cm tall. Stems whitish to pale pink. Inflorescences 1-flowered. Flowers 5-merous; calyx 3–4 mm long, the lobes 1–2 mm long; corolla 20–30 mm long, yellow to orange; anthers subsessile; ovary without glands. Capsules medially dehiscent. Seeds filiform. Fl (Jan, Jul); known from wet, open rocky places at Pic Matécho, near rapids at Saut Maïs, and in forest at base of Mont Galbao.

Voyria aurantiaca Splitg. PL. 62d

Herbs, 5–20 cm tall. Stems salmon to orange. Inflorescences 1- to many-flowered. Flowers 5-merous; calyx 5–10 mm long, the lobes 3–4 mm long; corolla 15–25 mm long, yellow or orange, the tube hairy adaxially; anthers subsessile; ovary without glands. Capsules dehiscing by 2 valves. Seeds subglobose. Fl (Jan, Feb, Mar, Jun, Jul, Aug, Oct, Sep, Nov, Dec); in marshy forest and in moist nonflooded forest.

Voyria caerulea Aubl. PL. 62e

Herbs, 5–10 cm tall. Stems bluish to white, sometimes deep purple. Inflorescences 2–6-flowered (rarely to 25-flowered); bracts enclosing flowers. Flowers 5(6)-merous; calyx 10–25 mm long, the tube longitudinally irregularly split, the lobes 1–2 mm long; corolla

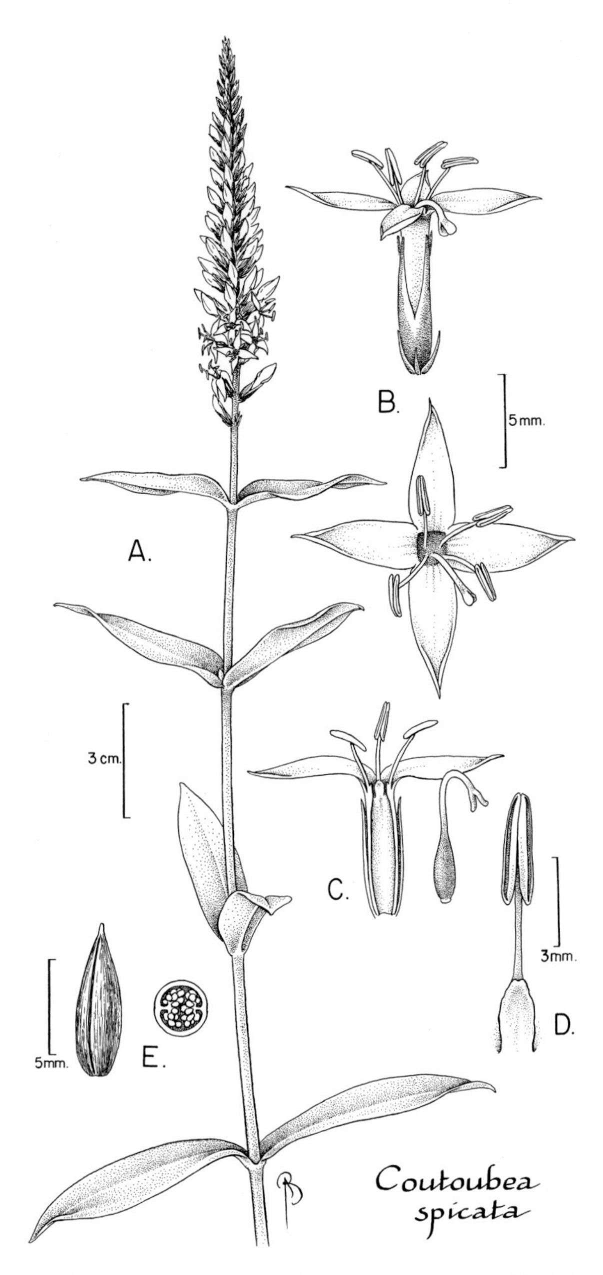

FIG. 134. GENTIANACEAE. *Coutoubea spicata* (A–D, *Mori et al. 20947*; E, *Skog 7111*). **A.** Apex of stem with terminal spike; note the leaf blade bases connate across node. **B.** Lateral (above) and apical (below) views of flowers. **C.** Medial section of flower (left) with pistil removed and pistil (right). **D.** Adaxial view of stamen. **E.** Lateral view (left) and transverse section (right) of fruit showing parietal placentation of seeds.

30–45 mm long, blue, purple, or whitish, the throat white; anthers with distinct filaments; ovary without glands. Capsules indehiscent. Seeds subglobose. Fl (Jan, Feb, Mar, May, Jun, Jul, Aug), fr (Jun, Sep); common, nonflooded moist forest.

Voyria corymbosa Splitg. PL. 63a,d

Herbs, 8–25 cm tall. Stems salmon to white. Inflorescences few- to many-flowered. Flowers 5(6)-merous; calyx 2–4 mm long, the lobes 1–2 mm long; corolla 8–20 mm long, the lobes white to salmon, the filaments absent to distinct, 0–5 mm long; ovary without glands. Capsules dehiscing by 2 valves. Seeds subglobose. Fl (Jan, Feb, Mar, Apr, May, Jul, Sep, Oct, Nov), fr (Jan, Sep); common, in nonflooded forest and along streams.

Voyria corymbosa was split by Maas and Ruyters (1986) into subsp. *corymbosa* and subsp. *alba* on features that no longer appear to hold up.

Voyria flavescens Griseb. PL. 63b

Herbs, to 10 cm tall. Roots tuberous, forming a dense, stellate clump. Stems cream-colored. Inflorescences 1-flowered. Flowers 5-merous; calyx ca. 5 mm long, the lobes ca. 1.5 mm long; corolla 12–13 mm long, the tube cream-colored, the lobes yellow adaxially, the throat orange; anthers with distinct filaments; ovary basally provided with 2 sessile glands. Capsules dehiscing by 2 valves. Seeds subglobose. Fl (Feb, Mar), in secondary vegetation near stream near old army camp just N of Eaux Claires.

Voyria rosea Aubl. PL. 63c,d

Herbs, 10–20 cm tall. Stems red to pinkish red. Inflorescences 1–6-flowered. Flowers 5-merous, smelling like those of *Convallaria* (Liliaceae *s. lat.*); calyx 6–15 mm long, the lobes 1.5–2 mm long; corolla 60–90 mm long, the lobes pink to red; anthers subsessile; ovary without glands. Capsules indehiscent. Seeds subglobose. Fl (Jan, Feb, Mar, Apr, May, Jul, Aug, Sep, Oct); seasonally common, in nonflooded moist forest.

Voyria spruceana Benth.

Herbs, 2–7 cm tall. Stems salmon-red. Inflorescences 1-flowered. Flowers 4–5-merous; calyx reddish, 4–6 mm long, winged to keeled, the lobes 2–3 mm long; corolla 12–16 mm long, yellow; stamens with hairy tail at base of thecae; subsessile ovary without glands. Capsules indehiscent. Seeds subglobose. Fl (Jan, Jul); in savanna forest on slope near Pic Matécho.

Voyria tenella Hook. PL. 64a

Herbs, 10–20 cm tall. Roots tuberous, forming dense, stellate clump. Stems whitish. Inflorescences 1-flowered, the buds nodding. Flowers 5-merous; calyx ca. 2 mm long, orange, the lobes 1 mm long; corolla 9–15 mm long, the tube and throat yellow to orange, the lobes blue to purple, caducous; anthers subsessile; ovary provided with 2 stalked glands. Capsules medially dehiscent. Seeds filiform. Fl (Feb, May, Jun, Jul), fr (Feb); in moist nonflooded forest and in submontane forest.

Voyria tenuiflora Griseb. PL. 64b

Herbs, 10–20 cm tall. Stems salmon to orange. Inflorescences 1-flowered. Flowers 5-merous; calyx 6–10 mm long, the lobes 2–6 mm long; corolla 25–40 mm long, white to yellow-orange at base, creamy white in upper part, the tube constricted in upper part, the lobes distinctly recurved; anthers subsessile; ovary without glands. Capsules indehiscent. Seeds subglobose. Fl (Feb, May); in moist nonflooded forest.

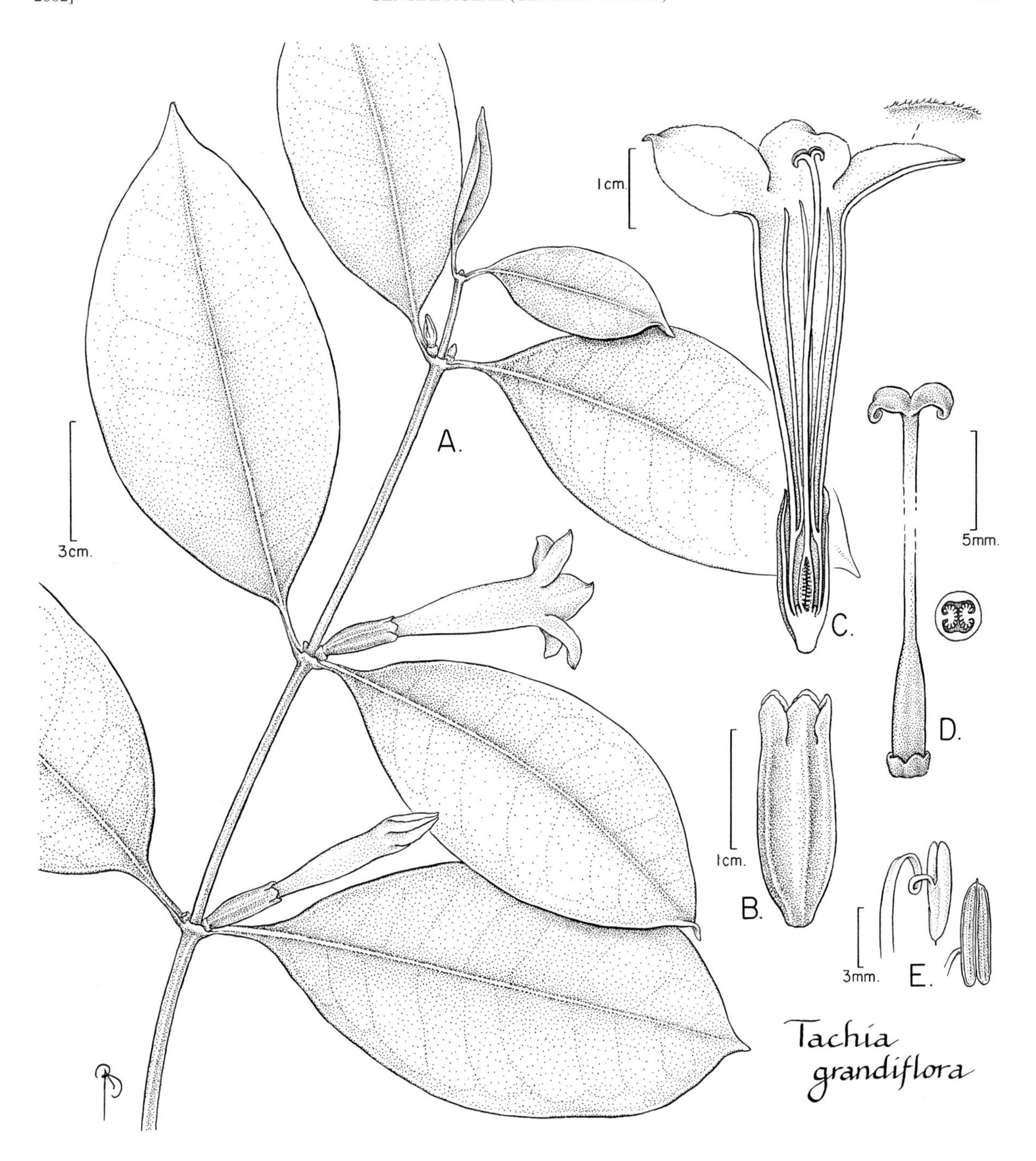

FIG. 135. GENTIANACEAE. *Tachia grandiflora* (A–D, *Mori et al. 22277*; E, *Dick 112* from Amazonas, Brazil). **A.** Upper part of stem with leaves, flower buds, flower, and old flower. **B.** Lateral view of keeled calyx. **C.** Medial section of flower (left) with detail of adaxial surface of corolla lobe (above). **D.** Lateral view of pistil (left) and transverse section of ovary (above). **E.** Two views of anthers.

VOYRIELLA (Miq.) Miq.

Saprophytic herbs. Stems markedly 4-angled. Leaves small and scale-like, without chlorophyll; petioles absent; blades not amplexicaulous. Inflorescences terminal, capitate, composed of many-flowered dichasia. Flowers actinomorphic, 5-merous; sepals free; corolla tubular, hardly exceeding sepals, caducous. Capsules indehiscent. Seeds subglobose.

Voyriella parviflora (Miq.) Miq. PL. 64c

Herbs, 5–10 cm tall. Stems white. Inflorescences many-flowered, globose, to 10 × 15 mm. Flowers: sepals 4–5 mm long; corolla 6–7 mm long, white; anthers subsessile; ovary without glands. Fl (Feb, Apr, May, Jul, Sep), fr (May); in moist nonflooded forest, often buried under leaf litter.

GESNERIACEAE (African Violet Family)

Christian Feuillet and Laurence E. Skog

Epiphytic or terrestrial herbs, subshrubs, shrubs, or vines. Stems with or without well developed internodes. Stipules lacking. Leaves simple, opposite, or rarely whorled in the Guianas; petioles present; blades of a pair equal to strongly unequal. Inflorescences of solitary flowers, racemes, or modified cymes. Flowers usually zygomorphic, rarely nearly actinomorphic, bisexual; calyx of 5 lobes, rarely 2 (*Codonanthe*); corolla 5-lobed; stamens 4 or 5, adnate to corolla, staminode present or absent, the anthers often connivent; ovary superior, 1-locular, the style simple; placentation parietal, the placentae 2. Fruits berries or dry or fleshy capsules, 2-valved. Seeds ± fusiform or oblong, numerous, very small.

Feuillet, C. & J. A. Steyermark. 1999. Gesneriaceae. *In* J. A. Steyermark, P. E. Berry, K. Yatskievych & B. K. Holst (gen. eds.), Flora of the Venezuelan Guayana **5:** 542–573. P. E. Berry, K. Yatskievych & B. K. Holst (vol. eds.). Missouri Botanical Garden Press, St. Louis.
Leeuwenberg, A. J. M. 1984. Gesneriaceae. *In* A. L. Stoffers & J. C. Lindeman, Flora of Suriname **V(1):** 592–650. E. J. Brill, Leiden.

1. Plants stoloniferous. . . . . *Episcia.*
1. Plants not stoloniferous.
   2. Stems with very short internodes. Leaves in apical rosette-like clusters. . . . . *Napeanthus.*
   2. Stems with obvious internodes. Leaves not in clusters.
      3. Plants 1–2 m tall when blooming, self-supporting.
         4. Plants terrestrial. Inflorescences lacking bracts. . . . . *Besleria.*
         4. Plants epiphytic. Inflorescences with pink to red bracts. . . . . *Drymonia coccinea.*
      3. Plants <1 m tall when blooming or vines.
         5. Plants terrestrial, not vines.
            6. Plants tuberous. Calyx campanulate, orange; corolla bright yellow. . . . . *Chrysothemis.*
            6. Plants lacking tubers. Sepals free or nearly free, green to red; corolla white or cream-colored.
               7. Stems shorter than leaves. . . . . *Paradrymonia densa.*
               7. Stems longer than leaves. . . . . *Nautilocalyx.*
         5. Plants vines or epiphytes.
            8. Leaves of a pair strongly unequal (the small one caducous, or only 1/2 to 1/10 as large).
               9. Plants glabrous. Leaf margins entire. . . . . *Drymonia psilocalyx.*
               9. Plants hirsute. Leaf margins serrate-dentate. . . . . *Columnea sanguinea.*
            8. Leaves of a pair equal or subequal (when subequal, the smaller one >1/2 as large).
               10. Stems hirsute, at least near apex.
                  11. Corollas yellow or orange, red hirsute. . . . . *Columnea calotricha.*
                  11. Corollas white, white hirsute. . . . . *Paradrymonia campostyla.*
               10. Stems glabrous to puberulent.
                  12. Inflorescences >5-flowered, pedunculate. . . . . *Drymonia coccinea.*
                  12. Inflorescences 1–3-flowered, epedunculate.
                     13. Flowering stems wiry. Fruits berries.
                        14. Plants often growing in epiphytic ant nest. Pedicels shorter than leaves. Corolla white to yellowish, pinkish, or light purple. . . . . *Codonanthe.*
                        14. Plants not growing in ant nest. Pedicels longer than leaves. Corollas orange-red. . . . . *Columnea oerstediana.*
                     13. Flowering stems fleshy. Fruits fleshy capsules. . . . . *Drymonia serrulata.*

BESLERIA L.

Herbs to subshrubs, terrestrial, to 2 m tall, without stolons or tubers. Stems erect, single or branched only at base, the internodes well developed. Leaves equal or subequal in a pair; blades with acuminate apices, the margins serrate. Inflorescences cymose, umbellate, paniculate, or fasciculate, ebracteate. Flowers: calyx connate only at base to 3/4 length, 5-lobed; corolla not spurred at base; stamens 4; staminode present, with a sterile anther. Fruits apiculate berries.

1. Inflorescences epedunculate. Calyx lobes connate >1/2 of length; corolla <50% longer than calyx. . . . . *B. insolita.*
1. Inflorescences pedunculate. Calyx lobes connate <1/2 length. Corolla about twice as long as calyx.
   2. Inflorescences congested-paniculate; pedicels 1–3 mm long. Calyx lobes connate only at base, the lobes orbicular or nearly so; corolla 5–6 mm long, white, partly purple in bud, the lobes pubescent adaxially. . . . . *B. flavovirens.*
   2. Inflorescences umbellate or subcymose, loose; pedicels 3–30 mm long. Calyx lobes connate 1/3–1/2 length, the lobes triangular; corolla 15–25 mm long, orange, pink, or red, the lobes glabrous. . . . . *B. patrisii.*

Besleria flavovirens Nees & Mart.

Subshrubs or herbs, 0.3–2 m tall. Stems simple or branched only at base. Leaves: petioles 1.5–6 cm long; blades oblong-lanceolate, 12–35 × 5–12 cm, glabrous adaxially, appressed-pubescent abaxially. Inflorescences congested-paniculate with numerous flowers; peduncle 5–40 mm long, irregularly branched; pedicels 1–3 mm long. Flowers small, aggregated; calyx lobes connate at base, orbicular or nearly so, 2–3 × 2–2.5 mm, greenish white, round or emarginate at apex; corolla zygomorphic, dorsally ventricose, white (partly purple in bud), the tube 3.5–4 mm long, the lobes spreading, the 2 dorsal ones 1 mm long, the 3 ventral ones 1.5–2 mm long, pubescent adaxially, rounded; stamens exserted; stigma capitate. Fruits globose berries, 4–6 mm diam., dark purple. Fr (Jan); common but infrequently collected, in or near running water, especially on boulders.

Besleria insolita C. V. Morton FIG. 136, PL. 64d

Subshrubs or herbs, 0.5–1.7 m tall. Stems simple or branched only at base. Leaves: petioles 2.5–12 cm long; blades elliptic or oblong-elliptic, 10–30 × 4–11 cm, sparsely strigose adaxially, strigillose abaxially. Inflorescences fasciculate, epedunculate, 1–6-flowered; pedicels 5–20 mm long. Flowers: calyx tube 7–13 mm long, bright yellow, the lobes connate for 2/3–3/4 length, triangular, 3–6 × 1–2 mm, yellow with white tips, acuminate; corolla actinomorphic, slightly longer than calyx, not ventricose, white, the tube 9–15 mm long, the lobes erect, subequal, 2 mm long, rounded; stamens included; stigma bilobed. Fruits subglobose berries, 7 × 6 mm, bright red. Fl (Jan, Feb, Jun, Mar, Aug, Oct, Nov), fr (Feb, Mar, Jul, Sep); common, in moist areas, especially near running water on boulders.

Besleria patrisii DC. [Syns.: *Besleria verecunda* C.V. Morton, *B. maasii* Wiehler] PL. 64e

Subshrubs or herbs, 0.3–2 m tall. Stems simple or branched only at base. Leaves: petioles 1–7 cm long; blades elliptic or oblong-elliptic, 5–25 × 1.5–9 cm, sparsely pubescent adaxially, strigose abaxially. Inflorescences umbellate or subcymose, 1–6-flowered; peduncle 1–6 cm long; pedicels 3–30 mm long. Flowers: calyx tube 5–7 mm long, green or yellow to red, the lobes connate for 1/3–1/2 length, triangular, 5–9 × 3–5 mm, acuminate; corolla actinomorphic, to 2× as long as calyx, not ventricose, orange, pink or red, the tube 17–25 mm long, the lobes erect, sub-equal, 2 mm long, rounded; stamens included; stigma bilobed. Fruits subglobose berries, 10–12 mm diam., pale green, whitish, yellow, orange, red or pink. Fl (Jan, Mar, Jul, Aug, Nov, Dec), fr (Jan, Mar, Dec); common, in or near running water, especially on boulders.

CHRYSOTHEMIS Decne.

Herbs, terrestrial with tubers, without stolons. Stems erect, mostly unbranched. Leaves equal or subequal in a pair, the internodes well developed; blades with acuminate apices, the margins crenate-serrate. Inflorescences axillary, umbellate, much shorter than leaves, bracteate. Flowers: calyx persistent; corolla actinomorphic; stamens 4; staminode absent. Fruits dry capsules.

Chrysothemis pulchella (Sims) Decne.

Herbs, 0.1–1 m tall. Leaves: blades elliptic, ovate or oblong-ovate, to 30 × 12 cm, sparsely pubescent adaxially, puberulous abaxially, the base decurrent onto petiole. Inflorescences with numerous flowers; peduncle 1–5 cm long, the bracts oblong or lanceolate, 4–15 × 1–6 mm; pedicels 1–3 cm long. Flowers: calyx campanulate, 5-angled, 10–20 mm long, puberulous or villose abaxially, orange or red, the lobes acuminate, dentate; corolla yellow with brownish-red markings, the tube 15–30 mm long, white-sericeous abaxially, the lobes 3–8 mm diam.; stamens included, the anthers all coherent or free, the filaments inserted at base of corolla tube; stigma bilobed. Fruits capsules, 6 mm diam., sparsely pubescent. Known from our area only by a sterile collection, flowering in August in other parts of southern French Guiana; on boulder in forest near Crique Cochon.

Description based on collections from French Guiana outside our area.

CODONANTHE Hanst.

Herbs to subshrubs, epiphytic, often associated with ant nests, without stolons or tubers. Stems with well developed internodes. Leaves equal in a pair; blades with obtuse to acuminate apices, the margins entire or serrulate. Inflorescences axillary, epedunculate, fasciculate. Flowers: corollas white, yellowish, pinkish, or light purple; anther cells separated by enlarged connective, partially dehiscent in middle or at base. Fruits berries.

1. Leaf blade margins often with few serrations toward apex. Calyx lobes 2, the dorsal lobe small, the ventral lobe large, round to 4-toothed. . . . . *C. calcarata.*
1. Leaf blade margins usually entire. Calyx with 5 subequal, free lobes. . . . . *C. crassifolia.*

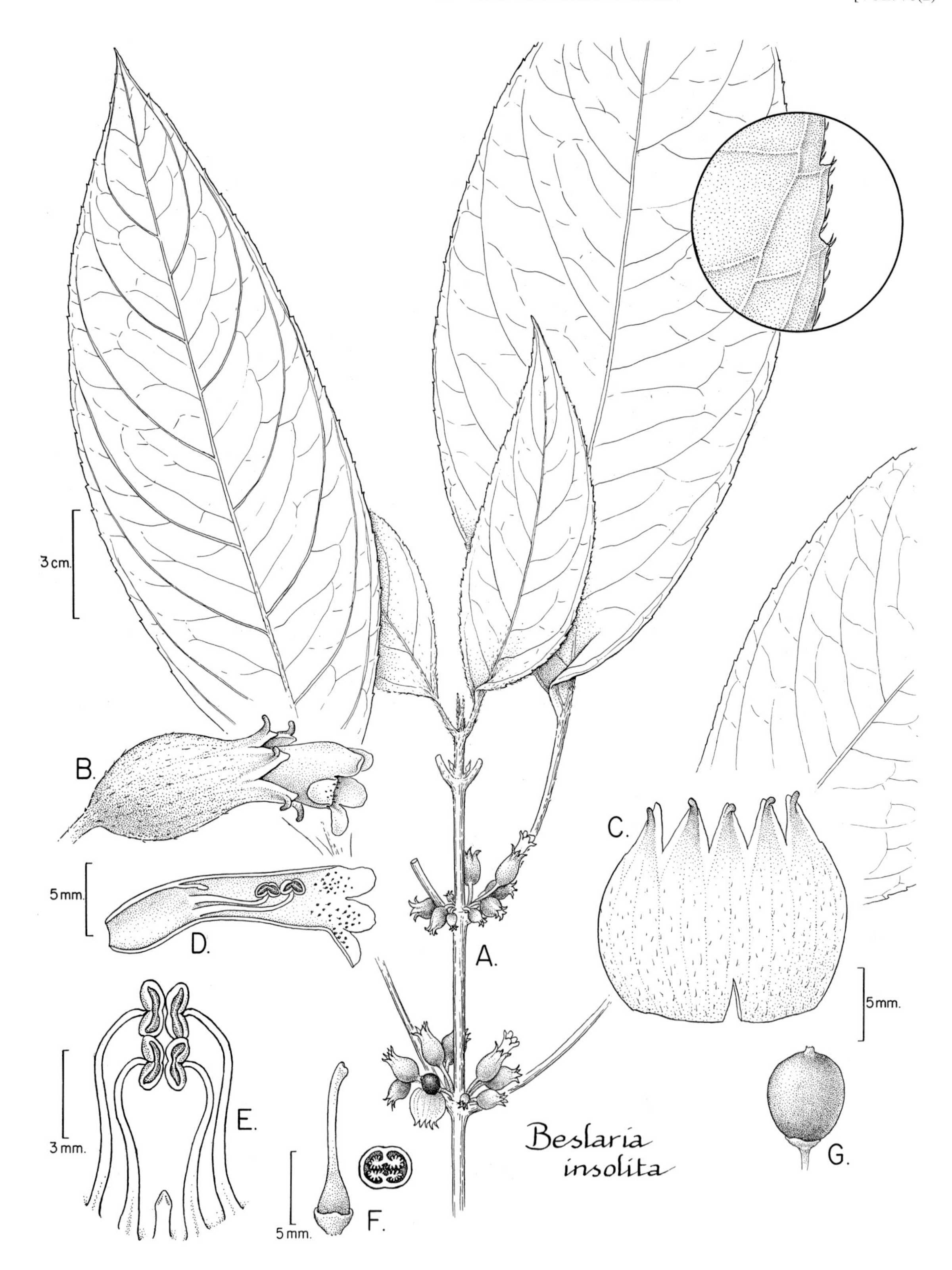

FIG. 136. GESNERIACEAE. *Besleria insolita* (*Mori & Gracie 21176*). **A.** Apex of stem with leaves, axillary inflorescences, and detail of ciliate leaf margin (right); note the spreading calyx subtending one of the lower fruits. **B.** Lateral view of whole flower; note the appendaged calyx lobes. **C.** Abaxial view of opened calyx from below fruit. **D.** Medial section of corolla with two stamens and staminode. **E.** Stamens and staminode; note longitudinal anther dehiscence. **F.** Pistil with cupular nectary at base (left) and transverse section of ovary (right). **G.** Mature fruit after abscission of calyx.

Codonanthe calcarata (Miq.) Hanst. PL. 65a

Epiphytes. Stems 0.15–1 m long. Leaves equal in a pair; petioles 1–10 mm long; blades oblong-elliptic, 1.5–12 × 0.5–5 cm, often fleshy, puberulous to glabrous, dark green adaxially, paler green to dark purple abaxially, the base cuneate, the apex acuminate, the margins mostly serrulate toward apex. Inflorescences 1–8-flowered. Flowers: calyx lobes 2, the dorsal one linear, 2–4 × <1 mm, curved around corolla spur, the ventral one larger, oblong or rectangular, 3–7 × 1.5–4 mm, green, the apex round to 4-dentate; corolla oblique, nearly transverse in calyx, spurred at base, glabrous abaxially, white, pinkish or light purple, with or without pink to purple markings, the tube 16–25 mm long, the lobes 3–8 mm wide; stamens included. Fruits globose berries, 8–12 mm diam., glabrous, purple to dark red. Fl (Oct, Nov, Dec); mostly in ant nests in forest on well drained and wet soils, occasionally on rocks.

Codonanthe crassifolia (Focke) C. V. Morton FIG. 137, PL. 65c

Epiphytes. Stems 0.15–2 m long. Leaves equal in a pair; petioles 2–15 mm long; blades oblong-ovate or oblong-elliptic, 2–8.5 × 0.5–4 cm, fleshy to succulent, glabrous, green or reddish, sometimes yellowish abaxially, the base cuneate or round, the apex acute or obtuse, the margins mostly entire, rarely obscurely sinuate. Inflorescences 1–4-flowered. Flowers: sepals 5, free, 2.5–9 × 0.5–1.5 mm, green, subequal, the dorsal one curved around corolla spur; corolla oblique, nearly transverse in calyx, spurred at base, puberulous or glabrous abaxially, white or yellowish, with or without pink markings abaxially, yellow in throat, the tube 12–18 mm long, the lobes 2–6 mm wide; stamens included. Fruits subglobose berries, 10–12 × 7–9 mm, nearly glabrous, bright red. Fl (Sep, Oct, Nov), fr (Feb, Nov); mostly in ant nests in forest on well drained and wet soils, occasionally on rocks.

COLUMNEA L.

Herbs to subshrubs, epiphytic, without stolons or tubers. Stems to 5 m long, with a few irregular branches, the internodes well developed. Leaves equal to strongly unequal in a pair; blades variable in size, shape, and texture. Inflorescences axillary, fasciculate, cymose, or of solitary flowers. Flowers: sepals 5, nearly free; corolla tubular, erect in calyx, slightly to strongly zygomorphic. Fruits berries.

1. Leaves of a pair equal or subequal.
    2. Stems fleshy. Leaves much longer than flowers. Corolla not markedly bilabiate. . . . . *C. calotricha.*
    2. Stems wiry. Leaves shorter than flowers. Corolla strongly bilabiate. . . . . *C. oerstediana.*
1. Leaves of a pair strongly unequal. . . . . *C. sanguinea.*

Columnea calotricha Donn. Sm. [Syn.: *Trichantha calotricha* (Donn. Sm.) Wiehler] FIG. 138

Herbs or subshrubs, epiphytic or saxicolous. Plants hirsute. Stems horizontal to erect or ascending, fleshy, hirsute. Leaves subequal in a pair; petioles 5–25 mm long; blades oblong-elliptic or lanceolate, 7–16 × 1–4 cm, dark green adaxially, pale green, purplish, or red abaxially, the base obliquely cuneate, the apex acute, the margins crenulate. Inflorescences axillary, fasciculate, epedunculate, 1–7-flowered; pedicel 3–10 mm long. Flowers: sepals free, subequal, 12–25 × 3–6 mm; corolla erect in calyx, bright yellow or orange, red-hirsute, the tube gibbous at base, 17–35 mm long, the lobes subequal, oblong, 3–4 × 2–3 mm, the apex acute, entire; stamens exserted at anthesis; stigma bilobed. Fruits ovoid berries, 10 × 5 mm, with scattered hairs, red. Fl (May, Nov); common, in canopy, occasionally on rocks.

Columnea oerstediana Oerst.

Herbs or subshrubs, epiphytic. Plants appressed-pubescent to glabrescent. Stems pendent, creeping or shortly erect, wiry, to 5–6 m long. Leaves equal in a pair: petioles to 2 mm long; blades ovate to oblong-elliptic, 1–3.7 × 0.5–2 cm, green adaxially, pale green abaxially, the base cuneate, rounded or subcordate, the apex obtuse to acute or acuminate, the margins entire, somewhat revolute. Inflorescences axillary, 1-flowered; pedicels 10–25 mm long. Flowers: sepals free, subequal, 10–18 × 3–11 mm; corolla erect in calyx, strongly bilabiate, pilose, orange-red, the tube gibbous at base, 20–40 mm long, the ventral lobe reflexed, 12–16 mm long, the upper lip erect with 2 lateral, spreading lobes, 5–10 mm long and 2 dorsal lobes fused into a galea, 10–20 × 7–15 mm, rounded to slightly retuse or cuspidate at apex; stamens exserted at anthesis; stigma bilobed. Fruits ovoid berries, 6–8 mm diam., pubescent, pink. Fl (Jan, Sep, Nov); common, in canopy, sometimes in lower strata.

Columnea sanguinea (Pers.) Hanst. [Syns.: *Columnea aureonitens* Hook., *Dalbergaria aureonitens* (Hook.) Wiehler, *D. sanguinea* (Pers.) Steud.]

Subshrubs, epiphytic, or saxicolous. Plants mostly hirsute or tomentose. Stems spreading, fleshy, tomentose at apex, glabrescent elsewhere. Leaves strongly unequal in a pair, the larger one with petioles 5–15 mm long, tomentose; blades oblique, lanceolate, 15–30 × 3–8 cm, dark green adaxially, pale green abaxially, the base obliquely cuneate, the apex acuminate, the margins serrate-dentate, the smaller blades stipule-like, 1/10 as large as larger one. Inflorescences axillary, fasciculate, epedunculate, 1–3-flowered, the bracts lanceolate, laciniate; pedicels 2–5 mm long. Flowers: sepals free or nearly so, subequal, 12–20 × 3–4 mm; corolla slightly oblique in calyx, white-, yellow-, or red-pubescent, the tube gibbous at base, 21–27 mm long, yellowish white, the lobes subequal, oval, 2.5–4 × 2–2.5 mm, yellow; stamens included; stigma bilobed. Fruits subglobose berries, 15 × 10 mm, sparsely pubescent, white. Fl (Jan, Nov), fr (Jan, Nov); common, on tree trunks or rocks.

DRYMONIA Mart.

Herbs or subshrubs (terrestrial or epiphytic), or vines, without stolons or tubers. Stems branched at base and sometimes also in younger parts, the internodes well developed. Leaves opposite, equal to strongly unequal in a pair; blades variable in size, shape, and texture. Inflorescences axillary, cymose, racemose, or of solitary flowers. Flowers: calyx lobes 5, free or connate at base; corolla spurred at base; stamens 4, included. Fruits fleshy capsules, becoming coriaceous, dehiscing by 2 valves, the valves spreading or recoiling to display colored interior.

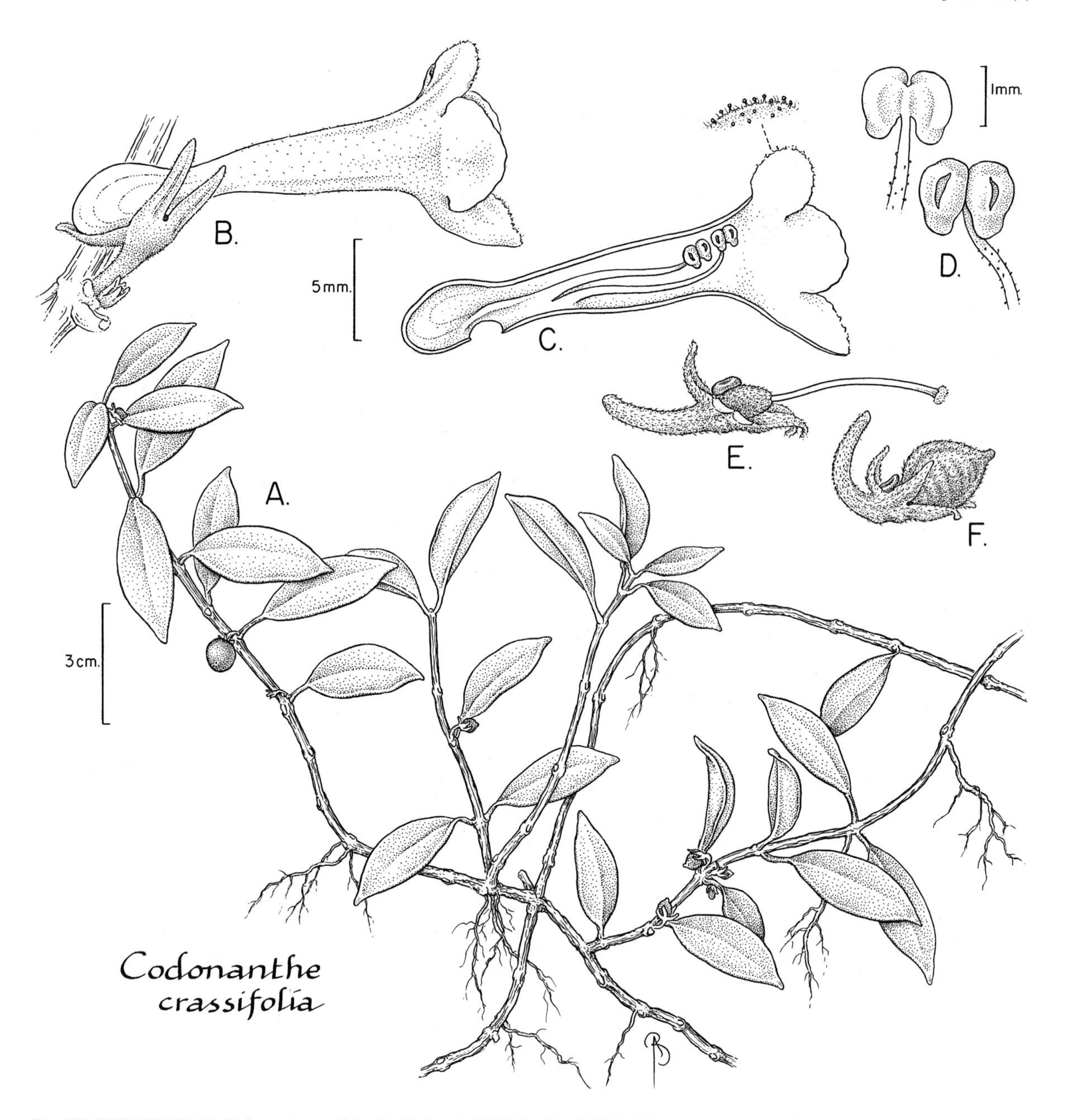

FIG. 137. GESNERIACEAE. *Codonanthe crassifolia* (A, *Mori et al. 21575*; B, *Mori 24016*; C–F, *Mori et al. 21610*). **A.** Part of stem showing adventitious roots from nodes on the stem, leaves, and fruit; note the absence of visible higher order venation in the leaf blades. **B.** Lateral view of flower; note shape of calyx and spurred corolla. **C.** Medial section of corolla showing adnate stamens and detail of margin. **D.** Abaxial (above left) and adaxial (above) views of stamens. **E.** Pistil on pedicel and subtended by three of the five sepals. **F.** Immature fruit.

1. Inflorescences with pink to red bracts. Corolla whitish or more often sulfur yellow. . . . . *D. coccinea.*
1. Inflorescences without bracts. Corolla white to peach or pink.
   2. Epiphytes. Leaves of a pair strongly unequal, the margins entire. . . . . *D. psilocalyx.*
   2. Climbers. Leaves of a pair equal, the margins serrulate. . . . . *D. serrulata.*

Drymonia coccinea (Aubl.) Wiehler [Syns.: *Alloplectus patrisii* DC., *A. coccineus* (Aubl.) Mart.]

FIG. 139, PL. 65d

Subshrubs, epiphytic or terrestrial, 0.6–2.5 m tall at maturity, creeping vines as juveniles. Stems sarmentose, branched or unbranched, puberulous at apex. Leaves of a pair equal to unequal, the larger one to 2.5× times as long as shorter one; petioles 0.5–6 cm long; blades fleshy, obliquely elliptic to oblong, 8–30 × 3–10 cm, fleshy, sparsely appressed-pubescent to glabrous on both surfaces, dark green adaxially, pale green to dark red abaxially, the base cuneate to almost decurrent, the apex acuminate, the margins entire

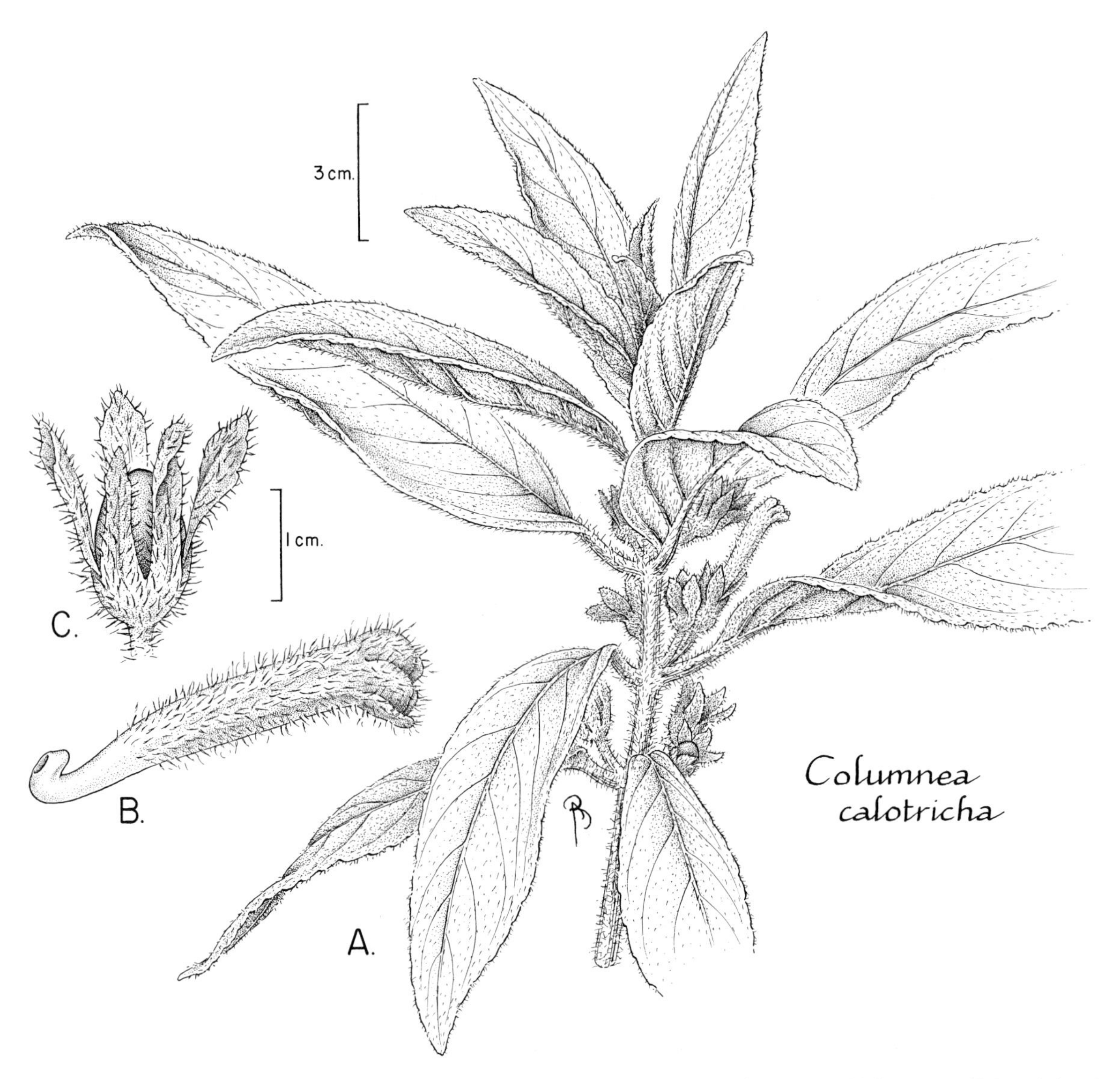

FIG. 138. GESNERIACEAE. *Columnea calotricha* (*Mori et al. 22216*). **A.** Apex of stem with leaves, inflorescences, and infructescences. **B.** Lateral view of corolla; note short adaxial spur. **C.** Fruit with persistent calyx.

or sometimes obscurely denticulate near apex. Inflorescences pendent, racemose, the axis branched, 6–15-flowered, the flowers congested, included in bracts; peduncle nearly absent to 6 cm long; bracts and bracteoles leafy, nearly orbicular, 1.5–4.5 cm wide, pinkish to red, occasionally yellowish-white. Flowers: sepals free, colored like bracts, 13–25 × 6–15 mm; corolla oblique in calyx, whitish or more often sulfur yellow, the tube 28–35 mm long, the lobes subequal, rounded, 4–10 mm wide, with or without red-brown dots; stigma capitate. Fruits subglobose capsules, 13–15 × 9–10 mm, fleshy, sparsely appressed-pubescent, whitish or yellow, the inside wall pink. Fl (Feb, Aug, Sep, Oct, Nov, Dec), common, often on trees at lower heights. *Crête poule* (Créole).

Drymonia psilocalyx Leeuwenb. [Syn.: *Drymonia psila* Leeuwenb.]

Herbs or subshrubs, epiphytic. Plants entirely glabrous except for some hairs at apex of ovary. Stems spreading, to 1.5 m long, succulent. Leaves succulent, brittle, strongly unequal in a pair, the larger leaf with petioles 10–30 mm long; blades narrowly elliptic, 10–24 × 2.5–6.5 cm, the base obliquely cuneate, the apex acuminate, the margins entire, medium green adaxially, much paler abaxially; shorter blades 1/10 to 1/2 as large. Inflorescences 1-flowered; pedicels 5–12 mm long. Flowers: calyx lobes connate at base, the dorsal one 1/3 as long as others, pale green; corolla spurred, oblique in calyx, white, with a large pale yellow to orange spot ventrally in tube, the tube 25–30 mm long, the lobes rounded, fimbriate, the ventral lobe larger than others, ca. 8 mm wide; stigma capitate. Fruits obliquely ovoid capsules, 18 × 10 mm, fleshy, dark violet, the apex acute, the inside wall reddish-brown. Seed mass black. Fl (Jan), fr (Nov, Dec); on tree trunks or main branches in forest.

Drymonia serrulata (Jacq.) Mart. [Syn.: *Drymonia cristata* Miq.]

Vines, herbaceous or suffrutescent. Stems branched. Leaves of a pair equal; petioles 6–30 mm long; blades elliptic, oblong or ovate, 4–19 × 2–8 cm, sparsely pubescent to glabrous on both surfaces,

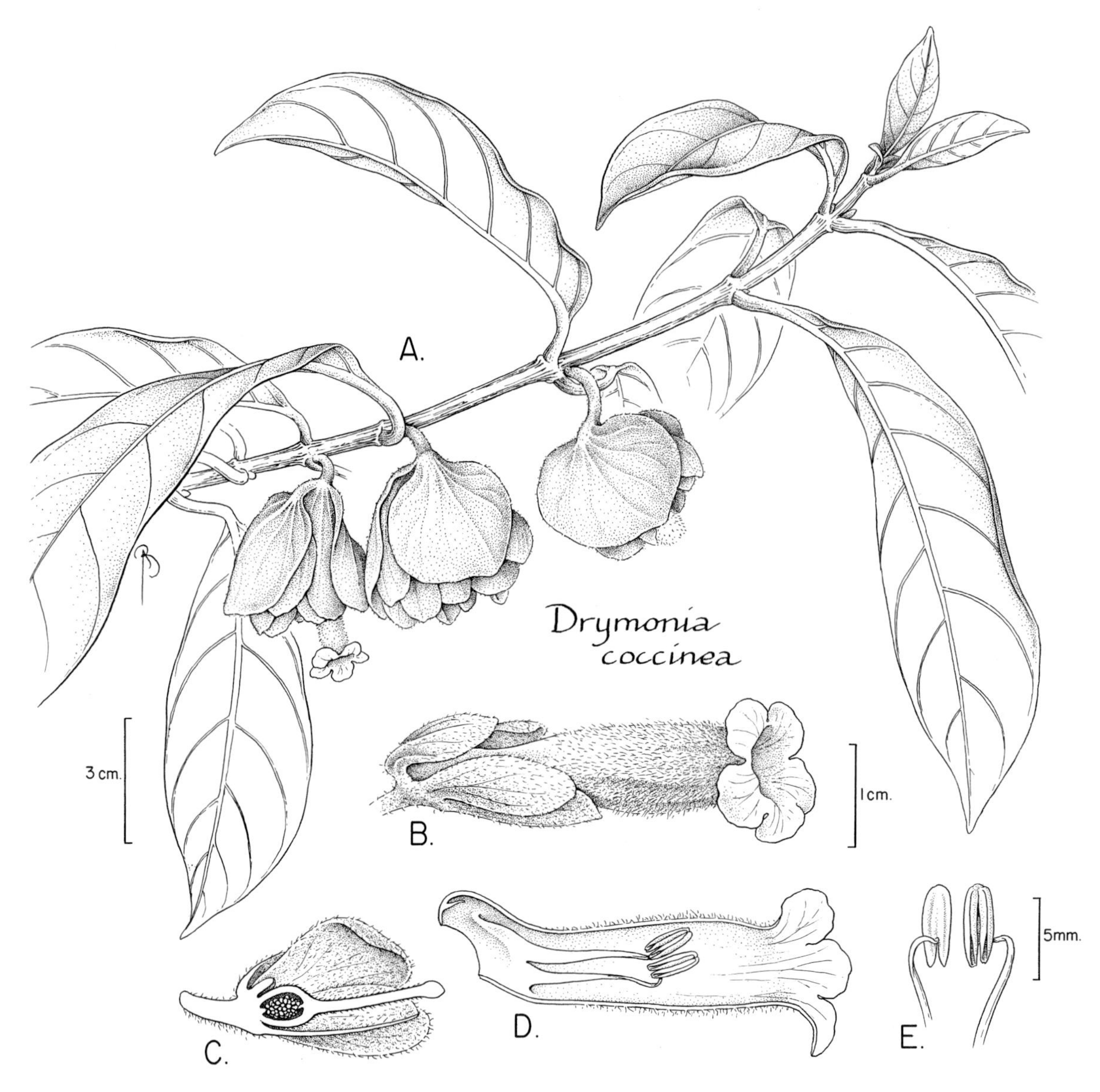

FIG. 139. GESNERIACEAE. *Drymonia coccinea* (*Mori et al. 20937*). **A.** Apex of stem with leaves and axillary inflorescences subtended by bracts. **B.** Lateral view of flower. **C.** Medial section of flower with corolla and androecium removed; note the superior ovary and the adaxial nectary. **D.** Medial section of the corolla and androecium; note the short spur on the corolla, reduced staminode, and the paired, but not connivent, fertile stamens. **E.** Stamens; note the basal dehiscence of the anthers.

dark green (sometimes with lighter pattern) adaxially, light green to purple abaxially, the base cuneate to rounded, sometimes oblique, the apex acute or acuminate, the margins serrulate to sinuate-dentate. Inflorescences axillary, epedunculate, 1–3-flowered; pedicels 5–25 mm long. Flowers: calyx lobes nearly free, 2–5 × 0.5–3 cm, accrescent in fruit, the dorsal lobe smaller; corolla spurred, white to peach, the tube 2–5 cm long, the lobes rounded, 1–1.5 cm wide, serrulate, reflexed, the ventral lobe ca. 2× as large as others, erose-fimbriate, not reflexed; stigma bilobed. Fruits globose to broadly ovoid capsules, 1–2 × 1–2 cm, fleshy, yellowish to purplish, the inside wall bright red-orange. Seed mass black. Fl (Feb, Mar); common, growing in areas of sun flecks on forest floor.

EPISCIA Mart.

Herbs, terrestrial or epiphytic, short, stoloniferous, without tubers. Stems decumbent, creeping, rooting at nodes, actively branching from lower nodes (2nd or 3rd node up), the internodes, except for stolons, not well developed. Leaves crowded, subequal in a pair; blades with obtuse to acute apices, the margins crenate-serrate. Inflorescences axillary, cymose. Flowers zygomorphic; sepals free; corolla spurred, oblique in calyx; stamens 4, included. Fruits fleshy capsules.

1. Corolla white, the lobes serrate to fringed. . . . . *E. sphalera.*
1. Corolla yellow, the lobes entire. . . . . *E. xantha.*

Episcia sphalera Leeuwenb.

Herbs, stoloniferous. Stems decumbent. Leaves equal or subequal in a pair; petioles hirsute, 0.5–4 cm long; blades elliptic or oblong-ovate, 2–7 × 1.5–4 cm, hirsute, the base rounded to subcordate, the apex obtuse or acute, the margins crenate-serrate. Infloresences axillary, cymose, 1–few-flowered, bracteate; peduncle 0–1 cm long; pedicels 20–50 mm long. Flowers: sepals free, 6–9 mm long, green; corolla spurred at base, obliquely oriented in calyx, white, the tube 20–25 mm long, the lobes subequal, broadly suborbicular, 6–10 mm wide, obtuse to rounded, serrate to fringed, spreading; stamens slightly exserted before filaments become contorted; stigma large, saucer-shaped or slightly bilobed. Fruits bivalved, fleshy capsules, laterally compressed, 6 × 4 × 3 mm. Seeds dark, the funicles gray. Fl (Sep), fr (Sep); on rocks or on well drained soil in forest.

Episcia xantha Leeuwenb. FIG. 140, PL. 65b

Herbs, stoloniferous. Stems decumbent. Leaves equal or subequal in a pair; petioles hirsute, 1–8 cm long; blades elliptic or ovate, 5–20 × 3–15 cm, bullate, hirsute, the base rounded, subcordate, or less often cuneate, the apex obtuse or acute, the margins crenate-serrate. Inflorescences axillary, cymose, 1–7-flowered, bracteate; peduncle 1.5–5 cm long; pedicels 3–12 mm long. Flowers: sepals free, 10–12 mm long, green; corolla spurred at base, obliquely oriented in calyx, pale to bright yellow, with or without brown-red markings in throat, the tube 15–18 mm long, the lobes subequal, broadly suborbicular, 5 × 6–7 mm, rounded, entire, spreading; stamens included; stigma large, capitate. Fruits bivalved, fleshy capsules, laterally compressed, 10 × 8–9 × 6 mm. Seed mass gray. Fl (Jan, Apr, May, Jun, Jul, Aug, Sep), fr (Aug, Sep, Oct); on well drained soil or decaying logs in damp forest understory.

NAPEANTHUS Gardner

Herbs or subshrubs, terrestrial, without stolons or tubers. Stems semi-woody, very short to long and decumbent, the internodes not well developed. Leaves in clusters at apex of stems, sessile or subsessile; blades oblong-spathulate. Inflorescences axillary, cymose or pseudo-umbellate, bracteate. Flowers actinomorphic or zygomorphic, small; corolla white, weakly attached to calyx; stamens 4–5. Fruits dry capsules.

1. Corolla tube 3–5 mm long, lobes round at apex. . . . . *N. jelskii.*
1. Corolla tube 10–15 mm long, lobes emarginate at apex. . . . . *N. macrostoma.*

Napeanthus jelskii Fritsch

Subshrubs, terrestrial, often hanging from rooting point. Stems simple. Leaves: petioles lacking or very short; blades 2–14 × 1–5 cm, the base gradually narrowed, the apex obtuse or rounded, the margins remotely serrate to entire. Inflorescences cymose with numerous flowers; peduncle filiform, 2.5–6 cm long; bracts lanceolate, 2–4 × 0.5–1 mm; pedicels 1–2.5 cm long. Flowers: calyx lobes connate to 1/3 of length, lanceolate, 3 × 1 mm (to 6 × 2 mm in fruit), green; corolla white, cup-shaped or rotate, actinomorphic, the tube 3–5 mm, the lobes rounded, about 2 mm long; stamens 5 (or 4?, with 5 subequal filaments in bud); stigma obscurely saucer-shaped. Fruits globose capsules, 1.5 mm diam. (not >1/3 length of calyx). Not documented from our area but to be expected.

Napeanthus macrostoma Leeuwenb. PL. 65e

Subshrubs, terrestrial, often hanging from rooting point. Stems simple. Leaves: petioles lacking or very short; blades 11–17 × 3.5–5.5 cm, the base gradually narrowed, the apex obtuse or rounded, the margins shallowly serrate to sinuate. Inflorescences subumbellate, with numerous flowers; peduncle stout, 2.5–6 cm long; bracts lanceolate, 8–12 × 2–3 mm; pedicels 0.5–1 cm. Flowers: calyx lobes connate to 1/4 length, lanceolate, 8 × 2 mm, green; corolla white, campanulate, actinomorphic, the tube 1–1.5 cm long, the lobes ovate-rectangular, 3 mm wide, emarginate; stamens 4; stigma obscurely saucer-shaped. Fruits globose capsules (not >1/3 length of calyx). Fl (Jan, May, Oct, Nov); occasional, in forest on boulders in very wet places or on creek banks.

NAUTILOCALYX Hanst.

Herbs, terrestrial, without stolons or tubers (mature plants of some species with tubers outside our area). Stems decumbent, occasionally short scandent, rooting at nodes, to 1.5 m tall, the internodes well developed. Leaves of a pair equal; blades with obtuse to acuminate apices, the margins crenate-serrate. Inflorescences axillary, cymose, fasciculate, or of solitary flowers. Flowers: corolla spurred, obliquely oriented in calyx; stamens 4, included. Fruits capsules.

1. Leaf blades not decurrent.
    2. Petioles ≥1/2 as long as leaf blades. . . . . *N. adenosiphon.*
    2. Petioles <1/3 as long as leaf blades. . . . . *N. pictus.*
1. Leaf blades decurrent onto petioles. . . . . *N. mimuloides.*

Nautilocalyx adenosiphon (Leeuwenb.) Wiehler [Syn.: *Episcia adenosiphon* Leeuwenb.]

Herbs, terrestrial, ca. 0.1–0.2 m tall. Stems decumbent, tomentose at apex, branching at rooted nodes. Leaves of a pair equal; petioles 1–3 cm long; blades ovate or nearly so, pale green adaxially, paler and glaucous abaxially, the base rounded or subcordate, the apex obtuse or acute, the margins crenate-serrate. Inflorescences axillary, of solitary flowers; pedicels 1.5 cm long. Flowers: calyx lobes connate at very base, ovate, about 12 × 5 mm, pale green; corolla spurred, white,

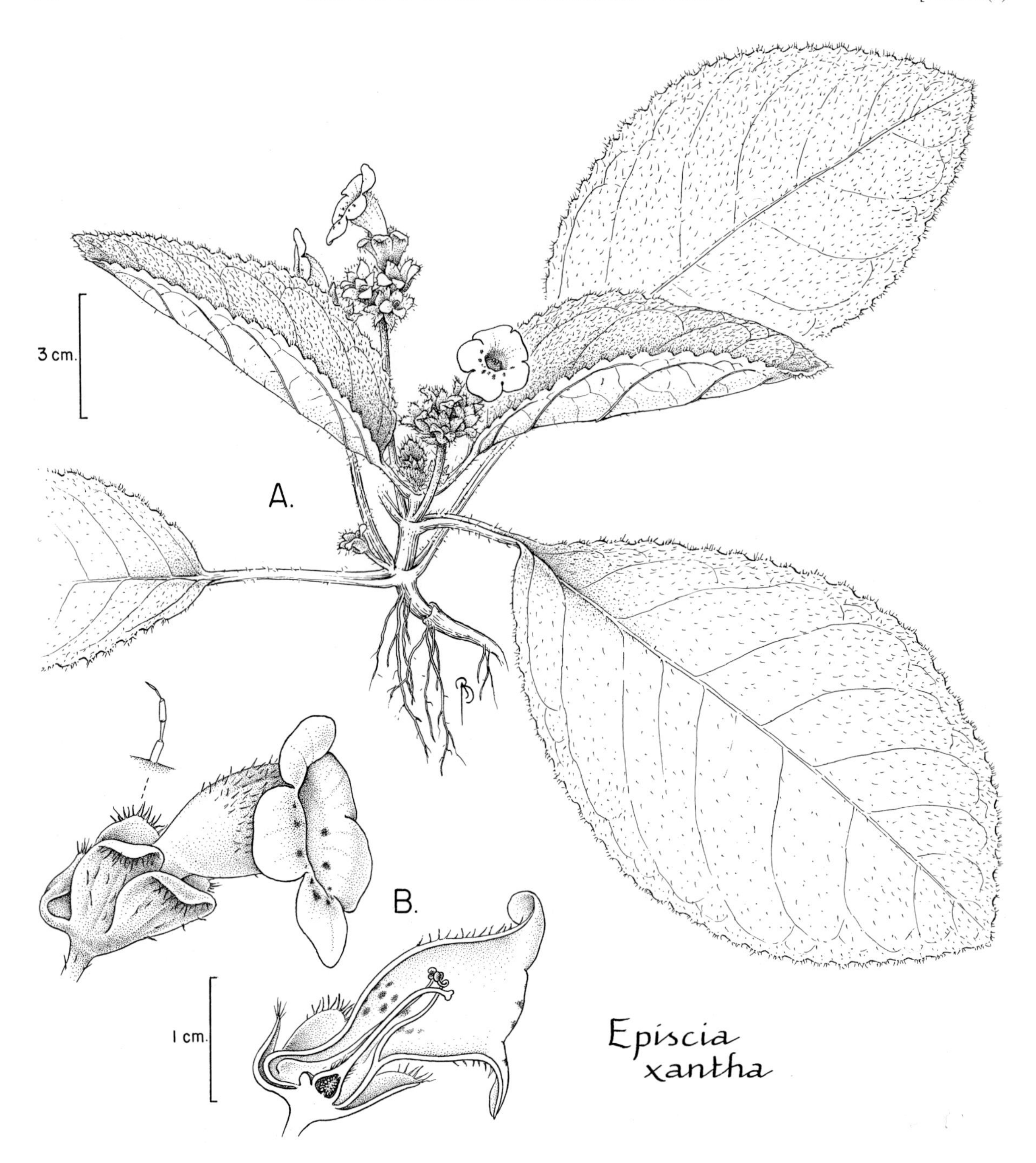

FIG. 140. GESNERIACEAE. *Episcia xantha* (*Mori et al. 22285*). **A.** Plant with leaves and inflorescences. **B.** Lateral view of flower (above), with detail of pubescence, and medial section (below) of flower; note the superior ovary, adaxial nectary, short spur, and fertile stamens with connivent anthers.

the tube 2.5–3 cm long, the lobes round, 7–8 mm wide, the margins entire; stigma bilobed. Fruits unknown. Fl (Aug); known by a single collection from along the Mana river just outside our area (*Cremers 7567*, CAY).

Nautilocalyx mimuloides (Benth.) C. V. Morton [Syn.: *Episcia mimuloides* Benth.] FIG. 141, PL. 66a,b

Herbs, terrestrial, 0.1–1.5 m tall, size and pubescence varying according to exposure to sun. Stems decumbent or short scandent, rooting at nodes. Leaves of a pair subequal; petioles 1–5(12) cm long; blades oblong, 5–19 × 2–8 cm, dark or pale green adaxially, paler abaxially, the base long decurrent, the apex acuminate, the margins crenate-serrate. Inflorescences cymose, 1–12-flowered; peduncle <2 cm long, the bracts 0.5–1.5 × 0.2–0.5 mm, acuminate, green; pedicels 1–3 cm long. Flowers: sepals free, 1–1.6 × 0.5–0.7 cm, green; corolla white, the tube 3–4.5 cm long, the lobes round, 4–6 mm wide, the margins entire; stigma bilobed. Fruits globose, dry capsules, 1 cm long. Fl (May, Jun, Aug, Oct); common, in forest gaps.

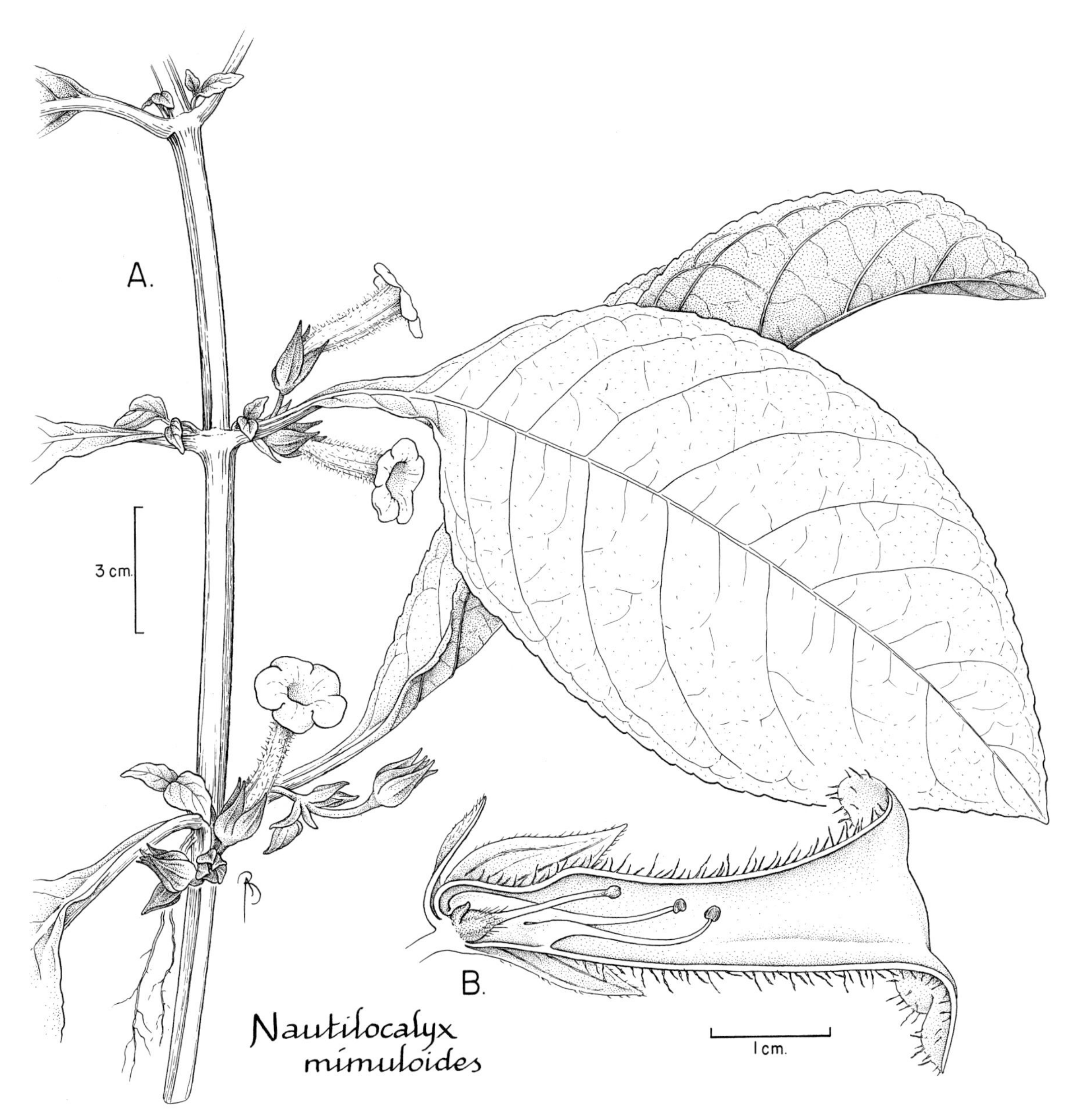

FIG. 141. GESNERIACEAE. *Nautilocalyx mimuloides* (*Mori et al. 22212*). **A.** Part of scandent stem with leaves, roots at lower node, and inflorescences; note the decurrent leaf blade bases. **B.** Medial section of flower; note the short spur, superior ovary, adaxial nectary, and separate fertile stamens.

Nautilocalyx pictus (Hook.) Sprague

Herbs, terrestrial, 0.1–1 m tall. Stems decumbent or short scandent, rooting at nodes. Leaves of a pair subequal; petioles 1–8 cm long; blades oblong-elliptic to oblong-lanceolate, 5–23 × 2–10 cm, often bullate, color very variable in a population, dark to pale green or dark red or bronze with or without paler/darker green or white or silver or red markings adaxially, pale green or reddish abaxially, the base cuneate, the apex acuminate, the margins crenate-serrate. Inflorescences fasciculate, 1–10-flowered; peduncle obsolete or very short, the bracts sepal-like; pedicels 0.3–1.5 cm long. Flowers: calyx lobes nearly free, 1.5–2.5 × 0.1–0.4 cm, green or purple; corolla white or yellowish, the tube 2.5–3.5 cm long, the lobes round, 4–7 mm wide, the margins serrulate; stigma bilobed. Fruits globose, dry capsules, 6–8 cm long. Fl (Mar, Apr, May, Sep, Nov), fr (Aug); common, in understory of nonflooded forest.

PARADRYMONIA Hanst.

Herbs (epiphytic or terrestrial) or vines, without stolons or tubers. Stems erect or ascending, the internodes either well developed or not. Leaves of a pair subequal; blades diverse in shape, size and texture. Inflorescences 1–3- to multi-flowered, cymose, bracteate. Flowers: corolla obliquely oriented in calyx, spurred, hirsute; stamens 4, included. Fruits capsules.

1. Herbs. Stems shorter than leaves, the internodes usually <1 cm long. . . . . *P. densa.*
1. Vines. Stems much longer than leaves, the internodes usually >3 cm long. . . . . *P. campostyla.*

Paradrymonia campostyla (Leeuwenb.) Wiehler [Syn.: *Drymonia campostyla* Leeuwenb.]

Vines, herbaceous to suffrutescent, to 3 m long. Stems much longer than leaves, rooted at nodes, the flowering stems free from substrate, hirsute near apex, the internodes usually >3 cm long. Leaves of a pair equal; petioles 0.6–4.5 cm long; blades often obliquely oblong-ovate or oblong-elliptic, 2.5–10 × 1–4.5 cm, hirsute or pilose on both surfaces, dark green adaxially, paler abaxially, the base cuneate or rounded, the apex acute or acuminate, the margins entire or obscurely repand-serrate. Inflorescences cymose, 1–3-flowered, subsessile; pedicels 0.5–2 cm long. Flowers: sepals free, long-acuminate, green; corolla white, white-hirsute abaxially, the throat yellow with purple lines adaxially, the tube 3–4.5 cm long, the lobes round, 6–12 mm wide, the margins entire; stigma bilobed. Fruits globose, 10 mm diam., hirsute. Fl (Feb), fr (Feb); common, in cloud forest.

Paradrymonia densa (C. H. Wright) Wiehler [Syn.: *Episcia densa* C. H. Wright]

Herbs, terrestrial, 0.3–0.6 m tall (mostly due to large leaves). Stems shorter than leaves, glabrous, the internodes usually <1 cm long. Leaves of a pair subequal; petioles (1)6–12(14) cm long; blades oblong-lanceolate or oblong-elliptic, 5–30 × 1–12 cm, glabrous or nearly so on both surfaces, sometimes strigose on veins abaxially, dark green adaxially, paler or reddish abaxially, the base cuneate to obliquely subcordate, the apex acuminate to obtuse, the margins entire to crenate-serrate. Inflorescences umbellate or cymose, the flowers mostly numerous, aggregated; peduncle to 1 cm long; pedicels 0.7–3 cm long. Flowers: calyx lobes connate for 1/3 length, purple abaxially, green adaxially, the apex acute; corolla white, white-hirsute abaxially, sometimes with purple or purple-tipped lobes, the tube 2.5–4.7 cm long, the lobes round, 5–7 mm wide, the margins entire; stigma capitate. Fruits subglobose, ca. 10 mm diam., with scattered hairs. Not collected from our area, but to be expected.

HERNANDIACEAE (Hernandia Family)

Scott A. Mori and John L. Brown

Trees or lianas. Leaves simple, alternate. Stipules absent. Inflorescences axillary, cymose, or dichotomously branched panicles. Flowers actinomorphic, unisexual (plants monoecious) or bisexual; tepals 3–10, in one or two whorls; stamens 3–5, the anthers opening by 2 valves, the filaments with basal nectaries in *Hernandia*; ovary inferior, 1-locular, with nectaries at summit in *Hernandia*; placentation apical, the ovule solitary, pendulous. Fruits dry drupes, sometimes with lateral or terminal wings but not in our area, sometimes enclosed in bladder-like cupule (*Hernandia*). Seeds without endosperm; embryo straight, the cotyledons 2, large, oily, folded, wrinkled, or lobed.

Kostermans, A. J. G. H. 1966. Hernandiaceae. *In* A. Pulle (ed.), Flora of Suriname **II(1):** 338–344. E. J. Brill, Leiden.
Kubitzki, K. U. 1976. Hernandiaceae. *In* J. Lanjouw & A. L. Stoffers (eds.), Flora of Suriname **II(2):** 485–486. E. J. Brill, Leiden.
Miller, J. S. & P. E. Berry. 1999. Hernandiaceae. *In* J. A. Steyermark, P. E. Berry, K. Yatskievych & B. K. Holst (gen. eds.), Flora of the Venezuelan Guayana **5:** 592–593. P. E. Berry, K. Yatskievych & B. K. Holst (vol. eds.). Missouri Botanical Garden Press, St. Louis.

1. Trees. Inflorescences bracteate. Flowers >2 mm diam., with nectaries. Fruits surrounded by inflated, bladder-like cupule. . . . . *Hernandia.*
1. Lianas. Inflorescences not bracteate. Flowers <2 mm diam., without conspicuous nectaries. Fruits not surrounded by inflated, bladder-like cupule. . . . . *Sparanttanthelium.*

HERNANDIA L.

Trees. Leaves: blades with 2–4 basal secondary veins, these almost as prominent as primary veins (midrib), the remaining secondary veins pinnately arranged along midrib, the higher order venation reticulate. Inflorescences axillary, appearing corymbose, the ultimate units 3-flowered cymes, the central flower pistillate, the lateral ones staminate, bracteate throughout, the bracts persistent; peduncle of lower inflorescences longer than those of upper. Flowers unisexual. Staminate flowers with 3 stamens, the filaments with basal nectaries at apex on each side. Pistillate flowers surrounded at base by cupules derived from bracts; nectaries usually 4. Fruits globose drupes, enclosed in bladder-like cupule with opening at apex. Seeds with embryo flattened, the cotyledons ± wrinkled.

Hernandia guianensis Aubl. FIG. 142, PL. 66c

Trees, 10–30 m tall. Trunks with buttresses. Leaves: petioles 3.5–11 cm long; blades narrowly ovate to widely ovate (almost cordate in some specimens), 10–18 × 5–10 cm, glabrous, the margins entire. Inflorescences pubescent. Flowers ca. 3–4 mm diam., pale green or greenish-white. Staminate flowers with 4–6, imbricate tepals. Pistillate flowers with 6–8 imbricate tepals; nectaries 4, globose, glabrous, very well developed; style pubescent, the stigma peltate. Fruits black, ribbed, at least when dry, ca. 3 × 2–2.5 cm, surrounded by bladder-like cupule with opening at apex. Fl (Feb); in nonflooded forest. *Bois banane, mirobolan* (Créole), *ventoza* (Portuguese).

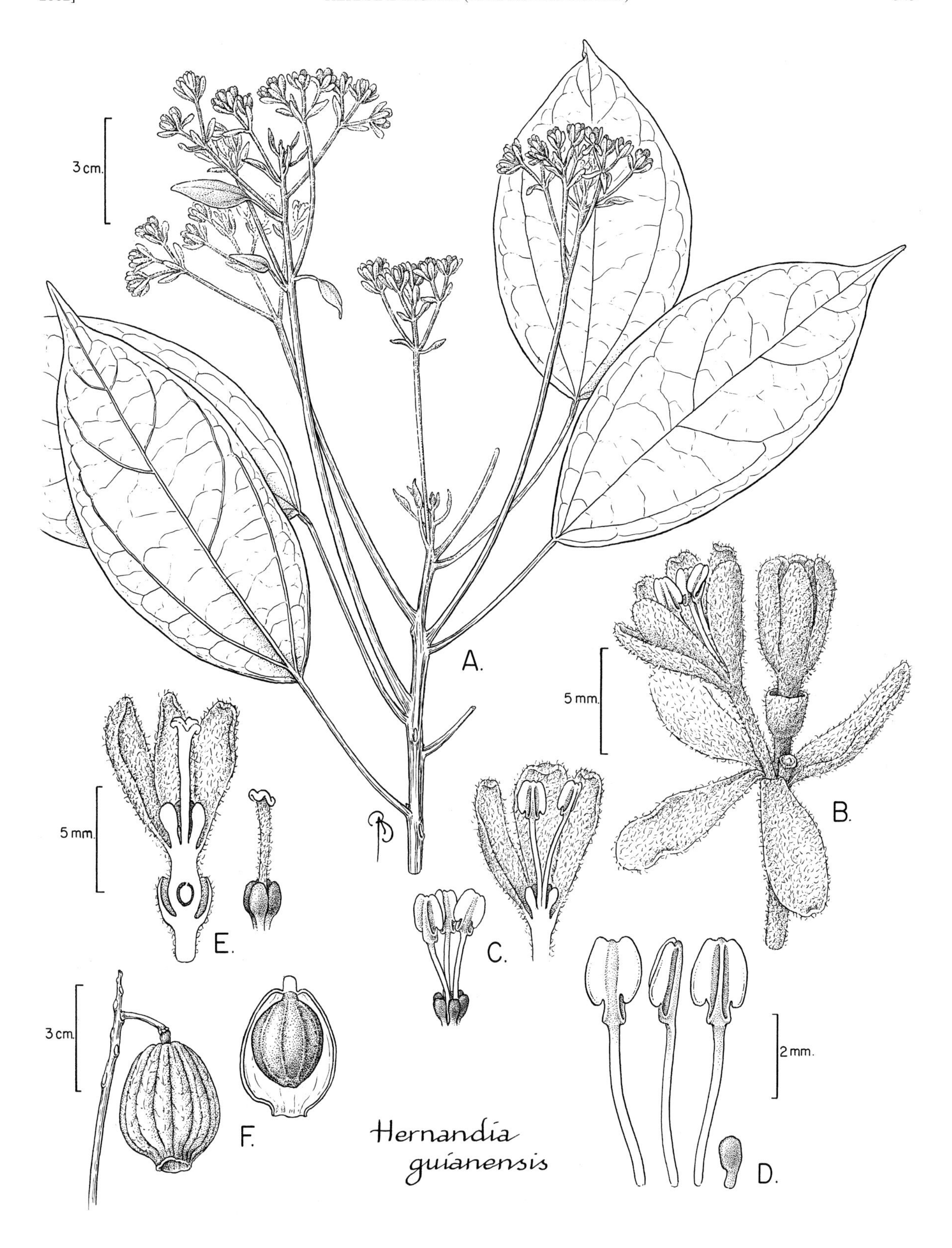

FIG. 142. HERNANDIACEAE. *Hernandia guianensis* (A, voucher information lost; B–E, *Steyermark 88867* from Venezuela; F, *Steyermark et al. 114687* from Venezuela). **A.** Part of stem with leaves and inflorescences. **B.** Part of inflorescence subtended by four bracts; note open staminate flower. **C.** Medial section of staminate flower (right) and androecium surrounded by glands (left). **D.** Abaxial (farthest left), lateral (second from left), and adaxial (third from left) views of stamens, and gland (near left). **E.** Medial section of pistillate flower (left) and pistil surrounded by glands (right). **F.** Calyx-enclosed fruit (left) and fruit with front half of calyx removed (right).

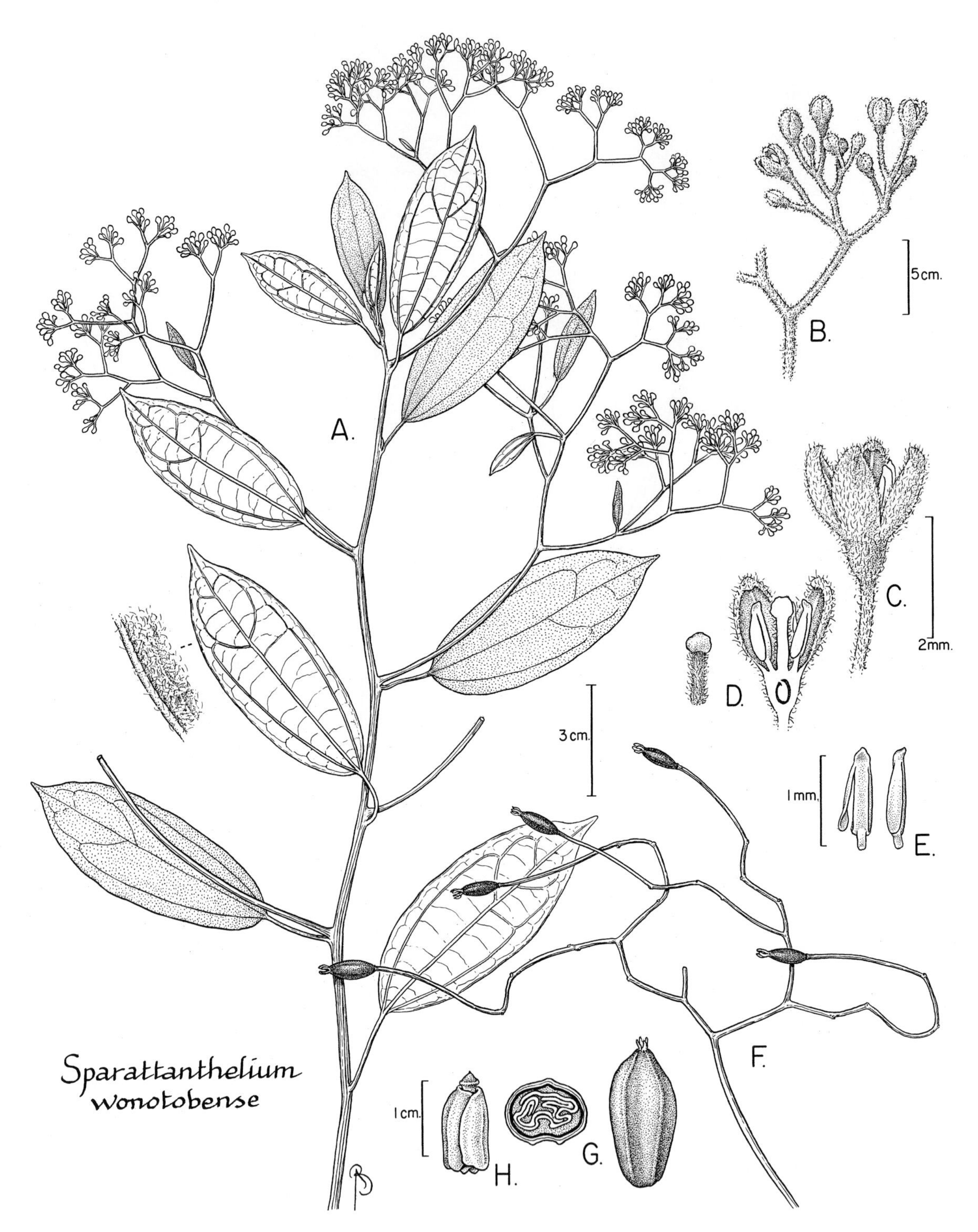

FIG. 143. HERNANDIACEAE. *Sparattanthelium wonotobense* (A–E, *Mori & Gracie 18986*; F–H, *Mori et al. 23044*). **A.** Stem with leaves, inflorescences, and detail of abaxial leaf margin. **B.** Part of inflorescence. **C.** Lateral view of flower. **D.** Medial section of flower (right) and style and stigma (left). **E.** Stamens showing flap-like opening of anther. **F.** Part of infructescence. **G.** Entire fruit (right) and transverse section of fruit (left) showing convoluted cotyledons. **H.** Embryo.

SPARATTANTHELIUM Mart.

Lianas. Leaves: blades with 2 basal secondary veins nearly as prominent as primary vein (midrib), the remaining secondary veins inconspicuous, pinnately arranged along midrib, the tertiary veins often arranged at right angles to midrib. Inflorescences axillary or subterminal, without bracts, appearing dichotomously branched, at least in our area. Flowers bisexual, usually <2 mm diam., at least in our flora; stamens 4–5, the filaments filiform, without nectaries; style cylindrical, straight, the stigma subcapitate. Fruits dry drupes, ovoid to ovoid-ellipsoid. Seeds with convoluted and folded cotyledons.

1. Leaf blades ± concolorous, sparsely pubescent abaxially. . . . . *S. guianense.*
1. Leaf blades distinctly discolorous, densely pubescent abaxially. . . . . *S. wonotoboense.*

Sparattanthelium guianense Sandwith

Leaves: petioles 1.5–3 cm long; blades elliptic to slightly oblanceolate, 6–18 × 2.5–6 cm, sparsely pubescent abaxially, ± concolorous, the margins entire. Description based on a single sterile specimen (*Granville 2695*, CAY).

Sparattanthelium wonotoboense Kosterm. FIG. 143

Leaves: petioles 1–3 cm long; blades elliptic, 5–8.5 × 2.2–3.6 cm, densely pubescent abaxially, distinctly discolorous, the margins entire. Inflorescences axillary, in dichotomously branched panicles. Flowers ca. 1.5 mm diam.; tepals 4, valvate; stamens 4, dehiscing by long valves opening from base and remaining attached at apex; style cylindric, the stigma capitate. Fl (Jun, Aug). *Liane basilic* (Créole).

HIPPOCRATEACEAE (Hippocratea Family)

Ara Görts-van Rijn

Lianas, shrubs with scandent branches, or occasionally slender trees, glabrous or some parts with simple hairs. Leaves simple, opposite. Inflorescences thyrsoid, corymbose, or dichotomously branched panicles or fasciculate on reduced peduncles; bracts minute. Flowers bisexual; sepals 5, not covering petals in bud, connate at base, equal or unequal; petals 5, free; nectary disc always present as low ring, saucer-like, or short-tubular, or cushion-like with or without flattened or upturned basal margin, or nectary disc discontinuous and forming 3 staminiferous pockets; stamens 3, inserted within disc, the anthers opening extrorsely by transversely or obliquely confluent clefts; ovary superior, 3-locular, the style subulate or absent, the stigmas obsolete or forming terminal shield, or stigmas 3, entire or bilobed; placentation axile, with 2–14 ovules per locule. Fruits either capsular (dehiscent), then consisting of 3 mericarps attached to swollen receptacle, or berries (indehiscent). Seeds 6 to many, in capsular fruits attached basally by conspicuous or reduced wing, in non-capsular fruits embedded in mucilaginous pulp.

Görts-van Rijn, A. R. A. 1994. Hippocrateaceae. *In* A. R. A. Görts-van Rijn (ed.), Flora of the Guianas Ser. A, **16:** 1–81, 141–148. Koeltz Scientific Books, Havlíčkův Brod, Czech Republic.

Mennega, A. M. W. & J. P. Hedin. 1999. Hippocrateaceae. *In* J. A. Steyermark, P. E. Berry, K. Yatskievych & B. K. Holst (gen. eds.), Flora of the Venezuelan Guayana **5:** 594–617. P. E. Berry, K. Yatskievych & B. K. Holst (vol. eds.). Missouri Botanical Garden Press, St. Louis.

1. Plants with flowers.
  2. Petals with adaxial transverse lines of hairs below apex. . . . . *Hippocratea.*
  2. Petals without adaxial transverse lines of hairs below apex.
    3. Nectary disc with 3 staminiferous pockets. . . . . *Cheiloclinium.*
    3. Nectary disc without staminiferous pockets.
      4. Nectary disc conical-truncate, cylindrical, or cushion-like, with or without flattened or upturned basal margin; stigmas obscure. . . . . *Salacia.*
      4. Nectary disc short tubular, annular, or annular-pulvinate with conspicuous, flattened basal margin; stigmas entire, bilobed, the style ending in minute stigmatic shield, or stigmas obsolete.
        5. Leaf blades scabrous. Flowers 9–12 mm diam. . . . . *Prionostemma*
        5. Leaf blades not scabrous. Flowers 1–8 mm diam.
          6. Petals erect; nectary disc short-tubular or annular, the rim often thickened, with pointed extensions between stamens; style ending in minute stigmatic shield. . . . . *Elachyptera.*
          6. Petals spreading; nectary disc short-tubular or annular, the rim without pointed extensions between stamens; style short, stigmas 3, entire or deeply bilobed, or stigmas absent.
            7. Filaments linear; nectary disc short-tubular; stigmas 3, obvious or stigmas absent. . . . . *Tontelea.*
            7. Filaments deltoid; nectary disc annular; stigmas 3, minute. . . . . *Pristimera.*
1. Plants with fruits.
  8. Fruits consisting of 3 dehiscent mericarps. Seeds winged or almost wingless.
    9. Leaf blades scabrous. Mericarps woody. . . . . *Prionostemma.*
    9. Leaf blades not scabrous. Mericarps coriaceous.
      10. Inflorescences tomentellous. Mericarps thick. . . . . *Hippocratea.*
      10. Inflorescences glabrous. Mericarps thin.
        11. Leaf blades brownish or olivaceous when dried; secondary veins in 4–6 pairs. Wing of seed reduced, smaller than body of seed, sometimes expanded laterally. . . . . *Elachyptera.*

11. Leaf blades greenish when dried; secondary veins in 6–8 pairs. Wing of seed much longer than body of seed. . . . . *Pristimera*.
8. Fruits berries. Seeds embedded in mucilaginous pulp. . . . . *Cheiloclinium*, *Salacia*, *Tontelea*.

CHEILOCLINIUM Miers

Lianas, shrubs, or slender trees. Leaf blades not scabrous. Inflorescences thyrsoid or dichotomously branched panicles, the flowers in cymes at end of ultimate axes. Flowers: petals spreading, without adaxial transverse line of hairs below apex; nectary disc discontinuous, forming 3 pockets, these surrounding stamens; filaments ligulate; ovary trigonous, the stigmas adnate to apex. Fruits berries (indehiscent), on thickened pedicels. Seeds embedded in mucilaginous pulp.

1. Shrubs or small trees. Inflorescences thyrsoid. Fruits usually with 3 basal, pale, longitudinal bands. . . . . *C. cognatum*.
1. Lianas. Inflorescences dichotomously branched. Fruits with 6–10 basal, longitudinal lines or ridges or fruits unknown.
    2. Flowers ca. 2.5 mm diam.; stigmas bilobed. Fruits with 6–10 basal, longitudinal lines or ridges. . . . . *C. hippocrateoides*.
    2. Flowers ca. 4.5 mm diam.; stigmas entire. Fruits unknown. . . . . *Cheiloclinium* sp. 1.

Cheiloclinium cognatum (Miers) A.C. Sm. PL. 67a

Small trees or shrubs. Leaves: blades narrowly elliptic to elliptic, ovate, or oblong, 8–20 × 2.5–7 cm, chartaceous, the base rounded to acute, the apex acuminate, the margins slightly crenate; secondary veins in 6–12 pairs. Inflorescences thyrsoid, 2–10 cm long; pedicels <1 mm long. Flowers ca. 5 mm diam., brownish-red or dark yellow with crimson margin, or orange, brown dotted; stigmas entire, linear. Fruits globose or ellipsoid, or somewhat trigonous, 3–5 × 3 cm, green to orange, usually with 3 pale, basal longitudinal bands. Fl (May, Jun), fr (Jan, Feb, Mar, Apr, May, Jul, Aug, Oct, Dec); common, in nonflooded forest.

Cheiloclinium hippocrateoides (Peyr.) A. C. Sm.

Lianas. Leaves: blades elliptic, 7–20 × 3–11 cm, chartaceous to coriaceous, the base obtuse or acute, decurrent onto petiole, the apex cuspidate, the margins entire or crenulate; secondary veins in 5–11 pairs. Inflorescences dichotomously branched panicles, to 6 cm long; pedicels to 2 mm long. Flowers ca. 2.5 mm diam., greenish white to yellow or orange; stigmas bilobed. Fruits ellipsoid, to 3 × 2 cm, yellow or orange, with 6–10 lines or ridges from base. Fl (Nov); rare, in nonflooded forest.

Cheiloclinium sp. 1 PL. 67b

Lianas. Leaves: blades elliptic or elliptic-oblong, 9–15 × 3.5–7 cm, the base acute, the apex acuminate, the margins slightly crenate; secondary veins in ca. 11 pairs. Inflorescences dichotomously branched panicles, 4–6 cm long; pedicels <1 mm long. Flowers ca. 4.5 mm diam., yellow-green; stigmas entire, linear. Fruits unknown. Fl (Sep); collected once (*Mori et al. 23785*, NY), in nonflooded forest.

ELACHYPTERA A.C. Sm.

Lianas, scandent shrubs, or slender trees with scandent branches. Leaf blades not scabrous. Inflorescences paniculate-corymbose. Flowers: petals without adaxial transverse line of hairs below apex; nectary disc short tubular or annular, the rim often thickened and with pointed extensions between stamens; filaments ligulate. Fruits capsular, consisting of 3 thin, coriaceous, divergent mericarps separately attached to swollen receptacle. Seeds attached by short wing.

Elachyptera floribunda (Benth.) A. C. Sm.

Lianas. Branches purplish brown or cinereous. Leaves: ovate-elliptic or elliptic, 5–10 × 2.5–5 cm, chartaceous or thin-coriaceous, brownish or olivaceous when dried, the base acute to rounded, the apex acute or acuminate, the margins entire or crenate; secondary veins in 4–6 pairs. Inflorescences 2–11 cm long; bracts fimbriate, minute. Flowers 1–2.7 mm diam., white, cream-colored, or purplish white; petals erect; style ending in minute shield. Mericarps elliptic or obovate, 3.5–6 × 2–3 cm, striate because of numerous fine ribs, the apex obtuse or emarginate. Seeds to ca. 3 cm long, the basal wing 8 mm long. Fl (Sep); rare, in nonflooded forest.

HIPPOCRATEA L.

Lianas with long, twining branches. Young parts puberulous. Leaf blades not scabrous. Inflorescences dichotomously branched panicles, the flowers solitary or in cymes at end of stems or arranged laterally along stems. Flowers: petals with adaxial transverse line of hairs below apex; nectary disc cushion-like to truncate-conical; filaments ligulate, somewhat broadened at base; style subulate; stigmas inconspicuous. Fruits with 3 thick mericarps, attached to swollen receptacle. Seeds attached by a wing.

Hippocratea volubilis L. FIG. 144; PART 1: FIG. 7

Leaves: blades elliptic, oblong, ovate, or obovate, 3–14 × 2–6 cm, the base rounded to acute, the apex acute or acuminate, the margins sinuate to crenate; secondary veins in 5–8 pairs. Inflorescences to 6 × 9 cm, brown to orange-puberulous. Flowers 4–8 mm diam., puberulous; petals greenish, cream-colored or yellow, with adaxial transverse line of hairs just below apex;

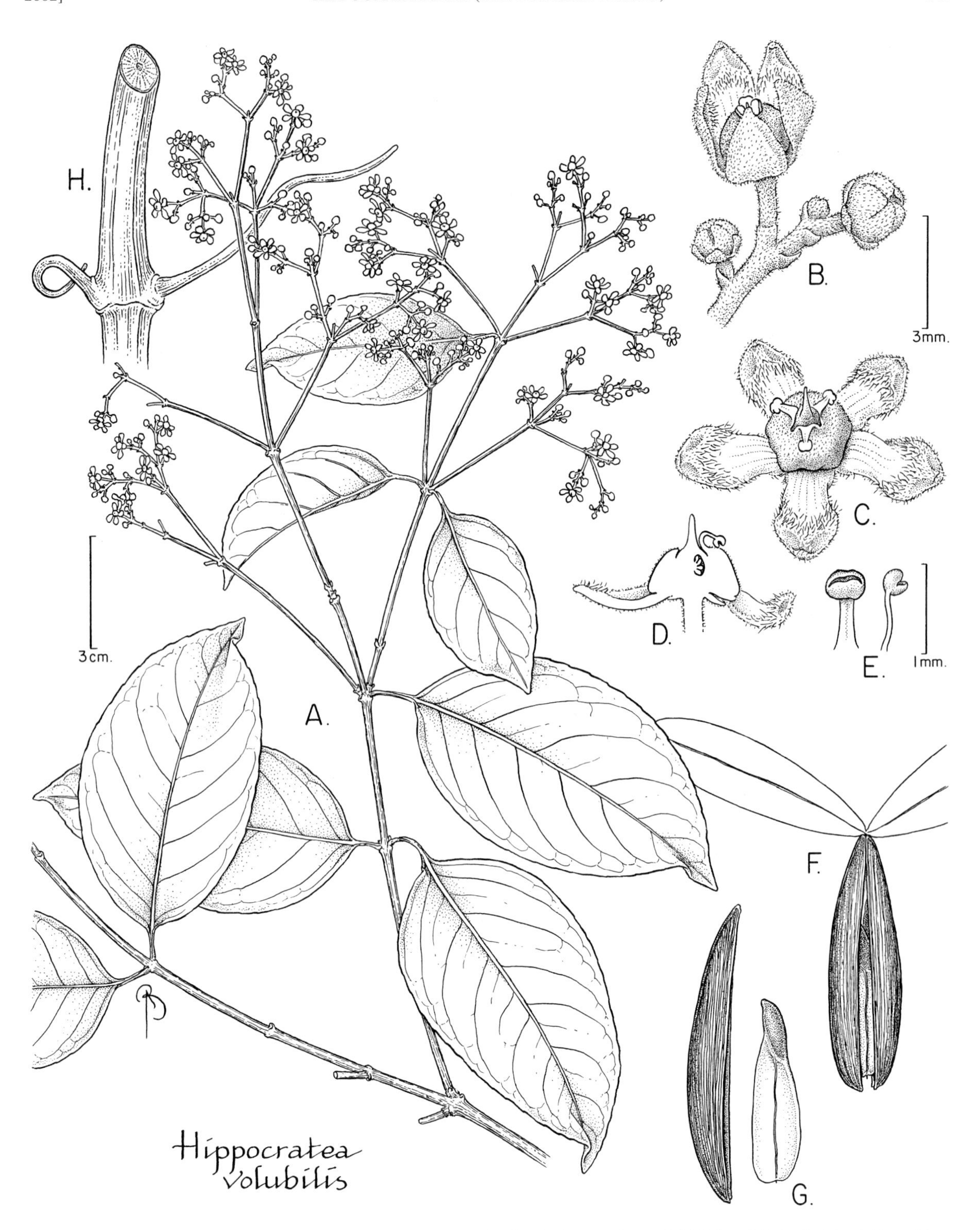

FIG. 144. HIPPOCRATEACEAE. *Hippocratea volubilis* (A–E, *Mori & Gracie 18947*; F, G, *Görts-van Rijn et al. 94*; H, from field sketch of plant in Puerto Rico). **A.** Stem with leaves and inflorescences. **B.** Part of inflorescence with flower buds and partially opened flower. **C.** Apical view of flower showing nectary disc, subulate style, and reflexed stamens. **D.** Medial section of flower. **E.** Abaxial (left) and lateral (right) views of stamen. **F.** Fruit showing one mericarp beginning to dehisce and arrangement of other two mericarps. **G.** Half of one mericarp of fruit (far left) and seed (near left). **H.** Part of stem showing node with tendrils.

nectary disc puberulous. Mericarps narrowly obovate, narrowly oblanceolate, or elliptic, 4–8 × 1.5–3 cm, densely, finely striped, the apex rounded or emarginate. Seeds 3–7.5 cm long, the basal wing 2–4 cm long. Fl (Jun, Sep), fr (Apr, May); common, in non-flooded forest.

PRIONOSTEMMA Miers

Lianas or shrubs. Branches (sub)opposite or alternate. Leaf blades scabrous. Inflorescences dichotomously branched panicles, tomentose. Flowers tomentellous; petals spreading, without adaxial transverse line of hairs below apex; nectary disc annular-pulvinate, with conspicuous, flattened, basal margin; filaments linear; style subulate; stigmas obsolete. Fruits with 3 woody, divergent mericarps on swollen receptacle. Seeds attached by large, basal wing.

Prionostemma aspera (Lam.) Miers PL. 67d; PART 1: FIG. 7

Lianas or shrubs with scandent or creeping branches. Bark with colorless or red exudate. Leaves: blades ovate or elliptic-oblong, 5–10 × 2.5–5 cm, thin-coriaceous, scabrous; secondary veins in 5–7 pairs. Inflorescences 3–12 cm long. Flowers fragrant, 9–12 mm diam.; sepals unequal; petals green to yellow; stamens white; stigmas orange. Mericarps green to brown, 5.5–8 × 4–6 cm, bluntly ribbed, the apex rounded or emarginate. Seeds with 4–6 cm long wing. Fl (Feb, Jun, Aug), fr (Feb, May, Aug); relatively common, in nonflooded forest.

PRISTIMERA Miers

Lianas, small shrubs, or trees with scandent branches. Leaf blades not scabrous. Inflorescences paniculate-corymbose, dichotomously branched. Flowers: petals spreading, without adaxial transverse line of hairs below apex; nectary disc annular; filaments deltoid; stigmas 3, minute. Fruits capsular, the mericarps 3, thin, coriaceous, divergent, attached to swollen receptacle. Seeds with large, papery wing, the wing much longer than body of the seed.

Pristimera nervosa (Miers) A. C. Sm. PL. 67c

Lianas or small trees. Stems dark-purple or maroon, the twigs grayish. Leaves: blades ovate or elliptic, 7–13(18) × 4–8 cm, chartaceous or membranous, bright green adaxially, paler abaxially, the base obtuse or acute, the apex obtusely short-acuminate, the margins undulate or crenate; secondary veins in 6–8 pairs, the tertiary veins orthogonally reticulate. Inflorescences 1–4 cm long; peduncles 0.5–2.5 cm long, the bracts fimbriate, persistent. Flowers 2–3 mm diam., fragrant; sepals fimbriate; petals slightly fleshy, thickened in middle, greenish-white or yellow; nectary disc annular; filaments deltoid; style with 3 minute stigmas. Mericarps flattened, thin-coriaceous. Fl (Oct); rare, in forest near stream.

SALACIA L.

Lianas, shrubs, or trees. Leaf blades not scabrous. Inflorescences paniculate, dichotomously branched, or fasciculate; peduncles reduced, thickened and flowers aggregated. Flowers: petals without adaxial transverse line of hairs below apex; nectary disc conical-truncate, cylindrical, or cushion-like, with flattened or upturned basal margin; filaments ligulate, the anthers opening by transversely or obliquely confluent clefts; style short; stigmas obscure. Fruits indehiscent, globose or ellipsoid, smooth to scrobiculate. Seeds embedded in mucilaginous pulp.

1. Inflorescences fasciculate, the flowers aggregated; peduncles reduced.
    2. Leaf blades narrowly ovate to ovate. Flowers 12–22 mm diam. Fruits smooth to slightly rugulose when dried. . . . . . *S. impressifolia.*
    2. Leaf blades elliptic-oblong. Flowers ca. 14 mm diam. Fruits strongly tuberculate when dried. . . . . . *S. juruana.*
1. Inflorescences paniculate, dichotomously branched, the flowers not aggregated; peduncles developed.
    3. Leaf blades chartaceous, 25–28 cm long, the apex acuminate; secondary veins in 10–14(18) pairs. Inflorescences paniculate, not dichotomously branched, to 25 cm long, the axes and pedicels very slender. . . . . . *S. insignis.*
    3. Leaf blades subcoriaceous to coriaceous, 5–19 cm long, the apex rounded, cuspidate, or emarginate; secondary veins in 6–10 pairs. Inflorescences dichotomously branched panicles, to 6 cm long, the axes and pedicels not very slender.
        4. Branches arising at various angles from stem, densely lenticellate. Inflorescences densely ferruginous-pubescent. Flowers 3–4 mm diam. . . . . . *S. amplectens.*
        4. Branches arising at right angles from stem, not densely lenticellate. Inflorescences glabrous or reddish tomentellous. Flowers 4–10 mm diam.
            5. Leaf blade base attenuate, the apex mucronulate. Inflorescences glabrous. Flowers 4–7 mm diam; sepals glabrous.. . . . . . *S. multiflora* subsp. *mucronata.*
            5. Leaf blade base acute, obtuse, or slightly cordate, the apex rounded, cuspidate, or short-acuminate. Inflorescences occasionally reddish tomentellous. Flowers 7–10 mm diam.; sepals occasionally reddish tomentellous. . . . . . *S. multiflora* subsp. *multiflora.*

Salacia amplectens A. C. Sm.

Lianas. Stems fluted, the lateral branches arising at various angles from main stem. Leaves: petioles 7–10 mm long; blades narrowly elliptic or oblong, 7–13 × 2–5 cm, subcoriaceous, the base rounded to attenuate, the apex cuspidate, the acumen 1.5 cm long; secondary veins in 6–9 pairs, flattened or impressed adaxially, inconspicuous abaxially. Inflorescences axillary or ramiflorous, dichotomously branched panicles, densely ferruginous-pubescent, the axes not very slender, to 6 cm long; pedicels not very slender, 1–3 mm long. Flowers 3–4 mm diam., green; sepals pubescent; nectary disc conical, with thin upturned basal ring, fleshy. Fruits ellipsoid or subglobose, rugulose, orange or yellow when ripe. Not known by collections from our area but to be expected.

Salacia impressifolia (Miers) A. C. Sm.

Lianas or shrubs. Leaves: petioles 8–21 mm long; blades narrowly ovate to ovate, 9–20 × 4–8 cm, subcoriaceous or coriaceous, the base rounded to slightly cordate, often somewhat falcate, the apex abruptly obtusely acuminate or rounded; secondary veins in 7–11 pairs, impressed or flat on both surfaces, occasionally slightly elevated abaxially, the tertiary veins impressed or obsolete on both surfaces. Inflorescences with peduncles reduced, the flowers aggregated; pedicels slender, 4–15 mm long, increasing in length in fruit. Flowers 12–22 mm diam., yellowish green, pale brown or orange; nectary disc cushion-like, flattened at base. Fruits globose, 3–4 cm diam., smooth or slightly rugulose when dried, grayish glaucous when unripe, dull orange when ripe. Fl (May); in nonflooded forest.

Salacia insignis A. C. Sm. Pl. 67e

Lianas. Stems sometimes reddish brown. Leaves: petioles 10–16 mm long; blades elliptic-oblong, 25–28 × 4–7.5 cm, chartaceous, the base acute, decurrent onto petiole, the apex abruptly acuminate, the acumen to 1 cm long; secondary veins in 10–14(18) pairs, prominent abaxially, the tertiary veins parallel, plane adaxially, prominulous abaxially. Inflorescences paniculate, not dichotomously branched, few together or solitary, the flowers loosely arranged at ends of branches, the axes very slender, to 25 cm long; pedicels very slender, to 9 mm long. Flowers 6–10 mm diam., whitish or greenish yellow; nectary disc cushion-like, flattened towards base. Fruits globose to ellipsoid, to 3 cm diam., smooth, lenticellate or rugulose, at first bluish- or grayish green then turning orange. Fl (Aug), fr (Jan, Apr, Aug); in nonflooded forest.

Salacia juruana Loes.

Trees, to 6–15 m tall or lianas. Leaves: petioles 10–20 mm long; blades elliptic-oblong, 12–29 × 7–15 cm, coriaceous, the base rounded or obtuse, the apex obtuse, rounded, or cuspidate; secondary veins in 7–10 pairs, prominulous or plane adaxially, prominent abaxially, the tertiary veins immersed or impressed. Inflorescences fasciculate, the flowers aggregated; peduncles reduced; pedicels slender, to 1.5 cm long, increasing to 3 cm long × 8 mm thick in fruit. Flowers ca. 14 mm diam., greenish or yellow, often with dark spot at base; nectary disc cushion-like, flattened at base. Fruits globose, to 6 cm diam., strongly tuberculate (when dried), at first bluish green then turning yellow. Fr (Apr); in nonflooded forest.

Salacia multiflora (Lam.) DC. subsp. **mucronata** (Rusby) Mennega

Lianas. Lateral branches often at right angles with main stem. Leaves: petioles 4–9 mm long; blades obovate, 5–12 × 2–6.5 cm, (sub)coriaceous, the base attenuate, the apex mucronulate, with callose mucro, or emarginate; secondary veins in 6–10 pairs. Inflorescences dichotomously branched, often on defoliated branches, to 6 cm long, glabrous, the axes and pedicels not very slender, to 6 cm long; pedicels 1–3.5 mm long, articulate. Flowers 4–7 mm diam., greenish or creamy white, often farinose-puberulent; sepals glabrous; nectary disc conical-truncate or cylindrical, the basal and upper part differently colored. Fruits ellipsoid, often apiculate, 3.5 × 2.5 cm, smooth or rugulose, brown or orange, often white speckled. Fl (May, Jun); in nonflooded forest.

Salacia multiflora (Lam.) DC. subsp. **multiflora**

Lianas. Lateral branches often at right angles with main stem. Leaves: petioles 4–9 mm long; blades obovate or elliptic-oblong, 6–19 × 3.5–9 cm, coriaceous or subcoriaceous, the base acute, obtuse, or slightly cordate, the apex rounded or cuspidate, often with glandular tip; secondary veins in 6–10 pairs. Inflorescences dichotomously branched, often on defoliated branches, to 6 cm long, the axes and pedicels not very slender, to 6 cm long; pedicels 1–3.5 mm long, articulate. Flowers 7–10 mm diam; sepals occasionally reddish tomentellous; nectary disc conical-truncate or cylindrical, the basal and upper part differently colored. Fruits obovoid or ellipsoid, to 7 × 4 cm. Fl (Mar), fr (May); in nonflooded forest.

TONTELEA Aubl.

Lianas, shrubs, or occasionally slender trees. Leaf blades not scabrous. Inflorescences thyrsoid or dichotomously branched panicles. Flowers: petals spreading, without adaxial transverse line of hairs below apex; nectary disc short-tubular; filaments linear, the anthers broader than long, the 2 cells distinct; ovary depressed-conical or subtrigonous, the style short, the stigmas 3, entire or deeply bilobed, or stigmas obsolete. Fruits berries, cylindrical, globose to ellipsoid; pedicels thickened. Seeds embedded in mucilaginous pulp.

1. Inflorescences from branches or stems. Fruits cylindrical, to 16 cm long. . . . . *T. cylindrocarpa.*
1. Inflorescences axillary. Fruits globose, ellipsoid, or ovoid, to 4 cm diam.
    2. Leaf blades coriaceous.
        3. Leaf blades with secondary veins in 5–7 pairs. Flowers 2.2 mm diam. . . . . *T. nectandrifolia.*
        3. Leaf blades with secondary veins in 7–11 pairs. Flowers 4–7 mm diam. . . . . *T. ovalifolia.*
    2. Leaf blades chartaceous, sometimes firmly so.
        4. Inflorescences laxly flowered, to 9 cm long; peduncles and axes slender. Stigmas 3, each deeply divided, the 6 lobes equal. . . . . *T. laxiflora.*
        4. Inflorescences semiglobose, rather densely flowered, 2–3.5 cm long; peduncles and axes robust. Stigmas entire or shallowly bilobed.
            5. Inflorescences not ceriferous. Flowers green or white; stigmas entire. . . . . *T. attenuata.*
            5. Inflorescences ceriferous. Flowers orange or yellow; stigmas shallowly bilobed. . . . . *T. sandwithii.*

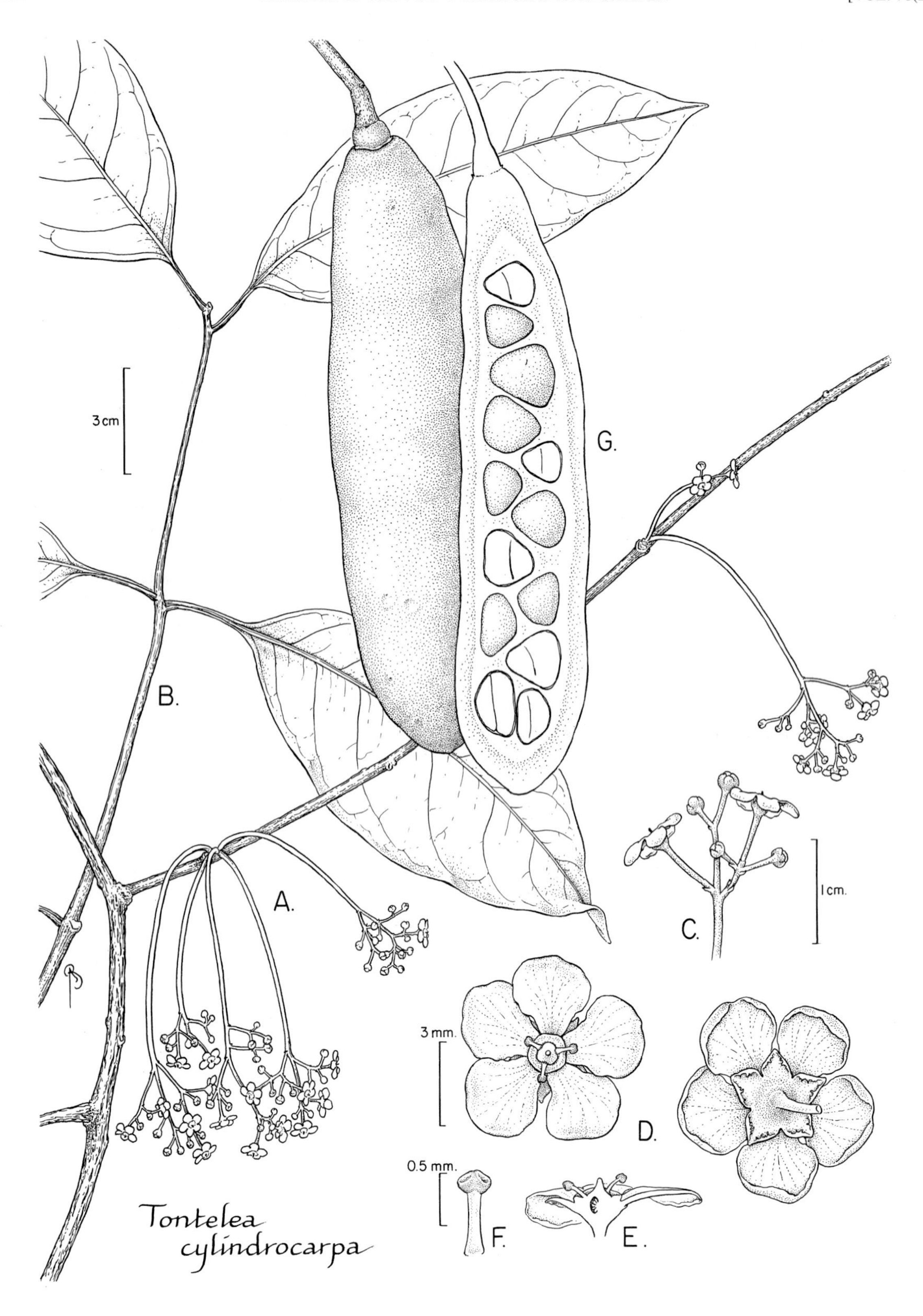

FIG. 145. HIPPOCRATEACEAE. *Tontelea cylindrocarpa* (A, B, G, *Görts-van Rijn et al. 110*; C–F, *Mori et al. 24194*). **A.** Part of stem with inflorescences. **B.** Apex of stem with leaves. **C.** Part of inflorescence. **D.** Apical (left) and basal (right) views of flower. **E.** Medial section of flower. **F.** Stamen. **G.** Medial section of fruit showing embedded seeds.

Tontelea attenuata Miers

Lianas or shrubs. Leaves: blades narrowly elliptic, elliptic-oblong to ovate, 8–19 × 2.5–7 cm, chartaceous, sometimes firmly so, the base cuneate, the apex acuminate; secondary veins in 9–11 pairs, plane to prominulous abaxially, the tertiary veins plane to prominulous abaxially. Inflorescences dichotomously branched panicles (branching close to base), semiglobose, 2–3 cm long, glabrous, the axes robust; pedicels to 1 mm long. Flowers fragrant, 3–4 mm diam., green or white; stigmas 3, entire, deltoid, spreading. Fruits subglobose or ellipsoid, to 4.5 × 2 × 2 cm, at first green then turning orange. Seeds 6. Fl (Sep, Oct, Nov); in nonflooded forest.

Tontelea cylindrocarpa (A.C. Sm.) A.C. Sm. FIG. 145, PL. 67f

Lianas or shrubs. Branches twining. Leaves: blades elliptic, 12–27 × 4–12 cm, chartaceous to thin-coriaceous, the base obtuse to attenuate, the apex acuminate or cuspidate; secondary veins in 4–7 pairs, prominent abaxially, the tertiary veins parallel, distinct abaxially. Inflorescences on branches or stem, thyrsoid-paniculate, solitary, or few together, 5–20 cm long, the axes slender; peduncles to 16 cm long; pedicels 5–7 mm long. Flowers 6–8 mm diam., greenish white or yellow; style subulate, the stigmas obsolete. Fruits cylindrical, pendent, to 16 × 5 cm, brown, often with purple hue or grayish blue. Seeds 20–24. Fl (May, Jul, Sep), fr (Mar, May, Sep, Oct, Nov); common and conspicuous because of gray, sausage-shaped fruits; in nonflooded forest.

Tontelea laxiflora (Benth.) A. C. Sm.

Lianas or occasionally slender trees. Leaves: blades narrowly elliptic to elliptic, ovate, or obovate, 8–16 × 3.5–5 cm, chartaceous to subcoriaceous, the base cuneate, the apex acuminate; secondary veins in 6–9 pairs, prominulous abaxially, the tertiary veins densely reticulate, prominulous abaxially. Inflorescences axillary, laxly and dichotomously branched panicles, to 9 cm long, the axes slender; peduncles <1 cm long; pedicels 3–15 mm long. Flowers 4–5 mm diam., fragrant, greenish or yellowish; stigmas 3, deeply divided, recurved, the 6 lobes equal, giving an umbrella-like appearance. Fruits ellipsoid, to 2.4 × 1.8 cm, lemon-yellow, yellow, or orange. Seeds ≤6. Fl (Jun), fr (Feb, Mar, May, Jul, Nov); common, in nonflooded forest.

Tontelea nectandrifolia (A. C. Sm.) A. C. Sm.

Lianas or epiphytic shrubs. Leaves: blades elliptic to oblong, 6–14 × 3–7 cm, coriaceous, the base acute, the apex rounded or bluntly acuminate; secondary veins in 5–7 pairs, plane or prominulous abaxially, the tertiary veins obsolete. Inflorescences thyrsoid-paniculate, 3–5 cm long, glabrous; pedicels slender, 1–2 mm long. Flowers 2.2 mm diam., greenish; nectary disc short tubular; style truncate, the stigmas obsolete. Fruits ovoid to globose, ca. 3 cm diam., green, slightly stipitate. Not yet collected in our area but to be expected.

Tontelea ovalifolia (Miers) A. C. Sm. subsp. **ovalifolia**

Lianas. Leaves: blades elliptic or somewhat obovate, 9–22 × 4–11 cm, coriaceous, the base obtuse or subacute, the apex rounded or obtuse, the margins entire or somewhat serrulate; secondary veins in 7–11 pairs, plane to prominulous abaxially, the tertiary veins immersed, obsolete abaxially. Inflorescences thyrsoid-paniculate, to 4 cm long, the axes robust; pedicels to 2 mm long. Flowers 4–7 mm diam.; stigmas 3, deltoid, spreading. Fruits ovoid to subglobose, 3.5 × 5 cm. Fl (Sep); in nonflooded forest.

Tontelea sandwithii A. C. Sm.

Lianas. Leaves: blades elliptic-oblong, 9–18 × 4–7.4 cm, chartaceous to firm; secondary veins in 6–10 pairs, these plane to slightly prominulous adaxially, more distinct abaxially, the tertiary veins obsolete. Inflorescences dichotomously branched panicles, 2–3.5 cm long, ceriferous, the axes robust; pedicels <1 mm long. Flowers ca. 4 mm diam., orange or yellow; stigmas shallowly bilobed. Fruits ovoid, 1.5–2 cm long, dark-brown, coriaceous. Fl (Sep); in nonflooded forest.

HUGONIACEAE (Hugonia Family)

Daniel Sabatier

Trees. Stipules present, small. Leaves simple, alternate; blades with both surfaces minutely punctate. Flowers actinomorphic, bisexual; sepals 5; petals 5, deciduous; stamens 10, alternately short and long, fused at base; ovary superior, globose, mostly 5-celled, the styles 5; placentation axile. Fruits drupes, the exocarp thin, fleshy, the endocarp woody.

This family is sometimes treated as part of the Linaceae.

Bakhuizen van den Brink, R. C., fil. 1966. Linaceae. *In* A. Pulle (ed.), Flora of Suriname **III(1):** 409–411. E. J. Brill, Leiden.
Lindeman, J. C. 1986. Linaceae. *In* A. L. Stoffers & J. C. Lindeman (eds.), Flora of Suriname **III(1–2):** 485–487. E. J. Brill, Leiden.
Ramírez, N., P. E. Berry & A. Jardim. 1999. Hugoniaceae. *In* J. A. Steyermark, P. E. Berry, K. Yatskievych & B. K. Holst (gen. eds.), Flora of the Venezuelan Guayana **5:** 618–623. P. E. Berry, K. Yatskievych & B. K. Holst (vol. eds.). Missouri Botanical Garden Press, St. Louis.

HEBEPETALUM Benth.

Trees. Branches and leaves glabrous. Stipules small, deciduous. Leaves: blades chartaceous, the margins entire to crenulate; secondary veins numerous, most loop-connected near margin, some reticulately connected. Inflorescences terminal, paniculate.

Hebepetalum humiriifolium (Planch.) Benth. FIG. 146

Understory trees, to 20 m tall. Trunks with plank buttresses, the bole angled. Bark rough, brown, with numerous, large, whitish lenticels. Leaves: petioles scarcely developed in seedlings to long and finely winged in adults; blades variable in shape, from linear in seedlings and saplings to elliptic-oblong in adults, often with two faint longitudinal lines, one on each side of the midrib, the

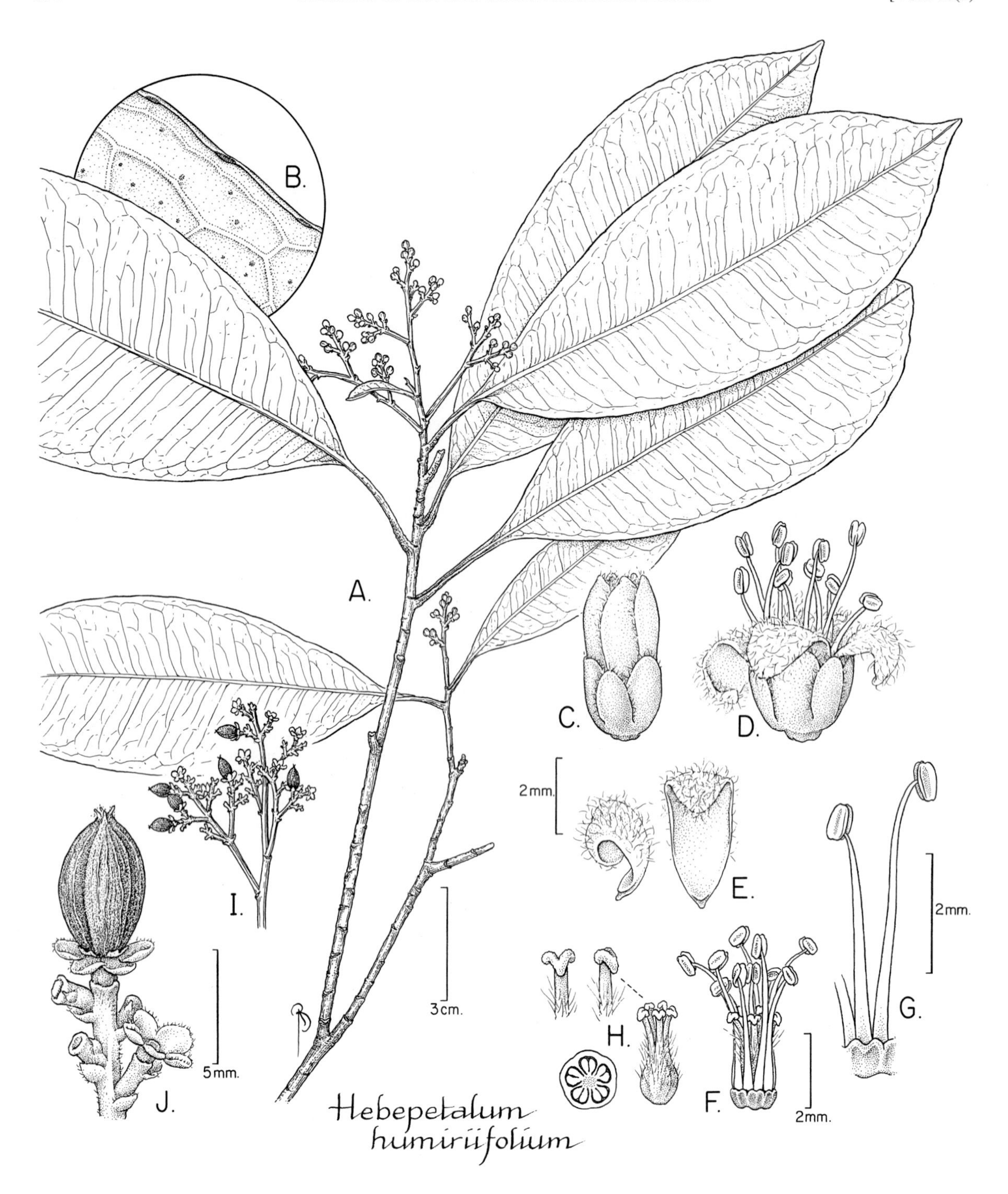

FIG. 146. HUGONIACEAE. *Hebepetalum humiriifolium* (A, *Mori et al. 19183*; B, I, J, *Oldeman B.676*; C–H, *Mori & Cardoso 17723* from Brazil). **A.** Stem with inflorescences in bud. **B.** Abaxial leaf surface showing inconspicuous punctations. **C.** Flower just before anthesis. **D.** Flower at anthesis. **E.** Petals. **F.** Flower with perianth removed. **G.** Stamens with basal nectaries. **H.** Pistil (right), detail of stigmas (above left), and transverse section of ovary (below left). **I.** Part of infructescence. **J.** Detail of infructescence with fruit.

base acute to cuneate in adults, the apex acuminate in adults, the margins serrate in seedlings and saplings, entire to crenulate in adults. Flowers yellow abaxially, white adaxially; petals covered with long, white hairs adaxially; stamens glabrous, shortly connate, with 5 bifid glands. Drupes globose, ca. 6 × 7 mm, ripening dark green to black, with whitish longitudinal lines. Fl (Aug), fr (Nov); scattered in our area, in nonflooded moist forest. *Bois-patagaïe* (Créole).

HUMIRIACEAE (Humiria Family)

Daniel Sabatier

Trees. Leaves simple, alternate; blades frequently punctate-glandulose near petiole or margin, the margins entire to crenate-serrate. Stipules very small, in pairs, soon deciduous or absent. Inflorescences terminal, subterminal, or axillary, equal to or shorter than leaves. Flowers actinomorphic, bisexual; calyx with 5 persistent sepals; corolla with 5 deciduous petals; stamens connate at base (monadelphous), either uniseriate in definite number (10 or 20) (*Humiria*, *Humiriastrum*, *Sacoglottis*) or pluriseriate in indefinite number (60–200) (*Vantanea*), the anthers with very thick, prolonged connective; intrastaminal disc present; ovary superior; placentation axile. Fruits drupaceous, the exocarp carnose, the endocarp very hard, woody, sculptured, with germinal dehiscence by valves or apertures, 1–7-locular, with 1–2 seeds per locule.

Bakhuizen van den Brink, R. C. fil. 1966. Humiriaceae. *In* A. Pulle (ed.), Flora of Suriname **III(1):** 412–421. E. J. Brill, Leiden.
Cuatrecasas, J. 1961. A taxonomic revision of the Humiriaceae. Contr. U. S. Natl. Herb. **35(2):** 25–214.
——— & O. Huber. 1999. Humiriaceae. *In* J. A. Steyermark, P. E. Berry, K. Yatskievych & B. K. Holst (gen. eds.), Flora of the Venezuelan Guayana **5:** 623–641. P. E. Berry, K. Yatskievych & B. K. Holst (vol. eds.). Missouri Botanical Garden Press, St. Louis.

1. Stamens numerous (60–200), the filaments flexuous, the anthers with 2 bilocular thecae. Endocarp longitudinaly furrowed, with lingulate valves. . . . . *Vantanea.*
1. Stamens less numerous (10 or 20), the filaments straight, the anthers with 2 unilocular thecae. Endocarp not longitudinally furrowed or with inconspicuous furrows.
   2. Stamens 10. Endocarp smooth or slightly tuberculate outside, without 5 apical apertures, filled with vacuolar resinous cavities inside. . . . . *Sacoglottis.*
   2. Stamens 20. Endocarp smooth outside, with 5 apical apertures, without resinous cavities inside.
      3. Thecae glabrous. Endocarp 20–30 mm long, with conspicuous elliptic-oblong subapical valves alternating with deep apical apertures. . . . . *Humiriastrum.*
      3. Thecae pilose. Endocarp 9–12 mm long, with inconspicuous narrow longitudinal valves alternating with small apical apertures. . . . . *Humiria.*

HUMIRIA J. St.-Hil.

Canopy trees. Flowers small; stamens 20, of 2 alternating sizes, the filaments straight, united for lower half, the free portion papillose, the anthers with 2, 1-celled, pilose thecae, these basal on thick connective. Fruits drupaceous, the exocarp black, juicy, the endocarp smooth outside, 5-locular, with 5 inconspicuous, narrow, longitudinal valves alternating with small apical apertures, without resinous cavities.

Humiria balsamifera (Aubl.) J. St.-Hil.

Canopy trees. Trunks with small, thick buttresses. Bark dark brown, deeply longitudinally fissured. Young stems winged, glabrous. Leaves: petioles nearly absent; blades variable in shape and size, from linear-elliptic in young trees to spatulate in older trees, usually 4–8 × 2.5–4 cm, the base auriculate clasping in young trees to attenuate in older trees, the apex often emarginate in older trees. Inflorescences axillary or terminal, paniculate-cymose, 4–7 cm long; peduncle winged or angular. Flowers white, buds not exeeding 7 mm long. Fruits small, ellipsoid-oblong, the endocarp to 12 × 5 mm. Common in savannas outside our area, rare in forest. Collected from nearby Nouragues and expected from our area. *Bois rouge* (Créole), *umiri* (Portuguese).

HUMIRIASTRUM (Urb.) Cuatrec.

Canopy trees. Trunks with buttresses. Stems hirtellous. Leaves: petioles short (1–2 mm long), hirtellous; blades with margins ± conspicuously crenate-serrate. Inflorescences terminal or subterminal, contracted paniculate-cymose, the axes hirtellous. Flowers small, buds not exceeding 3 mm long; stamens 20. Fruits with endocarp elliptic-oblong, smooth outside, with pronounced acute apex and 5 well developed apical apertures alternating with 5 conspicuous, elliptic-oblong, subapical valves, without resinous cavities.

Humiriastrum excelsum (Ducke) Cuatrec.

Canopy trees. Trunks with thick, steep buttresses. Bark light red-brown, smooth, finely lenticelate or peeling in irregular, large plates. Leaves: petioles 1–2 mm long, hirtellous; blades obovate-elliptic, 3–8 × 1.5–3 cm, chartaceous to subcoriaceous, glabrous, the base cuneate, the apex acuminate to cuspidate, the margins slightly crenate-serrate; secondary veins in 6–11 pairs, inconspicuous. Flowers whitish; ovary glabrous. Drupes yellow-brown to reddish when ripe, the endocarp 25–30 × 17–20 mm. Fl (Sep), fr (Aug, Sep); rare, in nonflooded forest.

SACOGLOTTIS Mart.

Canopy trees. Trunks with very small and thick buttresses or buttresses absent. Leaves: blades with acuminate to cuspidate apex, the margins ± deeply serrate-crenate. Inflorescences usually axillary, sometimes terminal, paniculate-cymose, shorter than leaves; pedicels long. Flowers small, buds not exceeding 5 mm; stamens 10, 5 longer than alternating ones, the filaments straight, the thecae baso-lateral, on thick connective. Fruits globose or oblong, with 1–3 developed locules, 1 seed per locule, the endocarp woody, smooth or slightly tuberculate, filled with resinous vacuolar cavities.

1. Bark sparsely covered with small red-brown lenticels, not cracked as below. Fruits globose to pyriform, 15–27 × 15–20 mm, ripening orange-red, the inner layer of exocarp fibrous. . . . . *S. cydonioides.*
1. Bark not covered with lenticels as above, deeply grid cracked. Fruits oblong, 25–30 × 12–15 mm, ripening orange-yellow, the inner layer of exocarp not fibrous. *S. guianensis* var. *guianensis.*

Sacoglottis cydonioides Cuatrec.

Canopy trees. Trunks with small thick, running buttresses. Bark smooth, light brown, sparsely covered with small, red-brown lenticels. Glabrous except on inflorescence axis. Leaves: petioles short, 5–7 mm long; blades 5–11 × 3–5 cm, thick-coriaceous, the apex acuminate, the base obtuse, the margins ± revolute, serrate. Inflorescences axillary or terminal. Flowers pale green; buds not exceeding 3.5 mm long; petals with upper half minutely hispidulous. Fruits globose to pyriform, 15–27 mm long, the outer layer of exocarp carnose, orange-red, the inner layer fibrous, the endocarp slightly tuberculate. Collected from nearby Nouragues in primary, nonflooded forest and expected from our area. *Bofo udu* (Bush Negro).

Sacoglottis guianensis Benth. var. **guianensis**

Canopy trees. Trunks unbuttressed. Bark rough, dark brown, grid cracked. Inflorescence axis and often terminal stems ± hirtellous to pubescent. Leaves: petioles 10–15 mm long; blades 8–12 × 3–5.5 cm, coriaceous, the apex acuminate to cuspidate, the margins slightly crenate to serrate-crenate. Inflorescences axillary. Flowers pale green; buds not exceeding 4.5 mm long; petals glabrous. Fruits ellipsoid-oblong, the exocarp carnose, orange-yellow when ripe, the endocarp to 20 × 12 mm, smooth or slightly tuberculate. Fr (Sep); in nonflooded moist forest.

VANTANEA Aubl.

Canopy trees. Trunks often with buttresses. Leaves: blades coriaceous, the margins entire. Inflorescences terminal or subterminal, paniculate-cymose, usually shorter than leaves, sometimes equal to leaves in length. Flowers with numerous (50–120) stamens, the filaments flexuous, the thecae 2, bilocular, basal on thick connective with acute tip. Fruits: endocarp woody, longitudinally furrowed, with ligulate valves, the surface rough or smooth.

1. Leaves with petioles not thickened at base; blades elliptic to oblong-elliptic, the apex acuminate. Flower buds ca. 3 cm long. Petals red; ovary and disc glabrous. . . . . *V. guianensis.*
1. Leaves with petioles often thickened at base; blades obovate-elliptic or oblanceolate. Flower buds 0.8–1.2 cm long. Petals white or greenish-white; ovary and disc pubescent.
    2. Leaf blades obovate-elliptic, 4–11 cm long. Fruits ca. 2.4–2.8 × 2.2–2.5 cm, the exocarp ca. 2 mm thick, the endocarp rough. . . . . *V. parviflora.*
    2. Leaf blades oblanceolate to obovate-elliptic, 8–15 cm long. Fruits ca. 5–7 × 4–5 cm, the exocarp ca. 5 mm thick, the endocarp smooth. . . . . *V. ovicarpa.*

Vantanea guianensis Aubl. FIG. 147; PART 1: FIG. 9

Trunks with steep, plank buttresses. Bark rough, scaly, densely lenticelate, dark reddish. Leaves: petioles flattened, not thickened at base, 6–12 mm long; blades elliptic to oblong-elliptic, ca. 6–14 × 3–6 cm, the apex shortly and obtusely acuminate. Inflorescences: axes glabrous. Flower buds ca. 3 cm long. Flowers: petals red; ovary glabrous. Fruits ovoid-oblong, ca. 6 × 4.5 cm, ripening red-brown, the exocarp carnose, 5 mm thick, the endocarp rough, with brain-like surface, deeply 5-furrowed. Fl (Aug, Sep), fr (Jun, Aug); scattered, in primary nonflooded forest.

Vantanea ovicarpa Sabatier

Trunks with small buttresses. Bark scaly, red-brown. Leaves: petioles sulcate, thickened at base, 8–10 mm long; blades oblanceolate to obovate-elliptic, ca. 8–15 × 4–7 cm, the apex rounded and emarginate, sometimes short acuminate. Inflorescences: axes puberulous. Flower buds ca. 12 mm long. Flowers: petals greenish-white, sometimes tinged with pink at apex; ovary and disc densely pilose. Fruits ovoid to ovoid-oblong, ca. 5–7 × 4–5 cm, dull green the exocarp ca. 5 mm thick, the endocarp smooth, 6- or 7-locular. Fl (Sep), fr (Jan, but present on ground in Sep); in nonflooded moist forest at ca. 500 m in our area.

Vantanea parviflora Lam.

Trunks with small, thick buttresses or buttresses absent. Bark rough, grid cracked in young trees, peeling in large plates in older trees. Leaves: petioles terete, often thickened at base, 5–15 mm long; blades obovate-elliptic, 4–11 × 2–5.5 cm, the apex obtuse or

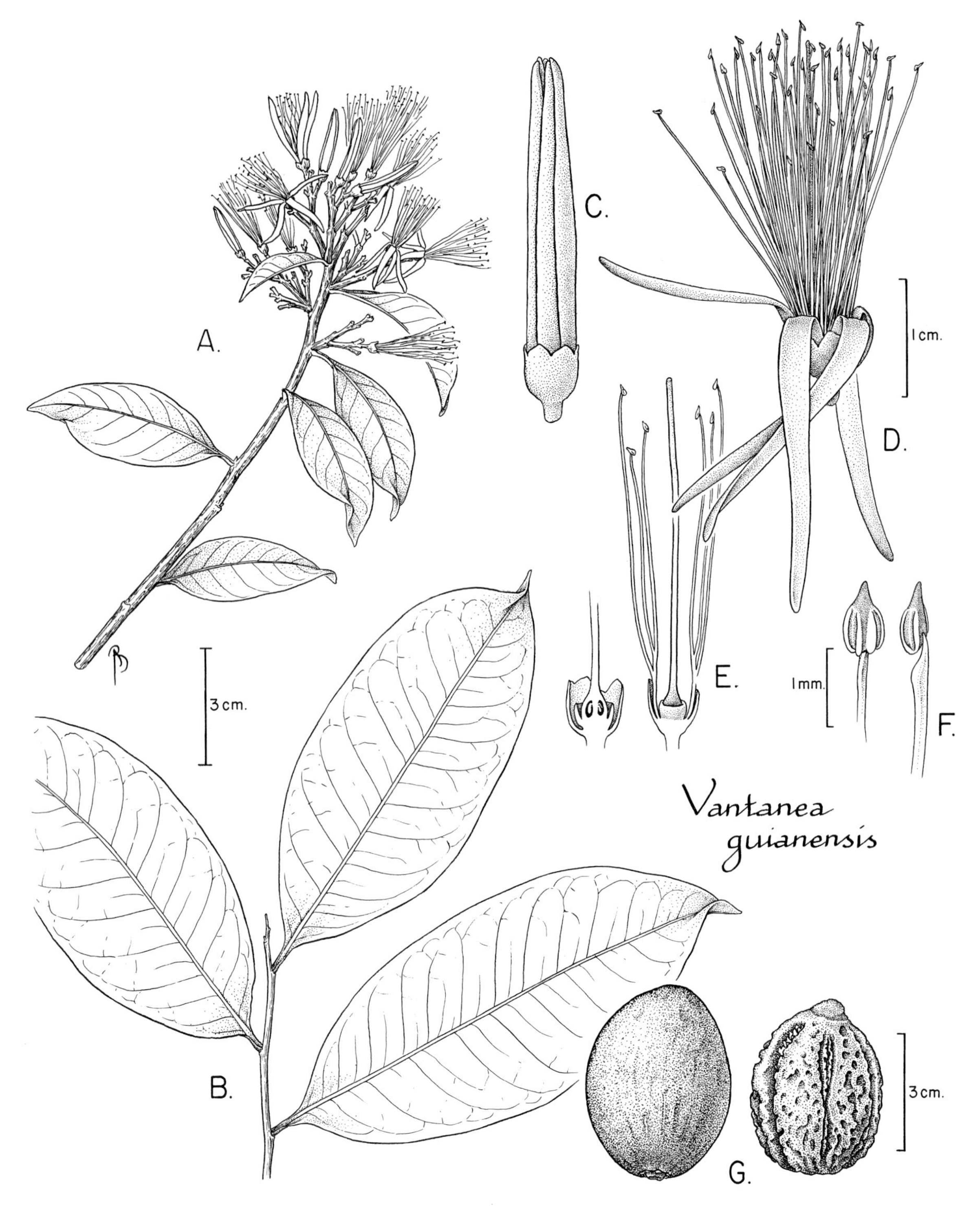

FIG. 147. HUMIRIACEAE. *Vantanea guianensis* (A, *Mori 18722*; B, *Mori & Boom 15303*; C–F, *Mori & Boom 14778*; G, *Mori & Gracie 18978*). **A.** Leafy apex of flowering stem. **B.** Leafy apex of vegetative stem. **C.** Lateral view of flower bud. **D.** Open flower showing numerous stamens. **E.** Medial section (left) of base of flower with corolla, androecium, and upper part of pistil removed; note the axile placentation of ovules, and medial section of flower (right) with corolla removed; note the connate filament bases and cupular disc. **F.** Adaxial (far left) and lateral (near left) views of stamens showing the expanded and prolonged connective. **G.** Entire fruit (left) and endocarp (right).

obtusely acuminate and emarginate. Inflorescences: axes puberulous. Flower buds ca. 0.8 cm long. Flowers: petals white; ovary villose. Fruits ellipsoid-globose, 2.4–2.8 × 2.2–2.5 cm, ripening gray-green, the exocarp thin, ca. 2 mm thick, the endocarp rough. Fl (Sep); common but at low density, in primary nonflooded forest. *Gris-gris rouge* (Créole).

ICACINACEAE (Icacina Family)

Adrian C. de Roon and Scott A. Mori

Trees, shrubs, or lianas. Leaves simple, alternate. Stipules absent. Inflorescences terminal, axillary, or supra-axillary, cymose or paniculate with cymose branches. Flowers usually articulate below calyx, small, actinomorphic, usually bisexual, infrequently unisexual; calyx fleshy, the lobes or teeth (4)5(6); petals (4)5(6), free or united at base, valvate, usually fleshy; stamens (4)5, alternating with petals, the anthers mostly 4-celled, rarely 2-celled, basi- or dorsifixed, with introrse or lateral dehiscence; ovary superior, 1-locular, rarely 2–3-locular, each locule with 2 apical, pendulous ovules. Fruits drupaceous. Seeds with small or large embryo, the endosperm copious.

Howard, R. A. & R. Duno de Stefano. 1999. Icacinaceae. *In* J. A. Steyermark, P. E. Berry, K. Yatskievych & B. K. Holst (gen. eds.), Flora of the Venezuelan Guayana **5:** 646–658. P. E. Berry, K. Yatskievych & B. K. Holst (vol. eds.). Missouri Botanical Garden Press, St. Louis.
Jansen-Jacobs, M. J. 1979. Icacinaceae. *In* A. L. Stoffers & J. C. Lindeman, Flora of Suriname **V(1):** 344–355. E. J. Brill, Leiden.
Roon, A. C. de. 1994. Icacinaceae. *In* A. R. A. Görts-van Rijn (ed.), Flora of the Guianas Ser. A; **16:** 82–109, 142–151. Koeltz Scientific Books, Havlíčkův Brod, Czech Republic.

1. Trees or shrubs.
   2. Flowers unisexual (pistillode present in staminate flowers). Fruits flattened, bicolored, the convex side black, concave side white at maturity. . . . . *Discophora.*
   2. Flowers bisexual. Fruits not flattened (*Emmotum* often compressed), not bicolored at maturity.
      3. Leaves with conspicuous peltate scales on both surfaces, the tertiary veins indistinct even with hand lens. Petals glabrous adaxially. . . . . *Dendrobangia.*
      3. Leaves without conspicuous peltate scales, the tertary veins distinct. Petals pubescent adaxially.
         4. Abaxial leaf surface usually with closely appressed, abundant, soft hairs. Inflorescences congested. Petals very densely cottony; ovary 2–3-locular. . . . . *Emmotum.*
         4. Abaxial leaf surface glabrous or subglabrous. Inflorescence spicate. Petals pubescent but not densely cottony; ovary 1-locular. . . . . *Poraqueiba.*
1. Lianas.
   5. Leaf blades pilose, especially abaxially. Petals glabrous adaxially. . . . . *Pleurisanthes.*
   5. Leaf blades glabrescent. Petals pilose adaxially.
      6. Inflorescences axillary. Anther connectives linear. Mature fruits purple at maturity. . . . . *Leretia.*
      6. Inflorescences terminal. Anther connectives triangular. Mature fruits yellow to orange at maturity. . . . *Casimirella.*

CASIMIRELLA Hassl.

Shrubs or lianas, often with large, tuberous rhizomes. Peltate scales with fringed margins absent. Inflorescences large, terminal or supra-axillary panicles. Flowers bisexual; calyx deeply 5-lobed, the lobes almost as long as petals; petals 5–6, free, tomentose adaxially, the apices recurved; stamens 5–6, the filaments glabrous, the anthers dorsifixed, the connective prolonged beyond thecae into triangular projection; ovary densely pubescent, 1-locular; style 1, rarely 2–3, the stigma small, capitate. Fruits peach-like drupes, the pericarp densely pubescent, the endocarp woody. Seed 1, the testa thin.

Casimirella ampla (Miers) R. A. Howard

Lianas. Leaves: petioles 8–10 mm long; blades narrowly ovate to lanceolate, 8–20 × 3–10 cm, chartaceous, glabrous, the base nearly acute to rounded, the apex obtuse to acuminate, the margins entire; secondary veins in 3–6 pairs. Inflorescences supra-axillary or terminal, to 20 cm long. Flowers ca. 4 mm diam.; calyx densely strigose, the lobes lanceolate, ca. 1.5 mm long; petals ovate-lanceolate, 3–4 × 1.5–2 mm, villose adaxially. Fruits globose to ovate, pubescent, ca. 5 cm diam, yellow to orange at maturity. Not yet collected in our area but to be expected.

DENDROBANGIA Rusby

Canopy to emergent trees. Trunks with well developed buttresses. Stems, leaves, inflorescences, and external parts of flowers covered with conspicuous peltate scales with fringed margins. Inflorescences axillary, short panicles with glomerules of sessile flowers. Flowers bisexual; calyx with 5 imbricate sepals; petals 5, fused at bases, glabrous adaxially, the lobes reflexed, the apices attenuate and differentiated from rest of lobe, glabrous adaxially; stamens 5, adnate to corolla, the anthers dorsifixed, the filaments short, filiform; ovary pubescent, 1-locular. Fruits drupes. Seed 1, with copious endosperm, the embryo minute.

Dendrobangia boliviana Rusby FIG. 148, PL. 68a

Trees, to 46 m tall. Trunks with steep, high, often branched buttresses. Bark smooth, with incomplete hoop marks when young, becoming rough and peeling when older, the slash orange with darker streaks. Leaves: petioles 12–25 mm long; blades widely elliptic to elliptic, 10–14(18) × 5–9.5 cm, chartaceous, drying dark brown, with abundant peltate scales on both surfaces, the base acute

Plates 65–72

Plate 65. GESNERIACEAE. **a.** *Codonanthe calcarata* (*Mori et al. 21603*), flowers. **b.** *Episcia xantha* (*Mori et al. 22285*, photographed in NY from plant grown from seed of this collection), apical view of flower. **c.** *Codonanthe crassifolia* (*Mori et al. 21610*), part of epiphytic plant with fruit. **d.** *Drymonia coccinea* (unvouchered), inflorescence with one open flower. **e.** *Napeanthus macrostoma* (*Mori et al. 21657*), flowers; note emarginate corolla lobes.

Plate 66. GESNERIACEAE. **a.** *Nautilocalyx mimuloides* (*Mori et al. 22212*), part of plant growing along ground with flowers. **b.** *Nautilocalyx mimuloides* (*Mori et al. 23049*), lateral view of flower. HERNANDIACEAE. **c.** *Hernandia guianensis* (*Prévost 3500*), fruits surrounded by inflated calyces. [Photo by M.-F. Prévost]

Plate 67. HIPPOCRATEACEAE. **a.** *Cheiloclinium cognatum* (Görts-van Rijn 107), flowers. **b.** *Cheiloclinium* sp. 1 (*Mori et al. 23785*), flowers. **c.** *Pristimera nervosa* (*Mori et al. 23137*), flower. **d.** *Prionostemma aspera* (*Görts-van Rijn 70*), three-parted fruit. **e.** *Salacia insignis* (*Mori et al. 23158*), flowers; note that calyx lobes do not show between petals. **f.** *Tontelea cylindrocarpa* (*Görts-van Rijn 110*), flower; note that calyx lobes show between petals.

Plate 68. ICACINACEAE. **a.** *Dendrobangia boliviana* (*Mori et al. 22879*), flowers and bud; note attenuate apices of corolla lobes. **b.** *Discophora guianensis* (*Acevedo-Rodríguez 5013*), bicolored fruits. **c.** *Poraqueiba guianensis* (*Mori et al. 23157*), flowers and buds. LACISTEMACEAE. **d.** *Lacistema grandifolium* (*Mori et al. 19161*), flowers and fruits. [Photo by S. Mori]

Plate 69. LAMIACEAE. **a.** *Hyptis lanceolata* (*Mori et al. 23800*), inflorescence with three flowers. **b.** *Hyptis pachycephala* (*Mori et al. 23743*), inflorescence with three flowers. **c.** *Scutellaria uliginosa* (*Mori et al. 22056*), flower, bud, and calyces. LAURACEAE. **d.** *Endlicheria punctulata* (*Mori et al. 22268*), immature fruits in cupules. **e.** *Aiouea guianensis* (*Mori & Gracie 21168*), immature fruits in cupules. **f.** *Licaria guianensis* (*Acevedo-Rodríguez 4962*), fruit in cupule.

Plate 70. LAURACEAE. **a.** *Endlicheria pyriformis* (*Mori & Gracie 23935*), flowers in staminate phase. **b.** *Nectandra hihua* (*Mori & Gracie 24212*), axillary inflorescence with red-violet axis. **c.** *Nectandra reticulata* (*Mori et al. 24188*), flowers; note papillose surface of petals. **d.** *Ocotea diffusa* (*Mori et al. 22948*), buds and flower. **e.** *Sextonia rubra* (*Mori et al. 21001*), flowers.

Plate 71. LAURACEAE. **a.** *Rhodostemonodaphne grandis* (*Mori et al. 23998*), buds and flowers. **b.** *Rhodostemonodaphne rufovirgata* (*Mori et al. 23832*), buds and flower. **c.** *Rhodostemonodaphne kunthiana* (*Mori et al. 22324*), flowers. **d.** *Rhodostemonodaphne saülensis* (*Mori et al. 23821*), flowers and buds.

Plate 72. LECYTHIDACEAE. **a.** *Corythophora amapaensis* (*Mori et al. 23342*), flowers, one with androecial hood pulled open to show reproductive structures. **b.** *Corythophora rimosa* subsp. *rubra* (*Mori et al. 23302*), flower and buds. **c.** *Corythophora amapaensis* (*Mori et al. 18675*), part of inflorescence with buds, open flowers, and immature fruits. **d.** *Corythophora rimosa* subsp. *rubra* (*Mori et al. 21007*), flower with androecial hood pulled open to show staminal ring on left and staminodes on right. **e.** *Couratari guianensis* (unvouchered), canopy tree flowering while leafless.

Codonanthe calcarata **a.**

Episcia xantha **b.**

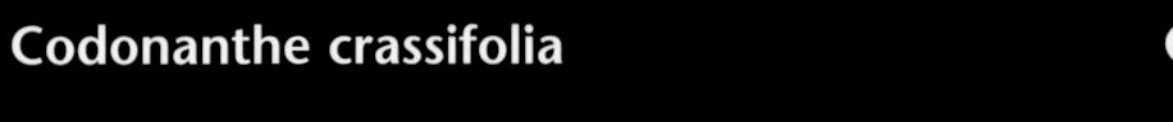

Codonanthe crassifolia **c.**

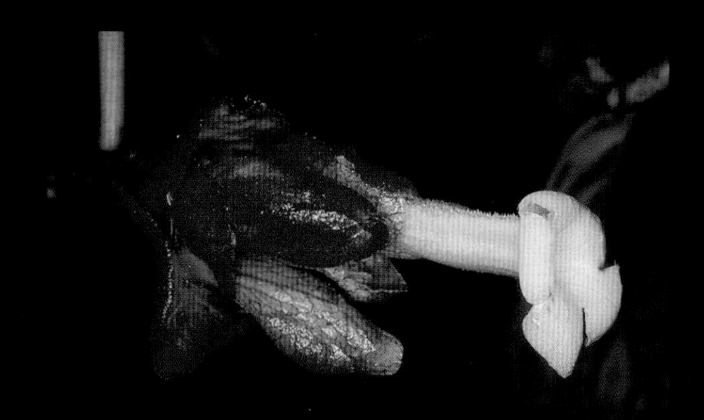

Drymonia coccinea **d.**

Napeanthus macrostoma **e.**

Nautilocalyx mimuloides

a.

b.

HERNANDIACEAE

Hernandia guianensis

c.

Plate 66

Cheiloclinium cognatum a.

Cheiloclinium sp. 1  b.

Pristimera nervosa c.

Prionostemma aspera d.

Salacia insignis e.

Tontelea cylindrocarpa f.

ICACINACEAE

Dendrobangia boliviana a.

Discophora guianensis b.

Poraqueiba guianensis c.

LACISTEMACEAE

 Lacistema grandifolium d.

LAMIACEAE

Hyptis lanceolata **a.**

Hyptis pachycephala **b.**

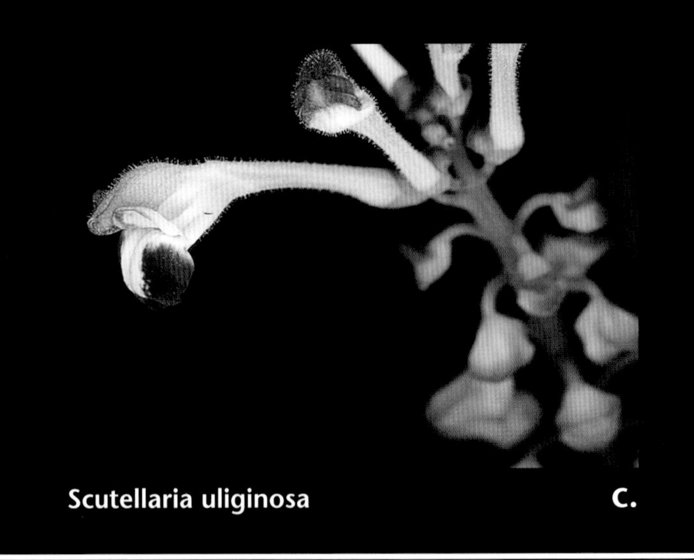

Scutellaria uliginosa **c.**

Endlicheria punctulata **d.**

LAURACEAE

Aiouea guianensis **e.**

Licaria guianensis **f.**

Nectandra hihua b.

Nectandra reticulata c.

Ocotea diffusa d.

Sextonia rubra e.

Plate 70

a.

Rhodostemonodaphne grandis

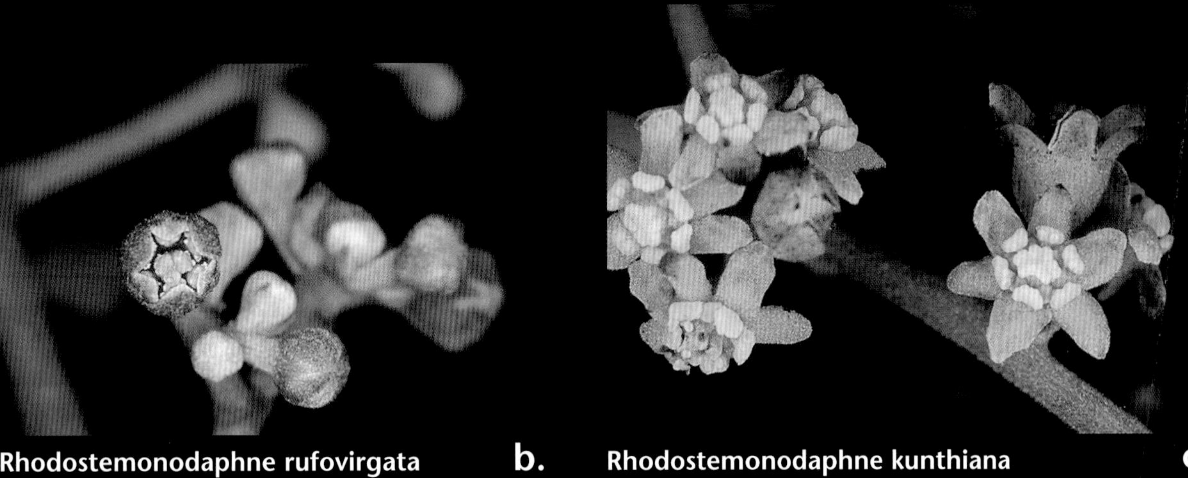

Rhodostemonodaphne rufovirgata b.

Rhodostemonodaphne kunthiana c.

d.

Rhodostemonodaphne saülensis

LECYTHIDACEAE

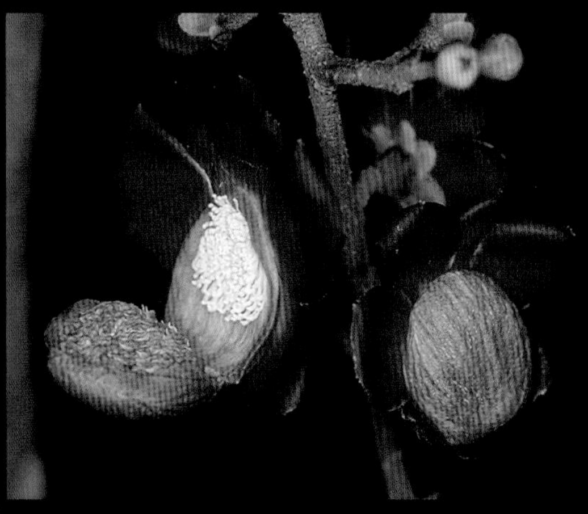

Corythophora amapaensis  a.

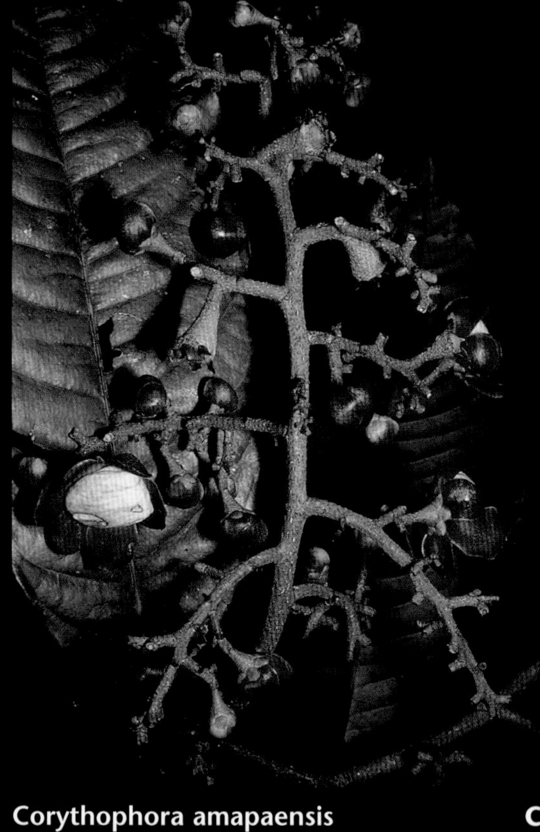

Corythophora amapaensis c.

Corythophora rimosa subsp. rubra  b.

Couratari guianensis e.

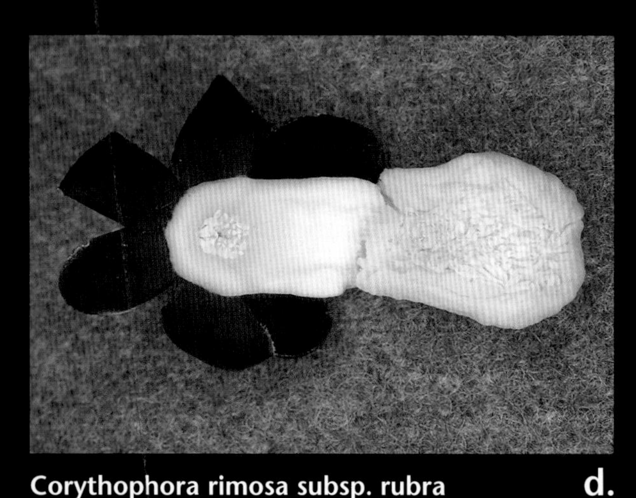

Corythophora rimosa subsp. rubra d.

Plate 72 (Lecythidaceae continued on Plate 73)

FIG. 148. ICACINACEAE. *Dendrobangia boliviana* (A, *Mori & Pipoly 15465*; B–E, *Mori et al. 22879*; F, G, *Mori & Pennington 17978*). **A.** Part of stem with leaves and inflorescences and detail of adaxial surface of young leaf showing stellate hairs. **B.** Detail of inflorescence. **C.** Oblique-apical (left) and lateral (right) views of flower. **D.** Opened corolla and pistil. **E.** Lateral (near right) and adaxial (far right) views of stamen. **F.** Part of stem with fruit. **G.** Medial section of fruit with intact seed.

to obtuse, the apex acuminate, the margins entire; secondary veins in 6–10 pairs, the tertiary veins indistinct. Inflorescences 3–4.5 cm long, glomerules of 3–5 flowers. Flowers ca. 2 mm diam.; corolla 3 mm long, white or yellowish-green. Fruits triangular in transverse section, 2 × 1 × 0.5 cm, yellow to orange, the mesocarp thin, fleshy, the endocarp thin, woody at maturity. Fl (Feb, Mar, Jul), fr (Mar, May, Jul); scattered, in nonflooded forest. *Méquoi* (Créole), *taapoutiki* (Bush Negro).

DISCOPHORA Miers

Small trees or shrubs. Peltate scales with fringed margins absent. Inflorescences axillary, short panicles. Flowers unisexual; calyx with 5 triangular, poorly developed teeth; petals 5, cucullate, with hooked apices, free, glabrous adaxially; nectary present. Staminate flowers: stamens 5, free, the filaments with densely woolly swelling just above middle, the anthers ovate, versatile, the thecae diverging at base; pistillode present. Pistillate flowers: ovary l-locular. Fruits drupes, flattened, ridged on convex side, smooth on concave side, the mesocarp thin, fleshy, the endocarp thin, crustaceous, scarcely woody. Seed with copious endosperm, the embryo minute.

Discophora guianensis Miers FIG. 149, PL. 68b

Small trees, to 10 m tall. Bark smooth, dirty white to gray. Leaves: petioles 10–25 mm long; blades oblong to narrowly oblong, 12–30 × 4–10 cm, chartaceous, glabrous, the base acute, the apex acuminate, the margins entire, finely revolute; secondary veins in 7–10 pairs. Inflorescences to 6 cm long, densely golden strigose, becoming glabrous and stout in fruit. Flowers ca. 2.2 mm diam.; petals 2–3 mm long, cream-colored. Fruits black, ribbed on convex side, white, smooth on concave side at maturity. Fl (Jan, Dec), fr (Feb, Mar, May); uncommon, in understory of non-flooded forest.

EMMOTUM Ham.

Trees. Peltate scales with fringed margins absent. Inflorescences axillary, short fascicled and often congested panicles. Flowers bisexual; calyx with 5 lobes; petals 5, free, densely woolly along midrib adaxially; stamens 5, free, the anthers basifixed to subdorsifixed; ovary (2)3-locular, the locules all on one side, ovary thus asymmetric as viewed in cross section. Fruits drupes, these often compressed, the endocarp with thick, woody, ornamented wall, usually with one, curved seed.

Emmotum fagifolium Ham.

Tree, to 30 m. Leaves: petioles 12–15 mm long; blades ovate to narrowly ovate, 7–18 × 4–8 cm, coriaceous, glabrous adaxially, with closely appressed, abundant, soft hairs abaxially, the base obtuse to rounded, the apex acuminate, the margins entire; secondary veins in 9 pairs, finely impressed adaxially, salient abaxially, the tertiary veins distinct. Inflorescences axillary, congested, to ca. as long as petiole, sericeous. Flowers ca. 6 mm diam.; calyx with 5 triangular teeth, ca. 5 × 6 mm; petals linear, reflexed at apex, 4–7 mm long, white, the midrib with woolly hairs along entire length adaxially, the adaxial margins free of hairs; filaments glabrous, 4 mm long, the anthers subdorsifixed; ovary pubescent, 3-locular, the style glabrous, the stigma punctiform. Fruits depressed, ribbed, ca. 1.5 cm diam. In nonflooded moist forest. *Bois-agouti* (Créole).

LERETIA Vell.

Lianas, or rarely shrubs or small trees with scandent branches. Peltate scales with fringed margins absent. Inflorescences axillary, large, pedunculate, much branched panicles. Flowers usually bisexual, sometimes unisexual by abortion; calyx 5-lobed; petals (4)5, free, the apex inflexed; stamens (4)5, the filaments filiform, the anthers dorsifixed, the connective prolonged into linear projection beyond thecae; ovary hirsute, with disc-like columniform base, 1-locular, the style excentric, glabrous, with two aborted styles at base. Fruits drupes, glabrescent, the mesocarp thin, scarcely fleshy, the endocarp very thin, scarcely woody. Seed 1; embryo with wrinkled cotyledons.

Leretia cordata Vell.

Lianas. Leaves: petioles 5–15 mm long; blades ovate to oblong to oblong to elliptic, 10–20(30) × 3–7(12) cm, chartaceous to coriaceous, sparsely pubescent when young, the base acute to obtuse, the apex acute, sometimes obtuse to rounded or apiculate to acuminate, the margins entire; secondary veins in 6–8 pairs. Inflorescences to 18 × 10 cm. Flowers ca. 5 mm diam.; calyx patelliform, the lobes 0.5 mm long, strigose; petals oblong-lanceolate, ca. 5 × 2 mm, lanate-villose adaxially except at base and inflated apex, white. Fruits oblong, nearly glabrous, ca. 2 × 1 cm, purple at maturity. Fl (May), fr (Sep).

PLEURISANTHES Baill.

Lianas. Stems frequently strap-shaped. Leaves with prominent reticulate venation. Inflorescences axillary, supra-axillary, or terminal, panicles or single or clustered, sometimes spiciform racemes, the rachis often flattened; pedicels not articulate. Flowers small, bisexual; calyx cupulate, 5-lobed; petals 4–5, free or connate at base, fleshy, the apex inflexed, the midrib frequently prominent, glabrous adaxially, strigose abaxially; stamens 4–5, free; ovary conical, pilose, 1-locular, the style minute to well developed, the stigma capitate; ovules 2, collateral. Fruits drupes.

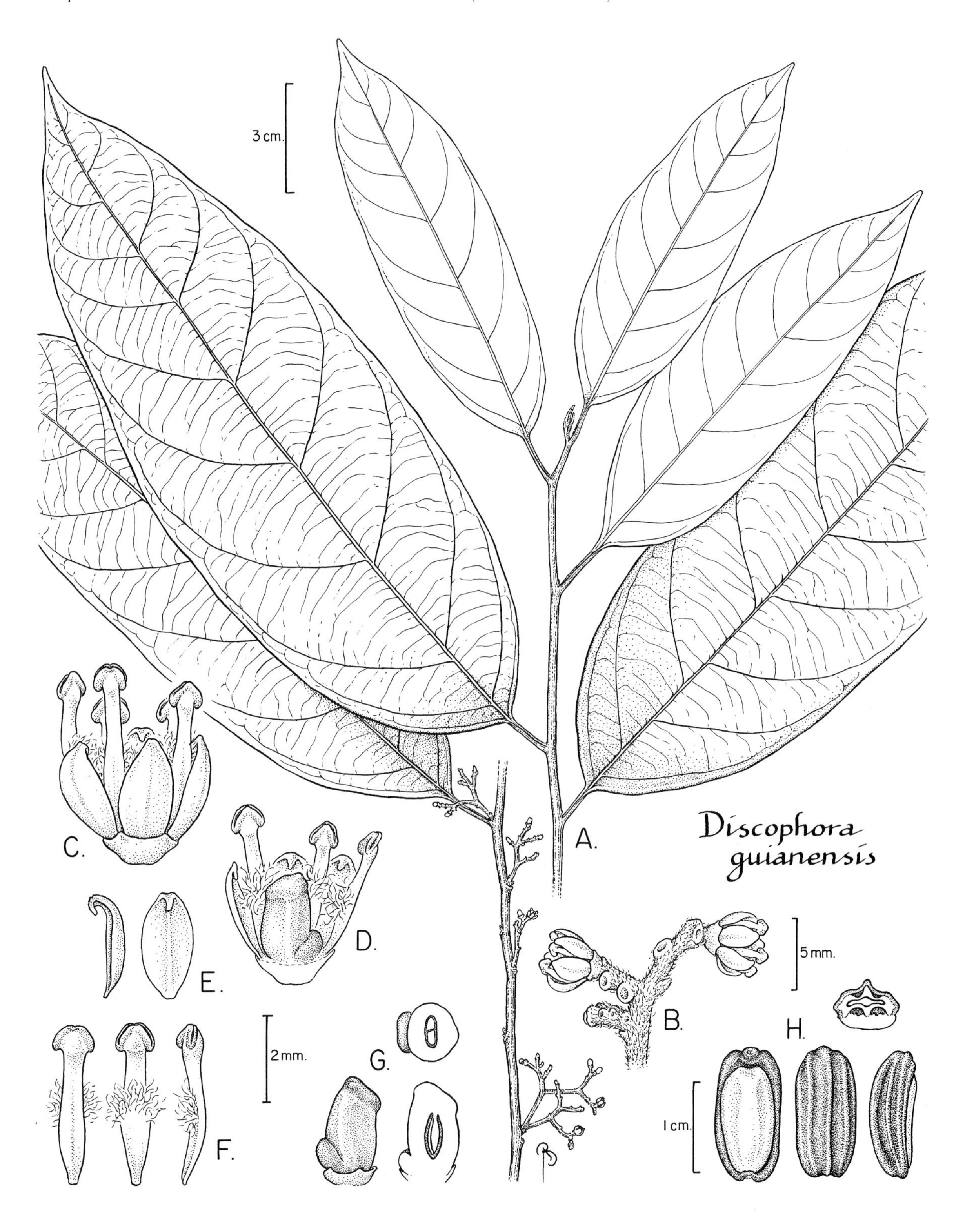

FIG. 149. ICACINACEAE. *Discophora guianensis* (A–G, *Mori & Pepper 24295*; H, *Mori et al. 22961*). **A.** Terminal part of stem with leaves (near left) and part of stem with inflorescences (far left). **B.** Detail of inflorescence. **C.** Lateral view of flower. **D.** Section of flower with two petals removed and showing intact pistil. **E.** Lateral (far left) and adaxial (near left) views of petals. **F.** Abaxial (far left), adaxial (middle left), and lateral (near left) views of stamens. **G.** Lateral view (below left), medial section (below right), and transverse section (above right) of pistil with gland. **H.** Three views of fruit (below) and transverse section of fruit (above right).

Pleurisanthes parviflora (Ducke) R. A. Howard

Leaves: petioles ca. 1 cm long, densely tomentose; blades elliptic-oblong to ovate-oblong, 8–16 × 3–6 cm, subcoriaceous, glabrous except for midrib adaxially, tomentose abaxially, the base rounded to subcordate, the apex acuminate, the margins entire, slightly revolute; secondary veins in 8–9 pairs. Inflorescences axillary, loosely paniculate, 5–15 cm long, the axes tomentose, the flowers single or in clusters of 2–6; pedicels 2–3 mm long. Flowers ca. 2.5 mm diam.; calyx with triangular lobes; petals oblong to lanceolate, 2–2.5 × 0.7–1 mm; stamens 5, the filaments ca. 2 mm long, the anthers oblong, ca. 1 mm long; ovary hirsute, the style glabrous. Mature fruits not known.

Not yet collected from our area but to be expected.

PORAQUEIBA Aubl.

Trees. Peltate scales with fringed margins absent. Inflorescences axillary, paniculate, branched from base; pedicels absent or nearly absent. Flowers bisexual; calyx with 5 imbricate lobes; petals free or fused at base, the midrib expanded and fleshy adaxially, often pubescent, the apex inflexed; stamens 5, flattened, free, the anthers basifixed; ovary glabrous, l-locular, with 2, apical, pendulous ovules. Fruits drupes, the mesocarp fleshy, often edible, the endocarp woody. Seed with minute, curved or nearly straight embryo.

Poraqueiba guianensis Aubl. PL. 68c

Understory to canopy trees. Trunks with or without buttresses. Bark smooth, with hoop marks and horizontally oriented lenticels, the slash reddish-orange to reddish-brown, with darker vertical streaks. Leaves: petioles 10–29 mm long; blades ovate to elliptic, 12–25 × 5–10 cm, coriaceous, glabrous, the base obtuse to rounded, the apex acuminate, the margins entire; secondary veins in 7–8 pairs, the tertiary veins departing midrib at nearly right angles, then bending abruptly downward to join secondaries. Inflorescences to 9 cm long, sericeous; pedicels 0.5 mm long. Flowers 5 mm diam.; calyx green, the lobes ovate, 1 mm long; corolla white, the petals oblong-ovate, ca. 3 × 1 mm, with thickened medial, densely pubescent ridge adaxially; stamens with flattened filaments, these wider towards apex; ovary with simple style, the stigma punctiform. Fruits ellipsoid, to 3–4 × 2 cm, green, drying blackish. Fl (Aug, Sep), fr (Nov); occasional in nonflooded forest. *Umari* (Portuguese).

LACISTEMATACEAE (Lacistema Family)

Scott A. Mori

Small trees. Leaves simple, alternate; venation pinnate. Stipules present, the stipule scars sometimes encircle young stem. Inflorescences clustered in leaf axils or along stem, catkin-like, the flowers sessile; bract broadly ovate, the bracteoles smaller. Flowers small, zygomorphic, bisexual; sepals 2–6, very rarely only one or absent, small, free; petals absent; nectary present and well developed or absent; stamen 1, usually bifid; ovary 1-locular, the style distinct or nearly absent, the stigmas 2–3; placentation parietal, the placentae 3, the ovules pendulous. Fruits capsules, splitting into 3 valves. Seeds 1, rarely 2–3 per fruit, arillate; endosperm fleshy, sparse, the embryo straight, with foliaceous cotyledons.

Raechal, L. J. 1999. Lacistemataceae. *In* J. A. Steyermark, P. E. Berry, K. Yatskievych & B. K. Holst (gen. eds.), Flora of the Venezuelan Guayana **5:** 676–677. P. E. Berry, K. Yatskievych & B. K. Holst (vol. eds.). Missouri Botanical Garden Press, St. Louis.
Vaandrager, G. 1966. Lacistemaceae. *In* A. Pulle (ed.), Flora of Suriname **I(1):** 258–261. E. J. Brill, Leiden.

1. Stamen scarcely bifid; style very short, nearly absent, the stigmas sessile. . . . . *L. aggregatum.*
1. Stamen markedly bifid; style well developed, the stigmas not sessile.
    2. Stems and leaves densely, silky pubescent. Leaf blades usually >12 cm long. Inflorescences with flowers scattered along rachis. Flowers with well developed, annular nectary; ovary glabrous. . . . . *L. grandifolium.*
    2. Stems and leaves merely pubescent. Leaf blades usually ≤12 cm long. Inflorescences with flowers tightly congested along rachis. Flowers without nectary; ovary pilose. . . . . *L. polystachyum.*

LACISTEMA Sw.

A single genus in our area, description same as family description.

Lacistema aggregatum (Bergius) Rusby

Small trees. Young stems pubescent. Leaves: petioles 4–5 mm long, pubescent; blades elliptic, 7.5–11.5 × 3–4.5 cm, glabrous; secondary veins in 5–6 pairs. Inflorescences mostly axillary, the flowers tightly congested along rachis. Flowers: stamen scarcely bifid; nectary well developed; ovary glabrous, the style essentially absent. Fl (Mar, Oct, Nov), fr (Oct); in forest.

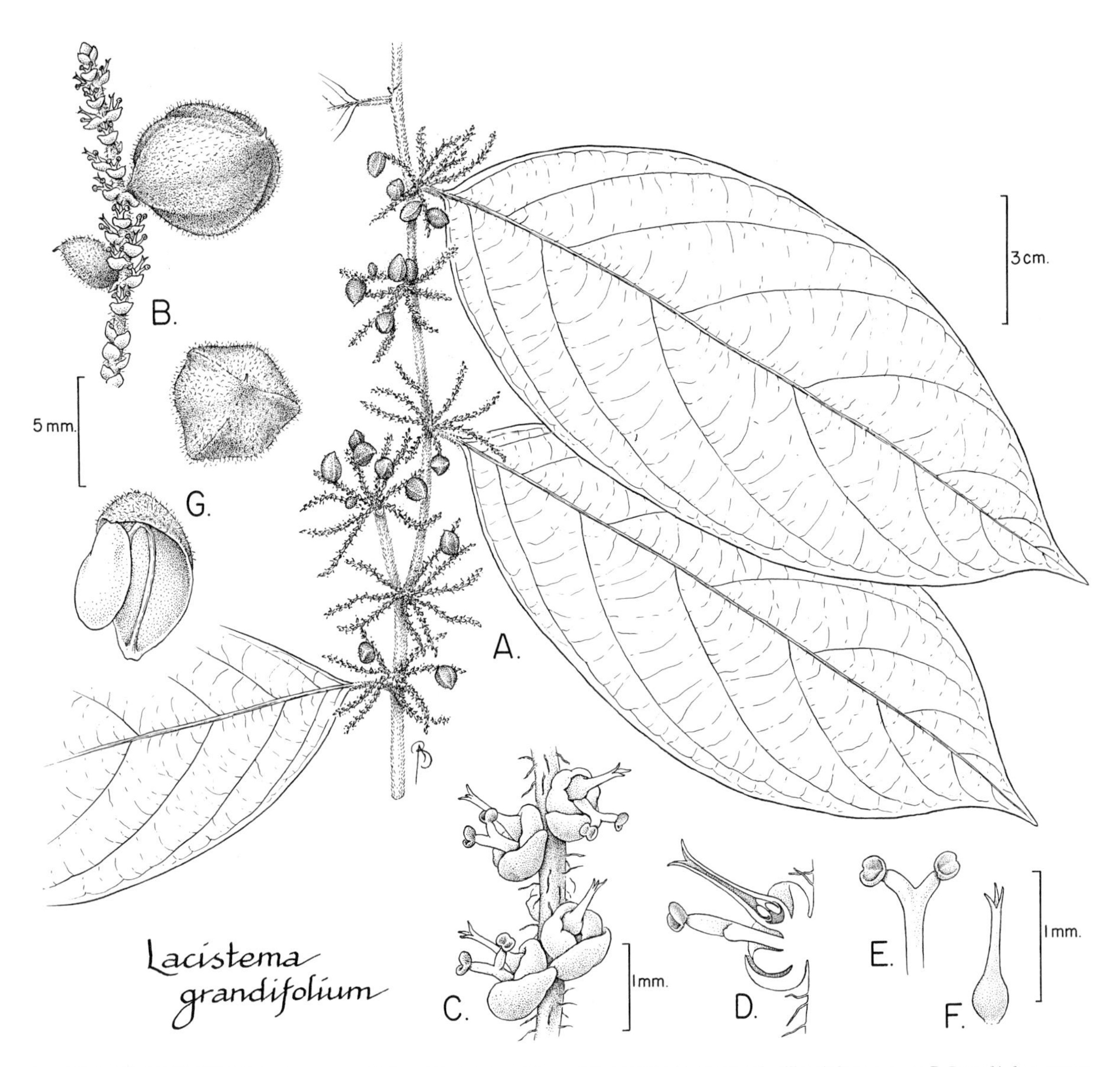

FIG. 150. LACISTEMATACEAE. *Lacistema grandifolium* (*Mori et al. 20983*). **A.** Part of stem with leaves and axillary infructescences. **B.** Part of infructescence showing developing fruit amidst flowers. **C.** Part of inflorescence showing much reduced flowers and bifid stamens. **D.** Medial section of flower. **E.** Bifid stamen. **F.** Pistil. **G.** Apical view of fruit (above) and carpel wall with single seed (below).

Lacistema grandifolium W. Schnizl. FIG. 150, PL. 68d

Trees, to 15 m tall. Young stems very densely, silky pubescent. Leaves: petioles 8–11 mm long, densely pubescent; blades elliptic, 14–23 × 5.5–10 cm, densely, silky pubescent, especially abaxially; secondary veins in 6–9 pairs. Inflorescences mostly from branches, the flowers scattered along rachis. Flowers: stamen markedly bifid; nectary well developed, annular, extrastaminal; ovary glabrous, the style well developed. Fruits red, inflated. Fl (Aug, Sep), fr (Aug, Sep); in nonflooded forest.

Lacistema polystachyum W. Schnizl.

Small trees, to 8 m tall. Young stems pubescent. Leaves: petioles 5–7 mm long, pubescent; blades narrowly elliptic, 9.5–12.5 × 2.5–4 cm, with scattered pubescence; secondary veins in 5–6 pairs. Inflorescences mostly axillary, the flowers tightly congested along rachis. Flowers: stamen markedly bifid; nectary not developed; ovary pilose, the style well developed. Fruits red. Seed surrounded by white arilloid. Fl (Apr, Aug, Sep, Oct, Nov), fr (May, Aug, Sep, Oct, Nov); scattered but common, in nonflooded forest.

LAMIACEAE (Mint Family)

Ray M. Harley

Herbs, shrubs, or rarely trees, usually glandular and aromatic. Stems often 4-angled. Stipules absent. Leaves usually simple, opposite. Inflorescences axillary and/or terminal, in axils of bracts or upper leaves, cymose, often much modified by reduction into false whorls (verticillasters) or heads (capitula), collectively usually forming a thyrse or by modification a "spike" or "raceme," bracts similar to or often reduced and markedly different from leaves, in capitulate inflorescences often forming involucre; bracteoles usually present. Flowers zygomorphic, bisexual, with functionally pistillate flowers also present in gynodioecious plants, or rarely flowers unisexual (plants dioecious); calyx tubular, usually 5-lobed or toothed, the lobes subequal or the calyx 2-lipped, frequently with three posterior (adaxial) and two anterior (abaxial) lobes or teeth, but often modified in various ways, sometimes one or both lips entire, the adaxial lip sometimes with decurrent wings along tube, or with a rounded scale-like appendage (*Scutellaria*); corolla gamopetalous, usually strongly zygomorphic (rarely seeming ± actinomorphic) and often 2-lipped, the adaxial (upper) lip weakly 2-lobed to entire, flat to concave or hooded, the abaxial (lower) lip usually ± 3-lobed, longer or shorter than adaxial, the median lobe flat to concave or boat-shaped and sometimes compressed; stamens 4, didynamous or subequal, sometimes reduced to 2, adnate to corolla, the connective sometimes elongate; staminodia often present, then 2; ovary superior, borne on simple or lobed disc, bicarpellate and 4-ovulate, deeply 4-lobed by development of false septa, the style usually gynobasic, shortly bifid at apex. Fruits of four (rarely fewer), usually dry nutlets, held loose in usually persistent calyx, the pericarp mucilaginous or not when wetted; endosperm scanty or none.

Pogostemon cablin (Blanco) Benth. is cultivated in the village, but there is no evidence that it escapes from cultivation.

Harley, R. M. 1999. Lamiaceae. *In* J. A. Steyermark, P. E. Berry, K. Yatskievych & B. K. Holst (gen. eds.), Flora of the Venezuelan Guayana **5:** 678–700. P. E. Berry, K. Yatskievych & B. K. Holst (vol. eds.). Missouri Botanical Garden Press, St. Louis.

Kostermans, A. J. G. H. 1966. Labiatae. *In* A. Pulle (ed.), Flora of Suriname **IV(1):** 334–353, 498. E. J. Brill, Leiden.

1. Calyx two-lipped, with rounded scale-like appendage (the scutellum) on adaxial side, the lips entire, closed in fruit. . . . . *Scutellaria.*
1. Calyx two-lipped or not, toothed at least abaxially, scale-like appendage absent, although adaxial lobe sometimes broad with decurrent wings.
   2. Calyx 8–10-toothed; corolla bright orange; stamens held under adaxial lip of corolla. . . . . *Leonotis.*
   2. Calyx 5-lobed or -toothed; corolla not orange; stamens divergent, or declinate toward abaxial lip of corolla.
      3. Calyx strongly 2-lipped, the adaxial lip broad, with margins decurrent onto calyx tube.. . . . . . *Ocimum.*
      3. Calyx not 2-lipped, ± equally 5-toothed, without broad adaxial lip as above.
         4. Viscid herbs. Inflorescences subtended by slender bracts <1 mm wide. Calyx teeth widely spreading in fruit, broadly triangular; corolla violet-blue. Nutlets concave on adaxial face, with involute, fimbriate margin. . . . . *Marsypianthes.*
         4. Nonviscid herbs. Inflorescences subtended by bracts usually >1 mm wide. Calyx teeth ± erect in fruit, subulate, rarely triangular; corolla white if calyx teeth triangular. Nutlets ovoid or flattened, not as above. . . . . *Hyptis.*

HYPTIS Jacq.

Often aromatic, perennial herbs or shrubs, rarely annual. Stems erect to prostrate. Inflorescences bracteate cymes, forming congested or rarely diffuse thyrsoid panicles, verticillasters, or capitula, these surrounded by an involucre of bracts and sometimes secondarily aggregated and forming thyrsoid panicles, pseudoracemes or spike-like inflorescences; bracteoles usually present. Flowers: calyx usually 10-veined, with 5 ± subequal teeth, the tube usually strongly accrescent in fruit, rarely inflated; corolla 2-lipped, enclosing stamens and forming explosive pollination mechanism, often white, sometimes pink-spotted on adaxial lip, otherwise usually lilac to purplish; stamens 4, declinate, the anthers 2-thecous, the thecae parallel and confluent; stylopodium absent or rarely present. Nutlets usually ovoid, but sometimes elongate or flattened; pericarp often mucilaginous when wet.

1. Flowers not in hemispherical or spherical capitula, clustered in few-flowered glomerules, on short peduncles in axils of reduced leaves, forming elongate, terminal, spike-like inflorescences. Calyx narrowly cylindrical, elongating conspicuously in fruit to 4.5–7 mm long, the teeth very short, subulate; corolla usually pale blue or lilac. . . . . *H. mutabilis.*
1. Flowers arranged in spherical or hemispherical capitula on peduncles in axils of scarcely reduced leaves. Calyx not as above; corolla white, often with pale pink spots.
   2. Calyx teeth broadly triangular. . . . . *H. pachycephala.*
   2. Calyx teeth subulate.
      3. Robust, erect herbs, ± subglabrous. Leaf blades lanceolate to elliptic-rhomboid. Inflorescences with involucral bracts lanceolate to linear. . . . . *H. lanceolata.*
      3. Lax, decumbent to ascending, but not erect, herbs, with weak spreading hairs. Leaf blades ovate to elliptic-lanceolate. Inflorescences with involucral bracts ovate. . . . . *H. atrorubens.*

Hyptis atrorubens Poit.

Decumbent to ascending, but not erect, ± odorless, often reddish-tinged, sparsely pubescent herbs. Stems often rooting at lower nodes. Leaves: petioles 5–9 mm long; blades ovate to lanceolate, 16–35 × 8–15(20) mm, the base truncate to long attenuate, the apex acute, the margins crenate-serrate. Inflorescences in axils of upper leaves, capitula 7–12 mm diam.; peduncles 4–9 mm long; bracts ovate, 3–4 mm wide, appressed. Flowers: calyx 3.5–4 mm long at anthesis, 7–9 mm long in fruit, the teeth finely subulate, 1.5–2 mm long; corolla 4.5–5 mm long, white with pink spots on upper or lower lip. Nutlets oblong-ovoid, 1 mm long, dark brown. Fl (Jun, Jul, Nov); in damp, open grassy places, damp forest margins, and by water. *Hortelão bravo* (Portuguese), *ti bombe noir, ti bombe rouge* (Créole), *trevo roxo* (Portuguese).

Hyptis lanceolata Poir. FIG. 151, PL. 69a

Robust, erect, aromatic, ± subglabrous herbs, 1–2 m tall. Stems sometimes woody at base, not rooting at lower nodes. Leaves: petioles difficult to distinguish from blade; blades elliptic-lanceolate or rhomboid-lanceolate and sublobate, 90–116 × 20–42 mm, the base long-attenuate, grading into petiole, the apex acute, the margins irregularly and bluntly serrate or doubly serrate. Inflorescences in leaf axils, capitula spherical, 10–12 mm diam., greenish; peduncles 10–20 mm long; bracts lanceolate to linear, to 1–3.5 mm wide, spreading at anthesis. Flowers: calyx 3 mm long at anthesis, 4.5 mm long in fruit, the teeth subulate, 1.5 mm long; corolla 3–3.5 mm long, white, the lobes often faintly spotted pink. Nutlets elliptic-oblong, 1 mm long, dark brown. Fl (Jun, Sep, Oct); along streams and as a weed in cultivated fields. *Mélisse sauvage* (Créole).

Hyptis mutabilis (Rich.) Briq.

Erect, aromatic, subglabrous or sparsely hairy, perennial herbs, 1–2.5 m tall. Stems quadrangular, the angles with small, retrorse prickles toward base, sometimes woody at base, not rooting at lower nodes. Leaves: petioles 9–35(52) mm long; blades ovate or ovate-rhomboid, 25–70(100) × 15–50(60) mm, the base broadly cuneate, the apex acute to acuminate, the margins doubly serrate. Inflorescences terminal, but in axils of upper reduced leaves, forming elongate, branched, spike-like panicles of many small glomerules, the glomerules to 15-flowered, 1.5–7 mm diam. at anthesis, 10 mm diam. in fruit; peduncles 3–5 mm long; bracts obovate or elliptic-lanceolate, to 1–2 mm wide, appressed, the veins prominent abaxially when dry. Flowers: calyx 1.75–2.5 mm long at anthesis, the teeth subulate, 0.5 mm long, the tube elongating to (4.5)5.5–7 mm long and becoming narrowly cylindrical in fruit; corolla 4.5–6 mm long, pale blue or lilac, the adaxial lip with darker lines, the tube slightly curved, white. Nutlets ellipsoid-oblong, 1.2–1.5 mm, smooth to finely rugulose, black or dark brown, smooth to finely rugulose. Fl (Jun); at forest margins and in weedy areas. *Radié crise* (Créole).

Hyptis pachycephala Epling PL. 69b

Erect, sparsely branched, odorless herbs, 0.5–1 m tall. Stems often densely villous, sometimes rooting at lower nodes. Leaves: petioles 4–9 mm, densely villous; blades elliptic-lanceolate, 85–155 × 18–52 mm, the base attenuate, the apex acuminate, the margins shallowly serrate. Inflorescences axillary from upper leaves, capitula (6)15–25 mm diam., small and apparently few-flowered at first, soon accrescent, forming showy, spherical-depressed heads; peduncles 5–25 mm long; outer bracts ovate-elliptic, to 2.5–5 mm wide. Flowers: calyx 4–4.5 mm long at anthesis, 7–7.5 mm long in fruit, the teeth broadly triangular with blunt apex, 0.5–1.25 mm long; corolla 4–5 mm long, pearly white, the adaxial lip with pink spots. Nutlets ovoid, subtrigonous, 1.5 mm long, pale brown with darker striae and dark brown coronate apex. Fl (Feb, Mar, May, Jun, Jul, Aug, Sep, Oct), fr (Aug); in wet areas in forest or at shady margins of forest clearings.

The central French Guianan populations are disjunct from the main area of distribution in western South America.

LEONOTIS (Pers.) R. Br.

Robust annual to perennial herbs or shrubs. Inflorescences of 2–11 dense, ± spherical, many-flowered verticillasters; bracts usually leaf-like; bracteoles numerous, spinescent. Flowers: calyx tubular, 10-veined, the teeth 8–10, usually spinescent, unequal, the posterior tooth usually much longer than others; corolla tubular, deeply 2-lipped, bright orange, or rarely yellow or white, the adaxial lip entire, hooded, almost as long as tube, the abaxial lip 3-lobed, soon withering; stamens 4, didynamous, inserted near throat of corolla and directed under adaxial lip, the anterior pair of stamens longer, the anthers 2-thecous, divaricate; style 2-lobed at apex, the anterior branch much longer than posterior. Nutlets 3-angled, glabrous.

Leonotis nepetifolia (L.) R. Br. FIG. 152

Erect, annual or short-lived perennial herbs, 1–3 m tall. Leaves: petioles 20–60 mm long; blades ovate, 40–100 × 25–75 mm, sparsely pubescent, the base truncate to cuneate, attenuate onto petiole, the apex acute, the margins deeply crenate to crenate-dentate. Inflorescences terminal, of 2–5 globose, widely spaced, many-flowered verticillasters 20–50 mm diam; bracts leaf-like, narrower than leaves; bracteoles linear, spine-tipped, to 10–12 mm wide. Flowers: calyx tube strongly curved just below throat, 7–11 mm long at anthesis, to 12–20 mm long in fruit, the teeth rigid, spinescent, the adaxial teeth 4–7 mm long; corolla 15–20 mm long, strongly 2-lipped, adaxial lip 7–9 mm long, twice as long as abaxial, bright orange. Nutlets trigonous, 3–3.5 mm long, dark brown. Fl (Mar, Jun, Oct); introduced weed of cultivated or disturbed ground. *Cordão de frade* (Portuguese), *gros pompon, pompon, pompon soda* (Créole).

MARSYPIANTHES Benth.

Usually viscid and aromatic perennial herbs or subshrubs. Stems procumbent to ascending or erect. Inflorescences axillary, in subglobose cymose, many- or rarely few-flowered capitula; pedicels short; bracts long, curved, linear to subulate. Flowers: calyx campanulate to broadly funnel-shaped; calyx teeth subequal, broadly lanceolate to deltate, wide spreading in fruit, not densely hairy in throat; corolla 2-lipped, with explosive pollination mechanism, the adaxial lip 2-lobed or emarginate; the abaxial lip 3-lobed, the outer lobes spreading, the

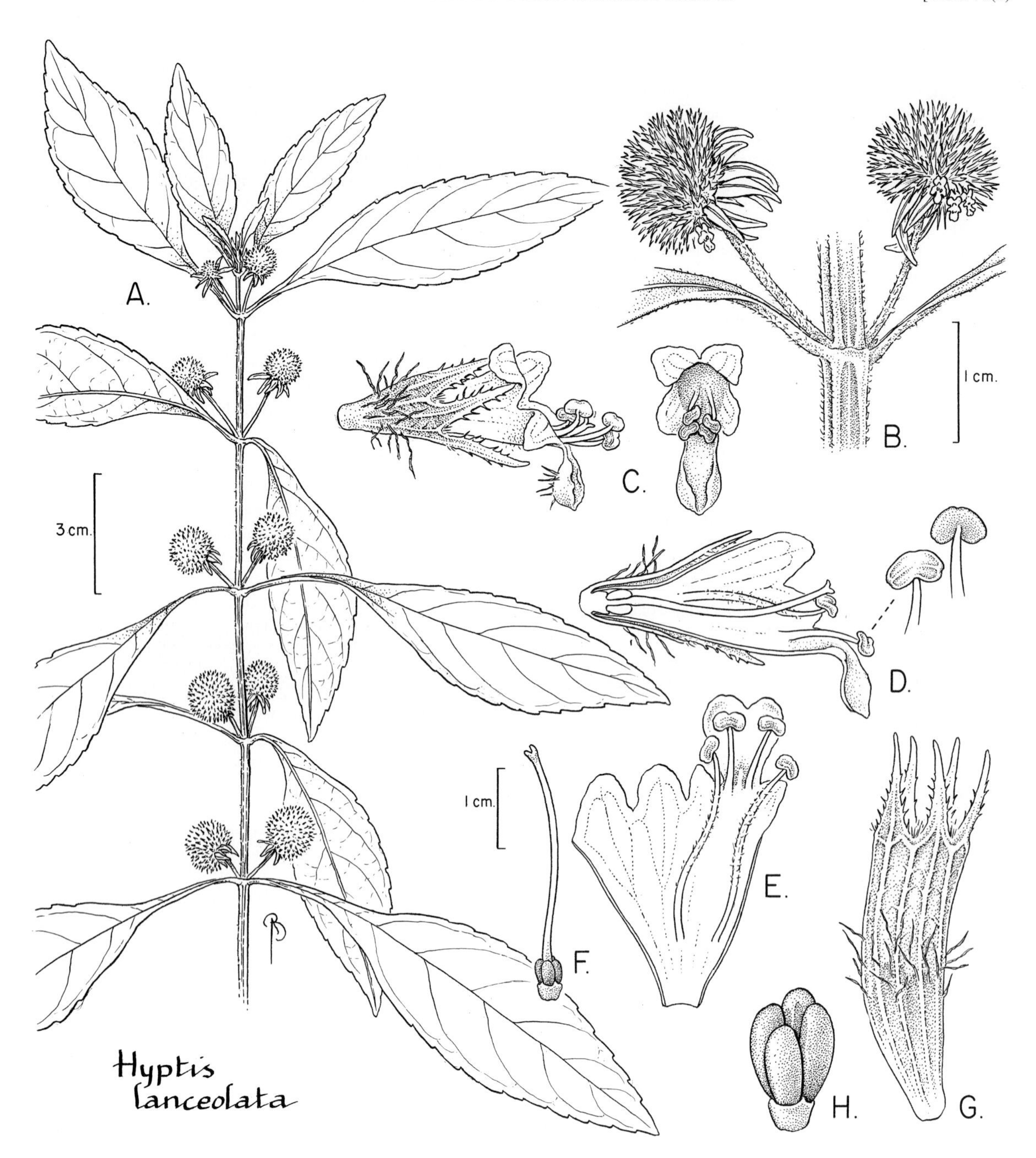

FIG. 151. LAMIACEAE. *Hyptis lanceolata* (A, B, G, H, *Mori et al. 23800*; C–F, *Mori et al. 22040*). **A.** Part of stem with leaves and inflorescences. **B.** Detail of node with inflorescences and petioles. **C.** Lateral (left) and apical (right) views of flower. **D.** Medial section of flower (left) and detail of anthers (above right). **E.** Corolla opened to show arrangement of stamens. **F.** Lateral view of pistil. **G.** Lateral view of calyx. **H.** Oblique-apical view of fruit comprised of four nutlets.

middle lobe saccate, hinged at base, the tube slender; stamens 4, declinate, the anthers 2-thecous, the thecae parallel and confluent. Nutlets ovate, cymbiform, smooth on abaxial surface, the adaxial surface concave with involute, fimbriate margin.

Marsypianthes chamaedrys (Vahl) Kuntze

Viscid herbs, 0.2–0.75 m tall, with spreading, weakly villous, usually gland-tipped hairs. Stems prostrate to ascending or erect. Leaves: petioles (3)6–15 mm; blades ovate to ovate-lanceolate, 15–36 × 7–18 mm, viscid, dull green, the base cuneate to truncate, slightly decurrent onto petiole. Inflorescences in leaf axils, capitula 10–15-flowered, 12–18 mm diam.; peduncles 15–30 mm long; bracts very narrowly elliptic to linear or filiform, 5–8 mm long, often bluish or purple-tinged. Flowers: calyx ca. 5 mm long at anthesis, the tube broadly funnel-shaped, the teeth broadly triangular, ca. 2 mm long, connivent in fruit at first, later spreading widely

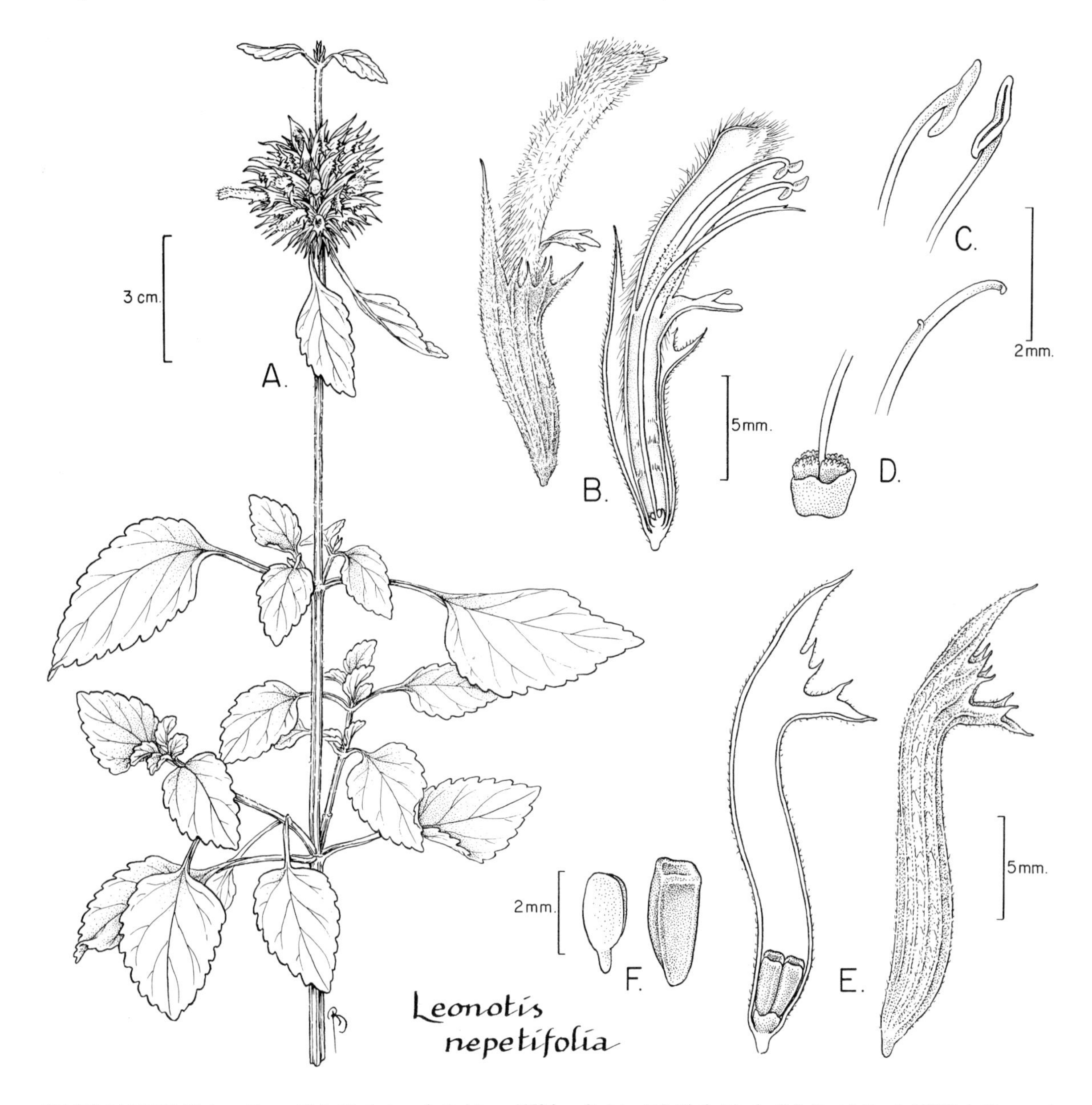

FIG. 152. LAMIACEAE. *Leonotis nepetifolia* (A, B, *Acevedo-Rodríguez 3878* from St. John, U.S. Virgin Islands; C–F, *Mori & Gracie 18956*). **A.** Upper part of stem with leaves and inflorescence. **B.** Lateral view (left) and medial section (right) of flower. **C.** Two views of anthers. **D.** Lateral view of ovary with base of style (left) and apical portion of style with stigma (right). **E.** Medial section (left) and lateral view (right) of fruiting calyx. **F.** Embryo (left) and nutlet (right). (Adapted with permission from P. Acevedo-Rodríguez, *Flora of St. John, U.S. Virgin Islands*, Mem. New York Bot. Gard. 78, 1996); note that C, D, F, and part of E have been modified.

and becoming strongly reflexed to expose nutlets; corolla ca. 5–6 mm long, violet-blue. Nutlets 2–3 mm long, adaxial surface smooth, pale, abaxial surface concave, with incurved fimbriate margin. Fl (Apr, Jun, Aug, Sep); in damp grassy areas, weedy, and cultivated fields. *Gadu paepina in Badu* (Créole), *sete sangrias* (Portuguese), *ti bombe blanc* (Créole).

OCIMUM L.

Annual to perennial aromatic herbs or small shrubs. Inflorescences in axils of bracts in upper part of stems, interrupted, spike-like, composed of verticillasters; bracteoles absent. Flowers: calyx strongly 2-lipped, usually deflexed in fruit on curved pedicel, the adaxial lip ovate-orbicular, larger than abaxial, entire, with margins decurrent onto calyx tube, the abaxial lip 4-toothed, the teeth acuminate; corolla strongly 2-lipped, the adaxial lip 4-lobed, the abaxial lip entire or toothed; stamens 4, declinate along abaxial lip of corolla, the filaments of upper pair appendaged in some species, the anthers 2-thecous. Nutlets ovoid, mucilaginous when wetted.

Ocimum campechianum Mill. [Syn.: *Ocimum micranthum* Willd.] FIG. 153

Aromatic, annual or short-lived perennial herbs, 0.15–0.75 m tall. Stems single, erect, from tap-root. Leaves: petioles 3–35 mm long; blades ovate, (12)20–60(110) × 5–30(65) mm, subglabrous, very pale green abaxially, the base attenuate, the apex acute to acuminate, the margins serrate. Inflorescences erect, verticils usually lax in fruit; bracts ovate, 2–10 mm; pedicels slightly flattened, in fruit 4–5 mm long, curving downwards. Flowers: calyx at anthesis to 2.5–4 mm long, pale green or purple-tinged, 7–10 mm long with margins of adaxial lip broadly decurrent almost to base of tube, deflexed in fruit; corolla 3–4 mm long, pink, the adaxial lip 4-lobed, the abaxial lip obovate; stamens exserted. Nutlets narrowly obovoid, 1.5–2 mm long, verruculose, dark gray-brown, with paler raised areas. Fl (Sep); in open grassy areas, roadsides, and forest clearings. *Alfavaca* (Portuguese), *basilic*, *basilic fombazin*, *basilic sauvage*, *framboisier*, *Grand basilic* (Créole), *remedio de vaqueiro* (Portuguese).

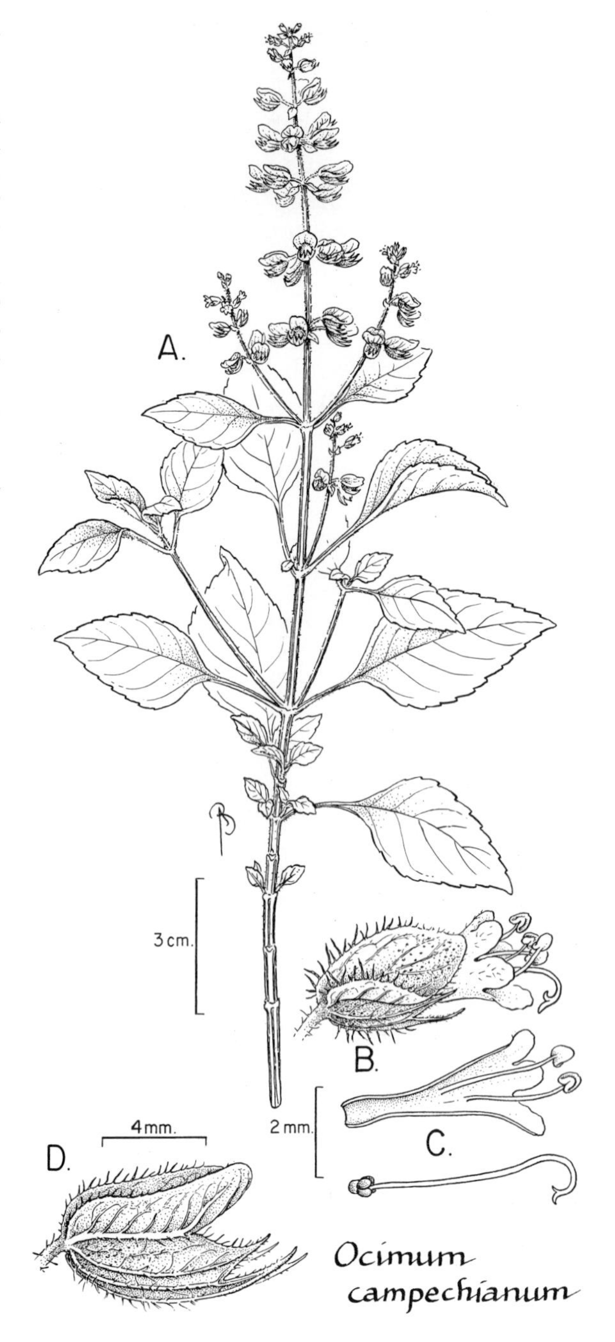

FIG. 153. LAMIACEAE. *Ocimum campechianum* (A–C, *Mori s.n.* from St. John, U.S. Virgin Islands; D, *Acevedo-Rodríguez 3771* from St. John, U.S. Virgin Islands). **A.** Upper part of stem with leaves and flowers. **B.** Lateral view of flower. **C.** Medial section of corolla showing adnate stamens (above) and pistil (below). **D.** Lateral view of calyx. (Reprinted with permission from P. Acevedo-Rodríguez, *Flora of St. John, U.S. Virgin Islands*, Mem. New York Bot. Gard. 78, 1996.)

SCUTELLARIA L.

Perennial, odorless herbs or subshrubs. Leaves usually petiolate, at least toward base of plant. Inflorescences racemes or spikes, rarely solitary; pedicels often curved. Flowers: calyx 2-lipped, accrescent in fruit, the tube ventricose-campanulate, the adaxial side usually with rounded, concave, scale-like appendage (the scutellum), the lips entire, rounded, closed in fruit, the adaxial lip of calyx eventually deciduous, the abaxial lip persistent; corolla 2-lipped, the tube long, usually sigmoid, or arcuate, dilated distally, the adaxial lip galeate; stamens 4,

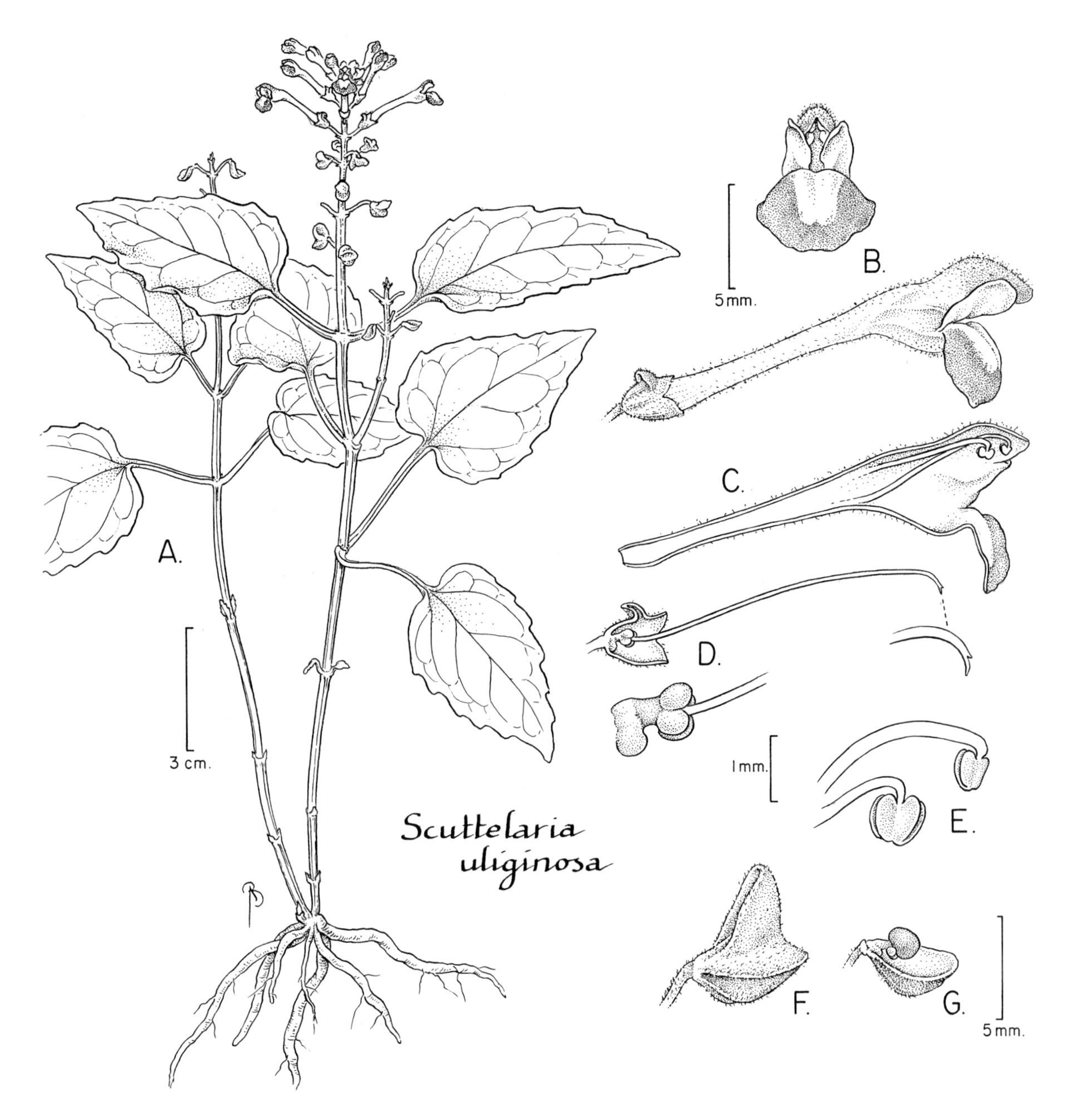

FIG. 154. LAMIACEAE. *Scutellaria uliginosa* (A, C, F, G, *Mori et al. 22323*; D, E, *Mori & Gracie 24218*). **A.** Plant. **B.** Apical (above) and lateral (below) views of flower. **C.** Medial section of corolla showing two adnate stamens. **D.** Medial section of flower with corolla (and stamens) removed and details of apex of style (lower right) and ovary (lower left). **E.** Two views of stamens. **F.** Lateral view of fruit. **G.** Dehisced fruit with one mature nutlet.

ascending under upper lip and included within it, the anterior pair longer, monothecous, the posterior pair with two convergent thecae; style branches unequal, the posterior branch reduced. Nutlets depressed-globose to ovoid, borne on elongate, peg-like gynophore.

Scutellaria uliginosa Benth. FIG. 154, PL. 69c

Erect herbs, to 0.35 m tall. Stems 1 to several from short rhizome bearing fascicle of fleshy, fusiform roots. Leaves: petioles 20–35 mm long, the uppermost often shorter; blades ovate to ovate-lanceolate, 48–80 × 24–48 mm, the base cordate, the apex acute to acuminate, the margins usually shallowly and broadly serrate to serrate-dentate. Inflorescences terminal racemes, the flowers opposite; pedicels 1–3 mm long; bracts linear. Flowers: calyx 2.5–3 mm, the lobes short, rounded, the scutellum erect, calyx in fruit 6–8 mm long, the scutellum spreading, conspicuous; corolla 15–20 mm long, the tube pale, slender, the adaxial lip not spreading, pale lilac or white, the abaxial lip bright purple. Nutlets ovoid, 2 mm long, verrucate, dark brown. Fl (Feb, Mar, May, Jun, Jul, Sep, Oct, Nov, Dec), fr (Jul, Aug, Sep, Nov); moist forest, especially along roads.

Plants from French Guiana appear to be intermediate between *S. purpurascens* Sw., from northern South America and the Caribbean, and *S. uliginosa* which extends southwards along the Atlantic forests of Brazil.

LAURACEAE (Avocado Family)

Henk van der Werff

Trees or shrubs. Slash of trunk often emitting slightly spicy aroma. Indument of simple hairs. Stipules absent. Leaves often emitting slight spicy aroma when crushed, alternate, infrequently whorled or opposite; blades with entire margins; secondary veins often decurrent along midrib. Flowers small, bisexual or unisexual (plants dioecious); tepals 6, equal, rarely outer 3 smaller than inner 3; stamens 3, 6, or 9, each with 2 or 4 locelli, opening by small flaps; ovary superior. Fruits one-seeded "berries," subtended by unmodified or swollen pedicel or by a cupule, the cupule variously modified, but often cup-shaped and red.

Jansen-Jacobs, M. J. 1976. Lauraceae. *In* J. Lanjouw & A. L. Stoffers (eds.), Flora of Suriname **II(2):** 451–484. E. J. Brill, Leiden.
Kostermans, A. J. G. H. 1966. Lauraceae. *In* A. Pulle (ed.), Flora of Suriname **II(1):** 244–337, 481–487. E. J. Brill, Leiden.
Werff, H. van der. 1991. A key to the genera of Lauraceae in the New World. Ann. Missouri Bot. Gard. **78:** 377–387.
——— & J. G. Rohwer. 1999. Lauraceae. *In* J. A. Steyermark, P. E. Berry, K. Yatskievych & B. K. Holst (gen. eds.), Flora of the Venezuelan Guayana **5:** 700–750. P. E. Berry, K. Yatskievych & B. K. Holst (vol. eds.). Missouri Botanical Garden Press, St. Louis.

1. Stamens 3, all 2-celled. . . . . *Licaria.*
1. Stamens 6 or 9, 2- or 4-celled.
    2. Flowers bisexual.
        3. Stamens with 4 locelli (+ flaps).
            4. Outer tepals smaller than inner ones. Leaves clustered at tips of branches. . . . . *Sextonia.*
            4. Tepals equal. Leaves evenly distributed along branches.
                5. Leaves tripliveined. Staminodia with cordate tips. . . . . *Cinnamomum.*
                5. Leaves pinnately veined. Staminodia lacking or stipitiform.
                    6. Tepals papillose adaxially, united at base and, in old flowers, falling off as one unit together with stamens. Stamens papillose, the locelli arranged in arc. . . . . *Nectandra.*
                    6. Tepals glabrous or pubescent but not papillose, free, falling individually or persisting in old flowers, a few stamens often present on cupule of young fruits. Stamens glabrous or pubescent but not papillose, the locelli arranged in 2 rows. . . . . *Ocotea.*
        3. Stamens with 2 locelli.
            7. Inflorescences few-flowered (rarely >7 flowers). Flowers red; tepals spreading at anthesis; inner 3 stamens with fused filaments. . . . . *Kubitzkia.*
            7. Inflorescences many-flowered. Flowers white, green, or yellow; tepals erect; inner 3 stamens with free filaments.
                8. Inflorescences and flowers with very short, mostly erect hairs, the indument sparse to moderate, but never covering flowers completely. Tepals quickly deciduous, leaving ovary fully exposed. . . . . *Beilschmiedia.*
                8. Inflorescences and flowers with appressed hairs or glabrous, the indument, when present, sparse to very dense, covering inflorescences and flowers completely. Tepals ± persistent, the ovary largely or completely enclosed in hypanthium.
                    9. Flowers densely pubescent, the surface usually completely covered by the indument. . . . . *Aniba.*
                    9. Flowers glabrous or very sparsely pubescent, the surface always visible. . . . . *Aiouea.*
    2. Flowers unisexual.
        10. Stamens with 2 locelli (locelli in pistillate flowers usually visible as thin spots on staminodia). . . . . *Endlicheria.*
        10. Stamens with 4 locelli.
            11. Tepals at anthesis mostly erect, rarely spreading. Locelli arranged in two rows; filament and anther well differentiated, the filament clearly narrower than anther. . . . . *Ocotea.*
            11. Tepals at anthesis spreading or erect. Locelli arranged in an arc; filament as wide as anther, the two not clearly differentiated. . . . . *Rhodostemonodaphne.*

Some fruiting or vegetative charactes which allow rapid identification include the following:

- Cupule cup-shaped and with double margin: *Licaria*, *Ocotea cujumary*, and *Kubitzkia*.
- Fruit not subtended by a cupule or persistent tepals: *Beilschmiedia*.
- Leaves opposite: *Licaria chrysophila* and *L. debilis*.
- Leaves clustered or whorled: *Endlicheria*, *Aniba*, and *Sextonia*.
- Leaves tripliveined: *Aiouea longepetiolata*, *Cinnamomum*, and some species of *Rhodostemonodaphne*.
- Outer tepals smaller than inner ones: *Aniba parviflora* and *Sextonia*.

AIOUEA Aubl.

Small or medium-sized trees. Leaves alternate, evenly distributed along branches. Inflorescences axillary, usually many-flowered. Flowers bisexual; tepals at anthesis mostly erect; stamens 3, 6, or 9, all 2-celled; staminodia 3, 6, or 9. Fruits seated in shallow cupule or on flat disc; pedicel usually swollen in fruit.

1. Petioles >2.5 cm long; leaf blades tripliveined. . . . . *A. longipetiolata.*
1. Petioles ≤1 cm long; leaf blades pinnately veined.
    2. Leaves shiny. Tepals much shorter than hypanthium; stamens 6, the locelli extrorse. . . . . *A. guianensis.*
    2. Leaves not shiny, opaque. Tepals as long as hypanthium; stamens 9, the outer 6 with locelli introrse. . . . *A. opaca.*

Aiouea guianensis Aubl. PL. 69e

Trees, to 20 m tall. Stems mostly glabrous. Terminal buds glabrous or nearly so. Leaves: petioles 5–10 mm long; blades elliptic, 9–20 × 4–8 cm, glabrous, shiny, the base acute, the apex obtuse or acute; venation pinnate, the midrib immersed adaxially. Inflorescences 2–3 together on short, leafless shoots, many-flowered. Flowers glabrous or pubescent; tepals much shorter than hypanthium; stamens 6, the locelli extrorse. Fruits round, seated on small disc, the pedicel swollen in fruit. Fr (Feb); rare, in nonflooded forest.

Identification of the single collection of this species from central French Guiana is based on fruiting material and is therefore tentative.

Aiouea longipetiolata van der Werff

Trees, to 35 m tall. Stems minutely brown-tomentellous, glabrescent with age. Terminal buds light-brown tomentellous. Leaves: petioles 27–35 mm long; blades elliptic or ovate, 10–17 × 4–7.5 cm, minutely brown-pubescent abaxially, the base obtuse or rounded, the apex acute; venation tripliveined, the midrib impressed adaxially. Inflorescences 8–14 cm long, brown-tomentellous. Flowers at base slightly tomentellous, otherwise glabrous; tepals much shorter than hypanthium, erect at anthesis; stamens 9. Fruits to 3 cm long, the cupule very shallow, gradually narrowed into pedicel. Fl (Aug, Sep, Nov), fr (Mar); occasional.

The long petioles and tripliveined leaves with minute, brown hairs on the abaxial surface are characteristic of this species.

Aiouea opaca van der Werff FIG. 155

Trees, to 12 m tall. Stems glabrous or with few appressed hairs. Terminal buds densely whitish pubescent. Leaves: petioles 6–9 mm long; blades elliptic, 12–18 × 6–9 cm, glabrous, opaque, the base acute, the apex acute or shortly acuminate; venation pinnate, the midrib impressed adaxially. Inflorescences ca. 16 cm long, glabrous or with a few hairs towards the flowers. Tepals sparsely appressed pubescent, as long as hypanthium; stamens 9. Fruits unknown. Fl (Aug, Sep); in nonflooded forest.

ANIBA Aubl.

Trees or shrubs. Leaves alternate, evenly distributed along stems or clustered; blades sometimes papillose, glabrous or pubescent abaxially. Inflorescences axillary, generally puberulous. Flowers small, bisexual; tepals 6, erect at anthesis, equal or unequal; stamens 9 or 6, 2-celled, generally with tomentellous filaments. Fruits seated in cup-shaped cupule, frequently with a few stamens persisting on cupule.

1. Terminal buds thick, brown tomentose-tomentellous. Leaves strongly clustered near tips of branches; blades 15–35 cm long, tomentulose abaxially. . . . . *A. williamsii.*
1. Terminal buds slender, inconspicuous. Leaves ± evenly distributed along stems; blades rarely >20 cm long, glabrous or nearly so.
    2. Abaxial leaf blade surfaces microscopically papillose. Outer 3 tepals smaller than inner 3 tepals. . . . *A. parviflora.*
    2. Abaxial leaf surfaces smooth, not microscopically papillose. All 6 tepals equal.
        3. Tepals pubescent adaxially; fertile stamens 6, representing outer 2 whorls. . . . . *A. kappleri.*
        3. Tepals glabrous or pubescent adaxially; fertile stamens 9.
            4. Secondary veins in 5–8 pairs. Tepals pubescent adaxially. . . . . *A.* aff. *jenmanii.*
            4. Secondary veins in 8–13 pairs. Tepals glabrous adaxially. . . . . *A. citrifolia.*

Aniba citrifolia (Nees) Mez

Trees, to 20 m tall. Stems slightly tomentellous at apex when young, but soon becoming glabrous. Leaves alternate, sometimes slightly clustered; blades elliptic, 9–20 × 3–6 cm, glabrous, the base acute, the apex acute. Inflorescences axillary, to 15 cm long. Tepals 6, equal, ca. 1.2 mm long, minutely pubescent abaxially, glabrous adaxially; stamens 9, 2-celled; pistil glabrous. Fruit cupules cup-shaped. Fruits to 1.5 cm long. Fr (Aug).

Aniba aff. **jenmanii** Mez

Trees, to 15 m tall. Stems appressed pubescent. Leaves alternate; blades elliptic, 10–20 × 5–8 cm, glabrous, the base acute, the apex acute or acuminate. Inflorescences axillary, to 6 cm long, (sparsely) appressed pubescent. Flowers yellow; tepals 6, equal, ca. 1 mm long, the adaxial surface pubescent, especially toward base; stamens 9; pistil glabrous. Fl (Aug); rare, in nonflooded forest.

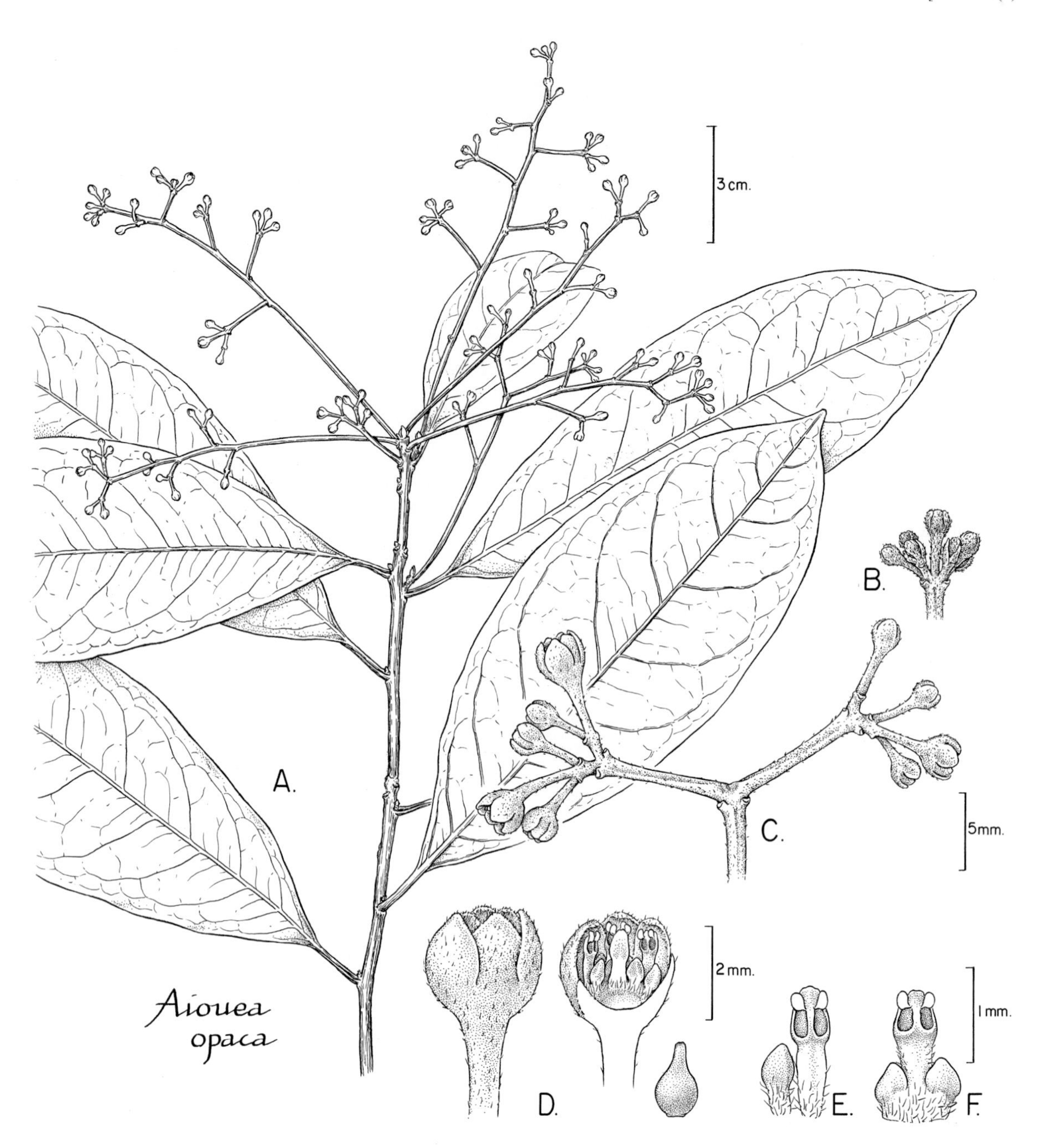

FIG. 155. LAURACEAE. *Aiouea opaca* (A, C–F, *Mori et al. 20928*; B, *Mori et al. 20735*). **A.** Stem with leaves and inflorescences. B-**C.** Details of inflorescence. **D.** Lateral view of flower (left), flower with three tepals and pistil removed (near right), and pistil that has been removed from flower (far right). **E.** Stamen of whorl II and adjacent staminode. **F.** Stamen of whorl III with basal glands attached to filament.

Diagnostic are the few secondary veins and the weakly raised reticulation of the leaves of this species. *Aniba jenmanii* is poorly known to me; the identification is based on comparison of written descriptions and is therefore tentative.

Aniba kappleri Mez

Trees, to 18 m tall. Stems slightly pubescent when young, soon glabrous. Leaves alternate; blades elliptic, 6–15 × 2–6 cm, glabrous, the base acute, the apex somewhat acuminate. Inflorescences axillary, ca. 10 cm long, slightly pubescent. Flowers tomentellous, the surface completely covered; tepals 6, equal, ca. 1.2 mm long; stamens 6, representing 2 outer whorls, the filaments densely pubescent; staminodia 3; pistil minutely pubescent. Fruit cupules cup-shaped. Fruits ca. 1.5 cm long. Fl (Jul, Aug); rare, in nonflooded forest.

Aniba parviflora (Meisn.) Mez

Trees, to 12 m tall. Stems rather densely but minutely tomentellous. Leaves alternate; blades elliptic, 10–20 × 4–8 cm, glabrous, but microscopically papillose abaxially, the base acute, the apex acute. Inflorescences axillary, to 8 cm long, minutely tomentellous. Flowers yellowish green, tomentellous; tepals 6, unequal, the outer 3 shorter and narrower than inner 3, the inner ones ca. 1.1 mm long, the outer ones ca. 0.7 mm long, the filaments densely pubescent; pistil pubescent or glabrous. Fruit cupules cup-shaped. Fruits to 3 cm long. Fl (Sep); rare, in nonflooded moist forest. *Bois de rose femelle* (Créole).

Aniba williamsii O.C. Schmidt

Trees, to 25 m tall. Stems thick (ca. 5 mm diam. 5 cm below terminal bud), brown, tomentose. Terminal buds brown-tomentose, thick, conspicuous. Leaves clustered at tips of branches; blades elliptic, 15–35 × 6–10 cm, firmly chartaceous, brown-pubescent abaxially, the base acute to rounded, the apex acute. Inflorescences clustered at tip of branches, to 15 cm long. Flowers tomentellous-tomentose; tepals 6, equal, ca. 1.2 mm long; stamens 9, 2-celled, the filaments tomentose; pistil pubescent. Fruit cupules cup-shaped, warty. Fruits to 2 cm long. Fl (May, Aug); rare, in non-flooded forest.

BEILSCHMIEDIA Nees

Medium-sized trees. Leaves alternate. Flowers small, bisexual; tepals 6; stamens 9 or 6; 2-celled; ovary 1-locular, with 1 ovule. Fruits drupes.

Beilschmiedia hexanthera van der Werff FIG. 156

Trees, to 20 m tall. Leaves alternate; blades 5–12 × 3–5 cm, glabrous adaxially, minutely appressed pubescent abaxially; secondary veins in 5–7 pairs, the higher order venation finely areolate-reticulate. Inflorescences axillary. Flowers bisexual, greenish yellow, sparsely pubescent; tepals 6, erect at anthesis, equal, ca. 1 mm long; stamens 6, 2-celled, the cells introrse; staminodia 3, columnar, each with 2 glands near base; pistil glabrous, the shallow receptacle densely pubescent adaxially. Fruits unknown. Fl (Aug); rare, in forest on well-drained soil.

CINNAMOMUM Presl

Shrubs or trees. Leaves alternate or opposite, frequently tripliveined, often with tufts of hairs in axils of lowermost secondary veins. Inflorescences axillary. Flowers bisexual; tepals 6, usually erect at anthesis; stamens 9, 4-celled; staminodia 3, well developed. Fruit cupules small, with a simple margin, the tepals frequently persisting in fruit.

Cinnamomum triplinerve (Ruiz & Pav.) Kosterm.

Trees, to 15 m tall. Stems glabrous or sparsely pubescent. Terminal buds appressed pubescent. Leaves alternate; blades elliptic, 8–15 × 4.5–6 cm, glabrous, except for tufts of hair in axils of basal secondary veins; venation tripliveined, the midrib impressed adaxially. Inflorescences ca. 10 cm long, glabrous or nearly so. Flowers yellow or green; tepals erect at anthesis, glabrous or pubescent. Fruit cupules shallowly cup-shaped, the tepals persisting on margin of cupule. Fl (Aug, Sep), fr (Feb); in forest and secondary vegetation.

Diagnostic are the tripliveined leaves with small tufts of hairs in the axils of the lowermost secondary veins.

ENDLICHERIA Nees

Trees or shrubs. Leaves alternate or whorled, pubescent or glabrous. Inflorescences axillary, usually many-flowered. Flowers unisexual (plants dioecious); tepals 6, equal, erect or spreading at anthesis. Staminate flowers with 9, 2-celled stamens and a rudimentary ovary. Pistillate flowers with 9 staminodia, pistil with conspicuous stigma, this usually raised above staminodia. Fruit cupules cup-shaped, with simple margin, the pedicel usually swollen in fruit.

1. Leaves whorled; blades with with short, erect hairs abaxially.. . . . . *E. melinonii.*
1. Leaves alternate; blades glabrous or densely appressed pubescent (sericeous) abaxially.
    2. Young stems densely pubescent. Leaf blades densely appressed pubescent abaxially. . . . . *E.* cf. *sericea.*
    2. Young stems glabrous or nearly so. Leaf blades glabrous or nearly so abaxially.
        3. Stems pale green. Leaf blades elliptic; reticulate venation raised adaxially. Tepals spreading at anthesis. . . . . *E. pyriformis.*
        3. Stems dark brown. Leaf blades obovate, reticulate venation immersed, not visible adaxially. Tepals erect at anthesis. . . . . *E. punctulata.*

Endlicheria melionii Benoist

Trees, to 10 m tall. Stems and terminal buds densely brown-tomentose. Leaves whorled at tip of branches; blades obovate or elliptic obovate, 10–18 × 3–6 cm, abaxial surface with erect hairs, the base gradually narrowed into petiole, the apex shortly acuminate. Inflorescences to 15 cm long, many-flowered, densely pubescent. Flowers pubescent abaxially, glabrous adaxially. Staminate

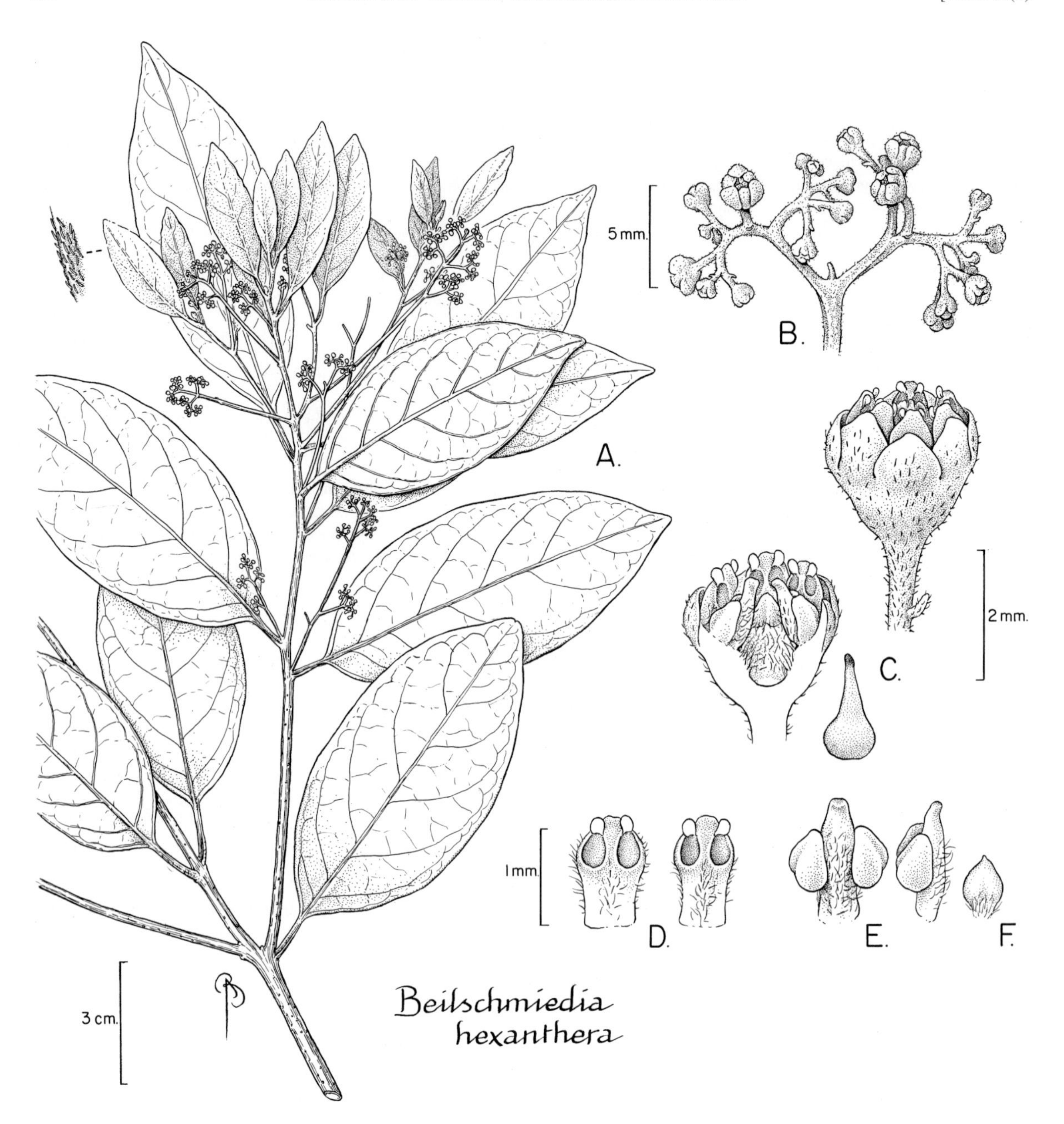

FIG. 156. LAURACEAE. *Beilschmiedia hexanthera* (*van der Werff et al. 12951*). **A.** Stem with leaves and inflorescences. **B.** Detail of inflorescence. **C.** Lateral view of flower (above), lateral view with three tepals and pistil removed (far left), and pistil (near left) that has been removed from flower. **D.** Stamens of whorls I and **II. E.** Staminodes representing whorl III; note glands attached to filaments. **F.** Staminode representing whorl IV.

flowers: tepals widely spreading at anthesis; stamens 9, 2-celled. Pistillate flowers: tepals spreading; staminodia 9; pistil large. Fruit cupules deeply cup-shaped. Fr (Sep); rare, in nonflooded forest.

In other parts of its range, this species occurs in periodically flooded forest. The obovate, whorled leaves and the erect indument on the abaxial leaf blade surface are diagnostic.

Endlicheria punctulata (Mez) C. K. Allen PL. 69d

Shrubs or small trees, to 5 m tall. Stems glabrous. Terminal buds appressed pubescent. Leaves alternate; blades obovate, 6–15 × 3–5 cm, glabrous, the base acute, the apex acuminate. Inflorescences to 4 cm long, few-flowered, glabrous. Flowers glabrous, white; tepals erect at anthesis. Staminate flowers: stamens 9, 2-celled. Pistillate flowers not seen. Fruit cupules shallowly cup-shaped, the pedicel slightly swollen. Fl (Sep), fr (May); nonflooded forest.

This species can be confused with *Ocotea cernua* which differs in its smaller, deeper fruit cupules and 4-celled stamens.

Endlicheria pyriformis (Nees) Mez PL. 70a

Shrubs or small trees, to 4 m tall. Stems glabrous. Terminal buds glabrous. Leaves alternate; blades elliptic, 10–24 × 3–7 cm, glabrous, pale green, the base acute, the apex acuminate. Inflorescences to 15 cm long, laxly flowered, glabrous, reddish. Flowers glabrous, green with red hue; tepals spreading at anthesis, minutely pubescent on adaxial surface. Staminate flowers: stamens short, the cells almost lateral. Pistillate flowers: staminodia often with opened locelli. Fruit cupules cup-shaped, the pedicel thickened. Fl (Jul, Aug, Sep), fr (Feb, Apr, Oct, Dec); rather common, in nonflooded forest.

Diagnostic are the pale green stems and leaves. It is very similar to *Ocotea diffusa* which has darker foliage and stems, flowers in February, and which has pubescent terminal buds.

Endlicheria cf. **sericea** Nees

Trees, 18 m tall. Stems densely pubescent. Terminal buds densely pubescent. Leaves alternate; blades elliptic, ca. 25 × 8 cm, densely appressed pubescent abaxially. Inflorescences to 8 cm long, pubescent. Tepals at anthesis erect. Fruit cupules thick, cup-shaped. Known only by a single sterile specimen from our area (*Mori et al. 23269*, NY); in nonflooded forest.

KUBITZKIA van der Werff

Trees, to 10 m tall. Leaves alternate; blades rather thinly chartaceous. Inflorescences axillary, few-flowered. Flowers bisexual; tepals 6, equal; stamens 9, 2-celled. Fruit cupules with double margin, the tepals persistent. Fruits drupes.

Kubitzkia mezii (Kosterm.) van der Werff

Small trees. Stems slender. Leaves alternate; blades elliptic, 6–11 × 2.5–4 cm, glabrous. Inflorescences axillary, few-flowered (rarely >7 flowers). Flowers 5–6 mm diam., red; tepals 6, equal, spreading at anthesis, pubescent adaxially; stamens 9, slender, the anther about as wide as filament, the 2 anther cells subapical, the long filaments densely pubescent, the filaments of inner 3 stamens (partly) fused into a cylinder. Fruit cupules with tepals persistent and reflexed on outer margin. Fruits ellipsoid, ca. 0.5 cm long. Fl (Sep), fr (Oct); rare, in nonflooded moist forest.

Red flower color is unusual for Lauraceae.

LICARIA Aubl.

Shrubs to canopy trees. Leaves alternate, rarely opposite. Inflorescences axillary, sometimes in axils of bracts at tips of branches and appearing terminal. Flowers bisexual; stamens 3, 2-celled; staminodia 0, 3, or 6, small. Fruit cupules deeply cup-shaped, with a double margin, but in some species the double margin scarcely visible, the small, dried stamens frequently persistant on cupules.

1. Leaves opposite.
    2. Stems sparsely appressed pubescent. Leaf blades subglabrous abaxially, the apex acuminate, the acumen 1–1.5 cm long. . . . . *L. debilis.*
    2. Stems densely pubescent (surface of young stems completely covered). Leaf blades densely appressed pubescent abaxially, the apex acute. . . . . *L. chrysophylla.*
1. Leaves alternate.
    3. Young stems densely pubescent, the pubescence completely covering stems.
        4. Leaf blades densely pubescent abaxially, the surface mostly covered by indument; secondary veins in 4–8 pairs, immersed on adaxial surface and not or scarcely loop-connected. . . . . *L. martiniana.*
        4. Leaf blades sparsely pubescent abaxially, the surface readily visible; secondary veins in 2–5 pairs, somewhat impressed on adaxial surface and clearly loop-connected. . . . . *L. guianensis.*
    3. Young stems glabrous or sparsely pubescent, the surface readily visible.
        5. Terminal buds glabrous. Tepals spreading. Infructescences to 1.5 cm long. . . . . *L. vernicosa.*
        5. Terminal buds pubescent or glabrous. Tepals erect. Infructescences (much) longer.
            6. Terminal buds pubescent. Leaf blades somewhat bullate. Flowers glabrous (but inflorescences pubescent); locelli roundish, on abaxial surface of anther. . . . . *L. subbullata.*
            6. Terminal bud glabrous. Leaf blades flat. Flowers pubescent; locelli slit-like, on adaxial surface of anther. . . . . *L. cannella.*

Licaria cannella (Meisn.) Kosterm.

Trees, 20 m or taller. Stems glabrous. Terminal buds glabrous. Leaves alternate, often grouped at tips of branches; blades elliptic to obovate-elliptic, 6–20 × 3–7 cm, glabrous, the margin often thickened and pale; secondary veins not or scarcely loop-connected. Inflorescences ± 5 cm long or shorter, pubescent, densely flowered. Flowers pubescent; tepals erect; stamens not surpassing tepals, 2-celled, the locelli slit-like, on adaxial surface of stamens. Fruit cupules deeply cup-shaped, to 4 cm long, the double margin easily visible. Fruits to 5 cm long. Fl (Aug, Sep); rare. *Bois canelle* (Créole).

The epithet is usually, but incorrectly, spelled "*canella*." An additional vegetative character is the dark color of the petiole, which contrasts with the grayish bark of the stems; this character can also occur in *L. subbullata*.

Licaria chrysophylla (Meisn.) Kosterm.

Trees, ca. 20 m tall. Stems densely pubescent. Terminal buds densely pubescent. Leaves opposite; blades narrowly elliptic, 6–17 × 2–6 cm, densely pubescent abaxially; secondary veins weakly loop-connected. Inflorescences to 6 cm long, densely pubescent. Flowers pubescent; tepals erect; stamens enclosed in flower, the locelli roundish, apical or nearly so. Fruits to 2 cm long, the cupules to 2.5 cm long, the double margin variously developed. Fl (Oct); rare, nonflooded forest.

Diagnostic are the opposite, rather narrow, densely pubescent leaves.

Licaria debilis (Mez) Kosterm.

Small trees, to 10 m tall. Stems (sparsely) pubescent. Terminal bud pubescent. Leaves opposite; blades (broadly) elliptic, 6–10 × 3–5 cm, subglabrous abaxially, the apex acuminate; secondary veins loop-connected. Inflorescences to 10 cm long, slender, sparsely pubescent. Flowers white, glabrous or with a few hairs; stamens about as long as tepals, the locelli ± terminal, roundish. Fruits to 2.5 cm long, the cupules cup-shaped, the double margin scarcely, if at all, noticeable. Fl (Aug, Sep); in nonflooded forest.

Diagnostic are the opposite, subglabrous, acuminate leaves and slender inflorescences.

Licaria guianensis Aubl. PL. 69f

Trees, to 25 m tall. Young stems and terminal buds densely pubescent. Leaves alternate; blades elliptic, 5–10 × 2–4 cm, sparsely pubescent abaxially, the apex acuminate; secondary veins clearly loop-connected. Inflorescences to 8 cm long, densely pubescent. Flowers yellowish, pubescent; tepals erect; stamens included, the locelli roundish, apical. Fruit cupules deeply cup-shaped, the double margin scarcely visible. Fruits ca. 1 cm long. Fl (Mar, Apr, Jul, Sep), fr (Jul); common, in nonflooded forest.

Diagnostic are the alternate, acuminate leaves with few secondary veins. See discussion under *L. martiniana*.

Licaria martiniana (Mez) Kosterm.

Trees, to 25 m tall. Stems densely pubescent. Terminal buds densely pubescent. Leaves alternate; blades elliptic, 7–15 × 2.5–5 cm, densely pubescent abaxially, the apex acute or slightly acuminate; secondary veins weakly loop-connected. Inflorescences to 6 cm long, densely pubescent. Flowers yellow, densely pubescent; tepals erect or incurved; stamens included, the locelli apical, roundish. Fruits ca. 1.5 cm long, the cupules cup-shaped, the double margin not clearly developed. Fl (Nov); nonflooded forest.

Licaria martiniana is close to *L. guianensis*, but differs in its larger, more coriaceous leaves, greater number of secondary veins which are not strongly loop-connected and the denser pubescence on the abaxial leaf blade surface, although the indument apparently wears off with age.

Licaria subbullata Kosterm.

Small trees, to 10 m tall. Stems sparsely pubescent. Terminal buds pubescent. Leaves alternate, elliptic or obovate-elliptic, ± bullate, 14–20 × 5–7 cm, glabrous, the apex shortly acuminate; secondary veins weakly loop-connected. Inflorescences to 6–8 cm long, gray-pubescent. Flowers glabrous, yellow, drying dark and contrasting with the gray-pubescent inflorescences; tepals erect, the outer 3 shorter than inner 3; stamens exserted; the locelli roundish, on abaxial surface of anthers. Fruit cupules cup-shaped, the inner margin longer than outer one, the outer one consisting of 6 spreading lobes. Fl (Jul, Sep); rare, in nonflooded forest.

Diagnostic are the subbullate leaves and pubescent inflorescences with glabrous flowers.

Licaria vernicosa (Mez) Kosterm.

Small trees, to 10 m tall. Stems glabrous. Terminal buds glabrous. Leaves alternate; blades elliptic or obovate, 8–14 × 4–6 cm, glabrous, the apex shortly acuminate; secondary veins weakly loop-connected. Inflorescences (ex descr.) to 1.5 cm long, sparsely pubescent. Flowers: tepals spreading, the inner 3 longer than outer 3, the locelli on abaxial surface of stamens, roundish. Fruits 1.5–2 cm long, the cupules cup-shaped, the infructescences very short and cupule close to stems. Fr (Sep); rare.

This is a poorly known species. It is vegetatively glabrous, as *L. cannella*, but has sepals spreading instead of erect as in that species.

NECTANDRA Rottb.

Medium to tall trees. Leaves alternate, glabrous or pubescent with simple hairs. Flowers bisexual, the adaxial surface of tepals and stamens densely papillose pubescent; tepals fused at bases, in old flowers falling as unit with stamens; stamens 9, all 4-celled, the cells arranged in an arc. Fruits drupes, seated in a cup-shaped cupule.

1. Young stems densely brown-tomentellous, the hairs curled, not appressed. Leaf blades with base with two reflexed lobes. . . . . *N. reticulata.*
1. Young stems densely or sparsely appressed pubescent. Leaf blades without reflexed lobes.
   2. Stems densely, but minutely appressed pubescent. Flowers 3–5 mm diam. . . . . *N. cissiflora.*
   2. Stems subglabrous or sparsely appressed pubescent. Flowers 3–8 mm diam.
      3. Flowers 3–5 mm diam.; tepals with adaxial surface pubescent near base, papillose distally; anthers without sterile tip, sparsely papillose. . . . . *N. purpurea.*
      3. Flowers 7–8 mm diam.; tepals with adaxial surface papillose, without hairs near base; anthers with sterile tip, distinctly papillose. . . . . *N. hihua.*

Nectandra cissiflora Nees

Trees, to 15 m tall. Young stems densely appressed pubescent, the surface covered by the indument. Leaves alternate; blades 12–18 × 4–7 cm, minutely appressed pubescent, the base flat. Inflorescences axillary. Flowers 3–5 mm diam., fragrant; anthers obtuse, with small sterile tip, densely papillose. Fruit cupules shallow. Fr (Oct, immature); uncommon.

Nectandra hihua (Ruiz & Pav.) Rohwer PL. 70b

Trees, to 20 m tall, but sometimes flowering when much smaller. Stems subglabrous or sparsely appressed pubescent, the surface almost entirely visible. Leaves alternate; blades 10–30 × 3–12 cm, nearly glabrous, the base flat or slightly inrolled. Inflorescences axillary. Flowers 7–8 mm diam., fragrant; anthers with distinct sterile tip, densely papillose. Fruit cupules cup-shaped. Fl (Sep, Oct); uncommon, in nonflooded forest.

Nectandra purpurea (Ruiz & Pav.) Mez

Trees, to 15 m tall. Young stems sparsely appressed pubescent, becoming glabrous with age. Leaves alternate; blades 12–22 × 5–10 cm, glabrous or nearly so, the base flat. Inflorescences axillary. Flowers 3–5 mm diam.; anthers without sterile tip, sparsely and inconspicuously papillose. Fruit cupules shallowly bowl-shaped. Fruits ellipsoid, to 2.5 cm long. Fl (Jan); apparently rare, in nonflooded forest.

Nectandra reticulata (Ruiz & Pav.) Mez PL. 70c

Tall trees, to 35 m tall. Young stems densely brown-pubescent. Leaves alternate; blades 14–35 × 5–11 cm, pubescent abaxially, the base with reflexed lobes. Inflorescences axillary. Flowers ca. 1 cm diam., fragrant, snow white; tepals spreading at anthesis, densely papillose inside; stamens 9, 4-celled, the anthers with large, sterile tip. Fruit cupules deeply cup-shaped. Fruits to 2 cm long. Fl (Sep, Oct, Dec), fr (Dec); in nonflooded forest.

OCOTEA Aubl.

Shrubs to tall trees. Leaves alternate, very rarely clustered at tips of branches, sometimes with domatia (tufts of hair in axils of secondary veins). Flowers small, usually green, yellow, or white, bisexual or unisexual; tepals erect or spreading at anthesis; stamens 9, all 4-celled; staminodia (3) present or absent, if present, stipitiform; pistil glabrous or rarely pubescent. Fruit cupules variable, ranging from small disc to deep cup, sometimes with tepals persisting, rarely with double margin.

1. Apparent petioles 3–5 cm long; leaf blades with base decurrent onto petiole. . . . . *O. tomentella.*
1. Apparent petioles <3 cm long; leaf blades with base variable, but not decurrent.
    2. Leaf blades with predominantly erect hairs.
        3. Flowers pedicellate, not arranged in tight clusters. Fruit cupules small, plate-like, the tepals not persistent. . . . . *O. fendleri.*
        3. Flowers sessile or nearly so, arranged in tight clusters. Fruit cupules cup-shaped, or, when rather shallow, with persistent tepals.
            4. Leaf blades chartaceous, the apex acuminate. Inflorescences (only pistillate known) shorter than leaves. Fruit cupules with persistent tepals, the cupule shallow. . . . . *O. scabrella.*
            4. Leaf blades firmly coriaceous to chartaceous, the apex acute. Inflorescences longer than or as long as leaves. Fruit cupules without persistent tepals, the cupule cup-shaped. . . . . *O. glomerata.*
    2. Leaf blades with appressed hairs or glabrous abaxially, rarely with curled hairs on midrib, but not on lamina.
        5. Leaf blades with base abruptly obtuse or rounded. . . . . *O. commutata.*
        5. Leaf blades with base acute or decurrent.
            6. Flowers and inflorescences glabrous.
                7. Leaf blades with secondary veins loop-connected. Pedicels ca. 3 mm long. Tepals spreading at anthesis. Fruit cupules very shallow, almost plate-like. . . . . *O. diffusa.*
                7. Leaf blades with secondary veins not loop-connected, fading near margin. Pedicels 1–1.5 mm long. Tepals erect at anthesis. Fruit cupules cup-shaped. . . . . *O. cernua.*
            6. Flowers and/or inflorescences pubescent.
                8. Pistils (sparsely) pubescent. Fruit cupules flat, plate-like, thick, appearing double-margined, the tepals often persisting as (reflexed) lobes on cupule. . . . . *O. floribunda.*
                8. Pistils glabrous. Fruit cupules cup-shaped or plate-like; if plate-like, thin, not double-margined and without persisting tepals.
                    9. Leaf blades (narrowly) oblong. Inflorescences and young stems densely brown-tomentellous. Fruit cupules deeply cup-shaped, with pronounced double margin. . . . . *O. cujumary.*
                    9. Leaf blades (broadly) elliptic. Inflorescences variously pubescent or glabrous. Fruit cupules deeply cup-shaped, but without a double margin or small, plate-like.
                        10. Young stems glabrous or sparsely pubescent, the surface readily visible.
                            11. Leaf blades with secondary veins immersed abaxially, the surface smooth.
                                12. Leaf blades coriaceous; secondary veins indistinct, usually in ≥8 pairs. . . . . *O. ceanothifolia.*
                                12. Leaf blades chartaceous; secondary veins readily visible abaxially, usually in 4–6 pairs. . . . . *O. splendens.*
                            11. Leaf blades with secondary veins raised abaxially, discernible to touch.

13. Leaf blades without domatia (inconspicuous axillary tufts of hairs or small depressions in axils of secondary veins). Receptacle in staminate flowers shallow, glabrous adaxially. Fruit cupules plate-like. . . . . *O. puberula.*
13. Leaf blades with domatia (inconspicuous tufts of hairs or depressions in axils of secondary veins). Receptacle in staminate flowers rather deep, pubescent adaxially. Fruit cupules (deeply) cup-shaped.
    14. Leaf blades elliptic to elliptic-obovate, 5–11 cm long. Fruit cupules 6–8 mm diam. . . . . *O. cinerea.*
    14. Leaf blades elliptic or broadly elliptic, 11–20 cm long. Fruit cupules ca. 16 mm diam. . . . . *O. subterminalis.*

10. Young stems densely pubescent, the indument completely covering surface.
    15. Fruit cupules cup-shaped. . . . . *O. canaliculata.*
    15. Fruit cupules plate-like.
        16. Terminal buds 1–1.5 cm long. Leaf blades widest above middle, oblanceolate to obovate. Fruit cupules very small, scarcely wider than swollen, lenticellate pedicel. . . . . *O. oblonga.*
        16. Terminal buds to 0.5 cm long. Leaf blades widest at or below middle. Fruit cupules small, plate-like, wider than scarcely swollen pedicel. . . . . *O. puberula.*

Ocotea canaliculata (Rich.) Mez

Trees, to 30 m tall. Stems densely tomentellous. Leaves alternate; blades elliptic, 10–15 × 3–5 cm, coriaceous, glabrous, domatia absent, the base acute, the apex acute; secondary veins raised abaxially. Inflorescences about as long as or longer than leaves, brown-tomentellous. Flowers unisexual, pubescent, white; tepals ± erect. Fruit cupules deeply cup-shaped, with a single margin, the tepals briefly persistent on young cupules. Fr (Nov, Dec); occasional, in nonflooded forest.

Diagnostic are the tomentellous stems, coriaceous leaves, and cup-shaped fruit cupules.

Ocotea ceanothifolia (Nees) Mez

Trees, to 25 m tall. Stems glabrous. Leaves alternate; blades elliptic, 10–20 × 3–7.5 cm, coriaceous, glabrous, domatia absent, the base acute, the apex acute or slightly acuminate; secondary veins immersed. Inflorescences not seen; infructescences much shorter than leaves. Flowers (ex descr.) unisexual, tomentellous. Fruit cupules cup-shaped, with single margin, the tepals deciduous. Fr (Sep, Oct, Nov); occasional, in nonflooded forest.

Diagnostic are glabrous leaves, stems, and coriaceous leaves with immersed, inconspicuous secondary veins. *Ocotea ceanothifolia* is a poorly known species and the identifications, based on fruiting specimens, are tentative.

Ocotea cernua (Nees) Mez

Small trees, to 8 m tall. Stems glabrous. Leaves alternate; blades elliptic, 10–15 × 4–6 cm, chartaceous, glabrous, domatia absent, the base acute, the apex acuminate; secondary veins weakly raised abaxially. Inflorescences shorter than leaves, glabrous. Flowers unisexual, glabrous, yellow; tepals erect. Fruit cupules cup-shaped, with single margin, the tepals not persistent. Fl (Jun), fr (Jan); occasional, in old secondary or nonflooded forest.

Diagnostic are the glabrous stems and flowers, and the acuminate leaves.

Ocotea cinerea van der Werff

Large trees, to 50 m tall. Stems sparsely and minutely appressed pubescent or glabrous. Leaves alternate; blades elliptic, 5–11 × 1.8–4 cm, glabrous or nearly so, inconspicuous tufts of hairs or small depressions frequently present in axils of secondary veins, the base acute or somewhat decurrent onto petiole, the apex acute or shortly acuminate. Flowers unisexual, sparsely puberulous abaxially; tepals spreading at anthesis. Fruit cupules shallowly cup-shaped, with single margin, the tepals not persistent. Fr (Nov); occasional, in nonflooded moist forest.

Diagnostic are the slightly obovate leaves with attenuate bases, the sparsely pubescent, unisexual flowers, and the small, shallowly cup-shaped cupules.

Ocotea commutata (Nees) Mez

Small trees, to 6 m tall. Stems glabrous or sparsely appressed pubescent. Leaves alternate; blades narrowly elliptic, 10–24 × 4–8 cm, chartaceous, glabrous, domatia absent, the base gradually narrowed, finally abruptly obtuse, truncate or cordate, the apex acute; secondary veins raised on abaxial surface. Inflorescences shorter than leaves, pubescent. Open flowers not seen; buds pubescent. Fruit cupules small, shallow, the tepals persisting as lobes, the pedicel swollen in fruit. Fl (Apr), fr (Aug, Sep, Oct, Nov); occasional, in forest and secondary vegetation.

Diagnostic are the narrowly obtuse, truncate, or cordate leaf base and the small cupules with persistent tepals.

Ocotea cujumary Mart.

Small or medium-sized trees, to 15 m tall. Stems densely brown tomentellous. Leaves alternate; blades ± oblong, 10–15 × 2–5 cm, firmly chartaceous, glabrous adaxially, sparsely appressed pubescent abaxially, domatia absent, the base acute, the apex acute or acuminate; secondary veins loop-connected, weakly raised abaxially. Inflorescences densely tomentellous, exceeding leaves. Flowers unisexual, pubescent; tepals erect at anthesis. Fruit cupules cup-shaped, with pronounced double margin, the tepals persisting on the outer margin, eventually wearing off. Fr (Aug, Sep); rare, in nonflooded forest.

Diagnostic are the long, tomentellous inflorescences, the oblong leaves with loop-connected secondary veins, and the double-margined fruit cupules.

Ocotea diffusa van der Werff FIG. 157, PL. 70d

Shrubs or small trees, to 12 m tall. Stems glabrous. Leaves alternate; blades elliptic, 10–15 × 3.5–5 cm, chartaceous, glabrous,

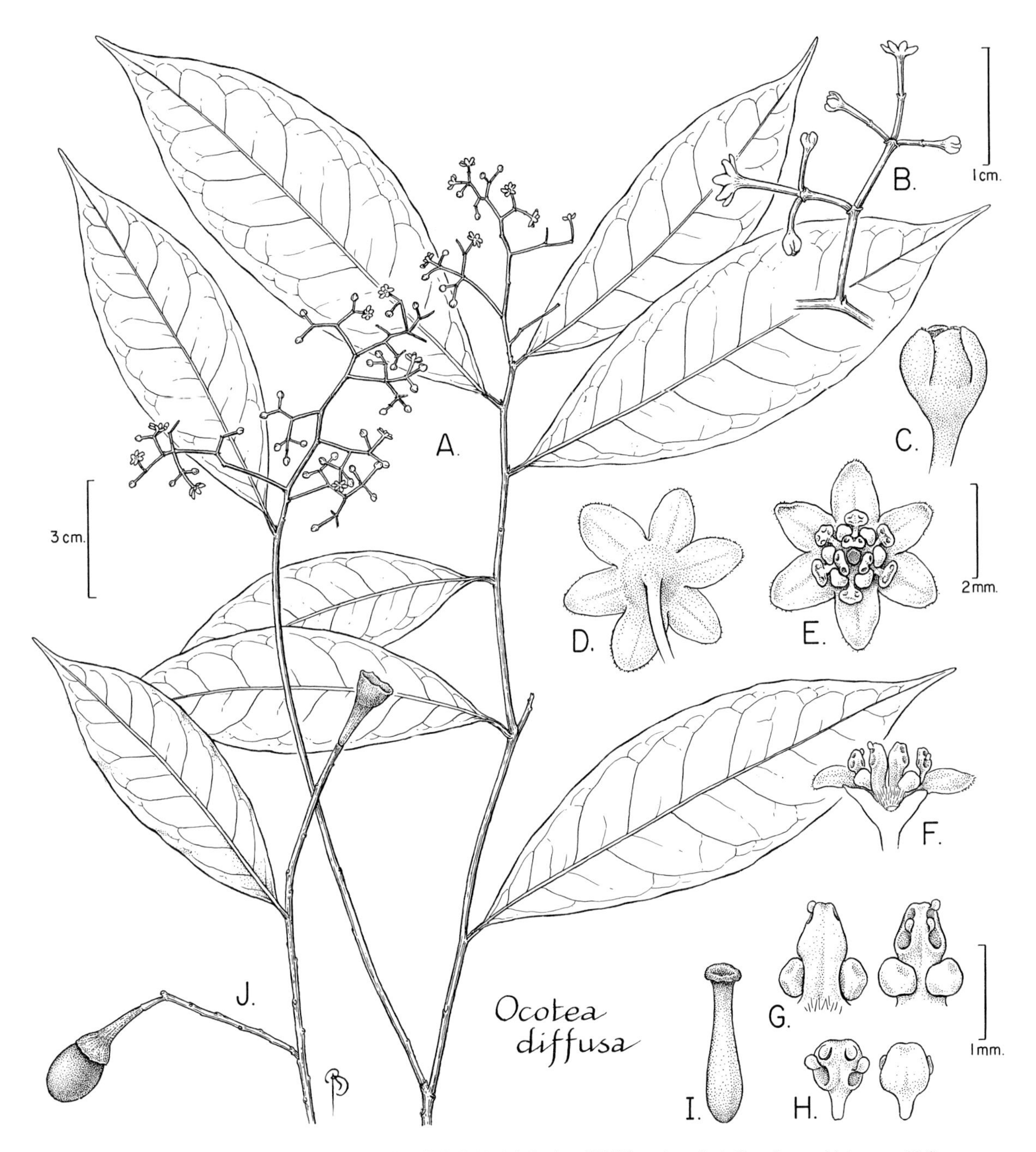

FIG. 157. LAURACEAE. *Ocotea diffusa* (A–I, *de Granville B.4750*; J, *Mori & Cardoso 17718* from Amapá). **A.** Part of stem with leaves and inflorescences. **B.** Detail of inflorescence. **C.** Lateral view of flower bud. **D.** Basal view of flower. **E.** Apical view of flower. **F.** Medial section of flower with pistil removed. **G.** Abaxial (near right) and adaxial (far right) views of staminodes showing glands at base. **H.** Adaxial (near right) and abaxial (far right) views of stamens. **I.** Lateral view of pistil. **J.** Part of stem with leaf, fruit, and cupule.

but small tufts of hairs in axils of lower secondary veins, the base acute, the apex acuminate; secondary veins slightly impressed adaxially, loop-connected. Inflorescences glabrous, shorter than or as long as leaves. Flowers unisexual, glabrous; tepals spreading at anthesis. Fruit cupules shallow, bowl-shaped, with simple margin, the tepals deciduous. Fl (Feb, Mar), fr (Oct); occasional, in nonflooded forest.

Diagnostic are the leaves with inconspicuous domatia, and the slightly impressed, loop-connected secondary veins. This species has been confused with *Endlicheria pyriformis*, which it resembles closely. It differs in having inconspicuous domatia, brown (instead of pale green) stems, and 4-celled anthers; it flowers in February–March, while *E. pyriformis* flowers in July-September.

Ocotea fendleri (Meisn.) Rohwer

Trees, 20 m tall. Stems densely brown-tomentellous. Leaves alternate; blades elliptic, 12–20 × 3–6 cm, firmly chartaceous, with short, erect hairs abaxially, domatia absent, the base narrowly acute or acute, the apex obtuse; secondary veins raised abaxially. Inflorescences not seen. Flowers unisexual, pubescent. Infructescences as long as leaves, pubescent. Fruit cupules, when very small, cup-shaped, at maturity, a thin, single-margined plate, leaving the fruit fully exposed. Fr (Aug, Sep); occasional, in nonflooded forest.

Diagnostic are the erect indument on the leaves and the small, plate-like cupule. This species can be confused with *O. glomerata*; in addition to characters mentioned in the key, *O. glomerata* differs in its sharply acute leaves.

Ocotea floribunda (Sw.) Mez

Trees, to 25 m tall. Stems minutely appressed pubescent, soon becoming glabrous. Leaves alternate; blades elliptic, 10–15 × 3–5 cm, chartaceous, glabrous or with some minute, appressed hairs abaxially, domatia absent, the base acute, the apex acute; secondary veins weakly raised abaxially. Inflorescences pubescent, shorter than leaves. Flowers unisexual, pubescent; tepals ± spreading at anthesis. Fruit cupules a thick, double-margined, flat plate, the tepals frequently persisting on outer margin. Fr (Aug, immature); rare, in nonflooded forest.

Diagnostic are the thick, flat cupules and the (sparsely) pubescent pistil or pistillode, especially the style.

Ocotea glomerata (Nees) Mez

Trees, to 20 m tall. Stems angular, densely pubescent, the hairs mostly erect. Leaves alternate; blades elliptic to ovate-elliptic, 10–20 × 3–7 cm, firmly chartaceous, pubescent with erect hairs abaxially, domatia absent, the base acute, the apex acute. Inflorescences densely pubescent, as long as or longer than leaves, the flowers clustered. Flowers unisexual, pubescent; tepals erect at anthesis. Fruit cupules cup-shaped, with a simple margin, the tepals persisting on very immature cupules, but falling off as cupule matures. Fl (Feb, Apr, Jun), fr (Aug, Dec); occasional, in nonflooded forest and secondary vegetation.

Diagnostic are the erect indument on the abaxial leaf surface, the cup-shaped fruit cupules, and the acute leaf apices. See also note under *O. fendleri*.

Ocotea oblonga (Meisn.) Mez

Trees, to 35 m tall. Stems brown-tomentellous. Leaves alternate; blades elliptic, 12–20 × 3–6 cm, chartaceous, appressed pubescent abaxially, domatia absent, the base acute, the apex acute or obtuse; secondary veins raised abaxially. Inflorescences pubescent, about as long as leaves. Flowers unisexual, pubescent. Fruit cupules a small disc, the pedicel swollen and lenticellate. Fl (May, Jun), fr (Sep, immature); rare, in nonflooded forest.

This species is very similar to *O. fendleri*, differing mostly in the pubescence on the abaxial leaf blade surface (appressed vs. erect).

Ocotea puberula (Rich.) Nees

Trees, to 25 m tall. Stems sparsely appressed pubescent, becoming glabrous. Leaves alternate; blades elliptic, 9–18 × 3–6 cm, chartaceous, glabrous or with some appressed hairs abaxially, domatia absent, the base acute, the apex acute; secondary veins raised abaxially. Inflorescences pubescent, as long as or shorter than leaves. Flowers unisexual, sparsely pubescent; tepals spreading at anthesis. Fruit cupules small, plate-like, with simple margin, the tepals not persistent, the pedicel thickened. Fl (Apr), fr (Jun, Jul, Aug); occasional, in nonflooded forest.

This is a widespread, variable and poorly defined species, best recognized by its plate-like fruit cupules, spreading tepals, and rather thin leaves.

Ocotea scabrella van der Werff

Small trees, to 8 m tall. Stems densely brown-pubescent, the hairs short, erect. Leaves alternate; blades broadly elliptic, 12–20 × 5–8 cm, chartaceous, with sparse, short, erect hairs abaxially, domatia absent, the base acute, the apex acuminate. Inflorescences pubescent, much shorter than leaves, the flowers nearly sessile, arranged in dense clusters. Flowers unisexual, pubescent; tepals spreading at anthesis. Fruit cupules shallow, thin, with simple margin, the tepals persistent. Fl (Sep), fr (Nov); occasional, in nonflooded forest.

Diagnostic are the very short, erect indument on the abaxial leaf surface, and the small cupules with persistent tepals. Fruiting specimens suggest that the inflorescences can be as long as the leaves.

Ocotea splendens (Meisn.) Baill.

Small trees, to 15 m tall. Stems puberulous when young, becoming glabrous with age. Leaves alternate; blades elliptic, 10–15 × 4–7 cm, firmly chartaceous, sparsely appressed pubescent or glabrous abaxially, domatia absent, the base acute, the apex acuminate; secondary veins weakly raised abaxially. Inflorescences as long as or longer than leaves, pubescent. Flowers unisexual, pubescent; tepals erect at anthesis. Fruit cupules rather deeply cup-shaped, with simple margin, the tepals not persistent. Fl (Aug, Sep), fr (Feb, May, Sep); occasional, in nonflooded forest.

Ocotea subterminalis van der Werff

Trees, to 18 m tall. Stems sparsely pubescent, soon becoming glabrous. Leaves alternate; blades (broadly) elliptic, 15–22 × 5–7 cm, chartaceous, glabrous except for domatia in axils of secondary veins, the base acute, the apex acuminate. Inflorescences sparsely pubescent, about as long as leaves or shorter. Flowers unisexual, sparsely pubescent; tepals erect at anthesis. Fruit cupules warty, deeply cup-shaped, the margin simple, the tepals not persistent. Fl (Nov), fr (Aug, Nov); rare, in nonflooded forest.

Diagnostic are the domatia, the yellow-green abaxial leaf surface and the warty fruit cupules.

Ocotea tomentella Sandwith PART 1: FIG. 9

Trees, to 40 m tall. Stems angular, tomentellous. Leaves alternate; blades elliptic, 15–25 × 6–10 cm, chartaceous, tomentellous abaxially, domatia absent, the base decurrent onto petiole and inrolled, the inrolled part 2–4 cm long, the apex acute or obtuse; secondary veins raised abaxially. Inflorescences pubescent, shorter or longer than leaves. Flowers unisexual, pubescent; tepals erect at anthesis. Fruit cupules deeply cup-shaped, the margin simple, the tepals not persistent. Fr (May, Dec); rare, in nonflooded forest.

Diagnostic are the leaves with decurrent, inrolled bases.

RHODOSTEMONODAPHNE Rohwer & Kubitzki

Santiago Madriñán

Trees or shrubs (sometimes scandent shrubs outside our area). Leaves alternate; tripli- to penniveined. Inflorescences thyrsoid, erect or pendulous. Flowers unisexual (plants dioecious); tepals equal, erect, spreading or reflexed at anthesis; stamens 9 (staminodial in pistillate flowers), mostly 4-celled (2(3)-celled in outer whorls of *R. morii*); staminodia of whorl IV present or lacking; pistillode much reduced or absent in staminate flowers. Fruits drupes, seated in thick cupule, the tepals deciduous or persistent.

1. Young stems with inconspicuous, minute hairs (puberulous), soon glabrescent.
   2. Leaf blades membranaceous, the tertiary and higher order veins conspicuous abaxially. . . . . *R. morii.*
   2. Leaf blades coriaceous, the tertiary and higher order veins barely visible abaxially. . . . . *R. elephantopus.*
1. Young stems with conspicuous, dense hairs, the indument persistent.
   3. Young stems with silver to golden, appressed, straight hairs (sericeous).
      4. Leaf blades with 3–6 pairs of secondary veins, these reddish when fresh, the basal most pair with more acute angles (acrodromous), tripliveined. Flowers reddish when fresh. . . . . *R. saülensis.*
      4. Leaf blades with 6–10 pairs of secondary veins, these yellow-green when fresh, the pairs all parallel (pinnate). Flowers yellowish when fresh. . . . . *R. grandis.*
   3. Young stems with red to brown to yellow, erect, straight to crisped hairs (pubescent to tomentose).
      5. Leaf blades with 7–13 pairs of secondary veins. . . . . *R. kunthiana.*
      5. Leaf blades with ≤6 pairs of secondary veins.
         6. Leaf blades yellowish to light green, with white, straight, erect, inconspicuous hairs throughout abaxially (puberulous). . . . . *R. leptoclada.*
         6. Leaf blades dark green to brown, with reddish brown hairs (at least on veins) abaxially.
            7. Leaf blades with white, crisped hairs abaxially (pubescent), only veins with reddish-brown hairs. . . . . *R. rufovirgata.*
            7. Leaf blades with brown, crisped, conspicuous hairs throughout abaxially. . . . *Rhodostemonodaphne* sp. 1.

Rhodostemonodaphne elephantopus Madriñán FIG. 158

Large trees, to 40 m tall. Trunks with low, thick buttresses. Stems canaliculate, inconspicuously silver puberulous, soon glabrescent. Leaves: petioles 2–13 × 0.8–1.8 mm; blades elliptic to slightly obovate, 2–9 × 1–3 cm, coriaceous, glabrous, the base decurrent, the apex obtuse to acuminate; venation pinnate, the midrib plane ad- and abaxially, the secondary veins in 4–9 pairs, inconspicuous, the tertiary and higher order veins scarcely visible abaxially. Inflorescences erect, glabrous. Flowers green when fresh; tepals spreading. Fl (Aug, Sep); occasional, in nonflooded forest.

This species is known only from central French Guiana. Diagnostic are the glabrous leaf blades with inconspicuous venation and the canaliculate petioles.

Rhodostemonodaphne grandis (Mez) Rohwer PL. 71a

Large trees, to 30 m tall. Stems silver to golden sericeous, the hairs persisting for one growth period. Leaves: petioles 19–44 × 2–3.8 mm; blades elliptic to ovate, 10–30(45) × 5–11(20) cm, coriaceous, glabrous adaxially, puberulous to sericeous, soon glabrescent or the hairs persistent, light green to whitish, the veins yellow-green abaxially when fresh, the base rounded to acute, plane, the apex rounded to acuminate; venation pinnate, the secondary veins in 6–10 pairs. Inflorescences erect, silver puberulous; pedicels reddish. Flowers: tepals yellowish when fresh, reflexed; stamens red. Fruit cupules tuberculate, tinged with red when fresh, the tepals caducous. Fruits first green, maturing black. Fl (Sep), fr (Mar); occasional, in nonflooded forest. *Baaka-apisi*, *cèdre jaune*, *cèdre noir* (Créole), *guéli-apisi* (Bush Negro).

The recently described *R. saülensis* has been segregated from this species. As now circumscribed, *R. grandis* is restricted to Surinam, French Guiana, and Amapá, Brazil. It is distinguished from *R. saülensis* by the higher number (6–10) and pinnate arrangement of the secondary veins and the yellowish flowers.

Rhodostemonodaphne kunthiana (Nees) Rohwer PL. 71c

Medium-sized trees, to 18 m tall (to 30 m outside our area). Trunks with low, spreading buttresses. Stems brown pubescent, the hairs persistent. Leaves: petioles 10–25(50) × 5–9 mm; blades elliptic, 10.5–23.5 × 5–12.5 cm, glabrous and shiny adaxially, sparingly hairy, dull and lighter abaxially, the base rounded, the apex obtuse to acuminate; venation pinnate, the secondary veins in 7–13 pairs, the tertiary veins scalariform. Inflorescences erect, brown pubescent; pedicels short. Flowers: short pedicellate; tepals brownish pubescent abaxially, cream-colored to yellow adaxially when fresh; stamens yellow-green. Fruit cupules with thin margin, reddish when fresh, the tepals caducous. Fruits shiny green. Fl (May, Jun), fr (Oct); occasional, in nonflooded moist forest. *Cèdre noir* (Créole).

This is a widely distributed species characterized by the number of secondary veins, the strongly scalariform tertiaries, and the shortly pedicellate, brown pubescent, cream-colored to yellow flowers.

Rhodostemonodaphne leptoclada Madriñán

Shrubs to small trees, 3–4(10) m tall. Stems yellow-brown tomentose, the hairs persistent, becoming dark brown. Leaves: petioles 5–12 × 0.8–1.2(2.2) mm; blades elliptic, (4.8)7.9–11(15.4) × (1.7)3.1–4(6.8) cm, coriaceous, plane (slightly bullate outside our area), glabrous except for tomentose midrib, when dry conspicuously shiny yellow adaxially, inconspicuously puberulous (conspicuously puberulous outside our area), opaque yellow-green abaxially, the base decurrent to cuneate, the apex acute to acuminate; venation tripliveined, the secondary veins in (3)4(6) pairs. Inflorescences slender, little-branched, pauciflorous, pendulous (?).

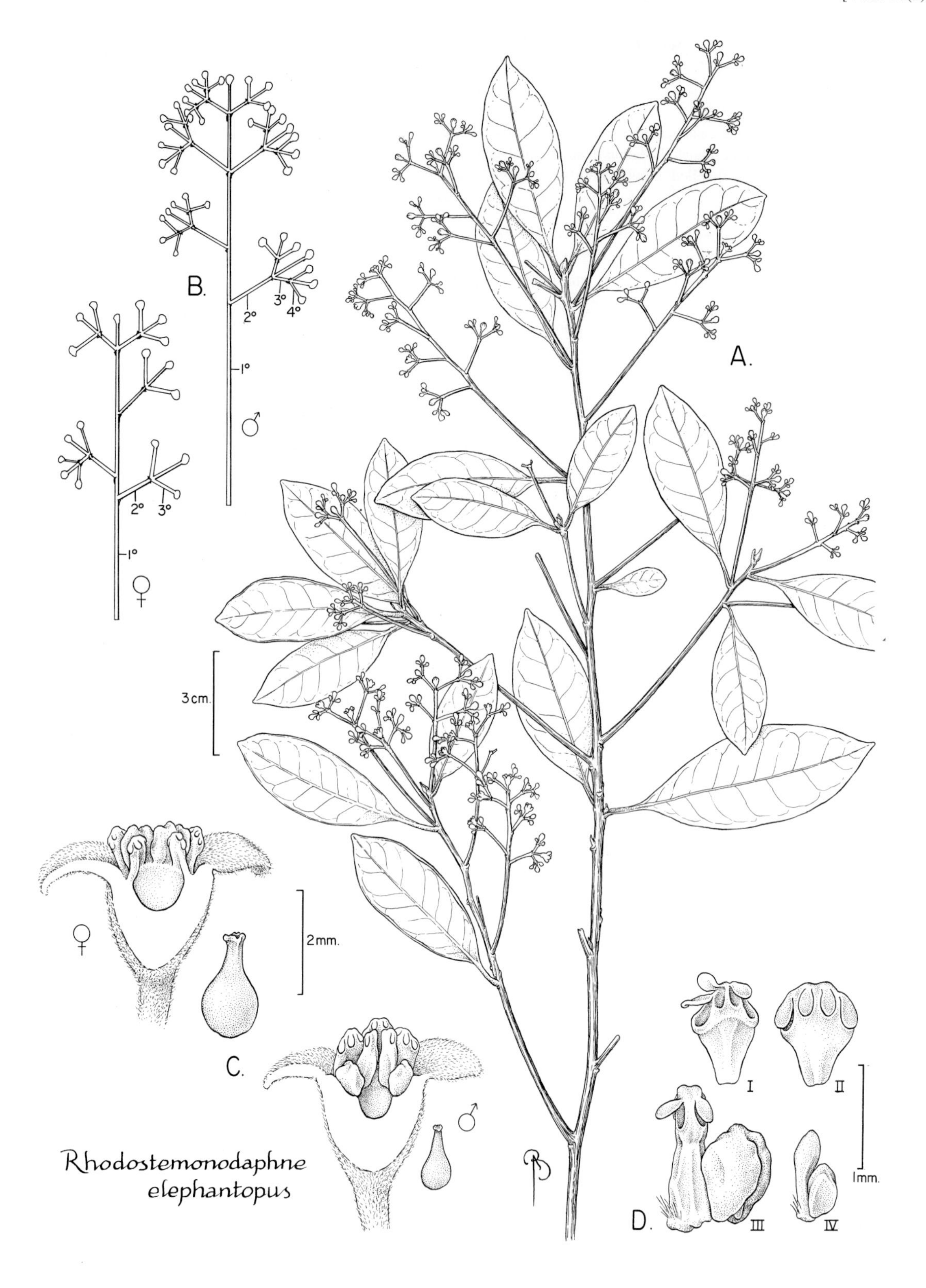

FIG. 158. LAURACEAE. *Rhodostemonodaphne elephantopus* (*Mori et al. 20774*). **A.** Stem with leaves and inflorescences. **B.** Staminate (right) and pistillate (left) inflorescence branching patterns. **C.** Medial section of pistillate flower (above) with pistil removed and shown at right and medial section of staminate flower (below) with pistillode removed and shown at right. **D.** Stamens of different series showing valvate anther dehiscence; note the lower stamens' glandular appendages.

Flowers green when fresh; tepals spreading. Fl (Aug, Sep); rare, in nonflooded moist forest.

Diagnostic are the slender stems, light yellow-green leaves, and few-flowered, reduced inflorescences.

Rhodostemonodaphne morii Madriñán FIG. 159

Medium-sized to large trees, to 35 m tall. Stems brown puberulous, soon glabrescent. Leaves: petioles (40)92–166 × 2–4.6(7.1) mm; blades elliptic, 4–16 × 2–7.1 cm, membranaceous, glabrous, olive green (drying dark green to black) adaxially, pubescent, shiny olive green to slightly glaucous abaxially, the base cuneate, the apex attenuate to acuminate; venation pinnate to slightly tripliveined, the secondary veins in (2)3–4 pairs, the tertiary and higher order veins conspicuous abaxially. Inflorescences pendulous, brown puberulous, profusely branched. Flowers cream-colored when fresh; tepals erect; anthers of outer whorls 2(3)-celled. Fruit cupules smooth to tuberculate, red when fresh, the margin slightly undulate, the tepals persistent. Fruits first green, maturing black. Fl (Sep, Oct), fr (May, Oct, Dec); common, in nonflooded moist forest.

The membranaceous, somewhat glaucous leaves and the erect tepals are characteristic of this species.

Rhodostemonodaphne rufovirgata Madriñán PL. 71b

Large trees, to 40 m tall. Stems reddish-brown tomentose, the hairs peristent and becoming dark brown. Leaves: petioles 13–25 × 1.3–2 mm; blades elliptic to lanceolate, (7.5)10–18(21) × 2.9–6.5 (8.7) cm, coriaceous, reddish-brown tomentose, soon glabrescent, turning dark green adaxially, white pubescent, the veins turning dark brown pubescent abaxially, the base decurrent to cuneate, the apex attenuate to acuminate; venation slightly tripliveined, the secondary veins in 4(5) pairs. Inflorescences erect, reddish-brown tomentose. Flowers yellow-brown when fresh; tepals erect at anthesis. Fl (Sep, Oct); common, in nonflooded moist forest.

Diagnostic are the reddish-brown tomentum on all the vegetative parts and the precocious growth in which all leaves, branches and inflorescences are produced before final stem elongation and lignification.

Rhodostemonodaphne saülensis Madriñán PL. 71d

Medium-sized to large trees, to 42 m tall. Stems silver to golden sericeous, the hairs persistent. Leaves: petioles 15–32 × 1.4–2.4 mm; blades elliptic to lanceolate, (9.7)14–17(21) × (2.8)4.2–6.5(7) cm, membranaceous, glabrous, shiny adaxially, puberulous, slightly glaucous, the veins distinctly reddish when fresh abaxially, the base decurrent and slightly inrolled, the apex attenuate to acuminate; venation tripliveined, acrodromous, the secondary veins in (3)4–5(6) pairs. Inflorescences erect, silver puberulous; pedicels reddish. Flowers: tepals reflexed, reddish, with lighter margins when fresh; stamens red. Fruit cupules with straight margin, red when fresh, the tepals caducous. Fruits yellow-brown, smooth to wrinkled. Fl (Aug, Sep, Oct), fr (Aug); common, in nonflooded moist forest.

This species, collected only in central French Guiana, was previously included as part of the *R. grandis* complex, but it can easily be separated from it by the lower number (3–6) and acrodromous arrangement of the secondary veins and the reddish flowers.

Rhodostemonodaphne sp. 1

Medium-sized trees, to 18 m tall. Stems brown tomentose, the hairs persistent. Leaves: petioles 13–18 × 3–4 mm; blades elliptic, 15–25 × 7–11 cm, coriaceous, glabrous, shiny, brown adaxially, brown tomentose abaxially, the base acute to decurrent, the apex rounded to acuminate, the margins often conspicuously revolute; venation tripliveined, the secondary veins in 3–4 pairs. Fruit cupules tuberculate at base, the margins thin, brown tomentose. Fr (May); collected only once (*Mori & Pennington 18052*, NY), in nonflooded moist forest.

This is a very distinctive individual characterized by brown tomentum on the vegetative parts and the fruits and by the large, broad leaves with revolute margins. The placement of this specimen in *Rhodostemonodaphne* is tentative until flowers are collected.

SEXTONIA van der Werff

Tall trees. Leaves clustered at apices of thick, corky branches; blades without domatia. Flowers bisexual, white; tepals unequal, the 3 outer ones smaller than 3 inner ones, erect at anthesis, not persisting in fruit; stamens 9, 4-celled, the outer 6 with cells arranged in shallow arc, the inner 3 with cells arranged in 2 pairs; staminodia 3; receptacle deep. Fruits drupes, about 1/3 to almost entirely covered by cupule.

Sextonia rubra (Mez) van der Werff [Syn.: *Ocotea rubra* Mez] PL. 70e; PART 1: PL. Ib as *Ocotea rubra*

Trees, to 40 m tall. Stems appressed pubescent when very young, soon becoming glabrous and corky. Leaves clustered at apices of branches; blades obovate, 10–20 × 3–6 cm, firmly chartaceous, glabrous, domatia lacking, the base acute, the apex obtuse or rounded. Inflorescences sparsely appressed pubescent, shorter than leaves. Flowers glabrous, white, fragrant; tepals unequal, the outer three smaller than inner three, erect at anthesis. Fruit cupules thick, cup-shaped, the margin simple, the tepals not persistent. Fl (Aug, Sep), fr (Nov, Dec); occasional, in nonflooded forest.

Diagnostic are the bark shedding in large, irregular plates (Plate Ib in Part 1) and the clustered, obovate leaves.

SPECIES UNIDENTIFIED TO GENUS

Lauraceae sp. I

Trees, 18 m. Stems glabrous. Leaves alternate; blades elliptic, 15–20 × 5–7 cm, glabrous, domatia absent, the base acute or slightly obtuse, the apex acuminate. Fruit cupules large (3 cm diam.), with double margin, the lower margin undulate. Fr (Aug, old); known by a single specimen from our area (*Mori et al. 23320*, NY).

Diagnostic are the large, thick, flat fruit cupules. It is probably an *Ocotea*, but *Endlicheria* is also possible.

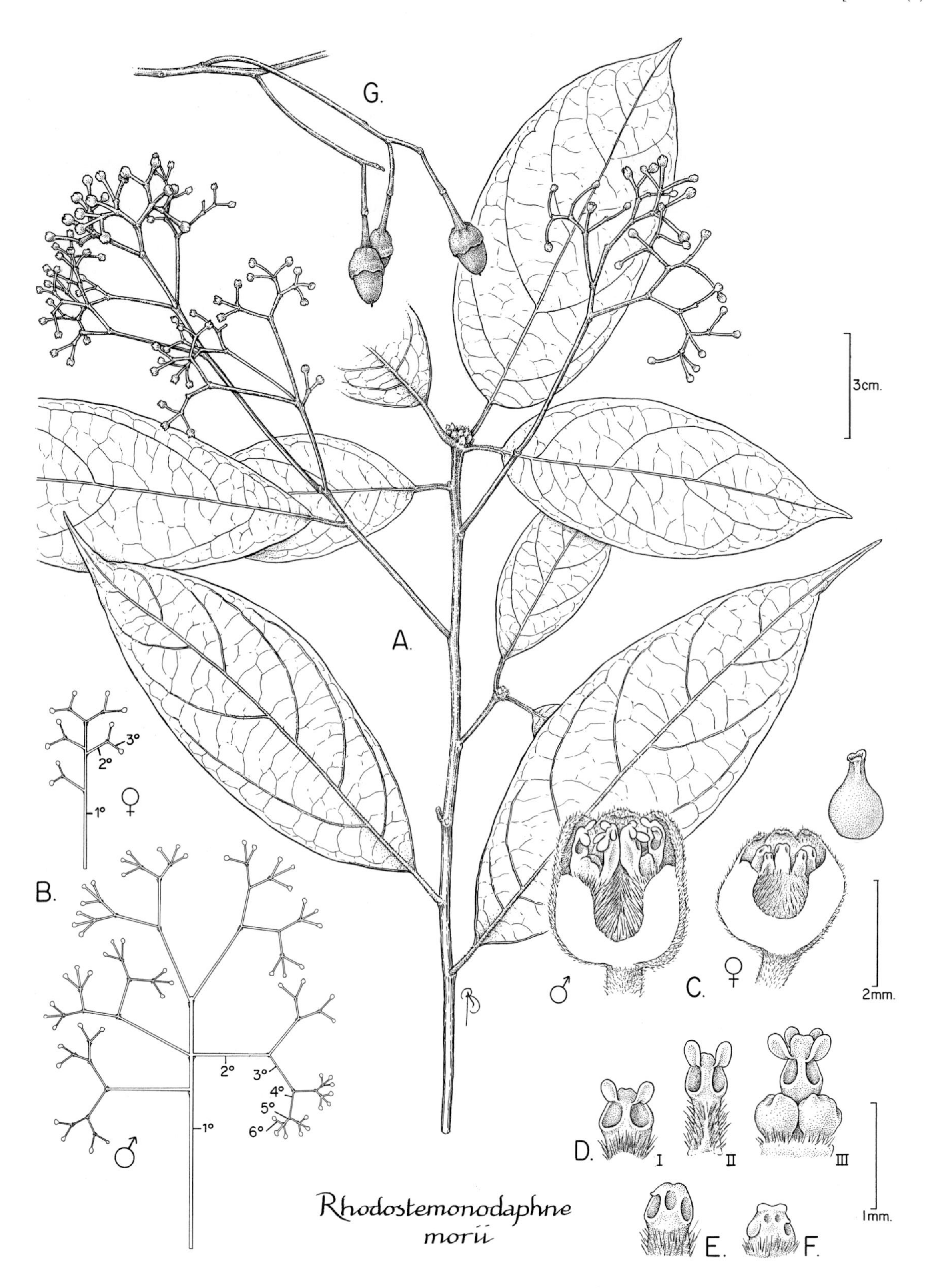

FIG. 159. LAURACEAE. *Rhodostemonodaphne morii* (A, C (staminate flower), D, *Mori & Prance 15044*; B, unvouchered diagram; C (pistillate flower), *Irwin 84564*; E, *Mori & Prance 15031*; F, G, *Mori & Boom 15122*). **A.** Stem with leaves and inflorescences. **B.** Branching patterns of male (above right) and female (below right) inflorescences. **C.** Medial sections of staminate (left) and pistillate (right) flowers with gynoecia removed, fertile pistil shown above. **D.** Fertile stamens of different series showing valvate anther dehiscence and appendages at base of stamen at right. **E.** Staminode. **F.** Staminode. **G.** Branch with three fruits. (Reprinted with permission from S. Madriñan, Brittonia 48(1), 1996.)

LECYTHIDACEAE (Brazil-nut Family)

Scott A. Mori

Trees. Bark usually fibrous. Leaves simple, alternate, large and clustered at branch ends in *Gustavia augusta*, more frequently medium-sized and scattered at branch ends. Stipules usually absent or very small and inconspicuous. Inflorescences terminal, axillary, or cauline, of simple racemes or spikes, paniculate arrangements of racemes or spikes, or infrequently fasciculate. Flowers actinomorphic in *Gustavia*, zygomorphic in remaining genera, bisexual; calyx nearly entire in *Gustavia augusta*, with 6 sepals in remaining species in our area; petals (5)6–8; stamens connate into symmetrical staminal ring in *Gustavia*, the ring prolonged on one side into strap-like structure curving over summit of ovary in remaining genera; ovary inferior, 2–6-locular; placentation axile, the ovules anatropous, inserted at base, middle, or apex of septum or from floor of locule. Fruits indehiscent and berry-like in *Gustavia*, dehiscent, circumscissile, woody capsules in remaining genera, the sepals often persisting or their scars evident as a calycine ring. Seeds with wings surrounding entire seed in *Couratari*, or without wings in remaining genera; aril present or absent, when present lateral or basal; embryo without cotyledons (*Corythophora*, *Eschweilera*, *Lecythis*), with leafy cotyledons (*Couratari*), or with plano-convex cotyledons (*Gustavia*).

Eyma, P. J. 1932. The Polygonaceae, Guttiferae and Lecythidaceae of Surinam. Meded. Bot. Mus. Herb. Rijks Univ. Utrecht **4:** 1–77.

———. 1966. Lecythidaceae. *In* A. Pulle (ed.), Flora of Suriname **III(1):** 119–155. E. J. Brill, Leiden.

Lindeman, J. C. & A. C. de Roon. 1986. Lecythidaceae. *In* A. L. Stoffers & J. C. Lindeman (eds.), Flora of Suriname **III(1–2):** 332–350. E. J. Brill, Leiden.

Mori, S. A. & collaborators. 1987. The Lecythidaceae of a lowland Neotropical forest: La Fumée Mountain, French Guiana. Mem. New York Bot. Gard. **44:** 1–190.

——— & G. T. Prance. 1990. Lecythidaceae — Part II. The zygomorphic-flowered New World genera (*Couroupita*, *Corythophora*, *Bertholletia*, *Couratari*, *Eschweilera*, & *Lecythis*). Fl. Neotrop. Monogr. **21(II):** 1–376.

——— & ———. 1992. Lecythidaceae. *In* A. R. A. Görts-van Rijn (ed.), Flora of the Guianas Ser. A, **12:** 1–144. Koeltz Scientific Books, Koenigstein.

——— & ———. 1999. Lecythidaceae. *In* J. A. Steyermark, P. E. Berry, K. Yatskievych & B. K. Holst (gen. eds.), Flora of the Venezuelan Guayana **5:** 750–779. P. E. Berry, K. Yatskievych & B. K. Holst (vol. eds.). Missouri Botanical Garden Press, St. Louis.

Prance, G. T. & S. A. Mori. 1979. Lecythidaceae - Part I. The actinomorphic-flowered New World Lecythidaceae (*Asteranthos*, *Gustavia*, *Grias*, *Allantoma*, & *Grias*). Fl. Neotrop. Monogr. **21:** 1–270.

1. Flowers actinomorphic, the anthers dehiscing by apical pores. Fruits indehiscent. . . . . *Gustavia.*
1. Flowers zygomorphic, the anthers dehiscing by lateral slits. Fruits dehiscent.
    2. Androecial hood with external flap. Ovary 3-locular. Seeds surrounded by wing. . . . . *Couratari.*
    2. Androecial hood without external flap. Ovary 2- or 4-locular. Seeds without wing.
        3. Ovaries 2-locular.
            4. Androecial hood flat, dorsiventrally expanded. Fruits campanulate. Seeds with basal aril. . . . *Corythophora.*
            4. Androecial hood not flat, forming at least one inward coil. Fruits globose, turbinate, or cup-shaped. Seeds usually with lateral aril. . . . . *Eschweilera.*
        3. Ovaries (3)4(5)-locular.
            5. Androecial hood forming complete coil. Seeds with lateral aril. . . . . *Eschweilera.*
            5. Androecial hood flat or forming partial coil. Seeds with basal aril. . . . . *Lecythis.*

CORYTHOPHORA R. Knuth

Canopy trees. Inflorescences terminal or axillary, racemes or paniculate arrangements of racemes. Flowers zygomorphic; sepals 6; petals 6; anthers dehiscing by lateral slits, the androecial hood dorsiventrally expanded, appendages not curved inwards; ovary usually 2-locular, the ovules attached to floor of locule or lower part of septum, the style not well-differentiated from summit of ovary. Fruits woody, dehiscent, campanulate. Seeds many per fruit, elongate; aril basal; embryo with undifferentiated cotyledons.

1. Bark not deeply fissured. Hypanthium ferruginous, squamulose, contrasting with dark brown to blackish sepals, the lobes well developed, greater than 2 mm wide; androecial hood markedly dorsiventrally expanded, the appendages fused together for most of length, antherless or with a few vestigial anthers; staminal ring with 194–230 anthers. . . . . *C. amapaensis.*
1. Bark deeply fissured. Hypanthium glabrous, smooth, same color and texture as sepals, the lobes poorly developed, <2 mm wide; androecial hood only slightly dorsiventrally expanded, appendages free, with anthers; staminal ring with 50–70 anthers. . . . . *C. rimosa* subsp. *rubra.*

Corythophora amapaensis S. A. Mori & Prance
FIG. 160, PL. 72 a,c; PART 1: FIG. 10

Trunks unbuttressed. Bark brown, nearly smooth, with shallow, vertical cracks, the outer bark thinner than inner bark. Leaves: petioles 8–20 mm long; blades elliptic to oblong, 14–26 × 5–10.5 cm, the margins crenulate; secondary veins in 17–23 pairs. Flowers 3.5–4 cm diam.; petals pinkish-red to dark red or purple; staminal ring with 194–230 anthers, the androecial hood flat, dorsiventrally expanded, dark red or purple, the appendages nearly entirely fused, antherless. Fruits 6–15 × 6–10 cm (excluding operculum), the pericarp 13–27 mm thick. Fl (Jul, Aug, Sep, Oct, Nov), fr (Dec, Apr);

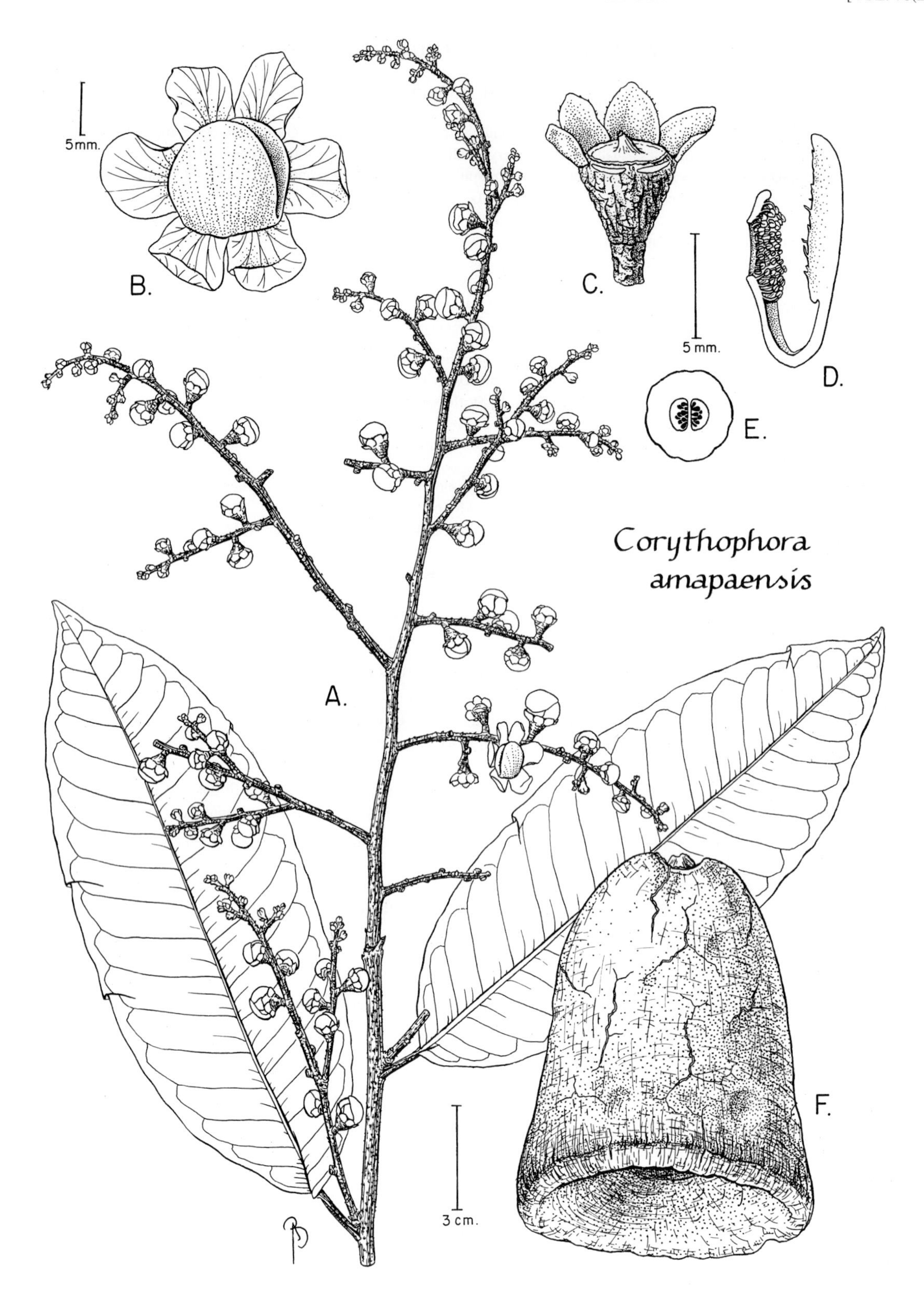

FIG. 160. LECYTHIDACEAE. *Corythophora amapaensis* (A, *Maguire et al. 47121* from Amapá, Brazil; B–F, *Cowan 38313* from Amapá, Brazil). **A.** Part of stem with leaves and inflorescences. **B.** Apical view of flower. **C.** Lateral view of flower with two sepals, perianth, and androecium removed. **D.** Medial section of androecium. **E.** Transverse section of ovary. **F.** Base of fruit. (Reprinted from S. A. Mori, Brittonia 33(3), 1981.)

occasional, in nonflooded forest on well-drained soil but usually in lower areas.

Corythophora rimosa W. A. Rodrigues subsp. **rubra** S. A. Mori PL. 72b,d; PART 1: PL. Vb

Trunks cylindric to base. Bark brown, with deep vertical fissures, the outer bark thicker than inner bark, the inner bark reddish. Leaves: petioles 7–19 mm long; blades elliptic, 4–6 × 2–9 cm, the margins undulate, minutely crenate; secondary veins in 12–18 pairs. Flowers ca. 2.5 cm diam.; petals red; staminal ring with 50–70 anthers, the androecial hood flat, only slightly expanded dorsiventrally, white, the appendages well developed, with yellow anthers. Fruits 8–2 × 8–11.5 cm (excluding operculum), the pericarp 18–28 mm diam. Fl (Jul, Aug, Sep), fr (Aug, Oct, Nov, Dec); relatively common in nonflooded forest, especially along ridges.

COURATARI Aubl.

Emergent or canopy trees. Inflorescences terminal or axillary, racemes or paniculate arrangments of racemes. Flowers zygomorphic, appearing with or without leaves; sepals 6; petals 6; anthers dehiscing by lateral slits, the androecial hood with well developed inner coil and extra external flap; ovary 3-locular, the style very short; ovules attached to junction of septum with floor of locule. Fruits dehiscent, cylindric or campanulate, the pericarp relatively thin-walled. Seeds many per fruit, surrounded by membranous wing; aril absent; embyro with leaf-like cotyledons.

1. Leaves entirely glabrous; blades ≤8.5 cm long. Flowering when leaves absent.
   2. Leaves: petioles slender, 12–20 mm long; blades with base usually acute to attenuate. Fruits 4–6 cm long (excluding stipe). . . . . *C. multiflora.*
   2. Leaves: petioles stout, 7–12 mm long; blades with base usually obtuse to rounded. Fruits 7–9 cm long (excluding stipe). . . . . *C. oblongifolia.*
1. Leaves with stellate hairs abaxially, especially along veins; blades ≥8 cm long. Flowering when leaves present or absent.
   3. Leaf blades with secondary veins nearly plane abaxially, pubescent mostly on midrib. Flowers with white petals and yellow hood; androecial hood echinate. Flowering when leaves present. . . . . *C. stellata.*
   3. Leaf blades with secondary veins very salient abaxially, pubescent throughout. Flowers, at least the hood, pink or purple; androecial hood not echinate. Flowering when leaves absent.
      4. Leaf blades with secondary veins impressed adaxially. Flowers entirely pink; stamens 50–65. Plants only found along streams. . . . . *C. gloriosa.*
      4. Leaf blades with secondary veins plane adaxially. Flowers with hood pink and petals greenish toward base and white toward apex; stamens 15–25. Plants in nonflooded moist forest. . . . . C. *guianensis.*

Couratari gloriosa Sandwith

Canopy trees. Trunks with low buttresses. Leaves: petioles stout, 15–20 mm long; blades elliptic or obovate-elliptic, 14–17 × 7.5–16 cm, with scattered stellate hairs over veins and surface abaxially, the base subcordate; secondary veins in 19–28 pairs, impressed adaxailly, very salient abaxially. Flowers when leaves absent; petals purple; androecial hood ca. 40 mm diam., not echinate, purple, the staminal ring with 50–65 stamens. Fruits cylindrical-campanulate, 8–9 (excluding stipe) × 3–3.8 cm, with persistent, slender pedicel. Fl (Feb), fr (Apr, Dec); locally common, found only along streams.

Couratari guianensis Aubl. emend. Prance PLS. 72e, 73a,b

Canopy trees. Trunks with well developed buttresses. Leaves: petioles stout, 15–25 mm long; blades obovate-oblong to elliptic, 8–19 × 4–10 cm, densely covered by stellate hairs over entire surface abaxially, the base rounded; secondary veins in 16–22 pairs, plane adaxially, very salient abaxially. Flowers when leaves absent; petals greenish toward base and white toward apex; androecial hood ca. 30 mm diam., not echinate, pink, the staminal ring with 15–25 stamens. Fruits cylindric, 12–17 (excluding stipe) × 6 cm, the pericarp usually markedly lenticellate. Fl (Aug, Sep), fr (empty fruits May, Sep, Dec); occasional, in nonflooded forest. *Ingui pipa*, *mahot cigare* (Créole).

Couratari multiflora (Sm.) Eyma

Canopy or emergent trees. Trunks unbuttressed or with broad flat buttresses. Leaves: petioles slender, 12–20 mm long; blades usually elliptic to oblong, 4.5–8 × 2–5 cm, entirely glabrous, the base usually acute to attenuate; secondary veins in 6–8 pairs. Flowers when leaves absent; petals pink; androecial hood ca. 10 mm diam., not echinate, pink, the staminal ring with 30–35 stamens. Fruits cylindrical, 4–6 (excluding stipe) × 2–3.5 cm, the pericarp thinner than in other species. Fl (Aug, apparently toward end of rainy season), fr (Sep); occasional, in nonflooded forest. *Ingui pipa*, *mahot cigare* (Créole), *tauari* (Portuguese).

Couratari oblongifolia Ducke & R. Knuth

Canopy or emergent trees. Trunks buttressed. Leaves: petioles stout, 7–12 mm long; blades elliptic to oblong, 6.5–8.5 × 3.5–4.5 cm, entirely glabrous, the base usually obtuse to rounded; secondary veins in 10–14 pairs. Flowers when leaves absent; petals pink; andreocial hood not echinate, the staminal ring with 10–14 stamens. Fruits cylindrical, 7–9 (excluding stipe) × 2.5–3.5 cm. Fr (Oct); rare, known only from two collections from nonflooded moist in our area.

Couratari stellata A. C. Sm FIG. 161, PL. 73c–e; PART 1: FIGS. 7, 10

Canopy to emergent trees. Trunks usually with well developed buttresses. Leaves: petioles stout, 5–12 mm long; blades elliptic to oblong-elliptic, 5–14 × 2.5–8 cm, longitudinal striations parallel to primary vein, with stellate hairs mostly restricted to surface of midrib abaxially, the base obtuse; secondary veins in 11–22 pairs.

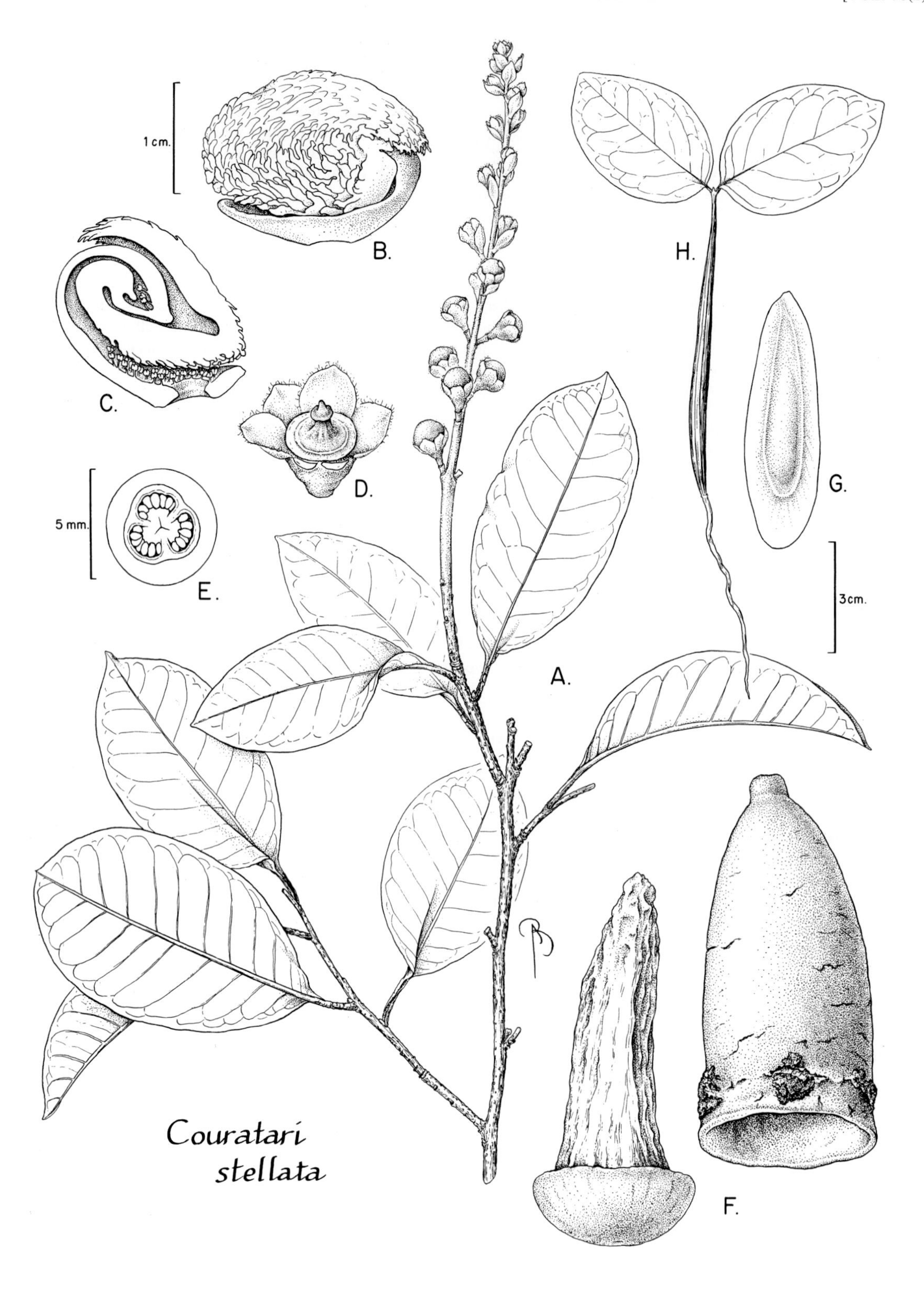

FIG. 161. LECYTHIDACEAE. *Couratari stellata* (A, *Heyde & Lindeman 23* from Surinam; B–E, *de Granville B.5380*; F, *Mori & Pipoly 15454*). **A.** Part of stem with leaves and inflorescence. **B.** Lateral view of androecium; note echinate surface and extra external flap. **C.** Medial section of androecium. **D.** Apical-oblique view of flower with two sepals, perianth, and androecium removed. **E.** Transverse section of ovary. **F.** Base of fruit (above) and operculum (left). **G.** Winged seed; note that the wing completely surrounds the seed. **H.** Seedling; note the somewhat winged stem. (Reprinted from S. A. Mori & G. T. Prance, Fl. Neotrop. Monogr. 21(2), 1990.)

Flowers when leaves present; petals usually white; androecial hood 15–20 mm diam., echinate externally, yellow, the staminal ring with 40–45 stamens. Fruits campanulate, 5–11 (excluding stipe) × 3–5 cm. Fl (Oct, Nov, Dec, Feb), fr (Mar, Apr, Dec); very common, in nonflooded moist forest.

ESCHWEILERA DC.

Understory, canopy, or emergent trees. Inflorescences terminal, axillary, or cauline, racemes, spikes, or paniculate arrangments of racemes or spikes. Flowers zygomorphic; sepals 6; petals (5)6; anthers dehiscing by lateral slits, the androecial hood coiled inwards, the innermost appendages modified into nectaries; ovary 2–4-locular; ovules attached to floor of locule. Fruits woody, dehiscent. Seeds 1 to many per fruit; aril usually lateral, absent or surrounding seed outside our area; embryo with undifferentiated cotyledons.

1. Petals 5(6), green. . . . . *E. piresii* subsp. *viridipetala.*
1. Petals 6, not green.
    2. Pedicels absent, the flowers tightly congested (i.e., with no space between flowers). . . . . *E. simiorum.*
    2. Pedicels present, or if absent, the flowers not tightly congested.
        3. Sepals usually 7–14 mm wide, the apex rounded. Bruised flowers, fruits, and seeds turning bluish-green.
            4. Bark not scalloped. Sepals as long as wide, the margins not undulate. Fruits truncate directly below calycine ring, the pericarp rough. . . . . *E. decolorans.*
            4. Bark scalloped. Sepals longer than wide, the margins undulate. Fruits tapered from calycine ring to pedicel, the pericarp smooth. . . . . *E. laevicarpa.*
        3. Sepals usually ≤7 mm wide, the apex acute or obtuse or 10–20 mm wide, the apex rounded (*E. grandiflora*). Bruised flowers, fruits, and seeds not turning bluish-green.
            5. Petals and androecial hood both the same shade of yellow. Fruits usually <3 cm diam. . . . . *E. parviflora.*
            5. Petals white or pink, androecial hood yellow or white tinged with pink or petals light yellow and hood darker yellow. Fruits usually >3 cm diam. (*E. apiculata* may have smaller fruits).
                6. Flowers ≤3 cm diam.
                    7. Bark deeply fissured, peeling in long, rectangular plates. Pedicels absent. . . . . *E. squamata.*
                    7. Bark not deeply furrowed nor peeling in rectangular plates. Pedicels present.
                        8. Bark not scalloped, nearly smooth, with well-defined hoop marks. Inflorescences unbranched, the pedicels 15–25 mm long. Fruits much wider than long. . . . . *E. collina.*
                        8. Bark scalloped, without hoop marks. Inflorescences branched, the pedicels 5–10 mm long. Fruits as narrow or narrow than long.
                            9. Inner bark bright pink. Largest leaf blades ≤10 cm long. Sepals often with transverse crease at base; androecial hood without crease between hood proper and anterior most appendages. Base of fruit attenuate. . . . . *E. apiculata.*
                            9. Inner bark yellowish or orangish-brown. Largest leaf blades usually >10 cm long. Sepals without transverse crease at base; androecial hood with marginal crease separating hood proper from anterior most appendages. Base of fruit rounded.
                                10. Leaves: petioles not penetrated by secretory canal; blades chartaceous. Pedicels slender. . . . . *E. micrantha.*
                                10. Leaves: petioles penetrated by secretory canal; blades coriaceous. Pedicels robust. . . . . *E. sagotiana.*
                6. Flowers usually >3 cm diam.
                    11. Sepals 10–20 mm wide, imbricate for most of length; style 4–8 mm long, oblique. . . . . *E. grandiflora.*
                    11. Sepals <10–20 mm wide, not imbricate or imbricate for <1/2 length; style <4 mm long, erect.
                        12. Trees of canopy, mostly >20 m tall at reproductive age. Trunks buttressed. Pedicels pubescent. Fruits globose, abruptly constricted below calycine ring, then rounded to base. . . . . *E. coriacea.*
                        12. Trees of understory, the largest to 22 m tall at reproductive age. Trunks not buttressed. Pedicels glabrous. Fruits turbinate, tapered from calycine ring to base.
                            13. Petals and androecial hood white, flushed with pink to entirely pink; sepals 4–7 mm wide. Fruits broadly turbinate. . . . . *E. pedicellata.*
                            13. Petals light yellow, androecial hood bright yellow; sepals 3 mm wide. Fruits narrowly turbinate. . . . . *E. chartaceifolia.*

Eschweilera apiculata (Miers) A. C. Sm.

Canopy trees. Trunks with cylindric, unbuttressed, sometimes with slight fluting toward base. Bark peeling in irregular plates, scalloped, the outer bark ca. 1 mm thick, the inner bark ca. 6 mm thick, bright pink when first cut. Leaves: petioles 5–12 mm long; blades usually elliptic, 6.5–10 × 3–6 cm; secondary veins in 7–11 pairs. Inflorescences paniculate arrangements of racemes; pedicels 3–8 mm long, puberulous. Flowers 2–2.5 cm diam.; sepals thick, 2–3.5 × 2–2.5 mm, not or only slightly imbricate, often with transverse crease at base; petals white; androecial hood yellow; ovary 2-locular. Fruits small, turbinate, 1.5–2.5 (excluding operculum and pedicel) × 2–3.5 cm, the base attenuate. Fl (Sep, Oct); fr (Apr, Nov); occasional, in nonflooded forest.

Eschweilera chartaceifolia S. A. Mori

Understory trees. Bark with vertically oriented lines of lenticels, the inner bark red. Leaves: petioles 4–6 mm long; blades elliptic to narrowly elliptic, 11–13 × 3.5–5 cm, chartaceous, the margins entire to serrulate toward apex; secondary veins in 8–10 pairs. Inflorescences racemes or weakly branched paniculate arrangements of racemes, 1–5 cm long; pedicels 7–10 mm long, glabrous. Flowers ca. 3 cm diam.; sepals 5 × 3 mm; petals light yellow; androecial hood bright yellow. Fruits narrowly turbinate, tapered from calycine ring to base, 3–5 × 4–5 cm, the operculum convex. Fl (Aug), fr (old Oct); rare, in nonflooded forest.

Eschweilera collina Eyma — PL. 74a,b

Canopy trees. Trunks cylindric. Bark nearly smooth, with occasional vertical cracks and well defined hoop marks, the outer bark bark thinner than inner bark. Leaves: petioles 7–12 mm long; blades widely elliptic to elliptic, 8–14(20) × 4–9 cm; secondary veins in 7–10 pairs, salient on both surfaces in dried leaves. Inflorescences racemes; pedicels 15–25 mm long, tapered into hypanthium, glabrous. Flowers ca. 3 cm diam.; sepals 2–4 × 3–5.5 mm, separate or only slightly imbricate; petals white; androecial hood yellow; ovary 2-locular. Fruits depressed turbinate, much wider than long, 2–6 (excluding operculum) × 3.5–6 cm, the calycine ring prominent, the operculum umbonate. Fl (Aug), fr (May, Aug, Sep); occasional, in nonflooded forest.

Eschweilera coriacea (DC.) S. A. Mori — PL. 74c

Canopy trees. Trunks usually with well developed buttresses. Bark brown to very dark brown, nearly black, ± smooth, not markedly fissured, peeling in irregular plates but not leaving dipple marks, the outer bark 0.5–1 mm thick, the inner bark 5–7 mm thick, yellowish-white. Leaves: petioles 8–12 mm long; blades elliptic to narrowly elliptic or narrowly obovate to oblanceolate, 10–26 × 4.5–9 cm; secondary veins in 10–15 pairs. Inflorescences usually paniculate arrangements of racemes; pedicels 10–22 mm long, pubescent. Flowers 3.5–5 cm diam.; sepals 4–9 × 3.5–7 mm, often tinged with red, the base gibbous, scarcely imbricate; petals white; androecial hood yellow; ovary 2-locular. Fruits depressed globose, often abruptly contracted directly below calycine ring, then rounded to base, 3–4(excluding operculum and pedicel) × 5–8 cm. Fl (Sep, Oct), fr (Feb, Mar, Apr, Jun, Jul); common, in nonflooded forest, especially in valley bottoms and lower on hillsides. *Mahot noir* (Créole, referring to its dark colored outer bark).

Eschweilera decolorans Sandwith

Canopy trees. Trunks unbuttressed or with slightly developed buttresses. Bark nearly smooth but sometimes with inconspicuous vertical cracks and vertically oriented lenticels, the outer bark 0.5–1 mm thick, the inner bark 7–14 mm thick, yellow. Leaves: petioles 6–12 mm long; blades widely elliptic to elliptic, 10–17 × 5–8.5 cm; secondary veins in 9–14 pairs. Inflorescences racemes; pedicels 10–15 mm long, glabrous. Flowers bruising bluish-green, drying nearly black, ca. 5 cm diam.; sepals 8–10 × 6–8 mm, strongly imbricate, as long as wide; petals white; androecial hood light yellow; ovary 3–5-locular. Fruits depressed globose, 4–5 × 7–8 cm, truncate below calycine ring, the pericarp rough, turning bluish-green when bruised. Fl (Aug, Sep), fr (Nov); occasional, in nonflooded forest.

Eschweilera grandiflora (Aubl.) Sandwith — PL. 74d,e

Understory trees. Trunks unbuttressed. Bark ± smooth, with conspicuous, vertically oriented lenticels, both outer and inner barks thin. Leaves: petioles 9–30 mm long; blades elliptic or narrowly oblong to oblong, 10–35 × 5–14 cm, often with longitudinal striations parallel to midrib; secondary veins in 12–19 pairs. Inflorescences racemes, the rachis angular, often zig-zagged; pedicels ca. 10 mm long, glabrous. Flowers 4.5–6 cm diam.; sepals 12–24 × 11–20 mm, the bases imbricate for most of length; petals white tinged with pink; androecial hood white or yellow on outside, yellow on inside; ovary 4-locular, the style well developed, oblique, 4–8 mm long. Fruits depressed globose, 2.5–4 × 5–6 cm, with persistent sepals attached near base. Fl (Jan, Sep, Oct, Nov, Dec), fr (Feb, immature); occasional, in nonflooded forest. *Weti loabi* (Bush Negro).

Eschweilera laevicarpa S. A. Mori

Canopy trees. Bark gray, smooth, scalloped, the irregular depressions left by sloughing of bark, the inner bark <1 mm thick, the inner bark 7–10 mm thick, light yellow. Leaves: petiole 5–10 mm long; blades elliptic, 9–14 × 4.5–6.5 cm; secondary veins in 9–11 pairs. Inflorescences racemes; pedicels ca. 8 mm long, glabrous. Flowers turning bluish-green when bruised; sepals 12–17 × 11–14 mm, imbricate, longer than wide, the margins undulate; petals white; androecial hood white outside, yellow inside; ovary 2-locular. Fruits cup-shaped, tapered from calycine ring to pedicel, turning bluish-green when bruised. Fl (Aug, Sep, Oct), fr (Jan, Oct, Dec); occasional, in nonflooded moist forest.

Eschweilera micrantha (O. Berg) Miers

Canopy trees. Trunks usually unbuttressed, sometimes slightly fluted toward base. Bark gray to dark brown, scalloped, the outer bark thinner than inner bark, the inner bark yellowish or orangish-brown. Leaves: petioles 5–9 mm long; blades elliptic, infrequently narrowly ovate, 10–21 × 4–7 cm, chartaceous; secondary veins in 9–12 pairs. Inflorescences paniculate arrangements of racemes; pedicels slender, 5–9 mm long, pubescent. Flowers 1.5–2.5 cm diam.; sepals 1.5–3 × 1.4–2.5 mm, the bases scarcely imbricate; petals white; androecial hood yellow, with crease separating hood proper from anterior most appendages; ovary 2-locular. Fruits cup-shaped, 2–3 (excluding operculum) × 4–5 cm, the base rounded. Fl (Oct, Nov), fr (Mar, Apr); relatively common, in nonflooded forest, especially along ridges. *Mahot blanc* (Créole), *weti loabi* (Bush Negro).

Eschweilera parviflora (Aubl.) Miers — PL. 74f

Understory or, less frequently, canopy trees. Trunks usually unbuttressed. Bark nearly smooth, sometimes dippled, often with vertically disposed lenticels and very shallow cracks, often with maroon cast, the outer bark thinner than the white to yellowish-white inner bark. Leaves: petioles 5–10 mm long; blades elliptic, 7.5–15.5 × 2.5–7.5 cm; secondary veins in 10–13 pairs. Inflorescences usually paniculate arrangements of racemes; pedicels 8–13 mm long, puberulous to pubescent. Flowers 2–2.5 cm diam., petals and hood both same shade of yellow; sepals 2.5–6 × 2–5.5 mm, the bases imbricate; petals yellow; androecial hood yellow; ovary 2-locular. Fruits smallest of family in area, 1.5–2 (excluding operculum) × 2–3 cm. Fl (Nov, Dec), fr (Apr, May, Sep); occasional, in nonflooded forest.

Eschweilera pedicellata (Rich.) S. A. Mori FIG. 162, PL. 75a

Understory trees. Trunks unbuttressed. Bark smooth, with vertically oriented lenticels, often with reddish or maroon cast, the outer bark ca. 1 mm thick, the inner bark 2–3 mm thick, yellowish-white. Leaves: petioles 8–10 mm long; blades usually elliptic to narrowly elliptic, 9–16 × 3.5–7 cm; secondary veins in 8–13 pairs. Inflorescences racemes; pedicels 15–30 mm long, glabrous. Flowers 3.5–5 cm diam.; sepals 6–11 × 4–7 mm, longer than wide, the bases imbricate for short distance; petals white tinged with pink to entirely pink; androecial hood white or yellow tinged with pink; ovary 2-locular. Fruits depressed turbinate to turbinate, tapered from calycine ring to base, 2–4 × 2.5–5 cm. Fl (Aug, Sep, Oct, Nov), fr (Mar, Aug, Oct, Nov, Dec); relatively common, in nonflooded forest. *Baikaaki* (Bush Negro), *mahot noir* (Créole).

Eschweilera piresii S. A. Mori subsp. **viridipetala** S. A. Mori FIG. 163

Canopy trees. Trunks with steep, thick buttresses. Bark fissured, the outer bark 1–2 mm thick, the inner bark 10 mm thick, red. Leaves: petioles 5 mm long; blades widely elliptic, 5–7 × 3–5 cm; secondary veins in 8–11 pairs. Inflorescences spikes; pedicels poorly developed. Flowers 1.5–2 cm diam.; petals 5(6), green; androecial hood white. Fruits broadly cup-shaped, the largest 2.5 (excluding operculum) × 4.5 cm. Fl (Sep); rare, in nonflooded forest.

Eschweilera sagotiana Miers

Canopy trees. Trunks unbuttressed or with poorly developed, low buttresses. Bark with small depressions, the inner bark yellowish or orangish-brown. Leaves: petioles 8–15 mm long; blades elliptic, 10.5–22 × 4–10 cm, coriaceous; secondary veins in 10–15 pairs. Inflorescences racemes; pedicels robust, 5–12 mm long, usually puberulous, penetrated by secretory canal. Flowers 2.5 cm diam.; sepals 3.5–4.5 × 2.5–3.5 mm, the bases not or scarcely imbricate; petals white; androecial hood yellow, with marginal crease separating hood proper from anterior most appendages; ovary 2-locular. Fruits cup-shaped, 2–3 (excluding operculum) × 4–5.5 cm, the base rounded. Fl (Sep, Nov, Dec), fr (Apr, Oct, Nov); rare, in nonflooded forest.

Eschweilera simiorum (Benoist) Eyma

Understory trees. Bark reddish-brown, with abundant, vertically oriented lenticels, the outer bark ca. 1 mm thick, the inner bark 2 mm thick, white. Leaves: petioles 15–21 mm long, glabrous; blades elliptic to narrowly elliptic or oblong to narrowly oblong, 21–34 × 8.5–15 cm; secondary veins in 15–22 pairs. Inflorescences spikes, the flowers tightly congested (i.e., no space between flowers); pedicels absent. Flowers ca. 7 cm diam.; sepals 11–16 × 12–19 mm, the bases stongly imbricate; petals white with tinges of rose or purple; hood of androecium yellow; ovary 4-locular. Fruits globose to turbinate or cylindric, with persistent, enlarged, but scarcely lignified sepals, 5–7 × 5–6 cm. Fr (May, Dec); rare, in nonflooded forest. *Man tapouhoupa* (Bush Negro).

Eschweilera squamata S. A. Mori PL. 75b; PART 1: PL. VIb

Canopy to emergent trees. Trunks with steep, high buttresses. Bark distinctly fissured, peeling in long, rectangular plates, the outer bark 5 mm thick, the inner bark 10 mm thick, grayish-yellow. Leaves: petioles 10–25 mm long; blades elliptic, 16–23 × 7–11 cm; secondary veins in 13–14 pairs. Inflorescences spikes or paniculate arrangements of spikes; pedicels absent. Flowers ca. 1.5 cm diam.; sepals 1–1.3 × 2–2.7 mm, the bases not imbricate; petals white; androecial hood yellow; ovary 2-locular. Fruits depressed globose, 4–6.5 (excluding operculum) × 5–11 cm. Fl (Aug, Nov, Dec), fr (Aug, Dec); occasional, in nonflooded forest.

GUSTAVIA L.

Understory trees. Inflorescences terminal, axillary, or cauline, of solitary flowers or racemes. Flowers actinomorphic; calyx entire or with 4–6 distinct sepals; petals 6–8; stamens >500, the filaments fused at bases into ring, free at apices, the anthers dehiscing by apical pores; ovary 4–6-locular; ovules >10, attached to expanded apical-axile placenta. Fruits indehiscent, ± globose. Seeds >1 per fruit in our area, without aril; funicle well developed, contorted and yellow or poorly developed and white; embryo with plano-convex cotyledons.

1. Largest leaf blades >24 cm long. Petals and androecium flushed with pink; calyx entire to weakly 4-lobed; ovary smooth. Seeds with yellow, contorted funicle. . . . . *G. augusta.*
1. Largest leaf blades ≤24 cm long. Petals and androecium entirely white or white flushed with yellow; calyx with 6 distinct sepals; ovary winged. Seeds without yellow, contorted funicle. . . . . *G. hexapetala.*

Gustavia augusta L. PL. 75d

Understory trees. Trunks not buttressed. Stems relatively thick, 3–9 mm diam., the leaves clustered at ends. Leaves: petioles absent to 40 mm long; blades narrowly obovate to oblanceolate, 16–48 × 4–13 cm; secondary veins in 14–22 pairs. Inflorescences terminal, axillary, or cauline, racemes. Flowers 9–20 cm diam.; calyx nearly entire, sometimes with 4 poorly developed rounded lobes; petals usually 8, white flushed with pink; filaments white at base, pink towards apex; ovary not winged. Fruits 3–7 (excluding pedicel) × 3–8 cm, without wings, without persistent sepals, light yellow at maturity. Seeds black, with yellow, contorted funicle. Fl (Nov, Jan), fr (Feb, Mar, Aug, Dec); infrequent, in nonflooded forest, common along rivers in other parts of its range. *Bois pian*, *camaca*, *cona-da-cona-dou* (Créole), *jeniparana*, *mau tapouhoupa*, *mucurão* (Portuguese).

Gustavia hexapetala (Aubl.) Sm. FIG. 164, PL. 75f; PART 1: FIG. 10

Understory trees. Trunks not buttressed. Stems 3–5 mm diam., the leaves scattered. Bark distinctly dippled. Leaves: petioles 2–17 mm

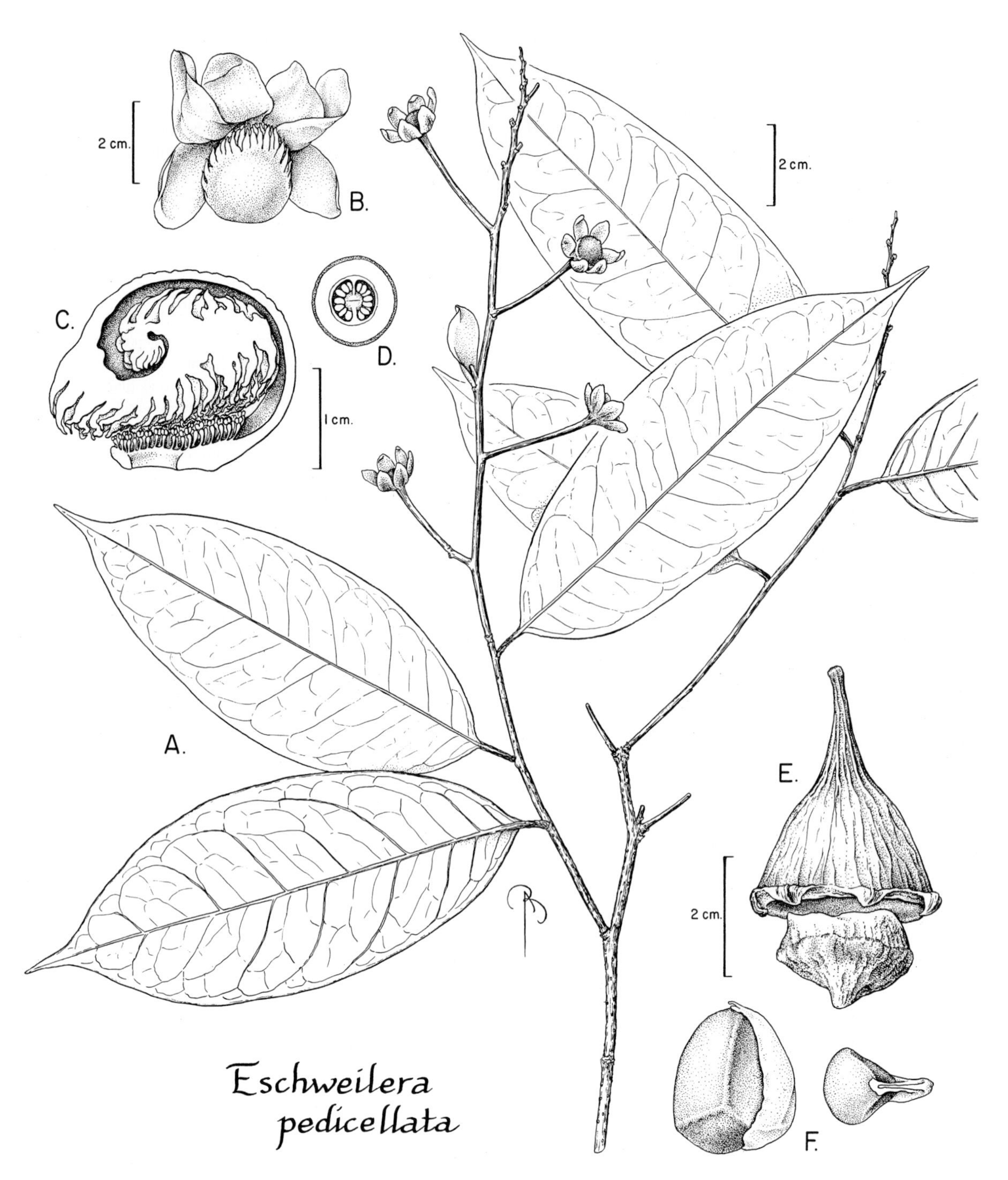

FIG. 162. LECYTHIDACEAE. *Eschweilera pedicellata* (A, *Mori & Boom 14817*; B, *Mori & Bolten 8432* from Surinam; C, D, *Mori & Bolten 8613* from Surinam; E, *Mori & Boom 15180*; F, *Mori & Bolten 8623* from Surinam). **A.** Part of stem with leaves and inflorescence. **B.** Apical view of flower. **C.** Medial section of androecium. **D.** Transverse section of ovary. **E.** Fruit base and operculum. **F.** Lateral view of seed and aril (left) and apical view of seed and aril (right). (Reprinted from S. A. Mori & G. T. Prance, Fl. Neotrop. Monogr. 21(2), 1990.)

long; blades usually oblanceolate or obovate, 9–24 × 3–13 cm; secondary veins in 9–13 pairs. Inflorescences terminal, racemes. Flowers 6–9 cm diam.; calyx with 6, well developed, triangular sepals; petals 6, white, flushed with yellow internally; filaments white, the staminal ring tinged with yellow on inside; ovary winged. Fruits 1–3 × 1–3.5 cm, markedly winged, with persistent sepals, yellow-orange at matuity. Seeds brown, with straight, poorly developed funicle. Fl (Jan, Apr, Jul, Aug, Sep, Dec), fr (Jan, Apr, May, Jul, Sep); very common, in nonflooded forest. *Pois puant* (Créole), *tapouhoupa* (Bush Negro).

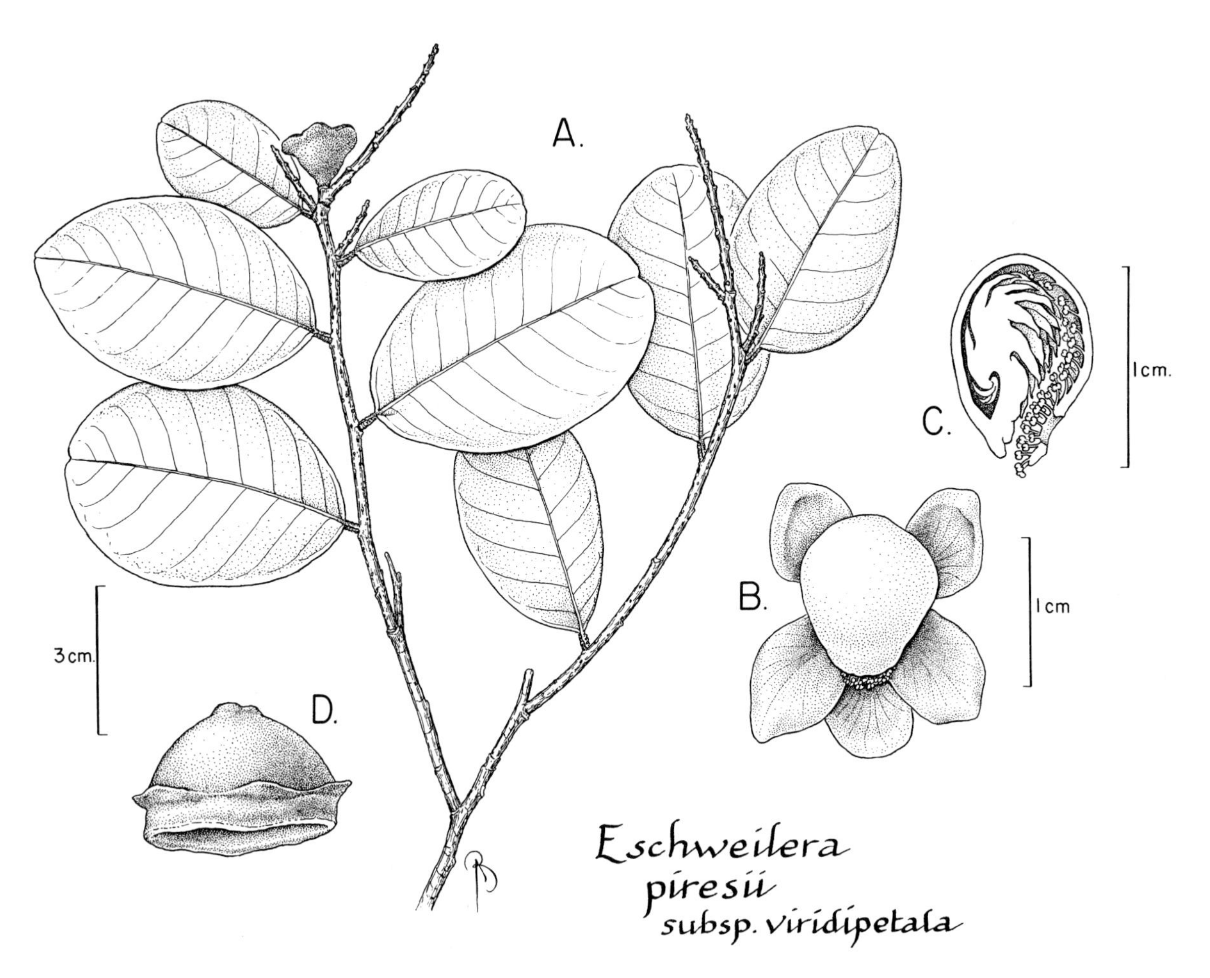

FIG. 163. LECYTHIDACEAE. *Eschweilera piresii* subsp. *viridipetala* (A, D, *Mori et al. 20800*; B, C, *Mori et al. 20805*). **A.** Apices of stems with leaves, old axes of infructescences, and fruit. **B.** Apical view of flower showing five petals and androecial hood. **C.** Medial section of androecium showing fertile stamens proximally and staminodial nectaries distally. **D.** Base of fruit. (Reprinted from S. A. Mori, Mem. New York Bot. 64, 1990.)

LECYTHIS Loefling

Canopy or emergent trees. Inflorescences terminal, axillary, or cauline, spikes, racemes, or paniculate arrangements of spikes or racemes. Flowers zygomorphic; sepals 6; petals 6; anthers dehiscing by lateral slits, the androecial hood flat, dorsiventrally expanded, or with appendages curved inwards but never forming complete coil, the appendages with or without anthers; ovary 4-locular, ovules inserted on lower part of septum. Fruits woody, dehiscent, globose, depressed globose, conical, or turbinate. Seeds >1 per fruit in our area; aril usually basal; embryo with undifferentiated cotyledons.

1. Pedicels and hypanthium rugose, less rugose in fruit.
    2. Pedicels ca. 2 mm long; hood of androecium dorsiventrally expanded, the anthers on ligular side of staminal ring yellow in contrast to white ones of remainder of ring. . . . . *L. corrugata* subsp. *corrugata*.
    2. Pedicels ≥5 mm long; hood of androecium not dorsiventral expanded, the anthers of staminal ring all white.
        3. Petals white, the hood with yellow anthers.
            4. Flowers 3–3.5 cm diam., the androecial hood white, sometimes tinged with blue. . . . . *L. persistens* subsp. *persistens*.
            4. Flowers 6–7 cm diam., the androecial hood reddish-orange. . . . . *L. persistens* subsp. *aurantiaca*.
        3. Petals pink or white tinged with pink, the hood without anthers.
            5. Largest leaf blades 8–10 cm long, chartaceous. Androecial hood nearly as wide as long, pink. Fruit with rugae inconspicous at maturity. . . . . *L. confertiflora*.
            5. Largest leaf blades 11–17 cm long, coriaceous. Androecial hood longer than wide, salmon colored. Fruit with rugae conspicuous at maturity. . . . . *L. idatimon*.

1. Pedicels and hypanthium smooth.
    6. Inner bark bright yellow. Abaxial leaf surface minutely papillate. Flowers opening at dusk, falling before daylight; petals green; androecium with >1000 stamens. . . . . *L. poiteaui.*
    6. Inner bark not bright yellow. Abaxial leaf surface not papillate. Flowers opening in early morning, falling in late afternoon; petals white or yellow; androecium with <500 stamens.
        7. Outer bark with deep, vertical fissures, markedly laminated. Petals yellow or white at base and purple or red at apex; style erect, with annular expansion toward apex; androecial hood flat, the appendages not curved inwards. Fruits to 20 × 20 cm.. . . . . *L. zabucajo.*
        7. Outer bark less markedly fissured and laminated. Petals white; style obliquely oriented; androecial hood not flat, the appendages curved inwards. Fruits seldom >10 × 10 cm.
            8. Leaf blades obovate. Hypanthium 10–15 mm diam. at anthesis. . . . . *L. holcogyne.*
            8. Leaves elliptic. Hypanthium ca. 5 mm diam. at anthesis.
                9. Flowers pedicellate. . . . . *L. chartacea.*
                9. Flowers sessile. . . . . *Lecythis* sp. A.

Lecythis chartacea O. Berg

Canopy trees. Trunks not buttressed. Bark shallowly fissured. Leaves: petioles 10–15 mm long; blades 9–14 × 4–6 cm; secondary veins in 8–12 pairs. Inflorescences weakly once-branched paniculate arrangements of racemes; pedicels present, smooth. Flowers ca. 1.5 cm diam.; petals white; androecial hood with appendages curved inwards, white on outside, yellow on inside, the appendages without anthers; hypanthium sulcate, ca. 5 mm diam. at anthesis, the style obliquely oriented. Fruits turbinate, 4–5 × 4 cm. Not yet collected from our area but to be expected.

Lecythis confertiflora (A. C. Sm.) S. A. Mori Pl. 75c

Canopy trees. Trunks not buttressed. Bark smooth, sometimes with vertical cracks. Leaves: petioles 5–10 mm long; blades elliptic to narrowly ellipitc, 6–10 × 3–6 cm; secondary veins in 8–10 pairs. Inflorescences racemes or once-branched paniculate arrangments of racemes; pedicels ca. 10 mm long, rugose. Flowers 2–2.5 cm diam.; petals pink to dark pink; androecial hood flat, nearly as wide as long, pink, the appendages without anthers; hypanthium rugose. Fruits narrowly conical, 4–5.5 (excluding operculum) × 2.5–3 cm, the pericarp nearly smooth, the rugae inconspicuous at maturity. Fl (Mar, Jul, Aug, Sep, Nov, Dec), fr (Mar, Apr, Sep, Nov, Dec); locally common, in forest along ridge tops. *Mahot blanc* (Créole), *weti loabi* (Bush Negro).

Lecythis corrugata Poit. subsp. **corrugata** Fig. 165, Pl. 75e

Canopy trees. Trunks usually not buttressed. Bark brown or grayish-brown with shallow vertical fissures. Leaves: petioles 10–15 mm long; blades elliptic, 8–12 × 4–6 cm; secondary veins in 7–9 pairs. Inflorescences racemes or once-branched paniculate arrangements of racemes; pedicels ca. 2 mm long, rugose. Flowers 2.5–3 cm diam.; petals pink; staminal ring with anthers on ligular side yellow in contrast to white ones of remainder of ring, the androecial hood flat, dorsiventrally expanded, pink, the appendages fused, not apparent; hypanthium rugose. Fruits turbinate, ca. 2–2.5 cm diam., rugose. Fl (Jan, Nov, Dec), fr (Mar, Apr); sporadic in our area, in nonflooded forest but usually in lower areas. *Lebi loabi* (Bush Negro), *mahot*, *mahot blanc*, *mahot noir*, *mahot rouge* (Créole), *weti loabi* (Bush Negro).

Lecythis holcogyne (Sandwith) S. A. Mori Pl. 76c

Canopy trees. Trunks not buttressed, often with swollen base. Bark shallowly fissured. Leaves: petioles 5–10 mm long; blades obovate, 6–13 × 2.5–7 cm; secondary veins in 7–10 pairs. Inflorescences spikes; pedicels absent. Flowers 2–3 cm diam.; petals white; androecial hood white tinged with yellow, the appendages curved inwards, the style obliquely oriented; hypanthium smooth to sulcate, 10–15 mm diam. at anthesis. Fruits broadly obovate, truncate at base, 6–7 × 5–7 cm. Fl (Feb, Dec), fr (Apr, May, Jun, Jul, Dec); uncommon, in nonflooded forest.

Lecythis idatimon Aubl.

Understory trees. Trunks buttressed. Bark smooth, with scattered vertical cracks. Leaves: petioles 9–15 mm long; blades narrowly elliptic to oblong, 8–17 × 3.5–7 cm, coriaceous; secondary veins in 12–16 pairs. Inflorescences racemes or weakly once-branched paniculate arrangements of racemes; pedicels 10–15 mm long, rugose. Flowers 2–2.5 cm diam.; petals white with tinges of pink or entirely pink; androecial hood flat, longer than wide, salmon colored, the appendages without anthers; hypanthium rugose. Fruits conical or turbinate, 2.5–3.5 (exluding operculum and pedicel) × 2–3.5 cm, rugose, the rugae conspicuous at maturity. Fl (Nov), Fr (Mar, Apr, Dec); locally common, in forest usually along ridge tops. *Lebi loabi* (Bush Negro), *mahot*, *mahot blanc*, *mahot rouge* (Créole), *oemanbarklak* (Bush Negro), *weti loabiu* (Bush Negro).

Lecythis persistens Sagot subsp. **aurantiaca** S. A. Mori Pl. 76a,b

Canopy trees. Trunks not buttressed. Bark smooth, with occasional vertical cracks. Leaves: petioles 13–18 mm long; blades elliptic, 12–18 × 6.5–9.5 cm; secondary veins in 14–18 pairs. Inflorescences racemes or once-branched paniculate arrangements of racemes; pedicels 25–40 mm long, rugose, with persistent bases 10–15 mm long left on rachises after flowers fall. Flowers 6–7 cm diam.; petals white; androecial hood flat, reddish-orange, the appendages with yellow anthers; hypanthium rugose. Fruits globose to depressed globose, usually >5 (including operculum) × 5 cm. Fl (Mar, Oct, Nov), fr (Jan, Apr, Dec); occasional, in nonflooded forest.

Lecythis persistens Sagot subsp. **persistens**

Canopy trees. Trunks not buttressed. Bark smooth, with occasional vertical cracks. Leaves: petioles 13–20 mm long; blades 11–21 × 5–8.5 cm; secondary veins in 13–18 pairs. Inflorescences racemes or once-branched paniculate arrangements of racemes; pedicels 8–15 mm long, rugose, the persistent pedicel bases ca. 5 mm long. Flowers 3–3.5 cm diam.; petals white, sometimes with bluish-tinge; androecial hood flat, white, sometimes tinged with blue, the appendages with yellow anthers; hypanthium rugose.

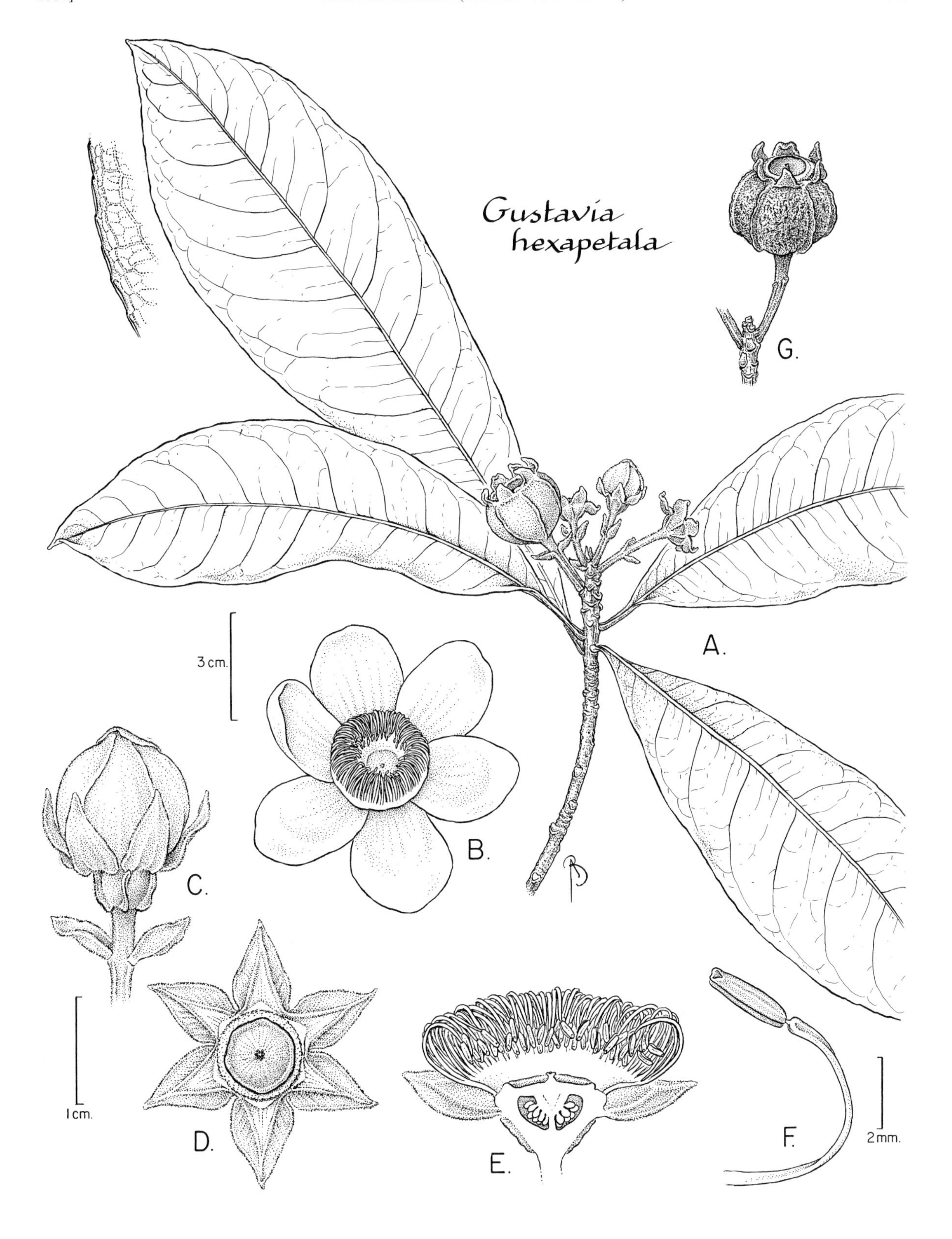

FIG. 164. LECYTHIDACEAE. *Gustavia hexapetala* (A, E, F, *Mori et al. 18676*; B, *Mori 15517*; C, D, *Mori & Bolton 8755*; G, *Jacquemin 2436*). **A.** Part of stem with leaves, flower bud, and immature fruits. **B.** Apical view of flower. **C.** Lateral view of flower bud. **D.** Apical view of flower after corolla and stamens have fallen. **E.** Medial section of flower. **F.** Stamen; note poricidal dehiscence. **G.** Lateral view of fruit.

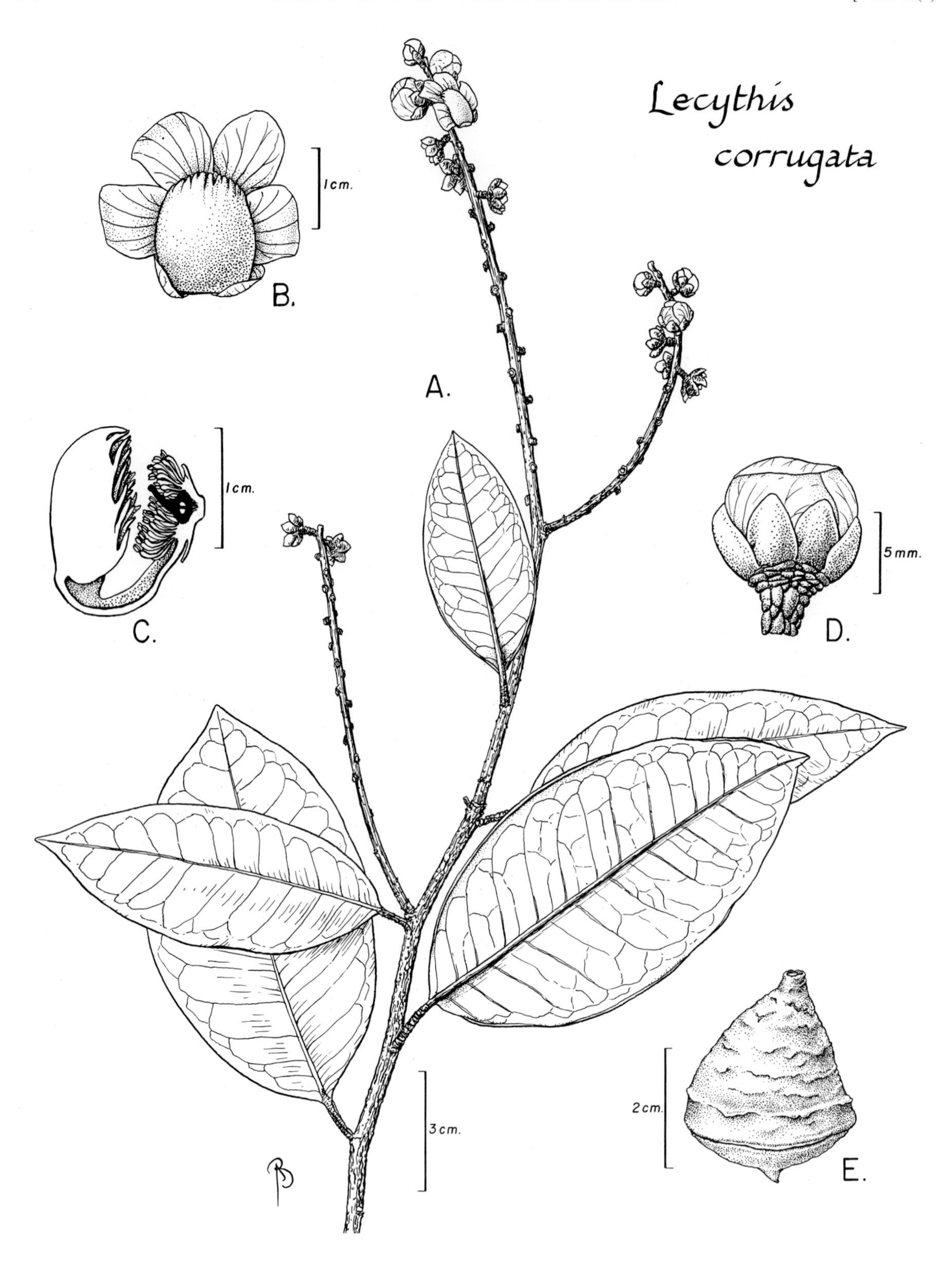

FIG. 165. LECYTHIDACEAE. *Lecythis corrugata* subsp. *corrugata* (A, *Mori & Bolten 8278* from Guyana; B–D, *Mori & Bolten 8682* from Surinam; E, *Mori & Bolten 8098* from Guyana). **A.** Part of stem with leaves and inflorescences. **B.** Apical view of flower. **C.** Medial section of flower with petals removed. **D.** Lateral view of bud; note rugose hypanthium and pedicel. **E.** Lateral view of nearly mature fruit.

Fruits turbinate, ca. 3–5 (including operculum) × 3–5 cm. Fl (Nov, Dec), Fr (Mar, Apr); occasional, in nonflooded forest. *Lebi loabi* (Bush Negro), *mahot, mahot blanc, mahot fer* (Créole).

Lecythis poiteaui O. Berg PL. 76d

Canopy trees. Trunks not buttressed. Bark with shallow, vertical fissures, the inner bark bright yellow. Leaves deciduous, appearing just before flowers; petioles 3–10 mm long; blades elliptic, 13–17 × 6–7 cm, minutely papillate abaxially; secondary veins in 16–20 pairs. Inflorescences spikes or racemes; pedicels absent to 5 mm long, smooth when present. Flowers open at dusk, fall before daylight, ca. 11 cm diam.; petals green; androecial hood flat, white, the proximal appendages with anthers, the distal ones without anthers; hypanthium smooth, glabrous. Fruits globose to depressed globose, 3.5–8 × 5–10.5 cm. Fl (Jan, Mar, Jul), fr (Apr); occasional, in nonflooded forest. *Mahot, mahot jaune, mahot rouge* (Créole), *meli* (Bush Negro).

Lecythis zabucajo Aubl. PART 1: FIG. 10, PL. VIc

Emergent trees. Trunks usually slightly buttressed. Bark brown to grayish-brown, with deep vertical fissures, the outer bark laminated, as thick or thicker than white inner bark. Leaves deciduous, flowering shortly after new leaf growth; petioles 3–10 mm long, blades 6–11.5 × 2–5.5 cm, the margins serrulate; secondary veins in 10–16 pairs. Inflorescences racemes, usually from young branches below leaves; pedicels 3–5 mm long, smooth. Flowers 4–5 cm diam.; petals yellow or white at base and purple or red at apex; staminal ring with >1000 stamens, the androecial hood flat, yellow or white tinged with yellow at apex, the proximal appendages with anthers, the distal ones without anthers; hypanthium smooth, the style erect, with annular expansion toward apex. Fruits globose or turbinate, 6–20 × 7.5–20 cm. Seeds with well developed aril at base. Fl (Aug), fr (Mar, Oct, Nov, Dec); occasional, in nonflooded forest. *Canari macaque* (Créole), *kouatapatou* (Bush Negro), *marmite de singe* (Créole).

The Pic Matécho population of this species, located at ca. 500 m alt., possesses flowers with petals that are white at the base and purple or red at the apex.

Lecythis sp. A

Canopy trees. Trunks not buttressed, often with swollen base. Bark vertically fissured. Leaves: petioles ca. 10 mm long; blades elliptic, 8–13 × 3–5.5 cm; secondary veins in 8–9 pairs. Inflorescences spikes; pedicels absent. Flowers ca. 1.5 cm diam.; petals white; appendages curved inwards; hypanthium smooth to sulcate, ca. 5 mm diam. at anthesis; the style obliquely oriented. Fruits unknown. Fl (Sep, Nov); known from only two collections, *Mori et al. 21580, 23953* (NY), from nonflooded forest along the Route de Bélizon north of Eaux Claires. These collections are intermediate between *L. chartacea* and *L. holcogyne*, having the small flowers of the former and the sessile flowers of the latter.

LENTIBULARIACEAE (Bladderwort Family)

Nathan P. Smith

Insectivorous herbs, terrestrial or aquatic, sometimes epiphytic. Roots absent. Stolons/rhizomes and rhizoids often present, tubers sometimes present. Insectivorous traps present (modified leaves), usually arising from stolons or rhizomes, thread-like and forked (*Genlisea*) or bladder-like (*Utricularia*). Stipules absent. Leaves simple, alternate, in basal rosettes (*Genlisea* and some *Utricularia*); blades linear, obovate, spathulate, or (in some *Utricularia*) divided. Inflorescences racemes, often scapose, few to many flowered; bracts and bracteoles often present. Flowers zygomorphic, bisexual; calyx 2- (*Utricularia*) or 5-lobed (*Genlisea*); corolla usually yellow to greenish-yellow or lavender to violet, bilabiate, the lower lip larger than upper lip, spurred; stamens 2; ovary superior, compound, bicarpellate, unilocular; placentation free-central/basal, the ovules usually numerous. Fruits capsules or less often indehiscent, dehiscent by 2–4 valves, the dehiscence irregular, or circumscissile. Seeds (1)-numerous.

Genlisea is unknown in our area; however, two species have been collected in French Guiana (Boggan et al., 1997). With further exploration of the rock outcrops found in the Pic Matécho area, other species of *Utricularia* are to be expected.

Taylor, P. G. 1967. Lentibulariaceae. *In* B. Maguire and collaborators (eds.), The botany of the Guayana Highland. Part VII. Mem. New York Bot. Gard. **17(1):** 201–228.

———. 1989. The genus *Utricularia* — a taxonomic monograph. Royal Botanic Gardens, Kew.

———. 1999. Lentibulariaceae. *In* J. A. Steyermark, P. E. Berry, K. Yatskievych & B. K. Holst (gen. eds.), Flora of the Venezuelan Guayana **5:** 782–803. P. E. Berry, K. Yatskievych & B. K. Holst (vol. eds.). Missouri Botanical Garden Press, St. Louis.

UTRICULARIA L.

Insectivorous herbs, terrestrial or aquatic, sometimes epiphytic. Roots absent. Stolons and rhizoids often present, tubers sometimes present. Traps bladder-like, often arising laterally on stolans and leaves. Leaves simple, alternate, in basal rosettes (many terrestrial species); blades obovate, spathulate, or divided. Inflorescences scapose racemes, these sometimes paniculate, erect, few- to many-flowered, often pubescent; scales often present on lower part of peduncle; bracts and bracteoles (when present) basal to pedicel, basifixed or fixed above the base. Flowers: calyx not fully developing until fruit ripe, the lobes 2; corolla generally yellow or lavender, bilabiate, the lower lip spurred, larger than upper lip, often with swelling at base; stamens inserted at junction of upper and lower lip. Terrestrial species are commonly found in wet habitats.

1. Bract and bracteole margins entire. Larger calyx lobe bifid; corolla mostly lavender. . . . . *U. calycifida.*
1. Bract margins fimbriate (bracteoles not apparent). Larger calyx lobe not bifid; corolla yellow. . . . . *U. hispida.*

Utricularia calycifida Benj.

Terrestrial herbs, to 30 cm tall. Rhizoids present, basal to peduncle. Stolons present. Traps minute, arising from stolons. Leaves basal to peduncle, rosulate, nearly glabrous; blades obovate to spatulate-oblanceolate, 3–13 × 1–4.5 cm, the base narrowly tapering to peduncle, the margins entire. Inflorescences scapose racemes, 1–2 per plant, to 30 cm tall, many flowered; scales sometimes present on lower peduncle, bract-like; pedicels 2–5 mm long; bract present, glabrous, basifixed, narrowly triangular, 2–3.5 mm long, the margins entire; bracteoles 2, glabrous, basifixed, lateral to bract, shorter and more narrow than bract. Flowers: calyx lobes unequal in size, the margins minutely denticulate, the larger lobe bifid, 5–6 × 6–9 mm, the smaller lobe 4–5 × 4–6 mm; corolla lavender with a central yellow patch surrounded by white at swelling of lower lip base, the upper lip 4 mm long, the lower lip 7 mm long, widest at apex, the apex slightly lobed, the spur 8 × 1 mm. Fl (Sep); in shade along streams flowing over granitic rock.

Utricularia hispida Lam.

Terrestrial herbs, to >40 cm tall. Traps unknown. Leaves unknown. Inflorescences paniculate racemes, to >40 cm, many-flowered; bract basifixed, triangulate to obovate, 0.5–1 mm long, the margins fimbriate, the mid-apex sometimes acuminate and extended; bracteoles not apparent; pedicels 4–9 mm long. Flowers: calyx lobes unequal in size, the larger lobe 1.5–2 × 1.5–2 mm (in fruit), the smaller lobe 1.5–2 × 1–1.5 mm (in fruit); corolla yellow. Capsules globose, ca. 2 mm diam. Seeds numerous, minute, angular. Fr (Sep); growing on or near open rock.

LOGANIACEAE (Logania family)

Bruno Bordenave

Trees, lianas (with axillary, hook-like tendrils), shrubs, subshrubs, or annual herbs. All parts of plants often bitter to taste and sometimes toxic. Stipules present as interpetiolar lines, these often well developed and forming an ocrea-like sheath. Leaves opposite; blades with entire margins; venation camptodromous (mostly eucamptodromous) or imperfectly acrodromous (*Strychnos*). Inflorescences axillary or terminal, cymose, corymbose, or thrysoid, condensed or elongate. Flowers actinomorphic, bisexual; corolla 4–5-merous, gamopetalous, infundibuliform to salverform (*Spigelia*); ovary superior, (1)2(5)-locular; placentation axile or parietal in unilocular taxa, the ovules numerous. Fruits septicidal capsules or berries. Seeds without wings or winged in *Antonia*; endosperm present.

Recent studies suggest that *Potalia* belongs to the Gentianaceae and that *Antonia*, *Spigelia*, and *Strychnos*, along with several other genera, should be treated as a separate family, the Strychnaceae (Struwe et al., 1994).

Raalte, M. H. van. 1966. Loganiaceae. *In* A. Pulle (ed.), Flora of Suriname **IV(1):** 103–110. E. J. Brill, Leiden.
Uittien, H. 1966. Loganiaceae. *In* A. Pulle (ed.), Flora of Suriname **IV(1):** 472–474. E. J. Brill, Leiden.

1. Annual herbs, subshrubs, or treelets <2 m tall.
   2. Treelets. Leaf blades >25 cm long. Flowers yellow. . . . . *Potalia.*
   2. Annual herbs or subshrubs. Leaf blades <20 cm long. Flowers white or pinkish-white.
      3. Leaves with petioles nearly absent. Inflorescences unbranched, secund spikes. Corolla 15–20 mm long. Fruits emarginate at apex but not shaped like a bishop's cap, the outer surface markedly verrucose to tuberculate. . . . . *Spigelia.*
      3. Leaves petiolate. Inflorescences branched dichotomously, cymose. Corolla ca. 3 mm long. Fruits shaped like a bishop's cap, the outer surface smooth. . . . . *Mitreola.*
1. Lianas, shrubs >2 m tall, or trees.
   4. Large trees. Leaf blades with eucamptodromous venation. Flowers clustered in small bracteate heads. Fruits dehiscent. Seeds winged. . . . . *Antonia.*
   4. Small trees, shrubs, or lianas. Leaf blades with acrodromous venation. Flowers not clustered in small, bracteate heads. Fruits indehiscent. Seeds not winged. . . . . *Strychnos.*

ANTONIA Pohl

A monotypic genus, description same as for *A. ovata*.

Antonia ovata Pohl FIG. 166, PL. 77a,b

Trees, 8–35 m tall. Trunks with dark gray bark spotted with white, buttresses present. Leaves: petioles canaliculate, 5–8 mm long, pubescent, the opposite petioles touching and surrounding stem; blades elliptic to oblong, 3–12 × 3–7 cm, coriaceous, glabrous adaxially, puberulous abaxially on veins, the base obtuse, the apex obtuse, the margins entire, often revolute; venation eucamptodromous, secondary veins in 3–5 pairs. Inflorescences terminal, cymose, the flowers in small, bracteate heads, the axes pubescent. Flowers small, yellow, fragrant; calyx with 5 scale-like, imbricate sepals; corolla with cylindrical tube 2–3 mm long, swollen at base, the petals linear,

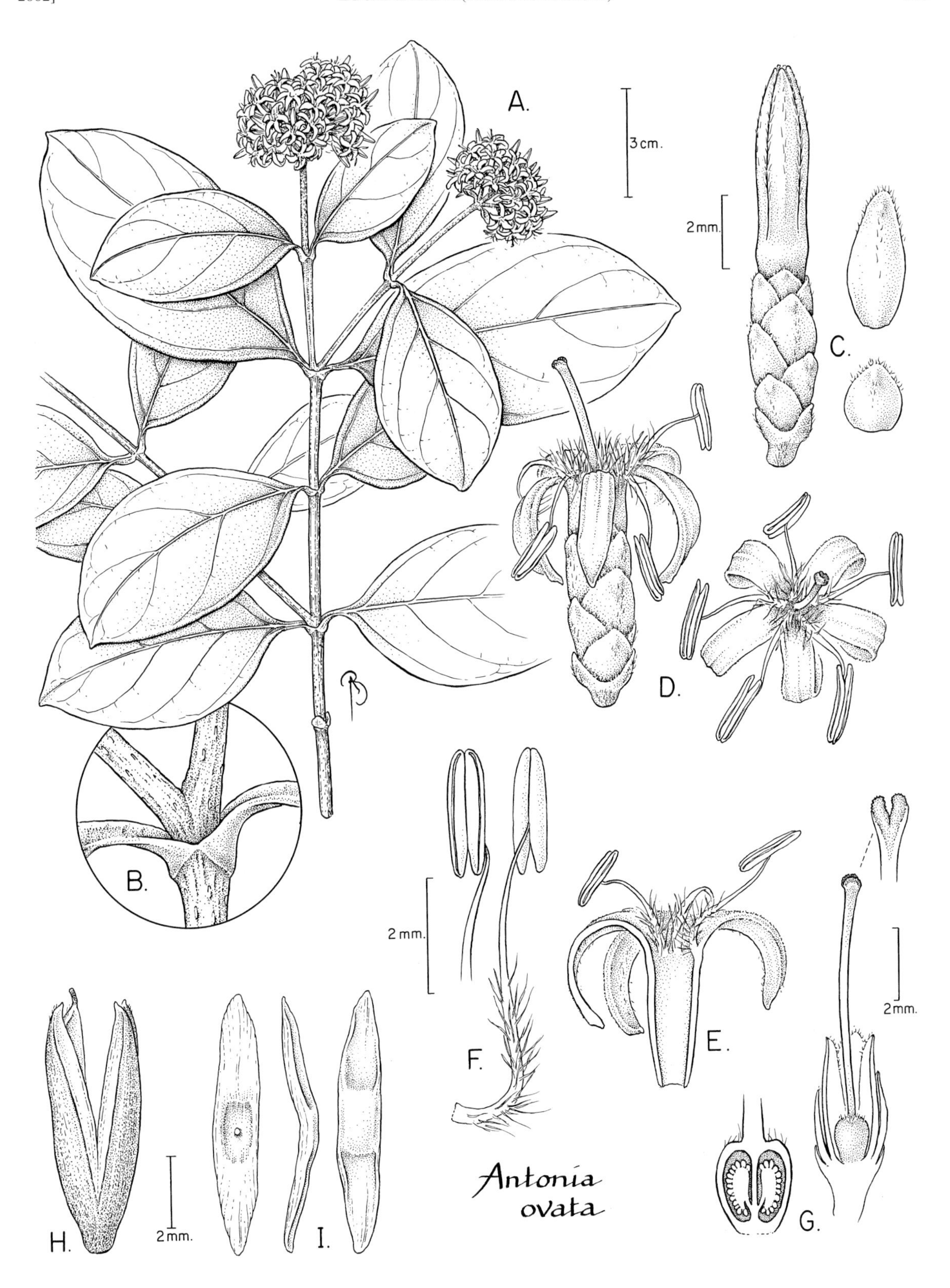

FIG. 166. LOGANIACEAE. *Antonia ovata* (*Mori et al. 24159*). **A.** Part of stem with leaves and inflorescences. **B.** Detail of node showing junction of petioles. **C.** Flower bud with sheathing bracteoles (left); inner bracteole (above right); outer bracteole (below right). **D.** Lateral (left) and apical (right) views of flower. **E.** Medial section of corolla with adnate stamens. **F.** Adaxial view of anther (above) and abaxial view of stamen (right). **G.** Pistil (right) with detail of stigma (above right) and medial section of ovary (left). **H.** Dehisced capsular fruit. **I.** Three views of seeds.

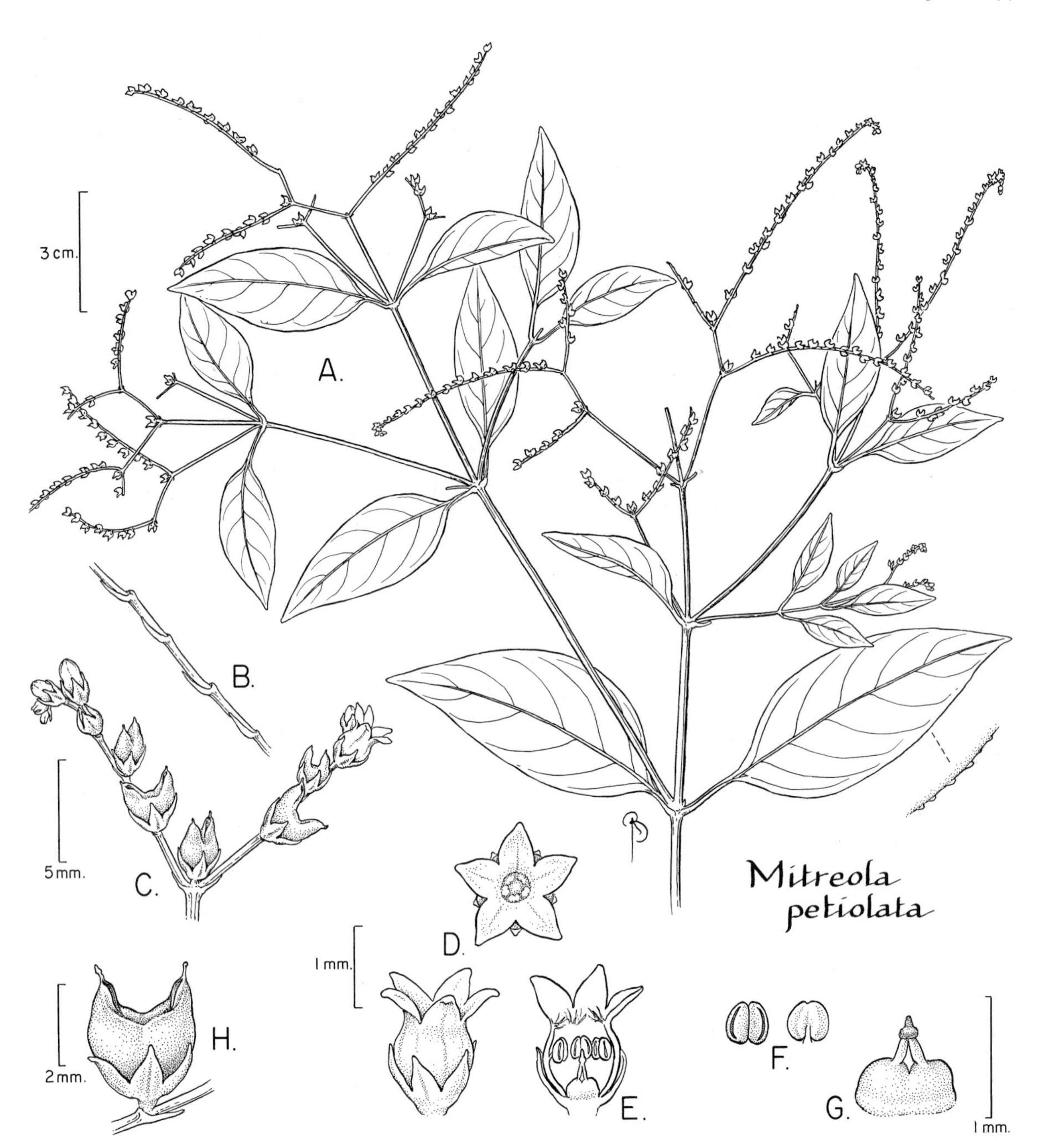

FIG. 167. LOGANIACEAE. *Mitreola petiolata* (*Mori et al. 22079*). **A.** Part of stem with leaves and infructescences and detail (lower right) of leaf margin. **B.** Detail of rachis of inflorescence. **C.** Detail of inflorescence with flowers, flower buds, and developing fruits. **D.** Apical (above right) and lateral view of flower (below). **E.** Lateral section of flower. **F.** Adaxial (left) and abaxial (right) views of anthers. **G.** Lateral view of pistil. **H.** Lateral view of dehisced fruit with subtending calyx.

acute, pubescent adaxially; stamens 5, the filaments pubescent at base; ovary puberulent, 2-locular, the style filiform, glabrous, the stigma bilobed. Fruits dehiscent, ovoid, the base cuneate. Seeds 1 per locule, with membranous wings at base and at apex. Fl (Sep, Nov), fr (Sep, Nov); in undisturbed nonflooded forest, bark and leaves used as a fish poison. *Bois-sabot*, *bois-sucre* (Créole).

MITREOLA R. Br.

Herbs or subshrubs. Inflorescences cymose, dichotomously branched; pedicels absent to poorly developed. Flowers: calyx with 5 sepals, these glandular adaxially; corolla white, twice as long as calyx, subcampanulate; stamens 5, inserted on lower half of corolla tube; styles 2, short, persistent, the stigma capitate. Fruits bifid, shaped like bishop's cap, with apical dehiscence, the outer surface smooth. Seeds elliptic-compressed, with reticulate-punctate seed coat; endosperm fleshy.

Mitreola petiolata (J. F. Gmel.) Torr. & A. Gray FIG. 167, PL. 77e

Herbs or subshrubs, to 1(2) m tall. Leaves: petioles 5–15 mm long; blades oblong to elliptic-lanceolate, 3–10 × 1.5–5 cm, membranous, the base decurrent onto petiole, the apex acuminate, the margins entire, plane; venation eucamptodromous, secondary veins in ca. 5–7 pairs. Inflorescences with flowers conspicuously on one side of axes. Flowers: corolla ca. 3 mm long, white. Fruits smooth. Fl (Jan, Mar, Oct), fr (Jan, Jul, Oct); road sides and along creek and river banks.

POTALIA Aubl.

Genus description same as for *P. amara*.

Potalia amara Aubl. FIG. 168, PL. 77c,d

Understory treelets or shrubs, 1–2 m tall, usually unbranched. Stipules forming sheath surrounding node. Leaves decussate; petioles 2.5–5 cm long; blades oblanceolate, usually 25–60 × 8–15 cm, membranous, the adaxial surface darker green than abaxial surface, the base long attenuate, decurrent onto petiole, the apex short acuminate, the margins entire, plane; venation eucamptodromous, secondary veins in ca. 12–15 pairs, the tertiary veins inconspicuous. Inflorescences terminal, cymose, the axes yellow or orange; pedicels 5–10 mm long; bracts and bracteoles ovate, 2–3 mm long. Flowers: calyx 4-merous, the sepals imbricate, ca. 5 mm long, the apex rounded; corolla yellow, the tube short, the lobes 10, convolute; stamens 10, inserted at middle of tube, alternate with petals; ovary 2-locular, the style swollen at base, filiform at apex, the stigma capitate. Fruits indehiscent, ovoid, the apex mucronate. Seeds numerous, angular. Fl (Aug, Sep, Oct, Dec), fr (Jul, Aug, Nov); scattered, in forest understory, an important medicinal plant in French Guiana (Grenand et al., 1987). *Anabi* (Portuguese), *mavévé grand bois* (Créole), *pau de cobra* (Portuguese).

Once considered a widely distributed species, *P. amara* has recently been split into several other species in such a way that *P. amara* is now restricted to the Guianas and Amapá, Brazil (Struwe et al., 1999, treated as part of Gentianaceae).

SPIGELIA L.

Herbs or subshrubs. Leaves with eucamptodromous venation. Inflorescences spicate. Flowers 5-merous; sepals with adaxial glands; corolla gamopetalous, the tube longer than lobes; stamens adnate to corolla; ovary 2-locular; placentation axile, the ovules numerous. Fruits dehiscent.

Spigelia multispica Steud. PL. 77f,g

Subshrubs, 0.5–1 m tall, usually unbranched. Stipules united and forming membranous sheath surrounding node. Leaves opposite at lower nodes, whorled at upper nodes subtending inflorescences; petioles absent or scarcely developed; blades elliptic to ovate, 3–15 × 2–5 cm, membranous, adaxial surface darker green than abaxial surface, the base decurrent onto petiole, the apex acuminate, the margins often finely revolute. Inflorescences terminal, 1–3 unbranched spikes per cluster, the flowers conspicuously on one side of axes. Flowers 5-merous, white to pinkish-white; sepals glandular adaxially, persisting on rachis after flowers or fruits have fallen; corolla salverform, 15–20 mm long, the tube white, the lobes 5, relatively small, valvate, pink to lavender; stamens inserted on upper part of tube, the filaments very short; ovary verrucose to tuberculate, the style persistent, the stigma hirsute. Fruits green, verrucose to tuberculate, the apex emarginate. Seeds polyhedral, the testa verrucose-reticulate. Fl (Jan, Feb, Mar, Apr, May, Aug, Sep, Oct, Nov), fr (Jan, Feb, Mar, Jul, Aug, Sep, Oct, Dec); common, in disturbed areas, especially along road sides through forest and along river banks. *Mavévé* (Créole).

STRYCHNOS L.

Usually lianas with woody, hook-like tendrils, less frequently shrubs or trees. Stipules usually represented only by stipular line running from base of one petiole to base of opposite petiole. Leaves opposite; venation acrodromous. Inflorescences terminal or lateral, cymose; pedicels with two bracts. Flowers: calyx small, 4–5-merous; corolla 4–5-merous, the throat and lobes often densely pubescent; stamens inserted in throat, alternate with petals; ovary 2-locular, rarely incompletely 2-locular, the style filiform, the stigma capitate or indistinctly 2-lobed; placentation usually with numerous ovules. Fruits indehiscent, globose, green, yellow, or orange berries. Seeds discoid or rarely globose; endosperm present.

Often all parts of plant are very bitter tasting; many of the species have alkaloids that provide part of the active ingredients for *curare* and other medicines.

1. Plants pubescent on vegetative parts, at least on petiole, midrib, or base of young leaf margins.
    2. Long hairs present.
        3. Leaf blades coriaceous. . . . . *S. cayennensis.*
        3. Leaf blades membranous.
            4. Leaf blades concolorous, the margins with hairs.
                5. Leaves pubescent mainly on petiole, midrib, and at base of margin; blades lanceolate, 3–7 cm long. . . . . *S. medeola.*
                5. Leaves pubescent on both surfaces of blade and margins; blades elliptic to oblanceolate, 6–20 cm long. . . . . *S. toxifera.*
            4. Leaf blades discolorous (dark green adaxially, pale abaxially), the margins without hairs. . . . . *S. tomentosa.*
    2. Only short hairs present, these erect or appressed.

6. Leaf blades membranous.
    7. Leaf blades 3–9 cm long. Inflorescences short racemes. . . . . *S. guianensis.*
    7. Leaf blades 7–30 cm long. Inflorescences clustered cymes. . . . . *S. peckii.*
6. Leaf blades coriaceous or subcoriaceous.
    8. Leaf blades with primary and secondary veins impressed adaxially, the higher order veins prominent on both surfaces. Mature fruits bluish-green. . . . . *S. panurensis.*
    8. Leaf blades with primary and secondary veins not impressed adaxially, the higher order veins not prominent on both surfaces. Mature fruits yellow or orange.
        9. Leaf blades with three primary veins. . . . . *S. cogens.*
        9. Leaf blades with five primary veins. . . . . *S. erichsonii.*

1. Plants glabrous on vegetative parts.
    10. Leaf blades membranous. . . . . *S. oiapocensis.*
    10. Leaf blades coriaceous.
        11. Scale-like cataphylls present at base of young shoots. . . . . *S. melinoniana.*
        11. Scale-like cataphylls not present at base of young shoots. . . . . *S. glabra.*

Strychnos cayennensis Krukoff & Barneby

Small trees, <9 m tall. Tendrils not apparent. Leaves: blades lanceolate, 7–12 × 2–4.5 cm, coriaceous, the base rounded, the apex with gradually tapering long acumen; primary veins 3. Inflorescences axillary, elongate, with 7–10 flowers, puberulent, the hairs short, appressed, pale yellow, the flowers not congested. Fruits ovoid, mucronate, small, ca. 2.5 cm diam., the pericarp thin, grayish-green, smooth. Seeds 1 or more, discoid. Fr (Feb); in understory of undisturbed nonflooded forest.

Strychnos cogens Benth. PART 1: FIG. 11

Lianas, often >20 cm diam. Tendrils axillary, hook-shaped, woody. Leaves: petioles densely pubescent, the hairs minute, rust colored; blades elliptic-ovate to narrowly lanceolate, 4–15 × 2–6 cm, subcoriaceous to coriaceous, discolorous, sparsely fulvus pubescent at first, becoming glabrescent with age except at junctions of secondary veins with midvein, with obscure punctations abaxially, the base obtuse to rounded, the apex acuminate; primary veins 3. Inflorescences axillary, few-flowered, congested cymes, 1.5 cm long, fulvus subsetulose, the hairs ascending; bracts ovate, puberulent distally, closely surrounding calyx. Flowers sessile. Fruits globose, small, to 2.5 cm diam., the pericarp very thin, shiny, smooth, yellow. Seeds 1–2, discoid, 15 × 11 × 3 mm. Fl (Mar); in primary forest.

Strychnos erichsonii Progel PL. 78b

Lianas, to 30 cm diam. at base. Bark orange-gray, with rough verrucose lenticels. Wood pale yellow. Tendrils terminal on branches or axes of inflorescence or axillary, hook-shaped, distally swollen. Leaves: petioles puberulent, with very short hairs, becoming glabrescent with age; blades variable, elliptic, lanceolate to suborbicular, 7–26 × 3–13 cm, coriaceous, shiny, glabrous except on midrib when young adaxially, dull, with punctations, puberulent with very short appressed, pale hairs abaxially; primary veins 5. Inflorescences axillary, thyrses, to 2.5 cm long, fulvus-puberulent with very short appressed hairs, sometimes appearing rust colored. Flowers: calyx ca. 1 mm long; corolla tube to 8 mm long, with golden woolly hairs on inside. Mature fruits globose, to 3.5 cm diam., the pericarp thin, shiny, smooth, pale yellow. Seeds 1 or more, discoid; endosperm horny. Fl (Jul), fr (Aug). *Dobouldoi, dobouldoi rouge, ledi dobouldoi* (Bush Negro), *liane canelle* (Créole), *ledi dobuldwa* (Bush Negro).

Large lianas reaching tops of tallest trees in primary forest, used as a stimulant and male aphrodisiac by Saramaka Bush Negroes (Grenand et al., 1987).

Strychnos glabra Progel

Lianas. Tendrils simple, axillary. Leaves: blades ovate to lanceolate, 4–13 × 3–7 cm, coriaceous, shiny, metallic in appearance, sometimes paler abaxially, glabrous, the base subcordate to obtuse, the apex rounded, often mucronate; primary veins 3. Inflorescences axillary, in short cymes of 5–7 flowers. Flowers: corolla tube woolly. Fruits oblong, mucronate, 2 cm long, the pericarp very thin, orange, crusty, glabrous. Seeds 1–2, discoid; endosperm horny. Fr (Jan); in primary nonflooded forest. *Curare* (French), *urari* (Portuguese).

Strychnos guianensis (Aubl.) Mart.

Small lianas. Tendrils axillary. Leaves: blades variable, ovate to oblanceolate, 3–9 × 1.5–4 cm, membranous, paler, with straight, appressed, very short hairs abaxially, the hairs most peristent on central vein, the base subcordate to cuneate, the apex rounded to acute, mucronulate, the margins often cilate toward base; primary veins 3(5). Inflorescences axillary, condensed in short racemes of 5–7 flowers. Flowers: corolla woolly in throat. Fruits oblong, mucronate, 2 cm long, the pericarp very thin, orange, crusty, glabrous. Seeds 1–2, discoid; endosperm horny. Fl (Aug), fr (Jan). *Curare* (French), *urari* (Portugese)

Strychnos medeola Progel

Small lianas. Tendrils small, axillary. Leaves: petioles densely puberulent, with many short and a few long hairs; blades lanceolate, relatively small, 3–7 × 1.5–4 cm, the base rounded, the apex long acuminate; primary veins 3. Inflorescences terminal, slender, few-flowered corymbs, covered with short, curved, rust colored hairs. Fruits ovoid, 1.5–3 cm diam., the pericarp very thin, dark yellow, shiny, smooth. Seeds 1–2. Fl (Jan), fr (Jun); in primary forest. *Matu bwa bâde* (Créole).

Strychnos melinoniana Baill. FIG. 169

Large lianas. Stems white, the young stems reddish-brown, with scale-like cataphylls at base of young shoots. All vegetative parts glabrous. Leaves: petioles drying black; blades elliptic, relatively large, 8–20 × 3–10 cm, coriaceous, sometimes glaucescent abaxially, drying grayish-green, the base obtuse to acute, the apex gradually narrowed to subacuminate; primary veins 3(5). Inflorescences terminal, congested cymes to 1.5 cm long, densely puberulent, the hairs short and pale, later becoming glabrescent. Flowers: corolla glabrous abaxially. Mature fruits ellipsoid, mucronate at apex, the pericarp very thin and fleshy. Fl (Nov), fr (Jul); in primary, nonflooded forest.

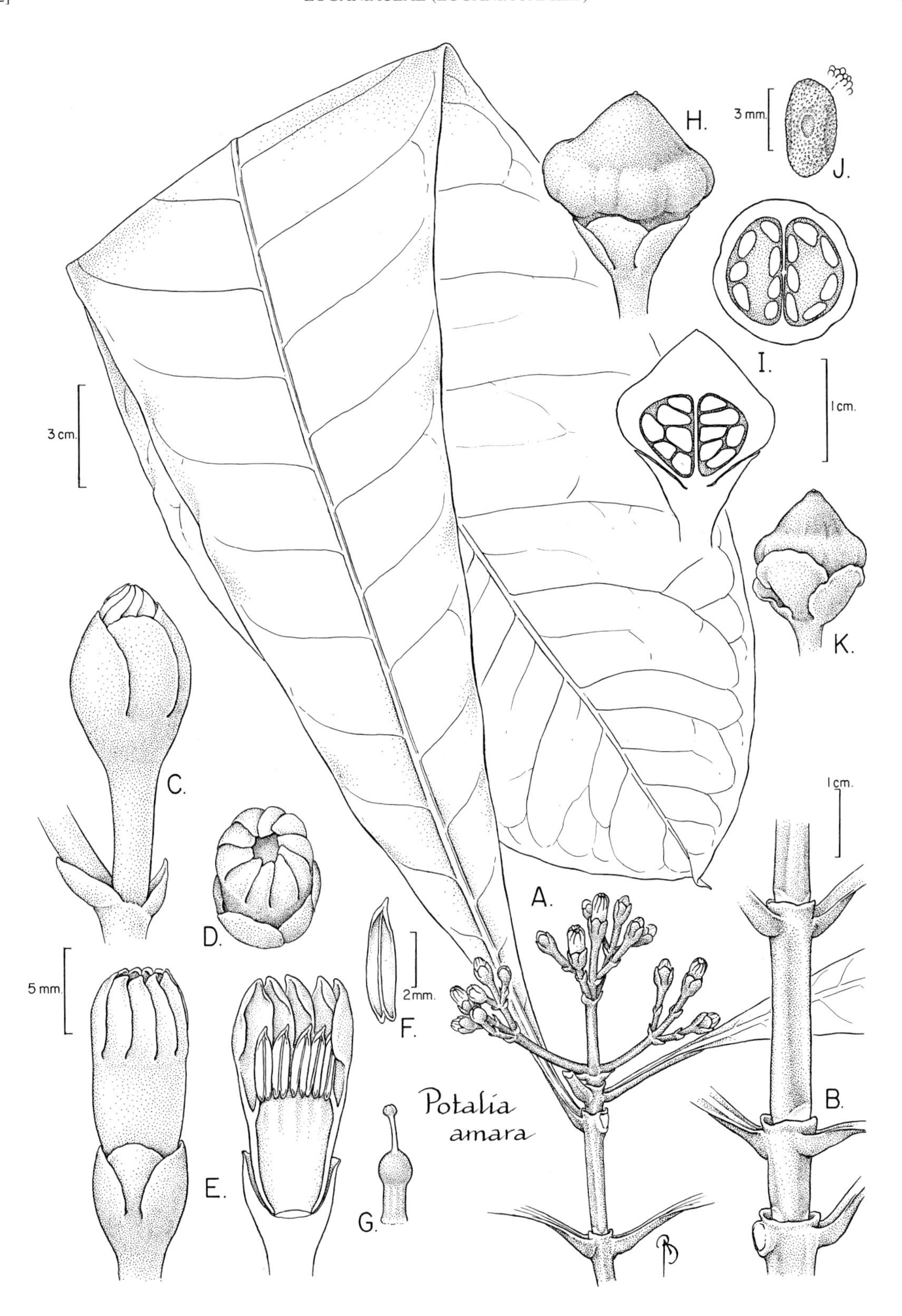

FIG. 168. LOGANIACEAE. *Potalia amara* (A, *Mori et al. 24074*; B, C, H–J, *Mori et al. 24073*; D, *Mori 23865*; E–G, *Skog & Feuillet 7067*). **A.** Apical part of stem showing leaf and inflorescence. **B.** Part of stem showing opposite arrangement of leaves and ocrea-like sheath. **C.** Lateral view of flower bud. **D.** Apical view of open flower. **E.** Lateral view of flower (left) and longitudinal section of flower (right). **F.** Adaxial view of stamen. **G.** Lateral view of pistil. **H.** Lateral view of fruit. **I.** Transverse section of fruit (above) and medial section of fruit (below). **J.** Seeds.

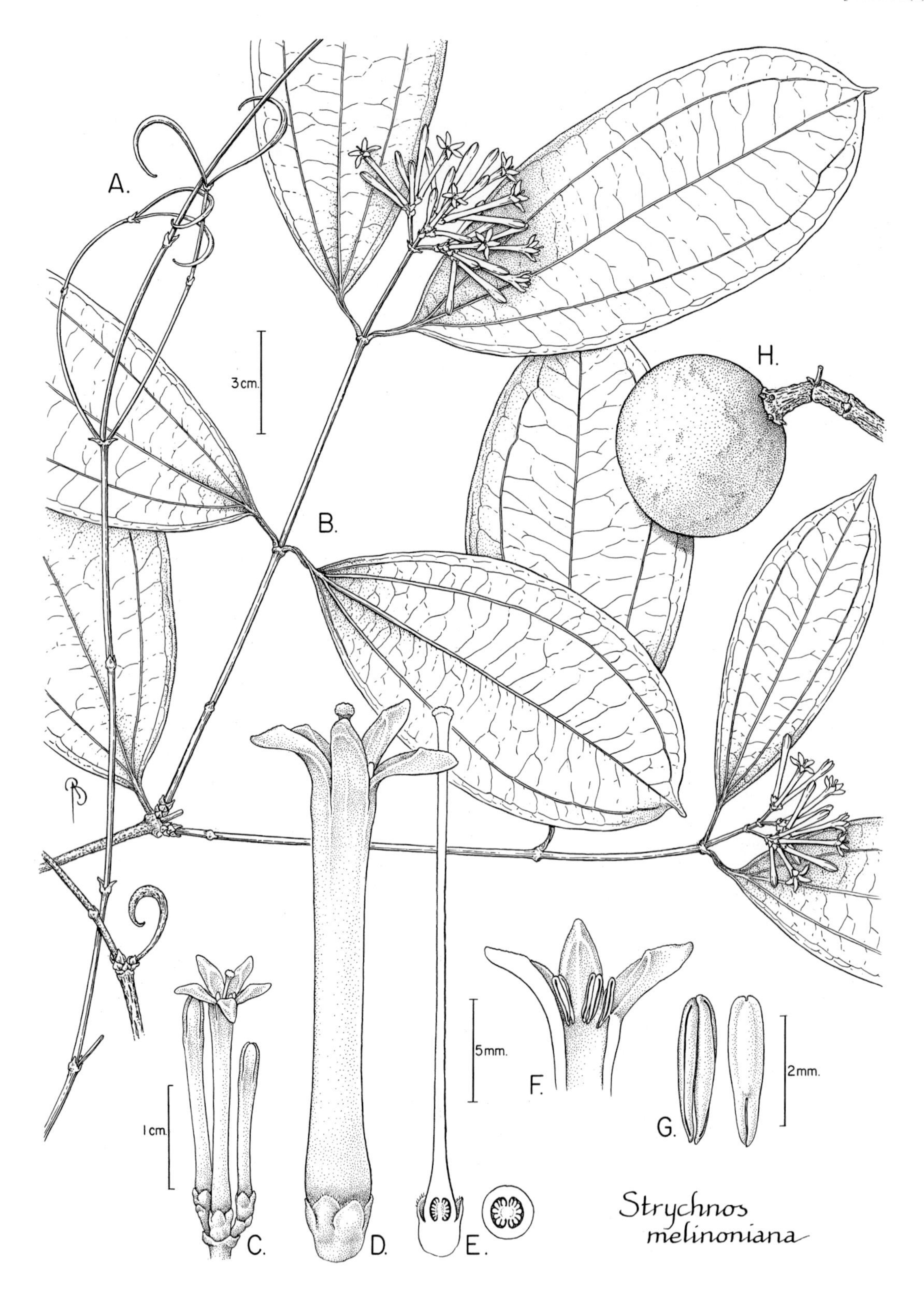

FIG. 169. LOGANIACEAE. *Strychnos melinoniana* (A–G, *Mori et al. 22839*; H, *Mori 18588*). **A.** Apical part of stem showing recurved tendrils. **B.** Part of stem with leaves and inflorescences. **C.** Detail of inflorescence showing buds and flower. **D.** Lateral view of flower. **E.** Medial section of pistil (left) and transverse section of ovary (right). **F.** Medial section of apical part of corolla showing adnate stamens. **G.** Adaxial (right) and abaxial (far right) views of anthers. **H.** Fruit.

Plates 73–80

Plate 73. LECYTHIDACEAE. **a.** *Couratari guianensis* (*Mori et al. 20973*), fallen flower. **b.** *Couratari guianensis* (*Mori et al. 20973*), androecial hood of flower in medial section showing stamens at bottom and nectar in innermost coil of hood. **c.** *Couratari stellata* (unvouchered), base of trunk with tall buttresses. **d.** *Couratari stellata* (*Mori et al. 22750*), flower. **e.** *Couratari stellata* (*Mori & Pipoly 15454*), dehisced fruit with its operculum, another operculum, and winged seeds. [Photo by S. Mori]

Plate 74. LECYTHIDACEAE. **a.** *Eschweilera collina* (*Mori & Gracie 19310* photographed in Amazonas, Brazil), lateral view of flower. **b.** *Eschweilera collina* (*Mori et al. 20594* from Amazonas, Brazil), flower with androecium sectioned to show stamens at base and staminodes in coil; normally the androecial hood is oriented vertically as seen in Plate 74a. **c.** *Eschweilera coriacea* (*Mori et al. 22651*), flowers, buds, and immature fruits. **d.** *Eschweilera grandiflora* (*Mori & Boom 15278*), flower and bud. [Photo by S. Mori] **e.** *Eschweilera grandiflora* (*Mori et al. 22988*), immature fruits; note persistent sepals. **f.** *Eschweilera parviflora* (*Mori et al. 21684*), flower and buds.

Plate 75. LECYTHIDACEAE. **a.** *Eschweilera pedicellata* (*Mori et al. 22679*), lateral view of flower. **b.** *Eschweilera squamata* (*Mori & Boom 15288*), fruit. [Photo by S. Mori] **c.** *Lecythis confertiflora* (*Mori et al. 20801*), flower. **d.** *Gustavia augusta* (*Phillippe 26974*), central part of flower showing staminal ring surrounding pistil. **e.** *Lecythis corrugata* subsp. *corrugata* (*Mori & Bolten 8682*, photographed in Surinam by S. Mori), flower and buds. **f.** *Gustavia hexapetala* (*Mori & Pipoly 15517*), flower and bud; note six petals. [Photo by S. Mori]

Plate 76. LECYTHIDACEAE. **a.** *Lecythis persistens* subsp. *aurantiaca* (*Mori et al. 22736*), flowers; compare with b. to note variation in intensity of flower color. **b.** *Lecythis persistens* subsp. *aurantiaca* (*Mori et al. 24724*), flowers and buds; note rugose pedicels; compare with a. to note variation in intensity of flower color. [Photo by S. Mori] **c.** *Lecythis holcogyne* (*Mori et al. 24785*), flower and buds. **d.** *Lecythis poiteaui* (*Mori & Snyder 24341*), nocturnal flower. [Photo by S. Mori]

Plate 77. LOGANIACEAE. **a.** *Antonia ovata* (*Mori et al. 24159*), flowers. **b.** *Antonia ovata* (*Mori et al. 24159*), infructescence; note bracts. **c.** *Potalia amara* (*Mori & Gracie 23865*), open flower and buds. **d.** *Potalia amara* (*Mori et al. 21574*), immature fruits. **e.** *Mitreola petiolata* (*Mori et al. 22079*), flowers and immature fruits. **f.** *Spigelia multispica* (*Mori et al. 24767*), flowers, buds, and immature fruits. **g.** *Spigelia multispica* (*Mori et al. 23113*), flower and buds.

Plate 78. LOGANIACEAE. **a.** *Strychnos panurensis* (*Mori & Boom 15328*), fruits. [Photo by S. Mori] **b.** *Strychnos erichsonii* (*Mori & Gracie 18609*), flowers and tendril. **c.** *Strychnos tomentosa* (*Mori et al. 23933*), flowers; note pubescence. **d.** *Strychnos tomentosa* (*Mori et al. 23933*), inflorescences and coiled tendril.

Plate 79. LORANTHACEAE. **a.** *Phthirusa pycnostachya* (*Mori et al. 23969*), inflorescences; note upward orientation of flowers. **b.** *Phthirusa pyrifolia* (*Mori et al. 23967*), inflorescences; note downward orientation of flowers. **c.** *Phthirusa stelis* (*Mori et al. 22874*), flower and buds. **d.** *Struthanthus syringifolius* (*Mori et al. 23848*), inflorescences. **e.** *Struthanthus* sp. (*Mori & Gracie 18342*), seedlings that have germinated on stem and leaves of *Trigonia villosa*.

Plate 80. LYTHRACEAE. **a.** *Cuphea carthagenensis* (*Mori et al. 22083*), flowers. MALPIGHIACEAE. **b.** *Byrsonima aerugo* (*Mori et al. 22196*), flowers and buds; note glands on calyces. **c.** *Byrsonima stipulacea* (*Mori et al. 23261*), fruits; note glands on calyces. **d.** *Hiraea fagifolia* (*Mori et al. 22924*), flowers and buds.

Couratari guianensis

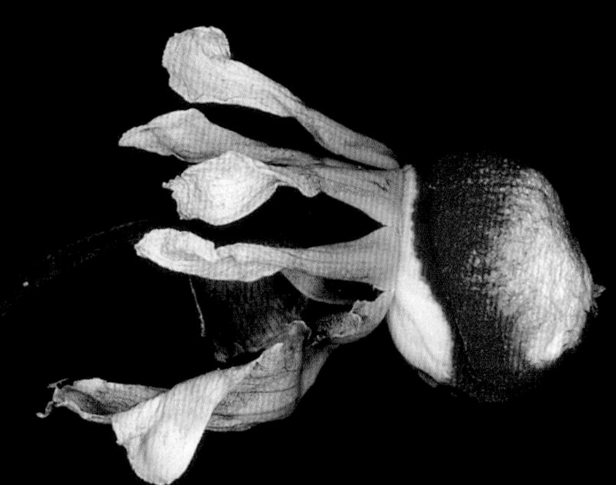

a.

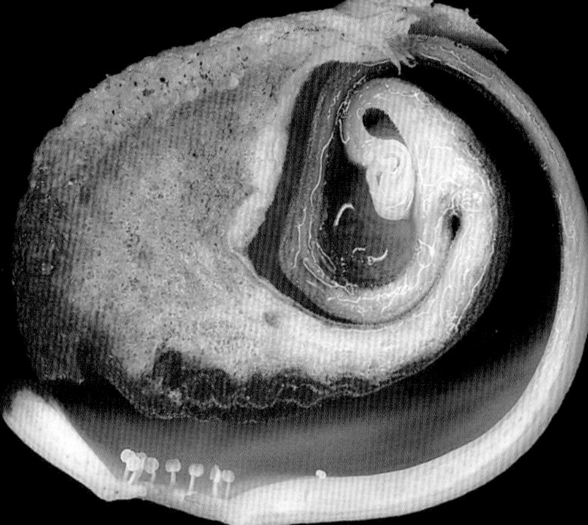

b.

c.

Couratari stellata

Couratari stellata

d.

e.

Couratari stellata

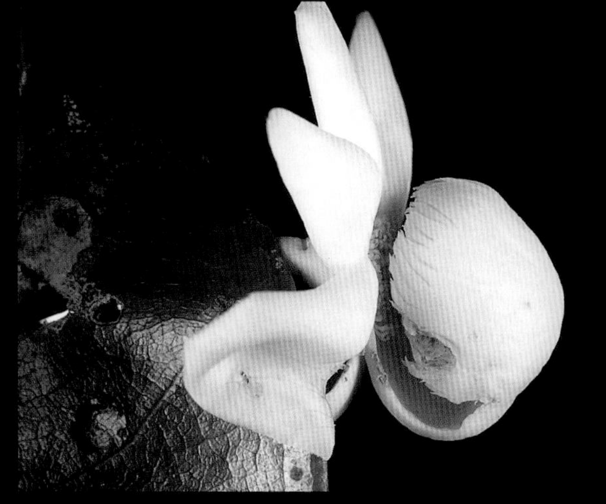

Eschweilera collina

a.

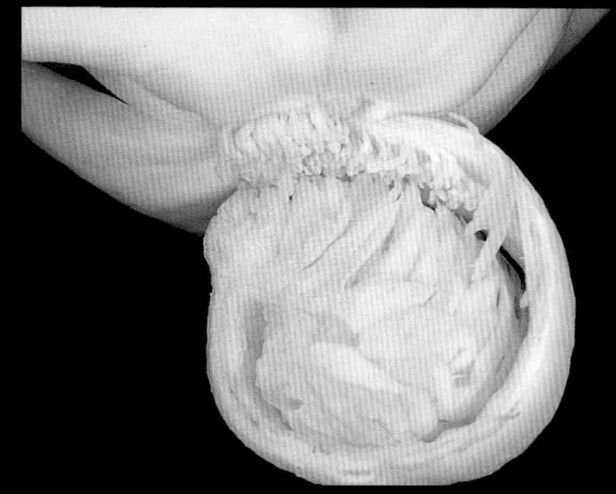

b.

Eschweilera coriacea c.

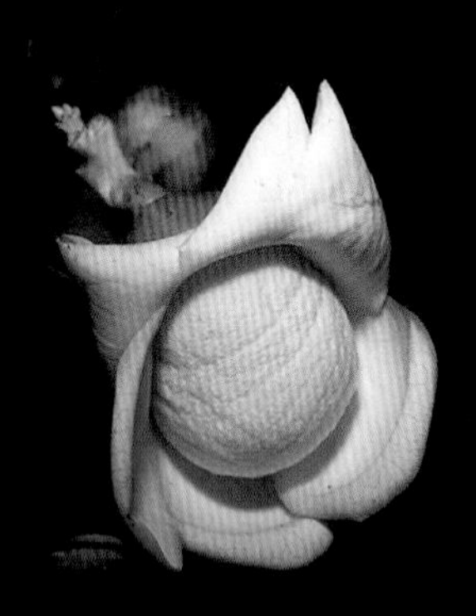

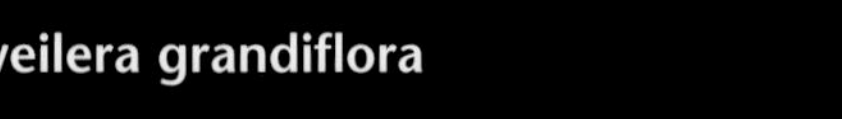

Eschweilera grandiflora d.

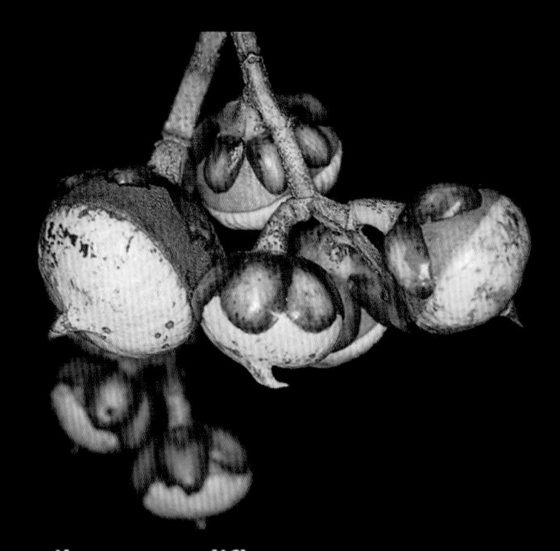

Eschweilera grandiflora e.

f.

Eschweilera parviflora

Plate 74

Eschweilera pedicellata **a.**

Eschweilera squamata **b.**

Lecythis confertiflora **c.**

Gustavia augusta **d.**

Lecythis corrugata subsp. corrugata **e.**

Gustavia hexapetala **f.**

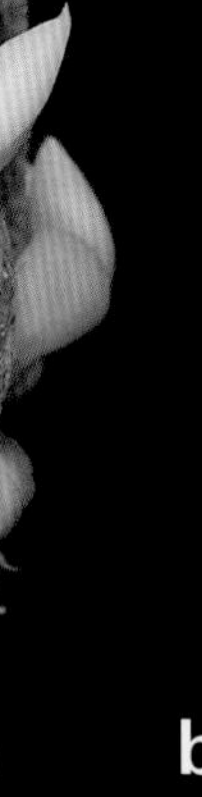

Lecythis persistens subsp. aurantiaca **b.**

Lecythis holcogyne **c.**

Lecythis poiteaui **d.**

Plate 76

Antonia ovata a.

Antonia ovata b.

Potalia amara c.

Potalia amara d.

Mitreola petiolata e.

Spigelia multispica f.

Spigelia multispica g.

Plate 77

Strychnos panurensis a.

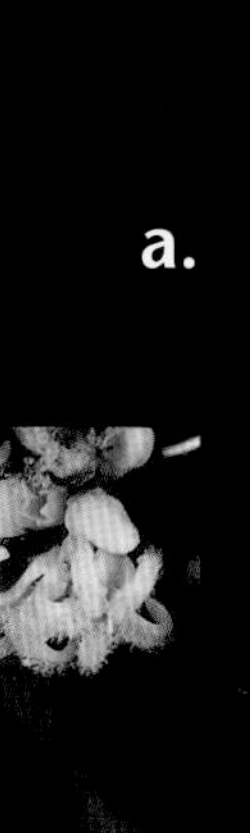

Strychnos erichsonii b.

Strychnos tomentosa c.

Strychnos tomentosa d.

Plate 78

Phthirusa pycnostachya a.

Phthirusa pyrifolia b.

Phthirusa stelis c.

Struthanthus syringifolius d.

Struthanthus sp. e.

Plate 79

LYTHRACEAE

Cuphea carthagenensis a.

MALPIGHIACEAE

Byrsonima aerugo b.

Byrsonima stipulacea c.

Hiraea fagifolia d.

Plate 80 (Malpighiaceae continued on Plate 81)

Strychnos oiapocensis Fróes

Erect, unarmed shrubs, sometimes scandent with age. Young stems glabrous, lustrous, becoming gray and verrucose with age. Tendrils absent. Leaves: blades ovate, 6–16 × 2.5–6.5 cm, membranous, glabrous, dull, drying olivaceous, the base rounded to cuneate, the apex acuminate; primary veins 3 or 5, the outer ones poorly developed when 5. Inflorescences terminal or subterminal, compact cymes; peduncle <5 mm long; pedicels concealed by glabrous bracteoles. Flowers glabrous except for inner part of tube. Fruits globose, ± 3.5 cm diam., the pericarp very thin, orange-red when mature, verrucose. Seeds 1–3, discoid. Fr (Mar, Sep); in understory of primary forest. *Dobouldoi* (Créole), *dobuldwa* (Bush Negro).

Strychnos panurensis Spruce & Sandwith PL. 78a

Erect shrubs as juveniles, lianas as adults. Leaves: petioles sparsely puberulent, with minute ascending hairs; blades elliptic to lanceolate, 5–2 × 4–7 cm, subcoriaceous to coriaceous, puberulent on central vein, with minute erect hairs abaxially and occasionally puberulent near base adaxially, the base rounded to cuneate, the apex abruptly short acuminate to caudate; primary veins 3(5), the primary and secondary veins impressed adaxially, the higher order veins prominent on both surfaces. Inflorescences axillary or terminal, elongate thyrses 2–7 cm long. Mature fruits globose, 2–2.5 cm diam., the pericarp thin, shiny, smooth, bluish-green. Seeds usually 1, discoid. Fl (Nov), fr (Dec); in primary forest.

Strychnos peckii B. L. Rob.

Large lianas, to 30 cm diam. Leaves: petioles rugose, densely puberulent when young, the hairs minute; blades elliptic to oblanceolate, 7–30 × 4–17 cm, membranous, drying ochre-yellow, puberulent abaxially, the hairs very short appressed, puberulent on central vein, especially when young, adaxially, the base rounded, the apex rounded to short acuminate; primary veins 3(5). Inflorescences axillary, clustered cymes, subglobose and densely flowered, fulvus-pulverulent. Fruits globose, large, to 7 cm diam., the pericarp ca. 5 mm thick, dull, reticulately marked, rugose and often warty; pedicels to 2 cm thick. Seeds many, discoid, 1.8 × 2.5 cm. Fl (Mar), fr (Apr); reaching top of tallest trees in nonflooded primary moist forest.

Strychnos tomentosa Benth. PL. 78c,d

Lianas. Stems pubescent. Tendrils axillary, distally swollen. Leaves: petioles 4 mm long, with spreading hairs; blades elliptic 6–10 × 4–5 cm, membranous, dark olive green adaxially, pale, densely fulvus-pubescent abaxially, the base rounded to obtuse, the apex mucronate; primary veins 3(5), impressed adaxially. Inflorescences terminal and subterminal, corymbose; pedicels to 0.5 cm long. Flowers: calyx lobes ovate to lanceolate; corolla tube 1.2 cm long, densely fulvus-pubescent abaxially, white lanate with short curled hairs in throat; anthers exserted, on short, distinct filaments; style glabrous. Fruits 6 cm diam., the pericarp 2 mm thick, shiny and tuberculate. Fl (Aug, Sep); in canopy of medium-sized to very tall trees in nonflooded forest. *Curare* (French), *urari* (Portuguese)

Strychnos toxifera Benth.

Lianas of intermediate size, to 10 cm diam. Leaves: petioles hirsute, with long, spreading hairs and short curved hairs; blades elliptic to oblanceolate, 6–20 × 3–8 cm, membranous, dull, densely pubescent on both surfaces when young, the pubescence persistent, especially on veins, the base subcordate to obtuse, the apex acuminate, finely pointed; primary veins 5. Inflorescences terminal cymes, the cymes corymbose, few-flowered, densely hirsute, the hairs long, rust colored. Fruits globose, large, to 7 cm diam., bluish-green, the pericarp thick, smooth, shiny. Seeds 10–15, discoid, ± 2.5 cm diam. Known only by sterile collections from our area; in primary forest.

One of the main ingredients used in the preparation of *curare* by the Amerindians of Guayana.

LORANTHACEAE (Showy Mistletoe Family)

Job Kuijt

Hemiparasitic shrubs on branches and trunks of woody plants, some with creeping roots. Pubescence generally absent, but young stems sometimes with scurfy lines. Stipules absent. Leaves simple, opposite. Inflorescences spikes, racemes, or paniculate arrangements of spikes or racemes, the flowers single or in triads. Flowers actinomorphic or nearly so, usually bisexual, less frequently unisexual; calyx reduced to small rim (calyculus); petals 6; stamens 6, dimorphic, adnate to petals; ovary inferior, 1-locular, the style and stigma simple; placentation basal, the ovules replaced by ovarian papilla. Fruits fleshy. Seeds 1 per fruit, without testa; embryo large, green, dicotylar (sometimes polycotylar in *Psittacanthus*), surrounded by sticky tissue; endosperm usually copious (endosperm absent in *Psittacanthus*).

Krause, K. 1966. Loranthaceae. *In* A. Pulle (ed.), Flora of Suriname **I(1):** 4–24. E. J. Brill, Leiden.
Lindeman, J. C. & A. R. A. Görts-van Rijn. 1966. Loranthaceae. *In* A. Pulle (ed.), Flora of Suriname **I(1):** 295–300. E. J. Brill, Leiden.

1. Leaves very fleshy. Buds fleshy; petals >1 cm long. Endosperm lacking. . . . . *Psittacanthus.*
1. Leaves not excessively fleshy. Buds not excessively fleshy; petals <1 cm long. Endosperm present.
    2. Flowers single in axils of paired, strap-shaped bracts fused with swollen axis, sunken in cavities, each cavity with 2 minute, strap-like bracteoles. . . . . *Oryctanthus.*
    2. Flowers in paired triads along inflorescence axis which is not swollen, each lateral flower of triad subtended by 1 bracteole.
        3. Inflorescences axillary or also terminal, especially the latter often branched. Petals usually orange to red; filaments absent or nearly so, the anthers basifixed. . . . . *Phthirusa.*
        3. Inflorescences axillary, unbranched. Petals usually greenish white; filaments slender, the anthers versatile. . . . . *Struthanthus.*

ORYCTANTHUS Eichler

Small parasitic shrubs with creeping roots radiating out only from primary haustorium, these roots bearing secondary haustoria. Young stems of some species with scurfy lines. Leaves: petioles present, often short, or absent; blades not or only slightly fleshy; venation pinnate to palmate. Inflorescences axillary, simple or in compound terminal groups, 1 or 3 spikes per axil, the flower-bearing part swollen, the flowers single in each scale-leaf axil, partly immersed in cavities, each cavity provided with 2 minute, strap-shaped bracteoles flanking single flower. Flowers bisexual; petals <2 mm long, usually dark wine red, rarely somewhat yellowish; stamens in 2 series, the anthers each with 4 small locules, tipped by small connectival horn. Fruits reddish or green, often with yellow base. Embryo massive, green, dicotylar, with swollen radicular end.

1. Leaf blades ovate to orbicular; venation palmate or nearly so. Inflorescences axillary and often in terminal panicles, the flowers and fruits not perpendicular to axis. Fruits with apex rounded. . . . . *O. alveolatus.*
1. Leaf blades mostly ovate; venation pinnate or nearly so. Infloresences strictly axillary, the flowers and fruits ± perpendicular to axis. Fruits with apex ± truncate. . . . . *O. florulentus.*

Oryctanthus alveolatus (Kunth) Kuijt

Plants variable in size. Stems terete even when young, lacking furfuraceous lines. Leaves: petioles short or absent; blades ovate to orbicular, to 12 × 12 cm; venation palmate, at least in broad-leaved forms. Inflorescences axillary and frequently also in open, leafless, terminal, panicle-like groups, the flowers and fruits not perpendicular to axis; peduncules short and stout or nearly absent. Fruits ovoid, the apex rounded, greenish-red, inclined at ca. 45°. Fl (Oct).

Oryctanthus florulentus (Rich.) Tiegh.

Sparsely branched, small hemiparasites. Stems angular when young, with brownish furfuraceous lines. Leaves: petioles short; blades broadly lanceolate to somewhat ovate, mostly ca. 4 × 2 cm; venation pinnate. Inflorescences to 4 cm long, often with 24 or more flowers per inflorescence, the flowers and fruits ± perpendicular to axis; peduncles <2 mm long. Fruits ellipsoid, the apex ± truncate, orange-red, oriented nearly perpendicular to infructescence axis. Fr (Aug); often in secondary vegetation. *Caca zozo* (Créole), *erva de passarinho* (Portuguese), *gui* (French).

PHTHIRUSA Mart.

Small leafy hemiparasites, glabrous at maturity, with creeping roots from just above primary haustorium and sometimes from leafy stems, all with inconspicuous secondary haustoria. Young stems with scurfy lines. Leaves petiolate to nearly sessile; blades not fleshy; venation pinnate. Inflorescences axillary and sometimes also terminal, the axillary ones sometimes, the terminal ones always branched, the flowers in paired triads, the two lateral flowers of triad subtended by one bracteole each. Flowers unisexual (plants dioecious) or bisexual; unisexual flowers with aborted organs of opposite sex; petals dark red to somewhat orange, <6 mm long; filaments absent or nearly so, the anthers with 4 locules and connectival horn. Fruits green, yellow, or orange. Embryo massive, green, dicotylar, with swollen radicular end.

1. Leaf blades with the apex rounded; venation palmate. . . . . *P. pycnostachya.*
1. Leaf blades with apex acute; venation pinnate.
   2. Adventitious roots from base of plants only. Inflorescences axillary, unbranched. Flowers bisexual. . . . . *P. pyrifolia.*
   2. Adventitious roots from base of plant and from stems. Inflorescences axillary and in terminal, branched groups, the axillary ones also branched. Flowers unisexual. . . . . *P. stelis.*

Phthirusa pycnostachya Eichler PL. 79a

Fairly large plants, not scandent. Stems terete, grayish with irregular brown lenticels, the internodes to 3.5 cm long, the adventitious roots not known. Leaves: petioles distinct, robust, 3 × 3 mm; blades broadly elliptic to nearly orbicular, to 7 × 6 cm, coriaceous, the base obtuse, the apex rounded; venation palmate, the veins well-defined, nearly reaching apex. Inflorescences 1–3 per axil, sometimes in terminal compound groups, to 2.5(8) cm long; peduncles ca. 2–50 mm long, terete, the flowers often in 20 or more pairs of densely crowded triads on rather fleshy axis or flowers spread out over 5 cm length of axis; triad peduncles at most 2.5 mm long, with small, persistent bracts and bracteoles; pedicels absent. Flowers usually not >1 mm long. Fruits ovoid, ca. 3.5 × 2.5 mm, the apex somewhat truncate. Fl (May, Sep), fr (Jun, Sep).

Phthirusa pyrifolia (Kunth) Eichler PL. 79b

Rather large plants, not scandent. Young stems keeled, bearing conspicuous scurfy lines without obvious lenticels even with age, the adventitious roots only from base. Leaves: petioles distinct, to 10 mm long, often somewhat compressed or keeled; blades lanceolate, mostly to 10 × 4–5 cm, the base obtuse to truncate, the apex usually acute; venation pinnate. Inflorescences 1 or more per leaf axil, rather open, the triads in 7 or more pairs; peduncles short or nearly absent. Flowers: petals dark wine-red to pale red. Fruits ellipsoid, ca. 8 × 5 mm, bright orange-red with yellow tip and dark, purple-green base at maturity. Fl (Jun, Sep), fr (Sep).

Phthirusa stelis (L.) Kuijt FIG. 170, PL. 79c

Plants scandent. Stems slightly angular and bearing scurfy, brownish lines when young, soon becoming ± terete and glabrous, attached by roots emerging from stem contact points with hosts, the adventitious roots from both base and stem. Leaves: petioles distinct, variable in length; blades ovate, 4–11 × 2–6 cm, the base often obtuse to rounded, the apex acute to acuminate; venation pinnate. Inflorescences axillary or terminal, racemose, either simple (axillary) or branched (terminal, but sometimes also in

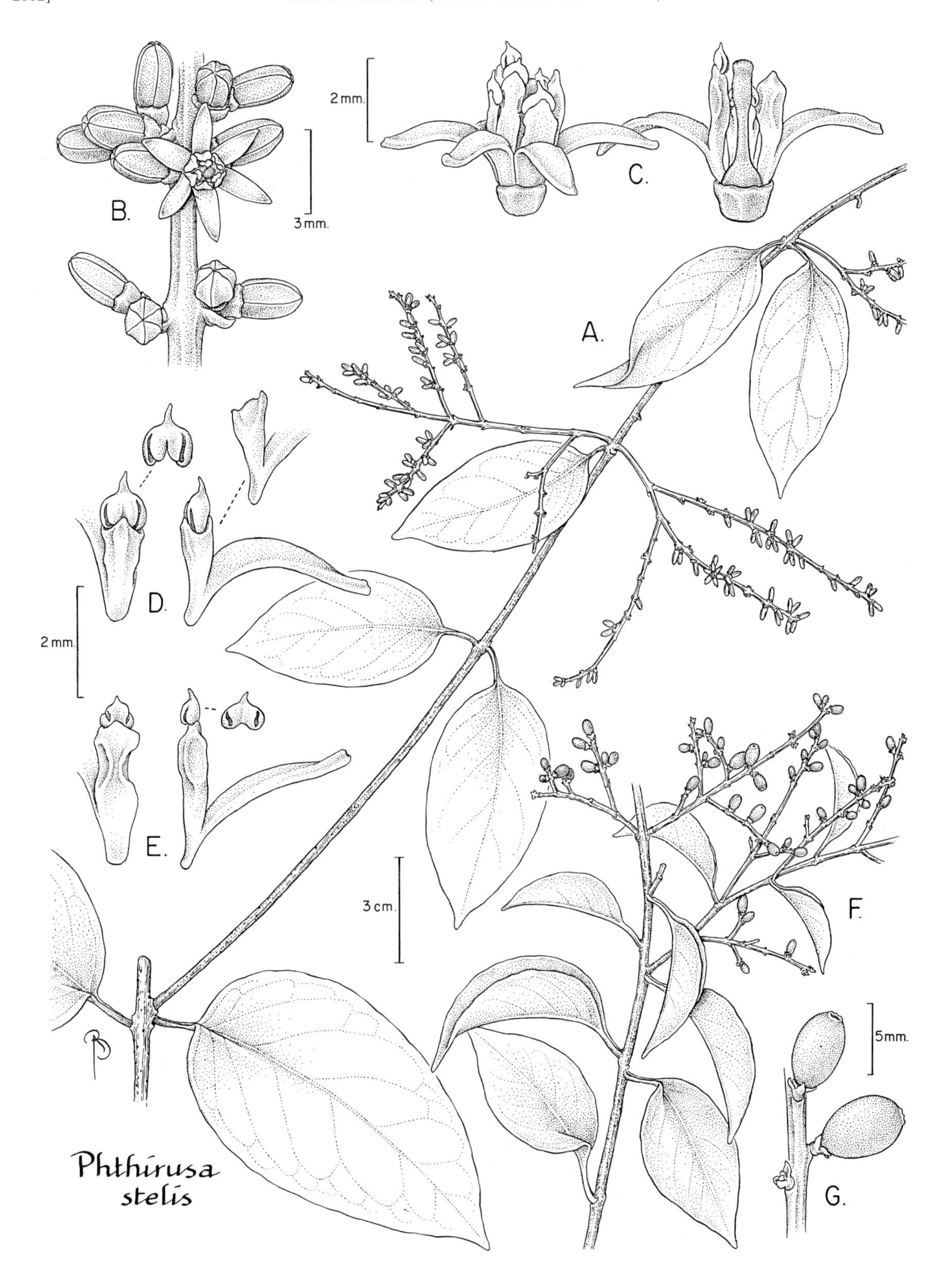

FIG. 170. LORANTHACEAE. *Phthirusa stelis* (A–E, *Mori et al. 22874*; F, G, *Mori & Gracie 18915*). **A.** Part of stem with leaves and staminate inflorescences. **B.** Detail of staminate inflorescence with open flower and flower buds. **C.** Lateral view of staminate flower (left) and lateral view with two tepals removed (right). **D.** Adaxial (left) and lateral (right) views of short stamens with details of anther (above left) and attachment of filament to base of tepal (above right). **E.** Adaxial (left) and lateral (right) views of long stamen with detail of anther (above right). **F.** Part of stem with leaves and infructescences. **G.** Detail of infructescence with immature fruits.

axils), bearing many pairs of pedunculate to nearly sessile triads; pedicels absent. Flowers variable in size, to 7 mm long in bud, staminate buds short-clavate, the pistillate buds rather acute and narrow. Fruits ellipsoid, to 9 × 5 mm, dull orange-red at maturity. Fl (Jan, Feb), fr (May, Jun).

PSITTACANTHUS Mart.

Large, stout hemiparasites, the primary haustorium very large, creeping roots absent. Stems without scurfy lines. Leaves: petioles short; blades fleshy; venation pinnate. Inflorescences at apex of leafy shoots, consisting of paired, pedunculate triads bearing pedicellate flowers, the flowers subtended by single bract or bracteole. Flowers bisexual; petals >10 mm long, usually reddish or yellowish; anthers linear, versatile; style long, nearly as long as petals. Fruits large one-seeded berries. Seeds occupying most of fruit; embryo green, consisting mostly of cotyledons with very short, rounded radicular end; endosperm absent.

1. Plants not dichotomously branched. Leaves with robust petioles and blades. Fruiting pedicels 2–4 mm thick. . . . . *P. corynocephalus*.
1. Plants probably dichotomously branched. Leaves without robust petioles and blades. Fruiting pedicels <2 mm thick.. . . . . *Psittacanthus* sp. A.

Psittacanthus corynocephalus Eichler

Plants coarse, not dichotomously branched. Leaves: petioles robust; blades ovate to nearly orbicular, very fleshy, the apex rounded. Inflorescences terminal, rarely in axils of distal leaves but then reduced in size; peduncles and pedicels stout. Flowers: buds to 12.5 cm long, very fleshy, the style straight, very long; anthers ≥2 cm long. Fruits large, purple berries; pedicels 2–4 mm thick. Fl (Jun); canopy.

Psittacanthus sp. **A** (Mori et al. 23867)

Plants not coarse, apparently dichotomously branched. Leaves: petioles rather slender; blades ovate, rather thin. Inflorescences and flowers unknown. Fruits large, red berries; pedicels rather slender, <2 mm thick. Fr (Sep); on branches of large tree.

STRUTHANTHUS Mart.

Fairly large, scandent, hemiparasites with haustoria-bearing roots both from stems and from near primary haustorium. Leaves petiolate, rather thin, venation pinnate. Inflorescences axillary, unbranched, the flowers in paired triads with lateral flowers subtended by 1 bracteole each. Flowers unisexual but aborted organs of opposite sex always present (plants dioecious); petals <6 mm long, usually greenish white; stamens with prominent, free filaments, the anthers versatile. Fruits orange, reddish, or blue. Seeds: embryo dicotylar, green, with swollen radicular end (Pl. 79e); endosperm copious.

1. Plants delicate. Mature leaf-bearing stems without lenticels. Leaf blades acute at base and apex. Inflorescences with 1–2 triad pairs. . . . . *S. dichotrianthus*.
1. Plants robust. Mature leaf-bearing stems with numerous lenticels. Leaf blades obtuse at base and apex (often with small acuminate tip). Inflorescences with 4–5 triad pairs. . . . . *S. syringifolius*.

Struthanthus dichotrianthus Eichler

Rather delicate plants. Stems with elongate, slender internodes lacking lenticels. Leaves: petioles slender; blades narrowly lanceolate, mostly 5 × 2 cm, the base acute, the apex acute. Inflorescences axillary, 1 to several per axil, the triads in 1–2, spreading pairs. Fl (Feb, Mar, May), fr (May).

Struthanthus syringifolius (Mart.) Mart. PL. 79d

Robust plants. Stems rough, with numerous minute lenticels. Leaves: petioles rather stout; blades broadly lanceolate to elliptic, the base obtuse, the apex obtuse but often with small acuminate tip. Inflorescences axillary, usually 1 per axil, the triads in 4–5 pairs oriented perpendicular to axis. Fl (Sep).

LYTHRACEAE (Loosestrife Family)

Maria Lúcia Kawasaki

Herbs, subshrubs, small shrubs, or trees (outside our area). Leaves simple, opposite. Stipules inconspicuous or absent. Inflorescences axillary, solitary, racemes, or appearing fasciculate. Flowers perigynous, actinomorphic (outside our area) or zygomorphic, bisexual, 4-, 6-, or 8-merous; petals often clawed or narrowed at base; stamens mostly twice as many as sepals or petals; ovary superior, 2-carpellate, 1-locular; placentation axile. Fruits commonly loculicidal capsules.

Görts-van Rijn, A. R. A. & M. J. Jansen-Jacobs. 1986. Lythraceae. *In* A. L. Stoffers & J. C. Lindeman (eds.), Flora of Suriname **III(1–2):** 493–497. E. J. Brill, Leiden.
Jonker, F. P. 1966. Lythraceae. *In* A. Pulle (ed.), Flora of Suriname **III(1):** 422–432. E. J. Brill, Leiden.

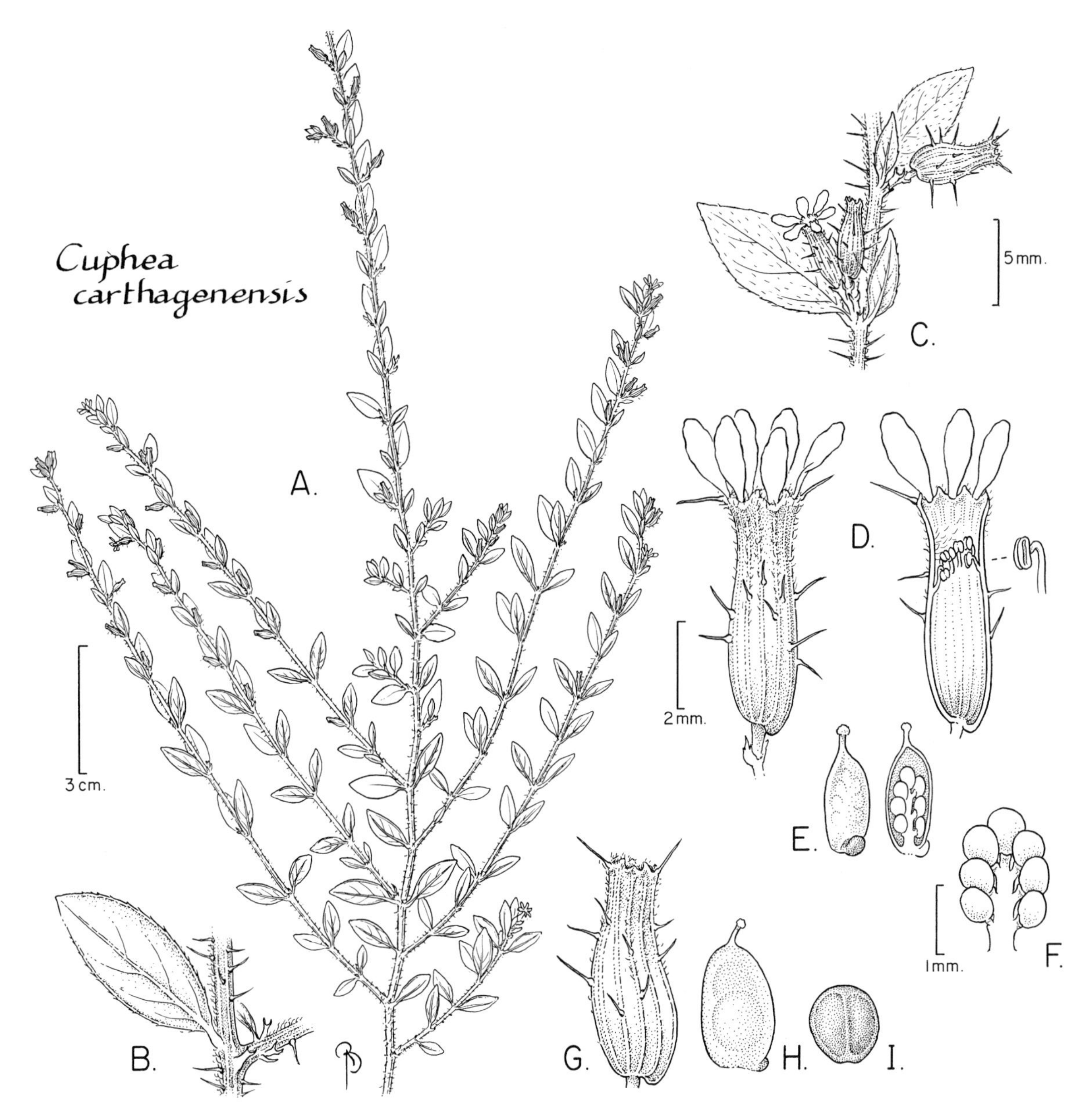

FIG. 171. LYTHRACEAE. *Cuphea carthagenensis* (A–F, *Mori & Gracie 18428*; G–I, *Mori et al. 22083*). **A.** Apical portion of plant with leaves, flowers, and fruits. **B.** Detail of stem with leaf and branch at node. **C.** Detail of stem with flower and fruits enclosed in calyces. **D.** Lateral view (left) and medial section of flower (near right) with enlargement of stamen (far right). **E.** Pistil with nectary disc at base (near right) and medial section of pistil (far right). **F.** Detail of ovules on placenta. **G.** Lateral view of fruit enclosed in calyx. **H.** Lateral view of fruit. **I.** Seed.

CUPHEA P. Browne

Herbs, subshrubs, or small shrubs. Flowers: calyx narrowly tubular, gibbous at base, 12-veined; nectary disc present; ovary asymmetric, oblong; ovules 2-several, the style short, included; stamens usually 11, included. Capsules thin-walled. Seeds dorsiventrally flattened, narrowly winged.

Cuphea carthagenensis (Jacq.) J. F. Macbr.
FIG. 171, PL. 80a

Subshrubs. Stems hispid, the hairs glandular. Leaves: petioles 1–2 mm long; blades elliptic to narrowly elliptic, 1–1.5 × 0.4–0.9 cm, heteromorphic, i.e., sometimes one leaf of node much larger than other, chartaceous, scabrous on both sides, the base cuneate, the apex acute. Inflorescences axillary or interaxillary; pedicels sparsely hirsute, to 2 mm long. Flowers: calyx 4–5 mm long; petals magenta or lilac, glabrous, ca. 2 mm long; ovary glabrous, ca. 2 mm long; stamens inserted at upper part of calyx-tube; nectary disc small, subglobose. Fruits ca. 4 mm long. Seeds suborbicular. Fl (Jun, Aug, Sep, Oct); common, in weedy places. *Radié raide* (Créole).

MALPIGHIACEAE (Malpighia Family)

William R. Anderson

Trees, shrubs, and lianas. Hairs unicellular, usually medifixed or submedifixed (malpighiaceous trichomes). Stipules usually present. Leaves usually opposite, often bearing large multicellular glands on petiole or blade or both; blades simple, usually entire, lobed in *Stigmaphyllon* spp. Flowers subtly to strongly zygomorphic, bisexual; sepals 5, eglandular or, most often, the lateral 4 or all 5 bearing (1)2 large multicellular abaxial glands; petals 5, distinct, clawed, alternating with sepals, imbricate, the innermost (flag) petal posterior and often different from lateral 4; stamens 10 (in ours), the anthers dehiscent by longitudinal slits; gynoecium superior, comprising (2)3 free to connate carpels, each fertile locule containing 1 pendent anatropous ovule, the styles 1 per carpel, distinct (connate in some species of *Bunchosia*). Fruits dry or fleshy, dehiscent or indehiscent, samaroid, nutlike, or drupaceous. Mature seeds without endosperm.

Anderson, W. R. 1981. Malpighiaceae. *In* B. Maguire (ed.), The Botany of the Guayana Highland — Part XI. Mem. New York Bot. Gard. **32:** 21–305.

The following references were not consulted in the preparation of this treatment. They are cited here at the request of the editor for the convenience of those who need to know what else has been published on the Malpighiaceae in the Guianas.

Görts-van Rijn, A. R. A. & M. J. Jansen-Jacobs. 1976. Malpighiaceae. *In* J. Lanjouw & A. L. Stoffers (eds.), Flora of Suriname **II(2):** 445–450. E. J. Brill, Leiden.

Jonker, F. P. 1966. Malpighiaceae. *In* A. Pulle (ed.), Flora of Suriname **II(1):** 478–480. E. J. Brill, Leiden.

Kostermans, A. J. G. H. 1936. Malpighiaceae. *In* A. Pulle (ed.), Flora of Suriname **II(1):** 146–243. J. H. De Bussy, Amsterdam.

Successful use of the keys to genera and species requires an understanding of what I mean by certain morphological terms. The most important of these are defined here; for more details, see my treatment of the Malpighiaceae of the Venezuelan Guayana (W. R. Anderson, 2001). The ancestral inflorescence of the Malpighiaceae was a raceme of cincinni, but in most genera the cincinni have been reduced to one-flowered units. Each flower is borne on a *pedicel*, whose base is defined by a joint; below the joint the stalk is called the *peduncle*, and the peduncle bears two *bracteoles*; the peduncle is subtended by a single *bract*. The peduncle has been lost in several evolutionary lines, in which case the pedicel is described as sessile, subtended then by a cluster of the bract and two bracteoles. The flower's plane of symmetry is defined by the *anterior sepal* and the *posterior petal*, which are often different from the *lateral sepals* and *lateral petals*, respectively. Of the three styles, the anterior lies on the plane of symmetry and is often unique, while the posterior two lie to the right and left of the plane of symmetry and are mirror images of each other.

1. Trees or shrubs. Fruits fleshy, unwinged, indehiscent.
    2. Leaf blades eglandular. Inflorescences terminal. Styles slender, subulate, the stigmas minute. Fruits containing 1 trilocular stone. . . . . *Byrsonima.*
    2. Leaf blades bearing abaxial glands. Inflorescences lateral, axillary. Styles stout, of uniform thickness, the stigmas large. Fruits containing 2 or 3 pyrenes, distinct or united in center.
        3. Inflorescences elongate pseudoracemes; bracteoles with one of each pair bearing 1 abaxial gland. Petals yellow; stamens ± alike; styles 2 (or 1 through fusion of 2), the stigmas terminal. . . . . *Bunchosia.*
        3. Inflorescences umbels of 2–4 flowers; bracteoles both eglandular. Petals pink or purplish; stamens heteromorphic; styles 3, the stigmas on internal angle at apex. . . . . *Malpighia.*
1. Lianas. Fruits dry, winged, breaking apart into 3 samaras (or fewer if 1 or 2 carpels abort).
    4. Specimens with flowers.
        5. Leaf blades moderately to deeply cordate at base.
            6. Styles bearing at apex a large rounded dorsal appendage 1.5–2 mm long, symmetrical on anterior style, unilateral on posterior styles. . . . . *Stigmaphyllon.*
            6. Styles rounded or truncate at apex, or extended at most into very short sharp dorsal hook.
                7. Stipules borne on stem between petioles, caducous or eventually deciduous, leaving wide scar. Inflorescences bearing many conspicuous leaflike bracts 8–25 mm long. . . . . *Tetrapterys.*
                7. Stipules borne on petiole, well above base, persistent. Inflorescences without leaflike bracts. . . . . *Hiraea.*
        5. Leaf blades cuneate to rounded or subcordate at base.
            8. Bracteoles globose-cymbiform, enclosing bud until flowers open, borne just below flower, the pedicel absent or to 2 mm long in fruit. . . . . *Mezia.*
            8. Bracteoles not globose or enclosing bud, separated from flower by well developed pedicel.
                9. Petals pink or lilac in bud and during flowering.
                    10. Leaf blades persistently velutinous adaxially. Bracts and bracteoles 6–9 mm long. Sepals 5–8.5 mm long, distally inflated with aerenchyma, the lateral 4 each bearing 1 large abaxial gland. . . . . *Jubelina.*

10. Leaf blades glabrate adaxially at maturity or with hairs persistent on midrib. Bracts and bracteoles 0.5–2 mm long. Sepals 1.8–5 mm long, not inflated, the lateral 4 each bearing 2 abaxial glands.
    11. Ultimate branches of inflorescence umbels of 4–6 flowers; pedicels sessile or subsessile, raised at most on peduncles 0.5 mm long; bracts and bracteoles deciduous. Styles with terminal capitate stigmas. . . . . *Banisteriopsis.*
    11. Ultimate branches of inflorescence pseudoracemes of 7–35 flowers; pedicels raised on peduncles 1–3(5) mm long; bracts and bracteoles persistent. Styles stigmatic on the internal angle, dorsally apiculate or short-hooked at apex. . . . . *Mascagnia.*

9. Petals yellow, or yellow and red, or yellow in bud turning red or orange during flowering.
    12. Ultimate branches of inflorescence bearing (2)4–40 flowers in pseudoracemes.
        13. Petals glabrous; styles with terminal stigmas. . . . . *Banisteriopsis.*
        13. Petals (at least the lateral 4) abaxially sericeous; styles stigmatic on internal angle. . . . *Tetrapterys.*
    12. Ultimate branches of inflorescence bearing 4–6(8) flowers in umbels.
        14. Stipules borne on petiole, well above base. . . . . *Hiraea.*
        14. Stipules borne on stem between petioles, or absent.
            15. Sepals becoming revolute at apex during flowering. . . . . *Heteropterys.*
            15. Sepals appressed during flowering. . . . . *Tetrapterys.*

4. Specimens with fruits.
    16. Samara with dorsal wing dominant, the nut bearing on its sides only short winglets or crests or quite smooth.
        17. Wing of samara with abaxial edge thickened, the veins diverging and branching from it toward thinner adaxial edge. . . . . *Heteropterys.*
        17. Wing of samara with adaxial edge thickened, the veins diverging and branching from it toward thinner abaxial edge.
            18. Petioles 25–95 mm long; blades moderately to deeply cordate at base. Styles stigmatic on internal angle, each bearing at apex a large rounded dorsal appendage 1.5–2 mm long, symmetrical on anterior style, unilateral on posterior styles. *Stigmaphyllon.*
            18. Petioles 3–15 mm long; blades cuneate to rounded or subcordate at base. Styles with terminal stigmas and without any sort of dorsal appendage at apex. . . . . *Banisteriopsis.*
    16. Samara with lateral wing(s) dominant, the dorsal wing smaller or reduced to winglet or crest, absent in a few species.
        19. Samara with 4 discrete lateral wings, 2 on each side.
            20. Stipules interpetiolar or borne on petiole between base and middle, or absent. . . . . *Tetrapterys.*
            20. Stipules borne on petiole near its apex. . . . . *Hiraea.*
        19. Samara with 1 continuous lateral wing, or 2, 1 on each side.
            21. Stipules well developed, borne on petiole well above base. Samaras butterfly-shaped with 2 discrete lateral wings. . . . . *Hiraea.*
            21. Stipules very small and interpetiolar or borne on very base of petiole, or absent. Samaras with lateral wing continuous to nearly distinct at base.
                22. Bracteoles globose-cymbiform, enclosing bud until flowers open, borne just below flower, the pedicel absent or to 2 mm long in fruit. . . . . *Mezia.*
                22. Bracteoles not globose or enclosing the bud, separated from flower by well developed pedicel.
                    23. Leaf blades persistently velutinous. Bracts and bracteoles 6–9 mm long. Sepals 5–8.5 mm long, distally inflated with aerenchyma, the lateral 4 each bearing 1 large abaxial gland. Samaras 55–80 × 35–60 mm. . . . . *Jubelina.*
                    23. Leaf blades thinly sericeous to glabrate at maturity. Bracts and bracteoles 0.5–1.3 mm long. Sepals 1.8–2.5 mm long, not inflated, the lateral 4 each bearing 2 abaxial glands. Samaras to 30 mm diam. . . . . *Mascagnia.*

BANISTERIOPSIS C. B. Rob.

Lianas (our species), vines, shrubs, or rarely small trees. Stipules small, distinct, interpetiolar. Inflorescences: peduncle usually absent or very short, but well developed in a few species. Flowers: anterior sepal eglandular, the lateral 4 (in ours) biglandular; petals yellow, pink, or white; anthers alike in some species but more commonly strongly dissimilar; styles 3, the stigmas terminal, without dorsal extensions.

Fruits breaking apart into 3 samaras, samara with dorsal wing dominant, thickened on adaxial (upper) edge, the veins terminating in thinner abaxial edge; much shorter winglets or crests present on sides of nut in some species.

Gates, B. 1982. *Banisteriopsis, Diplopterys* (Malpighiaceae). Fl. Neotrop. Monogr. **30:** 1–237.

1. Larger leaf blades 4.5–9 cm long. Branches of inflorescence terminating in umbels of 4–6 flowers; bracts and bracteoles deciduous. Petals pink. . . . . *B. schwannioides.*
1. Larger leaf blades 9.5–22 cm long. Branches of inflorescence terminating in pseudoracemes of 10–40 flowers; bracts and bracteoles persistent. Petals yellow.
    2. Leaves bearing several minute glands along margin of blade, plus 1 pair of larger glands at apex of petiole. Sepals appressed during flowering; lateral petals entire, denticulate, or dentate, posterior petal short-fimbriate; anthers all with connective not glandular-swollen; styles ± equal in length. . . . . *B. wurdackii.*
    2. Leaves usually bearing 2 large glands at juncture of blade and petiole and otherwise eglandular. Sepals becoming revolute during flowering; petals all fimbriate; anthers opposite 3 anterior sepals with connective glandular-swollen; anterior style much shorter than 2 posterior styles. . . . . *B. carolina.*

Banisteriopsis carolina W. R. Anderson

Lianas. Leaves: longer petioles 9–15 mm long, eglandular or biglandular at apex; larger blades elliptic or slightly ovate, 9.5–13.2 × 5–7.3 cm, initially thinly sericeous on both sides, at maturity glabrate adaxially and very thinly sericeous (apparently glabrate) abaxially, mostly bearing a pair of bulging glands at juncture of lamina and petiole and otherwise eglandular, the base obtuse to rounded, the apex abruptly short-acuminate. Inflorescence a slender axis 15–30 cm long, axillary to a full-sized vegetative leaf and bearing several pairs of much reduced leaves, each subtending 1 pseudoraceme of 15–35 flowers; bracts and bracteoles 0.7–1 mm long, persistent; pedicels 6–7.5 mm long, occasionally sessile but mostly raised on a peduncle 0.3–1 mm long. Flowers: sepals 2 mm long, becoming revolute during flowering; petals bright yellow, glabrous, the limb fimbriate all around margin with divisions glandular, at least proximally, on posterior 3 petals; anthers glabrous, the 3 opposite anterior sepals with connective glandular and much enlarged; anterior style ca. 2 mm long, bending forward at base and then erect, the 2 posterior styles 2.7–3 mm long, lyrate, bending strongly backward at base and then sigmoid-ascending. Fruits unknown. Fl (Feb); in nonflooded moist forest.

Banisteriopsis schwannioides (Griseb.) B. Gates FIG. 172

Lianas. Leaves: longer petioles 3–7 mm long, eglandular; larger blades ovate or elliptic to orbicular, 4.5–9 × 2.2–4.8 cm, tomentose to glabrate adaxially, persistently sericeous abaxially, bearing a pair of glands abaxially at base beside midrib and often 1–2 smaller pairs distally between midrib and margin, the base broadly cuneate or truncate to subcordate, the apex acuminate to apiculate. Inflorescences terminal or axillary, cymose, the ultimate branches umbels of 4–6 flowers; bracts and bracteoles 1.2–2 mm long, deciduous during or soon after flowering; pedicels 11–24 mm long, sessile or subsessile, the peduncle to 0.5 mm long. Flowers: sepals 3–5 mm long, appressed during flowering; petals pink, glabrous, dentate to fimbriate; anthers glabrous or sparsely pilose, the anterior 5 with enlarged glandular connectives, those opposite anterior-lateral sepals with connectives much exceeding locules; anterior style much thicker than posterior 2, the stigmas capitate. Samaras with dorsal wing 27–30 × 11–13 mm, the nut tuberculate to aculeate on sides. Fl (Nov); in nonflooded moist forest.

The Saül collections of this species have the leaf hairs more appressed than other populations, which all occur along the Amazon River in Brazil. Also, their petals are more deeply dissected and the anther locules are sparsely pilose, whereas they are glabrous in the Amazon. Perhaps eventually the French Guianan population will prove to be taxonomically recognizable, but for now it is best left in *B. schwannioides*, which it resembles in most characters.

Banisteriopsis wurdackii B. Gates

Lianas. Leaves: longer petioles 7–12 mm long, biglandular at apex; larger blades ovate to broadly elliptic, 10–22 × 4.5–11 cm, glabrous adaxially, sparsely sericeous abaxially, bearing minute glands on margins, the base cuneate to obtuse or rarely cordate, the apex short- to long-acuminate. Inflorescences axillary, paniculate, the ultimate branches pseudoracemes of 10–40 flowers; bracts and bracteoles 0.6–1 mm long, persistent; pedicels 6–12 mm long, subsessile, the peduncle 0.5–2 mm long. Flowers: sepals 1.2–1.6 mm long, appressed during flowering; petals glabrous, yellow, the lateral 4 entire or denticulate, the posterior short-fimbriate with basal fimbriae often glandular; anthers glabrous, the connectives not glandular-swollen; styles ± equal, the stigmas truncate. Samaras with dorsal wing 24–38 × 10–15 mm; nut bearing on each side a single wing parallel to areole, 4–10 × 2–4 mm, rarely absent. Fr (Dec); in nonflooded moist forest.

BUNCHOSIA Kunth

Trees or shrubs. Stipules small, distinct, borne on base of petiole. Leaves: petioles eglandular; blades usually bearing glands impressed in abaxial surface. Inflorescences elongate axillary pseudoracemes; bracts eglandular; bracteoles with 1 or both usually bearing 1(2) abaxial glands. Flowers: petals glabrous, yellow; anthers ± alike; carpels 2 (in our species) or 3; styles as many as carpels, distinct or partially to completely connate, stout, the large terminal stigmas subpeltate or apparently capitate. Fruits drupaceous, containing as many pyrenes as carpels in common fleshy exocarp, yellow, orange, or red at maturity, the pyrenes elongate, round or elliptic in transverse section, free from each other at maturity, with smooth, brittle, cartilaginous wall.

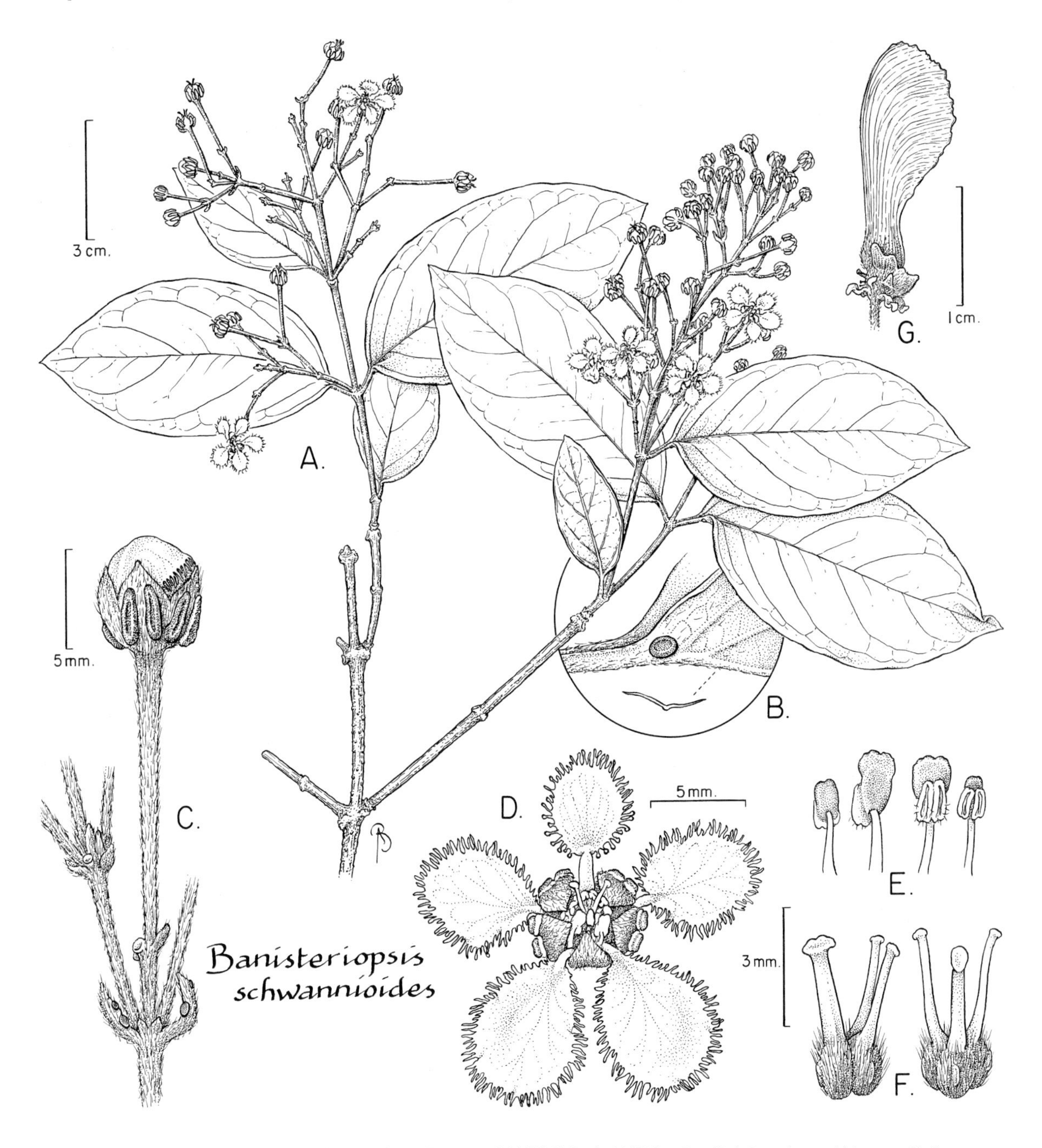

FIG. 172. MALPIGHIACEAE. *Banisteriopsis schwannioides* (A–F, *Mori et al. 21590*; G, *Ducke 11018* from Brazil). **A.** Part of stem with leaves and inflorescences. **B.** Detail of base of leaf showing abaxial gland and hair. **C.** Detail of cymose inflorescence showing one of the four flowers of the central umbel (in bud), the pedicels of three of the four flowers in the umbel on the left, and only the stalk of the umbel on the right; the bracts and bracteoles that subtend the pedicels are deciduous during or soon after flowering. **D.** Apical view of flower with posterior petal uppermost. **E.** Abaxial (left) and adaxial (right) views of anthers, the middle two from stamens opposite the anterior-lateral sepals, the outer two from stamens opposite the anterior-lateral petals. **F.** Lateral (left) and frontal (right) views of gynoecium; note that the anterior style is much thicker than the two posterior styles. **G.** Samara with large dorsal wing and outgrowths on side of nut.

1. Leaf blades nearly or quite glabrate at maturity, plane or slightly revolute at margin. Pseudoracemes usually bearing 20–40 or more flowers, 1 per bract. Style(s) glabrous, ca. 1.5 mm long. Dried fruits to 16 mm long, 16 mm diam., the apex rounded, the wall granulate. . . . . *B. decussiflora.*
1. Leaf blades thinly but persistently sericeous abaxially, undulate and crispate at margin. Pseudoracemes bearing 10–20 flowers, sometimes 2 (1 above the other) in axil of 1 bract. Style sericeous, ca. 3–3.5 mm long. Dried fruits 20–28 mm long, 15–20 mm diam., the apex beaked, the wall smooth. . . . . *B. glandulifera.*

Bunchosia decussiflora W. R. Anderson FIG. 173

Shrubs or trees, 3–12(25) m tall. Leaves: longer petioles 6–15 mm long; larger blades elliptic or slightly ovate or obovate, 13–23 × 4.5–8.5 cm, nearly or quite glabrate at maturity, the base cuneate, the apex acuminate, the margins plane or slightly revolute. Inflorescences usually 7–15 cm long, bearing 20–40 or more flowers, 1 per bract. Flowers: ovary bicarpellate, sericeous, the style usually 1 (formed from 2 completely connate) or occasionally 2 (ca. 1/2-connate), 1.5 mm long, glabrous. Fruits 9–16 × 9–16 mm (dried), orange to red, the apex rounded, the wall granulate. Fl (Jan, Jul); in forest.

Bunchosia glandulifera (Jacq.) Kunth

Shrubs or small trees, 2–6(8) m tall. Leaves: longer petioles 6–8 mm long; larger blades elliptic or ovate, 11–18 × 5.5–10(12) cm, thinly but persistently sericeous abaxially, the base cuneate to rounded and often slightly attenuate, the apex long-acuminate, the margins undulate and crispate. Inflorescences 4–11 cm long, bearing 10–20 flowers, sometimes 2 (1 above the other) in axil of same bract. Flowers: ovary bicarpellate, sericeous, the style 1 (formed from 2 completely connate styles), 3–3.5 mm long, sericeous. Fruits 20–28 × 15–20 mm (dried), orange to red, the apex with elongate beak, the wall smooth. Cultivated for the large edible fruits; not native or naturalized in our area.

BYRSONIMA Kunth

Trees or shrubs. Leaves, bracts, and bracteoles eglandular. Stipules intra- and epipetiolar, distinct or partially to completely connate. Inflorescences terminal, racemes of few-flowered cincinni or pseudoracemes (i.e., raceme of 1-flowered cincinni). Flowers: sepals all eglandular or all biglandular; petals yellow, white, pink, or red; anthers ± alike; styles 3, slender and subulate, the stigmas minute and apical or slightly internal. Fruits drupes, the thin flesh turning yellow, orange, red, purple, blue, or blue-black at maturity, the single stone with hard wall, trilocular.

1. Stipules distinct, 1.5–3 mm long. Inflorescence bracts 0.8–1.5 mm long. Petals white turning pink or red; anthers glabrous; ovary glabrous. Fruits 5–8 mm diam. (dried), glabrous. . . . . *B. densa.*
1. Stipules completely connate, at least 2.5 mm long. Inflorescence bracts 3.5–10 mm long. Petals yellow; anthers sericeous; ovary sericeous or short-velutinous. Fruits 10–18 mm diam. (dried), sericeous or tomentose to glabrate.
   2. Stipules 2.5–4.5 mm long, persistent on base of petiole. Leaf blades sericeous to glabrate abaxially, the hairs all sessile, straight, appressed, 2-branched. . . . . *B. aerugo.*
   2. Stipules (8)12–25 mm long, deciduous before leaves. Leaf blades velutinous abaxially, some hairs stalked and stellate (i.e., with more than 2 branches). . . . . *B. stipulacea.*

Byrsonima aerugo Sagot FIG. 174, PL. 80b

Trees, (4)15–45 m tall. Stems tightly sericeous. Stipules 2.5–4.5 mm long, completely connate, persistent. Leaves: longer petioles (15)20–40 mm long; larger leaf blades elliptic or slightly obovate, 11–22(25) × 4–9(11) cm, densely ferruginous-sericeous abaxially, the hairs sometimes patchily deciduous and blade eventually nearly glabrate, the hairs sessile, straight, appressed, the base cuneate or attenate, the apex acute or usually acuminate. Inflorescences: bracts 3.5–5 mm long; bracteoles 0.6–1.5 mm long. Flowers: petals yellow; anthers 2–3 mm long, sericeous, especially between locules, the connective equalling locules or extended beyond them to 0.6 mm; ovary densely sericeous. Fruits 10–13 mm diam. (dried), sericeous to glabrate, yellow. Fl (May), fr (Jun); in nonflooded moist forest.

This species is very close to *Byrsonima crispa* A. Juss., which is not known from the Guianas but extends from eastern Brazil through much of Amazonia. The only real difference between them is that the leaves of *B. crispa* are never densely sericeous as in *B. aerugo*; they have similar hairs, but they are very sparse, hardly visible without a lens. When the leaves of *B. aerugo* are glabrescent, as they tend to be in the populations in our area, the similarity of the two species is inescapable. *Byrsonima crispa* is much the older of the two names.

Byrsonima densa (Poir.) DC.

Trees, 5–30(40) m tall. Stems glabrous except hirsute in axils of stipules. Stipules 1.5–3 mm long, distinct, persistent. Leaves: longer petioles 7–12 mm long, sericeous to glabrate; larger blades elliptic or obovate, 9–12(13.5) × 3.8–5(7) cm, initially sericeous but soon nearly or quite glabrate, the base cuneate, the apex obtuse, acute, or abruptly short-acuminate. Inflorescences: bracts 0.8–1.5 mm long; bracteoles 0.5–1 mm long. Flowers: petals white turning pink or red; anthers 0.9–1.2 mm long, glabrous, the connective exceeding locules by 0.2–0.5 mm; ovary glabrous. Fruits 5–8 mm diam. (dried), green contrasting with subtending red sepals, glabrous. Fl (May), fr (Nov); in nonflooded moist forest.

Byrsonima stipulacea A. Juss. PL. 80c

Trees, 8–25 m tall. Stems velutinous. Stipules (8)12–25 mm long, amplexicaulous, completely connate, each pair deciduous independently of and often well before leaf. Leaves: longer petioles 12–27 mm long, velutinous; larger leaf blades elliptic or rhombic, 12–27 × 6–13 cm, velutinous (to glabrescent adaxially) with mixture of long, basifixed simple hairs, stalked stellate hairs, and sessile stellate hairs, the base cuneate, the apex acute or obtuse, sometimes acuminate or rounded. Inflorescences: bracts 6–10 mm long; bracteoles 2.5–5 mm long. Flowers: petals yellow; anthers 2.2–4.2 mm long, loosely sericeous on both sides, the connective usually exceeding locules by 0.4–1.1 mm; ovary densely short-velutinous with overlay of appressed hairs. Fruits 12–18 mm diam. (dried), tomentose to glabrate, orange-yellow. Fr (Dec, Aug); in nonflooded moist forest.

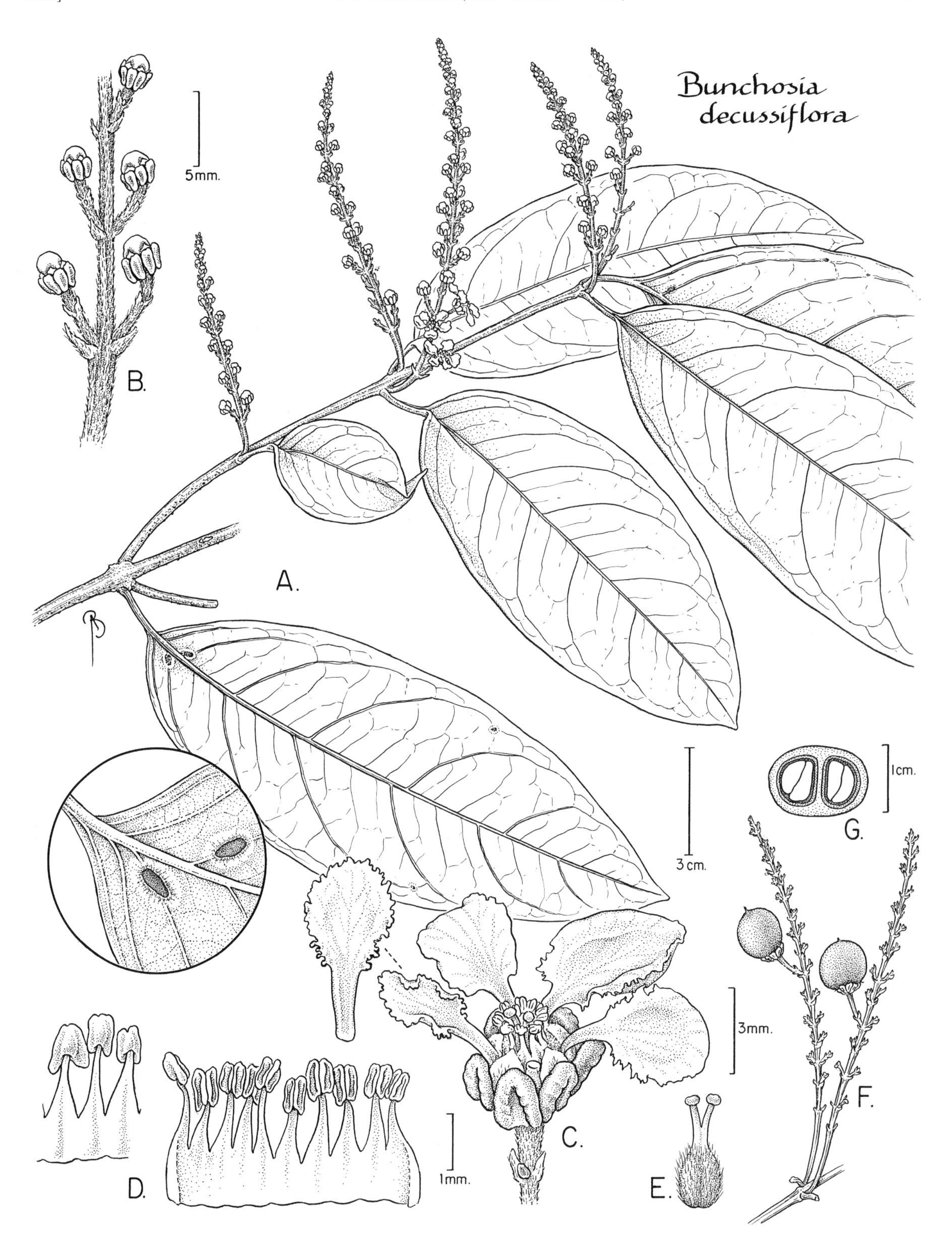

FIG. 173. MALPIGHIACEAE. *Bunchosia decussiflora* (A–E, *Cremers 7560* from Mana River, French Guiana; F, *Silva & Brazão 60764* from Brazil; G, *Cremers 6531* from between Sommet Tabulaire and Massif des Emerillons, French Guiana). **A.** Part of stem with leaves and inflorescences and detail of abaxial leaf base showing glands. **B.** Part of decussate pseudoraceme with flower buds. **C.** Oblique-lateral view of flower with posterior petal to left and also shown adaxially above left; one posterior-lateral petal has been removed; note gland on one bracteole. **D.** Adaxial view of opened androecium (right) and abaxial view of part of androecium with three stamens (left). **E.** Bicarpellate gynoecium, the two styles half-connate. **F.** Fruiting pseudoracemes. **G.** Transverse section of fruit, showing two large seeds, each enclosed in a thin, cartilaginous endocarp, the two pyrenes borne in thin common flesh.

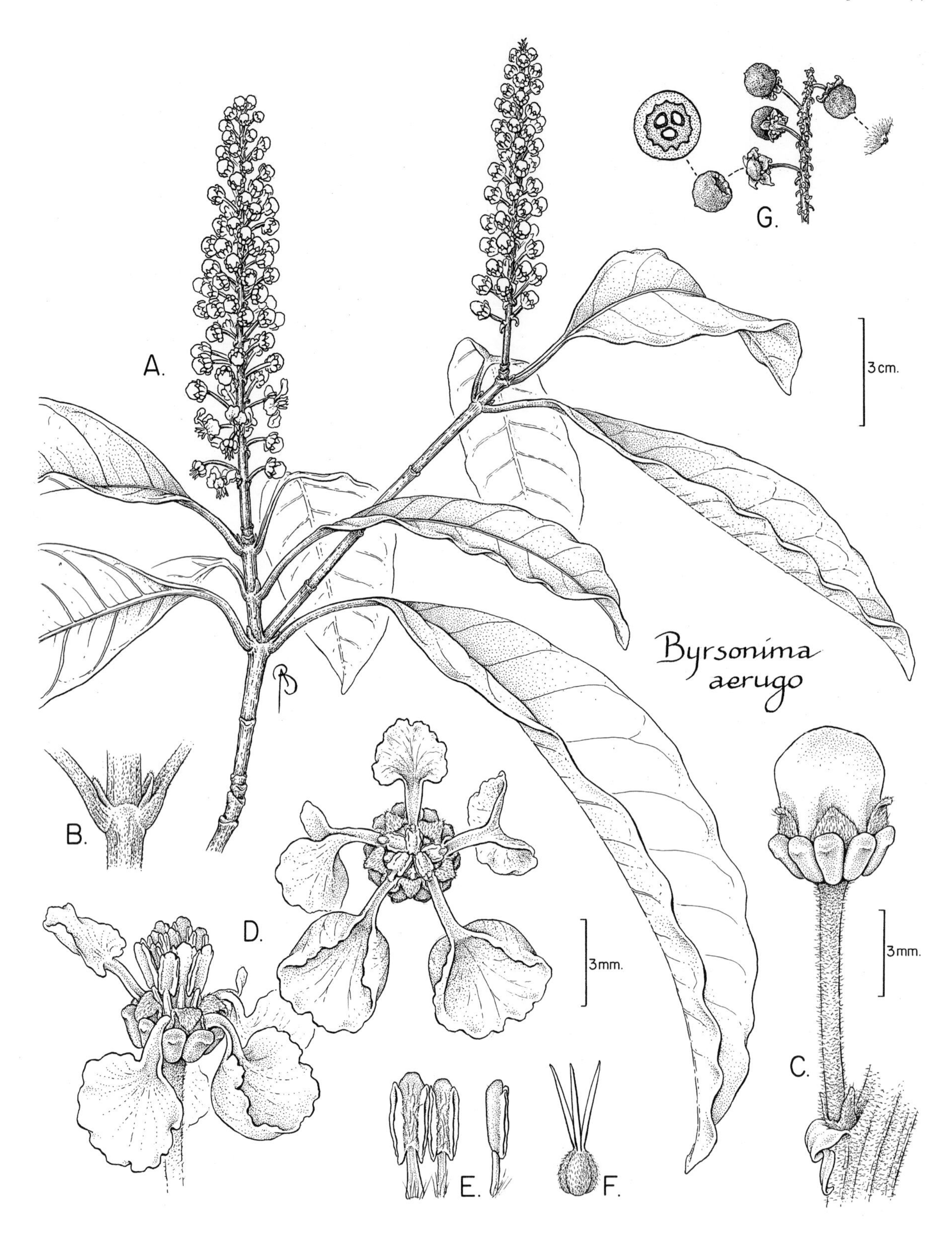

FIG. 174. MALPIGHIACEAE. *Byrsonima aerugo* (A–F, *Mori et al. 22196*; G, *Liesner 19586* from Venezuela). **A.** Stem with leaves and inflorescences. **B.** Node showing petiole bases and connate intrapetiolar stipules. **C.** Flower bud with sessile pedicel subtended by two short bracteoles and one long, reflexed bract; note oil glands on calyx. **D.** Lateral (left) and apical (right) views of flowers. **E.** Adaxial (left) and lateral (right) views of stamens. **F.** Gynoecium. **G.** Part of infructescence (right) with one fruit removed (near left), and transverse section of fruit (far left), showing three seeds in a common stony endocarp surrounded by fleshy mesocarp; detail of sericeous apex of fruit (above, far right).

HETEROPTERYS Kunth

Lianas (our species), vines, shrubs, or small trees. Stipules very small, distinct, triangular, borne on or beside base of petiole, or absent. Leaves usually bearing glands. Inflorescences umbels, corymbs, or pseudoracemes, these single or grouped in axillary or terminal racemes or panicles. Flowers: petals mostly yellow or pink; anthers ± alike; styles 3, the apex with large, usually internal stigma and dorsally rounded, truncate, acute, or hooked. Fruits breaking apart into 3 samaras, the samara with dorsal wing dominant, thickened on abaxial (lower) edge and usually bent upward, the veins terminating in thinner adaxial edge; much shorter winglets or crests present on sides of nut in some species.

Heteropterys oligantha W. R. Anderson
FIG. 175; PART 1: FIG. 6 as *H. siderosa*

Lianas. Stipules apparently absent. Leaves: longer petioles 4–10 mm long; larger leaf blades elliptic, 8–16.5 × 3–7.5 cm, bearing several abaxial glands scattered on proximal half between midrib and margin, the base cuneate to rounded, the apex abruptly short-acuminate. Inflorescences axillary panicles 3–5(10) cm long, always shorter than subtending leaf, with branches terminating in umbels of 4 flowers sometimes subtended by an additional pair of flowers. Flowers: sepals becoming revolute at apex during flowering, all eglandular or the anterior eglandular and the lateral 4 biglandular; petals abaxially sericeous on claw, yellow. Samaras falcate, the nut nearly horizontal and the wing strongly ascending; nut of samara cylindrical but laterally compressed, 5–8 mm long, smooth-sided; wing of samara 40–50 × 13–17 mm. Fl (Feb), fr (Apr); in moist forest.

HIRAEA Jacq.

Lianas, occasionally shrubby. Stipules borne on petiole, most often at or above middle, usually long and subulate. Leaves usually bearing glands; tertiary veins often strongly parallel. Inflorescences axillary, usually 1–several umbels of 4 or many flowers; pedicels sessile. Flowers: sepals all eglandular or anterior eglandular and lateral 4 biglandular; petals yellow or yellow with red markings (almost completely red during flowering in *H. morii*); anthers ± alike; styles 3, the apex with large internal stigma and dorsally rounded to prominently hooked. Fruits breaking apart into 3 samaras, the samara with largest wings lateral, usually with 2 discrete wings, then butterfly-shaped, the dorsal wing small, sometimes reduced to a crest or lost, intermediate winglets or slender projections rarely present.

1. Petioles 20–40 mm long; blades densely and persistently velutinous abaxially, the hairs Y- or T-shaped. . . . . *H. gracieana.*
1. Petioles 4–15 mm long; blades sericeous to glabrate abaxially at maturity, the hairs (if present) sessile, straight, appressed.
   2. Stipules 1–2 mm long. Leaf blade margins eglandular. Inflorescences 1–7 unbranched, 4-flowered umbels arranged vertically. Posterior petal eglandular. Samaras with dorsal wing absent or represented by rounded crest 0.2(0.5) mm high. . . . . *H. affinis.*
   2. Stipules 2.5–5.5 mm long. Leaf blade margins eglandular or, usually, bearing small glands distally. Inflorescences 1–2 mostly ternate cymes of 4-flowered umbels, sometimes only 1 of 3 umbels of cyme developing. Posterior petal glandular-dentate or glandular-fimbriate. Samaras with dorsal wing small but present, at least 1.5 mm high (samara not yet known for *H. morii*).
      3. Larger leaf blades 8–16 × 3–7 cm; secondary veins connected by cross-veins 1–1.5 mm apart. Pedicels 9–14 mm long. Petals yellow, sometimes turning reddish in age. . . . . *H. fagifolia.*
      3. Larger leaf blades 16–19 × 7.5–11 cm; secondary veins connected by cross-veins 2–4 mm apart. Pedicels 15–17 mm long. Petals red. . . . . *H. morii.*

Hiraea affinis Miq. PART 1: FIG. 6

Stems sericeous to glabrate. Stipules 1–2 mm long. Leaves: longer petioles 4–11 mm long; larger blades elliptic or obovate, 11–23 × (4)5–12 cm, apparently glabrous (actually often sparsely sericeous abaxially with tiny straight white appressed hairs), eglandular on margins, the base rounded or cordate, the apex obtuse, acute, or short-acuminate; secondary veins connected by parallel cross-veins mostly 2–5 mm apart. Inflorescences 1–7 unbranched, 4-flowered umbels arranged in vertical row in axil; pedicels (9)11–18 mm long, sericeous. Flowers: petals light yellow, all eglandular; styles with rounded dorsal hook at apex. Samaras with lateral wings 40–56 × 22–30 mm, the dorsal wing absent or represented by a rounded crest 0.2(0.5) mm high. Fr (May); in nonflooded moist forest.

Hiraea fagifolia (DC.) A. Juss. PL. 80d

Stems sericeous to glabrate. Stipules 2.5–5 mm long. Leaves: longer petioles (4)6–10(15) mm long; larger blades elliptic or somewhat ovate or obovate, 8–16 × 3–7 cm, soon glabrate except for sericeous abaxial midrib, eglandular on margins or bearing few small buttonlike glands distally, the base cuneate to rounded, the apex mostly acuminate (occasionally acute or obtuse); secondary veins connected by parallel cross-veins 0.5–1.5 mm apart. Inflorescences mostly cymes of 3 4-flowered umbels, sometimes only 1 of 3 umbels developing, sometimes 2 cymes present, one borne above the other, loosely sericeous; pedicels 9–14 mm long, sericeous. Flowers: petals yellow, sometimes turning reddish in age, the lateral 4 eglandular, the posterior glandular-dentate or glandular-fimbriate; styles with short acute or obtuse dorsal hook at apex. Samaras mostly butterfly-shaped, with unlobed membranous lateral wings

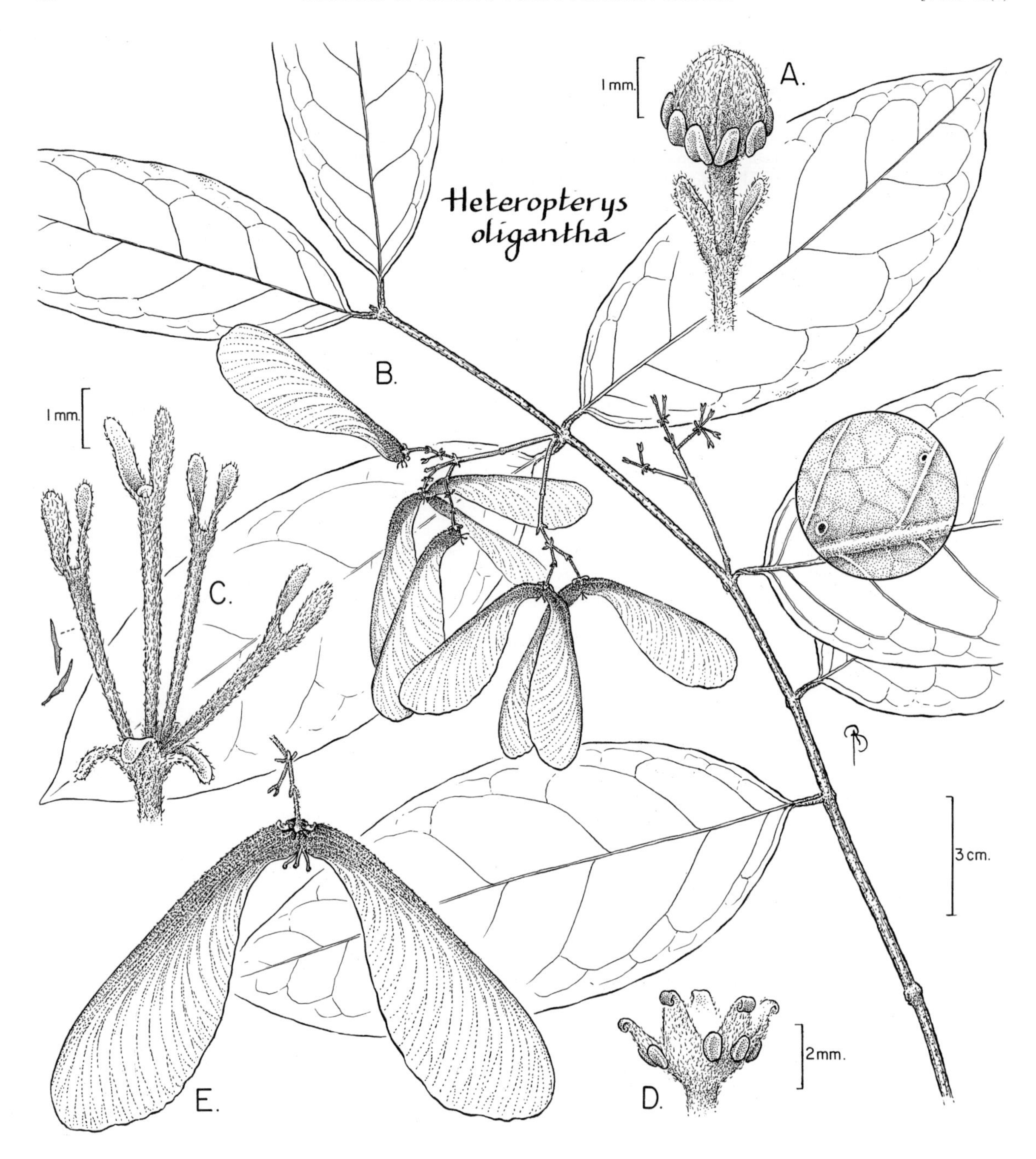

FIG. 175. MALPIGHIACEAE. *Heteropterys oligantha* (A, *Oldeman 2034*; B–E, *Mori & Pipoly 15549*). **A.** Lateral view of flower bud with sepals completely concealing petals. **B.** Part of stem with leaves and infructescences and detail of abaxial leaf surface showing dark bordered glands. **C.** Umbel of four peduncles subtended by reflexed bracts, each peduncle bearing at its apex two long, narrow, erect or spreading bracteoles with detail of malpighiaceous hairs. **D.** Lateral view of old flower with revolute sepals, four of the five sepals abaxially biglandular. **E.** Fruit comprising two samaras; the third carpel has aborted.

20–32 × 15–16 mm and dorsal wing 2–5 × 1.5–2 mm, but sometimes lateral wings coriaceous and irregularly lobed or reduced. Fl (Feb, Apr), fr (Apr); in moist forest.

Hiraea gracieana W. R. Anderson FIG. 176

Stems, leaves (at least abaxially), and inflorescences persistently velutinous, the hairs Y- or T-shaped. Stipules 3–5 mm long. Leaves: longer petioles 20–40 mm long; larger blades broadly elliptic or obovate, 19–34 × 11.5–18 cm, bearing small buttonlike glands distally on margins, the base broadly cuneate or rounded, the apex obtuse or rounded and abruptly short-acuminate and often apiculate; secondary veins connected by parallel cross-veins 2–6 mm apart. Inflorescences compact cymes of 3–7 4-flowered umbels, occasionally 2 cymes, one borne above the other; pedicels 16–25 mm long, persistently velutinous. Flowers: petals bright

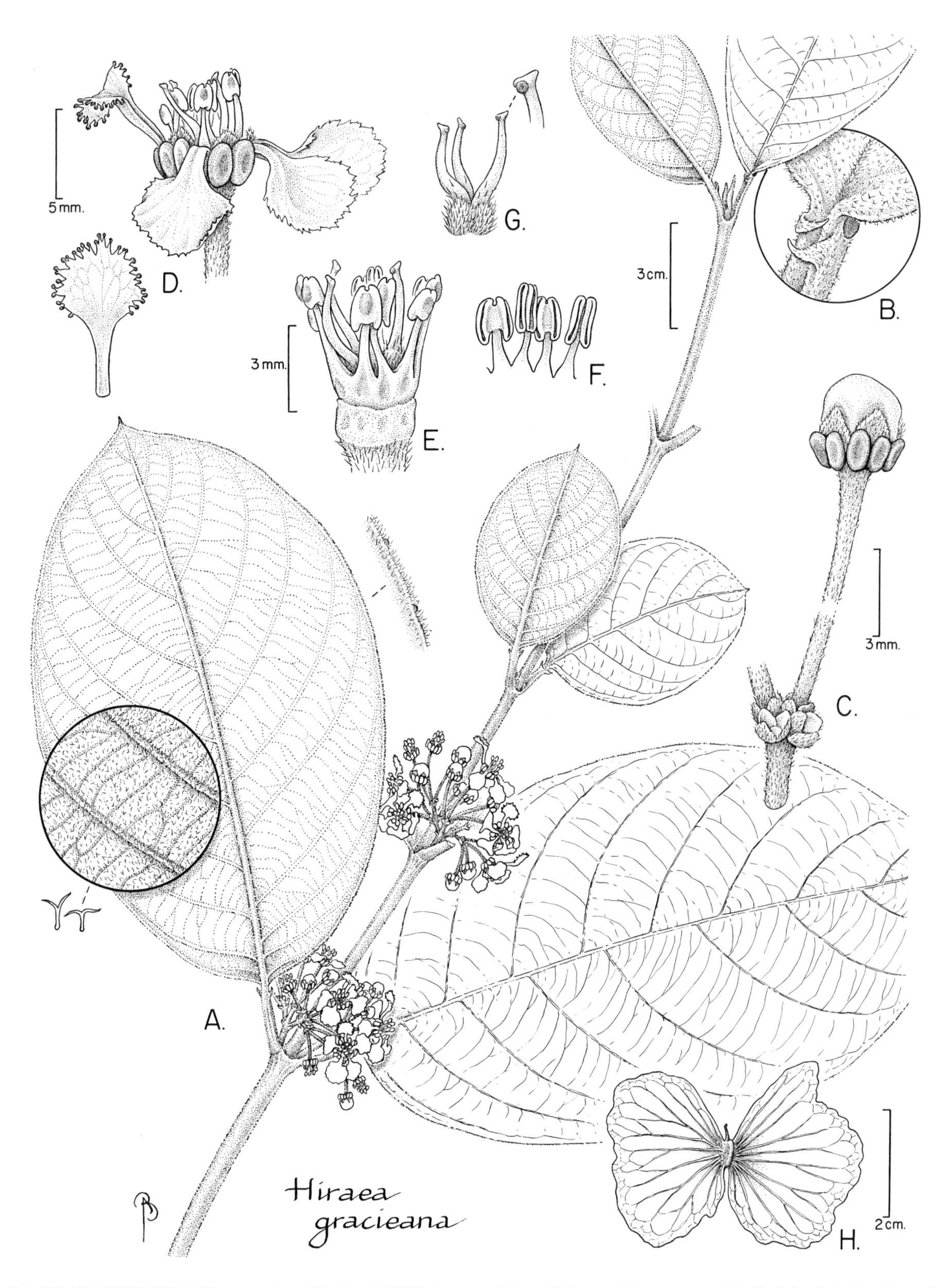

FIG. 176. MALPIGHIACEAE. *Hiraea gracieana* (*Mori et al. 22751*). **A.** Apex of stem with leaves, inflorescences, and details (left) of gland-bearing leaf margin (right) and abaxial leaf surface with stalked malpighiaceous hairs (left). **B.** Detail of apex of petiole with adaxial stipules and abaxial glands. **C.** Part of inflorescence showing one four-flowered umbel (two flowers removed), each pedicel subtended by two bracteoles and one bract. **D.** Lateral view of flower, the posterior petal uppermost (above) and adaxial view of posterior petal (below). **E.** Lateral view of flower with perianth removed; note connate bases of filaments. **F.** Adaxial view of part of androecium. **G.** Lateral view of gynoecium, the anterior style to the right and detail of apex of anterior style. **H.** Abaxial view of samara.(Reprinted from W. R. Anderson, Mem. New York Bot. Gard. 46(2), 1994.)

yellow, the lateral 4 eglandular, the posterior glandular-dentate; styles with a very short dorsal point at apex. Samaras with lateral wings 25–38 × 20–25 mm, the dorsal wing 3–5 × 1–1.5 mm. Fl (Nov), fr (Nov); in nonflooded moist forest.

Hiraea morii W. R. Anderson

Stems and leaf blades loosely sericeous to glabrate. Stipules 2.5–5.5 mm long. Leaves: longer petioles 8–15 mm long; larger blades obovate, 16–19 × 7.5–11 cm, bearing small buttonlike or short-cylindrical glands distally on margins, the base rounded, the apex short-acuminate to rounded; secondary veins connected by parallel cross-veins 2–4 mm apart. Inflorescences single cymes of 3 short-stalked, 4-flowered, side-by-side umbels with common stalk ± completely suppressed, sericeous to subvelutinous; pedicels 15–17 mm long, thinly sericeous, the hairs fusiform, short-stalked, ca. 0.5 mm long. Flowers: petals yellow in bud, red during flowering, the lateral 4 eglandular, the posterior glandular-dentate; styles with short acute dorsal projection at apex. Samaras unknown. Fl (Apr); in secondary vegetation.

JUBELINA A. Juss.

Lianas. Stipules small, triangular, borne on base of petiole. Leaves: petioles eglandular; blades usually bearing abaxial impressed glands; tertiary veins strongly parallel. Inflorescences decompound, the flowers ultimately borne in umbels of 4 or corymbs of 6; bracts and bracteoles large, pubescent on both sides. Flowers: calyx with anterior sepal eglandular and 1 large gland on each of 4 lateral sepals; petals pink or yellow; styles 3, the apex with large stigma on internal angle, dorsally truncate or short-hooked. Fruits breaking into 3 samaras, the samara with large lateral wings dominant, distinct at apex, confluent to nearly distinct at base, and smaller semicircular central dorsal wing.

Anderson, W. R. 1990. The taxonomy of *Jubelina* (Malpighiaceae). Contr. Univ. Michigan Herb. **17:** 21–37.

Jubelina rosea (Miq.) Nied. FIG. 177, PL. 81a

Stems, leaves, and inflorescence velutinous, the hairs Y-shaped. Leaves: longer petioles 7–21 mm long; larger blades obovate to broadly elliptic to rotund, 10–20 × 5–13 cm, bearing several glands abaxially in a row, the base cuneate or truncate, the apex rounded or short-acuminate. Inflorescences: bracts and bracteoles pink, 6–9 mm long; peduncles 1–3 mm long; pedicels 6–12 mm long. Flowers: sepals pink, 5–8.5 mm long, reflexed, distally inflated with aerenchyma; petals pink, the lateral 4 mostly abaxially sericeous, the posterior glabrous. Samaras roughly elliptic in outline, 55–80 × 35–60 mm, with corrugated and deeply lobed wing between large membranous outer lateral wing and central dorsal wing. Fl (Sep); in nonflooded moist forest.

MALPIGHIA L.

Shrubs or small trees. Stipules small, borne on stem between petioles. Leaves: petioles eglandular; blades usually bearing 2(10) glands abaxially. Inflorescences axillary corymbs or umbels; bracts and bracteoles eglandular; pedicels pedunculate. Flowers: petals pink, pale purple, or white; stamens 10, glabrous; ovary with 3 carpels usually completely connate, the 3 locules all fertile, the styles 3, the apex with large internal or subterminal stigma, dorsally rounded, truncate, or hooked. Fruit fleshy drupes (or berries), red or orange, with 3 pyrenes united in center or free at maturity but then usually retained in common fleshy exocarp, the hard wall of each pyrene showing rudimentary dorsal and lateral wings and sometimes rudimentary intermediate winglets or dissected outgrowths.

Malpighia emarginata DC.

Shrubs or small trees, 2–6 m tall, much branched with stiff branchlets. Leaves sometimes crowded in dense shoots with very short internodes, the same plants also bearing stems with well developed internodes; longer petioles (1)2–4 mm long; larger blades ovate, elliptic, or obovate, 2.5–10 × 1.4–5 cm, bearing 2(4) glands abaxially, sparsely sericeous or glabrate, the base cuneate or rounded, the apex most often rounded or obtuse, but sometimes emarginate and apiculate. Inflorescences umbels, sessile or raised on a stalk 1–3(5) mm long and containing 2–4 flowers. Flowers: calyx bearing 6–10 glands; petals pink or purplish, the lateral 4 with narrow abaxial keel; stamens heteromorphic, those opposite posterior-lateral petals with thicker filaments and larger anthers than others; styles dorsally truncate or apiculate at apex. Fruits to 17 × 22 mm, red. Cultivated for the edible fruit rich in vitamin C; not native or naturalized in our area. *Acerola* (origin unknown), *Barbados cherry* (English), *cerise de Cayenne* (Créole).

MASCAGNIA (DC.) Colla

Vines, mostly lianas. Stipules small, free, triangular, borne between petioles or on base of petiole. Leaves usually bearing glands. Inflorescences mostly axillary or terminal pseudoracemes, sometimes congested and reduced to form corymbs or umbels, single or grouped in panicles; floriferous peduncle usually well developed. Flowers: petals yellow or yellow and orange, pink, lilac, or white; stamens 10, the anthers ± alike; ovary with 3 carpels connate along central axis, all fertile, the styles 3, the apex with large internal stigma, dorsally rounded, truncate, acute, or short-hooked. Fruits breaking into 3 samaras, the samara with its largest wings lateral, a single wing continuous at base or at both base and apex, or 2 discrete wings, the dorsal wing small, sometimes reduced to a crest or lost.

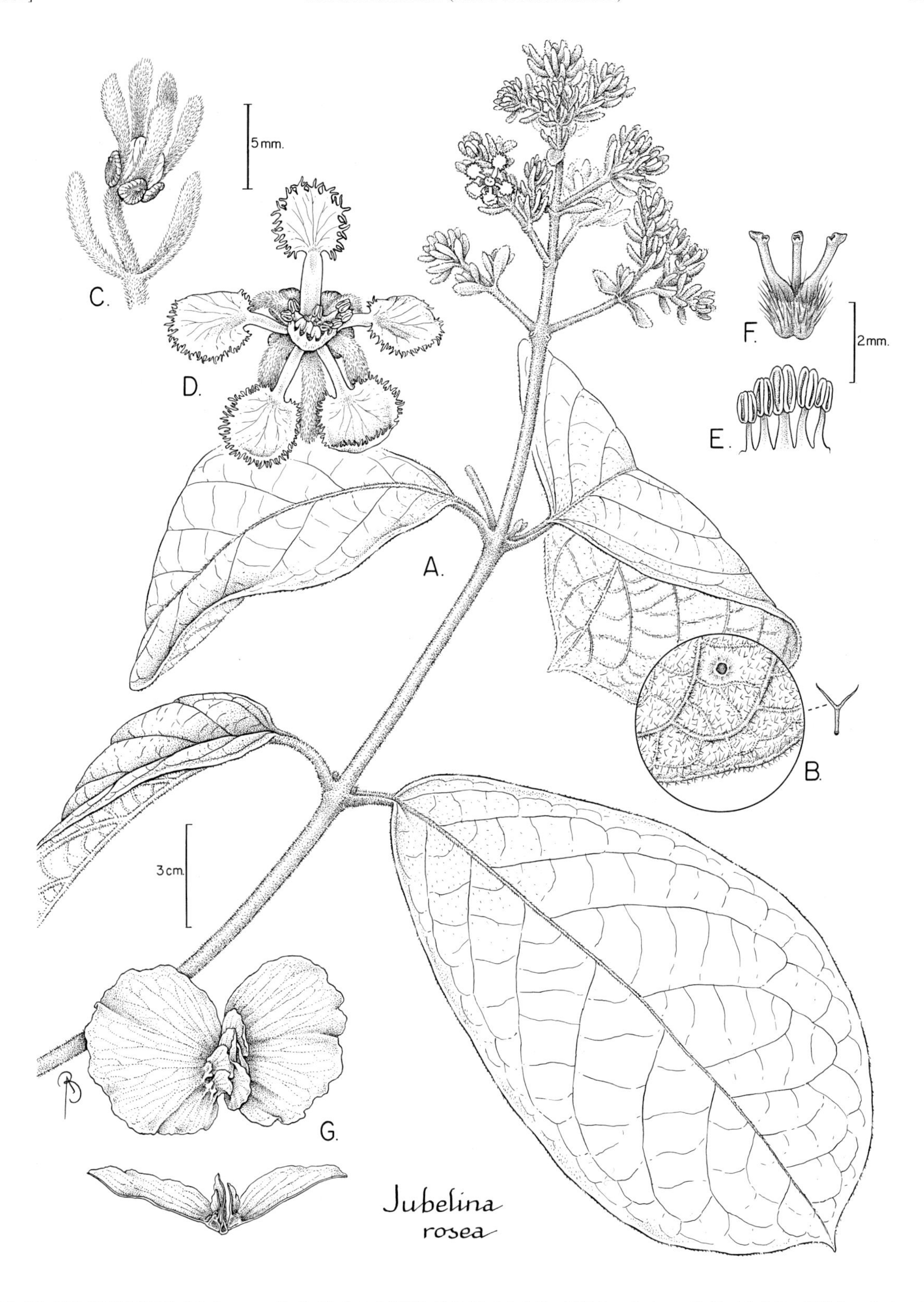

FIG. 177. MALPIGHIACEAE. *Jubelina rosea* (A–D, *Mori et al. 20929*; E, F, *Mori & Bolten 8551* from Surinam; G, *Mori & Bolten 8399* from Surinam). **A.** Apex of stem with leaves and inflorescence. **B.** Detail of abaxial leaf surface showing gland and stalked, bifurcate hair. **C.** Flower bud showing spatulate bracteoles and sepals, each lateral sepal bearing one large abaxial gland. **D.** Oblique-apical view of flower, the posterior petal uppermost. **E.** Adaxial view of part of the androecium, the central stamen opposite one of the posterior-lateral petals. **F.** Lateral view of gynoecium with anterior style in center. **G.** Abaxial view of samara (above) and medial section of samara (below).

Mascagnia divaricata (Kunth) Nied. FIG. 178, PL. 81b

Lianas. Leaves: longer petioles 10–19(24) mm long; larger blades narrowly to broadly ovate, 6–13 × 3.5–6.5(8) cm, thinly sericeous to glabrate at maturity, the base broadly cuneate to rounded, the apex gradually to abruptly acuminate. Inflorescences paniculate, velutinous, the hairs short and gray, the ultimate branches pseudoracemes of 7–35 flowers; bracts and bracteoles 0.5–1.3 mm long, persistent; pedicels 5–10(12) mm long, raised on peduncles 1–3(5) mm long. Flowers: sepals 1.8–2.5 mm long, the anterior eglandular, the lateral 4 biglandular; petals lilac or pinkish lilac, sometimes with yellow area at base of limb; anthers pilose, the connective not exceeding locules; styles dorsally apiculate or short-hooked at apex. Samaras subcircular, (18)20–30 × (18)20–28 mm, the lateral wing continuous at base and divided to halfway to nut at apex, the dorsal wing 4–6 × 1.5–4.5 mm. Fl (Mar, May), fr (Sep); in nonflooded moist forest.

MEZIA Nied.

Lianas, shrubs, or small trees. Stipules minute, interpetiolar, caducous, or absent. Leaves: petioles eglandular; blades usually bearing abaxial impressed glands. Inflorescences axillary and terminal, often decompound, the flowers ultimately borne in umbels of 4; bracts smaller than bracteoles; peduncle well developed; bracteoles borne just below flower, globose-cymbiform, large, the inner enclosing bud until flowers open, the outer enclosing bud and inner bracteole; pedicel absent or very short, to 2 mm long in fruit. Flowers: sepals narrowly oblong or spatulate, the anterior eglandular, the lateral 4 each bearing 2 large compressed glands, these distinct or partially to completely connate; petals yellow or the posterior petal yellow and red, stamens 10, dimorphic, the 5 opposite sepals differing from 5 opposite petals in size and shape, and sometimes in pubescence; styles 3, the apex with large internal stigma, dorsally truncate or short-hooked or pedaliform. Fruits breaking into 3 samaras, each bearing 2 large lateral wings, the wings distinct or more often confluent at base, a smaller dorsal wing, and often additional wings, winglets, or crests between them or outside lateral wings.

1. Petals abaxially loosely white-tomentose in center, the posterior petal without marginal glands; filaments tomentose, connate only at base; anthers tomentose at base. Samaras with single flat winglet between lateral and dorsal wings and a flat crest or winglet on nut outside lateral wing. . . . . *M. angelica.*
1. Lateral petals abaxially brown-sericeous in center, the posterior petal glabrous, glandular-fimbriate; filaments glabrous, connate for 1/2–2/3 of length; anthers glabrous. Samaras with space between lateral and dorsal wings filled by complex of irregular, ruffled, interconnected crests and winglets, the nut without crest or winglet outside lateral wing. . . . . *M. includens.*

Mezia angelica W. R. Anderson FIG. 179, PL. 81c

Lianas. Leaves: longer petioles 15–30 mm long; larger blades elliptic or somewhat ovate or obovate, 12–20(24) × 4.5–7.6(9.8) cm, sericeous to glabrate, the base cuneate to truncate, the apex abruptly short-acuminate. Inflorescences: bracts 3.5–6.5 mm long; bracteoles 5–7(8) mm long, the apex emarginate or bifid. Flowers: sepals with glands nearly or completely connate; petals abaxially loosely white-tomentose in center, the posterior petal distally yellow, proximally red and dentate or short-fimbriate, eglandular; filaments tomentose, especially distally, connate only for basal 0.5–1 mm, the anthers tomentose at base; styles pedaliform at apex (i.e., with a short, broad abaxial extension resembling from above the sole of a shoe). Samaras oblate, 55–80 × 45–60 mm, the lateral wing 25–34 mm wide, continuous at base, incised to nut at apex, the central dorsal wing 7–14 mm wide, with 1 flat winglet 3–7 mm wide present on each side of and parallel to central dorsal wing; nut bearing on each side a flat crest or winglet 1–9(15) mm wide, outside of and parallel to lateral wing. Fl (Sep), fr (Sep); in nonflooded moist forest.

Mezia includens (Benth.) Cuatrec.

Lianas. Leaves: longer petioles 14–27 mm long; larger blades elliptic, sometimes slightly ovate or obovate, 13–20 × 6–10.5 cm, sericeous to glabrate, the base cuneate to rounded or slightly attenuate, the apex abruptly short-acuminate. Inflorescences: bracts 3–4(5) mm long; bracteoles 6.5–12 mm long, the apex entire. Flowers: sepals with glands compressed but distinct; lateral petals abaxially brown-sericeous in center, the posterior petal yellow with red veins in center, glabrous, glandular-fimbriate all around margin; filaments glabrous, 1/2–2/3 connate, the anthers glabrous; styles acute or truncate at apex. Samaras subcircular, 70–100 × 50–90 mm, the lateral wing 30–45 mm wide, continuous at base, incised to nut at apex, the central dorsal wing 7–10 mm wide, the space between lateral and dorsal wings filled by complex of irregular, ruffled, interconnected crests and winglets; nut without crest or winglet outside lateral wing. Fr (Aug); in nonflooded moist forest.

STIGMAPHYLLON A. Juss.

Vines. Stipules interpetiolar, distinct, inconspicuous. Leaves: petioles usually long, bearing 2 large glands at apex; blades entire or lobed. Inflorescences dichasia of congested pseudoracemes, these usually corymbose or umbellate. Flowers: anterior sepal eglandular, the lateral 4 biglandular; petals yellow or yellow and red; anthers 10, very unequal; styles 3, the apex stigmatic on internal angle, dorsally truncate, hooked, or (most often) bearing a foliaceous appendage, symmetrical on anterior style, unilateral on posterior styles. Fruits breaking into 3 samaras, the samara with dorsal wing dominant, thickened on adaxial (upper) edge, the veins terminating in thinner abaxial edge, with much shorter winglets or crests present on sides of nut in some species.

Anderson, C. 1997. Monograph of *Stigmaphyllon* (Malpighiaceae). Syst. Bot. Monogr. **51:** 1–313.

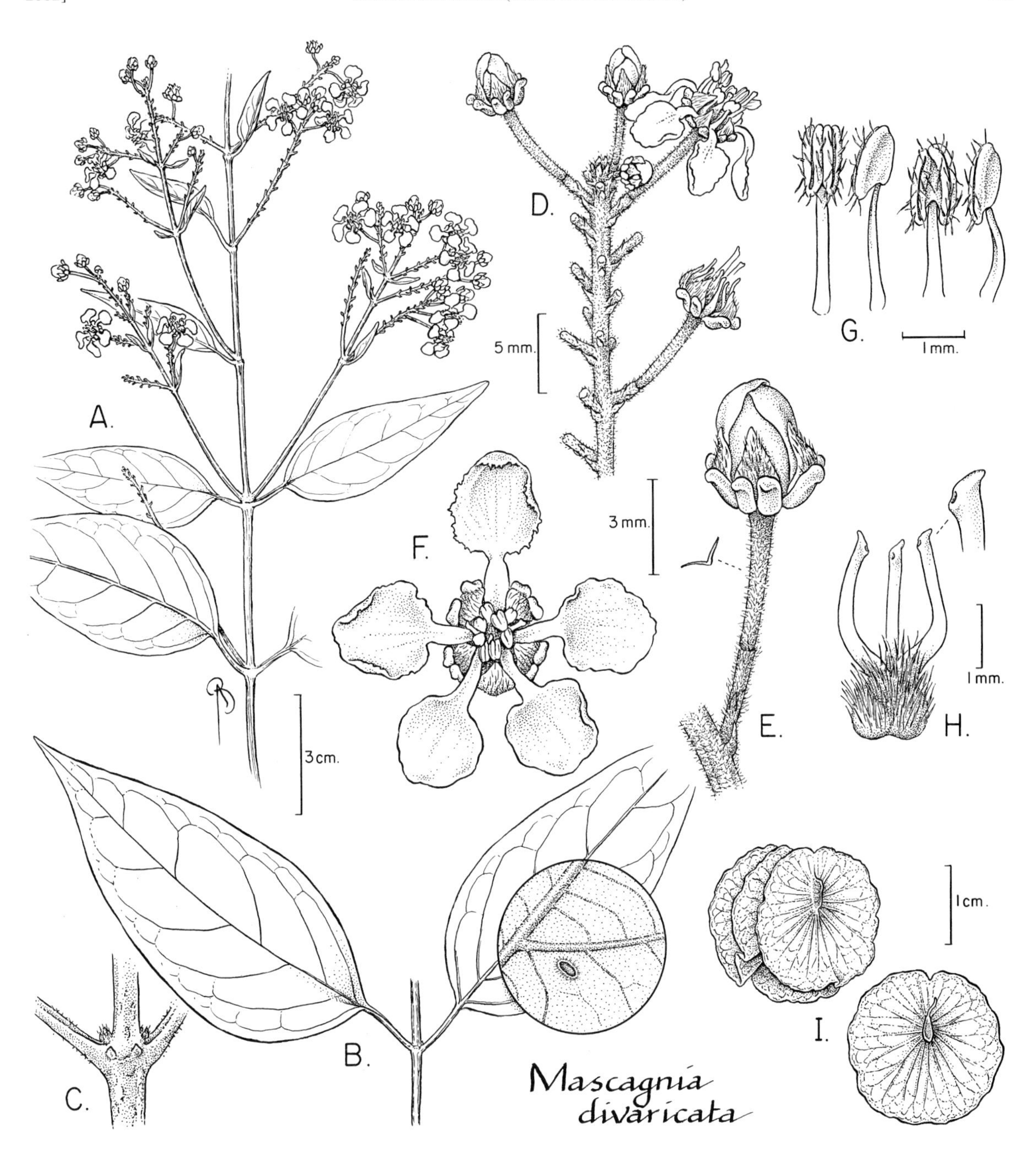

FIG. 178. MALPIGHIACEAE. *Mascagnia divaricata* (A–H, *Mori et al. 22205*; I, *Nee 38580* from Bolivia). **A.** Part of stem with leaves and inflorescences. **B.** Larger leaves and detail of abaxial surface with gland. **C.** Node showing small triangular, interpetiolar stipules. **D.** Distal portion of pseudoraceme with flower, buds, and old flower. **E.** Lateral view of flower bud showing bract on inflorescence axis, peduncle bearing bracteole near middle, pedicel beyond joint, and detail of hair. **F.** Apical view of flower, the posterior petal uppermost. **G.** Adaxial (far left and third from left) and lateral (second and fourth from left) views of stamens. **H.** Lateral view of gynoecium with anterior style in center (left) and detail of one style apex (above). **I.** Complete fruit with abaxial view of one samara (above) and adaxial view of one separated samara (right). (Reprinted from W. R. Anderson, Contr. Univ. Michigan Herb. 21, 1997.)

Stigmaphyllon sinuatum (DC.) A. Juss

FIG. 180, PL. 81f; PART 1: FIGS. 4, 6, PL. Xb

Vines. Leaves: longer petioles 25–95 mm long; larger blades broadly ovate to circular, 8–16 × 7.5–15 cm, ± densely and persistently sericeous abaxially, the hairs sessile, straight, tightly appressed, the base moderately to deeply cordate, the apex mostly obtuse to rounded. Flowers: styles all with large, rounded, horizontal or pendent folioles 1.5–2 mm long. Samaras 34–45 × 12–17 mm, the nut bearing an often dissected lateral crest to 2 mm wide. Fl (Aug, Nov, Feb), fr (Aug, Dec); in nonflooded moist forest and secondary vegetation.

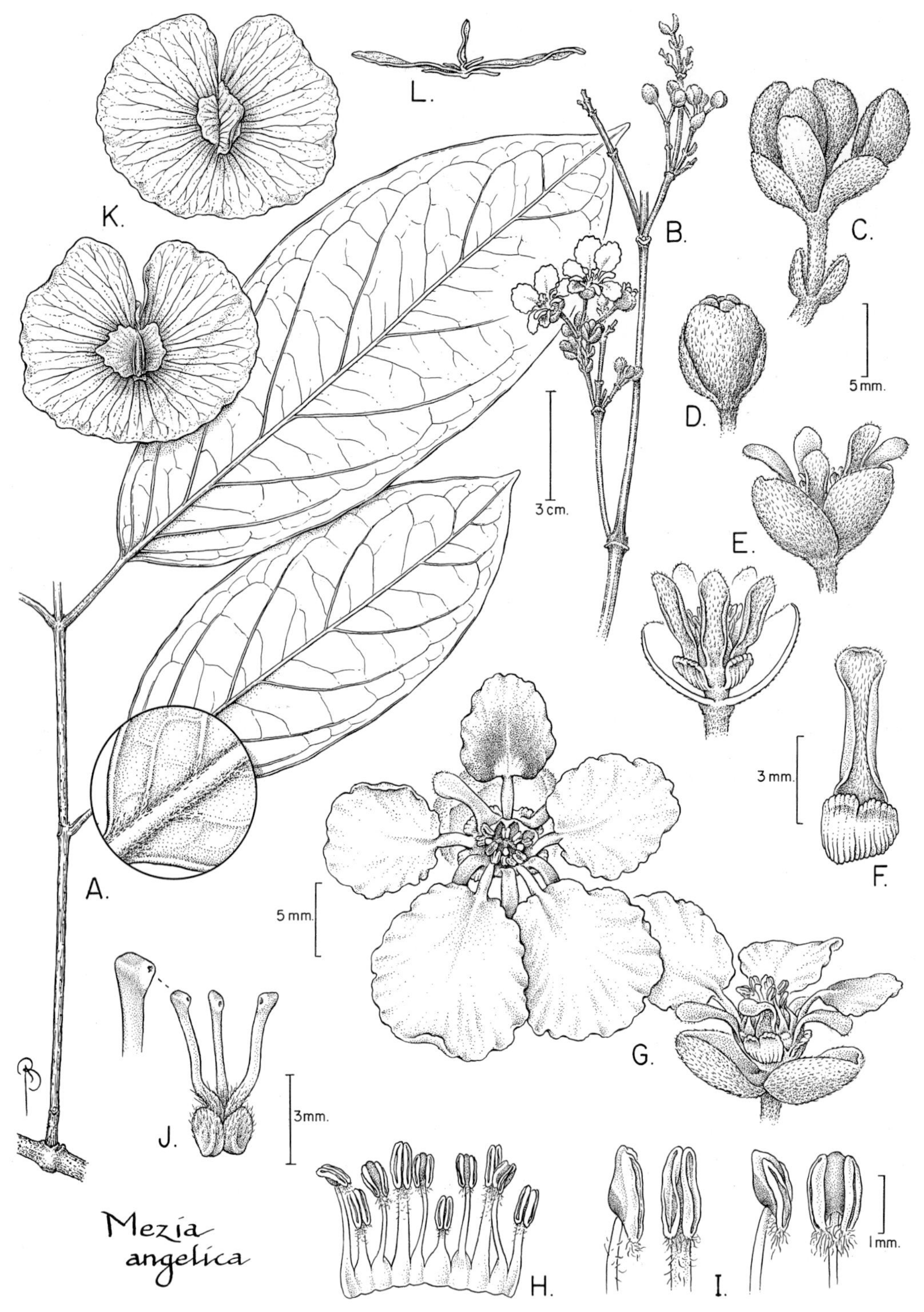

FIG. 179. MALPIGHIACEAE. *Mezia angelica* (A–J, *Mori et al. 20945*; K, L, *Silva 2830* from Brazil). **A.** Part of stem with leaves and detail of abaxial base of blade showing persistent hairs. **B.** Part of inflorescence with flowers and flower buds. **C.** Lateral view of umbel with four flower buds, each bud enclosed by two bracteoles and subtended by one bract, the stalk of the umbel bearing a pair of sterile bracts. **D.** Lateral view of flower bud about to open, with the two bracteoles being forced apart by the enlarging sepals. **E.** Lateral view of flower with petals removed, with the two subtending bracteoles intact (right) and with the bracteoles half cut away (below) to show sepals, with the eglandular anterior sepal in center. **F.** Abaxial view of one lateral sepal bearing a large double gland formed from two nearly connate glands. **G.** Apical view of flower (above left) with posterior petal uppermost and lateral view of flower (right) with two petals removed and the posterior petal to the right. **H.** Adaxial view of androecium opened with the shortest stamen (fifth from right) opposite the posterior petal. **I.** Lateral (far left) and adaxial (near left) views of anthers opposite petals and lateral (near right) and adaxial (far right) views of anthers opposite sepals. **J.** Lateral view of gynoecium with anterior style in center (right) and detail of one style-apex (above left). **K.** Abaxial (above) and adaxial (below) views of samaras. **L.** Medial section of samara showing large lateral wing, smaller dorsal wing (pointing up), single winglets between dorsal and lateral wings, and single winglets outside lateral wing.

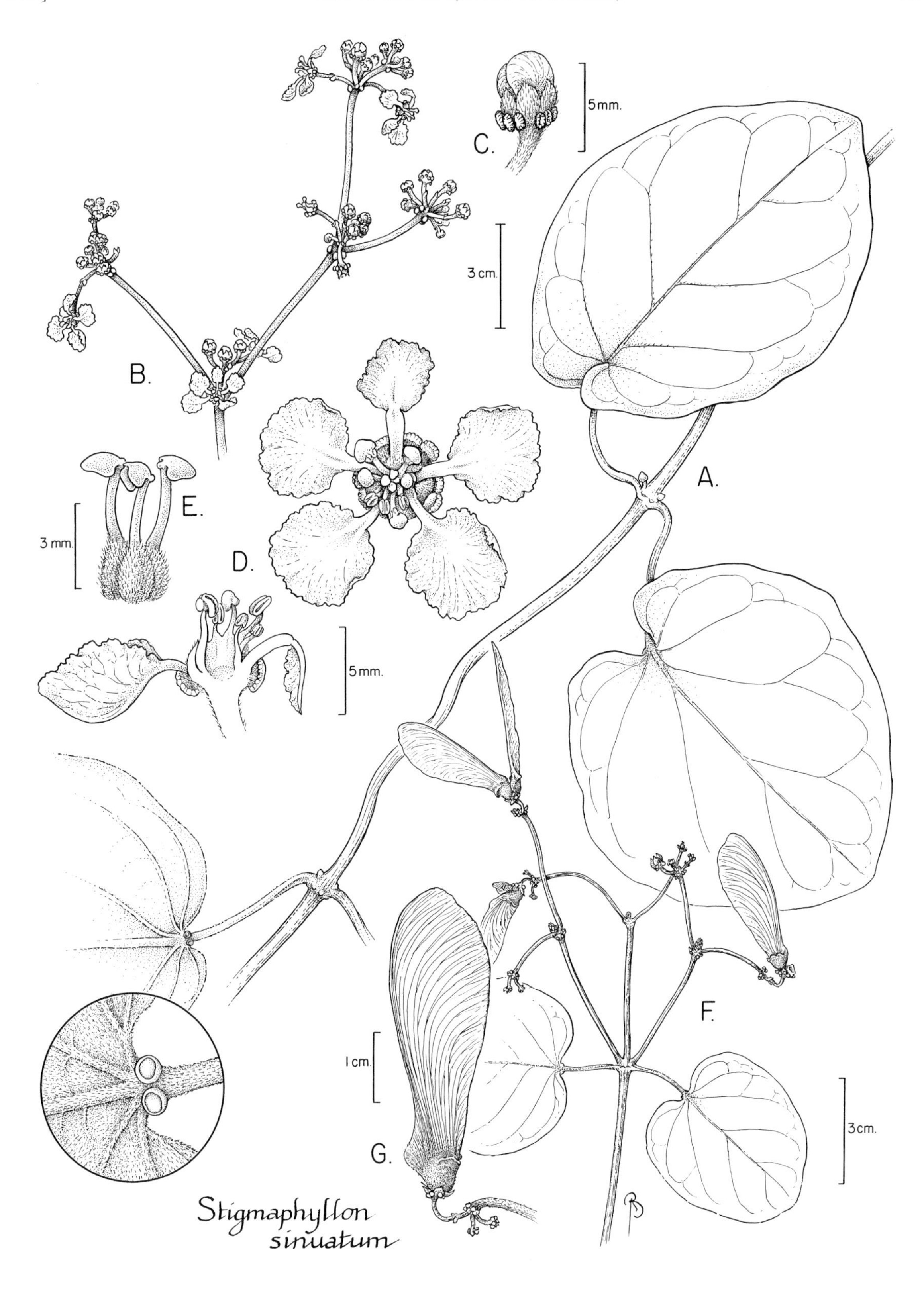

FIG. 180. MALPIGHIACEAE. *Stigmaphyllon sinuatum* (A, B, unvouchered field sketches; C–G, *Mori & Gracie 22849*). **A.** Part of leafy stem of liana showing detail (below) of glands at the apex of petiole on abaxial surface. **B.** Inflorescence. **C.** Lateral view of flower bud with paired glands at base of four lateral sepals. **D.** Apical view of flower (right) with posterior petal uppermost and longitudinal section of flower (below) with gynoecium removed and posterior petal to right; note unequal anthers. **E.** Gynoecium showing styles with large apical folioles. **F.** Apex of stem with infructescence. **G.** Samara.

TETRAPTERYS Cav.

Lianas, vines, or occasionally shrubs. Stipules small, interpetiolar or borne on petiole, or absent. Leaves usually bearing glands. Inflorescences umbels, corymbs, or pseudoracemes, these often grouped in panicles. Flowers: petals yellow or pink; stamens ± alike; styles 3, the apex with internal to apical stigma, dorsally smooth or truncate or short-hooked. Fruits breaking into 3 samaras, each samara with its largest wings lateral, usually 4 discrete wings, the dorsal wing smaller, sometimes reduced to crest or lost, intermediate winglets or outgrowths present between lateral and dorsal wings in some species.

1. Leaf blades eglandular or bearing glands on margin. Branches of inflorescence terminating in pseudoracemes of (2)4–8(10) flowers. Sepals revolute during flowering; petals (at least the lateral 4) abaxially sericeous. . . . *T. acutifolia.*
1. Leaf blades bearing glands on abaxial surface between midrib and margin. Branches of inflorescence terminating in umbels of 4–6 flowers. Sepals appressed during flowering; petals glabrous.
   2. Stipules distinct, persistent or eventually deciduous, leaving 2 tiny interpetiolar scars. Inflorescences with nonfloriferous bracts inconspicuous, narrowly lanceolate, mostly ≤3 mm long. Calyx glands (if present) usually becoming stalked in older flowers and fruit; styles slender, with small, discrete, nearly terminal stigmas. . . . . *T. mucronata.*
   2. Stipules connate in interpetiolar pairs, caducous or eventually deciduous, leaving a single large scar. Inflorescences containing conspicuous, often orbicular, foliaceous bracts, much smaller and thinner than vegetative leaves but much larger than floriferous bracts, these deciduous and usually absent from fruiting specimens. Calyx glands sessile; styles stout, the large stigmas internal and decurrent.
      3. Vegetative stems densely and ± persistently sericeous through first year of growth. Leaf blades abaxially persistently sparsely sericeous, the very short hairs evenly distributed over whole surface, bearing a cluster of 3–10 glands on each side of midrib abaxially. Sepals abaxially densely and persistently sericeous. Samaras with upper lateral wings at least 3 times as long as lower wings.. . . *T. glabrifolia.*
      3. Vegetative stems soon glabrescent to glabrate in first year of growth. Leaf blades abaxially soon glabrate or with some hairs persistent on midrib, bearing 0–1 gland on each side near base abaxially. Sepals abaxially glabrous or sparsely to moderately sericeous, often with marginal fringe of tiny hairs. Samaras with upper lateral wings to 2.5(3) times as long as lower wings.
         4. Leaf blades with abaxial glands at base, if present, 1–2.5 mm long. Some or all calyx glands asymmetrical, earshaped; limb of anterior-lateral petals 6–12 mm long. Nut of samara bearing several prominent outgrowths between lateral and dorsal wings, and usually a small rounded crest outside lateral wings, the samara with lateral wings subequal or the lower wings somewhat longer. . . . . *T. megalantha.*
         4. Leaf blades with abaxial glands at base, if present, to 0.5 mm long. All calyx glands symmetrical, elliptic or obovate; limb of anterior-lateral petals 4–5.5 mm long. Nut of samara smooth between lateral and dorsal wings (occasionally with 1 small outgrowth), and smooth outside lateral wings, the samara with upper lateral wings distinctly longer than lower. . . . . *T. crispa.*

Tetrapterys acutifolia Cav.

Lianas, sometimes shrubby when growing in open places without support. Stipules absent or minute (to 0.3 mm long), borne on petiole between base and middle. Leaves: longer petioles 4–7 mm long; larger leaf blades ovate or elliptic, 8–14 × 3–6.4 cm, eglandular or bearing several small glands on margins, the base cuneate to rounded, the apex acuminate to acute or obtuse. Inflorescence branches terminating in short, crowded pseudoracemes of (2)4–8(10) flowers, these often corymbose and sometimes approaching umbels when internodes very short. Flowers: sepals becoming revolute during flowering, the anterior eglandular, the lateral 4 biglandular, the glands sessile; petals abaxially sericeous (posterior petal sometimes nearly glabrous), yellow; styles stout, the stigma internal, dorsally rounded at apex. Samaras with lateral wings subequal, 7–15 mm long, the dorsal wing 2–3 mm wide; nut bearing several winglets or irregular outgrowths between lateral and dorsal wings. Fl (Feb).

Tetrapterys crispa A. Juss.

Lianas. Stems initially white- or tawny-sericeous, soon glabrescent to glabrate. Stipules connate in interpetiolar pairs, caducous. Leaves: longer petioles 12–22(27) mm long; larger blades elliptic or slightly ovate or obovate, 13–20 × (5.5)6–12 cm, eglandular abaxially at base or bearing a single gland to 0.5 mm diam. on each side and with distal row of tiny glands set in from margin, soon glabrate except for often sericeous abaxial midrib, the base cuneate to rounded or occasionally shallowly cordate, the apex short-acuminate. Inflorescence branches terminating in umbels of 4 flowers; reddish nonfloriferous bracts conspicuous in flower, 8–25 mm long. Flowers: sepals appressed during flowering, abaxially glabrous or with a few hairs beside glands and often bearing marginal fringe of tiny hairs, the anterior eglandular, the lateral 4 biglandular, the glands symmetrical, elliptic or obovate, sessile; petals glabrous, yellow turning orange with age, the limb of anterior-lateral petals 4–5.5 mm long; styles stout, the posterior two thicker than anterior one, the stigmas internal, large and decurrent. Samaras with upper lateral wings (15)22–38 mm long, always distinctly longer than lower wings (6)10–15(20) mm long, the dorsal wing 2–6 mm wide; nut smooth between lateral and dorsal wings (occasionally bearing 1 small outgrowth), without crests or winglets outside lateral wings. Fl (Aug, Feb), fr (Nov, Mar); in nonflooded moist forest and cloud forest.

There are several collections from central French Guiana that are closest to *T. crispa* but atypical in some characters. Their leaves are too small, and their samaras have too many outgrowths between the lateral and dorsal wings. I originally identified some of these plants as *T. discolor* (G. Mey.) DC., but on closer examination I have

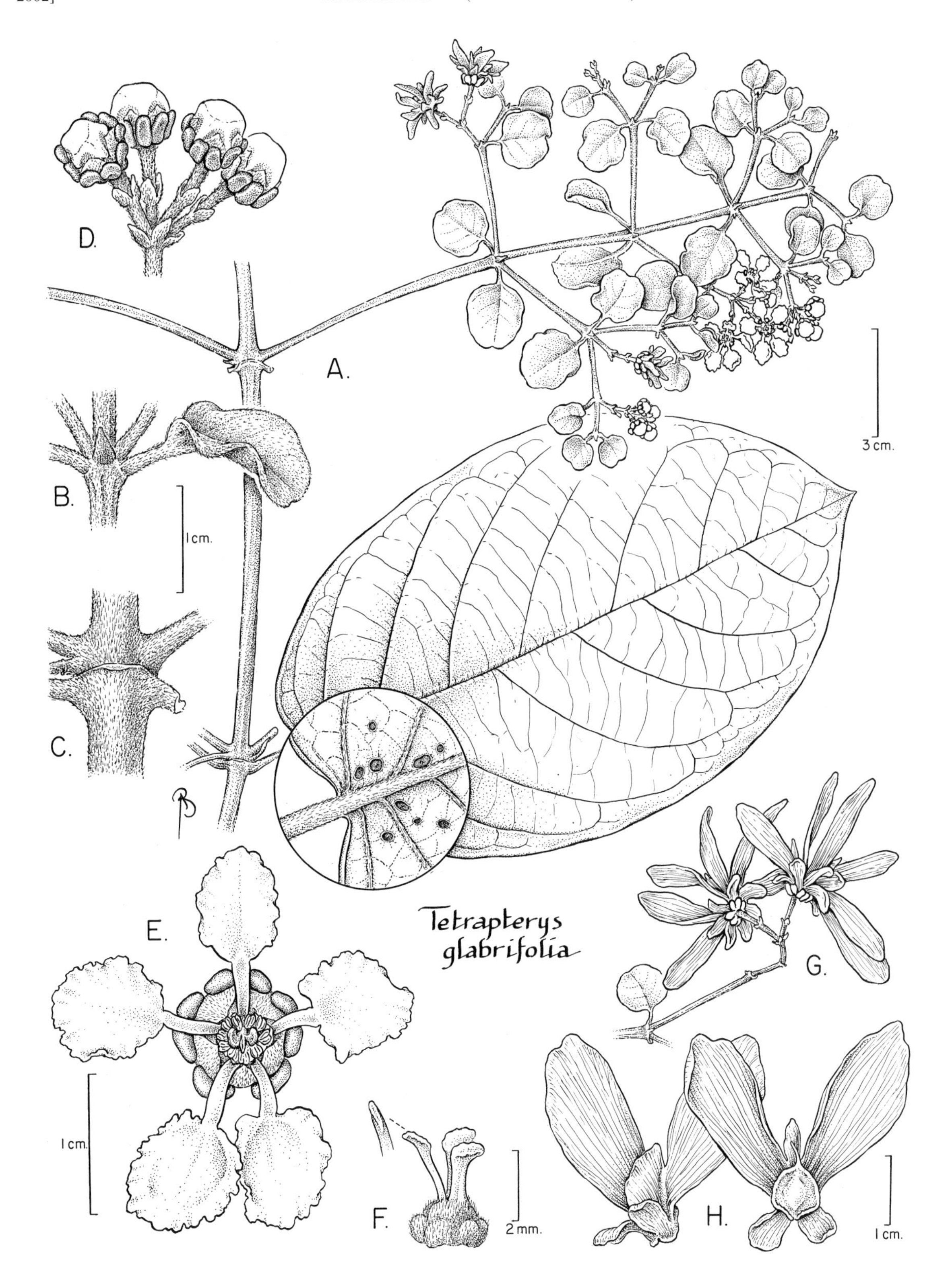

FIG. 181. MALPIGHIACEAE. *Tetrapterys glabrifolia* (A–F, *Mori et al. 22873*; G, H, *Mori & Pipoly 15630*). **A.** Part of stem with leaf and inflorescence and detail of abaxial surface of leaf showing impressed glands; note the rounded non-floriferous bracts in inflorescence. **B.** Node of inflorescence showing triangular interpetiolar stipule pair. **C.** Older node showing corky scar left after loss of stipule pair. **D.** Lateral view of umbel of four flower buds subtended by four bracts, each flower with a pair of bracteoles just below the peduncle-pedicel joint. **E.** Apical view of flower with the posterior petal uppermost. **F.** Lateral view of gynoecium (right) with anterior style to left and adaxial view of anterior style (above) showing decurrent stigma. **G.** Intact fruits, each comprising three samaras. **H.** Abaxial (left) and adaxial (right) views of separated samaras.

concluded that true *T. discolor* is not present in our area. I suspect that the intermediate plants in our area represent the results of hybridization between *T. crispa* and *T. megalantha*. The collections before me that seem intermediate are *Granville 8661* (CAY), *Mori & Pipoly 15609*, *Mori & Gracie 21095*, and *Mori & Pepper 24277* (all NY).

Tetrapterys glabrifolia (Griseb.) Small

FIG. 181, PL. 81d; PART 1: FIG. 6 as *T. crispa*

Lianas. Stems densely golden-sericeous, the hairs persistent at least during first year. Stipules connate in interpetiolar pairs, caducous or eventually deciduous. Leaves: longer petioles 13–21 mm long; larger blades elliptic to suborbicular or somewhat ovate or obovate, 11–25 × 7–15.5 cm, with a cluster of 3–10 glands on each side of midrib abaxially, these 0.3–0.7 mm diam. and distally few to many smaller glands (0.2–0.3 mm diam.) scattered over surface between midrib and margin, appearing nearly glabrous to naked eye but actually sparsely but persistently sericeous abaxially, the very short hairs evenly distributed over whole surface, the base broadly cuneate or rounded to cordate, the apex abruptly short-acuminate to rounded and apiculate. Inflorescence branches terminating in umbels of 4 flowers; reddish nonfloriferous bracts conspicuous in flower, 8–25 mm long. Flowers: sepals appressed during flowering, densely and persistently sericeous abaxially, the anterior eglandular, the lateral 4 biglandular, the glands elliptic or obovate, symmetrical or somewhat asymmetrical and earshaped, sessile; petals glabrous, yellow turning orange with age, the limb of anterior-lateral petals 6.5–8 mm long; styles stout, the posterior two notably thicker than anterior one, the stigmas internal, large and decurrent. Samaras with upper lateral wings 30–40 mm long (to 70 mm long in western Amazonia), the lower wings 8–11 mm long (to 17 mm in western Amazonia), the dorsal wing 4–7 mm wide; nut smooth between lateral and dorsal wings, without crests or winglets outside lateral wings. Fl (Feb, Apr), fr (Feb); in nonflooded moist forest.

This species is very close to *T. calophylla* A. Juss., which was described from French Guiana. The main, and perhaps only, difference between them is that in *T. calophylla* the leaf blade is obviously densely metallic-sericeous abaxially, whereas in *T. glabrifolia* the hairs are so sparse that the blade appears to the naked eye to be glabrous or nearly so. *Tetrapterys glabrifolia* was originally described as a variety of *T. calophylla*, which is a much older name.

Tetrapterys megalantha W. R. Anderson PL. 81e

Lianas. Stems initially thinly white-sericeous, soon glabrate. Stipules connate in interpetiolar pairs, caducous. Leaves: longer petioles (9)11–20 mm long; larger leaf blades elliptic or slightly ovate or obovate, (8)10–16 × 5–8.2 cm, eglandular abaxially at base or (often on same plant) bearing a single gland 1–2.5 mm long on each side and several smaller glands distally set in from margin, soon glabrate, the base cuneate or slightly decurrent to rounded, the apex abruptly short-acuminate. Inflorescence branches terminating in umbels of 4 flowers; reddish nonfloriferous bracts conspicuous in flower, 8–25 mm long. Flowers: sepals appressed during flowering, abaxially glabrous or occasionally loosely sericeous, often bearing marginal fringe of tiny hairs, the anterior eglandular, the lateral 4 biglandular, some or all glands asymmetrical, earshaped, sessile; petals glabrous, the limb yellow, the claw red, the limb of anterior-lateral petals 6–12 mm long; styles stout, the posterior two thicker than anterior one, the stigmas internal, large and decurrent. Samaras with lateral wings 14–25 mm long, the upper and lower wings subequal or the lower wings somewhat longer, the dorsal wing 4–7(10) mm wide; nut bearing several prominent outgrowths between lateral and dorsal wings, usually with a rounded crest 1–2 mm wide outside lateral wings. Fl (Dec, Jan), fr (Feb); in nonflooded moist forest.

Specimens from French Guiana and Surinam have smaller petals and less strongly asymmetrical calyx glands than those from Guyana, but we still have so few collections of this species in flower that it is difficult to know how much significance to attach to those differences. Specimens from the Sentier Botanique, just east of Eaux Claires, have unusually small leaves and unusually hairy sepals. For specimens that are intermediate between *T. megalantha* and *T. crispa*, see the discussion under *T. crispa*.

Tetrapterys mucronata Cav.

Lianas. Stipules minute, interpetiolar, distinct, persistent or deciduous. Leaves: longer petioles 5–14 mm long; larger leaf blades ovate or elliptic, 6–15 × 3–8 cm, 2 impressed abaxial glands at base and usually few to many smaller glands distally between margin and midrib, the base cuneate to rounded, the apex acuminate to obtuse. Inflorescence branches terminating in umbels of 4–6 flowers; nonfloriferous bracts inconspicuous, mostly narrowly lanceolate, ≤3 mm long. Flowers: sepals appressed during flowering, all eglandular or the lateral 4 biglandular, the glands becoming stalked in age; petals yellow, glabrous; styles slender, tapered distally, the small discrete stigma internal or nearly terminal, dorsally rounded at apex. Samaras with upper lateral wings 10–20 mm long, the lower wings 5–11 mm long, the dorsal wing 2–4 mm wide; nut usually bearing several aculeate outgrowths between lateral and dorsal wings. Fl (Sep); in nonflooded moist forest.

MALVACEAE (Mallow Family)

Laurence J. Dorr

Herbs, shrubs, or rarely small trees, often with stellate hairs. Leaves simple, alternate, entire, lobed, or dissected. Stipules persistent or caducous, sometimes inconspicuous. Inflorescences axillary or terminal, usually racemes or panicles or flowers fasciculate or solitary; involucel present or absent. Flowers: sepals 5, distinct or connate at base; petals 5, distinct, adnate to the staminal column at base; stamens numerous, monadelphous, the anthers monothecal; ovary superior, (3)5(many)-carpellate; placentation axile. Fruits loculicidal capsules, schizocarps of 5 to many mericarps, or rarely berries. Seeds 1 to numerous, glabrous, sparsely pubescent, or covered with long, fiber-like hairs.

Several species of Malvaceae are cultivated in central French Guiana, including *Abelmoschus moschatus* Medik. (*okra*, for edible fruit), *Gossypium barbadense* L. (*cotton*, for fiber), *Hibiscus rosa-sinensis* L. (*Chinese hibiscus*,for ornament), and *H. acetosella* Hiern (*false roselle*, for salad). None of them escapes from cultivation, and, therefore, these species are not included in the keys and descriptions.

Molecular data (Bayer et al., 1999) support an expanded concept of the Malvaceae, which would include genera herein assigned to Bombacaceae, Sterculiaceae, and Tiliaceae.

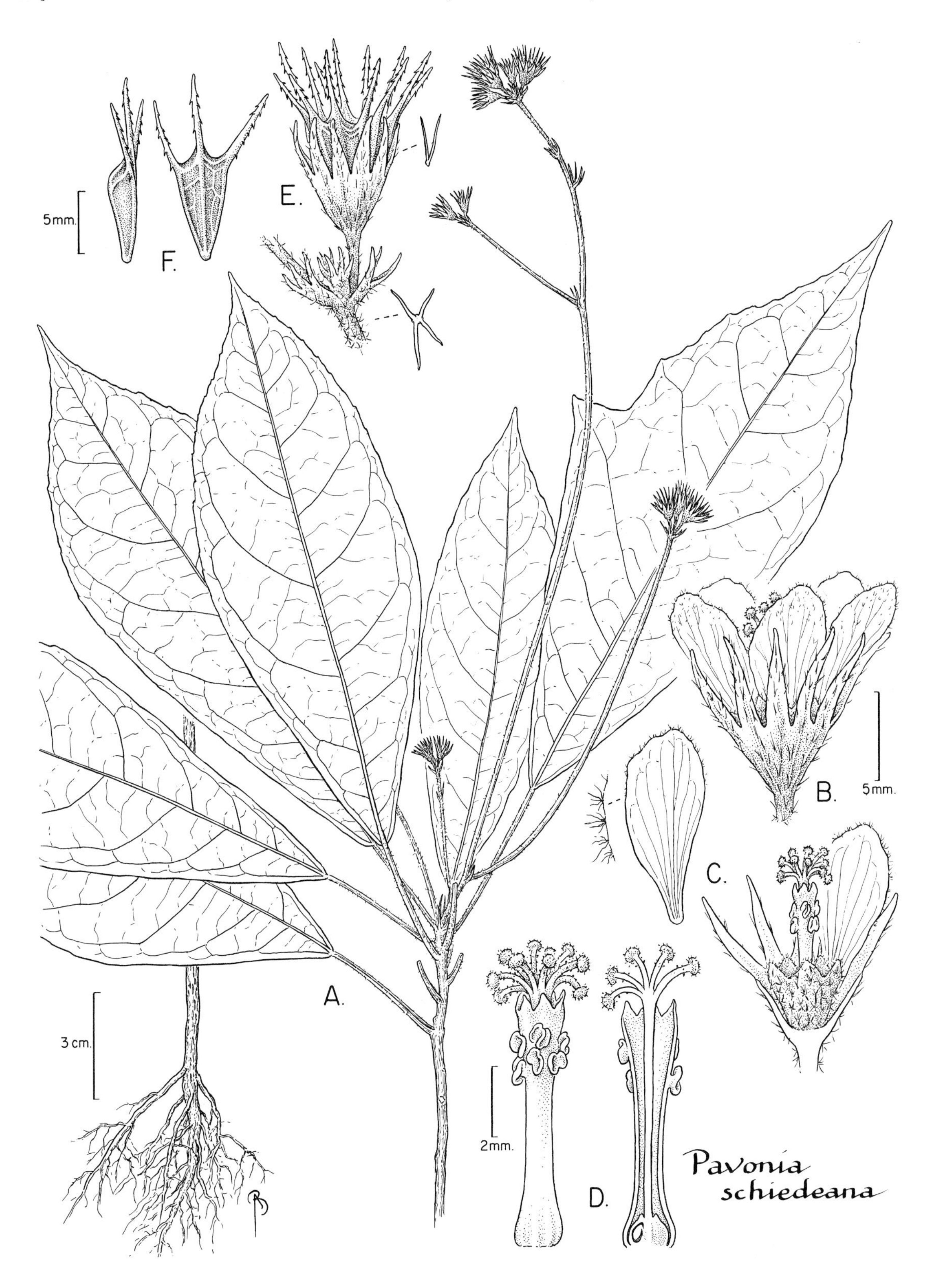

FIG. 182. MALVACEAE. *Pavonia schiedeana* (A–D, *Mori & Boom 15132*; E, F, *Mori et al. 22911*). **A.** Apex of stem with leaves and infructescences (right) and base of stem with roots (left). **B.** Lateral view of flower. **C.** Lateral view of flower with four petals and most of epicalyx removed to show calyx, monadelphous androecium, and 10-parted stigma (below right), and petal with detail of margin (above left). **D.** Monadelphous androecium and 10-parted stigma (left) and medial section of same (right). **E.** Part of schizocarp subtended by persistent calyx with details of hairs. **F.** Two views of individual mericarps.

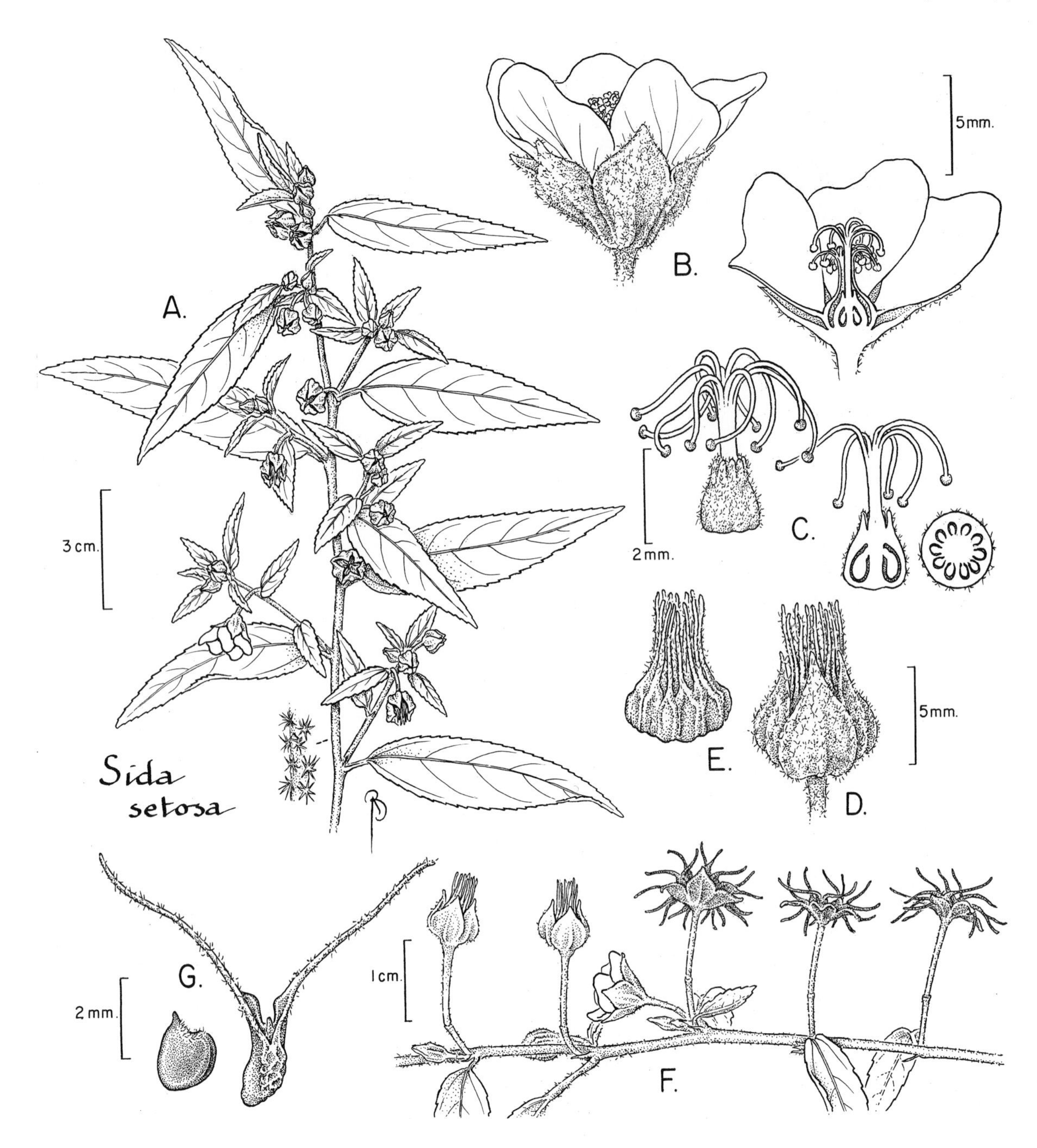

FIG. 183. MALVACEAE. *Sida setosa* (A–E, *Mori et al. 22333*; F, G, *Mori et al. 22055*). **A.** Apex of stem with leaves, flower buds, flowers, floral remains, and detail of stellate hairs on stem. **B.** Lateral view (left) and medial section (right) of flower. **C.** Lateral view (left) and medial section (near right) of pistil and transverse section of ovary (far right). **D.** Lateral view of immature schizocarp enclosed in calyx. **E.** Lateral view of immature schizocarp with calyx removed. **F.** Part of stem with a flower, immature, and mature fruits. The stem, shown horizontally here, is usually erect in nature. **G.** Mericarp (right) and seed (left).

Jansen-Jacobs, M. J. 1986. Malvaceae. *In* A. L. Stoffers & J. C. Lindeman (eds.), Flora of Suriname **III(1–2):** 259–276. E. J. Brill, Leiden.
Uittien, H. 1966. Malvaceae. *In* A. Pulle (ed.), Flora of Suriname **III(1):** 1–25, 433–435. E. J. Brill, Leiden.

1. Involucel present. Corollas white or pale lavender. . . . . *Pavonia.*
1. Involucel absent. Corollas yellow or pale yellow-orange. . . . . *Sida.*

PAVONIA Cav.

Prostrate perennial herbs, erect subshrubs, or shrubs, usually stellate-pubescent, sometimes viscid or glabrate. Leaves: blades ovate, elliptic, lanceolate, oblanceolate, or deltoid, the margins dentate or crenate. Inflorescences axillary, racemes, panicles, heads, or flowers solitary or paired; involucel present. Flowers: calyx 5-lobed; corolla small or large and showy, the petals white, lavender, purple, or yellow; androecium included or exserted; staminal column apically 5-toothed; styles 10, the stigmas capitate. Fruits schizocarps, the mericarps 5, indehiscent, usually ornamented. Seeds solitary, glabrous or slightly pubescent.

Pavonia schiedeana Steud. A. St.-Hil. & Naudin
FIG. 182, PL. 82a; PART 1: FIG. 8

Subshrubs, 0.15–0.45 m tall. Stems minutely stellate-pubescent. Stipules subulate. Leaves: petioles 1.5–5 cm long; blades elliptic, oblong, or obovate, occasionally with 1 lateral lobe, 6–18 × 3–6.5(8) cm, minutely stellate-pubescent on both surfaces (or glabrate adaxially), the base narrowly cuneate or truncate, the apex acute to acuminate, the margins crenate-dentate. Inflorescences terminal or on lateral axillary peduncles that may exceed the subtending leaf, condensed racemes or head-like; pedicels 0.5–5 cm long; involucel of ca. 8 bractlets, 4–6 × ca. 1 mm. Flowers: calyx 4–5 mm long; corolla white or pale lavender-white and fading to pink or pale lavender, the petals 6–9 mm long. Fruits 6–8 mm diam., glabrous, the mericarps 3-spined, the spines 2.5–5 mm long, retrorsely barbed. Seeds ca. 5 mm long, glabrous. Fl (Feb, Jul, Sep, Oct), fr (Jan, Feb, Jun, Sep); common, in forest understory, especially along trails and roads, epizoochorous.

SIDA L.

Herbs or subshrubs, erect or prostrate, glabrous or pubescent, sometimes viscid. Leaves: blades ovate, elliptic, rhombic, or linear, usually dentate. Inflorescences axillary or terminal, flowers solitary or in glomerules; involucel absent. Flowers: calyx 5-lobed, often 10-ribbed at base; corolla white, yellow, orange, rose, or purple, the petals sometimes with dark red basal spots; androecium included, apically antheriferous; styles 5–14, the stigmas capitate. Fruits schizocarps, glabrous or puberulent, the mericarps 5–14, indehiscent basally, indehiscent or dehiscent apically, usually with 2 apical spines or muticous. Seeds solitary, glabrous.

1. Leaf blades ± rhomboid, the apex acute to subobtuse. Mericarps muticous to spinescent, apically 2-spined (when spines present), the spines minute. *S. rhombifolia.*
1. Leaf blades lanceolate, lance-elliptic, or subrhombic, the apex narrowly acute. Mericarps spinescent, apically 2-spined, the spines to ca. 6 mm long. *S. setosa.*

Sida rhombifolia L.

Erect subshrubs, 0.4–1.5 m tall. Stems minutely stellate-pubescent. Stipules subulate. Leaves: petioles to 7 mm long; blades ± rhomboid, 2.5–9 × (0.5)1–3 cm, minutely stellate-pubescent ad- and abaxially, often glabrescent adaxially, ± discolorous, the base cuneate, the apex acute to subobtuse, the margins serrate distally. Inflorescences axillary, the flowers solitary; pedicels 1–3 cm long. Flowers: calyx 5–6 mm long; corolla yellow or yellow-orange, the petals 7–9 mm long, without dark basal spots; styles 10–14. Fruits oblate to conical, 4–5 mm diam., the mericarps 10–14, muticous to apically 2-spined (sometimes 1-spined through failure of dehiscence), the spines minute. Fl (Feb, May, Jun, probably year round), fr (Feb, Jun, probably year round); common, forest edges and along trails and roads. *Erva relogio*, *malva relogio* (Portuguese), *wadé-wadé* (Créole).

Sida setosa Colla FIG. 183, PL. 82b,c

Erect subshrubs or shrubs, 0.5–3 m tall. Stems minutely puberulent to glabrate. Stipules filiform. Leaves: petioles 0.8–1.5 cm long; blades lanceolate, lance-elliptic, or subrhombic, 7.5–11(14) × 2–5 cm, minutely puberulent to glabrate, ± discolorous, the base truncate to cuneate, the apex narrowly acute, the margins entire proximally, serrate distally. Inflorescences axillary, the flowers solitary, of several flowers, or sometimes subumbellate; pedicels 0.5–1 cm long. Flowers: calyx 6–8 mm long; corolla yellowish-orange, the petals 6–10 mm long, without dark basal spots; styles 10–12. Fruits conical, 6–7 mm diam., sparsely puberulent, the mericarps 10–12, ca. 8 mm long, apically 2-spined, the spines to ca. 6 mm long. Fl (May, Aug, Sep, Oct, probably year round), fr (May, Jun, Sep, Oct, probably year round); forest edges and weedy places, especially along roads.

MARCGRAVIACEAE (Shingle Plant Family)

Scott V. Heald, Adrian C. de Roon, and Stefan Dressler

Terrestrial, hemiepiphytic or epiphytic, shrubs or lianas, frequently with sprawling branches, juvenile stems often with adventitious roots (Pl. 14c), raphid cells and variously shaped sclereids frequent. Stipules absent. Leaves simple, alternate, distichous and dimorphic (*Marcgravia*) or spirally arranged (all other genera); petioles present or sometimes nearly absent, glabrous; blades frequently with abaxial glands, the margins entire, rarely crenate; primary vein ending in persistent or often deciduous, sclerified mucro. Inflorescences terminal, occasionally axillary or cauliflorous, erect or pendulous, racemes or pseudoumbels, sometimes pseudospikes; bracts transformed into variously shaped nectaries; bracteoles 2, sepaloid, persistent, sometimes lacking. Flowers bisexual, actinomophic, hypogynous; sepals 4–5, often unequal, free or nearly so, imbricate, persistent; petals (3)4–5, free or connate, imbricate or united into deciduous calyptra (*Marcgravia*); stamens 3 to many, the filaments free or sometimes connate, the anthers basifixed, dithecal, introrse, longitudinally dehiscent, the pollen yellow or often magenta; ovary superior, completely or incompletely 2–20-locular, the style distinct, sometimes reduced and then the stigma (sub)sessile; ovules mostly numerous. Fruits capsular and loculicidally and septifragously dehiscent from base (*Marcgravia*) or berry-like and irregularly dehiscent (other genera), (sub)globose, apiculate because of persistent style and stigma. Seeds hemispherical to mostly reniform, shiny reticulate, embedded in red pulp; endosperm scanty or lacking, the embryo straight.

Lanjouw, J. & P. F. Baron van Heerdt. 1966. Marcgraviaceae. *In* A. Pulle (ed.), Flora of Suriname **III(1):** 373–385. E. J. Brill, Leiden.

1. Leaves distichous, heterophylly between creeping sterile branches and fertile branches. Inflorescences umbelliform, the apical flowers abortive, only the nectaries developed; fertile flowers without nectaries; sepals 4; petals united into deciduous calyptra. . . . . . *Marcgravia.*
1. Leaves spirally arranged, not heterophyllous. Inflorescences umbelliform or racemes, all flowers fertile and provided with nectaries; sepals 5; petals 5, almost free or connate only at base, reflexed at anthesis.
   2. Inflorescences umbelliform. . . . . . *Marcgraviastrum.*
   2. Inflorescences racemes.
      3. Racemes 30–80 cm long, with 100–350 flowers; nectaries saccate, without auricles, inserted on pedicels below flower. Stamens 20–40. . . . . . *Norantea.*
      3. Racemes 10–30 cm long, with 15–30 flowers; nectaries spur-shaped with two wing-like auricles, inserted on pedicels at base of flower. Stamens 5. . . . . . *Souroubea.*

MARCGRAVIA L.

Climbing shrubs or lianas with dimorphic growth. Juvenile stems creeping, appressed to substrate (Fig. 184), attached by adventitious roots; leaves small, thin; petioles absent; blades ovate, the base often cordate, the margins crenate. Mature stems and fertile branches spreading, not rooting, often pendent, (sub)terete or angular, often provided with wart-like lenticels; leaves with petioles poorly developed; blades chartaceous to coriaceous, usually provided with variously shaped abaxial glands, the apex often with a drip-tip, the margins sometimes with blackish glands; secondary veins obscure to prominent when dry. Inflorescences (sub)umbelliform, terminal on normally leaved erect or hanging branches or axillary or cauliflorous on differently and deciduously leaved or bracteate hanging peduncles, central flowers abortive, only the nectariferous bracts (nectaries) well developed; pedicels present; rachis mostly contracted with fertile flowers and nectaries inserted at same point or sometimes the flowers and nectaries inserted at different points along rachis; nectaries hood-, pitcher- or dipper-shaped, connate with pedicels of abortive flowers. Fertile flowers borne erect or obliquely on pedicel; sepals 4, decussate; petals 4, connate into deciduous calyptra; stamens 6 to many, the filaments flattened, mostly filiform, the anthers ovate to triangular or sagittate; ovary 3–20-locular, the stigma capitate to umbonate. Fruits leathery capsules, loculicidally dehiscent from base. The seeds tiny, embedded in juicy red pulp, the testa reticulate, blackish purple.

1. Inflorescences terminal on normally leaved branches, with 14–26 flowers; rachis with interstices between fertile flowers and nectaries 2–7 mm long; nectaries 4–6.5 cm long. Calyptra 9–17 mm long; stamens 20–35. . . . . . *M. coriacea.*
1. Inflorescences, terminal, lateral, or cauliflorous, on long, hanging, deciduously leaved or bracteate peduncles, with 20–30 flowers; rachis without interstices between fertile flowers and nectaries; nectaries 0.5–6 cm long. Calyptra 4–10 mm long; stamens 9–20.
   2. Nectaries 0.5–1 cm long, the stalk 0.2–0.5 cm long. Calyptra 5–7 mm long; stamens 9–15. . . . . . *M. pedunculosa.*
   2. Nectaries 4–6 cm long, the stalk ca. 1 cm long. Calyptra ca. 9 mm long; stamens 14–20. . . . . . *Marcgravia* sp. A.

Marcgravia coriacea Vahl

Lianas, flowering branches stout, often angular and provided with lines of verrucose lenticels. Leaves: petioles 3–6 mm long; blades oblong to elliptic, 6–16 × 2–5 cm, (sub)coriaceous, with usually only two basal abaxial glands present, sometimes also with minute poriform glands near margin and blackish glands along margins, the base mostly rounded, sometimes obtuse or acute to cuneate or shortly attenuate, the apex acuminate when dry, the acumen ca. 1 cm long; secondary veins inconspicuous to prominent. Inflorescences with 14–26 fertile flowers, the interstices between flower and nectary insertions 2–7 mm long; nectaries 4–6 per inflorescence, cylindrical-saccate, somewhat clavate, bent outwards, 4–6.5 cm long, the stalk 5–15 mm long, the cup 3–5 cm long, the opening large, rounded, with a recurved rim; pedicels 2–3.5 cm long, verrucosely lenticellate; bracteoles small, inserted at base of calyx. Flowers ca. 1 cm diam.; outer sepals suborbicular, 2–3 × 4–6 mm, the inner sepals reniform, 2–3 × 7–12 mm, the calyptra subglobose to ovate-globose, 9–17 × 6–10 mm; stamens 20–35; ovary turbinate, 8–11-locular. Fruits ellipsoid to globose, 1–1.5 × 1–1.5 cm. Fr (Sep); common, especially in riverine forests.

Marcgravia pedunculosa Triana & Planch. FIG. 184i,j

Lianas, flowering branches subterete to angular, lenticellate. Leaves of fertile branches: petioles 4–8(10) mm long; blades oblong, 7–18 × 3–7 cm, chartaceous, usually with 1–2 inconspicuous, minute, poriform, abaxial glands on each side of leaf basally, the base acute to obtuse, the apex acuminate, the acumen 1.5–3 cm long; venation prominent on both sides when dry, secondary veins in 6–8 pairs. Inflorescences terminal, axillary, or cauliflorous, 15–30 cm long, hanging peduncles, provided with deciduous leaves, toward top of peduncle becoming smaller to bract-like, with 20–30 fertile flowers; pedicels 1–2.5 cm long; nectaries 4–10, cucullate, 0.5–1 cm long, the cup 3–6 mm long, the stalk 2–6 mm long. Flowers ca. 5 mm diam.; outer sepals ca. 1 × 2–3 mm, the inner sepals 1 × 3–4 mm, the calyptra ovoid to ovoid-conical, 5–7 × 4–5 mm; stamens 9–15, the filaments flattened, abruptly narrowed to anther, ca. 3 mm long, the anthers sagittate, ca. 3 mm long; pistil ca. 3 mm long, ovary 6–9-locular. Fruits 0.6–0.8 cm diam. Fl (Aug), fr (Dec).

Marcgravia sp. **A** FIG. 184a–h, PL. 82d,e

Lianas. Leaves: petioles ca. 4 mm long; blades elliptic to oblong, 6–11 × 2.5–4 cm, coriaceous, abaxial glands on each side of leaf basally, the base obtuse to rounded, the apex acute to acuminate; venation prominent on both sides when dry, secondary veins in 12–14 pairs. Inflorescences terminal, lateral, or cauliflorous, on long, hanging, deciduously leaved or bracteate peduncles, with 20–30 flowers; rachis without interstices between fertile flowers and nectaries; nectaries 4–6 cm long, the stalk ca. 1 cm long. Flowers: calyptra ca. 9 mm long; stamens 14–20. Fl (Aug, Sep), fr (Aug).

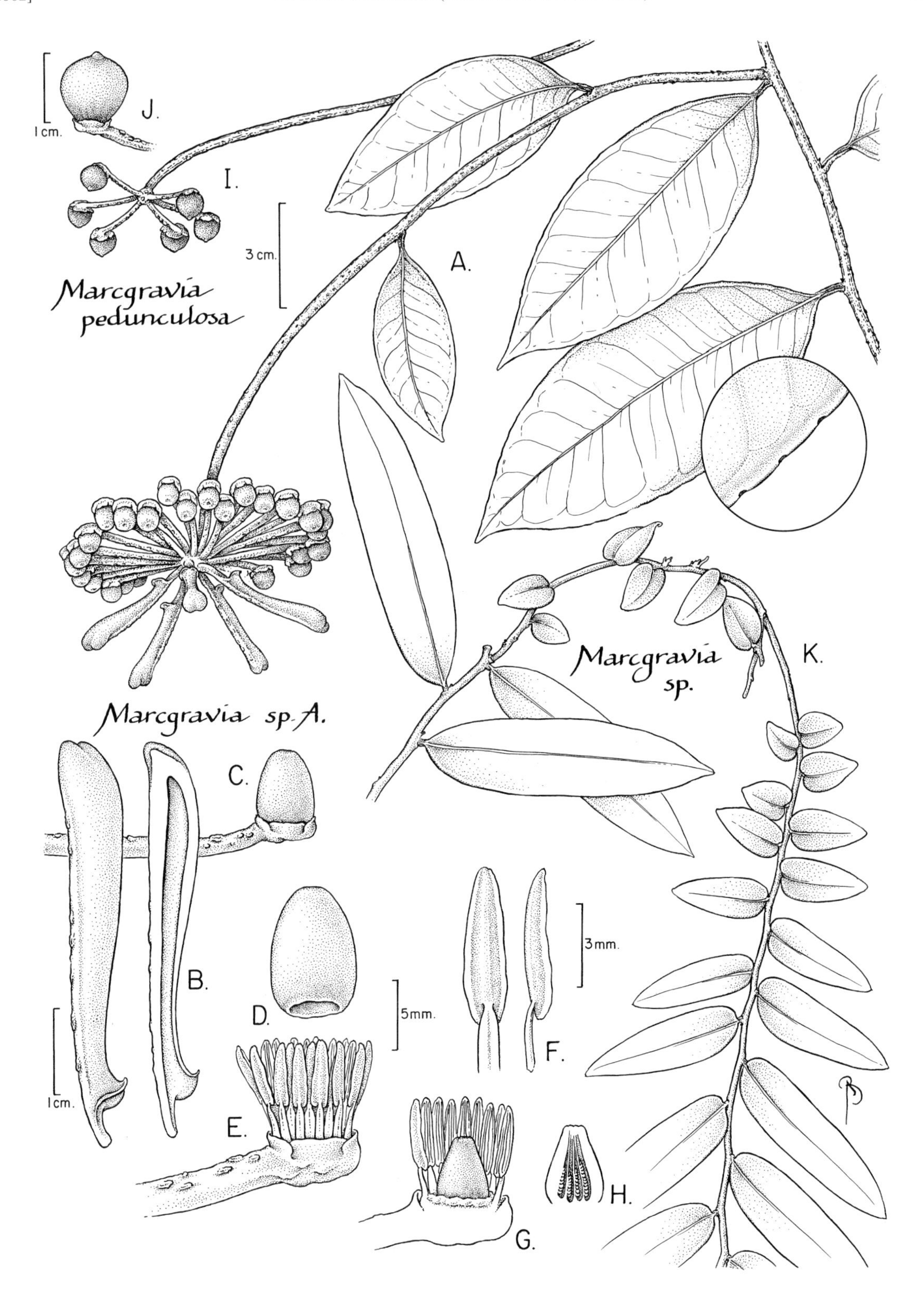

FIG. 184. MARCGRAVIACEAE. **A–H.** *Marcgravia* sp. A (*Mori et al. 23162*). **A.** Stem with leaves and inflorescence and detail of leaf margin showing glands. **B.** Lateral view (far left) and medial section (near left) of nectary bracts. **C.** Flower bud. **D.** Caducous calyptra of flower. **E.** Lateral view of flower at anthesis. **F.** Abaxial (far left) and lateral (near left) views of stamens. **G.** Lateral view of flower with half of stamens removed to show pistil. **H.** Medial section of ovary. **I, J.** *Marcgravia pedunculosa* (*Prance & Boeke 28112*) **I.** Infructescence. **J.** Lateral view of fruit. **K.** *Marcgravia* sp. (*Mori et al. 24087*). Stem from an unknown species showing juvenile (small) and mature leaves.

MARCGRAVIASTRUM (Szyszyl.) de Roon & S. Dressler

Terrestrial or (hemi)epiphytic shrubs or lianas. Leaves: petioles present; blades coriaceous, provided with numerous sclereids, often producing ciliate fracture when broken perpendicular to midrib, abaxial glands either 2–20 per leaf-half and arranged in fairly regular rows or very numerous (25–170) and arranged in rows and scattered over surface, the apex with conspicuous persistent or deciduous, sclerified mucro, the margins entire, often (slightly) revolute; secondary veins obscure to impressed adaxially, obscure to prominent abaxially when dry. Inflorescences contracted, umbell-like racemes, erect, with (2)5–25 flowers, the axis 0.5–3 cm long; pedicels slender to very stout; nectaries saccate, tubular or pouch-shaped, attached at lower quarter to third of pedicel; bracteoles appressed to calyx or rarely inserted some mm below calyx. Flowers 5-merous; sepals semi- or suborbicular to orbicular, coriaeous, the margins chartaceous; petals free to variously connate, coriaceous, strongly reflexed at anthesis, the margins often membranous; stamens (12)30–75, often adnate to base of corolla, the filaments free, linear, the anthers (sub)sagittate; ovary globose to turbinate, completely or incompletely 5–9-locular, the style conical to long cylindrical, the stigma capitate or radiate; ovules usually numerous. Fruits berry-like, (sub)globose, the pericarp hard leathery, reddish-green to red-violet, the mesocarp fibrous and pulpy. Seeds reniform, reticulate, shiny black or red-black.

Marcgraviastrum pendulum (Lanj. & Heerdt) Bedell

Sprawling shrubs or lianas. Leaves: petioles 3–7 mm long; blades oblong, 9–15 × 3–6 cm, abaxial glands 3–7 per side, arranged in row 1–7 mm from margin, the apex acute, obtuse, or mucronate, often retuse because of loss of mucro, the margins slightly revolute. Inflorescences with 6–15 flowers, the rachis stout, 1–2 cm long; pedicels slender, thickening toward apex, 4–7 cm long; nectaries sessile, inserted 1–1.5 cm from base of pedicel, bag-shaped, curved, pendent, 1.5–3 × 0.4–0.8 cm, greenish or yellowish with reddish tinge; bracteoles suborbicular, 4–6 × 7–9 mm. Flowers coriaceous, ca. 1.5 cm diam.; sepals semi- to suborbicular, 5–7 × 8–10 mm, green with rose margin; petals obovate-oblong, free or slightly connate at base, ca. 12 × 7 mm, green to pink abaxially, pale green adaxially; stamens 50–65, the filaments filiform, ca. 5 mm long, the anthers ca. 3 mm long; ovary turbinate, ca. 6 × 5 mm, the style 2–3 mm long. Fruits globose, ca. 2.5 cm diam. Seeds reniform, blackish.

Not known by collections from our area but to be expected; *Cremers 11628*, *Granville 10759*, *Granville & Cremers 11830* (all at CAY) collected from nearby areas.

NORANTEA Aubl.

Lianas or shrubs with sprawling branches. Stems often with long aerial roots. Leaves: petioles present or nearly absent; blades coriaceous, without sclereids, mostly with 2 glands at base, sometimes also with very minute abaxial glands along margins, the margins entire, slightly revolute; venation prominent when dry. Inflorescences long, erect to horizontally oriented racemes, with 100–350 flowers, the rachis stout, 25–100 cm long; pedicels 0.3–1.2 cm long, turned upward; nectaries inserted near middle of pedicel, saccate, stalked, bright red, orange-red, or purplish; bracteoles inserted at or at some distance from base of calyx. Flowers 5-merous; sepals suborbicular, imbricate; petals free or slightly connate at base, reflexed at anthesis; stamens 20–40, adnate to corolla, often falling as unit, the filaments curved, flattened at base, thickened and angular in section toward apex, the anthers linear, slightly falcate, laterally compressed; ovary ovate-conical, ca. 4 × 2–2.5 mm, incompletely (3)4–5(6)-locular; ovules 10–20 per locule. Fruits berry-like, the exocarp hard leathery, the mesocarp fibrous, pulpy. Seeds reniform, reticulate, reddish-black.

Norantea guianensis Aubl. FIG. 185, PL. 82f

Lianas. Leaves: petioles (0.5)1–2 cm long; blades obovate-oblong to oblanceolate or elliptic, 8–21 × 3.5–8.5 cm, coriaceous, with 2 glands at base, abaxial glands mostly lacking or, if present, few inserted apically and close to margin, these minute (visible only with magnification), the base acute, cuneate or attenuate, the apex obtuse to rounded; venation slightly prominent adaxially, prominent abaxially when dry. Inflorescences: nectaries cylindrically bag-shaped, 2.5–4 cm long, bright orange to red, the stalk (0.5)1–1.5 cm long. Flowers ca. 0.5 cm diam.; petals ovate to ovate-oblong, reflexed, 3–6 × 2–4 mm, reddish to purplish. Fruits globose, 0.5–1.5 cm diam., reddish at maturity. Fl (Jan, Feb, Dec), fr (Feb, Sep); very common, in moist lowland forest.

SOUROUBEA Aubl.

Terrestrial or (hemi)epiphytic shrubs and lianas. Leaves: petioles present; blades membranaceous to rigidly coriaceous, variously shaped sclereids frequent but sometimes lacking, abaxial glands variable in number, size and arrangement, often inserted in row parallel to margin. Inflorescences lax to dense racemes (5)10–35(50) cm long, with 15–60(100) flowers; pedicels slender to stout, (0.5)1–5 cm long; nectaries inserted at base of flowers, cup- or spur-shaped, hollow, often with 2 wing-like auricles. Flowers 5-merous, fragrant; sepals 5, coriaceous; petals (3)5, free or slightly connate, strongly reflexed; stamens (3)5, often adnate to base of petals, the filaments often connate basally, flattened, broad, abruptly narrowed apically, the anthers ovate to subglobose; ovary ovoid to pentagonal, 3–5 locular, the stigma sessile, often large, with radiating lobes. Fruits berry-like, (sub)globose, 0.5–2.5 cm diam., apiculate, reddish at maturity, the pericarp hard, leathery, the mesocarp fibrous and pulpy. Seeds reniform, shiny, reticulate.

Souroubea guianensis Aubl. FIG. 186, PL. 83

Shrubs or lianas. Stems often with long aerial roots. Leaves: petioles 0.5–1.5(2) cm long; blades variable in shape and size, often asymmetric, obovate to obovate-oblong or elliptic-oblong, 7–15(19) × 3–7(8) cm, abaxial glands 3–12 on each side of blade, <0.5 mm to >1 mm diam., scattered or in ± regular row, the base acute, cuneate, obtuse, or sometimes nearly rounded, the apex acute to obtuse or rounded, sometimes slightly acuminate; venation mostly invisible adaxially, prominent abaxially when dry. Inflorescences

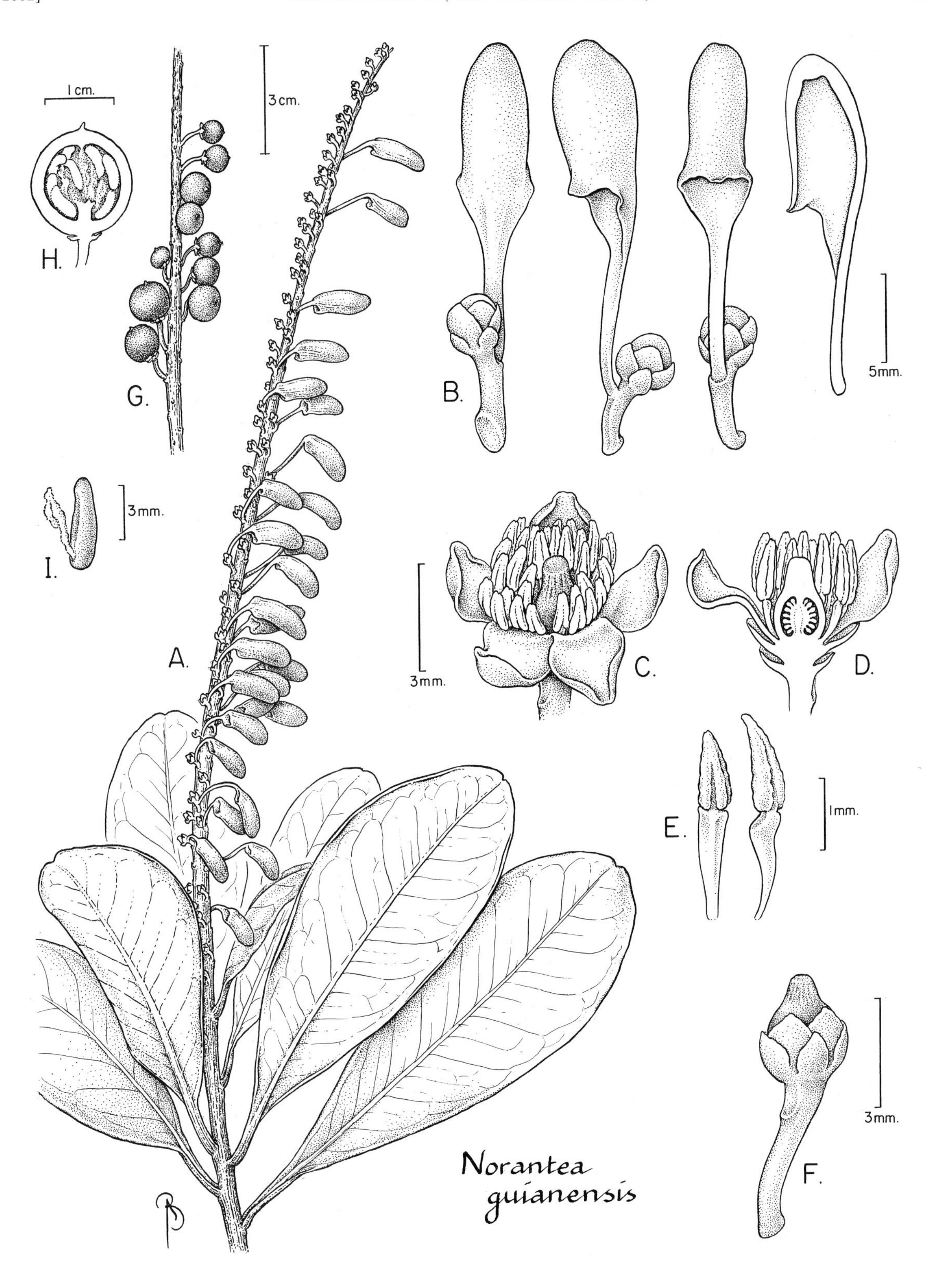

FIG. 185. MARCGRAVIACEAE. *Norantea guianensis* (A, F, *Mori et al. 22996*: B, *Mori et al. 8717*; C–E, *Lanjouw & Lindeman 1765* from Surinam; G–I, *Ek 842* from Guyana). **A.** Part of stem with leaves and inflorescence. **B.** Three views of flower buds and associated cup-like nectariferous bracts and lateral section of bract (far right). **C.** Oblique-apical view of flower. **D.** Medial section of flower. **E.** Two views of stamens. **F.** Immature fruit partially enclosed by calyx. **G.** Part of infructescence. **H.** Medial section of fruit. **I.** Seed with funicle.

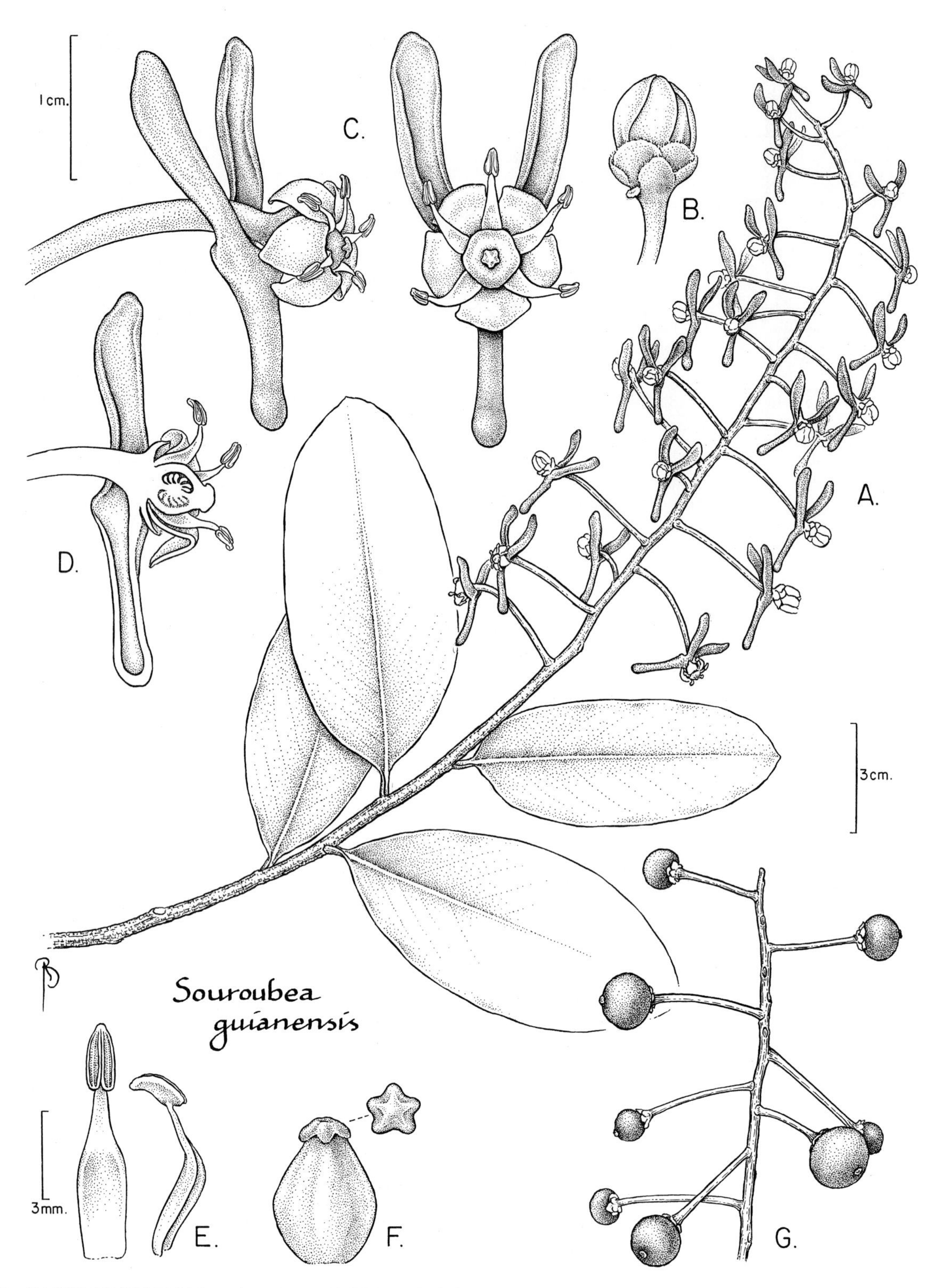

FIG. 186. MARCGRAVIACEAE. *Souroubea guianensis* (A–F, *Mori et al. 23766*; G, *Delgado 126* from Venezuela). **A.** Apex of stem with leaves and inflorescence. **B.** Flower bud with nectariferous bract removed. **C.** Lateral (left) and apical (right) views of flower with associated nectariferous bract. **D.** Medial section of flower and nectariferous bract. **E.** Two views of stamens. **F.** Lateral view of pistil (left) and apical view of stigma (above right). **G.** Part of infructescence.

laxly flowered, with mostly 15–30 flowers, the rachis 10–25(30) cm long; pedicels 1.5–3 cm long; nectaries spur-shaped, auriculate, 2–3 cm long, red, the spur as long as or slightly shorter than auricles. Flowers ca. 1 cm diam.; sepals suborbicular; petals obovate-oblong, 8–10 × 4–5 mm, basally connate to 1–3 mm, yellow, the apices sometimes orange; stamens 5, the filaments flattened, broad, ca. 5

mm long, narrowed toward anthers, basally connate, the anthers ovoid-globose, 1.5–2.5 mm long; ovary ovoid to subcylindrical, 2.5–4 mm long, 5-locular, the stigma large, sessile, 5-radiate. Fruits subglobose, ca. 1.5 cm diam. Fl (Sep), fr (Sep); in forest on poorly drained soils.

MELASTOMATACEAE (Melastome Family)

John Wurdack and Susanne Renner

Trees, shrubs, or herbs, usually terrestrial, sometimes lianas (*Adelobotrys*). Stipules absent (in the neotropics). Leaves simple, opposite (rarely whorled or pseudoalternate by abortion or early dehiscence of one member of each pair), homomorphic or heteromorphic; venation usually (except *Mouriri* and a few ericoid species in other genera) with 1 to several pairs of palmate primary veins arising at or above the blade base and ladder-like arrangement of secondary veins, pinnate in *Mouriri*. Inflorescences terminal, lateral, or cauline, usually cymose. Flowers actinomorphic, bisexual, perigynous or epigynous, (3)4–5(9)-merous; hypanthium bearing calyx lobes, petals, and stamens on a torus at or near base of calyx lobes; calyx usually open in bud (closed in bud and dehiscing irregularly in a few species of *Miconia* and *Bellucia*); petals free, right-contort, usually white to purple, usually spreading; stamens usually twice as many as petals (rarely the same fertile number with alternate staminodia), usually ± dimorphic, at least in size, usually glabrous, the anthers basifixed, ovate to subulate, (2)4-celled, opening by 1–2(4) terminal pores (rarely rimose), the connective sometimes prolonged below the thecae and variously appendaged adaxially and/or abaxially (see Figs. 190 & 197, illustrations of *Ernestia glandulosa* and *Tibouchina aspera*); ovary superior or more or less inferior, (1)2–5(15)-locular, placentation usually axile except in *Mouriri* where it is free central, usually multiovulate, the stigma punctiform or capitate. Fruits loculicidal capsules enclosed by persistent hypanthium if from superior ovary or berries if from inferior ovary. Seeds usually numerous and variously shaped (winged in *Acanthella*).

Gleason, H. A. 1966. Melastomaceae. *In* A. Pulle (ed.), Flora of Suriname **III(1):** 178–281. E. J. Brill, Leiden.

Görts-van Rijn, A. R. A. & M. J. Jansen-Jacobs. 1986. Melastomataceae. *In* A. L. Stoffers & J. C. Lindeman (eds.), Flora of Suriname **III(1–2):** 356–429. E. J. Brill, Leiden.

Lanjouw, J. 1966. Melastomaceae. *In* A. Pulle (ed.), Flora of Suriname **III(1):** 442–443. E. J. Brill, Leiden.

Renner, S. S. 1989. Systematic studies in the Melastomataceae: *Bellucia*, *Loreya*, and *Macairea*. Mem. New York Bot. Gard. **50:** 1–111.

Wurdack, J. J. 1993. Melastomataceae. *In* A. R. A. Görts-van Rijn (ed.), Flora of the Guianas Ser. A, **13:** 1–425. Koeltz Scientific Books, Koenigstein.

1. Ovaries superior. Fruits dehiscent capsules or berry-like and indehiscent (*Aciotis*).
   2. Stamen connective with adaxial (and sometimes also abaxial) appendages or not appendaged.
      3. Calyx lobes and petals 5.
         4. Fertile stamens 5, alternating with minute staminodia. . . . . *Rhynchanthera.*
         4. Fertile stamens 10, the two whorls more or less dimorphic, but the inner one never staminodial.
            5. Connective prolongation of larger anthers 2/3–3/4 as long as thecae. . . . . *Desmoscelis.*
            5. Connective prolongation of larger anthers 1/3 or less as long as thecae. . . . . *Tibouchina.*
      3. Calyx lobes and petals 4.
         6. Capsules or berry-like, 2-valved.
            7. Connective simply articulate with filament, not appendaged. Fruits capsules or thin-walled indehiscent and berry-like. . . . . *Aciotis.*
            7. Connective basally prolonged and adaxially bluntly bilobed, at least in the larger stamens. Fruits capsules. . . . . *Comolia.*
         6. Capsules, 3-valved (in our area).
            8. Stamen connectives with yellow adaxial appendages 2–3 mm long. *Ernestia.*
            8. Stamen connectives with purple adaxial appendages 0.3–0.7 mm long. . . . . *Nepsera.*
   2. Stamen connective with only abaxial appendages.
      9. Plants lianas. Ovary and fruit 5-locular. . . . . *Adelobotrys.*
      9. Plants herbs or subshrubs. Ovary and fruit usually 3-locular. . . . . *Macrocentrum.*
1. Ovaries inferior. Fruits berry-like.
   10. Leaf venation pinnate or only midrib evident (exceptionally, in *Mouriri sagotiana*, with prominent loop connected secondary veins). Anther connective with a concave abaxial gland. . . . . *Mouriri.*
   10. Leaves with 1 to several pairs of strongly developed primary veins diverging from midrib at or somewhat above blade base. Anther connective lacking an abaxial gland (occasionally with glandular hairs near or at base).
       11. Inflorescences terminal or pseudolateral by overtopping vegetative growth, but then solitary and not paired in upper leaf axils.
           12. Petals acute or acuminate at apex. . . . . *Leandra.*
           12. Petals rounded or retuse at apex.

13. Inflorescences in upper leaf axils or at defoliated nodes, rarely terminal and then always becoming pseudolateral. Abaxial leaf surface pubescence often gland-tipped. . . . . *Clidemia.*
13. Inflorescences in terminal panicles. Abaxial leaf surface pubescence amorphous or all dendritic-stellulate, lacking glandular hairs. . . . . *Miconia.*

11. Inflorescences lateral, paired in upper leaf axils or from branchlets below leaves.
14. Ant domatia present at petiole apex or blade base.. . . . . *Maieta.*
14. Ant domatia absent.
15. Flowers 6–8(9)-merous.
16. Flowers with 2 pairs of prominent long-persistent bracts. . . . . *Topobea.*
16. Flowers lacking prominent bracts.
17. Calyx irregularly dehiscent at anthesis; petals (20)22–24 mm long; ovary completely adnate to the hypanthium, the stigma fluted-capitate, 2.5–4 mm diam. . . . . *Bellucia.*
17. Calyx persistent; petals 7–9 mm long; ovary 1/10 adnate to the hypanthium, the stigma slightly expanded, 0.8–1 mm diam. . . . . *Clidemia octona* subsp. *guayanensis.*
15. Flowers 4–5(6)-merous.
18. Flowers in developed inflorescences with an obvious axis.
19. Small to large trees, <4–30 m tall. Petals adaxially with callus ridges. . . . . *Loreya.*
19. Shrubs. Petals lacking adaxial callus ridges. . . . . *Clidemia.*
18. Flowers solitary but aggregated in the upper leaf axils or on branchlets below leaves, the inflorescence axis not developed.
20. Hypanthium 1.7–3 mm long.
21. Petals rounded at apex, not cross-ridged within. . . . . *Clidemia conglomerata.*
21. Petals bluntly acute, cross-ridged within. . . . . *Henriettella.*
20. Hypanthium (3)4–15 mm long.
22. Anther tips spoon-shaped; stigma not or very slightly expanded. . . . . *Henriettea.*
22. Anther tips rounded or truncate; stigma capitate, to 2 mm diam. . . . . *Loreya.*

ACIOTIS D. Don

Herbs or subshrubs. Leaves homomorphic, at least in our area. Inflorescences terminal panicles. Flowers small, 4-merous; hypanthium terete or with inconspicuous longitudinal ridges; calyx lobes triangular, persistent; petals ovate to lanceolate, white, pink, or purplish; stamens 8, usually somewhat dimorphic in size, the anthers oblong to orbicular and 1-pored, the connective not or somewhat prolonged, not appendaged; ovary superior, 2-locular. Fruits 2-valved capsules or indehiscent and berry-like. Seeds numerous, more or less cochleate, tuberculate.

1. Largest leaf blades usually <6.5 cm long. Stamens slightly dimorphic, larger anthers <1.2 mm long, ovoid to almost spheroidal, the connective basally prolonged 0.1 mm or less. Placentae persistent on old infructescences. . . . . *A. acuminifolia.*
1. Largest leaf blades usually >6.5 cm long. Stamens slightly dimorphic, larger anthers >1.2 mm long, the connective basally usually distinctly prolonged. Entire capsule caducous.
2. Ovary apex glandular-puberulous. . . . . *A. indecora.*
2. Ovary apex glabrous.
3. Stems with spreading hairs >1.5 mm long. Leaf blades on both sides sparsely to moderately fine-setose, the hairs 1–1.5 mm long.. . . . . *A. caulialata.*
3. Stems glabrous or very inconspicuously puberulous on wings. Leaf blades on both sides glabrous or very sparsely and deciduously strigulose. . . . . *A. purpurascens.*

Aciotis acuminifolia (DC.) Triana [Syn.: *Aciotis aequatorialis* Cogn.] Pl. 84a

Herbs to 0.3 m tall, usually much branched. Stems narrowly 4-winged, fine-setose with eglandular hairs <1 mm long. Leaves: petioles 1–3.5 cm long; blades ovate to elliptic, 2–6.5(9.5) × 1.5–3.5(4.5) cm, the base obtuse, truncate, or cordulate, the apex acute, the margins inconspicuously ciliolate-serrulate; primary veins in 1(2) pairs. Flowers: hypanthium 2.5–3 mm long, with glandular hairs on the costae; petals pale pink to white, 2–3.5(4.5) mm long; stamens 2.2–3.5 mm long, anthers 0.5–1.2 mm long, ovoid or almost spheroidal, pink, the connective basally prolonged 0.1 mm or less; ovary apex glabrous. Fruits biloculicidal capsules; hypanthium caducous but placentae and sometimes ovary wall persistent on old infructescences. Fl (May, Jun, Oct); in weedy, often moist, habitats along roads.

Aciotis caulialata (Ruiz & Pav.) Triana [Syns.: *Aciotis alata* (Beurl.) Almeda, *A. rubricaulis* (DC.) Triana]

Herbs, 0.3–1 m tall. Stems conspicuously winged, with spreading eglandular hairs >1.5 mm long. Leaves: petioles 1–3(4.5) cm

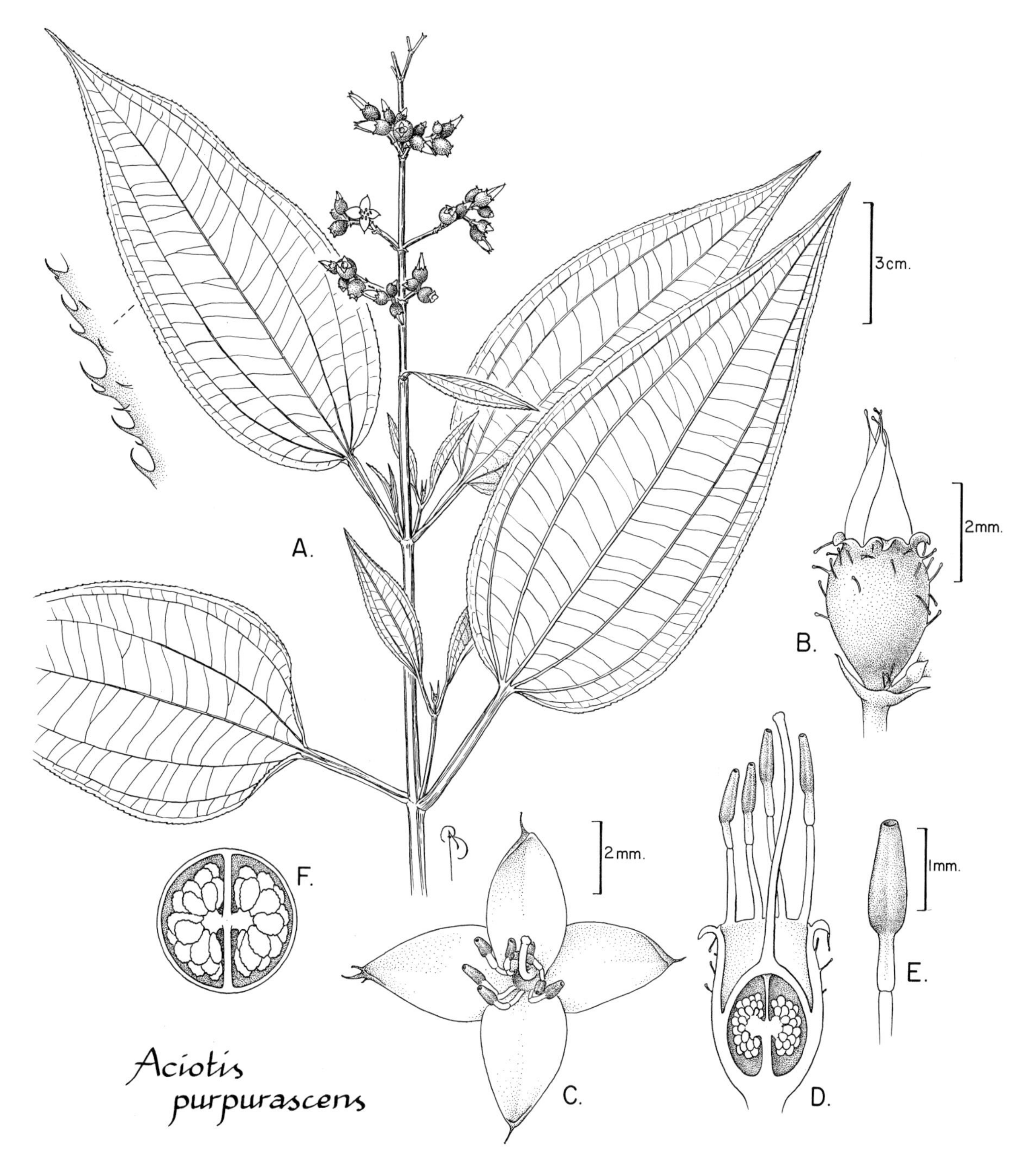

FIG. 187. MELASTOMATACEAE. *Aciotis purpurascens* (A, *Prance 28109*; B–F, *Mori & Gracie 19028*). **A.** Apex of stem with leaves, terminal inflorescence, and detail of fringed leaf margin (left). **B.** Flower bud. **C.** Apical view of flower. **D.** Medial section of flower with corolla removed; note the partly perigynous cupular hypanthium with calycine lobes and the axile placentation. **E.** Stamen showing jointed filament and poricidal dehiscence. **F.** Transverse section of fruit.

long, winged; blades oblong-ovate, 8–10(15) × 4–5(13) cm, on both sides sparsely to moderately fine-setose, the hairs 1–1.5 mm long, the base obtuse or cuneate, the apex acute, the margins ciliolate-serrulate. Flowers: hypanthium 2–3.5 mm long, hirsute with glandular and eglandular hairs; petals white, 4–5(8) mm long; stamens somewhat dimorphic, large anthers 1.2–1.8 mm long, the connective basally prolonged 0.5–0.7 mm, small anthers 0.7–1 mm long, the connective basally prolonged 0.7–0.8 mm; ovary apex glabrous. Fruits biloculicidal capsules, entire capsule caducous. Fl (Jun, Jul); in weedy habitats.

Aciotis indecora (Bonpl.) Triana [Syn.: *Aciotis laxa* (DC.) Cogn.]

Subshrubs, 1.5–2 m tall. Stems quadrangular, narrowly winged, at least in part with gland-tipped smooth hairs <1 mm long. Leaves: petioles 1.5–8 cm long; blades oblong-elliptic or ovate, 4.5–14 × 2–8.5 cm, the base obtuse, the apex shortly acuminate, the margins minutely ciliolate-serrulate with glandular and eglandular hairs; primary veins in 2–3 pairs. Flowers: hypanthium 1.5–2.5 mm long, densely glandular pubescent; petals white to pinkish, 1.5–2.5 mm long; large anthers 1.3–1.6 mm long, usually purple, the connective basally prolonged 0.2–0.3 mm, small anthers 0.8–1.1 mm long, the connective basally prolonged 0.2–0.3 mm; ovary apex glandular-puberulous. Fruits biloculicidal capsules, green at maturity and berry-like in aspect, entire capsule caducous Fl (Jan), fr (Mar); in forest and secondary vegetation.

Aciotis purpurascens (Aubl.) Triana FIG. 187

Subshrubs, 0.5–1 m tall. Stems 4-winged, glabrous or very inconspicuously puberulous on wings. Leaves: petioles 1–2 cm long; blades narrrowly ovate to oblong-ovate, 8–15 × 2.5–6 cm, both sides glabrous or very sparsely and deciduously strigulose, the base obtuse to cordulate, the apex narrowly acute to acuminate, the margins inconspicuously ciliolate-serrulate. Flowers: hypanthium 2–2.5 mm long; petals white or pinkish, 4–5 mm long; large anthers 1.2–1.6 mm long and narrow, the connective basally prolonged 0.7–0.9 mm, small anthers 0.9–1 mm long, the connective basally prolonged 0.8–1 mm; ovary apex glabrous. Fruits entirely caducous. Fl (Jun, Jul, Aug, Sep, Oct); in weedy habitats along trails and road and along streams.

ADELOBOTRYS DC.

Lianas. Stems often with adventitious roots on parts appressed to tree trunks, these usually not present on herbarium specimens. Leaves homomorphic or heteromorphic. Inflorescences terminal panicles. Flowers 5-merous; hypanthium terete or ribbed; calyx limb expanded, usually with indistinct oblate lobes, externally and/or internally with ± developed short projections or teeth; petals obovate, usually pink, the apex rounded or emarginate; stamens 10, sometimes heteromorphic, the anthers linear-subulate, with a small usually abaxially inclined pore, the connective basally not prolonged nor adaxially appendaged, abaxially with an acute basal spur and an elongate often bifid appendage ascending parallel to the anthers; ovary superior, 5-locular. Fruits capsules. Seeds numerous, narrowly lanceolate.

Adelobotrys adscendens (Sw.) Triana

Lianas climbing appressed to tree trunks to 8 m from ground. Stems and leaves with minute deciduous bifid hairs. Leaves: petioles 2.5–5.5 cm long; blades ovate to oblong-ovate, 8–16 × 5–12 cm, the base rounded to cordulate, the apex acute to short-acuminate, the margins deciduously fine-ciliolate; primary veins in 2–3 pairs. Inflorescence panicles 9–20 cm long, many-flowered; pedicels 3–6 mm long. Flowers: hypanthium 4–4.5 mm long, with deciduous bifid hairs; calyx 1.5–2 mm long; petals 8–10.5 mm long, pale pink; stamens dimorphic, anthers 5–7.5 mm or 4–4.5 mm long, the abaxial ascending appendage 2–3.5 mm long. Fl (Jan, Sep), fr (Jan, Jul); in nonflooded forest at all elevations.

BELLUCIA Raf.

Trees or shrubs (outside our area). Leaves homomorphic. Inflorescences axillary (in the Guianas). Flowers 6–8(9)-merous; hypanthium terete, globose; calyx calyptrate (in the Guianas), dehiscing in an irregular semicircle, drying as a hyaline membrane, often persistent after anthesis; petals obovate, clawed, fleshy with callus ridges adaxially, white or pink-flushed abaxially, the apex truncate or slightly emarginate; stamens isomorphic, laterally compressed, the anthers with blunt apices and two slightly introrse minute pores, the connective not prolonged but fleshy; ovary completely adnate to the hypanthium, 10–14(15)-locular, the stigma fluted-capitate, with 10–16 lobes. Fruits berries, yellow. Seeds numerous, ± ovoid.

Bellucia grossularioides (L.) Triana
FIG. 188, PL. 84c; PART 1: FIG. 9

Trees, 3–30 m tall. Leaves: petioles to 6 cm long; blades broadly ovate, 15–35 × 10–20 cm, coriaceous, glabrous, the base obtuse to rounded, the apex shortly acuminate, the margins entire or serrulate in young leaves. Inflorescences cymose, short, usually 2-flowered; pedicels 10–24 mm long. Flowers usually (6)7–8(9)-merous; hypanthium 8–10 mm long, glabrous; petals (20)22–24 mm long, white; anthers 5–6 mm long; ovary (12)13–14(15)-locular, the stigma 2.5–4 mm diam. Fruits globose, ca. 4.5 cm diam., yellow at maturity. Fl (Jun), fr (Jun, Jul, Aug); common in secondary vegetation along roads and wider trails, fruits are eaten and dispersed by animals (Charles-Dominique et al., 1981). *Araça de anta* (Portuguese), *bois mèle*, *bois-messe*, *caca Henriette*, *graine mèle* (Créole), *mandapuçu* (Portuguese), *mésoupou* (Créole).

CLIDEMIA D. Don

Erect (rarely scandent or radicant) shrubs. Leaves opposite or rarely pseudoalternate, rarely with ant domatia (outside the Guianas), homomorphic or slightly heteromorphic in size. Inflorescences lateral or pseudolateral (by overtopping vegetative growth) in upper leaf axils or at defoliated branchlet nodes. Flowers 4–5(8)-merous; hypanthium terete; calyx usually lobed and persistent, each lobe with an external tooth; petals oblong-obovate, the apex rounded, white to pink; stamens isomorphic or slightly dimorphic in size, the anthers usually slender, 1-pored, the connective not or only slightly prolonged, usually unappendaged or with a minute dorso-basal tooth; ovary partly to completely adnate to the hypanthium, (2)3–5(10)-locular. Fruits berries, globose, many-seeded, dark blue or black at maturity. Seeds ovoid to pyramidate, smooth to granulate.

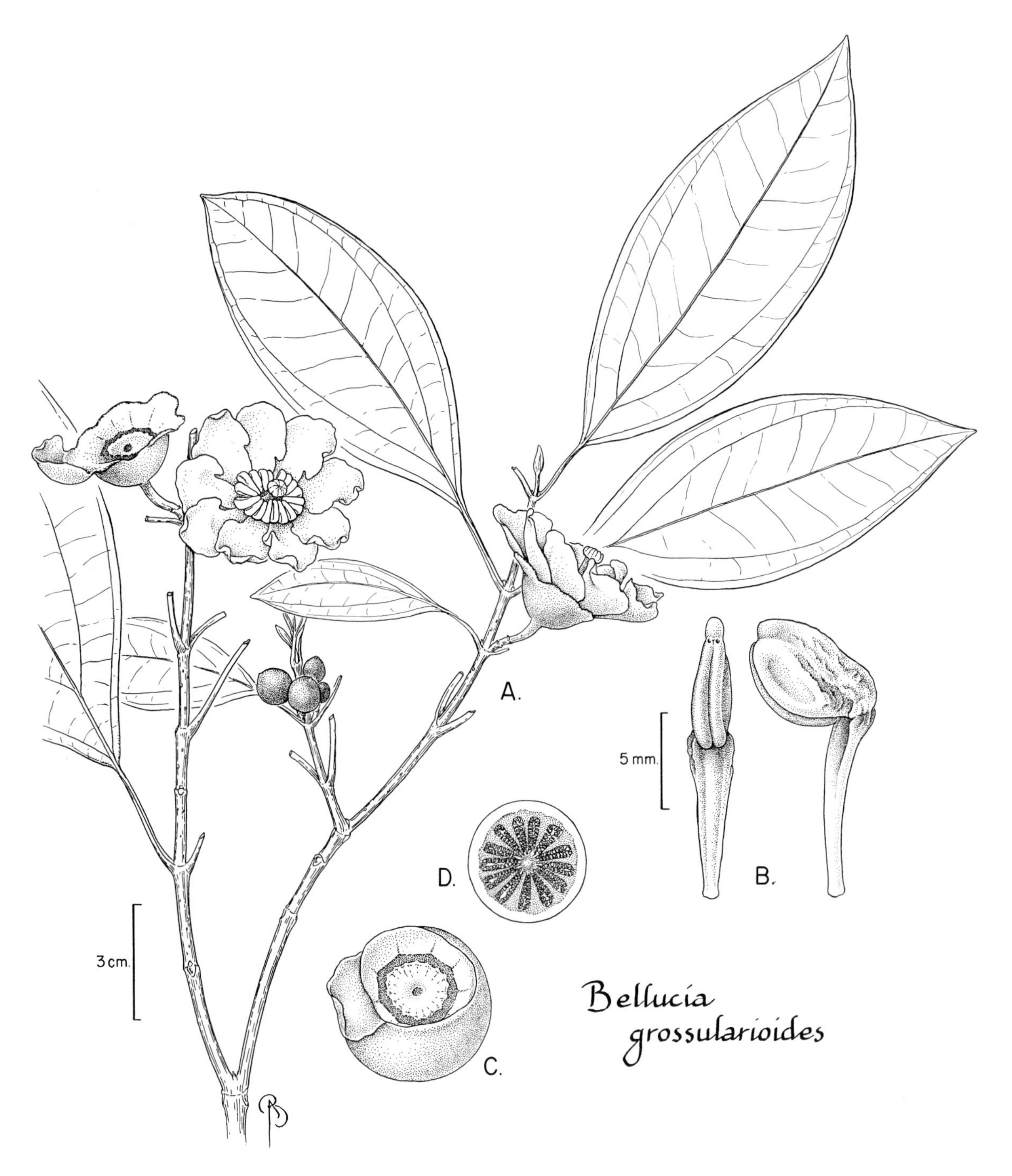

FIG. 188. MELASTOMATACEAE. *Bellucia grossularioides* (*Mori & Gracie 19012*). **A.** Stem with leaves, flowers in bud, at anthesis, and after anthesis. **B.** Adaxial (left) and lateral (right) views of stamens with double poricidal dehiscence. **C.** Fruit with a single persistent calyx lobe. **D.** Transverse section of fruit showing multiple locules.

1. Stems, branchlets, inflorescences, and hypanthia sparsely to moderately setose with eglandular hairs 4–10 mm long. . . . . *C. laevifolia.*
1. Stems, branchlets, inflorescences, and hypanthia glabrous or if pubescent with much shorter, often glandular hairs.
    2. Leaf base acute and long-decurrent; primary veins arising 2–4 cm above blade base. . . . . *C. septuplinervia.*
    2. Leaf base variously shaped but not decurrent; primary veins arising from blade base or to 1.5 cm above it.
        3. Flowers 4-merous.

4. Hypanthium densely setose to strigose.......... *C. rubra.*
4. Hypanthium resinous-granulose.......... *C. saülensis.*
3. Flowers 5–8-merous.
5. Flowers mostly 7–8-merous; ovary mostly 7–10-locular.......... *C. octona* var. *guyanensis.*
5. Flowers mostly 5-merous; ovary 3- or 5-locular.
6. Flowers in axillary sessile or very short-stalked heads.
7. Leaves 10–20 cm long; primary veins arising from blade base. Petals 5.2–5.5 mm long.......... *C. conglomerata.*
7. Leaves 2.5–12 cm long; inner pair of primary veins arising subalternately 0.3–1.5 cm above blade base. Petals ca. 3 mm long.......... *C. involucrata.*
6. Flowers in cymes or panicles.
8. Leaf bases inaequilateral and obtuse. Petals 5–6.5 mm long.......... *C. dentata.*
8. Leaf bases symmetric, rounded or cordulate. Petals 8–10 mm long.......... *C. hirta.*

Clidemia conglomerata DC.

Shrubs, 0.5–2(4) m tall. Young branchlets, petioles, and primary leaf veins moderately covered with appressed thick, flattened, shaggy hairs 1–2 mm long abaxially. Leaves: petioles 1–2.5 cm long; blades elliptic, 10–20 × 5–9 cm, chartaceous, sparsely and deciduously strigulose adaxially, sparsely setulose on the secondary veins and venules but glabrous on adaxial surface, sometimes purple-suffused abaxially, the base acute, the apex acuminate, the margins ± crenate-denticulate; primary veins in 2 pairs. Inflorescences in upper leaf axils, multiflorous heads; pedicels absent or rarely short, subtended by early caducous bracts 2–3 mm long. Flowers 5-merous; hypanthium ca. 1.8 mm long, densely strigose; calyx lobes ovate-oblong, 0.7–0.9 mm long, the subulate external teeth equalling or slightly exceeding the interior teeth; petals 5.2–5.5 mm long, white; anthers ca. 2.6 mm long, subulate, the pore minute, adaxially inclined, the connective not or slightly prolonged but not appendaged; ovary 5-locular. Fruits first green, lavender, purple, or black at maturity. Fl (Jan, Feb), fr (May, Aug); in understory of nonflooded forest.

Clidemia dentata D. Don

Shrubs, 1–3(6) m tall. Terete branches, primary leaf veins abaxially, and inflorescences moderately setose with fine smooth hairs 1–2 mm long. Leaves: petioles (0.5)1(2) cm long; blades ovate-elliptic to oblong-elliptic, 7–15 × 3–7 cm, chartaceous, sparsely loose-strigulose adaxially, sparsely appressed-setulose abaxially, the base obtuse, inaequilateral, the apex gradually acuminate, the margins obscurely ciliolate-denticulate; primary veins in 2 pairs. Inflorescences terminal but soon pseudolateral by overtopping growth, 3–15-flowered, 1–2(3) cm long; pedicels 3–5 mm long. Flowers 5-merous; hypanthium ca. 3 mm long, densely setose with smooth hairs 1–3 mm long; calyx 0.9–1.1 mm long, essentially truncate but with narrow, setose external teeth 1.5–3 mm long; petals 5–6.5 mm long, white; anthers stout-subulate, 2.8–3 mm long, the pore abaxially inclined, the connective basally very slightly prolonged but not appendaged; ovary 5-locular. Fruits dark blue. Fl (Nov), fr (Apr, May); along trails in nonflooded forest. *Radier macaque* (Créole).

Clidemia hirta (L.) D. Don Pl. 84b

Shrubs, 0.5–2(3) m tall. Terete branchlets, leaf veins and venules abaxially, inflorescences, and hypanthia moderately to sparsely fine-setose (often some of the shorter hairs gland-tipped) and sparsely stellulate-furfuraceous. Leaves: petioles 0.7–2 cm long; blades ovate to oblong-ovate, 5–16 × 3.5–8 cm, membranous, sparsely setose with fine smooth hairs adaxially, the base rounded to cordulate, the apex acute to shortly blunt-acuminate, the margins entire to crenulate, ciliate; primary veins in 2–3 pairs. Inflorescences terminal but becoming pseudolateral, 3–7-flowered, usually trichotomous near base, 2–3 cm long; pedicels 3–7 mm long, the minute bracteoles caducous. Flowers 5-merous; hypanthium 3–3.5 mm long; calyx lobes ovate, 0.2–0.6 mm long, the linear, setose external teeth 2–4 mm long; petals 8–10 mm long, white; anthers subulate, 3.5–4.5 mm long, the pore minute, abaxially inclined, the connective basally very slightly prolonged and with a small blunt dorso-basal tooth; ovary 5-locular. Fruits dark blue, tasty. Fl (Jun); in weedy areas. *Pixirica* (Portuguese), *radié macaque* (Créole).

Clidemia involucrata DC.

Shrubs, 1–3(4) m tall. Terete branchlets, primary leaf veins abaxially, and peduncles moderately setulose with fine smooth sometimes gland-tipped hairs. Leaves: petioles 0.4–2 cm long; blades elliptic-ovate, 2.5–12 × 1.5–5 cm, chartaceous, sparsely and finely appressed-setulose (a few hairs gland-tipped) on both surfaces, the base obtuse or acute, the apex acuminate, the margins ciliolate-serrulate; primary veins in 2–3 pairs, the inner pair diverging subalternately 0.3–1.5 cm above leaf base, the venules rather laxly reticulate. Inflorescences solitary in upper leaf axils, multibracteate heads; peduncle 0.5–1.5 cm long; pedicels lacking. Flowers 5-merous; hypanthium ca. 2.3 mm long, densely and finely appressed-setulose (a few hairs gland-tipped); calyx lobes ovate, ca. 0.8 mm long, the setulose external teeth ca. 1.5 mm long; petals ca. 3 mm long, white; anthers 3 mm long, oblong-subulate, slightly projecting adaxially below filament insertion, the pore minute, the connective not appendaged; ovary 3-locular. Fruits bluish-black at maturity. Fr (Jul); in nonflooded forest.

Clidemia laevifolia Gleason

Shrubs, 0.5–3 m tall. Terete branchlets sparsely setose with smooth eglandular hairs 4–10 mm long, as well as caducously stellulate-puberulous. Leaves: petioles 0.5–1.5 cm long; blades ovate-elliptic to oblong-elliptic, 6–12 × 4–7 cm, firm-chartaceous, glabrous or very sparsely and caducously fine-setose adaxially, abaxially sparsely fine-setose on veins but otherwise glabrous, often purplish-tinged abaxially, the base rounded, the apex shortly and abruptly acuminate, the margins entire, ciliate; primary veins in (1)2 pairs. Inflorescences terminal but becoming pseudolateral by overtopping vegetative growth, 3–5-flowered, usually trichotomous near base, 2–3 cm long; pedicels 1–5 mm long. Flowers 5-merous; hypanthium ca. 4(5.5) mm long, setose with eglandular hairs 3–6 mm long; calyx lobes barely or somewhat (to 1.5 mm) developed,

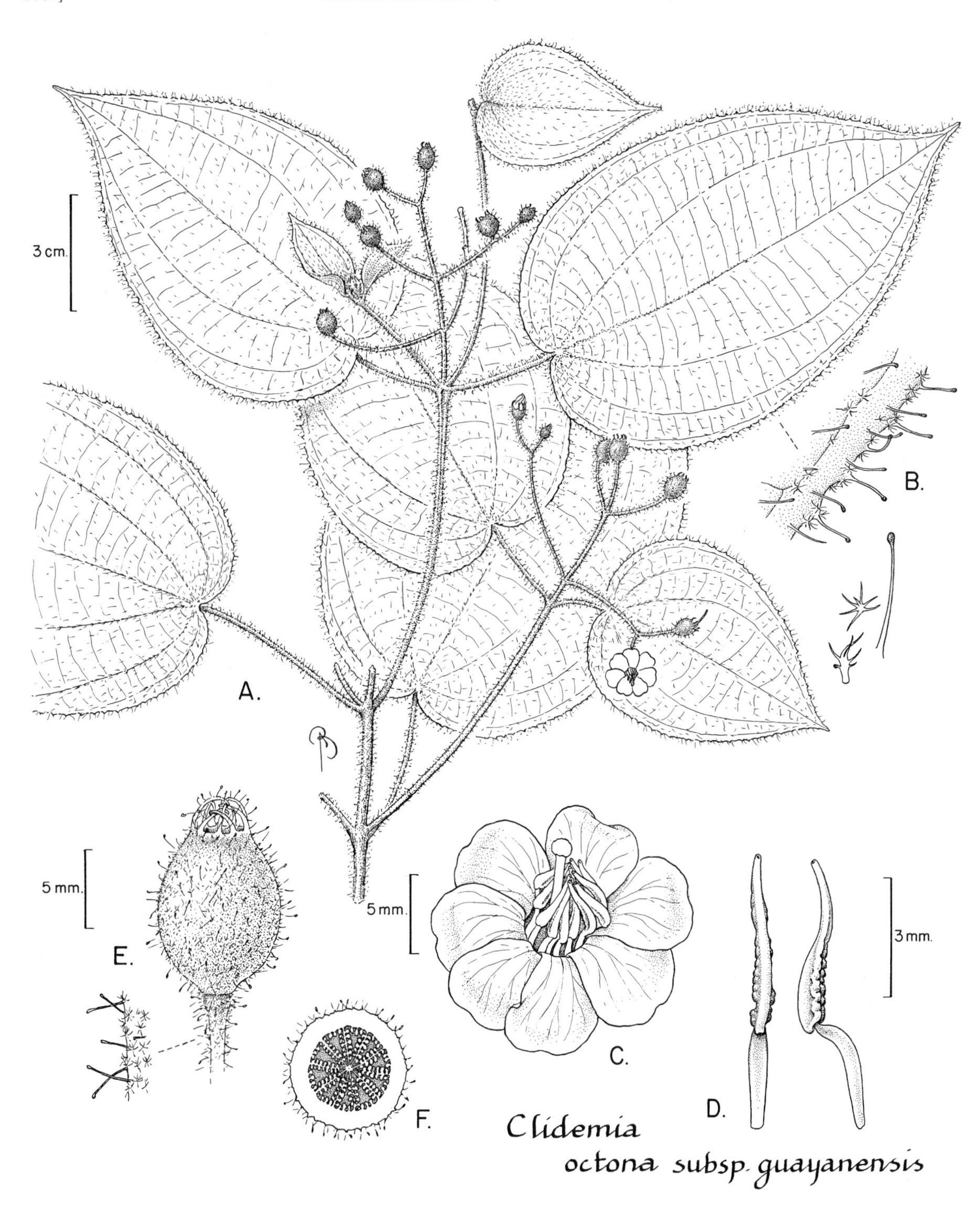

FIG. 189. MELASTOMATACEAE. *Clidemia octona* subsp. *guayanensis* (A, B, D–F, *Mori & Pipoly 15420*; C, *Mori et al. 18208*). **A.** Leafy stem with inflorescences becoming infructescences. **B.** Details of leaf margin (above) with glandular capitate and stellate hairs (below). **C.** Apical view of flower. **D.** Abaxial (left) and lateral (right) views of anthers showing single poricidal dehiscence. **E.** Fruit (right) with detail of pubescence (below). **F.** Transverse section of fruit with numerous locules.

each with a setose tooth 4.5–5(7) mm long; petals ca. 8(10) mm long, white to pale pink; anthers subulate, ca. 5(6) mm long, the pore abaxially inclined, the connective with an emarginate dorso-basal tooth 0.1–0.2 mm long; ovary 5-locular. Fruits dark blue. Fr (Apr); in nonflooded forest.

Clidemia octona (Bonpl.) L. O. Williams subsp. **guayanensis** Wurdack FIG. 189, PL. 84d

Shrubs, 1–4 m tall. Terete branchlets, inflorescences, and hypanthia sparsely glandular-setulose with smooth hairs 0.5–2 mm long, underlain densely by short-stipitate stellate hairs. Leaves: petioles 2–6 cm long; blades ovate, 10–21 × 7–14 cm, chartaceous, sparsely fine-setose with some of the hairs gland-tipped, the primary veins moderately puberulous with stipitate-stellate hairs adaxially, the veins and surface sparsely to moderately stellate-puberulous abaxially, the base cordate, the apex acute to short-acuminate, the margins denticulate; primary veins in (2)3(4) pairs. Inflorescences pseudolateral by overtopping growth, 21–35-flowered, 3–8 cm long; pedicels 3–5 mm long, with small deciduous bracteoles inserted shortly below hypanthium. Flowers (6)7–8-merous; hypanthium 4–4.5 mm long; calyx lobes 0.4–0.5 mm long, the external setose teeth 2–3 mm long; petals 7–9 mm long, white; anthers oblong-subulate, 3.5–5 mm long, the pore minute, the connective not prolonged but with a small blunt dorso-basal tooth; ovary (6)7–10-locular. Fruits lavender-blue or purple. Fl (Feb, Mar, Apr, Jun, Aug, Sep, Oct), fr (Jan, Mar, May, Jun, Aug, Sep, Oct); common, in weedy areas along trails and roads. *Feuille bome* (Créole).

Clidemia rubra (Aubl.) Mart. [Syn.: *C. sericea* D. Don] Pl. 84e

Shrubs, 0.7–2(3) m tall. Branchlets and abaxial primary leaf veins densely incurved-setose to strigose with fine smooth hairs 1–2.5 mm long. Leaves: petioles 0.2–1 cm long; blades ovate-oblong to elliptic, 6–13 × 3–6 cm, chartaceous, sparsely setulose adaxially, moderately fine-setose with hairs 1–2 mm long abaxially, the base rounded to cordulate, the apex acute to short-acuminate, the margins ciliate-serrulate; primary veins in 2–3 pairs. Inflorescences in upper leaf axils, short-stalked fascicles; pedicels nearly absent, with persistent, lanceolate bracteoles. Flowers 4-merous; hypanthium 3–4 mm long, densely fine-setose to strigose with some gland-tipped hairs; calyx lobes broadly ovate, ca. 0.6 mm long, the setose external teeth 0.6–0.9 mm long; petals 2–2.8 mm long, pale pink; anthers oblong-subulate, 3–3.8 mm long, the pore abaxially inclined, the connective very slightly prolonged but not appendaged; ovary 3–4-locular. Fruits dark blue, tasty. Fl (Feb, Jun, Aug), fr (Feb, Jun); in nonflooded forest. *Radié macaque*, *radier macaque* (Créole).

Clidemia saülensis Wurdack

Shrubs, 0.5–1.7 m tall. Terete branchlets and adaxial primary leaf veins densely fine-setose with slightly reflexed smooth hairs 3–4(5) mm long. Leaves: petioles 3–7 cm long; blades ovate, 10–19 × 5–11 cm, chartaceous, sparsely appressed-setose adaxially, rather sparsely setulose abaxially, the base cordate, the apex gradually acuminate, the margins ciliate-serrulate; primary veins in (2)3 pairs. Inflorescences from upper leaf axils or defoliated branchlet nodes, few-flowered, 2–3(5) cm long, the axes sparsely glandular-setulose; pedicels 3–6 mm long, the persistent subulate bracteoles inserted 0.8–1 mm below the hypanthium. Flowers 4-merous; hypanthium 3.5–4.2 mm long, densely resinous-granulose, essentially glabrous; calyx lobes rounded, 0.3 mm long, the setulose external teeth 0.7–0.8 mm long; petals 2.4–2.5 mm long, white; anthers oblong-subulate, 2.5–2.6 mm long, the pore abaxially inclined, the connective basally slightly prolonged but not appendaged; ovary mostly 4-locular. Fruits dark blue. Fl (Jan, Mar, May, Jun, Jul), fr (Jan, Feb, Mar, May, Jul, Sep); common, in weedy areas and on rock outcrops.

Clidemia septuplinervia Cogn.

Shrubs, 0.3–2(3) m tall. Terete young branchlets, primary and secondary leaf veins abaxially, inflorescences, and hypanthia moderately and minutely granulose-furfuraceous. Stems often rooting at nodes. Leaves: petioles 4–7 cm long; blades ovate to oblong-ovate, 13–28 × 5–15 cm, chartaceous, glabrous or very sparsely and deciduously strigulose adaxially, the base acute, abruptly long-decurrent, the apex gradually acuminate, the margins entire, appressed-ciliolate; venation of 3 pairs of primary veins, the inner pairs departing 2–4 cm above the blade base. Inflorescences usually paired at defoliated nodes or on stems, 3–5-flowered, 1(3) cm long; pedicels ca. 1 mm long, with obscure bracteoles. Flowers 4-merous; hypanthium 2.5–3 mm long, esetulose or very sparsely glandular-setulose; calyx lobes ovate, 0.3–0.4 mm long, the glandular-setulose external teeth 1.5–3 mm long; petals 1.9–2 mm long, white; anthers narrowly oblong, 1.3–1.9 mm long, the pore minute, abaxially inclined, the connective basally not or slightly prolonged but not appendaged; ovary 4-locular. Fruits first red, blue at maturity. Fl (Aug), fr (Jan, Feb, Mar, Apr, Jun, Aug, Sep, Dec); in wet or moist areas.

COMOLIA DC.

Shrubs. Leaves homomorphic. Inflorescences solitary or occasionally glomerate in upper leaf axils. Flowers 4-merous; hypanthium usually terete; calyx lobes persistent; petals obovate to oblong, pink to magenta; stamens somewhat dimorphic, anthers subulate to linear-subulate, the pore adaxially inclined, connectives basally somewhat prolonged, adaxially bluntly bilobed, abaxially unappendaged or with a blunt spur; ovary superior, 2(3–4)-locular. Fruits capsules. Seeds numerous, cochleate, foveolate.

Comolia villosa (Aubl.) Triana

Shrubs, 0.2–0.7 m tall. Young branchlets fine-setulose at nodes, moderately puberulous with glandular or eglandular hairs or glabrous in internodes. Leaves: petioles 0.2–0.5 cm long; blades elliptic to obovate-elliptic, 1–2.5 × 0.4–1.4 cm, chartaceous, thinly appressed-setose to glabrous adaxially, thinly setulose with glandular or eglandular hairs to glabrous abaxially, the base acute, the apex acute to obtuse, the margins denticulate; primary veins in 1–2 pairs. Inflorescences few-branched short cymes in upper leaf axils; pedicels ca. 1 mm long above bracteoles. Flowers: hypanthium 3–5 mm long, sparsely puberulous with glandular or eglandular hairs or glabrous; calyx lobes narrowly oblong, 1.5–2.5 mm long; petals 8–12 mm long, purple; anthers 5.5–6.5 mm long, the connective basally prolonged 1–2 mm or 0.5–1 mm and bilobed, abaxially unappendaged; ovary 2-locular. Fruits biloculicidal, globose. Savanna or in weedy, open areas.

DESMOSCELIS Naudin

Little-branched herbs or subshrubs. Inflorescences leafy panicles. Flowers 5-merous; hypanthium terete; calyx lobes lanceolate, persistent; petals obovate, the apex rounded to emarginate; stamens 10, dimorphic, the anthers oblong with a single pore, large stamens with connectives basally distinctly prolonged, adaxially with a pair of filiform lobes, small stamens with connectives not prolonged, adaxially with short, thick lobes; ovary superior, 5-locular, the apex densely fine-setose with deciduously gland-tipped hairs. Fruits capsules. Seeds numerous, cochleate, tuberculate.

Desmoscelis villosa (Aubl.) Naudin

Herbs or subshrubs, 0.5–1 m tall. Quadrangular stems moderately to densely fine-setose with hairs 2–7(10) mm long. Leaves: almost sessile, petioles 0.2–0.5 cm long; blades ovate to oblong-elliptic, (1.5) 3–5(6) × 1–2.5 cm, firm-chartaceous, moderately strigose to strigulose with fine hairs on both surfaces, the base rounded to cordulate, the apex acute, the margins entire; primary veins in 2–3 pairs. Inflorescences in upper bract axils of short lateral branches, usually solitary; pedicels 1.5–3 mm long. Flowers: hypanthium 3.5–5 mm long, moderately to densely fine-setose; calyx lobes 2–3 mm long; petals 6–9 mm long, pale purple; stamens dimorphic, large anthers 2.3–2.7 mm long, their connectives basally prolonged 1.5–2 mm, the adaxial lobes 1.8–2 mm long, short anthers 1.8–2 mm long, their connectives basally prolonged 0–0.1 mm, the adaxial lobes 0.3–0.5 mm long. Fruits loculicidal, globose. Often in open, moist, usually grassy areas.

ERNESTIA DC.

Shrubs or subshrubs (rarely radicant vines). Leaves homomorphic. Inflorescences axillary (*E. glandulosa*) or terminal panicles or cymes. Flowers 4-merous (in the Guianas); hypanthium terete; calyx lobes persistent; petals elliptic to obovate, pink or white; stamens somewhat to markedly dimorphic, anthers subulate, the pore small and adaxially inclined, the connective basally prolonged and adaxially bilobed with 2–3 mm long acute or aristate appendages, abaxially minutely tuberculate; ovary superior, 3-locular (sometimes 4-locular outside our area), glabrous (in the Guianas). Fruits capsules. Seeds numerous, cochleate, muriculate.

1. Inflorescences 1–2 cm long, lateral in upper leaf axils. Petals white. . . . . *E. glandulosa.*
1. Inflorescences 6–12 cm long, terminal. Petals lavender. . . . . *E. granvillei.*

Ernestia glandulosa Gleason FIG. 190, PL. 85a

Shrubs or subshrubs, 0.5–1.5 m tall. Quadrangular branchlets, leaves, inflorescences, and hypanthia moderately fine-setulose with gland-tipped hairs. Leaves: petioles 1–3(5.5) cm long; blades ovate-oblong, (6)11–16.5 × (2)3–5.5 cm, chartaceous, the base cordulate, the apex gradually acuminate, the margins denticulate; primary veins in 2–3 pairs. Inflorescences in upper leaf axils, 3–7-flowered, 1–2 cm long; pedicels 3–3.5 mm long. Flowers: hypanthium 2.7–3.3 mm long; calyx lobes linear-oblong, 2–3 mm long, at the apex sparsely setulose adaxially; petals 5–8 mm long, white; stamens dimorphic, large anthers 3–3.3 mm long, their connectives basally prolonged 0.6–0.7 mm and adaxial appendages 2 mm long, short anthers 2.7–3 mm long, their connectives prolonged 0.3–0.4 mm and their adaxial appendages 1 mm long; ovary 3-locular. Fruits globose, brown. Fl (Jan, Feb, Mar, May, Jul, Sep, Oct), fr (Jan, Jun, Jul, Oct, Nov); locally abundant, in disturbed areas, especially along trails and roads through forest.

Ernestia granvillei Wurdack

Shrubs, 1–2 m tall. Branchlets, primary leaf veins abaxially, and inflorescences densely glandular-setose. Leaves: petioles 3–5 cm long; blades narrowly to broadly ovate, 8–12 × 4–7 cm, chartaceous, sparsely puberulous with partly gland-tipped hairs 0.5–0.7 mm long on both surfaces, the base cordate, the apex narrowly acute, the margins inconspicuously ciliate-serrulate; primary veins in 3–4 pairs. Inflorescences panicles, 6–12 cm long; pedicels 2–3 mm long. Flowers: hypanthium 4–5 mm long, moderately puberulous with eglandular hairs ca. 0.1 mm long and glandular hairs 0.3–1 mm long; calyx lobes narrowly oblong, 3–3.6 mm long; petals 6.5–7.3 mm long, lavender; stamens dimorphic, large anthers 5.2–5.3 mm long, their connective basally prolonged 2.5–3.7 mm, the adaxial appendages ca. 3 mm long, short anthers ca. 4 mm long, their connectives 0.8–1 mm long, the adaxial appendages ca. 2.5 mm long; ovary 3-locular. Fruits globose, brown. Fl (Jul, Aug, Sep), fr (Jan); on talus slopes among large boulders.

HENRIETTEA DC.

Trees or shrubs. Leaves homomorphic. Inflorescences from defoliated nodes or in leaf axils, fasciculate; pedicels with basal, small bracteoles. Flowers 5-merous (in the Guianas); hypanthium terete, usually appressed strigose; calyx limb entire or lobed, persistent, usually with inframarginal external teeth; petals clawed, the apex obtuse to acute, with an external subapical tooth, white or pink; stamens isomorphic, anther tips attenuate or spoon-shaped, pore small and abaxially inclined, the connective usually slightly bilobed at base; ovary 5-locular, completely adnate to the hypanthium, the apex truncate or with short column. Fruits berries. Seeds numerous, ovoid to oblong-pyramidal, papillate.

1. Hypanthium 4–4.5 mm diam. and 5.5–8 mm long just after anthesis; calyx lobes 0.5–2 mm long; petals pink. . . . . *H. ramiflora.*
1. Hypanthium 6–8 mm diam. and 10–15 mm long just after anthesis; calyx lobes 4–5 mm long; petals white. . . . . *H. succosa.*

Henriettea ramiflora (Sw.) DC. FIG. 191, PL. 84f

Trees, (3)5–15(25) m tall. Young branchlets and primary leaf veins strigose. Leaves: petioles 0.8–1.6 cm long; blades elliptic to obovate-elliptic, 7–17 × 3.5–9 cm, sparsely and caducously strigulose adaxially, moderately to densely covered with stellate-based setulae abaxially, the base broadly to narrowly acute, the apex acute or shortly and abruptly acuminate; primary veins in 1 pair (excluding the tenuous marginal pair). Inflorescences: pedicels 2–5 mm long. Flowers: hypanthium 5.5–8 mm long and ca. 3 mm diam. (at or just after anthesis), densely strigulose; calyx 2–3 mm long (including the broadly ovate lobes 0.5–2 mm long) and sparsely

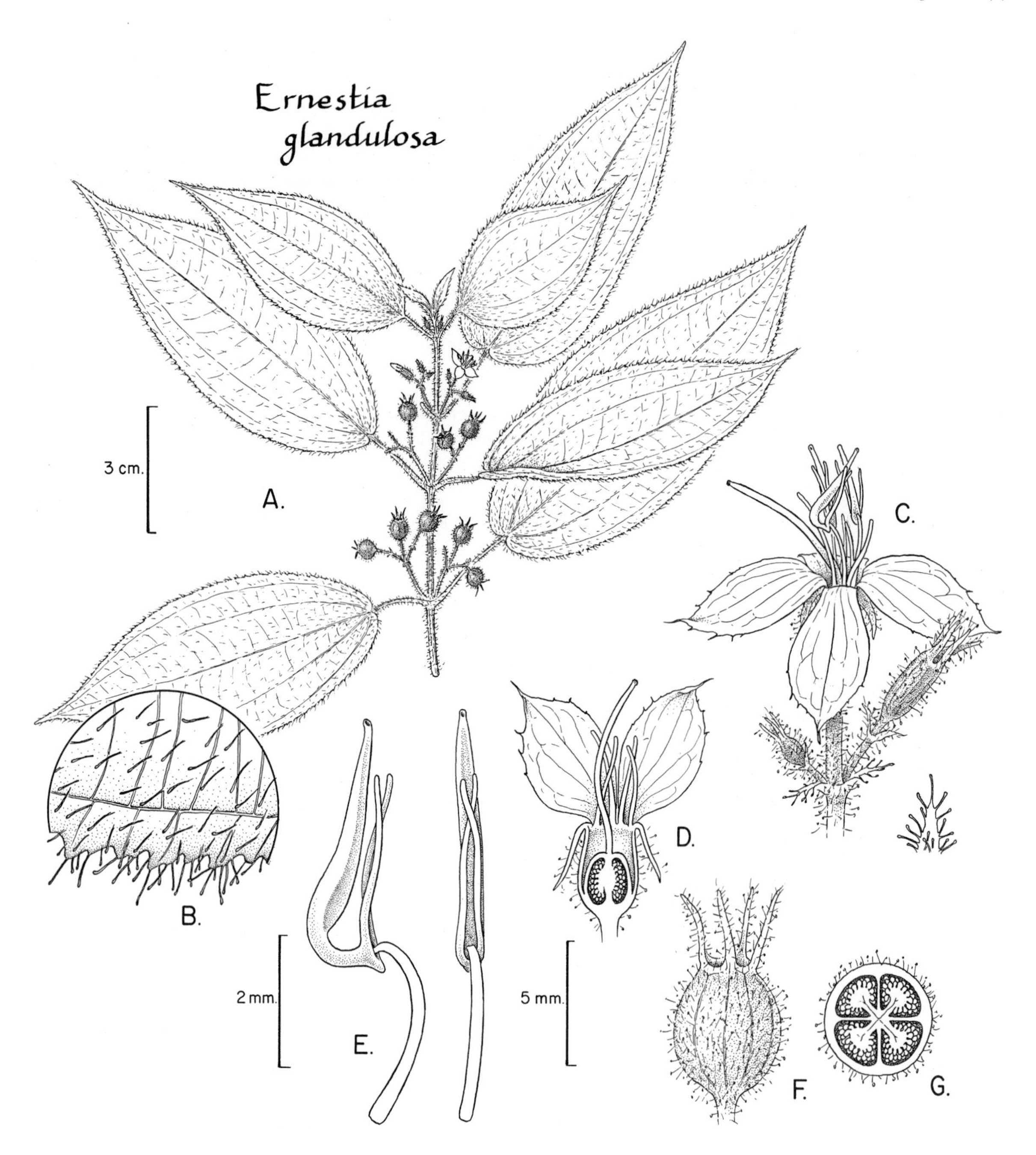

FIG. 190. MELASTOMATACEAE. *Ernestia glandulosa* (*Mori et al. 18207*). **A.** Apex of stem with leaves, inflorescences, and infructescences. **B.** Detail of crenulate leaf margin showing glandular pubescence. **C.** Lateral view of flower and two flower buds (left) with detail of bract (below). **D.** Medial section of flower showing cupular hypanthium and axile placentation. **E.** Lateral (left) and adaxial (right) views of stamens showing paired appendages and single poricidal dehiscence. **F.** Fruit. **G.** Transverse section of fruit.

strigulose adaxially, the external teeth to 1.2 mm long; petals (5)7–10 mm long, pink; anther tips distinctly spoon-shaped, 5.3–6(8) mm long. Fruits elongate, 2–3 cm long, first red, bluish-black, tasty at maturity. Fl (Aug); in secondary vegetation.

Henriettea succosa (Aubl.) DC.

Treelets, 3–12 m tall, the young branchlets and primary leaf veins densely appressed-setose. Leaves: petioles 1–2 cm long; blades elliptic to obovate-elliptic, 15–30 × 6–15 cm, sparsely short-setose adaxially, completely covered with dendritic or stellate-based setulae abaxially, the base broadly acute to obtuse, the apex abruptly short-acuminate; primary veins in 2 pairs. Inflorescences: pedicels 2–5 mm long. Flowers: hypanthium 10–15 mm long and 6–8 mm diam. (at or just after anthesis), densely appressed-setose with minutely roughened hairs 1–1.5 mm long; calyx lobes ovate-oblong, 4–5 mm long; petals 14–17 mm long, white; anther tips distinctly spoon-shaped, 7–9 mm long. Fruits elongate, 1.5–2.5 cm long, first red, dark blue, tasty at maturity. Fl (Jan); in secondary vegetation. *Caca Henriette* (Créole).

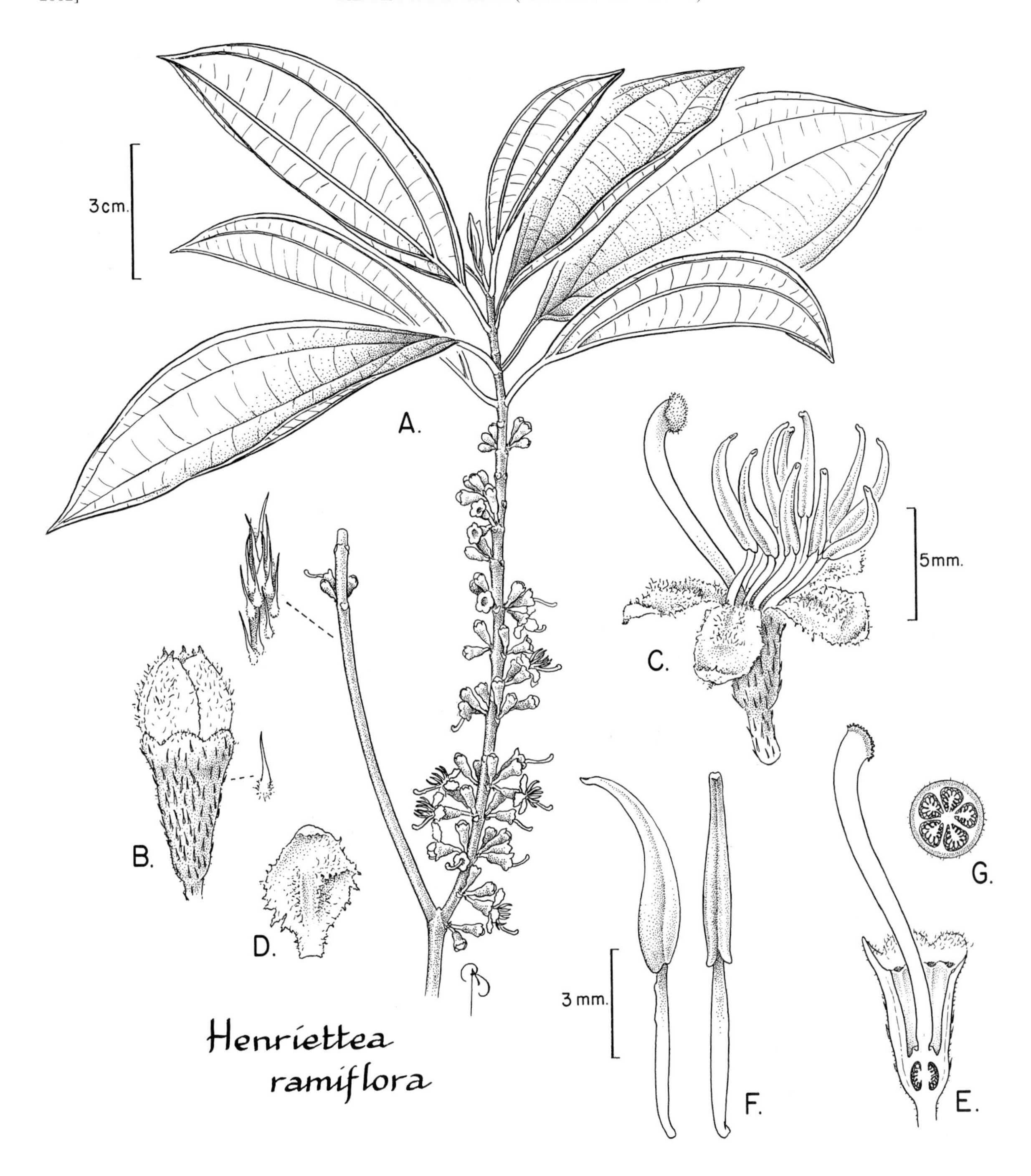

FIG. 191. MELASTOMATACEAE. *Henriettea ramiflora* (*Mori & Gracie 19212*). **A.** Leafy fertile stem with flowers; note detail of stem surface (left). **B.** Flower bud with detail of trichome (right). **C.** Open flower. **D.** Petal. **E.** Medial section of flower with corolla and androecium removed; note the campanulate hypanthium and parietal placentation. **F.** Lateral (left) and adaxial (right) views of stamens. **G.** Transverse section of ovary.

HENRIETTELLA Naudin

Shrubs or small trees. Leaves homomorphic. Inflorescences from defoliated branchlet nodes, fasciculate; pedicels with basal, mostly minute bracteoles. Flowers 5-merous (in our area); hypanthium terete, minutely resinous granulose or sparsely pubescent; calyx truncate or shortly lobed, glabrous adaxially, the external teeth minute; petals ovate to lanceolate, usually cross-ridged or appendaged adaxially, the apex bluntly acute, usually with an external mucro, white; stamens isomorphic, the anthers with 1–2 pores (in our area), the connective basally not or slightly prolonged and usually unappendaged; ovary completely adnate to the hypanthium, 5-locular (in our area). Fruits berries. Seeds numerous, obovate-angular.

1. Flower pedicels obsolete. . . . . *H. duckeana.*
1. Flowers on pedicels 4.5–8 mm long.
   2. Stems, petioles, and veins on adaxial leaf blade surface and entire abaxial leaf blade surface very soon glabrescent. . . . . *H. caudata.*
   2. Stems, petioles, and veins on adaxial leaf blade surface and entire abaxial leaf blade surface densely covered with hairs, these often bulbous based. . . . . *Henriettella* sp. A.

Henriettella caudata Gleason

Treelets or shrubs, 1.5–8 m tall. Young branchlets and petioles sparsely to moderately strigulose but very soon glabrescent. Leaves: petioles 2–5 cm long; blades oblong-elliptic to oblong-lanceolate, 15–25 × 5–9 cm, glabrescent with age, the base acute and slightly decurrent, the apex long acuminate; primary veins in 2 pairs, inserted 2–4 cm above blade base. Inflorescences: pedicels 4.5–6 mm long. Flowers: hypanthium ca. 2.2 mm long, sparsely covered with minute glands (only visible under high magnification); calyx ca. 0.3 mm long, truncate to barely 5-undulate; petals oblong-ovate, 2.6–3 mm long, granulose; anthers ca. 2.5 mm long, narrowly oblong, minutely 2-pored, the connective dorso-basally with a lanceolate tooth 0.2 mm long. Fruits globose, first red, dark blue at maturity. Fr (Jul). *Bois mêle, caca Henriette* (Créole).

Henriettella duckeana Hoehne Pl. 85b

Shrubs or small trees, 1–3(10) m tall. Stems denseley pubescent with simple, appressed hairs. Leaves: petioles 1–4 cm long; blades elliptic to broadly elliptic, 15–19 × 8–9 cm, pubescent on veins abaxially and to a lesser extent adaxially, the base acute, the apex long acuminate; primary veins in 2 pairs, inserted 0.5–4 cm above blade base. Inflorescences: pedicels absent. Flowers: hypanthium ca. 1.7 mm long, minutely resinous-granulose (visible under high magnification); calyx 0.8–1 mm long, truncate; petals elliptic-oblong, ca. 3.5 mm long; anthers oblong, ca. 2.2 mm long, retuse-pored, the connective not prolonged or appendaged. Fruits globose, first red, black at maturity. Fl (Feb), fr (Apr); in non-flooded forest.

Henriettella sp. A

Treelets. Stems densely pubescent, the hairs simple, appressed, markedly bulbous at base. Leaves: petioles 1.5–2 cm long; blades elliptic to broadly elliptic, 12–16 × 5.5–8 cm, pubescent throughout abaxially, on veins adaxially, the base acute to obtuse, the apex acuminate; primary veins in 2 pairs. Inflorescences: pedicels 6–8 mm long. Flowers: hypanthium ca. 3 mm long, with a few scattered hairs; calyx undulate and with minute apical teeth. Fruits globose. Fr (Sep, immature); known from a single collection (*Mori et al. 24238*, NY) from along the Route de Bélizon just N of Eaux Claires.

LEANDRA Raddi

Shrubs, subshrubs, or treelets (rarely climbing outside our area). Leaves homomorphic. Inflorescences usually terminal panicles. Flowers 4–5(6)-merous; hypanthium terete; calyx lobes with well developed persistent (in our area) external teeth; petals linear to ovate, the apex acute or acuminate, white to pink; stamens essentially isomorphic, anthers usually oblong, the minute pore usually adaxially inclined, the connective basally not or barely prolonged, usually unappendaged; ovary barely to mostly adnate to the hypanthium, 2–5-locular, glabrous or pubescent. Fruits berries. Seeds numerous, small, usually angular-pyramidate and smooth, occasionally ovate and tuberculate.

1. Flowers arranged in unilateral rows on inflorescence axes (most obvious in fruit).
   2. Leaf blade bases rounded to cordate. Flowers predominantly 6-merous. . . . . *L. solenifera.*
   2. Leaf blade bases acute, obtuse, or rounded but never cordate. Flowers 5-merous.
      3. Leaf blade margins serrate to doubly serrate. Ovary apex glandular-setulose. . . . . *L. agrestis.*
      3. Leaf blade margins entire to obscurely serrulate. Ovary apex glabrous. *L. divaricata.*
1. Flowers not arranged in unilateral rows on inflorescence axes.
   4. Ovary 5(6)-locular.
      5. Branchlet and inflorescence hairs flattened. Leaves with 1 pair of primary veins. . . . . *L. paleacea.*
      5. Branchlet and inflorescence hairs terete or stellulate. Leaves with 2(3) pairs of primary veins.. . . *L. rufescens.*
   4. Ovary 3(4)-locular.
      6. Leaves minutely strigulose. Anther connectives with an ascending abaxial appendage. . . . . . . *L. clidemioides.*
      6. Leaves pubescent, the hairs 1.5–2 mm long. Anther connectives lacking an abaxial appendage. . . . . *L. micropetala.*

Leandra agrestis (Aubl.) Raddi Pl. 85c,d

Shrubs, 0.5–1(2) m tall. Stems, primary leaf veins abaxially, inflorescences, and hypanthia moderately to densely fine-setose, the hairs lax. Leaves: petioles 1–3 cm long, blades elliptic to oblong-elliptic, (5)6–10(13) × (2)3–5(6) cm, both surfaces pubescent, the hairs spreading, the base broadly acute to obtuse, the apex gradually short-acuminate, the margins serrate to doubly serrate, distinctly ciliate; primary veins in 2(3) pairs. Inflorescences panicles, 3–5 cm long, rather few-flowered, sometimes densely congested, the flowers secund. Flowers 5-merous; hypanthium ca. 2.5 mm long; calyx lobes ovate, ca. 0.7 mm long, the setiferous external teeth 0.7–1 mm long; petals oblong-elliptic, 2.7–3 mm long, white; anthers 1.8–2 mm long; ovary 3-locular, the apex moderately glandular-setulose. Fruits globose, first red, purple at maturity. Fl (Jan, Feb, Mar, Jul, Aug), fr (Jan, Mar, May, Jun, Jul); in weedy habitats.

Leandra clidemioides (Naudin) Wurdack

Shrubs or treelets, 2–4 m tall. Stems, primary leaf veins, and inflorescences moderately to densely strigulose with obscurely roughened eglandular hairs. Leaves: petioles 1–3 cm long; blades oblong-elliptic to lance-elliptic, 7–20 × 4–8 cm, very sparsely and deciduously strigulose on both surfaces, the base acute, the apex gradually acuminate, the margins entire, distantly appressed-ciliolate; primary veins in 1(2) pairs. Inflorescences panicles 4–12 cm long, with a moderate number of flowers, the flowers not secund. Flowers 5-merous; hypanthium 1.5–2 mm long, sparsely to moderately strigulose; calyx lobes ca. 0.3 mm long and equaled or slightly (0.2 mm) exceeded by minute external teeth; petals narrowly oblong, 1–2 mm long, white; anthers slightly unequal in size, 1.5–2.5 mm long, the connective with an ascending blunt appendage abaxially; ovary 3-locular, the apex glabrous or very sparsely glandular-setulose. Fruits globose. Fl (Jan, Mar, May), fr (Mar); in forest.

Leandra divaricata (Naudin) Cogn.

Subshrubs, 0.3–1 m tall, often decumbent and rooting at the nodes. Stems, primary leaf veins abaxially, inflorescences, and hypanthia moderately long-strigulose or short-strigose (rarely setose) with fine smooth hairs. Leaves: petioles 2–4 cm long; blades elliptic to ovate-elliptic, 5–11 × 3–6 cm, sparsely fine-strigose adaxially, sparsely strigulose (rarely setulose) on secondary veins, sometimes glabrous on venules and surface abaxially, the base obtuse to broadly acute, the apex short-acuminate to acute, the margins entire to obscurely serrulate, appressed-ciliolate; primary veins in 2(3) pairs, the inner pairs inserted to 1.5 cm above blade base. Inflorescences panicles 3–9 cm long, the flowers secund. Flowers 5-merous; hypanthium 1.6–1.8 mm long, occasionally with some gland-tipped hairs; calyx lobes 0.1–0.2 mm long and semicircular, the oblong-ovate external teeth 0.5–0.7 mm long; petals oblong-lanceolate, 2–2.5 mm long, white; anthers 1–1.5 mm long, the connective basally inconspicuously prolonged; ovary 3-locular, the apex glabrous. Fruits globose, first red, black at maturity. Fl (Jan, Jun, Jul), fr (Jun, Jul, Sep, Nov); in light gaps in forest or in disturbed habitat along rivers.

Leandra micropetala (Naudin) Cogn.

Shrubs, 1–3 m tall, vegetatively similar to *L. rufescens*. Stems pubescent, the hairs ca. 3 mm long, somewhat deflexed. Leaves: petioles 0.5–2 cm long, densely fine-setose; blades moderately fine-setose with hairs 1.5–2 mm long adaxially. Inflorescences panicles, the axes moderately fine-setose with glandular and eglandular hairs, the flowers not secund. Flowers: hypanthium moderately fine-setose with eglandular hairs and sparsely fine-setose with gland-tipped hairs; calyx lobes semicircular, the external teeth minute (0.2 mm); petals 0.4–0.5 mm long, white; ovary predominantly 3(4)-locular, glabrous. Fruits globose, bright blue. Fl (Jan, Apr); on rock outcrops and in high-elevation forest.

Leandra paleacea Wurdack

Shrubs, 0.5–4 m tall. Stems and inflorescences densely strigose with flattened hairs. Leaves: petioles 0.5–1(1.5) cm long; blades elliptic, 7–10 × 2.5–4.5 cm, densely strigose along primary veins but otherwise glabrous adaxially, sparsely strigulose abaxially, the base narrowly acute to acute, the apex gradually short-acuminate, the margins entire, appressed-ciliolate; primary veins in 1 pair. Inflorescences panicles, 5–8 cm long, 30–35-flowered, the flowers not secund. Flowers 5(6)-merous; hypanthium 2.2–3.5 mm long, rather densely loose-strigose and sparsely glandular-setulose; calyx lobes ovate, 0.6 mm long, the setulose external teeth ca. 0.7 mm long; petals oblong, 1–1.5 mm long, white; anthers ca. 2 mm long; ovary 5(6)-locular, the apex setose. Fruits globose, first red, blue with white hairs at maturity. Fl (Jan), fr (Mar, Jun, Jul, Aug); often in disturbed habitats along trails and road.

Leandra rufescens (DC.) Cogn.

Shrubs, 1–3 m tall. Stems, primary leaf veins on both surfaces, and inflorescences densely setose with fine smooth (or very sparsely and obscurely roughened) hairs, hairs spreading, often slightly deflexed. Leaves: petioles 1–4 cm long; blades oblong-lanceolate to ovate-elliptic, 10–25 × 4–10 cm, sparsely appressed-setulose adaxially, moderately fine-setose abaxially, the base broadly acute to obtuse, the apex gradually acuminate, the margins undulate-denticulate to entire; primary veins in 2 pairs. Inflorescences panicles, 10–20 cm long, multiflorous, the axes often reddish, the flowers not secund. Flowers 5-merous; hypanthium 1.5–2 mm long, moderately fine-setulose (sometimes with a few gland-tipped hairs); calyx lobes ca. 0.1–0.2 mm long, the setulose external teeth 0.3–0.5 mm long; petals obovate-oblong, 1–1.3 mm long, white; anthers 1.5–1.7 mm long, subulate, the connective basally slightly prolonged but unappendaged; ovary inferior to almost superior, 5-locular, apically densely setulose. Fruits globose, first reddish-purple, dark blue with white hairs at maturity. Fl (Jan), fr (Mar, Jun, Jul, Aug); common, in gaps in nonflooded forests.

Leandra solenifera Cogn. FIG. 192, PL. 85f,g

Shrubs, 1–2 m tall. Stems, leaves, inflorescences, and hypanthia moderately to densely setulose with fine eglandular hairs intermixed with gland-tipped hairs 0.3–0.5 mm long. Leaves: petioles 3–6 cm long; blades ovate to widely ovate, 8–15 × 5–9 cm, the base rounded to cordate, the apex gradually acuminate, the margins clearly denticulate to serrulate, ciliate; primary veins in 3–4 pairs. Inflorescences panicles, 10–18 cm long, multiflorous, the flowers secund. Flowers 6-merous; hypanthium 2.4–2.7 mm long; calyx lobes ovate, 0.3–0.7 mm long, the external teeth minute; petals elliptic to ovate, 2.5–2.8 mm long, white to pink; anthers 1.2–2 mm long; ovary mostly 6-locular, the apex sparsely glandular. Fruits globose, first red, turning purple at maturity. Fl (Jan, Feb, Mar, Apr, May, Jun, Jul, Aug, Sep, Oct, Nov), fr (May, Jun, Jul, Aug, Sep, Nov); very common, in disturbed habitats everywhere.

LOREYA DC.

Small to large trees. Leaves homomorphic. Inflorescences from stems and branches below leaves, few-branched cymes or sessile clusters. Flowers 5(6)-merous; hypanthium terete, hemispherical; calyx truncate, the lobes vestigial; petals ovate, elliptic, or broadly triangular, fleshy, adaxially with callus ridges, often also with lateral flaps; stamens isomorphic, laterally compressed, the tips rounded or

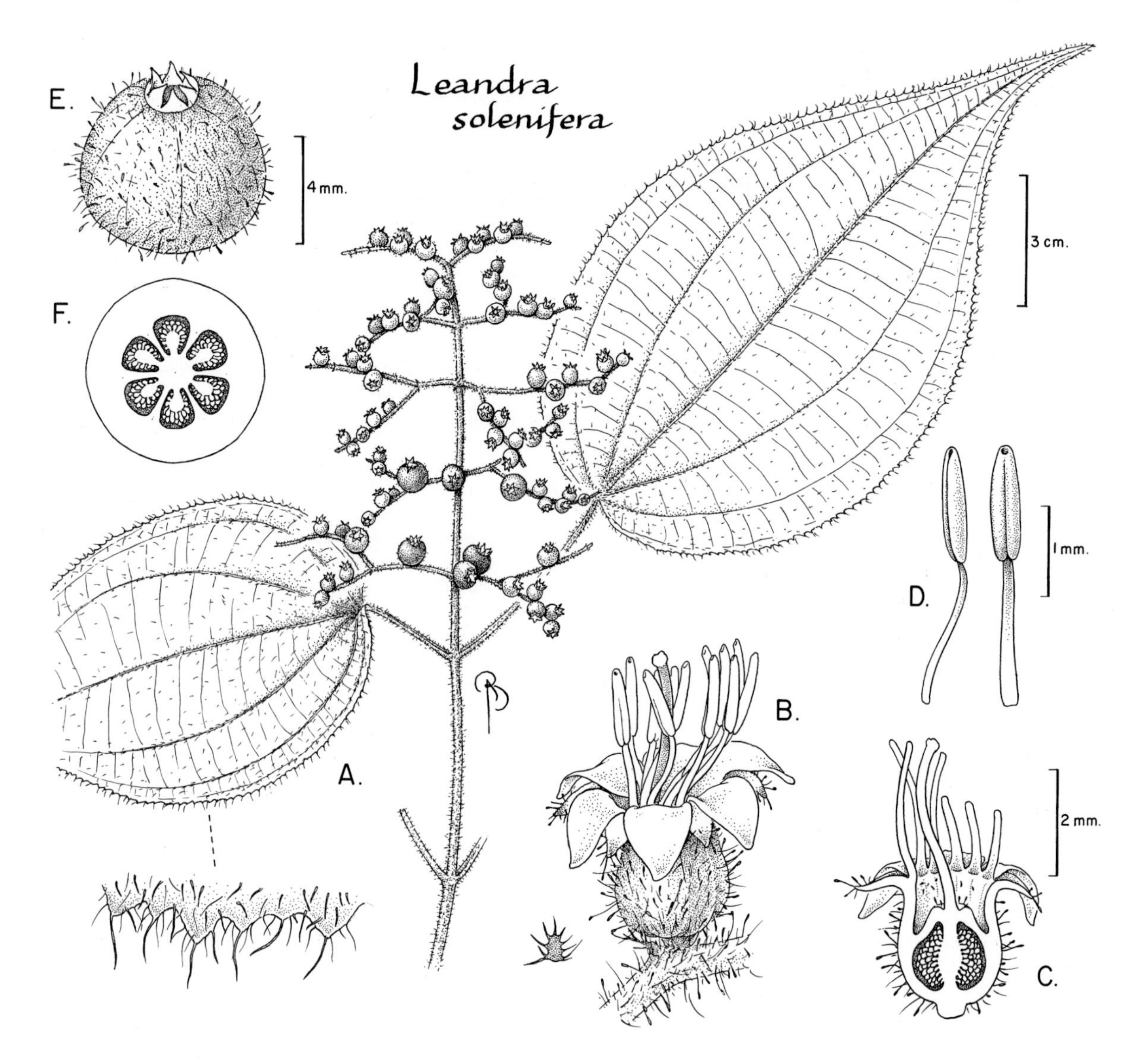

FIG. 192. MELASTOMATACEAE. *Leandra solenifera* (*Mori et al. 18205*). **A.** Apex of stem with leaves, infructescence, and detail of crenulate leaf margin with dense pubescence. **B.** Flower with detail of bract (below left). **C.** Medial section of flower showing axile placentation. **D.** Lateral (left) and adaxial (right) views of stamens with single poricidal dehiscence. **E.** Fruit with persistent calyx at apex. **F.** Transverse section of fruit.

truncate and 1- or 2-pored, the connectives not prolonged but fleshy and with a pronounced dorsi-basal callus; ovary completely adnate to the hypanthium, 5(6)- or 10-locular, apically glabrous, the stigma capitate with 5(6) lobes. Fruits berries. Seeds ovoid, numerous, with an elongate lateral hilum.

1. Stems densely pubescent, the hairs long and spreading. Leaf blades with shallowly dendiculate margins. Flowers in sessile clusters; anthers 1-pored. . . . . *L. mespiloides.*
1. Stems glabrous or, if hairs present, these short and appressed. Leaf blades with entire margins. Flowers in pedunculate cymes; anthers 2-pored.
    2. Leaf blades 9–15 × 4–8(12) cm. Hypanthium ca. (3)4 mm long; petals 5–6(7) mm long. . . . . *L. arborescens.*
    2. Leaf blades 15–26 × 13–23 cm. Hypanthium ca. 7 mm long; petals 13–15 mm long. . . . . *L. subrotundifolia.*

Loreya arborescens (Aubl.) DC. PL. 86a

Trees, to 30 m tall. Stems glabrous or with short appressed hairs. Leaves: petioles 1–3 cm long; blades ovate, elliptic, or obovate, 9–15 × 4–8(12) cm, glabrous except for puberulous veins abaxially, the base cuneate, the apex acuminate, bluntly acute, or rounded, the margins entire. Inflorescences cymes, 4.5–5 cm long; peduncles 5–20 mm long; pedicels 5–10 mm long. Flowers: hypanthium ca. (3)4 mm long; petals ovate, 5–6(7) mm long, white adaxially, pink abaxially; anthers 3–4.5 mm long, 2-pored; ovary 5-locular. Fruits ca. 8 mm diam., glabrescent, first red, blue or black at maturity. Fl (Aug, Sep); in nonflooded, old, secondary forest.

Plates 81–88

Plate 81. MALPIGHIACEAE. **a.** *Jubelina rosea* (*Mori et al. 20929*), flower. **b.** *Mascagnia divaricata* (*Mori et al. 22205*), flowers and buds; note glands on calyx. **c.** *Mezia angelica* (*Mori et al. 20945*), flower. **d.** *Tetrapterys glabrifolia* (*Mori et al. 22873*), flower; note calyx glands visible between clawed petals. **e.** *Tetrapterys megalantha* (*Mori et al. 21584*), winged fruit; note glands on calyx. **f.** *Stigmaphyllon sinuatum* (*Mori & Gracie 22849*), flower; note folioles on styles and calyx glands visible between clawed petals.

Plate 82. MALVACEAE. **a.** *Pavonia schiedeana* (*Mori et al. 23975*), flower. **b.** *Sida setosa* (unvouchered), flower and immature fruit. **c.** *Sida setosa* (unvouchered), mature and immature fruits. MARCGRAVIACEAE. **d.** *Marcgravia* sp. A (*Mori et al. 23162*), inflorescence in bud with nectar-bearing bracts below. **e.** *Marcgravia* sp. A (*Mori et al. 23162*), flower from manually-opened bud. **f.** *Norantea guianensis* (unvouchered), flowers and buds above and orange nectariferous bracts below.

Plate 83. MARCGRAVIACEAE. *Souroubea guianensis* (unvouchered, photographed in Brazil), flower and subtending red nectiferous bract.

Plate 84. MELASTOMATACEAE. **a.** *Aciotis acuminifolia* (*Mori & Gracie 19019*), flowers and buds; note winged, four-angled stem and punctations on leaves. **b.** *Clidemia hirta* (unvouchered), flower and immature fruits; note trichomes on calyces, stem, and leaf. **c.** *Bellucia grossularioides* (*Mori & Gracie 19012*), lateral view of flower showing deep pink color on abaxial surface of petals (white adaxially) and broadly flaring calyx; note bud completely enclosed by calyx. **d.** *Clidemia octona* subsp. *guayanensis* (*Mori & Gracie 19005*), flower; note eight petals and small beetle commonly found on flowers of this species. **e.** *Clidemia rubra* (*Mori & Gracie 18744*), flowers; note trichomes on stem and leaves. **f.** *Henriettea ramiflora* (*Mori & Gracie 19212*), flowers, buds, and immature fruits.

Plate 85. MELASTOMATACEAE. **a.** *Ernestia glandulosa* (Mori & Gracie 21146), lateral view of flower and buds. **b.** *Henriettella duckeana* (*Mori & Gracie 21163*), flowers and buds. **c.** *Leandra agrestis* (*Mori et al. 24758*), inflorescence; note dense white hairs on calyces. **d.** *Leandra agrestis* (*Mori & Gracie 18878*), infructescence with mostly immature fruits and one mature purple-black fruit; note pubescence on axes of infructescence and fruits. **e.** *Miconia aliquantula* (*Mori & Gracie 18886*), flowers and buds; note sessile leaves. **f.** *Leandra solenifera* (*Mori et al. 23802*), flower and buds. **g.** *Leandra solenifera* (*Mori & Gracie 18917*), part of infructescence with mostly immature fruit and single mature, black fruit.

Plate 86. MELASTOMATACEAE. **a.** *Loreya arborescens* (*Mori & Gracie 23202*), flower and buds. **b.** *Macrocentrum fasiculatum* (*Mori et al. 23757*, photographed in NY from plant grown from seeds of this collection), flowers. **c.** *Maieta guianensis* (*Mori et al. 22938*), flowers.

Plate 87. MELASTOMATACEAE. **a.** *Miconia affinis* (*Mori et al. 24158*), flowers. **b.** *Miconia ceramicarpa* var. *ceramicarpa* (*Mori et al. 22256*), apical view of plant showing leaves, flowers, immature salmon-colored fruits and mature pale blue-gray fruits. **c.** *Miconia cacatin* (*Mori & Gracie 18977*), flowers. **d.** *Miconia ceramicarpa* var. *ceramicarpa* (*Mori & Gracie 21138*), flowers and buds. **e.** *Miconia lateriflora* (*Mori & Gracie 18884*), infructescence with immature green fruits and mature white fruits on red axes. **f.** *Miconia sastrei* (*Mori & Gracie 21183*), flower and buds. **g.** *Miconia trimera* (*Mori & Gracie 18674*), immature fruits.

Plate 88. MELASTOMATACEAE. **a.** *Mouriri collocarpa* (*Mori et al. 24684*), flower and bud. [Photo by S. Mori] **b.** *Mouriri crassifolia* (*Mori et al. 22737*), flower and buds. **c.** *Tibouchina aspera* var. *asperrima* (*Mori & Gracie 18341*), flowers. **d.** *Topobea parasitica* (*Mori et al. 17349*, photographed in Amapá, Brazil), flower and bud. [Photo by S. Mori]

Jubelina rosea a.

Mascagnia divaricata b.

Mezia angelica c.

Tetrapterys glabrifolia d.

Stigmaphyllon sinuatum f.

Tetrapterys megalantha e.

MALVACEAE

Pavonia schiedeana a.

Sida setosa b.

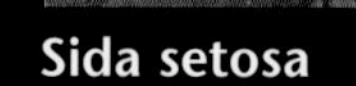

Sida setosa c.

MARCGRAVIACEAE

Marcgravia sp. A d.

Marcgravia sp. A e.

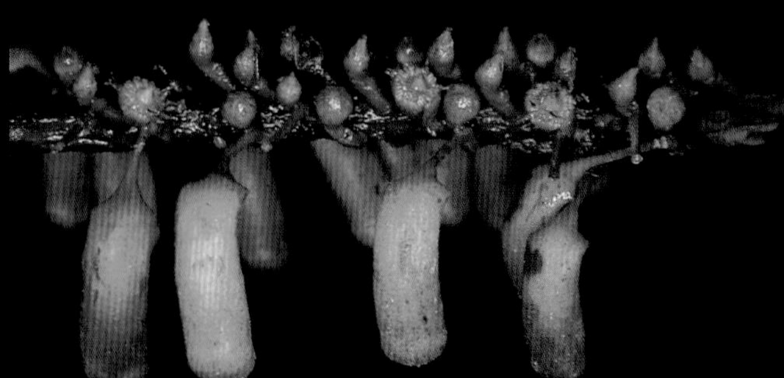

Norantea guianensis f.

Souroubea guianensis

Aciotis acuminifolia a.

Bellucia grossularioides c.

Clidemia hirta b.

Clidemia octona subsp. guayanensis d.

Clidemia rubra e.

Henriettea famiflora f.

Ernestia glandulosa a.

Henriettella duckeana  b.

Leandra agrestis c.

Leandra agrestis d.

Miconia aliquantula e.

Leandra solenifera f.

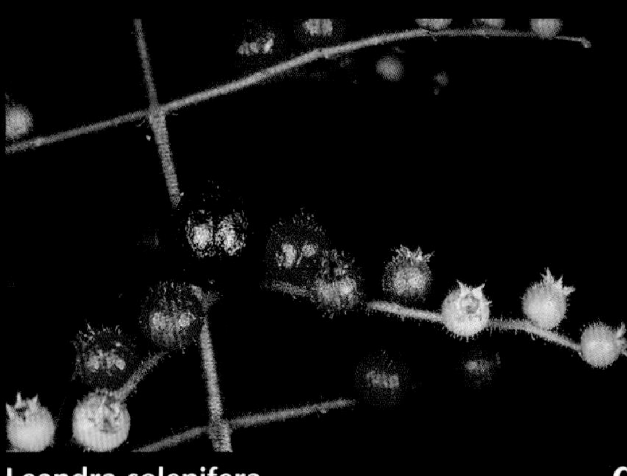

Leandra solenifera g.

Loreya arborescens a.

Macrocentrum fasiculatum

b.

Plate 86 Maieta guianensis c.

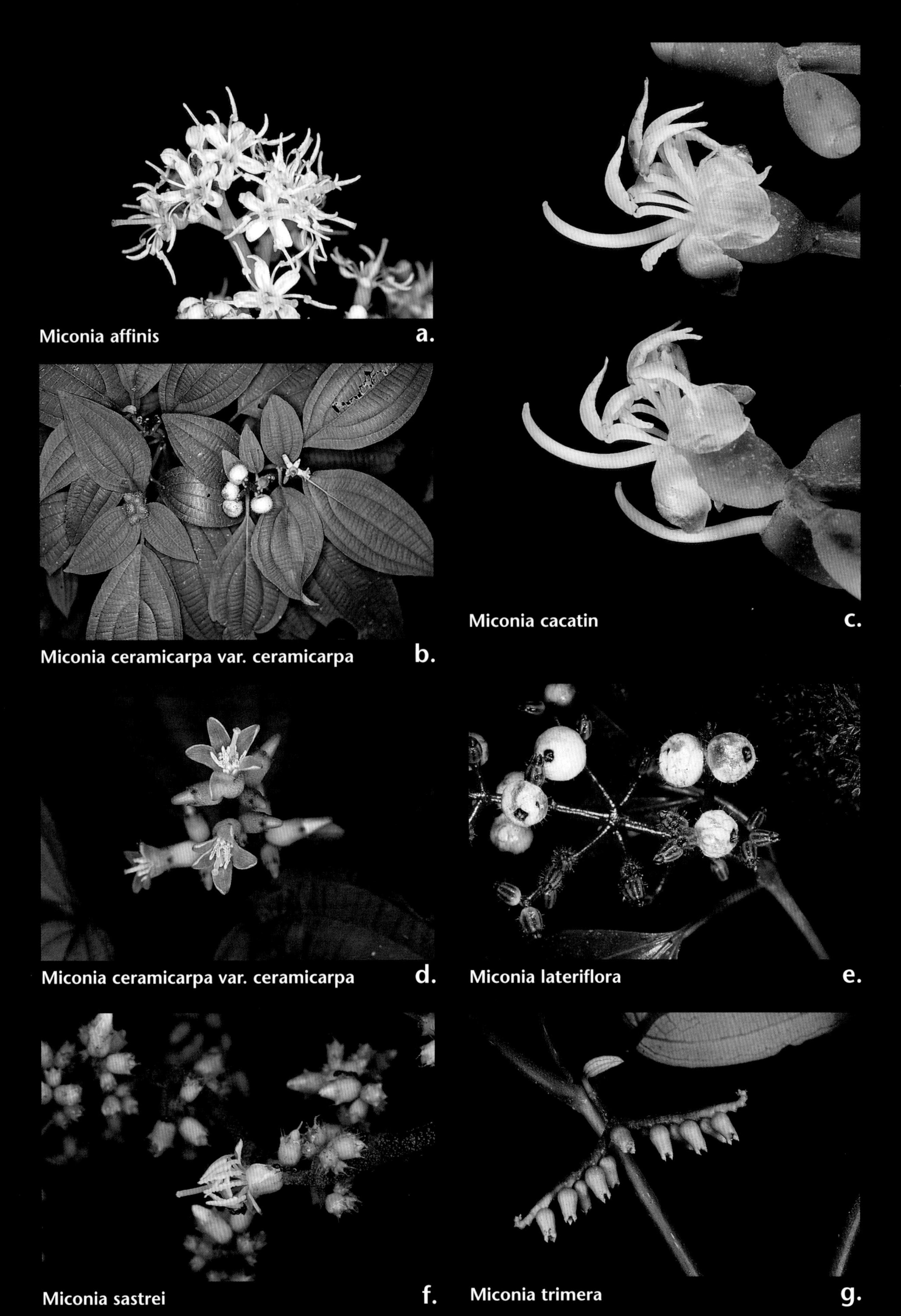

Miconia affinis a.

Miconia ceramicarpa var. ceramicarpa b.

Miconia cacatin c.

Miconia ceramicarpa var. ceramicarpa d.

Miconia lateriflora e.

Miconia sastrei f.

Miconia trimera g.

Plate 87

Mouriri collocarpa **a.**

Mouriri crassifolia **b.**

Tibouchina aspera
var. asperrima **c.**

Topobea parasitica **d.**

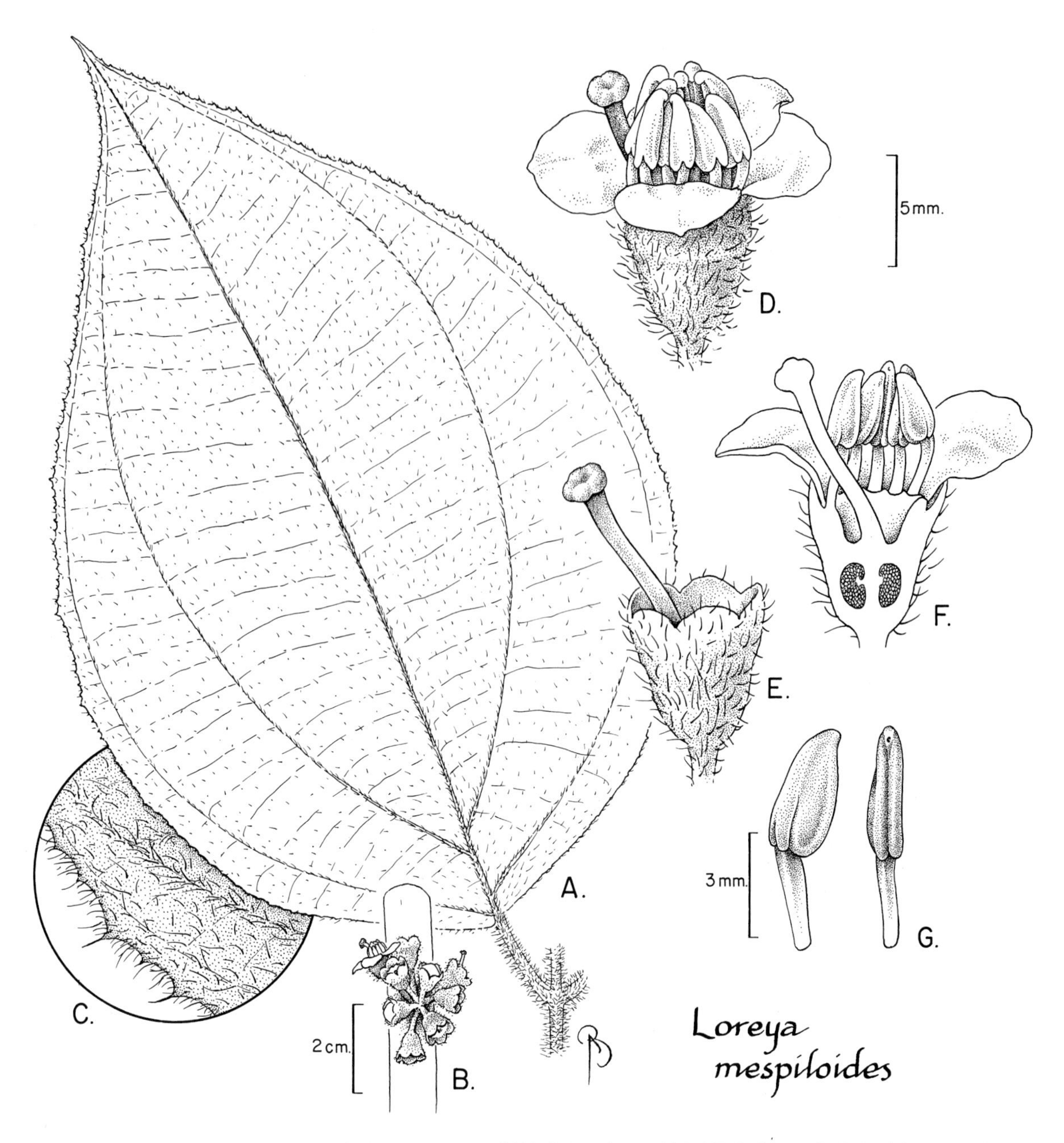

FIG. 193. MELASTOMATACEAE. *Loreya mespiloides* (*Mori & Gracie 18327*). **A.** Part of stem with leaf. **B.** Cauliflorous fasciculate inflorescence. **C.** Detail of serrulate leaf margin with pubescence. **D.** Lateral view of flower. **E.** Flower after anthesis and abscission of corolla and androecium. **F.** Medial section of flower showing axile placentation. **G.** Lateral (left) and adaxial (right) views of anthers with single poricidal dehiscence.

Loreya mespiloides Miq. FIG. 193

Treelets, usually <4 m tall in our area. Stems densely pubescent with long, spreading hairs. Leaves: petioles 2–6 cm long; blades broadly elliptic to ovate, 15–30(40) × 10–20 cm, moderately to densely strigose adaxially, sericeous abaxially, the base rounded to broadly acute, the apex acute to acuminate, the margins shallowly denticulate. Inflorescences from trunk and branches, clusters; peduncles absent; pedicels 10–15(20) mm long. Flowers: hypanthium (5)7–8 mm long, loosely to densely strigose, with basally expanded hairs; petals elliptic, 10–15 mm long, white or pink; anthers 5–7 mm long, 1-pored; ovary 5-locular. Fruits ca. 13–20 mm diam., densely puberulous, white or yellow. Fl (Jun, Sep), fr (Aug, Sep, Oct); common but scattered, in disturbed areas along roads, trails, and around gaps.

Loreya subrotundifolia (Wurdack) S. S. Renner [Syn.: *Bellucia subrotundifolia* Wurdack]

Small trees, 3–15 m tall. Stems glabrous. Leaves: petioles (2)3–4(6) cm long; blades orbicular, 15–26 × 13–23 cm, glabrous at maturity, the base rounded, truncate, or cordate, the apex obtuse or rounded, the margins entire. Inflorescences from trunk and branches, cymes, 3.5–4 cm long; peduncles 0.5–1 cm long; pedicels

5–10 mm long. Flowers: hypanthium ca. 7 mm long, sparsely puberulous to glabrescent; petals ovate or elliptic, 13–15 mm long, pink; anthers 4–6 mm long, 2-pored; ovary 5(6)-locular, the stigma ca. 2 mm diam. Fruits ca. 12 mm diam., glabrous, purple to dark blue at maturity. Fl (Oct), fr (Mar); in secondary vegetation.

MACROCENTRUM Hook. f.

Herbs or subshrubs. Inflorescences solitary at ends of leafy stems or terminal few-flowered cymes with sessile flowers. Leaves homomorphic or heteromorphic. Flowers 4–5-merous; hypanthium 8–10-costate; calyx lobes oblate to lanceolate, persistent; petals oblong-lanceolate to oblong, white or pinkish white, the apex cuspidate-acute to rounded; stamens essentially isomorphic, the anthers linear-subulate, minutely 1-pored, the connective basally slightly prolonged, adaxially unappendaged, abaxially with an acute descending tooth; ovary superior, usually 3-locular, with a truncate or collared apex. Fruits capsules, 8–10-ridged. Seeds numerous, ovoid-clavate, minutely roughened.

1. Leaves with 2 pair of primary veins. Flowers 4-merous. . . . . *M. latifolium.*
1. Leaves with 1 pair of primary veins. Flowers 4- or 5-merous.
   2. Flowers 4-merous.. . . . . *M. cristatum.*
   2. Flowers 5-merous. . . . . *M. fasciculatum.*

Macrocentrum cristatum (DC.) Triana

Herbs or subshrubs. Stems glabrous, with inconspicuous longitudinal ridges. Leaves homomorphic; petioles 1–1.5 cm long; blades ovate to oblong-ovate, 1.5–2 × 0.8–1.2 cm (outside our area), glabrous or very sparesly strigulose on both surfaces, the base obtuse to rounded, the apex acute to narrowly obtuse, the margins cilolate-serrulate; primary veins in 1 pair. Inflorescences few-flowered cymes; peduncles minute. Flowers 4-merous; hypanthium 2–2.6 mm long; calyx lobes oblate, 0.3–0.5 mm long; petals 6–10 mm long, white. Capsules 8-ridged, obpyramidal, 5–5.5 mm long Fl (Jan, Jun, Sep), fr (Jan); on exposed granitic bolders.

Macrocentrum fasciculatum (DC.) Triana PL. 86b

Subshrubs. Stems glabrous, narrowly or distinctly winged, sometimes rooting at lower nodes. Leaves homomorphic or heteromorphic; petioles 1–2 cm long; blades lanceolate to ovate, 3.5–5.5 × 1.5–3.5 cm, glabrous or sparsely strigulose adaxially, glabrous abaxially, the base obtuse, acute, or attenuate-acute, the apex acute, the margins ciliolate-serrulate; primary veins in 1 pair. Inflorescences 7–9-flowered cymes; peduncles short. Flowers 5-merous; hypanthium 2.5–3 mm long; calyx lobes oblate, 0.3–0.5 mm long, external projections minute or lacking; petals oblong-lanceolate, 6–9 mm long, white. Capsules 10-ridged, obpyramidal, 5–7 mm long. Fr (Jan, Aug, Sep); locally common, on grantic boulders.

Macrocentrum latifolium Wurdack

Subshrubs, sometimes radicant. Stems quadrangular, obscurely setulose at nodes but otherwise glabrous, rooting at lower nodes. Leaves homomorphic or subhomomorphic; petioles 1.5–3 cm long; blades ovate, 6–9 × 3–7 cm, adaxially sparsely to moderately but deciduously strigulose, abaxially glabrous, the base broadly obtuse to cordulate, the apex broadly acute, the margins minutely serrulate; primary veins in 2 pairs. Inflorescences few-flowered cymes; peduncles 2–6 cm long, bifurcate at apex. Flowers 4-merous; hypanthium 2.5–2.8 mm long; calyx lobes ca. 0.1 mm long and oblate, the external teeth minute; petals elliptic-oblong, 6.1–6.5 × 2–2.1 mm, white. Capsules 8-ridged, obpyramidal, ca. 7 mm long. Fl (Jul); in low forest on slopes.

MAIETA Aubl.

Shrubs, glandular-setose. Leaves strongly dimorphic, one of pair with ant domatia (formicaria), the other without domatia. Inflorescences in upper leaf axils, solitary or in short cymes. Flowers 5-merous; hypanthium terete (in the Guianas); calyx lobes with subulate external teeth; petals white, the apex rounded or emarginate; stamens isomorphic, anthers subulate, emarginate at apex and with an abaxially inclined pore, adaxially 2-lobed, the connective unappendaged; ovary 1/2–2/3 adnate to the hypanthium, 5-locular. Fruits berries. Seeds many, 0.5–0.8 mm long, pyramidate, minutely pebbled, with lateral raphe.

Maieta guianensis Aubl. FIG. 194, PL. 86c; PART 1: PL. VIIa

Shrubs, 0.5–1 m tall. Stems with long, spreading, glandular hairs. Leaves: petioles 0.2–0.5 cm long; blades elliptic to obovate-elliptic, largest 12–22 × 5–10 cm, with immersed basal ant domatia 1–3 cm long, smallest 2–9 × 0.6–3.5 cm, without domatia, both surfaces, but especially abaxial veins, with conspicuous spreading hairs, the base broadly acute to rounded, the apex shortly and abruptly acuminate, the margins undulate-denticulate. Inflorescences solitary or flowers in pairs from upper leaf axils. Flowers: hypanthium red; calyx red; petals white; stamens curved inwards, white; the stigma capitellate. Fruits globose, first red, bluish-black at maturity. Fl (Feb, Apr, Jun), fr (Jan, Feb, May, Jun, Jul, Sep, Oct); common, in understory of nonflooded forest everywhere.

MICONIA Ruiz & Pav.

Shrubs or trees (rarely lianas). Leaves homomorphic. Inflorescences in terminal (or pseudolateral outside our area) panicles. Flowers (3)4–6-merous; hypanthium terete; calyx usually regularly lobed (in the Guianas except in *M. diaphanaea* and *M. sastrei* where it dehisces rather irregularly), usually persistent in fruit; petals usually small, glabrous or granulose (rarely stellulate-puberulous), usually white (rarely

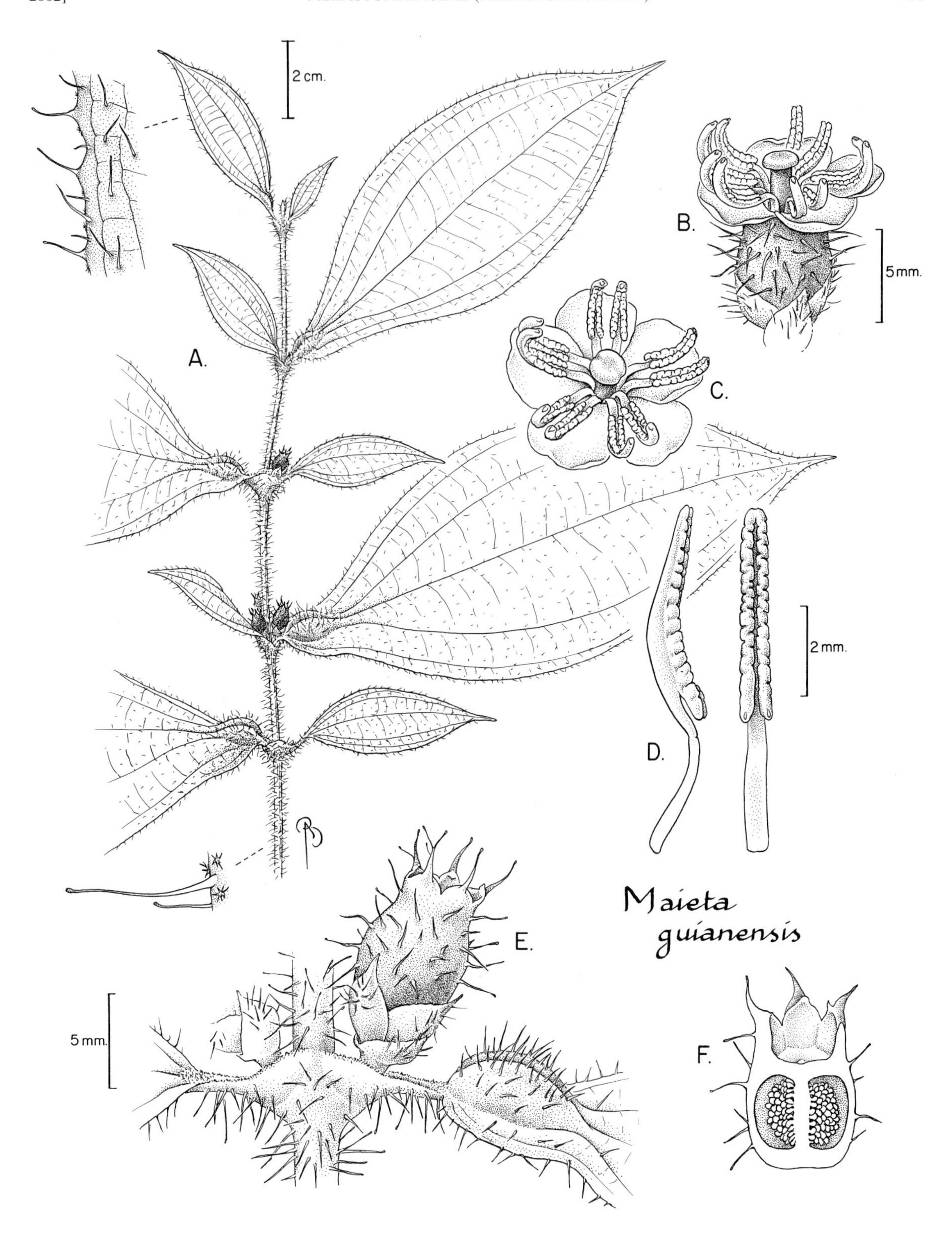

FIG. 194. MELASTOMATACEAE. *Maieta guianensis* (A, *Mori & Gracie 18991*; B–D, *Mori & Gracie 18901*; E, F, *Mori 15657*). **A.** Apex of stem with leaves and axile fruits; note details (left, above and below) of pubescence on leaf margin and stem. **B.** Lateral view of flower. **C.** Apical view of flower; note the paired stamens with single poricidal dehiscence. **D.** Lateral (left) and adaxial (right) views of stamens. **E.** Fruit in axile of leaf; note the swelling at leaf base which serves as an ant domatium. **F.** Medial section of fruit; note the axile placentation of ovules.

pink or yellow), the apex rounded to retuse; stamens usually slightly dimorphic (at least in size), anthers of various forms, usually 1-pored (rarely 2- or 4-pored or rimose), the connective simple or basally prolonged and appendaged; stigma not or somewhat expanded; ovary barely to completely adnate to the hypanthium, (2)3–5-locular. Fruits berries, globose at maturity, usually smooth, rarely ribbed. Seeds pyramidate to ovoid, usually smooth.

Key to artificial species groups

1. Leaf blades glabrous on surface abaxially (but often puberulous on veins, especially primaries). . . . . Group A.
1. Leaf blades pubescent on surface abaxially (but colorless indument sometimes so dense as to appear glabrous to unaided eye).
    2. Pubescence of abaxial leaf blade surface predominantly of discrete, smooth, unbranched hairs. . . . . Group B.
    2. Pubescence of abaxial leaf blade surface indiscrete or of dendritic, stellate, or lepidote hairs.
        3. Pubescence of abaxial leaf blade surface scattered, the surface visible between hairs. . . . . Group C.
        3. Pubescence of abaxial leaf blade surface dense, the surface not visible. . . . . Group D.

Key to Miconia *Group A*

1. Flowers 3-merous. . . . . *M. trimera.*
1. Flowers 4–5-merous.
    2. Flowers 4-merous.
        3. Leaf base cordulate. Hypanthium ca. 1.5 mm long. . . . . *M. aliquantula.*
        3. Leaf base acute to obtuse. Hypanthium ca. 3.5 mm long.. . . . . *M. lateriflora.*
    2. Flowers 5-merous.
        4. Flowers sessile on ultimate axes.
            5. Anther connectives basally prolonged and bilobate. . . . . *M. eriodonta.*
            5. Anther connectives neither prolonged nor appendaged.
                6. Leaves glabrous except for a few hairs abaxially; primary veins in 1 pair. Petals pink. . . . . *M. ciliata.*
                6. Leaves fine-setulose on primary veins abaxially; primary veins in 2 pairs. Petals white.. . . . *M. racemosa.*
        4. Flowers not sessile on ultimate axes.
            7. Petals 6–9 mm long, yellow. . . . . *M. cacatin.*
            7. Petals 1–4 mm long, white.
                8. Inflorescences racemiform or once-branched. . . . . *M. ceramicarpa* var. *ceramicarpa.*
                8. Inflorescences with tertiary branchlets.
                    9. Calyx apiculate-conic in bud, dehiscing irregularly at anthesis. . . . . *M. sastrei.*
                    9. Calyx open and lobed in bud.
                        10. Leaf blades with at least some primary veins arising above base.
                            11. Hypanthium narrowly campanulate, 4–6 × 1.8–2.2 mm. Fruits narrowly campanulate. . . . . *M. tillettii.*
                            11. Hypanthium globose, 2–3 × 2–3 mm. Fruits globose.. . . . . *M. prasina.*
                        10. Leaf blades with all primary veins arising from base.
                            12. Leaves 10–20 × 4–8 cm. . . . . *M. affinis.*
                            12. Leaves 5–12 × 1.5–4 cm.
                                13. Petals glabrous; anthers 1.1–1.4 mm long. . . . . *M. minutiflora.*
                                13. Petals granulose; anthers 1.7–1.9 mm long. . . . . *M. myriantha.*

Key to Miconia *Group B*

1. All primary leaf veins arising from base or from very near base. . . . . *M. bracteata*
1. At least some of primary leaf veins arising from above base.
    2. Petioles 0.5–3(4) cm long; blades abaxially densely setose. . . . . *M. nervosa.*
    2. Petioles >1 cm long; blades abaxially glabrous except for setose hairs on the midrib, primary and secondary veins. . . . . *M. ceramicarpa* var. *candolleana.*

Key to Miconia *Group C*

1. Branchlets 4-winged, the wings 2–3 mm wide. . . . . *M. alata.*
1. Branchlets terete to quadrate, not winged.

2. Flowers sessile on ultimate inflorescence axes.
 3. Ovary 3-locular. . . . . *M. dispar.*
 3. Ovary 5-locular. . . . . *M. eriodonta.*
2. Flowers not sessile.
 4. Hypanthium 5–6.5 mm long; anther connective with short-stalked glands at base. . . . . *M. tomentosa.*
 4. Hypanthium 1.6–2.7 mm long; anther connective without basal glands.
 5. Hypanthium 1.6–1.8 mm long. . . . . *M. diaphanea.*
 5. Hypanthium 2.5–2.7 mm long. . . . . *M. splendens.*

Key to Miconia *Group D*

1. Branchlets 4-winged, the wings 2–3 mm wide. . . . . *M. alata.*
1. Branchlets terete to quadrate, not winged.
 2. Leaf blade surface pubescence amorphous or arachnoid abaxially.
 3. Flowers 6-merous.
 4. Hypanthium ca. 3.5 mm long; stigma capitellate. . . . . *M. acuminata.*
 4. Hypanthium 5–7 mm long; stigma not or barely expanded. . . . . *M. holosericea.*
 3. Flowers 5-merous.
 5. Flowers sessile on ultimate inflorescence axes.
 6. Branchlets sharply 4-angled. . . . . *M. argyrophylla* subsp. *argyrophylla.*
 6. Branchlets obtusely quadrangular to subterete, but not sharply 4-angular.
 . . . . *M. argyrophylla* subsp. *gracilis.*
 5. Flowers not sessile on ultimate inflorescence axes. . . . . *M. gratissima.*
 2. Leaf blade surface pubescence of minute stellulate or lepidote-stellulate hairs abaxially.
 7. Hypanthium glabrous at anthesis. . . . . *M. mirabilis.*
 7. Hypanthium pubescent at anthesis, the hairs stellate or lepidote.
 8. Petioles absent; blades obovate-spatulate; innermost pair of primary veins arising to 30 cm above blade base. . . . . *M. plukenetii.*
 8. Petioles present; blades not obovate-spatulate; primary veins all arising from base.
 9. Petioles usually <1 cm long; blades 10–20 × 2–6 cm. Fruits ca. 3–4 mm diam. when dried. . . . . *M. chrysophylla.*
 9. Petioles usually >2 cm long; blades 20–26 × 8–12 cm. Fruits ca. 8 mm diam. when dried. . . . . *M. trailii.*

Miconia acuminata (Steud.) Naudin

Trees, (4)10–20 m tall. Young stems and lower leaf surfaces with appressed amorphous white indument, each node with an interpetiolar line. Leaves: petioles 1–1.5(2) cm long; blades elliptic to oblong-elliptic, 10–20 × 3–7 cm, the base acute, decurrent, the apex long-acuminate, the margins entire; primary veins in 2 pairs, all arising from or close to base. Inflorescences terminal or in upper leaf axils, panicles, 6–10 cm long; pedicels absent or present, with large caducous bracts, 5–8 mm long. Flowers predominantly 6-merous; hypanthium 3.5 mm long, densely stellulate-puberulous; calyx lobes oblong, ca. 3 mm long, falling at anthesis; petals obovate, 4.7–4.8 mm long, white; stamens slightly unequal in size, large anthers ca. 4.3 mm long, with an adaxial pore, small anthers ca. 3.8 mm long, with an abaxial pore; ovary 3–4-locular; stigma capitellate. Fruits first red, then dark blue. Fr (Mar).

Miconia affinis DC. PL. 87a

Shrubs or small trees, 3–6(15) m tall. Young stems rounded-quadrate, the nodes with interpetiolar ridge. Primary leaf veins abaxially, inflorescences, and hypanthia moderately appressed-puberulous with minute stellulate hairs. Leaves: petioles 0.5–2 cm long; blades oblong-elliptic to elliptic, 10–20 × 4–8 cm, glabrous adaxially, sparsely and deciduously stellulate-puberulous abaxially, the base broadly acute to obtuse, the apex abruptly acuminate, the margins entire or obscurely undulate-serrulate; primary veins in 1–2 pairs, all arising from base. Inflorescences panicles, 8–15 cm long, with tertiary branchlets and multiflorous; pedicels 0.5–2 mm long. Flowers 5-merous; hypanthium 1.3–1.7 mm long; calyx 0.7–0.8 mm long (including triangular lobes ca. 0.3 mm long), not persistent in fruit; petals narrowly obovate, 1.8–3 mm long, white; stamens dimorphic, anthers narrowly oblong, 2–2.6 mm or 1.2–2 mm long, 1-pored, the connective basally somewhat prolonged and with a cordiform or trilobed appendage; ovary 3-locular. Fruits first red, bluish-black at maturity. Fl (Feb, Sep, Oct, Nov, Dec), fr (Jan); in weedy habitats along forest roads and trails.

Miconia alata (Aubl.) DC.

Shrubs, 1–3 m tall. Stems strongly 4-winged, the wings 2–3 mm wide, each node with an interpetiolar line. Primary leaf veins abaxially, inflorescences, and hypanthia moderately stellate puberulous, with sessile or short-stalked hairs. Leaves: petioles absent; blades elliptic to pandurate-elliptic, 9–25 × 5–10 cm, rugulose, adaxially sparsely to moderately fine-setulose with smooth hairs, abaxially densely resinous-glandular and sparsely puberulous with sessile or short-stipitate stellate hairs, the base attenuate, the apex bluntly acute to obtuse, the margins entire, ciliolate; primary veins in 2(3) pairs, arising to 1–4 cm above base. Inflorescences panicles, 8–15(20) cm long, multiflorous; pedicels absent. Flowers 5-merous; hypanthium 2–2.5 mm long; calyx lobes 0.3–0.4 mm long, the apex triangular or rounded; petals oblong to obovate-oblong, 2–2.5 mm long, densely granulose, white; stamens slightly dimorphic, anthers subulate-oblong, 2–3 mm long, with a minute adaxial or slightly abaxial pore, the connective basally not prolonged, adaxially minutely bilobed, abaxially unappendaged; ovary 3-locular. Fruits first red, black at maturity. Fl (Nov), fr (Nov), in forest.

Miconia aliquantula Wurdack PL. 85e

Shrubs, 1–2.5 m tall. Stems, inflorescences, and hypanthia moderately puberulous with short-stipitate or sessile stellate hairs. Leaves: petioles 0.2–0.5 cm long; blades elliptic to oblong-elliptic, 10–20 × 5–8 cm, caducously stellate-puberulous adaxially, sparsely and deciduously stellate-puberulous along primary veins but otherwise glabrous abaxially, the base cordulate, the apex acuminate, the margins obscurely undulate-serrulate, appressed-ciliolate; primary veins in 1 pair (excluding tenuous marginal pair), arising from or slightly above base. Inflorescences panicles, 3–5(8) cm long, few-flowered; pedicels absent. Flowers 4-merous; hypanthium ca. 1.5 mm long; calyx lobes 0.2–0.3 mm long, the puberulous external teeth 0.2–0.3 mm long; petals obovate-oblong, 1–1.3 mm long, white; stamens isomorphic, anthers oblong-subulate, ca. 1.7 mm long, with a minute abaxially inclined pore, unappendaged; stigma not expanded; ovary 3-locular. Fruits mature dark blue. Fl (Jun), fr (Jul, Aug); in understory of nonflooded forest.

Miconia argyrophylla DC.

Shrubs or small trees, (1)3–5(8) m tall. Young stems, primary and secondary leaf veins abaxially, inflorescences, and hypanthia covered with a rusty rather deciduous stellulate pubescence underlain by persistent appressed arachnoid indument. Leaves: petioles 2–3 cm long; blades elliptic to oblong-elliptic, 11–28 × 4–13 cm, glabrous adaxially, completely covered with pale arachnoid tomentum and flocculose along midrib abaxially, the base acute to almost rounded, the apex acuminate, the margins entire, sometimes obscurely undulate; primary veins in 1 pair (excluding tenuous marginal pair), arising from or slightly above base. Inflorescences panicles 5–17 cm long, multiflorous; pedicels absent. Flowers 5-merous; hypanthium ca. 2 mm long; calyx lobes triangular, 0.2 mm long; petals narrowly obovate-oblong, 2–2.5 mm long, white; stamens dimorphic, anthers linear-oblong, 2–2.5 mm long, with a minute pore, the connective basally slightly prolonged, with a basal cordiform or trilobed appendage; ovary 3-locular. Fruits mature dark blue. Fl (Oct), fr (Jan); in forest.

Wurdack (1993), in the Flora of the Guianas, recognized two subspecies, subsp. *argyrophylla* with sharply 4-angled, robust stems (0.7–1 mm diam.) and subsp. *gracilis* Wurdack with obtusely quadrangular to almost terete, more slender stems (0.4–0.5 mm diam.).

Miconia bracteata (DC.) Triana [Syn.: *Miconia lappacea* (DC.) Triana].

Shrubs, (0.5)1–2(3) m. Stems, primary leaf veins on both sides, inflorescences, and hypanthia moderately to densely fine-setose with smooth or obscurely barbellate hairs, with a few fine gland-tipped ones intermingled in the inflorescence and hypanthia. Leaves: petioles 0.8–3.5 cm long; blades elliptic to oblong-elliptic, 8–21 × 4–12 cm, sparsely appressed-setulose or strigose adaxially, sparsely fine-setose abaxially, the base broadly acute to obtuse, the apex acuminate, the margins minutely serrulate to almost entire; primary veins in 2 pairs, arising from base. Inflorescences glomerulate-spicate or with short branches basally, 5–10 cm long; pedicels absent. Flowers 5-merous; hypanthium 2.5–3 mm long; calyx lobes oblong, 1–1.4 mm long, with subulate external teeth; petals oblong, 4.5–6 mm long, white; stamens almost isomorphic, anthers subulate, (2.5)3.5–5 mm long, with a minute pore, the connective basally slightly prolonged, with ± developed blunt abaxial tooth to 0.25 mm long; ovary 3–5-locular. Fl (Sep, Oct), fr (May, Jun, Aug, Nov); in forest understory and secondary vegetation, often along streams.

Miconia cacatin (Aubl.) S. S. Renner FIG. 195, PL. 87c

Small trees, 2–4(7) m tall. Young stems obtusely quadrangular. Young stems, petioles, primary leaf veins abaxially, inflorescences, and hypanthia essentially glabrous, sometimes with a few, very small (ca. 0.1 mm long), dendritic or stellate hairs especially associated with inflorescence axes and hypanthia. Leaves: petioles 6–10.5(13.5) cm long; blades elliptic-oblong, 21–36(65) × 8–12(18) cm, the base rounded to cordulate, the apex bluntly acute to obtuse, the margins entire; primary vein in 1 pair, arising from base, the secondaries widely spaced (ca. 1–1.5 cm apart). Inflorescences narrowly oblong panicles, 9–20 cm long, with a moderate number of flowers; pedicels 0–3 mm long, articulate at hypanthium base. Flowers 5-merous; hypanthium 6–9 mm long; calyx truncate, 1.2–1.4 mm long; petals obovate-suborbicular, 6–9 mm long, translucent yellow; stamens isomorphic, anthers subulate, ca. 5.5 mm long, with a small abaxially inclined pore, the connective basally prolonged 0.6–0.8 mm but not expanded; ovary 5-locular. Fruits maturing black. Fl (May, Jun), fr (Jul, Aug, Sep); common but scattered, in understory of nonflooded forest.

Miconia ceramicarpa (DC.) Cogn. var. **candolleana** Cogn.

Subshrubs, 0.4–1 m tall. Stems, primary leaf veins abaxially, inflorescences, and hypanthia sparsely to moderately appressed-pubescent. Leaves: petioles 1–3 cm long; blades elliptic-ovate to elliptic, 8–20 × 4–10 cm, moderately to densely fine strigose adaxially, moderately fine setose abaxially, midrib, primary veins and secondary veins abaxially setose, but surface otherwise glabrous, often purple or reddish-tinged abaxially, the base broadly acute to obtuse, the apex acute to short acuminate, the margins inconspicuously ciliolate-serrulate; primary veins in 2(3) pairs, arising from base to 3 cm above base. Inflorescences racemiform or once-branched, 2–9 cm long, 25–35-flowered; pedicels absent. Flowers 5-merous; hypanthium 2.5–3 mm long, the hairs patent at anthesis; calyx lobes oblate, 0.2–0.3 mm long; petals oblong to obovate-oblong, 2.5–3.5 mm long, white; stamens isomorphic, anthers narrowly oblong, 2–2.5 mm long, 1-pored, the connective basally not prolonged, dorso-basally with a descending tooth 0.2 mm long; ovary 3-locular. Fruits sparsely to moderately setulose with gland-tipped hairs, first pink, red at maturity. Fl (Dec), fr (Jan, Feb, Mar, Nov, Dec); less common than var. *ceramicarpa*, in understory of nonflooded forest.

Miconia ceramicarpa (DC.) Cogn. var. **ceramicarpa** PL. 87b,d

Similar to *M. ceramicarpa* var. *candolleana* except in the following. Leaves: blades sparsely strigose adaxially, strigose on secondary veins but glabrous on surface abaxially. Flowers: hypanthium with hairs appressed at anthesis. Fl (Jan, Feb, Apr, Jun), fr (Jan, Feb, Mar, Apr, May, Jun, Aug); common, along trails and in gaps in nonflooded forest.

Miconia chrysophylla (Rich.) Urb.

Trees or shrubs, (3.5)6–13(25) m tall. Young stems 2–4-angled. Young stems, abaxial leaf blade surfaces, inflorescences, and

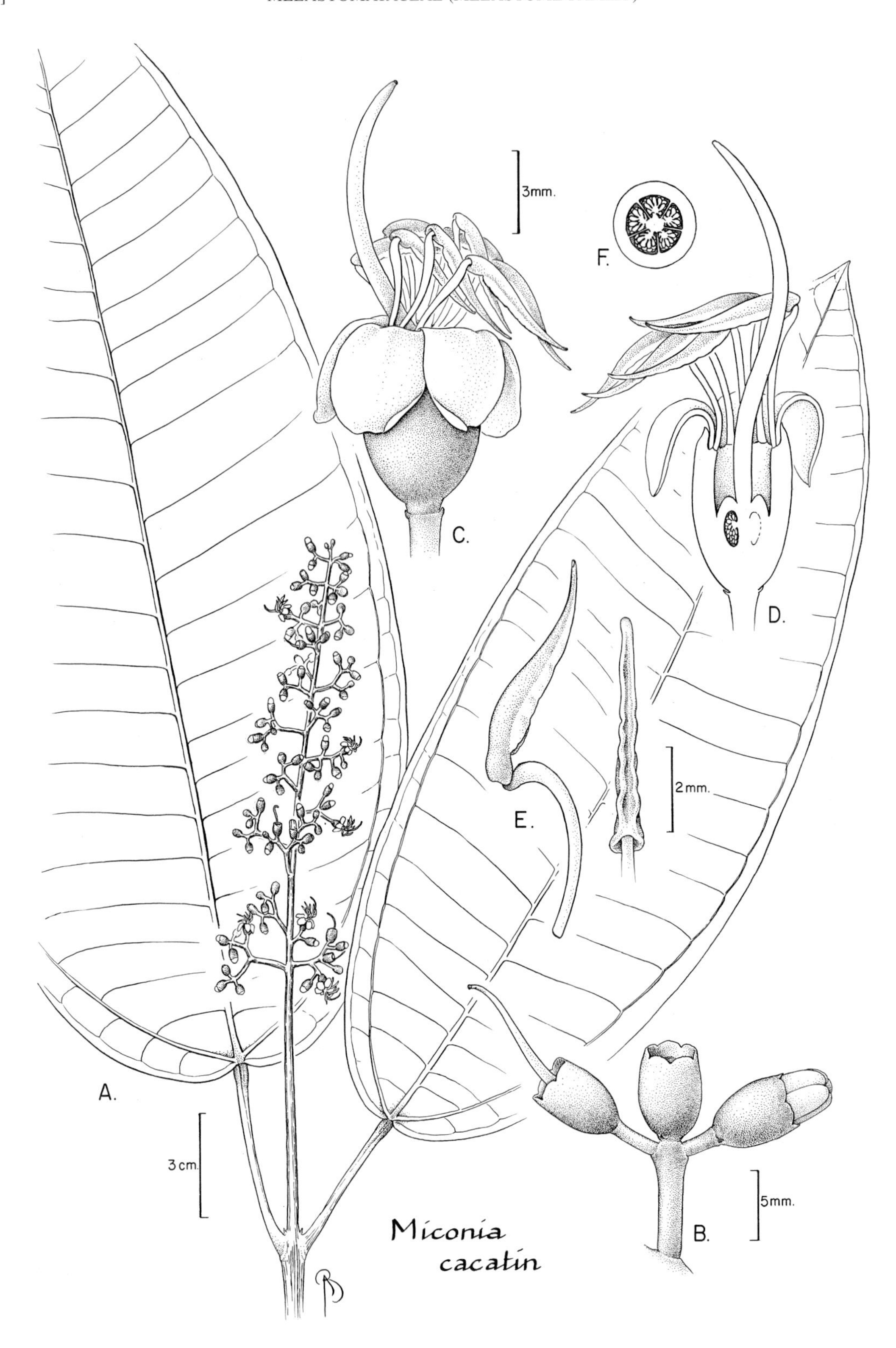

FIG. 195. MELASTOMATACEAE. *Miconia cacatin* (*Mori & Gracie 18977*). **A.** Apex of stem with leaves and terminal inflorescence. **B.** Part of inflorescence with flower in bud and two flowers past anthesis. **C.** Lateral view of flower. **D.** Medial section of flower showing axile placentation. **E.** Lateral (left) and adaxial (right) views of stamens with single poricidal dehiscence. **F.** Transverse section of ovary.

hypanthia completely covered with yellow-brown lepidote pubescence. Leaves usually 3–4-whorled; petioles 0.5–1 cm long; blades oblong to elliptic, 10–20 × 2–4(6) cm, sparsely and caducously lepidote-puberulous adaxially, the base acute, the apex acuminate, the margins undulate-serrulate to entire; primary veins in 1 pair, arising from base. Inflorescences panicles 6–15 cm long, multiflorous; pedicels absent. Flowers 5-merous; hypanthium ca. 1.2 mm long; the calyx ca. 0.4 mm long, scarcely 5-undulate; petals oblong-obovate, ca. 2 mm long, white; stamens slightly dimorphic in size, anthers oblong, rimose nearly to base, 0.7–0.9 mm long, the connective basally prolonged 0.4 mm, not appendaged; ovary 3-locular. Fruits dark blue, 3–4 mm diam. when dry.

Miconia ciliata (Rich.) DC.

Shrubs (rarely small trees), 0.5–2.5(4) m tall. Stem nodes and petiole apices fine-setulose. Leaves: petioles 0.4–2.5 cm long; blades elliptic to oblong, 4–20 × (0.7)1.5–7 cm, mature leaves glabrous except for a few hairs abaxially, the base obtuse to rounded, the apex acute, the margins obscurely serrulate and appressed ciliate; primary veins in 1 pair (excluding a tenuous marginal pair), arising from base. Inflorescences 4–10 cm long, with a moderate number of flowers, the flowers sessile along the axes. Flowers 5-merous; hypanthium 1.7–2 mm long; calyx shallowly (0.2 mm) lobed, 0.7–0.9 mm long; petals obovate-oblong, ca. 2.5 mm long, pink; stamens isomorphic, anthers oblong, ca. 1.3 mm long, with a minute adaxially inclined pore, the connective neither prolonged nor appendaged; ovary 3-locular. Fruits first red, dark blue or black at maturity. Fl (Jan).

Miconia diaphanea Gleason

Shrubs, 1–3(4) m tall. Stems, primary leaf veins abaxially, and inflorescences densely fine-setose with stipitate-stellate hairs 1–3 mm long. Leaves: petioles 0.5–1(2) cm; blades elliptic to oblong-elliptic, 15–30 × 8–13 cm, sparsely and caducously puberulous with stipitate-stellulate hairs adaxially, venules and surface sparsely to moderately setulose with short-stipitate stellate hairs abaxially, the base acute to obtuse, the apex acuminate, the margins obscurely to obviously crenulate-serrulate, deciduously ciliolate; primary veins in 2–3 pairs, arising to 6 cm from base. Inflorescences panicles, 4–7(12) cm long, multiflorous, the flowers in glomerules; pedicels absent. Flowers 5-merous; hypanthium 1.6–1.8 mm long, densely to moderately stellulate-puberulous; calyx ca. 0.9 mm long, closed in bud, dehiscing rather irregularly at anthesis into oblate lobes ca. 0.7 mm long; petals oblong, 1.7–2 mm long, white; stamens slightly dimorphic, anthers oblong-subulate, 1.4–1.7 mm long, with a minute abaxially inclined pore, the connective with a dorso-basal tooth 0.1–0.2 mm long; ovary 3(4)-locular. Fruits dark blue at maturity. Fl (Jul), fr (Sep).

Miconia dispar Benth.

Trees, (2)4–8(13) m tall. Stems obtusely quadrangular. Stems, petioles, and inflorescences densely puberulous with stellate-dendritic hairs 0.2–0.5(1) mm long. Leaves: petioles 2.5–5 cm long, deeply grooved; blades elliptic-oblong, 20–35 × 7–18 cm, glabrous adaxially, densely to moderately puberulous with short-stalked stellulate-dendritic hairs abaxially, the base broadly acute to obtuse, the apex abruptly short-acuminate, the margins obscurely undulate-serrulate; primary veins in 2 pairs, arising from base. Inflorescences narrowly oblong panicles, 20–30 cm long, multiflorous, the flowers sessile on short axes. Flowers 5-merous; hypanthium 1.3–1.5 mm long, densely stellulate-puberulous; calyx lobes ovate, ca. 1 mm long, equalled or somewhat exceeded by blunt external teeth; petals obovate-oblong, 1.5–2 mm long, white; stamens dimorphic, anthers narrowly oblong, 2–2.8 mm long, with an adaxially inclined pore, the connective basally slightly prolonged and with a large appendage surrounding the filament apex; ovary 3-locular. Fruits dark blue to black at maturity. Fr (Mar).

Miconia eriodonta DC.

Shrubs or small trees, 2–6 m tall. Stems terete, flowering branchlets obscurely 4-angular. Stems, inflorescences, and hypanthia densely puberulous with dendritic-stellulate hairs. Leaves: petioles 2–3 cm long; blades elliptic to oblong-elliptic, 15–20 × 7–9 cm, glabrous adaxially, sparsely and deciduously puberulous with dendritic-stellulate hairs on surface abaxially, the base broadly acute to narrowly obtuse, the apex short-acuminate, the margins entire; primary veins in 1 pair (excluding tenuous marginal pair), arising from base. Inflorescences panicles narrowly oblong, 15–20 cm long, multiflorous, the flowers crowded-sessile on short axes. Flowers 5-merous; hypanthium ca. 2 mm long; calyx ca. 1.2 mm long, barely 5-lobed, the external teeth minute (0.5 mm long); petals obovate-oblong, 3.8–4 mm long, white; stamens dimorphic, anthers ca. 3 mm or 2.3 mm long, with an abaxial or slightly adaxial pore, the connective basally prolonged 0.4 mm or 0.3 mm, with a cordate or bilobate basal appendage 0.3 mm long; stigma slightly capitellate, 0.4 mm diam.; ovary 5-locular. Fruits dark blue or black at maturity. Fr (Jun); in secondary vegetation.

Miconia gratissima Triana

Small trees, (2)4–8(10) m tall. Stems, inflorescences, and hypanthia moderately to densely covered with delicate dendritic or dendritic-stellulate hairs. Leaves: petioles 1–2(3) cm long; blades elliptic to lance-elliptic, (9)12–27(31) × (3.5)5–8(10) cm, glabrous adaxially, completely covered by appressed arachnoid pubescence abaxially, the base acute, the apex acuminate or acute; primary veins in 1 pair (excluding tenuous marginal pair), arising from base. Inflorescences panicles, 8–10(15) cm long, multiflorous; pedicels essentially absent to 3 mm long. Flowers very fragrant, 5-merous; hypanthium 7–8 mm long; calyx lobes oblong, caducous, 3.5–4.5 mm long, the thick external teeth ca. 1.5 mm long; petals obovate, 5–5.5 mm long, pinkish; stamens slightly dimorphic, anthers subulate, 7–8.5 mm long, the minute pore terminal or abaxially inclined (large stamens) or abaxially tilted (small stamens), the connective adaxially 2-lobed; ovary (4)5-locular. Fruits first red, dark blue at maturity. Fl (Mar).

Miconia holosericea (L.) DC.

Small trees, (2)6–8(10) m tall. Stems, leaf veins abaxially, inflorescences, and hypanthia densely covered with delicate stellulate or dendritic-stellulate hairs. Leaves: petioles (1)2–3 cm long; blades ovate-elliptic to oblong-elliptic, (11)15–25 × (5)8–12 cm, the base broadly acute to rounded, the apex short-acuminate, adaxially caducously stellate-puberulous, abaxially completely covered by appressed arachnoid pubescence, the margins entire; primary veins in 2 pairs, arising to 2 cm above base. Inflorescences terminal or axillary panicles, 3–8 cm long, the flowers glomerate; pedicels nearly absent. Flowers 6-merous; hypanthium 5–7 mm long; calyx lobes oblong, ca. 3 mm long, caducous at anthesis; petals obovate-oblong, 7–8 mm long, white to pale pink; stamens slightly dimorphic, anthers subulate, 6–8 mm long, the minute pore adaxially (large stamens) or abaxially (small stamens) inclined, the connective dorso-basally slightly thickened and with an inconspicuous obtuse spur, adaxially 2-lobed; ovary usually 4-locular; stigma not or barely expanded. Fruits first light yellow, eventually turning red, then black at maturity. Fr (Feb); in nonflooded forest. *Pied bois* (Créole).

Miconia lateriflora Cogn. PL. 87e

Shrubs, 1.5–3 m tall. Stems terete, glabrous or sparsely and caducously stellulate-furfuraceous. Leaves: petioles 0.8–2(3.5) cm; blades elliptic, 8–16(25) × 3.5–9(11) cm, glabrous on both surfaces, the base acute to obtuse, the apex short-acuminate, the margins entire or obscurely serrulate, sparsely ciliolate, the hairs caducous; primary veins in 1 pair, arising from base. Inflorescences panicles, small and 15–21-flowered, the flowers grouped in 3-flowered dichasia. Flowers 4-merous; hypanthium ca. 3.5 mm long, glabrous or very sparsely stellulate-furfuraceous; calyx lobes oblate, ca. 0.2 mm long, the acute external teeth 0.3–1.2 mm long; petals oblong-obovate, 1.6–1.8 mm long, white; stamens isomorphic, anthers subulate, 3.3–3.6 mm long, minutely 1-pored, the connective basally not prolonged but adaxially minutely bilobed; stigma not expanded; ovary 3–4-locular. Fruits sometimes sparsely glandular-setulose, dark blue at maturity. Fr (Apr, Jun); common, in understory of nonflooded forest and in secondary vegetation. *Basilic-grand-bois* (Créole).

Miconia minutiflora (Bonpl.) DC.

Small trees or shrubs, 5–8 m tall. Stems, primary leaf veins abaxially, inflorescences, and hypanthia (basally) very sparsely dendritic-puberulous but soon glabrescent, otherwise glabrous. Leaves: petioles 0.5–1 cm long; blades lanceolate to narrowly elliptic, 5–12 × 1.5–4 cm, the base acute to obtuse, the apex long-acuminate, the margins entire; primary veins in 1 pair, arising from base. Inflorescences-panicles, 5–15 cm long, multiflorous; pedicels 1–1.5 mm long. Flowers 5-merous; hypanthium 0.8–1 mm long; calyx lobes ca. 0.1 mm long; petals ovate-elliptic, ca. 2 mm long, white; stamens slightly dimorphic, anthers narrowly obcuneate, 1.1–1.4 mm long, with a broad pore, the connective basally prolonged 0.4–0.5 mm, with a cordiform or 3-lobed basal appendage; ovary 3-locular. Fruits dark blue at maturity. Fl (Apr), fr (Jun); in secondary vegetation, exhibits "big bang" flowering apparently triggered by heavy rains, the flowers are visited by numerous species of bees (Mori & Pipoly, 1984).

Miconia mirabilis (Aubl.) L. O. Williams

Trees, 5–15 m tall. Trunks markedly fluted. Stems, abaxial leaf blade surfaces, inflorescences, and bracts moderately to completely covered with minute stellulate or lepidote-stellulate hairs. Leaves: petioles 2.5–5 cm long; blades elliptic to ovate-elliptic, 10–15 × 4–7 cm, glabrous adaxially, densely stellate pubescent, occasionally with longer, barbellate hairs in primary vein axils abaxially, the base acute to rounded, the apex gradually acuminate, the margins entire; primary veins in 2 pairs, the innermost pair forming small domatia at junction with midrib. Inflorescences panicles, 9–12 cm long, multiflorous; pedicels 2–5 mm long. Flowers 5(6)-merous; hypanthium 3–4.2 mm long, glabrous; calyx 1–1.8 mm long, undulately lobed; petals narrowly obovate-oblong, 7–9 mm long, pruinose-granulose, white; stamens slightly dimorphic, anthers subulate, 6–8 mm long, with a small adaxially inclined pore, the connective basally not prolonged; ovary 3(5)-locular. Fruits red, dark purple to black at maturity. Fr (May, Sep); in nonflooded forest. *Caca Henriette* (Créole).

Miconia myriantha Benth.

Small trees or shrubs, 2.5–12 m tall. Vegetatively like *M. minutiflora* but leaf blades usually noticeably glaucous abaxially. Inflorescences as in *M. minutiflora*. Flowers: hypanthium 0.8–1 mm long; calyx lobes deltoid, 0.4–0.5 mm long; petals oblong-triangular, 2–2.3 mm long, granulose, white; stamens slightly dimorphic, anthers narrowly oblong, 1.7–1.9 mm long, the connective not or barely prolonged, with a dorso-basal cordiform or 3-lobed appendage; pistil and fruits as in *M. minutiflora*.

Miconia nervosa (Sm.) Triana

Shrubs or small trees, 1–5 m tall. Stems, primary leaf veins abaxially, and inflorescences densely strigulose to appressed-setose with fine eglandular hairs. Leaves: petioles 0.5–3(4) cm long; blades elliptic to ovate, 15–40 × 5–14 cm, sparsely fine-strigulose adaxially, abaxially sparsely to moderately appressed-setose with fine hairs, the base narrowly acute, the apex gradually acuminate, the margins obscurely undulate-serrulate; primary veins in 2–3 pairs, the innermost pairs arising to 10 cm above base. Inflorescences narrowly oblong, 4–12(17) cm long, the flowers glomerulate at ends of short lateral axes; pedicels absent. Flowers 5-merous; hypanthium 2.5–3.2 mm long, sparsely to moderately strigulose with fine smooth hairs; calyx lobes broadly triangular, 0.2 mm long; petals oblong, 3.5–4.5 mm long, white; stamens isomorphic, anthers oblong, 2.5–4 mm long, minutely 1-pored, the connective basally not or barely prolonged, not appendaged; ovary 3-locular. Fruits immature red, mature light to dark blue. Fl (Aug, Oct, Dec), fr (Aug, Oct); in secondary vegetation and along trails through nonflooded forest.

Miconia plukenetii Naudin

Trees, 4–10(16) m tall. Stems and inflorescences completely covered by tawny, sessile stellate/lepidote hairs. Leaves: sessile; blades obovate-spatulate, 30–60 × 10–25 cm, glabrous adaxially, densely stellate-lepidote-pubescent abaxially, the base auriculate, sometime clasping, the apex short-acuminate, the margins entire; primary veins in 2–3 pairs, the inner pair arising to 30 cm above blade base, this pattern unique in our area. Inflorescences panicles, 15–35 cm long, multiflorous; pedicels nearly absent. Flowers 5-merous; hypanthium 5–6 mm long, densely stellulate-puberulous; calyx 2.7–3.5 mm long, lobed to within ca. 1.5 mm of torus, stellulate/lepidote-pubescent adaxially; petals obovate-oblong, 5.5–7.9 mm long, white or pinkish; stamens slightly dimorphic, the filaments minutely glandular-puberulous, anthers subulate, 5.5–7.5 mm long, with a small adaxially inclined pore, the connective basally not prolonged but thickened at base and with a moderate number of short-stalked glands; ovary 3-locular. Fruit immature red, mature dark blue or black berries.

Miconia prasina (Sw.) DC.

Shrubs or small trees, 1.5–12 m tall. Stems, leaf veins abaxially, inflorescences, and hypanthia very sparsely to moderately but deciduously stellulate-puberulous, older plant parts glabrous. Leaves: petioles 0.3–2 cm long; blades elliptic to oblong or obovate-elliptic, 6–25 × 2.5–10 cm, glabrous on both surfaces, the base acute-attenuate, the apex short-acuminate, the margins entire to repand-denticulate; primary veins in 1–2 pairs, arising to 2.5 cm above base. Inflorescences panicles, 5–15 cm long, with tertiary branchlets and multiflorous; pedicels absent. Flowers 5-merous; hypanthium 2–3 × 2–3 mm; calyx lobes triangular, 0.1–0.3 mm long; petals obovate-oblong, 2–3 mm long, densely granulose, white; stamens slightly dimorphic, anthers narrowly oblong, 1.5–3 mm long, with a small adaxially inclined pore, the connective basally not or barely prolonged and adaxially bilobed; ovary 3-locular. Fruits 4–5 mm diam. when dry, first reddish, bluish-black at maturity. Fr (May); in nonflooded forest.

Miconia racemosa (Aubl.) DC.

Shrubs or small trees, (1)2–4(5) m tall. Stems glabrous along internodes, densely fine-setose at nodes. Leaves: petioles (1)2–6 cm long; blades obovate-elliptic to elliptic-oblong, 12–23 × 6–11 cm, caducously fine-setulose on primary veins, adaxially glabrous or sparsely barbellate-strigulose and stellate-puberulous at extreme margins, abaxially deciduously fine-setulose on primary veins, otherwise glabrous, the base broadly acute to obtuse, the apex acute to short-acuminate, the margins ciliolate-serrulate; primary veins in 2 pairs, arising from base. Inflorescences broadly oblong, with branches to 5 cm long. Flowers sessile, similar to those of *M. ciliata* but with white petals. Fruits maturing dark blue. *Radié macaque* (Créole).

Miconia sastrei Wurdack PL. 87f

Shrubs, 1–2 m tall. Stems terete. Stems, petioles, and primary leaf veins abaxially moderately fine-setose with smooth hairs (the hairs conspicuously patent, sometimes caducous) and (along with inflorescences) sparsely to moderately, caducously stellulate-puberulous. Leaves: petioles 2–3 cm long; blades elliptic, 12–21 × 5–10 cm, sparsely and deciduously setulose adaxially, glabrous on surface abaxially, the base obtuse to rounded or even cordulate, the apex acuminate, the margins entire to distantly crenulate, ciliolate; primary veins in 2 pairs, arising to 1 cm above base, often with domatia in vein axils. Inflorescences panicles, 5–8 cm long, with tertiary branchlets, but only a moderate number of flowers; pedicels absent. Flowers (4)5-merous; hypanthium ca. 2 mm long, glabrous or with scattered deciduous setae; calyx apiculate-conic in bud, ca. 1 mm long, hyaline, dehiscing irregularly into lobes at anthesis; petals obovate-oblong, 2.7–2.9 mm long, white; stamens isomorphic, anthers subulate, yellow, 2.2–2.8 mm long, with a minute abaxially inclined pore, the connective basally not prolonged but with a small dorso-basal tooth; ovary 3-locular. Fruits maturing dark blue. Fl (Feb, Mar, Apr, May, Jun, Oct), fr (Jun, Sep); common, usually in somewhat disturbed habitats along trails and roads through nonflooded forest.

Miconia splendens (Sw.) Griseb.

Shrubs or small trees, usually <6 m tall. Young stems rounded-quadrate, the nodes with interpetiolar ridge. Primary leaf veins abaxially, inflorescences, and hypanthia moderately appressed-pubescent with minute stellate-lepidote scales. Leaves: petioles 0.5–1.5 cm long; blades oblong-elliptic, 15–30 × 8–12 cm, glabrous adaxially, sparsely stellate-lepidote-pubescent abaxially, the epidermis visible between scales, the base acute to attenuate, the apex abruptly acuminate, the margins entire or obscurely undulate-serrulate; primary veins in 2 pairs, arising 0.5–2 cm above base. Inflorescences panicles, 15–20 cm long, multiflorous, with tertiary branchlets; pedicels 1–2.5 mm long. Flowers 5- merous; hypanthium 2.5–2.7 mm long; calyx ca. 1 mm long (including triangular lobes ca. 0.3 mm long), abaxial calyx teeth appearing as mere bumps, calyx not persistent in fruit; petals narrowly obovate, 2–3 mm long, white; stamens slightly dimorphic, the anthers narrowly oblong, 2.8–3 mm, 1-pored, the connective basally obtusely bilobed; ovary 3(4)-locular. Fruits first red, bluish-black at maturity. Fl (Aug), in nonflooded forest.

Miconia tillettii Wurdack

Shrubs or small trees, 1–3(8) m tall. Plants glabrous except for inflorescences. Leaves: petioles 1.5–5 cm long; blades elliptic to obovate, 10–30 × 4–10 cm, the base acute to cuneate, the apex short acuminate to acuminate, the margins entire; primary veins in 2 pairs, arising to 1–3 cm above base. Inflorescences panicles, 6–13 cm long, ample, sparsely and caducously furfuraceous, with tertiary branchlets; pedicels 1–1.2 mm long. Flowers 5-merous; hypanthium 4–6 × 1.8–2.2 mm, narrowly campanulate and 10-ridged; calyx lobes 0.5–0.6 mm long, the apex rounded, the apiculate external teeth 0.2–0.4 mm long; petals obovate-oblong, ca. 2 mm long, white; stamens slightly dimorphic, anthers subulate, curved, 2.3–2.6 mm long, with a minute pore, the connective basally not prolonged but adaxially united with the thecae for about 0.5–0.6 mm below the filament insertion; ovary 3-locular. Fruits campanulate, first red, dark blue at maturity. Fl (Aug), fr (May, Aug, Sep); common, in understory of nonflooded forest and in forest along streams.

Miconia tomentosa (Rich.) DC.

Small trees, 5–7 m tall. Stems, primary and secondary leaf blades abaxially, inflorescences, and hypanthia densely covered with dendritic-stellulate hairs. Leaves: petioles inconspicuous, ca. 5 mm long; blades broadly elliptic to pandurate, 20–40 × 10–25 cm, glabrous adaxially, moderately to densely pubescent abaxially, the hairs dendritic-stellate, the base attenuate, slightly auriculate, the apex acuminate, the margins entire; primary veins in 1 pair, arising to 4–8 cm above base. Inflorescences oblong panicles, 15–30 cm long, multiflorous; pedicels absent. Flowers 5-merous; hypanthium 5–6.5 mm long; calyx ovate-lobed, 1.5–2 mm long; petals obovate-oblong, 5–8 mm long, minutely pruinose-granulose, white to pink-purple; stamens slightly dimorphic, the filaments sparsely glandular-puberulous, anthers subulate, 7–8 mm long, with a minute adaxially inclined pore, the connective basally thickened, with numerous short-stalked glands; ovary 3(4)-locular. Fruits dark blue at maturity. *Poutsi-hô*. Fl (Oct); in nonflooded forest.

Miconia trailii Cogn.

Shrubs or small trees, 3–6 m tall. Stems, petioles, abaxial leaf blade surfaces, axes of inflorescences, and buds densely stellate-pubescent. Leaves: petioles 2–6 cm long; blades elliptic to oblong, 20–26 × 8–12 cm, the base obtuse to slightly rounded, the apex short-acuminate to acuminate, the margins entire to obscurely undulate-serrulate; primary veins in 1 pair (excluding tenuous marginal pair), arising from base. Inflorescences oblong panicles, 10–20 cm long, multiflorous; pedicels nearly absent. Flowers predominantly 6-merous; hypanthium 5–7 mm long; calyx lobes 2.5–3.5 mm long; petals 6–9 mm long; anthers 6–8 mm long, with abaxially inclined pores; ovary 4-locular. Fruits ca. 8 mm diam., purple to dark blue at maturity. Fl (Jan), fr (Aug); along road through nonflooded forest.

Miconia trimera Wurdack PL. 87g

Shrubs, 0.3–0.6 m tall. Stems terete, often rooting at lower nodes. Stems, leaves, inflorescences, and hypanthia sparsely and minutely puberulous. Leaves: petioles 2–6 cm long; blades ovate to elliptic, 8–14 × 5–7 cm, the base round to obtuse, sometimes asymmetric, the apex short acuminate to acuminate, the margins entire or obscurely serrulate, ciliolate; primary veins in 2 pairs, arising from base or to 1 cm above base. Inflorescences terminal or axillary, consisting of a peduncle and usually 2, less frequently 3, axes, the flowers sessile along axes. Flowers 3-merous; hypanthium ca. 3 mm long; calyx 0.4–0.5 mm long, the lobes oblate, their external teeth 0.6–0.7 mm long; petals oblong, ca. 2 mm long, white; stamens isomorphic, anthers oblong, 2.5 mm long, with a minute abaxially inclined pore, the connective basally not or barely prolonged, unappendaged; ovary 3-locular. Fruits purplish-blue at maturity. Fl (Feb, Aug, Sep, Nov), fr (Feb, May, Jun, Aug, Sep); common, in nonflooded forest.

MOURIRI Aubl.

Trees. Leaves homomorphic; venation pinnate, the secondary veins sometimes obscure. Inflorescences axillary, ramiflorous, or cauliflorous, small thyrses, dichasia, or solitary, sometimes umbelloid or fasciculate; pedicels often jointed. Flowers 5-merous (in our area); free hypanthium usually well developed, sometimes none; calyx often partly or completely enclosing petals in bud, splitting regularly into lobes at anthesis, the limb lobed or truncate; petals lanceolate to ovate, triangular, trullate, elliptic, or obovate, purple, pink, light blue, white, or yellow, rarely red; anthers straight or curved, dehiscent by two lengthwise slits, apical slits, or tear-shaped pores, the connective adaxially usually ± caudate, abaxially with an elliptic concave gland (when fresh producing a sticky exudate; drying shiny); ovary inferior, 1–5-locular; placentation free-central, axile, axile-basal, basal, or parietal, the locules close together to widely separated. Fruits berries or of 2–5 widely separated, 1-seeded lobes. Seeds 1–8 (in our area), sometimes shiny only on relatively small patch derived from original outer face of ovule, the rest of surface roughened, or seeds shiny all over except for rough hilum.

Mouriri (along with *Votomita*) is often treated as a separate family, Memecylaceae (Angiosperm Phylogeny Group, 1998; Renner, 1993).

1. Leaf blades with secondary veins visible, salient, connecting with each other to form well defined marginal veins. . . . . *M. sagotiana.*
1. Secondary veins invisible or, if visible, then not salient, without well-defined marginal veins.
    2. Largest leaf blades usually >12 cm long.
        3. Leaf blades narrowly ovate, the base cordate. Inflorescences mostly ramiflorous; peduncle <1 cm long. Petals light blue. . . . . *M. sideroxylon.*
        3. Leaf blades oblong, the base obtuse to rounded but not cordate. Inflorescences mostly axillary; peduncles 1–3 cm long. Petals mostly yellow. *M. crassifolia.*
    2. Largest leaf blades usually <10 cm long.
        4. Largest leaf blades <5 cm long. Buds not distinctly pointed. . . . . *M. duckeana.*
        4. Largest leaf blades >5 cm long. Buds distinctly pointed. . . . . *M. collocarpa.*

Mouriri collocarpa Ducke FIG. 196, PL. 88a

Trees, to 35 m tall. Trunks usually not buttressed, sometimes somewhat irregular at base. Bark rough, peeling in irregular, longish plates, the inner bark yellow, the wood hard. Leaves: petioles 3–5 mm long; blades ovate to elliptic, 5–11.5 × 2.3–5 cm, the base acute to obtuse, often decurrent onto petiole, the apex short acuminate to acuminate; secondary veins not visible, the marginal veins absent. Inflorescences axillary or in axils of leaf scars just below leaves; peduncle short. Flowers ca. 15 mm diam.; buds distinctly long-pointed; petals yellow. Fruits depressed globose, 3–5 cm diam., yellow, fragrant. Seeds flattened, ca. 1/3 smooth, the remaining surface rough. Fl (Sep, Nov, Dec), fr (May, Sep, Nov); common, in nonflooded forest.

Mouriri crassifolia Sagot PL. 88b

Trees, to 22 m tall in our area. Trunks unbuttressed or with incipient buttresses. Bark dark brown, with fine vertical fissures, the inner bark yellow or orange. Stems, especially when young, sometimes narrowly winged. Leaves: petioles absent or to 3 mm long; blades oblong, 10–19 × 4.5–6.5 cm, the base obtuse to rounded, the apex abruptly acuminate; secondary veins not visible, the marginal veins absent. Inflorescences mostly axillary; peduncle 1–3 cm long. Flowers 20–25 mm diam.; buds not distinctly long-pointed; petals yellow adaxially, yellow suffused with peach or red abaxially. Fruits ca. 2.5 cm diam., reddish-orange. Fl (Jan, Oct, Nov), fr (Mar); in nonflooded forest. *Bois de fer, bois-flèche, pâte d'amande* (Créole).

Mouriri duckeana Morley

Trees, to 30 m tall. Leaves: petioles 3–6 mm long; blades narrowly ovate or elliptic, 4–4.5 × 2.5–3 cm, the base obtuse to rounded, the apex acuminate; secondary veins scarcely visible, the marginal veins absent. Inflorescences axillary; peduncle absent. Flowers 10–15 mm diam.; buds not distinctly long-pointed; petals yellow to orange-yellow. Fruits ca. 1 cm diam. Fl (Sep); in nonflooded forest.

Mouriri sagotiana Triana

Small trees, to 4 m tall but to 15 m tall in other parts of French Guiana. Trunks without buttresses. Leaves: petioles 2–3 mm long; blades narrowly ovate to lanceolate, 4.5–10 × 2–3.5 cm, the base obtuse to rounded, the apex long acuminate; secondary veins salient and connecting with each other (anastomosing) several mm from leaf margin. Inflorescences axillary; peduncle absent. Flowers 5–7 mm diam.; buds not distinctly long-pointed; petals yellow. Fruits ca. 1–1.6 cm diam. Fl (Oct); in nonflooded forest. *Bois fer, bois-flèche* (Créole).

Mouriri sideroxylon Triana

Trees, to 30 m tall. Leaves: petioles nearly absent or to 3 mm long; blades narrowly ovate, 9–15 × 5–7 cm, the base cordate, the apex acuminate; secondary veins not visible. Inflorescences mostly ramiflorous, sometimes axillary; peduncle <1 cm long. Flowers ca. 8 mm diam.; buds pointed, but the apex not long-pointed; petals light blue. Fruits 1.5–2 cm diam. Fl (Apr); in nonflooded forest.

NEPSERA Naudin

Herbs or subshrubs. Stems 4-angled. Leaves homomorphic. Inflorescences terminal, lax, multiflorous panicles. Flowers 4-merous; hypanthium terete; calyx lobes ovate-subulate, persistent; petals oblong-lanceolate, the apex acute; stamens 8, slightly dimorphic in size, the anthers subulate, minutely 1-pored, the connective basally slightly prolonged (0.5–0.7 mm) and bilobed (lobes bluntly acute and 0.3–0.7 mm long), abaxially not appendaged; ovary superior, 3- locular, glabrous. Fruits capsules, 3-valved. Seeds numerous, cochleate, foveolate.

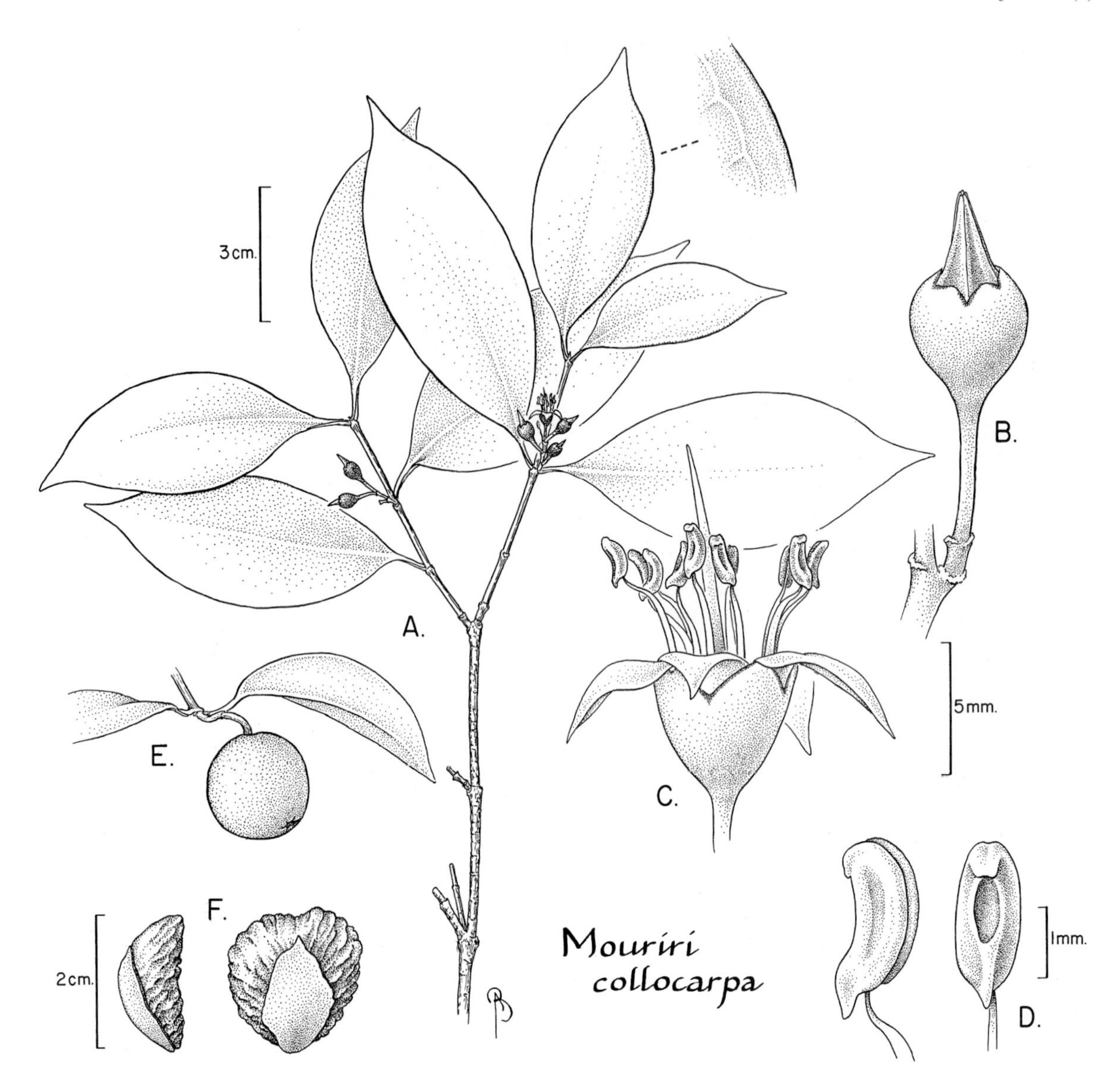

FIG. 196. MELASTOMATACEAE. *Mouriri collocarpa* (A, B, D, *Mori & Boom 15304*; C, *Mori & Bolten 8509* from Surinam; E, *Mori et al. 22293*; F, *Mori & Pennington 18055*). **A.** Stem with leaves, inflorescences, and detail of leaf margin (above right). **B.** Part of inflorescence with flower bud. **C.** Lateral view of flower. **D.** Lateral (far left) and adaxial (near left) views of stamens; note the elliptical concave gland on anther. **E.** Fruit on leafy stem tip. **F.** Seeds.

Nepsera aquatica (Aubl.) Naudin

Herbs or subshrubs, to 0.5–2 m tall. Stems, veins on abaxial leaf surfaces, inflorescences, and hypanthia sparsely and inconspicuously puberulous with partly gland-tipped hairs. Leaves: petioles 0.5–1 cm long; blades ovate to elliptic, 2–5 × 1.5–3 cm, glabrous or sparsely and deciduously strigulose adaxially, glabrous except for veins abaxially, the base rounded to cordate, the apex acute to acuminate, the margins serrulate; primary veins in 2–3 pairs, all arising from base. Inflorescences panicles to 30 cm long; pedicels 3–5 mm long. Flowers 5–7 mm diam.; petals 4.5–7 mm long, white; anthers 2.5–3 mm long, purple. In secondary vegetation.

RHYNCHANTHERA DC.

Shrubs or subshrubs. Leaves homomorphic. Inflorescences terminal, few- to many-flowered cymes forming thyrses. Flowers 5-merous; hypanthium globose or campanulate, terete, hirsute with glandular hairs to glabrescent; calyx lobes linear, subulate or narrowly triangular, often exceeding hypanthium, persisting; petals obovate, purplish red to magenta (rarely white); stamens 5, alternating with 5 staminodia, the antesepalous stamens sometimes of unequal length, with one stamen larger than other four, anthers usually subulate, the tips spoon-shaped with an adaxial or terminal pore, the connective basally greatly prolonged, adaxially with a simple or bilobed appendage; ovary superior, 3–5-locular, the style slender, declined or subsigmoid. Fruits capsules. Seeds numerous, oblong, angular, the hilum basal, the testa reticulate-foveolate.

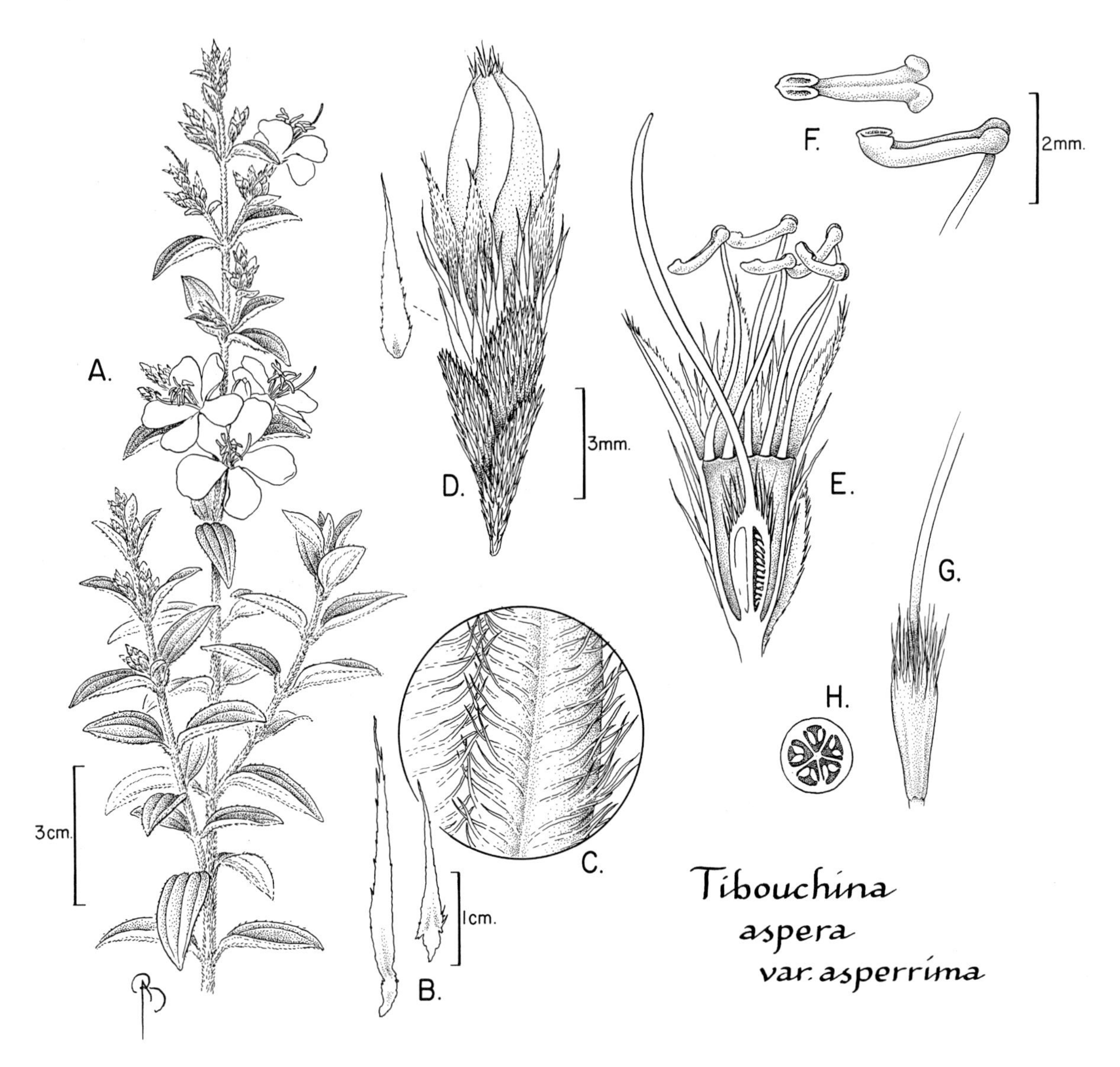

FIG. 197. MELASTOMATACEAE. *Tibouchina aspera* var. *asperrima* (A, *Mori & Gracie 18400*; B–H, *Mori & Gracie 18341*). **A.** Stem with leaves and inflorescences. **B.** Stem scales. **C.** Detail of adaxial leaf surface and margin. **D.** Flower bud subtended by bracts (right) with detail of scale (above left). **E.** Medial section of flower showing perigynous ovary and parietal placentation. **F.** Apical (above) and lateral (below) views of stamens; note the bilobed connective. **G.** Ovary with apical pubescence and part of style. **H.** Transverse section of ovary.

Rhynchanthera grandiflora (Aubl.) DC.

Subshrubs, to 1–2 m tall. Plants usually densely glandular-pubescent. Leaves: petioles 2.5–3.7 cm long; blades ovate, 4.5–8.2 × 2.4–3 cm, ± densely strigulose on both surfaces, the base cordate, the apex acute to acuminate, the margins crenulate; primary veins in 3–4 pairs, all arising from base. Inflorescences in axils of uppermost leaves, short cymes; pedicels ca. 1 mm long. Flowers large, 3–4 cm diam.; hypanthium 4–6 mm long, glandular-setulose; calyx lobes subulate, usually exceeding hypanthium, 6–15 mm long; petals ca. 20 mm long, reddish purple; anthers 5–6.5 mm long, the beak 3.5–4 mm long; ovary (3)5-locular. Usually found in humid places in savannas or open swampy habitats. Fl throughout the year in other areas. *Coquelicot* (Créole).

TIBOUCHINA Aubl.

Shrubs or infrequently herbs (outside our area). Leaves homomorphic. Inflorescences terminal, usually panicles. Flowers 5-merous (in the Guianas); hypanthium terete; calyx lobes persistent; petals obovate, usually magenta or pink; stamens sometimes dimorphic, anthers subulate, with a small ventro-terminal pore, the connective basally often prolonged, adaxially 2-lobed, abaxially usually unappendaged; ovary superior, 5-locular. Fruits capsules. Seeds numerous, cochleate, tuberculate.

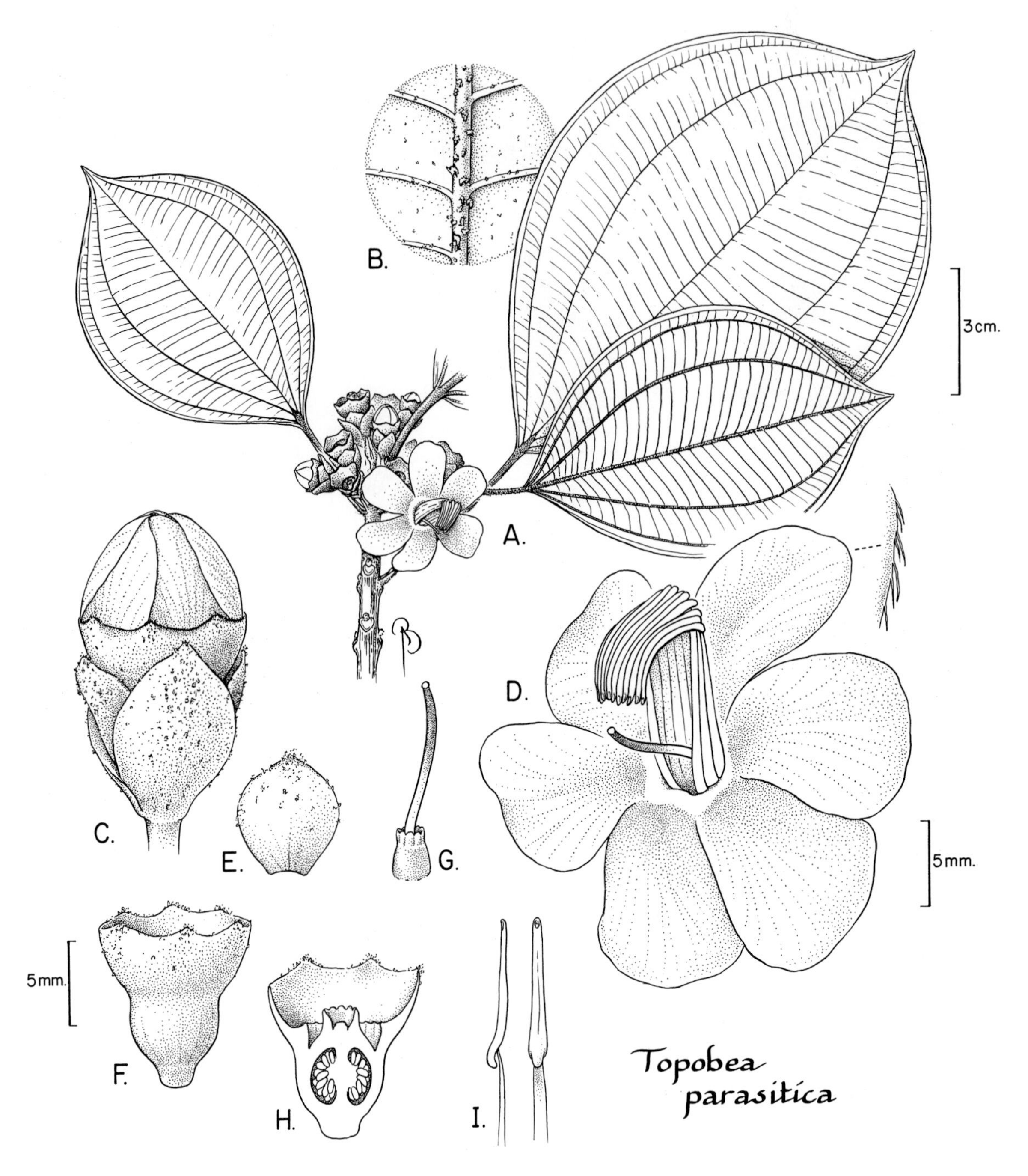

FIG. 198. MELASTOMATACEAE. *Topobea parasitica* (A–C, E–I, *Mori & Gracie 18724*; D, *Mori et al. 17349*). **A.** Apex of stem with inflorescences; note leaves with five arcuate veins from base to apex. **B.** Detail of abaxial leaf surface. **C.** Flower bud. **D.** Flower with detail of petal margin (right). **E.** Abaxial surface of inner bract. **F.** Immature fruit. **G.** Pistil surmounted by lobed cupular disc. **H.** Medial section of fruit; note axile placentation of seeds. **I.** Lateral (left) and adaxial (right) views of stamens with single poricidal dehiscence.

1. Leaf blades <3.5 cm wide. Flowers <2.5 cm diam. Plants native. . . . . *T. aspera* var. *asperrima*.
1. Leaf blades >5 cm wide. Flowers >2.5 cm diam. Plants cultivated. . . . . *T. multiflora*.

Tibouchina aspera Aubl. var. **asperrima** Cogn.

FIG. 197, PL. 88c

Shrubs, 0.3–2(3) m tall. Stems, primary leaf veins abaxially, and inflorescences with lanceolate flattened hairs (actually emergences of the epidermis, with stomates). Leaves: petioles 0.2–0.8 cm long; blades ovate to narrowly ovate, 2–8 × 1–3.5 cm, rigid-coriaceous, hairs oriented in longitudinal bands between primary veins adaxially, throughout abaxially, the base rounded, the apex acute, the margins entire, appressed ciliolate; primary veins in 2 pairs, all arising from base. Inflorescences terminal or axillary panicles, multiflorous. Flowers about 2 cm diam.; hypanthium 7–8 mm long; calyx lobes oblong-triangular to lanceolate, 5–5.5 mm long; petals 9–15 mm long, rose-purple; anthers 6–8 mm long, the connective prolonged 1–2 mm. Fl (Jun); in weedy habitats. *Cocorico* (Créole).

Tibouchina multiflora (Gardner) Cogn.

A showy ornamental with quadrate, winged stems, broad leaves (>5 cm wide), terminal, multiflorous inflorescences and large, purple-petaled flowers larger than 2.5 cm diam. Cultivated in the village. Fl (Aug).

TOPOBEA Aubl.

Shrubs or trees, sometimes vining, often epiphytic. Leaves homomorphic. Inflorescences axillary. Flowers 6-merous, surrounded by 2 pairs of prominent long-persistent bracts; hypanthium terete; calyx ± 6-lobed, usually persistent; petals pink (in the Guianas); stamens 12, isomorphic, anthers narrowly oblong, usually laterally coherent, with a dorso-apical pore, the connective usually with a dorso-basal spur; ovary at least partly inferior, 4–6-locular. Fruits rather large, yellow or red berries. Seeds numerous.

Topobea parasitica Aubl. FIG. 198; PL. 88d

Epiphytic shrub. Young stems caducously strigulose with roughened hairs. Leaves: petioles (2)3–8 cm long; blades broadly elliptic to orbicular, 10–18 × 5–12 cm, caducously roughened-puberulous on veins abaxially, the base obtuse, the apex abruptly short-acuminate, the margins entire; primary veins in (2)3 pairs, all arising from base. Inflorescences in clusters of 4–6 from upper leaf nodes; pedicels 0.5–1 cm long; bracts 12–15 mm long. Flowers: hypanthium ca. 6 mm long; calyx ca. 3.5 mm long, barely (0.5–0.7 mm) 6-lobed; petals broadly ovate, 15–17 mm long, the apex broadly obtuse, rounded, or truncate; anthers 10–12 mm long, the dorso-basal connective spur 0.7–1.5 mm long, blunt; ovary partly inferior, ca. 1/5–1/4 adnate to the hypanthium, (4?)6-locular. Fruits red at maturity. Fl (Feb, Aug); in large trees. *Gouman*, *vanille sauvage* (Créole).

MELIACEAE (Mahogany Family)

Terence D. Pennington

Canopy or understory trees or treelets. Latex absent. Bark exuding slightly spicy (*Guarea* and *Trichilia*) or garlic (*Cedrela*) aromas when cut. Indumentum of simple, malpighiaceous, or rarely stellate hairs. Stipules absent. Leaves spirally arranged, usually pinnately compound, with or without terminal leaflet, sometimes with terminal bud showing intermittent growth (*Guarea*), rarely trifoliolate; leaflets entire; venation eucamptodromous or brochidodromous. Inflorescences usually axillary, less frequently ramiflorous, cauliflorous, or terminal, usually thyrsoid or racemose, rarely fasciculate, paniculate, or spicate. Flowers actinomorphic, bisexual or unisexual; calyx 3–6-lobed or with 3–6 free sepals, the aestivation open or imbricate; petals 4–6, free or partly united, the aestivation imbricate or valvate; filaments rarely completely free, usually partly or completely united to form a staminal tube, with or without appendages, the anthers 7–12, attached apically on filament, on margin of staminal tube, or within throat of staminal tube; nectary (disc) intrastaminal, stipitate or annular; ovary superior, 2–10-locular, the locules uniovulate, biovulate, or multiovulate, the style simple, the stigma capitate or discoid; placentation axile. Fruits loculicidal, septifragal, or septicidal capsules. Seeds either winged and then attached to woody columella or unwinged and then usually with fleshy arillode or sarcotesta, rarely with corky sarcotesta; endosperm absent.

Pennington, T. D. & B. T. Styles. 1981. Meliaceae. Fl. Neotrop. Monogr. **28:** 1–470.
——— & A. R. A. Görts-van Rijn. 1984. Meliaceae. *In* A. L. Stoffers & J. C. Lindeman (eds.), Flora of Suriname **V(1):** 519–569. E. J. Brill, Leiden.

1. Locules with 1–2 ovules. Fruits loculicidal capsules.
   2. Leaves nearly always with terminal bud showing intermittent growth. Anthers attached within throat of staminal tube. . . . . *Guarea*.
   2. Leaves without terminal bud. Anthers attached at apices of filaments or on margin of staminal tube. . . . . . . *Trichilia*.
1. Locules with 3 to many ovules, these biseriate. Fruits septifragal or septicidal capsules.
   3. Bark fissured. Stamens 5, the filaments free but adnate to androgynophore below, the anthers attached apically on filaments. Fruits oblong or ellipsoid capsules with thinly woody valves. Seeds winged, flat. . . . . . *Cedrela*.
   3. Bark smooth. Stamens 10, completely united into tube, the anthers attached within throat of staminal tube. Fruits subglobose capsules with thickly woody valves. Seeds unwinged, angular. . . . . *Carapa*.

CARAPA Aubl.

Trees. Indumentum of simple hairs. Leaves paripinnate. Flowers 5-merous, unisexual (plants monoecious); sepals 5-lobed nearly to base, imbricate; petals 5, free, imbricate; staminal tube cup-shaped, divided at apex into 10 short appendages, the anthers 10, attached within throat of staminal tube and alternating with appendages; nectary well developed, cushion-shaped, surrounding base of ovary; ovary 5(6)-

locular, the locules with 3–8 biseriate ovules, the style short, the stigma discoid; pistillode similar to pistil but with rudimentary ovules. Fruits large, pendulous, subwoody, subglobose, septifragal capsules with 5 valves opening from base and apex simultaneously. Seeds large, angular, with thick, woody sarcotesta.

Carapa procera DC. PL. 89a; PART 1: FIG. 10

Trees, to 25 m tall. Trunks unbuttressed, the bole cylindric. Bark smooth, gray-brown, sometimes with incomplete hoop marks, the slash 7–10 mm thick, pink to dark red. Leaves crowded at ends of branchlets, to 1 m long, with extra-floral nectaries often present on rachis between pairs of leaflets and at leaflet tips; petioles prominently swollen at base; leaflets broadly oblong, to 35 × 10 cm, glabrous, the base rounded, the apex rounded, glabrous. Inflorescences axillary or terminal, to 80 cm long, widely branched thyrses. Flowers: calyx lobes rounded, 1–1.5 mm long; petals ovate to obovate, 4–8 mm long, greenish-white; staminal tube cup-shaped, 3–4.5 mm long. Capsules brown, lenticellate, 7–9(12) × 5–7(10) cm, the valves with median ridge. Seeds 3–4 cm long, 4–6 per valve, the sarcotesta dark brown. Fl (Oct, Nov, Dec), fr (Mar, Apr); common, in forest on well drained soil. *Andiroba* (Portuguese), *carapa* (Créole).

CEDRELA P. Browne

Trees. Bark fissured. Leaves paripinnate. Flowers 5-merous, unisexual (plants monoecious); calyx cup-shaped, irregularly dentate; petals 5, free, adnate to androgynophore below; anthers attached apically on filaments; ovary borne at apex of androgynophore, 5-locular, the locules with 10–14 biseriate ovules, the stigma discoid; pistillode more slender than pistil, with rudimentary ovules. Fruits woody, septicidal capsules, opening from apex by 5 valves; columella woody, broadly winged. Seeds with terminal wing, attached by seed end to apex of columella and winged toward base of capsule.

Cedrela odorata L. FIG. 199; PART 1: FIG. 7

Trees, to 20 m tall. Trunks buttressed, the bole cylindric. Bark grayish-brown, deeply fissured, slash pink, usually smelling of garlic. Leaves deciduous, 25–50 cm long; leaflets opposite to alternate, in 6–10 pairs, ovate-lanceolate, often falcate, 7–15 × 3–5 cm, the base acute to truncate, the apex acuminate, usually glabrous. inflorescences terminal, 20–40 cm long, much branched, the branches spreading horizontally. Flowers unisexual (plants monoecious); calyx cup-shaped, irregularly toothed; petals 7–8 mm long, imbricate, greenish-yellow; stamens 2–3 mm long. Capsules oblong or ellipsoid, pendulous, 2–3.5 cm long, brown or brownish-gray with prominent small whitish lenticels. Seeds 2–3 cm long including wing. Fl (Aug); apparently rare, in forest on well-drained soil. *Acajou, acajou de Guyane* (Créole).

GUAREA L.

Trees or treelets. Indumentum of simple hairs. Leaves nearly always pinnately compound with terminal bud with intermittent growth, rarely without terminal bud (*G. silvatica*). Inflorescences axillary, ramiflorous, or cauliflorous, narrow panicles, racemes, or spikes. Flowers unisexual (plants dioecious); calyx 3–7-lobed or margin truncate; petals 4–6, free, usually valvate, rarely slightly imbricate; staminal tube cylindrical, the margin entire, crenate, or lobed, the anthers 8–12, attached within throat of staminal tube; nectary stipitate, usually expanded into collar below ovary; ovary 4–8-locular, the locules with 1–2 superposed ovules, the stigma discoid. Fruits 2–8 valved loculicidal capsules, the valves 1–2 seeded. Seeds fleshy, with thin vascularized, frequently red or orange sarcotesta.

1. Leaves without terminal bud. Capsules 2–3-valvate, constricted between seeds. . . . . *G. silvatica.*
1. Leaves with terminal bud. Capsules 4–8-valvate, not constricted between seeds.
   2. Small, unbranched treelet with long (27–32 cm), strongly bullate leaflets. Inflorescences cauliflorous, subsessile. Capsules deep red, fleshy, with numerous convoluted and anastomosing ridges. . . . . *G. michel-moddei.*
   2. Trees, or if small treelets then leaflets not bullate and smaller. Capsules not deep red, not fleshy, without convoluted and anastomising ridges.
      3. Capsules 7–8-valvate, 3.5–6 × 3–4 cm, 2 superposed seeds in each valve. Flowers with pubescent ovary. . . . . *G. grandifolia.*
      3. Capsules 4(6)-valved, usually much smaller, if as large then valves 1-seeded or ovary glabrous.
         4. Flowers with glabrous ovary. Capsules with pale lenticels, 2 superposed seeds in each valve. . . . *G. kunthiana.*
         4. Flowers with pubescent ovary. Capsules not lenticellate, valves 1-seeded.
            5. Large, buttressed trees. Inflorescences to 30 cm long, stout. Capsules 4–6-valved, 2.5–3.2 cm long. . . . . *G. gomma.*
            5. Small, unbuttressed trees or treelets. Inflorescences much shorter, slender. Capsules 4-valved, <2.5 cm long.
               6. Capsules 1–1.5 cm long, depressed globose, globose, or pyriform, the valves often obscurely ribbed or tuberculate, puberulous intermixed with coarse long hairs. . . . . *G. pubescens*
               6. Capsules ca. 2 cm long, globose to pyriform, the valves smooth, finely puberulous without coarse, long hairs. . . . . *G. scabra.*

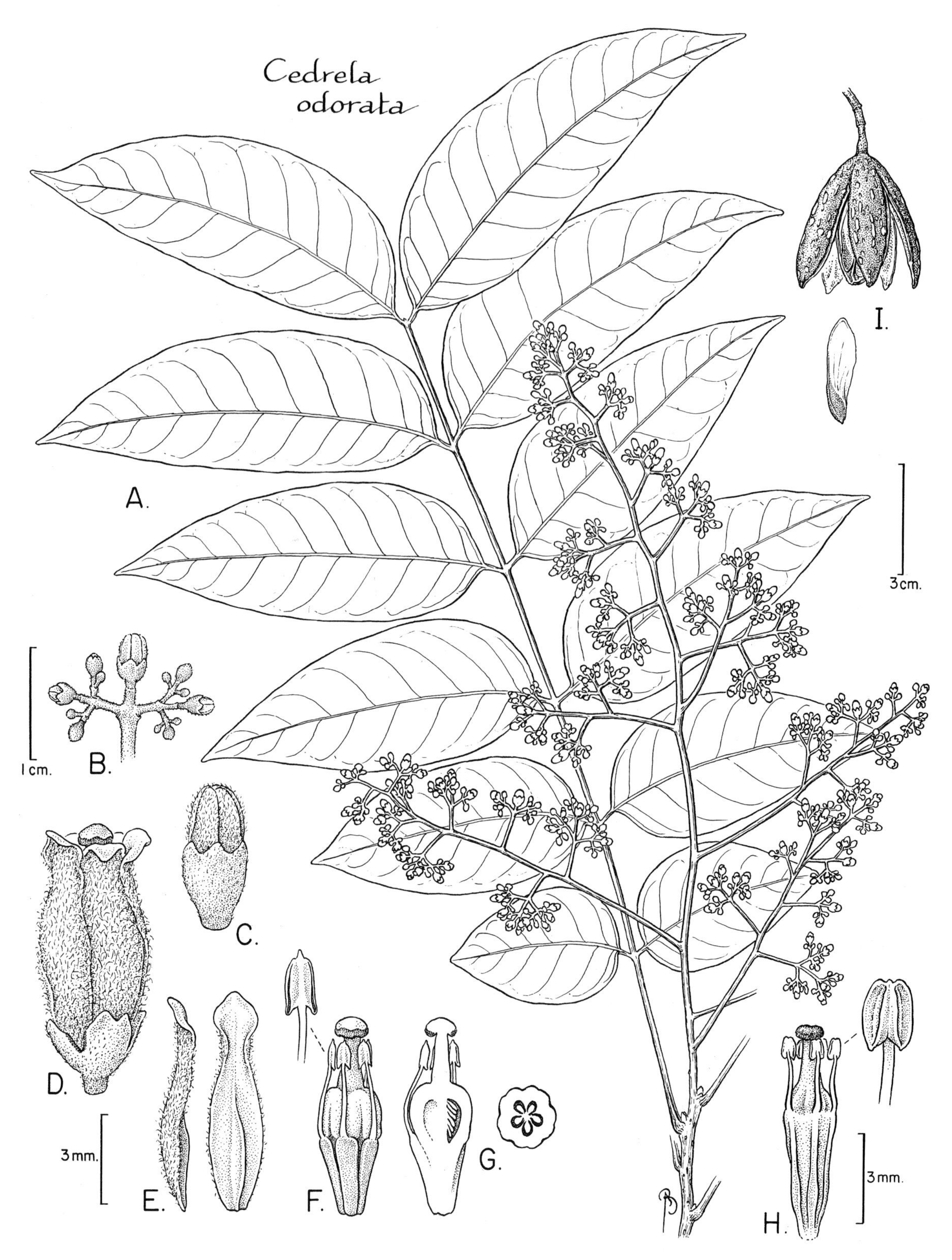

FIG. 199. MELIACEAE. *Cedrela odorata* (A–C, *Mori et al. 19208*; D–G, *J. de Bruijn 1031* from Venezuela; H, *Pennington & Sarukhán 9655* from Mexico; I, *Wurdack & Monachino 39721* from Venezuela). **A.** Part of stem with one leaf and terminal inflorescence. **B.** Part of inflorescence with flower buds at different stages. **C.** Flower bud showing cup-shaped calyx. **D.** Flower. **E.** Lateral view (near right) and adaxial view (far right) of petals. **F.** Pistillate flower with petals removed to show staminodes and pistil (right) and detail of staminode (above). **G.** Medial section of pistil (left) and transverse section of ovary (right). **H.** Staminate flower with petals removed to show stamens and pistillode (near right) and detail of stamen (above right). **I.** Capsular fruit (above) and winged seed (below).

Guarea gomma Pulle

Trees, to 35 m tall. Trunks buttressed to 1.5 m. Bark brown, scaling in large, irregular plates, the slash ca. 1 cm thick, reddish, with characteristic spicy aroma. Leaves with intermittent growth, with 9–12 pairs of leaflets; leaflets narrowly oblong or oblanceolate, to 22 × 6 cm, glabrous. Inflorescences axillary, to 30 cm long, the axes stout, the lateral branches short. Flowers: calyx cup-shaped, 2–3 mm long; petals 4–6, 8–12 mm long, cream-colored; staminal tube 6.5–8 × 2–4 mm, cream-colored; ovary 4–6-locular, puberulous. Capsules 2.5–3.2 cm long, 4–6-valved, globose to pyriform, red, smooth, finely puberulous, the valves 1-seeded. Seeds surrounded by bright orange sarcotesta. Fr (Jan, Jun, Aug, Oct); common, in forest on well-drained soils. *Bois jacquot* (Créole), *jatuauba preta* (Portuguese).

Guarea grandifolia DC. FIG. 200, PL. 89c

Emergent trees, to 40 m high. Trunks buttressed to 2 m high. Bark pale brown, scaling in irregular plates, the slash cream-colored to pinkish, to 1.5 cm thick. Stems massive, with dense terminal clusters of leaves. Leaves with intermittent growth, to 1.5 m long, with to 20 pairs of leaflets; leaflets broadly oblong, to 30 × 9 cm. Inflorescences axillary or ramiflorous, to 40 cm long, slender, pyramidal thyrses. Flowers: calyx cup-shaped, 2–6 mm long, reddish; petals 4–5, 10–14 mm long, cream-colored; staminal tube 9–11 × 3–5 mm, cream-colored; ovary 7–8-locular, pubescent. Capsules globose to pyriform, 7–8-valved, 3.5–6 × 3–4 cm, red, smooth or faintly lined, puberulous, the valves 2-seeded, the pericarp pure white adaxially, contrasting with bright orange seeds. Seeds pendulous on white funicles. Fl (Jun, Sep, Oct), fr (Apr, Oct); common, in forest on well-drained soil. *Carapa-oyac*, *encens* (Créole), *jatuauba preta* (Portuguese).

Guarea kunthiana A. Juss.

Small to medium-sized trees, to 26 m tall. Trunks cylindric, buttressed when larger, the buttresses to 40 cm tall. Bark grayish-brown, smooth, with some fine horizontal rings, the slash 4–5 mm thick, orange to reddish-brown with characteristic spicy aroma. Leaves with intermittent terminal growth, to 35 cm long; leaflets 2–6 pairs, broadly elliptic, 15–25 × 5–10 cm; secondary veins widely spaced, the higher order venation obscure. Inflorescences axillary, 3.5–25 cm long, broadly pyramidal thyrses. Flowers: calyx cup-shaped, 1.5–3 mm long, green or reddish; petals 4, 7–12 mm long, cream-colored; staminal tube 5.5–10 × 2–3 mm, cream-colored; ovary 4-locular, glabrous. Capsules ca. 3 × 2.5 cm, 4-valved, globose to ellipsoid, brown with conspicuous pale lenticels, glabrous, the valves with 2 seeds. Seeds superposed, bright orange. Fl (Apr, Nov), fr (May, Jun); common, in forest on well-drained soils and in disturbed areas.

Guarea michel-moddei T. D. Penn. & S. A. Mori
FIG. 201, PL. 89d,e

Unbranched treelets, to 4–5 m × 3–5 cm diam. Bark strongly suberized, thick, scaling. Leaves with intermittent terminal growth, in terminal cluster, spirally arranged, to 70 cm long; leaflets 4–7 pairs, oblong or oblanceolate, to 32 × 9 cm, strongly bullate; venation impressed adaxially, sharply prominent abaxially. Inflorescences cauliflorous, the flowers in small subsessile clusters. Flowers: calyx shallowly cup-shaped, 2 mm long, pink; petals 4, 8–9 mm long, creamy white; staminal tube ca. 7 × 3 mm, creamy-white; ovary 7-locular, densely pubescent. Capsules 4–4.5 cm diam., 7-valved, globose, covered with numerous deep, anastomosing and convoluted fleshy ridges, deep red-colored, the valves 1-seeded. Fl (Aug, Sep), fr (Jan, May); locally common, in understory of forest on well-drained soils.

Guarea pubescens (Rich.) A. Juss. PL. 89b

Understory treelets or trees, to 6 m. Leaves with intermittent terminal growth, to 40 cm long; leaflets 3–5 pairs, elliptic or obovate, to 30 × 7 cm, but usually much smaller, the apex often narrowly acuminate; tertiary venation fine, closely parallel, oblique. Inflorescences axillary or cauliflorous, variable in length, to 13 cm long, slender thyrses. Flowers: calyx inequally cup-shaped, 2–3 mm long, reddish; petals 4, 7–9 mm long, cream-colored, frequently tinged pink at apex; staminal tube 5–7 × 1.5–3 mm, cream-colored; ovary 4-locular, densely stiff-pubescent. Capsules 1–1.5 cm long, 4-valved, depressed globose, globose, or pyriform, the valves 1-seeded, often obscurely ribbed or tuberculate, dull reddish or purple. Fl (Mar, Apr, May, Aug, Sep, Oct), fr (Jan, Jun, Jul, Aug, Oct, Dec); common but scattered, in nonflooded forest. *Dangouti* (Bush Negro).

Guarea pubescens is represented in the vicinity of Saül by two distinct forms. The first is a small treelet with thick, suberous, fissured bark, leaves to 40 cm long, and leaflets to 30 × 7 cm . This form is cauliflorous with a small subsessile inflorescence and a depressed globose capsule with the valves shallowly ribbed. This form is represented by *Mori et al. 21652, 22028* and *Pennington & Mori 12141* (all NY). The second form is an understory tree lacking the suberized bark and with generally shorter leaves and shorter, broader leaflets. The inflorescence is a slender thyrse to 13 cm long. The fruit of this form is generally globose or pyriform and more or less smooth. The second form is represented by *Granville 10334, Hahn 3616, Mori & Gracie 18672, Mori & Pipoly 15626, Mori et al. 22213*, and *Phillippe et al. 26911, 27003* (all NY).

Guarea scabra A. Juss. PL. 90a

Trees, to 25 m tall. Bark peeling in irregular plates. Leaves with intermittent apical growth, to 35 cm long; leaflets in 4–7 pairs, oblong, to 16 × 5 cm, glabrous, the apex short acuminate. Inflorescences on stems below leaves, to 5 cm long. Flowers: calyx shallow, 1–1.5 mm long, red; petals 4, 10–12 mm long, pinkish; staminal tube 8–10 mm long, creamy white; anthers 8; ovary 4-locular, pubescent. Capsules ca. 2 cm long, globose to pyriform, red, 4-valved, the valves 1-seeded, smooth, finely puberulous. Fl (Sep), fr (Jan, Feb); in nonflooded forest.

Guarea silvatica C. DC. PL. 90b; PART 1: FIG. 10

Trees, to 25 m tall. Trunks buttressed to 75 cm, the bole cylindric. Bark brown, smooth or scaling in older trees. Leaves without terminal intermittent growth, to 35 cm long; leaflets in 3–4, opposite to alternate pairs, elliptic, to 18 × 6.5 cm, usually glossy on both surfaces. Inflorescences axillary, to 30 cm long, unbranched or with few lateral branches, the flowers arranged along axes in distinct cymose fascicles. Flowers: calyx shallow, 0.5–1 mm long; petals 4, 5–7.5 mm long, cream-colored; staminal tube 5–6 × 1.5–3 mm, cream-colored; ovary 2–3-locular, glabrous. Capsules 3.8–5 × 2.5–4 cm, 2(3)-valved, constricted between seeds, glabrous, the valves 1-seeded. Seeds 2.7–3 cm long, orange. Fr (Apr, Jun, Jul, Aug, Sep); forest on well-drained soil.

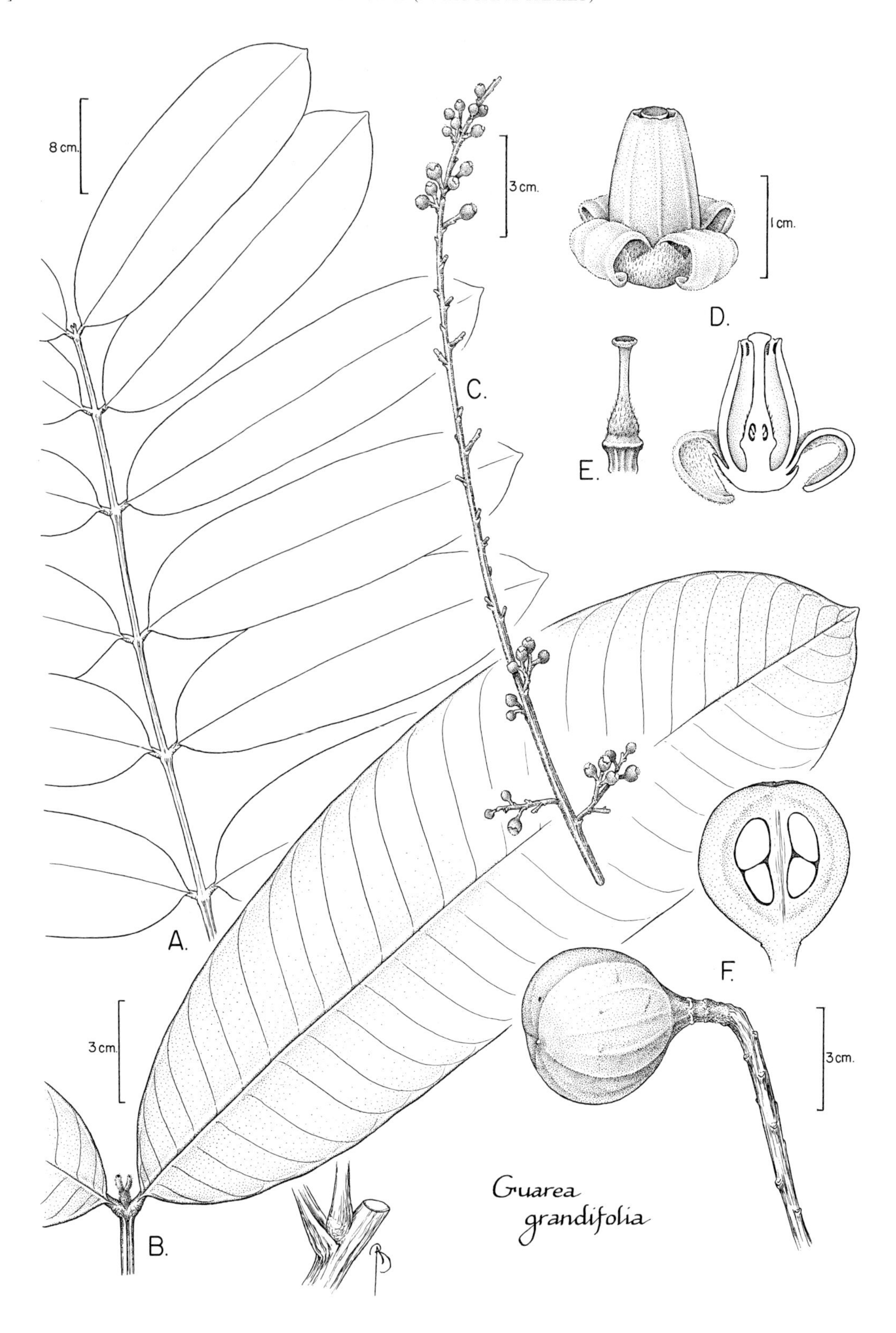

FIG. 200. MELIACEAE. *Guarea grandifolia* (A, B, F, *Mori et al. 15685*; C, *Mori & Boom 14845*; D, E, *Mori & Gracie 20828*). **A.** Part of leaf with base of rachis and part of stem (below). **B.** Indeterminate apex of rachis showing nascent leaflets and two expanded leaflets. **C.** Part of inflorescence. **D.** Lateral view of flower showing recurved petals and erect staminal tube. **E.** Pistil (left) surmounting gynophore and medial section (right) of flower showing axile placentation of ovules. **F.** Fruit on rachis (below) and medial section of fruit showing seeds (above).

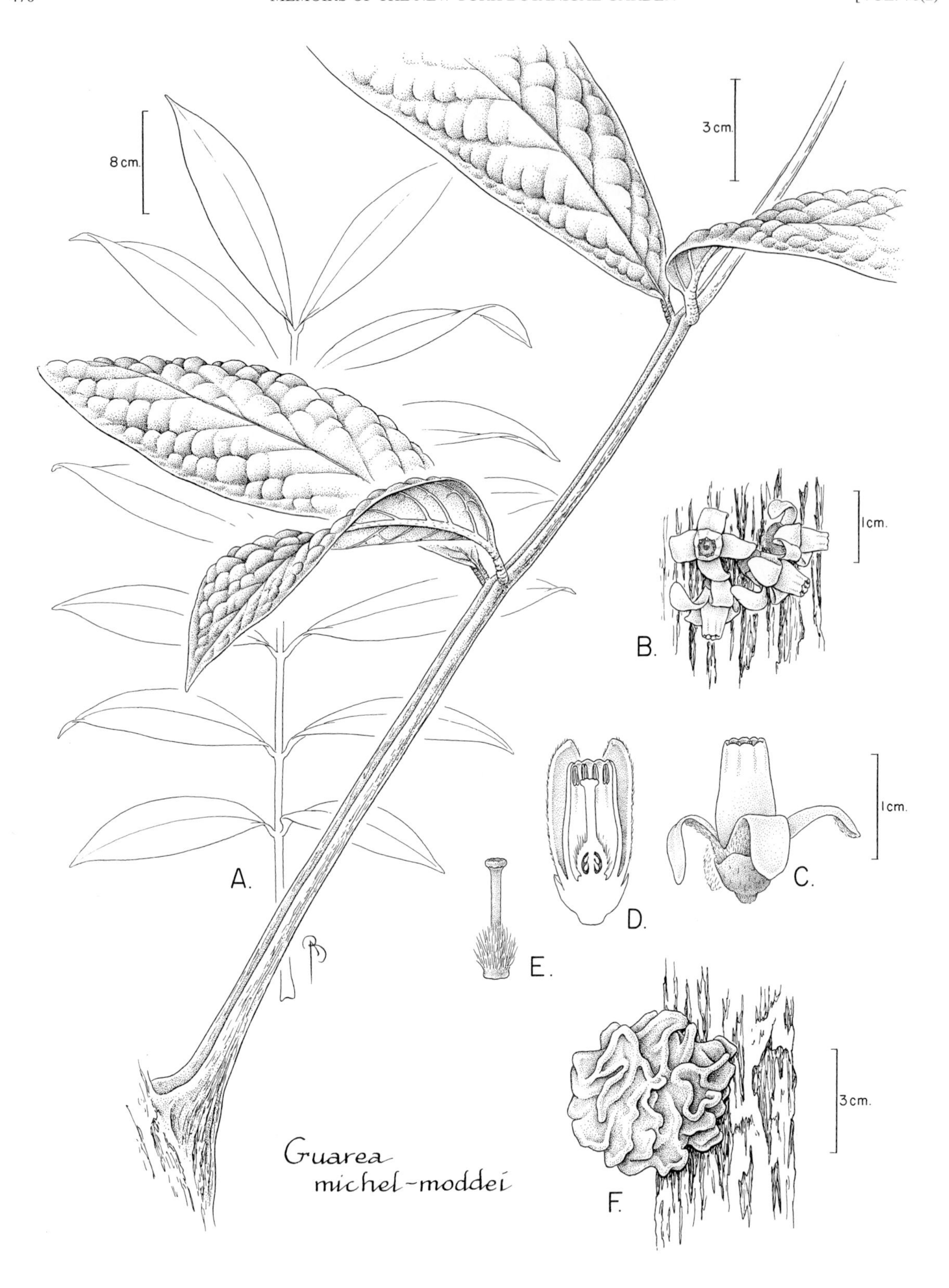

FIG. 201. MELIACEAE. *Guarea michel-moddei* (*Mori et al. 20963*). **A.** Part of stem and base of pinnately compound leaf with a diagram of the whole leaf in background. **B.** Cauliflorous flowers. **C.** Lateral view of whole flower showing spreading petals and erect staminal tube. **D.** Medial section of flower showing anthers inserted within the staminal tube, and ovary with axile placentation. **E.** Pistil. **F.** Fruit.

TRICHILIA P. Browne

Trees or treelets. Indumentum of simple, malpighiaceous, or rarely stellate hairs. Leaves pinnate or rarely trifoliolate. Flowers unisexual (plants dioecious). Flowers: calyx 4–6-lobed or sepals free; petals 4–5, free or partly united, imbricate or valvate; stamens 5–10, partly or completely fused into staminal tube, the anthers at apices of filaments or attached on margin of staminal tube; intrastaminal nectary present or absent, usually a fleshy annulus; ovary 2–3-locular, the locules with 1–2 ovules, the stigma usually capitate. Fruits 2–3-valved, loculicidal capsules, the valves 1–2-seeded. Seeds fleshy, partly or completely surrounded by fleshy arillode, or rarely by a sarcotesta.

1. Leaves trifoliolate (rarely with a few pinnate leaves on same plants). . . . . *T. pallida.*
1. Leaves pinnate.
    2. Petals free, aestivation imbricate; filaments partly or completely united.
        3. Indumentum of minute stellate hairs. . . . . *T. euneura.*
        3. Indumentum not of stellate hairs.
            4. Filaments 0.5–1.5 mm long, united only at base; petals 2–2.5 mm long. . . . . *T. micrantha.*
            4. Filaments 2–4.5 mm long, completely united into tube; petals 3–5.5 mm long. . . . . *T. septentrionalis.*
    2. Petals free or partly united, aestivation valvate; filaments completely united.
        5. Leaflets dimorphic, lowest pair(s) much reduced and often a different shape, or reduced to vestigial scales.
            6. Reduced basal leaflets represented by pair of subulate scales; leaflet base usually rounded. . . . *T. quadrijuga.*
            6. Reduced basal leaflets in 2–3 pairs, these orbicular or reniform; leaflet base tapering. . . . . *T. schomburgkii.*
        5. Leaflets monomorphic. . . . . *T. cipo.*

Trichilia cipo (A. Juss.) C. DC.

Understory treelets, to 10 m tall. Bark smooth, grayish-brown. Leaves pinnate, to 21 cm long; leaflets 7–11, alternate to subopposite, oblong to elliptic, to 15 × 4.5 cm, subglabrous. Inflorescences axillary, to 40 cm long, lax-branched, few- to many-flowered panicles. Flowers: calyx shallowly cup-shaped, 0.5–1.5 mm long; petals 4–5, 2–2.5 mm long, partially fused, valvate, greenish cream-colored; stamens 7–8, with filaments completely fused into tube; ovary 3-locular, the locules biovulate. Capsules 1.3–2 cm long, ellipsoid, smooth, appressed puberulous, grayish, 3-valved. Seeds 2, 1–1.5 cm long, partially surrounded by apical red arillode. Fl (Aug, Sep); in nonflooded forest. *Encens, encens blanc, encens petites feuilles* (Créole).

Trichilia euneura C. DC. PL. 90c,d

Understory trees, to 7(10) m tall. Leaves to 25 cm long; leaflets 5–7, alternate, oblong to broadly elliptic, to 15 × 7 cm, with scattered, minute stellate hairs abaxially. Inflorescences axillary, to 7 cm long, densely flowered thyrses. Flowers: calyx cup-shaped, 1–2 mm long; petals 5, free, imbricate, 4–5 mm long, yellowish-green; stamens 8 or 10, the filaments partly united into tube; ovary 3-locular, the locules with 2 collateral ovules. Capsules ellipsoid to obovoid, 2.7–3.8 × 1.5–1.6 cm, densely stellate puberulous, 3-valved. Seeds solitary, surrounded by orange sarcotesta. Fl (Sep, Oct), fr (Feb); occasional, in nonflooded forest at low elevations and in forest at higher elevations on Mont Galbao.

The higher elevation collection (*Boom 10825*, CAY, NY) has narrower leaflets and a much denser stellate indument than found in lowland collections.

Trichilia micrantha Benth.

Small to medium-sized trees, to 20 m tall. Trunks unbuttressed, the bole cylindric. Bark smooth, gray, the slash orange, 2 mm thick. Leaves imparipinnate, to 30 cm long; leaflets 9, opposite, elliptic or oblong-elliptic, to 16 × 6 cm, subglabrous, drying blackish. Inflorescences clustered around shoot apices in axils of undeveloped leaves, to 15 cm long, with spreading branches. Flowers: calyx shallow, to 0.75 mm long; petals 5, 2–2.5 mm long, free, imbricate, yellowish-green; stamens 10, the filaments fused at base, 0.5–1.5 mm long, pubescent above; ovary 3-locular, the locules uniovulate. Capsules broadly ellipsoid, 1–2.3 cm long, sericeous, yellowish, 3-valved. Seeds 1–3, surrounded by orange-red arillode. Fl (Apr, May); occasional, in nonflooded forest and in secondary forest.

Trichilia pallida Sw.

Medium-sized trees, to 18 m tall. Trunks with small buttresses, the bole angular or cylindric. Bark smooth, lenticellate, pale gray. Leaves trifoliolate (rarely a few pinnate), the rachis to 7 cm long; leaflets elliptic, to 15 × 7 cm, the lateral leaflets usually smaller, glabrous, drying pale green, the apex acuminate. Inflorescences axillary and ramiflorous, subfasciculate, to 2.5 cm long. Flowers: calyx shallow, 1–2 mm long; petals 4, 3–5.5 mm long, imbricate, cream-colored; stamens 8, 2–3.5 mm long, pubescent, the filaments partially fused; ovary 3-locular, the locules biovulate. Capsules ovoid to globose, 1–2 cm long, greenish-yellow, smooth, densely pubescent, valves 3. Seeds black, shiny, partially surrounded by red, fleshy arillode, usually only 1 developing. Fr (Apr); occasional, in nonflooded forest.

Trichilia quadrijuga Kunth

Trees, to 20 m tall. Trunks unbuttressed, the bole slightly fluted at base. Bark grayish-brown, scaling in long irregular pieces. Leaves pinnate, to 30 cm long; leaflets 9–11, with pair of linear to subulate vestigial leaflets 1–2 mm long attached 1–2.5 cm above base of petiole, alternate to subopposite, usually oblong or obovate, to 12 × 3–4 cm, abaxial surface with scattered, appressed, dibrachiate hairs, the base rounded. Inflorescences axillary, to 15 cm long, slender or broad branched thyrses. Flowers: calyx 0.5–1 mm long, shallow cup-shaped; petals 4–5, 3–4 mm long, free or partly fused, valvate, whitish-cream colored; stamens 8–10, the filaments completely united into tube 1–2 mm long; ovary 3-locular, the locules biovulate. Capsules ellipsoid or narrowly obovoid, 2–3 cm long, pinkish-red, 3-valved, granular-papillose with some appressed hairs. Seeds 1–2, surrounded by bright red arillode. Fl (Sep), fr (Mar); in nonflooded forest.

Trichilia schomburgkii C. DC.

Trees, to 20 m tall. Bark smooth, grayish. Leaves pinnate, to 25 cm long; leaflets 5–7, with 2–3 pairs of greatly reduced orbicular or reniform leaflets ("pseudostipules") clasping base of petiole, alternate to opposite, usually oblanceolate or oblong, to 22 × 8 cm, subglabrous, the base tapered, the apex attenuate. Inflorescences axillary, to 25 cm long, sparsely or profusely branched. Flowers: calyx shallow, cup-shaped, 0.5–1 mm long; petals 4–5, free or partly fused, 3.5–4.5 mm long, valvate, greenish-cream-colored; stamens 8, the filaments completely united into tube 2–3 × 1–2 mm long; ovary 3-locular, the locules biovulate. Capsules ovoid or ellipsoid, 2–3.5 cm long, smooth, grayish green, appressed puberulous, 3-valved. Seeds 1–2, 1.3–1.5 cm long, partially covered by apical arillode. Fl (Aug); in nonflooded forest.

Trichilia septentrionalis C. DC.

Small to medium-sized trees, to 15 m tall. Bark smooth, grayish-brown. Leaves imparipinnate, to 38 cm long; leaflets 7–9, opposite, broadly oblong, the terminal leaflet usually broadly elliptic, to 25 × 9.5 cm, subglabrous. Inflorescences axillary, erect, much branched panicles, to 30 cm long. Flowers: calyx cup-shaped, 1.5–2.5 mm long, of 5 free, strongly imbricate sepals; petals 5–7, free, 3–5.5 mm long, strongly imbricate, greenish cream-colored; stamens 8–10, the filaments completely united into tube 2–4.5 × 1–2.5 mm; ovary 3-locular, locules uniovulate. Capsules obovoid to ellipsoid, 1.3–3.2 cm long, smooth, densely puberulous, reddish. Seeds 1–2, 1–2.5 cm long, completely surrounded by red arillode. Fl (Aug), fr (Feb, Jul, immature); common, in nonflooded forest.

MENDONCIACEAE (Mendoncia Family)

Dieter C. Wasshausen

Herbaceous or suffrutescent twining (counterclockwise) shrubs or lianas (our species). Stipules absent. Leaves simple, opposite. Inflorescences in leaf axils, the flowers solitary or clustered; pedicels present; bracteoles (bracts as previously used by this and other authors) 2, large, flat or keeled, spathelike, variously shaped and vestured, valvate, often partially connivent or connate, green, often filled with liquid when enclosed around buds, widely spreading under fruit. Flowers zygomorphic, bisexual; calyx inconspicuous, annular or cupular; corolla sympetalous, ± hypocrateriform, whitish, greenish, or reddish, often with purplish markings adaxially, the tube cylindric to funnel-form, the limb subequally 5-lobed or bilabiate; stamens 4, didynamous, the anthers bilocular, with subequal locules; ovary superior. Fruits drupaceous, ovoid to ellipsoid. Seeds 1–2.

The Mendonciaceae are often included in the Acanthaceae, but they lack both the cystoliths and the specialized mechanism of seed-dispersal which characterize that family. Separation of the Mendonciaceae from the Acanthaceae is also supported by evidence from molecular biology (McDade et al., 2000).

Bremekamp, C. E. B. 1938. Acanthaceae. *In* A. Pulle (ed.), Flora of Suriname **IV(2):** 166–256. J. H. De Bussy, Amsterdam.

MENDONCIA Vand.

Characters the same as for the family in our area.

1. Bracteoles narrowly oblong, 8–9 mm wide. Corollas bright red or purplish. . . . . *M. hoffmannseggiana.*
1. Bracteoles oblong, oblong-elliptic to ovate, 8–16 mm wide. Corollas not bright red or purplish.
   2. Leaf blades glabrous or nearly glabrous. Bracteoles glabrous, ovate, 14–20 mm long. . . . . *M. glabra.*
   2. Leaf blades hirsute or sparsely pubescent. Bracteoles hirsute, oblong or oblong-elliptic to ovate, 20–41 mm long.
      3. Leaf blades narrowed, acuminate, or mucronulate at apex. Bracteoles oblong-elliptic to ovate, 20–25 × 8–16 mm, hirsute, the trichomes conspicuous, tawny, 3.5 mm long, the apex rounded or subobtuse. . . . . *M. bivalvis.*
      3. Leaf blades broadly rounded, abruptly acuminate to short-cuspidate at apex. Bracteoles oblong, 30–41 × 9–13 mm, yellowish-brown hirsute, the trichomes inconspicuous, 1 mm long, the apex rounded or obtuse and apiculate. . . . . *M. squamuligera.*

Mendoncia bivalvis (L. f.) Merr.

Leaves: blades ovate to oblong-ovate or elliptic, 6–15 × 3.5–4.5 cm, sparingly hirsute, the apex narrowed, acuminate or mucronulate. Inflorescences: pedicels densely pilose, the bracteoles oblong-elliptic to ovate, 20–25 × 8–16 mm, thin, veiny, tawny hirsute, the trichomes conspicuous, 3.5 mm long, the apex obtuse or rounded and mucronulate. Flowers 25 mm long; petals creamy white, tinged yellow distally. Fruits oblong to ovoid drupes, 15–18 × 9–12 mm, compressed, oblique, deep purple when ripe. Fl (Feb), fr (Feb); occasional, disturbed and open areas in forests.

Mendoncia glabra (Poepp. & Endl.) Nees PL. 90e

Leaves: blades elliptic-ovate, 6–10 × 2.5–5.5 cm, membranous, glabrous or nearly glabrous, the apex narrowed, acuminate, often mucronulate. Inflorescences: pedicels slender, glabrous, the bracteoles green, ovate, 14–20 × 12–18 mm, glabrous, the apex acute or slightly apiculate. Flowers 20–25 mm long; petals creamy white or white with purple splotches, or with dark purple veins at throat. Fruits ellipsoid to ovoid drupes resembling elongate olives 15 × 8–9 × 7 mm, dark purple when ripe. Fl (Feb), fr (Jan, Feb, Jul); occasional, in nonflooded forests.

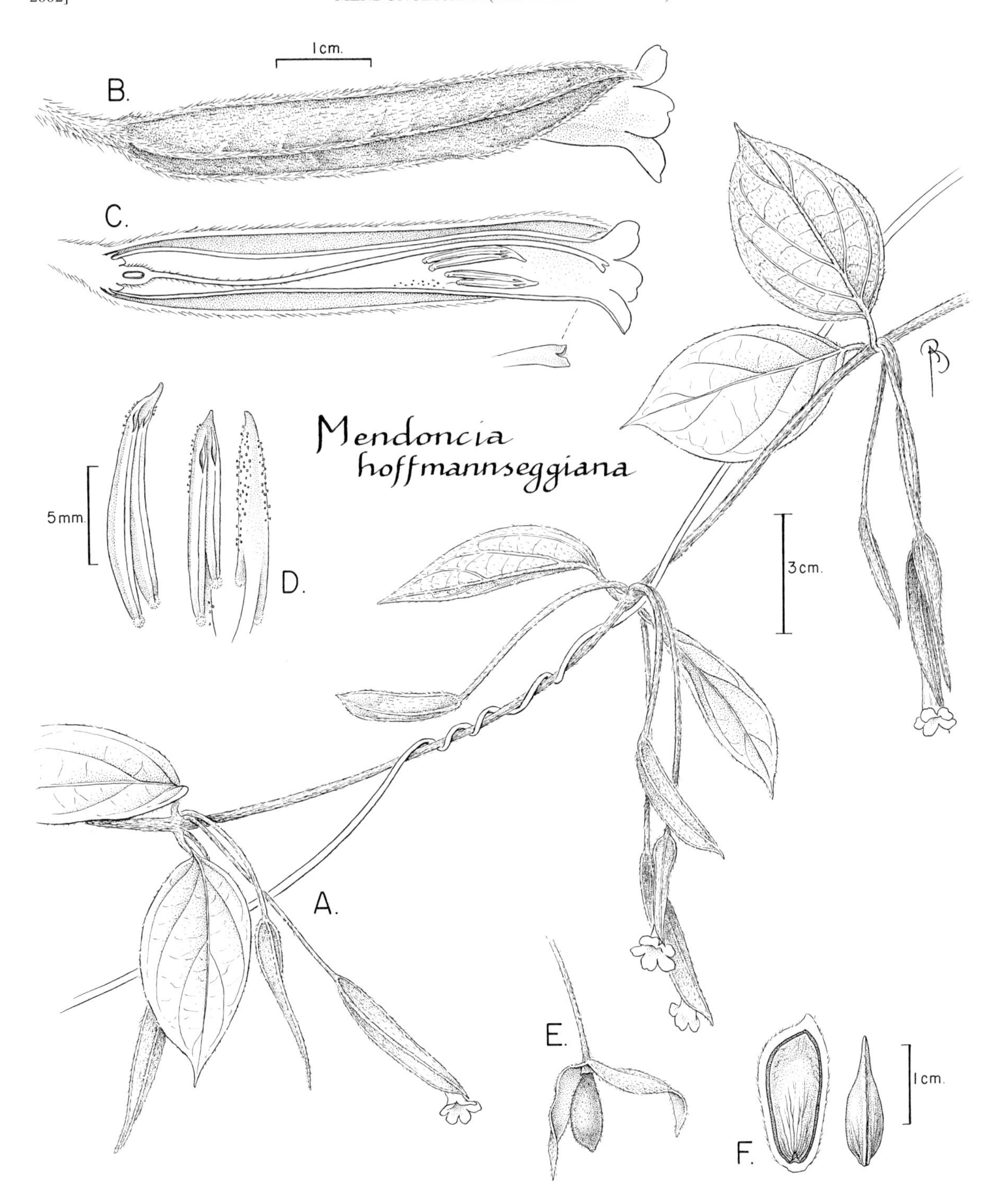

FIG. 202. MENDONCIACEAE. *Mendoncia hoffmannseggiana* (A–D, *Mori & Gracie 21202*; E, F, *Skog & Feuillet 5684*). **A.** Part of leafy stem with pendent inflorescences twining around another plant. **B.** Lateral view of flower. **C.** Medial section of flower. **D.** Anthers. **E.** Fruit subtended by two bracteoles. **F.** Medial section (near right) of fruit showing seed and seed (far right).

Mendoncia hoffmannseggiana Nees FIG. 202, PL. 90f

Leaves: blades narrowly to broadly elliptic, 5–12 × 3.7–8 cm, rather thin, minutely scabrous adaxially, yellow-brown hirsute abaxially, the apex abruptly acuminate, terminating in awn 2–4 mm long. Inflorescences: pedicels slender, densely yellowish-brown pilose, the bracteoles narrowly oblong, subfalcate, 30–40 × 8–9 mm, yellowish-brown hirsute, the trichomes inconspicuous, 1 mm long, the apex abruptly acute or acuminate. Flowers 35–55 mm long; petals bright red or purplish, the lobes 5–6 × 3 mm, emarginate, the lower part pinkish-red. Fruits obliquely obovate drupes, 15 mm diam., sparingly puberulous, turning from green to dark purple to blackish at maturity, succulent when ripe, the apex acute, tipped by part of persistent style. Fl (Jan, Feb, Mar, Aug), fr (Feb, Jul); common, at forest margins.

Mendoncia squamuligera Nees

Leaves: blades broadly elliptic to oblong, 8.5–10 × 5–6 cm, yellowish-brown hirsute, especially below, the apex abruptly acuminate to short-cuspidate. Inflorescences: pedicels pubescent; bracteoles oblong, 30–41 × 9–13 mm, rather firm, minutely punctate, yellowish-brown hirsute, the trichomes 1 mm long, the apex rounded or obtuse and apiculate. Flowers 50–60 mm long, the petals white or white with lilac streaks. Fruits oblong-obovate, lenticular-compressed drupes, 16–18 × 8–9.5 mm, brownish to purplish-black when ripe. Fl (Feb); occasional, in nonflooded forest.

MENISPERMACEAE (Moonseed Family)

Rupert C. Barneby

Lianas or treelets (*Abuta grandiflora*) in our area, some arborescent or subherbaceous outside our area. Stems terete or anomalously thickened laterally and becoming ribbonlike. Indumentum of simple trichomes or lacking. Stipules absent. Leaves simple, alternate; petioles commonly pulvinate at each end; blades with margins entire or seldom crenulate, or palmatilobate; primary and secondary venation either palmate or pinnate. Inflorescences unisexual (plants dioecious), either axillary, serially supra-axillary, or cauliflorous, the flowers bracteate, borne singly or fasciculate in spikes, racemes, or panicles. Flowers small, greenish, whitish or dull reddish; perianth commonly imbricate in cycles of 3; sepals 6 to many, free or partly united; petals 6 or lacking. Staminate flowers: androecium generally 6(3)-merous, or less often 12-merous and upward, or 1-merous, the filaments either free or united into column, the anthers bilocular, dehiscent by slits. Pistillate flowers: androecium staminodal or lacking; gynoecium apocarpous, the carpels commonly 3 (in *Sciadotenia* 6 or more), sessile on a torus or sometimes elevated on a gynophore; ovary superior, following fertilization commonly dilated on abaxial side and becoming hippocrepiform or annular but in some genera remaining symmetrical and straight, the stylar scar remaining sub-basal, the stigma linguiform; placentation submarginal, the ovules 2, but only 1 developing into seed. Fruits drupaceous monocarps; exocarp thin or coriaceous, red, orange, or yellow, the mesocarp mucilaginous or fibrous-mealy, the endocarp crustaceous, bony or woody, the interior wall either smooth or often engraved or ornamented with processes, the internal surface often intruded adaxially from various directions into cavity as a septum or as an externally concave condyle. Seeds with endosperm either continuous, ruminate, or lacking, or when lacking the horny cotyledons assuming storage function; embryo conforming to curvature of endocarp wall and to that of condyle when present, the radicle very short, the cotyledons either thick and horny, then pressed face to face, or vermiform, or foliaceous and then divaricate in one plane.

Generic and specific determination within the Menispermaceae is made more problematic by the lack of a modern circumtropical taxonomic revision. Consequently, some collections of this family are not yet identified to genus (see checklist at http://www.nybg.org/bsci/french_guiana). Continued botanical exploration will hopefully resolve the identity of these collections and will likely result in the addition of new records. Several species are likely to appear in the vicinity of Saül, though they have not yet been collected. These include *Cissampelos andromorpha* DC., *C. pareira* L., *C. tropaeolifolia* DC., *Disciphania unilateralis* Barneby, and *Orthomene schomburgkii* (Miers) Barneby & Krukoff.

Diels, L. 1966. Menispermaceae. *In* A. Pulle (ed.), Flora of Suriname **II(1):** 123–131. E. J. Brill, Leiden.

Jonker, F. P. 1966. Menispermaceae. *In* A. Pulle (ed.), Flora of Suriname **II(1):** 476–477. E. J. Brill, Leiden.

Ott, C. 1996. Verbreitung, Systematik, und Ökologie der Menispermaceae in Ecuador. Ph.D. dissertation, Johannes Gutenbergen-Universität, Mainz.

1. Plants staminate, in flower.
    2. Sepals imbricate, in generally >3 cycles of 3, strobiliform, silky-puberulent. *Sciadotenia.*
    2. Sepals imbricate or valvate, in 2 or 3 cycles of 3, not strobiliform, not silky-puberulent.
        3. Petals absent. . . . . *Abuta.*
        3. Petals present.
            4. Leaf-blades whitened abaxially with a dense coat of minute matted hairs. . . . . *Curarea.*
            4. Leaf-blades often glabrous abaxially, if pubescent thinly so, or loosely tomentulose.
                5. Innermost cycle of sepals united through 2/3 length into solid column. . . . . *Elephantomene.*
                5. Sepals either all free, or all shortly united into a cupule (*Disciphania moriorum*).
                    6. Leaf blades pinnately veined. Petals membranous, incurved, but not induplicate around opposed filament, not packed into a button-like pseudodisc. . . . . *Telitoxicum.*
                    6. Leaf blades palmately veined. Petals fleshy, induplicate around opposed filament, packed into a button-like pseudodisc [genera more easily distinguished by their fruits].
                        7. Flowers sessile; sepals of equal size. . . . . *Disciphania.*
                        7. Flowers pedicellate; sepals of unequal size, the outer set much smaller than the inner.
                            8. Petioles of longer leaves <2 cm long; blades widely obtuse. . . . . *Orthomene.*
                            8. Petioles of longer leaves >2 cm long, much >2 cm long if blade obtuse.
                                9. Leaf-blades broadly rounded at apex. . . . . *Caryomene.*
                                9. Leaf-blades at least shortly acuminate at apex.. . . . . *Anomospermum.*
1. Plants pistillate, in fruit [pistillate plants in flower are best identified by comparison in a herbarium].
    10. Seeds without endosperm, the cotyledons horny or fleshy, filling drupe-cavity.

11. Carpels of each flower 3. . . . . *Curarea*.
11. Carpels of each flower 9–15. . . . . *Sciadotenia*.
10. Seeds with endosperm, the cotyledons not horny or fleshy.
12. Drupes elevated on gynophores united at base into common column, free distally. . . . . *Elephantomene*.
12. Drupes sessile on torus.
13. Stylar scar terminal. Seed straight.
14. Endocarp longitudinally winged; endosperm not ruminate; cotyledons foliaceous. . . . . . *Disciphania*.
14. Endocarp not winged; endosperm ruminate; cotyledons linear. . . . . *Orthomene*.
13. Stylar scar lateral or almost basal. Seed incurved.
15. Stylar scar lateral to long axis of fruit. Seeds J-shaped in medial section, folded upon a condyle arising midway between pedicel and stylar scar and descending obliquely into seed-cavity. . . . . *Anomospermum*.
15. Stylar scar at or near base of drupe. Seeds inversely U-shaped in medial section, folded upon a condyle vertically intruded from base of drupe.
16. Inflorescences of solitary flowers. Interior wall of heavily ligneous endocarp armed with rows of compressed teeth interdigitating with lamellae of endosperm. Lamellae lacking integument. . . . . *Caryomene*.
16. Inflorescences racemosely several-flowered. Interior wall of endocarp unarmed. Lamellae of endosperm separated by membranous integuments.
17. Leaf-blades pinnately veined. . . . . *Telitoxicum*.
17. Leaf-blades 3–5-pliveined from insertion of petiole. . . . . *Abuta*.

ABUTA Aubl.

Lianas, some fertile when shrubs or treelets 1.5–4 m tall, but most flowering in forest canopy. Leaves glabrous, pilosulous, or tomentose; blades with apex obtuse to shortly acuminate, the margins entire or irregularly crenulate; venation with 3–5 principal veins arising from base. Inflorescences from year-old stems or from the leafless stem. Staminate inflorescences paniculate, the flowers borne singly along short secondary axes. Pistillate inflorescences racemose, several-flowered. Flowers of both sexes 6-sepalous; sepals in two whorls, the 3 outer ones usually much smaller than valvate inner ones; petals lacking. Staminate flowers: androecium 6-merous, the filaments either free or some connate. Pistillate flowers: gynoecium 3-merous, the carpels sessile on drum-shaped or globose receptacle, often only one maturing. Drupes sessile on torus, obovoid or oblong-obovoid, somewhat laterally compressed, the stylar scar remaining close to receptacle, the endocarp bent over an erect, basal condyle into an almost closed horseshoe, its interior wall smooth. Seeds: endosperm ruminate, its lamellae separated by membranous integuments; embryo vermiform.

1. Leaf blades glabrous. Drupes glabrous.
2. Commonly shrubs or treelets, 1.5–4 m tall in our area, but potentially high climbing lianas. Primary lateral veins of leaf blades divergent from midrib at very base of blade, the midrib (except for faint scalariform secondary venulation) simple. Inflorescences glabrous. . . . . *A. grandifolia*.
2. Canopy lianas. Primary lateral veins of leaf blades divergent from midrib a little beyond base of blade, the midrib giving rise beyond middle to 1–2 pairs of strong, incurved-ascending secondary veins. Inflorescences gray-puberulent or -pilosulous. . . . . *A. imene*.
1. Leaf blades abaxially either tomentose or strigose along principle veins. Drupes sericeous or velutinous, at least when young.
3. Leaf blades densely tomentose-velutinous overall abaxially. . . . . *A. rufescens*.
3. Leaf blades silky-strigose along principle veins, glabrous between them abaxially. . . . . *A. bullata*.

Abuta bullata Moldenke Pl. 91a

High-climbing lianas. Main stems to 20 cm diam, the young stems silky-strigose. Leaves olivaceous; petioles 2.5–9.5 cm long, silky-strigose; blades ample, ovate- or obovate-elliptic, 12–27 × 6.5–16 cm, glabrous except along abaxially strigulose primary veins, the base broadly cuneate or rounded, the apex abruptly shortly acuminate, the margins entire; venation with 3 principal veins arising from base, the midrib giving rise on each side to (2)3 major incurved-ascending and many scalariform secondary veins, these all impressed or depressed on blade's adaxial surface, prominent abaxially. Inflorescence axes silky-strigose. Staminate inflorescences: primary axes slender, ± 8–15 cm long, the secondary and tertiary axes scarcely 1 cm long; pedicels <2 mm long. Pistillate inflorescences: peduncles 1 cm long; rachises 1.5–3.5 cm long; pedicels 5–7 mm long. Flowers: sepals gray-silky, ± 1.5 mm long. Drupes plumply oblong-ellipsoid, 28–35 × 21–27 mm, the exocarp sericeous when young, glabrescent and yellow at maturity. Fr (Jan); in nonflooded forest.

Abuta grandifolia (Mart.) Sandwith Fig. 203, Pl. 91b,c

Commonly treelets, 1.5–4 m tall (potentially climbing into canopy), glabrous throughout. Leaves: petioles 1.5–6.5 cm long; blades broadly or narrowly elliptic, ovate-, or obovate-elliptic, the longer ones 10–30 × 4–10 cm, the base generally cuneate, the apex

FIG. 203. MENISPERMACEAE. *Abuta grandifolia* (A–E, *Mori et al. 23206*; F–I, *Mori et al. 23236*; J, K, *Mori et al. 19129*). **A.** Apex of leafy stem with pistillate inflorescence. **B.** Staminate inflorescence. **C.** Detail of staminate inflorescence. **D.** Apical view of staminate flower. **E.** Adaxial view of inner sepal. **F.** Androecium (left) and single stamen (right). **G.** Detail of pistillate inflorescence. **H.** Apical view of pistillate flower. **I.** Lateral view of pistillate flower with sepals removed (above) and medial section of carpel (right). **J.** Infructescence with two fruits. **K.** Seed (right) and medial section of seed (above).

shortly or obscurely acuminate, the margins entire; venation with 3 strong veins from base, the midrib simple except for weak scalariform secondary venules. Inflorescences from year-old and older stems, 1–5 cm long, the axes filiform; peduncles 0.5–1 cm long; rachises 2.5–3.5 cm long; pedicels 1.5–4 mm long. Flowers greenish, nigrescent when dried; longer sepals (of both sexes) ± 1.5 mm long. Drupes commonly 1, rarely 2 or 3 developing from each flower, obovoid or oblong-obovoid, obtuse, 19–23 × 10–13 mm, sessile, the exocarp yellow when ripe, lustrous black when dried. Fl (May, Aug), fr (Jan, Mar, Jul, Aug, Sep); in understory of non-flooded forest. *Abuta*, *abuta branca*, *abutua* (Portuguese), *bois bandé canelle*, *canelle sauvage* (Créole).

Abuta imene (Mart.) Eichler

Canopy lianas. Leaves: petioles 1.5–6.5 cm long; blades broad-elliptic, oblong-elliptic, or ovate, the larger ones 10–17 × 4–9 cm, the base either rounded or cuneate, the apex shortly acuminate, the margins entire; venation with 3 principal veins from base but the lateral primary veins divergent from midrib a little above base of blade and the midrib itself giving rise, near and beyond mid-blade, to 1 or 2 pairs of strong, incurved-ascending secondary veins. Inflorescences of both sexes gray-puberulent or -pilosulous, the primary axes stiffly ascending. Staminate inflorescences: primary axes 4–12 cm long, the secondary axes 1 cm long

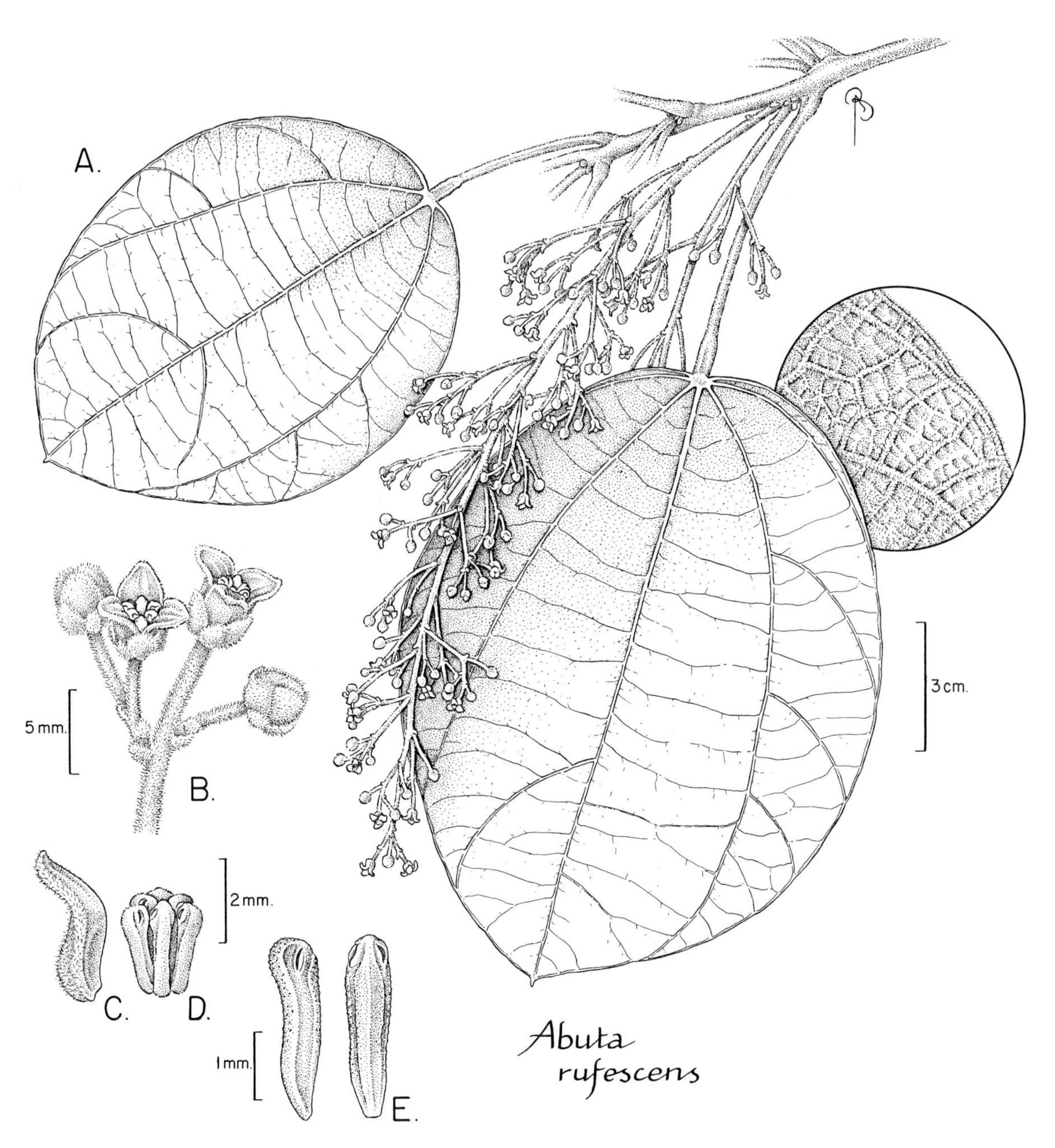

FIG. 204. MENISPERMACEAE. *Abuta rufescens* (*Mori et al. 23166*). **A.** End of stem with leaves and staminate inflorescence and insert of abaxial venation. **B.** Part of staminate inflorescence with two open flowers. **C.** Inner sepal. **D.** Androecium. **E.** Stamens.

or less, often recurved; pedicels often recurved. Pistillate inflorescences: peduncles 0.5–1 cm long; rachises 1.5–4.5 cm long; pedicels 2–4 mm long. Flowers: sepals gray-puberulent, ± 1–1.5 mm wide. Drupes commonly 1, rarely 2 or 3 from each flower, obovoid or oblong-obovoid, obtuse, 17–22 × 11–14 mm, sessile, the exocarp glabrous, yellow when ripe, lustrous black when dried. Fr (Feb); in nonflooded forest.

Abuta rufescens Aubl. FIG. 204, PL. 91d,e

Robust canopy lianas. Stems attaining 10 cm or more diam. Pubescence densely softly tomentulose or velutinous with short, lustrous white but in age tarnished or rufescent hairs on young stems, petioles, abaxial surface of leaf blades, ample inflorescences, perianth of all flowers, and drupes. Leaves: petioles of leaves associated with flowers 6–21 cm long; blades ample, suborbicular, mostly 8–15 cm diam., firm, those of shade-leaves and sapling-leaves to 30–40 cm diam., discolorous, smooth glabrous adaxially, firm, the base shallowly cordate, the apex obtuse or shortly acuminate, the margins irregularly crenulate; venation with 5 principal veins arising from base, immersed adaxially, deeply alveolate-reticulate, very prominent and tomentulose abaxially, the secondary veins incurved-ascending from midrib, 2–3 on each side, the connecting venules scalariform. Inflorescences: peduncles 1.5–3 cm long. Staminate inflorescences: primary rachises 11–22 cm long, the secondary axes 1–2.5 cm long; pedicels 4–8 mm long. Pistillate inflorescences 3–10 cm long. Flowers: sepals yellowish-green, the inner 3 sepals curved outward distally, 4–5 mm long, nearly twice as long as 6 outer ones. Drupes 24–29 × 15–19 mm, the exocarp densely white- or rufous-velutinous, orange when ripe; pedicels ± 1 cm long. Fl (Aug), fr (Sep); in nonflooded forest.

ANOMOSPERMUM Miers

Robust lianas flowering in canopy. Pubescence almost or entirely lacking. Leaves: petioles of longer leaves >2 cm long; blades stiffly coriaceous, the apex at least shortly acuminate; venation palmate. Inflorescences either axillary or shortly supra-axillary on contemporary or year-old stems; flowers of both sexes pedicellate. Staminate inflorescences pseudoracemose, the flowers either fasciculate at nodes or borne in short, few-flowered, sometimes subcymose clusters. Pistillate inflorescences 1-flowered, solitary or paired. Flowers: perianth of both sexes fleshy; sepals free, 2–3-seriate, the 3 inner ones longer, narrowly imbricate in bud; petals 6, uniform, fleshy, each induplicate and clasping a stamen or staminode, together forming a button-like pseudodisc; filaments free. Pistillate flowers: androecium staminodial; carpels 3. Drupes asymmetrically ovoid, sessile on torus, the stylar scar near midpoint of adaxial side, the exocarp coriaceous, green ripening orange, the mesocarp succulent, thin, the endocarp woody, either dimpled externally, then the dimples corresponding with ligneous projections intruded into copious ruminate endosperm, or smooth externally and internally. Seeds: embryo vermiform, threaded through the endosperm and with it evenly incurved through ± 3/4 circle over condyle descending obliquely into the cavity from near middle of their adaxial surface.

Anomospermum steyermarkii Krukoff & Barneby
PART 1: FIG. 10

Canopy lianas. Young stems black-purple, with short rufous pubescence. Leaves: petioles 2–7.5 cm long; blades ovate, 8–16 × 4.4–11 cm, with short rufous pubescence, especially on veins of abaxial surface, the base rounded or broad-cuneate, the apex shortly abruptly deltate-acuminate, the margins entire; venation with 3–5 principal veins arising from base, the midrib giving rise at and beyond mid-blade to 2–3 pairs of major incurved-ascending secondary veins. Staminate inflorescences scarcely known. Pistillate inflorescences 8–15 mm long. Flowers: 3 longer sepals (of either sex) ± 3 mm long; pseudodisc ± 4 mm diam. Drupes 31–35 × 25–30 × 20–24 mm; endocarp low-carinate around its abaxial periphery and dimpled in 2 vertical rows around each surface, the interior wall armed. Fr (Mar, Aug, Sep); in nonflooded forest.

CARYOMENE Barneby & Krukoff

Lianas, potentially high-climbing. Leaves: petioles of longer leaves >2 cm long, much >2 cm long if blade obtuse; blades ample, lustrous on adaxial surface and densely minutely papillate abaxially, the apex rounded; venation with 3–5 principal veins arising from base, finely reticulate. Staminate inflorescences unknown. Pistillate inflorescences axillary, 1(2)-flowered. Flowers unknown. Fruits massive pyriform drupes, sessile on torus, the stylar scar almost basal, the exocarp coriaceous, the mesocarp thin, pulpy, the endocarp extremely thick-walled, densely woody, folded over an erect condyle entering from the pedicel, the interior wall perforate and produced inward horizontally in the form of ligneous laminae inter-digitating with layers of ruminate endosperm. Seeds: endosperm laminae lacking integument; embryo vermiform, doubled onto itself into narrow hoop, the cotyledons linear, closely appressed, much longer than minute radicle.

Caryomene olivascens Barneby & Krukoff

High-climbing lianas, flowering in canopy. Leaves: petioles 5–17 cm long, those of shade-leaves much longer than those associated with flowers; blades broadly ovate, 10–23 × 8–13 cm, the base rounded or obscurely cordate, the apex depressed-deltate, the margins irregularly crenate and undulate; venation with 3–5 principal veins arising from base, the outermost pair sometimes weak, the midrib giving rise on each side, near and beyond mid-blade, to 2–3 strong incurved-ascending secondary veins, these all immersed on adaxial surface and abaxially prominent. Pistillate inflorescences 4 cm long. Drupes, including the solid, neck-like base, ± 4.5–5 × 3.5 × 3 cm, the exocarp shiny green turning yellow, black when dried, the mesocarp yellowish-white. Fr (Feb, May, Jul, Sep, Oct, Nov); in nonflooded forest.

CURAREA Barneby & Krukoff

Lianas. Leaves: blades notably discolorous, the abaxial surface gray- or white-felted; venation with 3–5 principal veins arising from base. Inflorescences either from contemporary leaf axils or from older defoliate stems. Staminate inflorescences cymose-paniculate. Pistillate inflorescences racemose. Staminate flowers: sepals 6, in two whorls, externally tomentulose, the outer 3 minute, the inner 3 valvate in bud; petals 3 or 6, obovate, shorter than inner sepals; stamens 6, the filaments free. Pistillate flowers: sepals as in staminate flower; petals 3; staminodia lacking; carpels 3, after fertilization each raised on drum-shaped carpophore. Drupes sessile on and deciduous from carpophore, oblong- or obovoid-ellipsoid, the stylar scar remaining subbasal, the endocarp hippocrepiform, its interior wall papery smooth, the condyle erect, basal. Seeds: endosperm lacking; embryo hippocrepiform, the cotyledons horny, filling drupe-cavity.

Curarea candicans (Rich.) Barneby & Krukoff

Lianas flowering in canopy or sometimes precociously fertile. Pubescence of minute matted hairs throughout, except for glabrous adaxial surface of leaves. Leaves: petioles 2.5–12 cm long; blades (those of shade leaves much larger than those associated with flowers), narrowly or broadly elliptic to ovate, the longer ones 9–28 × 5–11.5 cm, the base cuneate, the apex shortly acuminate to caudate, the margins entire; venation with 3–5 prominent veins arising from base, the primary nerves connected by subhorizontal secondaries. Staminate inflorescences 4–10 cm long, the secondary branchlets to ± 1.5 cm long; pedicels scarcely 2 mm long. Pistillate inflorescences simpler and shorter than staminate. Flowers: inner sepals ± 1.5 mm long; petals <1 mm long. Pistillate flowers: carpophores at maturity drum-shaped. Drupes oblong-ellipsoid, 17–27 × 11–21 mm, densely gray- or brownish-tomentulose. Fl (Nov); in nonflooded forest.

DISCIPHANIA Eichler

Slender lianas. Stems often corky. Leaves: blades membranous, either glabrous, setose, or pilosulous, the margins either entire or lobulate; venation palmate. Inflorescences axillary spikes. Flowers: perianth alike in both sexes, greenish; sepals 6, in two whorls, of subequal size, ± fleshy, either free or united into cupule; petals 6 (3 or lacking), fleshy, packed into a button-like disc, each induplicate around an opposed filament. Staminate flowers: androecium 3-merous, the filaments free, often dilated upward and incurved. Pistillate flowers: staminodia lacking; carpels 3. Drupes sessile on torus, grapelike, the stylar scar terminal, the exocarp thin, red when ripe, the mesocarp mucilaginous, the endocarp crustaceous or cartilaginous, straight (condyle lacking), dorsiventrally compressed, the broad surfaces low-convex, the narrow surfaces 3-keeled or -winged lengthwise. Seeds straight: endosperm thin, not lamellate; embryo straight, the cotyledons foliaceous.

1. Leaf blades cordate, 6–11.5 × 4–8 cm, the base not subtruncate, the margins entire. Staminate flowers when fully expanded nearly 3 cm diam. . . . . *D. moriorum*.
1. Leaf blades ovate, 4–7.5 × 2.5–5 cm, the base subtruncate, the margins generally entire but some produced on one or on both sides into a deltate lobe. Staminate flowers unknown but surely not as large. . . . *Disciphania* sp. A.

Disciphania moriorum Barneby FIG. 205, PL. 92a

Lianas. Young stems microscopically puberulent. Leaves: petioles 2.5–5 cm long; blades cordate, 6–11.5 × 4–8 cm, dark green, nigrescent when dry, scaberulous with short erect, basally thickened trichomes, the basal sinus wide and shallow, the apex acuminate, the acumen 1 cm long, the margins entire; primary venation palmate, of 5(7) veins, the midrib once forked on each side beyond mid-blade, the inner pair incurved-ascending well beyond midblade, all slenderly raised on each surface, the tertiary venules faint and distant. Staminate inflorescences axillary in upper leaf axils, spicate, 3–4-flowered; peduncles 3.5–6.5 cm long; rachises 3–12 mm long, the bracts ± 1 mm long, deflexed. Pistillate inflorescences unknown. Staminate flowers: perianth fleshy, nearly 3 cm diam., green when fresh but nigrescent on drying, glabrous; sepals 6, in 2 subequal cycles, ovate-oblong-elliptic, 12–14.5 × 7–8 mm, at base united into shallow cup, the free blades at full anthesis tightly reflexed behind pseudodisc; petals fleshy, cuneiform, ± 5–6 mm wide at distal edge, contiguous but not connate; stamens 3, the filaments obliquely claviform, ± 2.5 mm long, produced beyond inwardly leaning anthers into horizontal point. Fl (Sep); known only from two collections (probably from same plant) in low area in forest.

Disciphania sp. **A**

Slender lianas of unknown length. Fertile stems ± 1.5 mm diam., black-spotted, glabrous. Leaves: petioles 20–35 × 0.7–1 mm; blades ovate, 4–7.5 × 2.5–5 cm, thin-textured, minutely scabrous-puberulent, nigrescent, dull, the base subtruncate, the apex acuminate, the margins generally entire but occasionally produced on one or on both sides into a deltate lobe ± 4 mm long; venation faint, the primary veins 5, the midrib weakly 1–3-branched near and beyond mid-blade. Staminate inflorescences unknown. Pistillate inflorescences ± 6-flowered. Flowers not seen. Infructescences: peduncles 2.5 cm long; rachises ± 1–2 cm long. Drupes broadly ellipsoid, grapelike, ± 15 mm long, the exocarp red, the mesocarp mucilaginous, the endocarp black, in broad view 12–13 × 8–9 mm, the adaxial and abaxial keels narrow, entire, the lateral and intermediate wings lacerate. Fr (Feb, Nov); in nonflooded forest.

Known only from pistillate plants (*Mori et al. 21542*, CAY, NY and *Oldeman 2005*, CAY) collected in central French Guiana from the vicinity of Saül. The erratically lobed leaves resemble those of the Peruvian *D. heterophylla* Barneby, but the basal angles of the leaf blades are broadly rounded, not acute.

ELEPHANTOMENE Barneby & Krukoff

Lianas, sometimes of gigantic size. Main stems whitish. Leaves: blades coriaceous, abaxially tomentulose; venation palmate-pinnate. Inflorescences axillary or serially supra-axillary on new growth. Staminate inflorescences paniculate. Pistillate inflorescences racemose. Staminate flowers: sepals 6, in two whorls, the outer cycle short, the inner cycle united through 2/3 length into solid column bearing at its apex 3 small valvate lobes; petals 6, membranous; stamens 6, erect, presence or lack of fusion unknown. Pistillate flowers unknown. Drupes 3 (unless some abort), obovoid or oblong-obovoid, a little laterally compressed, elevated on stout gynophores united at base into common column, free distally, the stylar scar remaining close to receptacle, the exocarp hard, minutely pallid-velutinous, closely enveloping hard woody endocarp (mesocarp lacking), the endocarp bent over an erect, basal condyle into an almost closed horseshoe, its interior wall smooth. Seeds: endosperm ruminate; embryo vermiform.

A monotypic genus described from Saül and previously thought to be endemic, but recently discovered in far western Amazonian Brazil and Ecuador (Barneby, 1993).

Elephantomene eburnea Barneby & Krukoff FIG. 206

Lianas, attaining 40–50 m long. Main stems to 25 cm diam. at base. Leaves: petioles (2.5)3–12 cm long; blades ample, broadly ovate-orbicular, (7)9–20 × (6)8–18.5 cm, glabrous, lustrous on adaxial surface, densely minutely pallid- or brownish-tomentulose abaxially, the base widely rounded or sinuately subtruncate, the apex widely obtuse, the margins entire to irregularly crenulate; venation of palmate primary veins, the outer pair weak and short, the inner pair produced straight to leaf-margin and giving rise on distal side to 4–5 secondary veins, the costa pinnately 5–7-branched on each side upward from below mid-blade, all primary and secondary

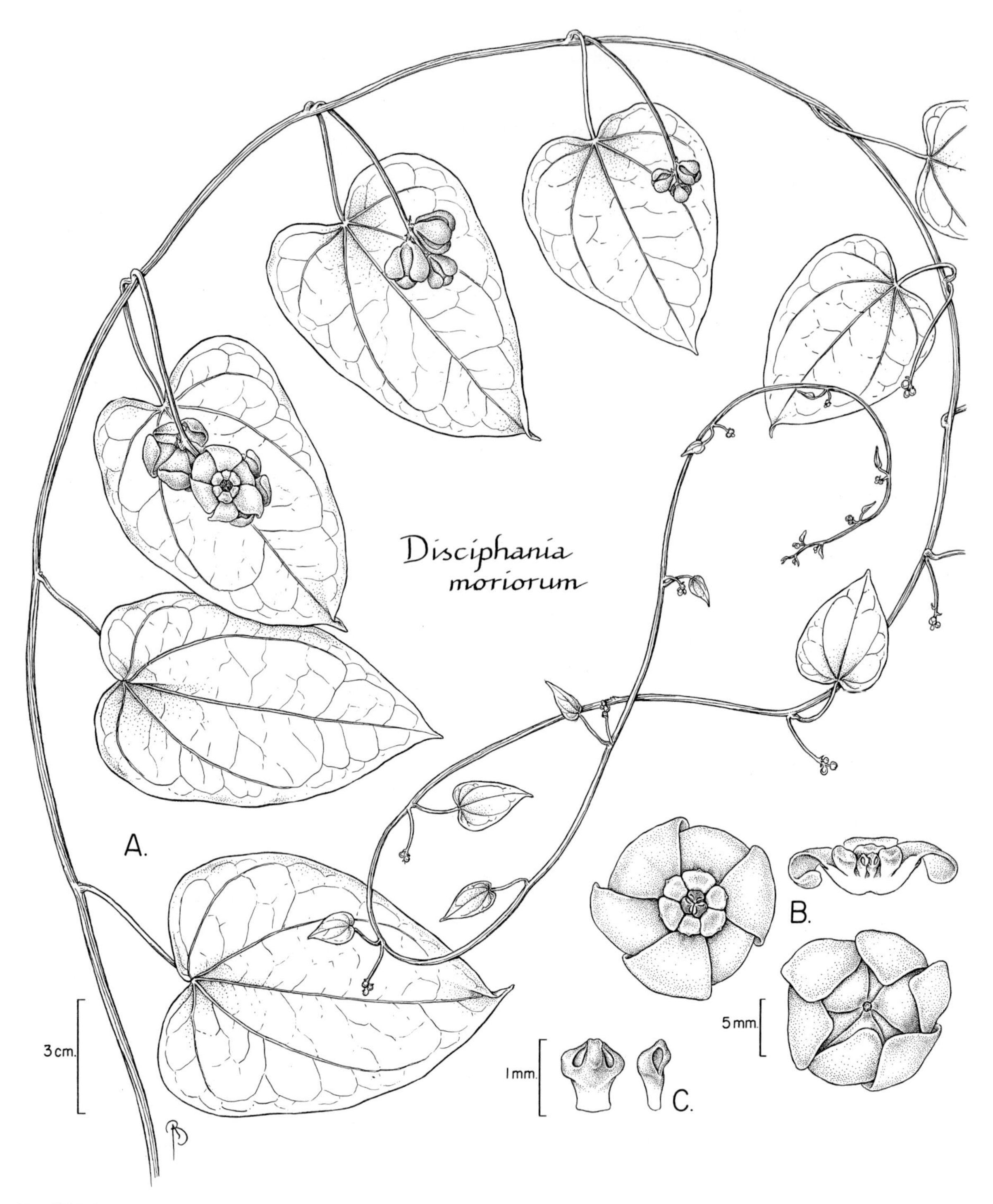

FIG. 205. MENISPERMACEAE. *Disciphania moriorum* (*Mori & Gracie 23980*). **A.** Apex of stem with axile inflorescences. **B.** Apical (left) and basal (right) views and medial section (above right) of staminate flowers showing two whorls of recurved sepals, thickened petals, and three stamens. **C.** Adaxial (left) and lateral (right) views of stamens. (Reprinted from R. C. Barneby, Brittonia 48(1), 1996.)

veins shallowly engraved on adaxial surface, cordlike abaxially. Staminate inflorescences: peduncles 0.5 cm long; primary rachises 17–27 cm long, the secondary axes 1.5 cm long or less; pedicels 2–5 mm long. Flowers: perianth light orange; sepals obtusely deltate, 1.5–1.8 mm long, the two cycles separated by a column ± 3.5 mm long; petals ± 0.6 mm long. Pistillate flowers: gynophores 1–2 cm long, united ± halfway into an inverted tripod. Drupes 45–57 × 25–30 mm, the exocarp light yellow-orange, the mesocarp fleshy, deeper yellow-orange. Fl (Jan, Aug), fr (May, Jul, Aug, Sep, Dec); in nonflooded forest.

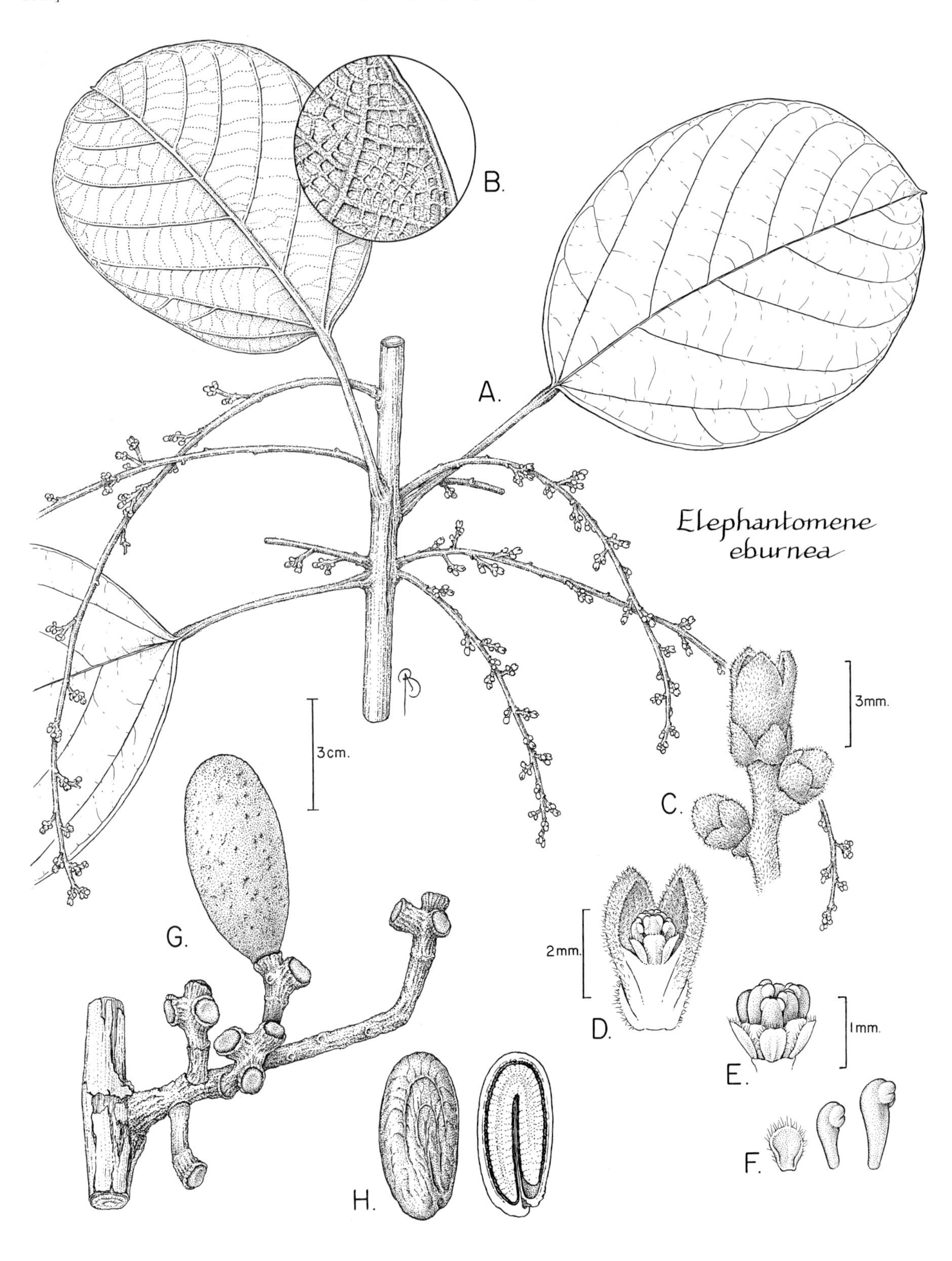

FIG. 206. MENISPERMACEAE. *Elephantomene eburnea* (A–F, *Mori et al. 23108*; G, *Mori et al. 18586*; H, *Mori et al. 23196*). **A.** Stem with leaves and staminate inflorescence. **B.** Abaxial view of leaf-blade showing venulation. **C.** Staminate stem of second order. **D.** Staminate flower, one large sepal removed. **E.** Staminate flower, all sepals removed. **F.** Petals and two stamens. **G.** Infructescence with fruit. **H.** Lateral view (near right) and medial section (far right) of endocarp.

ORTHOMENE Barneby & Krukoff

Robust lianas flowering in canopy. Pubescence almost or entirely lacking. Leaves: petioles of longer leaves <2 cm long; blades stiffly coriaceous, the apex broadly obtuse; venation palmate. Inflorescences either axillary or shortly supra-axillary on contemporary or year-old stems. Staminate inflorescences pseudoracemose, their flowers either fasciculate at nodes or borne in short, few-flowered, sometimes subcymose clusters. Pistillate inflorescences 1-flowered, solitary or paired; flowers of both sexes pedicellate. Flowers: perianth of both sexes fleshy; sepals free, 2–3-seriate, the 3 inner ones longer, narrowly imbricate in bud; petals 6, uniform, fleshy, each induplicate and clasping a stamen or staminode, together forming a button-like pseudodisc; filaments free. Pistillate flowers: androecium staminodial; carpels 3. Drupes symmetrically ellipsoid, sessile on torus, the stylar scar terminal; exocarp thinly coriaceous, smooth or verruculose, the endocarp smooth or externally pitted, laterally inverted into seed cavity in form of laminiform, vertical septum. Seeds straight; endosperm ruminate; embryo vermiform, the cotyledons linear.

Orthomene prancei Barneby & Krukoff FIG. 207, PL. 92c

Lianas. Stems of fertile branches densely sordid-pilosulous. Leaves: petioles of larger leaves 8–15 mm long; blades narrowly or broadly obovate, 5.5–9 × 3–5.5 cm, glabrate adaxially, thinly strigose abaxially, the base broadly cuneate, the apex rounded, mucronate, the margins entire; venation with 3 principal veins arising from base, the costa giving rise on each side to 3 incurved-ascending secondary veins, the marginal nerve cartilaginous. Inflorescence axes densely sordid-pilosulous. Staminate inflorescences axillary, solitary or paired, 1–3-flowered, subtended by abruptly diminished leaves toward apex of twining terminal branchlets; peduncles 2 cm long; rachises ± 1 cm long; longer pedicels 7–9 mm long. Pistillate inflorescences axillary, solitary, 1-flowered; pedicels ± 1 cm long. Staminate flowers: perianth green; sepals 9, 3-seriate, the two outermost series minute, 0.5–1 mm long, the inner series obovate-suborbicular, the largest ± 6.5–7 mm diam., imbricate in vernation, fleshy, the margins membranous; petals fleshy, compressed into pseudodisc, each in profile semi-obovate, 4 × 3 mm, induplicate around a stamen. Pistillate flowers unknown. Drupes ellipsoid, subcompressed, 24 × 12 × 10 mm; exocarp dry,-nigrescent, the endocarp ± 0.6 mm thick in section, finely incised-reticulate, its interior wall smooth, the septum laminar, ± 2 mm wide. Fl (Aug); in nonflooded forest.

A rare species, known by one staminate flowering collection from Saül and from one fruiting specimen from central Amazonian Brazil.

SCIADOTENIA Miers

Lianas, seldom of great length. Leaves: blades chartaceous, thinly pubescent or glabrate; venation with 3–5 principal veins arising from base, the secondary venation scalariform. Inflorescences either supra-axillary or cauliflorous, brown- or golden-silky-pilosulous. Staminate inflorescences very slender, spiciform or weakly paniculate; pedicels lacking. Pistillate inflorescences pendulous, 1-flowered. Flowers strobiliform; sepals in (3)4-several, 3-merous whorls, imbricate, the innermost whorl valvate in bud; petals 6, included. Staminate flowers: stamens 6, the filaments free, the connective exserted beyond anthers. Pistillate flowers: carpels 9–15, each in fruit elevated on a gynophore, the gynophores ± confluent at base. Drupes ultimately disjointing from gynophores, in profile obliquely pyriform, the stylar scar sub-basal; endocarp evenly incurved through ± 3/4 circle over an erect, basal condyle, its interior wall smooth. Seeds: endosperm lacking; embryo with cotyledons horny, filling drupe-cavity.

Sciadotenia cayennensis Benth. PL. 92b

Slender lianas, 3 m or more long. Stems wiry, densely and shortly sericeous when young. Leaves: petioles 2–3.5 cm long; blades commonly ovate, then the base rounded or subcordate, sometimes elliptic, the base cuneate, mostly 10–14 × 4.5–8.5 cm, the apex always acuminate, the margins entire; venation with 3 principal veins arising from base, the midrib giving rise to many horizontal secondary veins and also, well above mid-blade, to 1 pair of incurved-ascending ones. Staminate inflorescences 3–7 cm long. Pistillate inflorescences axillary from new growth or from leafless stems, solitary; peduncles 8–23 cm long, pliantly pendulous; rachises 1–2 cm long. Staminate flowers: innermost sepals ± 2.5 mm long. Pistillate flowers: perianth caducous; carpels 9–12. Drupes asymmetrically pyriform, 8–10 mm diam., appearing umbellate, each elevated on and finally disjointing from a longitudinally sulcate, silky-puberulent gynophore 7–15 mm long; exocarp orange when ripe, wrinkled, minutely velutinous. Fl (Feb), fr (Jan, Apr, Sep, Nov, Dec); in nonflooded forest.

TELITOXICUM Moldenke

Lianas, high-climbing. Leaves: blades coriaceous, glabrous; venation pinnate. Inflorescences either axillary or cauliflorous, puberulent or pilosulous. Staminate inflorescences paniculate; pedicels slender, mostly borne on short secondary branches. Pistillate inflorescences racemose. Flowers: perianth green, nigrescent upon drying; sepals 6, in two whorls, the 3 inner ones imbricate in bud, slightly larger than 3 outer; petals 6, free, membranous, concave, slightly shorter than sepals. Staminate flowers: stamens 6, the filaments free. Pistillate flowers: staminodia 6; carpels 3. Drupes sessile on torus, obovoid or oblong-obovoid, slightly laterally compressed, the stylar scar remaining close to receptacle; endocarp bent over an erect, basal condyle into an almost closed horseshoe, the interior wall of endocarp smooth; endosperm ruminate, its lamellae separated by membranous integuments; embryo vermiform.

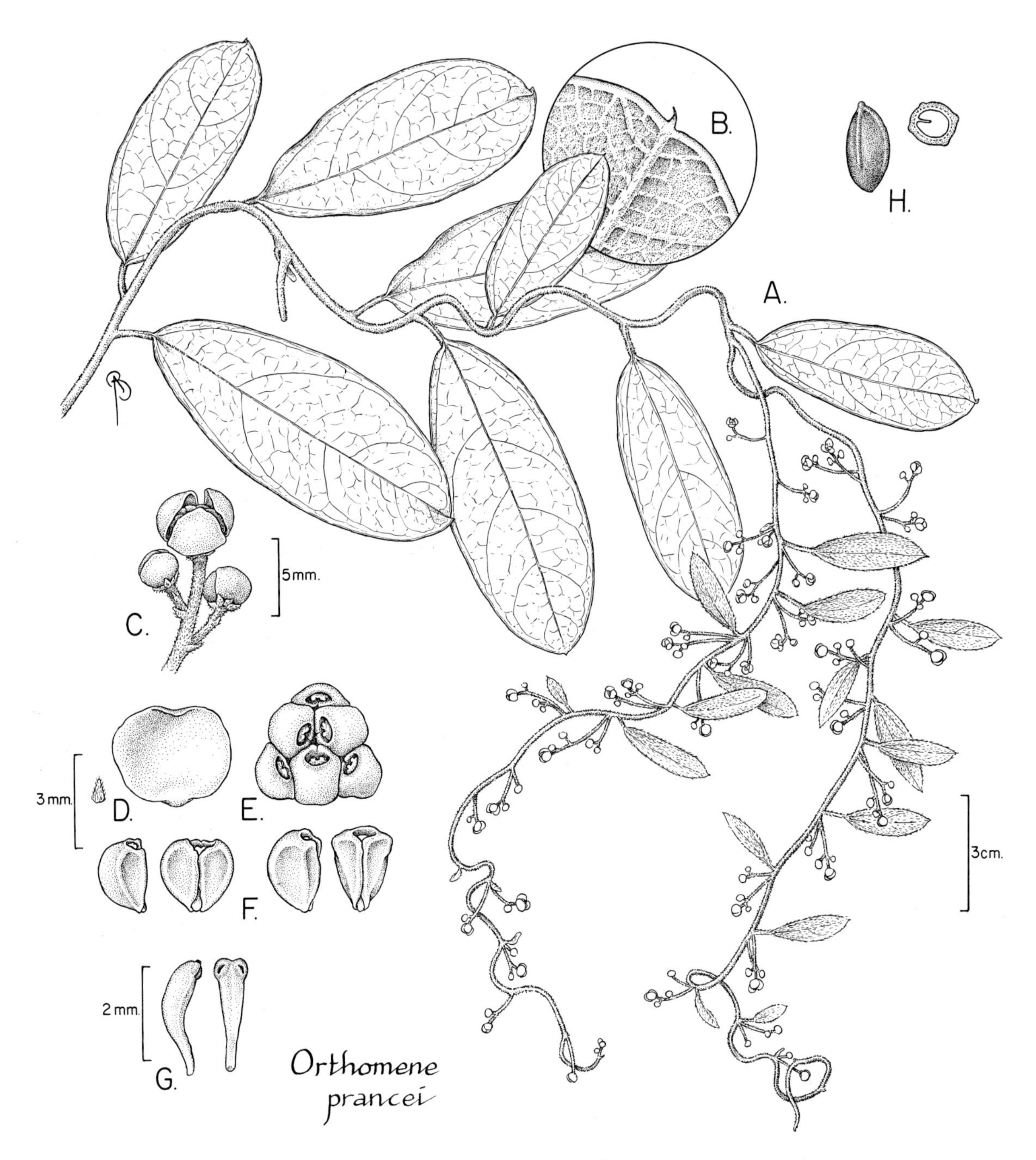

FIG. 207. MENISPERMACEAE. *Orthomene prancei* (A–G, *Mori et al. 23263*; H, *Prance et al. 5011* from Amazonas, Brazil). **A.** Leafy upper stem and staminate inflorescence. **B.** Abaxial view of leaf apex. **C.** Staminate buds and flower. **D.** Small outer (left) and larger inner sepal (right). **E.** Pseudodisc of 6 petals, apical view. **F.** Separated petals, each embracing a stamen. **G.** Stamen, lateral (near right) and adaxial (far right) views. **H.** Lateral view (left) and transverse section (above right) of fruit.

Telitoxicum inopinatum Moldenke

Canopy lianas. Pubescence generally lacking. Leaves: petioles 2–6 cm long; blades ovate or ovate-elliptic to oblanceolate, the longer ones 8.5–18 × 3.5–8 cm, the base either rounded or cuneate, the apex obscurely acuminate, the margins entire; pinnately veined, the secondary veins 5–8 from each side of midrib, immersed or shallowly depressed on adaxial surface, prominent abaxially. Staminate inflorescences ≤10 cm long, puberulent, the secondary axes ≤1.5 cm long, at some leaf nodes fasciculate with a solitary flower; pedicels filiform, 3.5–8 mm long. Pistillate inflorescences puberulent; peduncles 2–3 cm long; rachises 7–9 mm long. Flowers: sepals lanceolate, the inner ones ± 1.6–2.2 mm long; petals ± 1 mm long. Drupes puberulent, early glabrate, oblong-obovoid ± 3 × 1.6–2.2 cm, the stout fruiting pedicels, including globose receptacle, 9–12 mm long. Fr (Apr, Nov); in nonflooded forest.

MIMOSACEAE (Mimosa Family)

James W. Grimes

Shrubs, trees, lianas, vines, or herbs. Epidermal prickles present or lacking. Stipules present or lacking, when present quickly deciduous to persistent. Leaves alternate (opposite or whorled), compound, paripinnate or bipinnate; petioles pulvinate; blades if bipinnate, then pinnae 1 to many pairs, (sub)opposite (alternate); leaflets opposite or alternate, pulvinulate; venation pinnate, palmate-pinnate, palmate, or reduced to midvein. Inflorescences simple or compound, commonly axillary to contemporaneous leaves, or terminal or cauliflorous, either of capitula or spikes, or pseudoracemes or panicles of capitula or spikes; receptacles bracteate, the bracts deciduous or persistent; pedicels present or lacking, the bracteoles lacking. Flowers actinomorphic, those within any unit of inflorescence homomorphic, or heteromorphic andromonoecious (either the proximal or distal flowers staminate), or dimorphic because terminal 1 to few enlarged; perianth (3)4–5-merous; sepals fused; petals either united, or becoming free to base; stamens either as many, or twice as many as perianth segments, or numerous, free to base, or united with corolla at base into stemonozone, and sometimes united distal to stemonozone into a tube, the anthers dorsifixed, the connective sometimes with gland; ovary superior, unilocular, sessile or on a gynophore; placentation marginal. Fruits strongly laterally compressed to turgid and ± cylindric, inertly dehiscent through one or both sutures, or dehiscent and contorted, or indehiscent and variously modified, or craspedia. Seeds ± plump to strongly laterally compressed, sometimes persistent on funicle after dehiscence of fruit; pleurograms present or lacking, when present either complete or incomplete.

Barneby, R. C. & J. W. Grimes. 1996. Silk tree, guanacaste, monkey's earring: A generic system for the synandrous Mimosaceae of the Americas. Part I. *Abarema*, *Albizia*, and allies. Mem. New York Bot. Gard. **74(I):** 1–161.

——— & ———. 1997. Silk tree, guanacaste, monkey's earring: A generic system for the synandrous Mimosaceae of the Americas. Part II. *Pithecellobium*, *Cojoba*, and *Zygia*. Mem. New York Bot. Gard. **74(II):** 1–161.

Jansen-Jacobs, M. J. 1976. Mimosaceae. *In* J. Lanjouw & A. L. Stoffers (eds.), Flora of Suriname **II(2):** 611–653. E. J. Brill, Leiden.

Kleinhoonte, A. 1976. Mimosaceae. *In* J. Lanjouw & A. L. Stoffers (eds.), Flora of Suriname **II(1):** 258–331. E. J. Brill, Leiden.

1. Leaves pinnately compound, with gland between leaflets. Seeds surrounded by white, edible, sweet tasting sarcotesta. . . . . *Inga.*
1. Leaves usually bipinnately compound. Seeds not surrounded by white, edible, sweet tasting sarcotesta.
  2. Inflorescences of cauliflorous spikes of capitula. . . . . *Zygia.*
  2. Inflorescences of axillary capitula or spikes, or in terminal pseudoracemes, or panicles of capitula, or spikes.
    3. Vines, subshrubs, or herbs armed with epidermal prickles.
      4. Stamens 4, 5, 8, or 10.
        5. Leaves with 2 pairs of pinnae, petiolar gland present, leaflets in 3–4 pairs. Fruits with valves papery, framed by sutures, inertly dehiscent through both sutures.. . . . . *Piptadenia.*
        5. Leaves with 2–33 pairs of pinnae, when only 2 pairs, the leaflets in 1–2(3) pairs. Fruits with valves breaking away from replum into 1-seeded articles, these individually indehiscent or dehiscent. . . . . *Mimosa.*
      4. Stamens numerous (50–90). Fruits with valves continuous with replum, dehiscent through both sutures. . . . . *Acacia.*
    3. Trees or lianas, unarmed.
      6. Inflorescences of spikes.
        7. Stamens numerous, (12)14 or more, united into a tube. Fruits moniliform, dehiscent through adaxial suture. Seeds persistent on funicle. . . . . *Abarema*, in part.
        7. Stamens 10, free to the base. Fruits plano-compressed even when plump, indehiscent or dehiscent through adaxial suture. Seeds not persistent on funicle.
          8. Petiolar glands lacking; distal pinnae of some leaves modified into tendrils. Inflorescences dense one-sided pseudoracemes of spikes. Fruits craspedia, 4–8 cm wide. . . . . *Entada.*
          8. Petiolar glands present; distal pinnae of leaves not modified into tendrils. Inflorescences spicate, but not one-sided. Fruits dehiscent along adaxial suture, or indehiscent (not craspedia), ≤2.5 cm wide.
            9. Leaflets opposite along rachis, <25 mm long. Fruits flattened, dehiscent along one suture only. Seeds plano-compressed, the testa membranous, lacking pleurogram; endosperm lacking. . . . . *Pseudopiptadenia.*
            9. Leaflets either opposite or alternate, but when ≤25 mm long alternate. Fruits ± terete in transverse section, indehiscent. Seeds plump, the testa hard, with pleurogram; endosperm present. . . . . *Stryphnodendron.*
      6. Inflorescences of globose, biglobose, or clavate capitula.
        10. Capitula large, ≥1 cm wide in fruit. Flowers bisexual, with morphologically dissimilar neuter and/or staminate flowers basal to bisexual ones. . . . . *Parkia.*
        10. Capitula smaller, the receptacle <1 cm wide. Flowers all bisexual, or the pistillate-sterile one or few enlarged or coarser and at apex of capitulum.

11. Pinnae of 2–3 pairs per leaf; distal leaflets larger than proximal ones. Inflorescences terminal pseudoracemes of capitula. Fruits lomentiform, twisted through 90° between each seed. . . . . *Cedrelinga.*
11. Pinnae usually of >3 pairs per leaf, if of 2–3 pairs then distal leaflets equal in size or smaller than proximal ones. Inflorescences axillary, not pseudoracemes. Fruits not lomentiform, not twisted as above.
    12. Pinnae of 10–30 pairs per leaf, the leaflets 32–83 pairs per pinna. Fruits indehiscent, curved through 0.5 to 1.5 circles into annular or compressed-helicoid figure, the endocarp forming septa between seeds. . . . . *Enterolobium.*
    12. Pinnae of 1–14 pairs per leaf, the leaflets 3–29(33) pairs per pinna. Fruits dehiscent, not curved as above, the endocarp not forming septa between seeds.
        13. Leaflets of (16)18–29(33) pairs per pinna. Fruits follicular, the 3 layers of the walls irregulary coherent, the exocarp breaking into small tetragonal pieces, the mesocarp of transverse woody fibers, the endocarp coherent between seeds but not forming septa. Seeds not persistent on funicle. . . . . *Balizia.*
        13. Leaflets of 3–19 pairs per pinna. Fruits follicular or dehiscent through both sutures, the 3 layers of the walls coherent. Seeds persistent on funicle. . . . . *Abarema.*

ABAREMA Pittier

Canopy trees mostly without buttresses. Epidermal prickles lacking. Stipules herbaceous or sometimes lacking. Leaves alternate, bipinnate; petiole with glands between first pinna-pair and commonly between subsequent pairs; pinnae opposite; leaflets opposite; venation pinnate. Inflorescences of pedunculate capitula or short spikes, solitary, paired, or three or more in leaf-axils. Flowers hermaphroditic, subhomomorphic when spicate, when in capitula the terminal 1(few) dimorphic, i.e. usually wider, coarser and/or longer than peripheral ones; perianth greenish-white; stamens 12–28(36), white, united at base into tube. Fruits follicular or dehiscent through both sutures; valves contorting, internally scarcely and irregularly mottled to completely orange-red or red, the three layers of the walls coherent, the endocarp not forming septa between seeds. Seeds persistent on funicle, the testa white and/or translucent when mature and fresh; embryo usually aniline-blue and visible.

1. Inflorescences spicate racemes, the main axes 6–12 cm long. Fruits moniliform, the valves 4–8 mm thick over each seed. . . . . *A. curvicarpa* var. *curvicarpa.*
1. Inflorescences capitulate, or pseudoumbellate, the main axes <3 cm long. Fruits broad-linear, not moniliform, the valves ≤2 mm thick over each seed.
    2. Pinnae 1–2 pairs per leaf; leaflets 1–3(4) pairs per pinna; blades elliptic, obovate, ovate, or elliptic-ovate; midrib ± centric.
        3. Leaflets of all pinnae exactly 1 pair per pinna; pulvinules 1.5–3 mm long. . . . . *A. laeta.*
        3. Leaflets of the distal or only pair of pinnae 2–3 pairs per pinna; pulvinules 5–9 mm long. . . . . . *A. mataybifolia.*
    2. Pinnae 1–12 pairs per leaf; leaflets 3–12 pairs per pinna; blades rhombic or rhombic-elliptic; midrib oblique.
        4. Pinnae 6–12 pairs per leaf; leaflets 12–19 pairs per pinna, becoming smaller at distal end or at both ends of pinna. . . . . *A. barbouriana* var. *barbouriana.*
        4. Pinnae 1–7 pairs per leaf; leaflets 3–12 pairs per pinna, becoming larger at distal end of pinna.
            5. Pinnae 3–7 pairs per leaf; leaflets 7–12 pairs per pinna, becoming larger at distal end of pinna. . . . . *A. jupunba* var. *jupunba.*
            5. Pinnae 1–3 pairs per leaf; leaflets 3–6(7) pairs per pinna, not becoming larger at distal end of pinna.. . . . . *A. jupunba* var. *trapezifolia.*

Abarema barbouriana (Standl.) Barneby & J. W. Grimes var. **barbouriana**

Trees, to 30 m tall. Bark smooth, with small cankers and pock marks. Leaves: pinnae 6–12 pairs per leaf; leaflets 12–19 pairs per pinna, rhombic or rhombic elliptic, to 5–21 × 2–10 mm, becoming smaller at distal or at both ends of pinnae, almost always finely and densely silky villosulous beneath; midrib oblique. Inflorescences of capitula, the axis <3 cm long. Flowers: terminal flower of inflorescence differentiated from other flowers; stamens (12)14–28. Infructescences usually 1-fruited. Fruits sessile, curled 1–2.5 times into circle 3.7–4.4 cm diam., in profile broad-linear, (5)6–12 × (1)1.1–1.6 cm, the endocarp orange-red either overall or only in seed-cavities; dehiscence downward through both sutures, the valves coiling. Seeds (few seen) plump, ± 5–7 mm diam.; pleurogram either lacking or complete. Known by a single sterile collection from our area; apparently rare, in nonflooded forest.

Abarema curvicarpa (H. S. Irwin) Barneby & J. W. Grimes var. **curvicarpa** FIG. 208

Trees, to 30–45 m tall. Trunks sometimes with buttresses. Bark reddish brown, smooth, with scattered round and horizontally expanded lenticels. Leaves: pinnae 6–12 pairs per leaf; leaflets 14–26 pairs per pinna, equal in size (except for slightly broader furthest

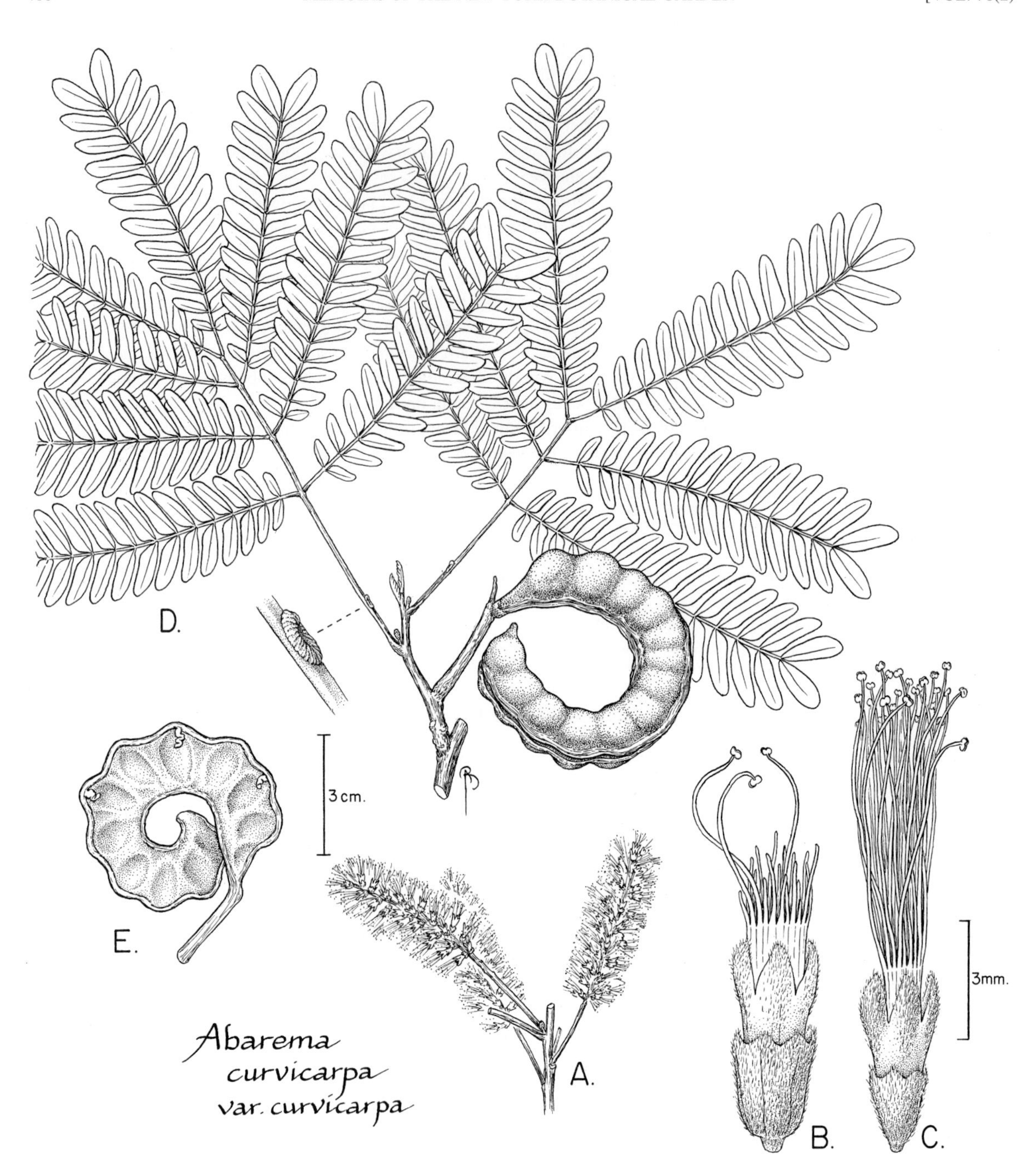

FIG. 208. MIMOSACEAE. *Abarema curvicarpa* var. *curvicarpa* (A–C, *Mori & Gracie 19172*; D, *Mori & Pipoly 15432*; E, *Sabatier 3051*). **A.** Spicate inflorescences. **B.** Distal flower with filament tubes connate into a tube. **C.** Proximal flower. **D.** Apex of stem with leaves, fruit, and detail of extrafloral nectary near rachis base (left). **E.** Part of dehisced fruit after dispersal of seeds.(Used with permission of the *Flora of the Guianas*.)

pair), obliquely oblong, elliptic-oblong or linear-oblong, 7–18 × 1.8–6 mm. Inflorescences compound, of racemes of spikes, the main axes 6–12 cm long, the secondary axes becoming 3.5–4.5 cm long. Flowers subhomomorphic, the distal one(s) slightly broader but their androecium not modified; stamens ± 28. Fruits turgid, moniliform, evenly incurved through 1–2 full spirals, ± 10–16 × 1.7–2.3 cm, the woody valves 4–8 mm thick where grossly dilated and domed over each seed; dehiscence tardy, primarily along adaxial suture. Seeds compressed-obovoid, in broad view ± 9 × 7 mm; pleurogram lacking. Fl (Aug), fr (Mar, Jun); common but scattered, in nonflooded forest.

Abarema jupunba (Willd.) Britton & Killip var. **jupunba**

Trees. Trunks sometimes with smooth buttresses. Bark smooth, scaly, gray-brownish. Leaves: pinnae 3–7 pairs per leaf; leaflets 7–12 pairs per pinna, the terminal pair largest, (1.3)1.6–6.5(7) × 1–3.6(4.4) cm. Inflorescences of capitula, (10)15–45-flowered, either compact or with one or more flowers downwardly displaced on peduncles. Flowers usually dimorphic, the terminal 1–3 ordinarily longer and broader and with androecium with more stamens, but these sometimes modified or lacking; stamens 14–25. Infructescences 1–2-

fruited per capitulum. Fruits estipitate or almost so, curled though 2/3 to nearly 2 circles, broad-linear in profile, 6–9.5 × 1–1.6 cm, dehiscence through both sutures, the valves coiling. Seeds plumply lentiform, as long or slightly longer than wide in broad view, 5–8 mm diam.; pleurogram lacking. Fl (Oct, Dec), fr (May); common but scattered. *Acacia* (Créole).

Abarema jupunba (Willd.) Britton & Killip var. **trapezifolia** (Vahl) Barneby & J. W. Grimes

Trees. Trunks sometimes with smooth buttresses. Bark smooth, scaly, gray-brownish. Leaves: pinnae 1–3 pairs per leaf; leaflets 3–6(7) pairs per pinna, the larger pairs (3.4)3.5–6.5(7) × 1.9–3.6 cm. Inflorescences of capitula, (10)15–45-flowered, either compact or with one or more flowers downwardly displaced on peduncles. Flowers usually dimorphic, the terminal 1–3 ordinarily longer and broader and with androecium with more stamens, but these sometimes modified or lacking; stamens 14–25. Infructescences 1–2-fruited per capitulum. Fruits estipitate or almost so, curled though 2/3 to nearly 2 circles, broad-linear in profile, 6–9.5 × 1–1.6 cm, dehiscence through both sutures, the valves coiling. Seeds plumply lentiform, as long or slightly longer than wide in broad view, 5–8 mm diam.; pleurogram lacking. Fl (Sep).

Abarema laeta (Benth.) Barneby & J. W. Grimes

Shrubs or small trees, to 6 m tall. Leaves: pinnae 1–2 pairs per leaf; leaflets exactly 1 pair per pinna; pulvinules 1.5–3 mm long; blades subequilaterally elliptic or (ob)ovate, 10–18(20) × 3–7(7.7) cm; midrib ± centric. Inflorescences of capitula, 12–28-flowered; axes <3 cm long; pedicels very short or lacking. Flowers dimorphic, the terminal one wider and coarser than peripheral ones; stamens 20–24. Infructescences 1(2)-fruited per capitulum. Fruits when short nearly straight but when longer recurved through to 3/4 circle, in profile oblong or broad-linear, (5)5.5–13(16) × 1.4–2.2 cm; dehiscence through both sutures, the valves coiling, crimson overall inside. Seeds plumply ellipsoid, in broad view 10–11 × 6–8 mm; pleurogram lacking. Fl (Aug, Sep); apparently rare, in nonflooded forest.

Abarema mataybifolia (Sandwith) Barneby & J. W. Grimes PL. 93a

Trees, 8–20 m tall. Bark gray-brown, smooth, finely cracked, soft. Leaves: pinnae 1–2 pairs per leaf; leaflets 2–3 pairs per pinna; pulvinules 5–9 mm long; blades subequilaterally ovate or elliptic-ovate, slightly larger distally, these 8–14 × 3–7.5 cm; midrib ± centric. Inflorescences of capitula, ± 15–20-flowered; pedicels present in peripheral flowers, absent in terminal one. Flowers dimorphic, the terminal one (few seen) somewhat coarser; stamens 18–28. Infructescences 1–2-fruited per capitulum. Fruits evenly recurved through half or nearly complete circle, in profile broad-linear, ± 10–20 × (1.3)1.8–2.6 cm; dehiscence through both sutures, the valves coiling, red-crimson overall inside. Seeds plumply lentiform-globose, in broader view 8–9 × 7–8 mm; pleurogram lacking. Fl (Sep, Nov), fr (Apr, May); common, in nonflooded forest.

ACACIA Mill.

Lianas. Epidermal prickles present. Stipules herbaceous and quickly deciduous. Leaves alternate, bipinnate; petioles and sometimes leaf-rachis between ultimate few to many pinna-pairs usually with stalked glands; leaflets opposite; venation of simple midrib, or rarely faintly palmately veined from petiolule. Inflorescences of axillary capitula, fasciculate, at any node the peduncles ± equal or very inequal in length. Flowers homomorphic, white; stamens numerous, ± free to the base. Fruits flattened, elliptic in outline; dehiscence inert, through both sutures, the valves continuous with replum. Seeds elliptic, round or obovate; testa ± hard, tan, light brown or dark brown, the pleurogram U-shaped.

Acacia tenuifolia (L.) Willd. var. **tenuifolia** PART 1: FIG. 10

Most commonly lianas, also prostrate to weakly erect shrubs. Leaves: pinnae 21–35 pairs per leaf; leaflets 43–107 pairs per pinna, elliptic, 2.3–3.8 × 0.3–0.64 mm, sensitive, folding along pinna-rachis. Inflorescences of panicles of white or cream-colored capitula; peduncles 2–5-fasciculate, maturing sequentially, the first formed always attaining greater size; pedicels lacking or rarely present, then to 0.25 mm long. Flowers tubular-campanulate; sepals 1.1–2 mm long; petals 1.85–2.5 mm long; stamens 50–90. Fruits elliptic, 10.2–12.5 × 1.9–3.1 cm, flat, the exocarp coarsely wrinkled, dark brown. Seeds (few seen mature) elliptic, ± 12.5 × 7.5 mm; testa shiny brown, the areole lighter in color, the pleurogram centric. Fl (Aug, Sep, Oct), fr (Sep, Oct, Nov); abundant, in nonflooded forest, especially conspicuous along forest margins, when in flower conspicuous everywhere. *Pois sucré*, *queue lézard* (Créole).

BALIZIA Barneby & J. W. Grimes

Canopy trees. Epidermal prickles lacking. Stipules linear-lanceolate or ligulate, caducous. Leaves alternate, bipinnate; petioles almost always with glands, leaf-rachis also often with one gland between furthest 1–2 pinna-pairs; pinnae opposite; leaflets opposite; venation pinnate. Inflorescences of pedunculate capitula, axillary, solitary or 2–3-fasciculate; pedicels present in peripheral flowers, lacking in terminal one. Flowers dimorphic, reddish, but stamens white, the peripheral ones mostly bisexual, the terminal one sterile; stamens numerous, united at base into tube. Fruits woody, plano-compressed, transversely fibrous, tardily follicular through adaxial suture, the 3 layers of the walls irregularly coherent, the exocarp breaking into small tetragonal pieces, the mesocarp of transverse woody fibers, the endocarp coherent between seeds but not forming septa. Seeds narrowly oblong, not persistent on funicles; testa either putty-colored and ± discolored within narrowly U-shaped pleurogram, or partly white and partly translucent; embryo green or anthocyanic.

Balizia pedicellaris (DC.) Barneby & J. W. Grimes FIG. 209

Trees, to 40 m tall. Leaves: pinnae 6–10(14) pairs per leaf, the rachis of distal, longer ones (4)5–9(10) cm; leaflets (16)18–29(33) pairs per pinna, narrowly oblong, lance-oblong or oblong-elliptic, 6–13.5(16) × 1.8–3.3(4) mm, the margins revolute. Inflorescences: peduncle (2.5)3–6.5(8) cm long; pedicels present in peripheral flowers (4)4.5–7.5 mm, lacking (or nearly so) in terminal flowers. Flowers: terminal 1 or more with sepals (3)3.4–4.6 mm long; petals (6.5)7–11 mm long. Infructescences 1–3-fruited per capitulum.

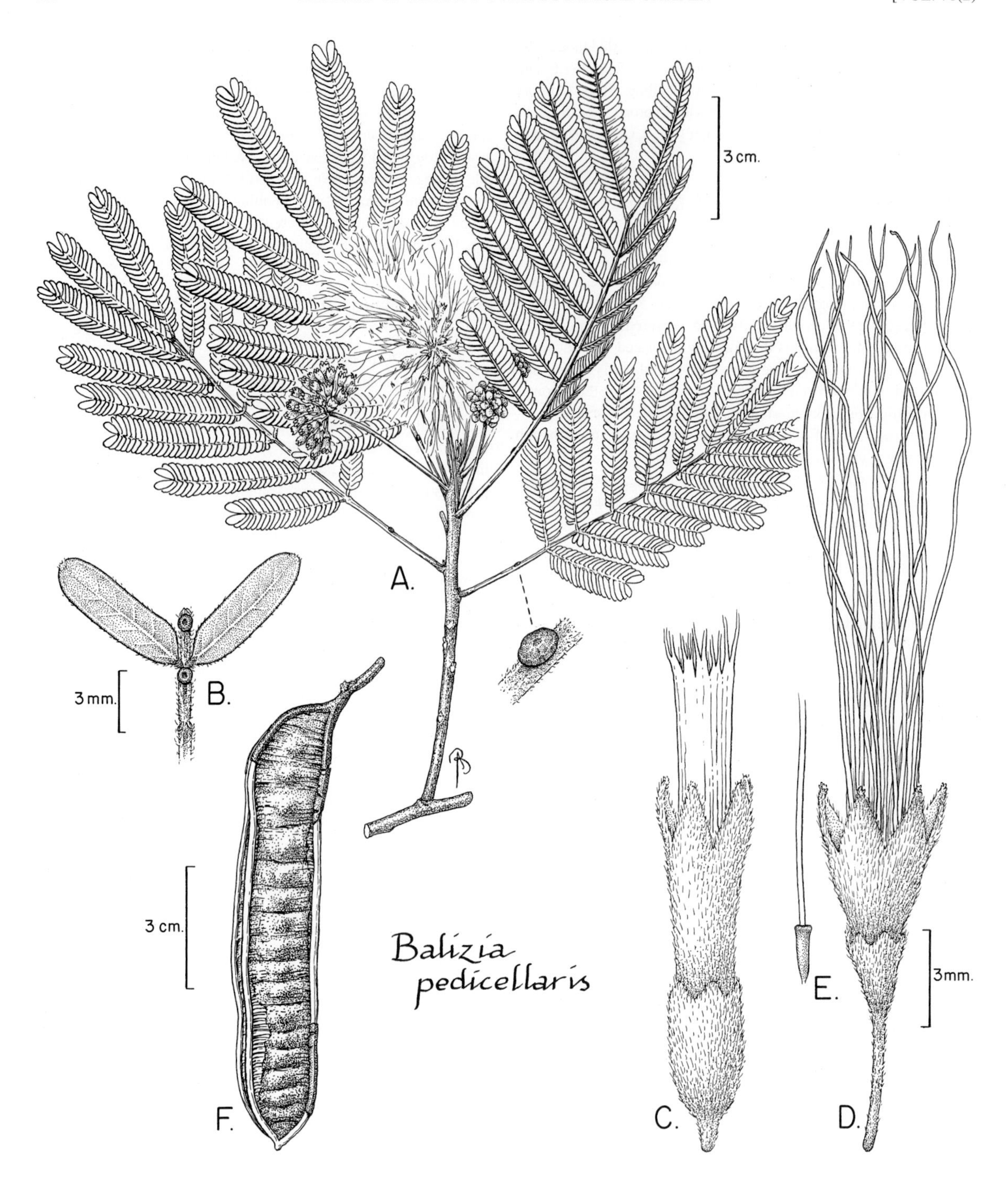

FIG. 209. MIMOSACEAE. *Balizia pedicellaris* (A–E, *Jansen-Jacobs 1858* from Guyana; F, *Woodherbarium Surinam 49* from Surinam). **A.** Apex of flowering stem showing bipinnately compound leaves and inflorescences; note detail of extrafloral nectary near midpoint of petiole (right). **B.** Apex of petiolule and lower part of leaflet showing extrafloral nectaries. **C.** Distal flower showing exserted staminal tube. **D.** Proximal flower. **E.** Base of pistil. **F.** Fruit.

Fruits estipitate, spreading or ascending from receptacle, 7–12.5(14) × (1.7)1.8–3.2 cm. Seeds compressed but plump, (6)7.5–9 × 3–4 mm in profile. Fl (Sep), fr (Apr, Sep); in forest on well-drained soil. *Bois macaque*, *cèdre marécage* (Créole).

CEDRELINGA Ducke

Trees. Trunks generally with buttresses. Epidermal prickles lacking. Stipules lacking. Leaves alternate, bipinnate; petioles with gland at or below insertion of each pinna-pair; pinnae opposite; leaflets opposite, asymmetrical, finely reticulate; midrib arched backward and posteriorly displaced from mid-blade. Inflorescences of terminal panicles or pseudoracemes of capitula; pedicels lacking. Flowers hermaphroditic, homomorphic; petals greenish-white, the lobes separate above the stemonozone, therefore at maturity lacking a tube free from the androecium; stamens numerous, white, united at base into a tube. Fruits papery lomentiform, twisted through about 90° between each seed-cavity. Seeds: testa thin, without an areole.

Cedrelinga cateniformis (Ducke) Ducke FIG. 210, PL. 93b

Canopy trees, to 30–60 m tall. Trunks with or without buttresses. Bark vertically furrowed, reddish. Leaves: pinnae 2–3 pairs per leaf, the distal ones a little longer, the rachis of longer ones 6–12(16.5) cm; leaflets ± accrescent distally, asymmetrically ovate or ovate-elliptic, the distal pair 7–12(13) × 3.5–5.7 cm. Inflorescences: primary axis 10–150 cm; peduncles 2–6 per node. Flowers brownish when dried. Infructescences 1-fruited per capitulum. Fruits pendulous, lomentiform, 2–7 dm long, in profile broad-linear but deeply constricted between each of the 2–5(6) very large plano-compressed 1-seeded articles, each segment 10–15(17) × (3.3)3.5–6 cm, the whole body twisted through ± 90° at each isthmus but plane and straight between them, indehiscent, but breaking apart between seed-cavities. Seeds disciform, in profile elliptic, 25–32 × 14–18 mm (orbicular 10–11 mm diam), ± 1.5 mm thick. Fl (Nov); rare, in nonflooded forest.

ENTADA Adans.

Lianas. Epidermal prickles lacking. Stipules small, caducous. Leaves alternate, bipinnate; petioles lacking glands; pinnae opposite; leaflets opposite. Inflorescences terminal, crowded pseudoracemes of mostly paired spiciform racemes, the racemes all ascending to vertical. Flowers small; stamens 10, free. Fruits plano-compressed, broad-linear craspedia, the persistent replum undulately constricted between seeds, the stipe 1.5–3 cm long, the exocarp brown, exfoliating when ripe from the pale tan endocarp, this breaking into transversely narrow-oblong 1-seeded papery segments. Seeds not persistent on funicles.

Entada polyphylla Benth.

Lianas. Leaves: pinnae 4–8 pairs per leaf, the median ones normally the longest, the rachis of one or both distal pinnae normally modified into a woody tendril; leaflets of longer pinnae 12–20 pairs per pinna; blades oblong, 11–22(23) × 3–7(8) mm. Inflorescences: main axes 1–3.5(4.5) dm; racemes subsessile, solitary or more usually paired, their axes (3)3.5–8(9) cm; pedicels 0.15–0.3 mm long. Fruits in profile broad-linear, nearly straight, flattened, 25–45 × 4–8 cm, the seed segments ± 2–3 cm wide. Bud (Sep); apparently rare, in nonflooded forest.

ENTEROLOBIUM Mart.

Trees. Trunks with narrow buttresses. Epidermal prickles lacking. Stipules small, mostly caducous. Leaves alternate, bipinnate; petiolar glands either cupular-patelliform, or mounded, or sunk in petiolar groove; pinnae opposite; leaflets opposite. Inflorescences axillary, capitulate; peduncles 3–9 per node. Flowers homomorphic or dimorphic, hermaphroditic, white; stamens numerous, united into a tube. Fruits sessile, in profile oblong or broad-linear decurved through 1/2–2 circles into a annular or compressed-helicoid figure, the exocarp thin, dark, the mesocarp thick, either dry mealy-fibrous or resinous-pulpy, the endocarp papery, pallid, inflexed to form complete, often partly membranous septa between seeds, indehiscent, the seeds released by weathering or excreted by herbivores. Seeds plumply compressed, ovoid-ellipsoid; testa hard, castaneous or fuscous, the pleurogram U-shaped.

Enterolobium schomburgkii (Benth.) Benth. FIG. 211

Trees, 10–50 m tall. Bark rough, peeling in flakes. Leaves: pinnae 10–30 pairs per leaf, the rachis of the longer ones at mid-stalk (2.2)3–6(6.7) cm; leaflets 32–83 pairs per pinna; blades linear or narrowly lance-linear, the larger ones 2.5–5(5.4) × 0.5–0.8 mm. Inflorescences: pedicels ± present in peripheral flowers, lacking in terminal one. Flowers dimorphic; sepals of peripheral flowers (2.3)2.5–3.3(3.5) mm; petals (3.2)3.5–5.2 mm; sepals of terminal flowers (2.2)2.4–3.8 mm; petals (4)4.5–8 mm. Fruits evenly recurved through 3/4–1½ circles into flattened spiral, the compressed body at middle of valves 5–12 × 1.8–3.3 cm. Bud (Oct), fr (Oct from ground); rare, in nonflooded forest. *Acacia franc* (Créole).

INGA Willd.

Odile Poncy

Small to large trees. Trunks often irregular, usually widened at base, often provided with low, rounded buttresses. Bark smooth, variable in color, lenticellate, the lenticels often horizontally oriented. Leaves paripinnately compound, alternate, the leaflets in pairs, the petiole and rachis winged or not winged, with sessile to stalked, generally cup-shaped glands on rachis between each pair of leaflets. Inflorescences from leaf axils, sometimes from axils of undeveloped leaves and therefore appearing as terminal panicles, or, less often, from branches, spicate or capitate. Flowers 5-merous; calyx and corolla greenish, yellowish, or whitish, rarely pinkish; calyx tubular or campanulate, dentate or shortly lobed; corolla tubular or funnel-shaped, the tube at least as long as lobes; stamens numerous, fused at bases into tube often exserted from corolla, the staminal tube white. Fruits indehiscent, usually broken open by animals. Seeds embedded in sweet, edible pulp, an outgrowth of integument; embryo with colored cotyledons.

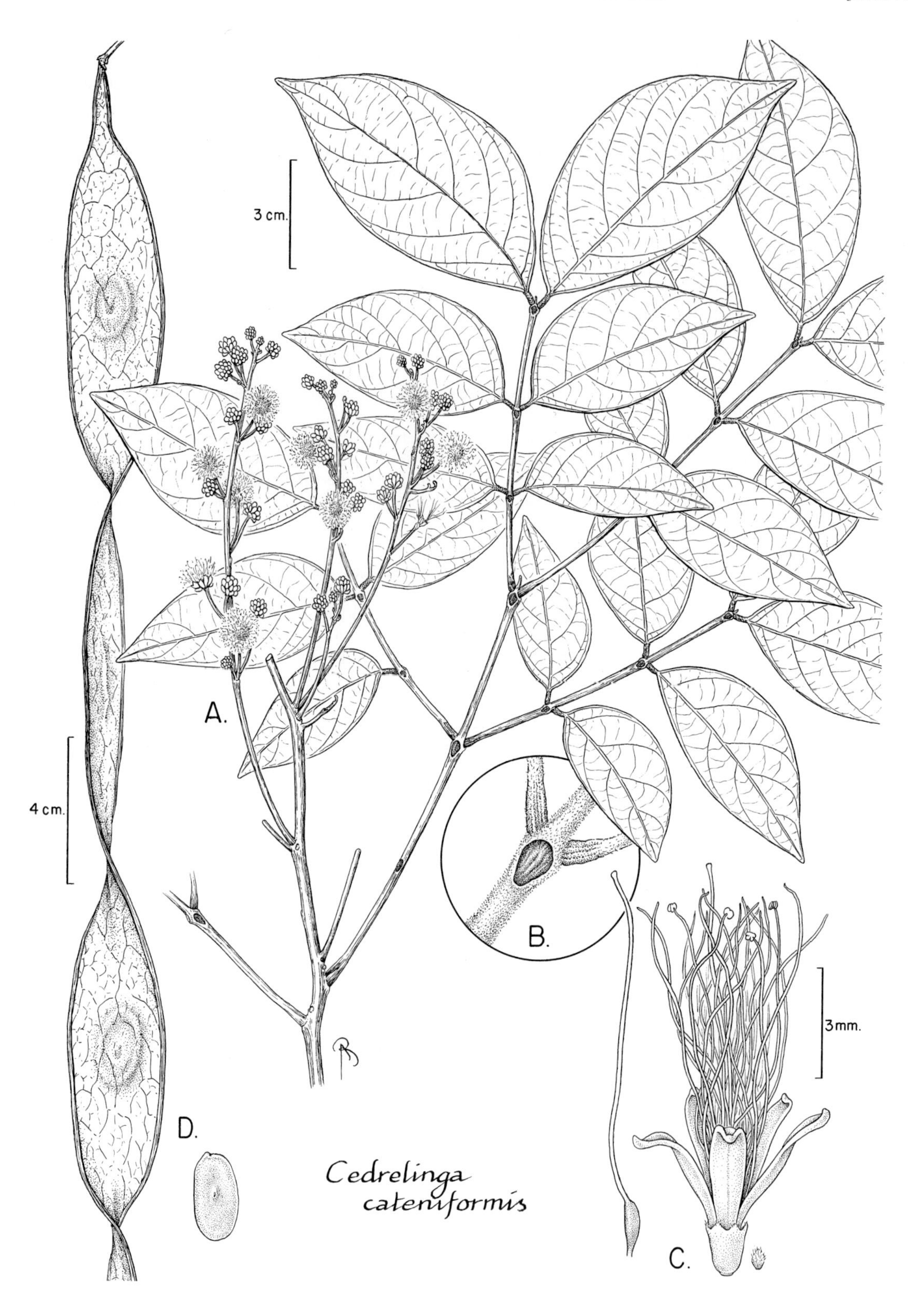

FIG. 210. MIMOSACEAE. *Cedrelinga cateniformis* (A, *SEF 8949* from Ecuador and *Vargas 146* from Peru; B, C, *SEF 8949*; D, *Neill 7131* from Ecuador). **A.** Part of stem with bipinnately compound leaf showing inflorescences and extrafloral nectaries near midpoint of rachis and below insertion points of pinna-pairs and leaflets. **B.** Detail of extrafloral nectary. **C.** Lateral views of pistil (left) and flower (right) with detail of calyx scale (right). **D.** Base of fruit (left) and seed (right).

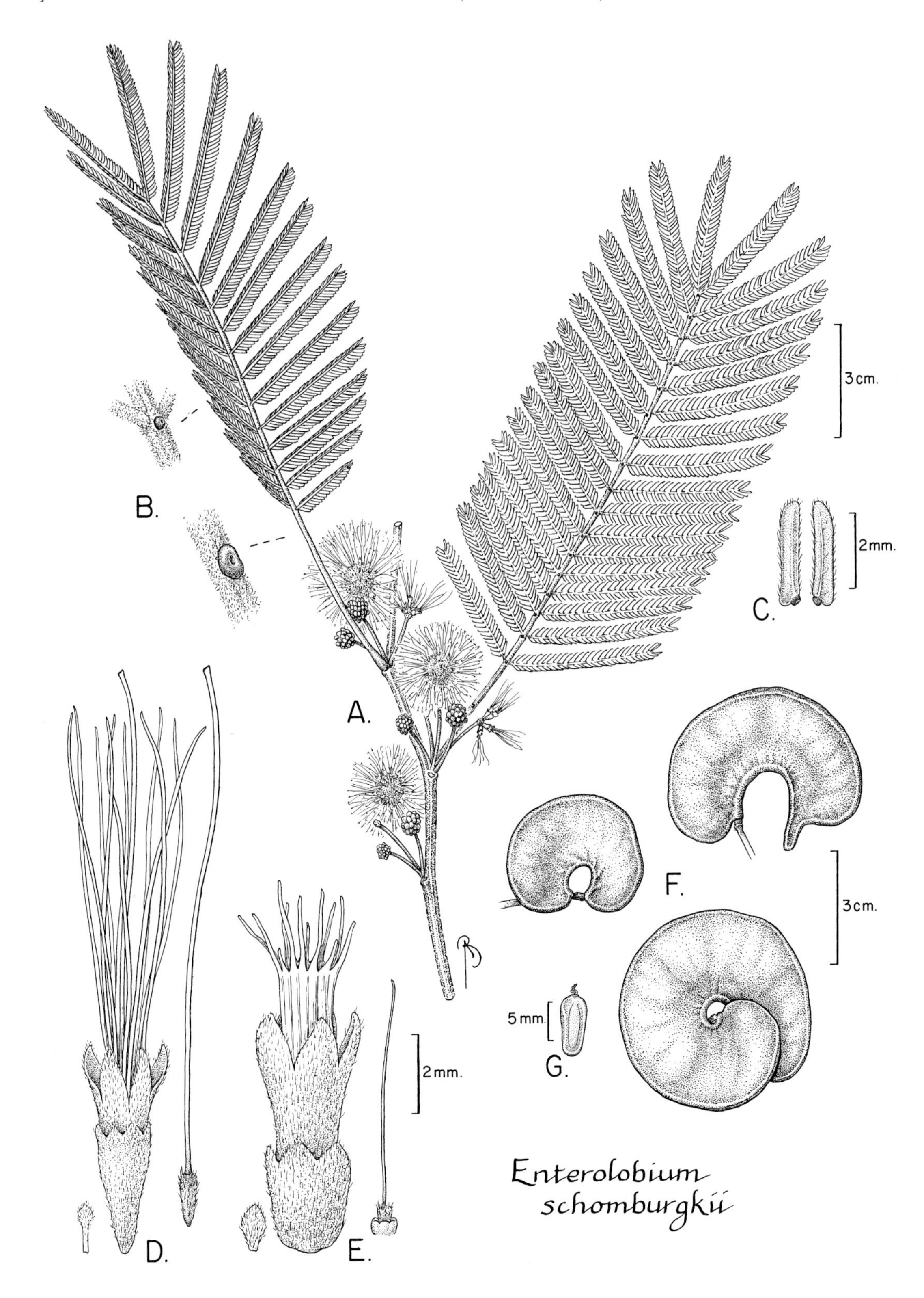

FIG. 211. MIMOSACEAE. *Enterolobium schomburgkii* (A–E, *N. T. Silva 1387* from Amazonia, Brazil; F, *M. G. Silva 4781* (left) from Mato Grosso, Brazil; *Gentry et al. 41781* (upper right) from Peru; *M. G. Silva & R. Souza 2474* (lower right) from Pará, Brazil; G, *M. G. Silva & R. Souza 2472* from Pará, Brazil). **A.** Part of stem with flowers, floral buds, and leaves. **B.** Detail of extrafloral nectaries on rachis between pinnae (above right) and on petiole at base of blade (below right). **C.** Two leaflets. **D.** Lateral view of terminal flower (near left), bracteole (far left), and pistil (right). **E.** Lateral view of peripheral flower (near left), bracteole (far left), and pistil (right). **F.** Three fruits showing variation in form. **G.** Seed. (Used with permission of the *Flora of the Guianas.)*

Pennington, T. D. 1997. The genus *Inga*. Botany. Royal Botanic Gardens, Kew.

Poncy, O. 1985. Le genre *Inga* (Légumineuses, Mimosoideae) en Guyane Française. Mém. Mus. Natl. Hist. Nat., Sér. B, Bot. **31:** 1–124, pls. I–XI.

Species of *Inga* are known as *pois sucré* in Créole.

1. Leaves with 3 or more pairs of leaflets (this part of key includes species having leaves with (2)3, or (2)3–4 pairs of leaflets (e.g., *I. fastuosa, I. paraensis, I. poeppigiana, I. sarmentosa*).
    2. Leaves with basically 3 pairs of leaflets (uniformly 3 pairs, but occasionally also (2)3 pairs, or 3(4)pairs on same tree).
        3. Leaf rachis winged, at least for part of length.
            4. Plants hairy except for calyx. Corolla >30 mm long. Fruits densely rusty pubescent.
                5. Stems and leaf rachis with green to whitish, not very dense hairs. Leaflets sparsely pubescent abaxially. Corolla 40–50 mm long. Fruits 1.5–2 cm wide. . . . . *I. poeppigiana.*
                5. Stems and leaf rachis with rusty, very dense hairs. Leaves densely rusty pubescent. Corolla ca. 40 mm long. Fruits 2–3 cm wide. . . . . *I. fastuosa.*
            4. Plants not hairy throughout. Corolla <30 mm long. Fruits glabrous.
                6. Leaf rachis segments terete, or marginate or partly winged. Calyx <2 mm long.
                    7. Leaf rachis terete or canaliculate; glands wider than rachis and inflorescences axillary, not on brachyblasts. . . . . *I. pezizifera.*
                    7. Leaf rachis marginate or slightly or partly winged; either glands narrower than rachis or inflorescences on brachyblasts.
                        8. Petioles <10 mm long; rachis segments conspicuously winged but only just below leaflet insertions. Fruits 2–3 cm wide. . . . . *I. bourgoni.*
                        8. Petioles 20–45 mm long; rachis segments marginate to narrowly winged at least in the distal third. Fruits 1–2 cm wide. . . . . *I. alba.*
                6. Leaf rachis segments (and petioles) conspicuously winged for entire length. Calyx ≥2 mm long.
                    9. Leaf rachis segment wings triangular or widely winged, but not elliptic. Corolla pubescent, always more pubescent than calyx.
                        10. Calyx not acuminate in bud; corolla 7–8 mm long. Fruits ca. 11 cm long, not woody. . . . . *I. acreana.*
                        10. Calyx acuminate in bud; corolla 17–20 mm long. Fruits 20–40 cm long, woody. . . . . *I. rhynchocalyx.*
                    9. Leaf rachis segment wings elliptic. Corolla sparsely pubescent, never more pubescent than calyx. . . . . *I. auristellae.*
        3. Leaf rachis not winged.
            11. Stipules very conspicuous, foliaceous, wider than long. . . . . *I. stipularis.*
            11. Stipules not conspicuous, not foliaceous, narrower than long (linear).
                12. Petioles and leaf rachis woody, with lenticels, the glands not conspicuous; leaflets coriaceous. Peduncle thick. Calyx ca. 10 mm long. Fruits thick, woody, rough.. . . . . *I. sarmentosa.*
                12. Petioles and leaf rachis not woody, without lenticels, the glands conspicuous; leaflets not coriaceous. Peduncle slender. Calyx 1.5–3 mm long. Fruits thin, not woody, smooth.
                    13. Petioles and leaf rachis puberulous. Inflorescences spikes (5–7 cm long); pedicels lacking or nearly lacking. . . . . *I. albicoria.*
                    13. Petioles and leaf rachis glabrous. Inflorescences capitate; pedicels present. . . . . *I. paraensis.*
    2. Leaves with (3)4 or more pairs of leaflets.
        14. Leaf rachis terete, neither winged, marginate, flattened, nor grooved.
            15. Stems, leaves, and fruits glabrous. Stipules very conspicuous, foliaceous (as wide as long or wider). . . . . *I. stipularis.*
            15. Stems, leaves, or fruits velvety pubescent. Stipules not foliaceous.
                16. Stipules persistent. Petioles thick, woody; leaflet blades 17–25 cm long, glabrous abaxially. Inflorescences from branches or trunk, subcapitate. Fruits with distinct, reticulate venation. . . . . *I. retinocarpa.*
                16. Stipules deciduous. Petioles slender, not woody; leaflet blades 6–14 cm long, pubescent abaxially. Inflorescences axillary, spicate. Fruits without distinct, reticulate venation.
                    17. Foliar glands cup-shaped, raised, usually wider than rachis; leaflets sparsely pubescent abaxially. Flowers with dense, short pubescence; calyx narrowly tubular. Fruits with margins ≤5 mm wide, slightly raised, the pericarp not rigid. . . . . *I. thibaudiana.*

17. Foliar glands discoid, sessile, not wider than rachis; leaflets densely velvety, rusty-brown abaxially. Flowers with dense long pubescence; calyx campanulate to cup-shaped. Fruits with margins to 20 mm wide, with prominent rounded crests, the pericarp rigid. . . . . *I. rubiginosa.*

14. Leaf rachis marginate to distinctly winged, flattened, or grooved, at least for part of length, i.e., not terete.

18. Leaf rachis segments narrowly and regularly winged for entire length, or marginate, or distinctly flattened, or grooved.

19. Leaf blades >6 cm long. Inflorescences spicate.

20. Leaves glabrous. Calyx ≤4 mm long; corolla <8mm long, sparsely pubescent. . . . *I. acrocephala.*

20. Leaves pubescent. Calyx >4 mm long; corolla >8 mm long, densely pubescent (velutinous). . . . . *I. suaveolens.*

19. Leaf blades <6 cm long. Inflorescences capitate.

21. Pedicels inconspicuous. Trees in nonflooded forest. . . . . *I. gracilifolia.*

21. Pedicels 8–10 mm long. Shrubs or treelets on granitic outcrops. . . . . *I. virgultosa.*

18. Leaf rachis segments widely winged for at least part of length.

22. Stipules persistent, ≥10 mm long, striate.

23. Stems and stipules densely rusty pubescent. . . . . *I. fastuosa.*

23. Stems and stipules glabrous to sparsely pubescent. . . . . *I. macrophylla.*

22. Stipules deciduous or ≤10 mm long, not striate.

24. Leaf rachis segments winged distally.

25. Leaflet blades asperous. Corolla ca. 18 mm long. Fruits tetragonal. . . . . *I. striata.*

25. Leaflet blades glabrous. Corolla ca. 4 mm long. Fruits cylindrical or flattened, not tetragonal. . . . . *I. alba.*

24. Leaf rachis segments distinctly winged for entire length.

26. Leaflet blades pubescent abaxially. Bracts >4 mm long, sheathing calyx at base. Calyx 5–6 mm long. Fruits cylindrical, with conspicuous longitudinal grooves. . . . . *I. edulis.*

26. Leaflet blades glabrous or scarcely pubescent. Bracts inconspicuous, <4 mm long, not sheathing calyx at base. Calyx 1–2 mm long. Fruits flattened, without conspicuous longitudinal grooves. . . . . *I. alata.*

1. Leaves with 1–2 pairs of leaflets (this part of key includes species with 2(3) pairs of leaflets on same tree, a variation occuring occasionally in *I. marginata* and *I. sarmentosa*).

27. Leaves uniformly with 1 pair of leaflets.

28. Petioles not winged. Inflorescences spicate. . . . . *I. nouragensis.*

28. Petioles winged. Inflorescences capitate. . . . . *I. nubium.*

27. Leaves with 2 pairs of leaflets (rarely (1)2(3) on same tree).

29. Leaf rachis segments terete or flattened, not winged.

30. Stems and leaves entirely glabrous.

31. Corolla ≤12 mm long. Fruits fleshy, ca. 2.5 cm wide, the pericarp smooth, yellow at maturity. . . . . *I. capitata.*

31. Corolla 13–17 mm long. Fruits woody, 3.5–4 cm wide, the pericarp rough, pale brown at maturity. . . . . *I. sarmentosa.*

30. Stems and leaves not entirely glabrous.

32. Inflorescences capitate; peduncle 1–2 cm long. . . . . *I. huberi.*

32. Inflorescences spicate; peduncle 4–10 cm long.

33. Petioles 30–105 mm long. Inflorescences cylindric, the rachises 4–6 cm long. Calyx ca. 7–8 mm long. Fruits woody, ca. 4.5 cm wide. . . . . *I. longipedunculata.*

33. Petioles 10–15 mm long. Inflorescences conical, the rachises 1–2 cm long. Calyx 3.5–5.5 mm long. Fruits coriaceous, ca. 2.5 cm wide. . . . . *I. leiocalycina.*

29. Leaf rachis not terete, marginate or winged.

34. Inflorescences capitate or umbellate.

35. Petioles woody; leaf rachis segments marginate, usually winged along distal half. Fruits 30–35 cm long. . . . . *I. graciliflora.*

35. Petioles not woody; leaf rachis segments winged along entire length. Fruits 7–10 cm long. . . . . *I. umbellifera.*

34. Inflorescences spicate.

36. Leaves densely velvety abaxially. Spikes <5 cm long. . . . . *I. brachystachys.*

36. Leaves glabrous abaxially. Spikes ≥5 cm long. . . . . *I. marginata.*

Inga acreana Harms

Small to medium-sized trees, to 15 m tall. Bark reddish gray, the lenticels elongate. Stems glabrous. Stipules oblanceolate, ca. 6 mm long, glabrous. Leaves: petioles 15–25 mm long, flattened and grooved, sometimes slightly winged; rachis segments 1.5–3.5(5) cm long, triangularly winged for entire length, ≤10 mm wide just below insertion of leaflets, the glands sessile, 1 mm diam.; leaflets 3(4) pairs; blades of first pair 5–7 × 2–3 cm, markedly smaller than others, glabrous on both surfaces, or puberulent when young. Inflorescences spicate; peduncle 3–4 cm long, slender; rachis ca. 1 cm long; bracts 1–2 mm long, scale-like; pedicels lacking. Flowers: buds not acuminate; calyx ca. 5 mm long, tubular, slightly pubescent, the teeth ca. 1 mm long, regular; corolla tubular, 7–8 mm long, pubescent, always more pubescent than sepals. Fruits flat, asymmetrical at base, ca. 11 × 2 cm, not woody, the pericarp glabrous, greenish brown when ripe, the valves shallowly rippled, sometimes lenticellate, prominent around seeds, the endocarp stringy, expanded between seeds. Seeds 10–13 per fruit, not contiguous; cotyledons 2 × 1–1.3 cm, dark blue-green. Fl (Jul), fr (Dec); uncommon, in nonflooded forest.

Inga acrocephala Steud.

Medium-sized trees, to 20 m tall. Bark gray to light brown, smooth, dipple marks occasionally present. Stems glabrous. Stipules spatulate, ca. 4 mm long, striate, glabrous, early deciduous. Leaves: petioles 25–40 mm long, glabrous; rachis segments 2–5 cm long, glabrous, not winged but flattened to slightly marginate, the glands orbicular, sessile; petiolules 2–3 mm long, dark; leaflets (3)4(5) pairs. Inflorescences axillary or on lateral short shoots, 2–5 spikes; peduncle 2–7 cm long, sparsely tomentose; rachis 1–2 cm long, bearing 20–30 crowded flowers; bracts 1–2 mm long, scale-like, persistent; pedicels lacking. Flowers: calyx 2–4 mm long, campanulate, sparsely pubescent, the teeth 5, unequal; corolla funnel-shaped, 5–7 mm long, sparsely pubescent. Fruits straight or slightly curved, 20–25 × 3–4 cm, the pericarp glabrous, woody-fibrous with many lenticels, bright green then brown at maturity. Seeds 12–16 per fruit; cotyledons ca. 2.5 × 1.5 cm, green. Fl (Jul), fr (Apr); in nonflooded forest.

Inga alata Benoist PART 1: FIG. 4

Medium-sized trees, to 30 m tall. Bark light gray-brown, the lenticels dark reddish-brown, dense, in horizontal rows, the slash often exuding amber-colored resin. Stems angular, velvety pubescent. Stipules conspicuous, to 8 × 4 mm. Leaves: petioles 15–30 mm long, thick, flattened to marginate; rachis segments 2.5–4 cm long, winged (the wing broadest distally, to 3 mm wide) for entire length, the glands conspicuous, sessile, thick, sometimes asymetrical; petiolules pubescent; leaflets 4(5) pairs; leaflet blades to 13–22 × 4.5–8 cm, smooth adaxially, sparsely pubescent abaxially. Inflorescences spicate, axillary or on short-shoots, 4–5 cm long; peduncle 1–3 cm long, slender; rachises ca. 1.5 cm long, 30–50-flowered; bracts minute, scale-like, not sheathing calyx, curved, persistent; pedicels lacking. Flowers glabrous or very slightly tomentose; calyx tubular, 1–2 mm long; corolla funnel-shaped, 5–7 mm long, the lobes sharp. Fruits flat, 15–25 × ca. 3 cm, glabrous, the pericarp dark green when ripe, the valves shallowly rippled, eventually verrucose, seeds often prominent. Seeds 8–14, not contiguous; cotyledons 1.5 × 1 cm, green. Known only by sterile specimens from our area; relatively common, in nonflooded forest.

Inga alba (Sw.) Willd. FIG. 212E–G

Canopy to emergent trees, to 40 m tall. Trunks commonly with simple, rounded buttresses. Bark reddish-brown, with wide pock marks, the lenticels small and dense, the slash often exuding reddish sap. Stems with short, rust-colored pubescence. Stipules <10 mm long, not striate. Leaves: petioles 20–45 mm long, terete or grooved; rachis segments 1.5–3.5(4) cm long, puberulous, the segments marginate to narrowly winged in the distal half, the wings to 2 mm wide (sometimes wider in young plants), the glands raised to stalked, sometimes wider than rachis (especially in young plants); leaflets (2)3(4) pairs; leaflet blades to 9.5–15.5 × 5–8 cm, glabrous, grayish abaxially. Inflorescences dense short spikes, on short-shoots, these axillary or from branches; peduncle 5–12 mm long, pubescent; bracts inconspicuous; pedicels lacking. Flowers: calyx 1 mm long, puberulous; corolla 4 mm long, minutely pubescent, white; staminal tube exserted 5–6 mm, the free filaments 5–6 mm long. Fruits cylindrical or flattened, 9–15 × 1–2 cm, the pericarp grayish green when ripe, with tiny reticulate ripples, glabrous. Seeds 8–12 per fruit. Fl (Aug, Oct), fr (Feb, Sep); relatively common, in nonflooded forest. *Bougouni* (Créole).

Inga albicoria Poncy

Trees, to 10 m tall in our area. Bark cream-colored, the lenticels darker, the slash green, then cream-colored. Stems reddish-brown, lenticellate. Stipules not conspicuous, linear. Leaves: petioles 10–20 mm long, slender, not woody, terete to marginate just below insertion of leaflets, puberulous; rachis segments 1.5–3 cm long, slender, not woody, terete to marginate just below insertion of leaflets, puberulous, the glands orbicular, raised, conspicuous; petiolules 1 mm long, glabrous; leaflets 3 pairs; leaflet blades to 10–12 × 3.5–5 cm, not coriaceous, glabrous. Inflorescences axillary, spikes; peduncle ca. 2 cm long; rachis 5–7 cm long, bearing over 50 widely spaced flowers; bracts ca. 1 mm long, some stalked and peltate; pedicels lacking or inconspicuous. Flowers: calyx ca. 1.5 mm long, pubescent; corolla ca. 5 mm long, glabrous, green to greenish-white; staminal tube exserted to scarcely 2 mm, the free filaments to 5 mm long. Fruits 15 × 22 cm, the pericarp thin, papery, smooth. Seeds 12–14 per fruit, not contiguous. Fl (Aug); apparently rare, in disturbed nonflooded forest.

Inga auristellae Harms

Small understory trees, to 8 m tall. Bark grayish, the lenticels sparse. Stems pubescent when young. Stipules narrowly rhomboidal, 4 mm long. Leaves: petioles ca. 10 mm long, densely pubescent, marginate or narrowly winged; rachis segments 2–3(4)cm long, pubescent, winged for entire length, the wings elliptic, wider below leaflet insertion, 1–3 mm wide, the glands 1 mm diam., raised; leaflets (2)3 pairs; leaflet blades to 6.5–11 × 2–4.5 cm, glabrous adaxially except for hairs along main veins and margins, sparsely pubescent abaxially. Inflorescences spicate, axillary; peduncle 1–2 cm long; rachis 1–2.5 cm long, 20–30-flowered; bracts spatulate, 1.5 mm long (to 3 mm on infructescences), persistent. Flowers: calyx tubular, ca. 3 mm long, sparsely pubescent; corolla tubular, ca. 7 mm long, sparsely pubescent; staminal tube as long as corolla. Fruits 11–16 × 1.5–2 cm, 5–10 mm thick when mature with seed, the pericarp thin, coriaceous, glabrous. Seeds 4–8 per fruit. Fl (May, Jul, Aug), fr (Jan, Feb, Mar, Jul, Oct, Dec); common, in undisturbed forest, often in temporarily inundated areas.

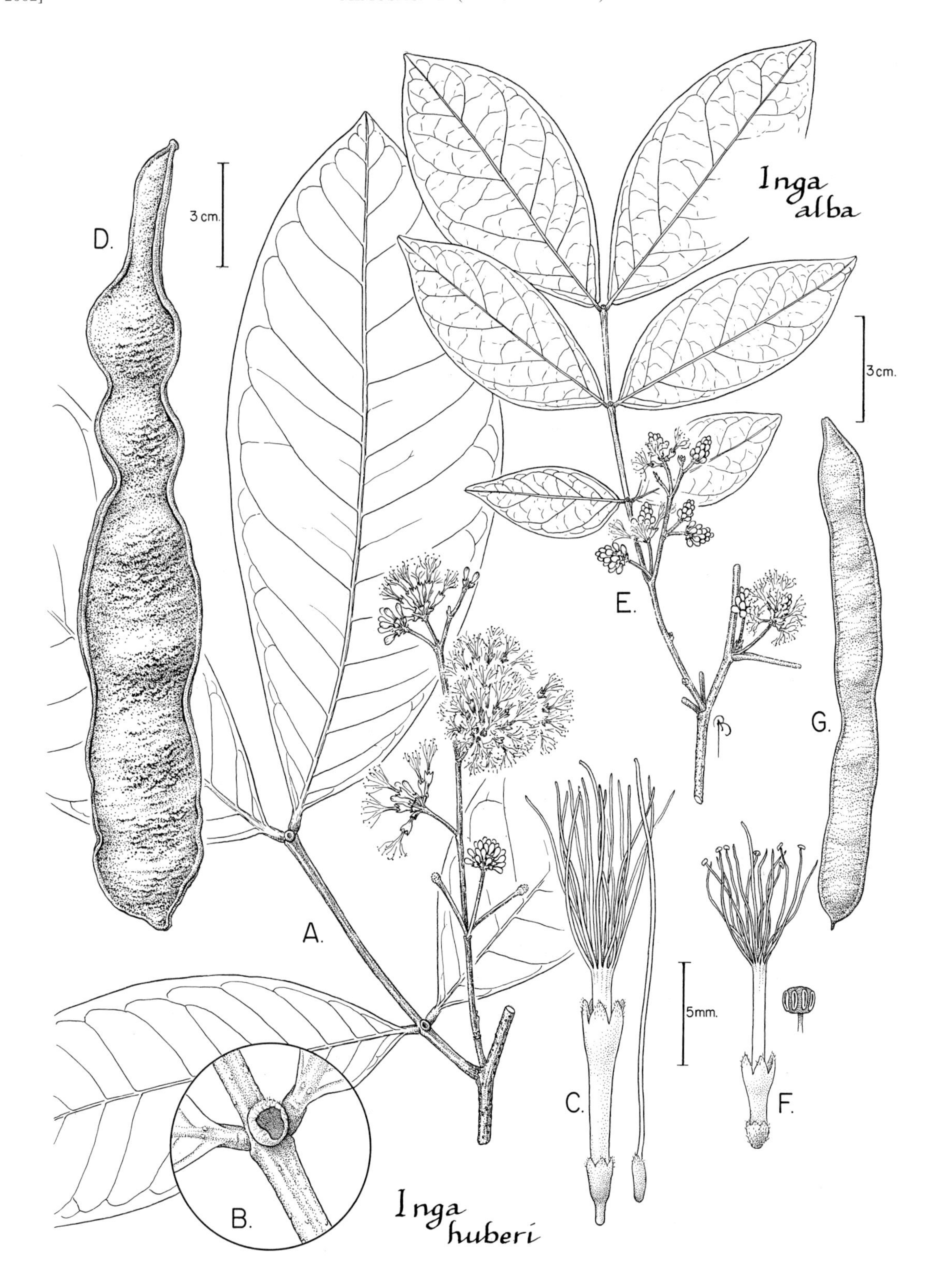

FIG. 212. MIMOSACEAE. **A–D.** *Inga huberi* (A–C, *Maguire 5444*; D, *Mori & Gracie 18903*). **A.** Part of stem with pinnately compound leaf and axillary inflorescence. **B.** Detail of extrafloral nectary at insertion point of leaflet pair. **C.** Lateral view of flower (near right) and pistil (far right). **D.** Fruit. **E–G.** *Inga alba* (E, F, *Marshall 191*; G, *French Guiana Forest Service 7667*). **E.** Part of stem showing pinnately compound leaf and inflorescences. **F.** Lateral view of flower (left) and anther (above). **G.** Fruit.

Inga bourgoni (Aubl.) DC.

Medium-sized trees, to 20 m tall. Bark reddish-brown, with hoop marks and pock marks, the lenticels in horizontal rows, the slash exuding sticky resin, the inner bark pink. Stems glabrous. Stipules conspicuous, linear or falcate, enlarged and rounded at apex, often persistent. Leaves glabrous; petioles reduced, <10 mm long, grooved; rachis segments 3–4 cm long, grooved, sometimes ridged around glands, conspicuously winged only below leaflet insertions, the wings 2 mm wide, the glands ca. 1 mm diam., narrower than rachis, sessile; leaflets 3 pairs; leaflet blades to 17–20 × 6.5–9 cm. Inflorescences axillary spikes; peduncle 0.5–1.5 cm long; rachis 2.5–5 cm long, 30–50-flowered; bracts scale-like, persistent; pedicels lacking. Flowers crowded; calyx 1 mm long, sparsely pubescent; corolla 5–6 mm long, sparsely pubescent, never more pubescent than calyx. Fruits 10–12 × 2–3 cm, cylindrical, fleshy when ripe, the pericarp glabrous, coriaceous, the sutures 4–5 mm wide, adorned with transverse, close, shallow ripples. Seeds 8–12 per fruit, contiguous. Fl (May, Jun); uncommon, in mature, nonflooded forest.

Inga brachystachys Ducke

Small to medium-sized understory trees, to 15 m tall. Stems slightly grooved, slightly pubescent. Leaves: petioles ca. 5 mm long, winged; rachis segments 1.5–2.5 cm long, narrowly to widely winged; nectaries orbicular; leaflets 2 pairs, sessile, the proximal ones (4)6–10(15) × (2)3–7 cm; blades glabrous adaxially, except the primary vein often brownish tomentose, densely velvety abaxially, the apex acute, or with a short, rounded acumen, the base acute, markedly asymmetrical. Inflorescences of 2–5 spikes, axillary to mature leaves or at defoliated nodes, or on lateral short shoots; peduncle slender, 1–2.5 cm long; rachis 1–1.5 cm long, 15–20-flowered; bracts inconspicuous, linear, ca. 0.5 mm long, pubescent. Flowers sessile, sparsely pubescent; calyx tubular, ca. 1.5 mm long, the teeth inconspicuous; corolla funnel-shaped, 5–7(10) mm long. Fruits flattened, 10–11 × 1.7–2 cm, the valves shallowly but conspicuously reticulate, sutures not thickened (mature fruit unknown). Fl (Nov); rare, in understory of undisturbed forest.

Inga capitata Desv.

Understory trees, usually to 12 m tall, occasionally taller. Trunks with occasional flying buttresses. Bark whitish to beige, sometimes with incomplete hoop marks. Stems lenticellate, glabrous. Stipules lanceolate, 5–11 × 1.5–2.5 mm, often persistent. Leaves: petioles ca. 15 mm long, often lenticellate, not winged; rachis 5.5–7.5 cm long, often lenticellate, not winged but sometimes slightly ridged around glands, the glands sessile, inconspicuous; leaflets 2 pairs; leaflet blades to 10–14 × 4.5–6.5 cm, glabrous, the base attenuate. Inflorescences spicate to subcapitate, axillary; peduncle to 6(8) cm long, thick, angular, lenticellate; bracts 3–4 × 1 mm, caducous. Flowers: calyx campanulate, ca. (4)8 mm long, sometimes persistent in fruit; corolla campanulate, (8)10–12 mm long, white, the lobes sometimes pinkish. Fruits when ripe subcylindrical, 14 × 2.5 cm, the pericarp fleshy, smooth, yellow at maturity. Seeds 8–12 per fruit; cotyledons black. Fl (Aug), fr (Apr, Aug); in undisturbed forest. *Pois sucré crapaud* (Créole).

Inga edulis Mart. FIG. 213A–D

Canopy trees, to 35 m tall. Bark light-colored, grayish to brownish, sometimes with incomplete hoop marks. Stems densely rusty pubescent, ribbed. Stipules 3–4 mm long, linear. Leaves: petioles 20–30 mm long, not winged; rachis segments ca. 3 cm long, pubescent, widely winged for entire length, the wings 5–8 mm wide, the glands conspicuous, transversely elliptic to kidney-shaped, 2–3 × 1–2 mm, sessile; leaflets (3)4(5) pairs; leaflet blades to 18 × 8 cm, conspicuously pubescent abaxially, especially pubescent on veins of both surfaces. Inflorescences spicate, often numerous in single axil; peduncle 2–4 cm long, rusty pubescent; rachis to 4 cm long, 30–40-flowered; bracts >4 mm long, sheathing calyx at base, caducous; pedicels lacking. Flowers: calyx tubular, 5–6 mm long, pubescent, striate; corolla 11–15 mm long, densely velutinous, creamy white or green; staminal tube slightly exserted. Fruits cylindrical, ca. 25 × 1 cm, with numerous, conspicuous, longitudinal grooves, with tightly appressed, light greenish-brown pubescence. Seeds 15–20 per fruit, longitudinally arranged; cotyledons brilliant black. Fl (Aug, Sep), fr (Jan, Feb, Jun); common, in disturbed areas.

Inga fastuosa (Jacq.) Willd. PART 1: FIG. 10

Small to medium-sized trees, to 20 m tall. Bark reddish-brown, the slash white. Stems, stipules, petioles, rachis and main veins densely rusty pubescent. Stipules lanceolate, 13–16 × ca. 7 mm, persistent, striate. Leaves: petioles ca. 15 mm long, rusty; rachis segments 3–6 cm long, widely winged, the wings 5–7 mm wide, the glands 1 mm diam., usually stalked; leaflets (2)3(4) pairs; leaflet blades to 14–21.5 × 6.5–13.5 cm, densely rusty pubescent. Inflorescences axillary, spicate; peduncle ca. 2 cm long, rachis ca. 1 cm long, densely pilose; bracts ca. 7 mm long, densely pilose, caducous; pedicels inconspicuous. Flowers: calyx 1.5–2 cm long, glabrous to sparsely pilose, striate; corolla ca. 40 mm long, pale yellow-green, densely velutinous, the hairs rusty-brown; staminal tube exserted ≥10 mm, the free filaments to 25–45 mm long, cream-colored. Fruits to 25 × 2–3 cm, twisted, densely hispid, the hairs rusty. Seeds to 25 per fruit; cotyledons brilliant black. Fl (May), fr (Jan, May, Oct); in disturbed areas.

Inga graciliflora Benth.

Understory trees, to 15 m tall. Stems minutely puberulous. Stipules scale-like, inconspicuous. Leaves: petioles 10–30 mm long, woody; rachis segments 3–6 cm long, woody, glabrous or very slightly pubescent, marginate or sometimes winged on distal half below insertion of leaflets, the glands ca. 2 mm diam., flattened, sessile; leaflets (1)2 pairs; leaflet blades to 23 cm long, with very tiny indumentum abaxially, mainly on veins. Inflorescences in axils of old leaves or from branches, capitate, the capitula 1–4, spherical or ovoid; peduncle ≤5 mm long, inconspicuously puberulous; bracts ca. 1 mm long, scale-like, inconspicuously puberulous, persistent; pedicels 2–3 mm long. Flowers: calyx 1 mm long, inconspicuously puberulous; corolla funnell-shaped, 4–5 mm long, inconspicuously puberulous; staminal tube exserted, twice as long as corolla. Fruits moniliform, to 30–35 × 2 cm, the pericarp glabrous, the sutures ribbed, the valves joining between seeds, with reticulate ornamentation. Seeds 8–14 per fruit; embryo dark green. Fl (Aug), fr (Jan); rare, in undisturbed forest.

Inga gracilifolia Ducke

Canopy or emergent trees, to 35 m tall. Bark reddish to brownish, the slash red, with red exudate. Stems reddish-brown, glabrous with age, slightly puberulous when young. Stipules linear, inconspicuous, usually deciduous. Leaves glabrous: petioles 10–15 mm long, slightly channelled, narrowly winged; rachis segments 1–1.5 cm long, slightly channelled, narrowly winged, the wings ca. 0.5 mm wide, the glands orbicular, small, raised; leaflets (4)5 pairs; leaflet blades to 4–6 × 1–1.5 cm, coriaceous, glabrous at maturity, glossy. Inflorescences capitate, solitary; peduncle 3–4 mm long,

Plates 89–96

Plate 89. MELIACEAE. **a.** *Carapa procera* (*Mori & Gracie 21528*), flowers and buds. **b.** *Guarea pubescens* (*Mori et al. 22028*), flowers and dehiscing fruit. [Photo by S. Mori] **c.** *Guarea grandifolia* (*Mori et al. 22645*), flowers and buds. **d.** *Guarea michel-moddei* (*Mori et al. 20963*), flowers on cauliflorous inflorescence. **e.** *Guarea michel-moddei* (*Pennington 12110*), two fruits on trunk. [Photo by S. Mori]

Plate 90. MELIACEAE. **a.** *Guarea scabra* (*Mori & Gracie 23886*), leaf with indeterminate apex between two leaflets. **b.** *Guarea silvatica* (*Mori et al. 24022*), dehisced fruits and seeds (some with red sarcotesta). **c.** *Trichilia euneura* (*Mori et al. 21989*), flowers and buds. **d.** *Trichilia euneura* (*Mori & Gracie 23931*), apical view of flower and buds. MENDONCIACEAE. **e.** *Mendoncia glabra* (*Mori & Gracie 21112*), flower. **f.** *Mendoncia hoffmannseggiana* (Mori & Gracie 21202), flower and bud.

Plate 91. MENISPERMACEAE. **a.** *Abuta bullata* (*Mori et al. 23444*, photographed in Sinnamary region of French Guiana), bullate leaves and cross-section of stem of liana; note bright yellow color of wood. **b.** *Abuta grandifolia* (*Mori et al. 24152*), flowers and buds. **c.** *Abuta grandifolia* (*Mori et al. 23206*), fruits. **d.** *Abuta rufescens* (*Mori et al. 23166*), inflorescence. **e.** *Abuta rufescens* (*Mori et al. 23166*), flowers.

Plate 92. MENISPERMACEAE. **a.** *Disciphania moriorum* (*Mori et al. 23980*), flower. **b.** *Sciadotenia cayennensis* (*Mori et al. 23906*), infructescence. **c.** *Orthomene prancei* (*Mori et al. 23263*), flower and buds.

Plate 93. MIMOSACEAE. **a.** *Abarema mataybifolia* (*Acevedo-Rodríguez 5029*), fruits, one dehisced to reveal seeds. **b.** *Cedrelinga cateniformis* (*Mori et al. 21547*), flower and buds. **c.** *Mimosa myriadenia* var. *myriadenia* (*Mori et al. 22893*), flowers. **d.** *Inga umbellifera* (*Mori et al. 24196*), inflorescence and base of leaf showing winged rachis and extrafloral nectary.

Plate 94. MIMOSACEAE. **a.** *Mimosa polydactyla* (*Mori et al. 24233*), inflorescence and infructescences of immature fruits. **b.** *Parkia nitida* (*Mori & Pipoly 15406*), inflorescence. [Photo by S. Mori] **c.** *Pseudopiptadenia psilostachya* (*Mori & Gracie 18649*), flowers and base of compound leaf with extrafloral nectary on petiole. MONIMIACEAE. **d.** *Siparuna decipiens* (*Mori et al. 24005*), pistillate flowers and buds. **e.** *Siparuna cuspidata* (*Mori et al. 22972*), fruits.

Plate 95. MONIMIACEAE. **a.** *Siparuna guianensis* (*Mori et al. 24170*), pistillate flowers. **b.** *Siparuna guianensis* (*Mori et al. 24170*), staminate flowers. **c.** *Siparuna poeppigii* (*Mori et al. 20762*), pistillate flowers. **d.** *Siparuna poeppigii* (*Mori et al. 22888*), fruits.

Plate 96. MORACEAE. **a.** *Brosimum rubescens* (*Mori et al. 24138*), abaxial surfaces of leaves, leaf buds, and immature fruit. **b.** *Clarisia illicifolia* (*Mori et al. 21593*), inflorescence. **c.** *Clarisia illicifolia* (*Mori et al. 22878*), infructescence. **d.** *Ficus gomelleira* (*Mori & Gracie 18284*), buttressed base of tree; note Scott Mori in tree to left of *Ficus*.

Carapa procera a.

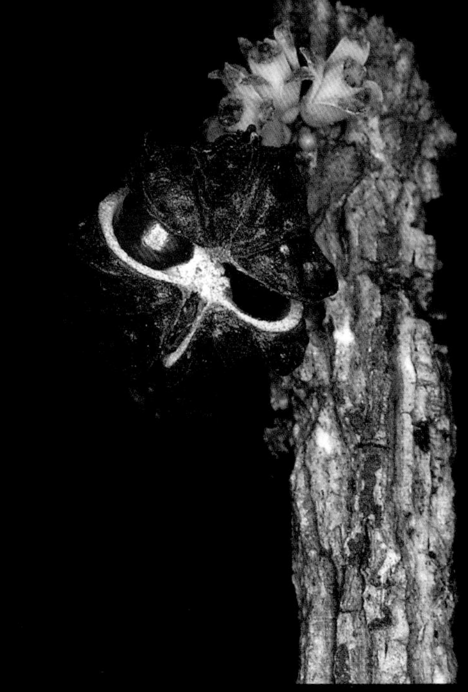

Guarea pubescens b.

Guarea grandifolia c.

Guarea michel-moddei d.

Guarea michel-modde e.

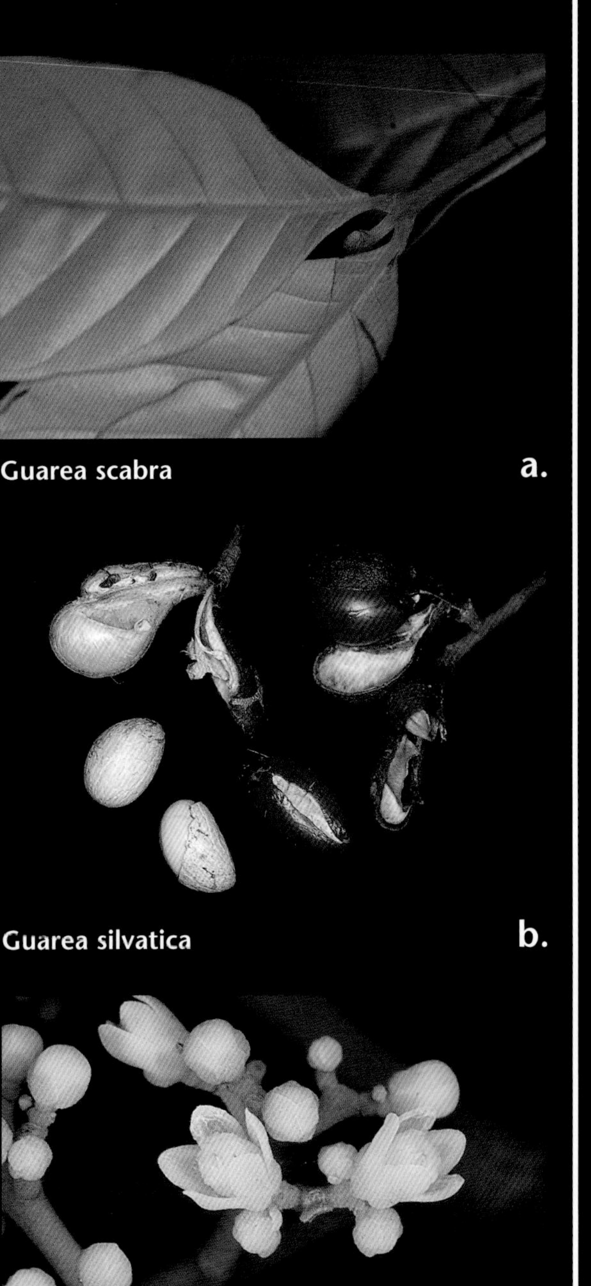

Guarea scabra a.

Guarea silvatica b.

Trichilia euneura c.

Trichilia euneura d.

Plate 90

MENDONCIACEAE

Mendoncia glabra e.

Mendoncia hoffmannseggiana f.

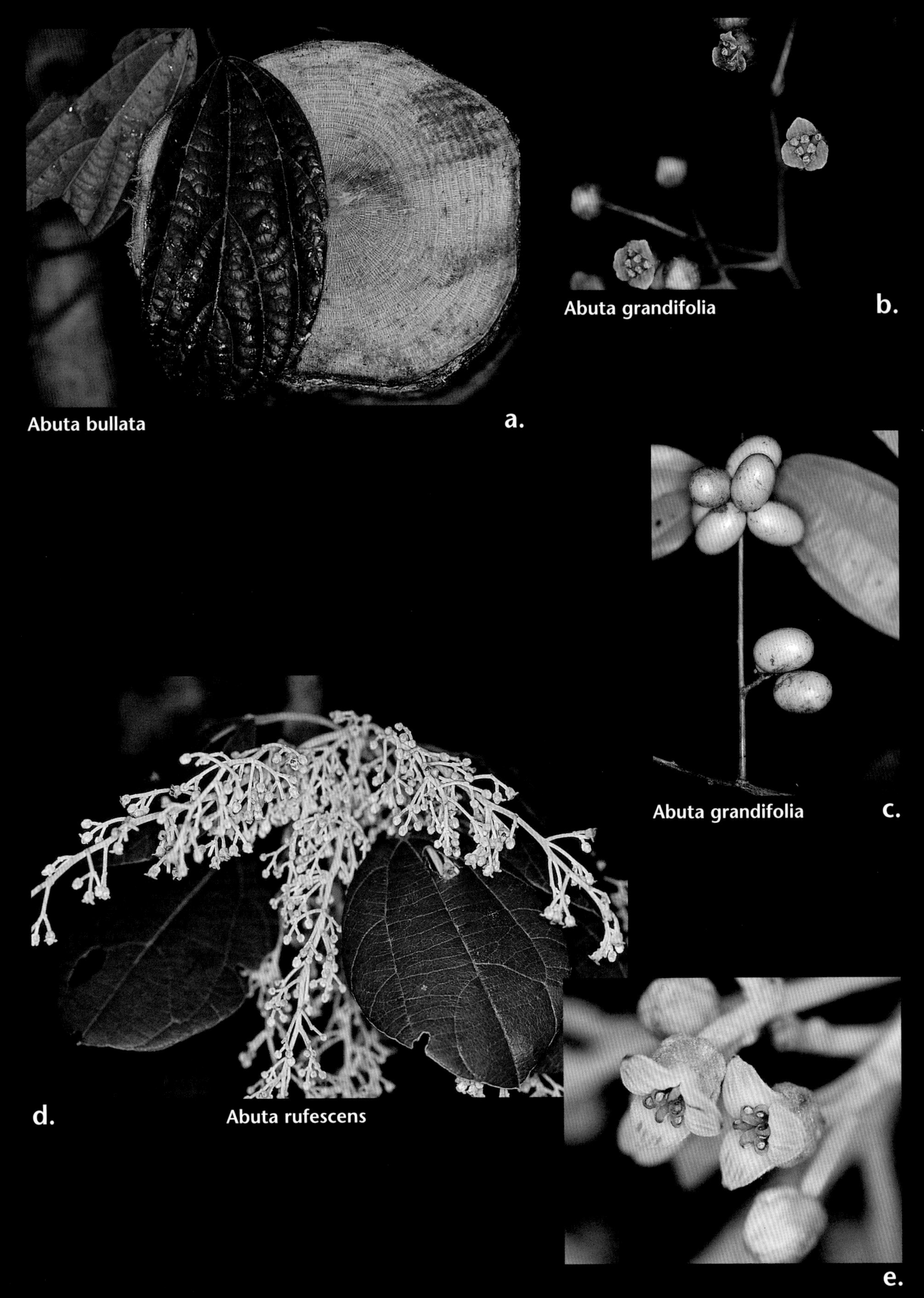

Abuta bullata a.

Abuta grandifolia b.

Abuta grandifolia c.

d. Abuta rufescens

e.

Plate 91

Disciphania moriorum **a.**

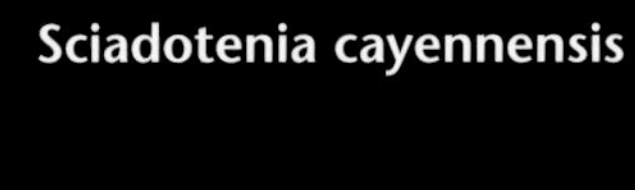

Sciadotenia cayennensis **b.**

Orthomene prancei **c.**

Plate 92

Abarema mataybifolia a.

Cedrelinga cateniformis b.

Mimosa myriadenia var. myriadenia c.

Inga umbellifera d.

Mimosa polydactyla a.

Pseudopiptadenia psilostachya c.

Parkia nitida b.

MONIMIACEAE

Siparuna decipiens d.

Siparuna cuspidata e.

Plate 94

Siparuna guianensis **a.**

Siparuna guianensis **b.**

Siparuna poeppigii 

Siparuna poeppigii **d.**

Plate 95

MORACEAE

Clarisia illicifolia **b.**

Brosimum rubescens **a.**

Clarisia illicifolia **c.**

Ficus gomelleira **d.**

Plate 96 (Moraceae continued on Plate 97)

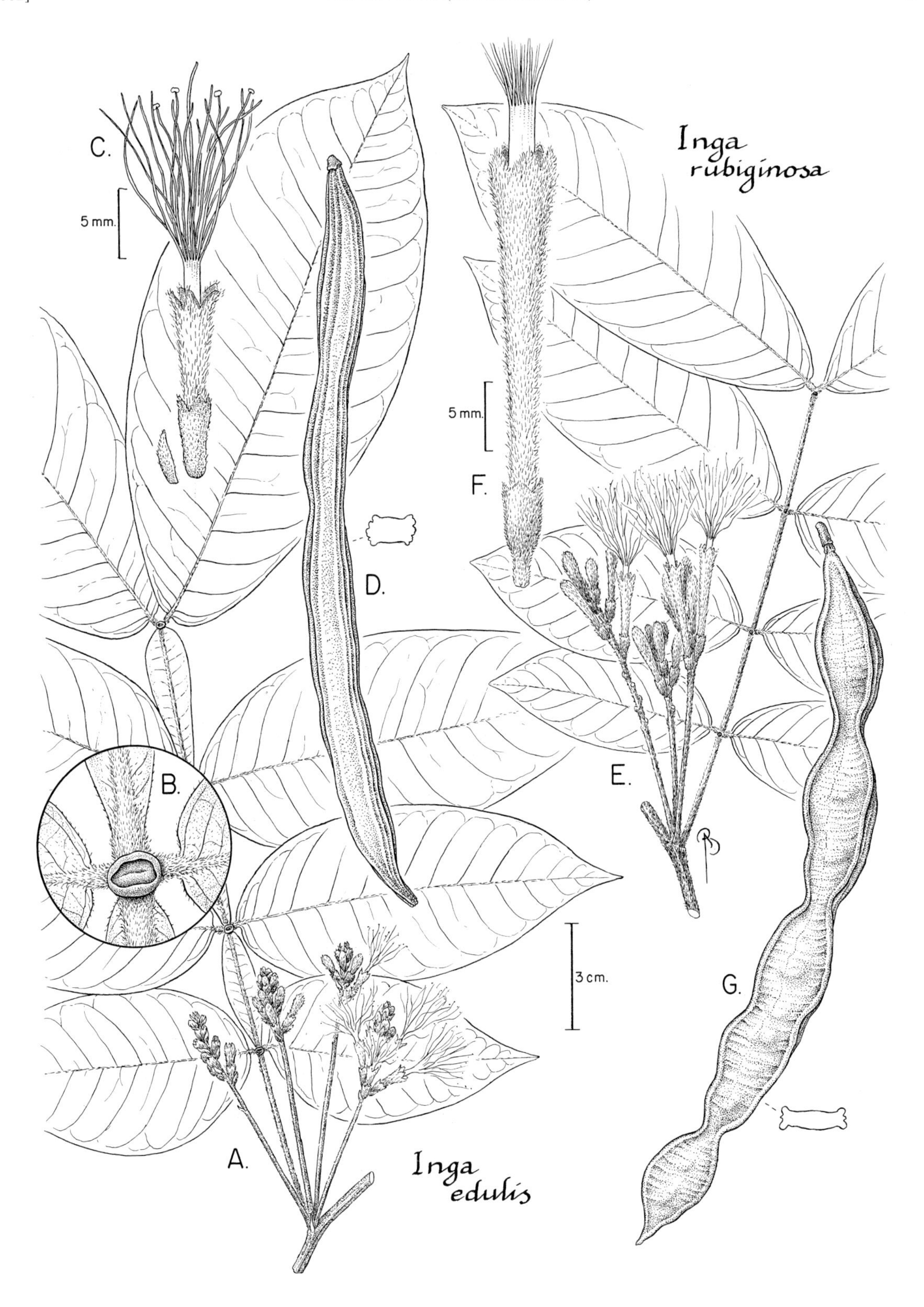

FIG. 213. MIMOSACEAE. **A–D.** *Inga edulis* (A–C, *Mori et al. 20844*; D, *Irwin 2699*). **A.** Part of stem showing pinnately compound leaf and axile inflorescences; note the winged rachis. **B.** Detail of extrafloral nectary at insertion of leaflet pair. **C.** Flower with bract (left). **D.** Fruit with diagrammatic transverse section (right). **E–G.** *Inga rubiginosa* (E, F, *Mori & Ek 20722*; G, *Oldeman 2243*). **E.** Part of stem with pinnately compound leaf and axile inflorescences; note the terete rachis. **F.** Flower showing base of filaments. **G.** Fruit with diagrammatic transverse section (right), this section wider when fruit is ripe.

slender; bracts extremely small (0.5 mm long), caducous; pedicels inconspicuous. Flowers: calyx 1 mm long; corolla narrowly tubular, then distally funnel-shaped, 5 mm long; staminal tube exserted. Fruits 18–25(>30) × 2–3 cm, flattened, the pericarp bright green, the valves with slightly prominent tiny reticulate ripples, the sutures not thickened, somewhat undulate. Seeds 13–18 per fruit, not contiguous, prominent. Known only by a single sterile collection from our area; apparently rare, in nonflooded forest.

Inga huberi Ducke FIG. 212A–D

Medium-sized to canopy trees, to 30 m tall. Trunks commonly with strongly irregular, even fluted buttresses. Bark dark, reddish-brown, sometimes dippled, the lenticels paler, the slash pink, exuding resin. Stems gray, puberulous, especially when young, lenticellate. Stipules linear, 3–4 mm long, velvety. Leaves: petioles 10–20 mm long, puberulous, not winged; rachis segments 2.5–3.5(4) cm long, puberulous, not winged, the glands conspicuous, cup-shaped, the proximal one often wider than distal one; leaflets 2 pairs; leaflet blades to 15.5–22 × 5.5–11 cm, glabrous to puberulous on veins. Inflorescences capitate, axillary to leaves or gathered on short-shoots; peduncle 1–1.5(2) cm long, slender, rusty pubescent, 20–30-flowered; bracts ca. 1.5 mm long, pubescent; pedicels present. Flowers: calyx campanulate, 2 mm long, puberulous; corolla 10 mm long, glabrous, pale green; staminal tube exserted ca. 5 mm, the free filaments to 7–9 mm long, white. Fruits flat, irregular in outline, 22–27 × 3.5–4 cm, the pericarp with short, rusty-brown pubescence, the margins raised, the valves with prominent network of oblique veins. Seeds 10–15 per fruit; cotyledons whitish to pale violet. Fr (Apr, Jun, Jul); occasional, in nonflooded forest in undisturbed and disturbed areas.

Inga leiocalycina Benth.

Medium-sized trees, to 25 m tall. Bark reddish-brown, the slash red, sometimes with white streaks. Stems reddish-brown, pubescent, lenticellate. Stipules narrowly spatulate, 3 mm long, puberulent. Leaves: petioles 10–15 mm long, pubescent, often flattened, not winged; rachis segments 2.5–3.5 cm long, densely pubescent, often flattened, not winged, the glands raised, conspicuous, usually wider than rachis; leaflets 2 pairs; leaflet blades to 12–15.5 × 4–5.5 cm, with scattered hairs on both surfaces, the veins more densely pubescent. Inflorescences axillary, of 2–5 spikes, conical; peduncle 4–8 cm long, slender, puberulous; rachis 1–2 cm long, bearing 20–30 congested flowers; bracts inconspicuous, ca. 1–3 mm long; pedicels lacking. Flowers: calyx tubular, 3.5–5 mm long, sparsely pubescent, irregularly cleft; corolla 6–8.5 mm long, velutinous, yellow-green; staminal tube exserted ca. 1 mm, the free filaments 5–7 mm long, white. Fruits to 30 × 2.5 cm, straight, constricted between seeds, coriaceous, the pericarp glabrous, dark green when ripe. Seeds 10–16 per fruit, not contiguous, prominent; cotyledons green. Fl (Aug, Sep), fr (Nov); common in undisturbed and disturbed forest.

Inga longipedunculata Ducke

Understory small to medium-sized trees, to 20 m tall. Bark gray, with incomplete hoop marks. Stipules linear, 2–3 mm long, early deciduous. Stems reddish-brown, glabrous. Leaves: petioles and rachis not winged, glabrous, woody when old; petioles 30–105 mm long, rachis segments 3–10.5 cm, not winged, glands raised, the distal one sometimes lacking; petiolules puberulous; leaflets (1)2(3) pairs; leaflet blades to 12–20 × 6.5–8.5 cm, sparsely pubescent adaxially, especially along veins. Inflorescences axillary, of very long spikes, cylindric; peduncle 8–10 cm long; rachis 4–6 cm long, the flowers widely spaced, the bracts ca. 1 mm long, acicular. Flowers: calyx 7–8 mm long, brown pubescent; corolla 12 mm long, yellow velutinous; free filaments ca. 10 mm long. Fruits to 45 × 4.5 cm, pericarp woody, thick, brown when ripe; sutures rounded, 1.5–2 cm wide. Seeds to 20 per fruit, not contiguous. Fl (Aug, Sep, Oct), fr (Jan, Feb, May, Nov); locally common, in undisturbed, nonflooded forest.

Inga macrophylla Willd. [Syn.: *Inga bracteosa* Benth.]

Small trees, to 16 m tall. Bark dark reddish-brown, the slash yellow. Stems angular, slightly grooved, glabrous to sparsely pubescent. Stipules narrowly ovate to lanceolate, 10–25 × 7–11 mm, glabrous, striate, persistent. Leaves: petioles (40)50–70 mm long, striate; rachis extending beyond distal pair of leaflets, the segments 3.5–7 cm long, slightly grooved, glabrous to puberulous, widely winged except for proximal part, the wings 3–6 mm wide, the glands raised to markedly stalked; leaflets 3 pairs; leaflet blades to 17–21 × 8–10.5 cm, scattered pilose hairs over both surfaces. Inflorescences of 2–4 axillary spikes; peduncle 6–10 cm long, thick, slightly grooved, glabrous to sparsely pilose; rachis 2–4 cm long; bracts 15–25 × 4 mm, lanceolate, striate, persistent. Flowers: calyx 25–30 mm long, glabrous, striate; corolla 45–50 mm long, densely velutinous, pale green; staminal tube exserted ca. 10 mm, the free filaments 30 mm long. Fruits to 40 × 4 cm, quadrangular in section when ripe, the pericarp pubescent only when very young, the sutures to 1.5 cm wide, winged, the faces shallowly rippled. Seeds to 20 per fruit. Fl (Aug, Sep, Oct, Dec), fr (Mar); locally common, in disturbed areas and in understory of primary forest.

Inga marginata Willd.

Small to medium-sized trees, to 20 m tall. Bark brown to orange-brown, sometimes with incomplete hoop marks, lenticels present. Stems glabrous, very sparsely lenticellate. Stipules striate, to 6 × 1 mm, caducous. Leaves glabrous; petioles 15–20 mm long, marginate to narrowly winged; rachis segments 2.5–4 cm long, marginate to narrowly winged, the wings 1–1.5 mm wide, widest below leaflet insertion, the glands 1 mm diam., sessile; leaflets (1)2 pairs; leaflet blades to 8.5–12.5 × 4–5 cm. Inflorescences of 1–2 axillary spikes; peduncle ca. 1 cm long; rachis 5–9 cm long, puberulous, the flowers widely separated; bracts 1–2 mm long, spatulate or acicular; pedicels lacking or short. Flowers: calyx 1–1.5 mm long, with scattered hairs (visible at 10×); corolla 3.2–4 mm long, glabrous; staminal tube exserted 2–3 mm, the free filaments 4–6 mm long. Fruits subcylindric, 13 × 1–1.6 cm at maturity, the pericarp glabrous, yellow, the margins slightly raised. Seeds 10–12 per fruit. Fl (Jun, Jul, Aug, Sep, Oct, Nov, Dec), fr (Jan, Feb, Sep, Nov); common, especially in disturbed areas.

Mori et al. 23276 (NY) and *Oldeman 2539* (CAY), which have been determined as *Inga* aff. *marginata*, differ from typical *I. marginata* by having conspicuous stipules, a more dense inflorescence, and both the petiole and rachis winged. These collections may represent a distinct species.

Inga nouragensis Poncy

Medium-sized to canopy trees, to 30 m tall. Bark cream-colored to brownish, with tightly spaced hoop marks. Stems grayish, irregular, the bark flaking off, internodes short, the young parts pubescent. Stipules inconspicuous. Leaves: petioles 12–15 mm long, marginate and grooved; rachis segments absent, the gland sessile, inconspicuous; leaflets 1 pair; leaflet blades often markedly asymmetrical and folded along midrib, 5–10(12) × 2.5–4(5) cm, glabrous or with very sparse short hairs; midrib conspicuously prominent adaxially. Inflorescences of 1–3 short axillary spikes; peduncle slender, 1.5–3 cm long; rachis 6–10 mm long, bearing ca. 25 congested flowers; bracts 1–2 mm long, scale-like; pedicels lacking. Flowers: calyx campanulate, 1–1.5 mm long, the lobes inconspicuous, pubescent at apex; corolla funnel-shaped, 4–6 mm long, glabrous; staminal tube exserted. Fruits (immature) 10–16 × 1.5–1.8 cm, the pericarp glabrous, the sutures not thickened. Seeds 8–10 per fruit. Fl (Aug), fr (Apr); in nonflooded, mature forest.

Inga nubium Poncy

Understory trees, to 10 m tall. Stems lenticellate. Stipules linear, scale-like, ca. 3 mm long. Leaves: petioles ca. 20 mm long, winged, the wings triangular, 5–8 mm wide; rachis segments absent, the glands raised, 1 mm diam.; leaflets 1 pair; leaflet blades to 6–8.5 × 3–3.5 cm, glabrous. Inflorescences axillary, capitate, solitary; peduncle 1–2.5 cm long, 20–30-flowered; bracts spatulate, 1.5–2 mm long, finely pubescent; pedicels 5 mm long. Flowers: calyx tubular, ca. 2 mm long, pubescent; corolla tubular, ca. 7 mm long, pubescent; staminal tube exserted ca. 1.5–2 mm, the free filaments 10 mm long. Fruits (immature) flat, 6–8 × 2.5 cm, the pericarp glabrous. Seeds 10 per fruit. Fl (Jan, Aug); in nonflooded forest.

Inga paraensis Ducke

Medium-sized to canopy trees, to 35 m tall. Bark beige to red brown, the lenticels conspicuous, in horizontal lines. Plants glabrous throughout. Stems angular, red brown, with pale lenticels. Stipules not conspicuous, linear. Leaves: petioles 10–20 mm long, flattened, shallowly grooved adaxially, not woody, glabrous; rachis segments 2–3.5 cm long, flattened, shallowly grooved adaxially, widening around glands, glabrous, the glands orbicular, conspicuous, appearing sunken; petiolules black, thickened; leaflets (2)3 pairs; leaflet blades to 7.5–14 × 3–7 cm, chartaceous. Inflorescences capitate; peduncle 6–8 cm long, slender, ca. 30-flowered; bracts linear, 2–3 mm long; pedicels ≥8 mm long. Flowers: calyx tubular, 3 mm long, insconspicuously lobed; corolla tubular, 10–11 mm long, the lobes 2 mm long. Fruits 18–30 × 2–2.5 cm, the pericarp thin, coriaceous, smooth, yellow at maturity. Seeds not contiguous, 10–22 per fruit, cotyledons green. Fr (Nov); in nonflooded, mature forest.

Inga pezizifera Benth.

Trees, to 25 m tall. Bark reddish brown. Stems with conspicuous, dense, clear-colored lenticels, glabrous. Stipules not foliaceous, caducous. Leaves glabrous; petioles 20–40 mm long, not winged (terete), often marginate; rachis segments terete to canaliculate adaxially, (3)4–5(7) cm long, not winged, the glands cup-shaped, to 2.5 mm diam., conspicuous, some, if not all, wider than rachis, raised; leaflets 3 pairs; leaflet blades to 12.5–15.5 × 5–7.5 cm; secondary veins in 8–9 pairs. Inflorescences axillary, spicate, solitary or clustered; peduncle 1.5–3 cm long, pubescent; rachis 1–1.5 cm long; bracts scale-like, ca. 1 mm long. Flowers minute; calyx ca. 1.5 mm long, pubescent; corolla 5.5–9 mm long, puberulous, green; staminal tube exserted 2–3 mm, the free filaments 9–10 mm long, white. Fruits flat, slightly to markedly curved, ca. 20 × 3–4 cm, the pericarp glabrous, subligneous, the sutures scarcely raised. Seeds 8–12 per fruit. Fl (May, Jun), fr (Feb, Apr, May); in nonflooded forest. *Bougouni* (Créole), *inga chichica* (Portuguese), *lebiweco* (Bush Negro).

Inga poeppigiana Benth.

Understory trees, to 10 m tall. Bark pale. Stems ribbed, irregular, young parts green, sparsely pubescent. Stipules lanceolate, 8–12 × 4 mm, striate, conspicuous, often persistent. Leaves with sparse hairs 1–2 mm long; petioles ca. 10 mm long, winged; rachis extending ca. 8 mm beyond distal leaflets, the segments 2–2.5 cm long, winged, the wings 4–6 mm wide, the glands circular, <1 mm diam., raised to very shortly stalked; leaflets (2)3 pairs; leaflet blades to 11 × 4.5 cm, supple, sparsely pubescent abaxially. Inflorescences axillary, spicate, solitary, with sparse hairs 1–2 mm long; peduncle ca. 1.2 cm long; rachis ca. 2 cm long, with ≤15 flowers; bracts ca. 5 mm long, channeled, resembling stipules but smaller; pedicels lacking. Flowers: calyx tubular, slightly swollen at base, ca. 20 mm long, glabrous, striate, green, the lobes irregular; corolla tubular, 40–50 mm long, sparsely pubescent, greenish-white; staminal tube exserted 2 cm, the free filaments 4 cm long. Fruits ca. 20 × 1.5–2 cm, straight, the pericarp densely rusty velutinous, the sutures not thickened, the valves thin, coriaceous. Seeds 14 per fruit, not contiguous, prominent. Fl (May); in undisturbed forest.

Inga retinocarpa Poncy

Understory treelets or trees, to 8 m tall. Bark beige or dark gray, warty. Stems hollow, covered with fine reddish-brown pubescence. Stipules linear, erect, 12 × 2 mm, persistent. Leaves very large, entirely glabrous; petioles terete, 150–180 mm long, thick, woody; rachis often extending beyond distal leaflets, the segments 6–8 cm long, thick, ligneous, terete, the glands sessile, inconspicuous; leaflets 3–4 pairs; leaflet blades to 17–25 × 5–6 cm, glabrous abaxially. Inflorescences from branches or trunk, 1–3 from same spot, subcapitate; peduncle ca. 1.5 cm long; bracts <1 mm long, triangular; pedicels short. Flowers greenish white; calyx minute, 1.5 mm long; corolla funnel-shaped, ca. 12 mm long, pubescent; staminal tube exserted 5 mm, the free filaments ca. 12 mm long. Fruits 8–14 × 1.5(2.5) cm, the pericarp not rigid, with distinct reticulate pattern, velutinous, yellowish-brown at maturity. Fl (Oct), fr (Mar, May); in nonflooded forest.

Inga rhynchocalyx Sandwith

Trees, to 30 m tall. Trunks with buttresses to 1 m tall. Young stems dark brown, densely lenticellate, glabrous. Stipules linear, 3–5 mm long, caducous. Leaves glabrous or inconspicuously pubescent when young; petioles ca. 30 mm long, winged; rachis segments 2.5–3.5(+) cm long, widely winged for entire length, the glands orbicular, 1 mm diam.; leaflets 3(4) pairs; leaflet blades to 13–18 × 6–7 cm, glabrous on both surfaces, glossy. Inflorescences axillary, subcapitate when immature, spicate at anthesis, solitary or geminate; peduncle 3–5 cm long, grooved, with inconspicuous sparse hairs; rachis subcapitate when young, to 4 cm long when in flower, 20–50(+)-flowered, the scars of flowers prominent and conspicuous; bracts linear, to 10 mm long, overtopping young buds; pedicels lacking. Flowers: buds acuminate; calyx tubular, ca. 10 mm long excluding acumen, glabrous, greenish to yellowish or cream-colored, the lobes acuminate, irregular, either only once-cleft and then unilabiate, or twice-cleft; corolla tubular, 17–20 mm long, pubescent, always more pubescent than calyx, cream-colored to white; staminal tube exserted. Fruits 20–40 × 3–4 cm, the pericarp woody, glabrous, the margins rounded around seeds. Fl (Aug); in nonflooded forest.

Inga rubiginosa (Rich.) DC. FIG. 213E–G

Trees, to 30 m tall. Bark cream-colored to light brown, with hoop marks. Stems densely rusty pubescent. Stipules deciduous. Leaves: petioles slender, 20–30 mm long, not woody; rachis segments slender, 2–3 cm long, not woody, densely pubescent, terete, the glands discoid, sessile, 1–2 mm diam., not wider than rachis; leaflets 4 pairs; leaflet blades to 6–14 × 3.5–6.5 cm, densely rusty-brown pubescent abaxially, with scattered pubescence adaxially. Inflorescences of 1–4 axillary spikes; peduncle 1–3 cm long, shallowly grooved; rachis 3–5 cm long, densely rusty pubescent, 10–20-flowered, the flowers loosely spaced. Flowers with long pubescence; calyx campanulate to cup-shaped, 3–5 mm long, nearly truncate to weakly lobed, velutinous, the hairs darker in color than those of corolla; corolla 20–27 mm long, velutinous, white to yellowish green; staminal tube exserted 2–5 mm, the free filaments 25–50 mm long. Fruits tetragonal at maturity, 21 × 2.2 cm, the pericarp rigid, velvety, brownish green, the venation not distinct, the margins prominently rounded, to 20 mm wide. Seeds to 20 per fruit. Fl (Aug, Sep); in undisturbed and disturbed forests.

Inga sarmentosa Harms

Medium-sized trees, 15–20 m tall. Trunks with low, running buttresses. Bark: slash red. Stems thick, glabrous, reddish-brown, the lenticels dense, verrucose. Stipules linear, falcate, 6 × 1.5 mm, often persistent. Leaves entirely glabrous; petioles ca. 10 mm long, thick, woody, lenticellate, not winged; rachis segments 4.5–5.5 cm long, thick, ligneous, lenticellate, not winged, the glands small (not conspicuous), sessile, immersed in rachis; petiolules 7 mm long, thick, blackish; leaflets (2)3 pairs; leaflet blades to 14–15.5 × 6–7.5 cm, coriaceous. Inflorescences axillary, often congested spikes, entirely glabrous; peduncle 2–6 cm long, thick; rachis ca. 2 cm long, the flowers crowded; bracts ca. 3 mm long; pedicels lacking or short. Flowers: calyx tubular, ca. 10 mm long, greenish, irregular; corolla tubular, 13–17 mm long, whitish, the lobes light purple; stamens and style pinkish. Fruits 5–13 × 3.5–4 cm, 2–3 cm thick, the pericarp thick, woody, the pericarp rough, lenticellate, pale brown at maturity, the sutures rounded. Fl (Jul), fr (May, Jun); apparently rare, in nonflooded forest.

Inga stipularis DC.

Small to medium-sized trees, to 20 m tall. Plants entirely glabrous. Bark gray, with incomplete hoop marks, the lenticels verrucose, the slash white. Stems irrgular, lenticellate. Stipules very conspicuous, suborbicular, often wider than long, 1–2 × 1–3 cm, persistent, foliaceous, especially when associated with inflorescences, the base asymmetric. Leaves: petioles 15 mm long, not winged; rachis segments terete, 2–5 cm long, the glands orbicular, ca. 2 mm diam., not wider than rachis, sessile; leaflets (2)3–4(5) pairs; leaflet blades to 10.5–17.5 × 4–7.5 cm, coriaceous. Inflorescences axillary spikes, sometimes in axil of undeveloped leaves at end of stems, appearing as terminal panicles; peduncle 2–4 cm cm long, thick; rachis 1–2 cm long, ca. 40-flowered, these congested, conspicuously scarred after flower abscission, nearly black when dried. Flowers: calyx tubular, 3–4 mm long, the lobes 5, acute; corolla ca. 8 mm long, pale green, the lobes pinkish; free filaments ca. 14 mm long, white. Fruits moniliform, 10–18 × 1–1.7 cm, the pericarp dark green at maturity, the valves coriaceous, joining between seeds, the sutures not raised. Fl (Feb, Apr, May), fr (Mar, Apr); common, in gaps and disturbed areas.

Inga striata Benth. [Syn.: *Inga nuda* Benth.]

Small trees, to 15 m tall in our area. Stems with reddish-brown pubescence, lenticellate. Stipules acicular, <10 × 1 mm, not striate. Leaves: petioles 35–40 mm long, not winged; rachis often extending beyond distal leaflets, the segments 3.5–4.5 cm long, pubescent, widely winged on distal half or two thirds, the wings wider on distal parts, to 2 mm, the glands circular, raised to stalked; petiolules ca. 2 mm long, pubescent; leaflets 4 pairs; leaflet blades to 14.5–19 × 5–8 cm, pubescent on both surfaces but more densely so on veins and abaxial surface; secondary veins in 14–17 pairs. Inflorescences axillary, spicate, >5 cm long; peduncle 1.5–2.5 cm long, angular, stout, pubescent; rachis ca. 3 cm long; bracts acicular, 2–3 mm long, caducous. Flowers: calyx tubular, 10 mm long, glabrous, striate due to salient venation; corolla tubular, 18 mm long, velutinous, white; free filaments to 25 mm long, white. Fruits tetragonal, 15–20 × 1.5 cm, the pericarp glabrous, the sutures raised, ribbed. Seeds 18 per fruit; cotyledons brown. Fl (Oct); in nonflooded forest.

Inga suaveolens Ducke

Small to medium-sized trees, 8–20 m tall. Trunks swollen at base. Bark smooth, with fine, closely spaced concentric striations, the slash with red bands separated by yellow bands. Stems reddish-brown, lenticellate, puberulous. Stipules linear, ca. 5 mm long, puberulent. Leaves: petioles 10–15 mm long, puberulous, narrowly winged distally; rachis segments 2–2.5 cm long, pubescent, marginate to slightly winged (more conspicuously so on distal parts), the glands raised, conspicuous, the proximal ones often transversely elliptic; petiolules pubescent; leaflets 4 pairs; leaflet blades 9.5–13.5 × 3–4.5 cm, with scattered hairs abaxially (visible at 10x). Inflorescences axillary, 1–4 spikes grouped together; peduncle 5.5–7.5 cm long, pubescent; rachis 1.5–2.5 cm long, bearing congested flowers; bracts ca. 2 × 1 mm, acicular. Flowers: calyx ca. 5 mm long, pubescent; corolla 11 mm long, velutinous; free filaments ca. 11 mm long. Fruits unknown. Fl (Aug, Sep, Oct); locally common, in mature forest.

Inga thibaudiana DC.

Small to canopy trees, to 25 m tall. Bark gray to cream-colored, the lenticels crowded, irregularly scattered. Stems densely pubescent, lenticellate. Stipules deciduous. Leaves: petioles slender, 15–25 mm long, not woody; rachis segments 2–3 cm long, densely pubescent, terete (winged, at least partly, in juvenile individuals), the glands cup-shaped, usually wider than rachis, raised; leaflets (4)5 pairs; leaflet blades to 8–10.5 × 3–4.5 cm, rough, sparsely pubescent adaxially, more pubescent and downy abaxially. Inflorescences of 1–5 axillary spikes; peduncle 2–3 cm long, pubescent; rachis 1–2.5 cm long, the flowers congested; bracts scale-like, ca. 1 mm long. Flowers with short pubescence; calyx narrowly tubular, 4–5 mm long, pubescent; corolla narrowly tubular, 14 mm long, velutinous; free filaments ca. 16 mm long. Fruits convex, straight, or distally curved, 18–21 × ca. 2 cm, the pericarp not rigid, densely velvety pubescent, greenish- to reddish-brown, the venation plane, the margins slightly raised, ≤5 mm wide. Seeds to 22 per fruit. Fl (Jul, Aug), fr (Jan, Feb, Apr, May, Jun, Jul); common, in undisturbed or disturbed forests in flooded or nonflooded areas.

Inga umbellifera (Vahl) Steud. PL. 93d

Shrubby to small trees, to 10 m tall. Stems glabrous, gray, the lenticels paler. Stipules linear, deciduous. Leaves: petioles 10–20 mm long, glabrous, not woody, winged for distal half; rachis segments 2–4 cm long, glabrous, winged along entire length, the wings 1.5 mm wide, sometimes spreading around glands; leaflets (1)2 pairs; leaflet blades to 11–16 × 4–7 cm. Inflorescences solitary in leaf-axils or on short-shoots, these 1.5–3 cm long, axillary or from branches, each unit of inflorescence umbellate; peduncle ca. 2 cm long, pubescent, the capitula 20–25-flowered; bracts linear, ca. 2 × 0.5 mm; pedicels 10 mm long, pubescent. Flowers: calyx 2–2.5 mm long, pubescent; corolla 7 mm long, glabrous, green; free filaments 14 mm long, white. Fruits subcylindrical at maturity, 7–10 × 2 cm, the pericarp smooth, bright yellow, the sutures not raised. Seeds 8–12 per fruit. Fl (May, Jul, Aug, Sep), fr (Feb); common, along rivers and in low forests on laterite or on granitic outcrops.

Inga virgultosa (Vahl) Desv.

Shrubs or treelets, often multi-stemmed, to 6 m tall, glabrous throughout. Bark grayish. Stems angular, verrucose, the lenticels brownish. Stipules linear, inconspicuous, persistent. Leaves: petioles ca. 5 mm long, marginate, channelled; rachis segments to 1 cm long, marginate, channelled, the glands <1 mm diam., circular or ovate; leaflets (3)4–6(10) pairs; leaflet blades to 3 × 1 cm, glossy, the base rhomboidal or nearly so. Inflorescences axillary, capitate, generally solitary; peduncle ca. 5 mm long, very slender; capitula 10–15-flowered; bracts scale-like, to 1 mm long; pedicels 8–10 mm long, slender. Flowers: calyx ca. 1 mm long; corolla funnel-shaped, ca. 4 mm long, the lobes 0.5 mm long, sharp; staminal tube not to shortly exserted. Fruits nearly cylindrical, variable in length but often <6 cm long, 2–2.5 cm wide, the base markedly asymmetrical, the pericarp glabrous, smooth, yellow at maturity, the sutures 2–3 mm wide, not raised. Seeds 4–10 per fruit. Fr (Jan); in low forest on Pic Matécho.

MIMOSA L.

Lianas, shrubs, subshrubs, or herbs. Epidermal prickles present. Stipules subulate to lanceolate, persistent, deciduous, or obsolete. Leaves bipinnate, alternate; petioles with or without glands; pinnae opposite; leaflets opposite, the first pair immediately beyond pinna-pulvinus often reduced and stipelliform; venation palmate or reduced to midrib. Inflorescences of spikes or capitula, solitary or fasciculate in leaf-axils, or terminal pseudoracemes or panicles; pedicels present or lacking. Flowers homomorphic; stamens 4, 5, 8, or 10. Fruits plano-compressed or turgid, the valves breaking away from replum into 1-seeded, dehiscent, or indehiscent articles. Seeds with testa pleurogrammic.

Barneby, R. C. 1991. Sensitivae censitae. A description of the genus *Mimosa* Linnaeus (Mimosaceae) in the New World. Mem. New York Bot. Gard. 65: 1–835.

1. Petioles with patelliform or broad-based glands. Stamens 5 or 10.
   2. Leaflets of longer pinnae 16–42 pairs per pinna, the largest <1 cm long. Inflorescences panicles of spikes. Stamens 5. . . . . *M. myriadenia* var. *myriadenia*.
   2. Leaflets of longer pinnae 1–3 pairs per pinna, the largest 2.5–12 cm long. Inflorescences panicles of capitula. Stamens 10. . . . . *M. guilandinae* var. *guilandinae*.
1. Petioles without glands. Stamens 4.
   3. Pinnae 3–5 pairs per leaf; leaflets of longer pinnae 28–65 pairs per pinna. . . . . *M. polydactyla*.
   3. Pinnae generally 2 pairs per leaf; leaflets of longer pinnae 11–27(36) pairs per pinnae. . . . . *M. pudica*.

Mimosa guilandinae (DC.) Barneby var. **guilandinae** FIG. 214

Lianas or sarmentose shrubs. Stems and leaf axes with prickles. Stipules triangular, subulate, caducous, or obsolete. Leaves: petioles with broad-based but narrow-pored glands, similar glands near tip of pinna-rachises; pinnae 2–3(4) pairs per leaf, the longest, further one 2–9.5 cm long; leaflets 1–2(3) pairs per pinna; blades obovate-suborbiculae or obovate elliptic, the largest 2.5–12 × 2–7.5 mm. Inflorescences nearly efoliate terminal panicles of globose capitula; peduncle 2–7 per node, rarely solitary. Flowers white or greenish-yellow; sepals 0.4–1 mm long; petals 1.3–2.3 mm long; stamens 10. Infructescences 1–3-fruited per capitulum. Fruits in profile oblong or broad-linear, 7–11 × 2.5–3 cm, plano-compressed, the replum smooth or with few recurved aculei, the valves breaking into individually dehiscent articles 5–7 × 15–35 mm. Fl (Aug), fr (Aug); often in weedy, open places.

Mimosa myriadenia Benth. var. **myriadenia** PL. 93c

Lianas, sometimes flowering as shrubs, 1–2 m tall. Stems and leaf axes with prickles. Stipules subulate, becoming dry and fragile in age. Leaves: petioles with shallowly patelliform glands, similar, smaller ones near ends of petioles and pinna-rachises; pinnae 5–33 pairs per leaf; rachises of longer ones 2.5–7.5 cm; leaflets 16–42 pairs per pinna; blades linear- to rhombic-oblong, the largest 3–9.5 × 0.6–3 mm. Inflorescences nearly efoliate panicles of spikelike racemes; peduncle 2–5 per node. Flowers white; sepals 0.5–0.7 mm long; petals 1.3–1.8 mm long; stamens 5. Infructescences 1–3-fruited per capitulum. Fruits pendulous, in profile linear-oblong, 5–10 × 1–1.8 cm, the articles individually dehiscent, 4–8 mm long. Fl (Jan, Feb); very common but conspicuous only when in flower, usually in disturbed areas. *Amourette, queue lézard* (Créole).

Mimosa polydactyla Willd. PL. 94a

Suffrutescent herbs, 50–200 cm tall. Stems with pair of prickles at each node. Stipules linear-attenuate or subsetiform, 4–8 × 0.25–0.6 mm, persistent. Leaves sensitive; primary axes without glands, 2–6(7) cm long; pinnae 3–5 pairs per leaf; rachises of longer ones 4–11 cm long; leaflets of longer pinnae 28–55(65) pairs per pinna; blades linear, the larger ones 4–9 × 0.7–1.8 mm, the base semicordate. Inflorescences axillary; peduncle 1–4 per node, 7–15 mm long, shorter than subtending petioles. Flowers whitish; sepals 0.2–0.4 mm long; petals 1.1–1.7 mm long; stamens 4, the filaments sometimes pinkish. Infructescences dense, spherical clusters. Fruits in profile narrowly oblong, 9–15 × 2.5–4.5 mm, the replum with straight setae 2–4 mm long, the articles individually indehiscent, 2.5–3.5 mm long. Fl (Feb, Jun, Aug, Sep, Oct), fr (Aug, Sep, Oct); common, in disturbed areas. *Juquiri* (Portuguese), *maillontre* (Créole), *malicia, malicia das mulheres* (Portuguese), *radié lan mort* (Créole), *sensitive* (French).

Mimosa pudica L.

Subshrubs, occasionally flowering as herbs. Stems with prickles below each node. Stipules lanceolate or linear-attenuate, persistent. Leaves sensitive; primary axes without glands, (1)2–5.5 cm long;

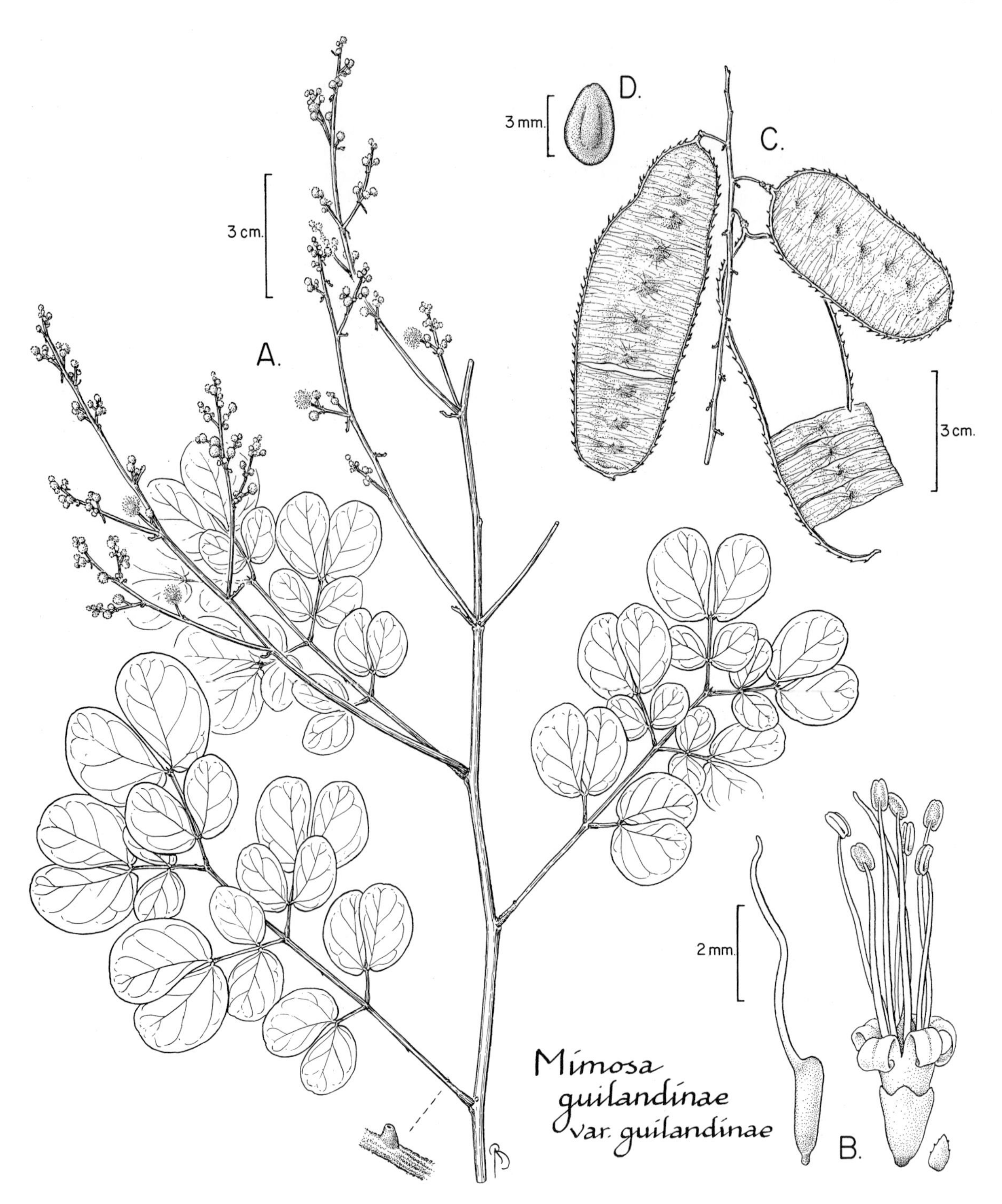

FIG. 214. MIMOSACEAE. *Mimosa guilandinae* var. *guilandinae* (A, B, *Feuillet 1554* from French Guiana outside our area; C, D, *Harley 24785* from French Guiana outside our area). **A.** Part of stem showing bipinnately compound leaves and inflorescences. **B.** Pistil (left) showing subapical attachment of style and flower (right) with bract (far right). **C.** Fruits showing partial dispersal of articles; note marginal prickles. **D.** Seed. (Used with permission of the *Flora of the Guianas*.)

pinnae 2 pairs per leaf, the rachises of longer ones 3–6.5(8) cm long; leaflets of longer pinnae 11–27(36) pairs per pinna; blades in outline linear-oblong, the longer ones (4)5–13(14) × 1–2.5 mm. Inflorescences of capitula, 1–7 per node; peduncle 1–3(3.5) cm long, the subtending leaf sometimes developing after flowers. Flowers whitish or pink-purple; sepals 0.1–0.2 mm long; petals 1.5–2.1 mm long; stamens 4. Infructescences several to many (30)-fruited per capitulum. Fruits in profile narrow-oblong, 8–15 × (2.7)3–4.5(5) mm, the replum hispid, with tapering setae 2–4(5) mm, the articles individually indehiscent, (2)2.5–4 mm long. *Juquiri*, *malicia*, *malicia das mulheres* (Portuguese), *radié lan mort* (Créole), *sensitive* (French).

PARKIA R. Br.

Trees. Epidermal prickles lacking. Stipules lacking. Leaves alternate, opposite or whorled; petioles with glands; pinnae opposite, subopposite, or alternate; leaflets opposite. Inflorescences simple or branched racemes of capitula; peduncles 1–2 per node, erect or pendent. Flowers either: 1) all bisexual and fertile, or 2) fertile or staminate flowers at base of inflorescence, the nectar-secreting ones at apex, or 3) neuter flowers at base, the staminate and nectar-secreting ones in middle of receptacle, and the fertile, hermaphroditic ones at apex; stamens 10, all fertile, filaments connate below into tube. Infructescences pendent or erect, the receptacle swollen, clavate. Fruits stipitate, strap-shaped in outline, sometimes falcate, dehiscent or indehiscent; endocarp sometimes mealy or gummy. Seeds in one or two rows, discoid, the testa hard, pleurogrammic.

Hopkins, H. C. F. 1986. *Parkia* (Leguminosae: Mimosoideae). Fl. Neotrop. Monogr. **43:** 1–124.

1. Capitula clavate or biglobose, with neuter staminodial flowers at base, modified nectar-secreting flowers in middle, and fertile flowers at apex. Fruits indehiscent.
    2. Leaves decussate (opposite and alternating in pairs at right angles along stem). Capitula erect. Fruits with valves densely velutinous. . . . . *P. decussata.*
    2. Leaves not decussate. Capitula pendent. Fruits with valves glabrate at maturity.
        3. Leaflets with midrib usually curved towards distal margin at apex. Peduncles alternate, usually solitary at nodes. Fruits with valves densely reticulate, depressed around widely spaced seeds. . . . . *P. reticulata.*
        3. Leaflets with midrib straight at apex. Peduncles opposite or subopposite, or several together at one node. Fruits with valves without distinct reticulations but corrugated over closely spaced seeds. . . . . . . *P. nitida.*
1. Capitula globose or oblate, lacking basal staminodial flowers. Fruits dehiscent.
    4. Inflorescences and infructescences pendent. Fruits dehiscent down adaxial suture only, the sutures with copious viscous gum, the valves glabrous. . . . . *P. pendula.*
    4. Inflorescences and infructescences erect. Fruits dehiscent down adaxial, and sometimes also abaxial, sutures, the sutures lacking gum, the valves covered with dense red-brown pubescence.
        5. Pinnae (17)30–45(55) pairs per leaf. Inflorescences sparsely branched or unbranched, proximal to whorl of leaves and not produced beyond foliage. Fruits with valves ligneous, ≤3 mm thick. . . . . . *P. velutina.*
        5. Pinnae 9–20 pairs per leaf. Inflorescences much branched, arising above leaves. Fruits with valves coriaceous or subligneous, ≤1 mm thick. . . . . *P. ulei* var. *surinamensis.*

Parkia decussata Ducke

Trees, to 40 m tall. Bark reddish-brown. Leaves bipinnate, opposite, decussate; pinnae 4–6(9) pairs per leaf; leaflets 21–35 pairs per pinna; blades oblong, (15)20–28 × (3)4–6 mm. Inflorescences simple racemes of capitula, the main axes ≤1 m long; peduncle 8–45 × 0.6–1 cm, erect; capitulum clavate, 5.5–7.5 cm long. Flowers nocturnal, of three types, the hermaphroditic ones with sepals 10–14 mm long; petals 11.5–15 mm long. Fruits widely undulately oblong, plano-compressed, indehiscent, (15)30–55 × 3.5–6 cm, the valves velutinous, the cavity with amber-colored sticky or crystalline gum. Seeds 19–22 × 7–9 mm, the testa black. Fr (Apr); in nonflooded forest.

Parkia nitida Miq. PL. 94b

Trees, to 40 m tall. Bark grayish. Leaves bipinnate, opposite or subopposite, not decussate; pinnae 3–9(12) pairs per leaf; leaflets 21–35(47) pairs per pinna; blades usually perpendicular to pinnarachis, oblong, (8.5)10–25 × 2.5–6.5 mm; midrib straight at apex. Inflorescences branched racemes of capitula; peduncle opposite or subopposite, 1–3 pairs per node, 2.5–16 cm long, pendent; capitula biglobose, 4.5–8 cm long. Flowers nocturnal, of three types, the hermaphroditic ones with sepals 6–10 mm long; petals 6.5–11 mm long. Fruits strap-shaped, often falcate, including stipe 20–40 cm long, indehiscent, the valves glabrous or glabrescent, without distinct reticulations but corrugated over closely spaced seeds, the cavity with amber-colored, sometimes cystalline gum. Seeds 17–22 × 8–11 × 5.5–8.5 mm, the testa black. Fl (Mar), fr (Jun, Jul, Sep); common, in nonflooded forest. *Bois bouchon* (Créole).

Parkia pendula (Willd.) Walp.

Trees, to 40 m tall. Bark gray or reddish, peeling in large rectangular flakes. Leaves alternate or whorled; petioles with small gland; pinnae (12)15–27 pairs per leaf, opposite or alternate; leaflets 45–96(112) pairs per pinna; blades linear, 3–6(11) × 0.5–1(1.5) mm. Inflorescences simple or branched racemes of capitula; peduncle usually 1 per node, alternate, (15)30–115 cm long, pendent; capitula globose, 3.8–4.9 × 3–3.4 cm. Flowers nocturnal, andromonoecious, of two types, the fertile, hermaphroditic flowers at base of receptacle, the nectar-secreting ones at apex; hermaphordite flowers with sepals 8–10.5 cm long; petals 9.5–12 mm long. Fruits strap-shaped, sometimes falcate, stipitate, 15–30 × 1.9–3 cm, dehiscent through adaxial suture, the valves glabrous, the adaxial suture much thickened and secreting large amounts of sticky amber-colored gum, the cavity without gum. Seeds released into gum, in one or partially two series, 7–11 × 4–6 × 2.5–3 mm, the testa dark with paler blotches. Fl (Aug), fr (Sep); apparently rare in our area, in nonflooded forest. *Acacia mâle, bois-ara* (Créole) *pau de arara, visgueira joerana* (Portuguese).

Parkia reticulata Ducke

Trees, 25–30 m tall. Trunks with buttresses, these often branched. Leaves alternate; pinnae 10–16 pairs per leaf, opposite or subopposite; leaflets 40–44 pairs, linear-oblong or sigmoid, 5–9 × 1–1.5 mm; midrib usually curved toward distal margin at apex. Inflorescences unbranched; peduncle ≤8 per cluster, alternate, usually solitary at nodes, pendent, 8–22 cm long; capitula biglobose, ± 5.5 cm long. Flowers nocturnal, hermaphroditic, of three kinds, the hermaphroditic flowers with sepals 5–5.5 mm long; petals ± 6.5 mm long. Fruits strap-shaped, stipitate, 28–35 × 3–3.8 cm, indehiscent, the valves usually glabrous, densely reticulate, depressed around widely spaced seeds. Seeds: testa unknown. Fl (Feb), fr (Mar, May from ground); in nonflooded forest.

Parkia ulei (Harms) Kuhlm. var. **surinamensis** Kleinhoonte

Trees, 35–50 m tall. Trunks with plank buttresses. Bark smooth. Leaves alternate or in short whorls; pinnae 9–20 pairs per leaf, opposite or subopposite; leaflets 30–55 pairs per pinna; blades narrowly oblong, 3–10 × 0.5–2 mm. Inflorescences erect, terminal racemes of capitula, arising above leaves; peduncles ≤15 per cluster, 0.4–1 cm long; capitula subglobose, 1–1.8 cm diam. Flowers diurnal, monomorphic, hermaphroditic; sepals 2.5–4 mm long; petals 3–5 mm long. Fruits strap-shaped, sometimes curved, stipitate, 9–36 × 3–4.7 cm, dehiscent along adaxial suture, tardily so along abaxial one, the sutures scarcely thickened, the valves with reddish-brown velutinous pubescence or subglabrous, coriaceous or subligneous, ≤1 mm thick, lacking gum. Seeds 11–18 × 5–12 × 4–5 mm, the testa black. Fl (May, Jul), fr (Jul); the only species of *Parkia* in our area with diurnal flowers, in nonflooded forest.

Parkia velutina Benoist PART 1, FIG. 10

Trees, 40 m tall. Bark smooth. Leaves in whorls at ends of stems; pinnae (17)30–45(55) pairs per leaf, opposite or subopposite; leaflets 40–110 pairs per pinna, linear to narrowly sigmoid, 5–11.5 × 1–2 mm. Inflorescences proximal to whorl of leaves, ascending, not produced beyond foliage, sparsely branched or unbranched; peduncles ≤15 per cluster, alternate, 1.5–4 cm long; capitula globose, 2.5–3.5 cm diam. Flowers nocturnal, monomorphic, hermaphroditic; sepals 7–8 mm long; petals 11.5–13 mm long. Fruits strap-shaped, sometimes curved, 18–50 × 4.7–6(8) cm, dehiscent along adaxial suture and sometimes along abaxial one, the valves with reddish-brown velutinous pubescence, ligneous, ≤3 mm thick, lacking gum. Seeds 18–20 × 9–10 × 5–6 mm, the testa black. Fr (Oct from ground); in nonflooded forest.

PIPTADENIA Benth.

Lianas. Epidermal prickles on stems and leaf axes. Stipules small, persistent, or apparently lacking. Leaves bipinnate, opposite; petioles with a gland, these often also present between some distal pairs of leaflets; leaflets with venation pinnate. Inflorescences of spikes either axillary to leaves or in exserted efoliate pseudoracemes or panicles of spikes; pedicels lacking. Flowers: stamens 10, the anthers (at least in bud) gland-tipped. Fruits stipitate or subsessile, broad-linear or oblong-elliptic, straight or almost so, plano-compressed, dehiscent through both sutures, the valves inertly separating to release seeds. Seeds plumply discoid, the testa hard, pleurogrammic.

Piptadenia floribunda Kleinhoonte FIG. 215

Leaves: pinnae 2 pairs per leaf, the rachis of further pair 4–7 cm long; leaflets 3–4 pairs per pinna, distally larger; blades of terminal pair 4–6 × 2.3–3.7 cm. Inflorescences: main axes 2–4 dm long, the secondary axes spicate, 2–4.5 cm long. Flowers bisexual and staminate, yellow; calyx campanulate, 0.6–0.8 mm long; petals 1.7–2 mm long. Fruits ± 10 × 3 cm, the valves papery, framed by straight sutures, the exocarp veinless. Seeds ellipsoid. Fl (Sep); apparently rare, in nonflooded forest.

PSEUDOPIPTADENIA Rauschert

Trees. Trunks buttressed. Epidermal prickles lacking. Stipules small and caducous or lacking. Leaves bipinnate, alternate; petioles with glands; pinnae opposite or subopposite, the distal pinnae not modified into tendrils; leaflets opposite or alternate; venation palmate. Inflorescences of shortly pedunculate spikes, solitary or paired in leaf-axils, or in terminal panicles. Flowers homomorphic; stamens 10, free to base. Fruits stipitate, in profile elongately linear, straight, decurved or randomly twisted, plano-compressed, dehiscence follicular, through adaxial suture, the valves glabrous. Seeds plano-compressed, in broad view oblong-elliptic to broad-linear, not persistent on funicles, the testa membranous, lustrous, translucent, widely winged, the pleurogram lacking; endosperm lacking.

1. Leaves with petiolar glands vertically elongate, mostly 4.5–14.5 mm long, ≥ width of petiolar groove; larger leaflets 7.5–11.5 × 2.5–4.5 mm. Fruits (1.4)1.8–2.4 cm wide. . . . . *P. psilostachya.*
1. Leaves with petiolar glands often obsolete or lacking, when present 2–6 mm long, sunk into petiolar groove; larger leaflets 5–9 × 1.4–2.3 mm. Fruits 0.9–1.5 cm wide. . . . . *P. suaveolens.*

Pseudopiptadenia psilostachya (DC.) G. P. Lewis & M. P. Lima. FIG. 216, PL. 94c

Trees. Leaves: petioles with vertically elongate, sessile glands (2)4.5–14.5 mm long, near or above mid-petiole, its rim ≥ width of groove of petiole, smaller glands at tip of primary rachises and yet smaller ones between furthest 1–2 pairs of leaflets; pinnae 6–10 pairs per leaf, the rachis of longer ones (6)7–11 cm long; leaflets 21–32 pairs per pinnae, oblong, straight or shallowly sigmoid, the longer ones near mid-rachis, 7.5–11.5 × 2.5–4.5 mm. Inflorescences of pedunculate spikes, solitary or fasciculate in leaf-axils or in short, terminal panicles, the ultimate axes 7–13 cm long. Flowers pale yellow; sepals 1.2–1.5 mm long; petals 2.7–3 mm long. Fruits pendulous, linear, straight or nearly so, rarely twisted, the stipe 1–2.5 cm long, the body (15)25–52 × (1.4)1.8–2.4 cm. Seeds 4.5–5.5 × 1–1.5 cm. Fl (Aug), fr (Jan); apparently rare, in nonflooded forest.

Pseudopiptadenia suaveolens (Miq.) J. W. Grimes

Trees, 35–60 cm tall. Trunks strongly buttressed. Leaves: petioles generally lacking glands, or when present sunk into adaxial sulcus, linear-elliptic, 2–6 mm long; pinnae 7–11(12) pairs per leaf, the rachis of longer ones 4.5–8.5 cm long; leaflets 23–33 pairs per pinna, the longer ones 5–9 × 1.4–2.3 mm. Inflorescences of pedunculate spikes, solitary or fasciculate in leaf-axils or in short terminal panicles, the ultimate axes 5–13 cm long. Flowers

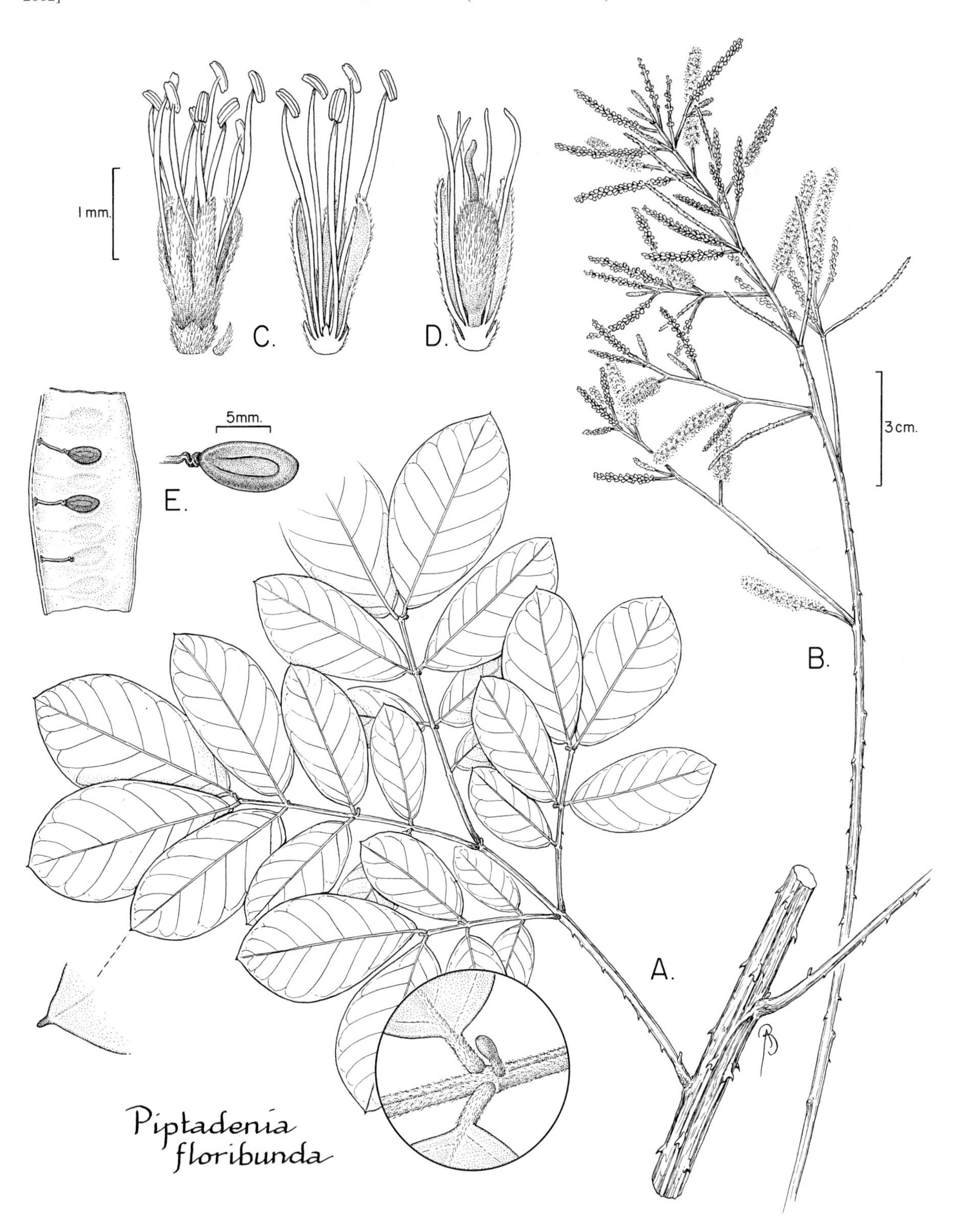

FIG. 215. MIMOSACEAE. *Piptadenia floribunda* (A, B, *Mori & Bolten 8463* from Surinam; C–E, *Rabelo 3097* from Amapá, Brazil). **A.** Part of stem showing armed ridges and bipinnately compound leaf with prickles and extrafloral nectaries near petiole base and below insertion points of leaflets; note detail of stalked nectary (left). **B.** Inflorescences. **C.** Lateral view (far left) and medial section (near left) of staminate flower with detail of bract (left). **D.** Pistillate flower with perianth and androecium partially removed. **E.** Part of dehisced fruit with seeds (left) and seed (right).

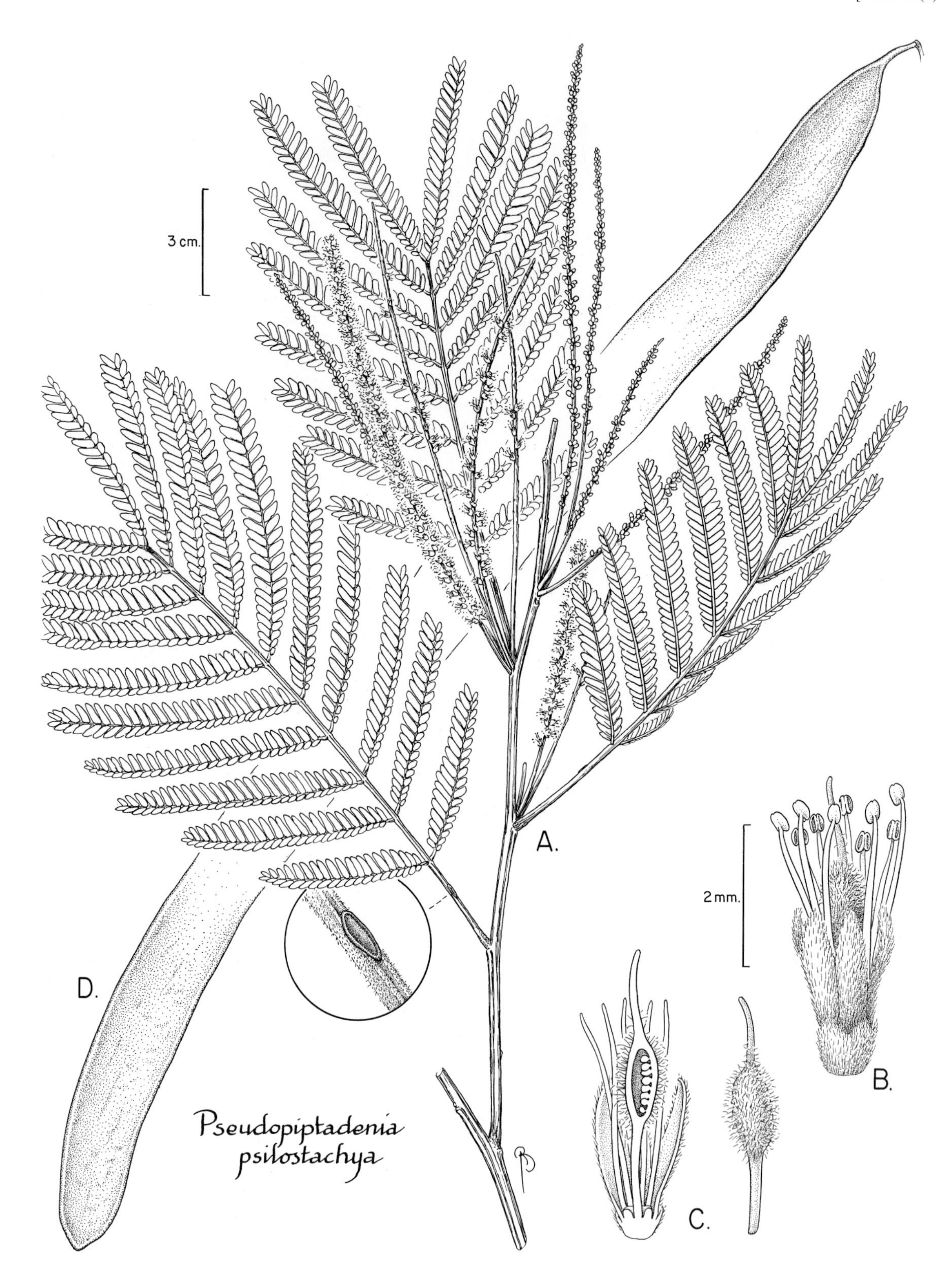

FIG. 216. MIMOSACEAE. *Pseudopiptadenia psilostachya* (A–C, *Irwin et al. 54449* from Surinam; D, *Service Forestier 7226*). **A.** Part of stem with bipinnately compound leaves and spicate inflorescences; note detail of extrafloral nectary near midpoint of petiole (left). **B.** Lateral view of flower. **C.** Medial section (left) of flower showing pistil surmounting gynophore and subtended by a nectary disc and pistil (right). **D.** Fruit.

yellow-green or pale yellow; sepals 0.7–1 mm long; petals 2.3–2.7 mm long. Fruits pendulous, linear, flattened, straight or nearly so, rarely twisted, the stipe 1–2.5 cm long, the body 20–42 × 0.9–1.5 cm. Seeds, including broad wing, 36–46 × 8–10 mm. Fl (Dec), fr (Jun), seedlings (Jun); common but scattered, in nonflooded forest.

STRYPHNODENDRON Mart.

Trees. Trunks smooth. Epidermal prickles lacking. Stipules lacking, or at least not seen. Leaves bipinnate, alternate; petioles with glands; pinnae opposite or alternate, the distal pinnae not modified into tendrils; leaflets alternate or opposite. Inflorescences of spikes, solitary or fasciculate in leaf-axils or in terminal panicles, the ultimate axes pedunculate; peduncle with two caducous bracts below flowers; pedicels lacking. Flowers homomorphic; stamens 10, the anthers gland-tipped in bud. Fruits (sub)sessile, linear-oblong, ± terete, straight or slightly falcate, ± pulpy, essentially indehiscent, the endocarp raised between seeds as partitions, the seeds released by animals or by weathering of the valves. Seeds plump, not persistent on funicles, the testa brown or black, hard, areolate, the pleurogram present; endosperm present.

1. Pinnae 7–13 pairs per leaf; leaflets alternate, 8–13 pairs per pinna, 10–23 × 4.5–10 mm. . . . . *S. guianense.*
1. Pinnae 1–6 pairs per leaf; leaflets opposite, 2–7 pairs per pinna, 55–150 × 18–75 mm.
   2. Pinnae 3–6 pairs per leaf; leaflets 4–7 pairs per pinna, pulvinules 1.5–3 mm long. Fruits incurved and twisted. . . . . *S. polystachyum.*
   2. Pinnae 1–3 pairs per leaf; leaflets 2–5 pairs per pinna, pulvinules 7–9 mm long. Fruits gently falcate but not twisted. . . . . *S. moricolor.*

Stryphnodendron guianense (Aubl.) Benth.

Leaves: primary axes of longer leaves 13–28 cm long; petioles 4–10(12) cm long, with glands below mid-petiole; pinnae either opposite or alternate, 7–13 pairs per leaf, the rachis of distal ones 4–14 cm long; leaflets alternate, 8–13 pairs per pinna; blades rhombic-oblong or oblong-obovate, the largest 10–23 × 4.5–10 mm. Inflorescences solitary spikes or 2–5-fasciculate in leaf-axils, the axes ascending or recurved; peduncle 8–35 mm long, the rachises 5.5–13.5 cm long. Flowers: sepals 0.5–0.8 mm long; petals 1.7–2.8 mm long, white. Fruits subsessile, broad-linear, straight or slightly falcate, compressed but plump, dehiscence very tardy along both sutures, but valves not gaping widely enough to release seeds. Seeds compressed-ovoid, in broad view 8–11 × 5–7 mm, the testa black. Fl (Dec), fr (Apr); in nonflooded forest, sometimes in disturbed areas such as along airport road.

Stryphnodendron moricolor Barneby & J. W. Grimes
FIG. 217

Leaves: primary axes of longer leaves 8–17 cm long; petioles 2–7 cm long, with glands between each pinna-pair; pinnae opposite, 1–3 pairs per leaf, the rachis of longer ones 10–25 cm long; leaflets opposite, 2–5 pairs per pinna; pulvinules 7–9 mm long; blades ovate or obovate-elliptic, the furthest, largest pair 10–15 × 4–7.5 cm. Inflorescences terminal efoliate panicles of spikes, the primary axes 3–5 dm long, the spikes 2–3-fasciculate; peduncle 3–6 mm long, the rachises 2–4.5 cm long. Flowers: sepals ± 0.5 mm long; petals ± 2 mm long, purple-red. Infructescences few-fruited per spike. Fruits sessile, broad-linear, gently falcate but not twisted, compressed but plump, 10–11 × 1.5–1.7 cm, indehiscent. Seeds in profile 10–11 × 5–6 mm, the testa brown. Fl (Nov), fr (Mar, Aug); rare, in nonflooded forest.

Stryphnodendron polystachyum (Miq.) Kleinhoonte

Leaves: primary axes of longer leaves 12–40 cm long; petioles 5–10.5 cm long, with glands below first pair of pinnae; pinnae 3–6 pairs per leaf, the rachis of longer ones 9–17 cm long; leaflets opposite, 4–7 pairs per pinna; pulvinules 1.5–3 mm long; blades subequilaterally ovate- or broadly lanceolate-acuminate, the largest 5.5–11 × 1.8–5.2 cm. Inflorescences of spikes, solitary or 2–3-fasciculate in leaf-axils; peduncle 3–7 mm long, the rachises 2.5–6.5 cm long. Flowers: sepals 0.7–0.9 mm long; petals 2–2.2 mm long, greenish; stamens red. Infructescences several-fruited per spike. Fruits sessile, broad-linear, the body compressed, incurved in nearly 3/4 circle and twisted, indehiscent, the fruit weathering on tree. Seeds compressed-obovoid, 6.5–8 × 4–5.5 mm, the testa brown. Known by a sterile collection from our area; in disturbed forest.

ZYGIA P. Browne

Understory trees. Epidermal prickles lacking. Stipules small, firm, caducous, usually striately veined. Leaves bipinnate, alternate; petioles with glands; pinnae opposite; leaflets opposite; blades with venation palmate at base of leaflet, pinnate distally. Inflorescences cauliflorous, the capitula spicate, on simple axes or paniculately branched; peduncles 1–4 per node, each fascicle subtended by deltate, deciduous bract. Flowers homomorphic; stamens numerous, united into tube at base. Infructescences 1–3-fruited per capitulum. Fruits falcately recurved or straight, dehiscence follicular, through adaxial suture; endocarp produced inward between seeds as incomplete or complete septa.

1. Inflorescences capitulate. Petals puberulent overall, 3.2–4.5 mm long. Fruits compressed but plump, 4–7 cm long. . . . . *Z. racemosa.*
1. Inflorescences spicate. Petals densely pilosulous overall, 5–7 mm long. Fruits bluntly tetragonal (not compressed), 8.5–11 cm long. . . . . *Z. tetragona.*

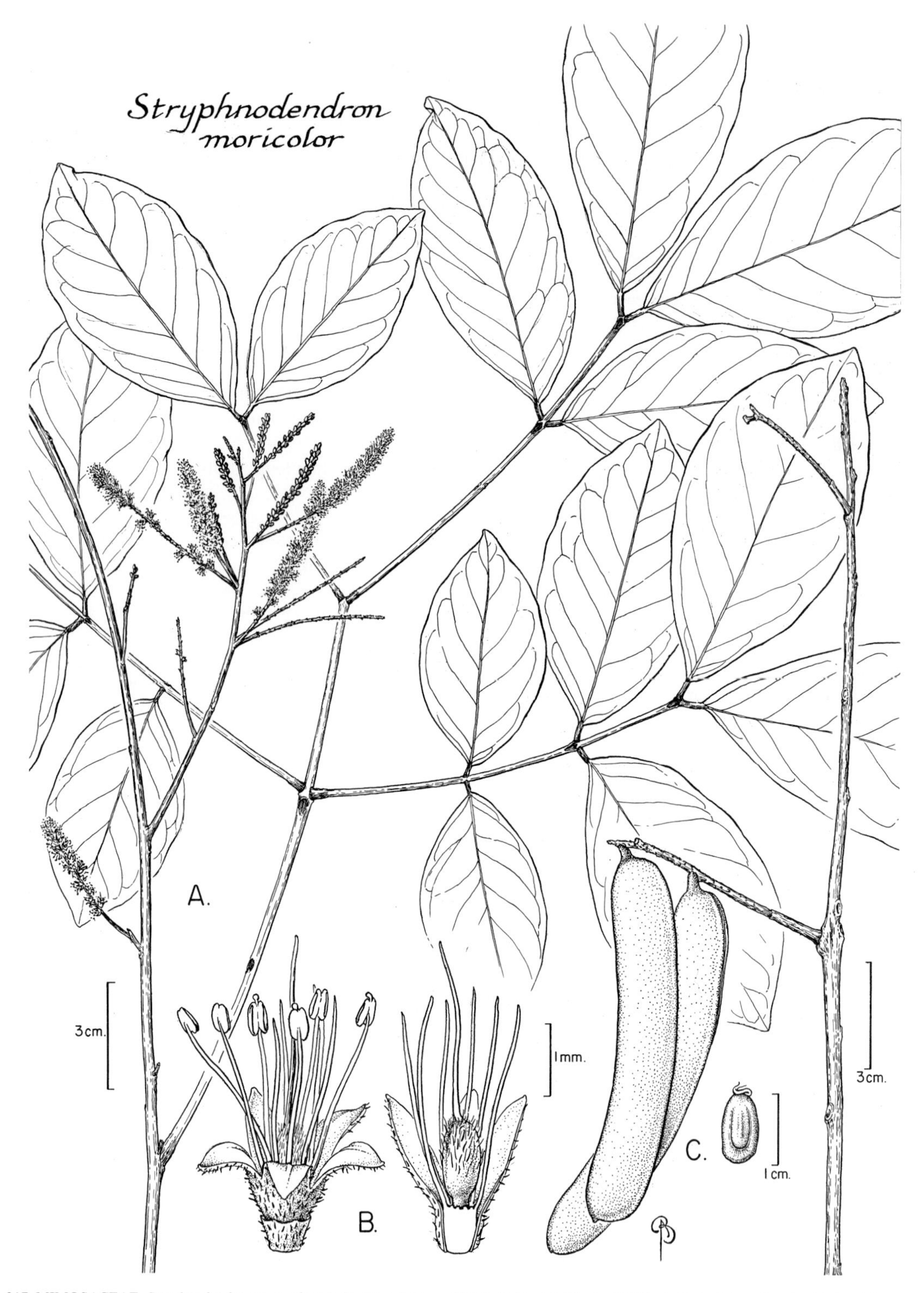

FIG. 217. MIMOSACEAE. *Stryphnodendron moricolor* (A, B, *Mori & Boom 15236*; C, *Mori & Pipoly 15407*). **A.** Part of stem showing bipinnately compound leaves and inflorescences. **B.** Lateral view of flower (left) and flower with part of perianth and androecium removed to show pistil; note the short gynophore. **C.** Stem with fruits and seed (below right). (Reprinted with permission from R. C. Barneby, Brittonia 48(1), 1996.)

Zygia racemosa (Ducke) Barneby & J. W. Grimes FIG. 218

Bark rough, peeling in irregular plates. Leaves: primary axes (4)5–14 cm long; petioles 3–24(28) mm long; pinnae (3)4–8(10) pairs per leaf, the rachis of penultimate, longest pair 5–9.5(11) cm long; leaflets appearing sessile against leaf-rachis, the larger blades at mid-petiole (9)10–19(20.5) × (3)3.3–6.5(7) mm. Inflorescences capitulate, either simple or branched at base, the primary axes

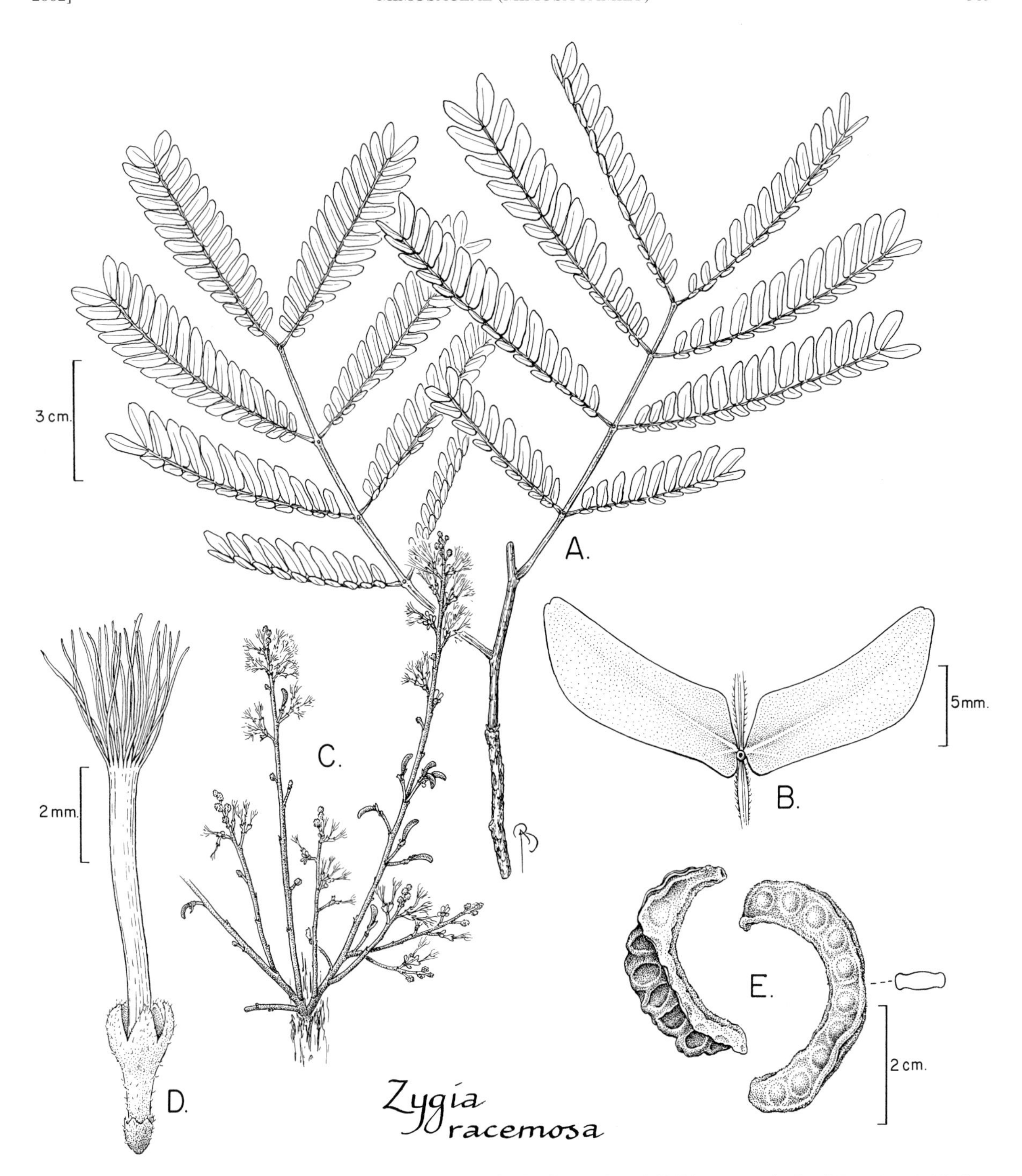

FIG. 218. MIMOSACEAE. *Zygia racemosa* (A, B, *Silva 1313* from Amazonian Brazil; C, *Daly et al. 1235* from Amazonian Brazil; D, *Daly et al. 1455* from Amazonian Brazil; E, *Lanjou & Lindeman 2217* from Surinam and *Woodherbarium Surinam 191* (left) from Surinam). **A.** Part of stem showing bipinnately compound leaves; note the extrafloral nectaries at insertion point of leaflets on rachis. **B.** Leaflet pair with associated nectary. **C.** Cauliflorous inflorescences. **D.** Flower. **E.** Fruits with diagrammatic transverse section of fruit (right).

(1)2–10(14) cm long; peduncle (1.5)3–11 mm long, the axes of the capitula 1–2(2.5) mm long. Flowers: sepals 0.7–1.1 × 0.7–0.9 mm; petals 3.2–4.3(4.5) mm long, puberulent overall, brown, salmon-pink, yellow, orange or whitish, but always brownish when dry. Fruits compressed but plump, in profile 4–7 × 0.7–1 cm, in transverse section I-shaped. Seeds in broad view 6–6.5 × 5 mm, the testa brown, papery, lacking pleurogram. Fl (Sep, Oct, Nov); common but scattered, in nonflooded forest. *Tamarin sauvage* (Créole).

Zygia tetragona Barneby & J. W. Grimes

Bark brown. Leaves: primary axes 7–12 cm long; petioles proper lacking, the first pair of pinnae attached just above pulvinus; pinnae 6–10(11) pairs per leaf, the rachis of longer ones (beyond mid-leaf) 4.5–8 cm long; leaflets sessile against leaf-rachis; larger blades 7–12 × 2–5 mm. Inflorescences of sessile spikes, these 8–13 mm long. Flowers: sepals 1.9–2 mm long; petals 5–7 mm long, densely pilosulous overall, white when fresh, drying brown; stamens with the filaments white at base, pink at apex. Fruits in profile 8.5–11 × 0.6 cm, in transverse section bluntly tetragonal. Seeds in broad view ± 10.5 × 3 mm, the testa brown, lacking pleurogram. Fl (Aug); collected once in our flora from granitic outcrop mountain, among very large boulders, at ca. 400 m elevation.

MONIMIACEAE (Monimia Family)

John D. Mitchell

Treelets to trees, infrequently shrub-like, cut trunk and crushed leaves often emitting lemon-like aroma. Indumentum simple, stellate, or lepidote. Stipules absent. Leaves simple, opposite or verticillate; blades with entire or toothed margins, often pellucid-punctate, the secondary veins often decurrent along midrib. Inflorescences usually axillary, cymose. Flowers relatively small and inconspicuous, unisexual (plants dioecious or monoecious), rarely bisexual outside our area, actinomorphic, perigynous, with a concave to cupulate, clavate, or urceolate receptacle (floral cup); perianth of usually 4 tepals, apically perforate, a velum (floral roof) present or absent; stamens few to numerous, often with short filaments, the anthers longitudinally or valvately dehiscent; gynoecium of several separate carpels, these often sunken in receptacle, with short or elongate style and terminal stigma; ovule solitary, subapical and pendulous (*Mollinedia*) or basal and erect (*Siparuna*), anatropous, crassinucellar, bitegmic or unitegmic. Fruit of separate drupaceous monocarps, these sometimes collectively enclosed in the enlarged floral cup to form aggregate pseudocarp (*Siparuna*).

Renner and Hausner (1997) and Renner (1999) separate *Siparuna* into a separate family, the Siparunaceae.

Jansen-Jacobs, M. J. 1976. Monimiaceae. *In* J. Lanjouw & A. L. Stoffers (eds.), Flora of Suriname **II(2):** 425–426. E. J. Brill, Leiden.
Lanjouw, J. 1966. Monimiaceae. *In* A Pulle (ed.), Flora of Suriname **II(1):** 471–472. E. J. Brill, Leiden.
Petter, F. M. 1966. Monimiaceae. *In* A Pulle (ed.), Flora of Suriname **II(1):** 107–112. E. J. Brill, Leiden.

1. Leaf margins dentate. Indumentum of simple trichomes only. Anthers longitudinally dehiscent. Ovules subapical and pendulous. Fruits not enclosed in floral cup. . . . . *Mollinedia.*
1. Leaf margins usually entire. Indumentum usually of lepidote to stellate (sometimes simple) trichomes. Anthers usually valvately dehiscent. Ovules basal and erect. Fruits enclosed in floral cup. . . . . *Siparuna.*

MOLLINEDIA Ruiz & Pav.

Treelets or small trees. Indumentum of simple trichomes. Leaves opposite; blades with dentate margins. Flowers unisexual (plants dioecious); receptacles densely pubescent; tepals 4. Staminate flowers with stamens 8–60 (outside our area), the anthers longitudinally dehiscent, hippocrepiform. Pistillate flowers with tepals caducous at anthesis; carpels 6–50 (outside our area), the styles short; ovules subapically pendulous. Fruits drupaceous monocarps, stipitate or sessile on thickened, discoid receptacle.

Mollinedia ovata Ruiz & Pav. [Syn.: *M. laurina* Tul.] FIG. 219

Treelets to small trees, to 6 m × 10 cm. Stems terete to slightly angled distally, glabrous to puberulous. Leaves: petioles 7–20 mm long; blades lanceolate or narrowly oblong, ovate, obovate, or elliptic, 13.6–22.5 × 4.8–7.2 cm, chartaceous to subcoriaceous, both surfaces glabrous to sparsely pubescent, the base obtuse, cuneate, or attenuate, the apex acute or acuminate, the margins subentire to dentate; midrib prominent abaxially, flattened or impressed adaxially, the secondary veins in 5–8 pairs, prominent abaxially, flattened to prominulous adaxially. Staminate inflorescences of one or more 3-flowered dichasia, subtended by lanceolate bracts, sparsely to densely pubescent; peduncle 4–25 mm long; pedicels 6.5–15 mm long. Staminate flowers campanulate, with widely spreading tepals at anthesis, 5–7.5 mm diam., the outer pair of tepals ca. 2 × 1.5 mm, the apex obtuse, the margins entire, the inner pair ca. 2.5 × 1.8 mm, with a laciniate, membranous appendage at apex; stamens 24–28, subsessile, the anthers ca. 1.3 mm long. Pistillate flowers urceolate, ca. 3.5 mm diam., the tepals caducous; carpels very numerous, the ovaries ovoid, densely pubescent, 1–1.4 mm long, the styles filiform, sigmoid. Fruits ovoid to ellipsoid monocarps, 13–18 × 7–10 mm, subtended by a somewhat repand, discoid receptacle. Fl (Nov), fr (Jan, Sep, Nov); cloud forest on Mt. Galbao.

SIPARUNA Aubl.

Treelets to large trees. Indumentum of lepidote to stellate, or occasionally simple trichomes. Leaves opposite or occasionally verticillate; blades usually with entire margins. Flowers strictly unisexual (plants monoecious or dioecious). Staminate flowers: tepals subequal, united basally to form an annulus, sometimes with a flat or strongly raised and variously differentiated, apically perforate velum adaxial to annulus sometimes present; stamens few to numerous, the anthers oblong, usually valvately dehiscent. Pistillate flowers often similar to staminate ones in form, with annulus and velum often present; carpels 4 to many, often with elongate, basally connate styles, these often exserted through

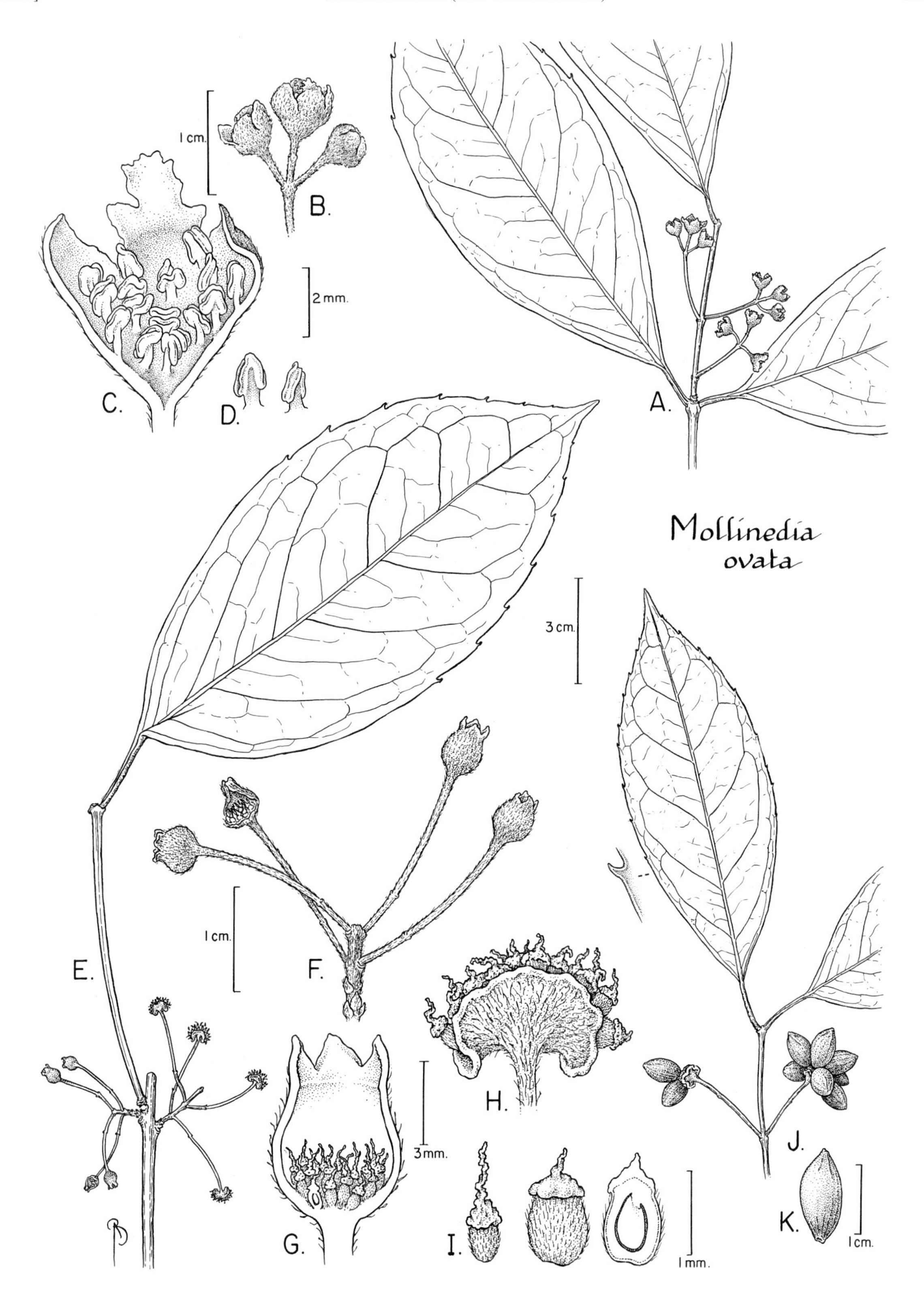

FIG. 219. MONIMIACEAE. *Mollinedia ovata* (A–D, *Broadway 8971* from Trinidad; E–I, *Wallnöfer & Tarin 13513*; J, K, *Granville et al. 8625*). **A.** Part of stem with leaves and staminate inflorescences. **B.** Detail of staminate inflorescence. **C.** Medial section of staminate flower. **D.** Stamens. **E.** Part of stem with leaf and pistillate inflorescences. **F.** Detail of pistillate inflorescence. **G.** Medial section of pistillate flower. **H.** Lateral view of receptacle with immature fruits. **I.** Immature fruits (near and middle right) with remnant stigmas and medial section of fruit (far right). **J.** Part of stem with leaves and infructescences and detail of leaf margin. **K.** Monocarp.

apical pore of velum if present; ovules basal and erect. Fruits drupelets, associated with brightly colored arilloids, enclosed in expanded floral cup which bursts open at maturity (the combination of expanded floral cup and enclosed drupelets is an aggregate fruit or pseudocarp).

1. Medium-sized to tall trees. Stems strongly 4–6-angled or sulcate, the cut stems, especially of saplings, exuding sap which oxidizes red. Leaves verticillate, the leaf apices usually emarginate. *S. pachyantha.*
1. Treelets to medium-sized trees. Stems not strongly angled or sulcate, the cut stems not exuding sap, or if sap exuded, then not oxidizing red. Leaves opposite, the leaf apices acute, obtuse, rounded, or acuminate.
   2. Staminate flowers with stamens wholly enclosed by floral cup (flowers closed except for apical pore). Pistillate flowers 4.5–6.5 mm long; tepals connate into conical apical tube. . . . . *S. decipiens.*
   2. Staminate flowers with stamens not wholly enclosed by floral cup. Pistillate flowers 1.2–3 mm long; tepals free or nearly so, or lacking.
      3. Tepals conspicuous, spreading. . . . . *S. poeppigii.*
      3. Tepals inconspicuous, small or lacking.
         4. Leaf blades with apex short to long acuminate. Indumentum stellate. Staminate flowers with 8–14 stamens. Pseudocarp relatively smooth, the indumentum sparsely stellate. . . . . *S. guianensis.*
         4. Leaf blades with apex abruptly long acuminate (i.e., cuspidate). Indumentum lepidote to stellate. Staminate flowers with 4–8 stamens. Pseudocarp tuberculate, the indumentum stellate-lepidote. . . . . *S. cuspidata.*

Siparuna cuspidata (Tul.) A. DC. [Syn.: *S. crassiflora* Perkins) FIG. 220, PL. 94e

Small to medium-sized trees, to 15 m × 20 cm. Indumentum stellate grading into lepidote. Stems terete, the smaller ones with dense stellate-lepidote hairs. Leaves opposite; petioles 5–10 mm long, stellate-lepidote-pubescent; blades chartaceous to subcoriaceous, narrowly elliptic, oblong or obovate, 8–30 × 2.5–12 cm, with sparse to densely stellate-lepidote hairs abaxially, glabrous or with sparse stellate-lepidote hairs adaxially, the base obtuse to cuneate, the apex abruptly long acuminate (i.e., cuspidate); midrib prominent abaxially, impressed to prominulous adaxially, the seconday veins in 5–8 pairs, prominent abaxially, flattened to prominulous adaxially. Inflorescences 0.7–2.5 cm long, the indumentum densely stellate-lepidote-pubescent, the staminate flowers in upper part and pistillate flowers in lower part; peduncle 1–5 mm long; pedicels of staminate flowers 1.5–4.5 mm long; pedicels of pistillate flowers subsessile to 0.6 mm long. Staminate flowers cup-shaped, 1.4–1.5 × 1.5–1.8 mm; tepals obscure; stamens 4–8, 0.5–0.6 mm long. Pistillate flowers ovoid, 1.2–2 × 0.8–1.8 mm; tepals minute or lacking, the velum conical with narrow apical pore; carpels 4 or more, the styles apically connate, the stigmas connate. Pseudocarps depressed globose, ca. 1 × 1.4 cm (dried), prominently tuberculate, stellate-lepidote-pubescent, green with irregular patches of reddish-purple to dark purple-black. Fr (Feb); rare, in nonflooded forest.

Siparuna decipiens (Tul.) A. DC. PL. 94d

Small to medium-sized trees, to 18 m × 17.8 cm. Trunks cylindrical or fluted, unbuttressed. Outer bark smooth to scaly, with vertical rows of lenticels, reddish-brown or gray, the inner bark cream- to buff-colored, fibrous, the slash without exudate. Indumentum lepidote. Stems terete to slightly flattened distally, the smaller ones with dense lepidote hairs. Leaves opposite: petioles 2.5–3 cm long, the indumentum lepidote; blades oblong, obovate, or ovate, 10.7–18 × 4.3–7.5 cm, chartaceous to coriaceous, with lepidote hairs on both surfaces, the base attenuate, the apex rounded to acute or acuminate; midrib very prominent abaxially, flattened adaxially, the secondary veins in 7–9 pairs, prominent abaxially, flattened adaxially. Inflorescences 3–5 cm long, the indumentum densely lepidote; peduncle 7–10 mm long; pedicels of staminate flowers 4–5 mm long; pedicels of pistillate flowers 2.5–6 mm long. Staminate flowers ovoid or globose, 2.2–2.5 × 1.7–2 mm; floral cup enclosing stamens, with only a minute apical pore; stamens 4–6, 0.7–0.8 mm long. Pistillate flowers lageniform, 4.5–6.5 × 2.2–3.5 mm; tepals connate into conical tube, the velum narrowly conical, surrounded by tepal cone, enclosing connate styles; carpels numerous, the styles connate, the stigmas connate. Pseudocarps subglobose or ovoid, slightly oblique, 1.5–2 × 1–1.8 cm (dried), the apex cuspidate, crowned by persistent cone of tepals. Fl (Aug, Sep, Oct), fr (Feb, Mar, Apr, Sep, Dec); very common, in nonflooded forest. *Vénéré, vulnéraie* (Créole).

Siparuna guianensis Aubl. PL. 95a,b

Small to medium-sized trees, to 5 m (to 20 m outside our area). Indumentum mostly stellate, with occasional simple trichomes. Stems terete or slightly flattened, the smaller stems glabrous to sparsely pubescent. Leaves opposite: petioles 5–8 mm long, pubescent; blades oblong, obovate, oblanceolate, or lanceolate, 12.5–21.5 × 4–7.5 cm, chartaceous to subcoriaceous, both surfaces glabrous to densely pubescent, the base rounded to cuneate, the apex short to long acuminate; midrib prominent abaxially, impressed to prominulous adaxially, the secondary veins in 10–14 pairs, prominulous to prominent ab- and adaxially. Inflorescences 1–1.5 cm long, stellate-pubescent, staminate and pistillate flowers in same or separate inflorescences; peduncle 2–5 mm long; pedicels of staminate and pistillate flowers similar in length, 1.5–3 mm long. Staminate flowers cup-shaped, 1.5–2.5 × 2–2.5 mm; tepals 4–6; stamens 8–14, the filaments ca. 0.5 mm long, the anthers ca. 0.2 mm long. Pistillate flowers ovoid, 1.5–3 × 1.3–2 mm; tepals 4–6, the velum conical, stellate-pubescent, somewhat obscured by tepals; carpels 5 or more, the styles connate except at apex where stigmas diverge. Pseudocarps globose, 0.8–1.5 cm diam., ± smooth, stellate-lepidote-pubescent, reddish. Fl (Jan, Sep, Nov), fr (Aug); second growth or along edges of primary nonflooded forest. *Caa-pitiu, capitiu* (Portuguese), *mavévé, vénéré, viniraie, viniré, viviré, vulnéraie* (Créole).

Siparuna guianensis is very similar to and difficult to distinguish from *S. poeppigii* in vegetative and fruiting characteristics (Renner & Hausner, 1997), but the differences in their flowers are striking, i.e., the large widely spreading tepals of the latter species versus the small straight tepals of the former.

Siparuna pachyantha A. C. Sm. [Syn.: *S. emarginata* R. S. Cowan]

Medium-sized to tall trees, to 40 m × 120 cm. Trunks cylindrical, with or without buttresses. Outer bark gray or gray-brown with

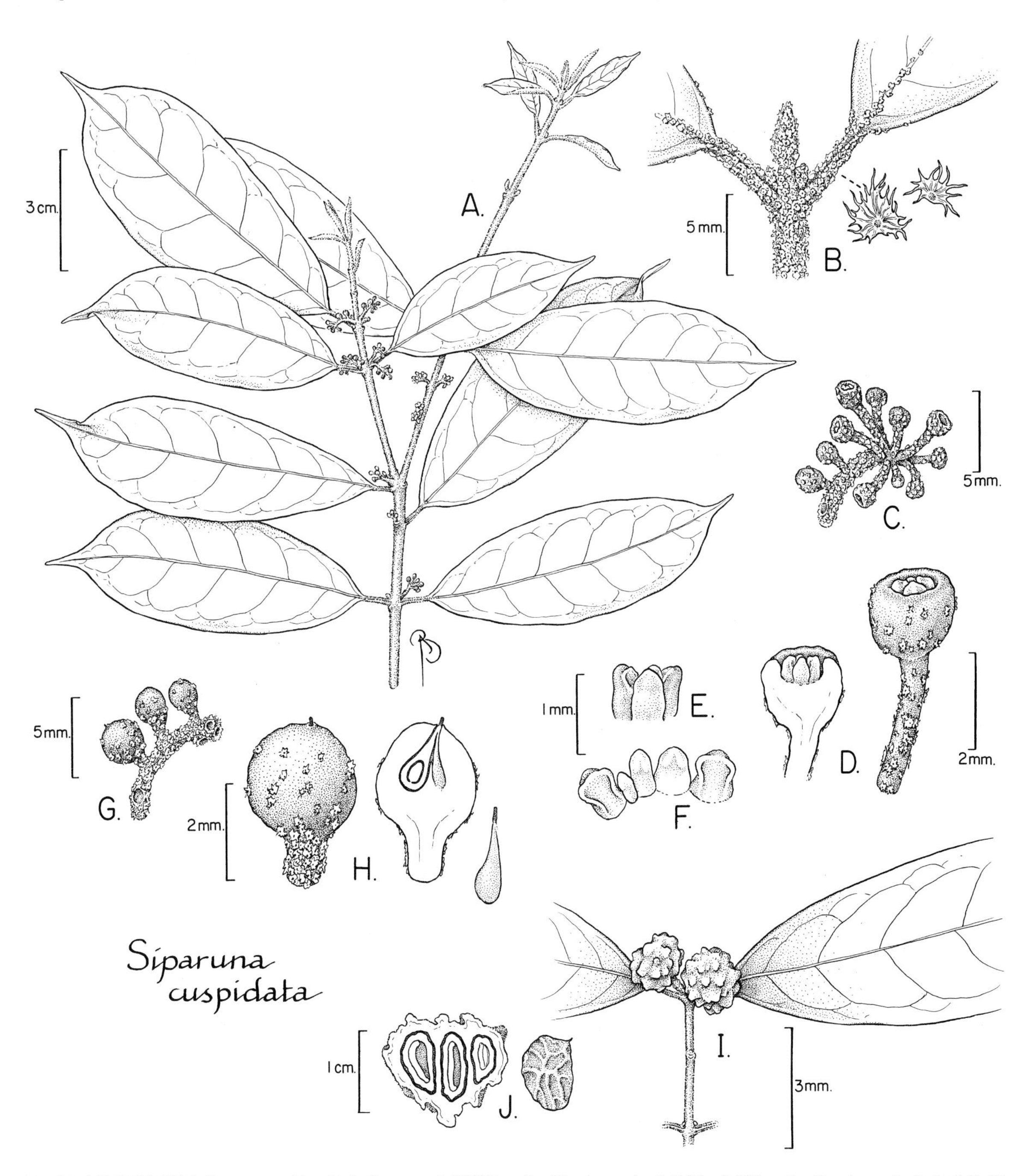

FIG. 220. MONIMIACEAE. *Siparuna cuspidata* (A, B, *Prance et al. 17832* from Brazilian Amazonia; C–F, *Silva 2432* from Brazilian Amazonia; G, H, *Kukle 77* from Brazilian Amazonia; I, J, *Mori et al. 22972*). **A.** Stem with leaves and axillary inflorescence. **B.** Detail of stem apex showing lepidote scales. **C.** Part of staminate inflorescence. **D.** Intact staminate flower (right) and medial section of staminate flower (left). **E.** Androecium of sessile staminodes and stamens. **F.** Opened androecium showing staminodes and stamens. **G.** Part of inflorescence showing pistillate flowers. **H.** Lateral view of pistillate flower (left), medial section of pistillate flower (near right), and carpel (far right). **I.** Part of stem with fruits. **J.** Medial section of fruit (left) and monocarp with reticulate surface (right).

shallow vertical fissures, the lenticels large, vertically oriented, the inner bark yellow with white streaks or yellow-orange becoming reddish. Indumentum stellate grading into lepidote. Stems distinctly 4–6-angled or sulcate, the indumentum densely stellate-lepidote, the exudate oxidizing red. Leaves verticillate: petioles 2–4.5 cm long, the indumentum stellate-lepidote; blades obovate, 10.7–60 × 5.1–23 cm, those of saplings much larger, chartaceous to subcoriaceous, with sparsely to densely stellate-lepidote hairs on both surfaces, the base attenuate, the apex usually emarginate, occasionally rounded or short cuspidate; midrib prominent abaxially, impressed adaxially, the secondary veins in 13–22 pairs, prominent abaxially, flattened adaxially. Inflorescences 6–14 cm long, the indumentum densely

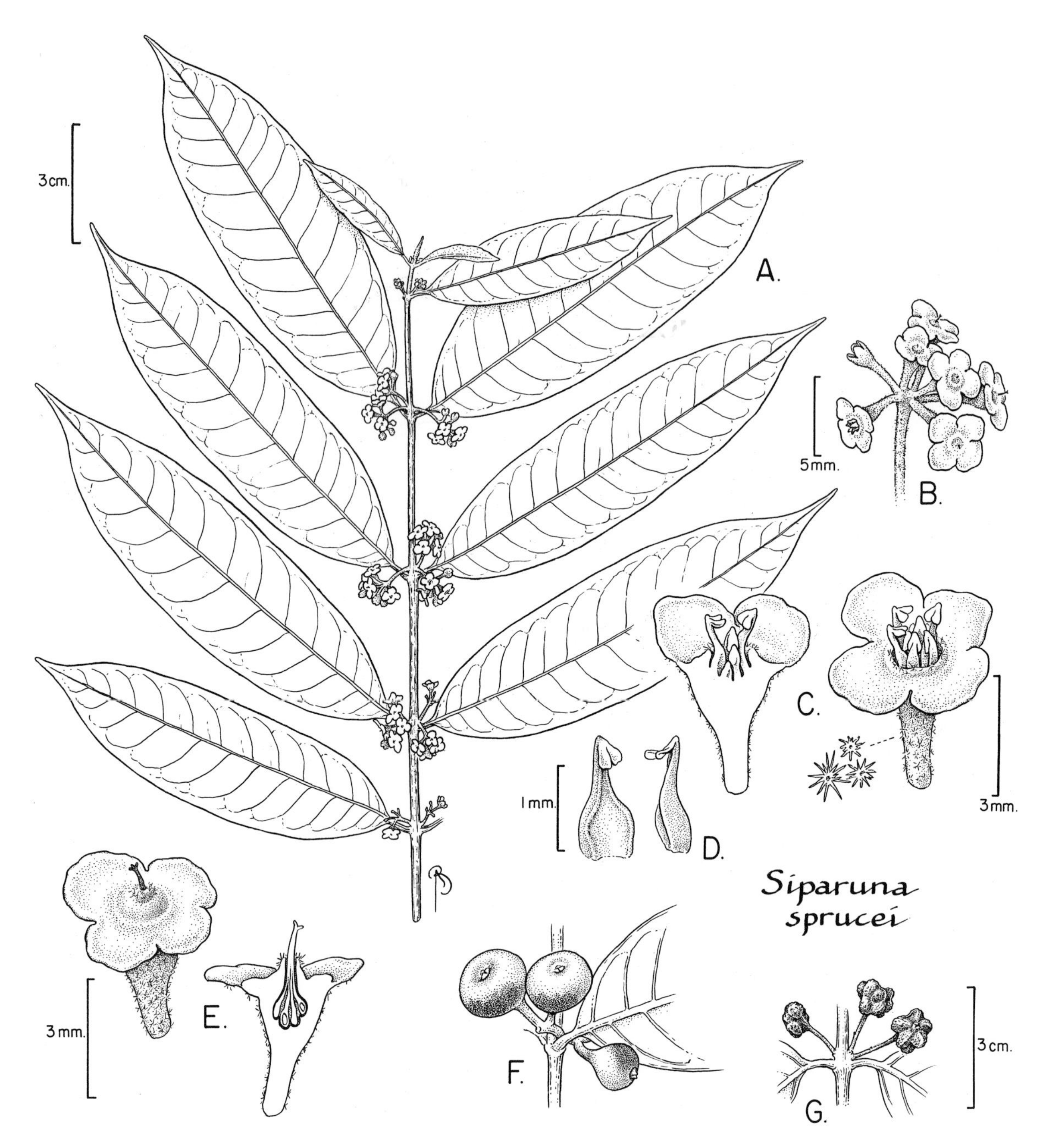

FIG. 221. MONIMIACEAE. *Siparuna poeppigii* (A–D, *Mori et al. 20762*; E, *Mori & Boom 15176*; F, G, *Mori et al. 22888*). **A.** Stem with leaves and staminate inflorescences. **B.** Part of staminate inflorescence. **C.** Oblique-apical view (right) and medial section (left) of staminate flowers with detail of stellate-lepidote hairs (below right). **D.** Stamens. **E.** Oblique-apical view (left) and medial section (right) of pistillate flowers. **F.** Fresh fruits drawn from photograph. **G.** Dried fruits drawn from herbarium specimen.

stellate-lepidote; peduncle 0–5.7 cm long; pedicels of staminate flowers 2–3 mm long; pedicels of pistillate flowers ca. 1 mm long. Staminate flowers clavate, 4–6 × 1.5–3.5 mm; hypanthium completely enclosing stamens except for apical pore; tepals obsolete; stamens 2, ca. 0.5–1 mm long. Pistillate flowers depressed globose, 4.5–6 × 4.5–6 mm; tepals not apparent, the velum conical, with a narrow apical pore; carpels ca. 6–10, the styles connate, the stigmas connate, not exserted through apical pore. Pseudocarps globose, ca. 2(4) × 1.3–1.7(3) cm (dried), the indumentum stellate-lepidote. Fl (Aug, Dec), fr (Jan, May, Aug, Oct); frequent, in nonflooded forest.

Siparuna poeppigii (Tul.) A. DC. [Syns.: *S. amazonica* (Mart.) A. DC., *S. sprucei* A. DC.] FIG. 221, PL. 95c,d

Treelets or small trees, to 6 m × 6 cm. Indumentum stellate, except for occasional simple trichomes. Stems terete, glabrous. Leaves opposite; petioles 4–10 mm long, usually glabrous; blades narrowly elliptic, oblong, or obovate, 11.5–16.7 × 3.2–6.5 cm, chartaceous, both surfaces glabrescent, the base rounded to cuneate, the apex short to long acuminate; midrib prominent abaxially, impressed to prominent adaxially, the secondary veins in 10–14 pairs, prominulous ab- and adaxially. Inflorescences 0.5–1.5 cm long, stellate-pubescent, staminate and pistillate flowers in same inflorescence; peduncle 7–10 mm long; pedicels of staminate and pistillate flowers similar in length, 1.4–2 mm long. Staminate and pistillate flowers similar in size and shape, campanulate, 2–3.5 × 2–4.4 mm diam.; tepals 4, widely spreading at anthesis, 1–2 × 1.3–2.3 mm. Staminate flowers: stamens 8(14), the filaments 0.5–0.7 mm long, the anthers 0.2–0.3 mm long. Pistillate flowers: velum conical, apically stellate; carpels 8–10, the styles connate except at tip where stigmas diverge. Pseudocarps globose, 1–1.5 cm diam. (dried), yellow-green to deep reddish-purple. Fl (Aug, Oct, Sep, Nov), fr (Feb, Mar, Apr, Oct, Dec); secondary growth or along margins of nonflooded forest. *Vénéré, vulnéraie* (Créole).

Specimens of this species are difficult to separate from those of *S. guianensis* without flowers (Renner & Hausner, 1997).

MORACEAE (Mulberry Family)

Cornelis C. Berg

Trees or shrubs, often hemiepiphytic in *Ficus*, herbs outside our area (*Dorstenia*). Latex usually milky, sometimes oxidizing reddish or brownish. Stipules present, often conspicuous, either entirely or partially encircling stem, usually leaving conspicuous scars. Leaves alternate, spirally arranged, or distichous, opposite only in *Bagassa*. Inflorescences racemes, spikes, heads, or with involucrate or non-involucrate receptacles, or with urn-shaped receptacles enclosing flowers in *Ficus* (syconium). Flowers actinomorphic, unisexual (plants dioecious or monoecious); perianth well developed to absent, when present uniseriate; tepals 4(6) or absent; stamens 1–4; ovary free or fused with perianth, 1-locular, the stigmas (1)2; placentation apical, the ovule solitary. Fruits drupe-like or achene-like, often enclosed by fleshy perianth (these pseudodrupes aggregated and solitary) or the fruits enclosed in fleshy receptacles (*Brosimum, Ficus, Trymatococcus*). Seeds with fleshy endosperm when small, without endosperm when large.

Berg, C. C. 1972. Olmedieae, Brosimeae (Moraceae). Fl. Neotrop. Monogr. **7:** 1–229.

———. 1992. Moraceae. *In* A. R. A. Görts-van Rijn (ed.), Flora of the Guianas Ser. A, **11:** 10–92, 192–222. Koeltz Scientific Books, Koenigstein.

——— & M. J. M. Dewolf. 1975. Moraceae. *In* J. Lanjouw & A. L. Stoffers (eds.), Flora of Suriname **V(1):** 173–299. E. J. Brill, Leiden.

1. Leaves opposite. . . . . *Bagassa.*
1. Leaves alternate.
   2. Stipules fully encircling stem, leaving annular scars.
      3. Leaf blades with waxy glandular spots abaxially. Inflorescences and infructescences figs, the flowers entirely enclosed at anthesis. . . . . *Ficus.*
      3. Leaf blades without waxy glandular spots. Inflorescences and infructescences not figs, the anthers and stigmas exposed at anthesis.
         4. Leaf blades pinnately incised. Cultivated, sometimes persisting at sites of old homesteads. . . . *Artocarpus.*
         4. Leaf blades entire or dentate, not pinnately incised. Native.
            5. Horizontal lateral branches not deciduous. Young parts often with hooked hairs. Stipules fused. Inflorescences (sub)globose to turbinate, not involucrate, with peltate bracts. . . . . *Brosimum.*
            5. Horizontal lateral branches deciduous. Young parts without hooked hairs. Stipules free. Inflorescences discoid (if not uniflorous), involucrate, with basally attached bracts.
               6. Staminate inflorescences not pedunculate. Pistillate infloresences one-flowered. . . . . . *Pseudolmedia.*
               6. Staminate inflorescences pedunculate. Pistillate inflorescences several- to many-flowered.
                  7. Leaf blades subglabrous, 5–17 cm long. . . . . *Naucleopsis.*
                  7. Leaf blades pubescent, usually >10–50 cm long. . . . . *Perebea.*
   2. Stipules not fully encircling the stem, scars not fully annular.
      8. Leaf blades with spinose apex or spinose marginal teeth. Inflorescences usually borne below leaves on older wood. . . . . *Clarisia.*
      8. Leaf blades without spinose apex or also spinose marginal teeth. Inflorescences not as above.
         9. Horizontal lateral branches deciduous. Young parts without hook-shaped hairs. Inflorescences unisexual, discoid, involucrate, with basally attached bracts.
            10. Leaf blades pubescent, dull adaxially when dry. . . . . *Helicostylis.*
            10. Leaf blades (sub)glabrous, ± shiny adaxially when dry. . . . . *Maquira.*
         9. Horizontal lateral branches not deciduous. Young parts often with hook-shaped hairs. Inflorescences bisexual or, if unisexual, then receptacle covered by peltate bracts.
            11. Inflorescences unisexual or bisexual, receptacle globose to turbinate to almost discoid, not crowned by staminate flowers, covered by peltate bracts. . . . . *Brosimum.*
            11. Inflorescences bisexual, receptacle cylindrical to subglobose, crowned by staminate flowers, not covered by peltate bracts. . . . . *Trymatococcus.*

ARTOCARPUS J. R. Forst. & G. Forst.

Trees. Hook-shaped hairs absent. Horizontal lateral branches not deciduous. Stipules fully encircling stem, free. Leaves spirally arranged; blades pinnately incised or entire, without waxy, glandular spots. Staminate inflorescences spikes, cylindric to clavate. Pistillate inflorescences globose. Flowers unisexual (plants monoecious); perianth tubular. Staminate flowers: stamen 1. Pistillate flowers fused; stigma 1. Fruits in large globose infructescences.

Artocarpus altilis (Parkinson) Fosberg

Trees, to 30 m tall. Stipules 10–25 cm long. Leaves 2–6 cm long; blades elliptic, pinnately incised, ca. 20–50(75) × 15–30(40) cm, coriaceous. Inflorescences solitary in leaf axils, unisexual. Staminate inflorescences 8–12 cm long. Pistillate inflorescences 8–12 cm diam. Pistillate infructescences to 30 cm diam., greenish, the surface rough in seeded form, smooth in seedless form. Fr (immature Jun, Sep); an Asian species cultivated for edible fruit (seedless form) and seeds, persisting near old homesteads and therefore possible to mistake for a native plant. *Arbre à pain* (seedless form, French), *breadfruit* (seedless form, English), *châtaignier* (seeded form, French), *fruta de pão* (Portuguese).

BAGASSA Aubl.

Trees. Hook-shaped hairs absent. Horizontal lateral branches not deciduous. Stipules lateral, free. Leaves opposite; blades without waxy and glandular spots; primary veins 3, arising from base, pubescent, the margins serrate. Staminate inflorescences spicate. Pistillate inflorescences globose. Flowers unisexual (plants dioecious). Staminate flowers: tepals 4; stamens 2. Fruits small, drupe-like, enclosed by fleshy perianth, greenish at maturity, aggregated into globose infructescences.

Bagassa guianensis Aubl. FIG. 222

Canopy or emergent trees, to 45 m tall. Leaf blades cordiform to ovate or suborbicular, 6–30 × 4–23 cm, subcoriaceous to chartaceous, abaxial surface scabrous in juvenile leaves, the margins unlobed or 3-lobed (when juvenile to 3-parted). Inflorescences in leaf axils, unisexual. Staminate inflorescences 4–13 cm long. Pistillate inflorescences 1–1.5 cm diam. Infructescences 2.5–3.5 cm diam., greenish. Fl (Aug, Dec); common, in secondary vegetation, especially along airport road. *Bagasse, bois vache, maman boeuf, odoun* (Créole), *tatajuba* (Portuguese).

BROSIMUM Sw.

Trees. Hook-shaped hairs present, especially on juvenile parts. Horizontal lateral branches not deciduous. Stipules fully encircling stem and fused or not fully encircling stem and free. Leaves alternate, distichous; blades without waxy, glandular spots, the margins entire. Inflorescences bisexual or unisexual; peduncule present; receptacle globose to turbinate (to almost discoid), covered by peltate bracts. Flowers unisexual (plants monoecious or dioecious). Staminate flowers normal (with distinct perianth and to 4 stamens) to reduced (without perianth and 1 stamen). Pistillate flowers 1 to several, embedded in the receptacle. Fruits surrounded by fleshy layer derived from the receptacle, the receptacle with remnant bracts. Seeds 1 to several per infructescence, large.

1. Stipules fully encircling stem, fused, usually 1–3 cm long.
    2. Leaf blades densely pubescent on all veins abaxially. . . . . *B. parinarioides* subsp. *parinarioides.*
    2. Leaf blades glabrous or sparsely pubescent only on main veins abaxially.
        3. Petioles with epidermis peeling off; blades mostly broadest at or below middle, usually >10 cm long. . . . . *B. utile* subsp. *ovatifolium.*
        3. Petioles with epidermis persistent; blades broadest at or above middle, usually <10 cm long. . . . . . *B. rubescens.*
1. Stipules not fully encircling stem, free, usually <1 cm long.
    4. Stipules with prominent fan-shaped venation. Leaf blades with midrib impressed adaxially. . . . . *B. lactescens.*
    4. Stipules without prominent fan-shaped venation. Leaf blades with midrib plane adaxially.
        5. Leaf blades hairy, dull adaxially when dry. . . . . *B. acutifolium* subsp. *acutifolium.*
        5. Leaf blades glabrous, ± shiny adaxially when dry. . . . . *B. guianense.*

Brosimum acutifolium Huber subsp. **acutifolium**

Canopy trees, to 35 m tall. Stipules free, 0.2–0.8 cm long, without prominent fan-shaped venation. Leaves: petioles 0.3–0.7 cm long, the epidermis persistent; blades oblong to lanceolate, convex (often folded in herbarium specimens), 3–20 × 2–7 cm, adaxial surface often ± scabrous, abaxial surface hirtellous to tomentose, the apex acuminate; midrib plane adaxially. Inflorescences unisexual, with densely white puberulous bracts. Pistillate inflorescences with one to several flowers. Staminate flowers reduced; perianth absent; stamens 1–3, small. Infructescences usually with 1 fruit, globose, 1.5–2 cm diam., yellow to orange at maturity. Fl (Sep), fr (Aug); infrequent, in forest on well-drained soil, the latex is the source of hallucinogenic compounds. *Bois mondan, lamoussé* (Créole), *mururé* (Portuguese).

A collection (*Allorge 396*) deviates from *B. acutifolium* subsp. *acutifolium* in having slightly more obovate leaves than other collections from central French Guiana.

Brosimum guianense (Aubl.) Huber

Canopy or understory trees or treelets. Stipules free, 0.2–0.5 cm long, without prominent fan-shaped venation. Leaves: petioles 0.2–0.6 cm long, the epidermis persistent; blades elliptic to obovate, 4–13 × 2–6 cm, subcoriaceous to chartaceous, adaxial surface glabrous, abaxial surface pubescent or glabrous, the apex rounded to

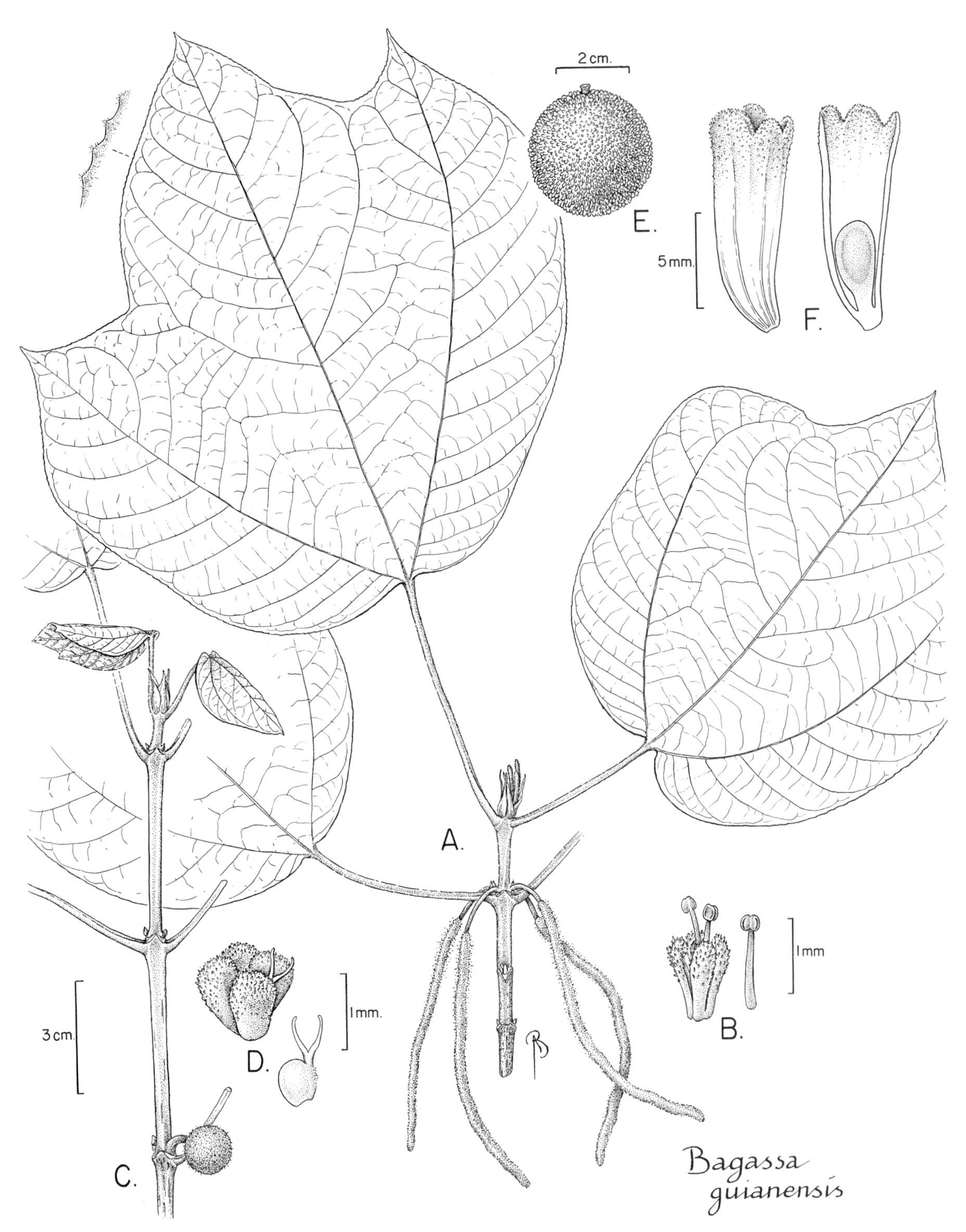

FIG. 222. MORACEAE. *Bagassa guianensis* (A, B, *Zarucchi 2667* from Brazil; C, D, *Silva 2553* from Brazil; E, F, *Mori & Boom 15346*). **A.** Apex of stem showing opposite leaves and spicate staminate inflorescences; note detail of leaf margin (above left). **B.** Staminate flower (left) showing tepals and stamen (right). **C.** Apex of stem showing globose pistillate inflorescences. **D.** Pistillate flower (left) showing tepals and pistil (right) showing subapical attachment of pistil. **E.** Infructescence. **F.** Lateral view (left) and medial section (right) of fruit showing expanded tepals.

acuminate; midrib plane adaxially. Inflorescences bisexual, usually with 1 to several pistillate flowers. Staminate flowers: perianth reduced; stamen 1, small. Infructescences subglobose, with 1 or more fruits, ca. 1.5 cm diam., reddish at maturity. Fl (Jan, May, Sep), fr (Feb, Mar); infrequent, in nonflooded forest. *Amourette*, *bois de lettre moucheté*, *lettre moucheté* (Créole), *pinde paya* (Bush Negro).

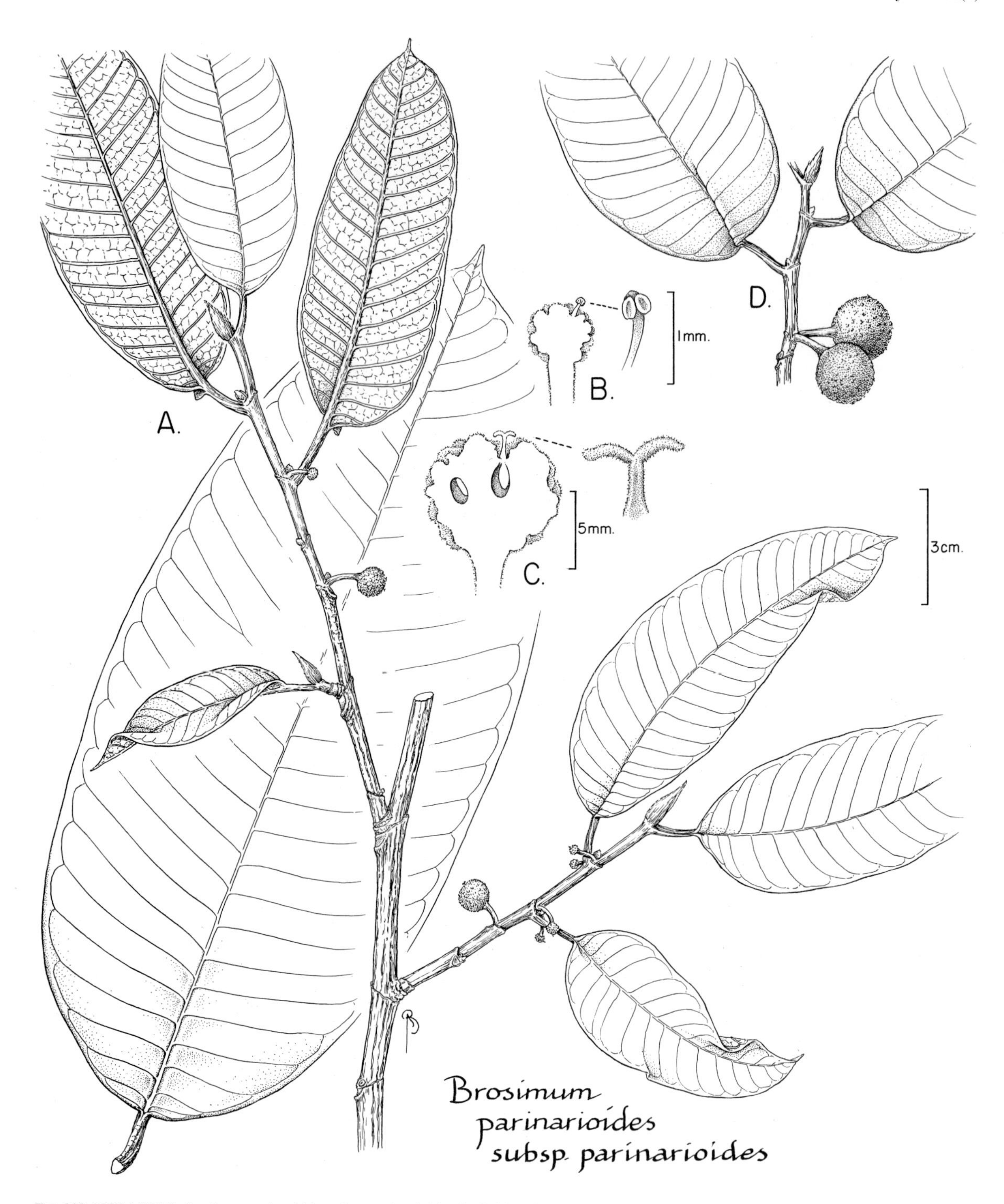

FIG. 223. MORACEAE. *Brosimum parinarioides* subsp. *parinarioides* (A–C, *Mori & Boom 15154*; D, *Mori & Boom 15283*). **A.** Part of stem showing leaves, inflorescences, and single large leaf (behind). **B.** Medial section of staminate inflorescence with detail of stamen (right). **C.** Medial section of pistillate inflorescence (left) with detail of style and stigma (right). **D.** Apex of stem with infructescences.

Brosimum lactescens (S. Moore) C. C. Berg

Canopy trees or emergents, to 50 m tall. Stipules free, 0.3–1(1.5) cm long, with prominent, fan-shaped venation. Leaves: petioles 0.3–0.5 cm long, the epidermis persistent; blades oblong to subovate, 5–20 × 2–6 cm, (sub)coriaceous, adaxial surface glabrous, abaxial surface ± pubescent, the apex acuminate to subacute; midrib impressed adaxially. Inflorescences unisexual. Pistillate inflorescences with 1 to several flowers. Staminate flowers well developed; stamens 2–4. Infructescences with 1 or more fruits, globose to discoid, 1–3 cm diam, red or yellow at maturity. Known only by sterile collections in our area; in forest on well-drained soil.

Brosimum parinarioides Ducke subsp. **parinarioides** FIG. 223

Canopy trees or emergents, to 40 m tall. Trunks broadened at soil level. Stipules fused, (0.5)1–2(2.5) cm long, with prominent, fan-shaped venation. Leaves: petioles 0.4–2 cm long, the epidermis peeling off; blades elliptic, 5–25 × 3–10 cm, coriaceous, adaxial surface glabrous, abaxial surface densely but minutely puberulous on veins, the apex acuminate to subacute; midrib almost plane adaxially. Inflorescences bisexual, usually with a single pistillate flower. Staminate flowers reduced; perianth absent; stamen 1, small. Infructescences usually with one fruit, globose, ca. 2–2.5 cm diam. Fl (Nov), fr (Nov); scattered, in nonflooded forest, often conspicuous because of the large number of deciduous leaves accumulating on the ground below crown, latex from this species has been used to adulterate *balata* (a latex usually derived from species of Sapotaceae). *Amaparana* (Portuguese), *mapa* (Créole), *murureran*a (Portuguese).

Brosimum rubescens Taub. PL. 96a; PART 1: PL. IXb

Canopy trees or emergents, to 40 m tall. Stipules fused, 1–2.5 (3.5) cm long, without prominent, fan-shaped venation. Leaves: petioles 0.2–1.3 cm long, the epidermis persistent; blades elliptic, 2–10 × 1–5(7) cm, coriaceous, adaxial surface glabrous, abaxial surface glabrous or sparsely hairy, the apex acuminate; midrib almost plane adaxially. Inflorescences bisexual, usually with 1 to several pistillate flowers. Staminate flowers: perianth ± reduced; stamens 1–3, small. Infructescences usually with 1 fruit, subglobose, 1–1.5 cm diam. Fl (Sep), fr (Nov); in nonflooded forest. *Bois d'arc* (wood was used to make bows), *bois de lettre*, *bois satiné*, *lettre rubané*, *satiné rubané* (Créole).

Brosimum utile (Kunth) Pittier subsp. **ovatifolium** (Ducke) C. C. Berg

Canopy trees or emergents, to 50 m tall. Stipules fused, (0.5)1–4.5 cm long, without prominent, fan-shaped venation. Leaves: petioles 0.3–2.5(3) cm long, the epidermis usually peeling off; blades elliptic to ovate, (5)10–20(26) × 3–10 cm, coriaceous, adaxial surface glabrous, abaxial surface glabrous or sparsely hairy, the apex acuminate; midrib almost plane adaxially. Inflorescences bisexual, usually with 1 pistillate flower. Staminate flowers: perianth ± reduced; stamen 1, small. Infructescences with 1 fruit, subglobose, 2.5–3 cm diam. Fl (May); in nonflooded forest.

CLARISIA Ruiz & Pav.

Treelets or shrubs. Hook-shaped hairs present or absent. Horizontal lateral branches not deciduous. Stems and roots often reddish (this needs to be confirmed for *C. ilicifolia*). Stipules lateral, free, not fully encircling stem. Leaves alternate, distichous; blades without waxy, glandular spots, the margins entire or pinnately incised. Inflorescences usually arising below leaves on older wood. Staminate inflorescences spicate. Pistillate inflorescences discoid-capitate, noninvolucrate, with 1 (outside our area) to many flowers. Flowers unisexual (plants dioecious). Staminate flowers small, sometimes not well defined one from the other; tepals 2–7; stamens 1–3. Pistillate flowers: perianth tubular. Fruits enclosed by a fleshy perianth, solitary (outside our area) or aggegated.

Clarisia ilicifolia (Spreng.) Lanj. & Rossberg PL. 96b,c

Understory treelets or shrubs. Hooked-shaped hairs always present. Leaves: petioles 0.3–2.5 cm long; blades elliptic to lanceolate, 10–30 × 3–10 cm, coriaceous, adaxial surface glabrous, abaxial surface sparsely pubescent, the apex acuminate, with a spinose tip, the margins spinose-dentate, entire or lobed; midrib prominent adaxially. Staminate infloresences usually to 1 cm long. Pistillate inflorescences with 6–20 flowers. Fruiting perianth ellipsoid, orange. Fl (Oct, Nov, Dec), fr (Jan, Feb, Mar, Jun); in understory of nonflooded forest. *Janita* (Portuguese).

FICUS L.

Free-standing trees or hemiepiphytes (stranglers) with aerial roots, sometimes becoming free-standing after "strangler" phase. Stipules fully encircling stem, free. Hook-shaped hairs absent. Horizontal lateral branches not deciduous. Leaves spirally arranged; blades with 1 or 2 waxy glandular spots abaxially, the margins entire. Inflorescences (figs or syconia) globose-urceolate receptacles, with 2 or 3 subtending (basal) bracts and a small apical opening (ostiole) barred by bracts, enclosing numerous flowers. Flowers unisexual (plants monoecious). Staminate flowers: stamens 1–2 (opening when seeds are mature). Pistillate flowers: tepals 2–4, the styles of different length, the stigmas 1 or 2. Fruits small achenes, enclosed in figs. Species of this genus are known as *figuier* and generally flower throughout the year.

Without opening the fig and carefully examining the contents, it is often difficult to tell if a fig is in flower or in fruit. A change in color, softening, and loss of latex in the fig are indications that fruits are ripe. Individuals of species of *Ficus* have their own flowering rhythms, in which flowering occurs year round, thereby providing the breeding sites for the pollinators which live for a few days only.

1. Free standing trees, without aerial roots. Leaf blades with 2 glandular spots in axils of basal secondary veins abaxially. Figs solitary in leaf axils.
    2. Stipules 4–10 cm long. Petioles with epidermis persistent; blades scabrous adaxially. . . . . . *F. insipida* subsp. *scabra*.
    2. Stipules 1.5–3.5 cm long. Petioles with epidermis peeling off; blades smooth or only slightly scabrous adaxially. . . . . *F. maxima*.
1. Usually hemiepiphytic trees or shrubs, with aerial roots. Leaf blades with 1 glandular spot at base of midrib abaxially. Figs in pairs in leaf axils or in clusters below leaves.
    3. Figs more than 2 together on short spurs below leaves, pedunculate.

4. Leaf blades lanceolate to linear, usually ≤2 cm wide. . . . . *F. leiophylla.*
4. Leaf blades elliptic to subobovate, usually >2 cm wide. . . . . *F. guianensis.*
3. Figs in pairs in leaf axils, sessile or pedunculate.
5. Leafy stems and stipules with brownish hairs. . . . . *F. gomelleira.*
5. Leafy stems and stipules with white hairs or glabrous.
6. Petioles >4 cm long; blades usually ca. 10–20 cm long, the base cordate. Figs 1–2.5 cm diam. . . . . *F. nymphaeifolia.*
6. Petioles usually <4 cm long; blades usually to 12 cm long or, if 10–17 cm long, then the base acute to obtuse and figs 0.8–1 cm diam.
7. Leaf blades ca. 10–17 cm long. Figs sessile, with ostiole surrounded by 3-lobed rim. . . *F. panurensis.*
7. Leaf blades usually to 12 cm long. Figs pedunculate, with ostiole not surrounded by rim.
8. Petioles usually ≤1 cm long; blades lanceolate. Figs 0.3–0.5 cm diam. . . . . *F. schumacheri.*
8. Petioles usually ≥1 cm long; blades elliptic to oblong. Figs 0.4–0.8(1) cm diam.
9. Leaf blades usually with subcordate to rounded base, usually drying brownish. Figs with ostiole not sunken in apex. . . . . *F. amazonica.*
9. Leaf blades usually with an acute to rounded base, usually drying greenish. Figs with ostiole sunken in apex. . . . . *F. pertusa.*

Ficus amazonica (Miq.) Miq.

Hemiepiphytic trees, to 20 m tall. Stems glabrous or minutely white-puberulous. Stipules 0.5–1(2.5) cm long, glabrous. Leaves: petioles (0.5)1–2.5(4.5) cm long, the epidermis not peeling off; blades elliptic to oblong, ca. 4–12 × 1.5–5 cm, smooth adaxially, glabrous abaxially, usually drying brownish, with glandular spot at base of midrib abaxially, the base cordate to rounded. Figs axillary, in pairs, 0.4–0.8(1) cm diam., greenish at maturity; peduncles 0.2–0.5 cm long; ostiole not surrounded by a rim, plane to slightly prominent. Fr (Oct).

Ficus gomelleira Kunth & Bouché FIG. 224, PL. 96d

Hemi-epiphytic shrubs or trees ("stranglers"), becoming free-standing trees to 40 m tall, larger trees with well developed buttresses. Stems brown-hairy. Stipules 0.5–2(3.5) cm long, brown-hairy. Leaves: petioles 1.5–5 cm long, the epidermis not peeling off; blades elliptic to subobovate, ca. 10–25 × 5–15 cm, smooth or ± scabrous adaxially, hairy abaxially, the base cordate to rounded, usually drying brownish, with glandular spot at base of midrib abaxially. Figs axillary, in pairs, 1–2.5 cm diam., pubescent, greenish at maturity; peduncles 0.5–1.5 cm long; ostiole surrounded by triangular rim. Fr (Mar, Aug); infrequent, in nonflooded forest but often on moist soils, the fig tree serving as a tourist attraction because of its large size to the SW of Saül belongs to this species.

Mori et al. 24961 (CAY) has sessile figs, subpersistent stipules, and scabrous adaxial leaf surfaces, features usually not found in *F. gomelleira.*

Ficus guianensis Desv. PL. 97a

Hemiepiphytic shrubs or trees ("stranglers"), becoming free-standing trees to 35 m tall. Stems glabrous or minutely white-puberulous. Stipules 0.3–1 cm long, glabrous or minutely white-puberulous. Leaves: petioles 0.5–3 cm long, the epidermis not peeling off; blades elliptic to subobovate, ca. 3–16 × 2–6 cm, smooth adaxially, glabrous or minutely brown-puberulous abaxially, drying brownish or greenish, with glandular spot at base of midrib abaxially, the base acute to rounded. Figs from below leaves, more than 2 together on short shoots, 0.3–1.2 cm diam., pink to red at maturity; peduncles 0.1–1 cm long; ostiole not surrounded by rim, plane to slightly prominent. Fr (Feb, May, Jun, Aug, Oct, Nov); common but scattered, in nonflooded moist forest.

Ficus insipida Willd. subsp. **scabra** C. C. Berg PL. 97b

Canopy (or emergent) trees, to 35 m tall. Stems white-hairy. Stipules 4–10 cm long, white-hairy. Leaves: petioles 2.5–6.5(8.5) cm long, the epidermis not peeling off; blades elliptic to oblong, 10–30 × 3–15 cm, scabrous adaxially, white-hairy abaxially, drying green, with glandular spots in axils of basal secondary veins abaxially, the base rounded to subcordate. Figs axillary, solitary, 1.5–2.5 cm diam., greenish at maturity; peduncles 0.7–1.5 cm long; ostiole not surrounded by rim, plane or prominent. Fr (Aug, Oct, Nov); common, in advanced secondary forest, especially along roads and in old gaps. *Apuí* (Portuguese), *bois figué* (Créole).

Ficus leiophylla C. C. Berg

Hemiepiphytic ("stranglers") or terrestrial shrubs or trees. Stems minutely white-puberulous. Stipules 0.3–1.7 cm long, minutely white-puberulous. Leaves: petioles 0.3–1.5(2.5) cm long, the epidermis not peeling off; blades lanceolate to linear, ca. 2.5–10(15) × 0.5–2(2.5) cm, smooth adaxially, subglabrous abaxially, drying brownish or greenish, with glandular spot at base of midrib abaxially, the base acute to obtuse. Figs from below leaves, more than 2 together on short shoots, 0.3–0.5 cm diam., greenish(?) at maturity; peduncles 0.2–0.3 cm long; ostiole not surrounded by rim, slightly prominent. Fr (Mar); rare, known by a single collection from our area.

Ficus maxima Mill.

Trees, to 25 m tall. Stems glabrous or minutely white-puberulous. Stipules 1.5–3.5 cm long, glabrous or minutely white-puberulous. Leaves: petioles 1–6 cm long, the epidermis peeling off; blades elliptic to oblong, 10–17 × 7–11 cm, smooth adaxially, white-puberulous to hispidulous abaxially, drying greenish, with glandular spots in axils of basal secondary veins abaxially, the base acute. Figs axillary, solitary, 0.8–1.8 cm diam., greenish at maturity; peduncles 0.6–2 cm long; ostiole not surrounded by rim, plane. Fr (Feb, Jul, Sep); in nonflooded forest but on poorly drained soil. *Caxinguba* (Portuguese).

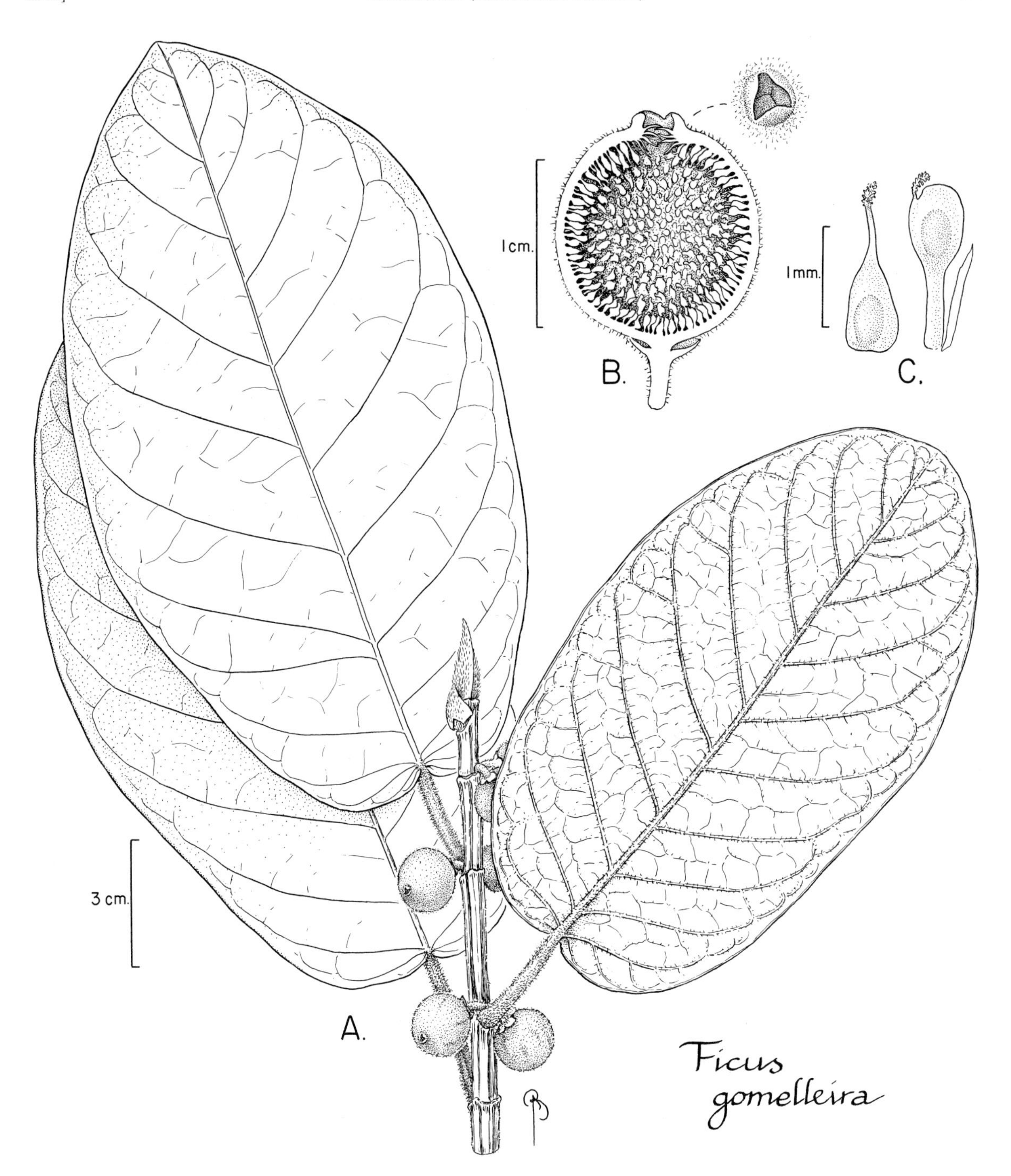

FIG. 224. MORACEAE. *Ficus gomelleira* (*Cid 310* from Brazil). **A.** Apex of stem showing leaves and inflorescences. **B.** Medial section of inflorescence (syconium) showing subtending bracts and pubescent surface; note the detail of apical view of osteole showing opening closed by bracts (right). **C.** Dimorphic pistils from two flowers: long-styled form (left) and short-styled form (right) showing one tepal.

Ficus nymphaeifolia Mill. PART 1: PL. IVc

Hemiepiphytic shrubs or trees ("stranglers"), becoming free-standing trees to 40 m tall, with well developed buttresses. Stems glabrous or minutely white-puberulous. Stipules 1–2.5 cm long, glabrous or minutely white-puberulous. Leaves: petioles 5–12 (20 cm long, the epidermis not peeling off; blades ovate to (broadly) elliptic, 10–25 × 7–18 cm, smooth adaxially, glabrous or minutely white-puberulous abaxially, drying brownish or greenish, with glandular spot at base of midrib abaxially, the base cordate. Figs axillary, in pairs, 1–2.5 cm diam., greenish at maturity; peduncles usually absent, sometimes to 0.4 cm long, the basal bracts to 0.5–0.8 cm long; ostiole not surrounded by rim, ± prominent. Fr (Nov, Dec); common but scattered, in nonflooded forest. *Bois figué*.

Ficus panurensis Standl.

Hemiepiphytic shrubs or trees ("stranglers"), becoming free-standing trees to 10 m tall. Stems white-hairy. Stipules ca. 1–2 cm long, sometimes peristent, white hairy. Leaves: petioles 0.7–2.5 cm long, the epidermis not peeling off; blades oblong to subobovate, 10–17 × 5–6.5, smooth adaxially, white-hairy or glabrous abaxially, drying brownish, with glandular spot at base of midrib abaxially, the base obtuse to rounded. Figs axillary, in pairs, 0.7–1 cm diam.; peduncles absent; ostiole surrounded by 3-lobed rim. Known by a single sterile collection from our area.

Ficus pertusa L. f.

Hemiepiphytic shrubs or trees ("stranglers"), becoming free-standing trees to 20 m tall. Stems glabrous or minutely white-puberulous. Stipules 0.5–1.5 cm long, glabrous or minutely white-puberulous. Leaves: petioles 1–2.5 cm long, the epidermis not peeling off; blades elliptic to oblong, 4.5–12.5 × 2.5–6.5 cm, smooth adaxially,, subglabrous abaxially, usually drying greenish, with glandular spot at the base of midrib abaxially, the base rounded to acute. Figs axillary, in pairs, 0.4–0.8 cm diam., greenish at maturity, often with red, purple or brownish spots; peduncles 0.3–0.4 cm long; ostiole not surrounded by rim, sunken. Fr (Jun, Aug); in nonflooded forest.

Ficus schumacheri (Liebm.) Griseb.

Hemiepiphytic shrubs or trees ("stranglers"), becoming free-standing trees to 25 m tall. Stems glabrous or minutely white-puberulous. Stipules 0.3–0.8 cm long, sometimes persistent, glabrous or minutely white-puberulous. Leaves: petioles 0.3–1(1.3) cm long, the epidermis not peeling off; blades lanceolate, 2.5–9 × 0.7–2 cm, smooth adaxially, subglabrous abaxially, usually drying greenish, with glandular spot at base of midrib abaxially, the base emarginate to acute. Figs 0.3–0.5 cm diam., greenish with red, purple or brownish spots at maturity; peduncles 0.1–0.2 cm long; ostiole not surrounded by rim, slightly prominent. Known by a single sterile collection from our area.

HELICOSTYLIS Trécul

Canopy trees. Hook-shaped hairs absent. Horizontal lateral branches deciduous. Stipules not fully encircling stem, free. Leaves alternate, spirally arranged on ± vertical branches, distichous on horizontal branches; blades without waxy, glandular spots, pubescent on both surfaces, dull adaxially when dry. Inflorescences on lateral branches, often several together on minute short-shoots, unisexual, discoid-capitate and involucrate with basally attached bracts; peduncle present; receptacle of staminate inflorescences ± concave beneath and involucre not covering flowers before anthesis. Flowers unisexual (plants monoecious or dioecious); tepals 4, partly fused. Staminate flowers: stamens 4. Pistillate flowers: stigmas band-shaped, often twisted. Fruits enclosed by fleshy perianths, these pseudodrupes yellow and aggregated in heads.

1. Leaf blades with hairs densely covering midrib abaxially, the base often rounded, the margins often dentate towards apex. Pistillate inflorescences with ca. 15–30 flowers. Perianth of staminate flowers with 4 deeply divided lobes. . . . . *H. pedunculata.*
1. Leaf blades with hairs less densely covering midrib abaxially, the base often acute, the margins usually entire. Pistillate inflorescences with 5–12 flowers. Perianth of staminate flowers with 4 shallow lobes.. . . . . *H. tomentosa.*

Helicostylis pedunculata Benoist

Canopy trees, to 25 m tall. Leaf blades elliptic to obovate, 10–28 × 5–12 cm, coriaceous to chartaceous, hairs densely covering midrib abaxially, the base often rounded, the margins usually dentate toward apex. Staminate inflorescences to 6 together or accompanying pistillate ones; receptacle 0.8–1 cm diam. Pistillate inflorescences with ca. 15–30 flowers; peduncle 0.5–2.5 cm long. Flowers unisexual (plants often monoecious); perianth with 4 deeply divided lobes. Infructescences discoid to hemispherical, 2–3 cm diam. Known by a single sterile collection from our area. *Satiné* (Créole).

Helicostylis tomentosa (Poepp. & Endl.) Rusby PL. 97c

Canopy trees, to 30 m tall. Leaf blades oblong to elliptic, 5–32 × 2–15 cm, subcoriaceous, ± densely hairy but not covering midrib abaxially, the base often acute, the margins usually entire. Staminate inflorescences to 15 together, rarely accompanying pistillate ones; receptacle 0.5–0.8 cm diam. Pistillate inflorescences with 5–12 flowers; peduncle to 0.5 cm long. Flowers unisexual (plants sometimes monoecious); perianth with 4 shallow lobes. Infructescences subglobose, 2.5–5 cm diam. Fl (Dec), fr (Feb, Dec); in nonflooded forest.

MAQUIRA Aubl.

Canopy or understory trees. Hook-shaped hairs absent. Horizontal lateral branches deciduous. Stipules not fully encircling stem, free. Leaves alternate, spirally arranged on vertical branches, distichous on horizontal branches; blades without way, glandular spots, subglabrous, ± shiny adaxially when dry. Inflorescences on horizontal branches on minute short-shoots, unisexual, discoid-capitate (or 1-flowered) and involucrate with basally attached bracts; peduncle present. Staminate inflorescences several together on small, short shoots, with involucre not covering flowers before flowering. Flowers unisexual (plants dioecious or monoecious). Staminate flowers: perianth 4-lobed to 4-parted; stamens 4. Pistillate flowers: perianth 4-lobed; ovary adnate to perianth, the stigmas lingulate. Fruits enclosed by fleshy perianths, these pseudodrupes free or fused in heads.

1. Leaf blades usually <15 cm long; tertiary venation reticulate. *M. guianensis* subsp. *guianensis.*
1. Leaf blades usually >15 cm; tertiary venation scalariform. . . . . *M. sclerophylla.*

Maquira guianensis Aubl. subsp. **guianensis** PL. 97d

Understory or canopy trees, to 10(30) m tall. Leaf blades oblong to elliptic, 5–18 × 3–6 cm, coriaceous, the margins entire. Staminate inflorescences to 3 together; peduncle to 1.5 cm long; receptacle 0.5–1.5 cm diam. Pistillate inflorescences sometimes accompanied by 1–2 staminate ones, to 1.2 cm long, with 5–35, free flowers; peduncule present; receptacle 0.5–2 cm diam. Infructescences (2)3–4 cm diam., with free, brown-velutinous perianths. Fl (Aug, Sep), fr (Feb, Apr); in nonflooded forest.

Maquira sclerophylla (Ducke) C. C. Berg FIG. 225

Canopy trees. Leaf blades elliptic to oblong, 13–30 × 5–15 cm, coriaceous, adaxial surface often somewhat scabrous, the margins entire. Staminate inflorescences to 4 together, to 1.2 cm long; peduncule present; receptacle 0.3–1.2 cm diam. Pistillate inflorescences sometimes accompanied by 1–2 staminate ones, to 0.8 cm long, with 2–4, partly fused flowers; peduncle present. Infructescences to 3(4) cm diam., with fused, nearly glabrous perianths. Fl (Sep), fr (Oct); common, in nonflooded forest.

NAUCLEOPSIS Miq.

Understory trees. Hook-shaped hairs absent. Horizontal lateral branches deciduous. Stipules fully encircling stem, free. Leaves spirally arranged on vertical branches, distichous on horizontal branches; blades without waxy glandular spots, subglabrous, the margins entire. Inflorescences unisexual, discoid to globose-capitate and involucrate with basally attached bracts. Staminate inflorescences often several together on minute short-shoots, the involucre covering flowers until flowering. Pistillate inflorescences with flowers fused together. Flowers unisexual (plants dioecious); tepals 3–7, free. Staminate flowers: stamens 2–4. Pistillate flowers: tepals with free parts conical to spine-like. Fruits enclosed by fleshy perianths, fused in heads.

Naucleopsis guianensis (Mildbr.) C. C. Berg PL. 97e

Understory trees, to 20 m tall. Stipules 0.7–2 cm long. Leaf blades elliptic to subobovate, 5–17 × 1.5–5.5 cm. Staminate inflorescences: peduncule present. Pistillate inflorescences (sub)sessile. Infructescences ca. 2–4 cm diam., yellow to orange. Fl (Aug, Sep, Oct), fr (Sep, Nov); common, in nonflooded forest.

PEREBEA Aubl.

Understory or canopy trees. Hook-shaped hairs absent. Horizontal lateral branches deciduous. Stipules fully encircling stem, free. Leaves alternate, spirally arranged on vertical branches, distichous on horizontal branches; blades without waxy glandular spots, pubescent, the margins enter or dentate. Inflorescences on horizontal branches on short spurs, unisexual, discoid-capitate and involucrate. Staminate inflorescences with involucre not covering flowers before flowering; peduncle present. Pistillate inflorescences with many flowers. Flowers unisexual (plants dioecious or monoecious). Staminate flowers: tepals 4, free or basally fused; stamens 2–4. Pistillate flowers: perianth tubular. Fruits enclosed by perianths, these pseudodrupes free or fused together in heads.

1. Hairs on stems, stipules, and blades appressed. Stipules 2–4 cm long. Inner involucral bracts ovate to triangular. . . . . *P. guianensis* subsp. *guianensis*.
1. Hairs on stems, stipules, and blades at least partly patent. Stipules 1–2.5 cm long. Inner involucral bracts narrowly triangular to subulate. . . . . *P. rubra* subsp. *rubra*.

Perebea guianensis Aubl. subsp. **guianensis** PL. 98a,b

Usually understory trees, to 20 m tall. Leaf blades oblong, 20–50 × 7–24 cm, coriaceous to chartaceous, the margins entire or dentate. Staminate inflorescences: peduncle 0.5–2 cm long; receptacle 0.5–1.5 cm diam. Pistillate inflorescences: peduncle 0.3–1 cm long; receptacle 0.8–1.5 cm diam. Pistillate flowers: stigmas lingulate. Infructescences discoid to hemispherical, 2.5–4 cm diam., red at maturity. Fl (Apr, Aug, Sep, Oct), fr (Jan, Feb, Apr, Aug, Sep, Oct); in nonflooded forest. *Abérémou* (Créole), *cauchorana* (Portuguese).

Perebea rubra (Trécul) C. C. Berg subsp. **rubra**

Canopy trees, to 30 m tall. Leaf blades oblong to elliptic to (sub)ovate, 10–28 × 4–12 cm, (sub)coriaceous, the margins entire to dentate. Staminate inflorescences: peduncle 1–3 cm long; receptacle 0.8–1.5 cm diam. Pistillate inflorescences: peduncle absent or 0.5 cm long; receptacle 0.8–1.2 cm diam. Pistillate flowers: stigmas filiform. Infructescences discoid, 1.5–2.5 cm diam. Fl (Feb), fr (Apr); in nonflooded forest.

PSEUDOLMEDIA Trécul

Canopy trees. Hook-shaped hairs absent. Horizontal lateral branches deciduous. Stipules fully encircling stem, free. Leaves alternate, spirally arranged on vertical branches, distichous on horizontal branches; blades without waxy, glandular spots, glabrous or sparsely pubescent, the margins entire to dentate toward apex. Inflorescences on horizontal branches, often 2 or 4 together, unisexual, sessile, capitate and involucate with basally attached bracts. Flowers unisexual (plants monoecious or dioecious). Staminate flowers indistinct; stamens free among "bracts" and covered by involucre until flowering. Pistillate inflorescences uniflorous. Pistillate flowers: perianth tubular; stigmas filiform. Fruits enclosed by fleshy perianths.

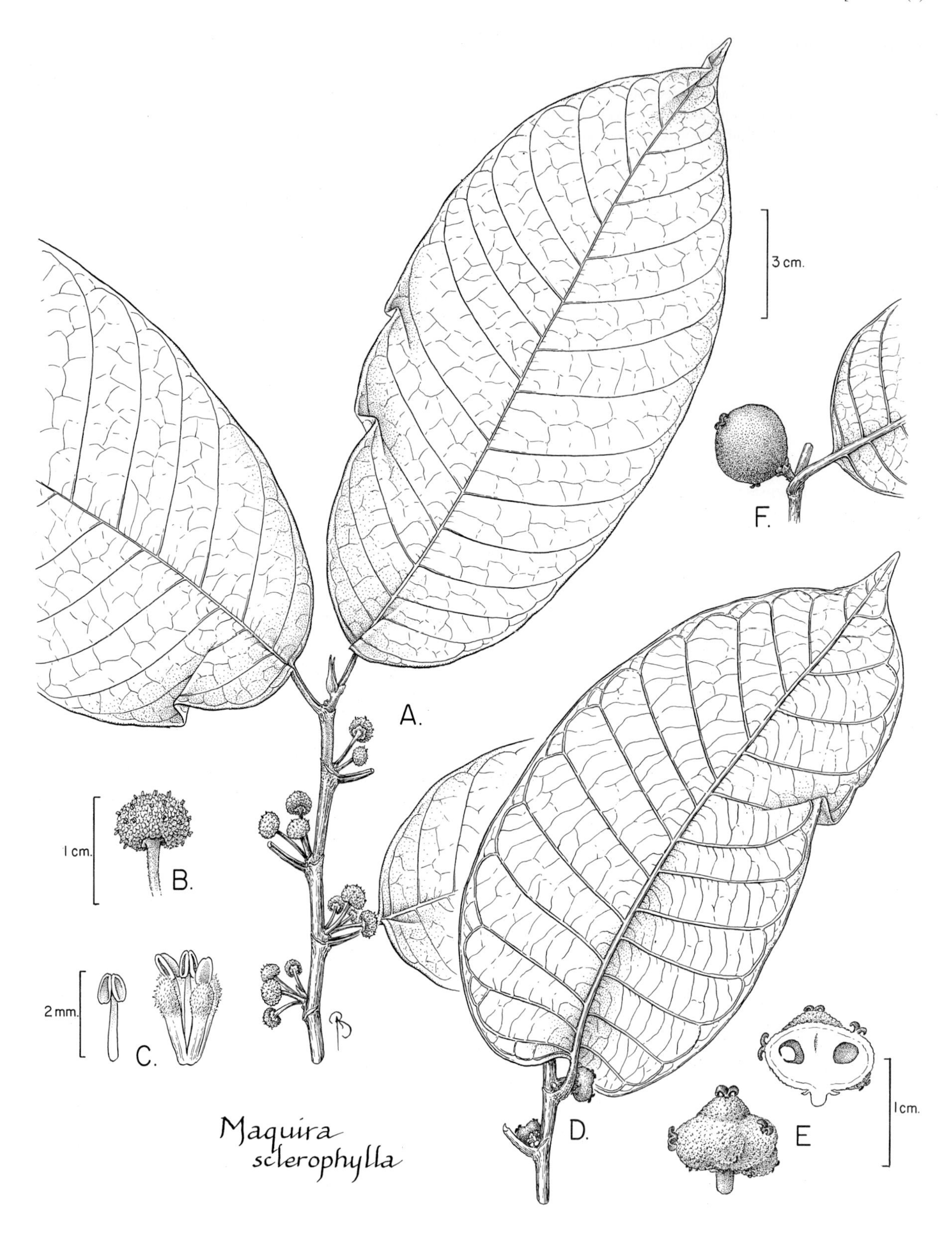

FIG. 225. MORACEAE. *Maquira sclerophylla* (A-C, *Mori et al. 22002*; D, E, *Silva 1354* from Brazil; F, *Silva 3707* from Brazil). **A.** Apex of stem showing leaves and staminate inflorescences. **B.** Staminate inflorescence. **C.** Staminate flower (right) and stamen (left). **D.** Part of stem showing leaves and pistillate inflorescences. **E.** Whole (below) and medial section (above) of pistillate inflorescences. **F.** Infructescence with a single fruit.

1. Branchlets and stipules with equally long appressed hairs. Terminal buds slender. Leaf blades with margins entire. Fruiting perianth puberulous, the hairs white. . . . . *P. laevigata.*
1. Branchlets and stipules with patent minute hairs intermixed with much longer patent to ± appressed hairs. Terminal buds often swollen. Leaf blades with margins often faintly dentate towards apex. Fruiting perianth hirtellous to subhirsute, the hairs yellow. . . . . *P. laevis.*

Pseudolmedia laevigata Trécul PL. 98c

Understory to canopy trees, to 35 m tall. Stems with equal length, appressed hairs, the terminal buds slender. Stipules with equal length, appressed hairs. Leaves: blades oblong to elliptic, 3–14 × 1–5 cm, coriaceous, the margins entire. Infloresences: involucral bracts with white hairs. Fruiting perianth red, edible, enclosing the fruit, puberulous, the hairs white. Fr (Apr, May); in nonflooded forest.

Pseudolmedia laevis (Ruiz & Pav.) J. F. Macbr.

Canopy trees, to 40 m tall. Stems with patent minute hairs mixed with longer to ± appressed hairs, the terminal buds swollen. Stipules with patent minute hairs mixed with longer to ± appressed hairs. Leaves: blades oblong, 5–20 × 3–8 cm, subcoriaceous, the margins often faintly dentate toward apex. Inflorescences: involucral bracts with yellowish hairs. Fruiting perianth red, enclosing the fruit, hirtellous to subhirsute, the hairs yellow. Known by a single sterile collection from our area.

TRYMATOCOCCUS Poepp. & Endl.

Canopy or understory trees. Hook-shaped hairs present, especially on young parts. Horizontal lateral branches not deciduous. Stipules partially encircling stem, free. Leaves alternate, distichous; blades without waxy, glandular spots. Inflorescences bisexual, receptacle cylindrical to turbinate, with 1 pistillate flower embedded in center and several staminate flowers borne on upper part of receptacle. Flowers unisexual (plants monoecious). Staminate flowers: tepals 3, partly fused; stamens 3. Pistillate flowers: stigmas filiform. Fruits enclosed by the fleshy receptacle, remnants of staminate flowers often persistent at apex.

1. Stems hirtellous to velutinous. Stipules 0.3–0.4 cm long. Leaf blades with tertiary venation partly scalariform. . . . . *T. amazonicus.*
1. Stems appressed-puberulous. Stipules ca. 0.2 cm long. Leaf blades with tertiary venation reticulate. . . . . . *T. oligandrus.*

Trymatococcus amazonicus Poepp. & Endl. [Syn.: *T. paraensis* Ducke]

Understory or canopy trees, to 30 m tall. Stems hirtellous to velutinous. Stipules 0.3–0.4 cm long. Leaves: blades elliptic to oblong, 2–12 × 1.5–5.5 cm; tertiary venation partly scalariform. Inflorescences narrowly turbinate to cylindrical, with 20 or more staminate flowers. Infructescences ca. 1.5 cm diam. Fr (Jun); apparently less common than *T. oligandrus*, in nonflooded forest.

Trymatococcus oligandrus (Benoist) Lanjouw PL. 98d

Understory trees, to 20 m tall. Stems appressed-puberulous. Stipules ca. 0.2 cm long. Leaves: blades elliptic to oblong, 1–12 × 0.5–4 cm; tertiary veins reticulate. Inflorescences cylindrical to narrowly turbinate, with 2–5 staminate flowers. Infructescences globose, ca. 1.5 cm diam. Fl (Aug, Sep), fr (Feb, May, Jul, Sep, Oct, Nov); common, in nonflooded forest. *Bois de lettre moucheté, satiné* (Créole).

MYRISTICACEAE (Nutmeg Family)

John D. Mitchell

Trees, often very aromatic. Bark exuding cream-colored or red-colored sap when cut. Indumentum simple, stellate, dendritic, bifid or malpighiaceous (T-shaped trichomes). Stipules lacking. Leaves evergreen, simple, alternate; blades with entire margins, often pellucid-punctate. Inflorescences usually axillary, cymose to racemose, sometimes ramiflorous or cauliforous. Flowers unisexual (plants dioecious or occasionally monoecious), actinomorphic, cupulate or campanulate; perianth uniseriate; tepals (2)3(5), basally connate, valvate. Staminate flowers with 3–20 stamens, the filaments united into a column, the anthers partially free or often laterally connate, dehiscence longitudinal, pistillode lacking. Pistillate flowers without staminodia; pistil monocarpellate; stigma simple or bilobed, sessile or subsessile; ovule 1, more or less basal, usually anatropous, bitegmic. Fruits bivalvate capsules, coriaceous to woody, dehiscent along 2 sutures. Seeds 1(2) per fruit, often with ruminate endosperm; aril well developed, entire to strongly laciniate; cotyledons cryptocotylar, often epigeal in Neotropics.

Jansen-Jacobs, M. J. 1976. Myristicaceae. *In* J. Lanjouw & A. L. Stoffers (eds.), Flora of Suriname **II(2):** 427–429. E. J. Brill, Leiden.
Ooststroom, S. J. van. 1966. Myristicaceae. *In* A. Pulle (ed.), Flora of Suriname **II(1):** 113–122, 473–475. E. J. Brill, Leiden.

1. Leaf blades with tertiary veins subparallel, nearly perpendicular to midrib. Seeds with black, red or purple spots. . . . . *Compsoneura.*
1. Leaf blades with tertiary veins reticulate. Seeds uniform in color.
    2. Abaxial leaf blade surfaces with stellate trichomes.

3. Slash of bark exuding cream-colored sap. Pedicels bracteolate. Anthers 12–20. Arils entire. . . . . *Osteophloeum*.
3. Slash of bark exuding red sap. Pedicels ebracteolate. Anthers 3(4–6). Arils deeply laciniate. . . . . *Virola*.
2. Abaxial leaf blade surfaces without stellate trichomes, but, if trichomes present, these simple or bifid (or T-shaped). . . . . *Iryanthera*.

COMPSONEURA Warb.

Trees. Sap usually red. Indumentum lacking. Leaf blades with tertiary veins subparallel, nearly perpendicular to the midrib. Plants dioecious. Flowers ebracteolate. Staminate flowers with staminal column obconical, truncate; stamens 4–11, the anthers adnate to column (in our area). Fruits oblong-ellipsoid, pericarp thin. Seeds with aril entire to laciniate. Seeds patterned with black, red or purple spots; endosperm not ruminate.

Compsoneura ulei Warb. FIG. 226

Treelets or small trees, 2–5(12) m tall. Leaves: petioles 6–15 mm long; leaf blades chartaceous to subcoriaceous, oblong, elliptic, obovate, or ovate, 11.4–20.5 × 4.7–2.7 cm, the base cuneate, acute to obtuse, the apex acuminate; midrib prominent abaxially, flattened to prominulous adaxially, the secondary veins in 4–7 pairs, arcuate, prominulous to prominent abaxially, prominulous adaxially, the tertiary veins prominulous on both surfaces. Inflorescences racemose to paniculate, often arising from short shoots, 1.5–3 cm long; bracts subtending pedicels ovate, 0.5 mm long, caducuous; pedicels 2.5–4 mm long. Flowers green to greenish yellow; tepals 3(4), divided to base, subdeltate, 1.7–2.2 × 2.2–2.5 mm, markedly papillate adaxially. Staminate flowers with staminal column 0.8–1 mm long, the anthers 9–14. Pistillate flowers with ovary subglobose, the stigma bifid. Fruits yellow to yellow-orange, 1.6–2 × 1–1.3 cm; aril red, entire to laciniate. Fr (Nov); rare, in nonflooded forest on La Fumée Mt.

IRYANTHERA Warb.

Trees. Sap red. Indumentum of bifid, T-shaped (malpighiaceous), or simple trichomes. Leaf blades with reticulately branching, somewhat obscure tertiary venation. Plants usually monoecious (in our area). Inflorescences often ramiflorous or cauliflorous. Flowers bracteolate. Staminate flowers with 3–4(6) anthers, these adnate to the staminal column. Fruits transversally ellipsoid; pericarp woody; aril entire or apically laciniate. Seeds concolorous; endosperm ruminate (in our area).

1. Flowers rotate; tepals divided to base of perianth. Staminate flowers with staminal column ca. 0.3 mm long. . . . . *I. sagotiana*.
1. Flowers urceolate; tepals divided 1/3 of way to base of perianth. Staminate flowers with staminal column ca. 2.5 mm long. . . . . *I. tessmannii*.

Iryanthera sagotiana (Benth.) Warb. FIG. 227, PL. 98e,f; PART 1: FIG. 10, PL. XVIc

Trees, to 30 m tall. Trunks cylindrical, unbuttressed. Outer bark brown to reddish-brown, shaggy, peeling in long plates; inner bark orange. Sap red. Indumentum usually absent on mature leaf blades, sparsely to densely ferruginous on branchlets, petioles, inflorescences, and flowers. Leaves: petioles 0.7–1.5 cm long; leaf blades chartaceous to subcoriaceous, oblanceolate to obovate or elliptic, 12–20 × 4–6.7 cm, the base attenuate, cuneate, or obtuse, the apex short to long acuminate; midrib prominent abaxially, prominulous adaxially, the secondary veins in 10–18 pairs, flattened abaxially, impressed to flattened adaxially. Pistillate inflorescences with thick primary and secondary rachises, arising from slender branchlets, fasciculate, 3–8 cm long; bracteole subtending perianth semicircular. Flowers rotate, yellow-green; tepals 3(4), divided to base of perianth, subdeltate, glabrous adaxially. Staminate flowers with staminal column slightly obconical, ca. 0.3 mm long, the anthers 3–4. Pistillate flowers with ovary somewhat obconical, truncate, ca. 0.7 mm long, glabrous, the stigma discoid. Fruits green, 1.8–2.5 × 2–3.1 cm; aril red. Fl(Jun, Aug, Sep, Oct), fr (Feb, Dec); fairly common, in nonflooded forest.

Iryanthera tessmannii Markgr. PL. 99a

Trees, to 30 m tall. Trunks cylindical. Outer bark reddish-brown, rough, peeling in long thin plates. Sap red. Indumentum very sparse or absent on mature leaf blades, densely ferruginous on inflorescences and flowers. Leaves: petioles 1.2–2 cm long; leaf blades chartaceous to subcoriaceous, oblanceolate or narrowly elliptic, 15–17 × 4–5.5 cm, the base attenuate, cuneate, or obtuse, the apex long acuminate or acute; midrib prominent abaxially, prominulous to prominent adaxially, the secondary veins in 13–15 pairs, flattened abaxially, impressed adaxially. Pistillate inflorescences with thick primary and secondary rachises, cauliflorous or ramiflorous. Staminate inflorescences with slender rachises arising from slender branchlets, fasciculate, 2.4–10 cm long; bracteole subtending perianth semicircular to ovate. Flowers narrowly urceolate, yellow-green; tepals 3(4), divided 1/3 of way to base of perianth, subdeltate, glabrous adaxially. Staminate flowers with staminal column swollen at base, ca. 2.5 mm long, the anthers 3–4. Pistillate flowers with ovary somewhat conical, glabrous, the stigma subpeltate, centrally concave. Fruits green, turning brown, 1–2.5 × 2–3.5 cm; aril red. Fl (Aug, Sep, Feb), fr (Aug, Nov); in nonflooded forest on La Fumée Mt. and vicinity of Eaux Claires.

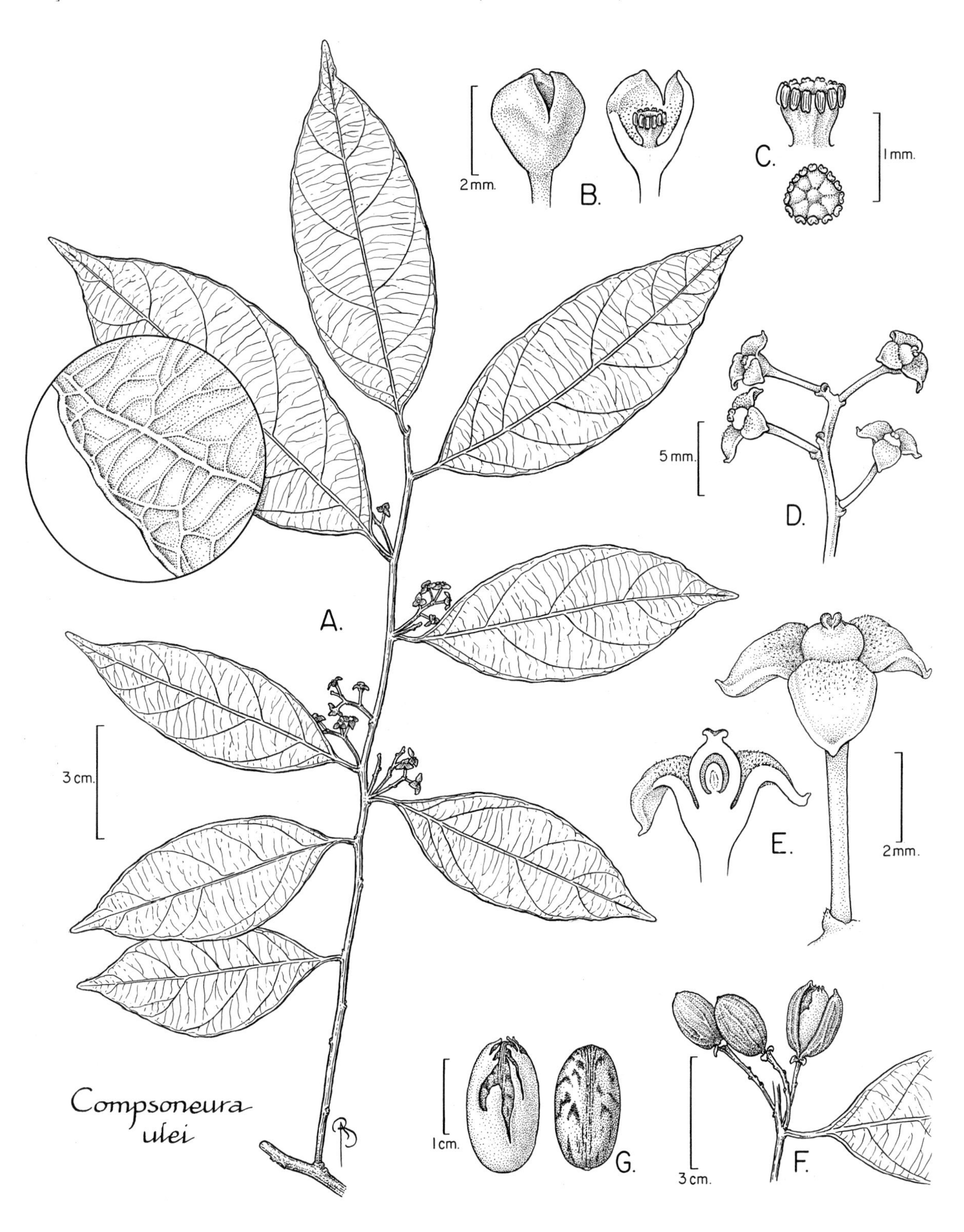

FIG. 226. MYRISTICACEAE. *Compsoneura ulei* (A, D, E, *Silva 928* from Amazonas; B, C, *Prance et al. 4649* from Amazonas; F, G, *Mori & Boom 15165*). **A.** Part of stem with leaves and inflorescences and detail of abaxial surface of leaf. **B.** Lateral view (left) and medial section (right) of staminate flower. **C.** Lateral view (above) and apical view (below) of androecium. **D.** Inflorescence of pistillate flowers. **E.** Medial section (left) and lateral view (right) of pistillate flower. **F.** Infructescence. **G.** Seed with (far left) and without (near left) aril.

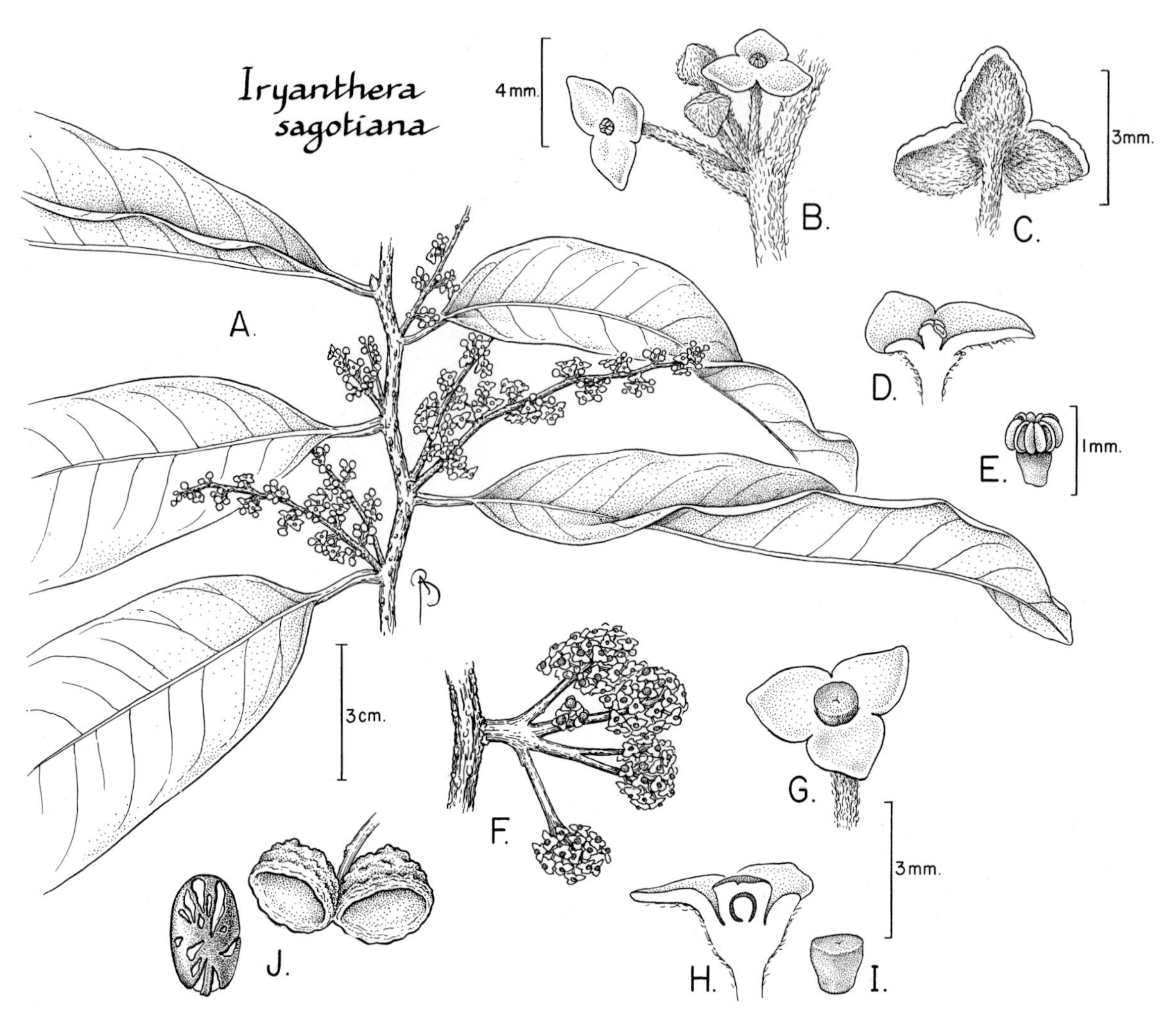

FIG. 227. MYRISTICACEAE. *Iryanthera sagotiana* (A, F–I, *Mori et al. 20912*; B–E, *Mori et al. 20755*; J, *Mori & Gracie 21117*; K, *Mori et al. 22951*). **A.** Part of stem showing leaves and staminate inflorescences. **B.** Part of staminate inflorescence. **C.** Basal view of staminate flower showing perianth. **D.** Medial section of staminate flower. **E.** Lateral view of androecium. **F.** Cauliflorous pistillate inflorescence. **G.** Oblique-apical view of pistillate flower. **H.** Medial section of pistillate flower showing basal placentation of ovule. **I.** Pistil. **J.** Fruit on pedicel. **K.** Dehisced fruit (right) and seed surrounded by aril (left).

OSTEOPHLOEUM Warb.

Trees. Sap usually cream-colored. Indumentum of stellate trichomes. Leaf blades with reticulately branching, somewhat obscure tertiary veins. Plants dioecious. Flowers bracteolate. Staminate flowers with staminal column apically conical, the anthers adnate to column, 12–20. Fruits transversally ellipsoid, the pericarp woody. Seeds concolorous; aril entire; endosperm ruminate.

Osteophloeum platyspermum (A. DC.) Warb.
FIG. 228, PL. 99b

Trees, to 44 m tall. Trunks cylindrical, unbuttressed. Outer bark brown, rough with vertical fissures; inner bark cream- to yellow-colored, the exudate usually cream-colored, sometimes clear to amber or slightly reddish-colored. Indumentum stellate, ferruginous, covering branchlets, young leaves, inflorescences, and flowers. Leaves: petioles 1.5–2 cm long; leaf blades oblanceolate or oblong, 10.3–17 × 3.5–5.4 cm, chartaceous to coriaceous, the abaxial surface papillate, the base usually attenuate, the apex obtuse, short acuminate, or slightly emarginate; midrib prominent abaxially, impressed adaxially, the secondary veins in 6–11 pairs, prominulous to prominent abaxially, impressed to flattened adaxially. Inflorescences paniculate, 2–5 cm long; bracts subtending pedicels semi-orbicular; bracteole subtending perianth semi-circular. Flowers ferruginous abaxially, greenish adaxially; tepals 3, divided about half-way to base, subdeltate, glabrous adaxially. Staminate flowers with staminal column 0.3–1 mm long. Pistillate flowers with ovary conical, densely stellate-pubescent, the stigma bifid. Fruits yellow-green, 1.5–1.7 × 2.4–2.6 cm. Fl (Sep, Oct, Nov); in nonflooded forest.

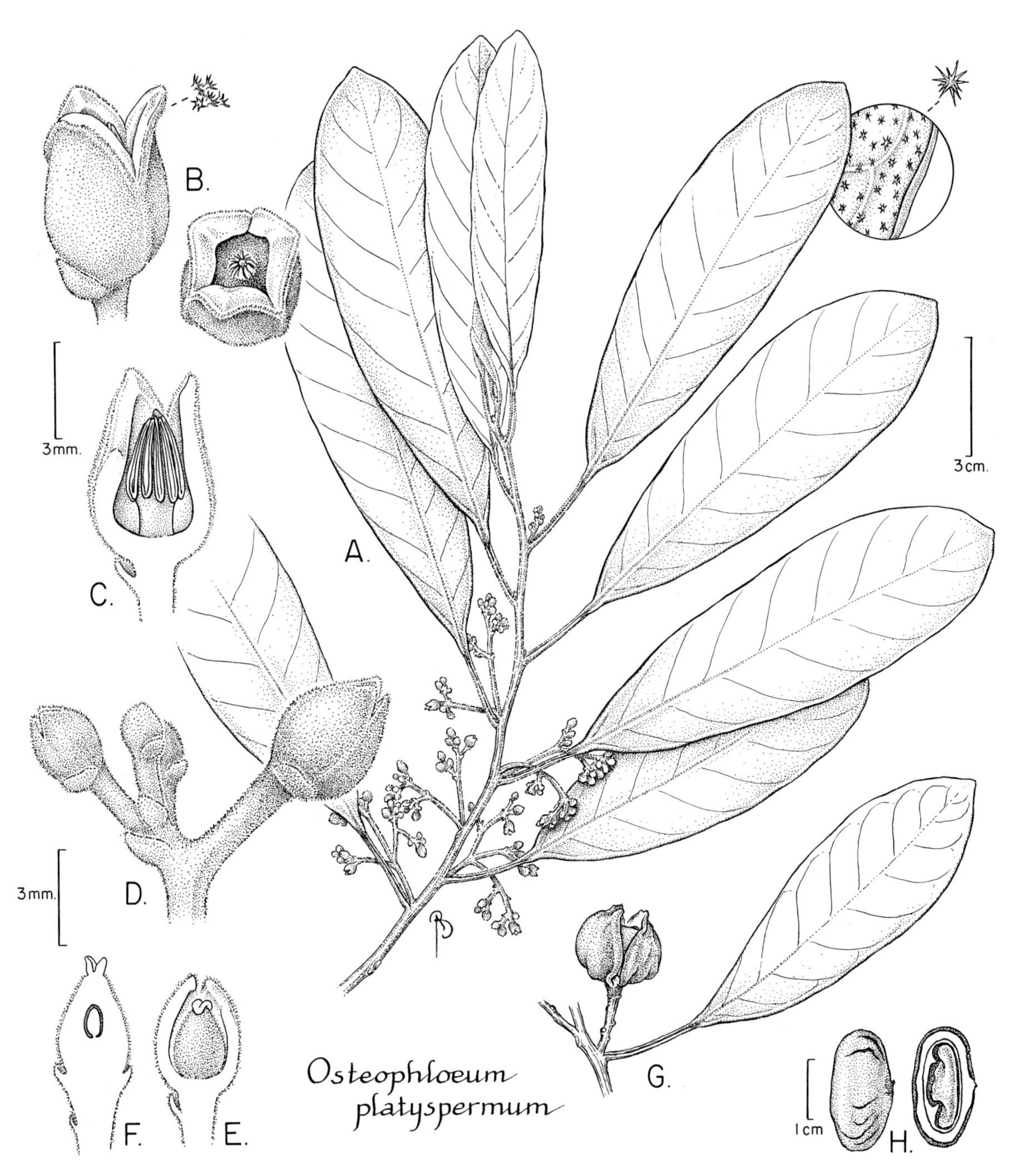

FIG. 228. MYRISTICACEAE. *Osteophloeum platyspermum* (A, *Mori et al. 20770*; B, C, *Mori et al. 22021* & photos *Mori & Gracie 23893*; D–F, *Prance et al.3139* from Amazonas; G, H, *Prance et al. 11117* from Roraima). **A.** Apex of stem showing staminate inflorescences, leaves, and detail of leaf showing revolute margin and stellate hairs on abaxial surface. **B.** Lateral view of staminate flower with detail of hairs on abaxial surface of perianth (left) and apical view of staminate flower (below right). **C.** Medial section of staminate flower. **D.** Part of inflorescence. **E.** Medial section of pistillate flower. **F.** Medial section of developing fruit showing position of ovule. **G.** Dehisced fruit. **H.** Lateral view of seed surrounded by aril (left) and medial section of fruit (right).

VIROLA Aubl.

Trees. Sap red. Indumentum of stellate and sometimes dendritic trichomes. Leaf blades with reticulately branching tertiary venation. Plants dioecious. Inflorescences clustered towards ends of branches. Flowers ebracteolate. Staminate flowers with anthers 3(4–6), adnate to staminal column. Fruits globose, ovoid, or ellipsoid, the pericarp woody. Seeds concolorous; aril laciniate, red; endosperm ruminate; germination epigeal, cryptocotylar.

1. Dendritic trichomes abundant on branchlets and inflorescences. Leaf bases cordate or truncate (occasionally rounded); secondary veins in 7–15 pairs. Anthers longer than staminal column, the connective apiculate, projecting beyond anthers. . . . . *V. sebifera.*
1. Dendritic trichomes absent or early caducous on branchlets and inflorescences. Leaf bases acute, attenuate, obtuse, or rounded (rarely subcordate); secondary veins in 17–27 pairs. Anthers shorter than or subequal to staminal column, the connective usually not apiculate, usually not projecting above anthers.
    2. Leaf blade bases obtuse or rounded, not decurrent. Tepals ≤2 mm long; staminal column <0.8 mm long; ovary subglobose, the stigma sessile. Fruits subglobose, 1.5–2 cm long.. . . . . *V. surinamensis.*
    2. Leaf blade bases attenuate to obtuse, decurrent onto petiole. Tepals ≥2 mm long; staminal column ≥0.8 mm long; ovary oblong-ovoid, the stigma subtended by short style. Fruits ellipsoid or ovoid, >2.5 cm long.
        3. Very large trees, with well developed buttresses. Leaf blades with dendritic and some stellate, early caducous trichomes abaxially, the adult blades nearly glabrous, with only a few vestigial clumps of trichomes, especially along midrib near base, the midrib not markedly impressed adaxially. Fruits 4–5(6.5) × 3.5–4 cm, glabrous. . . . . *V. kwatae.*
        3. Large trees, buttresses sometimes present but not as well developed. Leaf blades with stellate, persistent trichomes abaxially, the midrib impressed adaxially. Fruits 3.7–3.9 × 2.3–2.5 cm, ferruginous, stellate-pubescent. . . . . *V. michelii.*

Virola kwatae Sabatier PL. 99c,d

Trees, to 55 m tall. Trunks cylindric or angled, with plank buttresses to 3.5 m tall. Outer bark gray or blackish, vertically fissured. Sap red, often watery. Indumentum of both stellate and dendritic trichomes, adaxial surface of mature leaf blades usually glabrous, abaxial surface initially covered with mostly dendritic trichomes, soon becoming glabrous with only a few clumps of vestigial trichomes remaining, especially along midrib near base, the branchlets and petioles glabrescent, the inflorescences and flowers sparsely to densely golden-ferruginous or golden-brown stellate-pubescent. Leaves: petioles 0.6–1.5 cm long; blades oblanceolate or elliptic to oblong, 7–22(31) × 6(9) cm, chartaceous, densely papillate abaxially, the base attenuate to obtuse, decurrent onto petiole, the apex short to long acuminate; midrib prominent abaxially, shallowly impressed to flattened adaxially, the secondary veins in 15–27 pairs, prominulous to prominent abaxially, impressed to prominulous adaxially. Inflorescences 3.7 cm long. Flowers campanulate, yellow-orange; tepals 2–2.5 mm long (pistillate flowers slightly larger than staminate ones), sparsely stellate-pubescent, with simple trichomes present abaxially. Staminate flowers: staminal column subulate, 0.8–1 mm long, the anthers 4–5, 0.6–0.7 mm long, the connective usually not projecting above anther. Pistillate flowers: ovary oblong-ovoid, the style short, the stigma deeply cleft, slightly oblique. Fruits ovoid, 4–5(6.5) × 3.5–4 cm, glabrous when mature, greenish-yellow to yellow-orange to tan. Fl (Oct), fr (May, Sep); common, in nonflooded forest. *Yayamadou, yayamadou montagne* (Créole).

Virola michelii Heckel [Syn.: *V. melinonii* (Benoist) A. C. Sm.] PL. 99e

Trees, to 35 m tall. Trunks cylindric or angled, with low buttresses to 10 cm tall or unbuttressed. Outer bark gray, dark brown, or black, vertically fissured, the inner bark cream-colored to pink or orange. Sap red, often watery. Indumentum virtually of all stellate trichomes, adaxial surface of mature leaf blades usually glabrous, the branchlets, petioles, abaxial leaf blade surfaces, inflorescences, and flowers sparsely to densely golden-ferruginous. Leaves: petioles 1–1.5 cm long; blades oblanceolate or elliptic to oblong, 8.5–21 × 2.4–7 cm, chartaceous, densely papillate abaxially, the base attenuate or acute, decurrent onto petiole, the apex short- to long-acuminate; midrib prominent abaxially, impressed adaxially, the secondary veins in 17–22 pairs, prominulous to prominent abaxially, impressed to prominulous ab- and adaxially. Inflorescences 2.5–7.2 cm long. Flowers campanulate, yellow; tepals 2.2–3 mm long (pistillate flowers somewhat larger than staminate), sparsely stellate-pubescent adaxially. Staminate flowers: staminal column subulate, 1.2–1.5 mm long, the anthers 3, 0.6–0.7 mm long, the connective not projecting above anthers. Pistillate flowers: ovary oblong-ovoid, the style short, the stigma deeply cleft, slightly oblique. Fruits ovoid or ellipsoid, 3.7–3.9 × 2.3–2.5 cm, ferruginous, stellate-pubescent when mature. Fl (Jul, Aug, Oct), fr (Sep, Oct, Dec); less common than *V. kwatae*, in non flooded forest. *Yayamadou, yayamadou montagne* (Créole).

Virola sebifera Aubl. FIG. 229, PL. 99f

Trees, to 20 m tall. Outer bark reddish-brown, scaly. Sap red, somewhat watery. Indumentum densely stellate, ferruginous on all surfaces, adaxial leaf surface becoming glabrous with age, the dentritic trichomes abundant on branchlets, inflorescenes and flowers. Leaves: petioles 1–1.7 cm long; leaf blades chartaceous to subcoriaceous, lanceolate, oblong, narrowly elliptic, ovate, or obovate, 10–42 × 4.5–11.2 cm, with abaxial surface granulose-papillate, the base cordate, truncate, or occasionally rounded, the apex usually long acuminate or acute; midrib prominent abaxially, impressed to prominulous adaxially, the secondary veins in 7–15 pairs, prominent abaxially, impressed to prominulous adaxially. Inflorescences 8.5–18 cm long. Flowers funnel-shaped, golden-ferruginous; tepals 0.7–0.9 mm long, densely stellate-pubescent adaxially. Staminate flowers with staminal column short, stout, 0.2–0.3 mm long, the anthers 3(4), 0.8–1 mm long, the connective apiculate, projecting beyond anthers. Pistillate flowers with ovary subglobose, the stigma sessile, discoid, not deeply cleft, slightly oblique. Fruits blue-green, densely ferrigineous-brown pubescent when immature, globose, 1.5–1.8 cm diam. Fl (May), fr (Jul, Aug, Nov); rare, in nonflooded forest. *Djadja, ajamadou, yayamadou* (Créole).

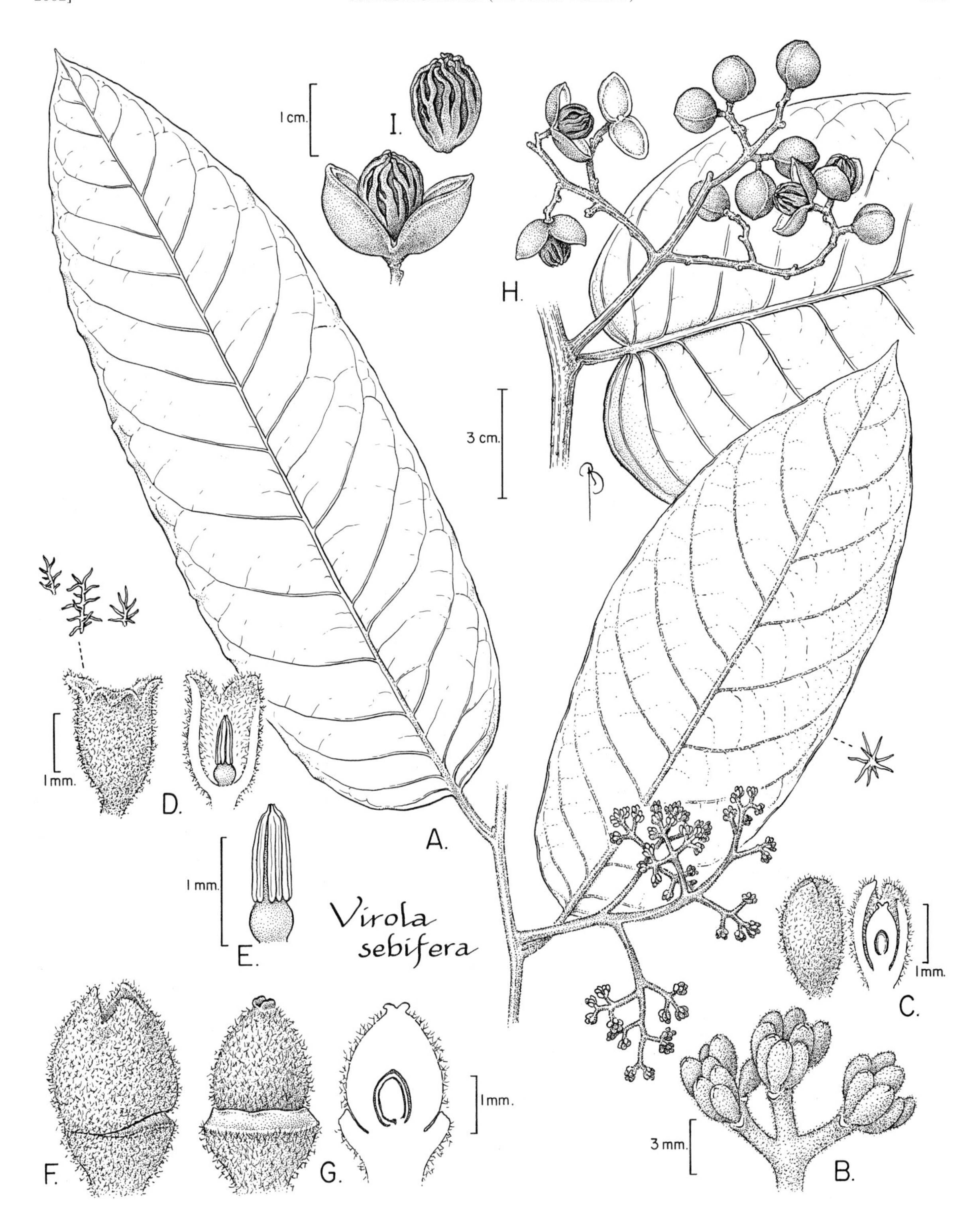

FIG. 229. MYRISTICACEAE. *Virola sebifera* (A–C, *Mori & Pennington 17947*; D, E, *Pipoly 11340* from Guyana; F, G, *Lleras et al. P17153* from Peru; H, I, *Mori & Boom 15188*). **A.** Part of stem with leaves and inflorescence showing enlargement of stellate hair from abaxial leaf surface. **B.** Part of inflorescence. **C.** Lateral view (far left) and medial section of pistillate flower (near left). **D.** Lateral view (left) showing enlargement of dendritic hairs on perianth and medial section (right) of staminate flower. **E.** Lateral view of staminate flower with perianth removed. **F.** Lateral view of developing fruit with perianth beginning to abscise. **G.** Lateral view of fruit after perianth has fallen (left) and medial section of immature fruit (right). **H.** Part of stem with infructescence. **I.** Lateral view of dehisced fruit showing seed (below) and seed covered with lanciniate aril (right).

Virola surinamensis (Rol.) Warb.

Trees, to 40 m tall. Trunks with steep high buttresses. Outer bark gray or brown, smooth to shallowly fissured; inner bark pink to reddish-brown. Sap red. Indumentum stellate (dendritic trichomes rarely present), usually absent on adaxial surface of mature leaf blades, sparsely to densely golden-ferruginous on abaxial leaf surface, branchlets, petioles, inflorescences and flowers. Leaves: petioles 0.4–0.7 cm long; leaf blades lorate to narrowly oblong, lanceolate or oblanceolate, 11.7–27 × 2.2–6 cm, chartaceous to subcoriaceous, the abaxial surface granulose-papillate, the base obtuse or rounded, the apex long acuminate or acute; midrib prominent abaxially, impressed adaxially, the secondary veins in 17–26 pairs, flattened to prominent abaxially, impressed adaxially. Inflorescences 3.8–13.5 cm long. Flowers campanulate, yellow; tepals 1–1.6 mm long, stellate-pubescent adaxially. Staminate flowers with staminal column subulate, 0.5–0.7 mm long, the anthers 3, 0.4–0.5 mm long. Pistillate flowers: ovary subglose, the stigma sessile, flaring, deeply cleft. Fruits subglobose, 1.5–2 cm diam. Fr(Jan, Mar, Jun); rare, in wet areas near streams. *Bicuiba* (Portuguese), *Dyadya*, *moussigot* (Créole), *ucúuba* (Portuguese), *wawichi*, *yayamadou*, *yayamadou-marécage* (Créole).

MYRSINACEAE (Myrsine Family)

John J. Pipoly III

Trees or shrubs. Stipules absent. Leaves simple, alternate or pseudoverticillate; blades with pellucid, black, orange, or red round punctations or punctate-lineations. Inflorescences terminal or lateral (axillary), panicles, racemes, corymbs, or fascicles. Flowers bisexual or functionally unisexual (then plants dioecious, or rarely monoecious or polygamous); sepals nearly free or short connate, persistent in fruit; petals fused or nearly free; stamens and staminodia opposite petals, the stamens free, or free from each other but adnate to petals, or, when connate by filaments to form a tube, the tube adnate to petals and stamens appearing epipetalous, the anthers dehiscent by longitudinal slits or apical pores; pistils and pistillodes unilocular; placentation basal, the placentae with several to many uni- or pluriseriate ovules. Fruits drupaceous. Seeds 1 per fruit.

Bottelier, H. P. 1986. Myrsinaceae. *In* A. Pulle (ed.), Flora of Suriname **IV(1):** 431–442. E. J. Brill, Leiden.

1. Stems and abaxial leaf blade surfaces furfuraceous. Inflorescences terminal or terminal and lateral. Anthers >3× longer than wide. . . . . *Ardisia.*
1. Stems and abaxial leaf blade surfaces glandular-papillate, stipitate-lepidote-pubescent, or glabrous, rarely furfuraceous. Inflorescences strictly lateral. Anthers ≤3× times longer than wide.
    2. Inflorescences racemes to spikes, or columnar or pyramidal panicles, the rachis green or greenish-brown in our area, not translucent. Petals not contorted and twisted in bud. . . . . *Cybianthus.*
    2. Inflorescences corymbs, the rachis pink, translucent. Petals contorted and twisted in bud. . . . . *Stylogyne.*

ARDISIA Sw.

Shrubs or trees. Stems furfuraceous. Leaves alternate; petioles present; blades furfuraceous, the margins entire or serrate. Inflorescences terminal or both terminal and lateral, normally paniculate with racemose, umbellate, or corymbose branches. Flowers bisexual, 4–5-merous; calyx lobes quincuncial or imbricate, almost free, or united to 1/4 length; corolla lobes quincuncial or imbricate, the lobes united to 1/3, or, more often, nearly free; stamens free, adnate at corolla base, shorter than anthers, the anthers long, sagittate, dorsifixed; pistil with long, slender style, the stigma punctiform; placentation of globose placentae, the ovules bi- to pluriseriate, numerous. Fruits globose, drupaceous, the style base persistent.

Ardisia guianensis (Aubl.) Mez [Syn.: *Icacorea guianensis* Aubl.] FIG. 230

Shrubs, to 2 m tall. Stems slender, subterete, finely furfuraceous. Leaves subsessile; petioles thick, 5 mm long; blades oblanceolate, obovate, or elliptic, 6–12 × 2.5–5 cm, membranaceous to chartaceous, the base subcuneate, the apex abruptly short acuminate, the margins entire to subentire to crenulate. Inflorescences terminal, tripinnately and pyramidally paniculate, to 9 cm long, finely furfuraceous, the flowers subcorymbose, numerous; pedicels 4–7 mm long. Flowers 4–5-merous; calyx lobes ovate, 1.3–1.4 mm long, membranaceous, punctate, with small, round, orange glands, the apex rounded, the margins hyaline, obscurely ciliolate or erose-ciliolate at first; corolla 3.5–5 mm long, the lobes elliptic to oblong, the apex obtuse; stamens 3–4.2 mm long, the filaments 1–1.5 mm long, the anthers linear, 2.4–3.3 mm long, apiculate, dehiscent by small flaring apical pores; ovary small, glabrous; ovules 5–14 in several series. Fruits globose, 5–6 mm diam. when dried. Fl (Jan, Feb, Nov, Dec), fr (Mar, Apr, May, Jun); common, in gaps and forest margins on well-drained soils at all elevations in our area.

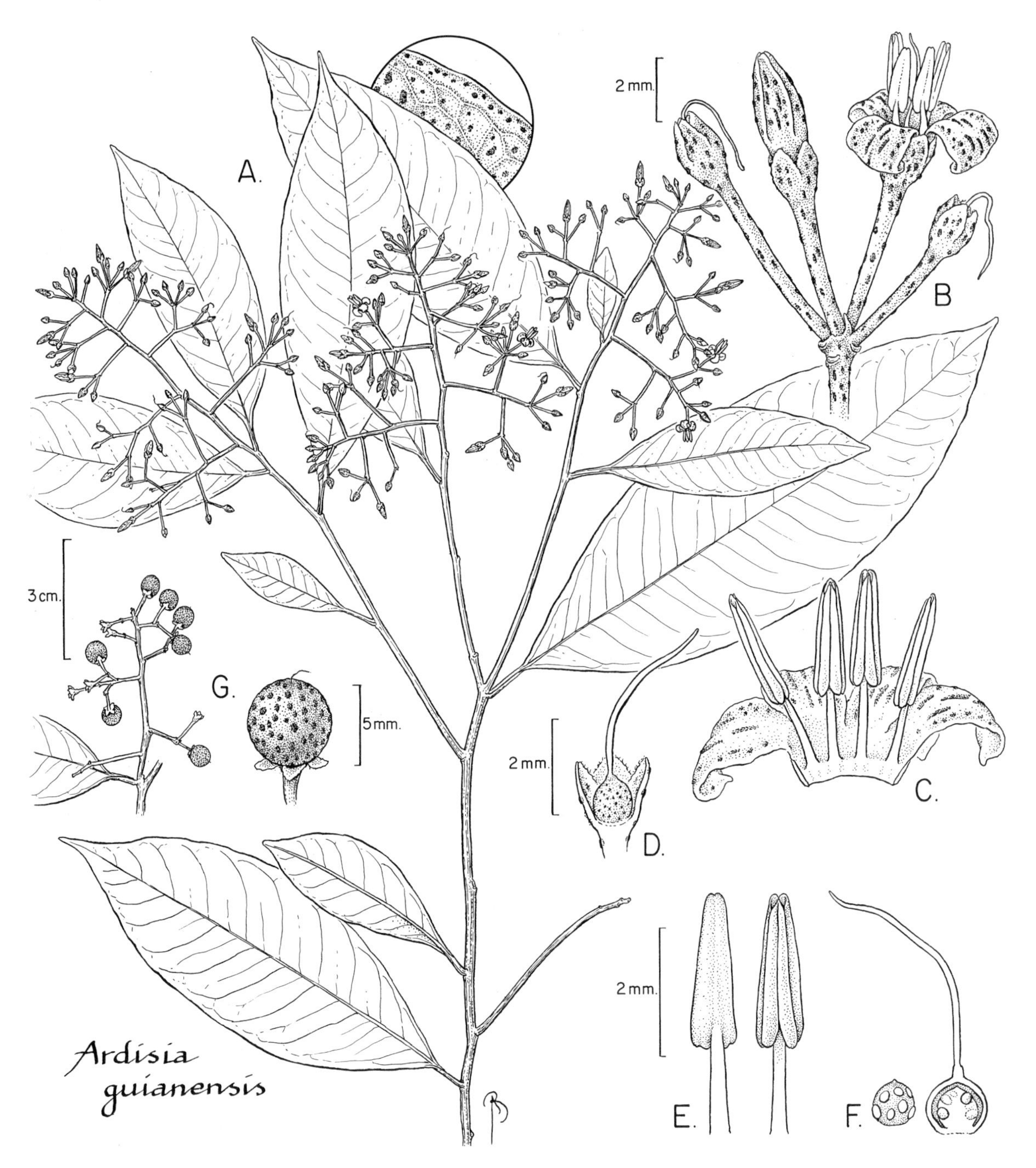

FIG. 230. MYRSINACEAE. *Ardisia guianensis* (A–F, *Mori et al. 22957*; G, *Mori & Pennington 17943*). **A.** Part of stem with leaves and inflorescences and detail of abaxial surface of leaf showing glandular punctations. **B.** Detail of inflorescence showing glandular punctations on flowers in various stages. **C.** Lateral view of corolla opened to show adnation of stamens. **D.** Lateral view of pistil in persistent calyx with two sepals removed. **E.** Abaxial (near right) and adaxial (far right) views of anthers. **F.** Placenta with ovules (near right) and lateral view of pistil with medial section of ovary (far right). **G.** Infructescence (left) and lateral view of fruit showing glandular punctations (right).

CYBIANTHUS Mart.

Shrubs or trees. Stems with pubescence variously lepidote, glandular-papillate, or furfuraceous. Leaves alternate, subopposite, or pseudoverticillate; petioles absent or present. Inflorescences lateral (axillary), simple racemes or spikes, panicles of racemose branches, or indeterminate umbels appearing racemose. Flowers functionally unisexual or bisexual, 3–6(7)-merous; calyx lobes valvate, or aberrantly contorted, basally connate 1/5–2/3 length; corolla lobes imbricate or valvate, basally connate 1/5–3/4 length; stamens and staminodia adnate to corolla tube, the filaments variously connate to form a tube, the staminal (or staminodial) tube adnate to corolla tube or, at times, developmentally fused with it (the stamens or staminodia appearing epipetalous), bearing fleshy lobes alternate with apically free portions of filaments or not; pistil with short style, the stigma capitate, capitate-lobate, or punctiform; ovules (1)2–5(7), uni- or biseriate. Fruits drupaceous, globose or depressed-globose, the style base persistent. Seeds 1(2).

1. Stem apices and/or abaxial leaf blade surfaces (at least at first) bearing minute furfuraceous scales or glandular papillae.
   2. Stems terete.
      3. Calyx lobe margins erose to erose-denticulate. . . . . *C. surinamensis.*
      3. Calyx lobe margins entire.
         4. Petioles canaliculate, not pulvinate, (0.8)1–1.3 cm long, glabrous; blades elliptic, (6)8–12(13.5) × (2.7)3–5 cm, the base acute. Pistillate inflorescences simple racemes. Calyx lobe margins glabrous, not bearing numerous, translucent, glandular scales. . . . . *C. leprieurei.*
         4. Petioles not canaliculate, pulvinate, (0.9)1.7–2.2 cm long, densely glandular-papillate; blades oblanceolate, 27.5–64 × 6.5–9.5(11) cm, chartaceous, the base gradually tapering, cuneate. Pistillate inflorescences columnar panicles with glomerulate branches. Calyx lobe margins glandular-ciliolate, bearing numerous, translucent, glandular scales. . . . . *C. prevostae.*
   2. Stems angular, or with narrow ridges.
      5. Leaf blades narrowly oblanceolate to narrowly elliptic, 2.5–5.5 cm wide, the apex long-acuminate. Calyx lobes black punctate, the margins somewhat erose apically. . . . . *C. microbotrys.*
      5. Leaf blades obovate to elliptic, 7.5–11 cm wide, the base decurrent on stem angles, the apex short acuminate to acute. Calyx lobes red punctate, the margins entire. . . . . *C. potiaei.*
1. Stem apices and/or abaxial leaf blade surfaces bearing ferruginous stipitate lepidote scales or rufous dendritic hairs, at least when young.
   6. Stem apices, calyces and abaxial leaf surfaces bearing ferruginous stipitate lepidote scales. . . . . *C. guyanensis* subsp. *multipunctatus.*
   6. Stem apices bearing rufous dendroid hairs, forming a tomentum, calyces and abaxial leaf blade glabrous or bearing small rufous stellate or dendritic hairs.
      7. Stems terete.
         8. Petioles marginate, 5–10(12) mm long; blades chartaceous, narrowly oblanceolate or oblong, 2–4(6) cm wide, conspicuously black or red punctate and punctate-lineate abaxially, the base and apex long-attenuate; tertiary veins inconspicuous. Pedicels obconic in fruit. . . . . *C. fuscus.*
         8. Petioles canaliculate, 10–20(30) mm long; blades thinly coriaceous, elliptic, (4)5.5–7(10.8) cm wide, inconspicuously pellucid-punctate abaxially, the base cuneate, the apex long-acuminate; tertiary veins prominently raised. Pedicels cylindrical in fruit. . . . . *C. resinosus.*
      7. Stems angular.
         9. Flowers erect. Calyx lobes narrowly ovate, apically obtuse to acute, densely and prominently brown punctate. Pedicels in staminate flowers 2.7–3 mm long, in pistillate flowers 0.7–1 mm long. Stamens with apical free portion of filaments 0.8–1 mm long, the anthers subglobose, apically broadly rounded, the pores not confluent. Pistil obnapiform.. *C. prieurii.*
         9. Flowers nodding. Calyx lobes widely ovate, apically rounded, densely and prominently black punctate. Pedicels in staminate flowers 2.1–2.7 mm long, in pistillate flowers 0.9–1.2 mm long. Stamens without free portion of filaments, the anthers quadrate, apically truncate, the pores confluent or nearly so. Pistil obturbinate. . . . . *C. venezuelanus.*

Cybianthus fuscus Mart.

Shrubs or small trees, to 3 m tall. Stems terete, rufous dendritic-tomentose, glabrescent. Leaves: petioles deeply canaliculate, marginate, 5–10(12) mm long, with basal pulvinus; blades very narrowly oblanceolate or oblong, (13)16–25(30) × 2–4(6) cm, chartaceous, glabrous adaxially, rufous papillate, conspicuously black punctate and punctate-lineate abaxially, the base long attenuate, fully decurrent onto petiole to pulvinus, the apex long attenuate, the margins entire, revolute, glabrous; tertiary veins inconspicuous. Staminate inflorescences lax racemes; pedicels 3.5–5 mm long. Pistillate inflorescences erect racemes; pedicels 1.5–4 mm long. Staminate flowers: calyx 0.8–1(1.8) mm long, the lobes very broadly ovate or linear-lanceolate, apically acute to acuminate to attenuate, the margins highly erose, densely ciliate; corolla 2–2.3 mm long, the lobes suborbicular divided, densely and prominently glandular-granulose; stamens 1 mm long, the filaments without an apical free portion (the stamens appearing epipetalous), the anthers very widely ovate, the base cordate, the apex acute, dehiscent by terminal pores, the pores confluent at anthesis; pistillode absent or highly reduced. Pistillate flowers as in staminate but calyx 0.7–0.9 mm long, the lobes oblate; corolla as in staminate but 1.4–1.8 mm long; staminodia resembling stamens but with antherodes smaller than anthers; pistil obconic, translucent, densely glandular-lepidote-pubescent; ovules 2–3. Fruits globose, 5–7 × 5–7 mm at maturity, densely and prominently punctate, with a few persistent translucent lepidote scales; pedicels obconic. Fl (Jan, Dec), fr (Mar, Apr, Jul); nonflooded moist forest and cloud forest.

Cybianthus guyanensis (A. DC.) Miq. subsp. **multipunctatus** (A. DC.) Pipoly [Syns.: *C. multipunctatus* A. DC., *Conomorpha multipunctata* (A. DC.) Miq.]

Shrubs or small trees, to 8 m tall. Stems terete, with densely lepidote pubescence, the scales stipitate, ferruginous. Leaves: petioles canaliculate, 0.5–1 cm long, the indumentum densely lepidote; blades narrowly obovate to narrowly elliptic, asymmetric, 8–12(15) × 2–3.5(6) cm, chartaceous or rarely membranaceous, the apex acuminate to abruptly acuminate, the acumen 0.5–1 cm long. Staminate inflorescences racemes or rarely subsessile pseudoverticils of 2–3 racemes; pedicels (0.3)0.5–1 mm long. Pistillate inflorescences simple racemes; pedicels (0.3)0.5–1 mm long. Staminate flowers: calyx 0.9–1 mm long, prominently brown or black punctate, the lobes ovate-triangular, apically attenuate to rounded; corolla 2–2.5 mm long, the lobes, ovate-triangular; stamens 1.1–1.4 mm long, the tube 0.6–0.8 mm long, the apically free filaments 0.6–0.9 mm long, the anthers ovate, apically acute, dehiscent by wide, sublaetrorse slits, conspicuously brown punctate; pistillode lageniform. Pistillate flowers as in staminate but calyx 0.8–0.9 mm long; corolla 1.8–1.9 mm long; staminodia 0.9–1 mm long, the filaments 0.7–0.8 mm long, the antherodes ovate, attenuate to an acute tip; pistil pyriform; ovules 2. Fruits 0.5–0.6 × 0.4–0.5 cm, the exocarp thick, black, inconspicuously pellucid punctate. Fl (Jan, Feb, Jun, Dec), fr (Jan, Mar, Apr, May, Jul, Aug); in nonflooded moist forest.

Cybianthus leprieurii G. Agostini [Syn.: *Weigeltia parviflora* Mez, non *Cybianthus parviflorus* Müll. Arg.]

Small trees, to 4(11) m tall. Stems terete, the indument densely and minutely rufous lepidote apically, glabrescent. Leaves: petioles canaliculate, (0.8)1–1.3 mm long, not pulvinate, glabrous; blades elliptic, (6)8–12(13.5) × (2.7)3–5 cm, chartaceous, glabrous and nitid adaxially, bearing numerous rufous hydropotes abaxially, the base acute, the apex acute or rarely subacuminate, the margins entire, flat. Staminate inflorescences bipinnate panicles; pedicels ca. 1 mm long, densely glandular papillate. Pistillate inflorescences simple racemes; pedicels as in staminate but to 2 mm long. Flowers unknown from our area (description from Mez, 1902). Staminate flowers: calyx 1.8–2 mm long, the lobes ovate, sparsely orange or pellucid punctate, the apex acute, the margins irregular, entire, glabrous, not bearing numerous, translucent, glandular scales; corolla ca. 2 mm long, the lobes ovate to suborbicular, the apex broadly rounded, often with subapical notch; stamens much shorter than petals, the filaments longer than anthers, the anthers subglobose, the connective epunctate; pistillode conical. Pistillate flowers unknown. Fruits depressed-globose, 5–7 mm diam., red, then purple-black at maturity. Not yet collected from our area but to be expected.

Cybianthus microbotrys A. DC. [Syn.: *Weigeltia microbotrys* (A. DC.) Mez]

Shrubs or small trees, to 3 m tall. Stems angular, the indument densely and minutely ferruginous furfuraceous-lepidote, glabrescent. Leaves: petioles slightly canaliculate, (0.7)1–1.5(2) cm long, glabrescent; blades narrowly oblanceolate to narrowly elliptic, (11.5) 12–23 × 2.5–5.5 cm, chartaceous, glabrous adaxially, bearing ferruginous hydropotes abaxially, the base attenuate, the apex long acuminate, the margins entire, flat. Staminate inflorescences subspicate racemes; pedicels 0.8–1 mm long, sparsely glandular-granulose. Pistillate inflorescences like staminate but pedicels 0.6–1 mm long, thickening in fruit. Flowers: calyx lobes black punctate, the margins somewhat erose apically. Staminate flowers: calyx 1–1.2 mm long, the lobes ovate, apically acute, prominently black punctate medially, the margins entire below, somewhat erose apically; corolla 2.2–2.4 mm long, the lobes suborbicular, apically obtuse, densely and prominently punctate and punctate-lineate abaxially, densely glandular-granulose adaxially basally and behind stamens, the margins entire, glabrous; stamens 1.6–1.8 mm long, the staminal tube sparsely glandular-granulose, the apically free filaments glabrous, the anthers oblate, apically and basally emarginate, the connective prominently black punctate dorsally; pistillode narrowly conic, hollow, the indument translucent glandular-lepidote. Pistillate flowers as in staminate but calyx 1–1.2 mm long; corolla 2.4–2.6 mm long; staminodia 1.7–1.8 mm long, the apical free filaments 0.7–0.8 mm long, the antherode vestigial, knob-like; pistil lageniform to obturbinate; ovules 2–3. Fruits globose, 6–8 × 6–8 mm, prominently pellucid punctate. Fl (Jan), fr (Jan, Mar, Apr, May, Jul, Sep); common, in all kinds of forest.

Cybianthus potiaei (Mez) G. Agostini [Syn.: *Weigeltia potiaei* Mez]

Trees, to 6 m tall. Stems angular, with narrow ridges, the indument very sparsely rufous glandular-lepidote (the scales from hydropotes), early glabrescent. Leaves pseudoverticillate; petioles deeply canaliculate, 8–15(20) cm long, only slightly thickened basally, the indument sparsely lepidote, glabrescent; blades obovate to elliptic, (11.5)16–25(30) × (4.7)5–10 cm, subcoriaceous to coriaceous, glabrous adaxially, bearing numerous rufous hydropotes (appearing like small lepidote scales) scattered over surface but more concentrated on midrib abaxially, the base decurrent on stem angles, the apex short acuminate to acute, the margins entire, flat. Staminate inflorescences pinnate panicles, columnar, the flowers in pseudoterminal glomerules at ends of side branches; pedicels 0.4–0.8 mm long, glandular-papillate. Pistillate inflorescences spikes; pedicels obsolete to <0.1 mm long. Flowers: calyx lobes red punctate, the margins entire. Staminate flowers 4-merous; calyx 0.6–0.8 mm long, the lobes suborbicular, apically obtuse to broadly rounded, densely and prominently red punctate medially, the margins irregular, glabrous, entire; corolla 1.7–1.9 mm long, the lobes broadly ovate to suborbicular, glabrous abaxially, sparsely glandular-granulose only above junction of corolla tube and lobes, prominently orange and pellucid punctate medially, the margins entire, irregular, glabrous; stamens 1.7–1.9 mm long, the filaments 1.6–1.7 mm long, the anthers oblate, apically and basally emarginate; pistillode subulate, hollow. Pistillate flowers as in staminate flowers but 2.6–2.8 mm long; calyx 2.1–2.3 mm long; corolla 1.8–2.1 mm long, the lobes oblong; staminodia 1.4–1.5 mm long, the filaments, 0.8–0.9 mm long, the antherodes obcordate, 0.3–0.4 × 0.3–0.4 mm, basally cordate, apically acute; pistil ellipsoid; ovules 3. Fruits slightly depressed-globose, 7–11 × 7–11 mm, densely and prominently orange punctate and punctate-lineate. Fl (Oct, Nov), fr (Jan); nonflooded moist forest.

Cybianthus prevostae Pipoly FIG. 231

Monoaxial trees, to 2.5 m tall. Stems terete, minutely rufous glandular-papillate. Leaves: petioles semiterete, not canaliculate, pulvinate, (0.9)1.7–2.2 cm long, densely glandular-papillate; blades oblanceolate, 27.5–64 × 6.5–9.5(11) cm, chartaceous, glabrous adaxially, bearing numerous hydropotes abaxially, the base gradually tapering, cuneate, the apex acute to rounded, the margins entire,

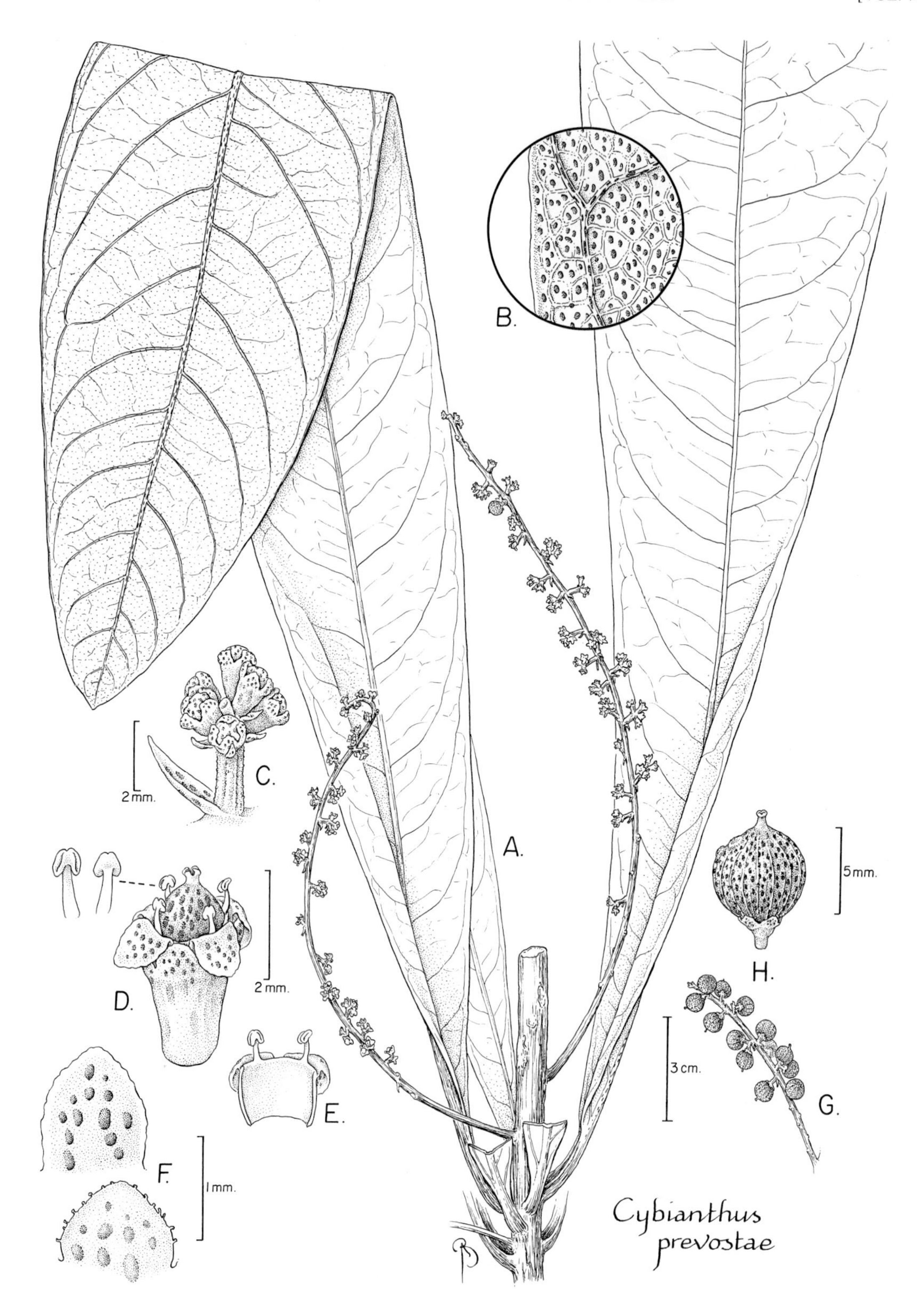

FIG. 231. MYRSINACEAE. *Cybianthus prevostae* (A, B, D–G, *Irwin et al. 48230* from French Guiana outside our area; C, *Granville 3395* from French Guiana outside our area; H, I, *Prévost & Grenand 972* from French Guiana outside our area). **A.** Part of stem with leaves and inflorescences. **B.** Detail of leaf surface showing punctations. **C.** Immature inflorescence. **D.** Flower (right) with abaxial (near upper left) and adaxial (far upper left) views of stamens. **E.** Adaxial view of part of gamopetalous corolla with adnate stamens. **F.** Sepal (below) and petal (above). **G.** Infructescence. **H.** Fruit. **I.** Fruit (above left) and see (below right). (Reprinted with permission from J. J. Pipoly, Brittonia 51(2), 1999.)

flat. Staminate inflorescences unknown. Pistillate inflorescences columnar panicles with glomerulate branches; pedicels stout, 1.4–1.7 × 2.6–2.8 mm, densely glandular-papillate. Flowers 2.2–2.5 mm long, grayish-green; calyx 2.1–2.3 mm long, the lobes suborbicular, apically obtuse, prominently black punctate medially, the margins entire, irregular, glandular-ciliolate, bearing numerous translucent glandular scales; corolla 2.2–2.5 mm long, the lobes suborbicular to very widely ovate, apically obtuse, the margins hyaline, entire, glabrous; staminodia 1.4–1.6 mm long, the filaments 0.4–0.5 mm long, the antherodes deeply cordate-sagittate, apically emarginate; pistil ellipsoid; ovules 3. Fruits globose, purple-black at maturity, (3.5)4.5–5.5 mm diam. when dried. Fl (Jan); in riparian forests.

Cybianthus prieurii A. DC. [Syns: *Peckia prieurii* (A. DC.) Kuntze, *Cybianthus nitidus* Miq., *Cybianthus subspicatus* Miq., *C. comatus* Mez, *C. viridiflorus* A. C. Sm.]

Trees, to 10 m tall. Stems angular, densely rufous dendritic-tomentose. Leaves: petioles trigonal but slightly canaliculate distally, 1–1.5 cm long; blades oblanceolate, 11.5–22 × (3.8)–7.6 cm, chartaceous, very sparsely rufous puberulent along midrib abaxially, glabrescent, inconspicuously pellucid punctate, the base acute, the apex acute to acuminate, the margins entire, flat. Staminate inflorescences erect, simple racemes, sparsely stellate puberulent, glabrescent; pedicels 2.7–3 mm long. Pistillate inflorescences as in staminate but pedicels 0.7–1 mm long. Staminate flowers erect, pale green; calyx 0.7–0.9 mm long, the lobes narrowly ovate, apically obtuse to acute, densely and prominently brown punctate, the margins opaque, glandular-ciliolate, erose apically; corolla 2–2.5 mm long, the lobes ovate, prominently red to brown punctate abaxially, glandular-granulose throughout adaxially, apically obtuse, the margins flat, glandular-granulose, subapically notched; stamens with apical free portion of filaments 0.8–1 mm long, the anthers subglobose, apically broadly rounded, the pores not confluent; pistillode obsolete. Pistillate flowers as in staminate but calyx 0.8–1 mm long; corolla 2.8–3 mm long; antherodes 0.3–0.4 mm long, the base subcordate, the apex obtuse; pistil obnapiform, the indument sparsely lepidote. Fruits globose, 5–7 mm diam., red at maturity. Fr (Jul, Aug); in forest at higher elevations.

Cybianthus resinosus Mez

Trees, to 15 m tall. Stems terete, densely ferruginous dendritic-tomentose at first, glabrescent. Leaves: petioles canaliculate, 10–20 (30) mm long, tapered, densely ferruginous dendroid-tomentose, glabrescent; blades elliptic, (11)15–21(26) × (4)5.5–7(10.8) cm, thinly coriaceous, nitid and glabrous ad- and abaxially, inconspicuously pellucid-punctate abaxially (not visible when dried), the base cuneate, decurrent onto petiole, the apex long-acuminate, the margins entire, irregular, flat; tertiary veins prominently raised. Staminate inflorescences lax, simple racemes, sparsely rufous stellate-puberulent; pedicels cylindrical, 2.1–1.7 mm long, sparsely puberulent, glabrescent. Pistillate inflorescences as in staminate. Staminate flowers nodding, grayish-brown; calyx 0.9–1 mm long, the lobes widely ovate, sparsely rufous stellate-puberulent, densely and prominently black punctate, apically acuminate, the margins hyaline, erose, short glandular-ciliate; corolla 1.6–1.8 mm long, the lobes very widely ovate, 1.2–1.5 mm long, apically obtuse, densely and prominently black punctate and glabrous abaxially, densely glandular-granulose and pusticulate adaxially, the margins irregular, glandular-granulose, entire; stamens 0.7–0.9 mm long, the anthers sessile, quadrate, 0.2–0.3 × 0.5–0.6 mm wide, apically truncate; pistillode cylindrical. Pistillate flowers as in staminate; calyx 1–1.2 mm long, the lobes 0.8–0.9 mm long; corolla 1.3–1.5 mm long; staminodia as in stamens but 0.4–0.6 mm long, the tube 0.3–0.5 mm long, the antherodes ca. 0.1 × 0.2–0.3 mm wide; pistil obturbinate; ovules 2. Fruits globose, 5–7 mm diam. at maturity, the exocarp black, juicy, edible at maturity; pedicels cylindrical. Fr (Mar); in forests on well-drained soils at higher elevations in our area.

Cybianthus surinamensis (Spreng.) G. Agostini [Syns.: *Salvadora surinamensis* Spreng. f., *Peckia surinamensis* (Spreng.) Kuntze, *Weigeltia surinamensis* (Spreng. f.) Mez, *W. myrianthos* A. DC., *Wallenia myrianthos* (A. DC.) Rchb., *Cybianthus myrianthos* (A. DC.) Miquel, *Weigeltia capitellata* Miq.]

Shrubs or trees, to 3(10) m tall. Stems terete, the indument rufous furfuraceous lepidote at first, glabrescent. Leaves: petioles thin, slightly canaliculate, (1)2–2.5(3) cm long, the indument sparsely rufous lepidote abaxially; blades narrowly oblanceolate to narrowly elliptic, (6.5)13–23 × (3.1)4–5(7.5) cm, chartaceous, densely and conspicuously black punctate and punctate-lineate abaxially, the base cuneate, not decurrent on stem, the apex long-acuminate. Staminate and pistillate inflorescences pinnately paniculate; staminate pedicels 1.5–2.5 mm long; pistillate pedicels 0.7–1.5 mm long. Staminate flowers: calyx 1.1–1.5 mm long, the lobes nearly free, widely ovate to suborbicular, sparsely brown punctate, sometimes sparsely glandular-papillate, the margins irregularly erose to erose-denticulate; corolla 2.4–2.6 mm long, the tube minute, 0.3 mm long, the lobes nearly free, widely ovate to suborbicular, 2.1–2.3 × 1.8–2 mm, sparsely brown punctate medially, the apex broadly rounded, the margins slightly irregular to very minutely erose; stamens 2 mm long, the tube swollen between portions of apically free filaments, the filaments 1–1.2 mm long, the anthers oblate, 0.3–0.4 × 0.5–0.6 mm, the base subcordate, the apex broadly rounded; pistillode conic, hollow. Pistillate flowers as in staminate but calyx 0.7–0.9 mm long; corolla unknown; pistil obnapiform; ovules 3. Fruits depressed-globose, 3–5 mm diam. Fl (Jan); riverine forests.

Cybianthus venezuelanus Mez [Syns.: *Peckia purpurea* Rusby, *Cybianthus brownii* Gleason]

Trees, to 5 m tall. Stems angular, densely rufous dendritic-tomentose. Leaves: petioles slightly canaliculate, (1.5)2–2.5 mm long, tapered, densely and minutely stellate rufous puberulent; blades oblanceolate to elliptic, (10)17–27(34) × (3)6–9(11) cm, chartaceous, rufous puberulent, smooth adaxially, glabrescent, sparsely rufous puberulent abaxially, the pubescence concentrated along midrib and secondary veins, inconspicuously pellucid punctate, the base acute, slightly decurrent onto petiole, the apex acute to acuminate, the margins entire, flat. Staminate inflorescences erect, simple racemes, densely rufous stellate-puberulent; pedicels cylindrical, 2.1–2.7 mm long, sparsely rufous stellate-puberulent, glabrescent. Pistillate inflorescences as in staminate but pedicels cylindrical, 0.9–1.2 mm long. Staminate flowers nodding, green; calyx 0.9–1.1 mm long, the lobes widely ovate, apically rounded, densely rufous stellate-puberulent, sparsely and prominently black punctate, the margins opaque,

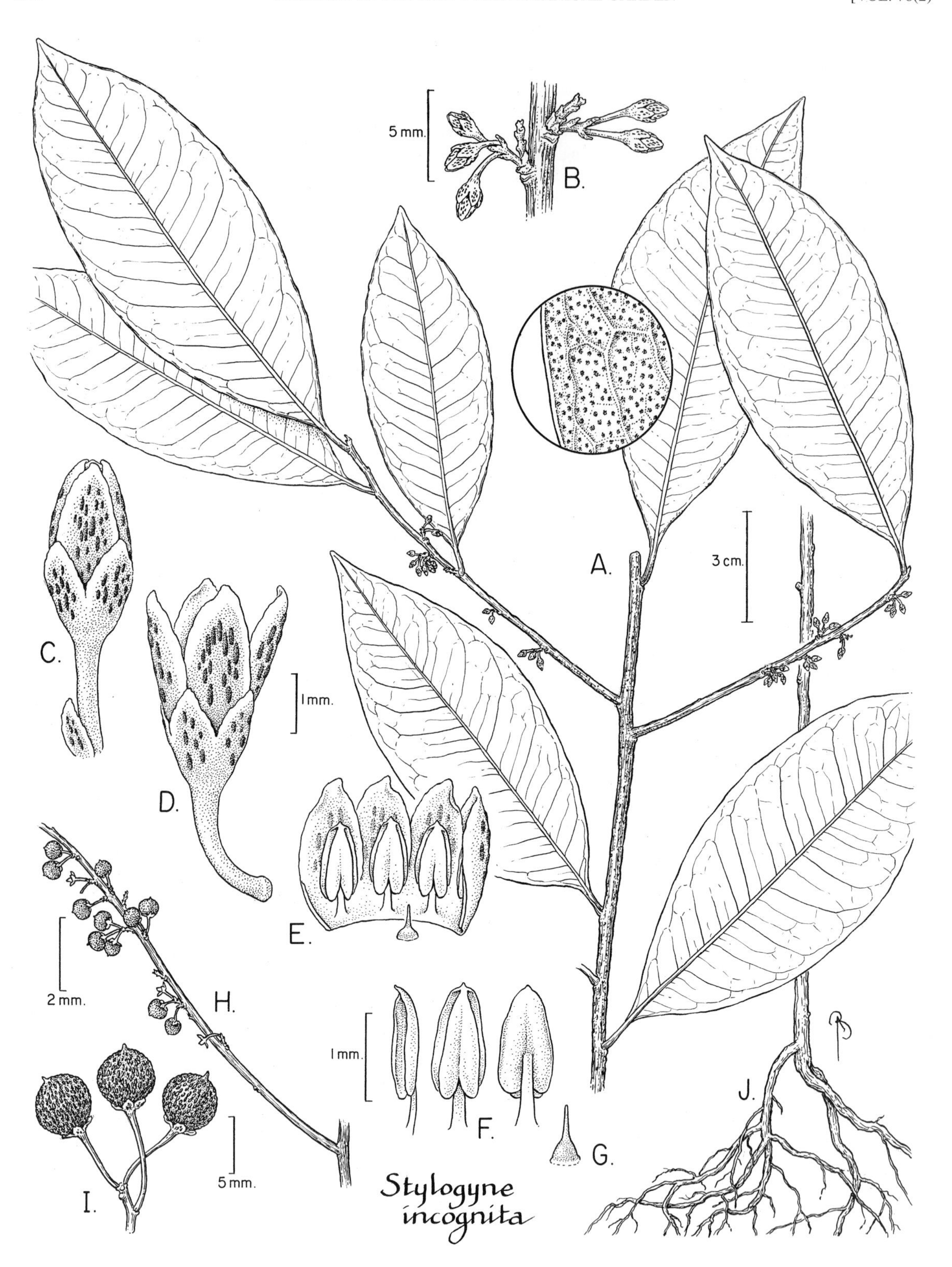

FIG. 232. MYRSINACEAE. *Stylogyne incognita* (A, B, D–G *Granville 8401*; C, *Oldeman B–4202*; H, J, *Granville 5434*; I, *Granville 4483*). **A.** Part of stem with leaves and inflorescences and detail of abaxial surface of leaf. **B.** Detail of inflorescences on stem. **C.** Lateral view of flower bud showing glandular punctations on abaxial surface of corolla, calyx, and bracteole. **D.** Lateral view of flower. **E.** Corolla opened to show arrangement of stamens and separate pistil. **F.** Lateral (far left), adaxial (near left), and abaxial (right) views of stamens. **G.** Lateral view of pistil. **H.** Part of stem with infructescences. **I.** Infructescence showing glandular punctations on fruits. **J.** Basal portion of stem and roots.

coarsely serrulate, densely glandular-ciliolate; corolla 1.6–1.8 mm long, the lobes widely triangular, apically acute, dorsally recurved, prominently and densely black punctate and glabrous abaxially, densely glandular granulose throughout adaxially, the margins slightly revolute, densely glandular-granulose; anthers apparently sessile, quadrate, 0.3–0.4 × 0.5–0.6 mm, the base truncate, the apex truncate, the connective prominently punctate dorsally, rufous glandular-papillate apically, the pores confluent or nearly so; pistillode obsolete or conical, hollow, glabrous. Pistillate flowers as in staminate but calyx 0.8–1.2 mm long; corolla 1.2–1.4 mm long; antherodes 0.2–0.3 mm long; pistil obturbinate; ovules 2–3. Fruits globose, 5–7(9) mm diam., black at maturity, the pericarp thick, juicy. Fl (Jan, Dec), fr (Mar, May, Sep); in nonflooded forest at all elevations.

STYLOGYNE A. DC.

Shrubs or small trees. Stems glabrous. Leaves alternate; petioles present. Inflorescences lateral or terminal, rarely lateral and terminal, corymbose or paniculate, the branches racemose or short umbellate-corymbose; peduncle absent to long; pedicels present. Flowers unisexual or bisexual, (4)5-merous; calyx lobes dextrorsely contorted, open in bud, free or short-connate basally, punctate and punctate-lineate; corolla lobes dextrorsely contorted, highly twisted in bud, short-connate, commonly lineate-punctate; stamens usually shorter than petals, the filaments slender, free from each other, adnate to petals, attached at petal base or above, the anthers oblong, subsagittate basally, dorsifixed or basifixed, twisted at anthesis, usually dehiscent by latrorse longitudinal slits, rarely first by apical pores then widening to slits; staminodia similar to stamens but reduced in size, the antherodes devoid of pollen; pistil obturbinate, the ovary ovoid, the style long, the stigma punctiform, subequaling or exceeding stamens; placentae ovoid or globose, basal, the ovules 3–5, uni- or rarely biseriate; pistillode lageniform, the ovary hollow, the style subcapitate, <1 mm long. Fruits drupaceous, the style base persistent. Seeds 1 per fruit.

Stylogyne incognita Pipoly FIG. 232

Subshrubs, to 1 m tall. Stems terete, glabrous. Leaves: petioles deeply canaliculate, 6–10 mm long, glabrous; blades elliptic (8)10.5–15(20) × (3.5)5–7.5 cm, chartaceous, smooth and somewhat nitid adaxially, prominently black punctate abaxially, the base acute, the apex acuminate, the acumen 6–16 mm long, the margins entire, flat. Staminate inflorescences lateral (axillary), corymbose, 8–10 mm long; pedicels 3–3.6 mm long, densely and conspicuously black punctate-lineate. Bisexual inflorescences as in staminate. Pistillate inflorescences unknown. Staminate flowers: calyx 1.4–1.7 mm long, the lobes ovate, medially thickened, densely and prominently red and brown punctate and punctate-lineate, the apex obtuse to broadly rounded, the margins irregular, hyaline, somewhat erose apically, essentially glabrous, but extremely minute glandular-ciliolate apically; corolla 2.6–2.9 mm long, the lobes elliptic to oblong, apically obtuse to rounded, asymmetric, prominently orange punctate-lineate dorsally, glabrous, the margins entire, glabrous; stamens 1.6–1.8 mm long, the filaments attached basally on corolla tube, the anthers oblong to lanceolate, basifixed just below middle, the base cordate, the apex obtuse but with prominent apiculus, the connective conspicuously brown punctate along edges; pistillode obnapiform. Bisexual flowers as in staminate but pistil 0.9–1.1 × 0.6–0.7 mm; placenta subglobose, the ovules 2, exposed. Fruits globose, (3.5)4–6(8) mm diam. when dried. Fl (Dec), fr (Jan, Mar, Jun, Jul, Aug); nonflooded moist forest.

MYRTACEAE (Myrtle Family)

Bruce K. Holst and Maria Lúcia Kawasaki

Trees and shrubs. Stipules absent. Leaves simple, opposite, glandular-punctate; secondary leaf veins usually anastamosing and forming a distinct, straight or looping, marginal vein (except *Psidium* in our area). Inflorescences paniculate, racemose, dichasial, spicate, or flowers solitary; bracteoles 2, attached at ovary base. Flowers actinomorphic, bisexual; hypanthium prolonged beyond ovary or not; calyx, if open in bud, with 4 or 5 lobes, these either distinct, partially connate and tearing irregularly or longitudinally at anthesis, or, if closed in bud, then calyptrate and circumscissile (*Calyptranthes*), or tearing irregularly (*Psidium*); petals 4 or 5, white, rarely absent; stamens many; ovary inferior, 2–5-locular; placentation axile, the ovules 2 to many per locule. Fruits berries, frequently crowned by persistent calyx lobes. Seeds 1 to many, globose, ellipsoid, or C-shaped; embryo of 3 distinct types: with fleshy, fused or free cotyledons and inconspicuous radicle (eugenioid); with leafy and folded cotyledons and well developed radicle equaling cotyledons in length (myrcioid); or with very small foliaceous cotyledons and much longer radicle (myrtoid).

The shape of the mature embryo is an important feature that facilitates identification to subtribe and genus. It can be observed by removing the seed coat. As far as is known, all fleshy-fruited Myrtaceae are nonpoisonous, though while some are delicious or unusual-tasting, not all are palatable. Most species in the family have bee-pollinated flowers, and seeds that are dispersed by animals, especially birds and monkeys.

Amshoff, J. H. 1951. Myrtaceae. *In* A. Pulle (ed.), Flora of Suriname **III(2):** 56–158. Royal Institute for the Indies, Amsterdam.

Landrum, L. R. & M. L. Kawasaki. 1997. The genera of Myrtaceae in Brazil: an illustrated synoptic treatment and identification keys. Brittonia **49(4):** 508–536.

Lindeman, J. C. 1986. Myrtaceae. *In* A. L. Stoffers & J. C. Lindeman (eds.), Flora of Suriname **III(1–2):** 518–546. E. J. Brill, Leiden.

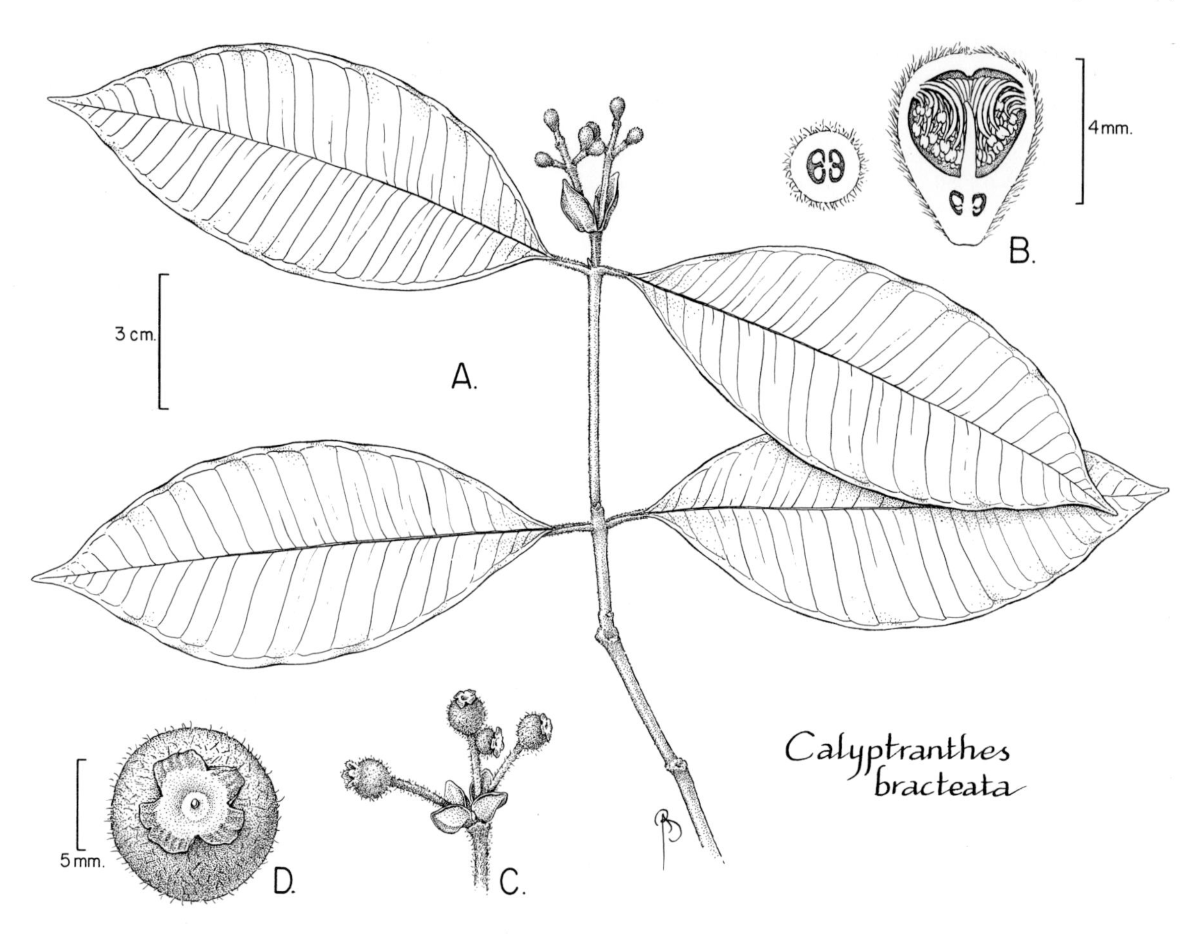

FIG. 233. MYRTACEAE. *Calyptranthes bracteata* (A, B, *de Granville 8936*; C, D, *de Granville 5482*). **A.** Apex of stem showing leaves and terminal inflorescences. **B.** Transverse section (left) and medial section (right) of flowers before anthesis showing axile ovule placentation in two locules and hypanthium prolonged above apex of pistil. **C.** Terminal infructescence showing persistent bracts. **D.** Apical view of fruit showing persistent perianth.

1. Inflorescences branched, paniculate. Embryo myrcioid.
   2. Calyx with 4 or 5 lobes (if calyx initially closed in bud, then leaves with corky or flaky petioles). . . . . *Myrcia.*
   2. Calyx calyptrate and circumscissile (petioles never corky or flaky). . . . . *Calyptranthes.*
1. Inflorescences not branched, racemose (the raceme frequently reduced and inflorescence appearing fasciculate or glomerate), dichasial, or flowers solitary. Embryo eugenioid or myrtoid.
   3. Young stems distinctly 4-angled. Calyx closed in bud, opening irregularly. Seeds many per fruit, the seed coat thick, bony; embryo myrtoid. . . . . *Psidium.*
   3. Young stems terete or flattened. Calyx open in bud, with 4 distinct lobes. Seeds 1 or 2 per fruit, the seed coat thin, membranaceous; embryo eugenioid.
      4. Hypanthium tube prolonged and narrowed above summit of ovary, circumscissile just above ovary after anthesis and deciduous along with calyx and androecium. . . . . *Myrciaria.*
      4. Hypanthium not prolonged above ovary, not circumscissile, calyx lobes usually persisting in fruit. . . . . *Eugenia.*

CALYPTRANTHES Sw.

Trees and shrubs. Branching usually bifurcate. Inflorescences in leaf axils, often paired panicles; bracteoles deciduous. Flowers: hypanthium prolonged beyond apex of ovary; calyx closed in bud, calyptrate, circumscissile at anthesis; ovary 2-locular, the ovules 2 per locule. Fruits globose, crowned by circular hypanthium scar. Seeds 1–2, the seed coat membranaceous; embryo myrcioid.

1. Inflorescence bracts early deciduous. Flower buds 2–3 mm long. Fruits glabrous. . . . . *C. amshoffae.*
1. Inflorescence bracts persistent. Flower buds ca. 5 mm long. Fruits brown-tomentose. . . . . *C. bracteata.*

Calyptranthes amshoffae McVaugh

Trees. Pubescence of silky-ferruginous hairs. Leaves: petioles 5–10 mm long; blades elliptic to narrowly elliptic, 12–20 × 5–8 cm, coriaceous, glabrous adaxially, pubescent abaxially, the base obtuse, the apex acuminate; midrib sulcate adaxially, the marginal veins 2, the innermost 1–4 mm from margin. Inflorescences with bracts early deciduous. Flower buds 2–3 mm long, pubescent. Fruits ca. 1.3 cm diam., glabrous. Fl (Nov), fr (Mar, Jul); in nonflooded forest on Mont Galbao.

Calyptranthes bracteata M. L. Kawas. & B. Holst FIG. 233

Shrubs or small trees. Pubescence of brown hairs. Leaves: petioles 5–10 mm long; blades elliptic to narrowly elliptic, 12–20 × 3.5–8 cm, coriaceous, glabrous adaxially, pubescent abaxially, the base obtuse, the apex acuminate; midrib sulcate adaxially, the marginal vein 1, 1–2 mm from margin. Inflorescences with bracts persistent. Flower buds ca. 5 mm long, pubescent. Fruits ca. 1.3 cm diam., brown-tomentose. Fl (Jan, Apr), fr (Apr, May); in non-flooded forest.

EUGENIA L.

Trees and shrubs. Inflorescences racemose, frequently reduced and appearing fasciculate or glomerate, or flowers solitary; bracteoles usually ovate to triangular and persistent. Flowers 4-merous; hypanthium not or only rarely prolonged beyond apex of ovary; calyx open in bud, with 4 distinct lobes or the lobes rarely partially connate; ovary usually 2-locular, the ovules (2)many per locule. Fruits usually crowned by calyx lobes. Seeds 1 or 2(5), the seed coat membranaceous; embryo eugenioid, the cotyledons fused.

1. Midrib of leaf blades convex, biconvex, plane, or forming a narrow, elevated line adaxially.
  2. Plants largely glabrous, or, at most, the floral parts ciliate or new growth with some hairs.
    3. Petioles of older leaves corky-flaky; larger leaf blades mostly >15 cm long. Fruits oblong, ellipsoid, or obovoid, the pedicels stout, woody.
      4. Leaf blades drying dark brown adaxially, olive-green or greenish brown abaxially. Calyx lobes in fruit spreading to slightly incurved, persistent. . . . . *E. latifolia.*
      4. Leaf blades drying dull or olive green ad- and abaxially. Calyx lobes in fruit tightly appressed to ovary summit, eventually deciduous. . . . . *Eugenia* sp. A.
    3. Petioles of older leaves not corky-flaky; larger leaf blades mostly <14 cm long. Fruits globose or turbinate, the pedicels not woody.
      5. Innermost marginal vein 1–2 mm from margin and ± less parallel to it. Pedicels 2–4 mm long. . . . . *Eugenia* sp. B.
      5. Innermost marginal vein 3–10 mm from margin and arching between secondaries. Pedicels 6–30 mm long.
        6. Stipule-like bracts not evident on new vegetative growth. Leaf blades mostly elliptic. Inflorescence bracts not imbricate, broadly ovate, ca. 1 mm long, discernible with difficulty; bracteoles ovate to triangular, persistent past anthesis. Fruits yellow to orange at maturity. . . . . *E. gongylocarpa.*
        6. Stipule-like bracts evident on new vegetative growth. Leaf blades mostly obovate. Inflorescence bracts imbricate, triangular-lanceolate, 2–4 mm long, puberulous, evident; bracteoles linear to subulate, deciduous before anthesis. Fruits red at maturity. . . . . *E. patrisii.*
  2. Plants noticeably pubescent, at least in part, the indument evident on flowers (particularly the ovary) and/or leaf blades abaxially.
    7. Bracteoles linear or subulate, deciduous before anthesis.
      8. Stipule-like bracts evident on new vegetative growth. Leaves mostly obovate. Inflorescence bracts imbricate, triangular-lanceolate, 2–4 mm long, puberulous. Ovary glabrous or sparsely pilose, the hairs reddish. Fruits red at maturity. . . . . *E. patrisii.*
      8. Stipule-like bracts not evident on new vegetative growth. Leaves mostly oblong or ovate. Inflorescence bracts not as above. Ovary densely pubescent, the hairs yellowish. Fruits golden yellow at maturity. . . . . *E. armeniaca.*
    7. Bracteoles ovate to triangular, persistent past anthesis and usually into fruit.
      9. Petioles of older leaves corky-flaky; blades lustrous, densely and minutely gray- to silvery-sericeous abaxially. . . . . *E. argyrophylla.*
      9. Petioles smooth; blades dull, glabrous or sparsely to densely ferruginous fading to tan-tomentose abaxially.
        10. Leaf blades 10–20 cm long; secondary veins in 7–10 pairs, widely spaced, the marginal vein 5–9 mm from margin and strongly arching between secondaries. Calyx lobes partially fused in bud, pale coppery-sericeous. Fruits ellipsoid. . . . . *E. feijoi.*
        10. Leaf blades 6–8.5 cm long; secondary veins in >15 pairs, not widely spaced, the marginal vein ca. 1 mm from margin and parallel to it. Calyx lobes free, ferruginous-tomentose to glabrescent. Fruits globose. . . . . *E. ferreiraeana.*
1. Midrib of leaf blades sulcate adaxially.

11. Inflorescence axis evident, >5 mm long.
    12. Racemes abbreviate, 0.5–1 cm long. Bracteoles linear-subulate, deciduous at or before anthesis. Calyx lobes lanceolate to narrowly triangular, 5–10 mm long, membranaceous, subfoliaceous. Fruits smooth when fresh, densely dark gland-dotted when dry, 3–5-seeded. . . . . *E. macrocalyx.*
    12. Racemes elongate, 7–15 cm long. Bracteoles ovate or triangular, persistent. Calyx lobes ovate, 2–3 mm long, rigid. Fruits strongly muricate-tuberculate, 1-seeded. . . . . *E. muricata.*
11. Inflorescence axis not evident or at most to 5 mm long, the inflorescence appearing fasciculate, glomerate, or flowers solitary.
    13. Flowers sessile.
        14. Leaf blades mostly obovate; marginal vein 2–3 mm from margin, the secondary and marginal veins flat or scarcely raised, barely visible abaxially. Calyx lobes pubescent. Fruits 3–4 cm diam. . . . . . . *E. morii.*
        14. Leaf blades elliptic; inner marginal vein 3–6 mm from margin, the secondary and marginal veins raised and forming distinct gridiron pattern abaxially. Calyx lobes glabrous. Fruits ca. 1 cm diam. or less. . . . . *E. coffeifolia.*
    13. Flowers distinctly pedicellate.
        15. Flowers glabrous.
            16. Principal marginal vein parallel to and 1–2 mm from margin. Fruits subglobose, the pedicels 7–30 mm long. . . . . *E. pseudopsidium.*
            16. Principal marginal veins strongly arched, 5–10 mm from margin. Fruits ellipsoid, the pedicels 3–12 mm long. . . . . *E. mimus.*
        15. Flowers or at least the ovary pubescent.
            17. Leaf blades glabrous to sparsely appressed-pubescent or sericeous abaxially, the abaxial surface never drying whitish.
                18. Hairs whitish or pale-coppery. Calyx lobes lanceolate or narrowly triangular, greenish; ovary appressed-pubescent with whitish or coppery hairs. Fruits glabrous, orange at maturity, smooth when fresh, densely dark gland-dotted when dry. Seeds 3–5 per fruit. . . . . . *E. macrocalyx.*
                18. Hairs dark-brown or dark coppery brown. Calyx lobes ovate to widely ovate, brownish; ovary densely brown- or coppery-brown tomentulose.
                    19. Petioles smooth when dry, broadly channeled. Bracteoles free, linear, deciduous prior to anthesis. . . . . *E. dentata.*
                    19. Petioles wrinkled when dry, the channel mostly obscured. Bracteoles connate, broadly ovate, persistent past anthesis. . . . . *Eugenia* sp. C
            17. Leaf blades sparsely to densely soft-pilose, or abaxial surface drying whitish.
                20. Leaf blades with secondary veins conspicuous in mature leaves, impressed adaxially, strongly elevated abaxially, the abaxial surface drying greenish, sparsely to densely and softly pilose. Bracteoles >7 mm long, linear, deciduous. . . . . *E. tetramera.*
                20. Leaf blades with secondary veins scarcely evident, flat or slightly elevated adaxially, plane or scarcely elevated abaxially, the abaxial surface drying whitish, often sparsely sericeous. Bracteoles ca. 1 mm long, ovate, persistent.
                    21. Leaf blades mostly elliptic to elliptic-oblong, the apex bluntly acuminate. Flowers borne in leafy axils, pendent below leaves; buds ca. 3 mm long. . . . . *E. albicans.*
                    21. Leaf blades mostly obovate to oblong, the apex sharply acuminate. Flowers borne at leafless nodes; buds 15–18 mm long. . . . . *E. spruceana.*

Eugenia albicans (O. Berg) Urb.

Small trees. Pubescence of whitish- or coppery-sericeous hairs. Leaves: petioles 2–5 mm long; blades elliptic to elliptic-oblong, 6–8 × 2.5–3.5 cm, chartaceous to coriaceous, discolorous, glabrous adaxially, sericeous abaxially, the base acute to obtuse, the apex bluntly acuminate; midrib sulcate adaxially, the marginal vein 1, ca. 1 mm from margin. Inflorescences much abbreviated racemes; pedicels 5–15 mm long, sericeous. Flowers: buds ca. 3 mm long, sericeous. Fl (Jan, Feb, Dec); in nonflooded forest.

A large-leaved variant with leaf blades to 14 × 7.5 cm (e.g., *Molina 1250*, CAY, SEL) occurs elsewhere in French Guiana.

Eugenia argyrophylla B. Holst & M. L. Kawas. FIG. 234

Shrubs. Pubescence of silvery hairs. Leaves: petioles 3–5 mm long, the older ones corky-flaky; blades elliptic to narrowly elliptic, 11–16 × 4.5–7.5 cm, chartaceous to coriaceous, lustrous, glabrous adaxially, densely and minutely gray- to silvery-sericeous abaxially, the base acute, the apex acuminate; midrib distinctly convex adaxially; marginal veins 2, the innermost 3–7 mm from margin. Inflorescences much abbreviated racemes; pedicels 3–6 mm long, sericeous. Flowers: buds ca. 3 mm long, sericeous. Fl (Jan); in nonflooded, moist forest.

Eugenia armeniaca Sagot PL. 100b

Small to medium-sized trees. Pubescence of yellowish hairs. Leaves: petioles 3–4 mm long; blades mostly oblong or ovate, 7–14 × 3–4.5 cm, chartaceous, glabrescent or persistently pilose abaxially, the base rounded, the apex acuminate; midrib shallowly biconvex adaxially, the marginal vein strongly arching between secondary veins, an additional outer, irregularly arching and less

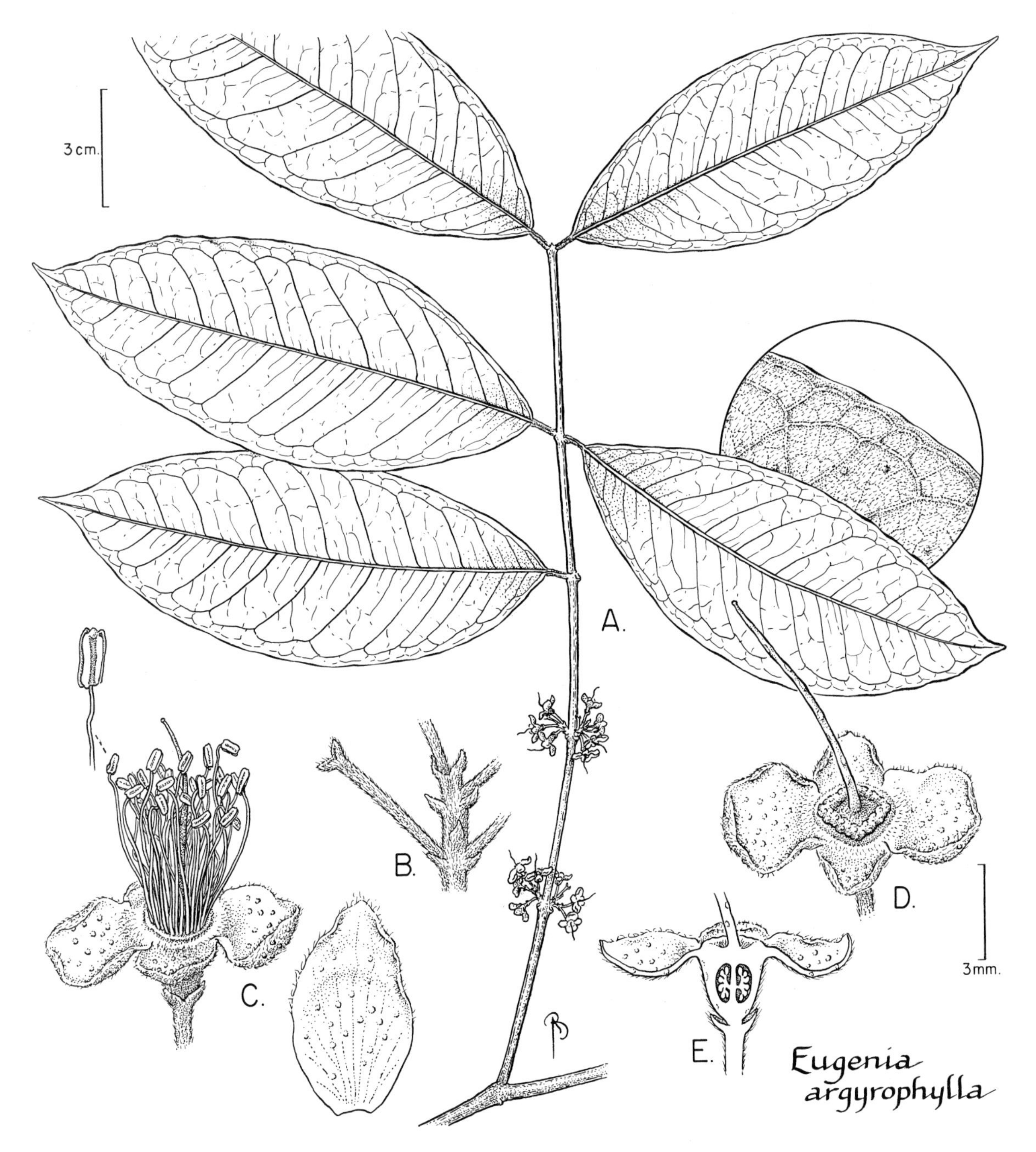

FIG. 234. MYRTACEAE. *Eugenia argyrophylla* (A, B, D, E, *Granville 10576*; C, *Granville 10504*). **A.** Part of stem with leaves and inflorescences. **B.** Detail of part of inflorescence showing bracts. **C.** Lateral view of flower with detail of stamen (left) and adaxial view of petal (right). **D.** Oblique- apical view of flower after stamens have fallen. **E.** Medial section of flower.

prominent marginal vein also present. Inflorescences abbreviated racemes, the axis 2–4(9) mm long; pedicels to 1 cm long, pubescent. Flowers: buds 4–5 mm long, densely pubescent. Fruits globose, large, ca. 3 cm diam., densely golden-yellow pubescent. Seeds 1 or 2. In nonflooded forest.

The only collections of *Eugenia armeniaca* (*Mori et al. 22741*, NY; *23330*, SEL) from our area are sterile. The strongly arching inner marginal vein, mostly oblong leaf blades, and yellowish indument are characteristic of this taxon.

Eugenia coffeifolia DC.

Shrubs or small trees. Pubescence of red hairs. Leaves: petioles 3–10 mm long; blades elliptic to narrowly elliptic, 8–15(17) × 3–5.5(6.5) cm, chartaceous to coriaceous, glabrous, the base acute, the apex acuminate; midrib sulcate adaxially, the marginal veins 2, the innermost 3–6 mm from margin, the secondary and marginal veins froming distinct gridiron pattern abaxially. Inflorescences much abbreviated; pedicels absent. Flowers: buds 1–2 mm long; calyx lobes glabrous; ovary pubescent, obscured by enveloping bracteoles. Fruits globose, ca. 1 cm diam. or less, sparingly pubescent. Fl (Aug, Oct, Nov), fr (Jan, Sep, Nov, Dec); common, in nonflooded forest.

Eugenia dentata (O. Berg) Nied.

Large trees. Bark reddish, exfoliating in plates. Leaves deciduous prior to flowering: petioles 5–8 mm long, smooth, broadly channeled; blades elliptic or less often slightly ovate or obovate, 6–9(12) × 2.5–3.5(5) cm, chartaceous, young blades thinly covered with slender, silky, pale coppery hairs, glabrescent, the base obtuse to acute, the apex abruptly short-acuminate, recurved; midrib sulcate adaxially, the marginal vein 2–4 mm from margin and slightly arching between the secondaries. Inflorescences short racemes, somewhat appressed-brown puberulous, the axis 4–5 mm long; pedicels 5–8 mm long, brown-puberulous, slender. Flowers: buds 7–9 mm long; calyx lobes membranous, thinly coppery brown sericeous; ovary irregularly ridged longitudinally, densely coppery brown tomentulose; bracteoles free, linear, deciduous before anthesis. Fruits ellipsoid, 5 × 3 cm, densely dark-brown coppery tomentulose (described as "yellow-orange" when fresh), lustrous. Seeds 1. Fl (Aug), fr (Jan); in nonflooded forest. *Goyavier sauvage* (Créole).

Eugenia feijoi O. Berg

Trees. Pubescence of ferruginous or pale coppery to tan hairs. Bark pinkish, peeling in irregular longitudinal plates. Leaves: petioles ca. 1 cm long, smooth to wrinkled when dry; blades elliptic, 10–20 × 4.5–6.5 cm, chartaceous, glabrous to sparsely appressed-pilose, the base acute to obtuse, the apex acuminate; midrib forming a narrow elevated line adaxially, the secondary veins widely spaced, the marginal veins 2, the outer one formed by lowermost secondary vein, the more prominent inner one formed by subsequent secondaries and arching between them at 5–9 mm from margin. Inflorescences much abbreviated racemes, 1 to several per node, few-flowered; pedicels ca. 1 cm long, sericeous. Flowers: buds 6–9 mm long, sericeous; calyx lobes connate nearly to apex, tearing regularly to form 4 lobes or irregularly to form 3; anthers gray. Fruits ellipsoid, abruptly tapered at both ends, 3–4 cm long, glabrescent. Fl (Sep); in nonflooded forest.

Eugenia ferreiraeana O. Berg

Small trees. Pubescence ferruginous, becoming tan or grayish with age. Leaves: petioles ca. 5 mm long; blades narrowly elliptic to lanceolate, 6–8.5 × 1.5–3 cm, coriaceous, glabrescent, the base acute, the apex acuminate; midrib convex adaxially, the marginal vein 1, ca. 1 mm from margin. Inflorescences much abbreviated racemes; pedicels 5–10 m long, tomentose. Flowers: buds ca. 3 mm long, tomentose. Fruits globose, ca. 8 mm diam., gray-tomentulose. Fl (Apr), fr (Sep); in nonflooded forest.

Eugenia gongylocarpa M. L. Kawas. & B. Holst
FIG. 235, PL. 100c

Small trees. Plants glabrous. Leaves: petioles 5–10 mm long; blades elliptic or some slightly obovate, 8–13 × 4–7 cm, chartaceous to coriaceous, the base acute to obtuse, the apex acute to acuminate, slightly recurved; midrib convex or biconvex adaxially, the marginal veins 2, strongly irregular and arched between secondaries, the innermost 5–10 mm from margin. Inflorescences much abbreviated racemes; pedicels 6–20 mm long. Flowers: buds ca. 6 mm long. Fruits globose, with slightly swollen base, ca. 2.5 cm diam., yellow to orange at maturity. Fl (Jun, Aug), fr (Feb, Mar, Apr, May, Jul, Aug, Oct, Dec); locally common, in nonflooded forest.

Eugenia latifolia Aubl.

Trees. Plants glabrous or nearly so. Leaves: petioles 5–10 mm long, stout, the older ones corky-flaky; blades oblong to elliptic, 15–22 × 6–9 cm, coriaceous, glabrous, drying dark brown adaxially, olive-green or greenish brown abaxially, the base obtuse to cuneate, the apex acute to shortly acuminate, slightly conduplicate; midrib broad, plane to shallowly impressed, the marginal veins 2, the innermost 3–9 mm from margin. Inflorescences abbreviated racemes, 1 to few per node, the main axes 1–5 mm long; pedicels initially thin, stout and woody in fruit, corky-flaky, 3–10 mm long, glabrous or nearly so. Flowers: buds 8–10 mm long. Fruits oblong to ellipsoid, 1.5–2.5 cm long. Fr (Sep); in nonflooded forest.

Eugenia macrocalyx (Rusby) McVaugh

Trees. Pubescence of whitish or pale coppery hairs. Leaves: petioles 5–15 mm long; blades elliptic to oblong, 7–14 × 3–6 cm, coriaceous to chartaceous, glabrous, the base acute to obtuse, the apex acute to acuminate, sometimes abruptly so; midrib sulcate adaxially, the higher order venation inconspicuous, the marginal veins 2, the innermost 2–3 mm from margin. Inflorescences usually short racemes 0.5–1 cm long or rarely longer; pedicels (6)20–30 mm long, glabrescent; bracteoles linear-subulate, deciduous at or before anthesis. Flowers: buds 5–6 mm long; calyx lobes subfoliaceous, lanceolate to narrowly triangular; ovary appressed-pubescent. Fruits oblong to pyriform, ca. 3 cm long, glabrous, orange, densely dark gland-dotted when dry. Seeds 3–5 per fruit. Fl (Feb), fr (Apr, Jun); in nonflooded forest.

Eugenia mimus McVaugh PL. 100d

Small trees. Plants glabrous. Leaves: petioles 5–10 mm long; blades elliptic to narrowly elliptic, 10–20 × 5–8 cm, chartaceous to coriaceous, the base acute to obtuse, the apex acute to acuminate; midrib sulcate adaxially, the marginal veins 2, the innermost 5–10 mm from margin, strongly arched between secondaries. Inflorescences much abbreviated racemes; pedicels 3–12 mm long. Flowers: buds ca. 5 mm long, glabrous. Fruits ellipsoid, ca. 2 cm long. Fr (Jan, May, Sep, Oct); in nonflooded forest.

Eugenia morii B. Holst & M. L. Kawas. FIG. 236

Small trees. Pubescence of reddish-brown hairs. Leaves: petioles 8–12 mm long; blades narrowly obovate to oblong, 9–15 × 3–5 cm, chartaceous to coriaceous, glabrous, the base acute to cuneate, the apex abruptly acuminate, reflexed; midrib narrowly sulcate adaxially, the marginal veins 1, 2–3 mm from margin, barely visible. Inflorescences much abbreviated, appearing glomerate; pedicels absent. Flowers: buds 2–3 mm long, pubescent. Fruits globose, 3–4 cm diam., dark purple, glabrous. Fl (Sep), fr (May); in nonflooded forest.

Eugenia muricata DC.

Trees. Pubescence of reddish hairs. Leaves: petioles 7–10 mm long; blades elliptic to narrowly elliptic, 10–20 × 3.5–8.5 cm, coriaceous, glabrous, the base acute to obtuse, the apex acuminate; midrib sulcate adaxially, the marginal vein 1, ca. 1 mm from margin. Inflorescences elongate racemes, 7–15 cm long, the axes pubescent; pedicels to ca. 1 cm long. Flowers: buds ca. 3 mm long, pubescent. Fruits globose, densely tuberculate, ca. 2 cm diam., glabrescent. Fl (Jun, Aug, Sep, Oct), fr (Jan, Mar, Aug, Sep); in nonflooded forest.

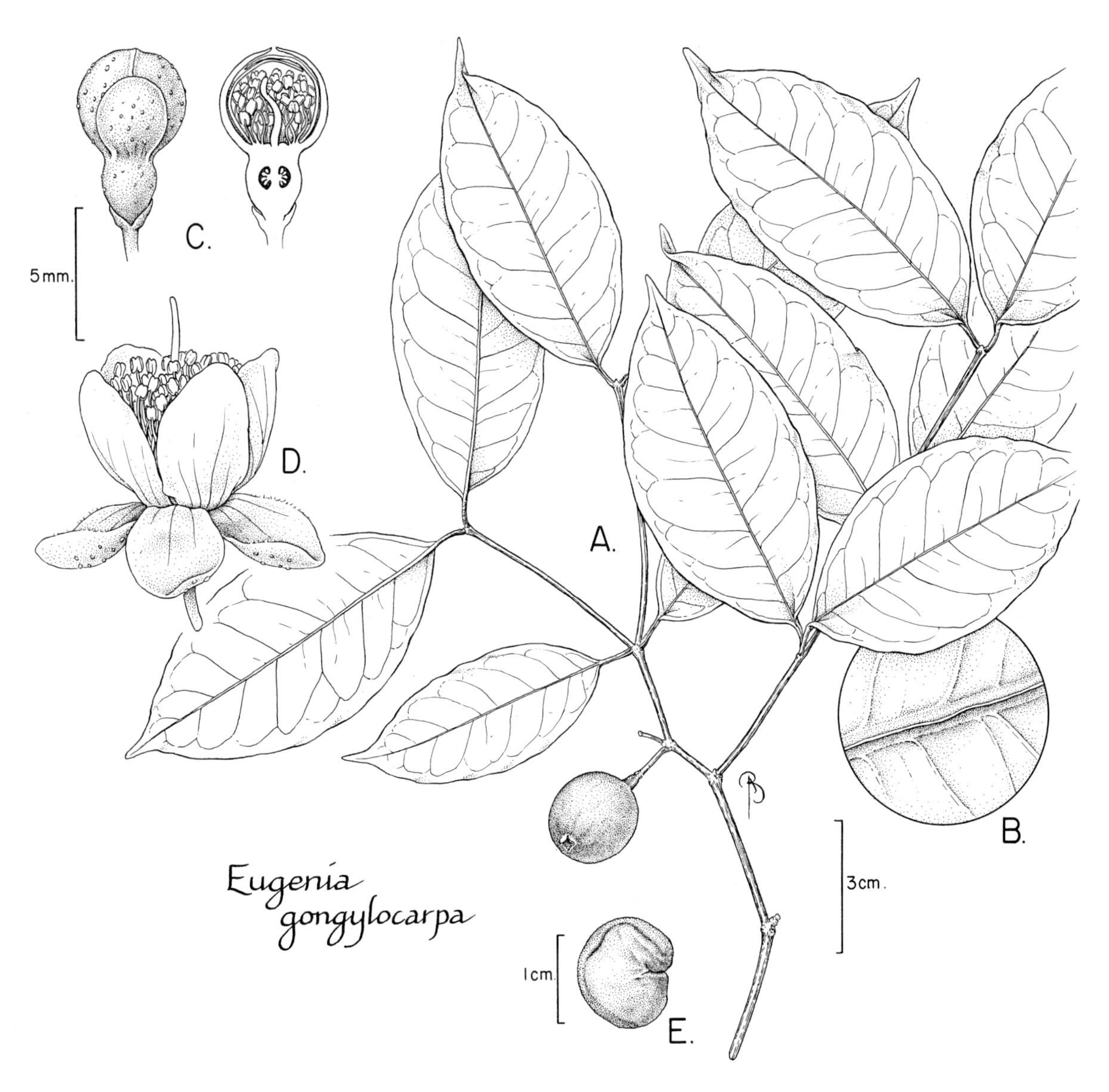

FIG. 235. MYRTACEAE. *Eugenia gongylocarpa* (A, B, E, *Mori & Pipoly 15554*; C, D, *Sabatier 3644*). **A.** Leafy stem and fruit showing apical perianth scar. **B.** Detail of canaliculate midrib on adaxial leaf surface. **C.** Lateral view (left) and medial section (right) of flower bud before anthesis showing axile placentation and persistent bracteoles. **D.** Flower at anthesis. **E.** Seed.

Eugenia patrisii Vahl PL. 100e

Trees, semideciduous. Pubescence of reddish hairs. Stipule-like bracts on new growth. Leaves: petioles ca. 5 mm long; blades obovate to elliptic or narrowly elliptic, 5–13 × 2.5–5 cm, chartaceous, glabrous, the base acute, the apex acuminate; midrib forming narrow, elevated line adaxially, the marginal veins 2, strongly arched between secondary veins, the innermost 3–8 mm from margin. Inflorescences much abbreviated, bracteate racemes; bracts imbricate, triangular-lanceolate, 2–4 mm long, puberulous; pedicels 10–30 mm long, puberulous to glabrous; bracteoles linear to subulate, deciduous before anthesis. Flowers: buds ca. 3 mm long, glabrous or sparsely pilose. Fruits globose, to ca. 1.5 cm diam., often of markedly varying diameters, bright red at maturity, glabrous. Fl (Sep, Oct, Dec), fr (Feb, Sep, Oct); in mossy forest at high elevations and in nonflooded forest at lower elevations.

Eugenia pseudopsidium Jacq.

Small trees. Plants glabrous. Leaves: petioles 5–10 mm long; blades elliptic to narrowly elliptic or oblong, 7–13(16) × 3–5.5(8) cm, chartaceous to coriaceous, the base acute to obtuse, the apex acuminate, often abruptly so; midrib sulcate adaxially, the marginal vein 1, 1–2 mm from margin. Inflorescences much abbreviated racemes; pedicels 7–30 mm long, slender. Flowers: buds ca. 3 mm long. Fruits subglobose, to 1.5 cm diam. Fl (Apr, Jun, Jul, Aug), fr (Jun, Jul, Aug, Sep, Oct, Dec); common, in nonflooded forest.

A collection from the summit of Mont Galbao, *Mori et al. 8774* (NY), apparently represents the same taxon. It has leaves similar to those of *E. pseudopsidium*, but the few immature fruits present have the calyx lobes relatively short and strongly inflexed, whereas typical collections of *E. pseudopsidium* have well developed and erect calyx lobes.

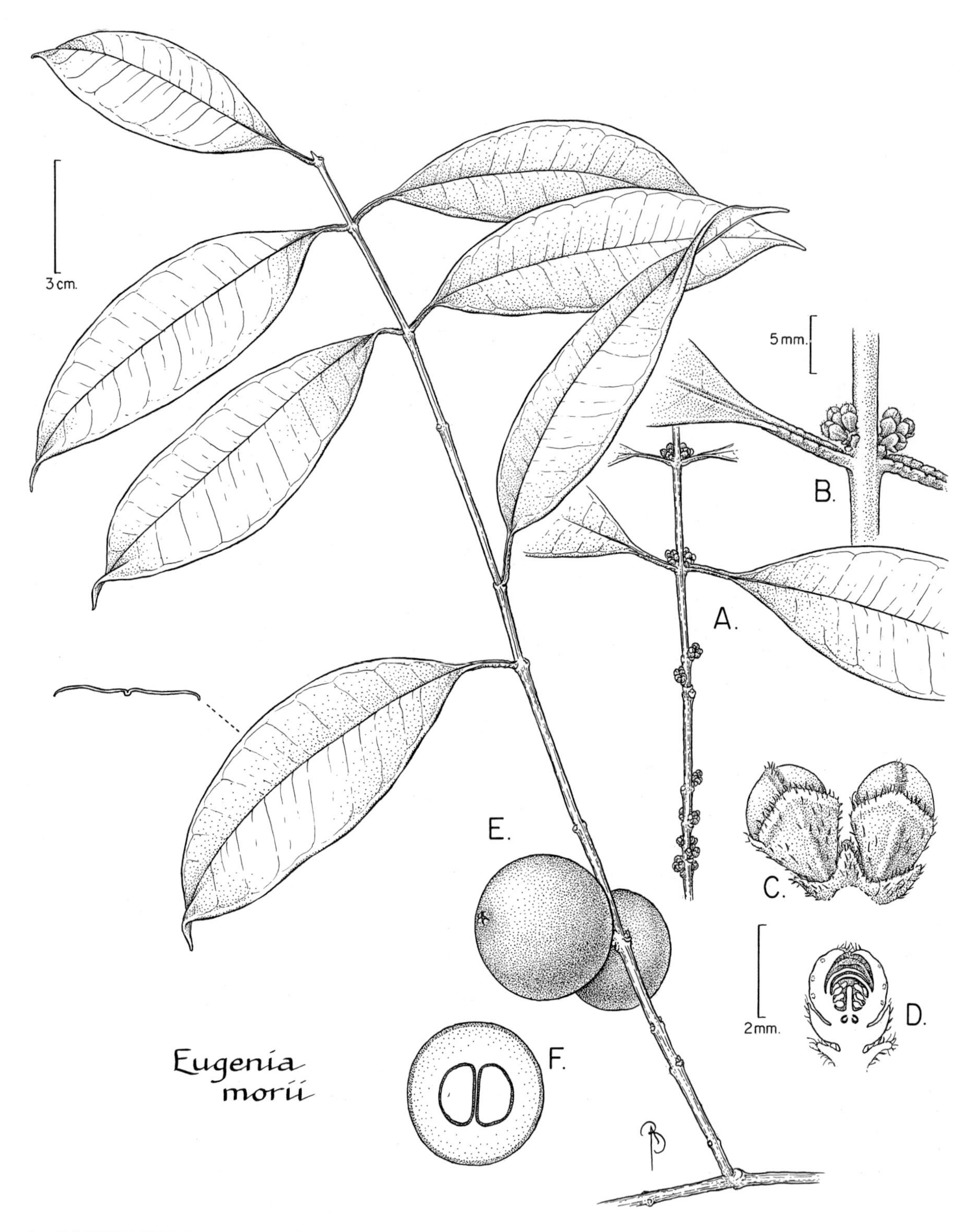

FIG. 236. MYRTACEAE. *Eugenia morii* (A–D, *Mori & Boom 14900*; E, F, *Mori et al. 22302*). **A.** Part of stem with leaf bases and inflorescences in bud. **B.** Detail of stem with inflorescences in axils of leaves. **C.** Lateral view of two flower buds. **D.** Medial section of flower in bud. **E.** Part of stem with leaves and fruits (right) and transverse section of leaf showing impressed midrib. **F.** Transverse section of fruit.

Eugenia spruceana O. Berg

Large trees. Pubescence of pale-coppery hairs, fading to ash or whitish color. Leaves: petioles 5–6 mm long; blades narrowly obovate to oblong, 9–13.5 × 3.5–4.7 cm, coriaceous, discolorous, glabrous adaxially, sparsely sericeous abaxially, the base acute to obtuse, the apex abruptly and sharply acuminate; midrib sulcate adaxially, the higher order venation scarcely evident, obscured by granular texture, the marginal vein 1, <1 mm from revolute margin. Inflorescences much abbreviated racemes, borne at leafless nodes; pedicels 7–10 mm long, sericeous. Flowers: buds 15–18 mm long; calyx lobes glabrescent; ovary sericeous. Fl (Jan); in moist forest.

This taxon, represented by *Mori et al. 23307* (NY, SEL, SP), resembles *Eugenia cucullata* Amshoff, but differs principally by having a markedly convex (vs. sulcate) midrib.

Eugenia tetramera (McVaugh) M. L. Kawas. & B. Holst

Trees. Pubescence of white or yellowish brown hairs. Bark reddish-brown and smooth or mottled. Leaves: petioles 5–15 mm long; blades elliptic, ovate, or obovate, 10–18 × 4–9 cm, coriaceous, puberulous to glabrous adaxially, softly pilose, especially on veins abaxially, the base acute to rounded, the apex acuminate; midrib sulcate adaxially, the secondary and principal marginal veins impressed adaxially, strongly elevated abaxially, the marginal veins 2, the innermost 3–8 mm from margin. Inflorescences much abbreviated racemes; pedicels 10–25 mm long, tomentose; bracteoles linear, 8–10 mm long, tomentose, deciduous. Flowers: buds ca. 10 mm long, densely pilose; ovary 2–4-locular. Fruits 3–4 cm diam., globose to subspherical-pyriform, densely pubescent, light orange to apricot-colored. Seeds several per fruit. Fl (Aug), fr (Aug); in nonflooded forest.

Eugenia sp. **A**

Trees. Plants glabrous. Leaves: petioles 6–9 mm long, stout, the older ones corky-flaky; blades oblong to elliptic, 11–19 × 5–9 cm, coriaceous, drying dull or olive green ad- and abaxially, the base obtuse to rounded, the apex shortly acuminate; midrib convex adaxially, the marginal veins 2, the innermost shallowly arching at 4–8 mm from margin. Flowers not known; bracteoles persistent. Infructescences 4-fruited, the axis thick, 5 mm long; calyx lobes ca. 5 mm long, tightly appressed to ovary summit, eventually deciduous; fruiting pedicels stout, woody, ca. 1 cm long. Fruits ellipsoid or obovoid, ca. 3 cm long, dark purple at maturity, borne ± upright along branches. Fr (Aug); in nonflooded forest.

Eugenia sp. **B**

Trees. Plants glabrous. Trunks angled. Bark orange-brown, scaly. Leaves: petioles 7–10 mm long; blades elliptic to narrowly elliptic or obovate, 8–11 × 3–4.5 cm, coriaceous, the base acute, the apex acuminate, slightly recurved; midrib convex adaxially, the marginal veins 2, the innermost 1–2 mm from margin. Inflorescences much abbreviated racemes; pedicels 2–4 mm long. Flowers: buds ca. 4 mm long. Fruits globose, ca. 2 cm diam., shiny red at maturity. Fl (Nov), fr (Feb); in nonflooded forest.

This taxon, represented by *Mori & Boom 15192* (NY, SEL) and *Mori 22945* (NY), belongs in a species complex that includes *E. dittocrepis* O. Berg and *E. citrifolia* Poir. which is in need of more study before species limits can be established.

Eugenia sp. **C** Pl. 100a

Large trees. Pubescence of lustrous dark brown hairs. Bark rough. Leaves: petioles 10–13 mm long, wrinkled when dry; blades elliptic to oblong, 12–15 × 5–6 cm, chartaceous, sparsely appressed-pubescent, probably glabrescent, the base acute to obtuse, the apex short-acuminate; midrib sulcate adaxially, the marginal veins 2, the innermost 2–6 mm from margin and arching between secondaries. Inflorescences abbreviated racemes, tomentulose, the axis 2–4 mm long; pedicels ca. 10–13 mm long. Flowers: buds 8–10 mm long; calyx lobes tomentulose. Fl (Feb); in nonflooded forest.

This taxon is based on a single flowering collection (*Mori et al. 22947*, NY, SEL). It appears to be related to *Eugenia brownsbergii* Amshoff.

MYRCIA Guill.

Trees and shrubs. Inflorescences paniculate or rarely racemose; bracts persistent or deciduous; bracteoles usually deciduous. Flowers 4 or 5-merous; calyx lobes free and distinct in bud, or rarely connate and then tearing irregularly and longitudinally; hypanthium prolonged or not beyond apex of ovary; ovary 2- or 3-locular, the ovules 2 per locule. Fruits crowned by calyx lobes or remnants of them. Seeds 1 or 2, the seed coat membranaceous; embryo myrcioid.

1. Flowers glabrous to rarely sparsely appressed-pubescent abaxially; hypanthium prolonged beyond apex of ovary; calyx lobes free or connate. Fruits globose.
    2. Midrib of leaf blades convex or plane adaxially.
        3. Leaf blades 10–22 cm long, the apex gradually acute to acuminate. Flowers 5-merous; buds 3–4 mm long; calyx lobes pubescent adaxially; ovary 3-locular. . . . . *M. gigas.*
        3. Leaf blades 9–12 cm long, the apex abruptly acuminate. Flowers 4-merous; buds ca. 2 mm long; calyx lobes glabrous adaxially; ovary 2-locular. . . . . *M. graciliflora.*
    2. Midrib of leaf blades sulcate adaxially, sometimes only shallowly so.
        4. Secondary veins slightly impressed adaxially. Calyx completely closed in bud. . . . . *M. rupta.*
        4. Secondary veins slightly elevated adaxially. Calyx open in bud.
            5. Petioles corky-flaky. Inflorescence bracts persistent past anthesis. . . . . *M. saxatilis.*
            5. Petioles smooth. Inflorescence bracts deciduous before anthesis. . . . . *M. platyclada.*
1. Flowers, particularly the ovary, densely appressed-pubescent abaxially; hypanthium not prolonged beyond apex of ovary; calyx lobes free. Fruits ellipsoid (or subglobose in *M. fenestrata* with bullate leaves).
    6. Petioles 5–10 mm long.
        7. Leaf blades abaxially and inflorescence axes mostly glabrous. . . . . *M. fallax.*
        7. Leaf blades abaxially and inflorescence axes densely pubescent. . . . . *M. magnoliifolia.*
    6. Petioles 1–2 mm long.

8. Leaf blades distinctly bullate. . . . . *M. fenestrata.*
8. Leaf blades flat.
    9. Leaf blades 0.5–1.5 cm wide, impressed-punctate adaxially (use 10× lens, the base obtuse; secondary and tertiary venation scarcely evident. . . . . *M. sylvatica.*
    9. Leaf blades 3.5–5 cm wide, not impressed-punctate adaxially, the base subcordate; secondary and tertiary venation distinct. . . . . *M. subsessilis.*

Myrcia fallax (Rich.) DC.

Shrubs or trees. Pubescence of yellowish white hairs. Leaves: petioles 5–10 mm long; blades elliptic to lanceolate, 8.5–16(18) × 2.5–5.5(7) cm, chartaceous to coriaceous, glabrescent, the base acute to obtuse, the apex acuminate; midrib sulcate adaxially, the marginal veins 2, the innermost ca. 2 mm from margin. Inflorescence axes mostly glabrous. Flowers 5-merous; buds 2–3 mm long, pubescent; hypanthium not prolonged beyond apex of ovary; ovary 2-locular, densely appressed-pubescent. Fruits ellipsoid, ca. 1 cm long, glabrescent. Fl (Jan, Feb, Jun, Oct, Dec), fr (Jan, Feb, May, Jul, Aug, Oct, Dec); often in disturbed vegetation along roads.

Myrcia fenestrata DC.

Small trees. Pubescence of yellowish white hairs. Leaves: petioles 1–2 mm long; blades narrowly elliptic, ovate or lanceolate, 10–13 × 3–4 cm, chartaceous to coriaceous, distinctly bullate, glabrous, the base obtuse, the apex acuminate; midrib sulcate adaxially, the marginal vein 1, ca. 1 mm from margin. Flowers 5-merous; buds ca. 2 mm long, pubescent; hypanthium not prolonged beyond apex of ovary; ovary 2-locular, densely appressed pubescent. Fruits subglobose, ca. 1 cm diam., glabrescent. Fl (Mar), fr (May); in nonflooded forest.

Myrcia gigas McVaugh FIG. 237

Trees. Plants mostly glabrous. Leaves: petioles 10–15 mm long; blades elliptic to narrowly elliptic or lanceolate, 10–22 × 4.5–8.5 cm, coriaceous, glabrous, the base acute, the apex gradually acute to acuminate; midrib convex to plane adaxially, the marginal veins 2, the innermost 3–7 mm from margin. Flowers 5-merous; buds 3–4 mm long, glabrous abaxially; calyx lobes pubescent adaxially; hypanthium prolonged beyond apex of ovary; ovary 3-locular. Fruits globose, ca. 1 cm diam. Fl (Aug, Sep), fr (Jan, Apr, Sep, Oct); in nonflooded forest.

Myrcia graciliflora Sagot

Trees. Plants glabrous. Leaves: petioles ca. 5 mm long; blades elliptic, 9–12 × 3.5–5 cm, chartaceous to coriaceous, glabrous, the base acute to cuneate, the apex abruptly acuminate; midrib convex adaxially, the marginal veins 2, the innermost 1–2 mm from margin. Flowers 4-merous; buds ca. 2 mm long; hypanthium prolonged beyond apex of ovary; ovary 2-locular. Fruits globose, 8–10 mm diam., glabrous. Fl (Aug, Sep, Dec), fr (Nov); in nonflooded forest.

Myrcia magnoliifolia DC.

Trees. Pubescence of yellowish white hairs. Leaves: petioles 5–7 mm long; blades elliptic, 11–19 × 5.5–8.5 cm, coriaceous, glabrous adaxially, pubescent abaxially, the base obtuse, the apex acute to shortly acuminate; midrib sulcate adaxially, the marginal veins 2, the innermost 2–3 mm from margin. Inflorescence axes densely pubescent. Flowers 5-merous; buds ca. 3 mm long, pubescent; hypanthium not prolonged beyond apex of ovary; ovary 2-locular, densely appressed-pubescent. Fruits ellipsoid. Fl (Jun, Nov); in nonflooded forest.

Myrcia platyclada DC.

Trees. Pubescence of reddish hairs. Leaves: petioles ca. 5 mm long; blades obovate, 5–9 × 2.5–4 cm, coriaceous, glabrous, the base attenuate, the apex acute, shortly acuminate to broadly rounded; midrib sulcate adaxially, the secondary veins slightly elevated adaxially, the marginal vein 1, ca. 1 mm from margin. Flowers 5-merous; buds ca. 2 mm long, glabrous; hypanthium prolonged beyond apex of ovary; ovary 2-locular. Fruits globose, ca. 1 cm diam., glabrous. Fl (May), fr (Feb, Sep, Oct, Nov); common, in nonflooded forest.

Myrcia rupta M. L. Kawas. & B. Holst FIG. 238

Small trees. Pubescence of reddish brown hairs. Leaves: petioles 6–10 mm long, corky-flaky; blades elliptic to narrowly elliptic, 11–17 × 4–7 cm, coriaceous, mostly glabrous, the base acute to obtuse, the apex abruptly acuminate; midrib sulcate and secondary veins slightly impressed adaxially, the marginal veins 2, the innermost 3–5 mm from margin. Flowers: buds 2–3 mm long, glabrous to sparsely appressed-pubescent; calyx completely closed in bud, irregularly 5-lobed at anthesis; hypanthium prolonged beyond apex of ovary; ovary 2-locular. Fruits globose, ca. 1 cm diam., glabrous. Fl (Jan, Nov, Dec), fr (Mar, Apr); in nonflooded forest.

Myrcia saxatilis (Amshoff) McVaugh

Shrubs. Pubescence reddish brown, turning gray with age. Leaves: petioles 1–2 mm long, corky-flaky; blades elliptic or obovate, 2.5–5 × 1–3 cm, coriaceous, glabrous, the base acute, the apex obtuse to rounded, the margins revolute; midrib sulcate adaxially, the secondary veins slightly elevated adaxially, the marginal vein 1, ca. 1 mm from margin. Inflorescence bracts persistent past anthesis. Flowers 5-merous; buds ca. 2 mm long, glabrous abaxially; hypanthium prolonged beyond apex of ovary; calyx lobes unequal, appressed-pubescent adaxially; ovary 2-locular. Fruits globose, 7–8 mm diam., glabrous. Fl (Jan); in nonflooded forest.

Myrcia subsessilis O. Berg

Small trees. Pubescence of yellowish white hairs. Leaves: petioles 1–2 mm long; blades ovate to narrowly ovate, 5–9 × 3.5–5 cm, coriaceous, glabrous, the base subcordate, the apex acute to acuminate; midrib sulcate adaxially, the marginal veins 2, the innermost 2–3 mm from margin. Flowers 5-merous; buds 2–3 mm long, pubescent; hypanthium not prolonged beyond apex of ovary; ovary 2-locular, densely appressed-pubescent. Fruits ellipsoid. Fl (Jan); in nonflooded forest.

Myrcia sylvatica (G. Mey.) DC.

Small trees. Pubescence of yellowish white hairs. Leaves: petioles 1–2 mm long; blades ovate to lanceolate, 2.5–5.5 × 0.5–1.5 cm, coriaceous, glabrous and impressed-punctate adaxially (use 10× magnification), glabrescent abaxially, the base obtuse, the apex acuminate-caudate; midrib sulcate adaxially, the marginal vein 1, ca. 1 mm from margin, the secondary and tertiary venation scarcely evident. Inflorescences largely axillary. Flowers 5-merous; buds ca.

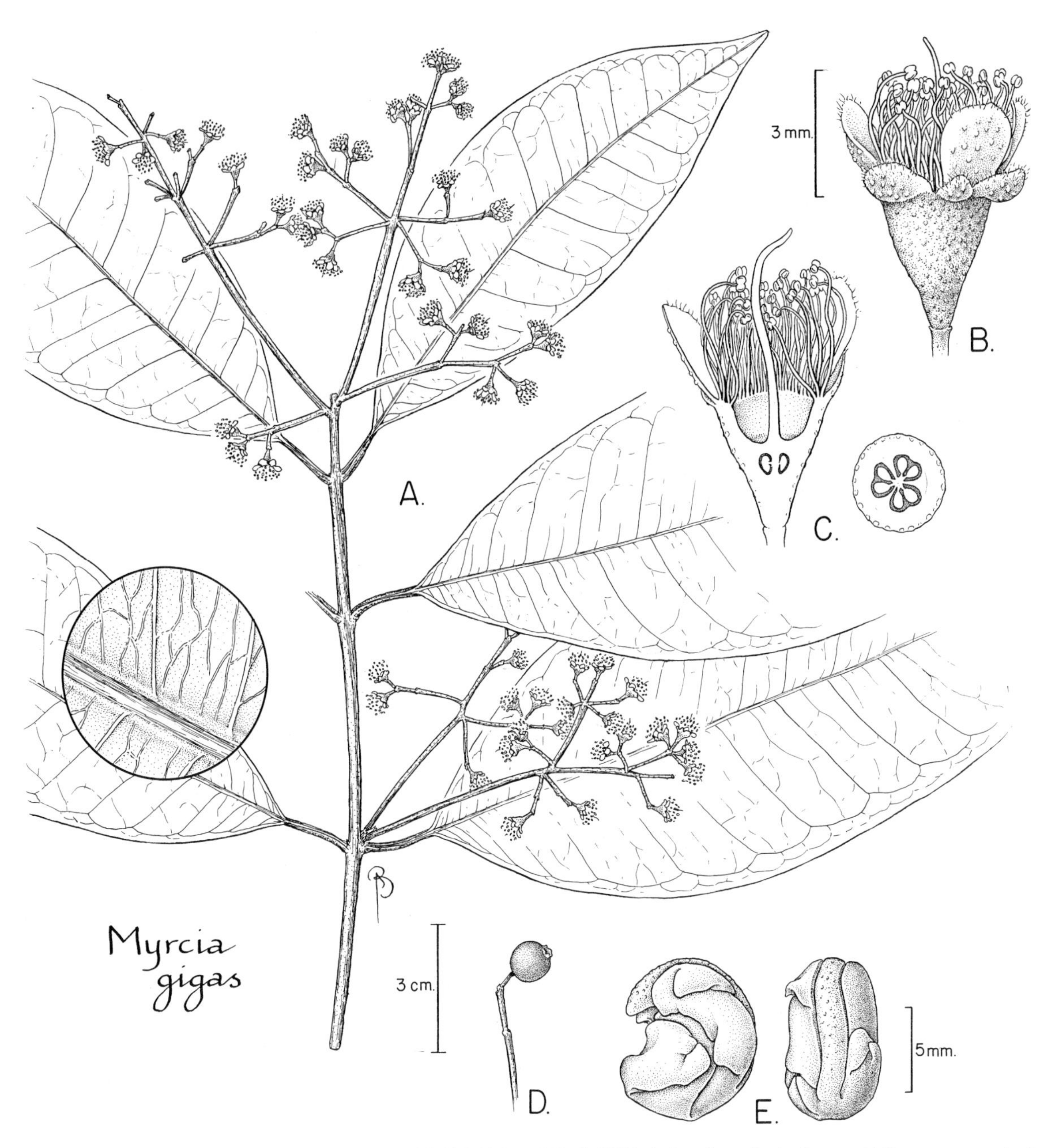

FIG. 237. MYRTACEAE. *Myrcia gigas* (A–C, *Mori & Boom 14966*; D, E, *Mori & Pipoly 15572*). **A.** Part of stem showing leaves and inflorescences; note detail of adaxial leaf surface (left). **B.** Lateral view of flower after dehiscence of bracteoles. **C.** Medial (left) and transverse sections of flower showing short hypanthium and axile placentation. **D.** Fruit showing apical hypanthium scar. **E.** Two views of myrcioid embryo showing folded cotyledons and prolonged radical.

2 mm long, pubescent; hypanthium not prolonged beyond apex of ovary; ovary 2-locular, densely appressed-pubescent. Fruits ellipsoid. Known only by sterile collections from our area; in non-flooded forest.

MYRCIARIA O. Berg

Trees and shrubs. Inflorescences in leaf axils or on trunks or older branches, much reduced racemes, appearing glomerate; bracteoles often connate, persistent; pedicels absent to poorly developed. Flowers 4-merous; hypanthium tube prolonged and narrowed above summit of ovary, circumscissile after anthesis at point of attachment with ovary, deciduous along with calyx and androecium; calyx open in bud; petals and stamens borne on hypanthium; ovary 2-locular, the ovules usually 2 per locule. Fruits globose, crowned by circular scar. Seeds 1 or 2, the seed coat membranaceous; embryo eugenioid.

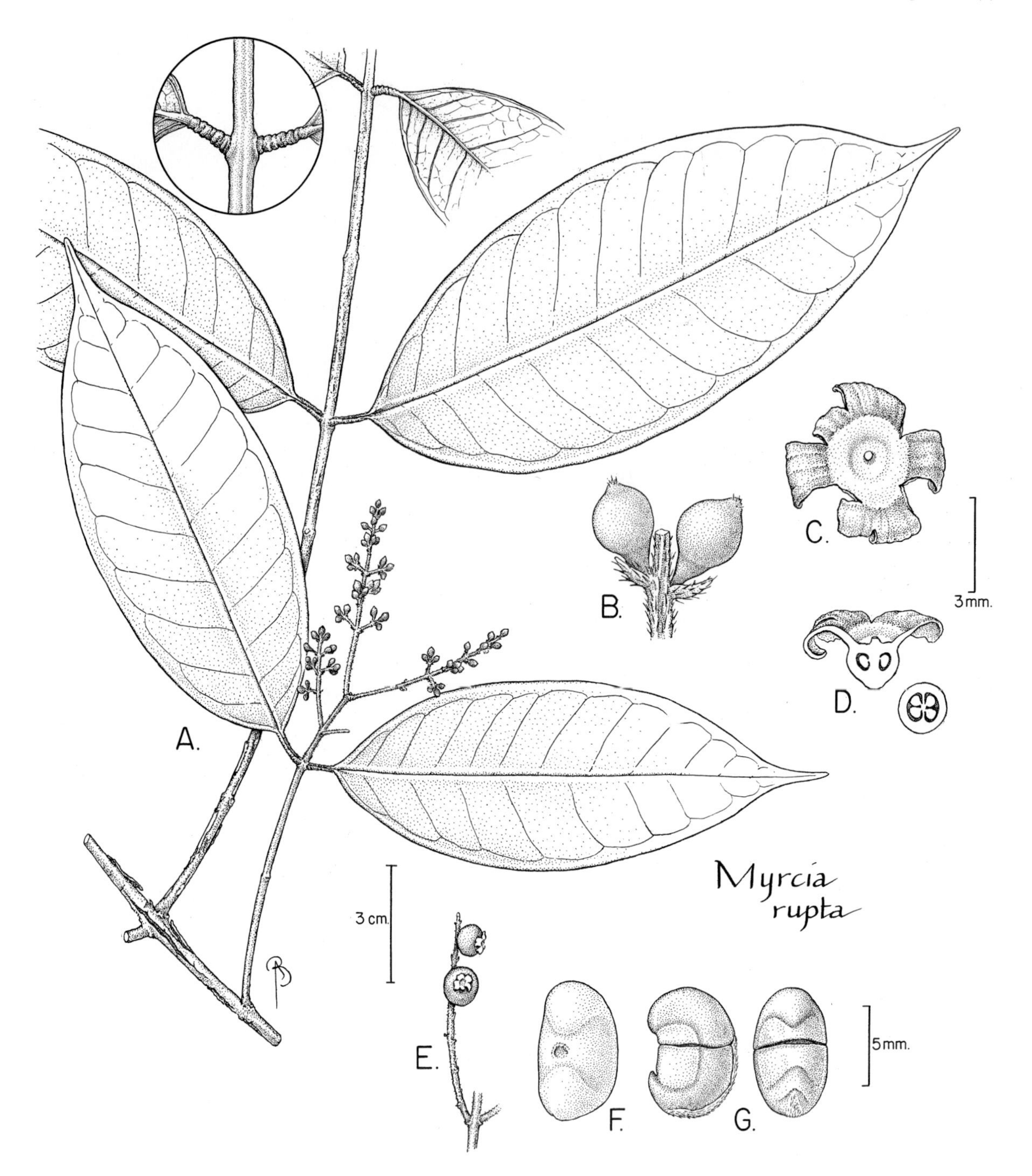

FIG. 238. MYRTACEAE. *Myrcia rupta* (A, B, *Mori et al. 8788*; C, *Cremers 11606*; D, *Oldeman B.4210*; E–G, *de Granville 2346*). **A.** Part of stem showing leaves and inflorescences; note detail of corrugated petioles (above). **B.** Part of inflorescence showing flowers buds. **C.** Apical view of immature fruit. **D.** Medial (above) and transverse (below) sections of immature fruits showing axile ovule placentation and two locules. **E.** Part of infructescence. **F.** Seed showing hylar scar. **G.** Two views of myrcioid embryos.

Myrciaria floribunda (Willd.) O. Berg

Small trees or shrubs. Young stems sparsely pilose. Leaves: petioles 4–6 mm long, sparsely pilose; blades elliptic, 5–7 × 1.5–2.5 cm, chartaceous, glabrous except along midrib, the base cuneate, the apex abruptly long-acuminate; midrib usually convex adaxially, the marginal vein 1, ca. 0.5 mm from margin. Flower buds ca. 3 mm long, glabrous. Fruits orange to red at maturity, glabrous. In non-flooded forest. *Goyavier* (Créole and French).

The only known collection from our area of this species (*Mori et al. 20970*, NY, SEL) is sterile. The leaf shape and fine, parallel secondary venation of *M. floribunda* are characteristic.

PSIDIUM L.

Trees and shrubs. Inflorescences dichasia or flowers solitary; bracteoles deciduous. Flowers 4- or 5-merous; calyx closed in bud, opening by irregular fissures; hypanthium prolonged beyond apex of ovary; ovary usually 3-locular, ovules many per locule. Fruits many-seeded, crowned by remnants of calyx lobes. Seed coat bony; embryo myrtoid.

Psidium guajava L.

Shrubs or small trees. Pubescence of whitish or yellowish white hairs. Young stems strongly 4-angled. Leaves: petioles 4–6 mm long; blades elliptic to narrowly elliptic, 8–12 × 3.5–6 cm, coriaceous, pubescent to glabrescent, the base obtuse, the apex obtuse to acute, apiculate; midrib sulcate adaxially, secondary veins usually noticeably parallel, approaching to 1–2 mm from margin, not forming distinct marginal vein. Inflorescences: pedicels 15–20 mm long, pubescent. Flowers: buds pyriform, 10–15 mm long, pubescent. Fruits ellipsoid to ovoid, ca. 6 × 4 cm, glabrous, green to yellowish green. Fl (Aug); cultivated. *Goiyave* (Créole), *goyavier* (French), *guava* (English).

NYCTAGINACEAE (Bougainvillea or Four O'clock Family)

Robert A. DeFilipps and Shirley L. Maina

Unarmed trees or shrubs. Leaves alternate, opposite or whorled, simple, entire, petiolate. Stipules absent. Inflorescences terminal or axillary, sometimes cauliflorous, paniculate or corymbose cymes. Flowers unisexual (sometimes bisexual in *Neea floribunda*), actinomorphic; perianth of 5 sepals connate to form 5-lobed tube, the lower portion persistent and accrescent to fruit; petals absent; stamens 5–11, free or fused at base; ovary superior, 1-locular, with basal placentation. Fruits indehiscent anthocarps adherent to indurate perianth-tube, sometimes ribbed. Seeds 1 per fruit.

Burger, W. 1983. Family No. 65. Nyctaginaceae. Flora Costaricensis. Fieldiana, Bot. n.s. **13:** 180–199.

1. Staminate flowers infundibuliform or campanulate; stamens exserted. Pistillate flower buds rounded or obtuse at apex. . . . . *Guapira.*
1. Staminate flowers urceolate; stamens deeply included. Pistillate flower buds acuminate at apex. . . . . *Neea.*

GUAPIRA Aubl.

Shrubs or trees. Trunks without buttresses. Leaves usually opposite; blades often pubescent abaxially. Inflorescences of noncauliflorous cymes; pedicels minutely 2- or 3-bracteolate at base. Flowers unisexual, often with abortive organs of other sex (plants dioecious). Staminate flowers: perianth infundibuliform or campanulate; stamens 6–10(11), exserted, unequal. Pistillate flowers: perianth tubular, rarely with staminodia.

1. Leaf blades glabrous on both surfaces at maturity. Inflorescence axes glabrous or glabrate. . . . . *G. eggersiana.*
1. Leaf blades rufous-puberulent abaxially, especially on veins, at maturity. Inflorescence axes rufous-puberulent.
   2. Staminate perianth rufous-pubescent to thick rufous-tomentose abaxially; stamens (7)8–10(11), usually >7. . . . . *Guapira* sp. **A.**
   2. Staminate perianth glabrous to rufous-puberulent, never approaching thick rufous-tomentose, abaxially; stamens (6)7. . . . . *G. salicifolia.*

Guapira eggersiana (Heimrl) Lundell

Shrubs or trees, 1.5–14 m tall. Young stems glabrous. Leaves usually opposite; petioles to 2 cm long; blades mostly elliptic or elliptic-lanceolate, 4.5–15 × 2.4(6.4) cm, glabrous on both surfaces at maturity, sometimes nitid, the apex obtuse, acute, or acuminate. Inflorescences: axes glabrous or glabrate. Staminate inflorscences corymbiform or subumbelliform. Staminate flowers: perianth glabrous at maturity, except for apically ciliate calyx-teeth; stamens 7–8. Pistillate flowers: perianth narrowly ellipsoid, subinfundibuliform or salverform, slightly constricted above, (2)3–5 × ca. 1 mm. Fl (Jan), fr (Jan); in nonflooded forest.

Guapira salicifolia (Heimerl) Lundell Pl. 101a

Shrubs or trees, 1.5–15 m tall. Young stems rufous-puberulent. Leaves opposite; petioles to 1.5 cm long; blades mostly elliptic or elliptic-lanceolate, 5–15 × 1.5–7 cm, rufous-puberulent abaxially, especially on veins at maturity, fuscous when dried, the apex subabruptly acuminate or long-acuminate. Inflorescences axes rufous-puberulent. Staminate inflorescences umbelliform. Staminate flowers: perianth glabrous to rufous-puberulent, never approaching rufous-tomentose, abaxially; stamens (6)7. Pistillate flowers: perianth subtubular, constricted above, (1.5)2–2.5 × 1 mm, the apex rounded. Fl (Nov, Dec), fr (Jul); uncommon, in nonflooded forest.

Pistillate flowering specimens of this species are superficially similar to *Neea ovalifolia* but can be distinguished as follows:

1. Leaf blades with adaxial surface dull, the abaxial surface rufous-puberulent along midrib. Pistillate flower buds rounded at apex; pistillate flowers (1.5)2–2.5 × 1 mm . . . . . . *Guapira salicifolia.*
1. Leaf blades with adaxial surface nitid, the abaxial surface glabrous. Pistillate flower buds tapering to long-acuminate at apex. Pistillate flowers 4 × 1.5 mm. . . . . *Neea ovalifolia.*

Guapira sp. A

Trees, 40 m tall. Young stems rufous-puberulent to rufous-tomentose. Leaves opposite to subopposite or alternate; petioles to 2 cm long; blades mostly elliptic to elliptic-lanceolate, sometimes approaching elliptic-obovate, 6–11 × 2.4–4.5 cm, rufous-puberulent, especially on veins, abaxially, dark brown when dried, the apex acute or short-acuminate to long-acuminate. Staminate inflorescences umbelliform, the axes rufous-puberulent. Staminate flowers: perianth campanulate to infundibuliform, rufous-pubescent to thick rufous-tomentose abaxially, ca. 0.45–0.50 × 0.38–0.51 cm, stamens (7)8–10(11), usually >7, to 1 cm long. Pistillate flowers not known. Fl (Sep); known by a single collection from nonflooded forest in our flora.

NEEA Ruiz & Pav.

Dioecious trees. Trunks with or without buttresses, or shrubs. Leaves opposite or alternate, sometimes pubescent abaxially. Inflorescences axillary or terminal, or sometimes cauliflorous; pedicels often minutely 1–3-bracteolate at base. Flowers usually unisexual (sometimes bisexual in *Neea floribunda*), often with abortive organs of other sex (plants dioecious). Staminate flowers: perianth urceolate; stamens 5–10, deeply included, unequal. Pistillate flowers: perianth urceolate, tubular-campanulate or tubular-cylindrical, with prominent sterile staminodia.

1. Leaves large, ca. 25–30(42) × 7.5–13.8 cm, when opposite, those of a pair equal in size.
    2. Trees, 4–18 m tall. Leaves glabrous. Inflorescences cauliflorous. . . . . *N. floribunda.*
    2. Shrubs or small trees, to 3(5) m tall. Leaves rufous-puberulent abaxially, especially on veins. Inflorescences not cauliflorous, terminal. . . . . *N. mollis.*
1. Leaves often smaller, 7–17(23) × 2–6.5 cm, when opposite, those of a pair sometimes unequal in size.
    3. Pistillate perianth with globose-campanulate limb, prominently constricted at throat. . . . . *N. constricta.*
    3. Pistillate perianth tubular-cylindrical, suburceolate, campanulate or indundibuliform, not constricted at throat.
        4. Leaves not nitid adaxially, often drying dark brown. Infructescence branches very thin, not subalate. Pistillate perianth flared at apex. . . . . *N. spruceana.*
        4. Leaves nitid adaxially, often drying greenish- or yellowish-brown. Infructescence branches thick and subalate. Pistillate perianth not flared at apex. . . . . *N. ovalifolia.*

Neea constricta J. A. Schmidt PL. 101b

Shrubs or trees, 3–12 m tall. Trunks sometimes fluted. Leaves opposite or sometimes alternate, those of a pair often markedly unequal in size; petioles to 2 cm long; blades elliptic, narrowly lanceolate or narrowly oblanceolate, to 17 × 6.2 cm, often smaller, glabrous, the apex acuminate. Inflorescences axillary or terminal, the axes rufous-puberulent. Staminate flowers: perianth urceolate, pale cream-colored with 5 magenta or crimson teeth; stamens 6. Pistillate flowers: perianth prominently constricted at throat into narrow lower portion and upper globose-campanulate limb. Fl (Sep), fr (Aug); seemingly rare, in nonflooded forest.

Neea floribunda Poepp. & Endl. FIG. 239, PL. 101d

Trees, 4–18 m tall. Trunks often with buttresses or fluted basally. Leaves opposite, sometimes alternate and opposite on same plant; petioles to 4 cm long; blades obovate, elliptic or oblanceolate, to 25–30(42) × (7.5)13.8 cm, glabrous, the apex obtuse to acuminate. Inflorescences usually cauliflorous, laxly branched cymes or panicles, the axes rufous-puberulent. Flowers unisexual or sometimes bisexual. Staminate flowers: perianth urceolate, greenish with ruffled lobes; stamens 6(9). Pistillate flowers: perianth tubular. Fl (Oct, Nov, Dec), fr (Feb, Nov); common, in nonflooded forest.

Neea mollis J. A. Schmidt PL. 101c

Shrubs or trees, 0.8–3(5) m tall. Trunks not buttressed. Branches densely rufous-puberulent. Leaves opposite or subopposite; petioles to 3 cm long; blades broadly elliptic, to 26.5 × 12.5 cm, glabrous adaxially, rufous-puberulent abaxially especially on veins, the apex obtuse to acuminate. Inflorescences terminal, the axes rufous-puberulent. Staminate flowers not seen. Pistillate flowers: perianth tubular-campanulate, green, constricted at apex of body of fruit. Fl (Feb, Jun, Aug, Oct), fr (May, Aug, Sep, Oct, Nov); occasional, in nonflooded forest.

Neea ovalifolia J. A. Schmidt

Shrubs (sometimes scandent) or trees, 3–21(30) m tall. Trunks sometimes buttressed. Leaves opposite or subopposite; petioles to 3 cm long; blades elliptic or oblanceolate (rarely ovate), to 13.5(23) × 6.5 cm, nitid adaxially, often drying greenish- or yellowish-brown, the apex acute or acuminate. Inflorescences axillary or terminal, the axes rufous-puberulent, thick, to 2 mm wide, subalate in fruit. Staminate flowers: perianth urceolate, green; stamens 6. Pistillate flowers: perianth tubular-cylindrical, 4 × 1.5 mm, the apex long-acuminate. Fl (Nov), fr (Jan, Feb, Aug, Sep, Nov); infrequent, in nonflooded forest.

Neea spruceana Heimerl

Shrubs or trees, 1–6 m tall. Leaves opposite or subopposite, those of a pair often unequal in size, sometimes verticillate in whorls of 3 or 4; petiole 0.4–1.5 cm or leaves subsessile; blades usually elliptic or oblong, sometimes ovate or oblanceolate, to 16.5(19.5) × 6 cm, dull adaxially and often drying dark brown, glabrous, the apex acuminate or long-acuminate. Inflorescences axillary or terminal, the axes sparsely spreading-rufous-puberulent, thin, not subalate in fruit. Staminate flowers: perianth urceolate, sometimes infundibuliform or tubular-ellipsoid, yellow or yellowish-green; stamens 5–6. Pistillate flowers: perianth suburceolate, campanulate, or infundibuliform, the limb flared at apex, with 5 lobes to 1 mm long. Fl (Sep, Dec outside our area); probably represented in our area by *Prévost 2106* (CAY).

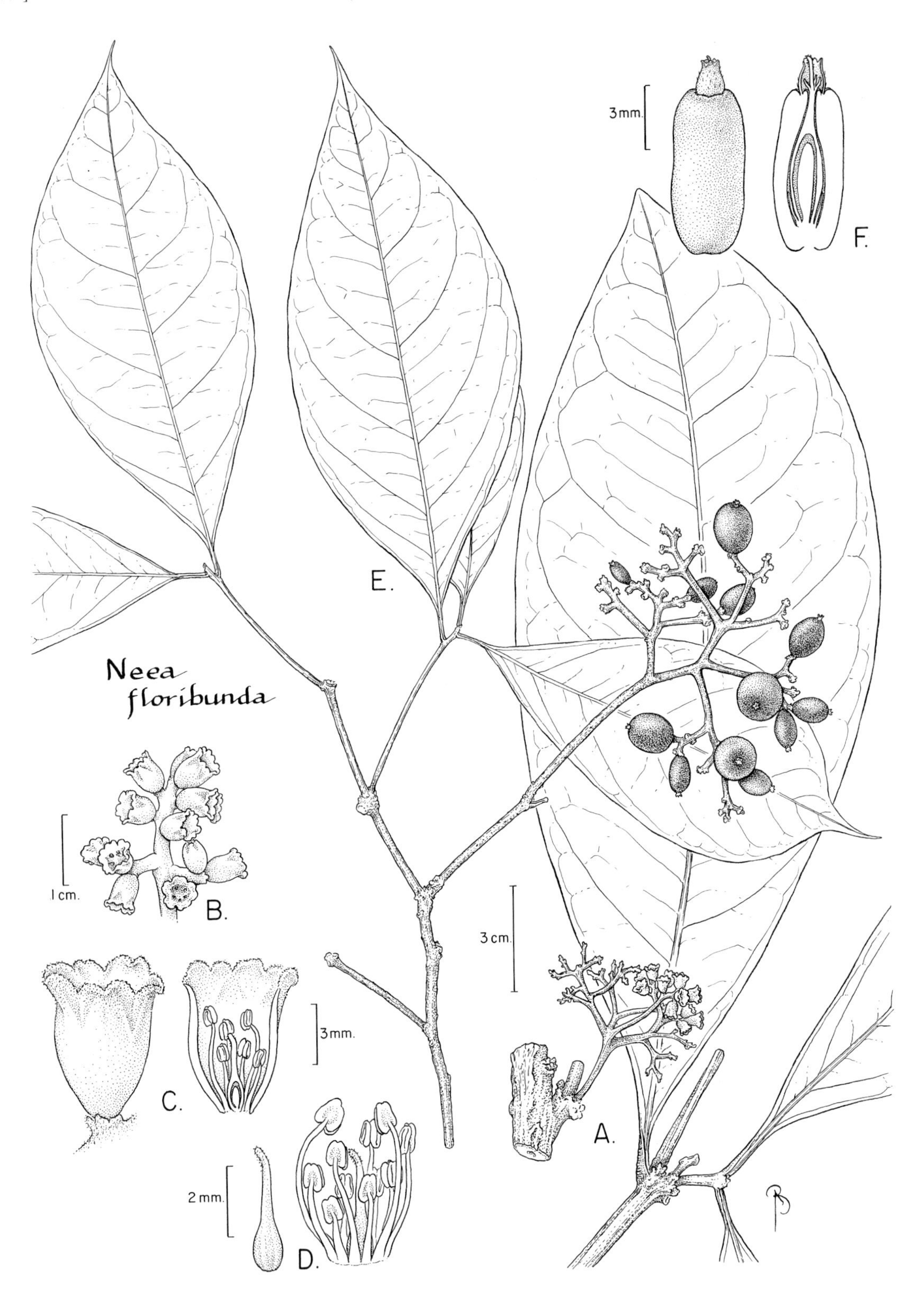

FIG. 239. NYCTAGINACEAE. *Neea floribunda* (A–D, *Mori et al. 22686*; E, *Mori & Gracie 21198*; F, *Mori & Boom 15193*). **A.** Part of stem with leaves (right) and inflorescence (left). **B.** Detail of inflorescence. **C.** Lateral view of flower (left) and Medial section of flower (right). **D.** Lateral views of pistil (left) and androecium surrounding pistil (right). **E.** Part of stem with leaves and infructescence. **F.** Fruit (far left) and medial section of fruit (near left).

NYMPHAEACEAE (Water-lily Family)

Scott A. Mori

Aquatic, perennial, rhizomatous, or tuber forming herbs. Leaves simple, alternate, usually floating but sometimes emergent because of drop in water level, sometimes with dimorphic floating and submerged leaves; petioles long; blades often peltate, mostly elliptic-sagittate to elliptic, the base cordate or sagittate, the margins entire or irregularly toothed, sometimes with upturned rim outside our area (*Victoria amazonica* (Poepp.) J. E. Sowerby). Inflorescences of solitary flowers, the flowers displayed at or above surface of water; pedicels long. Flowers actinomorphic, bisexual, large and showy, nocturnal or diurnal, often changing color from one day to next; sepals usually (3)4–9, free; petals numerous, free, often grading into stamens; stamens numerous; ovary superior or inferior, syncarpous or apocarpous, 5 to many locular, the stigmas sessile, as many in number as carpels. Fruits often spongy, berry-like, indehiscent or irregularly dehiscent, often pulled under water at maturity. Seeds often small and numerous, commonly operculate, often arillate; endosperm scanty, the perisperm copious, the cotyledons 2.

Cramer, J. 1979. Nymphaeaceae. *In* A. L. Stoffers & J. C. Lindeman (eds.), Flora of Suriname **V(1):** 370–384. E. J. Brill, Leiden.

NYMPHAEA L.

Rhizomes erect or horizontal, elongate stolons present or absent. Leaves floating, occasionally emergent or submersed; petioles glabrous or pubescent, with several larger air canals and many smaller ones; blades elliptic-sagittate to orbicular, the base cordate to sagittate, the apex acute, tapered to truncate; venation mostly radiate but with prominent midrib. Flowers diurnal or nocturnal, faintly to strongly aromatic, hypogynous to perigynous; sepals usually 4, rarely 3 or 5, the veins prominent or inconspicuous; petals 7–40, attached in several series, the transition from petals to stamens gradual or abrupt, white, blue, red, or yellow; stamens 20–700, multiseriate, the outer filaments broader and often petaloid, the inner filaments narrower and filiform, the connectives with or without terminal appendages, the anthers introrse; carpels 5–47, forming ring embedded in cup-shaped receptacular tissue, the upper surface of each carpel forming ray of stigmatic tissue, this ray usually terminating abaxially in an upwardly curved appendage; placentation laminar, the ovules numerous, anatropous. Fruits irregularly dehiscent, berry-like capsules, ripening under water. Seeds with membranous aril, this allowing seeds to float; embyro small, straight; endosperm scanty, the perisperm abundant.

Nymphaea glandulifera Rodschied

Aquatic herbs. Leaves floating, subpeltate; petioles ca. 45 cm long, spongy, glabrous, attached 1–2 mm from base of blade; blades broadly elliptic to nearly orbicular, sinus well developed, 8.5–9.5 cm long from sinus to apex × 15–16 cm wide, glabrous, uniformly green, the base cordate, the margins entire. Flowers ca. 7 cm diam. at anthesis; petals white. Seeds numerous, small, ovoid, ca. 1 mm long, pubescent. Fl (Apr); in slow flowing water near rapids.

OCHNACEAE (Ochna Family)

Claude Sastre

Trees, treelets, shrubs, or herbs. Stipules caducous or persistent. Leaves simple, alternate. Inflorescences axillary cymes (often much reduced), terminal panicles, or pseudoracemes. Flowers actinomorphic (except in species with partial fusion of calyx), bisexual; sepals usually free; petals 5, free; staminodia present or absent; stamens 5–10; ovary superior, with 3–10 carpels entirely united or united only at base, the style single, gynobasic or terminal, the stigma single; placentation axile or parietal, the ovules solitary or numerous per locule. Fruits either free and indehiscent monocarps on a torus (*Ouratea*) or capsules (*Sauvagesia*). Seeds exalbuminous and 1 per carpel in *Ouratea* or albuminous and numerous in *Sauvagesia.*

Görts-van Rijn, A. R. A. & M. J. Jansen-Jacobs 1986. Ochnaceae. *In* A. L. Stoffers & J. C. Lindeman (eds.), Flora of Suriname **III(1–2):** 455–463. E. J. Brill, Leiden.
Wehlburg, C. 1966. Ochnaceae. *In* A. Pulle (ed.), Flora of Suriname **III(1):** 328–341. E. J. Brill, Leiden.

1. Trees, treelets, or shrubs. Stipules entire. Petals yellow; stamens 10; carpels 5–8. Fruits of indehiscent monocarps. . . . . *Ouratea.*
1. Subshrubs or herbs. Stipules ciliate-glandular. Petals rose-colored or white; stamens 5; carpels 3. Fruits capsules. . . . . *Sauvagesia.*

OURATEA Aubl.

Trees, treelets, or shrubs, to 28 m tall. Stipules usually caducous, entire. Leaves: petioles short; blades obovate to ovate, chartaceous or coriaceous, the apex acuminate, the margins entire or serrate; secondary veins subopposite or alternate across midrib, subparallel, curving upward near margin and continuing almost as inframarginal veins. Inflorescences terminal, panicles or pseudoracemes. Flowers: sepals 5, free or united, coriaceous, green; petals 5, free, membranaceous, yellow; staminodia absent; stamens 10, yellow, the anthers (sub)sessile, the dehiscence poricidal; ovary with 5–8 carpels united only at base, the style gynobasic; placentation axile, the ovules 1 per carpel. Fruits of 1–4 dark purple to nearly black monocarps attached on ± conical to globose, red torus at maturity. Seeds ellipsoid, with two small, apical appendages, the surface smooth.

1. Trees, 6–28 m tall (one collection of *O. scottii* subsp. *scottii* has been described as a shrub). Secondary veins not clearly distinguishable from intersecondary veins. Carpels 6–8.
   2. Secondary and tertiary veins impressed on both surfaces (most distinctly abaxially). Sepals caducous at anthesis. . . . . *O. scottii* subsp. *scottii*.
   2. Secondary and tertiary veins plane or salient on both surfaces. Sepals persistent in young fruit.. . . . . *O. candollei*.
1. Shrubs or treelets, 0.6–5 m tall. Secondary veins clearly distinguishable from intersecondary veins. Carpels 5.
   3. Leaf blade margins entire or the teeth widely separated. Inflorescences distinctly paniculate. . . . . *O. leblondii*.
   3. Leaf blade margins serrate, the teeth closely spaced. Inflorescences pseudoracemose.
      4. Shrubs, 0.6–2.5 m tall. Leaf blades chartaceous, with 12–16 pairs of secondary veins, the tertiary veins distinct. Inflorescence axes puberulent. Flowers with free sepals. . . . . *O. erecta*.
      4. Treelets, 1.5–5 m tall. Leaf blades coriaceous, with 7–10 pairs of secondary veins, the tertiary veins indistinct. Inflorescence axes glabrous. Flowers with sepals united in 2–4 parts. . . . . *O. saülensis*.

Ouratea candollei (Planch.) Tiegh.

Trees, 10–28 m tall. Bark whitish-gray, rough, with vertical fissures and lines of lenticels, exfoliating in large flakes, the inner bark brick red, 3 mm thick. Leaves: petioles 2–4 mm long; blades narrowly elliptic to elliptic, 11–14 × 2.5–4 cm, coriaceous, the margins entire; secondary veins not clearly distinguishable from intersecondary veins, in numerous, difficult to count, pairs, the secondary and tertiary veins plane or salient on both surfaces. Inflorescences paniculate, 8–14.5 cm long, the axes glabrous; pedicels 5–7 mm long. Flowers: sepals basally fused, oblong, 5–7 × 2–3 mm, persistent in young fruit; petals obovate, 9 × 7 mm; stamens subsessile, 5 mm long; ovary with 6–8 carpels, yellow, the style 5 mm long. Monocarps 1–2, ovoid, 6–8 × 5–6 mm; torus with 1–3 lobes, 3–4 × 4 mm. Fl (Aug, Sep, Oct), fr (Feb); rare, in nonflooded forest. *Gaulette rouge* (Créole).

Ouratea erecta Sastre

Shrubs, 0.6–2.5 m tall. Leaves: petioles 3–5 mm long; blades narrowly obovate to oblanceolate, 10–23.5 × 2.5–8 cm, chartaceous, the margins serrate; secondary veins clearly distinguishable from intersecondary veins, in 12–16 pairs, the distinct tertiary veins perpendicular to secondary and intersecondary veins. Inflorescences pseudoracemose, rarely distinctly branched, the axes sparsely puberulent, red (in fruit); pedicels 8–10 mm long. Flowers: sepals free, lanceolate, 9 × 2.5–3 mm; petals narrowly obovate, 11 × 5 mm; stamens sessile, 8 mm long; ovary with 5 carpels, white, the style 6 mm long. Monocarps 1–2, oblong, 8 × 7 mm, turning from green to black; torus globose. Fl (Jan, Mar, Nov, Dec), fr (Jan, Feb, Mar, May, Dec); common, in nonflooded forest.

Ouratea leblondii (Tiegh.) Lemée [Syn.: *O. sagoti* (Tiegh.) Lemée]

Shrubs, 2–3 m tall. Bark blackish with white blotches. Leaves: petioles 4–6 mm long; blades ovate to elliptic, 10–21 × 5–9 cm, coriaceous, the margins entire or the teeth widely separated; secondary veins clearly distinguishable from intersecondary veins, in 6–8 pairs, impressed on both surfaces, faintly apparent on adaxial surface, the tertiary veins subperpendicular to secondary veins, impressed on both surfaces, faintly apparent on adaxial surface. Inflorescences paniculate, 8–20 cm long, the axes glabrous; bracts at base triangular, 5 × 2 mm at base; pedicels 1 cm long. Flowers: sepals free, ovate, 7–8 × 2–3 mm, persistent in young fruits; petals obovate, 9 × 7 mm; stamens sessile, 5 mm long; ovary with 5 carpels, the style 5 mm long. Fruits: mericarps 1–3, ovoid, 6–8 × 5–6 mm; torus compressed to subspherical, 3–7 × 5–7 mm. Fl (Jan, Apr), fr (Jan); common, in forest along rivers, occasional on granitic outcrops.

Ouratea saülensis Sastre FIG. 240

Unbranched shrubs or treelets, 1.5–5 m tall. Leaves: petioles 5–10 mm long; blades narrowly obovate to oblanceolate, 12–24.5 × 3.5–7 cm, coriaceous, the margins serrate; secondary veins clearly distinguishable from intersecondary veins, in 7–10 pairs, the intersecondary and tertiary veins very fine and difficult to distinguish. Inflorescences pseudoracemose, the axes glabrous, red (in fruit); pedicels 7–12 mm long. Flowers: sepals united in 2–4 parts, 8–9 × 3–5 mm; petals narrowly obovate, 10 × 5 mm; stamens sessile, 8 mm long; ovary with 5 carpels, the style 6 mm long. Monocarps 1–2, oblong, 10 × 5 mm, turning from green to black; torus globose to clavate, 12 × 10 mm, turning from pink to red. Fl (Sep), fr (Jan, Apr, Nov, Aug); occasional, in nonflooded forest.

Ouratea scottii Sastre subsp. **scottii** FIG. 241, PL. 102a

Trees, 6–20 m tall (*Oldeman B.4156* described as a shrub 2.4 m tall). Leaves: petioles 5–9 mm long; blades elliptic to narrowly elliptic, 6.5–17 × 2.5–5 cm, coriaceous, the margins entire; secondary veins not clearly distinguisable from intersecondary veins, in numerous, difficult to count, pairs, the secondary and tertiary veins impressed on both surfaces (most distinctly on abaxial surface). Inflorescences paniculate, generally markedly once-branched, the axes glabrous, red (in fruit); pedicels 7–9 mm long. Flowers: sepals free or united into 3–4 parts, oblong, 5–6 × 2–3 mm, caducous at anthesis; petals very widely obovate, 9 × 8 mm; stamens sessile, 5 mm long; ovary with 6–8 carpels, the style 5 mm long. Monocarps 1–3, ovoid, 6–8 × 5–6 mm; torus 1–3 lobed, subglobose, red. Fl (Aug, Sep, Oct), fr (Feb, Sep, Oct, Nov); common, in nonflooded forest.

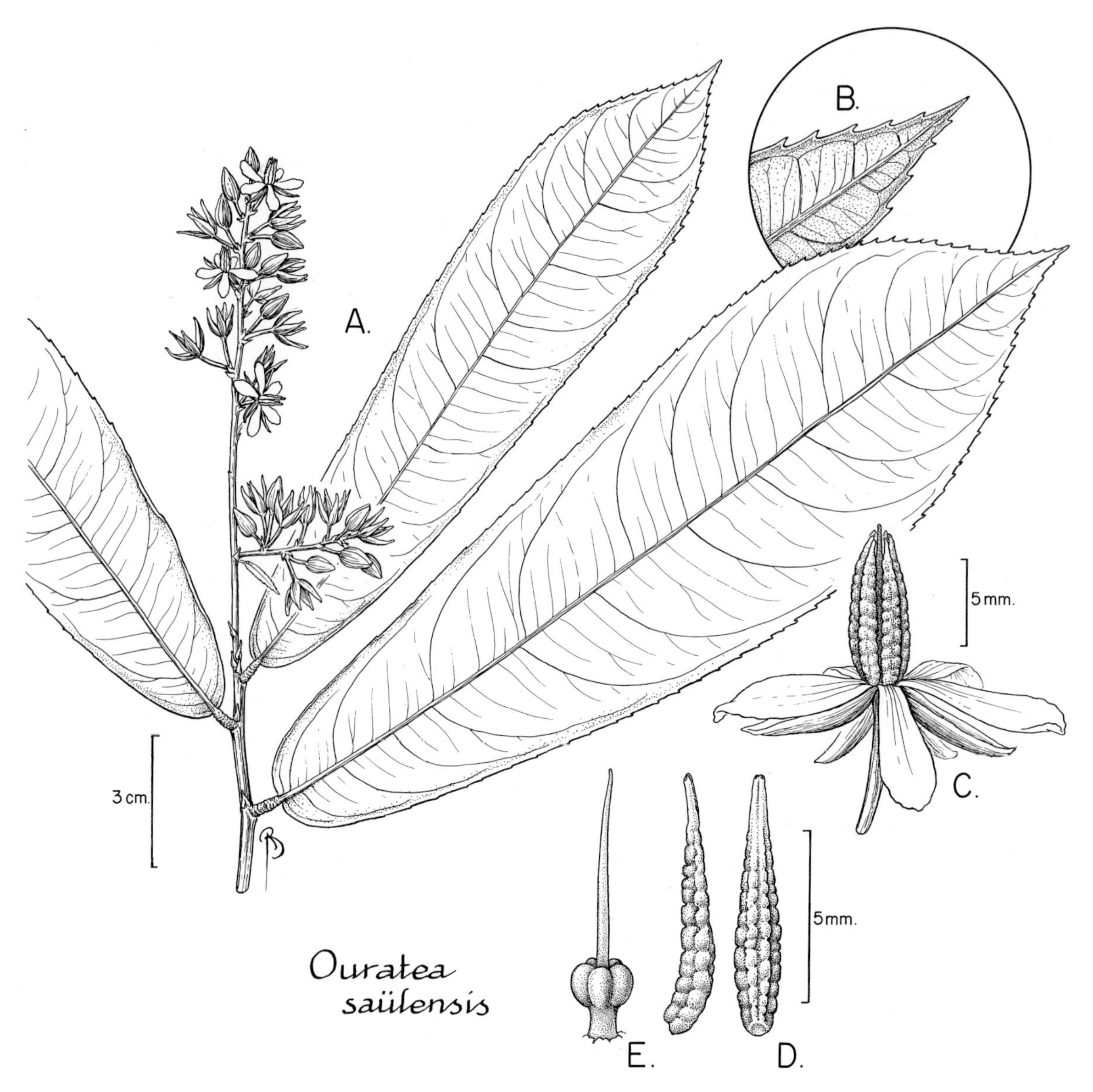

FIG. 240. OCHNACEAE. *Ouratea saülensis* (*Mori et al. 14889*). **A.** Apex of stem showing leaves and inflorescence. **B.** Detail of leaf apex. **C.** Lateral view of flower. **D.** Two views of stamens. **E.** Pistil surmounting short gynophore. (Reprinted with permission from C. Sastre, Brittonia 45(4), 1993.)

SAUVAGESIA L.

Subshrubs or perennial or annual herbs. Stipules persistent or caducous, ciliate-glandular. Leaves subsessile; blades linear, elliptic, or oblong to ovate, the apex acute or obtuse, the margins crenulate; secondary veins parallel, arcuate and ending as a marginal tooth. Inflorescences axillary or terminal, much reduced cymes or panicles. Flowers: sepals 5, chartaceous; petals membranaceous, rose-colored or white; staminodial whorls 1–2, the outer staminodia present or absent, if present, numerous, capitate, the inner staminodia 5, petaloid, forming pseudocorolla; fertile stamens 5, the anthers with longitudinal dehiscence; ovary with 3 united carpels, the style terminal; placentation parietal, the ovules numerous. Fruits 3-valved capsules. Seeds subglobose, not winged, the surface dimpled.

1. Annual herbs. Stipules caducous. Inflorescences terminal. Outer staminodia absent. . . . . *S. tafelbergensis.*
1. Subshrubs or perennial herbs. Stipules persistent. Inflorescences axillary. Outer staminodia present.
   2. Subshrubs. Stipules 10–12 mm long, with branched cilia along margins. Leaf-blades linear, 50–120 × 3–5 mm. . . . . *S. aliciae* subsp. *aratayensis.*
   2. Herbs. Stipules 5–8 mm long, with simple cilia along margins. Leaf-blades narrowly elliptic, 12–30 × 3–10 mm. . . . . *S. erecta.*

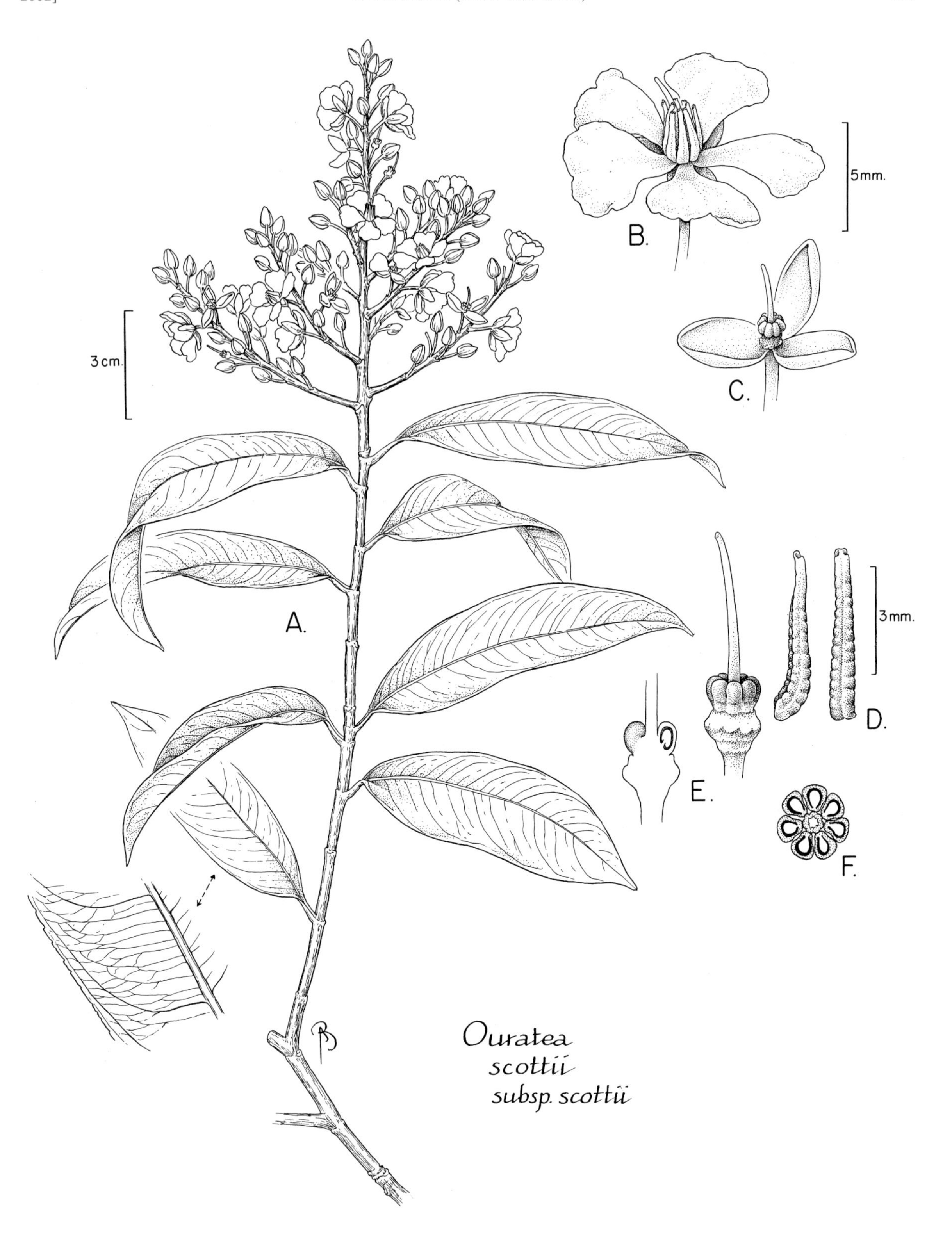

FIG. 241. OCHNACEAE. *Ouratea scottii* subsp. *scottii* (*Mori et al. 20956*). **A.** Part of stem showing leaves and terminal, paniculate inflorescence; note detail of leaf venation (below left). **B.** Oblique-apical view of flower. **C.** Flower with corolla and androecium removed. **D.** Two views of stamens. **E.** Intact (right) and medial section (left) of gynoecium and receptacle. **F.** Transverse section of ovary. (Reprinted with permission from C. Sastre, Brittonia 46(4), 1994.)

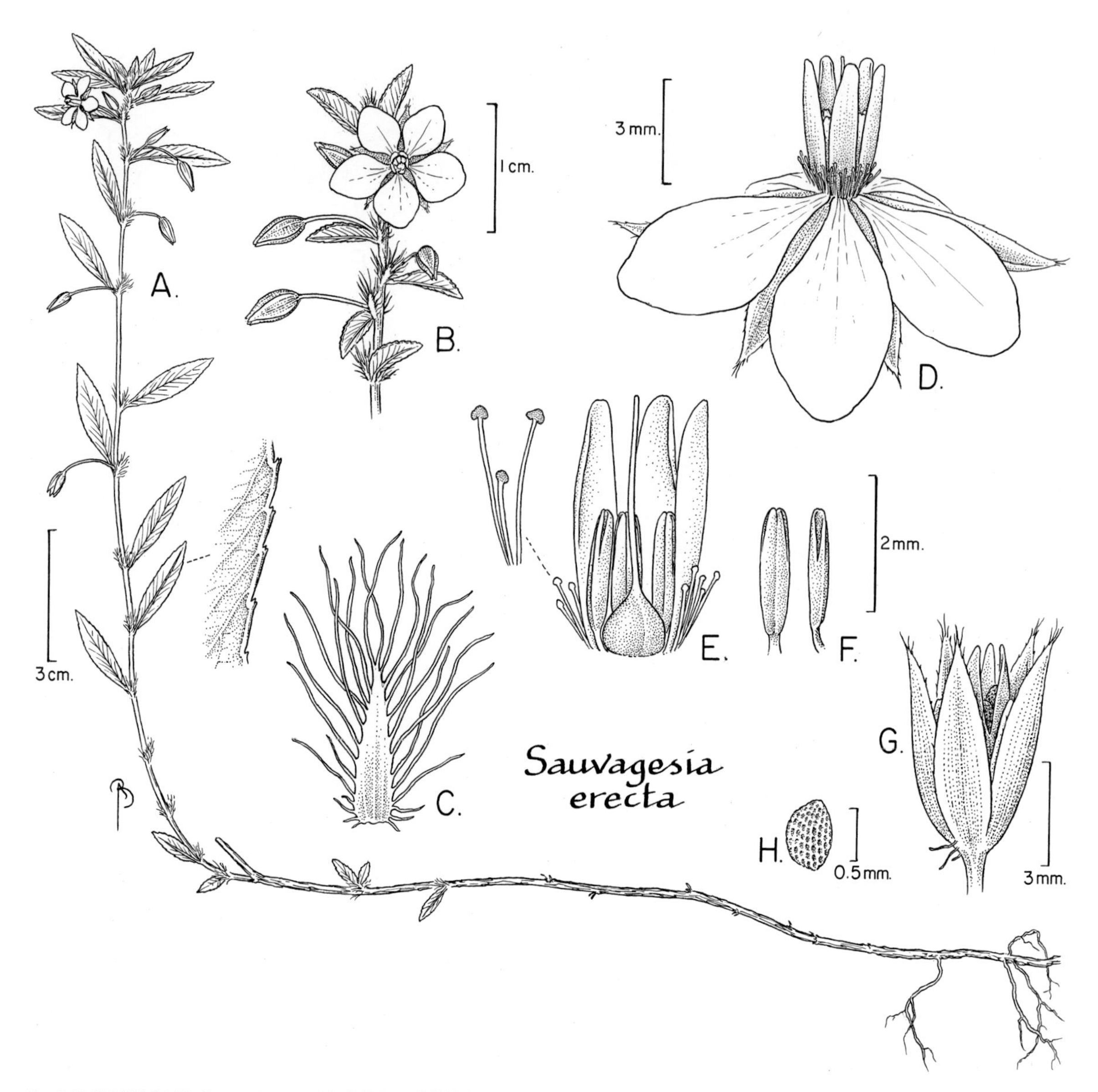

FIG. 242. OCHNACEAE. *Sauvagesia erecta* (A, C–F, *Pursell 8220* from Venezuela); B, *Mori 18387*; G, H, *Jansen-Jacobs 1292* from Guyana). **A.** Habit of plant and detail of abaxial surface of leaf. **B.** Distal portion of stem with leaves, calyces, and flower. **C.** Stipule. **D.** Lateral view of flower. **E.** Detail of androecium and gynoecium with several outer staminodia, two inner staminodia, and two stamens removed (near left) and three staminodia from outer whorl (far left). **F.** Adaxial (left) and lateral (right) views of stamens. **G.** Lateral view of fruit with persistent sepals. **H.** Seed.

Sauvagesia aliciae Sastre subsp. **aratayensis** Sastre

Subshrubs, 0.3–1.2 m tall. Stipules 10–12 mm long, the margins with branched and glandular cilia, some cilia nearly as long as stipule. Leaves subsessile; blades linear, 50–120 × 3–5 mm, the apex mucronate, the mucro between 2 small teeth. Infloresences axillary, much reduced cymes. Flowers: sepals ovate, 5–6 × 1–2 mm, green, the apex ciliate; petals obovate, 5–6 × 3–4 mm, white; outer staminodia in 3 cycles, 1–1.5 mm long, filiform, the inner staminodia subrectangular, petaloid, 3–4 × 1 mm, covering fertile parts of flowers; fertile stamens 2–2.5 mm long; ovary subconic, the style 3 mm long. Capsules 5–6 mm long. Fl (Jan); occasional, on granitic outcrops.

Sauvagesia erecta L. FIG. 242

Perennial, often repent, herbs, 0.09–0.7 m tall. Stipules 5–8 mm long, with simple cilia along margins. Leaves: blades narrowly elliptic, 12–30 × 3–10 mm. Inflorescences axillary, much reduced cymes. Flowers: sepals lanceolate, 4–7 × 1–2.5 mm, green, the apex often aristate, the margins sometimes glandular near base; petals widely to narrowly obovate, 5–7 × 2–4 mm, white to rose-colored; outer staminodia in 3 cycles, filiform, 1–2 mm long, the inner staminodia subrectangular, 2–4 × 1 mm, covering fertile parts of flower; fertile stamens 2–3 mm long; ovary subconic, the style 4 mm long. Capsules 5–6 mm long. Fl (Apr, Jun, Aug), but probably fl and fr year round; common, in disturbed

Plates 97–104

Plate 97. MORACEAE. **a.** *Ficus guianensis* (*Mori et al. 22986*), syconia. **b.** *Ficus insipida* subsp. *scabra* (*Mori & Mitchell 18783*), syconia. [Photo by S. Mori] **c.** *Helicostylis tomentosa* (*Mori et al. 22963*), fruits. **d.** *Maquira guianensis* subsp. *guianensis* (*Mori et al. 22891*), fruits. **e.** *Naucleopsis guianensis* (*Mori et al. 24020*), fruits.

Plate 98. MORACEAE. **a.** *Perebea guianensis* subsp. *guianensis* (*Mori et al. 24109*), inflorescence and infructescence. **b.** *Perebea guianensis* subsp. *guianensis* (*Mori et al. 22673*), inflorescences. **c.** *Pseudolmedia laevigata* (*Mori et al. 22279*), fruits. **d.** *Trymatococcus oligandrus* (*Mori et al. 23124*), flowers. MYRISTICACEAE. **e.** *Iryanthera sagotiana* (*Mori et al. 20912*), pistillate flowers. **f.** *Iryanthera sagotiana* (*Mori et al. 24004*), staminate flowers.

Plate 99. MYRISTICACEAE. **a.** *Iryanthera tessmannii* (*Mori et al. 21000*), staminate flowers. **b.** *Osteophloeum platyspermum* (*Mori et al. 23893*), flowers and buds. **c.** *Virola kwatae* (*Mori & Gracie 24227*), dehisced fruit with seed surrounded by red aril and seed with aril partially removed. **d.** *Virola kwatae* (*Mori et al. 23976*), transverse section of fruit showing red aril surrounding seed with ruminate endosperm. **e.** *Virola michelii* (*Mori et al. 23346*), pistillate flowers. **f.** *Virola sebifera* (*Mori & Boom 15188*), fruits, some of which have dehisced to reveal seeds surrounded by red arils. [Photo by S. Mori]

Plate 100. MYRTACEAE. **a.** *Eugenia* sp. C (*Mori et al. 22947*), flower and buds. **b.** *Eugenia armeniaca* (*Mori et al. 23330*), trunk with exfoliating red bark. **c.** *Eugenia gongylocarpa* (*Acevedo-Rodríguez 4959*), fruit. **d.** *Eugenia mimus* (*Acevedo-Rodríguez 5011*), immature orange fruits and single, mature black fruit. **e.** *Eugenia patrisii* (*Mori et al. 24023*), flowers.

Plate 101. NYCTAGINACEAE. **a.** *Guapira salicifolia* (*Mori et al. 22787*), flowers. **b.** *Neea constricta* (*Mori & Gracie 24205*), flowers. **c.** *Neea mollis* (*Mori et al. 23747*), immature fruits. **d.** *Neea floribunda* (*Mori et al. 22686*), flowers.

Plate 102. OCHNACEAE. **a.** *Ouratea scottii* subsp. *scottii* (*Mori et al. 20956*), flowers and buds. OLACACEAE. **b.** *Heisteria densifrons* (*Mori et al. 23713*), immature fruits. **c.** *Heisteria cauliflora* (*Mori et al. 22281*), immature and mature fruit. **d.** *Chaunochiton kappleri* (*Mori et al. 24118*), wind-dispersed fruit subtended by expanded calyx. **e.** *Heisteria cauliflora* (*Mori et al. 22880*), flower and buds.

Plate 103. OLACACEAE. **a.** *Heisteria scandens* (*Mori & Gracie 21164*), fallen fruits with persistent calyces. **b.** *Minquartia guianensis* (Mori & Gracie 23199), transverse section of fruit; note white latex. **c.** *Minquartia guianensis* (*Mori & Gracie 18885*), flowers and buds. ONAGRACEAE. **d.** *Ludwigia octovalvis* (*Mori et al. 22068*), flower, bud, and immature fruit; note calyx lobes visible between petals. OXALIDACEAE. **e.** *Oxalis barrelieri* (unvouchered, photographed in Trinidad), flowers.

Plate 104. PASSIFLORACEAE. **a.** *Dilkea johannesii* (*Mori et al. 23957*), flower. [Photo by S. Mori] **b.** *Passiflora amoena* (unvouchered, photographed in the Kaw Mountains, French Guiana), cauliflorous flowers and bud.

Ficus guianensis a.

Ficus insipida subsp. scabra

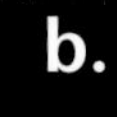

b.

Helicostylis tomentosa c.

Maquira guianensis subsp. guianensis d.

Naucleopsis guianensis e.

a.

Perebea guianensis subsp. guianensis

b.

Perebea guianensis subsp. guianensis

c.

Pseudolmedia laevigata

d.

Trymatococcus oligandrus

MYRISTICACEAE

e.

Iryanthera sagotiana

f.

Iryanthera sagotiana

Plate 98

Iryanthera tessmannii a.

Osteophloeum platyspermum  b.

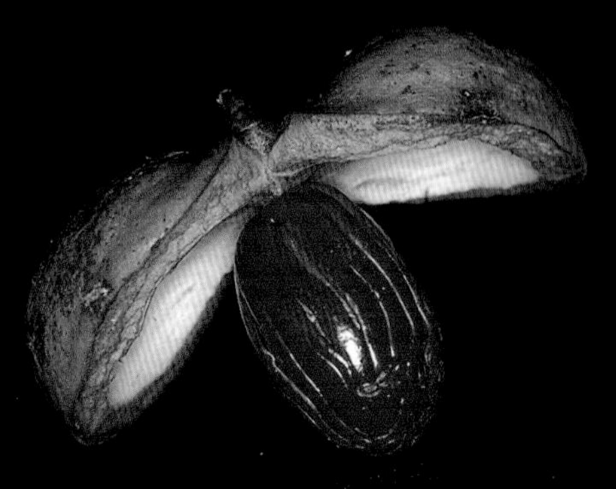

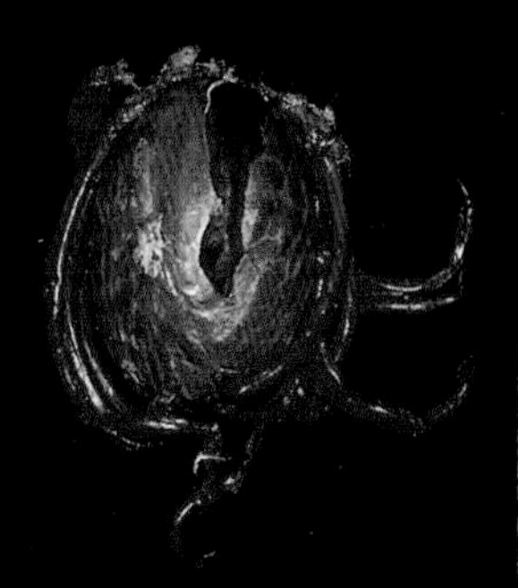

c. Virola kwatae

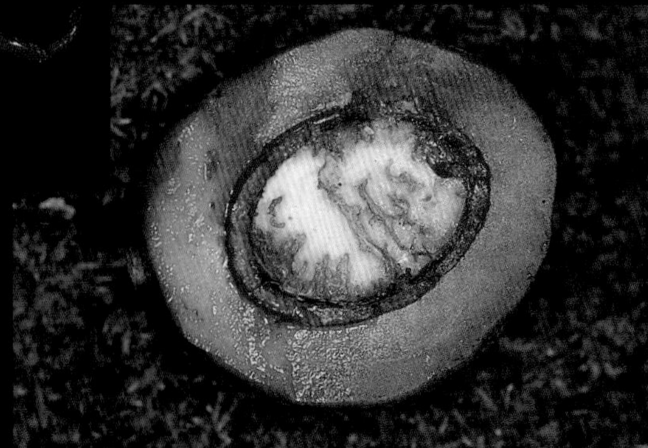

d.

Virola michelii e.

Virola sebifera f.

Eugenia sp. C a.

Eugenia armeniaca b.

Eugenia gongylocarpa c.

Eugenia mimus d.

Plate 100 Eugenia patrisii e.

Guapira salicifolia  a.

Neea constricta b.

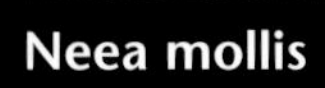

Neea mollis c.

Neea floribunda d.

OCHNACEAE

Ouratea scottii subsp. scottii a.

Heisteria densifrons b.

Heisteria cauliflora c.

OLACACEAE

Heisteria cauliflora e.

Chaunochiton kappleri d.

Plate 102

Heisteria scandens a.

Minquartia guianensis c.

Minquartia guianensis b.

ONAGRACEAE

Ludwigia octovalvis d.

OXALIDACEAE

e. Oxalis barrelieri

Dilkea johannesii **a.**

Passiflora amoena **b.**

areas, especially along airport landing field. *Erva de São Martinho* (Portuguese), *herbe St. Martin, zerb Saint Martin* (Créole).

Sauvagesia tafelbergensis Sastre

Annual herbs, 5–50 cm tall. Stipules 3–10 mm long, caducous, the margins with glandular cilia. Leaves: petioles 0–10 mm long; blades oblong, 15–90 × 7–14 mm, membranaceous, the apex apiculate. Inflorescences terminal, paniculate. Flowers: sepals ovate, 2–2.5 × 1 mm, the apex often aristate; petals obovate, 4 × 2 mm, acuminate, white; outer staminodia absent, the inner staminodia linguliform, 1.5 × 0.4 mm, not entirely covering fertile parts of flowers; fertile stamens 0.8–1 mm long; ovary subconic, the style 1 mm long. Fr (Sep); occasional, on granitic outcrops.

OLACACEAE (Olax Family)

Paul Hiepko

Trees, treelets, or erect, rarely scandent shrubs or lianas. Stipules absent. Leaves simple, alternate. Inflorescences axillary, rarely from stems. Flowers actinomorphic, generally bisexual, 3–7-merous; calyx usually cup-shaped, sometimes enlarged in fruit; petals free or connate, valvate, often hairy adaxially; disc sometimes present, rarely enlarged in fruit; stamens 1–3-seriate, sometimes in part staminodial; ovary superior, 2–5-locular near base (1-locular above); placentation of one pendulous ovule per locule. Fruits drupes with thin, fleshy pericarp, or pseudodrupes (external fleshy layer formed by accrescent calyx). Seeds 1; endosperm abundant; embryo minute, the cotyledons 2, 3, or 4.

Amshoff, J. H. 1966. Olacaceae. *In* A. Pulle (ed.), Flora of Suriname **I(1):** 262–272. E. J. Brill, Leiden.

Hiepko, P. 1993. Olacaceae. *In* A. R. A. Görts-van Rijn (ed.), Flora of the Guianas Ser. A, **14:** 3–35, 65–73. Koeltz Scientific Books, Koenigstein.

1. Stamens 3, staminodia 6. Fruits nearly covered by enlarged, fleshy calyx. . . . . *Dulacia.*
1. Stamens 5–15, staminodia absent. Fruits not covered by calyx, the calyx, however, sometimes enlarged.
    2. Stamens same number as petals.
        3. Calyx enlarged in fruit, forming large circular wing. . . . . *Chaunochiton.*
        3. Calyx not enlarged in fruit, the fruit ± covered by enlarged disc. . . . . *Cathedra.*
    2. Stamens double number of petals.
        4. Shrubs, treelets, lianas or trees without fenestrate trunks. Inflorescences fascicles. Petals free. Fruits with calyx covering basal part only or frill-like, expanded, suberect or reflexed.. . . . . *Heisteria.*
        4. Trees with fenestrate trunks. Inflorescences spikes. Petals united for lower half. Fruits without enlarged calyx. . . . . *Minquartia.*

CATHEDRA Miers

Shrubs or trees. Inflorescences axillary fascicles. Flowers: calyx cup-shaped, not or hardly enlarged in fruit; petals 5–6(7), free, barbate above middle adaxially; nectary disc hypogynous, much enlarged after anthesis to nearly covering fruit; stamens 5–6, epipetalous, the filaments flat, thickish; ovary 2-locular. Fruits pseudodrupes, covered halfway to almost completely by enlarged pericarp-like disc.

Cathedra acuminata (Benth.) Miers

Treelets or trees, (1.5)4–25 m × to 47 cm. Trunks with simple buttresses. Bark brown, smooth, the inner part orange. Leaves: blades lanceolate to oblong, slightly to strongly asymmetric, 6–15 × 2–7 cm; secondary veins in 4–6 pairs. Inflorescences fasciculate, usually in defoliate leaf axils, with 5–20(50) flowers. Flowers: petals 2–2.5 × 0.7–1 mm, fleshy, white to yellowish or greenish; nectary disc urceolate, 1 mm high at anthesis. Fruits 2 × 1.3–2 cm, ridged, covered by enlarged nectary disc except at very apex, yellow or orange-yellow. Fl (Sep), fr (Sep); in nonflooded forest.

CHAUNOCHITON Benth.

Glabrous trees. Inflorescences corymb-like panicles; peduncles short. Flowers: calyx cup-shaped, 5-dentate, small at anthesis, much enlarged in fruit; petals 5, free; stamens 5, epipetalous, the filaments filiform; ovary 5-ribbed lengthwise, 2-locular. Fruits drupes, ± globose, usually with 5–10 longitudinal, shallow ribs, the calyx accrescent, forming large, circular, papery wing.

Chaunochiton kappleri (Engl.) Ducke

PL. 102d; PART 1: FIG. 7

Trees, 6–30 m × 10–50(150) cm, deciduous when in flower. Bark red-brown, mottled with gray and whitish, with aroma of bitter almonds. Leaves: blades lanceolate- to oblong-ovate, 4–8 × 3–5 cm; secondary veins in 6–8 pairs. Inflorescences many-flowered panicles, to 5 cm long. Flowers: buds narrow-tubular

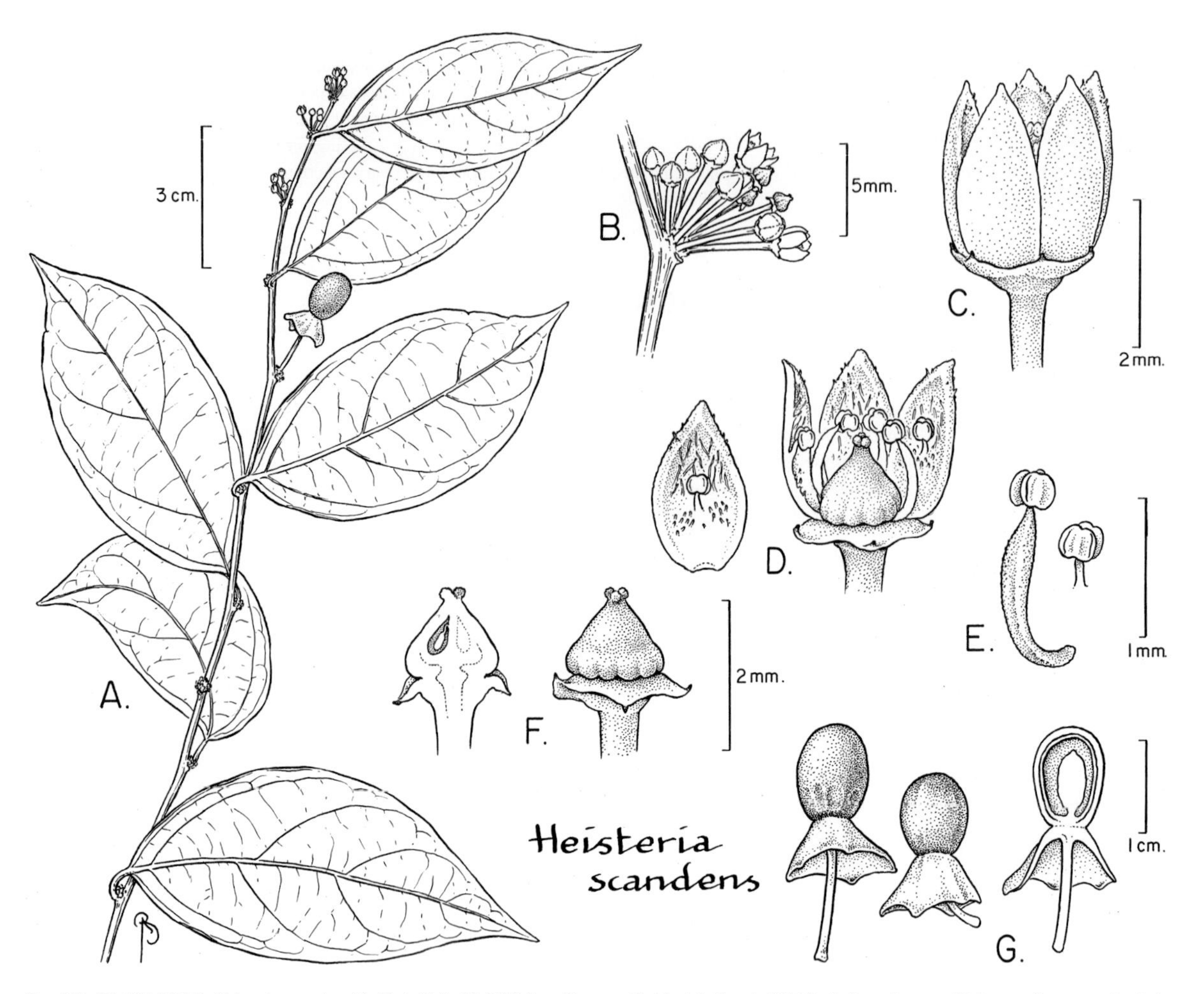

FIG. 243. OLACACEAE. *Heisteria scandens* (A–F, *A. C. Smith 2862* from Guyana; G, *Mori & Gracie 21164*). **A.** Part of stem with leaves, flowers and a fruit. **B.** Inflorescence. **C.** Lateral view of flower. **D.** Flower with two petals removed (right) and adaxial view of petal with adnate stamen (left). **E.** Two types of stamens. **F.** Lateral view of pistil (right) and medial section of pistil (left). **G.** Two fruits (left) and medial section of fruit with undeveloped seed (right).

with dilated subglobose apex; calyx 5-dentate; petals free, 7–10 × 0.5 mm, yellowish-greenish or whitish to purplish; filaments deep pink to crimson; ovary 1 mm long, the style red-violet. Fruits drupes, 8–10 mm diam., green, the calyx rotate, to 10 cm diam., the edge undulate. Fr (Sep, Nov); occasional, in nonflooded forest.

DULACIA Vell.

Trees. Inflorescences racemes or panicles. Flowers: calyx strongly enlarged in fruit, nearly entirely surrounding fruit; petals 6, connate ± half length; fertile stamens 3, adnate to fused bases of two petals, the filaments flat, usually hairy; staminodia 6, spatulate, usually hairy below, the upper half bifid and glabrous; ovary 3-locular. Fruits pseudodrupes, globose to ellipsoid, covered by enlarged, thin-fleshy calyx except for umbonate apex.

Dulacia guianensis (Engl.) Kuntze

Trees, 14–30 m × 15–40 cm. Trunks usually without buttresses. Bark rough, with strong unpleasant scent, the slash brownish with reddish streaks. Leaves: blades lanceolate to elliptic-oblong, slightly asymmetrical, 6–10 × 2–5 cm; secondary veins in (5)6(7) pairs. Inflorescences mostly 1–2-forked racemes or panicles. Flowers: petals fleshy, pale green or yellowish; upper part of ovary and lower half of style densely hairy. Fruits 2 × 1.5 cm, yellow or orange, with pleasant aroma. Fr (Aug); occasional, in nonflooded forest, the seeds reported to be edible.

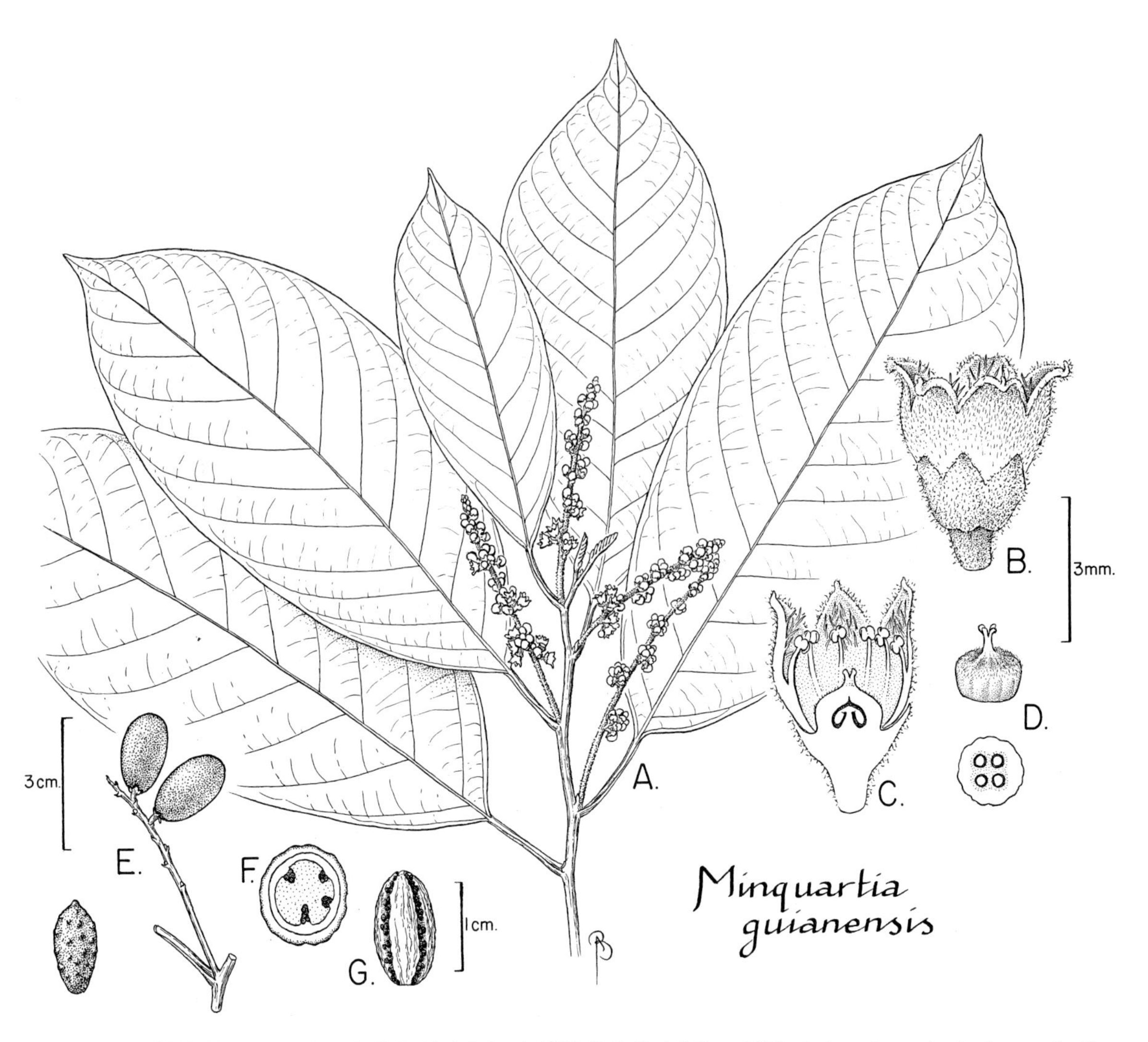

FIG. 244. OLACACEAE. *Minquartia guianensis* (A–D, *Mori & Gracie 18885*; E–G, *Mori & Boom 14760*). **A.** Apex of stem showing leaves and axillary inflorescences. **B.** Flower showing the cupular, lobed calyx and corolla. **C.** Medial section of flower showing the filament bases adnate to the corolla and the pendulous placentation from a free-central placenta. **D.** Pistil (above) and transverse section (below) of ovary showing the tetramerous stigma and compound ovary. **E.** Infructescence (right) and dried fruit (left). **F.** Transverse section of fruit. **G.** Lateral view of seed showing the canaliculate endocarp.

HEISTERIA Jacq.

Glabrous shrubs, treelets, or trees, rarely lianas. Inflorescences fascicles, with few to 20 flowers. Flowers: calyx 5 (6)-dentate or -lobed, usually markedly enlarged and brilliantly colored in fruit; petals 5(6), free, usually hairy adaxially; stamens usually 10 (12), some episepalous and others epipetalous; ovary 3-locular. Fruits drupes, globose to ellipsoid, subtended or partly covered at maturity by generally markedly enlarged, thin (i.e., slightly fleshy), shallowly to deeply 5(6)-lobed calyx.

1. Erect shrubs, treelets, or trees. Calyx in fruit not green or reflexed; pedicels ≤10 mm long in infructescences.
   2. Erect shrubs or treelets. Leaves 10–25 cm long.
      3. Calyx markedly enlarged, loosely enclosing fruit, lobed nearly to base, turning crimson red or purplish, sometimes pink. Drupes white to gray or yellowish at base, bluish at apex; pedicels 1–2 mm long in infructescences. . . . . *H. cauliflora.*
      3. Calyx only slightly enlarged under fruit (ca. 2 mm high), cup-shaped, only surrounding base of drupe, greenish or dull red. Drupes pale yellow to orange throughout; pedicels 1.5–4 mm long in infructescences. . . . . *H. densifrons.*
   2. Trees. Leaves 4.5–10 cm long. . . . . *H. ovata.*
1. Lianas or scandent shrubs. Calyx in fruit green, reflexed; pedicels 12–15(20) mm long in infructescences. . . . *H. scandens.*

Heisteria cauliflora Sm. PL. 102c,e

Shrubs or treelets, 1.5–6 m × 6 cm. Leaves: blades usually oblong-lanceolate, 12–25 × 3.5–9 cm; secondary veins of basal, curved-ascending pair and 10–14(20) distal pairs, departing from midrib at wide angle, then running straight to margin, ± parallel. Inflorescences fascicles, often from defoliate leaf axils, sometimes from stem; bracts numerous, forming small cushions. Flowers: calyx 5-lobed nearly to base; petals 5, 3 mm long, hairy adaxially; stamens 10. Drupes ellipsoid, 9–12 × 5–7 mm, white to gray or yellowish at base, bluish at apex, the enlarged calyx 5-lobed nearly to base, ca. 30 mm diam. when expanded, loosely enclosing drupes, pink, turning orange or usually red; pedicels 1–2 mm long. Fl (Feb, Apr), fr (Feb, May, Nov); in nonflooded forest.

Heisteria densifrons Engl. PL. 102b

Shrubs, treelets, or trees, 2–5 m × 5–15 cm. Leaves: blades ovate to oblong, 10–16 × 3–8 cm; secondary veins in 6–9 pairs, arcuately ascending. Inflorescences fascicles, with (1)3–6 flowers. Flowers: calyx shortly 5(6)-dentate, 2 mm diam.; petals 5(6), white, densely hispidulous-hairy adaxially; stamens 10(12). Drupes ellipsoid-oblongoid, 10–11 × 6–8 mm, pale yellow to orange throughout at maturity, the calyx becoming thickish, but only slightly enlarged and including only base of drupe, not deeply lobed, greenish to dull red; pedicels 1.5–4 mm long. Fl (Feb), fr (Feb, May, Jun, Aug, Oct); in nonflooded forest. *Mamayawé* (Créole).

Heisteria ovata Benth.

Trees in our area, to 20 m tall. Leaves: blades ovate to elliptic, 4.5–10 × 1.5–6 cm; secondary veins in ca. 6 pairs. Inflorescences axillary or along stems, fascicles; pedicels slender, 3–5 mm long. Flowers: calyx cup-shaped, acutely 5-lobed for 1/2 length; petals 5, somewhat pubescent adaxially; stamens 10. Drupes ovoid, 8–10 × 5–7 mm, the enlarged calyx 10–20 mm diam. when expanded, with undulate to lobed margin, red; pedicels 4–8 mm long. Fr (Aug); known from a single collection from our area; in nonflooded forest.

Heisteria scandens Ducke FIG. 243, PL. 103a

Lianas or scandent shrubs. Leaves: blades ovate to elliptic, 6–10 × 3–5 cm; secondary veins in 1(2), curved-ascending basal pair(s) and 3–4(5) upper short pairs. Inflorescences fascicles, with 8–20 (50) flowers; pedicels slender, 6–8 mm long. Flowers: calyx 5-lobed for 1/2 length; petals 5, laxly hairy or subglabrous adaxially; stamens 10, unequal in length, the filaments liguliform. Drupes globose or obovoid, slightly 10-sulcate, shiny, 9–12 mm diam., bright orange or cherry red at maturity, the enlarged calyx green, very shallowly lobed, strongly reflexed, 12–14 mm diam. when expanded; pedicels 12–15 (20) mm long. Fl (Dec), fr (Feb); occasional, in nonflooded forest.

MINQUARTIA Aubl.

Trees. Young parts covered with rusty pubescence. Trunks usually fenestrate, usually with buttresses. Inflorescences shortly pedunculate spikes, generally simple, rarely with a few, short branches; pedicels nearly absent. Flowers: calyx 5(6)-denticulate, persistent, not enlarging; petals (4)5–6(7), connate for lower half; stamens generally 10, sometimes 15, some episepalous and others epipetalous, the filaments filiform; ovary (3)4(5)-locular, hairy. Fruits drupes, often tuberculate externally, the pericarp with latex. Seeds with ruminate endosperm.

Minquartia guianensis Aubl.
FIG. 244, PL. 103b,c; PART 1: FIG. 10, PL. Ic

Trees, 10–30 m tall. Trunks with rather deep, long grooves, the older trees sometimes perforate, ± fenestrate. Bark brownish-gray, sometimes with sparse white latex when cut, the sapwood yellow, hard. Stems and all young parts usually ± rusty-puberulent, the hairs often stellate or dendritic. Leaves: blades oblong to elliptic, 8–25 × 3–12 cm, glabrous adaxailly, often puberulous, especially on veins abaxially; secondary veins in 7–13 pairs, rather straight-ascending, subparallel. Inflorescences spikes, 2–9 cm long, rusty-tomentellous. Flowers aromatic, cream-colored; petals connate into campanulate tube for 1–1.5 mm, the free lobes with long, erect hairs adaxially. Drupes ellipsoid or obovoid, 2–2.5 × 1.5 cm, yellowish-reddish initially, becoming purplish-black at maturity. Fl (Jun, Nov), fr (Aug, Sep, Dec); common, in nonflooded forest, the hard, heavy, extremely durable trunks are used for telephone poles and in construction. *Mecouart, méquoi, minquart, païcoussa rouge* (Créole).

ONAGRACEAE (Evening Primrose Family)

Maria Lúcia Kawasaki

Herbs or shrubs. Stipules small and inconspicuous, or absent. Leaves simple, alternate. Inflorescences of solitary flowers. Flowers actinomorphic, bisexual; calyx lobes usually 4 or 5; petals 4 or 5; stamens twice as many as calyx lobes, in two whorls; ovary inferior, usually 4- or 5-locular; placentation axile, with many ovules per locule. Fruits usually loculicidal capsules. Seeds numerous.

Jonker, F. P. 1942. Oenotheraceae. *In* A. Pulle (ed.), Flora of Suriname **III(2):** 13–34. J. H. De Bussy, Amsterdam.

Lindeman, J. C. 1986. Oenotheraceae or Onagraceae. *In* A. L. Stoffers & J. C. Lindeman (eds.), Flora of Suriname **III(1–2):** 507–510. E. J. Brill, Leiden.

Zardini, E. M. & P. H. Raven. 1991. Onagraceae. *In* A. R. A. Görts-van Rijn (ed.), Flora of the Guianas Ser. A, **10:** 1–44. Koeltz Scientific Books, Koenigstein.

LUDWIGIA L.

Herbs or shrubs. Inflorescences: bracteoles 2. Flowers: calyx lobes usually 4 or 5, persistent; petals 4 or 5, free, deciduous, usually yellow; stamens twice as many as calyx lobes; ovary 4- or 5-locular, the stigma capitate; ovules pluriseriate or uniseriate in each locule. Fruits loculicidal capsules.

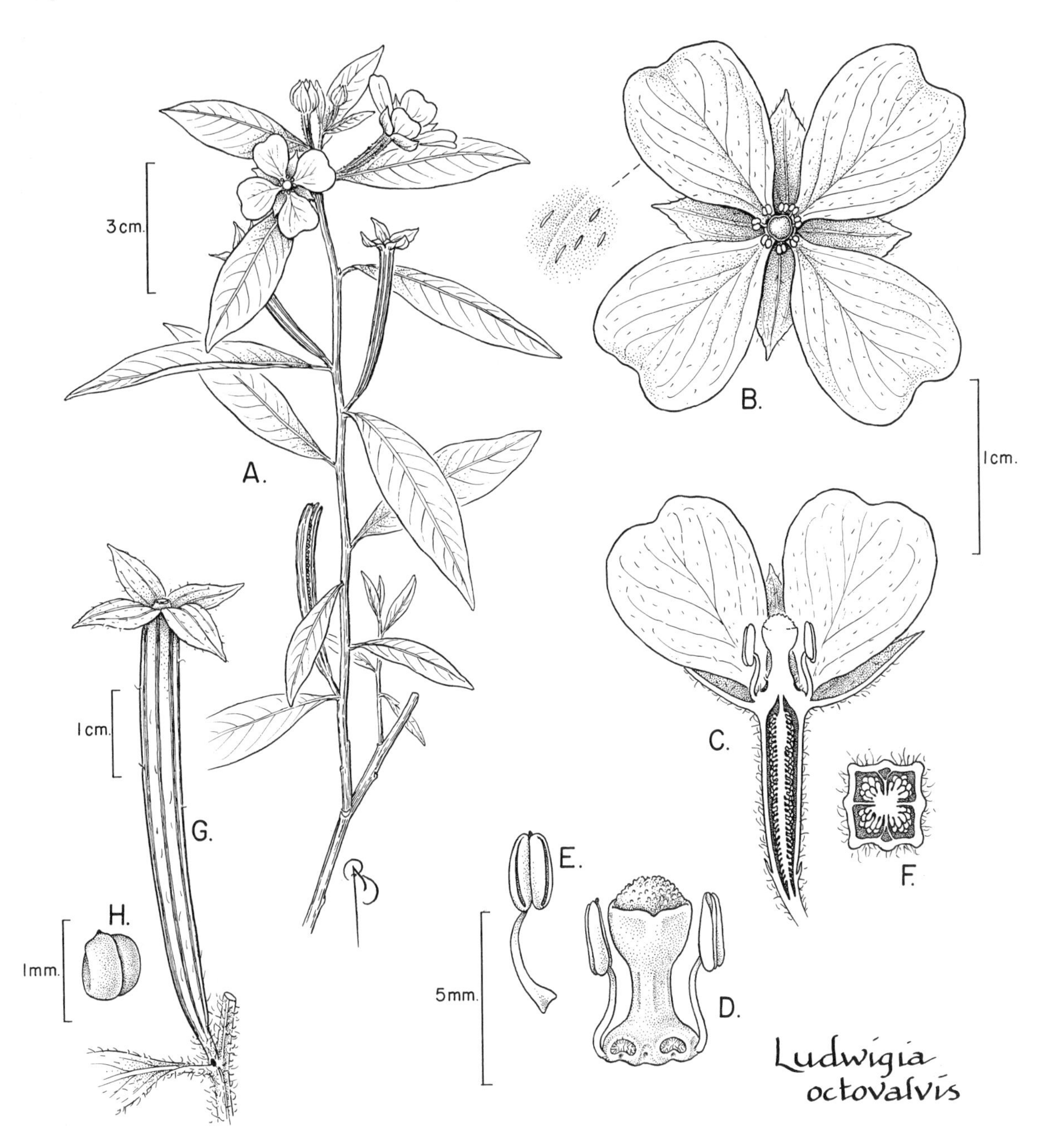

FIG. 245. ONAGRACEAE. *Ludwigia octovalvis* (*Fosberg 54099* from St. Croix, V.I.). **A.** Part of stem with leaves, flowers, and fruits. **B.** Apical view of flower with detail of petal surface. **C.** Medial section of flower. **D.** Detail of stamens and pistil. **E.** Stamen. **F.** Transverse section of ovary. **G.** Fruit. **H.** Seed.

1. Calyx lobes 4. Seeds pluriseriate in each locule.
   2. Bracteoles conspicuous, to 20 mm long. . . . . *L. foliobracteolata.*
   2. Bracteoles inconspicuous, to 1 mm long.
      3. Plants glabrous. Fruits quadrangular. . . . . *L. erecta.*
      3. Plants pubescent. Fruits terete. . . . . *L. octovalvis.*
1. Calyx lobes 5. Seeds uniseriate in each locule.
   4. Plants glabrous or nearly so. . . . . *L. dodecandra.*
   4. Plants distinctly pubescent.
      5. Leaf blades lanceolate, ≤1 cm wide. Calyx lobes 7–10 mm long. . . . . *L. leptocarpa.*
      5. Leaf blades elliptic to narrowly elliptic, 1.5–3 cm wide. Calyx lobes 3–5 mm long. . . . . *L. affinis.*

Ludwigia affinis (DC.) Hara

Herbs or shrubs. Plants with ferruginous or ochraceous pubescence. Leaves: petioles 2–10 mm long; blades elliptic to narrowly elliptic, 3–8 × 1.5–3 cm, pubescent, especially on veins, the base attenuate, the apex acute to acuminate. Inflorescences: pedicels 0–5 mm long, the bracteoles triangular, to 1 mm long. Flowers: hypanthium ca. 10 mm long, villous; calyx lobes 5, lanceolate, 3–5 mm long. Fruits terete, 20–30 × 2–4 mm, villous. Seeds uniseriate in each locule. Jul (fl). *Girofle d'eau* (Créole).

Description based on collections of this species from other localities archived at SP and determined by Raven & Zardini. The specimen from our area (*Granville 3137*, CAY) not examined.

Ludwigia dodecandra (DC.) Zardini & P. H. Raven

Herbs or shrubs. Plants glabrous or nearly so. Leaves: petioles 2–10 mm long; blades elliptic to narrowly elliptic, 3.5–12 × 1.5–4 cm, the base attenuate, the apex acute to acuminate. Inflorescences: pedicels 0–5 mm long, the bracteoles triangular, to 1 mm long. Flowers: hypanthium ca. 10 mm long; calyx lobes 5, lanceolate, 3–4 mm long. Fruits terete, 15–20 × 1.5–2 mm. Seeds uniseriate in each locule. Fl (Feb, Sep, Oct); young fr (Feb, Jun); in moist areas along streams. *Girofle d'eau* (Créole).

Ludwigia erecta (L.) Hara

Herbs. Plants glabrous. Leaves: petioles 2–15 mm long; blades narrowly elliptic to lanceolate, 4–9 × 1–2 cm, the base attenuate, the apex acute to acuminate. Inflorescences: pedicels 0–5 mm long, the bracteoles triangular, ca. 0.5 mm long. Flowers: hypanthium 5–10 mm long; calyx lobes 4, ovate to lanceolate, 2–4 mm long. Fruits quadrangular, 10–20 × 2–3 mm. Seeds pluriseriate in each locule. Fl (Aug, Sep), fr (Sep); in nonflooded forest, in disturbed area. *Girofle d'eau* (Créole).

Fruits described from other collections at SP from outside our area determined by E. Zardini.

Ludwigia foliobracteolata (Munz) Hara

Shrubs. Plants densely villous on very young stems, becoming glabrous with age. Leaves: petioles 5–50 mm long; blades elliptic, ovate, or oblanceolate, 5–25 × 2–8 cm, pubescent abaxially, especially on veins, the base cuneate, the apex acute to acuminate. Inflorescences: pedicels 10–20 mm long; bracteoles suborbicular to ovate, to 20 mm long. Flowers: calyx lobes 4, ovate, 10–14 mm long. Fruits quadrangular, 6–22 mm long. Seeds pluriseriate in each locule.

Not collected from our area, but to be expected. Description modified from Zardini and Raven (1991).

Ludwigia leptocarpa (Nutt.) Hara

Herbs or shrubs. Plants with yellowish pubescence. Leaves: petioles 2–10 mm long; blades lanceolate, 3.5–7 × 0.5–1 cm, pubescent, especially on veins, the base attenuate, the apex acute to acuminate. Inflorescences: pedicels 0–10 mm long, the bracteoles triangular, ca. 0.5 mm long. Flowers: hypanthium 10–15 mm long, villous; calyx lobes 5, lanceolate, 7–10 mm long. Fruits terete, 15–40 × 2.5–3 mm, villous. Seeds uniseriate in each locule. Fl (Oct), fr (Feb, Oct).

Description based on collections from other localities archived at SP and determined as this species by Raven & Zardini.

Ludwigia octovalvis (Jacq.) P. H. Raven FIG. 245, PL. 103d

Herbs or shrubs. Plants with yellowish to ochraceous pubescence. Leaves: petioles 2–10 mm long; blades lanceolate, 3.5–10 × 0.8–2 cm, pubescent, especially on veins, the base cuneate, the apex acuminate. Inflorescences: pedicels 0–5 mm long, the bracteoles triangular, to 1 mm long. Flowers: hypanthium 10–15 mm long, pubescent; calyx lobes 4, lanceolate, 5–8 mm long. Fruits terete, 20–30 × 2–4 mm, pubescent. Seeds pluriseriate in each locule. Fl (Aug, Sep, Oct), fr (Aug, Sep, Oct); nonflooded moist forest. *Girofle d'eau, giroflée, poivre sauvage* (Créole).

OXALIDACEAE (Wood-sorrel Family)

John D. Mitchell

Herbs or subshrubs (sometimes trees outside of our area). Stipules adnate to petiole or absent. Leaves 3(4)-foliolate (distinctly pinnate or palmate in some taxa outside of our area), alternate, or apparently basal. Inflorescences axillary, cymose, of solitary flowers, or sometimes umbelliform. Flowers actinomorphic, bisexual, pentamerous; calyx imbricate; corolla convolute, rarely imbricate; stamens usually 10, ± bicylic, the antepetalous cycle shorter, the filaments basally connate, the anthers longitudinally dehiscent; ovary superior, 5-locular, the styles 5, free, the stigmas capitate, often bifid; placentation axile, the ovules 1–5 per locule, epitropous, ± pendulous. Fruits loculicidal capsules. Seeds bitegmic, with basal aril breaking elastically to eject seed.

Görts-van Rijn, A. R. A. 1986. Oxalidaceae. *In* A. L. Stoffers & J. C. Lindeman (eds.), Flora of Suriname **III(1–2):** 516–517. E. J. Brill, Leiden.

Jonker, F. P. 1942. Oxalidaceae. *In* A. Pulle (ed.), Flora of Suriname **III(2):** 44–55. J. H. De Bussy, Amsterdam.

Lourteig, A. 1980. Oxalidaceae. *In* R. E. Woodson Jr. & R. W. Schery and collaborators (eds.), Flora of Panama. Ann. Missouri Bot. Gard. **67:** 823–850.

OXALIS L.

Herbs or subshrubs. Leaves 3(4)-foliolate, pinnate or palmate. The rest of description same as for family description.

Oxalis barrelieri L. FIG. 246, PL. 103e

Herbs to subshrubs, to 1 m tall, sparsely to densely pubescent. Roots fibrous, branched. Stems green to violaceous, leafy. Leaves: petioles 2–3.2 cm long, sparsely pubescent, the terminal petiolule 7–10 mm long, the lateral petiolules ca. 0.5 mm long; blades of terminal leaflet larger than lateral ones, the blades suborbicular to narrowly ovate, elliptic, oblong, or rarely obovate, 2.2–3.4 × 1–2.5

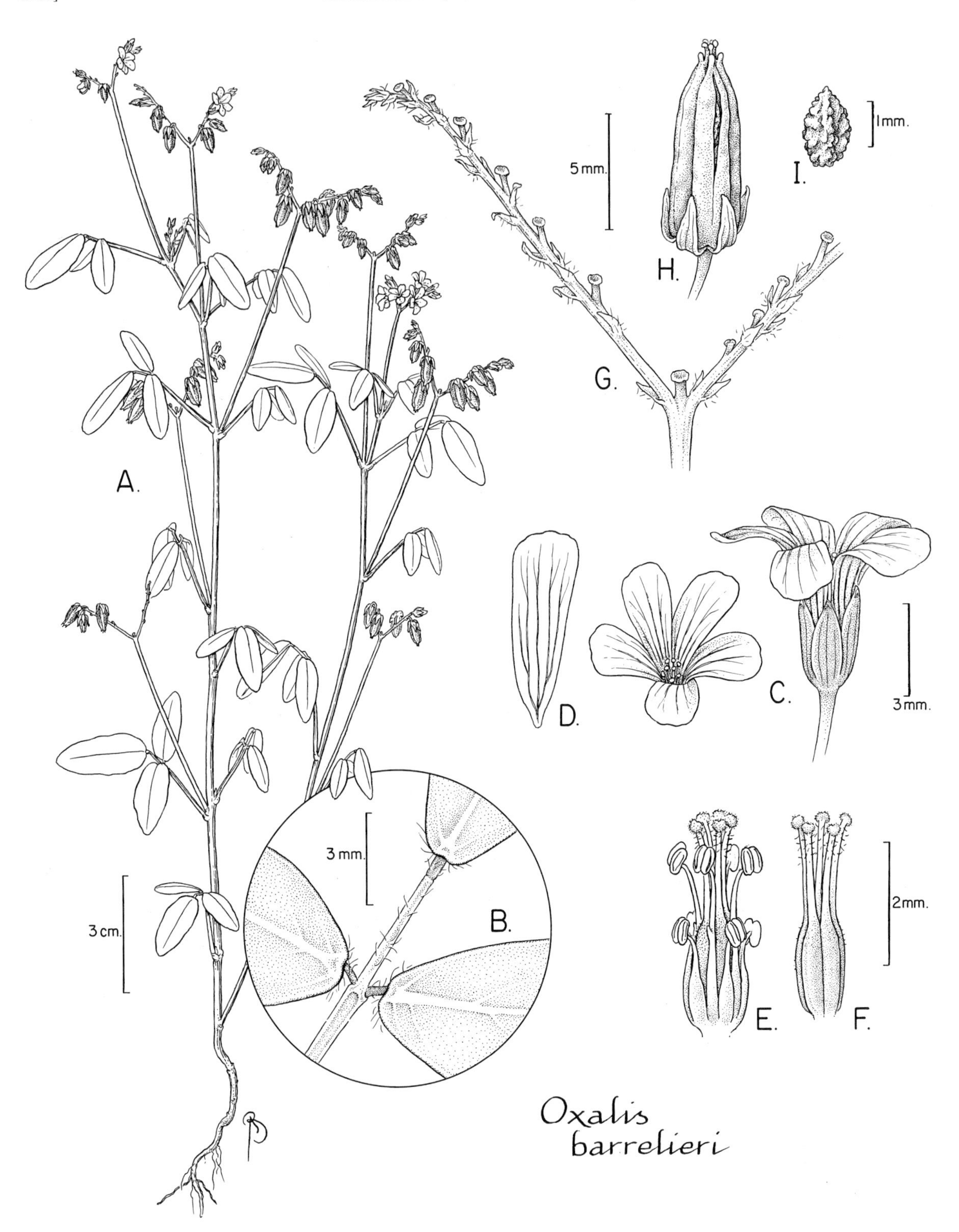

FIG. 246. OXALIDACEAE. *Oxalis barrelieri* (A, H, I, *Mori & Gracie 18266*; B–G, *Mori & Gracie 18431*). **A.** Plant with leaves, flowers, and fruits. **B.** Detail of part of leaf. **C.** Oblique-apical (left) and lateral (right) views of flower. **D.** Adaxial view of petal. **E.** Lateral view of flower with calyx and corolla removed showing dimorphic stamens. **F.** Lateral view of pistil. **G.** Infructescence after fruits have dehisced and fallen. **H.** Lateral view of fruit. **I.** Seed.

cm, chartaceous, sparsely pubescent abaxially, glabrous adaxially, the base rounded or cordate, the apex rounded or obtuse, the margins entire; midrib slightly prominent to prominent abaxially, flattened adaxially, the higher order venation obscure adaxially. Inflorescences bifid, 6–12.5 cm long; peduncle 5–7 cm long; pedicels 1.4–2.5 mm long. Flowers: sepals narrowly elliptic, oblong or ovate, 2.5–4 mm long; petals obovate, 6–7 mm long, pale pink to rose, yellow at throat; anthers yellow, 0.4–0.5 mm long. Capsules ovoid or obovoid-oblong, 5-angled, 6–7 mm long. Fl (Jun, Oct, Dec), fr (Jun, Oct, Dec); weed growing in disturbed areas around habitations and along roadsides. *Pâte dentrifice, trèfle à quatre feuilles* (Créole)

PASSIFLORACEAE (Passion-flower Family)

Christian Feuillet

Vines or lianas, climbing by axillary tendrils (infrequently shrubs or trees outside our area). Stipules present, these sometimes caducous. Leaves simple and unlobed, simple and palmately 2–9-lobed, or rarely palmately or pedately compound, alternate. Inflorescences axillary, cymose, rarely racemose, or fasciculate, 1- to several-flowered. Flowers actinomorphic, perigynous, bisexual; floral receptacle narrowly tubular to saucer-shaped; calyx with 4–5 sepals; corolla with petals as many as and alternate with sepals; corona of 1 to several series of filaments, membranes, or scales attached on hypanthium; operculum present or absent; limen present or absent; nectar ring present or absent; stamens 5–8(10); ovary superior, borne on gynophore (at least in our area), 3–4-carpellate, unilocular, the styles 3–4, at least partially distinct; placentation parietal. Fruits capsules or berries.

Jansen-Jacobs, M. J. 1986. Passifloraceae. *In* A. L. Stoffers & J. C. Lindeman (eds.), Flora of Suriname **III(1–2):** 452–454. E. J. Brill, Leiden.
Killip, E. P. 1966. Passifloraceae. *In* A. Pulle (ed.), Flora of Suriname **III(1):** 306–327. E. J. Brill, Leiden.

1. Leaves with glands on petiole or on blade or, if glandless, then lobed. Tendrils simple, generally not deciduous. Sepals 5; petals 5 (when present); stamens 5, borne on androgynophore, the anthers dorsifixed; styles 3. . . . . . *Passiflora.*
1. Leaves generally glandless, simple and unlobed. Tendrils 3-parted at apex, often deciduous. Sepals 4; petals 4; stamens 8, borne on base of floral receptacle, the anthers basifixed; style 1, dividing into 4 branches near middle. . . . . *Dilkea.*

DILKEA Mast.

Lianas or, outside our area, shrubs. Tendrils 3-parted at apex, often deciduous. Stipules caducous. Leaves simple, entire, alternate, generally glandless. Inflorescences usually axillary, rarely terminal, fasciculate, with small bracts, or of solitary flowers. Flowers: receptacle very small; calyx opening in 2 phases, forming temporary calyx tube, the sepals 4; corolla white, the petals 4, free to base; corona in 2 series; operculum missing; limen missing; stamens 8, the anthers basifixed; ovary 4-carpellate, the style dividing into 4 branches near middle. Fruits berries, globose or ovoid, apiculate, the pericarp coriaceous. Seeds 2–10, large, not flattened, bean-shaped.

Dilkea johannesii Barb. Rodr. FIG. 247, PL. 104a

Lianas, glabrous throughout. Leaves: blades elliptic to oblanceolate, 12–30 × 6–8 cm, the base shortly attenuate, the apex acuminate. Inflorescences of 1 flower or fasciculate with up to 5 flowers. Flowers: perianth white; calyx with sepals about 3 cm long; petals subequal to sepals; corona in 2 series, the outer filamentose, 2–2.5 cm long, the inner filiform at base, spatulate-inflated above middle, margined with floccose, crispate threads. Fruits globose, ca. 5 cm diam., yellow. Fl (Sep); in nonflooded forest.

PASSIFLORA L.

Vines or lianas (outside our area also shrubs or trees). Tendrils simple, usually not deciduous. Stipules present, sometimes caducous. Leaves simple and then unlobed or palmately 2–9-lobed or palmately or pedately compound, alternate, often with glands on petiole and/or blade. Inflorescences mostly cymose; peduncle usually missing in our area; bracts and bractlets attached on pedicel, often grouped in involucre near base of flower. Flowers: receptacle narrowly tubular to saucer-shaped; sepals 5; petals as many as and alternate with sepals or missing; corona of 1 to many series of filamentous or membranous appendages; stamens 5, borne on androgynophore, the anthers dorsifixed; ovary borne on short gynophore, 3-carpellate, the styles 3, distinct. Fruits capsules or berries.

1. Leaves compound. Inflorescences clearly pedunculate, 2-flowered, with a terminal tendril. . . . . *P. cirrhiflora.*
1. Leaves simple. Inflorescences racemose and more than 2-flowered or sessile with axillary tendril and 1–2 flowers.
   2. Petioles without glands.
      3. Plants glabrous or short pubescent. Leaves 2-lobed.
         4. Leaf blades with glands. Pedicels 0.5–1.5 cm long. Fruits indehiscent, purplish black. . . . . *P. vespertilio.*
         4. Leaf blades without glands. Pedicels 1–6 cm long. Fruits dehiscent at apex when ripe, reddish-green. . . . . *P. rubra.*
      3. Plants long hirsute. Leaves 3-lobed. . . . . *P. foetida* var. *hispida.*
   2. Petioles with glands.
      5. Leaves 3-lobed.
         6. Stipules round at apex, the margins red, glandular. Leaf blades coriaceous, medium to dark green adaxially, glaucous abaxially, the margins red, glandular. Flowers purplish-blue. . . . . *P. garckei.*

6. Stipules acuminate at apex, the margins not red or glandular. Leaf blades chartaceous, both sides pale glaucous green, drying black, the margins not red or glandular. Flowers pale pinkish lavender. . . . . *P. exura.*

5. Leaves not lobed.
   7. Stems 4-angled, 4-winged. . . . . *Passiflora* sp. A.
   7. Stems terete or 3- or 5-angled, not winged.
      8. Petiolar glands usually 2, rarely 4, mostly attached at or below middle, always lateral.
         9. Plants glabrous or finely puberulent. Leaf blade margins entire or sometimes emarginate in front of submarginal glands. Inflorescence bracts linear-lanceolate, greenish.
            10. Floral tube cylindric. . . . . *P. glandulosa.*
            10. Floral tube campanulate. . . . . *P. variolata.*
         9. Plants ferrugineous. Leaf blade margins serrate. Inflorescence bracts ovate, deep orange to red. . . . . *P. coccinea.*
      8. Petiolar glands 1–2, attached at apex, always adaxial.
         11. Leaf blades coriaceous, pubescent abaxially. Flowers white. . . . . *P. candida.*
         11. Leaf blades not coriaceous, glabrous. Flowers red or pink or greenish white with purple markings.
            12. Leaf blade margins slightly undulate. Flowers greenish white with purple markings. . . . . *P. plumosa.*
            12. Leaf blade margins flat. Flowers red or pink.
               13. Inflorescences ≤5 cm long. . . . . *P. amoena.*
               13. Inflorescences to ≥10 cm long. . . . . *P. saulensis.*

Passiflora amoena L. K. Escobar Pl. 104b

Lianas, glabrous throughout. Stems not conspicuously winged. Stipules setaceous, caducous. Leaves simple; petioles stout, 1–5 cm long, 2-glandular, the glands attached adaxially at apex; blades unlobed, elliptic to ovate, 9.5–13 × 4–7.5 cm, the base obtuse to rounded, the apex obtuse and abruptly acuminate to emarginate, the margins entire. Inflorescences cauline, racemose, <5 cm long, or occasionally flowers axillary; pedicels stout, 2–3.5 mm long; bracts <1.5 mm long. Flowers deep rich pink; floral tube narrowly tubular, 3–5 cm long; perianth 3.5–4 cm diam., the lobes spreading; corona in 3–5 series, the outermost filaments yellow, ca. 7 mm long, laterally compressed, dilated and bent outward 2 mm from base; operculum erect, borne at base of tube, ca. 5 mm long; limen lacking. Fruits ellipsoidal, 6–9 × 3–5 cm, with thin, coriaceous pericarp, green with pink bloom. Fl (Oct); in nonflooded forest.

Passiflora candida (Poepp. & Endl.) Mast.

Lianas, indumentum ferrugineous. Stems not conspicuously winged. Stipules caducous. Leaves simple; petioles stout, 2-glandular, the glands attached adaxially at apex; blades unlobed, broadly ovate or ovate-oblong, 8–18 × 7–14 cm (leaves of sapling or from stump shoots to 50 × 20 cm), coriaceous, the base rounded, the apex mostly short acuminate, the margins obscurely callous-denticulate. Inflorescences axillary, occasionally on wood, 1–2-flowered, sessile cymes. Flowers white; floral tube broadly funnel-shaped, 2.5–3.5 cm long; sepals and petals usually not spreading, 3–4 cm long; corona in 4 series, the outermost filaments yellow with brownish-red dots, ca. 25 mm long, liguliform, verrucose at margin, bearing short appendage near middle; operculum erect, borne 7 mm above base of tube, 4–5 mm long; limen lacking. Fruits ellipsoidal, 4–6 × 2–3 cm, with thick, coriaceous pericarp, green with rufescent indumentum. Fl (Oct); in nonflooded forest.

Passiflora cirrhiflora Juss.

Lianas, glabrous or puberulent throughout. Stems not conspicuously winged. Root system with proximal part fleshy. Stipules setaceous. Leaves compound; petioles to 10 cm long, 2-glandular, the glands attached ca. 1 cm above base; blades pedately 5–7-foliolate; leaflets with petiolules 2–5 mm long, the blades oblong, 5–8 × 2–4 cm, the base narrowed, the apex acute, acuminate, or long-aristate, the margins aristate-bidentate near base, otherwise entire. Inflorescences axillary, 2-flowered cymes, pedunculate, ending in tendril; pedicels 2–5 cm long; bracts linear-subulate, 1 cm long, 2-glandular at base, the bractlets similar to bracts. Flowers yellow, aging reddish brown or dull orange; floral tube broadly campanulate, 0.5–1 cm long; perianth to 8 cm diam., the lobes slightly spreading; corona in 3 series, the outer with liguliform filaments ca. 3 cm long, these verrucose or fimbriate at margin, yellow, with dark red at apex; operculum erect, plicate; limen tubular. Fruits subglobose, to 6 cm diam., green, with a whitish bloom. Seen in our area, but not collected.

Passiflora coccinea Aubl.
Fig. 248, Pl. 105a; Part 1: Figs. 4, 11, Pl. XIIb

Lianas, indumentum ferrugineous. Stems not conspicuously winged. Stipules narrowly linear, 4–6 mm long, the margins entire or minutely glandular-serrulate. Leaves simple; petioles to 3.5 cm long, 2-glandular at base; blades unlobed or obscurely 3-lobed, oblong, 6–20 × 3–12 cm, the base subcordate, the apex acute, acuminate, or almost obtuse, the margins double serrate or crenate. Inflorescences axillary, 1–2-flowered, sessile cymes; pedicels to 12 cm long; bracts ovate, 6–12 × 3.5–5 cm, concave, the margins glandular, crenate to sharply serrate, changing from green in bud to red or deep orange at anthesis to yellow-brown around fruit, enclosing flower buds as well as fruits, spreading for only a few hours at anthesis. Flowers scarlet or red; floral tube cup-shaped, 1.5–2 cm long; perianth to 12 cm diam., the lobes spreading; corona in 2 series, the outer filaments triangular-flattened, white, ca. 1 cm long, pointing inward to close tube; operculum recurved; limen membranous. Fruits ovoid, ca. 5 cm diam., pale yellow to orange. Fl (Jan, Feb, Jun, Aug, Sep); very common, in disturbed habitats, especially along airport road. *Liane serpent* (Créole), *maracujá poranga, maracujá do rato* (Portuguese), *Marie tambour, pomme liane sauvage* (Créole).

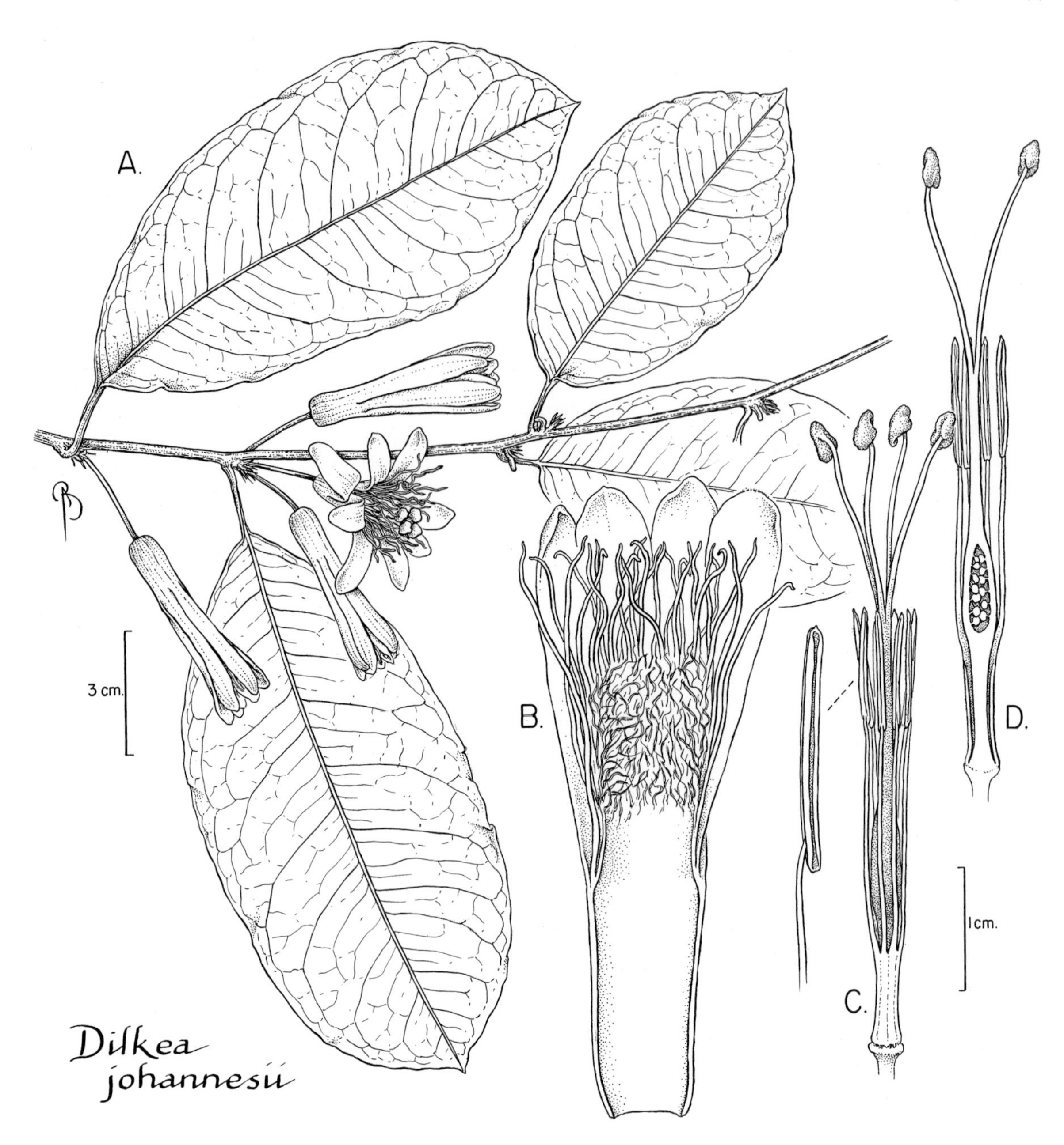

FIG. 247. PASSIFLORACEAE. *Dilkea johannesii* (*Mori et al. 23957*). **A.** Part of stem with leaves and flowers. **B.** Medial section of perianth. **C.** Androecium and gynoecium (right) and detail of lateral view of anther (left). **D.** Medial section of androecium and gynoecium.

Passiflora exura Feuillet

Lianas, glabrous throughout. Stems not conspicuously winged. Stipules triangular-falcate, 30 mm long, the base cordate on one side, the apex acuminate, the margins serrulate. Leaves simple; petioles 8–11 cm long, 2–4-glandular; blades 3-lobed, 8–11 cm long, the base cordate, subpeltate, the apices of lobes acute and shortly mucronate, the margins entire. Inflorescences axillary, 1–2-flowered, sessile cymes; pedicels 3.5 cm long; bracts ovate, 2.5–2.7 cm long, the base cordate, the apex acute and mucronate. Flowers pale pinkish lavender; floral tube campanulate; perianth ca. 13 cm diam., the lobes spreading; corona in several series, the outer filaments 3 cm long; operculum membranous, decurved; limen membranous. Fruits elliptic, 10 × 7.5 cm, with thick, spongy, orange-yellow mesocarp. Fl (Feb); in forest clearings.

Passiflora foetida L. var. **hispida** (DC.) Killip PL. 105b

Vines, the plants covered with long hirsute indumentum. Stems not conspicuously winged. Stipules semiclasping, deeply cleft into filiform, gland-tipped divisions. Leaves simple; petioles to 6 cm long, glandless; blades 3-lobed, the base cordate with well developed lateral lobes. Inflorescences axillary, 1–2-flowered, sessile cymes; pedicels to 6 cm long; bracts 3–4 cm long, 3–4-pinnatifid or

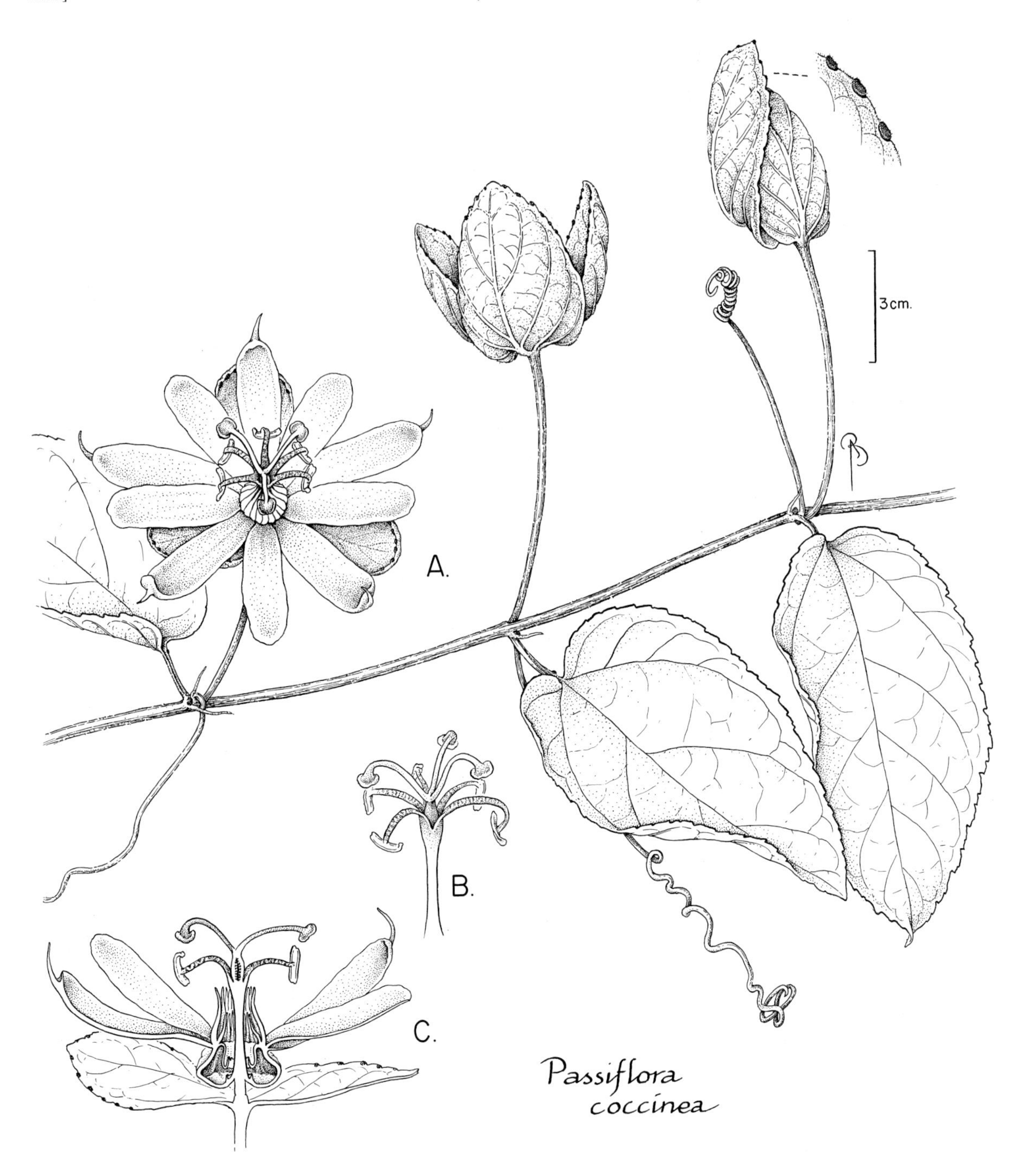

FIG. 248. PASSIFLORACEAE. *Passiflora coccinea* (*Mori et al. 20948*). **A.** Part of stem showing leaves, flower, stipules, and unbranched, axillary tendrils; note detail of glandular bract margin (right). **B.** Five-merous androecium and three-branched pistil surmounting androgynophore. **C.** Medial section of flower showing parietal placentation, androgynophore, and corona.

-pinnatisect, the segments filiform, gland-tipped. Flowers white, pink, or purple; floral tube cup-shaped; perianth to 2–5 cm diam., the lobes spreading; corona in several series, the outer filaments ca. 1 cm long; operculum annular, thick, triangular, 0.5 mm tall; nectar ring annular; limen erect, membranous, cup-shaped, denticulate at margin; ovary glabrous. Fruits spherical, 1–3 cm diam., yellow. Fl (Feb, Mar, Aug, Oct); common in open vegetation, especially along airport road.

Passiflora garckei Mast.

Lianas, glabrous throughout. Stems not conspicuously winged. Stipules semi-ovate or subreniform, 30–50 × 15–20 mm, the apex round, the margins red, glandular. Leaves simple; petioles to 10 cm long, 4–6-glandular; blades 3-lobed, 8–15 × 10–25 cm, coriaceous, the lobes oblong-lanceolate, the base subpeltate, truncate, or subcordate, the apices acute, acuminate or occasionally obtuse, the margins red, entire or minutely glandular-serrulate. Inflorescences axillary, 1–2-flowered, sessile cymes; pedicels to 6 cm long; bracts oblong-lanceolate, 6–10 × 4–6 mm, the base subcordate, the apex acuminate, borne 1–1.5 cm below flower. Flowers blue or purplish; floral tube campanulate, 6–8 mm long; perianth to 9 cm diam., the lobes spreading; corona in several series, the outer filaments filiform, ca. 3–3.5 cm long, blue-violet and yellowish; operculum deflexed at base, membranous, the upper 2/3 erect, filamentose; limen membranous. Fruits ovoid to spherical, to 5 cm diam., the pericarp thick, rigid, glaucous green. Known only by sterile collections in our area; in nonflooded wet forest.

Passiflora glandulosa Cav.

Lianas, glabrous throughout or finely puberulent on stem and abaxial leaf surface. Stems not conspicuously winged. Stipules linear-subulate or setaceous, caducous. Leaves simple; petioles to 2.5 cm long, 2-glandular at or below middle; blades unlobed, ovate-oblong to oblong-lanceolate, 6–15 × 4–10 cm, the base slightly cordate to slightly acute, the apex mucronulate, the margins entire or slightly emarginate in front of submarginal glands. Inflorescences axillary, 1–2-flowered, sessile cymes; pedicels to 8 cm long; bracts linear-lanceolate, 5–10 mm long, with 2–4 glands on margin at base. Flowers red; floral tube cylindric, 1.5–2.5 cm long; perianth to 11 cm diam., the lobes spreading; corona in 2 series, the outer filaments awl-shaped, ca. 1 cm long, white or pinkish; operculum recurved, attached 3 mm from base of tube; limen membranous. Fruits ovoid, 5–6 × 2.5–3 cm, yellow with white markings. Fl (Sep); in forest. *Pomme de liane sauvage* (Créole).

Passiflora plumosa Feuillet & Cremers

Lianas, glabrous throughout except ovary. Stems not conspicuously winged. Stipules linear, soon deciduous, about 1 mm long. Leaves simple: petioles 20–35 mm long, 2-glandular, the glands attached adaxially at apex; blades simple, unlobed, elliptic, 8–18 × 3–7.5 cm, the base rounded to oblique, the apex acute to acuminate, the margins minutely undulate. Inflorescences cauline, racemose, to 2 cm long; pedicels 7–8 mm long; bracts minute. Flowers greenish white; tube cylindric, 6–9 mm long; sepals about 16 × 5 mm, striped with purple; petals similar to the sepals; corona in 4 rows, the filaments of 3 outer rows white, with purple markings, flattened, decreasing toward inner ones from 6.5 to 3 mm long, the innermost row of white, recurved, branched, plumose filaments; operculum borne 4 mm from base of tube, erect, laciniate at apex. Fruits unknown. Known only by a sterile specimen (*Feuillet 2981*, CAY) in our area. Blooming in Aug just outside our area; in higher elevation forests.

Passiflora rubra L.

Vines, pubescent at least on abaxial leaf surface. Stipules linear-subulate, 5–7 mm long. Stems not conspicuously winged. Leaves simple; petioles 5–45 mm long, glandless; blades 2-lobed by abortion of middle lobe, 2–7 cm long along central midrib, 4–10 cm long along midribs of lobes, the base cordate, the apex of lobes usually acute, apiculate, the sinuses acute to truncate, the margins entire. Inflorescences axillary, 1–2-flowered, sessile cymes; pedicels 1–6 cm long; bracts lacking. Flowers greenish-white or pale yellow-green; floral tube wider than long, flat with short rim; perianth 2–6 cm diam., the lobes spreading; corona in 1–2 series, the outer filaments 1–1.5 cm long, white; operculum erect, plicate, borne on bottom of tube; nectar ring annular; limen annular. Fruits pyriform, 2.5–5 × 0.5–2 cm, with thick, spongy mesocarp, dehiscent for 1/3 to 2/3 length, reddish. Phenology not determined; uncommon, in disturbed forest.

Passiflora saulensis Feuillet

Liana, glabrous throughout. Stems not conspicuously winged. Leaves simple; petioles 2–3 cm long, 2-glandular, the glands attached adaxially at the apex; blades unlobed, ovate to suborbicular, 10–11 × 6–8 cm, the base rounded, the apex obtuse and slightly emarginate, the margins entire. Inflorescences cauline, racemose, 10–80 cm long; pedicels 18–24 mm long, articulate 6–8 mm from base; bracts minute. Flowers red; floral tube 2.5–4 cm long, not ventricose at base, funnel-shaped; perianth lobes 15–30 mm long; corona in 3 series, the outermost of flattened filaments, the apex rounded, slightly curved, 6 mm long; operculum attached 1.3 cm from base of tube, filamentose, 5–8 mm long; limen lacking. Fruits unknown. Fl (Sep); in mossy cloud forest.

Passiflora variolata Poepp. & Endl.

Vines or lianas, glabrous throughout or inconspicuously fine puberulent on stem and abaxial leaf surface. Stems not conspicuously winged. Stipules linear-subulate or setaceous, caducous. Leaves simple; petioles ca. 1 cm long, 2-glandular near base; blades unlobed, narrowly oblong, 5–10 × 2–5 cm, with small, submarginal glands, the base slightly cordate to slightly acute, the apex mucronulate, the margins entire. Inflorescences axillary, 2-flowered, sessile cymes; pedicels to 8 cm long; bracts linear-lanceolate, 5–8 mm long, with 2–4 glands on margin. Flowers orange red; floral tube broadly campanulate, 0.7 cm long; perianth to 9 cm diam., the lobes spreading; corona in 2 series, the outer tubular, ca. 1 cm long, including 2 mm long fringe of filaments, white; operculum recurved, 4–5 mm long, denticulate; limen membranous. Fruits spherical, 3–4 cm diam., sclerified, with a yellow and a green pole, line of color change clear, but very sinuous. Fr (Jun); in nonflooded forest.

Passiflora vespertilio L.

Vines, glabrous to slightly puberulent throughout. Stems not conspicuously winged. Stipules linear-setaceous, 3–5 mm long, sometimes caducous. Leaves simple; petioles glandless; blades 2-lobed by abortion of middle lobe, variable in shape, ranging from nearly as long as wide to >3x as wide as long, with glands between midribs of lateral lobes, the base rounded or truncate, the lobes divaricate, the apices acute to acuminate, the sinuses shallowly lunate or wanting, the upper margins often undulate. Inflorescences axillary, 1–2-flowered, sessile cymes; pedicels 0.5–1.5 cm long, 1–3 in fruit; bracts setaceous, 3 mm long, not verticillate, attached below middle of pedicel. Flowers white to yellowish-green; floral tube wider than long, ± flat, with short rim; perianth to 5 cm diam., the lobes spreading; corona in 2 series, the outer filaments liguliform, 1–1.5 cm long, greenish to yellowish-white; operculum erect, plicate, attached at bottom of tube, often dark purple-red; nectar ring annular; limen annular. Fruits spherical, 1–1.5(3.5) cm diam., the pericarp thin and soft, purplish-black. Fr (Mar, Aug); in forest gaps and open forest. *Marie tambour* (Créole).

Passiflora sp. A

Vines, glabrous throughout. Stems 4-angled, conspicuously 4-winged. Stipules narrowly elliptic. Leaves simple; petioles 2.5–3.5 cm long, 4-glandular near apex; blades unlobed, elliptic, 17–20 × 8–9 cm, the base rounded, the apex mucronulate, the margins entire. Inflorescences axillary, 1-flowered, sessile; pedicels to 5 cm long; bracts ovate, 4–5 cm long. Flowers seen only as young buds. Fruits unknown. Fl. buds (Sep); nonflooded forest.

This collection (*Mori et al. 23947*, NY) probably represents a new species related to *P. trialata* Feuillet & J. M. MacDougal but mature flowers are needed before it can be described.

PHYTOLACCACEAE (Pokeweed Family)

Robert A. DeFilipps and Shirley L. Maina

Herbs, subshrubs, or shrubs. Stipules present or absent. Leaves alternate, simple, entire, often petiolate. Inflorescences terminal or extra-axillary racemes, spiciform racemes, spikes or panicles; pedicels bracteate, sometimes bracteolate. Flowers bisexual, actinomorphic; tepals 4 or 5, free or often united below in a 4- or 5-lobed perianth, inconspicuous or petaloid; stamens 4–22, free or the filaments united at base, often borne on a hypogynous disc, the anthers 2-locular; ovary superior, 1- to 16-carpellate, the carpels free or partly to wholly connate, the styles, if present, as many as carpels, usually free, the stigma capitate, or sessile and penicillate; placentation basal in unicarpellate ovaries, axile in syncarpous ovaries, the ovules 1 per carpel. Fruits berries, achenes, capsules, samaras, or drupes. Seeds 1-numerous.

Stoffers, A. L. 1968. Phytolaccaceae. *In* A. Pulle & J. Lanjouw (eds.), Flora of Suriname **I(2):** 209–217. E. J. Brill, Leiden.

1. Plants with aroma of garlic or onion when crushed. Flowers sessile or subsessile; stamens 4; ovary 1-carpellate. Fruits retrorsely awned achenes. . . . . *Petiveria.*
1. Plants not with aroma of garlic or onion when crushed. Flowers distinctly pedicellate; stamens 5–14(22); ovary 2–16-carpellate. Fruits berries or drupes.
   2. Ovary 5–16-carpellate, the styles 10–16; stamens 9–14(22). Fruits smooth, many-seeded berries. . . . . *Phytolacca.*
   2. Ovary 2-carpellate, the style 1; stamens 5. Fruits echinulate, 1-seeded drupes. . . . . *Microtea.*

MICROTEA Sw.

Sprawling herbs, sometimes shrubby at base. Leaves petiolate or subpetiolate. Stipules absent. Inflorescences terminal or extra-axillary spikes or spiciform racemes; pedicels short, bracteate and 0- to 2-bracteolate. Flowers: tepals 5, shortly united below, persistent in fruit; stamens 5; ovary 2-carpellate, without hook-like awns, the style short to nearly absent, the stigmas 2, each sometimes 3-partite. Fruits small, thin walled, echinulate drupes. Seeds 1 per fruit, black, shiny.

Some authors place this genus in the Chenopodiaceae (Cronquist, 1981).

FIG. 249. PHYTOLACCACEAE. *Microtea debilis* (*Mori & Gracie 23818*). **A.** Habit and detail of leaf apex (upper right). **B.** Inflorescence. **C.** Lateral view of flower. **D.** Medial section of flower (right) and lateral view of ovary (below). **E.** Infructescence. **F.** Lateral view of fruit (left) and seed (right).

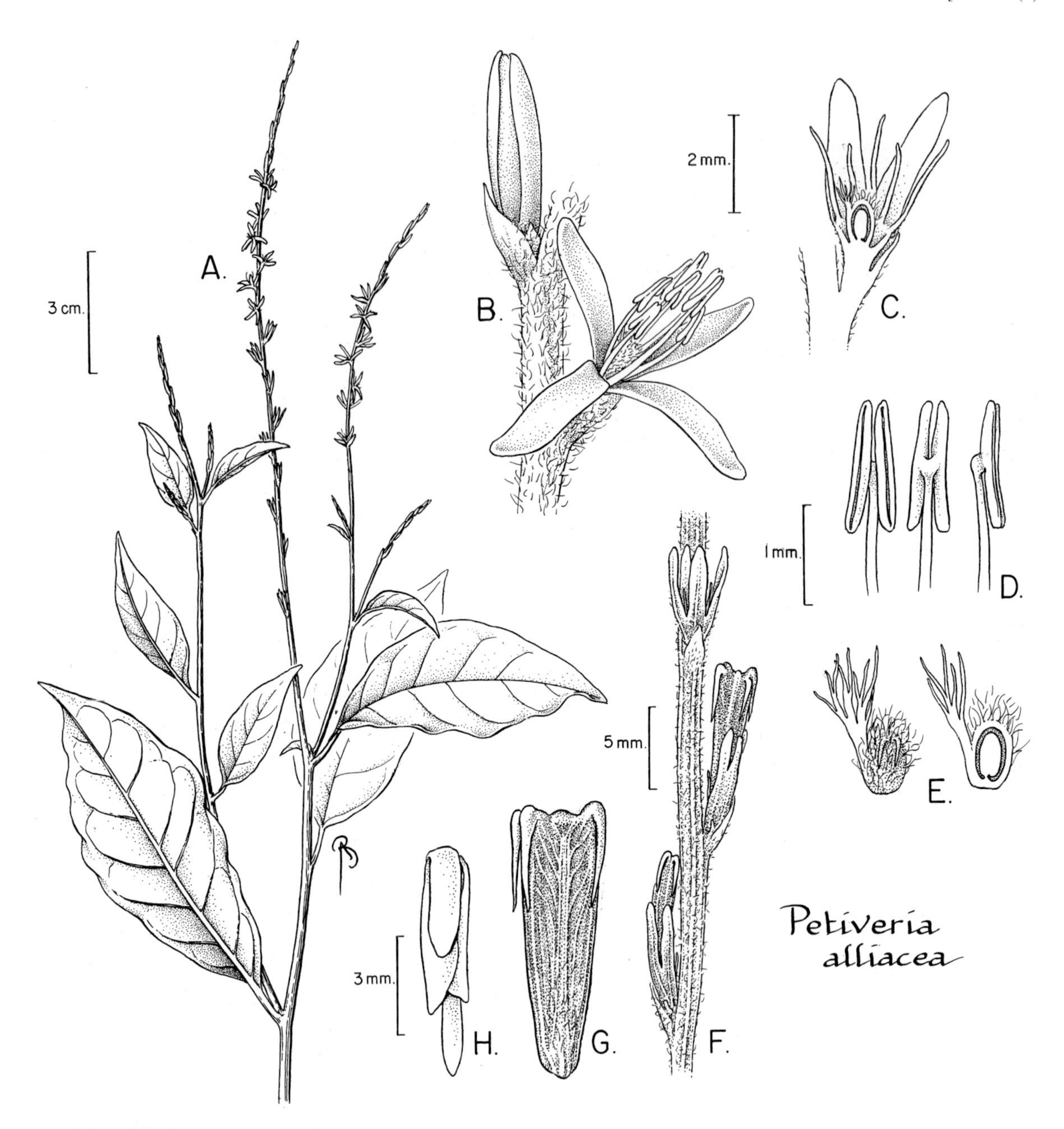

FIG. 250. PHYTOLACCACEAE. *Petiveria alliacea* (A, *Fishlock 133* from St. John, U.S. Virgin Islands; B, *Acevedo-Rodríguez 1863* from St. John, U.S. Virgin Islands; C–H, *Mori & Gracie 18246*). **A.** Upper part of plant with stems, leaves, and inflorescences. **B.** Detail of part of inflorescence. **C.** Medial section of flower with anthers fallen. **D.** Adaxial (far left), abaxial (middle left), and lateral (near left) views of anthers. **E.** Lateral view (left) and medial section (right) of pistil. **F.** Part of infructescence. **G.** Lateral view of fruit. **H.** Seed. (Reprinted with permission from P. Acevedo- Rodríguez, *Flora of St. John, U.S. Virgin Islands*, Mem. New York Bot. Gard. 78, 1996.)

Microtea debilis Sw. FIG. 249, PL. 105c

Annual, ascending to prostrate herbs, to 50 cm tall. Leaves: petioles to 3 cm long; blades (near base of plant) elliptic, ovate, or oblanceolate, to 8 × 3 cm, glabrous, the base broadly decurrent on to petiole. Inflorescences 10–25-flowered racemes, 1.5–4(5.5) cm long; bract seemingly only an enation, ovate, ca. 0.3 mm, the apex obtuse, green; bracteoles elliptic, scarious, ca. 0.8–1 mm long, often notched; pedicels 0.8–1.2 mm long. Flowers: tepals lanceolate or ovate, 0.5–1.1 mm long, white with green midrib; style short, the stigma 2-partite. Fruits spreading from inflorescence-axis, globose, 1–1.5 mm long, greenish, echinulate in honeycomb pattern. Seeds 1 per fruit, globose, ca. 1.3 mm diam., black, smooth, shiny. Fl (Aug, Sep). Weed of disturbed areas in non-flooded moist forest at Eaux Claires. *Alentou-case*, *entoucase* (Créole), *erva mijona* (Portuguese), *racine-pistache* (Créole).

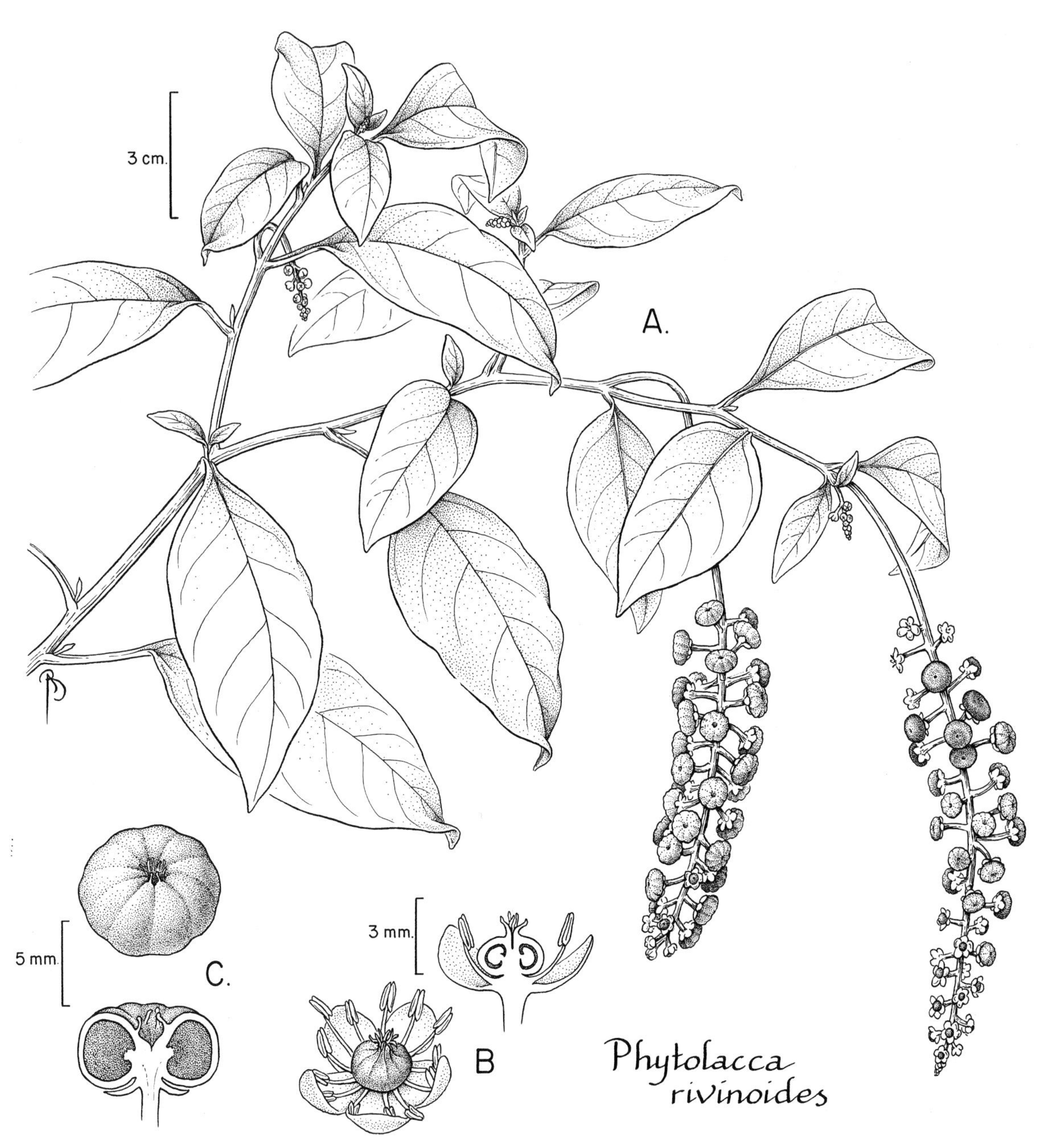

FIG. 251. PHYTOLACCACEAE. *Phytolacca rivinoides* (A, C, *Mori et al. 19065*; B, *Mori & Gracie 21186*). **A.** Apices of stems showing leaf- opposed racemose inflorescences. **B.** Oblique-apical view (left) and medial section (right) of flowers. **C.** Oblique-apical (above) and medial section (below) of fruits.

PETIVERIA L.

Herbs or subshrubs, slightly woody at base. Leaves shortly petiolate. Stipules present. Inflorescences elongate, extra-axillary or terminal spike-like racemes; pedicels distinct or scarcely developed, bracteate and bracteolate. Flowers: tepals 4, shortly united below, spreading in flower, persistent, enlarging and erect in fruit. Ovary 1-carpellate, with 4–6 retrorse hook-like awns; style absent; the stigma sessile, papillose, penicillate and laterally decurrent along ventral (flattened) side of ovary. Fruits 2-lobed achenes with 4–6 apical, retrorse awns. Seeds 1 per fruit, linear.

Petiveria alliacea L. FIG. 250; PART 1: FIG. 8

Perennial, erect herbs to 1–2 m tall, shrubby at base, with aroma of garlic or onion when crushed. Stems puberulent in lines between ribs. Leaves: petioles to 2 cm long; blades elliptic, obovate or ovate, to 16(20) × 7 cm, densely but very minutely pellucid-punctate, the abaxial surface glabrous or sparingly puberulent on veins, the base ± acute, the margins entire (but with minute bristle-tipped enations). Inflorescences laxly flowered spike-like racemes to 45 cm long; bracts lanceolate to deltate, 1–3 mm long, green puberulent; bracteole 1 mm long. Flowers weakly zygomorphic; tepals linear or oblong, 2.6–5 × 0.8–1 mm, prominently 3- to 5-veined, spreading and white or pink at anthesis, erect and greenish in fruit. Fruits linear or narrowly oblong, 8–10 mm long, cuneate, striate, puberulent, appressed to inflorescence-axis, with 4 apical, 2.5–4 mm long, retrorse awns. Fl (Jun), fr (Jun), common, in weedy habitats, epizoochorous. *Douvan-douvan, liane ail* (Créole), *mucura caa* (Portuguese), *ndongu ndongu* (Bush Negro), *radié pian* (Créole).

PHYTOLACCA L.

Herbs or shrubs. Leaves sessile or petiolate. Stipules absent. Inflorescences mostly terminal racemes or spikes; pedicels absent or present, 1–3-bracteate at base; bracteoles randomly attached along pedicel. Flowers: tepals 5, free, becoming reflexed and then deciduous in fruit; stamens 9–14(22); ovary 5- to 16-carpellate, unarmed, the carpels united, the styles 10–16, free, appressed-recurved. Fruits aggregations of one-seeded drupelets ("berries") or syncarpous 5–16-celled berries. Seeds 1 per locule, black, shiny.

Phytolacca rivinoides Kunth & Bouché FIG. 251, PL. 105d

Perennial, woody herbs or subshrubs to 3(5) m tall. Stems glabrous. Leaves: petioles to 7 cm long; blades ovate, elliptic, lanceolate, or oblong, to 21 × 7.5 cm, glabrous, calcium oxalate raphides in epidermis visible with lens, the base acute. Inflorescences many-flowered terminal or axillary racemes to 65(70) cm, simple (rarely with 1–2 long branches from near base); pedicels 5–15 mm long; bracts 1.5–4.2 mm long; bracteoles 1–2, usually attached randomly along pedicel but sometimes paired or subopposite, 0.4–1.5 mm long. Flowers: tepals ovate, 1.2–2.8 × 1.5 mm, white or pink, glabrous, becoming reflexed, deciduous in fruit. Fruits depressed-globose berries, 6–8 mm wide, dark purple or black, ribbed at carpel-margins. Seeds 8–12 per fruit, ovoid, biconvex, 1 × 1.3 mm. Weeds of disturbed areas. Fl (Feb, Aug), fr (Feb, Aug). *Bichouiac* (Créole).

PIPERACEAE (Pepper family)

Ara Görts-van Rijn

Herbs, subshrubs, shrubs, or treelets, sometimes scandent, rarely lianas. Stems of woody species usually with swollen nodes. Stipules present or absent. Leaves simple, alternate, opposite, or in whorls, when crushed often aromatic. Inflorescences terminal, axillary, or leaf-opposed (*Piper* and some species of *Peperomia*) spikes, with minute flowers, the spikes solitary or few together, or occasionally in panicles or umbels; floral bracts subtending flowers; pedicels absent but sometimes present outside our area. Flowers usually bisexual; calyx absent; corolla absent; stamens 2–5; ovary sessile or stipitate, unilocular, the stigmas 1–4; placentation basal, the ovule solitary. Fruits small drupes.

Kramer, K. U. & A. R. A. Görts-van Rijn. 1968. Piperaceae. *In* A. Pulle & J. Lanjouw (eds.), Flora of Suriname **I(2):** 414–421. E. J. Brill, Leiden.

Yuncker, T. G. 1968. Piperaceae. *In* A. Pulle & J. Lanjouw (eds.), Flora of Suriname **I(2):** 218–290. E. J. Brill, Leiden.

1. Small herbs, usually epiphytic or epilithic, occasionally terrestrial. Stems without swollen nodes. Leaves fleshy. Floral bracts round-peltate, glabrous. Stigma 1. . . . . *Peperomia.*
1. Larger herbs, subshrubs, shrubs, or treelets, occasionally scandent, trailing, or lianas. Stems with swollen nodes. Leaves not fleshy. Floral bracts various, glabrous or pilose, often fringed. Stigmas (2)3–4.
    2. Leaf blades not peltate. Spikes leaf-opposed, simple, solitary. . . . . *Piper.*
    2. Leaf blades peltate. Spikes axillary, in umbellate clusters. . . . . *Pothomorphe.*

PEPEROMIA Ruiz & Pav.

Small herbs, epiphytic, epilithic, or terrestrial, creeping, prostrate or erect. Stems without swollen nodes. Leaves fleshy. Spikes terminal, leaf-opposed, axillary, or occasionally paniculate, solitary or few together; floral bracts round-peltate, glabrous. Flowers numerous, usually loosely arranged; stamens 2; ovary sessile (in our area), the stigma single. Fruits globose, ellipsoid, cylindrical, sometimes with a slender beak, or apex obliquely scutellate or truncate.

1. Leaves peltate; blades with adaxial surface with erect, septate trichomes. . . . . *P. gracieana.*
1. Leaves not peltate; blades with adaxial surface glabrous or with trichomes but these not erect and septate.
    2. Leaves whorled, 2–4 at each node. . . . . *P. maguirei.*
    2. Leaves alternate, 1 at each node.
        3. Spikes in panicles. . . . . *P. pernambucensis.*
        3. Spikes solitary or few together.

4. Leaf blades small, 0.15–5 cm long, the apex emarginate, obtuse or rounded. Spikes to 6 cm long.
 5. Terrestrial, erect herbs. . . . . *P. pellucida.*
 5. Epiphytic or epilithic, repent or prostrate herbs.
  6. Leaf blades elliptic to obovate, with three pale, often reddish veins from base. . . . . *P. ouabianae.*
  6. Leaf blades orbicular, cordate, or somewhat elliptic, or slightly deltoid, without reddish veins from base.
   7. Leaf blades emarginate at apex. . . . . *P. emarginella.*
   7. Leaf blades obtuse to rounded at apex.
    8. Blades with veins obsolete. Peduncles shorter than spikes, without bracts. Fruits (sub)globose with oblique apex. . . . . *P. rotundifolia.*
    8. Blades with 3–5 distinct veins. Peduncles as long as or longer than spikes, with 1–2 bracts. Fruits ellipsoid with slender beak. . . . . *P. serpens.*
4. Leaf blades >5 cm, or if smaller, the apex acute or acuminate. Spikes 10–30 cm long.
 9. Leaf blades with waxy exudate present (obvious on at least dried leaves). Spikes usually 2–4 together. . . . . *P. macrostachya.*
 9. Leaf blades without waxy exudate. Spikes solitary, occasionally paired.
  10. Petioles glabrous; blades when dried not black-glandular dotted, the apex usually rounded, or obtuse, or occasionally acute. Fruits ellipsoid, the apex with slender beak. . . . . . *P. obtusifolia.*
  10. Petioles ciliate; blades when dried densely black-glandular dotted, the apex acute to acuminate. Fruits globose, the apex oblique with subapical stigma. . . . . *P. glabella.*

Peperomia emarginella (Sw.) C. DC.

Epiphytic or epilithic, small herbs. Stems repent, rooting at the nodes, glabrous. Leaves alternate; petioles 1–5 mm long, glabrous or with sparse hairs; blades almost orbicular, very small, 0.15–0.5 cm wide, bright green, thin fleshy, glabrous or with long, loose (red) hairs, the apex often emarginate, the margins slightly ciliate; veins 3 from base. Spikes terminal, solitary, white, to 1.5 cm long; peduncles pinkish, filiform, to 10 mm long. Fruits brown, ellipsoid. Known only by sterile collections from our area; mossy forest at summit of Mont Galbao.

Peperomia glabella (Sw.) A. Dietr.

Epiphytic herbs. Stems erect, hanging, or prostrate, rooting at nodes, red, brownish, or dark green, 5 mm thick when dried, with lines of hairs decurrent from nodes. Leaves alternate; petioles 5–10 mm long, ciliate; blades narrowly ovate to ovate, somewhat rhombic, 1.5–7 × 0.5–3.5 cm, thin fleshy, when dried with numerous black-glandular punctations, the base attenuate, the apex acute or acuminate, the margins ciliate toward apex; veins 3–5 from base. Spikes generally terminal, erect, solitary, green, 3–16 cm long; peduncles 5–15 mm long. Fruits globose with oblique apex. Fl (Jan, Jun, Jul, Aug, Oct, Nov, Sep, Dec), fr (Aug, Sep); common, at all elevations in our area. *Ti-mourou* (Créole).

Peperomia gracieana Görts — FIG. 252, PL. 106a

Terrestrial or sometimes epilithic herbs. Stems creeping, rooting at nodes, forming loose mats, slender, glabrous, some nodes bearing tiny stalked tubers. Leaves alternate, peltate; petioles 1–10 cm long, pale green; blades orbicular or widely elliptic, 1–3.5 × 1–4.5 cm, thin, almost pellucid when dried, adaxial surface with erect, septate trichomes; veins 6–8 from center. Spikes axillary, solitary, 4–6 cm long; peduncles ca. 15 mm long. Young fruits green, depressed globose, flat at apex. Fl (Sep), fr (Aug, Sep); in nonflooded forest.

Peperomia macrostachya (Vahl) A. Dietr.

Epiphytic herbs. Stems trailing or pendent, rooting at nodes, to 1 m or more long, 6 mm thick when dried, red speckled, especially at nodes, glabrous. Leaves alternate; petioles 5–20 mm long, red speckled, glabrous; blades broadly ovate, 4–11 × 1.5–6 cm, fleshy, coriaceous, waxy dendritic exudate obvious when dried, the base rounded, cordulate, or acute, the apex acute or acuminate, the margins somewhat ciliate; veins obscure. Spikes terminal, ascending, curved or pendent, usually 2–4 together but not in panicles, 20–30 cm long, yellow-green; peduncles to 15 mm long. Fruits cylindrical, the apex obliquely scutellate. Fl (Jan, Jul, Aug, Sep, Oct), fr (Aug); associated with ant nests.

Peperomia maguirei Yunck.

Epiphytic herbs. Fruiting stems erect, to 5 cm long. Leaves whorled, 2–4 at each node; petioles to 5 mm long, pubescent; blades elliptic-obovate, 1.2–2 × 0.5–1.2 cm, drying dull, coriaceous, glabrous except for scattered hairs on young leaves, the base acute, the apex rounded or emarginate; veins 3 from base. Spikes terminal, solitary, 2 cm long; peduncles 10 mm long, pubescent. Fruits globose with thick style. Fl (Jan, Aug, Sep); in nonflooded forest.

Peperomia obtusifolia (L.) A. Dietr.

Epiphytic, epilithic, or terrestrial herbs, with pungent aroma in fresh and dried specimens. Stems erect or ascending, rooting at nodes, 4 mm thick when dried, green, with or without red spots. Leaves alternate; petioles to 50 mm long, often red dotted, glabrous; blades elliptic-obovate or subspatulate, 2–16 × 2–7.5 cm, shiny, coriaceous, glabrous, the base cuneate, the apex rounded, obtuse, or occasionally acute; veins in 3–4 pairs arising from almost at base and 1 pair from midrib. Spikes terminal, leaf-opposed, or axillary, solitary or occasionally paired, 10–20 cm long; peduncles 10–80 mm long, often with red dots, minutely pubescent or glabrous. Fruits ellipsoid, the apex with slender, terminally hooked or curved beak. Fl (Jan, Mar, Apr, May, Jun, Sep), fr (Jul); common, in nonflooded forest at all elevations in our area. *Ti-mourou* (Créole).

Peperomia ouabianae C. DC.

Epiphytic or epilithic herbs. Stems repent, rooting at nodes, filiform, pubescent, the fruiting branches decumbent. Leaves

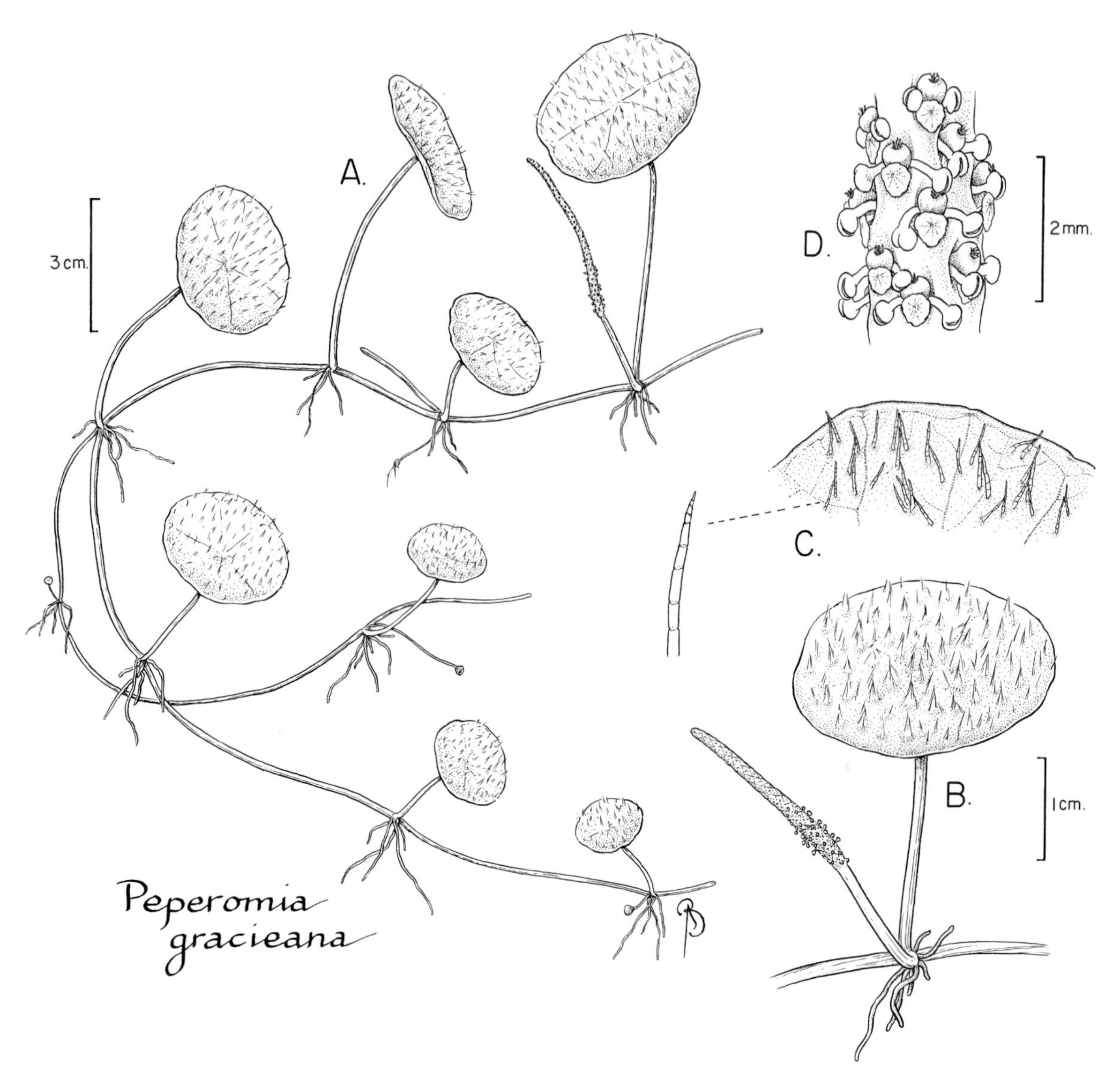

FIG. 252. PIPERACEAE. *Peperomia gracieana* (A, *Mori et al. 23264*; B–D, *Mori et al. 23773*). **A.** Part of repent stem with leaves and inflorescence; note roots at node. **B.** Part of stem with leaf and inflorescence. **C.** Part of adaxial leaf surface with detail of multicellular trichomes (left). **D.** Part of inflorescence showing numerous, minute flowers, each axillary to a peltate bract. (Reprinted with permission from A. R. A. Görts-van Rijn, Brittonia 50(1), 1998.)

alternate; petioles <5 mm long, sparsely villous; blades elliptic to obovate, 0.5–2 × 0.4–0.9 cm, dark green, villous adaxially, pale green, glabrous abaxially, the base acute to obtuse, the apex obtuse, the margins ciliate; veins 3 arising from base, these pale, sometimes reddish. Spikes terminal, solitary, 1.5–5.5 cm long; peduncles to 15 mm long, glabrous or slightly pubescent. Fruits ovoid, brown, warty, the apex oblique. Fl (Apr, May), fr (Sep); in nonflooded forest.

Peperomia pellucida (L.) Kunth

Terrestrial, delicate, annual herbs. Stems erect, fleshy, glabrous. Leaves alternate; petioles to 25 mm long; blades orbicular to deltoid, 1–3.7 × 1–3.5 cm, the base rounded, truncate, or subcordate, pale green, shiny, membranous when dried, pellucid, the apex obtuse or bluntly acute; veins 5–7 arising from base, indistinct. Spikes terminal, solitary, slender, 1–6 cm long; peduncles to ca. 10 mm long. Fruits brown, globose, finely longitudinally ridged. Fl (Jan, Feb, Nov); in forest along rivers, or in open forest understory. *Erva-de-jaboti* (Portuguese), *salade soldat* (Créole).

Peperomia pernambucensis Miq.

Terrestrial or epiphytic herbs. Stems simple, erect, fleshy, short, to 5 cm tall, glabrous. Leaves alternate; petioles to 25(50) mm long, glabrous; blades narrowly elliptic or oblanceolate, 15–30 × 4–12 cm, membranous, glabrous, the base cuneate, usually decurrent onto petiole, the apex usually acuminate; veins in 5–8 pairs arising from along midrib. Spikes terminal or axillary, to 20, in 4–8 cm long panicles; peduncles 50 mm long, densely velvety-puberulent. Fruits nearly cylindrical, verrucose, the apex truncate. Fr (Oct); in nonflooded forest.

Peperomia rotundifolia (L.) Kunth

Epiphytic or epilithic, small herbs. Stems prostrate when vegetative, erect when fertile, rooting at nodes, crisp-pubescent when young. Leaves alternate, the fertile branches often with progressively smaller leaves distally; petioles to 1 mm long, glabrous or pubescent; blades orbicular or somewhat elliptic, (0.5)1–1.2 cm wide, fleshy, pubescent, becoming glabrous, the apex obtuse to rounded, the margins often ciliate; veins obsolete. Spikes terminal, solitary, to 2 cm long; peduncles to 10 mm long, pubescent or glabrous, without bracts. Fruits (sub)globose, the apex oblique. Fl (Jan, Feb, May, Nov); in nonflooded forest. *Ti-mouron* (Créole).

Peperomia serpens (Sw.) Loudon PL. 106b

Epiphytic or epilithic herbs. Stems repent, crisp-pubescent. Leaves alternate; petioles to 25 mm long, crisp-pubescent; blades orbicular, cordate, or somewhat deltoid, slightly peltate, 1–2 cm diam., dark green, fleshy, crisp-pubescent, becoming glabrous, the apex obtuse to rounded, the margins ciliate toward apex; veins 3–5 arising from base, pale green. Spikes terminal or axillary, solitary, to 1.7 cm long; peduncles to 30 mm long, crisp-pubescent, with 1–2 bracts. Fruits ellipsoid, with a slender beak. Fl (Jan, Feb, May, Aug, Sep, Nov); common, in nonflooded forest at all elevations in our area, often seen on fallen branches or at bases of tree trunks. *Grand mourou, gros mourou, mourou, petit mouron, ti-mourou-grandes-feuilles, ti-mouron* (Créole).

PIPER L.

Shrubs, subshrubs, treelets, sometimes scandent or lianas, occasionally herbs. Stems with swollen nodes. Stipules sometimes present, the caducous. Leaves alternate; blades often glandular-dotted, the base symmetrical or asymmetrical; venation palmate or pinnate, the secondary veins originating from lower part of midrib or from throughout midrib. Spikes leaf-opposed, solitary; peduncles present; floral bracts triangular, rounded, peltate, or cucullate, glabrous or pilose at base, the margins often fringed. Flowers bisexual (except unisexual in *P. nematanthera* in our area); stamens 2–5; ovary sessile, the stigmas (2)3–4, sessile or on a style. Fruits globose, obovoid, trigonous, often depressed drupes; often dispersed by bats.

1. Leaf blades with venation palmate.
    2. Stems without spines. Leaf blades with veins not conspicuous. Fruits distinctly separate, glabrous or at most papillose, without disk. . . . . *P. amalago.*
    2. Stems often with spines opposite petiole. Leaf blades with veins very prominent abaxially. Fruits close together, papillate to puberulous with glabrous disk at apex. . . . . *P. reticulatum.*
1. Leaf blades with venation pinnate.
    3. Leaf blades large, 20–60(70) × (12)10–35 cm, with deeply lobed base, the lobes shorter or longer than petiole. Spikes (8)20–70 cm long.
        4. Stems densely pubescent. Fruits pubescent. . . . . *P. cernuum.*
        4. Stems glabrescent. Fruits glabrous. . . . . *P. obliquum.*
    3. Leaf blades smaller, 4.5–35 × 3–18 cm, without deeply lobed base, at most somewhat cordate. Spikes ≤20 cm long.
        5. Plants glabrous except for subnodal lines of pubescence on the stem or cilia on leaf blade margins.
            6. Plants glabrous except for cilia on leaf blade margins. Secondary veins in 10–16 pairs. Spikes 4–20 cm long. . . . . *P. augustum.*
            6. Plants glabrous, except for lines of hairs on the stem below nodes. Secondary veins in 8–12 pairs. Spikes 0.5–4 cm long.
                7. Leaf blades lanceolate to oblong or elliptic, 4–8.5 cm wide, the apices short-acuminate. Spikes 2.5–4 cm long. . . . . *P. anonifolium.*
                7. Leaf blades narrowly lanceolate, to 3 cm wide, the apices long-acuminate. Spikes 0.5–2 cm long. . . . . *P. eucalyptifolium.*
        5. Plants completely glabrous or variously pubescent, but not pubescent as described above.
            8. Leaf blade bases oblique, with one side (1)2–10 mm longer than other side (i.e., base asymmetrical).
                9. Usually treelets to 7 m tall. Stems glabrous. . . . . *P. arboreum.*
                9. Herbs, shrubs or occasionally treelets, sometimes scandent, to 4 m tall. Stems variously pubescent.
                    10. Leaves scabrous.
                        11. Leaf blades to 30 × 17 cm. Peduncles 15–20 mm long. Fruits hirsute. . . . . *P. tectonifolium.*
                        11. Leaf blades to 20 × 11 cm. Peduncles 5–10 mm long. Fruits papillose to puberulent.
                            12. Leaf blades somewhat scabrous; tertiary veins reticulate between secondary veins. Anthers dehiscing laterally. . . . . *P. dilatatum.*
                            12. Leaf blades strongly scabrous; tertiary veins parallel between secondary veins. Anthers dehiscing transversely.. . . . . *P. hispidum.*
                    10. Leaves not scabrous (sometimes somewhat scabrous in *P. tectonifolium* when dried).
                        13. Creeping subshrubs or herbs, to 0.2 m tall. Leaf blade apices obtuse. . . . . . *P. humistratum.*

13. Shrubs, treelets, or creeping herbs or shrubs, well over 0.2 m tall. Leaf blade apices acute to acuminate.
  14. Spikes pendent.
    15. Leaf blades shortly pubescent, especially abaxially, the base rounded, obtuse, or subacute. Floral bracts densely fringed. . . . . *P. avellanum.*
    15. Leaf blades ± less villous on both surfaces, the base obtuse to subcordate. Floral bracts not fringed. . . . . *P. brownsbergense.*
  14. Spikes not pendent.
    16. Creeping herbs or subshrubs, to 0.8 m tall. Secondary veins in 6–8 pairs. Spikes with unisexual flowers. Stamens long, exserted or recurved. . . . . *P. nematanthera.*
    16. Shrubs, subshrubs, or treelets, 1–5 m tall, sometimes scandent or even liana-like. Spikes with bisexual flowers, stamens not exserted.
      17. Leaf blade base cordate; secondary veins in 8–11 pairs. . . . . . . *P. demeraranum.*
      17. Leaf blade bases not markedly cordate; secondary veins in 3–7 pairs.
        18. Leaf blades broadly ovate; secondary veins originating from lower 1/2 of midrib. Spikes (6)10–20 cm long. . . . . *P. tectonifolium.*
        18. Leaf blades elliptic, oblong, or ovate; secondary veins originating from 1/2 to 3/4 of midrib. Spikes 8–12 cm long.
          19. Leaf blades pubescent abaxially only on veins. Spikes white, yellow, or green. . . . . *P. hostmannianum.*
          19. Leaf blades densely pubescent abaxially throughout. Spikes reddish. . . . . *P. trichoneuron.*

8. Leaf blade bases not obviously oblique, usually with both sides ± equal (i.e., the base symmetrical).
  20. Leaf blades scabrous. . . . . *P. amplectenticaule.*
  20. Leaf blades not scabrous (glabrous or variously pubescent).
    21. Stems and leaves glabrous or glabrescent.
      22. Fruits with persistent, distinct style. . . . . *P. crassinervium.*
      22. Fruits with sessile stigmas or style obsolete.
        23. Mature spikes 1–5 cm long.
          24. Leaves densely glandular-dotted. Secondary veins in 3–4 pairs. Spikes pendent. . . . . *P. nigrispicum.*
          24. Leaves not glandular-dotted. Secondary veins usually in 6–8(12) pairs. Spikes erect. . . . . *P. piscatorum.*
        23. Mature spikes usually >5 cm long.
          25. Leaf blades with secondary veins in 6–10 pairs. Spikes not apiculate. Fruits ridged or with appendages at apex.
            26. Petioles 12–17 mm long. Fruits with 4 appendages at apex, not ridged. . . . . *P. alatabaccum.*
            26. Petioles 5 mm long. Fruits without 4 appendages at apex, often ridged when dried. . . . . *P. bartlingianum.*
          25. Leaf blades with secondary veins in 4–6 pairs. Spikes usually apiculate. Fruits without ridges or appendages.
            27. Leaf blades not conspicuously glandular, the blade asymmetric. . . . . *P. aequale.*
            27. Leaf blades conspicuously glandular, the blade symmetric. . . . . *P. insipiens.*
    21. Stems and leaves variously and persistently pubescent.
      28. Leaf blades narrow, usually <3.5 cm wide; secondary veins in 6–10 pairs.
        29. Shrubs to 2 m. Leaf blade base acute, the apex long-acuminate. . . . . . . *P. angustifolium.*
        29. Creeping herbs or subshrubs to 0.3 m. Leaf blade base obtuse to rounded or subcordate, the apex acute. . . . . *P. consanguineum.*
      28. Leaf blades wider, usually ≥2.5 cm wide; secondary veins in 4–6(8) pairs.
        30. Stems glabrous. Leaves pilose. Spikes 5–8 cm long, not apiculate. . . . . *P. dumosum.*
        30. Stems and leaves variously pubescent. Spikes 1–9 cm long, apiculate or not apiculate (*P. cyrtopodon*).
          31. Leaf blade apices long-acuminate; secondary veins in 4–6 pairs. Floral bracts densely fringed.
            32. Leaf blades 8–16.5 × 3–6(7)cm. Spikes pendent, 1–5 cm long, green or yellow. . . . . *P. adenandrum.*

32. Leaf blades 22–28 × 7–11 cm. Spikes erect, 8–9 cm long, often reddish. . . . . *P. trichoneuron.*
31. Leaf blade apices acute or acuminate; secondary veins in 6–8 pairs. Floral bracts at most somewhat pilose at base.
33. Plants pilose, the hairs to 2 mm long. Leaf blade bases subcordate to rounded, the apices acuminate. Spikes pendent. . . . . *P. cyrtopodon.*
33. Plants hirsute, the hairs shorter. Leaf blade bases cuneate, obtuse, or rounded, the apices acute. Spikes erect. . . . . *P. paramaribense.*

Piper adenandrum (Miq.) C. DC.

Shrubs, 1.5 m tall. Stems crisp-pubescent or villous, the upper internodes densely glandular. Leaves not glandular-dotted; petioles 2–5(10) mm long, winged to middle; blades lance-elliptic or elliptic-ovate, 8–16.5 × 3–6(7) cm, glabrous or pubescent adaxially, (densely) crisp-pubescent or villous abaxially, the base nearly symmetrical, acute or obtuse, the apex long-acuminate, the margins inconspicuously ciliate; venation pinnate, the secondary veins in 5–6 pairs, originating from along 3/4 of midrib or from entire midrib. Spikes pendent, apiculate, 1–5 cm long, green to pale or dark yellow; peduncles 5–11 mm long, sometimes reddish tinged, crisp-pubescent or villous, the floral bracts densely fringed. Flowers: stigmas sessile. Fruits depressed globose, 2.5–3(4) mm diam., verruculose, separate at maturity. Fl (Aug, Oct, Dec), fr (Apr, May); common along trails, in nonflooded forest. *Ti-bombe* (Créole).

Piper aequale Vahl

Shrubs or treelets, 1–4 m tall. Stems glabrous. Leaves not conspicuously glandular-dotted; petioles 10–15(20) mm long; blades lanceolate to lance-elliptic or elliptic-ovate (broadly ovate on sterile branches), asymmetrical, 12–23 × 3.5–9 cm, glabrous, often pale or yellow, the base symmetrical, acute or obtuse, the apex acuminate; venation pinnate, the secondary veins in 4–6 pairs, originating from along 3/4 of midrib or along entire midrib. Spikes erect, apiculate or not, to 10 cm long, white or green; peduncles 10–15 mm long, the floral bracts fringed. Flowers: stigmas sessile. Fruits oblong to trigonous, without ridges or appendages, glabrous. Fl (Jan, Feb, Mar, Jul, Sep, Oct, Nov, Dec), fr (Feb, May); common, in disturbed moist and nonflooded forest.

Piper alatabaccum Trel. & Yunck.

Shrubs, 2–3 m tall. Stems glabrous. Leaves not conspicuously glandular-dotted; petioles 12–17 mm long; blades elliptic, 20–28 × 6–10 cm, glabrous, the base symmetrical, acute, the apex long-acuminate; venation pinnate, the secondary veins in 7–10 pairs, originating from along entire midrib. Spikes erect, not apiculate, 5–8 cm long; peduncles 10 mm long, the rachis puberulent to glabrous, the floral bracts papillate. Flowers: stigmas sessile. Fruits with 4 prominent, deltoid appendages near apex, not ridged, glabrous. Fl (Mar), fr (Dec); in nonflooded forest.

Piper amalago L.

Shrubs, 2–4(7) m tall. Stems nearly glabrous. Leaves not conspicuously glandular-dotted; petioles 5–10 mm long; blades lanceolate to rounded or subobovate, 8–11(15) × 2.5–6(10) cm, glabrous, the base symmetrical, acute, obtuse, or rounded, the apex acuminate; venation palmate, with 5–7 veins, these not very conspicuous. Spikes erect, not apiculate, 6–7 cm long; peduncles 8–15 mm long, the floral bracts cucullate, puberulent. Flowers: stigmas sessile. Fruits ovoid, distinctly separate, glabrous or at most papillate, without disk at apex. Phenology not recorded.

Piper amplectenticaule Trel. & Yunck.

Creeping herbs or subshrubs, to 4 m tall. Stems densely pubescent. Leaves not conspicuously glandular-dotted; petioles 2–10 mm long; blades elliptic to lanceolate-oblong, 7–15 × 3–6 cm, scabrous, glabrescent adaxially, appressed-pubescent abaxially, the veins densely so, the base nearly symmetrical, obtuse to rounded to subcordate, the apex short-acuminate; venation pinnate, the secondary veins in 4–5 pairs, originating from along lower 1/2 of midrib, prominulous abaxially, the tertiary veins widely reticulate. Spikes erect, apiculate, 2.5–4.5(8) cm long, green to white; peduncles 10–15 mm long, slender, glabrescent, the floral bracts densely fringed. Flowers: stigmas sessile. Fruits oblong or depressed globose, puberulent at apex. Fl (Feb, Jun), fr (Aug); common, along trails in nonflooded forest.

Piper angustifolium Lam.

Small shrubs, 1.5–2 m tall. Stems crisp-pubescent. Leaves not conspicuously glandular-dotted; petioles 1–4 mm long; blades narrowly lanceolate, 5–11 × 1–2.5 cm, glabrous, the base nearly symmetrical, acute, the apex long-acuminate; venation pinnate, the secondary veins in 6–10 pairs, originating from along entire midrib. Spikes erect, apiculate, <1 cm long; peduncles 2–4(7) mm long, the floral bracts cucullate, glabrous. Flowers: stigmas sessile. Fruits depressed globose or obovoid, glabrous. Fl (Aug, Nov), fr (Apr, Jul); common, in forest understory in nonflooded forest at all elevations in our area.

Piper anonifolium (Kunth) C. DC.

Shrubs, 1–3 m tall. Stems glabrous, except for subnodal lines of hairs. Leaves not conspicuously glandular-dotted; petioles 5–10 mm long, ciliately keeled; blades lanceolate to oblong or elliptic, 9–19 × 4–8.5 cm, drying silvery shiny, the base symmetrical, acute, the apex short-acuminate; venation pinnate, the secondary veins in 8–10 pairs, originating from along entire midrib. Spikes erect, apiculate, 2.5–4 cm long, white to green; peduncles 5–10 mm long, the floral bracts cucullate. Flowers: stigmas sessile. Fruits depressed globose, glabrous. Fl (Jan, Feb, May, Jun, Aug, Oct, Nov, Dec), fr (Mar, May, Dec); common, along trails or roads in nonflooded moist forest.

Piper arboreum Aubl.

Usually treelets, 2–7 m tall. Stems glabrous, the upper internodes warty. Leaves not glandular-dotted; petioles 5–10 mm long; blades elliptic-ovate or lanceolate-elliptic, 10–25 × 6–11 cm, glabrous, the base oblique, one side 5–10 mm longer than other, rounded, the apex acuminate; venation pinnate, the secondary veins in 8–10 pairs, originating from along entire midrib. Spikes erect, not apiculate, 5–10(15) cm long; peduncles 5–10 mm long, the floral bracts densely fringed.

Flowers: stigmas sessile. Fruits oblong, glabrous. Fl (Oct, Nov), fr (Feb, Sep); along trails in nonflooded forest or along streams.

Piper augustum Rudge

Shrubs or treelets, 2–3(8) m tall. Stems glabrous. Leaves glandular-dotted; petioles 5–10(40) mm long; blades elliptic-ovate, 15–35 × 9–15(18) cm, the base symmetrical, acute or obtuse, the apex acute, the margins ciliate; venation pinnate, the secondary veins in 10–16 pairs, strongly curved, originating from along entire midrib. Spikes pendent, not apiculate, 4–12(20) cm long, white; peduncles 10–20(30) mm long, red to brown, the floral bracts densely fringed. Flowers: stigmas sessile. Fruits oblong, glabrous. Fl (Feb, Mar, May, Jun, Aug, Sep, Oct), fr (Jan, Mar, Sep, Oct, Dec); common, often along streams but also in nonflooded moist forests at all elevations in our area.

Piper avellanum (Miq.) C. DC.

Shrubs, to 1.5 m tall. Stems densely short-pubescent. Leaves not conspicuously glandular-dotted; petioles 3–13 mm long, often conspicuously winged; blades lanceolate-elliptic, lanceolate-oblong, or oblanceolate, 7–18 × 3–7.5 cm, glabrous or sparsely short-pubescent adaxially, sparsely to densely pubescent abaxially, the base asymmetrical, one side 1–5 mm longer than other, rounded, obtuse or subacute, the apex short-acuminate or acute, the margins ciliate; venation pinnate, the secondary veins in 5–6(8) pairs, originating from along 3/4 of midrib, the tertiary veins irregularly reticulate. Spikes pendent, occasionally erect, apiculate or not, 3–5.5 cm long, white to green; peduncles slender, 5–10 mm long, pubescent, the floral bracts densely fringed. Flowers: stigmas sessile. Fruits trigonous or depressed globose, 2 mm diam., glabrous, papillose. Fl (May, Aug, Nov), fr (May, Jul); in nonflooded forest.

Piper bartlingianum (Miq.) C. DC. FIG. 253, PL. 106d

Shrubs, 1–4 m tall. Stems glabrous. Leaves not conspicuously glandular-dotted; petioles 5 mm long; blades elliptic-ovate or elliptic-oblong, 17–30 × 6–8(10) cm, the base symmetrical, acute; the apex acuminate, venation pinnate, the secondary veins in 6–8 pairs, originating from along entire midrib. Spikes erect, not apiculate, 10–12 cm long; peduncles 10 mm long, the rachis hirsute, the floral bracts cucullate, somewhat pilose at base. Flowers: stigmas sessile. Fruits globose, often somewhat ridged when dried, glabrous, separate at maturity. Fl (Jun, Oct, Dec), fr (Jan, Feb, Apr, May, Jun, Oct, Nov); common, along trails in nonflooded forest. *Poivrier* (Créole).

Piper brownsbergense Yunck.

Herbs or shrubs, 1–2 m tall. Stems with upper internodes drying finely ridged, somewhat (yellow-)glandular, villous, the hairs to 1 mm or more long. Leaves glandular-dotted; petioles 3–5(35) mm long; blades elliptic, 11–16 × 4–7.5 cm, villous on both surfaces, the base asymmetrical, one side 2–3 mm longer than other, obtuse to subcordate, the apex acute to acuminate; venation pinnate, the secondary veins in 5–7 pairs, originating from along entire midrib. Spikes pendent, hidden under leaves, to 5 cm long; peduncles 10–20 mm long, villous, the floral bracts cucullate, somewhat pilose at base. Flowers: stigmas sessile. Fruits depressed globose to obovoid, glabrous. Fr (Jan, May, Aug); locally common, on granitic outcrops. *Cèdre* (Créole).

Piper cernuum Vell.

Shrubs or treelets, 2–6(13) m tall. Stems densely pubescent. Leaves not glandular-dotted; petioles 40–90 mm long; blades broadly ovate to elliptic-ovate, 23–60 × 14–28 cm, glabrous or sparsely pubescent adaxially, densely pubescent abaxially, the base deeply unequally lobed, the apex acute; venation pinnate, the secondary veins in 5–8 pairs, originating from along 3/4 of midrib. Spikes pendent, not apiculate, 20–60 cm long; peduncles 10–30(50) mm long, densely pubescent, the floral bracts densely fringed. Flowers: stigmas sessile. Fruits obovoid, pubescent. Fl (Feb).

Piper consanguineum (Kunth) C. DC. PL. 106c

Herbs or subshrubs, 0.1–0.3 m tall. Stems sometimes creeping, crisp-pubescent. Leaves minutely glandular-dotted; petioles 2–12 mm long; blades lanceolate-oblong, (3.5)4.5–7 × 1–2(3.5) cm, glabrous adaxially, sometimes brown with white midrib, crisp-pubescent abaxially, the base symmetrical, obtuse to rounded or subcordate, the apex acute; venation pinnate, the secondary veins in 8–10 pairs, originating from along entire midrib. Spikes erect, apiculate, 1.5–2.5 cm long, yellow to green; peduncles 5 mm long, glabrescent, the floral bracts glabrous. Flowers: stigmas on short style. Fruits depressed globose, papillose, green with thickened border near apex, free from each other at maturity. Fl (Sep, Oct, Dec), fr (May, Jun); relatively common, in nonflooded forest.

Piper crassinervium Kunth

Shrubs, 1–4 m tall. Stems glabrous or glabrescent. Leaves not conspicuously glandular-dotted; petioles 12–25 mm long, sparsely pubescent; blades elliptic-ovate or ovate, 14–21 × 5.5–12.5 cm, glabrous, glabrescent, or slightly puberulous abaxially, the base symmetrical or nearly so, cuneate, obtuse, or truncate, the apex acuminate; venation pinnate, the secondary veins in 4–5 pairs, originating from along lower 1/2 of midrib. Spikes erect, not apiculate, 4–16 cm long; peduncles 7–15 mm long, glabrous or pubescent, the floral bracts densely fringed. Flowers: stigmas on distinct style. Fruits globose, with persistent, distinct style, glabrous. Fl (Jan); collected only once in our area, in open areas.

Piper cyrtopodon C. DC.

Shrubs, 1–3(4) m tall. Stems pilose, the hairs to 2 mm long. Leaves not conspicuously glandular-dotted; petioles 5 mm long; blades oblong-elliptic, 15–20 × 4–5.5 cm, glabrescent adaxially, pilose abaxially, especially on veins, the base nearly symmetrical, subcordate to rounded, the apex acuminate, the margins sparsely ciliate; venation pinnate, the secondary veins in 6–8 pairs, originating from along entire midrib. Spikes pendent, not apiculate, 3–3.5 cm long; peduncles ca. 10 mm long, pilose, the floral bracts cucullate, somewhat pilose at base. Flowers: stigmas sessile. Fruits ovoid, glabrous. Fl (Sep); in disturbed forest along Limonade Trail.

Piper demeraranum (Miq.) C. DC.

Shrubs, 1–2 m tall. Stems crisp-pubescent. Leaves minutely glandular-dotted; petioles 3–5(10) mm long; blades elliptic to oblanceolate, 15–26(35) × 5–9 cm, glabrous adaxially, crisp-pubescent abaxially, the base asymmetrical, with one side 1–4 mm longer than other, unequally cordate; the apex acuminate, venation pinnate, the secondary veins in 8–11 pairs, originating from along entire midrib, the tertiary veins widely reticulate. Spikes erect, not or slightly apiculate, 3.5–4.5 cm long, green to yellow; peduncles to 10 mm long, the floral bracts cucullate, densely fringed. Flowers: stigmas sessile. Fruits globose, puberulent at apex. Fr (May); uncommon, in nonflooded forest.

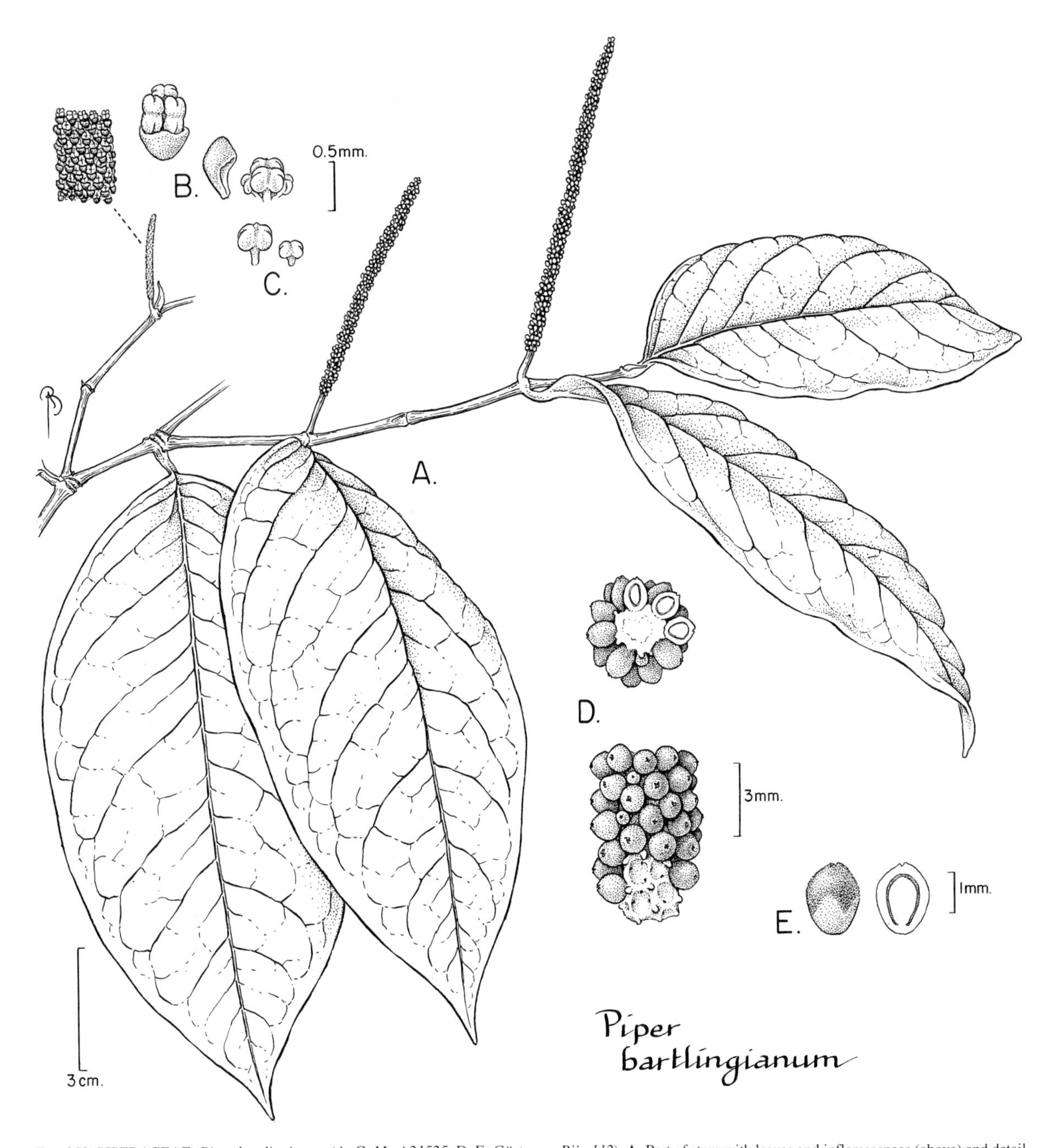

FIG. 253. PIPERACEAE. *Piper bartlingianum* (A–C, *Mori 21525*; D, E, *Görts-van Rijn 112*). **A.** Part of stem with leaves and inflorescences (above) and detail of inflorescence (upper left). **B.** Apical view of bract and anthers (above left), bract (near right), and flower showing four stamens (far right). **C.** Two stamens. **D.** Transverse section of infructescence (above) and detail of lateral view of infructescence (below). **E.** Fruit (near right) and medial section of fruit (far right).

Piper dilatatum Rich.

Shrubs, 2–3 m tall. Stems crisp-pubescent. Leaves not glandular-dotted; petioles 5–20 mm long, sparsely pubescent, often reddish tinged; blades rhombic to subobovate, 15–20 × 7–9 cm, somewhat scabrous, glabrous adaxially except for pubescent veins, sparsely pubescent abaxially, the base oblique, with one side 3–4 mm longer than other, unequally cordate, rounded, or obtuse, the apex acuminate; venation pinnate, the secondary veins in 5–6 pairs, originating from along lower 1/2 of midrib. Spikes erect, not apiculate, 7–8 cm long, creamy white; peduncles 5–10 mm long, puberulent, the floral bracts densely fringed. Flowers: anthers dehiscing laterally; stigmas sessile. Fruits trigonous, puberulent at apex. Fl (Feb, May, Jun, Aug, Oct); common, usually along roads and trails.

Piper dumosum Rudge

Shrubs or treelets, 2–4 m tall. Stems glabrous. Leaves glandular-dotted; petioles 5–20 mm long, pilose; blades lanceolate to elliptic (occasionally ovate), 18–26 × 5.5–12 cm, sparsely pilose on both surfaces, the base nearly symmetrical, usually acute, occasionally truncate to rounded when blades ovate, the apex short-acuminate;

venation pinnate, the secondary veins in 5–6 pairs, originating from along 3/4 of midrib. Spikes pendent, not apiculate, 5–8 cm long, green; peduncles 10–12 mm long, the floral bracts densely fringed, slender, pilose. Flowers: stigmas sessile. Fruits obovoid to trigonous, glabrous. Fl (Jan, May, Jul, Aug, Sep, Oct), fr (May, Nov); common, often along streams.

Piper eucalyptifolium Rudge

Shrubs or treelets, 2 m tall. Stems glabrous, except for subnodal lines of hairs. Leaves not conspicuously glandular-dotted; petiole 3–7 mm long, ciliately keeled; blades narrowly lanceolate, 6–16 × 1.5–3 cm, glabrous, the base symmetrical, acute, the apex long-acuminate; venation pinnate, the secondary veins in 8–12 pairs, originating from along entire midrib. Spikes erect, apiculate, 0.5–2 cm long; peduncles ca. 5 mm long, glabrescent, the floral bracts glabrous, cucullate. Flowers: stigmas sessile. Fruits depressed globose, glabrous. Fl (Jan, Nov), fr (Jan).

Piper hispidum Sw.

Shrubs, to 4 m tall, or infrequently liana-like. Stems hirsute to glabrescent. Leaves glandular-dotted; petioles 5–10 mm long, hirsute; blades elliptic or elliptic-ovate, 11–19 × 4–11 cm, strongly scabrous and glabrous adaxially, hirsute abaxially, the veins hirsute, grayish brown abaxially when dry, the base asymmetrical, with one side 2–5 mm longer than other, obliquely rounded or cuneate, the apex acuminate; venation pinnate, the secondary veins in 5–6 pairs, originating from along lower 1/2 of midrib, the tertiary veins transversely parallel, not conspicuously anastomosing near margin. Spikes erect, not apiculate, 8–14 cm long, white; peduncles to 10 mm long, hirsute, the floral bracts densely fringed. Flowers: anthers dehiscing transversely; stigmas sessile. Fruits oblong to trigonous, papillose to puberulent at apex. Fl (Jan, May, Jul, Aug, Sep, Oct, Nov, Dec), fr (Aug); common, in weedy habitats along streams and in nonflooded forest.

Piper hostmannianum (Miq.) C. DC.

Shrubs, or sometimes treelets, often scandent or even liana-like, 3 m or more tall. Stems crisp-pubescent. Leaves not glandular-dotted; petioles 5–10 mm long; blades elliptic or ovate, 13–22 × 6–10 cm, pubescent along veins, especially abaxially, the base oblique, with one side 3–5 mm longer than other, obtuse to acute, the apex acuminate; venation pinnate, the secondary veins in 3–4(6) pairs, originating from along 3/4 of midrib, departing midrib at wide angle, then recurving abruptly to run almost parallel to margin. Spikes erect, 10–12 cm long, white or yellow to green; peduncles 10 mm long, pubescent, the floral bracts densely fringed. Flowers: stigmas sessile. Fruits oblong or trigonous, glabrous or puberulent at apex. Fl (May, Aug, Sep, Oct), fr (May, Sep); common, in weedy areas in nonflooded forest.

Piper humistratum Görts & Kramer

Creeping subshrubs or herbs, to 0.4 m tall. Stems villous, the hairs sometimes white. Leaves not glandular-dotted; petioles 1–2(5.5) mm long, villous, the hairs sometimes white; blades elliptic, 9–15(18) × 5–7(8.5) cm, glabrous adaxially, villous abaxially, especially on veins, the base oblique, with one side 2–5 mm longer than other, rounded, the apex obtuse; venation pinnate, the secondary veins in 3–5 pairs, originating from along 3/4 of midrib or from along entire midrib. Spikes erect, occasionally pendent when in fruit, not apiculate, 2–3.5 cm long, green or yellow or reddish; peduncles 15–25 mm long, villous, the floral bracts glabrous. Flowers: stigmas sessile. Fruits oblong, glabrous, immersed. Fl (Sep, Nov); in understory of nonflooded forest, often on stream banks. *Feuille grage* (Créole).

Piper insipiens Trel. & Yunck.

Shrubs or treelets, 2–3 m tall. Stems glabrous. Leaves glandular-dotted; petioles (3)10–15 mm long; blades elliptic, ovate, lanceolate-oblong, 10–17 × 3–8 cm, glabrous, the base symmetrical, cuneate or obtuse, the apex acute to acuminate; venation pinnate, the secondary veins in 4–6 pairs, originating from along 3/4 of midrib, the tertiary veins widely reticulate. Spikes erect, apiculate, (3)7–12 cm long; peduncles (5)10–30 mm long, the floral bracts densely fringed. Flowers: stigmas sessile. Fruits obovoid to trigonous, glabrous, becoming free from each other. Fl (May), fr (May); along roads and trails in nonflooded forest.

Piper nematanthera C. DC.

Creeping subshrubs or herbs, to 0.8 m tall. Stems puberulent. Leaves not conspicuously glandular-dotted; petioles 5–10 mm long, puberulent to glabrescent; blades elliptic to subobovate, 9–17 × 3–6 cm, glabrescent adaxially, puberulent on veins, the base asymmetrical, with one side 2–5 mm longer than other, unequally cordate or subcordate, the apex long-acuminate; venation pinnate, the secondary veins in 6–8 pairs, originating from along entire midrib. Spikes horizontally oriented, not apiculate, unisexual or bisexual; peduncles 10–20 mm long, puberulent, glabrescent, the floral bracts glabrous. Staminate spikes ca. 1 cm long. Pistillate spikes 1–2 cm long. Staminate flowers: stamens long, exserted or recurved. Pistillate flowers: ovary ovoid, the stigmas sessile, rather long, recurved. Fruits obovoid to depressed globose, glabrous. Phenology not recorded.

Piper nigrispicum C. DC. PL. 106e

Shrubs, 2–3 m tall. Stems glabrous. Leaves densely glandular-dotted; petioles 4–8 mm long; blades ovate to elliptic-oblong, 8–15 × 3–6 cm, glabrous, the base nearly symmetrical, acute, cuneate, or obtuse, the apex acuminate; venation pinnate, the secondary veins in 3–4 pairs, originating from along 3/4 of midrib, the tertiary veins not conspicuous, widely reticulate. Spikes pendent, apiculate, 1–5 cm long, green to reddish; peduncles 5–13 mm long, slender, often reddish tinged, the floral bracts densely fringed. Flowers: stigmas sessile. Fruits trigonous, glabrous. Fl (Jan, Feb, May, Jun, Jul, Aug, Sep, Oct, Nov), fr (Jan, Feb, May, Aug, Sep, Oct); abundant, in disturbed areas in nonflooded forest.

Piper obliquum Ruiz & Pav.

Shrubs or treelets, 1.5–8 m tall. Stems glabrescent. Leaves not glandular-dotted; petioles 30–70(100) mm long, glabrous or densely pubescent, green to brown; blades elliptic-ovate to oblong, 20–60(70)

× (12)20–35 cm, glabrous adaxially, glabrous or sparsely pubescent abaxially, the veins sparsely pubescent, the base asymmetrical, with one side 10 mm longer than other, unequally lobed, the lobes shorter or longer than petiole, occasionally overlapping petiole, the apex acute to acuminate; venation pinnate, the secondary veins in 5–9 pairs, originating from along 3/4 of midrib. Spikes pendent, apiculate, (8)20–60(70) cm long, reddish when young; peduncles 20–50 mm long, glabrescent, the floral bracts cucullate, sometimes fimbriate. Flowers: stigmas sessile or on short style. Fruits oblong, glabrous. Fl (May, Oct, Nov), fr (Jan); in disturbed forest along streams and in nonflooded forest at high elevations.

Piper paramaribense C. DC.

Shrubs, creeping subshrubs or herbs, to 1 m tall. Stems hirsute. Leaves glandular-dotted; petioles 5 mm long; blades elliptic-oblong, 7–17 × 2.5–6 cm, appressed-hirsute on veins abaxially, dark green abaxially, the base slightly asymmetrical, with one side at most 1 mm longer than other, cuneate, obtuse, or rounded, the apex acute; venation pinnate, the secondary veins in 6–8 pairs, originating from along entire midrib, deeply impressed adaxially, salient abaxially. Spikes erect, apiculate, 2–4 cm long, yellow or green; peduncles 5–10 mm long, glabrescent, the floral bracts cucullate, somewhat pilose. Flowers: stigmas sessile. Fruits globose or obovoid, glabrous. Fl (Jan, Sep), fr (Mar); in nonflooded forest.

Piper piscatorum Trel. & Yunck.

Shrubs or treelets, 2 m tall. Stems glabrous. Leaves not glandular-dotted; petioles 5–10 mm long; blades elliptic-oblong, 8–16 × 3–6 cm, glabrous, the base symmetrical, acute, the apex acuminate; venation pinnate, the secondary veins in 6–8(12) pairs, originating from along entire midrib, prominent abaxially, conspicuously anastomosing. Spikes erect, not apiculate, 1–3 cm long; peduncles 5 mm long, glabrous, the rachis pubescent, the floral bracts cucullate, somewhat pubescent. Flowers: stigmas sessile. Fruits globose, glabrous. Fl (Aug), fr (May); uncommon, in nonflooded forest.

Piper reticulatum L.

Trees, to 10 m tall, or shrubs, sometimes scandent. Stems glabrous, often with spines opposite petioles. Leaves not glandular-dotted; petioles 7–12 mm long; blades ovate, usually broadly so, 14–30 × 7–18 cm, glabrous, often with aroma of anise, especially when crushed, the base symmetrical, rounded or obtuse, the apex acuminate; venation palmate, with 5–7(9) veins, the veins plane or impressed adaxially, very prominent abaxially, the secondary veins distinctly transversely cross-connecting with primary veins, prominent abaxially, the tertiary veins reticulate. Spikes 10–12(15) cm long, erect, white; peduncles slender, 10–20 mm long, the floral bracts glabrous or sparsely puberulous. Flowers: stigmas sessile. Fruits close together, papillate or puberulous, with glabrous disk at apex. Fl (Feb), fr (Jun, Sep); in disturbed area along the Route de Bélizon.

Piper tectonifolium Kunth

Shrubs, 1–3 m tall. Stems densely pubescent. Leaves not conspicuously glandular-dotted; petioles 15–25 mm long; blades broadly ovate, 17–30 × 10–17 cm, somewhat scabrous when dried, sparsely pubescent except veins densely pubescent abaxially, the base slightly oblique, with one side 2–3 mm longer than other, truncate, subcordate, or obtuse, the apex acuminate; venation pinnate, the secondary veins in 5–7 pairs, originating from along lower 1/2 of midrib. Spikes erect or horizontally oriented, not apiculate, (6)10–20 cm long; peduncles 15–20 mm long, pubescent, the floral bracts densely fringed. Flowers: stigmas sessile. Fruits trigonous, hirsute. Fl (Jan, Feb, Mar, Aug, Sep, Oct, Dec), fr (Sep); in nonflooded forest.

Piper trichoneuron (Miq.) C. DC.

Shrubs, subshrubs or treelets, 2 m tall. Stems densely pubescent. Leaves not glandular-dotted; petioles 10–20 mm long, densely pubescent; blades narrowly elliptic to narrowly oblong, 22–28 × 7–11 cm, glabrous adaxially, densely pubescent throughout abaxially, often white variegated, the base symmetrical or asymmetrical, with one side 1–3 mm longer than other, obtuse, the apex long-acuminate; venation pinnate, the secondary veins in 4–6 pairs, originating from lower 1/2–3/4 of midrib. Spikes erect, apiculate, 8–9 cm long, reddish; peduncles 10–20 mm long, pubescent, the floral bracts densely fringed. Flowers: stigmas sessile. Fruits oblong, pubescent. Fl (Feb, May, Jul), fr (Aug); common, along trails and roads in nonflooded forest.

POTHOMORPHE Miq.

Large herbs or subshrubs. Stems with swollen nodes. Leaves alternate, peltate. Spikes axillary, arranged in umbels on common peduncle, densely flowered. Flowers and fruits as in *Piper*.

Pothomorphe forms a natural clade closely related to *Piper* (Burger, 1971) and, therefore, I prefer to consider it as a separate genus and not as part of *Piper* as done by Tebbs (1993).

Pothomorphe peltata (L.) Miq. FIG. 254, PL. 107a

Shrubs or large herbs, to 3 m tall. Stems glabrous or sparsely and minutely pubescent. Leaves peltate, glandular-dotted; petioles attached above base of blade, 80–200 cm long; blades round-ovate, to 16–30 × 18–40 cm, glabrous except for short hairs on veins, the base rounded to cordate, the apex acute; 13–15 veins originating from base and 2 pairs originating from midrib. Spikes erect, numerous, 5–10 cm long; peduncles slender, 10–15 mm long, the floral bracts fringed. Fruits trigonous, glabrous. Fl (Jan, May, Jul, Aug, Oct), fr (Aug); common, in disturbed areas. *Caa-peba* (Portuguese), *feuille bomb*, *grand feuille bomb*, *petit bombe* (Créole).

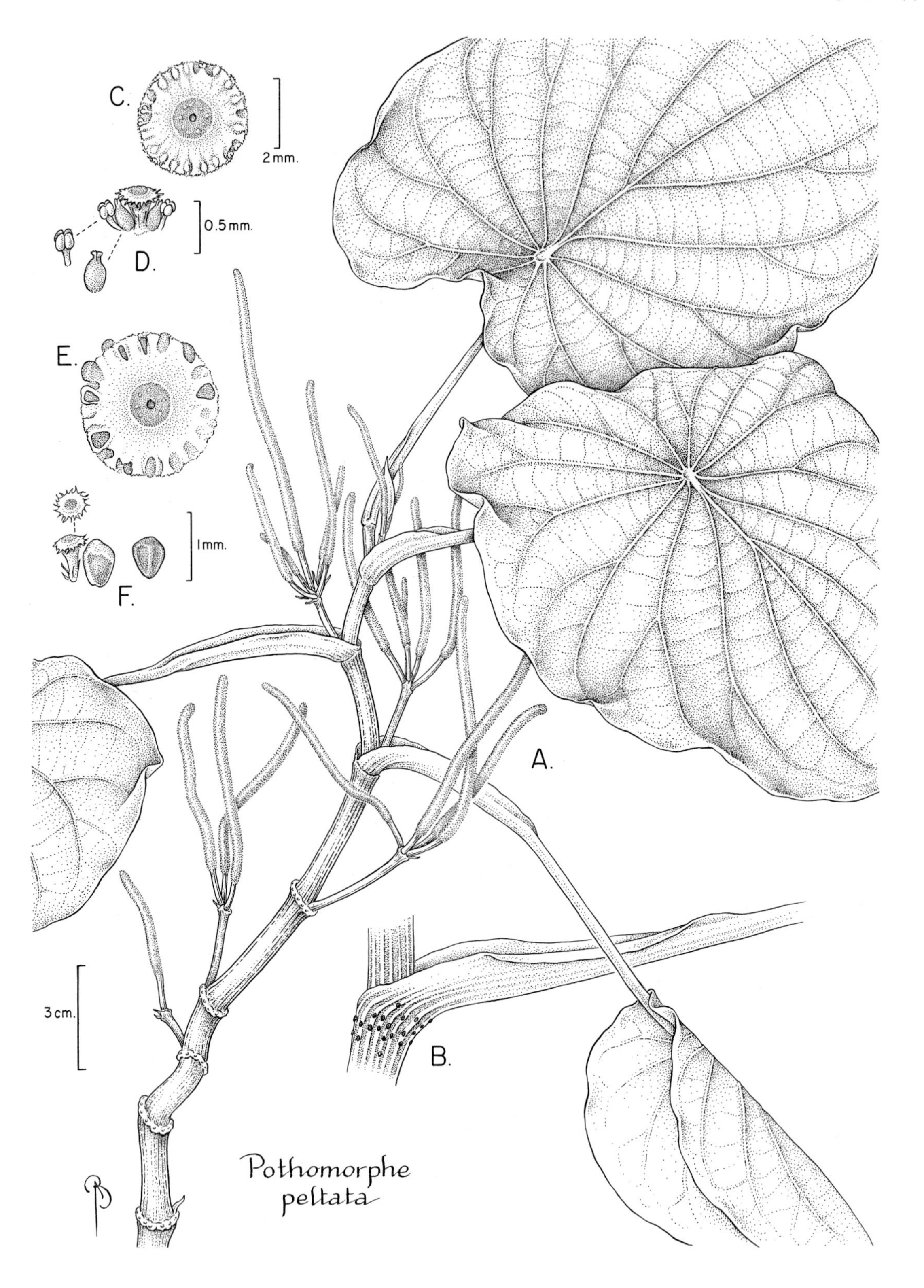

FIG. 254. PIPERACEAE. *Pothomorphe peltata* (*Görts-van Rijn 67*). **A.** Upper part of stem with leaves and inflorescences. **B.** Detail of petiole clasping stem; note glands. **C.** Transverse section of inflorescence. **D.** Lateral view of floral unit with peltate bract flanked by a flower on each side (above) and detail of pistil (near left) and stamen (far left). **E.** Transverse section of infructescence. **F.** Peltate bract (far left) with apical view of bract above, fruit (near left), and seed (above right).

PODOSTEMACEAE (Riverweed Family)

C. Thomas Philbrick

Tiny to large aquatic herbs, firmly attached to rocks, restricted to swift river currents, vegetative when submerged, flowering when exposed. Roots prostrate, branched or unbranched, dorsiventrally flattened or subterete. Stems arising from roots, often attached to rocks via haptera. Stipules absent or present. Leaves distichous, variable in form, simple, repeatedly forked, pinnate, repeatedly pinnate, or pinnatiform, sheathing or not, the ultimate divisions variable, linear, fimbriate, irregularly lobed. Inflorescences lateral or terminal, of solitary flowers or fascicled; pedicels elongating after anthesis. Flowers covered by a sac-like spathella, actinomorphic or zygomorphic, bisexual; tepals 2 to many, in a complete whorl or restricted to 1 side of ovary, scale-like, lanceolate, or filamentous, alternating with stamens, deciduous; stamens 1 to many, in a complete whorl or restricted to 1 side of ovary, the anthers basifixed or dorsifixed, introrse or extrorse, the base blunt to acute or sagittate, the apex blunt to acute; pollen 3-colpate or 3-sulcate, in monads; ovary superior, 2–3-carpellate, the carpels equal or subequal, the styles 2–3, apical, free or basally fused; placentae expanded and fleshy, the ovules numerous. Capsules similar in size to ovary, 2–3-valved, the valves equal to subequal, sometimes caducous, each to 7-ribbed or ribs lacking. Seeds numerous.

Van Royen, P. 1951. The Podostemaceae of the New World. Part I. Meded. Bot. Mus. Herb. Rijks Univ. Utrecht. **107:** 1–151.

APINAGIA Tul. em. Royen

Small to large herbs, trailing in water currents or prostrate when exposed. Stems branched or unbranched, terete or flattened and winged. Leaves united at base or free; petioles sheathed at base; blades variable in size and shape, pinnatisect, bipinnatisect, or variably lobed, often with tufts of hair-like filaments adaxially, the lobes blunt or acute; pinnativeined, palmativeined, or without obvious veins. Inflorescences solitary. Flowers with mature spathella (prior to rupture) clavate to nipple-shaped, the ruptured spathella tubular; tepals alternating with stamens; stamens 2–30, the anthers emarginate to sagittate at base, the thecae equal or unequal; ovary ellipsoidal, ovoid to obovoid, carpels 2, the stigmas 2, apical, linear, free or fused. Capsules 2-valved, each valve 3–4-ribbed.

1. Leaf blades linear. Tepals 15–20; stamens 25–30. . . . . *A. kochii.*
1. Leaf blades usually elliptic. Tepals 4–6; stamens 2–4.. . . . . *A. richardiana.*

Apinagia kochii (Engl.) P. Royen

Stems unbranched, to 10 cm long, nearly terete at base, angled to flattened upward, the internodes to 2.8 cm long. Leaves: petioles absent to 1.1 cm long, sometimes winged; blades linear, entire to pinatilobed, 5.6–24 × 1.4–4.1 cm, tufts of hair-like filaments linear, expanded near apex, to 2 mm long, the base triangular, the apex obtuse to acute, the margins with lobes rounded to crenate; midrib prominent to indistinct. Inflorescences axillary; pedicels 3–35 cm long. Flowers: tepals 15–20, lanceolate, 0.5–1 mm long; stamens 25–30, in complete whorl around ovary, 4–6 mm long, the anthers 1.5–2.5 mm long; ovary 2.5–4.5 mm long, the styles slightly cohering at base, about 1 mm long, subulate. Capsules 8-ribbed. In small stream.

Known by a single sterile collection (*Granville 5541*) from our area. Flowers and fruits not seen; description from Van Royen (1951).

Apinagia richardiana (Tul.) P. Royen

Stems 1–3 times branched or unbranched, to 28 cm long, nearly terete to angled at base, flattened upward, the internodes to 2.5 cm long. Leaves: petioles 0.9–1.4 cm long; blades elliptic or asymmetrically rectangular to rhombiform, pinnatilobed to pinnatisect, 1.6–6.9 × 0.9–4.5 cm, tufts of hair-like filaments 1.7–3.2 mm long, the base triangular, often extending below point of attachment, the apex obtuse, the margins with triangular to lanceolate lobes; veins prominent to distinct. Spathella (prior to rupture) 3.5–4.2 × 0.9–1.2 mm, the apex nipple-like. Inflorescences terminal or axillary; pedicels to 1.5 cm long. Flowers (only preanthesis flowers studied): tepals 4–6, lanceolate to slightly spatulate, confined to one side of ovary, sometimes 2 together lateral to stamens, 0.8–1.2 mm long, the apex acute to blunt; stamens 2–4, confined to one side of ovary, 1.9–2.1 mm long, the filaments 1.1–1.4 mm long, the anthers 0.5–1.3 × 0.3–0.5 mm, the base emarginate to sagittate, blunt; ovary 1.5–1.7 mm long, the styles free or slightly fused at base, 0.4–0.6 mm long, triangular, acute. Capsules 6-ribbed. Fl (Nov); in fast water of small stream.

Mature capsules not seen; description from Van Royen (1951).

POLYGALACEAE (Milkwort Family)

Arian Jacobs-Brouwer

Treelets, scandent scrubs, and lianas. Stems with xylem in closed ring. Leaves simple, alternate; petioles often with pairs of glands or spines at base; blades with entire margins; venation with primary vein raised abaxially, the secondary veins eucampto- to bronchidodromous, the intersecondary veins present, the tertiary veins reticulate. Inflorescences axillary or terminal, single- to many-flowered racemes or occasionally panicles; bracts 1 per flower, attached at base of pedicel; bracteoles 2 per flower, attached alongside bracts. Flowers nearly actinomorphic or zygomorphic, bisexual, often fragrant; sepals 5; petals 3 in zygomorphic flowers or 5 in nearly actinomorphic flowers; stamens 8, basally adnate to petals, free or united in staminal sheath, the anthers basifixed, poricidally dehiscent; ovary superior, usually 2(4–5)-locular, the style 1, straight or curved, the stigma terminal; placentation axile. Fruits 2-seeded dehiscent capsules, 1-winged samaras, or 2–5-seeded and drupe-like.

Görts-van Rijn, A. R. A. 1976. Polygalaceae. *In* J. Lanjouw & A. L. Stoffers (eds.), Flora of Suriname **II(2):** 518–523. E. J. Brill, Leiden.
Oort, A. J. P. 1966. Polygalaceae. *In* A. Pulle (ed.), Flora of Suriname **II(1):** 406–425. E. J. Brill, Leiden.

1. Leaf blades coriaceous; venation obscure. Corolla tubular. Fruits drupe-like. . . . . *Moutabea.*
1. Leaf blades membranous to subcoriaceous; venation distinct. Corolla not tubular, the petals free, or at most only basally connate. Fruits drupe-like, samaras, or dehiscent capsules.
    2. Shrubs or treelets. Fruits dehiscent capsules, the calyx persistent. . . . . *Polygala.*
    2. Lianas. Fruits 1-winged samaras or drupes, the calyx caducous.
        3. Leaves, especially when young, with 2 flat circular glands at apex of petioles and basal part of adaxial leaf blade. Flowers nearly actinomorphic; sepals 5, equal. Fruits drupe-like. . . . . *Barnhartia.*
        3. Leaves stipulate or with spines or glands at base of petiole. Flowers clearly zygomorphic; sepals 5, 2 forming petaloid wings. Fruits 1-winged samaras. . . . . *Securidaca.*

BARNHARTIA Gleason

Canopy lianas. Stipules absent. Leaves: blades subcoriaceous. Inflorescences terminal or axillary, compound racemes; pedicels poorly developed. Flowers nearly actinomorphic; sepals 5, free, equal, chartaceous; petals 5, one free, 4 connate in pairs, nearly equal; stamens in 3 groups; ovary 2-locular, with 1 ovule per locule, the style straight, the stigma 1. Fruits 2-seeded and drupe-like.

Barnhartia floribunda Gleason

Lianas. Stems flattened, 0.4–0.7 cm diam, with conspicuous dark gray lenticels, glabrous at base, rather densely pubescent toward apex. Leaves: petioles 5–10 mm long, sparsely pubescent, with 2 circular glands at apex; blades narrowly elliptic to elliptic, 9–14 × 2.5–5 cm, subcoriaceous, shiny, glabrous, with 2 circular glands at base adaxially, nearly glabrous abaxially, the base acute to cuneate, the apex acute to acuminate; venation prominent, the secondary veins eucampto-brochidodromous, in 7–10 pairs. Inflorescences terminal or axillary, 2–8 × 1–4 cm, densely to completely pubescent; pedicels 1–2 mm long; bracts persistent, elliptic to ovate; bracteoles persistent, broadly triangular. Flowers 6–8 mm long, fragrant; sepals elliptic to ovate, completely pubescent, the apex obtuse; petals white to pale yellow-orange; stamens 8, 2 adnate to free petal, 3 stamens adnate to each of 2 connate petals. Fruits globose, 0.7–1 cm diam., glabrous, yellow to orange at maturity. Seeds nearly ellipsoid, 5–7 × 3–5 mm, the apex and base with some red-brown hairs; aril absent. Fl (Dec), fr (Dec); apparently rare, collected only once from our area.

MOUTABEA Aubl.

Canopy lianas. Stems with inconspicuous small spines. Stipules absent. Leaves: blades thick-coriaceous. Inflorescences axillary, racemes; pedicels present. Flowers nearly actinomorphic; calyx lobes 5; corolla tubular, the lobes 5, nearly equal; stamens united in staminal sheath; ovary (2)4(5)-locular, with 1 ovule per locule in our area, the style straight, the stigma terminal. Fruits drupe-like, often orange at maturity. Seeds with thin aril.

Moutabea guianensis Aubl. FIG. 255, PL. 107b

Lianas. Stems curved, the inner surface of curve often characteristically wrinkled, very light gray to yellowish ochre. Leaves: petioles 5–11 mm long, glabrous, with 2 circular glands at base; blades usually obovate, sometimes ovate, 6–20 × 2.5–7 cm, markedly coriaceous, sparsely covered with circular glands abaxially, the base cuneate to acute, the apex obtuse, acute or emarginate; venation obscure except for midrib. Inflorescences axillary, racemes, 2–4 × 2.5–4 cm, sparsely pubescent; pedicels 2–2.5 mm long; bracts cup-shaped, with two circular glands at the base; pedicels 2–2.5 mm long; bracteoles absent. Flowers 15–25 mm long, fleshy, fragrant; calyx connate for 2/3 length, the lobes ± equal, dirty white; petals connate for 2/3 length, the lobes white; stamens adnate to corolla to 4/5–9/10 length, the filaments completely united into staminal sheath, the anthers in 2 groups of 4; ovary broadly transversely ovoid. Fruits globose, 2–4.5 cm diam., dark green when young, yellow to orange when ripe, edible. Seeds ovoid, laterally compressed, 16 × 8–13 mm, completely surrounded by whitish, thin, slimy aril. Fl (Feb, Aug), fr (Feb, Apr, Jun, Jul, Nov); common, in nonflooded forest. *Bois violet* (Créole), a name more commonly applied to species of *Peltogyne* (Fabaceae).

POLYGALA L.

Subshrubs, shrubs, or treelets. Leaves: petioles with triangular to rhombic or tubular glands; blades membranous to chartaceous. Inflorescences terminal or axillary, small racemes. Flowers zygomorphic; sepals free, the inner 2 forming petaloid wings, persistent in fruit; petals 3, free, the dorsal one forming helmet shaped keel; stamens united in staminal sheath; ovary 2-locular, ovules 1 per locule in our area, the style curved. Fruits dehiscent capsules, 2-locular. Seeds with aril.

1. Petioles with triangular to rhombic glands at base. Flowers 6.5–7 mm long. Fruits 2-winged. . . . . . . *P. membranacea.*
1. Petioles with tubular glands at base. Flowers 25–30 mm long. Fruits without wings. . . . . *P. spectabilis.*

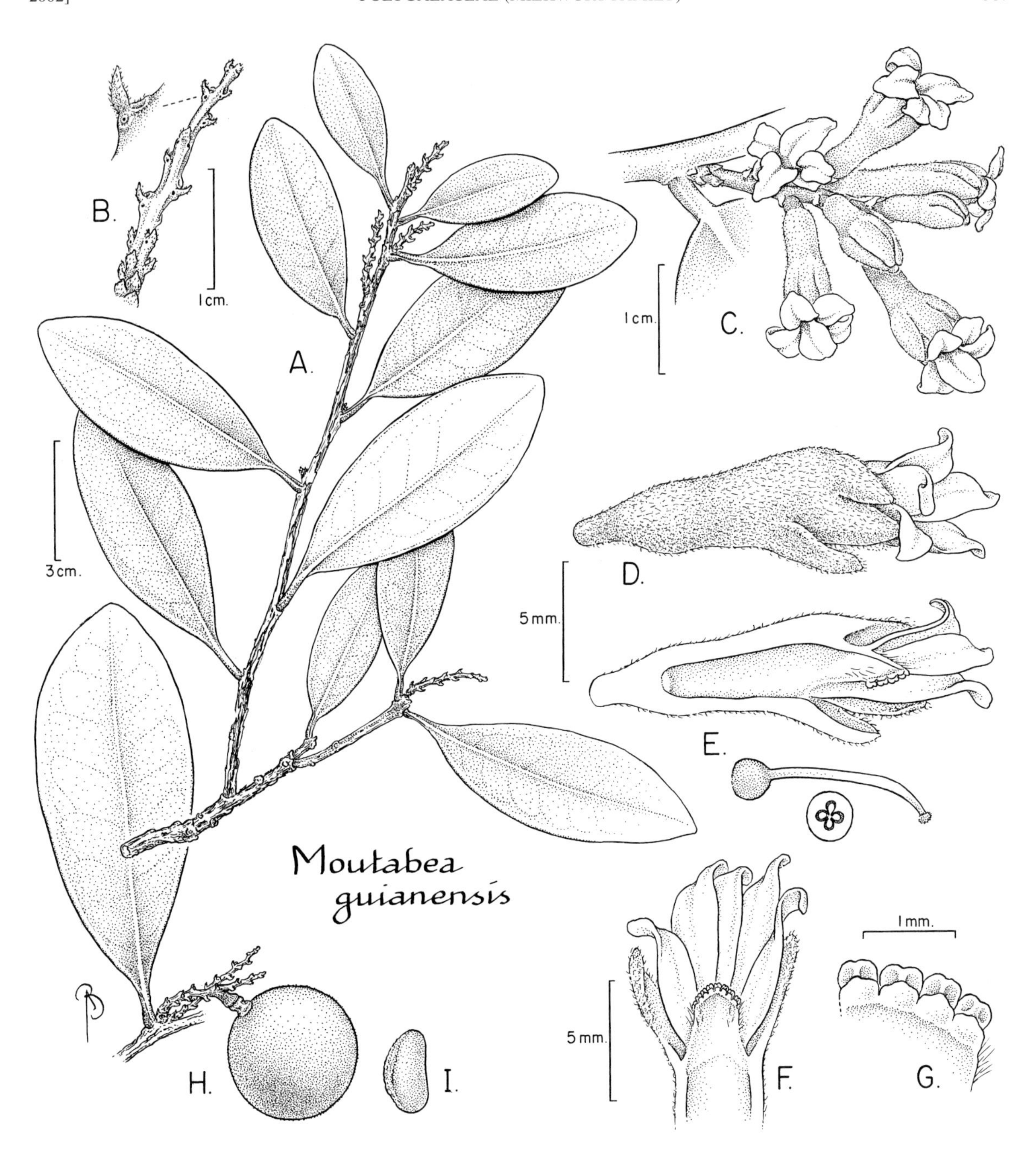

FIG. 255. POLYGALACEAE. *Moutabea guianensis* (A, B, D–G, *Mori et al. 19164*; C, *Mori 22145* from Kaw Mountains, French Guiana; H, *Mori & Hartley 18535*; I, *Mori & Pipoly 15512*). **A.** Part of stem with leaves and old inflorescence rachises. **B.** Rachis of inflorescence (right) with detail of bract subtended by a gland (above). **C.** Inflorescence. **D.** Lateral view of flower. **E.** Medial section of flower with pistil removed (above), pistil (near below), and transverse section of ovary (far below). **F.** Adaxial view of distal portion of sectioned flower showing fused stamens on column (staminal sheath). **G.** Adaxial view of one half of staminal sheath. **H.** Part of stem with leaf and fruit. **I.** Seed.

Polygala membranacea (Miq.) Görts

Shrubs or treelets, 0.3–2 m tall. Stems terete, 3–6 mm diam., glabrous, the branchlets rather densely to densely covered with appressed curved hairs toward apex, with a triangular to rhombic gland per internode. Leaves: petioles 2–3 mm long, slightly winged, with 2 triangular to rhombic glands at base; blades narrowly elliptic to elliptic, sometimes obovate, 5–18 × 1.4–6 cm, glabrous, the base obtuse, the apex acuminate; veins green to yellow, secondary veins eucampto-brochidodromous, in 6–10 pairs. Inflorescences axillary or terminal. Flowers 6.5–7 mm long; outer 3 sepals pale green, the wings green-yellow to white; corolla violet. Fruits 2-winged, transversely broadly ellipsoid to subglobose, 0.7–1.5 × 0.9–1.9 cm (including wings), pale green. Seeds cylindrical, 6–7.5 × 4.5–5 mm. Not yet collected from our area but to be expected.

Polygala spectabilis DC.

Treelets, to 3 m tall. Stems terete, 4–10 mm diam., the bark with vertical lines, the branchlets sparsely pubescent to densely pubescent with white curved hairs toward apex. Leaves: petioles 3–5 mm long, with 2 tubular glands at base; blades elliptic to broadly elliptic or obovate, 5–25 × 1.8–10 cm, sparsely pubescent with straight hairs, the base acute, the apex acute to acuminate, ending in sharp point; veins yellowish when dried, often purplish in young fresh leaves, secondary veins brochidodromous, in 7–9 pairs. Inflorescences terminal. Flowers 25–30 mm long; outer 3 sepals green, the wings bright purple, whitish with light purple, or light purple at base to greenish yellow toward apex, turning green in fruit; corolla white, dirty white, or pale purple, often with yellow outer margins. Fruits without wings, subglobose, 7–9 × 8–9 mm. Seeds subdeltoid, 4.5–5.6 × 2.4–3 mm. Fl (Jan, Aug), Fr (Jan); restricted to granitic outcrops.

SECURIDACA L.

Canopy lianas, or sometimes treelets, variously pubescent. Young stems greenish olive colored. Stipules only known in one species, reduced to glands or spines in other species. Leaves: blades chartaceous to subcoriaceous. Inflorescences of solitary flowers, many-flowered racemes, or sometimes panicles; pedicels present to nearly absent. Flowers zygomorphic; sepals free, the inner 2 forming petaloid wings, these not persistent in fruit; petals free, 3, the dorsal one forming helmet-shaped keel; stamens united in staminal sheath; ovary 1-locular, 1 ovule per locule, the style curved, the stigma terminal. Fruits 1-winged samaras. Seeds 1 per fruit, without arils.

1. Stipules present but caducous. Inflorescences of solitary flowers, not showy. Flower keel without crest. . . S. *uniflora*.
1. Stipules reduced to glands or spines. Inflorescences multiflowered, showy. Flower keel crested.
   2. Leaf blades obtuse at apex, densely covered with soft hairs adaxially. . . . . *S. volubilis*.
   2. Leaf blades acute to acuminate at apex, glabrous or nearly so adaxially.
      3. Stipules reduced to 2 circular glands. Leaf blades shiny grayish green adaxially. . . . . *S. pubescens*.
      3. Stipules reduced to 2 spines. Leaf blades light green adaxially. . . . . *S. spinifex*.

Securidaca pubescens DC. PL. 107c

Lianas or treelets. Secondary stems 3–5 mm diam., densely pubescent with short soft hairs. Stipules reduced to 2 circular glands. Leaves: petioles 5–8 mm long, densely pubescent; blades ovate to broadly ovate, rarely elliptic, 3–8.5 × 2.2–5.5 cm, shiny grayish green adaxially, sparsely pubescent only on midrib adaxially, densely pubescent abaxially, the base acute to obtuse, the apex acute to acuminate; secondary veins brochidodromous, in 4–5 pairs. Inflorescences multiflowered and showy. Flowers 10–12 mm long, fragrant; outer 3 sepals green with dark purple, the wings dark purple when young, bright to light purple when old; petals light purple or pink, the keel crested. Fertile fruits not seen from our area. Fl (Aug); collected only once from our area.

Securidaca spinifex Sandwith

Lianas. Stems 20–30 mm diam., somewhat flattened, the secondary stems densely to completely pubescent. Stipules reduced to 2 spines. Leaves: petioles 4–6 mm long, densely pubescent; blades elliptic, occasionally ovate or obovate, 4–12 × 2.5–6.5 cm, light green adaxially, very sparsely pubescent, mostly on midrib adaxially, densely pubescent abaxially, the base rounded, the apex acute to acuminate; secondary veins brochidodromous, in 7–9 pairs. Inflorescences multiflowered, showy, often with subtending leaves purple tinged. Flowers 8–11 mm long, entirely purple except for some white at base of petals; keel crested. Fruits to 2.4 cm long. the lower part ellipsoid, 1.2–1.6 × 0.8–1 cm, densely to completely pubescent, the primary wing 0.5–0.8 × 0.8–1 cm, the secondary wing nearly absent. Fl (Aug, Sep); rare, in forest gaps and in secondary vegetation.

No fruiting collections are known from the Guianas; the fruit description is based on *Mori & Cardoso 17724* (NY) and *Cowan 38309* (NY), both from Amapá, Brazil.

Securidaca uniflora Oort PART 1: FIG. 6

Slender liana. Stems flattened, green on one side, the secondary stems usually flattened, 2–5 mm diam., glabrous. Small stipules present but caducous, without glands. Leaves: petioles 2–3 mm long, glabrous, strongly wrinkled; blades usually ovate, sometimes elliptic, 2.5–8 × 1.2–4.5 cm, shiny grayish green adaxially, dull green abaxially, glabrous, the base rounded, the apex acute; secondary veins eucampto-brochidodromous, in 5–7 pairs. Inflorescences of solitary flowers, not showy. Flowers 5–8 mm long, with strong fragrance; wings white; keel without crest, white. Fruits to 7 cm long, the lower part ellipsoid to ovoid, 0.5–1 × 0.2–0.5 cm, the surface smooth, the primary wing membranous, 3.5–5.3 × 1.3–2 cm, the secondary wing usually short, but sometimes elongate, comma-shaped, to 1 cm long. Fl (Nov); known only by a single flowering collection from our area, in nonflooded forest.

The fruit description is based on *Larpin 736* (U) from French Guiana.

Securidaca volubilis L. emend. Oort
FIG. 256; PART 1: FIG. 6 as *S. diversifolia*

Lianas, densely to completely pubescent with very soft curved hairs. Stems terete, gray with light brown grooves, the secondary stems dark green, 3–5 mm diam. Stipules reduced to glands. Leaves: petioles 2–4 mm long, densely pubescent with 2 circular to tubular glands near base; blades elliptic, occasionally ovate, 3–5.5 × 1.5–3 cm, pubescent adaxially, more densely pubescent abaxially, the base rounded to cuneate, the apex obtuse; secondary veins eucampto-brochidodromous, in 5–7 pairs. Inflorescences multiflowered, showy. Flowers 9–12 mm long; outer 3 sepals green, the wings purple; keel purple, crested. Fruits 3.5–5 cm long, the lower part globose, 0.9–1.2 cm diam., ribbed, the primary wing 4–6 × 1.2–2 cm, with prominent parallel curved veins, the secondary wing comma shaped, 0.7–1 × 0.3–0.5 cm. Fl (Jun), fr (Feb).

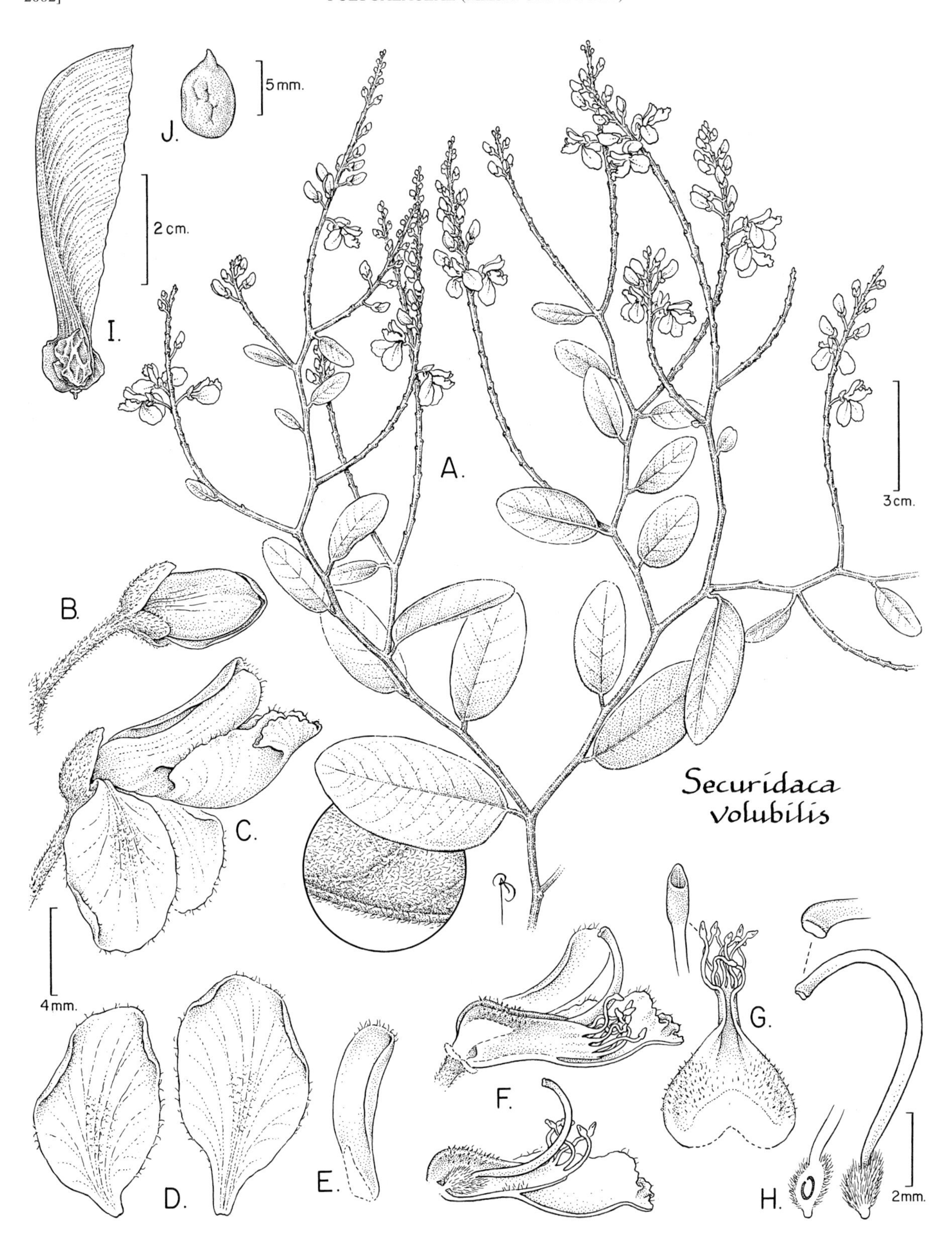

FIG. 256. POLYGALACEAE. *Securidaca volubilis* (A–H, *Mori & Gracie 18899*; I, J, *Mori & Gracie 21205*). **A.** Part of stem with leaves and inflorescences and detail of indument on abaxial surface of leaf. **B.** Lateral view of flower bud. **C.** Lateral view of flower. **D.** Adaxial views of two petaliod sepal wings. **E.** Adaxial view of an upper petal. **F.** Lateral view of flower with one upper petal and two wings removed (above) and lateral view of flower with two upper petals, two wings, and part of staminal sheath removed. **G.** Adaxial view of staminal sheath with detail of anther. **H.** Medial section of ovary (near right), lateral view of pistil (far right), and detail of stigma (above). **I.** Fruit. **J.** Seed.

POLYGONACEAE (Buckwheat Family)

John Brandbyge

Lianas in our area, annual or perennial herbs, shrubs, or trees elsewhere. Stipules commonly well developed and connate into tubular, persistent or deciduous, usually membranous to hyaline, often bilobed or fringed sheath (ocrea), this sometimes much reduced or lacking. Leaves simple, alternate in our area, seldom opposite or whorled elsewhere, the margins usually entire; venation usually pinnate. Inflorescences axillary or terminal, either on main stems or lateral shoots, paniculate, racemose, or spicate, but basically of clusters of dichasial or helicoid cymes, the partial inflorescences subtended by bracts, each flower subtended by a persistent membranous, tubular bracteole (ocreola); pedicels articulate. Flowers actinomorphic, relatively small, unisexual (plants polygamomonoecius or dioecious) or bisexual; perianth segments 2–6, basally connate into inconspicuous to conspicuous, green or colored floral tube, commonly in 2 similar to slightly dissimilar whorls of 3 segments each, sometimes in 1 whorl of 5 segments, mostly persistent into fruit; stamens (2)6–9, rarely more, the filaments free or basally connate, often adnate to perianth tube and forming distinct ring; annular nectary disc often present around base of ovary, or nectaries several and intrastaminal, the anthers usually versatile, introrse, opening by longitudinal slits; ovary superior, unilocular, the styles 1–3, distinct or proximally united, the stigmas filiform, peltate, or capitate, entire or variously fringed; placentation basal or shortly free-central, the ovule solitary. Fruits achenes or small nuts, trigonous or sometimes lenticular, mostly subtended or surrounded by accrescent perianth, this often fleshy and brightly colored. Seeds: embryo straight or curved, commonly excentric or peripheral, the endosperm well developed, ruminate in our area.

Eyma, P. J. 1932. The Polygonaceae, Guttiferae and Lecythidaceae of Surinam. Meded. Bot. Mus. Herb. Rijks Univ. Utrecht **4:** 1–77.

———. 1966. Polygonaceae. *In* A. Pulle (ed.), Flora of Suriname **I(1):** 49–71. E. J. Brill, Leiden.

Lindeman, J. & A. R. A. Görts-van Rijn. 1968. Polygonaceae. *In* A. Pulle & J. Lanjouw (eds.), Flora of Suriname **I(2):** 303–309. E. J. Brill, Leiden.

COCCOLOBA P. Browne

Lianas, shrubs (often with scrambling branches), or trees. Stems terete, commonly strongly striate, solid, or sometimes hollow except at nodes, the nodes occasionally swollen. Ocreae at first sheathing, later split along one or two sides, commonly coriaceous and persistent at base and membranous and irregularly breaking at apex, or entire ocrea membranous and caducous. Leaves alternate, persistent or deciduous, when deciduous, the young leaves usually turning black on drying, the leaves of adventitious or juvenile shoots commonly much larger and of different shape than those of normal shoots; petioles arising from base or from well above base of ocreae; blades glabrous, puberulous, or pilose, the margins entire; venation pinnate. Inflorescences: staminate flowers in 2–7-flowered partial inflorescences; pistillate flowers in 1- or rarely 2-flowered partial inflorescences; pedicels mostly enlarging in fruit. Flowers unisexual (plants polygamomonoecius or dioecious); perianth segments 5, the perianth 2–4 mm long, basally connate and forming 1–2 mm long campanulate tube, the lobes 1.5–2.5 mm long, imbricate in bud, often reflexed at anthesis; stamens 8, exserted in functionally staminate flowers, the filaments dilated at base into ring-like structure; staminodia not exserted; pistil exserted, the ovary trigonous, glabrous, the styles 3, the stigmatic surface often expanded; pistillode not exserted. Fruits trigonous achenes, often enclosed in expanded, fleshy, brightly colored perianth.

1. Leaf blades with 2 secondary veins on each side of midrib arising from same point near base, the secondary veins forming 45° angle with midrib, the tertiary veins salient adaxially when blades dry. . . . . *C. ascendens.*
1. Leaf blades with 1 secondary vein on each side of midrib arising from near base, the secondary veins forming >45° angle with midrib, the tertiary veins ± plane when blades dry.
    2. Petioles ca. 2 cm long, not winged; blades chartaceous. . . . . *C. excelsa.*
    2. Petioles usually >2 cm long, narrowly winged; blades coriaceous. . . . . *C. parimensis.*

Coccoloba ascendens Lindau

Lianas or high climbing shrubs, rarely shrubs or small trees. Ocreae 1–2 cm long, minutely puberulous to glabrous abaxially, apically acuminate. Leaves: petioles 2–2.5 cm long, glabrous; blades elliptic, broadly elliptic, ovate, or rarely obovate, (8)15–25 × (4)8–14(18) cm, coriaceous, glabrous, lustrous, especially adaxially, the base rounded to slightly cordate, the apex rounded, acute, short acuminate, or rarely emarginate; secondary veins in 8–9 pairs, 2 on each side of midrib arising from same point near base, the secondary veins forming 45° angle with midrib, the tertiary veins salient adaxially when blades dry. Inflorescences axillary or terminal, ± densely flowered, the rachis 6–15(20) cm long, glabrous or rarely pubescent; pedicels 1.5 mm long. Flowers: buds elongate, much exceeding bracteoles. Fruits (achenes plus surrounding perianth) ovoid to subglobose, (12)18–24 × (10)12–16 mm, the apex apiculate, greenish when young, turning red to deep blue and juicy when mature, the achenes trigonous, 10–16 × 8–12 mm, lustrous or dull, chesnut brown. Fl (Sep, Dec), fr (Dec, Mar); forests, often along rivers or in flooded forests, also found in secondary vegetation along roads.

Coccoloba excelsa Benth.

Lianas, or shrubs or small trees, then often leaning or scrambling. Ocreae 1–6 cm long, brownish puberulous or glabrate abaxially, apically long-acuminate. Leaves: petioles ca. 2 cm long, brownish puberulous, rarely glabrous; blades elliptic, 10–20 × 3.5–8 cm, chartaceous, brownish puberulous or rarely glabrous abaxially, the base obtuse, the apex ± acuminate; secondary veins in 8–11 pairs, 1 on each side of midrib arising from near base, the secondary veins forming >45° angle with midrib, the tertiary veins ± plane when blades

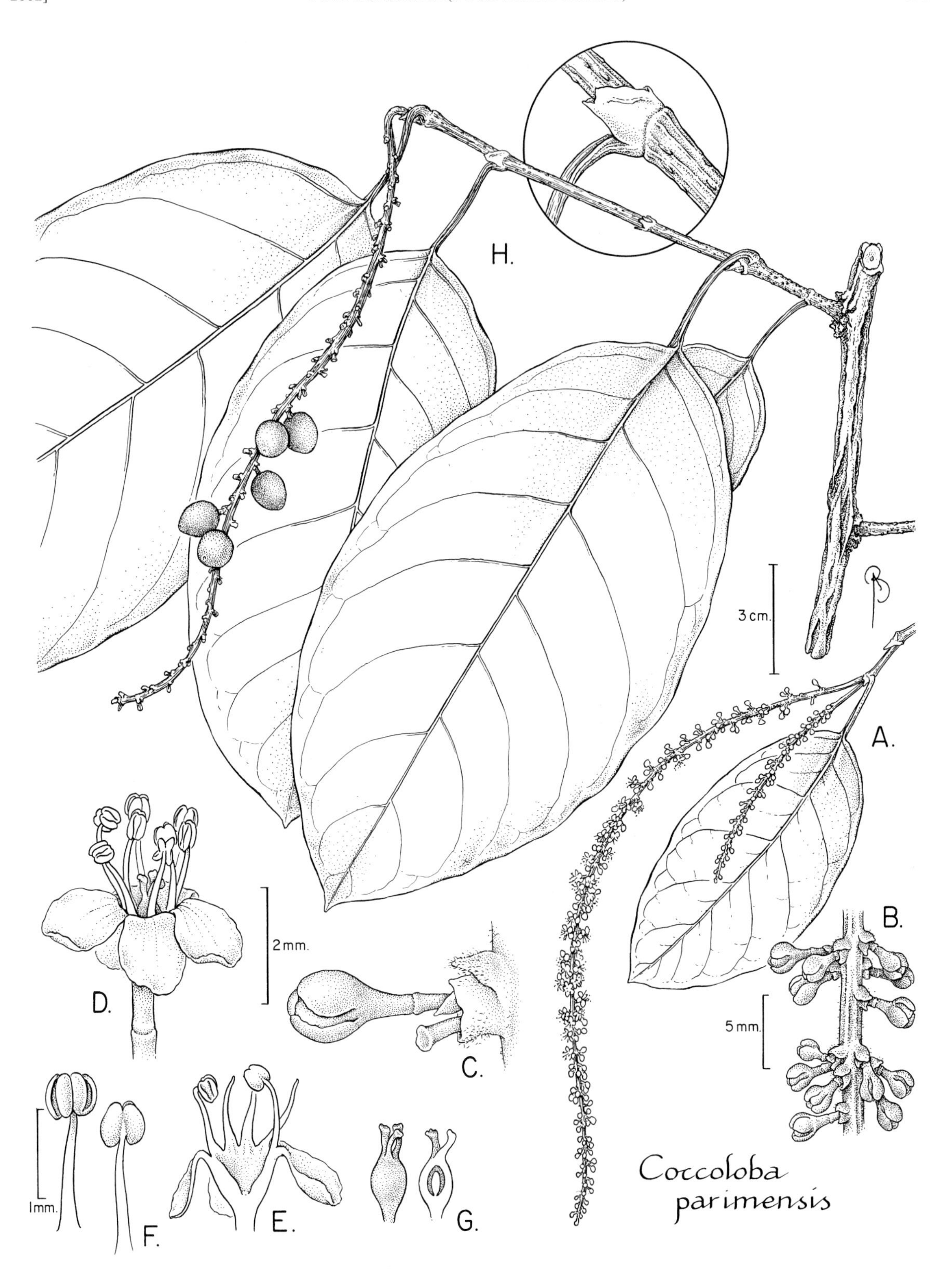

FIG. 257. POLYGONACEAE. *Coccoloba parimensis* (A–G, *Boom 8024* from Guyana; H, *Mori et al. 14993*). **A.** Inflorescence and leaf. **B.** Detail of inflorescence with buds. **C.** Bud in sheathing bract. **D.** Lateral view of flower. **E.** Medial section of flower with pistil removed. **F.** Adaxial (far left) and abaxial (near left) views of anthers. **G.** Pistil (far left) and medial section of pistil (right). **H.** Part of stem with leaves and infructescence with detail of ocrea at leaf node.

dry. Inflorescences axillary or sometimes terminal, ± loosely flowered, the rachis 5–13 cm long, puberulous; pedicels 2–3 mm long. Fruits (achenes plus surrounding perianth) ovoid to subglobose, 6–9 × 5–7.5 mm, green to wine-red when young, turning dark brownish to black, the achenes subglobose, 5–7.5 × 5–7 mm, chestnut to dark brown. Fl (Jul); collected once in our area, in nonflooded forest.

Coccoloba parimensis Benth. FIG. 257

Lianas or small trees, then often with scrambling branches. Ocreae 2–2.5 cm long, glabrate. Leaves: petioles 1.6–4 cm long, narrowly winged, glabrous; blades elliptic to broadly lanceolate, 8–25 × 6–14(20) cm, coriaceous, glabrous throughout, the base rounded to cuneate, the apex short acuminate; secondary veins in 8–12 pairs, 1 on each side of midrib arising from near base, the secondary veins forming >45° angle with midrib, the tertiary veins ± plane when blades dry. Inflorescences axillary or terminal, ± loosely flowered, the rachis (10)15–24 cm long, slightly puberulous to glabrate; pedicels 1.5–3 mm long. Fruits (achenes plus surrounding perianth) ovoid, 7–9(13) × 6–7(8.5) mm, the apex obtuse, dull green to black. Achenes ovoid, 6–8 × 6–6.5 mm, dull, pale brown. Fr (Sep); collected once in our area, in nonflooded forest on La Fumée Mountain Trail.

PROTEACEAE (Protea Family)

Vanessa Plana

Evergreen trees, treelets, or shrubs. Leaves simple or pinnately compound, sometimes heterophyllous, alternate, in verticils, or spirally arranged, rarely decussate. Stipules absent. Inflorescences axillary or lateral, racemose or paniculate with lateral conflorescences, the flowers paired, each pair subtended by persistent or caducous bracteole. Flowers actinomorphic or slightly to strongly zygomorphic, bisexual; perianth of 1 whorl, petaloid, 4-merous, the segments linear, valvate, recurving at anthesis, deciduous; stamens 4, opposite perianth segments, adnate to perianth, the anthers opening lengthwise; nectaries 4, hypogynous, fleshy, membranous, or scale-like, free or variously fused; ovary superior, sessile or shortly stipitate, 1-locular, the style straight or recurved, the stigma terminal or latero-apical. Fruits indehiscent or follicles, sometimes tardily dehiscent. Seeds 1 or 2, sometimes winged.

Mennega, A. M. W. 1966. Proteaceae. *In* A. Pulle (ed.), Flora of Suriname **I(1):** 154–157. E. J. Brill, Leiden.
———. 1968. Proteaceae. *In* A. Pulle & J. Lanjouw (eds.), Flora of Suriname **I(2):** 315–320. E. J. Brill, Leiden.

1. Leaves paripinnate, homophyllous. Nectary glands fleshy; style curved, the stigma latero-apical. . . . . *Euplassa.*
1. Leaves simple or imparipinnate in young, sterile plants (not apparent in herbarium specimens). Nectary glands either fleshy, scale-like, or membranous; style erect, the stigma terminal or not apparent.
   2. Leaves homophyllous; secondary vein angles with midrib ca. 45°. Nectary glands fused, membranous. Fruits indehiscent or tardily dehiscent follicles. Seeds fleshy, not flattened, not winged. . . . . *Panopsis.*
   2. Leaves heterophyllous (not apparent in herbarium specimens); secondary vein angles with midrib <45°. Nectary glands free, fleshy or scale-like. Fruits follicles, usually flattened. Seeds not fleshy, flattened, surrounded by broad wing. . . . . *Roupala.*

EUPLASSA Salisb.

Trees. Leaves pinnately compound, spirally arranged, paripinnate; leaflets frequently subopposite, in 3–6 pairs, the margins entire to broadly serrate; secondary vein angles with midrib ca. 45°. Inflorescences axillary or terminal, racemose or paniculate, solitary or in pairs; bracteole subtending flower pair narrowly to broadly triangular, minute, to 4 mm long; pedicels fused to differing heights, never free. Buds opening laterally. Flowers weakly to strongly zygomorphic; perianth segments all recurving at anthesis or frequently one remaining erect; anthers ovate, subsessile, housed in broad cavity at distal end of perianth segments; nectary glands fused, fleshy, lobed, or distinct, if the latter then heteromorphic; ovary shortly stipitate, glabrous or pubescent, the style curved, the stigma latero-apical. Fruits indehiscent, globose or subglobose, the pericarp coriaceous. Seeds 2 per fruit, ± flattened, not winged.

Euplassa pinnata (Lam.) I. M. Johnst. FIG. 258, PL. 107d

Leaves paripinnate, with 3–5 pairs of leaflets, the rachis 6–21.7 cm long; leaflets rarely subopposite, coriaceous. Inflorescences axillary, racemose; bracteoles persistent, the lowermost one unusually long, to 4 mm long; pedicels fused to half way. Buds with distinctive beak to 1 mm long. Flowers weakly zygomorphic; perianth segments keeled, recurving at anthesis; nectary glands entirely fused except for gap on ventral side; ovary square in transverse section, dark orange-red pilose, the style recurved at apex, the stigmatic surface ca. 2 mm long, the receptive area limited to tip of conical protuberance on stigma, often distinctly yellow. Fruits subglobose. Seeds not winged. Fl (Nov); tropical moist forest. *Bois grage, bois grage blanc, amoussaié blanc* (Créole).

PANOPSIS Salisb.

Trees, treelets, or shrubs. Leaves simple, spirally arranged, in verticils, or decussate; secondary vein angles with midrib ca. 45°. Inflorescences usually terminal, sometimes lateral or subterminal, branched or unbranched, subtended by small leaves; bracteoles small, caducous, ligulate; pedicels fused or free. Buds elongate, opening apically as result of elongation of style. Flowers essentially actinomorphic; perianth parts strongly recurved; stamens with free parts of filaments ribbon-like, free or attached on perianth segments below half way, the free parts of filaments ribbon-like; nectary glands fused, membranous, four-lobed; ovary stipitate, pubescent, the style terminal, erect, straight sided to clavate, the stigma small, circular; ovules 2, collateral, pendulous. Fruits indehiscent or tardily dehiscent follicles, either globose or fusiform and retaining pubescence. Seeds 1 per fruit, large, fleshy, not flattened, not winged.

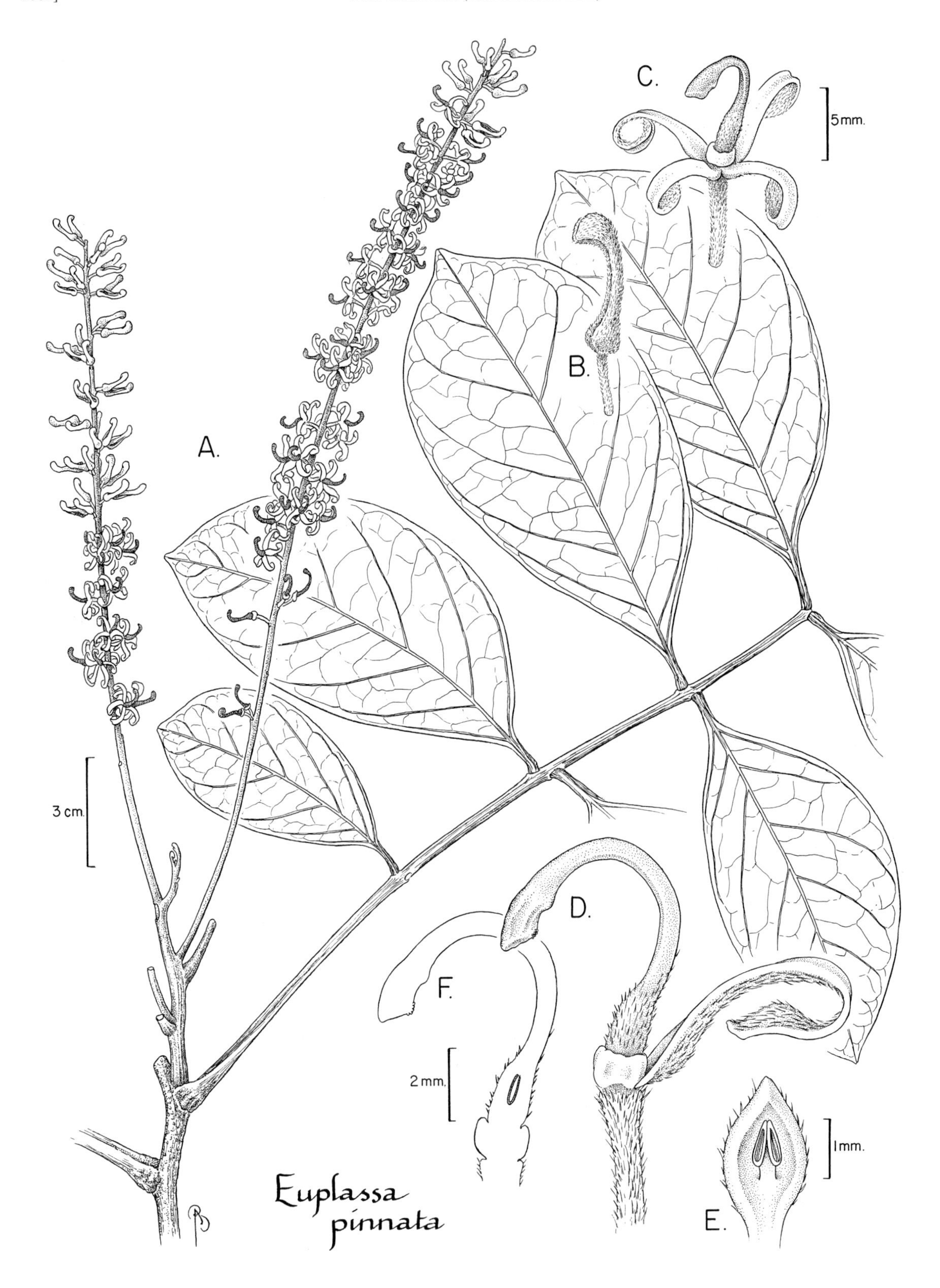

FIG. 258. PROTEACEAE. *Euplassa pinnata* (*Mori & Boom 15168*). **A.** Apex of stem showing leaf and inflorescences. **B.** Flower bud; note valvate aestivation of calyx. **C.** Oblique-apical view of flower. **D.** Lateral view of flower with one sepal attached. **E.** Stamen adnate to adaxial surface of sepal. **F.** Medial section of pistil.

Panopsis rubescens (Pohl) Rusby

Trees, less frequently treelets or shrubs. Leaves spirally arranged, rarely decussate, subcoriaceous to coriaceous. Inflorescences congested racemes, branched or unbranched, if branched with 1–6 lateral conflorescences; bracteoles minute, 1–1.5 mm long; pedicels free, very delicate, thread-like, 4–11(13) mm long, elongating to 15 mm long after anthesis. Flowers 3–6.5 mm long; filaments attached near base of perianth segments; nectary gland deeply lobed from 1/3 length to base; ovary indumentum long pilose, the style clavate, slightly drawn out, truncate at very tip with small central depression. Fruits resembling follicles, tardily dehiscent or indehiscent, densely velutinous. Seeds globose to fusiform. Known only by a sterile collection from our area; in forest on humid or wet soils to 500 m.

ROUPALA Aubl.

Trees or shrubs. Adult leaves simple, young sterile leaves pinnatifid to imparipinnate, alternate or spirally arranged; secondary vein angles with midrib <45°. Inflorescences axillary or terminal, racemose; bracteoles persistent or caducous, minute to several mm long, triangular; pedicels free to base, very rarely fused. Flowers actinomorphic; perianth segments all recurving at anthesis; filaments adnate to perianth segments, the anthers sessile or subsessile, linear-oblong; nectary glands free, fleshy or scale-like; ovary sessile, pubescent to glabrous, the style erect, elongate, subterete, the stigma a slit, terminal; ovules 2, collateral, pendulous. Fruits follicles, usually flattened. Seeds 2 per fruit, not fleshy, flattened, surrounded by broad wing.

Roupala montana Aubl.

Small trees, to 10 m tall. Bark usually light brown when young, red brown or red-gray after peeling of outer layer. Leaves spirally arranged, the adult leaves simple, the younger leaves imparipinnate, with at least 5 ranks of lateral leaflets, usually more. Inflorescences densly clustered racemes, 5–26 cm long, the indumentum pale brown to mid-brown throughout; bracteoles minute, orange pubescent; pedicels free. Flowers 7–9 mm long; anthers subsessile, 2–3 mm long; nectary glands fleshy, free; ovary with indumentum short to long sericeous. Fruits constricted at base, not significantly constricted at apex, light brown to dark brown, the indumentum adpressed pilose or velutinous when young, glabrescent to glabrous at maturity. Known only by a sterile collection from our area; a widespread species found in many different habitats. *Bois grage rouge* (Créole).

QUIINACEAE (Quiina Family)

Scott A. Mori

Trees or shrubs. Stipules interpetiolar, in pairs or solitary (*Touroulia*). Leaves simple or pinnately compound (*Froesia* and *Touroulia*), opposite or whorled; tertiary veins very fine, closely spaced, inconspicuous. Inflorescences compound racemose. Flowers actinomorphic to subactinomorphic (the sepals often unequal), unisexual or bisexual (plants polygamodioecious or dioecious); sepals 4–5, free or fused at base; petals 4–8, free; stamens numerous; ovary superior; placentation axile. Fruits berries or capsules. Seeds covered with a conspicuous indumentum or glabrous (*Froesia*); endosperm present or absent, the cotyledons thin or thick.

Görts-van Rijn, A. R. A. 1986. Quiinaceae. *In* A. L. Stoffers & J. C. Lindeman (eds.), Flora of Suriname **III(1–2):** 468–472. E. J. Brill, Leiden.
Lanjouw, J. & P. F. Baron van Heerdt. 1966. Quiinaceae. *In* A. Pulle (ed.), Flora of Suriname **III (1):** 355–365. E. J. Brill, Leiden.

1. Leaves pinnately compound. . . . . *Touroulia.*
1. Leaves simple.
    2. Leaves whorled. Ovary 6–14-locular. Fruits with distinct cavities as seen in transverse section of pericarp. . . . . *Lacunaria.*
    2. Leaves opposite or whorled. Ovary 1–2-locular. Fruits without distinct cavities as seen in transverse section of pericarp. . . . . *Quiina.*

LACUNARIA Ducke

Trees. Stipules in pairs, usually persistent, awl- or needle-shaped. Leaves simple, whorled; venation craspedidromous to eucamptrodromous, the secondary veins either extended beyond margins into teeth, these not conspicuously spine-like, or not extending beyond margins. Staminate and pistillate or bisexual inflorescences not markedly different in size. Flowers unisexual or bisexual; calyx 2–4-merous; corolla 4–8 merous; ovary 6–14-locular, with 2–4 ovules in each locule. Pistillate and bisexual flowers with styles poorly developed. Fruits berries, with very well developed cavities in pericarp, the pericarp usually longitudinally striate or furrowed. Seeds densely covered with hairs.

1. Leaf blades crenate toward apex, 10–18 cm long; secondary veins in 10–13 pairs. Inflorescences terminal. Stamens ca. 50; stigmas 8. . . . . *L. crenata.*
1. Leaf blades entire throughout, 15–24 cm long; secondary veins in 15–25 pairs. Inflorescences ramiflorous. Stamens ca. 18; stigmas 12–13. . . . . *L. jenmanii.*

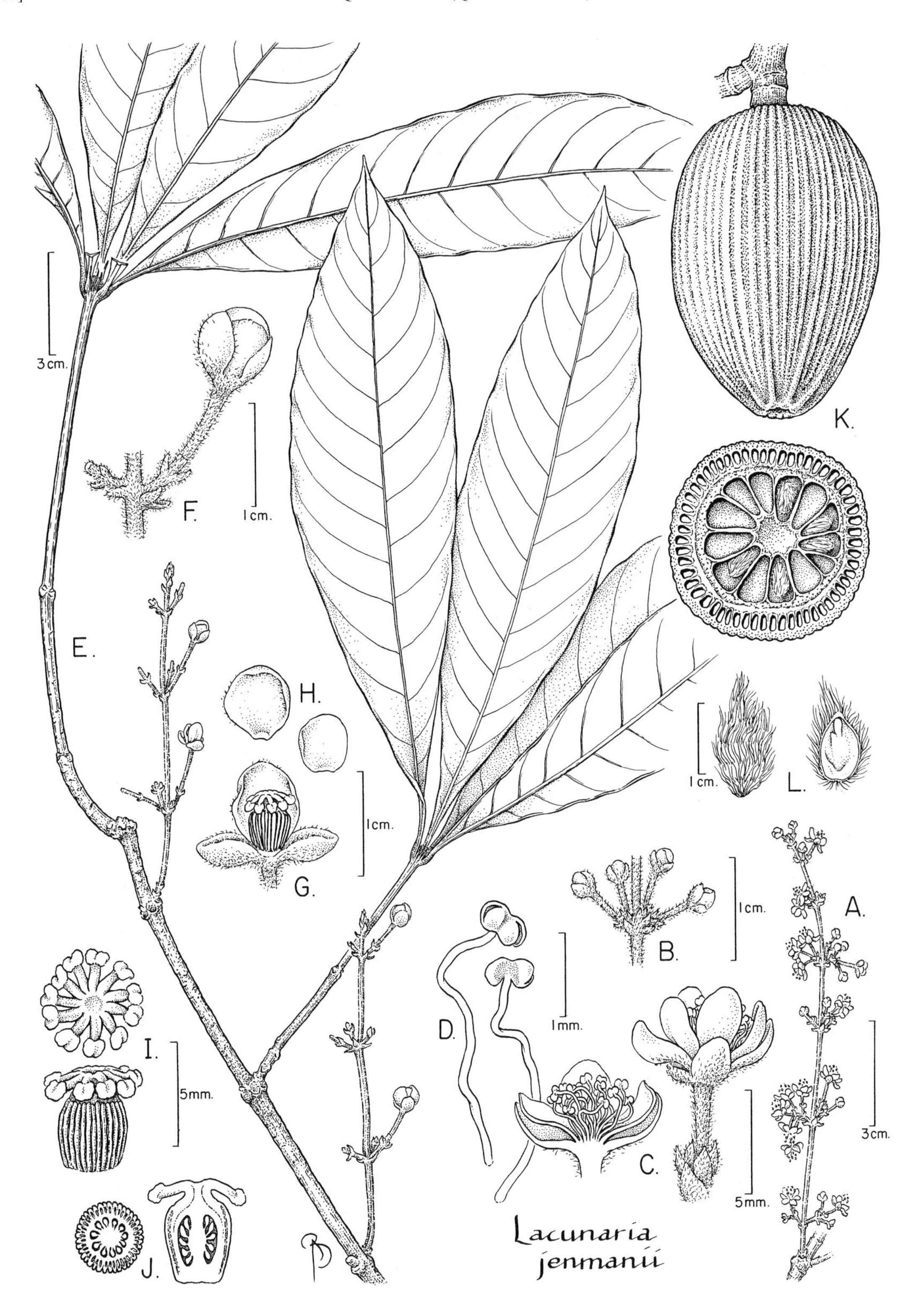

FIG. 259. QUIINACEAE. *Lacunaria jenmanii* (A, B, *Sabatier 1423*; C, D, *Mori & Werff 23280*; E–J, *Mori & Pipoly 15417*; K, L, *Mori & Hartley 18549*). **A.** Part of staminate inflorescence. **B.** Detail of staminate inflorescence. **C.** Medial section (left) and lateral view (right) of staminate flower. **D.** Two stamens. **E.** Part of stem with leaves and pistillate inflorescences. **F.** Detail of pistillate inflorescence. **G.** Lateral view of pistillate flower with three petals and one sepal removed. **H.** Adaxial views of two petals. **I.** Apical view of stigma (above) and lateral view of pistil (below). **J.** Transverse (left) and medial section (right) of pistil. **K.** Fruit (above) and transverse section of fruit (below); note cavities in pericarp. **L.** Seed (left) and medial section of seed (right).

Lacunaria crenata (Tul.) A. C. Sm. PL. 108a

Understory trees, to 23 m tall. Stipules needle-shaped. Leaves in whorls of 3–4, less frequently opposite; petioles nearly absent to 22 mm long, swollen to pulvinate at base; blades narrowly obovate to oblanceolate, less frequently elliptic, 10–18 × 3–7.5 cm, the base acute to cuneate, the apex acuminate, the margins crenate toward apex; secondary veins in 10–13 pairs. Inflorescences terminal, the axes ferruginous. Staminate flowers 9–12 mm diam.; sepals 4; petals 8; stamens ca. 50, the filaments to 1.5 mm long, the anthers subglobose, 0.4 × 0.6 mm. Pistillate flowers not known from our area. Fruits globose, with 8 separate, persistent stigmas. Fl (Nov, Dec), immature fr (Jan, May); in nonflooded forest.

One collection, *Mori et al. 22202* (NY), with the terminal inflorescences of this species but lacking the crenate leaf margins, probably represents this species, but was not used in writing the description.

Lacunaria jenmanii (Oliv.) Ducke FIG. 259, PL. 108b

Understory trees, 5–20 m tall. Stipules awl-shaped. Leaves usually in whorls of 4; petioles 12–20 mm long, swollen to pulvinate at base; blades elliptic, narrowly obovate, or oblanceolate, 15–24 × 6–10 cm, the base obtuse, acute, or narrowly cuneate, the apex acuminate, the margins entire; secondary veins in 15–25 pairs. Inflorescences ramiflorous, the axes ferruginous. Staminate flowers 5–7 mm diam.; sepals 4, with 2 smaller outer ones and 2 larger inner ones; petals 4; stamens ca. 18, the filaments 1.5–3 mm long in same flower, the anthers subglobose, 0.4 × 0.4 mm. Pistillate flowers ca. 12 mm diam; sepals 4?, petals ca. 8, usually white; stigmas 12–13, light colored, pad-like, laterally positioned. Fruits ovoid, to 9 × 7 cm. Fl (Mar, Apr, Aug, Sep), fr (Jan, Feb, Apr, Jul, Oct); in nonflooded forest.

QUIINA Aubl.

Trees. Stipules in pairs, usually persistent, awl-shaped, needle-shaped, or half-cordate, when half-cordate, the two together appearing cordate. Leaves simple, opposite or whorled; venation eucamptrodromous to brochidodromous, the secondary veins not extending beyond margins into teeth. Staminate and pistillate or bisexual inflorescences not markedly different in size. Flowers unisexual or bisexual; sepals (3)4(5); petals 3–6; ovary 1–2-locular, with 2 ovules in each locule. Pistillate and bisexual flowers with 2, well developed styles. Fruits berries, without well developed cavities in pericarp, the pericarp somewhat striate. Seeds densely covered with hairs.

1. Leaves whorled. . . . . *Q. obovata.*
1. Leaves opposite.
    2. Leaf blades 35–60 cm long.
        3. Stipules ovate to narrowly ovate. Leaf blades coriaceous; secondary veins in 18–21 pairs. . . . *Q. oiapocensis.*
        3. Stipules needle-shaped, awl-shaped, or half cordate. Leaves blades chartaceous; secondary veins in 54 pairs. . . . . *Quiina* sp. A.
    2. Leaf blades 8–22 cm long.
        4. Stipules needle-shaped or awl-shaped. Inflorescences axillary. Fruits nearly as long as wide. . . *Q. guianensis.*
        4. Stipules half-cordate. Inflorescences ramiflorous. Fruits distinctly longer than wide. . . . . *Q. sessilis.*

Quiina guianensis Aubl. PL. 108c

Understory trees, usually 5–10 m tall. Stipules needle-shaped or awl-shaped, 7–8 × 0.2–0.6 mm. Leaves opposite, subsessile; petioles 0–3 mm long; blades narrowly elliptic to elliptic, less frequently oblanceolate, 9–22 × 2.2–3.1 cm, the base rounded to nearly auriculate, the apex acuminate, the margins entire; secondary veins in 13–20 pairs. Inflorescences axillary, fasciculate, or racemose. Staminate flowers ca. 3 mm diam.; sepals 4; petals 4–5, yellow or cream-colored; stamens ca. 18, cream-colored, the filaments 0.8–0.9 mm long, the anthers 0.4 × 0.45 mm. Bisexual flowers unknown from our area. Fruits globose, orange. Seeds 2 per fruit, covered with golden hairs. Fl (Jan, Feb, Jun), fr (Apr, Jul); common, in nonflooded forest.

One collection, *Mori & Gracie 18365* (NY), has stipules intermediate in width between this species and *Q. sessilis*. In other respects, however, it more closely resembles *Q. guianensis*.

Quiina obovata Tul. PL. 108d

Understory trees, to 15 m tall. Stipules needle-shaped, 9.5–12 × 0.3–0.4 mm. Leaves whorled, 3–4 per node; petioles of adult leaves 12–22 mm long; blades of saplings and sprouts pinnatisect, 11–40 × 6.5–26 cm, those of adult leaves narrowly obovate to elliptic, 9.5–15.5 × 3.5–6 cm, the base cuneate to narrowly cuneate, the apex acute to acuminate, the margins entire; secondary veins in 17–18 pairs. Inflorescences axillary, racemose. Staminate flowers ca. 2 mm diam.; sepals 4; petals 4; stamens ca. 20. Bisexual flowers unknown from our area. Fruits ellipsoid, 15–18 × 8–10 mm, yellow-orange. Seeds 1 per fruit, covered with hairs. Fl (Aug, Oct), fr (Nov); common, in nonflooded forest.

Collections of this species from our area have previously been determined as *Q. pteridophylla* (Radlk.) Pires which has ovate stipules and occurs further south in eastern Amazonian Brazil.

Quiina oiapocensis Pires PL. 108e

Understory trees, to 15 m tall. Stipules ovate to narrowly ovate, 30–35 × 14–17 mm. Leaves opposite; petioles 60–70 mm long; blades narrowly obovate to oblanceolate, 38–39 × 12.5–15 cm, coriaceous, the base obtuse, acute, or cuneate, the apex obtuse, the margins entire; secondary veins in 18–21 pairs. Inflorescences axillary, the rachis 14–22 cm long. Flowers with 5 sepals as determined from fruits. Fruits fusiform, ferruginous, 20–24 × 7 mm. Seeds 2 per fruit, 15 × 5 mm, densely villose. Fr (Apr); uncommon, in nonflooded forest.

Quiina sessilis Choisy

Understory trees, usually 5–10 m tall. Stipules half-cordate, the two together cordate, appearing to have originally been one but now separated into two, 8–14 × 4–6 mm. Leaves opposite; petioles 4–6

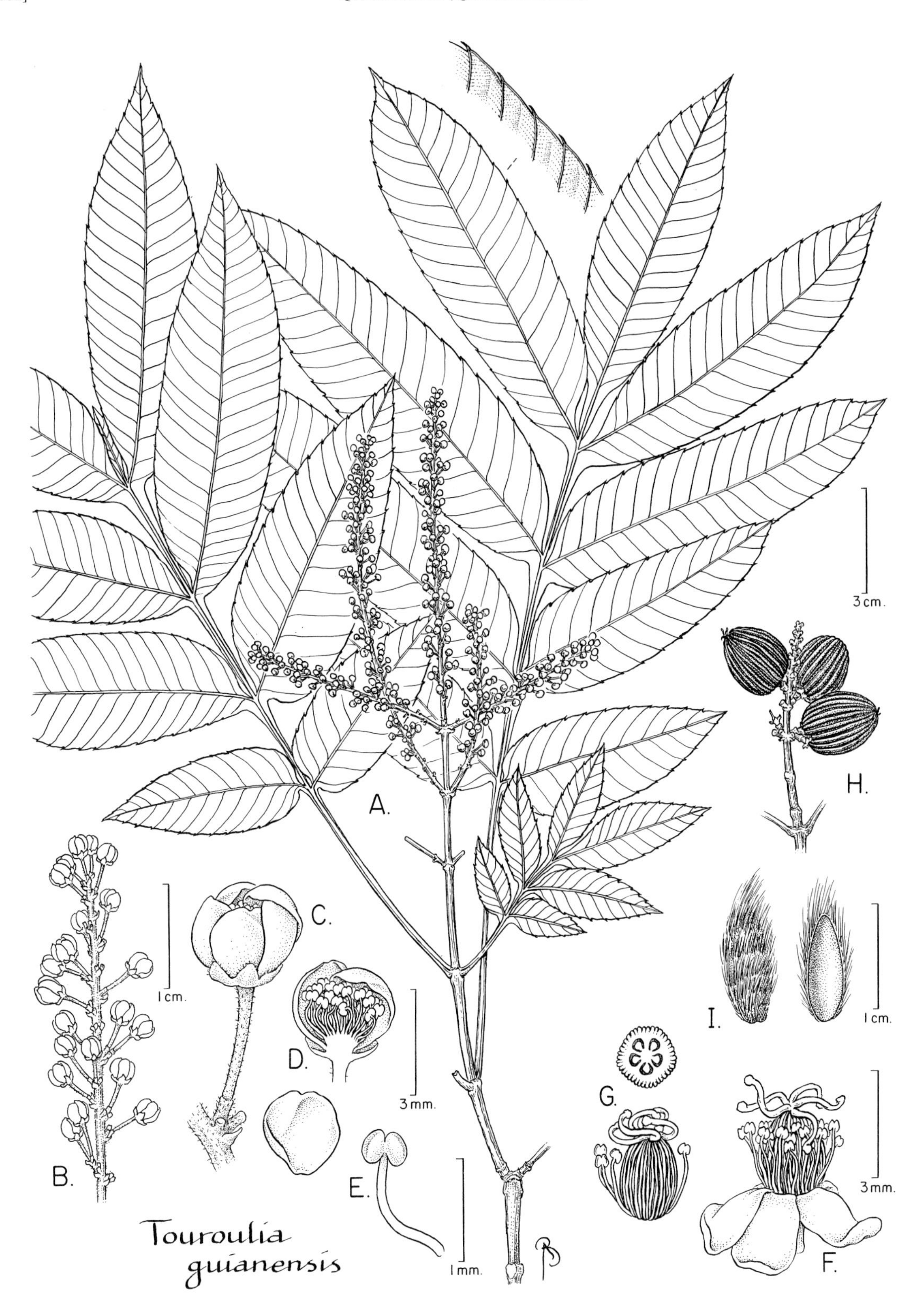

FIG. 260. QUIINACEAE. *Touroulia guianensis* (A–E, *Boom et al. 8692* from Brazil; F, G, *Blanco 534* from Venezuela; H, I, *Forest Department of British Guiana 4097* from Guyana). **A.** Part of stem with leaves and inflorescence in bud; detail of leaf margin above. **B.** Staminate inflorescence in bud. **C.** Staminate bud. **D.** Medial section of staminate flower (right) and adaxial view of a petal (below). **E.** Stamen. **F.** Lateral view of bisexual flower. **G.** Pistil from bud with a few stamens attached (below) and transverse section of ovary (above). **H.** Infructescence. **I.** Seed (near right) and longitudinal section of undeveloped seed (far right).

mm long; blades narrowly obovate to oblanceolate, 15–19 × 4.5–7 cm, the base rounded to nearly auriculate, the apex acuminate, the margins entire; secondary veins in 16–22 pairs. Inflorescences ramiflorous, racemose. Staminate flowers unknown from our area. Bisexual flowers with petals yellow in bud; ovary with two distinct styles. Fruits narrowly obovoid, 15–20 × 6–8 mm, the pericarp very woody and fibrous, difficult to cut into transverse sections. Seeds 1 per fruit, covered with hairs. Fl (Aug), fr (Oct); in nonflooded forest.

Quiina sp. A

Monopodial treelets, to 1.5 cm tall. Stipules needle-shaped, awl-shaped, or half cordate, 35–60 × 2–6 mm. Leaves opposite; petioles 10–20 mm long; blades oblanceolate, 50–70 × 9–16.5 cm, chartaceous, the base narrowly cuneate, the apex acute, the margins entire; secondary veins in 54 pairs. Known only from two sterile collections from understory plants (*de Granville 2296*, CAY and *Mori et al. 23338*, NY) which are probably not representative of adult plants.

TOUROULIA Aubl.

Trees. Stipules solitary, caducous but leaving conspicuous interpetiolar line as scar. Leaves imparipinnate, opposite; leaflets serrate; venation craspedodromous, the secondary veins extending beyond margin into spine-like teeth. Staminate inflorescences larger than bisexual ones. Flowers unisexual (plants dioecious or androdioecious) or bisexual; calyx 5-merous; corolla 5-merous. Pistillate and bisexual flowers with separate styles, the ovary 7-locular, with 1–2 ovules in each locule. Fruits berries, each locule containing a single seed. Seeds densely covered with long, villous hairs.

Touroulia guianensis Aubl. FIG. 260

Understory to canopy trees, 10–30 m tall. Leaves usually with 4 pairs of leaflets and a single terminal leaflet; leaflets sessile, usually narrowly oblong, less frequently narrowly elliptic or oblanceolate, 9–13 × 2–3.5 cm, the base decurrent onto rachis, the apex acuminate; secondary veins in 16–21 pairs. Inflorescences apparently with all staminate flowers or with all bisexual flowers, the staminate inflorescences larger than bisexual inflorescences. Staminate flowers 4–5 mm diam.; sepals imbricate; petals cucullate, pale yellow; stamens ca. 60, the filaments ca. 1 mm long, the anthers 0.3 × 0.45 mm. Pistillate flowers with petals pale yellow. Fruits ovoid, 2–2.5 × 2 cm, the exocarp finely and longitudinally sulcate. Seeds ellipsoid, 14 × 6 mm. Only two sterile collections known from our area; in nonflooded forest. *Bois macaque* (Créole).

RAFFLESIACEAE (Rafflesia Family)

Scott A. Mori

Minute, parasitic herbs on some species of Flacourtiaceae and legumes, vegetatively reduced to mycelium-like tissue penetrating tissue of host. Leaves simple, reduced to scales, or absent, whorled, opposite, or alternate. Inflorescences solitary, the flowers subtended by series of bracts. Flowers actinomorphic, unisexual (plants monoecious or dioecious), actinomorphic, minute to 10 mm diam. in Neotropical species, very large in Asian species of *Rafflesia*; tepals 4–10, caducous, petalloid and differentiated from bracts or not differentiated from bracts; stamens in 2 series around pistillode; ovary half-inferior or inferior, unilocular, the style distinct, the stigma capitate; placentation parietal, the placentae 4 and distinct or ovules not grouped in distinct placentae. Fruits berries.

This description based on *Apodanthes* and *Pilostyles*, the only genera known to occur in South America.

Gentry, A. H. 1973. Family 50A. Rafflesiaceae. *In* R. E. Woodson, R. W. Schery & collaborators (eds.), Flora of Panama. Ann. Missouri Bot. Gard. **60:** 17–21.

APODANTHES Poit.

Parasitic herbs on some species of Flacourtiaceae. Leaves absent. Inflorescences with flower subtended by several whorls of bracts, emerging from stems and trunks of hosts and appearing like cauline flowers of host. Pistillate flowers with 4, petalloid tepals, these inserted near top of ovary; ovary half-inferior, the placentae 4, protruding into locule.

Apodanthes caseariae Poit.

Inflorescences abundant along trunk and stems of host. Staminate flowers with numerous stamens. Pistillate flowers 4–6 × 2–3 mm; tepals white. Fl (Sep); common, in cloud forest on Mont Galbao.

RHABDODENDRACEAE (Rhabdodendron Family)

Scott A. Mori

Small trees or shrubs. Stipules small, obscure, caducous. Leaves simple, alternate, with small, lepidote scales and dark punctations adaxially, showing very small, numerous, pellucid punctations when viewed with hand lens against strong light. Inflorescences axillary

FIG. 261. RHABDODENDRACEAE. *Rhabdodendron amazonicum* (A, B, E–G, *Silva 1134* from Brazil; C, D, *Austin et al. 7067* from Brazil; H, I, *Maguire et al. 47095* from Brazil). **A.** Terminal part of stem with leaves and inflorescences. **B.** Leaf with detail of abaxial surface and lepidote scale. **C.** Part of inflorescence. **D.** Flower. **E.** Flower with most of anthers disarticulated. **F.** Medial section of flower (left) and pistil with gynobasic style (right). **G.** Adaxial (near right) and lateral (far right) views of anther. **H.** Part of infructescence. **I.** Medial section of fruit.

or supraxillary, infrequently terminal, racemose; pedicels with several scattered bracteoles. Flowers appearing actinomorphic but displaced style making them slightly zygomorphic; calyx with 5 lobes or lobes indistinct; corolla with 5, imbricate, caducous petals; stamens numerous; ovary superior, 1-locular, the style arising from one side of ovary, the stigma displaced to one side of style; placentation basal, the ovule single. Fruits drupes, subtended by cup-shaped receptacle. Seeds 1 per fruit, without endosperm; cotyledons thick, fleshy.

Prance, G. T. 1972. Rhabdodendraceae. Fl. Neotrop. Monogr. **11:** 1–22.

RHABDODENDRON Gilg & Pilg.

A monogeneric family. Description same as description for the family.

Rhabdodendron amazonicum (Benth.) Huber
FIG. 261, PL. 109a,b

Small trees, usually 2.5–15 m tall. Leaves: petioles 2–3.5 cm long; blades oblanceolate, the largest 34–37 × 4.5–9 cm, the base narrowly cuneate, decurrent onto petiole, the apex acuminate to very shortly and abruptly acuminate, the margins entire; secondary veins in 20–25 pairs, intersecondary veins present, making it difficult to count number of secondary veins, the tertiary veins inconspicuous. Flowers 15–18 mm diam.; corolla white or cream-colored, caducous, usually absent shortly after anthesis; stamens ca. 50, the anthers linear, ca. 6 × 0.3 mm, dehiscing laterally, readily disarticulating from filaments, the filaments 1.5 mm long; style 5 mm long, the stigmatic surface ca. 4 mm long. Fruits globose, ca. 10 mm diam, black or purple at maturity, subtended by red, cup-like receptacle. Fl (Nov), fr (May); scattered, in nonflooded forest forest.

RHAMNACEAE (Buckthorn Family)

Scott V. Heald

Trees or lianas, the latter with tendrils. Stipules lanceolate, pubescent, paired, caducous. Leaves simple, alternate; blades with or without glands at base, the margins entire or serrate. Inflorescences cymose or racemose. Flowers actinomorphic, bisexual; sepals 5, cuneate, thick, with valvate aestivation; petals 5 and clawed or lacking; stamens 5, opposite petals (alternate sepals when petals lacking) and ± enveloped by them; anthers introrse; nectary disc intrastaminal, surrounding and sometimes adnate to ovary; ovary superior or seemingly inferior by adnation of nectary disc, 3-locular, the style single at base, 3-cleft at apex, the stigmas 3; placentation basal, the ovules usually 1 per locule. Fruits capsules or drupes.

Lanjouw, J. 1966. Rhamnaceae. *In* A Pulle (ed.), Flora of Suriname **II(1):** 102–106. E. J. Brill, Leiden.

1. Lianas with axillary tendrils. Fruits capsular, strongly trigonal, with 3 narrow wings. . . . . *Gouania.*
1. Trees without tendrils. Fruits ellipsoid drupes without wings.. . . . . *Ziziphus.*

GOUANIA Jacq.

Lianas with axillary tendrils. Stems pubescent. Leaves: blades obovate to ovate, the margins serrate, the base symmetrical, sometimes with glands; secondary venation pinnate, arcuate. Inflorescences axillary and racemose or appearing terminal and paniculate. Flowers: petals 5, caducous; nectary disc fringed around style; ovary seemingly inferior by adnation of nectary disc; ovule 1 in each developing locule. Fruits capsular, strongly trigonal, with 3 narrow wings, fragmenting septacidally into 3 mericarps, the mericarps 2-winged, V-shaped.

Gouania blanchetiana Miq. [Syn.: *G. frangulifolia* (Schult.) Radlk.]
FIG. 262

Lianas. Stems with short, spreading pubescence. Tendrils transformed from branches, generally with 1 ± suppressed leaf, unbranched, coiled in one plane, infrequent on fertile shoots. Leaves: petioles 0.6–1.4 cm long, pubescent; blades narrowly obovate to narrowly ovate, 4.5–7.5 × 1.9–3.1 cm, pubescent in secondary vein axils abaxially, the base obtuse, occasionally slightly asymmetrical, with adaxial, caducous, horn-like glands,

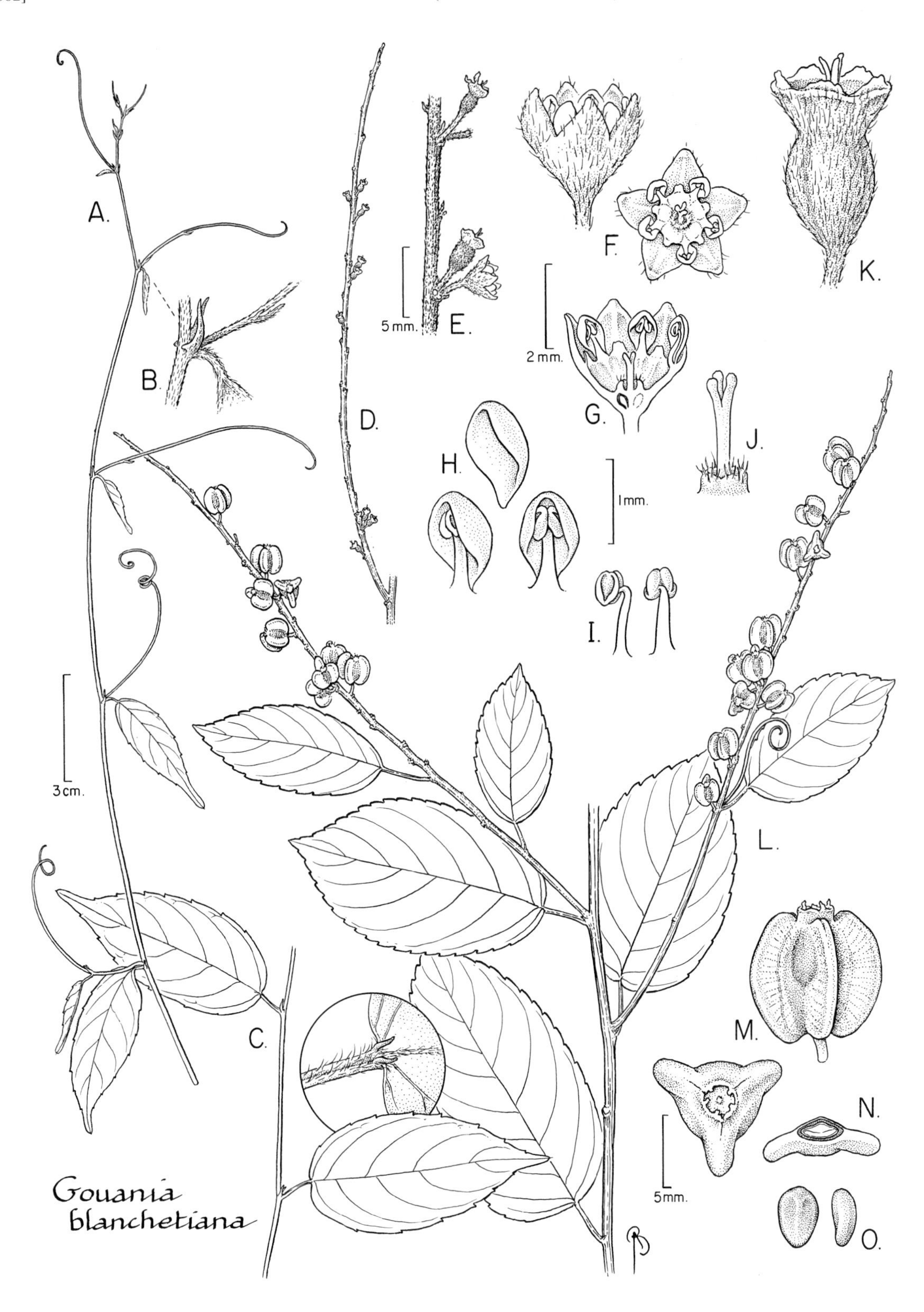

FIG. 262. RHAMNACEAE. *Gouania blanchetiana* (A–C, *Mori et al. 23714*; D–K, *Granville 8635*; L–O, *Mori & Gracie 21173*). **A.** Part of young liana with leaves and tendrils. **B.** Detail of node with stipules. **C.** Lower leaves of liana with detail (right) of stipels. **D.** Old inflorescence. **E.** Detail of inflorescence. **F.** Lateral (left) and apical (right) views of flowers. **G.** Medial section of flower. **H.** Petal (right) and two views of petals and stamens (below). **I.** Two views of stamens. **J.** Lateral view of style surrounded at base by intrastaminal disc. **K.** Lateral view of immature fruit. **L.** Part of liana with leaves and infructescences. **M.** Lateral (right) and apical (below) views of fruit. **N.** Transverse section of one carpel of fruit. **O.** Two views of seeds.

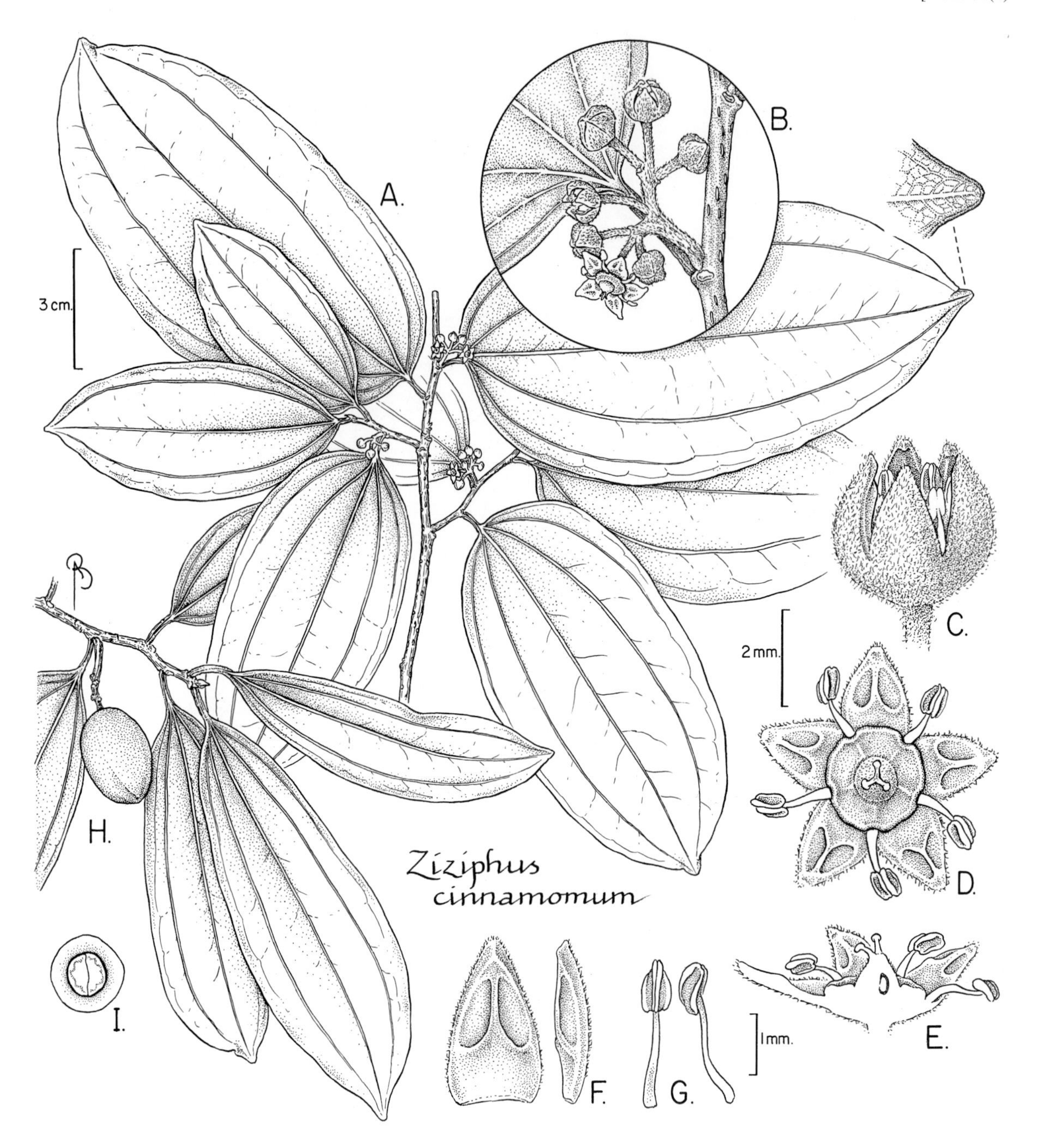

FIG. 263. RHAMNACEAE. *Ziziphus cinnamomum* (A–G, *Blanco 921* from Venezuela; H, I, *Mori et al. 22662*). **A.** Part of stem with leaves, inflorescences, and detail of leaf apex (right). **B.** Axillary inflorescence. **C.** Partially opened flower. **D.** Apical view of flower. **E.** Medial section of flower. **F.** Lateral (right) and adaxial (left) views of petals. **G.** Stamens. **H.** Part of stem with infructescence. **I.** Transverse section of fruit.

the apex acute with subapical, finger-like extension, the margins serrate. Inflorescences 7–25 cm long. Flowers: sepals 5, with slight medial ridge on adaxial surface; petals 5, clawed, caducous, initially nearly enveloping stamens. Fruits capsular, 0.8–1.1 cm long, strongly trigonal, with 3(4) narrow wings (0.4–0.6 cm wide). Seeds 3–4 mm long, brown, dorsally compressed, with slight adaxial ridge. Fl (Jan, Feb, Mar, Apr, Jun, Sep), fr (Feb, Apr); canopy and second growth lianas of lowland forest.

ZIZIPHUS Mill.

Trees. Stems pubescent, often with single thorns lateral to leaf axil, the thorns absent in our species. Leaves: petioles pubescent; blades glabrous or pubescent, the base asymmetrical, with or without glands on adaxial surface, the margins entire or serrate; 3-veined from base. Inflorescences axillary, cymose. Flowers: petals 5 or absent; nectary disc annular; ovary superior; ovule 1 in single developed locule. Fruits ellipsoid drupes, small (≤2 cm long), without wings.

Ziziphus cinnamomum Triana & Planch. FIG. 263, PL. 109c; PART 1: PL. VIa

Canopy trees. Stems thornless, with dense, very short, appressed pubescence. Bark slash with distinctive alternating reddish-brown and white bands. Leaves: petioles 0.8–1.1 cm long; blades elliptic to oblong, 9–14.7 × 3.4–7.3 cm, glabrous, the base asymmetrical, glands (Gentry, 1993) not evident in our material, the margins entire; venation of 3 salient veins, these arcuate from base. Inflorescences 0.8–1.2 cm long. Flowers: sepals with pronounced, medial ridge on adaxial surface; petals absent. Fruits dry drupes, 2.5–2.8 × 1.3–1.5 cm, the exocarp woody, yellow, the mesocarp sweet to taste, the septa papery. Seeds large, ca. 1.3 cm long, ± triangular in transverse section, the testa irregularly furrowed. Fr (Aug, Sep, Oct); rare, in nonflooded forest.

Fruits of this species are eaten by capuchin (*Cebus apella*) and spider (*Ateles paniscus*) monkeys. Although both monkeys disperse the seeds, capuchins are more efficient dispersers than spider monkeys (Zhang & Wang, 1995).

RHIZOPHORACEAE (Red Mangrove Family)

Scott A. Mori

Trees or shrubs. Stipules present, interpetiolar, persistent or caducous. Leaves simple, opposite, verticillate, or alternate. Inflorescences axillary, solitary, few-flowered cymes, fascicles, or infrequently racemes. Flowers actinomorphic, perigynous or epigynous, the hypanthium of epigynous flowers sometimes prolonged beyond ovary; sepals usually 4–8, valvate; petals as many as and alternate with sepals, free; stamens twice as many as petals or numerous, often inserted in pairs opposite petals, the filaments distinct or connate at base, attached to or around base of perigynous nectary disc, the disc sometimes wanting or the anthers sometimes sessile; ovary 2–6-carpelar, usually with as many locules as carpels, sometimes 1-locular, the style terminal, simple; placentation apical-axile, the ovules 2 per locule, pendulous. Fruits berries, drupes, or capsules. Seeds 1 per fruit or 1 per locule, sometimes arillate, with well developed endosperm.

Görts-van Rijn, A. R. A. & M. J. Jansen-Jacobs. 1986. Rhizophoraceae. *In* A. L. Stoffers & J. C. Lindeman (eds.), Flora of Suriname **III(1–2):** 511–515. E. J. Brill, Leiden.

Jonker, F. P. 1942. Rhizophoraceae. *In* A. Pulle (ed.), Flora of Suriname **III(2):** 35–43. J. H. De Bussy, Amsterdam.

Prance, G. T., M. Freitas da Silva, B. W. Alburquerque, I. de Jesus da Silva Araújo, L. M. M. Carreira, M. M. N. Braga, M. Macedo, P. N. da Conceição, P. L. B. Lisbôa, P. I. Braga, R. C. L. Lisbôa & R. C. Q. Vilhena. 1975. Revisão taxonômica das espécies amazônicas de Rhizophoraceae. Acta Amazonica **5(1):** 5–22.

CASSIPOUREA Aubl.

Small trees, infrequently shrubs. Stipules caducous. Leaves opposite. Inflorescences solitary or fasciculate. Flowers perigynous; calyx fused at base, the lobes 5, valvate; petals 5, laciniate, fimbriate, floccose; stamens numerous, the filaments distinct, inserted on outside of nectary disc and at base of hypanthium; ovary 2–4-locular. Fruits slightly fleshy, ovoid capsules. Seeds arillate.

Cassipourea guianensis Aubl. FIG. 264

Trees, 4–20 m tall. Stipule scars visible as faint line between leaf bases. Leaves: petioles 8–15 mm long, puberulous; blades elliptic to oblong, 6–15 × 2.5–6.5 cm, mostly glabrous, the base obtuse to rounded, the apex acuminate; venation brochidodromous, the secondary veins in 6–9 pairs. Inflorescences: pedicels nearly absent to ca. 5 mm long. Flowers: petals spatulate, white, very densely floccose, the margins markedly fimbriate; stamens ca. 25, the filaments 1.5–2.5 mm long, the anthers 1.5 mm long; nectary disc annular, membranous, 1.5 mm high; pistil pubescent, the stigma truncate. Fl (Jan, Mar, May), fr (Jan, May); uncommon, in nonflooded forest.

Prance et al. (1975) stated that the flowers of *C. guianensis* are sessile. However, the French Guianan collections range from subsessile to clearly pedicellate.

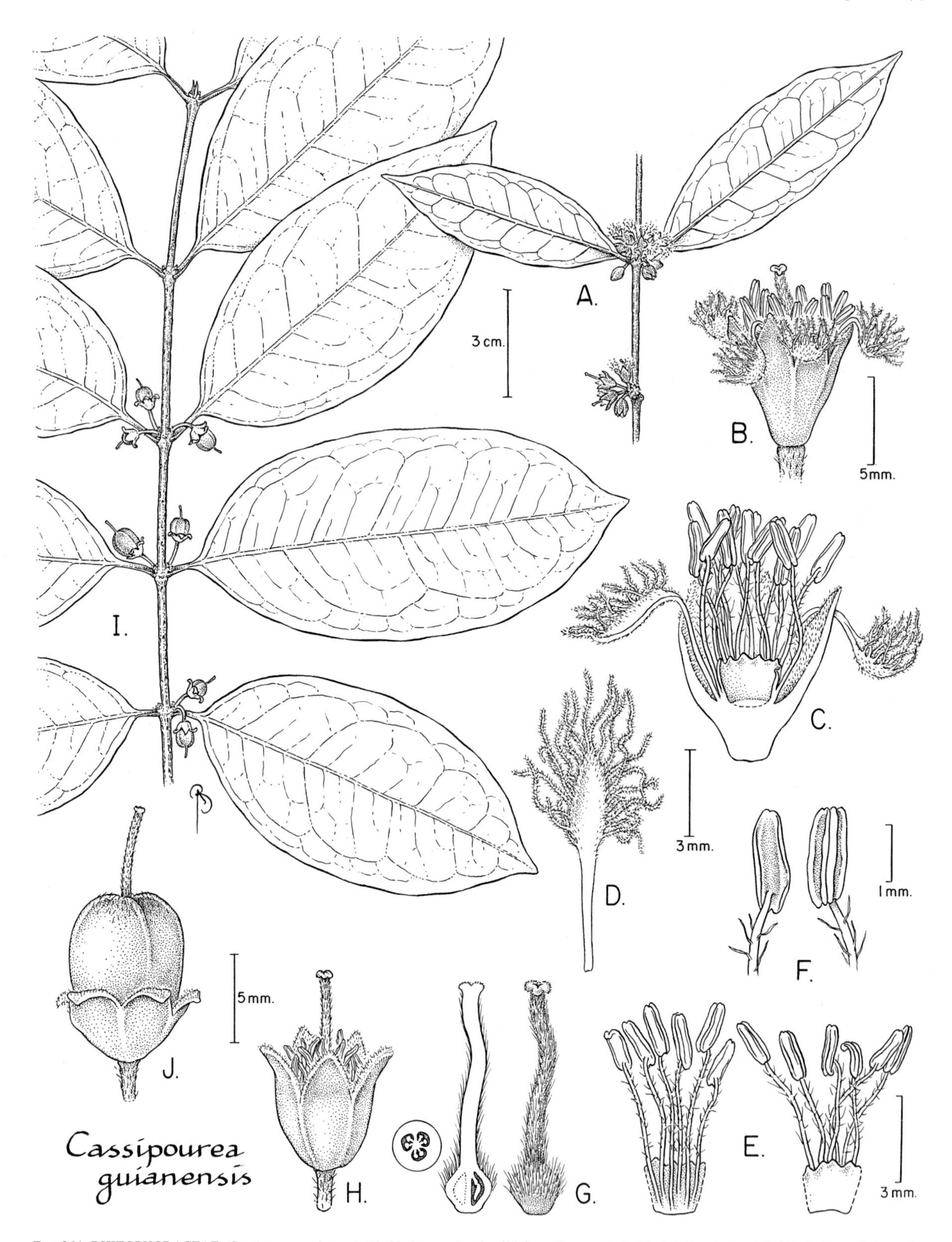

FIG. 264. RHIZOPHORACEAE. *Cassipourea guianensis* (A–H, *Jansen-Jacobs 693* from Guyana; I, J, *Mori & Pennington 18051*). **A.** Part of stem with leaves and inflorescences. **B.** Lateral view of flower. **C.** Medial section of flower with pistil removed. **D.** Adaxial view of petal. **E.** Abaxial (left) and adaxial (right) views of part of androecium and disc. **F.** Abaxial (left) and adaxial (right) views of anthers. **G.** Pistil (near left), medial section of pistil (middle left), and transverse section of ovary (far left). **H.** Lateral view of flower past anthesis, the petals have fallen. **I.** Part of stem with leaves and immature fruits. **J.** Immature fruit.

Plates 105–112

Plate 105. PASSIFLORACEAE. **a.** *Passiflora coccinea* (*Mori et al. 23293*), flower and bud still enclosed in bracts. **b.** *Passiflora foetida* var. *hispida* (*Mori & Mitchell 18693*), flower and fruit completely enclosed within hairy bracts. [Photo by S. Mori] PHYTOLACCACEAE. **c.** *Microtea debilis* (*Mori & Gracie 23818*), flowers, buds, and immature fruits. **d.** *Phytolacca rivinoides* (*Mori & Gracie 21186*), flowers on deep pink pedicels and rachis.

Plate 106. PIPERACEAE. **a.** *Peperomia gracieana* (*Mori et al. 23773*), peltate leaf and inflorescence. [Photo by S. Mori] **b.** *Peperomia serpens* (*Mori et al. 23057*), inflorescences. **c.** *Piper consanguineum* (*Görts-van Rijn 91*), infructescence; note light midribs of leaves. **d.** *Piper bartlingianum* (*Mori & Gracie 21157*), erect inflorescence. **e.** *Piper nigrispicum* (*Mori et al. 22053*), pendent inflorescence.

Plate 107. PIPERACEAE. **a.** *Pothomorphe peltata* (unvouchered), plant with peltate leaves and erect, white inflorescences. POLYGALACEAE. **b.** *Moutabea guianensis* (*Mori et al. 22145*, photographed in Kaw Mountains, French Guiana), flowers and buds. **c.** *Securidaca pubescens* (*Mori et al. 23725*), flowers and buds. PROTEACEAE. **d.** *Euplassa pinnata* (*Mori & Boom 15168*), flowers. [Photo by S. Mori]

Plate 108. QUIINACEAE. **a.** *Lacunaria crenata* (*Mori et al. 22202*), fruit cut to reveal seeds. **b.** *Lacunaria jenmani* (*Mori et al. 22062*), fruit. [Photo by S. Mori] **c.** *Quiina guianensis* (*Mori & Gracie 21159*), staminate flowers and buds. **d.** *Quiina obovata* (unvouchered), pinnately lobed juvenile leaves (adult leaves are simple). **e.** *Quiina oiapocensis* (*Mori & Pipoly 15499*), apex of stem with fruits; note stipules. [Photo by S. Mori]

Plate 109. RHABDODENDRACEAE. **a.** *Rhabdodendron amazonicum* (*Mori et al. 24695*), flowers and buds. [Photo by S. Mori] **b.** *Rhabdodendron amazonicum* (*Mori et al. 24695*), flower. [Photo by S. Mori] RHAMNACEAE. **c.** *Ziziphus cinnamomum* (*Mori et al. 22662*), fruit.

Plate 110. RUBIACEAE. **a.** *Amaioua guianensis* (*Mori et al. 22904*), flower, bud, and calyces. **b.** *Chimarrhis microcarpa* (*Mori et al. 20910*), flowers and buds. **c.** *Chimarrhis turbinata* (*Mori et al. 24239*), flowers and buds; note reflexed corolla lobes. **d.** *Diodia ocimifolia* (*Mori et al. 23716*), flowers; note angled stem. **e.** *Faramea guianensis* (*Mori & Gracie 23901*), inflorescence enclosed in bracts. **f.** *Coussarea violacea* (*Mori et al. 22765*), infructescence. **g.** *Chiococca alba* (unvouchered), flowers and buds.

Plate 111. RUBIACEAE. **a.** *Geophila cordifolia* (*Mori et al. 23719*), fruit. **b.** *Ferdinandusa paraensis* (*Mori & Gracie 22092*), flowers and bud. **c.** *Hamelia axillaris* (*Mori et al. 22080*), flower and immature fruits. **d.** *Hamelia axillaris* (*Mori et al. 22080*), infructescence with immature fruits. **e.** *Guettarda spruceana* (*Mori et al. 24780*), nocturnal flowers.

Plate 112. RUBIACEAE. **a.** *Isertia coccinea* (*Mori et al. 20888*), flower and buds. **b.** *Malanea macrophylla* (*Mori et al. 23308*), inflorescences. **c.** *Manettia reclinata* (*Mori et al. 23294*), flower and immature fruit. **d.** *Isertia spiciformis* (unvouchered, photographed in Sinnamary region of French Guiana), flower and buds. **e.** *Palicourea guianensis* (*Mori & Gracie 21131*), flowers and buds.

PASSIFLORACEAE (continued)

a.

Passiflora coccinea

PHYTOLACCACEAE

Passiflora foetida var. hispida

b.

Microtea debilis

c.

PIPERACEAE

Peperomia gracieana **a.**

Peperomia serpens **b.**

Piper consanguineum **c.**

Piper bartlingianum **d.**

Piper nigrispicum **e.**

Plate 106

Pothomorphe peltata a.

POLYGALACEAE

Moutabea guianensis b.

Securidaca pubescens c.

PROTEACEAE

Euplassa pinnata d.

Plate 107

QUIINACEAE

Lacunaria crenata a.

Lacunaria jenmani b.

Quiina guianensis c.

Quiina oiapocensis e.

Quiina obovata d.

RHABDODENDRACEAE

Rhabdodendron amazonicum **a.**

Rhabdodendron amazonicum **b.**

RHAMNACEAE

Ziziphus cinnamomum **c.**

Amaioua guianensis **a.**

Chimarrhis microcarpa **b.**

Chimarrhis turbinata **c.**

Diodia ocimifolia **d.**

Faramea guianensis **e.**

Coussarea violacea **f.**

Chiococca alba **g.**

Geophila cordifolia **a.**

Ferdinandusa paraensis **b.**

Hamelia axillaris **c.**

Hamelia axillaris **d.**

sertia coccinea a.

Malanea macrophylla b

Manettia reclinata c.

sertia spiciformis d.

Palicourea guianensis e

Plate 112 (Rubiaceae continued on Plate 113)

ROSACEAE (Rose Family)

John D. Mitchell

Trees or shrubs (sometimes herbs in extra-Guianan taxa). Stipules present. Leaves simple (compound in many taxa outside our area), alternate. Inflorescences axillary or terminal, racemose, umbellate, or solitary. Flowers actinomorphic, bisexual, perigynous, with distinct hypanthium, 4–5-merous; calyx imbricate; corolla imbricate; stamens (1)5–20(numerous), in ≥1 cycles, the filaments slender, usually distinct, attached to upper rim of hypanthium, the anthers usually longitudinally dehiscent; ovary superior (sometimes inferior in taxa outside our area), with 1 to many carpels; placentation axile, the ovules usually 1–2 per locule, often anatropous. Fruits drupes (various other fruit types found in extra-Guianan taxa).

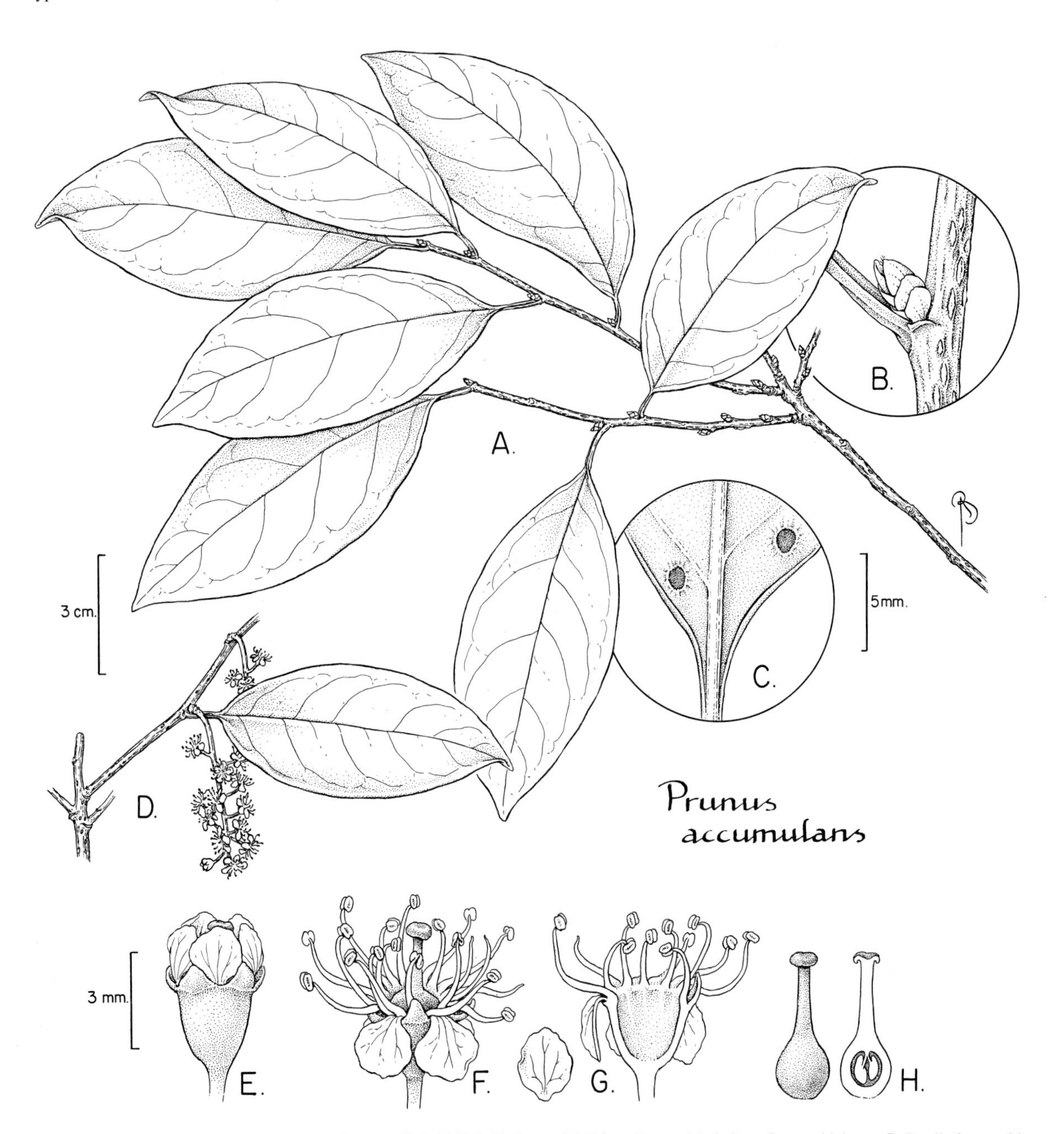

FIG. 265. ROSACEAE. *Prunus accumulans* (A–C, *Granville B.4646*; D–H, *Bernardi 565* from Venezuela). **A.** Part of stem with leaves. **B.** Detail of stem with flower bud in leaf axil. **C.** Detail of abaxial base of leaf blade showing glands. **D.** Part of stem with a leaf and inflorescences. **E.** Lateral view of flower bud beginning to open. **F.** Lateral view of flower (left) and adaxial view of petal (right). **G.** Medial section of flower with pistil removed. **H.** Pistil (far left) and medial section of pistil (right).

Kleinhoonte, A. 1966. Rosaceae. *In* J. Lanjouw & A. L. Stoffers (eds.), Flora of Suriname **II(2):** 426–456. E. J. Brill, Leiden.
Romoleroux, K. 1996. Rosaceae. *In* G. Harling & L. Andersson (eds.), Flora of Ecuador **56:** 1–152.

PRUNUS L.

Trees or shrubs. Leaves deciduous or evergreen; blades with glandular areas present near base, the margins entire; venation brochidodromous. Inflorescences axillary, racemose, with conspicuous bracts. Flowers: perianth 5-merous; stamens 15–30; ovary 1-locular, with 2 pendulous ovules, the style subterminal. Fruits drupes, the mesocarp thin, the endocarp bony. Seeds one per fruit.

Prunus accumulans (Koehne) C. L. Li & Aymard [Syn.: *Prunus myrtifolia* (L.) Urb. var. *accumulans* Koehne]
FIG. 265

Small to medium-sized trees, to 28 m × 150 cm. Bark gray with brown spots. Wood white. Stems reddish-brown, densely lenticellate. Stipules small, caducous, 7–11 mm long, glabrous Leaves: blades elliptic or ovate to lanceolate, 8–13 × 3.8–6 cm, chartaceous, glabrous, discolorous, the base obtuse to cuneate, decurrent onto petiole, the apex obtuse to acute, sometimes acuminate, the margins revolute; secondary veins in 5–8 pairs, arcuate, slightly prominent abaxially, impressed to flattened adaxially, the midrib prominent abaxially, impressed adaxially. Inflorescences racemose, 3–7 cm long; peduncle 5–15 mm long; pedicels 4–6.5 mm long. Flowers: hypanthium cupulate; calyx lobes deltoid; petals obovate, clawed, 1.5–2 mm long, white or cream-colored; stamens 15–30, the filaments unequal in length, 1.3–3 mm long; pistil with style 1.5–3 mm long, the stigma capitate-peltate. Drupes subglobose, 1.2–1.3 cm diam. (when dry), black when ripe. Rare, collected only once from our area, in nonflooded forest. *Bois noyo* (Créole), *marmelo bravo*, *viraru* (Portuguese).

RUBIACEAE (Coffee and Quinine Family)

Brian M. Boom and Piero G. Delprete

Shrubs to tall trees, less frequently lianas, vines, or erect or creeping herbs, mostly terrestrial, rarely epiphytic. Bark of shrubs and trees often containing quinine-related compounds. Stipules (Fig. 266) interpetiolar (intrapetiolar in *Capirona*), free, connate or sheathing at base, or rarely forming apical cap (*Amaioua*, *Duroia*), persistent or caducous, frequently with adaxial colleters, these secreting resinous compounds. Leaves simple, commonly opposite and decussate, rarely in whorls of 3–5 (only whorls indicated in descriptions); blades with entire margins. Inflorescences terminal, subterminal, or axillary, paniculate, cymose, capitate (often subtended by colorful bracts), or spicate, sometimes with foliose bracts (pherophylls) at base of secondary axes. Flowers usually actinomorphic, infrequently zygomorphic (e.g., flower buds bent at apex and stamens unequal in *Posoqueria*, or corolla with one-sided gibbous base in *Palicourea*), usually bisexual sometimes functionally unisexual (*Alibertia*, *Ibetralia*, *Posoqueria*, *Randia*), 4–5-merous (rarely 3- or 6–7-merous); calyx short-cupular or reduced to a wavy line, persistent or caducous, the lobes often minute to linear, rarely with a few flowers per inflorescence with one lobe expanded into calycophyll (*Capirona*); corolla gamopetalous, campanulate to hypocrateriform to tubular, actinomorphic, the aestivation valvate, contorted, or imbricate; stamens usually as many as corolla lobes, adnate to corolla tube, the anthers included or exserted, commonly opening by longitudinal slits, dorsifixed near base or middle, introrse; ovary inferior (rarely superior or half-superior, but not in our area), mostly 2-locular, rarely 5–7-locular (*Psychotria*), the style solitary, 2(4)-lobed or capitate, stylar pollen presentation sometimes present (e.g., *Ixora*); placentation axile (parietal in *Randia*) or with solitary ovules attached to roof or base of locules. Fruits baccate (many-seeded fleshy fruits), drupaceous (1–2–5-seeded fleshy fruits), capsular (loculicidal or septicidal), or fused to form fleshy syncarps (*Morinda*). Seeds with or without wings.

Bremekamp, C. E. B. 1966. Rubiaceae. *In* A. Pulle (ed.), Flora of Suriname **IV(1):** 113–298, 475–491. E. J. Brill, Leiden.
———. 1966. Remarks on the position, the delimitation, and the subdivision of the Rubiaceae. Acta Bot. Neerl. **15:** 1–33.
Delprete, P. G. 1999. Rondeletieae (Rubiaceae), Part I. Fl. Neotrop. Monogr. **77:** 1–226.
Kirkbride, J. H. 1977. Index to the Rubiaceae by Julian A. Steyermark in the Botany of the Guayana Highland by B. Maguire and collaborators. Phytologia **36:** 324–366.
Steyermark, J. A. 1964. Rubiaceae. *In* B. Maguire, J. J. Wurdak & Collaborators, Botany of the Guayana Highlands, Part V. Mem. New York Bot. Gard. **10:** 186–278.
———. 1965. Rubiaceae. *In* B. Maguire, J. J. Wurdack & Collaborators, Botany of the Guayana Highlands, Part VI. Mem. New York Bot. Gard. **12:** 178–285.
———. 1967. Rubiaceae. *In* B. Maguire, J. J. Wurdack & Collaborators, Botany of the Guayana Highlands, Part VII. Mem. New York Bot. Gard. **17:** 230–436.
———. 1972. Rubiaceae. *In* B. Maguire, J. J. Wurdack & Collaborators, Botany of the Guayana Highlands, Part IX. Mem. New York Bot. Gard. **23:** 227–832.
———. 1974. Rubiaceae. *In* T. Lasser & J. A. Steyermark, Flora de Venezuela **9:** 1–2070. Instituto Botánico, Caracas.

1. Epiphytic shrubs.
   2. Flowers showy; corolla >40 mm long. Fruits capsular, cylindrical, >50 mm long. Seeds numerous, fusiform, without longitudinal groove on ventral side, with a tuft of hairs on one end. . . . . *Hillia.*

2. Flowers inconspicuous; corolla 4.5 mm long. Fruits drupaceous, oblong, <10 mm long. Seeds 2, convex, with longitudinal groove on ventral side, without tuft of hairs. *Psychotria guadalupensis.*

1. Terrestrial shrubs, subshrubs, trees, lianas, vines, or herbs.
   3. Nonclimbing, suffrutescent, erect or creeping herbs.
      4. Fruits with 3 to many seeds.
         5. Stems glabrous. Inflorescences 1(3)-flowered, axillary. Corolla 2–2.2 mm long. . . . . *Oldenlandia.*
         5. Stems pubescent. Inflorescences 3- to many-flowered, axillary or terminal. Corolla 5–27 mm long.
            6. Stems procumbent. Corolla 4-lobed, 5–6 mm long, the aestivation valvate. Fruits baccate, metallic blue. . . . . *Coccocypselum.*
            6. Stems erect. Corolla 5-lobed, 12–18 mm long, aestivation contorted. Fruits capsular, brown. . . . . *Sipanea.*
      4. Fruits with 2 seeds.
         7. Flowers 5-lobed. Fruits fleshy.
            8. Stipules usually reflexed. Leaf blades cordate. Fruits slightly twisted at maturity, red to orange. . . . . *Geophila.*
            8. Stipules usually ascending or erect. Leaf blades not cordate. Fruits not twisted at maturity, blue. . . . . *Psychotria* (*P. ulviformis*, *P. variegata*).
         7. Flowers 4-lobed. Fruits dry.
            9. Fruits splitting into 2 indehiscent mericarps. . . . . *Diodia.*
            9. Fruits splitting into 2 mericarps, both dehiscent, or one dehiscent and the other indehiscent. . . . . *Spermacoce.*
   3. Vines, lianas, shrubs, or trees.
      10. Climbing or scrambling shrubs, vines, or lianas.
         11. Stems quadrangular, with paired, curved to coiled thorns. . . . . *Uncaria.*
         11. Stems terete, unarmed, or, if armed, the thorns not paired nor curved.
            12. Plants armed. Flowers 5-merous; corolla lobes contorted in bud. Placentation parietal. Fruits baccate, the pericarp coriaceous. Seeds embedded in gelatinous pulp.. . . . . *Randia.*
            12. Plants unarmed. Flowers 3–6-merous; corolla lobes valvate in bud. Placentation central or apical. Fruits drupaceous, baccate, or capsular, the pericarp membranous if fruit baccate or drupaceous, or thinly woody if capsular. Seeds not embedded in gelatinous pulp.
               13. Stems ≤2 mm thick, herbaceous. Anthers versatile. Fruits septicidal capsules. Seeds winged. . . . . *Manettia.*
               13. Stems ≥2 mm thick, woody (young stems subherbaceous). Anthers dorsifixed. Fruits drupaceous or baccate. Seeds not winged.
                  14. Fruits baccate, 3–5-locular. Seeds many per locule. . . . . *Sabicea.*
                  14. Fruits drupaceous, 2-locular. Seeds 1 per locule.
                     15. Leaf blades dark green and shiny adaxially, pale or silvery-green abaxially. Stamens inserted at mouth of corolla tube. Fruits oblong, terete. Seeds cylindrical. . . . . *Malanea.*
                     15. Leaf blades same shade of green ad- and abaxially. Stamens inserted at base of corolla tube. Fruits oblong-ellipsoid and terete, or suborbicular and laterally compressed. . . . . *Chiococca.*
      10. Trees or nonscrambling shrubs.
         16. Trees. Fruits capsular. Seeds winged.
            17. Flowers showy; corolla >1 cm long, the lobes contorted in bud.
               18. Stipules intrapetiolar, persistent, 2–4 cm long, pubescent. Inflorescences with a few flowers with one calyx lobe expanded and petaloid. Capsules septicidal. Seed wing ellipsoid, the margins irregularly fringed. . . . . *Capirona.*
               18. Stipules interpetiolar, deciduous, 0.8–1.5 cm long, glabrous. Inflorescences with all flowers with calyx lobes all equal and nonpetaloid. Capsules loculicidal. Seed wing bipolar, pointed at both ends, the margins entire. . . . . *Ferdinandusa.*
            17. Flowers inconspicuous; corolla <1 cm long, the lobes valvate or narrowly imbricate or open in bud.
               19. Inflorescences terminal, simple or branched spikes. Stamens inserted at base of corolla tube, the anthers versatile. Seed wings bipolar, tapered at both ends, the margins entire (not fringed). . . . . *Alseis.*
               19. Inflorescences axillary panicles (subterminal, 2 per node). Stamens inserted between corolla lobes, the anthers dorsifixed near base. Seed wings concentric (not bipolar), the margins irregularly fringed. . . . . *Chimarrhis.*

16. Trees, shrubs or subshrubs. Fruits baccate or drupaceous (with two pyrenes). Seeds not winged.
    20. Fruits with 1 seed per locule.
        21. Ovules pendulous from apex of ovary.
            22. Shrubs to treelets, 2–6 m tall. Corolla lobes valvate in bud. Fruits 2-locular.
                23. Plants unarmed. Corolla glabrous, campanulate, 6.5–14.5 mm long. Fruits oblong-ellipsoid and terete, or suborbicular and laterally compressed, 4–9 × 4–6 mm.. . . . . *Chiococca.*
                23. Plants armed with axillary thorns. Corolla pubescent, narrowly tubular, 30–35 mm long. Fruits oblong, terete, 13–15 × 2.5–4 mm. . . . . *Chomelia.*
            22. Trees, 6–20 m tall. Corolla lobes imbricate in bud. Fruits 4-locular.. . . . . *Guettarda.*
        21. Ovules erect from base of ovary.
            24. Corolla lobes contorted in bud.
                25. Stipules broadly acuminate, free at base. Inflorescences axillary. Calyx truncate, subtended by bracts. . . . . *Coffea.*
                25. Stipules narrowly acuminate or cuspidate or awned, connate at base. Inflorescences terminal in our species. Calyx denticulate or lobed, not subtended by bracts. . . . . . *Ixora.*
            24. Corolla lobes valvate in bud.
                26. Ovules attached to septum of ovary. . . . . *Morinda.*
                26. Ovules attached to base of ovary.
                    27. Fruits with 1 locule (by abortion of 1 locule, the ovaries 2-locular), or sometimes with 2-locules separated by thin septum.
                        28. Stipules usually connate at base, short-triangular, cuspidate, or aristate. Ovules solitary, not inserted on common basal column. Fruits wider than long, globose or reniform. . . . . *Faramea.*
                        28. Stipules usually free at base, ovate to triangular, truncate to obtuse (not cuspidate or aristate). Ovules connate, attached on common basal column. Fruits longer than wide, generally ellipsoid, rarely subglobose. . . . . . *Coussarea.*
                    27. Fruits with 2 or more locules separated by thick septum.
                        29. Inflorescence axes usually brightly colored red, orange, yellow, blue or purple. Corolla usually red, orange, yellow, pink, blue, violet, purple, or rarely white, or some combination of these colors, the tube laterally swollen at base (gibbous). . . . . *Palicourea.*
                        29. Inflorescence axes green, white, or pink. Corolla generally white, the tube not swollen basally.
                            30. Stipules usually with 2 lobes on each side, sometimes pectinate or truncate, rounded, triangular or prolonged apically (rarely scarcely developed). . . . . *Psychotria.*
                            30. Stipules usually laciniate, fringed, subulate, aculeiform, aristate, or dorsal surface sometimes with teeth or appendages. . . . . *Rudgea.*
    20. Fruits with several to many seeds per locule.
        31. Corolla lobes valvate in bud.
            32. Treelets to trees, 5–15(20) m tall. Corolla tube 30–70 mm long, reddish orange grading to yellow distally. . . . . *Isertia.*
            32. Shrubs, <4 m tall. Corolla tube <30 mm long, white or pale yellow.
                33. Stems with hollow internodes. Inflorescences axillary, congested umbels. Corolla pale yellow. Ovary 5-locular. . . . . *Patima.*
                33. Stems with solid internodes. Inflorescences terminal, elongate spikes. Corolla white. Ovary 2-locular. . . . . *Gonzalagunia.*
        31. Corolla lobes contorted or imbricate in bud.
            34. Fruits <10 mm long, with thin exocarp. Seeds minute, <3 mm long, the testa foveolate, usually tuberculate.
                35. Stems densely strigulose. Stipules connate basally, persistent. Ovary 2-locular. Fruits globose, 3–6 mm and dark blue at maturity. . . . . *Bertiera.*
                35. Stems glabrous to puberulent. Stipules free basally, deciduous. Ovary 4–5-locular. Fruits 3–6 mm diam. and orange at maturity, subglobose to oblong, or 6–8 × 3–6 mm and purple-black at maturity.
                    36. Inflorescences terminal. Corolla white. Fruits globose, 3–6 mm diam., orange at maturity.. . . . . *Bothriospora.*

36. Inflorescences axillary. Corolla yellow or orange. Fruits subglobose to oblong, 6–8 × 3–6 mm, purple-black at maturity. . . . . *Hamelia.*

34. Fruits >10 mm long, with coriaceous exocarp (except for *Alibertia myrciifolia* with fruits ca. 10 mm diam. and thin exocarp). Seeds large, >3 mm long, the testa smooth or fibrous.

37. Plants armed. . . . . *Randia.*

37. Plants not armed.

38. Flowers bisexual (both male and female organs present and functional in all flowers).

39. Corolla tubes <1.5 cm long, densely pubescent. *Genipa.*

39. Corolla tubes >2.5 cm long, glabrous.

40. Corolla zygomorphic, the buds curved to one side at apex; corolla tube 7–13 cm long; stamens unequal in length. . . . . *Posoqueria.*

40. Corolla actinomorphic, the buds straight; corolla tube ca. 30 cm long; stamens equal in length. . . . . *Tocoyena.*

38. Flowers functionally unisexual (with either male or female organs functional in same flower although both present).

41. Stipules not united to form a cap above apical bud, persistent.

42. Trees, 3–7 m tall. Trunks sulcate. Stipules 1.5–3 mm long. Leaf blades 3–15 × 1.5–8.5 cm. Flowers 3–4-merous, the corolla lobes 1.5–2.5 mm long. Fruits globose, with thin exocarp, ca. 1 cm diam. . . . . *Alibertia.*

42. Treelets, 1.5–4 m tall. Trunks not sulcate. Stipules ca. 5 mm long. Leaf blades 15–23 × 5.5–13 cm. Flowers 6–7-merous, the corolla lobes 10–14 mm long. Fruits oblong, with coriaceous exocarp, 2.5–4 cm diam. . . . . *Ibetralia.*

41. Stipules united to form conical cap above apical bud, the cap deciduous above abscission layer.

43. Staminate and pistillate flowers in cymose inflorescences. Fruits 2-locular, 1.2–1.5 cm. diam. . . . . *Amaioua.*

43. Staminate flowers in cymose inflorescences, pistillate flowers usually solitary. Fruits 1–4-locular, 4–7 cm diam.. . . . . *Duroia.*

ALIBERTIA A. Rich.

Shrubs or small trees, dioecious. Stipules interpetiolar, basally sheathing, persistent. Inflorescences terminal, capitate in staminate plants, of single flowers in pistillate plants. Flowers 3–4-merous, unisexual; corolla white or greenish-white, the lobes contorted in bud. Staminate flowers with dorsifixed anthers. Pistillate flowers: ovary 2–8-locular, the ovules 3 to numerous in each locule. Fruits berries, globose, the pericarp thin and firm, the pulp fleshy. Seeds large, subglobose, obtusely angled, vertically oriented.

Alibertia myrciifolia (K. Schum.) K. Schum. [Syns.: *Cordiera myrciifolia* K. Schum., *Alibertia triloba* Steyerm., *Alibertia uniflora* Standl.]

Treelets, 3–7 m tall. Trunks sulcate. Stems glabrous, with resinous exudate from buds, apical stipules, and inflorescences. Stipules sheathing at base, free portion triangular, 1.5–3 mm long. Leaves: blades elliptic to ovate, 3–15 × 1.5–8.5 cm. Staminate flowers usually 4-merous, pistillate flowers usually 3-merous; corolla tube 0.5–0.7 × 1–1.5 mm, green, the lobes 1.5–2.5 mm long. Berries globose, ca. 1 cm diam. Fl (Aug), fr (Jan), uncommon, in nonflooded forests.

ALSEIS Schott

Trees. Trunks fenestrate. Stipules interpetiolar, readily caducous. Inflorescences terminal, branched spikes, sometimes basally frondose. Inflorescences with sessile flowers on secondary axes. Flowers small, 5-merous; calyx lobes deciduous; corolla short, white, the lobes valvate or open in bud; stamens inserted at base of corolla tube, the anthers long exserted, versatile; ovary 2-locular, the ovules numerous in each locule. Fruits capsules, with septicidal dehiscence from apex. Seeds peltate, with bipolar wings tapered at both ends, the testa reticulate.

Alseis longifolia Ducke

Trees, to 25 m × 45 cm. Trunks fenestrate. Stems glabrous. Stipules narrowly triangular. Leaves: blades oblanceolate, ca. 18 × 6 cm. Inflorescences with axes ca. 10–20 cm long. Flowers: calyx lobes inconspicuous; corolla ca. 3 mm long. Fruits narrowly obconical, ca. 10 × 1.5 mm. Seeds 3–5 mm long. Fl (Mar), fr (Aug); in nonflooded forest. *Zolive* (Créole).

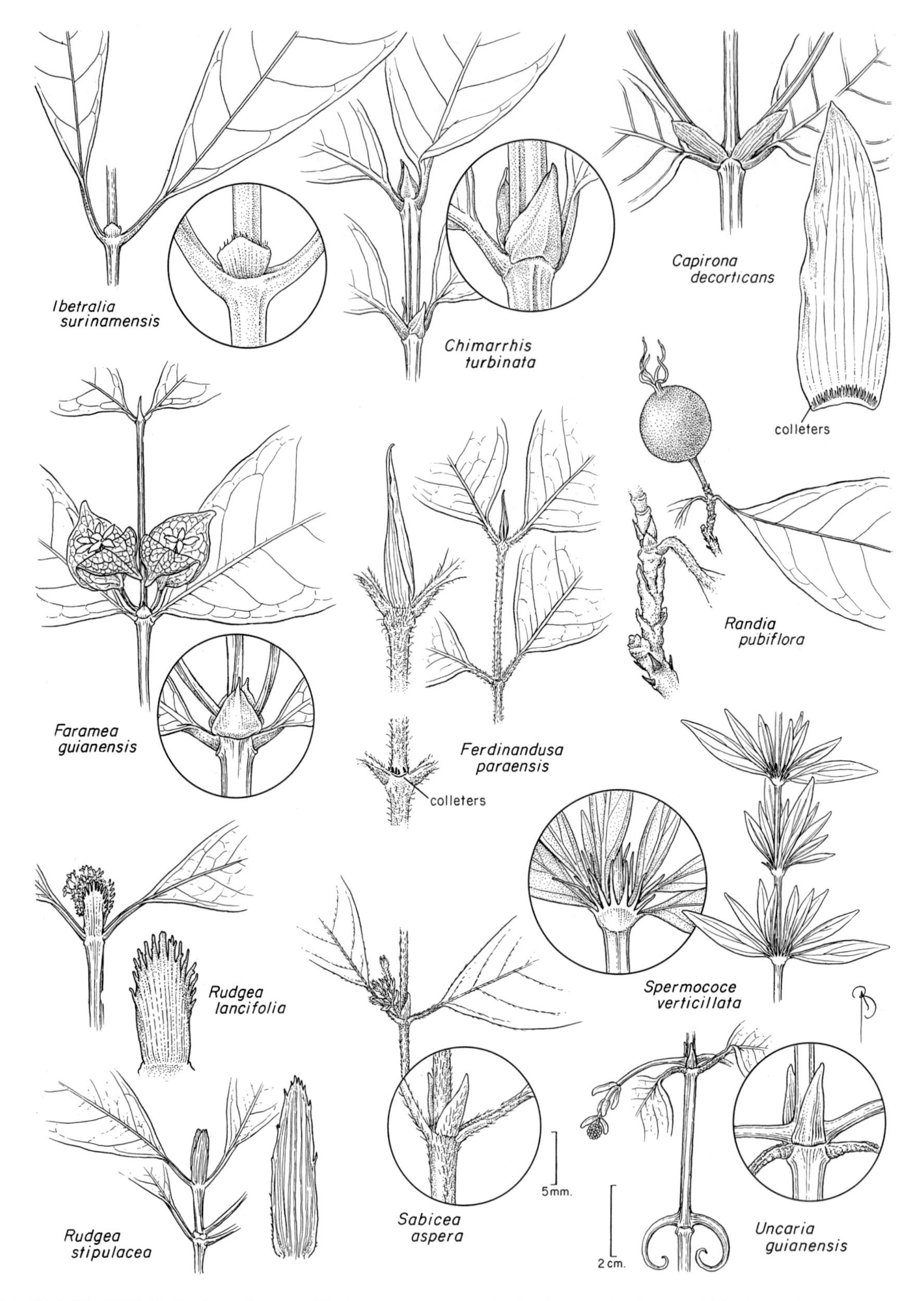

FIG. 266. RUBIACEAE. Details of several species of Rubiaceae showing stipules (see individual species plates for additional stipule illustrations). *Ibetralia surinamensis* (*Mori et al. 20884*), *Capirona decorticans* (*Granville 5548* from Cayenne, French Guiana), *Chimarrhis turbinata* (*Mori et al. 21567*), *Faramea guianensis* (*Boom 10762*), *Ferdinandusa paraensis* (*Boom & Mori 2349*), *Randia pubiflora* (*Mori 18746*), *Rudgea lancifolia* (*Granville 2754*), *Rudgea stipulacea* (*Mori & Pennington 17986*), *Sabicea aspera* (*Granville 3344* from Cayenne, French Guiana), *Spermacoce verticillata* (*Mori et al. 19068*), *Uncaria guianensis* with coiled spines (*Granville B.4427*).

AMAIOUA Aubl.

Trees, dioecious. Stipules interpetiolar, connate and forming apical cap, splitting irregularly, circumscissile, deciduous. Leaves usually aggregated at ends of stems. Inflorescences terminal, both staminate and pistillate flowers in cymose or umbelliform inflorescences. Flowers unisexual, 5–6-merous; calyx tubular; corolla sericeous abaxially, creamy white or greenish-gray, the lobes contorted in bud, ± falcate. Staminate flowers with stamens included, the anthers dorsifixed. Pistillate flowers: ovary 2-locular, the ovules numerous, in 2 horizontal rows per locule. Fruits berries, 2-locular. Seeds numerous, large, rounded or triangular, horizontally oriented.

Amaioua guianensis Aubl. PL. 110a

Trees, to 11 m × 15 cm. Trunks sulcate. Stems strigulose when young, glabrous when older. Stipule cap lanceolate, 2.2 × 1 cm, caducous. Leaves: blades elliptic, 10–26 × 3.5–14 cm. Inflorescences contracted. Staminate inflorescences few- to many-flowered; peduncules absent to 2 cm long. Pistillate inflorescences capitate or densely fasciculate, 4–21-flowered; pedicels and peduncle absent. Staminate flowers pedicellate; calyx 9–11 × 4–4.5 mm; corolla ca. 20 mm long. Pistillate flowers sessile; calyx 3–4 × 2.5 mm; corolla 9–13 mm long. Fruits oblong, 17–20 × 12–15 mm, reddish, drying black. Fl (Jan, Feb, Apr), fr (Mar, Apr, Jun); common, in understory of nonflooded forest. *Bois-négresse*, *qualité bois-négresse* (Créole).

BERTIERA Aubl.

Shrubs or small trees. Stems rounded and densely strigulose. Stipules interpetiolar, simple, connate basally, forming small sheath at base. Inflorescences terminal, paniculate. Flowers 5-merous; corolla white, the lobes usually pointed, contorted in bud; stamens inserted at throat of corolla, the anthers dorsifixed, conspicuously apiculate at apex; ovary 2-locular, with resinous dots, the ovules several to numerous in each locule. Fruits baccate, globose, 2-locular, dark blue, ± 10-costate when mature. Seeds angled, verrucose.

Bertiera guianensis Aubl.

Shrubs or treelets, 1.5–6 m tall. Stipules 5–11 mm long, the tubular portion connate, 4–6 mm long, the apical appendage narrowly acuminate, 3–8 mm long. Leaves: blades elliptic to oblong-elliptic, 10–17 × 2.5–7.5 cm. Inflorescences 5–18 × 1.5–9 cm. Flowers: calyx glabrous, the tube 0.2–0.3 mm long, the lobes 0.3–0.8 × 0.4–0.5 mm; corolla tube 2.5–3.5 × 1–1.5 mm, the lobes 1.2–2.7 × 0.7–1 mm. Fruits 3–6 mm diam., blue, drying black. Fl (Jan, Aug, Oct, Dec), fr (Jan, Feb, May, Jun, Aug, Sep, Oct, Nov, Dec); common, in weedy habitats. *Confiture macaque* (Créole).

BOTHRIOSPORA Hook. f.

Shrubs. Stems thin, terete. Stipules interpetiolar, free, deciduous. Inflorescences terminal, corymbose. Flowers 4–6-merous; calyx with persistent teeth; corolla shortly infundibuliform or subrotund, the lobes contorted in bud; stamens exserted, the anthers dorsifixed; ovary 4–5-locular, with numerous ovules, these horizontally oriented. Fruits baccate, globose, fleshy, with persistent calyces. Seeds numerous, minute, foveolate.

Bothriospora corymbosa (Benth.) Hook. f.

Shrubs, to 2 m tall. Stipules triangular, 10 × 3 mm. Leaves: petioles 0.8–1.2 cm long; blades elliptic, 5–7 × 3–3.5 cm. Inflorescences 4–6.5 cm long. Flowers: calyx 1–1.5 mm long; corolla 5–6 mm long, white. Fruits 2.5–3 mm diam. Fr (Jan); known from a single sterile collection (*de Granville 9001*, CAY) from our area.

CAPIRONA Spruce

Trees. Stipules intrapetiolar, large. Inflorescences terminal, paniculate, with large bracts surrounding flower buds. Flowers 5–6-merous, showy, zygomorphic; calyx unequally 5–6-dentate, some flowers with one lobe expanded and leaf-like (calycophyll), this pinkish or reddish; corolla with lobes contorted in bud; stamens inserted in lower part of corolla tube, the anthers included; ovary 2-locular, the ovules numerous in each cell, ascending, imbricate. Fruits capsular, septicidal. Seeds numerous, small, with irregular-ellipsoid wing, with margin irregularly fringed.

Andersson, L. 1994. *Capirona* in Rubiaceae (part 3). *In* G. Harling & L. Andersson (eds.), Flora of Ecuador 50: 3–7.

Capirona decorticans Spruce [Syn.: *Capirona leiophloea* Benoist, *Capirona surinamensis* Bremek.] FIG. 266 (stipules only)

Trees, to 15 m × 35 cm. Bark smooth, green to reddish-brown. Stems glabrous. Stipules lanceolate, 2–4 × 0.5–1 cm. Leaves: blades obovate to spatulate, 12–32 × 6–13 cm, glabrous. Inflorescences paniculate, glabrous. Flowers: calyx ca. 0.5 cm long, with one lobe expanded into calycophyll ca. 8 cm long; corolla ca. 2.5 cm long, reddish basally, pinkish-white apically. Fruits oblong, 32 × 8 mm, 10-costate, with persistent calyx lobes. Fl (Apr); known by a single fertile collection (*de Granville 5548*, CAY) from our area. *Bois de lettre à grandes feuilles*, *bois-palika* (Créole), *lisapau*, *mulatorana* (Portuguese), *mutende* (Bush Negro), *pau mulato* (Portuguese).

CHIMARRHIS Jacq.

Tall canopy trees. Trunks often irregular, older individuals frequently with large buttresses. Stipules interpetiolar, convoluted in bud, readily caducous (subgen. *Chimarrhis*) or persistent (subgen. *Pseudochimarrhis*). Leaves large, aggregated at end of stems; petioles present. Inflorescences axillary, subterminal, 2 per node, opposite on stems. Flowers small, protogynous; calyx tube short, persistent; corolla white to yellowish-white, the tube very short, the throat with internal ring of hair, the lobes curved, valvate (or narrowly imbricate) in bud; stamens inserted among corolla lobes, exserted after style maturity, the anthers dorsifixed near base; ovary 2-locular, style lobes exserted beyond closed corolla, the ovules numerous in each cell. Fruits capsular, dehiscing septicidally from apex. Seeds small, numerous, vertical, with concentric wing and irregularly fringed margin.

Delprete, P. G. 1999. *Chimarrhis*. *In* P. G. Delprete, Rondeletieae (Rubiaceae), Part I. Fl. Neotrop. Monogr. **77:** 137–187.

1. Trunks irregular but not sulcate. Stipules large, foliaceous, readily caducous. Leaf blades chartaceous. Capsules subglobose, 1.9–2.5 mm long, glabrous. . . . . *C. microcarpa*.
1. Trunks sulcate. Stipules small, not foliaceous, persistent. Leaf blades subcoriaceous. Capsules turbinate, 3.5–6 mm long, puberulent. . . . . *C. turbinata*.

Chimarrhis microcarpa Standl. [Syns.: *Chimarrhis microcarpa* Standl. var. *microcarpa*, non *Chimarrhis cymosa* Jacq. subsp. *microcarpa* Urb., *Chimarrhis longistipulata* Bremek.] FIG. 267, PL. 110b

Large canopy trees, to 35 m × 50 cm. Trunks irregular but not sulcate, with plank buttresses. Stipules triangular-lanceolate, (15)25–35 mm long, readily caducous. Leaves: blades widely elliptic to obovate, 10–20 × 5–10 cm, chartaceous, glabrous. Inflorescences paniculate. Flowers: calyx 0.5–0.75 mm long, glabrous; corolla 4–5 mm long, glabrous. Fruits subglobose, 1.9–2.5 mm long, glabrous. Fl (Apr, Jun, Jul, Sep, Oct), fr (Jan); common, in nonflooded forest.

Chimarrhis turbinata DC. [Syns.: *Pseudochimarrhis turbinata* (DC.) Ducke, *Pseudochimarrhis difformis* (Benoist) Benoist, *Bathysa difformis* Benoist, *Elaeagia brasiliensis* Standl.] FIG. 266 (stipules only), PL. 110c; PART 1: PL. Ia

Large canopy trees, to 35 m × 45 cm dbh. Trunks sulcate, without buttresses (Part 1: pl. Ia). Stipules triangular-lanceolate, glabrous to pubescent, 10–20 mm long, persistent. Leaves: blades obovate, 15–22 × 8–14 cm, subcoriaceous, glabrous. Inflorescences paniculate. Flowers: calyx ca. 3 mm long, puberulent; corolla 3.5–6 mm long, glabrous. Fruits turbinate, 3.5–6 mm long, puberulent. Fl (Aug, Sep, Oct), fr (Nov); infrequent, in nonflooded forest. *Bois-chapelle*, *bois-pagaïe* (Créole).

CHIOCOCCA P. Browne

Vines, lianas, or treelets. Stipules basally connate, acicular or shortly pointed apically, persistent. Inflorescences axillary panicles, simple or compound and opposite on stems. Flowers: calyx 4–6-lobed; corolla 4–5-lobed, the lobes valvate in bud, white; stamens 4–5, inserted at base of corolla tube; ovary 2-locular, the ovules solitary in each cell, pendulous from apex. Fruits drupaceous, fleshy, coriaceous, laterally compressed or terete, 2-locular, with one pyrene in each locule.

1. Lianas or scandent shrubs. Leaf blades lanceolate-ovate, 1.6–11 × 0.8–4.5 cm. Corolla funnel-shaped, 5–10 mm long, yellowish-white. Fruits suborbicular, laterally compressed, 4–6 × 4–6 mm.. . . . . *C. alba*.
1. Shrubs or treelets, 2–3 m tall. Leaf blades lanceolate-elliptic, 4.5–18 × 3–8 cm. Corolla mostly campanulate, 6.5–14.5 mm long, greenish-yellow. Fruits oblong-ellipsoid, terete, 7–9 × 4–5 mm. . . . . *C. nitida*

Chiococca alba (L.) Hitchc. FIG. 268, PL. 110g

Lianas or scandent shrubs. Stems glabrous. Stipules suborbicular-deltoid, glabrous, 0.3–1 × 1.1–2.5 mm. Leaves: blades lanceolate-ovate, 1.6–11 × 0.8–4.5 cm. Inflorescences 5–27-flowered. Flowers: calyx 1.5–2.8 mm long; corolla funnel-shaped, 5–10 mm long, yellowish-white. Fruits suborbicular, laterally compressed, 4–6 × 4–6 mm, glabrous, white. Fl (Jan, Feb, Mar, Apr, May, Aug, Sep), fr (Jun, Dec); common, in weedy habitats and in secondary forest.

Chiococca nitida Benth. [Syn.: *Chiococca erubescens* Wernham]

Shrubs or treelets, 2–3 m tall. Stipules glabrous, suborbicular to widely triangular, 1–3 × 3.5–4 mm. Stems glabrous. Leaves: blades lanceolate-elliptic, 4.5–18 × 3–8 cm. Inflorescences 5–21-flowered. Flowers: calyx 2.5–3.5 mm long; corolla mostly campanulate, 6.5–14.5 mm long, greenish-yellow. Fruits oblong-ellipsoid, terete, 7–9 × 4–5 mm, glabrous, white, turning dark purple. Fr (Oct, Dec); uncommon.

CHOMELIA Jacq.

Shrubs or treelets. Stems armed with axillary thorns. Stipules interpetiolar, triangular. Inflorescences axillary cymes. Flowers 4-merous; calyx tubular, with persistent teeth; corolla white, the lobes valvate in bud; stamens inserted at throat of corolla tube, included, the anthers dorsifixed, sessile; ovary 2-locular, the ovules one per locule, pendulous. Fruits drupaceous, 2-seeded. Seeds cylindrical, pendulous.

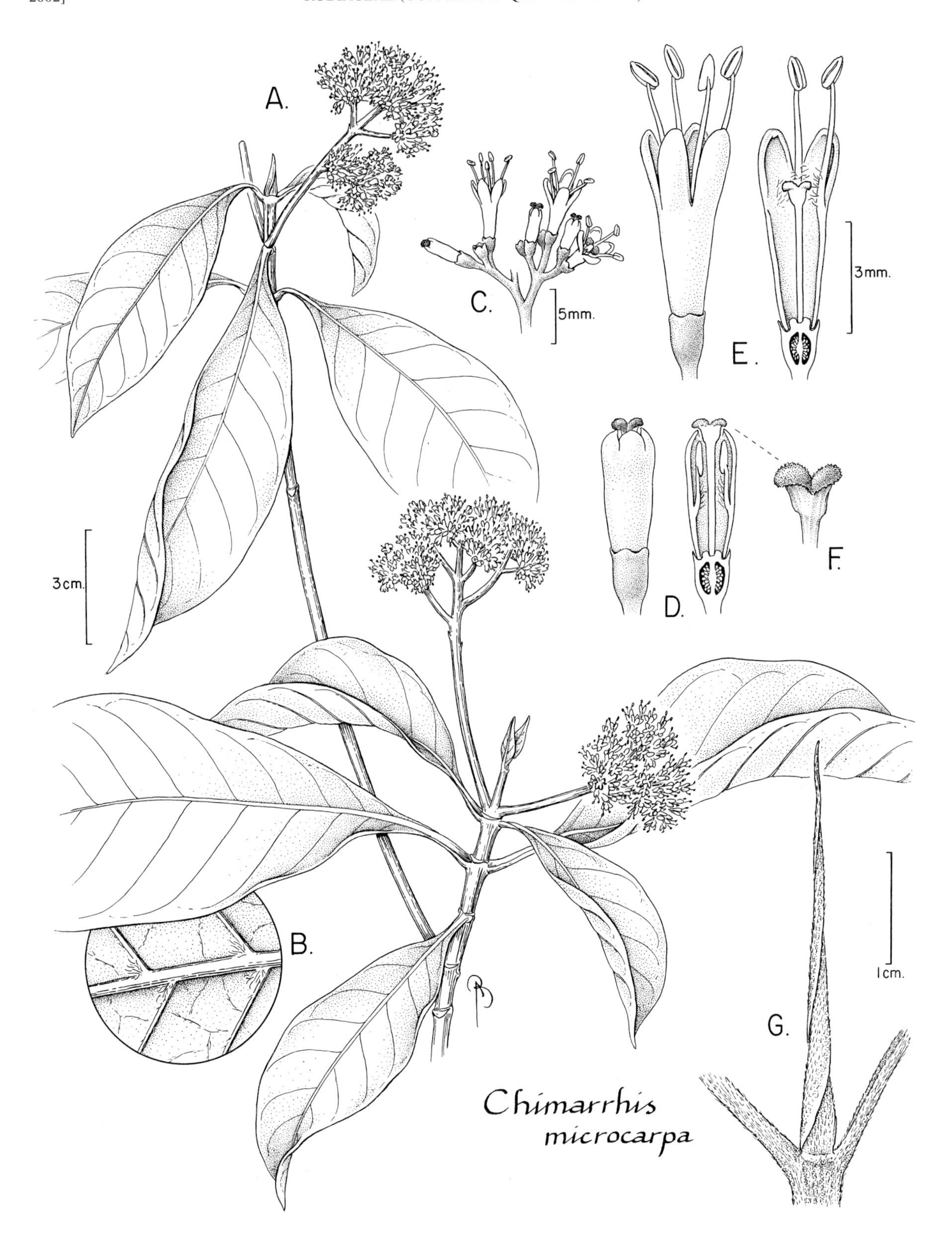

FIG. 267. RUBIACEAE. *Chimarrhis microcarpa* (A–F, *Mori et al. 20910*; G, *Boom & Mori 2037*). **A.** Apices of stems showing leaves and inflorescences. **B.** Detail of abaxial leaf surface showing tufts of pubescence in the vein axils. **C.** Part of inflorescence. **D.** Lateral (left) and medial section (right) of pin flowers. **E.** Lateral (left) and medial section (right) of thrum flowers. **F.** Detail of stigma. **G.** Stipule.

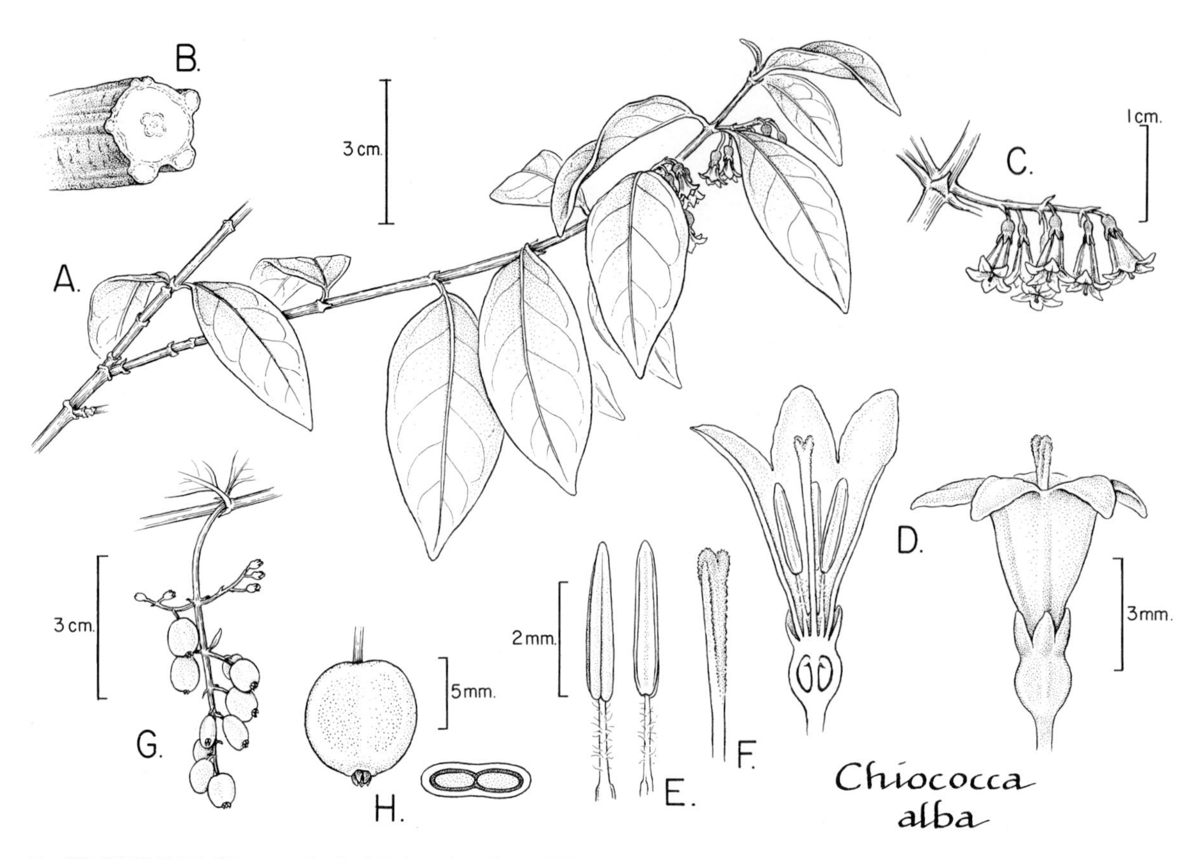

FIG. 268. RUBIACEAE. *Chiococca alba* (A, C–F, *Acevedo-Rodríguez 5077*; B, *Acevedo-Rodríguez 1914*; G, H, *Acevedo-Rodríguez 3821*, all from St. John, U.S. Virgin Islands). **A.** Part of stem with leaves and flowers. **B.** Transverse section of older stem. **C.** Detail of inflorescence. **D.** Medial section (left) and lateral view (right) of flower. **E.** Dorsal (left) and lateral (right) views of stamens. **F.** Detail of stigma. **G.** Infructescence. **H.** Lateral view (above) and transverse section (right) of fruit. (Reprinted with permission from P. Acevedo-Rodríguez, *Flora of St. John, U.S. Virgin Islands*, Mem. New York Bot. Gard. 78, 1996.)

Chomelia tenuiflora Benth. [Syn.: *Anisomeris tenuiflora* (Benth.) Pulle]

Shrubs or treelets, 3–6 m tall. Stems armed with axillary spines, densely pubescent when young. Stipules triangular, pubescent, 3.5–4.5 mm long. Leaves: blades obovate-elliptic to ovate-elliptic, 5–13 × 2.5–5.5 cm. Inflorescences 2–3-flowered. Flowers: calyx 3–4.5 mm long, pubescent; corolla 30–35 mm long, pubescent. Fruits cylindrical, 13–15 × 2.5–4 mm, sparsely pubescent. Fl (May, Nov). *Estrella*, *limaorana* (Portuguese).

COCCOCYPSELUM P. Browne

Herbs. Stems procumbent. Stipules interpetiolar, small, triangular, persistent. Inflorescences axillary (subterminal), capitate, 3–6-flowered, 1 cm wide at anthesis. Flowers 4-merous; corolla lobes valvate in bud, blue, lavender or rarely white; stamens dorsifixed; ovary 2-locular, the ovules numerous, horizontally arranged in each locule. Fruits baccate, oblong, spongy, blue, the pulp white. Seeds numerous, lenticular, brownish.

Coccocypselum guianense (Aubl.) K. Schum.

Creeping herbs. Stems 1–1.5 mm thick, pubescent. Stipules triangular, densely pubescent, 2–4 mm long. Leaves: blades ovate, 1.7–4.5 × 0.9–3 cm. Inflorescences 3–6-flowered. Flowers: calyces pubescent, 1.6–2.2 mm long; corolla 5–6 mm long, pale blue to lavender. Fruits 13–15 × 7–10 mm, spongy, metallic blue. Seeds many per locule, minute, angled or compressed. Fr (Jan); in understory of forest.

COFFEA L.

Shrubs. Stipules interpetiolar, triangular, broadly acuminate, persistent. Inflorescences axillary cymes. Flowers with pleasant fragrance; corolla white, the lobes 4–5, contorted in bud; stamens inserted at throat of corolla tube, the anthers dorsifixed; ovary 2-locular, ovules solitary in each locule. Fruits drupaceous, fleshy, ellipsoid to ovoid, with 2 pyrenes. Seeds with ventral groove.

Coffea arabica L.

Shrubs, 2–3 m tall. Stipules 4–6 mm long. Stems glabrous. Leaves: blades lanceolate-elliptic, 10–20 × 3–8 cm. Inflorescences 1–4-flowered. Flowers: calyx 5–6-lobed, subtended by bracts; corolla tube 17–19 mm long, the lobes 10–18 mm long. Fruits ca. 1.8 × 1.2 cm, red, drying black. Fr (Jun); cultivated but sometimes persisting around old homesteads and appearing native. The seeds are the source of coffee. *Café* (French, Spanish, Portuguese), *coffee* (English).

COUSSAREA Aubl.

Trees or shrubs, generally glabrous. Stipules interpetiolar, persistent, free (rarely connate) at base, ovate to triangular, the apex truncate to obtuse (not cuspidate or aristate). Inflorescences terminal, paniculate. Flowers 4-merous; corolla white, the throat glabrous adaxially, the lobes valvate in bud; stamens 4, the anthers dorsifixed; ovary 2-locular, with thin septum breaking down from above and ovary appearing unilocular, the ovules 1 per locule, ascendent, attached to a common basal column. Fruits drupaceous, fleshy or coriaceous, generally longer than wide (usually ellipsoid), with 1 seed by abortion of other ovule. Seeds suberect.

1. Inflorescences compoundly branched panicles, 5–20 cm long, with ≥20 flowers.
    2. Inflorescences epedunculate, trichotomously branched from base, ca. 19 cm long. Corolla ca. 13–15 mm long. Fruits 12–15 × 8–10 mm. . . . . *C. machadoana.*
    2. Inflorescences pedunculate, with one primary rachis, 6–11.5 cm long. Corolla ca. 16 mm long. Fruits 8–9 × 5–7 mm. . . . . *C. racemosa.*
1. Inflorescences simple panicles or condensed cymes, ≤5 cm long, with ≤20 flowers.
    3. Leaf blades with pit domatia (cavities) in axils of secondary veins abaxially. . . . . *C. micrococca.*
    3. Leaf blades without pit domatia (cavities).
        4. Small trees, to 10 m tall. Stipules triangular and short acuminate. Leaf blades oblanceolate to elliptic, drying blackish. Infructescences epedunculate to subsessile, the peduncle, when present, 1–3 mm long. Mont Galbao endemic. . . . . *C. granvillei.*
        4. Shrubs to treelets, to 3.5 m tall. Stipules truncate to suborbicular. Leaf blades oblong-elliptic or ovate-eliptic, drying greenish. Infructescences short pedunculate, the peduncle 3–15 mm long. Widespread.
            5. Young stems yellowish to pale green. Inflorescences 4–9-flowered cymes. Corolla 15–17 mm long. . . . . *C. amapaensis.*
            5. Young stems brownish. Inflorescences 8–15-flowered thrysoid panicles. Corolla 8–10.5 mm long. . . . . *C. violacea.*

Coussarea amapaensis Steyerm.

Shrubs to treelets, to 3.5 m tall. Stipules inconspicuous, truncate, persistent. Young stems yellowish to pale green. Leaves: blades oblong-elliptic, 5.5–10.5 × 2.2–4 cm, drying whitish-green; domatia absent. Inflorescences cymes, 4–9-flowered, rachis 3–20 mm long; peduncle 3–15 mm long. Flowers: calyx 1.5–2 mm long; corolla 15–17 mm long. Fruits 20 × 12 mm. Fl (Sep); uncommon in nonflooded forest.

Coussarea granvillei Delprete & B. M. Boom FIG. 269

Small trees, to 10 m tall. Outer bark green, smooth. Stems blackish, glabrous. Stipules triangular and short acuminate, 2 × 3 mm, persistent. Leaves: blades oblanceolate to elliptic, 8–17 × 3–7.5 cm, drying blackish; domatia absent. Inflorescences condensed cymes, (3)5–7-flowered; peduncle absent, or, when present, 1–3 mm long. Flowers unknown at anthesis for this species; calyx in flower buds to 1 mm long; corolla in flower buds 4-angular, to 14 mm long. Fruits sessile, ellipsoid, 14–20 × 12–14 mm. Fl bud (May), Fr (Jan, Mar, May); known only from Mont Galbao.

Coussarea machadoana Standl.

Trees, to 10 m × 12 cm. Trunks flattened. Stems pale green. Stipules connate at base, shortly apiculate, 4–5 mm long. Leaves: blades narrowly elliptic-oblong, 17–21 × 6–9 cm, drying pale olive-green, the apex caudate to cuspidate, the acumen ca. 1 cm long; domatia absent. Inflorescences compound panicles, ca. 19 cm long, trichotomously branched from base, densely flowered, the flowers aggregated in cymes or subumbels; peduncle absent; pedicels 1–2 mm long. Flowers: calyx 1 mm long; corolla 13–15 mm long. Fruits unknown. Fl (Mar); rare, in understory of nonflooded forest on Antenne Nord de La Fumée.

Coussarea micrococca Bremek.

Treelets, to 6 m tall. Stipules broadly ovate to deltoid, 3–4 × 3–4 mm, persistent. Young stems olive green. Leaves: blades ovate to elliptic, 6–16 × 2–7.5 cm, drying dark olive green; pit domatia present in axils of secondary veins, narrowly oblong, the opening 1–1.5 mm long. Inflorescences panicles, 3–4.5 cm long, with 6–7 flowers clustered at top of each axis; peduncle 2–3.5 cm long. Flowers: buds ca. 2 mm long; calyx ca. 0.5 mm long; corolla 6–7 mm long (corolla from *Oldeman B.694*, CAY). Fruits spherical, 4–5 mm diam., pale green (fruits from *Lim Sang LBB–16260*, NY). Fl (Jan); known by a single collection (*Granville 8679*, CAY, NY) from our area from the cloud forest of Mont Galbao.

Coussarea racemosa A. Rich. FIG. 270

Trees, to 16 m × 25 cm. Trunks without buttresses, angled. Outer bark smooth, gray, scaly, the inner bark orange, fibrous. Young stems pale yellowish-green. Stipules short, broad, widely rounded, 3–4 mm long. Leaves: blades elliptic to ovate, 12–19 × 5.2–8.5 cm, drying whitish green; domatia absent. Inflorescences pyramidal panicles, 6–11.5 cm long, many-flowered; peduncle 35–50 mm long, yellowish. Flowers nocturnal, heterostylous; calyx 2 mm long; corolla ca. 16 mm long. Fruits ellipsoid, 8–9 × 5–7 mm, white to pale green. Fl (Jul, Aug, Sep, Oct), fr (Jan, Mar, May, Jul, Nov); common, in forest, especially on ridges.

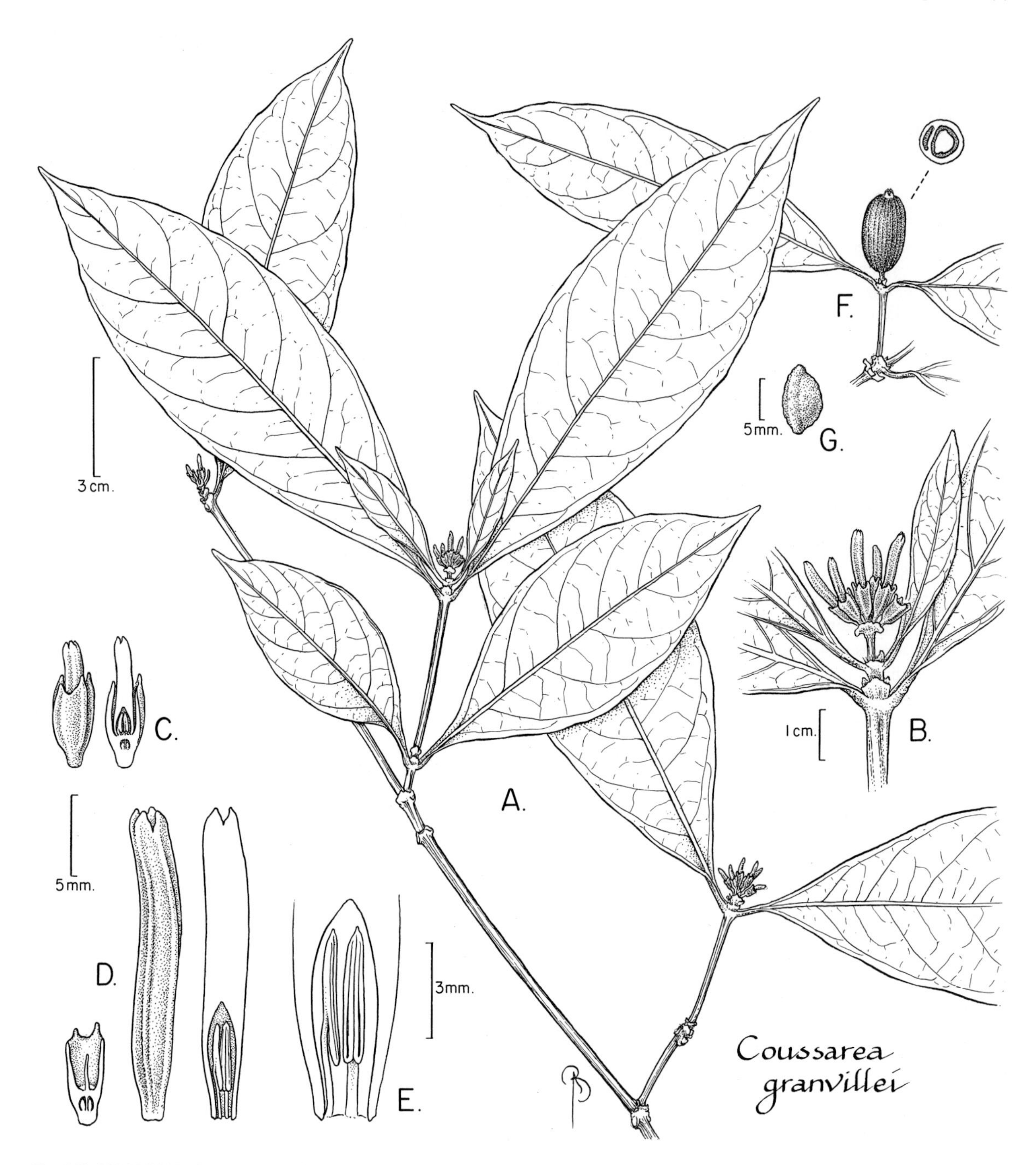

FIG. 269. RUBIACEAE. *Coussarea granvillei* (A–E, G, *Oldeman 3289*; F, *Granville 8601*). **A.** Part of stem with leaves and inflorescences. **B.** Detail of inflorescence with bases of leaves and stipules. **C.** Lateral view (far left) and medial section of flower bud (near left). **D.** Medial section of calyx and pistil (below), lateral view (near right) and medial section (far right) of corolla. **E.** Detail of stamens adnate to corolla. **F.** Part of stem with leaves and fruit (right), and transverse section of fruit (above right). **G.** Seed.

Coussarea violacea Aubl. [Syn.: *C. schomburgkiana* (Benth.) Benth. & Hook. f.] PL. 110f

Shrubs or small trees, to 3 m tall. Young stems brownish. Stipules suborbicular, widely rounded, 2–3 mm long. Leaves: blades ovate-elliptic, 10–12 × 4–6 cm, drying olive-green; domatia absent. Inflorescences trichotomous, thyrsoid panicles, 2–5 cm long, 8–15-flowered; peduncle 6–10 mm long. Flowers: calyx 1.5–2 mm long; corolla 8–10.5 mm long. Fruits oblong-ovoid, 22 × 11 mm, bright red-orange. Fr (Aug, Nov); uncommon, in nonflooded forest.

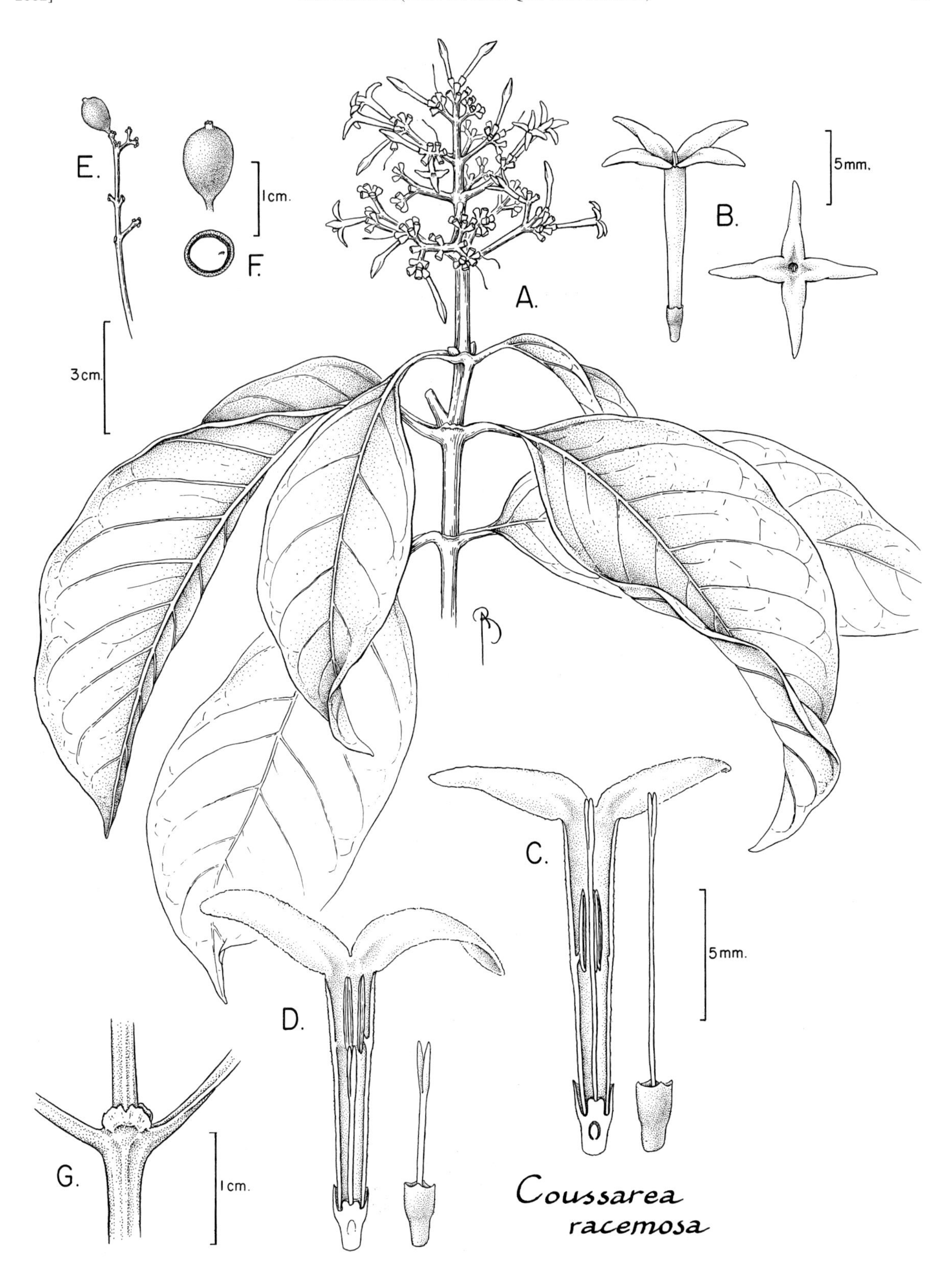

FIG. 270. RUBIACEAE. *Coussarea racemosa* (A, B, D, G, *Mori et al. 20953*; C, *Mori et al. 20754*; E, F, *Mori & Pennington 18007*). **A.** Apex of stem showing leaves and inflorescence. **B.** Lateral (left) and apical (right) views of flowers. **C.** Medial section (left) of pin flower and pistil after corolla dehiscence (right). **D.** Medial section (left) of thrum flower and pistil after corolla dehiscence. **E.** Part of infructescence. **F.** Lateral view (above) and transverse section (below) of fruit. **G.** Stipule.

DIODIA L.

Herbs or subshrubs. Stems round or quadrangular, thin. Stipules united with petioles by setose sheath, the setae filiform. Inflorescences axillary, densely capitate, each head subtended by a leaf. Flowers: calyx 2–4-lobed, the lobes persistent; corolla 4-lobed, valvate in bud, white; stamens equal in number to corolla lobes, inserted on corolla tube, the anthers dorsifixed, exserted; ovary 2-locular, the ovules 1 per locule, ascending, united to septum at middle. Fruits dry, splitting into 2 indehiscent mericarps. Seeds ellipsoid, convex dorsally, with longitudinal groove on ventral side.

1. Fertile axes with axillary inflorescences approximately equal size throughout, i.e., not distally reduced. Corolla 2.5–3.5 mm long. Fruits oblong-turbinate, 2.5–4 × 1.5 mm. . . . . *D. ocimifolia.*
1. Fertile axes with axillary inflorescences gradually reduced in size distally. Corolla 1.2–1.3 mm long. Fruits subglobose, 1–1.2 mm diam. . . . . *D. spicata.*

Diodia ocimifolia (Willd.) Bremek. PL. 110d

Erect or suberect herbs. Stems 0.3–1.5 m long. Stipular sheath 3–5 mm long, with 6–8 unequal reddish-brown setae. Leaves: blades lanceolate to lanceolate-elliptic, 3–9 × 0.6–3 cm. Inflorescences approximately equal size throughout. Flowers: calyx 1.5–2 mm long; corolla 2.5–3.5 mm long, white. Fruits stipitate, oblong-turbinate, 2.5–4 × 1.5 mm. Fl (Feb, Jun, Jul, Aug, Sep, Oct), fr (Feb, Aug); in weedy habitats.

Diodia spicata Miq.

Prostrate or erect herbs. Stems to 1.5 m long. Stipular sheath 3–3.5 mm long, with 7–8 setae. Leaves: blades lanceolate-elliptic, 4–4.8 × 1.1–1.3 cm, those subtending inflorescences progressively reduced apically. Inflorescences gradually reduced in size distally. Flowers: calyx 1.8 mm long; corolla 1.2–1.3 mm long, white. Fruits subglobose, 1–1.2 mm long. Fl (Sep, Oct), fr (Sep); in weedy habitats.

DUROIA L. f.

Trees, dioecious. Stipules united and forming conical cap above apical bud, the cap deciduous above circular slit (circumscissile). Leaves opposite or whorled. Inflorescences terminal. Staminate inflorescences cymose. Pistillate inflorescences of solitary flowers. Flowers unisexual, 5–6-merous; corolla lobes contorted in bud. Staminate flowers: stamens 5–6, inserted in tube, included, the anthers dorsifixed. Pistillate flowers: ovary 1–4-locular; placentation parietal, the ovules numerous. Fruits baccate, large, ovoid or globose, the pericarp thickened, woody. Seeds numerous, angled or prismatic, surrounded by gelatinous pulp.

1. Leaves 5 per node; blades cordate at base. . . . . *D. aquatica.*
1. Leaves 2 or 3–4 per node; blades decurrent, acute, or obtuse at base.
   2. Stems pubescent. Leaves opposite, 2 per node; blades pubescent abaxially, especially on midrib and veins. Fruits densely pubescent. . . . . *D. eriopila.*
   2. Stems glabrous. Leaves whorled, 3–4 per node; blades glabrous abaxially. Fruits glabrate to sparsely pubescent. . . . . *D. longiflora.*

Duroia aquatica (Aubl.) Bremek.

Trees, to 10 m × 13.5 cm. Trunks cylindrical. Outer bark flaky, soft, brown, the inner bark orange. Stems pubescent when young, glabrescent when older. Stipule cap pentagonal, acuminate, 3 cm long. Leaves in whorls of 5; petioles to 10 cm long; blades oblong-obovate, 35 × 15 cm, densely appressed pubescent on midrib and veins abaxially, the apex acuminate, the base cordate. Inflorescences terminal. Staminate inflorescences with ca. 24 flowers per sessile umbel. Staminate flowers 6-merous; calyx ca. 24 mm long; corolla ca. 50 mm long, densely pubescent abaxially. Pistillate flowers 10-merous; corolla tube wider than in staminate flowers. Fruits ovoid, 10 × 5.5 cm, woody, glabrescent. Fr (Mar, Apr, Jun); common but scattered, in nonflooded forest. *Confiture macaque* (Créole).

Duroia eriopila L. f. PART 1: PL. XIIIc

Trees, to 15 m × 10 cm. Stems densely pubescent. Stipule cap bifacial, obtuse, hirsute, ca. 1 cm long. Leaves opposite; petioles absent to 2 cm long; blades elliptic to obovate, 15–35 × 6–17 cm, densely hirsute abaxially, the base acute or decurrent, the apex acuminate. Inflorescences terminal. Staminate inflorescences of 10–30 flowers in sessile heads. Pistillate inflorescences with sessile flowers. Staminate flowers 6-merous; calyx 3–6.5 mm long; corolla 23–25 mm long. Pistillate flowers: calyx ca. 12 mm long; corolla 23–30 mm long, densely pubescent. Fruits globose, 5–7 cm diam., rust colored, densely pubescent. Fl (Jan, Sep, Oct), fr (Jan, May, Nov); in nonflooded forest. *Confiture macaque* (Créole).

Duroia longiflora Ducke

Trees, to 22 m × 29 cm. Trunks with steep rounded buttresses. Outer bark thin, scaly, brown, the inner bark orangish, fibrous. Stems 4-angular, glabrous. Stipule cap 4-angular. Leaves in whorls of 3–4; petioles 1–2 cm long; blades commonly obovate, 15–30 × 5–10 cm, glabrous. Inflorescences terminal. Staminate flowers: calyx ca. 3 mm long, densely sericeous; corolla 3.5–4 cm long, the tube greenish, the lobes white. Pistillate flowers unknown in our area. Fruits broadly obovoid, 4.5–5.5 × 4–4.5 cm, glabrate to sparsely pubescent. Fl (Oct, Nov), fr (Jun, Jul, Aug); in nonflooded forest.

FARAMEA Aubl.

Shrubs, subshrubs, or trees, generally glabrous. Stipules interpetiolar, persistent, usually connate at base, short-triangular, the apex cuspidate or aristate. Inflorescences terminal. Flowers usually 4-merous; corolla blue, white, or lavender, the lobes valvate in bud, the throat glabrous adaxially; stamens 4–6, inserted at throat or above middle of tube, the anthers dorsifixed; ovary 2-locular, 1-celled by abortion of one locule, or sometimes with 2 cells and quickly disappearing thin septum, the ovules 1 per locule, collateral, separate, erect or ascendent from ovary base. Fruits drupaceous, coriaceous, generally wider than long, with 1 functional locule and usually 1 seed by abortion of other seed (rarely with 2 seeds in our area). Seeds horizontally oriented, ventrally excavated, sometimes with vertical ridge on surface, the testa thin.

1. Inflorescences terminal.
   2. Inflorescences paniculate, the primary axes blue. Corolla blue. Fruits wider than long, subreniform. .... *F. multiflora* var. *salicifolia*.
   2. Inflorescences umbellate or corymbose, the primary axes green or yellowish. Corolla white. Fruits globose.
      3. Treelets to trees, to 8 m tall. Petioles 5–10 mm long; blades 8–17 × 2–6.5 cm. Corolla ca. 8 mm long. Fruits ca. 10 mm diam., yellow.. .... *F. corymbosa*.
      3. Subshrubs, to 0.5 m tall. Petioles ca. 3 mm long; blades 5.5–9 × 1.4–3.8 cm. Corolla ca. 17 mm long. Fruits ca. 7 mm diam., blue. .... *F. quadricostata*.
1. Inflorescences axillary.
   4. Leaf blades 2–4 × 0.8–2 cm. Corolla 3–4 mm long. .... *F. lourteigiana*.
   4. Leaf blades 10–23 × 4–8 cm. Corolla 15–17 mm long.
      5. Stipules ca. 8.5 mm long, with awn 3–4 mm long. Leaf blades elliptic to oblong, the base acute. Axillary inflorescences subtended by two pairs of light green, ovate-orbicular bracts. .... *F. guianensis*.
      5. Stipules 14–25 mm long, with awn 10–17 mm long. Leaf blades oblanceolate, the base cordate. Axillary inflorescences not subtended by bracts. .... *F. tinguana*.

Faramea corymbosa Aubl.

Treelets to trees, to 8 m × 10 cm. Stems glabrous. Stipules triangular, 1–4 mm long, terminating in erect awn 1–2 mm long. Leaves: petioles 5–10 mm long; blades lanceolate-elliptic to oblong-oblanceolate, 8–17 × 2–6.5 cm. Inflorescences terminal, umbellate, with 3–5 axes terminating in 5–15-flowered umbels; peduncle absent. Flowers: calyx 1.2–1.5 mm long; corolla 8–8.5 mm long, white. Fruits globose, ca. 10 mm diam., yellow. Fr (Apr, Sep); uncommon, in nonflooded forests.

Faramea guianensis (Aubl.) Bremek.
FIG. 266 (stipules only), PL. 110e

Shrubs, to 1.5 m tall. Stems glabrous. Stipules narrowly triangular, ca. 8.5 mm long, terminating in awn ca. 3 mm long. Leaves: petioles 1–2 mm long; blades elliptic to oblong, 11–16 × 4–5 cm. Inflorescences axillary, cymose, with ca. 7 flowers per cyme; bracts 4, ovate-orbicular, 2 larger outer ones (ovate, with reticate venation), and 2 inner smaller ones; peduncle 0.5–1 cm long. Flowers: calyx ca. 0.8 mm long; corolla ca. 15 mm long, white. Fruits slightly curved, turbinate, ca. 10 × 4 mm, grayish-white. Fl (Jan, Aug, Sep, Oct, Nov), fr (Mar, Oct); common, in nonflooded forest. *Petit ipéca* (Créole).

Faramea lourteigiana Steyerm.

Subshrubs, to 1 m tall. Stems glabrous. Stipules triangular, ca. 4–5 mm long, terminating in filiform awn 3–4 mm long. Leaves: petioles ca. 1 mm long; blades ovate to elliptic, 2–4 × 0.8–2 cm. Inflorescences axillary, subsessile, 1–5-flowered. Flowers: calyx 1.5 mm long; corolla 3–4 mm long, white. Fruits globose, ca. 6 mm diam., violet blue (drying pale blue). Fl (Mar, May, Jul, Sep), fr (Mar, Aug); common, in understory of nonflooded forest. *Bois-bandé, ti bois bandé* (Créole).

Faramea multiflora A. Rich. var. **salicifolia** (C. Presl) Steyerm. [Syn.: *F. salicifolia* C. Presl]

Shrubs or small trees, to 4 m tall. Stems glabrous. Stipules widely ovate to triangular, connate for 1–5 mm, terminating in awn 1.5–3 mm long. Leaves: petioles 4–10 mm long; blades variously shaped, usually narrowly elliptic, 7.5–17 × 1.5–6 cm. Inflorescences terminal, paniculate, ca. 4 × 4 cm, many-flowered, the axes dark blue basally grading to light blue apically, the secondary and tertiary axes white; peduncle absent. Flowers: calyx 1.5–2 mm long; corolla 11–13 mm long, light blue. Fruits wider than long, subreniform, 5–6 × 9–11 mm, dark orange to purple or bluish black. Fl (Jan, Mar, Apr, May, Aug, Nov), fr (Jun, Jul, Aug, Sep, Oct); common, in nonflooded forest.

Faramea quadricostata Bremek.

Subshrubs, to 0.5 m tall. Stems glabrous. Stipules triangular, ca. 5 mm long, awned. Leaves: petioles ca. 3 mm long; blades oblong to oblanceolate, 5.5–9 × 1.4–3.8 cm. Inflorescences terminal, corymbose; peduncle present. Flowers: calyx ca. 1 mm long; corolla ca. 17 mm long, white. Fruits globose, ca. 7 mm diam., blue, with persistent calyces. Fl (Mar), fr (May, Aug); uncommon, in nonflooded forest.

Faramea tinguana Müll. Arg.

Subshrubs, to 0.5 m tall. Stems glabrous. Stipules triangular, 14–25 mm long, with rigid awn 10–17 mm long. Leaves: petioles 2–4 mm long; blades oblanceolate, 10–23 × 4–8 cm, the base cordate. Inflorescences axillary, subsessile, few-flowered. Flowers: calyx ca. 2.7 mm long; corolla 16–17 mm long, white. Fruits globose, ca. 10 mm diam., purple to blue violet. Fl (Jan), fr (Feb, Apr, Aug, Sep, Oct, Nov); common, in nonflooded forest.

FERDINANDUSA Pohl

Trees. Stipules interpetiolar, readily caducous. Inflorescences terminal. Flowers: buds with narrow tube and globose, larger, apical expansion; corolla 4–5-lobed, the lobes contorted in bud, the apex notched; stamens usually 4, inserted at upper part of corolla tube, the anthers versatile; ovary 2-locular, the ovules imbricate, numerous or few in each locule. Fruits capsular, cylindrical or oblong, loculicidal. Seeds few or numerous, peltate, the seed wings bipolar, pointed at both apices, the margins of wings entire or lacerate.

Ferdinandusa paraensis Ducke
FIG. 266 (stipules only), PL. 111b

Trees, to 25 m × 40 cm. Trunks with base fluted or slightly buttressed. Outer bark grid-cracked, brown, the inner bark orange with yellow inclusions. Stems puberulous. Stipules 8–15 mm long, glabrous. Leaves: petioles 5–10 mm long; blades oblong to obovate-oblong to ovate, 8–18 × 4–8 cm. Inflorescences terminal, paniculate, frondose, with reduced leaf-like bracts (pherophylls). Flowers nocturnal; calyx 2–3 mm long, green, with 5 acute lobes; corolla 17 mm long, the tube green-white, the lobes white. Fruits oblong, ca. 3 cm long, with many seeds per locule. Seeds small, with bipolar wings. Fl (Sep, Oct), fr (Mar); infrequent, in nonflooded forest.

GENIPA L.

Trees. Stems glabrous. Stipules interpetiolar, simple, deciduous. Inflorescences terminal, cymose. Flowers large, bisexual, 5–6-merous; corolla densely pubescent abaxially and in throat, the lobes contorted in bud; stamens inserted on throat, exserted, the anthers dorsifixed; ovary 2-locular, the ovules numerous, horizontally oriented. Fruits baccate, globose or ovoid, glabrous, the pericarp firm, coriaceous, with several seeds per locule. Seeds large, laterally compressed, angular.

Genipa spruceana Steyerm. FIG. 271

Trees, to 30 m × 50 cm. Trunks with low, thick buttresses. Outer bark smooth, gray, the inner bark cream-colored with orange laminations. Stems glabrous. Stipules ovate-lanceolate or triangular-ovate, caudate, 9–20 mm long. Leaves: petioles 6–20 cm; blades elliptic-oblanceolate, 10–30 × 3–15 cm, membranous to coriaceous, glabrous, drying black. Inflorescences 5–10 × 5–10 cm; peduncle 1–3.5 cm long. Flowers: calyx 9–20 mm long; corolla cream-colored with yellow throat, the tube 6–13 mm long, the lobes 10–18 mm long. Fruits globose or ovoid, 6–6.5 × 4 mm, glabrous. Fl (Oct); rare in our area, in nonflooded forest. *Ti génipa* (Créole).

GEOPHILA D. Don

Herbs. Stems thin, creeping. Stipules interpetiolar, simple, membranous, usually reflexed. Leaves: petioles present; blades usually at least somewhat cordate, membranous. Inflorescences axillary or subterminal, few-flowered, subtended by involucre of bracts. Flowers usually 5-merous; calyx lobes persistent; corolla white, the lobes valvate in bud; stamens usually 5, the anthers dorsifixed; ovary 2-locular, the ovules 1 per locule, erect, basally attached. Fruits drupaceous, ovoid to globose, fleshy, red to orange, slightly twisted at maturity. Seeds with groove on ventral surface.

Andersson, L. 1999. *Geophila in* Rubiaceae (part 3). In G. Harling & L. Andersson (eds.), Flora of Ecuador **62:** 235–239.

1. Corolla 7–14 mm long. Fruits 9–10 mm long, glabrous. . . . . *G. repens.*
1. Corolla 5.5–9 mm long. Fruits 6–8 mm long, pubescent.
    2. Leaf blades 2.5–10 × 1.2–8.5 cm, pubescent ad- and abaxially. Inflorescences 7–17-flowered. Corolla 5.5–6 mm long. . . . . *G. cordifolia.*
    2. Leaf blades 1–4 × 0.6–3 cm, pubescent adaxially, glabrous abaxially. Inflorescences 3–6-flowered. Corolla ca. 9 mm long. . . . . *G. tenuis.*

Geophila cordifolia Miq. PL. 111a

Creeping herbs. Stipules ovate, 2–6 × 1.8–5 mm. Leaves: petioles 4–10 cm long, covered with crooked multicellular hairs; blades ovate-cordate to oblong-cordate to ovate-suborbicular, 2.5–10 × 1.2–8.5 cm, pubescent ad- and abaxially, the base deeply cordate. Inflorescences 7–17-flowered. Flowers: calyx 1.5 mm long, pubescent; corolla 5.5–6 mm long. Fruits 8 mm long, pubescent. Fl (Apr), fr (Jan, May, Jun, Aug, Sep, Nov); in shaded and somewhat open habitats, e.g., along forest trails.

Geophila repens (L.) I. M. Johnst. [Syn.: *Geophila herbacea* (L.) K. Schum.]

Creeping herbs. Stipules ovate, 1.5–2 × 1.5–2 mm. Leaves: petioles 0.5–8.5 cm long, with pubescence on one side; blades widely ovate-cordate, suborbicular-ovate-cordate or subreniform, 1.2–5.5 × 1.2–5 cm, glabrous to pubescent, the base long cordate. Inflorescences 2–4-flowered. Flowers: calyx 2–3.5 mm long; corolla 7–14 mm long. Fruits 9–10 mm long, glabrous. Fr (Jul, Sep); in shaded and somewhat open habitats, e.g., along forest trails.

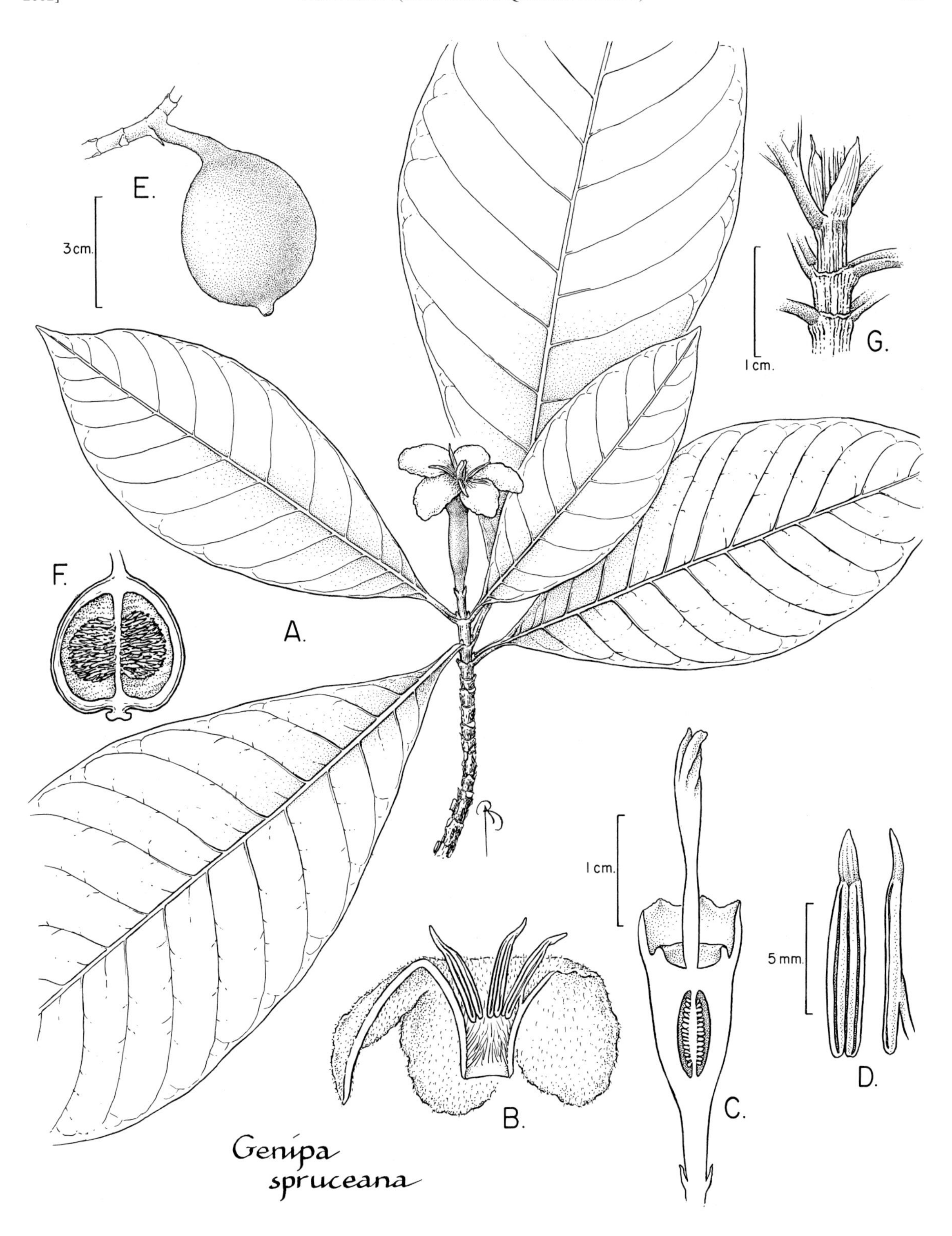

FIG. 271. RUBIACEAE. *Genipa spruceana* (A, E, G, *Mori & Boom 15089*; B–D, *Jansen-Jacobs et al. 1529* from Guyana; F, *Persson et al. 1959* from eastern French Guiana). **A.** Apex of stem showing leaves and axillary flower. **B.** Medial section of corolla with adnate stamens. **C.** Medial section of pistil. **D.** Two views of anthers showing the elongate connective. **E.** Fruit. **F.** Medial section of fruit. **G.** Stipule.

Geophila tenuis (Müll. Arg.) Standl.

Creeping herbs. Stipules ovate, 1.8–2.5 × 1–1.5 mm. Leaves: petioles 0.8–8 cm long, with pubescence on one side; blades triangular-ovate-cordate, 1–4 × 0.6–3 cm, pubescent adaxially, glabrous abaxially. Inflorescences 3–6-flowered. Flowers: calyx ca. 4 mm long; corolla ca. 9 mm long. Fruits 6–7 mm long, pubescent. Fl (Mar), fr (Mar, Jun, Sep); in shaded and somewhat open habitats, e.g., along forest trails.

GONZALAGUNIA Ruiz & Pav.

Shrubs. Stems usually elongate. Stipules interpetiolar, united basally into sheath, persistent. Inflorescences terminal, elongate spikes. Flowers 4-merous; calyx teeth persistent; corolla villous in throat, the lobes valvate in bud; stamens 4, the anthers dorsifixed. Fruits baccate, 2(4)-locular, each locule containing a pyrene with many seeds. Seeds minute, the testa slightly foveolate.

Gonzalagunia dicocca Cham. & Schltdl. [Syn.: *G. surinamensis* Bremek.] FIG. 272

Shrubs, 1–2.5 m tall. Stems densely pubescent. Stipules basally widely triangular, long-acuminate, 3–8 mm long, strigulose. Leaves: blades ovate to elliptic, 6.5–17.5 × 1.5–5.5 cm, membranous. Flowers heterostylous; calyx lobes triangular to suborbicular, 0.2–1.8 mm long; corolla hypocrateriform, 6–10 mm long, white. Fruits 2-locular, 2.5–3 mm diam., spongy, white to purple. Fl (Jan, Feb, Mar, Apr, Jun, Sep, Oct, Nov, Dec), fr (Jan, Feb, Apr, Jul, Aug, Sep, Oct, Nov, Dec); locally abundant, especially in somewhat open areas along roads through forest.

GUETTARDA L.

Trees. Stipules interpetiolar, ovate, shortly connate basally, the apex pointed. Inflorescences axillary, dichotomously branched cymes; peduncle present. Flowers 4-merous; calyx shortly 4-lobed; corolla pubescent abaxially, glabrous adaxially except at base, white to greenish-white; stamens inserted near top of tube; ovary 4-locular. Fruits drupaceous, oblong, the exocarp glabrous, coriaceous, the endocarp bony, with 4 pyrenes.

1. Leaf blades 7–13 × 3–7.5 cm, membranous, glabrous abaxially; tertiary venation reticulate. Corolla ca. 9 mm long. Fruits oblong, 5–7 mm diam. . . . . *G. acreana*.
1. Leaf blades (5)13–20 × (3)5–9.5 cm, thin-coriaceous, densely silver-pubescent abaxially; tertiary venation subparallel. Corolla 24–27 mm long. Fruits subglobose, 7–10 mm diam. . . . . *G. spruceana*.

Guettarda acreana K. Krause

Trees, to 20 m × 25 cm. Trunks fluted, with high, steep plank buttresses. Outer bark gray-green, smooth, the inner bark yellow. Stems sparsely pubescent when young, glabrous when mature. Stipules narrowly triangular, 3–5 mm long, glabrous abaxially, the apex not acuminate. Leaves: petioles 0.6–1.8 cm long; blades obovate-elliptic, ovate-oblong, or ovate-lanceolate, 7–13 × 3–7.5 cm, membranous, glabrous, the apex with acumen 1–1.5 cm long; tertiary venation reticulate. Inflorescences: peduncle 3–10 cm long. Flowers diurnal: calyx ca. 1.5 mm long; corolla ca. 9 mm long, sparsely pubescent abaxially, white. Fruits oblong, ca. 8–9 × 5–7 mm, glabrous. Fl (Mar, Apr, May), fr (Jun, Aug, Nov); common, in nonflooded forest.

Guettarda spruceana Müll. Arg. PL. 111e

Trees, to 20 m × 20 cm. Outer bark gray-green, smooth. Stems sparsely pubescent, with sparse lenticels when young. Stipules broadly triangular, 3–5 mm long, densely pubescent abaxially, the apex acuminate. Leaves: petioles 2.5–5 cm long; blades elliptic, (5)13–20 × (3)5–9.5 cm, thin-coriaceous, densely silver-pubescent abaxially, the apex with acumen 1.5–3 cm long; tertiary venation subparallel. Inflorescences: peduncle 7–14 cm long. Flowers nocturnal; calyx 2–2.5 mm long; corolla 24–27 mm long, densely pubescent abaxially, white. Fruits subglobose, 7–10 mm diam., densely velutinous-pubescent. Fl (Feb, Mar), fr (Mar); infrequent, in nonflooded forest.

HAMELIA Jacq.

Shrubs or treelets. Stipules interpetiolar, inconspicuous, deciduous. Leaves: blades membranous. Inflorescences terminal or axillary; pedicels absent, the flowers secund. Flowers 5-merous; calyx short, inconspicuous; corolla narrowly infundibular, yellow to orange to reddish-orange, the lobes imbricate in bud; stamens inserted at base of corolla tube, usually exserted; ovary 5-locular, the ovules numerous per locule. Fruits baccate, subglobose to oblong. Seeds minute, testa foveolate.

Elias, T. S. 1976. A monograph of the genus *Hamelia* (Rubiaceae). Mem. New York Bot. Gard. **26:** 81–144.

Hamelia axillaris Sw. PL. 111c,d

Shrubs or treelets, to 4 m tall. Stems glabrous to puberulent. Stipules 1.5–3 mm long, the base triangular, the apex acuminate. Leaves: petioles 0.5–4 cm long, blades elliptic or elliptic-ovate, 6–17 × 2.5–6.5 cm. Inflorescences axillary; peduncle 0.5–1.5 cm long. Flowers: calyx ca. 4 mm long; corolla 9–12 mm long, yellow. Fruits subglobose to oblong, 6–8 × 3–6 mm, glabrous, purple-black at maturity. Fl (Feb, Aug, Oct, Nov), fr (Feb, Aug, Oct, Nov, Dec); in somewhat disturbed areas, especially along trails and roads.

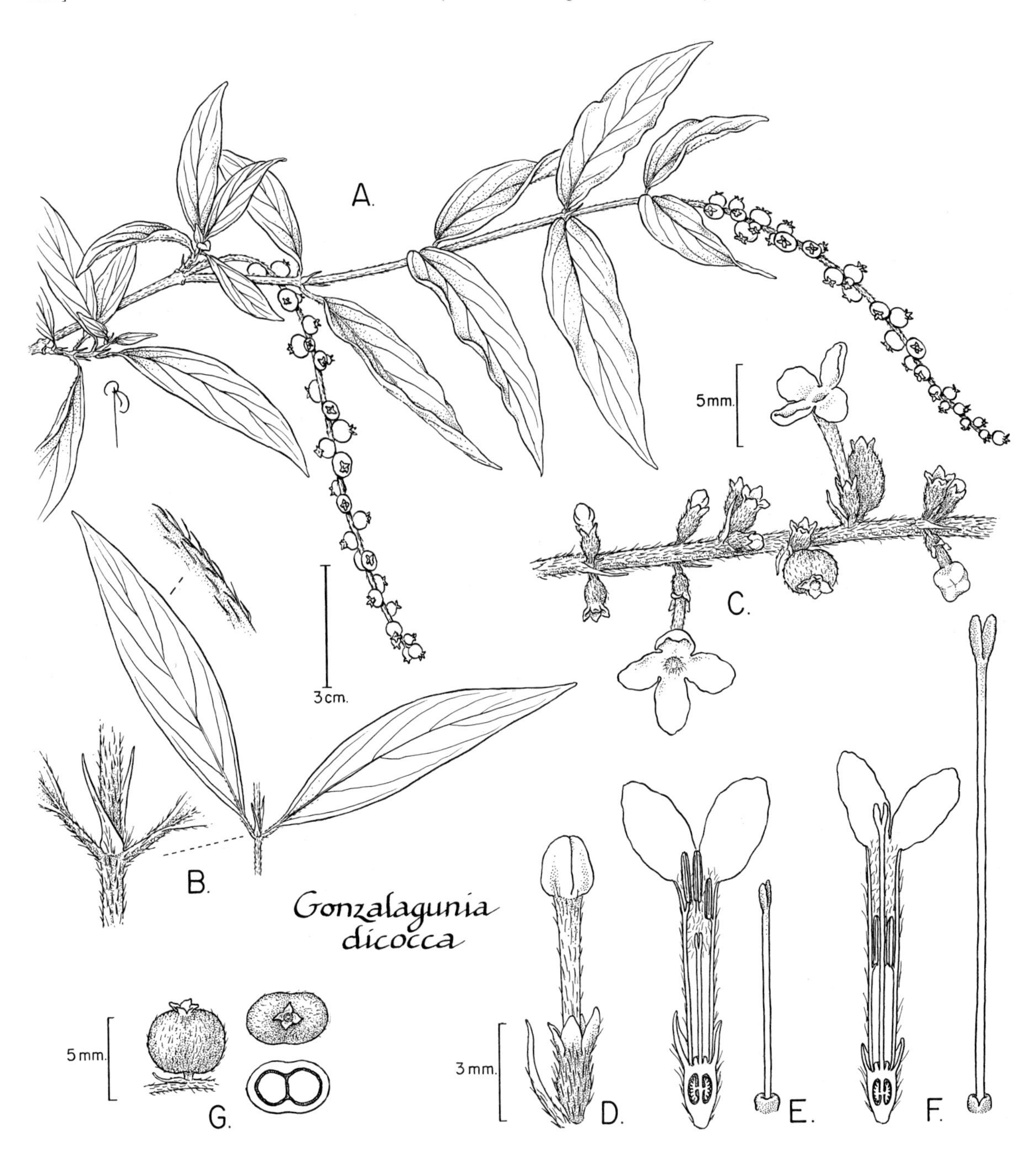

FIG. 272. RUBIACEAE. *Gonzalagunia dicocca* (A, B, *Mori et al. 14852*; C–E, *Mori et al. 22667*; F, *Andersson et al. 2024*; G, unvouchered field sketch). **A.** Part of stem with leaves and infructescences. **B.** Part of stem with leaves with detail of stipules (left) and leaf margin (above). **C.** Part of inflorescence. **D.** Lateral view of flower bud with bract. **E.** Medial section of short-styled flower (far left) and pistil of same (near left). **F.** Medial section of long-styled flower (left) and pistil of same (right). **G.** Lateral view (above left), apical view (above right), and transverse section (below right) of fruit.

HILLIA Jacq.

Epiphytic (rarely terrestrial) shrubs. Stems glabrous. Stipules interpetiolar, ligulate, caducous. Leaves: petioles present; blades fleshy, thickened. Inflorescences terminal, of solitary flowers. Flowers showy, (5)6-merous; corolla infundibuliform or salverform, the tube 4.5–11 cm long, white or green or yellowish-green. Fruits capsular, long, cylindrical, dehiscing septicidally and basipetally into 2 valves. Seeds numerous, fusiform, with tuft of "hairs" (exotestal extensions) at one end.

Taylor, C. M. 1994. Revision of *Hillia* (Rubiaceae). Ann. Missouri Bot. Gard. **81:** 571–609.

1. Leaf blades 9.5 × 2.5–8 cm; secondary veins usually in 5–6 pairs. Corolla infundibuliform, green to yellowish-green, the tube 45–65 mm long, the lobes 13–22 × 8–12 mm. Fruits 7.5–10 × 0.8–1.5 cm. . . . . *H. illustris.*
1. Leaf blades 4–10 × 2–5 cm; secondary veins usually in 3–4 pairs. Corolla salverform, white, the tube 45–110 mm long, the lobes 25–35 × 5–8 mm. Fruits 5.5–11 × 0.6–1.2 cm. . . . . *H. parasitica.*

Hillia illustris (Vell.) K. Schum. FIG. 273

Stipules ligulate, linear-oblong, or ligulate-lanceolate, 2.5–5 × 0.6–1.4 cm, the apex subobtuse to rounded. Leaves: blades elliptic-ovate to ovate-elliptic or lanceolate-elliptic, 9.5–15 × 2.5–8 cm, coriaceous, the apex acuminate; secondary veins in 5–6 pairs, scarcely visible. Flowers nocturnal: calyx 6-lobed; corolla infundibuliform, green to yellow-green, the tube 45–65 mm long, constricted to 0.6–0.8 mm wide at base, expanded to 23–42 mm wide at mouth, the lobes 6, widely ovate, 13–22 × 8–12 mm. Fruits 7.5–10 × 0.8–1.5 cm. Fl (Nov); scattered, in nonflooded forest at all elevations.

Hillia parasitica Jacq.

Stipules ligulate-oblong, 1.1–5.5 × 0.6–1.1 cm, the apex rounded or obtuse. Leaves: blades rhombic-elliptic to elliptic, 4–10 × 2–5 cm, coriaceous, the apex acuminate; secondary veins in 3–4 pairs, scarcely visible. Flowers nocturnal: calyx 6–9-lobed; corolla salverform, white, the tube 45–110 × 2.5–3 mm, the lobes 6, lanceolate, 25–35 × 5–8 mm, strongly reflexed. Fruits 5.5–11 × 0.6–1.2 cm. Known by only one collection from our area (*Granville 8563*, CAY).

IBETRALIA Bremek.

Treelets, dioecious. Stipules interpetiolar, basally sheathing at base, free portion broadly triangular, persistent. Inflorescences terminal. Staminate inflorescences corymbose. Pistillate inflorescences solitary. Flowers unisexual; corolla white or greenish-white, the lobes contorted in bud. Staminate flowers with dorsifixed anthers. Pistillate flowers: ovary 2–8-locular, the ovules numerous per locule. Fruits baccate, oblong, 2–8-locular, the exocarp coriaceous, the pulp fleshy, with tubular persistent calyx. Seeds large, vertically orientated, compressed or subglobose, obtusely angled.

According to recent molecular phylogenies (Claes Persson, 2000), this genus is better kept separate from *Alibertia.* It is closely related to *Kutchubaea.*

Ibetralia surinamensis Bremek. [Syns.: *Alibertia surinamensis* (Bremek.) Steyerm., *Alibertia dolichophylla* Standl.] FIG. 266 (stipules only)

Treelets, 1.5–4 m tall. Stems glabrous, with resinous exudate from young stipules (with adaxial colleters). Stipules sheathing at base, free portion broadly triangular, ca. 5 mm long. Leaves: blades obovate to oblanceolate, 15–23 × 5.5–13 cm. Staminate flowers usually 7-merous. Pistillate flowers usually 6-merous; corolla tube 8–12 × 3–4 mm, the lobes 10–14 mm long. Fruits oblong, 2.5–5 × 2.5–4 cm. Fl (Sep, Oct, Nov, Dec), fr (Jan, Mar, Apr, May, Sep, Nov, Dec); common, in cloud forests.

ISERTIA Schreb.

Understory treelets to trees. Trunks not buttressed. Stems slender, subterete or thick quadrangular. Stipules 4 and interpetiolar or 2 and intrapetiolar. Leaves: petioles present. Inflorescences terminal, thyrsiform-paniculate or thyrsiform-racemose. Flowers: calyx 4-lobed; corolla 6-lobed, the tube cylindrical, reddish-orange basally and yellow distally, with dense yellow pubescence at mouth, the lobes valvate in bud; stamens 5 or 6, inserted near mouth, the anthers dorsifixed; ovary 5–6-locular, the ovules many per locule. Fruits baccate, fleshy. Seeds numerous, minute, brownish, angular, the testa deeply foveolate.

Boom, B. M. 1984. A revision of the genus *Isertia* (Isertieae: Rubiaceae). Brittonia **36:** 425–454.

1. Stems quadrangular. Stipules 4, interpetiolar. Leaf blades floccose abaxially. Corolla tube 50–70 mm long. Fruits ca. 10 mm diam. . . . . *I. coccinea.*
1. Stems subterete. Stipules 2, intrapetiolar. Leaf blades glabrescent abaxially. Corolla tube 30–45 mm long. Fruits ca. 5 mm diam. . . . . *I. spiciformis.*

Isertia coccinea (Aubl.) J.F. Gmel. FIG. 274, PL. 112a

Treelets to trees, 5–15(20) m tall. Outer bark light gray-brown, the inner bark creamy white with orange inclusions, turning brown-black on exposure to air, with red resinous exudate, the sapwood white-yellow. Stems quadrangular, covered with brownish strigulose tomentum. Stipules 4, interpetiolar, cymbiform and triangular, 5–15 mm long. Leaves: blades obovate to elliptic, 12–25 × 9–13 cm, glabrous adaxially, floccose abaxially; secondary veins in 15–24 pairs. Inflorescences broadly ovoid to spherical, 8–13 × 5–11 cm. Flowers: calyx 4-lobed, 6–7 × 5–6 mm, red; corolla reddish-orange basally, grading to yellow distally, the tube 50–70 mm long, densely tomentose abaxially, with dense yellow pubescence around mouth adaxially. Fruits broadly ovoid, ca. 10 mm diam. Seeds 0.9–1 mm long. Fl (Jan, Apr, Jun, Sep, Oct), fr (Aug); common, in secondary vegetation, especially along roads and wide trails, visited and presumably pollinated by hummingbirds. *Bois pian* (Créole).

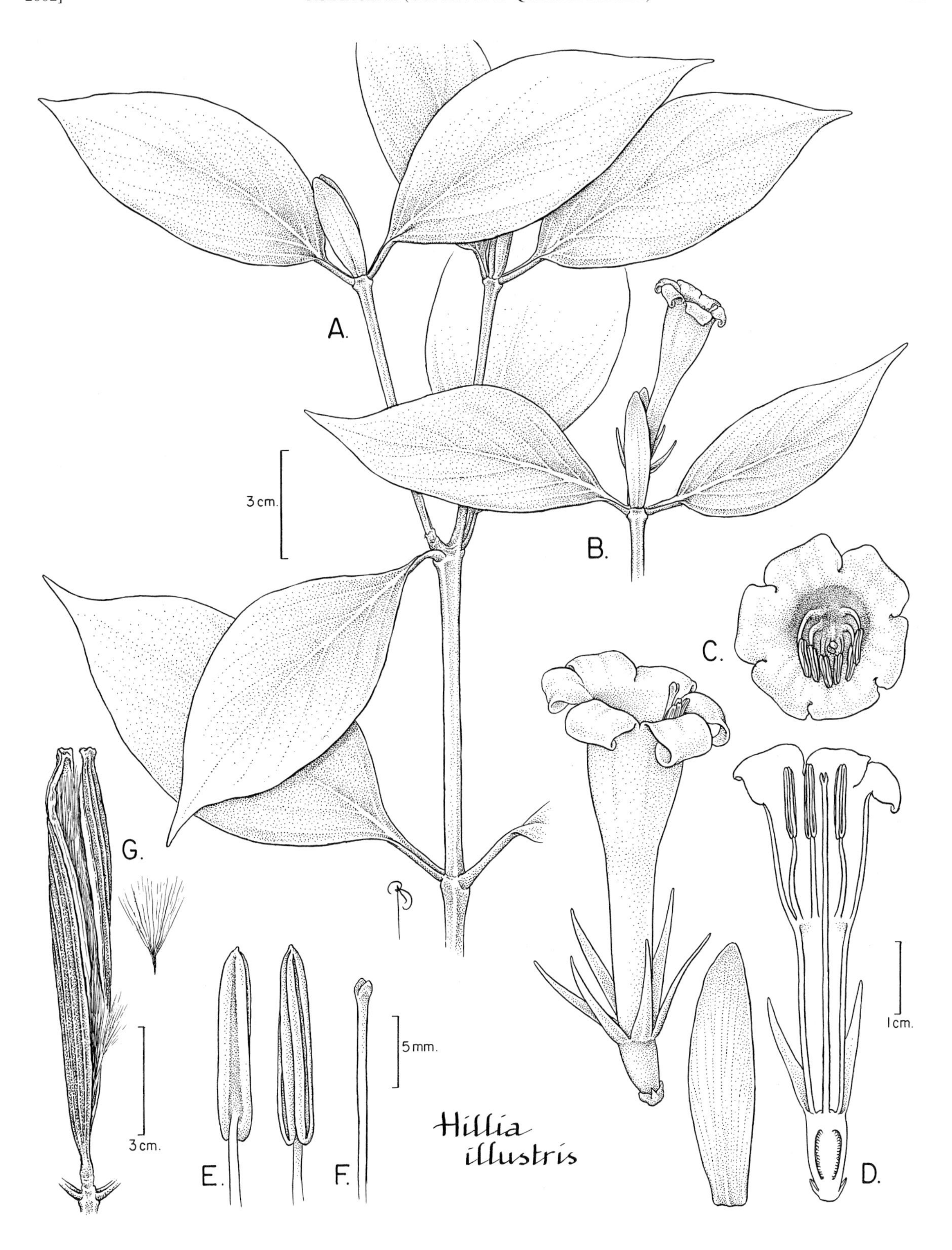

FIG. 273. RUBIACEAE. *Hillia illustris* (A, *Ek & Montfort 196*; B, D–F, *Mori & Boom 15259*; C, unvouchered photos from Amazonas, Brazil; G, *Gillespie 1551* from Guyana). **A.** Part of stem with leaves and stipules. **B.** Part of stem with leaves, stipules, and flower. **C.** Apical (right) and lateral (below left) views of flower. **D.** Medial section of flower (near left) and stipule (far left). **E.** Abaxial (near right) and adaxial (far right) views of anthers. **F.** Upper part of style. **G.** Lateral view of dehisced fruit (left) and seed (below).

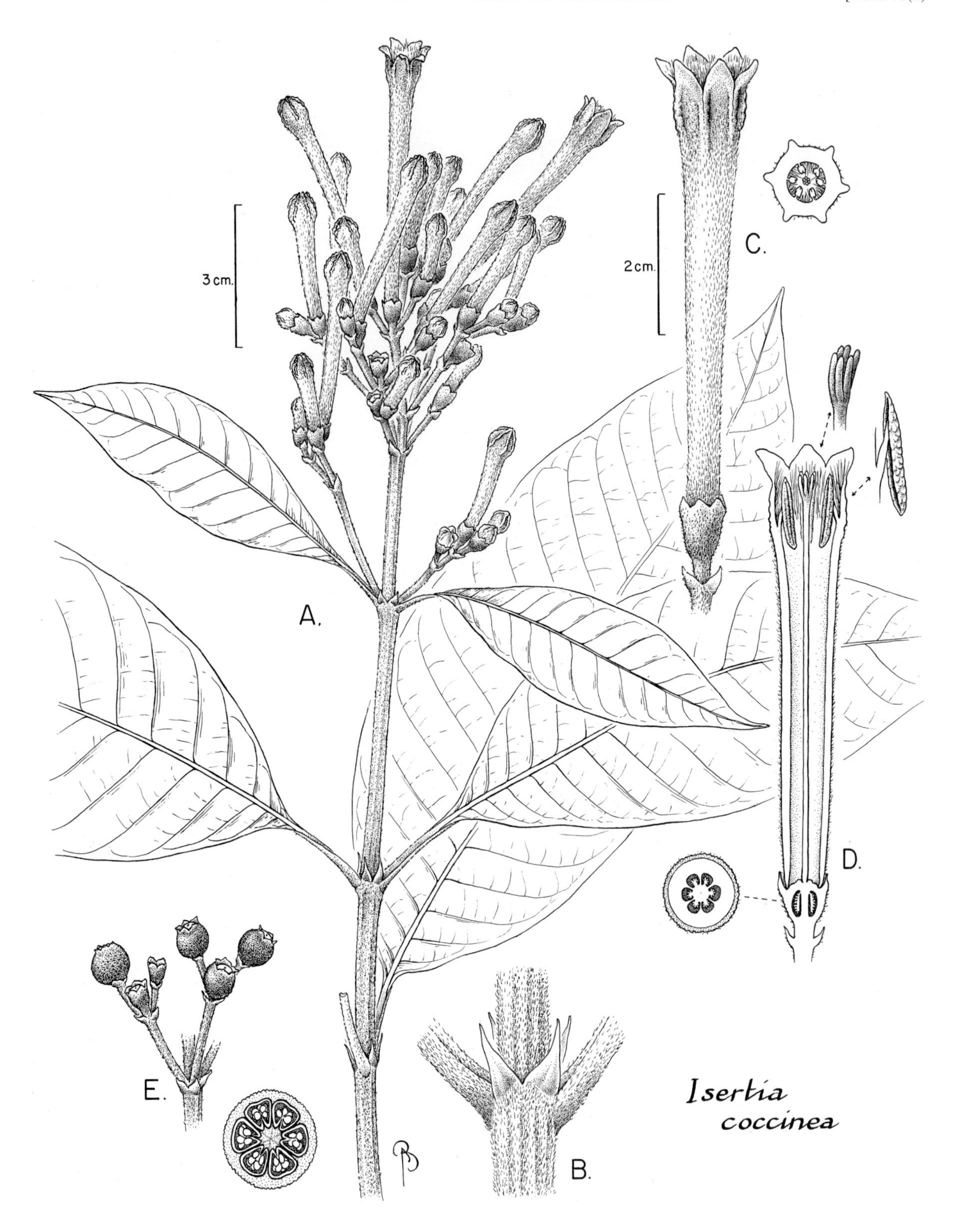

FIG. 274. RUBIACEAE. *Isertia coccinea* (A–D, *Mori et al. 20888*; E, *Marshall 177*). **A.** Apex of stem showing terminal and axillary inflorescences and leaves. **B.** Detail of node showing distinct, persistent stipules. **C.** Lateral view of flower (left) and transverse section (right) through corolla. **D.** Transverse section of ovary (left) and medial section of flower (right) showing the inferior ovary with details of stigma (above) and stamen (above right). **E.** Part of infructescence (right) and transverse section of fruit (far right).

Isertia spiciformis DC. [Syn.: *Isertia commutata* Miq., *Isertia pterantha* Bremek.] Pl. 112d

Treelets to trees, 5–10 m tall. Outer bark gray. Stems subterete, strigose to glabrescent. Stipules 2, intrapetiolar, 8–9 mm long, triangular. Leaves: blades obovate to ovate, 14–37 × 5–22 cm, glabrous adaxially, glabrescent abaxially; secondary veins in 19–21 pairs. Inflorescences narrowly ellipsoid, 5–19 × 5–11 cm. Flowers: calyx 4-lobed, ca. 3 × 3 mm, red; corolla reddish-orange basally, grading to yellow distally, the tube 30–45 mm long, strigulose abaxially, with dense yellow pubescence around mouth adaxially. Fruits oblong, ca. 5 mm diam. Seeds 1–1.3 mm long. Fl (May, Oct), fr (Feb, May, Jun, Oct), occasional, often associated with gaps in nonflooded forest.

IXORA L.

Shrubs or trees. Stems generally glabrous. Leaves opposite or whorled (3 per node in *I. piresii*). Stipules interpetiolar, simple, connate basally, generally persistent, the base wide, the apex acuminate or cuspidate. Inflorescences terminal. Flowers 4–5-merous; calyx denticulate or lobed; corolla salverform, the tube narrow, the lobes linear, lanceolate, contorted in bud; stamens inserted within corolla tube, the anthers dorsifixed; style exserted well beyond corolla tube, functioning in secondary pollen presentation; ovary 2-locular, the ovules solitary in each locule. Fruits drupaceous, globose, fleshy or coriaceous, with 2 pyrenes (rarely 1 pyrene by abortion of other). Seeds subglobose, with deep groove on ventral side.

1. Leaves whorled, 3 per node . . . . . . *I. piresii.*
1. Leaves opposite.
    2. Inflorescences sessile to very short pedunculate, the peduncle <5 mm long when present.
        3. Leaf blades 10–16.5 × 3–5 cm, the base acute. Corolla white. Native. . . . . . *I. aluminicola.*
        3. Leaf blades 3.5–8 × 1.8–5 cm, the base auriculate. Corolla orange-red. Cultivated. . . . . . *I. coccinea.*
    2. Inflorescences pedunculate, the peduncle 10–40 mm long.
        4. Leaf blades glabrous abaxially. Inflorescence axes, corolla, and calyces glabrate. Corolla apex obtuse in bud. . . . . . *I. graciliflora.*
        4. Leaf blades puberulent abaxially. Inflorescence axes, corolla, and calyces puberulent. Corolla apex acuminate in bud. . . . . . *I. pubescens.*

Ixora aluminicola Steyerm.

Shrubs, to 2 m tall. Stems glabrous. Stipules narrowly acuminate, 1–5 mm long, glabrous. Leaves opposite; blades lanceolate to oblong-elliptic, 10–16.5 × 3–5 cm, the base acute. Inflorescences 3–4-flowered; peduncle absent to <5 mm long. Flowers: calyx tube 1.5 mm long, the lobes 0.8–1 mm long; corolla white, the tube 5.5–6.5 mm long, the lobes 5.5 mm long. Fruits globose, 15–18 mm diam., glabrous, purple. Fl (Jan, Mar, Apr, Sep), fr (Jun, Jul, Aug, Sep); common, in nonflooded forests. *Bois-flèche*, *cèdre* (Créole).

Ixora coccinea L.

Shrubs, ca. 1 m tall. Stems glabrous. Stipules triangular, 4–5 mm long, with awn ca. 2 mm long. Leaves opposite, sessile; blades ovate to elliptic, 3.5–8 × 1.8–5 cm, coriaceous, the base auriculate to cordate, the apex acute. Inflorescences: condensed cymes; peduncle very short or absent. Flowers: calyx ca. 2 mm long, puberulous; corolla puberulous, orange-red to deep red, the tube 15–25 mm long, the lobes 10–14 mm long. Fruits subglobose. Fl (Sep); cultivated as an ornamental. *Buisson ardent* (Créole).

Ixora graciliflora Benth.

Shrubs, to 2 m tall. Stems puberulous when young, glabrescent when older. Stipules triangular, acuminate, ca. 3–5 × 2 mm. Leaves opposite; petioles 2–4 mm long; blades ovate-oblong, 8–20 × 3.5–7.5 cm, the base acute. Inflorescences terminal, large, laxly paniculate, glabrous; peduncle 1–3 cm long. Flowers: calyx 0.8 mm long, glabrous; corolla creamy-white, the tube 8–14 mm long, the lobes 4 mm long, glabrous. Fruits not seen. Fl (Mar); known from a single collection from our area (*Granville 2358*, CAY).

Ixora piresii Steyerm.

Single-stemmed shrubs, to 1.5 m tall. Stems glabrescent. Stipules triangular, ca. 3 mm long, with awn 1 mm long. Leaves 3 per node; petioles absent to 3 mm long; blades oblanceolate to oblong-oblanceolate, 29–33 × 8–8.5 cm, the base rounded. Inflorescences terminal, cymes of ca. 15 flowers, 4 × 6.5 cm (excluding peduncle), red; peduncle to 8 cm long. Flowers: calyx 0.8–1 mm long; corolla tube 16 mm long, the lobes 7 mm long, creamy-white. Fruits subglobose, 1 cm diam., puberulous, red. Fr (Jan); in forest at higher elevations.

Ixora pubescens Schult. & Schult. f.

Shrubs to treelets, to 3 m tall. Stems puberulous. Stipules narrowly triangular, subappressed, 4–9 mm long, puberulent, the apex acuminate. Leaves opposite; petioles 3–10 mm long; blades subelliptic to oblanceolate, 18–30 × 4.5–8.5 cm. Inflorescences terminal, laxly paniculate, 6–10 cm long; peduncle 3–4 cm long, puberulous. Flowers: buds reddish and yellow; calyx ca. 2 mm long, puberulous; corolla ca. 25 cm long, puberulous, the tube red, the lobes yellow. Fruits unknown from our area. Fl (Nov); known by two collections from our area (*Granville B.5185*, CAY and *Mori et al. 22807*, NY), in nonflooded forest.

MALANEA Aubl.

Lianas. Stipules interpetiolar, ligulate-oblong, deciduous. Leaves: petioles present; blades subcoriaceous. Inflorescences axillary, paniculate. Flowers 4-merous; calyx campanulate, shortly 4-lobed; corolla shortly infundibuliform, the mouth and lobes densely pubescent, the lobes valvate in bud; stamens 4, inserted at mouth of corolla tube, the anthers exserted, dorsifixed; ovary 2-locular, the ovules 1 per locule, pendulous. Fruits drupaceous, narrowly oblong, 1–2-locular (if 1-locular, by abortion of 1 locule). Seeds cylindrical.

Malanea macrophylla Griseb. [Syn.: *M. hypoleuca* Steyerm.] PL. 112b

Lianas. Stems unarmed, glabrous. Stipules ligulate-oblong, 11–25 × 5–11 mm, glabrous or strigulose, the apex rounded. Leaves: blades broadly ovate, ovate, elliptic, or obovate, 6.5–14.5 × 3.5–9 cm, glabrous, dark green and shiny adaxially, pale green or silvery pubescent abaxially; secondary veins in 5–8 pairs. Inflorescences 1–7.5 cm long, the rachis much branched, densely gray pubescent. Flowers: corolla infundibuliform, densely pubescent externally, pale yellow, the tube 1.8–4 mm long, 0.7–1 mm wide basally, 1.5–2.8 mm wide distally, the lobes 4, 2–2.5 × 1–1.5 mm, pubescent adaxially. Fruits linear-oblong, 8–10 × 3 mm, the endocarp bony. Fl (Aug), fr (Aug, Sep, Nov); common but scattered, in nonflooded forest.

MANETTIA Mutis

Subwoody climbers. Stems thin. Leaves small. Stipules interpetiolar, triangular. Inflorescences axillary. Flowers 4-merous; calyx usually with persistent lobes; corolla red or white, the lobes valvate in bud; stamens inserted at mouth of corolla, the anthers versatile; ovary 2-locular, the ovules numerous, imbricate, inserted on central axile placenta. Fruits capsular, septicidal, opening from apex. Seeds small, discoid, winged.

1. Inflorescences 5–24-flowered. Calyx lobes ca. 1 mm long; corolla ca. 5 mm long, greenish-white or white. Fruits oblong, ca. 5 × 3 mm. . . . . *M. alba.*
1. Inflorescences 1–3(5)-flowered. Calyx lobes ca. 5 mm long; corolla 16–30 mm long, red. Fruits obconic, ca. 8 × 6 mm. . . . . *M. reclinata.*

Manettia alba (Aubl.) Wernham

Stems slender, densely puberulent. Stipules triangular, 1 mm long, the apex acute. Leaves: petioles 5–12 mm long; blades elliptic-oblong to elliptic-lanceolate, 3.5–7 × 1.5–3.5 cm. Inflorescences 5–24-flowered; peduncle 1–2 cm long. Flowers: calyx 2.5–3 mm long, the lobes ca. 1 mm long; corolla 5 mm long, greenish-white or white. Fruits oblong, ca. 5 × 3 mm. Fr (Sep); uncommon, usually in forest edges or in forest gaps.

Manettia reclinata L. [Syn.: *M. coccinea* (Aubl.) Willd.] FIG. 275, PL. 112c

Stems slender, glabrous. Stipules broadly triangular, 0.5–1.5 mm long, the apex truncate, the margins denticulate. Leaves: petioles 3–10 mm long; blades lanceolate to ovate, 3–10 × 1–4 cm. Inflorescences 1–3(5)-flowered; peduncle absent to 3 cm long. Flowers: calyx 5–8 mm long, the lobes ca. 5 mm long; corolla 16–30 mm long, red. Fruits obconic, ca. 8 × 6 mm. Fl (Mar, May, Aug, Sep, Nov), fr (Jan, Aug, Sep); usually in disturbed areas, for example at forest edges. *Macoudia* (Créole).

MORINDA L.

Shrubs or small trees. Stipules interpetiolar, connate at base, persistent. Inflorescences terminal and/or axillary, small heads of sessile flowers. Flowers usually united at base, 4–5-merous; corolla white, the lobes valvate in bud; stamens inserted at throat of corolla tube, the anthers dorsifixed, usually with appendage at apex; ovary 2–4-locular, each locule with solitary, ascending ovule, or, when 2-locular, with 2 collateral ascending ovules, the ovules attached to septum of ovary. Fruits fleshy syncarps, each unit with 4 pyrenes. Seeds oblong or obovoid to reniform.

Morinda brachycalyx (Bremek.) Steyerm. [Syn.: *Appunia brachycalyx* Bremek.]

Shrubs or treelets, to 1.5 m tall. Stems quadrangular, glabrescent. Stipules triangular, 4 × 3 mm, the apex acuminate. Leaves: blades narrowly elliptic, elliptic, or narrowly lanceolate, (10)14–22 × 2.5–8 cm. Inflorescences: peduncles several, ca. 1.2 cm long. Flowers: calyx a short, entire rim, ca. 0.2 mm wide; corolla ca. 15 mm long, white. Fruits globose, fleshy, 0.5–1 cm diam., reddish purple. Fl (Jan, Mar, Apr, May, Jun, Sep), fr (May, Jun, Jul, Sep).

OLDENLANDIA L.

Glabrous herbs. Stipules interpetiolar, small, divided. Leaves small; blades with apex mucronate. Inflorescences axillary, 1–3-flowered. Flowers 4-merous; calyx deeply divided; corolla lobes obtuse, valvate in bud; stamens inserted at corolla throat, the filaments short, glabrous, exserted; ovary 2-locular, the ovules numerous. Fruits loculicidal capsules, 2-locular. Seeds minute, few or numerous.

Oldenlandia lancifolia (Schumach.) DC. [Syn.: *Hedyotis lancifolia* Schumach.]

Procumbent herbs. Stems glabrous. Stipules with 2 setiform appendages, 2–3 mm long. Leaves: petioles absent; blades linear or linear-lanceolate, 1–7 × 0.2–1.2 cm. Inflorescences axillary, 1(2) per node, commonly single-flowered (rarely 2–3-flowered); pedicels 0.5–3 cm long, filiform. Flowers: calyx 1–1.5 mm long; corolla white, the tube ca. 1 mm long, the lobes 1–1.2 mm long. Fruits depressed globose, 2–3 mm diam. Fr (Mar, Sep, Nov); in disturbed habitats.

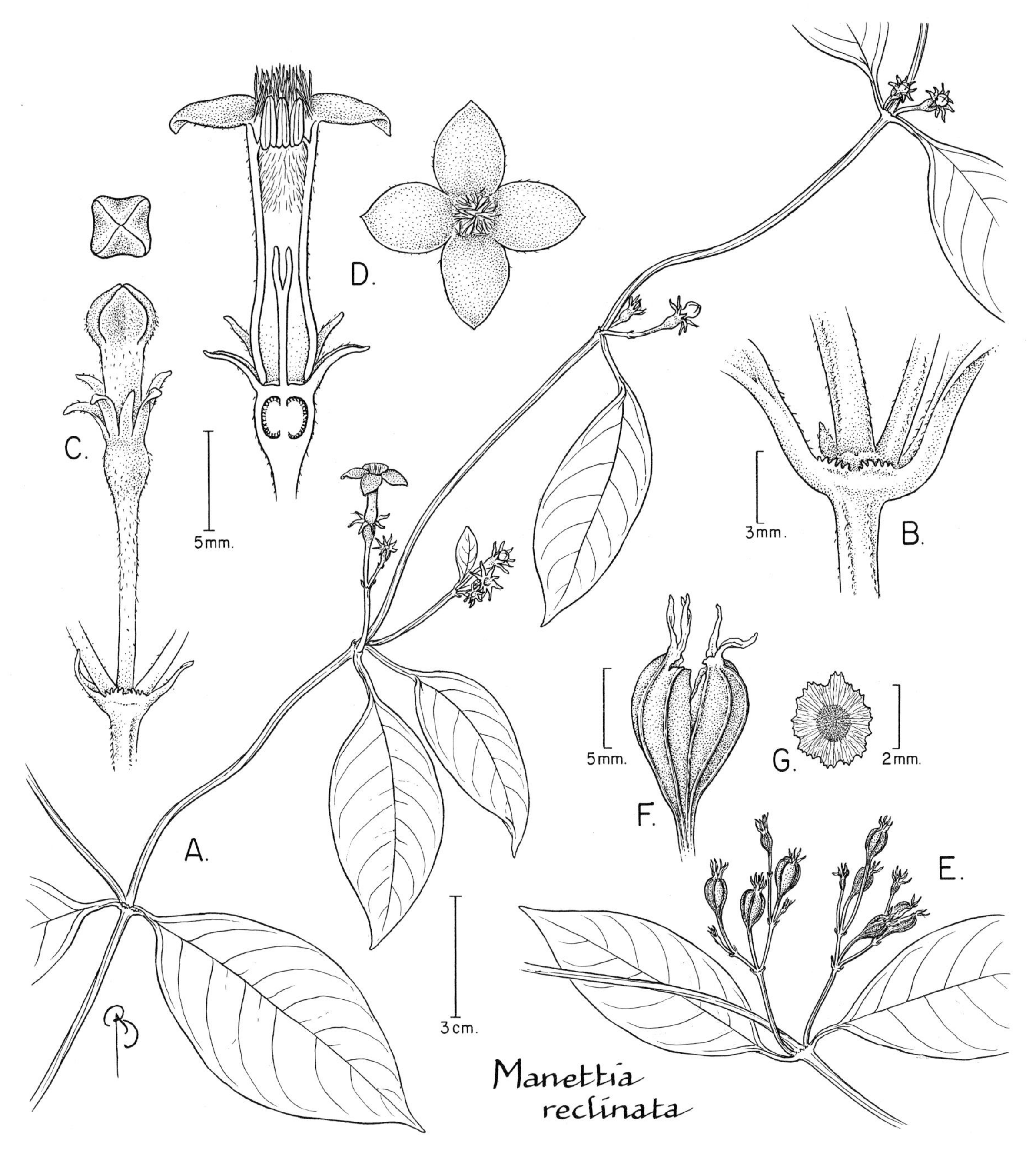

FIG. 275. RUBIACEAE. *Manettia reclinata* (A, C, D, *Mori et al. 22338*; B, *F. Hallé s.n.*, from photo taken in French Guiana outside our area; E–G, *Granville 648* from Grand Inini, French Guiana). **A.** Part of stem with leaves and inflorescences. **B.** Detail of node showing stipules and axillary bud. **C.** Lateral view of floral bud (right) and apical view of bud (above right). **D.** Medial section of flower (left) and apical view of flower (right). **E.** Part of stem with leaves and infructescences. **F.** Lateral view of dehiscing fruit. **G.** Seed.

PALICOUREA Aubl.

Trees or shrubs. Stipules interpetiolar, usually united below, bilobed or bicusped, glandular, persistent. Leaves usually opposite (3–4 per node in *P. quadrifolia*). Inflorescences terminal, the rachis red, orange, yellow, blue, purple, or some combination of these colors. Flowers 4–5-merous; calyx truncate; corolla usually yellow, orange, red, pink, blue, or violet (rarely white), the tube curved or swollen at base on one side, with ring of hairs near base inside, the lobes valvate in bud; stamens 4–5, the anthers dorsifixed; ovary 2–6-locular, the ovules solitary, ascending in each locule. Fruits drupaceous, fleshy, convex, usually deeply 5-costate, with 2–6 pyrenes. Seeds usually with depression on one side. Species of this genus are commonly pollinated by butterflies.

Taylor, C. M. 1997. Conspectus of the genus *Palicourea* (Rubiaceae: Psychotieae) with the description of some new species from Ecuador and Colombia. Ann. Missouri Bot. Gard. **84:** 224–262.

———. 1999. *Palicourea in* Rubiaceae (part 3). *In* G. Harling & L. Andersson (eds.), Flora of Ecuador **62:** 134–235.

1. Leaves whorled, 3–4 per node. . . . . *P. quadrifolia.*
1. Leaves opposite.
    2. Leaf blades pubescent throughout abaxially. . . . . *P. longistipulata.*
    2. Leaf blades glabrous or veins puberulent or short hirtellous abaxially.
        3. Shrubs 1–4 m tall.
            4. Stipule lobes 2–8 mm long. Inflorescences > 2× long as wide. Corolla tube pink, the lobes purple. . . . . *P. calophylla.*
            4. Stipule lobes ≤2 mm long. Inflorescences < 2× long as wide, often ± equally as long as wide. Corolla yellow or yellow-orange.
                5. Stipule lobes triangular-lanceolate, ca. 2 mm long, puberulent. Petioles 4–17 mm long. Inflorescences paniculate, generally with >20 flowers. Corolla 7–17 mm long, glabrous abaxially. . . . . *P. crocea.*
                5. Stipule lobes widely triangular, 1–2 mm long, glabrous. Petioles 2–8 mm long. Inflorescences generally corymbose, sometimes paniculate, with ≤20 flowers. Corolla 15–20 mm long, puberulent abaxially. . . . . *P. longiflora.*
        3. Trees to 12 m tall.
            6. Stems terete. Leaf blades short hirtellous along midrib abaxially. Inflorescence axes short hirtellous; peduncle 8–9 cm long. Corolla white. Rare. . . . . *P. brachyloba.*
            6. Stems quadrangular. Leaf blades glabrescent along midrib abaxially. Inflorescence axes puberulent; peduncle 8–15 cm long. Corolla yellow. Common. . . . . *P. guianensis.*

Palicourea brachyloba (Müll. Arg.) B. M. Boom [Syn.: *Psychotria brachyloba* Müll. Arg.]

Small trees, to 12 m tall. Stems terete, glabrous. Stipules 4–5 mm long, the lobes ovate, the apex rounded. Bark cracked, gray, with white flecks. Leaves opposite; petioles 10–20 mm long; blades lanceolate to oblanceolate, 14–30 × 6–12 cm long, short hirtellous along midrib abaxially. Inflorescences paniculate, ca. 8 × 6 cm, the axes yellow; peduncle 8–9 cm long. Flowers: calyx ca. 2 mm long; corolla 13–15 mm long, white. Fruits subglobose, 4–5 mm diam. Fl (Mar, Sep); rare.

Palicourea calophylla DC.

Shrubs, 2–4 m tall. Stems terete, glabrous. Stipule lobes lanceolate, 4–8 mm long. Leaves opposite; petioles 5–25 mm long; blades oblong-elliptic to lanceolate-elliptic or obovate, 15–30 × 4–10 cm, glabrous adaxially, puberulent on veins abaxially. Inflorescences thrysoid-paniculate, 6.5–10 × 2–4.5 cm, the axes dark pink or red; peduncle 2–9 cm long. Flowers: calyx 0.8 mm long; corolla tube pink, 12–13 mm long, the lobes purple. Fruits suborbicular, deeply 9-costate, ca. 7 × 6 mm, turning purple. Fl (Jan, May), fr (Jan, May, Aug, Sep); in nonflooded forest. *Wapa sec* (Créole).

Palicourea crocea (Sw.) Roem. & Schult.

Shrubs, 1–3 m tall. Stems glabrous, purplish. Stipules with vein 0.5–1 mm long, the lobes triangular-lanceolate, ca. 2 mm long, puberulent. Leaves opposite; petioles 4–17 mm long; blades elliptic-ovate to oblong-elliptic, 8–15 × 3–7 cm. Inflorescences paniculate, 3–11(16) × 2–7(10) cm, the axes purplish-red to pinkish-red to reddish-orange; peduncle 3.5–6 cm long. Flowers: calyx 1.8–2.5 mm long; corolla 7–17 mm long, yellow to yellow-orange, glabrous. Fruits suborbicular to ovoid, 8–10-costate, 4–10 × 3.5–8 mm, purple. Fl (Jan, Feb, Sep, Oct, Nov, Dec), fr (Jan, Apr, Jun, Aug); common, in understory of nonflooded forest, especially in somewhat disturbed areas.

Palicourea guianensis Aubl. PL. 112e

Trees, to 12 m × 10 cm. Stems quadrangular, glabrous. Stipules 6–8 mm long, the lobes suborbicular-ovate, glabrous, the apex rounded. Leaves opposite; petioles 1–4.5 cm long; blades elliptic-oblong to ovate-elliptic, 13–43 × 6–18 cm, green adaxially, green or purple abaxially. Inflorescences paniculate, 9–19 × 7–15 cm, the axes bright yellow, reddish at base; peduncle 8–15 cm long. Flowers: calyx yellow, 1.2–2.5 mm long; corolla 13–23 mm long, densely pubescent, bright yellow. Fruits suborbicular-obovoid, 4.5–6 × 2.5–4.5 mm. Fl (Feb, Mar, May), fr (Jul); common, mostly in secondary vegetation or disturbed forest margins. *Carapa* (Créole).

Steyermark (1972) recognized several subspecific ranks based on vestiture type which we consider to be normal variation over the wide geographic range of this species.

Palicourea longiflora (Aubl.) A. Rich. PL. 113a

Shrubs, 1–3 m tall. Stems glabrous. Stipules widely triangular, 1–2 mm long, glabrous, with vein 0.5–1 mm long. Leaves opposite; petioles 2–8 mm long; blades ovate to elliptic-oblong, 6–16 × 3.5–6.5 cm. Inflorescences corymbose, sometimes paniculate, 6–10 × 3–5 cm, pendulous, the axes orange; peduncle 3–8.5 cm long, purple. Flowers: calyx 1.2–2 mm long; corolla 15–20 mm long, yellow at base, purple apically. Fruits subglobose, 10-costate, 5 mm diam, green. Fl (Apr, May, Jul, Aug, Sep), fr (Apr, May, Sep); common, in understory of nonflooded forest.

Palicourea longistipulata (Müll. Arg.) Standl.

Shrubs, 2–4 m tall. Stems terete-quadrangular, pubescent. Stipules narrowly deltoid or lanceolate-linear, the lobes 3–17 mm long, with vein 2–4.5 mm long. Leaves opposite; petioles 4–10 mm long; blades oblong to oblong-elliptic, 12–23 × 3.5–11.5 cm, densely rust-brown pubescent abaxially. Inflorescences corymbose, 6–11 × 4–6 cm, the axes red to purple, with rust-brown pubescence; peduncle 5–10 cm long. Flowers: calyx 2–3 mm long,

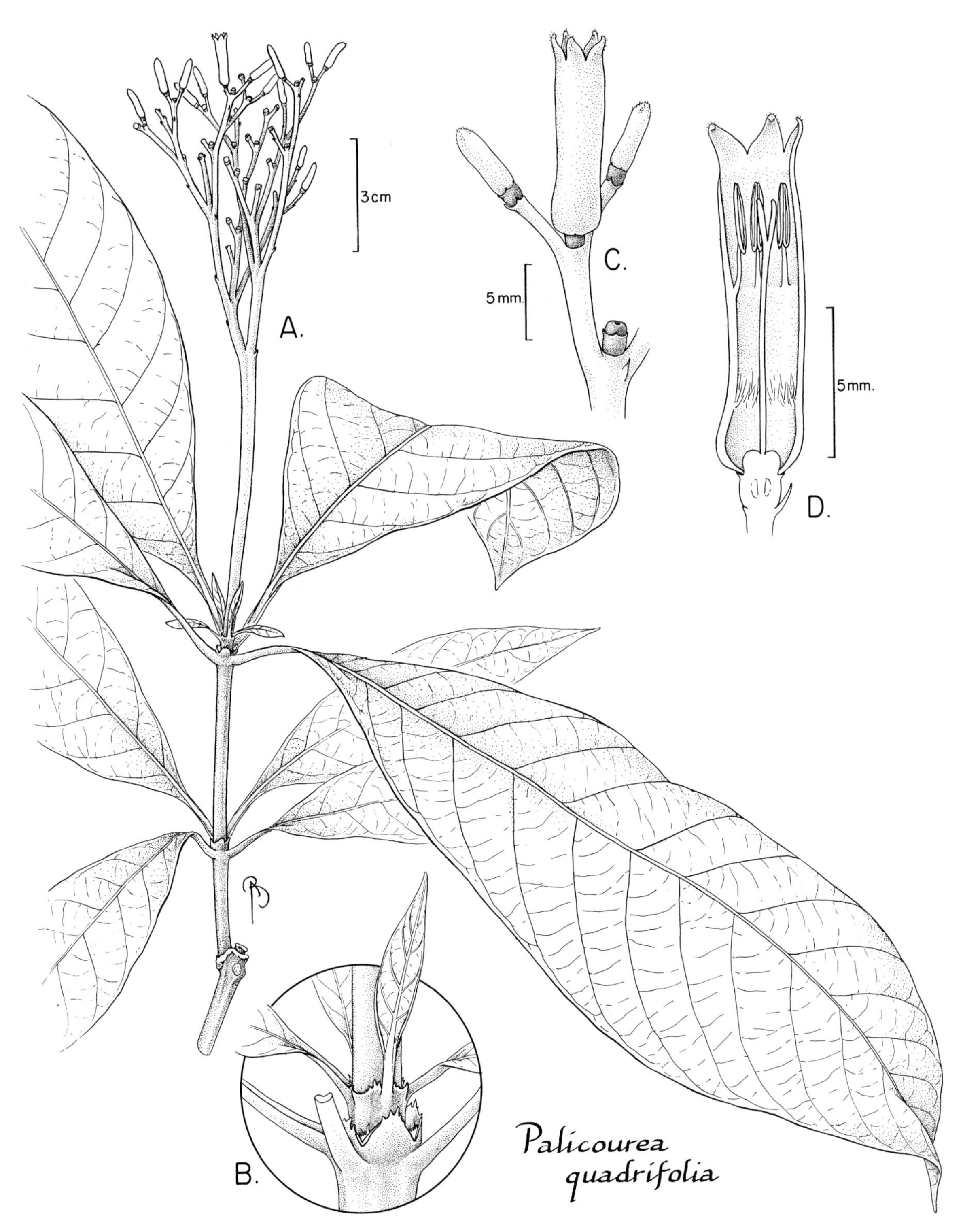

FIG. 276. RUBIACEAE. *Palicourea quadrifolia* (A, C, D, *Mori et al. 20978*; B, *Mori et al. 14934*). **A.** Apex of stem showing leaves and terminal inflorescence. **B.** Detail of two nodes showing the intrapetiolar, persistent stipules. **C.** Part of inflorescence showing open flower and flower buds. **D.** Medial section of flower showing slight basal swelling.

pubescent; corolla 8–13 mm long, the tube red, the lobes pubescent, yellow. Fruits widely ovoid, 8–10-costate, ca. 5 × 4 mm, hirtellous, blackish. Fl (Jan, Feb, Oct, Nov, Dec), fr (Mar), uncommon, in non-flooded forest.

Palicourea quadrifolia (Rudge) DC. FIG. 276

Shrubs, 2–3.5 m tall. Stems quadrangular, glabrous. Stipules glabrous, with vein 2 mm long, the lobes 2–3 mm long, apically rounded. Leaves 3–4 per node, whorled; petioles 10–20 mm long; blades ovate-elliptic to oblanceolate, 17–24 × 7–8.5 cm. Inflorescences paniculate, 5–8 × 6–10 cm, the axes bright yellow; peduncle 4–8 cm long. Flower: calyx, ca. 1–2 mm long, yellow-green; corolla ca. 12 mm long, variously greenish-white, yellow or red. Fruits subglobose, ca. 5 mm diam., glabrous, green. Fl (Jan, Feb, Jul, Sep, Dec), fr (Jul, Dec); common, in understory of nonflooded forest.

PATIMA Aubl.

Shrubs. Stems erect, terete, hollow at internodes. Stipules interpetiolar, connate basally, persistent. Inflorescences axillary, congested umbels, ebracteate. Flowers 5-merous; calyx tube truncate; corolla infundibuliform, the tube glabrous abaxially, densely villous at base adaxially, the lobes acuminate, valvate in bud; stamens with dorsifixed anthers; ovary 5-locular, the ovules many per locule, the ovules horizontally arranged on fleshy placentas. Fruits baccate, 5-locular, the calyces persistent. Seeds numerous, minute.

Patima guianensis Aubl.

Shrubs, ca. 1.5 m tall. Stipules triangular, ca. 5 mm long, glabrous, the apex acute. Leaves: petioles 3–6 cm long; blades oblanceolate, 32–50 × 10–15 cm. Inflorescences 3–15-flowered; pedicels ca. 1 cm long. Flowers: calyx 5–7 mm long; corolla ca. 20 mm long, pale yellow, pointed in bud. Fruits globose, ca. 1.2 cm diam., glabrous, with persistent calyces. Fl (Jan), fr (Jan, Jul); only known in our area from near Carbet Mais trail.

POSOQUERIA Aubl.

Trees or shrubs. Stems glabrous. Stipules interpetiolar, generally lanceolate.Leaves: blades coriaceous. Inflorescences terminal, subcorymbose, few- to many-flowered. Flowers 5-merous; buds curved to one side; calyx lobes auriculate or subimbricate basally, glandular adaxially; corolla showy, long-tubular, zygomorphic at apex, white (turning yellowish when old), the lobes contorted in bud; stamens unequal in length, inserted at mouth of corolla, four bending backward in two pairs, the single stamen depositing spherical mass of pollen onto pollinator (Fig. 277C); ovary 2-locular or incompletely 1-locular (by abortion of 1 locule), the ovules numerous, erect. Fruits baccate, (1)2-locular, globose, the exocarp thick-coriaceous. Seeds numerous, large, bony.

1. Shrubs or small trees, 1.5–5 m tall. Stipules lanceolate-triangular, ca. 6 mm long, usually deciduous. Leaf blades 7–10 × 3.5–5 cm, subcoriaceous. Inflorescences 3–5-flowered. . . . . *P. gracilis.*
1. Trees, 8–15 m tall. Stipules lanceolate-oblong, 5–14 mm long, usually persistent. Leaf blades 9–20 × 5–12 cm, coriaceous. Inflorescences 10–12-flowered. . . . . *P. latifolia.*

Posoqueria gracilis (Rudge) Roem. & Schult. [Syn.: *P. latifolia* (Rudge) Roem. & Schult. subsp. *gracilis* (Rudge) Steyerm.]

Shrubs or small trees, 1.5–5 m tall. Young stems glabrous, slender. Stipules lanceolate-triangular, ca. 6 mm long, usually deciduous, the apex acuminate. Leaves: blades oval to narrowly elliptic, 7–10 × 3.5–5 cm, subcoriaceous. Inflorescences, 3–5-flowered. Flowers nocturnal; calyx 3 mm long; corolla tube ca. 7 cm long, the lobes ca. 1 cm long. Fruits oblong-ovoid, 3.5 × 3 cm, yellow-orange streaked with white. Fl (fl), fr (Feb, Aug, Sep); infrequent.

Posoqueria latifolia (Rudge) Roem. & Schult. [Syn.: *Tocoyena longifolia* Kunth, *Posoqueria decora* DC.]

FIG. 277

Trees, 8–15 m × 25 cm. Trunks somewhat irregular. Young stems glabrous, thick. Stipules lanceolate-oblong, 5–14 mm long, usually persistent. Leaves: blades ovate to elliptic-oblong, 9–20 × 5–12 cm, thickly coriaceous. Inflorescences 10–12-flowered. Flowers nocturnal; calyx 3 mm long; corolla white, the tube 7–13 cm long, the lobes 1.3–2 cm long. Fruits globose, ca. 5 cm diam., yellow at maturity. Seeds surrounded by light yellow layer. Fl (Aug, Sep, Oct), fr (Jan, May, Aug, Sep, Nov); common but scattered, in nonflooded forest. The yellow pulp surrounding the seeds is eaten by monkeys. *Bois-canne*, *graine coumarou* (Créole).

PSYCHOTRIA L.

Trees, shrubs, herbs, or rarely epiphytic shrubs. Stipules interpetiolar, triangular to oblong-ovoid and drying reddish (most species of subgen. *Psychotria*) or with 2 short or elongate lobes on each side and not drying reddish (most species of subgen. *Heteropsychotria*). Inflorescences terminal or axillary, racemose, corymbose, or capitate and subtended by colorful bracts (species formerly placed in *Cephaelis*). Flowers 4–6-merous; corolla mostly white or creamy-white (rarely yellow or bluish), the lobes valvate in bud; stamens inserted at throat of corolla tube, the anthers dorsifixed; ovary usually 2-locular (rarely 5-locular), the ovules solitary in each locule, erect from base. Fruits drupaceous, fleshy, with 2(5) pyrenes, white, yellow, red, blue, purple, or black. Seeds convex, with longitudinal groove on ventral (rarely dorsal) surface.

Delprete, P. G. 2001. Notes on some species of *Psychotria* (Rubiaceae) occurring in the Guianas. Brittonia **53:** 396-404.

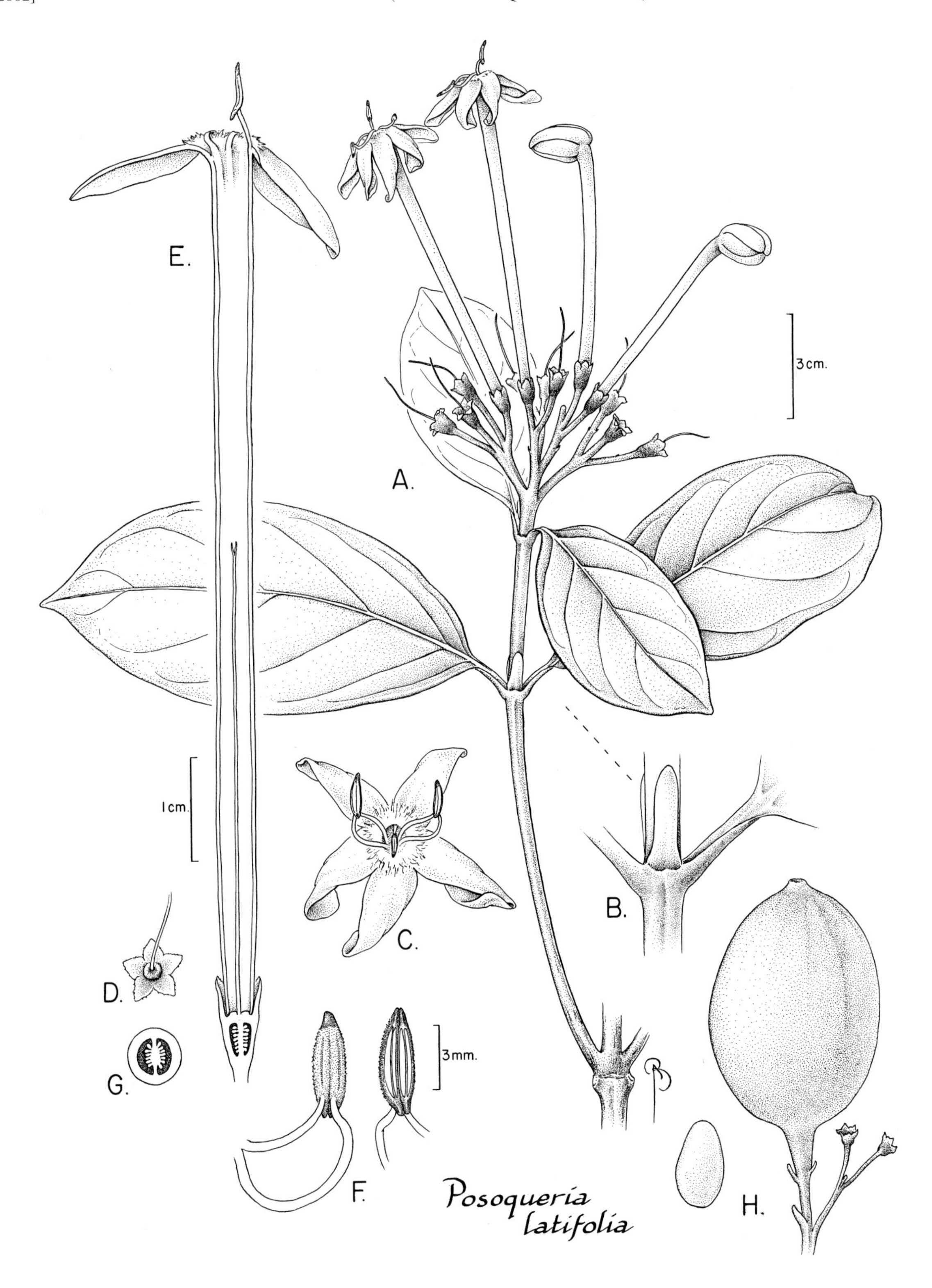

FIG. 277. RUBIACEAE. *Posoqueria latifolia* (A–G, *Mori et al. 20952*; H, *Mori & Boom 15228* and *Mori & Boom 14717*). **A.** Apex of stem showing leaves and terminal inflorescence. **B.** Detail of node showing stipule. **C.** Apical view of flower. **D.** Apical view of calyx and base of pistil. **E.** Medial section of flower. **F.** Abaxial (left) and adaxial (right) views of connate anthers and part of filaments. **G.** Transverse section of ovary. **H.** Part of infructescence (right) showing fruit and seed (left).

1. Plants epiphytic. . . . . *P. guadalupensis.*
1. Plants terrestrial.
    2. Trees, 10–25(30) m tall. . . . . *P. mapourioides.*
    2. Herbs, subshrubs, shrubs, or occasionally trees <10 m tall.
        3. Inflorescences axillary, or occasionally mixed axillary and terminal.
            4. Creeping herbs, the stems densely pubescent. . . . . *P. ulviformis.*
            4. Erect or prostrate shrubs or herbs, the stems glabrous or puberulent, but never densely pubescent.
                5. Shrubs. Petioles 2–13 mm long.
                    6. Stipule lobes broadly triangular, 3–5 mm long. Petioles 7–13 mm long; blades 8.5–17 × 2.5–9.5 cm. Inflorescences strictly axillary, capitate. Fruits 8–10 × 4.5–8 mm. . . . . *P. erecta.*
                    6. Stipule lobes lanceolate-linear, 1–2 mm long. Petioles 2–6 mm long; blades 4–9 × 1.5–4 cm. Inflorescences axillary or sometimes terminal, umbellate or corymbose. Fruits 3 × 3–5 mm. . . . . *P. officinalis.*
                5. Herbs, often succulent, base somewhat woody. Petioles 10–80 mm long.
                    7. Inflorescences capitate, sessile to subsessile. . . . . *P. lateralis.*
                    7. Inflorescences paniculate or cymose.
                        8. Leaf blade margins ciliate. Inflorescences with peduncle 0.4–1.1 cm long. Corolla 3–3.5 mm long. . . . . *P. saülensis.*
                        8. Leaf blade margins not ciliate. Inflorescences with peduncle 1.2–15 cm long. Corolla 4.5–7.5 mm long.
                            9. Stipules inconspicuous, 0.6–0.8 mm long. Inflorescences 1.5–2 × 2–3 cm. . . . . *P. microbracteata.*
                            9. Stipules conspicuous, 1–10 mm long. Inflorescences 2–15 × 1.5–14 cm.
                                10. Inflorescences usually of 3 main axes; peduncle stout, 2–4 mm thick. . . . *P. uliginosa.*
                                10. Inflorescences with >3 main axes; peduncle thin, 1–1.5 mm thick. . . . . *P. macrophylla.*
        3. Inflorescences terminal (sometimes appearing axillary, e.g., *P. officinalis*).
            11. Only terminal stipules persistent, the others deciduous, with fine brown hairs persistent at base of scars. Leaves drying dark purplish-brown (subgen. *Psychotria*).
                12. Terminal stipule deeply cut on each side into 2 narrowly linear-setaceous segments, these ferruginous-hirtellous. . . . . *P. perferruginea.*
                12. Terminal stipule not deeply cut on each side.
                    13. Petioles 0.5–5 mm long; blades 1.3–3.1 cm wide. Inflorescences 1.2–3.5 × 1.5–3 cm. Corolla 2–3 mm long. . . . . *P. borjensis.*
                    13. Petioles 3–25 mm long; blades 2–7.5 cm wide. Inflorescences 1–15 × 2–7 cm. Corolla 5–10 mm long.
                        14. Stipules lanceolate-oblong to ovate, 10–20 × 4–8 mm, the margins ciliolate, not white. Petioles 3–5 mm long. Flowers sessile, usually >50 per inflorescence. Corolla 5–6 mm long. . . . . *P. carthagenensis.*
                        14. Stipules broadly ovate, 7–9 × 6–8 mm, the margins not ciliolate, white. Petioles 3–12 mm long. Flowers pedicellate, usually <50 per inflorescence. Corolla (5)7–10 mm long. . . . . *P. cupularis.*
            11. Stipules persistent, these sometimes with erose margins or elongate teeth or lobes, without fine brown hairs at base of scars. Leaves usually drying green or sometimes various shades of reddish-brown (subgen. *Heteropsychotria*).
                15. Inflorescences evidently branched (not capitate or subcapitate [shortly branched]), with evident central rachis. Bracts absent, inconspicuous, or rarely conspicuous (*P. capitata*), never enclosing the inflorescence, mostly greenish or whitish.
                    16. Stems densely pubescent. Fruits 5-pyrenate, orange, turning black upon maturity. . . . . . *P. racemosa.*
                    16. Stems glabrous or sparsely pubescent (or densely retrorsely strigose, older stems becoming glabrate in *P. alloantha*). Fruits 2-pyrenate, blue, purple, green, white, or whitish-red upon maturity.
                        17. Plants herbaceous, often succulent apically, somewhat woody basally. . . . . *P. uliginosa.*
                        17. Plants woody.
                            18. Stipule lobes very conspicuous, elongate, 5–20 mm long.
                                19. Bracts subtending inflorescence axes conspicuous, usually 5–12 mm long. . . . . *P. capitata.*
                                19. Bracts subtending inflorescence axes not conspicuous (sometimes absent), usually <2 mm long. . . . . *P. microbotrys.*

18. Stipule lobes less conspicuous, shorter, 0.5–4(7) mm long.
    20. Main bracts of inflorescence axes ± conspicuous, 4.5–30 mm long.
        21. Inflorescences always terminal; bracts present at base of inflorescence, at apex of peduncle, and subtending lowest axes of inflorescences. . . . . *P. brachybotrya.*
        21. Inflorescences terminal or axillary; bracts absent at base of inflorescences and at apex of peduncle, but present on axes above their junction and subtending heads of flowers. . . . . *P. officinalis.*
    20. Main bracts of inflorescence axes ± inconspicuous or not developed, 0–2 mm long.
        22. Inflorescences of compact (1)3–26-flowered clusters at apex of each axis. . . . . *P. astrellantha.*
        22. Inflorescences with clusters of 1–3 flowers bunched together, or scattered along axes, or in forks of axes.
            23. Leaf blades with base rounded or obtuse. . . . . *P. pullei.*
            23. Leaf blades with base acute or acuminate.
                24. Fruits 5–10-costate, not didymous (subdidymous in *P. deflexa*), the 2 pyrenes appearing united without groove between them.
                    25. Shrubs to treelets, to 5 m tall. Tertiary venation subparallel. Stipular lobes triangular, 0.3–1 mm long. . . . . *P. paniculata*
                    25. Subshrubs to shrubs, 0.5–4 m tall. Tertiary venation sparsely reticulate. Stipular lobes narrowly triangular or awn-like, 1.5–7 mm long.
                        26. Subshrubs to shrubs, 0.5–1.5 m tall. Stipular lobes narrowly triangular or awn-like, 1.5–2 mm long. Inflorescences 3–6 cm long, unbranched or with basal axes to 5 mm long. Fruit not didymous, 5-costate. . . . . *P. pullei*
                        26. Subshrubs to shrubs, 0.5–4 m tall. Stipular lobes narrowly triangular, 3–7 mm long. Inflorescences (2)6–17 cm long, with basal axes to 20–45 mm long. Fruit subdidymous, 6(8)-costate. . . . . *P. deflexa.*
                24. Fruits smooth, didymous (bilobed, the two sides globose), the 2 pyrenes with conspicuous groove between them.
                    27. Flowers pedicellate. . . . . *P. acuminata.*
                    27. Flowers sessile. . . . . *P. bahiensis.*

15. Inflorescences capitate or subcapitate (shortly branched), without an evident central rachis. Bracts mostly conspicuous, green, white, orange, red, purple, enclosing at least part of inflorescence or fascicles, or inconspicuous and present at the base of inflorescence axes.
    28. Stems densely retrorsely strigose, becoming glabrate. Inflorescences subtended by many foliose, green, rhombic-spatulate, densely tomentellose bracts. Calyx densely hirsutulous; corolla elongate, salverform, pubescent, the lobes ligulate, recurved. Only known from Pic Matécho. . . . . *P. alloantha.*
    28. Stems glabrous or pubescent (not retrorsely strigose). Inflorescences and floral characters not as above.
        29. Inflorescences epedunculate to short-pedunculate, the peduncle <5 mm long.
            30. Leaf blades pubescent, or at least puberulent along midrib.
                31. Leaf blades pilosulous only along midrib abaxially.. . . . . *P. oblonga.*
                31. Leaf blades pubescent on surfaces of one or both sides, or on midrib.
                    32. Leaf blades pubescent on both surfaces, the midrib glabrous ad- and abaxially. . . . . *P. iodotricha.*
                    32. Leaf blades glabrous on both surfaces, the midrib spreading-hirsutulous ad- and abaxially.
                        33. Outer involucral bracts decussate, entire. Corolla tube glabrous throughout abaxially. . . . . *P. trichophoroides.*
                        33. Outer involucral bracts not decussate, irregularly lobed (rarely entire). Corolla tube pilose, at least distally abaxially.
                            34. Outer involucral bracts lanceolate, wider at base or at middle, the apex acute. . . . . *P. iodotricha.*

34. Outer involucral bracts oblong-pandurate, narrower at middle, the apex obtuse. . . . . *P. viridibractea.*
30. Leaf blades glabrous.
35. Stipular sheath terminating in many sharp awns 5–10 mm long. . . . . *P. pungens.*
35. Stipular sheath entire or terminating in only one linear lobe 1.5–2 mm long.
36. Stipular sheath entire. Outer involucral bracts rounded at apex, as long as wide or wider than long, 6–25 mm wide. . . . . *P. apoda.*
36. Stipular sheath terminating in one linear lobe 1.5–2 mm long. Outer involucral bracts narrowed at apex, longer than wide, 2–5.5 mm wide. . . . . *P. kappleri.*
29. Inflorescence pedunculate, the peduncle >5 mm long.
37. Peduncle glabrous.
38. Outermost involucral bracts basally connate for 2.5–20 mm from base.
39. Stipular sheath 1.5–3 mm long, the lobes triangular, 1–4 mm long, drying olive green. Outermost involucral bracts basally connate for 2.5–4 mm from base. . . . . *P. colorata.*
39. Stipular sheath <1 mm long, the lobes lanceolate, 5–6 mm long, drying creamy-white (hay color). Outermost involucral bracts basally connate for 8–20 mm from base. . . . . *P. urceolata.*
38. Outermost involucral bracts free or connate for only up to 1.5 mm from base.
40. Individual flowers in heads surrounded by 1–3 bracteoles.
41. Outermost involucral bracts 4–6 mm long, ovate, drying beige. . . . *P. kappleri.*
41. Outermost involucral bracts 12–45 mm long, linear or spatulate, drying olive green.
42. Leaves with 6–11 secondary veins each side, the tertiary veins sparsely reticulate. Bracts 4, subtending the inflorescence, decussate, spatulate, 23–45 × 8–25 mm.. . . . . *P. carapichea.*
42. Leaves with (14)25–45 secondary veins each side, the tertiary veins parallel (and parallel to secondary veins). Bracts 2–8, inserted on inflorescence axes, not decussate, linear, 12–30 × 2–4 mm. . . . . *P. ligularis.*
40. Individual flowers in heads not surrounded by 1–3 bracteoles. . . . . *P. platypoda.*
37. Peduncle puberulent or pubescent.
43. Entire inflorescences not subtended by showy bracts, nor with bracts inserted on axes. . . . . *P. moroidea.*
43. Inflorescences subtended by bracts, these often showy, or with bracts inserted on axes.
44. Stems pubescent to hirsute.
45. Outer involucral bracts united into tube.
46. Bracts red to orange-red. Corolla 14–16.5 mm long, yellow. Fruits 5.5–7 × 2.8–3 mm, blue. *P. poeppigiana.*
46. Bracts orange. Corolla 11–12 mm long, white. Fruits 13 × 8 mm, white. . . . . *P. granvillei.*
45. Outer involucral bracts free or united only to 3 mm from base.
47. Flowers in heads uniformly surrounded by 1–4 bracteoles.
48. Abaxial leaf blade surface softly pilose, the blades never purple abaxially, not striped adaxially. . . . . *P. callithrix.*
48. Abaxial leaf blade surface strigose pubescent throughout, or at least on veins, the blades often purple abaxially and gray or silvery-striped adaxially, or sometimes blades green thoughout and not striped. . . . . *P. variegata.*
47. Flowers in heads not uniformly surrounded by 1–4 bracteoles.
49. Erect subshrubs (basal nodes sometimes prostrate). Stems with dark brown or dark purple hairs (older internodes glabrate).
50. Stems and subtending bracts with dark purple hairs. Leaf blades with secondary veins in 6–9 pairs. Bracts narrowly lanceolate to linear, 13–22 × 1–2 mm. . . . *P. medusula*

50. Stems and subtending bracts with dark brown hairs. Leaf blades with secondary veins in 10–17 pairs. Bracts lanceolate, 5–12 × 2–3.5 mm. . . . . *P. bremekampiana.*

49. Prostrate to decumbent herbs to delicate subshrubs. Stems with pale purplish-blue hairs. . . . . *P. callithrix.*

44. Stems puberulous, glabrous, or glabrate.

51. Outermost involucral bracts narrowly ovate, apex obtuse to rounded. . . . . *P. oblonga.*

51. Outermost involucral bracts lanceolate, narrowly lanceolate, or linear, apex acute to acuminate.

52. Hypanthia pubescent.

53. Terminal internodes and abaxial midrib of leaf blades with hairs >1 mm long. Stipules pubescent, the lobes narrowly triangular to linear, ca. 5 mm long. Calyx lobes 2.5–3 mm long, longer than tube. . . . . *P. bremekampiana.*

53. Terminal internodes and abaxial midrib of leaf blades usually glabrous or with hairs <1 mm long. Stipules glabrous, the lobes shortly triangular, ≤2 mm long. Calyx lobes usually inconspicuous, 0.1–0.5 mm long, shorter than tube. . . . . *P. hoffmannseggiana.*

52. Hypanthia glabrous.

54. Bracts 3–17(20) per inflorescence, 3.5–8(11) mm long, the margins not as below. Corolla 4–7 mm long. . . . *P. hoffmannseggiana.*

54. Bracts 21–40 per inflorescence, 13–22 mm long, the margins conspicuously long setose-ciliate with dark purple hairs. Corolla 12–16 mm long. . . . . *P. medusula.*

Psychotria acuminata Benth.

Subshrubs or shrubs, 0.3–1 m tall. Stems sometimes rooting at nodes, glabrous. Stipules with basal sheath 1.5–2 mm long, puberulent, the lobes linear-lanceolate, 1.5–2 mm long, puberulent, persistent, the margins ciliate. Leaves: petioles 0.8–1.8 cm long, puberulent; blades widely ovate, oblong-elliptic, or ovate-elliptic, 9–20 × 3.5–8.5 cm, glabrous adaxially, glabrous or puberulent on midrib abaxially; secondary veins in 10–13 pairs. Inflorescences terminal, cymosely paniculate, 1.5–4.5 × 2.5–5.5 cm; peduncle 2–5 cm long, puberulent. Flowers: calyx ca. 1 mm long, usually truncate, puberulent, ciliolate; corolla 5-lobed, ca. 8 mm long, puberulent, white or whitish. Fruits bilobed (the two sides subglobose), smooth, 2-pyrenate, 3 × 5 mm, glabrous, white. Fl (Jan), fr (Feb, Aug); in understory of nonflooded forest.

Psychotria alloantha Steyerm.

Shrubs, 1–1.5 m tall. Stems densely retrorsely strigose, becoming glabrate. Stipules with basal sheath 4–5 mm long, densely strigillose, persistent, the lobes ligulate-lanceolate, 3–4 mm long, with obtuse apex. Leaves: petioles 6–8 mm long; blades oblong-elliptic, 11–19 × 4–9.5 cm, glabrescent or scarcely appressed pubescent adaxially, scarcely to moderately appressed pubescent abaxially, the midrib densely strigillose abaxially; secondary veins in 10–15 pairs. Inflorescences terminal, capitate (with short-cymose branching), 2.5–3 × 3.5–4 cm, conspicuously bracteate, the 10–12 outermost bracts rhombic-spatulate, 15–19 × 10–15 mm, densely tomentellose on both sides, green; peduncle (when present) to 7 mm long, densely strigose. Flowers: calyx 2.5 mm long, the lobes broadly triangular, ca. 0.8 mm long, densely hirsutulous, with a single squamella between base of lobes adaxially; corolla 5-lobed, ca. 20 mm long, the tube white, the lobes ligulate, recurved, pilosulous, yellow. Fruits ovoid, smooth, 2-pyrenate, blue. Fl (Jul); in forest understory, known only from Pic Matécho.

Psychotria apoda Steyerm. [Syn.: *Cephaelis violacea* (Aubl.) Sw.]

Subshrubs or shrubs, 0.4–3 m tall. Stems glabrous. Stipules with basal sheath 2–4.5 mm long, glabrous, persistent, the apex rounded, the margins entire (not lobed). Leaves: petioles 7–15 mm long, glabrous; blades oblong to oblong-elliptic, 8–19.5 × 2–9 cm, glabrous throughout; secondary veins in 10–17 pairs. Inflorescences terminal, capitate, hemispheric, 1.2–2.3 × 1.5–2 cm, densely flowered, with 4 pairs of involucral bracts, these 6–25 mm wide, purplish, glabrous; peduncle usually absent. Flowers: calyx 1.8–2.5 mm long, the teeth triangular, 0.2–1.2 mm long; corolla 5-lobed, 14–17 mm long, glabrous, white. Fruits ovoid, smooth, 2-pyrenate, 12 × 15 mm, white or blue. Fl (Jan, Mar, Nov, Dec); in nonflooded forest and along streams.

Psychotria astrellantha Wernham [Syn.: *Rudgea obscura* Sandwith]

Shrubs, 2–4 m tall. Stems glabrous. Stipules with basal sheath 0.6 mm long, puberulent, persistent, the lobes narrowly acuminate, 1–1.5 mm long (breaking off in old stipules). Leaves: petioles nearly absent to 4 mm long; blades oblong to obovate, 10–15 × 3–7 cm; secondary veins in 6–7 pairs. Inflorescences terminal, umbellately arranged in (1)3–26-flowered heads on axes to 1.5 cm long; peduncle 0.5–1 cm long, glabrous. Flowers: calyx 0.5–1 mm long, the teeth 0.1 mm long; corolla 5-lobed, ca. 2.4 mm long, white. Fruits ovoid, smooth, 2-pyrenate, 8 × 6 mm, green. Fl (Jan, Jul, Aug, Sep, Oct), fr (Jan, Sep); in understory of nonflooded forest.

Psychotria bahiensis DC.

Subshrubs or shrubs, 0.5–3 m tall. Stems glabrous or puberulent. Stipules with basal sheath short, glabrous, persistent, the lobes narrowly lanceolate to filiform, 0.5–3 mm long. Leaves: petioles 2–15 mm long, glabrous or puberulent; blades oblong-elliptic to lanceolate-elliptic, 6–21 × 2–10.5 cm, glabrous, the midrib sometimes puberulent abaxially; secondary veins in 6–13 pairs. Inflorescences terminal, cymose-paniculate to subumbellate, 0.7–3 × 1.3–5 cm; peduncle 0.4–3 cm long, puberulent to glabrous. Flowers sessile; calyx 1–1.5 mm long, puberulent to glabrous, dentate or subtruncate, the teeth less than 0.3 mm long; corolla 5-lobed, 4–8.5 mm long, glabrous to puberulent, white. Fruits bilobed (the two sides subglobose), smooth, 2-pyrenate, ca. 3 × 5 mm, glabrous, bluish to purplish. Fr (Sep); in nonflooded forest.

Psychotria borjensis Kunth

Shrubs, 1–3 m tall. Stems puberulent. Stipules lanceolate or ovate-lanceolate, 5–7 mm long, entire, glabrous or finely puberulent, drying reddish, readily caducous. Leaves: petioles 0.5–5 mm long, somewhat to densely tomentellose; blades oblanceolate to lanceolate-elliptic, 5–14 × 1.3–3.1 cm, glabrous; secondary veins in 11–15 pairs, minutely tomentose. Inflorescences terminal, umbellate, 1.2–3.5 × 1.5–3 cm; peduncle 0.6–2.8 cm long, glabrous to minutely puberulent. Flowers: calyx 1–1.5 mm long, minutely puberulent, the lobes subtriangular, 0.1–0.2 mm long, puberulent; corolla 5-lobed, 2–3 mm long, minutely puberulent, creamy- or greenish-white. Fruits ovoid-ellipsoid, slightly 10-costate, 3.5–5 × 3 mm, glabrous, orange to dark red. Fl (Jan, Sep), fr (Sep, Oct); infrequent in nonflooded forest and in secondary vegetation.

Psychotria brachybotrya Müll. Arg. FIG. 278

Subshrubs or shrubs, 0.8–1.5 m tall. Stems glabrous. Stipules 2–4 mm long, the lobes narrowly triangular, glabrous, persistent. Leaves: petioles 2–10 mm long, glabrous; blades ovate to widely elliptic, 9.5–17 × 2.5–7.5 cm, glabrous; secondary veins in 7–9 pairs. Inflorescences terminal, racemosely branched, 1–1.8 × 1.2–1.8 cm, with bracts subtending each group of flowers 4.5–9 × 2–3 mm; peduncle 0.5–4 cm long; rachis turning blue at fruit maturity. Flowers: calyx 1.2 mm long, puberulent to pilosulous, the lobes 0.2–0.3 mm long, shortly ciliolate; corolla 5-lobed, 2.2–4.8 mm long, pilosulous to glabrous, white. Fruits ovoid-ellipsoid, slightly bilobed, 8–10-costate, 3–10 × 3–8 mm, 2-pyrenate, glabrous, purple. Fl (Feb, Mar, May), fr (Feb, Jun, Jul, Aug, Sep); common, in nonflooded forest.

Psychotria bremekampiana Steyerm. [Syns.: *Cephaelis hirta* (Miq.) Bremek., *Petagomoa hirta* Miq.]

Erect subshrubs (basal nodes sometimes prostrate), to 0.5 m tall. Stems sparsely to densely hirsute, with dark brown hairs >1 mm long, older internodes glabrate, or rarely glabrous, sometimes rooting at basal nodes. Stipular sheath very short, the lobes narrowly triangular to linear, ca. 5 mm long, ciliate. Leaves: petioles absent to 3 mm long; blades oblong, 6–10 × 1.7–3 cm, tomentose (the hairs yellow to pale brown); secondary veins in 10–17 pairs. Inflorescences terminal, few-flowered heads, the subtending bracts lanceolate, 5–12 × 2–3.5 mm, dark red, hirsute, with brown hairs, the apex acute, the margins ciliate; peduncle ca. 1.5 cm long, ciliate. Flowers: calyx ca. 5 mm long, ciliate, the lobes lanceolate, 2.5–3 mm long, longer than tube; corolla 5-lobed, ca. 8 mm long, ciliolate, white. Fruits ovoid, smooth, ca. 10 × 8 mm, 2-pyrenate, fleshy, densely pubescent, blue, with persistent calyces. Fl (Jan, Feb, Aug); in nonflooded forest.

Psychotria callithrix (Miq.) Steyerm. [Syns.: *Gamotopea callithrix* (Miq.) Bremek., *Cephaelis callithrix* Miq., *Psychotria callithrix* var. *tontaneoides* (Britton & Standl.) Williams & Cheesman, *Cephaelis tontaneoides* (Britton & Standl.) Standl., *Evea tontaneoides* Britton & Standl., *Psychotria oiapoquensis* Steyerm.] FIG. 279

Prostrate to decumbent herbs to delicate subshrubs. Stems rooting at nodes, sparsely to densely hirsute, with purplish-blue hairs. Stipular sheath 1 mm long, the lobes linear, 3 mm long. Leaves: petioles 5–8 mm long, hirsute, with purplish-blue hairs; blades narrowly ovate to lanceolate to ovate, 3.5–5.5 × 0.8–1.8 cm, glabrescent adaxially, pubescent abaxially, the midrib densely hirsute with purplish-blue hairs on both sides; secondary veins in 8–10 pairs. Inflorescences terminal, heads of few flowers, subtended by lanceolate bracts, these 6.5 mm long; peduncle 0.7–3 cm long, hirsute, with purplish-blue hairs. Flowers: calyx hirsute, the lobes 1.5 mm long; corolla 4-lobed, 7 mm long, hirsute, white. Fruits ovoid-ellipsoid, ca. 7 mm diam., smooth, 2-pyrenate, sparsely pubescent, bright blue. Fl (Mar, Jun, Aug, Oct), fr (Jun, Jul, Oct); common, in nonflooded forest.

Psychotria capitata Ruiz & Pav.

Shrubs to treelets, 0.3–3 m tall. Stems glabrous. Stipular sheath 3–8 mm long, the lobes 5–15 mm long, glabrous to scarcely puberulent. Leaves: petioles 2–20 mm long, glabrous to pubescent; blades lanceolate to oblong-elliptic or ovate, 6–25 × 1.7–8.5 cm, glabrous, the midrib and secondary veins sometimes pubescent abaxially; secondary veins in 12–24 pairs. Inflorescences terminal, paniculate-cymose, 2–8 × 2–7.5 cm, the axes white to lavender, the bracts subtending axes conspicuous, lanceolate, 5–12 mm long, green to violaceous; peduncle 1.5–11.5 cm long, glabrous to pubescent, green to white. Flowers: calyx 1–1.8 mm long, the lobes 0.1–0.6 mm long, entire to ciliolate; corolla 4–5-lobed, 5–11 mm long, glabrous, white or cream-colored. Fruits globose-ovoid, 8-costate, 2-pyrenate, 5–6 mm diam., glabrous, blackish-blue or blackish-purple, rarely white. Fr (Jan, Apr); usually found in riverine vegetation. *Bois d'Inde* (Créole).

Psychotria carapichea Delprete [Syns.: *Psychotria ligularis* (Rudge) Steyerm. var. *carapichea* (Poir.) Steyerm., *Cephaelis guianensis* (Aubl.) Standl., *Callicocca guianensis* (Aubl.) Gmel., *Uragoga guianensis* (Aubl.) Pulle, *Cephaelis involucrata* Willd., *Psychotria galbaoensis* Steyerm.]

Subshrubs or shrubs, 1–3 m tall. Stems single, glabrous. Stipules with basal sheath extremely reduced, glabrous, the lobes broadly ovate, 3–6 mm long, persistent. Leaves: petioles (3)8–25 mm long; blades oblanceolate to elliptic, (7.5)11–20 × (2.5)3.5–7.5 cm; secondary veins in 6–11 pairs, the tertiary veins sparsely reticulate. Inflorescences terminal, capitate; bracts subtending inflorescence 4, decussate, spatulate (the basal stalk of the bracts oblong, the free portion narrowly to broadly ovate), 23–45 × 8–25 mm, the adaxial surface green, abaxial surface white to greenish-white; peduncle 1.7–3.3 cm long, glabrous. Flowers: calyx lobes absent (margin undulate) or to 0.2 mm long, glabrous; corolla 11–14 mm long, glabrous, white. Fruits oblong-ovoid, smooth (slightly costate when dry), 2-pyrenate, 7–9 × 4–6 mm, orange to orange-red. Fl (Apr), fr (Mar, Apr, May, Jul, Aug); frequent, in nonflooded forest.

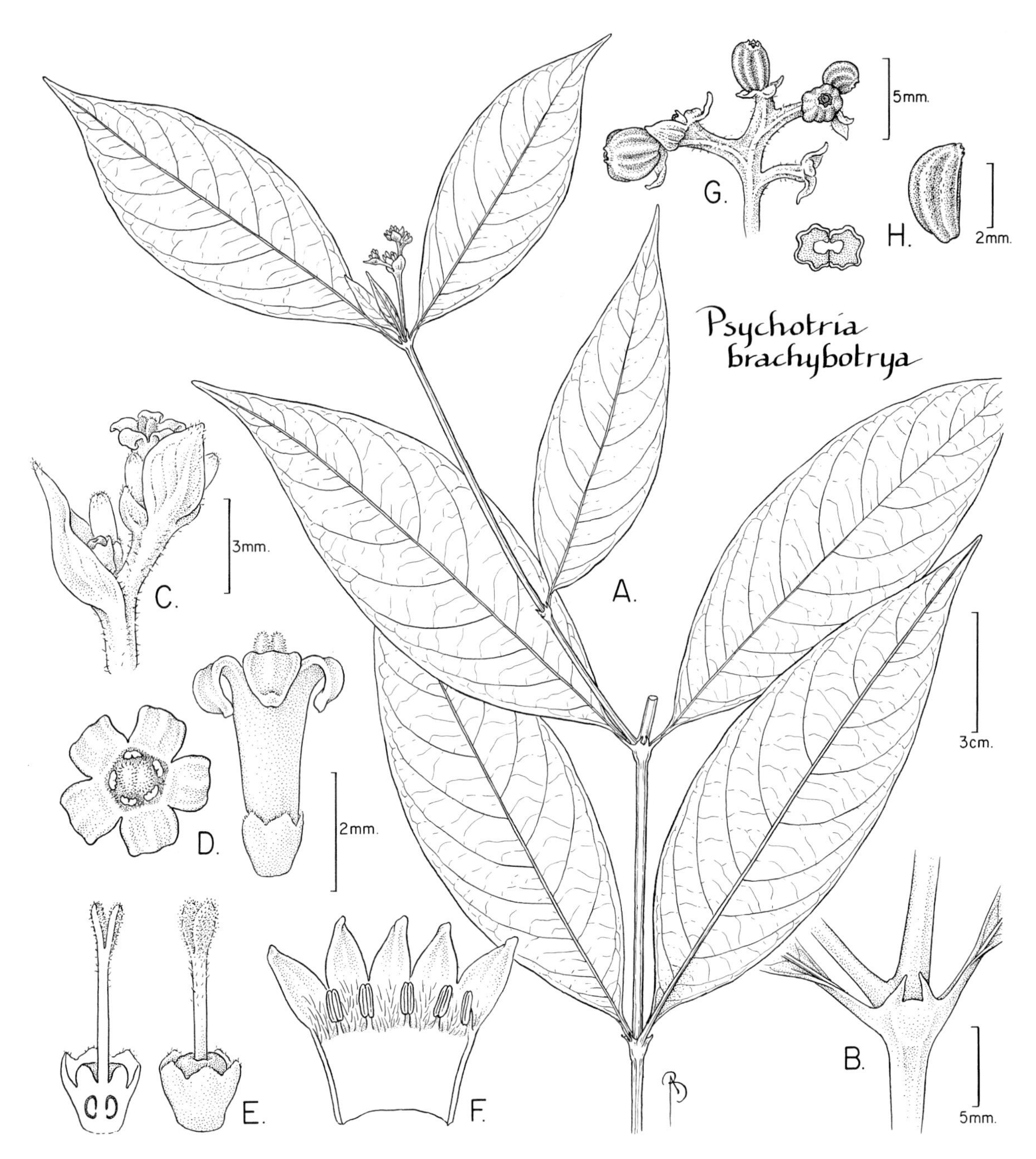

FIG. 278. RUBIACEAE. *Psychotria brachybotrya* (A, B, *Mori et al. 22900*; C–F, *Mori & Pennington 17941*; G, H, *Boom 10710*). **A.** Part of stem with leaves and immature inflorescence. **B.** Stipules. **C.** Lateral view of part of inflorescence. **D.** Apical (left) and lateral (right) views of flower. **E.** Medial section (far left) and lateral view (near left) of pistil. **F.** Corolla opened to show adnate stamens. **G.** Part of infructescence. **H.** Transverse section of fruit (left) and lateral view of fruit segment (right).

Psychotria carthagenensis Jacq.

Shrubs or treelets, to 5 m tall. Stems glabrous. Stipules lanceolate-oblong to ovate, 10–20 × 4–8 mm, glabrous or puberulent, drying reddish, readily caducous, the margins ciliolate. Leaves: petioles 3–5 mm long, glabrous to sparsely puberulent; blades obovate-elliptic to oblanceolate-elliptic to broadly elliptic, 6.5–21 × 2–11.5 cm, glabrous; secondary veins in 8–14 pairs. Inflorescences terminal, ovoid to subhemispheric in outline, >50-flowered, 5–15 × 2.5–7 cm; peduncle 0.7–2.5–3 cm long, glabrous. Flowers sessile; calyx 1.5–1.8 mm long, puberulent, the teeth short, widely triangular, to 0.2 mm long; corolla 5-lobed, 5–6 mm long, puberulent, white or greenish-white. Fruits oblong, slightly 10-costate, 2-pyrenate, 4.5–6.5 × 2.5–5 mm, glabrous, orange or reddish-orange. Fl (Feb, Aug, Sep, Oct, Nov, Dec), fr (Feb, Jan, Mar, Aug); common, in understory of nonflooded forest.

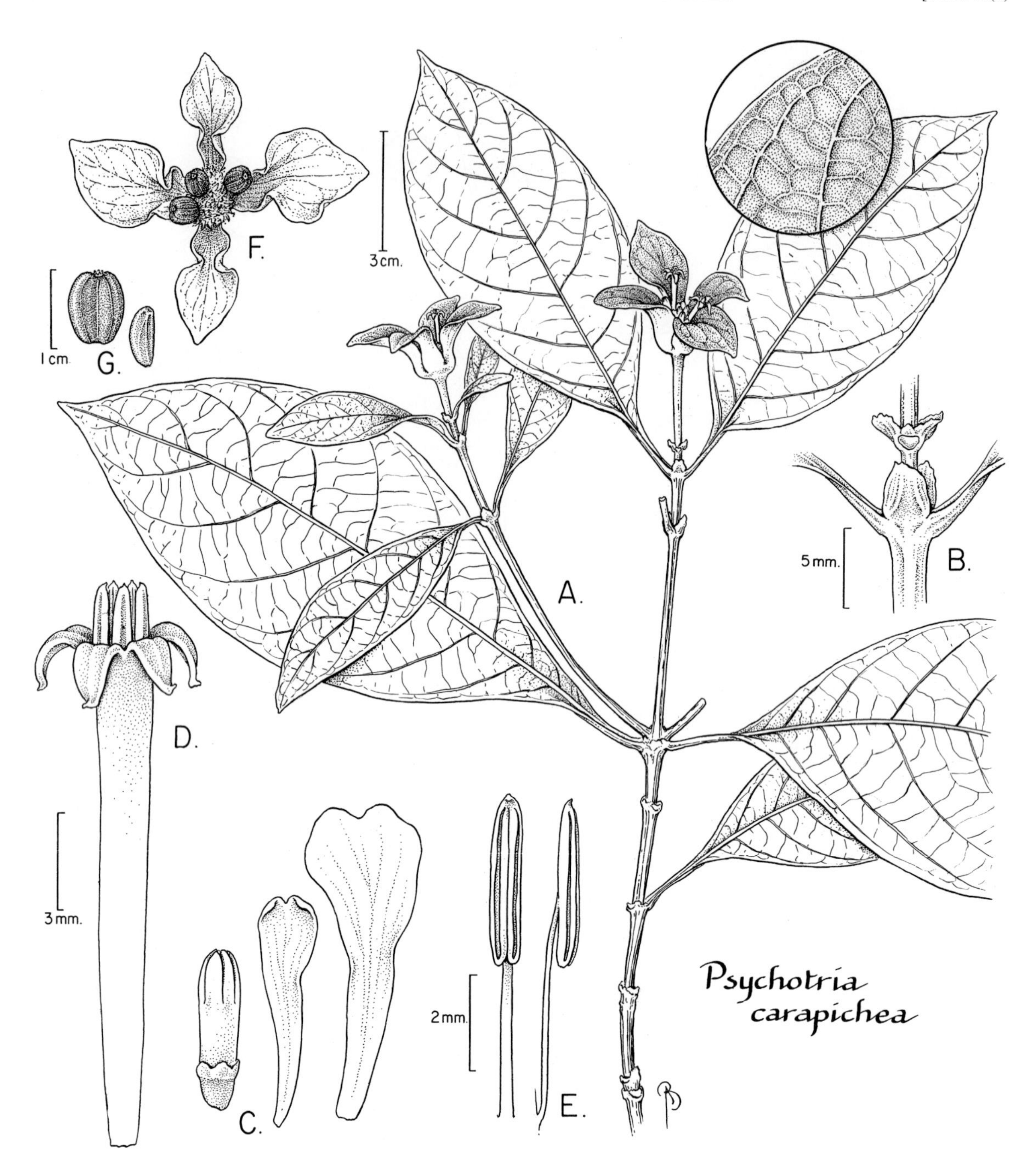

FIG. 279. RUBIACEAE. *Psychotria carapichea* (A–E, *Mori & Pipoly 15511*; F, G, *Marshall & Rombold 102*). **A.** Part of stem with leaves and inflorescences. **B.** Stipules. **C.** Flower bud (left) and two bracts (near and far right). **D.** Lateral view of corolla with projecting anthers. **E.** Adaxial (far left) and lateral (near left) views of stamens. **F.** Apical view of infructescence surrounded by bracts. **G.** Fruit (above) and seed (right). (Reprinted with permission from P. G. Delprete, Brittonia 53, 2001).

Psychotria colorata (Roem. & Schult.) Müll. Arg. [Syn.: *Cephaelis amoena* Bremek.]

Subshrubs or shrubs, 0.5–1.5 m tall. Stems glabrous. Stipules with sheath 1.5–3 mm long, glabrous, the lobes narrowly triangular, 1–4 mm long. Leaves: petioles 2–15 mm long; blades oblong-elliptic to ovate-oblong, 7–20 × 2.5–8 cm, surfaces glabrous, the midrib glabrous or sparsely pubescent, the secondary veins glabrous; secondary veins in 8–15 pairs. Inflorescences terminal, subhemispheric heads, 1.2–2 × 2.2–4 cm, the two outermost bracts subtending inflorescences 2–3.5 × 2–3.5 cm, connate in basal 2.5–4 mm, glabrous, purple; peduncle 0.5–5.8 cm long, glabrous, purple. Flowers: calyx 1.5–3.5 mm long, glabrous, the lobes 0.2–2 mm long; corolla 5-lobed, 10.5–14 mm long, glabrous, white to purplish. Fruits ovoid-oblong, 10-costate, 2-pyrenate, 5–5.5 × 3–3.5 mm, glabrous, blue or purplish. Fl (Feb, Mar, Dec), fr (Mar); in gaps and disturbed vegetation.

Psychotria cupularis (Müll. Arg.) Standl.

Shrubs or treelets, to 5 m tall. Stems glabrous to puberulent. Stipules broadly ovate, 7–9 × 6–8 mm, glabrous or puberulent, readily caducous, drying reddish rust-brown, with white margins, not ciliolate. Leaves: petioles 3–12 mm long, glabrous to puberulent; blades narrowly elliptic to oblanceolate, (6)7–21 × (2.5)3–7.5 cm, glabrous; secondary veins in 5–10 pairs. Inflorescences terminal, corymbosely paniculate, 4–7.5 × 2–7 cm; peduncle 1.8–3 cm long, usually densely puberulent. Flowers pedicellate; calyx 1.5–2.5 mm long, subtruncate or slightly lobed, the teeth to 0.2 mm long, puberulent; corolla 5-lobed, (5)7–10 mm long, puberulent, white. Fruits globose, 8–10-costate, 2-pyrenate, 5–7 × 3.5–5 mm, orange to red. Fl (Jul), fr (Jul, Aug, Sep, Nov); common, in nonflooded forest.

Psychotria deflexa DC.

Subshrubs or shrubs, 0.5–4 m tall. Stems glabrous. Stipules with basal sheath 2 mm long, the lobes narrowly lanceolate to linear, 3–7 × 0.5–4 mm, glabrous. Leaves: petioles 2–15 mm long, glabrous; blades lanceolate-oblong to ovate-elliptic, 7–20 × 2.7–10.5 cm, glabrous; secondary veins in 5–10 pairs. Inflorescences terminal, paniculate, (2)6–17 × 2–5 cm; peduncle 3–7.5 cm long, slightly puberulent. Flowers: calyx 0.5–1.1 mm long, the teeth widely triangular, 0.1–0.3 mm long, glabrous; corolla 5-lobed, 3.5–5 mm long, glabrous, white or creamy white. Fruits subglobose, slightly bilobed, 6(8)-costate, 2-pyrenate, ca. 3 mm diam, purple, blue, or white. Fl (Mar), fr (Feb, Jun, Aug, Sep, Oct); frequent, in nonflooded forest.

Psychotria erecta (Aubl.) Standl. & Steyerm.

Subshrubs or shrubs, 0.5–1.8 m tall. Stems glabrous to puberulent. Stipules with basal sheath to 1 mm long, the lobes broadly triangular, 3–5 mm long, pubescent. Leaves: petioles 7–13 mm long, glabrous or pubescent; blades elliptic-oblong to lanceolate-oblong, 8.5–17 × 2.5–9.5 cm, glabrous adaxially, glabrous or strigulose abaxially; secondary veins in 5–9 pairs. Inflorescences axillary, capitate, 3–7-flowered, the involucral bracts ca. 2 mm long; peduncle absent to 7 mm long, strigulose. Flowers: calyx truncate, ca. 4 mm long, glabrous to scarcely strigulose; corolla 5–6-lobed, ca. 7 mm long, glabrous, white. Fruits narrowly ovoid, slightly costate, 2-pyrenate, 8–10 × 4.5–8 mm, glabrous to scarcely strigulose, blackish-purple. Fl (Jan, Sep), fr (Feb, Mar, May, Sep); in nonflooded forest at all elevations.

Psychotria granvillei Steyerm. PL. 113c

Shrubs, 0.6–1.5 m tall. Stems pubescent. Stipules with basal sheath 5–8 mm long, densely pubescent, the lobes narrowly triangular, 5 mm long, the apex fringed. Leaves: petioles 8–17 mm long, pubescent; blades lanceolate to oblong-elliptic, 10–17.5 × 4.3–6 cm, pubescent ad- and abaxially; secondary veins in 11–12 pairs. Inflorescences terminal, hemispheric, many-flowered heads, 1.5 × 2.2 cm, the heads subtended by orange (sometimes with green acuminate tips) involucral bracts united into basal tube, the external pair suborbicular to ovate, 3.3 × 2.4 cm; peduncle erect, 3–5 cm long, pubescent. Flowers: calyx 4 mm long, densely pubescent, the lobes 1.5 mm long, glabrous, the margins ciliate; corolla 11–12 mm long, white. Fruits obovate, 13 × 8 mm long, white. Fl (Jan, Feb, Mar, Apr), fr (May, Jun); common, in nonflooded forest.

Psychotria guadalupensis (DC.) R. A. Howard

Epiphytic shrubs, to 1 m tall. Stems subquadrangular when young; older stems terete, glabrous, rooting at basal nodes. Stipules sheathing, cupular, truncate, 1.5–3 mm long, membranous, glabrous. Leaves: petioles 3–6 mm long, glabrous; blades ovate to lanceolate-oblong, 4–7 × 1.5–3.3 cm, glabrous; secondary veins in 4–5 pairs. Inflorescences axillary, cymose, trichotomously branched, 2–3.5 × 2–7 cm; peduncle 1.5–3 cm, glabrous. Flowers: calyx ca. 1.5 mm long, glabrous, the lobes ca. 0.4 mm long; corolla 4.5 mm long, glabrous, creamy white. Fruits oblong, 3.5 × 2 mm, bright red. Fl (Sep), fr (Jan, Sep); known only from forest on summit of Mont Galbao in our area.

Psychotria hoffmannseggiana (Roem. & Schult.) Müll. Arg.

Subshrubs to shrubs, 0.3–2 m tall. Stems glabrous to puberulous (hairs <1 mm long). Stipules with basal sheath truncate, 4–5 mm long, the lobes shortly triangular, to 2 mm long, glabrous. Leaves: petioles 2–8 mm long, glabrous or puberulent; blades elliptic to elliptic-ovate, 3.5–13 × 1.2–4 cm, glabrous adaxially, glabrous to sparsely puberulent abaxially; secondary veins in 6–14 pairs. Inflorescences terminal, at first one capitate head, then expanding to several heads in course of development, the heads 1–3 × 1.5–2 cm, with bracts inserted at base of fascicles, these 3–17(20) per inflorescence, narrowly lanceolate, white or green adaxially, green abaxially; peduncle 0.5–2.5 cm long, puberulent. Flowers: calyx 1–2 mm long, the lobes 0.1–0.5 mm long; corolla 5-lobed, 4–7 mm long, slightly puberulent to glabrous, white. Fruits ovoid-ellipsoid, 8–10-costate, 2-pyrenate, 3 × 4.5 mm, glabrous, blue or purple. Not yet collected in our area but to be expected.

Psychotria iodotricha Müll. Arg. PL. 113e

Erect subshrubs, 0.3–2 m tall (sometimes basal nodes procumbent). Stems rooting at basal nodes, densely hirsute, with black or dark purple hairs. Stipules with basal sheath 1.5 mm long, densely pubescent, the lobes lanceolate, 3–7 mm long. Leaves: petioles absent to 5 mm long, densely hirsute; blades lanceolate to ovate-oblong, 4.5–11 × 1.2–4 cm, glabrous to pubescent adaxially, densely pubescent abaxially; secondary veins in 8–14 pairs. Inflorescences terminal, hemispheric heads, 6–10 × 4–12 mm, usually with 4–10 flowers, subtended by bracts, the outer involucral bracts not decussate, lanceolate, 2–3.5 × 3–4 mm, the apex acute; peduncle absent. Flowers: calyx 6–9 mm long, the lobes 3.5–6.5 mm long, pubescent; corolla 5-lobed, 12–15 mm long, white. Fruits ovoid to subglobose, 8-costate, 12–13 × 12–15 mm, blue to purple. Fl (Nov, Dec), fr (Mar, Jul); in nonflooded forest.

Psychotria kappleri (Miq.) Benoist

Shrubs, ca. 1.5 m tall. Stipules with basal sheath 1 mm long, the lobes linear, 1.5–2 mm long. Leaves: petioles 0.7–1 cm long; blades lanceolate-oblong, 5–10 × 1.2–3.2 cm, glabrous; secondary veins in 6–11 pairs. Inflorescences terminal, capitate, the bracts free at base, 4–6 × 2–5.5 mm, glabrous, reddish or purplish (drying beige); peduncle absent to ≤6 mm long. Flowers surrounded by 1–3 bracteoles; calyx very short, the lobes triangular, 1 mm long, pubescent; corolla 4 mm long, glabrous, white. Fruits ovoid to fusiform, smooth, 2-pyrenate, ca. 3 mm long (when green). Fl (Mar, Sep, Oct), fr (Apr); in nonflooded forest.

Psychotria lateralis Steyerm.

Herbs, 20–50 cm tall. Stems decumbent, rooting at basal nodes, succulent, glabrous, purple-brown. Stipules sheathing, suborbicular, 4 × 6 mm, the margins irregularly erose. Leaves: petioles 1–3 cm long; blades elliptic-oblanceolate to oblong-elliptic, 11–19 × 3.5–5.5 cm, the midrib reddish-violet; secondary veins in 10–12 pairs. Inflorescences axillary, sessile, capitate, subhemispheric, 10 × 15 mm, subtended by purple or red involucral bracts, 13 × 9–12 mm. Flowers: calyx tube 5 mm long, red, the lobes 2–5 mm long, the margin ciliate; corolla 5-lobed, white. Fruits ovate, costate, 5–5.5 × 3.5–4 mm, red or reddish-white. Fl (Feb, Aug, Dec), fr (Jan, Feb, Mar, May, Jun, Aug, Nov); common, in nonflooded forest.

Psychotria ligularis (Rudge) Steyerm. [Syns.: *Psychotria ligularis* var. *ligularis*, *Schradera ligularis* Rudge]

Subshrubs or shrubs, 0.2–1.5 m tall. Stems single, glabrous. Stipules with basal sheath extremely reduced, glabrous, the lobes broadly triangular to broadly ovate, 2–5 mm long, persistent. Leaves: petioles 3–20 mm long; blades oblanceolate to elliptic, (4)11–18 × (1.5)2.5–5.5 cm; secondary veins in (14)25–45 pairs, the tertiary veins parallel to secondary veins. Inflorescences terminal, capitate to subcapitate (shortly branched); bracts 2–8, inserted on inflorescence axes, not decussate, linear, 12–30 × 2–4 mm, green throughout; peduncle 1.5–4.5 cm long, glabrous. Flowers: calyx lobes 0.3–0.5 mm long, glabrous; corolla 7–8 mm long, glabrous, white. Fruits oblong-ovoid, smooth (slightly costate when dry), 2-pyrenate, 6–10 × 4–6 mm, orange to orange-red. Fr (Mar, Oct); in our area, known by only two collections (*Andersson 1989*, NY, and *de Foresta 705*, CAY); in understory of nonflooded forest.

Psychotria macrophylla Ruiz & Pav. *sensu lato*

Herbs to subshrubs, 0.7–2.5 m tall. Stems erect, succulent, glabrous to pubescent. Stipules triangular-ovate to lanceolate, 1–10 × 1.5–2.5 mm, rigid, glabrous, persistent. Leaves: petioles 2–8 cm long, glabrous, sometimes reddish-purple; blades variably oblong to oblanceolate-elliptic, 14–25(28) × 2.5–9(12) cm, glabrous; secondary veins in 13–30 pairs. Inflorescences axillary, paniculate, widely oblong-ovoid, 2–11 × 1.5–3.5 cm (in our area), many-flowered; peduncle 1–5.5 cm long, glabrous to densely puberulent. Flowers: calyx 1–1.5 mm long, glabrous to densely puberulent, the lobes <0.5 mm long; corolla 4–5-lobed, 4.5–7.5 mm long, glabrous to puberulent, white or greenish-white. Fruits globose to oblong, 8–10-costate, 4–6 × 4 mm, white, red, yellow, or orange. Fr (Nov, Dec); in understory of nonflooded forest.

This species is extremely variable in leaf shape and size. Study of specimens from throughout its range may reveal that collections treated under this name represent a complex of several species.

Psychotria mapourioides DC.

Trees, 10–25(30) m tall. Trunks angled or cylindrical. Bark smooth, lenticellate, with buttresses. Stems glabrous. Stipules obovate, 8–23 × 4–15 mm, readily caducous, drying reddish, the apex obtuse or rounded, glabrous. Leaves: petioles 0.7–4 cm long, glabrous; blades obovate-elliptic to oblanceolate or lanceolate-elliptic, 12–28 × 3–12 cm, generally glabrous; secondary veins in 6–15 pairs. Inflorescences terminal, paniculate, 4–10 × 5–12 cm; peduncle 2.8–13 mm long, glabrous to minutely puberulent. Flowers: calyx subtruncate, 1–2.5 mm long; corolla 5-lobed, 5–9 mm long, glabrous or densely puberulent, white. Fruits subglobose, 5–6-costate, 2-pyrenate, 5–7.5 × 4–5 mm, glabrous, red. Fl (Feb, Mar, Apr, Aug, Sep, Dec), fr (Feb, May, Aug, Sep, Oct, Nov); common, in nonflooded forest.

Psychotria medusula Müll. Arg. [Syns.: *Psychotria blepharophylla* (Standl.) Steyerm., *Cephaelis blepharophylla* Standl, *Uragoga glabra* (Aubl.) Kuntze, *Tapogomea glabra* Aubl.] PL. 113b

Erect subshrubs, 25–75 cm tall (basal nodes sometimes prostrate). Stems single, rooting at nodes, sparsely pubescent, glabrate in older internodes, or rarely glabrous. Stipules with basal sheath 4 mm long, the lobes narrowly lanceolate, 9–13 mm long, with prominent dark purple hairs (drying black). Leaves: petioles 1–1.5 cm long, pubescent; blades ovate to elliptic, 10–17 × 5–8 cm, the margins with dark purple hairs (drying black); secondary veins in 6–9 pairs. Inflorescences terminal, few-flowered (<20) pseudoheads, 1–2 × 2–3 cm, subtended by 21–40 narrowly lanceolate to linear bracts, 13–22 × 1–2 mm, with prominent, numerous dark purple hairs (drying black); peduncle 5–10 mm long, with dark purple hairs (drying black). Flowers: calyx 2 mm long, the lobes 1 mm long; corolla 5-lobed, 12–16 mm long, lilac, mauve, or white basally, violet apically, ciliate. Fruits ovoid-ellipsoid, 8–10-costate, 2-pyrenate, 5 × 4 mm, blue. Fl (Jan, Feb, Mar, Jun, Sep, Oct, Dec), fr (Jan, Apr, Jun, Jul, Aug); common, in understory of nonflooded forest.

Psychotria microbotrys Standl.
PL. 113d; PART 1: PL. VIIIc as *P. surinamensis*

Shrubs, 1.7–3 m tall. Stems glabrous. Stipules widely ovate, 11–20 mm long, the lobes narrowly lanceolate, (2)4–10 mm long, glabrous, persistent. Leaves: petioles 1–3.5 cm long, glabrous; blades elliptic-oblong to elliptic obovate, 14–27 × 7–13 cm, glabrous; secondary veins in 11–15 pairs. Inflorescences terminal, paniculate, sometimes widely oblong-pyramidal in outline, 3–5 × 3–6.5 cm, the bracts subtending inflorescence axes inconspicuous (sometimes absent), to 2 mm long; peduncle 3.5–7 cm long, puberulent to glabrous. Flowers: calyx lobes widely rounded; corolla 3–3.5 mm long, glabrous. Fruits subglobose, 10-costate, ca. 3 mm diam., blue or whitish-red. Fl (Feb, Mar, Apr), fr (Jan, May, Jun, Jul); common, in undertory of nonflooded forest.

Psychotria microbracteata Steyerm.

Herbs, 30–60 cm tall, glabrous. Stipules inconspicuous, often reduced to minute, interpetiolar membrane, 0.6–0.8 mm long, persistent. Leaves: petioles 1.5–2.7 cm long, glabrous; blades lanceolate-elliptic, 10–16 × 3.5 cm, glabrous; secondary veins in 10–12 pairs. Inflorescences axillary, cymosely branched, 1.5–2 × 2–3 cm; peduncle 1.7–7 cm long, glabrous. Flowers: calyx ca. 2 mm long, the lobes ca. 0.4 mm long; corolla 6–6.5 mm long, glabrous, greenish-white. Fruits elliptic-oblong, 6–7 × 4.5–5 mm, white. Fl (Jan, Mar, Sep, Nov), fr (May, Jun, Jul, Sep); endemic to Mont Galbao, in forest.

Psychotria moroidea Steyerm.

Shrubs, 0.3–1.5 m tall. Stems glabrous. Stipules with basal sheath truncate, ca. 2 mm long, the lobes reduced to linear teeth 1–2 mm long, persistent. Leaves: petioles 2–3 mm long; blades rhomboid or obovate, 8–15 × 3–5 cm; secondary veins in 7–10 pairs. Inflorescences terminal, globose heads, ca. 6 mm diam.; peduncle 0.5–1.5 cm long. Flowers: calyx ca. 1 mm long, the lobes ca. 0.4 mm long; corolla ca. 2 mm long, greenish-white, (flowers from *Cowan 39028*, NY, from outside our area). Fruits globose-ovoid, smooth, 2-pyrenate, ca. 0.5 cm diam., red. Fl (Jan, Feb, Mar, Dec), fr (Mar, Apr, May, Aug); in nonflooded forest.

Psychotria oblonga (DC.) Steyerm. PL. 113f

Subshrubs, 25–80 cm tall. Stems branched, sometimes prostrate, glabrous. Stipules with basal sheath truncate, ca. 1 mm long, the lobes reduced to setaceous awns 1.5–2.5 mm long, persistent. Leaves: petioles ca. 5 mm long, glabrous to puberulent; blades ovate to elliptic, 6–10 × 2.5–4 cm, the midrib pilosulous abaxially; secondary veins in 8–15 pairs. Inflorescences terminal, sessile to subsessile, single-flowered, subtended by narrowly ovate bracts, the bracts 13–15 mm long, obtuse to rounded at apex, glabrous; peduncle absent to >5 mm long. Flowers: calyx green; corolla 13 mm long, glabrous, white. Fruits obovate, smooth, 2-pyrenate, 22–30 × 10–15 mm, spongy, lavender. Fl (Jan, Jun, Sep, Dec), fr (Mar, Jun); common, in nonflooded forest. *Gaulette noir* (Créole).

Psychotria officinalis (Aubl.) Sandwith

Shrubs, 0.3–1.5 m tall. Stems glabrous. Stipules with basal sheath truncate or subtruncate, 1–2 mm long, the lobes lanceolate-linear, 1–2 mm long, glabrous, persistent. Leaves: petioles 2–6 mm long, usually glabrous; blades lanceolate-elliptic, 4–9 × 1.5–4 cm, glabrous; secondary veins in 6–9 pairs. Inflorescences terminal (sometimes appearing axillary) axillary, umbellate or corymbose, a shortly branched head during anthesis, expanding to several heads after anthesis, 0.5–2.5 × 1–4 cm, the fascicles subtended by narrowly lanceolate bracts, green or reddish-green; peduncle 1.3–3.2 cm long, puberulent. Flowers: calyx 1.3 mm long, the lobes <0.3 mm long; corolla 5-lobed, 4–5 mm long, glabrous, white. Fruits globose to obovoid, 8-costate (when dry), 2-pyrenate, 3 × 3–5 mm, usually glabrous, shiny blue to purple. Fl (Jan), fr (Jul, Aug, Sep); common, in nonflooded forest.

Psychotria paniculata (Aubl.) Raeusch.

Shrubs to treelets, to 5 m tall. Stems glabrous. Stipules with basal sheath 2–3 mm long, the lobes reduced to two triangular teeth, 0.3–1 mm long, glabrous. Leaves: petioles 7–15 mm long, glabrous; blades oblong-elliptic to ovate-elliptic, 9.5–18 × 4–7.5 cm, glabrous; secondary veins in 7–11 pairs, the tertiary venation densely subparallel (perpendicular to secondary veins). Inflorescences terminal, paniculate, 5.5–7.5 × 4–5.5 cm; peduncle 3–4 cm long, glabrous. Flowers: calyx 0.2–0.3 mm long, with undulate margins or extremely reduced teeth, glabrous; corolla 5-lobed, ca. 5 mm long, glabrous, white. Fruits globose, 10-costate, 2-pyrenate, 4–4.5 mm diam. (fruit from *Cremers et al. 7682*, CAY). Fl (Feb); in our area known from a single collection (*Mori & Gracie 21148*, NY), in nonflooded forest.

Psychotria perferruginea Steyerm.

Shrubs, 1.4 m tall. Stems glabrous. Stipules 3.5–5 mm long, each terminal stipule with two narrowly linear-setaceous segments, these ferruginous-hirtellous, persistent. Leaves: petioles 0.5–1.5 cm long, glabrous; blades oblong-elliptic, 9–14.5 × 3–5 cm, glabrous. Inflorescences terminal, 2.5–3 × 4–5 cm; peduncle 3.8–4 cm long, glabrous. Flowers: calyx 2 mm long, glabrous, the lobes 0.5 mm long, ciliolate; corolla 5-lobed, 4–5 mm long, glabrous, white. Fruits unknown. Fl (Dec); in nonflooded forest, known only from the type collection (*Granville B.4640*, CAY) from Mont La Fumée.

Psychotria platypoda DC. PL. 113g

Subshrubs or shrubs, 1–3 m tall. Stems glabrous. Stipules with basal sheath truncate, 1–2 mm long, the lobes linear-lanceolate, 0.5–3.5 mm long, glabrous, caducous. Leaves: petioles 2–12 mm long, glabrous; blades oblong-elliptic to ovate-elliptic, 8–19 × 3–7.5 cm, glabrous; secondary veins in 6–16 pairs. Inflorescences terminal, single compact heads of 16–70 flowers, 1.2–2 × 0.7–0.9 cm at anthesis, slightly larger in fruit, the bracts subtending inflorescences free at base (inserted at base of axes), purple to green; peduncle 5–17 mm long, glabrous or puberulent, reddish. Flowers: calyx 1.2–1.3 mm long, glabrous, the lobes 0.2 mm long; corolla 7.5–8.5 mm long, glabrous, white to pink. Fruits subglobose, 10-costate, 2–4 × 3.5–4 mm, glabrous, dark blue to violet. Fl (Jan, Mar, Apr, Jul, Sep), fr (May, Jun, Jul, Aug, Sep); common, in somewhat disturbed areas in all kinds of forest.

Psychotria poeppigiana Müll. Arg. [Syn.: *Cephaelis tomentosa* (Aubl.) Vahl] PL. 114a,b

Subshrubs or shrubs, 0.6–3.5 m tall. Stems pubescent. Stipules with basal sheath densely villous, 4–5 mm long, the lobes linear-lanceolate, 4–11 mm long, densely pubescent, persistent. Leaves: petioles 0.6–5 cm long, densely pubescent; blades elliptic to widely ovate or lanceolate, 7–30 × 2.5–11.5 cm, pubescent throughout; secondary veins in 8–18 pairs. Inflorescences terminal, capitate, multi-flowered, 1–2.5 × 1.5–3.5 cm, subtended by prominent involucral bracts, the external pair connate at base, red to orange-red; peduncle 0.6–12 cm long, densely pubescent. Flowers: calyx 3–4 mm long, the lobes 0.6–2.2 mm long, glabrous; corolla 5-lobed, 14–16.5 mm long, yellow. Fruits obovate-oblong, 5.5–7 × 2.8–3 mm, blue. Fl (Jan, Feb, May, Aug, Sep, Dec), fr (May, Jun, Sep); very common, in slightly disturbed areas in all kinds of forest. *Radié-zoré* (Créole).

Psychotria pullei Bremek.

Subshrubs to shrubs, 0.5–1.5 m tall. Stems glabrous. Stipules with basal sheath extremely reduced, the lobes narrowly triangular-linear, 1.5–2 mm long, glabrous, persistent. Leaves: petioles 2 mm long; blades lanceolate, 3–14 × 2–5.5 cm; secondary veins in 6–11 pairs. Inflorescences terminal, paniculate, 3–6 cm long, the rachis filiform, usually unbranched, or with extremely reduced basal axes, these to 5 mm long; peduncle to 1–2.5 cm long. Flowers: calyx ca. 0.3 mm long, puberulous, the lobes 0.2 mm long; corolla 4-lobed, 3 mm long, white (or red according to Bremekamp). Fruits globose, slightly 5-costate, 2-pyrenate, purple. Fl (Jan); known only from the cloud forest of Mont Galbao in our area .

Psychotria pungens Steyerm.

Shrubs, ca. 1 m tall. Stems glabrous. Stipules persistent, with basal sheath truncate, 6–9 mm long, glabrous, with 5–6 long-spinulose segments per side, 5–10 mm long, the apical stipules forming conical cap topped by cluster of 12–13 spinulose awns. Leaves: petioles 2–5 mm long, glabrous; blades narrowly oblong-oblanceolate to oblong-elliptic, 9–13 × 2.5–4.5 cm; secondary veins in 9–11 pairs. Inflorescences terminal, congested heads, 10–18-flowered, 2–2.5 × 1.5–1.8 cm, subtended by spinulose, green involucral bracts, these oblong-oblanceolate, 16–17 × 3–4.5 mm, glabrous, the margins ciliate. Flowers: calyx 0.5–0.6 mm long, the lobes 3.5–5 mm long, glabrous, the margins ciliate; corolla unknown. Fruits oblong, smooth, subdidymous, 2-pyrenate, 9–10 × 5–6 mm, glabrous. Fr (Mar, Sep); in our area known from forests on the lower slopes of Mont Galbao.

Psychotria racemosa (Aubl.) Raeusch. PL. 114c

Subshrubs to shrubs, 0.5–3 m tall. Stems densely pubescent. Stipules with basal sheath extremely reduced, to 2 mm long, densely puberulent, the lobes narrowly lanceolate, 5–12 mm long, ciliolate, persistent. Leaves: petioles 0.3–1.5 cm long; blades oblong-elliptic to ovate-oblong, 8.5–18 × 2.5–8.5 cm, glabrous adaxially, glabrescent abaxially; secondary veins in 11–15 pairs. Inflorescences terminal, paniculate-racemose, oblong to subhemispheric in outline, 1.5–3.5 × 2–5 cm at anthesis, slightly larger upon fruiting (4–6 cm long), puberulent; peduncle 0.5–3.2 cm long, densely puberulent. Flowers: calyx 1.5–2 mm long, the lobes usually 0.3–0.4 mm long; corolla 3.5 mm long, puberulent, white to creamy white. Fruits subglobose, prominently 5-costate, 5-pyrenate, 3.5–4 × 5–6 mm, glabrous, first orange, turning black at maturity. Fl (Jan, Feb), fr (Feb, Mar, May, Jun, Jul, Aug, Sep, Oct, Nov); common, in nonflooded forest.

Psychotria saülensis Steyerm.

Herbs, 0.25–1 m tall. Stems hollow, the basal portion prostrate, rooting at nodes. Stipules suborbicular to extremely reduced, to 8 mm long, membranous, persistent. Leaves: petioles 3.5–5 cm long, glabrous; blades widely elliptic to ovate-elliptic, 20–22 × 7–9 cm, membranous, drying olive green, the margins conspicuously ciliate; secondary veins in 9–10 pairs. Inflorescences axillary, cymose-paniculate, 1–2.5 × 1.5–2 cm, the rachis filiform; peduncle 4–11 mm long. Flowers: calyx 0.5–1 mm long, the margins undulate or with 5 minute lobes; corolla 4–5-lobed, 3–4 mm long, white. Fruits subglobose, subbilobed, smooth, 2-pyrenate, 6.5 × 5 mm, red. Fl (Jan), fr (May); known only from forest on Mont Galbao.

Psychotria trichophoroides Müll. Arg.

Herbs to subshrubs, to 70 cm tall. Stems prostrate-decumbent, branching distally, rooting at nodes. Stipules with basal sheath 2–3 mm long, the lobes narrowly triangular, 6–11 mm long, conspicuously ciliate, persistent. Leaves: petioles 2–3 mm long, conspicuously spreading pubescent; blades elliptic, 5–8 × 2–3 cm, glabrous except for conspicuously spreading pubescent midrib, drying blackish-brown; secondary veins in 8–11 pairs. Inflorescences terminal, sessile to subsessile, capitate, subtended by many ciliate bracts, these lanceolate, 8–20 mm long, pale green, drying black, the apex acute; peduncle absent to 4 mm long. Flowers: calyx ca. 0.5 mm long; corolla 5-lobed, 6–7 mm long, white. Fruits unknown. Fl (Dec); known by a single collection from nonflooded forest in our area.

Psychotria uliginosa Sw.

Herbs, 0.5–1.5 m tall. Stems erect, thick, fleshy, glabrous, rooting at basal nodes. Stipules widely suborbicular or oblong-lanceolate, 3–6 mm long, glabrous, persistent. Leaves thick to almost fleshy; petioles 3–7.5 cm long, glabrous; blades elliptic-obovate to obovate-oblong, 9.5–32 × 5–12.5 cm, glabrous; secondary veins in 11–17 pairs. Inflorescences axillary or subterminal, paniculate, 2–3.5 × 1.5–3 cm at anthesis, to nearly twice as large at fruit maturity; peduncle stout, 2–4 mm thick, 7.5–15 mm long at anthesis (17–35 mm long at fruit maturity). Flowers: calyx subtruncate, 2–3 mm long; corolla ca. 5 mm long, white. Fruits ovoid to oblong-ovoid, 7–8 × 4.5–6 mm, purple. Fl (Mar, May, Sep, Dec), fr (Sep, Nov, Dec); in understory of forests.

Psychotria ulviformis Steyerm. [Syn.: *Cephaelis alba* (Aubl.) Wernham]

Herbs. Stems prostrate, densely pubescent, rooting at nodes and internodes. Stipules with basal sheath 2–3.5 mm long, membranous, each side split into two narrowly lanceolate lobes, 2–4.5 mm long, these strongly ciliate, with white hairs, persistent. Leaves: petioles 4–32 mm long, densely hirsute to villose-ciliolate; blades oval to broadly oval-oblong, 3–9 × 1.5–6.5 cm, puberulent to shortly pubescent throughout, lavender-maroon abaxially; secondary veins in 8–12 pairs. Inflorescences axillary, subhemispheric, 1–2.5 × 1–2.5 cm, multi-flowered, with 3 principal axes; peduncle 1.5–4 cm long, glabrous to villose. Flowers: calyx 1.5–1.7 mm long, the lobes 0.6–0.7 mm long, strongly ciliolate, with lavender hairs; corolla 5-lobed, 7 mm long, white. Fruits subglobose, 4–9 × 4–7 mm, glabrous, bluish. Fl (Jan, Feb, Mar, Apr, Aug); frequent, in nonflooded forest. *Feuille grage, herbe-grage, radié grage, radié serpent, radier serpent* (Créole), *zãzãpatu* (Bush Negro), *zégron léphan* (Créole).

Psychotria urceolata Steyerm.

Shrubs, 1.5–2.5 m tall. Stems glabrous. Stipules with basal sheath extremely reduced (<1 mm long), the lobes lanceolate, 5–6 mm long, drying creamy white (hay colored), persistent, the apex irregularly fringed. Leaves: petioles 3–8 mm long; blades oblong-elliptic, 9.5–18 × 4–7.5 cm, midrib and/or secondary veins puberulent or pilosulous; secondary veins in 7–9 pairs. Inflorescences terminal, capitate, subtended by purple involucral bracts basally fused into basal urceolate involucre, the bracts 2.5–3 × 2.2–2.5 cm; peduncle erect, 3–4.5 cm long. Flowers: calyx 5 mm long, the lobes 1.5–2 mm long; corolla 2.2–2.7 mm long, white. Fruits ellipsoid, ca. 7 × 4 mm, blue (from *Granville 2912*, CAY, from outside our area). Fl (Jan, Mar); in nonflooded forest.

Psychotria variegata Steyerm. PL. 115a; PART 1: PL. VIIIb

Herbs. Stems prostrate or procumbent, densely hirsute, the hairs multicellular, lavender. Stipules with basal sheath to 2–3 mm long, membranous, the lobes setose, 4–5 mm long, persistent, the margins ciliolate, the hairs multicellular, lavender. Leaves: petioles 3–10 mm long, densely hirsute; blades ovate to oblong-ovate to oblong-lanceolate, 2.5–5.7 × 1–3 cm, strigose pubescent throughout, or at least on veins, often gray or silvery-striped adaxially, purple or green abaxially, or sometimes blades green throughout; secondary veins in 8–11 pairs. Inflorescences terminal, subhemispheric heads, 8–10 × 12–17 mm, subtended by bracts, the two external bracts ovate to oblong-lanceolate, 8–9 × 3–4 mm, hirsute; peduncle usually 4–9 cm long, densely hirsute. Flowers: calyx 2–2.1 mm long, the lobes 1.1 mm long, densely setose-ciliate; corolla 5-lobed, 6–7.5 mm long, white. Fruits oblong-ovate to subglobose, smooth, 2-pyrenate, 4–7 × 2–7 mm, pilose, blue. Fl (Jan, Jun, Aug, Sep, Nov), fr (Mar, May, Jun, Jul, Aug, Sep, Dec); common, sometimes as ground-cover, in partly opened and disturbed areas in forests. *Herbe jacques* (Créole).

Psychotria viridibractea Steyerm.

Shrubs. Stems quadrangular, densely hirsute when young, less so or glabrous when older. Stipules with basal sheath 3–4 mm long, densely hirsute, the lobes acicular, 3–4.5 mm long, persistent. Leaves: petioles 4–5 mm long, densely hirsute; blades lanceolate-elliptic, 6.5–11 × 2.3–4.2 cm, glabrous except for spreading-hirsutulous veins; secondary veins in 9–10 pairs. Inflorescences terminal, subcapitate, broadly oblong, 5–17 × 4–7 mm, 3-flowered, the 4 outer involucral bracts oblong-pandurate, 19 × 7 mm, pale green, the apex obtuse; peduncle 2–4 mm long, densely hirsute. Flowers: calyx 1–2 mm long, glabrous, the lobes densely ciliolate; corolla apically pubescent. Fruits unknown. Fl (Mar); known only from the types collected from the forest on Mont Galbao.

RANDIA L.

Shrubs, trees, or lianas, dioecious. Stipules interpetiolar, imbricate toward apices of stems, frequently connate or subconnate basally, simple, persistent. Leaves opposite or fasciculate. Inflorescences axillary or terminal, the staminate flowers in terminal clusters, the pistillate flowers solitary. Flowers unisexual, 5-merous; corolla lobes contorted in bud; anthers dorsifixed; ovary 1(2)-locular, the ovules numerous, in double or triple rows, horizontally oriented on 2 fleshy parietal placentas. Fruits baccate, 1(2)-locular, globose or oval, the pericarp coriaceous, crowned by persistent calyx. Seeds many, horizontally oriented, compressed, obtusely angled, embedded in gelatinous pulp.

Determinations of the two species collected in this area were made by Claes Gustafsson (GB), who is working on a treatment of the South American species of this genus.

1. Lianas. Leaf blades pubescent. Calyx lobes narrowly triangular, stiff, drying cream-colored. . . . . *R. asperifolia.*
1. Shrubs or small trees. Leaf blades glabrous. Calyx lobes linear, lax, drying pale green. . . . . *R. pubiflora.*

Randia asperifolia (Sandwith) Sandwith [Syn.: *Basanacantha asperifolia* Sandwith]

Lianas. Stems terete, densely hirtellous, armed with spines 3–5 mm long. Stipules triangular, ca. 5 mm long, imbricate at apices of stems. Leaves: petioles 6–13 mm long; blades broadly elliptic to obovate-elliptic, 7–18.5 × 3–10 cm, pubescent throughout. Inflorescences 2–5-flowered cymes. Flowers: calyx 5–6 mm long; corolla tube 2.5–3.2 cm long, green, the lobes elliptic to obovate-elliptic, 1.2 cm long, greenish-white. Fruits not known from our area. Fl (Nov); in nonflooded forest and streamside habitats.

Randia pubiflora Steyerm. FIG. 266 (stipules only)

Shrubs or small trees, to 4 m tall. Stems densely strigulose when young, becoming glabrous when mature, armed with spines 4–14 mm long. Stipules triangular, 3–4 mm long, the apex cuspidate. Leaves: petioles 2–11 mm long, densely strigulose; blades elliptic-ovate to lanceolate-elliptic, 3.5–11 × 2.2–6 cm, glabrous, the midrib and secondary veins strigulose abaxially. Inflorescences 1–3-flowered cymes. Flowers unisexual: calyx 5–16 mm long, densely strigulose; corolla tube 2–3 cm long, the lobes 1–1.6 cm long. Fruits round to ellipsoid, 2–3 cm diam., yellow to orange. Fr (Feb, Jun, Aug, Oct); in secondary vegetation and nonflooded forest.

The identity of this species is still uncertain but it is not *Randia armata* DC. nor *R. nitida* (Kunth) DC. The variation throughout its range indicates that it might represent part of a complex of several species (Claes Gustafsson, pers. comm.).

RUDGEA Salisb.

Shrubs or trees. Stipules interpetiolar, usually cartilaginous at margins, laciniate, subulate, aculeiform, aristate, or dorsal surface sometimes with teeth or appendages. Leaves opposite. Inflorescences terminal or axillary. Flowers 5-merous; calyx divided to base; corolla white, with lobes valvate in bud; stamens inserted at throat or on lower part of tube, the anthers dorsifixed; ovary 2-locular, with 1 ascending ovule in each locule. Fruits drupaceous, with 2 pyrenes, each pyrene containing one seed. Seeds with narrow groove in central cavity.

1. Inflorescences axillary. . . . . *R. stipulacea.*
1. Inflorescences terminal.
    2. Treelets to trees, 4–17 m tall. Inflorescences pedunculate. Corolla puberulent. . . . . *R. lancifolia.*
    2. Shrubs or treelets, 1–6 m tall. Inflorescences epedunculate or pedunculate. Corolla glabrous.
        3. Stipules >5 mm long. Inflorescences capitate.
            4. Stipules foliose, with fringed margins, 10–15 mm long. Leaf blades 14–23 cm long, dark green adaxially, pale green abaxially. Inflorescences subtended by foliose, ovate bracts, ca. 15–20 × 10 mm, with fringed margins. . . . . *R. bremekampiana.*
            4. Stipules not foliose, deeply fimbriate, 7–10 mm long. Leaf blades 9–12 cm long, the same shade of green ad- and abaxially. Inflorescences subtended by extremely reduced, narrowly lanceolate bracts, 1–1.5 × 0.5 mm, with entire margins. . . . . *R. guyanensis.*
        3. Stipules 2–5 mm long. Inflorescences not capitate.
            5. Stipules fimbriate, ferruginous-puberulent. Inflorescences paniculate-thyrsoid, >18-flowered, with 15–25 filiform axes, ferruginous-puberulent. . . . . *Rudgea* sp. A.
            5. Stipules suborbicular or broadly ovate. Inflorescences paniculate, 6–12(18)-flowered, with 2–5 axes (not filiform), glabrous.
                6. Stipules suborbicular, 2–3 mm long, glabrous, with dorsal oblong-ligulate appendage 1.5–2 mm long, the apex rounded. Petioles 5–6 mm long; blades drying brownish. Peduncle 3–8 cm long. . . . . *R. standleyana.*
                6. Stipules broadly ovate, ca. 5 mm long, pubescent, without dorsal appendage, the apex fringed. Petioles 8–10 mm long; blades drying pale green to pale yellowish-green. Peduncle ca. 0.5 cm long. . . . . *Rudgea* sp. B.

Rudgea bremekampiana Steyerm.

Shrubs, to 4 m tall. Stipules obovate-suborbicular, 10–15 × 5–10 mm, the margins fringed. Leaves: petioles nearly absent to 15 mm long; blades elliptic-oblanceolate to oblong-oblanceolate, 14–23 × 4.5–8 cm, dark green adaxially, pale green abaxially. Inflorescences terminal, capitate, hemispherical, densely 20–45-flowered, 15 × 25 mm, subtended by foliose bracts, the bracts ovate, ca. 15–20 × 10 mm, with fringed margins; peduncle absent to 4 cm long, glabrous. Flowers: calyx 6 mm long; corolla 5-lobed, 22 mm long, white. Fruits ovoid, ca. 1 cm diam., color unknown. Fl (Jan, fr (Aug); known from the forests of Mont Galbao and Sommet Tabulaire (just south of our area).

Rudgea guyanensis (A. Rich.) Sandwith

Shrubs, ca. 2 m tall. Stems terete or decussately compressed. Stipules oblong-ovate, sheathing at base, 7–10 × 3–5 mm, the apex long-fimbriate, the laciniate appendages 2–3 mm long. Leaves: petioles 3–5 mm long; blades elliptic, 9–12 × 3.5–6 cm. Inflorescences terminal, capitate, 9–17-flowered, subtended by extremely reduced, narrowly lanceolate bracts, 1–1.5 × 0.5 mm, with entire margins; peduncle 0.9–2 cm long, glabrous. Flowers: calyx ca. 2.5 mm long, membranous; corolla 19–25 mm long, white. Fruits ovoid, 5–7 × 4 mm, pale orange. Fl (Aug); in nonflooded forest.

Rudgea lancifolia Salisb. [Orthographic variant: *R. lanceaefolia* Salisb.] FIG. 266 (stipules only)

Treelets to trees, 4–17 m tall. Trunks angled. Outer bark smooth, scaly, brown, the inner bark yellow. Stems quadrangular. Stipules ovate, 4–10 mm long, pectinate-setaceous. Leaves: petioles 5–15 mm long; blades lanceolate, 20–23 × 9–10 cm, the midrib puberulent abaxially. Inflorescences terminal, paniculate, drying black; peduncle 2.5–10 cm, sparsely puberulent. Flowers: calyx ca. 4 mm long; corolla ca. 5 cm long, puberulent, greenish white. Fruits oblong, ca. 1 cm long, orange or whitish, turning purple. Fl (Dec), fr (Jul); in nonflooded forest.

Rudgea standleyana Steyerm.

Shrubs or treelets, 1–6 m tall. Stems glabrous. Stipules suborbicular, 2–3 × 2–3 mm, glabrous, with dorsal oblong appendage 1.5–2 mm long, the apex rounded. Leaves: petioles 5–6 mm long, glabrous; blades oblong-elliptic, 6.2–16 × 2–5 cm, glabrous. Inflorescences terminal, panicles, 4.5–6 × 5–6 cm, 6–12(18)-flowered; peduncle 3–8 cm long. Flowers: calyx 1–1.5 mm long; corolla 50–60 mm long, white. Fruits oval or oblong, ca. 1.5 × 1 cm, orange. Fl (Jan, Dec), fr (Jan, Mar, Sep); frequent, in forest on Mont Galbao.

Rudgea stipulacea (DC.) Steyerm. FIG. 266 (stipules only)

Shrubs, 1–3 m tall. Young stems terete or decussately compressed, glabrous. Stipules oblong-lanceolate, 14–15 × 4–6 mm, glabrous, striate, minutely fringed at apex. Leaves: petioles 10–15 mm long, glabrous; blades oblong-lanceolate, oblong-elliptic, or obovate, 15–28 × 4.5–13 cm, glabrous. Inflorescences axillary, long-pedunculate, shortly branched to openly paniculate, 3–1.5 × 3.5–5.5 cm, the secondary axes 3–50 mm long, glabrous; peduncle winged or angled, 6–16 cm long. Flowers: calyx ≤1.5 mm long; corolla 15–25 mm long, glabrous, white or pale green. Fruits subglobose, 7–10 × 8–10 mm, black violet to dark blue. Fl (Mar, Apr, May, Jul), fr (Jun, Jul, Aug); in nonflooded forest.

Rudgea sp. **A** (sp. nov.?)

Shrubs, 1–1.5 m tall. Young stems terete, ferruginous-puberulent. Stipules fimbriate, ca. 3.5 mm long, the sheath extremely reduced, ferruginous-puberulent, with 4–5 aristae ca. 3 mm long. Leaves: petioles 10–20 mm long; blades elliptic to broadly obovate, (10) 10–16 × (2.5)4–7.5 cm, glabrous, the apex acuminate, drying reddish-brown. Inflorescences terminal, paniculate-thyrsoid, many-flowered, ferruginous-puberulent, with 15–25 filiform axes, each terminating in a few-flowered cluster; peduncle 15–25 mm long. Flowers: calyx 0.1–0.2 mm long; corolla 3.5–4.5 mm long, glabrous, white. Fruits unknown. Fl (Jan); known from a single collection (*Granville 8729*, CAY, NY) from Mont Galbao.

Rudgea sp. **B** (sp. nov.?)

Shrubs, 1–1.5 m tall. Stems glabrous, pale green. Stipules broadly ovate, ca. 5 mm long, pubescent, the apex and margins fringed, with several extensions. Leaves: petioles 8–10 mm long; blades ovate to oblanceolate, 10–18 × 3–5.5 cm, glabrous, the apex caudate, drying pale green to pale yellowish-green. Inflorescences terminal, paniculate, 9–11-flowered; peduncle ca. 5 mm long, glabrous. Flowers unknown. Fruits ovoid, ca. 15 × 10 mm, yellowish-pale green. Fr (Mar); known from three fruiting collections (*Granville 1536* at CAY and NY, *2386* at CAY, *8912* at CAY) from the forests of Mont Galbao.

SABICEA Aubl.

Lianas. Stems woody (young stems subherbaceous), generally pubescent. Stipules interpetiolar, generally triangular, persistent. Leaves opposite. Inflorescences axillary, few-flowered, umbelliform, condensed panicles. Flowers: calyx 3–6-lobed, the lobes persistent; corolla 4–6-lobed, the lobes valvate in bud; stamens 4–6, inserted in tube, the anthers dorsifixed; ovary 3–5-locular; placentation central, the ovules many per locule, horizontally inserted. Fruits baccate, fleshy, the pericarp membranous. Seeds small, numerous, ovoid or angled, foveolate.

Andersson, L. 1999. *Sabicea. In* G. Harling & L. Andersson (eds.), Flora of Ecuador **62:** 101–114.
Wernham, H. F. 1914. A monograph of the genus *Sabicea*. British Museum (Natural History), London.

1. Stems, stipules, and leaf midribs sparsely appressed-pubescent. Corolla 10–13 mm long. . . . . *S. aspera.*
1. Stems, stipules, and leaf midribs densely spreading-hirsute. Corolla 8–10 mm long. *S. villosa.*

Sabicea aspera Aubl. FIG. 266 (stipules only)

Lianas. Stems (2)3–4 mm thick, sparsely appressed-pubescent. Stipules narrowly triangular, (4.5)7–7.5 × 2–2.5 mm, sparsely appressed-pubescent, the apex acute. Leaves: petioles 5–12 mm long, appressed-pubescent; blades ovate to narrowly ovate to narrowly elliptic, 5–12 × 2–4.5 cm, the primary and secondary venation sparsely appressed-pubescent. Inflorescences sessile to short-pedicellate, few-flowered, contracted panicles; peduncle mostly 2–4 mm long or absent. Flowers: hypanthium densely appressed-pubescent; calyx (4)5-lobed, the lobes narrowly triangular, 2–3 mm long, sparsely appressed-pubescent; corolla 10–13 mm long, sparsely erect pubescent, white. Fruits subglobose, 4–6 mm diam., red. Fl (Jan); open areas in cloud forest.

Sabicea villosa Roem. & Schult. [Syns.: *S. hirsuta* Kunth var. *adpressa* Wernham, *S. villosa* var. *adpressa* (Wernham) Standl.]

Lianas. Stems (2)3–4 mm thick, densely spreading-hirsute. Stipules narrowly triangular, 5–13 × 2.5–3 mm, densely spreading-hirsute, the apex acute. Leaves: petioles 5–12(15) mm long, densely spreading-hirsute; blades elliptic to broadly elliptic, 5.5–12(19.5) × 2.5–5(9.3) cm, the primary and secondary venation densely spreading-hirsute. Inflorescences sessile to short-pedicellate, few-flowered contracted panicles; peduncle to 3 mm long, or absent. Flowers: hypanthium densely spreading-pubescent; calyx (4)5-lobed, the lobes narrowly triangular, 2–3 mm long, sparsely to densely pubescent; corolla 8–10 mm long, sparsely erect pubescent, white. Fruits unknown from our area (globose, red, sparsely pubescent; from *Gillespie 1013*, NY, from Guyana). Fl (Jan, Mar), fr (Mar); in disturbed areas at low elevation.

SIPANEA Aubl.

Herbs to subshrubs. Stems thin, erect, pubescent. Stipules interpetiolar, generally triangular, persistent. Inflorescences terminal. Flowers: calyx 4–5-lobed, the lobes elongate; corolla lobes 5, contorted in bud; stamens 5, included, the anthers linear, dorsifixed; ovary 2-locular, the ovules numerous in each locule, horizontally inserted. Fruits loculicidal capsules, 2-locular. Seeds numerous, small, angled, the testa foveolate, reticulate.

Sipanea wilson-brownei R. S. Cowan

Herbs to subshrubs, 0.5–1.5 m tall. Stems reddish, appressed-puberulent. Stipules triangular to lanceolate, 2–5 mm long, puberulent. Leaves: petioles 5–13 cm long, pubescent; blades ovate-elliptic, elliptic-ovate, or elliptic-lanceolate, 3.5–7 × 1–3.5 cm. Inflorescences terminal, usually dichotomously branched (or rarely unbranched), the axes each 3–6-flowered, the central flower solitary; peduncle 2.5–5 cm long, pubescent. Flowers sessile; calyx lobes 3.5–8 mm long, glabrous except at margins; corolla 12–18 mm long, the tube reddish-purple to lilac, the lobes pink to white. Fruits oblong, ca. 5 × 4 mm, moderately hirsute or hispid. Fl (Jan, Jul), fr (Jan); in nonflooded forest.

SPERMACOCE L. [Syn.: *Borreria*]

Herbs to subshrubs. Stems terete or quadrangular, thin. Stipules united with petiole by basal setose sheath, the setae filiform. Inflorescences various, usually many-flowered, the flowers sessile to shortly pedicellate, surrounded by elongate fimbrillae. Flowers: calyx 2- or 4(5)-lobed; corolla 4(5)-lobed, the lobes valvate in bud, white or blue; stamens 4(5), usually exserted, the anthers dorsifixed; ovary 2-locular, with 1 ascendent ovule in each locule. Fruits dry, splitting from apex to form two cocci united at base, both cocci dehiscent (*Borreria* G. Mey. *sensu stricto*) or one dehiscent and the other indehiscent (*Spermacoce sensu stricto*). Seeds oblong, convex dorsally, with longitudinal groove, finely granulate.

Borreria is treated as a synonym of *Spermacoce* by many workers. (Adams,1999; Burger & Taylor, 1993; Fosberg et al., 1993; Govaerts, 1996; Verdcourt, 1975, 1976).

1. Erect herbs to subshrubs, to 1.5 m tall.
    2. Stipules 3–4 × 2–4 mm, with 4–9 setae, 2–10 mm long. Leaf blades with abaxial surface (or at least midrib) pubescent. Calyx lobes 4 (or rarely 5). . . . . *S. capitata.*
    2. Stipules 1.5–2 × 2–3 mm, with 5–6 setae, 1–5 mm long. Leaf blades with abaxial surface glabrous. Calyx lobes 2 or 4 (one pair smaller than other pair). . . . . *S. verticillata.*
1. Creeping herbs.
    3. Stems weakly winged. Leaf blades 1–3 × 0.2–1.5 cm. . . . . *S. ocymoides.*
    3. Stems conspicuously winged. Leaf blades (1)2.5–6 × (0.5)1.5–2 cm.
        4. Stems and leaves pubescent. Corolla white. Plants of secondary vegetation along airstrip. . . . . *S. latifolia.*
        4. Stems and leaves glabrous. Corolla pale blue. Plants of open areas within primary forest at summit of Mont Galbao. . . . . *S. alata.*

Spermacoce alata Aubl. [Syn.: *Borreria alata* (Aubl.) DC.] PL. 115c

Creeping herbs. Stems conspicuously winged, 20–50 cm long, rooting at nodes. Stipules fimbriate, 2 × 4–5 mm, glabrous, with 6–8 setae, 3–8 mm long. Leaves: blades lanceolate to ovate-elliptic, (1.5)2.5–6 × (0.8)1.5–2 cm, glabrous. Inflorescences terminal or on main or lateral branches, subhemispheric, subtended by 4 involucral bracts. Flowers: calyx 6–8 mm long; corolla 9–10 mm long, pale blue. Fruits ca. 2 mm long. Fl (Jan, Sep); in open areas within primary forest at summit of Mont Galbao.

Spermacoce capitata Ruiz & Pav. [Syn.: *Borreria capitata* (Ruiz & Pav.) DC.]

Erect subshrubs or herbs, to ca. 1 m tall. Stems quadrangular. Stipules fimbriate, 3–4 × 2–4 mm, pubescent, with 4–9 brown setae, 2–10 mm long. Leaves: blades lanceolate, 0.7–6 × 0.2–1 cm, abaxial surface (or at least midrib) pubescent. Inflorescences terminal or axillary, subhemispheric, subtended by 2–5 involucral bracts. Flowers: calyx lobes 4 (or rarely 5), 3.5–5 mm long; corolla 3.8–6.5 mm long, white. Fruits narrowly turbinate, 2–2.3 mm long. Fl (Jun); in sunny, disturbed habitats, especially along airstrip.

Spermacoce latifolia Aubl. [Syn.: *Borreria latifolia* (Aubl.) K. Schum.]

Creeping herbs. Stems quadrangular, conspicuously winged, pubescent. Stipules fimbriate, the basal vein 1–2.5 × 4 mm, densely pubescent (the hairs 1.5 mm long), with 5–7 setae, these 1.5–6 mm long. Leaves: blades elliptic-ovate to ovate, 1–5 × 0.5–2 cm, pubescent. Inflorescences axillary, sessile. Flowers: calyx 3–4 mm long; corolla 4.5–6.5 mm long, white. Fruits 2.5–4 mm long, pubescent. Fl (Jun); in sunny, disturbed habitats.

The varieties and forms that were described by Steyermark (1972) are considered to be within the morphological variation of a widespread species.

Spermacoce ocymoides Burman f. [Syns.: *Borreria ocymoides* (Burman f.) DC., *Spermacoce prostrata* Aubl., *Spermacoce gracilis* Ruiz. & Pav., *Borreria parviflora* G. Mey.]

Creeping annual herbs. Stems sharply quadrangular, weakly winged, 0.1–0.5 m long, sometimes with a few thin roots at basal nodes. Stipules fimbriate, 1.8–3.5 mm long, with ca. 7 unequal setae, 0.5–3 mm long. Leaves: blades elliptic, 1–3 × 0.2–1.5 cm, membranous. Inflorescences terminal and axillary, at last (1)2–3 terminal nodes. Flowers: calyx 2–2.5 mm long; corolla 0.5–1 mm long. Fruits urceolate, 0.6–1 mm long, membranous. Fl (Aug, Sep); in disturbed, sunny habitats and open areas within nonflooded forest.

Spermacoce verticillata L. [Syns.: *Borreria verticillata* (L.) G. Mey., *Borreria podocephala* DC.]

FIG. 266 (stipules only), PL. 115b

Erect subshrubs or herbs, to 1.5 m tall. Stipules fimbriate, 1.5–2 × 2–3 mm, with 5–6 white setae. Leaves: blades lanceolate, 1.4–5.2 × 0.1–1 cm, glabrous. Inflorescences terminal and axillary, at last (1)2–3 terminal nodes, subtended by 1–4 involucral bracts. Flowers: calyx lobes 2 or 4 (one pair smaller than other), 3–3.2 mm long; corolla 1.2–3 mm long, white. Fruits widely oblong to subglobose, 1.5–2 mm long. Fl (Jun, Aug, Sep, Oct), fr (Jun); in disturbed habitats.

TOCOYENA Aubl.

Trees or shrubs. Stipules interpetiolar, triangular, persistent. Inflorescences terminal. Flowers: calyx 5–6-dentate, persistent; corolla large, showy, fragrant, white or creamy white distally and yellow at base (yellow after anthesis), the lobes 5, contorted in bud; stamens inserted at mouth of corolla, exserted; anthers dorsifixed; ovary 2-locular, with many ovules per locule. Fruits baccate, with many seeds, the pericarp thick-coriaceous. Seeds subcompressed, horizontal, embedded in fleshy pulp.

Tocoyena longiflora Aubl.

Treelets, to 3.5 m × 2 cm. Stipules triangular, ca. 1 cm long. Leaves: petioles 15–30 mm long; blades oblanceolate, ca. 40 × 15 cm, glabrous. Inflorescences terminal, with ca. 11 flowers. Flowers nocturnal; calyx ca. 10 mm long, the lobes ca. 2.5 mm long; corolla tube ca. 30 cm long, glabrous, the lower half yellow, the upper half white, lobes ca. 1 cm long, white. Fruits glabrous. Fl (Jun), fr (Aug, Sep); in nonflooded forest.

UNCARIA Schreb.

Lianas, climbing by means of paired, curved to coiled thorns. Stipules interpetiolar, simple, ovate. Inflorescences axillary, dense globose heads. Flowers white in bud, turning yellowish-orange at anthesis; calyx truncate or 5-lobed; corolla hypocrateriform, pubescent abaxially, the lobes 5, valvate; stamens 5, inserted on tube; ovary fusiform, 2-locular. Fruits capsules, dehiscing into two valves. Seeds numerous, imbricate, with bipolar wings.

Andersson, L. 1994. *Uncaria in* Rubiaceae (part 3). *In* G. Harling & L. Andersson (eds.), Flora of Ecuador **50:** 106–109.

1. Thorns strongly curved to coiled, glabrous. Leaf secondary and tertiary veins glabrous abaxially. Flowers pedicellate; calyx 3.5–4 mm long. . . . . *U. guianensis*.
1. Thorns slightly curved, sparsely short-pubescent. Leaf secondary and tertiary veins appressed short-pubescent abaxially. Flowers sessile; calyx 0.1–0.3 mm long. . . . . *U. tomentosa*.

Uncaria guianensis (Aubl.) J. F. Gmel.

FIG. 266 (stipules only); PART 1: FIG. 11

Lianas. Stems quadrangular, glabrous. Thorns strongly curved to coiled, glabrous. Leaves: blades ovate to lanceolate-elliptic, 7–10 × 3.5–5 cm; secondary and tertiary veins glabrous, the secondary veins in 5–6 pairs. Inflorescences heads (2)2.5–3 cm diam. at anthesis. Flowers pedicellate; calyx 5-lobed, 3.5–4 mm long, puberulent, the lobes 0.2–0.6 mm long; corolla tube 4–5 mm long, densely antrorse sericeous, the lobes 5, oblong-ovate, 1.5–3 mm long. Fruits fusiform, 1.2–2 cm long (*Clarke 6441*, NY). Fr (Apr); uncommon, known only from our area by one collection from secondary forest. *Piquant guadeloupe* (Créole).

Uncaria tomentosa (Roem. & Schult.) DC. [Syns.: *Uncaria surinamensis* Miq., *U. tomentosa* (Roem. & Schult.) DC. var. *dioica* Bremek.]

Lianas. Stems quadrangular, glabrous. Thorns slightly curved, sparsely short-pubescent. Leaves: blades ovate to lanceolate-elliptic, 7–13.5 × 2.5–7 cm; secondary and tertiary veins appressed short-pubescent, the secondary veins in 5–6 pairs. Inflorescences, flowers and fruits unknown from our area (descriptions of inflorescence and flower from *Aristeguieta & Zabala 7000*, NY, and fruit from *Schunke Vigo 13489*, NY). Inflorescences heads, 1.5–2 cm diam. at anthesis. Flowers sessile: calyx truncate or 5-lobed, 0.1–0.3 mm long, puberulent, the lobes absent (margin undulate) to extremely reduced; corolla tube 4–6 mm long, short-pubescent, the lobes 5, oblong-ovate, ca. 1.5 mm long. Fruits fusiform, 6–11 mm long. Rare, known only from two sterile collections from primary nonflooded forest.

RUTACEAE (Rue Family)

Jacquelyn A. Kallunki

Shrubs or trees. Leaves pinnate, 3-foliolate, or simple, alternate, often aromatic, with often pellucid, glandular dots on blade. Inflorescences terminal, axillary, or lateral, cymose, variously arranged. Flowers actinomorphic, bisexual or unisexual (dioecious in *Zanthoxylum*); calyx 5-parted (large and pink-red in *Erythrochiton*); petals 5, free (*Esenbeckia*, *Zanthoxylum*), coherent at anthesis (*Conchocarpus*, *Ticorea*), or connate (*Erythrochiton*); stamens 5; nectary disc intrastaminal; ovary superior, 2-or 5-carpelled, the carpels free or axially connate (*Ticorea*); placentation axile, the ovules 2 per locule. Fruits of 1–5 separate mericarps or 5-parted capsules (*Esenbeckia*), each carpel loculicidally dehiscent along axile suture, in all but *Zanthoxylum* with a bony, elastic, separating endocarp that ejects the seed. Seeds 1 or 2 per carpel.

Cowan, R. S. 1967. Rutaceae of the Guayana Highland. Mem. New York Bot. Garden **14(3):** 1–14.

1. Trunks with prickles. Leaves pinnate. Flowers unisexual. Seeds globose, shiny black. . . . . *Zanthoxylum.*
1. Trunks without prickles. Leaves 3-foliolate or simple. Flowers bisexual. Seeds ovoid, elliptic, or oblong, brown.
    2. Leaves 3-foliolate; blades with circular pits (domatia) in axils of secondary veins abaxially, the glandular dots blackish. . . . . *Ticorea.*
    2. Leaves 1-foliolate; blades without domatia, the glandular dots visible or not but not blackish.
        3. Leaf blades oblanceolate, the base long-cuneate. Inflorescences lateral, perennating, cauline with age, bearing flowers on short-shoots. Calyx 30–38 mm long, red to pink-red in flower; corolla 43–75 mm long. . . . . *Erythrochiton.*
        3. Leaf blades elliptic to narrow-ovate, the base acute. Inflorescences terminal or lateral but not perennating, bearing flowers on rachises with normally elongate internodes. Calyx 1–5 mm long, green or presumed to be green in flower; corolla 2–20 mm long.
            4. Leaf blades 15–33 cm long, the glandular dots visible on both surfaces, pellucid especially in older leaves. Inflorescences terminal, long-pedunculate, several-branched dichasia. Petals 20 mm long. Fruits of 1–5 free mericarps. *Conchocarpus.*
            4. Leaf blades 7–15 cm long, the glandular dots barely visible on abaxial surface, not pellucid. Inflorescences lateral, axillary dichasia. Petals 2 mm long. Fruits 5-parted capsules. . . . . . . *Esenbeckia.*

CONCHOCARPUS J. C. Mikan

Shrubs or trees. Leaves alternate, 1-foliolate; petioles ± swollen at apex. Inflorescences terminal, several-branched dichasia. Flowers: calyx cupular, 5-dentate; petals initially coherent into tube, ultimately free, creamy white, densely pubescent on both surfaces; filaments initially coherent to corolla by a dense pubescence but ultimately free; nectary disc thin-cupular, glabrous, shorter than ovary; ovary of 5, sparsley pubescent carpels, free except at apex where they join style. Fruits of 1–5, free mericarps, each without longitudinal ridges, rounded on back in transverse section. Seeds smooth, glabrous, brown.

Conchocarpus guyanensis (Pulle) Kallunki & Pirani [Syn.: *Almeida guyanensis* Pulle] FIG. 280

Shrubs or trees, to 8 m tall. Leaves: blades elliptic to narrow-ovate, 15–33 × 4.8–12 cm, glandular dots visible on both surfaces, pellucid especially in older leaves, the base acute, the apex long-acuminate. Inflorescences few-branched, long-pedunculate dichasia; peduncle 13–23 cm long. Flowers: calyx 4–5 mm long, presumed to be green, with rounded teeth, persistent below fruit; petals narrow-oblanceolate, ca. 20 × 3.5 mm wide; filaments barbate at midlength adaxially, the anthers lanceolate, 4–5 mm long. Fruiting mericarps 14–17 mm high, 6–7 mm deep, glabrous. Fl (Apr, Nov), fr (Mar, Jul, Sep, Oct); in nonflooded forest.

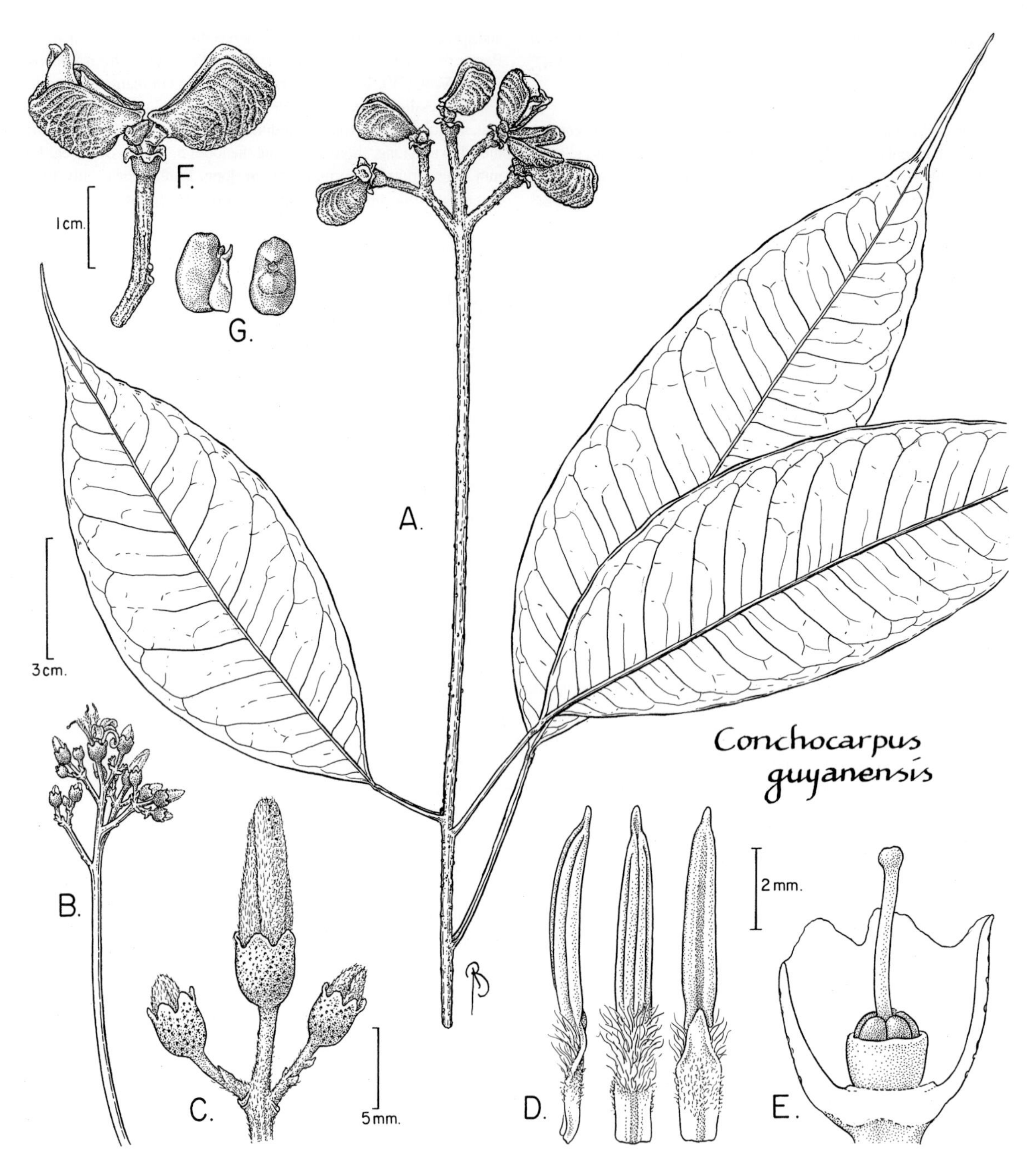

FIG. 280. RUTACEAE. *Conchocarpus guyanensis* (A, F, G, *Prévost 309*; B–E *Sabatier 1066*). **A.** Part of stem with leaves and infructescence. **B.** Inflorescence. **C.** Detail of inflorescence in bud. **D.** Lateral (near right), adaxial (middle right), and abaxial (far right) views of stamens. **E.** Lateral view of nectary disc and gynoecium surrounded by part of calyx. **F.** Dehisced mericarps. **G.** Lateral (above left) and adaxial (above right) views of seed.

ERYTHROCHITON Nees & Mart.

Shrubs or trees, unbranched or few-branched. Leaves alternate, 1-foliolate, often clustered at apex of trunk or branch; petioles ± swollen at apex. Inflorescences lateral but not obviously axillary. Flowers: calyx tubular, variously lobed; petals connate for ca. 1/2 length, glabrous on both surfaces, white; filaments adnate to corolla; nectary disc tubular, glabrous, taller than ovary; ovary of 5, glabrous carpels free except at apex where they join style and sometimes at base. Fruits of 5, free mericarps, each longitudinally ridged where it abuts adjacent carpels, angular on back in transverse section. Seeds ovate, tuberculate, tomentulose, pale brown.

Kallunki, J. A. 1992. A revision of *Erythrochiton sensu lato* (Cuspariinae, Rutaceae). Brittonia **44:** 107–139.

Erythrochiton brasiliensis Nees & Mart. FIG. 281

Shrubs or trees, to 12 m tall. Leaves: blades oblanceolate 15–58 × 3.5–8.8 cm, glandular dots obscure, more easily seen on abaxial surface of younger leaves, not obviously pellucid at any stage, the base long-cuneate, the apex acuminate to acute. Inflorescences cauline, initially pedunculate dichasia, in age often perennating, branched, 8–73 cm long; peduncle and rachis 3–4-angled; pedicels present, the flowers on short-shoots. Flowers: calyx 30–38 mm long, red to pink-red, irregularly splitting as fruit develops, ultimately deciduous from mature fruit; petals 43–75 mm long, the tube ca. 35 mm long, the lobes spreading, ca. 21 × 7.5 mm; anthers white, narrow-oblong to lorate, 5–7 mm long. Fruiting mericarps 12–16 mm high, 7–9 mm deep, glabrous or sparsely puberulent, red when fresh. Fl (Feb, Oct); on rocky hills.

ESENBECKIA Kunth

Shrubs. Leaves alternate, 1-foliolate; petioles eared at apex. Inflorescences lateral, axillary. Flowers: calyx of 5, free sepals; petals free, glabrous adaxially, pubescent abaxially, greenish white; filaments free; nectary disc thick-cupular, pubescent, equalling ovary; ovary of 5, connate, pubescent carpels. Fruits 5-parted capsules, round in transverse section. Seeds elliptic or oblong, smooth, glabrous, brown.

Kaastra, R. C. 1982. Pilocarpinae (Rutaceae). Fl. Neotrop. Monogr. **33:** 1–198.

Esenbeckia cowanii Kaastra FIG. 282

Shrubs, 3 m tall. Leaves: blades elliptic, 7–15 × 3–4.7 cm, glandular dots barely visible on abaxial surface, not pellucid, the base acute, the apex acute or acuminate. Inflorescences axillary dichasia, the subtending leaf sometimes reduced and inflorescence appearing internodal; peduncle 0.5–2 cm long. Flowers: calyx ca. 1 mm long, with acute sepals, persistent below fruit; petals narrow-ovate, ca. 2 × 1 mm; filaments glabrous, the anthers cordate, apiculate, ca. 0.6 mm long. Fruits 18–20 × 16–20 mm, appressed-pubescent. Fr (Aug); on granitic outcrops with large boulders.

TICOREA Aubl.

Shrubs or trees. Leaves alternate, 3-foliolate; petioles not swollen at apex (at least in dried specimens); blades with circular pits (domatia) in axils of secondary veins on abaxial surface. Inflorescences terminal. Flowers: calyx cupular, 5-dentate; petals coherent into tube at anthesis, ultimately free, pubescent on both surfaces, white; filaments initially coherent to corolla by pubescence but ultimately free; nectary disc cupular, short-pubescent on rim, equal to ovary in height; ovary of 5, glabrous carpels connate axially but free laterally. Fruits of 1–5 mericarps, initially ± connate axially, separating from each other when developing unequally and at dehiscence, each with one longitudinal dorsal ridge and two lateral ridges where abutting adjacent carpels. Seeds oblong, smooth, glabrous, brown.

Kallunki, J. A. 1998. Revision of *Ticorea* Aubl. (Rutaceae, Galipeinae). Brittonia **50(4)**: 500–513.

Ticorea foetida Aubl. FIG. 283; PART 1: PL. XIIc

Shrubs or trees, 1.5–14 m tall. Leaves: leaflet blades narrow-obovate or elliptic, the terminal ones 11–30 × 4.7–11 cm, the laterals 7–25.5 × 2.4–10.5 cm, glandular dots blackish, visible on both surfaces, not pellucid, the base acute, the apex long-acuminate. Inflorescences few-branched, long-pedunculate dichasia; peduncle 13.5–30 cm long. Flowers: calyx 2.5–3.9 mm long, with acute teeth, persistent below fruit; petals linear, 17–24 × 1.5–2.2 mm wide; anthers elliptic, 4–5.3 mm long, sterile at base above point of attachment to filament, with bifid basal appendages 0.6–1.2 mm long. Fruiting mericarps 10–12 mm high, 7–10 mm deep, short-pubescent, glabrescent except for adjacent lateral faces. Fl (Jan, Feb, Mar, Jul), fr (Mar, May, Jun, Jul, Oct); locally common, in understory of nonflooded forest, sometimes in secondary vegetation.

ZANTHOXYLUM L.

Dioecious trees. Trunks prickly. Leaves alternate, pinnate; leaflets opposite to alternate. Inflorescences terminal. Flowers: sepals free; petals free, white or yellowish white, glabrous; stamens free (absent in pistillate flowers); nectary disc in pistillate flowers cushion-like, subtending ovary, absent in staminate flowers; ovary in pistillate flowers of 2 free carpels, each with or without style but joined to single peltate stigma, in staminate flowers rudimentary, conical, tapering to 2 styles but lacking stigma. Fruits of 1, globose, 1-seeded mericarp. Seeds ± globose, smooth, glabrous, shiny black.

1. Leaves glabrous. . . . . *Zanthoxylum* sp. A.
1. Leaves stellate-pubescent.
    2. Leaflets densely stellate-pubescent abaxially, the margins entire, the largest ones 7.5–16.5 cm long, the petiole and rachis 17–60 cm long. Rudimentary ovary in staminate flowers and nectary disc in pistillate flowers densely pubescent. . . . . *Z. ekmanii.*
    2. Leaflets with scattered stellate hairs abaxially, the margins crenate, the largest ones 3.3–6.9 cm long, the petiole and rachis 3–14 cm long. Rudimentary ovary in staminate flowers and nectary disc in pistillate flowers glabrous. . . . . *Z. rhoifolium.*

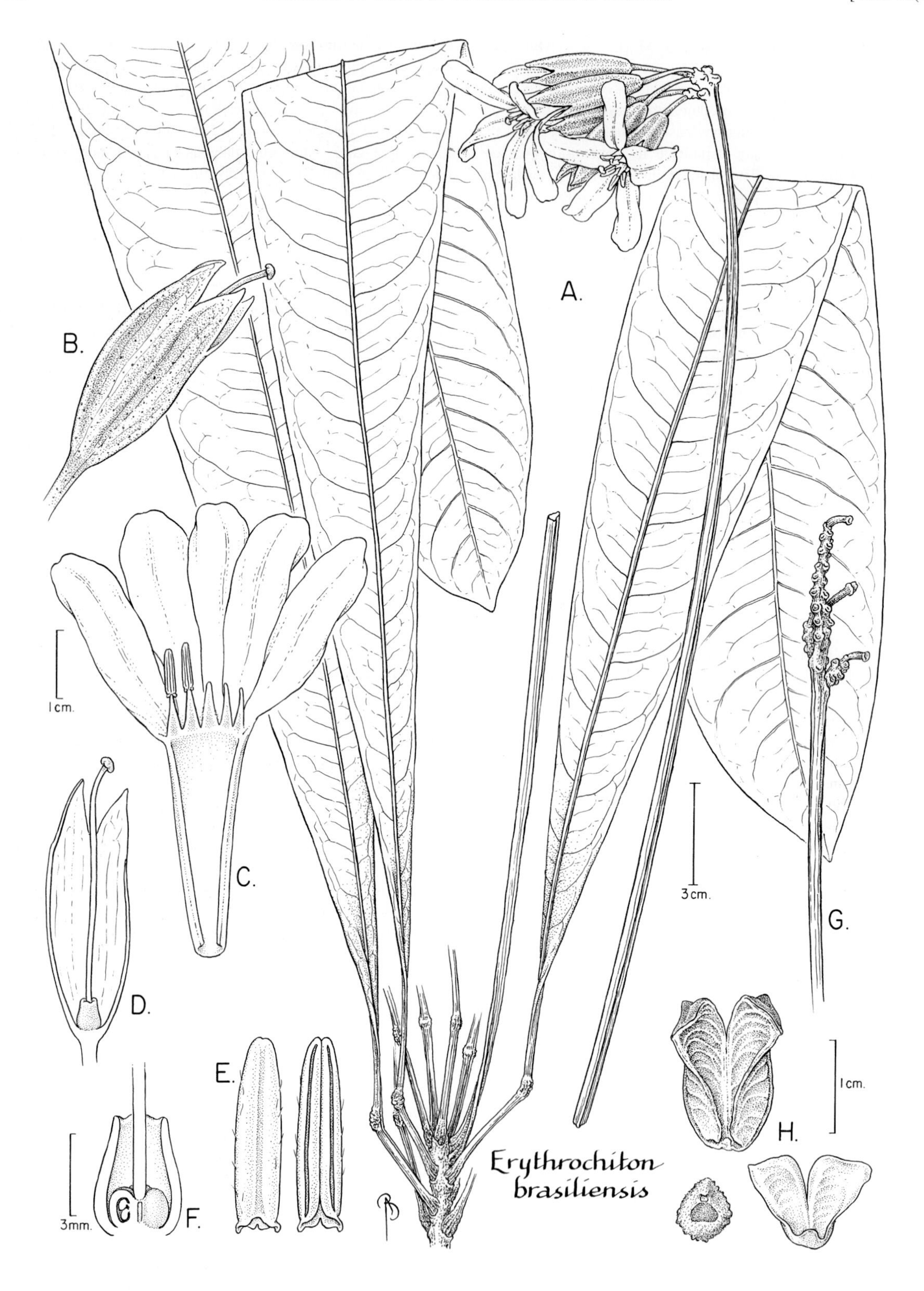

FIG. 281. RUTACEAE. *Erythrochiton brasiliensis* (A, *Oldeman 2047* and *Kallunki 609* from Brazil; B–F, *Kallunki 712* from Brazil; G, H, *Granville 7476*). **A.** Apex of stem with leaves and inflorescence. **B.** Lateral view of calyx with exserted style. **C.** Corolla opened to show adnate stamens (three anthers fallen). **D.** Lateral view of nectary disc and style surrounded by part of calyx. **E.** Abaxial (near right) and adaxial (far right) views of anthers. **F.** Medial section of disc, ovary (with one carpel shown in medial section), and base of style surrounded by part of calyx. **G.** Old inflorescence. **H.** Dehisced mericarp (left), endocarp (below right), and seed (below left).

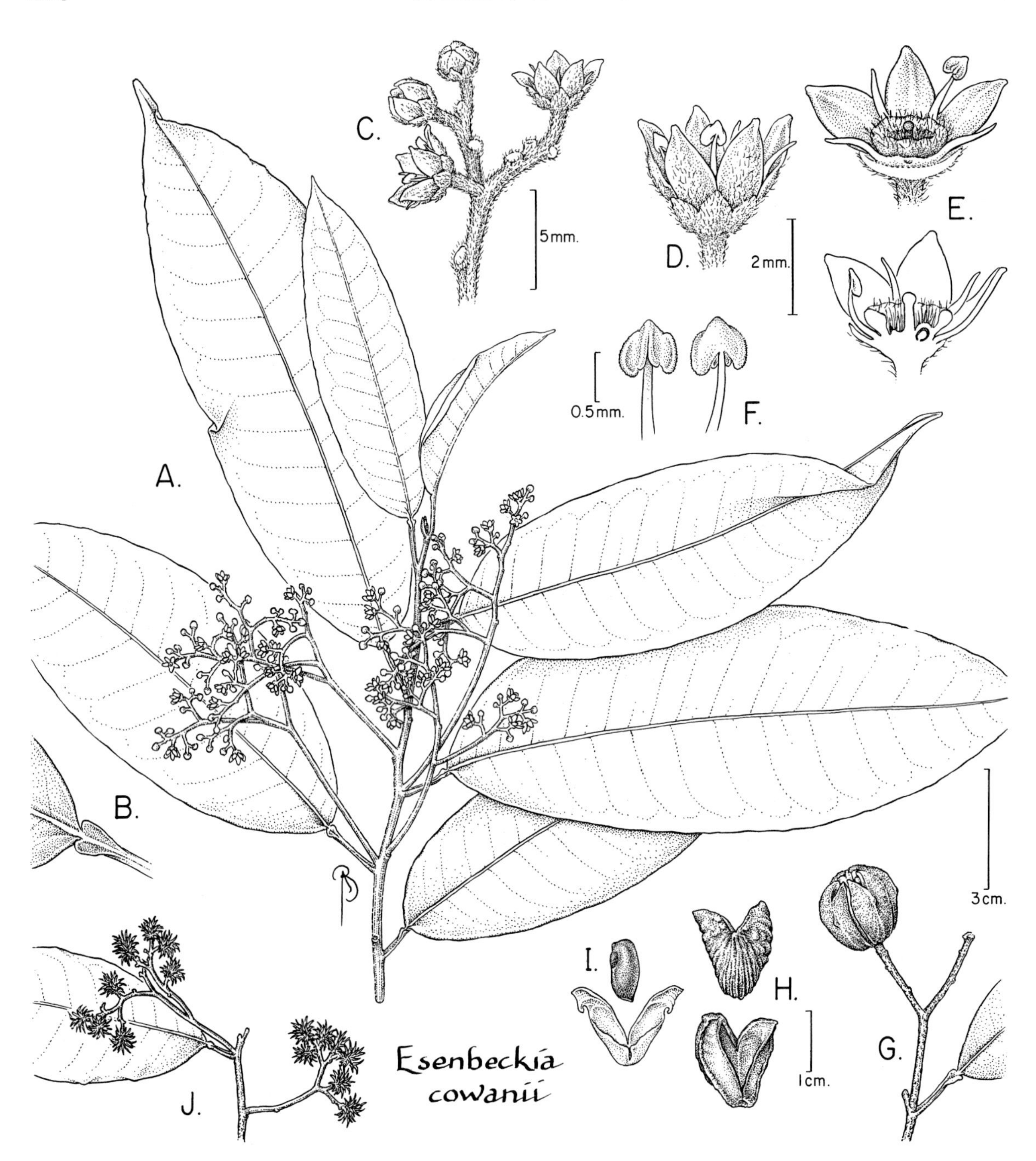

FIG. 282. RUTACEAE. *Esenbeckia cowanii* (A–F, *Cid et al. 3130* from Acre, Brazil; G, *Mori et al. 23260*; H, I, *A. Rosas Jr. et al. 259* from Acre, Brazil; J, *Cowan 38758* from Kaw Mountains, French Guiana). **A.** Part of stem with leaves and inflorescences. **B.** Junction of blade and petiole showing flared petiole apex. **C.** Detail of inflorescence. **D.** Lateral view of flower. **E.** Lateral view of flower with two petals removed (above) and medial section of flower (below); note the well developed disc. **F.** Adaxial (far left) and abaxial (near left) views of stamens. **G.** Fruit on part of infructescence. **H.** Abaxial (above left) and adaxial (below left) views of dehisced mericarp. **I.** Seed (right) and endocarp (below right). **J.** Part of stem with leaf base and inflorescences of galled flowers.

Zanthoxylum ekmanii (Urb.) Alain FIG. 284, PL. 115d

Trees, 8–30 m tall. Leaflets 10–20; petioles and rachis 17–60 cm long; blades elliptic, largest ones 7.5–16.5 × 3.4–5.5 cm, stellate hairs on both surfaces, glabrescent adaxially, persistently and densely pale-stellate abaxially, glandular dots blackish adaxially, obscured by pubescence abaxially, pellucid, the base rounded (obtuse), the apex usually short-acuminate (obtuse), the margins ± entire. Inflorescences panicles. Flowers: sepals 5, ca. 0.25 mm long, acute, short-ciliate, persistent below fruit; petals narrow-ovate, 1.2–1.7 × ca. 0.6–0.8 mm. Staminate flowers: anthers deeply cordate, ca. 0.5 mm long; rudimentary ovary densely pubescent. Pistillate flowers: nectary disc densely pubescent; each carpel lacking a style, the stigma sessile. Fruiting mericarps ca. 4 mm diam., glabrous, with many small glandular dots. Fl (Mar, Apr), fr (Mar, Jun); in secondary or disturbed vegetation, nonflooded forest.

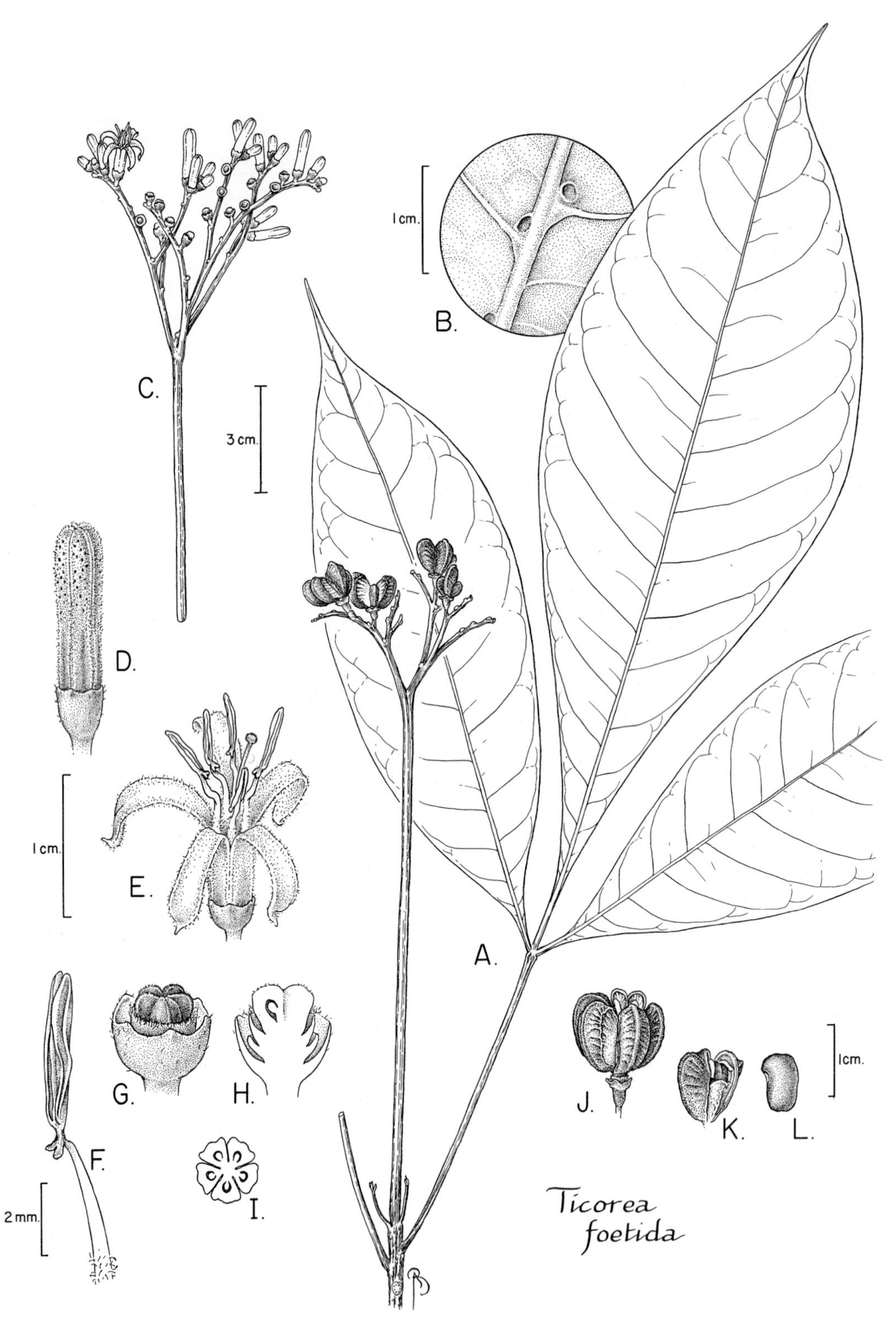

FIG. 283. RUTACEAE. *Ticorea foetida* (A–C, *Mori & Pipoly 15376*; D–I, *Loizeau & Loizeau 620*; J, *Mori & Pennington 17934*; K, L, *Mori & Hartley 18459*). **A.** Apex of stem with leaf and infructescence. **B.** Detail of abaxial leaflet surface showing domatia in axils of secondary veins. **C.** Inflorescence. **D.** Lateral view of flower bud. **E.** Lateral view of flower. **F.** Apex of filament with anther. **G.** Lateral view of flower with corolla removed showing the 5-merous, apocarpous gynoecium. **H.** Medial section of flower with corolla and androecium removed. **I.** Transverse section of ovary. **J.** Lateral view of fruit of distinct, dehiscent mericarps. **K.** Dehisced mericarp showing part of bony endocarp and seed. **L.** Seed. (Reprinted with permission from J. A. Kallunki, Brittonia 50(4), 1998.)

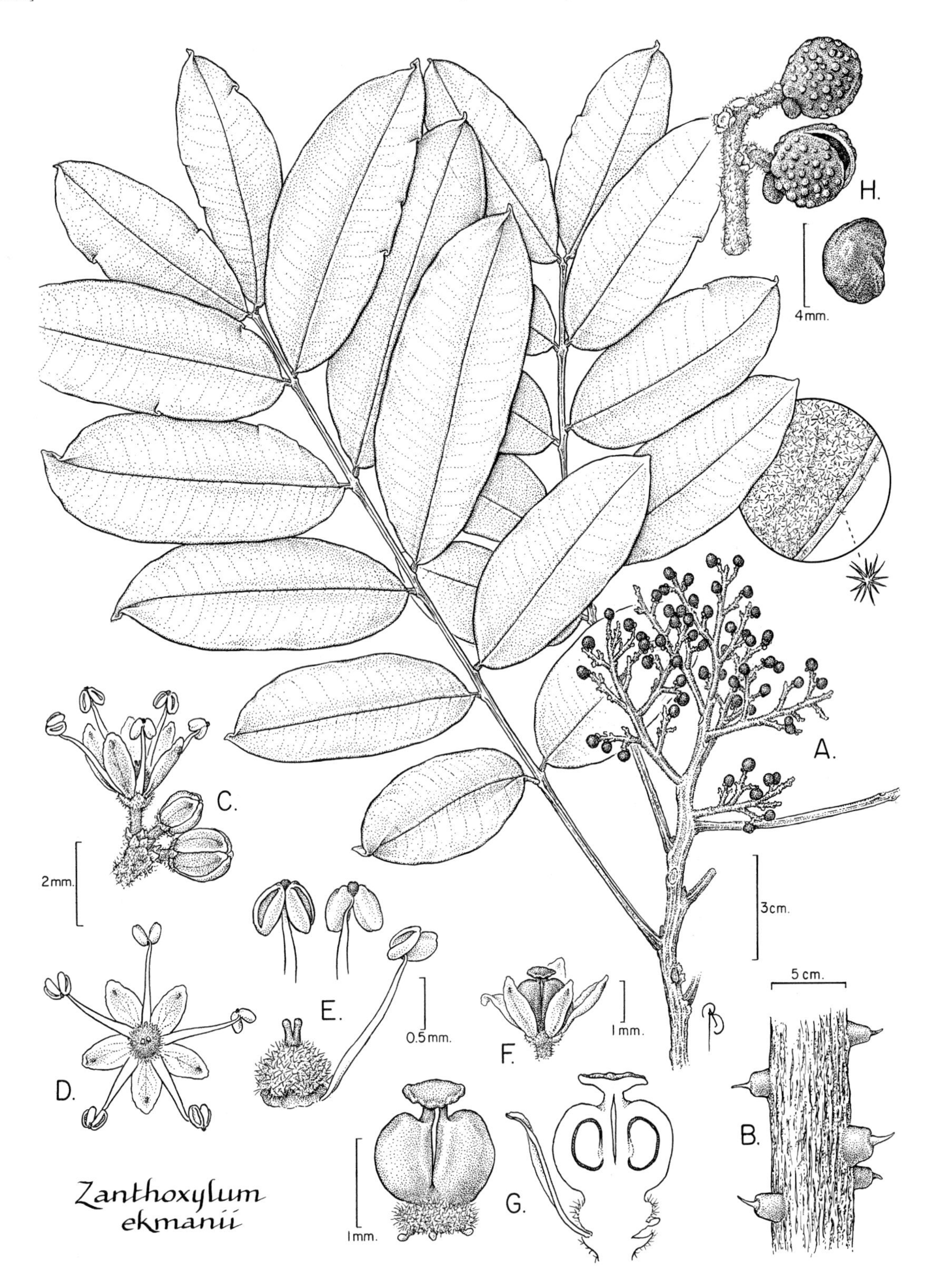

FIG. 284. RUTACEAE. *Zanthoxylum ekmanii* (A, *Mori & Gracie 19011*; B, *Mori et al. 21661*; C–E, *Mori & Pipoly 15603*; F, G, *Bernardi 2125* from Venezuela; H, *Rabelo et al. 2311* from Brazil). **A.** Part of stem with leaves and infructescence and detail of abaxial surface of leaf showing enlargement of stellate hair. **B.** Prickles on trunk. **C.** Detail of inflorescence. **D.** Apical view of staminate flower. **E.** Pistillode and one stamen (below) and apical portion of two filaments with adaxial (above left) and abaxial (above right) views of anthers. **F.** Lateral view of pistillate flower. **G.** Lateral view (left) and medial section of pistil (right). **H.** Two fruits on infructescence (left) and seed (below).

Zanthoxylum rhoifolium Lam.

Trees, 9–35 m tall. Leaflets 5–14; petioles and rachis 3–14 cm long; blades elliptic, largest ones 3.3–6.9 × 1.7–2.2 cm, with scattered stellate hairs on both surfaces, glandular dots usually visible in sinuses and on both surfaces, pellucid, the base obtuse, the apex usually acute, the margins crenate. Inflorescences panicles. Flowers: sepals 5, ca. 0.2 mm long, acute, short-ciliate, persistent below fruit; petals narrow-ovate, 1.3–1.5 × ca. 0.5 mm. Staminate flowers: anthers deeply cordate, ca. 0.6–0.75 mm long; rudimentary ovary glabrous. Pistillate flowers: nectary disc glabrous; each carpel with very short style, the stigma sessile. Fruiting mericarps ca. 3 mm diam., glabrous, with a few relatively large, circular glands. Fl (Feb, Dec), fr (Mar); in secondary or disturbed vegetation. *Bois piquant, bois zépine* (Créole), *tamanqueira* (Portuguese), *zépini titefeuille* (Créole).

Zanthoxylum sp. A

Treelets, 4 m tall. Leaflets 10; petioles and rachis 56 cm long; blades elliptic, largest ones in the only known leaf 18 × 7 cm, glabrous, glandular dots visible in sinuses and on both surfaces, pellucid; the base acute, the apex acuminate, the margins crenulate. Known only from a single sterile collection (*Mori 21661*, NY).

SAPINDACEAE (Soap-berry Family)

Pedro Acevedo-Rodríguez

Trees, shrubs, lianas, or less often herbaceous vines. Stems of lianas usually with multiple steles, and very often with white, milky latex. Stipules minute to large, present only in climbing species. Leaves pinnately or ternately compound, rarely simple or unifoliolate, alternate, spirally arranged, rarely opposite; leaf rachis of most arborescent species with terminal, rudimentary leaflet. Inflorescences axillary, terminal, or cauliflorous, racemose, paniculate, or spicate thyrses; peduncles angular to terete, bracts usually inconspicuous; pedicels articulate near base or not articulate. Flowers actinomorphic or zygomorphic, bisexual or unisexual (plants dichogamous, monoecious, or dioecious); sepals 4–5, distinct or less often connate; petals 4–5, distinct, usually white or yellowish, with adnate adaxial appendage, or petals auriculate (involute basal margins); appendages petaloid, simple, bifurcate, or hood-shaped with glandular apex, variously pubescent; nectary disc extrastaminal, annular or unilateral, entire or lobed; stamens (4–7)8(10), the filaments equal or unequal in length, free or connate at base, the anthers dorsifixed or basifixed, introrse, opening by longitudinal slits; ovary superior, of (2)3(5) connate carpels, with same number of locules, the style usually present, the stigmas elongate or capitate; placentation axial, the ovules 1 or 2 per locule. Fruits capsules, schizocarps, or indehiscent and baccate, winged or not, sometimes echinate. Seeds usually one per locule, sometimes arillate, or with fleshy testa.

Kramer, K. U. 1976. Sapindaceae. *In* J. Lanjouw & A. L. Stoffers (eds.), Flora of Suriname **II(2):** 487–511. E. J. Brill.
Uittien, H. 1966a. Sapindaceae. *In* A. Pulle (ed.), Flora of Suriname **II(1):** 345–396. E. J. Brill, Leiden.

1. Lianas or vines, often bearing pair of coiled tendrils at base of inflorescence rachis. Stems usually producing milky sap when cut.
    2. Fruits 3-winged schizocarps, splitting into 3, winged, indehiscent mericarps. Seeds exarillate.
        3. Leaves biternate. Thyrses racemose or paniculate. Flowers zygomorphic, with unilateral nectary disc modified into 2–4 glands. Mericarp wing proximal. . . . . *Serjania.*
        3. Leaves trifoliolate. Thyrses with umbellate units. Flowers actinomorphic, with annular nectary disc. Mericarp wing distal. . . . . *Thinouia.*
    2. Fruits dehiscent (capsules), the locules opening to expose seeds. Seeds arillate. . . . . *Paullinia.*
1. Shrubs or trees, without tendrils. Stems not producing milky sap when cut.
    4. Leaves trifoliolate or unifoliolate, the terminal leaflet well developed. . . . . *Allophylus.*
    4. Leaves pinnate, the terminal leaflet rudimentary.
        6. Fruits indehiscent (baccate). Seeds completely covered by fleshy, juicy testa. . . . . *Talisia.*
        6. Fruits dehiscent (capsular). Seeds with basal, fleshy arillode.
            7. Petals without appendages. Capsules coriaceous, the cocci 2(3), one usually larger. . . . . *Pseudima.*
            7. Petals with adaxial appendages. Capsules usually woody, less often coriaceous, the cocci indistinguishable, or if distiguishable, then of similar sizes.
                8. Ovary with two carpels. . . . . *Vouarana.*
                8. Ovary with three carpels.
                    9. Leaflets serrate, serrate-dentate, or rarely entire. Calyx with imbricate sepals. . . . . *Cupania.*
                    9. Leaflets usually entire. Calyx with valvate sepals. . . . . *Matayba.*

ALLOPHYLUS L.

Small trees or understory shrubs. Stipules wanting. Leaves alternate, trifoliolate or less often unifoliolate, usually long-petiolate. Inflorescences racemose or paniculate thyrses. Flowers zygomorphic, functionally pistillate or staminate, calyx 4(5)-merous, with bilateral or dorsiventral-transverse symmetry; petals 4, auriculate or with bifid appendage; nectary disc unilateral, semiannular, or 4-lobed; stamens 8, the filaments unequal, connate at base into short tube; ovary 2(3)-locular, with single ovule per locule. Fruits indehiscent, baccate, green, yellow, orange, or red, of 1–2 ellipsoid or ovoid monocarps, one usually rudimentary.

1. Leaves unifoliolate. . . . . *A. leucoclados.*
1. Leaves trifoliolate.
   2. Inflorescences unbranched (racemose). Pistillate flowers with stipitate ovary (androgynophore). . . . . *A. latifolius.*
   2. Inflorescences branched (at least with a basal branch). Pistillate flowers with sessile ovary.
      3. Leaflets elliptic, oblong, or oblanceolate, the margins entire or nearly so (obscurely serrate). Inflorescences tomentulose. . . . . *A. robustus.*
      3. Leaflets obovate to oblong, the margins repand to obtusely serrate. Inflorescences puberulent. . . . . *A. angustatus.*

Allophylus angustatus (Triana & Planch.) Radlk.

Trees, to 15 m tall, glabrous (except for puberulent inflorescences and young stems). Leaves trifoliolate, seldom unifoliolate; petioles terete, 2–4 cm long; leaflets obovate to oblong, 11–21 × 5–9.5 cm, chartaceous, the base nearly cuneate, or sometimes nearly falcate on lateral ones, the apex obtusely acuminate, the margins repand to obtusely serrate. Inflorescences paniculate thyrses, with 1–3 branches when axillary, or paniculate when terminal, 3–13 cm long; cincinni sessile or nearly so, 3-flowered. Flowers fragrant; calyx with dorsiventral-transversal symmetry, light green, puberulent, the sepals 4, concave, rounded, ciliate, outer sepals and lower inner sepal 1.5–1.7 mm long, upper inner sepal 2–2.2 mm long, notched at apex; petals white, spatulate, ca. 2.2 mm long, auriculate, sericeous on upper half; nectary disc divided into 4 glands; filaments glabrous to puberulent; ovary appressed pubescent. Monocarps obovoid, glabrous, costate at base, 1.2 cm long. Fl (Aug), fr (Aug, Nov); in nonflooded forest.

Allophylus latifolius Huber Pl. 115e

Trees, 3–7 m tall. Stems tomentulose, becoming glabrous with age. Leaves trifoliolate; petioles terete, appressed pubescent, 3.1–7 cm long; leaflets elliptic, lanceolate, or oblanceolate, 8–15 × 3.1–6.7 cm, chartaceous, glabrous except for sparingly pubescent veins abaxially, the base obtuse to nearly cuneate on terminal leaflets, obtuse-rounded on lateral ones, the apex long acuminate to cuspidate, the margins obtusely serrate to entire. Inflorescences racemose thyrses, 7–15 cm long; cincinni sessile, 2–3-flowered. Flowers: calyx with bilateral symmetry, light green, glabrous or puberulent, the sepals 4, concave, rounded, ciliate, 1–1.2 mm long, outer sepals slightly smaller than the inner ones; petals white, spatulate, ca. 1.2 mm long, auriculate, sericeous on upper half; nectary disc divided into 4 glands; filaments villosulous; ovary sericeous. Pistillate flowers with short androgynophore bearing staminodes and gynoecium. Monocarps obovoid, 1.2–1.5 cm long, green, glabrous. Fl (May, Aug), fr (Apr, Aug, Nov); in nonflooded forest.

Allophylus leucoclados Radlk.

Understory shrubs. Stems tomentulose, becoming glabrous with age. Leaves unifoliolate; petioles 6–12 mm long; leaflets elliptic, 12.5–21 × 4.5–7.5 cm, membranous, glabrous, the base acute, the apex acuminate to shortly so, the margins serrate. Inflorescences racemose thyrses, 2.5–3.5 cm long, tomentulose; cincinni sessile. Monocarps obovoid, 1–1.2 cm long, yellow, glabrous. Fr (May); known from our area by a single collection, in nonflooded forest.

Allophylus robustus Radlk.

Shrubs, 3–4 m tall. Stems tomentulose, becoming glabrous with age. Leaves trifoliolate; petioles terete, tomentulose, 6–13 cm long; leaflets elliptic, oblong or oblanceolate, 13–25 × 5.2–9.5 cm, chartaceous, glabrous except for tomentulose prominent veins abaxially, the base attenuate into elongate petiolule, acute to nearly cuneate on terminal leaflets, unequal on lateral leaflets, the apex short acuminate to obtuse, mucronate, the margins entire or obscurely serrate. Inflorescences paniculate thryses (1 or 2 branches from base), 8–12 cm long, tomentulose; cincinni sessile. Flowers: calyx with dorsiventral-transversal symmetry, puberulent, the sepals 4, concave, rounded, ciliate, ca. 1 mm long; nectary disc divided into 4 glands; ovary sericeous. Monocarps obovoid, 1.2–1.5 cm long, green, puberulent, the venation prominent. Fl (Sep, Dec), fr (Feb, Mar, May); in nonflooded forest.

CUPANIA L.

Small to large trees. Stipules wanting. Leaves alternate, pinnate with rudimentary distal leaflet; leaf rachis cylindrical or angular; petiolules usually short and slender; leaflets alternate or opposite, the margins serrate, dentate, or rarely entire. Inflorescences axillary and terminal, paniculate thyrses, with flowers in compound or simple dichasial cymes. Flowers actinomorphic, bisexual or unisexual (plants polygamous or dioecious); sepals 5, imbricate, usually tomentose; petals 5, with a pair of short, marginal appendages; nectary disc annular, usually 5-lobed; stamens (4, 6) 8, exserted, the filaments of equal length, free to base, the anthers dorsifixed; ovary 3-locular, each locule with single ovule. Capsules 2- or 3-locular, woody or coriaceous. Seeds 2–3, with fleshy arillode at base.

1. Petiolules 1–3 mm long; leaflets elliptic, chartaceous, the margins dentate-crenate or less often entire. . . . . *C. scrobiculata* subsp. *guianensis.*
1. Petiolules 5–10(15) mm long; leaflets elliptic, broadly elliptic, oblong, or less often obovate, coriaceous or nearly so, the margins entire or repand-dentate. . . . . *C. scrobiculata* subsp. *scrobiculata.*

Cupania scrobiculata Rich. subsp. **guianensis** (Miq.) Acev.-Rodr. [Syn.: *Cupania scrobiculata* Rich. f. *guianensis* (Miq.) Radlk., *Cupania guianensis* Miq.]

Trees, 12–20 m tall, young stems and inflorescence axes tomentulose. Leaves with 4–8 alternate leaflets; petioles plus rachis 15 cm long, slightly angular, furrowed, tomentulose; petiolules 1–3 mm long; leaflets elliptic, 7–13.5 × 4.1–5.5 cm, chartaceous, glabrous except for puberulent veins abaxially, the base unequal, obtuse-rounded, the apex acuminate, the margins dentate-crenate, or less often entire. Inflorescences 15–22 cm long. Flowers: sepals tomentulose, oblong, 1.8–2 mm long. Capsules 3-locular, tomentose, wrinkled, orange to reddish orange, the stipe 4–5 mm long, the cocci wider than long, nearly ellipsoid, 5–7 mm wide. Fr (Feb, Apr, May); in nonflooded forest.

Cupania scrobiculata Rich. subsp. **scrobiculata** [Syns.: *Cupania scrobiculata* Rich. f. *reticulata* (Cambess.) Radlk., *Cupania reticulata* Cambess.]

Trees, 5–9 m tall; young stems and inflorescence axes tomentose. Leaves with 2–7, alternate leaflets; petioles plus rachis (7)14–22(29) cm long, terete, striate, puberulent to tomentulose; petiolules 5–10(15) mm long; leaflets elliptic, widely elliptic, oblong or obovate, 7–22 × 3.5–9.2 cm, coriaceous or nearly so, glabrous or sometimes puberulent along veins abaxially, the base obtuse, unequal, the apex obtusely to abruptly acuminate, the margins entire to repand-dentate. Inflorescences 10–20 cm long. Flowers: sepals tomentulose, oblong, 2–2.2 mm long. Capsules 3-locular, wrinkled, tomentose, orange to reddish orange, the stipe 4–5 mm long, the cocci ascending, nearly ellipsoid, 1.5–2 cm long. Fl (Nov), fr (Feb, Mar); in nonflooded forest.

MATAYBA Aubl.

Small to large trees. Stipules wanting. Leaves alternate, pinnate, with rudimentary distal leaflet; leaflets alternate to opposite; petiolules usually short and slender; leaf rachis cylindrical or angular; leaflets entire. Inflorescences axillary and terminal, paniculate or racemose thyrses, with flowers in lateral dichasia, monochasia, cincinni, or congested cymes. Flowers actinomorphic, bisexual or unisexual (plants polygamous-dioecious); sepals 5, valvate, usually tomentose; petals 5, with a pair of short, marginal appendages; nectary disc annular, usually 5-lobed; stamens (4–6)8, exserted, the filaments free, tomentose, of equal length, the anthers dorsifixed; ovary 3-locular, each locule with single ovule. Capsules 2- or 3-locular, woody or coriaceous. Seeds with fleshy arillode at base.

1. Leaflets 4–8 cm long, with secondary venation inconspicuous, the apex abruptly acuminate to caudate. Capsules globose-trigonous. . . . . *M. peruviana.*
1. Leaflets (6)8–25(32) cm long, with conspicuous secondary venation, the apex obtuse to short-acuminate. Capsules bi- or trilobed.
    2. Leaflet apex usually retuse, the abaxial surface with prominent brownish venation. . . . . *M. arborescens.*
    2. Leaflet apex not retuse, the abaxial surface with prominent yellowish venation. . . . . *M. purgans.*

Matayba arborescens (Aubl.) Radlk.

Small trees, to 10 m tall. Stems tomentulose, becoming glabrous with age. Leaves with (4)6–8, alternate leaflets; petioles plus rachis 7–10 cm long, slightly angular; leaflets elliptic or oblong-elliptic, 8–17 × 4–6.2 cm, chartaceous, glabrous, the base acute to obtuse, the apex obtuse, retuse, or less often abruptly acuminate into obtuse, retuse acumen, the margins crenate; venation prominent and brownish abaxially. Inflorescences paniculate thyrses, 10–25 cm long. Flowers: sepals tomentulose, triangular, ca. 1 mm long; petals ca. 1.5 mm long, whitish, the appendage sericeous; nectary disc puberulent; ovary appressed pubescent. Capsules bilocular (third locule rudimentary) or less often 3-locular, glabrous, reddish, the stipe 3–5 mm long, the cocci nearly ovoid, ca. 1 cm long, the apex retuse. Seeds usually 2 per fruit, ellipsoid, ca. 7 mm long, dark brown to black. Fr (Sep, Nov); in nonflooded forest.

Matayba peruviana Radlk. [Syn.: *Matayba oligandra* Sandwith]

Trees, to 20 m tall; young stems, leaf rachis and inflorescence axes tomentulose. Leaves with 10–12, subopposite or alternate leaflets; petioles plus rachis 7–29 cm long; leaflets oblong-elliptic, 4–8 × 2–3.5 cm, coriaceous or nearly so, pubescent or glabrescent, barbate in axils of veins abaxially, the base unequal, the apex abruptly acuminate to caudate, the margins entire; secondary venation inconspicuous. Inflorescences paniculate thyrses, 6–18 cm long. Flowers: sepals puberulent, triangular, 0.7–1 mm long; petals 1–1.5 mm long, whitish, the appendage appressed pubescent; nectary disc glabrous; ovary puberulent, trilocular. Capsules globose-trigonous, 8–10 mm long, glabrous, reddish, the stipe 6–8 mm long, the apex apiculate. Seeds usually one per fruit, dark brown, ellipsoid. Fr (May); in nonflooded forest.

Matayba purgans (Poepp. & Endl.) Radlk.

Small trees, 4 to 12 m tall; young stems and inflorescence axes tomentulose. Stems slightly furrowed. Leaves with (2)4(6), alternate or subopposite leaflets; petioles plus rachis 8–16(28) cm long, finely furrowed; leaflets elliptic, widely elliptic or oblong-elliptic, (6)10–22(32) × (3)5–10(15) cm, subcoriaceous, glabrous, the base obtuse, sometimes unequal, the apex abruptly acuminate into obtuse tip, the margins entire; venation yellowish, prominent abaxially, midrib slightly furrowed adaxially. Inflorescences racemose thryses, clustered at leaf axils, 2–3.5 cm long. Flowers: sepals tomentulose, triangular, ca. 1 mm long; petals ca. 0.7 mm long, whitish, obovate, the appendage sericeous; nectary disc glabrous; filaments villous; ovary puberulent. Capsules trilobed, 8–10 mm long, glabrous, reddish, the stipe 6–8 mm long, the apex apiculate. Seeds usually one or two per fruit, dark brown, ellipsoid. Fr (Apr, Dec); in nonflooded forest.

PAULLINIA L.

Lianas or vines. Stems with single or multiple steles in transverse section, often producing milky sap. Stipules minute to foliaceous, persistent or deciduous. Leaves alternate, trifoliolate, 5-pinnately foliolate, biternate, triternate, or bipinnate. Inflorescences axillary, terminal, or cauliflorous, racemose, spicate, or paniculate thyrses, with flowers in lateral cincinni or drepania, with pair of coiled tendrils at base when axillary. Flowers zygomorphic; calyx 4–5-merous, the sepals distinct, or the two dorsal ones connate to varying degrees; petals 4, distinct; appendages hood-shaped, with fleshy, yellowish apex; nectary disc unilateral, modified into 2 or 4 glands; stamens 8, the filaments of unequal length; ovary 3-locular, the locules with single ovule. Fruits dehiscent, septifragal (marginicidal) capsules, smooth, winged or echinate. Seeds usually globose, with fleshy testa.

1. Leaf rachises winged.
   2. Petioles winged.
      3. Transverse sections of stem nearly terete, with single stele. Inflorescences axillary. . . . . *P. pachycarpa.*
      3. Transverse sections of stem trigonous, with large, central stele and 3 smaller, peripheral steles. Inflorescences cauliflorous. . . . . *P. alata.*
   2. Petioles not winged.
      4. Leaflets mostly glabrous, barbate only in leaf vein axils, the margins bidentate at apex, not ciliate. . . . . *P. lingulata.*
      4. Leaflets pilose along midrib abaxially, the margins glandular-serrate-ciliate. . . . . *P. plagioptera.*
1. Leaf rachis not winged (seldom marginate in *P. tricornis*).
   5. Leaves trifoliolate.
      6. Stems glabrous. Stipules deltoid.. . . . . *P.* cf. *anomophylla.*
      6. Stems hispidulous. Stipules subulate. . . . . *Paullinia* sp. A.
   5. Leaves pinnately 5-foliolate.
      7. Plants markedly hirsute. . . . . *P. rubiginosa.*
      7. Plants glabrous or tomentose.
         8. Leaf rachises glabrous or nearly so; venation reticulate. . . . . *P. tricornis.*
         8. Leaf rachis puberulent or tomentose; venation clathrate or nearly so.
            9. Leaflets caudate or cuspidate at apex.
               10. Leaflets cuspidate, the midrib pubescent abaxially, the vein axils not barbate. . . . . . *P. clathrata.*
               10. Leaflets caudate, the midrib glabrous abaxially, the vein axils barbate. . . . . *P. capreolata.*
            9. Leaflets acuminate, shortly acuminate, apiculate, or abruptly obtuse at apex.
               11. Stipules entire. Leaflets glabrous to sparsely pubescent abaxially.
                  12. Sepals 5. Capsules not winged. . . . . *P. latifolia.*
                  12. Sepals 4 (two of sepals connate for most of length forming larger "sepal" retuse at apex). Capsules winged. . . . . *P. fibulata.*
               11. Stipules deeply dissected. Leaflets tomentose abaxially. . . . . *P. stellata.*

Paullinia alata (Ruiz & Pav.) G. Don

Lianas, usually many-branched from base. Stems trigonous, glabrous, not producing milky sap; transverse section of stem with large, central stele and 3 peripheral smaller ones; old stems deeply, obtusely costate, bearing no leaves. Stipules subulate, ca. 3 mm long. Leaves pinnate, 5-foliolate; petioles and rachis broadly winged; petiolules 3–5 mm long; leaflets elliptic to oblong-elliptic, 10–15 × 5–6 cm, subcoriaceous, glabrous except for barbate vein axils, the base attenuate to obtuse, the apex obtusely acuminate, the margins repand-dentate; tertiary venation reticulate. Inflorescences cauliflorous, racemose thryses, congested in short fascicles. Capsules pyriform, 1–1.2 cm long, bright red. Fl (May, Jul); along roadside at border of nonflooded forest.

Paullinia cf. **anomophylla** Radlk.

Lianas. Stems terete, glabrous, smooth, slightly striate, not producing milky sap; transverse section of stem with single stele. Stipules deltoid, minute. Leaves trifoliolate; petioles glabrous, terete, not winged, 3.5–11 cm long; petiolules pulvinate, glabrous, the lateral ones 0.5–1.5 cm long, the distal one 2–4 cm long; leaflets elliptic or oblong-ovate, 8–16 × 3.6–8.4 cm, chartaceous, glabrous, the base acute to obtuse, the apex cuspidate, the margins crenate to repand-dentate; venation reticulate, the tertiary venation inconspicuous. Known from a single sterile collection (*Acevedo-Rodríguez 5006*, NY) from our area; in nonflooded forest on gentle slopes.

Identification of this species as *Paullinia anomophylla* is provisional. The terete stem of this single collection differentiates it from typical *P. anomophylla*, which has triangular stems.

Paullinia capreolata (Aubl.) Radlk.

Lianas. Stems terete, glabrous, lenticellate, producing milky sap; transverse section of stem with single stele. Stipules early deciduous. Leaves pinnate, 5-foliolate; petioles and rachis terete, not winged, puberulent; petiolules 3–8 mm long, puberulent; leaflets elliptic to obovate, 6–10 × 2.2–4.2 cm, chartaceous, glabrous except for barbate vein axils, the base acute to attenuate, the apex caudate, the margins ciliate, crenate to remotely serrate on upper 1/3 of blade; tertiary venation clathrate, yellowish. Inflorescences axillary or terminal, racemose thryses. Capsules nearly globose, crustose. Known from a single sterile collection (*Acevedo-Rodríguez 5034*, NY) from nonflooded forest along roadside in our area.

Paullinia clathrata Radlk.

Lianas. Stems terete, tomentose; transverse section of stem with single stele. Stipules early deciduous. Leaves pinnately 5-foliolate; petioles and rachis terete, not winged, pubescent; petiolules 6–10 mm long, the terminal one to 3.5 cm long; leaflets elliptic to obovate, 10–15 × 5–6 cm, chartaceous, sparsely, appressed pubescent abaxially, the base attenuate to obtuse, the apex cuspidate, the margins remotely serrate on upper 1/3 of blade; tertiary venation clathrate. Inflorescences axillary or terminal, acemose thyrses. Capsules nearly globose. Known from sterile collections (*Acevedo-Rodríguez 5035*,

NY, *Gentry 63071*, NY) from nonflooded forest on gentle slopes in our area.

Paullinia fibulata Rich.

Lianas. Stems terete, glabrescent; transverse section of stem with single stele. Leaves pinnate, 5-foliolate; petioles and rachis not winged, terete, tomentulose; petiolules 3–6 mm long; leaflets oblong-elliptic to obovate, 6–10 × 3.2–6 cm, coriaceous, sparsely pubescent, especially along veins, abaxially, the base obtuse to rounded, the apex apiculate, the margins repand-dentate; tertiary venation clathrate. Inflorescences axillary, racemose thryses. Flowers: sepals 4. Capsules obovoid to pyriform, tomentose, with 3 ascending wings on upper half, the cocci greenish yellow, the wings red, the margins crenate. Fr (Aug); border of nonflooded forest along roadsides and trails.

Paullinia latifolia Radlk.

Lianas. Stems terete, tomentulose, becoming glabrous and lenticellate with age; transverse section of stem with single stele. Stipules deltoid, minute. Leaves pinnate, 5-foliolate; petiole and rachis not winged, terete, tomentulose; petiolules 3–6 mm long; leaflets oblong-elliptic, to obovate, 14–16 × 7–8.7 cm, chartaceous to coriaceous, glabrous to sparsely pubescent abaxially, the base acute, obtuse or rounded, the apex acuminate to obtusely apiculate, the margins remotely glandular-serrate; tertiary venation clathrate. Inflorescences axillary or sometimes cauliflorous, racemose thryses. Flowers: sepals 5. Capsules nearly globose, 6-ridged, ferruginous tomentose, orange, the stipe elongated. Fl (Nov), fr (Apr, Nov); in nonflooded forests.

Paullinia lingulata Acev.-Rodr.

Lianas. Stems nearly terete, minutely tomentulose; transverse section of stem with single stele. Stipules minute, early deciduous. Leaves biternate; petioles not winged, puberulent; rachis narrowly winged; leaflets elliptic or oblong-elliptic, 3–8 × 1.5–3 cm, chartaceous, mostly glabrous except barbate in vein axils axils abaxially, the base attenuate to obtuse, the margins bidentate below lingulate apex; tertiary venation reticulate, inconspicuous. Inflorescences axillary, racemose thryses, densely flowered. Fl (Jan, Feb), fr (Feb); in nonflooded forest.

Paullinia pachycarpa Benth.

Lianas. Stems terete, finely striate, glabrous; transverse section of stem with single stele. Stipules early deciduous, leaving scar to 5 mm wide. Leaves pinnate, with 3 pinnae, the lower pair of pinnae trifoliolate; petioles and rachis broadly winged, glabrous; leaflets oblong to oblanceolate, 8–20 × 3–6.6 cm, chartaceous, glabrous, the base attenuate on terminal leaflets, abruptly attenuate on lateral ones and obtuse on basal ones, the apex cuspidate, the margins remotely serrate on upper half of blade; tertiary venation reticulate. Inflorescences axillary, paniculate thryses, densely flowered. Flowers: calyx canescent, greenish white, the sepals 4, the outer sepals rounded, much smaller than inner ones. Capsules depressed globose, wrinkled, 6–9-ridged, ferruginous sericeous. Fl (May); in nonflooded forest.

Paullinia plagioptera Radlk.

Lianas. Stems terete, glabrous, lenticellate; transverse section of stem with single stele. Stipules minute, early deciduous. Leaves pinnate, 5-foliolate; petioles marginate, pubescent; rachis pubescent, narrowly winged, ciliate; leaflets elliptic or oblong-elliptic, 4–9 × 2–3.2 cm, chartaceous-coriaceous, glabrous except for pilose midrib, the base obtuse to rounded, unequal on lateral leaflets, the apex acuminate, the margins broadly glandular-serrate-ciliate; tertiary venation reticulate. Inflorescences racemose thryses, with sessile cincinni. Capsules obovoid, 8–12 mm long, with 3 divaricate to ascending wings, ca. 1 cm long on upper half. Fr (Feb); at disturbed nonflooded forest edge.

Paullinia rubiginosa Cambess.

FIG. 285, PL. 115f; PART 1: PL. XVb

Lianas. Stems 4–5-costate, covered with reddish hispidulous hairs; transverse section of stem with single stele. Stipules persistent, to 1.5 cm long, deeply dissected, with ovate outline. Leaves pinnate, 5-foliolate; petioles and rachis terete, not winged, hispid; leaflets elliptic to lanceolate, 8–21 × 3.2–7.5 cm, chartaceous, hispidulous especially along margins, the base cuneate on terminal leaflets, obtuse to rounded on lateral ones, the apex narrowed into long, fine cusp, the margins remotely denticulate; tertiary venation clathrate. Inflorescences paniculate thryses, densely flowered, with long, overlapping, hispid bracts. Capsules elliptic or oblanceolate, 3-winged, densely hispid, red. Fl (Sep), fr (Jan, Feb, May, Aug); in nonflooded forest.

Paullinia stellata Radlk. PART 1: FIG. 11

Lianas. Stems 4–5-costate, puberulent; transverse section of stem with single stele. Stipules persistent, 3–4 mm long, deeply dissected with ovate outline. Leaves pinnate, 5-foliolate; petioles and rachis terete, not winged, tomentulose; leaflets elliptic, oblong-elliptic to oblanceolate, 7–14.5 × 3.5–7 cm, chartaceous, tomentose abaxially, the base acute on terminal leaflets, obtuse to rounded on lateral ones, the apex acuminate and mucronate, the margins remotely denticulate or entire; tertiary venation finely clathrate. Inflorescences racemose thyrses, densely flowered, with non-overlapping, bracts. Capsules depressed globose, golden, ferruginous tomentose. Fl (May); in nonflooded forest in our area.

Paullinia tricornis Radlk. PL. 116a

Lianas. Stems terete, lenticellate, glabrous, producing white sap, the transverse section with single stele. Stipules minute (2–3 mm long). Leaves pinnate, 5-foliolate; petioles not winged, terete, furrowed, glabrous; rachis not winged or marginate, glabrous or nearly so; leaflets elliptic to lanceolate, 4–14 × 2.8–7 cm, coriaceous, with minute, resinous dots, glabrous except for barbate or barbulate vein axils, the base attenuate to obtuse, the apex acuminate to long acuminate, the margins remotely serrate; tertiary venation reticulate. Inflorescences paniculate thyrses, densely flowered. Flowers: calyx canescent, greenish white, the sepals 5, outer sepals rounded, much smaller than inner ones. Capsules ferruginous tomentose, 3-winged. Fl (Jun), fr (Sep); in nonflooded forest.

Paullinia sp. A

Stems terete, hispidulous; transverse section with single stele. Stipules subulate. Leaves trifoliolate; distal leaflet much larger than lateral ones, with scattered hairs with swollen bases abaxially, the apex acuminate, the base acute to obtuse, the margins serrate. Known from a sterile collection of a juvenile plant from Pic Matécho (*Granville 4989*, CAY).

Because adult plants differ from juveniles, it is not possible to identify this species with certainty.

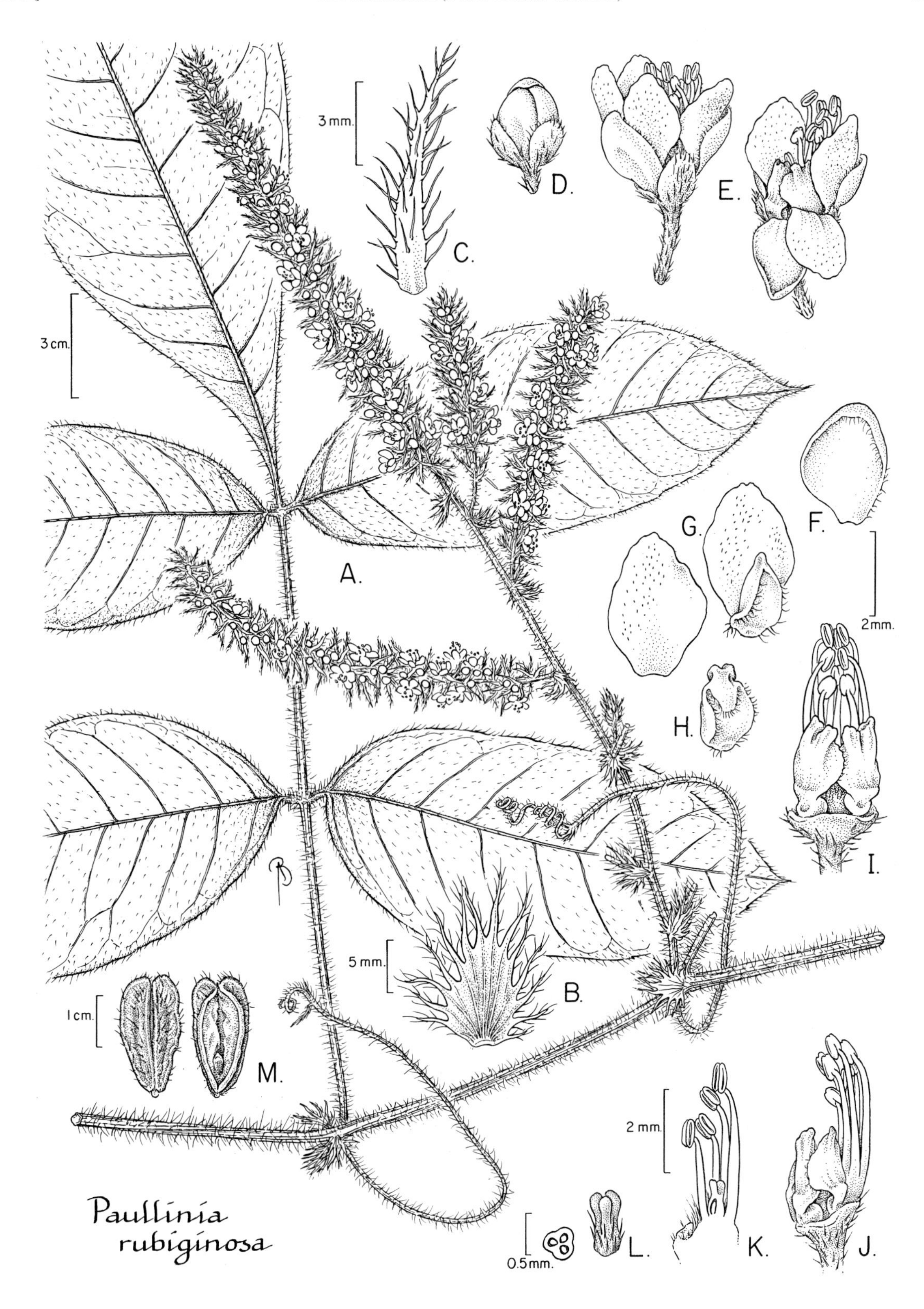

FIG. 285. SAPINDACEAE. *Paullinia rubiginosa* (A–L, *Mori & Gracie 23857*; M, *Mori & Gracie 21152*). **A.** Part of liana with leaves, tendrils, and inflorescence. **B.** Stipule. **C.** Bract. **D.** Lateral view of flower bud. **E.** Lateral view of flower (left) and lateral view of flower with one petal removed (right). **F.** Adaxial view of sepal. **G.** Abaxial view of petal (left) and adaxial view of petal with appendage (right). **H.** Adaxial view of petal appendage. **I.** Lateral view of flower with sepals and petals removed; note petal appendages inserted on nectary disc glands. **J.** Lateral view of flower with sepals and petals removed. **K.** Medial section of flower with sepals and petals removed. **L.** Lateral view (near left) and transverse section (far left) of pistillode. **M.** Fruit (left) and dehisced fruit (right).

PSEUDIMA Radlk.

Small to large trees. Stipules wanting. Leaves alternate, pinnate, with rudimentary distal leaflet; leaflets alternate or subopposite; petiolules short and slightly enlarged at base; leaf rachis angular, furrowed; leaflets with entire margins. Inflorescences axillary or terminal, racemose or paniculate thyrses, the flowers in compound dichasia. Flowers actinomorphic, bisexual or unisexual (plants polygamous-dioecious); sepals 5, imbricate, concave, the external 2 shorter; petals 5, longer than sepals, without appendages; nectary disc annular, 5-lobed; stamens 8 or 10, shorter than petals, the filaments villous, filiform, of equal length, the anthers basifixed; ovary 2(3)-locular, the locules with single ovule. Fruits dehiscent, 2(3)-locular, coriaceous capsules, the cocci equally developed or one of them smaller. Seeds large, arillate.

Pseudima frutescens (Aubl.) Radlk.

PL. 116b; PART 1: FIG. 10, PL. XVc

Unbranched trees, 3–10 m tall. Bark light brown, smooth. Stems furrowed to terete, the pith hollow (sometimes housing ants) toward apex. Juvenile leaves simple; adult leaves even-pinnate; leaflets 10–20, alternate; petioles plus rachis 8–75 cm long; leaflets oblong, lanceolate, elliptic or oblanceolate, 5–22(30) × 3–6.5(7.3) cm, chartaceous, glabrous, the base acute to obtuse, the apex acuminate, the margins entire or crenulate. Inflorescences paniculate thyrses, 30–45 cm long, densely flowered. Flowers: calyx greenish to light-yellow, sericeous, the sepals 12–15 mm long; petals yellowish. Capsules bilocular, red, each locule splitting in half. Seeds 2 per capsule, large, black, with white arillode on lower half. Fl (Jun, May), fr (Jun, Apr, Aug, Sep, Oct); scattered but common, in nonflooded forest.

SERJANIA Mill.

Lianas or vines of forest canopy or disturbed areas. Transverse section of stem with single or multiple steles, often producing milky sap. Stipules minute to small, early deciduous or persistent. Leaves alternate, trifoliolate, pinnate and 5-foliolate, biternate, or triternate, or seldom bipinnate. Inflorescences axillary or terminal, racemose or paniculate thryses, the flowers on lateral cincinni. Flowers zygomorphic; sepals 4 or 5, distinct, or 2 connate; petals 4, distinct, the appendage hood-shaped, with fleshy, yellowish apex; nectary disc unilateral, 2–4-glandular; stamens 8, the filaments of unequal length, the anthers dorsifixed; ovary 3-locular, the locules with single ovule. Fruits schizocarps, splitting into 3 samaroid mericarps with proximal wing. Seeds lenticular to globose, exarillate.

1. Transverse sections of stem with ≥3 peripheral steles.
   2. Stems terete, 8–10-striate, the transverse sections with 8 or 10 peripheral steles, disposed in circle around central stele. Leaf rachises not winged. . . . . *S. pyramidata.*
   2. Stems trigonous, 5-ribbed, the transverse sections with 3 peripheral steles, disposed in triangle around central cylinder. Leaf rachises winged. . . . . *S. paucidentata.*
1. Transverse sections of stem with single stele.
   3. Plants glabrous. Stems pentagonous. Leaflets 5–15 cm long, the apex cuspidate to caudate. Fruits 3–5 cm long. . . . . *S. grandifolia.*
   3. Plants pilose to setulose. Stems nearly terete. Leaflets 2–7.5 cm long, the apex acuminate or obtuse. Fruits 2.5–3 cm long.
      4. Plants setulose, herbaceous. Leaflets membranous, the margins deeply serrate to lobed, the apex acuminate. . . . . *S. setulosa.*
      4. Plants pilose, woody. Leaflets subcoriaceous, the margins obtusely serrate, the apex obtuse to acute. . . . . *S. didymadenia.*

Serjania didymadenia Radlk.

Lianas. Stems nearly terete, pilose, the transverse section with single stele, the pith usually hollow. Stipules inconspicuous. Leaves biternate; petioles and rachis furrowed, slender, pilose; leaflets ovate, 2–4 × 1.5–2.5 cm, subcoriaceous, sparsely pilose along veins, punctate abaxially, the base attenuate to nearly cuneate on terminal leaflets, obtuse to subcordate on lateral ones, the apex obtuse to acute, the margins obtusely serrate, mucronate. Flowers: sepals ca. 2 mm long. Mericarps 2.5–3 cm long, the wing membranous. Fr (May); secondary forest along airstrip.

Serjania grandifolia Radlk.

FIG. 286, PL. 116c; PART 1: FIG. 6

Lianas, with abundant milky sap. Stems pentagonous, glabrous, the transverse section with single stele, the pith usually hollow. Stipules inconspicuous. Leaves biternate; petioles and rachis furrowed, stout, glabrous; leaflets elliptic, 5–15 × 3–7 cm, chartaceous, glabrous, the base attenuate or acute on terminal leaflets, obtuse to nearly rounded on lateral ones, the apex cuspidate to caudate, the margins entire or repand-dentate. Flowers: anthers pinkish. Mericarps 3–5 cm long, reddish when mature, the cocci globose, the wings membranous. Fl (Jan), fr (Feb); in nonflooded forest on gentle slopes.

Serjania paucidentata DC.

Lianas, not producing milky sap. Stems trigonous, with 5 sharp ribs, glabrous, the transverse section with central stele surrounded by 3 triangular, peripheral steles. Stipules inconspicuous. Leaves biternate; petioles furrowed, marginate, stout, glabrous; rachis winged; leaflets elliptic, 6.5–16 × 3.7–6 cm, chartaceous, glabrous, the base long to short attenuate, the apex long apiculate, the margins bidentate below apex. Mericarps 2.4–2.7 cm long, pinkish tinged, the cocci globose, pubescent. Fl (Nov); along borders of nonflooded forest.

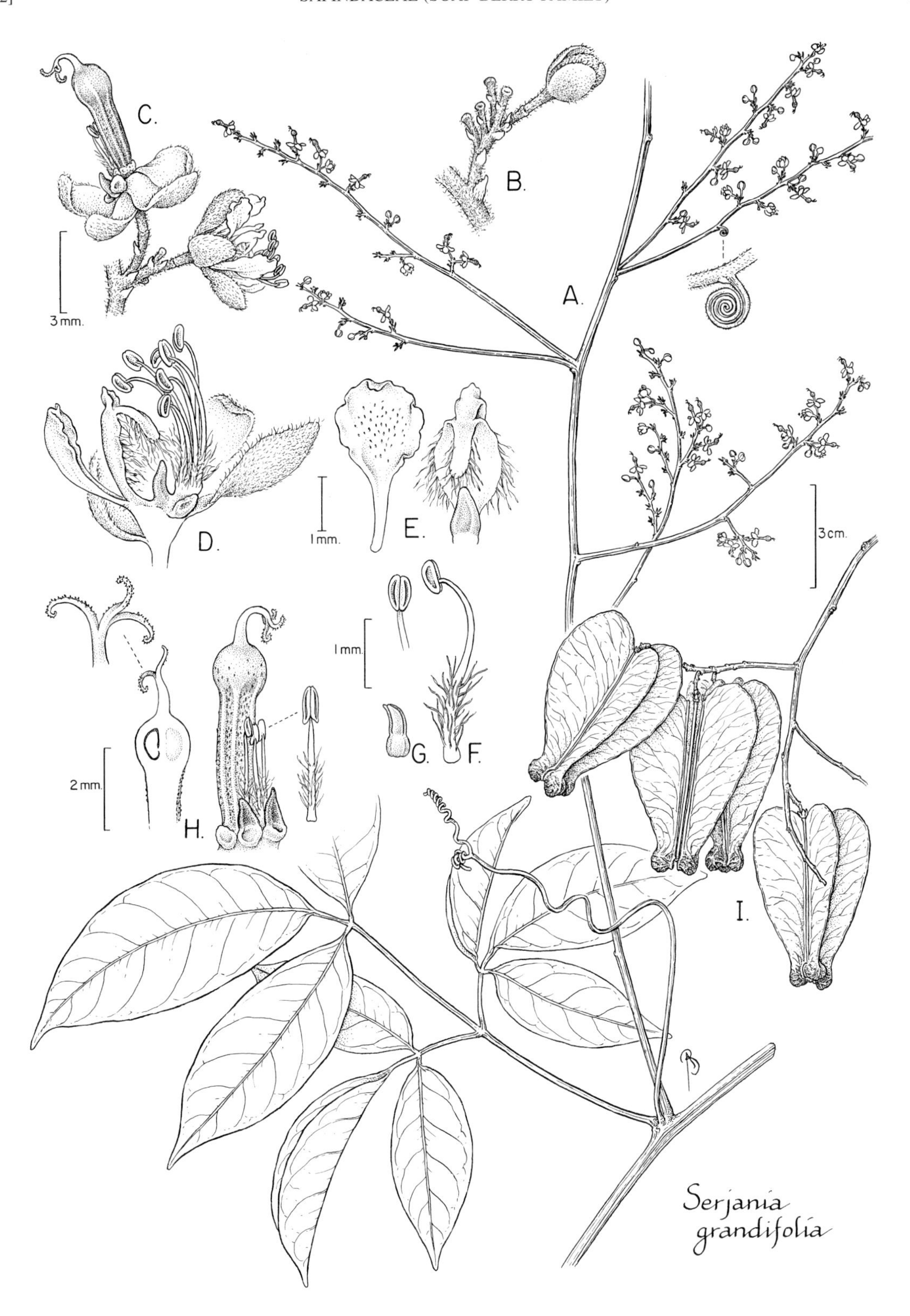

FIG. 286. SAPINDACEAE. *Serjania grandifolia* (A–H, *Mori & Granville 8916* from Route de l'Est, French Guiana; I, *Mori & Gracie 21151*). **A.** Part of vine with leaf, tendril, inflorescence, and detail of tendril. **B.** Detail of inflorescence cincinnus with flower bud, bracts, and pedicel bases. **C.** Detail of inflorescence cincinnus with one functionally pistillate flower and one functionally staminate flower. **D.** Lateral view of functionally staminate flower. **E.** Adaxial view of petal with appendage removed (left) and adaxial view of petal appendage and subtending gland with petal removed. **F.** Lateral view of stamen (near left) and detail of abaxial view of apical portion of stamen (upper left). **G.** Lateral view of pistillode. **H.** Medial section of pistil (near left) and detail of stigma (upper left), pistil with a few staminodes and glands (near right) and detail of staminode (far right). **I.** Winged schizocarps.

Serjania pyramidata Radlk.

Lianas, producing milky sap. Stems terete, 8–10-striate, tomentulose to glabrous, the transverse section with 8–10 peripheral, small steles surrounding large central one. Stipules inconspicuous. Leaves biternate; petioles and rachis furrowed, tomentulose or pilose, slender; leaflets oblong, ovate, or elliptic, 4.5–12 × 2.7–6 cm, chartaceous, sparsely pilose to glabrous, the base cuneate in terminal leaflets, obtuse in lateral ones, the apex long acuminate to cuspidate, the margins deeply serrate, usually on upper half of blade. Mericarps 2.5–2.8 cm long, the cocci slightly flattened. Fl (Jan, Feb, Mar), fr (Feb, Mar, May); in remnant forest along roadsides.

Serjania setulosa Radlk.

Herbaceous vines, to 2 m long, producing scanty milky sap. Stems nearly terete, striate, setulose, the transverse section of stem with single stele, the pith usually hollow. Stipules minute. Leaves biternate; petioles and rachis furrowed, slender, setulose to pilose; leaflets ovate, 3–7.5 × 2–3 cm, membranous, sparsely setulose, the base abruptly attenuate in terminal leaflets, obtusely to shortly attenuate in lateral ones, the apex acuminate, mucronate, the margins deeply serrate to lobed. Known from a single sterile collection (*Acevedo-Rodríguez 4992*, US) from nonflooded forest understory.

TALISIA Aubl.

Understory treelets to large trees. Stipules wanting. Leaves alternate, pinnate, the terminal leaflet rudimentary; leaflets opposite or alternate; petiolules usually swollen; leaflets with entire margins. Inflorescences axillary, terminal, or cauliflorous, paniculate thyrses, with flowers in lateral dichasia. Flowers actinomorphic, functionally staminate or pistillate (plants dichogamous); sepals 5, connate at base, usually shorter than petals; petals 5, distinct, erect or reflexed, the appendage petaloid, simple or bifid, erect and villous, or seldom wanting (the petal then with involute margins at base); nectary disc annular; stamens 5–8, the filaments of equal or unequal length, free, villous or glabrous, the anthers basifixed; ovary 3-locular, each locule with a single ovule. Fruits indehiscent, baccate, ellipsoid or globose, the pericarp coriaceous to woody. Seeds usually 1, less often 2–3 per fruit, the testa fleshy, usually edible.

1. Trees, ≥8 m tall, the lateral branches robust.
  2. Inflorescences and calyces with long, glandular hairs. . . . . *T. hemidasya.*
  2. Inflorescences and calyces without glandular hairs.
    3. Secondary veins of leaflets inconspicuous. Inflorescences paniculate thyrses. . . . . *T. simaboides.*
    3. Secondary veins of leaflets prominent. Infloresences simple thyrses.
      4. Leaflets hirsute to pilose along primary and secondary veins abaxially. . . . . *T. pedicellaris.*
      4. Leaflets glabrous abaxially.
        5. Leaves with 6–8(10) leaflets; venation prominent abaxially. Inflorescences congested in leaf axils or cauliflorous, tomentulose. . . . . *T. clathrata* subsp. *canescens.*
        5. Leaves with 2–4 leaflets; venation slightly prominent abaxially. Inflorescences solitary or congested in leaf axils, not cauliflorous, puberulent. . . . . *T. praealta.*
1. Unbranched shrubs or treelets, without robust lateral branches.
  6. Leaflets minutely hirsute abaxially. Fruits ovoid to globose-ovoid, the pericarp woody. . . . . *T. mollis.*
  6. Leaflets glabrous, sparing pubescent, or puberulent abaxially. Fruits ellipsoid to globose, trigonous-ellipsoid, or pyriform, the pericarp coriaceous or woody.
    7. Petiolules drying much darker than leaflets. Calyx 6–7 mm long, velvety pubescent; petals ca. 9 mm long, ferruginous sericeous-tomentose abaxially. Fruits velvety pubescent when young. . . . . . *T. megaphylla.*
    7. Petiolules not drying noticeably darker than leaflets. Calyx ≤3 mm long, minutely pilose or puberulent; petals to 7.5 mm long, glabrous or appressed pubescent abaxially. Fruits glabrous or puberulent.
      8. Stamens 6–8. . . . . *T. guianensis.*
      8. Stamens 5(6).
        9. Petals appressed pubescent abaxially.
          10. Leaflets with minute, scale-like punctations abaxially; rachis triangular at base, sharply angled toward apex. . . . . *T. longifolia.*
          10. Leaflets smooth abaxially; rachis nearly terete at base, obtusely angled toward apex. . *T. carinata.*
        9. Petals glabrous abaxially.
          11. Nectary disc glabrous; filaments glabrous. . . . . *T. macrophylla.*
          11. Nectary disc hispidulous to tomentose; filaments tomentose. *T. sylvatica.*

Talisia carinata Radlk.

Treelets, 2–4 m tall, unbranched or with few ascending, sympodial branches on upper part. Bark brownish, lenticellate. Stems striate, becoming lenticellate, glabrous or minutely pubescent, reddish brown. Leaves even- or odd- pinnate, spirally arranged toward apex of stem; leaflets (5)10–18(22), alternate or opposite; petiole plus rachis (20)24–65(120) cm long, terete toward base, obtusely angled toward apex, glabrous; petiolules pulvinate, swollen; blades elliptic, oblong-elliptic, or seldom nearly lanceolate, (4.5)6.5–18(25) × 1.7–4.3(8.2) cm, chartaceous, the base usually unequal, obtuse-acute or attenuate, the apex acuminate to cuspidate; primary and secondary veins prominent and glabrous abaxially, impressed adaxially. Inflorescences axillary, inserted toward apex of stems, paniculate thyrses, 10–30 cm long, sparsely flowered, pilose. Flowers: calyx pilose, 1.5–2 mm long, the sepals ovate; petals cream-colored, appressed-pubescent abaxially, 2.5–3 mm long, the appendage as long as petal, sericeous, white; nectary disc hispidulous at apex; stamens 5, the filaments pilose. Fruits ellipsoid to pyriform, ca. 2 cm long, glabrous, turning from green to yellow-orange, the apex apiculate, the pericarp coriaceous, granulose. Seeds ellipsoid, with creamy or yellow, fleshy testa. Fl (Jul, Aug), fr (Jan, Feb, Mar, Aug, Sep); common, in nonflooded forest. *Bois flambeau, gaulette indien, petite gaulette, tit-gaulette* (Créole), *touliatan, tuliata* (Amerindian).

Talisia clathrata Radlk. subsp. **canescens** Acev.-Rodr.

Trees, 8–15 m tall, branched on upper part. Bark grayish to brown, smooth to slightly rough. Stems minutely pubescent, becoming terete and lenticellate. Leaves even-pinnate or seldom odd-pinnate; leaflets 6–8(10), opposite to alternate; rachis plus petiole 13–21(30) cm long, terete, striate, puberulent; blades elliptic, lanceolate or oblanceolate, 7.5–19 × 2.7–5.5 cm, chartaceous, glabrous, the base obtuse or acute, sometimes slightly unequal, the apex acuminate to caudate; venation impressed adaxially, prominent abaxially. Inflorescences simple thyrses, 2.5–7 cm long, congested in leaf axils or cauliflorous, striate, tomentulose. Flowers: calyx light green, minutely canescent, the sepals 2.5–3 mm long, ovate; petals reflexed, ca. 4.5 mm long, whitish, the appendage as long as petal, woolly tomentose; stamens 8. Fruits ellipsoid, ca. 1.5 cm long, densely, minutely, ferruginous tomentulose, rusty-brown, the apex apiculate, the pericarp slightly woody, smooth. Seeds ellipsoid. Fl (May), fr (Jun, Aug); in nonflooded forest.

Talisia guianensis Aubl.

Slender treelets, 2.5–10 m tall, unbranched or with few ascending, sympodial branches on upper part. Bark dark brown, lenticellate. Stems nearly terete, glabrous toward apex. Leaves even-pinnate, spirally arranged toward apex of stems; leaflets 16–22, opposite or alternate; rachis plus petiole 24–60(100) cm long, terete, or slightly striate toward apex, glabrous, dark brown, glossy; blades oblong, oblong-elliptic, or elliptic, 8–18(44) × 2.1–6(13.5) cm, nearly coriaceous, glabrous and glossy on both surfaces, the base unequal, obtuse-acute, abruptly narrowing into long, pulvinate petiolule, the apex acuminate; midrib yellowish, prominent abaxially. Inflorescences terminal or axillary, inserted toward apex of stems, paniculate thyrses, 10–25 cm long, sparsely flowered, glabrous or minutely puberulent, pinkish tinged. Flowers: calyx pinkish tinged, puberulent, the sepals 2–3 mm long, ovate, ciliate at margins; petals whitish, 6–7.5 mm long, the appendages as long as petals, sericeous; nectary disc hispidulous; stamens (6)8. Fruits ellipsoid-ovoid, 2–2.5 cm long, glabrous, orange-yellow at maturity, the apex apiculate, the pericarp coriaceous, granulose. Seeds 1–3 per fruit, nearly ellipsoid, with yellowish fleshy testa. Fl (Sep, Nov), fr (Nov); in nonflooded forest.

Talisia hemidasya Radlk. [Syn.: *T. glandulifera* Steyerm.]

Trees, 12–18 m tall, with robust lateral branches. Bark grayish, thin, smooth, peeling in irregular plates, the inner bark reddish. Stems terete, tomentose, with numerous glandular hairs toward apex. Leaves even-pinnate, spirally arranged near apex of stem; leaflets (6)10–14, opposite to alternate; rachis plus petiole 15–30(50) cm long, terete, or slightly angular toward apex, slightly striate, densely covered with soft, erect hairs, 0.2–0.3 mm long, sometimes with scattered setaceous or glandular hairs; blades oblong-elliptic, 6–23.5 × 2–5.7(8.5) cm, chartaceous, glabrous, except for minutely pubescent midrib, adaxially, glabrous except for minutely pubescent (sometimes with glandular hairs) midrib abaxially, the base obtuse-acute, unequal, the apex acuminate; venation flattened or seldom slightly impressed adaxially, prominent abaxially. Inflorescences axillary, inserted toward apex of stems, paniculate thryses, 6–31 cm long, fulvo-tomentose, with scattered glandular hairs. Flowers: calyx tomentose, with scattered glandular hairs, the sepals ca. 3 mm long, ovate; petals whitish, 5–6 mm long, the appendages as long as petals, sericeous; stamens 8. Fruits globose-ovoid, ca. 2 cm long, densely hirsutulous, yellow to orange-yellow at maturity, the apex apiculate, the pericarp slightly woody, smooth. Seeds nearly globose, with thin, fleshy, orange-yellow testa. Fl (Sep, Oct), fr (May, Sep, Nov); common, in nonflooded forest.

Talisia longifolia (Benth.) Radlk.

Slender small trees or treelets, (3)9–10 m tall, unbranched or with few ascending, sympodial branches on upper part. Bark brown, with numerous black, protruding lenticels, the inner bark orangish. Stems furrowed, glabrous toward apex. Leaves even-pinnate, spirally arranged toward apex of stems; leaflets 8–16, opposite or alternate; rachis plus petiole 25–53(100) cm long, glabrous, triangular at base, trullate and sharply angled toward apex; petiolule elongate, pulvinate; blades elliptic or oblong, 11.5–30(54) × 3–8.5(14) cm, coriaceous, glabrous, with minute, scale-like punctations abaxially, the base unequal, obtuse-acute or attenuate, the apex acute to acuminate; midrib prominent on both surfaces. Inflorescences terminal or axillary, inserted toward apex of stems, paniculate thyrses, 19–38 cm long, sparsely flowered, puberulent and striate. Flowers: calyx yellowish green, puberulent, 2–2.5 mm long, the sepals ovate, the margins ciliate; petals cream-colored to yellowish, 3–4 mm long, appressed pubescent to puberulent abaxially, the appendages as long as petals, sericeous; stamens 5. Fruits trigonous-ellipsoid or ellipsoid-ovoid, ca. 2 cm long, puberulent, orange-yellow at maturity, the apex apiculate, the pericarp coriaceous, granulose. Seeds ellipsoid. Known only from sterile collections in our area; in nonflooded forest.

Talisia macrophylla (Mart.) Radlk. [Syn.: *Talisia allenii* Croat]

Slender small trees or unbranched shrubs, 8–10 m tall. Bark grayish with raised, dark lenticels. Stems terete, minutely lenticellate, dull, minutely pubescent. Leaves odd-pinnate, spirally arranged on upper part of stem; leaflets 9–13, alternate or opposite; petiolules 3–15 mm long, pulvinate and swollen at base; rachis plus petiole 45–72 cm long, terete, slightly angled toward apex, glabrous or minutely pubescent; blades oblong, oblong-elliptic or oblong-obovate, 11.5–42 × 4.5–13 cm, coriaceous, glabrous, the base unequal, obtuse-acute to attenuate, the apex abruptly acuminate into obtuse acumen; primary and secondary veins prominent abaxially, impressed adaxially. Inflorescences axillary, inserted toward apex of stem, paniculate thyrses, 20–30 cm long, sparsely flowered, minutely pilose. Flowers: calyx greenish, minutely pilose, 1.2–1.5 mm long, the sepals ovate-triangular, ca. 1 mm long; petals cream-colored, glabrous, 4.5–5 mm long, the appendage as long as petal, sericeous, white; nectary disc glabrous; stamens 5, the filaments glabrous. Fruits green, ellipsoid, apiculate, ca. 2 cm long, glabrous, the pericarp coriaceous, granulose. Fl (Feb, Apr), fr (May, Nov); in nonflooded forest.

Talisia megaphylla Radlk. PL. 117a

Slender trees or unbranched shrubs, 4–10 m tall. Stems furrowed, glabrous, lenticellate, the pith hollow, housing ants toward apex. Leaves odd- or even-pinnate, spirally arranged toward apex of stem; leaflets 11–15, opposite or alternate; rachis plus petiole 65–83 cm long, terete, striate, slightly angled toward apex, glabrous, yellowish brown, glossy; petiolules brownish, pulvinate, 7–14 mm long, drying much darker than leaflets; blades oblanceolate to oblong-elliptic, 17–43 × 4.5–10.5 cm, chartaceous, glabrous or sparingly pubescent abaxially, the base unequal, obtuse-acute to obtuse-cuneate, the apex abruptly acuminate; midrib yellowish, the venation plane adaxially, prominent abaxially. Inflorescences terminal or axillary, inserted toward apex of stem, paniculate thyrses, 20–42 cm long, sparsely flowered, velvety pubescent. Flowers: calyx 6–7 mm long, velvety pubescent, dull pale orange, the sepals 5–5.5 mm long, ovate; petals ca. 9 mm long, papillate adaxially, ferruginous sericeous-tomentose abaxially, the appendage as long as petal, sericeous, white; stamens 8. Fruits oblong-ellipsoid to nearly globose, 2 cm long, velvety pubescent, becoming glabrous, orange-yellow, the apex apiculate, the pericarp thickened, coriaceous, smooth. Seeds ellipsoid. Fl (Jul, Aug), fr (Oct, Nov); in nonflooded forest. *Bete* (Amerindian).

Talisia mollis Cambess. FIG. 287

Slender trees, 4–10 m tall, unbranched or with few ascending, sympodial branches on upper part. Bark grayish to dark brown, smooth or lenticellate. Stems terete, densely covered with soft, erect, yellow hairs 0.5–0.8 mm long, the pith sometimes hollow (housing ants) toward apex. Leaves even-pinnate, spirally arranged toward apex of stems; leaflets 10–16, opposite to alternate; rachis plus petiole 22–100 cm long, terete, or angular toward apex, slightly striate, pubescence same as stems; blades oblong to lanceolate, 9–25(42) × 2.2–8.5(12) cm, chartaceous, glabrous except for tomentulose midrib adaxially, pubescence same as stem but less dense abaxially, the base obtuse, unequal, the apex acuminate; venation impressed adaxially, prominent abaxially. Inflorescences terminal or axillary, inserted toward apex of stems, paniculate thyrses, 30–60 cm long, densely flowered, the rachis hirsute. Flowers: calyx light green, minutely hirsute, sometimes with a few glandular hairs, the sepals 3–4 mm long, oblong; petals whitish, 4–4.5 mm long, the appendages as long as petals, sericeous; stamens 8. Fruits ovoid to globose-ovoid, 2–2.5 cm long, densely to sparsely hirsutulous, yellow to orange-yellow at maturity, the apex apiculate, the pericarp woody, smooth, thickened. Seeds ellipsoid, with fleshy, yellowish to orange-yellow testa. Fl (Jan, Sep, Oct, Nov), fr (Jan, May); in nonflooded forest.

Talisia pedicellaris Radlk.

Trees, 12–14 m tall, with stout lateral branches on upper one-third of trunk. Bark brown or reddish brown, rough, thin, peeling in irregular plates, the inner bark reddish brown. Stems terete, puberulent, becoming lenticellate. Leaves even-pinnate; leaflets 4–6, opposite or alternate; rachis plus petiole 4.5–11 cm long, terete, minutely pubescent; petiolules slender, puberulent; blades ovate, widely oblong, elliptic, or widely elliptic, 4–14.5 × 1.8–6.6 cm, chartaceous, glabrous except for minutely hirsute or pilose primary and secondary prominent veins abaxially, the base obtuse, acute, to nearly rounded, sometimes slightly unequal, the apex long acuminate to caudate. Inflorescences from branches, inserted toward apex of stems, solitary or clustered simple thyrses, 4–8 cm long, puberulent. Flowers: calyx pale green, puberulent, the margins ciliate, the sepals 2–2.5 mm long, oblong to ovate; petals 2–2.5 mm long, whitish, the margins ciliate, involute at base, the appendages wanting; stamens 8. Fruits ellipsoid, apiculate, 1.8–2 cm long, glabrous, the pericarp coriaceous, granulose. Seeds ellipsoid. Fl (Aug); in nonflooded forest.

Talisia praealta (Sagot) Radlk.

Trees, to 27 m tall. Trunks buttressed. Bark light brown to grayish, rough, with numerous horizontal rows of lenticels, peeling in large irregular plates, the inner bark reddish brown. Stems terete, striate, minutely pubescent. Leaves even-pinnate; leaflets 2–4, opposite or subopposite; rachis plus petiole 1.5–5.5 cm long, slightly flattened adaxially, puberulent; petiolules slender, 2–6 mm long; blades elliptic, 5–12.5 × 1.9–5.5 cm, chartaceous, glabrous, the base acute or obtuse, decurrent onto petiolule, the apex obtuse or seldom shortly acuminate and retuse; midrib impressed adaxially, prominent abaxially, the secondary veins slightly prominent abaxially. Inflorescences axillary, solitary or congested simple thyrses, 2–4 cm long, puberulent. Flowers: calyx ferruginous tomentose, the sepals ca. 2.5 mm long, oblong; petals ca. 2.2 mm long. Fruits ellipsoid, minutely tomentose, light green, the pericarp nearly woody, smooth. Fr (Sep); in nonflooded forest.

Talisia simaboides Kramer

Trees, to 30 tall. Trunks buttressed. Bark grayish brown, smooth, with lenticels in horizontal rows, the inner bark orange or brown. Stems furrowed, minutely pubescent, becoming terete and lenticellate with age. Leaves even- or odd-pinnate; leaflets 5–8, alternate or subopposite; rachis plus petiole 4.5–20.5 cm long, terete, striate; petiolules slender, glabrous; blades oblong, elliptic, lanceolate or nearly obovate, 4–14 × 1.5–4.5 cm, chartaceous to subcoriaceous, minutely puberulent abaxially, especially along prominent midrib, the base obtuse or acute, decurrent onto elongate petiolule, the apex obtuse to shortly acuminate, usually retuse; venation inconspicuous. Inflorescences terminal, paniculate thyrses, 9–15 cm long, minutely, velvety pubescent. Flowers: calyx minutely pubescent, the sepals ovate, 2–2.5 mm long; petals lanceolate, ca. 3 mm long, appressed pubescent, whitish, the appendages as long as petals, sericeous, white; stamens 8. Known only by sterile collections from our area; in nonflooded forest.

Talisia sylvatica (Aubl.) Radlk. [Syns.: *Talisia micrantha* Radlk., *T. reticulata* Radlk. PL. 116d

Slender trees or treelets, 2–8 m tall, unbranched or with few ascending, sympodial branches on upper part. Stems minutely pubescent, dull, becoming lenticellate. Leaves even-pinnate, spirally arranged toward apex of stem; leaflets 8–14, alternate or opposite; rachis plus petiole (27)39–58(89) cm long, terete, obtusely angled toward apex of stem, puberulent; petiolules pulvinate and swollen; blades elliptic or oblong-elliptic, 7.4–26(40) × 3.5–9(11.2) cm, chartaceous, the base usually unequal, obtuse-acute to attenuate, the apex acuminate to long acuminate; primary and secondary veins prominent, puberulent abaxially, impressed adaxially. Inflorescences axillary, inserted toward apex of stems or seldom on leafless stems, paniculate thyrses, 15–30 cm long, densely flowered, the axes pinkish tinged, pilose. Flowers: calyx 1.7–2 mm long, pilose, pinkish tinged, the sepals ovate; petals cream-colored, glabrous, 3.5–4 mm long, the appendages as long as petals, sericeous, white; nectary disc hispidulous to tomentose at apex; stamens 5, the filaments tomentose. Fruits ellipsoid, ca. 2 cm long, puberulent, turning from green to yellow-orange, the apex apiculate, the pericarp coriaceous, granulose. Seeds ellipsoid, with creamy yellow fleshy testa. Fl (May, Jun), fr (Feb, Mar, Apr, May, Sep, Oct, Nov, Dec); in nonflooded forest. *Bois flambeau*, *gaulette indien* (Créole), *tepu*, *touliatan*, *tepuime* (Amerindian).

Plates 113–120

Plate 113. RUBIACEAE. **a.** *Palicourea longiflora* (*Mori et al. 24009*), immature fruits; note brightly colored peduncle and pedicels. **b.** *Psychotria medusula* (*Mori et al. 22115*), fresh lavender flowers, old orange flowers, and buds. **c.** *Psychotria granvillei* (*Mori et al. 22969*), inflorescence. **d.** *Psychotria microbotrys* (*Mori & Gracie 21148*), flowers and buds. **e.** *Psychotria iodotricha* (*Mori et al. 21554*), flower. **f.** *Psychotria oblonga* (*Mori & Gracie 18990*), flower and stipules. **g.** *Psychotria platypoda* (*Mori & Gracie 18889*), infructescence of immature fruits; note red-purple bracts.

Plate 114. RUBIACEAE. **a.** *Psychotria poeppigiana* (*Mori et al. 22895*), inflorescence. **b.** *Psychotria poeppigiana* (*Mori & Gracie 18861*), infructescence. **c.** *Psychotria racemosa* (*Mori et al. 18206*), infructescences.

Plate 115. RUBIACEAE. **a.** *Psychotria variegata* (*Mori et al. 18179*), infructescence. **b.** *Spermacoce alata* (*Mori & Gracie 22142*, photographed at Stoupan, French Guiana), flowers. **c.** *Spermacoce verticillata* (*Mori et al. 22084*), inflorescences. RUTACEAE. **d.** *Zanthoxylum ekmanii* (*Mori et al. 21661*), trunk with prickles. SAPINDACEAE. **e.** *Allophylus latifolius* (*Mori et al. 23316*), flowers and buds. **f.** *Paullinia rubiginosa* (*Mori & Gracie 23857*), flowers and buds; note brown bracteoles.

Plate 116. SAPINDACEAE. **a.** *Paullinia tricornis* (*Mori et al. 23894*), infructescence; note white aril subtending black seed. **b.** *Pseudima frutescens* (*Mori et al. 22666*), fruits; note white arils subtending black seeds. **c.** *Serjania grandifolia* (*Mori et al. 22875*), winged fruits. **d.** *Talisia sylvatica* (*Mori & Gracie 18928*), part of inflorescence.

Plate 117. SAPINDACEAE. **a.** *Talisia megaphylla* (*Mori et al. 23121*), flowers and buds. SAPOTACEAE. **b.** *Chrysophyllum argenteum* subsp. *auratum* (*Mori et al. 24189*), flowers and buds. **c.** *Chrysophyllum cuneifolium* (*Mori et al. 23371*), cauliflorous inflorescence with flowers and buds.

Plate 118. SAPOTACEAE. **a.** *Chrysophyllum cuneifolium* (*Mori et al. 22340*), fruit and transverse sections of fruits. **b.** *Ecclinusa lanceolata* (unvouchered), white latex dripping from slash of trunk. **c.** *Micropholis obscura* (*Mori et al. 20955*), flowers and buds; note nectar in flowers. **d.** *Pouteria cayennensis* (*Mori & Gracie 23200*), flower and buds; note style protruding from buds. **e.** *Pouteria speciosa* (*Mori & Gracie 21170*), fruit and seed showing large, pale hilar scar.

Plate 119. SCROPHULARIACEAE. **a.** *Scoparia dulcis* (*Mori et al. 22075*), flowers. **b.** *Scoparia dulcis* (*Mori et al. 22075*), upper part of plant with inflorescences. **c.** *Capraria biflora* (unvouchered, photographed in Florida), flower and bud. SIMAROUBACEAE. **d.** *Simaba guianensis* subsp. *guianensis* (*Mori et al. 24690*), fruit of three drupaceous monocarps. [Photo by S. Mori]

Plate 120. SOLANACEAE. **a.** *Cyphomandra endopogon* subsp. *guianensis* (*Mori & Gracie 19054*), inflorescence with fresh lavender flower, old white flower, and buds. **b.** *Lycianthes pauciflora* (*Mori & Gracie 18907*), fruits with reflexed calyx lobes. **c.** *Markea sessiliflora* (*Mori et al. 22739*), flower. **d.** *Markea coccinea* (*Mori et al. 23779*), flower and calyx after corolla has fallen. [Photo by S. Mori]

Palicourea longiflora a.

Psychotria medusula b.

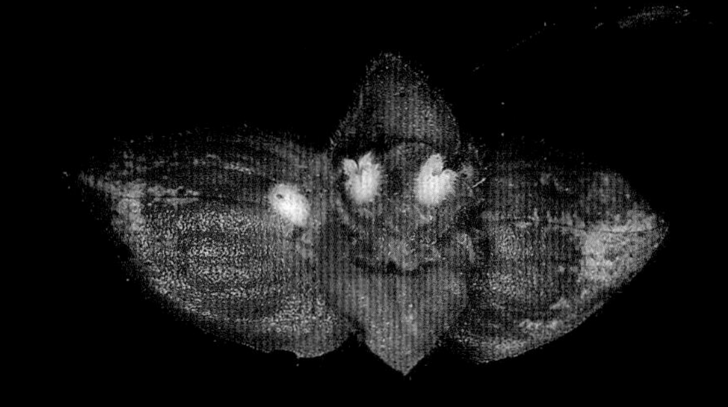

Psychotria granvillei c.

Psychotria iodotricha e.

Psychotria microbotrys d.

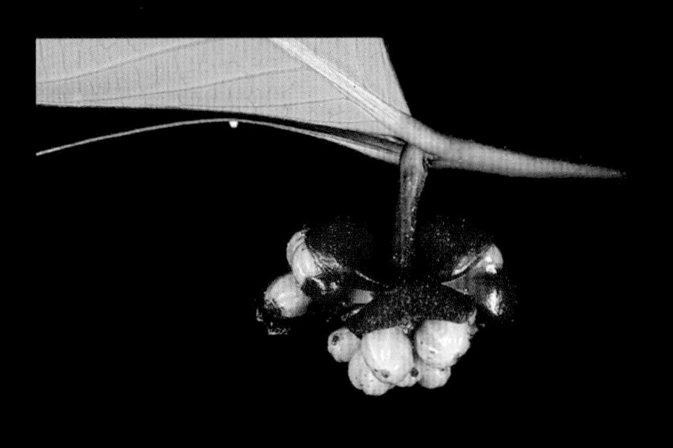

Psychotria platypoda g.

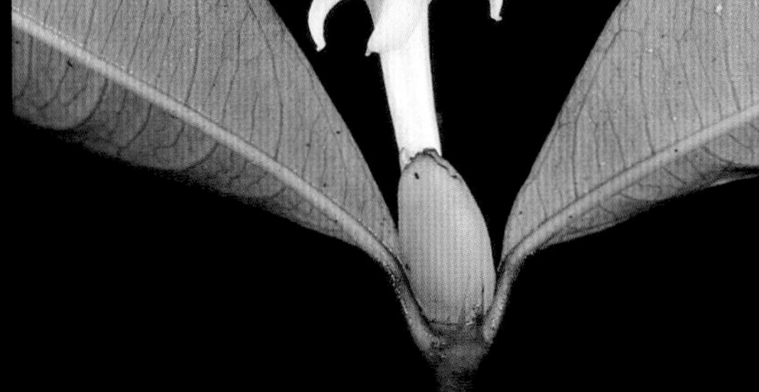

Psychotria oblonga f.

Psychotria poeppigiana

a.

Psychotria poeppigiana

b.

Psychotria racemosa

c.

Psychotria variegata a.

Spermacoce verticillata c.

Spermacoce alata b.

RUTACEAE

Zanthoxylum ekmanii d.

SAPINDACEAE

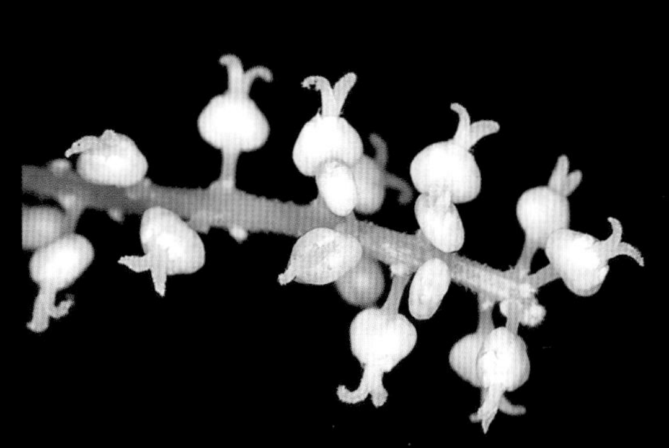

Allophylus latifolius e.

Paullinia rubiginosa f.

Plate 115

Paullinia tricornis a.

Pseudima frutescens b.

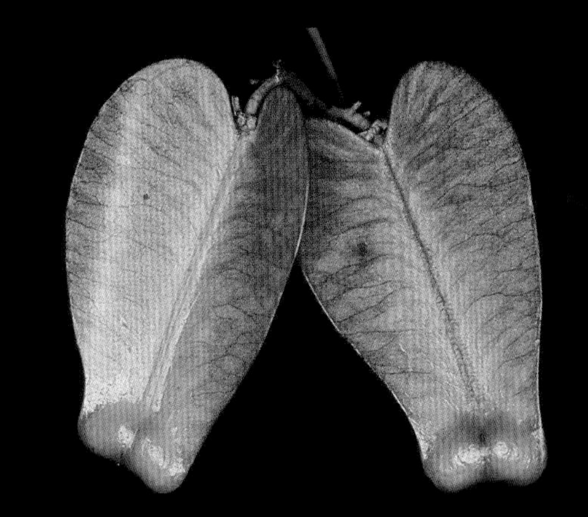

Serjania grandifolia c.

Talisia sylvatica d.

Plate 116

Talisia megaphylla

a.

SAPOTACEAE

Chrysophyllum argenteum subsp. auratum

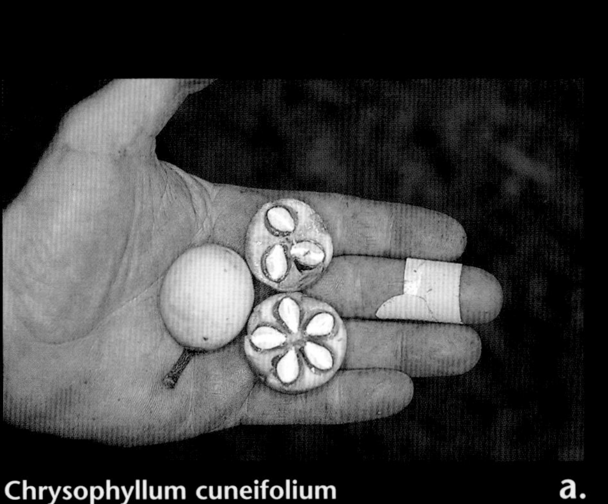

Chrysophyllum cuneifolium **a.**

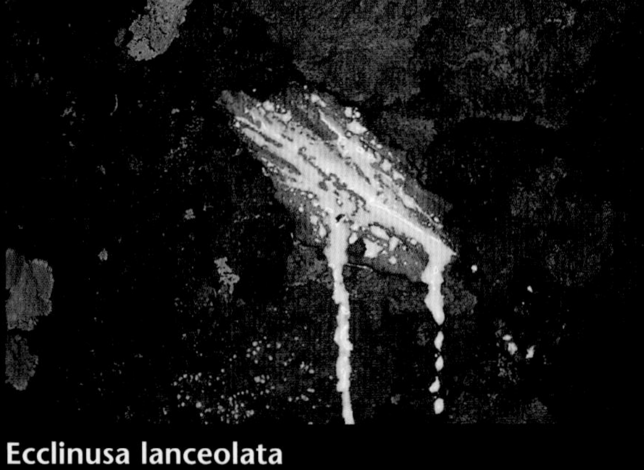
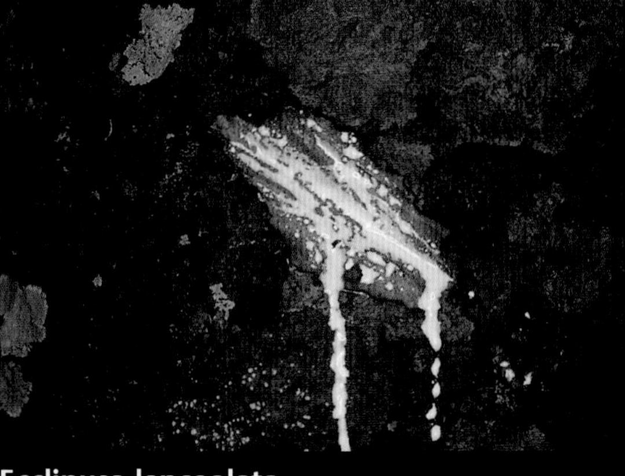

Ecclinusa lanceolata **b.**

Micropholis obscura **c.**

Pouteria cayennensis **d.**

Pouteria speciosa **e.**

Plate 118

SCROPHULARIACEAE

a.

Capraria biflora c.

Scoparia dulcis b.

SIMAROUBACEAE

Simaba guianensis subsp. guianensis d.

Plate 119

Cyphomandra endopogon subsp. guianensis a.

Lycianthes pauciflora b.

Markea sessiliflora c.

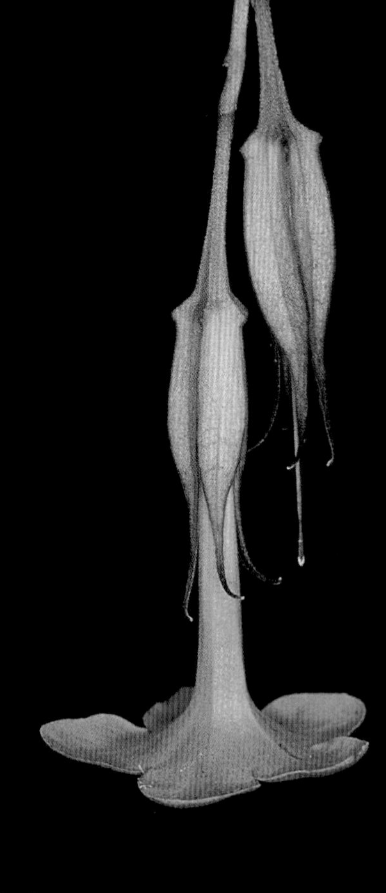

Markea coccinea d.

Plate 120 (Solanaceae continued on Plate 121)

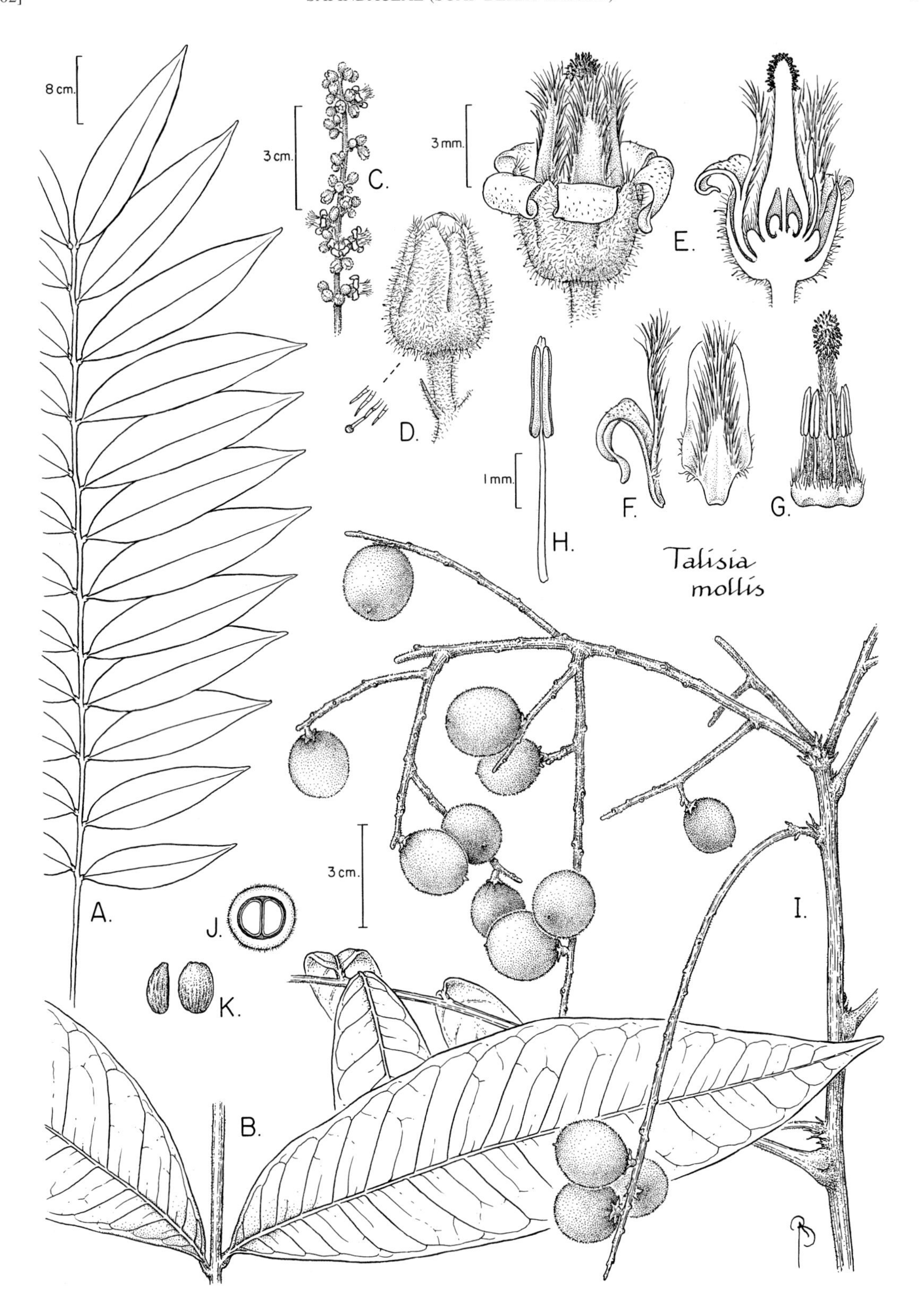

FIG. 287. SAPINDACEAE. *Talisia mollis* (A, B, I–K, *Acevedo-Rodríguez et al. 4953*; C–H, *Mori et al. 21607*). **A.** Leaf. **B.** Detail of leaf showing two subopposite leaflets. **C.** Part of inflorescence. **D.** Lateral view of flower bud with detail of hairs. **E.** Lateral view (left) and medial section (right) of flower. **F.** Lateral (near right) and adaxial (far right) views of petals with appendages. **G.** Lateral view of pistillate flower with petals and appendages removed. **H.** Adaxial view of stamen. **I.** Part of stem with infructescences. **J.** Transverse section of fruit. **K.** Two views of seeds.

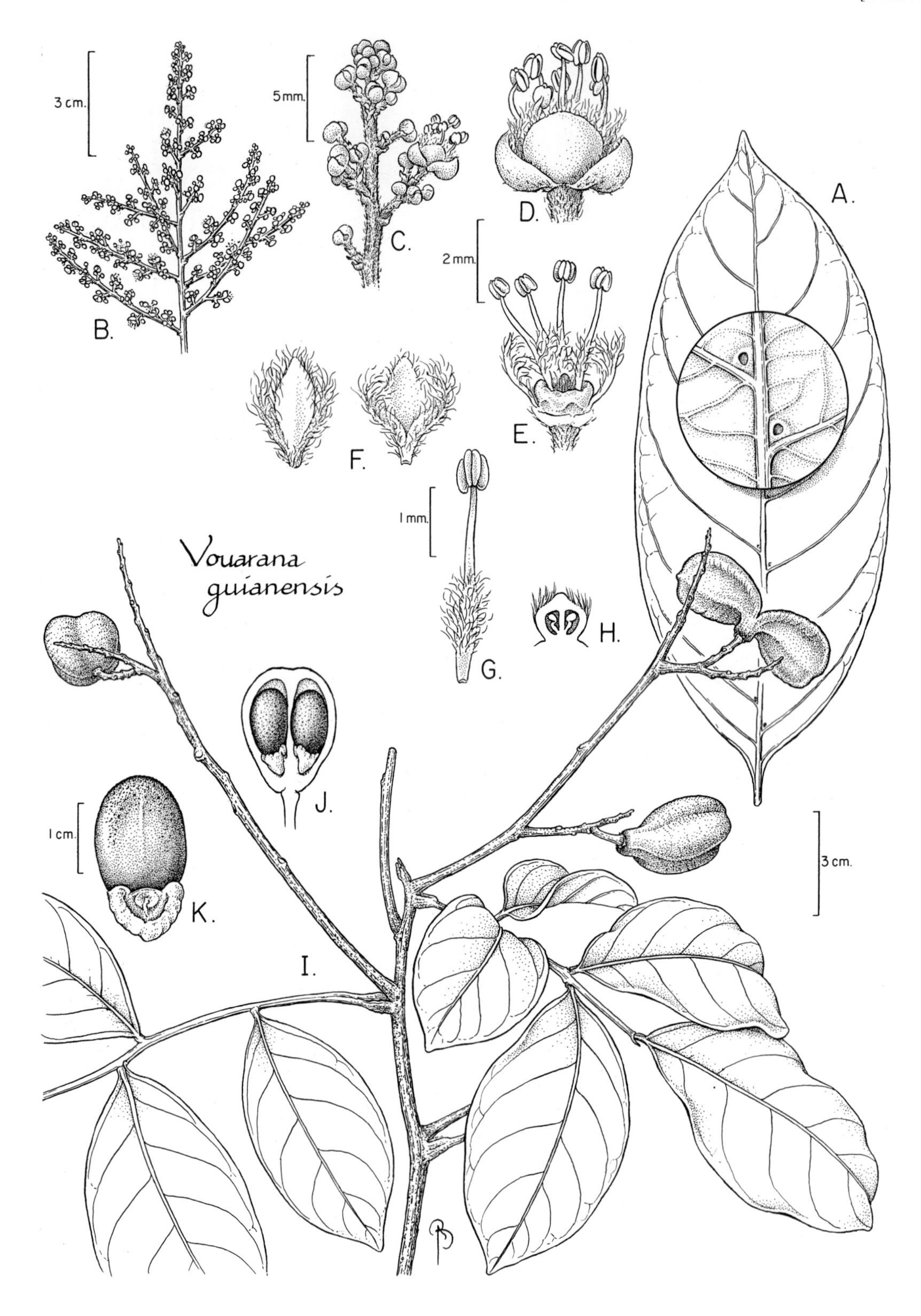

FIG. 288. SAPINDACEAE. *Vouarana guianensis* (A–H, *Sang LBB 16232* from Surinam; I–K, *Mori et al. 22318*). **A.** Abaxial surface of leaflet with enlarged portion showing domatia. **B.** Apical part of inflorescence. **C.** Detail of inflorescence. **D.** Lateral view of flower. **E.** Lateral view of flower with sepals and two petals removed. **F.** Abaxial (left) and adaxial (right) views of petals. **G.** Stamen. **H.** Medial section of ovary showing stalked ovules. **I.** Part of stem with leaves and fruits. **J.** Medial section of fruit with two seeds. **K.** Seed with basal arillode.

THINOUIA Triana & Planch.

Lianas of forest canopy or disturbed areas. Stems terete or trilobed, lenticellate, becoming warty with age, the transverse section of stem with single stele, the older stems with many cortical steles. Stipules minute, early deciduous. Leaves alternate, trifoliolate or less often biternate; petioles and rachises not winged; leaflets coriaceous. Inflorescences axillary, umbellate or racemose thyrses, with lateral cincinni, seldom bearing coiled tendrils. Flowers actinomorphic; calyx cup-shaped, the sepals 5, distinct; petals 5, spatulate, distinct, auriculate or with petaloid appendage; nectary disc annular; stamens 6–8, the filaments of equal length, free; ovary 3-locular, each locule with a single ovule. Fruits schizocarps, splitting into 3 samaroid mericarps with distal wing. Seeds subglobose, exarillate.

Thinouia myriantha Triana & Planch.

Lianas, not producing milky sap. Stems terete, glabrous. Leaves trifoliolate; petioles 5.5–12 cm long, slender, not winged; rachis slender, not winged; leaflets 6.5–14(18) × 3.5–7(10) cm, chartaceous, glabrous except for barbate vein axils, the apex long apiculate, the margins serrate, the terminal leaflets elliptic, with obtuse to rounded base, the lateral leaflets ovate, asymmetrical, with cuneate to cordate base. Inflorescences axillary or terminal, racemose thryses with lateral umbellate conglomerates of long-peduncled cincinni. Flowers: calyx cup-shaped, 0.7–1 mm long, pubescent; petals cuneiform, ca. 0.7 mm long, the appendages longer than petals. Mericarps 4–5.5 cm long, the cocci slightly flattened, glabrous. Fl (May), fr (Feb); along disturbed forest margins.

VOUARANA Aubl.

Medium to large trees of moist forests. Leaves alternate, pinnate, with rudimentary terminal leaflet; leaflets opposite or subopposite, usually foveolate at vein axils. Inflorescences axillary or terminal, paniculate thyrses with flowers in lateral dichasia. Flowers slightly zygomorphic, functionally unisexual (plants polygamous-monoecious); sepals 4–5, distinct, concave, as long as petals, the two outer ones smaller; petals 4–5, rhombic, auriculate; nectary disc annular and lobed; stamens 6–8, the filaments subulate and hirsute, free, the anthers dorsifixed; ovary 2-locular, each locule with a single ovule. Fruits woody capsules, pyriform, 1–2-seeded. Seeds ellipsoid, with triangular, white or yellowish arillode at base.

Vouarana guianensis Aubl. FIG. 288; PART 1: PL. XVIa

Trees, 12–15 m × to 25 cm diam. Bark smooth, brown, the inner bark reddish. Stems furrowed, puberulent. Leaves odd-pinnate; leaflets 5–9, alternate; rachis plus petiole 8–25 cm long, slightly flattened and angular, striate, puberulent; petiolules slender, puberulent; leaflets oblong or elliptic, 7–17 × 3.2–7.5 cm, chartaceous, glabrous or minutely puberulent abaxially, especially along prominent midrib, the vein axils foveolate, the base obtuse or acute, decurrent onto elongate petiolule, the apex obtuse to shortly acuminate, the margins entire. Inflorescences terminal, 9–25 cm long, the axes minutely pubescent. Flowers: calyx puberulent, the sepals ovate, 1.5–1.7(2) mm long, the margins ciliate; petals as long as sepals, the margins ciliate. Capsules woody, ovate, compressed, brown. Seeds brown, with trigonous, yellowish aril at base. Fr (May); in nonflooded forest.

SAPOTACEAE (Sapodilla Family)

Terence D. Pennington

Canopy or emergent (rarely understory) trees. Latex sticky, white, rarely yellow, present in bark, stems, leaves, and fruit. Indumentum of malpighiaceous hairs, rarely plants glabrous. Stipules usually minute or absent, large stipules only in *Ecclinusa*. Leaves simple, usually spirally arranged and clustered at branch ends, less frequently alternate and distichous, rarely opposite or verticillate; blades with entire margins; venation eucamptodromous or brochidodromous. Inflorescence axillary, ramiflorous, or cauliflorous, fasciculate. Flowers actinomorphic, bisexual, less frequently unisexual; calyx single whorl of 4–5 free or slightly united imbricate sepals or of 2 whorls of 3 sepals each and then the outer whorl valvate (*Manilkara*); corolla gamopetalous, the lobes 4–6, mostly entire, partly or completely divided into 3 segments in *Manilkara*; stamens 4–6, adnate to corolla tube (free from corolla in *Pouteria coriacea* and *P. laevigata*), opposite corolla lobes, included or less frequently exserted from corolla, the anthers usually extrorse; staminodes present or absent, when present in single whorl alternating with stamens or inserted in corolla tube sinus, simple or divided; annular nectary disc rarely present, then surrounding base of ovary; ovary superior, 1–6(9)-locular, the locules uniovulate, the style simple, included or exserted; placentation axile. Fruits berries or rarely drupes, the pericarp coriaceous or fleshy. Seeds 1 to several per fruit, globose or elliptic to strongly laterally compressed, typically with hard, smooth, shiny, dark brown testa separating easily from pericarp, less frequently testa roughened or wrinkled and adhering to pericarp, with prominent, roughened, adaxial or rarely basal hilar scar, the scar sometimes covering most of seed surface; endosperm present or absent; embryo with thin, foliaceous cotyledons and exserted radicle or with thick, plano-convex cotyledons with included radicle.

Eyma, P. J. 1966. Sapotaceae. *In* A. Pulle (ed.), Flora of Suriname **IV(1):** 354–399. E. J. Brill, Leiden.
Pennington, T. D. 1990. Sapotaceae. Fl. Neotrop. Monogr. **52:** 1–770.

1. Bark markedly fissured. Calyx of 2 whorls of 3 sepals each, the outer whorl valvate; corolla lobes usually divided to base into 3 segments. Seeds usually with small, narrow, basi-ventral scar. . . . . *Manilkara.*
1. Bark various, usually not markedly fissured. Calyx of a single whorl of 4–5 imbricate sepals; corolla lobes simple. Seeds with lateral (adaxial) or basal (*Diploön*), elongate, narrow, or broad scar.
   2. Stipules present, well developed, leaving conspicuous scar. Flowers sessile. . . . . *Ecclinusa.*
   2. Stipules absent (small stipules present in *Pouteria* aff. *flavilatex*). Flowers nearly always pedicellate.
      3. Leaves usually alternate and distichous; venation finely striate because of closely parallel secondary and higher order venation.
         4. Corolla cyathiform or tubular; staminodes present; stamens included or exserted. Seeds with endosperm, the scar adaxial. . . . . *Micropholis.*
         4. Corolla rotate; staminodes absent; stamens exserted. Seeds without endosperm, the scar basal. . . . . . *Diploön.*
      3. Leaves spirally arranged (alternate in *Sarcaulus brasiliensis* and *Chrysophyllum argenteum* and opposite in some *Pouteria eugeniifolia*), the venation not finely striate.
         5. Staminodes nearly always present, as many as corolla lobes.
            6. Leaves spirally arranged. Corolla and staminodes not carnose; stamens and staminodes not inflexed against style, the anthers included except in *Pouteria eugeniifolia*. . . . . *Pouteria.*
            6. Leaves alternate and distichous or weakly spirally arranged. Corolla and staminodes carnose; stamens and staminodes inflexed against style, the anthers exserted. . . . . *Sarcaulus.*
         5. Staminodes absent (minute staminodes present in *Chrysophyllum pomiferum*).
            7. Corolla rotate; stamens inserted at base of corolla lobes, exserted. . . . . *Pradosia.*
            7. Corolla cyathiform or tubular; stamens inserted within corolla tube, included.
               8. Abaxial leaf surface slightly glaucous. Ovary 2–4-locular. Seeds without endosperm. . . . . *Pouteria ambelaniifolia.*
               8. Abaxial leaf surface not glaucous. Ovary 5-locular. Seeds with endosperm. . . . . . *Chrysophyllum.*

CHRYSOPHYLLUM L.

Trees. Leaves alternate, distichous, or spirally arranged. Flowers unisexual or bisexual; calyx single whorl of 5 imbricate sepals, these frequently ciliate; corolla globose to cylindrical, the tube shorter or longer than lobes, the lobes 5; stamens 5, inserted on corolla tube, included; staminodes usually absent (minute staminodes present in *C. pomiferum*); ovary 5-locular, the style included. Fruits several-seeded berries. Seeds laterally compressed, with smooth, shiny or roughened testa, the seed scar narrow (broader and heart-shaped in *C. argenteum*), adaxial; endosperm present; embryo with thick, flat, or foliaceous cotyledons, the radicle exserted.

1. Leaves alternate, distichous. Seeds with broad, heart-shaped, basi-ventral scar.
   2. Leaf blades elliptic, sparsely to densely sericeous abaxially. . . . . *C. argenteum* subsp. *auratum.*
   2. Leaf blades oblong, glabrous abaxially. . . . . *C. argenteum* subsp. *nitidum.*
1. Leaves spirally arranged. Seeds with narrow, adaxial scar.
   3. Seed coat rough, not shiny, adherent to pericarp.
      4. Petioles 3–5 mm long; blades oblanceolate, glabrous, long tapering into rounded or truncate base. . . . . *C. cuneifolium.*
      4. Petioles ≥10 mm long; blades broadly oblanceolate or obovate, lower surface rufous-brown sericeous, not tapering into rounded or truncate base. . . . . *C. prieurii.*
   3. Seed coat smooth, shiny, free from pericarp.
      5. Sepals accrescent in fruit. Seed scar adaxial, extending around base. . . . *C. sanguinolentum* subsp. *sanguinolentum.*
      5. Sepals not accrescent in fruit. Seed scar adaxial, not extending around base.
         6. Leaf blades 6–14 × 2.5–5.5 cm. Fruits 2.5–3 cm long.
            7. Leaf blades oblanceolate, the apex acute; secondary veins slightly arcuate and convergent. Fruits hard-skinned, with pale lenticellate area around apex. . . . . *C. pomiferum.*
            7. Leaf blades elliptic, the apex caudate; secondary veins strongly arcuate and convergent. Fruits soft-skinned, without pale lenticellate area at apex. . . . . *C.* aff. *venezuelanense.*
         6. Leaf blades 15–32 × 6–12 cm. Fruits 3.5–5 cm long. . . . . *C. lucentifolium* subsp. *pachycarpum.*

Chrysophyllum argenteum Jacq. subsp. **auratum** (Miq.) T. D. Penn. PL. 117b

Trees, to 20 m tall. Trunks often with buttresses. Bark grayish, fissured, the slash with abundant white latex. Leaves: blades elliptic to broadly elliptic, 9–12 × 4–5.5 cm, sparsely to densely sericeous abaxially. Inflorescences mostly axillary. Flowers: corolla cylindric, 4–6 mm long, sericeous, light green or greenish-white. Fruits ellipsoid, 2–3 cm long, glabrous, purplish-black. Seeds solitary, with broad, basi-ventral scar. Fl (Feb, Jun, Jul, Aug, Sep, Oct), fr (Feb, Apr, Oct); common, in wide range of forest types. *Bois rouge* (Créole), *niam-boka* (Bush Negro), *wilapila* (Amerindian).

Chrysophyllum argenteum Jacq. subsp. **nitidum** (G. Meyer) T. D. Penn.

Trees, to 7 m tall. Leaves: blades oblong, 6–10 × 3–4 cm, glabrous abaxially. Inflorescences mostly axillary. Flowers: corolla cylindric, ca. 5 mm long, sericeous, yellow-green. Fl (Sep); apparently rare, in nonflooded forest.

Chrysophyllum cuneifolium (Rudge) A. DC.
PLS. 117c, 118a; PART 1: PL. XIVa,b

Trees, to 15 m tall. Trunks with low, thin, sometimes branched buttresses, the bole sometimes fluted. Bark light brown, with fine vertical cracks, the slash buff-colored. Leaves: petioles 3–5 mm long; blades oblanceolate, 11–22 × 3.3–7 cm, glabrous or sparsely pubescent on midrib abaxially, the apex acute to narrowly attenuate or acuminate; venation eucamptodromous, the secondary veins in 12–19 pairs. Inflorescences ramiflorous and cauliflorous to ground level. Flowers: sepals fringed-ciliate; corolla 4–4.5 mm long, greenish-white. Fruits obovate, the style persistent, 3.5 × 2.5 cm, edible, the young fruits with easily rubbed off indument. Seeds (1)5 per fruit, laterally compressed, the scar 1.5–2 mm wide, extending around base of seed, the testa rough, adherent to pulp. Fl (Aug), fr (Mar, May, Jun); scattered, in nonflooded forest. *Couata bealy* (Bush Negro), *guilapele* (Amerindian), *kwatabobi* (Bush Negro), *oulapele*, *quilapele*, *wilapele* (Amerindian), *zolive* (Créole).

Chrysophyllum lucentifolium Cronquist subsp. **pachycarpum** Pires & T. D. Penn.

Trees, to 30 m tall. Trunks with simple, spreading buttresses to 80 cm high, the bole cylindric. Bark brown, scaling and forming dipples, the slash cream-colored, with sticky, white latex. Leaves: blades broadly oblanceolate, 15–32 × 6–12 cm, glabrous, the apex obtuse to rounded; venation eucamptodromous, the secondary veins in 14–15 pairs. Inflorescences mostly from stems below leaves. Flowers: corolla 2–3 mm long, pale green. Fruits edible, subglobose, 3.5–5 cm long, glabrous, yellow. Seeds to 5 per fruit, laterally compressed with smooth shiny testa, the scar 1.5–3 mm wide. Fr (Oct); apparently rare, in nonflooded forest.

A possibly distinct species but presently determined as *C. lucentifolium* is represented by *Boom & Mori 1825*, *1916*, *1956*, *1959*, *1969*, *2004*, *2056* (all at NY), and *Pennington & Mori 12188* (K, NY). It differs from *C. lucentifolium* in the larger leaves to 32 × 12 cm, the fewer (11–13) secondary veins, and the acuminate leaf apex. The globose, hard-skinned fruits are orange (ca. 3.5 cm diam.), contain several laterally compressed seeds ca. 2 cm long, and posseses a wrinkled testa. The seed scar is 1–2 mm wide.

Chrysophyllum pomiferum (Eyma) T. D. Penn.

Trees, to 38 m tall. Trunks often buttressed. Bark grayish to red-brown, fissured, the slash exuding milky, white latex. Leaves: blades oblanceolate, 6–12 × 2.5–4 cm, glabrous, the apex acute to slightly attenuate. Inflorescences axillary and from stems below leaves. Flowers: corolla 3–4 mm long, greenish-white; staminodes present, minute. Fruits edible, globose, 2.5–3 cm long, smooth, glabrous, yellow, with pale lenticellate area at apex. Seeds several per fruit, laterally compressed, the scar 1–1.5 mm wide. Known only by sterile collections in our area; in nonflooded forest. *Balata jaune d'oeuf* (Créole).

Chrysophyllum prieurii A. DC.

Trees, to 31 m tall. Trunks with steep buttresses to 75 cm high, the bole cylindric, the crown sparse, the branches ascending. Bark mid-brown, finely and shallowly fissured, the ridges peeling into small, friable pieces, the slash pinkish-brown, streaked, with sticky, white latex. Leaves: blades broadly lanceolate to obovate, 9–13 × 5.4–7.5 cm, with reddish-brown, sericeous indumentum on abaxial surface, the hairs patent when young. Inflorescences from branches. Flowers: sepals fringed; corolla 2.5–3 mm long, yellowish-white. Fruits ellipsoid, 3.5–5 cm long, smooth. Seeds several(5), 2–3 cm long, laterally compressed, the scar 1.5–3 mm wide. Fl (Sep, Oct, Nov), fr (Sep); apparently rare, in nonflooded forest. *Balata jaune d'oeuf*, *balata poirier*, *jaune d'oeuf à grandes feuilles* (Créole), *malobi*, *pepe boiti* (Bush Negro), *pepe poirier*, *zolive* (Créole).

Chrysophyllum sanguinolentum (Pierre) Baehni subsp. **sanguinolentum** PART 1: FIG. 10

Trees, to 35 m tall. Trunks with steep, simple buttresses to 75 cm high, the bole often fluted. Bark reddish or rusty-brown, smooth in young individuals, becoming scaly with conspicuous dipples, the slash pink or red, with copious, sticky, white latex. Leaves: blades broadly oblanceolate, 15–35 × 6–13 cm. Inflorescences mostly from branches. Flowers: corolla 3–5 mm long, greenish, the lobes exceeding tube. Fruits edible, globose, 3–6 cm long, pale brown, rough-skinned, with accrescent sepals. Seeds several per fruit, laterally compressed, the scar 1–1.5 mm wide, extending around base of seed. Fr (Mar, May, Nov); common, in nonflooded forest. Latex formerly used as a substitute for rubber. *Balata singe rouge* (Créole), *mama-doosou*, *suitiamini* (Bush Negro).

Chrysophyllum aff. **venezuelanense** (Pierre) T. D. Penn.

Treelets, to 5 m tall. Leaves: blades elliptic, 10–15 cm long, glabrous, the apex caudate. Inflorescences on stems below leaves. Fruits globose, 2–2.5 cm diam, soft-skinned, glabrous, yellow-orange. Seeds several, ca. 1.5 cm long, strongly laterally compressed, with narrow adaxial scar. Fr (Sep); represented by a single collection from our area (*Mori & Gracie 21199*, NY).

Differs from typical *C. venezuelanense* in the smaller leaves with a caudate apex and the smaller fruit.

DIPLOÖN Cronq.

Trees. Trunks with buttresses. Bark thin, scaly. Stipules absent. Leaves alternate, distichous; venation brochidodromous, the secondaries joining and forming submarginal vein, the intersecondary veins long, usually extending to margin. Inflorescence axillary or from stem below leaves. Flowers bisexual; calyx with single whorl of 4–5, imbricate, free sepals; corolla rotate, with short tube and 4–5 larger lobes; stamens inserted at top of corolla tube, exserted; staminodes absent; ovary 1-locular, with 2 basal ovules. Fruits berries. Seeds solitary, broadly ellipsoid, with basal scar; endosperm absent; embryo with plano-convex cotyledons, the radicle included.

Diploön cuspidatum (Hoehne) Cronq.

Trees, to 20 m tall. Trunks with steep, narrow buttresses to 1.5 m tall. Bark dark brown, scaling in small, thin, irregular pieces, the slash ca. 2 mm thick, creamy brown, with white latex. Leaves: blades oblong-elliptic, 6.5–10 × 2–4 cm, glabrous, the apex caudate; secondary veins parallel, straight, in 17–21 pairs. Flowers: corolla greenish-white, 2.5–3 mm long, the tube ca. 0.5 mm long. Fruits 1.5–2 cm long, ripening reddish to black. Fr (Feb); known from a single collection from our area.

ECCLINUSA Mart.

Understory or canopy trees. Trunks unbuttressed. Bark dippled or peeling in irregular plates. Stipules present, well developed, falling but leaving conspicuous scar. Leaves spirally arranged, clustered; venation eucamptodromous, the tertiary veins oblique, numerous, close, parallel. Inflorescences axillary or from stem below leaves, the flowers sessile. Flowers unisexual (plants monoecious or dioecious); calyx with single whorl of 5, free, imbricate sepals; corolla campanulate or short tubular, the lobes 5, usually exceeding tube in length, simple; stamens 5, included, the anthers extrorse; staminodes absent; ovary 5–9-locular. Fruits berries. Seeds 1 to several per fruit, broad or laterally compressed, with adaxial scar extending around base; endosperm absent; embryo with plano-convex cotyledons, the radicle included.

1. Indumentum of young shoots and leaves fine, closely appressed.
    2. Stipules ca. 5 mm long. Leaf blades 12–16 × 3.5–4.5 cm. . . . . *E. guianensis.*
    2. Stipules 10–20 mm long. Leaf blades 30–40 × 9–17 cm. . . . . *E. lanceolata.*
1. Indumentum of young shoots and leaves of crisped or spreading hairs, not appressed. . . . . *E. ramiflora.*

Ecclinusa guianensis Eyma

Trees, to 20 m tall. Trunks unbuttressed, the bole cylindric, the crown small and dense. Bark dull grayish-brown, smooth or scaling in minute, rectangular pieces, with some incomplete hoop marks, the slash pale brown, with copious, flowing, white latex. Leaves: blades elliptic or oblong-elliptic, 12–16 × 3.5–4.5 cm, subglabrous, the apex narrowly attenuate. Flowers unisexual (plants monoecious), fragrant; corolla 2–2.5 mm long, with 5 lobes, greenish-white. Fruits globose, 1.5–3 cm long, yellow, puberulous. Known from a single sterile collection from our area; in nonflooded forest. *Bagasse, balata blanc* (Créole), *bataballi* (Bush Negro), *mapa* (Créole).

Ecclinusa lanceolata (Mart. & Eichler) Pierre
FIG. 289, PL. 118b

Trees, to 22 m tall. Trunks usually unbuttressed, the bole cylindric, the crown narrow, with widely spreading branches. Bark dark brown or black, strongly dippled, the slash cream-colored to orange-brown, laminated, to 1 cm thick, with copious, sticky, white latex. Leaves densely clustered; blades oblanceolate, 30–40 × 9–17 cm, subglabrous, the apex obtuse to obtusely cuspidate. Flowers unisexual (plants monoecious); corolla ca. 3–3.5 mm long, with 5–7 lobes, greenish-white. Fruits edible, globose, 5–6 cm diam., puberulous or glabrous, orange-yellow. Fl (Aug, Sep), fr (Feb, Mar); common, in nonflooded forest.

Ecclinusa ramiflora Mart.

Trees, to 20 m tall. Trunks unbuttressed, the bole cylindric. Bark dark brown, reddish-brown, or dark gray, scaling, with conspicuous dipples, the slash cream-colored to pinkish-brown, with copious, sticky, white latex. Leaves: blades oblanceolate, 15–30 × 5–9 cm, pubescent abaxially, the apex narrowly attenuate. Flowers unisexual (plants dioecious); corolla ca. 3.5 mm long, with 5 lobes, white toward base, green at apex. Fruits globose, 2.5–5 cm diam., densely pubescent, yellow. Fl (Aug, Sep, Oct), fr (Mar); common, in nonflooded forest.

MANILKARA Adans.

Canopy or emergent trees. Bark deeply fissured, the inner bark red or pink. Branching sympodial, the shoot apex often covered with transparent, varnish-like resin. Leaves usually densely clustered; venation brochidodromous. Inflorescences axillary or from branches just below leaves, solitary or fasciculate. Flowers bisexual; calyx of 2 whorls of 3 sepals each, the outer whorl valvate; corolla lobes 6, each lobe usually divided to base into 3 segments (entire or only partly divided in *M. zapota*), the median segment erect, clawed, clasping stamen, the lateral segments widely spreading; stamens 6, inserted at top of corolla tube, the filaments free or partially fused with staminodes; staminodes 6, well developed, entire or divided, alternating with stamens; ovary 6(12)-locular, the style exserted. Fruits 1- to several-seeded berries. Seeds strongly laterally compressed, with hard shiny testa, the hilar scar basi-ventral or adaxial, narrow; endosperm present; embryo with foliaceous cotyledons, the radicle exserted.

1. Corolla tube exceeding lobes, the lobes undivided or only partly divided at apex. Cultivated. . . . . *M. zapota.*
1. Corolla tube much shorter than lobes, the lobes divided nearly to base into 3 segments. Native.
    2. Leaf blades with fine, closely appressed, yellowish indumentum abaxially.
        3. Petioles 3.5–5 cm long; blades broadly ovate or elliptic, 14–20 × 5.5–9 cm; secondary and higher order venation easily visible on abaxial surface. Ovary puberulous. . . . . *M. huberi.*
        3. Petioles <1 cm long; blades oblanceolate, 7–8 × 2.5–3 cm; secondary and higher order venation inconspicuous on abaxial surface. Ovary glabrous. . . . . *M. paraensis.*
    2. Leaf blades glabrous abaxially. . . . . *M. bidentata.*

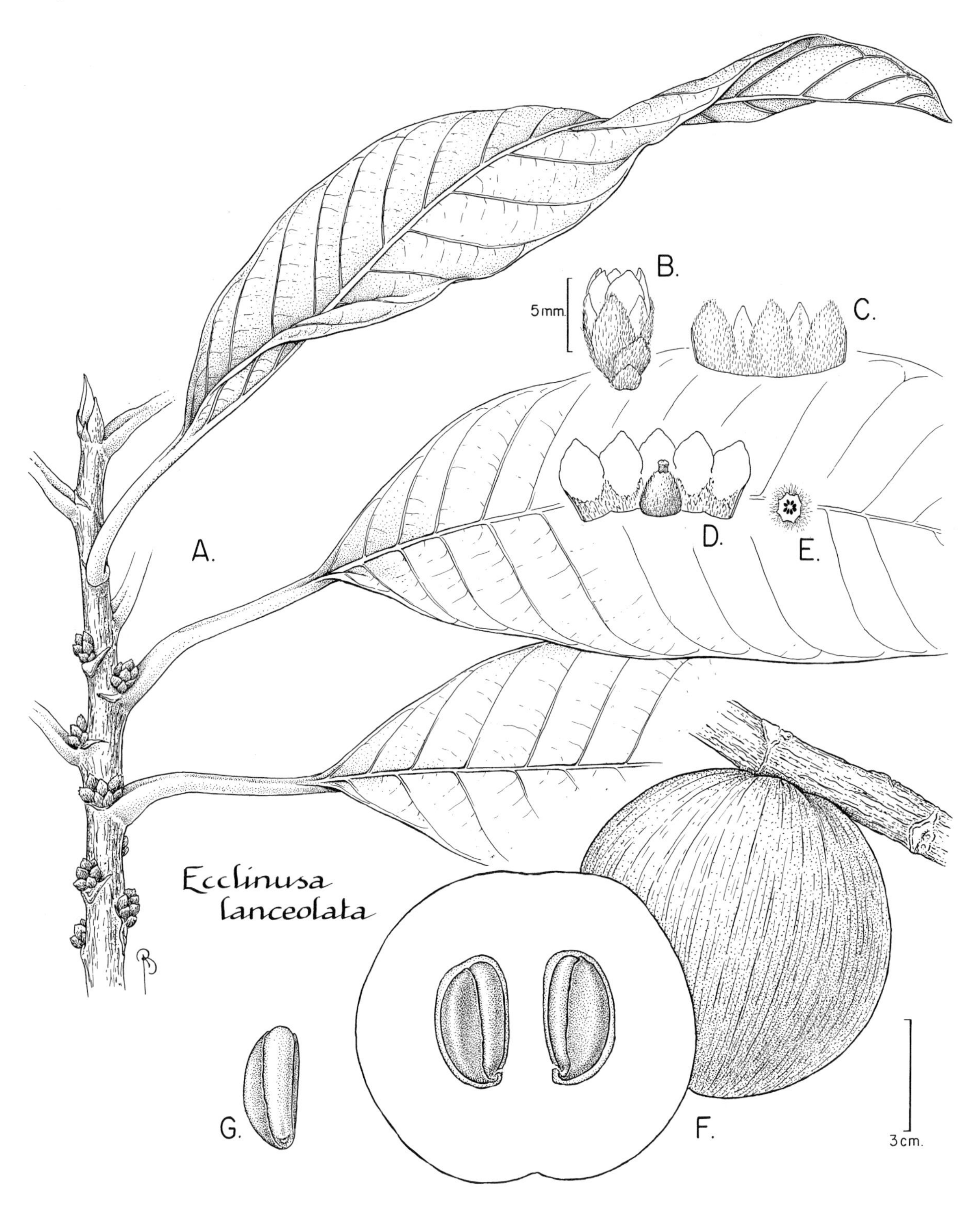

FIG. 289. SAPOTACEAE. *Ecclinusa lanceolata* (*Mori et al. 20897*). **A.** Apex of stem showing leaves, stipule scars, and immature inflorescences in leaf axils. **B.** Lateral view of pistillate flower. **C.** Abaxial view of opened calyx. **D.** Adaxial view of opened corolla and pistil of pistillate flower. **E.** Transverse section of ovary. **F.** Fruit on stem (above) and medial section of fruit (left). **G.** Seed; note hilar scar.

Manilkara bidentata (A. DC.) A. Chev.

Trees, to 30 m tall. Trees unbuttressed, the bole cylindric. Bark dark brown, fissured, with ridges peeling from trunk, the slash deep red, exuding copious, sticky, white latex. Leaves: petioles 2.5–3 cm long; blades widely elliptic to oblong-oblanceolate, 13–21 × 6.5–8.5 cm, glabrous abaxially, the apex acute or obtuse; venation inconspicuous. Flowers scented; corolla white or cream-colored,

3.5–5 mm long. Fruits ellipsoid or globose, 2–3 cm long, yellow to reddish-orange. In nonflooded forest. Known by a single sterile collection from our area. *Balata franc* (Créole).

Manilkara huberi (Ducke) Chev.

Trees, to 45 m tall. Trunks with simple, steep, thick buttresses to 2.5 m high, the crown round, widely spreading. Bark grayish- to blackish-brown, deeply fissured and grid-cracked, the slash pink with whitish streaks, exuding copious, sticky, white latex. Leaves: petioles 3.5–5 cm long; blades broadly oblong or elliptic, 14–20 × 5.5–9 cm, with closely appressed, yellowish indumentum forming fine covering abaxially, the apex obtuse or rounded; venation conspicuous abaxially. Flowers: corolla 4.5–5.5 mm long, greenish-white to yellowish. Fruits broadly ovoid to globose, 2.5–3 cm long, yellowish-green. Seedling leaves with silvery-white lower surface. Fl (Mar), young fr (Jul); scattered, in nonflooded forest.

Manilkara paraensis (Huber) Standl. PART 1: PL. Vc

Unbuttressed trees, to 30 m tall. Crown small and finely divided. Bark grayish-brown, deeply fissured, the slash deep red, exuding copious, sticky, white latex. Leaves: blades oblanceolate, 7–8 × 2.5–3 cm, with dense yellowish to grayish appressed indumentum forming fine covering abaxially, the apex obtuse or rounded. Flowers 3.5–4 mm long, yellowish-white. Fruits globose or ellipsoid, pale yellowish-green. Known by a single sterile collection from our flora, in forest on ridge tops.

Manilkara zapota (L.) Royen

Trees, to 10 m tall. Crown broad, dense, dark, with conspicuous sympodial branching, young individuals often with whorled branches. Bark grayish, fissured and grid-cracked in older individuals, the inner bark pink, exuding copious, sticky, white latex. Leaves densely clustered at shoot apices; blades elliptic, 7–10 × 2.5–3.5 cm, glabrous abaxially, the apex acute or narrowly attenuate. Flowers scented; corolla 8–9 mm long, greenish-white or cream-colored. Fruits broadly ovoid or elliptic, 5–7 cm long, pale brown, with rough, scaly pericarp. Seeds several per fruit. Fl (Jun); native of Mexico, cultivated in village for edible fruit. *Sapotille*, *sapotillier* (French).

MICROPHOLIS Pierre

Canopy or understory trees. Leaves mostly alternate and distichous, finely striate because of closely parallel secondary and higher order venation (higher order venation reticulate in *M. guyanensis*). Inflorescences axillary or on smaller branches, fasciculate, the fascicles sometimes on short, densely scaly shoots. Flowers unisexual or bisexual; calyx single whorl of 4–5 free, imbricate sepals; corolla cyathiform or tubular, the lobes 4–5, erect or reflexed, simple; stamens 4–5, inserted at corolla lobe sinuses; staminodes present; small annular nectary disc surrounding ovary base sometimes present; ovary 4–5-locular. Fruits 1-seeded berries. Seeds laterally compressed, the testa smooth or transversely wrinkled, shiny, the seed scar adaxial, narrow, extending length of seed; endosperm present; embryo with foliaceous cotyledons, the radicle exserted.

1. Stamens exserted.
    2. Leaf blades with secondary veins widely spaced, prominent, submarginal vein prominent. . . . . *M. submarginalis.*
    2. Leaf blades with secondary veins closely spaced, not prominent, submarginal vein inconspicuous.
        3. Buttresses convex. Corolla ca. 4 mm long. . . . . *M. obscura.*
        3. Buttresses not convex. Corolla 10–13 mm long. . . . . *M. cayennensis.*
1. Stamens included.
    4. Young stems golden woolly tomentose. Leaf blades densely sericeo-pubescent abaxially. . . . . *M. porphyrocarpa.*
    4. Young stems finely appressed puberulous or glabrous. Leaf blades finely appressed puberulous or glabrous abaxially.
        5. Bark smooth. Leaf blades usually finely appressed puberulous, the hairs buff or golden-yellow; higher order venation reticulate. . . . . *M. guyanensis.*
        5. Bark scaling, fissured, or smooth. Leaf blades ± glabrous; higher order venation, when visible, parallel with secondaries.
            6. Leaf blades 5–9 × 2–4 cm. . . . . *M. venulosa.*
            6. Leaf blades 11–19 × 4–8 cm.
                7. Emergent trees. Leaf blades coriaceous; venation steeply ascending. . . . . *M. melinoniana.*
                7. Understory trees. Leaf blades not coriaceous; venation shallowly ascending.
                    8. Leaf blades with secondary and tertiary veins not distinguishable from one another. Flowers 4-merous. Fruits 4-winged or ribbed. . . . . *M. acutangula.*
                    8. Leaf blades with secondary veins more pronounced than tertiary veins. Flowers 4–5-merous. Fruits not winged or ribbed.
                        9. Young stems with golden to ferruginous indumentum. Corolla 1.25–3 mm long. . . . *M. venulosa.*
                        9. Young stems not golden to ferruginous. Corolla 5–6.5 mm long.
                            10. Petioles 2–4 mm long. Pedicels 5–25 mm long. Corolla tube appressed puberulous abaxially. Fruit 2.5–3.5 cm long, the apex acute. . . . . *M. longipedicellata.*
                            10. Petioles 0.7–1.1 cm long. Pedicels 5–7 mm long. Corolla glabrous. Fruits ca. 5 cm long, the apex rostrate. . . . . *M. mensalis.*

Micropholis acutangula (Ducke) Eyma

Understory trees, to 18 m tall. Trunks with poorly developed buttresses. Bark smooth or vertically cracked, the slash reddish or orange with some white latex. Leaves: blades broadly oblong to elliptic, 11–16 × 4–6 cm, chartaceous, glabrous, the apex acuminate to caudate. Inflorescences axillary; pedicels 3–9 mm long. Flowers unisexual (plants monoecious); corolla 3–6 mm long, yellowish green; stamens included. Fruits narrowly obovoid or ellipsoid, 2.5–3 cm long, 4-winged or ribbed, glabrous, purple or black. Known only by sterile collections from our area; frequent, in nonflooded forest.

Micropholis cayennensis T. D. Penn.

Trees, to 30 m tall. Trunks with steep, simple buttresses to 1.5 m high, the bole cylindric, the crown dense, spreading. Bark rich brown, scaling in thin, longitudinal-rectangular pieces, the slash cream-colored, streaked with brown, with small amount of sticky white or yellowish latex. Leaves: blades elliptic to oblanceolate, 15–25 × 6–10 cm, glabrous or abaxial surface with sparse, finely appressed hairs. Inflorescences on smaller branches, fasciculate, with many flowers. Flowers: corolla 10–13 mm long, white, the lobes reflexed; stamens exserted. Fruits ellipsoid, 5 × 3 cm, glabrous. Fl (May, Sep), fr (Feb); in nonflooded forest.

Micropholis guyanensis (A. DC.) Pierre

Trees, to 34 m tall. Trunks with steep, simple or branched, plank buttresses, the bole cylindric. Bark gray-brown, ± smooth or finely vertically cracked, the slash pink or orange, exuding sticky, white latex sometimes darkening on exposure. Leaves: blades elliptic to oblanceolate, 6–20 × 2.5–6 cm, finely appressed puberulous with buff to golden-yellow hairs abaxially, the apex attenuate to acuminate; higher order venation reticulate. Inflorescences at first axillary, persisting as short, densely scaly shoots with terminal cluster of flowers. Flowers unisexual (plants dioecious); corolla 2.5–3 mm long, greenish cream-colored, the lobes erect; stamens included. Fruits ellipsoid, 1.5–2.5 cm long, puberulous or glabrous, purplish-black. Fl (Jul, Aug, Sep, Nov), fr (Jan, Mar, Sep); common, in nonflooded forest. *Balata blanc* (Créole).

Micropholis longipedicellata Aubrév.

Understory trees, to 15 m tall. Trunks unbuttressed. Bark thinly scaling, reddish-gray to dark brown, the slash orange-yellow, exuding small amount of white latex. Leaves: blades broadly elliptic, 5.5–13 × 2–5.5 cm, chartaceous, glabrous, the apex acuminate to caudate. Inflorescences axillary; pedicels 5–25 mm long. Flowers unisexual (plants dioecious?); corolla ca. 5 mm long, pale green; stamens included. Fruits ellipsoid, 2.5–3.5 cm long, yellow, the pedicel to 4 cm long. Fr (Aug); in poorly drained forest.

Micropholis melinoniana Pierre FIG. 290

Emergent trees, to 45 m × 120 cm. Trunks with well developed, steep, high, branched, running buttresses, the bole cylindric. Bark grayish-brown, scaling, the slash ca. 2 cm, orange, with white sticky latex. Leaves: blades oblong to broadly elliptic, 11–19 × 4–8 cm, coriaceous, glabrous, the apex acute to acuminate. Inflorescences axillary. Flowers unisexual; corolla 2.5–3.5 mm long, greenish-white; stamens included. Fruits broadly ellipsoid, 4–7 cm long, light orange, glabrous. Fl (Aug), fr (Jan, Apr, May, Dec); in nonflooded forest.

Micropholis mensalis (Baehni) Aubrév.

Small understory trees, to 15 m tall. Bark with slash with small amount of sticky white latex. Leaves: blades oblong-elliptic, 14–17 × 5–6 cm, glabrous. Inflorescences axillary or from below leaves. Flowers: corolla 5–6.5 mm long, whitish. Fruits ellipsoid, ca. 5 cm long, smooth, glabrous, whitish, the apex rostrate. Known only by a sterile collection from our area; in nonflooded forest.

Micropholis obscura T. D. Penn. PL. 118c; PART 1: PL. Va

Trees, to 35 m tall. Trunks with large, thin, convex buttresses to 3 m high, the bole cylindric or slightly fluted, the crown compact, of steeply ascending branches. Bark dark or reddish-brown, shaggy, coarsely scaling in large irregular plates, the slash brown or yellow, with brown streaks, with slight exudate of white or cream-colored latex. Leaves: blades elliptic or oblanceolate, 7–15 × 3.5–6 cm, glabrous, the apex narrowly acuminate; venation obscure. Inflorescences arising from below leaves on smaller branches. Flowers bisexual?, scented; corolla 4 mm long, the lobes reflexed; stamens exserted. Fruits ellipsoid, 2–3 cm long, glabrous, yellowish. Fl (Aug, Sep); scattered, in nonflooded forest. *Zolive* (Créole).

Micropholis porphyrocarpa (Baehni) Monach.

Trees, to 33 m tall. Trunks unbuttressed or with steep, simple, high buttresses, the bole usually somewhat fluted. Bark brown or orange-brown, scaling in small, thin, irregular pieces, the slash orange to pale cream-colored, laminated, with yellowish-white or white, sticky latex. Leaves: blades oblanceolate, oblong, or narrowly elliptic, 8–15 × 3.5–5 cm, densely yellowish to golden sericeo-pubescent abaxially, the apex acuminate to obtuse. Inflorescences axillary; pedicels 6–7 mm long. Flowers unisexual (plants dioecious); sepals ca. 3 mm long. Fr (May, immature); scattered, in nonflooded forest.

The Saül population of this species differs by having pedicellate instead of sessile flowers.

Micropholis submarginalis Pires & T. D. Penn.

Trees, to 20 m × 20 cm. Trunks unbuttressed. Bark slightly flaking, dark brown, the slash exuding scarce white latex. Leaves: blades ellliptic, 19–30 × 7.5–9.5 cm, the apex acute or shortly attenuate; venation brochidodromous, secondary veins parallel, impressed adaxially, in 25–30 pairs, the submarginal vein conspicuous. Corolla shortly tubular, 4–5 mm long, creamy white; stamens exserted. Fruits ellipsoid, ca. 2 cm, long, yellow. Known only by a sterile collection from our flora; in nonflooded forest.

Micropholis venulosa (Mart. & Eichler) Pierre

Trees, to 40 m tall. Trunks with large, steep buttresses to 3 m high, the bole cylindric or slightly fluted. Bark gray-brown, scaling, with small dipples or finely fissured and grid-cracked, the slash reddish-brown, with plentiful milky white latex. Leaves: blades elliptic or lanceolate, 5–9 × 2–4 cm, glabrous, the apex usually caudate. Inflorescences axillary. Flowers unisexual (plants monoecious); corolla 1.5–3 mm long, pale greenish, the lobes erect. Fruits broadly ellipsoid, 1.5–2.5 cm long, glabrous, yellowish. Fl (Jun), fr (Apr, May); in nonflooded forest. *Zolive grand bois* (Créole).

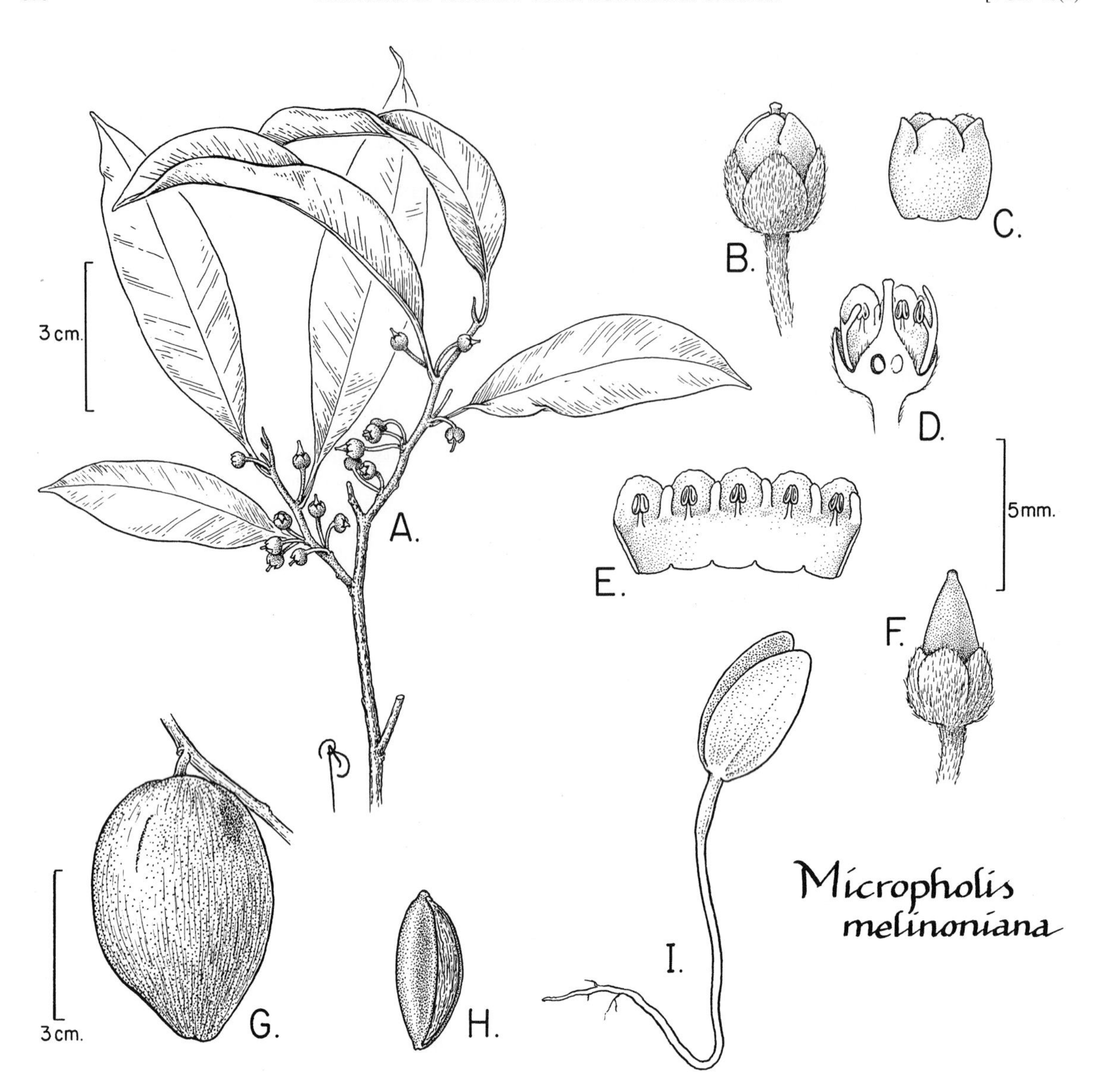

FIG. 290. SAPOTACEAE. *Micropholis melinoniana* (A–F, *Mori & Ek 20730*; G, *Mori 15561*; H, I, *Mori et al. 15560*). **A.** Apex of stem with leaves and inflorescences. **B.** Lateral view of flower; note exserted stigma. **C.** Lateral view of corolla. **D.** Medial section of flower. **E.** Adaxial view of opened corolla showing the alternating stamens and staminodes. **F.** Lateral view of immature fruit. **G.** Fruit. **H.** Seed; note hilar scar on right. **I.** Seedling.

POUTERIA Aubl.

Trees. Stipules usually absent (present only in *P.* aff. *flavilatex*). Leaves spirally arranged and often clustered at branch tips, rarely opposite; venation eucamptodromous or brochidodromous, not striate. Flowers bisexual or unisexual; calyx in single whorl, with 4–6, free, imbricate sepals; corolla cyathiform or tubular, the lobes 4–6, erect or slightly spreading, simple; stamens 4–6, inserted within corolla tube, rarely free, included or rarely exserted; staminodes 4–6, inserted in corolla lobe sinuses; ovary 1–6-locular. Fruits 1- to several-seeded berries, coriaceous or fleshy, often very large, the pericarp often edible. Seeds globose, ellipsoid, or laterally compressed, the testa smooth and shiny, the seed scar adaxial, narrow or broad, sometimes covering most of seed surface; endosperm usually absent; embryo with thick, plano-convex or rarely foliaceous cotyledons, included or rarely exserted.

There are six species of *Pouteria* that are so poorly known that they are not included in the key. Descriptions of these species follow at the end of the treatment of this genus.

1. Latex yellow. Small stipules present. . . . . *P.* aff. *flavilatex*.
1. Latex white. Stipules absent.
    2. Flowers 4-merous (sepals, corolla lobes, stamens, staminodes, and locules all 4); corolla lobes and staminodes fringed ciliate.

3. Bark reddish-brown, peeling profusely in large papery scales, the total bark thickness ca. 2 mm. . . . . *P. decorticans.*
3. Bark not peeling in large papery scales, >2 mm thick.
    4. Abaxial leaf blade surface with erect or spreading indumentum. . . . . *P. hispida.*
    4. Abaxial leaf blade surface with closely appressed indumentum or glabrous.
        5. Corolla 2.5–4 mm long. Fruits tapering into stipe. Seed scar covering almost all seed surface. . . . . *P. filipes.*
        5. Corolla exceeding 4.5 mm long. Fruits not tapering into stipe. Seed scar not covering >1/3 seed surface.
            6. Petioles 5.5–8 cm long; blades 30–42 cm long. . . . . *P. torta* subsp. *glabra.*
            6. Petioles 1–4 cm long; blades 10–22 cm long.
                7. Leaf blades with secondary veins in 6–8 pairs, the tertiary veins numerous, parallel, the higher order vein reticulum absent. Fruits narrowly ellipsoid. Seeds strongly laterally compressed. . . . . *P. singularis.*
                7. Leaf blades with secondary veins in 8–22 pairs, the tertiary veins few oblique to reticulate, the higher order reticulum conspicuous. Fruits ellipsoid to globose. Seeds not strongly laterally compressed.
                    8. Large trees. Trunks with steep narrow buttresses and fluted boles. Fruits tomentose (sometimes glabrous with age). . . . . *P. guianensis.*
                    8. Small, secondary forest trees. Trunks unbuttressed. Fruits glabrous. . . . . *P. caimito.*

2. Flowers not entirely 4-merous, the sepals 5 or more, if 4, then ovary 2-locular, or corolla lobes, stamens, staminodes, and locules 6–10; corolla lobes and staminodes usually not fringed ciliate.
    9. Sepals 4, the corolla lobes, stamens, and staminodes 6; locules 6–10, the style long, equalling corolla or exserted.
        10. Corolla 5.5–7 mm long. . . . . *P. brachyandra.*
        10. Corolla 9.5–15 mm long. . . . . *P. venosa* subsp. *amazonica.*
    9. Sepals 4–5, the corolla lobes, stamens, and staminodes 4–5; locules 1–5, the style exserted or not.
        11. Leaves often opposite. Corolla rotate; stamens inserted at top of corolla tube or at base of lobes, exserted. . . . . *P. eugeniifolia.*
        11. Leaves nearly always spirally arranged. Corolla cyathiform or tubular; stamens variously inserted, always included.
            12. Seeds with endosperm; embryo with thin, foliaceous cotyledons.
                13. Bark finely vertically cracked. Stamens free. Fruits 6–9 cm long. . . . . *P. laevigata.*
                13. Bark fissured or scaling. Stamens inserted in upper half or top of corolla tube. Fruits 2.5–5 cm long.
                    14. Young shoots, young leaves, and inflorescences densely short-tomentose, with crisped ferruginous hairs; corolla ca. 6.5 mm long. . . . . *P. cayennensis.*
                    14. Young shoots, young leaves, and inflorescences glabrous or only with sparse, minute, appressed indumentum; corolla 3.5–4.5 mm long. . . . . *P. oblanceolata.*
            12. Seed without endosperm; embryo with thick, plano-convex cotyledons.
                15. Flowers 5-merous; corolla tubular, 3–18 mm long, the tube exceeding lobes, the lobes and staminodes papillose, the style long, usually exserted.
                    16. Leaf blades ≤2 times as long as wide, the base cordate. Seed scar confined to adaxial surface. . . . . *P. maxima.*
                    16. Leaf blades ca. 3 times as long as wide, the base tapering. Seed scar covering most or all of seed.
                        17. Leaf blades 2.5–4.5 cm wide, with golden-brown, minute, appressed indumentum abaxially; secondary veins in 20–25 pairs. . . . . *P. rodriguesiana.*
                        17. Leaf blades 4.5–10 cm wide, glabrous; secondary veins in 8–13 pairs. . . . . . *P. speciosa.*
                15. Sepals, corolla lobes, and stamens 4–6; locules 1–2 (5 in *P. engleri* and *P. melanopoda*); corolla 1.5–4(6.5) mm long, usually cyathiform, the tube shorter or longer than lobes, the lobes and staminodes not papillose, the style short, included.
                    18. Leaf blades slightly glaucous on lower surface.
                        19. Higher order venation parallel to secondaries and descending from margin. Corolla lobes, stamens, and staminodes 4. . . . . *P. egregia.*
                        19. Higher order venation obscure. Corolla lobes and stamens 5.
                            20. Fruits sessile, ca. 5 × 4 cm, with a hard skin, the base truncate. . . . . *P. putamen-ovi.*
                            20. Fruits pedicellate, not exceeding 3 × 3 cm, with a soft skin, the base tapered.

21. Corolla 5.5–6.5 mm, the tube exceeding lobes. Young fruits oblate, mature fruits globose. Seeds with scar 9 mm wide. . . . . *P. ambelaniifolia.*
21. Corolla not >4 mm long, the tube equalling or shorter than lobes. Fruits ellipsoid. Seeds with scar 2–3 mm wide.
    22. Bark light brown, deeply fissured. Leaf blade apex rounded or obtuse; midrib sunken on adaxial surface. . . . . *Pouteria* sp. A.
    22. Bark gray, smooth. Leaf blade apex acuminate; midrib raised on adaxial surface. . . . . *P. cuspidata* subsp. *dura.*

18. Leaf blades not glaucous; venation eucamptodromous or brochidodromous.
    23. Higher order leaf blade venation finely areolate. Ovary 1-locular.
        24. Leaf blades with sparse, fine, appressed indumentum abaxially (as seen with hand lens). . . . . *P. gongrijpii.*
        24. Leaf blades glabrous.
            25. Petioles 1–1.5 cm long; blades 8–15 cm long. Corolla 1.5–2.5 mm long, the lobes ciliate. . . . . *P. reticulata.*
            25. Petioles 1.5–2.5 cm long; blades 15–28 cm long. Corolla ca. 3.5 mm long, the lobes not or scarcely ciliate. . . . . *P. retinervis.*
    23. Higher order leaf blade venation not finely areolate or, if so, then ovary 2 (5)-locular.
        26. Stamens free or inserted in lower half of corolla tube.
            27. Sepals, corolla lobes, stamens, staminodes 4; stamens free; ovary 2-locular, the style not exserted, slender.. . . . . *P. coriacea.*
            27. Sepals, corolla lobes, stamens 5, staminodes 0–5; stamens inserted in lower half of corolla tube; ovary 5-locular, the style long exserted, stout.. . . . . *P. engleri.*
        26. Stamens insesrted near top of corolla tube.
            28. Ovary 5-locular. . . . . *P. melanopoda.*
            28. Ovary 2-locular.
                29. Venation eucamptodromous.
                    30. Leaf blades 15–28 × 7–12 cm; secondary veins in 14–18 pairs, parallel, more or less straight, the fine areolate vein reticulum conspicuous on both surfaces (with lens). . . . . *P. retinervis.*
                    30. Leaf blades 11–17 × 5–7 cm; secondary veins in 7–11 pairs, convergent and arcuate, the higher order venation lax, open reticulum. . . . . *P. bilocularis.*
                29. Venation mostly brochidodromous.
                    31. Leaf blades 10.5–14 cm long, the apex rounded or obtuse; intersecondary veins long. Fruits ca. 4 cm long. Seed scar ca. 8 mm wide. . . . . *P. virescens.*
                    31. Leaf blades 8–11 cm long, the apex obtuse to shortly attenuate; intersecondary veins short or absent. Fruits 2.5–3 cm long. Seed scar 4.5–7 mm wide. . . . . *P. jariensis.*

Pouteria ambelaniifolia (Sandwith) T. D. Penn.

Trees, to 32 m tall. Trunks with steep buttresses. Bark pale gray to gray brown, with vertical cracks, peeling in irregular plates, the slash streaked with orange, exuding small amount of white latex. Leaves: blades broadly oblong to oblong-elliptic, 8–12 × 3–5 cm, glabrous, the apex obtuse to rounded. Inflorescences axillary and from stem below leaves. Flowers: corolla cyathiform, 5.5–6.5 mm long, greenish-yellow. Fruits oblate when young, globose at maturity, ca. 3 cm long, smooth, glabrous, yellow. Seeds ellipsoid, with broad scar ca. 9 mm wide. Fl (Aug), fr (Sep, Nov); apparently rare, in nonflooded forest.

Pouteria bilocularis (H. Winkl.) Baehni

Trees, to 35 m tall. Trunks buttressed to 1.5 m, the bole irregular. Bark grayish-brown, finely vertically cracked and scaling in thin irregular plates, the slash deep reddish, with some sticky white latex. Leaves: blades elliptic, 11–17 × 5–7 cm, the apex acuminate; secondary veins in 7–11 pairs, convergent, arcuate. Inflorescences axillary or on stems below leaves. Flowers: corolla 1.75–2.25 mm long, greenish-yellow. Fruits ellipsoid, 3–3.5 cm long, smooth, glabrous, tough-skinned, yellow, the apex rounded; endocarp soft, gelatinous, adherent to seed. Seeds ellipsoid, laterally compressed, 2–2.5 cm long, the testa shiny, wrinkled, scar adaxial, 2–4 mm wide. Fr (May); rare, in nonflooded forest.

Pouteria brachyandra (Aubrév. & Pellegr.) T. D. Penn.

Trees, to 20 m tall. Trunks with steep, narrow buttresses. Bark scaling in irregular plates, the slash thin, reddish-pink with white latex. Leaves: blades oblanceolate, 13–18 × 4–6.5 cm, glabrous.

Inflorescences axillary. Flowers: corolla tubular, 5.5–6 mm long, greenish-white. Fruits unknown for this species. Rare, in nonflooded forest. Known only by three sterile collections from our area.

Pouteria caimito (Ruiz & Pav.) Radlk.

Trees, to 10 m tall. Trunks of larger individuals outside French Guiana with fluted bole. Bark reddish-brown, peeling in thin sheets, the slash cream-colored, exuding some white latex. Leaves: blades oblanceolate, 11–15 × 3–4.5 cm, glabrous. Inflorescences mostly from stems below leaves. Flowers: corolla 5–7 mm long, pale green. Fruits ellipsoid or ovoid, 3–6 cm long, glabrous, yellow, with translucent white, flesh. Seed scar 2–6 mm wide. Fr (Jan, Feb, Aug). *Wilaka* (Amerindian), *zolive* (Créole).

Pouteria cayennensis (A. DC.) Eyma Pl. 118d

Trees, to 46 m tall. Trunks with steep, branched buttresses to 2 m high, the bole fluted or cylindric, the crown large, spreading. Bark brown to dark gray, slightly vertically cracked or scaling and becoming dippled, the slash pinkish or straw-colored, exuding abundant, sticky white latex. Leaves: blades broadly oblanceolate, 9–12 × 4.5–6 cm, with some scanty ferruginous indumentum along midrib abaxially, soon glabrous, the apex rounded. Inflorescences from mostly below leaves. Flowers: sepals with ferruginous indumentum; corolla broadly tubular, ca. 6.5 mm long, cream-colored. Fruits ellipsoid, ca. 3–5 cm long, yellow, the apex rounded, the mesocarp tough, pinkish, the endocarp 3–4 mm thick, transparent, gelatinous and adherent to seed coat. Seed scar 2–4 mm wide. Fl (Aug, Sep), fr (Jul); apparently rare, in nonflooded forest.

Pouteria coriacea (Pierre) Pierre

Trees, to 15 m tall. Trunks of larger individuals with poorly developed buttresses. Bark reddish-brown, smooth, the slash red, with white latex. Leaves: blades elliptic, 10–16 × 3.5–5.5 cm, the apex acuminate; secondary veins in 6–8 pairs, strongly arcuate. Inflorescences mostly axillary. Flowers: corolla 1.75–3 mm long, red. Fruits globose, ca. 3 cm diam., glabrous, yellow. Seeds 1–2, oblong, ca. 2 cm long, the testa shiny, strongly wrinkled, the scar adaxial, 5 mm wide. Fl (Aug), fr (Feb, Jun, Aug, Nov); frequent, in nonflooded forest.

Pouteria cuspidata (A. DC.) Baehni subsp. **dura** (Eyma) T. D. Penn.

Trees, to 22 m tall. Trunks with well developed buttresses. Bark smooth, gray, the slash reddish brown, with white latex. Leaves: blades elliptic to oblanceolate, 9–13 × 3–4 cm, coriaceous, subglabrous, the apex acuminate; venation obscure, slightly impressed on both surfaces. Inflorescences axillary. Flowers: corolla 1.75–3.5 mm long, cream-colored. Fruits oblong, 2.5–3 cm long, subglabrous, yellow. Seeds solitary, ellipsoid, 1.5–2.5 cm long, laterally compressed, the testa smooth, shiny, the scar adaxial, 1–3 mm wide. Fl (Sep), fr (Feb, immature); in nonflooded forest.

Pouteria decorticans T. D. Penn. Part 1: Pl. Vd

Trees, to 25 m tall. Trunks unbuttressed or with small, simple, or branched buttresses to 50 cm high, the bole fluted or irregular in transverse section. Bark reddish-brown, peeling in large papery scales, the slash orange, pink, or cream-colored, thin, the total thickness ca. 2 mm, exuding sticky, white latex. Leaves: blades oblanceolate, 13–23 × 4–9 cm, paler green abaxially, the apex acuminate to caudate. Inflorescences from stems below leaves; pedicels absent. Flowers: corolla ca. 10 mm long, greenish-white. Fruits ovoid or ellipsoid, 3–3.5 cm long, appressed puberulous. Seed scar ca. 2 mm wide. Fl (May), fr (Mar, May); common and conspicuous because of shaggy bark, in nonflooded forest. *Zolive* (Créole).

Pouteria egregia Sandwith

Trees, to 33 m tall. Trunks of larger trees with simple or branched, running buttresses, extending to 3 m and to 2.5 m high, the bole cylindric or slightly fluted at base. Bark pale buff-brown, vertically cracked, peeling in thin, narrow strips, the slash orange-streaked, with plentiful, sticky, white latex. Leaves: blades elliptic, 6–11 × 2.5–4 cm, gray-green to glaucous abaxially, the apex obtuse to acuminate. Inflorescences axillary and from stem below leaves. Flowers: corolla cyathiform, ca. 2 mm long, greenish-white. Fruits ellipsoid to globose, 2–2.5 cm long, yellow to orange, glabrous, the apex rounded. Seeds ellipsoid, 1.4–1.8 cm long, the scar 3.5–8 mm wide. Fl (Aug, Sep), in nonflooded forest.

Pouteria engleri Eyma

Trees, to 40 m tall. Trunks with simple, plank buttresses to 2 m high, the bole cylindric. Bark reddish-brown to gray-brown, smooth to rough and flaking, the slash orange to pinkish, with sticky, white latex. Leaves: blades broadly oblanceolate to obovate, 9–11 × 3–6 cm, glabrous, the apex attenuate to rounded; secondary veins in 6–8 pairs, the tertiary veins mostly horizontal. Inflorescences axillary and from stems below leaves. Flowers: corolla ca. 4 mm long, yellowish-green; style long, exserted, persistent. Fruits ellipsoid, 2–5 cm long, yellow-orange, the apex rounded. Seeds solitary, slightly laterally compressed, the testa shiny, wrinkled, the scar 2 mm wide. Fr (Apr, immature); uncommon, in nonflooded forest.

Pouteria eugeniifolia (Pierre) Baehni

Trees, to 32 m tall. Trunks with steep, simple or branched buttresses, merging into fluted bole, the buttresses sometimes extending several meters from trunk. Bark pale to rich brown, peeling in large, irregular sheets, the slash pinkish-brown, sometimes streaked with orange, exuding some whitish latex. Leaves opposite or spirally arranged; blades oblong or oblong-elliptic, 8–11 × 2–3.5 cm, slightly glaucous abaxially, the apex acuminate. Inflorescences mostly axillary. Flowers scented; corolla rotate, 2.5–3 mm long, yellowish-green; stamens exserted. Fruits subglobose, ca. 2 cm long, russet colored. Seed scar 3–5 mm broad. Fl (Aug), fr (May, Nov, immature), in nonflooded forest.

Pouteria filipes Eyma

Trees, to 40 m tall. Trunks with buttresses to 5 m high, the slash white, with sparse, milky, white latex. Leaves: blades oblanceolate, 6–11 × 3–4.5 cm, finely and sparsely appressed puberulous adaxially, sometimes forming thin covering abaxially. Inflorescences axillary or from stems below leaves. Flowers: corolla 2.5–4 mm long, yellowish. Fruits ellipsoid, tapering at base into stipe ca. 2 cm long, 4.5–5 cm long (including stipe), subglabrous. Seed scar covering most of seed surface. Known only by two sterile collections from our area; in nonflooded forest from 400–650 meters.

Pouteria aff. **flavilatex** T. D. Penn.

Trees, to 30 m × 80 cm. Trunks with simple, steep, slightly concave buttresses to 2 m high. Bark grayish- or reddish-brown, scaling in thin, irregular pieces, the slash yellowish-brown, 5–8 mm thick, slowly exuding watery, yellow latex. Leaves: blades oblanceolate, 14–18 × 6–8 cm, the base rounded, the apex rounded; venation eucamptodromous, secondary veins in 8–12 pairs. Flowers not seen. Fruits (immature) ellipsoid, 3–4 × 2–2.5 cm, densely tomentose, yellowish-brown. Seeds solitary. Immature fr (Feb); in forest on ridges and hillsides. Known only by a single collection from our area.

This population differs from typical *Pouteria flavilatex* in lacking the appressed indumentum of the abaxial leaf blade surface.

Pouteria gongrijpii Eyma

Trees, to 27 m tall. Trunks with small buttresses to 50 cm high, the bole slighty fluted, the crown slender. Bark reddish-brown, peeling in thin, irregular plates, the slash creamy brown, with small amount of sticky, white latex. Leaves: blades oblanceolate, 13–20 × 4–7 cm, abaxial surface usually with some fine, appressed indumentum (as seen with hand lens), the apex acuminate. Inflorescences axillary and from stem below leaves. Flowers with strong, sweet scent; corolla 2.5–3 mm long, greenish cream-colored. Fruits ellipsoid, ca. 3 cm long, orange, the apex rounded. Seed solitary, not laterally compressed, ca. 2 cm long, the testa smooth, adherent to pericarp, the scar ca. 4 mm wide. Fl (May, Oct, Nov, Dec), fr (May); in nonflooded forest. *Akoinsiba* (Bush Negro), *balata poirier* (Créole), *niamboka*, *pepe boiti* (Bush Negro), *zolive* (Créole).

Pouteria guianensis Aubl.

Trees, to 30 m tall. Trunks with steep, thick buttresses merging into fluted bole, the crown narrow or broad, of ascending branches. Bark reddish-brown, finely vertically cracked or fissured, the ridges peeling in long, thin, narrow, friable strips, the slash pale brown or cream-colored, exuding small amount of sticky, white latex. Leaves: blades broadly oblanceolate, 10–22 × 4–8 cm, the apex attenuate to obtusely cuspidate. Inflorescences axillary or from stems below leaves. Flowers: corolla 6–12 mm long, pale green. Fruits globose, 3–5 cm long, tomentose at first, often subglabrous at maturity, orange-yellow. Seed scar 6–10 mm broad. Fl (Sep, Nov, Dec), fr (Feb, Apr, May, Dec); relatively common, in nonflooded forest and in forests in lower areas on poorly drained soils. *Balata kamwi* (Amerindian), *zolive* (Créole).

Pouteria hispida Eyma

Trees, to 36 m tall. Trunks with steep, rounded, simple or branched buttresses to 5 m high, the crown small, round, the branches spreading. Bark dark- or reddish-brown, peeling in thin, rectangular or irregular pieces, the slash cream-colored to pale orange or pinkish, exuding abundant, sticky, white latex. Leaves: blades broadly oblanceolate or elliptic, 17–40 × 6–15 cm, softly pubescent abaxially, the apex attenuate to rounded. Flowers: corolla 4–5 mm long, greenish-white. Fruits globose, 3.5–8 cm long, velutinous, yellowish or orange. Seed scar 2–3.5 mm wide. Fl (Aug, Sep, Oct, Nov), fr (May); common, in nonflooded forest.

Included under this name is a small-leaved, subglabrous form (leaves ca. 13 × 5.5 cm) represented by *Mori et al. 19119*, *20882*, and *20896*, NY. The flowers and fruits of this form do not differ from typical *P. hispida*, but further study may prove it to be a distinct species.

Pouteria jariensis Pires & T. D. Penn.

Trees, to 26 m tall. Trunks with steep, simple buttresses to 1.5 m high, the bole fluted. Bark gray-brown, peeling in thin slender pieces, the slash pink, with white, sticky latex. Leaves: blades elliptic or oblanceolate, 8–11 × 3.5–5 cm, the apex obtuse to shortly attenuate; intersecondary veins short or absent. Inflorescences axillary and from stems below leaves. Flowers bisexual?; corolla 3–4 mm long, greenish-yellow. Fruits ellipsoid to obovoid, 2.5–3 cm long, yellow or orange, smooth, velutinous, the apex rounded or obtuse. Seeds 1–2 per fruit, ellipsoid, the scar 4.5–7 mm wide. Fl (Nov); in nonflooded forest. *Zolive* (Créole).

Pouteria laevigata (Mart.) Radlk.

Trees, to 37 m tall. Trunks unbuttressed, the bole cylindric, the crown small, spreading. Bark pale brown, finely vertically cracked, the slash cream-colored to pale brown, exuding slow flowing, creamy white latex, the total bark thickness ca. 20 mm. Leaves in dense, terminal clusters; blades oblanceolate, 8–15 × 3.5–5 cm, glabrous, the apex rounded. Inflorescences mostly from stems below leaves; pedicels 2–3 mm long. Fruits globose, 6–9 cm diam., bright yellow, with few to many brown or black lenticels, the flesh rough, cream-colored or pinkish cream-colored. Seeds with rough testa (not shiny) adherent to layer of greenish, gelatinous pulp, the scar 3–4 mm wide. Fr (May, Jun); in nonflooded forest.

Pouteria maxima T. D. Penn.

Trees, to 40 m tall. Trunks buttressed, the buttresses steep, to 8 m, massive, some branched near base, the bole fluted. Bark grayish-brown, profusely scaling in rectangular or irregular plates, the slash pinkish, with abundant sticky, white latex. Leaves alternate and distichous; blades ovate to broadly elliptic, 10–20 × 10–12 cm, glabrous, the base cordate, the apex shortly attenuate; venation eucamptodromous, secondaries in 12–14 pairs, widely spreading, ± parallel. Inflorescences mostly on stems below leaves. Flowers: corolla ca. 6 mm long, greenish cream-colored. Fruits ellipsoid, 4–5 cm long, orange, hard-skinned, the pericarp fibrous, 2–3 mm thick, the endocarp gelatinous. Seeds ca. 2.5 cm long, laterally compressed, the testa smooth, shiny, the scar ca. 0.5 mm wide. Fr (May); in nonflooded forest.

Pouteria melanopoda Eyma

Trees, to 36 m tall. Trunks with simple or branched, steep or slightly concave buttresses to 2.5 m high, the bole usually cylindric. Bark reddish-brown, peeling profusely in small or large, thin, irregular pieces exfoliating from base of trunk, the slash cream-colored or pink, sometimes striate, with sparse to copious, white latex. Leaves: blades broadly oblanceolate, 17–25 × 6–10 cm, often with some residual brown indumentum at base of midrib abaxially, the apex attenuate to obtusely cuspidate. Inflorescences mostly from stems below leaves. Flowers: corolla 3–3.5 mm long. Fruits obovoid when young, ca. 2 cm long, mature fruit unknown. Fr (May, immature); in nonflooded forest.

Pouteria oblanceolata Pires

Trees, to 35 m tall. Trunks with small, simple or branched buttresses, the bole fluted towards base, cylindric higher up. Bark dark blackish or reddish brown, often black-orange, shallowly and narrowly fissured, the slash pink, with some sticky, white latex. Leaves: blades broadly oblanceolate, 9–30 × 3–11 cm, glabrous, the apex obtuse; midrib prominent adaxially. Inflorescences from stems mostly below leaves. Flowers: corolla 3.5–4.5 mm long, greenish-yellow. Fruits narrowly obovoid, 2.5–3.5 cm long, glabrous except for tuft of hairs at base, the apex rounded. Seeds laterally compressed, shiny, the scar 2–3 mm wide. Fl (Sep); in nonflooded forest.

Pouteria putamen-ovi T. D. Penn.

Trees, to 35 m tall. Trunks with steep, simple buttresses to 1 m high. Bark grayish-brown, sparsely peeling in thin plates, the slash brown, with small amount of sticky, white latex. Leaves: blades oblong or oblong-oblanceolate, 11–15 × 3–5 cm, the apex acuminate; secondary veins in 8–10 arcuate pairs. Inflorescences mostly from stem below leaves. Flowers not known from our area. Fruits 1-seeded, ellipsoid to globose, 5 × 4 cm, orange, the apex rounded. Seeds ca. 3 cm long, ellipsoid, with smooth, shiny, hard testa, the scar broad, covering ca. 1/3 of surface. Fr (Apr, May, Jun, Sep, Nov); in nonflooded forest.

Differing from the western Amazonian *P. putamen-ovi* in leaf shape.

Pouteria reticulata (Engl.) Eyma subsp. **reticulata**

Trees, to 37 m tall. Trunks with steep buttresses to 2 m, the bole usually fluted, the crown of steeply ascending branches. Bark brown to pale gray, peeling in large, thin, long, irregular plates, the slash orange, with some sticky, white latex. Leaves: blades oblanceolate or elliptic, 8–15 × 3–6.5 cm, glabrous, the apex obtusely cuspidate to acuminate. Inflorescences axillary or occasionally clustered on leafless axillary shoots, or from branches below leaves. Flowers: corolla 1.5–2.5 mm long, the lobes ciliate, greenish-white. Fruits ovoid or ellipsoid, 2–3 cm long, yellow or orange, smooth, glabrous. Seeds solitary, slightly laterally compressed, 1.5–2 cm long, the scar 2–5 mm wide. Fl (Aug, Oct), Fr (Feb, May, Sep, Oct, Nov); relatively common, in nonflooded forest.

Pouteria retinervis T. D. Penn.

Trees, to 22 m tall. Trunks with steep, slender buttresses to 40 cm high, the bole cylindric, the crown spreading. Bark dark brown, vertically cracked and peeling profusely in thin, longitudinal, friable strips, the slash laminate, with alternating bands of orange and white, with small amount of white latex. Leaves: blades elliptic, 15–28 × 7–12 cm, glabrous; higher order vein reticulum conspicuous on both surfaces (as seen with hand lens). Inflorescences mostly axillary. Flowers: corolla ca. 3.5 mm long, greenish-white, the lobes not or scarcely ciliate. Fl (Nov); in nonflooded forest.

Pouteria rodriguesiana Pires & T. D. Penn.

Trees, to 31 m tall. Trunks with steep, simple buttresses to 1 m high, the bole often fluted, the crown narrow. Bark reddish-brown, peeling in small rectangular pieces, the slash pinkish, with some sticky, white latex. Leaves: blades elliptic or oblong, 9–16 × 2.5–4.5 cm, abaxial surface golden-brown, the apex acuminate; secondary veins widely spreading, in 20–25 pairs. Inflorescences mostly axillary. Flowers: corolla shortly tubular, 3–4 mm long, greenish-white. Fruits ovoid to globose, 4–5 cm long, glabrous, the apex rounded. Seeds globose, the scar covering most of seed surface. Fl (Sep, Nov); in nonflooded forest.

Pouteria singularis T. D. Penn.

Trees, to 20 m tall. Trunks unbuttressed. Bark smooth, the slash pink, exuding white latex. Leaves: blades oblanceolate, 10–13 × 4–5.5 cm, glabrous. Inflorescences axillary. Flowers: corolla ca. 6 mm long, greenish-white. Fruits narrowly ellipsoid, 3.5–4 cm long, yellow, the surface mealy-granular. Seed scar 2.5–3.5 mm wide. Known by only two sterile collections in our area, in nonflooded forest.

Pouteria speciosa (Ducke) Baehni — Pl. 118e

Trees, to 40 m tall. Trunks unbuttressed, the bole cylindric or slightly fluted at base, the young growth with coppery-brown indumentum. Bark yellowish-brown, peeling to form pronounced dipples near base, the slash very thick (total thickness to 4.5 cm), dirty orange to pink, with scarce, slow-flowing, white latex. Leaves clustered at branch ends; blades broadly oblanceolate, 20–30 × 4.5–10 cm, glabrous, the apex rounded; secondary veins in 8–13 pairs. Inflorescences axillary or from stems below leaves. Flowers scented; sepals ferruginous puberulous; corolla cylindric, slightly broader at middle, 15–18 mm long, the tube whitish cream-colored, the lobes pale green. Fruits edible, ovoid-oblong, 10–12 cm long, velutinous, the hairs purplish-brown, becoming glabrous with age. Seeds solitary, ovoid or ellipsoid, 6–9 cm long, with thick bony testa, the scar very rough, covering entire seed except for shiny, abaxial strip. Fl (May, Aug), fr (Feb, Aug); common but scattered, in nonflooded forest.

Pouteria torta (Mart.) Radlk. subsp. **glabra** T. D. Penn.

Trees, to 20 m tall. Trunks with small buttresses, irregular, the crown long and slender, with sympodial branching. Bark pale gray-brown, peeling in irregular, narrow pieces, the slash pale yellowish-brown, with scant, white latex. Leaves: petioles 5.5–8 cm long; blades 30–42 × 8–13 cm, glabrous; venation with conspicuous fine reticulum abaxially. Inflorescences mostly clustered on smaller branches. Flowers: corolla 7–16 mm long. Fruits ellipsoid to globose, 3–6 cm long, covered with irritating, caducous hairs, yellow-orange. Seed scar 3–9 mm wide. Fr (Mar, Sep, Oct, Dec); apparently rare, in nonflooded forest.

Pouteria venosa (Mart.) Baehni subsp. **amazonica** T. D. Penn.

Trees, to 30 m tall. Trunks with small buttresses, the base of bole fluted. Bark dark brown, peeling in small flakes, the slash pink, soft, laminated, with copious white latex. Leaves: blades 9–15 × 3–6 cm, glabrous, the apex obtuse or obtusely cuspidate. Inflorescences axillary or from stems below leaves. Flowers: corolla broadly tubular, 9.5–15 mm long, pale green. Fruits globose, 3–4 cm long, velutinous at first, eventually glabrous, orange or yellow. Seed scar covering 1/3–1/2 of seed surface. Fl (Sep), fr (Jun, Aug), in nonflooded forest. *Bruchi-soke*, *mongui-soke*, *niamboka* (Bush Negro).

Pouteria virescens Baehni

Trees, to 27 m tall. Trunks with steep, slender buttresses to 2 m high, the bole strongly fluted, the crown small, slender. Bark deep reddish-brown, shallowly and narrowly fissured, the slash creamy brown, with small amount of white latex. Leaves: blades broadly oblanceolate, 10.5–14 × 3.5–6.5 cm, the apex rounded or obtuse; intersecondary veins long. Inflorescences axillary, from stems below leaves. Flowers unisexual (plants dioecious); corolla 3–5 mm long, yellowish-white; stamens absent in pistillate flowers. Fruits broadly ellipsoid, ca. 4 cm long, velutinous, the apex obtuse. Seeds 2 per fruit, ellipsoid, the scar ca. 8 mm wide. Fr (Dec); apparently rare, in nonflooded forest.

Pouteria sp. **A**

Trees, to 35 m tall. Trunks with steep, narrow, rounded buttresses. Bark light brown, deeply fissured, the slash laminated, streaked with white. Leaves: blades oblanceolate to oblanceolate-oblong, 6–9 × 2.5–3 cm, markedly coriaceous, the apex obtuse or rounded; midrib sunken on adaxial surface. Fruits ellipsoid, ca. 3 × 2.5 cm, finely appressed puberulous to glabrous, yellow. Seeds solitary, laterally compressed, ca. 1.5–2 cm, the testa smooth, shiny, the scar 2–3 mm wide. Fr (Mar); known only from a single collection (*Mori & Pipoly 15422*, NY) made at the junction of Antenne Nord with the main La Fumée Mountain trail.

The leaf and seed morphology indicate a relationship with *P. elegans* (A. DC.) Baehni, a widespread species common in periodically flooded forests of Amazonia.

The following species of *Pouteria* are so poorly known that they are not included in the key.

Pouteria sp. 1

Trees. Bark smooth, brown, the slash 10 mm thick, orange, with sticky, white latex. Leaves clustered at shoot apices; blades oblanceolate, 3.5–7.5 × 1.8–3 cm, glabrous, the apex rounded or emarginate; venation brochidodromous, secondaries in 8–9 pairs, slightly arcuate and ascending, fine higher order reticulum visible with lens (*Mori & Boom 15310*, NY).

Pouteria sp. 2

Trees, to 25 m tall. Trunks with low, thin, widely spreading buttresses, bole cylindric. Bark brown, with fine vertical cracks, peeling in small rectangular plates, the slash red or orange, 2 mm thick, with white, sticky latex. Leaves: blades elliptic to obovate, 17–23 × 7–10.6 cm, subglabrous adaxially, sparsely stiff ferruginous, pubescent abaxially, the apex abruptly acuminate; venation eucamptodromous or brochidodromous towards apex, the secondaries in 12–13 pairs, parallel, strongly arcuate, slightly bullate with venation impressed adaxially and very prominent abaxially (*Boom & Mori 2138*, *Mori & Hartley 18487*, both at NY).

Pouteria sp. 3

Unbuttressed trees, to 20 m tall. Trunks with cylindric bole. Bark brown, grid-cracked, lenticellate, with some incomplete hoops, the slash 1–1.5 cm thick, cream-colored, with sticky, white latex. Leaves: petioles 4–5 cm long; blades broadly oblanceolate or obovate, 35–45 × 12–15 cm, glabrous, the apex rounded; venation eucamptodromous, the secondaries in 19–24 pairs, straight, parallel (*Boom & Mori 1908*, *2381*, both NY).

Pouteria sp. 4

Trees, to 14 m tall. Trunks unbuttressed. Bark dark reddish brown, scaling in small, thin, irregular pieces, the slash thin, pink, with small amount of sticky, white latex. Leaves: petioles 2–3.5 cm long; blades oblong or oblong-elliptic, 17.5–22 × 5–7.5 cm, sparse, more or less appressed, brown indumentum on veins abaxially, the apex shortly acuminate; venation brochidodromous with distinct submarginal vein, the secondaries in 21–24 pairs, shallowly ascending, straight, parallel (*Pennington & Mori 12126*, NY).

Pouteria sp. 5

Trees, to 30 m tall. Trunks with bole fluted toward base. Bark shallowly fissured and reticulately cracked, grayish-brown, the slash pale orange, changing to dark orange on exposure, with white latex. Leaves: blades obovate, 8.5–15 × 5–9 cm, coriaceous, glabrous, the apex truncate; venation eucamptodromous, secondaries 7–8 pairs, arcuate, parallel, the higher order venation reticulate, prominent abaxially. Fruits borne on branches below leaves, sessile, subglobose, 7 × 6 cm (immature), glabrous, shiny green, containing copious sticky, white latex (*Mori & Pipoly 15423*, NY).

Pouteria sp. 6

Trees, to 20 m tall. Trunks with simple, symmetrical, plank buttresses, the bole cylindric. Bark orange-brown, scaling in small, thin pieces, the slash 4 mm thick, flesh-colored. Leaves: blades elliptic to obovate, 11–13 × 4–5 cm, the apex acuminate; venation eucamptodromous, secondaries in 11–13 pairs, ± parallel, slightly arcuate, the intersecondaries well developed, the higher order venation obscure. Fruits broadly oblong, 5.5 × 3.5 cm (immature), glabrous, green, with copious white latex, the apex rounded (*Mori & Boom 15199*, NY).

PRADOSIA Liais

Canopy to emergent trees. Leaves spirally arranged or verticillate; venation eucamptodromous. Inflorescences often ramiflorous or cauliflorous. Flowers bisexual; calyx single whorl of 5 imbricate sepals; corolla rotate, the lobes 5; stamens inserted at base of corolla lobes, exserted, the filaments long, geniculate below apex, strongly narrowed below insertion of anther; staminodes absent; ovary 5-locular, the style short. Fruits drupes, often slightly asymmetric; endocarp thin, cartilaginous. Seed solitary, with smooth, shiny testa and full-length adaxial scar; endosperm thin sheath; embryo with plano-convex cotyledons, the radicle exserted.

1. Leaves lax, spirally arranged; blades ≤15 cm long, glabrous; secondary veins in ca. 10 pairs. . . . . *P. ptychandra*.
1. Leaves densely clustered in whorls of 5–7; blades ≥15 cm long, pubescent on veins abaxially; secondary veins in >20 pairs. . . . . *P. verticillata*.

Pradosia ptychandra (Eyma) T. D. Penn.

Trees, to 25 m tall. Trunks with buttresses absent or poorly developed, the bole cylindric, the crown slender, with spreading branches. Bark grayish-brown, smooth to dippled at base, often with vertical rows of lenticels, the slash cream-colored, with orange streaks, with slow flowing, sticky, white latex. Leaves: blades oblanceolate, 10–15 × 3.5–6 cm, glabrous, the apex attenuate. Inflorescences from branches and trunk. Flowers: corolla ca. 5 mm long, wine-red. Fruits mostly ellipsoid, 3.5–4 cm long, orange-yellow, the inner pericarp gelatinous, transparent, sweet, edible. Seed solitary, ca. 2.5 cm long, the scar 8–9 mm wide. Fr (Apr, May), common, in nonflooded forest. *Balata pomier* (Créole), *kimboto*, *kwatabobi*, *malobi-weti* (Bush Negro), *zolive* (Créole).

Pradosia verticillata Ducke

Trees, to 35 m tall. Trunks with short, stout buttresses to 0.5 m high, the bole cylindric, the crown broad, with massive stems and dense terminal clusters of leaves. Bark pale buff-brown, peeling in large, irregular, thin sheets, the trunk becoming dippled, the slash orange-brown, with some sticky, white latex. Leaves: blades oblanceolate, 15–27 × 5–9 cm, the apex attenuate; secondary veins in 22–26 pairs. Inflorescences on stems and branches. Flowers: corolla ca. 4 mm long, violet-black. Fruits narrowly ovoid, asymmetric, 3.5–5 cm long, glabrous. Known only by two sterile collections from our area, in nonflooded forest.

SARCAULUS Radlk.

Understory or canopy trees. Leaves mostly alternate and distichous or only weakly spirally arranged. Flowers unisexual (plants dioecious); calyx single whorl of 5 imbricate sepals; corolla globose or cyathiform, the tube strongly carnose, equalling lobes, the lobes 5, subvalvate or slightly imbricate; stamens 5, inserted at top of corolla tube, exserted, the filaments short and swollen, the anthers inflexed; staminodes 5, thick, carnose, inflexed towards style; ovary 2–5-locular. Fruits fleshy berries. Seeds laterally compressed, the scar adaxial, extending around base; endosperm absent; embryo with plano-convex cotyledons, the radicle included.

Sarcaulus brasiliensis (A. DC.) Eyma subsp. **brasiliensis**

Trees, to 20 m tall. Trunks unbuttressed, the crown small, with horizontally spreading branches. Bark gray-brown, slightly scaly, the slash pinkish red, with plentiful, sticky, white latex. Leaves widely spaced, mostly alternate and distichous; blades elliptic or oblong-elliptic, 8.5–14 × 3–3.5 cm, the apex narrowly attenuate. Inflorescences axillary and from along stem; pedicels reflexed in bud. Flowers: corolla globose, 2.5–3.5 mm long, yellowish-white. Fruits edible, subglobose or ellipsoid, 2–3 cm long, smooth, subglabrous, yellow. Seeds 1–2 per fruit, laterally compressed, ca. 1.5 cm long, the scar 2–4 mm wide. Fl (Jun); in nonflooded forest.

SCROPHULARIACEAE (Figwort Family)

Noel H. Holmgren

Annual, biennial, or perennial herbs, vines, or shrubs. Stipules absent. Leaves simple or compound, alternate, opposite, whorled, or sometimes all basal. Inflorescences spikes, racemes, thyrsoid panicles, or flowers solitary or fascicled in leaf axils; bracts foliaceous to reduced; bractlets present in some. Flowers zygomorphic, rarely actinomorphic (*Scoparia*), bisexual; calyx of 5 or 4 (2 or 1) distinct or variously united segments; corolla sympetalous, 5- or 4-lobed; stamens 4 or 2, rarely 5 (as in *Capraria biflora*), adnate to corolla, some reduced and present as staminodes; anthers 2(1)-celled; ovary superior, 2-locular, the style slender and entire with single often capitate stigma or forked with 2 stigmatic lobes; placentation axile. Fruits capsules, septicidal or loculicidal or both (4-valved), some irregularly poricidal apically, rarely berries. Seeds usually many and small, rarely few; endosperm usually well developed; embryo small.

The results of a recent molecular genetic study (Olmstead & Reeves, 1995) of representatives of the families and major tribes in the subclass Asteridae, using DNA sequences in the chloroplast genes *rbcL* and *ndhF*, have implied radically different relationships between members of the traditional Scrophulariaceae and members of the Myoporaceae, Callitrichaceae, Hippuridaceae, Plantaginaceae, and Orobanchaceae. The authors have recombined these to form three presumed monophyletic families (Scrophulariaceae, Plantaginaceae, and Orobanchaceae). Although they did not use the taxa treated below, by inference they would be placed in the Plantaginaceae. However, such a study should be regarded as an untested hypothesis because their cladistic analyses were based on an insufficient sample size of taxa (seven species in seven genera) and did not take morphological characters into account.

Holmgren, N. H. & U. Molau. 1984. 177. Scrophulariaceae. *In* G. Harling & B. Sparre (eds.), Flora of Ecuador **21:** 1-–188.
Pennell, F. W. 1920. Scrophulariaceae of Colombia—I. Proc. Acad. Nat. Sci. Philadelphia **72:** 136–188.

1. Calyx segments parted to or near base.
   2. Leaves opposite or whorled. Calyx segments 4; corolla 1.5–2.7 mm long. . . . . *Scoparia.*
   2. Leaves alternate. Calyx segments 5; corolla longer than 4 mm. . . . . *Capraria.*
1. Calyx segments united into tube, sometimes tearing apart with expansion of fruit. *Lindernia.*

CAPRARIA L.

Erect branched herbs. Leaves simple, alternate; petioles absent; blades usually with serrate margins. Inflorescences (1)2–4-flowered fascicles. Flowers: calyx 5-parted to near base; corolla campanulate to subrotate, 5-lobed, the lobes subequal, the adaxial 2 external in bud; stamens 4 or 5, all fertile; style clavate, the stigmas united. Capsules dehiscing septicidally and secondarily loculicidally at apex, glandular-punctate, the placental mass remaining free standing. Seeds numerous, reticulate.

Capraria biflora L. FIG. 291, PL. 119c

Coarse, erect, branched herbs, to 2 m tall. Sparsely hirsute to densely hispid or villous on vegetative parts. Leaves: blades oblanceolate, 3–11.5 × 0.6–2(2.5) cm, the base narrowly cuneate, the apex acute, the margins serrate distally, sometimes subentire; veins prominent. Inflorescences: pedicels 7–17 mm long, ascending, hispid-villous, sometimes densely so. Flowers: calyx 3.5–6 mm at anthesis, 4.5–7.5 mm in fruit, usually longer than capsule, the segments linear to narrowly lanceolate; corolla campanulate, (5)8–11 mm long, subregular, the tube 2–3.5 mm long, the lobes 4–7 mm long, ovate, pubescent adaxially, white, the apex obtuse, mucronate; anthers 0.9–1.1 mm long; style 3–4(5) mm long. Capsules broadly ovoid, 4–6 mm long, the apex emarginate to rounded apically. Seeds conical-cylindrical, 0.4–0.6 mm long, reticulate, in longitudinal rows, light brown. Fl (Aug, Sep), fr (Aug, Sep); in moist disturbed places along roadsides and river banks. *Chá de Marajó* (Portuguese), *thé pays* (Créole).

LINDERNIA All.

Small herbs. Leaves simple, opposite or sometimes 3-whorled; petioles absent or present; blades with margins denticulate to subentire. Inflorescences solitary axillary flowers. Flowers: calyx 5-parted to near base or united below into tube; corolla bilabiate, the abaxial lip with 3 spreading lobes, the adaxial lip short, narrow, erect, 2-lobed, external in bud, the palate with 2 hairy ridges; stamens 4, all fertile or 2 fertile and 2 sterile; style branched, the stigmas flattened. Capsules septicidal, the plate-like septum persisting with placental mass. Seeds numerous, smooth or transversely lined or shallowly pitted.

1. Pedicels 7–25 mm long. Calyx 3–4.2 mm long. Capsules globose, 2.8–4.6 mm long. . . . . *L. crustacea.*
1. Pedicels 1–3(4.5) mm long. Calyx 4.5–7 mm long. Capsules ellipsoid-fusiform, 6–12 mm long. . . . . *L. diffusa.*

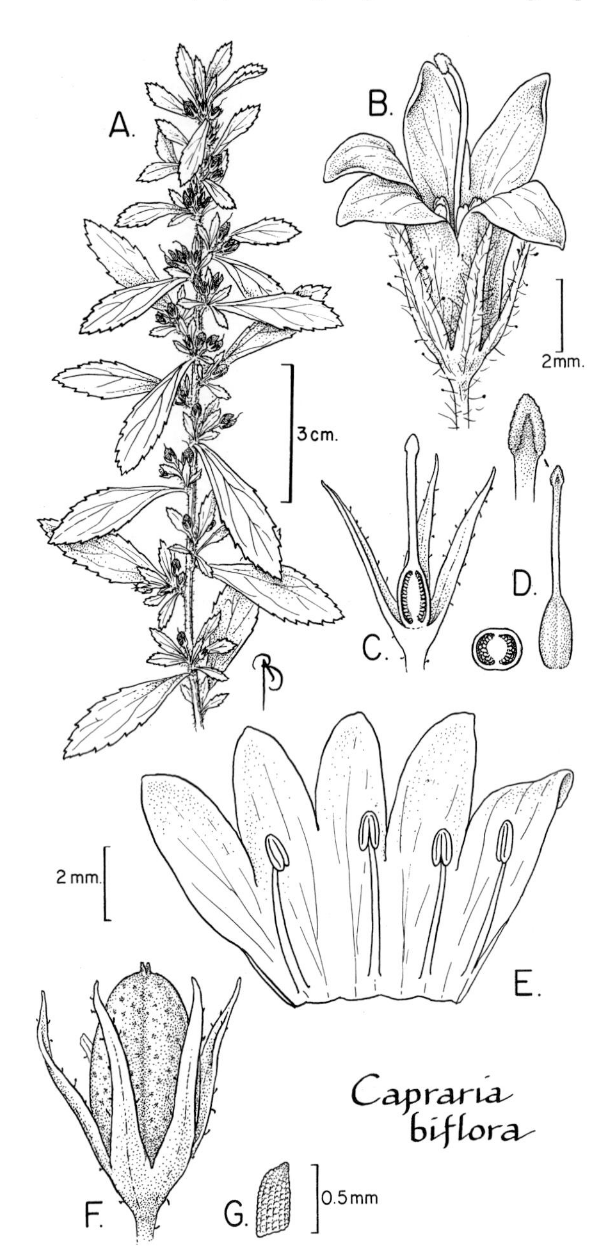

FIG. 291. SCROPHULARIACEAE. *Capraria biflora* (A, B, E, *Acevedo-Rodríguez 633* from St. John, U.S. Virgin Islands; C, D, F, G, *Prévost 777*). **A.** Upper part of stem with leaves and fruits. **B.** Lateral view of flower. **C.** Medial section of pistil in calyx with two sepals removed. **D.** Lateral view of pistil (right) with detail of stigma (above) and transverse section of ovary (below). **E.** Corolla opened to show adnate stamens. **F.** Lateral view of fruit in calyx. **G.** Seed. (Reprinted with permission from P. Acevedo-Rodríguez, *Flora of St. John, U.S. Virgin Islands*, Mem. New York Bot. Gard. 78, 1996.)

Lindernia crustacea (L.) F. Muell. FIG. 292

Small, prostrate, often much branched annuals. Stems 5–15 cm long, quadrangular. Vegetative parts glabrous to sparsely pubescent. Leaves: petioles 1–7 mm long; blades broadly ovate, 8–14(16) × 6–13 mm, the base widely cuneate to truncate, the apex obtuse to rounded, the margins crenate to crenate-denticulate. Inflorescences: pedicels 7–15(25) mm long, sparsely pubescent. Flowers: calyx 3–4.2 mm long, united below, the lobes 0.6–1.5(2) mm long, eventually splitting to base as fruit expands, appressed-pilose on veins and margins; corolla 5–7 mm long, the tube curved forward, ± dorsiventrally flattened, the tube and throat violet with broad, yellow line inside throat abaxially, the lobes pale blue with purple markings at rim of throat, the palate glabrous; stamens 4, the posterior pair with unbranched, incurved filaments, the anther cells ca. 0.35 mm long, divaricately spreading, the anterior pair with branched filaments, the longer incurved branch antheriferous, the shorter, straight branch a sterile knob. Capsules globose, 2.8–4.6 mm diam., the apex obtuse. Seeds subspherical, ca. 0.4 mm long, weakly pitted, pale yellowish-brown. Fl (Jun, Aug, Oct, Nov), fr (Jun, Aug, Sep, Oct); common, in weedy, open areas. *Douradinha do campo* (Portuguese), *petite griffe*, *ti mignonette* (Créole).

Lindernia diffusa (L.) Wettst.

Small, prostrate annuals. Stems 5–16 cm long, spreading-pilose, rooting at lower nodes. Leaves: petioles 1–2(3) mm long; blades broadly ovate to orbicular, 12–22 × 9–18 mm, pilose on veins and margins, the surfaces glabrous to sparsely pubescent with appressed-ascending hairs, the base rounded, the apex rounded, the margins crenate to crenate-serrate. Inflorescences: pedicels 1–3(4.5) mm long. Flowers: calyx 4.5–7 mm long, united below, the lobes lanceolate, 2–4 mm long, the apex acute or acuminate; corolla 6–9 mm long, white, the adaxial lip sometimes pink-tinged; stamens 4, the anthers yellow. Capsules ellipsoid-fusiform, 6–12 mm long, finely puberulent, the style base enlarged and persistent. Seeds 0.4–0.5 mm long, reticulate-pitted, yellowish. Known from a single collection (*Raynal-Roques 18643*, P) from our area; wet roadsides, ditch banks, and savannas in other parts of its range.

SCOPARIA L.

Erect, branched herbs. Leaves in whorls of 3 or 4, sometimes opposite, petiolate, glandular-punctate. Inflorescences (1)2- or 3-flowered axillary fascicles. Flowers: calyx 4-parted to near base; corolla subregular, rotate, 4-lobed by fusion of 2 adaxial lobes, the lobes longer than tube, the adaxial lobe external in bud, densely pilose within on all sides (in our species); stamens 4, subequal, exserted, the anthers 2-celled;

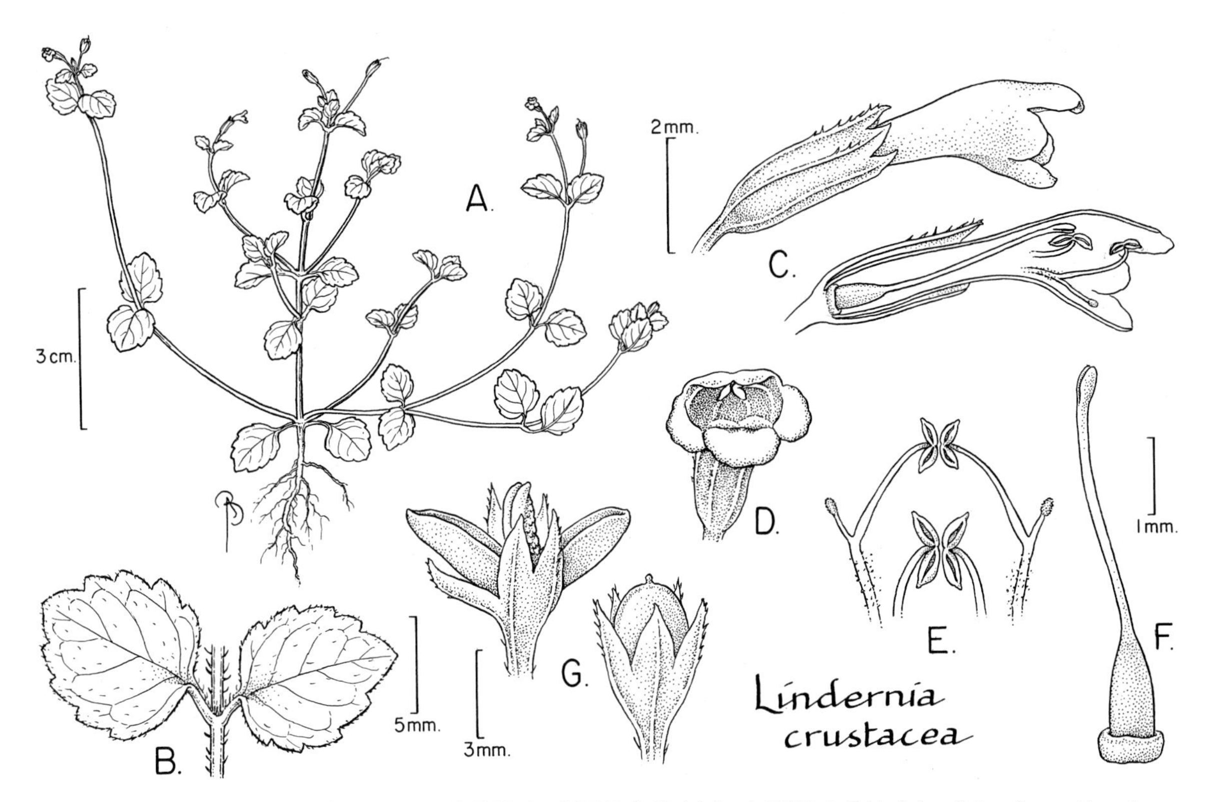

FIG. 292. SCROPHULARIACEAE. *Lindernia crustacea* (A–F, *Mori et al. 23042*; G, *Mori & Gracie 18453*). **A.** Habit of plant. **B.** Part of stem with two leaves. **C.** Lateral view of flower (above) and longitudinal section of flower (below). **D.** Oblique-apical view of flower. **E.** Stamens, the upper pair with appendages. **F.** Lateral view of pistil. **G.** Lateral view of intact (below right) and dehisced (above left) fruits.

style exserted, the stigmas united, capitate, sometimes slightly bifid. Capsules septicidal, secondarily loculicidal, the placental mass remaining free-standing. Seeds numerous, reticulate.

Scoparia dulcis L. PL. 119a,b

Erect herbs, 40–100 cm tall. Stems branched above, stems longitudinally ribbed. Vegetative parts minutely puberulent, at least around the nodes. Leaves: petioles 4–10 mm long; blades of main leaves rhombic-ovate, 2–3(4.5) × 0.9–2 cm, the base cuneate, the apex rounded, the margins irregularly doubly crenate-serrate, slightly revolute. Inflorescences: pedicels 3.5–7 mm long, ascending-spreading. Flowers: calyx 2–3 mm long, the segments 4, broadly (ob)lanceolate, 3–5-veined, glandular-ciliate distally; corolla 1.5–2.7 mm long, the lobes ca. 1 mm long, white or sometimes pale blue or pinkish-white, the apex rounded; anthers 0.5–0.7 mm long; style 1.5–1.7 mm long. Capsules globose, 2.3–4 mm long, the suture line furrowed, the apex emarginate. Seeds conical, 0.2–0.4 mm long, yellowish-brown, the apex truncate. Fl (Oct), fr (Oct); in disturbed habitats, especially along river banks and roadsides. *Balai doux*, *ti balai* (Créole), *vassourinha* (Portuguese).

SIMAROUBACEAE (Tree-of-heaven Family)

Wm. Wayt Thomas

Trees or shrubs. Stipules absent. Leaves imparipinnately (rarely paripinnately) compound, alternate; leaflets alternate or opposite. Inflorescences paniculate or racemose. Flowers actinomorphic, bisexual or unisexual (plants dioecious); calyx and corolla 5-merous; stamens 10 (*Simaba* and *Simarouba*) or 5 and opposite petals (*Picramnia*), the filaments appendaged in *Simaba* and *Simarouba*; gynoecium borne on disk or gynophore, the carpels 5 and essentially free with common style and divergent stigmas in *Simaba* and *Simarouba*, or 2–4 and united in *Picramnia* with very short style and divergent stigmas. Fruits a drupaceous monocarp (1 or rarely 2–5 per flower) derived from separate carpels in *Simaba* and *Simarouba*, or a berry in *Picramnia*.

Recent research has revealed that *Picramnia* is unrelated to members of the Simaroubaceae *sensu stricto* and best placed in the segregate family Picramniaceae (Fernando & Quinn, 1995).

Jansen-Jacobs, M. J. 1979. Simaroubaceae. *In* A. L. Stoffers & J. C. Lindeman, Flora of Suriname **V(1):** 319–330. E. J. Brill, Leiden.

1. Plants with unisexual flowers (plants dioecious). Leaflets usually chartaceous, the bases usually oblique, often strongly so. Inflorescences pendent. Fruits 2–4-seeded, berries. . . . . *Picramnia.*

1. Plants with bisexual or unisexual (plants dioecious) flowers. Leaflets often subcoriaceous, the bases usually symmetrical and cuneate or slightly oblique. Inflorescences erect. Fruits single-seeded, drupaceous.
   2. Plants with bisexual flowers. Leaflets opposite, usually 3–7(29). . . . . *Simaba.*
   2. Plants with unisexual flowers (dioecious). Leaflets alternate to subopposite, usually 11–23. . . . . *Simarouba.*

PICRAMNIA Sw.

Small trees, treelets, or shrubs. Leaves imparipinnate; leaflets alternate, usually chartaceous, often darkening upon drying, the base of lateral leaflets often oblique. Inflorescences pendent, simple racemes, either of separate flowers or glomerules of flowers, or compound thyrses. Flowers unisexual (plants dioecious); stamens or staminodia (if present) 5, the filaments unappendaged; intrastaminal disk entire or (4)5-lobed. Staminate flowers with pistillodium usually conical and pubescent. Pistillate flowers with ovary ovoid, glabrous to densely pubescent, the carpels 2(4), the style very short, the stigmas divergent. Fruits berries, obovoid to ellipsoid, glabrous to glabrate and shiny or short-pubescent, the calyx persistent, the sessile stigmas persistent. Seeds 2–4 per fruit.

1. Leaflets 7–9, usually glabrous, the largest 4–8 cm wide. Inflorescences lateral (cauliflorous), or lateral and terminal. . . . . *P. latifolia.*
1. Leaflets 9–19(23), usually puberulent or pubescent, the largest 2–5 cm wide. Inflorescences terminal or subterminal.
   2. Leaflet apex rounded to slightly acuminate. Fruits finely pubescent, the hairs cylindrical to subclavate. . . . . *P. guiananensis.*
   2. Leaflet apex narrowly acute to caudate. Fruits glabrous to puberulent, the hairs tapering apically. . . . . *P. sellowii* subsp. *spruceana.*

Picramnia guianensis (Aubl.) Jans.-Jac.

Small trees or shrubs, 2–6 m tall. Leaves 9–26 cm long; petioles 0.3–3 cm long; leaflets 11–19(23), ovate to elliptic, glabrous and shiny adaxially, glabrous to sometimes puberulent abaxially, the base usually oblique, obtuse to attenuate, the apex rounded to slightly acuminate, the margins ciliate; larger leaflets 5.5–9 × 2–4 cm. Inflorescences terminal or subterminal racemes, solitary. Staminate racemes 10–16 cm long. Pistillate racemes ca. 12 cm long. Staminate flowers cream-colored; petals 1.5–2 mm long; stamens ca. 2.5 mm long. Pistillate flowers yellowish-green; petals 0.5–0.8 mm long; staminodia not evident; ovary densely pubescent. Fruits 0.9–1.7 × 0.6–1.3 cm long, obovoid to globose, yellow to orange, densely short-pubescent, the hairs cylindrical to subclavate.

A rare and inconspicuous species endemic to the lowland forests of the Guianas and adjacent Brazil; it has not yet been collected around Saül, but has been collected nearby and is likely to occur in our area.

Picramnia latifolia Tul.

Small trees or shrubs, 3–15 m tall. Leaves 20–41 cm long; petioles 3–5 cm long; leaflets 7–9, elliptic to ovate or rarely obovate, glabrous although sometimes puberulent on primary veins and margins, the base usually oblique, obtuse to attenuate, the apex acuminate; larger leaflets 9–18 × 4–8 cm. Inflorescences cauline or terminal racemes, solitary or in fascicles. Staminate racemes 7–23 cm long. Pistillate racemes 12–60 cm long. Staminate flowers yellowish; petals 1–1.2 mm long; stamens 1.8–2.2 mm long. Pistillate flowers whitish; petals 0.8–1.2 mm long; staminodia petaloid; ovary densely to sparsely pubescent. Fruits 0.5–1.6 × 0.4–0.8 cm, ellipsoid to obovoid, glabrate or glabrous, orange or red becoming black at maturity. Fl (Aug, Dec), fr (Oct, Dec); uncommon, in nonflooded forest.

Picramnia sellowii Planch. subsp. **spruceana** (Engl.) Pirani

Small trees or shrubs, 1.5–15 m tall. Leaves 18–40 cm long; petioles 1–2.5 cm long; leaflets 9–19, elliptic or narrowly so to ovate, glabrous to glabrate although usually puberulent on larger veins and margins, the primary and secondary veins usually sulcate adaxially, the base oblique, obtuse to acute, the apex narrowly acute to caudate; larger leaflets 7–16 × 2.5–5 cm. Inflorescences terminal thyrses, usually with 1–5 branches. Staminate thyrses 10–65 cm long. Pistillate thyrses 10–40 cm long. Staminate flowers yellowish-green to cream-colored; petals 0.7–1.8 mm long; stamens 1.5–3 mm long. Pistillate flowers yellowish-green to cream-colored; petals 1.5–2 mm long; staminodia present or absent; ovary glabrous to sparsely pubescent. Fruits 1.4–2 × 0.6–1.4 cm long, ellipsoid, puberulent (the hairs tapering apically) to glabrous, red becoming black at maturity, the apex apiculate. Fl (Oct); in nonflooded and periodically flooded forest.

SIMABA Aubl.

Shrubs or trees. Leaves with leaflets opposite or subopposite. Inflorescences usually erect, paniculate, puberulent; pedicels usually subtended by a small pendulous gland. Flowers bisexual; sepals and petals (4)5; stamens 10, the filaments with adaxial, slender, usually pubescent, adnate or distally free appendage; gynophore cylindrical; ovary globose to depressed globose, of (4)5 separate carpels joined by common style. Fruits drupaceous monocarps, usually 1–5 from each flower.

1. Leaves 17–29 leaflets, the apices tipped with a small, hard gland. Petals 25–30 mm long. . . . . *S. cedron.*
1. Leaves with 3–23 leaflets, the apices not tipped with a small, hard gland. Petals 4–9 mm long.
   2. Leaves with (5)11–23 leaflets; leaflets usually with base oblique. Style in developed flowers 3.5–4.6 mm long. . . . . *S. polyphylla.*
   2. Leaves with 3–7(13) leaflets; leaflets usually with base symmetric. Style in developed flowers 0.2–2.9 mm long.

3. Trees, 6–18(30) m tall. Leaflets usually elliptic to obovate or ovate. Style 0.8–2.9 mm long. . . . . *S. guianensis* subsp. *ecaudata*.
3. Shrubs, 0.3–2 m tall. Leaflets usually narrowly elliptic. Style 0.2–0.5 or 1.3–1.5 mm long. . . . . *S. guianensis* subsp. *guianensis*.

Simaba cedron Planch.

Trees or treelets, unbranched or seldom branched, (1)2–10 m tall. Leaves usually grouped near apex, 45–120 cm long (including petiole and terminal leaflet), 17–29 leaflets per leaf; petioles 10–35 cm long; petiolules absent to poorly developed; leaflets opposite or subopposite, elliptic or oblong to narrowly oblong, 8–18 × 2.5–6(8) cm, the base obtuse to rounded, often strongly oblique, the apex abruptly acuminate, terminating in hard, brown gland. Inflorescences terminal or subterminal, paniculate or racemose, 50–115 cm long, the flowers fasiculate along the axes. Flowers: calyx cupuliform, the margin almost entire; petals narrowly oblong, 25–30 × 2.5–5 mm, sparsely pubescent, pale green to cream-colored; stamens 20–25 mm long, the staminal appendage 16–17 mm long, the free portion 4–5 mm long, liguliform, pilose; gynophore 3 mm long, the stigma indistinct. Drupaceous monocarps obovoid, 5–8(10) × 3–4.5 cm, brown to ferruginous, minutely puberulent. In nonflooded forest and forest edges. *Pau para tudo* (Portuguese), *wan édè* (Bush Negro).

This widespread species is known from French Guiana, but has not yet been collected from our area.

Simaba guianensis Aubl. subsp. **ecaudata** (Cronq.) Cavalcante [Syn.: *Simaba nigrescens* Engl.]

Trees, 6–18(30) m tall. Leaves with 3–7(13) leaflets, 9–22 cm long (including petiole and terminal leaflet); petioles 1.4–5.1 cm long, the larger leaflets ± sessile or with petiolules to 1 cm long; leaflets elliptic to slightly obovate or ovate, 3.1–12 × 1.3–5.6 cm, the bases cuneate to narrowly so, the apices rounded to acute or acuminate or occasionally caudate. Inflorescences terminal or from uppermost leaf axils, the main axis at anthesis 2–9(13) cm long; pedicels 1–3.5 mm long. Flowers: sepals 0.5–0.7(1) mm long, puberulent; petals narrowly elliptic, (4.2)4.5–5.9 × 1.2–2.1 mm, puberulent abaxially, usually white, occasionally pale yellow, cream-colored, or reddish-green; stamens with filaments 3.6–5.6 mm long, the staminal appendage 2.2–2.6 mm long, distally ciliate; gynoecium 0.3–0.8 × 0.8–1.3 mm, pilose, the style 0.8–2.9 mm long, lobed at apex. Drupaceous monocarps ellipsoid, lenticular, 1.2–1.4 × 0.7–0.8 mm, orange to green when ripe, the epicarp thin and rugose or ± fleshy and smooth, glabrous or sometimes puberulent. Fl (Aug); uncommon, in nonflooded forest.

Simaba guianensis Aubl. subsp. **guianensis** PL. 119d

Shrubs, 0.3–2 m tall. Leaves with 3–7 leaflets, 8.5–30 cm long (including petiole and terminal leaflet); petioles (0.9)1.8–9 cm long, the larger leaflets ± sessile or with petiolules to 2.5 cm long; leaflets narrowly elliptic to elliptic or occasionally narrowly obovate, 4.3–19 × 1.6–6.4 cm, the bases attenuate to narrowly cuneate, occasionally oblique, the apices caudate to acuminate. Inflorescences terminal or from uppermost leaf axils, slender, the main axis at anthesis 0.8–6 cm long; pedicels 1–4 mm long. Flowers: sepals 0.4–0.8 mm long, puberulent; petals narrowly elliptic to elliptic, 4–5 × 1.2–2 mm, puberulent abaxially, cream-colored to yellow, yellowish-orange or white; stamens with filaments 2.2–3.2 mm long, the staminal appendages 1.5–2.2 × 0.5–0.7 mm, distally ciliate; gynoecium 0.3–1 × 0.6–1.3 mm, oblate to transversely ellipsoid, short tomentose, the style 0.2–0.5 or 1.3–1.5 mm long (short- and long-styled flowers). Drupaceous monocarps ellipsoid, lenticular, 1–1.6 × 0.6–1.1 cm, rugose and yellow to orange when ripe, the epicarp thin, fleshy and sometimes slightly winged. Fl (Aug), fr (Sep, Nov); common, in nonflooded forest.

Research in progress suggests that the two subspecies recognized here may deserve status as distinct species.

Simaba polyphylla (Cavalcante) W. W. Thomas [Syn.: *Simaba guianensis* subsp. *polyphylla* Cavalcante]

Trees or large shrubs, 3–40 m tall. Leaves with (5)11–23 leaflets (including petiole and terminal leaflet); petioles 5–12 cm long, the rachis (including petiole) 11–35 cm long; petiolules to 3 mm long or leaflets sessile; terminal leaflet elliptic or narrowly so to slightly obovate or rarely ovate, (3.6)5–6–15 × (1.2)1.8–5.5 cm, the base usually oblique (terminal leaflet symmetrical and cuneate), the apex attenuate, often abruptly so, with rounded tip, subcoriaceous, glabrous or very sparsely puberulent adaxially, glabrous or short tomentose abaxially. Inflorescences terminal, paniculate, stout, the main axis 2–10 cm long, 2–3.5(5) mm diam. basally, glabrous to puberulent, lower lateral branches often longer than main axis; pedicels 2.5–5 mm long, puberulent, often subtended by small pendulous glands. Flowers: sepals 5, (0.5)0.7–0.8 mm long, pubescent; petals 5, narrowly ovate to elliptic, 4.6–9 × 1.8–2.2 mm wide, cream-colored; stamens with filaments 3.1–5 mm long, the staminal appendage 1.9–2.3 mm long, free portion 1–2 mm long, pilose, even on face; gynoecium 0.5–1.2 mm high, 1–1.6 mm broad, pubescent, the style 3.5–4.6 mm long. Drupaceous monocarps ellipsoid, lenticular, 1.4–1.8 × 0.8–1.3 cm wide, puberulent to glabrate, black to brown or olive-brown, the epicarp thin, slightly fleshy, rugose when dry. Known only by a sterile specimen (*Boom & Mori 1611*, NY) from our area; in nonflooded forest. *Pau para tudo, serve para tudo* (Portuguese), *wan édé* (Bush Negro).

SIMAROUBA Aubl.

Medium-sized to large trees. Leaves imparipinnate, leaflets alternate to subopposite. Inflorescences erect or arching, paniculate, the staminate inflorescences larger and with more flowers than pistillate ones. Flowers unisexual (plants dioecious); sepals and petals 5; stamens 10, the filaments with an adaxial, pubescent, distally free appendage; gynophore cylindrical; ovary globose to depressed globose, of (4)5 separate carpels joined by common style. Fruits drupaceous monocarps, 1(5) from each flower.

Simarouba amara Aubl. FIG. 293

Trees, 10–30 m tall. Bark with bitter taste. Leaves with 11–23 leaflets, 25–50 cm long (including petiole and terminal leaflet); petioles 5–11 cm long; larger leaflets elliptic to oblanceolate, 6.5–10 × 4–3.7 cm, the base cuneate, the apex rounded to retuse, sometimes with short mucronate tip. Inflorescences terminal. Staminate inflorescences 20–35 cm long. Pistillate inflorescences 10–15 cm long. Flowers greenish; sepals 0.7–0.8 mm long, depressed triangular, puberulent; petals 2.5–3.2 mm long. Staminate

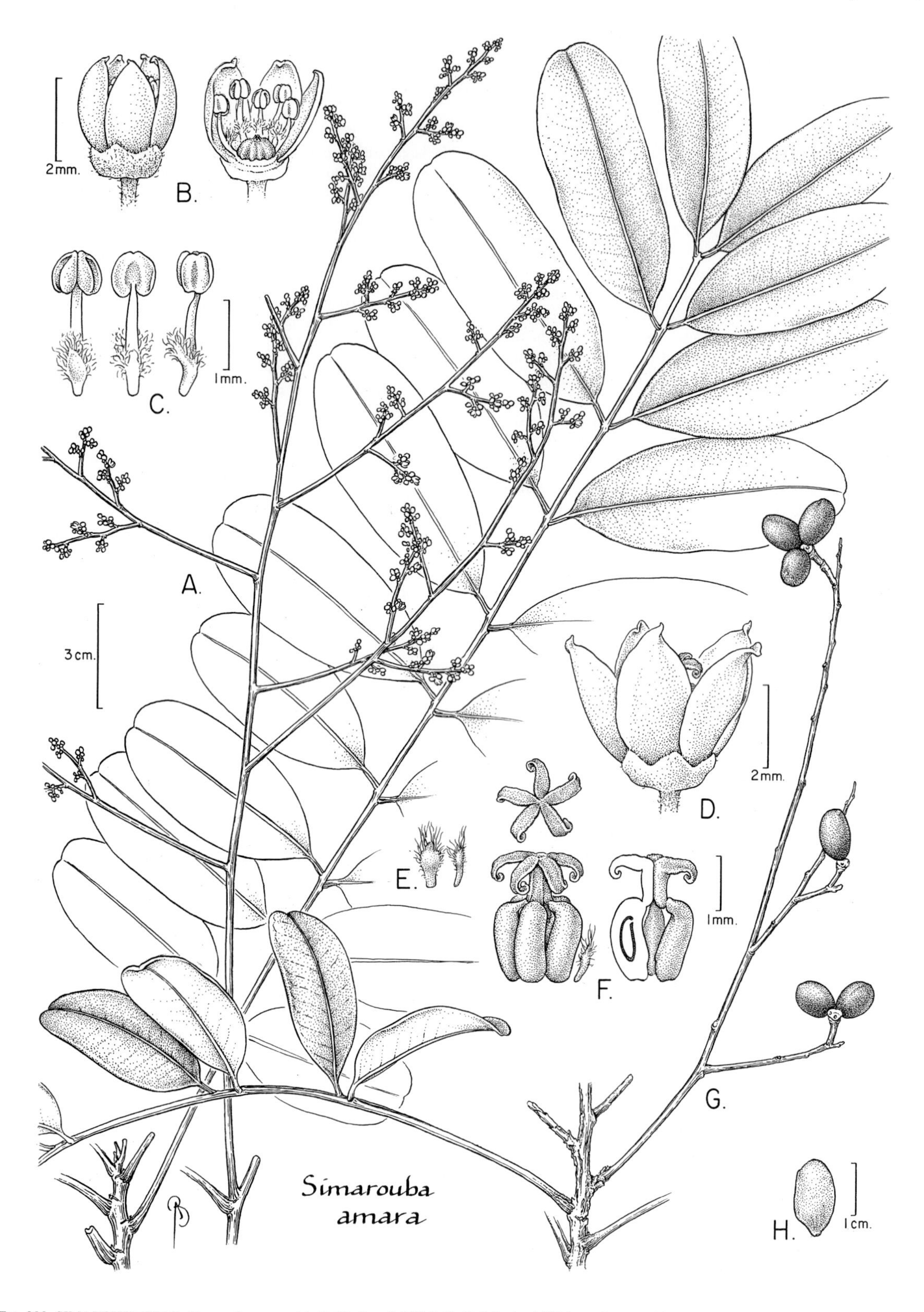

FIG. 293. SIMAROUBACEAE. *Simarouba amara* (A–C, *Mori et al. 21570*; D–F, *Sabatier 1159* from Cayenne, French Guiana; G, H, *Mori & Gracie 21141*). **A.** Compound leaf and inflorescence. **B.** Lateral view (left) and lateral view with two petals and part of calyx removed (right) of staminate flower. **C.** Stamens with pubescent appendages. **D.** Lateral view of pistillate flower. **E.** Staminodia. **F.** Lateral view of pistil (far left); note free carpels with united style and separated stigmatic lobes, and a staminode (near left); same view with several carpels removed and one carpel in medial section (right), and apical view of stigmatic lobes (above left). **G.** Part of infructescence and base of leaf; note multiple monocarps from a single flower. **H.** Seed.

flowers with stamens 1.6–2.5 mm long, the basal appendage of filament ± 0.5 mm long, tomentose. Pistillate flowers with staminodia ± 0.6 mm long, the basal appendage of filament ± 0.5 mm long and tomentose; ovary 1.4 × 1.8 mm, stigma branches 5, each ± 0.5 mm long. Drupaceous monocarps 1.2–1.6 × 0.7–1 cm, lenticular, ellipsoid, fleshy, smooth, glabrous, green becoming black upon maturity. Fl (Nov), fr (Feb); scattered, in nonflooded forest. *Assoumaripa* (Bush Negro), *marupa* (Portuguese), *simarouba* (Créole).

SOLANACEAE (Nightshade Family)

Michael Nee

Herbs, shrubs, lianas, vines or trees, terrestial, rarely epiphytic. Stipules absent. Leaves simple or less commonly compound, basically alternate, but often geminate, with one larger than other. Flowers actinomorphic or zygomorphic, bisexual; calyx 5-lobed, or more rarely truncate or spathe-like; corolla 5-lobed; stamens 5, less commonly reduced to 4 or 2, borne on corolla tube; ovary superior, rarely semi-inferior, 2-carpellate and usually 2-locular; placentation axile. Fruits berries or septicidal capsules. Seeds usually numerous.

1. Anthers dehiscing by apical pores, sometimes later with longitudinal slits; nectary disc absent.
    2. Calyx truncate, with 5 or 10 appendages emerging from below rim. . . . . *Lycianthes*.
    2. Calyx 5-lobed, or if nearly truncate, without appendages from below rim.
        3. Inflorescence from bifurcations of stem. Anthers with enlarged connective. . . . . *Cyphomandra*.
        3. Inflorescences extra-axillary, i.e., lateral on stem between or opposite leaves (only axillary in *Solanum anceps*). Anthers without enlarged connective. . . . . *Solanum*.
1. Anthers dehiscing by longitudinal slits; nectary disc present at base of ovary.
    4. Stamens 4, didynamous, included in narrow corolla tube. Fruits fleshy capsules. . . . . *Brunfelsia*.
    4. Stamens 5, ± equal, exserted from wide corolla tube. Fruits juicy berries.
        5. Calyx truncate, without lobes or appendages. Berries extremely pungent ("hot") but edible. . . . . . . *Capsicum*.
        5. Calyx lobed. Berries not pungent.
            6. Perennial epiphytic shrubs, with soft-woody stems and branches. Flowers several in an inflorescence. Fruits visible in accrescent but apically open calyx. . . . . *Markea*.
            6. Annual terrestrial herbs. Flowers solitary, axillary. Fruits completely enclosed in inflated calyx tube. . . . . *Physalis*.

BRUNFELSIA L.

Shrubs or treelets. Leaves simple, alternate; blades glabrous or with simple hairs, the margins entire. Inflorescences cymes or reduced to single flower. Flowers: corolla hypocrateriform, slightly zygomorphic; filaments inserted in upper part of corolla tube, the anthers in two pairs; style equalling filaments, the stigma held between two pairs of anthers, anthers longitudinally dehiscent. Fruits fleshy capsules, tardily dehiscent. Seeds few to numerous, angular.

1. Leaves drying dark green or blackish. Fruits baccate (actually fleshy capsules). *B. guianensis*.
1. Leaves drying yellowish green. Fruits capsules. . . . . *B. martiana*.

Brunfelsia guianensis Aubl. FIG. 294

Shrubs or treelets, 1–4 m tall, glabrous. Leaves: petioles 3–8 mm long; blades obovate to elliptic, 6–15 × 2–6.5 cm, the apex abruptly acuminate, drying dark green to blackish. Inflorescences axillary or terminal, with 1–2 flowers; pedicels 2–5 mm long. Flowers: calyx ovoid-campanulate, 7–10 × 4–7 mm; corolla white, the tube 20–26 × 1–1.5 mm, gradually dilated above middle, the limb spreading, 15–22 mm wide; stamens didynamous, included in tube, one pair of filaments 4–5 mm long, the other pair 2–4 mm long, the anthers orbicular-reniform, 1 mm diam. Fruits fleshy capsules (baccate), globose to ovoid, 2.5–4 cm diam., yellow when ripe. Seeds 7–10, 10–13 mm long, black, in molasses-like pulp. Fl (Jan, Dec), fr (Mar, Apr, May, Jun, Aug, Sep, Dec); common but scattered, in nonflooded forest. *Graine macaque* (Créole), *manaca* (Portuguese).

One collection (*Mori et al. 23756*, NY) has leaves that match *B. guianensis* but apparently a capsular fruit like *B. martiana*, its placement here is provisional.

Brunfelsia martiana Plowman

Shrubs, to 3 m tall, glabrous. Leaves: petioles 1–4 mm long; blades oblong, 10–25 × 4–8 cm, the base cuneate to obtuse, the apex acuminate, drying yellowish green. Inflorescences terminal or axillary, the axis to 10 mm long, with 1–7 flowers; pedicels 3–6 mm long. Flowers: calyx tubular-campanulate, 8–12 × 3–8 mm; corolla 20–24 × 1–3 mm, the limb 15–22 mm wide; stamens included in tube, didymous, the longer pair 4 mm long, the shorter 3 mm long, the anthers orbiculate-reniform, 1 mm diam. Fruits capsules. Fr (May); apparently rare, in nonflooded forest.

CAPSICUM L.

Herbs or small shrubs. Leaves simple, alternate or in pairs; blades with margins entire or shallowly lobed. Inflorescences axillary, solitary or fasciculate. Flowers: calyx truncate and entire, sometimes with 5 or 10 appendages emerging from below rim; corolla subrotate to

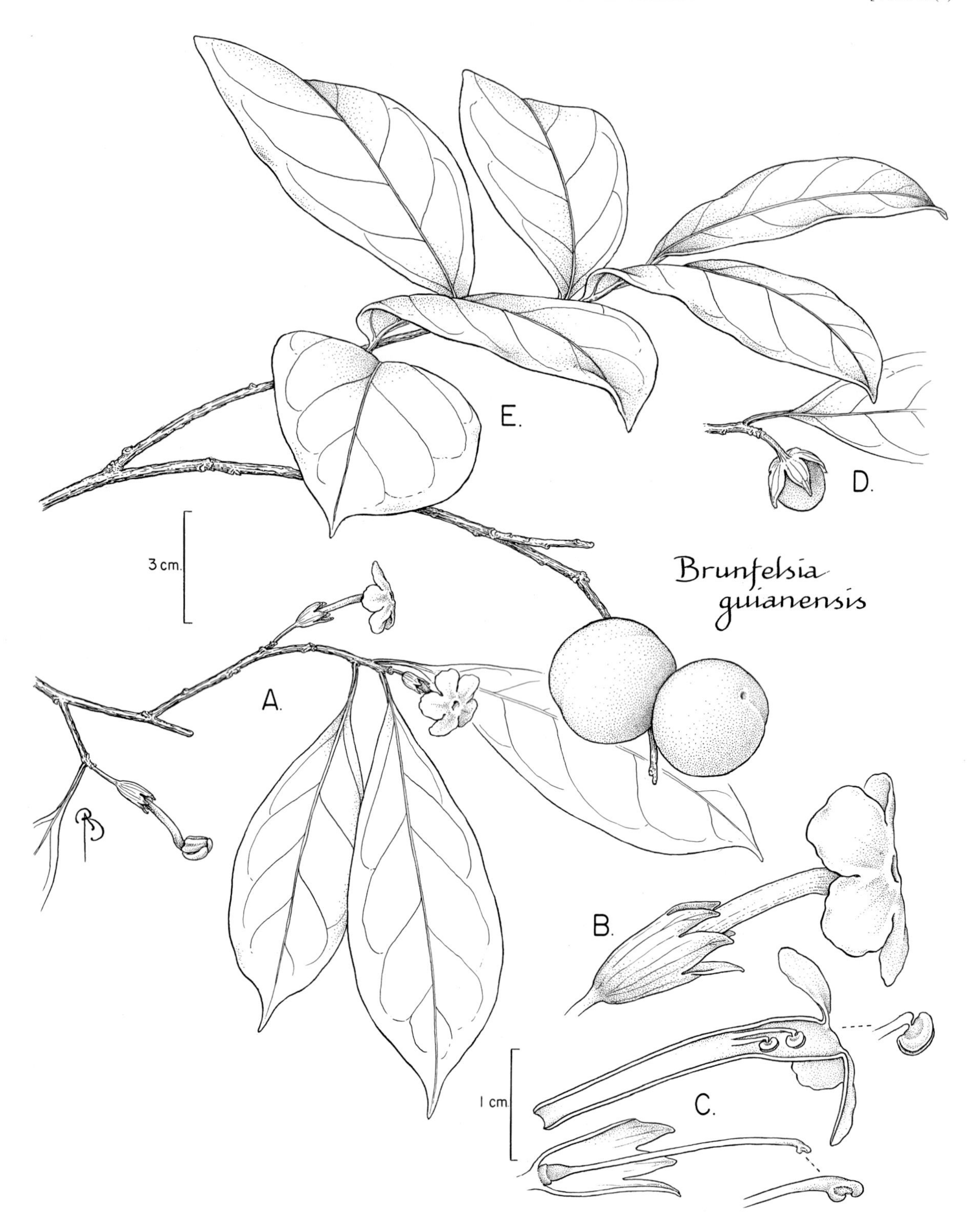

FIG. 294. SOLANACEAE. *Brunfelsia guianensis* (A–C, *Mori et al. 8767*; D, *Boom 10805*; E, *Mori & Gracie 18857*). **A.** Stem with leaves a flowers. **B.** Lateral view of flower. **C.** Medial section of corolla with detail of stamen (above) and pistil surrounded by one-half of calyx with detail of stigma (below). **D.** Immature fruit. **E.** Stem with leaves and two mature fruits.

campanulate, 5-lobed; stamens 5, equal, the anthers longitudinally dehiscent. Fruits berries, bright red, yellow or orange, sometimes purple, juicy to almost dry, bland to intensely pungent. Seeds numerous or few, flattened, reniform.

The peppers cultivated in local gardens seem to be mostly *C. chinense* Jacq., which was derived from the wild types of *C. frutescens*.

Capsicum frutescens L.

Herbs or small shrubs, to 2 m tall, nearly glabrous. Leaves: petioles 3–11 mm long; blades ovate, 3–8 × 1.5–4 cm, with a few hairs on major veins ad- and abaxially. Inflorescences axillary at leaf insertion and branch point, fasciculate, with 1–2 flowers; pedicels erect, 1 cm long and slender in flower, 2 cm long and distinctly enlarged at summit at junction with ribbed calyx in fruit. Flowers: calyx truncate, without appendages, 2 mm long in fruit; corolla nodding, ca. 5 mm wide, pale white to greenish yellow; anthers blue or green, ca. 1 mm long. Fruits held erect, bright red when ripe, narrowly ovoid, 1.5 × 0.5 cm, much larger in cultivated plants, very pungent. Seeds 3 mm long. Fl (Sep), fr (Sep); weed along roadsides and other weedy places, tolerated or planted in gardens for the edible fruits. *Piment* (Créole), *pimenta* (Portuguese), *poivre de Cayenne* (French).

CYPHOMANDRA Sendtn.

Soft-wooded shrubs to small trees, usually with a single stem, then with divergent branching above, glabrous or with simple or branched hairs. Leaves simple, often dimorphic, ill-smelling, those of stem often lobed, those of fertile branches entire. Inflorescences from bifurcations of stem, but often appearing axillary or extra-axillary, simple or branched scorpioid cymes. Flowers actinomorphic; corolla 5-lobed; stamens equal, the anthers usually tapered, dehiscent by apical pores, the connective expanded and usually of different color. Fruits juicy berries. Seeds numerous, flattened, reniform.

Cyphomandra is considered to be part of *Solanum* by some authors (Bohs, 1995).

Bohs, L. 1994. *Cyphomandra* (Solanaceae). Fl. Neotrop. Monogr. **63:** 1–175.

Cyphomandra endopogon Bitter subsp. **guianensis** Bohs Fig. 295, Pl. 120a

Shrubs or small trees, 2–15 m tall, the branches densely puberulent. Leaves dimorphic, those of stem entire or 3–5-lobed, those on fertile branches (almost all herbarium material) entire; blades elliptic to ovate, 4–25 × 4–15 cm, puberulent, especially abaxially, the base obtuse to deeply cordate, usually oblique. Inflorescences usually simple, with 20–45 flowers, the rachis hanging, 3–40 cm long; pedicels 20–35 mm long. Flower buds narrowly oblong, purplish; corolla lobed to near base, lavender when first open, fading to white on second day, the lobes narrowly oblong, 15–22 × 2–3 mm, the margins densely ciliate; anthers 8–10 mm long, yellow or purplish, the connective strongly expanded, especially in lower part. Fruits ellipsoid or globose berries, 4–5 × 3–5 cm, green, the mature color not known. Seeds numerous, 5–6 mm long. Fl (Jul, Aug, Sep, Oct, Nov), fr (Feb, Sep, Oct, Nov, Dec); common, in disturbed forest or secondary growth; pollinated by male euglossine bees (Gracie, 1993). *Kuzu, mavévé chien* (Créole).

LYCIANTHES (Dunal) Hassl.

Shrubs, vines, lianas, or perennial herbs from enlarged root, glabrous or pubescent with simple, branched, or stellate hairs. Leaves simple, generally alternate at base of plant and paired above, simple; blades with entire margins. Inflorescences axillary, with one to numerous fasciculate flowers. Flowers: calyx truncate and entire, usually with 5 or 10 appendages emerging from below rim; corolla deeply to very shallowly 5-lobed; stamens 5, the filaments equal, or 1 or 3 longer than rest, the anthers ellipsoid or oblong, dehiscent by apical pores. Fruits globose or conical berries, rarely drupaceous. Seeds flattened, reniform.

Lycianthes pauciflora (Vahl) Bitter [Syns.: *Solanum pauciflorum* Vahl, *Solanum geminatum* Vahl, *Lycianthes geminata* (Vahl) Bitter, *Solanum guianense* Dunal, *Lycianthes guianensis* (Dunal) Bitter] Fig. 296, Pl. 120b

Lianas, sometimes appearing as scandent shrubs, the stem with small, sessile, stellate hairs. Leaves simple, single or paired; petioles 8–13 mm long; blades elliptic, 7.5–15 × 4–7.5 cm, glabrous except on midrib adaxially, more tomentose with minute, sessile, stellate hairs abaxially, the base acute to rounded, the apex attenuate. Inflorescences axillary, with 1–3 flowers. Flowers: calyx 12 mm wide at maturity, with 10 appendages 2–3 mm long; corolla rotate, white with greenish midribs, probably opening at night (thus with few good flowering collections in herberia). Fruits globose berries, 2 cm diam., turning red at maturity. Seeds numerous, 3 mm long. Fl (Apr), fr (Feb, Mar, Apr, May, Jun); primary or secondary forest.

MARKEA Rich.

Epiphytic or hemiepiphytic shrubs or lianas, often growing in ant nests. Leaves simple; blades coriaceous, the margins entire. Inflorescences terminal, solitary or paniculate, racemose or corymbose; peduncles very short to very long. Flowers small to large; calyx often accrescent and sometimes showy; corolla campanulate to tubular, the tube longer than lobes; stamens 5, equal, the anthers longitudinally dehiscent; ovary superior to partly inferior. Fruits juicy or coriaceous berries. Seeds numerous, small, somewhat flattened.

1. Flowers on long flagellate peduncle to nearly 30 cm long; corolla orange to red. *M. coccinea.*
1. Flowers solitary to racemose, the peduncle and axis together to 7 cm long; corolla yellowish, creamy white, or greenish.
    2. Leaves with petioles 20–30 mm long; blades attenuate at base. Pedicels 16–24 mm long. . . . . *M. longiflora.*
    2. Leaves with petioles 8–10 mm long; blades truncate to subcordate at base. Pedicels nearly obsolete. . . . . *M. sessiliflora.*

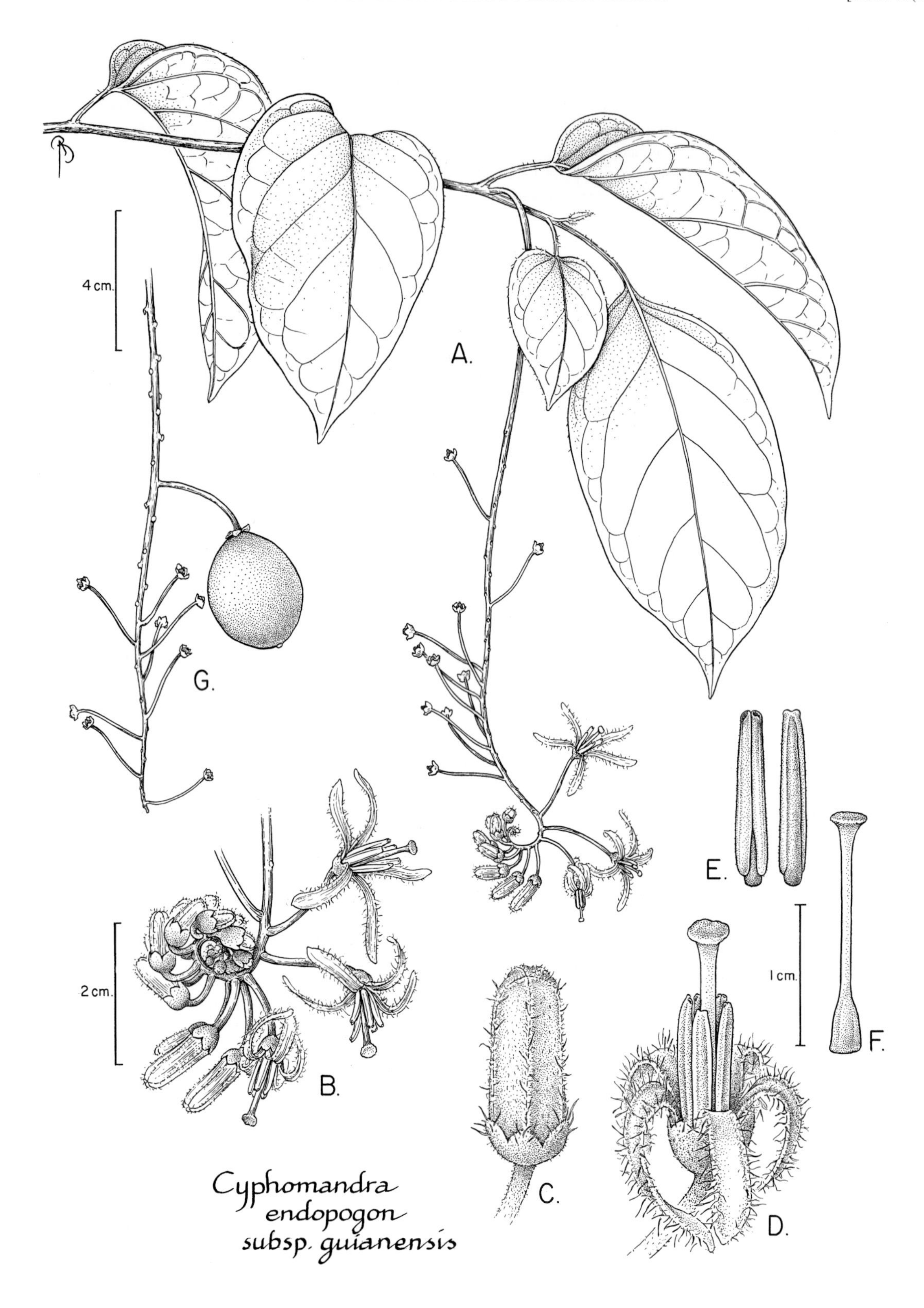

FIG. 295. SOLANACEAE. *Cyphomandra endopogon* subsp. *guianensis* (A–F, *Mori et al. 19054*; G, *Mori & Prance 15028*). **A.** Apex of stem showing leaves and pendulous inflorescence. **B.** Apex of inflorescence. **C.** Lateral view of flower bud. **D.** Lateral view of flower. **E.** Adaxial (left) and abaxial (right) views of stamens showing poricidal dehiscence. **F.** Lateral view of pistil. **G.** Part of infructescence showing fruit.

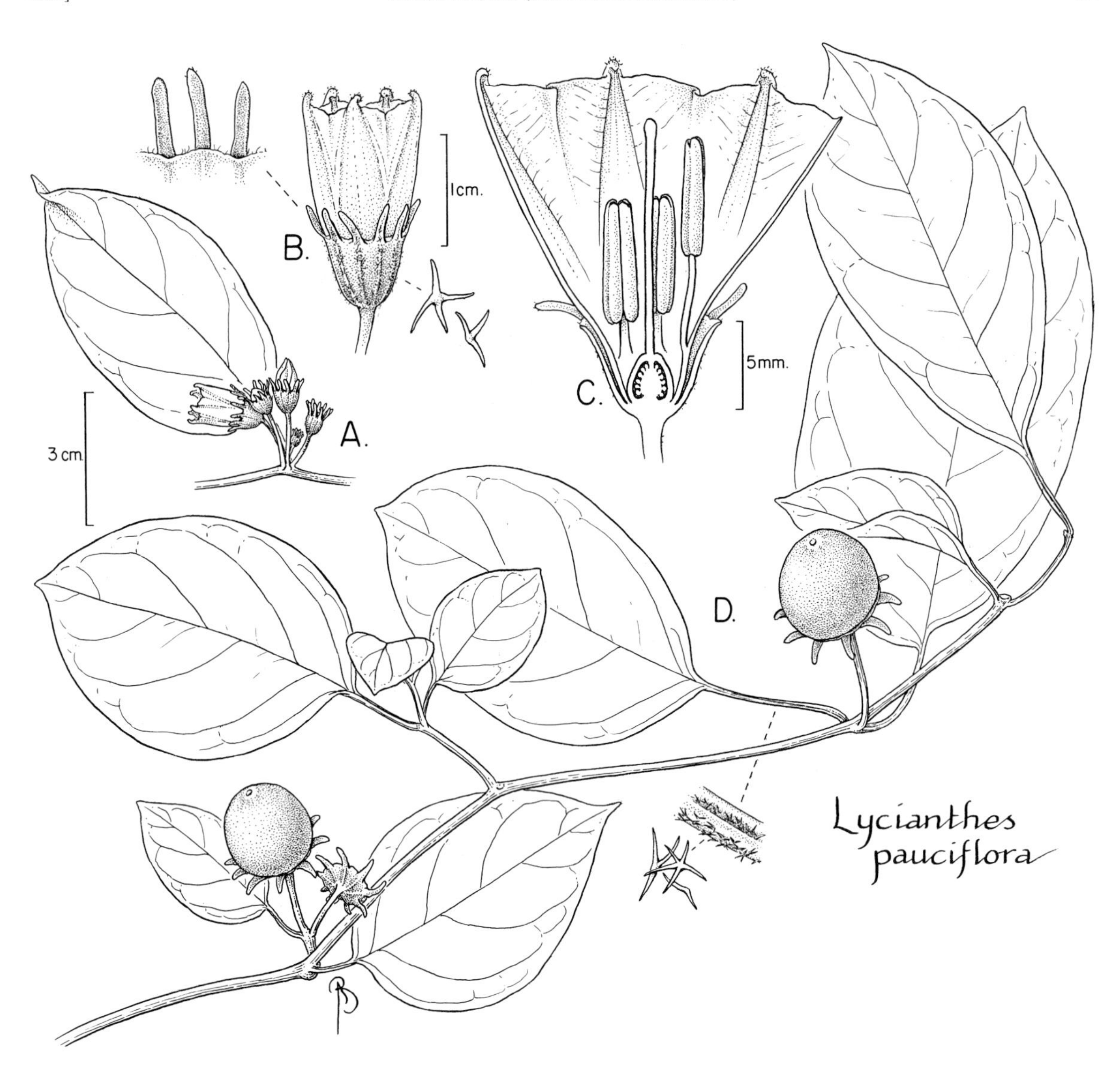

FIG. 296. SOLANACEAE. *Lycianthes pauciflora* (A, *Mori & Cardoso 17637* from Amapá, Brazil; B, C, *Cowan & Soderstrom 2017* from Guyana; D, *Acevedo-Rodríguez et al. 5015*). **A.** Axillary inflorescence and leaf. **B.** Flower with detail of calyx (left) and stellate hairs of calyx (right). **C.** Medial section of flower. **D.** Stem with leaves, infructescences, and detail of stellate hairs (below).

Markea coccinea Rich. FIG. 297, PL. 120d

Epiphytic shrubs with lianoid branches, glabrous. Leaves: petioles 4–7 mm long; blades oblanceolate, 5–5.5 × 12–15 cm long, subcoriaceous, glabrous, the base acute, the apex apiculate. Inflorescences axillary, or appearing terminal, simple or few-branched at apex, to 30 cm long, the flowers few; pedicels 16–20 mm long. Flowers: calyx lobed nearly to base, the lobes lanceolate, ca. 30 × 4 mm, the apex long-attenuate; corolla orange to red, the tube 50 × 5 mm, the limb 45 mm wide, the lobes ovate; free part of filaments 2.5 mm long, attached in villous zone 8 mm above base of corolla tube, the anthers 4 mm long with apices barely exserted from corolla tube. Fruits berries, ca. 1 cm diam. Fl (Aug, Sep, Oct, Dec), fr (Apr, Oct); in nonflooded forest.

Markea longiflora Miers [Syn.: *Markea camponoti* Dammer]

Epiphytic vining shrubs, the young parts minutely puberulent. Stems tan, corky. Leaves: petioles 20–30 mm long; blades oblanceolate, 16–25 × 5–8 cm, subcoriaceous, glabrous, the base attenuate, the apex attenuate. Inflorescences terminal, racemose, 1–7 cm long; pedicels 16–24 mm. Flowers: calyx deeply lobed, the lobes oblong-lanceolate, 32 × 9 mm, pale green or creamy white with purple veins, the apex long-attenuate; corolla creamy white with purple veins, the tube 70–100 mm long, the limb 15–20 mm wide; free part of filaments 16 mm long, the anthers 8 mm long. Fruits oblong yellow berries, 20 × 10 mm. Fl (Nov), fr (Dec); in nonflooded forest.

Markea sessiliflora Ducke [Syn.: *Markea porphyrobaphes* Sandwith] PL. 120c

Epiphytic shrubs, glabrous. Stems tan, corky-cracked. Leaves often in clusters of 2 or 3; petioles 8–10 mm long, corky-cracked; blades elliptic to obovate, 10–17 × 5–9 cm, coriaceous, drying dark adaxially and dark brown abaxially, the base truncate to subcordate at base. Inflorescences terminal, soon obviously lateral, simple or few-branched, to 4 cm long, the flowers one to several, the pedicels

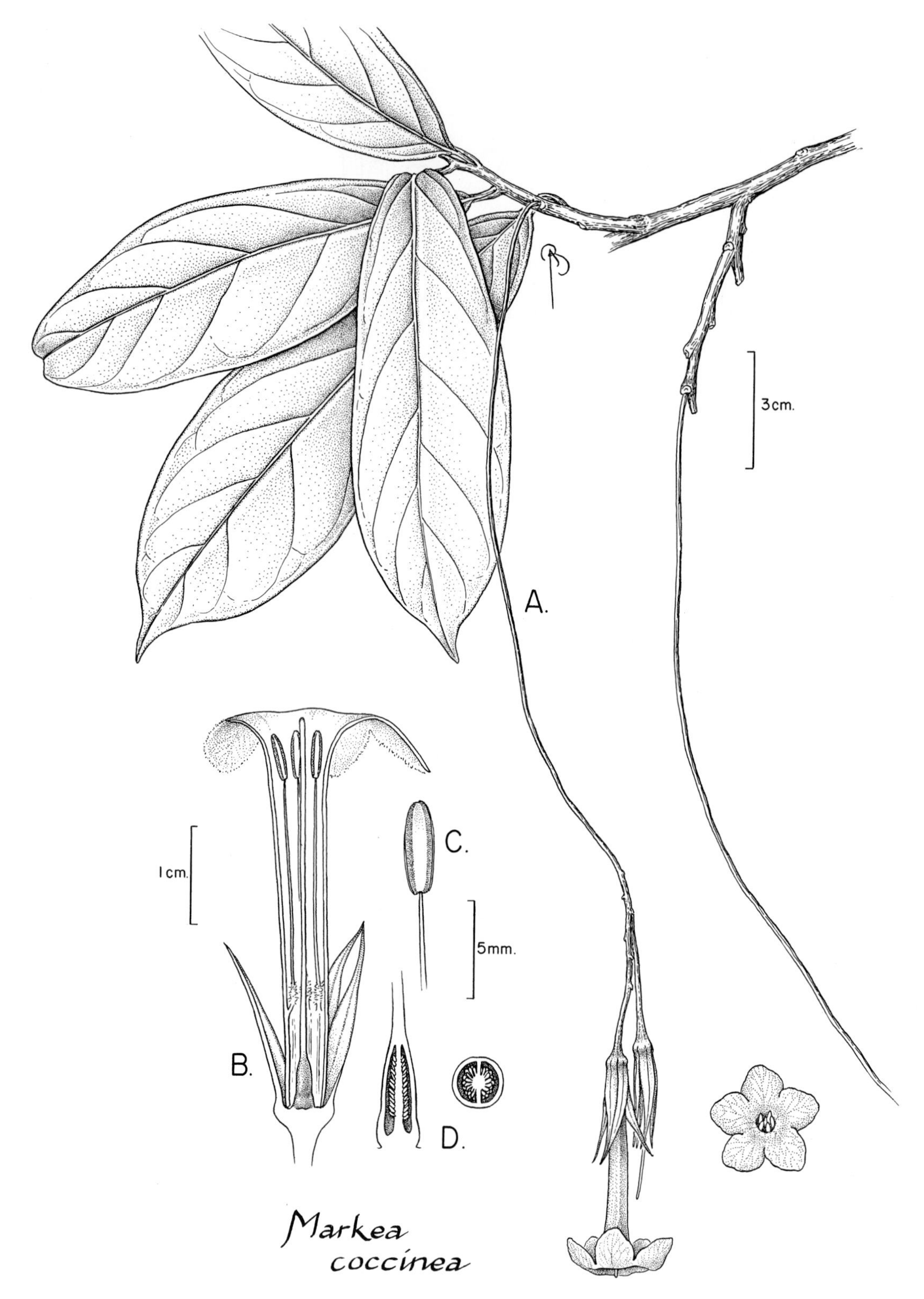

FIG. 297. SOLANACEAE. *Markea coccinea* (*Mori et al. 20899* and *Mori et al. 20749*). **A.** Part of leafy stem showing inflorescence with lateral (left) and apical (below right) views of flower. **B.** Medial section of flower. **C.** Anther and part of filament. **D.** Medial (left) and transverse (right) sections of ovaries.

nearly obsolete. Flowers: calyx light green, deeply lobed, the lobes 25 × 5 mm; corolla light yellow or creamy white, turning brown with age, the tube 50 × 15 mm, the lobes to 7 mm long; free part of filaments 13–30 mm long, attached in villous zone 13 mm above base of corolla tube, the anthers 8–11 mm long, included. Fruits berries. Fl (Jan, Nov, Dec); in nonflooded forest.

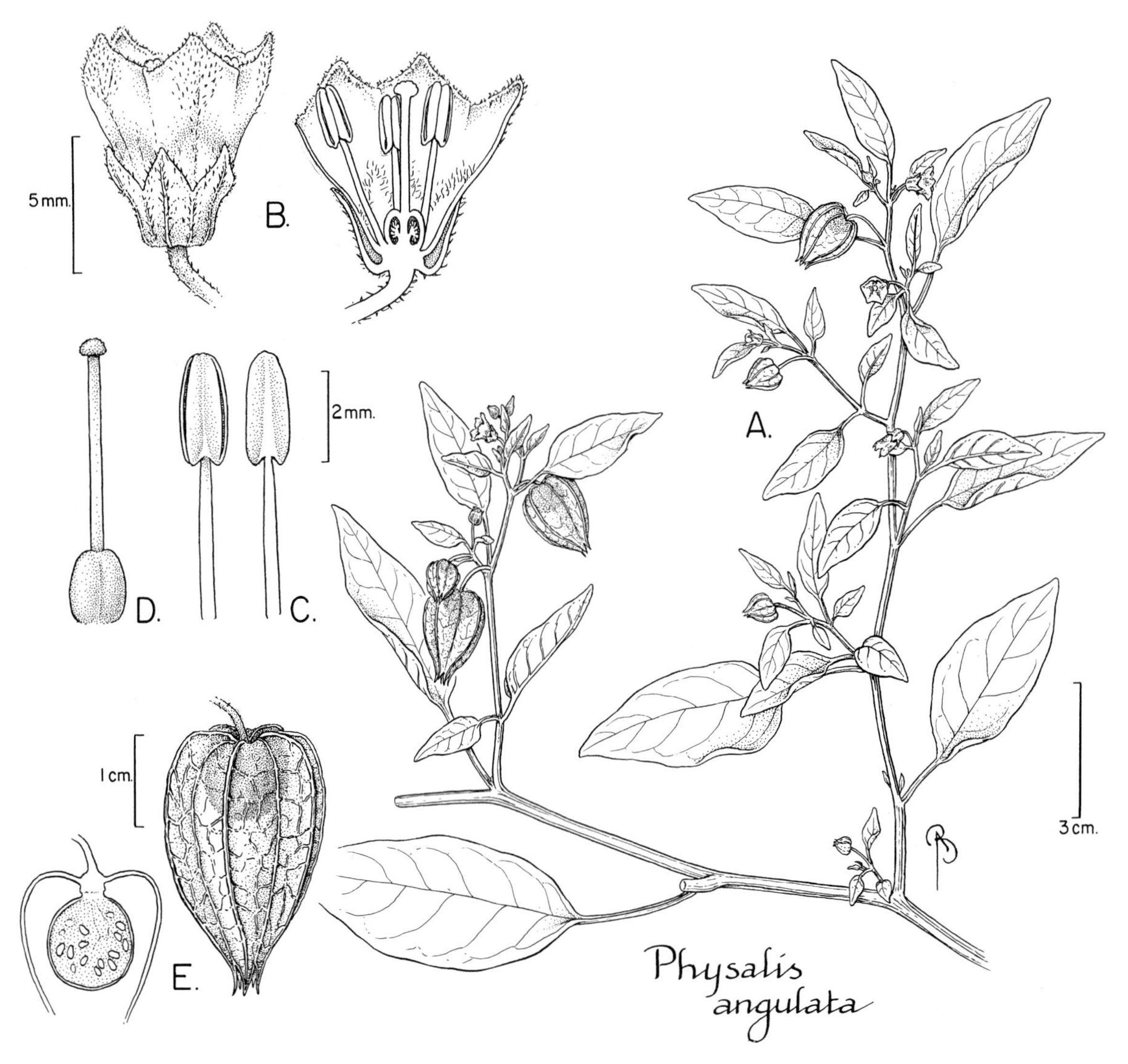

FIG. 298. SOLANACEAE. *Physalis angulata* (*Mori & Gracie 23923*). **A.** Part of stem with leaves, flowers, and fruits. **B.** Lateral view (left) and medial section (right) of flowers. **C.** Stamens. **D.** Pistil. **E.** Fruit surrounded by inflated calyx (right) and medial section of calyx and fruit (left).

PHYSALIS L.

Herbs, rarely shrubs, annual or perennial from rhizomes, glabrous or with simple or branched hairs. Leaves simple, alternate or paired; blades with margins entire to coarsely dentate or angular. Inflorescences axillary, of single or few fascicled flowers; pedicels present. Flowers: calyx 5-lobed, the tube expanding greatly after anthesis, inflated around fruit; corolla usually campanulate-rotate, often with 5 large spots or 5 groups of small spots in throat; stamens 5, equal when fully developed, the anthers longitudinally dehiscent. Fruits juicy or dry berries, the surface often glutinous. Seeds numerous, flattened, reniform.

1. Plants glabrous or with only a few very short hairs on younger parts. Calyx in fruit nearly terete, with 10 ± equal ribs; corolla without clearly demarcated spots. . . . . *P. angulata.*
1. Plants, especially lower part of stem, pilose villous with hairs to 2 mm long. Calyx in fruit strongly 5-angled; corolla with 5 clearly demarcated dark brown spots near base within. . . . . *P. pubescens.*

Physalis angulata L. FIG. 298

Annual, tap-rooted herbs, 20–100 cm tall, glabrous or with a few minute incurved hairs on younger parts. Leaves: blades ovate to ovate-lanceolate, 4.5–10 × 3–6 cm, dentate. Inflorescences with pedicels 8–11 mm long in flower, 10–25 mm long in fruit. Flowers: calyx 3–4 mm long in flower, much inflated and nearly terete in fruit, with 10 ± equal ribs, with short, incurved hairs on veins; corolla rotate-campanulate, 6–12 × 7–12 mm when fully expanded, yellowish, spots absent or obscure. Fruits juicy berries, completely enclosed by the accrescent calyx tube, 10–12 mm diam. Seeds 1.6–1.7 mm long. Fl (Sep, probably year round), fr (Sep, probably year round); weeds of disturbed soil in full sun and in gardens.

Physalis pubescens L. PL. 121b

Annual, tap-rooted herbs, 40–80 cm tall, lower stem strongly villous with spreading hairs 1–2 mm long. Leaves: blades ovate, 3–8 × 1.5–6 cm, pubescent, the hairs usually <0.5 mm long, the margins dentate or sinuate-dentate. Inflorescences with pedicels 3–7 mm long in flower, 5–15 mm long in fruit. Flowers: calyx 3.5–6 mm long in flower, much inflated and strongly 5-angled in fruit, 2–3.5 × 1.5–3 cm; corolla rotate-campanulate, yellow, with 5 clearly demarcated dark brown spots near base within. Fruits yellow or yellowish orange, juicy berries, 10–18 mm diam., completely enclosed in calyx. Seeds 1.3–1.5 mm long. Fl (Feb, Aug, probably year round), fr (Nov, probably year round); weeds of disturbed soil in full sun and in gardens. *Batoto* (Créole), *camapu* (Portuguese), *graine pok*, *herbe à cloques*, *zerb à cloques* (Créole).

SOLANUM L.

Herbs, shrubs, trees, lianas, or vines, unarmed or armed with prickles, glabrous or pubescent with simple to branched or stellate hairs, often glandular. Leaves simple, pinnately lobed or parted, to pinnately compound, alternate, solitary or paired and then frequently unequal. Inflorescences axillary or extra-axillary and sometimes opposite leaves, appearing terminal or lateral, cymes, simple or branched. Flowers: calyx 5-lobed, campanulate, often accrescent; corolla very shallowly to deeply 5-lobed, the lobes usually spreading, white, yellow, or some shade of lavender or purple; stamens 5, equal or more rarely unequal, the anthers oblong or attenuate, initially dehiscent by apical pores, sometimes later by longitudinal slits. Fruits berries, rarely capsules, usually with 2 locules. Seeds numerous, flattened and reniform, or rarely angular.

Solanum is sometimes circumscribed to include *Cyphomandra* (Bohs, 1995), which differs by anthers having an expanded connective. *Solanun anceps* is the species in this flora most closely related to *Cyphomandra*.

1. Plants always unarmed. Anthers oblong, with two terminal pores which later sometimes extend as longitudinal slits.
  2. Hairs branched or stellate.
    3. Lianas. Petioles coiling. Stamens with one filament longer than others. . . . . *S. pensile.*
    3. Erect herbs, shrubs, treelets, or small trees. Petioles straight. Stamens with filaments all equal in length.
      4. Leaves only somewhat lighter green abaxially. Inflorescences erect, held above foliage, the distinct peduncle longer than flowering portion.
        5. Young stems and abaxial surfaces of leaves densely yellowish-tomentose, the hairs mostly stalked, stellate. Leaf blades lanceolate to narrowly elliptic, 2–3 cm wide. . . . . *S. asperum.*
        5. Young stems and abaxial surfaces of leaves more sparsely and not as yellowish-tomentose, the hairs sessile, stellate. Leaf blades obovate or elliptic, 4–10 cm wide. . . . . *S. rugosum.*
      4. Leaves strongly bicolorous, green adaxially and whitened abaxially. Inflorescences lateral, ± concealed beneath associated pair of leaves, the peduncle shorter than flowering portion. . . . . *S. schlechtendalianum.*
  2. Hairs simple or absent.
    6. Inflorescences axillary. Fruits conical, tuberculate. . . . . *S. anceps.*
    6. Inflorescences extra-axillary, opposite or subopposite pair of leaves, or terminal. Fruits globose, smooth.
      7. Herbs, sometimes with a slightly woody stem. Inflorescences very short racemose but appearing umbellate, simple. Berries turning black. . . . . *S. americanum.*
      7. Shrubs or lianas. Inflorescences racemose, simple or several-times branched. Berries green or green basally and cream-colored apically, rarely seen turning purple or black.
        8. Lianas. Inflorescences terminal. Filaments unequal. . . . . *S. pensile.*
        8. Erect shrubs to treelets. Inflorescences opposite or subopposite pair of leaves. Filaments equal.
          9. Petioles 3–7 mm long; major leaf blades 1.5–3.7 cm wide, the margins undulate when dry. . . . . *S. morii.*
          9. Petioles 5–30 mm long; major leaf blades 3.5–12 cm wide, the margins plane or nearly so when dry.
            10. Leaves drying olive-green. Pedicels to 1.5 cm long in fruit. . . . . *S. oppositifolium.*
            10. Leaves drying dark green or blackish. Pedicels 1.3–2.2 cm long in fruit.
              11. Pedicels lignified in fruit and greatly swollen at apex. . . . . *S. leucocarpon.*
              11. Pedicels slender in fruit, not swollen at the apex. . . . . *Solanum* sp. A.
1. Plants armed, prickles present, at least on lower part of plant. Anthers tapered, with two small terminal pores.
  12. Plants pilose-tomentose, the stellate hairs with long midpoint (3–5 mm long). Leaf blades ovate, the base rounded to deeply cordate. . . . . *S. velutinum.*
  12. Plants tomentose, the stellate hairs with short midpoint. Leaf blades ovate or narrower, the base not cordate.
    13. Leaves essentially sessile. Anthers 3–4 mm long. Fruits 0.6–1 cm diam. . . . . *S. jamaicense.*
    13. Leaves distinctly petiolate. Anthers 6–14 mm long. Fruit 1–4 cm in diam.

14. Lianas or shrubs. All prickles on stem, petiole and abaxial leaf blade surface recurved "cat-claws," rarely (in *S. costatum*) with straight prickles on adaxial leaf blade surface. Corolla blue-purple or violet.
  15. High climbing lianas. Plants with stem and abaxial leaf blade surface nearly glabrous. . . . . *S. coriaceum*.
  15. Shrubs, treelets, or high-climbing lianas. Plants with stem and abaxial leaf blade surface densely stellate-tomentose.
    16. Lianas, sometimes appearing shrub-like when young. Inflorescences 2–5 times branched. . . . . *S. rubiginosum*.
    16. Shrubs or treelets with whorled, horizontal branches. Inflorescences simple-racemose. . . . . *S. costatum*.
14. Shrubs. Prickles mixed recurved and straight. Corolla white. . . . . *S. torvum*.

Solanum americanum Mill. [Syn.: *S. nodiflorum* Jacq.]

Short-lived perennial herbs, sometimes with slightly woody stem, 0.3–1.5 m tall, nearly glabrous, or with some short, incurved hairs. Leaves paired; blades ovate, 5–10 × 2–5 cm, the margins entire to sinuate-dentate. Inflorescences simple, very short racemose but appearing umbellate, on a peduncle 2–3 cm long, with 4–8 flowers; pedicels 3–6 mm in flower, 5–10 mm in fruit. Flowers: corolla white, 3–5 mm long; anthers 1.4–2.2 mm long. Fruits globose, soft, juicy berries, 5–8 mm diam., at first green, becoming black at maturity. Seeds numerous, 1–1.4 mm long. Known from a single collection from our area, flowering and fruiting year-round in French Guiana; ubiquitous weed of disturbed soil and in gardens. *Agouman*, *agouman-alaman*, *alaman* (Créole), *erva moura*, *pimenta de galinhas* (Portuguese).

Solanum anceps Ruiz & Pav. [Syn.: *Bassovia sylvatica* Aubl., non *Solanum sylvaticum* Dunal]

Herbs or subshrubs, 1–1.5 m tall, glabrous except for minute simple hairs on youngest parts. Leaves alternate; petioles 25–45 mm long; blades elliptic, 18–26 × 8–13 cm, usually purple abaxially, the base acute, the apex acuminate, the margins entire. Inflorescences axillary, simple, racemose, to 3 cm long, several-flowered. Flowers: corolla deeply lobed, white, the lobes recurved, 2.5 mm long, the anthers 1.3–1.5 mm long. Fruits conical green berries, 15 × 10 mm, the surface tuberculate. Seeds few, 2.5 mm long. Fl (Jan, Feb, Mar, May, Jul, Aug, Sep, Oct, Dec), fr (Jan, Feb, Mar, May, Jul, Aug, Sep, Oct, Dec); forest understory herb.

Solanum asperum Rich.

Shrubs or treelets, 1–6 m tall, the young stems and leaves densely yellowish tomentose, the hairs mostly stalked. Leaves: blades lanceolate to narrowly elliptic, 10–17 × 2–3 cm, greenish with sparse stellate hairs adaxially, densely tomentose with yellowish, stellate hairs abaxially. Inflorescences erect, 7–15 cm long, the common peduncle 4–13 cm, then several times branched. Flowers: corolla 9–12 mm wide, white; anthers 0.6–1.5 mm long; ovary tomentose. Fruits globose green berries, 8–10 mm diam., pubescent. Seeds numerous, 1.4–2 mm long. Fl (Feb, Mar, Aug, Nov, Dec), fr (Feb, Nov); common weed, in secondary vegetation. *Radier-sable* (Créole).

Solanum coriaceum Dunal

Lianas, to 20 m long or more and scrambling into canopy, with a few stellate hairs on young parts, soon glabrate except for outside of corolla, the stem with recurved prickles. Leaves: petioles 5–10 mm long, usually with recurved prickles; blades oblong, 6–18 × 3–9.5 cm, subcoriaceous, usually glabrous or sometimes tomentose, unarmed and lustrous adaxially, armed with recurved prickles on midrib and sometimes with a few stellate hairs abaxially, the margins entire or coarsely angular, slightly revolute. Inflorescences simple or rarely once-branched, 8–15 cm long, racemose. Flowers: corolla deeply lobed, densely and closely tomentose with minute stellate hairs abaxially, 35–40 mm wide, blue-purple or violet; anthers 10–14 mm long; ovary tomentose. Fruits globose, glabrous, shiny yellow berries, 30–40 mm diam. Seeds 3.5–4.5 mm long. Fl (Aug), fr (Jan, Feb, Aug); forest edges and secondary growth.

Solanum costatum M. Nee

Shrubs or small trees, to 4 m tall, the branches whorled and horizontal, the stems with recurved prickles, densely ferruginous-tomentose, the stems, petioles, and abaxial midribs of leaf blades with recurved "cat-claw" prickles, the adaxial side of leaf blades sometimes with straight prickles. Leaves paired, very unequal; petioles 1–1.3 cm long; blades of major leaves ovate to elliptic, 5–7.5 × 15–18 cm, the minor blades ca. 1/3 this size, strongly bicolorous, dark and glabrous except for sessile stellate hairs on major veins adaxially, densely pale-tomentose abaxially. Inflorescences extra-axillary, lateral, racemose, to 2.5 cm long; pedicels 10–11 mm long. Flowers: calyx 4 mm long, lobed about half way, the lobes attenuate, accrescent; corolla deeply lobed, purple, the lobes narrowly triangular, 20 × 2 mm; anthers 10–11 mm long. Fruits globose berries, to at least 1 cm diam., glabrous. Seeds 4 mm long. Fl (Dec); known from a single collection from our area, in openings in forest.

Solanum jamaicense Mill.

Shrubs, erect or more commonly with spreading or weak branches, to 3 m tall, tomentose, armed with recurved prickles. Leaves usually paired, with one about half the size of other; petioles nearly absent; major blades rhombic to ovate, 7–18 × 3–15 cm, tomentose ad- and abaxially, the base cuneate, the margins angular-sinuate, the base cuneate, the apex acute. Inflorescences lateral, sessile to short-pedunculate, simple, racemose, with up to 15 flowers; pedicels 7–12 mm long. Flowers: corolla 8–11 mm wide, deeply 5-lobed, white; anthers 3–4 mm long. Fruits globose juicy berries, 6–10 mm diam., bright red-orange. Seeds 1.8–2.2 mm long. Fl (Aug, Oct), fr (Aug, Oct); weeds of brushy, sunny, disturbed second growth and roadsides. *Cordichier blanc* (Créole).

Solanum leucocarpon Dunal [Syn.: *Solanum surinamense* Steud.]

Shrubs or treelets, 1–6 m tall, mostly glabrous. Leaves paired; petioles 12–30 mm long; major blades elliptic, 10–19 × 4–10.5 cm, the base rounded, the apex acute, the smaller ones orbiculate and much smaller, both glabrous or sometimes pubescent abaxially, ill-smelling when crushed, drying dark green or blackish. Inflorescences opposite leaves, simple, 1–4 cm long, with 5–15 flowers; pedicels 13–17 mm long, deflexed and lignified in fruit and greatly swollen near apex. Flowers: corolla deeply lobed, 15–28 mm diam., white; anthers 3.5–6 mm long. Fruits globose, glabrous berries, 10–15 mm diam., green, turning yellowish green. Seeds numerous, 3–3.5 mm long. Fl (Feb), fr (Feb); in forest understory. *Bitawiri* (Bush Negro), *bitayouli*, *mavévé*, *mavévé chien* (Créole).

This is a common species in the Guianan-Amazon forest, but has only been collected once (*Granville B5462*, CAY) from our area. The description is, therefore, based mostly on specimens from outside our area. A related but distinct and undescribed species is treated as *Solanum* sp. A below.

Solanum morii S. Knapp PL. 121a

Shrubs, 1–3 m tall, the young parts with curly simple hairs. Leaves paired, one much smaller than other; petioles 3–7 mm long; major blades lanceolate, 7–15 × 1.5–3.7 cm, with a few curly hairs along major veins abaxially, the apex long-attenuate, the margins undulate when dry. Inflorescences opposite or subopposite leaf pair, racemose, 5–10 mm long. Flowers: corolla deeply lobed, ca. 8 mm wide, white, the lobes 5 × 1.5 mm; anthers 2.5 mm long. Fruits globose berries, 15 mm diam., initially green at base and cream-colored at apex, purple when ripe. Seeds 3 mm long. Fl (Jan, Feb, Aug, Sep, Oct, Nov, Dec), fr (Jan, Mar, Aug, Sep, Nov); in forest along trails or in secondary vegetation.

Solanum oppositifolium Ruiz & Pav. [Syns.: *Solanum schizopodium* Sendtn., *Solanum viliflorum* Sendtn.]

Shrubs or treelets, 1–5 m tall, the young stems and leaves puberulent with simple hairs. Leaves paired, with one slightly smaller than other; petioles 5–11 mm long; major blades obovate to elliptic, 7–30 × 3.5–12 cm, drying olive-green, glabrous adaxially and often glabrous abaxially, the apex acute to acuminate. Inflorescences opposite leaves, simple or several-times branched, 1–12 cm long, with to 50 flowers; pedicels 4–6 mm long at anthesis, 0.6–1.5 cm long at fruiting. Flowers: corolla deeply lobed, 8–9 mm wide, white or greenish white; anthers 2–2.5 mm long. Fruits globose berries 10–15 mm diam., green, glabrous or nearly so. Seeds 3–4 mm long. Fl (Feb, Jun, Jul, Sep, Oct, Nov, Dec), fr (Jun, Jul, Oct); in forest, probably usually along streams.

Solanum pensile Sendtn. [Syns.: *Solanum ipomoea* Sendtn., *Solanum scandens* L. f., *Solanum laetum* Miq., *Solanum sempervirens* Dunal, *Solanum miquelii* C. V. Morton]

Lianas, glabrous or with small branched hairs. Leaves alternate, simple; petioles 20–45 mm long, at least some coiling around supports; blades ovate, 7–14 × 4–8 cm, the base obtuse to subcordate, the margins entire. Inflorescences terminal, to 15 cm long, several times branched; pedicels 7–8 mm long in fruit. Flowers: calyx very shallowly lobed; corolla deeply lobed, purple, the lobes lanceolate to linear, 15 × 3 mm; 4 filaments 2.5 mm long, the fifth 6.5 mm long, the anthers 8.5 mm long. Fruits globose berries, 12 mm diam., turning purple. Seeds flattened, 4 mm long. Fl (Jan), fr (Apr); in forest.

Solanum rubiginosum Vahl

Lianas, sometimes appearing shrub-like when young, densely ferruginous-tomentose with sessile, stellate hairs with short midpoints, the stems, petioles and abaxial midribs of leaf blades with uniformly recurved "cat-claw" prickles, these sometimes lacking on flowering specimens. Leaves alternate; petioles 1–3.5 cm long; blades elliptic, 7–14 × 3.5–7 cm, strongly bicolorous, dark green with sparse stellate hairs adaxially, densely yellowish tomentose abaxially, the veins more ferruginous, the base rounded, the apex attenuate, the margins entire or occasionally sinuate. Inflorescences terminal, with 2–5 branches, to 12 cm long; pedicels 7–10 mm long in fruit. Flowers: calyx deeply lobed, the lobes triangular, 6 mm long, appressed to fruit; corolla deeply lobed, ca. 20 mm wide, purple, the lobes lanceolate, 10 × 3 mm; anthers 8 mm long. Fruits globose berries, softly and finely tomentose, 15 mm diam. Seeds flattened, 3.5 mm long. Fl (Aug, Oct), fr (Aug, Oct, Nov); usually at edges or in gaps of forests.

Solanum rugosum Dunal FIG. 299, PL. 121c,d

Shrubs or small trees, 1–9 m tall. Leaves: obovate to elliptic, 15–25 × 4–10 cm, dark green and scabrous with sparse, sessile, stellate hairs adaxially, more tomentose abaxially. Inflorescences erect, 8–18 cm long, the common peduncle 5–15 cm, much branched above. Flowers: corolla 1.4–1.6 cm, white; anthers 2.2–3.2 mm long; ovary glabrous to pubescent. Fruits globose berries, 9–11 mm diam., glabrescent, green, yellow only when fully ripe. Seeds 1.5–2.1 mm long. Fl (Jun, Aug, Sep, Dec), fr (Jun, Aug); common, weed of secondary vegetation, especially along roadsides.

Solanum schlechtendalianum Walp.

Shrubs, 1–4 m tall, densely white-tomentose with stellate hairs. Leaves ill-smelling when crushed, in pairs, similar in shape but one much smaller than other; petioles 2–6 mm long; major blades ovate to ovate-elliptic, 7–15 × 3.5–7 cm, sparsely tomentose and dark green adaxially, very densely tomentose and whitened abaxially. Inflorescences lateral (terminal only when very young); peduncle 1.5–2.5 cm long, shorter than flowering branches, ± concealed beneath associated pair of leaves; pedicels 6–8 mm long, nodding in flower, but erect in fruit and putting fruit in contact with underside of associated leaf. Flowers: corolla 8 mm wide, white; anthers 2.5 mm long; ovary tomentose. Fruits globose berries, usually seen green, but purple when fully ripe, 7–8 mm diam., nearly glabrous. Seeds 2–2.5 mm long. Fl (Feb, Jul, Aug, Oct), fr (Feb, Jun, Jul, Aug, Oct); in disturbed forest and secondary vegetation.

Solanum torvum Sw. [Syns.: *Solanum ficifolium* Ortega, *Solanum daturifolium* Dunal]

Shrubs, 1–2.5 m tall, tomentose with stellate hairs, vigorous young plants with straight or recurved prickles or both, older flowering plants often unarmed. Leaves in pairs, with one 1/3–1/2 the size of other; petioles 15–65 mm long; major blades ovate, 10–20 × 6–15 cm, tomentose ad- and abaxially, the margins sinuate or with about 3 pairs of lobes per side, especially deeply lobed in young plants or vigorous shoots, rarely entire on older plants. Inflorescences lateral, simple or with 2–4 cymose branches, 3–6 cm long, the peduncle 0–2 cm long; pedicels with mixed simple and stellate hairs, both gland-tipped. Flowers: corolla ca. 25 mm wide, white; anthers 6–6.5 mm long; ovary glabrous or sparsely glandular. Fruits globose berries, 12–14 mm diam., green. Seeds numerous, 2.5 mm long. Fl (Feb, May, Jun, Aug, Sep, Oct), fr (Feb, May, Jun, Aug, Sep, Oct); common, weeds of secondary vegetation and disturbed areas. *Aubergine sauvage*, *tomate sauvage* (Créole).

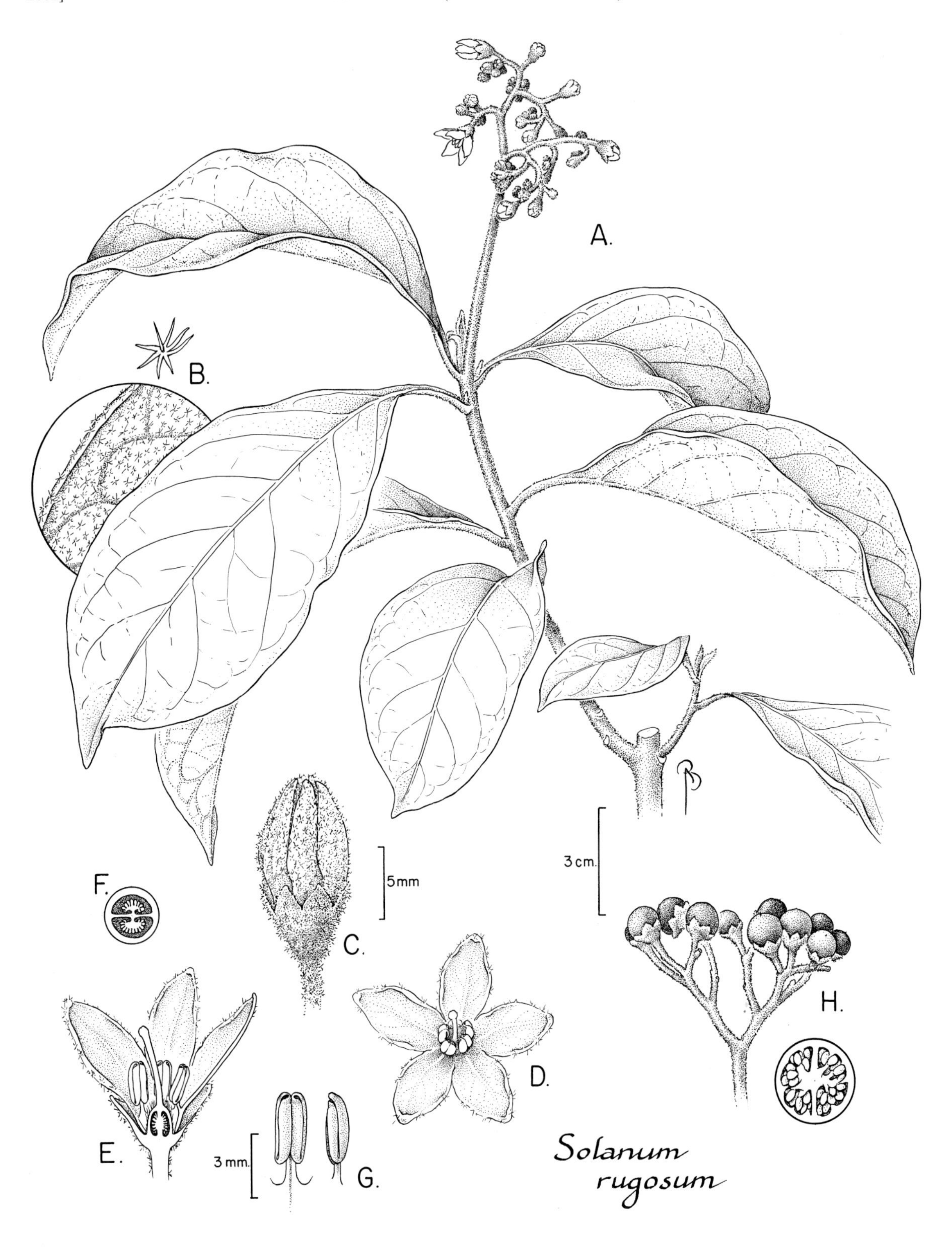

FIG. 299. SOLANACEAE. *Solanum rugosum* (*Mori et al. 21050*). **A.** Part of stem showing leaves and inflorescence. **B.** Detail of abaxial leaf surface showing stellate pubescence and single trichome (above) with rays and midpoint. **C.** Flower bud. **D.** Apical view of flower at anthesis. **E.** Medial section of flower. **F.** Transverse section of ovary. **G.** Adaxial (far left) and lateral (left) views of stamens. **H.** Infructescence (above) and transverse section of fruit (below).

Solanum velutinum Dunal

Shrubs, 2–4 m tall, densely ferruginous, pilose-tomentose with sessile or short-stipitate, porrect, stellate hairs with very long midpoint 3–5 mm long, armed with recurved spines on stems and midribs of abaxial leaf blade surface. Leaves alternate; petioles 10–50 mm long; blades ovate, 9–18 × 4.5–11.5 cm, densely pilose-tomentose ad- and abaxially, sticky, the base rounded to deeply cordate, the apex acuminate. Inflorescences lateral, simple, racemose, few-flowered, the peduncle 1–4.5 cm long. Flowers: calyx lobes lanceolate, 3 × 9 mm, spreading away from fruit; corolla, deeply lobed, 20 mm broad, white or pale purple; anthers 6.5 mm long. Fruits globose berries, 10–15 mm diam., glandular-puberulent with simple glandular hairs ca. 1 mm long, green (when ripe). Seeds 3.5 mm long. Fl (Feb, Aug), fr (Feb, Jul, Aug, Oct, Nov); secondary vegetation and edges of forests.

Solanum sp. **A**

Shrubs, to 2 m tall, the young parts minutely puberulent, soon glabrate. Leaves paired, one 1/4–1/3 the size of other; petioles 20–25 mm long; major blades elliptic, 17–20 × 5.5–7.5 cm, drying dark green or blackish, the base acute, the apex attenuate. Inflorescences opposite leaf pair, simple, to 1 cm long; pedicels 15–22 mm long. Flowers: corolla ca. 5 mm long, white; anthers ca. 2 mm long. Fruits globose berries, green, 10 mm diam. Seeds 4 mm long. Fl (Feb, Jul, Aug, Sep, Nov, Dec), fr (Aug, Sep, Nov, Dec); in montane forest on Mont Galbao.

This species most closey resembles *S. leucocarpon* and is based on *Boom 10807*, *10859* (both at NY); *Cremers 10847* (CAY); *Hahn 3607* (NY, US); and *Mori 8742*, *23112* (both at NY).

STERCULIACEAE (Cacao Family)

Laurence J. Dorr

Trees, treelets, shrubs, or herbs (rarely lianas), usually stellate-pubescent, but also with simple hairs or rarely peltate scales; mucilage present. Stipules present. Leaves simple or digitately compound, alternate; petioles usually present; blades with margins entire or lobed. Inflorescences axillary or terminal, sometimes cauliflorous (or ramiflorous) cymose, paniculate, umbelliform, or more complex, or flowers solitary; bracteoles present, sometimes with involucel of 3 or 4 distinct bractlets. Flowers actinomorphic, bisexual or unisexual; sepals (2)3–5(8), usually valvate, connate (rarely distinct), often glandular at base; petals usually 5, free, sometimes adnate at base to staminal tube, sometimes clawed, sometimes cucullate and with terminal appendage, or absent; stamens 5–15(45), in two whorls, the outer staminodial, free or more usually monadelphous in tubular column, often on androgynophore, sterile in pistillate flowers, the anthers 2- or 3-thecate; ovary superior, with 2–5(60) syncarpous carpels, rarely secondarily apocarpous or 1-carpellate, the styles equal in number to carpels, distinct or variously connate; placentation generally axile. Fruits various, often capsules, follicles or schizocarps, or coriaceous or woody berries, dehiscent or indehiscent.

Molecular data (Bayer et al., 1999) suggest that this family should not be maintained and that the genera treated here are best referred to an expanded Malvaceae.

Jansen-Jacobs, M. J. 1986. Sterculiaceae. *In* A. L. Stoffers & J. C. Lindeman (eds.), Flora of Suriname **III(1–2):** 283–292. E. J. Brill, Leiden.
Uittien, H. 1966. Sterculiaceae. *In* A. Pulle (ed.), Flora of Suriname **III(1):** 34–48, 437. E. J. Brill, Leiden.

1. Inflorescences not cauliflorous. Flowers staminate or bisexual; apetalous. Fruits apocarpous. . . . . *Sterculia*.
1. Inflorescences cauliflorous (or ramiflorous). Flowers bisexual; petals present. Fruits syncarpous.
    2. Leaves digitately compound. . . . . *Herrania*.
    2. Leaves simple.
        3. Fruits large, 8–20 cm long, smooth, indehiscent. Seeds many, surrounded by fleshy pulp.. . . . . . . *Theobroma*.
        3. Fruits small, 2–2.5 cm long, spiny, dehiscent. Seeds solitary, not surrounded by fleshy pulp. . . . . . . *Byttneria*.

BYTTNERIA Loefl.

Erect, decumbent, or scandent shrubs, subshrubs, small trees or lianas, armed or unarmed, glabrous or pubescent with simple, stellate or glandular hairs (sometimes in combination). Stipules caducous. Leaves simple; blades entire or lobed, sometimes with 1–5 foliar nectaries on the principal veins abaxially. Inflorescences axillary or subterminal, sometimes appearing cauliflorous, cymes or dichasia, or umbellate. Flowers bisexual; sepals 5, basally connate; petals 5, unguiculate, cucullate, and ligulate; staminal tube short, with 5 sessile or short-stipitate anthers alternating with 5 staminodes, the anthers 2-thecate; ovary 5-carpellate, syncarpous, sessile, each locule 2-ovulate (one ovule aborting), the styles simple, the stigmas capitate or 5-lobed. Fruits small, woody, globose, dehiscent capsules armed with spines. Seeds solitary, glabrous.

Byttneria morii L. C. Barnett & Dorr FIG. 300, PL. 122a

Small trees, to 10 m tall; stems sparingly stellate-pubescent, becoming glabrate. Outer bark brown, flaking in places, finely fissured with lenticels in vertical fissures, the inner bark thin, with fascicles of reddish-brown fibers broken by rows of pale yellowish parenchyma. Leaves: petioles to 1.5 cm long; blades elliptic to obovate, 12–30.5 × (2.5)4.5–12 cm, with scattered 2–3-armed hairs on midrib and minute glandular excrescences abaxially, the midrib with single foliar nectary near base abaxially, the base cuneate to narrowly cuneate, the apex acuminate to long acuminate, the margins entire. Inflorescences axillary (seemingly cauliflorous when in axils of fallen leaves), glomerate. Flowers: calyx reflexed at anthesis; petal claw ca. 1.5 mm long, maroon, the apical portion forming hood of 2 adaxial teeth and 2 abaxial wings that encloses stamens, petal lamina cylindric, narrowly elliptic, ca. 3–4 mm long, with simple hairs, maroon basally, cream-colored distally. Capsules woody, ellipsoidal, 2–2.5 × 2 cm, surface minutely stellate-pubescent, with scattered spines. Seeds fusiform, to ca. 2 cm long, striate. Fl (Jun, Jul, Aug), fr (Jul, Aug, Sep); scattered, in nonflooded forest.

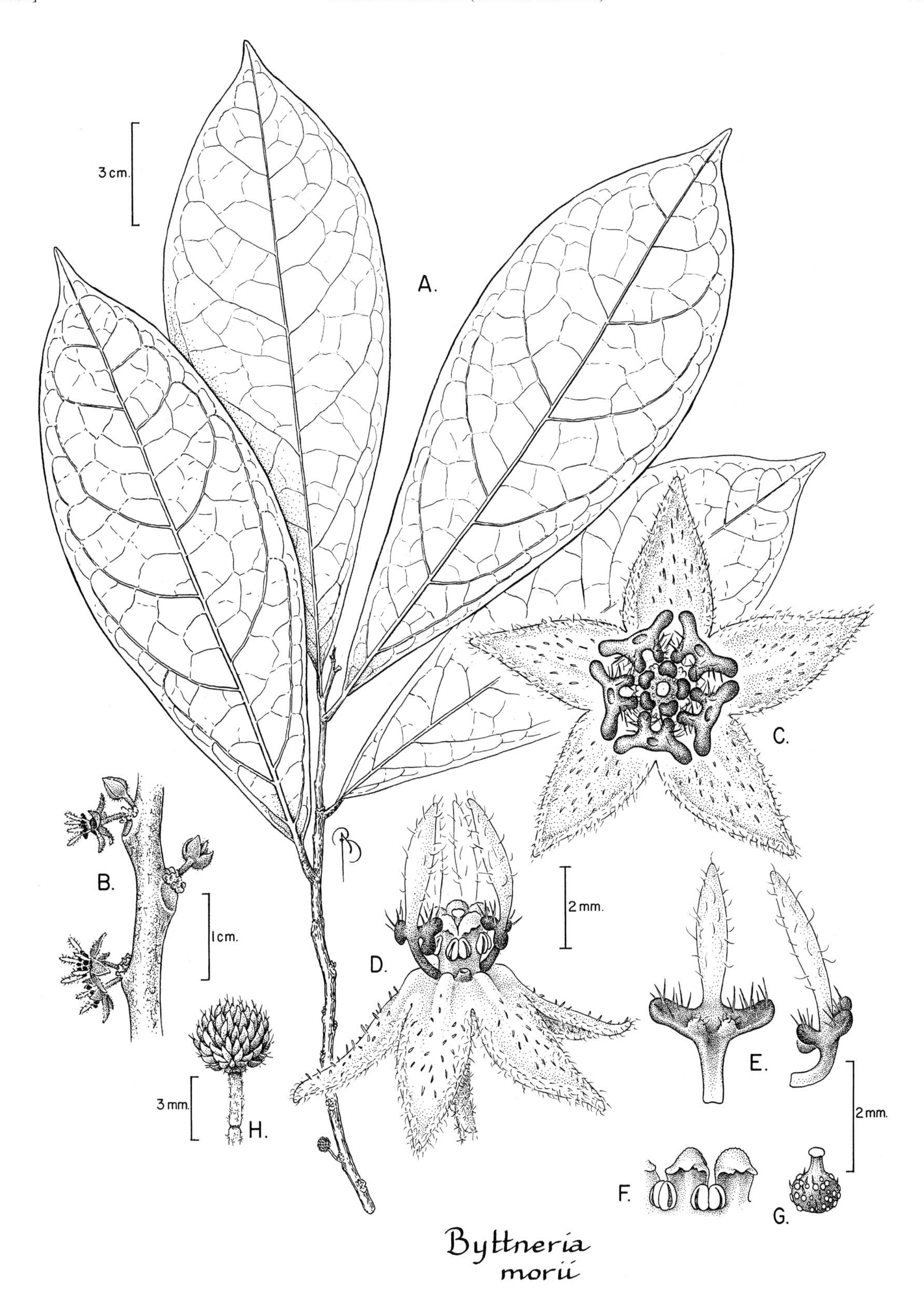

FIG. 300. STERCULIACEAE. *Byttneria morii* (A, H, *Mori 18721*; B–G, *Mori et al. 19160*). **A.** Apex of leafy stem showing immature fruit near base. **B.** Part of stem with ramiflorous inflorescences. **C.** Apical view of flower with the petal lobes removed, showing petal appendages. **D.** Lateral view of flower with one petal removed. **E.** Adaxial (left) and lateral (right) views of petals showing ant-mimicking appendages. **F.** Abaxial view of part of staminal tube. **G.** Pistil. **H.** Immature fruit.(Reprinted with permission from L. C. Barnett & L. J. Dorr, Brittonia 42(4), 1990.)

HERRANIA Goudot

Shrubs or treelets; stellate-pubescent. Stipules usually caducous, sometimes persistent. Leaves usually large, digitately compound, with 4–9 leaflets; leaflets inarticulate, the margins entire or not; venation pinnate. Inflorescences cauliflorous, fasciculate; bracteoles present. Flowers bisexual; calyx patelliform or cymbiform, 3–4(5)-lobed at anthesis; petals 5, cucullate, apically inflexed, with conspicuously elongate ligule; androphore or androgynophore absent; androecium of 5 fertile stamens alternating with 5 staminodes, connate into short tube, the filaments short, minutely 3-branched, each branch with one 2-thecate anther; staminodes petaloid; gynoecium 5-carpellate, syncarpous, the ovary sessile, with many ovules in two rows in each locule, the styles filiform, the stigmas obscurely 5-lobed. Fruits moderately large, indehiscent, dry berries, with 10 longitudinal ribs. Seeds numerous, flattened-ovoid, angular, embedded in pulpy tissue.

Herrania kanukuensis R. E. Schult.

Treelets, to 6 m tall, the stems densely reddish-brown stellate-pubescent. Bark brownish-black, somewhat striate. Stipules caducous. Leaves: petioles 14–52 cm long; blades (3)5–6-digitate; leaflets sessile, unequal, obovate, the base abruptly cuneate, the apex abruptly acuminate, the margins almost entire, but slightly undulate to sinuate or occasionally incised near apex, the major vein endings conspicuously (albeit minutely) mucronate, the central (largest) leaflets 25–47 × 11–20 cm, sparingly stellate-pubescent adaxially, moderately and uniformly stellate-pubescent abaxially, the principal veins densely reddish-brown stellate-pubescent ad- and abaxially. Flowers: calyx 3-lobed, the sepals ca. 14 × 6 mm, purple; petals 5, dark purple, the hood elongate to obovate, ca. 8–12 × 4–6(10) mm, ligules ca. 40(60) mm long. Fruits pedunculate, ovoid to fusiform, 10-ribbed, 8–11 × 3.5–4.5 cm, densely and minutely stellate-pubescent, the base rounded, the apex shortly cuspidate, the pericarp thin, yellow when ripe. Seeds ca. 60 or more, triangular or angular-ovate in outline. Fl (Oct), fr (Jul, Oct, Nov); understory in old second growth.

STERCULIA L.

Subcanopy or canopy trees, to 40 m tall. Trunks often buttressed. Stipules bractlike, caducous or rarely persistent. Leaves evergreen or deciduous, heterophyllous; mature leaf blades simple or palmately compound, the margins entire or 3–5-lobed, the immature leaf blades often palmately lobed or entire and palmately lobed. Inflorescences axillary or terminal, not cauliflorous, panicles or racemes. Flowers staminate or bisexual, with sexual parts borne on an elongate androgynophore; calyx 5-lobed, the sepals sometimes with internal appendages; petals absent; stamens 10, 12 or 15, sessile or shortly stipitate, the anthers 2-thecate, the anther locules irregularly arranged; staminodes absent; gynoecium 5-carpellate (vestigial in staminate flowers), secondarily apocarpous; ovary sessile, each locule with 2–20 ovules, the styles coherent, recurved, the stigmas capitate or obscurely 5-lobed. Fruits pin-wheel shaped, 1–5 woody follicles, sessile or stalked, beaked or not, dehiscent, densely minutely velutinous externally, hispid or glabrous internally. Seeds ellipsoid or obovoid, 2–20 per follicle, pendent from follicular suture.

Sterculia sp. A, represented by *Boom & Mori 14712* (NY), is not included in the key because it is known in our area only by a collection from a juvenile tree.

1. Terminal vegetative buds glabrous, shiny. Calyx interior villous, lacking vermiform hairs. . . . . *S. frondosa.*
1. Terminal vegetative buds pubescent, not shiny. Calyx interior always with vermiform hairs.
   2. Leaf blades oblong to elliptic, broadest at or below the middle; tertiary and higher order veins scarcely discernable abaxially. Inflorescence branches with short appressed stellate hairs only. Mature follicles stalked, shortly beaked. . . . . *S. pruriens* var. *pruriens.*
   2. Leaf blades ovate to obovate or oblanceolate, broadest above the middle; tertiary and higher order veins salient abaxially. Inflorescence branches with short appressed stellate hairs and scattered long simple, bifurcate or stellate hairs. Mature follicles ± sessile, not beaked. . . . . *S. villifera.*

Sterculia frondosa Rich. FIG. 301, PL. 122b

Large trees, to 25 m tall. Trunks cylindric, not buttressed. Bark reddish-brown, lenticellate, the slash friable, dark reddish-brown. Stipules caducous. Terminal vegetative buds 10–20 mm long, subulate, glabrous, shiny. Leaves: petioles 4–7.5 cm long, glabrous; blades obovate to oblanceolate, 10–26 × 4–9.5 cm, glabrous and lustrous adaxially, minutely stellate-pubescent (hairs scarcely visible at 10×) and dull abaxially, the base cuneate, the apex obtuse, the margins entire, often mucronate, rarely emarginate; tertiary and higher order venation visible, but not strongly salient. Inflorescences erect, 12–20 cm long, minutely grayish stellate-pubescent. Flowers fragrant: calyx short campanulate, the lobes 4–6 mm long, lanceolate, pink to red, with a darker red throat, the exterior densely stellate-pubescent, the interior villous, the appendage obscured by hairs. Mature follicles stalked, ca. 4–8 × 2–6 cm, laterally compressed, shortly beaked, the exterior densely reddish-brown puberulous, the wall ca. 5–7 mm thick, the interior hirsute with straw-colored hairs. Seeds solitary, brownish-black. Fl (Jul, Aug), fr (Apr, May, old fruits often persisting), seedling (Apr); nonflooded forest. *Kobé* (Bush Negro).

Seedlings have 3-lobed leaves, with long acuminate lobes and rounded sinuses, and are glabrous ad- and abaxially.

Sterculia pruriens (Aubl.) K. Schum. var. **pruriens**

Large trees, to 35 m tall. Trunks with poorly developed, low, spreading, thick buttresses. Bark (white to) gray, peeling in irregular plates, with scattered lenticels, the slash of outer bark white, the slash of inner bark flesh-colored to dark red. Stipules caducous. Terminal vegetative buds 3–5 mm long, ovoid, heterotrichous, with dense layer of short hairs interspersed with longer hairs, not shiny. Leaves: petioles 5–16 cm long, minutely pubescent, pulvinate at apex; blades oblong to elliptic, 14.5–28 × 6–16 cm, glabrous adaxially (densely pubescent when young), densely reddish-brown pubescent with small stellate hairs abaxially, the base obtuse to

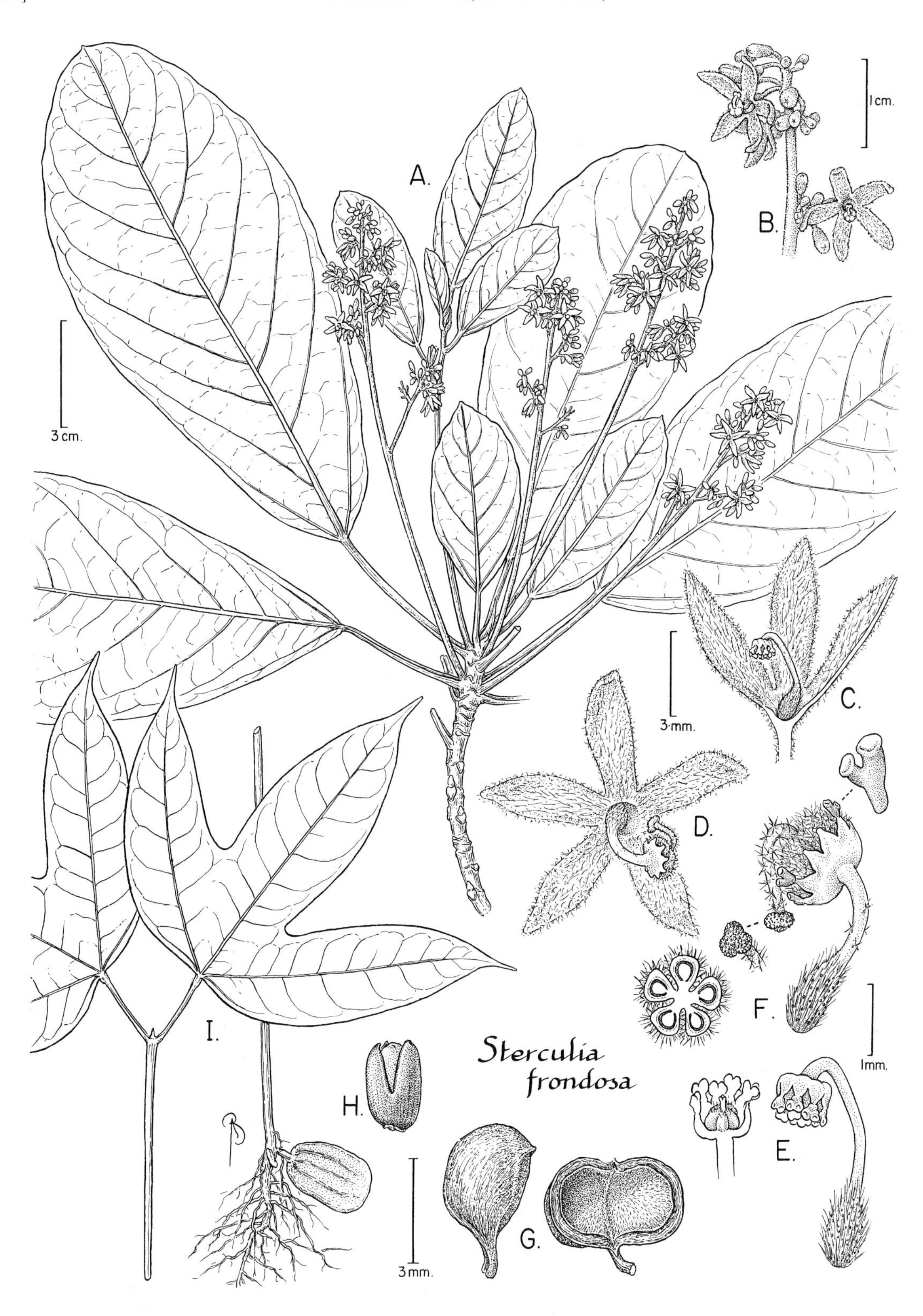

FIG. 301. STERCULIACEAE. *Sterculia frondosa* (A–G, *Mori et al. 23117*; H, *Mori & Pipoly 15496a*; I, *Mori 18183*). **A.** Part of stem with leaves and inflorescences. **B.** Detail of inflorescence. **C.** Lateral view of staminate flower with two perianth units removed. **D.** Apical view of pistillate flower. **E.** Androgynophore of staminate flower (right) with section of apex (left) showing pistillode. **F.** Androgynophore of pistillate flower (right) with details of staminode (above), and stigma (below), and transverse section of ovary (left). **G.** Intact follicle (left) and dehisced follicle (right). **H.** Seed. **I.** Seedling showing three-lobed leaves on apex (left) and cotyledon at base of stem (right).

truncate, sometimes slightly cordate, the apex acute to acuminate, the margins slightly revolute; tertiary and higher order venation poorly developed and scarcely discernable. Inflorescences ± erect, lax, 8.5–17 cm long, minutely reddish-brown pubescent, appearing with growth of newer and smaller (ca. 6.5–9.5 × 3.5–5 cm) leaves. Flowers: calyx short campanulate, the lobes lanceolate to ovate, 8–10 × 3–4 mm, the exterior and apex of interior yellow, the interior red (older flowers turning entirely red), the exterior densely stellate-pubescent, the interior stellate- and glandular-pubescent, with short multi-cellular vermiform hairs below calyx appendage and ring of papillae around calyx mouth. Mature follicles stalked, ca. 4.5 × 6.5 cm, rounded, shortly beaked, the exterior densely brownish velutinous, the interior not seen. Seeds not seen. Fl (Sep, Oct, Nov), fr (Nov); in nonflooded forest. *Kobé* (Bush Negro), *mahot cochon* (Créole), *tacacazeiro* (Portuguese).

With some doubt, juvenile leaves (*Gentry et al. 63087*, NY, and *63093*, MO) are associated with this species. They are palmately 3-lobed with moderate to deep sinuses, base truncate, apex short to long acuminate, glabrous adaxially, pubescent abaxially with sessile, 4-rayed stellate hairs, the rays ± equal in length, appressed (scarcely discernible at 10×).

Sterculia villifera Steud. PART 1: FIG. 10

Large trees, to 40 m tall. Trunks cylindric to fluted at base, not buttressed. Stipules caducous. Bark smooth, gray, with lighter vertically oriented lenticels, the slash of outer bark thick, with orange streaks, the slash of inner bark orange or creamy orange, with vertical streaks. Terminal vegetative buds 10–15 mm long, lacrimiform, heterotrichous with dense layer of minute hairs interspersed with long simple hairs, not shiny. Leaves: blades obovate to oblanceolate, 12–27 × 5–10 cm, minutely stellate-pubescent adaxially (immature leaves glabrous), densely stellate-pubescent abaxially, the base truncate to slightly cordate, the apex obtuse (immature leaves mucronulate), the margins entire; tertiary and higher order veins salient. Inflorescences erect, ca. 8–15 cm long, appearing with growth of leaves; peduncles and pedicels with short, appressed stellate hairs and scattered, long simple, bifurcate or stellate hairs. Flowers: calyx short campanulate, the lobes narrowly triangular, 6–8 mm long, the exterior red, densely tomentose, the interior pubescent, densely golden pubescent above appendage, red with long multi-cellular vermiform hairs below appendage. Mature follicles ± sessile, ellipsoidal, beakless, ca. 3–4 × 2.5 cm, the exterior glabrous, the interior densely pubescent with long, simple hairs that readily break. Seeds not seen from our area. Fl (Aug); in nonflooded forest. *Kobé* (Bush Negro).

The juvenile leaves are palmately 3-lobed with deep sinuses, base truncate, apex long acuminate, glabrous adaxially except for scattered hairs along midrib and secondary veins, pubescent with sessile, multi-rayed stellate hairs abaxially, the hairs with individual rays appressed (scarcely discernible at 10×).

Sterculia sp. A

Immature trees, to 6 m tall. Juvenile trees heterophyllous, palmately lobed and entire leaves on same plant but segregating by shoot; blades to ca. 55 × 40 cm, weakly rugose, glabrous adaxially, pubescent with long-stalked multi-rayed stellate hairs abaxially, the hairs with individual rays ascending (scarcely discernible at 10×), the base cordate, the apex long acuminate, the margins entire and ± sinuate. Flowers, fruits, and seeds unknown in our area. In nonflooded forest.

A single, sterile collection (*Mori & Boom 14712*, NY) is known from central French Guiana. The long-stalked multi-rayed hairs (not found in any other species of *Sterculia* in central French Guiana) suggest that this collection is the same as an imperfectly known, undescribed species that has been collected elsewhere in French Guiana, Guyana, Venezuela, and Brazil.

THEOBROMA L.

Small to medium-sized trees. Stems stellate-pubescent, becoming glabrate. Stipules caducous. Leaves simple; blades coriaceous, the margins entire. Inflorescences axillary or cauliflorous, 1- to many-flowered, fasciculate cymes. Flowers bisexual; sepals 5, basally connate; petals 5, cucullate, with oblong to elliptical or discoid apical appendage (lamina); androphore or androgynophore absent; androecium of 5 fertile stamens alternating with 5 staminodes, connate into short staminal tube, the filaments short, minutely 2–3-branched, each branch with one 2-thecate anther; staminodes distinct, petaloid or linear; gynoecium 5-carpellate, syncarpous, the ovary sessile, with many ovules in two rows in each locule, the styles 5, free or partially connate, the stigmas acute. Fruits large, indehiscent, fleshy, drupe-like berries. Seeds numerous, ellipsoid, ovoid or amygdaloid, surrounded by thick, pulpy tissue filling single cavity created when at maturity the original carpel walls decay.

1. Leaf blades glabrous abaxially (except for pubescent or puberulous midrib). Filaments 2-antheriferous. Fruits glabrous. . . . . *T. cacao.*
1. Leaf blades softly velutinous abaxially. Filaments 3-antheriferous. Fruits densely and softly stellate pilose-velutinous. . . . . *T. velutinum.*

Theobroma cacao L.

Small trees, to 10 m tall. Leaves: petioles to 2 cm long, pulvinate at both ends, puberulous to pubescent with simple and scattered stellate hairs; blades obovate-elliptic to oblong-elliptic, 12–40 × 4.5–13.5 cm, plane, glabrous ad- and abaxially except for pubescent or puberulous midrib, the base rounded to obtuse, slightly asymmetrical, the apex acute to caudate-acuminate, the margins entire. Inflorescences cauliflorous, many-flowered. Flowers: sepals lanceolate or oblong-lanceolate, to 8 mm long, reddish; petals white, the petal hoods obovate, to 4 × 5 mm, 3-veined with two very conspicuous, purple lateral veins, the petal lamina spatulate, 1.5–2.5 mm long, yellowish, the apex attenuate; staminodes linear subulate, 4–6 mm long, with conspicuous purple median vein; filaments 2-antheriferous. Fruits 10–20 × 10 cm, variable in size and shape, but generally oblong to fusiform, glabrous, terete or ± pentagonal in transverse section. Seeds ca. 20–40 per pod, the surrounding pulp white and sweet. Fl (year round?), fr (year round?); cultivated

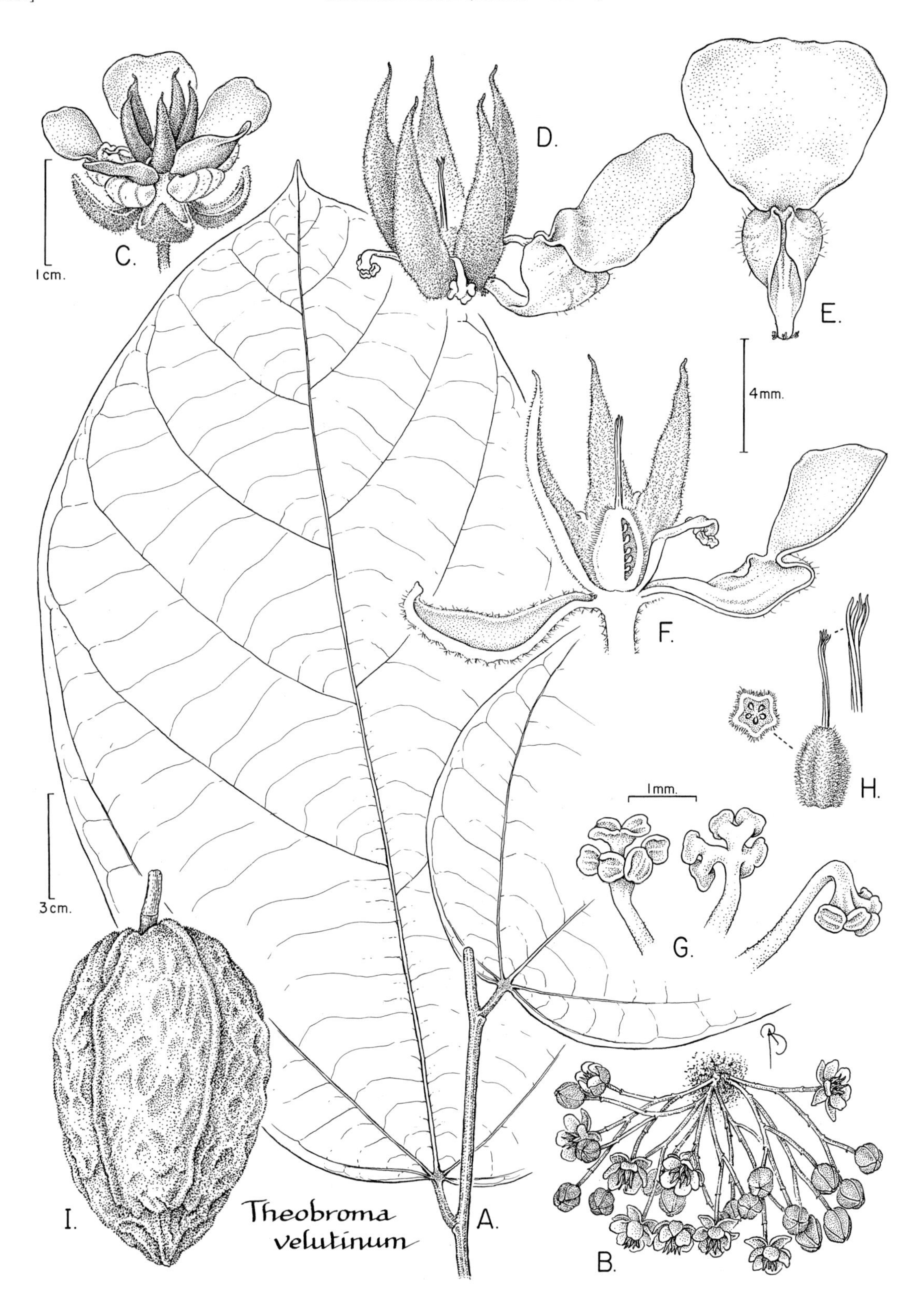

FIG. 302. STERCULIACEAE. *Theobroma velutinum* (A, *Mori et al. 15098*; B, C, I, *Mori et al. 24183*; D–H, *Mori et al. 22735*). **A.** Leaves attached to part of stem. **B.** Cauliflorous inflorescence. **C.** Lateral view of flower. **D.** Lateral view of flower with some petals removed showing style in center of five staminodes; two stamens are shown at left and center and one petal with stamen inserted at right. **E.** Adaxial view of petal showing pocket (or hood) into which the stamen is inserted. **F.** Medial section of flower; note stellate pubescence on abaxial surface of sepals. **G.** Three views of anthers. **H.** Lateral view of pistil (near left), transverse section of ovary (far left), and detail of style (above). **I.** Lateral view of fruit; note pubescence.

for the chocolate-producing seeds and persisting from cultivation around old homesteads. *Cacao* (Créole).

Theobroma velutinum Benoist FIG. 302, PL. 122c

Small trees, to 14 m tall. Bark gray, flecked with white, the wood cream-colored. Leaves: petioles to 2 cm long, not pulvinate, densely stellate-pubescent; blades oblong-ovate or ovate-oblong, 28–40 × 15–24 cm, slightly and broadly bullate, glabrous adaxially, softly velutinous with dense, subappressed stellate hairs abaxially, the base broadly rounded, ± symmetrical, the apex long acuminate, the margins entire or slightly sinuose. Inflorescences cauliflorous, many-flowered. Flowers: sepals lanceolate-oblong, ca. 8–10 mm long, wine-red or dark reddish-brown; petals red, the petal hoods, obovate-oblong, to 7 × 3 mm, 3-veined, the petal lamina subtrapezoidal, 6–7.5 mm long, light wine-red, the apex slightly sinuate; staminodes lanceolate-subulate, ca. 6 mm long, dark winered; filaments 3-antheriferous. Fruits ellipsoid, 8–9 × 6–6.5 cm, densely and softly stellate pilose-velutinous, 5-ribbed, the ribs thick and prominent. Seeds ca. 25–30 per pod. Fl (Apr, Sep, Oct, Nov), fr (May, Aug); scattered, in nonflooded forest. *Cacahuette sauvage* (Créole), *cacao* (Créole and Portuguese), *cacao sauvage* (Créole).

STYRACACEAE (Storax Family)

Peter W. Fritsch, Scott A. Mori, and John L. Brown

Trees or shrubs. Vestiture of stellate hairs or peltate scales. Stipules absent. Leaves simple, alternate; venation pinnate. Inflorescences lateral or pseudoterminal, cymose or racemose. Flowers actinomorphic, bisexual or rarely unisexual (plants gynodioecious); hypanthium adnate to ovary at various levels; calyx synsepalous, cup-like, (0)2–9-toothed or -lobed; corolla 4–5(8)-lobed or -parted, imbricate or valvate in bud, sympetalous; stamens generally twice (rarely equal) number of corolla lobes, essentially uniseriate, adnate to corolla, the anthers oblong to linear, basifixed, 2-locular, longitudinally and introrsely dehiscent; ovary usually 1-locular but 2–4(5) septate proximally, partly to completely inferior, the style filiform, hollow, the stigma terminal, truncate or minutely lobed; placentation essentially axile (rarely nearly basal), the ovules (1)4–9(ca. 30) in each locule, anatropous. Fruits dehiscent (loculicidal capsules) or indehiscent (dry, drupaceous, or baccate), with persistent calyx. Seeds 1–3(ca. 50), globose to fusiform, rarely winged, the testa thin to indurate, brown; endosperm cellular, oily, abundant; embryo straight, the cotyledons flattened or nearly terete.

STYRAX L.

Trees. Older stems with outer epidermal layer fibrous, dull brown or more often gray, the inner epidermal layer dull maroon, the buds superposed. Leaves: blades with margins entire. Inflorescences lateral or pseudoterminal, unbranched or once-branched, pubescent; bracteoles present. Flowers bisexual, fragrant; hypanthium adnate to base of ovary; calyx minutely 5-toothed, pubescent; corolla 5-lobed, tomentose, white or pale pink, the lobes longer than tube, valvate in bud, reflexed; stamens 10, the free portion of filaments distinct or connate into short tube, the ventral side of distinct portion sometimes auriculate, with a dense mass of stellate hairs, the anthers bright yellow; ovary 1-locular but 3-septate proximally, pubescent; placentation axile, with placental obturators present, the ovules 5–8 per carpel. Fruits drupaceous, ± ellipsoid, purplish, the mesocarp juicy, the thin endocarp adnate to seed at maturity. Seeds 1(3) per fruit, ellipsoid, the testa indurate, smooth except for 3(6) longitudinal grooves, the hilum broad.

The generic description is based on characters found in northeastern South America, especially from our area.

1. Young stems and petioles lepidote. Leaf blades abaxially glabrous or nearly so. Corolla 13–20 mm long. . . . . *S. glabratus.*
1. Young stems and petioles with stellate-tomentose indument. Leaf blades abaxially tomentose. Corolla 10–13 (to 15 in *S. sieberi*) mm long.
   2. Abaxial surface of principal veins of leaf blades easily visible through the pubescence. Anther sacs tapered at apex, the longer arms of hairs nearest distal end of stamen filament 2.3–4 mm long. . . . . *S. sieberi.*
   2. Abaxial surface of principal veins of leaf blades obscured by pubescence or nearly so (rarely surface of principal veins visible in *S. discolor*). Anther sacs not tapered at apex, the longer arms of hairs nearest distal end of stamen filament 0.6–1.4 mm long.
      3. Leaf blades with secondary veins in 8–10 pairs, these often with domatia in axils. Lower pedicels 2–6 mm long. Longer arms of hairs nearest distal end of stamen filament 0.6–0.8 mm long. Drupes 13–16 × 10–12 mm. . . . . *S. pallidus.*
      3. Leaf blades with secondary veins in 4–7 pairs, these without domatia in axils. Lower pedicels 5–8 mm long. Longer arms of hairs nearest distal end of stamen filament 1–1.4 mm long. Drupes 10–13 × 6–7 mm. . . . . *S. discolor.*

Styrax discolor M. F. Silva

Trees, to 30 m tall. Trunks buttressed. Bark not rough, the inner bark red. Young stems thinly brown-stellate-tomentose. Leaves: petioles 6–11 mm long, stellate-tomentose; blades elliptic, 10–15.5 × 3.5–7 cm, abaxial surface covered with minute grayish white stellate hairs and often also a lesser and variable amount of small rust-colored stellate hairs, the principal veins nearly always

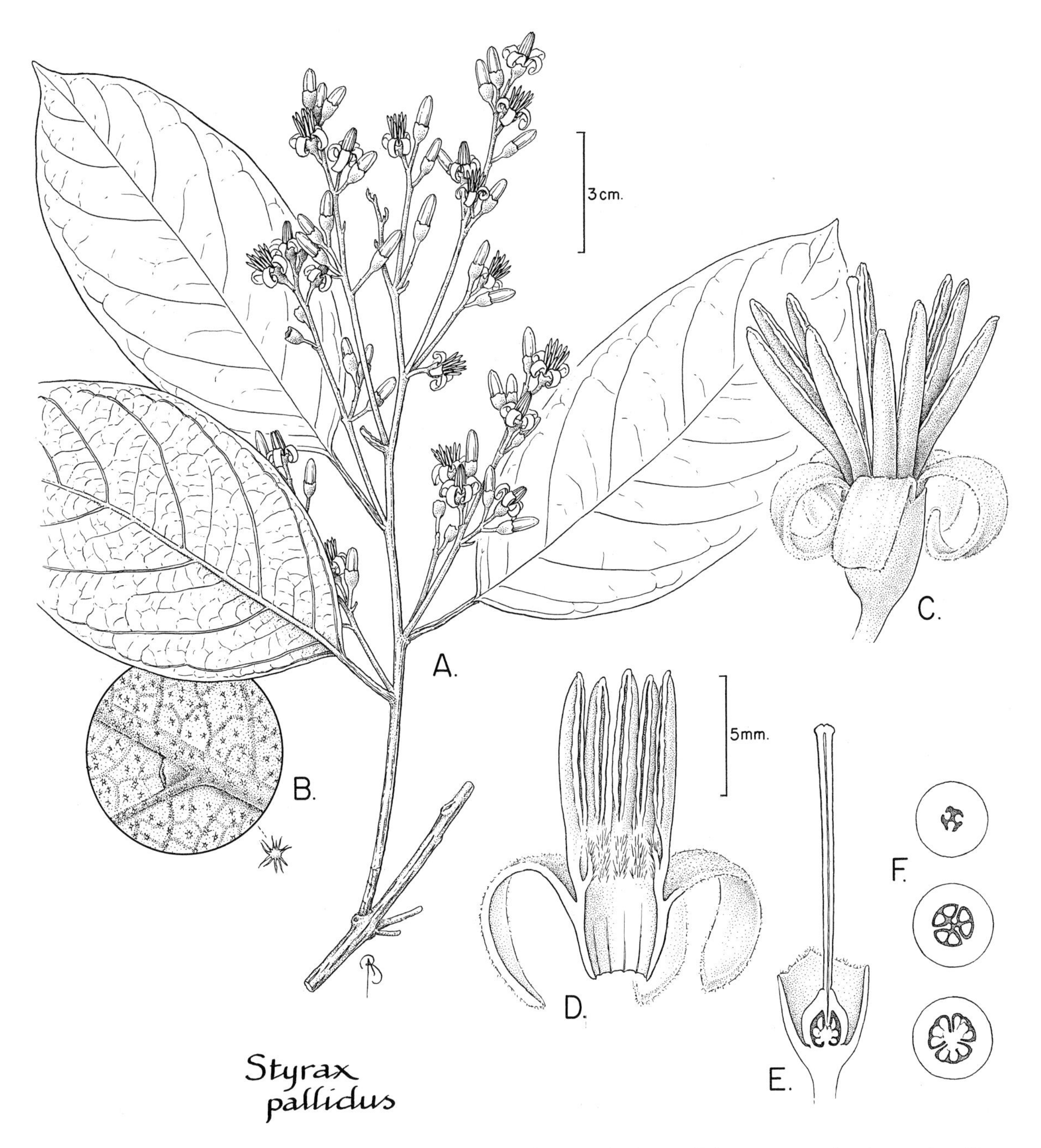

FIG. 303. STYRACACEAE. *Styrax pallidus* (*Mori & Gracie 18925*). **A.** Part of stem showing leaves and inflorescences; note the valvate corollas of flower buds. **B.** Abaxial leaf surface showing stellate pubescence and stellate hair below. **C.** Lateral view of flower. **D.** Medial section of flower with gynoecium removed showing the connate filament bases adnate to corolla. **E.** Medial section of flower with the corolla and androecium removed; note axile placentation. **F.** Transverse sections of ovary through apex (above), middle (middle), and base (below), showing the incomplete partitioning of locules.

bearing both types of hairs, the surface of these veins obscured by pubescence or nearly so (rarely the minute hairs absent on principal veins and thus the surface visible through pubescence), domatia absent; secondary veins in 4–7 pairs. Inflorescences racemose, unbranched, 3–4 cm long; longer peduncles 2–9 mm long; lower pedicels 5–8 mm long. Flowers: calyx 4–5 × 5–6 mm, light brown- to olive-green-stellate-tomentose, sometimes with a few rust-colored hairs toward base, the tomentum thin; corolla 10–12 mm long, white to pale pink, the lobes 7–9 × 2 mm; stamen filaments not auriculate, the longer arms of hairs nearest distal end of filament 1–1.4 mm long, the anthers 5.5–6.5 mm long, the sacs not tapered at apex. Drupes 10–13 × 6–7 mm, subtended by calyx 4–6 mm long. Known only by sterile collections from our area; rare, in forest.

Styrax glabratus Schott

Trees, to 20 m tall. Bark not rough, the inner bark creamy yellow. Young stems sparsely to densely, greenish gray, lepidote. Leaves: petioles 5–16 mm long, lepidote-pubescent; blades elliptic or oblanceolate, 9–19 × 2.5–7 cm, abaxially glabrous or nearly so except for scattered scales along midrib, often with domatia at junction of secondary veins and midrib, these 1.2–2.7 mm long; secondary veins in 6–9 pairs. Inflorescences racemose, unbranched or once-branched, 2–6 cm long; longer peduncles 3–9(19) mm long; lower pedicels 2–13 mm long. Flowers: calyx 4–7 × 4–6 mm, grayish green lepidote-pubescent, sometimes with a few rust-colored scales toward base, the vesture thin; corolla 13–20 mm long, white to pale pink, the lobes 8–13 × 2.5–3 mm; stamen filaments not auriculate, the longer arms of hairs nearest distal end of filament 1.2–2 mm long; anthers 5–8 mm long, the sacs tapered at apex. Drupes 13–18 × 7–9 mm, subtended by calyx 4–8 mm long. Known only by a sterile collection from our area; rare, in forest.

Styrax pallidus A. DC. FIG. 303, PL. 122d

Trees, to 40 m tall. Trunks buttressed. Bark not rough, the inner bark dark red, contrasting with pearl-white sapwood. Young stems thinly gray- to brown-stellate-tomentose. Leaves: petioles 9–19 mm long, stellate-tomentose; blades elliptic, ovate-elliptic, or oblong-elliptic, 8.5–16.5 × 3.5–8 cm, abaxial surface and veins covered with minute grayish white stellate hairs, usually also with minute golden-yellow to orange stellate hairs scattered on surface and principal veins, the surface of these veins obscured by the pubescence or nearly so, often with domatia at junction of secondary veins and midrib, these 1–1.8 mm long; secondary veins in 8–10 pairs. Inflorescences cymose, unbranched or once-branched, the longer ones 5–13 cm long; longer peduncles 11–53 mm long; lower pedicels 2–6 mm long. Flowers: calyx 3.5–4.5 × 4–5 mm, grayish green-stellate-tomentose, sometimes with a few golden-yellow or orange hairs toward base, the tomentum thin; corolla 10–13 mm long, white to pale pink, the lobes 8–10 × 1.5–2 mm; stamen filaments slightly auriculate, the auricles nearly obscured by pubescence, the longer arms of hairs nearest distal end of filament 0.6–0.8 mm long, the anthers 5–6 mm long, the sacs not tapered at apex. Drupes 13–16 × 10–12 mm, subtended by calyx 2–4 mm long. Fl (Jun, Oct); rare, in nonflooded forest.

Styrax sieberi Perkins

Trees, to 30 m tall. Trunks buttressed. Bark rough, the inner bark red, contrasting with yellow sapwood. Young stems thinly to thickly tawny, rust-colored-, golden-yellow-, or brown-stellate-tomentose. Leaves: petiole 4–13 mm long, stellate-tomentose; blades ovate, elliptic, or oblong-elliptic, 9.1–27 × 4.1–15 cm, abaxial surface covered with minute white stellate hairs and also a lesser and variable amount of larger white to golden-yellow or orange, somewhat stiff stellate hairs, the principal veins bearing only larger hairs, the surface of these veins easily visible through pubescence, often with inconspicuous (<0.3 mm long) domatia at junction of secondary veins and midrib; secondary veins in 6–12 pairs. Inflorescences racemose, unbranched or once-branched, 2–7 cm long (2–3 cm long in our area), the racemes often congested; longer peduncles 0–26 mm (0–7 mm in our area); lower pedicels 1–6 mm long. Flowers: calyx 3–5 × 3–5 mm, grayish green-, tawny-, or golden-yellow stellate-tomentose, the tomentum thin to thick; corolla 10–15 mm long, white, the lobes 7–12 × 1.5–3 mm; stamen filaments not auriculate, the longer arms of hairs nearest distal end of filament 2.3–4 mm long, the anthers 4.5–6 mm long, the sacs tapered at apex. Drupes 8–14 × 5–8 mm, subtended by calyx 2–5 mm long. Fr (Nov); apparently rare, in nonflooded forest.

SYMPLOCACEAE (Sweetleaf Family)

Scott A. Mori and John L. Brown

Shrubs or trees. Stipules absent. Leaves simple, alternate; blades with margins entire to serrate, usually with black glands; secondary venation pinnate. Inflorescences axillary. Flowers actinomorphic, bisexual, rarely unisexual; sepals 5; petals 5–8(15), often strap-shaped, fused to one-half length; stamens 5–100, adnate to corolla, the filaments markedly constricted at apex, the anthers small, globose; ovary half inferior to inferior, 2–6 locular, the stigma capitate, often divided into shallow lobes; placentation axile, the ovules pendulous from apex of locule. Fruits most commonly single-stoned drupes, crowned by persistent calyx lobes. Seeds with abundant endosperm, the embryo straight or curved, with very short cotyledons.

SYMPLOCOS Jacq.

A monogeneric family. Genus description same as for family description.

Symplocos martinicensis Jacq. FIG. 304

Trees, to 25 m tall. Stems grayish. Leaves: petioles 8–10 mm long; blades oblanceolate, 9–12 × 3–4 cm, sparsely pubescent, the hairs simple, the margins distinctly crenate; secondary veins in 5–8 pairs. Inflorescences spikes. Flowers oblong in bud; sepals rounded, ca. 1.5 × 2 mm, pubescent, the margins fimbriate; petals ca. 12 × 2.5 mm, white; stamens ca. 40, unequal in length, the staminal tube white; ovary half-inferior, 5-locular, the nectary disc pubescent, the stigma with 5, shallow lobes. Fl (Oct, Nov), rare, in nonflooded forest.

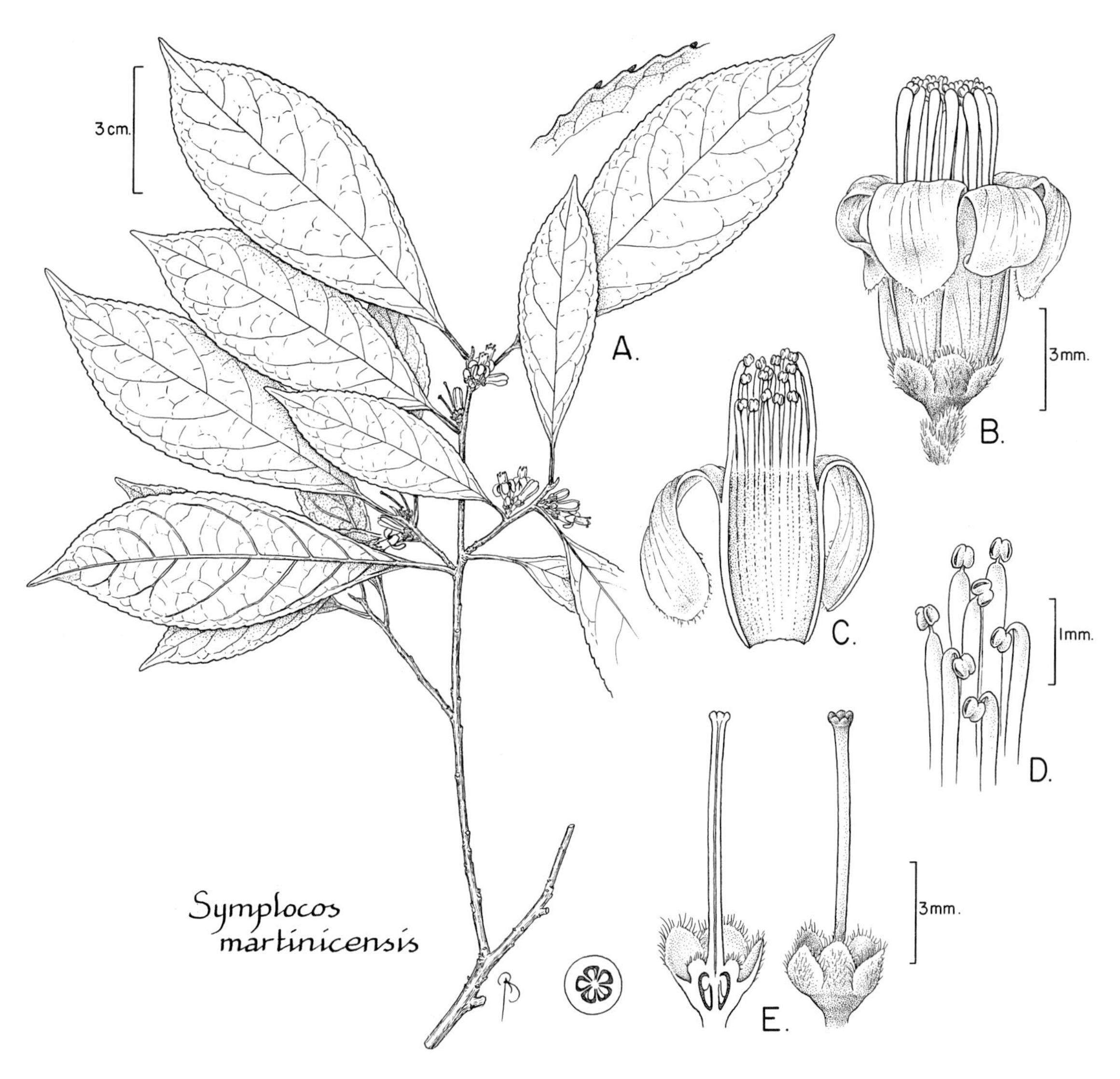

FIG. 304. SYMPLOCACEAE. *Symplocos martinicensis* (*Mori et al. 15037*). **A.** Part of stem showing leaves and axillary inflorescences; note detail of leaf margin (above). **B.** Flower. **C.** Medial section of flower with gynoecium removed; note the connate filament bases adnate to the corolla. **D.** Apices of stamens. **E.** Transverse section (left) and medial section (middle) and part (right) of flower with corolla and androecium removed.

THEOPHRASTACEAE (Theophrasta Family)

Bertil Ståhl

Shrubs, treelets, or small trees. Stipules absent. Leaves simple, alternate, often pseudoverticillate. Flowers actinomorphic, bisexual or unisexual, (4)5-merous; calyx persistent; corolla sympetalous, the lobes imbricate in bud; staminodes fused with corolla, alternating with lobes; stamens isomerous, opposite corolla lobes; ovary superior, 1-locular; placentation free central. Fruits berries, the pericarp coriaceous or woody. Seeds few to many, ± imbedded in juicy, sweet pulp.

Lindeman, J. 1979. Theophrastaceae. *In* A. L. Stoffers & J. C. Lindeman (eds.), Flora of Suriname **V(1):** 367–369. E. J. Brill, Leiden.
Ståhl, B. 1989. A synopsis of Central American Theophrastaceae. Nordic J. Bot. **9:** 15–30.

CLAVIJA Ruiz & Pav.

Unbranched or sparsely branched shrubs or treelets. Leaves often large, clustered in 2–4 pseudowhorls toward stem apices. Inflorescences cauliflorous, appearing along stem but usually most abundant among and just beneath foliage, racemose; pedicels present. Flowers often unisexual (plants mostly dioecious); corolla subrotate, fused to ca. 1/3 of length, firm with waxy texture, pale orange to orangish-red, the lobes

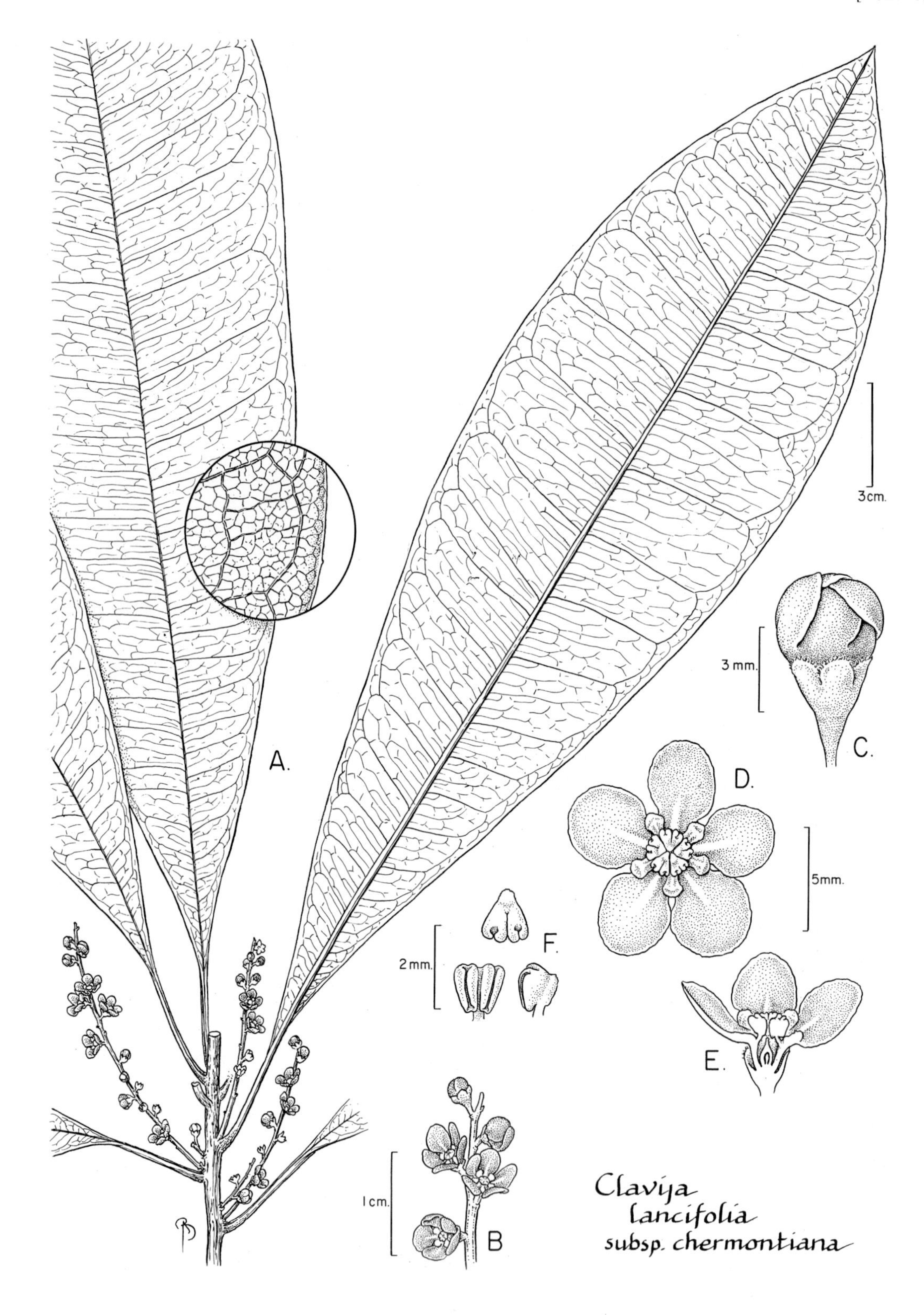

FIG. 305. THEOPHRASTACEAE. *Clavija lancifolia* subsp. *chermontiana* (*Mori & Ek 20731*). **A.** Part of stem with leaves and inflorescences and detail of leaf venation. **B.** Inflorescence. **C.** Lateral view of flower bud. **D.** Apical view of flower. **E.** Medial section of flower. **F.** Apical (above left), abaxial (below, far left), and lateral (below, near left) views of anthers.

suborbicular, somewhat unequal in size; staminodes gibbous, obovoid to oblong, adnate to corolla at mouth of tube; filaments connate at base only (always in pistillate flowers) or fused into tube (always in staminate flowers). Fruits subglobose, yellow or orange, the pericarp usually thin and somewhat elastic when fresh, brittle when dry. Seeds rather large, subspherical, completely imbedded in orange pulp.

Ståhl, B. 1991. A revision of *Clavija* (Theophrastaceae). Opera Bot. **107:** 5–77.

Clavija lancifolia Desf. subsp. **chermontiana** (Standl.) B. Ståhl FIG. 305, PL. 122e

Understory unbranched or sparsely branched shrubs or treelets, to 3 m high. Leaves: petioles 1–8 cm long; blades oblanceolate, sometimes elliptic or narrowly obovate, 14–45(53) × 3.5–12(14) cm, coriaceous, glabrous, the base attenuate, the apex acute or short-acuminate, the margins entire or serrulate. Racemes in staminate or bisexual plants to 20 cm long with 10–40 flowers, in pistillate plants to 4 cm long with 5–12 flowers. Flowers 5-merous. Fruits 1–3 cm diam. Seeds 2–12, 5–13 mm long, brown. Fl (Jan, May, Aug, Sep), fr (Aug, Sep); uncommon, in low-lying areas of nonflooded forest.

THYMELAEACEAE (Mezereum Family)

Maria Lúcia Kawasaki and Scott A. Mori

Shrubs or small trees. Bark peeling in strips. Stipules absent. Leaves simple, alternate, sometimes opposite or subopposite outside our area. Flowers actinomorphic, unisexual (plants dioecious); perianth usually uniseriate (small and scale-like petals sometimes present and then perianth biseriate); sepals usually 4, petaloid, fused at base; stamens twice as many and opposite to sepals, the filaments short or obsolete; staminodes present or absent; nectary disc often present; ovary superior, 1–2-locular; placentation apical, the ovules pendulous, 1 per locule. Fruits usually drupes. Seeds 1.

1. Leaf blades auriculate at base; secondary veins in 12–15 pairs. Inflorescences lateral, from stems, unbranched to scarcely branched. Stamens 8, the filaments absent or poorly developed; pistillate flowers often with staminodes. . . . . *Daphnopsis.*
1. Leaf blades cuneate at base; secondary veins in ca. 8 pairs. Inflorescences terminal, branched. Stamens 4, the filaments well developed; pistillate flowers without staminodes. . . . . *Schoenobiblus.*

DAPHNOPSIS Mart.

Shrubs or small trees. Inflorescences terminal or axillary, racemose, cymose, paniculate, or fasciculate, the ultimate axes usually umbelliform or capitulate. Flowers unisexual (plants dioecious), 4-merous; petals absent. Staminate flowers: stamens 8, usually not exserted, in two cycles, the distal cycle adnate to sepals, the proximal cycle adnate to hypanthium, the filaments absent or poorly developed. Pistillate flowers often with staminodes. Fruits drupes, the pericarp thin.

Daphnopsis sp. **1**.

Shrubs, 2 m tall. Leaves: petioles 3–6 mm long, glabrous; blades oblanceolate to widely oblanceolate, 22–27 × 7–8.5 cm, chartaceous, glabrous, somewhat striate abaxially when dry, the base auriculate, the apex acuminate to long acuminate, the margins entire; secondary veins in 12–15 pairs, the tertiaries reticulate. Inflorescences lateral, from stems, fasciculate to very short racemose, unbranched to scarcely branched. Flowers white (fide *de Granville B.5402*, CAY), but flowers not with specimens. Fruits ovate, ca. 10 × 6 mm, the apex pointed, glabrous, first green then red at maturity. Fl (Dec), fr (Jan from *de Granville 6377* (CAY) from just outside our area); collected once in our area on Mont Galbao.

The cauline position of the inflorescences and the remnant staminodes as well as the overall gestalt of these collections indicate that these collections represent a species of *Daphnopsis*, a genus heretofore not reported for French Guiana (Woodson & Schery, 1958; Boggan et al., 1997). Our determination, however, needs confirmation with more complete flowering collections.

SCHOENOBIBLUS Mart.

Shrubs or trees. Inflorescences terminal, paniculate, the ultimate axes umbelliform. Flowers unisexual (plants dioecious), 4-merous; petals absent. Staminate flowers: stamens 4, exserted, in a single cycle, adnate, opposite sepals, the filaments well developed. Pistillate flowers without staminodes. Fruits drupes, the pericarp thin.

Schoenobiblus daphnoides Mart. FIG. 306

Shrubs. Leaves: petioles 0.5–1 cm long; blades narrowly elliptic, lanceolate, or oblanceolate, 13–23 × 3.5–6 cm, chartaceous, glabrous, the base cuneate, the apex acuminate, the margins entire; secondary veins in ca. 8 pairs. Inflorescences terminal, paniculate, the axes yellowish-tomentose, glabrescent; pedicels of staminate flowers ca. 5 mm long, of pistillate flowers (in fruit) ca. 10 mm long. Staminate flowers: sepals 3–4 mm long, reflexed, yellowish-tomentose; stamens 4. Pistillate flowers: sepals 4, reflexed, ca. 4 mm long, yellowish-tomentose. Fruits elliptic, 10–15 × 8–10 mm, glabrous, orange. Seeds 1 per fruit. Fl (Jan), fr (May, Jul); in understory of nonflooded forest and in secondary vegetation.

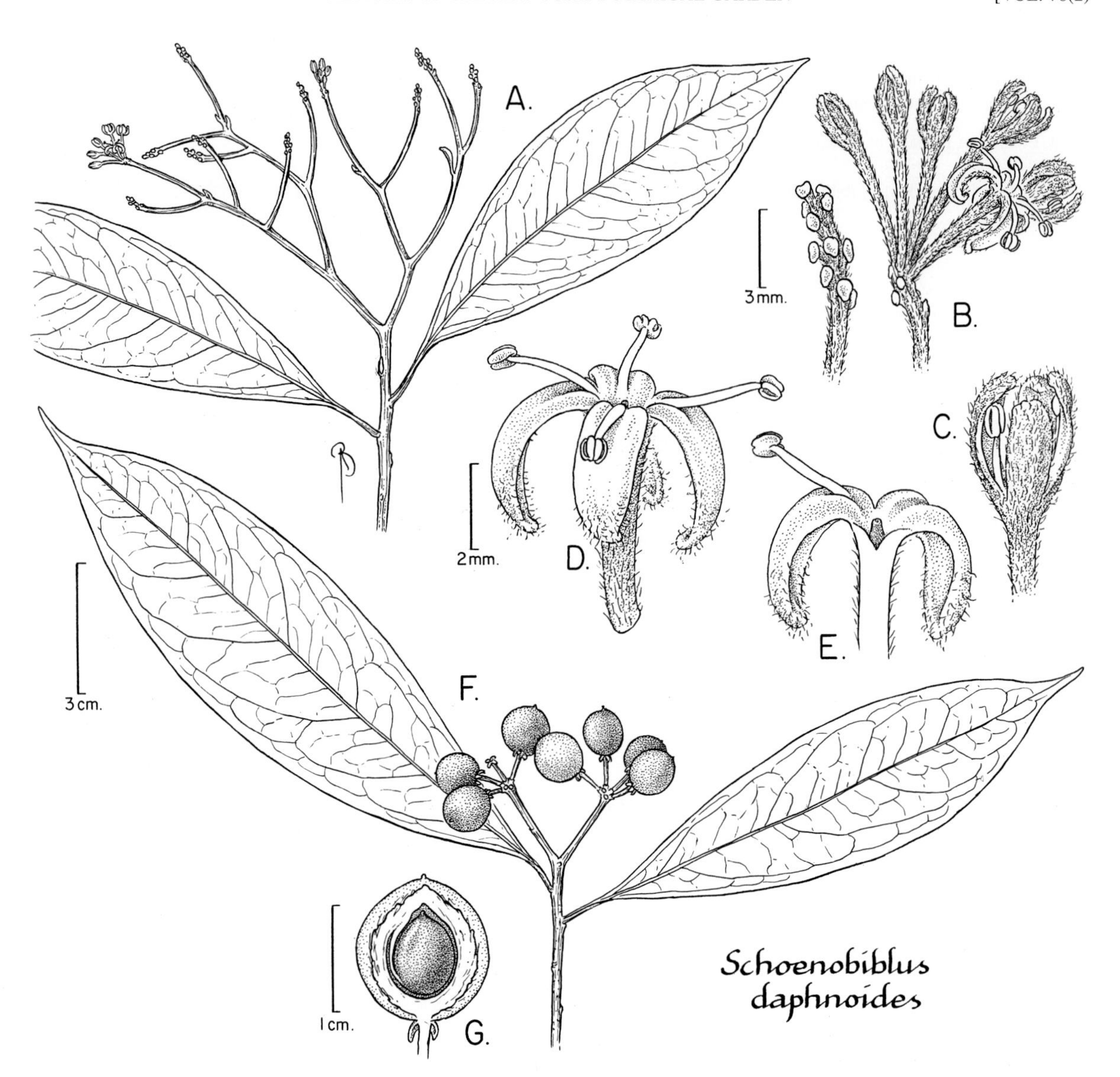

FIG. 306. THYMELAEACEAE. *Schoenobiblus daphnoides* (A–E, *Granville 2657*; F, G, *Mori & Hartley 18461*). **A.** Part of stem with leaves and inflorescences. **B.** Detail of part of inflorescence (near left) and inflorescence after flowers have fallen (far left). **C.** Lateral view of staminate flower bud. **D.** Lateral view of staminate flower. **E.** Section of staminate flower with two perianth lobes and three stamens removed; note pistillode. **F.** Part of stem with leaves and infructescence. **G.** Longitudinal section of fruit with seed intact.

TILIACEAE (Linden Family)

Laurence J. Dorr

Trees, shrubs, or rarely suffrutescent herbs, usually stellate-pubescent, but also with simple hairs or rarely peltate scales; mucilage present. Stipules persistent or caducous. Leaves simple, alternate; petioles present; blades with entire, often toothed or lobed margins. Inflorescences axillary or terminal cymes or panicles or flowers axillary and solitary; involucre sometimes present. Flowers: sepals usually 4–5, valvate, free or sometimes connate at base; petals 4–5, free, or absent, often glandular at base; androgynophore present or absent; stamens (5)numerous, free or shortly connate into 5 or 10 fascicles, the anthers bithecal; ovary superior, syncarpous (rarely apocarpous), (1)2–5(10)-carpelled; placentation axile (or intruded parietal). Fruits nut-like or drupes (rarely berries), capsules or schizocarps, dehiscent or indehiscent, sometimes with spines or wings. Seeds winged or not.

Molecular data (Bayer et al., 1999) suggest that this family should not be maintained and that the genera treated here are best referred to an expanded Malvaceae.

Jansen-Jacobs, M. J. 1986. Tiliaceae. *In* A. L. Stoffers & J. C. Lindeman (eds.), Flora of Suriname **III(1–2):** 293–300. E. J. Brill, Leiden.

——— & W. Meijer. 1995. Tiliaceae. *In* A. R. A. Görts-van Rijn (ed.), Flora of the Guianas Ser. A, **17:** 5–49, 68–73. Koeltz Scientific Books, Koenigstein.

Uittien, H. 1966. Tiliaceae. *In* A. Pulle (ed.), Flora of Suriname **III(1):** 49–57, 438–441. E. J. Brill, Leiden.

1. Flowers subtended by involucre. Capsules ovoid or oblong, without bristles or spines, opening by 4–5 valves. Seeds winged.
   2. Involucre of free bractlets. Anther thecae parallel, opening by longitudinal slits. *Luehea.*
   2. Involucre not of free bractlets (cup-shaped with 6–9 teeth or lobes). Anther thecae diverging above middle, opening initially by oblique apical pores, later dehiscing to base.. . . . . *Lueheopsis.*
1. Flowers without involucre. Capsules globose, subglobose, or disk-shaped, covered with bristles or conical spines, opening simply or by broad, short teeth (or appearing to be indehiscent). Seeds not winged. . . . . *Apeiba.*

APEIBA Aubl.

Trees, to 30 m tall. Stems, petioles, and inflorescences hispid, with simple cinnamon-colored hairs and shorter stellate hairs, or densely stellate-pubescent (sometimes minutely so). Leaves: blades elliptic to obovate, membranaceous to rugose or bullate, the base cordate to truncate, the apex acute to acuminate, the margins entire to coarsely serrate; secondary veins in 4–12 pairs. Inflorescences axillary, cymose; pedicels present, subtended by caducous bracts. Flowers: sepals (4)5; petals (4)5, shorter than sepals; stamens numerous, shorter than petals, arranged in whorls, the filaments of outer whorl usually fused at bases, the anthers linear, opening by longitudinal slits, with sterile appendage projecting beyond thecae; ovary superior, subglobose or disc-shaped, 5–10-carpelled, the style simple, slender, hollow, exceeding stamens in length, the stigma simple, slightly expanded, coarsely toothed. Fruits capsules, opening simply or by broad, short teeth (or appearing indehiscent), typically covered by spines or bristles, these either pubescent or glabrous. Seeds not winged, imbedded in pulp.

1. Leaf blade margins minutely serrate. Stems, petioles, and inflorescences hispid, densely covered with conspicuous, simple and short stellate hairs. Capsules with long, slender, flexible, pubescent spines to 2(3) cm long.
   2. Leaves bullate. Petals white. . . . . *A. albiflora.*
   2. Leaves smooth to slightly rugose. Petals bright yellow. . . . . *A. tibourbou.*
1. Leaf blade margins entire to sinuate. Stems, petioles, and inflorescences stellate-pubescent (or appearing glabrous to naked eye). Capsules with short, stout, conical, or fragile, thin, glabrous spines to 6 mm long.
   3. Leaves discolorous, grayish-white abaxially; secondary veins in 5–7 pairs, the basal pair extending half or less the length of leaf. Flowers 5-merous. Fruit covered with stout, conical spines. . . . . *A. petoumo.*
   3. Leaves concolorous, green abaxially; secondary veins in 3–4 pairs, the basal pair extending more than half the length of leaf. Flowers 4–5-merous. Fruits covered with fragile, thin spines. . . . . *A. glabra.*

Apeiba albiflora Ducke [sometimes considered a synonym of *Apeiba tibourbou* Aubl.] PL. 123a; PART 1: FIG. 10

Trees, to 25 m tall. Stems, petioles, and inflorescences hispid, with cinnamon-brown simple hairs mixed with shorter stellate hairs. Trunks with steep, thin, plank buttresses. Bark smooth, lenticellate, the slash white with yellow streaks, oxidizing yellow. Stipules ± persistent. Leaves: petioles to 1.5 cm long; blades narrowly elliptic, 12–25 × 4–10 cm, bullate, stellate-pubescent on veins adaxially, densely stellate-tomentose abaxially, the base cordate, the apex acuminate, the margins minutely serrate; secondary veins in 7–9 pairs. Flowers: sepals 5, yellow; petals 5, white; outer whorl of stamens ± sterile, the anthers with simple or bifurcate sterile extension. Fruits globose, ca. 3 cm diam. (excluding spines), the spines slender, flexible, pubescent, to 2(3) cm long. Fl (Oct), fr (Jul, Aug, Sep); often in gaps in nonflooded forest. *Peigne singe rouge* (Créole).

Apeiba glabra Aubl. PL. 123b

Trees, to 23 m tall. Stems, petioles, and inflorescences densely and minutely stellate-pubescent. Trunks with steep, high, thin cylindrical buttresses. Stipules caducous. Bark gray, smooth with large, prominent lenticels, the slash creamy white. Leaves: petioles to 1.5 cm long; blades elliptic to slightly obovate, 12–16 × 4–6 cm, membranous, glabrate adaxially, minutely stellate-pubescent with hairs scattered on veins and veinlets abaxially, the base obtuse to oblique, the apex acute to long acuminate, the margins entire; secondary veins in 3–4 pairs, the basal pair extending more than half the length of leaf. Flowers: sepals 4–5, yellowish-white; petals 4–5, yellow; anthers with simple or rarely emarginate or bifurcate sterile extension. Fruits globose, ± compressed, ca. 4–7 cm diam. (excluding spines), the spines fragile, thin, glabrous, to 6 mm long. Fl (Apr, May, Jun), fr (Jun); in secondary vegetation. *Bois bouchon* (Créole).

Apeiba petoumo Aubl. [Syns.: *Apeiba echinata* J. Gaertn., *A. membranacea* Benth., *A. aspera* Aubl., in part] FIG. 307; PART 1: FIG. 10, PL. XIIa

Canopy trees, to 30 m tall. Stems, petioles, and inflorescences densely and minutely stellate-pubescent. Trunks with thick, spreading buttresses. Bark brown or gray-brown, smooth with vertically oriented lenticels, the slash white, yellow or yellow-brown, oxidizing orange. Stipules caducous. Leaves: petioles to 3 cm long; blades elliptic to obovate, 8–23.5(30) × 4–11(13) cm, discolorous, membranous, glabrate adaxially, heterotrichous, with dense felt-like covering of grayish stellate hairs interspersed with cinnamon-colored stellate hairs abaxially, with conspicuous tufts of cinnamon colored hairs in axils of basal veins abaxially, membranous, the base truncate to subcordate, the apex acuminate, the margins entire or sinuate; secondary veins in 5–7 pairs, the basal pair extending half or less the length of leaf. Flowers: sepals 5, yellow; petals 5, yellow; anthers with simple, rounded or truncate sterile extension. Fruits disc-shaped, 4.5–6 cm diam. (excluding spines), the spines stout, conical, glabrous, 1–3 mm long. Fl (Sep, Oct, Nov), fr (Apr, Jun, Jul, Sep, Oct); often in disturbed habitats in nonflooded forest. *Bois bouchon*, *bois calou* (Créole), *cassawa-pau* (Bush Negro).

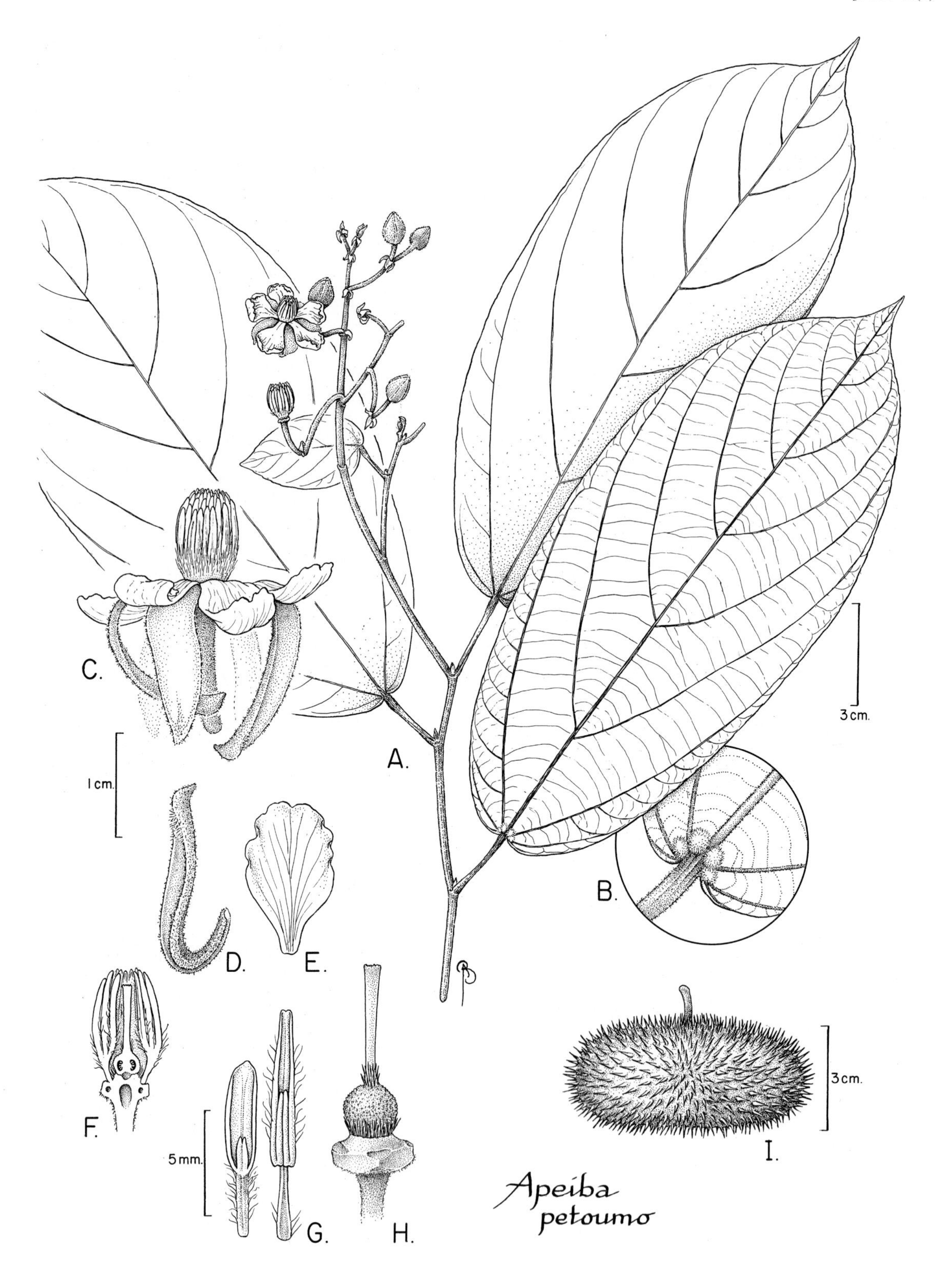

FIG. 307. TILIACEAE. *Apeiba petoumo* (*Mori 21523*). **A.** Part of stem with leaves and inflorescence. **B.** Detail of abaxial base of leaf blade showing tufts of hairs in vein axils. **C.** Lateral view of flower. **D.** Lateral view of sepal. **E.** Adaxial view of petal. **F.** Medial section of flower with calyx and corolla removed. **G.** Stamens from outer whorl (far left) and inner whorl (near left). **H.** Lateral view of pistil. **I.** Lateral view of fruit.

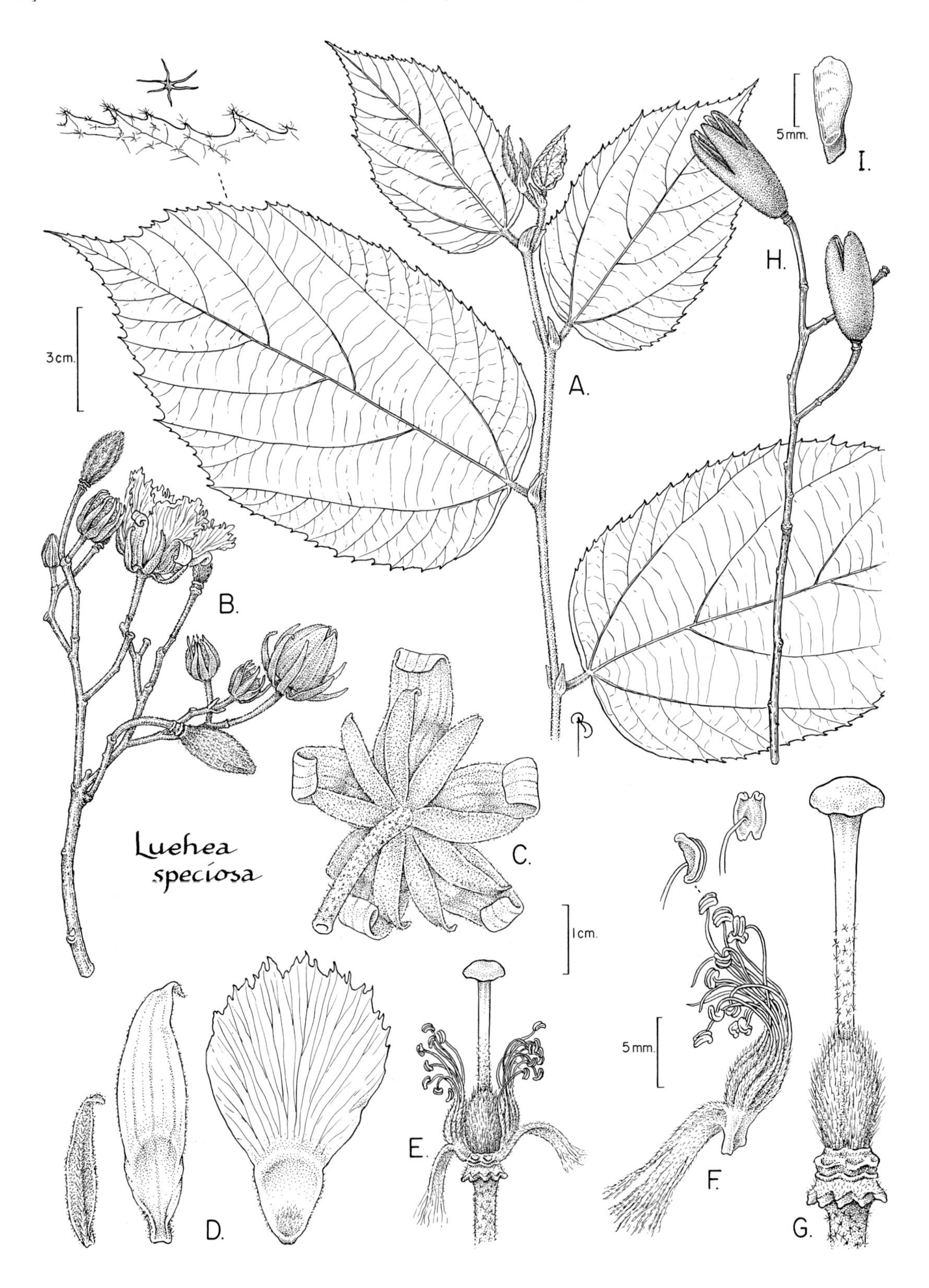

FIG. 308. TILIACEAE. *Luehea speciosa* (A, C–G, *Mori et al. 23884*; B, *Steyermark 88701* from Venezuela; H, I, *Smith 3087* from Guyana). **A.** Part of stem with leaves and detail of leaf margin and stellate hair. **B.** Inflorescence. **C.** Abaxial view of flower showing epicalyx. **D.** Petal (right), sepal (near left), and unit of epicalyx (far left). **E.** Lateral view of pistil and two fascicles of stamens. **F.** Fascicle of stamens with pubescent staminodes. **G.** Lateral view of pistil. **H.** Part of infructescence. **I.** Winged seed.

Apeiba tibourbou Aubl.

Small, low-branched trees, to 10(14) m tall. Stems, petioles, and inflorescences densely hispid with cinnamon-brown simple hairs mixed with shorter stellate hairs. Bark gray, smooth, with large prominent lenticels, the slash white or creamy white. Stipules ± persistent. Leaves: petioles to 3 cm long; blades elliptic, 10–30(35) × 6–12(17) cm, smooth to slightly rugose, sparingly pubescent with simple and stellate hairs mostly confined to major veins adaxially, densely stellate-pubescent with simple hairs mostly confined to major veins abaxially, the base cordate, the apex acuminate, the margins minutely serrate; secondary veins to 12 pairs. Flowers: sepals 5, light yellow to brown; petals 5, bright yellow; anthers with simple (sometimes shallowly bifurcate) sterile extension. Fruits globose, compressed, 4–7 cm diam. (excluding spines), the spines slender, flexible, pubescent, to 1 cm long. Fl (Jun, Aug), fr (Aug, Sep); in secondary vegetation. *Bois bouchon* (Créole), *pente de macaco* (Portuguese).

LUEHEA Willd.

Trees or tall shrubs. Stems, when young, petioles, and inflorescences densely stellate-tomentose, glabrate with age. Stipules caducous. Leaves: blades elliptic to ovate or oblong, glabrate or sparingly stellate-pubescent on veins adaxially, densely pubescent with arachnoid or stellate or both types of hairs abaxially, the base acute or obtuse to cordate, the apex usually acuminate, the margins serrate to dentate; secondary veins in 3–6 pairs. Inflorescences axillary or terminal panicles or cymes, sometimes flowers solitary. Flowers showy, with involucre of 5–9 free bractlets, these sometimes ± coherent at base; sepals 5; petals 5, white, violet, or rose-colored; stamens numerous, obscurely or very shortly coalescent in 5 phalanges; staminodes numerous, attached to base of each stamen phalange, fused or free, sometimes as long as stamens; anther thecae parallel, opening longitudinally; ovary (4)5-carpelled, the style simple, the stigma capitate. Capsules woody, imperfectly (4)5-valved. Seeds numerous, winged.

Luehea speciosa Willd. FIG. 308

Trees, to 12(25) m tall. Stems, when young, petioles, and inflorescences densely pubescent with ferruginous-colored stellate hairs. Leaves: petioles ca. 1 cm long; blades elliptic to obovate, 10–20 × 4.5–14 cm, subcoriaceous, glabrate or sparsely pubescent with scattered stellate hairs on veins adaxially, ferruginous-colored stellate hairs (mostly on veins) interspersed with whitish or grayish arachnoid hairs abaxially, the base unequally rounded to subcordate or cordate, the apex shortly acuminate, the margins coarsely serrate. Inflorescences axillary or terminal panicles; involucral bracts (8)9, narrowly lanceolate, to 2.5 cm long. Flowers fragrant; petals large, to ca. 4.5 cm long, white; stamens in 5 fascicles; staminodes fused for ca. 1/2 length. Capsules ovoid or oblong, 2.5–3.5 × 1.5–1.8 cm, dehiscing for ca. half length by 4–5-valves, stellate-tomentose. Fl (Feb), fr (Sep), rare in our area, in nonflooded forest.

LUEHEOPSIS Burret

Large trees. Stems, when young, petioles, and inflorescences densely covered with stellate hairs. Stipules caducous. Leaves: blades elliptic to obovate or nearly orbicular, sparingly stellate-pubescent to glabrate adaxially, arachnoid or stellate or both types of hairs abaxially, the base obtuse to truncate, the apex obtuse to acute or short acuminate, the margins entire to sinuate or serrate; secondary veins in 4–6 pairs. Inflorescences in axillary or terminal panicles; involucre cup-shaped, with 6–9, triangular to lanceolate teeth. Flowers: sepals 5; petals 5, white, pink, or rose-violet; stamens numerous, the filaments fused at base into mostly glabrous tube, the anther thecae diverging above middle, opening first by oblique, apical pores, later dehiscing to base; staminodes arising from top of tube; ovary 5-carpelled, the style simple, dilated at apex, the stigma 5-lobed. Capsules woody, imperfectly 5-valved. Seeds small, winged.

Lueheopsis rugosa (Pulle) Burret

Trees, 12–15(35) m tall. Stems, when young, petioles, and inflorescences densely covered with rusty-brown hairs. Leaves: petioles to 1.5 cm long; blades broadly ovate to nearly orbicular, 11–16 × 8–10 cm, stiffly rugose, subcoriaceous, glabrate, with scattered stellate hairs on veins adaxially, densely covered with arachnoid and stellate hairs on veins abaxially, the base truncate, the apex obtuse to acute (rarely emarginate), the margins entire to slightly wavy; secondary veins in 4–5 pairs. Flowers subtended by involucre, the teeth 6–9, triangular, to 2 mm long; petals rose-violet. Capsules ovoid, 1–2 × 3–4 cm, dehiscing to ca. one-half length by 5 loculicidal valves, densely dark rusty-brown stellate-pubescent. Fl (Feb); a poorly known species in our area.

TRIGONIACEAE (Trigonia Family)

Scott A. Mori

Lianas in our area, shrubs, rarely trees outside our area. Stipules interpetiolar, small, often connate, caducous. Leaves simple, usually opposite, alternate in *Trigoniodendron* of Malaysia. Inflorescences thyrses, panicles, or racemes. Flowers zygomorphic, bisexual; sepals 5, unequal, fused at base; petals 5, the two anterior ones forming a keel, the posterior one (standard) saccate, the lateral ones (wings) spatulate; stamens 5–8, the filaments fused at base; staminodes present or absent; stamens with anthers introrse, longitudinally dehiscent; nectary glands present, usually opposite standard; ovary superior, usually 3-locular, infrequently 4-locular or unilocular; placentation axile or parietal. Fruits septacidal capsules in our area, 3-winged samaras in *Humbertodendron* of Madagascar. Seeds covered by long hairs in our area, without endosperm.

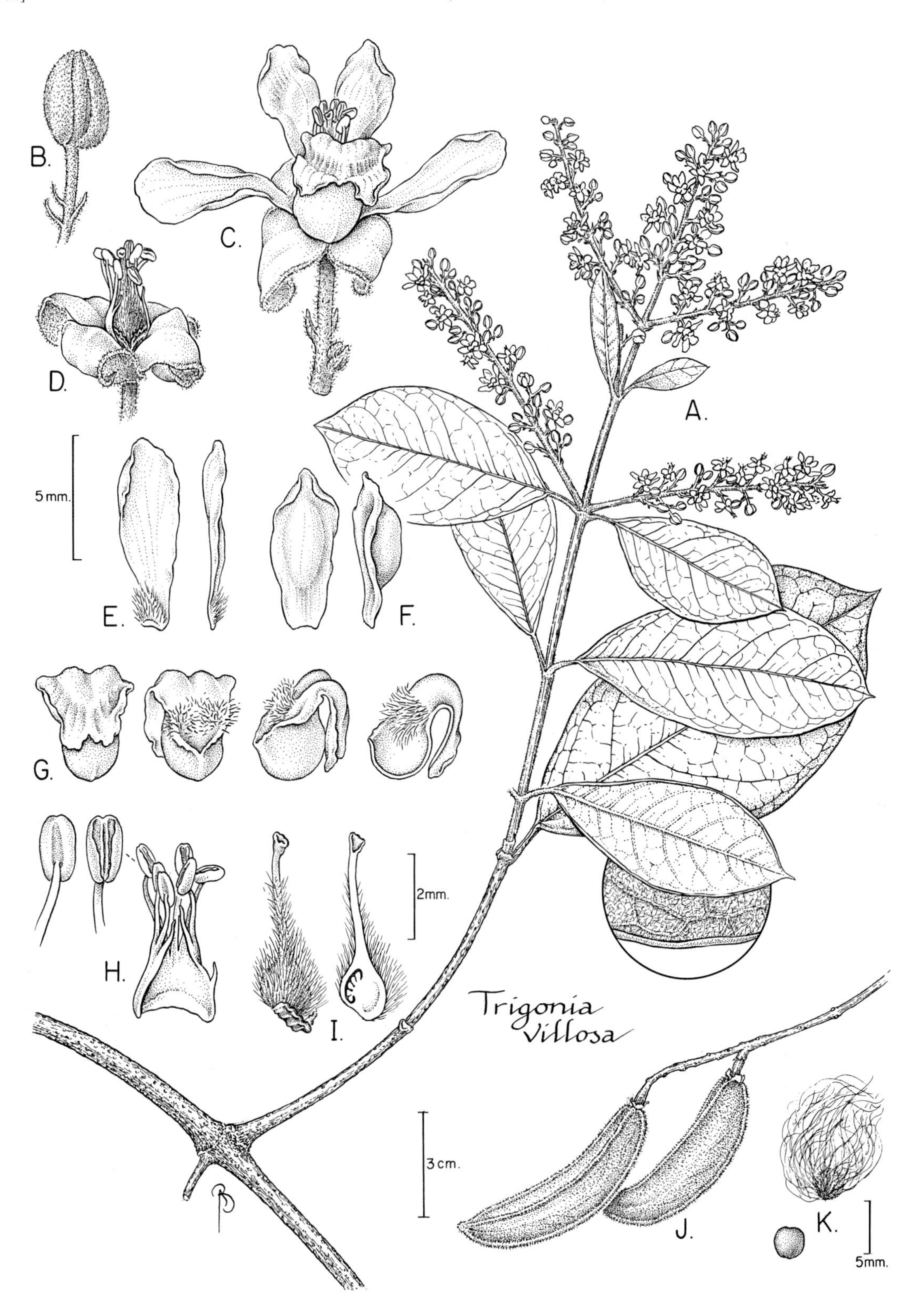

FIG. 309. TRIGONIACEAE. *Trigonia villosa* (A–I, *Mori & Mitchell 18702*; J, *Oldeman 2041*; K, *Tillet et al. 43907* from Guyana). **A.** Part of stem with leaves and inflorescences and detail of abaxial surface of leaf. **B.** Lateral view of floral bud. **C.** Oblique-apical view of flower. **D.** Flower after petals have fallen. **E.** Adaxial (near right) and lateral (far right) views of a lateral petal. **F.** Adaxial (far left) and lateral (near left) views of anterior petal. **G.** Three views and medial section (far right) of saccate posterior petal. **H.** Abaxial (far upper left) and adaxial (near upper left) detailed views of anthers and entire androecium (right). **I.** Lateral view (left) and medial section (right) of pistil with basal nectary. **J.** Fruits. **K.** Seed with hairs (above) and with hairs removed (below left).

Lindeman, J. 1986. Trigoniaceae. *In* A. L. Stoffers & J. C. Lindeman (eds.), Flora of Suriname **III(1–2):** 550–551. E. J. Brill, Leiden.
Lleras, E. 1978. Trigoniaceae. Fl. Neotrop. Monogr. **19:** 1–73.
———. 1998. Trigoniaceae. *In* A. R. A. Görts-van Rijn & M. J. Jansen-Jacobs (eds.), Flora of the Guianas Ser. A, **21:** 49–60, 87–98. Royal Botanic Gardens, Kew.
Stafleu, F. A. 1951. Trigoniaceae. *In* A. Pulle (ed.), Flora of Suriname **III(2):** 173–177. The Royal Institute for the Indies.

TRIGONIA Aubl.

Lianas in our area, shrubs, scandent shrubs, or treelets. Leaves opposite. Inflorescences with flowers ultimately in cymes or cincinii. Flowers papilionaceous (pea-like); staminodes 0–4; nectary glands usually 2, lobed, sometimes laciniate. Fruits trigonous, septacidal capsules.

1. Leaf blades glabrous to slightly puberulous abaxially. Flowers 1.5–2 mm diam. Fruits 8–14 mm long, glabrous to puberulous. . . . . *T. laevis* var. *microcarpa.*
1. Leaf blades densely villous abaxially. Flowers 4–5 mm diam. Fruits 40–90 mm long, densely pubescent. . . . . . *T. villosa.*

Trigonia laevis Aubl. var. **microcarpa** Sagot Pl. 123c

Lianas. Leaves: petioles 5–10 mm long, glabrous to puberulous; blades 8.5–10 × 3.5–6 cm, mostly glabrous abaxially; venation plane to salient adaxially. Inflorescences with rachises slender, these 8–20 cm long. Flowers small, 1.5–2 mm diam.; petals white to greenish-white. Fruits nearly as long as broad, 8–14 × 8–14 mm, glabrous to puberulous. Fl (Oct), fr (Feb); in nonflooded forest. *Liane câble* (Créole).

Trigonia villosa Aubl. Fig. 309

Lianas. Leaves: petioles 8–15 mm long, densely pubescent; blades 7–12.5 × 4–7 cm, densely villous abaxially; venation often impressed adaxially. Inflorescences with rachises stout, these 6–7 cm long. Flowers 4–5 mm diam.; petals mostly white, some with either yellow or purple spots even in flowers of same collection. Fruits longer than broad, 40–90 × 15–24 mm, densely pubescent. Fl (Apr, May, Aug), fr (Jan, Feb, Sep); in nonflooded forest.

TURNERACEAE (Turnera Family)

Michel Hoff

Shrubs, treelets, or small trees, rarely herbs. Pubescence of simple, glandular, or stellate hairs. Stipules small or absent. Leaves simple, alternate; petioles present or rarely absent, often with 1–2 pairs of glands; blades with entire, serrulate, dentate, or crenulate margins, often with 1–2 glands inserted toward base. Inflorescences usually axillary, rarely terminal, the flowers solitary or in few-flowered clusters; peduncle or pedicels free from or adnate to petiole; bracts 2, opposite or alternate, rarely absent. Flowers actinomorphic, bisexual, often heterostylous; calyx with quincuncial aestivation, the sepals 5, free or fused into tube to 1/2 length, often provided with glands adaxially; corolla with contorted aestivation, the petals 5, alternating with sepals, free or adnate to floral tube, unguiculate, the base of petals often provided with membranous, fimbriate corona; nectary disc entire, free or adnate to floral tube, reduced to epipetalous appendages, or absent; stamens 5, alternate with petals, adnate or free from floral tube, the filaments free or connate, the anthers subbasifixed or dorsifixed, with 2 thecae, the dehiscence longitudinal, introrse; ovary superior, 3-carpellate, 1-locular, the styles 3, filiform, simple or rarely bifid, connivent basally, the stigmas very short fimbriate or penicillate; placentation parietal, the placentae 3, the ovules 1 to numerous, anatropous. Fruits globose to linear loculicidal capsules. Seeds obovoid to oblong-cylindric, straight or curved, striate, reticulate, or rarely tuberculate, with membranous white aril.

Bremekamp, C. E. B. 1966. Turneraceae. *In* A. Pulle (ed.), Flora of Suriname **III(1):** 342–354. E. J. Brill, Leiden.
Görts-van Rijn, A. R. A. 1986. Turneraceae. *In* A. L. Stoffers & J. C. Lindeman (eds.), Flora of Suriname **III(1–2):** 464–467. E. J. Brill, Leiden.

TURNERA L.

Shrubs or treelets. Leaves often with glands on petiole and at base of blade; blades entire or dentate. Inflorescences with peduncles free or sometimes fused to petioles. Flowers homo- or heterostylous; sepals fused at base into short tube; petals generally longer than sepals, yellow or sometimes white, ± unguiculate, the claw adnate to floral tube; stamens shorter than perianth, inserted toward base of calyx, the filaments sometimes adnate to claw of petals and often connate for variable length, the anthers basifixed or subbasifixed, introrse; ovary sessile, the styles 3, simple, the stigmas generally penicillate. Capsules with 3 valves. Seeds with longitudinal, unilateral, membranous aril shorter than or sometimes slightly longer than seed.

Turnera glaziovii Urb. Fig. 310, Pl. 123d

Shrubs or treelets, 0.4–3.5 m tall. Stems pubescent, 2–4 mm diam. Stipules absent. Leaves: petioles 0.5–2.5 cm long, narrowly winged for most of length, usually with 1–2 pairs of crateriform glands along margin; blades oblanceolate to narrowly oblanceolate, (8)14–25 × (1.5)2.5–5 cm, the base long and narrowly cuneate, the apex acuminate, the margins serrulate, with 1 to several crateriform

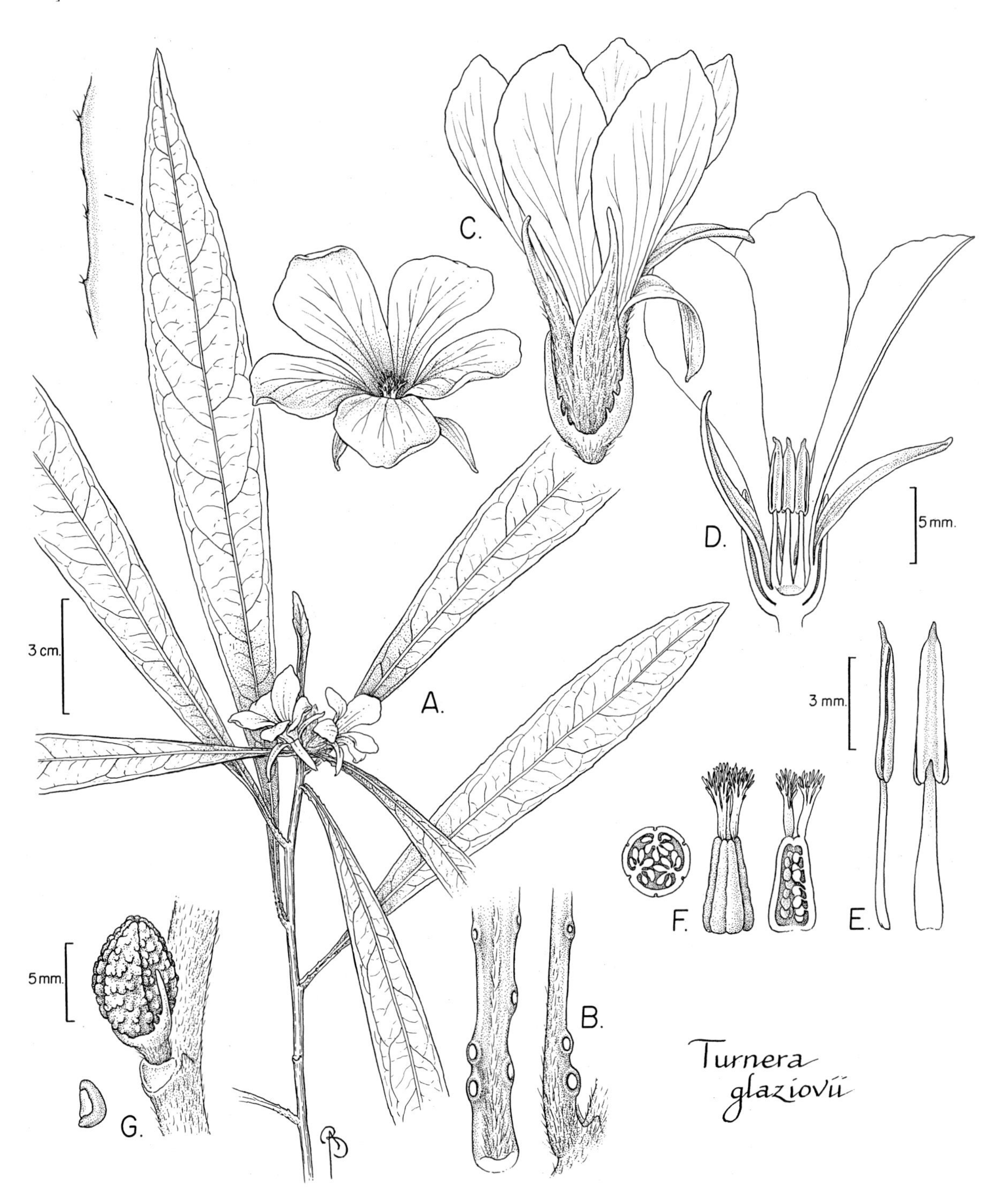

FIG. 310. TURNERACEAE. *Turnera glaziovii* (*Mori et al. 22774*). **A.** Apex of stem with leaves and inflorescences. **B.** Lateral (right) and adaxial (left) views of petioles showing glands. **C.** Lateral (right) and apical (left) views of flowers. **D.** Medial section of flower with pistil removed. **E.** Lateral (left) and adaxial (right) views of stamens. **F.** Pistil (center), medial section of pistil (right), and transverse section of ovary (left). **G.** Fruit (right) and seed (left).

glands along margin toward base; secondary veins in 6–10 pairs. Inflorescences in axils of leaves, usually of 1 to several flowers. Flowers 2–3 cm long; calyx yellow, the lobes narrowly linear; petals oblong to oblanceolate, (20)30–35 × 5–10(15) mm, glabrous, bright yellow; stamens 7–10 mm long, the filaments white, expanded at base, 7 mm long in short-styled plants, 4 mm long in long-styled plants, the anthers yellow, linear to oblong, apiculate; ovary conical, smooth, the styles free, yellow, pilose, the stigmas penicillate. Capsules ovoid, 0.7–1 × 0.4 cm, tuberculate, brown to reddish-brown. Seeds 4 × 2 mm, finely pubescent, pellucid-translucent, yellow-orange, very finely alveolate, with membranous white aril. Fl (May, Oct, Nov); infrequent, in somewhat open areas in nonflooed forest.

ULMACEAE (Elm Family)

Cornelis C. Berg

Trees, shrubs, or lianas, usually unarmed, sometimes with axillary thorns. Stipules present, lateral, free, caducous or persistent. Leaves simple, alternate and distichous, opposite only in *Lozanella* (outside our area). Inflorescences axillary or ramiflorous, solitary (especially pistillate inflorescences), cymose, or fasciculate. Flowers actinomorphic, bisexual or functionally unisexual (plants dioecious, monoecious, or polygamous); perianth uniseriate; tepals (3)4–5, free or fused at base; stamens 3–8, free; ovary superior, 1-locular, the styles 2, free or fused at base, the stigmas 2, sometimes branched into 4 ultimate divisions (*Celtis*); placentation apical, the ovule solitary. Fruits drupes in our area. Seeds with straight or curved embryo; endosperm sparse or absent.

Berg, C. C. 1992. Ulmaceae. *In* A. R. A. Görts-van Rijn (ed.), Flora of the Guianas Ser. A, **11:** 3–9, 192–222. Koeltz Scientific Books, Koenigstein.

Mennega, A. M. W. & A. R. A. Görts-van Rijn. 1968. Ulmaceae. *In* A. Pulle & J. Lanjouw (eds.), Flora of Suriname **I(2):** 301–302. E. J. Brill, Leiden.

Ooststroom, S. J. van. 1966. Ulmaceae. *In* A. Pulle (ed.), Flora of Suriname **I(1):** 47–48. E. J. Brill, Leiden.

1. Lianas or scandent shrubs, with axillary thorns. Stigma with 4 ultimate divisions. . . . . *Celtis.*
1. Trees, without axillary thorns. Stigma with 2 ultimate divisions.
    2. Leaf blades coriaceous, glabrous, the margins entire; venation with secondary veins distinct from midrib, pinnately arranged. Plants of primary forest.. . . . . *Ampelocera.*
    2. Leaf blades chartaceous, pubescent, the margins serrate; venation with basal secondary veins and midrib ± equal, the two dominant secondaries arising from base. Plants of secondary habitats.. . . . . *Trema.*

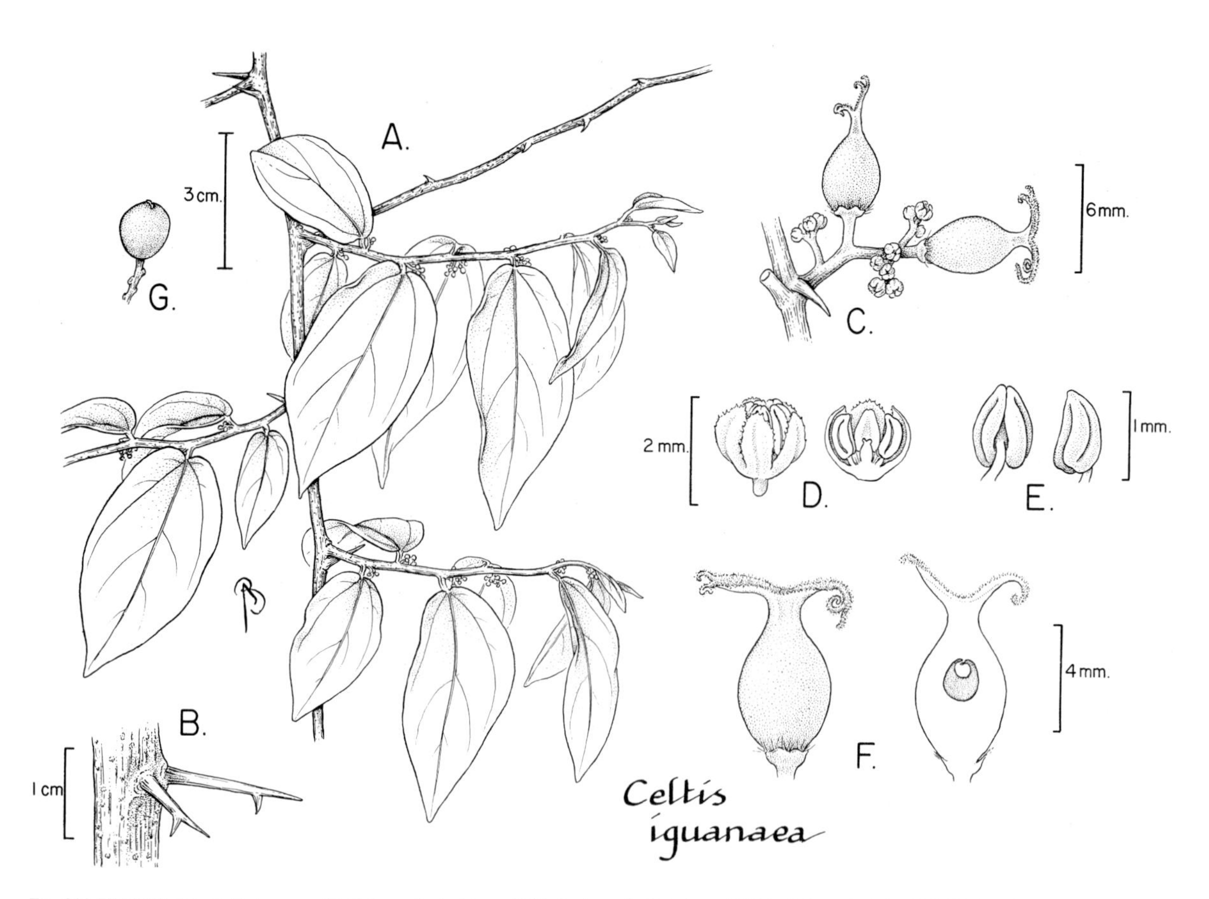

FIG. 311. ULMACEAE. *Celtis iguanaea* (A–F, *Acevedo-Rodríguez 2691*; G, *Acevedo-Rodríguez 2021*, both from St. John, U.S. Virgin Islands). **A.** Part of stem with leaves, flowers, and stipular spines. **B.** Detail of stem with stipular spines. **C.** Inflorescence with two pistillate flowers. **D.** Lateral view (left) and medial section (right) of staminate flower. **E.** Abaxial (left) and lateral (right) views of stamen. **F.** Lateral view (left) and medial section (right) of immature fruits. **G.** Fruit. (Reprinted with permission from P. Acevedo-Rodríguez, *Flora of St. John, U.S. Virgin Islands*, Mem. New York Bot. Gard. 78, 1996.)

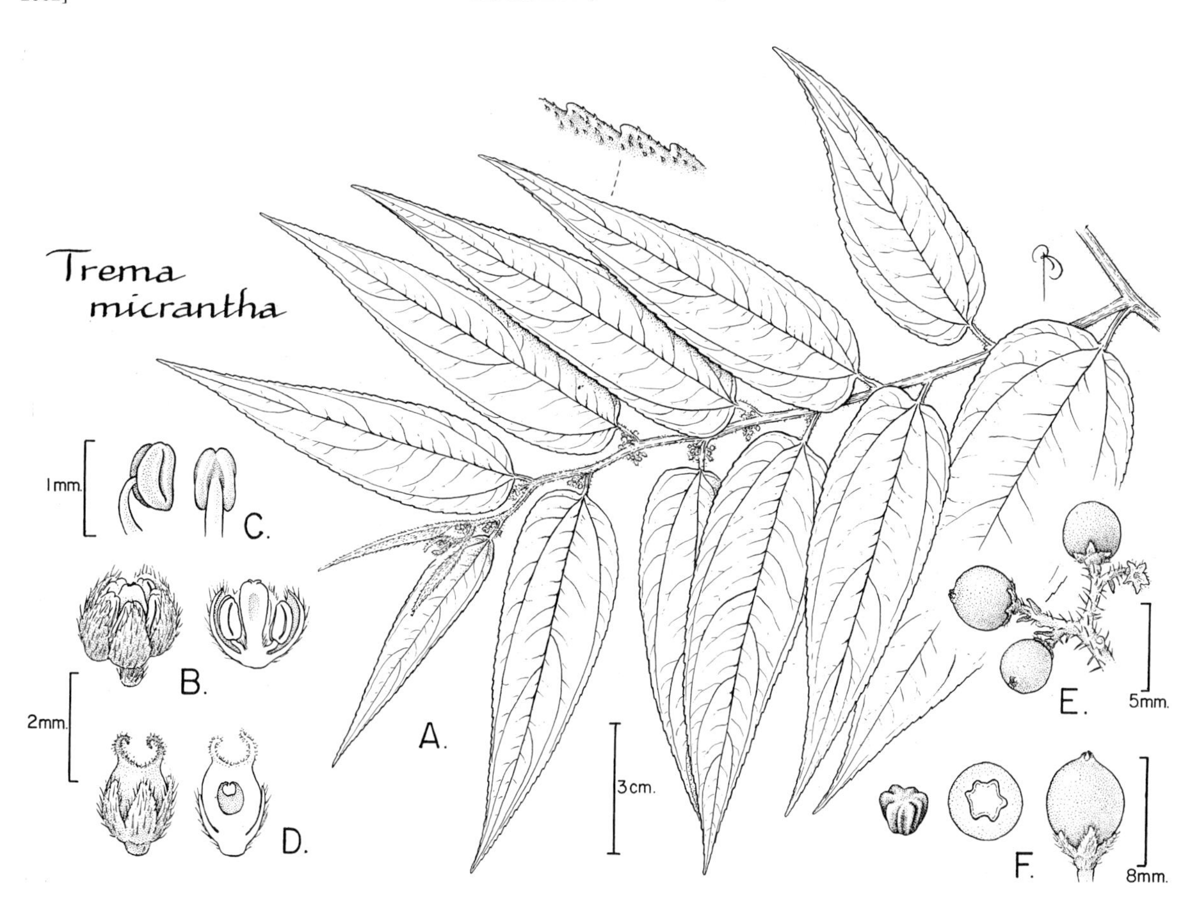

FIG. 312. ULMACEAE. *Trema micrantha* (A, D, *Acevedo-Rodríguez 619* from St. John, U.S. Virgin Islands; B, C, *Prance 29314* from St. John, U.S. Virgin Islands; E, F, *Acevedo-Rodríguez 3841* from St. John, U.S. Virgin Islands). **A.** Part of stem with leaves and flowers and detail of leaf margin (right, above). **B.** Lateral view (left) and medial section (right) of staminate flower. **C.** Lateral (far left) and abaxial (near left) views of stamen. **D.** Lateral view (far left) and medial section (near left) of pistillate flower. **E.** Infructescence. **F.** Stone (far left), transverse section of fruit (near left), and lateral view of fruit (right). (Reprinted with permission from P. Acevedo-Rodríguez, *Flora of St. John, U.S. Virgin Islands*, Mem. New York Bot. Gard. 78, 1996)

AMPELOCERA Klotzsch

Small or large trees, unarmed. Leaves: blades with base nearly symmetric to somewhat asymmetric, the margins entire or coarsely dentate; venation pinnate or palmate. Inflorescences sparsely to much branched, cymose or fasciculate, often with bisexual flowers toward apex and functionally staminate flowers toward base. Flowers bisexual or functionally staminate (then plants monoecious); tepals 4–5, usually united at base; stamens (3)8(16), the anther connective prolonged beyond apex of anther; style branches 2, free to base, the stigma not branched. Fruits drupes, yellow or orange. Seeds globose.

Todzia, C. A. 1989. A revision of *Ampelocera* (Ulmaceae). Ann. Missouri Bot. Gard. 76: 1087–1102.

Ampelocera edentula Kuhlm. PL. 124a

Canopy or understory trees, to 25 m tall. Stipules caducous. Leaves: petioles 5–12 mm long, the epidermis irregularly peeling in dried specimens; blades elliptic to oblong to subovate, 4–20 × 2–9 cm, coriaceous, glabrous, the adaxial surface smooth, the base slightly asymmetrical, rounded, the apex acuminate, the margins entire; secondary veins pinnately arranged, in 4–8 pairs. Inflorescences to 2 cm long. Flowers: tepals yellow-green; stamens usually 8, white. Fruits 1–1.5 cm long. Fl (Sep), fr (Oct); uncommon, in nonflooded forest.

CELTIS L.

Lianas or scandent shrubs (trees or shrubs outside our area), sometimes armed along stems. Leaves: blades with base symmetric to asymmetric, the margins serrate, especially toward apex; venation with 2 secondary veins nearly equal to midrib arising from base and several other pinnately arranged secondary veins toward apex. Staminate inflorescences cymose or fasciculate. Pistillate inflorescences solitary or few-flowered fasciculate. Flowers bisexual, staminate, or functionally pistillate, the plants monoecious or polygamous; tepals (4)5, free or slightly fused at base; stamens 5, the anther connective not prolonged beyond apex of anther; style branches 2, the stigma branched, with 4 ultimate divisions. Staminate flowers with pistillode. Pistillate flowers with staminodes. Fruits drupes.

Celtis iguanaea (Jacq.) Sarg. FIG. 311

Lianas or scandent shrubs, armed with curved axillary thorns on scandent stems, these ± straight and sometimes branched on erect stems. Stipules 0.2–0.5 cm long. Leaves: petioles ca. 5 mm long, the epidermis not peeling irregularly in dried specimens; blades ovate to elliptic to oblong, 3–14 × 2–6.5 cm, chartaceous, pubescent to sub-glabrous, the adaxial surface smooth, the base symmetric to slightly asymmetric, obtuse to rounded, the apex acuminate, the margins serrate toward apex. Inflorescences to 4 cm long. Flowers bisexual or functionally pistillate; stigmas white. Fruits 1–1.5 cm long, orange or red. Fl (Dec), fr (Feb); mostly in secondary growth.

TREMA Lour.

Trees, unarmed. Leaves: blades with base markedly asymmetric, the margins serrate for entire length; venation with 2 secondary veins nearly equal to midrib arising from base and several other pinnately arranged secondary veins toward apex. Inflorescences solitary, cymose, or fasciculate. Flowers bisexual or functionally unisexual (plants dioecious or monoecious); tepals (4)5, free; stamens (4)5, the anther connective not prolonged beyond apex of anther; style branches 2, the stigmas not branched. Staminate flowers with pistillode. Pistillate flowers without staminodes. Fruits small, ovoid to subglobose drupes.

Trema micrantha (L.) Blume FIG. 312, PL. 124b,c

Trees, to 15(20) m tall. Stipules subulate, 0.2–0.5 cm long. Leaves: petioles 5–10 mm long, the epidermis not peeling irregularly in dried specimens; blades narrowly ovate, 4–18 × 1–6 cm, chartaceous, pubescent, the adaxial surface scabrous, the base markedly asymmetric, rounded to subcordate, the apex long acuminate, the margins serrate along entire length. Inflorescences to 2 cm long. Fruits globose, 0.2–0.3 cm diam., orange or reddish. Fl (Feb, Aug, Sep, Oct, possibly year round), fr (Feb, Aug, Sep, Oct, possibly year round); pioneer tree, common in secondary vegetation, especially along airport road, the fruits are eaten by birds. *Bois l'homme, bois l'orme, bois ramier* (Créole).

URTICACEAE (Nettle Family)

Cornelis C. Berg

Herbs, subshrubs, shrubs, or treelets. Plants sometimes armed with stinging hairs or prickles. Stipules present or occasionally obsolete, lateral, free or fused, if fused then intrapetiolar. Leaves simple, alternate or opposite, always with variously shaped cystoliths in epidermal cells, these especially conspicuous in dried material. Inflorescences unisexual or bisexual, branched, often in pairs in leaf axils. Flowers unisexual (plants dioecious or monoecious); perianth uniseriate, the tepals 3–5. Staminate flowers actinomorphic; stamens 3–5, inflexed in bud, the pollen released explosively at anthesis. Pistillate flowers zygomorphic or actinomorphic; ovary free from perianth, 1-locular, the stigma 1; placentation basal, the ovule solitary. Fruits small achenes, enclosed or not enclosed by persistent perianth.

Berg, C. C. 1992. Urticaceae. *In* A. R. A. Görts-van Rijn (ed.), Flora of the Guianas Ser. A, **11:** 125–139, 192–222. Koeltz Scientific Books, Koenigstein.

——— & M. J. M. De Rooij. 1975. Urticaceae. *In* J. Lanjouw & A. L. Stoffers (eds.), Flora of Suriname **V(1):** 300–318. E. J. Brill, Leiden.

1. Treelets, shrubs, or subshrubs. . . . . *Urera.*
1. Herbs.
    2. Leaves opposite, without stinging or glandular hairs. . . . . *Pilea.*
    2. Leaves alternate, with stinging or glandular hairs.. . . . . *Laportea.*

LAPORTEA Gaudich.

Herbs, with stinging or glandular hairs. Stipules fused at base. Leaves alternate, spirally arranged to distichous; blades with pinnate venation. Inflorescences bisexual, ± paniculate, with flowers in loose glomerules. Flowers unisexual (plants monoecious). Staminate flowers: tepals (4)5; stamens (4)5. Pistillate flowers slightly zygomorphic; tepals 4, unequal, fused at base, with 2–5 stinging or glandular hairs on larger tepals; stigma tongue-shaped. Achenes not enclosed by perianth.

Laportea aestuans (L.) Chew FIG. 313

Herbs, to 2 m tall. Stipules 0.2–1.2 cm long, fused for ca. 1/2 length. Leaves: petioles usually 5–10 cm long; blades ovate, 5–20 × 4–17 cm, papyraceous, often reddish in color, the margins conspicuously dentate. Inflorescences to 25 cm long. Pistillate flowers with either stinging or glandular hairs on larger tepals. Achenes ca. 2 mm long. Fls (Aug, Sep, Nov, probably year round); common weed in disturbed areas, especially in village. *Zouti, zouti rouge* (Créole).

PILEA Lindl.

Herbs, without stinging or glandular hairs. Stipules fused, sometimes obsolete. Leaves opposite, one leaf at a node often smaller than other; blades with venation pinnate, subtripliveined, or single-veined. Inflorescences unisexual or bisexual, branched to nearly unbranched.

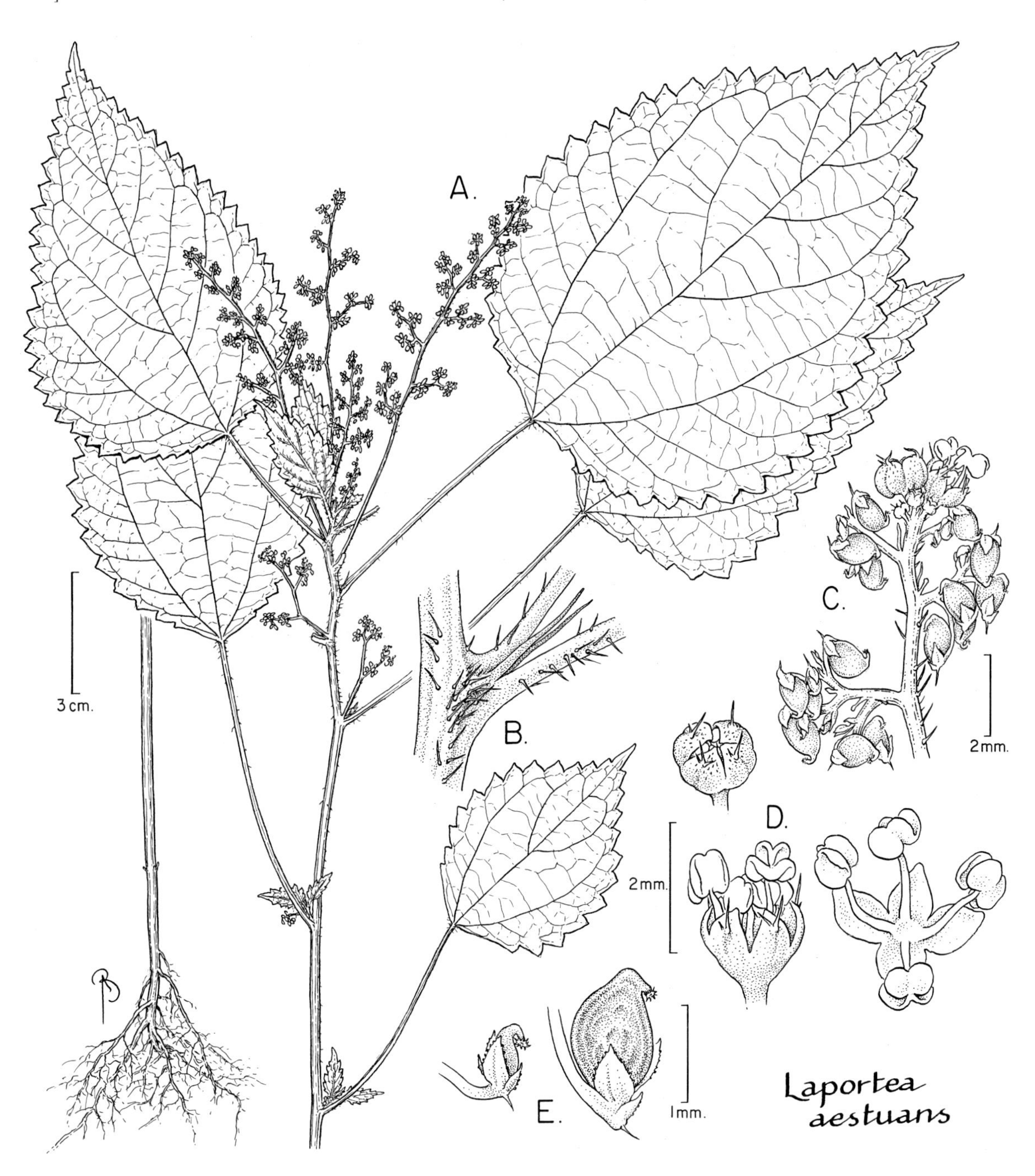

FIG. 313. URTICACEAE. *Laportea aestuans* (*Mori & Gracie 18657*). **A.** Apex of stem with leaves and inflorescences (below) and roots and base of stem (below left). **B.** Detail of stem showing stinging hairs. **C.** Infructescence with one flower still at anthesis. **D.** Apical view of flower bud (above), lateral (below left) and apical (below right) views of staminate flowers at anthesis. **E.** Immature (left) and mature (right) fruits in perianths.

Flowers unisexual (plants dioecious or monoecious). Staminate flowers: tepals (3)4; stamens (3)4. Pistillate flowers zygomorphic; tepals 3, basally fused, unequal in size, the middle one the largest; stigma capitate-penicillate. Achenes not enclosed by perianth.

1. Leaf pairs markedly unequal.
    2. Leaf blades subtripliveined to pinnately veined, to 50 mm long. . . . . *P. imparifolia.*
    2. Leaf blades single-veined, to 5 mm long. . . . . *P. microphylla.*
1. Leaf pairs ± equal. . . . . *P. pubescens.*

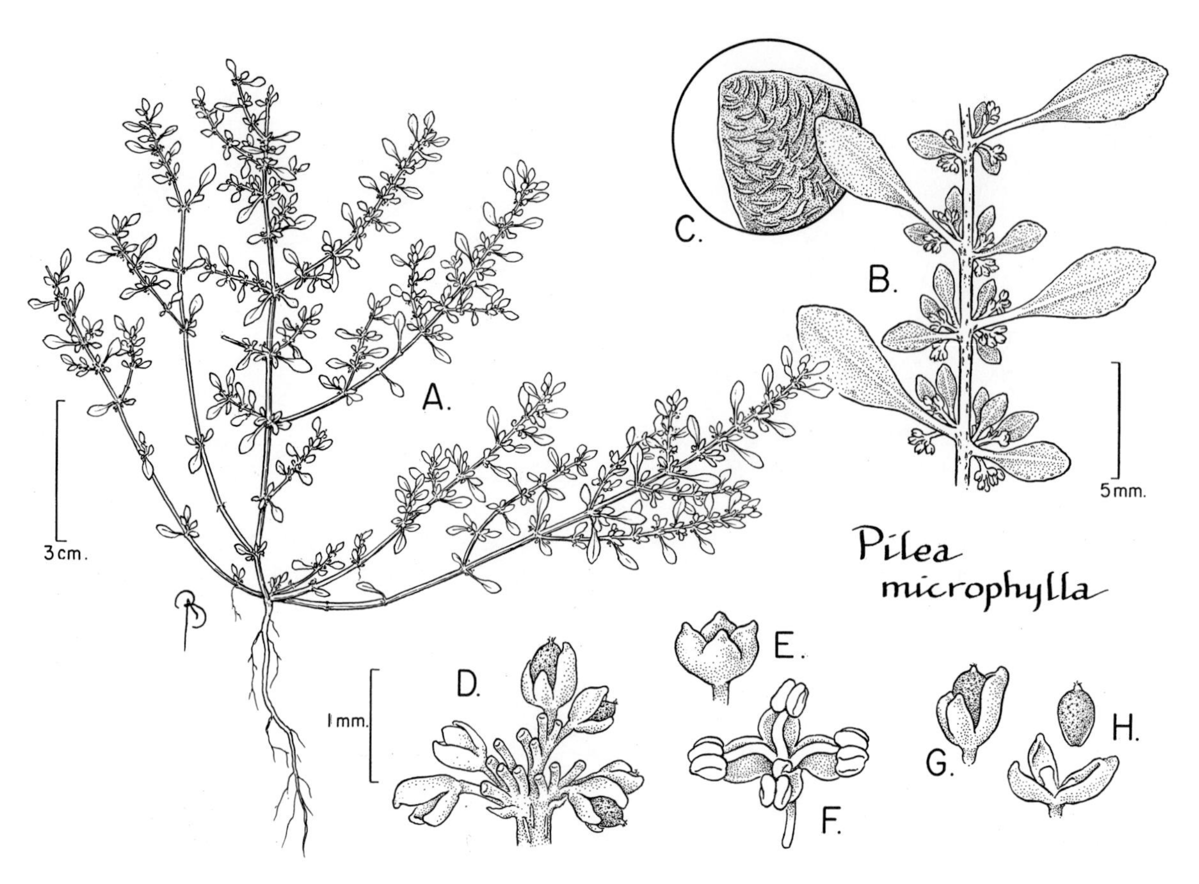

FIG. 314. URTICACEAE. *Pilea microphylla* (*Mori et al. 24155*). **A.** Plant. **B.** Detail of leafy stem with inflortescence. **C.** Detail of adaxial surface of dried leaf tip showing cytoliths. **D.** Infructescence. **E.** Lateral view of bud of staminate flower. **F.** Apical view of staminate flower. **G.** Lateral view of fruit and perianth. **H.** Lateral view of fruit (left) and perianth (below left).

Pilea imparifolia Wedd.

Herbs, with trailing stems and ascending branches. Stipules minute, 0.5–0.7 mm long, persistent. Leaves unequal at each node, one larger than other; blades oblong or obovate, 10–50 mm long, papyraceous, pubescent, the margins crenate-dentate or entire; venation conspicuous, subtripliveined to pinnate. Inflorescences hardly branched, with 2–8 flowers. Fl (Jan), in forest understory on rocks or on bases of tree trunks.

Pilea microphylla (L.) Liebm. FIG. 314, PL. 124d

Herbs, often prostrate, ascending stems to 30 cm tall. Stipules obsolete. Leaves unequal at each node, one larger than other; blades obovate to elliptic to oblong to subovate, 2–5 mm long, papyraceous, glabrous, the margins entire; venation inconspicuous, only midrib conspicuous. Inflorescences often subsessile, with to ca. 15 flowers. Fl (May, Aug, Sep, possibly year round), fr (Aug); common, weed in disturbed areas, especially in the village and around homesteads.

Pilea pubescens Liebm.

Herbs, to 30 cm tall, with creeping or ascending stems. Stipules 0.5–3 mm long, persistent. Leaves ± equal at each node; blades ovate to suborbicular, 5–65 × 3–40 mm, papyraceous, pubescent, the margins crenate to subentire; venation conspicuous, with secondary veins in 2–6 pairs. Inflorescences distinctly pedunculate, branched, to 3.5 cm long. Fl (Jan); in forest understory.

URERA Gaudich.

Treelets, shrubs, or subshrubs, with or without stinging hairs. Stipules fused. Leaves alternate, spirally arranged to distichous; blades with venation pinnate. Inflorescences paniculate, with flowers solitary or in glomerules. Flowers unisexual (plants dioecious). Staminate flowers: tepals 4–5. Pistillate flowers actinomorphic to zygomorphic; tepals 4, basally fused, equal or unequal; stigma capitate-penicillate. Achenes partly enclosed by whitish, juicy fruiting perianth.

1. Plants without stinging hairs and prickles. Leaf blades not lobed, with crenate-dentate margins. . . . . . . *U. caracasana.*
1. Plants with stinging hairs and prickles. Leaf blades pinnately lobed, the margins entire. . . . . *U. laciniata.*

Urera caracasana (Jacq.) Griseb. PL. 124e

Shrubs or treelets, to 5(10) m tall, unarmed. Stipules 0.5–1.5 cm long, caducous. Leaves: petioles 5–17 cm long; blades ovate to widely ovate, not lobed, ca. 10–35 × 8–25 cm, pubescent, the adaxial surface ± scabrous, the base subcordate to cordate, the margins crenate-dentate; secondary veins in 5–9 pairs. Inflorescences in leaf axils or below leaves, to 20 cm long. Fl (Feb, Jun, Sep, Dec, possibly year round), fr (Sep); in secondary growth and forest gaps. *Cansanção* (Portuguese), *zouti montagne* (Créole).

Urera laciniata Wedd.

Subshrubs or shrubs, to 4 m tall, armed with stinging hairs and prickles on stems, petioles, and leaf blades. Stipules 0.3–0.5 cm long. Leaves: petioles 11–15 cm long; blades widely ovate, pinnately lobed, ca. 10–25 × 10–25 cm, pubescent, the adaxial surface almost smooth, the base truncate to cordate, the margins not crenate-dentate; secondary veins in 5–7 pairs. Inflorescences in leaf axils, to 20 cm long. Known only by a single sterile specimen from our area.

VERBENACEAE (Verbena Family)

Marion J. Jansen-Jacobs

Herbs, shrubs, trees, or lianas. Stipules absent. Leaves simple or palmately compound (*Vitex*), mostly opposite, sometimes whorled. Flowers homostylous or heterostylous, ± zygomorphic, sometimes actinomorphic (*Aegiphila*), 4–5-merous; calyx tubular-campanulate, persistent, often enlarged in fruit; corolla with slender tube abruptly expanded into flat limb to funnel-shaped, mostly weakly 2-lipped; stamens 4(5), often didynamous, or 2 (*Stachytarpheta*); ovary superior, 2–4-locular, each locule with 1 ovule, the style 1, the stigma capitate or bifid. Fruits drupaceous with 1–4 stones, fleshy with enlarged calyx (*Aegiphila, Citharexylum, Vitex*) or calyx not enlarged (*Lantana*), or drupaceous and enclosed in calyx with winged lobes (*Petrea*), or dry schizocarps enclosed in calyx (*Lippia, Priva, Stachytarpheta*).

Jansen-Jacobs, M. J. 1988. Verbenaceae. *In* A. R. A. Görts-van Rijn (ed.), Flora of the Guianas Ser. A, **4:** 1–114. Koeltz Scientific Books, Koenigstein.

Moldenke, H. N. 1940. Verbenaceae. *In* A. Pulle (ed.), Flora of Suriname **IV(2):** 257–321. J. H. De Bussy, Amsterdam.

1. Leaves palmately compound. . . . . *Vitex*.
1. Leaves simple.
   2. Leaf blades with entire margins.
      3. Trees. Leaves with cup-shaped glands at apex of petiole. . . . . *Citharexylum*.
      3. Small trees, shrubs, or lianas. Leaves without glands at apex of petiole.
         4. Stems not scabrous, not lenticellate or with few lenticels. Leaf blades with scattered, disc-shaped glands abaxially. Inflorescences cymes or terminal panicles. Flowers 4-merous. Fruiting calyx cup-shaped. . . . . *Aegiphila*.
         4. Stems scabrous, lenticellate. Leaf blades without glands. Inflorescences racemes. Flowers 5-merous. Fruiting calyx with 5 wings. . . . . *Petrea*.
   2. Leaf blades with dentate or serrate margins.
      5. Inflorescences capitate racemes. Fruits fleshy drupes or dry schizocarps enclosed in calyx.
         6. Stems usually with small spines present. Fruits fleshy drupes, not enclosed in calyx. . . . . *Lantana*.
         6. Stems with small spines absent. Fruits dry schizocarps, enclosed in calyx. . . . . *Lippia*.
      5. Inflorescences in long racemes. Fruits dry schizocarps, enclosed in calyx.
         7. Stems with hirsute hairs. Flowers half-immersed in pits on inflorescence rachis. Fruiting calyx not inflated, without uncinate hairs, not sticking to animal fur or clothing. . . . . *Stachytarpheta*.
         7. Stems with uncinate hairs. Flowers not immersed in pits on inflorescence rachis. Fruiting calyx inflated, with uncinate hairs, sticking to animal fur or clothing. . . . . *Priva*.

AEGIPHILA Jacq.

Shrubs, small trees, or lianas. Leaves simple, decussate; blades with some scattered disc-shaped glands on abaxial surface, the margins entire. Inflorescences axillary cymes or terminal panicles. Flowers heterostylous, actinomorphic, ca. 1 cm long, 4-merous, subglabrous, greenish white to yellow; stamens 4; ovary 4-locular, the style bifurcate. Fruits drupaceous, with 1–4 stones, the exocarp fleshy, the fruiting calyx enlarged, cup-shaped.

1. Flowers in axillary cymes.
   2. Leaf blades strigose abaxially. . . . . *A. integrifolia*.
   2. Leaf blades densely villous or lanate abaxially. . . . . *A. villosa*.
1. Flowers in terminal panicles.

3. Leaf blades membranous, the base cuneate or acute. . . . . *A. membranacea.*
3. Leaf blades coriaceous, the base rounded or broadly cuneate.
    4. Leaf blades pubescent abaxially. . . . . *A. racemosa.*
    4. Leaf blades subglabrous abaxially. . . . . *A. laevis.*

Aegiphila integrifolia (Jacq.) B. D. Jacks.

Shrubs to small trees, to 8 m tall. Stems pithy, tetragonal, densely strigose-sericeous. Leaves at same node often unequal in size; blades 7–30 × 3–12 cm, dark-punctate, strigose, the base attenuate, decurrent onto petiole; secondary veins in 10–12 pairs. Inflorescences axillary, many-flowered cymes; peduncle 2–5 cm long. Fruits globose, about 0.8 cm long, orange-red. Fl (Mar), fr (Jul); in secondary forest.

Aegiphila laevis (Aubl.) J. F. Gmel.

Shrubs, to 4 m tall, or lianas. Stems subterete, minutely pubescent. Leaves at same node equal in size; blades 10–16 × 4–8 cm, coriaceous, dark punctate and subglabrous, the base rounded or broadly cuneate; secondary veins in 7–8 pairs. Inflorescences terminal, many-flowered panicles; peduncle 1.5–3 cm long. Fruits depressed globose, 1–1.2 cm long, yellow-orange. Fr (Jun); in secondary forest and at forest margins.

Aegiphila membranacea Turcz.

Shrubs to small trees, to 7 m tall, or lianas. Stems subterete, with erect hairs of different lengths. Leaves at same node equal in size; blades 8–22 × 3–7 cm, membranous, dark punctate, subglabrous, the base cuneate or acute; secondary veins in 6–9 pairs. Inflorescences terminal, many-flowered panicles; peduncle 2–3.5 cm long. Fruits oblong-ellipsoid, 0.6–1 cm long, green to yellow-orange. Fl (Jan, Jun), fr (Mar, May, Jun); in secondary forest.

Aegiphila racemosa Vell.

Shrubs to small trees, to 4 m tall, or lianas. Stems subterete, densely appressed pubescent. Leaves at same node equal in size; blades 6–15 × 3–6 cm, coriaceous, pubescent, the base rounded or broadly cuneate; secondary veins in 6–8 pairs. Inflorescences terminal, many-flowered panicles; peduncle 3–5 cm long. Fruits ellipsoid, 0.7–1.2 cm long, green to orange. Fr (May); in secondary forest.

Aegiphila villosa (Aubl.) J. F. Gmel. FIG. 315, PL. 125a

Shrubs to small trees, to 7 m tall. Stems pithy, tetragonal, densely yellowish-white villous or lanate. Leaves at same node equal in size; blades 15–50 × 5–20 cm, densely villous or lanate, the base attenuate; secondary veins in 10–14 pairs. Inflorescences axillary, many-flowered cymes; peduncle 1–2 cm long. Fruits globose, ca. 0.8 cm long, green to yellow . Fl (Feb, Mar, May, Jun, Jul, Sep), fr (Jun, Sep); in secondary forest. *Bois calou*, *bois tabac*, *feuille tabac*, *tabac sauvage* (Créole).

CITHAREXYLUM L.

Trees or shrubs. Leaves simple, opposite or in whorls of 3; petioles with one or more pairs of glands on petiole or base of blade, the margins entire. Inflorescences axillary racemes or terminal panicles of racemes. Flowers homostylous, ± zygomorphic, 1–1.5 cm long, 5-merous, glabrous, hairy in throat, white or yellow; stamens 5; ovary 4-locular, the stigma capitate. Fruits drupaceous, with 2 stones, the exocarp fleshy, the fruiting calyx enlarged and funnel-shaped.

Citharexylum macrophyllum Poir. FIG. 316; PART 1: FIG. 4

Trees, to 20 m tall. Stems hollow, subglabrous. Leaves mostly in whorls of 3; petioles with pair of large, cup-shaped glands at apex; blades 10–25 × 4–11 cm, subglabrous; secondary veins in 7–13 pairs. Inflorescences terminal, panicles, to 30 cm long. Fruits obovoid, 1–1.5 cm long, green to orange-red. Fl (Feb, Nov), fr (Jun, Jul); in secondary forest. *Bois-côtelette* (Créole).

LANTANA L.

Shrubs or herbs. Leaves simple, opposite or in whorls of three; blades with dentate margins. Inflorescences axillary, capitate racemes. Flowers homostylous, somewhat zygomorphic, 4–5-merous; stamens 4; ovary 2-locular, the stigma capitate. Fruits fleshy drupes, with 1–2 stones, the fruiting calyx not enlarged.

Lantana camara L.

Shrubs, to 3 m tall, mostly armed with small spines, aromatic. Stems tetragonal, sparsely hairy, the pubescence of glandular globules and hairs. Leaves: blades 3–10 × 2–6 cm, strigose, with glandular globules; secondary veins in 5–8 pairs. Inflorescences axillary capitules; peduncle 2–8 cm long, the bracts ca. 0.5 cm long. Flowers: corolla ca. 1 cm long, variously colored, white to pink or orange to red. Fruits subglobose, 0.3–0.5 cm long, blue or black. Fl (Feb, Apr, Jun, Aug, Oct, Nov), fr (Aug); common, in secondary vegetation, especially along roads and wider trails. *Cambara de cheiro* (Portuguese), *corbeille d'or* (Créole), *erva chumbinho* (Portuguese), *herbe des putains*, *marie-crabe*, *thé indien*, *vervein*, *zerb des putains* (Créole).

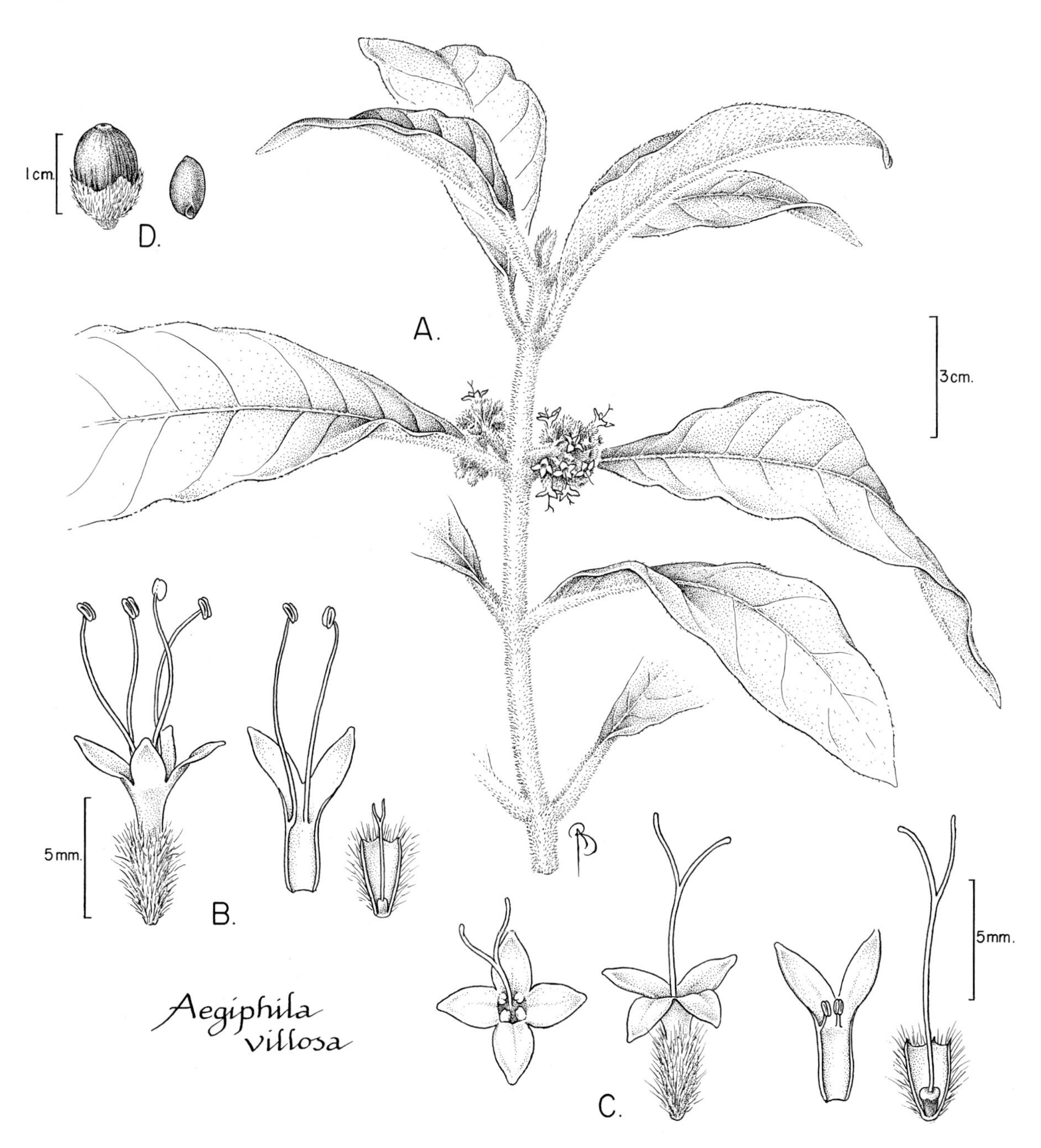

FIG. 315. VERBENACEAE. *Aegiphila villosa* (A–C, *Mori et al. 22175*; D, *Mori & Gracie 19014*). **A.** Apex of stem showing leaves and axillary inflorescences. **B.** Lateral view of short-styled flower (left) and medial sections of corolla (middle) and calyx (right) showing ovary. **C.** Apical (far left) and lateral (middle left) views of long-styled flowers and medial sections of corolla (middle right) and calyx with pistil (far right). **D.** Fruit (left) with persistent calyx and seed (right).

LIPPIA L.

Shrubs or herbs. Leaves simple, opposite or in whorls of three; blades with dentate margins. Inflorescences axillary, capitate racemes. Flowers homostylous, somewhat zygomorphic, 4–5-merous; stamens 4; ovary 2-locular, the stigma capitate. Fruits dry, 2-parted schizocarps, the fruiting calyx enclosing fruit.

Lippia alba (Mill.) N. E. Br.

Herbs or small shrubs, to 2 m tall, aromatic. Stems somewhat tetragonal, sparsely hairy, the hairs mixed with glandular globules. Leaves: blades 2–7 × 1–4 cm, strigose, with glandular globules; secondary veins in 6–7 pairs. Inflorescences axillary capitules; peduncle 0.5–1 cm long, the bracts ca. 0.4 cm long. Flowers: corolla ca. 0.5 cm long, pink, blue, or white. Fruits subglobose, 0.1–0.2 cm long, pale brown. Fl (May), fr (May); in open areas, cultivated as a spice or for medicinal use. *Erva cidreira* (Portuguese), *mélisse de calme* (Créole), *piepiepo* (Portuguese).

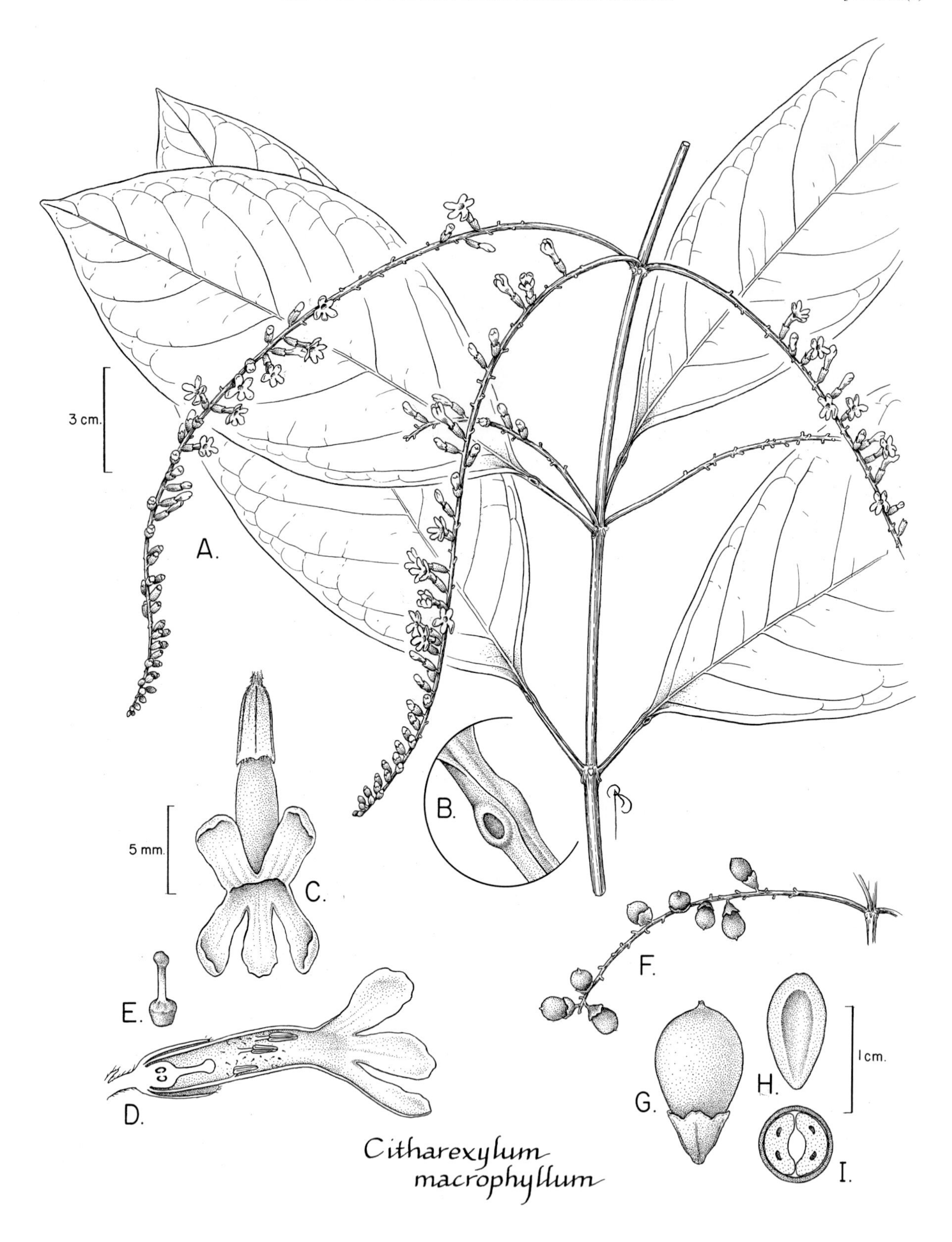

FIG. 316. VERBENACEAE. *Citharexylum macrophyllum* (A–E, *Mori et al. 8236* from Surinam; F–I, *Mori & Gracie 18912*). **A.** Part of stem showing leaves and axillary inflorescences. **B.** Detail of gland at apex of petiole. **C.** Oblique view of flower. **D.** Medial section of flower. **E.** Pistil. **F.** Part of stem and infructescence. **G.** Fruit with persistent calyx. **H.** Stone showing concave interior surface. **I.** Transverse section of fruit showing two stones.

Plates 121–128

Plate 121. SOLANACEAE. **a.** *Solanum morii* (*Mori & Gracie 23866*), counter-shaded fruits. **b.** *Physalis pubescens* (*Mori & Gracie 23741*), flower and immature fruit surrounded by inflated calyx. **c.** *Solanum rugosum* (*Mori & Gracie 19007*), inflorescence. **d.** *Solanum rugosum* (*Mori & Gracie 19007*), flowers and buds.

Plate 122. STERCULIACEAE. **a.** *Byttneria morii* (*Mori & Gracie 19160*), flowers; note ant-mimicking outgrowths. [Photo by S. Mori] **b.** *Sterculia frondosa* (*Mori et al. 23117*), flower. **c.** *Theobroma velutinum* (*Mori et al. 24183*), flowers. STYRACACEAE. **d.** *Styrax pallidus* (*Mori & Gracie 18925*), flowers and bud. THEOPHRASTACEAE. **e.** *Clavija lancifolia* subsp. *chermontiana* (*Mori & Ek 20731*), flowers and bud. [Photo by S. Mori]

Plate 123. TILIACEAE. **a.** *Apeiba albiflora* (*Mori et al. 23986*), fruit. **b.** *Apeiba glabra* (*Mori & Gracie 18922*), flower. TRIGONIACEAE. **c.** *Trigonia laevis* var. *microcarpa* (*Mori et al. 21997*), flowers and immature fruit. TURNERACEAE. **d.** *Turnera glaziovii* (*Mori et al. 22774*), flower.

Plate 124. ULMACEAE. **a.** *Ampelocera edentula* (*Mori et al. 23970*), inflorescence. **b.** *Trema micrantha* (*Mori et al. 23290*), part of stem, flowers, and immature fruits; note asymmetric leaf bases. **c.** *Trema micrantha* (*Mori et al. 23290*), flower and immature fruit. URTICACEAE. **d.** *Pilea microphylla* (*Mori & Gracie 23959*), part of stem. **e.** *Urera caracasana* (*Mori & Gracie 24225*), inflorescence.

Plate 125. VERBENACEAE. **a.** *Aegiphila villosa* (*Mori et al. 22175*), flowers. **b.** *Petrea bracteata* (*Mori et al. 20804*), flowers and bud; note corolla-like calyx. **c.** *Vitex triflora* (*Mori et al. 23160*), flower. **d.** *Stachytarpheta cayennensis* (*Mori & Gracie 18743*), flowers. VIOLACEAE. **e.** *Noisettia orchidiflora* (unvouchered), flowers. **f.** *Rinorea amapensis* (*Mori et al. 20803*), flowers.

Plate 126. VISCACEAE. **a.** *Phoradendron crassifolium* (*Mori et al. 21678*), inflorescences. **b.** *Phoradendron crassifolium* (*Mori et al. 22778*), flowers. **c.** *Phoradendron racemosum* (*Mori et al. 23287*), inflorescence. **d.** *Phoradendron racemosum* (*Mori et al. 23287*), flowers. VITACEAE. **e.** *Cissus verticillata* (*Mori et al. 23066*), flowers and buds.

Plate 127. VOCHYSIACEAE. **a.** *Erisma uncinatum* (*Mori et al. 21585*), flowers; note single style and stamen. **b.** *Qualea mori-boomii* (*Mori et al. 21643a*), flower. **c.** *Qualea rosea* (*Mori et al. 24682*), flower; note single style and stamen. [Photo by S. Mori]

Plate 128. VOCHYSIACEAE. **a.** *Ruizterania albiflora* (*Mori et al. 21011*), flower; note single style and stamen. **b.** *Qualea rosea* (*Mori & Gracie 18600*), node showing pair of glands and abaxial leaf bases. **c.** *Vochysia surinamensis* (*Mori et al. 23844*), immature fruits.

SOLANACEAE (continued)

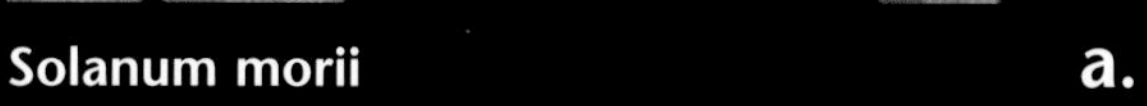

Solanum morii **a.**

Physalis pubescens

b.

Solanum rugosum **c.**

Solanum rugosum **d.**

STERCULIACEAE

Byttneria morii a.

Sterculia frondosa b.

Theobroma velutinum c.

STYRACACEAE

d. Styrax pallidus

THEOPHRASTACEAE

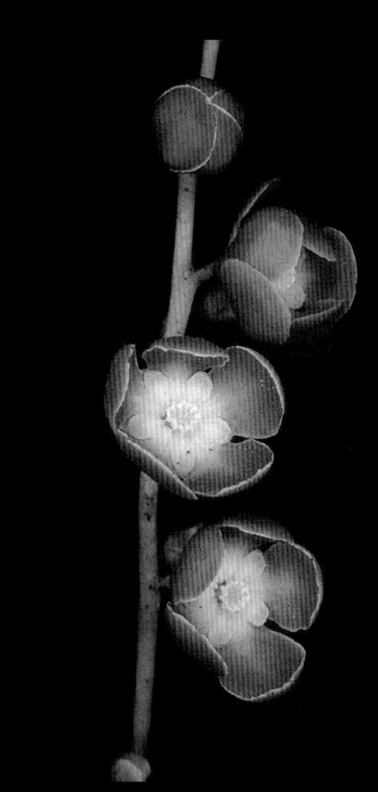

e. Clavija lancifolia subsp. chermontiana

Plate 122

TILIACEAE

Apeiba albiflora a.

Apeiba glabra b.

TRIGONIACEAE

Trigonia laevis var. microcarpa c.

TURNERACEAE

Turnera glaziovii d.

Ampelocera edentula **a.**

b. Trema micrantha

c.

URTICACEAE

Pilea microphylla **d.**

Urera caracasana **e.**

Aegiphila villosa a.

Petrea bracteata b.

Vitex triflora c.

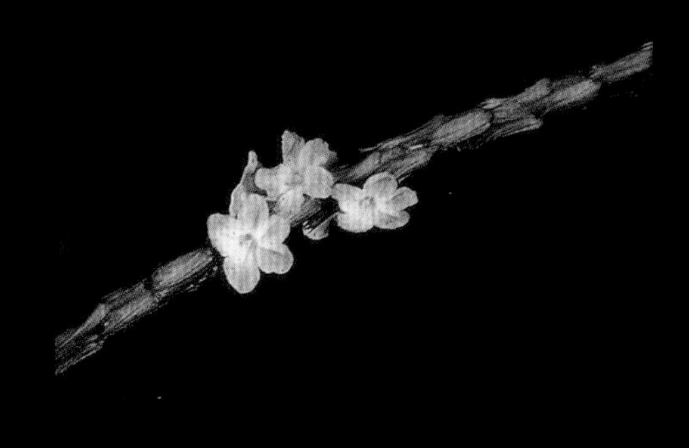

Stachytarpheta cayennensis d.

VIOLACEAE

Noisettia orchidiflora e.

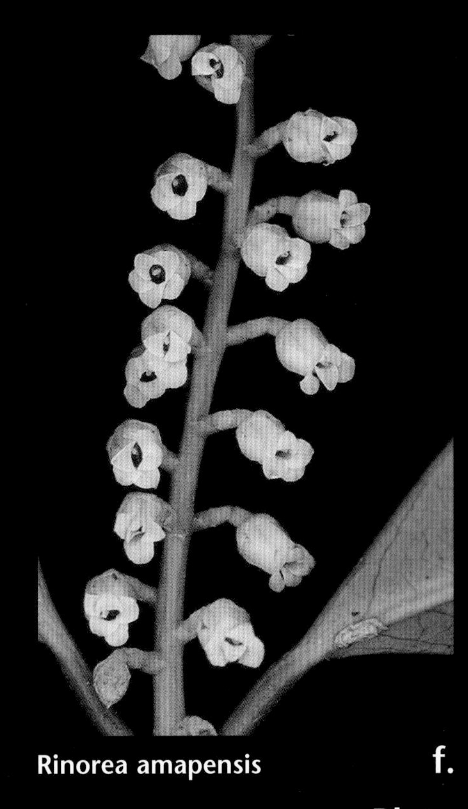

Rinorea amapensis f.

VISCACEAE

Phoradendron crassifolium a.

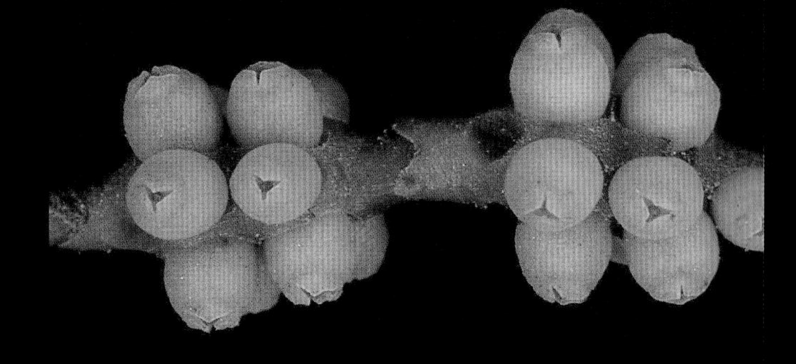

Phoradendron crassifolium b.

Phoradendron racemosum c.

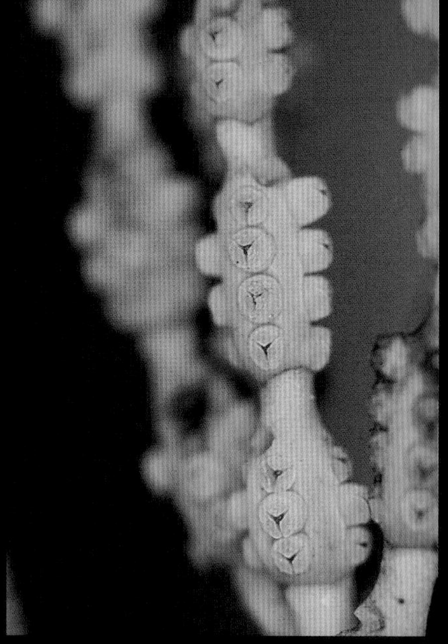

Phoradendron racemosum d.

VITACEAE

Cissus verticillata e.

Plate 126

Erisma uncinatum a.

Qualea mori-boomii b.

Qualea rosea c.

Ruizterania albiflora a.

Qualea rosea b.

Plate 128 Vochysia surinamensis c.

PETREA L.

Lianas. Leaves opposite, simple; blades with entire margins. Inflorescences axillary or terminal, racemes. Flowers mostly homostylous, slightly zygomorphic, 5-merous; stamens 4; ovary 2-locular, the stigma capitate. Fruits drupaceous, with 2 stones, enclosed in calyx tube, the lobes forming wings on fruit.

1. Leaf blades sparsely hirsute on veins abaxially, bullate, the veins impressed adaxially.. . . . . *P. bracteata.*
1. Leaf blades glabrous, not bullate, the veins prominulous adaxially. . . . . *P. volubilis.*

Petrea bracteata Steud. PL. 125b

Lianas. Stems subterete, scabrous, lenticellate. Leaves: blades 10–26 × 3–9 cm, bullate, scabrous, subglabrous adaxially, sparsely hirsute on veins abaxially; secondary veins in 8–13 pairs, the veins impressed adaxially. Inflorescences racemes, to 50 cm long. Flowers: calyx lavender, 2–3.5 cm long in fruit, the lobes 1.5–2.5 cm long in fruit; corolla 2–2.5 cm long, deeper lavender with purplish lines. Fl (Sep); in forest. *Liane gris* (Créole).

Petrea volubilis L. [Syn.: *P. kohautiana* C. Presl]

Lianas. Stems subterete, scabrous, distinctly lenticellate. Leaves: blades 5–22 × 3–11 cm, not bullate, somewhat scabrous, glabrous; secondary veins in 7–12 pairs, the veins prominulous adaxially. Inflorescences racemes, to 50 cm long. Flowers: calyx violet or blue, 2.5–3 cm long in fruit, the lobes 1.5–2.5 cm long; corolla 1.5–2 cm long, violet or blue. Fl (Feb); at forest margins and along rivers. *Liane gris* (Créole).

Petrea kohautina was considered a distinct species in the Flora of the Guianas (Jansen-Jacobs, 1988).

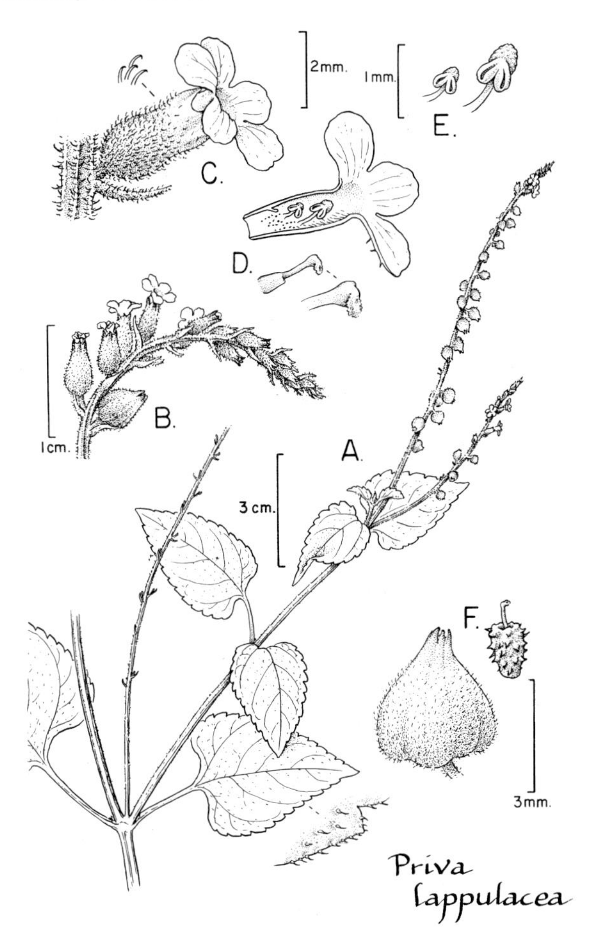

FIG. 317. VERBENACEAE. *Priva lappulacea* (*Acevedo-Rodríguez 2825* from St. John, U.S. Virgin Islands). **A.** Part of stem with leaves and inflorescences (right) and detail of abaxial surface leaf margin (below). **B.** Part of inflorescence. **C.** Lateral view of flower with subtending bracteole (near above) and detail of indument (far above). **D.** Medial section of corolla showing adnate stamens (above), pistil (near right), and detail of stigma (far right). **E.** Dimorphic stamens. **F.** Inflated calyx containing fruit (below left) and immature fruit (right). (Adapted with permission from P. Acevedo-Rodríguez, *Flora of St. John, U.S. Virgin Islands*, Mem. New York Bot. Gard. 78. 1996).

PRIVA Adans.

Herbs. Leaves opposite, simple; blades with margins dentate-serrate. Inflorescences terminal, racemes. Flowers homostylous, somewhat zygomorphic, 5-merous; stamens 4; ovary 4-locular, the stigma unequally 2-lobed. Fruits dry 2-parted schizocarps, the exocarp echinate or ridged, the fruiting calyx inflated, enclosing fruit.

Priva lappulacea (L.) Pers. FIG. 317; PART 1: FIG. 8

Herbs, to 1 m tall. Stems tetragonal, sparsely covered with uncinate hairs. Leaves: blades 2–9 × 1–6 cm, sparsely strigose; secondary veins in 3–5 pairs. Inflorescences racemes, 5–15 cm long. Flowers: calyx in fruit ovoid, inflated, 0.5–0.7 cm long, covered with uncinate hairs, green; corolla ca. 0.5 cm long, white to pale violet. Fruits ornamented with ridges and spines. Fl (Aug), fr (Aug); in weedy places, the fruiting calyx with included fruit sticks to animal fur or human clothing and is thereby dispersed. *Vingt-quatre heures* (Créole).

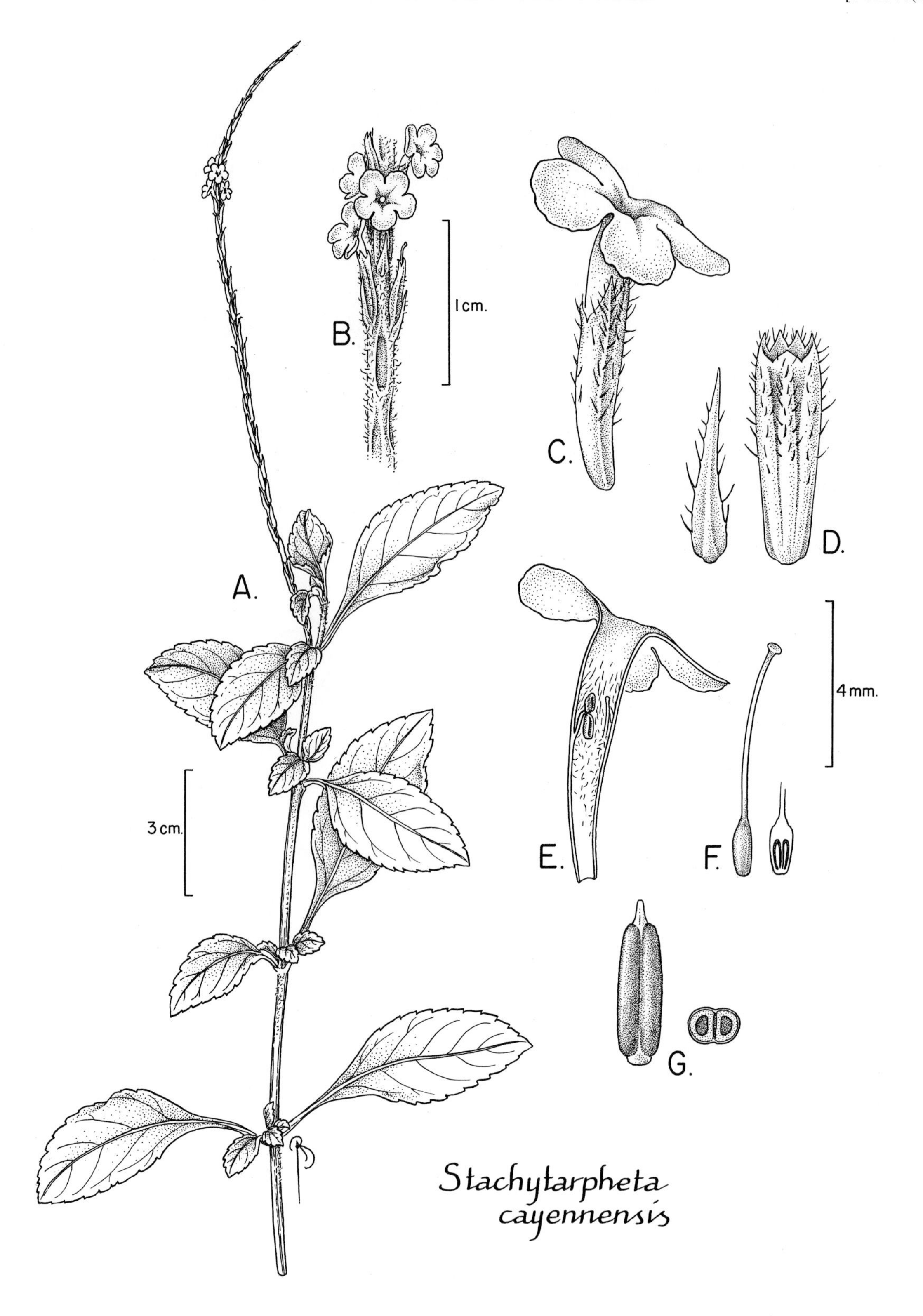

FIG. 318. VERBENACEAE. *Stachytarpheta cayennensis* (A–F, *Mori et al. 21059*; G, *Mori & Gracie 18743*). **A.** Apex of stem showing leaves and axillary inflorescence. **B.** Part of inflorescence showing half-immersed flowers in pits of rachis. **C.** Flower. **D.** Bract (left) and calyx (right). **E.** Medial section of corolla and stamen showing divaricate thecae on left side of corolla tube and staminode on right side. **F.** Lateral view (left) and medial section (right) of part of pistil. **G.** Lateral view (left) and transverse section (right) of fruit.

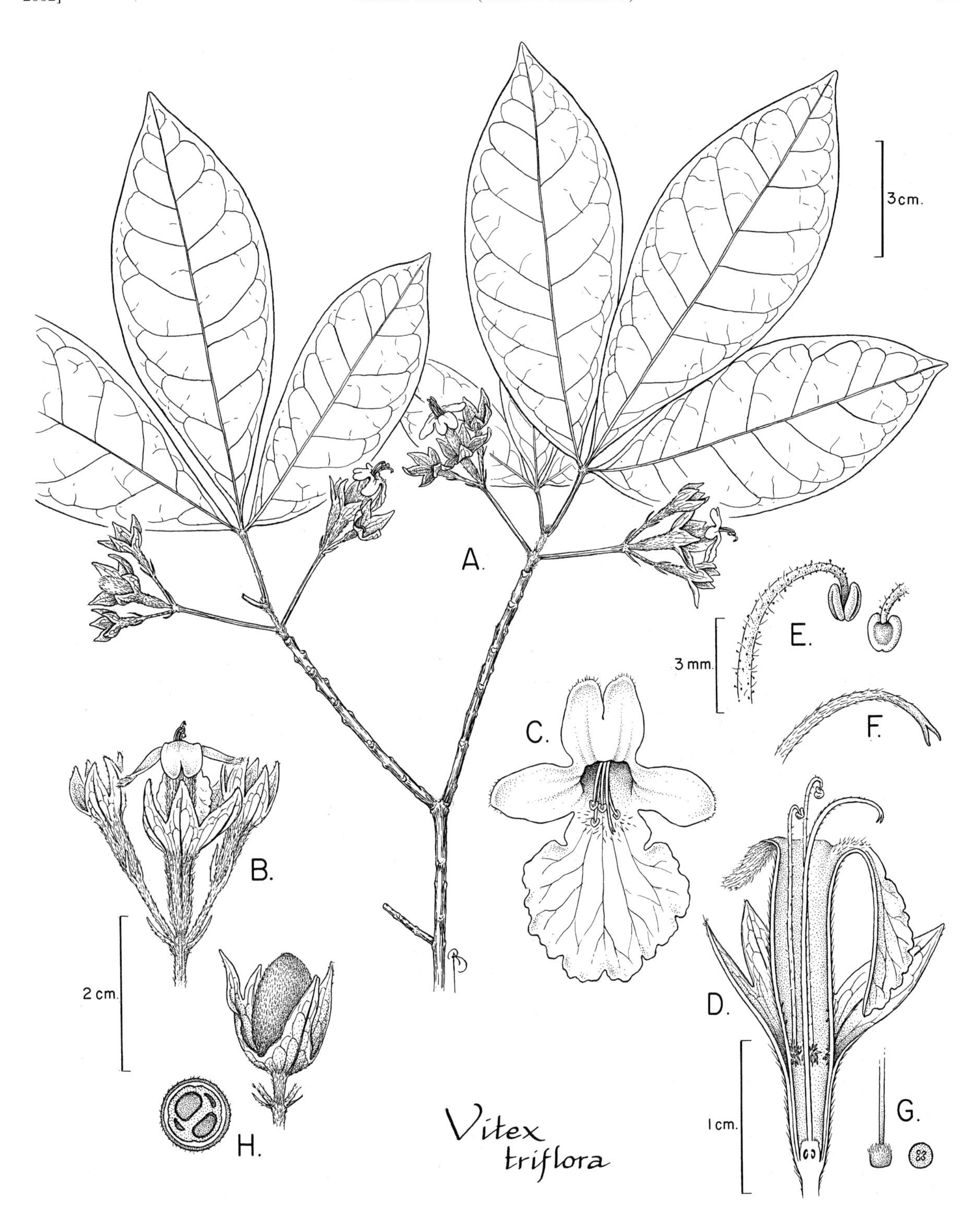

FIG. 319. VERBENACEAE. *Vitex triflora* (A–G, *Mori et al. 19194*; H, *Cowan 38820*). **A.** Apices of stems showing 3-foliolate leaves and inflorescences. **B.** Apex of inflorescence showing terminal flower at anthesis. **C.** Apical view of flower. **D.** Medial section of flower. **E.** Apices of stamens. **F.** Apex of pistil. **G.** Base of pistil (left) and transverse section (right) of ovary. **H.** Whole (right) and transverse section (left) of fruit.

STACHYTARPHETA Vahl

Herbs or shrubs. Stems pithy, somewhat tetragonal. Leaves opposite, simple; blades with dentate margins. Inflorescences terminal, racemes, the flowers sessile, half-immersed in pits on rachis. Flowers homostylous, ± zygomorphic, 5-merous; stamens 2, the anthers with divaricate thecae; staminodes 2; ovary 2-locular, the stigma capitate. Fruits dry, 2-parted schizocarps, ± smooth, enclosed in calyx.

1. Pits in inflorescence as wide as rachis, the rachis 1–2 mm thick; bracts 3–5 × 1 mm. . . . . *S. cayennensis.*
1. Pits in inflorescence narrower than rachis, the rachis 2–4 mm thick; bracts 5 × 2 mm. . . . . *S. jamaicensis.*

Stachytarpheta cayennensis (Rich.) Vahl FIG. 318, PL. 125d

Herbs or small shrubs. Stems sparsely hirsute. Leaves: blades 2–10 × 1.5–4.5 cm, sparsely strigose; secondary veins in 4–6 pairs. Inflorescences racemes, to 30 cm long, the rachis 1–2 mm thick, the pits as wide as rachis; bracts 3–5 × 1 mm. Flowers: corolla ca. 0.7 cm long, pale blue, pale mauve, or lilac, with a white throat. Fruit oblongoid, 0.3 cm long. Fl (Feb, Mar, Jun, Aug, Sep, Nov), fr (Jun, Aug, Sep); in weedy places. *Rinchão* (Portuguese), *verveine, verveine queue de rat* (Créole).

Stachytarpheta jamaicensis (L.) Vahl

Herbs or small shrubs. Stems sparsely hirsute. Leaves: blades 3–10 × 1.5–4.5 cm, sparsely strigillose; secondary veins in 4–6 pairs. Inflorescences racemes, to 50 cm long, the rachis 2–4 mm thick, the pits narrower than rachis; bracts 5 × 2 mm. Flowers: corolla ca. 1 cm long, blue or deep violet, with a white throat. Fruits oblongoid, 0.3 cm long. Fl (Feb, Jun, Nov); in weedy places. *Gros verveine, verveine* (Créole).

VITEX L.

Trees or shrubs. Leaves decussate, palmately compound, 3–7-foliolate; leaflet blades with entire margins. Inflorescences axillary cymes or terminal panicles. Flowers homostylous, zygomorphic, 5-merous; stamens 4; ovary 4-locular, the style shortly bifid. Fruits drupaceous, with 1–4 stones, the exocarp fleshy, the fruiting calyx somewhat enlarged.

1. Leaves 5-foliolate, villous. Corolla ca. 1.5 cm long. . . . . *V. guianensis.*
1. Leaves 3-foliolate, subglabrous. Corolla 2.5–3 cm long. . . . . *V. triflora.*

Vitex guianensis Moldenke

Canopy trees, to 30 m tall. Trunks ± buttressed or fluted at base. Stems yellowish-brown villous when young. Leaves 5-foliolate; leaflet blades 5–16 × 2–7 cm, villous; secondary veins in 9–14 pairs. Inflorescences axillary, cymes; peduncle 0.5–1.5 cm long. Flowers: corolla ca. 1.5 cm long, blue or purple with whitish lobes. Fruits subglobose, 2 cm diam., sericeous. Fr (Jun); rare, in forest.

Vitex triflora Vahl FIG. 319, PL. 125c

Understory tree, to 15 m tall. Trunks without buttresses. Stems pithy, golden-brown pubescent when young. Leaves 3-foliolate; leaflet blades 6–30 × 2.5–9 cm, subglabrous; secondary veins in 7–12 pairs. Inflorescences axillary, cymes; peduncle 1.5–6 cm long. Flowers: corolla 2.5–3 cm long, blue or purple with paler upper lip and darker colored lower lip. Fruits ovoid-subglobose, 1.5–2 cm diam., pubescent, green or yellow when young, turning to red or dark purple. Fl (May, Aug, Oct, Nov), fr (Aug, Oct); common, in nonflooded forest.

VIOLACEAE (Violet Family)

Willem H. A. Hekking and John D. Mitchell

Trees or treelets, subshrubs, or herbs. Stipules persistent or deciduous. Leaves simple, alternate or opposite, distichous or polystichous; domatia present or absent. Inflorescences terminal or axillary, fasciculate, thyrsoid, (pseudo)racemose, cymose, mono- or dichasial, or 1–2-flowered. Flowers strictly actinomorphic (*Rinorea, Gloeospermum, Leonia*), subactinomorphic (*Paypayrola*), or strictly zygomorphic (*Hybanthus, Noisettia*), usually bisexual; sepals 5, usually free; petals 5, free; stamens (3)5, the filaments free or ± connate and forming a tube, the anthers usually introrse, usually with abaxial glands or spur-like nectaries, the connective often prolonged into scale-like appendage; ovary superior, sessile, usually 3(5)-locular, the style solitary, the stigma simple or lobed; placentation parietal, the ovules 1-numerous per locule, anatropous, bitegmic, crassinucellar. Fruits capsular, dehiscing into 3 valves or sometimes indehiscent, globose, baccate (*Gloeospermum*) or nut-like (*Leonia*). Seeds 3 to numerous, (sub)globose or pyriform.

Hekking, W. H. A. 1988. Violaceae. Part I -*Rinorea* and *Rinoreocarpus*. Fl. Neotrop. Monogr. **46:** 1–207.

1. Trees or treelets. Flowers actinomorphic or subactinomorphic, i.e., all petals equal in length. Fruits capsular or not capsular (indehiscent and globose).
    2. Flowers subactinomorphic, i.e., the innermost petal with emarginate apex, all other petals ligulate. Fruits capsular. . . . . *Paypayrola.*
    2. Flowers strictly actinomorphic, i.e., all petals ligulate. Fruits capsular or not capsular (indehiscent and globose).
        3. Fruits dehiscent (capsular). Seeds not mucilaginous, with a caruncle and an areola. . . . . *Rinorea.*
        3. Fruits indehiscent (baccate or nut-like). Seeds mucilaginous, without a caruncle or an areola.
            4. Leaves polystichous. Inflorescences >5 cm long. Petals yellow; anther connective not scale-like. Pericarp somewhat woody. . . . . *Leonia.*
            4. Leaves distichous. Inflorescences <5 cm long. Petals white; anther connective scale-like. Pericarp usually coriaceous. . . . . *Gloeospermum.*

1. Herbs or subshrubs. Flowers zygomorphic, i.e., the lower petal longer than other petals. Fruits capsular.
    5. Leaves alternate; petioles conspicuous. Petals orange, pink, or vermillion, the lower petal spurred. . . . . . *Noisettia.*
    5. Leaves opposite; petioles nearly absent. Petals white to violet, the lower petal not spurred. . . . . *Hybanthus.*

GLOEOSPERMUM Triana & Planch.

Trees or treelets. Stipules deciduous. Leaves alternate, distichous. Inflorescences axillary, solitary, cymose, mono- or dichasial. Flowers strictly actinomorphic; petals whitish; stamens 5, the filaments usually partly or completely fused into filament tube, occasionally free, the anthers longitudinally dehiscent, the connective scale-like, usually laminar, occasionally minute, subulate; ovary subglobose, trapezoid, or conical, usually tapering into style, the style sometimes inflated; ovules 1–22 per locule. Fruits indehiscent, globose, baccate or nut-like, the pericarp usually coriaceous, occasionally somewhat woody. Seeds ± globose, mucilaginous, without areola and caruncle.

Gloeospermum sphaerocarpum Triana & Planch. FIG. 320

Small trees or treelets, 2–6 m tall. Bark gray. Stems pilosulous to glabrous. Leaves: petioles 4–7 mm long; blades usually elliptic, 4.5–22 × 1.2–7 cm, chartaceous, the base cuneate to obtuse, the apex long acuminate, the margins entire to serrulate. Inflorescences 6.5–12 mm long, pilosulous or glabrous; peduncles 0–4.3 mm long; pedicels 1–2 mm long. Flowers: sepals 1.7–2.3 × 1–3 mm, the margins ciliate; petals 4.5–6 × 2–3 mm; filaments partly to completely fused into filament tube, the apical free part of filaments 0.1–0.8 mm long, the filament tube 0.2–1.3 mm long, the anthers 1–1.8 × 0.5–0.8 mm, the connectives laminar, 0.2–1.5 × 0.2–0.5 mm, variable in form, entire to laciniate, colorless, transparent; ovary 1–2 × 0.7–1.5 mm, glabrous, the style 2.5–3.5 mm long, exceeding stamens by 0.7–1.5 mm (excluding the connective); ovules 3–12 per locule. Fruits apiculate berries, 1.5–4.5 cm diam. Seeds 9–15, 6–8 mm diam. Fl (Dec, Jan), fr (Dec, Jan); in understory of nonflooded forest and submontane forest (Mont Galbao).

HYBANTHUS Jacq.

Herbs or subshrubs. Stipules somewhat persistent. Leaves opposite (often alternate outside our area). Inflorescences solitary or racemose. Flowers strictly zygomorphic; petals with the lower one larger than others, the lateral petals longer than ligulate upper ones, both lower and lateral petals differentiated into a claw and a limb; stamens 5, the filaments free (in our area), those of 2 adaxial ones appendaged by a biglobular abaxial gland, the anthers longitudinally dehiscent, the connective scale-like, laminar; ovary subglobose to ovoid, the style curved near base; ovules 2 per locule (sometimes very numerous outside our area). Fruits capsular, dehiscing into 3 valves. Seeds pyriform, with an areola and caruncle.

Hybanthus oppositifolius (L.) Taub.

Subshrubs, to ca. 1 m tall. Stems hispidulous to glabrous. Leaves: petioles nearly absent; blades lanceolate to narrowly ovate, 1.5–8 × 0.2–1.8 cm, glabrous on midrib abaxially, hispidulous to glabrous on midrib adaxially, the base acute to obtuse, the apex acuminate, the margins serrate. Inflorescences axillary, solitary or sometimes accompanied by cleistogamous or reduced inflorescence; pedicels 5–12.5 mm long, hispidulous to glabrous. Flowers: sepals narrowly ovate, 2–3.5 × 0.5–1 mm, hispidulose, the apex acuminate; petals white to violet, the lower one 9 mm long, the claw symmetric, 2.5–4 × 0.5–1 mm, the limb deltoid, 3–5 mm wide, with a yellow nectar guide at base, the lateral petals 4–5.5 mm long, the claw asymmetric, 3–3.5 × 1.5–1.8 mm, the limb elliptic, 1.5–2 × 1–1.3 mm, the upper petals 3–4 mm long; filaments free, 0.2–0.5 mm long, the anthers 0.7–1 × 0.6–0.8 mm, the connective 0.7–1 × 0.5–0.8 mm, orange-brown; ovary 0.8–1 × 0.7–0.8 mm, the style scarcely exceeding stamens in length, 1.2–2 mm long, the stigma clavate; ovules 2 per locule. Fruit valves 3.5 × 1–1.5 mm. Seeds 2 per valve, 1.5–2 mm diam. Not collected, but to be expected from our area.

LEONIA Ruiz & Pav.

Trees or treelets. Stipules deciduous. Leaves alternate, polystichous. Inflorescences axillary and/or cauliflorous, solitary or 2–8-fasciculate, thyrsoid or cymose. Flowers strictly actinomorphic; petals all ligulate, yellow; stamens (3)5, the filaments fused into tube, the anthers apically dehiscent, extrorse, the connective not scale-like, inconspicuous; ovary subglobose to trapezoid, 3–5-locular, the style stipitate, the stigma indistinct; ovules 5–15 per locule (more numerous outside our area). Fruits globose, indehiscent, nut-like, the pericarp somewhat woody. Seeds subglobose, mucilaginous, without areola and caruncle.

Leonia glycycarpa Ruiz & Pav. var. **glycycarpa** FIG. 321

Trees, to 16 m × 20 cm. Bark smooth, light brown, with indistinct hoop marks, the slash yellow, with orange inclusions in inner bark. Stems glabrous, purplish with white, callose lenticels (when dried). Leaves: blades narrowly to broadly elliptic, 14–21 × 4.2–9 cm, glabrous, the base cuneate to rounded, the apex acuminate, the margins entire; venation brochidodromous, the secondary veins in 3–6 pairs, arcuate. Inflorescences 7–17 cm long, hispidulous; peduncles of lateral cymules 5–17 mm long; pedicels 3–5 mm long, hispidulous. Flowers with lemon aroma; sepals hemispherical, 1.2–1.8 × 0.5–2 mm, pilosulous, the apex rounded, the margins ciliate; petals oblong-elliptic, 3–6 × 1.5–3 mm, glabrous, the apex obtuse; filament tube 0.5–1 mm tall; ovary 4–5-locular, 1–1.8 × 0.7–1 mm, the style ca. 0.3 mm long. Fruits 3–3.5 cm diam., grayish, brownish, or whitish, the pericarp 4–6 mm thick. Seeds ca. 15, 8–15 × 5 mm. Fl (Aug, Sep); in nonflooded forest.

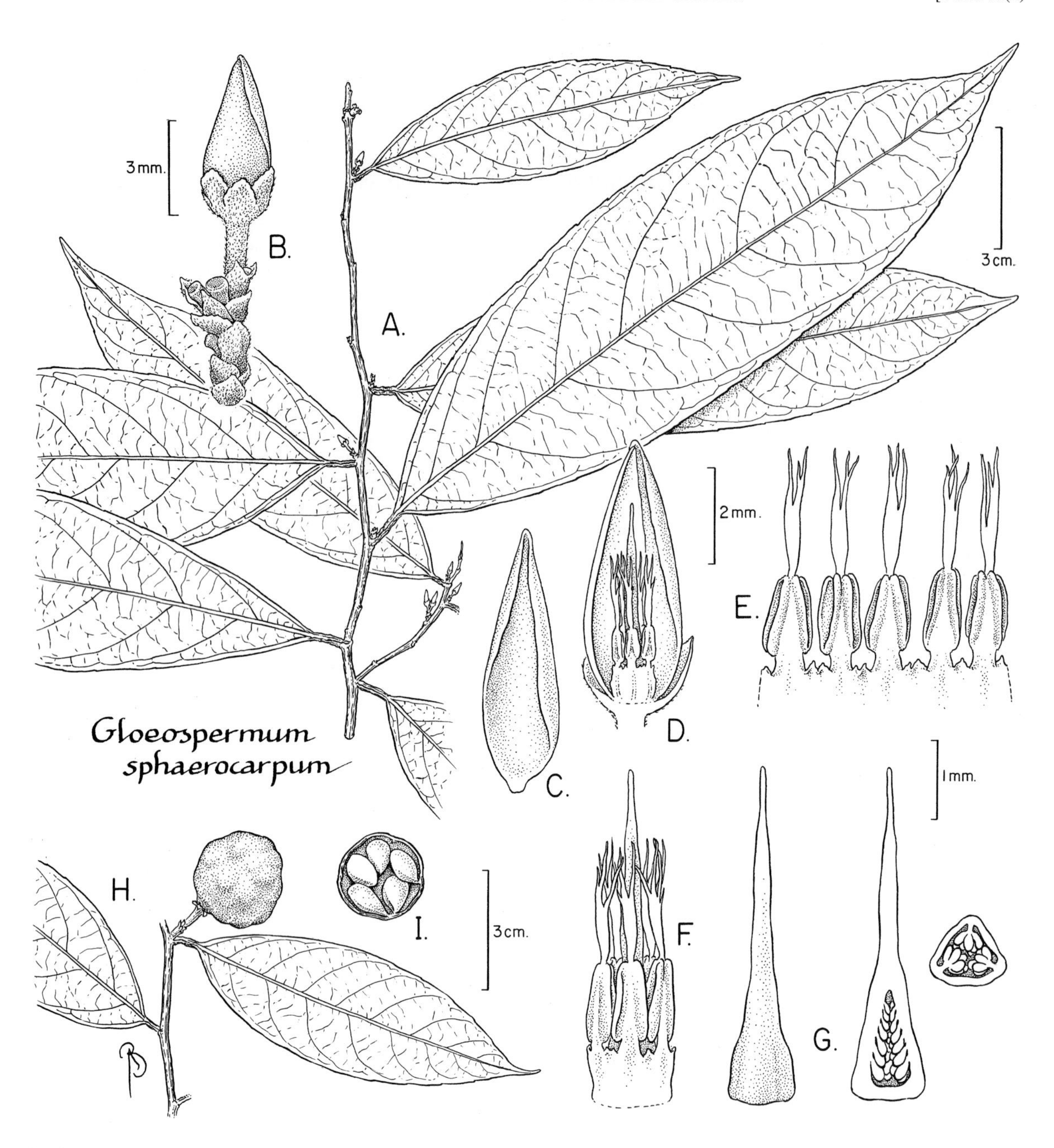

FIG. 320. VIOLACEAE. *Gloeospermum sphaerocarpum* (A, *Prance et al. 25478* from Amazonas, Brazil; B–G, *Granville 2672*; H, *Steyermark & Rabe 96234* from Venezuela; I, *Granville 8622*). **A.** Part of stem with leaves and flower buds. **B.** Lateral view of inflorescence with flower bud. **C.** Adaxial view of petal. **D.** Lateral view of flower with two petals removed. **E.** Adaxial view of upper portion of androecium opened out; note laciniate connectives on apices of anthers. **F.** Lateral view of androecium surrounding style. **G.** Lateral view (left) and medial section (near right) of pistil and transverse section of ovary (far right). **H.** Part of stem with leaves and dried fruit. **I.** Transverse section of fruit showing parietal attachment of seeds.

NOISETTIA Kunth

Subshrubs. Stipules deciduous. Leaves alternate. Inflorescences fasciculate or solitary. Flowers strictly zygomorphic; petals usually orange to red, the lower petal larger than others, differentiated into a claw and a limb, the lateral petals larger than upper ones; stamens 5, the filaments free, those of the two associated with the lower petal with a conical gland and a long spur, the spurs free from each other, the connective scale-like, laminar; ovary subglobose to trapezoid, 3-locular, the style sigmoid at base, the stigma trumpet shaped; ovules 4–5 per locule. Fruits capsular, dehiscing into 3 valves. Seeds pyriform, with areola and caruncle.

FIG. 321. VIOLACEAE. *Leonia glycycarpa* var. *glycycarpa* (A, *Phillippe et al. 26969*; B–F, *Jansen-Jacobs et al. 3073* from Guyana; G, *Jansen- Jacobs et al. 2199* from Guyana). **A.** Stem with leaves and inflorescences. **B.** Inflorescence. **C.** Detail of inflorescence. **D.** Lateral view of flower (right) and lateral view of flower with two petals and half of calyx removed (left). **E.** Lateral (right), adaxial (below left), and abaxial (below right) views of stamens. **F.** Lateral view (near right) and medial section (far right) of pistil. **G.** Lateral view (left) and medial section (below) of fruit.

Noisettia orchidiflora (Rudge) Ging. FIG. 322, PL. 125e

Subshrubs, to ca. 1.5 m tall. Stems glabrous. Leaves: petioles 1.5–2.5 cm long; blades narrowly to broadly elliptic, 3.3–16 × 1–3.5 cm, the base decurrent, the apex acuminate, the margins serrulate, hispidulous to glabrous abaxially. Inflorescences axillary; pedicels 5–18 mm long, strigillose. Flowers: sepals narrowly ovate, 3.8–4.3 × 0.5–0.8 mm, the apex gradually acuminate; petals orange to pink or vermillion (spur of lower petal whitish), the claw essentially symmetric, ca. 2 × 1–1.5 mm, the spur 8 × 1 mm, the limb of lower petal orbicular, ovate, or rhomboidal, 5–5.5 mm wide, with a yellow nectar guide at base, the lateral petals obovate, somewhat asymmetric, 3.5–4 × 1–1.8 mm, the upper petals ovate, symmetric, 1.5–2 × 0.5–1.3 mm; filaments 0.3–0.5 mm long, those of the two stamens associated with the lower petal appendaged by a conical gland, 0.2–0.3 mm long and by a free spur 7 mm long, the anthers 0.6–0.7 × 0.5 mm, the connective 0.5–0.8 × 0.6–1 mm, orange-brown; ovary 0.7–1 mm diam., the style 1.5 mm long, exceeding stamens by 0.5 mm. Fruit valves 8 × 3–4 mm. Seeds 4–5 per valve, ca. 1 mm diam. Fl (Jan, Apr, May, Jun, Jul, Oct); in primary and secondary non-flooded forest at all elevations in area.

PAYPAYROLA Aubl.

Small trees or treelets. Stipules deciduous. Leaves alternate, polystichous. Inflorescences axillary or cauliflorous, solitary or narrowly thyrsoid to (pseudo)racemose . Flowers subactinomorphic; petals yellow in our area, also white or pink outside our area, the innermost one with the base tapering into a claw, the apex emarginate, the lateral and upper ones ligulate; stamens 5, the filaments fused into a tube, the anthers longitudinally dehiscent, introrse, the connective minutely scale-like, subulate to nearly absent; ovary 3-locular, conical to lageniform, ± tapering into the style, the stigma trumpet-shaped; ovules 4–6 per locule. Fruits capsular, dehiscing into 3 valves. Seeds pyriform, with areola and caruncle.

Paypayrola guianensis Aubl. FIG. 323

Trees or treelets, to 6 m tall. Stems pilosulous to glabrescent. Leaves: blades obovate, 8.5–35 × 2.5–14 cm, glabrous, the base obtuse, the apex rounded or acuminate, the margins entire or with a few teeth; venation eucamptodromous, the secondary veins in 5–9 pairs, strongly arcuate. Inflorescences 3–6 cm long, the central axis pilosulous; pedicels nearly absent. Flowers: sepals broadly ovate, 2–3 × 1.5–2 mm; petals 10–18 × 1–2.5 mm; filament tube 2 mm high, the anthers ca. 1 mm diam., the connective scale-like, subulate, ca. 0.2 mm long; ovary 3 × 1 mm, glabrous, the style 2.5–3 mm long, emergent above filament tube. Fruit valves 2.3–4 × 1–1.5 cm. Seeds 4 per valve, 8–10 × 4–5 mm. Fl (Jul), fr (Jul); in non-flooded forest.

RINOREA Aubl.

Trees or treelets. Stipules deciduous. Leaves opposite or alternate (outside our area), polystichous; hairy tuft domatia sometimes present in secondary vein axils. Inflorescences racemes, occasionally 1–3-fasciculate or solitary. Flowers strictly actinomorphic; corolla often urceolate; petals all ligulate, whitish or yellow; stamens 5, the filaments free or fused into a tube, the anthers introrse, the connective scale-like, laminar, entire to fringed, orange-brown, brown, or yellow; ovary 3-locular, (sub)globose, occasionally trapezoidal, or conical, the style usually filiform, erect or slightly curved, the stigma obtuse or truncate: ovules 1–4 per locule. Fruits capsular, dehiscing into 3 valves. Seeds globose, with an areola and caruncle.

1. Leaf blades with base asymmetric, strongly oblique, obtuse, cordate, or auriculate. Seeds glabrous. . . . . . *R. neglecta.*
1. Leaf blades with base usually symmetric, not oblique, rounded to attenuate to cuneate. Seeds pubescent.
    2. Leaf blades with midrib puberulous adaxially. . . . . *R. pubiflora.*
    2. Leaf blades with midrib glabrous adaxially.
        3. Stems without numerous lenticels. Leaf blades with hairy tufts (domatia) present in secondary veins abaxially. Ovary without "spiny" indumentum. . . . . *R. amapensis.*
        3. Stems with numerous whitish, callose lenticels. Leaf blades with hairy tufts (domatia) lacking in secondary veins abaxially. Ovary "spiny" strigose.. . . . . *R. riana.*

Rinorea amapensis Hekking PL. 125f

Trees or treelets, 1–4(5) m tall. Stems pilosulous to densely pilose. Leaves: petioles 2–7 mm long; blades elliptic to obovate, 2.5–15 × 2–5 cm, glabrous on midrib and with hairy tuft domatia in secondary vein axils abaxially, glabrous adaxially, the base symmetric, rounded to cuneate, the apex acuminate, the margins entire to serrulate or crenulate; secondary veins in (7)9–11(13) pairs. Inflorescences racemes or racemose, 5–8.5 cm long, brownish to whitish, pilosulous; pedicels 4–4.5 mm long. Flowers: sepals 1.2–2.3 × 1.3–1.8 mm, whitish, pilosulous abaxially; petals 3–4.3 × 1.8 mm, brownish, pilosulous abaxially; filaments 0.5–0.8 mm long, free, the abaxial glands nearly always free, the anthers 1.2–1.4 × (0.4)0.6–0.8 mm; ovary subglobose, 0.7–1.3 × 0.7–0.8 mm, brownish pilose, the style 2–2.5 mm long, exceeding stamens by 0.2–0.5 mm. Fruit valves 0.7–3.3 × 0.2–1 cm. Seeds 1–2 per valve, 5–7 mm diam., erect, brownish, pilosulous, purplish when dried. Fl (Aug, Nov, Dec); in understory of nonflooded forest.

Rinorea neglecta Sandwith

Small trees or treelets, 1–4(7) m tall. Outer bark gray, the slash white, the wood white. Stems ferruginous hispid (to hispidulous) to glabrescent. Leaves: petioles 1–7 mm long; blades (narrowly) obovate to elliptic, 3–18 × 1.5–7.5 cm, ferruginous strigose on midrib and occasionally with hairy tuft domatia in secondary veins abaxially, puberulous to glabrescent adaxially, the base distinctly oblique, obtuse, cordate, or auriculate, the apex acuminate, the

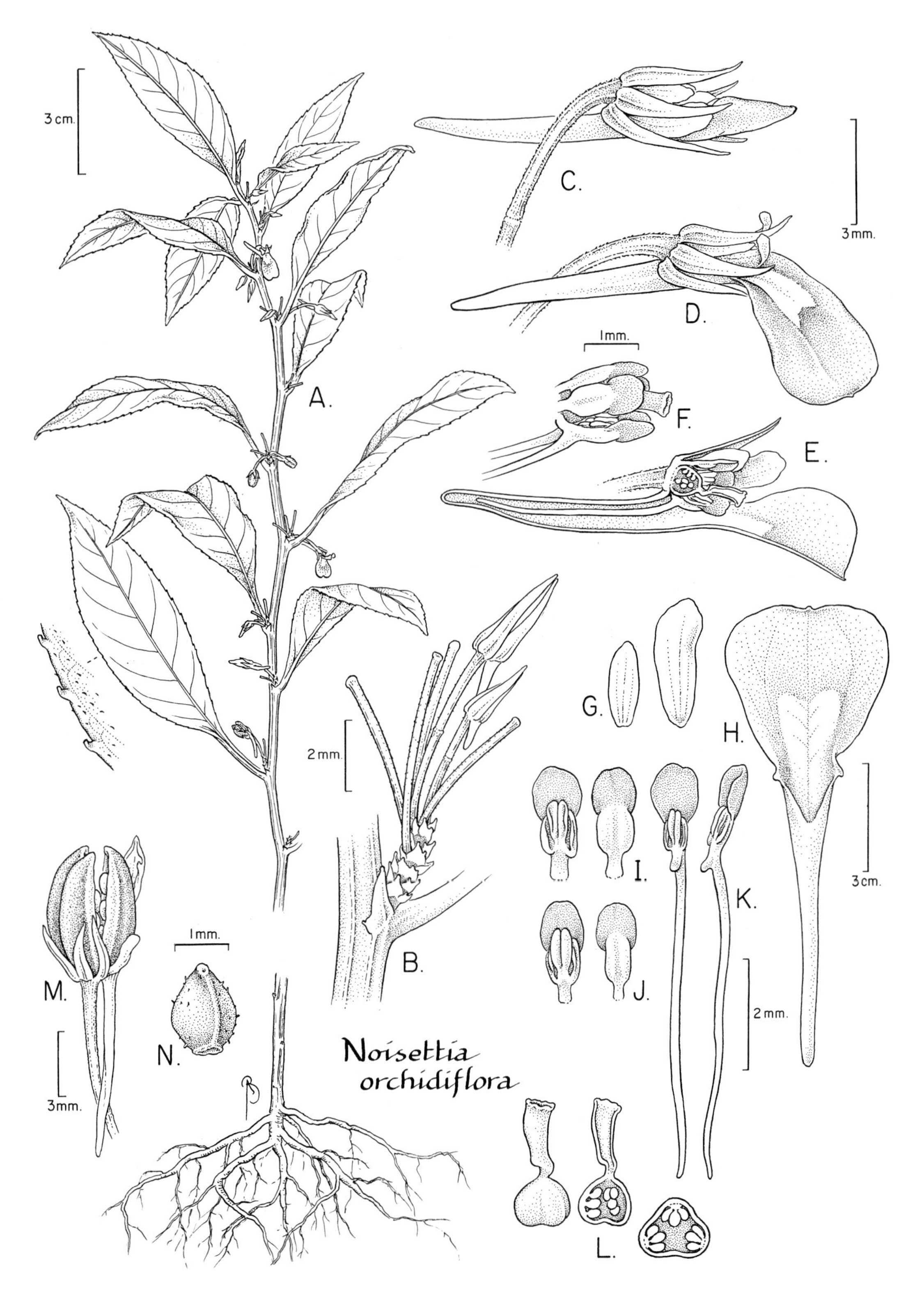

FIG. 322. VIOLACEAE. *Noisettia orchidiflora* (A, B, D, H, *Mori et al. 22184*; C, E–G, I–L, *Görts-van Rijn et al. 75*; M, N, *Mori & Pennington 17950*). **A.** Upper and lower parts of plant (near left and below) and detail of adaxial surface of leaf margin (far left). **B.** Node with inflorescence fascicle and stipule. **C.** Lateral view of flower bud. **D.** Lateral view of flower. **E.** Medial section of flower. **F.** Lateral view of anthers and stigma with upper part of style. **G.** Upper (left) and lateral (right) petals. **H.** Lower petal. **I.** Upper (left) and abaxial (right) views of lateral anther. **J.** Adaxial (left) and abaxial (right) views of anther of stamen associated with upper petals. **K.** Adaxial (left) and lateral (right) views of anther of stamen associated with lower petal. **L.** Lateral view (above left), medial section (above), and transverse section (right) of pistil. **M.** Fruit. **N.** Seed.

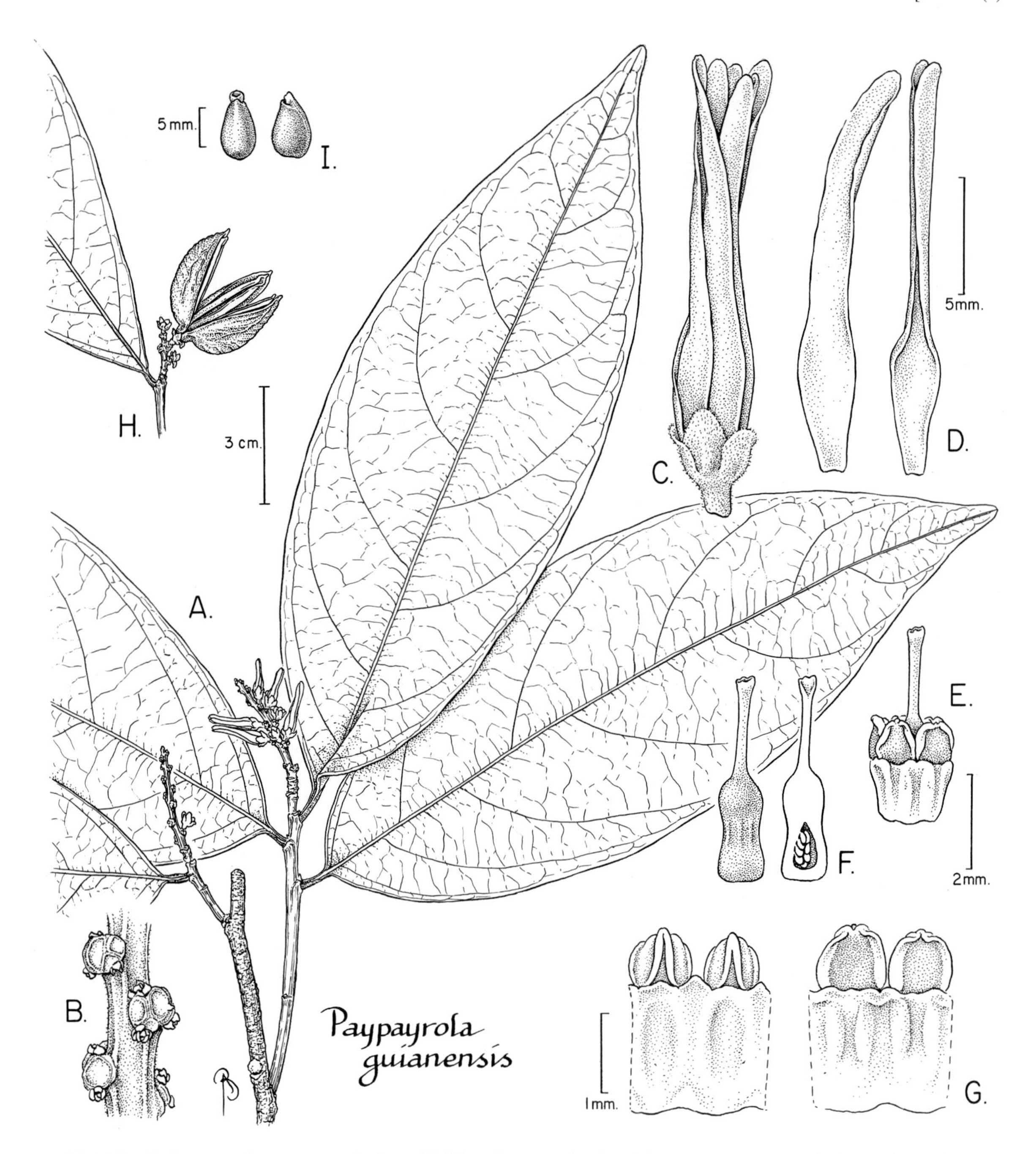

FIG. 323. VIOLACEAE. *Paypayrola guianensis* (A, B, *Cowan 38907* from Cayenne region, French Guiana; C–I, *Granville 3052*). **A.** Part of stem with leaves and inflorescences. **B.** Detail of inflorescence rachis with flowers removed. **C.** Lateral view of flower. **D.** Two petals, the innermost (near left) with emarginate apex. **E.** Lateral view of anther column surrounding pistil. **F.** Lateral view (far left) and medial section (near left) of pistil. **G.** Adaxial (far left) and abaxial (near left) views of portion of anther tube. **H.** Dehisced fruit at apex of stem. **I.** Two views of seeds.

margins serrulate; secondary veins in 9–14 pairs. Inflorescences racemes, 1–6 cm long, ferruginous hispidulous; pedicels 2–7 mm long. Flowers: sepals 1.5–2 × 0.7–1.5 mm, ferruginous strigose abaxially; petals 3.5–5 × 1.5–2.3 mm, glabrous; filaments 0.5–1 mm long, free, the abaxial glands free, the anthers 1.2–1.5 × 0.4–0.6 mm; ovary subglobose to conical, ca. 1 mm diam., golden to whitish pilose, the style 2–3 mm long, exceeding stamens by 0.2–0.8 mm. Fruit valves 1–1.8 × 0.5–0.8 cm. Seeds 1 per valve, 5–6 mm diam., glabrous. Fr (Dec, Jan, Feb); in nonflooded forest.

Rinorea pubiflora (Benth.) Sprague & Sandwith

Small trees or treelets, (0.25)1–6 m tall. Outer bark smooth, gray, the wood white. Stems ferruginous pubescent. Leaves: petioles (2)4–15(20) mm long; blades elliptic, ovate, or obovate, 3.5–20 × 1.5–8 cm, sparsely ferruginous strigose on midrib abaxially, puberulous adaxially, with hairy tufts (domatia) usually present in secondary vein axils abaxially, the base symmetric, rounded to cuneate, the apex acuminate, the margins entire to serrate or crenate;

secondary veins in (6)8–11(13) pairs. Inflorescences racemes 1–11.5 cm long, ferruginous, puberulous, or hispidulous; pedicels 2–6 mm long. Flowers: sepals 1.5–2.3 × 1.7–3 mm, ferruginous hispidulous abaxially; petals 3.2–6.5 × 1.2–2.8 mm, ferruginous strigose abaxially; filaments 0.5–1.5 mm long, free, the abaxial glands free, the anthers 1.6–1.8 × 1 mm; ovary subglobose, 0.5–1.3 mm diam., ferruginous strigose, the style 2–3.3 mm long, equalling or slightly exceeding stamens. Fruit valves 1.5–3.5 × 0.5–1 cm. Seeds 2–4 per valve, 4–8 mm diam., brownish puberulous. Fl (Jan, Feb, Jul, Aug, Sep, Oct, Nov, Dec), fr (Jan, Apr, May, Aug, Sep, Oct, Nov); common, in secondary and primary nonflooded forest.

Rinorea riana Kuntze

Small trees or treelets, 1–4(8) m tall. Outer bark gray with white patches, the wood yellow to chartreuse. Stems with numerous, whitish, callose lenticels, sparsely puberulous to glabrescent. Leaves: petioles (1)2.5–10(15) mm long; blades (narrowly)ovate to elliptic, (4.5)8–24.5 × (1)3–9 cm, glabrous, hairy tufts (domatia) absent, the base rounded to attenuate, the apex short to long attenuate, the margins entire to serrate or crenate; secondary veins in 8–13 pairs. Inflorescences racemes, 1.2–14.5 cm long, sparsely ferruginous pilosulous; pedicels 1.7–4.5 mm long. Flowers: sepals 1–2 × 0.5–1.8 mm, ferruginous to golden pilose or pilosulous abaxially; petals 3–5 × 1.2–2 mm, drying to chocolate brown; filaments 0.5–0.8 mm long, free, the abaxial glands free, the anthers 1.5–2 × 0.5–1 mm; ovary subglobose, 0.5–1.5 mm diam., pubescence of two types, "spiny" strigose and densely hispidulous, maroon, the style 1.2–2.5 mm, exceeding the stamens by 0.2–0.5 mm. Fruit valves 1.5–2.8 × 0.5–1 cm. Seeds 1–2 per valve, 4–10 mm diam., ferruginous hirtellous. Fl (Jan, Aug), fr (Jan, Feb, Mar, May, Jun, Aug); in nonflooded forest at all elevations.

VISCACEAE (Mistletoe Family)

Job Kuijt

Brittle, branching, glabrous hemiparasites lacking creeping roots. Stipules lacking. Leaves simple, opposite, of various shapes and sizes, including some scale-leaves, especially at base of lateral branches and on inflorescences. Inflorescences spikes subtended by 1 to several pairs of scale-leaves, followed by several fertile internodes with sessile flowers arranged serially above fertile scale-leaves. Flowers minute, actinomorphic, unisexual (plants monoecious in our area, both sexes present on each inflorescence); perianth lobes 2–4; stamens 3–4, adnate to tepals; ovary inferior, 1-locular. Fruits berries, small, fleshy. Seeds one per fruit, without testa; embryo dicotylar, green, embedded in endosperm and sticky tissue.

Krause, K. 1966. Loranthaceae. *In* A. Pulle (ed.), Flora of Suriname **I(1):** 4–24. E. J. Brill, Leiden.

PHORADENDRON Nutt.

Lateral branches with 1 or more pairs of scale leaves (basal cataphylls) inserted at various distances above subtending leaf. Leaves often brittle, in some species with 1 or more pairs of scale leaves (intercalary cataphylls) placed between successive pairs of expanded leaves. Inflorescences of various numbers of sterile internodes (below) followed by fertile internodes. Flowers mostly 3-parted, partly sunken in inflorescence axis but not individually subtended by foliar organs. Fruits white, yellowish, or red.

1. Stems with successive leaf pairs separated by ≥3 pairs of intercalary cataphylls subtending inflorescences. . . . . *P. crassifolium.*
1. Stems without successive leaf pairs separated by fertile intercalary cataphylls.
    2. Lateral stems with ≥3 pairs of basal cataphylls, with one pair of inconspicuous, sterile, intercalary cataphylls above each foliar node. Leaf blades with acute apex. . . . . *P. piperoides.*
    2. Lateral stems with 1–2 pairs of basal cataphylls, intercalary cataphylls lacking. Leaf blades with obtuse to rounded apex.
        3. Stems without dichotomous branching. Leaf blades >3× as long as wide; abaxial midrib not strongly raised.. . . . . *P. perrottetii.*
        3. Stems with dichotomous branching. Leaf blades ≤ 2× as long as wide; abaxial midrib strongly raised.. . . . . *P. racemosum.*

Phoradendron crassifolium (DC.) Eichler

FIG. 324, PL. 126a,b

Stout, percurrent hemiparasites. Stems terete; basal cataphylls in 1–2 pairs, rather close to subtending leaves, the successive leaf-pairs on all stems separated by several pairs of intercalary cataphylls, the lower pair(s) sterile, the upper several pairs fertile. Leaves: petioles not well-defined to 7 mm long; blades ovate, broadly lanceolate, elliptic, or narrowly elliptic, ca. 2× as long as wide, 6–14 × 2.5–7 cm, coriaceous; venation palmate. Inflorescences 1–3 per cataphyll, additional inflorescences in axils of foliar leaves, each with 2 to several pairs of crowded, sterile, scale leaves followed by 5–9 short, fertile internodes, to 3.5 cm long. Flowers biseriate, 3–7 above each fertile bract, in 2 pistillate series plus staminate apical flower. Berries yellowish to white, or reddish brown or orange when exposed to sun. Fl (May, Oct, Nov), fr (May, Sep, Nov); scattered but common.

Phoradendron perrottetii (DC.) Eichler

Large, percurrent hemiparasites. Stems terete or slightly keeled; basal cataphylls mostly in 1 pair, the intercalary cataphylls lacking. Leaves: petioles to 10 mm long; blades ovate, often curved, 3× as long as wide, thin, the apex rounded; venation of ± equivalent, large veins from base. Inflorescences to 8 cm long, either without or with

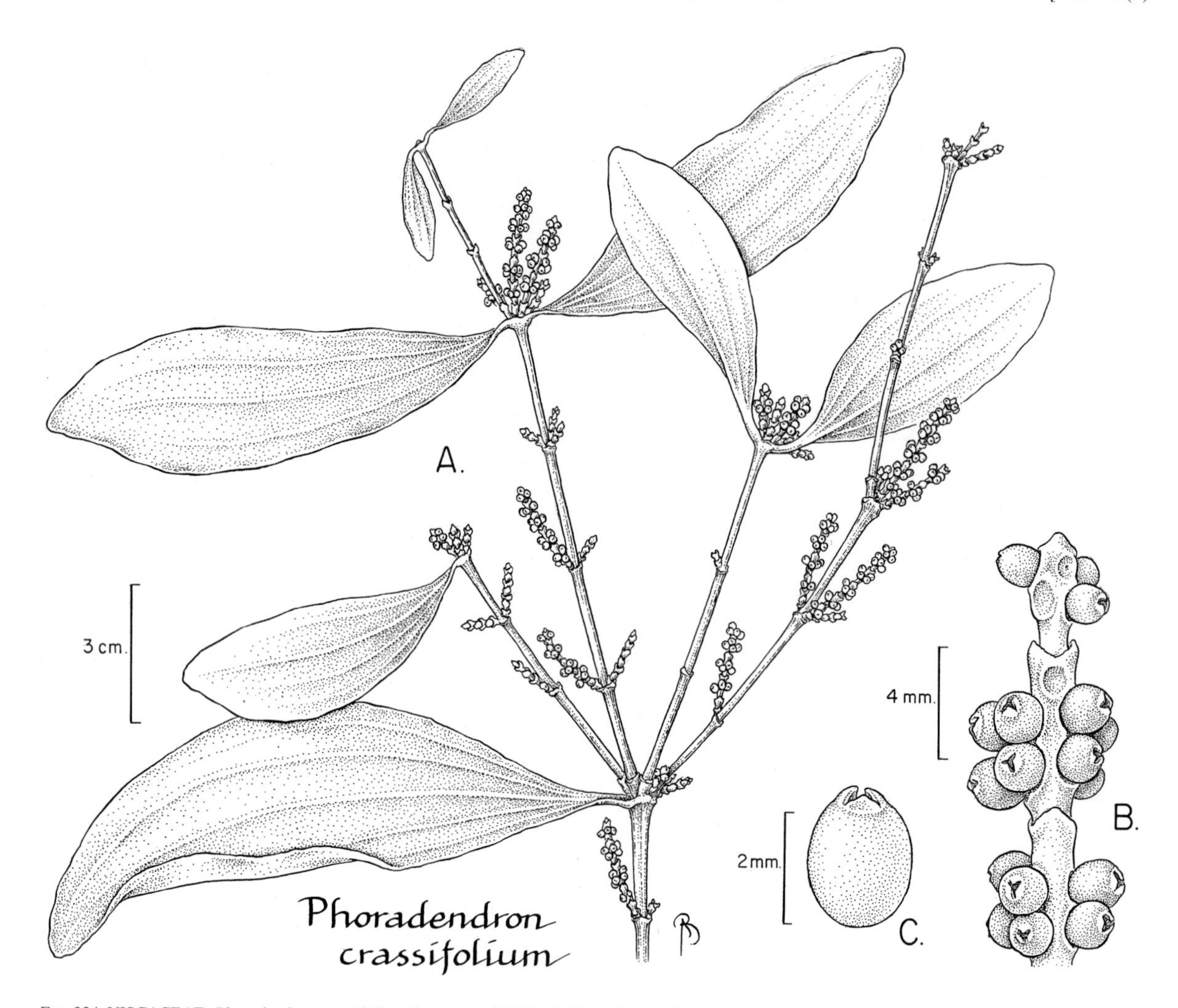

FIG. 324. VISCACEAE. *Phoradendron crassifolium* (*Mori et al. 23977*). **A.** Part of stem with leaves and inflorescences. **B.** Detail of inflorescence. **C.** Lateral view of pistillate flower.

1–2 pairs of sterile cataphylls, the fertile internodes several, each with numerous, crowded flowers arranged in 3 series. Fr (Jun); collected only once in our area.

Phoradendron piperoides (Kunth) Trel.

Leafy, branched, percurrent hemiparasites of medium size. Stems terete; basal cataphylls of lateral branches mostly in several pairs spaced along lower part, the intercalary cataphylls in 1 pair, very low and inconspicuous, sterile. Leaves: petioles ca. 5 mm long; blades lanceolate, mostly <10 × 5 cm, 2× as long as wide, the base acute, the apex acute; midrib prominent. Inflorescences with (0 to) several pairs of basal cataphylls and 5–8 fertile internodes above. Flowers biseriate, to 15 above each fertile bract, the staminate flowers few, either above or among pistillate ones above each bract. Fruits small, yellow or reddish. Fl (Feb), fr (Nov). *Bois lait* (Créole).

Phoradendron racemosum (Aubl.) Klug & Urb. PL. 126c,d

Large, dichotomously branched hemiparasites, new growth terminating in 2 leaves each subtending an inflorescence or new growth. Stems terete; basal cataphylls in 2 pairs, one pair very low and inconspicuous, the other more prominent, ca. 5 mm above base. Leaves: petioles 10–12 mm long; blades broadly ovate to elliptic, to 13 × 10 cm, nearly as wide as long, the base obtuse to rounded, the apex obtuse to rounded; midrib prominent. Inflorescences with 2–3 sterile internodes proximal to 3–4 fertile internodes. Flowers biseriate, 7–11 above each bract, distribution of sexes unknown. Fruits rather small, whitish. Fl (Aug).

VITACEAE (Grape Family)

Scott V. Heald

Lianas. Stems terete or ± quadrangular with short wings. Stipules persistent or caducous. Tendrils leaf-opposed, dichotomously branched. Leaves simple or trifoliolate, alternate; blades with margins serrulate. Inflorescences borne in leaf-opposed or seldom terminal, compound umbels. Flowers actinomorphic, bisexual, minute; calyx indistinctly lobed; corolla 4–5-merous, with valvate aestivation; petals free (*Cissus*)

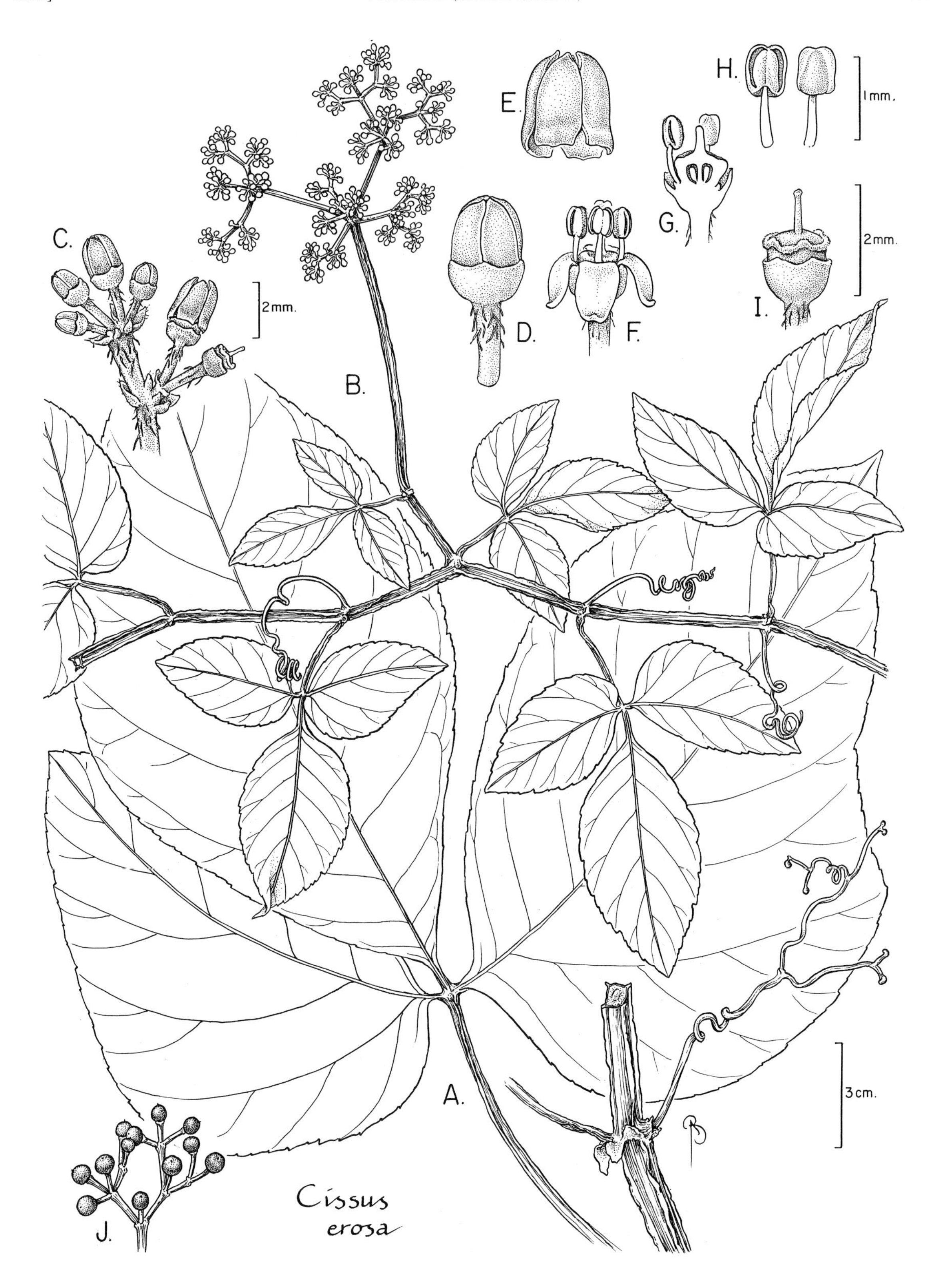

FIG. 325. VITACEAE. *Cissus erosa* (A, *Mori et al. 23902*; B, *Mori & Gracie 18392*; C–J, *Mori et al. 21022*). **A.** Leaf (near right) and part of stem with branching tendril (far right). **B.** Part of stem with leaves and inflorescence. **C.** Detail of inflorescence. **D.** Lateral view of flower bud. **E.** Calyptrally dehiscent corolla. **F.** Lateral view of flower showing non-calyptrally dehiscent corolla and extrose dehiscence of anthers. **G.** Medial section of flower with petals removed. **H.** Abaxial (near right) and adaxial (far right) views of anthers. **I.** Developing ovary surrounded by calyx. **J.** Infructescence.

or distally fused and calyptrate (*Vitis*); stamens 4–5, petal-opposed; nectary disc intrastaminal; pistil single, the ovary superior, 2-locular; placentation axile at base of ± incomplete septum, the ovules 3–4. Fruits berries. Seeds 1–2.

Görts-van Rijn, A. R. A. 1979. Vitaceae. *In* A. L. Stoffers & J. C. Lindeman (eds.), Flora of Suriname **V(1):** 335–343. E. J. Brill, Leiden.

CISSUS L.

Lianas. Leaves simple or trifoliolate. Inflorescences flat topped. Flowers 4-merous; petals free, spreading; nectary disc 4-lobed. Seeds 1 per fruit.

1. Stems ± quadrangular, often with narrow wings, the nodes not or only slightly swollen. Stipules persistent. Leaves trifoliolate. Flowers red. . . . . *C. erosa.*
1. Stems terete, without wings, the nodes swollen. Stipules caducous. Leaves simple. Flowers yellowish-green. . . . . *C. verticillata.*

Cissus erosa Rich. FIG. 325; PART 1: FIG. 11

Stems ± quadrangular, with narrow wings, the nodes not or only slightly swollen. Stipules persistent. Leaves trifoliolate; blades 2.6–12.3 × 1–5.3 cm on fertile stems, much larger on sterile stems, the margins with minute teeth spaced 1.5–5(7) mm apart. Peduncles, inflorescence rachises, and flowers red. Fruits ca. 7 mm diam. Fl (Jun, Aug, Sep, Oct); common, in weedy habitats and in disturbed areas in forest. *Liane-ravet* (Créole).

Cissus verticillata (L.) Nicolson & C. E. Jarvis [Syn.: *Cissus sicyoides* L.] PL. 126e

Stems terete, the nodes swollen. Stipules caducous, leaving triangular scar. Leaves simple; blades 7.2–8.5 × 3.4–5 cm on fertile stems, the margins with minute teeth spaced 3–13 mm apart. Peduncles, inflorescence rachises, and flowers yellowish-green. Fruits ca. 7 mm diam. Fl (Aug, Sep), fr (Sep); common, in weedy habitats.

This species is occasionally infected with the smut *Mycosyrinx cissi* (Poir.) G. Beck, which causes a witches' broom of the inflorescence in which the peduncles and inflorescence rachises are elongate and the flowers are transformed into narrow, hollow bodies that release spores. Until the dimorphism between infected and uninfected plants was recognized, infected plants were known as *Viscum verticillatum* L. and uninfected plants were known as *Cissus sicyoides* L. (Nicolson & Jarvis, 1984).

VOCHYSIACEAE (Vochysia Family)

Scott A. Mori

Trees. Leaves simple, opposite or whorled. Stipules present, sometimes small and inconspicuous or even absent, sometimes replaced by extrafloral nectaries. Inflorescences racemes or panicles. Flowers zygomorphic, bisexual; calyx with 5 lobes, one sepal often conspicuously spurred; petals 0–3; stamen 1; ovary superior or inferior (*Erisma*), 1(*Erisma*)–3-locular; placentation apical (*Erisma*) or axile. Fruits either loculicidal capsules and without persistent calyx lobes or indehiscent with persistent, wing-like calyx lobes (*Erisma*). Seeds 3 (*Vochysia*) to numerous per locule and winged or solitary and not winged (*Erisma*).

Lindeman, J. 1986. Vochysiaceae. *In* A. L. Stoffers & J. C. Lindeman (eds.), Flora of Suriname **III(1–2):** 552–562. E. J. Brill, Leiden.
Marcano-Berti, L. 1998. Vochysiaceae. *In* A. R. A. Görts-van Rijn & M. J. Jansen-Jacobs (eds.), Flora of the Guianas Ser. A, **21:** 1–44, 87–98. Royal Botanic Gardens, Kew.
Stafleu, F. A. 1951. Vochysiaceae. *In* A. Pulle (ed.), Flora of Suriname **III(2):** 178–203. The Royal Institute for the Indies.

1. Ovary inferior, 1-locular. Fruits indehiscent, with persistent winged calyx. Seeds one per fruit, not winged. . . . . . *Erisma.*
1. Ovary superior, 3-locular. Fruits dehiscent, without persistent calyx lobes. Seeds more than one per fruit, winged.
   2. Secondary veins widely spaced, >5 mm apart. Petals 0 or 3, yellow, when present smaller than larger calyx-lobe. . . . . *Vochysia.*
   2. Secondary veins very closely spaced, <1 mm apart. Petals 1, blue or white tinged with red or yellow, larger than larger calyx-lobe.
      3. Leaf blades with minutely apiculate apex (the apiculus appearing as extension of midrib). Inflorescences axillary, the flowers few. Anther basifixed, bearded. . . . . *Ruizterania.*
      3. Leaf blades without apiculate apex. Inflorescences terminal, the flowers numerous. Anther dorsifixed, not bearded. . . . . *Qualea.*

ERISMA Rudge

Canopy to emergent trees. Trunks with well developed buttresses. Stems and leaves with stellate pubescence. Extrafloral nectaries not developed. Leaves opposite or whorled; secondary veins widely spaced, <2 per centimeter. Inflorescences terminal, paniculate. Flowers: petal single, obcordate, larger than larger calyx-lobe; anther dorsifixed, not bearded; ovary inferior, 1-locular. Fruits indehiscent, the calyx lobes expanded into persistent wings. Seeds 1 per locule, without wings.

Kawasaki, M. L. 1998. Systematics of *Erisma* (Vochysiaceae). Mem. New York Bot. Gard. **81:** 1–40.

1. Buttresses branched. Pubescence on stems yellowish-brown. Leaf blades with petioles <10 mm long; blades with rounded base; secondary veins in 12–18 pairs. Inflorescences with persistent bracts. Flowers with pale yellow petal, the spur cylindric, only slightly recurved. . . . . *E. floribundum.*
1. Buttresses unbranched. Pubescence on stems gray. Leaf blades with petioles >10 mm long; blades with cuneate base; secondary veins in 5–8 pairs. Inflorescences without persistent bracts. Flowers with purple petal, the spur markedly curved inwards (uncinate). . . . . *E. uncinatum.*

Erisma floribundum Rudge PART 1: PL. IVa

Canopy trees. Trunks with symmetric, branched buttresses. Bark smooth, with abundant, small, white lenticles. Stems with yellowish-brown pubescence. Leaves opposite or in whorls of 4; petioles <10 mm long; blades 9–13 × 4–6 cm, the base rounded; secondary veins in 12–18 pairs. Inflorescences with persistent, orbicular, white bracts. Flowers: calyx with spur cylindric, only slightly recurved; petal pale yellow. Fl (May, Nov), fr (Feb); rare, in nonflooded forest.

Erisma uncinatum Warm. PL. 127a; PART 1: FIG. 7

Canopy to emergent trees. Trunks with steep, thick, unbranched buttresses. Stems with gray pubescence. Bark rough, peeling in irregular plates. Leaves opposite or in whorls; petioles >10 mm long; blades 11–16 × 5–7.5 cm, the base cuneate; secondary veins in 5–8 pairs. Inflorescences without persistent bracts, the axes purple. Flowers: calyx with spur markedly curved inward (uncinate); petal purple, this often with white base and white line down middle, pubescent; anther brown; stigma green. Fruits 7–8 cm long (including wing). Seeds with cryptocotylar germination, the seedlings with acicular stipules ca. 2 mm long. Fl (Sep, Nov), fr (Feb, Apr); occasional, in nonflooded forest. *Jaboty* (Portuguese).

QUALEA Aubl.

Canopy to emergent trees. Trunks with buttresses. Stems and leaves glabrous or with simple hairs. Extrafloral nectaries developed in stipular region. Leaves opposite; secondary veins closely spaced, >10 per centimeter. Inflorescences terminal, paniculate, the flowers numerous. Flowers: petal single, obcordate or obovate, larger than largest calyx-lobe; anther dorsifixed, not bearded; ovary superior, 3-locular. Fruits dehiscent, the valves separating from central column. Seeds >1 per fruit, >1 per locule, winged.

1. Leaf blades with apex rounded, then abruptly acuminate. Inflorescences with axes blue-violet. Buds blue-violet; petal deep blue, 15–20 mm wide; anther pubescent. . . . . *Q. mori-boomii.*
1. Leaf blades with apex acuminate. Inflorescences with axes not blue-violet. Buds not blue-violet; petal white with red spot in center, ca. 40 mm wide; anther glabrous. . . . . *Q. rosea.*

Qualea mori-boomii Marc.-Berti PL. 127b

Canopy to emergent trees. Trunks with poorly developed, steep buttresses. Bark smooth, finely lenticellate, peeling in irregular plates toward base. Extrafloral nectaries flat, not markedly raised above surface of stems. Leaves: blades 5.5–12 × 3.5–6 cm, not glaucous abaxially, the apex rounded, then abruptly acuminate; intersecondary veins well developed. Inflorescences mostly terminal, the axes blue-violet, the flowers numerous. Flowers: buds blue violet; petal blue, 15–20 mm wide. Fls (Nov); scattered, in nonflooded forest.

Similar to *Q. coerulea* Aublet but the buds and flowers, especially the spurred sepal, are 2–4 times larger.

Qualea rosea Aubl. PLS. 127c, 128b; PART 1: FIG. 4

Canopy to emergent trees. Trunks often with buttresses. Bark reddish-brown, smooth, sometimes finely and reticulately cracked, with incomplete hoop marks, lenticellate. Extrafloral nectaries, when present, cylindric, markedly raised above surface of stems. Leaves: blades 8–13 × 3.5–6 cm, glaucous abaxially, the apex acuminate; intersecondary veins not well developed. Inflorescences mostly terminal, the axes not blue-violet, the flowers numerous. Flowers: buds with white petals; petal white, with central red spot, ca. 40 mm wide. Fruits cylindric, ca. 7 × 4 cm, the exocarp often separates from endocarp when fruit dries. Seeds 10 × 4 mm, the wing 40 × 15 mm. Fl (Oct, Nov), fr (Apr); common, in nonflooded forest. *Gonfolo* (Créole).

RUIZTERANIA Marc.-Berti

Canopy trees. Trunks with buttresses. Stems and leaves glabrous or with simple hairs. Extrafloral nectaries in position of stipules, crateriform. Leaves opposite; secondary veins closely spaced, >10 per centimeter. Inflorescences axillary, the flowers few. Flowers: petal single, round to obovate, larger than largest calyx lobes; anther basifixed, bearded; ovary superior, 3-locular. Fruits dehiscent, the valves separating from central column. Seeds >1 per locule, winged.

Ruizterania albiflora (Warming) Marc.-Berti PL. 128a

Canopy trees. Trunks with thick, steep, running buttresses. Bark smooth, with elliptic, white lenticels. Extrafloral nectary in position of stipules, crateriflorm. Leaves: blades 5–9 × 2.5–3 cm, not glaucous abaxially, the apex minutely apiculate (the apiculus appearing as extension of midrib), often slightly emarginate with apiculus in middle; intersecondary veins well developed, the midrib extending beyond apex, forming terminal hair. Flowers: petals white, with central yellow spot, ca. 20–25 mm wide. Fls (Sep); rare, in nonflooded forest. *Gonfolo* (Créole).

VOCHYSIA Aubl.

Canopy, less frequently, emergent trees. Stems and leaves glabrous or with simple hairs. Extrafloral nectaries not developed, stipules caducous or persistent. Leaves opposite or whorled; secondary veins widely spaced, 1–2 per centimeter. Flowers entirely yellow; petals 0–3, when present smaller than largest calyx lobes in our flora; stamen with adaxial groove into which style and stigma fit, the anther basifixed; ovary superior, 3-locular. Fruits dehiscent, the valves not separating from central column. Seeds 1 per locule, winged.

1. Stems and leaves glabrous.
    2. Leaves opposite, chartaceous; secondary veins ca. 10 mm apart. . . . . *V. guianensis.*
    2. Leaves in whorls of 3, coriaceous; secondary veins ca. 5 mm apart. . . . . *V. surinamensis.*
1. Stems and leaves pubescent.
    3. Stipules triangular. Leaf blades >10 cm long; secondary veins in 24–27 pairs, the marginal vein well developed. Flowers with spurred sepal 10–11 mm long; petals lacking. . . . . *V. neyrattii.*
    3. Stipules acicular. Leaf blades <10 cm long; secondary veins in 8–12 pairs, the marginal vein not well developed. Flowers with spurred sepal ca. 4–7 mm long; petals 3. . . . . *V. tomentosa.*

Vochysia guianensis Aubl. PART 1: FIG. 7

Canopy trees. Trunks with rounded, spreading, branched buttresses. Stems glabrous. Bark rusty-orange, papery, readily flaking off. Stipules triangular, caducous. Leaves opposite; petioles ca. 20 mm long; blades 11–14 × 4.5–5.5 cm, chartaceous, glabrous, the apex obtuse to rounded, the very apex often slightly emarginate; secondary veins in 10–14 pairs, the marginal vein not well developed. Flowers glabrous; calyx with spur longer than largest sepal, the spur ca. 8 mm long; petals 3; anther 4 mm long. Fruits 5 × 1.5 cm. Seeds ca. 5 cm long, the seed itself densely pilose. Fl (Nov, Dec), fr (Apr); rare, in nonflooded forest. *Acacia mâle*, *couali*, *couari*, *kouali* (Créole), *quaruba branca*, *rabo de tucano* (Portuguese).

Vochysia neyratii Normand

Canopy to emergent trees. Trunks sometimes with low, poorly developed buttresses. Bark smooth or scalloped, sometimes with hoop marks. Stems pubescent. Stipules triangular, caducous. Leaves opposite; petioles 20–30 mm long; blades 12–18 × 5–7.5 cm, pubescent, the apex acuminate; secondary veins in 24–27 pairs, the marginal vein well developed. Flowers pubescent; calyx with spur shorter than largest sepal, the spur 10–11 mm long; petals 0; anthers 9 mm long. Fl (Jan, Feb, Nov), fr (Nov); occasional, in nonflooded forest.

Vochysia surinamensis Stafleu FIG. 326, PL. 128c

Canopy trees. Trunks without or with low spreading buttresses. Bark smooth, sometimes peeling in fine irregular plates, sometimes with distinct hoop marks. Stems glabrous. Stipules triangular, caducous. Leaves in whorls of 3; petioles 15–20 mm long; blades 8–15 × 4–7 cm, coriaceous, glabrous, the apex obtuse, rounded, or truncate, the very apex slightly to markedly emarginate; secondary veins in 14–18 pairs, the marginal vein not well developed. Flowers glabrous; calyx with spur nearly as long as largest sepal, the spur ca. 7 mm long; petals 3; anthers 4.5 mm long. Fl (Jul, Aug, Sep), fr (Aug, Sep); occasional, in nonflooded forest.

Vochysia tomentosa (G. Meyer) A. P. DC.

Canopy trees. Trunks with symmetrical, rounded, simple buttresses. Bark gray, rough, with prominent lenticels. Stipules acicular, persistent, ca. 3 mm long. Stems pubescent. Leaves opposite; petioles ca. 5 mm long; blades 4–7 × 1.5–2.5 cm, pubescent, the apex acuminate; secondary veins in 8–12 pairs, the marginal vein not well developed. Flowers pubescent; calyx with spur slightly shorter than largest sepal, the spur 4–7 mm long; petals 3; anther 7 mm long. Fl (Dec, Jan), fl (Jul); rare, in nonflooded forest.

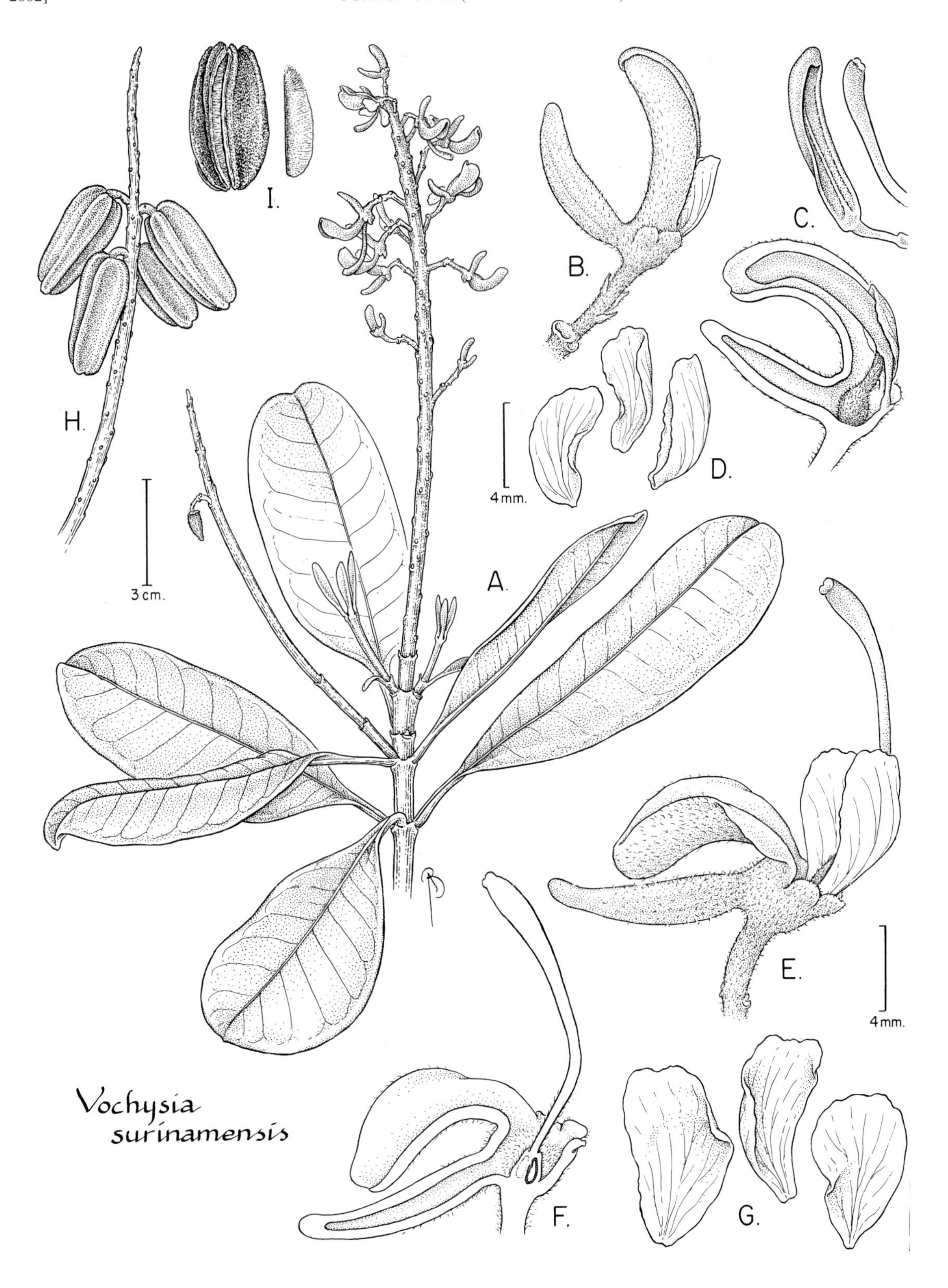

FIG. 326. VOCHYSIACEAE. *Vochysia surinamensis* (A, *Mori et al. 20980*; B–D, *Bunting 4271* from Venezuela; E–G, *Mori & Hartley 18538*; H, *Mori et al. 23844*; I, *Holst & Liesner 2674* from Venezuela). **A.** Part of stem with leaves and inflorescence. **B.** Lateral view of flower. **C.** Stamen (near right), style (far right), and longitudinal section of flower showing base of pistil, stamen surrounding upper part of style, and petal surrounding base of stamen (below), all within medial section of flower with two petals removed. **D.** Three views of petals. **E.** Lateral view of flower with expanded pistil after stamen has fallen. **F.** Medial section of flower; note spurred calyx-lobe. **G.** Three petals. **H.** Part of infructescence with immature fruits. **I.** Fruit (above left) and winged seed (above right).

Additions and Corrections to Part 1

The scale in Figure 2 is not correct. The bar should have been 3.4 cm long to indicate 10 km. As published, the scale bar gives the impression that the area covered by the *Guide* is smaller than it really is.

Since the publication of Part 1, six species of ferns new to our area have been collected:

Asplenium hostmannii Hieron. (add to Aspleniaceae, Part 1: 69–72)

Plants epiphytic. Rhizomes short, erect, with brown to black scales. Fronds pinnate, 30–50 × 6–8 cm; petiole brownish green, slender, covered at base with scales similar to those of rhizome; blade lanceolate, pinnatifid toward apex; pinnae horizontal, in 15–25 pairs, 15–40 × 5–10 mm, petiolulate, glabrous, asymmetric at base; veins arcuate. Sori linear. In lowland forest.

This species is vouchered by *Mori et al. 23780* (CAY).

This species fits into the key modified from that on page 69 as follows:

7. Stem short-creeping. Leaves closely spaced; petiole brown to black, shiny. . . . . *A. laetum.*
7. Stem erect. Leaves fasciculate; petiole brownish-green, dull.
    7′. Basal pinnae with auricle at base. . . . . *A. auritum.*
    7′. Basal pinnae without auricle at base. . . . . *A. hostmannii.*

Lindsaea cultriformis K. U. Kramer (add to Dennstaedtiaceae, Part 1: 79–85)

Fronds pinnate or bipinnate, with terminal pinna similar to others; petiole as long as lamina, castaneous to atropurpureous at base, abaxially rounded; blades with 3–5 pinnae to side, the pinnae subopposite or alternate, short stalked to sessile, 5–10 cm long, the rachises narrowly winged; pinnules lanceolate, slightly falcate, 2–3 times longer than wide, mostly with 3–5 shallow incisions, 10–30 per side, mostly subopposite, the upper ones alternate, 5–12 × 3–5 mm, the basal ones slightly shorter, the upper ones distinctly reduced in size, confluent into lobed or toothed acuminate apex to 2 cm long. Sori interrupted, mostly 2-veined, the basal ones 3–4-veined. In understory of slope forest.

Boom & Mori 1648 (CAY, NY) represents the first collection of this species for French Guiana.

This species fits into the key modified from that on page 80 as follows:

5. Pinnules slightly falcate, the apex acute to acuminate.
    5′. Blades nearly always pinnate; pinnules 5–6 times longer than wide, the upper pinnules often reduced in size. Sori continuous. . . . . *L. surinamensis.*
    5′. Blades pinnate or bipinnate; pinnules 2–3 times longer than wide, the upper pinnules distinctly reduced in size. Sori interrupted. . . . . *L. cultiformis.*
5. Pinnules not falcate, the apex round to quadrangular.

Grammitis mollissima (Fée) Proctor (add to Grammitidaceae, Part 1: 101–103)

Rhizomes small, erect. Fronds 5–10, erect, pinnatifid, incised to rachis, 3–15 × 0.6–1.2 cm, the triangular to ovate lobes adnate, 3–6 mm long, with brownish, 1 mm long hairs; veins simple, terminating in small hydathode; petiole 2–3 cm long. Sori inserted singly at end of each ultimate segment. Understory, in lowland forest.

This species is represented by *Boudrie 3285* (CAY) and *Mori et al. 23229* (CAY).

This species fits into the key modified from that on page 101 as follows:

4. Blade 2–35 mm wide, with reddish hairs on midrib, with 6–10 veinlets. . . . . *G. taxifolia.*
4. Blade 3–15 mm wide, glabrous or pubescent, with 1–5 veinlets.
    5. Blade glabrous, 10–40 × 0.6–1.5 cm. . . . . *G. suspensa.*
    5. Blade hirsute, 3–15 × 0.6–1.5 cm.
        6. Fronds not incised to rachis; petiole 2–3 cm long. Sori inserted along midvein of ultimate segment. . . . . *G. staheliana.*
        6. Fronds incised to rachis; petiole 0.4–0.6 cm long. Sori inserted at end of each ultimate segment. . . . . *G. mollissima.*

Elaphoglossum longicaudatum Mickel (add to Lomariopsidaceae, Part 1: 116)

Plants epiphytic. Rhizomes compact, 3–6 mm diam., covered with linear-lanceolate, red-brown, long acuminate scales 2–3 mm long. Sterile fronds with petiole 1–3 cm long, covered with small, stellate, orange scales; blades oblanceolate, 12–32 × 1.5–3.5 cm, covered with dense, linear-lanceolate, 1–2 mm long scales, the margins with hair-like teeth, the veins indistinct, free, simple, or 1-forked. Fertile fronds ca. 1/2 length of sterile fronds; petiole 3 cm long; blades linear, 8 × 0.7 cm. In lowland forest.

Mori et al. 21568 (CAY) represents the first collection of this species for French Guiana.

This species fits into the key modified from that on page 116 as follows:

3. Rhizome 3–10 mm diam.
   4. Rhizome scales orange. Fronds pendent, linear, 50–200 cm long. . . . . *E. herminieri.*
   4. Rhizome scales not orange. Fronds erect, ellptic or oblanceolate, <35 cm long.
      5. Sterile fronds with petiole 1–5 cm long.
         5′. Blades elliptic, with brown scales. . . . . *E. luridum.*
         5′. Blades oblanceolate, with orange scales. . . . . *E. longicaudatum.*
      5. Sterile fronds with petiole 10–15 cm long.

Pecluma hygrometrica (Splitg.) M. G. Price (add to Polypodiaceae, Part 1: 130–131)

Plants epiphytic. Rhizomes short-creeping, covered with orangish, flexuous scales, 1 mm long. Fronds 8–12 cm long; petiole blackish-brown, with brown pubescence, 1–2.5 cm long; blade elliptic to oblong, 7–10 × 2–4 cm, the midrib blackish with same hairs as petiole; segments linear, 2–4 mm wide, with white pubescence, with entire, ciliate margins; veins bifurcate, scarcely visible. Sori round, at ends of veins. In lowland forest.

Boudrie 3297 (CAY) represents the first collection of this species for French Guiana.

A new key to the species of *Pecluma* in our area follows:

1. Rhizome scales orangish. Fronds ≤12 cm long; segments 2–4 mm wide. . . . . *P. hygrometrica.*
1. Rhizome scales brown or black. Fronds ≥15 cm long; segments 2–7 mm wide.
   2. Fronds 15–30 cm long; segments 2–3 mm wide; petiole black. . . . . *P. plumula.*
   2. Fronds 20–90 cm long; segments 3–7 mm wide; petiole brown. . . . . *P. pectinata.*

Thelypteris biolleyi (H. Christ) Proctor (add to Thelypteridaceae, Part 1, pp. 152–157)

Plants terrestrial. Rhizomes woody, creeping or suberect. Fronds monomorphic, to 100 cm long, with to 15 pairs of pinnate-pinnatifid pinnae, the segments reduced in size at base; pinnae 12–13 × 1.5–2 cm, incised 1/3–1/2 distance to midrib; veins with 1 pair united at sinus; petiole and rachis covered with dense, sessile, stellate and acicular hairs, especially on young parts. In understory of lowland forest.

Boudrie 3295 and *3296* (both CAY) represent the first collections of this species for French Guiana.

This species fits into the key modified from that on page 152 as follows:

7. Lateral pinnae in 8–15 pairs, the segments reduced in size at base.
   8. Blades lacking stellate hairs between veins. . . . . *T. abrupta.*
   8. Blades with sessile stellate hairs between veins.
      8′. Rachis lacking acicular hairs. . . . . *T. nephrodioides.*
      8′. Rachis with acicular hairs, especially on young parts. . . . . *T. biolleyi.*
7. Lateral pinnae in 15–25 pairs, the segments not reduced in size at base. . . . . *T. gemulifera.*

Three orchids, *Beloglottis* sp. (*S. A. Mori 24129* at NY), *Scaphyglottis bifida* (Rchb.f.) Schweinf. (*J. P. Bikaeff s.n.* at NY), and *Sigmatostalix amazonica* Schltr. (*S. A. Mori 24996* at CAY), represent new taxa for the flora. All three were determined by Eric Christenson.

Rapatea ulei Pilg. (*Granville et al. 14070*) is now known to be common in swampy areas in the vicinity of Pic Matécho.

On page 31, the leaves of *Parabignonia steyermarkii* (Bignoniaceae) are opposite, not alternate.

On pages 32 and 37, *Bunchosia* sp. should be *B. decussiflora.*

On page 34, *Peperomia* sp. should be *P. gracieana.*

On page 36, *Tachigali* sp. B should be *T.* aff. *paniculata.*

On page 38, *Mezia* sp. should be *Mezia angelica* and *M. includens* and the following sentence should be in the plural.

On page 42, *Diospyros matheriana* should be *D. ropourea.*

On page 44, *Simaba guianensis* should be *S. guianensis* subsp. *guianensis* and *S. nigrescens* should be *S. guianensis* subsp. *ecaudata.*

On page 46, *Diospyros melinonii* should be *D. caprifolia.*

On page 72, *Asplenium perkinsii* Jenman should be *A. rutaceum* (Willd.) Mett. This change should also be made in the key to species of *Asplenium*, p. 69.

On page 87, the key to *Cyclodium* should be modified as follows:

2. Pinnae in (15)20–30 pairs, the midrib with ovate scales on abaxial surface. . . . . *C. guianense.*
2. Pinnae in 9–15(24) pairs, the midrib without ovate scales on abaxial surface. . . . . *C. inerme.*

On page 132, *Pleopeltis revoluta* (Willd.) A. R. Sm. should be *P. astrolepis* (Liebm.) E. Fourn. This change should also be made in the key to species of *Pleopeltis.*

On page 135, Michel Boudrie has pointed out that a much easier way to tell *Polypodium lasiopus* from *P. sororium* is by the presence of white hairs on the abaxial leaf surface of the former.

On page 179, *Monstera* sp. nov. should be *Monstera dubia* (Humb., Bonpl. & Kunth) Engl. & K. Krause.

On page 185, *Philodendron scandens* K. Koch & Sellow should be *P. hederaceum* (Jacq.) Schott. This change should also be made in the key to species of *Philodendron* on page 180.

On page 215, in the generic description of *Syagrus*, the inflorescences should be described as interfoliar, not infrafoliar.

On page 266, *Dioscorea trifida* L. f., not L.

On page 273 the correct abbreviation for the authors of *Heliconia lourteigiae* is Emygdio & E. Santos not Mello & E. Santos.
On page 381, craspedium, not crespidium.
On page 395, in the definition of "stemonozone," adnation should be replaced by fusion.
On page 396, syngenesious, not syngynesious.
On page 412, *Andropogon bicornis*, not *Axdropogon bicornis*.
On page 418, *Pseudima frutescens*, not *Pseudima fructescens*.
On page 420, *Vouarana guianensis*, not *Vourana guianensis*.
Figure 6, *Securidaca diversifolia* should be *Securidaca volubilis*; *Tetrapterys crispa* should be *T. glabrifolia*; *Heteropterys siderosa* should be *H. oligantha*.
Figure 9, *Sloanea* sp. (*Mori et al. 15681*) should be *Sloanea* sp. B.
Figure 10, *Apeiba aspera* should be *Apeiba petoumo*.
Figure 21D, E. *Asplenium perkinsii* Jenman should be *A. rutaceum* (Willd.) Mett.
Figure 61G, H. *Pleopeltis revoluta* (Willd.) A. R. Sm. should be *P. astrolepis* (Liebm.) E. Fourn.
Figure 73, The names *Thelypteris decussata* and *T. gemmulifera* were switched. E-G = *T. gemmulifera* and I-K = *T. decussata*.
Plate Ib, *Ocotea rubra* should be *Sextonia rubra*.
Plate IIa, *Sloanea* sp. (*Mori et al. 21665*) should be *S. echinocarpa*.
Plate VIIIc, *Psychotria surinamensis* should be *Psychotria microbotrys*.
Plate XIIa, *Apeiba aspera* should be *Apeiba petoumo*.

Glossary (supplement)

This is a supplement to the glossary presented in Part 1. See pp. 375–397 of Part 1 for an introduction to the glossary as well as for the definition of many other terms used to describe the vascular plants of central French Guiana. An important, well-illustrated source for the definition of terms used to describe the vegetative features of tropical plants has recently been provided by Keller (1996). A useful reference for defining foliar trichome types is Webster et al. (1996).

acrodromous. Referring to a type of leaf venation in which two or more primary or strongly developed secondary veins arch upward from or near the base; perfect acrodromous refers to a type of acrodromous leaf venation in which the primary or strongly developed secondary veins converge near the apex, e.g., *Strychnos* (Loganiaceae); imperfect acrodromous refers to a type of acrodromous leaf venation in which the primary or strongly developed secondary veins do not converge at the apex (Hickey, 1979).
actinodromous. Referring to a type of leaf venation in which three or more primary veins diverge radially from a single point (Hickey, 1979).
aculeate. Covered with aculei (prickles).
acumen. Apex.
andromonoecious. See glossary in part 1. In the Mimosaceae, referring to the presence of staminate and pistillate flowers in the same cluster.
annotinous. Referring to structures, e.g., inflorescences, arising from branches of last year's growth.
anomalous phloem. See anomalous secondary growth. See definition, Part 1: 378.
apophysis. A projection or protuberance.
areolate. With areoles; in mimosoid and caesalpinioid legumes, referring to seeds with a pleurogram.
areola. A flattened spot, circular or elliptic in outline, often on the apex of the seeds of some Violaceae (e.g., *Rinorea*).
areole. The smallest area of an organ, e.g., leaf or fruit, enclosed by united veins.

basal stoppers. Valves in the woody pericarp of *Parinari* that allow for the escape of the seedling.
bilabiate capitulum (head). Inflorescence of Asteraceae with at least some bilabiate flowers. Restricted to tribe Mutisieae and not recorded for our area.
bilabiate flower. A zygomorphic flower with two distinct lips, e.g., species of Asteraceae Tribe Mutisieae which are not yet recorded from our area.
bitegmic. Referring to ovules with two integuments.
brachyblast. In *Inga* (Mimosaceae), referring to a short, leafless lateral axis of limited growth that bears inflorescences (e.g., *I. alba*).

calycophyll(s). Expanded sepal or sepals of some Rubiaceae that function to attract pollinators (semaphylls) or aid in wind dispersal (pterophylls). See pterophyll and semaphyll.
calyculate. Referring to species of Asteraceae (common in the Senecioneae) possessing a secondary series of bracteoles subtending the primary phyllaries.
calyx tube. In Ericaceae with inferior ovaries, the proximal portion of the calyx fused with the ovary wall. Same as hypanthium as used for other plant families.
carpopodium. Differentiated base of the cypsela (Asteraceae), commonly zygomorphic and occasionally sculptured
cat-claw tendril. A tendril with the distal part divided into three equal, conspicuously recurved, spiny parts, e.g., in *Macfadyena unguis-cati* and *Parabignonia steyermarkii*.
chaff. Same as pale. See definition of pale, Part 1: 389.
clathrate. Lattice-like in appearance.
clavuncle. Same as style head, Part 1: 395.
conflorescence. The aggregation of several inflorescences into one, e.g., some species of Proteaceae.
corpus. Body, e.g., the main part of the style head (=clavuncle) in Apocynaceae.
craspedium. Misspelled as crespedium, Part 1: 381.
craspedidromous. Referring to a type of leaf venation in which the secondary veins terminate at the margin (Hickey, 1979).
crassulate. Thick.
cruciform (cruciate). Cross-shaped.

dichasium (plural = dichasia). A cymose inflorescence in which each axis produces a terminal flower and two more-or-less opposite lateral axes.
dicliny. Possessing separate staminate and pistillate flowers, see dioecious and monoecious.

dichogamy. Referring to flowers with the stamens shedding pollen and the stigmas receptive at different times, see protandrous and protogynous.
dimidiate. Divided into two halves; referring to an organ divided into two halves such that the smaller half is almost lacking, e.g. the leaflet blades of *Dimorphandra* which are sometimes divided into two halves distinctly different in size by the midrib.
dimorphic. See glossary, Part 1. In the Mimosaceae, referring to different flower shapes in the same inflorescence.
disciform capitulum (head). Inflorescence of Asteraceae with marginal actinomorphic, pistillate flowers and central disk flowers.
discoid capitulum (head). Inflorescence of Asteraceae with only disk flowers.
double margin. In the Bignoniaceae, referring to a calyx with the apex folded over to give the impression that two calyx margins are present, e.g., species of *Amphilophium*.
drepanium. Referring to a cymose inflorescence shaped like a sickle.

enantiosylous. Referring to flowers whose styles protrude alternately to the right and to the left of the main axis.
ericoid. Like members of the Ericaceae in one feature or another, e.g., used to describe the small leaves without typical venation in some species of Melastomataceae.
eugenioid embryo. A type of embryo found in the Myrtaceae in which the cotyledons are thick, separate, and plano-convex (like those of a bean) and the hypocotyl is a short protrusion, or the cotyledons are fused partially or completely in a single mass and the hypocotyl is not distinguishable (Landrum & Kawasaki, 1997).

floret. The individual flower of Asteraceae, together forming the capitulum or head characteristic of the family. See definition of floret as used for Poaceae, Part 1: 384.
frondose. Leafy, used to describe inflorescences bearing numerous, leaf-like bracts, a condition especially common in the Rubiaceae.

glomerule. A small, generally spherical, densely packed cluster of capitula in the Asteraceae.

hapter (plural = haptera). Disk-like or irregularly formed lateral outgrowths of roots (rarely shoots) that affix plants of many Podostemaceae to the substrate. Same as holdfast.
holdfast. Same as hapter.
hermaphrodite. See glossary, Part 1. In the Mimosaceae, referring to the presence of both staminate and pistillate flowers in the same inflorescence.
heterogamous. Capitula of Asteraceae in which two types of flowers are present. See alternative definition, Part 1: 385.
heteromorphic. See glossary, Part 1. In the Mimosaceae, referring to inflorescences with flowers of different sexes in the same inflorescence.
homogamous. Capitula of Asteraceae in which one type of flower is present. See alternative definitions, Part 1: 385.
humifuse. Spread out over the ground, e.g., the stems of *Desmodium axillare*.
hydathode. A structural modification, usually in leaves, that permits the release of water through an opening in the epidermis.
hydropote. In the Myrsinaceae, a multicellular, epidermal structure, often of leaves, serving for the absorption of water and mineral salts. When young, hydropotes consist of a basal stalk and a group of cap cells, but at maturity the superficial structure breaks off and the remaining depression often regulates water as a hydathode.
hypanthium. See glossary, Part 1. In Ericaceae with inferior ovaries, the proximal portion of the calyx fused with the ovary wall. Same as calyx tube.

interpetiolar glands. Excretory structures located between the petioles of some plants, e.g., species of Bignoniaceae.
involucrum. An adherent group of hairs arising from the style head in the Apocynaceae which often forms a ring that closes off the corolla. Sometimes called the "ring."

jugum (plural = juga). A pair of leaflets on a pinnately compound leaf.

ligulate capitulum (head). Inflorescence of Asteraceae with only ligulate flowers. Restricted to tribe Lactuceae and not recorded for our area.
ligulate flower. A zygomorphic flower with a strap-like corolla found in the Asteraceae tribe Lactuceae. Differring from a ray flower by having five instead of the usual three apical corolla lobes.
limen. In the Passifloraceae, a ring or a cup-shaped membrane more or less closely surrounding the base of the gynophore which may not be present (Jørgenson et al., 1984).

marginicidal dehiscence. See septifragal dehiscence.
midpoint. Erect, central cell(s) of a stellate hair of *Solanum*, the basal cells are called rays.
myrcioid embryo. A type of embryo found in the Myrtaceae in which the cotyledons are normally thin, leafy, and folded into a bundle, and the narrow, cylindrical hypocotyl is about the same length as the cotyledons and encircling them (Landrum & Kawasaki, 1997).
myrtoid embryo. A variable type of embryo found in the Myrtaceae in which the hypocotyl is the same length or much longer than the cotyledons; in those genera with hard seed coats the embryo is C-shaped; in those genera with membranous or submembranous seed coats it is common for the hypocotyl to be greatly swollen and sometimes for the whole embryo to form a spiral (Landrum & Kawasaki, 1997).

napiform. Turnip-shaped.
nectar ring. In the Passifloraceae, a low narrow ring situated below the operculum which may or may not be present (Jørgenson et al., 1984).
nitid. Shiny or lustrous.

obturamenta (singular = obturamentum). See basal stoppers.
ocreola. A persistent membranous, tubular bracteole (ocreola) surrounding the flowers of species of Polygonaceae.
operculum. In the Passifloraceae, normally a membranous structure of various forms situated below the corona (Jørgenson et al., 1984).
ovariodisc. In the Burseraceae, referring to a single, parenchymatous structure representing the ontogenetic fusion of the nectary disc and the pistillode.

palate. In the Scrophulariaceae, the space at the throat and limb of the corolla used by pollinating insects as a landing platform.
paraphyllidium (plura = parphyllidia). Minute appendage resembling very reduced leaves or stipules inserted on leaf or pinna rachis just above the pulvinule, e.g., species of *Mimosa*.
patelliform. Shaped like a knee-cap, e.g., the glands of some species of *Diospyros*.

pedaliform. Resembling the sole of a foot in shape.

peduncle. See glossary, Part 1. In the Mimosaceae, referring to the stalk of the capitulum which is actually a secondary peduncle.

percurrent. In the Viscaceae, the longitudinal pattern of shoot growth through continued activity of the apical meristem. This contrasts with dichotomous growth also found in the family.

perulate. Referring to buds covered with scales.

pherophyll. Leaf-like bract found in frondose inflorescences.

phloem arm. A segment of cross-shaped or star-shaped phloem as seen in transverse section in some lianas, especially species of Bignoniaceae.

placental obturators. A massive outgrowth of the placenta. In the Styracaceae, the micropyle of each ovule opens upon an obturator; the obturators may or may not be coalescent.

plinerved. See pliveined.

pliveined. Referring to leaf blade venation in which the midvein is accompanied by several nearly equal secondary veins arising at or near the base; found, for example, in many Ericaceae and some Euphorbiaceae.

primary veins. Same as principal veins (see glossary in part 1). Used by some authors for what we call secondary or lateral veins (see definitions in glossary, Part 1). All the nearly equal first order veins of most species (excluding species of *Mouriri* and *Tovomita*) of Melastomataceae.

prominent. Raised above the surface, e.g., the veins of a leaf.

prominulous. Diminutive of prominent.

pseudostaminodia (singular = pseudostaminodium). In the Amaranthaceae, referring to sterile, filamentous, entire to fimbriate processes as long as or longer than the stamens and alternating with them.

pseudostipules. 1) Leaves (e.g., in some species of Aristolochiaceae) or leaflets (e.g., in some species of *Trichilia*) that resemble stipules; 2) the bud scales of the axillary buds of some Bignoniaceae that resemble stipules, see Plate Ixc, Part 1 (*Cydista aequinoctialis*) in which the pseudostipules are covered by small glands visited by ants. The pseudostipules are erroneously called stipules in the legend of this plate.

pseudosyncarpous. In the Annonaceae, referring to fruits that are seemingly connate, but in reality are free or almost free (e.g., species of *Duguetia*)

pterophyll. Expanded sepal or sepals (usually white or green) of some Rubiaceae that function to aid in wind dispersal. See calycophyll.

pulverulent. Appearing dusty or powdery.

radiate capitulum (head). Inflorescence of Asteraceae with marginal ray flowers and central disk flowers.

sarmentose. Referring to a plant that produces long, slender runners.

semaphyll. Expanded sepal or sepals, usually brightly colored, of some Rubiaceae that function to attract pollinators. See calycophyll.

septifragal dehiscence. Referring to a type of fruit dehiscence in which the valves break away from the septa, e.g., in species of *Paullinia*.

spathella. Sac-like covering of the young flowers of Podostemaceae.

stipule cap. In some Rubiaceae (e.g., *Duroia*), the structure formed by the fusion of the stipules. It covers the apical buds and usually falls off as soon as the bud develops.

stipule persistence. In the Rubiaceae, the persistency of the stipules provides useful characters for identifying genera and subgenera (Delprete, 1999). Readily caducous stipules are present only when the leaves are in bud; caducous stipules are present during leaf development but fall before the leaves fall; and persistent stipules are still present after the leaves of their node fall.

suture. A line of fusion (for example of the valves of the carpels of an ovary) or the line of dehiscence (e.g., the lines along which anthers or fruits open).

thalloid. A generalized term used to describe the prostrate axis of a plant when demarcation into stem, leaf, and root is not obvious. Use of "thalloid" in Podostemaceae does not imply homology with the vegetative bodies of lower plants, e.g., some species of liverworts.

vaginate. Possessing a sheath, e.g., used to describe the petiole base of some species of *Piper*.

vermifuge. Referring to a substance that expels intestinal worms, e.g., extracts of *Senna alata*.

vestiture. The epidermal outgrowths or coverings of a plant.

Acknowledgments

The *Guide to the Vascular Plants of Central French Guiana* would not have been possible without the collaboration of many botanists. Therefore, we dedicate this book to those botanists whose collections and knowledge of the neotropical flora made this *Guide* possible. We are especially grateful to the 78 specialists who contributed treatments of the groups in which they specialize to this *Guide*. These botanists, who comprise a major portion of the plant specialists currently working in Neotropical botany, are shown in Plates 1–5.

The writing of a detailed inventory of the plants of any geographic region is an expensive undertaking. We are thankful to the following funding agencies for financial support of many aspects of this undertaking, including field work, herbarium study (including computerization), preparation of manuscripts (including the cost of illustration), and the expense of the publication of the color plates: the Beneficia Foundation (general support), Bristol-Myers Squibb Foundation (computerization), the Oliver S. and Jennie R. Donaldson Charitable Trust (botanical illustration), the Eppley Foundation for Research (salary support), Dorothy Salant and the G.A.G. Charitable Coorporation (general support), the Harriet Ford Dickenson Foundation (general support), the National Geographic Society (three field trips), the National Science Foundation (BSR–9024530) (general support), the Andrew W. Mellon Foundation (salary support), the Rockefeller Foundation (one field trip), and the Rhulen Family Foundation (botanical illustration). We are particularly appreciative of the support of the Beneficia Foundation, the Eppley Foundation for Research, and the Andrew W. Mellon Foundation for the support that made it possible for us to continue with this project after funds from the initial National Science Foundation grant were depleted. The senior editor is grateful to the Harriet Ford Dickenson Foundation for making it possible for him to hold the Nathaniel Lord Britton Curator of Botany chair at the New York Botanical Garden. We are grateful to WWF Guianas for financial support of the publication of the second part of the *Guide to the Vascular Plants of Central French Guiana.*

We are especially grateful to Beth Mitchell, wife of J. D. Mitchell, for her long-term financial support and for her encouragement to us over the course of this and other botanical projects.

Another source of funding has been through the neotropical ecotour program of the Institute of Systematic Botany coordinated by Carol Gracie, and led by Mori, Gracie, and J. D. Mitchell. The publication of the color photographs was partially supported with funds generated by this program. In addition to those individuals acknowledged in Part 1 for their earlier contributions, we are grateful to Sheila and Thane Asch, John Bernstein, Diana Davenport, Susan Fredericks, Gerhard Haas, Anne Hubbard, Katie Lee, Bob and Carol Russell, and Hazel Tuttle for their additional support for the art work by Bobbi Angell and for the publication of the color plates.

Many of the individuals who have helped us with our field work remain the same as cited in Part 1 of this book. However, we would be remiss if we did not give special thanks to: the Allinckx family at Eaux Claires who have taken us into their family on our expeditions to central French Guiana since 1990; to Joep and Marijka Moonen for logistical support and friendship in Cayenne; to Françoise Crozier, Venise Isidore, and Lucien Aboukrat for their help with specimens during our visits to the Herbier de Guyane; to the successive directors of the Centre IRD (ex ORSTOM) de Cayenne, who have allowed us and many of the specialists who have worked on this *Guide* to use the collections of the Herbier de Guyane; and to our special friends Bernard Hermier, Marie-Françoise Prévost, and Gerard Tavakilian who helped make our trips to French Guiana so pleasant.

We are fortunate to be able to illustrate this *Guide* with the exquisite art of Bobbi Angell. Her skill as an artist, knowledge of botany, and patience with a project spanning over nearly eleven years are sincerely appreciated. Once again, we thank Pedro Acevedo-Rodríguez for allowing us to use some of the illustrations prepared by Bobbi Angell for his *Flora of St. John, U.S. Virgin Islands*. We also thank Rupert Barneby, Jim Grimes, and the Smithsonian Institution for permitting the use of several illustrations of Mimosaceae prepared by Bobbi for treatments to be published in the *Flora of the Guianas*. These illustrations are attributed to *Flora of St. John, U.S. Virgin Islands* and the *Flora of the Guianas* in the figure legends.

The management of our collections over most of the course of this project has been meticulously handled by Sarah Hunkins and Eileen Whalen. We thank them for their careful typing of labels, for the distribution of specimens, and for many other aspects of specimen handling. We are also grateful to Jill Appel, Tony Kirchgessner, and the Computer Services Department of The New York Botanical Garden (directed by Nancy Steger) for conversion of the French database (AUBLET) of specimens upon which this guide was originally based as well as for making the database available for consultation on the Internet <http://www.nybg.org/bsci/hcol/french_guiana>. We thank Myrna Álvarez, Sharol Charles, Anthony Giordano, Eliza Habegger, Edmond Hecklau, John Janovec, Laura Marx, Oliver Pihlar, Ruth Rosenthal, and especially Nathan Smith for their assistance with the laborious task of recording specimen data for insertion into the database as well as for help with numerous other time consuming tasks. We are grateful to Myrna Álvarez for the many hours she spent in the preparation of the index and her willingness to help in many other ways.

We are grateful to C. C. Berg (additions to the Aids to Identification for Cecropiaceae and Moraceae), P. E. Berry (for his review of the entire manuscript), J. M. Cardiel (*Acalypha*), T. Daniel (for his review of the entire manuscript), M. Fleury (for advice on the origins of common names), P. Hiepko (Menispermaceae), M. L. Kawasaki (Melastomataceae), G. Morillo (Apocynaceae), A. van Prossdij (Hernandiaceae), L. Struwe (Gentianaceae and Loganiaceae), and D. Wasshausen (Begoniaceae) for their reviews of various parts of the manuscript. We are grateful to Paul Berry, Michel Boudrie (especially for ferns), Beat Fischer, and Robert Voss for pointing out errors in Part 1.

We thank Tom Zanoni for his suggestions based on his careful review of several versions of the manuscript and for copyediting the entire manuscript; the editor of the Memoirs of The New York Botanical Garden, William R. Buck, for his many corrections, useful suggestions, and punctual handling of the editorial process; and Joy E. Runyon for her patience and for the many hours of careful work that she put into the production of this book. Their efforts greatly improved the final product.

This is publication number 92 in the Studies on the Flora of the Guianas series of the Flora of the Guianas Project.

Literature Cited

Al-Shehbaz, I. A. & R. A. Price. 1998. Delimitation of the genus *Nasturtium* (Brassicaceae). Novon **8:** 124–126.

Alverson, W. S. 1994. New species and combinations of *Catostemma* and *Pachira* (Bombacaceae). Novon **4:** 3–8.

———**, B. A. Whitlock, R. Nyffeler, C. Bayer & D. A. Baum**. 1999. Phylogeny of the core Malvales: evidence from *ndh*F sequence data. Amer. J. Bot. **86:** 1474–1486.

Anderson, W. R. 2001. Malpighiaceae. *In* J. A. Steyermark, P. E. Berry, K. Yatskievych & B. K. Holst (gen. eds.), Flora of the Venezuelan Guayana. P. E. Berry, K. Yatskievych & B. K. Holst (vol. eds.). Missouri Botanical Garden Press.

Angiosperm Phylogeny Group. 1998. An ordinal classification of the families of flowering plants. Ann. Missouri Bot. Gard. **85:** 531–553.

Barneby, R. C. 1993. Two taxonomic equations relevant to the flora of Saül, French Guiana. Brittonia **45(3):** 235–236.

Baum, D. A., W. S. Alverson & R. Nyffeler. 1998. A durian by any other name: taxonomy and nomenclature of the core Malvales. Harvard Pap. Bot. **3(2):** 315–330.

Bayer, C., M. F. Fay, A. Y. De Bruijn, V. Savolainen, C. M. Morton, K. Kutitzki, W. S. Alverson & M. W. Chase. 1999. Support for an expanded family concept of Malvaceae within a recircumscribed order Malvales: a combined analysis of plastid *atp*B and *rbc*L DNA sequences. Bot. J. Linn. Soc. **129:** 267–303.

Bedell, H. G. 1985. A generic revision of the Marcgraviaceae. I., The *Norantea* complex. Ph.D. dissertation, University of Maryland, College Park.

Boggan, J., V. Funk, C. Kelloff, M. Hoff, G. Cremers & C. Feuillet. 1997. Checklist of the plants of the Guianas. Ed. 2. Smithsonian's Biological Diversity of the Guianas Program **30:** 1–238.

Bohs, L. 1995. Transfer of *Cyphomandra* (Solanaceae) and its species to *Solanum*. Taxon **44:** 583–587.

Brummit, R. K. & C. E. Powell. 1992. Authors of plant names. Royal Botanic Gardens, Kew.

Burger, W. 1971. Family 41. Piperaceae. Flora Costaricensis. Fieldiana Bot. **35:** 5–218.

———. 1999. *Spermacoce*. In: D. H. Lorence, A nomenclator of Mexican and Central American Rubiaceae. Monogr. Syst. Bot. Missouri Bot. Gard. **73:** 166–170.

——— **& C. M. Taylor.** 1993. *Spermacoce* (Rubiaceae). *In* W. C. Burger (ed.), Flora costaricensis. Fieldiana, Bot. n.s. **33:** 1–333.

Charles-Dominique, P., M. Atramentowicz, M. Charles-Dominique, H. Gérard, A. Hladik, C. M. Hladik & M. F. Prévost. 1981. Les mammifères frugivores aboricoles nocturnes d'une forêt Guyanaise: Inter-relations plantes-animaux. Rev. Ecol. (Terre et Vie) **35:** 341–435.

Cronquist, A. 1981. An integrated system of classification of flowering plants. Columbia University Press, New York.

Delprete, P. G. 1999. Rondeletieae (Rubiaceae), Part I. Fl. Neotrop. Monogr. **77:** 1–226.

Eichler, A. W. 1871. Flacourtiaceae as Bixaceae Tribus Flacourtieae. Flora Brasiliensis **13(1):** 421–516.

Elias, T. S. & G. T. Prance. 1978. Nectaries on the fruit of *Crescentia* and other Bignoniaceae. Brittonia **30:** 175–181.

Fernando, E. S. & C. J. Quinn. 1995. Picramniaceae, a new family, and a recircumscription of Simaroubaceae. Taxon **44:** 177–181.

Fosberg, F. R., M.-H. Sachet & R. L. Oliver. 1993. Rubiaceae. Flora of Micronesia 5. Smithsonian Contr. Bot. **81:** 44–135.

Gentry, A. H. 1982. Bignoniaceae. Flora de Venezuela **8(4):** 1–433.

———. 1992. Bignoniaceae—Part II (Tribe Tecomeae). Fl. Neotrop. Monogr. **25(II):** 51–105.

———. 1993. A field guide to the families and genera of woody plants of northwest South America (Colombia, Ecuador, Peru), with supplementary notes on herbaceous taxa. Conservation International, Washington, DC.

——— **& W. Morawitz**. 1992. *Jacaranda*. *In* A. H. Gentry, Bignoniaceae—Part II (Tribe Tecomeae). Fl. Neotrop. Monogr. **25(II):** 51–105.

Gibbs, P. E., J. Semir & N. D. da Cruz. 1988. A proposal to unite the genera *Chorisia* Kunth and *Ceiba* Miller (Bombacaceae). Notes Roy. Bot. Gard. Edinburgh **45(1):** 125–136.

Gill, G. E., R. T. Fowler & S. A. Mori. 1998. Pollination biology of *Symphonia globulifera* (Clusiaceae) in central French Guiana. Biotropica **30(1):** 139–144.

Govaerts, R. 1996. *Spermacoce* (Rubiaceae), pp. 14–19, in R. Govaerts, World Checklist of Seed Plants Vol. **2(1)**. Continental Publishing, Antwerp, Belgium.

Gracie, C. 1991. Observation of dual function of nectaries in *Ruellia radicans* (Nees) Lindau (Acanthaceae). Bull. Torrey Bot. Club 11**8(2):** 188–190.

———. 1993. Pollination of *Cyphomandra endopogon* var. *endopogon* (Solanaceae) by *Eufriesea* spp. (Euglossini) in French Guiana. Brittonia **45(1):** 39–36.

Grayum, M. H. & B. E. Hammel. 1999. Taxonomic concepts in flux. Cutting Edge **VI(1):** 1–2.

Grenand, P., C. Moretti & H. Jacquemin. 1987. Pharmacopées traditionnelles en Guyane. ORSTOM Mém. **108:** 1–569.

Hallé, F., R. A. A. Oldeman & P. B. Tomlinson. 1978. Tropical trees and forests. Springer Verlag, New York.

Hansen, B. 1980. Balanophoraceae. Fl. Neotrop. Monogr. **23:** 1–80.

Hickey, L. J. 1979. Revised classification of the architecture of dicotyledonous leaves. Pages 25–39 *in* C. R. Metcalfe & L. Chalk (eds.), Anatomy of the dicotyledons. Vol. 1. Ed. 2. Clarendon Press, Oxford.

Jansen, R. K. 1985. The systematics of *Acmella* (Asteraceae - Heliantheae). Syst. Bot. **8:** 1–115.

Jansen-Jacobs, M. J. 1988. Verbenaceae. *In* A. R. A. Görts-van Rijn (ed.), Flora of the Guianas Ser. A, **4:** 1–114. Koeltz Scientific Books, Koenigstein.

Jørgenson, P. M., J. E. Lawesson & L. B. Holm-Nielsen. 1984. A guide to collecting passionflowers. Ann. Missouri Bot. Gard. **71:** 1172–1174.

Judd, W. S., C. S. Campbell, E. A. Kellogg & P. F. Stevens. 1999. Plant systematics: A phylogenetic approach. Sinauer Associates, Sunderland, MA.

Keller, R. 1996. Identification of tropical woody plants in the absence of flowers and fruits: a field guide. Birkhäuser Verlag, Boston.

Landrum, L. R. & M. L. Kawasaki. 1997. The genera of Myrtaceae in Brazil: an illustrated synoptic treatment and identification keys. Brittonia **49(4):** 508–536.

Maas, P. J. M. & P. Ruyters. 1986. *Voyria* and *Voyriella* (Saprophytic Gentianaceae). Fl. Neotrop. Monogr. **41:** 1–93.

McDade, L. A., S. E. Masta, M. L. Moody & E. Waters. 2000. Phylogenetic relationships among Acanthaceae: Evidence from two genomes. Syst. Bot. **25(1):** 106–121.

Mori, S. A., C. A. Gracie & J. D. Mitchell. 1998. Le centre de la Guyane française: une expérience touristique unique et rude. J. Agric. Trad. Bot. Appl. **40:** 299–310.

———**, G. Cremers, C. Gracie, J.-J. de Granville, M. Hoff & J. D. Mitchell**. 1997. Guide to the vascular plants of central French Guiana. Part 1. Pteridophytes, gymnosperms, and monocotyledons. Mem. New York Bot. Garden **76(1):** 1–422.

——— **& J. J. Pipoly**. 1984. Observations on the big bang flowering of *Miconia minutiflora* (Melastomataceae). Brittonia **36(4):** 337–341.

Nesom, G. L. 1994. Review of the taxonomy of *Aster sensu lato* (Asteraceae: Astereae), emphasizing the New World species. Phytologia **77:** 141–297.

Nicolson, D. H. & C. Jarvis. 1984. *Cissus verticillata*, a new combination for *C. sicyoides* (Vitaceae). Taxon **33(4):** 726–727.

Olmstead, R. G. & P. A. Reeves. 1995. Evidence for the polyphyly of the Scrophulariaceae based on chloroplast *rbc*L and *ndh*F sequences. Ann. Missouri Bot. Gard. **82:** 176—193.

Polhill, R. M. 1981. Sophoreae. *In* R. M. Polhill & P. H. Raven (eds.), Advances in Legume Systematics **1:** 212–230. Royal Botanic Gardens, Kew.

Persson, C. 2000. Phylogeny of neotropical *Alibertia* group (Rubiaceae), with emphasis on the genus *Alibertia*, inferred from *ITS* and *5S* ribosomal sequences. Amer. J. Bot. **87:** 1018–1028.

Prance, G. T., M. Freitas da Silva, B. W. Alburquerque, I. de Jesus da Silva Araújo, L. M. M. Carreira, M. M. N. Braga, M. Macedo, P. N. da Conceição, P. L. B. Lisbôa, P. I. Braga, R. C. L. Lisbôa & R. C. Q. Vilhena. 1975. Revisão taxonômica das espécies amazônicas de Rhizophoraceae. Acta Amazonica **5(1):** 5–22.

Renner, S.S. 1993. Phylogeny and classification of the Melastomataceae and Memecylaceae. Nordic J. Bot. **13:** 519–540.

———. 1999. Circumscription and phylogeny of the Laurales: evidence from molecular and morphological data. Amer. J. Bot. **86(9):** 1301–1315.

——— **& G. Hausner**. 1997. Siparunaceae, Monimiaceae. *In* G. Harling & L. Andersson (eds.), Flora of Ecuador **59:** 1–125.

Ribeiro, J. E. L. da S., M. J. G. Hopkins, A. Vicentini, C. A. Sothers, M. A. da S. Costa, J. M. de Brito, M. A. D. de Souza, L. H. P. Martins, L. G. Lohmann, P. A. C. L. Assunção, E. da C. Pereira, C. F. da Silva, M. R. Mesquita & L. C. Procópio. 1999. Flora da Reserva Ducke. Guia de identificação das plantas vasculares de uma floresta de terra-firme na Amazônia Central. INPA, Manaus.

Simmons, M. P. & J. P. Hedin. 1999. Relationships and morphological character changes among genera of Celastraceae *sensu lato* (including Hippocrateaceae). Ann. Missouri Bot. Gard. **86:** 723–757.

Sleumer, H. O. 1980. Flacourtiaceae. Fl. Neotrop. Monogr. **22:** 1–499.

Smith, C. E., Jr. 1954. The New World species of *Sloanea* (Elaeocarpaceae). Contr. Gray Herb. **175:** 1–114.

Steyermark, J. A. 1972. Rubiaceae. *In* B. Maguire, J. J. Wurdack & collaborators, Botany of the Guayana Highlands, Part IX. Mem. New York Bot. Gard. **23:** 227–832.

——— **& W. D. Stevens**. 1988. Notes on *Rhodognaphalopsis* and *Bombacopsis* (Bombacaceae) in the Guayanas. Ann. Missouri Bot. Gard. **75:** 396–398.

Struwe, L. & V. A. Albert. 1998. *Lisianthus* (Gentianaceae), its probable homonym *Lisyanthus*, and the priority of *Helia* over *Irlbachia* as its substitute. Harvard Pap. Bot. **3(1):** 63–71.

———**, V. A. Albert & B. Bremer**. 1994. Cladistics and family level classification of the Gentianales. Cladistics **10:** 175–206.

———**, P. J. M. Maas, O. Pihlar & V. A. Albert**. 1999. Gentianaceae. *In* J. A. Steyermark, P. E. Berry, K. Yatskievych & B. K. Holst (gen. eds.), Flora of the Venezuelan Guayana **5:** 474–542. P. E. Berry, K. Yatskievych & B. K. Holst (vol. eds.). Missouri Botanical Garden Press, St. Louis.

Tebbs, M. C. 1993. Revision of *Piper* (Piperaceae) in the New World 3. The taxonomy of *Piper* sections *Lepianthes* and *Radula*. Bull. Brit. Mus. (Nat. Hist.), Bot. **23(1):** 1–50.

Thomas, V. & Y. Dave. 1992. Structure and biology of nectaries in *Tabebuia serratifolia* Nichols (Bignoniaceae). Bot. J. Linn. Soc. **109:** 395–400.

Van Royen, P. 1951. The Podostemaceae of the New World. Part I. Meded. Bot. Mus. Herb. Rijks Univ. Utrecht **107:** 1–151.

Verdcourt, B. 1975. Studies in the Rubiaceae–Rubioideae for the "Flora of Tropical East Africa": I. Kew Bull. **30:** 247–326.

———. 1976. *Spermacoce*. Pp. 339–374 *in* R. M. Polhill (ed.), Flora of Tropical East Africa, Rubiaceae. Part 1. Crown Agents for Oversea Governments and Administrations, London.

Webster, G. L., M. J. Del-Arco-Aguilar & B. A. Smith. 1996. Systematic distribution of foliar trichome types in *Croton* (Euphorbiaceae). Bot. J. Linn. Soc. **121:** 41–57.

Woodson, R. E., Jr. & R. W. Schery. 1958. Thymelaeaceae. *In* Flora of Panama. Part VII(Fasc. 2). Ann. Missouri Bot. Gard. **45(2):** 93–97.

Wurdack, J. J. 1993. Melastomataceae. *In* A. R. A. Görts-van Rijn (eds.), Flora of the Guianas Ser. A, **13:** 1–425. Koeltz Scientific Books, Koenigstein.

Zardini, E. M. & P. H. Raven. 1991. Onagraceae. *In* A. R. A. Görts-van Rijn (ed.), Flora of the Guianas Ser. A, **10:** 1–44. Koeltz Scientific Books, Koenigstein.

Zhang, Shu-Yi & Li-Xin Wang. 1995. Fruit consumption and seed dispersal of *Ziziphus cinnamomum* (Rhamnaceae) by two sympatric primates (*Cebus apella* and *Ateles paniscus*) in French Guiana. Biotropica **27(3):** 397–401.

Index to Common Names

This index includes Amerindian, English, Bush Negro (including Aluku, Paramaka, Saramaka, etc.), Créole, French, Portuguese (usually names from the Amerindians of Brazil but now incorporated in Brazilian Portuguese), and commercial names. The scientific names for corresponding common names in this index are only provided to species. Some of the common names may more specifically apply to a subspecies or variety of the species, which can be determined by consulting the text.

Abérémou *Perebea guianensis* (Moraceae)
Abriba sauvage *Rollinia cuspidata, Rollinia exsucca* (Annonaceae)
Abuta *Abuta grandifolia* (Menispermaceae)
Abuta branca *Abuta grandifolia* (Menispermaceae)
Abutua *Abuta grandifolia* (Menispermaceae)
Acacia *Abarema jupunba* (Mimosaceae)
Acacia franc *Enterolobium schomburgkii* (Mimosaceae)
Acacia mâle *Parkia pendula* (Mimosaceae); *Vochysia guianensis* (Vochysiaceae)
Acajou *Cedrela odorata* (Meliaceae)
Acajou de Guyane *Cedrela odorata* (Meliaceae)
Acariquara branca *Geissospermum argenteum* (Apocynaceae)
Acerola *Malpighia emarginata* (Malpighiaceae)
African Violet Family Gesneriaceae
Agouman *Solanum americanum* (Solanaceae)
Agouman-alaman *Solanum americanum* (Solanaceae)
Akoinsiba *Pouteria gongrijpii* (Sapotaceae)
Alaman *Solanum americanum* (Solanaceae)
Alentou-case *Microtea debilis* (Phytolaccaceae)
Alfavaca *Ocimum campechianum* (Lamiaceae)
Algodão-do-praia *Emilia sonchifolia* (Asteraceae)
Amandier sauvage *Buchenavia nitidissima* (Combretaceae)
Amaparana *Brosimum parinarioides* (Moraceae)
Amarante *Peltogyne paniculata* (Caesalpiniaceae)
Amaranth Famiy Amaranthaceae
Amourette *Mimosa myriadenia* (Mimosaceae); *Brosimum guianense* (Moraceae)
Anabi *Potalia amara* (Loganiaceae)
Anani *Symphonia globulifera* (Clusiaceae)
Andiroba *Carapa procera* (Meliaceae)
Angelique *Dicorynia guianensis* (Caesalpiniaceae); *Swartzia panacoco* (Fabaceae)
Angouchi *Terminalia amazonica* (Combretaceae)
Apa *Eperua falcata* (Caesalpiniaceae)
Apuí *Ficus insipida* (Moraceae)
Araça de anta *Bellucia grossularioides* (Melastomataceae)
Arbre à pain *Artocarpus altilis* (Moraceae)
Arouma *Dialium guianense* (Caesalpiniaceae)
Arua peludo *Cordia nodosa* (Boraginaceae)
Assacu *Hura crepitans* (Euphorbiaceae)
Assoumaripa *Simarouba amara* (Simaroubaceae)
Aubergine sauvage *Solanum torvum* (Solanaceae)
Avocado Family Lauraceae

Baaka-apisi *Rhodostemonodaphne grandis* (Lauraceae)
Bagasse *Ambelania acida* (Apocynaceae); *Bagassa guianensis* (Moraceae); *Ecclinusa guianensis* (Sapotaceae)
Baikaaki *Eschweilera pedicellata* (Lecythidaceae)
Balai doux *Scoparia dulcis* (Scrophulariaceae)
Balanophora Family Balanophoraceae
Balata blanc *Ecclinusa guianensis, Micropholis guyanensis* (Sapotaceae)
Balata franc *Manilkara bidentata* (Sapotaceae)
Balata jaune d'oeuf *Chrysophyllum pomiferum, Chrysophyllum prieurii* (Sapotaceae)
Balata kamwi *Pouteria guianensis* (Sapotaceae)
Balata poirier *Chrysophyllum prieurii, Pouteria gongrijpii* (Sapotaceae)
Balata pomier *Pradosia ptychandra* (Sapotaceae)
Balata singe rouge *Chrysophyllum sanguinolentum* (Sapotaceae)
Barbados cherry *Malpighia emarginata* (Malpighiaceae)
Basilic *Ocimum campechianum* (Lamiaceae)
Basilic fombazin *Ocimum campechianum* (Lamiaceae)
Basili-grand bois *Miconia laterifolia* (Melastomataceae)
Basilic sauvage *Ocimum campechianum* (Lamiaceae)
Batablli *Ecclinusa guianensis* (Sapotaceae)
Batoto *Physalis pubescens* (Solanaceae)
Bean Family Fabaceae
Begonia Family Begoniaceae
Bellflower Family Campanulaceae
Bete *Talisia megaphylla* (Sapindaceae)
Bichouiac *Phytolacca rivinoides* (Phytolaccaceae)
Bicuiba *Virola surinamensis* (Myristicaceae)
Bìíudu *Eperua falcata* (Caesalpiniaceae)
Birthwort Family Aristolochiaceae
Bita udu *Geissospermum argenteum* (Apocynaceae)
Bitawiri *Solanum leucocarpon* (Solanaceae)
Bitayouli *Solanum leucocarpon* (Solanaceae)
Bladdernut Family Lentibulariaceae
Blueberry Family Ericaceae
Buttersweet Family Celastraceae
Bofo udu *Sacoglottis cydonioides* (Humiriaceae)
Bois-acorbeau *Swartzia polyphylla* (Fabaceae)
Bois-agouti *Emmotum fagifolium* (Icacinaceae)
Bois-ara *Parkia pendula* (Mimosaceae)
Bois banane *Hernandia guianensis* (Hernandiaceae)
Bois-bandé *Faramea lourteigiana* (Rubiaceae)
Bois bandé canelle *Abuta grandifolia* (Menispermaceae)
Bois-blanc *Jacaranda copaia* (Bignoniaceae)
Bois-blanchet *Terminalia amazonica* (Combretaceae)
Bois bouchon *Parkia nitida* (Mimosaceae); *Apeiba glabra, Apeiba petumo, Apeiba tibourbou* (Tiliaceae)
Bois-calou *Apeiba aspera* (Tiliaceae); *Aegiphila villosa* (Verbenaceae)
Bois canelle *Licaria cannella* (Lauraceae)
Bois-canne *Posoqueria latifolia* (Rubiaceae)
Bois-canon *Cecropia obtusa, Cecropia sciadophylla, Pourouma bicolor, Pourouma guianensis, Pourouma melinonii, Pourouma minor, Pourouma saülensis* (Cecropiaceae)
Bois-canon mâle *Cecropia sciadophylla, Pourouma bicolor, Pourouma guianensis, Pourouma minor* (Cecropiaceae)
Bois capayou *Copaifera guianensis* (Caesalpiniaceae)
Bois-chapelle *Chimarrhis turbinata* (Rubiaceae)
Bois-charbon *Diospyros capreifolia* (Ebenaceae)
Bois-corbeau *Swartzia amshoffiana* (Fabaceae)
Bois-côtelette *Citharexylum macrophyllum* (Verbenaceae)
Bois-crapaud *Taralea oppositifolia* (Fabaceae)
Bois d'arc *Brosimum rubescens* (Moraceae)
Bois dartre *Senna alata* (Caesalpiniaceae); *Vismia cayennensis, Vismia gracilis, Vismia guyanensis, Vismia latifolia, Vismia sandwithii* (Clusiaceae)
Bois de fer *Mouriri crassifolia* (Melastomataceae)
Bois de lettre *Brosimum rubescens* (Moraceae)

Bois de lettre à grandes feuilles *Capirona decorticans* (Rubiaceae)
Bois de lettre moucheté *Brosimum guianense*, *Trymatococcus oligandrus* (Moraceae)
Bois de rose femelle *Aniba parviflora* (Lauraceae)
Bois-diable *Hura crepitans* (Euphorbiaceae)
Bois d'Inde *Psychotria capitata* (Rubiaceae)
Bois di vin *Aparisthmium cordatum*, *Hyeronima alchorneoides* (Euphorbiaceae)
Bois fer *Mouriri sagotiana* (Melastomataceae)
Bois figué *Ficus insipida*, *Ficus nymphaeifolia* (Moraceae)
Bois flambeau *Aspidosperma marcgravianum* (Apocynaceae); *Talisia carinata*, *Talisia sylvatica* (Sapindaceae)
Bois-flèche *Mouriri crassifolia*, *Mouriri sagotiana* (Melastomataceae); *Ixora aluminicola* Steyermark (Rubiaceae)
Bois fourmis *Cordia nodosa* (Boraginaceae)
Bois gaulette *Licania micrantha*; *Parinari excelsa* (Chrysobalanaceae)
Bois gelet *Quararibea duckei* (Bombacaceae)
Bois grage *Euplassa pinnata* (Proteaceae)
Bois grage blanc *Euplassa pinnata* (Proteaceae)
Bois grage rouge *Roupala montana* (Proteaceae)
Bois jacquot *Guarea gomma* (Meliaceae)
Bois lait *Bonafousia siphilitica* (Apocynaceae); *Mabea piriri*, *Sagotia racemosa* (Euphorbiaceae); *Phoradendron piperoides* (Viscaceae)
Bois la Saint-Jean *Schefflera morototoni* (Araliaceae)
Bois lélé *Quararibea duckei* (Bombacaceae); *Mabea piriri*, *Mabea speciosa* (Euphorbiaceae)
Bois lélé blanc *Quararibea duckei* (Bombacaceae)
Bois lézard *Bonafousia disticha* (Apocynaceae)
Bois l'homme *Trema micrantha* (Ulmaceae)
Bois l'orme *Trema micrantha* (Ulmaceae)
Bois macaque *Aspidosperma album* (Apocynaceae); *Balizia pedicellaris* (Mimosaceae); *Touroulia guianensis* (Quiinaceae)
Bois mèle *Bellucia grossularioides*, *Henriettella caudata* (Melastomataceae)
Bois-messe *Bellucia grossularioides* (Melastomataceae)
Bois mondan *Brosimum acutifolium* (Moraceae)
Bois-négresse *Amaioua guianensis* (Rubiaceae)
Bois noyo *Prunus accumulans* (Rosaceae)
Bois pagaïe *Aspidosperma carapanauba*, *Aspidosperma marcgravianum* (Apocynaceae); *Swartzia polyphylla* (Fabaceae); *Chimarrhis turbinata* (Rubiaceae)
Bois-palika *Capirona decorticans* (Rubiaceae)
Bois-patagaie *Aspidosperma album* (Apocynaceae); *Hebepetalum humiriifolium* (Hugoniaceae)
Bois pian *Jacaranda copaia* (Bignoniaceae); *Gustavia augusta* (Lecythidaceae); *Isertia coccinea* (Rubiaceae)
Bois piquant *Zanthoxylum rhoifolium* (Rutaceae)
Bois-puant *Senna occidentalis* (Caesalpiniaceae)
Bois ramier *Croton matourensis* (Euphorbiaceae); *Trema micrantha* (Ulmaceae)
Bois rouge *Humiria balsamifera* (Humiriaceae); *Chrysophyllum argenteum* (Sapotaceae)
Bois-sabot *Antonia ovata* (Loganiaceae)
Bois-Saint-Michel *Croton draconoides* (Euphorbiaceae)
Bois satiné *Brosimum rubescens* (Moraceae)
Bois sip *Cordia exaltata* (Boraginaceae)
Bois-sucre *Antonia ovata* (Loganiaceae)
Bois tabac *Aegiphila villosa* (Verbenaceae)
Bois vache *Bagassa guianensis* (Moraceae)
Bois violet *Peltogyne paniculata* (Caesalpiniaceae); *Taralea oppositifolia* (Fabaceae); *Moutabea guianensis* (Polygalaceae)
Bois zépine *Zanthoxylum rhoifolium* (Rutaceae)
Borage Family Boraginaceae
Bougainvillea Family Nyctaginaceae
Bougouni *Inga alba*, *Inga pezizifera* (Mimosaceae)
Bouton d'or *Synedrella nodiflora*, *Tilesia baccata* (Asteraceae)
Brazilnut Family Lecythidaceae
Breadfruit *Artocarpus altilis* (Moraceae)
Bruchi-soke *Pouteria venosa* (Sapotaceae)
Buckthorn Family Rhamnaceae
Buckwheat Family Polygonaceae
Buisson ardent *Ixora coccinea* (Rubiaceae)

Caa-peba *Pothomorphe peltata* (Piperaceae)
Caa-pitiu *Siparuna guianensis* (Monimiaceae)
Caca chien *Hymenaea courbaril* (Caesalpiniaceae)
Caca Henriette *Bellucia grossularioides*, *Henriettea succosa*, *Henriettella caudata*, *Miconia mirabilis* (Melastomataceae)
Caca poule *Catharanthus roseus* (Apocynaceae)
Cacahuette sauvage *Theobroma velutinum* (Sterculiaceae)
Cacao *Theobroma cacao*, *Theobroma velutinum* (Sterculiaceae)
Cacao Family Sterculiaceae
Cacao sauvage *Theobroma velutinum* (Sterculiaceae)
Caca poule *Catharanthus roseus* (Apocynaceae)
Cacau *Theobroma cacao* (Sterculiaceae)
Caca zozo *Oryctanthus florulentus* (Loranthaceae)
Cachiman épineux *Annona muricata* (Annonaceae)
Cactus Family Cactaceae
Cadrio *Asclepias curassavica* (Asclepiadaceae)
Caesalpinia Family Caesaliniaceae
Café *Coffea arabica* (Rubiaceae)
Café zerb pian *Senna hirsuta*, *Senna obtusifolia*, *Senna occidentalis* (Caesalpiniaceae)
Caiaté *Omphalea diandra* (Euphorbiaceae)
Caja *Spondias mombin* (Anacardiaceae)
Cajou *Anacardium occidentale* (Anacardiaceae)
Caju *Anacardium occidentale* (Anacardiaceae)
Cajueiro *Anacardium occidentale* (Anacardiaceae)
Calebassier *Crescentia cujete* (Bignoniaceae)
Camaca *Gustavia augusta* (Lecythidaceae)
Camapu *Physalis pubescens* (Solanaceae)
Cambará amarela *Tilesia baccata* (Asteraceae)
Cambara de cheiro *Lantana camara* (Verbenaceae)
Canari macaque *Lecythis zabucajo* (Lecythidaceae)
Canella Family Canellaceae
Canelle sauvage *Abuta grandifolia* (Menispermaceae)
Caper Family Capparaceae
Cansanção *Urera caracasana* (Urticaceae)
Capitiu *Siparuna guianensis* (Monimiaceae)
Carapa *Carapa procera* (Meliaceae); *Palicourea guianensis* (Rubiaceae)
Carapa-oyac *Guarea grandifolia* (Meliaceae)
Caroba *Jacaranda copaia* (Bignoniaceae)
Carrapateira *Ricinus communis* (Euphorbiaceae)
Carrapicho de agulha *Bidens cynapiifolia* (Asteraceae)
Carrot Family Apiaceae
Cashew *Anacardium occidentale* (Anacardiaceae)
Cashew Family Anacardiaceae
Cassawa-pau *Apeiba aspera* (Tiliaceae)
Cassialata *Senna alata* (Caesalpiniaceae)
Castanha de cotia *Omphalea diandra* (Euphorbiaceae)
Castor bean *Ricinus communis* (Euphorbiaceae)
Cauchorana *Perebea guianensis* (Moraceae)
Caxinguba *Ficus maxima* (Moraceae)
Cecropia Family Cecropiaceae
Cèdre *Licania guianensis* (Chrysobalanaceae); *Ixora aluminicola* (Rubiaceae); *Piper brownsbergense* (Piperaceae)
Cèdre jaune *Rhodostemonodaphne grandis* (Lauraceae)
Cèdre marécage *Balizia pedicellaris* (Mimosaceae)
Cèdre noir *Rhodostemonodaphne grandis*, *Rhodostemonodaphne kunthiana* (Lauraceae)

Cèdre remy *Tachigali melinonii* (Caesalpiniaceae)
Cèdre-sam *Cordia alliodora* (Boraginaceae)
Centorel *Coutoubea spicata* (Gentianaceae)
Cerise de Cayenne *Malpighia emarginata* (Malpighiaceae)
Chá de Marajó *Capraria biflora* (Scrophulariaceae)
Chaine d'enfant *Cyathula prostrata* (Amaranthaceae)
Chardon-bénit *Eryngium foetidum* (Apiaceae)
Châtaignier *Artocarpus altilis* (Moraceae); *Sloanea* spp. (Elaeocarpaceae)
Chawari *Caryocar glabrum*, *Caryocar villosum* (Caryocaraceae)
Chinese hibiscus *Hibiscus rosa-sinensis* (Malvaceae)
Cipó catinga *Mikania guaco* (Asteraceae)
Cipó d'alho *Mansoa alliacea* (Bignoniaceae)
Citronnelle blanc *Aspidosperma marcgravianum* (Apocynaceae)
Coca Family Erythroxylaceae
Cocoa-plum Family Chrysobalanaceae
Cocorico *Tibouchina aspera* (Melastomataceae)
Codio *Asclepias curassavica* (Asclepiadaceae)
Coeur dehors *Diplotropis purpurea* (Fabaceae)
Coffee *Coffea arabica* (Rubiaceae)
Coffee Family Rubiaceae
Composite Family Asteraceae
Concombre *Cayaponia ophthalmica*, *Gurania reticulata* (Cucurbitaceae)
Cona-da-cona-dou *Gustavia augusta* (Lecythidaceae)
Confiture macaque *Bertiera guianensis*, *Duroia aquatica*, *Duroia eriopila* (Rubiaceae)
Connarus Family Connaraceae
Copahu *Copaifera guianensis* (Caesalpiniaceae)
Copaïba *Copaifera guianensis* (Caesalpiniaceae)
Copal *Hymenaea courbaril* (Caesalpiniaceae)
Copal du Brésil *Hymenaea courbaril* (Caesalpiniaceae)
Copalier *Copaifera guianensis* (Caesalpiniaceae)
Coquelicot *Rhynchanthera grandiflora* (Melastomataceae)
Corbeille d'or *Lantana camara* (Verbenaceae)
Cordão de frade *Leonotis nepetifolia* (Lamiaceae)
Cordichier blanc *Solanum jamaicense* (Solanaceae)
Corossol *Annona muricata* (Annonaceae)
Corossol sauvage *Rollinia exsucca* (Annonaceae)
Corossol-grand-bois *Rollinia exsucca* (Annonaceae)
Corossolier *Annona muricata* (Annonaceae)
Cotton *Gossypium barbadense* (Malvaceae)
Couali *Vochysia guianensis* (Vochysiaceae)
Couari *Vochysia guianensis* (Vochysiaceae)
Couata bealy *Chrysophyllum cuneifolium* (Sapotaceae)
Coui *Crescentia cujete* (Bignoniaceae)
Coumarounda *Dipteryx odorata* (Fabaceae)
Counami *Clibadium sylvestre* (Asteraceae)
Coupaia *Jacaranda copaia* (Bignoniaceae)
Coupawa *Copaifera guianensis* (Caesalpiniaceae)
Coupaya *Jacaranda copaia* (Bignoniaceae)
Courbaril *Hymenaea courbaril* (Caesalpiniaceae)
Couzou *Bonafousia disticha* (Apocynaceae)
Cramentin *Justicia pectoralis* (Acanthaceae)
Cramentine rouge *Pachystachys coccinea* (Acanthaceae)
Crêpe denne *Heliotropium indicum* (Boraginaceae)
Crèque dinde *Heliotropium indicum* (Boraginaceae)
Cresson *Nasturtium officinale* (Brassicaceae)
Crête-coq *Heliotropium indicum* (Boraginaceae)
Crête d'Inde *Heliotropium indicum* (Boraginaceae)
Crête poule *Drymonia coccinea* (Gesneriaceae)
Crista de galo *Heliotropium indicum* (Boraginaceae)
Crodio *Asclepias curassavica* (Asclepiadaceae)
Crotane *Croton matourensis* (Euphorbiaceae)
Cumahy *Lacmellea aculeata* (Apocynaceae)
Cumaru *Dipteryx odorata* (Fabaceae)
Cumaru roxo *Dipteryx odorata* (Fabaceae)
Cupiúba *Goupia glabra* (Celastraceae)
Curare *Strychnos glabra*, *Strychnos guianensis*, *Strychnos tomentosa* (Loganiaceae)
Curimbo *Tanaecium nocturnum* (Bignoniaceae)
Custard-apple Family Annonaceae

Dangouti *Guarea pubescens* (Meliaceae)
Dartrier *Senna alata* (Caesalpiniaceae)
Diambarana *Coutoubea spicata* (Gentianaceae)
Dibo *Hirtella racemosa* (Chrysobalanaceae)
Dichapetalum Family Dichapetalaceae
Digo *Senna occidentalis* (Caesalpiniaceae)
Dillenia Family Dilleniaceae
Djadja *Virola sebifera* (Myristicaceae)
Djago *Vatairea paraensis* (Fabaceae)
Djãgo *Vataireopsis surinamensis* (Fabaceae)
Dobouldoi *Strychnos erichsonii*, *Strychnos oiapocensis* (Loganiaceae)
Dobouldoi rouge *Strychnos erichsonii* (Loganiaceae)
Dobuldwa *Strychnos oiapocensis* (Loganiaceae)
Dogbane Family Apocynaceae
Douradinha do campo *Lindernia crustacea* (Scrophulariaceae)
Douvant-douvan *Petiveria alliaceae* (Phytolaccaceae)
Douvant-douvant *Mansoa alliacea* (Bignoniaceae)
Dur bois *Hirtella racemosa* (Chrysobalanaceae)
Dyadya *Virola surinamensis* (Myristicaceae)

Ébène blanc *Tabebuia capitata* (Bignoniaceae)
Ébène souffré *Tabebuia capitata*, *Tabebuia serratifolia* (Bignoniaceae)
Ébène verte *Tabebuia capitata*, *Tabebuia serratifolia* (Bignoniaceae)
Ebènier de Guyane *Tabebuia serratifolia* (Bignoniaceae)
Ebony Family Ebenaceae
Éféa *Hevea guianensis* (Euphorbiaceae)
Elaeocarpus Family Elaeocarpaceae
Elm Family Ulmaceae
Encens *Protium demerarense*, *Protium heptaphyllum*, *Protium opacum*, *Protium pilosum* (Burseraceae); *Guarea grandifolia*, *Trichilia cipo* (Meliaceae)
Encens blanc *Protium opacum*, *Protium pilosum* (Burseraceae); *Trichilia cipo* (Meliaceae)
Encens grands bois *Protium opacum* (Burseraceae)
Encens petites feuilles *Trichilia cipo* (Meliaceae)
Encens rouge *Protium demerarense*, *Protium plagiocarpium* (Burseraceae)
Encens tites feuilles *Protium aracouchini* (Burseraceae)
Entoucase *Microtea debilis* (Phytolaccaceae)
Erva chumbinho *Lantana camara* (Verbenaceae)
Erva cidreira *Lippia alba* (Verbenaceae)
Erva-de-jaboti *Peperomia pellucida* (Piperaceae)
Erva de passarinho *Oryctanthus florulentus* (Loranthaceae)
Erva de São Martinho *Sauvagesia erecta* (Ochnaceae)
Erva mijona *Microtea debilis* (Phytolaccaceae)
Erva moura *Solanum americanum* (Solanaceae)
Erva relogio *Sida rhombifolia* (Malvaceae)
Espèce de concombre *Cayaponia ophthalmica* (Cucurbitaceae)
Estrella *Chomelia tenuiflora* (Rubiaceae)
Evening primrose Family Onagraceae

False roselle *Hibiscus acetosella* (Malvaceae)
Faux-gaïac *Dipteryx odorata* (Fabaceae)
Faux simarouba *Jacaranda copaia* (Bignoniaceae)
Femelle mamayawé *Unonopsis stipitata* (Annonaceae)
Feuille bomb *Potomorphe peltata* (Piperaceae); *Clidemia octona* (Melastomataceae)

Feuille grage *Piper humistratum* (Piperaceae); *Psychotria ulviformis* (Rubiaceae)
Feuille singe rouge *Combretum rotundifolium* (Combretaceae)
Feuille tabac *Aegiphila villosa* (Verbenaceae)
Feuille trèfle *Aristolochia trilobata* (Aristolochiaceae)
Fève tonka *Dipteryx odorata* (Fabaceae)
Figuier *Coussapoa angustifolia* (Cecropiaceae)
Figwort Family Scrophulariaceae
Flambeau rouge *Aspidosperma album* (Apocynaceae)
Flacourtia Family Flacourtiaceae
Four o'clock Family Nyctaginaceae
Framboisier *Ocimum campechianum* (Lamiaceae)
Frankincense Family Burseraceae
Fromager *Ceiba pentandra* (Bombacaceae)
Fruta de pão *Artocarpus altilis* (Moraceae)

Gadu paepina in Badu *Marsypianthes chamaedrys* (Lamiaceae)
Gagnac *Dipteryx odorata*, *Taralea oppositifolia* (Fabaceae)
Gagnac rivière *Taralea oppositifolia* (Fabaceae)
Gaiac *Dipteryx odorata* (Fabaceae)
Gaïac *Taralea oppositifolia* (Fabaceae)
Gaïac de l'eau *Taralea oppositifolia* (Fabaceae)
Galibi *Senna chrysocarpa* (Caesalpiniaceae)
Gaulette *Couepia caryophylloides*, *Couepia parillo*, *Hirtella bicornis*, *Hirtella hispidula*, *Hirtella racemosa*, *Hirtella silicea*, *Hirtella tenuifolia*, *Licania* spp., *Licania alba*, *Licania apetala*, *Licania canescens*, *Licania caudata*, *Licania heteromorpha*, *Licania kunthiana*, *Licania laevigata*, *Licania membranacea*, *Licania micrantha*; *Parinari* spp. (Chrysobalanaceae); *Connarus fasciculatus* (Connaraceae)
Gaulette azon *Licania alba* (Chrysobalanaceae)
Gaulette blanc *Hirtella glandulosa*, *Licania heteromorpha* (Chrysobalanaceae)
Gaulette couepi *Licania apetala* (Chrysobalanaceae)
Gaulette fourmi *Hirtella physophora*, *Hirtella silicea* (Chrysobalanaceae)
Gaulette gris-gris *Couepia habrantha*, *Licania caudata* (Chrysobalanaceae)
Gaulette indien *Licania heteromorpha* (Chrysobalanaceae); *Talisia carinata*, *Talisia sylvatica* (Sapindaceae)
Gaulette-marécage *Licania micrantha* (Chrysobalanaceae)
Gaulette noir *Couepia caryophylloides*, *Licania canescens*, *Licania membranacea* (Chrysobalanaceae); *Psychotria oblonga* (Rubiaceae)
Gaulette petites feuilles *Couepia parillo*, *Hirtella bicornis* (Chrysobalanaceae)
Gaulette rouge *Hirtella tenuifolia*, *Licania canescens*, *Licania heteromorpha* (Chrysobalanaceae); *Ouratea candollei* (Ochnaceae)
Gentian Family Gentianaceae
Ginseng Family Araliaceae
Girofle d'eau *Ludwigia affinis*, *Ludwigia dodecandra*, *Ludwigia erecta*, *Ludwigia octovalvis* (Onagraceae)
Giroflée *Ludwigia octovalvis* (Onagraceae)
Goiyave *Psidium guajava* (Myrtaceae)
Gomme-gutte de la Guyane *Vismia cayennensis* (Clusiaceae)
Gonfolo *Qualea rosea*, *Ruizterania albiflora* (Vochysiaceae)
Goué-goué *Terminalia amazonica* (Combretaceae)
Gouman *Topobea parasitica* (Melastomataceae)
Goupi *Goupia glabra* (Celastraceae)
Goyavier *Myrciaria floribunda*, *Psidium guajava* (Myrtaceae)
Goyavier sauvage *Eugenia dentata* (Myrtaceae)
Graine biche *Ambelania acida* (Apocynaceae)
Graine coumarou *Posoqueria latifolia* (Rubiaceae)
Graine en bas feuille *Phyllanthus amarus*, *Phyllanthus urinaria* (Euphorbiaceae)
Graine-hocco *Terminalia amazonica* (Combretaceae)
Graine macaque *Brunfelsia guianensis* (Solanaceae)
Graine mèle *Bellucia grossularioides* (Melastomataceae)
Graine oko *Terminalia amazonica* (Combretaceae)
Graine pok *Physalis pubescens* (Solanaceae)
Grand basilic *Ocimum campechianum* (Lamiaceae)
Grand feuille bomb *Pothomorphe peltata* (Piperaceae)
Grand mourou *Peperomia serpens* (Piperaceae)
Grand ricin *Ricinus communis* (Euphorbiaceae)
Grão de galo *Cordia nodosa* (Boraginaceae)
Grape Family Vitaceae
Graviola *Annona muricata* (Annonaceae)
Griffe-chat *Marcgravia coriacea* (Marcgraviaceae)
Griffe chatte *Macfadyena unguis-cati* (Bignoniaceae)
Griffes de chat *Macfadyena unguis-cati* (Bignoniaceae)
Gris-gris *Couepia caryophylloides*, *Hirtella glandulosa*, *Hirtella racemosa*, *Licania alba*, *Licania canescens*, *Licania caudata*, *Licania majuscula* (Chrysobalanaceae), *Parinari* spp., *Parinari excelsa* (Combretaceae)
Gris-gris noir *Licania majuscula* (Chrysobalanaceae)
Gris-gris-rouge *Couepia caryophylloides* (Chrysobalanaceae); *Vantanea parviflora* (Humiriaceae)
Gros mourou *Peperomia serpens* (Piperaceae)
Gros pompon *Leonotis nepetifolia* (Lamiaceae)
Gros verveine *Stachytarpheta jamaicensis* (Verbenaceae)
Guava *Psidium guajava* (Myrtaceae)
Guéli-apisi *Rhodostemonodaphne grandis* (Lauraceae)
Gui *Oryctanthus florulentus* (Loranthaceae)
Guilapele *Chrysophyllum cuneifolium* (Sapotaceae)

Haricot-blanc *Phaseolus lunatus* (Fabaceae)
Haricot pigeon *Cajanus cajan* (Fabaceae)
Herbe à chat *Hebeclinium macrophyllum* (Asteraceae)
Herbe à cloques *Physalis pubescens* (Solanaceae)
Herbe aiguille *Bidens cynapiifolia* (Asteraceae)
Herbe canard *Zornia latifolia* (Fabaceae)
Herbe chapentier *Justicia pectoralis* (Acanthaceae)
Herbe des putains *Lantana camara* (Verbenaceae)
Herbe-grage *Psychotria ulviformis* (Rubiaceae)
Herbe jacques *Psychotria variegata* (Rubiaceae)
Herbe Saint Martin *Sauvagesia erecta* (Ochnaceae)
Hernandia Family Hernandiaceae
Hévéa *Hevea guianensis* (Euphorbiaceae)
Hippocratea Family Hippocrateaceae
Hog plum *Spondias mombin* (Anacardiaceae)
Hortelão bravo *Hyptis atrorubens* (Lamiaceae)
Hugonia Family Hugoniaceae
Huile de ricin *Ricinus communis* (Euphorbiaceae)
Humiria Family Humiriaceae

Icacina Family Icacinaceae
Indian almond Family Combretaceae
Indigo *Senna occidentalis* (Caesalpiniaceae)
Inga chichica *Inga pezizifera* (Mimosaceae)
Ingui pipa *Couratari guianensis*, *Couratari multiflora* (Lecythidaceae)

Jaboty *Erisma uncinatum* (Vochysiaceae)
Jajamadou *Virola sebifera* (Myristicaceae)
Janita *Clarisia ilicifolia* (Moraceae)
Jatobá *Hymenaea courbaril* (Caesalpiniaceae)
Jatuauba preta *Guarea gomma*, *Guarea grandifolia* (Meliaceae)
Jaune d'oeuf à grandes feuilles *Chrysophyllum prieurii* (Sapotaceae)
Jeniparana *Gustavia augusta* (Lecythidaceae)
Jicama *Pachyrrhizus erosus* (Fabaceae)
Juquiri *Mimosa polydactyla*, *Mimosa pudica* (Mimosaceae)
Jutaí *Hymenaea courbaril* (Caesalpiniaceae)

Kktri *Ceiba pentandra* (Bombacaceae)
Kapok Family Bombacaceae
Kete poule *Gurania reticulata* (Cucurbitaceae)
Kimboto *Pradosia ptychandra* (Sapotaceae)
Kobé *Sterculia frondosa, Sterculia pruriens, Sterculia villifera* (Sterculiaceae)
Kouali *Vochysia guianensis* (Vochysiaceae)
Kouatapatou *Lecythis zabucajo* (Lecythidaceae)
Kudzu-vine *Pueraria phaseoloides* (Fabaceae)
Kumãti udu *Aspidosperma album* (Apocynaceae)
Kuzu *Cyphomandra endopogon* (Solanaceae)
Kwatabobi *Chrysophyllum cuneifolium, Pradosia ptychandra* (Sapotaceae)

Lacistema Family Lacistemataceae
Lamoussaié blanc *Euplassa pinnata* (Proteaceae)
Lamoussé *Pseudoxandra cuspidata* (Annonaceae); *Cordia bicolor, Cordia nodosa* (Boraginaceae); *Brosimum acutifolium* (Moraceae)
Lamoussé-fourmi *Hirtella physophora* (Chrysobalanaceae)
Lamoussé fourmis *Cordia nodosa* (Boraginaceae)
Langue de boeuf *Elephantopus mollis* (Asteraceae)
Langue-poule *Eclipta prostrata* (Asteraceae)
L'arbre sensible *Pfaffia glomerata* (Amaranthaceae)
Lavandeira *Catharanthus roseus* (Apocynaceae)
Lébène *Tabebuia serratifolia* (Bignoniaceae)
Lebi loabi *Lecythis corrugata, Lecythis idatimon, Lecythis persistens* (Lecythidaceae)
Lebiweco *Inga pezizifera* (Mimosaceae)
Ledi dobouldoi *Strychnos erichsonii* (Loganiaceae)
Ledi dobuldwa *Strychnos erichsonii* (Loganiaceae)
L'encens *Protium opacum* (Burseraceae)
L'encens petites feuilles *Protium pilosum* (Burseraceae)
Lettre moucheté *Brosimum guianense* (Moraceae)
Lettre rubané *Brosimum rubescens* (Moraceae)
Liane ail *Mansoa alliacea, Mansoa standleyi* (Bignoniaceae); *Petiveria alliacea* (Phytolaccaceae)
Liane basilic *Sparattanthelium wonotoboense* (Hernandiaceae)
Liane câble *Trigonia laevis* (Trigoniaceae)
Liane canelle *Strychnos erichsonii* (Loganiaceae)
Liane caoutchouc *Pacouria guianensis* (Apocynaceae)
Liane chasseurs Dilleniaceae spp.
Liane du lait *Bonafousia siphilitica* (Apocynaceae)
Liane gris *Petrea bracteata, Petrea volubilis* (Verbenaceae)
Liane noyau *Tanaecium nocturnum* (Bignoniaceae)
Liane noyo *Tanaecium nocturnum* (Bignoniaceae)
Liane panier *Cydista aequinoctialis* (Bignoniaceae)
Liane papaye *Omphalea diandra* (Euphorbiaceae)
Liane-ravet *Cissus erosa* (Vitaceae)
Liane serpent *Passiflora coccinea* (Passifloraceae)
Liane trèfle *Aristolochia trilobata* (Aristolochiaceae)
Lima bean *Phaseolus lunatus* (Fabaceae)
Limaorana *Chomelia tenuiflora* (Rubiaceae)
Linden Family Tiliaceae
Lingua de vaca *Elephantopus mollis* (Asteraceae)
Lipstick-tree family Bixaceae
Lisapau *Capirona decorticans* (Rubiaceae)
Loofah sponge *Luffa acutangula* (Cucurbitaceae)
Logania Family Loganiaceae
Loosestrife Family Lythraceae
Loseille bois *Begonia glabra* (Begoniaceae)

Macoudia *Manettia reclinata* (Rubiaceae)
Madlomé *Euphorbia hirta, Euphorbia thymifolia* (Euphorbiaceae)
Madlomé rouge *Euphorbia thymifolia* (Euphorbiaceae)
Mahot *Lecythis corrugata, Lecythis idatimon, Lecythis persistens, Lecythis poiteaui* (Lecythidaceae)
Mahot blanc *Eschweilera micrantha, Lecythis confertiflora, Lecythis corrugata, Lecythis idatimon, Lecythis persistens* (Lecythidaceae)
Mahot cigare *Couratari guianensis, Couratari multiflora* (Lecythidaceae)
Mahot cochon *Sterculia pruriens* (Sterculiaceae)
Mahot fer *Lecythis persistens* (Lecythidaceae)
Mahot jaune *Lecythis poiteaui* (Lecythidaceae)
Mahot noir *Eschweilera pedicellata, Lecythis corrugata* (Lecythidaceae); *Tachia grandiflora* (Gentianaceae)
Mahot rouge *Lecythis corrugata, Lecythis idatimon, Lecythis poiteaui* (Lecythidaceae)
Maillontre *Mimosa polydactyla* (Mimosaceae)
Mal-manmanyaret *Fusaea longifolia* (Annonaceae)
Mâle bois-canon *Pourouma bicolor, Pourouma guianensis, Pourouma minor* (Cecropiaceae)
Mâle manmanyaret *Fusaea longifolia* (Annonaceae)
Malicia *Mimosa polydactyla, Mimosa pudica* (Mimosaceae)
Malicia das mulheres *Mimosa polydactyla, Mimosa pudica* (Mimosaceae)
Malnommée *Euphorbia thymifolia* (Euphorbiaceae)
Malobi *Chrysophyllum prieurii* (Sapotaceae)
Malobi-weti *Pradosia ptychandra* (Sapotaceae)
Malva relogio *Sida rhombifolia* (Malvaceae)
Mama-doosou *Chrysophyllum sanguinolentum* (Sapotaceae)
Maman boeuf *Bagassa guianensis* (Moraceae)
Maman-manioc *Manihot esculenta* (Euphorbiaceae)
Mamanyawé *Guatteria punctata* (Annonaceae)
Mamão *Carica papaya* (Caricaceae)
Mamayawé *Anaxagorea dolichocarpa, Cremastosperma brevipes, Duguetia calycina, Duguetia pycnastera, Duguetia riparia, Fusaea longifolia, Guatteria punctata, Unonopsis perrottetii, Unonopsis rufescens, Unonopsis stipitata* (Annonaceae); *Heisteria densifrons* (Olacaceae)
Mamayawé-piment *Duguetia calycina, Duguetia inconspicua, Duguetia surinamensis, Duguetia yeshidan* (Annonaceae)
Mamona *Ricinus communis* (Euphorbiaceae)
Man tapouhoupa *Eschweilera simiorum* (Lecythidaceae)
Manaca *Brunfelsia guianensis* (Solanaceae)
Mandapuça *Bellucia grossularioides* (Melastomataceae)
Mandioca *Manihot esculenta* (Euphorbiaceae)
Manger lapin *Tilesia baccata* (Asteraceae)
Mango *Mangifera indica* (Anacardiaceae)
Mangosteen Family Clusiaceae
Manguier *Mangifera indica* (Anacardiaceae)
Mani *Symphonia globulifera* (Clusiaceae)
Manil *Moronobea coccinea, Symphonia globulifera* (Clusiaceae)
Manil chêne *Symphonia globulifera* (Clusiaceae)
Manioc *Manihot esculenta* (Euphorbiaceae)
Manioc blanc *Manihot esculenta* (Euphorbiaceae)
Manioc jaune *Manihot esculenta* (Euphorbiaceae)
Manioc sauvage *Manihot esculenta* (Euphorbiaceae)
Mapa *Brosimum parinarioides* (Moraceae); *Ecclinusa guianensis* (Sapotaceae)
Maracujá do rato *Passiflora coccinea* (Passifloraceae)
Maracujá poranga *Passiflora coccinea* (Passifloraceae)
Mara-sacaca *Connarus perrottetii* (Connaraceae)
Maria-congo *Geissospermum argenteum, Geissospermum laeve, Geissospermum sericeum* (Apocynaceae)
Marie-crabe *Lantana camara* (Verbenaceae)
Marie poil *Diospyros ropourea* (Ebenaceae)
Marie tambour *Passiflora coccinea, Passiflora vespertilio* (Passifloraceae)
Marimari *Senna multijuga* (Caesalpiniaceae)

Marmelo bravo *Prunus accumulans* (Rosaceae)
Marmite de singe *Lecythis zabucajo* (Lecythidaceae)
Marupa *Simarouba amara* (Simaroubaceae)
Marupa falso *Jacaranda copaia* (Bignoniaceae)
Mata-cachorro *Connarus perrottetii* (Connaraceae)
Matapasto *Senna alata* (Caesalpiniaceae)
Matu bwa bâde *Strychnos medeola* (Loganiaceae)
Mau tapouhoupa *Gustavia augusta* (Lecythidaceae)
Mavévé *Spigelia multispica* (Loganiaceae); *Siparuna guianensis* (Monimiaceae); *Solanum leucocarpon* (Solanaceae)
Mavévé chien *Cyphomandra endopogon*, *Solanum leucocarpon* (Solanaceae)
Mavévé grand bois *Potalia amara* (Loganiaceae)
Mavévé sucrier *Banara guianensis* (Flacourtiaceae)
Mecouart *Minquartia guianensis* (Olacaceae)
Médecinier béni *Jatropha gossypiifolia* (Euphorbiaceae)
Médecinier rouge *Jatropha gossypiifolia* (Euphorbiaceae)
Meli *Lecythis poiteaui* (Lecythidaceae)
Mélisse de calme *Lippia alba* (Verbenaceae)
Mélisse sauvage *Hyptis lanceolata* (Lamiaceae)
Méquoi *Dendrobangia boliviana* (Icacinaceae); *Minquartia guianensis* (Olacaceae)
Mezereum Family Thymelaeaceae
Mésoupou *Bellucia grossularioides* (Melastomataceae)
Mignonette *Drymaria cordata* (Caryophyllaceae)
Milkweed Family Asclepiadaceae
Milkwort Family Polygalaceae
Minquart *Minquartia guianensis* (Olacaceae)
Mint Family Lamiaceae
Mirobolan *Hernandia guianensis* (Hernandiaceae)
Mistletoe Family Viscaceae
Mombin *Spondias mombin*, *Spondias purpurea* (Anacardiaceae)
Mombin blanc *Tapirira guianensis* (Anacardiaceae)
Mombin faux *Tapirira guianensis* (Anacardiaceae)
Mombin fou *Tapirira guianensis* (Anacardiaceae)
Mombin sauvage *Tapirira guianensis* (Anacardiaceae)
Mongui-soke *Pouteria venosa* (Sapotaceae)
Morning-glory Family Convolvulaceae
Morototo *Schefflera morototoni* (Araliaceae)
Mourou *Peperomia serpens* (Piperaceae)
Moussigot *Virola surinamensis* (Myristicaceae)
Moutouchi *Dussia discolor*, *Pterocarpus rohrii* (Fabaceae)
Moutouchi de marécage *Dussia discolor* (Fabaceae)
Moutouchi de montagne *Diplotropis purpurea* (Fabaceae)
Moutouchi rubanée *Pterocarpus rohrii* (Fabaceae)
Mucura caa *Petiveria alliacea* (Phytolaccaceae)
Mucurão *Gustavia augusta* (Lecythidaceae)
Mulatorana *Capirona decorticans* (Rubiaceae)
Mururé *Brosimum acutifolium* (Moraceae)
Mururerana *Brosimum parinarioides* (Moraceae)
Mustard Family Brassicaceae
Mutende *Capirona decorticans* (Rubiaceae)
Myrtle Family Myrtaceae

Ndongu ndongu *Petiveria alliacea* (Phytolaccaceae)
Nettle Family Urticaceae
Niam-boka *Chrysophyllum argenteum* (Saptaceae)
Niamboka *Pouteria gongrijpii*, *Pouteria venosa* (Sapotaceae)
Nightshade Family Solanaceae
Noix de cajou *Anacardium occidentale* (Anacardiaceae)
Noyau sauvage *Momordica charantia* (Cucurbitaceae)

Ochna Family Ochnaceae
Odoun *Bagassa guianensis* (Moraceae)
Oemanbarklak *Lecythis idatimon* (Lecythidaceae)
Oficial de sala *Asclepias curassavica* (Asclepiadaceae)
Okra *Abelmoschus moschatus* (Malvaceae)
Olax Family Olacaceae
Olho de boi *Mucuna urens* (Fabaceae)
Ortie *Amaranthus caudatus* (Amaranthaceae)
Ouabé *Omphalea diandra* (Euphorbiaceae)
Oulapele *Chrysophyllum cuneifolium* (Sapotaceae)

Pacouri *Platonia insignis* (Clusiaceae)
Païcoussa *Aspidosperma marcgravianum* (Apocynaceae)
Païcoussa rouge *Minquartia guianensis* (Olacaceae)
Palétuvier *Tovomita brasiliensis* (Clusiaceae)
Palétuvier grand-bois *Chrysochlamys membranacea* (Clusiaceae)
Palma-christi *Ricinus communis* (Euphorbiaceae)
Panchi mouti *Copaifera guianensis* (Caesalpiniaceae)
Papaya *Carica papaya* (Caricaceae)
Papaya Family Caricaceae
Papaye *Carica papaya* (Caricaceae)
Papaye biche *Ambelania acida* (Apocynaceae)
Passion-flower family Passifloraceae
Patate douce *Ipomoea batatas* (Convolvulaceae)
Pâte d'amande *Merremia dissecta* (Convolvulaceae); *Mouriri crassifolia* (Melastomataceae)
Pâte dentrifice *Oxalis barrelieri* (Oxalidaceae)
Pau d'arco *Tabebuia serratifolia* (Bignoniaceae)
Pau de arara *Parkia pendula* (Mimosaceae)
Pau de chicle *Lacmellea aculeata* (Apocynaceae)
Pau de cobra *Potalia amara* (Loganiaceae)
Pau de leite *Ambelania acida* (Apocynaceae)
Pau lacre *Vismia cayennensis* (Clusiaceae)
Pau mulato *Capirona decorticans* (Rubiaceae)
Pau para tudo *Simaba cedron*, *Simaba polyphylla* (Simaroubaceae)
Peigne singe rouge *Combretum rotundifolium* (Combretaceae); *Apeiba albiflora* (Tiliaceae)
Pente de macaco *Apeiba tibourbou* (Tiliaceae)
Pepe boiti *Chrysophyllum prieurii*, *Pouteria gongrijpii* (Sapotaceae)
Pepe poirier *Chrysophyllum prieurii* (Sapotaceae)
Pepper Family Solanaceae
Pepino do mato *Ambelania acida* (Apocynaceae)
Persil diable *Bidens cynapiifolia* (Asteraceae)
Pervenche de Madagascar *Catharanthus roseus* (Apocynaceae)
Petit bomb *Pothomorphe peltata* (Piperaceae)
Petit ipèca *Faramea guianensis* (Rubiaceae)
Petit mouron *Peperomia serpens* (Piperaceae)
Petit quinine *Drymaria cordata* (Caryophyllaceae)
Petite gaulette *Hirtella physophora* (Chrysobalanaceae); *Talisia carinata* (Sapindaceae)
Petite gaulette fourmi *Hirtella physophora* (Chrysobalanaceae)
Petite gaulette rouge *Hirtella racemosa*, *Hirtella tenuifolia* (Chrysobalanaceae)
Petite griffe *Lindernia crustacea* (Scrophulariaceae)
Petite madlomé *Euphorbia hirta* (Euphorbiaceae)
Pião roxo *Jatropha gossypiifolia* (Euphorbiaceae)
Pied bois *Miconia holosericea* (Melastomataceae)
Pied mangue *Mangifera indica* (Anacardiaceae)
Piepiepo *Lippia alba* (Verbenaceae)
Pigeon pea *Cajanus cajan* (Fabaceae)
Piment *Capsicum frutescens* (Solanaceae)
Pimenta *Capsicum frutescens* (Solanaceae)
Pimenta de galinhas *Solanum americanum* (Solanaceae)
Pimenta de nambu *Erythroxylum citrifolium* (Erythroxylaceae)
Pinde paya *Brosimum guianense* (Moraceae)
Pindia udu *Vismia cayennensis* (Clusiaceae)

Pink Family Caryophyllaceae
Piquant guadeloupe *Uncaria guianensis* (Rubiaceae)
Pistache sauvage *Desmodium barbatum* (Fabaceae)
Pixirica *Clidemia hirta* (Melastomataceae)
Poirier *Tabebuia capitata, Tabebuia insignis* (Bignoniaceae)
Pois congo *Cajanus cajan* (Fabaceae)
Pois d'Angola *Cajanus cajan* (Fabaceae)
Pois d'Angole *Cajanus cajan* (Fabaceae)
Pois puant *Gustavia hexapetala* (Lecythidaceae)
Pois-savon *Phaseolus lunatus* (Fabaceae)
Pois-sept-ans *Phaseolus lunatus* (Fabaceae)
Pois sucré *Acacia tenuifolia, Inga* spp. (Mimosaceae)
Pois sucré crapaud *Inga capitata* (Mimosaceae)
Poivre de Cayenne *Capsicum frutescens* (Solanaceae)
Poivre sauvage *Ludwigia octovalvis* (Onagraceae)
Poivrier *Piper bartlingianum* (Piperaceae)
Pokeweed Family Phytolaccaceae
Pomme cajou *Anacardium occidentale* (Anacardiaceae)
Pomme de liane sauvage *Passiflora glandulosa* (Passifloraceae)
Pomme liane sauvage *Passiflora coccinea* (Passifloraceae)
Pommier cajou *Anacardium occidentale* (Anacardiaceae)
Pompon *Leonotis nepetifolia* (Lamiaceae)
Pompon soda *Leonotis nepetifolia* (Lamiaceae)
Poué blanc *Macoubea guianensis* (Apocynaceae)
Poutsi-hô *Miconia tomentosa* (Melastomataceae)
Protea Family Proteaceae
Pumpkin *Cucurbita moschata* (Cucurbitaceae)
Purple-heart *Peltogyne paniculata* (Caesalpiniaceae)

Qualité bois-négresse *Amaioua guianensis* (Rubiaceae)
Quaruba branca *Vochysia guianensis* (Vochysiaceae)
Quebra-pedras *Phyllanthus amarus* (Euphorbiaceae)
Queue du singe rouge *Combretum rotundifolium* (Combretaceae)
Queue lézard *Acacia tenuifolia, Mimosa myriadenia* (Mimosaceae)
Quilapele *Chrysophyllum cuneifolium* (Sapotaceae)
Quiina Family Quiinaceae
Quinine Family Rubiaceae

Rabo de galo *Heliotropium indicum* (Boraginaceae)
Rabo de tucano *Vochysia guianensis* (Vochysiaceae)
Racine-pistache *Microtea debilis* (Phytolaccaceae)
Radié albumine *Cyanthillium cinereum* (Asteraceae)
Radié capiaï *Bonafousia siphilitica* (Apocynaceae)
Radié chancre *Maprounea guianensis* (Euphorbiaceae)
Radié commandeur *Rolandra fruticosa* (Asteraceae)
Radié crise *Hyptis mutabilis* (Lamiaceae)
Radié divin *Justicia secunda* (Acanthaceae)
Radié du sang *Justicia secunda* (Acanthaceae)
Radié François *Ageratum conyzoides* (Asteraceae)
Radié grage *Mikania guaco* (Asteraceae)
Radié grage *Psychotria ulviformis* (Rubiaceae)
Radié-jaunâtre *Tilesia baccata* (Asteraceae)
Radié la fièvre *Eryngium foetidum* (Apiaceae)
Radié lan mort *Mimosa polydactyla, Mimosa pudica* (Mimosaceae)
Radié macaque *Clidemia hirta, Clidemia rubra* (Melastomataceae)
Radié macaque *Miconia racemosa* (Melastomataceae)
Radié maringouin *Chromolaena odorata, Hebeclinium macrophyllum* (Asteraceae)
Radié pian *Petiveria alliacea* (Phytolaccaceae)
Radié pisser *Synedrella nodiflora* (Asteraceae)
Radié raide *Cuphea carthagenensis* (Lythraceae)
Radié serpent *Mikania guaco* (Asteraceae); *Psychotria ulviformis* (Rubiaceae)
Radié zoré *Psychotria poeppigiana* (Rubiaceae)
Radier camphre *Unxia camphorata* (Asteraceae)
Radier macaque *Clidemia dentata, Clidemia rubra* (Melastomataceae)
Radier maringouin *Chromolaena odorata* (Asteraceae)
Radier pisser *Synedrella nodiflora* (Asteraceae)
Radier-sable *Solanum asperum* (Solanaceae)
Radier serpent *Psychotria ulviformis* (Rubiaceae)
Rafflesia Family Rafflesiaceae
Raguet maringouin *Chromolaena odorata* (Asteraceae)
Raquette *Opuntia cochenillifera* (Cactaceae)
Razié pisser *Synedrella nodiflora* (Asteraceae)
Red mangrove Family Rhizophoraceae
Remedio de vaqueiro *Ocimum campechianum* (Lamiaceae)
Rhabdodendron Family Rhabdodendronaceae
Rinchão *Stachytarpheta cayennensis* (Verbenaceae)
Riverweed Family Podostemaceae
Rose Family Rosaceae
Roseate periwinkle *Catharanthus roseus* (Apocynaceae)
Roucou *Bixa orellana* (Bixaceae)
Roucou sauvage *Sloanea* spp. (Elaeocarpaceae)
Roucouyer *Bixa orellana* (Bixaceae)
Rue Family Rutaceae

Sablier *Hura crepitans* (Euphorbiaceae)
Saint John *Justicia secunda* (Acanthaceae)
Saint-Martin *Ormosia flava* (Fabaceae)
Saint Martin blanc *Glycydendron amazonicum* (Euphorbiaceae)
Saint-Martin gris *Platymiscium ulei* (Fabaceae)
Salade Madame Hector *Emilia sonchifolia* (Asteraceae)
Salade soldat *Peperomia pellucida* (Piperaceae)
Salade tortue *Begonia glabra* (Begoniaceae)
Salade toti *Begonia glabra* (Begoniaceae)
Sandbox tree *Hura crepitans* (Euphorbiaceae)
Sapodilla Family Sapotaceae
Sapotille *Manilkara zapota* (Sapotaceae)
Sapotillier *Manilkara zapota* (Sapotaceae)
Satiné *Helicostylis pedunculata, Trymatococcus oligandrus* (Moraceae)
Satiné rubané *Brosimum rubescens* (Moraceae)
Sensitive *Mimosa polydactyla, Mimosa pudica* (Mimosaceae)
Serve para tudo *Simaba polyphylla* (Simaroubaceae)
Sete sangrias *Marsypianthes chamaedrys* (Lamiaceae)
Showy mistletoe Family Loranthaceae
Simarouba *Simarouba amara* (Simaroubaceae)
Soapberry Family Sapindaceae
Soari Family Caryocaraceae
Sorossi *Momordica charantia* (Cucurbitaceae)
Spanish plum *Spondias purpurea* (Anacardiaceae)
Spurge Family Euphorbiaceae
Squash Family Cucurbitaceae
Storax Family Styracaceae
Suitiamini *Chrysophyllum sanguinolentum* (Sapotaceae)
Sumauma *Ceiba pentandra* (Bombacaceae)
Suspiro *Asclepias curassavica* (Asclepiadaceae)
Sweet potato *Ipomoea batatas* (Convolvulaceae)
Sweetleaf Family Symplocaceae

Taapoutiki *Dendrobangia boliviana* (Icacinaceae)
Taba taba *Emilia sonchifolia* (Asteraceae)
Tabac sauvage *Aegiphila villosa* (Verbenaceae)
Tabaco bravo *Chelonanthus alatus* (Gentianaceae)
Tacacazeiro *Sterculia pruriens* (Sterculiaceae)
Tachi *Tachigali melinonii, Tachigali paraënsis* (Caesalpiniaceae)
Takamala *Chrysophyllum prieurii* (Sapotaceae)
Tamanqueira *Zanthoxylum rhoifolium* (Rutaceae)

Tamarin sauvage *Zygia racemosa* (Mimosaceae)
Tapereba *Spondias mombin* (Anacardiaceae)
Tapouhoupa *Gustavia hexapetala* (Lecythidaceae)
Tassi *Tachigali melinonii, Tachigali paraënsis* (Caesalpiniaceae)
Tatajuba *Bagassa guianensis* (Moraceae)
Tatapirica *Tapirira guianensis* (Anacardiaceae)
Tauari *Couratari multiflora* (Lecythidaceae)
Tepu *Talisia sylvatica* (Sapindaceae)
Tepuime *Talisia sylvatica* (Sapindaceae)
Tête-nègre *Rolandra fruticosa* (Asteraceae)
Thé indien *Lantana camara* (Verbenaceae)
Thé pays *Capraria biflora* (Scrophulariaceae)
Theophrasta Family Theophrastaceae
Ti balai *Scoparia dulcis* (Scrophulariaceae)
Ti bois bandé *Faramea lourteigiana* (Rubiaceae)
Ti-bombe *Piper adenandrum* (Piperaceae)
Ti bombe blanc *Marsypianthes chamaedrys* (Lamiaceae)
Ti bombe noir *Hyptis atrorubens* (Lamiaceae)
Ti bombe rouge *Hyptis atrorubens* (Lamiaceae)
Ti génipa *Genipa spruceana* (Rubiaceae)
Timignonette *Drymaria cordata* (Caryophyllaceae)
Ti mignonette *Lindernia crustacea* (Scrophulariaceae)
Ti-mouron *Peperomia rotundifolia, Peperomia serpens, Peperomia glabella, Peperomia obtusifolia* (Piperaceae)
Ti-mourou-grandes-feuilles *Peperomia serpens* (Piperaceae)
Tit-gaulette *Talisia carinata* (Sapindaceae)
Tobitoutou *Schefflera morototoni* (Araliaceae)
Tomate sauvage *Solanum torvum* (Solanaceae)
Tonka *Dipteryx odorata* (Fabaceae)
Topa *Clibadium sylvestre* (Asteraceae)
Topa noir *Clibadium sylvestre* (Asteraceae)
Touliatan *Talisia carinata, Talisia sylvatica* (Sapindaceae)
Tree-of-heaven Family Simaroubaceae
Trèfle *Aristolochia trilobata* (Aristolochiaceae)
Trèfle à quatre feuilles *Oxalis barrelieri* (Oxalidaceae)
Trevo cumaru *Justicia pectoralis* (Acanthaceae)
Trevo roxo *Hyptis atrorubens* (Lamiaceae)
Trigonia family Trigoniaceae
Trumpet-creeper Family Bignoniaceae
Tuliata *Talisia carinata* (Sapindaceae)
Turnera Family Turneraceae
Tytiudu *Aspidosperma album* (Apocynaceae)

Ucúuba *Virola surinamensis* (Myristicaceae)
Umiri *Humiria balsamifera* (Humiriaceae)
Umari *Poraqueiba guianensis* (Icacinaceae)
Urari *Strychnos glabra, Strychnos guianensis, Strychnos tomentosa* (Loganiaceae)
Urucu *Bixa orellana* (Bixaceae)

Vanille sauvage *Topobea parasitica* (Melastomataceae)
Vaquinha *Maprounea guianensis* (Euphorbiaceae)
Vassourinha *Scoparia dulcis* (Scrophulariaceae)
Vénéré *Siparuna decipiens, Siparuna guianensis, Siparuna poeppigii* (Monimiaceae)
Ventoza *Hernandia guianensis* (Hernandiaceae)
Verbena Family Verbenaceae
Verveine *Lantana camara, Stachytarpheta cayennensis, Stachytarpheta jamaicensis* (Verbenaceae)
Verveine queue de rat *Stachytarpheta cayennensis* (Verbenaceae)
Vervine blanc *Elephantopus mollis* (Asteraceae)
Vervine crabe *Sphagneticola trilobata* (Asteraceae)
Vervine sauvage *Elephantopus mollis* (Asteraceae)
Vingt quatre heures *Cyanthillium cinereum* (Asteraceae); *Priva lappulacea* (Verbenaceae)
Viniraie *Siparuna guianensis* (Monimiaceae)
Viniré *Siparuna guianensis* (Monimiaceae)
Violet Family Violaceae
Viraru *Prunus accumulans* (Rosaceae)
Visgueira joerana *Parkia pendula* (Mimosaceae)
Viviré *Siparuna guianensis* (Monimiaceae)
Vochysia Family Vochysiaceae
Vulnéraie *Siparuna decipiens, Siparuna guianensis, Siparuna poeppigii* (Monimiaceae)

Wacapou *Recordoxylon speciosum, Vouacapoua americana* (Caesalpiniaceae)
Wadé-wadé *Sida rhombifolia* (Malvaceae)
Wan édé *Simaba cedron, Simaba polyphylla* (Simaroubaceae)
Wapa *Eperua falcata* (Caesalpiniaceae)
Wapa sec *Palicourea calophylla* (Rubiaceae)
Watercress *Nasturtium officinale* (Brassicaceae)
Water-lily Family Nymphaeaceae
Wawichi *Virola surinamensis* (Myristicaceae)
Weti loabi *Eschweilera grandiflora, Eschweilera micrantha, Lecythis confertiflora, Lecythis corrugata* (Lecythidaceae)
Weti loabiu *Lecythis idatimon* (Lecythidaceae)
Wilaka *Pouteria caimito* (Sapotaceae)
Wilapele *Chrysophyllum cuneifolium* (Sapotaceae)
Wilapila *Chrysophyllum argenteum* (Sapotaceae)
Wood-sorrel Family Oxalidaceae

Yam bean *Pachyrrhizus erosus* (Fabaceae)
Yaoui *Duguetia calycina* (Annonaceae)
Yayamadou *Virola kwatae, Virola michelii, Virola sebifera, Virola surinamensis* (Myristicaceae)
Yayamadou-marécage *Virola surinamensis* (Myristicaceae)
Yayamadou montagne *Virola kwatae, Virola michelii* (Myristicaceae)

Zzpatu *Psychotria ulviformis* (Rubiaceae)
Zégron léphan *Psychotria ulviformis* (Rubiaceae)
Zépini titefeuille *Zanthoxylum rhoifolium* (Rutaceae)
Zerb à cloques *Physalis pubescens* (Solanaceae)
Zerb canard *Zornia latifolia* (Fabaceae)
Zerb carême *Tilesia baccata* (Asteraceae)
Zerb charpentier *Justicia pectoralis* (Acanthaceae)
Zerb des putains *Lantana camara* (Verbenaceae)
Zerb Saint Martin *Sauvagesia erecta* (Ochnaceae)
Zerb vin *Justicia secunda* (Acanthaceae)
Zerb zaiguille *Bidens cynapiifolia* (Asteraceae)
Zerbe chat *Hebeclinium macrophyllum* (Asteraceae)
Zerb'grage *Mikania guaco* (Asteraceae)
Zieu bourrique *Mucuna urens* (Fabaceae)
Zognon sauvage *Clusia grandiflora, Clusia nemorosa* (Clusiaceae)
Zolive *Alseis longifolia* (Rubiaceae); *Chrysophyllum cuneifolium, Chrysophyllum prieurii, Micropholis obscura, Pouteria caimito, Pouteria decorticans, Pouteria gongrijpii, Pouteria guianensis, Pouteria jariensis, Pradosia ptychandra* (Sapotaceae)
Zolive grand bois *Micropholis venulosa* (Sapotaceae)
Zouti *Laportea aestuans* (Urticaceae)
Zouti montagne *Urera caracasana* (Urticaceae)
Zouti rouge *Laportea aestuans* (Urticaceae)

Index to Scientific Names

Names in **boldface** indicate taxa described in this volume; those in normal type indicate names referred to in other contexts. Page numbers in **boldface** indicate primary page references, and asterisks (*) indicate pages with illlustrations. As plates are not on numbered pages, they are referred to by their plate numbers, which follow the page references for a taxon.

Abarema, **485**
  barbouriana, **485**
  curvicarpa, **485**, 486*
  jupunba, **486**
  laeta, **487**
  mataybifolia, **487**; Pl. 93a
Abelmoschus moschatus, 428
Abuta, **475**
  bullata, **475**; Pl. 91a
  grandifolia, 474, **475**, 476*; Pl. 91b,c
  imene, **476**
  rufescens, **477***; Pl. 91d,e
Acacia, **487**
  tenuifolia, **487**
Acalypha, **268**
  diversifolia, **268**; Pl. 53a,b
Acanthaceae, **31**, 472
Acanthella, 437
Aciotis, **438**
  acuminifolia, **438**; Pl. 84a
  aequatorialis, 438
  alata, 438
  caulialata, **438**
  indecora, **439**
  laxa, 439
  purpurascens, **439***, 440
  rubricaulis, 438
Acmella, **97**
  brachyglossa, 97
  ciliata, 97
  radicans, 97
  uliginosa, **97**, 98*; Pl. 21a
Acosmium, 167, **169**, 298, 301
  praeclarum, **169**
Adelobotrys, 437, **440**
  adscendens, **440**
Adenocalymna, **121**
  saülense, **121**, 121*; Pl. 23a,b
Aegiphila,7, **725**
  integrifolia, 6, **726**
  laevis, **726**
  membranacea, **726**
  racemosa, **726**
  villosa, **726**, 727*; Pl. 125a
Ageratum, **97**
  conyzoides, **97**, 99*; Pl. 21b
Aiouea, **371**
  guianensis, **371**; Pl. 69e
  longipetiolata, 370, **371**
  opaca, **371**, 372*
Alchornea, **268**
  triplinervia, **269**
Alchorneopsis, **269**
  floribunda, **269**, 270*; Pl. 53d
Alibertia, 606, **609**, 624
  dolichophylla, 624
  myrciifolia, 6, **609**
Alibertia (*continued*)
  surinamensis, 624
  triloba, 609
  uniflora, 609
Allophylus, **656**
  angustatus, **657**
  latifolius, **657**, Pl. 115e
  leucoclados, **657**
  robustus, **657**
Alloplectus
  patrisii, 338
  coccineus, 338
Almedia guyanensis, 649
Alseis, **609**
  longifolia, 9, **609**
Alternanthera, **40**
  bettzickiana, 40
  ficoidea
    var. bettzickiana, 40
    var. spathulata, 40
  halimifolia, **40**
  paronychoides var. bettzickiana, 40
  sessilis, **40**
  tenella, **40**
  tenella 'bettzickiana', 40
Amaioua, 606, **611**
  guianensis, **611**; Pl. 110a
Amaranthaceae, **40**
Amaranthus, **41**
  blitum, **41**
  caudatus, **41**
Ambelania, **70**
  acida, **70**; Pl. 14a
Ampelocera, **721**
  edentula, **721**; Pl. 124a
Amphilophium, 119, **121**, 749
  paniculatum, **121**; Pl. 23c
Anacardiaceae, **43**
Anacardium, **43**
  occidentale, **44**
  spruceanum, **44**, 45*
Anartia, **70**
  meyeri, **70**
Anaxagorea, **54**
  dolichocarpa, **54**, 55*
Andropogon bicornis, 748
Anemopaegma, **121**
  ionanthum, 9, **123**
  oligoneuron, **123**
Anguria
  bignoniacea, 240
  leptantha, 242
  spinulosa, 240
  subumbellata, 240
  triphylla, 244
Aniba, 370, **371**
  citrifolia, **371**
Aniba (*continued*)
  jenmanii, **371**
  kappleri, **372**
  parviflora, 370, **373**
  williamsii, **373**
Anisacanthus, **32**
  secundus, **32**
Anisomeris tenuiflora, 614
Annona, 53, **54**
  ambotay, **54**
  haematantha, **56**
  muricata, 7, **56**
  prevostiae, **56**, 57*
Annonaceae, 7, **53**, 750
Anomospermum, **478**
  steyermarkii, **478**
Anthodiscus, 191
Antonia, **398**
  ovata, **398**, 399*; Pl. 77a,b
Aparisthmium, **269**
  cordatum, 9, **269**, 271*; Pl. 53e
Apeiba, **713**
  albiflora, **713**; Pl. 123a
  aspera, 713, 748
  echinata, 713
  glabra, **713**; Pl. 123b
  membranacea, 713
  petuomo, **713**, 714*; 748
  tibourbou, **716**
Aphelandra, 31, **32**
  aurantiaca, **32**
Apiaceae, **67**
Apinagia, **585**
  kochii, **585**
  richardiana, **585**
Apocynaceae, 7, 9, **69**, 248
Apodanthes, **598**
  caseariae, **598**
Appunia brachycalyx, 628
Araceae, 6
Araliaceae, **84**
Ardisia, **532**
  guianensis, **532**, 533*
Aristolochia, **87**
  bukuti, **87**, 88*; Pls. 17c, 18a, c
  cremersii, **89**
  didyma, **89**
  iquitensis, **89**
  stahelii, **89**
  trilobata, **89**; Pl. 18b
Aristolochiaceae, **87**, 750
Arrabidaea, 7, **123**
  candicans, 7, **124**
  cinnamomea, **124**
  fanshawei, **124**
  florida, **124**
  inaequalis, **124**; Pl. 23d

Arrabidaea (*continued*)
japurensis, **125**
nigrescens, 9, **125**; Pl. 23e
oligantha, **125**
patellifera, 8, **125**; Pl. 23f
trailii, **125**; Pl. 24a, b
triplinervia, 7, **125**
Artocarpus, **516**
altilis, **516**
Asclepiadaceae, **89**
Asclepias, **90**
curassavica, **90**, 91*; Pls. 19, 20a
Aspidosperma, 69, **70**
album, **71**, 72*; Pl. 14b
carapanauba, **71**
cruentum, **71**
excelsum, **71**
macrophyllum subsp. morii, 73
marcgravianum, **71**
morii, 73
oblongum, **73**
sandwithianum, **73**
schultesii, **73**
spruceanum, **73**
Aspleniaceae, 746
Asplenium
hostmannii, **746**
perkinsii, 747, 748
rutaceum, 748
Aster, 113
laevis, 113
Asteraceae, **94**, 748, 749, 750
Asteridae, 683
Astronium, 43, **44**
ulei, **44**, 46*; Pl. 10a,b
Asystasia, **32**
gangetica, **33**

Bagassa, 515, **516**
guianensis, 6, **516**, 517*
Balanophoraceae, **116**
Balboa, 212
Balizia, **487**
pedicellaris, **487**, 488*
Banara, **320**
guianensis, **320***; Pl. 60e
Banisteriopsis, **411**
carolina, **412**
schwannioides, **412**, 413*
wurdackii, **412**
Barleria, **33**
lupulina, **33**
Barnhartia, **586**
floribunda, 7, **586**
Basanacantha asperifolia, 645
Bassovia sylvatica, 697
Batesia, **169**
floribunda, **169**; Pl. 34a,b
Bathysa difformis, 612
Bauhinia, **169**
guianensis, **170**; Pl. 34c,d
outimouta, **170**; Pl. 34f

Bauhinia (*continued*)
siqueirae, **170**
surinamensis, **170**; Pl. 34e
Begonia, **117**
glabra, **117**, 118*; Pl. 22e
prieurii, **117**; Pl. 22f
Begoniaceae, **117**
Beilschmiedia, 370, **373**
hexanthera, **373**, 374*
Bellucia, 437, **440**
grossularioides, **440**, 441*; Pl. 84c
subrotundifolia, 451
Beloglottis sp., 747
Bertiera, **611**
guianensis, **611**
Besleria, **334**
flavovirens, **335**
insolita, **335**, 336*; Pl. 64d
maasii, 335
patrisii, **335**; Pl. 64e
verecunda, 335
Bidens, **98**
alba var. radiata, 100
cynapiifolia, **100***
pilosa, **100**
Bignoniaceae, 7, 8, 9, **118**, 749, 750
Bixa, **139**
orellana, 7, **139**
Bixaceae, 7, **139**
Blechum, **33**
brownei, 33
pyramidatum, **33**
Bocoa, **301**
viridiflora, **301**
Bombacaceae, **139**, 140, 428
Bombacopsis, 142
Bonafousia, **73**
angulata, **74**
disticha, **74**, 75*
macrocalyx, **74**
sananho, **74**; Pl. 14c
siphilitica, **74**; Pl. 14c
Boraginaceae, 8, **145**
Borreria, 647
alata, 647
capitata, 647
latifolia, 648
ocymoides, 648
parviflora, 648
podocephala, 648
verticillata, 648
Boswellinae, 152
Bothriospora, **611**
corymbosa, **611**
Brassicaceae, **151**
Brosimum, 515, **546**
acutifolium subsp. **acutifolium**, **516**
guianense, **516**
lactescens, **518**
parinarioides subsp. **parinarioides**, 518*, **519**
rubescens, **519**; Pl. 96a
utile subsp. **ovatifolium**, **519**

Brunfelsia, **689**
guianensis, **689**, 690*
martiana, **689**
Buchenavia, **224**
grandis, **224**
nitidissima, **225**
parvifolia, **225**
Bufforrestia candolleana, 6
Bunchosia, 7, 410, **412**
decussiflora, 7, **414**, 415*, 747
glandulifera, **414**
sp., 747
Burseraceae, **151**
Bursereae, 152
Burserinae, 152
Byrsonima, **414**
aerugo, **414**, 416*; Pl. 80b
crispa, 414
densa, **414**
stipulacea, **414**; Pl. 80c
Byttneria, **700**
morii, **700**, 701*; Pl. 122a

Cactaceae, **165**
Caesalpiniaceae, **167**, 169, 298
Cajanus, **301**
cajan, **301**, 302*
Calea solidaginea subsp. deltophylla, 94
Callichlamys, **126**
latifolia, **126**
Callicocca guianensis, 638
Callitrichaceae, 683
Calophyllum madruno, 217
Calyptranthes, 539, **540**
amshoffae, **541**
bracteata, **540***,541
Campanulaceae, **183**
Canarieae, 151, 152
Candolleodendron, **301**
brachystachyum, **301**
Canellaceae, **184**
Capirona, 606, **611**
decorticans, **611**
leiophloea, 611
surinamensis, 611
Capparaceae, **186**
Capparis, **186**
leprieurii, **186**, 187*; Pl. 38e
Capraria, **683**
biflora, **683**, 684*; Pl. 119c
Capsicum, **698**
chinense, 690
frutescens, 690, **691**
Carapa, **465**
procera, **466**; Pl. 89a
Carica, **188**
papaya, **188**, 189*
Caricaceae, 6, **188**
Cardiopetalum, **56**
surinamense, **56**
Carpotroche, **321**
crispidentata, **321**, 322*; Pl. 60d

Caryocar, 191
glabrum subsp. **glabrum, 191**, 192*; Pl. 39b
villosum, 191
Caryocaraceae, 191
Caryomene, 478
olivascens, 478
Caryophyllaceae, 193
Casearia, 321, 323
Sect. Casearia, 323
Group Decandrae, 327
Group Singulares, 323
Sect. Crateria, 327
acuminata, 322; Pl. 61a, b
bracteifera, 321, 324*, 325; Pl. 61c
combaymensis, 323, 325
commersoniana, 323; Pl. 61d
javitensis, 323
negrensis, 323
pitumba, 323, 325*; Pl. 61e,f
rusbyana, 323
singularis, 324, 325, 326*
sylvestris var. **sylvestris, 325**
ulmifolia, 327; Pl. 61g
sp. A, 327
sp. B, 327
Casimirella, 358
ampla, 358
Cassia, 170
cowanii var. **cowanii, 171**
fastuosa, 171
grandis, 171
spruceana, 171; Pl. 35a
Cassipourea, 603
guianensis, 603, 604*
Catalpa, 119
Catharanthus, 74
roseus, 74
Cathedra, 559
acuminata, 559
Catostemma, 139, **140**
commune, 140
Cavendishia, 262
callista, 263
duidae, 263
Cayaponia, 7, **237**
jenmanii, 237, 238*
ophthalmica, 237; Pl. 47a
racemosa, 237
rigida, 237
Cecropia, 194
obtusa, 194, 195*
sciadophylla, 194; Pl. 39c
Cecropiaceae, 6, **194**
Cedrela, 465, **466**
odorata, 466, **467***
Cedrelinga, 488
cateniformis, 489, 490*; Pl. 93b
Ceiba, 139, **140**
pentandra, 140, 141*; Pl. 28
Celastraceae, 199
Celtis, 720, **721**
iguanaea, 720*, **722**
Centropogon, 183
cornutus, 183, 184*; Pl. 38a,b
Centrosema, 302
vexillatum, 302
Cephaelis, 632
alba, 644
amoena, 640
blepharophylla, 642
callitrhix, 638
guianensis, 638
hirta, 638
involucrata, 638
tomentosa, 643
tontaneoides, 638
violacea, 637
Chamaesyce
hirta, 280
hyssopifolia, 281
thymifolia, 281
Chaunochiton, 559
kappleri, 559; Pl. 102d
Cheiloclinium, 348
cognatum, 348; Pl. 67a
hippocrateoides, 348
sp. 1, 348; Pl. 67b
Chelonanthus, 329
alatus, 329, 330*; Pl. 62b
longistylus, 329
purpurascens, 329
uliginosus, 329
Chenopodiaceae, 571
Chimarrhis, 612
subgen. Pseudochimarris, 612
cymosa, 612
subsp. microcarpa, 612
longistipulata, 612
microcarpa, 612, 613*; Pl. 110b
var. microcarpa, 612
turbinata, 610*, **612**; Pl. 110c
Chiococca, 612
alba, 612, 614*; Pl. 110g
erubescens, 612
nitida, 612
Chomelia, 612
tenuiflora, 614
Chorisia, 140
Chromolaena, 101
odorata, 101, 102*
Chrysobalanaceae, 202
Chrysochlamys, 212, **213**
membranacea, 213
Chrysophyllum, 670
argenteum
subsp. **auratum, 670**; Pl. 117b
subsp. **nitidum, 671**
cuneifolium, 671; Pls. 117c, 118a
lucentifolium subsp. **pachycarpum, 671**
pomiferum, 671
prieurii, 671
sanguinolentum subsp. **sanguinolentum, 671**
venezuelanense, 671
Chrysothemis, 335
pulchella, 335
Cinnamodendron, 185
tenuifolium, 185*, **186**; Pl. 38c,d.
Cinnamomum, 370, **373**
triplinerve, 7, **373**
Cissampelos
andromorpha, 474
pareira, 474
tropaeolifolia, 474
Cissus, 742
erosa, 741*, **742**
sicyoides, 742
verticillata, 742; Pl. 126e
Citharexylum, 725, **726**
macrophyllum, 726, 728*
Clarisia, 519
ilicifolia, 519; Pl. 96b, c
Clavija, 709
lancifolia subsp. **chermontiana**, 710*, **711**; Pl. 122e
Cleome, 186
Clibadium, 101
sylvestre, 101
Clidemia, 440
conglomerata, 442
dentata, 8, **442**
hirta, 442; Pl. 84b
involucrata, 442
laevifolia, 8, **442**
octona subsp. **guayanensis, 8**, 443*; Pl. 84d
rubra, 444; Pl. 84e
saülensis, 444
septuplinervia, 444
sericea, 444
Clitoria, 303
sagotii var. **sagotii, 303**
ternatea, 303
Clusia, 212, **213**
colorans, 215
flavida, 214
grandiflora, 214
leprantha, 214; Pl. 41c
melchiorii, 214
nemorosa, 214
obovata, 215
octandra, 215
palmicida, 215, 216*
panapanari, 215; Pl. 42a
platystigma, 215
purpurea, 214
scrobiculata, 215; Pl. 42b
Clusiaceae, 8, **212**
Clytostoma, 119, **126**
binatum, 9, **126**; Pl. 25a
Cnestidium, 228
guianense, 228, 229*; Pl. 4a,b
Coccocypselum, 614
guianense, 614
Coccoloba, 590
ascendens, 590
excelsa, 590

Coccoloba (*continued*)
- **parimensis**, 591*, **592**

Codonanthe, 334, **335**
- **calcarata**, **337**; Pl. 65a
- **crassifolia**, **337**, 338*; Pl. 65c

Coffea, **614**
- **arabica**, **615**

Columnea, **337**
- aureonitens, 337
- **calotricha**, **337**, 339*
- **oerstediana**, **337**
- **sanguinea**, **337**

Combretaceae, **224**

Combretum, 224, **227**
- **laxum**, **227**; Pl. 44c
- **rotundifolium**, 225*, **227**; Pl. 44a,b

Commelinaceae, 6

Comolia, **444**
- **villosa**, **444**

Compsoneura, **526**
- **ulei**, **526**, 527*

Conceveiba, **269**
- **guianensis**, **270**, 272*; Pl. 53f

Conchocarpus, **649**
- **guyanensis**, **649**, 650*

Condylocarpon, **75**
- **amazonicum**, **76**
- **pubiflorum**, **76**

Connaraceae, **227**

Connarus, 227, **228**
- **fasciculatus** subsp. **fasciculatus**, **228**, 230*
- **perrottetii** var. **perrottetii**, **228**

Conocarpus, 224

Conomorpha multipunctata, 535

Convolvulaceae, 7, **231**

Copaifera, **171**
- **guianensis**, **171**

Cordia, **146**
- **alliodora**, 8, **146**; Pl. 31a
- **bicolor**, **146**
- **exaltata**, **146**
- **goeldiana**, **147**
- hirta, 147
- **lomatoloba**, **147**
- **naidophila**, **147**
- **nervosa**, **147**
- **nodosa**, 146, **147**, 148*
- **sagotii**, **147**; Pl. 31b
- **schomburgkii**, **147**

Cordiera myrciifolia, 690

Corythophora, **385**
- **amapaensis**, **385**, 386*; Pl. 72a,c
- **rimosa** subsp. **rubra**, **387**; Pl. 72b,d

Cosmos, **101**
- **caudatus**, **101**, 103*

Couepia, **202**
- **caryophylloides**, **203**
- **guianensis**, **203**
- **habrantha**, **203**
- **joaquinae**, **203**
- **obovata**, **203**
- **parillo**, **203**, 204*

Couma, 69, **76**
- **guianensis**, **76**

Couratari, 385, **387**
- **gloriosa**, **387**
- **guianensis**, **387**; Pls. 72e, 73a,b
- **multiflora**, **387**
- **oblongifolia**, **387**
- **stellata**, **387**, 388*; Pl. 73c-e

Coussapoa, **194**
- **angustifolia**, 6, **194**; Pl. 39e
- **latifolia**, **196***; Pl. 39d

Coussarea, **615**
- **amapaensis**, **615**
- **granvillei**, **615**, 616*
- **machadoana**, **615**
- **micrococca**, **615**
- **racemosa**, **615**, 617*
- schomburgkiana, 616
- **violacea**, **616**; Pl. 110f

Coutoubea, **329**
- **spicata**, **331**, 332*; Pl. 62a

Cremastosperma, **56**
- **brevipes**, **56**; Pl. 11c,d

Crepidospermum, **152**
- **goudotianum**, **152**

Crescentia, 118, 119, **126**
- **cujete**, **126**

Crotalaria, **303**
- **anagyroides**, **303**

Croton, **273**
- **draconoides**, **273**, 274*
- **matourensis**, 6, **273**
- **palanostigma**, **273**
- **schiedeanus**, **273**

Cucurbita moschata, 236

Curcurbitaceae, **236**

Cupania, **657**
- guianensis, 657
- reticulata, 658
- **scrobiculata** subsp. **guianensis**, **657**
  - f. guianensis, 657
- **scrobiculata** subsp. **scrobiculata**, **658**
  - f. reticulata, 657

Cuphea, **409**
- **carthagenensis**, 6, **409***; Pl. 80a

Curarea, **478**
- **candicans**, **478**

Curatella, 250

Cyanthillium, **104**
- **cinereum**, **104**

Cyathula, **41**
- **prostrata**, **41**, 42*

Cybianthus, **533**
- brownii, 537
- comatus, 537
- **fuscus**, **534**
- **guyanensis** subsp. **multipunctatus**, **535**
- **leprieuri**, **535**
- **microbotrys**, **535**
- multipunctatus, 535
- myrianthos, 537
- nitidus, 537
- parviflorus, 535

Cybianthus (*continued*)
- **potiaei**, **535**
- **prevostae**, **535**, 536*
- **prieurii**, **537**
- **resinosus**, **537**
- subspicatus, 537
- **surinamensis**, **537**
- **venezuelanus**, **537**
- viridiflorus, 537

Cyclodium, 747

Cydista, 118, 119, **127**
- **aequinoctialis**, 7, 9, **127***, 750

Cymbopetalum, **56**
- **brasiliense**, **57**; Pl. 11e

Cyphomandra, **691**, 696
- **endopogon** subsp. **guianensis**, **691**, 692*; Pl. 120b

Dacryodes, **153**
- **cuspidata**, **153**
- **roraimensis**, **153**

Dalbergaria
- aureonitens, 337
- sanguinea, 337

Dalbergia, **303**

Dalechampia, **273**
- **brevicolumna**, **275**
- **dioscoreifolia**, **275**, 276*
- **fragrans**, **277**
- **heterobractea**, **277**
- **stipulacea**, **277**; Pl. 54a,b
- **tiliifolia**, **277**

Daphnopsis, **711**
- **sp. 1**, **711**

Davilla, **250**
- aspera, 250
- **kunthii**, **250**
- pilosa, 250
- **rugosa** var. **rugosa**, **250**

Decaphalangium, 212

Dendrobangia, **358**
- **boliviana**, **358**, 359*; Pl. 68a

Dennstaedtiaceae, 746

Derris, **304**
- **pterocarpa**, **304**

Desmodium, **304**
- **adscendens**, **304**
- **axillare**, **304**, 749; Pl. 56d
- **barbatum**, **305**
- **incanum**, **305**; Pl. 56e,f
- **wydlerianum**, **305**

Desmoscelis, **444**
- **villosa**, **445**

Dialium, **172**
- **guianense**, **172**; Pl. 35c

Dichapetalaceae, **247**

Dichapetalum, **247**
- **pedunculatum**, **247**
- **rugosum**, **247**, 248*

Dicorynia, **172**
- **guianensis**, **172**; Pl. 35b

Dicranostyles, **231**
- **ampla**, **231**; Pl. 45d

Dicranostyles (*continued*)
guianensis, **231**, 232*
integra, **232**
Didymopanax morototoni, 87
Dilkea, **566**
johannesii, **566**, 568*; Pl. 104a
Dilleniaceae, **250**
Dimorphandra, **172**, 749
macrostachya subsp. **glabrifolia**, **172**
multiflora, **173**
pullei, 173
Dioclea, **305**
macrocarpa, **305**, 306*; Pl. 57a-d
virgata, **307**; Pl. 57e,f
sp. A, **307**
Diodia, **618**
ocymifolia, **618**; Pl. 110d
spicata, **618**
Diospyros, **254**, 749
capimnensis, 255
capreifolia, **255**
caprifolia, 747
carbonaria, **255**; Pl. 51b
cauliflora Blume, 256
cauliflora Mart., 256
cavalcantei, **255**; Pl. 51d
cayennensis, **255**; Pl. 51c
dichroa, **256**
duckei, 255
guianensis, 256
ierensis, 255
maritima, 256
martinii Benoist, 256
martinii Amshoff, 256
matheriana A. C. Sm., 256, 747
matheriana Auct., 256, 747
melinonii, 255
praetermissa, 256
ropourea, **256**, 257*, 747; Pls. 50c,d, 51a
tetrandra Hiern, **256**
tetrandra Span., 256
vestita Benoist, **256**
vestita Bakh., 256
Diploön, **671**
cuspidatum, **672**
Diplotropis, **307**
purpurea, **307**, 308*; Pl. 57a
Dipteryx, **307**
odorata, **307**
Disciphania, **479**
heterophylla, 479
moriorum, **479**, 480*; Pl. 92a
unilateralis, 474
sp. A, **479**
Discophora, **360**
guianensis, **360**, 361*; Pl. 68b
Distictella, 119, **127**
elongata, 9, **128**
Dodecastigma, **277**
integrifolium, **278**, 279*; Pl. 54c
Doliocarpus, 250, **251**
brevipedicellatus subsp. **brevipedicellatus**, **251**, 252*; Pl. 50a
Doliocarpus (*continued*)
dentatus subsp. **latifolius**, **251**
guianensis, **251**
paraensis, **251**
surinamensis, 251
Dorstenia, 515
Drymaria. **193**
cordata, **193***
Drymonia, **337**
campostyla, 344
coccinea, **338**, 340*; Pl. 65d
cristata, 339
psila, 339
psilocalyx, **339**
serrulata, **339**
Drypetes, **278**
variabilis, **278**, 280*; Pl. 54d
Duguetia, 53, **58**, 750
cadaverica, **58**, 59*; Pl. 11f
calycina, **58**
eximia, **58**
granvilleana, **58**
inconspicua, **58**
pycnastera, **60**
riparia, **60**
surinamensis, **60**; Pl. 12a
yeshidan, **60**
Dulacia, **560**
guianensis, **560**
Duroia, 606, **618**, 750
aquaticam, **618**
eriopila, **618**
longiflora, **618**
Dussia, **309**
discolor, **309**; Pl. 58b

Ebenaceae, **254**
Ecclinusa, **672**
guianensis, **672**
lanceolata, **672**, 673*; Pl. 118b
ramiflora, **672**
Eclipta, **104**
alba, 104
prostrata, **104**, 105*
Elachyptera, **348**
floribunda, **348**
Elaeagia brasilienensis, 612
Elaphandra, **104**
moriana, **104**
Elaeocarpaceae, 6, 7, **258**
Elephantomene, **479**
eburnea, **479**, 482*
Elaphoglossum longicaudatum, **746**
Elephantopus, **106**
mollis, **106**, 107*; Pl. 21d
Elizabetha, **173**
leiogyne, **173**
princeps, **173**
Emilia, **106**
fosbergii, 106
sonchifolia var. **sonchifolia**, **106**; Pl. 21c
Emmotum, **360**
fagifolium, **360**
Endlicheria, 370, **373**
melinonii, **373**
punctulata, **374**; Pl. 69d
pyriformis, **374**; Pl. 70a
sericea, **375**
Entada, **489**
polyphylla, 9, **489**
Enterolobium, **489**
schomburgkii, **489**, 491*
Eperua, **173**
falcata, **173**
grandiflora, **173**
schomburgkiana, **173**
Epiphyllum, **165**
phyllanthus, **165**; Pl. 33d-f
Episcia, **340**
adenosiphon, 341
densa, 344
mimuloides, 342
sphalera, **341**
xantha, **341**, 342*; Pl. 65b
Erechtites, **106**
hieracifolia, **107**
Ericaceae, **262**, 748, 749, 750
Eriotheca, 139, **141**
globosa, **141**
longitubulosa, **141**
Erisma, 742, **743**
floribundum, **743**
uncinatum, **743**; Pl. 127a
Ernestia, **445**
glandulosa, 437, **445**, 446*; Pl. 85a
granvillei, **445**
Eryngium, **67**
foetidum, **67**, 68*
Erythrochiton, 649, **650**
brasiliense, **651**, 652*
Erythroxylaceae, **263**
Erythroxylum, **264**
citrifolium, **264**
coca, 264
kapplerianum, **264**
macrophyllum, **264**, 265*; Pl. 53a,b
mucronatum, **264**
Eschweilera, 385, **389**
apiculata, **390**
chartaceifolia, **390**
collina, **390**; Pl. 74a,b
coriacea, **390**; Pl. 74c
decolorans, **390**
grandiflora, **390**; Pl. 74d,e
laevicarpa, **390**
micrantha, **390**
parviflora, **390**; Pl. 74f
pedicellata, **391**, 392*; Pl. 75a
piresii subsp. **viridipetala**, **391**, 393*
sagotiana, **391**
simiorum, **391**
squamata, **391**; Pl. 75b
Esenbeckia, 649, **651**
cowanii, **651**, 653*
Eugenia, **541**
albicans, **542**

Eugenia (*continued*)
argyrophylla, **542**, 543*
armeniaca, **542**; Pl. 100b
citrifolia, 547
coffeifolia, **543**
cucullata, 547
dentata, **544**
dittocrepis, 547
feijoi, **544**
ferreiraeana, **544**
gongylocarpa, **544**, 545; Pl. 100c
latifolia, **544**
macrocalyx, **544**
mimus, **544**
morii, **544**, 546*
muricata, **544**
patrisii, **544**; Pl. 100e
pseudopsidium, **545**
tetramera, **546**
sp. A, **547**
sp. B, **547**
sp. C, **547**; Pl. 100a
Eupatorium
macrophyllum, 108
odoratum, 101
Euphorbia, 266, **278**
cyathophora, **278**
hirta, **280**
hyssopifolia, **281**
thymifolia, **281***; Pl. 54e
Euphorbiaceae, 6, 7, 8, 9, **266**, 750
Euplassa, **592**
pinnatam, **592**, 593*; Pl. 107d
Evea tontaneoides, 638

Fabaceae, 169, **298**
Faramea, **619**
corymbosa, **619**
guianensis, 610*, **619**; Pl. 110e
lourteigiana, **619**
multiflora var. **salicifolia**, **619**
quadricostata, **619**
salicifolia, 619
tinguana, **619**
Ferdinandusa, **620**
paraensis, 610* **620**; **Pl. 111b**
Ficus, 7, 515, **519**
amazonica, **520**
gomelleira, **520**, 521*; Pl. 96d
guianensis, **520**; Pl. 97a
insipida subsp. **scabra**, **520**; Pl. 97b
leiophyllam, **520**
maxima, **520**
nymphaeifolia, **521**
panurensis, **522**
pertusa, **522**
schumacheri, **522**
Flacourtiaceae, 7, 8, **319**
Forsteronia,**76**
acouci, **76**; Pl. 14g
guyanensis, **77**, 78*; Pls. 14f, 15a
Froesia, 594
Fusaea, **60**
longifolia, **60**, 61*; Pl. 12b

Gamotopea callitrhix, 638
Garcinia, 212, **217**
madruno, **217**; Pl. 42c
Geissospermum, **77**
argenteum, **77**
laeve, **77**, 79*
sericeum, **77**
Genipa, **620**
spruceana, **620**, 621*
Gentianaceae, 5, **328**, 329, 398, 401
Geophila, **620**
cordifolia, **620**; Pl. 111a
repens, **620**
tenuis, **622**
Gesneriaceae, 6, **334**
Gloeospermum, 732, **733**
sphaerocarpum, **733**, 734*
Glycydendron, **281**
amazonicum, **281**
Gonzalagunia, **622**
dicocca, **622**, 623*
surinamensis, 622
Gossypium barbadense, 428
Gouania, **600**
blanchetiana, **600**, 601*
frangulaefolia, 600
Goupia, **199**
glabra, **199**, 200*; Pl. 40a
Goupiaceae, 199
Grammitidaceae, 746
Grammitis mollissima, 746
Guapira, **551**
eggersiana, **551**
salicifolia, **551**; Pl. 101a
sp. A, **552**
Guarea, 465, **466**
gomma, **468**
grandifolia, **468**, 469*; Pl. 89c
kunthiana, **468**
michel-moddei, **468**, 470*; Pl. 89d,e
pubescens, **468**, Pl. 89b
scabra, **468**; Pl. 90a
silvatica, **468**; Pl. 90b
Guatteria, **60**
blepharophylla, **61**
chrysopetala, 62
foliosa, **62**; Pl. 12e
oblonga, **62**
punctata, **62**; Pl. 12c,d
sp. A, **62**
Guatteriopsis blepharophylla, 61
Guettarda, **622**
acreana, **622**
spruceana, **622**; Pl. 111e
Gurania, **240**
bignoniacea, **240**; Pl. 47b
lobata, **240**; Pl. 47e
reticulata, **240**; Pl. 47c
spinulosa, 240
subumbellata, 239*, **240**; Pl. 48a,c
Gustavia, 385, **391**
augusta, 385, **391**; Pl. 75d
hexapetala, **391**, 395*; Pl. 75f
Gynocraterium, 31, **33**
guianense, **34**

Hamelia, **622**
axillaris, **622**; Pl. 111c,d
Hasseltia, **327**
floribunda, 7, **327**
Havetia, 212
flavida, 214
Havetiopsis, 212
flavida, 214
flexilis, 214
Hebeclinium, **108**
macrophyllum, **108**, 109*; Pl. 21e
Hebepetalum, **353**
humiriifolium, **353**, 354*
Hedyotis lancifolia, 628
Heisteria, **561**
cauliflora, **562**; Pl. 102c,e
densifrons, **562**; Pl. 102b
ovata, **562**
scandens, **562**
Heliconia lourteigiae, 748
Helicostylis, **522**
pedunculata, **522**
tomentosa, **522**; Pl. 97c
Heliotropium, **149**
indicum, **149**
Helmontia, **240**
cardiophylla, **241**
leptantha, 241*, **242**; Pl. 48f
Helosis, **117**
cayennensis var. **cayennensis**, 116*, **117**; Pl. 22a-d
Henriettea, **445**
ramiflora, 8, **445**, 447*; Pl. 84f
succosa, 8, **446**
Henriettella, **447**
caudata, **448**
duckeana, **448**; Pl. 85b
sp. A, **448**
Hernandia, **344**
guianensis, **344**, 345*; Pl. 66c
Hernandiaceae, **344**
Herpetacanthus, **34**
rotundus, **34**
Herrania, **702**
kanukuensis, **702**
Heteropterys, **417**
oligantha, **417**, 418*, 748
siderosa, 417, 748
Hevea, **282**
guianensis, **282**, 283*; Pl. 54f
Hibiscus
acetosella, 428
rosa-sinensis, 428
Hillia, **623**
illustris, **624**, 625*
parasitica, **624**

Himatanthus, 69, **77**
  articulatus, 80
  speciosus, 80; Pl. 15b
Hippocratea, 348
  volubilis, 348, 349*
Hippocrateaceae, 199, **347**
Hippuridaceae, 683
Hiraea, 417
  affinis, 417
  fagifolia, 417; Pl. 80d
  gracieana, 418, 419*
  morii, 420
Hirtella, 203
  bicornis var. **pubescens, 205**
  davisii, 205
  glandulosa, 205
  hispidula, 205
  physophora, 205
  racemosa, 205, 206*
  silicea, 205; Pl. 40b
  suffulta, 205
  tenuifolia, 205
Huberodendron, 139, **142**
  swietenioides, 142, 143*; Pl. 29
Hugoniaceae, 353
Humiria, 355
  balsamifera, 355
Humiriaceae, 355
Humiriastrum, 355
  excelsum, 355
Hura, 282
  crepitans, 282, 284*; Pls. 54g, 55a,b
Hybanthus, 732, **733**
  oppositifolius, 733
Hydrocotyle, 67
Hyeronima, 282
  alchorneoides var. **alchorneoides, 284**
  oblonga, 285, 286*
Hymenaea, 174
  courbaril, 174, 175*
Hymenolobium, 309
  petraeum, 309
Hyptis, 364
  atrorubens, 365
  lanceolata, 365, 366*; Pl. 69a
  mutabilis, 365
  pachycephala, 365; Pl. 69b

Ibetralia, 606, **624**
  surinamensis, 610*, **624**
Icacinaceae, 358
Icacorea guianensis, 532
Ichthyothere davidsei, 94
Icica altissima, 161
Inga, 489, 748
  acreana, 494
  acrocephala, 494
  alata, 494
  alba, 494, 495*
  albicoria, 494
  auristellae, 494
  bourgoni, 496
  brachystachys, 496

Inga (*continued*)
  bracteosa, 498
  capitata, 496
  edulis, 496
  fastuosa, 496
  graciliflora, 496
  gracilifolia, 496
  huberi, 495*, **498**
  leiocalycina, 498
  longipedunculata, 498
  macrophylla, 498
  marginata, 498
  nouragensis, 498
  nubium, 499
  nuda, 500
  paraensis, 499
  pezizifera, 499
  poeppigiana, 499
  retinocarpa, 499
  rhynchocalyx, 499
  rubiginosa, 497*, **499**
  sarmentosa, 500
  stipularis, 500
  striata, 500
  suaveolans, 500
  thibaudiana, 500
  umbellifera, 500; Pl. 93d
  virgultosa, 501
Ipomoea, 233
  batatas, 7, **233**
  batatoides, 233; Pl. 45e
  phyllomega, 233
  quamoclit, 233; Pl. 46a
  squamosa, 233; Pl. 46c
Irlbachia
  alata, 329
    subsp. alata, 329
    subsp. longistyla, 329
  purparascens, 329
Iryanthera, 526
  sagotiana, 526, 528*; Pl. 98e,f
  tessmannii, 526; Pl. 99a
Isertia, 624
  coccina, 624, 626*; Pl. 112a
  commutata, 627
  pterantha, 627
  spiciformis, 627; Pl. 112d
Ixora, 606, **627**
  aluminicola, 627
  coccinea, 627
  graciliflora, 627
  piresii, 627
  pubescens, 627

Jacaranda, 128
  copaia, 119, **128**, 129*; Pl. 25c,d
    subsp. spectablis, 128
Jacaratia, 188
  spinosa, 6, **188**, 190*; Pl. 39a
Jarilla, 188
Jatropha, 285
  gossypiifolia, 285

Jubelina, 420
  rosea, 420, 421*; Pl. 81a
Justicia, 34
  cayennensis, 34
  pectoralis, 34*
  potarensis, 34
  secunda, 34
  sprucei, 34

Kubitzkia, 370, **375**
  mezii, 375
Kutchubaea, 624

Lacistema, 362
  aggregatum, 362
  grandifolium, 363*; Pl. 68d
  polystachyum, 363
Lacistemataceae, 362
Lacmellea, 80
  aculeata, 80; Pl. 15c-e
Lactuceae, 749
Lacunaria, 594
  crenata, 596; Pl. 108a
  jenmani, 595*, **596**; Pl. 108b
Laetia, 327
  procera, 327
Lamiaceae, 363
Lantana, 725, **726**
  camara, 726
Laportea, 722
  aestuans, 722, 723*
Lauraceae, 7, **370**
  sp. I, 383
Laxoplumeria, 69, **80**
  baehniana, 80
Leandra, 448
  agrestis, 448; Pl. 85c, d
  clidemioides, 449
  divaricata, 449
  micropetala, 449
  paleacea, 449
  rufescens, 449
  solenifera, 449; 450*; Pl. 85f, g
Lecythidaceae, 385
Lecythis, 385, **393**
  chartacea, 394
  confertiflora, 394; Pl. 75c
  corrugata subsp. **corrugata, 394**, 396*; Pl. 75e
  holcogyne, 394; Pl. 76c
  idatimon, 394
  persistens
    subsp. **aurantiaca, 394**; Pl. 76a,b
    subsp. **persistens, 394**
  poiteaui, 397; Pl. 76d
  zabucajo, 397
  sp. A, 397
Lentibulariaceae, 397
Leonia, 732, **733**
  glycycarpa var. **glycycarpa, 733**, 735*
Leonotis, 365
  nepetifolia, 365, 367*

Lepidagathis, **36**
    alopecuroidea, **36**
Leretia, **360**
    cordata, **360**
Licania, **205**
    alba, **207**; Pl. 40d
    albiflora, **208**
    amapaensis, **208**
    apetala, **208**
    canescens, **208**
    caudata, **208**
    discolor, **208**; Pl. 40c
    fanshawei, **208**
    glabriflora, **208**
    granvillei, **208**, 209*
    guianensis, **208**
    heteromorpha var. **heteropmorpha**, 209
    kunthiana, **209**; Pl. 41a
    laevigata, **210**
    laxiflora, **210**
    majuscula, **210**
    membranacea, **210**
    micrantha, **210**
    octandra, **210**; Pl. 41b
Licaria, 370, **375**
    cannella, **375**
    chrysophylla, 370, **375**
    debilis, 370, **376**
    guianensis, **377**; Pl. 69f
    martiniana, **376**
    subbullata, **376**
    vernicosa, **376**
Linaceae, 353
Lindernia, **683**
    crustacea, **684**, 685*
    diffusa, **684**
Lindsaea cultriformis, 746
Lippia, **727**
    alba, **727**
Lisianthus uliginosus, 329
Loganiaceae, 5, 329, **398**, 748
Lomariopsidaceae, 746
Loranthaceae, **405**
Loreya, **449**
    arborescens, **450**; Pl. 86a
    mespiloides, **451***
    subrotundifolia, **451**
Lozanella, 720
Ludwigia, **562**
    affinis, **564**
    foliobracteolata, **564**
    dodecandra, **564**
    erecta, **564**
    leptocarpa, **564**
    octovalvis, 563*, **564**; Pl. 103d
Luehea, **716**
    speciosa, 715*, **716**
Lueheopsis, **716**
    rugosa, **716**
Luffa acutangula, 236
Lundia, 118, 119, **128**
    corymbifera, **130**
Lycianthes, **691**
    geminata, 691
    guianensis, 691
    pauciflora, **691**; 693*; Pl. 120b
Lysianthus
    alatus, 329
    Purpurascens, 329
Lythraceae, 6, **408**

Maba
    cauliflora, 256
    melinonii, 255
    mellinonii, 255
Mabea, **285**
    argutissima, 287
    piriri, **285**
    salicoides, 8, **287**
    speciosa, **287**
    subsessilis, 8, **287**
Macfadyena, 118, 119, **130**
    unguis-cati, 7, 9, **130**, 132*, 748; Pl. 25e
Machaerium, 303, **309**
    altiscandens, **310**
    aureiflorum, **310**
    floribundum, **310**
    paraënse, **310**
    quinata, **310**
Macoubea, **80**
    guianensis, **80**; Pl. 16a-c
Macrocentrum, **452**
    cristatum, **452**
    fasciculatum, **452**, Pl. 86b
    latifolium, **452**
Macrolobium, **174**
    bifolium, **174**
Macropharynx, **81**
    spectabilis, **81**, 82*
Magnoliopsida, **10**
Maieta, **452**
    guianensis, **452**, 453*; Pl. 86c
Malanea, **627**
    hypoleuca, 628
    macrophylla, **628**; Pl. 112b
Malpighia, **420**
    emarginata, **420**
Malpighiaceae, 7, **410**
Malvaceae, **428**, 700, 712
Mandevilla, **81**
    rugellosa,7, **81**; Pl. 16e
Manettia, **628**
    alba, **628**
    coccinea, 628
    reclinata, **628**, 629*; Pl. 112c
Mangifera, 43, **44**
    indica, **44**, 47*
Manihot, **287**
    quinquepartita, **287**, 289*
    esculenta, **287**
    sp. A, **290**
Manilkara, 669, **672**
    bidentat, **673**
    huberi, **674**
Manilkara (*continued*)
    paraensis, **674**
    zapota, **674**
Mansoa, 7, 118, **130**
    alliacea, 9, **130**
    standleyi, **131**
Maprounea, **290**
    guianensis, 7, **290**
Maquira, **522**
    guianensis subsp. **guianensis**, **523**; Pl. 97d
    sclerophylla, **523**, 524*
Marcgravia, 431, **432**
    coriacea, **432**
    pedunculosa, **432**, 433*
    sp. A, **432**, 433*; Pl. 82d,e
Marcgraviaceae, **431**
Marcgraviastrum, **434**
    pendulum, **434**
Margaritaria, **290**
    nobilis, **290**, 291*
Maripa, **234**
    glabra, 7, **234**, 235*; Pl. 46b
    scandens, **234**; Pl. 46e
Markea, **691**
    camponoti, 693
    coccinea, **693**, 694*; Pl. 120d
    longiflora, **693**
    porphyrobaphes, 693
    sessiliflora, **693**; Pl. 120c
Marsypianthes, **365**
    chamaedrys, **366**
Martinella, **131**
    obovata, **131**
Martiodendron, **175**
    parviflorum, **175**; Pl. 36a,b
Mascagnia, **420**
    divaricata, **422**, 423*; Pl. 81b
Matayba, **658**
    arborescens, **658**
    olygandra, 658
    peruviana, **658**
    purgans, **658**
Matelea, **90**
    gracieae, **90**, 92*; Pl. 20a,b
    palustris, **91**
Matisia, 139, **142**
    ochrocalyx, **142**, 144*; Pl. 30a
Mayna, **328**
    odorata, **328**; Pl. 61h
Maytenus, **199**
    floribunda, **201***
    myrsinoides, **201**
    oblongata, **201**
Melanpodium camphoratum, 115
Melastomataceae, 7, 8, **437**, 749, 750
Meliaceace, 8, **465**
Melothria, **242**
    fluminensis, 243
    guadalupensis, 243
    pendula, 242*, **243**
Memora, 9, 119, **131**
    flavida, **132**

Memora (*continued*)
moringiifoli, **132**; Pl. 26a
racemosa, **132**; Pl. 26b
tanaeciicarpa, **132**
Mendoncia, **472**
bivalvis, **472**
glabra, **472**; Pl. 90e
hoffmannseggiana, **473***; Pl. 90f
squamulifera, **474**
Mendonciaceae, 31, **472**
Menispermaceae, **474**
Merremia, **234**
dissecta, **236**
glaber, 236
macrocalyx, **236**; Pl. 46d
Metalepis, **94**
albiflora, 93*, **94**
Mezia, **422**
angelica, **422**, 424*, 747; Pl. 81c
includens, **422**, 747
sp., 747
Miconia, **452**
acuminata, 8, **455**
affinis, **455**; Pl. 87a
alata, 8, **455**
aliquantula, **456**; Pl. 85e
argyrophylla
subsp. **argyrophylla**, 8, **456**
subsp. gracilis, 456
bracteata, **456**
cacatin, **456**, 457*; Pl. 87c
ceramicarpa
var. **candolleana**, **456**
var. **ceramicarpa**, **456**; Pl. 87b,d
chrysophylla, 8, **456**
ciliata, **458**
diaphanea, 8, **458**
dispar, 8, **458**
eriodonta, 8, **458**
gratissima, 8, **458**
holosericea, 8, **458**
lappacea, 456
lateriflora, **459**; Pl. 87e
minutiflora, **459**
mirabilis, 7, 8, **459**
myriantha, **459**
nervosa, **459**
plukenetii, 8, **459**
prasina, **459**
racemosa, **459**
sastrei, 7, 8, **460**; Pl. 87f
splendens, **460**
tillettii, **460**
tomentosa, 8, **460**
traillii, 8, **460**
trimera, **460**
Micropholis, **674**
acutangula, **675**
cayennensis, **675**
guyanensis, **675**
longipedicellata, **675**
melinoniana, **675**
mensalis, **675**
Micropholis (*continued*)
obscura, **675**; Pl. 118c
porphyrocarpa, **675**
submarginalis, **675**
venulosa, **675**
Microtea, **571**
debilis, 571*, **572**; Pl. 105c
Mikania, **108**
congesta, 108
cordifolia, 108
gleasonii, **108**
guaco, **109**
micrantha, 108
microptera, **110**
parviflora, **110**
Mimosa, **501**, 749
guilandinae var. **guilandinae**, **501**, 502*
myriadenia var. **myriadenia**, **501**; Pl. 93c
polydactyla, **501**; Pl. 94a
pudica var. **tetrandra**, **501**
Mimosaceae, 9, **484**, 748, 749, 750
Minquartia, **562**
guianensis, 8, 561*, **562**; Pl. 103b,c
Mitreola, **400**
petiolata, 400*, **401**; Pl. 77e
Mollinedia, **510**
ovata, **510**, 511*
Momordica, **244**
charantia, 243*, **244**; Pl. 48d,e
Monimiaceae, **510**
Monopteryx, **310**
inpae, **310**; Pl. 58c
Monstera
dubia, 6, 747
sp. nov., 747
Moraceae, 6, 7, **515**
Morinda, 606, **628**
brachycalyx, **628**
Moronobea, **217**
coccinea, 7, **217**; Pl. 42d,e
Mouriri, 437, **461**, 750
collocarpa, **461**, 462*; Pl. 88a
crassifolia, **461**; Pl. 88b
duckeana, **461**
sagotiana, **461**
sideroxylon, **461**
Moutabea, **586**
guianensis, **586**, 587*; Pl. 107b
Mucuna, **311**
urens, **311**
Mutisieae, 748
Mycosyrinx cissi, 742
Myoporaceae, 683
Myrcia, **547**
fallax, **548**
fenestrata, **548**
gigas, **548**, 549*
graciliflora, **548**
magnoliifolia, **548**
platyclada, **548**
rupta, **548**, 550*
Myrcia (*continued*)
saxatilis, **548**
subsessilis, **548**
sylvatica, **548**
Myrciaria, **549**
floribunda, **550**
Myristicaceae, 8, **525**
Myrsinaceae, **532**, 749
Myrtaceae, **539**, 749

Napeanthus, **341**
jelskii, **341**
macrostoma, **341**; Pl. 65e
Nasturtium, **151**
officinale, **151**
Naucleopsis, **523**
guianensis, **523**; Pl. 97e
Nautilocalyx, **341**
adenosiphon, **341**
mimuloides, **342**, 343*; Pl. 66a,b
pictus, 6, **343**
Nectandra, **376**
cissiflora, **376**
hihua, **377**; Pl. 70b
purpurea, **377**
reticulata, **377**; Pl. 70c
Neea, **552**
constricta, **552**; Pl. 101b
floribunda, 551, **552**, 553*; Pl. 101d
mollis, **552**; Pl. 101c
ovalifolia, 551, **552**
spruceana, **552**
Nepsera, **461**
aquatica, **462**
Noisettia, 732, **734**
orchidiflora, **736**, 737*; Pl. 125e
Norantea, **434**
guianensis, **434**, 435*; Pl. 82f
Nyctaginaceae, **551**
Nymphaea, **554**
glandulifera, **554**
Nymphaeaceae, **554**

Ochnaceae, **554**
Ocimum, **367**
campechianum, **368***
micranthum, 368
Ocotea, **377**
canaliculata, **378**
ceanothifolia, **378**
cernua, **378**
cinerea, 7, **378**
commutata, **378**
cujumary, 370, **378**
diffusa, 7, **378**, 379*; Pl. 70d
fendleri, **379**
floribunda, **380**
glomerata, **380**
oblonga, **380**
puberula, **380**
rubra, 383, 748

Ocotea (*continued*)
scabrella, **380**
splendens, **380**
subterminalis, 7, **380**
tomentella, **380**
Odontadenia, **81**
perrottetii, **81**, 83*; Pl. 16d
puncticulosa, **81**; Pl. 17a
Oedematopus, 212
obovatus, 215
octandrus, 215
Olacaceae, 8, **559**
Oldenlandia, **628**
lancifolia, **628**
Omphalea, **290**
diandra, **290**, 292*; Pl. 55d-f
Onagraceae, **562**
Opuntia, 165, **166**
cochenillifera, **166**
Oreopanax, **84**
capitatus, **84**, 85*
Ormosia, **311**
flava, **311**
Orobanchaceae, 683
Orthomene, **482**
prancei, **482**, 483*; Pl. 92c
schomburgkii, 474
Oryctanthus, **406**
alveolatus, **406**
florulentus, **406**
Osteophloeum, **528**
platyspermum, **528**, 529*; Pl. 99b
Ouratea, **554**
candollei, **555**
erecta, **555**
leblondii, **555**
sagotii, 555
saülensis, **555**, 556*
scottii subsp. **scottii**, **555**, 557*; Pl. 102a
Oxalidaceae, **564**
Oxalis, **564**
barrelieri, **564**, 565*; Pl. 103e
Oxandra, **62**
asbeckii, **62**; Pl. 13a

Pachira, 139, **142**
insignis, **142**; Pl. 30b
macrocalyx, **145**
Pachyrrhizus, **311**
erosus, **311**
Pachystachys, **36**
coccinea, **36**, 37*; Pl. 9b
Pacouria, **83**
guianensis, 9, **84**
Palicourea, 606, **629**
brachyloba, **629**
calophylla, **630**
crocea, **630**
guianensis, **630**; Pl. 112e
longiflora, **630**; Pl. 113a
longistipulata, **630**
quadrifolia, **631***
Panopsis, **592**
rubescens, **594**
Parabignonia, 118, 119, **133**
steyermarkii, 9, **133**, 747
Paradrymonia, **343**
campostyla, **344**
densa, **344**
Paragonia, 119, **133**
pyramidata, 7, 9, **133**
Parinari, 202, **210**
excelsa, **212**
montana, 211*, **212**
Parkia, **503**
decussata, **503**
nitida, **503**; Pl. 94b
pendula, **503**
reticulata, **503**
ulei var. **surinamensis**, **504**
velutina, **504**
Passiflora, **566**, 749
amoena, **567**; Pl. 104b
candida, **567**
cirrhiflora, **567**
coccinea, **567**, 569*, Pl. 105a
exura, **568**
foetida var. **hispida**, **568**; Pl. 105b
garckei, **569**
glandulosa, **570**
plumosa, **570**
rubra, **570**
saülensis, **570**
trialata, 570
variolata, **570**
vespertilio, **570**
sp. A, **570**
Passifloraceae, **566**
Patima, **632**
guianensis, **632**
Paullinia, **658**, 750
alata, **659**
anomophylla, **659**
capreolata, **659**
clathrata, **659**
fibulata, **660**
latifolia, **660**
lingulata, **660**
pachycarpa, **660**
plagioptera, **660**
rubiginosa, **660**, 661*; Pl. 115f
stellata, **660**
tricornis, **660**; Pl. 116a
sp. A, **660**
Pausandra, **291**
fordii, **291**
marginii, **291**
Pavonia, **430**
schiedeana, 429*, **431**; Pl. 82a
Paypayrola, 732, **736**
guianensis, **736**, 738*
Peckia
prieurii, 537
purpurea, 537
surinamensis, 537
Pecluma hygrometrica, 747
Peltogyne, **175**
paniculata subsp. **pubescens**, 175
Peperomia, **574**
emarginella, **575**
glabella, **575**
gracieana, **575**, 576*, 747; Pl. 106a
macrostachya, **575**
maguirei, **575**
obtusifolia, **575**
ouabianae, **575**
pellucida, **576**
pernambucensis, **576**
rotundifolia, **577**
serpens, **577**; Pl. 106b
sp., 747
Perebea, **523**
guianensis subsp. **guianensis**, **523**; Pl. 98a,b
rubra subsp. **rubra**, **523**
Pereskia, 165
Petagoma hirta, 638
Petiveria, **573**
alliacea, 572*, **573**
Petrea, 725, **729**
bracteata, **729**; Pl. 125b
kohautiana, 729
volubilis, **729**
Pfaffia, **41**
glomerata, **42**
Phaseolus, **312**
lunatus, **312**
Philodendron, 747
hederaceum, 747
scandens, 747
Phoradendron,**739**
crassifolium, **739**, 740*; Pl. 126a,b
perrottetii, **739**
piperoides, **740**
racemosum, **740**; Pl. 126c,d
Phthirusa, **406**
pycnostachya, **406**; Pl. 79a
pyrifolia, **406**; Pl. 79b
stelis, **406**, 407*; Pl. 79c
Phyllanthus, **293**
amarus, **293**
urinaria, **293**; Pl. 56a
Physalis, **695**
angulata, **695***
pubescens, **696**; Pl. 121b
Phytolacca, **574**
rivinoides, 573*, **574**; Pl. 105d
Phytolaccaceae, **571**
Picramnia, 685, **686**
guianensis, **686**
latifolia, **686**
sellowii subsp. **spruceana**, **686**
Picramniaceae, 685
Pilea, **722**
imparifolia, **724**
microphylla, **724***; Pl. 124d
pubescens, **724**
Pilostyles, 598

Pinzona, 250, **251**
  coriacea, **251**, 253*; Pl. 50b
Piper, 574, **577**, 750
  adenandrum, **579**
  aequale, **579**
  alatabaccum, **579**
  amalaqo, **579**
  amplectenticaule, **579**
  angustifolium, **579**
  anonifolium, **579**
  arboreum, **579**
  augustum, **580**
  avellanum, **580**
  bartlingianum, **580**, 581*; Pl. 106d
  brownsbergense, **580**
  cernuum, **580**
  consanguineum, **580**; Pl. 106c
  crassinervium, **580**
  cyrtopodon, **580**
  demeraranum, **580**
  dilatatum, **581**
  dumosum, **581**
  eucalyptifolium, **582**
  hispidum, **582**
  hostmannianum, **582**
  humistratum, **582**
  insipiens, **582**
  nematanthera, **582**
  nigrispicum, **582**
  obiquum, **582**
  paramaribense, **583**
  piscatorum, **583**
  reticulatum, **583**
  tectonifolium, **583**
  trichoneuron, **583**
Piperaceae, 8, **574**
Piptadenia, **504**
  floribunda, **504**, 505*
Pithecoctenium, 119, **133**
  crucigerum, 9, **133**; Pl. 26c
Plantaginaceae, 683
Platonia, **217**
  insignis, **217**; Pl. 42f
Platymiscium, **312**
  ulei, **312**; Pl. 58d
Pleonotoma, **134**
  albiflora, **134**
  dendrotricha, **134**
Pleopeltis
  astrolepis, 747, 748
  revoluta, 747, 748
Pleurisanthes, **360**
  parviflora, **362**
Plukenetia, **293**
  polyadenia, **293**; Pl. 56b
Plumeria, 69
Poaceae, **749**
Pochotoa, 142
Podandrogyne, 186
Podostemaceae, **585**, 750
Poecilanthe, **312**
  effusa, **313**
  hostmannii, **313**
Pogostemon cablin, 364
Poinsettia cyathophora, 278
Polygala, 8, **586**
  membranacea, **587**
  spectabilis, **588**
Polygalaceae, 7, 8, **585**
Polygonaceae, **590**, 749
Polylychnis, **36**
  fulgens, **36**, 38*; Pl. 9c
Polypodiaceae, 747
Poypodium
  lasiopus, 747
  sororium, 747
Poraqueiba, **362**
  guianensis, **362**; Pl. 68c
Posoqueria, 606, **632**
  decora, 632
  gracilis, **632**
  latifolia, **632**
    subsp. gracilis, 632
Potalia, 5, 329, 398, **401**
  amara, **401**, 403*; Pl. 77c,d
Pothomorphe, **583**
  peltata, 8, **583**, 584*; Pl. 107a
Pourouma, 6, **196**
  bicolor
    subsp. **bicolor**, 197
    subsp. **digitata**, 197
  guianensis subsp. **guianensis**, **197**
  melinonii subsp. **melinonii**, **197**
  minor, 194, **197**; Pl. 39f
  saülensis, **197**
  tomentosa subsp. **maroniensis**, 197, **198***
  villosa, **197**
Pouteria, **676**
  ambelaniifolia, **678**
  bilocularis, **678**
  brachyandra, **678**
  caimito, **679**
  cayennensis, **679**; Pl. 118d
  coriacea, 669, **679**
  cuspidata subsp. **dura**, **679**
  decorticans, **679**
  egregia, **679**
  engleri, **679**
  eugeniifolia, **679**
  filipes, **679**
  flavilatex, **679**
  gongrijpii, **680**
  guianensis, **680**
  hispida, **680**
  jariensis, **680**
  laevigata, 669, **680**
  maxima, **680**
  melanopoda, **680**
  oblanceolata, **680**
  putamen-ovi, **680**
  reticulata subsp. **reticulata**, **681**
  retinervis, **681**
  rodriguesiana, **681**
  singularis, **681**
  speciosa, **681**
Pouteria (*continued*)
  torta subsp. **glabra**, **681**
  venosa subsp. **amazonica**, **681**
  virescens, **681**
  sp. 1, **682**
  sp. 2, **682**
  sp. 3, **682**
  sp. 4, **682**
  sp. 5, **682**
  sp. 6, **682**
  sp. A, **681**
Pradosia, **682**
  ptychandra, **682**
  verticillata, **682**
Prionostemma, **350**
  aspera, **350**; Pl. 67d
Pristimera, **350**
  nervosa, **350**; Pl. 67c
Priva, 725, **729**
  lappulacea, **729***
Proteaceae, 7, **592**, 748
Protieae, 151, 152
Protium, **153**, 154
  sect. Icicopsis, 151, 153, 154, 155, 157
  apiculatum, **157**; Pl. 32a
  aracouchini, 154, **175**; Pl. 32b
  cuneatum, **157**; Pl. 32c
  decandrum, **157**
  demerarense, **158**; Pl. 32d
  divaricatum subsp. **fumarium**, **158**
  firmum, 157
  guianense, **158**
  heptaphyllum subsp. **heptaphyllum**, **158**
  hostmannii, **158**
  inodorum, **158**
  insigne, 160
  morii, **158**, 159*
  neglectum, 161
  occultum, **158**; Pl. 33a
  octandrum, 158
  opacum subsp. **rabelianum**, **158**
  pallidum, **160**; Pl. 32f
  pilosum, **160**; Pl. 32g
  plagiocarpium, 153, **160**
  robustum, 153, **160**
  sagotianum, 153, **160**
  subserratum subsp. **subserratum**, **160**
  tenuifolium, 153, **161**
  trifoliolatum, **161**
Prunus, **606**
  accumulans, 605*, **606**
  myrtiflora var. accumulans, 606
Pseudima, **662**
  frutescens, **662**, 748; Pl. 116b
Pseudocatalpa, 119
Pseudochimarrhis, 612
  difformis, 612
  turbinata, 612
Pseudolmedia, **523**
  laevigata, **525**; Pl. 98c
  laevis, **525**
Pseudopiptadenia, **504**

Pseudopiptadenia (*continued*)
    psilostachya, **504**, 506*; Pl. 94c
    suaveolens, **504**
Pseudoxandra, **62**
    cuspidata, **62**
Psidium, 539, **551**
    guajava, **551**
Psiguria, **244**
    triphylla, **244**, 245*; Pl. 49a
Psittacanthus, 405, **408**
    corynocephalus, **408**
    sp. A, **408**
Psychotria, 606, **632**
    subgen. Heteropsychotria, 632, 634
    subgen. Psychotria, 632, 634
    acuminata, **637**
    alloantha, **637**
    apoda, **637**
    astrellantha, **637**
    bahiensis, **637**
    blepharophylla, 642
    borjensis, **638**
    brachybotrya, **638**, 639*
    brachyloba, 630
    bremekampiana, **638**
    callithrix, **638**, 640*
        var. tontaneoides, 638
    capitata, **638**
    carapichea, **638**
    carthagenensis, **639**
    colorata, **640**
    cupularis, **641**
    deflexa, **641**
    erecta, **641**
    galbaoensis, 638
    granvillei, **641**; Pl. 113c
    guadalupensis, **641**
    iodotricha, **641**; Pl. 113e
    kappleri, **641**
    lateralis, **641**
    ligularis, **641**
        var. carapichea, 638
        var. ligularis, 642
    macrophylla, **642**
    mapourioides, **642**
    medusula, **642**; Pl. 113b
    microbotrys, **642**, 748; Pl. 113d
    microbracteata, **642**
    moroidea, **642**
    oblonga, **643**; Pl. 113f
    officinalis, **643**
    oiapoquensis, 638
    paniculata, **643**
    perferruginea, **643**
    platypoda, **643**; Pl. 113g
    poeppigiana, **643**; Pl. 114a,b
    pullei, **643**
    pungens, **643**
    racemosa, **643**; Pl. 114c
    saülensis, **644**
    surinamensis, 642, 748
    trichophoroides, **644**
    uliginosa, **644**

Psychotria (*continued*)
    ulviformis, **644**
    urceolata, **644**
    variegata, **644**; Pl. 115a
    viridibracteata, **644**
Pterocarpus, **313**
    rohrii, **313**; Pl. 58e,f
Pueraria, **313**
    phaseoloides, **313**
Pulchranthus, **38**
    congestus, **39**
    variegatus, **39**; Pl. 9f
Pyrostegia, 119

Qualea, **743**
    mori-boomii, **743**; Pl. 127b
    rosea, **743**; Pls. 127c, 128b
Quapoya, 212
Quararibea, 139, **145**
    duckei, **145**; Pl. 30c
    spatulata, **145**; Pl. 30d
Quiina, **596**
    guianensis, **596**; Pl. 108c
    obovata, **596**; Pl. 108d
    oiapocensis, **596**; Pl. 108e
    pteridophylla, 596
    sessilis, **596**
    sp. A, **598**
Quiinaceae, **594**

Rafflesia, 598
Rafflesiaceae, **598**
Randia, 606, **645**
    asperifolia, **645**
    pubiflora, 610*, **645**
Rapatea ulei, 747
Rauvolfia, 69
Recordoxylon, **176**
    speciosum, **176**
Renggeria, 212
Rhabdodendraceae, 8, **598**
Rhabdodendron, **600**
    amazonicum, 599*, **600**; Pl. 109a,b
Rhamnaceae, **600**
Rheedia
    acuminata, 217
    kappleri, 217
    madruno, 217
    spruceana, 217
Rhipsalis, **166**
    baccifera, **166**
Rhizophoraceae, **603**
Rhodognaphalopsis, 142
Rhodostemonodaphne, 370, **381**
    elephatopus, **381**, 382*
    grandis, **381**; Pl. 71a
    kunthiana, **381**; Pl. 71c
    leptoclada, **381**
    morii, **383**, 384*
    rufovirgata, **383**; Pl. 71b
    saülensis, **383**; Pl. 71d
    sp. 1, **383**

Rhynchanthera, **462**
    grandiflora, **463**
Rhynchosia, **314**
    minima, **314**
    phaseoloides, **314**, 315*; Pl. 59b,e
Richeria, **293**
    grandis, **294**
    laurifolia, 294
    racemosa, 294
Ricinus, **294**
    communis, **294**
Riencourtia pedunculosa, 94
Rinorea, 732, **736**, 748
    amapensis, **736**; Pl. 125f
    neglecta, **736**
    pubiflora, **738**
    riana, **739**
Rolandra, **110**
    fruticosa, **110**, 111*
Rollinia, **63**
    cuspidata, **63**
    elliptica, **63**; Pl. 13b
    exsucca, **63**, 64*; Pl. 13c
Ropourea guianensis, 256
Rorippa, 151
    nasturtium-aquaticum, 151
Rosaceae, **605**
Roupala, **605**
    montana, 7, **594**
Rourea, **228**
    pubescens var. **pubescens**, **228**
    surinamensis, **228**; Pl. 45c
Rubiaceae, 9, **606**, 748, 749, 750
Rudgea, **645**
    bremekampiana, **646**
    guyanensis, **646**
    lanceaefolia, 646
    lancifolia, 610*, **646**
    standleyana, **646**
    stipulacea, 610*, **646**
    sp. A, **646**
    sp. B, **646**
Ruellia, **39**
    rubra, **39**; Pl. 9e
    saülensis, **39**
Ruizterania, **743**
    albiflora, **744**; Pl. 128a
Rutaceae, **649**
Ryania, **328**
    speciosa, 8, **328**

Sabicea, **646**
    aspera, 610*, **646**
    hirsuta var. adpressa, 647
    villosa, **647**
        var. adpressa, 647
Sacoglottis, 355, **356**
    cydonioides, **356**
    guianensis var. **guianennsis**, **356**
Sagotia, **294**
    racemsa, **294**, 295*; Pl. 56c
Salacia, **350**

Salacia (*continued*)
amplectens, **350**
impressifolia, **351**
insignis, **351**; Pl. 67e
juruana, **351**
multiflora
subsp. **mucronata**, **351**
subsp. **multiflora**, **351**
Sapindaceae, **656**
Sapium, **294**
argutum, **294**
glandulosum, **295**
klotzschianum, 295
lanceolatum, 295
montanum, 294
paucinervium, 296*, **287**
Sapotaceae, **669**
Sarcaulus, **682**
brasiliensis subsp. **brasiliensis**, **683**
Satyria, **263**
cerander, 262, 263
Sauvagesia, 554, **556**
aliciae subsp. **aratayensis**, **558**
erecta, **558***
tafelbergensis, **559**
Scaphyglottis bifida, 747
Schefflera, **84**
decaphylla, **84**
morototoni, 86*, **87**
paarensis, 84
Schlegelia, 7, 9, 118, 119, **134**
fuscata, **134**; Pl. 26d,e
paraensis, **135**, 136*
Schoenobiblus, **711**
daphnoides, **711**, 712*
Schradera ligularis, 642
Sciadotenia, 474, **482**
cayennensis, **482**; Pl. 92b
Sclerolobiuim, 180
albiflorum, 181
amplifolium, 180
guianensis, 181
melinonii, 181
paraënse, 181
Scoparia, 683, **684**
dulcis, **685**; Pl. 119a,b
Scrophulariaceae, **683**
Scutellaria, 364, **368**
purpurascens, 369
uliginosa, **369***; Pl. 69c
Securidaca, **588**
diversifolia, 588, 748
pubescens, 8, **588**; Pl. 107c
spinifex, **588**
uniflora, **588**
volubilis, 8, **588**, 589*, 748
Selenicereus, **166**
extensus, **166**
Selysia, **244**
prunifera, **244**, 245*
Senna, **176**
alata, **177**, 750
chrysocarpa, **177**, 178*
Senna (*continued*)
hirsuta var. **hirsuta**, **177**
latifolia, **177**
lourteigiana, **178**; Pl. 36c
multijuga, **179**
obtusifolia, **179**
occidentalis, **179**
quinquangulata, **179**
reticulata, **179**
Serjania, **662**
didymadenia, **662**
grandifolia, **662**, 663*; Pl. 116c
paucidentata, **662**
pyramidata, **664**
setulosa, **664**
Sextonia, 370, **383**
rubra, **383**, 748; Pl. 70e
Sida, **431**
rhombifolia, **431**
setosa, 430*, **431**; Pl. 82b,c
Sigmatostalix amazonica, 747
Simaba, 685, **686**
cedron, **687**
guianensis
subsp. **ecaudata**, **687**, 747
subsp. **guianensis**, **687**, 747; Pl. 119d
subsp. polyphylla, 687
nigrescens, 687, 747
polyphylla, **687**
Simarouba, 685, **687**
amara, **687**, 688*
Simaroubaceae, **685**
Sipanea, **647**
wilson-brownei, **647**
Siparuna, **510**
crassiflora, 512
cuspidata, **512**, 513*; Pl. 94e
decipiens, **512**; Pl. 94d
emarginata, 512
guianensis, **512**; Pl. 95a,b
pachyantha, **512**
poeppigii, **512**, 514*; Pl. 95c,d
Siparunaceae, 510
Sloanea, 7, **258**
brachytepala, **258**
brevipes, **259**; Pl. 52d
echinocarpa, **259**, 748
grandiflora, **259**, 260*
guianensis, **259**
latifolia, 6, **259**
laxiflora, **259**
synandra, **259**; Pl. 52a
tuerckheimii, **261**; Pl. 52b,c,f
sp. 748
sp. A, **261**
sp. B, **262**, 748
Solanaceae, **689**, 749
Solanum, 691, **696**
americanum, **697**
anceps, 696, **697**
asperum, **697**
coriaceum, **697**
costatum, **697**
Solanum (*continued*)
daturifolium, 698
ficifolium, 698
geminatum, 691
guianense, 691
ipomoea, 698
jamaicense, **697**
laetum, 698
leucocarpon, **697**
miquelii, 698
morii, **698**; Pl. 121a
nodiflorum, 697
oppositifolium, **698**
pauciflorum, 691
pensile, **698**
rubiginosum, **698**
rugosum, **698**, 699*; Pl. 121c,d
scandens, 698
schizopodium, 698
schlechtendalianum, **698**
sempervirens, 698
surinamense, 697
sylvaticum, 697
torvum, **698**
velutinum, **700**
viliflorum, 698
sp. A, **700**
Souroubea, **434**
guianensis, **434**, 436*; Pl. 83
Sparattanthelium, **346**
guianense, **347**
wonotoboense, 346*, **347**
Spermacoce, **647**
alata, **647**; Pl. 115c
capitata, **647**
gracilis, 648
latifolia, **648**
ocymoides, **648**
prostrata, 648
verticillata, 610*, **648**; Pl. 115b
Sphagneticola, **110**
trilobata, **111**, 112*
Sphyrospermum, **263**
buxifolium, **263**
Spigelia, 398, **401**
multispica, **401**; Pl. 77f,g
Spilanthes, 97
uliginosa, 97
Spondias, **44**
lutea, 46
mombin, **46**, 48*; Pl. 10b
purpurea, **46**, 48*
Stachytarpheta, 725, **731**
cayennensis, 730*, **732**; Pl. 125d
jamaicensis, **732**
Sterculia, **702**
frondosa, **702**, 703*; Pl. 122b
pruriens var. **pruriens**, **702**
villifera, **704**
sp. A, **704**
Sterculiaceae, 428, **700**
Stigmaphyllon, 410, **422**
sinuatum, **423**, 425*; Pl. 81f

Stizophyllum, 8, 9, **135**
inaequilaterum, **135**; Pl. 27a
riparium, 8, **135**
Struthanthus, **408**
dichotrianthus, **408**
syringifolius, **408**, Pl. 79d
Strychnaceae, 398
Strychnos, 5, 398, **401**, 748
cayennensis, **402**
cogens, **402**
erichsonii, **402**; Pl. 78b
glabra, **402**
guianensis, **402**
medeola, **402**
melinoniana, **402**, 404*
oiapocensis, **405**
panurensis, **405**; Pl 78a
peckii, **405**
tomentosa, **405**; Pl. 78c,d
toxifera, **405**
Stryphnodendron, **507**
guianense, **507**
moricolor, **507**, 508*
polystachyum, **507**
Stylogyne, **539**
incognita, 538*, **539**
Styracaceae, **706**, 750
Styrax, **706**
discolor, **706**
glabratus, **708**
pallidus, 707*, **708**; Pl. 122d
sieberi, **708**
Swartzia, 301, **314**
amshoffiana, **314**; Pl. 59
arborescens, **316**
panacoco var. **sagotii**, **316**; Pl. 59c
polyphylla, **316**, 317*; Pl. 59d
sp. A, **316**
sp. B, **316**
Sweetia, 169
praeclara, 169
Syagrus, 747
Symphonia, **218**
globulifera, **218**, 219*; Pl. 43a
Symphyotrichum, **112**, 113
laeve, **113**
Symplocaceae, **708**
Symplocos, **708**
martinicensis, **708**, 709*
Synedrella, **113**
nodiflora, **113**, 114*; Pl. 21f

Tabebuia, **135**
capitata, 8, **137**
insignis, **137**
serratifolia, 7, 8, **137**, 138*
Tabernaemontana
angulata, 74
disticha, 74
marcrocalyx, 74
meyeri, 70
sananho, 74
siphilitica, 74
Tabernaemontana (*continued*)
undulata, 70
Tachia, **331**
grandiflora, **331**, 333*; Pl. 62c
Tachigali, **180**
amplifolia, **180**; Pl. 37b
bracteolata, **180**
guianensis, **181**
melinonii, **181**
paniculata, **181**, 747
paraënsis, **181**
sp. B, 747
sp. C, **181**
Talisia, **664**
allenii, 665
carinata, **664**
clathrata subsp. **canescens**, **665**
glandulifera, 665
guianensis, **665**
hemidasya, **665**
longifolia, **665**
macrophylla, **665**
megaphylla, **665**; Pl. 117a
micrantha, 666
mollis, **666**, 667*
pedicellaris, **666**
praealta, **666**
reticulata, 666
simaboides, **666**
sylvatica, **666**; Pl. 116d
Tanaecium, 118, **137**
nocturnum, 9, **137**; Pl. 27b
Tapirira, **47**
bethanniana, **49**, 50*
guianensis. **40**, 50*
marchandii, 49
myriantha, 49
obtusa, **49**, 50*; Pl. 10e
peckoltiana, 49
Tapogomea glabra, 642
Tapura, **247**
amazonica var. **amazonica**, **247**
capitulifera, **247**
guianensis, **248**, 249*
Taralea, **316**
oppositifolia, **318**
Teliostachya alopecuroidea, 36
Telitoxicum, **482**
inopinatum, **483**
Terminalia, 224, **227**
amazonia, **227**
guyanensis, 226*, **227**; Pl. 44d,e
Tetracera, 250
Tetragastris, 151, **161**
altissima, **161**, 162*; Pl. 33b
panamensis, **163**
paraensis, 163
phanerosepala, 161
pilosa, 160
stevensonii, 163
trifoliolata, 161
Tetrapterys, **426**
acutifolia, **426**
Tetrapterys (*continued*)
calophylla, 428
crispa, **426**, 748
discolor, 426
glabrifolia, 427*, 428, **748**; Pl. 81d
megalantha, **428**; Pl. 81e
mucronata, **428**
Thelypteridaceae, 747
Thelypteris
biolleyi, **747**
decussata, 748
gemmulifera, 748
Theobroma, **704**
cacao, **704**
velutinum, 705*, **706**; Pl. 122c
Theophrastaceae, **709**
Thinouia, **669**
myriantha, **669**
Thymelaeaceae, **711**
Thyrsodium, **49**
guianense, **49**, 52*
puberulum, 51*, 52*, **53**
spruceanum, **53**
spruceanum, 52*, **53**; Pl. 10g
Tibouchina, 437, **463**
aspera var. **asperrima**, 468*, **464**; Pl. 88c
multiflora, **465**
Ticorea, 649, **651**
foetida, **651**, 654*
Tilesia, **113**
baccata var. **baccata**, 113, **115***
Tiliaceae, 428, **712**
Tocoyena, **648**
longiflora, 632, **648**
Tontelea, **351**
attenuata, **353**
cylindrocarpa, 352*, **353**; Pl. 67f
laxiflora, **353**
nectandrifolia, **353**
ovalifolia subsp. **ovalifolia**, **353**
sandwithii, **353**
Topobea, **465**
parasitica, 464*, **465**; Pl. 88d
Tournefortia, **149**
bicolor, **149**
maculata, **149**
paniculata var. **spigeliiflora**, **149**, 150*
syringaefolia, 149
ulei, **151**; Pl. 31d
Touroulia, 594, **598**
guianensis, 597*, **598**
Tovomita, **218**, 750
brasiliensis, **218**, 220*
brevistaminea, **221**
calodictyos, **221**
grata, **221**
longifolia, **221**
macrophylla, **221**
obovata, **221**
richardiana, 221
schomburgkii, **221**; Pl. 43b,c
tenuiflora, **222**

Tovomitopsis, 212
membranacea, 213
Tragia, **297**
lessertiana, 297*, **298**
Trattinnickia, 151, **163**
boliviana, **163**
cuspidata, 163
lawrancei var. boliviana, 163
rhoifolia, **163**, 164*
subsp. sprucei, 163
Trema, **722**
micrantha, 721*, **722**; Pl. 124b,c
Trichantha calotricha, 337
Trichilia, 465, **471**, 750
cipo, **471**
euneura, 8, **471**; Pl. 90c,d
micrantha, **471**
pallida, **471**
quadrijuga, **471**
schomburgkii, **472**
septentrionalis, **472**
Trigonia, **718**
laevis var. **microcarpa**, **718**; Pl. 123c
villosa, 717*, **718**
Trigoniaceae, **716**
Trigoniodendron, 716
Trymatococcus, 515, **525**
amazonicus, **525**
oligandrus, **525**; Pl. 98d
paraensis, 525
Turnera, **718**
glaziovii, **718**, 719*; Pl. 123d
Turneraceae, **718**
Tynanthus, 119, **137**
polyanthus, 9, **139**
pubescens, 9, **139**; Pl. 27c

Ulmaceae, **720**
Uncaria, **648**
guianensis, 610, **648**
tomentosa, **649**
var. dioica, 649
surinamensis, 649
Unonopsis, **63**
perrottetii, **63**
rufescens, **64**, 65*; Pl. 13d
stipitata, **65**; Pl. 13e,f
Unxia, **115**
camphorata, **115**
hirsuta, 115
Uragoga
glabra, 642
Uragoga (*continued*)
guianensis, 638
Urera, **724**
caracasana, **725**; Pl. 124e
laciniata, 7, **725**
Urticaceae, 7, **722**
Utricularia, **397**
calycifida, **398**
hispida, **398**

Vantanea, **356**
guianensis, **356**
ovicarpa, **356**
parviflora, **356**
Vatairea, **318**
paraensis, **318**
Vataireopsis, **318**
surinamensis, **318**; Pl. 60
Verbenaceae, 6, 7, **725**
Vernonia cinerea, 104
Victoria amazonica, 554
Vigna, **318**
vexillata, **318**
Vinca rosea, 74
Virola, **529**
kwatae, 8, **530**; Pl. 99c,d
melinonii, 530
michelii, 8, **530**; Pl. 99e
sebifera, **530**, 531*; Pl. 99f
surinamensis, **532**
Violaceae, 732
Viscaceae, **739**, 750
Viscum verticillatum, 742
Vismia, **222**
amazonica, 222
cayennensis, **222**
glaziovii, 8, 222
gracilis, **222**
guianensis, 8, **222**; Pl. 43e
latifolia, **8**, 224
ramuliflora, **224**
sandwithii, 8, 223*, **224**; Pl. 43d
Vitaceae, **740**
Vitex, 725, **732**
guianensis, **732**
triflora, 731*, **732**; Pl. 125c
Vochysia, 742, **744**
guianensis, **744**
neyratii, **744**
surinamensis, **744**, 745*; Pl. 128c
tomentosa, **744**
Vochysiaceae, **742**
Vouacapoua, **181**
americana, 182*, **183**; Pl. 37d,e
Vouarana, **669**
guianensis, 668*, **669**, 748
Voyria, 328, **331**
aphylla, **331**
aurantiaca, **331**; Pl. 62d
caerulea, **331**; Pl. 62e
corymbosa, **332**; Pl. 63a,d
flavescens, **332**; Pl. 63b
rosea, **332**; Pl. 63c,d
spruceana, **332**
tenella, **332**; Pl. 64a
tenuiflora, **332**; Pl. 64b
Voyriella, 328, **333**
parviflora, **334**; Pl. 64c

Wallenia myrianthos, 537
Wedelia
paludosa, 111
trilobata, 111
Weigeltia
capitellata, 537
microbotrys, 535
myrianthos, 537
parviflora, 535
potiaei, 535
surinamensis, 537
Wulfia, 113
baccata var. baccata, 113
stenoglossa, 113

Xylopia, **67**
cayennensis, 66*, **67**
frutescens, **67**
longifolia, 67
nitida, **67**; Pl. 13g,h
Xylosma, 319, **328**
benthamii, **328**

Zanthoxylum, 649, **651**
ekmanii, **653**, 655*; Pl. 115d
rhoifolium, **656**
sp. A, **656**
Ziziphus, **603**
cinnamomum, 602*, **603**; Pl. 109c
Zornia, **319**
latifolia, **319**; Pl. 60c
Zygia, **507**
racemosa, **508**, 509*
tetragona, **509**